H₃C—CH₃	H₃C—F	H₃C—Cl	H₃C—Br
H₃C—I	H₃C—OH	H₃C—O—CH₃	H₂O
H₃C—NH₂	H₃C—NH—CH₃		H₃C—CN
H₃C—NO₂	H₃C—CF₃	CH₃CH₂CH₂CH₂F	
CH₃CHO	CH₃C(O)CH₃	CH₃CO₂H	CH₃CO₂CH₃
	CH₃S(O)CH₃	H₃C—SH	

H_3C-CH_3 H_3C-F H_3C-Cl H_3C-Br H_3C-I H_3C-OH $H_3C-O-CH_3$ H_2O H_3C-NH_2 $H_3C-NH-CH_3$ H_3C-CN H_3C-NO_2 H_3C-CF_3 $CH_3CH_2CH_2CH_2F$ CH_3CHO $CH_3C(O)CH_3$ CH_3CO_2H $CH_3CO_2CH_3$ $CH_3S(O)CH_3$ H_3C-SH

Modern Physical Organic Chemistry

Modern Physical Organic Chemistry

Eric V. Anslyn

UNIVERSITY OF TEXAS, AUSTIN

Dennis A. Dougherty

CALIFORNIA INSTITUTE OF TECHNOLOGY

University Science Books
www.uscibooks.com

University Science Books
www.uscibooks.com

Production Manager: *Christine Taylor*
Manuscript Editor: *John Murdzek*
Designer: *Robert Ishi*
Illustrator: *Lineworks*
Compositor: *Wilsted & Taylor Publishing Services*
Printer & Binder: *Edwards Brothers, Inc.*

This book is printed on acid-free paper.

Library of Congress Cataloging-in-Publication Data

Anslyn, Eric V., 1960–
    Modern physical organic chemistry / Eric V. Anslyn, Dennis A. Dougherty.
        p. cm.
    Includes bibliographical references and index.
    ISBN 978-1-891389-31-3 (alk. paper)
    1. Chemistry, Physical organic.   I. Dougherty, Dennis A., 1952–   II. Title.

QD476.A57   2004
547′.13—dc22

                                                                2004049617

Printed in the United States of America
10   9   8   7   6   5   4

Abbreviated Contents

Contents

vii

PART III
ELECTRONIC STRUCTURE:
THEORY AND APPLICATIONS

CHAPTER 14: Advanced Concepts in Electronic
           Structure Theory   807

Intent and Purpose   807

Highlights

xix

The twentieth century saw the birth of physical organic chemistry—the study of the inter-relationships between structure and reactivity in organic molecules—and the discipline matured to a brilliant and vibrant field. Some would argue that the last century also saw the near death of the field. Undeniably, physical organic chemistry has had some difficult times. There is a perception by some that chemists thoroughly understand organic reactivity and that there are no *important* problems left. This view ignores the fact that while the rigorous treatment of structure and reactivity in organic structures that is the field's hallmark continues, physical organic chemistry has expanded to encompass other disciplines.

In our opinion, physical organic chemistry is alive and well in the early twenty-first century. New life has been breathed into the field because it has embraced newer chemical disciplines, such as bioorganic, organometallic, materials, and supramolecular chemistries. Bioorganic chemistry is, to a considerable extent, physical organic chemistry on proteins, nucleic acids, oligosaccharides, and other biomolecules. Organometallic chemistry traces its intellectual roots directly to physical organic chemistry, and the tools and conceptual framework of physical organic chemistry continue to permeate the field. Similarly, studies of polymers and other materials challenge chemists with problems that benefit directly from the techniques of physical organic chemistry. Finally, advances in supramolecular chemistry result from a deeper understanding of the physical organic chemistry of intermolecular interactions. These newer disciplines have given physical organic chemists fertile ground in which to study the interrelationships of structure and reactivity. Yet, even while these new fields have been developing, remarkable advances in our understanding of basic organic chemical reactivity have continued to appear, exploiting classical physical organic tools and developing newer experimental and computational techniques. These new techniques have allowed the investigation of reaction mechanisms with amazing time resolution, the direct characterization of classically elusive molecules such as cyclobutadiene, and highly detailed and accurate computational evaluation of problems in reactivity. Importantly, the techniques of physical organic chemistry and the intellectual approach to problems embodied by the discipline remain as relevant as ever to organic chemistry. Therefore, a course in physical organic chemistry will be essential for students for the foreseeable future.

This book is meant to capture the state of the art of physical organic chemistry in the early twenty-first century, and, within the best of our ability, to present material that will remain relevant as the field evolves in the future. For some time it has been true that if a student opens a physical organic chemistry textbook to a random page, the odds are good that he or she will see very interesting chemistry, but chemistry that does not represent an area of significant current research activity. We seek to rectify that situation with this text. A student must know the fundamentals, such as the essence of structure and bonding in organic molecules, the nature of the basic reactive intermediates, and organic reaction mechanisms. However, students should also have an appreciation of the current issues and challenges in the field, so that when they inspect the modern literature they will have the necessary background to read and understand current research efforts. Therefore, while treating the fundamentals, we have wherever possible chosen examples and highlights from modern research areas. Further, we have incorporated chapters focused upon several of the modern disciplines that benefit from a physical organic approach. From our perspective, a protein, electrically conductive polymer, or organometallic complex should be as relevant to a course in physical organic chemistry as are small rings, annulenes, or non-classical ions.

We recognize that this is a delicate balancing act. A course in physical organic chemistry cannot also be a course in bioorganic or materials chemistry. However, a physical organic chemistry class should not be a history course, either. We envision this text as appropriate for many different kinds of courses, depending on which topics the instructor chooses to emphasize. In addition, we hope the book will be the first source a researcher approaches when confronted with a new term or concept in the primary literature, and that the text will provide a valuable introduction to the topic. Ultimately, we hope to have produced a text that will provide the fundamental principles and techniques of physical organic chemistry, while also instilling a sense of excitement about the varied research areas impacted by this brilliant and vibrant field.

Eric V. Anslyn
Norman Hackerman Professor
University Distinguished Teaching Professor
University of Texas, Austin

Dennis A. Dougherty
George Grant Hoag Professor of Chemistry
California Institute of Technology

Acknowledgments

Many individuals have contributed to the creation of this textbook in various ways, including offering moral support, contributing artwork, and providing extensive feedback on some or all of the text. We especially thank the following for numerous and varied contributions: Bob Bergman, Wes Borden, Akin Davulcu, Francois Diederich, Samuel Gellman, Robert Hanes, Ken Houk, Anthony Kirby, John Lavigne, Nelson Leonard, Charles Lieber, Shawn McCleskey, Richard McCullough, Kurt Mislow, Jeffrey Moore, Charles Perrin, Larry Scott, John Sherman, Timothy Snowden, Suzanne Tobey, Nick Turro, Grant Willson, and Sheryl Wiskur. Scott Silverman has provided numerous corrections and suggestions.

A very special thanks goes to Michael Sponsler, who wrote the accompanying Solutions Manual for the exercises given in each chapter. He read each chapter in detail, and made numerous valuable suggestions and contributions.

Producing this text has been extraordinarily complicated, and we thank: Bob Ishi for an inspired design; Tom Webster for dedicated efforts on the artwork; Christine Taylor for orchestrating the entire process and prodding when appropriate; John Murdzek for insightful editing; Jane Ellis for stepping up at the right times; and Bruce Armbruster for enthusiastic support throughout the project.

Finally, it takes a pair of very understanding wives to put up with a six-year writing process. We thank Roxanna Anslyn and Ellen Dougherty for their remarkable patience and endless support.

A Note to the Instructor

Our intent has been to produce a textbook that could be covered in a one-year course in physical organic chemistry. The order of chapters reflects what we feel is a sensible order of material for a one-year course, although other sequences would also be quite viable. In addition, we recognize that at many institutions only one semester, or one to two quarters, is devoted to this topic. In these cases, the instructor will need to pick and choose among the chapters and even sections within chapters. There are many possible variations, and each instructor will likely have a different preferred sequence, but we make a few suggestions here.

In our experience, covering Chapters 1–2, 5–8, selected portions of 9–11, and then 14–16 creates a course that is doable in one *extremely fast-moving* semester. Alternatively, if organic reaction mechanisms are covered in another class, dropping Chapters 10 and 11 from this order makes a very manageable one-semester course. Either alternative gives a fairly classical approach to the field, but instills the excitement of modern research areas through our use of "highlights" (see below). We have designed Chapters 9, 10, 11, 12, and 15 for an exhaustive, one-semester course on thermal chemical reaction mechanisms. In any sequence, mixing in Chapters 3, 4, 12, 13, and 17 whenever possible, based upon the interest and expertise of the instructor, should enhance the course considerably. A course that emphasizes structure and theory more than reactivity could involve Chapters 1–6, 13, 14, and 17 (presumably not in that order). Finally, several opportunities for special topics courses or parts of courses are available: computational chemistry, Chapters 2 and 14; supramolecular chemistry, Chapters 3, 4, and parts of 6; materials chemistry, Chapters 13, 17, and perhaps parts of 4; theoretical organic chemistry, Chapters 1, 14–17; and so on.

One of the ways we bring modern topics to the forefront in this book is through providing two kinds of highlights: "Going Deeper" and "Connections." *These are integral parts of the textbook that the students should not skip when reading the chapters* (it is probably important to tell the students this). The Going Deeper highlights often expand upon an area, or point out what we feel is a particularly interesting sidelight on the topic at hand. The Connections highlights are used to tie the topic at hand to a modern discipline, or to show how the topic being discussed can be put into practice. We also note that many of the highlights make excellent starting points for a five- to ten-page paper for the student to write.

As noted in the Preface, one goal of this text is to serve as a reference when a student or professor is reading the primary literature and comes across unfamiliar terms, such as "dendrimer" or "photoresist." However, given the breadth of topics addressed, we fully recognize that at some points the book reads like a "topics" book, without a truly in-depth analysis of a given subject. Further, many topics in a more classical physical organic text have been given less coverage herein. Therefore, many instructors may want to consult the primary literature and go into more detail on selected topics of special interest to them. We believe we have given enough references at the end of each chapter to enable the instructor to expand any topic. Given the remarkable literature-searching capabilities now available to most students, we have chosen to emphasize review articles in the references, rather than exhaustively citing the primary literature.

We view this book as a "living" text, since we know that physical organic chemistry will continue to evolve and extend into new disciplines as chemistry tackles new and varied problems. We intend to keep the text current by adding new highlights as appropriate, and perhaps additional chapters as new fields come to benefit from physical organic chemistry. We would appreciate instructors sending us suggestions for future topics to cover, along with particularly informative examples we can use as highlights. We cannot promise that they will all be incorporated, but this literature will help us to keep a broad perspective on where the field is moving.

Given the magnitude and scope of this project, we are sure that some unclear presentations, misrepresentations, and even outright errors have crept in. We welcome corrections and comments on these issues from our colleagues around the world. Many difficult choices had to be made over the six years it took to create this text, and no doubt the selection of topics is biased by our own perceptions and interests. We apologize in advance to any of our colleagues who feel their work is not properly represented, and again welcome suggestions.

We wish you the best of luck in using this textbook.

Modern Physical Organic Chemistry

MOLECULAR STRUCTURE
AND THERMODYNAMICS

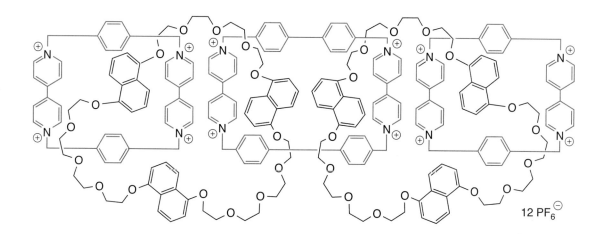

12 PF$_6^{-}$

Introduction to Structure and Models of Bonding

Intent and Purpose

There are three goals for Chapter 1. The first is to review simple notions of chemical bonding and structure. This review is meant for readers who have a knowledge of atomic and molecular structure equivalent to that given in introductory chemistry and organic chemistry textbooks. In this review, concepts such as quantum numbers, electron configurations, valence-shell electron-pair repulsion (VSEPR) theory, hybridization, electronegativity, polar covalent bonding, and σ and π bonds, are covered in an introductory manner. A large fraction of organic chemistry can be understood and predicted based upon these very simple concepts in structure and bonding. However, the second goal of the chapter is to present a more advanced view of bonding. This is known as qualitative molecular orbital theory (QMOT), and it will lay the foundation for Chapter 14, where computational methods are discussed. This more advanced approach to bonding includes the notion of group orbitals for recurring functional groups, and an extension of molecular orbital theory called perturbational molecular orbital theory that will allow us to make rational predictions as to how bonding schemes arise from orbital mixing. We show these bonding models first with stable molecules, and then apply the lessons to reactive intermediates. By covering stable structures alongside reactive intermediates, it should be clear that our standard models of bonding predict the reactivity and structure of all types of organic structures, stable and otherwise. Showing such a correlation is the third goal of the chapter.

A recurrent theme of this chapter is that organic functional groups—olefins, carbonyls, amides, and even simple alkyl groups such as methylene and methyl—can be viewed as having transferable orbitals, nearly equivalent from one organic structure to another. We will describe several of these molecular orbitals for many common organic functional groups. In all the discussions there is a single unifying theme, that of developing models of bonding that can be used to explain reactivity, structure, and stability, as a preparation for future chapters.

You may be aware that modern computational methods can be used to describe the bonding in organic molecules. Why, then, should we develop simple descriptive theories of bonding? With the advent of universally available, very powerful computers, why not just use quantum mechanics and computers to describe the bonding of any molecule of interest? In the early twenty-first century, it is true that any desktop computer can perform sophisticated calculations on molecules of interest to organic chemists. We will discuss the methodology of these calculations in detail in Chapter 14, and we will often refer to their results during our discussions in this and other chapters. However, for all their power, such calculations do not necessarily produce *insight* into the nature of molecules. A string of computer-generated numbers is just no substitute for a well-developed *feeling* for the nature of bonding in organic molecules. Furthermore, in a typical working scenario at the bench or in a scientific discussion, we must be able to rapidly assess the probability of a reaction occurring without constantly referring to the results of a quantum mechanical calculation. Moreover, practically speaking, we do not need high level calculations and full molecular orbital theory to understand most common reactions, molecular conformations and structures, or kinetics and thermodynamics. Hence, we defer detailed discussions of sophisticated calculations and full molecular orbital theory until just before the chapters where these methods are essential. Also, as powerful as they are, calculations are still severely limited in their ability to address large systems such as proteins, nucleic acids, or conducting polymers. This limitation is even more severe when solvation or solid state issues become critical. Therefore, it is still true—and will be true for some time—that descriptive models of bonding that are readily applicable to a wide range of situations are the best way to attack complex problems. The models must be firmly rooted in rigorous theory, and must stand up to quantitative computational tests. Two such models are developed in this chapter.

1.1   A Review of Basic Bonding Concepts

In this section we present a number of basic concepts associated with chemical bonding and organic structure. Most of this material should be quite familiar to you. We use this section to collect the terminology all in one place, and to be sure you recall the essentials we will need for the more advanced model of bonding given in Sections 1.2 and 1.3. For most students, a quick read of this first section will provide an adequate refresher.

1.1.1   Quantum Numbers and Atomic Orbitals

Every molecule is made up from the nuclei and electrons of two or more atoms via bonds that result from the overlap of atomic orbitals. Hence, the shapes and properties of atomic orbitals are of paramount importance in dictating the bonding in and properties of molecules. The **Bohr model** of atoms had electrons moving in specific orbits (hence the term **orbitals**) around the nucleus. We now view the shapes and properties of atomic orbitals as they are obtained from basic quantum mechanics via solution of the Schrödinger equation. The solutions to the Schrödinger equation are termed **wavefunctions**, and in their most common implementation these wavefunctions correspond to atomic or molecular orbitals.

The atomic orbital wavefunctions come in sets that are associated with four different quantum numbers. The first is the **principal quantum number**, which takes on positive integer values starting with 1 ($n = 1, 2, 3, \dots$). An atom's highest principal quantum number determines the **valence shell** of the atom, and it is typically only the electrons and orbitals of the valence shell that are involved in bonding. Each row in the periodic table indicates a different principal quantum number (with the exception of d and f orbitals, which are displaced down one row from their respective principal shells). In addition, each row is further split into **azimuthal quantum numbers** ($m = 0, 1, 2, 3, \dots$; alternatively described as $s, p, d, f, \dots$). This number indicates the angular momentum of the orbital, and it defines the spatial distribution of the orbital with respect to the nucleus. These orbitals are shown in Figure 1.1 for $n = 2$ (as with carbon) as a function of one of the three Cartesian coordinates.

The shapes given in Figure 1.1 are a schematic representation of the orbitals in regions of space around the nucleus. For $n = 1$, only a 1s atomic orbital is allowed. The highest electron density is at the atomic nucleus, with decreasing density in all directions in space at increas-

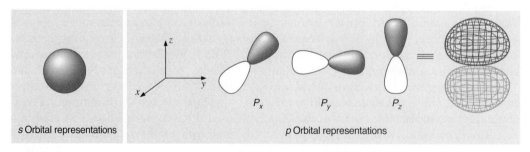

| s Orbital representations | p Orbital representations |

Figure 1.1
The general shape of s and p atomic orbitals for carbon. These cartoons are the schematics that chemists typically sketch. Shown also is a more realistic representation for the p orbital produced by quantum mechanical calculations.

ing distances from the nucleus. We pictorially represent such a population density as a sphere.

The principal quantum number 2 has s and p orbitals. The 2s orbital is similar to the 1s orbital, but has a spherical surface in three-dimensional space where the electron density goes to zero, called a **node**. A node is a surface (a sphere for s orbitals, a plane for p orbitals) that separates the positive and negative regions of a wavefunction. There is zero probability of finding an electron at an orbital node. The spherical node of a 2s orbital cannot be seen in the representation of Figure 1.1. In reality, this nodal surface in the 2s orbital has little impact on bonding models, and again, we pictorially represent this orbital as a sphere, just as with a 1s orbital.

A p orbital can orient along three perpendicular directions in space, defined to be the x, y, and z axes. The 2p orbitals have a **nodal plane** that contains the nucleus and is perpendicular to the orbital axis. As such, the electron density is zero at the nucleus. The population density of a p orbital reaches a maximum along its axis in both the negative and positive spatial directions, and then drops off. This population density is shown as a dumbbell-like shape.

The directionality of an orbital in space is associated with a third set of quantum numbers called **magnetic**. For the p orbitals the magnetic quantum numbers are $-1, 0$, and 1, each representing one of the three different orthogonal directions in space (see the three 2p orbitals in Figure 1.1). The 2s and 2p orbitals make up the valence shell for carbon. Later in this chapter, we will examine metals, which contain d orbitals. The magnetic quantum numbers for d orbitals are $-2, -1, 0, 1$, and 2.

The **phasing** of the atomic orbitals shown in Figure 1.1 (color and gray / clear) is solely a result of the mathematical functions describing the orbitals. One color indicates that the function is positive in this region of space, and the other color indicates that the function is negative. It does not matter which color is defined as positive or negative, only that the two regions are opposite. There is no other meaning to be given to these phases. For instance, the probability of finding an electron in the differently phased regions is the same. The probability is defined as the **electron density** or **electron distribution**. It is specifically related to the square of the mathematical function that represents the orbitals.

The fourth and final quantum number, m_s, is associated with the spin of an electron. Its value can be $+\frac{1}{2}$ or $-\frac{1}{2}$. An orbital can only contain two electrons, and their spin quantum numbers must be oppositely signed (termed **spin paired**) if the electrons reside in the same orbital. Because electrons have wave-like properties, these waves are overlapping in space when the electrons are in the same orbital. However, because the electrons are negatively charged and have particle character also, they tend to repel each other. As a result, their movements are actually correlated, so as to keep the like charges apart. **Correlation** is the ability of an electron to feel the trajectory of another electron and therefore alter its own course so as to minimize Coulombic repulsions and keep the energy of the system to a minimum.

1.1.2 Electron Configurations and Electronic Diagrams

The **electron configuration** of an atom describes all the atomic orbitals that are populated with electrons, with the number of electrons in each orbital designated by a superscript. For example, carbon has its 1s, 2s, and 2p orbitals each populated with two electrons. Hence, the electron configuration of carbon is $1s^2\,2s^2\,2p^2$. This is the **ground state** of carbon, the most stable form. Promotion of an electron from an atomic orbital to a higher-lying atomic orbital produces a higher energy **excited state**, such as $1s^2\,2s^1\,2p^3$.

In an **electronic diagram** the atomic orbitals are represented by horizontal lines at different energy levels, where the higher the line on the page the higher the energy. Symbols are placed near the lines to indicate which orbitals the lines are meant to represent. The arrows represent electrons, and their direction indicates the relative spin of the individual electrons. Several rules are used to decide how these lines (orbitals) are populated (filled) with electrons. The **aufbau principle** (from German for "building-up") states that one populates the lower energy orbitals with electrons first. Furthermore, only two electrons can be in each or-

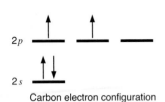

Carbon electron configuration

bital, and when they are in the same orbital they must be spin paired (a result of the **Pauli principle**). **Hund's rule** tells us how to handle the population of **degenerate orbitals**, which are orbitals that have the same energy. We singly populate such orbitals sequentially, and all electrons in singly-occupied orbitals have their spins aligned.

Carbon has six electrons, two in the 1s orbital, and four valence electrons that occupy the 2s and 2p valence orbitals. Based on the rules briefly reviewed here, the lowest energy electronic diagram of the valence shell of carbon is as shown in the margin.

The familiar **octet rule**, which states that atoms are most stable when their valence shell is full, suggests that carbon in a molecule will take on four more electrons from other atoms so as to possess an octet of electrons and thereby attain a noble gas configuration. The number of bonds that an atom can make is called its **valence number**. If each bond that carbon makes is created by the donation of a single electron from an adjacent atom's atomic orbitals, carbon will make four bonds. Carbon is said to have a valence of four. This valence is by far the most common bonding arrangement for C. When carbon has fewer than four bonds it is in a reactive form, namely a carbocation, radical, carbanion, or carbene. When a similar analysis is done for N, O, and F, it is found that these atoms prefer three, two, and one bond(s), respectively.

1.1.3 Lewis Structures

G. N. Lewis developed a notation that allows us to use the valence electrons of atoms in a molecule to predict the bonding in that molecule. In this method, the electrons in the valence shell of each atom are drawn as dots for all atoms in the molecule (see examples below). Bonds are formed by sharing of one or more pairs of electrons between the atoms, such that each atom achieves an octet of electrons. In an alternative to the electron dot symbolism, we can draw a line to represent a bond. A single bond is the sharing of two electrons, while double and triple bonds involve the sharing of four and six electrons, respectively. Despite its simplicity, this notation can be used to accurately predict the number of lone pairs that an atom will have and whether that atom will use single, double, or triple bonds when incorporated into specific molecules.

A few examples of Lewis structures

The problem with Lewis dot structures is that they provide no insight into molecular shapes, orbitals, or distributions of electrons within molecules. Instead, they are only useful for predicting the number of bonds an atom forms; whether the atom has lone pairs; and whether single, double, or triple bonds are used. Once an atom is found to have an octet using a Lewis analysis, no further insight into the structure or reactivity can be obtained from the Lewis structure. We have to turn to more sophisticated molecular structure and bonding concepts to understand structure and reactivity.

1.1.4 Formal Charge

Often it is convenient to associate full charges with certain atoms, even though the charges are in fact delocalized among the atoms in the molecule, and the overall molecule may be neutral. Such charges derive from the Lewis structure, and these full charges on atoms are called **formal charges**, denoting they are more of a formality than a reality.

The formula generally given in introductory chemistry textbooks for calculating the formal charge is formal charge = number of valence electrons − number of unshared electrons

$-\frac{1}{2}$ the number of shared electrons. In organic chemistry, it is easier to just remember a few simple structures. For example, the oxygen in water has two bonds to hydrogen and is neutral. In contrast, the oxygen in the hydronium ion (H_3O^+) bonds to three hydrogens and is positive; and the oxygen of hydroxide (OH^-) has only one bond and has a formal negative charge. This series can be generalized. Whenever oxygen has an octet of electrons and has one, two, or three bonds it is negative, neutral, or positive, respectively. More generally, whenever an atom has an octet and has one bond more than its neutral state it is positive; when it has one bond fewer it is negative. Hence, a nitrogen atom having two, three, or four bonds is negative, neutral, or positive, respectively; similarly, a carbon having an octet and three, four, or five bonds is negative, neutral, or positive, respectively. Although formal charge can be rapidly evaluated in this manner, you should not take the charge on a particular atom too literally, as demonstrated in the following Going Deeper highlight.

Going Deeper

How Realistic are Formal Charges?

Formal charge is more or less a bookkeeping tool. For the tetramethylammonium ion, for example, we draw a positive charge on the nitrogen because it is tetravalent. However, it is now possible to develop very accurate descriptions of the electron distributions in molecules using sophisticated computational techniques (Chapter 14). Such calculations indicate that a much more reasonable model for the tetramethylammonium ion describes the N as essentially neutral. The positive charge resides on the methyls, each carrying one-fourth of a charge. What is going on here? Looking ahead to Section 1.1.8, we know that N is more electronegative than C, so it should have more negative charge (less positive charge) than C. Indeed, in trimethylamine there is a substantial negative

charge on the N. On going from trimethylamine to tetramethylammonium the N does become more positive than in a neutral molecule. It is just that it goes from partial negative to essentially neutral, rather than from neutral to positive, as implied by the formal charge symbolism. Beyond bookkeeping, formal charge is really only useful for indicating the charge on the *molecule*, not on individual atoms.

Formal charge on quaternary ammonium

1.1.5　VSEPR

Once we have a basic idea of the bonds to expect for organic structures, the next key issue is the three-dimensional shape of such structures. We now introduce two important concepts for rationalizing the diverse possibilities for shapes of organic molecules: VSEPR and hybridization.

The **valence-shell electron-pair repulsion (VSEPR) rule** states that all groups emanating from an atom—whether single, double, or triple bonds, or lone pairs—will be in spatial positions that are as far apart from one another as possible. The VSEPR method does not consider singly occupied orbitals to be groups (see below for the reason). VSEPR is purely a theory based upon the notion that the electrostatic repulsions between entities consisting of two or more electrons dictate molecular geometries.

This rule can be applied to carbon when it is bonding to either four, three, or two other atoms. Acetylene has a **linear** arrangement of the C–C triple bond and the C–H bond, because a 180° angle places these two groups as far apart as possible. When three groups are attached to an atom, such as the three hydrogens of CH_3^+, the geometry is **trigonal planar**

Geometries based upon VSEPR

with 120° H–C–H bond angles. Finally, in a molecule such as methane, a **tetrahedral** arrangement of the four bonds places them as far apart in space as possible (H–C–H angles of 109.5°).

The geometries for acetylene, methyl cation, and methane correspond to the bond angles for idealized linear, trigonal, and tetrahedral systems. While these geometrically perfect angles do appear in simple molecules like these, in most organic molecules where different groups are attached to the various atoms, measurable deviations from these ideals are observed. However, we will still loosely refer to the carbons as tetrahedral or trigonal, even though we don't expect angles of exactly 109.5° or 120°, respectively.

The VSEPR model provides a simple way to understand such deviations from perfection. Since the geometries derived from VSEPR are based solely upon maximizing the distance between electron pairs, it makes sense that the geometries would also depend upon the "sizes" of the electron pairs. A central tenant of VSEPR is that lone pairs *behave as if* they are larger than bonded pairs. Always keep in mind that VSEPR is not based on any first principles analysis of electronic structure theory. It is a simple way to rationalize observed trends. It is debatable whether a lone pair of electrons actually is larger than a bonded pair of electrons plus the associated atoms. In fact, it is not even clear that size is a well defined concept for a lone pair. The point is, in VSEPR we consider lone pairs to be larger than bonded pairs because that approach leads to the right conclusions. This view allows us to rationalize the fact that the H–X–H angles in ammonia and water are smaller than 109.5°. Both systems are considered to have four groups attached to the central atom because, as stated earlier, lone pairs count as groups in VSEPR. Since a lone pair is larger than a bonding pair, the N–H bonds of ammonia want to get away from the lone pair, causing contractions of the H–N–H angles. The effect is larger in water, with two lone pairs.

The VSEPR rule uses a common principle in organic chemistry to predict geometry, that of **sterics**, a notion associated with the through-space repulsion between two groups. **Steric repulsion** arises from the buttressing of filled orbitals that cannot participate in bonding, where the negative electrostatic field of the electrons in the orbitals is repulsive. The reason that singly occupied orbitals are not considered to be groups in VSEPR is that they can participate in bonding with doubly occupied orbitals. Intuitively, we expect larger groups to be more repulsive than smaller groups, and this is the reasoning applied to the lone pairs in ammonia and water. Likewise, due to sterics, we may expect the central carbon in 2-methylpropane to have an angle larger than 109.5°, and indeed the angle is larger than this value (see margin).

107°   104.5°

Perturbations from a perfect tetrahedral angle

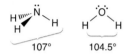

110.6°

Perturbation from the tetrahedral angle

1.1.6 Hybridization

It was stated earlier that CH_3^+ prefers bond angles of 120°, and methane prefers bond angles of 109.5°. How do we achieve such bond angles when the s and p atomic orbitals are not oriented at these angles? The s orbitals are spherical and so have no directionality in space, and the p orbitals are oriented at 90° angles with respect to each other. We need a conceptual approach to understand how s and p atomic orbitals can accommodate these experimentally determined molecular bond angles. The most common approach is the idea of hybridization, first introduced by Pauling.

Pauling's assumption was that bonds arise from the overlap of atomic orbitals on adjacent atoms, and that the better the overlap the stronger the bond. **Orbital overlap** has a quantitative quantum mechanical definition (given in Chapter 14). In a qualitative sense, overlap can be thought of as the extent to which the orbitals occupy the same space. However, if there are regions of overlap with matched and mismatched phasing, the contributions to the overlap have opposite signs and will cancel. The more space occupied where the phasing reinforces, the larger the overlap. When the opposite phasing in the various areas completely cancels, there is no overlap. For example, consider the arrangements of the s and p orbitals shown in the margin. The top shows how the s and p occupy some of the same space, but the phasing completely cancels: zero overlap is the result. Any movement of the s orbital to the side increases overlap, until the greatest overlap, shown for the bottom arrangement, takes advantage of the directionality of a p orbital.

No overlap between s and p orbital

Increasing overlap

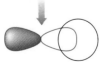

Most overlap

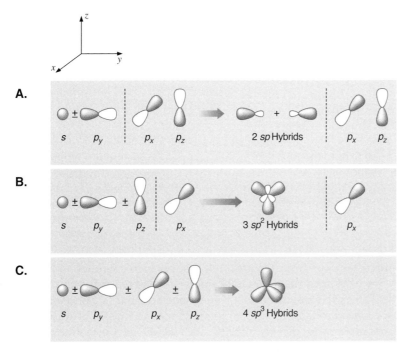

Figure 1.2
Forming hybrid orbitals. Combining an *s* orbital with one, two, or three *p* orbitals produces the familiar **A.** *sp*, **B.** *sp²*, and **C.** *sp³* hybrid orbitals.

Pauling also argued that orbitals with directionality would give stronger bonds because the overlap would be higher. Pauling suggested that to achieve orbitals with directionality, mixtures of atomic orbitals on the same atom are formed in a process known as hybridization. **Hybridization** is the method of adding and subtracting atomic orbitals on the same atom. Remember that orbitals are mathematical solutions to the Schrödinger equation, and that the addition and subtraction of mathematical equations is just an exercise in algebra. It is a perfectly valid operation to add orbitals as long as one also does the corresponding subtraction. Qualitatively, we use the positive and negative phasing along with the three-dimensional shapes of the orbitals to visualize what the result of adding and subtracting the orbitals would be. For example, Figure 1.2 **A** shows the result of combining a 2*s* and one of the 2*p* orbitals, in this case the $2p_y$ orbital. Each of the resultant orbitals has a large lobe on one side of the atom and a small lobe on the other side, and therefore has greater directionality than the original orbitals. The addition leads to an orbital with directionality along the negative *y* axis, and the subtraction leads to an orbital with directionality along the positive *y* axis. The two lobes have different phasing. The combination of an *s* orbital with a single *p* orbital creates what are called **sp hybrid orbitals**. Note that these two new orbitals point 180° apart, as is found in acetylene. Hence, the carbons in acetylene are considered to be *sp* hybridized in order to accommodate the experimentally determined geometry. This leaves two pure *p* orbitals: p_x and p_z.

This addition and subtraction can be carried further to give **sp²** and **sp³ orbitals**. Figure 1.2 **B** shows this addition and subtraction for *sp²*. Note that the new orbitals are now all 120° apart. The remaining *p* orbital is perpendicular to the *sp²* hybrid orbitals. It makes good sense that the mixing of this last *p* orbital with the *sp²* hybrids would lead to new hybrids that are above and below the plane formed by the *sp²* hybrids. In this case four identical orbitals called *sp³* hybrids are the result (Figure 1.2 **C**). Each points toward the corners of a tetrahedron. In organic molecules one of these hybridization states—*sp*, *sp²*, or *sp³*—is invoked as appropriate when explaining the linear, trigonal planar, or tetrahedral geometry of an atom, respectively.

The geometries for acetylene, methyl cation, and methane correspond to the bond angles for the different hybridization states sp, sp^2, and sp^3, respectively. Again, most organic molecules display measurable deviations from these ideals, but we still loosely refer to the atoms as sp, sp^2, or sp^3 hybridized, even though we don't expect angles of exactly 180°, 120° or 109.5°.

Hybridization provides an alternative "explanation" to VSEPR for such deviations from ideal angles. In going from pure sp to sp^2, sp^3, and pure p, the angles go from 180° to 120°, 109.5°, and 90°. Thus, decreasing s character leads to decreasing bond angles. We could say that in ammonia the N–H bonds have lost s character from N relative to a pure sp^3 N, because the angle is smaller than the perfect tetrahedral angle. In fact, we can quantify this analysis with a simple relationship. We define a **hybridization index**, i (Eq. 1.1). Here, the observed bond angle θ is used in the equation to solve for i.

$$1 + i \cos \theta = 0 \qquad\qquad\qquad \text{(Eq. 1.1)}$$

We then define the hybridization as sp^i. For example, since by definition the tetrahedral angle is the arc $\cos(-\frac{1}{3})$ (~109.5°), perfect tetrahedral angles imply $i = 3$. For ammonia, we conclude that the N hybrids that bond to H are $sp^{3.4}$, and in water the bonds to H are formed by sp^4 hybrids. That is, in water the orbitals that make up the O–H bonds are 80% p in character and 20% s, versus the 75:25 mixture implied by sp^3. The lone pairs must compensate, and they take on extra s character in NH_3 and H_2O. We will see that this notion of non-integral hybridizations is more than just an after-the-fact rationalization, and has experimental support (see the following Connections highlight). It can have predictive power. However, we must first introduce another important bonding concept: electronegativity.

Connections

NMR Coupling Constants

The view of variable hybridization has some experimental support. The magnitude of $^{13}C-^1H$ NMR coupling constants is expected to be proportional to the amount of carbon s character in the bond, because only s orbitals have density at the carbon nucleus and can affect neighboring nuclear spin states. In several systems, a clear correlation has been observed between NMR coupling constants and percent s character, as predicted from the geometry and the associated hybridization index.

For example, in cyclic alkanes, the smaller the ring, the larger the p character that would be expected in the hybrid orbitals used to form the C–C bonds, because p orbitals better accommodate smaller bond angles. Correspondingly, the C–H bonds would have higher s character.

This correspondence is indeed seen from an analysis of the C–H coupling constants given below. The smaller rings have the larger coupling constants.

Ring system	$J_{^{13}C-^1H}$
Cyclopropane	161
Cyclobutane	134
Cyclopentane	128
Cyclohexane	124
Cycloheptane	123
Cyclooctane	122

Ferguson, L. N. (1973). *Highlights of Alicyclic Chemistry, Part 1*, Franklin Publishing Company, Inc., Palisade, MI.

1.1.7   A Hybrid Valence Bond/Molecular Orbital Model of Bonding

There are two dominant models for considering bonding in organic molecules: valence bond theory (VBT) and molecular orbital theory (MOT). While often viewed as competing theories, VBT and MOT actually complement each other well, and our ultimate model for bonding will borrow from both theories. VBT was developed first. The idea, as originally put forth by Heitler and London and expanded by Pauling, was that the binding energy between two atoms arises primarily from exchange (resonance) of electrons between the two atoms in a bond. The starting point for VBT has one electron on each atom that contributes to an **electron pair bond**. It is this assignment of electrons primarily to individual atoms—or, more precisely, to individual orbitals on atoms—that is the hallmark of VBT. Bonding is, in effect, viewed as a perturbation of this arrangement. That is, when two atoms are brought together, each electron is permitted to interact with either nucleus, and this produces bond energies in adequate agreement with the experimental values. Hence, the conclusion was that bonds consist of two electrons in the region between two nuclei.

In VBT a molecule is formed by adjacent atoms sharing electrons. As suggested by the name, the electrons that are involved in bonding are those from the atoms' valence shells. Each atom donates one electron to the bond, and the resulting electron pair is considered to be mostly localized between the two adjacent atoms. This localization of the electrons is exactly the impression of bonding that is given by a Lewis structure. Furthermore, localization of the electrons between the atoms would require orbitals that point in the appropriate directions in space. It is this kind of reasoning that led Pauling to develop hybrid orbitals, an essentially valence bond concept. In essence, VBT nicely encompasses the topics discussed to this point in the chapter. However, one other notion is required by VBT—that of resonance. As discussed in Section 1.1.10, if more than one Lewis dot structure can be drawn for a molecule, then VBT states that the actual molecule is a hybrid of these "canonical forms".

Creating Localized σ and π Bonds

The most common model for bonding in organic compounds derives from VBT and the hybridization procedure given previously. **Sigma bonds** (σ bonds) are created by the overlap of a hybrid orbital on one atom with a hybrid orbital on another atom or an s orbital on hydrogen (Figures 1.3 **A** and **B**, respectively). **Pi bonds** (π bonds) are created by the overlap of two p orbitals on adjacent atoms (Figure 1.3 **C**). Specifically, σ bonds are defined as having their electron density along the bond axis, while π bonds have their electron density above and below the bond axis. The combination of the two orbitals on adjacent atoms that creates **in-phase** interactions (signs of the orbitals are the same) between the two atoms is called the **bonding orbital**. The combination that results in **out-of-phase** interactions (signs of the orbitals are opposite) is called the **antibonding orbital**. The bonding orbital is lower in energy than the antibonding orbital. There are also orbitals that contain lone pairs of electrons, which are not bonding or antibonding. These are called **nonbonding orbitals**. In standard neutral organic structures, only the bonding orbitals and nonbonding orbitals are occupied with electrons. Recall that an alkene functional group has a single σ and a single π bond between the adjacent carbons, whereas an alkyne has a single σ and two π bonds between the carbons. The number of bonds between two atoms is called the **bond order**.

The creation of bonding and antibonding orbitals is actually a molecular orbital theory notion. Therefore, the orbitals of Figure 1.3 are in effect **molecular orbitals**. We will have much more to say in Section 1.2 about how to linearly mix orbitals to create bonding and antibonding molecular orbitals. However, you may recall molecular orbital mixing diagrams from introductory organic chemistry, such as that shown in the margin for the π bond in ethylene. These diagrams give a picture of how chemists visually create bonding and antibonding orbitals via mixing. The mixing to derive the molecular orbitals gives both a plus

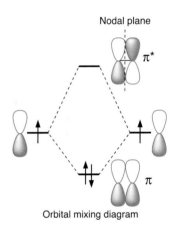

Nodal plane

π*

π

Orbital mixing diagram

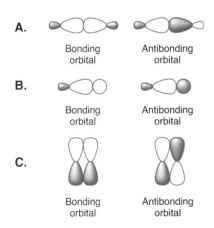

A.
Bonding orbital   Antibonding orbital

B.
Bonding orbital   Antibonding orbital

C.
Bonding orbital   Antibonding orbital

Figure 1.3
A. Combination of two hybrid orbitals on adjacent atoms gives bonding and antibonding orbitals. Population of the bonding orbital with two electrons creates a σ bond. **B.** Combination of a hybrid and a 1s orbital on hydrogen gives a bonding and antibonding pair. Population of the bonding orbital with two electrons creates a σ bond. **C.** Combination of two p orbitals on adjacent atoms also gives a bonding and antibonding set. Population of the bonding orbital with two electrons creates a π bond.

and minus combination of the starting atomic orbitals. Note, therefore, that the orbitals given in Figure 1.3 are actually derived from a hybrid VBT/MOT approach to bonding. One creates discrete, localized bonds between adjacent atoms as pictured with VBT, but it is done using the linear combination ideas of MOT.

The simple molecular orbital mixing diagram given above serves to illustrate many concepts and terms used with molecular orbital theory. The bonding molecular orbital (MO) is **symmetric** with respect to a mirror plane that resides between the two carbons making the bond, while the antibonding MO is **antisymmetric**. Furthermore, a nodal plane exists in the antibonding MO between the two atoms making the π bond, which means that populating this orbital with electrons leads to a repulsive interaction between the atoms.

The picture of σ and π bonds that consist of bonding and antibonding molecular orbitals that reside primarily between adjacent atoms is the standard that organic chemists most commonly use. The reactivity of the vast majority of organic compounds can be nicely modeled using this picture, and it forms the starting point for the electron pushing method of presenting organic reaction mechanisms (see Appendix 5). Hence, this theory of bonding is extremely important in organic chemistry.

1.1.8 Polar Covalent Bonding

Once the geometry of a molecule has been established, the next crucial feature for predicting the reactivity is its charge distribution. Notions such as VSEPR and hybridization control shape and structure. Here we discuss how electronegativity is the primary determinant of the charge distribution in a molecule, with hybridization playing a secondary but still important role.

Covalent bonds predominate in organic chemistry. In our simple theory of bonding, the two electrons in the bond are shared between the two adjacent atoms, as implied by a Lewis dot structure and the σ and π bonds discussed previously. Very few, if any, organic structures can be considered to have ionic bonds. However, whenever a carbon forms a bond to any atom or group not identical to itself, the bond develops some polar character; there is a positive end and a negative end to the bond. This charge separation means the sharing of electrons is unequal. A covalent bond that has an unequal sharing of the bonding pair of electrons is called **polar covalent**. Pauling argued that introducing polarity into a bond strengthens it, and we will see in Chapter 2 that trends in bond strengths generally support this view.

Electronegativity

To predict the charge distribution in an organic molecule, we need to examine the **electronegativity** of the atoms in the molecule. Pauling originally developed this important concept and described it as "the power of an atom in a molecule to attract electrons to itself". Pauling assigned values to various atom types by examining bond dissociation energies of molecules. As such, the Pauling electronegativity scale depends upon *molecular* properties, and is not an intrinsic property of the atoms. The Pauling scale is most commonly used, and is given in all introductory chemistry textbooks.

Mulliken defined an electronegativity scale that is derived from the average of the ionization potential and electron affinity of an atom, and therefore is solely an *atomic* property. The **ionization potential** is the energy required to remove an electron from an atom or molecule. Hence, this number reflects the affinity of an atom for the electrons it already has. The **electron affinity** is the amount of energy released or required to attach another electron to an atom or molecule. Hence, this number reflects the affinity of the atom for an additional electron. Using these values is a logical basis for determining the ability of an atom to attract electrons toward itself. Along with the electronegativity scales of Pauling and Mulliken, comparable scales have been developed by Nagle, Allen, Sanderson, Allred-Rochow, Gordy, Yuan, and Parr. Suffice it to say that the electronegativity of an atom is a difficult concept to put a precise number on, and that the use of different scales is appropriate for different applications. Table 1.1 compares the Pauling and Mulliken electronegativity scales, showing that the two are similar. We should always remember that the key issue is the *rela-*

Table 1.1
Electronegativities of Atoms According to
the Scales of Pauling and Mulliken*

Atom	Pauling	Mulliken
H	2.1	3.01
B	2.0	1.83
C	2.5	2.67
N	3.0	3.08
O	3.5	3.22
F	4.0	4.44
Cl	3.0	3.54
Br	2.8	3.24
I	2.5	2.88
Li	1.0	1.28
Na	0.9	1.21
K	0.8	1.03
Mg	1.2	1.63
Ca	1.0	1.30
Al	1.5	1.37
Si	1.8	2.03
P	2.1	2.39
S	2.5	2.65

*Pauling, L. (1960). *The Nature of the Chemical Bond and the Structure of Molecules and Crystals; an Introduction to Modern Structural Chemistry,* 3d ed., Cornell University Press, Ithaca, NY, and Allen, L. C. "Electronegativity is the Average One-Electron Energy of the Valence-Shell Electrons in Ground-State Free Atoms." *J. Am. Chem. Soc.,* **111**, 9003 (1989).

tive electronegativities of two moieties—we mainly want to know which is the more electronegative of the two, and whether the difference is relatively large or small. In most situations, all the various electronegativity scales lead to the same predictions.

The major factor influencing electronegativity is the energy of the orbitals that the atom uses to accept electrons. As one moves left-to-right across the periodic table, the valence orbitals become lower in energy within the same row. Going down a column, the atoms get bigger and the valence orbitals are higher in energy. This would mean that He has the lowest energy valence orbitals, but He cannot accept any more electrons because it is a noble gas. The atom with the lowest energy valence orbitals that is not a noble gas is F, which is the most electronegative element. In fact, a useful way to estimate electronegativities if Table 1.1 is unavailable is simply to recall that F is the most electronegative element, and moving left or down in the periodic table progressively diminishes electronegativity.

When an atom with a higher electronegativity than carbon forms a bond to carbon, the electrons in the bond will reside more toward the electronegative atom, producing a partial negative charge on this atom and a partial positive charge on carbon. Conversely, when the atom has a lower electronegativity, the carbon will possess a partial negative charge. These partial charges are denoted $\delta+$ and $\delta-$ (see a few examples below). The larger the difference between the electronegativities for atoms in a bond, the more the bond is polarized. The magnitude of the polarization can be gauged using the values of Table 1.1. For example, us-

Partial charges

ing the Pauling scale, a C–F bond is more polar than a C–Cl bond, because the electronegativity differences for the atoms in these bonds are 1.43 and 0.61, respectively. A bond whose electronegativity difference is 1.7 is considered to be 50% ionic and 50% covalent, and so bonds with differences greater than this are considered to be ionic. In this view, bonds of Li, Na, or K to F, Cl, Br, or I are all ionic.

Some conclusions that can be drawn from the data of Table 1.1 may seem a bit counterintuitive. For example, the electronegativity difference between C and I in both scales is actually smaller than the difference between C and H. Therefore, a C–I bond is predicted to have a smaller charge polarization than a C–H bond, based solely upon electronegativities. This polarization may come as a surprise, since iodide is a good leaving group in S_N2 reactions. Often the electronegativity of iodine is erroneously invoked to explain such reactions. As we will see below, electronegativity is not the whole story. Polarizability is also important, especially when rationalizing reactivity trends. Also note that the electronegativity difference between C and O is smaller than between P and S, or P and O, on the Mulliken scale. Hence, S–O and P–O bonds are more polarized than C–O bonds.

Electrostatic Potential Surfaces

In a polar bond, one end is designated as $\delta+$ and the other as $\delta-$, and this designation is often adequate for discussing simple molecules such as methyl chloride. However, in more complex molecules there will be many different types of bonds with differing degrees of polarity, and the overall molecule will reflect the sum of these and any interactions they experience. The simple $\delta+/\delta-$ symbolism is no longer adequate, so we need an alternative way to view the charge distribution in complex organic molecules. In recent years, many scientists have found that plots of electrostatic potential surfaces are quite useful in this regard. Such plots are given in Appendix 2, which contains a gallery of representative electrostatic potential surfaces for prototype organic structures. In these pictures, red represents negative electrostatic potential whereas blue represents positive electrostatic potential. A green color is a region that is essentially neutral.

Let's examine for a moment the electrostatic potential surface for methyl acetate (see Appendix 2). In our $\delta+/\delta-$ method, we would denote methyl acetate as shown in the margin. The carbonyl carbon is partially positive, and so is the methyl group. However, both oxygens would be denoted as negative. This notation is less than optimum, however, because we have no idea which oxygen is more negative. Yet, a quick glance at the electrostatic surface shows that the carbonyl oxygen is more negative (you may predict this fact from resonance; see Section 1.1.10). Thus, an electrostatic potential surface can provide deeper insight into the electronic distribution in a molecule than a simple $\delta+/\delta-$ picture.

What exactly are these surfaces? First, they result from a full quantum mechanical calculation of the electronic structure of the molecule. Note that we do not use electronegativities or hybridizations or bond dipoles or any of the descriptive features of our bonding model in such calculations. These are *a priori* quantum mechanical calculations, and their output enables the charge distribution in a molecule to be computed.

What are we actually showing in such plots? First, we give a surface to the molecule. The surface is very similar to a **van der Waals surface**, the surface that would be obtained by considering each atom to be a sphere with a radius equal to its van der Waal radius (Table 1.5). For technical reasons, though, the surface is more typically an **isodensity surface**, meaning a surface with a constant electron density, such as 0.002 electrons/$Å^2$. The distinction is small. Next, we color the surface according to **electrostatic potential** (i.e., red for negative and blue for positive).

What is electrostatic potential and how is it determined? Imagine taking a very small sphere with a charge of +1 and rolling it around the isodensity surface. At each point, we ask whether the sphere is attracted to or repulsed by the surface and what the energetic magnitude of the interaction is. The magnitude of the interaction is the electrostatic potential. Thus, the plots have units not of charge but of energy (we will use kcal/mol). We are not plotting partial charge; we are plotting electrostatic potential, although the two will track

Partial charges

each other. As a result, these plots are extremely useful for visualizing the charge distribution in organic molecules. We encourage the student to consult Appendix 2 frequently while reading this text.

There are definitely some caveats to the interpretation of electrostatic surface potentials, and some are given in the next Going Deeper highlight. An important caveat is to appreciate that these are electrostatic potential surfaces for the *ground states* of the molecules. They show the charge distribution absent any external perturbation. When a chemical reaction occurs, we expect a substantial reorganization of charge. For example, when an anionic nucleophile adds to the carbonyl of acetone, we expect a very different electrostatic potential surface for the transition state than in the picture of acetone shown in Appendix 2. Since transition states control reactivity, it is risky to use these electrostatic potential surfaces to predict or rationalize reactivity. They can be helpful in this regard, but caution is in order.

Going Deeper

Scaling Electrostatic Surface Potentials

Electrostatic potential surfaces are very useful guides to charge distributions, and they are now commonly shown in introductory organic texts. However, some caution is warranted when interpreting them. Most important is to pay attention to the energy scale associated with any particular structure. That is, what value of positive electrostatic potential does it take to achieve the maximum in blue, and what negative electrostatic potential will maximize the red? Realize that in the analysis of any electrostatic potential surface, positive or negative potentials larger in magnitude than the arbitrarily set range will simply be the most intense blue or red, and will not be distinguished from any other value over the limit.

There are two ways to scale these plots. Some electrostatic potential surface presentations take the direct results of the calculation and use as the maxima the maximum values for plus and minus electrostatic potential in the molecule. This appears to be the most common approach in introductory texts. In this approach, we could end up with a range such as +57.29 kcal/mol to −36.43

kcal/mol. A potential problem with this approach is that the color scale is "linear" from plus to minus, so the zero electrostatic potential color—an important benchmark—will be different for this structure than for some other structure with a range such as +27.22 to −49.83. For this reason, we avoid such presentations in Appendix 2 and present electrostatic potential surfaces with a symmetrical range of electrostatic potentials, such as ±25 kcal/mol or ±50 kcal/mol. In this way, zero electrostatic potential is always the same green color. However, *it is also very important to be aware of the range of electrostatic potentials being plotted*. A plot of the benzene electrostatic potential surface with a ±25 kcal/mol range will look different from a plot with a ±50 kcal/mol range. It is especially risky to compare two structures with different ranges plotted. Whenever possible, we will provide comparisons with the same range, and we will always make it clear what range of electrostatic potentials is used for any figure. The student always needs to be aware of electrostatic potential range when interpreting and comparing such plots.

Inductive Effects

We have seen that when carbon bonds to an electronegative element like O, N, Cl, or F, a bond polarization develops, making the C $\delta+$ and the heteroatom or halogen $\delta-$. We might expect a functional group containing electronegative atoms to also be electron withdrawing (see examples in the margin). The phenomenon of withdrawing electrons through σ bonds to the more electronegative atom or group is called an **inductive effect**. It is an effect that we will often cite to explain trends in thermodynamics or reactivity throughout this book (an interesting twist to induction is given in the next Going Deeper highlight). The inductive effect is what gives rise to bond polarizations, polarizations within molecules, and bond and molecular dipole moments.

A similar but separate phenomenon is a **field effect**. This is a polarization in a molecule that results from charges that interact through space, rather than through σ bonds, and it can influence the structure and reactivity of other parts of the molecule. We will see systems later in which both field and inductive effects seem to be operative.

$$\overset{\delta\oplus}{CH_3}-\overset{\delta\ominus}{NO_2} \quad \overset{\delta\oplus}{CH_3}-\overset{\delta\ominus}{CF_3}$$

Partial charges due to induction

Going Deeper

1-Fluorobutane

Consider the specific case of 1-fluorobutane. We expect a large bond dipole, with C1 positive, and the F negative. Moving along the σ chain we would expect the magnitude of the charges to progressively diminish, with some polarization at C2, and less at C3, etc.

Inductive effects have been extensively investigated experimentally, and some conflicting trends are seen. Advanced quantum mechanical calculations of the sort described in Chapter 14 are able to assign partial charges to atoms or groups of atoms, and while there is some debate as to the best way to do this, all methods produce similar trends. In the case of 1-fluorobutane the F atom carries a large partial negative charge, on the order of −0.44. As expected, the CH_2 group at C1, as a unit, carries a comparable positive charge, creating a very large C–F bond dipole. What about the CH_2 at C2? Instead of being just part of a simple C–C bond, it is paired to a carbon with a substantial positive charge. At the same time, it is two atoms away from a very electronegative element—fluorine.

Perhaps surprisingly, the C2 CH_2 shows a small *negative* charge, on the order of −0.03. Rather than a progressively diminishing positive charge as we move down the chain, a charge alternation is seen. At some levels of theory, this charge alternation continues down the chain, but the magnitude of the charges at C3 and C4 are so small as to be inconsequential.

This result is fairly general in computational studies, but it is not in line with experimental observations of inductive effects. As we will see in Chapter 5 and other places, inductive effects on thermodynamics and kinetics do not usually show an alternation pattern. For example, for a linear alkanoic acid, adding a strongly electronegative element like F to the alkyl chain always increases the acidity of the carboxylic acid functional group, and the effect is always stronger the closer the F is to the incipient carboxylate. This trend is an example of the danger of directly extrapolating ground state electronic structure features to reactivity patterns.

Magnitude of the delta charges
diminishes but alternates

Group Electronegativities

It is often convenient to consider groups that make up particular portions of a molecule as having their own electronegativity. For example, we would expect a CF_3 group to affect the charge distribution in a molecule via induction much more than a CH_3 group, but if we consider only carbon electronegativities, the two are the same. Table 1.2 lists some group electronegativity values that were derived to be comparable to the Pauling scale for atoms. We find that a methyl group is essentially the same as a C, whereas the CF_3 group has an electronegativity similar to that for O. Alkenyl and alkynyl groups are quite electronegative, as are nitro and cyano groups. Finally, a full positive charge, such as that associated with a protonated amine, has the highest group electronegativity.

Table 1.2
Group Electronegativities, Scaled to be Compatible with the Pauling Scale*

Group	Electronegativity
CH_3	2.3
CH_2Cl	2.8
$CHCl_2$	3.0
CCl_3	3.0
CF_3	3.4
Ph	3.0
$CH=CH_2$	3.0
$C\equiv CH$	3.3
$C\equiv N$	3.3
NH_2	3.4
NH_3^+	3.8
NO_2	3.4
OH	3.7

*Wells, P. R. "Group Electronegativities." *Prog. Phys. Org. Chem.*, **6**, 111 (1968).

Hybridization Effects

The relative electronegativities of C and H, a critical issue in organic chemistry, has in fact been the topic of some debate. In Table 1.1 we see that the Pauling scale describes C as more electronegative than H, while the Mulliken scale gives the opposite ordering. It is now believed that the cause of this discrepancy is a hybridization effect. Since *s* orbitals have substantial density at the nucleus while *p* orbitals have a node at the nucleus, the more *s* character in a hybrid orbital, the closer to the nucleus the electrons in that hybrid tend to be. Because electronegativity describes an atom's ability to attract electrons to itself, sp^2 hybrids should be more electronegative than sp^3 hybrids, and this is indeed the case. The data in Table 1.2 are completely consistent with this view. The electronegativities are $C \equiv CH > CH = CH_2 > CH_3; sp > sp^2 > sp^3$.

What does this mean, then, about the relative electronegativities of C and H? A good deal of evidence, including molecular quadrupole moments discussed below, points to the conclusion that an sp^2 C is more electronegative than H, while an sp^3 C and H have very similar electronegativities. In a sense, both Pauling and Mulliken were right. A great many observations, especially those involving noncovalent interactions (Chapters 3 and 4), can be understood from this simple statement.

Taken together, electronegativity and hybridization provide an appealing rationalization of many structural trends. For example, the smaller bond angles in ammonia and water vs. methane discussed previously are nicely explained. There is a competition as to whether the central atom (O or N) should place more *s* character in the hybrid orbital that contains the lone pair(s) or the hybrid orbital used to make bonds to the hydrogen atoms. Since the lone pair electrons are not shared with another atom, an electronegative element prefers greater *s* character in its own lone pair orbitals, because it can better keep these electrons to itself. This effect places more *p* orbital character in the bonds to hydrogen, which in turn reduces the H–N–H or H–O–H bond angles relative to methane. The effect is more pronounced for O because it is more electronegative.

As another example, let's consider methyl fluoride. The H–C–F angle is contracted, and as a result the H–C–H bonds are slightly expanded. The H–C–F contraction is due to the fact that F is the more electronegative substituent. The F prefers to bond to a carbon hybrid that has more *p* character, because it is easier to withdraw electrons from a *p* orbital on carbon than an *s* orbital on carbon. It is often said that *s* orbitals have better **electron penetration** to the nucleus than *p* orbitals, suggesting again that it is harder to withdraw electrons from *s* orbitals. If the carbon uses more *p* character in a hybrid to bond to F, more *s* character will be devoted to the hybrids that comprise the H–C–H bonds. It is difficult to imagine a rationalization of this result using VSEPR, because F is larger than H, and may be expected by VSEPR to open up the H–C–F angle. For the most part, organic structures are better rationalized using hybridization and electronegativity arguments than VSEPR.

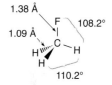

Geometry of CH_3F

1.1.9 Bond Dipoles, Molecular Dipoles, and Quadrupoles

One goal of our discussion of electronegativity was to delineate the relative electron withdrawing nature of an atom, group, or orbital. The term "relative" is important, because the exact numbers associated with atom and group electronegativities are not used on a day to day basis when practicing organic chemistry. Instead, the trends and relative electron donating and accepting abilities are of paramount importance. Now, however, we will be considering bond dipoles, and we need to get more quantitative. The exact charges on the atoms in the bonds need to be known, or the electronegativity numbers associated with atoms need to be used.

Bond Dipoles

When two atoms of differing electronegativites are bonded, one end of the bond will be $\delta+$ and the other will be $\delta-$. This analysis leads to the notion of a **bond dipole** as the local moment that is associated with a polar covalent bond. A **moment** reflects the electrostatic force that would be exerted by a charge on a neighboring charge. The dipole moment pro-

Bond dipole moments

Pulled in this direction

Opposite ends of dipoles attract

vides a means of comparing which bonds are more polar and evaluating the relative force that a dipole exerts on neighboring charges or dipoles. Certainly, examination of the electrostatic potential surfaces of Appendix 2 reveals bond dipoles. If we know the partial charges on the atoms of the bond, we can calculate a bond dipole.

A dipole moment (μ) is given in units of electrical charge times distance (Eq. 1.2), where q is charge and r is distance. It is usually expressed in units of Debye (D, where $1D = 10^{-18}$ esu cm). "Esu" stands for "electrostatic unit", and the charge of an electron or proton is negative or positive 4.80×10^{-10} esu, respectively.

$$\mu = q \times r \qquad \text{(Eq. 1.2)}$$

For example, a bond that has a 0.2 positive and negative charge on the opposite ends with a separation of 1.54 Å (1 Å = 10^{-8} cm) would have an associated bond dipole of $(0.2)(4.80 \times 10^{-10} \text{ esu})(1.54 \times 10^{-8} \text{ cm}) = 1.47 \times 10^{-18}$ esu cm = 1.47 D. Chemists consistently use a symbolism in which the positive end of the dipole is represented by a cross, along with an arrow that points in the direction of the negative end of the dipole.

Recognizing polar covalent bonds and bond dipoles in organic molecules is a great aid to predicting chemical reactivity. Species with partial or full negative charges should be attracted to the $\delta+$ region in a molecule or the positive end of the bond dipole. Conversely, positively charged species would be attracted to the $\delta-$ region of the molecule or the negative end of the bond dipole (see Chapter 10 for the use of these guidelines in predicting reactivity). Such attractions are crucial in controlling weak, noncovalent interactions such as solvation and molecular recognition (see Chapters 3 and 4). Just like electrostatic potential surfaces, molecular and bond dipoles reflect ground states. While the polarization patterns described here can provide valuable clues to reactivity, it is also crucial to consider how bond polarity affects transition states when discussing reactivity.

Molecular Dipole Moments

While the bond dipoles we have just described are, in a sense, conceptualizations, related to our notions of electronegativity, the **molecular dipole** is a well-defined, intrinsic property of a molecule. A molecule has a dipole moment whenever the center of positive charge in the molecule is not coincident with the center of negative charge. This separation of charged centers feeds into Eq. 1.2 also, making it possible to calculate the molecular dipole moment.

Table 1.3 lists experimentally determined dipole moments for select molecules. The numbers tell the relative separation of charges within the molecules, thereby giving an idea of the intensity of the electric field around the molecule. They also give a sense as to how

Table 1.3
Molecular Dipole Values*

Compound	Molecular dipole (D)	Compound	Molecular dipole (D)
CCl_4	0.0	CH_3COCH_3	2.9
$CHCl_3$	1.0	CH_3COOH	1.7
CH_2Cl_2	1.6	CH_3COCl	2.7
CH_3Cl	1.9	CH_3COOCH_3	1.7
CH_3F	1.8	C_6H_5Cl	1.8
CH_3Br	1.8	$C_6H_5NO_2$	4.0
CH_3I	1.6	1-Butene	0.34
CH_3OH	1.7	1-Propyne	0.80
CH_3OCH_3	1.3	cis-2-Butene	0.25
CH_3CN	4.0	cis-1,2-Dichloroethene	1.9
CH_3NO_2	3.4	Tetrahydrofuran	1.6
CH_3NH_2	1.3	Water	1.8

**Handbook of Chemistry and Physics*, CRC Press, Inc., Boca Raton, FL (1979).

strongly an approaching molecule or charge can differentiate one end of the molecule from the other or, alternatively, how favorable a potential electrostatic interaction can be. For example, all approaches to a molecule with a molecular dipole of zero, such as tetrachloromethane, encounter essentially the same electric field. This argument ignores both polarization effects and higher moments such as quadrupoles. In contrast, the electric field felt by a molecule approaching a structure with a dipole of 4.0, such as acetonitrile, is quite different, depending upon the direction of approach. Note that the electrostatic potential surfaces of several small molecules in Appendix 2 provide a clear way to visualize molecular dipole moments.

An often informative exercise is to analyze a molecular dipole as a vector sum of bond dipoles. Examples of this analysis are shown in Figure 1.4. Note that in high symmetry cases all the local bond dipoles cancel and the overall molecule has no molecular dipole. Thus, the absence of a molecular dipole does not rule out the existence of bond dipoles, and the presence of bond dipoles does not guarantee the existence of a molecular dipole.

Several trends emerge from examining Table 1.3. The more chlorines attached to methane, from CH_3Cl to CCl_4, the lower the dipole. This trend might at first seem counterintuitive, because we are progressively adding polar bonds to the system. However, it can be understood as a consequence of vector mathematics, in which the individual bond dipoles increasingly cancel as the number of chlorines increases. The incorporation of nitro or cyano groups into molecules results in very large molecular dipoles when there are no other bond dipoles to cancel them. An important feature of dipole moments is illustrated by the fact that the dipole moments of CH_3Br and CH_3F are the same. We would expect a much larger charge polarization in the C–F bond compared to the C–Br bond, and this is so. However, the C–Br bond is longer than the C–F bond, and even though the charge separation is smaller, the distance is larger. The two phenomena both affect the molecular dipole, and coincidentally lead to the same dipole moment for the two compounds.

Molecular Quadrupole Moments

In a complete description of a molecule's charge distribution, the dipole moment is just one term in a series: monopole, dipole, quadrupole, octupole, hexadecapole, etc. A monopole is just a point charge—the dominant term for ions. For neutral molecules organic chemists usually truncate the series after the dipole. However, the quadrupole moment of a molecule can often be quite important. As such, we take a moment here to remind you about some basic electrostatics.

A **quadrupole** is simply two dipoles aligned in such a way that there is no net dipole (if there was a dipole, we'd have a dipole, not a quadrupole). Interestingly, the multipole expansion follows a familiar topological pattern. Monopoles look like s orbitals (spheres); dipoles look like p orbitals (a + end and a – end); quadrupoles look like d orbitals; octupoles like f orbitals, etc. The analogy between multipoles and orbitals is given just to illustrate phasing properties; orbitals do not have polar character.

Figure 1.5 illustrates this point. The most common quadrupole has the two dipoles side-by-side pointing in opposite directions, giving four charge regions (two + and two −) and the topology of a d_{xy} orbital. This topology is also the arrangement in a quadrupole mass spectrometer. However, there is an alternative arrangement—two end-to-end dipoles point-

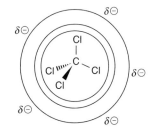

Electric field around CCl_4

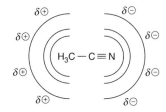

Electric field around acetonitrile

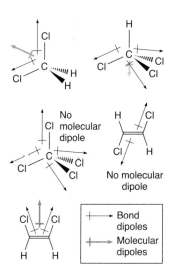

Figure 1.4
Examples of analyzing molecular dipoles as a sum of bond dipoles.

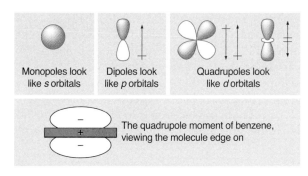

Monopoles look like s orbitals

Dipoles look like p orbitals

Quadrupoles look like d orbitals

The quadrupole moment of benzene, viewing the molecule edge on

Figure 1.5
The topological relationships between atomic orbitals and electrostatic moments, and the quadrupole moment of benzene.

ing in opposite directions—the topology of a d_{z^2} orbital. Actually, this arrangement is more important in organic chemistry, because it is present in benzene (see below). The multipole expansion series—monopole, dipole, quadrupole, etc.—is *not* a perturbation series. It is not true that quadrupoles are somehow intrinsically weaker than dipoles in electrostatic interactions. In fact, in some important organic molecular recognition phenomena, quadrupoles prove to be stronger than dipoles (see Chapter 3).

The most common and important quadrupole moment in organic chemistry is that of benzene. Experimental measurements have determined that benzene has a large quadrupole moment, with a charge distribution as in Figure 1.5 (see Appendix 2). Just as with molecular dipole moments, we can rationalize a molecular quadrupole moment as a sum of bond dipoles. In this case, we add six $C^{\delta-}$–$H^{\delta+}$ dipoles to get the molecular quadrupole. The existence of a large, permanent quadrupole moment in benzene is *unambiguous proof* that an sp^2 C is more electronegative than H. We must have six $C^{\delta-}$–$H^{\delta+}$ dipoles to explain the effect. Note that cyclohexane has a negligible quadrupole moment, indicating that an sp^3 C and H have similar electronegativities.

1.1.10   Resonance

The bonding model we have developed thus far is quite "classical", relying on fairly simple notions, such as Lewis structures. Some structures, however, cannot be adequately described by a single Lewis structure. In these cases, two or more Lewis structures are drawn, and the actual molecule is a hybrid or mixture of these **resonance structures**. The superposition of two or more Lewis structures to describe the bonding in a molecule is called **resonance** (also known as **mesomerism** in very old literature).

A classic example of resonance occurs for acetate. Two structures showing different positions for the double bond and the negative charge are possible. In this case the two structures are identical, and the charge on each oxygen is $-\frac{1}{2}$. Another familiar case is benzene, which also involves two equivalent structures, so that the C–C bond length is appropriate for a bond order of 1.5. All six bonds are equivalent and are represented by two equivalent resonance structures (called **Kekulé structures**).

Kekulé structures

Acetate resonance structures

Resonance structures are not separate molecules that are interconverting. There is really only one structure, which is best thought of as a hybrid of the various resonance structures. The two C–O bonds of acetate are equivalent and the negative charge is distributed equally between the two oxygens. Often, one symbolizes a combination of the resonance structures by a single structure meant to describe the hybrid.

Although the examples of acetate and benzene are ones in which the resonance structures are equivalent, this is not usually the case. For example, *p*-nitrophenol and methylvinylketone also have reasonable resonance structures, but they are significantly different in the arrangement of the bonds, lone pairs, and charges.

The picture of resonance implies that the electrons are covering a larger number of atoms than given by any one resonance structure, and this is defined as **delocalization**. Generally speaking, the more resonance structures that a molecule has and the more reasonable these structures are, the more stable the molecule. The energy of stabilization imparted by resonance is called the **resonance energy** or **delocalization energy**. The reason that a molecule with resonance structures is considered more stable is the effect of delocalization. As we will see in Chapter 14, the more spread out the orbital that electrons occupy, the lower the energy of those electrons. This is related to a calculation often covered in physical chemistry classes, called the "particle-in-a-box" calculation, which we briefly review in the following Going Deeper highlight.

Acceptable resonance structures

Going Deeper

Particle in a Box

The manner in which one finds the energy of electrons is to solve the Schrödinger equation ($H\Psi = E\Psi$). As a very simple example, imagine an electron in a one-dimensional "box". Here, the potential energy of the electron is zero if the electron is within the box, but is infinite at the edges and beyond the edges of the box. The potential energy cannot be infinite, so the electron is confined to the region within the box. The solutions (E_n) to the Schrödinger equation for this scenario are very simple, and take the form shown (consult any undergraduate physical chemistry textbook to see how the solutions are derived). The possible energies are quantized (n is an integer, 1, 2, 3, . . .), with the length of the box (L) in the denominator (m is the mass of the electron and h is Planck's constant). As the box gets bigger, the energy decreases.

The "box" is an analogy to an orbital. With an orbital the electrons have their greatest probability in certain regions of space. The lesson is that if the electrons are allowed to occupy a larger amount of space, their energy decreases. Specifically, the kinetic energy of the electrons drops, which will be a key issue we discuss in Chapter 14. Because resonance yields a picture of bonding that spreads the electrons out in space, it is a stabilizing concept.

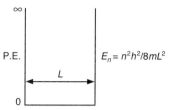

$$E_n = n^2h^2/8mL^2$$

Parameters for the particle in a box

In order to consider what are appropriate resonance structures for a molecule, we first draw all the possible Lewis structures. In these structures only electrons are allowed to "move around". *The positions of the nuclei never change*. The Lewis structures can have the maximum number of electrons appropriate for each atom (for example, eight for second-row atoms) or fewer electrons. Next, judgment must be made about which resonance structures are reasonable, and Figure 1.6 gives some guidance. Factors that contribute to making a particular resonance structure acceptable include having a noble gas configuration for the atoms, a maximum number of covalent bonds, a minimum number of like charges, close proximity of unlike charges, and placement of negative charges on electronegative atoms. Not all these guidelines need to be met to make a reasonable resonance structure. Many reasonable structures may contribute only a little to the true electronic structure of the molecule, depending upon just how reasonable they are. Yet, the identification of all reasonable resonance structures imparts information about polarity and polarizations in a molecule. As already mentioned, it is generally true that the larger the number of reasonable resonance structures associated with a molecule, the more stable it is. In addition, resonance is especially favorable when it involves two or more equivalent resonance structures (as, for example, with acetate and benzene), and when all second-row atoms have a full complement of eight valence electrons.

Reasonable resonance structures   Unreasonable resonance structures in shading

Figure 1.6
Reasonable and unreasonable resonance structures.

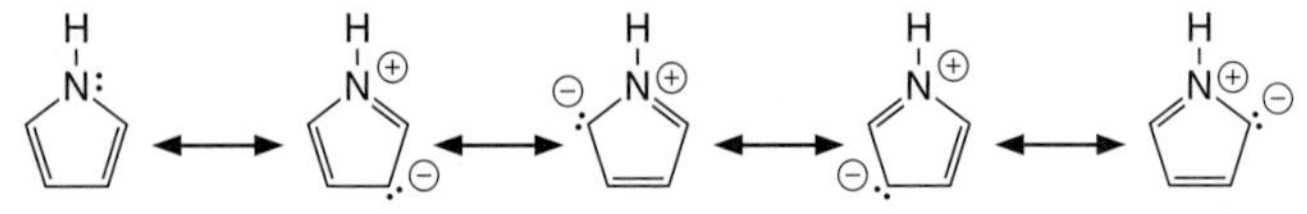

Resonance structures of pyrrole

Often we will see an atom adopt a nonstandard hybridization in order to maximize resonance. This most commonly arises when an atom has a lone pair of electrons that is in conjugation with (directly bonded to) a π system. For example, consider pyrrole, shown above. The nitrogen atom would seem at first glance to best be described as an sp^3 hybrid, because four groups are attached to it: two N–C bonds, an N–H bond, and a lone pair. However, in order to accommodate the resonance structures shown, four of which have a double bond between N and a C, the lone pair must be in a p orbital, not an sp^3 hybrid orbital. This requirement makes the nitrogen atom sp^2 hybridized, which is experimentally supported by the fact that the N is trigonal planar. Such resonance effects on hybridization are common and should be routinely looked for when assigning hybridization to various atoms. Another example is the hybridization of N in an amide, which should be classified as sp^2. However, as shown in the following Connections highlight, the validity of resonance in an amide has recently come into question.

1.1.11   Bond Lengths

The simple bonding model we have developed thus far can rationalize many geometrical features. We have already discussed bond angles, and most organic chemists know what standard angles are for organic structures. It is generally true that it is more difficult to distort bond lengths from standard values than it is to bend angles. Nevertheless, it is worthwhile to know some standard bond lengths, as a significant deviation from these values is a clear indication of substantial strain in a molecule or some non-standard bonding situation (see Chapter 2). In addition, we need to know bond lengths to evaluate bond and molecular dipoles.

Several trends can be gleaned from the series of average bond lengths listed in Table 1.4.

Table 1.4
Typical Bond Lengths of Some Covalent Bonds*

Bond		Length (Å)		Bond		Length (Å)
Single bonds				**Double bonds**		
$C(sp^3)$–$C(sp^3)$		1.53–1.55		$C(sp^2)$–$C(sp^2)$	alkenes	1.31–1.34
$C(sp^3)$–$C(sp^2)$		1.49–1.52		$C(sp^2)$–$C(sp^2)$	arenes	1.38–1.40
$C(sp^2)$–$C(sp^2)$	conjugated	1.45–1.46		$C(sp^2)$–$O(sp^2)$	aldehydes and ketones	1.19–1.22
	nonconjugated	1.47–1.48		$C(sp^2)$–$O(sp^2)$	esters	1.19–1.20
$C(sp)$–$C(sp)$		1.37–1.38		$C(sp^2)$–$O(sp^2)$	amides	1.225–1.24
$C(sp^3)$–$O(sp^3)$	ethers	1.42–1.44		$C(sp^2)$–$N(sp^2)$	imines	1.35
$C(sp^3)$–$N(sp^3)$	amines	1.46–1.48		**Triple bonds**		
C–F		1.39–1.43				
C–Cl		1.78–1.85		$C(sp)$–$C(sp)$	alkynes	1.17–1.20
C–Br		1.95–1.98				
C–I		2.15–2.18				
$C(sp^3)$–H		1.09–1.10				
$C(sp^2)$–H		1.075–1.085				
$C(sp)$–H		1.06				
N–H		1.00–1.02				
O–H		0.96–0.97				

*Allen, F. H., *et al.* "Tables of Bond Lengths Determined by X-ray and Neutron Diffraction. Part 1. Bond Lengths in Organic Compounds." *J. Chem. Soc., Perkin Trans. II*, S1–S19 (1987).

Connections

Resonance in the Peptide Amide Bond?

A foundation of structural biology and bioorganic chemistry is the peptide bond, the link between consecutive amino acids in a protein. Interestingly, the modern view of this prototype structure is evolving, and as such it provides an excellent example of various rationalizations of structure and bonding.

The **peptide bond** is an amide formed between a primary amine and a carboxylic acid. The most important structural features are (a) the amide group is planar, with a substantial barrier to rotation (typically in the range of $\Delta G^{\ddagger} = 15–20$ kcal/mol); and (b) there is a significant preference for the Z conformer (termed "trans" in the protein literature), which places the N–H and C=O bonds trans to each other. Another key feature of amides is their strong propensity toward hydrogen bonding. Two key protein secondary structural elements—the α-helix and the β-sheet—depend on the amide group to act as both a hydrogen bond donor and a hydrogen bond acceptor.

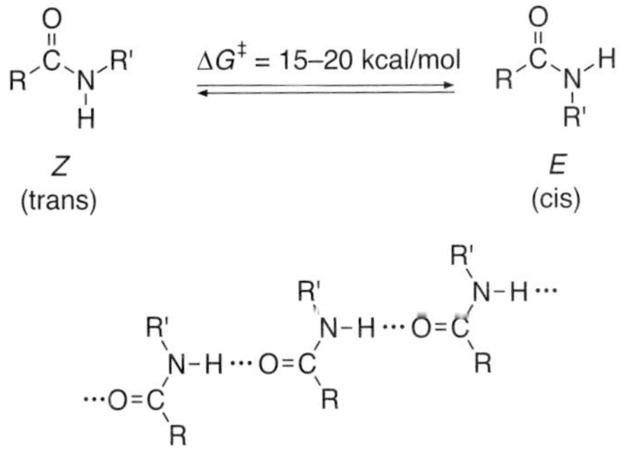

Features of amides

In the traditional model of an amide, resonance is the key concept. As shown below, one can write a reasonable resonance structure for an amide that places a double bond between the C and the N (structure **B**). This "double-bond character" leads to a planar structure, and hindered rotation about the C–N bond. To the extent that electrostatic interactions control hydrogen bond strengths (see Chapter 3), the charges implied by resonance structure **B** suggest strong hydrogen bonding in amides, as is observed.

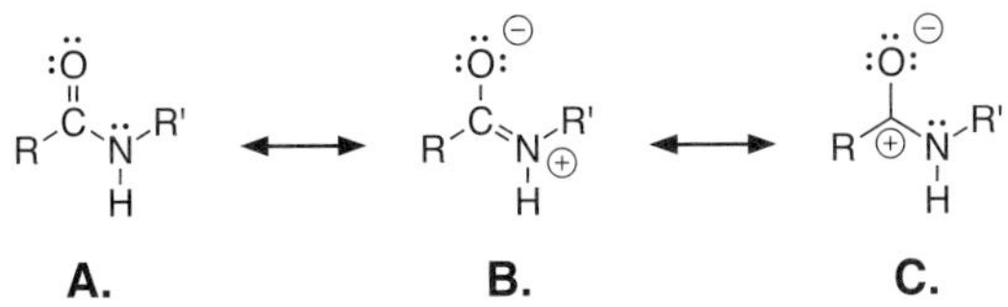

Amide resonance structures

More recent work, however, has suggested a refinement of the traditional model. Modern computational methods allow detailed analyses of molecular and electronic structure. These tools were used to look in detail at amide rotation. The classical resonance model predicts C–N double-bond character and C–O single-bond character in a planar amide. Since resonance is destroyed on rotating 90° about the C–N bond, we would expect the C–N bond to lengthen and the C–O bond to shorten in the perpendicular form. Calculations do show that the C–N bond lengthens upon rotation, but the C–O bond length is essentially unchanged. Also, examination of the charges calculated on the atoms of a typical amide do not support structure **B**. The N of an amide is not significantly more positive than the N of an amine, for which such resonance is not possible. Also, the O of an amide is not significantly more negative than the O of a ketone. Finally, the C of an amide is quite positive, just like the C of a ketone.

These observations suggested that the simple two-structure resonance model was inadequate. Instead three structures seem necessary, as shown. Both resonance structures **B** and **C** are considered to be of major importance for the rotation barrier. In this view, the role of the oxygen is to polarize the C–O bond, introducing a large $\delta+$ on carbon (resonance structure **C**). The N does not develop a significant $\delta+$, despite the implications of resonance structure **B**. Instead, this model emphasizes dipole interactions—the C–O dipole is aligned for a favorable interaction with the N–H bond dipole. Note that a $\delta+$ on N would not be consistent with this picture.

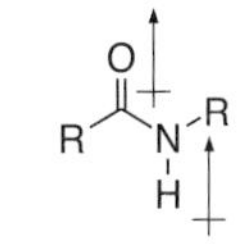

Amide bond dipoles

This refined resonance model was inspired by high-level calculations, and it reflects changing views on the importance of bond dipoles (i.e., $C^{\delta+}–O^{\delta-}$) in structure and reactivity. Bond dipoles are also important in rationalizing the Z–E preference. There are two strong bond dipoles in a secondary amide: $C^{\delta+}–O^{\delta-}$ and $H^{\delta+}–N^{\delta-}$. As shown, the Z form aligns these two dipoles favorably (the positive end near the negative end), but the E form would produce an unfavorable alignment. Finally, the strong propensity of the peptide bond toward hydrogen bonding is readily understood based on the charge distribution implied by the resonance and bond dipole arguments.

Wiberg, K. B. "The Interaction of Carbonyl Groups with Substituents." *Acc. Chem. Res.*, **32**, 922–929 (1999).

One obvious trend is that multiple bonds tend to be shorter than single bonds. The bond lengths in arenes are between C–C and C=C bond lengths, just as predicted by resonance theory. Bonds involving atoms with larger atomic radii (further down in a column and / or to the left within a row in the periodic table) are longer.

The state of hybridization affects bond lengths; more *s* character in the hybrid decreases the bond length. This effect is another manifestation of the fact that *s* orbitals lie closer to the nucleus than *p* orbitals. This last effect is sometimes under appreciated. For example, the fact that the C2–C3 bond of 1,3-butadiene is much shorter than a typical single bond in an alkane, 1.48 Å vs. 1.54 Å, might lead to the conclusion that there is some double-bond character in C2–C3 (see the resonance structure in the margin, which is sometimes taught in introductory organic chemistry). However, Table 1.4 shows that this is primarily a hybridization effect, as systems that are not planar (nonconjugated and hence have no double-bond character) but are sp^2–sp^2 single bonds, show comparably short bond lengths. Further support that the conjugation in butadiene has only a small effect on the C2–C3 bond length comes from the rotation barrier about this bond, which is only on the order of 4–5 kcal/mol (see Chapter 2 for an expanded discussion of rotational barriers).

Bond lengths can also be understood with reference to the radii of the constituent atoms. There are actually three different kinds of radii that we use to understand molecular dimensions: **covalent, ionic**, and **van der Waals**. The covalent radius is half the distance between two identical atoms bonded together. For example, the C–C bond length in ethane is 1.54 Å, and therefore the covalent radius of C is 0.77 Å. The ionic radius of an atom is its size as determined in the crystal lattice of various salts. The radius depends upon the charge on the ion, and is smaller as the positive charge increases and larger as the negative charge increases. Finally, the van der Waals radius of an atom is the effective size of the electron cloud around an atom when in a covalent bond, as perceived by an atom to which it is *not* attached. The measurement comes from an analysis of crystal packing, and effectively sets the steric size of an atom or a group. Table 1.5 shows just a few covalent, ionic, and van der Waals radii. There is some debate as to the exact values of some of these, especially the van der Waals radii, but the basic trends are clear. Using these values molecular surface areas and molecular volumes can be calculated.

Butadiene resonance structures

Table 1.5
Covalent, Ionic, and van der Waals Radii of Select Atoms (Å)*

Atom	Covalent	VDW	Ion	Ionic
C	0.77	1.68		
H	0.30	1.11	H⁻	2.08
N	0.70	1.53		
O	0.66	1.50		
F	0.64	1.51	F⁻	1.36
Cl	0.99	1.84	Cl⁻	1.81
Br	1.14	1.96	Br⁻	1.95
I	1.33	2.13	I⁻	2.16

*Pauling, L. (1960). *The Nature of the Chemical Bond and the Structure of Molecules and Crystals; an Introduction to Modern Structural Chemistry,* 3d ed., Cornell University Press, Ithaca, NY.

1.1.12   Polarizability

An important property of molecules that we have yet to discuss is polarizability. Since electrons are charged particles they respond to electric fields. In particular, electrons in molecules are mobile to varying degrees, and so their positions can shift to differing extents in response to an applied electric field. The electron cloud is said to become polarized in response to the electric field. The ability of the electron cloud to distort in response to an external field is known as its **polarizability**. Upon distortion, a dipole is typically induced in the molecule, adding to any permanent dipole already present.

Table 1.6
Atomic and Molecular Polarizabilities (α, cm^3/10^{-24})*

Atomic polarizabilities

H	0.6668											He	0.205
		C	1.76	N	1.10	O	0.802	F	0.557				
				P	3.13	S	2.90	Cl	2.18				
								Br	3.05				
								I	4.7 (or 5.35)				

Selected molecular polarizabilities

CH_4	2.6	NH_3	2.21	H_2O	1.45	H_2S	3.8
CO_2	2.91	CS_2	8.8	CF_4	3.84	CCl_4	11.2
C_2H_2	3.6	C_2H_4	4.25	C_2H_6	4.45	CH_3OH	3.23
Benzene	10.32	Cyclohexene	10.7	Cyclohexane	11.0		

*CRC Handbook of Chemistry and Physics, D. R. Lide (ed.), CRC Press, Inc., Boca Raton, FL (1990–1), pp. 10-193–10-209.

We define the polarizability of a molecule as the magnitude of the dipole induced by one unit of field gradient, which works out to be in units of volume. Often, the larger the volume occupied by the electrons, the more polarizable those electrons. Since an electric field gradient is directional, it can encounter a molecule or bond in different ways depending upon the relative orientation between the field and the molecule or bond. Hence, polarizabilities are often broken down into one longitudinal (along the bond or molecular axis) and two transverse (perpendicular to the axis) values. Table 1.6 lists some atomic and molecular polarizabilities, for which the directional components have been averaged. Just as with permanent dipoles, the induced dipoles that arise from such polarizations can be analyzed as molecular dipoles or as bond dipoles. We can thus speak of molecular/atomic polarizabilities, as well as of bond polarizabilities.

Several trends are evident in Table 1.6. First, as we move left to right across a row of the periodic table, polarizability decreases. This trend is clear in atomic polarizabilities (C > N > O > F) and molecular polarizabilities (CH_4 > NH_3 > H_2O). Electronegativity plays an important role. Atoms that hold on to their electrons tightly are not polarizable.

Polarizability has profound consequences. Water is a very polar molecule—it has a large dipole moment. However, methane is much more polarizable than water, and alkanes in general are among the most polarizable of molecules. Thus, while aqueous media provide a very polar environment, alkanes and other hydrophobic environments are more polarizable.

The second clear trend in Table 1.6 is that as we move down a column in the periodic table, polarizability increases substantially (S > O, P > N, and H_2S > H_2O). This observation explains an earlier apparent contradiction. According to Table 1.1, I is not significantly more electronegative than C, yet C–I bonds are much more reactive than C–Cl bonds in reactions like S_N2 and E2. In such instances, $C^{\delta+}$–$X^{\delta-}$ polar bonds are often invoked, but C–I bonds are not very polar. However, dramatic changes in molecular and electronic structure occur in the course of a reaction. A model for just the ground state of the reactant is often inadequate. When an anionic nucleophile approaches a C–X bond for an S_N2 reaction, it can induce a large bond dipole, especially if X is highly polarizable. Thus, the large polarizability of I makes up for the lower electronegativity, and the C–I bond is more reactive than C–Cl.

Another perhaps surprising point is revealed when considering the polarizabilities of the simple hydrocarbons given in Table 1.6. Because of their greater reactivity, most chemists might assume that alkenes are more polarizable than alkanes, but the opposite is true. Once again, alkanes are among the most polarizable of molecules. In particular, ethane is significantly more polarizable than ethylene. Perhaps even more surprisingly, cyclohexane is more polarizable than benzene. While perhaps counterintuitive, these trends are consistent with electronegativity arguments. Alkenes and arenes, with the more electronegative sp^2 carbons, are less polarizable than alkanes with only sp^3 carbons.

1.1.13 Summary of Concepts Used for the Simplest Model of Bonding in Organic Structures

The material presented to this point has been mostly a review of concepts presented in introductory organic chemistry classes. Only in a few cases was the discussion extended further. This review sets the stage for us to go deeper into bonding and its effects on structure, and a deeper analysis is what we present in the remainder of this chapter. First, however, it is instructive to summarize the concepts used in the simplest picture of bonding in organic structures; many of these concepts extend nicely into our second approach to bonding.

Hybridization.   Atoms contribute consistent sets of hybrid atomic orbitals to the bonding in a molecule. A carbon with four ligands is sp^3 hybridized, a C with three ligands is sp^2 hybridized, and a C with two ligands is sp hybridized.

σ and π Bonds.   Hybrid orbitals on two separate atoms overlap in the region between the adjacent atoms to create σ bonds. The overlap of p orbitals on adjacent atoms creates π bonds. The localized bonds consist of bonding and antibonding molecular orbitals, of which only the bonding orbitals are populated with electrons. This arrangement of bonds is a VBT notion, with discrete and localized bonds between adjacent atoms, but the orbitals are created using MOT mixing notions.

Resonance.   For certain molecules, a single valence bond structure cannot properly describe the bonding, so resonance is involved. Resonance structures show various arrangements of electrons within a structure, where each contributes to the bonding arrangement in that molecule. Resonance structures can also be used to suggest subtle features of the electronic structure of functional groups.

Electronegativity and bond polarization.   Electronegative elements pull electron density toward themselves. This introduces polarity into bonds, resulting in bond dipoles and molecular dipoles.

Induction.   Electronegativity and polarization effects can be felt a few bonds away through an inductive effect.

Polarizability.   Polarizability reflects the extent to which electrons can be perturbed from the standard bonding arrangement in response to the approach of another molecule. This effect is important in understanding solvent properties and many reactivity patterns.

Sterics.   Two or more groups tend to stay away from each other due to adverse electrostatic repulsion between filled orbitals.

We draw upon each of these notions as necessary to explain various aspects of bonding, reactivity, and structure. We will add solvation effects later in the book. For particular reactions or structures, we may have to refine and/or combine these notions to get the optimal model for the molecule. In some cases we may even need to completely re-think and modify these foundations. However, there is often a tendency in physical organic chemistry to become quite focused on the "exceptions to the rule". We should keep in mind that the vast majority of organic structures are well described by this simple model.

1.2   A More Modern Theory of Organic Bonding

In Section 1.1 we described the basic bonding patterns of organic molecules and the properties of different types of localized bonds. The concepts introduced were mostly fairly classical valence bond concepts. They have definite predictive power, and that is an important measure of the value of any model. However, the picture we are about to present also has the same predictive power, but can better explain certain structural issues and experimental observations. In addition, understanding chemical reactivity poses a serious challenge, and this second approach to bonding is very good at predicting the reactivity patterns of organic molecules.

The highest extent of rigor that can be given for bonding is presented in the advanced quantum mechanical analysis that was developed in the first half of the 20th century. Modern calculational methods now provide accurate representations of the molecular orbitals not only of stable molecules, but also of reactive intermediates and even transition states. These powerful computational methods will be described in detail in Chapter 14. Our goal in this chapter is to develop an *understanding* of chemical bonding, and the detailed numerical output of a quantum mechanical calculation is not enough. Here, we will show that certain key concepts and trends that result from the output of such calculations lead to a more rigorous descriptive model of organic bonding than given in the first part of this chapter. Molecular orbital theory (MOT) forms the core of this second model. However, we will see that certain key concepts from the valence bond theory based model for bonding will still be very useful (e.g., sterics, induction, and polarizability), and in fact the distinctions between the two theories are often blurred. Therefore, we again will present a hybrid VBT/MOT model, but now more tilted toward MOT.

Although the starting points for VBT and MOT are different, and the initial mathematics used to explain them are different, both theories can be shown to give similar results when correction factors are added and their respective mathematics are solved completely in a manner consistent with quantum mechanical rules. In addition, straightforward and theoretically justifiable mathematical operations can, in most cases, transform the fully or partially delocalized molecular orbitals we are about to examine into localized orbitals that clearly resemble our VBT notions of discrete and localized bonds. MOT and VBT are not as far apart as they may seem. Hence, one theory is not necessarily *more correct* than the other. Also, there is nothing wrong with a hybrid approach to bonding. Both MOT and VBT are approximations to the "true" answer—a full solution to the Schrödinger equation. A combination of the two is no less approximate, as long as it is thoughtfully applied.

1.2.1   Molecular Orbital Theory

In contrast to VBT, "full-blown" MOT considers the electrons in molecules to occupy molecular orbitals that are formed by linear combinations (addition and subtraction) of *all* the atomic orbitals on *all* the atoms in the structure. In MOT, electrons are not confined to an individual atom plus the bonding region with another atom. Instead, electrons are contained in MOs that are highly delocalized—spread across the entire molecule. MOT does not create discrete and localized bonds between neighboring atoms. An immediate benefit of MOT over VBT is its treatment of conjugated π systems. We don't need a "patch" like resonance to explain the structure of a carboxylate anion or of benzene; it falls naturally out of the delocalized nature of the MOs. The MO models of simple molecules like ethylene or formaldehyde also lead to bonding concepts that are pervasive in organic chemistry.

On the other hand, the fully delocalized view of MOT that is so useful for certain molecules like benzene is not so useful in other organic molecules. For example, if we want to anticipate the reactivity of 3-heptanone, we really don't need to consider orbitals that span all seven carbons plus the oxygen. Chemical experience tells us that the action is at the carbonyl, and we want to be able to focus on the π system and lone pairs. In order to do so, we will use an MO approach to establish certain fundamental bonding principles by studying small prototype molecules—ethylene for learning about C=C bonds, formaldehyde for C=O, acetic acid for carboxylic acids, etc. This analysis gives us the MOs for these functional groups. Fortunately, analysis of the computed MOs of complex systems shows that the MOs of the smaller model molecules appear with only minor modifications in the larger molecules. In other words, we can concentrate our analysis on the molecular orbitals of functional groups to understand structure and reactivity. Thus, a key concept we present below is that of **group orbitals**—orbitals that are delocalized only over a defined group of atoms. If we need to understand the molecular orbitals of the entire molecule, we combine these group orbitals to create molecular orbitals for the entire molecule, and the rules for this orbital mixing procedure are presented in the following discussions.

To create group orbitals or delocalized molecular orbitals, we need to understand how to combine atomic orbitals properly. Therefore, the starting point in developing our second model of organic bonding is a set of rules that lead by inspection to group orbitals and molecular orbitals. This procedure is called qualitative molecular orbital theory (QMOT).

1.2.2   A Method for QMOT

The protocol we will follow was developed by many workers, including Hoffmann and Salem. An especially succinct statement of the procedure embodied in a series of rules was provided by Gimarc, and is presented in Table 1.7. Below we give several examples of using rules 1–13, where rules 14 and 15 become more important in future chapters. Several MOT and quantum mechanical concepts go into the origin of these rules. For example, the orbital mixing follows the well-developed rules of perturbation theory, which we will often discuss (glance back to Section 1.1.7 for our first quick look at mixing). Furthermore, the symmetry of the molecule (see the next Connections highlight) guides the creation of the MOs, a notion that has a quantum mechanical origin discussed in Chapters 14 and 15.

Table 1.7
The Rules of QMOT*

1. Consider valence orbitals only.

2. Form completely delocalized MOs as linear combinations of s and p AOs.

3. MOs must be either symmetric or antisymmetric with respect to the symmetry operations of the molecule.

4. Compose MOs for structures of high symmetry and then produce orbitals for related but less symmetric structures by systematic distortions of the orbitals for higher symmetry.

5. Molecules with similar molecular structures, such as CH_3 and NH_3, have qualitatively similar MOs, the major difference being the number of valence electrons that occupy the common MO system.

6. The total energy is the sum of the molecular orbital energies of individual valence electrons.

7. If the two highest energy MOs of a given symmetry derive primarily from different kinds of AOs, then mix the two MOs to form hybrid orbitals.

8. When two orbitals interact, the lower energy orbital is stabilized and the higher energy orbital is destabilized. The out-of-phase or antibonding interaction between the two starting orbitals always raises the energy more than the corresponding in-phase or bonding interaction lowers the energy.

9. When two orbitals interact, the lower energy orbital mixes into itself the higher energy one in a bonding way, while the higher energy orbital mixes into itself the lower energy one in an antibonding way.

10. The smaller the initial energy gap between two interacting orbitals, the stronger the mixing interaction.

11. The larger the overlap between interacting orbitals, the larger the interaction.

12. The more electronegative elements have lower energy AOs.

13. A change in the geometry of a molecule will produce a large change in the energy of a particular MO if the geometry change results in changes in AO overlap that are large.

14. The AO coefficients are large in high energy MOs with many nodes or complicated nodal surfaces.

15. Energies of orbitals of the same symmetry classification cannot cross each other. Instead, such orbitals mix and diverge.

*Adapted, with modifications, from Gimarc, B. M. (1979). *Molecular Structure and Bonding: The Qualitative Molecular Orbital Approach*, Academic Press, New York.

Connections

A Brief Look at Symmetry and Symmetry Operations

It is clear from Table 1.7 that symmetry plays an important role in QMOT because the concept is used in rules 3, 4, 7, and 15. A full explanation of symmetry operations, point groups, and their relationships to orbitals is beyond the scope of this book (see the reference at the end of this highlight for an excellent discussion). However, we will look at proper and improper rotations in Chapter 6. To understand the QMOT examples discussed below, you should be aware of the symmetries inherent in only two geometries: trigonal planar with all three groups the same, and tetrahedral where one group is different than the other three (also called **pyramidal**). Hence, we discuss these two systems briefly.

A trigonal planar structure with three equivalent ligands belongs to the point group called D_{3h}. The structure possesses a C_3 axis perpendicular to the plane of the molecule that passes through the central atom. It also possesses three C_2 axes perpendicular to the central C_3 axis, one along each M–H bond vector (only one is shown to the right). To possess a C_3 **axis** and a C_2 **axis** means that the molecule can be rotated along this axis by 120° and 180°, respectively, returning exactly the same structure with atoms returned to identical positions in space. A D_{3h} structure also possesses three internal mirror planes, called σ **planes**. They contain the C_3 axis and lie along each M–H bond (only one is shown), where reflection of the atoms through these mirror planes likewise gives back the same structure.

The MH_3Y structure belongs to the point group called C_{3v}. This structure possesses one C_3 axis, no C_2 axes, but does have three σ planes (one defined by each H–M–Y plane; only one is shown).

Rules 3 and 7 address the symmetry properties of the orbitals. With respect to each symmetry operation, an orbital must be symmetric or antisymmetric. For an orbital to be symmetric, it must be unchanged by the sym-

metry operation. To be antisymmetric, all signs (phases) of the orbital must be reversed by the symmetry operation. Rule 4 speaks of "high symmetry" and "less symmetric structures". While these are not precisely defined terms, their meaning in context is usually clear. Certainly, the D_{3h} structure is of higher symmetry than the C_{3v}. Rule 15 is not one we are going to need until Chapter 15, so it will be addressed at that time. The use of these rules and the notions of symmetry will hopefully become evident with the examples given below, and with the practice given in the end-of-chapter Exercises.

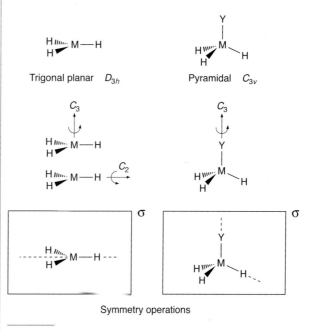

Symmetry operations

Cotton, F. A. (1971). *Chemical Applications of Group Theory*, 2nd ed., Wiley Interscience, New York.

1.2.3 Methyl in Detail

To illustrate the procedures of QMOT, we will "build" the MOs of planar CH_3 from scratch. We are choosing a reactive species such as this, because we want a simple structure as an easy starting point. To begin with, we will not worry as to whether we have methyl cation, anion, or radical, for, as we will describe later, the orbital diagram is not overly sensitive to such distinctions. For now, let's just make the MOs of planar CH_3; we'll put the electrons in later. The starting point is planar CH_3, because pyramidal CH_3 is of lower symmetry (rule 4).

Planar Methyl

Because of rule 1, the orbitals we have to work with for methyl are simple: three hydrogen $1s$ orbitals, a carbon $2s$ orbital, and three carbon $2p$ orbitals. Next, we form the delocalized MOs using the other rules, and those are shown in Figure 1.7. How this is done may at first seem mysterious, but with a little practice it should become clear. For example, using

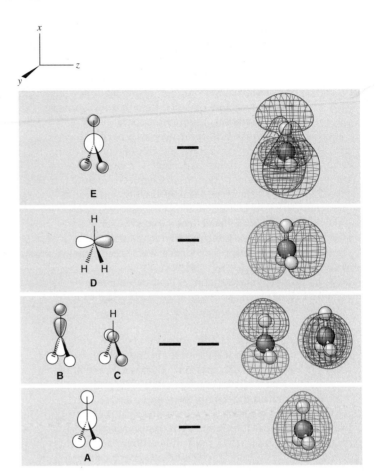

Figure 1.7
The orbitals of a planar methyl group, created using the rules of Table 1.7.

rules 2 and 3, we mix the carbon $2s$ orbital with the three hydrogen atomic orbitals (AOs) in-phase to produce orbital **A**. Mixing just means adding and subtracting, and we do this visually. Orbital **A** is symmetric with respect to the C_3 axis in the molecule. Our goal here is to focus on the low-lying, bonding MOs, so we mix these s orbitals in phase, in a bonding manner. If we mix out of phase, we create a high energy, antibonding orbital, shown as **E** in Figure 1.7. Next come the carbon p orbitals. The p_z AO cannot mix with any of the hydrogens, as they all lie on the node of this orbital. We thus have a molecular orbital that is just an atomic p orbital, called a nonbonding orbital, and labeled as **D** in Figure 1.7. The p_x and p_y AOs can mix with the hydrogens to give favorable interaction patterns, as shown in **B** and **C** in the figure. We would call orbitals **A, B, C,** and **D** the group orbitals for planar methyl.

To create the group orbitals for planar methyl we used a total of seven atomic orbitals. There must be a conservation of the number of orbitals, meaning that the number of molecular orbitals created must equal the number of atomic orbitals we start with. Yet, in Figure 1.7 we only show five orbitals. Orbital **E** and the two MOs that we do not show are called **virtual orbitals**, meaning that they exist but are not typically populated with electrons (see the discussion of adding electrons below). We do not draw two of them because they are not involved in bonding schemes, except with excited electronic states.

We also show the relative energies of these MOs in Figure 1.7, with energy increasing as we go from the bottom of the diagram to the top. Orbitals **A–C** are stabilized by mixing with the H AOs, but **D** is not. The reason **A** lies below **B** and **C** is that carbon $2s$ AOs are lower in energy than $2p$ AOs, and this carries over to the MOs. MOs **B** and **C** are of equal energy—they are degenerate. In this high symmetry system, the x and y axes are equivalent, and so p_x and p_y are degenerate, as are the MOs derived from them. If you are uncomfortable with this symmetry argument, please simply accept this degeneracy as fact for now.

Figure 1.7 also shows the actual MOs of CH_3^+, obtained from a quantum mechanical calculation. As you can see, our simple qualitative analysis has predicted the essence of the full calculation quite well. You don't actually see the individual s and p atomic orbitals, but you see the results of mixing them. Deciphering the origin of the calculated orbitals is a major goal of our analysis. We want our QMOT model to faithfully capture the essence of the actual computed results, and we will be constantly checking our qualitative reasoning against quantitative calculations to be sure we are getting things right. If the QMOT rules prove effective in reproducing calculated structures, it gives us confidence to use the method to predict other orbitals, too.

The Walsh Diagram: Pyramidal Methyl

Once we have the orbitals for a fragment, we can predict how geometrical distortions will affect the MOs, again using the rules from Table 1.7. We make the MOs of a pyramidal methyl group by distorting the planar system and following the changes as indicated in rule 4. A diagram that follows orbital energies as a function of angular distortions is called a **Walsh diagram**, named after A. D. Walsh, the theoretician who first devised the approach. Figure 1.8 shows the Walsh diagram for CH_3, along with the results of a calculation on pyramidal methyl.

We begin with the planar system we just developed, and it is reproduced on the left side of Figure 1.8. On pyramidalization, the energy of **A** will change very little. Perhaps a slight

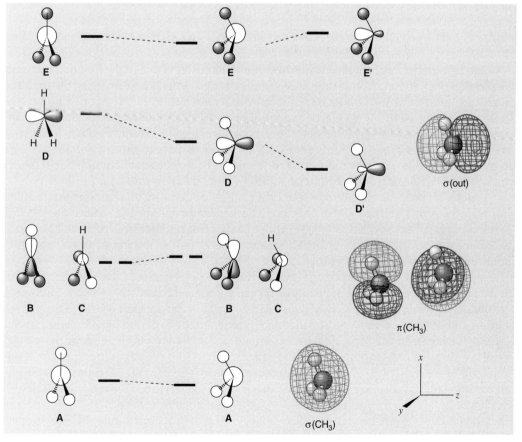

Figure 1.8
The Walsh diagram for CH_3. Beginning with planar CH_3 on the left and moving right, pyramidalization lowers orbital **A** slightly, raises **B** and **C** more so, and lowers **D** substantially. Also shown is a secondary mixing between **D** and **E** to produce **D'** and **E'**, the final MOs of pyramidal CH_3. Atomic orbitals are drawn to show phase relationships only; they are not to scale. Also shown are the computed MOs for the system, and the conventional labeling of the group orbitals for methyl.

energy lowering will occur because the hydrogens are closer to each other and can interact favorably. The **B/C** pair will rise in energy on pyramidalization significantly more than **A** is lowered because of the greater directionality of the bonding. In the planar form the p orbitals can point more directly at the hydrogens, maximizing overlap. This overlap diminishes as the system distorts, and this destabilizes the orbitals. Even with distortion, the **B/C** pair of orbitals remains degenerate.

The biggest effect by far of pyramidalization is on orbital **D**. A formerly nonbonding orbital becomes significantly bonding. The hydrogens have moved off the node of the p orbital, allowing favorable bonding interactions to develop. Thus, the energy of **D** is considerably lower in pyramidal vs. planar methane.

Let's look at the bonding in the planar and pyramidal forms. Orbitals **A–C** are strongly C–H bonding, whether in the planar or pyramidal forms. They can contain six electrons. In a VB model, we would want three C–H bonds, each with two electrons, for a total of six C–H bonding electrons. The two models agree. With QMOT, we still have three C–H bonds, described by three occupied MOs that are strongly C–H bonding.

We are not done with the QMOT analysis, because we have not yet considered rule 7. MOs **D** and **E** have the same symmetry, but one is based on a carbon $2p$ orbital and the other on a $2s$ orbital. Thus, rule 7 tells us to mix these two orbitals, and Figure 1.8 shows the result as **D′** and **E′**. Now **D′** looks more like a lone pair orbital, but contrary to simple Lewis diagrams, there is still significant C–H bonding in this lone pair orbital. Note also that **D′** is very much like an sp^n hybrid at carbon; it contains contributions from both the $2s$ and the $2p_z$ atomic orbitals. Orbital **E′** also looks like a hybrid, but points in the opposite direction from **D′**, just as is expected when one creates sp hybrids (they point in opposite directions on the same atom).

"Group Orbitals" for Pyramidal Methyl

One of the most important concepts in organic chemistry is that of the functional group, a collection of atoms and bonds that shows a consistent reactivity pattern in a wide range of molecular environments. This same notion carries over to electronic structure. Certain collections of atoms contribute a consistent set of orbitals to any molecule in which they are present. To a considerable extent these groups will correspond to the familiar functional groups, but there will be some differences. We will refer to these as **group orbitals**, a collection of partially delocalized orbitals that is consistently associated with a functional group or similar collection of atoms in molecules.

You are already familiar with a few orbitals that can be considered as group orbitals from our first VBT/MOT model of bonding. The localized bonding and antibonding C–C σ and π orbitals given in Section 1.1.7 are group orbitals. We call them group orbitals because we can use these orbitals to describe any σ and π bond, hence describing any alkane C–C bond and any C=C functional group.

As mentioned earlier, one of the goals of introducing a second model of bonding is to define the group orbitals of many types of functional groups, and many will be examined as we go along. However, the analysis just presented actually defined group orbitals also, but for a group not normally considered a functional group—methyl. We are used to thinking of a carbonyl or an olefin as a functional group, but not a methyl or methylene group. However, there are group orbitals for these groups, and we will find they play an important role in understanding many aspects of organic structure and reactivity. We will use these group orbitals to model the bonding in any molecules that contain methyl groups (see Section 1.3).

For convenience, the group orbitals for methyl are given specific descriptive names, as shown in Figure 1.8. We begin with a low-lying, C–H bonding orbital derived from the carbon $2s$ orbital. It is of σ **symmetry**, possessing no nodes along the z axis (the axis that would bond CH_3 to anything else) and is called $\sigma(\mathbf{CH_3})$. Next comes the C–H bonding orbitals that are derived from carbon $2p$ orbitals. There is a nodal plane that passes through the carbon and contains the z axis in each group orbital of this type, so it is sensible to consider them to be of π **symmetry**. They are called $\pi(\mathbf{CH_3})$ (a degenerate pair). Next comes another orbital of σ-type symmetry. It points away from the hydrogens, and so it is called $\sigma(\mathbf{out})$.

Putting the Electrons In—The MH₃ System

As stated in rule 5 of Table 1.7, the Walsh diagram we just developed applies not just to CH_3, but to any MH_3, where M is a main group element. It is thus the Walsh diagram for BH_3 and NH_3, too. The only difference is the number of valence electrons. Since B, C, and N have three, four, and five valence electrons, respectively, and the three H's contribute a total of three electrons, we must deposit into Figure 1.8 six, seven, or eight electrons, depending on whether we are considering BH_3, CH_3, or NH_3, respectively. Because neutral CH_3 is a radical, we'll discuss the ramifications of Figure 1.8 on methyl when we look at the geometries of reactive intermediates (see Section 1.4).

Let's focus first on the even electron systems, BH_3 and NH_3. The six valence electrons of BH_3 occupy MOs **A–C** of either the planar or pyramidal form. Now, consider which geometry, planar or pyramidal, the molecule will prefer. For BH_3, converting the planar form to the pyramidal form will be *destabilizing*. MO **A** is slightly stabilized, but **B** and **C** will be significantly destabilized. Rule 6 states that we will model the energy as the sum of the one-electron energies—that is, as the sum of the orbital energies for occupied orbitals, including a factor of two if the MO is doubly occupied, but only a factor of one for singly occupied MOs. The net effect of pyramidalization is thus destabilizing, and so BH_3 is predicted to be planar. Note that since **D** or **D'** is empty in BH_3, it has no influence on the geometry.

In contrast, with eight valence electrons, NH_3 is predicted to be pyramidal. Now orbital **D** is occupied in the planar form and **D'** in the pyramidal form, and the substantial stabilization of **D'** associated with pyramidalization outweighs all other considerations. Our simple model has thus made a clear prediction: BH_3 should be planar, while NH_3 should be pyramidal. Moreover, the prediction is correct and is identical to the prediction from VSEPR. However, the MOT model is rooted in the most modern theories of bonding.

1.2.4   The CH₂ Group in Detail

The methyl group was just a start. Let's now use the same procedure for the CH_2 group. Our goals are the same as they were with CH_3: define the group orbitals and examine how electron population leads to differing structures. Figure 1.9 shows the Walsh diagram for methylene, where we consider the relationship between the linear and bent forms.

The Walsh Diagram and Group Orbitals

Beginning with the linear structure, we have a low-lying, CH bonding orbital (**A**) derived from the carbon $2s$ orbital. Now, only one p orbital can bond to the hydrogens in the linear form, leaving a degenerate pair of nonbonding p orbitals. On going from linear to bent, orbital **A** is slightly stabilized due to the H•••H overlap. **B** is destabilized more, as C–H overlap drops, just as in the case with CH_3. However, the dominant change is the drop in **C**, as it goes from nonbonding to bonding. Orbital **D** remains an isolated p orbital, so it is nonbonding. As in the CH_3 diagram, we expect a secondary mixing, this time between **C** and **E**, resulting in an important hybrid orbital, **C'**.

The actual MOs of the bent form are also shown in Figure 1.9. These constitute the group orbitals for CH_2. Similar descriptors for the group orbitals are used as with methyl. We find a σ(**CH₂**), now a single π(**CH₂**), a σ_out(**CH₂**), and a lone p orbital. Together, the σ(CH₂) and the π(CH₂) define the C–H bonding orbitals. The σ_out(CH₂) and the p orbital can be used to make bonds to other groups (see Section 1.3).

Putting the Electrons In—The MH₂ System

The diagram given in Figure 1.9 applies to all MH_2 molecules, the most prominent of which is water. Because neutral CH_2 is a carbene, we'll discuss the ramifications of Figure 1.9 on the geometry of CH_2 as a discrete molecule when we analyze reactive intermediates (see Section 1.4). Here, let's briefly consider only H_2O. Water has eight valence electrons, and so both **C'** and **D** are doubly occupied. Thus, water will prefer a bent structure, in order to take advantage of the large stabilization of **C'** that occurs.

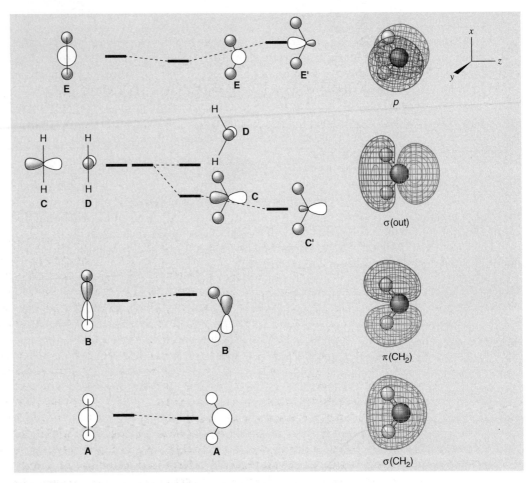

Figure 1.9
The Walsh diagram for CH_2. The linear form is shown on the left, and it is converted to the bent form. Also shown is a secondary mixing between the **C** and **E** orbitals to make **C′** and **E′**. The computed MOs and the standard group orbital designations are also given.

We all know that water has two lone pairs on the oxygen, but you may be wondering where they are. In this bonding scheme the lone pairs are best thought of as MOs **C′** and **D**. In the MO model, the two lone pairs of water (and those of all ethers or other dicoordinate oxygens) are *not* equivalent sp^3 hybrids. One is of σ symmetry (**C′**), has some O–H bonding character, and lies significantly lower in energy than the other because it has contributions from an oxygen s orbital. The higher lying lone pair is a pure p orbital, and thus has π symmetry. This model is supported by experiment. Water in the gas phase has two ionizations corresponding to the two different oxygen lone pairs, and they are separated by 2.2 eV. In an isolated water molecule the lone pairs are not identical!

What, then, of the often invoked model in which water is treated as if it has a "tetrahedral" structure, with two equivalent lone pairs? As shown in Figure 1.10, the two equivalent lone pairs can be obtained by taking in-phase and out-of-phase combinations of the **C′** and **D** orbitals. Is this model also viable for water? For properties such as dipole moment, polarizability, etc., either bonding model is acceptable. That is because these properties depend only on the total electron density of the molecule, which is the same for either model of bonding.

However, the environment that a molecule finds itself in also influences the orbital description. Our analysis to this stage has been for a single isolated MH_2 system. Orbitals **C′** and **D** could not mix since they have differing symmetry. However, as soon as another mole-

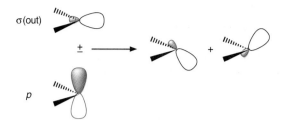

Figure 1.10
Mixing the two MOs of water that correspond to the lone pairs produces two equivalent lone pair hybrid orbitals.

cule approaches a water molecule, it lowers the symmetry of the system. Now the **C'** and **D** orbitals can mix in MOT and two sp^3-like lone pair orbitals are obtained (as in Figure 1.10). This would happen in bulk water.

In summary, with QMOT and the Walsh diagrams for CH_3 and CH_2, we made use of simple notions: s orbitals are lower in energy than p orbitals; strong overlap stabilizes an orbital, and weakening that overlap raises the orbital energy. In the end, though, we mixed atomic orbitals to create delocalized molecular orbitals on the groups. Mixing orbitals is a common tool, and it follows a precise protocol. As mentioned earlier, the QMOT rules were devised in part on a mixing protocol. It is now time to describe that procedure in detail. This will allow us to further develop our repertoire of group orbitals, and to begin the analysis of more standard functional groups, namely alkenes and carbonyls.

1.3   Orbital Mixing—Building Larger Molecules

The essence of orbital mixing is stated in rules 8 and 9 of Table 1.7. The rules are depicted pictorially in Figure 1.11, where we schematically mix orbitals of two separate molecules. You should already be familiar with the idea that mixing two orbitals produces an in-phase, bonding combination, and an out-of-phase, antibonding combination (Figure 1.3 and Section 1.1.7). The former is stabilized (lowered in energy), whereas the latter is destabilized (raised in energy). Figure 1.11 includes some further detail. The figure shows two mixings: Figure 1.11 **A** describes the mixing of degenerate orbitals, and Figure 1.11 **B** shows the mixing of two orbitals that start at different energies. The essential features of the two are the same.

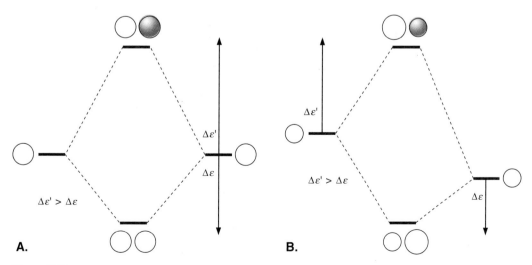

Figure 1.11
Orbital mixing. Simple spherical orbitals are shown to illustrate the concept. **A.** The first-order, degenerate mixing. **B.** The second-order, nondegenerate mixing.

A key aspect of orbital mixing is that the antibonding combination is raised in energy more than the bonding combination is lowered in energy. This difference in energy is true for both mixings in Figure 1.11. If both original orbitals are doubly occupied, the resulting two orbitals will also be doubly occupied, and because two electrons are raised in energy more than the other two are lowered, the net interaction is destabilizing. *The mixing of filled orbitals on two separate molecules is always destabilizing.* This four-electron interaction is referred to as **closed shell repulsion**. A different outcome arises if we have only two electrons in the system. The two electrons can come from one in each starting orbital or two in one starting orbital with the other empty. Now the two electrons end up in the lower orbital, and the mixing is stabilizing. *It is always favorable to mix a filled orbital with an empty orbital or to mix two singly occupied orbitals.* These two aspects of orbital mixing are universal, and they will be used throughout this text.

Figure 1.11 **B** illustrates an additional feature of the nondegenerate mixing, which is summarized in rule 9. In this situation, the lower energy, bonding orbital will have a larger contribution from the original orbital that started out lower in energy. The high energy, antibonding orbital will have a larger contribution from the higher-lying original orbital. As shown in Figure 1.11 **B**, mixing produces a **polarization** of the resulting orbitals. This will be a very important feature in subsequent orbital mixings.

We will see in Chapter 14 that the mixings just described can be treated quantitatively using **perturbation theory**. The mixing of Figure 1.11 **A**, involving degenerate orbitals, is called a **first-order perturbation**, while the nondegenerate mixing of Figure 1.11 **B** is a **second-order perturbation**. With the quantitative tools of perturbation theory, we can provide more detailed analyses of some of the issues addressed here, including, for example, why the orbitals **B** and **C** of Figure 1.7 are degenerate.

In reality, for most situations encountered in organic chemistry, the qualitative descriptions of orbital mixing presented here are adequate. We do need, however, one aspect of perturbation theory that is not evident from the simple analysis given so far. It is generally true that the first-order, degenerate mixing is stronger than the second-order mixing. Also, when comparing second-order mixings, the larger the energy gap between the initial pair of orbitals, the *smaller* the mixing interaction. This statement is known as the **energy gap law**, and it is given as rule 10 in Table 1.7. Note that the terms "extent of mixing", or "mixing interactions", refer to the energy changes that occur with mixing. The statement that first-order mixings are stronger than second-order is really just an aspect of the energy gap law, since first-order systems have the smallest possible energy gap: zero. Finally, rule 11 states that the more the interacting fragments overlap, the stronger the mixing interaction. Hence, overlap and energy gap both determine the extent of mixing.

1.3.1   Using Group Orbitals to Make Ethane

In Section 1.2.3 we built the orbitals of CH_3 (actually MH_3) using simple reasoning and symmetry. This construction gets more difficult to do as the molecules get bigger. It is not at all obvious what the orbitals of methylcyclohexane or acrolein should look like. However, there is a way to gain considerable insight into the bonding of larger molecules. The strategy is to build up larger molecules by combining small fragments whose MOs we understand, using the orbital mixing rules we have just developed. We are now in a position to illustrate this strategy by combining two methyl fragments to make ethane.

We want to create an orbital mixing diagram that combines the orbitals of two CH_3 groups to make ethane. We should use the MOs of pyramidal methyl, as this is the geometry appropriate to ethane. The diagram is set up as in Figure 1.12. We need consider only the degenerate (first-order) mixings, because they will be the strongest. The $\sigma(CH_3)$ and $\pi(CH_3)$ orbitals are primarily C–H bonding; they do not point out into the C–C bonding region. As such, the overlap in each case should be small, and the mixing interaction energy should be small (rule 11). Thus, these are shown as not especially strong mixings in Figure 1.12.

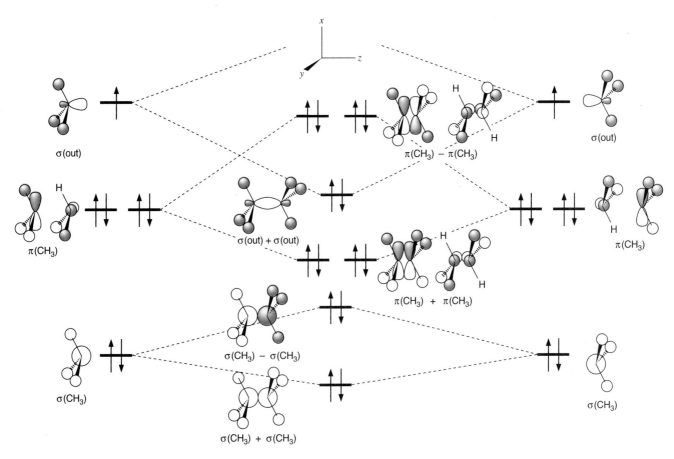

Figure 1.12
The orbital mixing diagram for the formation of ethane from two pyramidal CH_3 groups.
The computed MOs are shown in Figure 1.13.

The interesting mixing is of the pair of σ(out) orbitals. Because of the hybridization that resulted from rule 7 (Table 1.7) discussed previously, the σ(out) orbitals point out into the C–C bonding region. They overlap very well, and so the mixing interaction is quite strong, as shown in the figure. The mixing is strong enough that the [σ(out) + σ(out)] MO drops below the out-of-phase combination of the π(CH_3) orbitals. It is difficult to predict how far this orbital will drop and its relative placement with respect to the other orbitals with only the rules of Table 1.7, so this result must just be accepted at this stage of the discussion.

Now we add electrons. As noted above, CH_3 has seven valence electrons. Ethane thus has 14 valence electrons, and as we place them into Figure 1.12, we see that the **highest occupied molecular orbital (HOMO)** of ethane is a degenerate pair of orbitals that can best be described as C–H bonding, but slightly C–C antibonding.

All the combinations derived from σ(CH_3) and π(CH_3) orbitals are occupied. There are six such orbitals with 12 electrons, and ethane has six C–H bonds. Again, the accounting implied by the MO treatment is completely consistent with our simpler views of bonding, consisting of localized C–C and C–H σ bonds. It may at first seem odd that the out-of-phase combinations of the σ(CH_3) and π(CH_3) orbitals are occupied. Shouldn't these be antibonding MOs? They are C–C antibonding, but they are still C–H bonding, and this is enough to make them overall bonding. The [σ(out) + σ(out)] MO is the major C–C bonding orbital, and it has two electrons in it.

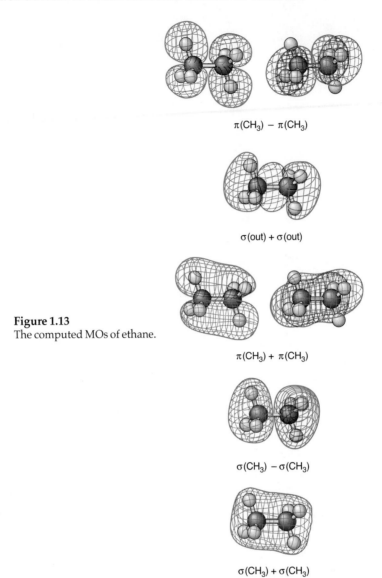

Figure 1.13
The computed MOs of ethane.

$\pi(CH_3) - \pi(CH_3)$

$\sigma(out) + \sigma(out)$

$\pi(CH_3) + \pi(CH_3)$

$\sigma(CH_3) - \sigma(CH_3)$

$\sigma(CH_3) + \sigma(CH_3)$

For comparison, we show the calculated MOs of ethane in Figure 1.13. Prior to our analysis these might have seemed to be fairly strange orbitals, but now we can see they match up quite nicely with the orbitals we have derived. This highlights the power of group orbitals; their combinations nicely rationalize the origins of orbitals derived from sophisticated calculations. Moreover, experience has shown that the C–C σ bonding orbital of ethane, [σ(out) + σ(out)] in Figures 1.12 and 1.13, transfers consistently to all alkane fragments in organic molecules. Thus, whenever we see a C–C single bond in a molecule, we can anticipate an orbital of this type. Likewise, we can anticipate orbitals of π-like character for the C–H bonding interactions.

1.3.2 Using Group Orbitals to Make Ethylene

The standard bonding picture for ethylene is viewed as being made from two sp^2 hybridized carbons, and it contains a C–C double bond comprised of a σ bond and a π bond. In MOT, however, we don't hybridize, and we don't presume bonding arrangements. Instead, we build ethylene with no assumptions by just combining two CH_2 groups. Let's see how well this construction turns out.

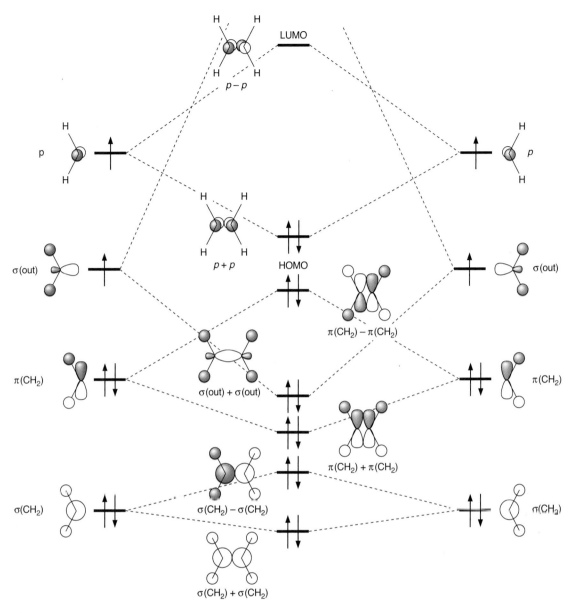

Figure 1.14
Orbital mixing diagram for the formation of ethylene by mixing two CH_2 groups. The computed MOs are shown in Figure 1.15.

The creation of ethylene from two CH_2 groups is shown in Figure 1.14. The actual MOs are shown in Figure 1.15. Hopefully by now the origins of all the MOs are clear. As with ethane, the MOs derived from $\sigma(CH_2)$ and $\pi(CH_2)$ make four MOs that are primarily C–H bonding, just as we need for ethylene. The interesting interactions involve $\sigma(out)$ and p.

The hybrid $\sigma(out)$ orbital is strongly directional, pointing along the C–C bond. We expect a strong interaction. The p orbital does not point across the C–C bond. As such, the $(p + p)$ interaction should be weaker than $[\sigma(out) + \sigma(out)]$, because of poorer overlap. The combination of these interactions produces the C–C double bond of ethylene. Thus, $[\sigma(out) + \sigma(out)]$ is the major σ bond component of the double bond. The $(p + p)$ mixing produces the π bond of ethylene.

A CH_2 group has six valence electrons, so ethylene has 12. When we place 12 electrons into Figure 1.14, we see that the HOMO is the π orbital. The **lowest unoccupied molecular**

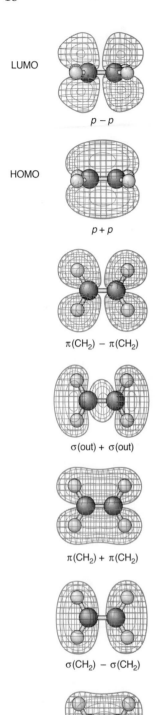

LUMO

$p - p$

HOMO

$p + p$

$\pi(CH_2) - \pi(CH_2)$

$\sigma(out) + \sigma(out)$

$\pi(CH_2) + \pi(CH_2)$

$\sigma(CH_2) - \sigma(CH_2)$

$\sigma(CH_2) + \sigma(CH_2)$

Figure 1.15
The computed MOs of ethylene.
Note the molecule is shown in
different orientations for
different MOs.

orbital (LUMO) is the out-of-phase combination of p orbitals, making it an antibonding π^* orbital. The standard model of bonding in ethylene has been reproduced. We simply follow the rules of orbital mixing, as described in Table 1.7.

1.3.3   The Effects of Heteroatoms—Formaldehyde

In our rules for orbital mixing, rule 5 states that similar molecules have similar MO diagrams. This is essentially true, but there are important differences that we must consider. Formaldehyde and ethylene are **isoelectronic**; they have the same number of valence electrons and the same types of valence orbitals. Thus, we can expect similar MOs for formaldehyde and ethylene, but with some changes (more properly termed "perturbations") introduced by the oxygen. Experience has shown that the primary consequence of introducing heteroatoms into a hydrocarbon system is to alter orbital energies, as stated in rule 12.

A convenient guideline for orbital energies to be used in mixing diagrams is the set of valence state ionization energies given in Table 1.8. We can see the trends expected from our earlier discussions of electronegativity. The key point is that *electronegative elements have relatively low-lying atomic orbitals*. The larger the ionization energy, the harder it is to remove the electron, so the energy of the orbital is lower. For the particular case at hand, it is clear from Table 1.8 that the atomic orbitals that O contributes to formaldehyde start at much lower energies than those of C or H. As a result, the mixings are no longer degenerate as they were for ethylene, and we must consider the consequences of the second-order perturbation rules for orbital mixing.

The mixing diagram for formaldehyde is shown in Figure 1.16, and the resulting MOs are given in Figure 1.17. The precise energies where we place the oxygen p orbitals are not crucial. However, we must place the oxygen p orbitals below the CH_2 p orbital. Orbitals that are analogous to all the key orbitals of ethylene are created in Figure 1.16, including both the σ and π orbitals of the double bond. Although mixing rule 9 now predicts polarizations in all the orbitals, let's first focus on the π and π^* orbitals.

Table 1.8
Valence State Ionization Energies*

Orbital		$-H_{ii}$ (eV)
H	1s	13.6
C	2s	21.4
	2p	11.4
N	2s	26.0
	2p	13.4
O	2s	32.3
	2p	14.8
F	2s	40.0
	2p	18.1
Si	3s	17.3
	3p	9.2
P	3s	18.6
	3p	14
S	3s	20
	3p	11
Cl	3s	30
	3p	15

*Technically, these are not experimentally determined numbers, but rather they are parameters (H_{ii} values) used in a popular computational method termed extended Hückel theory (see Chapter 14, p. 834).

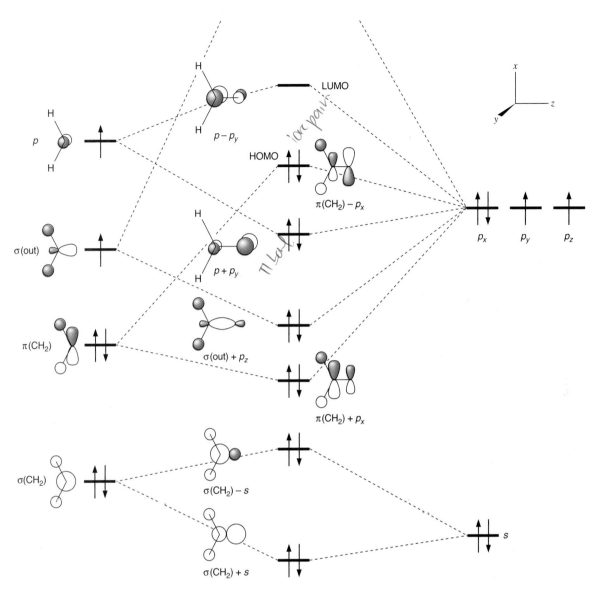

Figure 1.16
The orbital mixing diagram of $CH_2{=}O$ from CH_2 plus an oxygen atom. The final MOs are given in Figure 1.17.

Since the building blocks for π and π^* are isolated p orbitals, there is no ambiguity about the relative energies of the two initial orbitals: the oxygen orbital lies below the carbon orbital. As such, there is also no ambiguity about the expected polarizations. The lower energy MO, the π orbital, must be polarized toward the oxygen. Similarly, the higher energy MO, π^*, must be polarized toward carbon. The actual MOs show that this is indeed the case. When you examine the π MO of formaldehyde in Figure 1.17, the expected polarization may not be evident. However, you must remember that the atomic orbitals of electronegative elements are smaller; both the covalent and the van der Waals radii of oxygen are smaller than those of carbon (see above). So, the π MO of formaldehyde does have a larger contribution from the oxygen (numerical rather than graphical descriptions of the MO make this clear). The expected polarization toward carbon is very clear in the π^* MO.

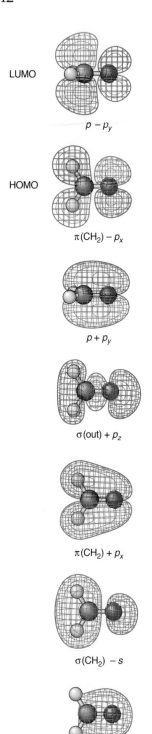

LUMO

$p - p_y$

HOMO

$\pi(CH_2) - p_x$

$p + p_y$

$\sigma(out) + p_z$

$\pi(CH_2) + p_x$

$\sigma(CH_2) - s$

$\sigma(CH_2) + s$

Figure 1.17
The computed MOs of
formaldehyde. Note the
molecule is shown in different
orientations for different MOs.

We still have 12 electrons to allocate. An interesting feature of Figure 1.16 is that when we insert our 12 valence electrons, we find that the LUMO is π^*, as expected. However, the HOMO is not π, but rather a lone pair, which we describe as $\pi(CH_2)$ minus the oxygen p_x orbital (see Figure 1.16). It is a π-type lone pair with a significant out-of-phase mixture of the $\pi(CH_2)$ orbital. The second lone pair lies below the HOMO, and is σ-type, formed from mixing $\sigma(out)$ with p_z. Neither of these orbitals are "pure" lone pair orbitals. The higher energy one is significantly C–H bonding, and the lower energy one is significantly C–O bonding. MOT does not always lead to a simple correspondence to our classical views. The situation with formaldehyde is directly analogous to what we saw previously with the water molecule. The canonical (fully delocalized) MOs of oxygen-containing molecules predict the existence of two distinct lone pairs. You can combine these orbitals to produce orbitals that resemble the classic sp^2 hybridization for the oxygen.

Looking ahead, it should be clear that the MO diagram for this prototype carbonyl has significant implications for predicting and rationalizing reactivity patterns. Nucleophiles will interact with the LUMO preferentially at the larger coefficient, the one on carbon. Note that this π^* MO "leans away" from the bonding region, as is typical for antibonding orbitals. The π^* orbital is perpendicular to the plane of the molecule but is slightly tilted toward the CH_2 and O groups. It is better represented in a cartoon as shown below. This "tilting" of the orbital has ramifications for reactivity that we will return to in Chapter 10. The nature of π^* is completely consistent with what we expect for nucleophilic addition to a carbonyl, namely attack at C from a "backside" direction. Also, the polarization of the HOMO toward oxygen has implications for reactivity. For example, we expect protonation on oxygen, not carbon.

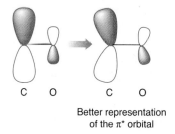

C   O   C   O

Better representation
of the π^* orbital

As in other cases discussed so far, our QMOT model of a simple carbonyl is completely compatible with more conventional bonding models. In order to explain the polarization we have just discussed, models emphasizing VBT concepts would focus on the greater electronegativity of O vs. C. This electronegativity difference should polarize the bond, as shown below. Alternatively, we could invoke a resonance argument. While a single Lewis structure is certainly a good model for a carbonyl, the resonance interaction shown is not unreasonable. Although it creates charge and disrupts octets, it places a negative charge on a highly electronegative element. Certainly the two forms would not contribute equally to the structure of a carbonyl. The C=O form is preferred—perhaps the structure is best thought of as a 4:1 or a 10:1 mixture of the two—the exact details are unimportant. However, even a small contribution from the ionic form has significant implications for carbonyl chemistry. It implies a partial negative charge on O, and a partial positive charge on C. Again, the reactivity patterns of a carbonyl, protonation at O and nucleophilic addition to C, are completely consistent with this model.

$O \; \delta\ominus$
$\delta\oplus$

Resonance in a carbonyl

The group orbitals of an olefin would be those we derived for ethylene in Figure 1.14. The most important ones will be the π and π^* MOs, designated as the HOMO and LUMO in Figures 1.14 and 1.15. Similarly, the group orbitals of a simple carbonyl compound, an aldehyde or ketone, are those we developed for formaldehyde (Figures 1.16 and 1.17). Now we should consider at least four different MOs: π and π^* along with the two lone pairs.

1.3.4 Making More Complex Alkanes

In alkanes, CH_2 and CH_3 groups will contribute orbitals such as in Figures 1.8 and 1.9. If we consider a linear alkane, $CH_3(CH_2)_nCH_3$, we would describe the bonding as follows. For the terminal methyl groups, the $\sigma(CH_3)$ and two $\pi(CH_3)$ orbitals would be used for C–H bonding. The methyl $\sigma(out)$ orbital would be used to make a C–C bond. The resulting orbital would be the C–C bonding orbital discussed for ethane.

The CH_2 group is more challenging to consider because it must make two C–C bonds. To do this, we use the CH_2 $\sigma(out)$ and the p orbitals. You don't have to, but it is convenient to make linear combinations of these two, as we did for water in Figure 1.10 (in fact, our discussion of orbital mixing in water can be extended to any MH_2). This linear combination creates orbitals similar to the sp^3 hybrids available to bond to two other groups. We just mix $\sigma(out)$ and p to give orbitals that point in the correct direction to make CH_2 a tetrahedral center. A linear combination of group orbitals to give orbitals reminiscent of our standard hybridization notions *will always be possible*. The important point is that the orbitals developed by the QMOT approach predict the delocalized orbitals found by high level calculations, where hybridization is not usually the result. We will show experimental evidence in Chapter 14 that supports the fact that hybridization does not actually occur in the standard sp^3, sp^2, and sp manner.

Once we see how the C–C bonds are made, we just assume them directly. So, in a linear alkane, we have a backbone of C–C bonding orbitals. The terminal methyl groups make C–C bonds with their respective $\sigma(out)$ orbitals. In addition, each CH_3 group contributes three C–H bonding orbitals, one $\sigma(CH_3)$ and two $\pi(CH_3)$, while each CH_2 contributes two C–H bonding orbitals, one $\sigma(CH_2)$ and one $\pi(CH_2)$.

This model of alkane bonding is useful, but admittedly, it really does not produce any new insights compared to the more conventional model emphasizing sp^3 hybridized carbons and simple localized σ bonds. Frankly, if we only ever considered alkanes, there would be no need for CH_2 and CH_3 group orbitals. However, we will now show that in more interesting molecules, the use of group orbitals provides very valuable insights.

1.3.5 Three More Examples of Building Larger Molecules from Group Orbitals

All functional groups have a set of group orbitals associated with them. Appendix 3 shows the full MOs of a number of representative small molecules. These can be considered to be the group orbitals of the analogous functional group in larger molecules. From the basic building blocks we have presented so far, the MOs for these essential functional groups of organic chemistry can be understood. For example, the MOs of a carboxylic acid (Appendix 3) can be developed from an orbital mixing diagram, combining the π MOs of formaldehyde with the appropriate AOs of an oxygen atom. Then, a methyl ester can be "prepared" by mixing in a $\pi(CH_3)$ orbital. When considering the reactivity of a specific functional group, it will be useful to look at its group orbitals in Appendix 3 as a guideline. Let's do such a mixture for a few additional cases.

Propene

In most instances alkyl groups are spectators in organic reactivity, and the simple VBT model based on hybridization is adequate. However, for a CH_2 or CH_3 that is adjacent to a conventional π system, such as an alkene, a carbocation, or a carbonyl, important interactions between the alkyl group and the π system may occur. In these instances the $\pi(CH_2)$ or $\pi(CH_3)$ group orbitals are quite useful.

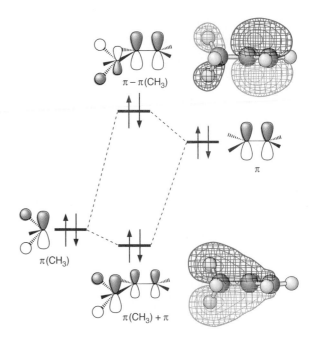

Figure 1.18
An orbital mixing diagram for propene,
along with the computed MOs of the
molecule.

As an example, Figure 1.18 shows an orbital mixing diagram for propene. We are going to focus only on the π system, so we start with the ethylene π MO of Figures 1.14 and 1.15. In conventional views of a methyl group, there really is no way for the CH_3 to interact with the π system of propene. However, it is not just a coincidence that two of the C–H bonding orbitals for CH_3 are called π(CH_3). These groups orbitals are of the correct symmetry to interact with conventional π orbitals of the sort seen in alkenes and carbonyls. In our orbital mixing diagram, the two degenerate π(CH_3) group orbitals are low lying in energy; they are C–H σ bonding, and so are lower than a C–C π bond. In any particular conformation of propene, it will always be true that one of the π(CH_3) group orbitals will be able to mix with the ethylene π MO. This mixture is a standard second-order mixing, producing two new propene MOs, as in Figure 1.18. Let's consider the consequences of this mixing.

Compared to ethylene, the π bond of propene is (a) slightly higher in energy, and (b) slightly delocalized onto the methyl group. Note the clear prediction that the olefin π orbital and the π(CH_3) group orbital should be *out of phase* in the "π" orbital, the HOMO of propene. This prediction is confirmed by the actual MO (Figure 1.18). In addition, the mixing is confirmed by experiments. The predicted elevation in orbital energy is fully supported by the fact that the ionization energy of propene (9.73 eV) is measurably lower than that of ethylene (10.51 eV), meaning that the HOMO of propene is higher in energy. The π(CH_3) group orbital has been lowered in energy, a factor not usually of importance in understanding reactivity. An important lesson of this analysis is that there are orbitals of π symmetry in simple alkyl groups such as CH_3 and CH_2, and these can interact with the more conventional π systems of alkenes, carbonyls, and so on.

The mixing we have just discussed is *not* **hyperconjugation**, which is the mixing of C–H and C–C bonding orbitals with adjacent empty or partially empty π and p orbitals (see Section 1.4.1). However, a similar mixture can be done with the π* orbital of the alkene that does represent hyperconjugation, and you are asked to show this in an Exercise at the end of this chapter.

Note that we have mixed filled orbitals in Figure 1.18, which may seem like a violation of the previous statement that such mixings are always destabilizing. Indeed, this interaction *is* destabilizing because four electrons go into the resulting orbitals. However, it is enforced by the close juxtaposition of the π(CH_3) orbital with the π bond due to the presence of the C–C bond between the methyl and the alkene of propene.

There is a distinction in mixing filled orbitals that must be kept straight. If we are considering a possible reaction between two species, and we want to know if favorable or unfavor-

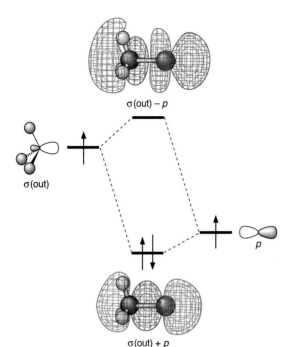

$\sigma(\text{out}) - p$

$\sigma(\text{out})$

p

$\sigma(\text{out}) + p$

Figure 1.19
Orbital mixing diagram for CH_3Cl, with computed orbitals shown. $\sigma(\text{out}) + p$ is the bonding orbital. $\sigma(\text{out}) - p$ is the LUMO, σ^* antibonding orbital.

able orbital interactions develop during the reaction, we clearly want to maximize two-electron mixings and minimize four-electron mixings. However, we are doing something different in our analysis of propene. Really, we are performing an after-the-fact analysis of the molecule, trying to understand the results of a full quantum mechanical calculation on the system. The molecule was not *really* formed by mixing fragment orbitals; we have simply found that such analyses lead to the true MOs. For a destabilizing mixing to occur within the molecule, such as that shown in Figure 1.18 for propene, there must be other σ and π bond stabilizing interactions that more than compensate, thus stabilizing the molecule overall.

Methyl Chloride

As another example of the use of orbital mixing strategies for understanding structure and reactivity, we consider methyl chloride as a prototype of an organic molecule singly bonded to an electronegative element. We could construct a complete diagram, using all the CH_3 group orbitals and the AOs of chlorine, but we can also focus only on the C–Cl bond as the most interesting feature of the system.

The C–Cl bond is formed by the mixing of $\sigma(\text{out})$ of the CH_3 with a chlorine p orbital. The mixing diagram is shown in Figure 1.19. Both electronegativity (Table 1.1) and orbital energy arguments (Table 1.8) allow us to place the Cl orbitals below the carbon orbitals. This means that the polarizations should have the bonding combination polarized toward Cl, and the antibonding combination polarized toward C. Actual calculated MOs are shown in Figure 1.19. The most interesting MO is the LUMO. It is the prototype σ^* orbital. It is C–Cl antibonding, in that there is a nodal surface that splits the C–Cl bond in two. A great deal of orbital character lies outside the region between the C and the Cl, and is especially accentuated at the carbon end of the molecule. There is considerable orbital density on the "backside" of the carbon, and hence a cartoon of the σ^* orbital is better represented as shown below rather than as given for generic σ^* orbitals shown in Figure 1.3.

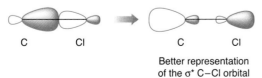

C   Cl   C   Cl

Better representation
of the σ^* C–Cl orbital

This σ^* orbital leads to predictions about reactivity. In the S_N2 reaction, a nucleophile will react with CH_3Cl. As we will see, nucleophiles have high-lying filled orbitals that can mix with low-lying empty orbitals. The σ^* orbital is the appropriate empty orbital on CH_3Cl, and the fact that it is substantially polarized toward the backside of the carbon atom is nicely consistent with the observed inversion of configuration (backside attack) seen in S_N2 reactions. We can view this as a group orbital for a C–Cl functional group that can be considered present in any alkyl chloride.

Butadiene

If we are going to build up larger π systems than are present in simple alkenes and carbonyls, we again need to linearly combine the appropriate group orbitals. This is done for the π system of butadiene in Figure 1.20, and calculated orbitals are given. The π and π^* orbitals of ethylene are our starting point, and they mix in the manner shown to give four molecular orbitals. These resulting orbitals can be considered to be the π MOs of butadiene, but also the four group orbitals for any diene. Note that the simple schematic mixing of the ethylene π and π^* orbitals does not predict the relative contributions of each p orbital to a given MO, and we will have to wait until Chapter 14 to see why. Yet, the simple mixture gets the nodal properties perfectly correct.

1.3.6 Group Orbitals of Representative π Systems: Benzene, Benzyl, and Allyl

We have focused our discussion to this point on group orbitals for small alkyl fragments, σ systems, and a few simple π systems. Yet, as shown in Appendix 3, a significant fraction of important functional groups in organic chemistry have complex π systems. To build many complex π systems by combining small π systems, we need to know the group orbitals of some representative small π systems other than just ethylene. We now give the MOs of three other essential building blocks of organic chemistry: benzene, benzyl, and allyl. In evaluating organic structure and reactivity, these units show up repeatedly, and so it is worthwhile to present them here.

Figure 1.21 presents the MOs of benzene, both in symbolic form and as produced by an accurate quantum mechanical calculation. Also shown is the "HOMO" of benzyl—that is, the singly occupied orbital of benzyl radical, the empty orbital of benzyl cation, and the doubly occupied HOMO of benzyl anion. To a good approximation, the MO has the same form for all three structures. Here, we show the orbitals from the top, so that the p orbitals that make up the molecular orbitals appear only as spheres.

Figure 1.22 shows the group MOs of allyl. The most distinctive feature is that the middle MO has a node through the central carbon, and so is formed from only C1 and C3. This is the MO that is empty in allyl cation, singly occupied in allyl radical, and doubly occupied (HOMO) in allyl anion.

The HOMOs of allyl and benzyl are completely consistent with resonance models for the two (shown below). Allyl resonance places the charge (* = +, −, or •) only on C1 and C3, where the orbital has finite coefficients. In benzyl, there is activity only at the benzylic carbon and the ortho and para positions of the ring. Once again, we emphasize that the MOT approach does not require the "patch" fix of resonance to get the π properties correct.

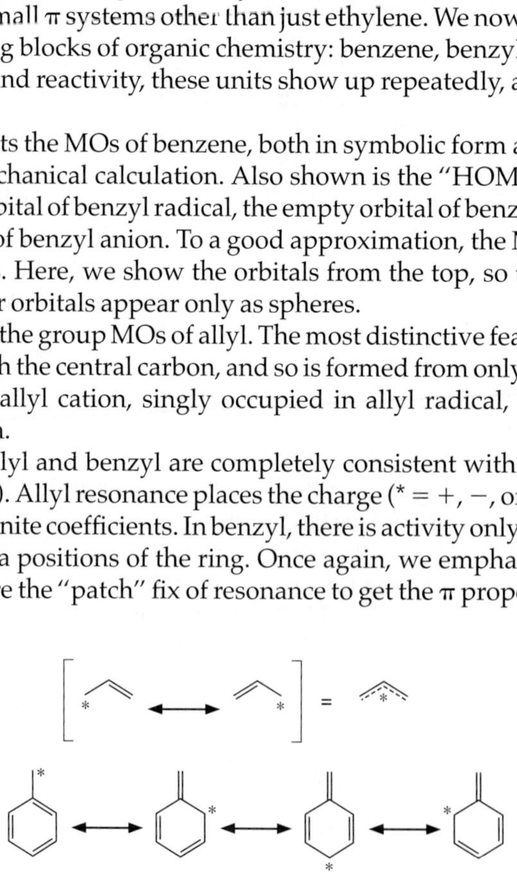

Resonance in allyl and benzyl

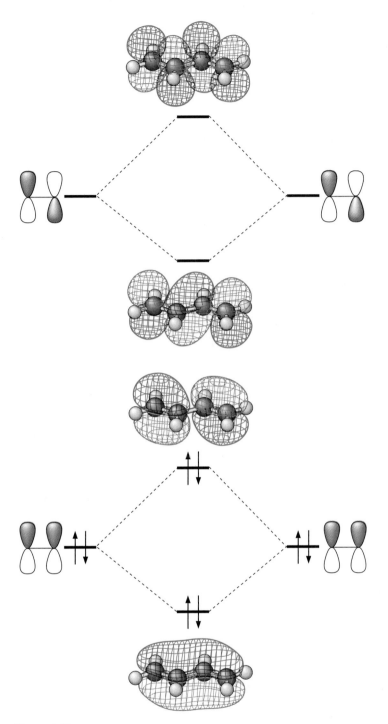

Figure 1.20
The mixing diagram for creating the π MOs of butadiene from ethylene group
orbitals and the calculated orbitals for the π system of butadiene.

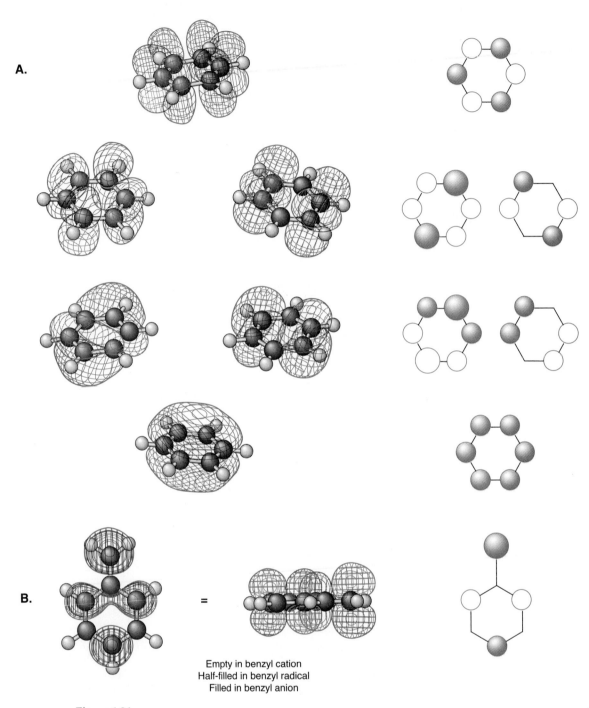

Figure 1.21
A. The computed MOs of benzene and conventional representations.
B. Two views of the key MO of a benzyl group and the conventional representation.

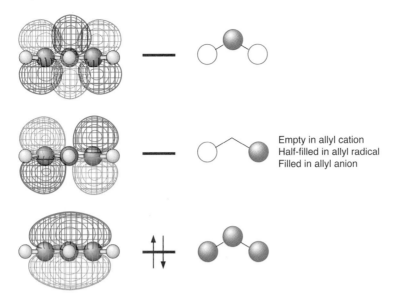

Empty in allyl cation
Half-filled in allyl radical
Filled in allyl anion

Figure 1.22
The MOs of an allyl fragment.

It is an interesting exercise to form the MOs of allyl using an orbital mixing strategy, and such an analysis is shown in Figure 1.23. Here, we show the *p* orbitals viewed from the top, so all we see is a circle. For allyl, it is simplest to first mix C1 and C3. We can form the in-phase and out-of-phase combinations of *p* orbitals, but to a good approximation, these nonadjacent orbitals do not overlap significantly, so there is no energy split. Now we bring in C2. It can mix only with the in-phase combination of C1 and C3, because it lies on the node of the out-of-phase combination. The new mixing produces two new MOs, while the C1/C3 out-of-phase combination comes across unchanged. The result is the allyl pattern of Figure 1.22.

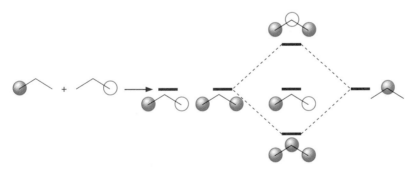

Figure 1.23
An orbital mixing approach to forming the MOs of an allyl fragment.

1.3.7 Understanding Common Functional Groups as Perturbations of Allyl

One of the most useful insights to be derived from the group orbitals of allyl is a picture of the group orbitals for the carboxylic acid, ester, and amide functional groups. These functional groups have π systems that can be viewed as isoelectronic with allyl anion. The resonance structures shown in the margin highlight this concept. Appendix 3 shows the actual π systems for these functional groups, where it is clear that the allyl nature is retained. The three π MOs of the functional groups are polarized in various ways due to the addition of the heteroatoms, yet the general nodal properties of allyl are retained.

Allyl anion analogs

1.3.8   The Three Center–Two Electron Bond

Starting from the classic picture of bonds between adjacent atoms, we often need other concepts to fully explain the bonding in a given molecule. As explained, resonance is such a concept, required to explain the bonding in molecules with delocalized π systems. Another bonding situation that requires a special treatment for VBT, but not with MOT, is the three center–two electron bond. As the name implies, the three center–two electron bond is associated with electron-deficient systems, those for which there are not enough valence electrons to make conventional two center–two electron bonds among all the atoms. The boranes and related structures use the three center–two electron bond extensively, creating highly bridged structures. In organic chemistry, the most common electron-deficient species are carbocations. The three center–two electron bond also figures prominently in carbocation chemistry, so we describe it briefly here.

The starting point for understanding the three center–two electron bond is the chemistry of boron-containing compounds. Simple boron compounds are isoelectronic with carbocations: BH_3 has the same number of valence electrons and orbitals as CH_3^+. We already noted that a QMOT treatment of the two structures produces qualitatively similar results, so BH_3 has the same planar structure as CH_3^+. As you may recall from introductory chemistry, though, BH_3 in the gas phase dimerizes to diborane, B_2H_6, and this molecule has the unconventional structure shown in Figure 1.24 **A**. The molecule has 12 valence electrons, eight of which are used to make the four conventional B–H bonds. The remaining four valence electrons are partitioned, two apiece, to the two B–H–B bridging systems, forming a pair of three center–two electron bonds.

The stability of this well-established bonding pattern can be understood from simple MO arguments. The B–H–B arrangement in diborane is another three-orbital mixing problem, like allyl. In fact, the mixing diagram we developed for allyl translates directly to the

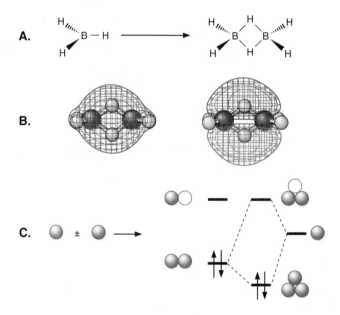

Figure 1.24
Three center–two electron bonds. **A.** Borane dimerizes to produce diborane. **B.** The calculated MOs primarily responsible for the bonding of the bridging hydrogens in diborane. Since there are two B–H–B bridges, there are two bonding orbitals. Shown are the in-phase (left) and out-of-phase (right) combinations of the fully symmetric B–H–B orbital combinations. **C.** A qualitative picture for how a single three center–two electron bond is constructed. Only the lowest of the three combinations, the fully in-phase MO, is occupied. The two MOs of part **B** are in-phase and out-of-phase combinations of this MO, because diborane has two three center–two electron bonds.

three center–two electron bond. In Figure 1.23, the open and filled circles were meant to be the "tops" of p orbitals, showing their phase relationships. However, they could just as easily be s orbitals or sp^3 hybrids. A recurring theme of QMOT, which we have already seen when discussing CH_2 and CH_3, is that the distinction between σ and π systems is not as clear-cut as introductory texts would suggest, and so we should not be surprised that the MO patterns in a σ system such as we are developing here are topologically identical to those of an analogous π system. Although formally σ bonding is involved, the mixing for the B–H–B system (Figure 1.24 **C**) still produces three molecular orbitals in a manner analogous to Figure 1.23. Since only one of the three orbitals is highly stabilized, the optimal arrangement involves only two electrons. It is a three-atom system involving only one occupied MO, a **three center–two electron bond**.

Note that it is generally better to make two, conventional, two center–two electron bonds rather than a single three center–two electron bond. However, boranes are electron deficient; there just are not enough valence electrons to make the requisite number of conventional bonds. The system does the best it can by forming a pair of three center–two electron bonds.

There is absolutely no doubt that three center–two electron bonding is an important component of the electronic structures of a wide range of boron-containing compounds. Given the isoelectronic relationship to CH_3^+, it should come as no surprise that similar bonding should arise in carbocations, and indeed it does (see the next section). There was initially considerable resistance to this notion; the tetravalent carbon was sacrosanct. However, it is now clear that three center–two electron bonds do contribute to many carbocation structures. The three center–two electron bond allows a wide range of exotic looking, highly-bridged structures, often referred to as non-classical carbocations (see Chapters 2, 11, and 14 for discussions). While a detailed description of the bonding in some of the more elaborate systems may seem complicated, at its core is the special stability of the three center–two electron bond.

1.3.9 Summary of the Concepts Involved in Our Second Model of Bonding

The more modern theory of bonding extends the notion of bonding and antibonding orbitals that describe localized σ and π bonds to increasingly delocalized orbitals. All organic functional groups can be envisioned as consisting of a set of delocalized orbitals that are combined to build up the whole molecule. These "group orbitals" will be very useful in future analyses, and as described in this chapter, they nicely predict the structures of stable molecules and reactive intermediates. The new concepts that were introduced in developing this second model of bonding are summarized as follows:

QMOT. A set of rules for predicting how orbitals will mix to make delocalized molecular orbitals for molecular groups or molecules.

Orbital mixing. MOs associated with particular bonds can be formed by mixing orbitals from the bonding partners according to the rules of QMOT. When such mixing occurs, the destabilization of the higher orbital is greater than the stabilization of the lower orbital, and in the case of the nondegenerate second-order mixing, predictable orbital polarizations result. This mixing can be between atomic orbitals on adjacent atoms, or between group orbitals on adjacent groups.

Transferable, partially localized MOs—group orbitals. Functional groups such as olefins, ketones, amides, aromatics, etc., contribute a set of transferable orbitals to a molecule. For example, a simple olefin always has π and π^* orbitals. Even CH_3 and CH_2 groups contribute partially delocalized molecular orbitals to molecules, and importantly, these include $\pi(CH_3)$ and $\pi(CH_2)$ group orbitals that can interact strongly with conventional π systems.

1.4   Bonding and Structures of Reactive Intermediates

In this chapter we have focused on the molecular and electronic structures of organic molecules and on the theories of bonding that explain them. A major goal of physical organic chemistry, however, is to explain reactivity, and a key concept in reaching that goal is the reactive intermediate. A **reactive intermediate** in organic chemistry is typically some form of carbon that either does not have the requisite four bonds, or has less than an octet of electrons, or is highly strained (see Chapter 2). Heteroatom analogs of these structures are also important. For example, carbanions do not have four bonds, while radicals, carbocations, and carbenes are also lacking an octet. Rather than disperse our discussion of the structures of these prototype reactive intermediates throughout the text, we group them here. This allows us to directly connect the structures to the theoretical models of bonding we have developed and to compare and contrast the properties of the basic classes of reactive intermediates. Our notions of bonding should be applicable to both stable structures and reactive structures. The structures of stable organic molecules are mostly straightforward. However, interesting issues arise in describing reactive intermediates that will allow us to further develop many of the models described thus far in this chapter. This further development includes the QMOT model given in Sections 1.2 and 1.3, but also other concepts such as electronegativity and resonance. This is the first of two preliminary descriptions of reactive intermediates. In the following chapter, which emphasizes thermodynamic stabilities, we will evaluate quantitatively the features that stabilize or destabilize reactive intermediates. We begin our analysis of reactive intermediates here by showing how the Walsh diagrams we have already developed provide a solid foundation for understanding the structural features of reactive intermediates.

We generated the group orbitals of CH_2 and CH_3 so we could use them as fragments to build up and understand larger molecules. However, each of these is also a well-defined species in its own right, and more importantly, they are the fundamental structures for a range of reactive intermediates we will see throughout this textbook. Here we will consider the extent to which the simple Walsh diagrams can help us anticipate the fundamental features of reactive intermediates that will allow us to understand their reactivities. A key issue for these structures is whether the planar or pyramidal (linear or bent) form is preferred. A basic assumption of the group orbital concept is that the essential features of an orbital diagram will not change as we alter atom types or the number of electrons. Thus, the basic orbital diagrams for the prototype structures (methyl cation, methyl radical, and methyl anion) are all the same to a good approximation. The primary difference is just the number of electrons we put into the diagram of Figure 1.8. We will see that this distinction is enough to yield very different properties for the three types of reactive intermediates.

1.4.1   Carbocations

From the standpoint of molecular and electronic structure, carbocations present a much greater variety and more significant conceptual challenges than carbanions, radicals, or carbenes. Many aspects of carbocation chemistry have been at times quite controversial. Here, we describe the bonding models for basic carbocation structures. First, however, we must explain some nomenclature.

A **carbocation** is a structure with a positive charge that is associated primarily with a carbon center or, in the case of delocalized systems, a collection of carbons. Heteroatom substitution is allowed, and it will inevitably lead to some positive charge being on the heteroatoms. There are two types of carbocations—carbenium ions and carbonium ions. **Carbenium ions** are trivalent species with a formula of R_3C^+. **Carbonium ions** encompass two related types of structures. First are pentavalent species of the general formula R_5C^+. While not common in solution phase chemistry, such pentavalent carbons are common in the gas phase. Also referred to as carbonium ions are carbocations that have an important contribution from three center–two electron bonding. The nomenclature will become more clear as we present examples below.

This distinction between the two types of carbocations was championed by Olah, and is consistent with other nomenclature. Just as hydronium and ammonium ions represent protonated forms of water and ammonia, carbonium ions can be considered to contain protonated forms of methane and other typical, tetravalent carbon compounds. Likewise, carbenium ions can be thought of as protonated carbenes ($:CH_2 + H^+$ gives CH_3^+), consistent with the nomenclature system. In the older literature, only the term carbonium ion is used in connection with what is more precisely called a carbenium ion. We feel the more modern nomenclature has value, so we use it in this book.

Carbenium Ions

The prototype carbenium ion is methyl cation, CH_3^+. It has six valence electrons (one from each of three H's and three from C, given the + charge). These six electrons will fill the lowest three MOs of Figure 1.8, with the p_z orbital remaining empty. Since pyramidalization destabilizes the **B/C** pair in Figure 1.8 more than it stabilizes the single **A** MO, CH_3^+ will prefer a planar structure. The lowering of the **D** orbital is inconsequential, because there are no electrons in it. The preference for planarity is therefore substantial. Calculations suggest that a CH_3^+ that is pyramidalized to have tetrahedral angles is less stable than the planar form by about 30 kcal/mol. We will see experimental evidence in support of this in Chapter 2. This is completely consistent with the classic view of sp^2 hybridization and what we know about the planar structure of CH_3^+.

This very simple analysis predicts the correct geometry, but as we now describe, there are a lot of complications to carbenium structures that require more complex pictures of bonding than is apparent from this simple analysis, and even just the substitution of an alkyl group will start to perturb the analysis.

Alkyl substitution stabilizes carbocations, and we will see in Chapter 2 that the order of stabilities for carbenium ions is $3° > 2° > 1° >$ methyl. Why should this be so? The reason is an orbital mixing phenomenon. The difference between methyl and ethyl cations is the CH_3 group, and as we have emphasized in this chapter, there are two orbitals on a methyl group with π symmetry. Just as with propene (Figure 1.18), in any given conformation of the ethyl cation one of these $\pi(CH_3)$ orbitals will be of the right symmetry to mix with this empty p orbital. The difference between this situation and the mixing in propene is that the $\pi(CH_3)$ is filled and the p orbital is empty, and it is always highly favorable to mix a filled orbital with an empty orbital.

The orbital mixing diagram for ethyl carbenium is shown in Figure 1.25 **A**, along with the resulting HOMO and LUMO. The empty orbital (LUMO) of ethyl carbenium is delocalized onto the methyl group. As such, we expect a significant amount of the positive charge on the CH_3 of ethyl cation, and high-level calculations suggest a charge of roughly +0.17 associated with the CH_3. Also, there should be geometry changes. Examination of the filled orbital shows a sort of "π bond" in ethyl carbenium, and we might expect this to lead to a shortening of the C–C bond. Indeed, calculations predict a bond length of 1.425 Å, significantly shorter than a typical bond between an sp^3 and an sp^2 carbon (1.51 Å; Table 1.4). Support for this bond shortening comes from the crystal structure of the t-butyl cation [$(CH_3)_3C^+$], a result made possible by the development of stable ion media (see Chapter 2). The C–C bonds in this cation are 1.442 Å long, again shorter than a typical sp^3–sp^2 bond. We would also expect a lengthening of the C–H bonds of the CH_3, due to the removal of electrons from the bonding $\pi(CH_3)$ molecular orbital, and this too is borne out by calculations. Another interesting effect is the distortion of the methyl C–H bond that aligns with the empty orbital. As shown in Figure 1.25 **A**, this bond "leans" toward the empty orbital, reducing the C–C–H angle to 95°.

The orbital mixing interaction we have just described is simply the MOT version of what in VBT terms would be called **hyperconjugation**. Hyperconjugation is often depicted with structures such as that shown in the margin, where a neighboring bond formed from a carbon sp^3 hybrid leans toward the carbocation center. Figure 1.25, on the other hand, gives a more sophisticated picture. The essence of hyperconjugation can also be viewed with the

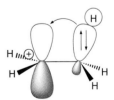

Hyperconjugation

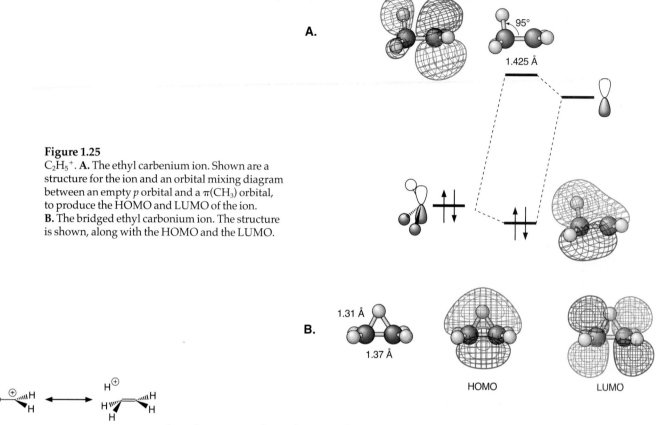

Figure 1.25
$C_2H_5^+$. **A.** The ethyl carbenium ion. Shown are a structure for the ion and an orbital mixing diagram between an empty p orbital and a $\pi(CH_3)$ orbital, to produce the HOMO and LUMO of the ion. **B.** The bridged ethyl carbonium ion. The structure is shown, along with the HOMO and the LUMO.

No-bond resonance

no-bond resonance form shown in the margin, and it successfully recapitulates all the major features of the MOT analysis.

Interplay with Carbonium Ions

The structure we have discussed for the ethyl cation is appealing and makes sense. In addition, as shown in Eq. 1.3, it is well-established that there is a very rapid scrambling of the five hydrogens of the ethyl cation both in solution and in the gas phase (we will discuss these rearrangements in Chapter 11). The scrambling of hydrogens in ethyl cation suggests some role for the bridged structure shown in the margin. The issue is whether this symmetrical, bridged structure should be considered as a transition state or an intermediate. The bridged species is now considered a carbonium ion, because it has a hypervalent hydrogen. The bonding situation is directly analogous to that of diborane discussed previously. The C–H–C unit involves a three center–two electron bond.

Bridged ethyl cation

$$\text{(Eq. 1.3)}$$

The structure of the bridged ion (be it transition state or intermediate) is intriguing, and a calculated structure is shown in Figure 1.25 **B**. To a good approximation, it is simply ethylene that has been protonated directly in the middle of the π bond. As expected, the C–C bond has elongated (1.37 Å vs. 1.31–1.34 Å for a typical olefin; see Table 1.4), and the bridging C–H bonds are quite long. The HOMO and LUMO of the cation look very much like the HOMO and LUMO of ethylene. At the same time, they are not really very different from the corresponding orbitals of the "conventional" hyperconjugated ethyl cation in Figure 1.25 **A**.

Our discussion of ethyl cation illustrates many issues that are universal in carbocation chemistry. As already mentioned, carbenium ions are especially prone to rearrangement. Often, several different but similar structures can equilibrate rapidly via such rearrangements. This rapid equilibration is a manifestation of the fact that the potential energy surfaces of carbocations are often very flat, involving several related structures that are similar in energy and that interconvert via processes with very small barriers. This is a challenging situation. Many of the most contentious debates in physical organic chemistry have boiled down to efforts to sort out very subtle structural and energetic differences on a given carbocation potential energy surface. If we look at the structures and MOs of the two forms of ethyl cation shown in Figure 1.25, the differences are certainly small. In addition, we might expect that such situations would be strongly influenced by context effects, such as solvent, counterion, temperature, exact substitution pattern, and so on.

In the specific case of ethyl cation, high-level theory actually favors the bridged structure as the lowest in energy by several kcal/mol. In fact, many calculations indicate that the classical non-bridged structure is best thought of as a transition state for the hydrogen scrambling reaction. However, in other cases the balance may be tilted in the other direction to favor the carbenium ion. For example, the *t*-butyl cation $[(CH_3)_3C^+]$ is a classical hyperconjugated carbenium ion. It is planar at the cationic carbon with 120° bond angles. Our goal at this point is simply to alert you to the potential for such complications, and to keep the big picture in mind; carbocation potential energy surfaces are often flat, with several possible isomers that differ very little in structure and energy.

Allyl and benzyl cations are the prototype delocalized carbenium ions (look at Figures 1.21 **B** and 1.22). Conventional π delocalization does not convert a carbenium ion to a carbonium ion in this case. No hypervalent atoms are involved in any of the resonance structures for allyl or benzyl. The same orbitals that carry the negative charge in allyl and benzyl anion carry the positive charge in allyl and benzyl cation.

Carbonium Ions

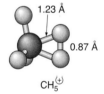

CH$_5^{(+)}$

The prototype carbonium ion, CH_5^+, is readily generated in a mass spectrometer. Also, with very strong acids like FSO_3H, protonated alkanes can be generated, making CH_5^+ an important model compound. MH_5 compounds are well known, the prototype being the trigonal bipyramidal PH_5. However, CH_5^+ does not adopt such a highly symmetric geometry. A representative structure is shown in the margin. The easiest way to think about this structure is as a pyramidal CH_3^+ bonded to a molecule of H_2. This makes a three center–two electron bond of the sort we have described for BH_3. The CH_3^+ contributes an empty σ_{out} orbital that interacts with the filled σ bonding orbital of H_2, making the three center–two electron system. The H_2 fragment has a slightly elongated H–H bond distance of 0.87 Å (vs. 0.746 Å for the isolated molecule), while the C–H bonds involving these hydrogens are substantially elongated. The positive charge is distributed around the entire structure. The molecular structure given for CH_5^+ is reasonable for discussion purposes, but as the next Going Deeper highlight describes, it does not represent the true "nature of the beast".

Going Deeper

CH$_5^+$—Not Really a Well-Defined Structure

We have just described the structure given for CH_5^+ as "representative" because the situation is actually fairly complicated. The CH_5^+ ion undergoes a rapid scrambling process in which all five hydrogens become equivalent. Estimates are that the barrier to this process is less than 1 kcal/mol, making CH_5^+ a highly fluxional molecule. The potential energy surface for this carbocation is extremely flat. In fact, some very high-level calculations have led to the conclusion that there is effectively *no barrier* to the fluxional process, in which case the notion of "molecular structure" is poorly defined. This small ion may be a quantum mechanical species, for which the classical model of a

rigid geometrical object is simply not applicable. Not surprisingly, spectroscopic characterization of such a structure is challenging. However, a very sophisticated gas phase infrared (IR) spectroscopy study reported in 1999 found a complex collection of lines that have been assigned to CH_5^+. The complexity of the spectrum and its dissimilarity to spectra of other simple molecules like CH_3^+ or CH_4 seems consistent with the non-standard description of CH_5^+. However, full assignment of the lines was not possible. Stay tuned.

White, E. T., Tang, J., and Oka, T. "CH$_5^+$: The Infrared Spectrum." *Science*, **284**, 135 (1999). Marx, D., and Parrinello, M. "Structural Quantum Effects and 3-Center 2-Electron Bonding in CH$_5^+$." *Nature*, **375**, 216 (1995).

Cyclopropylcarbinyl cation

Norbornyl cation

As discussed, an R_5C^+ species can be considered to be a bridging structure with a hypervalent carbon. The bridged picture of ethyl cation has a hypervalent hydrogen. Many other carbocations are also bridged species best defined as carbonium ions. Some of these compounds are called **non-classical cations**. These are defined as containing bridged structures with three center–two electron bonds and a hypervalent atom. Although ethyl cation fits this description, the term is normally associated with structures where a C–C σ bond bridges to create a hypervalent carbon. Two classic examples are cyclopropylcarbinyl cation and norbornyl cation, shown in the margin. We will discuss these structures much more in Chapters 2, 11, and 14.

1.4.2 Carbanions

Now let's consider the prototypical carbanion, CH_3^-, with a total of eight valence electrons. The **D** orbital of Figure 1.8 is now filled, and the stabilization it gains by pyramidalization dominates the Walsh diagram. This stabilization is enhanced by further mixing to produce **D′**. Thus, simple carbanions should be pyramidal, again consistent with experimental observations.

As noted previously, a basic tenet of the group orbital concept is that the MOs of MH_3 are essentially the same, regardless of the identity of M, the only difference being the number of valence electrons. We have already noted the isoelectronic relationships between BH_3 and CH_3^+, and between NH_3 and CH_3^-, and so, again, Figure 1.8 predicts that NH_3 should be pyramidal and BH_3 should be planar, as is observed experimentally.

The prototype structure of a simple, pyramidal carbanion has a lone pair of electrons in the σ_{out} orbital of Figure 1.8. An equivalent way to think about the structure is as an sp^3 hybrid, placing the lone pair in an sp^3-like orbital. In most cases the barrier to inversion at a carbanion center is small, roughly 1–2 kcal/mol for simple systems. Thus, on most reaction time scales, carbanions behave as if they were effectively planar (Eq. 1.4).

$$R_3\overset{\cdots}{\underset{R_2}{\text{—}}}\!\!\!\!\ominus\quad\rightleftharpoons\quad R_3\overset{R_2}{\underset{\cdots}{\text{—}}}\!\!\!\!R_1\ominus \qquad\qquad \text{(Eq. 1.4)}$$

Cyclopropyl anion

Several factors can significantly raise the **inversion barrier** at a carbanion center. One is incorporation into a small ring. Consider the inversion of the cyclopropyl anion. In the pyramidal ground state, the lone pair occupies a hybridized orbital and the carbanion center would be roughly sp^3 hybridized. In the planar transition state for inversion, the lone pair is in a p orbital, and the carbanion center is essentially sp^2 hybridized. In both the ground and transition states, the C–C bonds must be bent to accommodate the 60° bond angles of the cyclopropane ring. This is more difficult when a carbon is sp^2 hybridized (favoring 120° angles) than when the carbon is sp^3 hybridized (favoring 109.5° angles). This effect raises the energy of the planar transition state for inversion in the small ring relative to an open system. As a result, cyclopropyl anions often react as if they were stably pyramidal.

Electronegative substituents attached to the anionic center can also substantially raise the inversion barrier. This electronegativity effect is general. While ammonia has an inversion barrier of ~5 kcal/mol, NF_3 has a barrier of ~50 kcal/mol. This high barrier can be understood by recalling that electronegative substituents prefer to bond to orbitals with greater amounts of p character, because p orbitals in general are easier to withdraw electrons from than s orbitals. Thus, electronegative substituents preferentially stabilize the sp^3 ground state over the sp^2 transition state, raising the inversion barrier.

An alternative rationalization can be developed by referring to the CH_3 Walsh diagram. On going from NH_3 to NF_3 we replace H by the highly electronegative F. As noted in the discussion of formaldehyde (Section 1.3.3), electronegative atoms lower the energies of *all* the MOs to which they contribute. Returning to the Walsh diagram (Figure 1.8), the effect of fluorine on the planar structure is to lower all the orbital energies except **D**, which has no contribution from attached fluorines. However, with the pyramidal structure, all the orbitals are lowered in energy, including **D**, because they all have contributions from the fluorines. This stabilizes more electrons in the pyramidal structure of an anion because the **D** orbitals are populated, and thus raises the inversion barrier.

Connections

Pyramidal Inversion: NH₃ vs. PH₃

The pyramidal inversion barrier, E_{inv}, is a crucial property of any pyramidal tricoordinate molecule. As discussed previously, NH_3 is pyramidal, but the barrier is so low (~5 kcal/mol) that amines act planar at room temperature. PH_3, however, has a very high E_{inv}, on the order of 35 kcal/mol, so phosphines are stable pyramids at conventional temperatures.

To rationalize this difference between NH_3 and PH_3, we focus on the key interaction between **D** and **E** to generate **D′** and **E′** (Figure 1.8). This greatly stabilizes **D′** and is a key factor favoring the pyramidal form.

The magnitude of the stabilization increases as the initial energy gap between **D** and **E** decreases. Two factors make this gap *smaller* for PH_3 than for NH_3. First, P is less electronegative than N, so **D** will be higher-lying in PH_3, thus diminishing the gap. Second, the $2s$ and $2p$ orbitals in the first row of the periodic table have similar sizes, but beyond the first row the s orbitals are significantly more contracted than the p orbitals. Consequently, at distances where the phosphorus $3p$ orbitals interact strongly with the H $1s$ orbitals, the P $3s$ orbitals interact relatively weakly, making **E** less antibonding in PH_3 than in NH_3 and further diminishing the initial **D/E** gap.

Substituents that stabilize a carbanion by π delocalization will favor the planar structure. This preference is because the interaction with the π substituent will be greatest in this form, where we have a pure p orbital rather than a σ(out) orbital to participate in the delocalization. Given the small intrinsic preference for a pyramidal structure, we would expect the delocalization (resonance) effect to win out, making the anionic carbon planar. Examples of such substituents that will lead to a planar neighboring anionic carbon are cyano, nitro, and carbonyl. Resonance structures make this argument clear.

Allyl and benzyl anions are also planar. The HOMOs of these anions are shown in Figures 1.21 **B** and 1.22, respectively, and we expect significant negative charge only on those atoms that have significant contribution in the HOMO.

1.4.3 Radicals

Methyl radical has seven valence electrons. The crucial orbital that is stabilized by pyramidalization, σ_{out}, is only singly occupied (Figure 1.8), and the stabilization it provides is to some extent offset by the destabilization of other, doubly occupied orbitals. It is difficult to predict the geometry of this species based on the Walsh diagram, and it is best to say that no obvious preference for planar or pyramidal geometry can be predicted. This, too, is "consistent" with experiment, as simple radicals show only a very weak preference for the planar structure, and simple substitution can produce pyramidal radicals. The net result is that the parent methyl radical, $CH_3{}^\bullet$, is planar, but the energy cost for distorting away from planarity is small.

In fact, extensive study has revealed that the intrinsic preference for planarity in methyl radical is so small that essentially all other localized radicals are not planar. Two factors favor pyramidalization in radicals. The first is an electronic effect of the sort discussed previously for anions. $CF_3{}^\bullet$ is very strongly pyramidal for the same reason as discussed above for NF_3. Electronegative substituents prefer bonding to an sp^3 hybrid rather than an sp^2, and this preference completely shifts the balance toward pyramidalization in the case of the radical system. The second reason that radicals are pyramidal is a conformational effect, and this is discussed in Chapter 2.

Just as with allyl and benzyl cations and anions, allyl and benzyl radicals are stabilized by delocalization. The crucial molecular orbital, the **singly occupied molecular orbital (SOMO)**, is essentially the same as the MO that is doubly occupied for the respective anions (see Figures 1.21 **B** and 1.22).

There are also reactive intermediates known as **radical cations**. The geometries of such species can also be understood using our notions of bonding. One example is the one-electron oxidation of an alkene, where the electron is removed from the π orbital. For all alkenes besides ethylene (see Chapter 2 for a discussion of the radical cation of ethylene), oxidation retains a planar structure in the alkene. However, the mixing of the alkyl group's π-like group orbitals with the now singly occupied π orbital becomes even more pro-

Stabilized carbanions

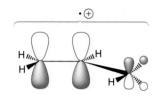

Radical cation of propene

nounced than we saw for propene (Figure 1.18). Therefore, the SOMO is even more delocalized onto the neighboring R group, as shown in the margin for the propene radical cation, where a bracket is used to denote one electron and a formal positive charge in the MO.

1.4.4   Carbenes

Lastly, let's consider carbenes, neutral :CR_2 species, the prototype of which is the molecule methylene, :CH_2. For the reactive intermediates we have considered so far—cations, anions, and radicals—the Walsh diagram for CH_3 provides the starting point for the discussion. For carbenes, the relevant electronic structure questions can be considered by referring to the Walsh diagram for CH_2 in Figure 1.9. We begin with the linear form on the left side of Figure 1.9. Methylene has six valence electrons, and we can place two electrons each in the **A** and **B** MOs, leaving two valence electrons to occupy the degenerate **C/D** pair. This leads to a unique feature of carbenes vs. the other reactive intermediates we have studied. Each carbene is in fact two reactive intermediates, differentiated by the spin state of the system. If the spins are aligned, then the spin state of the system is $S = \frac{1}{2} + \frac{1}{2} = 1$, and the multiplicity is $m_s = 2S + 1 = 3$. This is a **triplet** state. If we have opposing spins, $S = \frac{1}{2} + (-\frac{1}{2}) = 0$ and $m_s = 1$. This is a **singlet** state. These two states are expected to have substantially different molecular and electronic structures and to show distinct reactivities. The Walsh diagram of Figure 1.9 provides an excellent starting point for considering these issues.

Hund's rule predicts the high-spin, triplet state should be preferred at the linear geometry, and indeed it is. As the H–C–H angle contracts, a gap opens up between **C** and **D**. If we keep one electron in each MO, this distortion should be mildly stabilizing. With small bending, there is no large benefit to pairing two electrons into **C**, and the triplet is still preferred. However, when the angle becomes small enough, the lower energy of orbital **C** will overcome the electron repulsion energy, and a singlet with both electrons in **C** will become the ground state.

It is impossible to unambiguously predict the absolute ground state of methylene with simple models such as Walsh diagrams. As we will see in Chapter 14, understanding and predicting spin preferences requires more advanced treatments of electronic structure than we are providing here. However, some predictions are still possible. For example, the triplet state should have a wider H–C–H angle than the singlet. This is indeed the case; the angle is 136° for the triplet and 105° for the singlet. Experimentally it turns out that the triplet is the global ground state in methylene, by approximately 9 kcal/mol. All simple dialkyl carbenes have triplet ground states.

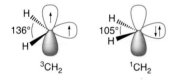

Carbene electron configurations

The Walsh diagram provides a satisfying analysis of the electronic structure of carbenes, and the essential features of the system can be summarized by the simple representations shown in the margin. The triplet has two electrons in two very different orbitals, what we have called p and σ(out). We might expect its reactivity to be similar to that of radicals, and indeed this is the case. The singlet state is quite different. It contains a lone pair of electrons in an MO [σ(out)] that is reminiscent of an sp^2 hybrid. It also contains an empty p orbital, just like a simple carbenium ion. Its reactivity patterns should be quite different from radicals, and we will see that they are (Chapter 10).

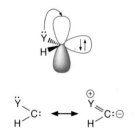

Resonance in singlet carbenes

While simple carbenes have a triplet ground state, appropriate substituents can reverse this preference, such that some substituted carbenes show a large energetic preference for the singlet. The bonding and structural model we have developed provides excellent guidance as to how we might create a carbene that has a singlet ground state. The most effective way is to interact with the empty **D** orbital of the singlet, as shown in the margin. Carbenes with lone-pair donating substituents such as N, O, and halogens can have singlet ground states because of such an interaction (see the next Going Deeper highlight for an example). The effect can be quite large; in difluorocarbene the singlet lies below the triplet by ~50 kcal/mol! Detailed theoretical studies reveal a linear correlation between the singlet–triplet gap and the electron pair donating ability of the attached substituent(s). As is typical, a resonance model also nicely rationalizes the stabilizing effect of donating substituents on carbenes.

Going Deeper

Stable Carbenes

We generally think of carbenes as extremely reactive species, and for the most part they are. In recent years, however, clever application of the concepts described here has led to some remarkable new carbenes. The breakthrough occurred in 1991 when Arduengo and co-workers at DuPont reported the synthesis and *isolation* of carbene *i*. Two factors contribute to the stability of this type of carbene. First is the steric bulk of the R group. The first example had R = adamantyl, a large, aliphatic ring system. Later examples included heavily substituted aromatics as the R group. The second effect is electronic. Two potent electron donors are attached to the carbenic center. As just discussed, these should greatly stabilize the singlet state, and indeed these types of carbenes have a singlet ground state. Remarkably, these molecules can be crystallized, and an x-ray structure reveals an N–C–N angle of 102° at the carbene center, in excellent agreement with the expectation for a singlet carbene. Samples of *i* are stable for years, as long as they are protected from air. Many derivatives have been made, and extensive physical characterization has provided detailed insights into carbene electronic structure. These stable carbenes are not solely theoretical curiosities. They have recently found use as ligands for an important class of ruthenium-based olefin metathesis catalysts that have profoundly influenced synthetic organic chemistry (see Chapter 12). Thus, research into basic reactive intermediates can lead to fundamental insights and useful new materials.

i

A stable carbene

Arduengo, A. J., III. "Looking for Stable Carbenes." *Acc. Chem. Res.*, **32**, 913–921 (1999).

1.5   A Very Quick Look at Organometallic and Inorganic Bonding

One theme of this textbook is to consistently tie organic chemistry to organometallic chemistry, which is just one of the current chemical subdisciplines where the tools of physical organic chemistry are often applied. In this regard, it is useful in this chapter to develop a simple bonding model for organometallic and inorganic complexes, and not just look at organic bonding. Here we examine a model analogous to the first VBT/MOT model of organic bonding given in Section 1.1. We will leave an examination of structure in organometallic systems to Chapter 12, and we will examine more complex MOT ideas about bonding in metal-containing systems in Chapter 14.

The same kind of localized and discrete σ and π bonds often associated with organic compounds can be assigned to the bonding in organometallic compounds. One simple concept for visualizing bonding in metal-containing systems is to make direct analogies to such bonding in organic systems. For example, let's examine the shapes and nodal properties of the d orbitals shown in Figure 1.26. First, notice that the d_{z^2} orbital is directional along the z axis, as is a hybrid orbital along an axis, and is therefore "sigma-like". In fact, we refer to it as having **sigma symmetry** along the z axis. Likewise, the $d_{x^2-y^2}$ orbital is aligned along the x and y axes, and is therefore considered to have sigma symmetry when viewed down these axes. Hence, when forming discrete localized bonds to ligands, the d_{z^2} orbital and the $d_{x^2-y^2}$ orbital can make sigma bonds that are placed along the z or the x and y axes, respectively. We essentially use these orbitals just as we use hybrid or σ(out) orbitals on carbon.

Now let's examine the analogies that can be drawn between the rest of the d orbitals and p orbitals. When citing down the x axis, the d_{xz} and d_{xy} orbitals look like two orthogonal p orbitals. Likewise, citing down the y axis, both the d_{yz} and d_{xy} look like p orbitals. Therefore, these orbitals are considered to have π **symmetry**. When creating double bonds to ligands,

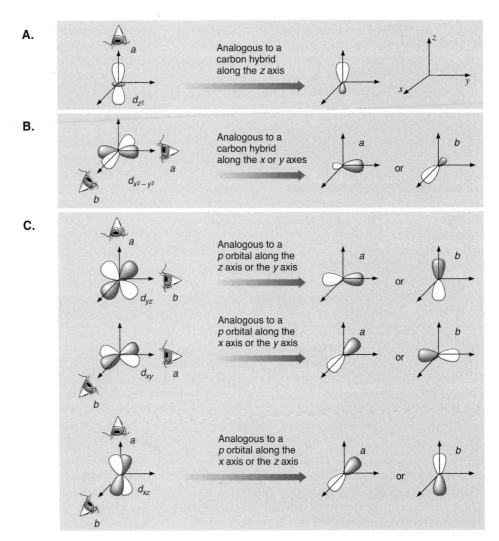

Figure 1.26
Schematic representations of d orbitals and their analogous carbon orbitals. **A.** Looking along the positive z axis (designated as an eyeball), the carbon hybrid along the z axis appears the same as a d_{z^2} orbital. They are referred to as having σ symmetry, analogous to σ(out) orbitals or the standard hybrid orbitals for making σ bonds to organic groups. **B.** Looking along the positive x or y axes, the $d_{x^2-y^2}$ orbital looks like a carbon hybrid. **C.** The d_{yz}, d_{xy}, and d_{xz} orbitals appear as p orbitals, depending upon the line of sight. They are referred to as having π symmetry. The a and b descriptors on each orbital refer to the line of sight for the corresponding eyeballs.

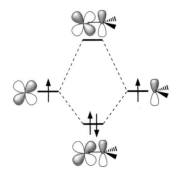

Figure 1.27
Orbital mixing diagram for the combination of a π-symmetry d orbital with a carbon p orbital to create a metal–carbon π bond.

these orbitals on the metals are used exactly as we use p orbitals on carbons. This combination is very clear in Figure 1.27, where an orbital mixing diagram is given for a π bond between a metal and a carbon, as would be present in an organometallic complex that has an M=CR_2 functional group.

Lastly, there is even a hybrid orbital approach to modeling the bonding in organometallic and inorganic complexes. Using the same concepts presented in Section 1.1.6 for the linear combinations of s and p orbitals to create hybrids, we can mix d orbitals with the s and p orbitals. For example, if you start with an sp^2 hybridization state, and mix the remaining p_z orbital with the d_{z^2} orbital, a hybridization state know as dsp^3 is obtained. As shown in Figure 1.28, the bonds to the apical positions are pd hybrid orbitals, while the equatorial positions are sp^2 hybrid orbitals. Such hybridization is appropriate for trigonal bipyramidal complexes.

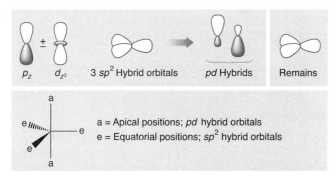

Figure 1.28
Hybridization for trigonal bipyramidal geometries starts with the central atom sp^2 hybridized. The orbitals aligned along the equatorial positions remain as sp^2 hybrids. However, the orbitals along apical positions are pd hybrids.

The mixing of an s, three p's, and the d_{z^2} and $d_{x^2-y^2}$ orbitals leads to d^2sp^3 hybridization, which is appropriate for octahedral complexes. For almost all bonding geometries in inorganic and organometallic chemistry, hybrid orbitals are useful, and this very brief introduction to bonding using metals will be enough to take us a long way in understanding structure and reactivity.

Summary and Outlook

This chapter represents just a beginning. It is only a first look into bonding, with hints at structure and reactivity. We presented two models for bonding, a classical one and a more modern approach, and showed that they can be used to understand stable organic structures and reactive intermediates. Hopefully this chapter has refreshed your memory, and has sparked an interest in you to learn how these notions of bonding can be put to use. We do exactly that in the next several chapters where the focus is more upon structure. Furthermore, after analyzing structure, we can take the bonding concepts and look at reactivity. Structure and reactivity actually take up approximately two-thirds of this book, and not until we have to look at some very specialized reactions (pericyclic and photochemical) will we need to develop a more sophisticated theory of bonding. With our current qualitative models we can go a long way in our analysis of topics in organic chemistry.

For many students and professors, this chapter may completely suffice as a review of bonding, because it is sufficient for almost all organic transformations. For other students and professors, however, it may be desirable to now go directly to Chapter 14, where the concepts introduced herein are discussed more quantitatively and modern methods in computational electronic structure theory are covered. This is a decision to be made on an individual basis. However, it should be appreciated that to the best of the authors' abilities, we took the topics of this chapter only to a depth that is routinely used when thinking about organic structure and bonding by a non-expert in quantitative methods. Our intention is for Chapter 14 to stand alone, so it can be covered at any point during the course to learn more advanced concepts and quantitative methods.

Exercises

1. State the hybridization of the non-hydrogen atoms in the following structures.

A.   **B.**   **C.**   **D.**   **E.**

2. Assuming that each atom in the following structures has an octet of electrons, identify which compounds have an atom that has a formal charge. Identify what and where that charge is. In the compounds without formal charges, identify any significant bond polarizations by writing $\delta+$ and $\delta-$ near the appropriate atoms.

A.   **B.**   **C.**   **D.**   **E.**

3. For the bond polarizations in Exercise 2, draw a dipole arrow for the polarized bonds.

4. Draw the σ and σ^* molecular orbitals for discrete and localized bonds formed between hybridized carbons and heteroatoms (a heteroatom is any atom besides C and H), as well as heteroatoms and hydrogen, in the following structures. Indicate any polarization these bonds may have in your diagram by drawing the shapes of the discrete bonding and antibonding orbitals in a manner indicative of this polarization.

A.   **B.**   **C.**

5. Draw the π and π^* molecular orbitals for the discrete and localized π bonds in the following structures. Indicate any polarization these bonds may have in your diagram through the shapes of the discrete bonding and antibonding orbitals.

A.   **B.**   **C.**   **D.**

6. Show any plausible resonance structures for the molecules given in Exercises 1 and 5. If there are no plausible resonance structures, indicate this (note that the molecule corresponding to letter **A** is the same in each problem).

7. Discuss the hybridization in the C–C and C–H bonds of cyclopropane, given that the H–C–H bond angle is 118°.

8. Carbon tetrachloride has neither a dipole moment nor a quadrupole moment, but it does have an octopole moment. Provide a simple description of this moment.

9. Formaldehyde has a fairly large dipole moment of 2.33 D, but CO has a small dipole moment of 0.11 D. Use resonance and electronegativity arguments to explain these results.

10. Consider bond dipoles to predict which conformer of formic acid should have the higher dipole moment, **A** or **B**.

A.   **B.**

11. Pauling proposed the following correlation between electronegativity difference and percent ionic character in a bond: Ionic character $= 100 \times [1 - e^{-(X_A - X_B)^2/4}]$ where X_A is the electronegativity of A. Calculate the percent ionic character in HF, HCl, HBr, and HI. The most polar bond for the elements in Table 1.1 would be in KF. What is its percent ionic character?

12. Which should have a larger dipole moment, formic acid (the higher dipole conformer of Exercise 10) or formamide?

13. The electronegativities of the –C≡CH and –C≡N groups are the same (Table 1.1), but the dipole moments of $CH_3C\equiv CH$ and $CH_3C\equiv N$ are 0.78 D and 3.92 D, respectively. Explain why the dipole moments are so different.

14. Draw a molecule that has no dipole moment, but has a quadrupole moment with the topology of a d_{xy} orbital, rather than the d_{z^2} topology.

15. Predict the hybridization state, sp^n, for the carbon hybrid orbitals used for the C–H bonds of CH_3Cl (H–C–H angle of 110.5°), and the C–H bonds of ethylene (H–C–H angle of 117.3°).

16. Rationalize the differences between the following dipole moments for water, dimethyl ether, and ethylene oxide (oxirane):

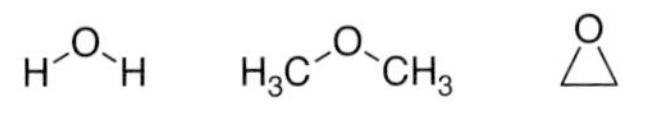

H⌒O⌒H	H_3C⌒O⌒CH_3	△ (O)
1.85	1.30	1.89

17. Convince yourself that the C–H bonds of cyclohexane would add up to give a quadrupole moment, as they do in benzene, if carbon were significantly more electronegative than hydrogen. Contrast this quadrupole moment to that in benzene.

18. Geometries for small molecules such as methyl fluoride can be determined from microwave spectroscopy, whereas various forms of diffraction—x-ray, electron, or neutron diffraction—can be applied to larger molecules. Shown below are bond lengths and angles for various methyl halides determined by various methods. Rationalize the trend in the C–X bond lengths (angstroms). Also rationalize the H–C–H and H–C–X bond angles using VSEPR. Does an argument based upon electronegativities also explain the trends in bond angles?

Molecule	C–H bond length	C–X bond length	H–C–H bond angle	H–C–X bond angle
CH_3F	1.09	1.385	110.2	108.2
CH_3Cl	1.096	1.781	110.52	108.0
CH_3Br	1.11	1.939	111.12	107.14
CH_3I	1.096	2.139	111.5	106.58

19. The calculated H–C–H bond angle in methyl radical is 120°, the calculated H–C–F angle in CH_2F radical is 115°, and the calculated F–C–F angle in CHF_2 is 112°. Rationalize why the structures become more pyramidal with increasing fluorine substitution.

20. Describe and draw (using cartoons) how the three group orbitals for a pyramidal methyl group (Figure 1.8) can be linearly combined to give orbitals similar to the individual σ bonds formed from sp^3 hybridized carbon atoms and hydrogen 1s orbitals.

21. Draw mixing diagrams using σ(out) orbitals that compare the formation of the C–C σ bond in ethane, created by the approach of two pyramidal methyl groups, to the formation of the C–F σ bond in CH_3F. Use the same energy axis for both plots. Highlight the differences and explain how this diagram would rationalize reactivity differences in alkanes and alkyl fluorides with nucleophiles.

22. Rationalize the differences between the π orbitals of butadiene and acrolein (the MOs are shown in Appendix 3). The results of Figure 1.20 should be a useful starting point.

23. Sketch the highest occupied MO of propene and compare and contrast this MO with the analogous MO of acetaldehyde ($CH_3CH=O$). Consider the methyl C–H bonding in your answer.

24. Sketch the π* MO of propene. Consider the methyl C–H bonding in your answer.

25. The H–P–H bond angle in PH_3 is nearly 90°, whereas that for NH_3 is only slightly contracted from the idealized 109.5°. Explain this difference using a VSEPR argument, and then use a hybridization/electronegativity argument.

26. Draw an orbital mixing diagram for a metal–methyl single bond using the d_{z^2} orbital on the metal and a σ(out) orbital on the methyl group.

27. Sketch the π and lone pair group orbitals for the following organometallic group, called a Fischer carbene. (*Hint:* Use the group orbitals of formic acid as your guide.)

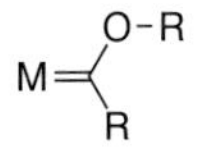

28. Use an orbital mixing diagram to rationalize the nature of the HOMO of methylamine shown in Appendix 3.

29. Starting with two allyl groups, use orbital mixing to predict all the orbitals for the π system of hexatriene.

30. Create the group orbitals for an MH system, where M is a second row element such as C or N.

31. Use the appropriate group orbitals and the QMOT rules in Table 1.7 to create the molecular orbitals of protonated formaldehyde ($CH_2=OH^+$), starting with methylene and OH.

32. Sketch the π orbitals for allyl anion and for an enolate anion, and discuss the differences.

33. Combine the lone pair orbitals on formaldehyde to achieve what resemble the classic sp^2 hybrid lone pairs.

34. Draw the virtual orbitals that were not shown in Figure 1.8 for the planar and pyramidal forms of CH_3 and in Figure 1.9 for the linear and bent forms of CH_2.

35. If there are two types of ligands bound to a trigonal bipyramidal metal atom, will the more electronegative ligands prefer the apical or equatorial positions? Explain your answer.

Further Reading

Much of the material in this chapter is review, and if you are not familiar with the material in Section 1.1, you should consult your introductory texts in general chemistry, organic chemistry, and/or physical chemistry. Selected references to more advanced topics are given below.

Qualitative Molecular Orbital Theory, Perturbation Theory, and Group Orbitals

Albright, T. A., Burdett, J. K., and Whangbo, M.-H. (1985). *Orbital Interactions in Chemistry*, John Wiley & Sons, New York.

Borden, W. T. (1975). *Modern Molecular Orbital Theory for Organic Chemists*, Prentice–Hall, Englewood Cliffs, NJ.

Gimarc, B. M. (1979). *Molecular Structure and Bonding: The Qualitative Molecular Orbital Approach*, Academic Press, New York.

Jorgensen, W. L., and Salem, L. (1973). *The Organic Chemist's Book of Orbitals*, Academic Press, New York. (A classic book that is unfortunately out of print—get one if you can!)

Valence Bond Theory, Hybridization, Electronegativity, and Resonance

Pauling, L. (1960). *The Nature of the Chemical Bond and the Structure of Molecules and Crystals; an Introduction to Modern Structural Chemistry*, 3d ed., Cornell University Press, Ithaca, NY.

Wheland, G. W. (1995). *Resonance Theory in Organic Chemistry*, John Wiley & Sons, New York.

Epiotis, N. D. (1983). *Unified Valence Bond Theory of Electronic Structure*, Springer–Verlag, Berlin.

Bodrowicz, F. W., and Goddard, W. A., III in *Modern Theoretical Chemistry, Methods of Electronic Structure Theory*, H. F. Schaefer III (ed.), Plenum Press, New York, 1977, Vol. 3, Chapter 4.

A Useful Compilation of Standard Bond Lengths

Allen, F. H., Kennard, O., Watson, D. G., Brammer, L., Orpen, A. G., and Taylor, R. "Tables of Bond Lengths Determined by X-ray and Neutron Diffraction. Part 1. Bond Lengths in Organic Compounds." *J. Chem. Soc., Perkin Trans. II*, S1–S19 (1997).

CHAPTER 2

Strain and Stability

Intent and Purpose

When organic chemists consider the reactivity of a new molecule, or propose a potential synthesis of a new target, or attempt to predict the lowest energy conformation of a new structure, a rapid evaluation of strains and stabilizing effects, in part, leads to the answer. The main goal of this chapter is to bring the student "up-to-speed" in this thought process. The focus is on the energetics associated with structure. In that regard, a major goal of the chapter is to explore the energetic consequences of deviations from the standard geometrical parameters described in Chapter 1.

To start off, we review some basic concepts of thermochemistry (i.e., strain and stability), and the quantities used to measure them (i.e., Gibbs free energy, enthalpy, and entropy). Since strain and stability can be tied to bond strengths, a logical first topic is various trends in bond dissociation energies. In this discussion, we link stretching vibrational modes to bond homolysis, and discuss the fact that all internal motions of molecules are dictated by internal forces that are represented by potential surfaces. Next, the overall stability of a compound is defined; chemists routinely use the heat of formation as a number that can be compared from structure to structure. We then show that the energetics of basic organic molecules can be estimated using a surprisingly simple model called thermochemical group increments. We end our discussion of thermochemistry by considering basic reactive intermediates—radicals, carbocations, and carbanions. As in Chapter 1, we combine our discussion of stable molecules and reactive intermediates in order to emphasize the similarities and the differences.

Next, we extend the discussion of internal motions and consider more subtle variations from standard structural parameters, such as eclipsed vs. staggered ethane, and the variations that occur with ring puckering. This is a field known as conformational analysis. In addition, dramatic deviations from standard bonding parameters, such as in cyclopropane, are covered. This leads to the idea of strain energy, a very important concept in physical organic chemistry. We also consider molecules in which special bonding arrangements lead to novel structural and energetic consequences, often associated with increased stability. These include topics such as aromaticity, the anomeric affect, and others that can be rationalized by orbital interactions. We also explore the structures of some of the "exotic" molecules that chemists have made in an attempt to push the limits of structure and bonding. Strain, stability, and conformational effects all have ramifications on reactivity. However, except for a few Connections highlights that emphasize our current discussions, the effects on reactivity are left to the chapters in Part II of this book.

Finally, we will show that the relationship between deviation from standard bonding parameters and energy can be put on a quantitative basis using the molecular mechanics method. While this simple method has limited theoretical justification, it is remarkably useful in predicting the structures and energies of a wide range of organic structures. Almost all practicing organic chemists now avail themselves of this method, making it a must topic for organic chemistry in the early 21st century. Too often this method is used as a "black box" computational tool, and the goal of this section of the chapter is to present the basic tenets and the strengths, but also the weaknesses, of the method.

2.1   Thermochemistry of Stable Molecules

Structure and energy are intimately related. Associated with any structure is a "total energy" or "internal energy"—numbers that are of minimal value in isolation, but are quite telling when compared to the same number for another molecule. For organic molecules, a large number of experimental energies of different kinds are known, and systematic variations are seen that clearly relate structure to energy. In this chapter we discuss the types of energies involved, and we present correlations that show how useful the energies are in comparing similar structures. Importantly, we are leading up to a precise definition of a key concept in organic chemistry—strain energy (see Section 2.1.7). Furthermore, certain bonding arrangements lead to a stabilization of a chemical structure, and these should be considered, along with strain, when analyzing a molecule's structure and reactivity.

2.1.1   The Concepts of Internal Strain and Relative Stability

What do chemists mean by **strain**? We are referring to a structural stress within a molecule that is not present in some reference compound. This results in more internal energy within the strained molecule relative to the reference. It is important to always know what the reference is, in order to fully understand the strain that is being discussed. All thermodynamic values, strain energy being only one example, are relative; *there must always be a reference state*. In organic chemistry, strain is typically associated with a conformational distortion or nonoptimal bonding situation relative to standard organic structures. The reference structure may be a completely different chemical compound that lacks the particular strain, or a different conformation where the strain is relieved.

What is **internal energy**? It is the energy held or stored within a molecule. Part of this energy can be released when given an outlet such as a chemical reaction. In that sense, it is analogous to potential energy within a compressed spring, or a brick raised above the ground. Hence, examining a common everyday structure, such as a spring, as an analogy to internal energy can be quite informative.

The introduction of strain into an organic molecule is perfectly analogous to the stretching or compressing of a spring. If the relaxed state of the spring is the reference (the most stable arrangement), then stretching or compressing leads to a strained form of the spring, where the spring now has stored potential energy (PE) (increased internal energy; see Figure 2.1 **A**) that can be released by returning to the relaxed form. *Fundamentally, all energy is related to the ability of a system to do work.* A stretched spring has more internal energy because it can do work by returning to the reference state, perhaps pulling a block in the process. Chemicals can do work via chemical reactions, such as the force created by an explosion of TNT.

Because of this analogy to potential energy, chemists often write reaction coordinate diagrams and conformational analysis plots with the energy axis labeled as PE or just E (Figure 2.1). We refer to the diagrams as potential energy diagrams (or surfaces), which implies some function that constrains the different internal motions of the molecule or any possible chemical reactions. This concept nicely conveys the notion of higher internal energy for one structure on the diagram relative to another, where that energy is capable of being released. Very often, however, we are more explicit in defining the energy, using Gibbs free energy or enthalpy (see below).

Molecules can also occupy different quantized energy states associated with various internal vibrational and electronic states. For example, absorption of a photon of infrared radiation can lead to a higher energy molecule that has an excited vibrational mode (see a brief discussion below), or absorption of UV light can lead to a higher energy electronic state (see Chapter 16). These phenomena also add potential energy to the molecules. Chemists do not typically associate these higher energy states with strain, but there are many ties between the two concepts, and we comment upon them in this chapter where appropriate.

To obtain a strained chemical structure or higher quantized energy state, energy needs to be added, just as with a spring. The energy can come from a collision, from a photon, or can be placed in the molecule during a synthetic procedure. However, there is also thermal

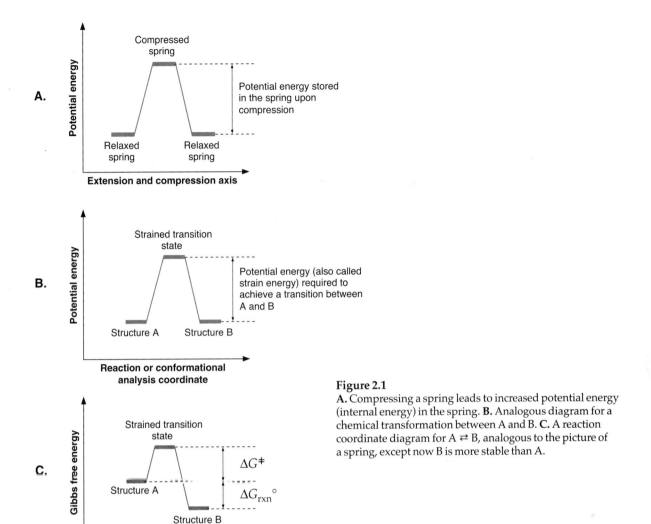

Figure 2.1
A. Compressing a spring leads to increased potential energy (internal energy) in the spring. **B.** Analogous diagram for a chemical transformation between A and B. **C.** A reaction coordinate diagram for A ⇄ B, analogous to the picture of a spring, except now B is more stable than A.

energy in the solution, which is measured by temperature. Various quantized energy states, or strained chemical structures, can be populated depending upon the temperature. The Boltzmann distribution (covered in Chapter 7) tells us how to predict this based upon the temperature. We compare the energy required to get to the higher energy state, on a per mole basis, to RT (where R is the gas constant and T is the temperature in kelvin). If the energy required to obtain the higher energy state or strained form of a molecule is lower than or comparable to RT, then this state will be significantly populated at that temperature without the addition of more energy. This will be important when we examine small barriers to conformational changes, such as rotation of the methyl groups in ethane.

In summary, when we place a molecule higher on a potential energy diagram, we are implying that it has more internal energy than the more stable reference compounds placed lower on that plot. Chemists use several terms to describe this situation, stating that these structures are higher in energy, less stable, and/or strained. Before looking at some types of strain or stabilizing factors, let's recall a few of the basics of thermodynamics. The goal here is to gain an understanding of the forms of potential energy that chemists use to discuss chemical stability and strain.

2.1.2   Types of Energy

When we compare the internal energy of one molecule to another as a means of understanding stability and strain, the analysis focuses upon a chemical reaction, a conformational difference, or a difference in population of quantized energy states. In this regard, we always examine the relationship between two or more states. Our experiments tell us the extent to which one state is preferred over the other.

Gibbs Free Energy

It is the change in **Gibbs free energy** between two different chemical states (compounds or conformations) that determines the position of the equilibrium between these states. We denote the Gibbs free energy change for any process as $\Delta G°$. Here, the $\Delta G°$ of a transformation is best thought of as the difference in stability of two different compositions of an ensemble of molecules at standard states and at constant pressure. A more precise definition is given in Section 3.1.5, where we show that Gibbs free energy is akin to a driving force (or a potential energy) for a spontaneous change in composition.

Eq. 2.1 presents the relationship between the equilibrium constant, K_{eq}, and the free energy change for any chemical process, $\Delta G°$ (the ° symbolizes we are considering standard states; see Section 3.1.5 for a discussion of standard states and insight into the origin of Eq. 2.1). Eq. 2.2 gives the ratio of species A and B at equilibrium for a simple transformation that interconverts A and B ($A \rightleftarrows B$). The same equations are used to describe a full-fledged reaction or a simple interconversion of two conformations. Figure 2.1 **C** shows a reaction coordinate diagram for the interconversion of A and B, where B has less internal energy. In this chapter we consider only thermodynamics, so we are concerned with the relative energies of A and B—the $\Delta G°$. In Chapter 7 we will discuss the pathway interconverting A and B and the meaning of $\Delta G^{\ddagger}$.

$$\ln K_{eq} = \frac{-\Delta G°}{RT} \qquad \text{(Eq. 2.1)}$$

$$K_{eq} = \frac{[B]}{[A]} \qquad \text{(Eq. 2.2)}$$

Recall that the square brackets in the equilibrium relationship given in Eq. 2.2 designate concentration, measured in mol/L. We are using kcal/mol as the standard unit of energy in this text. The unit kJ/mol is also commonly used, and the conversion is **1 kcal/mol = 4.184 kJ/mol**.

When the Gibbs free energy of B is lower than A, the transformation of A to B is **exergonic**. If we start with excess A, a spontaneous change in the composition of the solution will occur to create B. Conversely, when the free energy of B is higher than A, the reaction is **endergonic**. Use these terms when examining a Gibbs free energy diagram. Different terms ("exothermic" and "endothermic"; see below) are used when the energy axis is enthalpy. It is also important to remember that when we say "a spontaneous change will occur" we are implying nothing about how long it will take for that change to occur. Thermodynamics only tells us the direction in which the change will occur; we need kinetics (Chapter 7) to provide a time scale for the transformation.

A useful relationship to remember is that every 1.36 kcal/mol in $\Delta G°$ is worth a factor of 10 in an equilibrium constant at 298 K (see Table 2.1). Eq. 2.1 tells us that temperature effects are important also. A transformation that has a $\Delta G°$ of –1.0 kcal/mol at 298 K has a 85:15 ratio of B and A, but at –78 °C we will see a 93:7 ratio.

The free energy has two components, the **enthalpy** ($\Delta H°$) and the **entropy** ($\Delta S°$), related by the Gibbs–Helmholtz equation (Eq. 2.3). Typically, enthalpy is measured in kcal/mol, and entropy in **entropy units (eu)**, which are equivalent to cal/mol•K. Therefore, for entropy to be an energy value, it must be multiplied by the temperature. This has an important consequence. Changes in temperature affect the free energy between A and B, and therefore

Table 2.1
Composition of an A/B Mixture as a Function of Gibbs Free Energy Difference and Temperature

$\Delta G°$	K	% B	% A
At 298 K			
0	1	50	50
−1.36	10	90.9	9.1
−2.72	100	99	1
−4.08	1000	99.9	0.1
−0.5	2.33	70	30
−1.0	5.44	85	15
At 195.15 K (−78 °C)			
−1.0	13.2	93	7

the equilibrium constant. Combining Eq. 2.1 and Eq. 2.3, we obtain Eq. 2.4, showing how an equilibrium constant is influenced by temperature.

$$\Delta G° = \Delta H° - T\Delta S° \qquad \text{(Eq. 2.3)}$$

$$\ln K_{eq} = -\frac{\Delta H°}{RT} + \frac{\Delta S°}{R} \qquad \text{(Eq. 2.4)}$$

Enthalpy

The change in **enthalpy** is defined as the change in heat between two different compositions of an ensemble of molecules at constant pressure if no work is done. We have already stated that the energy of any system is defined as its capacity to do work. The energy between a starting and final state changes as work is done. However, the energy of a system also changes due to variations in temperature, where now the energy difference between two states has been transferred in the form of heat. This is what is most relevant to chemistry. Chemical reactions often perform work, such as combustion reactions of gasoline that drive the pistons in car motors. But, in the laboratory, most chemical transformations are not set up to do work, so instead the reactions release or absorb energy by changing the temperature of the reaction vessel. In relatively rare but interesting cases, the energy is released in the form of a photon, and a few examples of this will be given in Chapter 16.

In chemistry, a change in heat is accompanied by a change in bonding between two states (both intramolecular and intermolecular bonding). Hence, the easiest way to view a change in enthalpy is to associate it with changes in bond strengths, and we will explore bond strengths extensively in this chapter. In fact, all kinds of strains and issues of stability can be related to bond strengths. In this chapter we will explicitly define several forms of strain: bond angle strain, torsional strain, steric strain, etc. Each of these strains results from the weakening of the bonds in the molecule, which makes the molecule less stable. That is why this chapter is primarily focused upon enthalpy considerations.

For any reaction or change in conformation we define a **heat of reaction** (designated $\Delta H_{rxn}°$, but normally just written as $\Delta H°$). This value reflects the difference in heat between two compositions of a solution. We will look at several heats of reaction in this book (such as heats of formation, combustion, hydrogenation, etc.).

When the conversion of A to B releases energy, most often in the form of heat, we say the reaction is **exothermic** ($\Delta H°$ is negative). If the conversion requires an uptake of energy, which can cool the solution, we say the reaction is **endothermic** ($\Delta H°$ is positive). For exothermic and endothermic reactions we place the energy of B lower or higher than the energy of A, respectively, on a reaction coordinate diagram (or conformational analysis plot) with enthalpy as the energy axis.

Entropy

The **entropy** of a system is a measure of the disorder of that system. Increased heat always leads to more molecular disorder. The easiest way to consider entropy in chemistry is to associate it with all the kinds of molecular and atomic movements, referred to as **degrees of freedom**. The more degrees of freedom, and the more "loose" these are, the greater (and more favorable) the entropy. There are three different kinds of degrees of freedom: **translational, rotational**, and **vibrational**. Translational and rotational refer to the translation of the molecule throughout space and the tumbling of the molecule, respectively. Vibrational entropy is much more complex. Here we refer to every kind of internal motion of the molecule, such as bond stretches, bond rotations, and various forms of bond angle vibrations. The more freely that a bond rotates, a bond angle bends, or a bond stretch occurs, the more favorable the entropy. In general, the more kinds of motions and the more unconstrained those motions are, the more favorable the entropy.

There is also a statistical viewpoint from which to examine the entropy of a system. The more **configurations** that a system can exist in, the more favorable the entropy. By configurations, we mean in part different geometries—for example, conformations (see the discussion of conformational analysis given below). The more conformations a molecule can have, the more disordered it is. For conformations of the same energy, the entropy increases as $R\ln(n)$ where n is the number of identical conformations. The relationship is more complex for configurations that are not at the same energy.

There are approximate values that we can give to the entropies of certain degrees of freedom. These are convenient numbers to remember, because they can give quick insight into entropy changes that occur between two states. The translational entropy of a small molecule at 1 M concentration is around 30 eu. The more concentrated the solution, the lower the entropy because the movement of the molecule is more restricted (see Section 3.1.5 for further discussion). This means that the loss of translation of a molecule during a reaction will cost around 30 eu. The rotational entropy is also high, around 30 eu for a common organic molecule. It is independent of concentration. Lastly, high frequency vibrations (as found with strong covalent bonds) make no significant contributions to the entropy of the molecule, but low frequency vibrations can contribute a few entropy units (see Section 2.3.2 on ring vibrational modes). Internal bond rotations contribute on average around 3 to 5 eu. This means that the restriction of a freely rotating bond during a chemical transformation will cost around 3 to 5 eu. Examples of this are given in the following Going Deeper highlight, which gives the entropy differences between linear and cyclic alkanes.

While entropy is certainly important for significant changes in chemical structure (such as a cyclization), often when comparing two similar structures, the *difference* in entropies, $\Delta S°$, will be fairly small. For this reason, the majority of the discussions of thermochemistry focus on $\Delta H°$. Often, convincing arguments about the relative stabilities of two or more structures can be made by considering $\Delta H°$ values alone, because they can display large variations among closely related structures. However, we must always remember that it is $\Delta G°$ that sets the equilibrium constant. Especially when differences in $\Delta H°$ are small, we may have to consider entropy. In addition, as we will see in Chapters 3 and 4, entropy effects are often very important in molecular recognition and solvation phenomena, and so we will discuss entropy much more in those chapters.

2.1.3 Bond Dissociation Energies

We have already stated that enthalpy is best considered in chemistry as being manifest in bond strengths. A valuable thermodynamic quantity that has been extensively studied is the **bond dissociation energy (BDE)**. Recall from introductory organic chemistry that BDE is defined as $\Delta H°$ for the process shown in Eq. 2.5, and a wide variety of techniques has been developed to determine BDEs directly or indirectly. This is a gas phase reaction, all species are in their standard states, and the bond cleavage is always **homolytic** (producing two radicals). The BDE provides a rigorous definition of a **bond strength**.

$$R - R \rightarrow R^\bullet + R^\bullet \quad \Delta H° = BDE \qquad \text{(Eq. 2.5)}$$

Going Deeper

Entropy Changes During Cyclization Reactions

In this chapter we will examine the enthalpy changes for bond rotations in linear chemical structures, and the preferences for particular conformations of single bonds. We will also examine the relative enthalpies of conformations of cyclic systems. However, it is the comparison of the entropies of these two kinds of systems, linear and cyclic, that leads to the conclusion that there is on average a 3 to 5 eu change for the freezing of a bond rotation. At right are shown a series of differences in entropies between linear and cyclic alkanes. Note that the entropy difference becomes less favorable as the ring gets bigger, with the exception of cyclohexane, which is the least favorable relative to the open chain form. As the rings become bigger, more bond rotations must be frozen to form the ring. As we'll show later, cyclohexane is an exception to the general trends in cycloalkanes. If you want to just remember one number when considering how much entropy is lost when a bond rotation is frozen, most chemists use 4.5 eu.

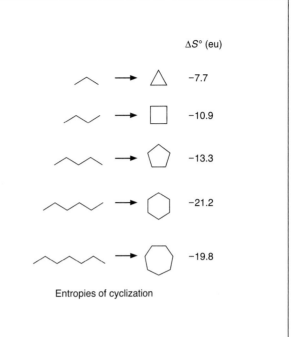

Entropies of cyclization

Page, M. I., and Jencks, W. P. "Entropic Contributions to Rate Accelerations in Enzymic and Intramolecular Reactions and the Chelate Effect." *Proc. Natl. Acad. Sci, USA,* **68**, 1678–1683 (1971).

Since BDE is a meaningful representation of the strength of a particular bond, comparing the BDEs of various bonds allows us to consider which are the strong and weak bonds in a molecule. The sum of all the BDEs in one molecule relative to the sum in another molecule with the same number and types of bonds will tell us which molecule is more stable, assuming there are no large entropy changes. Although we must always keep in mind the distinction between thermodynamic and kinetic stability, in many instances the thermodynamic BDE can be used to predict the reactivity of a molecule. This is especially true of thermal and photochemical reactions, in which bond homolysis is often a key initial step. In addition, BDEs are reflective of the nature of the chemical bond, so our models of chemical bonding should be able to rationalize variations in bond dissociation energies.

Table 2.2 lists a number of prototypical BDEs, and several trends are evident. As with all thermochemical data, the values are temperature dependent, and depend upon whether they are measured in the gas phase or solution (Table 2.2 gives gas phase values for room temperature). For CH_3–X bonding, there are clear trends, the first being F > OH > NH_2. Electronegativity is the major influence operating here. We noted in Chapter 1 that introducing some polarity into a covalent bond can strengthen the bond. The CH_3–X BDE trend supports this, in that the larger the electronegativity difference, the stronger the bond. This is the main effect in the first-row series involving F, O, N, and C. Recall also from Table 1.4 that bond lengths follow the inverse trend: C–F < C–O < C–N < C–C. Data such as these lead to the general view that shorter bonds are stronger bonds. Similar trends are seen in H–X bond strengths: O–H > N–H > C–H.

In considering the methyl–halogen bonds, CH_3X, the trend is F > Cl > Br > I. The same electronegativity effect discussed above contributes to this trend. In addition, as we move down the periodic table, the valence orbitals of X get progressively larger. The larger orbital size leads to a size mismatch with the carbon valence orbitals, and this weakens the bond by decreasing orbital overlap. We noted in Chapter 1 that orbital mixing decreases with smaller overlap and larger energy gaps. Hence, the C–I bond is much weaker than the C–C bond, even though C and I have comparable electronegativities.

Hybridization and resonance also contribute significantly to bond strengths. A C(sp)–H bond is stronger than a C(sp^2)–H bond, which is stronger than the analogous C(sp^3)–H bond.

Table 2.2
Some Specific Bond Dissociation Energies (in kcal/mol)*

Bond	BDE	Bond	BDE	Bond	BDE
H–H	104.2 (104.2)	$CH_2=CH–H$	110 (110.7)	$CH_3–CH_3$	90.4 (90.1)
$CH_3–H$	105.1 (105.0)	$C_6H_5–H$	110.9 (112.9)	$CH_3–F$	109.9 (115)
$CH_3CH_2–H$	98.2 (101.1)	$HC≡C–H$	132 (131.9)	$CH_3–Cl$	84.6 (83.7)
$(CH_3)_2CH–H$	95.1 (98.6)	$C_6H_5CH_2–H$	88 (89.7)	$CH_3–Br$	70.9 (72.1)
$(CH_3)_3C–H$	93.2 (96.5)	$CH_2=CHCH_2–H$	86.3 (88.8)	$CH_3–I$	57.2 (57.6)
$c(CH_2)_3–H$	106.3	$CH_3C(O)–H$	86 (88.1)	$CH_3–OH$	92.3 (92.1)
$c(CH_2)_4–H$	96.5	HO–H	119 (118.8)	$CH_3–NH_2$	84.9 (85.2)
$c(CH_2)_5–H$	94.5	$CH_3O–H$	104.4 (104.6)	$CH_3–SH$	74
$c(CH_2)_6–H$	95.5	$NH_2–H$	107.4 (107.6)	$CH_3–SiH_3$	88.2
(cyclopentenyl)–H	82.3	$CH_3S–H$	90.7 (87.4)	$CH_3–SiMe_3$	89.4
		HO–OH	51	$CH_3–GeMe_3$	83
(cyclopentadienyl)–H	71.1	$CH_3O–OCH_3$	37.6 (38)	$CH_3–SnMe_3$	71
		$HOCH_2–H$	94 (96.1)	$CH_3–PbMe_3$	57
(cyclohexadienyl)–H	73	$H_2C=CH_2$	(174.1)	$CH_3–OCH_3$	(83.2)
(cyclopropylmethyl)–H	97.4	$HC≡CH$	(230.7)	$CH_3–C_2H_5$	(89.0)
(cyclopropenyl)–H	90.6	$H_2C=O$	(178.8)	$CH_3–CH(CH_3)_2$	(88.6)
$CH_3–CH=CH_2$	(101.4)	$CH_3–C_6H_5$	(103.5)	$CH_3–C(CH_3)_3$	(87.5)
$C_6H_5–C_6H_5$	(118)	$CH_3–CH_2C_6H_5$	(77.6)	$CH_3–CH_2CH=CH_2$	(76.5)

*The bond of interest is shown in color. Values are from two sources. Numbers not in parentheses are from McMillen, D. F., and Golden, D. M. "Hydrocarbon Bond Dissociation Energies." *Ann. Rev. Phys. Chem.*, **33**, 493 (1982). Numbers in parentheses are from a recent attempt to provide the most current estimates, reconciling variations among results obtained from different methods. Blanksby, S. J., and Ellison, G. B. "Bond Dissociation Energies of Organic Molecules." *Acc. Chem. Res.*, **36**, 255–263 (2003).

Remember from Chapter 1 that more *s* character in a hybrid orbital makes the group more electronegative and decreases the bond length, leading to this trend in BDE.

Some interesting BDEs are associated with bonds to oxygen. An O–H bond is one of the strongest covalent bonds. As shown in the following Connections highlight, this has substantial consequences in biology. An interesting contrast is seen with the peroxides of Table 2.2. The O–O bonds of peroxides in general are quite weak, making this type of bond very susceptible to thermally induced homolysis. We will see this effect in practice when we study radical reactions in Chapters 10 and 11.

Before leaving this topic, let's consider BDEs in an absolute rather than a relative sense. C–C and C–H bonds generally have BDEs well in excess of 85 kcal/mol. As described in an upcoming Going Deeper highlight, this makes alkanes incredibly stable. In contrast, dimethyl peroxide has a BDE of only 38 kcal/mol. The roughly 50 kcal/mol difference in thermodynamic stability, if it translates to a comparable difference in kinetic stability, would correspond to a factor of 10^{37} difference in reactivity!

Using BDEs to Predict Exothermicity and Endothermicity

Since bond strengths are measured in enthalpy units, we can use bond strength differences between reactants and products to predict the exothermicity or endothermicity of a reaction. This is an exercise routinely done in introductory organic chemistry classes, so we only briefly review it here with one example. Consider the reaction given in Eq. 2.6. We only need to look at the differences in the BDEs for those bonds that change between reactants and products. The reactants possess the stronger bonds, which means they are the more stable structures, and hence the reaction is endothermic. The important point is that we can

Connections

A Consequence of High Bond Strength: The Hydroxyl Radical in Biology

One of the strongest bonds in chemistry is the O–H bond of water, with a BDE of 119 kcal/mol. Since thermodynamics often presages reactivity in radical chemistry, it should not be surprising to hear that the hydroxyl radical, HO•, is extremely reactive. That is, a reaction such as

$$HO• + RH \rightarrow H_2O + R•$$

should be exothermic for almost any organic compound RH, and it is generally very fast, too.

The implications of this observation for biological chemistry are profound. Water is plentiful in a biological system, but the generation of HO• is a high energy process. If HO• were to form, it would rapidly abstract a hydrogen from almost any organic molecule. This includes carbohydrates, proteins, and, perhaps most significantly, DNA. Hydroxyl radical does indeed react rapidly with DNA, a major reaction pathway being abstraction of a H from a C–H bond in a deoxyribose moiety. This creates a potentially lethal (to the cell) lesion in the genetic material. As such, we would expect any process that can generate hydroxyl radicals from water to have severe consequences for a biological system, and indeed this is the case. The most common source of biological hydroxyl radicals is ionizing radiation. Exposure of a living system to high energy radiation (γ rays, x rays, high energy electrons) leads to ionization of water and ultimately the production of many toxic oxygen-based species, including HO•. From that point on, some quite lethal chemistry can ensue, all because of the great strength of an OH bond.

Foote, C. S., Valentine, J. S., Greenberg, A., and Liebman, J. F. (1995). *Active Oxygen in Chemistry. Structure Energetics and Reactivity in Chemistry*, Vol. 2, Chapman & Hall, New York.

predict whether a reaction is exothermic or endothermic simply by examining which side of the reaction arrow has the stronger bonds.

$$CH_3–H + H–OH \longrightarrow CH_3–OH + H–H$$

$$105.1 \quad + \quad 119 \qquad\qquad 92.3 \quad + \quad 104.2 \qquad\qquad\qquad \text{(Eq. 2.6)}$$

$$\Delta H° = (105.1 + 119) - (92.3 + 104.2) = 27.6 \text{ kcal/mol}$$

A variety of experimental techniques has been employed to measure BDEs. Often, different techniques produce slightly different values, and over time, different workers have produced detailed analyses that choose the best value for a given BDE. We present the results of two such analyses in Table 2.2, one from 1982 and one from 2003. Where comparisons are available, there are small but significant differences. Certainly, the trends in BDEs are the same for the two. In general, though, when comparing two different BDEs it is best to use values from the same data set.

2.1.4 An Introduction to Potential Functions and Surfaces—Bond Stretches

For a simple molecule, the homolytic cleavage of a bond that defines the BDE is a chemical reaction that can be related to a **normal mode** within the molecule. Normal modes are the vibrational degrees of freedom that a molecule has—namely, stretches, bends, wags, torsions, etc. Each has its own **fundamental frequency**. The structures of molecules are not

Going Deeper

The Half-Life for Homolysis of Ethane at Room Temperature

The ~90 kcal/mol C–C BDE of ethane sets a lower limit to the activation energy for the thermally induced homolysis of the molecule. In Chapter 7 we will introduce the Arrhenius equation, which can be used to calculate rate constants from activation energies:

$$k = Ae^{-E_a/RT}$$

If we assume an Arrhenius pre-exponential factor (A) of 10^{13} (a common value for a unimolecular process), the half-life for homolysis of ethane at 25 °C would be approximately 10^{44} years. Our universe is postulated to have been around for at most only 10^{10} years. Thus, hydrocarbons are thermally very stable!

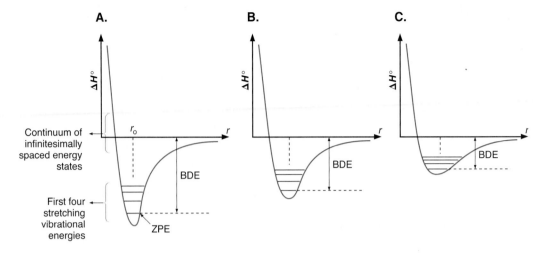

Figure 2.2
Morse potentials for bonds of varying strength. The rungs represent the quantized vibrational states.
A to **B** to **C** represents increasingly weaker bonds, with correspondingly lower frequency ZPE vibrations.
As the bond becomes weaker, the Morse potential becomes more shallow and the difference between the
energies of the vibrational states decreases.

static, but are dynamic, instead. Each fundamental frequency is described by a **potential function**, a mathematical function that describes how the potential energy changes along the coordinate for the internal motion. These functions all derive from the bonding forces that hold the atoms together in the molecule. We do not show many of these mathematical functions, but instead we simply draw graphs of the functions, called **potential surfaces**. The goal is to give a qualitative insight into how potential functions describe each and every molecular motion, such as bond stretches, bends, and conformational changes. Let's start with one of the simplest potential functions, the Morse potential, which describes bond-stretching vibrations.

Chemists pictorially tie the strength of a bond to its length using a **Morse potential**. A Morse potential is a plot of energy (specifically enthalpy) as a function of distance between the atoms. Figure 2.2 shows several such plots. Chapter 14 discusses the origin of the shape of these curves, noting that they arise from internal forces that cause bonds to form (see Section 14.1.5). The energy minimum represents a balance between the overlap of atomic orbitals, which is stabilizing, and nuclear repulsions, which are destabilizing. At very short distances between the atoms (small r on this curve), the nuclear repulsions dominate, and the atoms are repelled apart. At very long atomic distances (long r) the atoms move freely with respect to each other. The shape of this curve is indicative of an **anharmonic oscillator**.

Bonds vibrate in a manner constrained by the Morse potential. The bonds stretch and contract. The energies of the vibrations are not continuous, but are instead quantized as given by Eq. 2.7, where v is the frequency of the bond vibration (this equation is actually for a harmonic oscillator; see the next Going Deeper highlight). The energies are drawn on the Morse potential as rungs. The lowest energy vibration ($n = 0$) is defined as the **zero point energy (ZPE)**, which will be a very important notion that we will return to when isotope effects are discussed in Chapter 8. One result of the vibrational energies being quantized is that the bond never actually has the energy represented by the lowest point on the Morse potential surface, but instead has the ZPE. Higher energy vibrations are the other rungs drawn on the Morse potential. As with any series of quantized energy states, absorption of a photon of light corresponding to the energy difference between the states can form the basis of a spectroscopic method (see the discussion of IR spectroscopy given below). Population of a higher energy vibration leads to what is called an excited vibrational state of the molecule. For most compounds at ambient temperature the vibrational states for bond stretching above the ZPE state are not significantly populated, and we show this in a Going Deeper highlight on page 76 after having defined force constants. Because the atoms vibrate, their

Going Deeper

The Probability of Finding Atoms at Particular Separations

As already stated, the Morse potential is our first example of a potential surface that describes a particular motion. The bond vibrates within the constraints imposed by this potential. One may ask, "At any given moment, what is the probability of having a particular bond length?" This is similar to questions related to the probability of finding electrons at particular coordinates in space, which we will show in Chapter 14 is related to the square of the wavefunction that describes the electron motion. The exact same procedure is used for bond vibrations. We square the wavefunction that describes the wave-like nature of the bond vibration. Let's explore this using the potential surface for a **harmonic oscillator** (such as with a normal spring), instead of an anharmonic oscillator (Morse potential). For the low energy vibrational states, the harmonic oscillator nicely mimics the anharmonic oscillator.

Shown below is a potential surface for the vibrations of a harmonic oscillator. We are using this potential to model bonds, and we draw the quantized energy states again as rungs. For both bonds in low energy vibrational states and for springs in the macroscopic world, the frequencies of the vibrations do not change. Every spring possesses its own fundamental frequency. The energy of the vibration changes due to increases in the amplitude of the wave that describes each vibrational state. On the rungs are drawn the wavefunctions that describe these vibrations. Just as with electron wavefunctions (Section 1.1), as energy increases the number of nodes (zeroes in the function) increases. The second diagram shows the same picture except with the square of the wavefunctions, thus indicating probabilities. Let's look at the probability of bond lengths for three different vibrational states, the ZPE, the next highest, and a very high energy state.

The lowest energy vibration has the highest probable bond length centered right at the traditional bond length (r_o). These are the bond lengths reported in Table 1.4. The probability function decreases toward the short and long bond lengths confined by the potential surface. The second vibrational state is quite interesting. The probability is high at short and long bond lengths relative to r_o. This means that there are actually two bond lengths with high probabilities in the second quantized state. However, the average distance between the atoms is still r_o. With an anharmonic oscillator that describes real bonds, the average bond lengths actually increase with each increasing energy level (see Figure 2.2).

Lastly, the high energy state shown starts to approach what is observed in our macroscopic world for springs. The square of the wavefunction finds distances between the atoms that are most probable at short and long bond lengths. This is exactly what is found for two balls connected by a spring. During a vibration, the two balls spend most of their time at the short distance and long distance, because the balls slow down when they are changing direction at the extreme compression and extension of the spring. The balls are moving most rapidly at the average distance between them, and hence the probability of finding them at this distance is low.

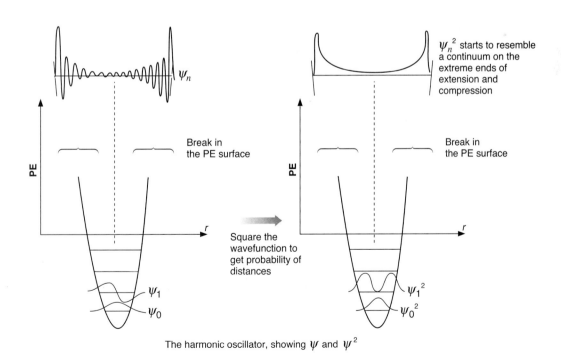

The harmonic oscillator, showing ψ and ψ^2

positions move with time relative to each other, meaning that the bond length varies with time. The manner in which one predicts the probability of finding a particular bond length requires quantum mechanics, and is discussed qualitatively in another Going Deeper highlight on page 75.

$$e_n = (n + \tfrac{1}{2}) h\nu \qquad \text{(Eq. 2.7)}$$

It makes good sense that the bond stretching vibration is the internal motion within the molecule that leads to homolysis. Imagine the extreme of a bond stretch—if it is large enough the bond just breaks. The energy axis is relative to the free dissociated atoms, where enthalpy is defined as zero, just as implied in Eq. 2.5 for the reaction that defines BDE. Hence, the energy that needs to be put into the bond to fully break it (the BDE) is shown as an arrow that starts at the bottom rung (ZPE) of the curve and ends at the reference state with the atoms at an infinite distance.

The shape of the Morse potential is indicative of the strength of the bond, and the looseness of the associated vibration (hence, its entropy). A deeper potential represents a stronger bond because the BDE is larger. For otherwise similar bonds, the stronger bonds have higher vibrational frequencies, meaning that the potential surfaces are more narrow (Figure 2.2 **A**). A narrow potential surface indicates a **stiff vibration**. As a bond gets weaker, the Morse potential is more shallow and the width becomes greater. Now the frequency of vibration is lower, and the vibration is considered to be a **loose vibration** (Figures 2.2 **B** and **C**). As noted earlier, the concept of describing a vibration as stiff or loose has consequences for the entropy of the compound, and the entropy of reaction, which we will return to in Chapters 7 and 8.

As described in the above Going Deeper highlight, the trends in vibrational states can be easily rationalized by modeling a bond as a spring that connects the two atoms. Eq. 2.8 gives the vibrational frequency of a spring with two masses (m_1 and m_2) at the ends, where m_r is the reduced mass of m_1 and m_2 $[m_r = (m_1 m_2)/(m_1 + m_2)]$. A very similar equation is used to model angle bending motions. This equation gives the frequency, ν, for a harmonic oscillator. The force constant (k) for the spring reflects the strength of the bond, where stronger bonds have larger force constants. Note that a larger k gives a larger frequency to the vibration. Recall from your everyday experience that the fundamental frequency of a stiff spring is large, whereas a loose spring vibrates slowly. Bonds behave in much the same way for the lower energy vibrations. The value of ν from this equation is used in Eq. 2.7 to calculate the energy of the stretching vibration.

$$\nu = \frac{1}{2\pi} \sqrt{\frac{k}{m_r}} \qquad \text{(Eq. 2.8)}$$

Going Deeper

How do We Know That $n = 0$ is Most Relevant for Bond Stretches at $T = 298$ K?

Previously we commented that vibrational states higher than $n = 0$ are not important for examining bond dissociation energies at ambient temperatures. We know this by calculating the energy difference between the states with $n = 0$ and $n = 1$. Below we list a table of force constants for bond stretches. Let's focus on the force constant for a C–C bond. Using this force constant in Eq. 2.8 to calculate the fundamental frequency, and using Eq. 2.7 to calculate the energy difference between the $n = 0$ and $n = 1$ vibrations, we obtain 3.22 kcal/mol. That gap gives a preference for

the low energy state of 233 to 1 at 298 K, meaning that only a small fraction of the molecules exist in the $n = 1$ energy level.

Type of Bond	Force Constants for a Stretch (mdyne/Å)
C–H in a typical methyl	4.7
C–C in an alkane	4.5
C=C in an alkene	11
C≡C in an alkyne	16

Infrared Spectroscopy

The trends discussed above can all be observed in infrared (IR) spectra. For example, the typical frequencies for C–C, C=C, and C≡C bond stretches are 450–500 cm^{-1}, 1617–1640 cm^{-1}, and 2260–2100 cm^{-1}, respectively. This comparison shows that the stronger bond has the higher frequency vibration. Because the reduced masses are all the same, the force constants must increase in this series (as shown by the table in the previous Going Deeper highlight).

Bond stretches are not the only vibrations that are quantized. Bond angle bends, molecular wags, and scissoring motions are also quantized. There are separate potential functions and surfaces that describe each of these vibrational modes. Just as with stretching vibrations, there are only certain energies that these other vibrations can have, and hence excitation from one energy level to another also gives absorption bands in the IR spectrum. These vibrations typically are at lower frequency, and are found at smaller cm^{-1} in IR spectra. For example, Figure 2.3 shows all the normal modes (vibrational degrees of freedom) of formaldehyde, along with the IR absorption values. For the purposes of this textbook, it is simply important to recognize that all vibrational degrees of freedom are quantized, and that each motion is defined by a potential surface that describes it as stiff or loose, giving insight into the relative entropy of one vibration to another.

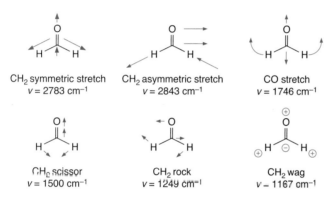

Figure 2.3
IR frequencies for the degrees of freedom of formaldehyde.

2.1.5   Heats of Formation and Combustion

For any molecule, the **total energy** is the energy required to convert the molecule into isolated atoms (breaking all bonds) and then remove all electrons from those atoms. This is typically a very large number (tens of thousands of kcal/mol), and while it can be produced by *ab initio*, quantum mechanical calculations, it is not of much value to practicing chemists. Furthermore, as with all energies, total energy is relative to some standard state taken for the isolated atoms. Hence, it is not known in an absolute sense.

A more useful energetic quantity associated with a particular molecule is its heat of formation. Recall our discussion of enthalpy, where energy was related to the ability of a system to perform work or release or absorb heat. To define the energy of a molecule, we therefore need a reaction that can release or absorb heat, and a reference state to which we can relate all molecules. In theory, any reaction that all organic molecules can undergo could be used. However, chemists have settled upon one particular reaction, the heat of formation.

The **heat of formation** of a hydrocarbon is defined as $\Delta H°$ for the process of Eq. 2.9—the formation of the molecule from its constituent elements. If the elements are in their standard states (graphite for C; H_2 at 1 atm for H), we have the standard heat of formation, symbolized as $\Delta H_f°$. All enthalpy values are typically expressed on a per mole basis. Furthermore, all molecules are referenced to the elements, which are defined to have a heat of formation of zero. If other elements are involved in the molecule, then we include the elements in their

Going Deeper

Potential Surfaces for Bond Bending Motions

All internal degrees of freedom can be described by potential surfaces, which are considered to be loose or stiff. As just one more example, let's compare the out-of-plane bending motions of a CH_2 group in ethylene and methane (we do this in preparation for the discussion of secondary isotope effects given in Chapter 8). The out-of-plane bend for ethylene sends the hydrogens toward the π bond, while the same bend in methane sends the hydrogens toward other C–H bonds. It is less sterically demanding to push the hydrogens toward a π bond, and hence this bend in ethylene is easier (described by a looser potential surface as shown). As with all degrees of freedom, the energies of the out-of-plane bending motions are quantized, which could be written on these potential surfaces as rungs in a manner analogous to the energies of the stretching vibrations placed on a Morse potential.

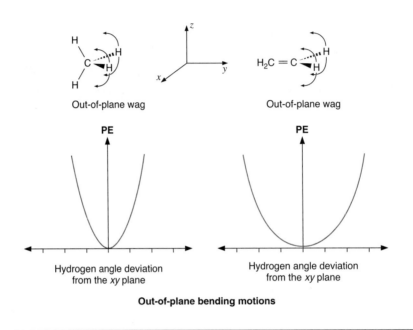

Out-of-plane bending motions

standard states on the left-hand side of Eq. 2.9. One must also designate a temperature, and unless explicitly noted otherwise, all ΔH_f° values will refer to 298 K. The correct designation is then $\Delta H_{f\ 298}^\circ$, but typically we will assume 298 K and not write it.

$$n\,C\,(\text{graphite}) + (\tfrac{m}{2})H_2\,(1\text{atm}) \longrightarrow C_nH_m + \text{heat} \tag{Eq. 2.9}$$

Heats of formation are typically not easy to measure directly. Certainly Eq. 2.9 represents a reaction that is difficult to perform and isolate a single product! However, for many organic molecules—especially for hydrocarbons, thanks to the petroleum industry—the **heat of combustion** has been determined accurately. This is defined as ΔH° for the reaction of Eq. 2.10—the "burning" of the molecule. Since ΔH_f° for CO_2 and H_2O are known (–94.05 and –57.80 kcal/mol, respectively), it is easy to convert a heat of combustion to a heat of formation, and this is how most experimental ΔH_f° values are actually obtained.

$$C_nH_m + (n + \tfrac{m}{4})O_2 \longrightarrow n\,CO_2 + (\tfrac{m}{2})H_2O + \text{heat} \tag{Eq. 2.10}$$

Students often confuse the trends in heats of formation and combustion. When comparing two or more structures that are isomers, a larger negative heat of formation means that the compound is more stable. The reaction that creates the more stable organic molecule from the elements releases more energy, and so there must be less internal energy within that organic molecule (the molecule is less strained). In contrast, a large negative heat of combustion means the compound is more unstable, more strained. In combustion, the compound is releasing its internal energy in the form of heat. The more unstable the structure, the more heat is released when it burns.

Many heats of formation values are available, and it is worthwhile to spend a few moments to consider some differences among them. For example, the heat of formation of cyclohexane is −29.9 kcal/mol. What is the meaning of this number? The fact that it is negative means that Eq. 2.9 is exothermic. In other words, a mole of cyclohexane is more enthalpically stable than the equivalent amount of graphite plus H_2. A further insight arises when we compare cyclohexane's heat of formation to another molecule with the same molecular formula, such as methylcyclopentane ($\Delta H_f^\circ = -25.5$ kcal/mol). Clearly, methylcyclopentane is less stable than cyclohexane.

The ΔH_f° values become even more powerful when we compare molecules of different molecular formulas. Keep in mind that this comparison must always be tied back to the elements. Let's consider the heat of formation of benzene plus three H_2 (ΔH_f° for $H_2 = 0$ kcal/mol by definition), and compare it to cyclohexane. Benzene is an especially "stable" molecule, so it may surprise some that ΔH_f° for benzene is +19.82 kcal/mol. That is, benzene is thermodynamically unstable relative to its constituent elements (graphite plus H_2). Therefore, relative to the elements, cyclohexane is actually the more stable structure.

We must always keep clear in our minds the difference between thermodynamic stability, kinetic stability, and reactivity. **Thermodynamic stability** involves the position of a particular equilibrium, and so it is set by ΔG°, as emphasized earlier in this chapter. Assuming for the moment that entropy effects are small, we would conclude that the equilibrium in Eq. 2.9 lies well to the right for cyclohexane and well to the left for benzene. Benzene does not spontaneously convert to graphite and H_2, however, because it is kinetically stable. **Kinetic stability** involves the rates of reactions, and so is controlled not by ΔG°, but by $\Delta G^\ddagger$ (see Figure 2.1 **C** and Chapter 7). A molecule is kinetically stable if there are substantial activation barriers to all available reaction paths. We know from experience that there must be a very large kinetic barrier to the conversion of benzene to its elements; that is, $\Delta G^\ddagger$ must be very large (going from right to left) for the "reaction" of Eq. 2.9 when benzene is the organic molecule.

2.1.6 The Group Increment Method

As already stated, knowing the heat of formation of a molecule tells us the position of the equilibrium of Eq. 2.9, not something that is of great value to most organic chemists. Heats of formation become useful when we compare values for similar molecules. Then, our reference state in effect becomes not the elements C and H, but rather some standard organic molecule that we can compare other molecules to. When we start to compare heats of formation for a number of hydrocarbons, definite patterns arise. From these patterns we can develop several notions that are very valuable in considering the thermodynamic stabilities of organic molecules. It is best to see how this happens by just looking at the patterns.

Table 2.3 shows experimental ΔH_f° values for a series of n-alkanes. After some initial

Table 2.3
Values for the Heats of Formation of
Simple Linear Alkanes (in kcal/mol)

Structure	ΔH_f°	$\Delta\Delta H_f^\circ$
$CH_3–CH_3$	−20.24	—
$CH_3–CH_2–CH_3$	−24.84	4.60
$CH_3–CH_2–CH_2–CH_3$	−30.15	5.31
$CH_3–CH_2–CH_2–CH_2–CH_3$	−35.00	4.85
$CH_3–CH_2–CH_2–CH_2–CH_2–CH_3$	−39.96	4.96
$CH_3–CH_2–CH_2–CH_2–CH_2–CH_2–CH_3$	−44.89	4.93
$CH_3–CH_2–CH_2–CH_2–CH_2–CH_2–CH_2–CH_3$	−49.81	4.92

variation, there is a clear pattern. Averaging over a large dataset reveals that every additional CH_2 group adds -4.93 kcal/mol to ΔH_f°. This pattern continues, and suggests the notion of a **group increment**—a consistent contribution to the ΔH_f° of a molecule by a particular grouping. The symbolism for the CH_2 group we have been discussing is $C–(H)_2(C)_2$—a carbon attached to two hydrogens and two alkyl carbons. Only this kind of a CH_2 has a group increment of -4.93 kcal/mol. The CH_2 of 2-butanone makes a different contribution to ΔH_f° because it is attached to one alkyl carbon and one carbonyl carbon (although the deviation from -4.93 kcal/mol is small; see below). This notion of a transferable energy increment associated with a chemical group is nicely consistent with the bonding model discussed in Chapter 1, in which a group such as CH_2 also contributes a transferable set of orbitals.

Once we have a CH_2 group increment, it is simple to determine the group increment for a CH_3 group in an alkane. We take ΔH_f° for *n*-hexane, for example, subtract $4(-4.93$ kcal/mol) for the four CH_2's, and divide the remainder by 2, because there are two CH_3's. When this is done and averaged over a number of alkanes, a value of -10.20 kcal/mol is obtained for the CH_3 group increment, designated as $C–(H)_3(C)$. We can now calculate *a priori* ΔH_f° for *any* linear alkane. For $CH_3(CH_2)_{n-2}CH_3$, ΔH_f° is predicted to be exactly $(n-2)(-4.93$ kcal/mol) $+ 2(-10.20$ kcal/mol).

There are enough experimental determinations of ΔH_f° for alkanes that group increments for an alkane CH and an alkane quaternary carbon (C) can be derived, and these are given in Table 2.4. Similarly, ΔH_f° values for a wide variety of organic molecules with a considerable diversity of functionality have been determined. This allows group increment values to be established for many types of groups, and hence, ΔH_f° values for a wide range of molecules to be *estimated*. Table 2.4 lists a small subset of all available group increment values. References to more complete collections are given at the end of the chapter. These group increments are sometimes referred to as **Benson increments**, after Sidney Benson, who was the prime developer of the concept. Note the manner in which a group increment is designated—enough attached atoms must be given to unambiguously identify the type of group. However, some bonding partners are implicit; for example, a C_d (an olefinic C) must have a C_d partner, so that is not designated.

As an example, Figure 2.4 shows how to calculate ΔH_f° for isobutylbenzene. The process is straightforward, and it gives surprisingly good results. We will use the group increment method sporadically in later chapters, and soon we will introduce some additional features of the method. At this point, however, we state a few general observations, and a few caveats.

First, the quality of a group increment is only as good as the quality of the experimental thermodynamic data that were used to obtain it. For the basic hydrocarbon groups, vast quantities of thermodynamic data are available, and so the group increments are quite reliable. On the other hand, there are published group increment values for some fairly obscure functional groups, such as diazenes and oximes. Much less data are available for such compounds, and so considerable caution is in order when using these group increments. Furthermore, the Benson group method ignores interactions between groups, which sometimes contribute to heats of formation. There are examples where the Benson group values fail, not

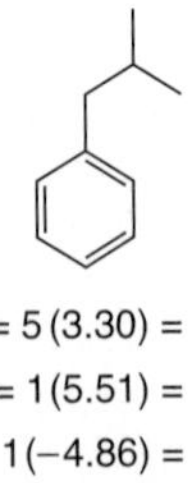

Figure 2.4
Sample calculation of ΔH_f° using group increments, shown for isobutylbenzene.

5 C_B–(H) = 5 (3.30) =	16.50
1 C_B–(C) = 1 (5.51) =	5.51
1 $C–(C_B)(C)(H)_2$ = 1 (−4.86) =	−4.86
1 $C–(H)(C)_3$ = 1 (−1.90) =	−1.90
2 $C–(H)_3(C)$ = 2 (−10.20) =	−20.40
	−5.15 kcal/mol

Experimental: −5.15 ± 0.34 kcal/mol

Table 2.4
Group Increments (in kcal/mol) for Fundamental Groupings*

Group	ΔH_f°	Group	ΔH_f°	Group	ΔH_f°
C–(H)$_3$(C)	–10.20	C–(O)(C$_d$)(H)$_2$	–6.5	C–(O)$_2$(C)$_2$	–18.6
C–(H)$_2$(C)$_2$	–4.93	C$_B$–(O)	–0.9	C–(O)$_2$(C)(H)	–16.3
C–(H)(C)$_3$	–1.90	O–(C)$_2$	–23.2	C–(O)$_2$(H)$_2$	–16.1
C–(C)$_4$	0.50	O–(C)(H)	–37.9	C–(N)(H)$_3$	–10.08
C$_d$–(H)$_2$	6.26	O–(C$_d$)$_2$	–33.0	C–(N)(C)(H)$_2$	–6.6
C$_d$–(H)(C)	8.59	O–(C$_d$)(C)	–30.5	C–(N)(C)$_2$(H)	–5.2
C$_d$–(C)$_2$	10.34	O–(C$_B$)$_2$	–21.1	C–(N)(C)$_3$	–3.2
C$_d$–(C$_d$)(H)	6.78	O–(C$_B$)(C)	–23.0	C$_B$–(N)	–0.5
C$_d$–(C$_d$)(C)	8.88	O–(C$_B$)(H)	–37.9	N–(C)(H)$_2$	4.8
C$_d$–(C$_B$)(H)	6.78	C–(CO)(C)$_3$	1.58	N–(C)$_2$(H)	15.4
C$_d$–(C$_B$)(C)	8.64	C–(CO)(C)$_2$(H)	–1.83	N–(C)$_3$	24.4
C$_d$–(C$_d$)$_2$	4.6	C–(CO)(C)(H)$_2$	–5.0	N–(C$_B$)(H)$_2$	4.8
C$_B$–(H)	3.30	C–(CO)(H)$_3$	–10.08	N–(C$_B$)(C)(H)	14.9
C$_B$–(C)	5.51	C$_B$–(CO)	9.7	N–(C$_B$)(C)$_2$	26.2
C$_B$–(C$_d$)	5.68	CO–(C)$_2$	–31.4	N–(C$_B$)$_2$(H)	16.3
C$_B$–(C$_B$)	4.96	CO–(C)(H)	–29.1	N$_I$–(H)	16.3
C–(C$_d$)(C)(H)$_2$	–4.76	CO–(H)$_2$	–26.0	N$_I$–(C)	21.3
C–(C$_d$)$_2$(H)$_2$	–4.29	CO–(C$_B$)$_2$	–25.8	N$_I$–(C$_B$)	16.7
C–(C$_d$)(C$_B$)(H)$_2$	–4.29	CO–(C$_B$)(C)	–30.9	CO–(N)(H)	–29.6
C–(C$_B$)(C)(H)$_2$	–4.86	CO–(C$_B$)(H)	–29.1	CO–(N)(C)	–32.8
C–(C$_d$)(C)$_2$(H)	–1.48	CO–(O)(C)	–35.1	N–(CO)(H)$_2$	–14.9
C–(C$_B$)(C)$_2$(H)	–0.98	CO–(O)(H)	–32.1	N–(CO)(C)(H)	–4.4
C–(C$_d$)(C)$_3$	1.68	CO–(O)(C$_d$)	–32.0	N–(CO)(C)$_2$	—
C–(C$_B$)(C)$_3$	2.81	CO–(O)(C$_B$)	–36.6	N–(CO)(C$_B$)(H)	0.4
C–(O)(C)$_3$	–6.6	CO–(C$_d$)(H)	–29.1	N–(CO)$_2$(H)	–18.5
C–(O)(C)$_2$(H)	–7.2	O–(CO)(C)	–43.1	N–(CO)$_2$(C)	–5.9
C–(O)(C)(H)$_2$	–8.1	O–(CO)(H)	–58.1	N–(CO)$_2$(C$_B$)	–0.5
C–(O)(H)$_3$	–10.08	C$_d$(CO)(C)	7.5		
C–(O)(C$_B$)(H)$_2$	–8.1	C$_d$–(CO)(H)	5.0		

C$_d$ = double bond; C$_B$ = benzene carbon; N$_I$ = imine nitrogen.

*Data are from Benson, S. W. (1976). *Thermochemical Kinetics: Methods for the Estimation of Thermochemical Data and Rate Parameters*, 2d ed., John Wiley & Sons, New York.

because there is insufficient thermochemical data, but because we do not have enough data regarding interactions between different substituents.

Along with group increments for ΔH_f°, there are analogous group increments for ΔS_f°. In our discussion of entropy above, we noted that bond rotations and other low energy molecular motions contribute to the entropy of a molecule, and hence it may be expected that alkyl and functional groups would contribute specific entropies. Since ΔG° determines the equilibrium constant, ΔS_f° group increments are potentially useful. In practice, however, they are much less frequently used, in part because the entropy differences between molecules are often small, and in part because fewer ΔS_f° group increments are available.

Group increments for heat capacity (ΔC_P°) are also available. Recall that **heat capacity** reflects the variation of ΔH° with temperature (see Section 4.1.1). Heat capacity group increments are essential when considering temperatures that are considerably different from 298 K. This is a very important issue, for example, in chemical engineering, where it is often crucial to know how much heat will be liberated by a particular process—especially if it is being run on a 10,000-gallon scale! In more typical physical organic chemistry, however, heat capacity group increments are not commonly used. Both the ΔS_f° and heat capacity group increments are applied in the same manner as the ΔH_f° group increments, and special texts on the topic are available.

2.1.7 Strain Energy

Now that we have established a definition of the stability of a compound—its heat of formation—we can return to the notion of strain. We introduced strain at the very outset of the chapter, defining it as an internal stress within a molecule due to some bonding or structural/conformational distortion. Now that we have a way to predict the stability of a molecule without any bonding distortions, we can give a very precise definition of strain energy.

The group increments method suggests that we can predict ΔH_f° for any organic molecule—a powerful tool, indeed. Let's consider some more examples. Table 2.5 shows calculated and experimental ΔH_f° values for simple cycloalkanes. The calculated ΔH_f° for cyclo[n] is just $n(-4.93\,\text{kcal/mol})$. For cyclohexane the result is quite good, but then things start to go downhill. The error for cyclopentane is considerable, while cyclobutane and cyclopropane are completely off.

Table 2.5
Calculated vs. Observed ΔH_f° Values
(in kcal/mol) for Selected Molecules

	ΔH_f° calculated by group increments	ΔH_f° experimental
Cyclohexane	–29.6	–29.9
Cyclopentane	–24.7	–18.3
Cyclobutane	–19.7	+6.7
Cyclopropane	–14.8	+12.7

The problem here is that the smaller rings are strained—cyclopentane to some extent, and cyclobutane and cyclopropane to a much greater extent. The strain arises, at least in part, from the significant distortion of C–C–C angles from the standard value. The group increment method does not account for strain. It was parameterized on simple, acyclic alkanes, where no bonding distortions occur, so it fails for these molecules. In fact, the values in Table 2.4 are often referred to as strain-free group increments. One significant result of the group increments approach is that we can now unambiguously define strain energy. **Strain energy** *is the difference between the experimental ΔH_f° for a molecule and the value of ΔH_f° calculated using strain-free group increments.*

It is tempting to dissect the total strain energy into components, such as angle strain, torsional strain, ring strain, etc., and we discuss all these in this chapter. In principle, this is a dangerous practice. The quantum mechanics that defines bonding does not naturally lead to such a partitioning, so often fairly arbitrary distinctions must be made. In practice, however, this is a useful strategy, but one that should be used cautiously.

2.2 Thermochemistry of Reactive Intermediates

As we did with electronic structure in Chapter 1, in this chapter on molecular structure we will consider reactive intermediates alongside stable molecules. There are many parallels and interconnections, but also some unique features for reactive intermediates. The same notions of strain and stability can be applied to molecules even if they are reactive, because structural issues are fundamentally the same regardless of whether an atom has an octet or not. Before looking at thermochemistry, however, we must consider a general issue for reactive intermediates. We consider a reactive intermediate to be "unstable", although for all types of reactive intermediates there are special cases that are considered "stable". These are imprecise terms, and in the next section we will seek to clarify their meaning.

2.2.1 Stability vs. Persistence

A recurring theme in physical organic chemistry, and especially in the study of reactive intermediates, is the notion of stability. We describe one structure as stable, but another as reactive. Chemists often describe a reactive intermediate as "stabilized". What do we mean by this? What does "stable" imply? To be stable, does a species have to live for minutes, hours, or days? Furthermore, under what conditions are we considering stability? Are there any other species present that could react with our "stable" species? An especially lucid discussion of this point was provided by K. U. Ingold, leading him to invoke a quote from Humpty Dumpty in *Alice in Wonderland*: "When I use a word it means just what I choose it to mean— neither more nor less". Actually, this is an appropriate description of the typical usage of the term "stable" in physical organic chemistry. The term is used by chemists in any way that is appropriate to the system under analysis, but it is *always* used to denote lower internal energy relative to a reference system.

Similarly, it is difficult to exactly define what it means to be unstable (reactive). A molecule that spontaneously reacts all by itself at 0 degrees Kelvin in interstellar space is certainly unstable. But in a more practical discussion, instability is again always a relative term, where the compound under analysis is unstable relative to some reference system. Yet, the reference system must be studied under the same experimental conditions as the compound whose stability is being questioned, because as we now describe, the experimental conditions affect the reactivity of compounds.

Stability implies some sort of minimal lifetime, but for all reactive intermediates the lifetime will be extremely dependent on the conditions of observation. For example, a free radical that might be indefinitely long lived under an inert atmosphere could have a very short lifetime in the presence of oxygen. Also, we will often encounter two related reactive intermediates, both of which have transient existences, but only one of which is stabilized with regard to the other. To clarify this situation, we follow the suggestion of Ingold and distinguish between **stability** and **persistence**. **Persistent** simply means long lived, and is a term associated with *kinetics*. A persistent structure has relatively high activation barriers for any reaction. Of course, we must agree upon the definition of "long", and that will depend on the circumstances. In some instances a lifetime of seconds will qualify as persistence, while in others we will need a lifetime of days or more before we call the structure persistent. Persistence is a notion that is *very* context dependent. Whether a reactive intermediate is long lived or not will depend crucially on whether there is anything around for the reactive intermediate to react with. Simple variables such as temperature and concentration will also affect the persistence of a reactive intermediate. So, by definition, we can describe a reactive intermediate as persistent only with regard to a well-defined set of conditions.

In contrast, stability is an intrinsic property of a reactive intermediate. We define a structure as stable or **stabilized** if it is *thermodynamically* (Gibbs free energy) more stable than some reference structure, as discussed above. Here, our focus is upon electronic stability rather than sterics, because for most reactive intermediates their electronics dominates their reactivity. For example, the benzyl cation is stabilized, because it is thermodynamically more stable than its reference, the methyl cation. However, under typical conditions the benzyl cation is not expected to be persistent. Fundamentally, stability is a thermodynamic notion, while persistence is a kinetic one. Stability is intrinsic to a structure, while persistence is very much context sensitive. We will do our best to keep these distinctions clear in the following sections. In the chemical literature, however, such precision in terminology is not always maintained.

2.2.2 Radicals

More so than any other area of reactive intermediate chemistry, thermodynamics is a valuable predictor of radical reactivity. There will always be exceptions, and we can never totally ignore kinetics, but as a first step, we will in most instances turn to the thermochemistry of a given situation to predict or rationalize radical reactivity patterns. In addition, there is a vast collection of relevant thermodynamic data for radicals—the collection of BDEs discussed in Section 2.1.3.

BDEs as a Measure of Stability

Consider Eq. 2.11, a simple variant of Eq. 2.5. This defines the BDE for a generic R–H bond. When comparing Eq. 2.11 for a variety of organic molecules, the contribution to ΔH_f°

from H• cancels out. As such, the relative BDEs for a series of R–H bonds provide an excellent collection of relative stabilities of the R• radicals. We are indeed considering stability, not persistence, here, as we are discussing the relative thermodynamics of a series of reactive intermediates.

$$R - H \longrightarrow R^\bullet + H^\bullet \quad \Delta H^\circ = BDE \qquad \text{(Eq. 2.11)}$$

Returning to Table 2.2 and focusing on C–H BDEs, a good deal of information on radical stabilities can be obtained. The clear trend in BDEs of methane > ethane > propane > isobutane leads directly to the series of radical stabilities: 3° > 2° > 1° > methyl. The overall effect is substantial, with over 10 kcal/mol in stabilization for t-butyl radical relative to methyl radical. The reason for this trend is hyperconjugation, which we will explore in depth when discussing carbocations (see below). Based upon BDEs, vinyl and phenyl radicals are substantially *less* stable than alkyl radicals.

Allyl and benzyl radical are substantially stabilized, as anticipated from the resonance structures (see Section 1.3.6). Comparing the BDEs of propene and toluene to an appropriate reference such as ethane suggests resonance stabilization energies of 12.4 and 14.1 kcal/mol, respectively. An alternative way to estimate allyl stabilization is to consider allyl rotation barriers (Eq. 2.12). Rotating a terminal CH_2 90° out-of-plane completely destroys allyl resonance, and so the transition state for rotation is a good model for an allylic structure lacking resonance. For allyl radical the rotation barrier has been determined to be 15.7 kcal/mol, in acceptable agreement with the direct thermochemical number.

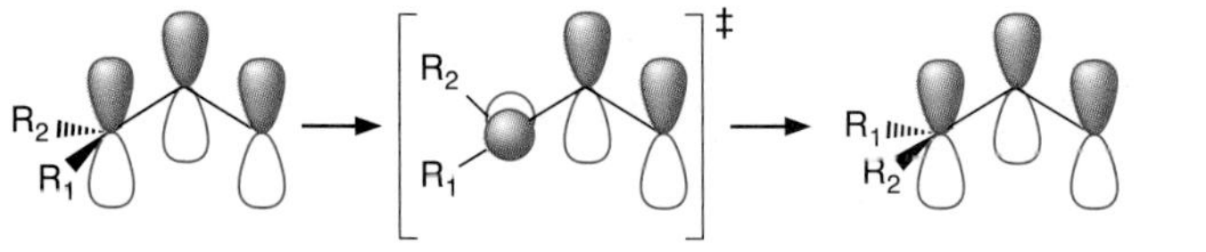

$$\text{(Eq. 2.12)}$$

The reasoning behind the above analysis is somewhat circular. BDEs are used to determine the stability of radicals, and the stability of a radical is used to rationalize trends in BDEs. BDE is really only the energy it takes to break a bond, which as with any other process, depends solely on the properties of the initial and final states of the system. BDE values are therefore a measure of the relative stability of the radical products compared to the organic reactants, not just the stability of the radicals. We have to assume that other factors, such as the stability of R–H (or R–X), are similar in the series for the BDE to reflect solely the radical stability. This assumption actually holds true reasonably well for C–H and C–C bonds. However, the BDE values for the C–Cl bonds in methyl chloride, ethyl chloride, isopropyl chloride, and t-butyl chloride are 84.1, 84.2, 85.0, and 83.0 kcal/mol, respectively. There is no trend here. Clearly other factors are involved, and these BDE values are poor measures of radical stability.

Radical Persistence

The issue of stability vs. persistence of free radicals is an important one that dates back to the birth of the field. In 1900 Gomberg prepared the triphenylmethyl or trityl radical according to Eq. 2.13. Under appropriate conditions, the free radical persists in solution indefinitely at room temperature. This initially controversial result was arguably the birth of reactive intermediate chemistry, and it spurred volumes of work. The trityl radical is in equilibrium with a dimer that, for decades, was assumed to be hexaphenylethane. However, nuclear magnetic resonance (NMR) and ultraviolet (UV) studies in 1968 revealed that the actual dimer was the unsymmetrical structure shown in Eq. 2.13, in which one trityl center added to the para position of a ring of another radical.

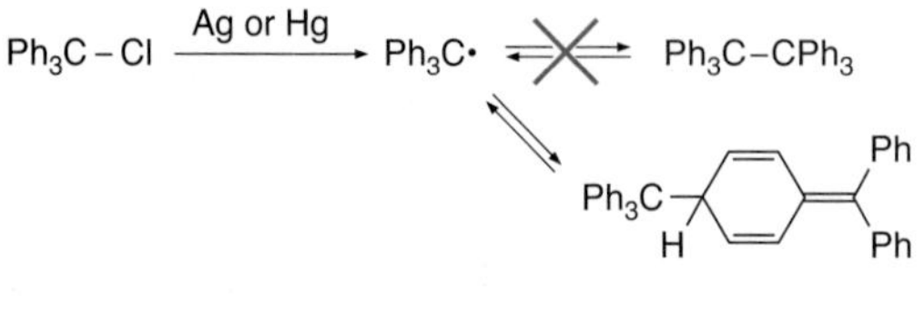

$$\text{(Eq. 2.13)}$$

We know that a phenyl group stabilizes an adjacent radical due to resonance (benzyl, for example), and three phenyls should provide more stabilization than two. This is indeed the case; trityl radical is certainly stabilized. However, it is not possible for all three aromatic rings of trityl to be in perfect conjugation with the radical center. Simple model building will convince you that a fully planar structure would experience severe steric clashes. The trityl radical adopts the nonplanar, propeller-like structure shown in Figure 2.5. In this structure, each ring stabilizes the radical considerably, but the overall stabilization is not three times the benzyl stabilization.

Why, then, is trityl so much more persistent than a typical radical? Methyl and most other radicals generated under similar conditions would have a fleeting existence. A clue comes from the dimerization chemistry. Apparently, the simple C–C bond formation we would expect for other radicals does not happen with trityl. Instead, radical coupling occurs in an unusual way that destroys the aromaticity of one ring. Another clue comes from the observation that placing *t*-butyl groups in the para positions of all the rings greatly enhances the persistence of the radical; dimerization is completely suppressed (Eq. 2.14). A para *t*-butyl group wouldn't significantly stabilize the trityl radical, so why does it make the structure more persistent?

$$\left(\diagdown\!\diagup\!\!\!-\!\!\bigcirc\!\!-\right)_3 C\cdot \longrightarrow \text{No dimers}$$

(Eq. 2.14)

Ultimately, extensive study showed that the stabilization provided by the phenyl rings contributes little to the *persistence* of trityl. The major factor influencing the persistence of trityl and other radicals is sterics. Figure 2.5 shows that the system is quite crowded. The twisting of the rings seriously occludes the radical center, and this is why trityl is persistent. Reagents cannot easily reach the radical center.

Once it was appreciated that steric bulk was the key to producing persistent radicals, a wide range of long lived structures could be prepared. Table 2.6 shows a collection of radicals and their half-lives under equivalent conditions. Trityl would have a half-life of ~5 minutes under these conditions. Very long lived radicals can be prepared without any significant stabilization. Note that the last two entries in Table 2.6 could be viewed as persistent but *de*stabilized radicals, in that both have sp^2 hybridized radical centers. The lesson from free radicals that steric protection to reactivity is much more important to persistence than electronic stabilization carries over to a wide range of reactive structures.

The extent to which a species is persistent very much depends on the environment. The data in Table 2.6 refer to fluid solution, at room temperature, with radical concentrations of 10^{-5} M. Under other conditions, different lifetimes would be observed. For example, samples of methyl radical have been prepared in which the radical is embedded in glass. These

Figure 2.5
The trityl radical. Shown are two views of a calculated structure for the radical. The radical center is shown in color.

Table 2.6
The Persistence of Various Radicals*

R·	$t_{1/2}$, 25 °C, 10^{-5} M	R·	$t_{1/2}$, 25 °C, 10^{-5} M
CH₃·	20 µs	Me₃Si, SiMe₃ C–CH Me₃Si, SiMe₃	> 110 days
(sec-butyl-type radical)	1 min	(di-t-butyl phenyl radical)	6 ms
(tetramethylcyclohexyl radical)	4.2 min		
(*t*-Bu)₃C·	8.4 min	Me₃Si–C=C, SiMe₃ CF₃	1 min
(Me₃Si)₃C·	2.3 days		

TMS = trimethylsilyl.
*Griller, D., Ingold, K. U. "Free Radical Clocks." *Acc. Chem. Res.*, **13**, 317 (1980).

samples are indefinitely stable at room temperature. This is because diffusion is impossible, negating all possible bimolecular paths. Oxygen and any other reactive species cannot come into contact with the radical, and the methyl radical has no available photochemical reactions.

When steric effects are combined with stabilizing effects such as delocalization, radicals that are very "stable" can be prepared. For example, both galvinoxyl and diphenylpicrylhydrazyl (DPPH) are commercially available free radicals that can be handled with no special precautions (see margin).

Group Increments for Radicals

If we know ΔH_f° for R–H, and the BDE, then we can derive ΔH_f° for R• since ΔH_f° for H• is known. In addition, a number of spectroscopic techniques have directly produced values of ΔH_f° for a wide variety of free radicals. Because we have a large collection of ΔH_f° values for radicals, we can derive group increment values for organic free radicals. Table 2.7 lists selected results of this effort. The symbolism is as before, and all the caveats concerning group increments for neutrals apply even more so for radicals. Nevertheless, a fairly broad range of group increments for radicals is available. Again, comparable values for ΔS_f° are also available. To calculate the heat of formation of an organic radical, we will typically combine increments from Tables 2.7 and 2.4, as appropriate.

The major value of free radical group increments is in the prediction of stabilities of proposed radical and biradical intermediates in various thermal and photochemical reactions. For example, we might want to determine whether an observed thermal rearrangement occurs homolytically, via biradical intermediates, or by a concerted, pericyclic process. One valuable piece of information is ΔH_f° of the proposed biradical intermediate. If it is too high for the biradical to lie on the reaction path, then the biradical route can be rejected. We will see examples of this type of analysis in Chapter 15.

Galvinoxyl

Diphenylpicrylhydrazyl (DPPH)

Some stable radicals

Table 2.7
Group Increment Values for Free Radicals (kcal/mol)*

Radical	ΔH_f°	Radical	ΔH_f°
$[^\bullet C–(C)(H)_2]$	35.82	$[C–(O^\bullet)(C)(H)_2]$	6.1
$[^\bullet C–(C)_2(H)]$	37.45	$[C–(O^\bullet)(C)_2(H)]$	7.8
$[^\bullet C–(C)_3]$	38.00	$[C–(O^\bullet)(C)_3]$	8.6
$[^\bullet C–(H_2)(C_d)]$	23.2	$[C–(CO_2^\bullet)(H)_3]$	−47.5
$[^\bullet C–(H)(C)(C_d)]$	25.5	$[C–(CO_2^\bullet)(H)_2(C)]$	−41.9
$[^\bullet C–(C)_2(C_d)]$	24.8	$[C–(CO_2^\bullet)(H)(C)_2]$	−39.0
$[^\bullet C–(C_B)(H)_2]$	23.0	$[^\bullet N–(H)(C)]$	(55.3)
$[^\bullet C–(C_B)(C)(H)]$	24.7	$[^\bullet N–(C)_2]$	(58.4)
$[^\bullet C–(C_B)(C)_2]$	25.5	$[C–(^\bullet N)(C)(H)_2]$	−6.6
$[C–(C^\bullet)(H)_3]$	−10.08	$[C–(^\bullet N)(C)_2(H)]$	−5.2
$[C–(C^\bullet)(C)(H)_2]$	−4.95	$[C–(^\bullet N)(C)_3]$	(−3.2)
$[C–(C^\bullet)(C)_2(H)]$	−1.90	$[^\bullet C–(H)_2(CN)]$	(58.2)
$[C–(C^\bullet)(C)_3]$	1.50	$[^\bullet C–(H)(C)(CN)]$	(56.8)
$[C_d–(C^\bullet)(H)]$	8.59	$[^\bullet C–(C)_2(CN)]$	(56.1)
$[C_d–(C^\bullet)(C)]$	10.34	$[^\bullet N–(H)(C_B)]$	38.0
$[C_B–C^\bullet]$	5.51	$[^\bullet N–(C)(C_B)]$	42.7
$[C–(^\bullet CO)(H)_3]$	−5.4	$[C_B–N^\bullet]$	−0.5
$[C–(^\bullet CO)(C)_2(H)]$	2.6		
$[C–(^\bullet CO)(C)(H)_2]$	−0.3		

C_d = double bond; C_B = benzene carbon; N_I = imine nitrogen. Values in parentheses are highly approximate.

*Data are from Benson, S. W. (1976). *Thermochemical Kinetics: Methods for the Estimation of Thermochemical Data and Rate Parameters*, 2d ed., John Wiley & Sons, New York.

2.2.3 Carbocations

There is a wealth of information on gas phase ion thermodynamics because of the power of mass spectrometry and ion cyclotron resonance techniques. Before we discuss carbocation, and subsequently carbanion, stabilities, keep in mind that ionic structures are much more sensitive to environmental influences than radicals. The polarity, nucleophilicity, and hydrogen bonding ability of the solvent are important influences, as are the nature of the counterion. As such, thermodynamic information is a less reliable predictor of reactivity for carbocations and carbanions than it is for radicals. Nevertheless, gas phase thermodynamics is an excellent starting point, defining the *intrinsic* stabilities of ions. Any deviation in trends between gas phase and solution studies is likely a consequence of solvation effects, a theme we will visit many times throughout this book.

Hydride Ion Affinities as a Measure of Stability

A common and very valuable measure of gas phase carbocation stability is the **hydride ion affinity (HIA)**, defined as $\Delta H°$ for the reaction in Eq. 2.15. This is simply the heterolytic analogue of the homolytic cleavage associated with the bond dissociation energy (BDE) just discussed. Just as a larger BDE implies a less stable radical, a larger HIA implies a less stable carbocation. And, just as with the BDE, the usefulness of the HIA is that it provides a number that can be compared directly for cations of dissimilar structure. This is less true of other measures of cation stability.

$$RH \longrightarrow R^{\oplus} + H^{\ominus} \quad \Delta H° = HIA \qquad \text{(Eq. 2.15)}$$

HIA values reflect differences in the energies of the initial and final systems under analysis, not just carbocation stabilities (see Section 2.2.2 for similar reasoning related to radicals). However, in the case of HIAs, factors other than cation stability are not as influential as with radicals and carbanions (see below), and therefore the HIA values track very nicely carbocation stability. This is in part because the differences between the HIA values for various cations are much larger than the differences between BDEs for various bonds, and thus the HIA trends are less sensitive to other factors.

Table 2.8 lists HIA values for representative carbocations as determined *in the gas phase*. If we know $\Delta H_f°$ for RH, we can obtain $\Delta H_f°$ for the cation R^+, because H^- contributes a constant $\Delta H_f°$ value of 34.2 kcal/mol to all systems. Note also that not all the values in Table 2.8 were obtained using the same experimental techniques, and as such there is some uncertainty in the numbers. However, the trends in the data are quite reliable.

The data of Table 2.8 present many familiar, and perhaps some not so familiar, trends. All introductory courses describe a carbocation stability sequence based on reactivity patterns that is 3° > 2° > 1° > methyl, and the data of Table 2.8 support that finding. Before quantifying the trend, however, an additional effect must be considered. Consider the series of 1° cations: ethyl, 1-propyl, and 1-butyl. The larger cations are more stable. The effect is more pronounced in the series of 3° cations derived from *t*-butyl cation. Adding extra carbons increases stability at first, and then the effect levels off. This pattern is seen in other gas phase ion data. This effect is observed because a naked ion in the gas phase is desperate for solvation of any kind. A vacuum has the lowest possible dielectric constant, so anything is an improvement. In effect, adding carbons to a small ion slightly ameliorates this situation. It is as if a small amount of solvation is being provided by the neighboring carbons. It is not a large effect, and it is pretty much saturated by the time we reach a seven-carbon system, but any comparison between ions should take this effect into account. Thus, to compare a 2° ion to a 3° ion in the gas phase, we should not compare isopropyl to *t*-butyl, as the latter has one more carbon.

We did not have to worry greatly about this issue in our discussion of free radical stabilities in the gas phase. A neutral free radical in the gas phase is not desperate for solvation like an ion is. As such, adding "spectator" carbons to a free radical does not alter its stability significantly.

Table 2.8
Gas Phase Hydride Ion Affinities
(HIA, in kcal/mol) for Selected
Carbocations (in kcal/mol)*

Structure	HIA	Structure	HIA
$CH_3^{\oplus}$	312	[structure]	256
$C_2H_5^{\oplus}$	273	[structure]	236
$CH_3CH_2CH_2^{\oplus}$	266	[structure]	225
$CH_3CH_2CH_2CH_2^{\oplus}$	265	[structure]	225
[structure]	265	[structure]	248
[structure]	246	[structure]	225
[structure]	247	[structure]	212
[structure]	231	$HC \equiv C-CH_2^{\oplus}$	270
[structure]	229	$H_2C=\overset{\oplus}{C}H$	287
[structure]	228	$H_2C=\overset{\oplus}{C}-CH_3$	258
[structure]	227	[structure]$-CH_2^{\oplus}$	234
[structure]	226	[structure]	220
[structure]	224	[structure]	287
[structure]	231	[structure]	249
[structure]	225	[structure]	230
$NH_2CH_2^{\oplus}$	218	[structure]	218
$HOCH_2^{\oplus}$	243	[structure]	201
$FCH_2^{\oplus}$	290	[structure]	258
$NCCH_2^{\oplus}$	318		

*An excellent source of gas phase ion data is the fol-
lowing series of volumes: Bowers, M. T. (ed.) (1979).
Gas Phase Ion Chemistry, Academic Press, New York,
and subsequent years.

Taking the carbon number into effect, we find that 2° ions are indeed more stable than 1° ions by 18–20 kcal/mol, and 3° ions are more stable than 2° ions by about 17 kcal/mol. These are very large numbers—significantly larger than analogous comparisons for radicals. We will see in Chapter 11 that experimental differences for reactions in solution that are considered to involve carbocations are nowhere near this large. For reasons discussed in Chapter 3, solvation always attenuates ionic effects, so the gas phase data always represent the absolute maximum any ionic effect can show.

The origin of the stabilization due to alkyl substitution was discussed in Section 1.4.1. It can be described as a mixing of filled $\pi(CH_3)$ or $\pi(CH_2)$ orbitals with the empty p orbital associated with the cation, or equivalently as a hyperconjugative interaction. The former model nicely rationalizes why the comparable trend seen in radical chemistry is of smaller magnitude. With a free radical, the mixing is a filled orbital with a singly occupied orbital. Such three-electron interactions are less stabilizing than the two-electron interaction seen with carbocations.

Many other observations based on Table 2.8 are consistent with expectation and experience. Allyl cation is substantially more stable than 1-propyl. Perhaps surprisingly, though, allyl is less stable than 2-propyl. In the gas phase, therefore, the 2°/1° distinction is more important than allylic resonance. Substituting just one end of the allyl with a methyl is enough to make the delocalized system more stable than the localized 2° systems.

Increasing delocalization, as in the cyclohexadienyl cation, is further stabilizing. This structure is simply protonated benzene, and it serves as a model for the intermediate in electrophilic aromatic substitution (see Section 10.18). Similarly, benzyl ion is quite stable for a formally 1° ion. Aromaticity effects are clear, as in the much greater stability of the six π electron tropylium ion vs. the four π electron cyclopentadienyl ion (see Section 2.4.1 for a discussion of aromaticity).

Hybridization effects are quite evident. Cations derived from sp^2 carbons are much less stable than those from sp^3 carbons (based on the instabilities of vinyl and phenyl cations). Recall from Chapter 1 that hybridization affects carbon electronegativity in the following order: $sp > sp^2 > sp^3$. The more electronegative the atom attached to the cationic center, the higher the energy of the cation. As another example, the substantially diminished stabilization of the propargyl cation ($HC\equiv C-CH_2^+$) relative to allyl can be explained as a hybridization effect arising from the electron withdrawing alkynyl group. Carbons with sp hybridization are expected to be inductively electron withdrawing more so than sp^2, and this should destabilize propargyl vs. allyl.

Heteroatom effects are interesting. If we compare the $X–CH_2^+$ ions, using ethyl as a reference, we see that NH_2 is very strongly stabilizing, OH less so, while F is destabilizing. This tracks with the expected π donating ability of the atoms. In the case of F, it is only weakly π donating, but is strongly σ withdrawing, and the net effect is destabilizing. It should not be surprising that the highly electron withdrawing cyano group greatly destabilizes a cation, as would other, similar groups.

We noted in Chapter 1 that carbocations prefer to be planar at the cationic center. From HIA data, we can estimate the penalty for pyramidalization. In the 1-adamantyl cation, ring constraints prevent the ion from achieving planarity. Consistent with this, we find that the difference in HIAs between the 1-adamantyl cation and the isomeric 2-adamantyl cation is only 9 kcal/mol. This relatively small gap between a 3° cation and a 2° cation is consistent with the destabilization due to pyramidalization of the 3° ion. In support of this, the x-ray crystal structure of the 3,5,7-trimethyl-1-adamantyl cation has been determined. As anticipated, the cationic center lies out of the plane formed by its three neighboring CH_2 carbons by a full 0.212 Å, a substantial amount for an sp^2 center.

While the gas phase data just described are very informative, we also have a significant amount of thermodynamic data on carbocations in solution. This is due to the collection of data in **stable ion media** for carbocations developed so brilliantly by Olah and co-workers, leading to the receipt of the Nobel Prize for Chemistry in 1994. In a stable ion medium the ions have no possibilities for reactions that quench the ionic charge. While carbocations are reactive, if we remove potential reaction partners we will generally not see dimerization or other reactions in which two carbocations react with each other. The most common reaction

1-Adamantyl

2-Adamantyl

3,5,7-Trimethyl-1-adamantyl

paths for a carbocation are capture by a nucleophile and reaction with a base to form an olefin (see Chapter 11). If we can create an environment that is devoid of nucleophiles and bases, carbocations can be long lived. We can thus envision a stable ion medium. Given our discussion in Section 2.2.1, we are really not making the ion any more stable; we are instead making it persistent. It would perhaps be more precise to refer to "persistent ion media", but the "stable ion media" moniker has stuck, and we will use it here.

For the most part the stable ion media developed by Olah are based on antimony pentafluoride, SbF_5. This is a very powerful Lewis acid, and exposure of an alkyl halide to it produces a carbocation with the $Sb_2F_{10}X^-$ counterion, as in Eq. 2.16. The counterion to R^+ is very large, and the negative charge is dispersed over a large number of atoms. As such, $Sb_2F_{10}X^-$ is a *very* poor nucleophile and a *very* weak base. Inert solvents such as SO_2 or SO_2ClF can be added as diluents for the SbF_5. In addition, powerful protonic acids such as $HF–SbF_5$ and $FSO_3H–SbF_5$ allow the generation of carbocations from alcohol and olefin precursors (see Chapter 5). Another advantage of these systems is that they are often fluid to very low temperatures. This enables spectroscopic studies, most especially NMR, to be performed at low enough temperatures to suppress most carbocation rearrangements.

$$RX + SbF_5 \longrightarrow R^{\oplus} + Sb_2F_{10}X^{\ominus} \qquad \text{(Eq. 2.16)}$$

In a remarkable achievement, Arnett was able to perform calorimetric measurements in stable ion media, allowing the thermodynamics of carbocations to be studied. Arnett measured both $\Delta H°$ for the reaction in Eq. 2.16 and the solvation energy for a series of RX compounds. This allowed a determination of $\Delta H_f°$ for R^+ plus $Sb_2F_{10}X^-$ in the stable ion medium for a series of carbocations. When these were compared with comparable data from the gas phase, an extraordinarily good correlation between the values in the two different environments was seen. Each kcal/mol of relative stabilization in the gas phase was matched by a kcal/mol of relative stabilization in stable ion media. For example, the difference in stabilities between 2-butyl and t-butyl cations was determined to be 14.5 ± 0.5 kcal/mol in stable ion media, in good agreement with the gas phase difference of 16 kcal/mol. These data provide compelling support for the argument that what is being observed under stable ion media conditions truly is a carbocation that is essentially devoid of complications from solvation or counterion effects. As such, direct comparisons of data from stable ion media to gas phase data (be it experimental or computational) are valid.

An interesting analysis of HIAs can be made for the so-called non-classical carbocations. We saw in Section 1.4.1 that these are compounds with a bridging C–C σ or π bond to a carbocation center, thereby creating a hypervalent C. A good deal of evidence indicates that such bridged structures are more stable than classical carbocations in the gas phase and in stable ion media. The gas phase HIA data support the view that bridging imparts stability. As shown in Table 2.8, the 2-methyl-2-norbornyl cation has an HIA of 225 kcal/mol, a typical value for an eight-carbon 3° cation, and it is generally accepted that this is a conventional carbenium ion. However, the HIA of 2-norbornyl is 231 kcal/mol, only 6 kcal/mol higher than the analogous 3° system. We discussed above that a typical 2°/3° energy difference is 17 kcal/mol. The substantially smaller value in this case indicates a special stabilization of the 2° ion, consistent with the notion that there is something unique about its structure.

Lifetimes of Carbocations

There are less data related to carbocation lifetimes as compared to radical lifetimes. Yet, some extensive studies by Mayr, Richards, and others have provided much insight into substituent effects on their lifetimes. In general, the lifetimes are extremely short in water. For example, Toteva found that the t-butyl carbocation has a lifetime of only 10^{-12} s in water. Hence, although we consider tertiary carbocations stable, they are clearly not persistent in this medium. Secondary carbocations are even more reactive toward addition of water, and many secondary derivatives undergo concerted hydrolysis in water that avoids formation of the carbocation reactive intermediate. The primary 4-methoxybenzyl carbocation intermediate of solvolysis of 4-methoxybenzyl chloride exists for several nanoseconds in water, but relatively little is known about the lifetimes of more unstable primary benzyl carbocations, because their neutral precursors show a strong preference for concerted hydrolysis that avoids formation of the primary carbocation intermediate.

2.2.4 Carbanions

By analogy to the previous sections, we should consider Eq. 2.17 to evaluate the stabilities of carbanions. Because of the universal importance of acid–base chemistry, Chapter 5 is entirely devoted to this subject. There we will discuss both carbon acids (where the negative charge on A$^-$ is primarily associated with a carbon) and heteroatom acids. Here we briefly mention trends associated with carbon acids with a goal of defining the essential nature of carbanions. Selected $\Delta H°$ values for the reaction of Eq. 2.17 in the gas phase for carbon acids are presented in Table 2.9.

$$HA \longrightarrow A^{\ominus} + H^{\oplus} \qquad \text{(Eq. 2.17)}$$

A great number of acidities have also been measured in solution. In such cases, the common value reported is the pK_a. Recall (or skip ahead to Chapter 5) that pK_a is the negative log of the acid dissociation constant in water. As such, a smaller pK_a value implies a stronger acid, and each pK_a unit represents a factor of 10 in equilibrium acidity. Table 2.10 lists a number of pK_a values for carbon acids.

Our starting point for discussion is alkanes, with pK_a values in the range of 45–70. Remember, a pK_a of 50 implies an equilibrium constant of 10^{-50} for the dissociation of the acid! Deprotonation at an sp^2 center (pK_a 40–45), as in ethylene or benzene, is favored over an sp^3 center, and the effect is even larger with an sp center (pK_a 20–25). A negative charge is more stable in a hybrid orbital with more s character, because s orbitals (unlike p) have finite density at the positively charged nucleus. Allyl anion and benzyl anion are greatly stabilized due to a resonance effect. An anion that contains a cyclic array of six electrons is aromatic. Four electrons is antiaromatic, hence the 46-unit difference in pK_a values for cyclopentadiene vs. cyclopropene (Eq. 2.18 vs. Eq. 2.19). Electron withdrawing and/or delocalizing substituents such as carbonyl, cyano, and nitro produce an expected lowering of the carbon acid pK_a.

$$\text{Cyclopentadiene} \xrightarrow[pK_a = 15–16]{-H^{\oplus}} \text{Six } \pi \text{ electrons} \qquad \text{(Eq. 2.18)}$$

$$\text{Cyclopropene} \xrightarrow[pK_a = 61]{-H^{\oplus}} \text{Four } \pi \text{ electrons} \qquad \text{(Eq. 2.19)}$$

The analysis of pK_a values to make comparisons of carbanion stability is a standard approach. Yet, as discussed earlier with regard to the use of BDEs and HIAs to analyze radical and carbocation stabilities, other factors can affect the values. This is a more important caveat when evaluating pK_a values than with BDEs or HIAs. As will be discussed in Chapter 5, solvation effects can dramatically change pK_a values, and sometimes in different solvents some pK_a values actually reverse in order. Hence, when using pK_a values to compare carbanion stabilities, it is very important to compare the values under as similar a set of conditions as possible.

Carbanion stable ion media exist. Carbanions are generally less reactive than carbocations to begin with, and it is often quite straightforward to directly observe carbanions under relatively conventional conditions. The major requirement for having a carbanion last for a significant time period is for the pH of the medium to be above the pK_a of the conjugate acid of the carbanion. When working in an aprotic solvent, this means that the pK_a of the solvent must be significantly higher than the pK_a of the conjugate acid of the carbanion. Under these conditions the carbanion will often be persistent. This acid–base chemistry notion is discussed extensively in Chapter 5.

Table 2.9
Gas Phase Values of $\Delta H°$ for Eq. 2.17 (in kcal/mol)

Methane	416.6
Ethane	420.1
Ethylene	407.5
Propene	390.8
Benzene	400.7
Fluoromethane	409
Chloromethane	396
Bromomethane	392
Iodomethane	386
Acetonitrile	373.5
Acetone	370.0
Toluene	377.0
Nitromethane	357.6

Table 2.10
pK_a Values of Selected Carbon Acids*

Ethane	50
Cyclohexane	45
$(CH_3)_3CH$	71
Ethylene	44
Benzene	43 or 37
Acetylene	24
Phenylacetylene	19.9
PhCH_3	41.2
$CH_2=CHCH_3$	43
Ph$_2CH_2$	33.0
Ph$_3CH$	31.5
Cyclopentadiene	16.0
Cyclopropene	61
CH_3COCH_3	20.0
$CH_3COCH_2COCH_3$	8.84
CH_3NO_2	10.2
CH_3CN	25.0
CH_3SOCH_3	28.5

*The student should realize that the very large values are approximate.

2.2.5 Summary

The thermochemistry of the three most common reactive intermediates—radicals, carbocations, and carbanions—is conveniently analyzed using heats of very similar reactions. Homolysis of C–H bonds is used for radicals and then heterolysis of C–H bonds to create hydride or proton is used to analyze carbocations or carbanions, respectively. Just as we use heats of formation and combustion to analyze the stabilities of standard organic compounds, these three simple reactions and their associated enthalpies are very good starting points whenever you are considering the relative stabilities of reactive intermediates. Now that we have covered measurements of stability and strain for both normal and reactive organic compounds, it is time to look at pathways for interconversion between structures, and their impact on strain and stability. This next discussion will further tie structure and energetics together.

2.3 Relationships Between Structure and Energetics— Basic Conformational Analysis

We will now discuss at some length the many ways in which deviations from standard bonding parameters lead to energetic destabilization of a molecule. We will focus on "stable" structures (i.e., not on reactive intermediates), but the notions we develop here also apply to reactive intermediates. We first explore acyclic systems, wherein molecular motions directly lead to strained forms. Note that we are not yet considering conventional chemical reactivity. We will be considering **conformers**, or **conformational isomers**. Recall that conformers are stereoisomers that interconvert by rotation around single bonds (see Chapter 6 for definitions of stereochemical concepts). These isomers are not to be confused with **constitutional isomers**, where the molecular formula is the same, but the atoms are arranged differently.

In this section we discuss both stable conformers and the transition states that interconvert them—a topic generally referred to as **conformational analysis**. While this is in some ways an unconventional placement of this topic in a physical organic textbook, it follows naturally from our discussion concerning strain and stability, and ties in well with the notion of potential surfaces and molecular motions. For example, a conformer that is less stable than the ground state is, by definition, strained. In addition, the paths that interconvert conformers derive from particular vibrational degrees of freedom. So, here we are discussing strained organic molecules in general, and vibrational modes that interconvert them. After considering the acyclic systems, we examine cyclic ones. There can be strains inherent to the ring size (small and large), as well as strains introduced upon achieving differing conformations. Many of these topics are covered in all introductory organic chemistry texts, and so in some places our discussion will be brief.

The terminology associated with conformational analysis is in many ways akin to the terminology used to discuss stereochemistry—namely, cis, trans, anti, syn, etc. In Chapter 6 there is a glossary of terms used in conformational analysis and stereochemistry, so if a term is used below that you are unfamiliar with, look to the end of Chapter 6 for a definition.

2.3.1 Acyclic Systems—Torsional Potential Surfaces

In considering acyclic systems we distinguish between two types of strain. The first is distortion of the ground state, generally resulting from an accumulation of bulky substituents around a C–C bond. This leads to bond lengths and angles that deviate substantially from the standard values (we examine some extremes in Section 2.5). The second type of strain concerns conformers, in which the strained form is manifest in the transition state of a rotation or in a less stable conformer. We begin with the latter type, where the focus is on rotations around C–C bonds. In these discussions, the familiar term "sterics" is used. As defined in Chapter 1, **sterics**, a **steric relationship**, or **steric strain**, are all terms that imply a buttressing of chemical groups that is destabilizing to the molecule.

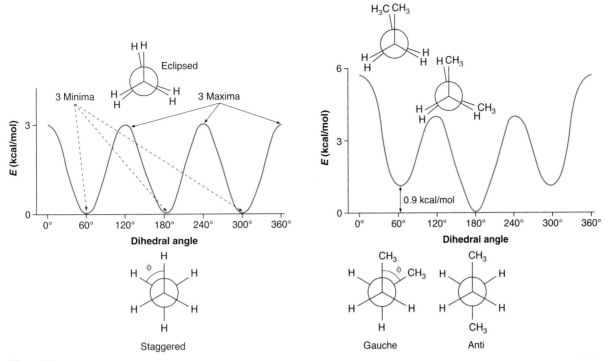

Figure 2.6
Left: Torsional itinerary for ethane, showing the three-fold nature of the barrier and the eclipsed and staggered forms. Right: Torsional itinerary for butane. The anti–gauche interconversion crosses a barrier of ca. 3.4 kcal/mol, while the direct gauche–gauche barrier is 5–6 kcal/mol.

Ethane

Recall the use of Newman projections from introductory organic chemistry, in which one sights directly down a C C bond and draws the back atom as a large circle and the front atom as a point. The x axis of Figure 2.6 is the **dihedral angle** between two arbitrarily chosen hydrogens, one on the front and one on the rear carbon. We usually think of these as the projection angles in a Newman projection. Alternatively, the A–B–C–D dihedral angle is defined as the angle between the A–B–C plane and the B–C–D plane. This is, in fact, how computer programs evaluate dihedral angles.

The staggered form of ethane is more stable than the eclipsed. The energy difference between the two—the **rotation barrier**, E_{rot}—is 3 kcal/mol. Thus, rotation about even the simplest C–C bond is not "free". Instead, it is hindered by a barrier of 3 kcal/mol. However, the barrier is so small that the rotations are very fast, as shown in the next Going Deeper highlight.

Remember, the eclipsed form is a transition state; the staggered conformation is the only stable form. Interconversion between one staggered form and another is a torsional motion, where the methyls twist relative to each other. When the torsional vibration achieves enough energy to traverse the transition state, the staggered conformers interconvert.

What does it physically mean when we state that "the staggered form is 3 kcal/mol

---

Going Deeper

How Big is 3 kcal/mol?

As an activation barrier, 3 kcal/mol is quite small. Let's calculate how fast the bond in ethane is rotating. As in the final Going Deeper highlight of Section 2.1.3, we use the Arrhenius equation with an A value of 10^{13}.

The rate constant at 298 K for a 3 kcal/mol barrier is 6×10^{10} s^{-1}. This corresponds to a half-life of 10 picoseconds. Rotation about simple C–C bonds is extremely rapid!

more stable than the eclipsed form"? As ethane molecules undergo the conversion of staggered to eclipsed back to staggered conformations, on average 3 kcal per Avogadro's number of molecules is absorbed from the ensemble and then released back to the ensemble. The energy that it takes to undergo the rotation/distortion can come from a collision between ethanes. Here, some of the kinetic energy of the ethanes involved in the collision is converted to the potential energy stored in an ethane as it achieves the eclipsed transition state. However, the energy can also arise from a redistribution of the energy between various vibrational modes within the ethanes, meaning that a collision is not required to achieve the transition state.

Where is this potential energy stored at the transition state? It is placed in the form of a **torsional strain**. As the hydrogens become eclipsed, a distortion of the C–H and C–C bonds occurs (albeit small). As these bonds go back to their normal lengths and angles in the staggered form, the energy required for the distortions is released to the ensemble of molecules or into another vibrational mode of the same molecule.

The transfer of energy back and forth between vibrational modes (stretches, bends, wags, etc.) and torsional modes implies that a torsion is a vibration, and we have already defined it as such. Moreover, because the other vibrations are quantized, it makes sense that torsional modes are quantized also, as we describe in the next Going Deeper highlight. The

---

Going Deeper

Shouldn't Torsional Motions be Quantized?

The bond stretching motions and bond bending motions discussed in Section 2.1.4 only had certain energies. Bond rotations are just another example of an internal degree of freedom within molecules. Therefore, why are we not explicitly discussing quantized energy states for bond rotations?

All internal degrees of freedom are indeed quantized. Below we show a schematic representation of this notion in the context of torsions. We define a torsional motion as one methyl group rotating relative to the other, but if the energy of this motion is below 3 kcal/mol, the barrier to the next staggered conformation is not surpassed. Superimposed on a portion of the torsional potential surface for ethane are energy states for this torsional motion. Only certain energies for this single torsion motion are actually allowed.

As evidence of the fact that torsional motions are quantized, excitations between the energy levels of a torsion can be observed in the microwave spectra of organic molecules. In fact, a large fraction of the lessons taught in this chapter about conformational preferences were obtained from the microwave spectra of the compounds.

The low energy motions appear somewhat harmonic, but the torsional motion quickly becomes anharmonic for the higher energy states. A continuum of energy states exists near and above the top of the barrier, meaning that multiple energy levels are possible. If the continuum of states is thermally populated, the rotations appear similar to any rotation that occurs in the macroscopic world.

To see if enough energy is present to occupy the continuum of states for ethane, we compare the 3 kcal/mol barrier to RT, which is 0.6 kcal/mol at 298 K. Although 0.6 is less than 3, it is certainly of the same order of magnitude. A significant fraction of the ethane molecules are in energy states at or near the continuum. This is why we view bond rotations as nearly "free rotors", and quantized effects on bond rotations are not relevant for conventional organic chemistry at ambient temperatures.

Hansen, G. E., and Dennison, D. M. "The Potential Constants of Ethane." *J. Chem. Phys.*, **20**, 313–326 (1952).

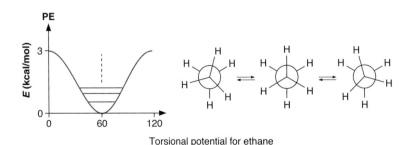

Torsional potential for ethane

picture of atomic motion within a molecule that arises is quite complex, where stretches, bends, wags, and torsions can exchange energy and be excited to various extents, depending on collisions and the temperature.

Butane—The Gauche Interaction

Figure 2.6 also shows the torsional profile for rotation about the C2–C3 bond of butane. Now there are two types of staggered forms—the **gauche** and **anti** conformers—and two types of eclipsed transition states. The anti conformer is more stable than the gauche by 0.9 kcal/mol. We could say that gauche butane has a strain energy of 0.9 kcal/mol. The conventional explanation of this result is to invoke a steric interaction between the methyl groups in the gauche form, termed a **vicinal repulsion** because it is between groups attached to adjacent carbons. The gauche form is chiral, and there are two enantiomeric forms on the complete torsional itinerary. This means there is a statistical factor of two favoring the gauche form when considering thermodynamic parameters. That is, if we want to know the relative proportions of gauche and anti butane at equilibrium, we must take into account the fact that there are two gauche forms but only one anti. The 0.9 kcal/mol number is an enthalpy. The statistical factor in favor of the gauche shows up as $R\ln2$ in the entropy (see the discussion of entropy in Section 2.1.2). Thus, for butane we would calculate the Gibbs free energy using Eq. 2.20. The result is that butane is roughly 70% anti and 30% gauche at 298 K.

$$\Delta G^\circ = \Delta H^\circ - T\Delta S^\circ = 0.9 \text{ kcal/mol} - T(R\ln2) \qquad \text{(Eq. 2.20)}$$

Just as with ethane, the differing forms of butane possess quantized torsional motions, now of the methyl groups relative to each other, and of the methyl hydrogens relative to the propyl group. Thermal population, collisions, or microwave excitation can lead to torsional energy states possessing the proper energy to overcome the barriers and interconvert the conformers. The transition state for interconverting gauche and anti forms is 3.4 kcal/mol above the anti, while the direct gauche-to-gauche interconversion must cross a higher barrier of 5–6 kcal/mol. This is a general result of conformational analysis—the easier path for interconverting two gauche forms is usually through the anti form, rather than crossing the direct gauche–gauche barrier.

While gauche and anti are unambiguous terms for butane, for more complex structures the more systematic **Klyne–Prelog system** is often used. Following the guidelines in Figure 2.7, gauche butane is synclinal (+ or –, depending on which enantiomer), anti butane is antiperiplanar, and eclipsed butane is synperiplanar.

The notions of eclipsing and gauche interactions are applicable not only to stable organic structures, but also to reactive intermediates. In fact, as described in the next Going Deeper highlight, the geometry of radicals is influenced by a conformational effect similar to those discussed for standard alkanes.

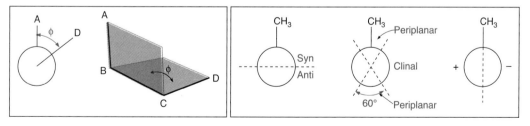

Figure 2.7
Dihedral angles. Left: Definition of a dihedral angle, ϕ, both in Newman projection and as the angle between two planes. Right: The Klyne–Prelog system for describing conformations about a single bond. We view the placement of the group on the front atom as being in regions of space called anti/syn, and clinal/periplanar relative to a reference group (here a methyl) on the rear atom.

Going Deeper

The Geometry of Radicals

In Chapter 1 we noted that radicals prefer planar structures, but the barrier to pyramidalization is very low. In fact, there is a conformational effect that leads to pyramidalization. Computational studies by Padden-Row and Houk have shown that even the simplest alkyl radicals are pyramidalized, and methyl radical is really the only totally planar radical. At right are shown three examples. The interpretation of these structures invokes simple conformational analysis. In each case, pyramidalization occurs in the direction that minimizes eclipsing interactions.

For ethyl and isopropyl the effect is not large. Theory predicts, however, that the *t*-butyl is substantially pyramidalized, with a C–C–C angle of 117.3°. For *t*-butyl and the other alkyl radicals, the barrier to pyramidal inversion is quite small. One measurement suggests a barrier of roughly 0.45 kcal/mol for *t*-butyl radical. So, under most experimental conditions, simple alkyl radicals can be

Pyramidalization of radical centers

treated as if they were planar. Nevertheless, when considering the reactions of radicals and the potential effects of substituents, there really is no strong bias toward planarity at the radical center, and in fact, the bias is toward pyramidalization.

Padden-Row, M. N., and Houk, K. N. "Origin of the Pyramidalization of *tert*-Butyl Radical." *J. Am. Chem. Soc.*, **103**, 5046 (1981).

Table 2.11
A Few Common Corrections to Standard Group Increments (kcal/mol)

Gauche alkane	0.8
Cis alkene	1.0
Ortho correction	0.6

2-Methylpentane

3-Methylpentane

3-Methylpentane

The 0.9 kcal/mol destabilization of gauche butane relative to anti is not typically thought of as a strain energy, but by our definition, that's exactly what it is. The group increments method, as developed above, ignores such interactions, and so is applicable to anti butane, not gauche (in fact, there is a subtle ambiguity about how group increments handle gauche interactions; see Exercise 32 at the end of the chapter). However, the group increments method is not only used to define strain-free ΔH_f° values. A number of correction factors have been developed to allow the method to predict accurate ΔH_f° values even for strained molecules, and a few of the most common are given in Table 2.11. The first such correction is the gauche correction. On considering a large number of structures, a value of 0.8 kcal/mol has been settled on as the group increments correction for a gauche interaction. Let's see how this correction is used.

Consider calculating ΔH_f° for 2-methylpentane and 3-methylpentane. Using the values from Table 2.4, a "strain-free" estimate of −42.36 kcal/mol is obtained for both molecules. However, the experimental fact is that 2-methylpentane is more stable than 3-methylpentane by 0.66 kcal/mol. This is a gauche effect. Draw a Newman projection sighting down the C2–C3 bond of 2-methylpentane. There is necessarily at least one gauche interaction. So, the best estimate of ΔH_f° for 2-methylpentane using group increments is −42.36 + 0.8 = −41.56 kcal/mol. Now consider 3-methylpentane. There are two gauche interactions—one along the C2–C3 bond and one along the C3–C4 bond (see margin). So, the best estimate of ΔH_f° for 3-methylpentane using group increments is −42.36 + 2(0.8) = −40.76 kcal/mol. Now the group increment method correctly "predicts" that the 2-methyl isomer is more stable. Table 2.12 lists the number of gauche interactions that occur across all bond types. When using group increments to get the best possible estimate of ΔH_f° for a molecule, the appropriate number of gauche corrections should be used.

Another common group increment correction is for cis double bonds. If an alkene is cis, a 1.0 kcal/mol penalty is added to ΔH_f°, while no such increment is applied to the trans isomer. This increment is a fairly crude estimate, so it does not apply to all alkenes. In particular, if one of the olefin substituents is a *t*-butyl or similar group, the increment should become considerably larger. Similarly, two substituents in an ortho arrangement on a benzene ring would need a correction, called the ortho correction in Table 2.11. Once again, this increment would be group dependent, and hence the value given here is a very crude estimate.

Table 2.12
Number of Gauche Interactions in Variously
Substituted C–C Bonds

	For X–Y bond	
X	Y	Number of gauche
R$_2$CH	CH$_3$	0
	CH$_2$R	1 or 2
	CHR$_2$	2 or 3
R$_3$C	CH$_3$	0
	CH$_2$R	2
	CHR$_2$	4
	CR$_3$	6

Barrier Height

The barrier to conformational interconversion in butane is larger than ethane due to the larger steric repulsions present in the transition states, which possess the eclipsing interactions. Table 2.13 shows some other measured rotation barriers around C–C bonds. For simple alkyl substituents, the rotation barriers around CH$_3$–X bonds generally increase as X gets larger, as may be expected. The barriers to rotation along C–N and C–O bonds are lower than analogous C–C bonds, indicating that lone pairs act smaller than C–H bonds. In the halogen series, it may at first seem surprising that the larger halogens do not necessarily give larger barriers. This is because the C–X bond length increases with higher atomic number, and the longer the bond, the further the halogen is from the methyl group.

Table 2.13
A Few CH$_3$–Y Rotation Barriers
for the Bonds Indicated*

Compound	Barrier height (kcal/mol)
CH$_3$–CH$_3$	2.9
CH$_3$–CH$_2$CH$_3$	3.4
CH$_3$–CH(CH$_3$)$_2$	3.9
CH$_3$–C(CH$_3$)$_3$	4.7
CH$_3$–CH$_2$F	3.3
CH$_3$–CH$_2$Cl	3.7
CH$_3$–CH$_2$Br	3.7
CH$_3$–CH$_2$I	3.2
CH$_3$–NH$_2$	2.0
CH$_3$–OH	1.1

*Lowe, J. P. "Barriers to Internal Rotation About Single Bonds." *Prog. Phys. Org. Chem.,* **6**, 1 (1968).

Barrier Foldedness

The ethane and butane molecules both have a three-fold rotation profile. That is, on rotating 360° about the C–C bond, the system goes through three maxima and three minima (Figure 2.6). Not all single bond rotations are three-fold, however. For example, rotation about the CH$_3$–C$_6$H$_5$ bond in toluene produces a six-fold barrier. Sketch out a rotational diagram if this is not clear. Whenever a three-fold rotor (methyl) opposes a two-fold rotor (phenyl), we'll see a six-fold barrier, because $3 \times 2 = 6$. The C–N bond in nitromethane also has a six-fold barrier, and the Cp–Co–(CO)$_3$ system (Cp = cyclopentadienyl) can be considered to have a fifteen-fold barrier (5 × 3). Generalizing, the foldedness (F) of a barrier involving an n-fold rotor and an m-fold rotor is given by Eq. 2.21, where q is the number of bonds eclipsed in the transition state. Thus, the ethane barrier is $(3 \times 3)/3 = 3$.

$$F = (n \bullet m)/q \qquad \text{(Eq. 2.21)}$$

A general attribute of n-fold barriers where n is large is that E_{rot} tends to be small. For example, E_{rot} for nitromethane is 0.006 kcal/mol, and for toluene it is < 0.1 kcal/mol. One way to rationalize the low barriers is to appreciate that for toluene to have a 3 kcal/mol barrier like ethane but still be six-fold, the energy of the system would have to change very rapidly in response to only small changes in torsional angle. This rapid fluctuation in energy is an unreasonable situation.

Tetraalkylethanes

Consider 2,3-dimethylbutane (tetramethylethane). As shown in Figure 2.8, there are two conformers, anti and gauche (defined now by the H–C–C–H dihedral angle). The anti form has two classical gauche–butane interactions, while the gauche form has three. So, in direct analogy to butane, we would expect a 70:30 mixture of anti to gauche. The experimental result, however, is that 2,3-dimethylbutane exists as a 1:2 mixture anti:gauche—a statistical mixture (the gauche form is still chiral) implying $\Delta H° = 0$! What has gone wrong?

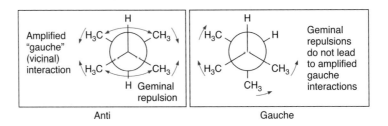

Figure 2.8
In tetramethylethane, geminal repulsions between methyls cause expansion of the C–C–C angle, leading to enhanced gauche butane-type interactions in the anti form only. This repulsion counters the fact that there are more gauche interactions in the gauche conformer (3) than in the anti (2). The net effect is that the two conformers have very nearly equal enthalpies.

One error in the above analysis is the assumption that the steric relationships present in n-butane will carry over unchanged in 2,3-dimethylbutane. It is true that there are gauche Me•••Me relationships in 2,3-dimethylbutane, and these are expected to be destabilizing. However, there are also other interactions, including a **geminal repulsion**, a steric interaction between methyls attached to the same carbon (Figure 2.8). The geminal repulsion will occur in addition to the vicinal repulsion.

The consequences of this geminal repulsion are different for gauche and anti-2,3-dimethylbutane. As shown in Figure 2.8, the geminal repulsion in the anti form exacerbates the vicinal repulsion, further destabilizing that form. No such effect exists for the gauche form, where the geminal repulsion does not seriously influence the vicinal repulsion. Thus, geminal repulsion preferentially destabilizes the anti form. The net effect is that the two forms have essentially the same enthalpies, and a 1:2 mixture (anti:gauche) results from the statistical arguments discussed above.

This geminal repulsion is a general result, and the preference for the gauche form of R_2CHCHR_2 molecules increases as the steric size of R increases. Thus, an important conclusion of this analysis is that all tetraalkylethanes are expected to show a preference for the gauche conformer. For a molecule with a very large R group, such as 1,1,2,2-tetrakis(t-butyl)-ethane, the effect is so large that *only* the gauche form is viable.

1,1,2,2-Tetrakis(t-butyl)ethane

These problems illustrate the danger of quantitatively extrapolating notions of conformational analysis based on very simple systems such as butane to more complicated structures. This caveat is true both in thinking about structures in general and in applying group increments correction factors. The correction factors are the best possible estimate, but we must be wary of their use. Another important example of this concerns *n*-pentane and related structures, which we consider next.

The g+g– Pentane Interaction

A subtle, but important conformational effect is the **g+g– pentane** interaction, which gives rise to what is commonly called *syn*-**pentane** strain. Recall that there are two possible gauche forms around an internal bond in a linear alkane, and the + / – designation of Figure 2.7 is used to differentiate them. For *n*-pentane the all-anti conformer is the most stable. However, in some situations gauche interactions will develop. A common example is in a reaction transition state, when reactants are held in a cyclic array.

As shown in Figure 2.9, the conformation with consecutive gauche bonds of opposite handedness—the g+g– structure—forms a nearly cyclic structure (Eq. 2.22). In fact, the geometry is very similar to that of the envelope conformation of cyclopentane (see below). This form brings the two methyl groups very close together, and an adverse H•••H steric repulsion develops. Compared to the g–g– form, the g+g– is less stable by about 3 kcal/mol, a very significant effect. A simple but flawed analysis of these two structures would suggest they should be very similar in energy, because both have two gauche interactions. The g+g–pentane interaction is another important example of the potential vulnerability of simple additivity arguments.

$$R\diagup\diagdown\diagup\diagdown R' \rightleftharpoons \underset{\text{R)(R'}}{\diamond} \qquad\qquad \text{(Eq. 2.22)}$$

g+g– pentane

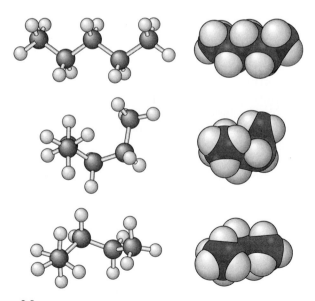

Figure 2.9
Conformations of *n*-pentane. Top: All anti. Middle: g–g–. Bottom: g+g–.
Note the close H•••H contact in the bottom structure.

Going Deeper

Differing Magnitudes of Energy Values in Thermodynamics and Kinetics

Conformational analysis is a good context in which to point out the differences in thermodynamic and kinetic phenomena for the same energy value. We just stated that 3 kcal/mol is a "very significant effect". However, we also noted in a previous Going Deeper highlight

that a 3 kcal/mol kinetic barrier is extremely small, resulting in bond rotations on ps time scales. This analysis highlights a stark difference between thermodynamic vs. kinetic effects. A 3 kcal/mol thermodynamic difference represents over a 100-fold preference for the lower energy conformation, yet a 3 kcal/mol barrier is insignificant at ambient temperature.

Allylic ($A^{1,3}$) Strain

Closely related to the g+g– pentane interaction is **allylic strain** or **$A^{1,3}$ strain** (also known simply as **A strain**). As shown in Eq. 2.23, this interaction can develop in a simple olefin between a substituent on one end of the olefin, and an allylic substituent on the other end. It is, in a sense, a more restricted version of the g+g– pentane interaction, and the energetic penalty is comparable. The $A^{1,3}$ interaction is more commonly invoked—especially in reaction transition states—perhaps because only one C–C bond rotation needs to be constrained to achieve the effect (vs. two for the g+g– pentane interaction). This has an effect on the conformations of cis alkenes, leading to a preference for the more linear structure.

$$ \text{(Eq. 2.23)} $$

allylic strain

2.3.2 Basic Cyclic Systems

The cycloalkanes have long served as important prototypes for conformational analysis. They are also building blocks for many important organic molecules, and so an understanding of their shape preferences is crucial for many studies. The essentials of cycloalkane conformational analysis are covered in introductory organic chemistry, but we review and expand upon them here.

Cyclopropane

Since any three points define a plane, we do not need an experiment to tell us that the carbon framework of cyclopropane is planar, the only saturated carbocycle for which this is true. The defining feature of cyclopropane is the C–C–C angle of 60°. As discussed in some detail in Chapter 14, the bonding in cyclopropane is unique, not really amenable to conventional bonding models, and so here we simply summarize some important structural features of the molecule.

The C–C bonds of cyclopropane are short—1.51 Å vs. normal C–C bonds of 1.54 Å. Also, the H–C–H angle is opened up quite a bit to 115° vs. a value of 106° for the H–C–H angle of the CH_2 group of propane. In addition, the C–H bonds of cyclopropane are more acidic than normal alkanes. All these observations are consistent with the rehybridization expected in cyclopropane. To make the C–C–C bond angles smaller, we use more p character. This leaves excess s character in the C–H bonds, and greater s character leads to greater acidity.

The strain energy of cyclopropane is 27.5 kcal/mol. The majority of the strain results from the deviation of bond angles from their normal values, but there is also expected to be a significant contribution from the eclipsing C–H interactions across C–C bonds forced by the planar structure.

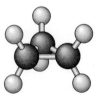

Cyclopropane

Cyclobutane

If cyclobutane were planar, the C–C–C angles would be 90°. However, the planar conformation produces perfect C–H eclipsing along the C–C bonds, an unfavorable situa-

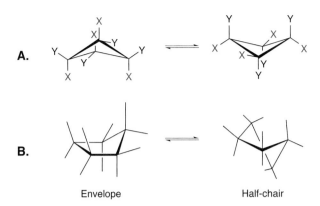

Figure 2.10
Aspects of cycloalkane conformational analysis. **A.** The two puckered conformations of cyclobutane. Note how the substituents (X and Y) exchange positions when puckered forms interconvert. **B.** Left: The envelope form of cyclopentane—the conformer contains a mirror plane in the plane of the page. Right: The half-chair form of cyclopentane, looking down the two-fold rotation axis.

tion. As a result, the molecule puckers (Figure 2.10 **A**), relieving eclipsing but reducing the C–C–C angles to about 88°. The angle between the two C–C–C planes is 28°. Inversion of the ring to interconvert the two puckered forms proceeds over a very small barrier (1.45 kcal/mol), and so in many circumstances one can think of cyclobutane as effectively planar (see Chapter 6 for a more thorough discussion). The inversion is sometimes called a **butterfly motion**.

The C–C bond lengths of cyclobutane are 1.55 Å—slightly elongated. The strain energy is 26.5 kcal/mol, essentially identical to that of cyclopropane. The strain results primarily from angle bending, but again residual torsional strain (the eclipsings are not completely relieved by puckering) plays a role.

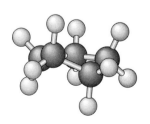

Cyclobutane

Cyclopentane

Planar cyclopentane would have a C–C–C angle of 108°, very close to the tetrahedral angle. However, as with cyclobutane, planar cyclopentane would suffer from a large number of C–H eclipsing interactions, and so the planar form is not a viable conformation. Again, the molecule puckers, even though the puckering compresses the C–C–C bond angles (Figure 2.10 **B**). Cyclopentane has 6.2 kcal/mol of strain energy, most of it due to torsional (eclipsing) effects, but some due to angle strain.

Cyclopentane can distort from the planar form to relieve eclipsings in two different ways. One way is to move just one carbon out of the plane, producing the **envelope** conformation, which retains a mirror plane of symmetry (Figure 2.10 **B**). This conformer resembles the g+g– pentane geometry discussed above, but now an adverse H•••H contact is replaced by a bond (see margin).

Alternatively, cyclopentane can twist about C–C bonds, producing the **half-chair** form, which has a two-fold rotation axis. These two forms are very nearly equal in energy, and they interconvert very rapidly (the barrier is < 2 kcal/mol) by a process termed **pseudorotation**. The rapid interconversion of envelope and half-chair forms makes all hydrogens (and carbons) equivalent, producing a time-averaged structure that is equivalent to the planar form. Note that the interconversion of envelope and half-chair forms does not require passing through the fully planar form, which lies ~5 kcal/mol above these structures. Substituents can cause a cyclopentane to prefer one conformation over the other, or distort this relatively flexible ring to intermediate conformations. An example of where the conformational dynamics of cyclopentane has important biological ramifications is in the conformations of DNA, as described in the following Connections highlight.

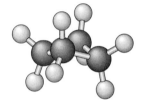

Cyclopentane vs. g+g– pentane

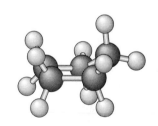

Cyclopentane envelope conformation

Cyclopentane half-chair conformation

Cyclohexane

Cyclohexane is a benchmark of conformational analysis, and it is an important building block of many natural and synthetic products. Ironically, though, cyclohexane is much more the exception than the rule for carbocycles. Cyclohexane is unique in that it is predominantly found in a single, relatively rigid conformation, the chair form (Figure 2.11 **A**).

Chair cyclohexane is best appreciated by a Newman projection sighting down two C–C bonds (see below). This form is quite compatible with standard bond and torsional angles. The actual structure is slightly flattened from ideal, such that the C–C–C–C torsional angle is ~55°. In the chair form, cyclohexane is generally considered to have a strain energy of 0 kcal/mol. From a Newman projection it is easy to see that each CH_2 contains two different types of hydrogens, termed **equatorial** and **axial** (labeled H_e and H_a, respectively, in the picture). The axial and equatorial positions interchange via a ring flipping process (see Figure 2.11 **A**).

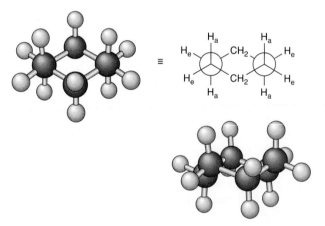

Cyclohexane chair conformation

Like the hydrogens, a substituent can occupy two different positions—axial or equatorial. Consider a methyl substituent (Figure 2.11 **B**). If we place a CH_3 in the axial position at C1 and sight down the C1–C2 bond, the methyl is gauche to one of the carbons of the ring. Based on the gauche butane interaction, we might expect this interaction to destabilize the structure by 0.9 kcal/mol. A similar interaction is also evident if we sight down the C1–C6 bond. In contrast, if we place a CH_3 in the equatorial position, no such gauche interactions develop. Thus, we might expect axial methylcyclohexane to be less stable (or strained) relative to the equatorial conformer by $2 \times 0.9 = 1.8$ kcal/mol. In this case, such a simple analysis

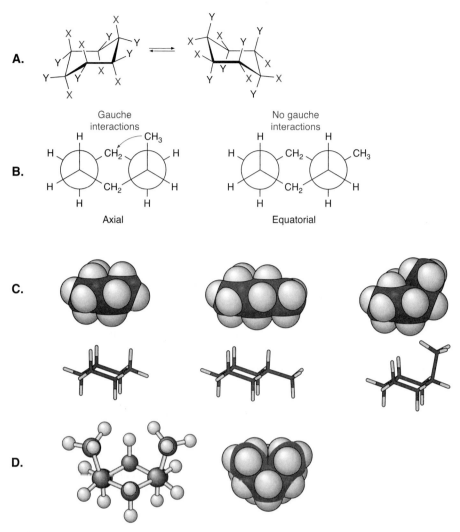

Figure 2.11

Aspects of cyclohexane conformational analysis. **A.** Interconverting chair forms of cyclohexane, with axial and equatorial locations labeled. In the left structure letters x are axial while letters y are equatorial. Note that the chair flip moves axial substituents to the equatorial position and vice versa. **B.** Newman projections down the C1–C2/C5–C4 bonds of methylcyclohexane. In the axial form, there is a gauche butane interaction between the methyl and C3. **C.** Views of cyclohexane, equatorial methylcyclohexane, and axial methylcyclohexane. Note that chair cyclohexane is a relatively disk-shaped molecule, and an equatorial methyl does little to disrupt this shape. In contrast, an axial substituent puts a "kink" into the structure. Also evident is the steric interaction between one methyl hydrogen and the two axial hydrogens on C3 and C5. **D.** *Cis*-1,3-dimethylcyclohexane in the diaxial conformation. Note the significant steric interaction between the 1,3-diaxial substituents.

holds up, producing a very good approximation to the actual situation. As shown in Figure 2.11 **C**, equatorial and axial substituents are in substantially different environments, and influence the overall shape of the molecule quite differently. In the equatorial form, the molecule retains the overall disk shape of the parent cyclohexane, but an axial substituent puts a decided "kink" in the structure.

For most substituents the equatorial position is preferred, and the magnitude of this preference ($\Delta G°$) is called the cyclohexane *A* **value**. A large number of cyclohexane *A* values have been determined (Table 2.14). The *A* value is one measure of the **steric size** of a substituent, in that larger groups *tend* to have larger *A* values (for another measure of steric size,

Table 2.14
Selected Cyclohexane *A* Values (in kcal/mol)*

Group	*A* value	Group	*A* value
D	0.006	CH_3	1.74
F	0.25–0.42	C_2H_5	1.79
Cl	0.53–0.64	$CH(CH_3)_2$	2.21
Br	0.48–0.67	$C(CH_3)_3$	4.7–4.9
I	0.47–0.61	CF_3	2.4–2.5
OH	0.60–1.04	C_6H_5	2.8
OCH_3	0.55–0.75	C_6H_{11}	2.2
OC_6H_5	0.65	CH_2Br	1.79
$OCOCH_3$	0.68–0.87	$Si(CH_3)_3$	2.5
$OSi(CH_3)_3$	0.74	$CH=CH_2$	1.5–1.7
NH_2	1.23–1.7	CHO	0.56–0.8
$N(CH_3)_2$	1.5–2.1	$COCH_3$	1.0–1.5
NO_2	1.1	CO_2^-	2.0
SH	1.21	CO_2H	1.4
SO_2CH_3	2.50	CO_2CH_3	1.2–1.3

*Eliel, E. L., Wilen, S. H., and Mander, L. N. (1994). *Stereochemisty of Organic Compounds*, John Wiley & Sons, New York.

see the next Going Deeper highlight). For example, we have the trend $CH_3 < C_2H_5 < CH(CH_3)_2 << C(CH_3)_3$. However, it seems hard to understand the fact that *A* values of Cl, Br, and I are essentially the same, when clearly there are large differences in their van der Waals radii (Table 1.5). This is a good illustration of the fact that *with all measurements of steric size, context is crucial*. Some substituents with large *A* values may, in another structural context, appear to be relatively small. In the cyclohexane context, the longer bond length of C–I compensates for the larger van der Waals radius of I, minimizing the adverse steric interaction of the gauche form. However, in another context, I will appear to be larger than CH_3.

Going Deeper

Alternative Measurements of Steric Size

A common alternative reference system for steric size is the rotation barrier in substituted biphenyls, as shown to the right. Here, the substituents point toward each other, and they must clash considerably during biphenyl rotation. For such a system, we see a perhaps more reasonable size sequence: $Br > CH_3 > Cl > NO_2 > COOH > OCH_3 > F > H$. Nevertheless, when considering the apparent steric sizes of simple substituents such as CH_3, Cl, Br, and I, almost any sequence of sizes can be observed, depending on the exact structural arrangement of the system being used to probe size. As such, some caution is in order when trying to anticipate the "size" of a substituent in a new context.

Many other attempts have been made to develop **steric parameters**, universal gauges of the steric demands of substituents. Perhaps the best known of these are due to Taft (see Section 8.4.1). However, given the strong context effects of steric interactions, the value of such terms is variable. In addition, with the ready availability of molecular mechanics calculations (see Section 2.6) for a wide range of structures, the usefulness of such parameters has diminished.

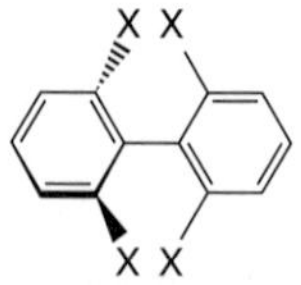

Twisted biphenyl

For an overview of efforts to evaluate steric size, see Förster, H., and Vögtle, F. "Steric Interactions in Organic Chemistry: Spatial Requirements of Substituents." *Angew. Chem. Int. Ed. Eng.*, **16**, 429–441 (1977).

Before moving on further, let's review some nomenclature from introductory organic chemistry that is often confusing. When two or more groups are attached to a cyclohexane ring, they can either be cis or trans to each other. The definitions can be seen from a view of a flat cyclohexane ring where the groups are either on the same side of the ring (cis) or on opposite sides (trans). This usage is distinct from the axial/equatorial terminology. Hence, cis and trans groups can occupy various axial and equatorial positions depending upon the substitution pattern. A few examples are given to the right.

A values are frequently additive when the axial substituents are trans. The relative stabilities of various conformations can be obtained by combining *A* values. When the *A* values are additive, one can use the values as a tool to measure other energies, as described in the next Connections highlight. Such additivity can break down, however, for cis arrangements. A common example is the **1,3-diaxial interaction**. When two substituents are in a cis-1,3 relationship in a cyclohexane, they experience a strong interaction if both are axial, as shown in Figure 2.11 **D**. This destabilizing interaction is, in fact, just a cyclic version of the g+g− interaction discussed above. The energetic consequence of this interaction can be substantial. For the case of the two methyl groups shown, the 1,3-interaction contributes 3.7 kcal/mol of strain to the diaxial form.

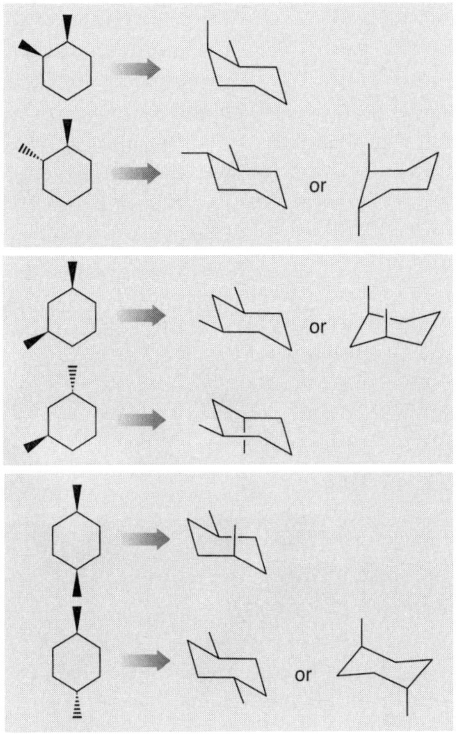

Dimethyl cyclohexanes

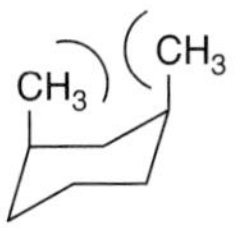

1,3-Diaxial interaction

Connections

The Use of *A* Values in a Conformational Analysis Study for the Determination of Intramolecular Hydrogen Bond Strength

The formation of intramolecular hydrogen bonds can stabilize particular conformations of cyclohexane rings. The measurement of the strengths of intramolecular hydrogen bonds can be performed by examining the equilibria between different cyclohexane ring conformers. For example, the equilibrium shown to the right involves conformations missing (*i*) and containing (*ii*) an intramolecular hydrogen bond. Form *i* is destabilized by two axial hydroxyls while form *ii* is destabilized by an axial isopropyl. The $\Delta G_{eq}°$ of the interconversion can be broken down into the *A* values and the Gibbs free energy of the hydrogen bond as written.

$$\Delta G_{eq}° = A_{ipr} + \Delta G_{HB}° - 2A_{OH}$$

The $\Delta G_{eq}°$ was experimentally determined using IR spectroscopy. Using the *A* values of Table 2.14, the $\Delta G_{HB}°$ was determined to be −1.9 kcal/mol.

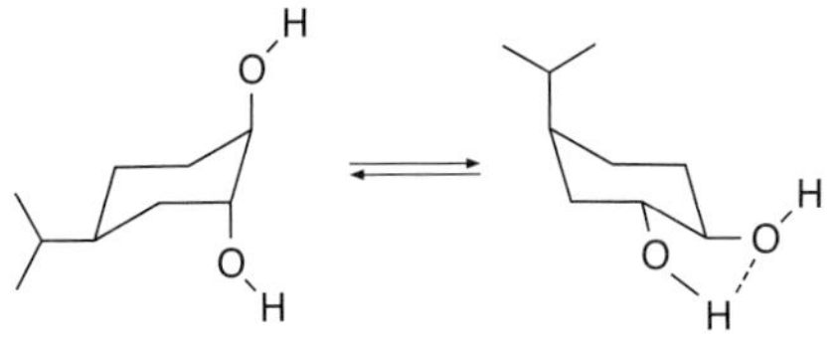

Evaluating an intramolecular hydrogen bond

Huang, C.-Y., Cabell, L. A., and Anslyn, E. V. "Molecular Recognition of Cyclitols by Neutral Polyaza–Hydrogen Bonding Receptors: The Strength and Influence of Intramolecular Hydrogen Bonds Between Vicinal Alcohols." *J. Am. Chem. Soc.*, **116**, 2778–2792, (1994).

As in other cycles, the different types of hydrogens (axial and equatorial) interconvert by a conformational process. What is unique about cyclohexane is that the barrier for this process is ~10.8 kcal/mol, much higher than for other carbocycles. Thus, in many circumstances, it is important to distinguish between the axial and equatorial positions. Although this barrier is high for a cycloalkane, it is still low enough to be very easily traversed at ambient temperatures. In fact, the ring flipping process is fast on the NMR time scale. (For a discussion of the NMR time scale, see the next Going Deeper highlight.)

The path by which the chair forms interconvert has been extensively studied, and it is summarized in the potential energy diagram given in Figure 2.12. After passing through a **half-chair** transition state, the molecule enters the **twist boat** conformer. This conformer is flexible, like cyclobutane and cyclopentane, and undergoes rapid interconversions among equivalent forms via a **boat** transition state, a process known as **pseudorotation**. The boat

---

Going Deeper

The NMR Time Scale

One of the most powerful tools for investigating conformational analysis and a range of other phenomena discussed in this text is dynamic NMR spectroscopy. This is because the time scale associated with NMR spectroscopy is in a range that is well matched to the rates of conformational processes. All spectroscopies have an associated time scale (see Table 16.1 for a more comprehensive list). The time scale is set by the Heisenberg uncertainty principle, one expression of which is shown in the equation below. For spectroscopy, ΔE is the energy gap between the two states that are connected by an absorption. For example, UV spectroscopy involves relatively large energies, implying that Δt must be very small, and indeed UV absorptions can be thought of as essentially instantaneous. However, the energies separating the various spin configurations probed in NMR spectroscopy are extraordinarily small. If ΔE is small, Δt must be relatively large, making the NMR time scale relatively slow.

$$\Delta E \bullet \Delta t \approx \pi \sqrt{2}$$

So, what is the NMR time scale? Actually, although we don't often think of it this way, common NMR terminology directly defines the time scale. When two peaks are separated by 100 Hz, they are separated by 100 s^{-1}, the units of a first-order rate. The reciprocal of such a rate is time; in this case, it is 0.01 s or 10 ms. The NMR time scale is typically in the ms range.

In particular, we are considering the case of two nuclei that display distinct NMR signals, but that interconvert through some dynamic process in the molecule. If the rate of interconversion is fast on the NMR time scale (larger than the reciprocal of their peak separation in Hz), we will see one peak. The result is that we see **coalesced peaks** (two peaks have become one) in an NMR spectrum because of a dynamic process.

For two singlets of equal intensity that show a chemical shift separation of $\Delta \nu$ under conditions in which there is no exchange, the rate of interconversion required to achieve coalescence is

$$k_{\text{coal}} = 2.22 \bullet \Delta \nu$$

Note that 2.22 is simply $\pi/\sqrt{2}$. For more complex processes, this equation is not exact, but it is still a useful guideline. Detailed spectral fitting programs allow quantitative analysis of such systems.

Consider again two singlets separated by 100 Hz as a common separation. At coalescence, $k = 222$ s^{-1}. Whether we see two signals now depends on the magnitude of the barrier to the interconversion process and the temperature. Using the Arrhenius equation we find that at room temperature coalescence corresponds to $E_a \cong 15$ kcal/mol, and this is a useful number to remember. If E_a for a process is 15 kcal/mol, and the chemical shift separation is about 100 Hz, we expect two signals at temperatures significantly below room temperature ($k << 222$ s^{-1}) and one signal at temperatures significantly above room temperature ($k >> 222$ s^{-1}).

We noted that the barrier to ring-flipping in chair cyclohexane is about 10.8 kcal/mol. Since this is substantially less than 15 kcal/mol, a compound such as *trans*-1,3-dimethylcyclohexane shows only one methyl peak in the ^{1}H NMR spectrum. However, if we lower the temperature, the rate constant for interconversion slows, eventually reaching a point where two methyl signals can be seen. Cyclohexane conformational dynamics are well suited to the NMR time scale. Considering the range of temperatures and chemical shift differences accessible, the range of activation barriers that can be probed with conventional NMR spectrometers is 5–20 kcal/mol—a very convenient range for conformational analysis.

Interconversion of 1,3-dimethylcyclohexane

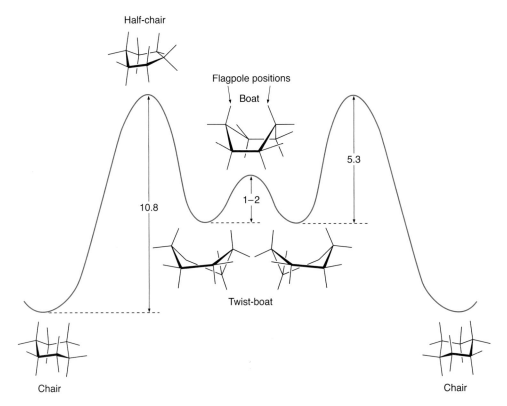

Figure 2.12
Conformational interconversions in cyclohexane. Relative energies are given in kcal/mol. The 1–2 kcal/mol barrier for interconversion of twist-boat forms is an estimate, as this process has never been directly observed. Note that because of the small barrier, one can expect many twist-boat–twist-boat interconversions before escape from this local minimum back to the chair manifold.

form is destabilized by, among other things, a long range steric interaction called the **flag-pole** interaction. Eventually, the system escapes the twist boat regime via another half-chair transition state to the "flipped" chair. At no time during the chair flip process are all six carbons in one plane, as this would be a very highly strained structure.

The value of cyclohexane as a workhorse of conformational analysis is undeniable. With its relatively rigid structure, a structure that can be made more rigid by substitution or ring fusion, cyclohexane has been a platform for innumerable studies of conformational effects on reactivity. Cyclohexane is also an important component of many biological structures, such as steroids and pyranose forms of sugars (see the next Connections highlight). Figures 2.11 **C** and **D** emphasize the shape of cyclohexane and simple derivatives, an important aspect when considering biological recognition processes.

Larger Rings—Transannular Effects

As soon as we progress beyond cyclohexane, we return to the more typical model of carbocycles. Several conformers are close in energy, and they often interconvert over relatively small barriers. The situation rapidly becomes very complicated and a domain primarily for the aficionado of conformational analysis. Here we describe a few general features of these systems.

The larger rings—$(CH_2)_n$, where $n > 6$—do not have large strain energies of the sort seen in the small rings. Perhaps surprisingly, though, they are not strain free. Table 2.15 illustrates this point. These observations make it possible to group cycloalkanes into small rings ($n = 3$ and 4), common rings ($n = 5$, 6 and 7), medium rings ($n = 8$–12), and large rings ($n \geq 13$).

Table 2.15
Strain Energies and Group Increment Correction
Factors for Cycloalkanes, $(CH_2)_n$ (in kcal/mol)

n	Strain energy	Group increment		n	Strain energy
3	27.5	27.6		10	12.4
4	26.3	26.2		11	11.3
5	6.2	6.3		12	4.1
6	0.1	0		13	5.2
7	6.2	6.4		14	1.9
8	9.7	9.9		15	1.9
9	12.6	12.8		16	2.0

Connections

Ring Fusion—Steroids

One way to "lock" a cyclohexane into one particular chair is by fusing a second ring onto it. A common building block is decalin, which combines two cyclohexanes. There are two ways to fuse two cyclohexanes, termed cis and trans. In the trans form, the second ring is joined to the first by two equatorial connections. The trans form is quite rigid. No chair flipping is possible, and so substituents around the ring can be unambiguously assigned as axial or equatorial based simply on their cis or trans relationships to the ring fusion. This assignment has frequently proven to be a useful strategy for designing stereochemical probes of reaction mechanisms.

The cis ring fusion must involve one axial and one equatorial C–C bond. This requirement has several important consequences. First, *cis*-decalin is less stable than *trans*-decalin by about 3 kcal/mol. Second, the cis form is flexible, as the axial and equatorial linkages can interconvert by a chair flipping process.

The third key difference between *cis*- and *trans*-decalin relates to shape. The trans form retains and extends the relatively flat, disk shape of chair cyclohexane, but the cis form forces a kink into the system. Fused cyclohexanes are common in biological structures, and because molecular shape is always crucial in biological recognition processes, the cis vs. trans ring fusion issue is quite important.

Steroids are important fused-ring cyclohexanes. Most steroids, such as cholesterol, have all trans fusions or have olefins at a fusion (as in cholesterol or testosterone) or aromatic rings (as in estrone). Either way, the flat, *trans*-decalin shape is maintained, making these generally lipophilic molecules adopt an elongated, disk-like structure. A major role of cholesterol is to insert into and thereby stabilize cell membranes, and no doubt the molecular shape is crucial to this function. There are exceptions, such as cholic acid (a component of bile) which adopts one cis ring fusion. This cis ring fusion alters the molecular shape considerably, and also creates an interesting juxtaposition of the three hydroxyls.

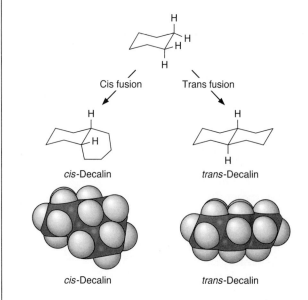

Cis fusion   Trans fusion

cis-Decalin   *trans*-Decalin

cis-Decalin   *trans*-Decalin

Cholesterol

Cholic acid

Triangle is 2.2 Å
on a side

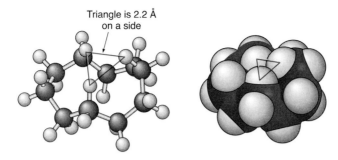

Figure 2.13
A low energy conformation of cyclodecane shown in ball-and-stick and
space-filling (CPK) representations. Note the close H•••H transannular
contact—a similar interaction occurs on the "back" side of the ring.

There is significant strain in the medium-ring compounds. The larger rings show diminishing strain with increasing n, and for $n \geq 17$ the rings become so large that they are barely distinguishable from long chains.

What is the origin of the strain in medium-ring compounds? The strain primarily results from a novel feature of these systems that is well illustrated by cyclodecane (Figure 2.13). It is not obvious at first, but examination of models shows that in closing up $(CH_2)_{10}$ into a cycle, some carbons that lie across the ring from one another will end up being fairly close to one another. This **transannular strain** cannot be avoided without substantial torsional and/or bond angle distortion. As such, medium-sized rings are strained. Also, the transannular carbons can react with one another in ways that would not normally be expected, because of their proximity. Again, a number of interesting reaction mechanisms depend on this effect.

Group Increment Corrections for Ring Systems

Except for cyclohexane and the very large cyclic alkanes, all cyclic alkanes are strained to some extent. Any strain that primarily arises from non-standard bond angles is called **angle strain** (or **Baeyer strain**). The strain is often further broken down into **small angle** and **large angle strain**. However, as discussed, the strain in cyclic systems also arises in part from torsional strain. In the context of rings, this is often called **Pitzer strain**.

In the group increments method, we assumed that a CH_2 always makes a constant contribution to ΔH_f° for a molecule. We also saw that a small ring such as cyclobutane leads to a substantial failure for the group increment method, because of its strain energy. Just as we assumed that a gauche butane interaction will make a consistent 0.8 kcal/mol contribution to ΔH_f°, it is reasonable to assume that a cyclobutane will do the same. With this in mind, a series of correction terms for common ring systems has been developed, with the goal of obtaining accurate ΔH_f° values for cyclic systems. Representative values are given in Table 2.15. Note that these are not identically equal to the accepted strain energies for the parent ring system, although they are quite close. The group increment correction for a cyclobutane is based on ΔH_f° values for a number of structures, and represents an average value that gives the best agreement with the range of experimental data. In contrast, the strain energy of cyclobutane is specific to the parent compound. With these new correction terms, it is now possible to predict ΔH_f° values for strained ring systems, by first adding up all the basic group increments [C–(H)$_2$(C)$_2$, etc.] and then adding appropriate ring strain correction values.

Ring Torsional Modes

The internal motions that we have discussed for the interconversion of conformers of cyclobutane, cyclopentane, cyclohexane, and those alluded to with the larger rings, all derive from vibrational modes. Just as high energy bond torsions lead to bond rotations, the interconversion between ring conformers is predicated on high energy ring torsional motions. With rings, more than one bond torsion is required (if only one bond rotated the ring would break). The concerted bond rotations lead to the interconverting ring conformers. For cyclo-

Spiro[2.2]pentane

Spiro[5.4]decane

Bicyclo[3.2.1]octane

Bicyclo[4.3.0]nonane

Bicyclo[2.2.1]heptane
norbornane

Figure 2.14
Nomenclature of bicyclic
systems. Note that, if we define
m as the sum of the numbers
within the brackets, there are
m + 1 atoms in the ring system
for spiro compounds, and *m* + 2
atoms in the ring system for
bicyclic compounds.

butane and cyclopentane, as well as the larger ring systems, the barriers to interconversion between conformers are very low, and little energy needs to be put into the concerted torsional vibrational modes to interconvert the conformers. The barriers are all near *RT*. The potential surfaces that describe the torsional motions of the atoms in these rings are very shallow and there is significant motion of the ring atoms. These low energy vibrational modes therefore impart favorable entropy to the ring systems, causing the ΔS_f° for these systems to be more positive than conformationally locked rings.

Cyclohexane is, as we have already alluded to, the exception to the norm. The potential energy diagram in Figure 2.12 has substantial barriers, and the concerted torsional motions that lead to the interchange between a chair and twist boat are relatively stiff. Therefore, cyclohexane has a significantly higher energy torsional motion than other rings, a motion that involves the ring carbons moving toward and away from a half-chair conformation. Once again, the energy of this motion would be quantized, but for practical purposes, this quantum mechanical effect has little relevance.

Bicyclic Ring Systems

Complex, multi-ring systems figure prominently in many aspects of organic chemistry, from exotic tests of theory to equally exotic natural products. We will encounter such systems throughout the text. Here we mention some fundamental issues inherent to multi-ring systems, beginning with a bit of nomenclature.

When a molecule has two rings and the two share only one carbon in common, the system is termed **spiro**, and the nomenclature is fairly straightforward (Figure 2.14). More typically, the two rings have two carbons in common, and these are called the **bridgehead carbons**. There will be three different paths between the two common carbons, and the bicycle is named by listing the lengths of these paths. Thus, decalin is bicyclo[4.4.0]decane. The systematic nomenclature of more complex multi-ring structures builds on this system, and rapidly becomes unwieldy. We'll avoid it as much as possible here. See http://www.chem.qmw.ac.uk/iupac/fusedring/ for all the gory details.

Figure 2.15 shows ring strains for a number of cyclic and polycyclic systems. As much as possible, the data are consistent with each other, but these are not all absolutely firm numbers. Care must be taken not to overinterpret the precise values, but the basic patterns can be expected to hold.

Is the strain in a bicyclic system simply the sum of the individual ring strains? Perhaps surprisingly, this is very often the case for simple systems. For example, the strain in bicyclo[3.1.0]hexane would be expected to be the sum of the strains of cyclopentane (6.2 kcal/mol) and cyclopropane (27.5 kcal/mol), which is 33.7 kcal/mol. The actual value is 33.9 kcal/mol, in excellent agreement. Significant exceptions occur for the smallest rings. Thus, for both possible ways of combining two cyclopropanes the expected strain energies are 2 × 27.5 = 55 kcal/mol. However, for spiro[2.2]pentane and bicyclo[1.1.0]butane, the ring strains are in the 65–66 kcal/mol range, roughly 10 kcal/mol more than expected. For these systems, there is extra strain induced by the ring fusion event. The effect is significantly diminished in bicyclo[2.1.0]pentane, and with larger systems the additivity approach is usually effective. In an amazing example, the strain energy of cubane (Figure 2.15) is predicted to be 6 × 26.3 = 158 kcal/mol, and the experimental value is ~166 kcal/mol! It is also true that polycyclic ring systems (if they do not contain cyclopropane and cyclobutane rings) are often very well treated by the molecular mechanics method (see below), making an analysis of ring strain straightforward.

Cycloalkenes and Bredt's Rule

Incorporating an olefin into a small ring increases the strain of the system; cyclopropene, for example, is very highly strained (Figure 2.15). The effect is consistent with expectation, in that the smaller rings enforce compressed C–C–C angles, and olefins prefer larger angles than alkanes. As such, it should be more destabilizing to have an olefin in a ring than an acyclic system.

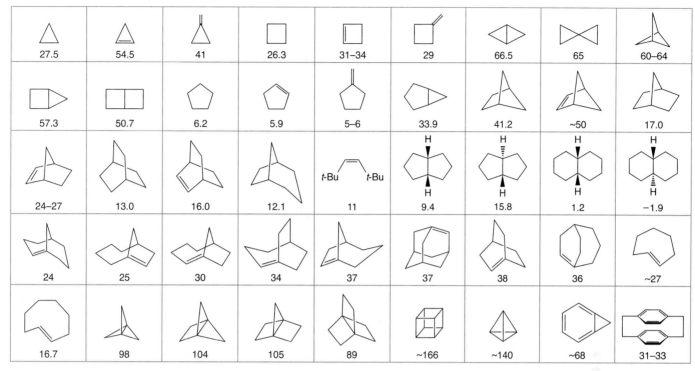

Figure 2.15
A potpourri of strained molecules and their associated strain energies (in kcal/mol).

When discussing such systems we automatically put the olefin in a cis geometry—that is, *trans*-cyclopropene seems like a pretty unreasonable structure. But what about *trans*-cyclohexene or *trans*-cyclopentene? The smallest cycle into which a trans olefin can be embedded has been a subject of considerable study. It has long been known that *trans*-cyclooctene is relatively stable, with a strain energy of ~16 kcal/mol. *trans*-Cycloheptene has been prepared and experimentally characterized at low temperatures, and it has an estimated strain energy of 27 kcal/mol. The olefin in this structure is substantially distorted, as indicated in Figure 2.16.

A related issue is whether it is possible to incorporate an olefin at the bridgehead position of a bicyclic system. Extensive investigations have shown that there are substantial lim-

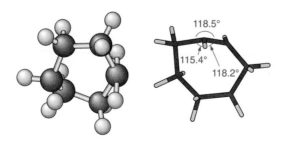

Figure 2.16
Two views of a calculated structure for *trans*-cycloheptene. The view on the left, sighting directly down the C=C bond, shows the substantial twisting in the carbon framework. Note, however, that the olefin is not only twisted; a significant amount of pyramidalization occurs, as indicated by the structure on the right. The extent of pyramidalization is indicated by the three valence angles. At a planar center, these three will sum to 360°, which is clearly not the case here.

its to such efforts. **Bredt's rule** was an attempt to rationalize which bicyclic systems could tolerate a bridgehead olefin. Bredt's rule considered bicyclo[*a.b.c*]alkanes, and focused on the minimal value of $S = a + b + c$ that allowed a bridgehead olefin. Early work suggested $S = 9$ was the limit, but this limit was then dropped to 8 and then to 7. However, it was eventually realized that better predictions could be made by simply considering the size of the ring that contains the trans olefin. A series of bridgehead olefins is shown in Figure 2.15, and it can be seen that a bridgehead olefin must be trans in one ring of a bicyclic system. To a good approximation, the stability of the bridgehead olefin follows the pattern for the simpler trans cyclic olefins. Systems in which the olefin can be trans in an eight-membered ring are less strained than those in a seven-membered ring and so on. There is variation, but the basic trend holds. Although the analysis has changed somewhat, olefins that are especially strained because they are at a bridgehead position are still referred to as **anti-Bredt olefins**. Comparable studies of cycloalkynes have been made. The results are similar: cyclooctyne is the smallest isolable system, but even cyclopentyne has been trapped as a transient intermediate.

Summary of Conformational Analysis and Its Connection to Strain

Structural distortions give rise to strains within molecules, all of which are manifest in weaker bonds. The strains arise from both static and dynamic processes. For example, the ring strain of cyclopropane cannot be relieved by any dynamic motion, and is therefore static. Similarly, strains such as are inherent in anti-Bredt olefins and fused small rings are static, and are not relieved by any particular conformational change. Yet, for cyclobutane a small pucker arises that alleviates some of the strain. In contrast, for linear butane, the all staggered form has no strain, and strain is only introduced during a dynamic process, that of bond rotation. Similarly, cyclohexane is also not strained. Yet ring interconversion motions have higher barriers in cyclohexane because significant strain is introduced during the coupled torsions that lead to ring flipping. All these examples show that some strains are inherent in a structure, and cannot be relieved by a molecular motion, while others are only partially relieved by a conformational change. On the other hand, some molecules are not strained, and strain only arises during a conformational change. These lessons from organic chemistry are applicable to any other field of chemistry—organometallic, inorganic, biochemistry, and polymer chemistry.

2.4   Electronic Effects

Thus far, our analysis of structure and energetics has been focused primarily on saturated hydrocarbons and simple olefins, where steric clashes and angle constraints imposed by ring systems dominate both structure and energetics. With that as a foundation, we can now consider more complex systems involving unsaturated groups and heteroatoms. We will focus on both stabilizing and destabilizing interactions involving such groups, and we will discuss some conformational biases associated with them. Along with steric interactions, we will need to invoke orbital mixing arguments to rationalize many observations.

2.4.1   Interactions Involving π Systems

Substitution on Alkenes

We have already seen that there is a steric strain in cis olefins relative to trans, resulting in corrections to the group increments for olefins. There are also stabilizing effects associated with increased substitution of the alkene by alkyl groups. Recall from introductory organic chemistry that increased substitution on an alkene leads to a stabilization of one alkene relative to another. One way to gauge the stability of an olefin is to evaluate its **heat of hydrogenation** ($\Delta H_{\mathrm{hyd}}°$). This $\Delta H°$ is for Eq. 2.24, the reaction of an olefin with H_2 to produce the analogous alkane, and this reaction is usually quite exothermic. A typical example—

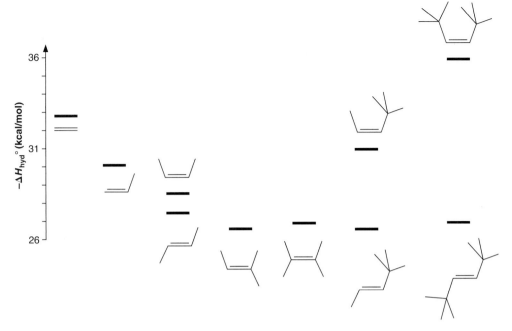

Figure 2.17
Heat of hydrogenation values ($\Delta H°$) for several alkenes. The effect of alkyl substitution is evident. Also, the stability of trans double bonds remains relatively constant, but significant destabilization of the cis alkenes is seen as the R group size increases. Derived from data in Turner, R. B., Jarrett, A. D., Goebel, P., and Mallon, B. J. "Heats of Hydrogenation. IX. Cyclic Acetylenes and Some Miscellaneous Olefins." *J. Am. Chem. Soc.* **95**, 790 (1973).

trans-2-butene—is shown, and for most conventional olefins $\Delta H_{hyd}°$ is near the –27.75 kcal/mol value shown.

$$\text{trans-2-Butene} + \text{H}_2 \longrightarrow \text{Butane}$$
$$\Delta H_f° \qquad -2.58 \qquad 0 \longrightarrow -30.37$$
$$\Delta H° = \Delta H_{hyd}° = -27.75 \text{ kcal/mol}$$

(Eq. 2.24)

Figure 2.17 shows a collection of heats of hydrogenation for various substituted olefins. The stabilization due to alkyl substitution is evident. This stabilization is a sizeable effect, such that *cis*-2-butene is more stable than propene, despite the adverse steric interaction. The tetrasubstituted olefin is less stable than the trisubstituted, but less so than expected given the destabilizing cis interactions. One way to rationalize the stabilization of olefins by alkyl substitution is an orbital mixing phenomenon of the sort we invoked for propene in Chapter 1. There is some mixing of a filled $\pi(\text{CH}_3)$ orbital with the π^* orbital, resulting in the LUMO shown in the margin. This is a filled–empty mixing, and so it must be stabilizing. The contribution of the $\pi(\text{CH}_3)$ orbital here is much smaller than in the HOMO of propene (Figure 1.18), in a clear manifestation of the energy gap law. The olefin π^* MO is higher in energy than the π MO, and so the energy gap to the $\pi(\text{CH}_3)$ orbital is greater and the mixing is less.

Figure 2.17 also shows several cis–trans comparisons. As the alkyl group increases in size, the energy difference increases. With two *t*-butyl groups on an olefin, the cis–trans difference is 10 kcal/mol. Because of these trends, the simple cis alkene group increment of Table 2.11 must be used with caution.

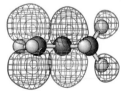

LUMO of propene

Conformations of Substituted Alkenes

The most obvious conformational effect in alkenes is that they are planar. The planarity is a contra-steric effect that derives from the geometry of the π bond. Here, all the groups around the alkene, including the alkene carbons, are **coplanar**. However, simple substituted π systems have additional conformational features to consider. For example, propene shows a 2 kcal/mol preference for the eclipsed over the staggered form (Eq. 2.25), with the staggered form being the transition state for rotation.

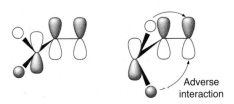

Eclipsed (preferred)   Staggered   (Eq. 2.25)

The origin of the preference for eclipsed over the staggered is subtle. As just discussed above, in Chapter 1 we showed how a π(CH₃) orbital will mix with the classical olefin π orbital to give the HOMO shown in Figure 1.18. To first order, this interaction is the same for the eclipsed and staggered forms. However, there is a secondary interaction between C1 and C3 that can discriminate between the two conformers. This interaction, shown below, can only occur for the staggered form, and it is necessarily out of phase and is thus destabilizing. This destabilization of the staggered form has been put forth as the cause of the conformational preference in propene. In 1-butene, the eclipsed form is still preferred, and surprisingly there is a slight preference for the methyl group to be in the same plane as the alkene.

Adverse
interaction

Secondary orbital interaction in propene

Preferred conformation
of 3-pentanone

A similar bias is seen in carbonyl compounds. Acetaldehyde shows a 1.0 kcal/mol preference for the eclipsed form over the staggered (Eq. 2.26), with the latter again being a transition state for rotation. In propanal the conformation having the methyl group eclipsed with the carbonyl (Eq. 2.27) is preferred by ~1 kcal/mol. Similarly, for 2-butanone and even 3-pentanone, the conformations with the methyls eclipsed with the carbonyl are preferred. It is generally accepted that electronic interactions are controlling these preferences, but the effects are subtle.

Eclipsed (preferred)   Staggered   (Eq. 2.26)

(Eq. 2.27)

Preferred

The 1-methylallyl cation displays an interesting conformational effect that is similar to those discussed above, and is another nice example of how the group orbitals given in Chapter 1 are useful. As shown in Figure 2.18, the structure with the methyl "in" (s-cis) is less stable than the one with the methyl "out" (s-trans), the energy difference being a surprisingly large 5–7 kcal/mol. One way to rationalize this effect is to examine the HOMO of the ion, shown in Figure 2.18. This orbital is the out-of-phase combination of a π(CH₃) orbital and the allyl HOMO. With the methyl tucked in, as in the s-cis form, there is some overlap between

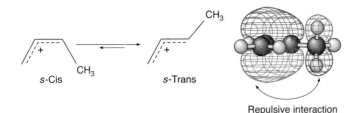

Repulsive interaction

Figure 2.18
The geometrical isomers of 1-methylallyl cation, and the HOMO of the
s-cis form.

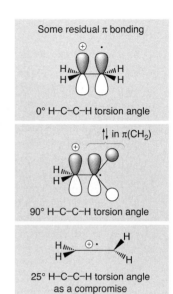

Some residual π bonding

0° H–C–C–H torsion angle

↑↓ in π(CH₂)

90° H–C–C–H torsion angle

25° H–C–C–H torsion angle
as a compromise

Ethylene radical cation

Conjugation

Cross conjugation

π(CH₃) and the *p* orbital at C3. This overlap is necessarily out of phase and so is destabilizing to the HOMO and the molecule. Such an interaction is absent in *s*-trans, thus rationalizing the conformational preference.

Neighboring group orbitals also influence the conformations of radical cations. With ethylene, oxidation of the π bond leads to a structure with a 25° dihedral angle. This dihedral angle is due to a competition between residual π bonding (preferring a 0° angle) and the ability of an occupied π(CH₂) group orbital to mix with and stabilize an adjacent empty *p* orbital on the other CH₂ group (preferring a 90° dihedral angle).

Conjugation

Direct attachment of alkenes, without any intervening atoms (**conjugation**), leads to π molecular orbitals that are delocalized across all the *sp²* hybridized carbons (see butadiene in Appendix A). The term conjugated is normally associated with π bonds that are arranged in a line or loop, such as with butadiene, α,β-unsaturated ketones, or benzene. A **cross-conjugated** system is a term used to define conjugation with the π bond arranged in a branched fashion, such as with the structures shown to the side. Either form of conjugation leads to delocalization, which stabilizes the π system. Estimates of the energetic stabilization due to conjugation vary, depending on the reference structure. Simple analyses, such as comparing hydrogenation energies of butadiene vs. 1-butene, neglect the stabilizing effect of the ethyl group in 1-butene and so tend to underestimate the stabilization. Recent estimates put the conjugative stabilization at 8 kcal/mol for butadiene and 9 kcal/mol for butadiyne. The effects are typically not as large as they are in reactive intermediates, where allylic stabilization is worth 10–15 kcal/mol.

Conjugation favors a planar structure, and for a prototype compound such as 1,3-butadiene, there are two choices, termed *s*-trans and *s*-cis (the *s* signifies geometry around a single bond). Due to sterics, the *s*-trans conformation is preferred, and the barrier to interconversion is near 4 kcal/mol. In fact, the *s*-cis is not actually present at all. It is a transition state between conformations that are referred to as skew. In the skew conformation, the C1–C4 repulsion is slightly relieved. This distortion from planarity is further evidence that the conjugation in molecules such as butadiene is not an energetically strong effect.

$$s\text{-Trans} \quad \rightleftharpoons \quad s\text{-Cis} \quad \vert \quad \text{Skew} \tag{Eq. 2.28}$$

As with conjugated dienes, α,β-unsaturated ketones prefer planar conformations. In propenal the *s*-trans conformation is strongly preferred (~5 kcal/mol, Eq. 2.29), while in 3-buten-2-one the *s*-trans is preferred by a factor of about three (Eq. 2.30). This preference is reversed in 4-methyl-3-penten-2-one (Eq. 2.31), where the *s*-cis conformation dominates by about a factor of three (you are asked to rationalize these trends in the Exercises at the end of the chapter).

$$\text{Preferred} \rightleftharpoons \tag{Eq. 2.29}$$

$$\text{(Eq. 2.30)}$$

$$\text{(Eq. 2.31)}$$

Connections

A Conformational Effect on the Material Properties of Poly(3-Alkylthiophenes)

Polythiophene is a classic conducting organic polymer, often used to study the mechanism of electrical conductivity (see Chapter 17 for a discussion of conducting polymers). The mechanism of conduction relies upon effective conjugation of the π system along the polymer and good stacking between adjacent polymer chains. In the standard random polymerization syntheses that start with 3-alkylthiophenes as the monomers, one achieves a variety of linkages (head-to-tail, head-to-head, and tail-to-tail, shown to the right). The head-to-head linkage creates a steric interaction that leads to a conformation along the polymer chain that disrupts the conjugation and would impede stacking between adjacent polymer strands. However, a more directed synthesis can produce a polymer that has pure head-to-tail linkages throughout. This direct synthesis removes the adverse steric interactions that caused twisting along the backbone, and the conductivity of thin films of this polymer is higher by a few orders of magnitude. This example shows that careful control of conformational effects can have dramatic effects on material properties.

McCullough, R. D., Tristram-Nagle, S., Williams, S. P., Lowe, R., and Jayaraman, M. "Self-Orienting Head-to-Tail Poly(3-alkylthiophenes): New Insights on Structure–Property Relationships in Conducting Polymers." *J. Am. Chem. Soc.*, **115**, 4910–4911 (1993).

Aromaticity

Many students are undoubtedly familiar with the **Hückel 4n + 2** rule for predicting aromaticity. Any hydrocarbon or heterocycle with 4n + 2 electrons in a fully conjugated cyclic π system is considered **aromatic**. Remember, it is the number of *electrons*, not the number of atoms that defines a system as aromatic or not. An aromatic system is more stable than expected when compared to similar structures. Benzene is the paradigmatic example of an aromatic system. As such, several aspects of the chemistry of benzene derivatives are considered hallmarks of aromatic character—namely, a pattern of substitution, not addition, with electrophilic reagents, and an unusual resistance to oxidation.

Any 4n + 2 system is aromatic, and hence cyclopropenyl cation, cyclopentadienyl anion, cycloheptatrienyl cation (called tropylium ion), pyrrole, and furan are all aromatic. They are all unusually stable, as described for cyclopropenyl cation in the following Connections highlight. Furthermore, they are all planar, like benzene. One would hope for a good theoretical justification for aromaticity, unifying the similar character of all these compounds. Indeed, aromaticity has been studied extensively with electronic structure theory methods. But, as we discuss in Section 14.5.1, there is still significant debate as to the origin of aromaticity. We leave a detailed presentation of the theory to that section of the book.

Aromaticity is not restricted to completely conjugated rings. There are rare cases where the geometry of the molecule allows an orbital overlap such that the compound can be aromatic, although there is a saturated center somewhere in the ring. These structures are called

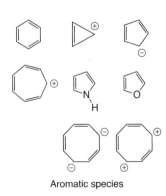

Aromatic species

homoaromatic. For example, homotropylium ion has been extensively investigated in this regard. Furthermore, even trishomocyclopropenium ion is considered to be aromatic, although it also can be viewed as a non-classical carbocation. These structures are also considered **homoconjugated**, meaning that conjugation arises between atoms that are not formally σ bonded together (see Section 11.5.11 for an example of homoconjugation affecting reactivity).

Homotropylium ion

Trishomocyclopropenium ion

In some cases, a resonance structure is required to see an aromatic system. The increased stability associated with an aromatic system is found for the structure, although the compounds do not appear aromatic unless the resonance structure is considered. Azulene, which can be drawn as a cyclopentadienyl anion fused to a cycloheptatriene cation, and cyclopropenone, which can be written as possessing a cyclopropenyl cation, are two examples (see margin).

Azulene

A long standing issue in physical organic chemistry is the quantitative extent to which benzene is stabilized by aromaticity. Thermochemical strategies have frequently been employed for such analyses. For benzene the task is relatively straightforward. The standard analysis is to compare the heat of hydrogenation of cyclohexene to that of benzene. One way to set up the analysis is to simply calculate $\Delta H°$ for the process of Eq. 2.32. This equation shows what is known as an **isodesmic** reaction, because the numbers and kinds of C's and H's are equal on both sides of the equation; for example, there are 12 sp^3 and six sp^2 carbons on each side of the equation. Using experimental values for $\Delta H_f°$, $\Delta H°$ is found to be –35.5 kcal/mol, so the reaction is quite exothermic. It is generally accepted that benzene and cyclohexane are strain free, but cyclohexene has a strain energy of 1.4 kcal/mol, and with this correction we conclude that the aromaticity of benzene is worth nearly 32 kcal/mol. This thermochemical value agrees well with those determined by other approaches.

Cyclopropenone

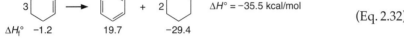

$$3 \quad \xrightarrow{\hspace{1cm}} \quad + \quad 2 \qquad \Delta H° = -35.5 \text{ kcal/mol}$$

$\Delta H_f° \quad -1.2 \qquad\qquad 19.7 \qquad -29.4$

(Eq. 2.32)

Antiaromaticity, An Unusual Destabilizing Effect

In contrast to the stability of aromatic systems, a planar π system with $4n$ electrons is generally observed to be unstable, and is called **antiaromatic**. The instability of such systems is quite severe, as shown with one example in the Connections highlight on the next page. In Section 14.5.6 we will show that such systems can have biradical character, and

---

Connections

Cyclopropenyl Cation

Cyclopropenyl cation is stable enough to be isolated and stored in a bottle, albeit at –20 °C. It can be prepared by chloride abstraction from 3-chlorocyclopropene, as shown below. The ^{1}H NMR spectrum in nitromethane solvent shows a singlet at a very downfield shift: 11.2 ppm.

$$\text{Cl} \quad \xrightarrow{\text{SbCl}_5} \quad \oplus \quad + \quad \text{SbCl}_6^{\ominus}$$

Generating cyclopropenyl cation

One can compare the reactivity of cyclopropenyl cation to allyl cation as a means of estimating the stability imparted by aromaticity. In the following equilibrium, it

was found that cyclopropenyl cation has an equilibrium constant 10^{13} more favorable than allyl cation—quite an astounding effect!

Cyclopropenyl vs. allyl cation

Breslow, R., and Groves, J. T. "Cyclopropenyl Cation. Synthesis and Characterization." *J. Am. Chem. Soc.*, **92**, 984 (1970).

hence are expected to be unusually reactive. The paradigmatic example of an antiaromatic system is cyclobutadiene.

Estimating the destabilization imparted by the antiaromaticity of cyclobutadiene is more of a challenge than estimating the stability imparted to benzene. It was not until late 1999 that $\Delta H_f°$ for this prototype structure was determined by Snyder and Peters. Using photoacoustic calorimetry, a $\Delta H_f°$ value of 114 ± 11 kcal/mol was determined for cyclobutadiene. Using this and other relevant data, we find that Eq. 2.33, the direct analogue of Eq. 2.32, is *endothermic* by 46 kcal/mol ± 11 kcal/mol. It is not obvious exactly how to incorporate strain effects, but to first order it is a good approximation to say that Eq. 2.33 is strain neutral. This analysis leads to the conclusion that the antiaromaticity of cyclobutadiene is more substantial than the aromaticity of benzene. This difference is especially true when we consider these values on a per carbon basis.

$$2\ \square \longrightarrow \square + \square \qquad \Delta H° = 45.7 \text{ kcal/mol}$$

$$\Delta H_f° \quad 37.5 \qquad\quad 6.7 \quad 114$$

(Eq. 2.33)

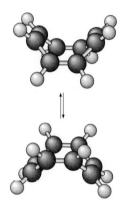

Cyclooctatetraene "tub"

Cyclobutadiene does not adopt a perfectly square structure. It rapidly interconverts between two rectangular forms. The distortion from a square arises from a **pseudo-Jahn–Teller effect**, which leads to a lowering of the energy of the system. This dynamic process (Eq. 2.34) has a very low activation barrier, estimated to be 5–10 kcal/mol.

$$\square \rightleftharpoons \square$$

(Eq. 2.34)

Cyclooctatetraene (COT) is also antiaromatic when planar. To avoid the antiaromaticity and severe bond angle strain, the ring puckers into a **tub** shaped conformation (shown in the margin). The barrier to "flipping" the tub is 13.7 kcal/mol. Interestingly, addition of two electrons or removal of two electrons would create aromatic systems, and indeed both reactions are known and lead to planar structures.

NMR Chemical Shifts

Aromatic systems possess diagnostic ^{1}H NMR chemical shifts. The circulation of electrons within the MOs that are above and below the plane of the ring (**ring current**) creates magnetic fields, giving rise to **anisotropy**. Anisotropy means that the magnetic field strength felt by the hydrogens varies as a function of the orientation of neighboring bonds. In an aromatic system, the ring current leads to a magnetic field that reinforces the applied field in the region of the protons, and shifts the proton resonances downfield (higher ppm), normally in

---

Connections

Cyclopropenyl Anion

When one deprotonates propene, it is the methyl hydrogens that are the most acidic. Deprotonation creates the resonance stabilized allylic anion. When the analogous reaction is attempted with cyclopropene, a vinylic hydrogen is the one removed. Deprotonation of the CH$_2$ group in cyclopropene (Eq. 2.19) would create an antiaromatic anion, an undesirable effect, and this reversal in acidities provided early support for the notion of destabilization due to antiaromaticity.

Schipperijn, A. J. "Chemistry of Cyclopropene. Preparation and Reactivity of the Cycloprop-1-enyl Anion in Liquid Ammonia." *Recl. Trav. Chim. Pays-Bas*, **90**, 1110–1112 (1971).

Cyclopropenyl vs. allyl anion

the range of 6 to 8 ppm. Antiaromatic systems show upfield shifts, although it is not clear that this shift is due to a ring current effect. The downfield shift due to aromaticity is defined as a **diamagnetic shift**, while a upfield shift is known as a **paramagnetic shift**.

Polycyclic Aromatic Hydrocarbons

Fused benzene rings also possess the reactivity characteristics of aromatic systems. For example, naphthalene and anthracene undergo substitution reactions with electrophiles instead of addition reactions. Furthermore, using an analysis similar to the isodesmic analysis given above for benzene, one can calculate stabilization energies due to aromaticity of 61 and 84 kcal/mol for naphthalene and anthracene, respectively. Not all polycyclic systems, however, are simple aromatic molecules. Phenanthrene undergoes addition upon treatment with bromine. Apparently, the reactivity associated with aromaticity is not as evident in a ring when the other rings in the compound can be identified as isolated (non-fused) benzene rings.

Naphthalene

Anthracene

Phenanthrene

Large Annulenes

An **annulene** is a cyclic, fully conjugated hydrocarbon, denoted with the nomenclature [n]annulene where n is the number of carbon atoms in the ring. Hence, [6]- and [8]annulene are benzene and cyclooctatetraene, respectively. The aromaticity of these compounds has been extensively investigated. As the ring becomes bigger, there is a lower stabilization imparted by conforming to the $4n + 2$ rule. As a general rule, it is believed that 22 electrons approaches the limit of aromaticity, and there is no stabilization for larger aromatic systems. An example of an intermediate case is [18]annulene. The structure is planar, and its NMR spectrum shows a strong diamagnetic ring current indicative of aromatic character. As the ring gets smaller, an effect analogous to transannular strain arises, which does not allow the structure to be planar. However, if the compound is bridged, thereby removing the offending hydrogens, it can become aromatic. An example is 1,6-methano[10]annulene. Further discussion of larger annulenes is given in Section 14.5.1.

[18]-Annulene

1,6-Methano[10]annulene

Connections

Porphyrins

The stabilization imparted by aromaticity has been exploited by nature in the use of the porphyrin ring system. Shown to the right is the basic porphyrin skeleton. If we exclude two nitrogens and alkenes not directly involved in creating a fully conjugated cyclic system, we can see a porphyrin as a ($4n + 2$) system. Extensive investigations have been conducted to explore the extent of stabilization imparted to these structures due to their aromatic character. It is generally believed that the incorporation of heteroatoms increases the stabilization, allowing these large rings to have significantly more stabilization than the analogous all-carbon annulenes.

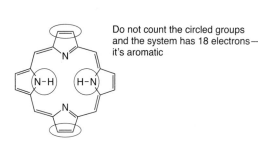

Do not count the circled groups and the system has 18 electrons— it's aromatic

Franck, B., and Nonn, A. "Novel Porphyrinoids for Chemistry and Medicine by Biomimetic Synthesis." *Angew. Chem. Int. Ed. Eng.*, **34**, 1795 (1995).

2.4.2 Effects of Multiple Heteroatoms

A basic tenet of organic conformational analysis is that *the lessons learned from the extensively studied hydrocarbon systems will carry over more or less unperturbed to systems with heteroatoms.* The conformational analysis of methyl ethyl ether should not be too different from that of *n*-butane. However, in certain cases, when multiple heteroatoms are superimposed on a hydrocarbon framework in close proximity, a number of novel "effects" arise, which often stabilize otherwise unstable conformations. These effects, such as the anomeric effect and the gauche effect, all have similar origins that can be easily understood using the bonding models of Chapter 1. Let's start by analyzing bond length effects.

Bond Length Effects

One simple difference in the conformations of pure hydrocarbons and heterocyclic rings results from the fact that C–heteroatom bonds are of different lengths than C–C bonds. Bonds to O and N are shorter, often causing increased steric strain. Bonds to S are significantly longer.

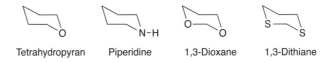

Tetrahydropyran   Piperidine   1,3-Dioxane   1,3-Dithiane

Dioxane diaxial interaction

These differences are reflected in the *A* values for groups attached to cyclohexane analogues. Table 2.16 shows a number of *A* values for groups attached to various positions on heterocycles relative to cyclohexane. Note the dramatic difference between the 2- and 5-positions on 1,3-dioxane. This difference reflects the shorter C–O bond lengths, making the 1,3-diaxial interactions more repulsive.

Table 2.16
A **Values ($\Delta G°$ in kcal/mol) for Three Groups in Particular Positions in Heterocyclic Systems**

	Cyclohexane	Tetrahydropyran	1,3-Dioxane		1,3-Dithiane	
		2-Position	2-Position	5-Position	2-Position	5-Position
Group						
Methyl	1.8	2.9	4.0	0.8	1.8	1.0
i-Pr	2.1		4.2	0.7	1.5	0.8
t-Bu	> 4.5			1.4	> 2.7	

Orbital Effects

Recall that the introduction of electronegative elements such as F, O, and N has a general effect of lowering the energies of *all* MOs to which they make a significant contribution. Especially important are the low-lying empty MOs (often σ* orbitals). In addition, heteroatoms introduce lone pair MOs, filled orbitals with very little bonding character that are relatively high-lying in energy, even though they are associated with an electronegative element. Many lone pair orbitals also tend to be relatively "localized", presenting a large electron density at one site for orbital mixing. This combination, high-lying filled MOs and low-lying empty MOs, is perfect for the always stabilizing two center–two electron interaction discussed in Chapter 1.

It is useful to think of this situation as a **donor–acceptor interaction**. The high-lying filled orbital donates electrons to the low-lying empty orbital, producing a stabilizing interaction. However, this interaction is *not* electron transfer nor the kind of donor–acceptor interactions often discussed in excited-state phenomena (Section 3.2.4). It is simply orbital

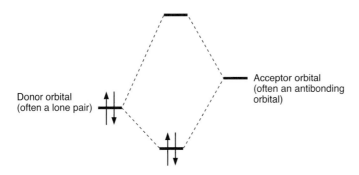

mixing (see above). No charge–transfer bands are seen in UV/vis spectroscopy and no highly polarized states are seen. It is still a "covalent" bonding situation, but there are some special orbital mixing possibilities.

Within this framework, then, it is useful to classify the donor and acceptor capabilities of certain kinds of groups. Useful sequences are shown in the margin.

The trends are fairly standard. Lone pairs are better donors than bonding pairs because they are at higher energy. Amongst lone pairs two effects dominate. First, donor ability increases as electronegativity decreases; and second, donor ability increases as you move down a column of the periodic table. These trends are consistent with the bonding models we developed in Chapter 1.

For acceptor MOs (these are empty σ^* orbitals), the trends are shown in the margin. Again, electronegativity and periodic table effects are evident. Note that moving down a column of the periodic table makes a C–X bond both a better donor and a better acceptor. What is dominating here is the polarizability of the X atom (see Section 1.1.12). Polarizability is not explicitly treated in simple orbital mixing models, because it, by definition, involves the reorganization of electron density (and hence of orbital shapes) that occurs in response to an interaction. We simply have to treat such effects as an extra layer on top of the simple orbital mixing models.

Given this information, we would predict that molecules would adopt conformations that maximize interactions between good donor orbitals and good acceptor orbitals (i.e., that maximize the mixing of filled with empty orbitals). The only remaining issue is the preferred geometry of the interacting orbitals, which is a bit counterintuitive. As shown in Figure 2.19, *the optimal arrangement places the donor orbital anti to the C–X bond that is acting as the acceptor.* The reason for this arrangement is the unique nodal character of a σ^* orbital. Figure 2.19 shows the interaction of a generic lone pair donor with a σ^* orbital, the latter modeled

Donors

Lone pairs > bonding pairs
$C:^- > N: > O: (p) > O: sp^2 > F:$
$I: > Br: > Cl: > F:$
C–H > N–H > O–H > F–H
C–Cl > C–C > C–H > C–F
C–I > C–Br > C–Cl > C–F
C–S > C–C > C–N > C–O

Acceptors

C–F > C–O > C–N > C–C
C–I > C–Br > C–Cl > C–F
C–S > C–O
C–P > C–N

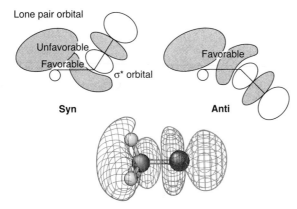

Figure 2.19
Preferred geometry for the interaction of a donor (shown as a lone pair) with an acceptor σ^* orbital. The σ^* orbital is modeled after the LUMO of CH_3Cl, shown at the bottom.

after the LUMO of CH_3Cl. There is considerable σ^* orbital density on the backside of the carbon, and the interaction of this density with the lone pair orbital is more extensive in the anti arrangement. Also, while it is difficult to portray in the picture, detailed orbital analyses reveal that the syn orientation experiences both favorable and unfavorable interactions. The net effect is that the anti arrangement is preferred. We are now ready to predict molecular shapes.

We begin with a simple system that very nicely illustrates the key principles. Consider (fluoromethyl)amine, FCH_2NH_2. As shown in Eq. 2.35, this system is perfectly set up for a donor–acceptor interaction. The preferred conformation puts the nitrogen lone pair (donor) anti to the C–F bond, optimizing the donor–acceptor interaction. This is really an optimal case, and the conformational preference is substantial.

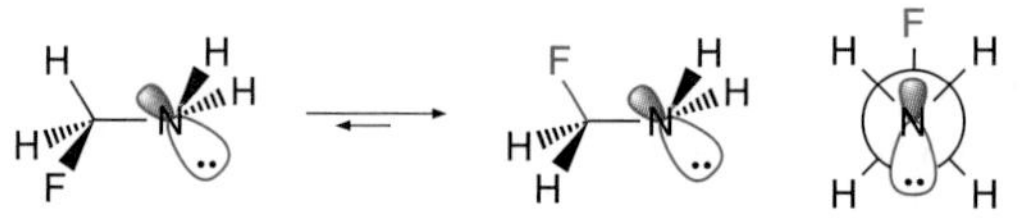

(Eq. 2.35)

Another simple system is 1,2-difluoroethane, in which the conformation with the fluorines gauche is preferred over the anti by 1.8 kcal/mol, in what would appear to be contrary to conventional steric arguments (Eq. 2.36). Donor–acceptor analysis explains the result. Aligning the fluorines anti places a poor donor (C–F bond) anti to a good acceptor (C–F bond). However, having the fluorines gauche places the two good acceptor bonds (C–F) anti to C–H bonds (Eq. 2.37). While C–H bonds are not especially strong donors, they are better donors than C–F bonds, and so the gauche conformation is preferred. In this case a favorable orbital interaction stabilizes what would otherwise be a strained structure.

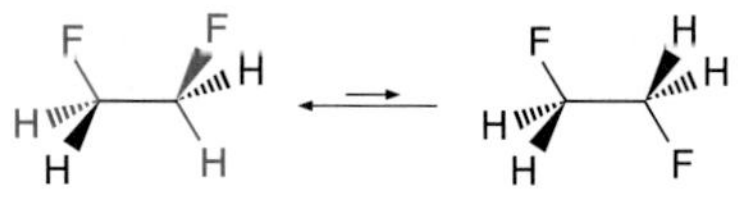

(Eq. 2.36)

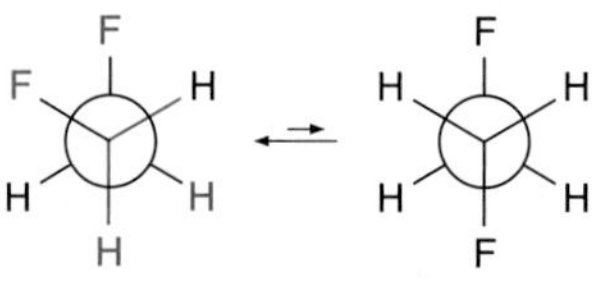

(Eq. 2.37)

A classic example of donor–acceptor interactions is seen in hydrogen peroxide, which also introduces an additional effect that can arise when several polar bonds are present in a molecule. In H_2O_2, solely steric arguments predict that the preferred conformation should have an H–O–O–H dihedral angle of 180°. This steric argument is augmented by a second effect. In the anti conformation the two large O–H bond dipoles are aligned anti to one another, often a significantly stabilizing effect. However, the anti conformer is opposed by donor–acceptor effects (see structures below).

The O–H bond is an excellent acceptor, and the best donor is an O lone pair. We know that such an oxygen has two types of lone pairs, a σ(out)-type orbital that is roughly an sp^2 hybrid and a pure p orbital (see Section 1.3.3, and water in Appendix 3). The p-type lone pair is higher in energy, and so by the energy gap law we expect it to mix with the acceptor orbital. As shown below, this mixing would favor a 90° dihedral angle. The final geometry reflects a compromise among the various interactions, producing a dihedral angle of ~120°. A similar effect might be expected for S–S bonds, and as the following Connections highlight shows, the preferred angle is 90°.

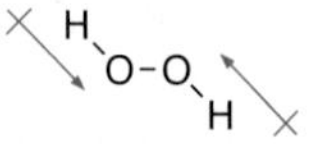

Steric preference 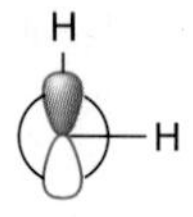   Donor–acceptor   Compromise
                           preference

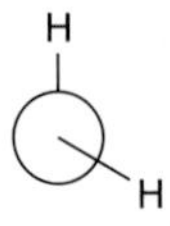

Connections

Protein Disulfide Linkages

The conformational preferences of dialkyl disulfides are similar to those of hydrogen peroxide. The dihedral angle is ~90° in a typical molecule such as dimethyl disulfide, perhaps because the dipole effect is smaller (S is less electronegative than O). Disulfides are common components of protein structures, formed by linking the sidechains of the amino acid cysteine. Invariably, such disulfide linkages in proteins are approximately gauche. Just like gauche butane, a gauche disulfide is chiral, and so exists in two enantiomeric forms. In the context of a protein, which is always chiral, the two disulfide gauche forms

are now diastereomeric rather than enantiomeric (see Chapter 6 for definitions). Thus, any protein that has a single disulfide can exist in two diastereomeric forms, differing in the geometry around the C–S–S–C bond. If there are n disulfides, 2^n diastereomers are expected (assuming there is no global symmetry in the protein).

The two enantiomeric forms
of a simple disulfdide

A particularly important conformational phenomenon that can be explained using the types of arguments developed here is the **anomeric effect** of carbohydrate chemistry. The anomeric effect can be defined as a contrasteric bias toward the axial (α) glycosidic linkage at the acetal carbon over the equatorial (β). This preference results from aligning the exocyclic C–O bond anti to a lone pair of the oxygen in the ring (see below). Such conformational preferences are extremely important in carbohydrate chemistry, so much so that the central carbon involved (C1 of a sugar) is often referred to as the **anomeric carbon**. Since the formation of such acetals and ketals is generally reversible, it is a simple matter to equilibrate axial and equatorial groups at anomeric centers and directly determine which is the more stable form without resorting to calorimetry or other more complicated procedures.

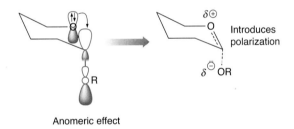

Anomeric effect

Figure 2.20 shows four examples of the anomeric effect. In each case, the large group on the anomeric carbon of the pyranoside prefers the axial position. The magnitude of the preference depends upon the group and the substituents on the ring. However, it is also influenced by the polarity of the solvent. For the third entry in Figure 2.20, the axial preference is larger in carbon tetrachloride than in acetonitrile. One might have expected the donor–acceptor interaction to be enhanced in the more polar solvent due to the polarization implied by the effect (see above). However, the opposite is found. The preferred conformation of the spirocycle shown as the last entry of Figure 2.20 is the one where each C–O bond is antiperiplanar to an oxygen lone pair orbital.

The simple model developed here provides a convenient way to explain and predict variations in structure seen in more complex systems. There is, however, some controversy concerning the anomeric effect. While most chemists accept that a donor–acceptor interaction of the kind shown above exists, there are clearly other factors. For example, an axial arrangement of the exocyclic C–O bond cancels dipoles, a potentially favorable effect. Such an effect is expected to be most important in low polarity solvents, perhaps explaining the solvent effect shown as the third entry of Figure 2.20.

Because of their common physical origin, the various donor–acceptor effects discussed here have been collectively called the **gauche effect**. The best conformation of a molecule has the maximum number of gauche interactions between adjacent lone pairs and / or polar

Dipole cancellation

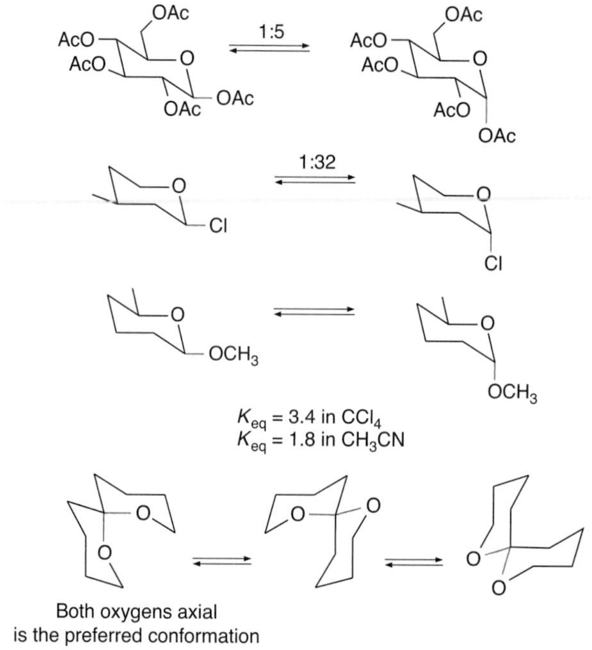

Figure 2.20
Top: Three glycosidic systems for which the large group prefers the axial position. Bonner, W. A. "The Acid-Catalyzed Anomerization of the D-Glucose Penta Acetates. A Kinetic Thermodynamic and Mechanistic Study." *J. Am. Chem. Soc.* **73**, 2659 (1951). Anderson, C. B., and Sepp, D. T. "Conformation and the Anomeric Effect in 2-Halotetrahydropyrans."*J. Org. Chem.* **32**, 607 (1967). Eliel, E. L., and Giza, C. A. "Conformational Analysis. XVIII. 2-Alkoxy- and 2-Alkylthiotetrahydropyrans and 2-Alkoxy-1,3-Dioxanes. Anomeric Effect." *J. Org. Chem.* **33**, 3754 (1968). Bottom: Another example of a strong conformational bias introduced by the anomeric effect.

bonds. Thus, in FCH_2CH_2F the polar bonds are gauche, and in a peroxide or a hydrazine (R_2NNR_2) the lone pairs are gauche. It is called the gauche effect, but its origin is the preference for having lone pairs anti to acceptor σ^* orbitals, rather than anti to one another.

An argument based upon an analysis of the relative orientation (stereochemistry) of orbitals is called a **stereoelectronic effect**. The placement of a lone pair orbital antiperiplanar to a polarized acceptor bond is just our first example. We will see stereoelectronic effects on reactivity in several places in this book.

Now that we have examined organic conformational analysis and various effects that lead to strain and stability, let's look at some structures where chemists have put these notions to the test.

2.5   Highly-Strained Molecules

One of the fundamental goals of physical organic chemistry has been to establish the limits of our models for structure and energetics. How long can a C–C bond be? How much angle strain can a molecule tolerate? How crowded can a structure be? Such questions have defined many brilliant research efforts and have produced a fantastic array of bizarre and wonderful structures. Here we present a collection of representative highly-strained molecules, with an emphasis on the structural concepts that are being tested.

2.5.1   Long Bonds and Large Angles

Typical C–C bond lengths were noted in Chapter 1, and while there is considerable variation, a C–C bond ≤ 1.59 Å is not considered exceptional. Many compounds with bonds >

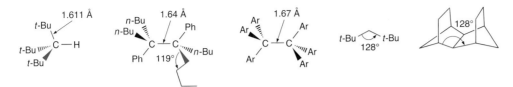

Figure 2.21
Simple structures in which excessive steric crowding leads to long bonds and/or expanded angles.

1.6 Å are now known, the primary strategy being to increase the steric demands around the bond (Figure 2.21). It is generally true that a long bond is a weak bond, and indeed many structures with long C–C bonds are thermally labile.

Several hexasubstituted ethanes with very long central bonds are known. For example, the diphenyltetrakis(*n*-butyl)ethane of Figure 2.21 has a central bond of 1.64 Å. A prototype of this family is hexaphenylethane. We discussed in Section 2.2.2 the fact that the triphenylmethyl radical does not dimerize to hexaphenylethane, but instead makes the unsymmetrical dimer of Eq. 2.13. Finally, in 1986 a true hexaphenylethane was observed and structurally characterized. It has a very long C–C bond of 1.67 Å [Figure 2.21, where Ar = 2,6-di(*t*-butyl)phenyl].

Steric repulsions also provide the primary strategy for creating expanded C–C–C angles, and we mentioned this briefly in Chapter 1. Even a simple molecule like di(*t*-butyl)methane has a greatly expanded central angle (Figure 2.21). In such a structure it is interesting to consider whether sp^3 is really the relevant hybridization for the central carbon (recall the variable hybridization discussion of Chapter 1). Certainly, with an angle of 128°, the bonding must be different from that of a typical CH_2.

2.5.2   Small Rings

Deviations of bond angles in the opposite sense—values much smaller than 109.5°—are routinely seen in small-ring compounds. We've seen that cyclopropane is highly strained for such a small molecule, and fusion of two rings to produce bicyclo[1.1.0]butane leads to ~65 kcal/mol of strain (Figure 2.15). The ultimate concatenation of cyclopropane rings is tetrahedrane, with an estimated strain energy of 140 kcal/mol. After decades of effort from many groups, Maier succeeded in synthesizing the tetra(*t*-butyl) derivative of this structure. Remarkably, this molecule is completely stable at room temperature.

Tetrahedrane   [1.1.1] Propellane

Another surprising observation is the considerable stability of [1.1.1]propellane, first synthesized by Wiberg in 1982. Along with cubane (another very strained but very persistent molecule synthesized by Eaton in 1964), [1.1.1]propellane and tetra(*t*-butyl)tetrahedrane illustrate an important concept. Typically, we expect a very strained molecule to be "unstable" or "reactive"—requiring very low temperatures or special conditions for characterization—and this is usually the case. However, simply having a very large amount of strain does not guarantee that a molecule will be reactive. The molecule must have a kinetically viable path to release the strain. The molecules are unstable, but persistent. Alternatively, we say the molecules are thermodynamically unstable, but kinetically stable.

Connections

From Strained Molecules to Molecular Rods

A major goal of modern materials chemistry is the development of molecular-scale analogues of the gates and switches that comprise modern computer chips and electronic devices—so-called **molecular electronics** or **molecular devices**. For this dream to succeed, basic structural building blocks that allow precise arrangements and positioning of molecular structures will be useful. The finding that [1.1.1]propellane was stable, and in fact readily synthesized in relatively large quantities, surprised the entire organic chemistry community. The unusual bonding in this structure suggested novel reactivity patterns, and indeed that has been found to be true. Under a variety of conditions, the central bond breaks and C–C bonds are formed between bridgehead carbons of separate propellanes. Michl and others have shown that this process can be controlled to produce rigid linear structures termed **staffanes**. Such structures could be one component of

a collection of "molecular Tinker Toys®" that may prove useful in rationally building molecular-scale devices.

Staffanes

Mazieres, S., Raymond, M. K., Raabe, G., Prodi, A., and Michl, J. "[2]Staffane Rod as a Molecular Rack for Unraveling Conformer Properties: Proposed Singlet Excitation Localization Isomerism in anti,anti,anti-Hexasilanes." *J. Am. Chem. Soc.*, **119**, 6682–6683 (1997).

For example, in cubane, homolysis of a C–C bond releases only a fraction of the total strain of the molecule and produces a biradical that has nowhere else to go (Eq. 2.38). The two newly formed radicals are trapped in an arrangement in which they are simply staring at each other—the most sensible reaction is reforming the broken bond. A concerted pericyclic process (Eq. 2.39) that might rearrange several bonds and thereby release much more strain, is forbidden by the orbital symmetry rules (see Chapter 15). However, if given a pathway, we might expect cubane to react very violently (see the Connections highlight below). The interplay between kinetics and thermodynamics is a recurring theme in all of chemistry, and it will be discussed in greater detail in Part II of this text.

$$\text{(Eq. 2.38)}$$

$$\text{(Eq. 2.39)}$$

Connections

Cubane Explosives?

We noted above that cubane has a strain energy of roughly 166 kcal/mol, but the structure is quite persistent. Rapid decomposition of cubane might be expected to release a great deal of energy, and when this release of energy is coupled with the fact that cubane has a higher density as a solid than almost any other hydrocarbon, the potential for cubane-based explosives and/or propellants is clear. Most organic explosives contain a number of nitro groups, including compounds such as TNT, RDX, HMX, and CL-20, which is perhaps the most powerful non-nuclear explosive known. For such structures, combustion leads to the release of a great deal of energy and a number of small volatile molecules, such as CO_2 and N_2, enhancing the explosive power. Imagine, then, the potential energy

stored in a molecule such as octanitrocubane, with its high density, huge strain, and very large NO_2/carbon ratio. This compound would be a potent material, and it has been the object of long-standing (and careful!) synthetic efforts. In 2000, Eaton and co-workers succeeded in making this remarkable structure. Meeting the synthetic challenge led to a new challenge. It turns out that octanitrocubane did not crystallize with quite the high density that theory predicted. So, now the quest is to find the alternative crystal form that will have the desired high density.

Eaton, P. E. "Cubanes: Starting Materials for the Chemistry of the 1990s and the New Century." *Angew. Chem. Int. Ed. Eng.*, **31**, 1421–1436 (1992). Zhang, M.-X., Eaton, P. E., and Gilardi, R. "Hepta- and Octanitrocubanes." *Angew. Chem. Int. Ed. Eng.*, **39**, 401–404 (2000).

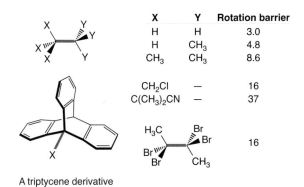

| TNT | RDX | HMX | CL-20 | Octanitrocubane |

2.5.3 Very Large Rotation Barriers

Along with very strained molecules, a related goal has been the development of structures in which severe steric interactions are present in the transition state for a C–C rotation, but not (or much less so) in the ground state. This steric interference would produce very large rotation barriers. In some cases, E_{rot} is so large that different conformers can be separated and remain stable at room temperature (such structures are termed **atropisomers**—see Chapter 6).

Again, steric bulk is the primary strategy, and quite substantial barriers can be achieved using just this strategy. A more clever approach uses the unique shape of structures such as triptycene to point substituents directly along the C–C bond and force a gearing-type interaction that can lead to very large bond rotation barriers. A spectacular example is the dimethylbitriptycyl derivative in Figure 2.22, with $E_{rot} > 54$ kcal/mol! With this strategy even hindered rotation around a C–C≡C–C bond can be seen, as the ditriptycene acetylene of Figure 2.22 gives a rotation barrier of 15 kcal/mol. Furthermore, if the hindered bond rotation can be coupled to another motion, one can envision controlled "gearing", as described in the next Going Deeper highlight.

X	Y	Rotation barrier
H	H	3.0
H	CH₃	4.8
CH₃	CH₃	8.6
CH₂Cl	—	16
C(CH₃)₂CN	—	37

(H₃C, Br, Br, Br, Br, CH₃ structure) 16

A triptycene derivative

Figure 2.22
Structures with very large rotation barriers. Values are in kcal/mol.

Rotation barrier   >54   15

Going Deeper

Molecular Gears

Often we think of alkyl groups as generic steric placeholders. However, in some contexts, the precise shape of the group can have important consequences. It has long been appreciated that in the right context, certain groups could fit together as the cogs of a gear, creating systems with potentially novel static and dynamic behaviors. One such system is hexaisopropylbenzene. The molecule adopts a perfect cyclic gearing array, in which each isopropyl is firmly locked in with its neighbor. Each methine H (shown in color) is tucked into the small space between geminal methyl groups of the adjacent isopropyl. The barrier to any kind of rotation of an isopropyl group is ≥ 22 kcal/mol, and molecular mechanics calculations (see below) suggest that simultaneously reversing the sense of the gearing (making all isopropyls point in the opposite direction) has a barrier on the order of 35 kcal/mol. Note that this is an extreme example of the context dependence of efforts to rate the relative steric sizes of groups. One can place six isopropyls around a benzene because of the potential for gearing. However, converting isopropyl to t-butyl in this context would have disastrous consequences, and hexakis (t-butyl)benzene is a very highly-strained, as yet unknown structure.

Perhaps the ultimate gearing system is based on triptycene units. We saw in Figure 2.22 how two triptycenes facing each other directly can lead to very high rotation barriers. When we attach two triptycenes to a central CH_2 group, we now must interlace the ring systems in order to avoid severe steric clashes. The result is that we see **correlated rotation** of the triptycenes in a gearing fashion. This correlated motion produces novel stereochemical phenomena, and also constitutes a molecular realization of a simple mechanical object, the bevel gear.

A triptycene gear

Siegel, J., Gutierrez, A., Schweizer, W. B., Ermer, O., and Mislow, K. "Static and Dynamic Stereochemistry of Hexaisopropylbenzene: A Gear-Meshed Hydrocarbon of Exceptional Rigidity." *J. Am. Chem. Soc.*, **108**, 1569–1575 (1986). Iwamura, H., and Mislow, K. "Stereochemical Consequences of Dynamic Gearing." *Acc. Chem. Res.*, **21**(4), 175–182 (1988).

Hexaisopropylbenzene

2.6   Molecular Mechanics

We've introduced the concept of strain as the energetic penalty that results from distorting a structure from normal bonding parameters. For example, consider angle distortion at a typical sp^3 carbon, where the "normal" angle is 109.5°. If the angle is compressed to 108°, strain is introduced, and the energy of the system should go up by some amount. If the angle is further compressed to 107°, the energy should go up more; 106° even more, and so on. It is reasonable to assume that the further we distort from the ideal, the larger the strain energy, and qualitative observations bear this out.

Can we put this analysis on a quantitative basis? Could we develop some sort of equation that relates the extent of distortion to the energy of the molecule? We can, and the method is called **molecular mechanics**. Here we will lay out the basic tenants of molecular mechanics and provide a description of its strengths and weaknesses. The method is now quite common and easily implemented for sizable molecules on a standard personal computer. It is a powerful aid to experimentalists in all fields of organic chemistry, as well as in molecular-scale studies of biology and materials science. It should be appreciated from the start, however, that the method has significant limitations and is susceptible to misuse.

2.6.1 The Molecular Mechanics Model

The fundamental concept of molecular mechanics is embodied in Eq. 2.40. That is, the total energy of a system can be represented as a sum of individual energies, one related to bond stretching, one for angle bending, one for torsional effects, one for nonbonded interactions, and perhaps many more. It is important to appreciate from the start that *there is no theoretical justification for this model*. If we look at the quantum mechanics of molecular structure, as embodied by the Schrödinger equation (Chapter 14), there is no "bond stretching" term. Molecular mechanics is completely and solely justified on empirical grounds—it is valid only to the extent that it works. Hence, the term **empirical force field** is sometimes used as a more realistic synonym for molecular mechanics. We will return to this point below after we define some terms.

$$
\begin{aligned}
E_{\text{tot}} &= E_{\text{bond}} + E_{\text{angle}} + E_{\text{torsion}} + E_{\text{nonbond}} + \ldots \\
&= E_r \quad + E_\theta \quad + E_\phi \quad + E_{\text{nb}} \quad + \ldots
\end{aligned}
\qquad \text{(Eq. 2.40)}
$$

We begin by defining the individual terms of the equation for E_{tot}, as well as presenting some discussion of the nature of the various parameters. The total energy, E_{tot}, produced by a molecular mechanics calculation is also referred to as the **steric energy**. It is not to be confused with strain energy, a very different quantity, as we will elaborate below.

The individual terms in Eq. 2.40 can each be viewed as a potential function, and they have the same mathematical forms as those for stretches, bends, and torsions that we discussed earlier in this chapter. It is important to remember, however, that the parameters used in the equations that describe the real degrees of freedom of molecules do not necessarily have any relation to the parameters used in the equations of the molecular mechanics method. Moreover, whereas the potential surfaces that describe the vibrational degrees of freedom in molecules derive from the forces that hold the atoms together, the potential functions in molecular mechanics are derived simply to get the right answer.

Bond Stretching

The standard equation for bond stretching is Eq. 2.41, where r is the length of the bond being evaluated, k_r is analogous to a force constant, and r_o is the "natural" bond length.

$$
E_r = \frac{k_r}{2}(r - r_o)^2
\qquad \text{(Eq. 2.41)}
$$

This equation is a classical Hooke's law potential function, and the plot of E_r vs. r (Figure 2.23) is the parabola found for the harmonic oscillator (see the Going Deeper highlight entitled "Probability of Finding Atoms at Particular Separations" on page 75). Note that k_r and r_o are *parameters*—they are obtained by a fitting process described in more detail below. They are *not* "experimental" force constants or bond lengths of any sort. They are parameters that have the *form* of a force constant and a length. For example, in the popular MM3 force field, r_o for a C–C single bond is 1.5247 Å—not at all a standard C–C bond length (recall Table 1.4).

We need a pair of parameters (k_r and r_o) for *each type* of bond in a molecule. That is, C–C single bonds have one such pair, C–C double bonds have another, C–H bonds have another, C–O another, etc. Every *type* of bond in a molecule has its own set of parameters. It is *not* the case that every bond in a molecule has its own set of parameters. In some force fields (we will call a particular implementation of the molecular mechanics method a **force field**), further distinctions are made. For example, a RCH_2–CH_2R bond might have a different parameter pair than a RCH_2–CHR_2 bond. Almost all modern force fields would differentiate between single bonds that are $C(sp^2)$–$C(sp^3)$ (as in toluene) vs. $C(sp^3)$–$C(sp^3)$ (as in an alkane). This differentiation can greatly increase the number of parameters.

We know from experiment that a Hooke's law function is a poor representation of a real covalent bond. The actual potential surface is something more like a Morse potential (Figure 2.2). When r is fairly close to r_o, a parabola is a good approximation of a Morse potential.

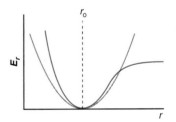

Figure 2.23
Hooke's law (parabola; black) vs. a Morse potential (color; see also Figure 2.2) to describe a bond stretching interaction.

However, at greater values of r—when a bond is stretched—the approximation is quite poor. For this reason, many force fields add a cubic term to the stretching potential function (Eq. 2.42).

$$E_r = \frac{k_r}{2} (r - r_o)^2 + k_r' \, (r - r_o)^3 \qquad \text{(Eq. 2.42)}$$

This expansion introduces another parameter (k_r'), but it does improve the force field. For the highest possible precision in calculations of organic molecules, such additional terms are usually included. However, in a force field for proteins or nucleic acids (see below), structures that rarely deviate substantially from standard bonding parameters, cubic terms are often unnecessary.

Angle Bending

A similar equation holds for angle bending (Eq. 2.43), where θ is the value of the angle being evaluated, k_θ is analogous to a force constant, and θ_o is the "natural" bond angle.

$$E_\theta = \frac{k_\theta}{2} (\theta - \theta_o)^2 \qquad \text{(Eq. 2.43)}$$

Again, there is a pair of parameters for each kind of angle. As with bond stretching, this parabolic-type function is often not optimal, and so a cubic term is added (Eq. 2.44).

$$E_\theta = \frac{k_\theta}{2} (\theta - \theta_o)^2 + k_\theta' \, (\theta - \theta_o)^3 \qquad \text{(Eq. 2.44)}$$

Torsion

The simplest form for a torsional potential function is Eq. 2.45, where n is the foldedness of the barrier, and $B = \pm 1$. If $B = +1$, then the staggered form of the bond is preferred, whereas if $B = -1$, the eclipsed form of the bond is preferred.

$$E_\phi = \frac{k_\phi}{2} [B + \cos(n\phi)] \qquad \text{(Eq. 2.45)}$$

When do we ever want $B = -1$? We want it for C–C double bonds, as in ethylene or benzene! Remember, molecular mechanics knows nothing about π bonds or molecular orbitals. We have to explicitly tell it that a double bond wants to be planar (i.e., eclipsed with a two-fold barrier). Again, every particular torsion type has its own set of parameters.

More modern force fields have found that an expanded torsional equation is beneficial (Eq. 2.46).

$$E_\phi = V_1 [B + \cos\phi] + V_2 [B + \cos(2\phi)] + V_3 [B + \cos(3\phi)] \qquad \text{(Eq. 2.46)}$$

That is, each torsion is treated as having one-fold, two-fold, and three-fold components. The subtleties for more complicated systems can be better treated in this way. For example, inspecting the torsional itinerary for butane (Figure 2.6) shows that it is not a perfect three-fold system, as ethane is. One way to accommodate the deviations is to add non-three-fold terms. This addition introduces still more parameters.

Nonbonded Interactions

Generally the most important component of any molecular mechanics force field is the nonbonding potential function. The traditional form is the **Lennard–Jones "6–12" potential** (Eq. 2.47), where ε and r^* are parameters that depend on the identities of the two interacting atoms and r is the distance between the atoms.

$$E_{nb} = \varepsilon \left[\left(\frac{r^*}{r} \right)^{12} - \left(\frac{r^*}{r} \right)^6 \right]$$ (Eq. 2.47)

When $r = r^*$, $E_{nb} = 0$. When $r > r^*$, E_{nb} goes slightly negative for a while—that is, there is a nonbonded *attraction* rather than a repulsion. This attraction is illustrated in Figure 2.24. This subtle feature can often be quite important. The parameter r^* is a cutoff distance, inside which a nonbonded interaction becomes repulsive. But what is the nature of ε? It is a parameter that defines the "hardness" of a nonbonding interaction. A large value of ε implies the energy goes up steeply as r becomes less than r^*—the interaction is "hard". A smaller value of ε gives a less steep rise and a "softer" interaction. This distinction is illustrated in Figure 2.24. Table 2.17 shows nonbonding parameters for one particular force field. As we would expect, r^* increases in the order H•••H < H•••C < C•••C. What is less obvious is why this particular force field makes a C•••C interaction much harder than an H•••H interaction, with H•••C softer still. Apparently, this particular combination gives the best fit to experimental data. The message again is that the parameters of a molecular mechanics force field are just that—parameters. They do not necessarily reflect any kind of experimental reality.

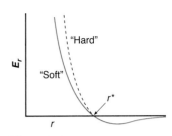

Figure 2.24
The Lennard–Jones "6–12" potential function, with examples of both a hard and a soft potential.

Table 2.17
Selected Nonbonding Parameters
From a Particular Force Field

Nonbonding pair	r^* (Å)	ε (Arbitrary units)
H•••H	3.20	2.8
H•••C	3.35	2.1
C•••C	3.85	6.6

Cross Terms

For hydrocarbons and simple organics, the force field we have defined so far is often sufficient. However, some force fields also include "cross terms". For example, a stretch–bend term couples bond lengthening with angle bending. It could, for example, make it easier to stretch a bond if the bond is also involved in a distorted angle. Such terms usually make only small contributions to the total energy.

Electrostatic Interactions

In polar molecules, including proteins and nucleic acids, coulombic interactions between charged groups and/or partial charges on atoms can become quite significant. Coulombic interactions are treated by an equation of the form Eq. 2.48, where q_i = the charge on atom i (usually a partial charge), ε = the dielectric constant of the medium, r_{ij} = distance between atoms i and j, and N = the number of atoms. This equation is simply Coulomb's law. Again, the charges are parameters that are specific to each particular kind of atom. One of the biggest challenges of the molecular mechanics method is to obtain the optimal set of charges.

$$E_{elec} = \sum_{j=1}^{N} \sum_{i>j}^{N} \frac{q_i q_j}{\varepsilon_{ij} r_{ij}}$$ (Eq. 2.48)

Hydrogen Bonding

In some force fields hydrogen bonding is handled simply by the electrostatic term just introduced. In others, there is an explicit equation for hydrogen bonds. One form for such an equation is Eq. 2.49.

$$E_{HB} = \sum_{j=1}^{N_H} \sum_{i>j}^{N_H} \left(\frac{C_{ij}}{r_{ij}^{12}} - \frac{D_{ij}}{r_{ij}^{10}} \right)$$ (Eq. 2.49)

This equation is a derivative of a Lennard–Jones potential function. In this equation N_H is the number of hydrogen bonds, while C and D are parameters depending on the type of hydrogen bond. In this approach we have to explicitly define all the hydrogen bonds in advance so this equation can be applied to them. When the simple electrostatic approach is used, hydrogen bonds need not be defined explicitly.

The Parameterization

Two things define a particular force field—the set of potential functions and the values of the parameters. The options in the first case are, for example, whether to include cubic terms in bond stretching or angle bending; whether to use the torsional equation with one-, two-, and three-fold terms; how to handle electrostatics and hydrogen bonding; etc. Once these decisions are made, it remains to determine values for all the parameters included in the various equations. There will be scores if not hundreds of such parameters for a moderately complete force field. Among the most widely used force fields are **MM#**, where # = 1, 2, and 3 delineates a version of the molecular mechanics (MM) parameters developed by Allinger and co-workers. Others are **AMBER, CHARMM**, and **UFF** (universal force field), the latter including a treatment for molecules possessing main group elements.

Where do the parameters come from? Fundamentally, they result from a fitting procedure, in which many types of experimental data are used. Structural information is crucial. There is a large database of experimentally determined structures for organic molecules, and a good force field should be able to reproduce them. So, parameters are adjusted to properly reproduce experimental structures.

However, energies are just as crucial, and these are sometimes harder to come by. For hydrocarbons and simpler organics, there is a large database of heats of formation, and, hence, strain energies, and these are valuable in parameterization. Other energies include rotation barriers and conformational differences. A competent force field should reproduce the butane torsional profile of Figure 2.6, and should obtain the A values for many cyclohexane substituents. Due to the similarity to real molecular vibrational modes, IR vibrations should be a valuable source for a force field, but in practice few modern force fields use them in their parameterization.

The value of a force field is directly proportional to the quality of its parameterization, and that in turn depends completely on the quantity and quality of experimental structural and energetic data that are available. Thus, good force fields for hydrocarbons exist because there is a wealth of experimental data on such systems. Another issue is that the factors that determine structure and energetics in hydrocarbons are fairly simple, in part because the electrostatic and hydrogen bonding terms are not very relevant. As structures become more complex, with more and more polar groups, parameterization becomes more difficult.

A recent boon to force field development has been the success of modern, *ab initio* quantum mechanical methods in predicting the properties of molecules (see Chapter 14 for a thorough description of these methods). These computational methods can now provide reliable data on small prototype systems for which experimental data are unavailable, and then force fields can be developed based on the quantum mechanical calculations. This is a valuable approach, but it is limited in that many interesting systems are too large to be treated by the quantum mechanical methods.

Heat of Formation and Strain Energy

After all the parameters are obtained, we can now do a molecular mechanics calculation. What is the result? A structure for the molecule is obtained by minimizing the total energy. This is a straightforward task in principle. Since Eq. 2.40 constitutes an analytical expression relating energy and geometry, we can use the derivatives of this equation to assist us in geometry optimization.

The other outcome from a molecular mechanics calculation is a value for E_{tot} (Eq. 2.40). However, E_{tot} is not a particularly useful quantity. It is just a number, obtained by adding up a collection of equations. We want to minimize E_{tot} to obtain the best possible geometry, but the actual value of the number does not directly relate to *any* experimental quantity. As such, *another* set of parameters must be developed that converts E_{tot} to the heat of formation. Once we have the heat of formation, we can obtain the strain energy in the usual way.

There are some instances in which E_{tot} is useful. If we are comparing *stereoisomers*, values of E_{tot} provide useful *relative* energies. That is because stereoisomers will always have identical contributors to E_{tot}, both in terms of the equations and the parameters involved. Since all structures along a torsional path are stereoisomers (conformers), E_{tot} can be used to determine rotation barriers. Note that E_{tot} cannot be used for constitutional isomers, such as *n*-butane vs. isobutane. That is because different parameters are likely involved, such as a CH_3–CHR_2 k and r_o in isobutane vs. the CH_3–CH_2R k and r_o for *n*-butane. Only after E_{tot} values for these two structures are converted to heats of formation can energy comparisons be made.

In general, then, the molecular mechanics method produces ΔH_f° values. In principle, the information to derive ΔS_f° is embedded in the method, but in practice the method is not nearly accurate enough to produce meaningful ΔS_f° values.

2.6.2   General Comments on the Molecular Mechanics Method

1. There is no theoretical justification for the method.

That is, nothing that we know about chemistry justifies dissection of the total energy of a molecule into separable components as implied by Eq. 2.40. The only justification for the method is that it works—not always, but often.

2. There is no unique, optimal force field.

A number of different workers have developed molecular mechanics force fields, often with different goals in mind. Since there is no theoretical basis for the method, there is no reason to think that one particular approach is intrinsically superior to another. Some force fields are better at some things than others.

3. Because of points 1 and 2, it is risky to attach significance to the individual energy terms of Eq. 2.40.

Consider the following hypothetical, but quite plausible, results from two different force fields evaluating the same molecule. They get the same geometry for the molecule, but the energies look quite different.

Force field #1 might produce:

$$E_{tot} = E_r + E_\theta + E_\phi + E_{nb} + E_{other}$$

$$= 5 + 30 + 2 + 7 + 2 = 46 \, kcal/mol$$

Heat of formation = –37 kcal/mol; strain energy = 45 kcal/mol

Force field #2 might produce:

$$E_{tot} = E_r + E_\theta + E_\phi + E_{nb} + E_{other}$$

$$= 27 + 12 + 16 + 43 + 12 = 110 \, kcal/mol$$

Heat of formation = –37 kcal/mol; strain energy = 45 kcal/mol

Remember, there is a force-field specific set of parameters that converts E_{tot} to ΔH_f°, and so two different force fields can get very different values for E_{tot} but the same ΔH_f°. In this example, the two force fields are equally good—both get the same geometry and the same heat of formation.

What about interpreting the individual terms of E_{tot}? Force field #1 predicts most of the strain comes from angle bending, while #2 predicts nonbonding interactions and bond stretching are most important. Which is correct? Neither! These terms have no meaning because molecules do not partition their total energy into neat compartments.

Force field #1 has chosen to make bond stretching easy, but angle bending hard; #2 makes nonbonding contacts quite adverse, and bond stretching difficult. The differing terms compensate for each other. As long as the developers of the individual force fields did their parameterization jobs well, either force field can get useful results.

4. *Because the energy expressions are all analytical, geometry optimization can be quite efficient.*
Inherent to the molecular mechanics method is a set of analytical expressions for the total energy of a system. It is a simple matter to derive the first and second derivatives of the energy expression. The first derivatives define the forces on the molecule. At a minimum there are no forces—the system is "at rest". Thus, geometry optimization involves minimizing the first derivatives—a process that can be much more efficient than just randomly searching for a minimum. Furthermore, at a minimum, all second derivatives are positive.

5. *Generally, getting a good geometry is easier than getting reliable relative energies.*
Figure 2.25 illustrates this point. Basically, it is easier to find the bottom of a well than it is to know whether a nearby well is higher- or lower-lying.

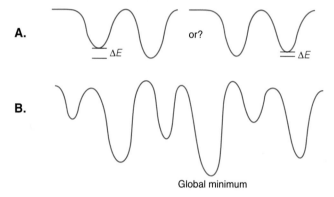

Figure 2.25
Two aspects of force field minimization. **A.** Two different force fields will usually find the same minima (geometries), but may differ in their relative energies. **B.** It is always difficult to be sure that the minimum you have found is the global minimum.

6. *Finding the global minimum can be challenging.*
It is a simple matter to know you are at the bottom of a well—in a true energy minimum. But how can you be sure it is the lowest possible structure, the **global minimum** (Figure 2.25)? In fact, you cannot be sure. There is no general, reliable solution to the global minimum problem. Just because a geometry optimization has produced a stable structure does not mean that a more stable structure cannot be found. The molecular mechanics method is especially susceptible to this problem. The more complicated the system, the more likely there are multiple minima.

There are many strategies for avoiding or at least minimizing the global minimum problem. These range from exhaustive search approaches, to ways to "kick" a structure out of a local minimum and into the global minimum. The user of the molecular mechanics approach needs to be aware of this potential pitfall.

7. *The greater the number of polar atoms and/or functional groups, the less reliable the results.*
There is a huge database of structural and thermodynamic data on hydrocarbons. However, the number of structures that contain an ester plus an aryl ether plus a dialkylamine for which we have accurate structural and heat of formation data is small (probably zero). So, parameterization of the force field is weaker for this type of structure, and the molecular mechanics method must be expected to be less reliable. Also, polar groups in

close proximity within a molecule can produce special "effects", such as the gauche and anomeric effects discussed earlier, that molecular mechanics knows nothing about (unless we add new parameters). The bottom line is always that we must be careful when applying the molecular mechanics method to systems that differ substantially from the structures on which the method was parameterized.

8. *The molecular mechanics method generally evaluates structures in the gas phase—in the absence of solvent.*

Most chemistry, however, is done in a solvent. Again, the difference between the gas phase and solution is expected to be greatest for polar molecules. This difference can be corrected by explicitly evaluating solvation, and we will discuss strategies for this in Chapter 3.

9. *The molecular mechanics method is* much *faster than quantum mechanical methods.*

We have gone to some length to point out the weaknesses of the molecular mechanics methods. However, when applied carefully, the method can produce very useful results. And, most importantly, the method is *much* faster than any quantum mechanical method will ever be. As such, for many systems it is the only game in town. This speed and the applicability to experimentally interesting systems are the method's greatest assets.

2.6.3 Molecular Mechanics on Biomolecules and Unnatural Polymers—"Modeling"

The molecular mechanics method just described was developed with organic chemistry in mind—that is, for "small" molecules with 10 to perhaps 50 "heavy" (i.e., non-hydrogen) atoms. However, the temptation to apply the method to biological macromolecules proved irresistible, and **modeling**, as it is often called, is now a standard tool. In this section we highlight some of the major differences in molecular mechanics as it is applied to macromolecules vs. small organic molecules. Typically, a number of simplifications are made in order to make the calculations more manageable, and hence applicable to very large molecules.

The force fields used for biopolymers (and unnatural polymers) are typically simplified versions of the general force field described above. For example, proteins and nucleic acids rarely have C–C bonds that are substantially elongated from normal values or valence angles that are greatly expanded or contracted. Typically, biomolecules achieve their complexity by concatenation of fairly ordinary organic structures, not by distorting molecules from their usual structural parameters. This allows simplifications to be made.

First, the cubic terms in bond stretching and angle bending are rarely included. In fact, some biopolymer force fields keep all bond lengths and bond angles at fixed, standard values; only dihedral angles and nonbonded contacts matter.

Second, the **united atom** or **extended atom** approach is quite common. There are a large number of C–H, N–H, and O–H bonds in proteins and nucleic acids, and varying their bonds lengths and bond angles is usually unimportant. In many instances, they are just steric placeholders. Thus, it is reasonable to remove them completely. For example, a CH_2 becomes a single, united atom—a sphere with a van der Waals radius much larger than a normal C. If every CH_3, CH_2, and CH is replaced by single, united atoms (with different kinds of united atoms for methyl, methylene, and methine), the number of bond stretching, angle bending, and, most importantly, nonbonding terms that must be evaluated drops substantially. This united atom approach is not a terrible approximation for carbons, although usually the sphere is centered at the carbon, while it should be offset toward the hydrogens somewhat. It is less attractive for NH and OH centers, and most force fields do not make this approximation.

The electrostatic and/or hydrogen bonding terms are especially important in force fields focused on biopolymers, because of the crucial role of hydrogen bonds and ion pair interactions in these systems. Unfortunately, these are often the most controversial, and least tested aspects of a force field.

In general, because of these and perhaps other approximations, force fields used for biopolymers are often considerably less accurate than the small molecule force fields. Since many of these methods come as part of larger modeling packages, which also include a number of visualization and analysis tools, it is often difficult to determine which approximations are being made at any one time, so caution is in order. Nevertheless, when used properly, valuable results can be obtained.

2.6.4   Molecular Mechanics Studies of Reactions

Since it is a non-quantum mechanical method, molecular mechanics is not intrinsically well suited to treating reaction mechanisms other than "reactions" that are simply conformational changes. That is, it would be completely unreasonable to study a bond-breaking process using a standard molecular mechanics package, because the method was not at all parameterized to treat bond-broken structures. Similarly, we might expect that an insufficient data base would exist to allow the development of reliable molecular mechanics parameters for reactive intermediates. Nevertheless, in some specific cases the method has been applied successfully to the evaluation of reaction mechanisms.

The first successes came with carbocation rearrangements. Schleyer and co-workers have studied multistep rearrangements of polycyclic hydrocarbons under strong acid conditions. For example, exposure of a hydrocarbon to excess $AlBr_3$ leads to reversible hydride abstractions such that a carbocation can be formed at essentially any carbon. The cations can then undergo [1,2]-carbon shifts. Since these are equilibrating conditions, thermodynamic predictions can be of value, and it was reasoned that perhaps the relative stabilities of neutral hydrocarbons that could be formed would make it possible to predict whether they could be involved in a rearrangement path. A classic example is shown in Eq. 2.50. Strong acid can isomerize the readily available tetrahydrodicyclopentadiene to adamantane, a ring system that is difficult to prepare by a conventional route. By assuming that hydride abstraction was possible from any carbon, and that all [1,2]-shifts were possible, molecular mechanics was used to evaluate the stabilities of potential intermediates. In this way, a path that was progressively downhill thermodynamically was developed, as shown in Eq. 2.50. The high speed of the molecular mechanics method was essential here.

$$\Delta H_f^\circ \quad -12.3 \qquad\qquad -10.9 \qquad\qquad -16.7 \qquad\qquad -20.2 \qquad\qquad -21.1 \qquad\qquad -32.6 \qquad\qquad\qquad\qquad\text{(Eq. 2.50)}$$

Tetrahydro-
dicyclopentadiene   Adamantane

These rearrangements are not solely of academic interest. The facile synthesis of the adamantyl ring system made possible the development of 1-aminoadamantane, known also as Symmetrel®, which has been used for the treatment of influenza A virus and perhaps Parkinson's disease.

Another interesting strategy for applying the molecular mechanics approach to reaction mechanisms has been proposed by Houk and co-workers. The reaction considered is simple radical addition to an olefin (Eq. 2.51). Using quantum mechanical computational methods of the kind described in Chapter 14, the detailed structures and energetics of the transition states for the addition of simple radicals, such as methyl and ethyl, to prototype olefins were characterized. From this information, a set of molecular mechanics parameters for the *transition state* of the reaction was developed. These parameters were then merged with a force field for conventional molecules. This made it possible to predict the relative energies of transition states for a series of radical addition reactions. The most interesting cases are cyclization reactions, in which R• and the olefin are part of the same molecule (Eq. 2.52). Such reactions can build in ring strain in the transition state, and molecular mechanics is quite good at predicting ring strain. Once a reasonable model of the transition state was in place, the relative strains of various cyclization transition states could be evaluated, allowing successful prediction of relative reaction rates. Such a merger of quantum mechanical computational methods with molecular mechanics is likely to see increasing use in coming years.

$$R\cdot \;+\; H_2C=CH_2 \longrightarrow \; R\diagdown\!\!\diagup\cdot \qquad\qquad\text{(Eq. 2.51)}$$

$$\qquad\qquad\qquad\qquad\qquad\text{(Eq. 2.52)}$$

Summary and Outlook

Topics covered in this chapter include thermochemistry, strain, stability, potential surfaces and functions, vibrational states, conformational analysis, and molecular mechanics. The unifying theme for all of these is structure and energetics. Let's briefly review some of the key lessons. The bonding forces that hold molecules together dictate their molecular structures and energetics. In that regard, a sum of the bond strengths of one molecule relative to another gives a good estimate of relative stability, and therefore a table of BDEs makes a good reference when predicting stabilities. Another excellent method is the group increments approach. The group increments method provides a way to estimate ΔH_f° values for a wide range of structures, and it leads to a quantitative definition of strain energy. It is important to keep clear the two different uses of the group increment method. If the goal is to obtain the best possible estimate for ΔH_f°, then all corrections—gauche, cis olefin, ring, etc.—are applied as appropriate. On the other hand, if the goal is to determine a strain energy, these corrections are not made. Just the basic group increments are combined, and the sum is subtracted from the true ΔH_f° to obtain a strain energy.

For all the basic classes of reactive intermediates, thermodynamic data that allow valuable comparisons of relative stabilities are available: BDEs for radicals, HIAs for cations, and pK_a values for anions. The trends in relative stabilities of reactive intermediates are generally well treated by the bonding model developed in Chapter 1.

We also covered the fact that molecular structures are dynamic, not static. Multiple degrees of vibrational freedom exist—namely, stretches, bends, torsions, etc. Each is quantized and the motions are constrained by a potential surface. For most organic chemistry purposes, only the quantization of bond stretches becomes relevant. The torsional degrees of freedom, when possessing enough energy, lead to the interconversion of conformers, both in acyclic and cyclic systems. The study of these interconversions is called conformational analysis.

Lastly, we showed that structure and energetics can be calculated using a method called molecular mechanics. The equations and force constants used in this method are similar to, but not identical to, those in the potential functions that describe real molecular vibrations and structure. The method, when properly parameterized, can predict structure, give strain energies, and calculate heats of formation.

Given these lessons about structure and energetics, we can now turn our attention to multiple topics in advanced organic chemistry. In the next few chapters we examine forces that hold pairs or ensembles of molecules together—the intermolecular forces involved in solvation and molecular recognition. Predicting solvation and binding phenomena relies on the same principles for intermolecular bonding—an examination of the enthalpy and entropy of the particular interaction. Entropy will play a much larger role in these chapters than it has in this chapter. We then turn our attention to acid–base chemistry, where thermodynamics is of paramount importance, and many of the lessons presented here will be recalled. After that, stereochemistry is covered, where the insights into molecular structure and conformational analysis given here will be essential to a complete understanding of stereochemical principles. Taking an even longer look forward, we will see in the chapters on kinetics and mechanisms that the interconversion of one molecule into another, a chemical reaction, is in fact the excitation and coupling of the kinds of vibrational modes discussed herein. Hence, this chapter is one that has very "long legs", being essential for many of the future chapters in this book.

Exercises

1. Estimate the percentage of the twist-boat conformation of cyclohexane present at 25 °C, assuming that the entropies of the chair and twist-boat are identical.

2. Sketch a complete (360°) torsional itinerary for toluene.

3. Describe the hybridization at the central carbon of di(*t*-butyl)methane (Figure 2.21) relative to the central carbon of propane.

4. Other strategies for determining the energetic consequences of aromaticity in benzene have been advanced. For example, one criticism of the analysis given in the chapter is that benzene contains only sp^2–sp^2 bonds, but in the cyclohexene reference the olefinic carbons are attached to sp^3 carbons. One possible solution would be to use CH groups from 1,3-butadiene ($\Delta H_f^\circ = 26.3$ kcal/mol) as a reference. Derive an aromaticity value for benzene using this approach, compare it to the value determined in the text, and comment on which seems more appropriate.

5. The C=C double bond of *trans*-cyclooctene is relatively short: 1.33 Å vs. 1.347 Å for *trans*-2-butene. Provide a rationalization for this.

6. Bulky substituents prefer the equatorial to the axial position in cyclohexane. Nevertheless, the equilibrium shown lies to the right. Provide an explanation for this.

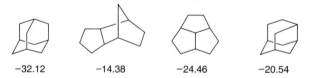

7. Suggest several different thermochemical strategies to evaluate the aromaticity of naphthalene. Comment on any differences among the values you obtain, and also on the degree of aromaticity in naphthalene vs. that in benzene. Some potentially useful data are given below; other useful data may be in the text or could be calculated by group increments.

	ΔH_f° (kcal/mol)
Naphthalene	36
trans-Decalin	−43.5
Tetralin	6.22
Butadiene	26.3

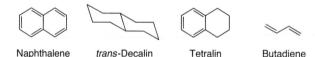

| Naphthalene | *trans*-Decalin | Tetralin | Butadiene |

8. Adamantane has been described as a "thermodynamic sink" because it is the most stable of all the $C_{10}H_{16}$ molecules. Shown below are adamantane (far left) and several isomers with their molecular mechanics computed heats of formation (in kcal/mol). Calculate the strain energy of each. Are your results consistent with the thermodynamic sink notion?

$$-32.12 \qquad -14.38 \qquad -24.46 \qquad -20.54$$

9. Explain why the cyclohexane derivative shown prefers the conformation with the methyl group axial (**B**) rather than equatorial (**A**).

A. ⇌ **B.**

10. Predict which isomer is preferred, and briefly explain why.

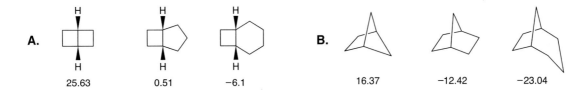

A.   **B.**

11. Sketch a Newman projection of what you think would be the preferred conformation of hydrazine (H_2NNH_2), and briefly explain your choice.

12. Series **A** and **B** can be viewed as two different ways to annulate an ethano bridge onto four-, five-, and six-membered rings. Using the ΔH_f° values given (in kcal/mol), calculate the strain energy for each compound. Briefly discuss whether the trend seen in each series is consistent with expectations based on additivity of ring strain.

A.

| 25.63 | 0.51 | −6.1 |

B.

| 16.37 | −12.42 | −23.04 |

13. Bicyclopentyl shows a strong preference for the conformation with the highlighted hydrogens anti. Rationalize this result within the context of the conformational preferences of other, similarly substituted alkanes.

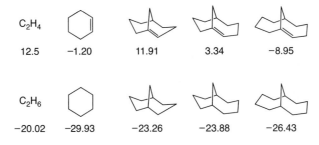

14. The experimental ΔH_f° value for C_{60} (see Figure 13.10 for a picture) is 634.8 kcal/mol. Based on this, is C_{60} better thought of as an aromatic molecule or a collection of C–C double bonds? The C_{BF}–$(C_{BF})_3$ group increment, with a value of 1.5, will be of use. A C_{BF} is a fused benezoid carbon, and C_{BF}–$(C_{BF})_3$ would therefore apply to the C's in graphite or C_{60}.

15. Olefin strain energy has been defined as the difference between the strain energies of an olefin and the corresponding saturated hydrocarbon. Generally, the olefin is more strained than the alkane. Given the experimental heats of formation (in kcal/mol) below, calculate the olefin strain for the olefins shown. Comment briefly on the implications of your findings. For some systems, the olefin is actually *less* strained than the alkane. These have been termed **hyperstable olefins**. Are any of the olefins in this set hyperstable?

C_2H_4

| 12.5 | −1.20 | 11.91 | 3.34 | −8.95 |

C_2H_6

| −20.02 | −29.93 | −23.26 | −23.88 | −26.43 |

16. Usually cyclohexane A values are reported as ΔG° values, but for some substituents, ΔH° and ΔS° values are also available. Such values are shown below for methyl and isopropyl (equatorial to axial interconversion). Consider the two chair conformers of *cis*-1-methyl-4-isopropylcyclohexane. Calculate the percentage of each form present at (a) 300 K, (b) 100 K, and (c) 75 K.

	ΔH° (kcal/mol)	ΔS° (cal/mol•K)
CH$_3$	1.75	0
CH(CH$_3$)$_2$	1.52	−2.31

17. Shown below are the values of ΔH_f° (in kcal/mol) for hexamethylbenzene, hexamethyl Dewar benzene, and hexamethylprismane. Determine the strain energy for each compound. You will need to make estimates for some group increments. Justify your choices for estimates.

−24.0	25.5	67.2

18. Draw a reasonable representation of the three-dimensional shape of cholic acid (see the Connections highlight on page 108). Comment on the relationship among the three hydroxyls and speculate how this may be important in the biological activity of the molecule.

19. If you are a physical organic chemist interested in conformational analysis, one thing you might want to do is design a molecule that locks a cyclohexane into the boat form. Consider a molecule in which a one-carbon bridge links the "flagpole" positions of a boat cyclohexane. What do you think the strain energy of such a molecule would be? (*Hint:* The answer is given in this chapter.)

20. Predict the preferred conformation of fluoromethanol, FCH_2OH, around the C–O bond and briefly rationalize your choice.

21. Consider a hypothetical explosive process in which one mole of octanitrocubane is converted to $8\ CO_2$ and $4\ N_2$. Estimate how much energy would be liberated. Useful heat of formation data: $\Delta H_f^\circ = -94.05$ kcal/mol for CO_2; $\Delta H_f^\circ = -19.3$ kcal/mol for CH_3NO_2. (*Hint:* You will need to make some approximations.)

22. How many grams of octanitrocubane would be needed in the process in Exercise 21 to heat 1 liter of water from 25 °C to 50 °C. (*Hint:* You need the heat capacity of water.)

23. Use the group increments of Table 2.4 to substantiate or refute the following statements:
(a) Branched alkanes are more stable than linear alkanes.
(b) For alkenes in a linear chain, an internal double bond is more stable than a terminal double bond.
(c) Hydrogenation of olefins is generally more exothermic than hydrogenation of analogous carbonyls.

24. Tetraalkylethanes are gauche, and the gauche preference increases as the alkyl group gets larger. However, 1,1,2,2-tetraphenylethane is anti. Suggest why this might be so.

25. How do you rationalize the fact that the cyclohexane A value for phenyl (2.8) is bigger than that for isopropyl (2.21)?

26. Provide an explanation for why cyclopentene is *less* strained than cyclopentane (Figure 2.15).

27. Show how a blind implementation of a cubic bond stretching term (Eq. 2.42) can lead to the **long bond catastrophe**, in which the force field fails for very long bonds. How would you fix this problem?

28. Explain why molecular mechanics is appropriate to calculate the differences in dipole moments between two similar molecules, but is inappropriate to calculate the differences in polarizability.

29. A molecular mechanics calculation with a particular force field gets the heat of formation of spirononane correct, but not for [5.3.5.3]fenestrane. Why?

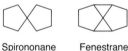

Spirononane   Fenestrane

30. Consider the various bicyclic alkanes for which strain energies are given in Figure 2.15. For which one will homolytic cleavage of a single C–C bond release the largest amount of strain? (*Hint:* Consider the strain energy of the ring-opened biradical to be equivalent to that of the analogous cycloalkane.)

31. X-ray crystallography reveals that tetracyclohexyldiphosphine (Cy_2P–PCy_2, where Cy = cyclohexyl) adopts a gauche-like conformation, with the dihedral angle between the P lone pairs being ~90°. This has been rationalized by invoking electronic interactions involving the P lone pairs (the gauche effect). Suggest an alternative explanation. [*Hint:* Tetracyclohexyldisilane ($Cy_2SiHSiHCy_2$) adopts a very similar conformation.]

32. The Benson group increments that we discuss herein (Table 2.4) are derived from $\Delta H_f°$ data of the sort in Table 2.3. However, a little thought will convince you that, for example, n-octane at 298 K will have a significant proportion of gauche conformers. This in a sense compromises the group increment approach developed above. Using a gauche correction of 0.9 kcal/mol, and ignoring any non-additive effects (such as the g+g– pentane interaction), what fraction of n-octane molecules will have one or more gauche bonds at 298 K in the gas phase? An alternative set of group increments has been developed that takes this effect into account, such that the C–(H)$_2$(C)$_2$ group increment more correctly reflects what one would expect for an all-trans alkane. While these alternative group increments do give better agreement with experiment, they have only been developed for the simplest hydrocarbon groups. As such, the Benson group increments, which are more broadly applicable, are the most commonly used. Predict whether the value for C–(H)$_2$(C)$_2$ in this modified type of group increment should be less than or greater than –4.93.

33. Use the thermochemical data of the various tables in this chapter to predict the bond dissociation energy for the O–H bond of ethanol. Does your value make sense, considering the comparable value for methanol given in Table 2.1?

34. Rationalize the following trends.

R	s-trans	s-cis
CH$_3$	0.7	0.3
CH$_2$CH$_3$	0.55	0.45
iPr	0.3	0.7
t-Bu	near 0	~1

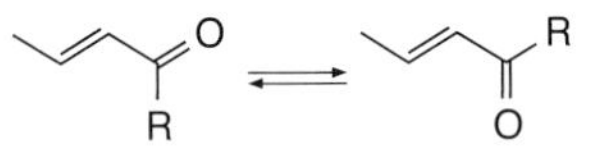

35. Predict the most stable conformation of the following molecule, called Kemp's triacid.

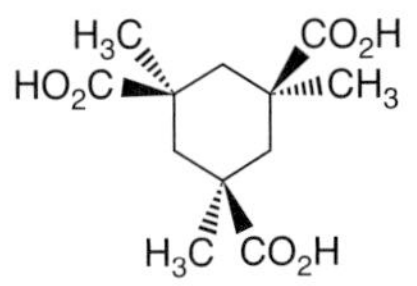

36. Calculate the $\Delta H_{rxn}°$ for the following reaction using group increments, and then using BDEs for C–C bonds. Ignore changes in the strengths of the C–H bonds. How do the values compare?

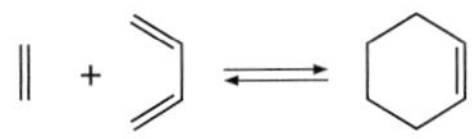

37. Draw on the same plot your predictions as to the relative shapes of the potential surfaces that describe the angle bending modes shown in the following molecules. On the x axis, place the minima of the plots (the preferred bond angles of approximately 109.5°, 120°, and 180°, respectively) all at the same place so that they all overlay at the energy minima. Explain your answers.

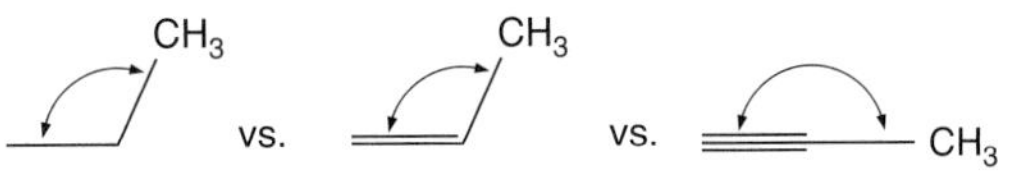

38. Draw on the same plot your predictions as to the relative shape of the Morse potentials for C–C, C–O, and C–F bonds. Explain your answers.

39. In the Going Deeper highlight entitled "How Do We Know That $n = 0$ is Most Relevant for Bond Stretches at $T = 298$ K?" (page 76), we calculated the energy gap between the $n = 0$ and $n = 1$ vibrational states for a C–C bond (modeled by a harmonic oscillator) with a force constant of 4.5 mdyne/Å. The answer was 3.22 kcal/mol. Confirm this number with your own calculation (1 dyne = g cm/s^2). Does this energy gap properly correspond to the IR stretching frequencies of common C–C bonds?

40. Given that the A value for a methyl group on cyclohexane is 1.8 kcal/mol, draw the potential surface for the ring interconversion that takes the methyl group from an equatorial to an axial position. What is your prediction as to the relative energies for the different twist-boat conformations, and how does this affect your prediction as to the lowest energy pathway for the ring interconversions? (*Hint:* Molecular models will be helpful for this problem.)

41. We stated in this chapter that many heats of formation are known due to the fact that many heats of combustion have been measured by the petroleum industry. Given that the heat of combustion of 1-butene is –649.5 kcal/mol, calculate its heat of formation. (*Hint:* The heats of formation of CO_2 and H_2O are –94.05 and –68.32 kcal/mol, respectively.) How does this compare to what you get using group increments?

42. After examination of Figure 2.17, what would you calculate as a proper group increment correction for the placement of two *t*-butyl groups cis in an alkene? Assume no strain correction needed for the alkane you need to consider to answer the question.

43. Draw a potential energy surface for rotation along the Cp–Mn (Cp = cyclopentadienyl) vector in the following organometallic complex. Do you expect the barrier to rotation to be large?

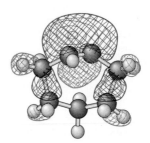

44. The **Thorp–Ingold effect** (also called the **gem–dimethyl effect**) is an effect on reactivity in cyclizations due to geminal methyl groups. Draw the differing conformations of 3,3-dimethylpentane, and predict the lowest energy conformation. Use the terms syn, anti, clinal, and periplanar to define the relationships between the main chain carbons. Make a guess as to what the Thorp–Ingold effect is.

45. Acetaldehyde shows the same conformational preference for the eclipsed over the staggered form as does propene, but the magnitude of the preference is reduced to 1.2 kcal/mol, vs. 2.0 kcal/mol for propene. Provide a rationalization for this observation.

46. A tertiary radical such as $(CH_3)_3C\bullet$ is more stable than an analogous primary radical such as $CH_3CH_2CH_2CH_2\bullet$. Yet, in Table 2.7, the $[\bullet C-(C)(H)_2]$ group increment is more stabilizing (actually, less destabilizing) than the $[\bullet C-(C)_3]$ group increment. Explain this.

47. In Chapter 1 we discussed hybridization extensively, including how distortions from normal bonding arrangements imply altered hybridizations. Shown below is the HOMO of *trans*-cycloheptene (see Figure 2.16 for other views of this molecule). Focusing only on the π bond, discuss how the shape of this MO does or does not reflect the expected hybridization changes for this strained molecule.

48. Instead of a hyperconjugative effect, some have described the stabilization of cations by alkyl substitution as due to the generic electron donating ability of alkyl groups. In this light, discuss the HIA value for 2-methylallyl shown.

HIA = 248

49. Using group increments and HIA data, provide a best estimate of the relative stabilities of 3-phenyl-2-butyl cation and the corresponding phenonium ion.

Phenonium ion

50. Use your best estimates of pK_a values, ring strains, and bond strengths to determine whether the equilibrium shown should indeed favor the alkoxycyclopropane as shown.

51. One of the most useful free radical rearrangements is the ring closure of the 5-hexenyl radical to form cyclopentylcarbinyl rather than cyclohexyl radical. Use group increments to estimate the relative energies of the 5-hexenyl, cyclopentylcarbinyl, and cyclohexyl radicals.

| 5-Hexenyl | Cyclopentylcarbinyl | Cyclohexyl |

52. We noted in the chapter that the BDE values for the C–Cl bonds in methyl chloride, ethyl chloride, isopropyl chloride, and *t*-butyl chloride are 84.1, 84.2, 85.0, and 83.0 kcal/mol, respectively. There are other factors that make the trend in BDEs a poor measure of radical stability. Discuss what these factors could be.

Further Reading

Strain and Strain Energies

The primary source of strain energies for this chapter is: Wiberg, K. B. "The Concept of Strain in Organic Chemistry." *Angew. Chem. Int. Ed. Eng.*, **25**, 312–322 (1986).

Another useful source of data, along with an extensive discussion of the topic, is: Greenberg, A., and Liebman, J. F. (1978). *Strained Organic Molecules. Organic Chemistry*, Vol. 38, Academic Press, New York.

Conformational Analysis and Related Topics

Eliel, E. L., Wilen, S. H., and Mander, L. N. (1994). *Stereochemistry of Organic Compounds*, John Wiley & Sons, New York.

An Overview of the Various Attempts to Analyze the Steric Size of Substituents

Forster, H., and Vogtle, F. "Steric Interactions in Organic Chemistry: Spatial Requirements of Substituents." *Angew. Chem. Int. Ed. Eng.*, **16**, 429–441 (1977).

The Complete Guide to Group Increments

Benson, S. W. (1976). *Thermochemical Kinetics: Methods for the Estimation of Thermochemical Data and Rate Parameters,* 2d ed., John Wiley & Sons, New York.

Molecular Mechanics

Burkert, U., and Allinger, N. L. (1982). *Molecular Mechanics*. ACS monograph, 177, American Chemical Society, Washington, D.C.

Osawa, E., and Musso, H. "Application of Molecular Mechanics Calculations to Organic Chemistry." *Top. Stereochem.*, **13**, 117 (1982).

Osawa, E., and Musso, H. "Molecular Mechanics Calculations in Organic Chemistry: Examples of the Usefulness of This Simple Non-Quantum Mechanical Model." *Angew. Chem. Int. Ed. Eng.*, **22**, 1 (1983).

Molecular Mechanics and Carbocation Rearrangements

Saunders, M., Chandrasekhar, J., and Schleyer, P. v. R. "Rearrangements of Carbocations" in *Rearrangements in Ground and Excited States*, P. de Mayo (ed.), Academic Press, New York, 1980, Vol. 1, Chapter 1, pp. 1–53.

The Antiaromaticity of Cyclobutadiene

Deniz, A. D., Peters, K. S., and Snyder, G. J. "Experimental Determination of the Antiaromaticity of Cyclobutadiene." *Science*, **286**, 1119–1122 (1999).

CHAPTER 3

Solutions and Non-Covalent Binding Forces

Intent and Purpose

The first goal of this chapter is to examine how molecular properties manifest themselves in the properties of condensed phases. The forces that hold molecules together in solutions and solids derive from the individual molecules that make up the aggregate. Several solvent scales for determining polarity and internal cohesion are presented. Next, we focus the discussion on the properties of solutes (entities dissolved) in solutions, including information on diffusion. Our goal is to set the stage for examining reactions that take place in solution. Therefore, a discussion of the thermodynamics of solutions and the driving force for reactions in solutions is given. The solvation forces for solutes are much the same forces that constitute solute–solute interactions. Hence, after examining solvation, we explore binding forces as a lead into the next chapter on molecular recognition and supramolecular chemistry. Chapters 3 and 4 will set the stage for Chapter 9 on catalysis, which will rely heavily upon a discussion of binding forces. We can discuss the binding forces involved in solvation, molecular recognition, and supramolecular chemistry, without examining kinetics and mechanisms, because we are concerned with systems that are under thermodynamic control. Finally, this chapter ends with an examination of modern computational methods for modeling solvation. Our intent is to give the student a sufficient background in the properties of solutions to rationally design experiments that probe reaction mechanisms and molecular recognition phenomena.

3.1 Solvent and Solution Properties

In Chapters 1 and 2 we covered molecular polarizabilities, dipoles, and conformations. We are now ready to explore how these properties dictate the properties of solvents, the interactions of solutes with the solvent, and the interactions between solutes. Since the vast majority of reactions performed by organic chemists occurs in solution, the choice of solvent can play an extremely important role in controlling the reactions. We need to choose solvents that not only solubilize the reactants, but also accelerate the desired reaction and/or impede undesirable reactions. Moreover, we can change the solvent to probe reaction mechanisms and look for the existence of various intermediates (see Grunwald–Winstein scales in Chapter 8). Finally, the interactions between the molecules of a solvent, and the interactions between solvent and solute, are some of the same interactions that occur between enzyme and substrate, antibody and antigen, and synthetic receptors and various target molecules—all topics of the next chapter.

Molecules "stick" together using combinations of forces that chemists have categorized as follows: ion pairs, dipole—dipole, dipole–induced-dipole, hydrogen bonding, van der Waals/London dispersion forces, solvophobic forces, Lewis acid–base interactions, metal coordination, and charge–transfer interactions. Each of these interactions is covered in various places in this book. As with many definitions and classifications used in chemistry, there is considerable overlap with some of these terms, and often molecules stick together using combinations of these interactions. Most common solvents interact with other solvent molecules or solutes using dipole–dipole, hydrogen bonding, and London dispersion forces. All three topics are discussed later in this chapter.

Before exploring the forces that cause solvents to stick together, it is instructive to give a general picture of the structure of liquids. **Liquids** are best described by a state of rapidly changing molecular order, which retains a high degree of cohesive interactions between the molecules.

3.1.1   Nature Abhors a Vacuum

As with all chemical phenomena, enthalpy and entropy determine the free energy of the system and hence the system's structure. The weak binding interactions that hold solvents together are all related to enthalpy, and in general they lower the free energy of the liquid state due to negative enthalpy contributions. Yet, entropy has a very large influence on solvent structure also. The entropy of most solvents is relatively large and positive compared to the solid state. This large entropy is due to the substantial freedom of movement of the solvent molecules relative to molecules in a crystal lattice.

Liquids prefer not to have empty spaces, leading to the common dictum, "Nature abhors a vacuum". The creation of a bubble in a solvent is very costly, because there are fewer configurations for the entire ensemble of molecules to adopt. As such, the tendency of liquids to fill space is fundamentally an entropy effect. Enthalpy is also significant, because bubbles increase the surface area at the expense of intermolecular attractive forces. Yet, in some cases enthalpy can become more favorable with a more open structure, such as ice relative to liquid water.

Liquids have structures in between gases (complete randomness) and crystals (highly ordered). The average location of the individual molecules in a solvent is expressed in terms of a radial distribution function, $g(r)$. This function relates the probability of finding another molecule at a particular distance r from each molecule. Figure 3.1 shows a schematic representation of $g(r)$ for a liquid and a perfect crystal. There are definite distances separating each molecule in the crystal, and hence there are predictable and reproducible distances at which each molecule in the crystal will be found relative to each other molecule. These repetitive distances are what lead to the diffraction of x rays in single-crystal crystallography. This repetitive nature is referred to as **long range order**. Such a high degree of order is not found in a liquid. There is a good probability of finding a layer of nearest neighbor solvent molecules around each individual solvent molecule, but the distances to the molecules in the second, third, etc., layers becomes less certain. This drop off in repetitiveness is called **short range order**.

The forces that hold liquids together are the same as those that hold molecular solids together. However, on raising the temperature of a system, these forces become less able to compete with thermal energy, and so we transition from a system with long range order to one with only short range order. We will discuss these intermolecular forces in considerable detail in this chapter. However, first we consider efforts to characterize solvents on a more macroscopic scale, emphasizing the bulk properties of the liquid.

3.1.2   Solvent Scales

Each of the binding forces that hold solvent molecules together plays a role in determining the bulk properties of the solvent. By bulk properties, we are not referring to the microscopic interactions between the individual solvent molecules, but instead to the properties that the solvent displays as a whole. For example, boiling points and melting points, the solubilizing behavior to solutes, surface tension, and refractive index are all bulk solution properties.

Solvents can be classified as protic or aprotic, and as polar or nonpolar. A **protic solvent** has a hydrogen atom attached to a heteroatom, such as O, N, or S, and can form hydrogen bonds with a solute molecule as well as with other solvent molecules. An **aprotic solvent** lacks a hydrogen on a heteroatom, and therefore cannot act as a donor.

Creating a definition of a polar solvent is a more difficult task. Phenomenologically, a **polar solvent** can be described as a solvent that can solubilize salts or molecules with large

A.

Distance to first
shell of molecules

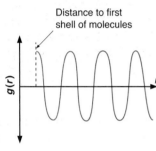

B.

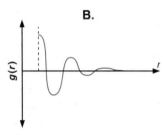

Figure 3.1
A. Schematic representation of the radial distribution function $g(r)$ for a typical solid.
B. Schematic representation of $g(r)$ for a typical liquid. After a few solvent spheres, there is no longer any spatial correlation to another solvent molecule. The origin on the y axis represents a 50% chance of finding another solvent molecule.

permanent dipoles, while a **nonpolar solvent** is one that does not. There are shades of gray to this definition, because certain organic ions can be solubilized in very nonpolar solvents, and not all polar solvents dissolve all common salts or molecules with large permanent dipoles. Solvents whose individual molecules have large dipole moments are often quite polar. When no hydrogen bond donor is present, they are called **dipolar aprotic** solvents, and include *N,N*-dimethylformamide (DMF), dimethylsulfoxide (DMSO), and hexamethylphosphoramide (HMPA). Protic solvents are also often quite polar, being able to solubilize many salts via hydrogen bonding. Lastly, although CCl_4 and liquid Xe are certainly not considered polar, they are often good solvents because they are quite polarizable.

Dielectric Constant

Most often chemists examine the **dielectric constant** (ε) of a solvent to determine whether it is polar or nonpolar (Table 3.1), with higher ε values reflecting greater polarity. The dielectric constant is a bulk property, measured by determining the effect of an intervening solvent on the electric field between two oppositely charged plates. The capacitance on the plates is measured, telling the extent to which the solvent screens the opposite charges on the plates from feeling each other. The electric field generated by the charges on the plates orients the solvent molecules to oppose the applied field. Large molecular dipoles, large molecular polarizabilities, and hydrogen bonding sites on the solvent molecules combine to give large dielectric constants, and hence the ε values correlate with our definition of polarity.

Table 3.1
Various Solvent Scales*

Solvent	ε	Z	$E_T(30)$	π^*	α	β
Formamide	111	83	57	0.97	0.71	0.48
Water	78	95	63	1.1	1.17	0.47
DMSO	47	71	45	1.0	0.00	0.76
DMF	37	69	44	1.0	0.00	0.76
Acetonitrile	36	71	46	0.75	0.19	0.40
Methanol	33	84	55	0.60	0.93	0.66
HMPA	29	63	41	0.87	0.00	1.05
Ethanol	25	80	52	0.54	0.83	0.75
Acetone	21	66	42	0.71	0.08	0.43
Isopropanol	20	76	48	0.48	0.76	0.84
t-Butyl alcohol	12	71	43	0.41	0.42	0.93
Pyridine	13	64	40	0.87	0.00	0.64
Methylene chloride	9	64	41	0.82	0.13	0.10
THF	8		37	0.58	0.00	0.55
Acetic acid	6	79	52	0.64	1.12	0.45
Ethyl acetate	6		38	0.55	0.00	0.45
Chloroform	5		35	0.27	0.20	0.10
Diethyl ether	4		34	0.27	0.00	0.47
Benzene	2	54	34	0.59	0.00	0.10
Carbon tetrachloride	2		32	0.28	0.00	0.10
n-Hexane	2		31	−0.04	0.00	0.00

*Data taken from the following sources: Riddick, J. A., Bunger, W. B., and Sakano, T. K. (1986). *Organic Solvents; Physical Properties and Methods of Purification*, 4th ed. (Techniques of Chemistry, Vol. II), Wiley–Interscience, New York. Kosower, E. M. (1968). *An Introduction to Physical Organic Chemistry*, John Wiley and Sons, Inc., New York. Kosower, E. M. "The Effect of Solvent on Spectra. I. A New Empirical Measure of Solvent Polarity: Z-Values." *J. Am. Chem. Soc.*, **80**, 3253 (1958). Reichardt, C. (1988). *Solvents and Solvent Effects in Organic Chemistry*, 2nd ed., VCH, Weinheim. Kamlet, M. J., Abboud, J.-L. M., Abraham, M. H., and Taft, R. W. "Linear Solvation Energy Relationship. 23. A Comprehensive Collection of Solvatochromic Parameters, TM*, K, and 2, and Some Method for Simplifying the Generalized Solvatochromic Equation." *J. Org. Chem.*, **48**, 2877 (1983).

Throughout this chapter the ε parameter will be used in various equations that describe binding forces (such as Eq. 3.1, below). Mathematically, it is defined as the ratio of the permittivity of the medium (ε_μ) to the permittivity of a vacuum (ε_o). Hence, $\varepsilon = \varepsilon_\mu / \varepsilon_o$. Therefore, it is a dimensionless parameter, which is often referred to as the **relative permittivity** (also known as the dielectric constant).

The dielectric constant gives insight into how well the solvent screens electrostatic forces. Solvents with high dielectric constants more effectively screen the attractive or repulsive forces between ions and the ends of dipoles. The partial charges on the polar solvent molecules interact with and diminish the effective charges on solutes and hence diminish the attractive or repulsive forces between charges on solutes.

The solvent with the highest dielectric constant is formamide, with water running second. Formamide has a large dipole, has hydrogen bonding capabilities, and is more polarizable than water. These three factors combine to give formamide the highest dielectric constant. Comparing water and methanol reveals a significant difference, indicating a significant decrease in polarity caused by replacing a single hydrogen of water with even the smallest organic fragment (methyl). Completely organic structures such as benzene and carbon tetrachloride have very little ability to mediate the forces between charges and so are nonpolar solvents.

The screening effect manifests itself in the equations that describe the electrostatic energies between full and partial charges. As a first example, **Coulomb's law**, which describes the attractive or repulsive potential energy (E) between two charges q_1 and q_2 at a distance r (Eq. 3.1), has ε in the denominator. Thus, the larger the dielectric constant, the lower the interaction energy between the two charges. We will return to an analysis of this equation when ion pairs are discussed (Section 3.2.1).

$$E = \frac{q_1 q_2}{4 \pi \varepsilon \varepsilon_o r} \qquad \text{(Eq. 3.1)}$$

Other Solvent Scales

Many other scales have been developed to measure the polar nature of solvents and other specific properties (Table 3.1). These scales make for handy reference when choosing a solvent for a particular purpose. Most of the other scales are based upon the solvatochromism of the solvent. **Solvatochromism** is the change in shape, intensity, and/or position of the UV/vis or emission spectrum of a chromophore or fluorophore induced by the solvent. The most extensively used scales are the Z scale and the $E_T(30)$ scale.

The Z scale is based upon the spectrum of N-ethyl-4-methylcarboxypyridinium iodide (Eq. 3.2). On excitation, this ion undergoes a charge–transfer transition to form the neutral radical species shown. The excited state thus has a much smaller dipole than the ground state. In a polar solvent, the ground state is therefore preferentially stabilized relative to the excited state, and the energy of the light required for the excitation increases (shorter wavelength). The Z parameters are correlated to the $\lambda_{\max}$ (nm) for excitation via Eq. 3.3. This parameter finds water the most polar solvent, with formamide similar to methanol.

$$\text{(Eq. 3.2)}$$

$$Z = 2.859 \times 10^4 / \lambda_{\max} \qquad \text{(Eq. 3.3)}$$

The $E_T(30)$ scale is based upon the spectrum of the pyridinium betaine shown in Eq. 3.4, which upon excitation leads to a less polar excited state due to a charge redistribution. Again, more polar solvents lead to a higher energy excitation (lower $\lambda_{\max}$). One limitation is that the presence of any acids that can protonate the phenoxide of the betaine negate the activity. Similar to the Z scale, the $E_T(30)$ scale lists water as the most polar.

$$\text{(Eq. 3.4)}$$

A scale known as π^* is based upon several different dyes, not just one as with the Z and $E_T(30)$ scales, and gives a good measure of the extent to which the solvent stabilizes ionic or polar species. The scale is best viewed as a measure of non-specific electrostatic solvation. Once again water wins, but formamide, DMSO, and DMF all run a close second.

Finally, scales to determine the hydrogen bonding ability of a solvent have also been developed. The α scale is a measure of the solvent's ability to act as a hydrogen bond donor to a solute, while the β scale is a measure of the solvent's ability to act as a hydrogen bond acceptor from a solute. The acceptor and donor ability can be correlated to other similar non-hydrogen bonding interactions. The α scale derives from a measurement of the UV/vis spectrum of 4-nitroaniline, which is sensitive to hydrogen bond donation from the NH_2 group. The β scale is much more complex, being derived from studies of a number of dyes in protic solvents, subtracting away effects of polarity and polarizability. Water is the best at hydrogen bond donation, with acetic acid a close second, but many solvents are better than water at accepting a hydrogen bond. The better hydrogen bond accepting solvents are those with strongly polarized bonds to oxygen, such as DMSO, DMF, and HMPA. Alcohols are also better than water at accepting a hydrogen bond. Ethyl acetate and diethyl ether are similar to water in hydrogen bond accepting ability.

The various solvent scales can be used to determine which property of a solvent has the greatest influence on reactivity or any other physical/chemical phenomena. An example of their use in a common reaction is given in the following Connections highlight, and we will also showcase their use in a Connections highlight concerned with the hydrophobic effect in the next chapter.

Connections

The Use of Solvent Scales to Direct Diels–Alder Reactions

The rates, regiochemistry, and stereochemistry of Diels–Alder reactions are affected by the solvent, and are often correlated to solvent polarity scales. In Chapter 15, we will cover orbital interactions that dictate the dominant regioisomers of Diels–Alder reactions similar to that given below. The diene A is considered to be a nucleophile and

the methyl vinyl ketone B an electrophile, and preferential orbital mixing gives the pseudo-para isomer predominately, an effect known as normal electronic demand.

For this particular reaction, the pseudo-para/meta regioselectivity did not correlate with the polarity scales ε, Z, or $E_T(30)$. However, a plot of log(para/meta) versus α, the hydrogen bond donor ability, was linear with increasing pseudo-para product for larger α values. The conclusion is that the electrophilic activation of methyl vinyl ketone by a hydrogen bond from the solvent reinforces the normal electronic demand, further accentuating the orbital interactions.

Cativiela, C., Garcia, J. I., Mayoral, J. A., and Salvatella, L. "Solvent Effects on endo/exo- and Regio-Selectivities of Diels–Alder Reactions of Carbonyl-Containing Dienophiles." *J. Chem. Soc., Perkin Trans.*, **2**, 847 (1994).

Heat of Vaporization

The **heat of vaporization** ($\Delta H_{vap}°$) of a solvent is the amount of energy required to vaporize the solvent per gram or mole of solvent at the boiling point. It is a direct measure of the energy required to overcome the attractive forces between the adjacent solvent molecules (Table 3.2). Water has the highest value for such a small molecule, indicating the greatest cohesive forces per surface area. Nonpolar solvents such as benzene and chloroform have quite low values until their surface area becomes large, as with decane.

Table 3.2
Heats of Vaporization of Some Common Solvents at 1.0 atm (cal/g) and δ Parameters*

Solvent	$\Delta H_{vap}°$	δ
Water	540	23.4
Methanol	263	14.3
Ethanol	204	12.7
Acetone	125	9.6
Benzene	94	9.2
Chloroform	59	
Methane	122	
Decane	575	

*Atkins, P. (1998). *Physical Chemistry,* 6th ed., W. H. Freeman and Company, New York. Abraham, M. H. "Solvent Effects on Transition States and Reaction Rates." *Prog. Phys. Org. Chem.,* **11**, 1 (1974).

Another informative solvent parameter that is similar to the heat of vaporization is the **cohesive energy density** (D). This energy is the mean potential energy of attraction between the solvent molecules within a given sample. In other words, it is the energy of cohesion per unit volume of solvent, and is defined by the molar heat of vaporization divided by molar volume ($D = \Delta H_{vap}°/V$). The cohesive energy density (D) of the solvent gives insight into how difficult it is to create a bubble of a given volume, such as an empty space that a solute would need to occupy. Therefore, D has been found to be related to the solubility of solutes, and **solubility parameters** (δ) are defined, where $D = \delta^{1/2}$ (Table 3.2).

Surface Tension and Wetting

Table 3.3
A Few Surface Tension Values (γ, mN/m)*

Solvent	γ
Water	72.8
Methanol	22.6
Benzene	28.9
Hexane	18.4
Mercury	472

*Atkins, P. (1988). *Physical Chemistry,* 6th ed., W. H. Freeman and Company, New York.

The **surface tension** is another measure of the internal cohesive forces within a solvent. All liquids tend to adopt shapes that minimize their surface area, because this leads to the maximum number of molecules in the bulk interacting with their neighbors. At the surface of a solution the solvent molecules cannot have the normal number of intermolecular interactions because these molecules are at an interface with air.

Table 3.3 lists the surface tensions (γ) of a few solvents. Solvents with a high surface tension require the greatest energy to increase their surface area, and will tend to minimize their exposed surface the most. Solvents with low cohesive forces will have a low surface tension and less of a driving force to minimize exposed surface area. Eq. 3.5 expresses this idea, where the incremental amount of work (energy, ∂w) that is needed to change the surface area of a solvent drop is equal to the surface tension times a incremental change in surface area ($\partial \sigma$). Mercury has an astounding surface tension of 472 relative to water's 73. Have you ever broken a mercury thermometer? The mercury metal beads up immediately on almost any surface, reflecting the very high surface tension.

$$\partial w = \gamma \partial \sigma \qquad \text{(Eq. 3.5)}$$

The ability of the solvent to adhere to a surface is called **wetting**. When there is sufficient attraction between the solvent molecules and the surface such that the solvent spreads over the surface and does not have a propensity to bead, we consider the surface wetted. When the energy of interaction between the surface and the solvent is similar to (or greater than) that of the solvent molecules with themselves, the solvent will spread out and wet the surface. For example, a drop of water on glass spreads to some extent, and wets the surface due to hydrogen bonds formed between the water molecules and the Si–OH groups on the glass. Conversely, when water is placed on a teflon surface it beads up, and does not wet the surface. The C–F teflon surface does not make strong interactions with the water molecules, and hence the water prefers to stick to itself.

A phenomenon related to wetting is **capillary action**. This phenomenon is the tendency of liquids to rise up the interior of narrow bore tubes. Liquids that adhere to the interior wall of the tubes will creep up the inside, having the effect of curving the surface of the liquid within the tube, creating a **meniscus**. A meniscus will form in a tube, but also between any two surfaces. A force results which pulls on the edges of the tube or surfaces toward the interior. A fascinating use of this force for assembling small objects has recently been reported, and is discussed in the following Connections highlight.

Connections

The Use of Wetting and the Capillary Action Force to Drive the Self-Assembly of Macroscopic Objects

Recently, capillary action has been used to self-assemble macroscopic objects. Objects of various shapes were cut from polydimethylsiloxane, a polymer that is not wettable by water but is wetted by fluorinated hydrocarbons. Designated surfaces were then made wettable by water by using controlled oxidation. These objects were then floated at an interface between perfluorodecalin ($C_{10}F_{18}$) and water. When two non-oxidized surfaces (wettable by $C_{10}F_{18}$) approached each other within a distance of approximately 5 mm, they moved into contact, which with time created an ordered, self-assembled pattern of the objects. The movement and self-assembly was driven by the solvent adhesive forces that produce the capillary action, thereby leading to an elimination of the curved menisci between non-oxidized surfaces. One such pattern is shown to the right.

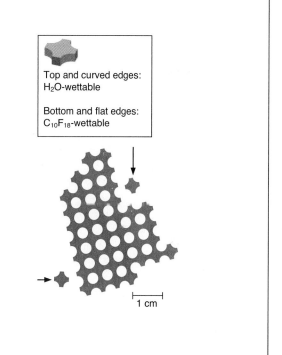

Top and curved edges: H_2O-wettable

Bottom and flat edges: $C_{10}F_{18}$-wettable

1 cm

Bowden, N., Terfort, A., Carbeck, J., and Whitesides, G. M. "Self-Assembly of Mesoscale Objects into Ordered Two-Dimensional Arrays." *Science*, **276**, 233 (1997).

Water

Water is becoming more and more important in the field of organic chemistry. The first reason for this is that bioorganic chemistry often explores chemical phenomena that occur in water, and thus the kinetics and thermodynamics of catalytic reactions and molecular recognition interactions are increasingly being studied in water. In addition, there is a strong push in the chemical industry to move away from the use of large amounts of organic solvents, and when possible, to perform chemical reactions in water so that there is less organic chemical waste (an example of **green chemistry**). Hence, understanding the properties of water is important to our understanding of nature, and may prove invaluable in helping our ecology.

Water is often thought of as a "special" solvent, with singular properties. Rather than having "special" properties, it is at the extreme limit of most solvent properties. For example, water has either the highest value or close to the highest value in the different polarity and hydrogen bond donor solvent scales discussed previously. However, water is not among the best hydrogen bond acceptors, having a lower β value than DMF, DMSO, and alcohols. Early life had to learn to deal with these extreme properties, and evolved to take advantage of them. The structures of proteins, nucleic acids, and cell membranes, as well as many other biological molecules, strictly depend upon water being the solvent.

The strong intermolecular forces in water, as evidenced by the high surface tension and heat of vaporization, are a direct result of the large charge polarization in the O–H bonds, leading to large dipole–dipole attractions and hydrogen bonding properties. The result is the high attractive force between the individual water molecules. Because of the tetrahedral geometry of water, each water molecule has the potential to hydrogen bond with four neighboring water molecules, thus being capable of making more intermolecular interactions than any other solvent. Specifically, in liquid water at 0 °C, each water molecule makes on average 3.4 hydrogen bonds, with an average O to O distance of 2.90 Å at 15 °C.

Thus, upon the melting of ice, which is fully hydrogen bonded (four per molecule), only about 15% of the hydrogen bonds are broken. Liquid water has considerable ice-like short range order but no long range order. **Flickering clusters** is a term that has been used to describe liquid water, implying short lived ice-like regions. The fluidity of these regions is imparted by the extremely rapid rate at which the hydrogen bonds are broken and formed. The half-life of each hydrogen bond in liquid water is only about 10^{-10} to 10^{-11} s. A similar, although even less ordered structure, is expected for other hydrogen bonding solvents, such as alcohols and thiols.

Recall that a polar solvent dissolves salts and molecules with large permanent dipoles. Thus, most crystalline salts and ionic compounds dissolve in water, as do many organic structures that have dipole moments and/or hydrogen bonding capabilities. The organics include sugars, alcohols, and various carbonyl containing structures. The ability of water to align its dipole and hydrogen bond to these organics leads to their solubility.

The picture of rapid fluxuation in water and other liquids leads to the general phenomenon that liquids take up more space that solids (water is an exception—ice expands relative to liquid water). Most liquids fill only about 55% of the space they occupy. This has interesting ramifications, one of which is on the design of molecular receptors, as discussed in the following Going Deeper highlight.

Going Deeper

The Solvent Packing Coefficient and the 55% Solution

In the next chapter we are going to cover molecular recognition phenomena—how solute molecules "stick together". There, binding forces, complementarity and preorganization will be important issues in the design of molecular receptors. However, a very simple postulate has recently been put forth by Rebek to guide the design of molecular receptors, and it is solely related to solvent packing. It is called the **55% solution**.

Organic liquids only occupy a certain percentage of space. The volume of filled space by a solvent is defined as its packing coefficient (PC), and is another bulk solvent property and parameter. It is a ratio of the sum of the van der Waals volumes for a solvent (V_W) to the given volume of space (V).

$$PC = V_W / V$$

Water has the largest PC (0.63), while most organic solvents vary between 0.6 and 0.5, with a mean near 0.55. In other words, most organic solvents fill just over 50% of the space they occupy.

Rebek postulates that one should design a molecular receptor for a target molecule where the target fills approximately 55% of the volume within the interior of the receptor. This would create a system with a volume-optimized binding behavior that is not significantly different from the bulk solvent. A suitable target for a receptor is one that has the right shape to fit the receptor, but also has a PC of around 55%.

Mecozzi, S., and Rebek, J., Jr. "The 55% Solution: A Formula for Molecular Recognition in the Liquid State." *Chem. Eur. J.*, **4**, 1016–1022 (1998).

3.1.3 Solubility

Most reactions that occur in solution require that the reactants be soluble. In general, reactions occur within homogeneous solutions. A **homogeneous solution** is one where there are no precipitates, solids, or different phases. In contrast, a **heterogeneous solution** has solids present or different phases. Solubility is a complex phenomenon having both a thermodynamic and a kinetic component. In general, if the solute can make more favorable interactions with the solvent than the interactions formed with itself in a crystal, the solute will dissolve. This discussion is a simple thermodynamic analysis, but very often the "practical" solubility is limited by the rate at which the solute crystal can break apart, losing molecules into the solution. Such kinetic considerations are hard to predict, and are usually just empirical observations. Therefore, we focus below on the thermodynamic aspects of solubility, and we discuss the mobility of solutes. However, in all these discussions it is important to remember that the practicalities faced in a laboratory are often more complex than the presentation given, often frustrating the chemist when he or she is working to dissolve a particular reactant.

General Overview

If a solute is to dissolve in a solvent, a reduction of the Gibbs free energy of the system must occur (see Section 3.1.5 for the mathematical description). There are several elements that can be considered separately as contributing to the free energy change, even though they do not occur separately during dissolution. First, a cavity must be created in the solvent. The creation of a cavity will be entropically disadvantageous (see Section 3.1.1), but also enthalpically unfavorable because it leads to fewer solvent–solvent interactions. The higher the cohesive energy of the solvent per volume, the greater the cost of creating a cavity. This is reflected in the δ solvent parameters discussed above. The second consideration for solubility is that the solute has to separate from the bulk solute (dissolve), leading to fewer solute–solute interactions. There is an enthalpic price to pay here, because intermolecular solute–solute interactions are breaking. Third, the solute must occupy the cavity created in the solvent. This leads to solvent–solute interactions, which are enthalpically favorable. Lastly, there is the entropy of mixing, which is favorable because the solute crystal and pure solvent taken together are more ordered than the co-mixture of solvent and solute. The first two considerations (the solvent–solvent and solute–solute interactions) can be tied to the heats of vaporization of the solvent and solute, which correlate with their respective internal cohesivenesses. The last two considerations (the enthalpy and entropy of solvent–solute interactions) give the energy gained upon solvation. All these contributors taken together constitute what is called the **solvation energy**. If the solvent and solute have strong intermolecular interactions, often similar to the kinds of interactions formed between the solvent molecules themselves, high solubility will be the result. This leads to the familiar paradigm, "Like dissolves like".

The solvation energies for many solutes have been measured (we give some in Section 3.2.2), and can be found in standard references such as the *CRC Handbook*. However, the energy value that is more useful is the **free energy of transfer** (ΔG_{tr}). This value measures the free energy for transferring a dilute solute from one solvent to another. Therefore, this number does not include the solute–solute interactions, but only focuses upon differential solvation between two solvents. Any solvent can be chosen as the reference, and Table 3.4 gives a few values for the salt $Et_4N^+I^-$ and for *t*-BuCl in several solvents relative to methanol. The values indicate that the only solvent better than methanol for solubilizing the salt is water, whereas the only solvent worse than methanol for solubilizing the organic structure is water, too. The values strongly reinforce the "like-dissolves-like" paradigm.

In a solution, the solute and surrounding solvent molecules exert an attractive force on one another. This leads to aggregation of the solvent around the solute, often causing the solute to act larger than it's intrinsic size (see the discussion of diffusion below). The region of solvent around the solute whose structure is significantly different than bulk solvent is called the **cybotactic region**. The size of the cybotactic region varies depending upon the dielectric constant of the solvent and the nature of the solute. Charged or highly polar solutes

Table 3.4
$\Delta G_{tr}°$ Values (in kcal/mol) Relative to Methanol*

Solvent	$\Delta G_{tr}°$ ($Et_4N^+I^-$)	$\Delta G_{tr}°$ (*t*-BuCl)
Water	−1.79	5.26
Ethanol	2.51	−0.29
Isopropanol	5.0	−0.34
t-Butanol	8.29	−0.53
DMSO	0.19	−0.12
CH_3CN	0.59	−0.45
Acetone	3.49	−0.95
Benzene	26.0	−1.22

*Janz, G. J., and Tomkins, R. P. T. (1972). *The Nonaqueous Electrolytes Handbook*, Academic Press, New York.

A.

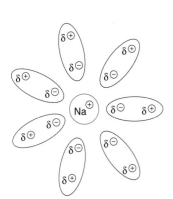

B.

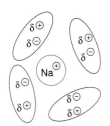

Figure 3.2
Solvent shape can affect solubility. **A.** A good arrangement, and **B.** a poor arrangement.

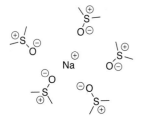

Solvation of sodium cation by DMSO

orient high dielectric solvents in the immediate vicinity of the solute due to the strong solvation. However, the ordering rapidly drops off with distance because the high dielectric solvents mediate the electric field of the solute. In low dielectric solvents, the cybotactic region around charged and polar molecules is larger because the electric fields extend further in space. Interestingly, with charged and polar solutes, the density of the cybotactic region is larger than the density of the bulk solvent, because the solvation forces pull the solvent in close to the solute. This leads to a phenomenon known as **electrostriction**, giving a reduction in volume.

Shape

The shape of the individual molecules in a solvent has a large influence on the solvent's ability to solubilize solutes. For example, molecules with their dipole along the long molecular axis can nicely solubilize an ion because several solvent molecules can approach the ion (Figure 3.2). However, when the dipole is along the short axis, solvation is not very effective because fewer molecules can approach the ion.

Using the "Like-Dissolves-Like" Paradigm

As stated, "like dissolves like" is the guiding principle when considering solubility properties. Solutes with full or partial charges dissolve well in solvents with full or partial charges. When attempting to dissolve a highly charged or polar molecule, we start by trying the highly polar solvents, typically those with the higher dielectric constants. Conversely, when dissolving an organic structure with little polarity, we start with solvents of low polarity. Recall from Section 3.1.2 that the concept of polarity was difficult to define, but it is directly related to dipole moments, hydrogen bonding capabilities, and polarizability.

Hydrogen bonding plays an important role in solubility. Solvents capable of being hydrogen bond donors and/or acceptors are very good at solubilizing solutes that can also form hydrogen bonds. Most polar organic molecules and those that have hydrogen bonding sites will dissolve in one or more of the following solvents: THF, acetonitrile, DMSO, DMF, and HMPA. Even though water is very polar, most polar organic structures will not dissolve unless they possess full positive and/or negative charges, or are small molecules (such as acetone and THF). Conversely, nonpolar solutes tend to dissolve best in lower polarity solvents, such as ether, ethyl acetate, or toluene.

We can examine some of the solvent scales to predict solubility. HMPA, DMF, and DMSO all have very large hydrogen bond accepting β values. This means they are good hydrogen bond acceptors, but also that they can coordinate to positive charges well. Hence, these solvents can often be used to solubilize alkali metal salts of common organic molecules due to their solvation of the cations. HMPA, DMF, and DMSO have hydrogen bond donating α solvent values of 0.0, meaning that they have no ability to donate a hydrogen bond, and therefore cannot readily stabilize negative charges. Indeed, these solvents supply little to no solvation to anions. We will return to this effect when we explore the nucleophilicity of anions in various solvents in Chapter 8.

Connections

Solvation Can Affect Equilibria

Some compounds change their polarity and hydrogen bonding capabilities in rearrangement processes. A prototypical example is tautomerization. One of the most well studied tautomerizations is the interconversion of 2-hydroxypyridine and 2-pyridone. The equilibrium between these two tautomeric forms is sensitive to the solvent, where the equilibrium is shifted to the tautomer most stabilized by solvation. 2-Hydroxypyridine is more stable in the gas phase, but 2-pyridone can be stabilized by polar solvents. The equilibrium constants in different solvents are given below.

Although 2-hydroxypyridine has an OH group capable of hydrogen bonding, it is 2-pyridone that is better stabilized in the high polarity solvents. You are asked in the end-of-chapter Exercises to explain this dichotomy.

Wong, M. W., Wiberg, K. B., and Frisch, M. J. "Solvent Effects. 3. Tautomeric Equilibria of Formamide and 2-Pyridone in the Gas Phase and Solution. An ab initio SCRF Study." *J. Am. Chem. Soc.*, **114**, 1645 (1992).

Tautomerization

Solvent	K_{eq}
Gas phase	0.40
Cyclohexane	1.7
Chloroform	6.0
CH_3CN	148
Water	910

3.1.4 Solute Mobility

The ability of an enzyme to bind its substrate, a carbonyl to condense with an amine, or a Pd catalyst to couple two alkenyl halides, all depends upon the reactants encountering each other in solution. The rate of the encounters depends upon the mobility of the solutes. Thus, before exploring reactivity (Part II of this book) or the structures of molecular complexes (Chapter 4), it is best to understand how molecular encounters occur. Here we present a brief introduction into the molecular details and mathematics of diffusion and molecular encounters.

Diffusion

The **diffusion** of a molecule through a solvent is best described as a "random walk". The molecule collides with solvent molecules, changing direction and speed with each collision. Each little step (jostling) is smaller even than atomic sizes, because there is little space in a solvent for the solute to hop around in. Yet, the speed at which molecules diffuse is relatively rapid (see below). Adding up all the random motions leads to what is referred to as **Brownian motion**.

Molecules with charges or dipoles diffuse slower in polar solvents. This slower diffusion is because polar molecules are well solvated in polar solvents, and hence must shed and interchange solvent molecules as they diffuse, or they must take the solvent with them. Shedding the solvent is costly. However, dragging the solvent is also costly because it results in increased friction due to the larger size of the entity that is moving. The friction that a solute feels as it diffuses through a solvent is related to its size, shape, and the viscosity of the solvent. This friction enters into the equations for translation in solution and determines how much solute molecules slow down in each step of the random walk.

Fick's Law of Diffusion

Diffusion of a solute in a solvent is caused by a **concentration gradient**. A thermodynamic driving force (F) exists for diffusion of a solute toward a uniform concentration of the solute, which is achieved throughout the solvent at equilibrium. However, on a microscopic level, even after bulk equilibrium has been achieved, a solute has a driving force for Brownian motion. This is because incremental movements (∂x) take the solute to areas of incrementally different solute concentration (∂c). The driving force (at constant pressure and temperature) for the diffusion of a solute in an ideal solution is given by Eq. 3.6, where c is concentration and x is a one-dimensional axis in space. After differentiation we get Eq. 3.7.

$$F = -RT\left(\frac{\partial \ln c}{\partial x}\right) \qquad \text{(Eq. 3.6)}$$

$$F = \frac{-RT}{c}\left(\frac{\partial c}{\partial x}\right) \qquad \text{(Eq. 3.7)}$$

The solute will achieve a steady **drift speed** (s) determined by the thermodynamic driving force, and the viscous drag from the solvent. The **solute flux** (J, the number of particles passing through a given area of space per unit time) is the drift speed times the concentration (Eq. 3.8). Further, the flux is determined by the diffusion coefficient (D, a proportionality constant that takes into account the nature of both the solute and the solvent) times the concentration gradient (Eq. 3.9, which is called **Fick's law** of diffusion). Combining Eqs. 3.7, 3.8, and 3.9 gives Eq. 3.10 for the diffusion speed or rate.

$$J = sc \qquad \text{(Eq. 3.8)}$$

$$J = -D\left(\frac{\partial c}{\partial x}\right) \qquad \text{(Eq. 3.9)}$$

$$s = \frac{DF}{RT} \qquad \text{(Eq. 3.10)}$$

To calculate the speed (rate) at which a solute will diffuse through a solution, we need to know the driving force for the diffusion, and the diffusion coefficient for the solute in the particular solvent. The diffusion coefficient depends upon the shape of the solute and the specific kinds of interactions it has with the solvent. Further, the viscosity of the solvent itself affects the diffusion coefficient. Table 3.5 shows several diffusion coefficients for different kinds of species in different solvents. In general, standard rate constants for diffusion of a solute through a solvent are on the order of 10^8 to 10^9 s^{-1}. Therefore, **diffusion controlled reactions** occur on a timescale of ns.

Several interesting trends arise from the diffusion coefficients given in Table 3.5. There is a large number for H$^+$ in water, meaning that this ion moves the fastest of all species in water. This is due to a hopping mechanism, whereby the H$^+$ diffuses by transfer between waters instead of as a single intact H_3O^+ molecule diffusing through the water. Similarly, OH$^-$ migrates quite rapidly, via deprotonation of a neighboring water molecule. In general, smaller molecules with little surface area diffuse rapidly through organic solvents. However, large biological molecules, such as the enzymes ribonuclease, lysozyme, and the oxygen carrying protein hemoglobin, diffuse quite slowly. Finally, collagen, a long polypeptide, diffuses very slowly due to its string-like shape.

Correlation Times

Correlation times for common organic molecules can be thought of as rotational diffusion times. The correlation times indicate the time it takes for the molecular orientation to be randomized relative to the starting orientation. A common organic molecule rotates in solvents very much in the same manner that it diffuses. Constant and continual collisions ran-

Table 3.5
Diffusion Coefficients (D)*

Solute	D (10^{-9} m^2 s^{-1})
H$^+$ in water	9.3[a]
I$_2$ in hexane	4.1[a]
Na$^+$ in water	1.33[a]
Sucrose in water	0.52[a]
H$_2$O in water	2.3[a]
CH$_4$ in CCl$_4$	2.9[a]
OH$^-$ in water	5.3[a]
Cl$^-$ in water	2.0[a]
Ribonuclease in water	0.12[b]
Lysozyme in water	0.10[b]
Serum albumum in water	0.059[b]
Hemoglobin in water	0.069[b]
Collagen in water	0.0069[b]

*Atkins, P. (1998). *Physical Chemistry*, 6th ed., W. H. Freeman and Company, New York.
[a] At 298 K.
[b] At 293 K.

domly rotate the molecules. Small molecules, especially those that are close to spherical, can rotate more freely within a cluster of solvent molecules, and hence they have very low correlation times.

3.1.5   The Thermodynamics of Solutions

Now that we have a basic understanding of solvents and solutes, let's examine the thermodynamics of solubility in more detail. The concepts involved lead directly to the thermodynamics of reactions. The second section of this book delves into the kinetics and mechanisms of organic transformations, which are highly dependent upon the nature of the solvent and the reactants. Hence, many of the topics discussed above will be revisited in these discussions. However, because the thermodynamics of solutions affects reactions and molecular recognition (the topic of the next chapter), it makes sense to discuss the thermodynamics of reactions here also. Therefore, in this section we explore the thermodynamic driving force for solubility and chemical reactions.

Our goal is to answer the following question: "Why do chemical transformations spontaneously occur?" As with all concepts in chemistry, a quick and easy answer is, "Because the energy of the system decreases". The *details* of this answer are what is fascinating to chemists.

There are three key tenets of thermodynamics that are important to an understanding of solubility and chemical reactions that we want to review here. The first is the concept of the chemical potential (μ), the second is that all energies are relative (recall Section 2.1), and the third is the manner in which the total Gibbs free energy of a solution varies as a function of composition.

The **Gibbs free energy** (**GFE**, G) is the energy of an entire system at constant pressure. It is an important parameter, as the difference between two GFEs is what most chemists use as the benchmark for the difference in stabilities of two systems. In the analysis given below, our system is a solution of solvent and solutes that can undergo a change in composition. Since energies are relative, we need a reference point to which we relate the energies of the molecules that we are studying. This naturally leads to the fact that the GFEs that we are interested in are differences in energy (ΔG). Let's see how all these concepts are developed mathematically.

Chemical Potential

Recall from your physical chemistry courses that the stability of an ideal gas is in part related to the volume that the gas occupies. The entropy of a gas is proportional to $nR(\ln V)$. This discussion is a simple statistical analysis, stating that the number of ways to arrange a set number of gas molecules (n) with a volume V increases with larger V. Here, the entire ensemble of gas molecules is considered to be more stable when V increases. It is important to note that the chemical structures of the individual gas molecules themselves have *not* become more stable just because they occupy a larger volume.

Ideal gases are not very relevant to most organic chemistry research. Instead, we need to analyze solutions. Further, our goal is to analyze reactions in solutions. Reactions are controlled by the stability of the entire solution when reactants are mixed, not just the stability of the individual reactants. Hence, our analysis needs to focus upon solutions as a prelude to understanding reactions. In essence, we need to understand how the stability of a solution varies as a function of the addition of reactants. Let's start by analyzing the addition of a single reactant, herein called a solute as we have done throughout this chapter.

For a solute dissolved in a solvent, the entropy of the solution becomes larger as the solute is diluted, an effect that lowers the overall Gibbs free energy of the solution. This is analogous to increasing the volume for a gas. The favorable entropy can be derived from the statistical mechanics of mixing. The solute has more ways to occupy the vessel when it is dilute.

The GFE of the solution also includes the energy of the individual solute and solvent molecules. All the normal enthalpy and entropy factors associated with structure and energy given in the last two chapters (bond strengths, strains, solvation, degrees of freedom, etc.) are considered. Hence, the GFE of the solution is a complicated sum of terms reflecting the stability of the solvent, the solute, solvation, and importantly, the entropy of mixing the solute with the solvent.

To determine the GFE of a solution, a term called the **chemical potential** (μ) of the solute is defined. The chemical potential of A (μ_A) is the extent to which the GFE of the solution (G_t, where t stands for *total*) will change due to a change in the amount of solute A (Eq. 3.11, where n_A is the number of moles of A). The chemical potential therefore tells us how the stability of a solution changes as a function of composition, where the solution will spontaneously evolve toward greater stability (lower G_t). Hence, μ_A is the link between energy and spontaneous changes in composition, such as solutes dissolving and chemical reactions occurring. For a single solute, it is convenient to think of the chemical potential of the solute as the driving force for dissolving more A into the solution, or precipitating A out of solution. This changes for each specific amount of A already dissolved. Energy is not force, but driving force gives a good mental image.

$$\mu_A = \frac{\partial G_t}{\partial n_A} \qquad \text{(Eq. 3.11)}$$

More precisely, chemical potential is analogous to potential energy. The higher potential energy of a compressed spring relative to a relaxed spring tells us that a spontaneous change will occur when the spring is released. Similarly, a higher chemical potential for a solution with a particular amount of A dissolved tells us that the concentration of A will spontaneously increase or decrease if given a chance.

The total GFE of the solution for any particular amount of A dissolved is represented by Eq. 3.12. This takes into account the chemical potential of the solvent also (μ_S). The chemical potential of the solvent would be the change in GFE of the solution as a function of the moles of solvent molecules in the solution (an equation analogous to Eq. 3.11).

$$G_t = n_A \mu_A + n_S \mu_S \qquad \text{(Eq. 3.12)}$$

Remember, energy is relative. To determine the magnitude of the chemical potential that drives a change in the composition of the solution, we need a reference state—defined for a particular amount of A dissolved in the solvent. The chemical potential of A would therefore be the chemical potential at this reference state (μ_A°) plus a correction for changing the system away from this state (Eq. 3.13). The $RT \ln(a_A)$ term is the correction to the chemical potential at conditions different than the reference state.

$$\mu_A = \mu_A^\circ + RT \ln(a_A) \qquad \text{(Eq. 3.13)}$$

The **activity** of A (a_A) is used in the correction because we are concerned with the amount of A in the solution (n_A) that affects the entropy of mixing. In our analysis, we define the activity as in Eq. 3.14, where γ is the activity coefficient (see Section 5.2.4 for a more thorough discussion of activities). Activity is "like" concentration but without units. Here $[A]_o$ is a reference concentration, set to 1 M (see discussion later in this section). Activity coefficients reflect the fact that solutes undergo non-ideal behavior, such as aggregation, which decreases the number of particles in solution. The activity gives the number of particles of the solute in the solution that affect the entropy of mixing. You may recall from a course in quantitative analysis that the activity coefficients for dilute solutions of ions can be estimated using **Debye–Hückel theory**, which uses interionic forces to estimate aggregation state. For now, realize that the value of the activity of a compound approaches the concentration of that compound as the compound's molarity goes to zero.

$$a_A = \frac{\gamma[A]}{[A]_0} \qquad \text{(Eq. 3.14)}$$

Let's look at some of the ramifications of Eq. 3.13. Figure 3.3 shows a plot of the total Gibbs free energy of solution as a function of the activity of A. The slope at any point along the curve is the chemical potential of the solution. When no A has been added to the solution there is an infinite driving force to dissolve A in the solvent. This tells us that all molecules will dissolve in all solvents at least to some extent. If solid A is added to the solvent, a spontaneous evolution will take place causing some A to dissolve. When the GFE is at a minimum, there is no longer any potential energy in the solution to be released when dissolving more A. When the activity of A is 1, Eq. 3.13 tells us that $\mu_A = \mu_A^\circ$. Yet, the slope that corresponds to μ_A° is for an arbitrary point along the curve, defined by whatever we choose as the standard state. Therefore, we now have to define a standard state. The **standard state** is taken as the concentration of A being a molarity or molality of one (we use molarity here). Therefore $[A]_o = 1$ M in Eq. 3.14 and the activity of A is simply $\gamma[A]/1$ M, which has no units.

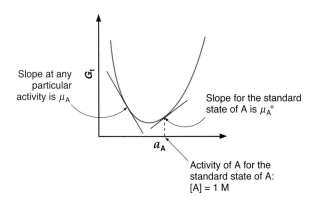

Slope at any particular activity is μ_A

Slope for the standard state of A is μ_A°

a_A

Activity of A for the standard state of A: [A] = 1 M

Figure 3.3
Plot of the total GFE as a function of the activity of a solute A. The slope at each point in the curve is the chemical potential of solute A (μ_A). There is one specific slope that is defined as the reference point. This is the slope for the activity of A when $[A] = 1$ M (μ_A°).

To summarize, the Gibbs free energy (stability) of a solution has multiple factors associated with it. First, there is the intrinsic stability of the solvent, the intrinsic stability of the solute, and the resulting solvation upon their interaction. Yet, there is also an important factor related to the mixing of the solvent and solute. We combine all these factors into the notion of the total GFE. The change in total GFE as a function of composition is used to determine if changes in the solution will occur, a notion called the chemical potential. We need a reference point for the chemical potential, which is defined as the change in Gibbs free energy of the solution as a function of composition for concentrations of solute at a molarity of 1. We can now tie this analysis to the driving force for reactions.

The Thermodynamics of Reactions

To analyze the thermodynamics of a simple reaction as given in Eq. 3.15, we compare the stability of a solution of A and a solution of B. The analysis does not tell us if there is a plausible pathway connecting A and B, but only whether A or B will dominate at equilibrium and to what extent. We start by writing an equation for B that is identical to Eq. 3.13. We then subtract the equations for B and A to achieve Eq. 3.16.

$$A \rightleftharpoons B \qquad \text{(Eq. 3.15)}$$

$$\mu_B - \mu_A = \mu_B{}^\circ - \mu_A{}^\circ + RT\,[\ln(a_B) - \ln(a_A)] \qquad \text{(Eq. 3.16)}$$

Since μ_A and μ_B are slopes themselves, it can be shown that $\mu_B - \mu_A$ is the slope of the GFE of the solution when plotted against a parameter called the extent of a reaction that converts A to B (you are asked to show this in the end-of-chapter Exercises). The **extent of reaction** is designated by ξ, and starts at 0 with the mole fraction of B being zero (activity equals zero also), and ends at 1, which signifies that all of A was converted to B. Since the μ's are akin to driving forces for changing the composition of the solution with respect to each single solute, a difference in μ's for individual solutes must be the driving force for interchanging the composition of the solution by interchanging those solutes. In other words, this difference is the potential energy stored in a solution ready to be released when the reaction occurs, in this case A to B. This analysis is expressed by Eq. 3.17, and is normally designated as ΔG_{rxn}.

$$\mu_B - \mu_A = \frac{\partial G_{rxn}}{\partial \xi} = \Delta G_{rxn} \qquad \text{(Eq. 3.17)}$$

If we now define $\mu_B{}^\circ - \mu_A{}^\circ$ as $\Delta G_{rxn}{}^\circ$, we obtain Eq. 3.18. This equation allows us to relate the driving force (ΔG_{rxn}) for interconverting A and B to their activities. When ΔG_{rxn} is negative, increasing the amount of B results in a lowering of the solution's GFE, and is a thermodynamically favorable process that will occur spontaneously. In fact, if no B is present, there is infinite driving force to form some B. Conversely, when ΔG_{rxn} is positive, B will revert to A, and if only B is present, there is an infinite driving force to create some A. Figure 3.4 should make this clear; the solution will always spontaneously evolve in the direction that lowers the total GFE of the solution.

$$\Delta G_{rxn} = \Delta G_{rxn}{}^\circ + RT \ln\left(\frac{a_B}{a_A}\right) \qquad \text{(Eq. 3.18)}$$

Figure 3.4
A plot of the total GFE of the solution as a function of the extent of interconversion of A and B (ξ). The slope at each ξ is ΔG_{rxn}, which is the driving force for achieving equilibrium. When equilibrium is achieved, ΔG_{rxn} is zero.

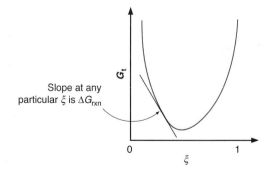

However, defining $(\partial G_{rxn}/\partial \xi)$ as ΔG_{rxn} is a bit confusing. We normally think of a ΔG term as a difference between two energy states. In this case you might think it would represent the difference in energy of the solution at any point ξ and the minimum possible energy. This, however, is incorrect. Instead, ΔG_{rxn} is the slope of the function that relates the change in GFE of the solution to the extent of the reaction. It is the driving force for achieving the minimum GFE for the solution at a particular composition of A and B, and is zero when equilibrium has been achieved. This is similar to the notion that μ was akin to a driving force for lowering the energy of a solution upon dissolving a solute.

The $\Delta G_{rxn}°$ is a term of paramount importance. Understanding the meaning of this term is one of the prime goals of this discussion. Unlike ΔG_{rxn}, $\Delta G_{rxn}°$ *does* truly reflect a difference in Gibbs free energies between two particular compositions of a solution. The reason is as follows. We defined $\Delta G_{rxn}°$ as $\mu_B° - \mu_A°$, the difference in the chemical potentials of A and B in their respective standard states. The total GFE (G_t) for an ideal solution of A and B in solvent S would be expressed by Eq. 3.19.

$$G_t = n_A \mu_A + n_B \mu_B + n_S \mu_S \qquad \text{(Eq. 3.19)}$$

Now we define $\Delta G_{rxn}°$ to be a per mole quantity. Hence, when a solution of one mole of A is considered at its standard state we get $G_A° = \mu_A° + n_S \mu_S$, and when one mole of B is considered at its standard state we get $G_B° = \mu_B° + n_S \mu_S$. Therefore, if we solve for $\mu_B° - \mu_A°$, we find it is a difference of two Gibbs free energies, $G_B° - G_A°$ (with the assumption that μ_S does not change with composition). Therefore, $\Delta G_{rxn}°$ is the difference in the stability of a solution of one mole of A in its standard state and a solution of one mole of B in its standard state. Stated in another way, it is the energy difference for the conversion of one mole of A to one mole of B, both at their standard states. This takes into account the intrinsic stabilities of the solutions of the two separate solutes when at their standard states, which includes the stabilities of the solutes themselves.

Since activities are commonly assumed to be close to concentrations, Eq. 3.18 reduces to the more familar Eq. 3.20, where $Q = [B]/[A]$. This equation is simply another (and approximate) way of expressing Eq. 3.18. It gives the driving force that exists for a reaction to occur when the concentrations of B and A do not reflect the difference in the intrinsic Gibbs free energies of their respective solutions at standard states ($\Delta G_{rxn}°$).

$$\Delta G_{rxn} = \Delta G_{rxn}° + RT \ln Q \qquad \text{(Eq. 3.20)}$$

After the Gibbs free energy of the solution has been minimized, ΔG_{rxn} is zero. Now the ratio of B to A does reflect the intrinsic stabilities of separate solutions of B and A at standard states. Equilibrium is said to have been achieved, and B and A are at their equilibrium concentrations. At this point Q is defined as K_{eq}, the equilibrium constant. When equilibrium has been achieved, we can rearrange Eq. 3.20 to Eq. 3.21. K_{eq} therefore reflects a ratio of B to A which is indicative of the intrinsic stabilities of A and B, the relative solvation of A and B, and the entropies of mixing A, B, and the solvent, at the standard states of A and B.

$$\Delta G_{rxn}° = -RT \ln K_{eq} \qquad \text{(Eq. 3.21)}$$

You might be wondering how we measure $\Delta G_{rxn}°$ values if the standard state experimental conditions are never used. In fact, it is physically impossible to convert one mole of A completely to a mole of B both at their standard states. Hence, the notion of the standard state is a bit esoteric. Importantly, our analysis had this in mind. Once equilibrium has been achieved, regardless of the actual concentrations involved, the manner in which we have set up our analysis leads to GFE values that reflect the intrinsic stabilities of solutions of A and B at their standard states.

One important ramification of our analysis needs to be mentioned at this stage. With an equilibrium where both the forward and reverse reactions are unimolecular, the composition of the solution can be directly determined by the K_{eq}. In other words, for all total concentrations of reactants and products in the reaction flask, the ratio of reactants and products at equilibrium is given by K_{eq}. This is quite different for equilibria that involve reactions of different molecularity in the forward and reverse reactions. When the molecularities of the forward and reverse reactions are different, the composition of the solution at equilibrium changes depending upon the total concentration of reactants and products, even though the value of K_{eq} does not change. Look ahead to Section 4.1.1 to see this.

Lastly, have you ever wondered why a reaction of A going to B that is exothermic does not just totally convert to all B? After all, if B is more stable, why doesn't all A just completely become B? Inherent in the question is the fact that the term exothermic relates to enthalpy. An exothermic reaction means that the intrinsic stability of solvated B is greater than the stability of solvated A. However, in our discussion of the thermodynamics just above, we analyzed solutions of A and B mixed together, and focused upon the total Gibbs free energies of the solution to describe the reaction. We found an infinite driving force for creating A or B when starting with pure B or A, respectively. Since it is the stability of the overall solution that dictates reactions, not just the stability of the solutes themselves, there will always be some of A and B present independent of how large the endothermicity or exothermicity of the reaction. The fundamental reason for this is the entropy of the solution, which is always more favorable when some of A and B are present, regardless of the stabilities of A and B.

Since it is the $\Delta G_{rxn}°$ that controls any equilibrium, and we have now found that part of this $\Delta G°$ depends upon the mixing of solutions, how do we determine just the stability of the reactants and products independent of the mixing? In the last chapter we focused upon enthalpy changes to determine the stability of organic structures. Therefore, we would like to calculate whether a reaction is exothermic or endothermic to make this determination. Hence, we need $\Delta H°$ values.

Calculating $\Delta H°$ and $\Delta S°$

We will spend a significant amount of the next chapter analyzing methods to measure equilibrium constants, from which the standard GFE of the reaction can be derived using Eq. 3.21. Yet, a lot of chemical insight derives from measuring $\Delta H°$ and $\Delta S°$. This can be quite easily done using a **van't Hoff analysis**. By substituting the Gibbs free energy equation, $\Delta G° = \Delta H° - T\Delta S°$, into Eq. 3.21 and rearranging, we get Eq. 3.22. A plot of $\ln K_{eq}$ versus $1/T$ gives a $\Delta H°$ value from the slope and a $\Delta S°$ value from the intercept. Hence, by measuring K_{eq} values at a variety of temperatures, the enthalpy and entropy of reaction can be determined, and we show one example in the following Connections highlight. A straight line is obtained in a van't Hoff analysis only if the heat capacity ($\Delta C_P°$) of the solution does not change (see Chapter 4). Curvature in a van't Hoft plot indicates that $\Delta C_P° \neq 0$.

$$\ln K_{eq} = -\frac{\Delta H°}{RT} + \frac{\Delta S°}{R} \qquad \text{(Eq. 3.22)}$$

In summary, it is clear that the Gibbs free energy of solutions plays a pivotal role in the thermodynamics of chemical reactions. In the last chapter we examined the stability of various organic structures, which is part of this total Gibbs free energy. Now, in this chapter, we found that the nature of the solvent, the resulting interactions with the solutes (solvation), and the simple act of mixing solutes and solvents, are also part of the total GFE. It is now appropriate to explore the interactions between solvents and solutes, and between solutes themselves, in detail. Once we understand these interactions, we can put together all the concepts—chemical structure and stability, solvation, and total Gibbs free energy—and start to explore some reactions.

3.2   Binding Forces

Now that we have a background into the structure of solvents, insight into polarity parameters, and solute mobility, it is time to explore the forces that hold the solvent molecules together. The same interactions that hold solvent molecules together are those that cause

Connections

A van't Hoff Analysis of the Formation of a Stable Carbene

In Chapter 11 we will discuss the structure and reactivity of carbenes. These are traditionally extremely unstable structures, where carbon only has six electrons. However, there are cases of stable carbenes, typically possessing resonance structures with stabilizing features such as zwitterionic and aromatic character. For example, for moderately large R, carbene A can be isolated and does not dimerize to a tetraaminoethylene derivative. Yet, carbene B dimerizes irreversibly, presumably due to the lack of additional aromatic stability.

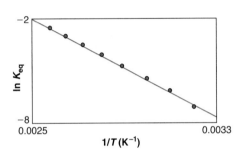

A.

B.

Stable carbenes

As a means to measure the double bond strength in a tetraamino-substituted ethylene, the structure shown to the right was synthesized. This compound does exist in

equilibrium with the carbene shown. The equilibrium constants for this transformation were determined as a function of temperature, and the van't Hoff plot shown gave $\Delta H° = 13.7 \, kcal/mol$ and $\Delta S° = 30.4$ eu. The bond strength for the double bond (13.7 kcal/mol) is exceptionally low relative to normal C=C bond strengths (approximately 160 kcal/mol; see Chapter 2).

Reversible carbene formation

Liu, Y., Lindner, P. E., and Lemal, D. M. "Thermodynamics of a Diaminocarbene–Tetraaminoethylene Equilibrium." *J. Am. Chem. Soc.*, **121**, 10626 (1999).

---

solutes to dissolve, and are responsible for solute–solute interactions and molecular recognition. Often, these binding forces are present within the same molecule, such as intramolecular hydrogen bonding. Hence, we examine the binding forces all together, and do not necessarily focus upon intermolecular or intramolecular interactions. The interactions can be as simple as the electrostatic attraction between a small cation and a small anion, or as complex as those associated with the multi-component enzyme assemblies that initiate gene expression. Hence, we use the term solute to refer to any species dissolved in a solvent, from a simple ion to a complex biomolecule.

In most cases the binding forces discussed herein are weak. Therefore, in reading the following sections, it may at times seem that we are discussing such weak phenomena that the forces are insignificant. On the contrary, we will demonstrate that cooperativity among many weak interactions can be quite powerful. It is an accumulation of many weak interactions that leads to large binding forces between solutes (molecular recognition). This is pervasively true both in chemical biology and in materials chemistry, and it is a phenomenon we will consistently observe.

3.2.1 Ion Pairing Interactions

Oppositely charged ions attract each other strongly. In the gas phase the "binding" between a simple cation and a simple anion can be worth well over 100 kcal/mol. The major contributor to the binding is an **electrostatic interaction**. We will be discussing electrostatics extensively in this section, and it is important to be clear on its usage here. By an electrostatic interaction, we mean a strictly Coulombic attraction or repulsion between charges or partial charges that existed prior to the interaction and remain unchanged in the interaction. The last restriction is not universally applied. Some would first bring two molecules together, allow their charge distributions to rearrange in response to each other's presence, and then consider the Coulombic interaction of these altered charge distributions to be an electrostatic interaction. This is not an unreasonable type of analysis. Here, however, we will retain the "static" of electrostatic, and we will consider binding that results from rearranged charge distributions to be *not* a strictly electrostatic effect.

An **ion pair** is defined to exist when a cation and anion are close enough in space that the energy associated with their electrostatic attraction is larger than the thermal energy (RT) available to separate them. This means that the ions stay associated longer than the time required for Brownian motion to separate non-interacting species. We have already examined the energy between two charges at a distance r (Eq. 3.1), and found it to depend inversely upon the dielectric constant. Hence, the extent of ion pairing will also depend upon the dielectric constant of the solvent. The inverse correlation with dielectric constant is imperfect, because other interactions between the cation and anion can be involved. The dielectric constant of the solvent does not take into account the specific coordinating ability or hydrogen bonding ability of the solvent toward particular cations or anions. Further, the size and shape of the anions and cations will influence their energy of attraction. The solvent molecules also become very organized when surrounding a cation or anion, which is entropically costly, whereas the surface around an ion pair is smaller and the solvation requirements lower. Thus, ion pair formation can be viewed as a competition with ion solvation as a means to lower the Gibbs free energy of the solution. Since most organic reactions are performed in solvents of relatively low dielectric constant, ion pairing is a common phenomenon for charged reactive intermediates (carbocations and carbanions).

Since Coulomb's law (Eq. 3.1) includes the dielectric constant of the medium (ε), we expect the energetics of an ion pair to be medium dependent. On moving from the gas phase ($\varepsilon = 1$) to an organic solvent ($\varepsilon < 10$), the energy of an ion pair is still expected to be quite significant. However, in water, with $\varepsilon = 78$, the interaction should be substantially attenuated. In other words, we do not expect oppositely charged ions to bind tightly to one another in water. Sodium chloride, for example, is dissociated in water. Note that we do not expect zero binding energy in water, only a relatively small binding energy.

Ionic interactions become stronger with **polyions**. A polyion is a polymer of repeating ionized units. For example, a dilute solution of sodium acetate in water is completely dissociated, while polyacrylate [$-(CH_2CHCO_2^-)_n-$] has a substantial fraction of the sodium ions bound to the polymer. This polymer is referred to as a **weak electrolyte**, in contrast to sodium acetate, which is a **strong electrolyte**. The large negative charge density on the polymer leads to a greater fraction of the sodiums being held in the vicinity of the polymer. A biological example of this is DNA and RNA, which are repeating units of negative phosphate diesters. These structures are well known to have large numbers of cations closely associated with the strands.

Salt Bridges

We have already noted previously that in molecular recognition, and especially in biological recognition, large effects often result from the accumulation of a large number of small effects. Thus, it becomes quite important to distinguish "no interaction" from a weak interaction, and 0.1 kcal/mol from 1.0 kcal/mol. Not surprisingly, when small distinctions are controlling, some debate and even controversy can arise.

The controversy can be quite intense when it is in the context of the **salt bridge**. A salt bridge is an ion pair between two side chains of a protein. The anion is a carboxylate (from Asp or Glu) and the cation is an ammonium (RNH_3^+, from Lys) or a guanidinium [$RNHC(NH_2)_2^+$, from Arg]. To what extent do salt bridges contribute to protein stability? There is no simple answer. We should anticipate that context would be important. If the salt bridge is on the surface of the protein, the dielectric constant should be close to that of pure water. Such an "exposed" salt bridge might contribute very little to protein stability. Again, ammonium acetate is dissociated in water, so an Asp•••Lys salt bridge should be weak or negligible in bulk water. Alternatively, the salt bridge might be somewhat or completely buried in the interior of the protein. Here the question of dielectric constant becomes complex. Often an "effective dielectric" constant anywhere in the range of 4–37 is ascribed to the interior of a protein. However, we are no longer in a relatively homogeneous medium like in a pure solvent, and so any such approximation must be considered fairly crude.

The consensus from a large number of studies of salt bridges is that they can contribute to protein stability, but there is considerable variation. Typically, a surface-exposed salt bridge is worth around anywhere from 0 to 2 kcal/mol, and a buried salt bridge can be worth up to 3 kcal/mol, with some exceptional cases being worth more. These are small effects, but again, molecular recognition is controlled by interactions that are individually small but add up to a large effect.

Another issue in considering the contribution of a salt bridge to protein stability, and one that must be considered whenever thermodynamic issues are discussed, is the appropriate reference state. Stability, whether we are talking about a protein fold or a reactive intermediate, is *always* a relative term. We noted this earlier in this chapter, in Chapter 2, and we will return to it often throughout this book. The following Going Deeper highlight presents the problem of defining an appropriate reference state.

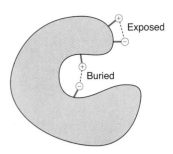

Exposed and buried salt bridges

Going Deeper

The Strength of a Buried Salt Bridge

What kind of an experiment would determine the strength of a buried salt bridge? It might seem that the sensible thing to do would be to measure the stability of the protein, a straightforward process involving merely heating the protein and watching it "unfold", with and without the salt bridge. In this experiment, the stability of the protein is defined as the difference in stability of the unfolded and folded states. It is a simple matter nowadays to alter protein structure in controlled ways.

What does "without the salt bridge" mean? Do we simply remove the amino acid side chains that make the salt bridge? This would leave a hole in the protein, and as we noted at the start of this chapter, nature abhors a vacuum. This seems like an unfair reference state. Recall that the interiors of proteins have low dielectric microenvironments. Perhaps a more sensible reference state would be to replace the two ionic side chains with two "greasy" side chains of comparable size. Now we are asking a question more like, "Which is more stable, a salt bridge or a hydrophobic contact in the interior of a protein?" Studies have been performed to address just this question, and often the outcome is that the protein is more stable with the hydrophobic pair than with the salt bridge. The conclusion would now be that *the salt bridge destabilized the protein*! Clearly, the choice of reference state influences the conclusions—a very important lesson for any thermodynamic experiment.

Hendsch, Z. S., and Tidor, B. "Do Salt Bridges Stabilize Proteins? A Continuum Electrostatic Analysis." *Protein Sci.*, **3**, 211–226 (1994).

3.2.2 Electrostatic Interactions Involving Dipoles

Just as full opposite charges attract each other, oppositely charged ends of dipoles attract each other. This leads to a rough alignment of the dipoles such that positively charged ends interact with negatively charged ends. Because solvents are not completely ordered, there is considerable disorder in this alignment. Yet, this attraction is one of the forces that holds solvent molecules together and raises boiling points. The dipoles do not have to be between solvent molecules, but can also be between solutes and solvents, and between two solutes.

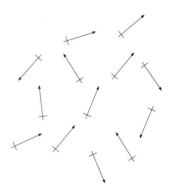

Dipoles aligned to some degree

Ion–Dipole Interactions

When a charged solute is dissolved in a solvent with a dipole moment, the electric field associated with the charge exerts a force on the dipole, orienting the oppositely charged end of the dipole toward the charge. For a dipole whose orientation is fixed in space, the potential energy of the interaction varies as the inverse squared distance r between the charge and dipole (Eq. 3.23, where ε is the dielectric constant of the solvent and μ is the dipole moment;

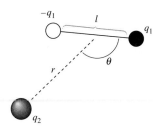

Ion–dipole alignment parameters

$\mu = q_1 l$). Thus, the ion–dipole energy falls off more rapidly than the attraction between two oppositely charged ions (Eq. 3.1). This equation holds for r significantly larger than l.

$$E = \frac{\mu q_2 \cos \theta}{4 \pi \varepsilon_0 r^2} \qquad \text{(Eq. 3.23)}$$

The attractive force can be quite large for a polar solvent molecule in direct contact with an ion. This is part of the large exergonic physical change when solid salts dissolve in water. The entropy of mixing also favors dissolution (see Section 3.1.5). Table 3.6 shows several heats of hydration (equivalent to the heat of solution for water as solvent) for various ions, salts, and a few organic structures.

Important solvation trends are evident in considering the simple ions. A clear trend in hydration energies emerges, with $Li^+ > Na^+ > K^+ > Rb^+$. The smaller the ion, the greater the hydration energy. This trend is an indication of a largely electrostatic effect. If we consider these ions as spheres of charge, the smaller ion has the same total charge as a larger ion, but it is distributed over the surface of a smaller sphere. Thus, the charge per unit area is larger, and so Coulombic interactions are stronger. Whenever a trend correlating ionic radius and interaction energy appears, we should suspect a strong electrostatic component to the interaction. The same trend is seen with the simple halogen anions. Consistent with this electrostatic analysis, divalent cations have *much* larger hydration energies than monovalent cations.

The hydration energies for simple salts are more difficult to interpret because they arise from a composite of many phenomena (see the description of solubility in Section 3.1.3), but a few trends are evident. The ionic radius trend discussed above is evident when comparing the chloride salts of Li^+, Na^+, and K^+—that is, it is more exothermic to solvate the smaller cations when keeping the anion constant. With the hydroxide salts, however, the exact opposite trend is found. With respect to solvating the anion, the sodium or tetramethylammonium salts of chloride, bromide, and iodide are better solvated the smaller the anion, again due to increased dipolar attraction with the smaller anion. Interestingly, the dissolution of some salts is endothermic, and indeed when NH_4Cl or NH_4NO_3 dissolves in water, the solution cools.

A Simple Model of Ionic Solvation—The Born Equation

The solvation energies of many simple ions are known, especially the hydration energies. As discussed above, a universal trend is that hydration strongly depends on the radius of the ion, with the smaller ions being better solvated. The **Born equation** (Eq. 3.24) attempts to put this kind of trend on a more quantitative basis. It is a simple correlation involving the dielectric constant, the ionic radius, and the charge of the ion. Plugging in the appropriate values reveals that for a monovalent ion in water at 298 K, the Born solvation energy, $E_{sol} = -164/a$, in kcal/mol, when a is in Å.

$$E_{sol} = -(1 - 1/\varepsilon)(q^2/8\pi\varepsilon_0 a), \quad \text{where } a \text{ is the radius of the ion} \qquad \text{(Eq. 3.24)}$$

Such a model is too simple, because it ignores the highly specific kinds of solute–solvent interactions discussed later, such as hydrogen bonds. But, it is not as bad as you may expect. For example, a chemist may consider NH_4^+ and K^+ as quite different (the former can hydrogen bond, etc.). However, simple modeling will convince you that their ionic radii are actually quite similar, and indeed, as shown in Table 3.6, their hydration energies are also quite similar. Also, Na^+ and Ca^{2+} have similar ionic radii, but the divalent ion has roughly quadruple the hydration energy, consistent with the q^2 term in Eq. 3.24.

One interesting implication of the Born equation concerns long range solvation of an ion by a solvent with a dipole such as water. We can concede that very close to an ionic solute—within the first two or even three solvation shells—such a simple model might be inadequate because it neglects specific effects. But what about further out? It is probably quite

Table 3.6
Heats of Solution of Various Compounds in Water*

Structure	Hydration energy (kcal/mol)[‡]	Ionic radius (Å)
A. Ions		
Li^+	−122	0.60
Na^+	−98	0.95
K^+	−81	1.33
Rb^+	−76	1.48
Cs^+	−71	1.69
Mg^{2+}	−476	0.65
Ca^{2+}	−397	0.99
Zn^{2+}	−485	
Sr^{2+}	−346	
Ba^{2+}	−316	
F^-	−114	1.36
Cl^-	−82	1.81
Br^-	−79	1.81
I^-	−65	2.16
NH_4^+	−80	
Me_3NH^+	−59	
$CH_3CO_2^-$	−80	
B. Salts		
LiOH	−5.6	
NaOH	−10.6	
KOH	−13.7	
LiCl	−8.8	
NaCl	0.93	
KCl	4.1	
NaBr	−0.14	
NaI	−1.8	
NH_4NO_3	6.1	
NH_4Cl	3.5	
$N(CH_3)_4Cl$	0.97	
$N(CH_3)_4Br$	5.8	
$N(CH_3)_4I$	10.1	
C. Simple Molecules		
NH_3	−7.3	
CH_3OH	−5.1	
Acetone	−3.8	
CH_3COOH	−6.7	
Benzene	−0.9	
n-Octane	2.9	

*Burgess, M. A. (1978). *Metal Ions in Solution*, John Wiley & Sons, New York.
[‡]Negative values represent an exothermic process.

acceptable. Then, what fraction of the total solvation energy of an ion such as K^+ is due to just long range interactions with the dielectric of the medium? To answer this question, we simply treat the ion as a very large ion, and plug the distance into the Born equation. For example, it is a simple matter to show that over 19 kcal/mol of solvation for a monovalent ion comes from water molecules that are ≥ 8.5 Å from the ion (see the end-of-chapter Exercises). This is actually quite a large number, and is an important factor to be considered when discussing aqueous solvation of ions.

Dipole–Dipole Interactions

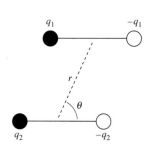

Dipole–dipole alignment parameters

Similar to the attraction between a dipole and a charge, interactions between dipoles on solutes and solvents can be attractive or repulsive. The force between two dipoles depends upon their relative orientation and, if the dipoles are fixed in space, the interaction energy falls off as a function of the inverse distance between the dipoles to the third power. Therefore, dipole–dipole interactions are very sensitive to the distance between the dipoles. Eq. 3.25 gives the energy between two fixed dipoles that are in the same plane and parallel, where ε is the dielectric constant of the medium and the μ's are the two respective dipole moments. If they are not parallel and in the same plane, the equation simply gets more complicated. Further, this is a simplification where r is significantly longer than the dipole length l ($\mu_1 = q_1 l_1$). The angle for which the two dipoles feel no attractive or repulsive force has an important use in spectroscopy, as discussed in the following Going Deeper highlight.

$$E = \frac{-\mu_1 \mu_2 (3\cos^2\theta - 1)}{4\pi\varepsilon\varepsilon_o r^3} \qquad \text{(Eq. 3.25)}$$

Going Deeper

The Angular Dependence of Dipole–Dipole Interactions—The "Magic Angle"

An interesting feature of Eq. 3.25 is the $3\cos^2\theta - 1$ term. Consider the value of θ required to make the magnitude of a dipole–dipole interaction go to zero [arc cos $(1/\sqrt{3})$]. This corresponds to ~54.7°. For *any* pair of dipoles, their interaction energy is zero if they are aligned at this angle. This is a familiar angle to spectroscopists and is referred to as the "magic angle". Why is it magic? In NMR spectroscopy, the nuclear spins can be treated as dipoles, as can the external magnetic field of the spectrometer. As such,

in a solid sample (remember, Eq. 3.25 refers to *fixed* dipoles, not rapidly tumbling dipoles as in a free solution), each nuclear spin will experience a *different* interaction with the external magnetic field depending on the precise angle between the field and the nuclear moment, producing extraordinary complexity in the spectra. To remove this, the NMR tube is tilted relative to the external magnetic field at the magic angle. This trick, coupled with rapidly spinning the tilted tube, removes this complexity. The spinning causes signals from any spins not aligned with the rotation axis to average and cancel.

3.2.3 Hydrogen Bonding

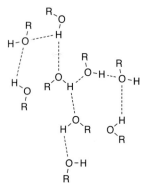

Network of hydrogen bonds in an alcohol

Hydrogen bonding is another very important binding force. While detailed, quantum mechanical analyses of hydrogen bonds can be complex, for weak to moderate hydrogen bonds a solely electrostatic model is adequate for most purposes. Such a model describes a **hydrogen bond** as a Coulombic interaction between a polar donor bond ($Dn^{\delta-}-H^{\delta+}$) and an acceptor atom ($:Ac^{\delta-}$). We use this simple model in all the discussions given below until short–strong hydrogen bonds are considered. Since the hydrogen bond is a simple Coulombic interaction, any partial negative charge can accept a hydrogen bond, not just electronegative atoms, but even π systems (as we will show later). The next Connections highlight indicates just how unusual hydrogen bond acceptors can become.

One of the most common examples of hydrogen bonds are those formed in liquid alcohols. Most OH groups make a hydrogen bond to an oxygen of an adjacent alcohol, thereby creating a network of hydrogen bonds. In liquid alcohols there is a rapid interchange of the hydrogen bonds, with the molecules oriented imperfectly with their neighbors.

Connections

An Unusual Hydrogen Bond Acceptor

If hydrogen bonds are essentially electrostatic in origin, then any region of a molecule with a partial negative charge should act as a hydrogen bond acceptor. Can hydrogens be hydrogen bond acceptors in some circumstances?

In Chapter 12 we will explore organometallic systems known as metal hydrides. A typical example is $LiAlH_4$. Similar to the hydrogens attached to Al, hydrogens attached to most transition metals possess partial negative charges. Hence, metal hydrides might be hydrogen bond acceptors. Indeed, a few such examples exist. One in particular is the iridium complex shown to the right, where a very short interaction (1.8 Å) between the metal hydride and the hydrogen atom of an appended alcohol was found in the crystal structure.

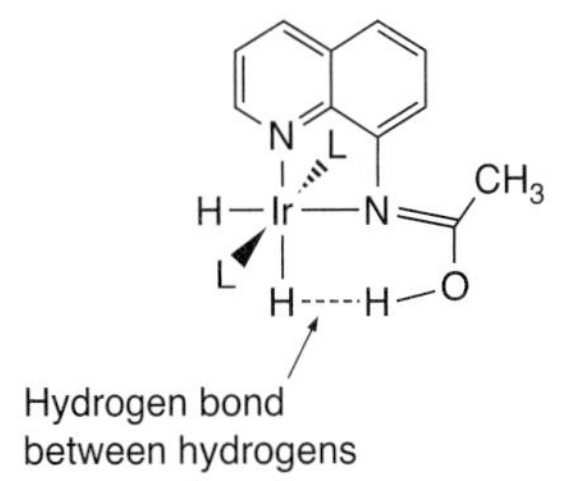

Hydrogen bond between hydrogens

Lee, J. C., Jr., Peris, E., Rheingold, A. L., and Crabtree, R. H. "An Unusual Type of H–H Interaction. Ir–H---HO and Ir–H---NH Hydrogen Bonding and its Involvement in σ-Bond Metathesis." *J. Am. Chem. Soc.*, **116**, 11014 (1994).

Geometries

Since electrostatic considerations dominate for most hydrogen bonds, the geometry of the hydrogen bond is not a major contributing factor to strength (data supporting this is given in the next Connections highlight). Still, the optimal geometry has a collinear arrangement of the three atoms involved, even though significant deviations from linearity can be tolerated. In cyclic systems, nine-membered rings containing hydrogen bonds give the most linear arrangement, and have been shown to be optimum (see the Connections highlight below). In addition, the Dn–H bond axis generally coincides with the imagined axis of a specific lone pair of :Ac. As discussed in Chapter 1, the hybridization of atoms and the directionality of lone pairs can be debated. Figure 3.5 shows a few representative geometries for hydrogen bonding. When there is only one lone pair, as with RCN: or :NH_3, we expect a linear geometry. With two lone pairs, VSEPR theory can help rationalize the observed angles. For water, with an H–O–H angle of $\sim 104°$, we expect a nearly tetrahedral arrangement, and the 55° angle of Figure 3.5 is consistent with this.

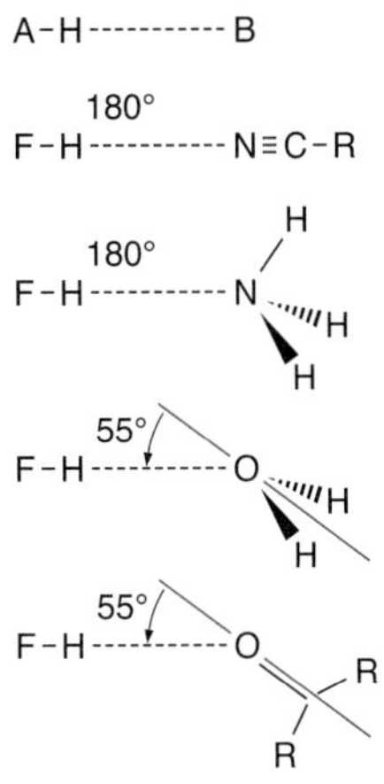

Figure 3.5
Hydrogen bonding. Shown are experimentally determined geometries for prototype hydrogen bonding complexes, showing the alignment of the donor with the putative lone pair acceptor.

Connections

Evidence for Weak Directionality Considerations

For a carbonyl compound, the hydrogen bond should be in plane and at an angle consistent with $\sim sp^2$ hybridization of the O—hence, an angle of 120°. However, as we have already alluded to, geometry is not so important in an electrostatic interaction, and even the directionality of the lone pairs is debatable. In support of this view, studies of hundreds of crystal structures analyzing the hydrogen bonding angles between carbonyls and various donors are consistent with diffuse lone pairs. As shown below, the H•••O=C angles range from 0° to 90° (as defined in

the picture), with a maximum at 40° (close to the expected angle for a carbonyl lone pair). However, a considerable number of hydrogen bonds are oriented along other angles, including the axis of the C=O bond ($\phi = 90°$).

Taylor, R., Kennard, O., and Versichel, W. "Geometry of the N H–O=C Hydrogen Bond. 1. Lone-pair Directionality." *J. Am. Chem. Soc.*, **105**, 5761–5766 (1983). Murray-Rust, P., and Glusker, J. P. "Directionality Hydrogen-Bond to sp^2 and sp^3 Hybridized Oxygen Atoms and its Relevance to Ligand–Macromolecular Interactions." *J. Am. Chem. Soc.*, **106**, 1018–1025 (1984). For a review, see Hubbard, R. E. "Hydrogen Bonding in Globular Proteins." *Prog. Biophys. Molec. Biol.*, **44**, 97 (1984).

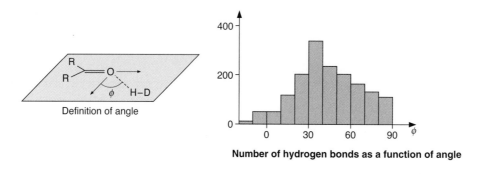

Definition of angle

Number of hydrogen bonds as a function of angle

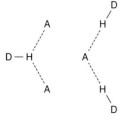

Bifurcated hydrogen bonds

Since directionality is not a dominant factor in the strength of normal hydrogen bonds, it is not surprising that there are a multitude of bridging hydrogen bonding geometries. Structures such as those shown in the margin are referred to as **three-center hydrogen bonds**, and also frequently as **bifurcated hydrogen bonds**. In cases where the two donors or the two acceptors are part of the same molecule, the term **chelated hydrogen bond** is sometimes used.

Connections

Intramolecular Hydrogen Bonds are Best for Nine-Membered Rings

In Chapter 2 we examined the stabilities of various rings, and found that the transannular effect raises the energy of rings with sizes beyond six carbons. However, using variable temperature NMR and IR studies, it has been determined that nine-membered rings are best for intramolecular hydrogen bonds between terminal amides (as shown to the right). In methylene chloride, the enthalpy of the hydrogen bonded state is 1.4 to 1.6 kcal/mol more favorable than the open chain structure, while the open chain structure is entropically favored by 6.8 to 8.3 eu. The enthalpic preferences for the hydrogen bonded state are significantly smaller for larger and smaller rings. The reason for the preference of a nine-membered ring derives

from lower torsional strains present in the hydrocarbon linker between the amides when a nine-membered ring is formed.

Nine-membered ring optimal
for hydrogen bonding

Gellman, S. H., Dado, G. P., Liang, G.-B., and Adams, B. R. "Conformation-Directing Effects of a Single Intramolecular Amide–Amide Hydrogen Bond: Variable-Temperature NMR and IR Studies on a Homologous Diamide Series." *J. Am. Chem. Soc.*, **113**, 1164–1173 (1991).

Now that we have discussed the electrostatic origin and geometries of normal hydrogen bonds, let's explore those factors that accentuate the electrostatic attraction. These include electronegativity, resonance, polarization, and solvent effects. The goal is to understand trends in hydrogen bond strengths, because actual bond dissociation energies for hydrogen bonds in solution are hard to come by. We start by analyzing why hydrogen bond strengths are difficult to determine.

Strengths of Normal Hydrogen Bonds

Hydrogen bonding can be a potent force for molecular recognition, but it should come as no surprise that context effects can be substantial. For example, the strength of a hydrogen bond depends upon both the nature of the donor and the acceptor, and the microenvironment of the hydrogen bond. Since the microenvironment of the hydrogen bond strongly affects its strength, hydrogen bond enthalpies cannot be transferred from one situation to another as can the bond dissociation energies for covalent bonds.

Thermochemical studies to determine hydrogen bond strengths have been performed, but systematic studies are not as extensive as those involving covalent bonds. Difficulties arise in measuring hydrogen bond strengths (enthalpies) because intermolecular interactions are influenced by significant entropic considerations, thereby making the measurement of association Gibbs free energies not easily related to simple enthalpies of the hydrogen bonds. Even the enthalpies of association of a Dn–H and an :Ac molecule cannot be directly related to the strength of the hydrogen bond, because the Dn–H and :Ac were to some extent solvated to start, and these solvation interactions influence the enthalpy of association. Very often the strengths of hydrogen bonds are determined by examining conformational equilibria, where one conformation possesses the hydrogen bond, and another conformation does not (see the Connections highlight in Section 2.3.2, and the one below about solvent scales and hydrogen bonds). Otherwise, measurements are made in the gas phase or very nonpolar solvents, where the solvation issue is nonexistent or less severe. On rare occasions, and in very clear-cut cases, one can determine hydrogen bond strengths when the association constant of two almost structurally identical molecules with a receptor can be determined, wherein one molecule can make the hydrogen bond and one cannot. The difference in Gibbs free energies of binding can roughly be equated to the intrinsic enthalpy of the hydrogen bond.

In general, hydrogen bond strengths are roughly broken into three catagories. Those of 15 to 40 kcal/mol are considered to be very strong, those in the range of 5 to 14 kcal/mol are moderate, and those between 0 and 4 kcal/mol—the most common hydrogen bonds—are weak. Consistent with the electrostatic model, there is a general trend that the hydrogen bond is stronger if one or both of the partners is charged, meaning that the electrostatic nature significantly increases due to large Coulombic attraction.

i. Solvation Effects

Probably the factor that most influences the strength of a hydrogen bond formed between a Dn–H and :Ac is the solvent. In the next section we tabulate a few hydrogen bond strengths for the gas phase or nonpolar solvents, which vary from 5 to 10 kcal/mol. However, a value of 0.5 to 1.5 kcal/mol is generally used as the strength of a hydrogen bond in the interior of a protein that is dissolved in water (see the α-helix Going Deeper highlight on page 176). If the hydrogen bond is not in the interior of the protein, it is best considered to be worth 0 kcal/mol, because water provides fierce hydrogen bonding competition. When one of the components, either the donor or acceptor, is charged, the strength increases substantially, and some researchers quote 4.0 to 4.5 kcal/mol. This is a bit larger than the 3 kcal/mol we gave for a buried salt bridge (see Section 3.2.1 on salt bridges). These numbers are not fully consistent, which just goes to show the rough nature of the values, and the considerable work in this area that is still needed.

The solvent dramatically influences the strength of hydrogen bonds because the donor and acceptor are solvated prior to formation of the Dn–H • • •:Ac hydrogen bond. Many polar solvents can form hydrogen bonds themselves, meaning that the donor and acceptor al-

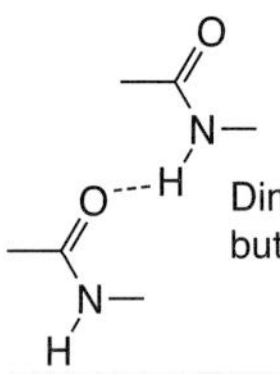

Dimer exists in CCl$_4$
but not in dioxane

Secondary amide dimers

ready possess hydrogen bonds prior to their combination. Hence, if the hydrogen bonds between Dn–H, :Ac, and the solvent S are essentially the same in strength, it is a "wash" to undergo the reaction shown in Eq. 3.26. Such a solvent is referred to as a **competitive solvent**. When the solvent is nonpolar and cannot form hydrogen bonds, the Dn–H• • •:Ac interaction more effectively influences the thermodynamics of Eq. 3.26, making the hydrogen bond appear stronger. Therefore, the most important factor for determining strength is a solvent's ability to form hydrogen bonds. For example, the dimerization of N-methylacetamide occurs in carbon tetrachloride, but is nearly nonexistent in the solvent dioxane, which has the same dielectric constant, because dioxane can accept hydrogen bonds. Since the solvent influences the strength of hydrogen bonds so dramatically, it is not surprising that the ability to form hydrogen bonds correlates to various solvent parameters, and an example of this is given in the following Connections highlight.

$$D-H \bullet\bullet\bullet S \;+\; A \bullet\bullet\bullet H-S \;\rightleftharpoons\; D-H \bullet\bullet\bullet A \;+\; S \bullet\bullet\bullet H-S$$

(Eq. 3.26)

Connections

Solvent Scales and Hydrogen Bonds

Since the polarity and hydrogen bonding capabilities of a solvent are of paramount importance in determining the strengths of hydrogen bonds, we might expect a correlation with solvent parameters. Indeed, such correlations have been found. In one specific case, the intrinsic $\Delta G°$ for the intramolecular hydrogen bond in the substituted cyclohexane shown to the right was plotted against several different solvent parameters. The best linear fit was a combination of the $E_T(30)$ and β values, where the β value of the solvent dominated the correlation. Recall that the β value is a measure of the hydrogen bond accepting ability of the solvent, whereas the $E_T(30)$ value correlates general polarity. The conclusion is that as the polarity of the solvent increases, the strength of the intramolecular hydrogen bond decreases, but that this is a secondary effect compared to the hydrogen bond accepting ability of the solvent. A higher hydrogen bond accepting ability in the solvent significantly decreases the free energy of formation of the intramolecular hydrogen bond.

Intramolecular hydrogen bond

Beeson, C., Pham, N., Shipps, G. Jr., and Dix, T. A. "A Comprehensive Description of the Free Energy of an Intramolecular Hydrogen Bond as a Function of Solvation: NMR Study." *J. Am. Chem. Soc.*, **115**, 6803–6812 (1993).

---

ii. Electronegativity Effects

The electrostatic model predicts that for a neutral donor, the larger the partial charge on H, the stronger the hydrogen bond. Indeed, hydrogen bonding strengths to a variety of acceptors follow the trend for donors, HF > HCl > HBr > HI. Note that the hydrogen bond strength is not following the strength of the acid for these donors (see Section 5.4.5 for acid strengths), but instead the charge on hydrogen. However, when we contrast hydrogens attached to the same kind of atom, the stronger acids have a larger charge on the hydrogen, and therefore are the better hydrogen bond donors. Therefore, we expect the trend CF_3CO_2H > CCl_3CO_2H > CBr_3CO_2H > CI_3CO_2H, which follows the trend in acid strength (see Chapter 5).

For the acceptor, we see trends such as H_2O > H_3N > H_2S > H_3P. We would anticipate that electronegativity on the acceptor atom is a double-edged sword. It increases the $\delta-$ on the atom, which is good for hydrogen bonding, but it makes the element less willing to share its electrons, which is bad for hydrogen bonding. As such, bonds to F are quite polar, but F is a very poor hydrogen bond acceptor (i.e., a poor electron donor). Hydrogen bonds involving F as the acceptor are actually rare. The poor hydrogen bonding seen with S and P is likely due to the very diffuse nature of the lone pairs in third row elements, which makes them poor acceptors. Examples of some of the trends we have discussed above are given in Table 3.7 for gas phase and very nonpolar solvents.

Table 3.7
Values of $\Delta H°$ for Some Selected Hydrogen Bonds*

Hydrogen bond	Compounds involved	Medium	Strength (kcal/mol)
O–H•••O=C	Formic acid/formic acid	Gas phase	–7.4
O–H•••O–H	Methanol/methanol	Gas phase	–7.6
O–H•••OR$_2$	Phenol/dioxane	CCl$_4$	–5.0
O–H•••SR$_2$	Phenol/n-butyl sulfide	CCl$_4$	–4.2
O–H•••SeR$_2$	Phenol/n-butyl selenide	CCl$_4$	–3.7
O–H•••sp^2 N	Phenol/pyridine	CCl$_4$	–6.5
O–H•••sp^3 N	Phenol/triethylamine	CCl$_4$	–8.4
N–H•••SR$_2$	Thiocyanic acid/n-butyl sulfide	CCl$_4$	–3.6

*Jeffrey, G. A. (1998). *An Introduction to Hydrogen Bonding (Topics in Physical Organic Chemistry)*, Oxford University Press, Oxford.

iii. Resonance Assisted Hydrogen Bonds

As already noted, hydrogen bonds are very sensitive to their context. Solvent and electronegativity effects likely play the largest roles in modulating their strength. However, several other factors can be identified as major contributors. The most frequently cited factors are resonance and polarization enhancement, although more recently another factor called "secondary hydrogen bonds" has found wide acceptance.

Resonance assisted hydrogen bonds are those that benefit from a particular resonance structure of the donor or acceptor. For example, the intramolecular hydrogen bond of o-nitrophenol is known to be exceptionally strong, and is enhanced by the resonance structure shown below. Such an interaction might just as well be considered as hydrogen bond assisted resonance; it is just a case of semantics. Amides in linear chains, as found in protein α-helices (Appendix 4), are also postulated to benefit from such an interaction, and even the base pairs in the DNA helix are often considered to possess such an interaction. The following Connections highlight gives some data that supports the notion of resonance assisted hydrogen bonding.

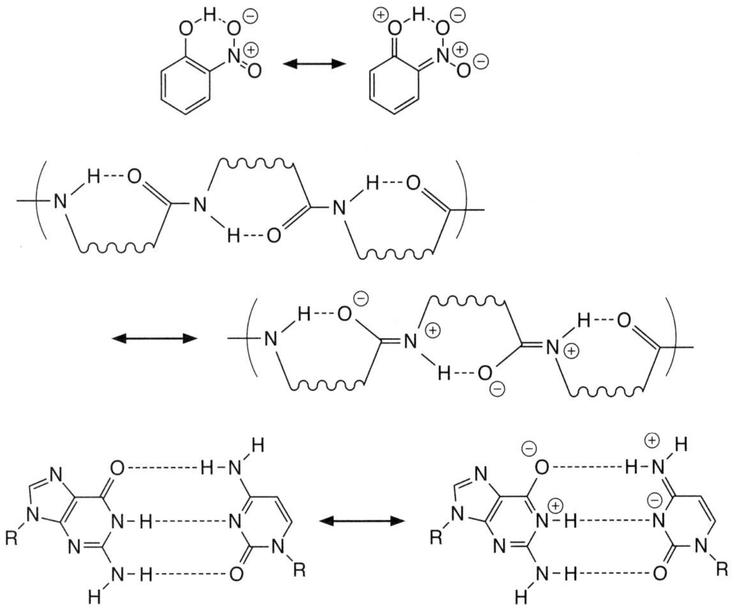

Examples of resonance assisted hydrogen bonding

Connections

The Extent of Resonance can be Correlated with Hydrogen Bond Length

A correlation has been found between a parameter that measures the extent of resonance delocalization and hydrogen bond length in β-diketone enols. The greater the contribution of the ionic resonance structures for chains of β-diketones shown below, the closer are the bond lengths d_1, d_2, d_3, and d_4.

To measure the relative contribution of the two resonance structures, a parameter called Q was defined as $Q = d_1 - d_2 + d_3 - d_4$. As the ionic resonance structure becomes more important, the parameter Q becomes smaller. In an examination of 13 crystal structures and a single neutron diffraction study of β-diketone enols, as well as several other intermolecular hydrogen bonded chains, a correlation was found between parameters such as Q and hydrogen bond distance (defined as the intermolecular O–O distance). Smaller O–O distances (meaning a stronger hydrogen bond) correlate well with lower Q values, meaning more resonance delocalization.

Gilli, G., Bertolasi, V., Feretti, V., and Gilli, P. "Resonance-Assisted Hydrogen Bond. III. Formation of Intermolecular Hydrogen-Bonded Chains in Crystals of β-Diketones and its Relevance to Molecular Association." *Acta. Cryst.*, 564–576 (1993).

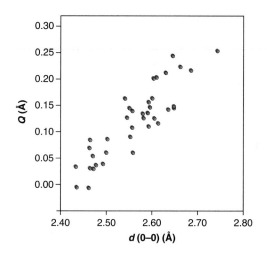

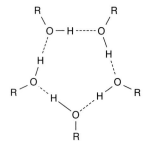

Cyclic structure formed from hydrogen bonding

Definitions of bond lengths used to calculate Q

iv. Polarization Enhanced Hydrogen Bonds

Polarization enhanced hydrogen bonds (also known as **cooperative hydrogen bonds**) are similar in concept to resonance enhanced hydrogen bonds. This phenomenon arises when there are neighboring hydrogen bonding groups that assist the polarization in the Dn–H bonds, making them better donors. Consider the water trimer shown in Eq. 3.27. Stabilization of the partial charges on the hydrogens and oxygens of the already formed dimer occurs when the third water makes a hydrogen bond.

(Eq. 3.27)

The best evidence that such a concept is important in hydrogen bonding arises from *ab initio* calculations. The strengths of hydrogen bonds have been calculated for alcohols in a cyclic arrangement, such as the pentamer of an alcohol shown in the margin with all cooperative hydrogen bonds. The strengths are found to increase from 5.6 kcal/mol for a cyclic trimer, to 10.6 kcal/mol for a cyclic pentamer, and 10.8 kcal/mol for a cyclic hexamer. However, some evidence also comes from crystal structures, and the following Connections highlight describes evidence from oligosaccharide structures.

Connections

Cooperative Hydrogen Bonding in Saccharides

Chains of cooperative hydrogen bonds are commonly seen in crystal structures of mono- and oligosaccharides. Shown below is a picture of the crystal structure of *p*-nitrophenyl α-maltohexaoside. A long running chain of hydrogen bonds can be identified along the 2,3-vicinal diol portion of the pyranosides, which orients one monomer with respect to the next.

Hindricks, W., and Saenger, W. "Crystal and Molecular Structure of the Hexasaccharide Complex (*p*-Nitrophenyl α-Maltohexaoside)BaI₃·27H₂O. *J. Am. Chem. Soc.*, **112**, 2789–2796 (1990).

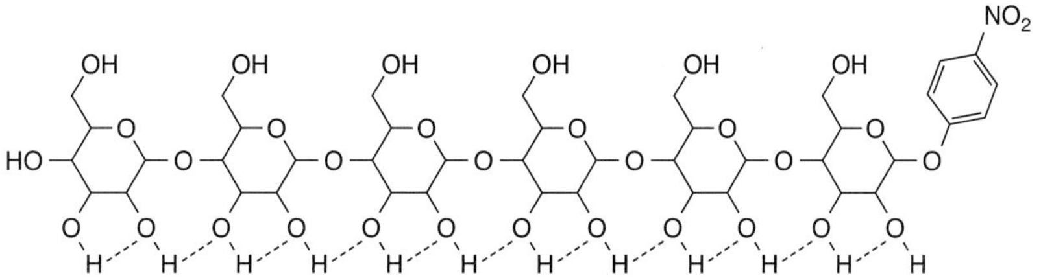

Intramolecular hydrogen bonding in oligosaccharides

---

v. Secondary Interactions in Hydrogen Bonding Systems

Since the microenvironment near hydrogen bonds greatly influences their strength, it makes sense that the proximity of other hydrogen bonds would also have an influence. In fact, when there are hydrogen bonds adjacent to one another, secondary interactions can arise which can either reinforce or weaken the primary hydrogen bonds. For example, the dimerization of two carboxylic acids yields two hydrogen bonds. However, there are also two "transannular" repulsive interactions between the hydrogen bonded species. Electrostatic arguments nicely rationalize these. In this system, the hydrogens are $\delta+$, the oxygens $\delta-$, and so the H•••H and O•••O interactions are repulsive. In contrast, when the donors are on one structure, and the acceptors on the other, the primary hydrogen bonds are supported by the secondary interactions.

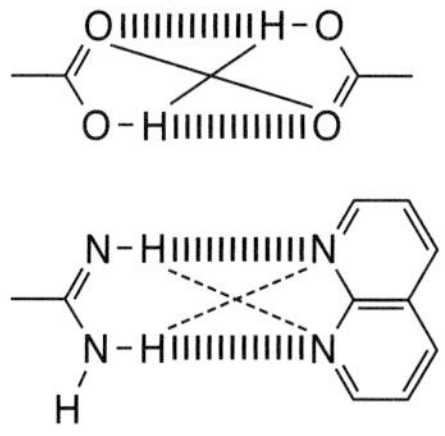

Primary hydrogen bonds (IIII)
Secondary hydrogen bonds (----)
Repulsive interactions (——)

vi. Cooperativity in Hydrogen Bonds

If hydrogen bonds are so weak in water, why is it that they can create such complex and diverse three-dimensional molecular architectures? As we will note in our discussion of the hydrophobic effect (see below), the major driving force for molecular associations in water is nonpolar binding derived from a release of water from around nonpolar surfaces. This means that organic molecules will tend to non-selectively aggregate with other organic molecules in water due to the hydrophobic effect. This non-specific association can contribute to making hydrogen bonds significant in water. A significant part of the reason that simple hydrogen bonds do not lead to strong association in water is the entropic penalty that must be paid for freezing the motions of the two partners. This $\Delta S°$ penalty is typically not adequately compensated by the favorable $\Delta H°$ for the interaction, remembering that the *net* $\Delta H°$ might be quite small (Eq. 3.26). However, if two large molecules are already brought together because of the hydrophobic effect, the entropy penalty has been partially pre-paid (local conformations must still be restricted to form the hydrogen bond). In this situation, it is more likely that hydrogen bonding could contribute to the overall association.

Hydrophobic association is generally non-specific, but selectivity can be imparted to organic association in water by hydrogen bonds, and especially by arrays of hydrogen bonds. As with a salt bridge, we might expect that an isolated hydrogen bond on the surface of a protein would contribute little to protein stability. Once again we find a significant context effect because the force is weak to start, and we need a reference point to determine the strength of the interaction (see the next Going Deeper highlight). However, a spectacular example of hydrogen bonding in protein structure is the α-helix (Appendix 4). We noted in

Chapter 1 that an amide functionality of the sort found in a typical peptide bond has excellent hydrogen bonding capability, both as a donor and an acceptor. In an α-helix a continuous stretch of the protein has all the amide hydrogen bonding potential completely satisfied. This creates a regular structure in the protein that nature exploits extensively. Why is this hydrogen bonding successful in water? One factor is the way the amides are to some extent shielded by the α-helix structure, making the microenvironment more "organic like". This partially desolvates the amides, making competition by water less of a factor. Another important issue, though, is **cooperativity**. The repeating structure of the α-helix reinforces itself. Once a few hydrogen bonds are formed, the system naturally propagates and each hydrogen bond reinforces the next. This can be viewed as an entropic effect. The first few hydrogen bonds pay most of the entropic cost, making it more and more favorable to continue the stretch of hydrogen bonding.

Going Deeper

How Much is a Hydrogen Bond in an α-Helix Worth?

Hydrogen bonding is the key feature that holds together the α-helix of protein secondary structure. To quantify such an interaction, though, is more difficult than it may seem. We have already noted the problems associated with placing values on hydrogen bond strengths. However, through a clever combination of organic chemistry and molecular biology, Schultz and co-workers were able to obtain a good estimate of the magnitude of the key hydrogen bond of the α-helix. Perhaps surprisingly, the protein synthesis machinery, the ribosome, can be coaxed into incorporating an α-hydroxy acid instead of an α-amino acid into a specific site in a protein. As shown in the picture to the right, this replaces the usual amide of the protein backbone with an ester, which disrupts the hydrogen bonding in the α-helix. By removing an NH and replacing it with O, one hydrogen bond of an α-helix would be lost. However, it is also true that an amide carbonyl is a much better hydrogen bond acceptor than an ester carbonyl, and so the backbone substitution should also weaken a second hydrogen bond. By studying a well-defined helix in a protein of known stability, and by placing esters at the beginning, middle, and end of the helix, it was possible to dissect out the contributions of these various factors. The substitution of an ester for an amide

destabilized the α-helix by 1.6 kcal/mol. Perhaps surprisingly, the weakening of the carbonyl as an acceptor was determined to have a larger effect (0.89 kcal/mol) than the deletion of the NH (0.72 kcal/mol).

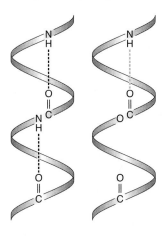

Koh, J. T., Cornish, V. W., and Schultz, P. G. "An Experimental Approach to Evaluating the Role of Backbone Interactions in Proteins Using Unnatural Amino Acid Mutagenesis." *Biochemistry*, **36**, 11314–11322 (1997).

Vibrational Properties of Hydrogen Bonds

In Section 2.1.4 we described the vibrational properties and potential wells of covalent bonds. Any bond possesses thermal motion, even at absolute zero, due to the zero point vibrational state. For a Dn–H bond, formation of a hydrogen bond to :Ac restricts the motion of the hydrogen atom because the hydrogen is now restrained by two bonds rather than one. Using infrared spectroscopy to measure the vibrational frequencies of the Dn–H bond is therefore a good experimental tool for characterizing hydrogen bonds. The vibrational frequencies of both the Dn–H bond and the H•••:Ac bond can often be observed.

When hydrogen bonds are formed, the single well potential that describes the covalent Dn–H bond is converted to an energy surface with two minima, reflecting the addition of the Ac•••H bond (Figure 3.6 **A**). The second minimum describes transfer of the hydrogen from the donor to the acceptor. In a typical weak hydrogen bond, there is a significant energy bar-

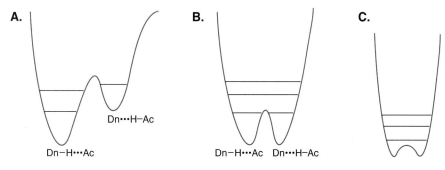

Figure 3.6
Potential energy plots for the vibrational states of various hydrogen bonds.
A. A normal hydrogen bond, **B.** a low-barrier hydrogen bond, and **C.** a no-barrier
hydrogen bond.

rier between the preferred Dn–H● ● ●:Ac form and the less favorable Dn–● ● ●H–Ac form. In
addition, the zero-point energies for both are well below the barrier.

There are characteristic vibrational modes that can be observed in the infrared spectra
that are diagnostic of the double well potential and hence hydrogen bonds. Table 3.8 shows
the stretches and bends found for normal hydrogen bonds such as those described by Figure
3.6 **A**. We find new frequencies for the in-plane and out-of-plane bends of the Dn–H bond,
but also new stretching and bending modes for the hydrogen bond itself. In keeping with the
picture that the bond between the Dn and H atom is weakened upon formation of a hydrogen bond, the Dn–H stretch moves to lower frequency, accompanied by an increase in intensity and band width. In support of the picture that the hydrogen atom is now held between
two atoms, the bending frequencies move to higher values.

Table 3.8
**Characteristics Vibrational Modes for Normal
Hydrogen Bonds, R–Dn–H ● ● ● Ac***

Vibrational modes	Frequencies (cm^{-1})
Dn–H stretch	3700–1700
Dn–H in-plane bend	1800–1700
Dn–H out-of-plane bend	900–400
H● ● ●Ac bond stretch	600–50
H● ● ●Ac bond bend	< 50

*Jeffrey, G. A. (1998). *An Introduction to Hydrogen Bonding (Topics in
Physical Organic Chemistry)*, Oxford University Press, Oxford.

Short–Strong Hydrogen Bonds

There are some important properties of hydrogen bonds that are evident from the double well potential of Figure 3.6 **A**. Imagine a case for which placing the hydrogen on either
the donor or the acceptor is of equal energy. Further, if the distance between the heteroatoms
is made short, often around 2.4 to 2.5 Å, the barrier to transfer of the hydrogen bond between
the donor and acceptor becomes close to the zero-point energy of the vibration that holds
the H atom in the complex (Figure 3.6 **B**). Hence, when the energies of the Dn–H● ● ●Ac and
Dn● ● ●H–Ac forms become essentially equal and the distance between Dn and Ac is short,
the barrier either becomes very low or completely disappears. These hydrogen bonds are referred to as **low-barrier hydrogen bonds** (LBHB) or **no-barrier hydrogen bonds** (Figures
3.6 **B** and **C**). When the barrier to transfer drops completely below or is very close to the zero-
point energy, the hydrogen moves in quite a wide potential well, and on average is centered
between the donor and acceptor atom. The wide potential well is accompanied by a lower

force constant for the stretching vibration, thereby having an interesting ramification on isotope effects. Both the low-barrier and no-barrier hydrogen bonds are referred to as **short–strong hydrogen bonds**.

The model that emerges from this analysis is that we can expect a LBHB in a Dn–H● ● ●:Ac system whenever the Dn and Ac atoms are very close and the pK_a values of Dn–H and H–Ac$^+$ are close, because this puts the two potential wells at nearly equal energies (see Section 5.2.1 for a discussion of pK_a values). If :Ac is anionic, as is often true for LBHBs, then it is the pK_a values of Dn–H and H–Ac that must be close. We are not saying that some "special" stabilization occurs when the pK_a values are close, just that this creates the strongest hydrogen bond. The closer the pK_a values, the stronger the hydrogen bond.

The low-barrier and no-barrier hydrogen bonds possess considerable degrees of electron sharing between the hydrogen atom and the donor and acceptor atoms. In this regard, the bond is a **three center–four electron bond**, and it has a considerable amount of covalent character. Hence, the directionality of these bonds is much more important than for traditional hydrogen bonds, with linear Dn● ● ●H● ● ●Ac geometries being strongly preferred.

The dependence of hydrogen bond strength upon bond length for a series of hydrogen bonds in the gas phase is shown in Figure 3.7. For a series of O–H● ● ●O hydrogen bonds, the energy of the hydrogen bond is plotted as a function of the O● ● ●O distance. The plot is decidedly non-linear. Consider a hydrogen bond with an O● ● ●O distance of 2.52 Å. It would have a hydrogen bond energy of less than 10 kcal/mol. Now consider the consequence of shrinking the hydrogen bond to 2.45 Å. For a very modest contraction of 0.07 Å, the hydrogen bonding energy goes up to more than 25 kcal/mol. This would now be a short–strong hydrogen bond.

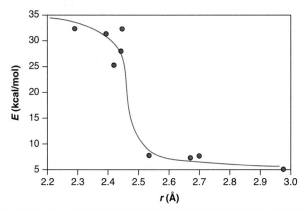

Figure 3.7
Hydrogen bond strengths as a function of heteroatom distances in the gas phase. See the first reference for short-strong hydrogen bonds at the end of the chapter.

Compounds proposed to possess low-barrier hydrogen bonds

The prototypical short–strong hydrogen bond is bifluoride [F–H–F]$^-$, which has a F–F distance of 2.25 Å and a bond strength of 39 kcal/mol. Table 3.9 shows a handful of other hydrogen bond strengths for short–strong hydrogen bonds.

In solution, very short distances between oxygen heteroatoms are observed in β-diketo enols and some diacid monoanions. Shown in the margin are just a few structures possessing hydrogen bond lengths consistent with low-barrier character.

At present, short–strong hydrogen bonds are well documented in the gas phase, and theoretical studies support their existence, but there is still some controversy as to the significance of the phenomenon in high polarity solvents. If they do occur in water, they have the potential to profoundly influence molecular recognition phenomena and enzymology. This point is addressed further in the following two Connections highlights.

Table 3.9
Strengths of Short–Strong Hydrogen Bonds*

Hydrogen bond	Strength (kcal/mol)[‡]	Hydrogen bond	Strength (kcal/mol)[‡]
$F^- \bullet\bullet\bullet HF$	39	$F^- \bullet\bullet\bullet HO_2CCH_3$	21
$Cl^- \bullet\bullet\bullet HF$	22	$F^- \bullet\bullet\bullet HOCH_3$	30
$Br^- \bullet\bullet\bullet HF$	17	$F^- \bullet\bullet\bullet HOPh$	20
$I^- \bullet\bullet\bullet HF$	15	$F^- \bullet\bullet\bullet HOH$	23
$CN^- \bullet\bullet\bullet HF$	21	$H_3N \bullet\bullet\bullet H-NH_3^+$	24

*Jeffrey, G. A. (1998). *An Introduction to Hydrogen Bonding (Topics in Physical Organic Chemistry)*, Oxford University Press, Oxford.
[‡]Values were determined in the gas phase by ion cyclotron resonance.

Connections

Proton Sponges

Probably the most common use of molecular geometries that enforce a very short heteroatom–heteroatom distance is in the creation of "proton sponges". These are fused-ring aromatic diamines where the amines are oriented in such a way as to cooperatively bind a single proton. Three examples of the conjugate acids of proton sponges are shown to the right. The first has a pK_a of 12.1 and the second has a pK_a of 16.1, while the third has a pK_a of 13.9. Therefore, the second compound is 10,000 times less acidic than the first. Since the substitution of the methoxy groups in the para position did not give the four orders of magnitude decrease in the acidity of the parent compound, it must be the steric compression from the *o*-methoxy groups that makes the center compound the least acidic. This shows how important it is to enforce the short distances between the heteroatoms to achieve the short–strong hydrogen bonds.

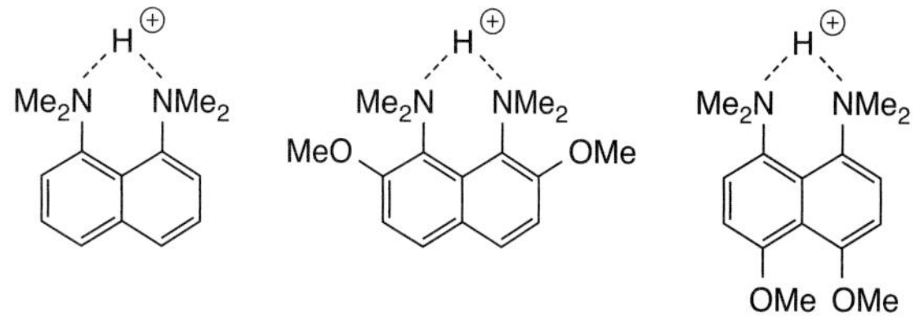

Compounds referred to as "proton sponges"

Staab, H. A., Kriéger, C., Hieber, G., and Oberdorf, K. "1,8-Bis(dimethylamino)-4,5-dihydroxynaphthalene, a Neutral, Intramolecularly Protonated 'Proton Sponge' with Zwitterionic Structure." *Angew. Chem. Int. Ed. Eng.*, **36**, 1884–1886 (1997).

Connections

The Relevance of Low-Barrier Hydrogen Bonds to Enzymatic Catalysis

Other than just gaining a basic understanding of the phenomenon of hydrogen bonds, why is the discussion of short–strong hydrogen bonds significant? Consider a substrate bound to the active site of an enzyme (or any other catalyst). As discussed in greater detail in Chapter 9, enzymes achieve their rate acceleration by preferential binding of the transition state of the reaction. Since the rate accelerations are often quite dramatic, this *preferential* binding must be substantial. The problem is that the enzyme also binds the substrate (the ground state), and on going from the ground state to the transition state, the geometry changes are often small, and no new hydrogen bonds are produced. However, if a very small binding change can lead to a very large increase in hydrogen bonding energy, we have the ideal situation for preferential binding of the transition state. Based on this, then, the role of the enzyme is to create a microenvironment in which

the necessary change in pK_a of the substrate relative to the transition state can occur. The postulate would be that the pK_a of the transition state is becoming closer to the pK_a of the functional group on the enzyme making contact with the transition state. It is well established that a properly designed protein environment can substantially alter pK_a values (see Chapter 5), and so this is an attractive mechanism for enzymatic catalysis.

Many studies have looked for low-barrier hydrogen bonds at enzyme active sites, with decidedly mixed results thus far. Currently, the question still remains as to whether LBHBs are important in many systems or are just a novelty associated with specialized hydrogen bonds in the gas phase. Stay tuned!

Gerlt, J. A., and Gassman, P. G. "Understanding the Rates of Certain Enzyme-Catalyzed Reactions: Proton Abstraction from Carbon Acids, Acyl-Transfer Reactions, and Displacement Reactions of Phosphodiesters." *Biochemistry*, **32**, 11943–11952 (1993). Cleland, W. W., and Kreevoy, M. M. "Low-Barrier Hydrogen Bonds and Enzymatic Catalysis." *Science*, **264**, 1887–1890 (1994).

In summary, hydrogen bonds are among the most important of the binding forces, yet for the most part they are purely electrostatic in nature. Although several factors determine their strength, such as resonance, geometry, and the nature of the donor and acceptor, it is the solvent that plays the largest role. In competitive solvent systems, a series of hydrogen bonds is required to impart a defined structure. The creation of artificial systems that possess various hydrogen bonding capabilities that mimic natural systems is an active area of modern physical organic chemistry. The following Connections highlight shows a recent example of exploiting hydrogen bonding for structural purposes in a totally unnatural system.

Connections

β-Peptide Foldamers

A universal feature of proteins is that they fold into well-defined, three-dimensional structures, partially due to hydrogen bonding (see Chapter 6). This is crucial to the proper functioning of living systems, but it is also a very interesting phenomenon. It is perhaps surprising that it has not been a long-standing goal of physical organic chemistry to learn how to make artificial systems that do the same thing. What would it take to build organic molecules that spontaneously fold into well-defined shapes? In recent years, this fundamentally interesting question has begun to attract the attention of physical organic chemists.

The targets of such research have been termed **foldamers**, and are defined as any polymer or oligomer with a strong tendency to adopt a specific, compact conformation. Taking a lead from nature's best known "foldamer",

researchers have used amide hydrogen bonding analogous to that seen in the α-helix (Appendix 4) to create well-defined, unnatural folds. A good deal of success has been obtained by Seebach and Gellman with β-peptides, polypeptides that use β-amino acids instead of the α-amino acids of biology. Oligomers of appropriate β-amino acids will fold into well-defined structures. As with the α-helix, the major organizing force is the chains of amide hydrogen bonding. This opens up many new opportunities for the rational design of organic molecules with well-defined structures and properties.

Gellman, S. H. "Foldamers: A Manifesto." *Acc. Chem. Res.*, **31**, 173–180 (1998). Seebach, D., Beck, A. K., and Bierbaum, D. J. "The World of β- and γ-Peptides Comprised of Homologated Proteinogenic Amino Acids and Other Components." *Chem. Biodiversity*, **1**, 1111–1239 (2004).

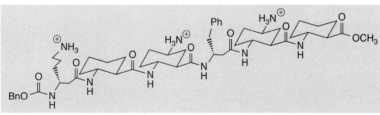

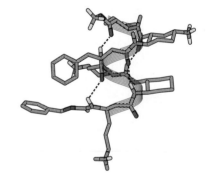

β-Amino acid foldamer

3.2.4   π Effects

In our discussions of ion pairing, dipole interactions, and normal hydrogen bonding, electrostatic factors played a dominant role. In fact, most binding forces have simple electrostatic attractions at their origin (see the hydrophobic effect, below, for an exception). Therefore, regions of negative charge, no matter what their nature, will in general be attracted to regions of positive charge, no matter what their nature. It is the character of the partners that leads to our definitions and discussions of the forces.

One region of negative charge associated with a large number of molecules derives from π systems, whether in aromatic structures or simple alkenes. The existence of such regions leads us to expect π systems to be involved in a variety of molecular recognition phenomena. These interactions can be surprisingly strong, or at times, exceedingly weak; it is once again a matter of context. Three general π binding forces are discussed here: the cation–π interaction, the polar–π interaction, and π donor–acceptor interactions.

Cation–π Interactions

Another non-covalent binding force that is comparable in strength to a salt bridge or a hydrogen bond (depending on the context!) is the **cation–π interaction**. This is the non-covalent interaction between a cation and the face of a simple π system such as benzene or ethylene. Only in recent years has it begun to be appreciated that this interaction can be quite strong and can make significant contributions to molecular recognition phenomena in both biological and synthetic systems. Figure 3.8 shows that in the gas phase the interaction can be quite strong—the Li⁺•••benzene interaction is comparable to even the strongest hydrogen bond. Before we discuss context and solvation effects, we need to develop a physical model for the interaction.

The clear trend of Figure 3.8—$Li^+ > Na^+ > K^+ > Rb^+$—is reminiscent of the hydration trends we discussed in Section 3.2.2. The hydration trends were rationalized with an electrostatic and size model, and an electrostatic model of the cation–π interaction has also proven to be quite powerful. How can we develop an electrostatic model with benzene as one of the partners?

The electrostatic model of water binding to an ion can be described as an ion–dipole interaction (Section 3.2.2). The cation interacts with the negative end of the large permanent dipole moment of water. Benzene has no dipole moment, but it does have a large, permanent quadrupole moment. Recall from our discussion in Chapter 1 that a quadrupole moment is simply two dipoles aligned in such a way that there is no net dipole. The quadrupole moment of benzene is of the form in which two dipoles are aligned end-to-end.

Recall also that the quadrupole moment of benzene arises because an sp^2 C is more electronegative than H. This creates six $C^{\delta-}$–$H^{\delta+}$ bond dipoles, and under the symmetry of benzene, they add up to a quadrupole moment. Similarly, the four $C^{\delta-}$–$H^{\delta+}$ bond dipoles in ethylene combine to make a substantial quadrupole in that molecule. This argument has

A.

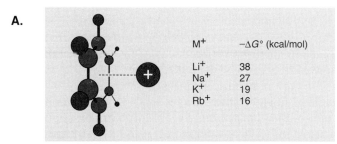

M⁺	$-\Delta G°$ (kcal/mol)
Li⁺	38
Na⁺	27
K⁺	19
Rb⁺	16

B.

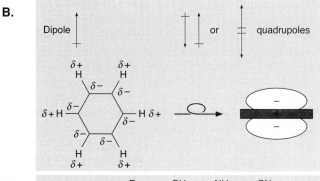

C.

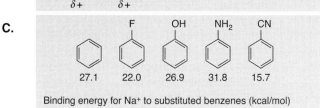

Binding energy for Na⁺ to substituted benzenes (kcal/mol)

Figure 3.8
The cation–π interaction. **A.** The basic nature of the interaction and binding energies for simple cations to benzene (gas phase experimental numbers). **B.** The relationship between dipoles and quadrupoles, and an illustration of six bond dipoles giving rise to a molecular quadrupole. Note that the left image is top down on the benzene, while the right image is edge on. **C.** Substituent effects on the cation–π interaction. These are calculated values. See also the analogous electrostatic potential surfaces in Appendix 2.

nothing to do with aromaticity, and so is not unique to benzene and its derivatives. While the emphasis in molecular recognition studies has been on benzene and its derivatives, ethylene and acetylene derivatives can participate in exactly the same way. Another important point is that the multipole expansion—pole, dipole, quadrupole, octapole, . . . —is *not* a perturbation series. Terms do not get progressively "smaller" as we move along the series. There is no reason that a quadrupole cannot bind an ion electrostatically just as well as a dipole, and to first order that is what is going on in the cation–π interaction. Another way to visualize the quadrupole moment of benzene is by viewing the electrostatic potential surfaces of the molecules. As shown in Appendix 2, the electrostatic potential surface of benzene is negative on the face of the ring and positive along the edge. Again, it is evident that cations should be attracted to the face. The same is true for alkenes and alkynes, as shown in the electrostatic potential surfaces for these molecules.

Once we accept the existence of quadrupole moments and appreciate that they can bind ions in the same way that dipole moments can, we should not be surprised by any of the "π effects" of this section. The only surprise is the large magnitude of the effects. For example, water binds K^+ in the gas phase with $\Delta H° = -18$ kcal/mol, an interaction we would describe to first order as that between the dipole of water and the ion. Benzene binds K^+ in the gas phase with $\Delta H° = -19$ kcal/mol. Clearly, a quadrupole can compete with a dipole!

As with other strongly electrostatic interactions, we would expect the cation–π interaction to be strongest in the gas phase, slightly weakened in organic solvents, and significantly attenuated in aqueous solvent. This is true to some extent, but the weakening of the interaction on moving into water is much less than we might expect. For example, the methylammonium•••acetate ion pair is worth ~120 kcal/mol in the gas phase, but ≤ 2 kcal/mol in water. On the other hand, the methylammonium•••benzene cation–π interaction is worth only ~19 kcal/mol in the gas phase, but is ~5 kcal/mol in water. Apparently, water is much less effective at attenuating a cation–π interaction than an ion pair or a hydrogen bond.

There appear to be two reasons for the retained strength of the cation–π interaction in water. First, remember that one component of the cation–π interaction, the benzene, is hydrophobic. So, to cover one face of it with an ion might be favorable in water (see the discussion of the hydrophobic effect given below).

The second issue is more subtle and complex, but relates back to our earlier discussion of Born solvation and the substantial long range solvation that water exerts on an ion (Section 3.2.2). This long range solvation arises because water molecules will tend to align their dipoles for a favorable interaction with the ion. At long distances these waters are not locked into a particular orientation. On average, however, there is a tendency for the water dipoles to be found more often in the favorable rather than the unfavorable dipole orientation. Now consider an ion pair at close contact. What should a water molecule that is 8–10 Å away do with its dipole? Many waters will be essentially equidistant from the two ions, and it will not be possible to achieve a favorable interaction with one ion without simultaneously achieving an unfavorable interaction with the other ion. It is as if forming the ion pair neutralized the charges, or at least that is what the more distant solvent molecules must feel. On the other hand, when a cation binds to benzene, there is no charge neutralization—the system remains a full cation regardless of the separation between the interacting partners. Full "Born" solvation is possible.

The electrostatic potential surfaces of simple aromatics also nicely rationalize the substituent effects on the cation–π interaction (Figure 3.8 **C**). These effects are not what might be immediately expected. Usually we think of phenol as electron rich, and so it is a bit surprising that it is not a better cation–π binder than benzene. However, the electrostatic potential surfaces fully support this result and the other results of Figure 3.8. To a considerable extent, the cation–π interaction is more affected by the inductive influence of a substituent than by π donation.

In summary, although less well known than ion pairs and hydrogen bonds, cation–π interactions contribute significantly to molecular recognition. They are very common in protein structures (Lys/Arg interacting with Phe/Tyr/Trp), and many binding sites for cationic ligands use cation–π interactions (see the example given in the next Connections highlight).

Synthetic receptors such as cyclophanes can substantially exploit the cation–π interaction in binding (see Section 4.2.5). Also, in crystal packing and many catalytic systems, cation–π interactions can be important players.

Connections

A Cation–π Interaction at the Nicotine Receptor

Acetylcholine (ACh, $Me_3N^+CH_2CH_2OC(O)CH_3$) is a common neurotransmitter. Every time you move a muscle voluntarily it is because this small, cationic molecule is released from a nerve terminal, drifts across the synapse, and binds to a specific neuroreceptor. The same process also occurs in the brain, and interestingly, nicotine is able to fool the neuroreceptor and elicit a physiological response. For this reason, the receptor is called the nicotinic acetylcholine receptor (nAChR), and the first step of nicotine addiction is nicotine binding to this receptor in the brain. The nAChR is a complex, integral membrane protein, and no crystal structure is available. However, a cation–π interaction is involved in binding ACh to the receptor. To prove this, the electrostatic model of the cation–π interaction was invoked. In particular, at a specific tryptophan residue of the receptor, successive fluorination was used to modulate the cation–π interaction. Fluorine has a predicable and *additive* effect on the quadrupole moment, and hence the cation–π binding ability, of simple aromatics. At the receptor, the tryptophan of interest was successively replaced with monofluoro-, difluoro-, trifluoro-, and tetrafluorotryptophan, and ACh binding was measured. A linear free energy relationship was seen between cation–π binding ability of the aromatic and the effectiveness of ACh at the modified receptor (see Chapter 8 for a discussion of linear free energy relationships). This effect was seen at only one specific tryptophan, establishing a cation–π interaction between the quaternary ammonium group of ACh and this aromatic group in the protein.

Zhong, W., Gallivan, J. P., Zhang, Y., Li, L., Lester, H. A., and Dougherty, D. A. "From *ab initio* Quantum Mechanics to Molecular Neurobiology: A Cation–π Binding Site in the Nicotinic Receptor." *Proc. Natl. Acad. Sci. (USA)*, **95**, 12088–12093 (1998).

Polar–π Interactions

Water binds cations electrostatically by aligning its large permanent dipole moment appropriately. Benzene binds cations electrostatically by aligning its large permanent quadrupole moment appropriately. Does this mean that benzene is a polar molecule? The most sensible answer is "yes". Typically, to say a molecule is polar is to say it has a substantial, permanent dipole moment. But why shouldn't a quadrupole moment count just as much as a dipole? If a molecule can bind ions strongly through a predominantly electrostatic interaction, it should be considered to be polar. Benzene is polar—it's just quadrupolar rather than dipolar. However, benzene is not a polar solvent and is, in fact, hydrophobic, too. This emphasizes a clear distinction between molecular phenomena and bulk, condensed phase phenomena. The two are not always tightly coupled.

If benzene is a polar molecule, it should experience molecular phenomena besides just cation binding, similar to what other polar molecules do. Water binds water well, and benzene binds water, too. The binding energy between benzene and water is 1.9 kcal/mol in the gas phase, and the geometry is as expected with the water hydrogens (the positive end of the water dipole) pointed into the benzene ring (see margin). Similarly, ammonia binds to benzene with 1.4 kcal/mol of binding energy in the gas phase. In a nonpolar solvent such as cyclohexane, the binding between the NH_2 group of aniline and the face of benzene is worth 1.6 kcal/mol.

Such interactions have been called hydrogen bonds to benzene. However, this seems to be pushing the hydrogen bond designation a bit far. A preferable term is a **polar–π** interaction, to indicate that a conventionally polar molecule is interacting with the quadrupole moment of a π system. Any hydrogen bond donor, such as an amide NH or an alcohol OH, will experience a favorable electrostatic interaction with the face of a benzene ring because of the large bond dipole associated with the hydrogen bond donor. Although weaker than a cation–π interaction, these polar–π interactions are also observed in protein structures, and are important contributors to solid state packing interactions.

π Hydrogen bonds

Connections

The Polar Nature of Benzene Affects Acidities in a Predictable Manner

The polar nature of benzene can influence reactivity in predictable ways. For example, the substituted benzoic acid shown to the right has a substantially perturbed pK_a value of 6.39 (X = Y = H), compared to 4.2 for benzoic acid itself. This is consistent with the negative electrostatic potential on the faces of the neighboring phenyls destabilizing the ionized carboxylate, thereby shifting the pK_a to a higher value. Substituents X and Y influence the pK_a further in ways consistent with this model (see end-of-chapter Exercise 4 on predicting these pK_a shifts).

Chen, C. T., and Siegel, J. S. "Through Space Polar–π Effects on the Acidity and Hydrogen Bonding Capacity of Carboxylic Acids." *J. Am. Chem. Soc.*, **116**, 5959–5960 (1994).

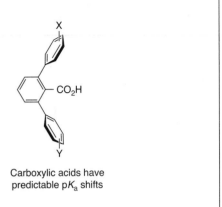

Carboxylic acids have predictable pK_a shifts

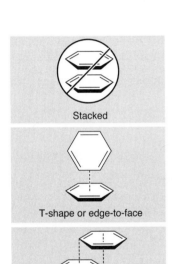

Stacked

T-shape or edge-to-face

Displaced or slip stacked

π–π Stacking geometries

Aromatic–Aromatic Interactions (π Stacking)

One of the most misused terms in molecular recognition is **π stacking**. Generally, it is an ill-defined concept that would seem to imply that it is somehow favorable to stack two π systems on top of each other. However, the electrostatic potential surface of benzene clearly shows that this is not the case. To directly stack two benzenes on top of one another will lead to an adverse electrostatic repulsion.

Nevertheless, simple aromatics do experience favorable interactions with each other. For simple systems like benzene, the **T-shaped** or **edge-to-face** geometry is better than stacking. This geometry places a region of negative electrostatic potential (the face of the ring) in contact with a region of positive electrostatic potential (the edge). In the gas phase, this is the preferred geometry, with a $\Delta H°$ of roughly –2 kcal/mol. Even in water, where we might expect the hydrophobic effect to favor the stacked form (see the discussion of the hydrophobic effect below), the T-shaped and displaced stacks are two of several structures that are preferred over the stacked arrangement.

In some more complicated structures the T-shaped geometry cannot be obtained. In these cases, then, it is best to form a **displaced** or **slipped stack**. This still aligns regions of positive electrostatic potential with regions of negative electrostatic potential. This type of "π stacking" is energetically favorable. There is also a favorable hydrophobic component to the slipped stack interaction (if water is the solvent—see below) such that slipped stacking becomes increasingly important for larger arenes such as naphthalene or anthracene. We prefer the term **aromatic–aromatic interaction** (or **π–π interaction**, because aromaticity is not really the issue here) to π stacking, because it does not imply the direct overlap of regions of negative electrostatic potential.

Note that the benzene–benzene interaction, especially in the T-shaped geometry, is just the logical extension of the notion that benzene is a polar molecule, like water. Thus, if water binds water electrostatically, which it does, benzene should bind benzene.

The Arene–Perfluoroarene Interaction

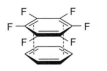

Arene–perfluoroarene stacking

While H is less electronegative than an sp^2 C, F is more electronegative than an sp^2 C. Because of this, it turns out that hexafluorobenzene (C_6F_6) has a quadrupole moment that is roughly equal in magnitude but opposite in sign to that of benzene. This means that regions of negative electrostatic potential in benzene are regions of positive electrostatic potential in C_6F_6, and so on. See the electrostatic potential surface in Appendix 2. One implication of this is that benzene and hexafluorobenzene should experience a *favorable stacking interaction,* which can be viewed as a quadrupole–quadrupole interaction. This is indeed the case, and the most dramatic manifestation is reflected in the solid state properties of the systems. Benzene melts at 5.5 °C and forms a herringbone structure in the solid state that maximizes the T-shaped interaction. Hexafluorobenzene melts at 4.0 °C and has the same crystal structure. However, a 1:1 mixture of the two melts at 24 °C and has a totally new crystal structure that emphasizes perfect stacks of alternating benzene–hexafluorobenzene molecules. It is rare that a mixture is higher melting than either pure compound, and this result is a potent testimony to the power of electrostatic interactions involving π systems. It turns out this interaction is general, such that almost any simple arene will stack with the analogous perfluoroarene in the solid state to form a mixed crystal of exceptional stability. An example of using this interaction in materials chemistry is given in the following Connections highlight.

Connections

Use of the Arene–Perfluorarene Interaction in the Design of Solid State Structures

One of the most challenging goals of modern physical organic chemistry is the rational design of solid state packing patterns—so-called **crystal engineering**. Many phenomena, most notably non-linear optics and magnetism (see Chapter 17), are most commonly observed in solids. These and other more mundane, but very important properties, like solubility and processability, depend strongly on the exact packing pattern in the crystal. Progress has been slow. It has been considered a "scandal" that, with modern theoretical methods and substantial computational power, we still cannot predict the most basic property of an organic molecule—namely, its melting point.

As the x-ray crystallography of small molecules has become fairly routine, a large database of structures has developed. From this, certain patterns of favorable packing patterns have emerged. As a potential organizing principle for the field, the notion of a **supramolecular synthon** has been proposed (see the next chapter for a discussion of supramolecular chemistry). This is a recurring, supramolecular motif (also known as a non-covalent interaction) that appears frequently in molecular crystal structures and encourages structural order. Many of the synthons involve hydrogen bonding and/or metal coordination, while others involve related electrostatic interactions. One novel interaction that has been established as a way to design solids is the arene–perfluoroarene interaction.

As an example of the use of a supramolecular synthon in materials design, we consider solid state diacetylene polymerization (see to the right). Single crystals of some diacetylene derivatives can be photopolymerized to produce long conjugated chains within the crystal. Because of their extensive conjugation, such polymerized diacetylenes have novel optical and electrical properties. For polymerization to occur, the diacetlyene must crystallize in a specific geometry that is conducive to polymerization—the potential reactive centers must be near each other and aligned properly. An interesting system would be diphenyldiacetylene (mp = 87 °C), but it crystallizes in a form that is not conducive to photopolymerization. The same is true of perfluorodiphenyldiacetylene (mp = 114 °C). However, a 1:1 mixture of the two diacetylenes (mp = 152 °C) does crystallize in the proper form because of the arene–perfluoroarene supramolecular synthon, and photopolymerization is possible. Photopolymerization can also be seen in pure crystals of phenyl (pentafluoro)phenyl diacetylene (mp = 124 °C), which nicely crystallizes into a stacked structure. Other examples of solid state engineering through the arene–perfluoroarene supramolecular synthon have also been seen.

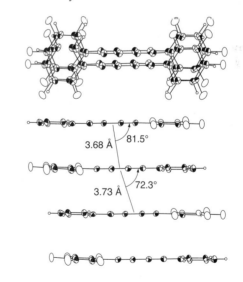

3.68 Å   81.5°

3.73 Å   72.3°

Coates, G. W., Dunn, A. R., Henling, L. M., Dougherty, D. A., and Grubbs, R. A. "Phenyl–Perfluorophenyl Stacking Interactions: A New Strategy for Supermolecule Construction." *Angew. Chem. Int. Ed. Eng.*, **36**, 248 (1997).

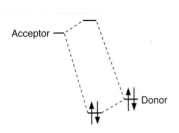

Donor–acceptor orbital mixing

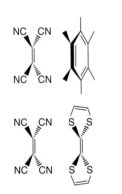

Donor-acceptor dimers

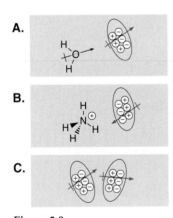

Figure 3.9
Examples of interactions involving induced dipoles. The ellipsoid represents a nonpolar molecule, and the colored arrow represents the induced-dipole.
A. Dipole–induced-dipole,
B. ion–induced-dipole, and
C. induced-dipole–induced-dipole.

π Donor–Acceptor Interactions

The last binding force that we examine which, at least in part, has its origin in electrostatic attractions is the π donor–acceptor interaction. A **donor–acceptor interaction** occurs between any two molecules, or regions of a molecule, where one has a low energy empty orbital (**acceptor**) and the other a high energy filled orbital (**donor**). When these two orbitals are aligned properly, some extent of **charge transfer** can occur from the donor to the acceptor. This is a stabilizing interaction. We examined in Section 2.3 several examples of orbital mixings that were important for the conformations of hydrocarbons that contain heteroatoms. A donor–acceptor interaction in that context was defined as a lone pair (or a σ or π bond) that could donate toward a low-lying empty orbital, possibly an antibonding orbital (recall the anomeric effect). A donor–acceptor binding interaction is another weak force that can be used to impart structure and hold compounds together (see the following Connections highlight).

The systems we are considering here differ in two ways from the simple orbital mixing described in Chapter 1. First, the donor and acceptor are not part of the same molecule. Second, the energy gap between the interacting orbitals is much smaller, leading to a stronger interaction. To achieve this, the partners in a π donor–acceptor interaction are generally heavily substituted, one with electron withdrawing groups and one with electron donating groups. For example, tetracyanoethylene is an excellent acceptor, and it forms complexes with electron rich systems such as hexamethylbenzene and tetrathiafulvalene.

Generally, a large extent of charge transfer leads to colors. For example, tetracyanoethylene and hexamethylbenzene form a complex that is deep purple. No new bonds are formed, however, as each partner can be re-isolated intact. Further, tetracyanoethylene and tetrathiafulvalene crystallize as an almost black solid. The complexes formed between the donor and acceptor are referred to as **charge–transfer complexes**. The color arises from an absorbance of light that promotes an electron from the donor to the acceptor (we will return to this in Chapter 16)—the full charge transfer occurs in the excited state, while only "orbital mixing" occurs in the ground state. The absorbance found in the UV/vis spectrum that is indicative of this electron transfer is called the **charge–transfer band**. It is the presence of this charge–transfer band that most clearly distinguishes this type of interaction from the others involving arenes discussed above. For simple systems, no charge–transfer band is seen in a cation–π interaction or an arene–perfluoroarene interaction, and so the electrostatic model is emphasized over the orbital mixing/charge–transfer model. When color appears on complexation, though, the orbital mixing model takes precedence. The true situation is a continuum, with varying degrees of both effects occurring in differing systems. However, it is important to note that the electron transfer that gives rise to the optical effect contributes little to nothing energetically to the association of the donor and acceptor. It is the orbital mixing in the ground state that drives the association.

3.2.5   Induced-Dipole Interactions

Thus far, in discussing some of the primary binding forces, we have emphasized an electrostatic model. The underlying principle is simply to match regions of positive charge with regions of negative charge. We did this because such a simple model is in fact quite successful in making qualitative predictions about the geometries of interactions between molecules and the relative strengths of nonbonding interactions. If, however, we want a fully *quantitative* model of such interactions, we must go beyond electrostatics. It is certainly true that when a cation moves close to an anion, the electronic wavefunctions of the two change in response to each other's presence, and this change is termed a polarization. This will certainly enhance the interaction, and the same will happen in hydrogen bonding, dipole interactions, or π interactions. In such a case, no fundamentally new effects arise from consideration of such polarization—we simply get a better quantitative picture of the interaction. However, the perturbation of the wavefunction of a nonpolar molecule by a polar one leads to electrostatic attractions that otherwise would not have existed (Figure 3.9 **A**).

Connections

Donor–Acceptor Driven Folding

One of the first studies of foldamers centered on molecules that form reproducible secondary structures due to π donor–acceptor interactions. Stringing together and alternating aromatic donors and acceptors in the short oligomer shown below led to the well-defined secondary structure that is shown schematically. The oligomer was called an **aedamer**, *a*romatic *e*lectron *d*onor–*a*cceptor. There is also a significant hydrophobic effect driving the condensed and stacked arrangement in water. X-ray crystallography of a co-crystal of the monomeric donors and acceptors confirmed the preference for an alternating structure, and UV / vis analysis showed the spectroscopic changes indicative of the stacking arrangement. This is an excellent example of the use of a small molecular binding force to create a large ordered structure.

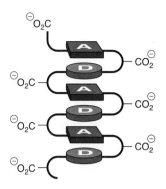

Folded structure of
aedamer in solution

Lokey, S. L., and Iverson, B. L. "Synthetic Molecules that Fold into a Pleated Secondary Structure in Solution." *Nature*, **375**, 303–305 (1995).

Linear aedamer

Ion–Induced-Dipole Interactions

Consider bringing a small cation near a molecule of ethane. Electrostatically, we expect essentially no interaction because ethane has neither a dipole nor a quadrupole. However, ethane is a fairly polarizable molecule—it can readily adjust its electron distribution to create a favorable interaction with the ion. The ethane will move some valence electrons toward the cation, leaving behind a region of depleted electron density (Figure 3.9 **B**). In so doing, we establish a dipole in ethane, where one did not exist before. This **ion–induced-dipole** interaction is weak—certainly weaker than the interaction of an ion with a permanent dipole. But the interaction is not negligible, and the fact is that a cation would rather bind to ethane than bind to nothing at all. The interaction energy is described by Eq. 3.28. Not surprisingly, the polarizability of the neutral molecule, α, is involved (see Chapter 1). The distance dependence is now r^{-4}, which means that the energy of interaction falls off more quickly than the interactions we have seen before.

$$E = \frac{-q^2\alpha}{(4\pi\varepsilon\varepsilon_o)^2 r^4}$$

(Eq. 3.28)

Dipole–Induced-Dipole Interactions

We now consider what happens when a polar molecule, one with a permanent dipole moment μ, approaches a nonpolar but polarizable molecule, producing a **dipole–induced-dipole** interaction. To understand this interaction, we start with an examination of the electric field generated by a dipole. It is the sum of the fields generated by each partial point charge on the ends of the dipole. The field felt along the axis of the dipole at a distance r from the center of the dipole is given by Eq. 3.29.

$$E_{\text{field}} = \frac{2\mu}{4\pi\varepsilon\varepsilon_o r^3}$$ (Eq. 3.29)

The size of the induced dipole in the polarizable molecule is $\mu = \alpha E_{\text{field}}$. If we combine this expression with Eq. 3.25, the dipole–dipole potential energy equation (where we drop the $3\cos^2\theta - 1$ term, because we are considering only aligned dipoles), we obtain Eq. 3.30 for the potential energy of a dipole–induced-dipole interaction (the subscript 1 refers to the molecule with the permanent dipole and subscript 2 is for the polarizable molecule). The important point is that the potential energy of a dipole–induced-dipole interaction varies with inverse distance to the sixth power, and hence is exceedingly sensitive to distance.

$$E = \frac{-\mu_1\alpha_2 E_{\text{field}}}{4\pi\varepsilon\varepsilon_o r^3} = \frac{-2\mu_1^2\alpha_2}{(4\pi\varepsilon\varepsilon_o)^2 r^6}$$ (Eq. 3.30)

Induced-Dipole–Induced-Dipole Interactions

We can take this one step further and create an **induced-dipole–induced-dipole** interaction. Consider bringing two molecules of ethane together (Figure 3.9 **C**). If one molecule instantaneously generates a dipole and the other does the same, a net attraction can develop. The more polarizable the atoms or molecules involved in these interactions, the larger the attraction. Although these forces are exceedingly small relative to hydrogen bonds and dipole–dipole interactions, they cannot be ignored. In fact, if there is a large surface area for the two molecules to interact, these forces can become considerable (see the heat of vaporization of decane, Table 3.2). They cause common alkanes to condense together into liquids. The induced-dipole–induced-dipole concept is one way to describe what are also known as the **van der Waals** or **London dispersion** forces.

An alternative way to think of the induced-dipole–induced-dipole interaction is as an electron correlation effect. The motions of valence electrons on the two interacting molecules are correlated. That is, as electrons on one molecule move to the "right", electrons on the other molecule also move to the "right". We simply note here that because van der Waals interactions are a consequence of electron correlation, simple molecular orbital theories are not able to quantitatively model these weak interactions.

The derivation of the potential energy for London dispersion forces is quite involved, and usually such interactions are not quantitatively modeled by equations of the sort we have been presenting here. Typically, the empirically derived Lennard–Jones "6–12" potential discussed in Chapter 2 or a related function is used. To a first approximation, as with the dipole–induced-dipole, the energy of interaction can be considered to drop off with an r^{-6} dependence.

Summarizing Monopole, Dipole, and Induced-Dipole Binding Forces

The induced-dipole binding forces discussed here can be compared to the permanent dipolar binding forces discussed in Section 3.2.2. One of the most important comparisons is how the energies of interaction vary as a function of distance. Table 3.10 tallies the distance dependence as a function of the type of interaction.

Table 3.10
Comparison of the Distance Dependence of the Energy of Interaction for Various Binding Interactions

	Monopole	Dipole	Induced-dipole
Monopole	$1/r$	$1/r^2$	$1/r^4$
Dipole		$1/r^3$	$1/r^6$
Induced-dipole			$1/r^6$

3.2.6 The Hydrophobic Effect

Up to this point all the binding forces we have discussed have electrostatic attractions as their origin, or at least as a major component. The last binding force we consider—the hydrophobic effect—is a deviation from this theme. The hydrophobic effect drives the association of organics together in water. As we noted above, simple organics such as alkanes have little attraction for each other (only dispersion forces). There is no permanent electrostatic attraction between alkanes. The precise physical origin of the hydrophobic effect has been intensely investigated and is still debated. We will not settle that debate here. Instead, we present some phenomenology and a model that provides a useful way to think about the effect.

Earlier we noted the many exceptional properties of water as a solvent. As much as what does dissolve in water, what doesn't dissolve has a profound effect on molecular recognition phenomena. We all know that "oil and water do not mix". This is the simplest statement of the **hydrophobic effect**—the observation that hydrocarbons and related "organic" compounds are insoluble in water. The hydrophobic effect is the single most important component in biological molecular recognition. It is the strongest contributor to protein folding, membrane formation, and in most cases, small molecule binding by receptors in water. As such, it is essential for organic chemists to have some sense of this crucial phenomenon.

Aggregation of Organics

From the outset we should distinguish two different manifestations of the hydrophobic effect. One is the low solubility of hydrocarbons in water, which is studied by considering $\Delta G°$ for the transfer of an organic molecule from the gas phase or hydrocarbon solution to water. The other manifestation is the tendency of organics to associate or aggregate in water, typically probed by measuring $\Delta G°$ of association and/or binding constants. While the physical origins of the two must ultimately be related, often we see conflicting conclusions from the two different types of studies. To some extent this is due to the differing reference states and types of measurements made.

Much of the essential physical chemistry of the hydrophobic effect has emphasized the transfer of small organics from the gas phase to water. As we have said, hydrocarbons have very low solubilities in water. While this is the characteristic feature of the hydrophobic effect, other thermodynamic effects are seen, including unusual entropy effects and often large heat capacity effects. To a very good approximation, $\Delta G°$ of transfer scales with surface area of the hydrocarbon that is exposed to water on dissolution. The exact scaling factor is debated and appears to depend on context. Values as low as 15 cal/mol in $\Delta G°$ for every $Å^2$ of exposed aliphatic or aromatic hydrocarbon and as high as 75 cal/mol•$Å^2$ are reported, but a more typical range is 30–50 cal/mol•$Å^2$. If we settle on 40 cal/mol•$Å^2$, and assume a surface area of 29 $Å^2$ for a CH_2 in an alkane, then every additional CH_2 adds 1.2 kcal/mol of destabilization in a hydrophobic effect.

The hydrophobicity of organic groups can also be measured by the partitioning of organic molecules between a nonpolar solvent, typically n-octanol, and water. We define the **hydrophobicity constant** π for an organic group R as in Eq. 3.31, where P_o is the partitioning of an organic molecule between octanol and water without R, and P is the partitioning of the organic structure with R attached. Small organic R substituents are found to make constant and additive contributions to the hydrophobicity of a molecule (Table 3.11). This reinforces our view that the hydrophobicity arises simply from the surface area of the group, and is not dramatically affected by the environment.

$$\pi = \log\left(\frac{P}{P_o}\right)$$

(Eq. 3.31)

Given the 30–50 cal/mol•$Å^2$ value, one would expect that once they are in water, hydrocarbons should minimize their exposed surface area. They can do this in two ways: shape changes and aggregation. As an example of the first, consider n-butane in water. Not surprisingly, gauche butane is a more compact structure than anti butane. We would expect a

Table 3.11
Some Values of π and the Incremental Gibbs Free Energy of Transfer from n-Octanol to Water*

R group	π	$\Delta G°$ (kcal/mol)
$-CH_3$	0.5	0.68
$-CH_2CH_3$	1.0	1.36
$-CH_2CH_2CH_3$	1.5	2.05
$-CH(CH_3)_2$	1.3	1.77
$-CH_2Ph$	2.63	3.59

*Leo, A., Hansch, C. et al. "Partition Coefficients and Their Uses." *Chem. Rev.*, **71**, 525–616 (1971).

Gauche butane reduces
exposed surface area

shift in the conformational equilibrium for n-butane in water, and indeed this is seen. The effect is small, but enough to change the 70:30 anti:gauche equilibrium mixture seen in the gas phase or in liquid butane to 55:45 in water. We expect this to be a general effect for any flexible organic molecule in water, and for larger molecules that can experience more substantial changes in surface area as a result of conformational changes, the effect could be quite large. In fact, just such an effect is the primary driving force for protein folding.

Figure 3.10 shows how the hydrophobic effect can also drive aggregation. The exposed hydrocarbon surface area will always be diminished when two organics aggregate. Because $\Delta G°$ is always favorable for such aggregation, the process is spontaneous in water. The spontaneous aggregation of organic groups in water was likely a key event in the development of primitive forms of life and/or their precursors (see further discussions of spontaneous self-assembly in the next chapter).

Because most pure hydrocarbons barely dissolve in water, aggregation has more typically been probed by studying **amphiphilic** molecules—structures that have both a hydrophobic region and a polar (**hydrophilic**) region (Figure 3.10). Such molecules are also often referred to as **surfactants**. Consider a long chain aliphatic carboxylic acid such as stearic acid. The polar carboxylate end is quite hydrophilic and the long alkyl chain is hydrophobic. The tail is **lipophilic**, a synonym for hydrophobic. The result is the spontaneous formation of a **micelle**, a roughly spherical structure with the hydrocarbon tails facing inward and the polar carboxylates on the surface. These structures form only above a certain concentration of the surfactant, known as the **critical micelle concentration**. This is a good example of the spontaneous self-assembly of a simple molecule into a more complex, partially ordered larger structure—a **supermolecule**. It would be very difficult to "rationally" build a large system with a hydrophobic core and a polar surface using the standard strategies of organic synthesis. However, when the building block is designed properly, the system puts itself together. As we will see in the next chapter, this kind of process has inspired chemists to try to learn the rules of self-assembly. The goal is the design and synthesis, by self-assembly, of beautiful, complex systems.

The spherical picture of a micelle shown in Figure 3.10 should not be taken too literally. A micelle is dynamic at many levels, as shown by a large number of physical organic studies. Individual surfactants can depart from and return to micelles on a microsecond timescale, while stepwise dissolution of micelles and reassembly occurs on the millisecond timescale. A long standing debate is the extent to which water penetrates into the hydrophobic core—that is, how perfect is the barrier between oil and water? It is now generally agreed that water penetrates fairly deeply, perhaps halfway down the hydrocarbon chain. For example, an olefin halfway down the hydrocarbon chain can react with polar reagents.

In nature, the more common amphiphiles are **phospholipids**. These are derivatives of glycerol (1,2,3-trihydroxypropane), in which two alcohols form esters with long chain carboxylic acids. The third alcohol forms a phosphate ester, and the phosphate then makes another ester with a simpler alcohol. This creates structures such as phosphatidyl choline, phosphatidyl serine, and phosphatidyl ethanolamine (see next page). The polar group can

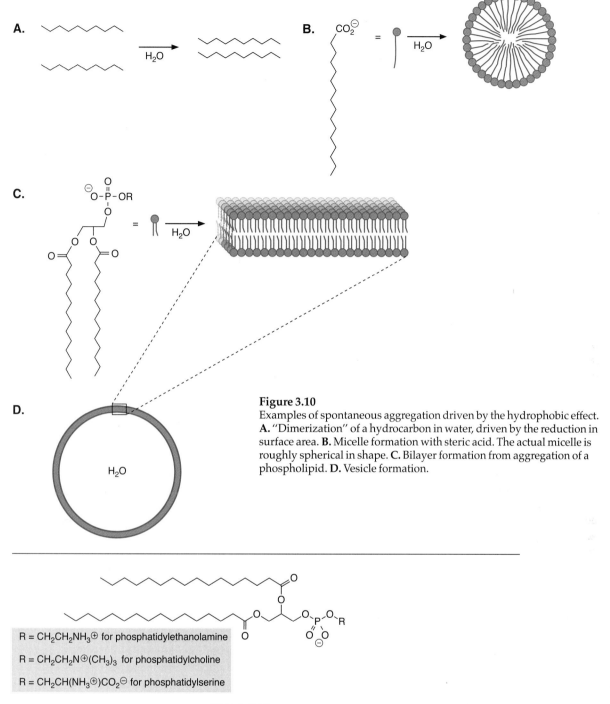

Figure 3.10
Examples of spontaneous aggregation driven by the hydrophobic effect.
A. "Dimerization" of a hydrocarbon in water, driven by the reduction in
surface area. **B.** Micelle formation with steric acid. The actual micelle is
roughly spherical in shape. **C.** Bilayer formation from aggregation of a
phospholipid. **D.** Vesicle formation.

R = $CH_2CH_2NH_3^{\oplus}$ for phosphatidylethanolamine

R = $CH_2CH_2N^{\oplus}(CH_3)_3$ for phosphatidylcholine

R = $CH_2CH(NH_3^{\oplus})CO_2^{\ominus}$ for phosphatidylserine

Phospholipids

be either anionic (phosphatidyl serine) or **zwitterionic** (having both a cation and an anion)
as in phosphatidyl choline or ethanolamine.

Because of their different shape in terms of the polar vs. hydrophobic groups, phospholipids do not form micelles. Instead, they can spontaneously assemble to form **bilayers** and
ultimately, **vesicles** (Figure 3.10 **C** and **D**). Vesicles are not nearly as dynamic as micelles.
Further, there is a clear demarcation between inside and outside with vesicles. We can imagine that such vesicles could form very small reaction vessels and, ultimately, primitive precursors of life.

The size of the head group relative to the tail of a surfactant has a significant effect on whether micelles or vesicles are formed. Soaps, detergents, and other single-tail amphiphiles have polar head groups that are wide (when including solvation) relative to the width of the nonpolar tails. The best way to achieve close-packing of such cone-shaped structures is an object with a high radius of curvature, a micelle. Conversely, the head group and tail widths are more nearly equivalent in double chain species like most lipids, leading to a cylindrical shape. Close-packing of cyclinders leads to aggregates with a low radius of curvature, like bilayer structures. This geometric analysis provides a conceptual framework that can be easily extended to other shapes for designing aggregates driven by the hydrophobic effect.

The Origin of the Hydrophobic Effect

What is the physical origin of the hydrophobic effect? Several factors are involved. First is the high cohesive energy or, equivalently, the high surface tension of water. The water–water interaction is very strong. As such, there is a significant penalty for creating a cavity in water. This must occur in order to dissolve a hydrocarbon solute, because some water–water interactions are broken (recall our discussion of solvation in Section 3.1.3). Second, water and hydrocarbons fail the "like-dissolves-like" test. Hydrocarbons are nonpolar, water is very polar, and therefore very little binding occurs between the solute and solvent to make up for the lost interactions between the solvent. Moreover, hydrocarbons are polarizable and water is not. So, water would much rather interact with water, and hydrocarbons would rather interact with hydrocarbons (the latter effect is smaller, as evidenced by the lower cohesive energies/surface tensions of organic liquids). All these factors are enthalpy considerations, and indeed these factors are important, but a recurring observation concerning the thermodynamics of the hydrophobic effect suggests entropy is a factor, too.

As we have already noted, hydrocarbons aggregate in water. If two molecules of hydrocarbon are placed in water, $\Delta G°$ is favorable (< 0) for the (non-covalent) aggregation. Surprisingly, though, it is often observed that $\Delta H°$ for the aggregation is small and perhaps even unfavorable (> 0). Necessarily, $\Delta S°$ is favorable (> 0), leading to the conclusion that *hydrophobic association is often entropy driven*. This is certainly counterintuitive. We would expect a process in which two or more molecules are brought together to be entropically unfavorable. To rationalize these thermodynamic observations, the model shown in Figure 3.11 is often invoked.

In our discussion, we compare the *water structure* before and after aggregation of the organic structures. First, as just stated above, water has a very high cohesive energy. Still, liquid water is dynamic and is not maximally hydrogen bonded. The perfect, rigid structure with four hydrogen bonds per water molecule is only seen in solid ice. While ice has a lower enthalpy than water due to more hydrogen bonds, it is entropically disfavored due to the increase in order. In the model of Figure 3.11, it is proposed that water in contact with a hydrophobic surface becomes more "ice-like". As stated, water in contact with an organic molecule loses favorable water–water contacts. To compensate, it strengthens its remaining water–water contacts, making them more ice-like. The local water structure becomes more rigid, and the strengths and number of individual water hydrogen bonds around the solute increase. This increase in the number and strength of hydrogen bonds can compensate for the lost hydrogen bonds due to the presence of the cavity created by the organic entity, and may even be enthalpically favorable. However, and most importantly, due to the increased ice-like nature of the waters around the organic, the entropy has significantly decreased. The near equal enthalpy of the water before and after dissolution of the organic, along with the clearly worse entropy, taken together lead to the low solubility of the organic structure. This is an example of enthalpy–entropy compensation, where decreased enthalpy leads to decreased entropy also.

Now let's analyze the same situation with two organic structures that dimerize. In essence, due to the lower exposed organic surface area upon dimerization, all the negative aspects discussed in the previous paragraph are diminished. When the two hydrophobic molecules associate, the hydrocarbon surface area exposed to water decreases, diminishing the

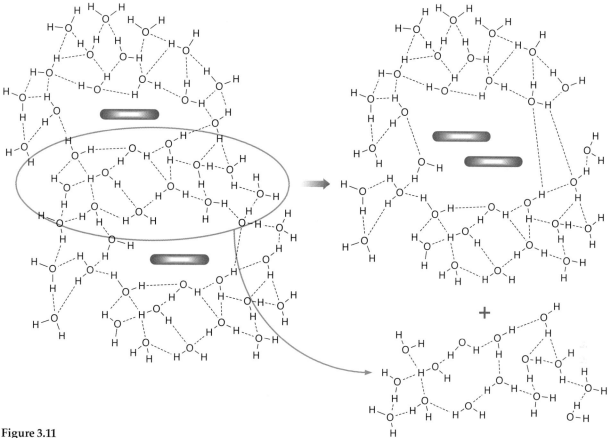

Figure 3.11
A model for the hydrophobic effect. Water near the surface of a hydrocarbon is ordered. Reducing surface area by dimerization frees some of the ordered water, producing a favorable entropy for hydrophobic aggregation.

amount of ice-like water. The release of ice-like water from around the organic structures upon dimerization leads to more "normal water" with the associated regular hydrogen bonds, which can result in either an unfavorable enthalpy change or a close-to-zero enthalpy change. Importantly, however, there is an accompanying increase in the disorder of the water. The association liberates a number of water molecules from the more constrained ice-like state, and so association is *entropically favorable*. The net effect is that the $T\Delta S°$ term outweighs the $\Delta H°$ term, producing a favorable $\Delta G°$. Hydrophobic association is entropy driven.

The discussion above demonstrates that there are some hallmarks of hydrophobically driven association of organic structures. One is a favorable entropy. However, another is a change in heat capacity during the binding, and in fact, this is often a more reliable indicator of the hydrophobic effect than entropy. In the next chapter we discuss the mathematical relationship used to measure a change in heat capacity (ΔC_P). For now, recall that the **heat capacity** of a solution measures the amount of energy the solution absorbs per unit change in temperature. Because there is a significant change in heat capacity associated with the hydrophobic effect, the entropy dominated signature we discussed above for the hydrophobic effect is most commonly observed near ambient temperature, but not necessarily at higher temperatures. At higher temperatures enthalpy effects commonly start to dominate the driving force for the hydrophobic effect. The extent of change of the heat capacity depends upon the surface area involved in the hydrophobically driven association. If the fraction of hydrophobic surface area exposed to water is diminished upon association of one or more entities, a negative change in heat capacity will occur.

Table 3.12
$\Delta S°$ and $\Delta C_P°$ of Association of Biological Receptors
and Their Substrates in Water at 298 K*

System	ΔS (cal/K•mol)	ΔC_P (cal/K•mol)
Aldolase and hexitol-1,6-diphosphate	34	−401
Heart LDH and NAD⁺	3.5	−84
tRNA ligase and isoleucine	19.7	−430
Avidin and biotin	1.3	−24
Hemoglobin and haptoglobin	−73	−940

*Blokzijl, W., and Engberts, J. B. F. N., "Hydrophobic Effects. Opinions and Facts." *Angew. Chem. Int. Ed. Engl.*, **32**, 1545–1579 (1993).

Table 3.12 shows some entropy and heat capacity changes for the binding of several biological structures with small organic molecules. Although other binding forces besides the hydrophobic effect must be involved in each of these cases, the hydrophobic effect is certainly a large fraction of the driving force. Note that the change in heat capacity is always negative, whereas the entropy is not always favorable.

The "classical" model shown in Figure 3.11 is just one of several viable views of the hydrophobic effect. However, it is simple, and depicts many of the unusual features, such as unfavorable $\Delta H°$ and favorable $\Delta S°$ values, and the overall dependence on surface area. Perhaps the biggest weakness of the model is that it ignores any possible attraction between the organic fragments—an enthalpic contribution that should be primarily due to van der Waals/dispersion forces. This should be a small but not entirely negligible effect. It is certainly not strong enough, nor directional enough, to justify such terms as the "hydrophobic bond", which should not be used. The classical model is essentially a **solvophobic effect**. Hydrocarbons associate in water not because they are attracted to each other, but rather because they are repulsed by the solvent—it is simply lower in energy for the water to get away from them. As with the other binding forces we have discussed herein, solvophobic effects lead to structural ordering, and the next two highlights give examples in natural and unnatural systems.

Going Deeper

The Hydrophobic Effect and Protein Folding

An essential feature of proteins is that they spontaneously fold into well-defined, three-dimensional structures. The single most important contributor to protein folding is the hydrophobic effect. It is imperative that amino acids such as leucine and valine, which have hydrophobic side chains, bury those side chains in the core of the protein, away from the aqueous environment of the cell. This **hydrophobic collapse** is a key early event in the process of converting a disordered chain of amino acids into a well-defined, properly folded protein. As a result, protein folding typically shows the thermodynamic hallmarks of the hydrophobic effect, including a favorable entropy (even though the folded protein is more ordered than the unfolded) and large negative heat capacity changes.

Dill, K. A. "Dominant Forces in Protein Folding." *Biochemistry*, **29**, 7133 (1990).

3.3   Computational Modeling of Solvation

In Chapter 2 we described the molecular mechanics approach to computing the structures and energies of organic molecules in the gas phase. There are also quantum mechanical methods for achieving the same goals, and these are discussed in some detail in Chapter 14. But, of course, most chemistry occurs in solution, and theorists, therefore, have made great

Connections

More Foldamers: Folding Driven by Solvophobic Effects

Another foldamer strategy involves oligo(phenylene ethynylene) structures that fold into helical conformations, creating tubular cavities. The folding is driven primarily by solvophobic effects—the nonpolar aromatic portions want to get away from the polar solvent, while the polar ethylene oxide side chains are exposed. Favorable aromatic–aromatic interactions may also be involved. These helical structures resemble a common protein motif—the α/β-barrel—and are also promising scaffolds for future study.

Nelson, J. C., Saven, J. G., Moore, J. S., and Wolynes, P. G. "Solvophobically Driven Folding of Nonbiological Oligomers." *Science*, **277**, 1793–1796 (1997).

Foldamer structure

$R = -(CH_2CH_2O)_3CH_3$

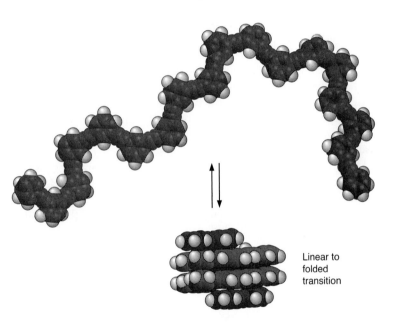

Linear to folded transition

efforts to model solvation phenomena. This is distinct from the empirical scales such as $E_T(30)$ discussed earlier. We are now considering efforts to provide a detailed theoretical description of solvents and solvent–solute interactions. This is a vast and evolving field, and a detailed treatment is beyond the scope of this text. Nevertheless, the future of physical organic chemistry will involve more and more modeling of solvents and solvent–solute interactions (solvation), and so we present an overview of the various strategies here.

The modeling of a solvent—a liquid phase—is especially challenging. In the gas phase, the molecules can be treated as isolated species that are easily modeled using quantum mechanics (Chapter 14) or molecular mechanics (Chapter 2). Modeling a solid is certainly challenging, but at least in the crystalline state there is periodic order, which in principle, simplifies the problem. Still, accurate computer modeling of solids is a major challenge.

In some ways, though, a liquid is the most challenging medium. It is a condensed phase, like a solid, and so is inherently a **many-body problem**. However, there is no long range periodic order (recall Figure 3.1). Also, liquids are by their very nature dynamic, and any model that does not take this into account will likely be inadequate. The challenges are clear, and there are two fundamentally different strategies to modeling solutions. In **continuum (or implicit) models**, the solvent is treated as a homogeneous medium that surrounds the solute molecule. Computationally, this is implemented as a fairly simple set of adjustments to the basic molecular mechanics (or quantum mechanics) model. In **explicit solvation models**, a large number of individual solvent molecules are added to a single solute molecule, and the entire system is treated by molecular mechanics. These methods have the advantage of being closer to physical reality, and being more easily interpreted. However, these benefits are achieved at the price of an enormous increase in computational complexity.

3.3.1   Continuum Solvation Models

The simplest continuum model includes the dielectric constant of the medium in evaluating electrostatic terms in molecular mechanics calculations. Recall that Eq. 3.1 (for simple electrostatic interactions) included a dielectric term (ε). Such a scaling of electrostatic interactions by the solvent dielectric constant is in principle useful and is theoretically justifiable. Note that for molecules dissolved in a solvent, the charges (q_i) are partial charges associated with each atom of the molecule that must be obtained by some other method. In principle this is a viable strategy, but in practice it has little impact on calculations.

More advanced continuum models are based on parameterized, atom-specific terms that scale with the exposed surface area. In a molecular mechanics based approach, the amount of atomic surface (the sphere defined by an atom's van der Waals radius) that is exposed to solvent is determined for each particular atom in a molecule. Then, an equation that includes parameters related to the type of atom and to the specific solvent calculates a solvation term. These terms are summed over all atoms in the molecule. Such approaches blend into the molecular mechanics method quite naturally, without an overly burdensome increase in computation time.

An especially interesting model, termed the **generalized Born** model, has been developed primarily for water as a solvent. We will describe it briefly here, because it nicely illustrates in a quantitative way some of the topics we have discussed in this chapter. The approach is a parameterized method that produces G_{solv}, the solvation free energy for a molecule or ion. First, G_{solv} is divided into three terms (Eq. 3.32).

$$G_{solv} = G_{cav} + G_{vdW} + G_{pol}$$

(Eq. 3.32)

The G_{cav} term represents the energy cost for forming a cavity in the solvent. As we noted above, this is a substantial effect for water as solvent because of its high cohesive energy. It will be less important but still significant for other solvents. The G_{vdW} term is a solute–solvent van der Waals term, accounting for the weak dispersion forces discussed above. Finally, G_{pol} is the solute–solvent electrostatic polarization term, which accounts for the interactions of charges on the solute with the solvent. It is assumed that for an alkane solute, $G_{pol} = 0$, and because the solvation energies of alkanes scale with exposed surface area, we arrive at Eq. 3.33.

$$G_{cav} + G_{vdW} = \Sigma s_i (SA)_i$$

(Eq. 3.33)

Here, s_i is a parameter for each atom type (in the spirit of molecular mechanics) and SA is the solvent accessible surface area for atom i.

What about G_{pol} for an ion in water? We need to consider two types of interactions. The first is the interaction between solute ions, which should be modeled by Coulomb's law. The other is the interaction of an ion with the solvent, and this can be modeled by the Born equation, as mentioned in Section 3.2.2. These two equations are, to some extent, of a similar form, and so can be combined to give Eq. 3.34.

$$G_{pol} = -\tfrac{1}{2}\,(1 - 1/\varepsilon)\sum_i \sum_j (q_i q_j / f_{GB})$$

(Eq. 3.34)

where ε = the dielectric constant, q_i is the charge on atom i, and f_{GB} (the generalized Born function) is $(r_{ij}^2 + a_{ij}^2\, e^{-D})^{0.5}$, where $a_{ij} = (a_i a_j)^{0.5}$ and $D = r_{ij}^2 / (2a_{ij})^2$ and a_i is the radius of ion i

Admittedly, it is not completely obvious where f_{GB} comes from. It is an intuitive combination of Coulomb's law and the Born equation. However, it does reduce to the Born equation in the limit of $r = 0$ (i.e., only one ion is present), and it is purely Coulombic if $r >> a$. The bottom line is this method works well, as shown in Table 3.13. The results are really quite remarkable, and they span the entire range from hydrocarbons to polar organics to ions. Importantly, because the calculation of solvation energy follows very much the form of a molecular mechanics calculation, this method can be easily added to any force field. Also, calculating the solvation adds an insignificant amount of time to the calculation. Perhaps more important for our purposes, this approach shows that useful results can be obtained by considering such effects as cavitation, surface area, and electrostatics.

Table 3.13
Comparison of Experimental Aqueous Solvation Energies with Those Calculated by the Generalized Born Model*

Solute	G_{solv} (kcal/mol)	
	Experimental	Calculated
Methanol	–5.1	–6.2
Acetone	–3.8	–3.2
Acetic acid	–6.7	–6.5
Benzene	–0.9	–1.0
n-Octane	+2.9	+2.9
NH_4^+	–80	–91
Me_3NH^+	–59	–63
$CH_3CO_2^-$	–80	–83

*Still, W. C., Tempczyk, A. et al. "Semianalytical Treatment of Solvation for Molecular Recognition and Dynamics." *J. Am. Chem. Soc.*, **112**, 6127–6129 (1990).

A potentially significant improvement of this generalized Born approach involves coupling this model with high-level quantum mechanical calculations of the charge distribution of the solute molecule. As discussed in considerable detail in Chapter 14, it is now routinely possible to calculate the full wavefunctions for typical organic molecules using so-called *ab initio* methods. One outcome of such calculations is a detailed and accurate charge distribution for the molecule. It is now possible to use the quantum mechanical charge distribution, rather than the much cruder molecular mechanics charges, to evaluate the electrostatic component of the solvation energy. It is even possible to calculate the perturbation to the molecular charge caused by the solvent and vice versa. This leads to the so-called self-consistent field (SCF) calculation, directly analogous to the SCF methods described in detail in Chapter 14. These are developing methodologies, but they do hold considerable promise as tools for evaluating the effects of solvation on structure and reactivity.

3.3.2 Explicit Solvation Models

A great deal of work has been expended to develop explicit solvent models within the molecular mechanics approach. Water has been the most extensively studied solvent because of its obvious importance for biology, and a popular approach is the TIP4P model (*t*ransferable *i*ntermolecular *p*otentials with a *4* *p*oint charge model). In this approach, a water molecule is treated as three van der Waals spheres (two hydrogens and one oxygen) with four centers of partial charge—two positive charges on the hydrogens and two negative charges at "tetrahedral" locations on the oxygen. Another popular model is TIP3P, which has two positive charges that are compensated by a single negative charge on the oxygen. Each water molecule is held rigidly—there is no optimization of bond lengths or bond angles.

Similar models exist for other solvents, such as CH_2Cl_2, THF, etc. In each instance, the solvent molecules are treated as rigid—that is, their internal geometries are not optimized. Molecular mechanics-type calculations are now done to evaluate interactions between the solute and the many solvent molecules.

A single solute molecule is placed in a box that is then filled with solvent molecules. The box has **periodic boundary conditions**, meaning that if a solvent molecule exits the box on the right, an image solvent molecule enters on the left to take its place. It is as if the box is just one of a lattice of boxes.

How big should the box be? If it is a cube, and we want to put a moderately-sized solute molecule in it, a box with 5 Å sides would be too small—solute molecules might protrude out of the box. A 100 Å box would be much better, but really very large in terms of computation. For small organic solutes, a cube with 20 Å sides is often adequate. It is a simple matter to calculate that 267 water molecules will fit into a $20 \times 20 \times 20$ Å box. If the solute is ethane, for example, it would take the place of two waters, based on its size. Thus, our calculation would be on a box with 265 water molecules and one ethane.

What do we do with such a system? Do we "optimize" its geometry? Not really. Liquid systems are dynamic. An "optimized" geometry is simply a snapshot of what is a constantly changing, equilibrating system. Even if we could obtain an optimized structure (image the possibilities for false and/or non-global minima!), it would not really tell us what we want to know about the system. To get a feeling for a liquid system, we need to evaluate its properties as an average over a particular period of time. In this way, meaningful thermodynamic properties of a liquid system can be obtained.

There are two different ways to execute this averaging: Monte Carlo methods and molecular dynamics methods. Both methods are commonly used, and both have particular advantages and disadvantages. We will briefly lay out the basics of these two methods below. A thorough derivation of these two fairly complex procedures is beyond the scope of this book. Our goal is to provide some familiarity, so modern work in the field can be intelligently read.

3.3.3 Monte Carlo (MC) Methods

The Monte Carlo (MC) method starts with a particular arrangement of all the particles (solute and solvent molecules) in the system—a configuration. Then, a three-step procedure is applied.

i. Calculate the energy;

ii. Move a randomly chosen particle a random distance, in a random direction; and

iii. Recalculate the energy and return to step ii.

It is from step *ii* that the method derives its name—the process of choosing random numbers is as if dice were thrown at a casino.

This is statistical mechanics, so classical terms such as free energy (G), density (ρ), pressure (P), temperature (T), volume (V), enthalpy (H), and entropy (S) will be relevant. In principle, if enough configurations are evaluated, the Monte Carlo method will produce an **average energy** that is meaningful. In practice, however, an unrealistically large number of configurations (perhaps hundreds of millions) would have to be evaluated before the average would become meaningful.

This problem can be circumvented by biasing the "randomness" of step *ii*, introducing **importance sampling**. This causes the method to favor "good" configurations over bad. The most important approach to importance sampling is the **Metropolis method** (Monte Carlo is a city, but Metropolis is a person's name). Steps *i* and *ii* are the same as above, followed by:

iii. Recalculate the energy.

    a) If the energy (E) goes down, keep the new structure.
    b) If E goes up, generate a random number p, such that $0 < p < 1$:
        If $p < e^{-(\Delta E/RT)}$, keep the new structure.
        If $p > e^{-(\Delta E/RT)}$, discard the new structure and return to the original (and count it again).

iv. Return to step ii.

This approach biases the sampling toward low energy structures. It can be shown that Metropolis sampling produces averages that are meaningful from a statistical mechanics viewpoint. Another sampling bias usually introduced is to favor moving solvent molecules that are closer to, rather than farther from, the solute molecule.

With these approaches, the Monte Carlo method becomes a feasible, but still large, calculation. For example, to evaluate a simple solute like ethane in water, we might first evaluate 10^6 configurations just to let the system "settle down" (i.e., equilibrate). Then, we would average over $2–4 \times 10^6$ configurations to consider the solvation.

An interesting feature of such sampling methods is that the final average energy is in fact a $\Delta G°$ value, even though a molecular mechanics force field is used to evaluate the energies of each configuration. How can a method based on molecular mechanics (which evaluates $\Delta H°$) produce a $\Delta G°$? Remember that $\Delta S°$ is innately a statistical term (recall the discussion of the two conformers of gauche butane in Chapter 2). Thus, by averaging over a very large number of configurations, statistical biases for particular arrangements will factor in naturally, and so $\Delta G°$ will emerge from the calculation. Since equilibrium constants are in fact determined by $\Delta G°$, not $\Delta H°$, this is a very useful feature.

3.3.4 Molecular Dynamics (MD)

The molecular dynamics (MD) method provides an alternative strategy for generating the large number of configurations of solute and solvent necessary for meaningful liquid simulations. Instead of randomly generating structures as in the Monte Carlo method, we take advantage of the fact that molecular mechanics methods provide not only energies but also forces, via the first derivatives of the force field equations. The method proceeds as follows.

We begin with a system in an initial state, such as a solute and many solvent molecules. We calculate the molecular mechanics energy and also the forces on the molecules via the derivatives of the force field equations. Unless the system is at an absolute minimum with respect to all degrees of freedom—an unlikely situation for an initial configuration—there will be finite forces on the system. We now simply apply Newton's classical equations of motion and let the system accelerate along the trajectories established by the forces. After a set amount of time, we stop and consider the new structure as a new configuration to be averaged, and compute its energy. We then proceed along the dynamics trajectory for another time step and repeat the process. After enough steps, this will generate an ensemble of structures that is comparable to one generated by Monte Carlo methods.

How long should each time step be? Experience has shown that this must be a very brief time—on the order of 1–2 femtoseconds (fs = 10^{-15} s). Allowing the structure to follow any one trajectory for longer times will carry the system into unrealistic geometries because the molecular mechanics method is imperfect—these are not "true" forces. How many steps are enough? The more the better. Realistically, it would be useful to run a simulation long enough to "see" a conformational interconversion take place, such as a chair–chair interconversion in cyclohexane, but this often is unrealistic. Using the Arrhenius equation ($k =$

$Ae^{-E_a/RT}$; see Chapter 7), and sensible activation parameters ($E_a = 10.8$ kcal/mol; $\log A = 13$), $k = 10^{13} \times e^{-(10,800/1.987 \cdot 298)} = 1.2 \times 10^5$, then $t(\frac{1}{2}) = 5.8 \times 10^{-6}$ s ≈ 6 μs. The even longer ms timescale is an appropriate one when considering protein folding and unfolding. With a 1 fs step time, we would need 6×10^9 configurations! This is three orders of magnitude more than is typically generated in a Monte Carlo simulation, and is currently unfeasible computationally. Typically, the lengths of the trajectories studied are in the nanosecond range, and this is often enough to get meaningful thermodynamic data, but not enough to directly "see" a structural change.

3.3.5   Statistical Perturbation Theory/Free Energy Perturbation

We introduce here one more extremely useful molecular mechanics based technique: **perturbation methods**. Although somewhat advanced, the method is so powerful that students of modern organic chemistry should know of it. The fact is that the explicit solvation methods only became really meaningful for experimentalists when the perturbation methods discussed here were introduced. We will provide only a very brief introduction. Note the method is equally compatible with MC and MD methods.

Suppose we want to calculate the aqueous solvation energy of organic molecule **A**. One approach would be to first fully equilibrate a box of TIP4P water molecules and obtain the average energy. We could then introduce one molecule of solute **A** and obtain another average energy. We could then subtract the two energies, and obtain the solvation energy. In practice, this is unfeasible for two reasons. First, the perturbation of dropping an **A** molecule into an equilibrated box of water is substantial, and it would take a long time to be sure we reach a real equilibrium. More seriously, we would be subtracting two very large numbers (the energies of systems with hundreds of molecules) to obtain a relatively small number— always a risky procedure. In practice, this just does not work.

Actually, experimentalists are rarely interested in absolute solvation energies. We want *relative* solvation energies. We noted this when we discussed heats of transfer of solutes between two solvents in Section 3.1.3. How much more or less soluble is **B** than **A**? If we really need an absolute energy for **B**, we start with another molecule (say **A**) whose experimental solvation energy is known. We then determine the *relative* solvation energy of **B**, and then combine it with the experimental number for **A** to get the absolute solvation energy for **B**. A recently developed method termed **statistical perturbation theory**, SPT (equivalently termed **free energy perturbation**, FEP), can answer this kind of relative energy question quite well. The essence of SPT is the Zwanzig equation (Eq. 3.35),

$$\Delta G = G_j - G_i = -kT \ln \langle \exp[(H_j - H_i)/kT] \rangle_i \qquad \text{(Eq. 3.35)}$$

where G_i is the free energy of state i, etc., and "$\langle \ \rangle_i$" means averaging over configurations generated for state i.

According to Eq. 3.35, the free energy difference between two states can be obtained from a collection of enthalpy differences generated by MC or MD for configurations that follow a smooth perturbation of one state into the other. As long as the perturbation on going from state j to state i is small, and as long as a proper averaging is done (as in Monte Carlo and MD methods), the free energy *difference* between the two states is obtained.

So, to get the relative solvation for **A/B**, we equilibrate **A**, incrementally permute (morph) it to **B** and apply the above equation. There are two important issues. First, how do we morph molecules? Actually, in the molecular mechanics method, this is not difficult. Consider **A** = ethane and **B** = methanol. To convert ethane to methanol, we simply change all the bond lengths, bond angles, and *molecular mechanics terms*, such as van der Waals radii, partial charges, etc., from those for ethane to those for methanol.

The second issue arises from the phrase, "as long as the perturbation on going from state j to state i is small", given above. Jumping straight from ethane to methanol, believe it or not, is much too dramatic. Just like dropping a molecule into the pure solvent system was too severe, the solvent system will just have too much trouble readjusting to this dramatic perturbation, and the method fails. We need a smaller perturbation. So, we go from ethane to a molecule that is 95% ethane and 5% methanol. This is a small perturbation for sure, but what does it mean? Remember, we are not dealing with real molecules, but rather with sets of molecular mechanics parameters and equations. It is actually no problem to simply *scale* the molecular mechanics terms to create a mythical system that is 95% ethane and 5% methanol. We are not saying there are many solute molecules, 95% of which are ethanes and 5% of which are methanols. There is only one solute molecule, and its geometry and molecular mechanics terms are a 95:5 weighted average of those for ethane and methanol. This perturbation is small enough that we can obtain an accurate $\Delta G°$ value by Monte Carlo or MD methods. Then, we permute the 95:5 to a 90:10, and so on until we get to our endpoint of 100% methanol. Adding up all the $\Delta G°$'s for the individual steps gives us the free energy change we seek between the initial and final states. Basically, we are simply permuting one molecule to another with small enough changes so that the solvent can keep up with them. The molecular mechanics method is well suited to this.

The bottom line is that SPT methods are very successful. The ethane/methanol relative solvation energy is obtained with essentially experimental accuracy. Once the concept is established, much more than just relative solvation energies can be obtained, as indicated in the following Going Deeper highlight. The method is computationally intensive—the study described above would require 21 full MC or MD runs—but the results are often worth it.

Going Deeper

Calculating Drug Binding Energies by SPT

A common situation in the pharmaceutical industry is as follows. A successful inhibitor (I_1) of some protein (P) has been developed, and a crystal structure of the inhibitor–protein complex is obtained. The inhibitor is not optimal, however, and one would like to design molecules that bind more tightly to the protein. It is very difficult to *a priori* calculate binding energies for small molecules to large proteins. The SPT method, however, is perfect for this kind of problem. Consider the following thermodynamic cycle:

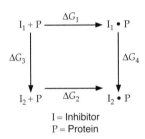

I = Inhibitor
P = Protein

Thermodynamic cycle used in SPT

We know ΔG_1 by measurement. We want to know ΔG_2, where I_2 is a molecule that is proposed, but perhaps not even synthesized yet. It is easy to see that $\Delta G_1 - \Delta G_2 = \Delta G_3 - \Delta G_4$. Note that ΔG_3 and ΔG_4 are easily obtained by SPT. ΔG_3 is just the relative solvation energy of the two inhibitors, as in the ethane/methanol example in the text (the protein, P, does not even figure into the calculation of ΔG_3.) Similarly, ΔG_4 can be readily obtained from SPT by permuting I_1 as it is bound to the protein to I_2 in its molecular mechanics calculated geometry for binding to the protein. Thus, from two SPT runs that might be expected to be quite reliable, we can get $\Delta G_1 - \Delta G_2$ and, because we know ΔG_1, we obtain ΔG_2. In principle, this could be done for many compounds, and the information could be used to decide which new inhibitors are worth the effort of synthesis and testing.

Summary and Outlook

We have discussed solvent structure, solvation, the thermodynamics of solutions, several binding forces, and finally computational methods to model solvation. We found that the molecular structures of solvent molecules are the origin of the bulk solvent properties. The interaction of the solvent with solutes determines solvation properties, which are combined with the intrinsic stability of the solvents and solutes, and the entropy of mixing, to give the total Gibbs free energy of a solution. It is this total Gibbs free energy of a solution that drives the dissolving of a solute, and any spontaneous chemical transformation. The solvation properties can be analyzed as separate binding forces: ion pairing, hydrogen bonding, dipole interactions, π interactions, and the hydrophobic effect. We will return to these concepts of solvation, solvent properties, and binding forces, when we examine reaction mechanisms and catalysis. However, our next goal is to show how the combination of several binding forces in the design of synthetic receptors leads to the fields of molecular recognition and supramolecular chemistry. Hence, it is time to explore how the incorporation of distinct binding forces in the design of multiple chemical entities can lead to the controlled assembly of large molecular aggregates from several small molecule precursors.

Exercises

1. Chloroform shows a significant binding interaction with benzene, but carbon tetrachloride does not. Predict the preferred geometry for the interaction and describe the physical nature of the attraction between the two molecules.

2. Show how we know that 267 water molecules fill a 20 Å × 20 Å × 20 Å box.

3. Benzene is a polar molecule, but not a polar solvent. In light of the cation–π interaction and other molecular recognition effects involving benzene that we have discussed above, explain why KCl is soluble in water but not in benzene (there are at least three reasons).

4. Predict a trend for electron donating and accepting substituent effects in the Connections highlight entitled "The Polar Nature of Benzene Affects Acidities in a Predictable Manner". Explain your predictions.

5. Use a strictly electrostatic argument to rationalize the fact that the binding energy of ammonia to benzene is less than that of water to benzene.

6. We stated in the text that for a monovalent ion in water at 298 K, the Born solvation energy, E_{sol} equals $-164/a$ in kcal/mol ($\varepsilon_o = 8.854 \times 10^{-12}\ C^2/J{\bullet}m$). Show that this is so.

7. We state in the text that over 19 kcal/mol of solvation energy for a monovalent ion comes from water molecules that are ≥ 8.5 Å from the ion. Show that this is so.

8. The ΔC_P (cal/K${\bullet}$mol) for water is 18; for ice it's 9. Do these data provide a simple explanation for the heat capacity effects generally seen in hydrophobic associations?

9. In a G${\bullet}{\bullet}{\bullet}$C base pair of DNA, there are three hydrogen bonds formed between the bases. As the hydrogen bonds are in close proximity, there is a good opportunity for secondary interactions (Section 3.2.3). Jorgensen has analyzed this system in general. Consider all possible arrangements of three hydrogen bonds (e.g., three donors on one partner with three acceptors on the other), and the various ways of having two plus one. Determine whether the secondary interactions are stabilizing or destabilizing for each set. Where does the G${\bullet}{\bullet}{\bullet}$C pair fall?

Jorgensen, W. L., and Pranata, J. "Importance of Secondary Interactions in Triply Hydrogen-Bonded Complexes: Guonine-Cytosine vs. Uracil-2,6-diamino Pyridine." *J. Am. Chem. Soc.,* **112**, 2008–2010 (1990).

10. Using the data given in Section 3.2.5, and a 40 cal/mol Å value for the hydrophobic effect, calculate the difference in surface area for anti and gauche butane. Given an estimated surface area for anti butane of 127 Å², estimate the surface area of gauche butane.

11. In reference to the discussion of Section 3.1.5, what is the driving force to form some of B when pure A is first added to the solvent?

12. Arrange the following compounds in order of increasing hydrogen bond donating ability toward methylamine. Rationalize your answer.

13. Arrange the following compounds in order of increasing hydrogen bond accepting ability from methanol. Rationalize your answer.

14. Calculate the energy of attraction in a vacuum for the following arrangement of dipoles (look back to Chapter 1 for bond dipoles and bond lengths). What stops the two molecules from simply collapsing together? (1 Debye = 3.33564×10^{-30} C•m.)

15. If one drop each of 1 M solutions of NaCl and sucrose were added to separate 1 L portions of water without stirring, which would more quickly form a homogeneous solution? Why?

16. Why is $e^{(-\Delta E/RT)}$ used as a criterion for importance sampling in Monte Carlo calculations? Why aren't the endothermic steps just discarded?

17. Why is water a better hydrogen bond donor than methanol, whereas methanol is a better hydrogen bond acceptor (see Table 3.1)?

18. What force(s) is (are) responsible for the higher heat of vaporization of acetone compared to benzene? What force(s) is (are) responsible for the higher heat of vaporization of benzene compared to chloroform?

19. Why is a lack of solvation an important factor in forming a low-barrier hydrogen bond?

20. List all the possible driving forces for π stacking found in DNA duplexes. Why is it possible for these π systems to stack on top of one another, while herein we noted that benzene does not do this?

21. We noted in the discussion of donor–acceptor interactions that the charge transfer seen in the UV/vis spectrum is not a significant factor in the binding force. When might you expect charge transfer to become a significant factor in the binding force?

22. The C–N bond rotation barriers in amides are generally lower in the gas phase than in solution. For example, the barrier in dimethylformamide is on average 1.5 kcal/mol lower in the gas phase than in the solution phase. There are at least two possible explanations. What are these?

23. Why are there no correlation times reported for spherical cations such as Na^+?

24. On average, the diffusion coefficients for lithium salts are smaller than for sodium salts. Explain.

25. In the following heterocyclic compounds, the keto form dominates over the enol form in solution. Suggest a reason for this.

Further Reading

Solvent Structure

Henderson, D. in *Physical Chemistry. An Advanced Treatise*, H. Eyring, D. Henderson, and W. H. Jost (eds.), Academic Press, New York, 1971, Vol. 8, pp. 377, 414.

Rawlinson, J. S. (1969). *Liquids and Liquid Mixtures*, Butterworth, London.

Kohler, F. (1972). *The Liquid State*, Verlag Chemie, Weinheim.

McKonald, I. R., and Singer, K. "Computer Experiments on Liquids." *Chem. in Brit.*, **9**, 54 (1973).

Solvent Scales

Kamlet, M. J., and Taft, R. W. "The Solvatochromic Comparison Method. I. The β-Scale of Solvent Hydrogen-Bond Acceptor (HBA) Basicities." *J. Am. Chem. Soc.*, **98**, 377 (1976).

Burden, A. G., Collier, G., and Shorter, J. "Influence of Aprotic Solvents on the O–D Stretching Band of Methan[²H]ol." *J. Chem. Soc. Perkin II Trans.*, 627 (1976).

Kosower, E. M. (1968). *An Introduction to Physical Organic Chemistry*, Wiley, New York, p. 293.

Reichardt, C. "Empirical Parameters of the Polarity of Solvents." *Angew. Chem. Int. Ed. Engl.*, **29**, 4 (1965).

Reichardt, C. (1979). *Solvent Effects in Organic Chemistry*, Verlag Chemie, Weinheim.

Taft, R. W., and Kamlet, M. J. "The Solvatochromic Comparison Method. 2. The α-Scale of Solvent Hydrogen-Bond Donor (HBD) Acidities." *J. Am. Chem. Soc.*, **98**, 2886 (1976).

Kamlet, M. J., Abboud, J.-L., Jones, M. E., and Taft, R. W. "Linear Solvation Energy Relationships. Part 2. Correlation of Electronic Spectral Data for Aniline Indicators with Solvent π and β Values." *J. Chem. Soc. Perkin II Trans.*, 342 (1979).

Kirkwood, J. G. "Theory of Solutions of Molecules Containing Widely Separated Charges With Special Application to Zwitterions." *J. Chem. Phys.*, **2**, 351 (1934).

Onsager, L. "Electric Moments of Molecules in Liquids." *J. Am. Chem. Soc.*, **58**, 1486 (1936).

The Structure of Water

Bills, J. L., and Snow, R. L. "Molecular Shapes and the Pauli Force. An Outdated Fiction." *J. Am. Chem. Soc.*, **97**, 6340 (1975).

Hall, M. B. "Valence Shell Electron Pair Repulsions and the Pauli Exclusion Principle." *J. Am. Chem. Soc.*, **100**, 6333 (1978).

Bartell, L. S., and Barshad, Y. Z. "Valence Shell Electron–Pair Repulsions: A Quantum Test of a Naive Mechanical Model." *J. Am. Chem. Soc.*, **106**, 7700 (1984).

Thermodynamics of Solutions

Pigogene, I., and Defuy, R. (1954). *Chemical Thermodynamics*, Longmans, London.

Benson, S. W. (1960). *Foundations of Chemical Thermodynamics*, McGraw–Hill, New York.

Caldin, E. F. (1961). *An Introduction to Chemical Thermodynamics*, Oxford University Press, Oxford.

Guggenheim, E. A. (1967). *Thermodynamics*, North Holland, Amsterdam.

Smith, E. B. (1977). *Basic Chemical Thermodynamics*, Oxford University Press, Oxford.

Ion Pairing

Janz, G. J., and Tomkins, R. P. T. (1972). *The Non-Aqueous Electrolytes Handbook*, Academic Press, New York.

Coplan, M. A., and Fuoss, R. M. "Single Ion Conductance in Nonaqueous Solvents." *J. Phys. Chem.*, **68**, 1177 (1964).

Greenacre, G. C., and Young, R. N. "Ion-Pairing of Substituted 1,3-Diphenylallyl Carbanions With Alkali-Metal Cations." *J. Chem. Soc. Perkin II Trans.*, 1661 (1975).

Szwarc, M. (ed.) (1972). *Ions and Ion-Pairs in Organic Reactions*, Wiley, New York.

Szwarc, M. "Ions and Ion Pairs." *Acc. Chem. Res.*, **2**, 87 (1969).

Robbins, J. (1972). *Ions in Solution*, Oxford University Press, Oxford.

Burley, J. W., and Young, R. N. "Ion Pairing in Alki-Metal Salts of 1,3-Diphenylalkenes. Part II. The Determination of Equilibrium Constants From Absorption Spectra." *J. Chem. Soc. Perkin II Trans.*, 835 (1972).

Szwarc, M. (ed.) (1974). *Ions and Ion Pairs in Organic Reactions*, Wiley, New York, Vol. 2.

Hydrogen Bonding

Umeyama, H., and Morokuma, K. "The Origin of Hydrogen Bonding. An Energy Decomposition Study." *J. Am. Chem. Soc.*, **99**, 1316–1332 (1977).

Legon, A. C. "Directional Character, Strength, and Nature of the Hydrogen Bond in Gas-Phase Dimers." *Acc. Chem. Res.*, **20**, 39–46 (1987).

Pimentel, G. S., and McLellan, A. L. (1960). *The Hydrogen Bond*, Freeman, San Francisco.

Hadzi, D. (ed.) (1959). *Hydrogen Bonding*, Pergamon, London.

Hamilton, W. C., and Ibers, J. A. (1968). *Hydrogen Bonding in Solids*, Benjamin, New York.

Covington, A. K., and Jones, P. (1968). *Hydrogen-Bonded Solvent Systems*, Taylor and Francis, London.

Vinogradov, S. N., and Linnell, R. H. (1971). *Hydrogen Bonding*, Van Nostrand, New York.

Emsley, J. "Very Strong Hydrogen Bonding." *Chem. Soc. Rev.*, **9**, 91 (1980).

Symons, M. C. R. "Water Structure and Reactivity." *Acc. Chem. Res.*, **14**, 179 (1981).

Fersht, A. R., Shi, J.-P., Knill-Jones, J., Lowe, D. M., Wilkinson, A. J., Blow, D. M., Brick, P., Carter, P., Waye, M. M. Y., and Winter, G. "Hydrogen Bonding and Biological Specificity Analyzed by Protein Engineering." *Nature*, **314**, 235–238 (1985).

Cox, J. P. L., Nicholls, I. A., and Williams, D. H. "Molecular Recognition in Aqueous Solution: An Estimate of the Intrinsic Binding Energy of an Amide–Hydroxyl Hydrogen Bond." *J. Chem. Soc. Chem. Commun.*, 1295–1296 (1991).

Short–Strong Hydrogen Bonds

Hibbert, F., and Emsley, J. "Hydrogen Bonding and Reactivity." *Adv. Phys. Org. Chem.*, **26**, 255–379 (1990).

Frey, P. A., Whitt, S. A., and Tobin, J. B. "A Low-Barrier Hydrogen Bond in the Catalytic Triad of Serine Proteases." *Science*, **264**, 1927–1930 (1994).

Warshel, A., Papazyan, A., and Kollman, P. A. "On Low Barrier Hydrogen Bonds and Enzyme Catalysis." *Science*, **269**, 102–104 (1995). Responses by Cleland, Kreevoy, and Frey.

Scheiner, S., and Kar, T. "The Nonexistence of Specially Stabilized Hydrogen Bonds in Enzymes." *J. Am. Chem. Soc.*, **117**, 6970–6975 (1995).

Shan, S., Loh, S., and Herschlag, D. "The Energetics of Hydrogen Bonds in Model Systems: Implications For Enzymatic Catalysis." *Science*, **272**, 97–101 (1996).

π Effects

Ma, J. C., and Dougherty, D. A. "The Cation-π Interaction." *Chem. Rev.*, **97**, 1303–1324 (1997).

Meyer, E. A., Castellano, R. K., and Diederich, F. "Interactions with Aromatic Rings in Chemical and Biological Recognition." *Angew. Chem. Int. Ed. Eng.*, **42**, 1210–1250 (2003).

π Donor–Acceptor Interaction

Pearson, R. G. "Symmetry Rules for Chemical Reactions." *Acc. Chem. Res.*, **4**, 152 (1971).

Pearson, R. G. "Orbital Symmetry Rules for Unimolecular Reactions." *J. Am. Chem. Soc.*, **94**, 8287 (1972).

Klopman, G. (ed.) (1974). *Chemical Reactivity and Reaction Paths*, Wiley, New York, p. 55.

Fleming, I. (1976). *Frontier Orbitals and Organic Chemical Reactions*, Wiley, London.

Levin, C. C. "A Qualitative Molecular Orbital Picture of Electronegativity Effects on XH_3 Inversion Barriers." *J. Am. Chem. Soc.*, **97**, 5649 (1975).

Hydrophobic Effect and Heat Capacity Changes

Blokzijl, W., and Engberts, J. B. F. N. "Hydrophobic Effects. Opinions and Facts." *Angew. Chem. Int. Ed. Eng.*, **32**, 1545–1579 (1993).

Sturtevant, J. M. "Heat Capacity and Entropy Changes in Processes Involving Proteins." *Proc. Natl. Acad. Sci, USA*, **74**, 2236–2240 (1977).

Orchin, M., Kaplan, F., Macomber, R. S., Wilson, R. M., and Zimmer, H. (1980). *The Vocabulary of Organic Chemistry*, Wiley–Interscience, New York, pp. 255–256.

Singh, S., and Robertson, R. "The Hydrolysis of Substituted Cyclopropyl Bromides in Water. IV. The Effect of Vinyl and Methyl Substitution on Cp." *Can. J. Chem.*, **55**, 2582 (1977).

Robertson, R. "The Interpretation of $\Delta Cp^{\ddagger}$ for S_N Displacement Reactions in Water." *Tetrahedron Letters*, **17**, 1489 (1979).

Muller, N. "Search for a Realistic View of Hydrophobic Effects." *Acc. Chem. Res.*, **23**, 23 (1990).

Computational Modeling of Solvation

Jorgensen, W. L. "Free Energy Calculations: A Breakthrough for Modeling Organic Chemistry in Solution." *Acc. Chem. Res.*, **22**, 184–189 (1989).

Cramer, C. J., and Truhlar, D. G. "Implicit Solvation Models: Equilibria, Structure, Spectra, and Dynamics." *Chem. Rev.*, **99**, 2161–2200 (1999).

Molecular Recognition and Supramolecular Chemistry

Intent and Purpose

We have seen how intermolecular forces such as hydrogen bonds and dipole–dipole interactions determine the properties of liquids and the strengths of solute–solvent interactions. These same forces also lead to **molecular recognition**—the specific, non-covalent association between a receptor molecule and a particular substrate. Nature is a master of molecular recognition. Whether it is an enzyme–substrate, antibody–antigen, or neuroreceptor–neurotransmitter pair, large molecules bind smaller molecules tightly and specifically, and binding of this type is at the heart of most biological processes. In addition, large molecules interact with other large molecules to form elaborate multisubunit complexes. In each case a large number of weak interactions cooperate to produce a substantial effect.

Inspired by nature's prowess, physical organic chemists have attempted to mimic the binding events of biological systems. Immediately, certain key questions arise. What forces will be most useful in such systems? What is the role of solvation? How do we design an artificial receptor? Can synthetic catalysts with enzyme-like properties be built? By characterizing such model systems, physical organic chemists hope to learn how nature's receptors work. Also, the possibility of designing useful new catalysts or sensors by mimicking natural systems is quite real.

Our overview of molecular recognition and supramolecular chemistry cannot be exhaustive. The fields are exploding at an astounding rate. Our focus will be upon the energetics of different binding interactions, the fundamental origins of these interactions, and the guiding principles for creating complex ensembles of molecules. Further, the analysis of affinity constants and the $\Delta H°$ and $\Delta S°$ of binding, as well as the manner in which they are measured, are covered. We will see that solvation is a factor of paramount importance in molecular recognition phenomena, and therefore a discussion of molecular recognition is a natural outgrowth of the concepts of solvation presented in the previous chapter. Toward the end of the chapter the lessons from molecular recognition are used in a discussion of the art of constructing large three-dimensional architectures.

The guiding principles for molecular recognition that we present should lead to an introductory understanding of the manner in which the molecules of life associate and perform functions and tasks. This understanding is a major focus of current physical organic chemistry, and therefore this chapter is quite relevant to research that many students will encounter in their graduate careers.

4.1 Thermodynamic Analyses of Binding Phenomena

Several different binding forces were discussed in the previous chapter. In many cases, approximate values were placed upon the strengths of these binding forces. However, we also noted discrepancies in the numbers, debate about the significance or origin of the effects, and that multiple small effects are used to generate large overall effects. Due to the relevance to biological systems and the need to control these forces to design functional molecules and molecular aggregates, physical organic chemists over the last 30 years have devoted significant research into these topics. Since it is difficult to get exact quantitative numbers on individual small effects between a solute and solvents (see the discussion of the difficulties of measuring hydrogen bond strengths in Section 3.2.3), the study of binding forces has primarily been explored through solute–solute interactions. As a means of distinguishing the solutes, we commonly refer to one as the **host** and the other as the **guest**. Typically, the host is the larger molecule, and it encompasses the smaller guest molecule. This nomenclature is mainly used with synthetic systems, whereas the more familiar terms of enzyme and substrate, antibody and antigen, are the analogs to host and guest in biological systems. By making rational changes in the structures of the host and/or guest, we hope to gain a better understanding of the individual binding forces between them and their interactions with the solvent. This is a classic approach to understanding natural phenomena, known as structure–activity studies (see Chapter 8 for a thorough discussion relative to reactivity). The primary data collected to gain this understanding are the thermodynamics of binding.

A knowledge of the thermodynamics of binding is essential if we are to rationally and confidently design synthetic systems that perform functions and tasks. Often the goal of molecular recognition with synthetic systems is to bind (sequester) a desired molecule from solution and interact with it in some manner, such as performing a reaction (catalysis; see Chapter 9) or giving off a signal indicative of its presence (sensing). However, we first must have confidence in how to bind the guest, and past precedent with studies of binding constants and design principles for artificial binding sites are crucial to efforts to create our new designs. Before we explore such basic binding studies, we need to analyze how the binding constants listed in the following sections are obtained. This provides an excellent opportunity to return to and expand upon the thermodynamic lessons given in Section 3.1.5.

Although our presentation is within the context of molecular recognition, it is very important to realize that the principles are very general. Hence, the following sections can be used to understand the thermodynamics of any two species that combine to form one species in a reversible equilibrium: acid + base, metal + ligand, Lewis acid + Lewis base, etc. We are about to go through an extensive discussion of the thermodynamics relevant to any one of these equilibria. Do not get "bogged down" in the discussions and lose sight of the goal—namely, to obtain an understanding of the thermodynamic origin of binding events and how to measure and interpret the thermodynamic parameters (K_a, $\Delta G°$, $\Delta H°$, and $\Delta S°$).

4.1.1 General Thermodynamics of Binding

Using our notation of host (H) and guest (G), we can express the binding equilibrium as in Eq. 4.1, the binding constant as in Eq. 4.2 or 4.3, and $\Delta G°$ as in Eq. 4.4. Note that K_a, the **association constant** (Eq. 4.2), and K_d, the **dissociation constant** (Eq. 4.3), are simply reciprocals of each other. Historically, chemists have tended to use K_a, with units of inverse concentration and the trend that bigger numbers imply stronger association. Biochemists, though, have generally favored K_d, with units of concentration and the trend that smaller numbers imply stronger association. The student needs to be comfortable with both approaches. The units of K_a and K_d are traditionally expressed as M^{-1} and some concentration (mM, μM), respectively, but as the next Going Deeper highlight notes, there are caveats associated with these units.

Equations 4.1–4.4 explicitly ignore the solvent, even though the discussion in the last chapter indicates that the solvent can have a dramatic influence on the magnitude of binding forces. We do not need the solvent explicitly written as part of these equations because $\Delta G°$ for the association reflects the stability of solvated H and G relative to solvated H•G and released solvent. As such, binding constants and related thermodynamic quantities should always be tabulated as being measured in a particular solvent, as well as at a particular temperature.

$$H + G \rightleftharpoons H•G \qquad \text{(Eq. 4.1)}$$

$$K_a = \frac{[H \bullet G]}{[H][G]} \qquad \text{(Eq. 4.2)}$$

$$K_d = \frac{[H][G]}{[H \bullet G]} \qquad \text{(Eq. 4.3)}$$

$$\Delta G° = -RT \ln(K_a) \qquad \text{(Eq. 4.4)}$$

In Section 3.1.5 we explored the thermodynamics of a simple reaction involving a single reactant A going to a single product B. The ratio of A and B at equilibrium reflects their intrinsic stabilities at their standard states. The intrinsic stabilities are expressed as $\Delta G°$ values, which are indicative of the energy it would take to convert one mole of A to B if A started at its standard state and B ended at its standard state. The ratio of A to B at equilibrium is constant for any initial concentration of the reactant ($[A]_o$). For example, for a reaction with an equilibrium constant K of 10 and an initial concentration $[A]_o$ of 11 mM, at equilibrium there will be 10 mM B and 1 mM A, because $10/1 = 10$. If the initial concentration of A was 2.2 μM, then at equilbrium there will be 2 μM B and 0.2 μM A, because $2/0.2 = 10$. We now explore how a reaction such as that given in Eq. 4.1 differs.

For a reaction as in Eq. 4.1, the ratios of [H], [G], and [H•G] are not constant for different initial concentrations of H and G. This is because the numerator of Eq. 4.2 is a concentration to the first power, but the denominator is related to concentration squared, and vice versa for Eq. 4.3. Let's examine some scenarios of various concentrations to delineate the trends. In our analysis we use an association constant of 10 M^{-1}, reflecting an exergonic reaction. However, for the sake of the following argument, let's assume that the reaction is also exothermic, a fact that we would not know unless we measured $\Delta H°$.

Assume a binding constant of 10 M^{-1}. Let's start the reaction with 10 mM H and G. At equilibrium we have [H] and [G] = 9.2 mM and [H•G] = 0.84 mM (you have to use the quadratic equation to get these numbers; see the end-of-chapter Exercises). Less than 10% of H and G are in the complexed form.

Now start with H and G both at 1000 mM. At equilibrium [H] and [G] = 270 mM and [H•G] = 730 mM. Under these conditions over 70% of H and G are part of the complex H•G.

Figure 4.1 puts our analysis in a pictorial fashion for an exergonic reaction between H and G. Just looking at this figure, we might conclude that under any conditions, a mixture of H and G will be mostly in the form of H•G. But we just established that this is not so if the initial concentrations are, for example, 10 mM. To get out of this conundrum we must recall the true meaning of the diagram in Figure 4.1. $\Delta G_{rxn}°$ reflects the change in Gibbs free energy for the conversion of one mole of H and G to one mole of H•G *if H and G started in their standard states and H•G ended in its standard state*. Yet, we never really work in the laboratory at standard states. Depending on whether we work at high concentration or low concentration, the dominant species in the flask might be H and G or H•G.

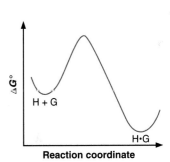

Figure 4.1
A normal reaction coordinate diagram for an exergonic binding process.

Going Deeper

The Units of Binding Constants

Given the definition of K_a from Eq. 4.2, we find that the binding constant has units of M^{-1}. Other equilibria can have different units. Putting units on K_a values is the most common convention used by chemists. Yet, we must remember that all equilibrium constants are really defined by ratios of activities, which are dimensionless values (see Section 3.1.5). True binding constants are therefore dimensionless. This should not come as a surprise, because the fact that $\Delta G° = -RT \ln(K_a)$ requires that the quantity we are taking the natural log of must not have any units (you can't take the ln of a unit!). Chemists, therefore, routinely give binding constants units only because we most commonly work with concentrations. Yet, this assumes that the concentrations are similar enough to the activities that it is acceptable to take the natural log of these equilibrium constants.

What do we mean by the low and high concentrations that make H and G or H•G dominant? Low or high relative to what? The most convenient reference point is the dissociation constant K_d (Eq. 4.3). When the initial concentrations of H and G are the same and are both below K_d, free H and G will dominate at equilibrium. When the initial concentrations of H and G are above K_d, then H•G will dominate. Finally, when the concentrations of free H and G equal K_d, then $[H•G] = K_d$. These relationships are why most biochemists (and now many other chemists) tabulate and think in terms of dissociation constants instead of association constants.

How can free H and G dominate for certain experimental conditions, even for exothermic reactions? There is a very simple mathematical analysis that leads to this conclusion. Let's imagine diluting a solution of H, G, and H•G that is at equilibrium but with arbitrary concentrations of these species. To see what happens to the relative concentrations of these three species upon dilution, we take Eq. 4.2 and express the concentrations as moles (n_X) and volume V, as shown in Eq. 4.5. The volume V is the same for all species, because H, G, and H•G are all in the same flask, which leads to a single V in the numerator. Hence, if we keep increasing V (dilution), the product $n_H n_G$ must go up faster than $n_{H•G}$ to keep the entire term constant. This simply means that dilution always increases free H and G relative to H•G.

$$K_a = \frac{\left(\dfrac{n_{H•G}}{V}\right)}{\left(\dfrac{n_H}{V}\right)\left(\dfrac{n_G}{V}\right)} = \left(\frac{Vn_{H•G}}{n_H n_G}\right) \qquad \text{(Eq. 4.5)}$$

The above discussion was a mathematical analysis, yet a thermodynamic analysis is probably more satisfying. Another vantage point from which to understand how free H and G can dominate the binding equilibrium, even for an exothermic reaction, is to analyze entropy. Remember that the Gibbs free energy of the entire solution controls the reaction, not just the stabilities of H and G relative to H•G. Recall from Section 3.1.5 that the entropy of a solution is greater when solutes exist in a larger volume (are more dilute). Hence, diluting any solution of H, G, and H•G is a stabilizing act. However, the entropy of a solution of H, G, and H•G will increase more rapidly upon dilution when creating free H and G, because there are two species that are leading to an entropic stabilization (H and G) relative to the one H•G. Therefore, upon dilution, an H•G complex will continually dissociate to H and G due to the favorable entropy of the entire solution.

We have discussed this phenomenon in the context of a host–guest complex, but as we mentioned earlier, it is true of any structure that can dissociate. We can define any two species that stick together to be a host–guest complex. For example, an acid HA will predominately dissociate to H^+ and A^- when its concentration is below K_a, where K_a is the acid dissociation constant. A confusing but historically sanctioned bit of symbolism defines an acid *dissociation* constant as K_a, not K_d. The subscript "a" stands for "acid", not "association". Here you might consider A^- the host for H^+. A metal–ligand complex ML will dissociate to M and L when the complex is diluted to a concentration below $1/K_a$, where K_a is the affinity constant for M and L (now following standard symbolism). Hence, this is a general concept to keep in mind for all molecular coordination phenomena.

The Relevance of the Standard State

The concept that the entropy of a solution of reactants (H and G together) becomes more favorable faster than the entropy of a solution of products (H•G) upon dilution means that the Gibbs free energy of the solution is changing upon dilution. How can this be, because $\Delta G°$ diagrams are not concentration dependent? Further, because we use the values of $\Delta G°$ to compare the favorability of one reaction to another, is this fair if the entropy due to the mixing of H, G, and H•G with the solvent is also part of the total Gibbs free energy of the reaction?

The answers to these questions can be understood by analyzing the reference state. The key is to remember that the $\Delta G°$ value is for a single set of conditions, those of the standard state. Hence, one cannot analyze Figure 4.1 to draw conclusions about the relative concentrations of species in solution, except for when they are in the standard state. Further, the entropies of the solution of solutes that are being considered in $\Delta G_{rxn}°$ are for standard states, and when we deviate from the standard state the entropies of the solution and hence the ratios of H, G, and H•G change. Yet, the $\Delta G_{rxn}°$ value also includes the inherent stabilities of H, G, and H•G, along with their solvation. To compare different systems we make the reasonable approximation that the entropy resulting from dissolution in a solvent is the same for any solute, regardless of its identity. This term cancels out in comparing two reactions. Thus, we can compare one $\Delta G_{rxn}°$ to another and be confident that it is telling us about the relative inherent stabilities of H, G, and H•G for two different systems. These stabilities take into account the bond strengths, strains, solvation, and degrees of freedom of H, G, and H•G. Hence, we use $\Delta G_{rxn}°$ values as a way of comparing the favorability of one reaction to another, allowing us to draw conclusions about the strengths of interactions involved within H, G, and H•G.

Since the key to understanding $\Delta G°$ diagrams is the realization that they are relevant only to the standard state, we can ask what will happen if we change our definition of the standard state. This is very seldom done, but it can be quite instructive to do so. During this exercise, it is important to state right up front that any mathematical analysis will not affect the intrinsic stabilities of H, G, and H•G. Yet, as you will see, it does effect what we define as an exergonic or endergonic reaction.

Let's once again look at some thermodynamic relationships, essentially repeating much of what we did in Section 3.1.5. For H, G, and H•G, we write an equation analogous to Eq. 3.13 and shown for H only below (Eq. 4.6). The ΔG_{rxn} of the reaction would then be Eq. 4.7, which can be simplified to Eq. 4.8, where $\Delta G_{rxn}°$ is defined as $G_{H•G}° - G_H° - G_G°$. We find that the potential for a spontaneous change in composition to occur is the difference in the Gibbs free energies of H, G, and H•G in their standard states plus a term that reflects the mismatch between the experimental ratios of H, G, and H•G and their intrinsic ratios at standard state. Once equilibrium has been achieved ($\Delta G_{rxn} = 0$), we write Eq. 4.9. Since the activities are defined as in Eq. 3.14, a little bit of algebra results in Eq. 4.10, and then Eq. 4.11 upon substituting in K_a. Here we use $K_a = (\gamma_{H•G}[H•G]/\gamma_H[H]\gamma_G[G])$, but chemists also normally drop the activity coefficients.

$$\mu_H = \mu_H° + RT \ln(a_H) \tag{Eq. 4.6}$$

$$\Delta G_{rxn} = \mu_{H•G} - \mu_H - \mu_G = \mu_{H•G}° - \mu_H° - \mu_G° + RT \ln(a_{H•G}) - RT \ln(a_H) - RT \ln(a_G) \tag{Eq. 4.7}$$

$$\Delta G_{rxn} = \Delta G_{rxn}° + RT \ln\left(\frac{a_{H•G}}{a_H a_G}\right) \tag{Eq. 4.8}$$

$$\Delta G_{rxn}° = -RT \ln\left(\frac{a_{H•G}}{a_H a_G}\right) \tag{Eq. 4.9}$$

$$\Delta G_{rxn}° = -RT \ln\left[\left(\frac{\gamma_{H•G}[H•G]}{\gamma_H[H]\gamma_G[G]}\right)\left(\frac{[H]_0[G]_0}{[H•G]_0}\right)\right] \tag{Eq. 4.10}$$

$$\Delta G_{rxn}° = -RT \ln\left[K_a\left(\frac{[H]_0[G]_0}{[H•G]_0}\right)\right] \tag{Eq. 4.11}$$

Let's first assume a K_a of 10^6 M^{-1} or equivalently a K_d of 10^{-6} M. Normally, we let the standard state concentrations be 1 M. If we do this and use Eq. 4.4, we get a $\Delta G_{rxn}°$ of -8.16 kcal/mol, reflecting an exergonic reaction. However, we are now going to change the standard state.

Let's see what happens if the standard state for H, G, and H•G are all set to 10^{-12} M, which will lead to a Gibbs free energy for reaction that we call ΔG_{rxn}^*. The reference state we choose has nothing to do with the size of the K_a that we will experimentally measure, which is a value that depends solely upon the ratios of H, G, and H•G that we find at equilibrium, and so a K_a of 10^6 M^{-1} is still correct for our new ΔG_{rxn}^* analysis. Yet, now with a different reference state ($[H]_o$, $[G]_o$, and $[H \bullet G]_o = 10^{-12}$ M in Eq. 4.11) the ΔG_{rxn}^* is now found to be $+8.16$ kcal/mol, reflecting an endergonic reaction! This tells us that there is an 8.16 kcal Gibbs free energy change to convert one mole of H and G both starting at 10^{-12} M to one mole of H•G at 10^{-12} M. This is what we would predict for these dilute species, because indeed H and G are preferred when dilute.

We chose this standard state for demonstration purposes because it is far below the K_d for this reaction and would therefore be an experimental condition that leads to a preference for free H and G relative to the complex H•G. A ΔG^* diagram would show H•G higher in energy than H and G, meaning that given this reference state, free H and G are preferred. Hence, the key to understanding $\Delta G°$ diagrams is to note that they reflect the exergonicity or endergonicity of the reaction *in the defined reference state*. These diagrams do not necessarily reflect the relative ratios of the reactants and products under your experimental conditions.

The Influence of a Change in Heat Capacity

In discussing binding interactions of the sort considered throughout Chapters 3 and 4, we see many binding constants, K_a, which in turn lead directly to $\Delta G°$. It is also common to dissect $\Delta G°$ into $\Delta H°$ and $\Delta S°$ terms, in hopes of gaining some physical insight into the nature of the binding interactions. Chemists commonly assume that $\Delta H°$ and $\Delta S°$ do not change with temperature, and we have assumed this in all our thermodynamic analyses thus far. However, this is often not true for many types of binding interactions. Instead, we find that $\Delta H°$ and $\Delta S°$ values do change with temperature, a clear indication that the heat capacities of the reactants and products are different. Let's see how this happens.

The enthalpy of any substance increases as the temperature increases at constant pressure via Eq. 4.12. The slope of a graph of enthalpy as a function of temperature is the **heat capacity** of the substance [$C_{P(A)}°$, where "A" designates the substance and the "°" means standard state]. The heat capacity is the amount of energy that a substance absorbs as a function of temperature. It relates to all the different ways the substance can store internal energy at constant pressure. For example, vibrational and rotational modes can absorb thermal energy, and a compound that has more such modes will be expected to have a larger heat capacity.

$$\frac{\partial H_A}{\partial T} = C_{P(A)}° \qquad \text{(Eq. 4.12)}$$

For any substance, then, we can describe enthalpy as a function of temperature. It is convenient to define a reference temperature, let's say 298 K, leading to Eq. 4.13, where $H_A°'$ is the enthalpy of A at 298 K.

$$H_A° = H_A°' + (T - 298)\, C_{P(A)}° \qquad \text{(Eq. 4.13)}$$

To get $\Delta H_{rxn}°$ for the conversion of A to B, we write an equation analogous to Eq. 4.13 for B, and then substraction gives Eq. 4.14. $\Delta C_P°$ is the difference in heat capacities between B and A. Since $298\Delta C_P°$ is a constant, we can incorporate it into $\Delta H_{rxn}°'$, leading to a new term defined as $\Delta H_o = \Delta H_{rxn}°' - 298\Delta C_P°$ (Eq. 4.15, where we now drop the "rxn" subscripts for simplicity). This equation gives the $\Delta H°$ for the reaction as a function of temperature relative to a reference enthalpy ΔH_o. The important point is that the $\Delta H°$ for the reaction is found to be temperature dependent if the heat capacities of the solvated reactants and products are different. For some reactions, such as thermal isomerizations, we might expect $\Delta C_P°$ to be close to zero, and an assumption of a $\Delta H°$ that does not vary with temperature is acceptable.

$$\Delta H_{rxn}° = \Delta H_{rxn}°' + (T - 298)\, \Delta C_P° \qquad \text{(Eq. 4.14)}$$

$$\Delta H° = \Delta H_o + T\Delta C_P° \qquad \text{(Eq. 4.15)}$$

Now what about the entropy? It makes good sense that the entropy of a system should increase with temperature, because any form of increased movement leads to more disorder. Once again, the amount of energy that the system absorbs as a function of temperature (the heat capacity) should affect how much the disorder of the system changes as a function of temperature. The more energy a system can absorb per unit change in temperature, the larger the change in entropy as a function of temperature. Eq. 4.16 gives the entropy of substance A at any final temperature (T_f) relative to the entropy at some initial temperature (T_i), let's say again 298 K.

$$S_A^\circ(T_f) = S_A^\circ(T_i) + C_{P(A)}^\circ \ln\left(\frac{T_f}{T_i}\right) \qquad \text{(Eq. 4.16)}$$

Similar mathematical substitutions as above produce Eq. 4.17 and Eq. 4.18, where $\Delta S_o = \Delta S^{\circ\prime} + \Delta C_P^\circ \ln(1/298)$.

$$\Delta S_{rxn}^\circ = \Delta S_{rxn}^{\circ\prime} + \Delta C_P^\circ \ln\left(\frac{T}{298}\right) \qquad \text{(Eq. 4.17)}$$

$$\Delta S^\circ = \Delta S_\circ + \Delta C_p^\circ \ln T \qquad \text{(Eq. 4.18)}$$

Given these new relationships between ΔH° and ΔS° for reactions as a function of temperature, we can consider how K_a changes as a function of temperature. Recall Eq. 3.22, which leads to a van't Hoff plot. Combining this with Eqs. 4.15 and 4.18 gives Eq. 4.19. When fitting experimental data of $R \ln K_a$ vs. $1/T$ using this equation, we have to adjust three variables to obtain the best fit (ΔH_o, ΔS_o, and ΔC_P). Then, to get ΔH° or ΔS° at a particular temperature, Eqs. 4.15 and 4.18 are used. We will use these relationships later in the chapter when examining the hydrophobic effect, but they are of general utility for any reaction where there is a difference in heat capacity between the reactants and products.

$$R \ln K_a = -\Delta H_\circ\left(\frac{1}{T}\right) + \Delta C_P^\circ \ln T + (\Delta S_\circ - \Delta C_P^\circ) \qquad \text{(Eq. 4.19)}$$

Cooperativity

The last thermodynamic concept that we want to describe here is the energetic consequence of forming multiple binding interactions in a single molecular recognition event, or any binding event. We have already discussed that nature uses multiple small effects to create a large effect. Each individual binding force, dipole attraction, hydrogen bonding, or π effect, is often quite small in magnitude, but when combined they can lead to a substantial affinity between two molecules. The large effect is not a simple sum of all the small effects, though, because there is often a further interaction called **cooperativity**. **Positive cooperativity** arises when the Gibbs free energy of binding is more negative than the sum of all the Gibbs free energy changes for each individual binding interaction. **Negative cooperativity** occurs when the Gibbs free energy of binding is more positive than the sum of the individual parts. Another form of cooperativity observed with biological molecules that bind more than one molecule is discussed in a Going Deeper highlight in Section 4.1.2.

Figure 4.2 shows a schematic binding event that occurs between a host H and guest molecule called A–B. The host has two independent binding regions, one for region A and one for region B of the guest. These separate binding regions can be thought of as consisting of single binding interactions, such as a single hydrogen bond, or they can represent several binding interactions.

The binding energy for molecule A–B to the host can be broken into three parts (Eq. 4.20): the intrinsic Gibbs free energy (GFE) for binding of A (ΔG_A^i), the intrinsic GFE for binding of B (ΔG_B^i), and what is known as the connection GFE (ΔG^s). The **connection GFE** is the extent to which the binding of A–B differs from the sum of the individual interactions of

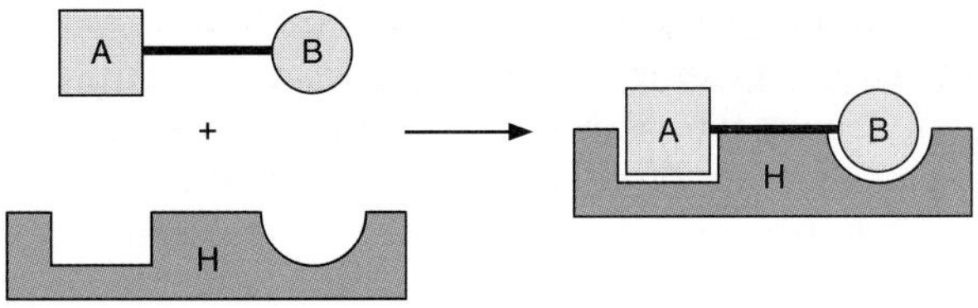

Figure 4.2
Binding of a molecule A–B to a host H
that has completely independent binding
regions for the separate entities A and B.

A and B. The favorable portion of the connection GFE is often postulated to be mostly entropic (the following Connections highlight notes that enthalpy can be commonly involved also). To a first approximation, the losses in translational and rotational entropy upon binding A, B, and A–B to the host are all the same. This is because translational entropy has only a small dependence upon size, and differences in rotational entropy are not very large for molecules of moderate size. In essence, the entropy required for binding B while A is bound is partly paid by linking A and B. Likely there is some loss in the entropy of internal bond rotations in A–B upon its binding relative to separate A and B, but this is small compared to translational and rotational entropy (see Section 2.1.2). Hence, in binding A and B separately to the host, the translational/rotational entropy is paid twice, while in binding A–B to the host it is only paid once. The binding of A–B will be more favorable than $\Delta G_A{}^i + \Delta G_B{}^i$ by an amount that corresponds to the loss of translational and rotational entropy of a single molecule. This additional favorable binding found when A and B are tied together is often referred to as the **chelate effect** in inorganic chemistry (for example, ethylene diamine binds metals better than two separate amines), and the term now finds widespread use in organic chemistry.

$$\Delta G_{AB}{}^\circ = \Delta G_A{}^i + \Delta G_B{}^i + \Delta G^s \qquad \text{(Eq. 4.20)}$$

Now we need to define the intrinsic binding energies of A and B. The **intrinsic binding energy** of any group X is the additional binding that this group imparts to the binding of the rest of the molecule if no differences in strain or entropy result upon interaction of X with the host. This is expressed by Eq. 4.21. Substituting B for X in this equation, doing a similar substitution of A for X in an equivalent equation involving B subscripts, and using the results in Eq. 4.20 leads to Eq. 4.22. Now we see that a positive Gibbs free energy of connection is representative of favorable positive cooperativity, and a negative Gibbs free energy of connection is representative of unfavorable and negative cooperativity.

$$\Delta G_X{}^i = \Delta G_{AX}{}^\circ - \Delta G_A{}^\circ \qquad \text{(Eq. 4.21)}$$

$$\Delta G^s = \Delta G_B{}^\circ + \Delta G_A{}^\circ - \Delta G_{AB}{}^\circ \qquad \text{(Eq. 4.22)}$$

There are enthalpic considerations that can negate the positive cooperativity that results from a favorable GFE of connection. For example, in the extreme case, if A is connected to B in such a way that once A binds, B cannot even reach its binding site on the host, then all the intrinsic binding energy for B cannot be gained. More commonly, conformational changes in A–B or the host may be necessary to bind A–B that are not necessary in the binding of separate A and B. Further, strains may be introduced or relieved in the binding of A–B that were not present in binding A and B separately. When unfavorable enthalpy effects overwhelm the favorable entropy effects, ΔG^s is negative and negative cooperativity occurs. An interesting example of how cooperatively can be influenced by enthalpy considerations is given in the following Connections highlight.

There is an interesting practical use of Eq. 4.22 that needs to be mentioned. It is common practice to multiply the K_a values for binding A and B to the host as a means of estimating the expected binding constant of A–B if no positive or negative cooperativity occurs. Any difference between the real K_a of A–B and this estimated value is considered the cooperativity. However, using $\Delta G° = -RT \ln K$ and Eq. 4.22 leads to Eq. 4.23. This gives an empirical way to measure the cooperativity. One experimentally determines K_A, K_B, and K_{AB}, and using these values in Eq. 4.23 gives the Gibbs free energy of connection.

$$\Delta G^s = RT \ln \left(\frac{K_{AB}}{K_A K_B} \right)$$

(Eq. 4.23)

There are two other forms of cooperativity often found in molecular recognition events. The cooperativity phenomena discussed above involved extra binding energy for A–B above and beyond the energy of binding A and B separately. Another form of cooperativity arises when binding A to one portion of the receptor enhances the binding of B at another site. In some manner the binding of A affects the affinity of a remote site for the binding of B. This is called **allosteric cooperativity**. This is commonly found in nature, and a Connections highlight concerned with hemoglobin (on page 219) discusses such an example.

Another form of cooperativity is called multivalency. **Multivalency** arises when multiple receptors tethered together (often on a surface, such as a biological cell) bind multiple **ligands** (a term used analogous to guests) that are tethered together. These are very often all

---

Connections

Cooperativity in Drug Receptor Interactions

One of the most challenging problems in modern medicine is the increasing resistance of infectious bacteria to long used antibiotics. This has spurred the development of new generations of antibiotics which are effective against resistant strains of bacteria. A mainstay of the new generation of antibiotics is vancomycin. In addition to its medicinal importance, this novel structure has served as a prime target of the synthetic organic community and as a valuable platform for unraveling key issues in molecular recognition.

Vancomycin (in black to the right) and related structures disrupt bacterial cell wall synthesis by binding to the C terminus of a protein that will be incorporated into the cell wall. The protein ends in the sequence D-alanine– D-alanine (color), an interesting use of the unnatural configuration for an amino acid. The antibiotics make numerous hydrogen bonding contacts to this dipeptide (see the drawing to the right), as well as hydrophobic contacts with the methyl groups of the D-Ala's. By systematically modifying the structure of small polypeptide mimics of the protein sequence and studying the binding of these analogs, Williams and co-workers aimed to discover the magnitude of, for example, an individual amide–amide hydrogen bond. In early efforts, one hydrogen bond was deleted (for example, #2 in the drawing). The deletion led to a substantial drop in the binding of the drug, suggesting the hydrogen bond was worth almost 5 kcal/mol in water. Further analysis, though, revealed that removing hydrogen bond #2 weakened the remaining interactions, so that the drop in binding overestimated the strength of the single hydrogen bond. Similar results were seen as the other interactions were deleted. It becomes apparent that, working in the other direction, as each new binding

interaction is added, the pre-exisiting interactions become stronger. This is related to the entropy–enthalpy compensation discussed in Chapter 3 and in the next section. Each additional interaction leads to a tighter complex, further restricting vibrational and translational entropy and thereby strengthening all the interactions.

Vancomycin

D–Ala–D–Ala

Detailed analysis of this system led to values of 0–2 kcal/mol for an amide–amide hydrogen bond, as well as an estimate of ca. 50 cal/mol·Å^2 for the hydrophobic effect. Such values are well in line with other estimates. These studies provide an excellent example of the use of a natural system to provide detailed, quantitative insights into the fundamentals of molecular recognition and the effect of cooperativity.

Williams, D. H., and Westwell, M. S. "Aspects of Weak Interactions." *Chem. Soc. Rev.*, **27**, 57–63 (1998).

the same ligand. Tethering all the ligands together via covalent or non-covalent means can lead to substantial increases in binding of the tethered ligands to the tethered receptors. The issues discussed above that give rise to positive cooperativity are certainly involved here.

Enthalpy–Entropy Compensation

When studying the $\Delta H°$ and $\Delta S°$ of binding as a means to try to improve the affinity of a receptor, an often frustrating feature arises, generally referred to as **enthalpy–entropy compensation**. Whether looking at one binding event over a range of conditions or a series of related binding interactions, we often find that as the binding becomes more enthalpically favorable, it becomes more entropically unfavorable. The net effect can be that $\Delta G°$, which determines the equilibrium constant, remains fairly constant. Why does this happen?

Long considered mysterious, and perhaps some consequence of the unique features of water, in recent years an interesting, general model for the effect has emerged. Remember these are inherently weak associations, much less than the strength of a covalent bond. As such, there is a considerable range in how "tight" or "locked in" a binding event is. Consider a fairly weak association, one with a small, favorable $\Delta H°$. The ligand will bind at the receptor binding site, but not very tightly—any pair-wise interaction will be weak, and there still will be considerable flexibility in both ligand and receptor, meaning considerable **residual motion**. Now make the interaction stronger—that is, make $\Delta H°$ more favorable. Individual interactions will become stronger, and the receptor will grab onto the ligand more tightly. But this will restrict motions of both the receptor and the ligand more significantly, making $\Delta S°$ less favorable. In this view, enthalpy–entropy compensation is a natural consequence of the fact that stronger binding is tighter binding, and so as $\Delta H°$ becomes more favorable, $\Delta S°$ becomes less favorable. It is very important to keep such issues in mind when considering the design of better receptors or better ligands, the latter being a central aspect of modern drug design. We shall discuss enthalpy–entropy compensation again in Chapter 8 when linear free energy relationships are examined.

4.1.2   The Binding Isotherm

To this point considerable discussion has been given to $\Delta G°$ and K_a values, as well as enthalpy, entropy, and heat capacity. We have noted that these parameters are the primary data used to decipher the binding forces of molecular recognition, and to test our postulates for creating molecules that associate in predictable ways. Now that we have some understanding of these thermodynamic parameters, we can delve into the manner in which they are determined. We would like, however, to once again stress that although we present this discussion in the context of molecular recognition, the mathematics and experimental methods we are discussing here are completely general for any two different compounds that are involved in an equilibrium to form a single structure.

Almost all experimental methods to measure binding constants for any kind of reaction rely on the analysis of a binding isotherm. A **binding isotherm** is the theoretical change in the concentration of one component as a function of the concentration of another component at constant temperature. We measure the concentrations using some experimental method (NMR, UV/vis spectroscopy, etc.), and fit the experimental data to the theoretical binding isotherm. Here we discuss only one-to-one binding, such as that between a single host (H) and guest (G), but binding isotherms have been generated for equilibria involving significantly more complex stoichiometries.

Normally the experiment is performed holding the concentration of one species relatively constant (either strictly constant or with slight dilutions during the experiment), while varying the concentration of the other species. For example, when holding the concentration of the host constant, it is common to employ a relationship such as Eq. 4.24, where $[H]_o$ is the initial concentration of H. This equation is derived from Eq. 4.2 using the relationship that $[H]_o = [H \bullet G] + [H]$. If we can measure the H$\bullet$G concentration as a function of the concentration of free G in solution, we can calculate K_a. Most commonly, some parameter related to the concentration of H$\bullet$G is plotted along the y axis (see examples below), not actually [H$\bullet$G] itself.

$$[H \bullet G] = \frac{[H]_o \, K_a \, [G]}{1 + K_a \, [G]}$$

(Eq. 4.24)

Importantly, [G] in Eq. 4.24 is not the amount of G that is added to the solution ($[G]_o$), but instead it is free G. Since the experimentalist controls $[H]_o$ and $[G]_o$, in a graphical analysis it is one of these values that would be used for the x axis. Routinely, the analysis is done for many concentrations of G_o with H_o constant, where $[G]_o = [G] + [H \bullet G]$. A plot of experimentally determined $[H \bullet G]$s as a function of many $[G]_o$ values gives the curve to be analyzed.

Since $[G]_o$ is not in Eq. 4.24, whereas [G] is, we must make some assumptions about [G] or solve for [G] in terms of $[G]_o$. If $[G]_o \gg [H]_o$, we can make the assumption that $[G] = [G]_o$, because at all points in the isotherm only a small fraction of G is converted to H•G. This is the basis of many methods for determining binding constants, and one in particular called the Benesi–Hildebrand method that is covered in an upcoming Going Deeper highlight. However, when this assumption is not valid, Eq. 4.25 can be used to relate [G] and $[G]_o$. Now for each $[G]_o$ value one solves this quadratic to get [G], and that particular [G] is used in Eq. 4.24 to generate the isotherm. To calculate [G] from this equation we must first know K_a, which is the goal of our analysis. Hence, we first guess a K_a to calculate [G] values for different $[G]_o$ values, and then iteratively change K_a until the theoretical isotherm matches the experimental data (see the next section for examples).

$$K_a[G]^2 + (K_a[H]_o - K_a[G]_o + 1) \, [G] - [G]_o = 0$$

(Eq. 4.25)

A plot of [H•G] versus $[G]_o$ is given in Figure 4.3. A hyperbolic relationship is found where the concentration of [H•G] approaches $[H]_o$ with increasing concentrations of $[G]_o$. This is called **saturation behavior**. At high concentrations of G_o, most of H is converted to the H•G complex and is considered saturated with G. We can see this using some simple mathematics.

Assume a binding constant of $10 \, M^{-1}$. If we start the reaction with 100 mM H and 10 mM G, at equilibrium [H] = 95.1 mM, [G] = 5.1 mM, and [H•G] = 4.9 mM. About half of the initial G is part of H•G, but we have barely changed the concentration of H (you should check the math for yourself).

If we start the analysis with 100 mM H and 1000 mM G, instead, at equilibrium [H] = 10 mM, [G] = 910 mM, and [H•G] = 90 mM. Now most of the host is in the form of H•G.

What is a high enough concentration of G to give saturation? High relative to what? Just as in our analysis of dilution (Section 4.1.1), the concentration is relative to the K_d for the

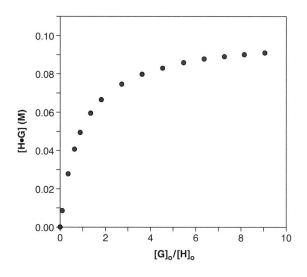

Figure 4.3
A theoretical binding isotherm as a function of $[G]_o$ for a system with $K_a = 10 \, M^{-1}$ and $[H]_o = 100 \, mM$.

equilibrium, but now it also depends upon $[H]_o$. In the two numerical examples above, we choose $[H]_o$ to be the same as K_d.

Eq. 4.26, obtained from Eq. 4.3 by substitution of $[H] = [H]_o - [H \bullet G]$, gives quick insight into predicting saturation behavior, especially when $[H]_o = K_d$. For example, when $[H]_o = K_d$ and $[G] << K_d$, the denominator of Eq. 4.26 can be approximated by K_d, and consequently we find that $[H \bullet G] = [G]$. Under these conditions the host is not saturated but the concentration of $H \bullet G$ tracks with how much G has been added. Conversely, when $[H]_o = K_d$ and $[G] >> K_d$, the denominator can be approximated by $[G]$, and $[H \bullet G] = [H]_o$, meaning that the host is now saturated. If we now use a low $[H]_o$ relative to K_d, it will take more G to saturate H, and if we start with a high $[H]_o$ relative to K_d, it will take less G to saturate H.

$$[H \bullet G] = \frac{[H_o][G]}{K_d + [G]} \qquad \text{(Eq. 4.26)}$$

Such analyses can be performed for a series of host–guest complexes with differing binding constants (K_a). With a higher binding constant, it takes less G to saturate H than for a complex that has a lower binding constant. To see this, let's plot several binding isotherms as a function of different K_a values but the same host concentration. We now plot $[G]_o/[H]_o$ as the x axis (Figure 4.4). A very distinct shape to the curves is evident. Each curve is still hyperbolic, has its highest extent of curvature at a 1:1 ratio of guest to host, and is saturating to the same level. These three points are clear for the two curves with the higher K_a values. However, when $1/K_a$ is significantly below H_o, the curve is quite flat, and as alluded to above, it takes more and more guest to saturate the host. For an optimal analysis, it is a good practice to perform the experiment at a $[H]_o$ near K_d. Recall that all these binding isotherms are for the same $[H]_o$. Hence, we want to make it quite clear that the shape of the isotherm is only indicative of the association constant when coupled with a knowledge of how $[H]_o$ compares to K_d. This is just another example of the effects discussed in Section 4.1.1, where we showed that the composition of the solution varies as a function of the starting concentrations of the reactants.

In the discussions above we focused upon measuring $[H \bullet G]$ as a function of $[G]_o$. In practice, it is arbitrary which compound we call H and which we call G, so we can plot $[H \bullet G]$ as a function of either the added concentration of host or guest while keeping the other constant. Remembering that this is a completely general discussion, you can plot against either of the two components that come together to create a complex.

Figure 4.4
Binding isotherms for varying K_a with $[H]_o = 0.1$ M.
■ $K_a = 1$ M^{-1} (hence $K_d = 1$ M and is above $[H]_o$),
● $K_a = 10$ M^{-1} (hence $K_d = 0.1$ M and is equal to $[H]_o$), and
▲ $K_a = 100$ M^{-1} (hence $K_d = 0.01$ M and is below $[H]_o$).

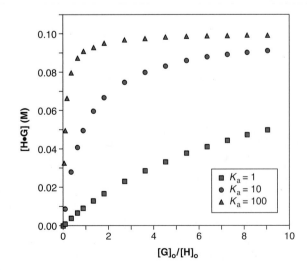

Going Deeper

The Hill Equation and Cooperativity in Protein–Ligand Interactions

An interesting cooperativity in binding is seen with some proteins that have multiple binding sites for the same ligand. The classic example is hemoglobin, which has four binding sites for O_2, but many other examples exist. If the four binding sites are independent of one another, such a system has no real advantage over four copies of a molecule with one binding site. However, if the binding sites can communicate with one another, either directly or, more commonly, indirectly, then an important kind of cooperativity can develop. If binding of the ligand at one site makes the subsequent binding event more favorable, we have positive cooperativity. Alternatively, the binding can inhibit further binding, producing negative cooperativity. This is a different kind of negative and positive cooperativity than that discussed in Section 4.1.1, where we analyzed how adding binding sites can assist binding. Here, several binding sites communicate with each other via an **allosteric effect**.

What is the biological advantage of positive cooperativity? Consider the binding isotherm for a cooperative system, shown in the graph. We define a saturation value, S, as the extent to which all binding sites are occupied,

varying from 0 to 1. In a cooperative system, as we increase ligand concentration, binding becomes more and more favorable, because the first binding site is filling up, followed by the second, which is a stronger binder. One way to model such a system is with the **Hill equation**, which considers the degree of saturation as a function of ligand concentration, [L]; EC_{50}, the ligand concentration necessary to achieve half-maximal binding; and an exponent, n_H, called the **Hill coefficient**, which measures the degree of cooperativity. The graph shows three situations, all with the same EC_{50}, but with differing degrees of cooperativity, as evidenced by the Hill coefficient. Note how in the cooperative systems, the dose–response curve is much steeper around EC_{50}. This means that relatively small changes in ligand concentration can produce large changes in response. Now consider a system in which ligand binding produces a signal, but only when all binding sites are occupied. Cooperativity has allowed nature to shape the dose–response curve to make a system more sensitive to small changes in ligand concentration, and this is just what you would want from a signaling system. This strategy is used in a range of biological signaling pathways.

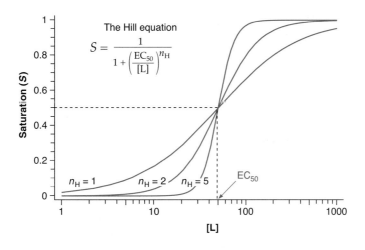

The Hill equation

$$S = \frac{1}{1 + \left(\dfrac{EC_{50}}{[L]}\right)^{n_H}}$$

4.1.3 Experimental Methods

Given the general shape, caveats, and lessons from the 1:1 binding isotherm noted above, we can now consider how one implements an experimental study. We need methods to determine [H•G] as a function of $[G]_o$ or $[G]_o/[H]_o$. In practice, because [H•G] is being measured from some chromatographic, electrochemical, or spectroscopic method, there are proportionality constants that relate the experimental signals to [H•G]. Most commonly we do not know what these proportionality constants are, and we do not know K_a. Therefore, theoretical binding isotherms for various K_a's and proportionality constants are compared to the experimental data, and the "best fit" between experiment and theory gives the K_a.

A key issue in such studies is the timescale of the measurement. Remember that the total binding energy for a host–guest complex is much less than a covalent bond energy in all but the most exceptional cases (see Section 4.3.3 on carcerands and carceplexes for exceptions). As a result, under most conditions, the complex is constantly forming and breaking up. These are dynamic systems. To emphasize this, we show in Eq. 4.27 an alternative way to represent K_d and K_a as a ratio of rate constants. These expressions make sense. A tight binding constant should have a slow dissociation rate—a small value of k_{off} relative to k_{on}. This would make K_d smaller and K_a larger, as expected.

$$H + G \underset{k_{off}}{\overset{k_{on}}{\rightleftharpoons}} H \bullet G$$

$$K_d = \frac{k_{off}}{k_{on}} \quad K_a = \frac{k_{on}}{k_{off}}$$

(Eq. 4.27)

We can get a feeling for these rate constants by making some reasonable assumptions. For small molecules, k_{on} will often be near the diffusion rate constant k_d. This occurs for "open" hosts binding to small guests, bases binding to protons, and unsaturated metal complexes binding to ligands. We can make the assumption that k_d is on the order of 10^9 M^{-1} s^{-1}. This means that for a reasonably strong complex with $K_d = 10$ μM, k_{off} must be on the order of 10^3 s^{-1}, meaning our complex has a lifetime ($\tau = 1/k_{off}$) on the order of a millisecond. If our probe method is faster than this time scale, we can expect to see individual signals for H, G, and H•G. Alternatively, a slower probe will see only a time average of the various species. That is, signals we might associate with G will be a weighted average of the spectrum of free G and the spectrum associated with H•G. This will complicate the analysis, but typically the problem can be solved. Below we will briefly discuss the several common approaches to evaluating binding constants, although any method that produces distinguishable signals for H and/or G vs. H•G can be used. References that provide detailed discussions of this topic are given at the end of the chapter.

UV/Vis or Fluorescence Methods

In Chapter 16 we analyze UV/vis (absorption) and fluorescence (emission) spectroscopies. For the present discussion we need consider only two issues. First, both methods are quite rapid; absorption and related phenomena occur on ps or faster time scales, and fluorescence emission requires ns or less. Thus, we are in the rapid observation regime, where individual signals from the various species can be observed. Second, for both methods it is a straightforward matter to relate the concentration of a species to the intensity of the signal associated with it (for one way, see the Going Deeper highlight below). An advantage of these approaches is that they can be quite sensitive (especially fluorescence). A disadvantage is that they are "guest limited"—only guests with desirable spectroscopic properties are amenable to study.

NMR Methods

It is often observed that the ^{1}H NMR signals of a guest shift considerably on binding to a host. This is especially true for cyclophane hosts (see below), which have aromatic "walls" and so can produce large shielding effects. NMR has thus been a powerful tool for evaluating host–guest complexes. However, in most cases the NMR time scale is slow compared to on and off rates for the complex (for an explanation of the NMR time scale, see the Going Deeper highlight in Chapter 2). This means that we cannot see a separate signal for the host–guest complex. The signal observed for "G" will actually be an averaged signal for G and H•G. The consequence is that the chemical shift of the G in the complex H•G is an additional unknown that must be solved for in our analysis of the experimental shifts as a function of concentration changes. This complicates the analysis, and increases the error bars associated with values of K_d determined by NMR. Nevertheless, NMR has been the most

---

Going Deeper

The Benesi–Hildebrand Plot

It is common and convenient in host–guest chemistry for one component of the study (host or guest) to be transparent in the absorption/emission range being probed. If we say, for example, the host is transparent, this means we only have to consider G and H•G (as always, the roles of H and G can be reversed). Using Beer's law (A = εbc, see Chapter 16), this leads to the equation shown for the absorbance of the system (a comparable equation for fluorescence can be obtained).

$$A = \varepsilon_{HG}b[H•G] + \varepsilon_{G}b[G]$$

When we define $A_o = \varepsilon_G b[G]_o$ as the first absorbance value (that obtained before addition of any H); $\Delta\varepsilon = \varepsilon_{H•G} - \varepsilon_G$ as the difference in extinction coefficients between the two species involved; and $\Delta A = A - A_o$

as the change in absorbance on adding an incremental amount of H, we get the following equation:

$$\Delta A = b[H•G]\Delta\varepsilon$$

After substituting the binding isotherm ([H•G] = [[G]$_o K_a$[H]/(1 + K_a[H])]) for [H•G] and assuming that [H] = [H]$_o$, we get the first equation below. The assumption is only valid when [H]$_o$ >> [G]$_o$. A double reciprocal plot of 1/ΔA as a function of 1/[H]$_o$ gives a line from which $\Delta\varepsilon$ can be calculated from the intercept, and K_a can be calculated from the slope. This is called the Benesi–Hildebrand method, and it has seen extensive use in both artificial and biological molecular recognition.

$$\Delta A = b\Delta\varepsilon[G]_o K_a[H]_o/(1 + K_a[H]_o)$$
$$1/\Delta A = 1/b\Delta\varepsilon[G]_o[H]_o K_a + 1/b\Delta\varepsilon[G]_o$$

---

common tool used for evaluating artificial (non-biological) host–guest system.

A benefit of the NMR method is that along with K_a, the fitting produces the NMR spectrum of the H•G complex. Since NMR spectra can provide valuable structural insights (much more so than UV/vis or fluorescence spectroscopy), NMR titrations produce important additional insights into the binding event, often providing considerable detail on the precise binding geometry. This is especially true when hydrogen bonding is the primary binding force, as the existence or absence of hydrogen bonds can be established directly. Let's briefly look at the mathematics used with the NMR technique.

If, for example, our experiment involves monitoring the chemical shift of the host upon addition of various concentrations of guest, Eq. 4.28 would hold. The value of the observed resonance (δ_{obs}, chemical shift, ppm) is a weighted average of the chemical shifts of the free host (δ_H) and the host–guest complex ($\delta_{H•G}$). The observed shift is weighted by the mole fractions of H ($X_H = [H]/[H]_o$) and H•G ($X_{H•G} = [H•G]/[H]_o$). Since $X_H = 1 - X_{H•G}$, we can rearrange Eq. 4.28 to Eq. 4.29. Once again, we use the standard binding isotherm (Eq. 4.24) for [H•G], but now we also divide by [H]$_o$ to get Eq. 4.30. After defining $\delta_{obs} - \delta_H = \Delta\delta$ and ($\delta_{H•G} - \delta_H$) = $\Delta\delta_{tot}$, we obtain Eq. 4.31. This equation once agains predicts a hyperbolic shape to the observed isotherm.

$$\delta_{obs} = \delta_H X_H + \delta_{H•G} X_{H•G} \qquad \text{(Eq. 4.28)}$$

$$\delta_{obs} - \delta_H = X_{H•G}(\delta_{H•G} - \delta_H) \qquad \text{(Eq. 4.29)}$$

$$X_{H•G} = \frac{K_a[G]}{1 + K_a[G]} \qquad \text{(Eq. 4.30)}$$

$$\Delta\delta = \frac{\Delta\delta_{tot} K_a[G]}{1 + K_a[G]} \qquad \text{(Eq. 4.31)}$$

Isothermal Calorimetry

The NMR and UV/vis techniques discussed above are used to determine a single binding constant K_d, from which $\Delta G°$ can be determined via Eq. 4.4. To determine the $\Delta H°$ and $\Delta S°$

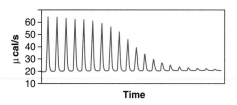

Figure 4.5
A plot of the heat released per time for a solution of the enzyme RNase as a function of injections of 2'-cytidine monophosphate. Early in the titration there are large changes in the concentration of H•G that occur during the spontaneous evolution to equilibrium, while toward the end of the titration there are small changes in H•G concentration, and hence little change in heat released. Wiseman, T., Williston, S., Brandts, J. F., and Lin, L.-N. "Rapid Measurement of Binding Constants and Heats of Binding Using a New Titration Calorimeter." *Anal. Biochem.*, **179**, 131–137 (1989).

values, a van't Hoff analysis would need to be employed (see Section 3.1.5). Recall that in a van't Hoff analysis, K_a values at different temperatures are measured, and a plot of $1/T$ versus $\ln K_a$ gives the thermodynamic parameters $\Delta H°$ and $\Delta S°$. However, there is an experimental technique that gives K_a and $\Delta H°$ all in one analysis, thereby allowing one to solve easily for $\Delta G°$ and $\Delta S°$ using Eq. 4.4 and $\Delta G° = \Delta H° - T\Delta S°$. It is referred to as isothermal calorimetry.

Upon association of a host and a guest, heat will either be released or absorbed. In isothermal calorimetry, a measurement of the change in heat is made for a series of additions of guest to host (Figure 4.5). The heat released or absorbed in each step is measured by comparing the temperature changes that occur between a sample cell where guest is added to host, and a reference cell where the guest solution is added to a "blank" solution (an identical solution, but lacking host). Additions of guest into the two cells are monitored until addition of guest no longer results in temperature differences between the cells. Eq. 4.32 relates the total heat (Q) generated or absorbed for all the additions of guest, where V is the volume of the vessel. Using the binding isotherm (Eq. 4.24) for [H•G] gives Eq. 4.33. The reason that the heat released or absorbed can be directly related to $\Delta H°$ for the reaction is given in the following Going Deeper highlight.

$$Q = V\Delta H°[\text{H•G}] \qquad \text{(Eq. 4.32)}$$

$$Q = \frac{V\Delta H° K_a[\text{H}_o][\text{G}]}{1 + K_a[\text{G}]} \qquad \text{(Eq. 4.33)}$$

In this way, we produce a binding titration just like we did from an NMR or a UV/vis study. This allows us to determine K_a and $\Delta H°$. From K_a we get $\Delta G°$, and then we can solve for $\Delta S°$. As promised, a single titration has produced all the quantities of interest. Isothermal calorimetry is thus a very powerful technique for evaluating binding. It does require fairly sophisticated instrumentation that is dedicated to such measurements. Also, fairly large quantities of material are sometimes needed, especially for hydrophobic binding interactions which can have fairly small $\Delta H°$ values, and thus release or absorb relatively little heat.

4.2 Molecular Recognition

Molecular recognition presents special challenges for physical organic chemistry, and defines one of the frontiers of the field. We will present an overview of the basic strategies and some key results of studies in molecular recognition. This is a large and rapidly growing field, and we cannot hope to cover all systems. Several comprehensive reviews are noted at the end of the chapter.

We have already explored the reasons why K_a and K_d values are the primary data obtained in our analyses—that is, to gain insight into weak binding forces. Molecular recognition events can be quite potent—binding interactions of tens of kcal/mol are common—and the consequences in a biological context or in a designed sensor can be quite dramatic. However, when we try to analyze the physical underpinnings of an event—the stock-in-trade of

Going Deeper

How are Heat Changes Related to Enthalpy?

To understand how the heat released or absorbed can be related to $\Delta H°$, we must delve once again into some thermodynamic relationships. First, we can ask why heat is released or absorbed. In our discussion of spontaneous changes in composition in Section 3.1.5, we analyzed how the Gibbs free energy of a solution evolves to a minimum upon addition of solutes to a solution. This is exactly what happens when one adds an aliquot of a guest to a solution of host. Immediately after addition, the system is not at equilibrium, and it spontaneously evolves to change the composition to that of the equilibrium composition of host, guest, and host–guest complex. The Gibbs free energy of the solution is minimized, but the enthalpy of the solution may either increase or decrease in this reaction depending upon whether the reaction is endothermic or exothermic.

To relate the enthalpy change to heat, we must better understand what enthalpy consists of. The enthalpy of a solution equals the internal energy of the solution (U) plus a term that includes the pressure and the volume of the solution (see right). The change in internal energy of a solution (dU), such as that which might occur during the evolution of the system to the lowest Gibbs free energy, is the heat released or absorbed (dq) plus how much work (dw) the solution performs during this evolution. We now write an equation describing the change in enthalpy of the solution during the change in composition from the initial state (i) to the final state (f). When the solution does no work ($dw = 0$) and there is no pressure or volume change, the change in heat is simply the change in enthalpy.

Hence, the heat evolved or absorbed during the spontaneous change that ensues while a solution evolves toward equilibrium is simply the enthalpy change for the solution.

$$H = U + PV$$
$$dU = dq + dw$$
$$dH = dq + dw + (P_f V_f - P_i V_i)$$
$$dH = dq$$

Here dH is not a molar quantity because it is the heat evolved for addition of an aliquot of G to H. It needs to be scaled to the molar quantity that we define at standard state, which gives the following equation. Here, $\Delta[\text{H}\bullet\text{G}]$ is the change in moles of H•G that occurs immediately after each injection of G, and V is the volume of the solution. Taken together, they represent a scalar (a proportionality constant) that reduces $\Delta H°$ to a quantity that represents the heat change for the number of moles actually changing during the addition of each aliquot of guest.

$$dq = \Delta H° \, (V \bullet \Delta[\text{H}\bullet\text{G}])$$

The total heat released for all additions of guest (Q, the sum of dq's after each injection) is given by Eq. 4.32. Hence, we find that the total heat released is directly proportional to the total amount of H•G formed and the heat of reaction ($\Delta H°$).

---

physical organic chemistry—we find that big effects result from the accumulation of a large number of relatively small effects. Weaker interactions such as those involved here can be difficult to analyze. We are not going to see classical substituent effects, where adding one phenyl group increases the rate of a reaction by 10^3. We will often be at the edge of detectability, in a region where interpretation is challenging. Also, weak interactions are especially sensitive to environment. An interaction that is potent in chloroform might be nonexistent in DMSO, or even in 95% chloroform / 5% DMSO. And then there are the special complexities of water, a key solvent for molecular recognition, and the difficulties in deciphering the origin of the hydrophobic effect. The student should appreciate going into this discussion that hard and fast quantitative rules will not emerge. Trends are clear, and unifying principles exist, but many challenges remain. Nevertheless, the importance of such interactions cannot be denied. Our ability to understand biology at a chemical level and to design materials with desired properties depends to a great degree on our ability to understand and control the weak interactions of molecular recognition.

The analysis of molecular recognition is a natural extension of our understanding of solvent–solvent, and solute–solvent interactions. In analyzing the driving force for association of any solute we must consider two factors: differential solvation effects and the interactions between the two solutes. For example, consider Eq 4.34. Upon association of A and B, their specific intermolecular interactions are replacing those that A and B made with the solvent. Similarly, the replacement of solvent molecules around A and around B by the interactions between A and B leads to release of solvent molecules, which in turn solvate each other. All these interactions are part of the total Gibbs free energy of the solution. Hence, we can relate all molecular recognition phenomena to differential solvation between the reactants and products. If the solvent makes less favorable interactions (either enthalpically or entropically) with A and B than those gained when A and B associate and the released solvent associates, then the complex A–B will be preferred over free A and B. Therefore, in the following discussions, we will be focusing upon the differences between the binding forces between A and B with solvent and A and B together, but also often analyzing interactions between the released solvent molecules, as we did with the hydrophobic effect.

$$\text{(Eq. 4.34)}$$

4.2.1 Complementarity and Preorganization

Before exploring various systems that have the primary binding forces at their origin, it is instructive to analyze two guiding principles of molecular recognition: complementarity and preorganization. **Complementarity** is the concept of using a host that has the proper structure to complement the structure of the guest. As a crude analogy, one does not put a circular block into a square hole. **Preorganization** is the notion of having the receptor be complementary *prior* to the binding event, a notion that is quite energetically favorable. It costs energy to deform a square hole to fit a circular block, and this energy cost is subtracted from the overall binding energy. It is much better for binding to preorganize the hole to the shape of a circle. Our exploration of these two principles is in the context of crown ethers, the molecules that sparked the birth of "molecular recognition", although the principles are applicable to all kinds of hosts and guests.

Crowns, Cryptands, and Spherands—Molecular Recognition with a Large Ion–Dipole Component

A watershed observation was reported by Pedersen in 1967 based on studies performed at DuPont Central Research. Pedersen noted that the cyclic polyether called 18-crown-6 formed remarkably tight complexes with simple cations like K^+. The "crown" nomenclature arose because of the shape of the complex, with six oxygens binding the cation in a characteristic macrocyclic shape (Figure 4.6). The seminal observations of Pedersen were brilliantly exploited by other workers, most notably Cram and Lehn, and these three shared the 1987 Nobel Prize in chemistry for this work. "Host–guest" chemistry was born, with the crown ether serving as a molecular host to the cationic guest.

It is not surprising that one ether oxygen should interact favorably with K^+. The lone pairs on oxygen or, alternatively, the dipole associated with a C–O–C unit, will have a significant electrostatic attraction to a cation (see the ion–dipole interaction in Section 3.2.2). The same forces are involved when a simple salt dissolves in water. What, then, is special about the crown ether?

Consider the interaction of K^+ with a single molecule of water (Eq. 4.35), the oxygen of which is similar to the oxygen of a crown ether. High pressure mass spectrometry studies establish that the binding interaction is enthalpically favorable, with $\Delta H° \approx -18$ kcal/mol in the gas phase. However, there is a significant entropy price for freezing out the motions of two particles, so that $\Delta S°$ is considerably negative (in this case ≈ -22 eu). The net effect is that $\Delta G°$ is only -11.5 kcal/mol, perhaps not as favorable as we might expect. With each additional water we add, $\Delta H°$ becomes slightly less favorable (the K^+ is now not as desperate to find a binding partner), but the $\Delta S°$ penalty persists for each step. To get six ether or water

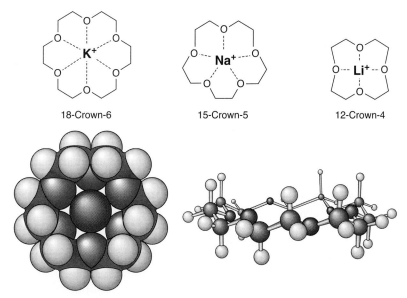

Figure 4.6
Several views of representative crown ethers binding alkali metal ions, along with two views of 18-crown-6.

oxygens around the K⁺, we have to freeze out the motions of seven particles—a severe entropic penalty.

$$K^{\oplus} + H_2O \rightleftharpoons K^{\oplus} \bullet H_2O \qquad \text{(Eq. 4.35)}$$

Now consider the complexation of K⁺ by 18-crown-6. We get six favorable O•K⁺ interactions, but now we only have to restrict the movement of two molecules. The $\Delta H°$ component has not changed substantially, while $\Delta S°$ clearly has, and hence $\Delta G°$ will be more favorable. As is always true in thermodynamics, you can't get something for nothing. The entropy penalty has to be paid somewhere. It is paid *during the synthesis* of 18-crown-6 (it is difficult to close the macrocyclic ring) instead of during the complexation process. As a result, $\Delta G°$ for the complexation is much more favorable. This is a general theme, in which the synthesis of an artificial receptor involves steps that pay the entropy price of preorganizing a binding site, thereby making complexation more favorable.

Another important insight from the crown ether work was the **template effect** in synthesis (Figure 4.7). Crown ethers are macrocycles, and the synthesis of macrocycles is often

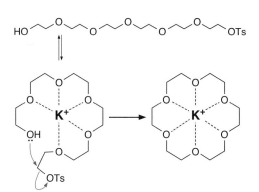

Figure 4.7
The template effect in the synthesis of crown ethers.

challenging, because of the statistical difficulty of getting the ends of a long chain together so they can react to make a ring. However, with the crown ethers it was discovered that if a potential guest for the final macrocycle is present during the ring closure reaction, the yield of the ring closure improved significantly. For example, having a K^+ salt in the reaction mixture improves the synthesis of 18-crown-6. The acyclic polyether is being coaxed by the K^+ into a form that resembles the macrocycle and so is more amenable to cyclization. This template effect has found good use in the crowns / cryptands field, and is also adaptable to other types of ring closures.

Besides being an important scientific advance, this kind of success in molecular recognition has practical uses, too. Salts such as KCN can be dissolved in organic solvents when an equivalent of 18-crown-6 is present. In such a system the K^+ is encapsulated by the crown, but the CN^- is in a sense "naked", lacking any significant ion pairing or solvation, making it an especially potent nucleophile (see the discussion of nucleophilicity in Chapter 8). In another example, the oxidant $KMnO_4$ can be dissolved in benzene when 18-crown-6 is present, and this "purple benzene" has unique oxidizing powers. The key advance of the crowns, then, was the development of a way to bring a normally water soluble salt into a less polar medium. This is done by creating a very polar "microenvironment" that stabilizes the cation. Essentially nothing is done to stabilize or solvate the anion, but the salt will still dissolve. Another modern use of crown ethers is in the creation of molecular devices, and an example of a transport agent is given in the next Connections highlight.

A key advance in the crown ether field was the development of a range of crowns, such as 12-crown-4, 15-crown-5, etc. Table 4.1 shows a series of binding constants between various metals and crowns. Now the concept of complementarity is evident. There is a trend, but not a perfect one, that the smaller crowns have selectivity for the smaller cations, while the larger crowns target the larger cations preferentially.

Connections

Using the Helical Structure of Peptides and the Complexation Power of Crowns to Create an Artificial Transmembrane Channel

The predictable helical nature of many peptides leads to the possibility of using an α-helix as a scaffold to create chemical devices. One such device builds upon the molecular recognition properties of crown ethers discussed here. When every third amino acid in a 21 amino acid long peptide has an appended benzo-21-crown-7 entity, the crown ethers stack to form a column. This structure aligns within lipid bilayers spanning one side to another. The affinity of the crown ethers for K^+ leads to conduction of this cation from one side of the membrane to the other via migration through the channel created by the stacked crowns.

Meillon, J.-C., and Voyer, N. A. "Synthetic Transmembrane Channel Active in Lipid Bilayers." *Angew. Chem. Int. Ed. Eng.*, **36**, 967–969 (1997).

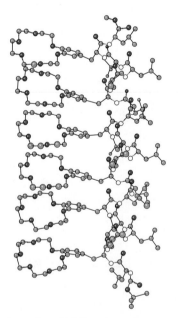

An artificial ion channel

Table 4.1
Log K_a Values for Various Crown-Like Structures and Cations*

Host	Cation	Conditions	logK_a
12-Crown-4	Li$^+$	MeOH (anion Cl$^-$)	<0.5
	Na$^+$	MeOH (anion Cl$^-$)	1.73
	K$^+$	MeOH (anion Cl$^-$)	0.86
15-Crown-5	Li$^+$	MeOH (anion Cl$^-$)	1.21
	Na$^+$	MeOH (anion Cl$^-$)	3.42
	K$^+$	MeOH (anion Cl$^-$)	3.38
18-Crown-6	Li$^+$	CDCl$_3$ (anion picrate)	5.63
	Na$^+$	CDCl$_3$ (anion picrate)	6.11
	K$^+$	CDCl$_3$ (anion picrate)	>11.0
	Li$^+$	MeOH/C$_6$H$_6$ 8/2 (Cl$^-$)	<0.5
	Na$^+$	MeOH (anion Cl$^-$)	4.32
	K$^+$	MeOH (anion Cl$^-$)	6.15
[2,1,1]-Cryptand	Li$^+$	MeOH (0.05 M Et$_4$NClO$_4$)	7.90
	Na$^+$	MeOH (0.05 M Et$_4$NClO$_4$)	6.64
	K$^+$	MeOH (anion nitrate)	2.36
[2,2,1]-Cryptand	Li$^+$	MeOH (anion nitrate)	4.69
	Na$^+$	MeOH (0.05 M Et$_4$NClO$_4$)	9.71
	K$^+$	MeOH (0.05 M Et$_4$NClO$_4$)	8.40
[2,2,2]-Cryptand	Li$^+$	MeOH (anion nitrate)	2.46
	Na$^+$	MeOH (0.05 M Et$_4$NClO$_4$)	7.8
	K$^+$	MeOH (0.05 M Et$_4$NClO$_4$)	10.49
Spherand	Li$^+$	D$_2$O saturated CDCl$_3$ (anion picrate)	>16.7
	Na$^+$	D$_2$O saturated CDCl$_3$ (anion picrate)	10.0

*There is considerable variation in these values depending upon the conditions and the technique used to measure the binding constants. To the best of our ability, we choose similar conditions and techniques. However, for a very comprehensive tabulation of binding constants see Izatt, R. M., Pawlak, K., and Bradshaw, J. S. "Thermodynamic and Kinetic Data for Macrocycle Interactions with Cations and Anions." *Chem. Rev.,* **91**, 1721–2085 (1991).

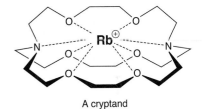

A cryptand

A triazacrown

Many variants of crown ethers have been developed, with the goal of altering selectivity or increasing binding. The **cryptands**, tricyclic compounds that fully encapsulate or entomb (hence the name *crypt*and) the cation, provide tight and selective binding in which the guest really is substantially encapsulated. More so than for the crown ethers, Table 4.1 shows a correlation between the binding constant and how well the cavity size matches the cation size for the cryptands.

18-Crown-6 can bind not only K$^+$, but also ammonium ions of the form RNH$_3$$^+$, although the selectivity is minimal. In contrast, the triazacrown shown in the margin selects ammoniums over K$^+$, making it a specific receptor for that functional group.

An important lesson from these early studies was the value of space-filling models (also known as Corey–Pauling–Koltun or CPK models) in designing new systems. In particular, when designing hosts that are intended to have cavities or voids, only space-filling models, versus various forms of ball-and-stick or wire-frame models should be trusted. Whether using physical models or computer images, the message is the same.

Although 18-crown-6 binds K$^+$ well, we want to be able to quantify this process and learn how to maximize it. A central theme that we have already mentioned is preorganization. We described above the significant entropic advantage of a macrocycle such as 18-crown-6 compared to six molecules of water. However, as noted in Chapter 2, we expect an 18-membered macrocycle to be conformationally quite flexible. This flexibility is lost on complexation, with a concomitant entropic penalty, and so 18-crown-6 is not truly opti-

A spherand

"Cleft"

"Tweezer"

mal. This is confirmed by crystal structures, which clearly show that in the absence of a complexing cation, 18-crown-6 adopts a collapsed structure quite different from the binding, crown shape. Thus, the molecule must reorganize itself to bind a cation, and there must be an energetic penalty for this. The preorganization is imperfect.

Cram showed that in better designed systems, in which the six oxygens are held in the optimal binding position even in the absence of a cation, much higher affinities can be achieved. The more closely the structures of the free and bound host are matched, the better the binding. An example of such a structure is the **spherand** shown in the margin. Now, crystal structures of host with and without guest are essentially superimposable—that is, the host is perfectly preorganized. The consequences of such preorganization can be profound. For a spherand of the type shown, the benefit of preorganization can be as much as 16 kcal/mol in the binding of a simple ion (use Table 4.1 to compare the binding of Li$^+$ to 12-crown-4 and the spherand). Recalling from Chapter 2 that every 1.36 kcal/mol is worth a factor of 10 in an equilibrium constant at room temperature, the 16 kcal/mol corresponds to almost 12 orders of magnitude in a binding constant!

Tweezers and Clefts

The crown ether structures nicely highlight the advantage of having a macrocycle for molecular recognition studies. Several other macrocyclic structures will be examined below with regards to specific kinds of binding interactions. However, other general shapes have also found significant utility. The terms **tweezer** and **cleft** are now quite common in the literature. A cleft often resembles "half" of a macrocycle, and involves a well-defined but open-ended surface that can bind a guest. A tweezer is a chemical structure with two or more binding sites that converge on a single guest. Often a tweezer will contain a cleft. Inherent in these structures is a "binding site"—quite analogous to the clefts and cavities of natural systems such as enzymes and antibodies. A commonly used guiding principle for molecular design that arose from these kinds of molecular receptors is that of **convergence**. The binding entities of the host are preorganized to point toward (converge on) a common area, creating the binding site.

Now that a few general design principles have been described, we explore the use of several binding forces in the creation of synthetic receptors, and the lessons learned from these systems with regards to the forces involved. We emphasize again that it will become clear that context is very important for these weak interactions. The energetic importance of the non-covalent interactions considered here will vary considerably depending on the surrounding environment, and a major goal of this section will be to develop an appreciation of these context effects.

4.2.2   Molecular Recognition with a Large Ion Pairing Component

One way to examine ion pairing is to measure the equilibrium constants for ion pair formation. Table 4.2 shows the association constants for the tetra(n-butyl)ammonium and perchlorate ions in a variety of solvents. Exceptionally strong ion pairing takes place in the low

Table 4.2
Association Constants (K_a, M^{-1}) for Tetra(n-butyl)ammonium and Perchlorate in Various Solvents*

Solvent	Dielectric constant (ε)	K_a
Benzene	2	3×10^{17}
o-Dichlorobenzene	10	93,400
Isopropanol	18	1950
Acetone	21	200
Acetonitrile	36	53

*Janz, G. J., and Tomkins, R. P. T. (1972). *The Non-Aqueous Electrolytes Handbook*, Academic Press, New York.

Going Deeper

Preorganization and the Salt Bridge

We discussed in Chapter 3 how the salt bridge, the ion pair between a side chain carboxylate and a side chain cation (ammonium or guanidinium) in a protein often contributes very little to protein stability. A recent study shows that if one of the components of the pair is immobilized, the energetic contribution of the salt bridge is significant. In particular, a water-exposed Lys–Glu salt bridge was evaluated and found to contribute essentially nothing to protein stability. However, in the same protein, the interaction between a different Glu side chain and the

N-terminal $-NH_3^+$ of the protein was found to contribute 1.5 kcal/mol to protein stability. The N-terminus of the protein is highly structured, and so the process of protein folding immobilized the terminal ammonium. This removed the entropic cost of freezing out bond rotations that would accrue when Lys contributes the cation to the salt bridge. Again, preorganization is a key to molecular recognition.

Strop, P., and Mayo, S. L. "Contribution of Surface Salt Bridges to Protein Stability." *Biochemistry*, **39**, 1251–1255 (2000).

---

dielectric solvents, but it becomes significantly weaker as the dielectric increases. Whereas benzene leads to an association constant of 10^{17}, acetonitrile only gives an association constant of 53, and in water the salt is totally dissociated. The influence of microenvironment upon the strength of ion pairing is further discussed in the two Connections highlights presented here.

Recent studies of molecular recognition driven primarily by electrostatic interactions have focused upon deciphering the roles of enthalpy and entropy during the binding event. Certain trends may be expected. In the discussion of crown ethers, a large enthalpic driving force for coordination of the ether oxygens to the cation was revealed, with preorganization of the binding sites further accentuating the binding. With electrostatic binding, similar considerations should be evident. However, with the large electrostatic driving force between a full anionic and cationic charge, even larger enthalpy considerations may be expected, as is evident from the variation of K_a with solvent dielectric given in Table 4.2. Yet, some surprises "pop up" from time to time (see the Connections highlight below).

Connections

A Clear Case of Entropy Driven Electrostatic Complexation

Using the host shown to the right, with varying X and R groups, the thermodynamic parameters for the complexation of sulfate (SO_4^{2-}) in methanol were determined using isothermal calorimetry. Surprisingly, all the enthalpies of complexation were positive (endothermic), while large favorable entropies of complexation were found (between 4.9 and 17.2 eu). Since both the guanidinium and sulfate anions are well solvated in methanol, the positive enthalpies of complexation reflect the endothermic reorganization of the solvent shells upon complexation. Even with the electrostatic attraction between the host and guest, the overall solvation is worse for the complex. Yet, the release of solvent upon coordination of the host and guest must dominate, given the large positive entropy. The study reveals an important lesson. Our instinct is to design receptors incorporating more favorable enthalpic interactions, and this can indeed be quite a successful

strategy. However, we must not forget about solvent release and differential solvation. This lesson is another example that all the factors implied by Eq. 4.34 should be considered during a binding event.

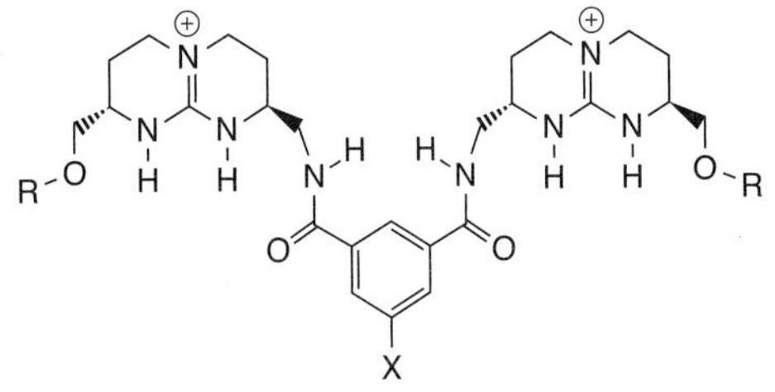

A sulfate binder

Berger, M., and Schmidtchen, F. P. "The Binding of Sulfate Anions by Guanidinium Receptors is Entropy-Driven." *Angew. Chem. Int. Ed. Eng.*, **37**, 2694–2697 (1998).

Connections

Salt Bridges Evaluated by Non-Biological Systems

We have noted several times that it is a challenging task to evaluate the contribution of an ion pair or salt bridge to a binding constant in aqueous media. We expect the exact value to depend strongly on the context of the interaction. In an effort to minimize context effects, Schneider evaluated a large number of ion pairs in fairly simple complexes in water. Sample structures are shown. Perhaps surprisingly, on considering about 40 ion pairs, a roughly constant increment of 1.2 ± 0.2 kcal/mol per salt bridge was seen. This is consistent with other studies, and provides a nice confirmation of the protein studies using quite simple model systems.

Schneider, H.-J., Schiestel, T., and Zimmermann, P. "The Incremental Approach to Noncovalent Interactions: Coulomb and van der Waals Effects in Organic Ion Pairs." *J. Am. Chem. Soc.*, **114**, 7698–7703 (1992).

Simple ions to probe ion pairing

4.2.3 Molecular Recognition with a Large Hydrogen Bonding Component

In the gas phase, and in organic solvents, hydrogen bonding is a potent, reliable force for molecular recognition. Because there is a significant directional component involved, hydrogen bonding can be used to orient molecules and build quite specific structures, and we'll see examples of its use in designed complexes below. One reason hydrogen bonding has seen extensive use in this regard is because it is easier to create organic molecules with hydrogen bonding abilities that are soluble in organic solvents than it is to exploit full electrostatic attraction with ion pairs, given the low solubility of ionic compounds. However, due to the competitive nature of water and alcohols, these solvents have received much less attention for studing hydrogen bond driven molecular recognition than have solvents such as chloroform and acetonitrile.

Representative Structures

A large fraction of the molecular complexes that have been synthesized to study hydrogen bond driven molecular recognition have involved rigid molecular scaffolds. Discussed below are just a few examples of the kinds of structures and designs that have been used in this manner. Admittedly, much of this work was performed to show that chemists can create structures that are complementary to reasonably complex guests, rather than to specifically study effects directly relevant to a biological process. However, several general lessons were gained that can be used to design receptors for practical purposes in low dielectric media, and these and other examples demonstrated that chemists can create receptors that function using hydrogen bonds.

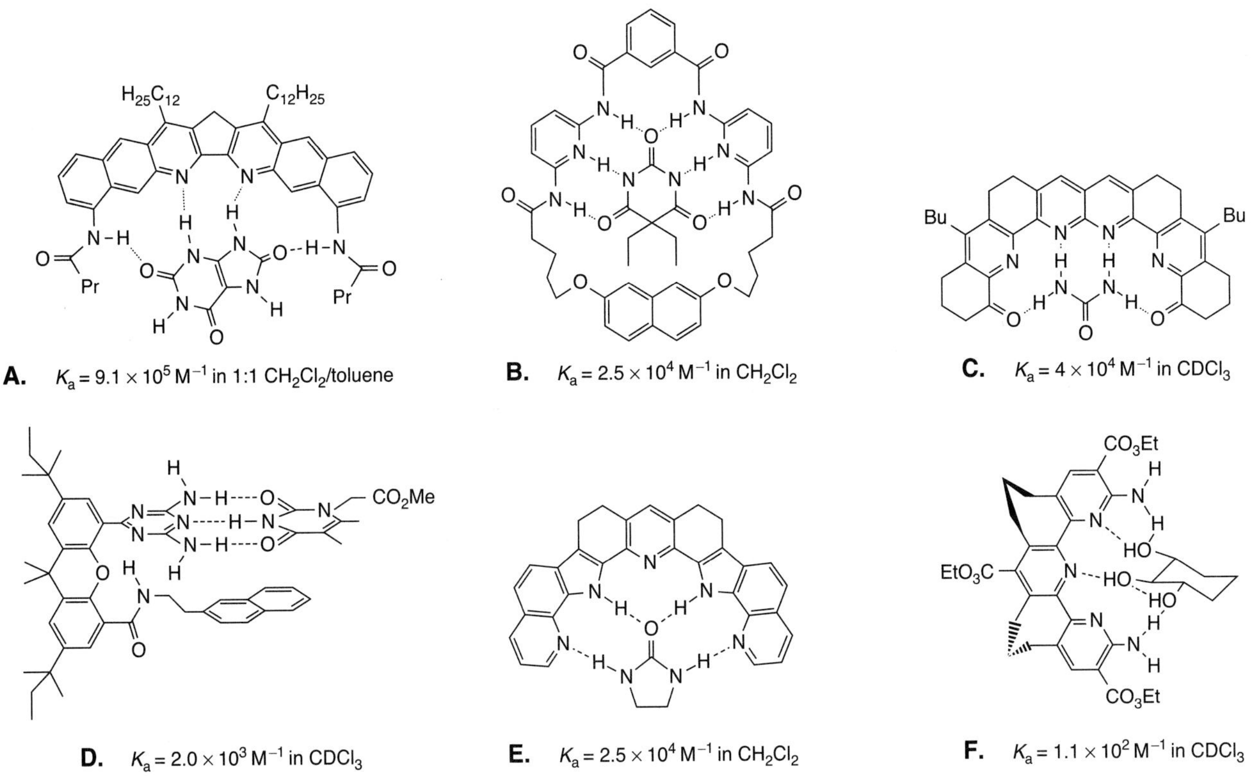

A. $K_a = 9.1 \times 10^5 \, M^{-1}$ in 1:1 CH$_2$Cl$_2$/toluene

B. $K_a = 2.5 \times 10^4 \, M^{-1}$ in CH$_2$Cl$_2$

C. $K_a = 4 \times 10^4 \, M^{-1}$ in CDCl$_3$

D. $K_a = 2.0 \times 10^3 \, M^{-1}$ in CDCl$_3$

E. $K_a = 2.5 \times 10^4 \, M^{-1}$ in CH$_2$Cl$_2$

F. $K_a = 1.1 \times 10^2 \, M^{-1}$ in CDCl$_3$

Figure 4.8
Examples of hydrogen bond driven molecular recognition in low dielectric media. **A.** Kelly, T. R., and Maguire, M. P. "A Receptor for the Oriented Binding of Uric Acid Type Molecules." *J. Am. Chem. Soc.,* **109**, 6549 (1987). **B.** Chang, S.-K., Engen, D. V., Fan, E., and Hamilton, D. A. "Hydrogen Bonding and Molecular Recognition: Synthetic, Complexation, and Structural Studies of Barbiturate Binding to an Artificial Receptor." *J. Am. Chem. Soc.,* **113**, 7640 (1991). **C.** Bell, T. W., and Liu, J. "Hexagonal Lattice Hosts for Urea. A New Series of Designed Heterocyclic Receptors." *J. Am. Chem. Soc.,* **110**, 3673 (1988). **D.** Park, T. K., Schroeder, J., and Rebek, J., Jr. "New Molecular Complements to Imides. Complexation of Thymine Derivatives." *J. Am. Chem. Soc.,* **113**, 5125 (1991). **E.** Hung, C.-Y., Höphner, T., and Thummel, R. P. "Molecular Recognition: Consideration of Individual Hydrogen-Bonding Interactions." *J. Am. Chem. Soc.,* **115**, 12601 (1993). **F.** Huang, C.-Y., Cabell, L. A., and Anslyn, E. V. "Molecular Recognition of Cyclitols by Neutral Polyaza–Hydrogen-Bonding Receptors: The Strength and Influence of Intramolecular Hydrogen Bonds Between Vicinal Alcohols." *J. Am. Chem. Soc.,* **116**, 2778 (1993).

Several examples are given in Figure 4.8. The complex **A** has a very high K_a, and was one of the earliest examples of an array of hydrogen bonds between a host and a guest that are displayed in such a manner as to be complementary. Both the concepts of preorganization and complementarity are evident in this complex. Similarly, complex **B** with barbital as the guest is an excellent example of how simple synthetic construction can lead to multiple interactions between a host and a guest. Likewise, due to its high complementarity, complex **C** solubilizes urea in chloroform, in which it is normally completely insoluble. Complex **D** is just one example of a myriad of complexes formed between synthetic hosts and thymine derivatives. Interestingly, by varying the number of hydrogen bonds in complexes such as **E**, good estimates of the strength of hydrogen bonds in chloroform have been made. These range from 1.2 to 1.7 kcal/mol, but do not take into account secondary interactions (see Chapter 3). Lastly, with an analysis of complex **F**, it became evident that multipoint hydrogen bonding is much less effective for the recognition of arrays of hydroxyls on cyclitols and carbohydrates, in part due to the intramolecular hydrogen bonds within these guests. In essence, these guests are internally solvating themselves.

All these examples illustrate how the use of small binding forces such as hydrogen bonds can result in good binding affinities when several are added together. Yet, it is important to emphasize that low dielectric organic solvents were used. In all cases the studies were done in **non-competitive solvents**—solvents that themselves are neither hydrogen bond donors nor acceptors (solvents with low α and β values; see Table 3.1). Removing any competition from the solvent maximizes the host–guest hydrogen bonding interactions. Recalling Eq. 4.34 shows why the strategies are successful. The solvation of the separate hosts and guests is low, and the creation of hydrogen bonds only occurs upon interaction of the host and guest. This would not be the case in alcohols or water as the solvent. Since water is such a potent hydrogen bond donor and acceptor, it competes effectively with any particular hydrogen bond that might be designed into a molecular recognition system. As a result, the design and characterization of synthetic molecular recognition systems based on hydrogen bonding has been restricted, with few exceptions, to organic solvents such as chloroform. Even addition of a small amount (eg., 10%) of a solvent like DMSO, a potent hydrogen bond acceptor, can be enough to completely disrupt a hydrogen bonding system.

Molecular Recognition via Hydrogen Bonding in Water

We just stated that synthetic molecular recognition with hydrogen bonding as a major binding force is only possible in non-competitive solvents, and that is certainly the case for the structures given in Figure 4.8 and related systems. But, let's not lose perspective. Hydrogen bonding is successfully exploited in aqueous media by biological systems. Recall the α-helix and β-sheet of protein secondary structure and the double helix of nucleic acid structure (however, the role of hydrogen bonding in the DNA duplex is currently debated, as shown in the Connections highlight given at the end of this discussion). In proteins and nucleic acids, hydrogen bonding is used to augment associations that have a significant hydrophobic component, and there is cooperativity in the formation of many hydrogen bonds. Can a similar combination of effects, in which hydrophobic and perhaps ion pairing interactions lead to a strong association and hydrogen bonding leads to selectivity, be developed for a synthetic molecular recognition system? The answer is "yes", as shown in a series of synthetic systems designed to recognize specific sequences of DNA.

As shown in Figure 4.9, the edge of the DNA double helix in the so-called "minor

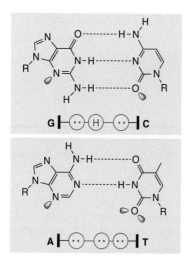

Figure 4.9
Molecular recognition by hydrogen bonding in the minor groove of DNA. The polyamide makes specific hydrogen bonding contacts with the various base pairs, allowing sequence recognition. For example, when hydroxypyrrole pairs with pyrrole, a T•A base pair is recognized; when imidazole pairs with pyrrole, a G•C pair is recognized. Side views of the lone pairs within the minor groove are designated as circles with double dots, and side views of hydrogens in the minor groove are designated as circles with H inside. Dervan, P. B., and Edelson, B. S. "Recognition of the DNA Minor Groove by Pyrrole–Imidazole Polyamides." *Curr. Op. Struct. Biol.*, **13**, 284 (2003).

groove" exposes a variety of hydrogen bond donors and acceptors within a shallow hydrophobic cleft, with structural patterns that vary depending on the base pair involved. Building off the structure of the natural product distamycin, Dervan and co-workers developed a series of polyamides based on three small heterocyclic building blocks: methylpyrrole, hydroxymethylpyrrole, and imidazole. The minor groove of DNA is wide enough to accommodate a stacked pair of such small heterocycles. When the heterocycles are put in the correct sequence and with the right spacers, a hairpin turn develops, and the organic structures nestle into the minor groove. The binding is tight and the sequence specificity remarkable, allowing simple pairing rules for targeting any desired sequence. There is cooperativity here —hydrophobic interactions drive the association; but many hydrogen bonds are formed directing the site of association. Also, when the polyamide "lays down" in the minor groove, the newly formed hydrogen bonds are very much protected from solvent, making competition from water nearly impossible. Thus, when well-designed, multiple hydrogen bonding contacts are formed, the competition with water and the intrinsic entropy penalty of association can be overcome. The ability to rationally design small organic molecules that bind specifically to any desired DNA sequence could be of considerable value, and work is underway to see if polyamides of the sort shown could be used as a form of gene therapy.

Connections

Does Hydrogen Bonding *Really* Play a Role in DNA Strand Recognition?

Probably the most well known example of how hydrogen bonds can direct selectivity is the base pairing in DNA and RNA. Guanine (**G**) pairs with cytosine (**C**) while adenine (**A**) pairs with thymine (**T**) or uridine (**U**). The accepted reason for this specific pairing is the three hydrogen bonds in the first combination, and two hydrogen bonds in the second combination. It is this specificity that has always been thought to guide the fidelity of DNA replication, where **A** is always inserted opposite to **T** and vice versa, while **G** is inserted opposite to **C** and vice versa.

However, this reasoning has recently been brought into question. In an ingenious study, 2,4-difluorotoluene (**F**) was used as a thymine isostere in a DNA replication experiment. A strand of DNA where **F** was one of the "bases" at a particular site was created. Surprisingly, **A** is efficiently inserted opposite to **F** during replication with almost as high an efficiency as that opposite **T**, while **C**, **T**, and **G** are selected against very effectively. Recall that fluorine is generally considered to be a *very* poor hydrogen bond acceptor (if it is an acceptor at all), and the interaction shown between the H of **F** and the N of **A** would be much weaker than a conventional hydrogen bond. The assertion is that shape and steric fit are the more important factors in the selection of bases during the replication process, and not the ability to form hydrogen bonds. More work is needed to delineate the role of shape versus hydrogen bonding in DNA replication, but the ramifications could be far reaching.

Moran, S., Ren, R. X.-F., Rummey, S., IV, and Kool, E. T. "Difluorotoluene, a Nonpolar Isotere for Thymine, Codes Specifically and Efficiently for Adenine in DNA Replication." *J. Am. Chem. Soc.*, **119**, 2056–2057 (1997).

4.2.4 Molecular Recognition with a Large Hydrophobic Component

In order to learn about molecular recognition in a biologically relevant context and to tackle the diversity of sizes and shapes of organic guests, many workers have attempted to develop water soluble organic molecules with well-defined, hydrophobic binding sites. Solubility is a challenge, but two systems have proven to be especially valuable: cyclodextrins and cyclophanes. In some sense these structures are the inverse of crowns and cryptands, which are typically organic soluble with a polar core. We are now considering water soluble molecules with a polar surface (for solubility) and a hydrophobic core.

Cyclodextrins

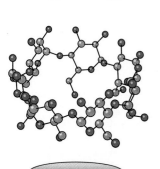

Cyclodextrin

Cyclodextrins are cyclic oligomers of glucose obtained by the enzymatic degradation of amylose. α-, β-, and γ-cyclodextrin have six, seven, and eight glucoses in a ring, respectively. The toroid of sugars elaborates a central cavity and, perhaps surprisingly, this cavity is fairly hydrophobic. Organic molecules, especially small aromatics such as substituted benzenes, show significant binding affinities toward cyclodextrins in water. Because of their ready availability and their potential for functionalization (a significant synthetic challenge, but one that was eventually overcome), cyclodextrins serve as an ideal platform for developing concepts of molecular recognition.

When the size and shape complementarity of host and guest are well matched, substantial binding affinities can be seen. Table 4.3 list several binding constants of various guests to cyclodextrins in water. Notice a few trends. First, anthracene has the highest binding affinity to the medium-sized β-cyclodextrin. Smaller hydrophobic entities such as benzoic acid and cyclohexanol bind best to the smaller cyclodextrin. However, the large aromatic compound pyrene has the highest affinity to the largest cyclodextrin. Where detailed thermodynamic measurements have been made, binding to cyclodextrins shows all the hallmarks of hydrophobic association, including relatively modest $\Delta H°$ values and favorable $\Delta S°$ values.

Table 4.3
Log(K_a) for Various Hydrophobic Guests with Cyclodextrins in Water (298 K)*

Guest	(α-CD)	(β-CD)	(γ-CD)
Anthracene	1.87	3.31	2.35
Benzoic acid	2.96	2.57	2.10
Cyclohexanol	1.81	2.70	—
1-Hexanol	2.95	2.34	—
Pyrene	2.17	2.69	3.05
Ferrocene carboxylate	—	2.85	—

*Saenger, W. "Cyclodextrin Inclusion Compounds in Research and Industry." *Angew. Chem. Int. Ed. Eng.,* **19**, 344–362 (1980).

Cyclophanes

Cyclodextrins were important in establishing many principles of **biomimetic chemistry**—a phrase coined by Breslow to describe efforts to invent new substances and reactions that imitate biological chemistry. However, cyclodextrins are limited by the inability to modify cavity size. If a guest of interest does not fit into α-, β-, or γ-cyclodextrin, it cannot be studied. Also, the hydrophobic cavity of cyclodextrins is not easily modified so as to evaluate different binding forces. With these features in mind, a number of groups turned to **cyclophanes**, which are simply cyclic compounds with aromatic rings as part of the cycle (Figure 4.10). When water-solubilizing groups are attached (see several examples on the next page), adjustable sizes and shapes could be designed and a much wider range of guests became amenable to study. In addition, several other receptor designs similar to cyclophanes and cyclodextrins have been created, and a few are shown in the Going Deeper highlight at the end of this section.

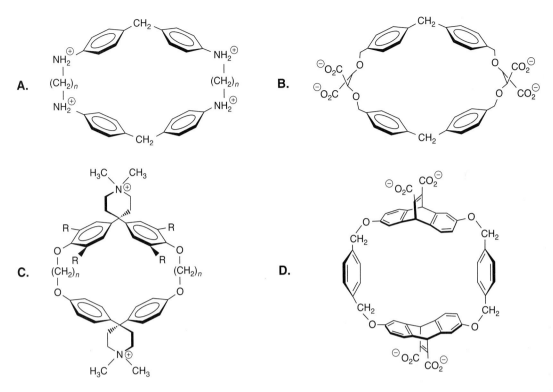

Figure 4.10
Representative examples of cyclophanes studied for molecular recognition in water. **A.** Lai, C.-F., Odashima, K., and Koga, K. "Synthesis and Properties of Water-Soluble Bis-Paracyclophanes." *Tetrahedron Lett.*, **26**, 5179–5182 (1985). **B.** Dhaenens, M., Lacombe, L., Lehn, J.-M., and Vigneron, J.-P. "Binding of Acetylcholine and Other Molecular Cations by a Macrocyclic Receptor Molecule of Speleand Type." *J. Chem. Soc. Chem. Commun.*, 1097–1099 (1984). **C.** Diederich, F. "Complexation of Neutral Molecules by Cyclophane Hosts." *Angew. Chem. Int. Ed. Eng.*, **27**, 362–386 (1988). **D.** Petti, M. A., Shepodd, T. J., Barrans, R. E., Jr., and Dougherty, D. A. "'Hydrophobic' Binding of Water-Soluble Guests by High- Symmetry, Chiral Hosts. An Electron-Rich Receptor Site with a General Affinity for Quaternary Ammonium Compounds and Electron-Deficient π Systems." *J. Am. Chem. Soc.*, **110**, 6825–6840 (1988).

With well-defined host–guest systems available, fairly precise measurements of hydrophobic association processes can be made (in contrast to studies of micelle or vesicle formation, where many particles come together to form an ill-defined structure). Early systems (Figure 4.10 **A**) established basic trends. Generally, cyclophane cavities are much more hydrophobic than cyclodextrin cavities, and hydrophobicity is the key determinant of binding ability. As long as the structure can fit into the cavity, the larger, and thus the more hydrophobic, the guest, the stronger the binding. Also, aromatic guests generally seem to be preferred over alkanes, contrary to solubility (hydrophobicity) trends. This presumably reflects the increased binding possibilities available to aromatics via cation–π and polar–π interactions. Even with such simple systems, fairly strong binding can be seen with $K_d < 1$ mM.

Along with shape complementarity and hydrophobicity, preorganization is also important in cyclophane binding. In several systems it was demonstrated that increasing rigidity in the "linker" region [the $(CH_2)_4$ region of host **A** and related structures] increases binding affinity.

We stated that an advantage of cyclophanes and other models is that they allow more detailed dissection of the structural features that promote molecular recognition. Let's examine a couple of cyclophane studies in detail—first, one that again accentuates how complex the hydrophobic effect really is. Figure 4.11 shows the results of a study to determine $\Delta H°$ and $\Delta S°$ for the association reaction of the cyclophane given in Eq. 4.36 and methylquin-

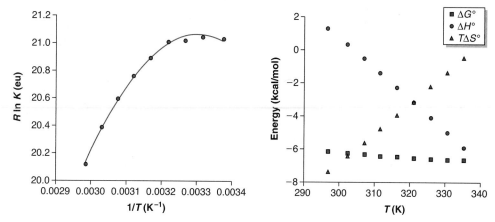

Figure 4.11
Determining the themodynamic properties of the binding event shown in Eq. 4.36 in water. **A.** The van't Hoff plot. Note the curvature, implying a significant heat capacity change. **B.** Variation of $\Delta G°$, $\Delta H°$, and $T\Delta S°$ with temperature. Modified from Stauffer, D. A., Barrans, R. E., Jr., and Dougherty, D. A. "Concerning the Thermodynamics of Molecular Recognition in Aqueous and Organic Media. Evidence for Significant Heat Capacity Effects." *J. Org. Chem.*, **55**, 2762–2767 (1990).

oline. The first thing to notice is that the van't Hoff plot is curved. Since the slope of the plot is related to $\Delta H°$, a curved plot (i.e., one with a changing slope) must indicate that $\Delta H°$ changes over the temperature range studied (see Section 4.1.1). Thus, a curved van't Hoff plot indicates $\Delta C_P° \neq 0$, but we would expect this for a process with a large hydrophobic component (recall our first discussion of the hydrophobic effect in Section 3.2.6). Fitting $\ln K$ vs. $1/T$ now requires the complicated Eq. 4.19, which does fit the data quite well.

$$(Eq.\ 4.36)$$

With all the relevant thermodynamic data in hand, we can hope to learn something about the mechanism of the association reaction. Figure 4.11 also shows the variation of $\Delta G°$, $\Delta H°$, and $T\Delta S°$ with temperature for this simple association process. We see that $\Delta G°$ is relatively insensitive to temperature. Remember, though, that $K_a = e^{-\Delta G/RT}$, so the association constant still changes with temperature. Note, however, the substantial variations in $\Delta H°$ and $T\Delta S°$ with temperature. These lead to some puzzling conclusions. At high temperatures we would conclude that this process is almost completely "enthalpy driven" ($|\Delta H°| \gg |T\Delta S°|$). However, if we lower the temperature only 40 K, suddenly the process becomes "entropy driven" ($|T\Delta S°| \gg |\Delta H°|$). This is not uncommon—it is often observed that enthalpy dominates hydrophobic association at high temperatures, while entropy is more important at low temperatures. Certainly, though, this makes detailed mechanistic interpretation challenging. Perhaps the most we can conclude is that the hydrophobic effect is complicated.

The second cyclophane study we examine sheds light on the "special" nature of water. Is water absolutely unique as a solvent for molecular recognition? Or, is the hydrophobic ef-

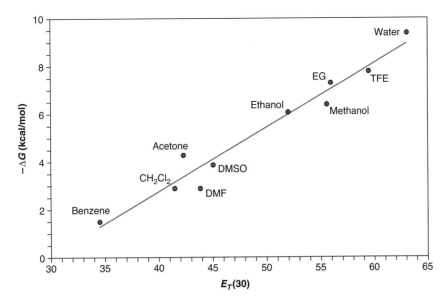

Figure 4.12
A plot of $-\Delta G^\circ$ of binding versus $E_T(30)$ for the binding of pyrene to the cyclophane shown in Eq. 4.37. Note that water is only a slight outlier. TFE = 2,2,2-trifluoroethanol; EG = ethylene glycol. Smithrud, D. B., and Deiderich, F. "Strength of Molecular Complexation of Apolar Solutes in Water and in Organic Solvents is Predictable by Linear Free Energy Relationships: A General Model of Solvation Effects on Apolar Binding." *J. Am. Chem. Soc.*, **112**, 339 (1990).

fect just one end of a continuum of **solvophobic** effects? To address this question, the binding of the cyclophane given in Eq. 4.37 with pyrene was studied across a very wide range of solvents (Figure 4.12). Both the guest and the host are essentially devoid of functionality that might lead to specific binding effects (i.e., no hydrogen bonds, salt bridges, etc., are possible). To a good approximation, binding affinity should be determined by solvophobicity and any innate attraction between host and guest, with the latter assumed to be solvent independent. Surprisingly, a steady continuum of binding affinities across the range of solvents was found. In fact, ΔG° correlated well with the simple solvent polarity parameter $E_T(30)$. In this particular system, water is not unique, it just defines the endpoint of a continuum.

$$\text{(Eq. 4.37)}$$

This experiment provides considerable food for thought, but extrapolation to other systems must be done cautiously. The host–guest interactions in Eq. 4.37 are almost exclusively aromatic–aromatic, while more typical studies of the hydrophobic effect involve aliphatic groups. As noted earlier and below, aliphatics and aromatics have some considerable differences in their properties, mostly related to the quadrupole moment of the aromatics. Pyrene especially may not be representative of a typical hydrophobic molecule. In other manifestations of the hydrophobic effect, water does seem unique. For example, formamide is quite close to water in the $E_T(30)$ scale, and would therefore be expected to be close to water in the pyrene binding experiment of Figure 4.12. Nevertheless, micelles and vesicles are not found at all in formamide.

Going Deeper

Calixarenes—Important Building Blocks for Molecular Recognition and Supramolecular Chemistry

As physical organic chemists design ever more complicated systems for studying molecular recognition and supramolecular chemistry, it is very useful to have synthetically accessible building blocks that readily lend themselves to elaboration into complex structures. Certain types of cyclophanes have proven to be amenable to large scale synthesis and adaptable to molecular recognition studies. We highlight two such structures here.

The **calixarenes** (from the Greek *calix*, meaning chalice) are a family of structures that have seen extensive use in many contexts. Shown below is the remarkable, one-step synthesis of the prototype calixarene. This condensation reaction is a variant of the phenol–aldehyde condensation that produces bakelite, the first synthetic polymer. Early workers spent considerable effort to optimize this synthesis, biasing the product toward the calixarene

rather than the polymer. Like the cyclodextrins, calixarenes come in differing sizes, with anywhere from four to eight aromatics in the ring. The tetramer shown is the most easily prepared and isolated, and is the basis of most advanced systems.

Extensive studies of the conformational preferences of calixarenes have established certain shape preferences, and a view of the cone-shaped form that has found the most use in molecular recognition studies is shown below. It should be remembered, though, that without additional structural constraints that favor this form, a simple calixarene is conformationally mobile and exists in several different forms.

A related construct is the resorcinol analog of a calixarene. It is also a bowl-shaped molecule that can be used in many ways. In such structures there is a conformational constraint from the additional hydrogen bonding motif shown.

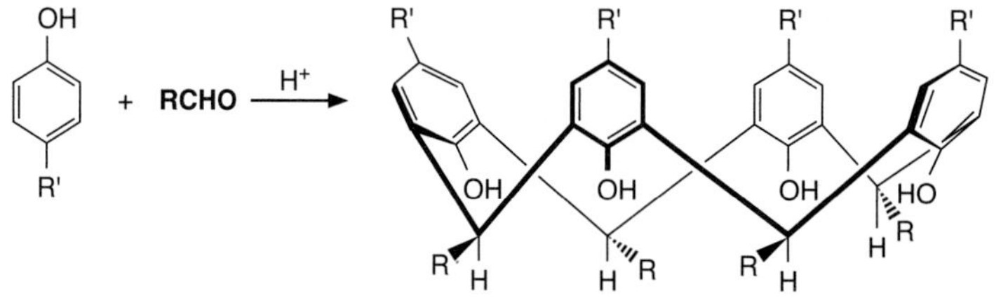

Calixarene

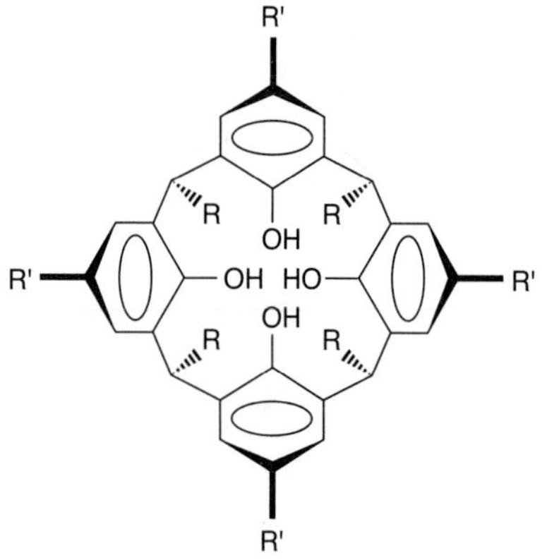

A calixarene cone conformation

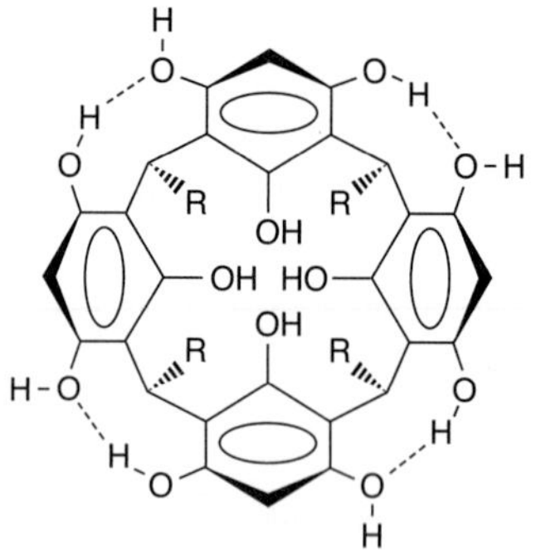

A resorarene analog of a calixarene

A Summary of the Hydrophobic Component of Molecular Recognition in Water

After our discussion here and in Chapter 3, the student may feel somewhat confused about the hydrophobic effect. Welcome to the club! Perhaps one source of the confusion is that such a broad range of phenomena is rationalized by invoking a "hydrophobic effect". Is it sensible to use the same explanation for such diverse observations as the insolubility of ethane in water, the formation of vesicles by phosphatidylcholine, the binding of pyrene by a cyclophane, and protein folding? Indeed, the origins of these phenomena are related.

Focusing on biological molecular recognition, some overall conclusions remain. Certainly bilayer formation is driven by the hydrophobic effect. In addition, most bimolecular associations in biology—be they protein–small molecule, protein–protein, protein–nucleic acid, etc.—have a substantial hydrophobic component. In the most studied cases—small molecules binding to proteins and protein folding—hallmarks of the "classical" hydrophobic effect are seen. These include large heat capacity effects and a strong entropy component. Even if we are uncomfortable that a precise physical model for the hydrophobic effect is unavailable, we can certainly anticipate that hydrophobic effects will be important whenever organic molecules interact with water, and that the magnitude of the effect will scale with the amount of hydrocarbon surface area exposed to water.

4.2.5   Molecular Recognition with a Large π Component

Cyclophanes are ideally suited to probing and quantifying the various binding interactions involving π forces discussed in Chapter 3. Early studies of cyclophane binding consistently found that aromatic guests were significantly preferred over aliphatic guests, while efforts to prepare hosts that were primarily aliphatic were generally unsuccessful. Systematic studies came to reveal that there are various π interactions that were favoring aromatics in molecular recognition. Therefore, layered on top of the hydrophobic effect that is the primary driving force are the π interactions, which enhance binding, provide selectivity, and promote specific binding orientations. In fact, as the next Connections highlight hints, these secondary interactions are commonly exploited in natural receptors.

Connections

Aromatics at Biological Binding Sites

We have noted above that the hydrophobic effect is a dominant force in chemical biology, both in protein folding and in protein–ligand interactions. As such we might expect an overrepresentation of hydrophobic residues at protein binding sites. This is true, but hydrophobicity is not the only issue. The simple hydrophobic amino acids such as Leu, Ile, and Val are not overly common at biological binding sites. However, it has repeatedly been observed that the aromatic amino acids Phe, Tyr, and Trp are much more likely to appear at binding sites than other residues. For example, nature's generic binding site is found in antibod-

ies, where a wide range of diverse binding sites are produced on a common scaffold. Structural analysis of antibody binding sites shows that both tyrosine and tryptophan are roughly five times more likely to appear at an antibody binding site than they are found in proteins in general. In fact, in one study, over 35% of the residues that contributed to antigen binding were tyrosine or tryptophan. This is no doubt due to their ability to make π interactions as well as present hydrophobic surface area.

Mian, I. S., Bradwell, A. R., and Olson, A. J. "Structure, Function and Properties of Antibody Binding Sites." *J. Mol. Biol.*, **217**, 133–151 (1991).

Cation–π Interactions

Cyclophanes have figured most prominently in studies of the cation–π interaction. More recently, structural studies of the cation–π interaction have derived from crown ether hosts (see the next Connections highlight). Qualitatively, the effect is perhaps best described when the guest shown in the margin binds with cyclophanes. The *t*-butyl and trimethylammonium groups are essentially isosteric, the only real difference between the two being that one is neutral and one has a positive charge. In all cyclophanes, ^{1}H NMR is a powerful tool for studying binding. Any guest will experience a large upfield shift in its NMR spectrum on binding, because its protons will point into the faces of the aromatic rings of the cyclophane, experiencing the shielding associated with the aromatic ring current. When this guest and a cyclophane host (Figure 4.10 **D**) are mixed in aqueous media, the *trimethylammonium* protons shift upfield the most. The hydrophobic effect favors binding the *t*-butyl group. Calculations suggest that the aqueous solvation energy of the NMe_3^+ group is better than that of the *t*-butyl group by 60 kcal/mol! Nevertheless, the hydrophobic cavity of the cyclophane

t-Butyl-trimethylammonium guest

prefers the charged group, establishing the viability of the cation–π interaction as a binding force in aqueous media.

Several efforts have been made to quantify the cation–π interaction through cyclophane studies. As with salt bridges and hydrogen bonding, it is difficult in water to sort out the quantitative contribution to binding of one particular host–guest interaction, because usually several interactions are occurring and, more importantly, solvation–desolvation phenomena are so complex. A study of a large number of relatively simple systems that can experience a cation–π interaction concluded that a typical interaction between an organic cation and a single aromatic ring contributes about 1 kcal/mol to ΔG° of binding. Another interesting comparison concerns the binding of the two guests shown to the side to the cyclophane of Figure 4.10 **D**. As much as possible, these two guests have the same size, shape, and hydrophobic surface area. Like the *t*-butyl/trimethylammonium comparison, the major difference is that one is neutral and the other is a cation, and the cation is better solvated than the neutral. The 2.5 kcal/mol preference for the cationic guest gives a sense of the extent to which cation–π interactions can influence binding in an optimal system. Well controlled studies of cation–π binding in proteins also lead to the conclusion that, in an appropriate system, a cation–π interaction can contribute 2–3 kcal/mol of binding energy.

The perfluoroarene effect has also been probed using cyclophane systems. In particular the interaction of a cation with an arene and a fluorinated arene has been probed in several systems. An interesting comparison between the π effects found for standard arenes and perfluoroarenes can be made simply by examining the electrostatic surface maps shown in Appendix 2 for benzene and hexafluorobenzene. The charges derived from the quadrupole moments for benzene and hexafluorobenzene are opposite on the faces of the arenes, negative for benzene (red in Appendix 2) and positive for hexafluorobenzene (blue in Appendix 2). As expected, fluorination turns an attractive cation–π interaction into a repulsive cation–fluoroarene interaction, and the difference can be on the order of several kcal/mol.

−ΔG°   5.9   8.4
(kcal/mol)

Connections

Combining the Cation–π Effect and Crown Ethers

Earlier in this chapter we noted the selectivity of 18-crown-6 for K+ relative to other crown ethers of different sizes. Shown below is a diaza-18-crown-6 receptor possessing phenol groups as arms (lariats). This crown has a similar selectivity for K+, but also binds this cation with some degree of a cation–π effect. The accompanying crystal structure (selected portions) shows that the phenyl rings stack above and below the plane formed by the crown ring, placing the electron rich π system and negative ends of the benzene quadrupole directly oriented at the cation. This is a beautiful structural confirmation of the binding of alkali metals we discussed in our original introduction of the cation–π binding force.

DeWall, S. L., Barbour, L. J., and Gokel, G. W. "Cation–π Complexation of Potassium Cation with the Phenolic Sidechain of Tyrosine." *J. Am. Chem. Soc.*, **121**, 8405–8406 (1999). (The iodide counterion is shown as a dark sphere.)

Crown and cation–π complexation

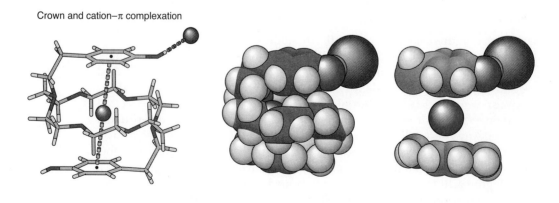

Polar–π and Related Effects

Cyclophane hosts are in principle ideal for exploiting the various weak interactions involving aromatics, such as polar–π interactions (polar bonds such as NH and OH interacting with the face of the aromatic) and aromatic–aromatic interactions (T-shaped and/or slipped-stack). An aromatic guest slipping into a cavity formed by a cyclophane host can hardly avoid aromatic–aromatic interactions. Often a cyclophane forms a box-shaped cavity, and an aromatic guest will naturally point its edges toward one of the walls of the box, forming T-shaped interactions. Similarly, slipped-stack interactions can be expected between the face of the guest and other walls of the host. We expect aromatic–aromatic interactions to be pervasive in cyclophane binding, and this is one reason why aromatic guests are generally better than aliphatics for cyclophane hosts.

It is precisely this pervasive nature, along with the intrinsically weak magnitude of such interactions, that makes quantifying polar–π interactions quite challenging. In aqueous media, where a hydrophobic effect is superimposed on any aromatic–aromatic interaction, quantitative analysis is difficult. The same is true of other polar–π interactions in which NH or OH bonds interact favorably with the face of an aromatic. Typically, the NH or OH groups are very well solvated by water. Given the weak nature of the polar–π interaction, it is difficult to overcome this favorable solvation. As such, documented examples in which such interactions are established to contribute significantly to binding in aqueous media are rare. Instead, confirmation of their importance has relied more on statistical arguments. Analysis of a large number of protein crystal structures reveals that NH and OH bonds tend to point toward the faces of aromatic ring systems.

In organic solvents polar–π interactions are often easier to quantify, just as was the case with hydrogen bonds. We noted in Section 3.2.4 an example of perturbation of the pK_a of a carboxylic acid through a polar–π effect. Another example is given in the Connections highlight below. Opportunities for such interactions also occur in many of the "hydrogen bonding" receptors noted above. These are weak but potentially important and fairly pervasive interactions that should always be considered in studies of molecular recognition.

4.2.6 Summary

Sections 4.2.1–4.2.5 described systems designed to probe the origin and strength of the types of binding forces discussed in the previous chapter: ion–dipole, ion pairing, hydrogen bonding, solvophobic effects, cation–π, and polar–π. Although the list of binding forces discussed is large, it is not comprehensive. For example, we did not cover dipole–dipole binding (an important factor in alignment in materials; see the discussion of liquid crystals in Chapter 17). Regardless of the length of the list of possible binding forces, the origin of all the forces derives primarily from electrostatics and solvation/desolvation phenomena. Electrostatic considerations dominate the discussion of ion–dipole, ion pairing, hydrogen bonding, cation–π, polar–π, and dipole–dipole interactions. In contrast, solvation/desolvation phenomena dominate the discussion of solvophobic effects. Yet, the entropically favorable shedding of solvent is often also part of molecular recognition events that have strong electrostatic components. For example, the Going Deeper highlight of Section 4.2.2 described entropy driven ion pairing complexation. Therefore, although electrostatic considerations give a good foundation for understanding molecular recognition, we must never forget Eq. 4.34, where the enthalpy and entropy of host–guest solvation and solvent release are important contributing factors to the overall affinities between hosts and guests.

Connections

A Thermodynamic Cycle to Determine the Strength of a Polar–π Interaction

As we have already discussed in this chapter, one of the goals of studying synthetic receptors is to quantify a particular binding force. By comparing the Gibbs free energy of binding for two very similar complexes, one possessing and one lacking the binding force, we can estimate the strength of that binding force. However, it is often challenging to design a system that precisely isolates one particular interaction. In the case analyzed below, a clever strategy was implemented to overcome this limitation. This "thermodynamic cycle" approach was originally developed for studying biological receptors using site-directed mutagenesis. However, it adapts very well to synthetic receptors.

The goal of the study was to measure the strength of an NH–π interaction (the "attractive" interaction in complex **A**). Comparing the Gibbs free energy of association of complex **A** and **B** could give an estimate of its strength.

However, complex **A** also has a repulsive interaction between the pyrrole NH of interest and the nearby amide NH of the binding partner, and this interaction is lacking in complex **B**. Hence, the strength of the NH–π interaction would be underestimated by comparing just **A** and **B**. To address this, a comparable study was made on complexes **C** and **D**, one of which has and one of which lacks the offending interaction. The Gibbs free energies of association were determined for all four complexes. With the accompanying equation (you are asked to derive it in Exercise 5 at the end of the chapter), a $\Delta G°(NH–π)$ value of 1.1 kcal/mol was determined. This is a nice example of measuring the strength of a molecular recognition process, coupled with a good insight into the different interactions present in the complexes under study.

$$\Delta G°(NH–\pi) = \Delta G_A° - \Delta G_B° - \Delta G_C° + \Delta G_D°$$

Adams, H., Harris, K. D. M., Hembury, G. A., Hunter, C. A., Livingstone, D., and McCabe, J. F. "How Strong is a π-Facial Hydrogen Bond?" *J. Chem. Soc. Chem. Commun.*, 2531–2532 (1996).

Thermodynamic cycle

Going Deeper

Molecular Mechanics/Modeling and Molecular Recognition

Model building has proven to be an indispensable aid in efforts to design new systems for molecular recognition. While physical models of the CPK type were initially used, now computer-based models are more common. Given this, it is often an irresistable temptation to try to make quantitative predictions about a given system using molecular mechanics. This should be done with caution for at least two reasons. First, simple molecular mechanics ignores the solvent, and we have gone to great lengths to emphasize that weak interactions are extremely sensitive to context. The importance of a hydrogen bond, for example, will almost always be substantially overestimated by molecular mechanics. Second, weak attractions can occur between simple groups that are near van der Waals contact. These are the induced-dipole–induced-dipole interactions discussed in Chapter 3. In molecular mechanics they arise from the nonbonded term as the slightly attractive region of the Lennard–Jones potential at distances just beyond van der Waals contact (see Figure 2.24). However, this is one of the most poorly parametrized aspects of the molecular mechanics methodology—that is, there are few quantitative studies of such interactions that are relevant to the kinds of systems considered here. As such, the molecular mechanics method is in general poorly suited to quantitative predictions concerning molecular recognition. It can serve a qualitative role, much like physical models can, but anything beyond that is risky.

4.3 Supramolecular Chemistry

We noted above that one reason the hydrophobic effect is so important is that it provides a way for the spontaneous assembly of many molecules into a more complex, (partially) ordered structure. In pure aqueous media, neither hydrogen bonding, ion pairing, nor the cation–π interaction can accomplish this alone. Nevertheless, nature self-assembles more complicated structures than just micelles and vesicles. Especially impressive are structures such as tobacco mosaic virus, in which exactly 2130 identical copies of a protein spontaneously assemble around a strand of RNA to make a 3000 Å long virus; or the ribosome, in which over 50 different proteins and thousands of nucleotides of RNA co-assemble into a remarkable molecular machine that executes protein synthesis.

While the hydrophobic effect is a necessary contributor to these remarkable assembly processes, the high architectural specificity is achieved by using the more directed interactions—the hydrogen bond, the ion pair, and the cation–π interaction. To first order, the hydrophobic effect provides the initial driving force for assembly and pays the entropy cost of bringing together many molecules, and then the more specific interactions define the precise structure.

Because of their intellectual and practical importance, and their aesthetic beauty, chemists have longed to mimic such large, multimolecular complexes. The phrase **supramolecular chemistry**—chemistry beyond the molecule—has been coined to describe this field. Certainly, the host–guest systems we discussed above are simple examples of supramolecules, and the molecular recognition forces we've enumerated will provide the driving force for assembly. But, when a large number of molecules assembles into one complex system, new issues arise. Here we will summarize some of the general issues associated with supramolecular chemistry, and then we will present some examples of especially successful or informative systems.

A key point in considering supramolecular chemistry is that we are now doing organic synthesis under **thermodynamic control** rather than **kinetic control** (see Chapter 7 for a full explanation of these terms). Synthetic organic chemistry has been fabulously successful in assembling complex structures, generally in a stepwise fashion. With few exceptions, the reactions of organic synthesis are under kinetic control. They are irreversible, and the product is determined by the relative energies of all possible *transition states*. A major goal of physical organic chemistry has always been to develop a detailed understanding of the mechanisms of such reactions, so that predictions can be made and syntheses can be designed.

In supramolecular chemistry, however, reactions are generally under thermodynamic control. The components are mixed, and the final product is the one with the most stable *ground state*. This requires that all reactions be reversible under the reaction conditions, so that an inherently dynamic system can find the lowest energy structure. As such, newly formed bonds must be relatively weak, so they can break and reform repeatedly under the reaction conditions until the **thermodynamic sink** (the lowest energy structure) is found. This is supramolecular chemistry in its purest form. In some systems, a mixture of weak and strong interactions may be involved, ultimately trapping the system in one state.

As we will see, supramolecular chemistry allows the synthesis of very large, complex, but well-defined organic structures. And, often, the synthesis involves just a few steps! While the accomplishments of modern synthetic organic chemistry are impressive indeed, structures with molecular weights in the thousands require long difficult routes (except many polymers), and generally produce only small amounts of product. Many have argued that if we are to design "smart" molecules for molecular electronics or targeted drug delivery, we will have to design them so that they "synthesize themselves". Nature has shown that this is possible, and if physical organic chemists can master this art, the potential payoff is tremendous.

Another general feature of supramolecular chemistry research is that thus far, few successes have been achieved in aqueous media. Early studies meant to establish the principles of the field have emphasized organic solvents, in which the more directional forces such as hydrogen bonding are stronger and can mediate spontaneous self-assembly into complex systems. Certainly, though, a long term goal of modern physical organic chemistry must be to design systems that use the hydrophobic effect in combination with other forces and spontaneously assemble into complex, well-defined systems.

4.3.1 Supramolecular Assembly of Complex Architectures

One goal of supramolecular chemistry is to design molecules that spontaneously assemble into complex systems with well-defined structures. Lehn has noted an analogy to computer programming—namely, the molecules must have information programmed into their designs that directs the assembly of the supramolecular system. The chemistry then reads out the information embedded in the design and builds the complex. Many examples have been developed in recent years in which many components are simply mixed together in a flask, and this "intelligent" molecular system assembles itself.

Self-Assembly via Coordination Compounds

Figure 4.13 shows an example of self-assembly using metals to guide the association. A flat, hexaazatriphenylene unit provides three bipyridyl-type ligands at the corners of a triangle. The quaterpyridyl (four pyridines in a row) provides another pair of bipyridyl binding sites. Cu^+ binds bipyridyls in a tetrahedral arrangement, so it seemed possible that adding Cu^+ would assemble these two units, and indeed this happens as shown. It really is remarkable to add eleven components (two triphenylenes, three quaterpyridyls, and six coppers) into a solution and then just wait while a beautiful, complex structure "makes itself". Imagine the effort required to make an analogous structure with all covalent bonds by conventional synthetic methods. We recognize that purists may object to the presence of metals in this and other supramolecular systems, but we see no value in such a limited view. As emphasized in this textbook, physical organic chemistry reaches across all of chemistry, biology, and materials science in its effort to understand how molecules behave.

A key in supramolecular assembly is to have some weak, reversibly formed bonds. In this way, "mistakes" can be corrected. Figure 4.13 shows an example. When assembling many particles into one supramolecular structure, misconnections and blind alleys will result. If the linkages are reversible, however, the mistakes can be undone. And if the design is clever enough that the desired product is the thermodynamic sink, eventually the goal will be reached. In many examples of supramolecular assembly, it seems likely that the final product is, in a sense, formed irreversibly. Once the entire system is in place, the weak bonds can reinforce each other. For example, with the "correct" complex in Figure 4.13, if one bond

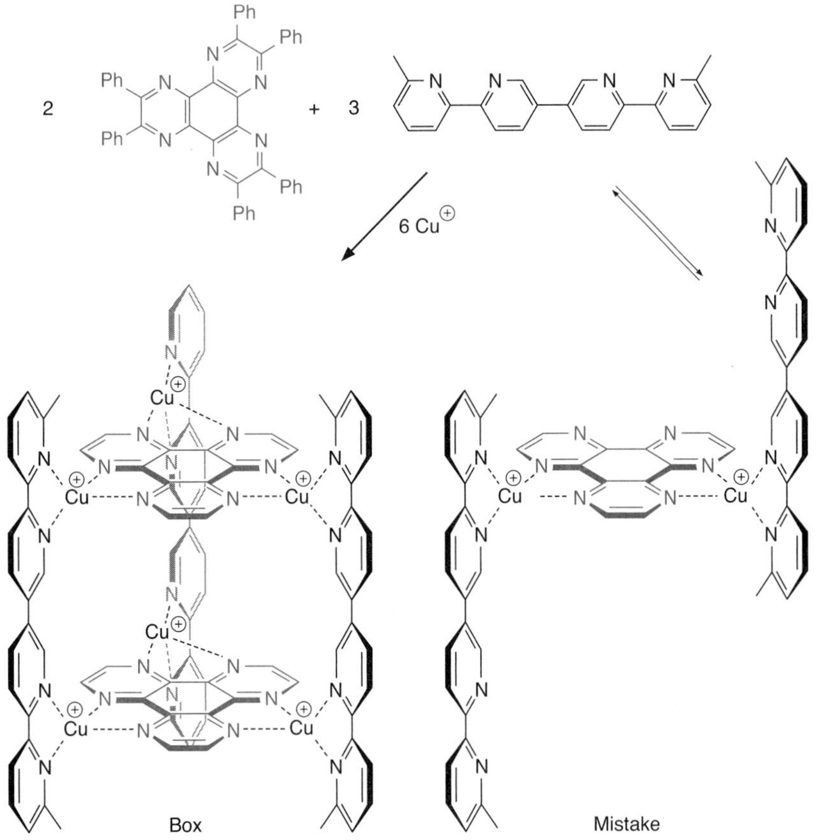

Figure 4.13
Supramolecular assembly of a complex "box" from 11 separate components by spontaneous self-assembly. Note that the phenyl groups of the precursor hexaphenylhexaazatriphenylene are omitted in the complex for clarity, and the back quaterpyridine is gray. The complex on the right is a "mistake" that is corrected by equilibration. Baxter, P., Lehn, J. M., Decian, A., and Fischer, J. "Multicomponent Self-Assembly—Spontaneous Formation of a Cylindrical Complex from 5 Ligands and 6 Metal-Ions." *Angew. Chem. Int. Ed. Eng.*, **32**, 69–72 (1993).

is broken, we have a dangling edge, but the whole system does not unravel. Unless one or more other bonds also break quickly, the most likely path is to simply repair the initial lesion. This kind of cooperativity in holding the structure together is reminiscent of the hydrogen bonding pattern of the α-helix in proteins.

Self-Assembly via Hydrogen Bonding

Another common motif used for supramolecular assembly is the hydrogen bond. For reasons discussed above, most studies are performed in a non-competitive solvent such as chloroform. In this regard, one motif that has found extensive use is the association of melamines and isocyanuric acids (Figure 4.14). "Tapes" and "ribbons" are formed from this association in the solid state. Building upon this motif, very complex three-dimensional structures have been assembled from relatively simple monomers. For example, a hexameric melamine structure (given in Figure 4.14) and six isocyanuric acids self-assemble into a stacked three-dimensional supramolecule. No partially formed aggregates are observed when only three equivalents of isocyanuric acid are mixed with the hexameric melamine structure. Instead, only the fully assembled structure and free monomers can be observed. This indicates an extremely high level of positive cooperativity among the binding sites of the hexameric melamine structure. Upon binding each individual isocyanuric acid to the hexameric melamine structure, the subsequent binding of other isocyanuric acids is enhanced.

Consider the various forces of molecular recognition involved in these systems, as well as the other design principles that favor controlled assembly. One key issue is directionality.

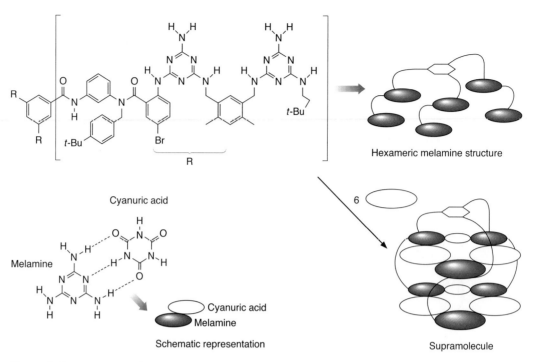

Hexameric melamine structure

Cyanuric acid

Melamine

Cyanuric acid
Melamine

Schematic representation

Supramolecule

Figure 4.14
The creation of a hexameric melamine compound gives very high positive cooperativity for the creation of a supramolecule containing six isocyanuric acids. Mathias, J. P., Seto, C. T., Simanek, E. E., and Whitesides, G. M. "Self-Assembly through Hydrogen Bonding: Preparation and Characterization of Three New Types of Supramolecular Aggregates Based on Parallel Cyclic CA₃–M₃ 'Rosettes'." *J. Am. Chem. Soc.*, **116**, 1725–1736 (1994).

If you want to assemble a "box", you need bidentate ligands and connectors that support a 90° angle. For a helix, a structure compatible with twisting must be designed. Also, preorganization is evident in several systems, where units are linked together beforehand in order to bias the assembly process. Another strategy has been termed **peripheral crowding**. More generally, this involves strategic placement of sterically imposing groups that can *disfavor* formation of one type of product, leading the system to assemble in the desired way. An example of this is the methyl groups at the end of the quaterpyridyl group of Figure 4.13. Self-assembly will be an essential component of other interesting supramolecular systems discussed below. The rules of supramolecular chemistry are emerging, and we can anticipate even more spectacular, self-assembled supramolecules in the near future.

4.3.2   Novel Supramolecular Architectures—Catenanes, Rotaxanes, and Knots

Organic chemists have always been fascinated by cyclic molecules and elaborate arrays of complex ring systems. As such, the notion of looping two rings through each other has long been a goal. Such a structure is termed a **catenane** (Figure 4.15). If there are only two rings, it is a [2]catenane; three rings make a [3]catenane, etc. A catenane presents a novel bonding situation with the potential for novel properties. It is a supramolecule held together because of the topology rather than exclusively binding of the sort we have discussed here. Related structures are **rotaxanes**, which have a dumbbell-like unit plus one or more rings entrapped on the "bar" of the dumbbell (Figure 4.15), and **knots**, single ring systems with novel topologies (see Chapter 6). The notions of molecular recognition and supramolecular chemistry have revolutionized efforts to prepare such systems. We will show several examples of such systems here and also in Chapter 6, where we discuss the novel stereochemical features associated with them.

Early attempts at preparing catenanes relied on a more or less statistical threading approach. By cyclizing very long chains (with over 30 carbons) to make very large rings, there

A [2]catenane

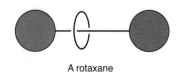

A rotaxane

Figure 4.15
Definitions of catenanes and rotaxanes.

was some chance that one chain would be looping through one pre-formed ring when it closed, and a catenane would result. Brilliant studies in the 1950s established the feasibility of this approach, and produced the first catenanes. The yields were extremely low, however, so catenanes remained rare curiosities.

The breakthrough came when it was realized that the concepts of molecular recognition and supramolecular chemistry could be put to use to make catenane synthesis a rational and efficient processes. Two independent strategies—the use of metal complexation chemistry and π donor–acceptor forces in organic systems—were exploited to make novel structures.

The system exploiting π donor–acceptor effects is shown in Figure 4.16. Early host–guest studies along the lines of those discussed above established that a cyclophane crown could bind cationic guests such as the bipyridinium compound (also known as paraquat,

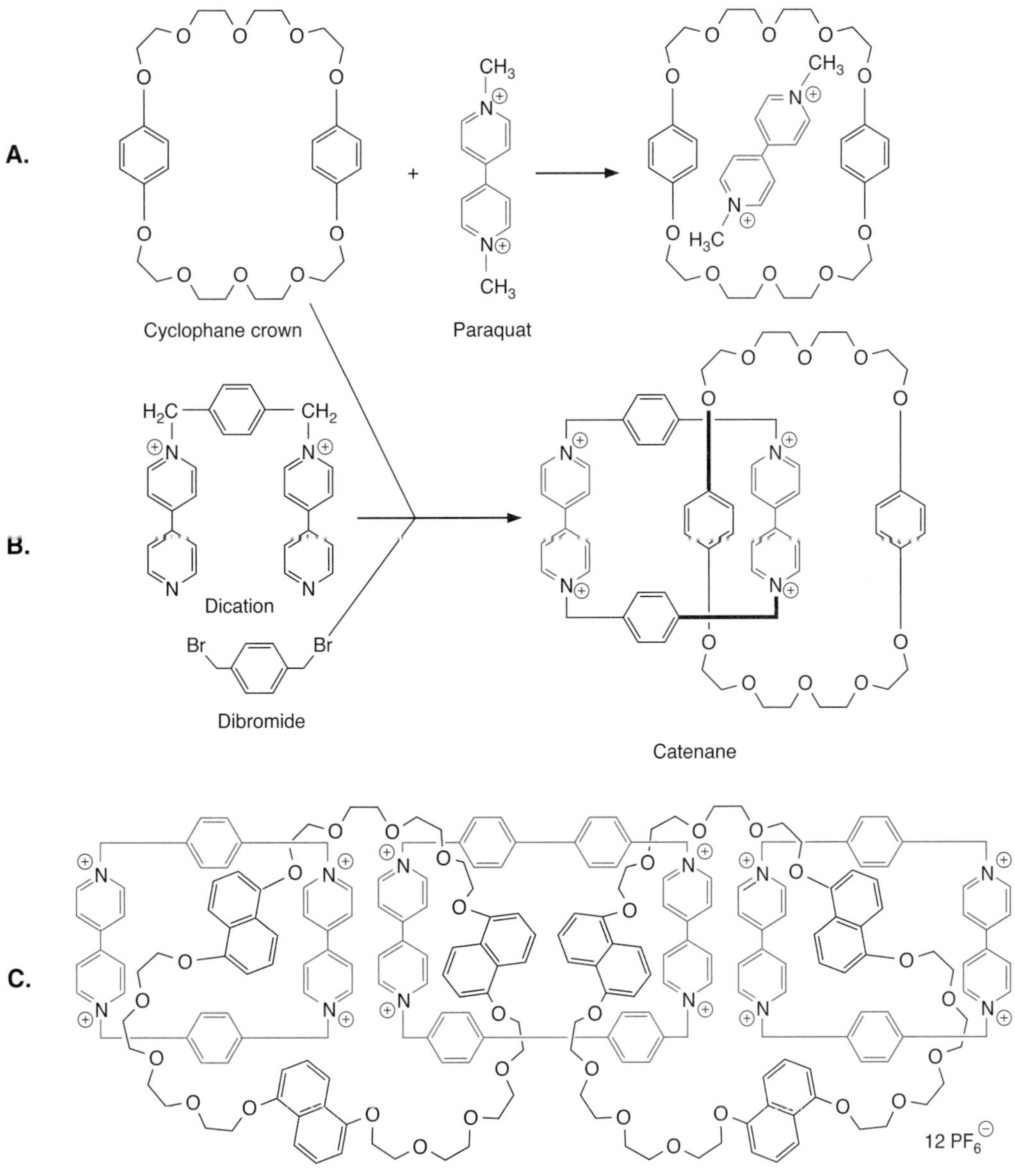

Figure 4.16
A. An example of a host–guest complex between paraquat and a polyether cyclophane.
B. Catenane formation using the molecular recognition forces involved in the complex in part A.
C. "Olympiadane", a complex catenane made by such a threading procedure. Fyfe, M. C. T., and Stoddart, J. F. "Synthetic Supramolecular Chemistry." *Acc. Chem. Res.*, **30**, 393–401 (1997).

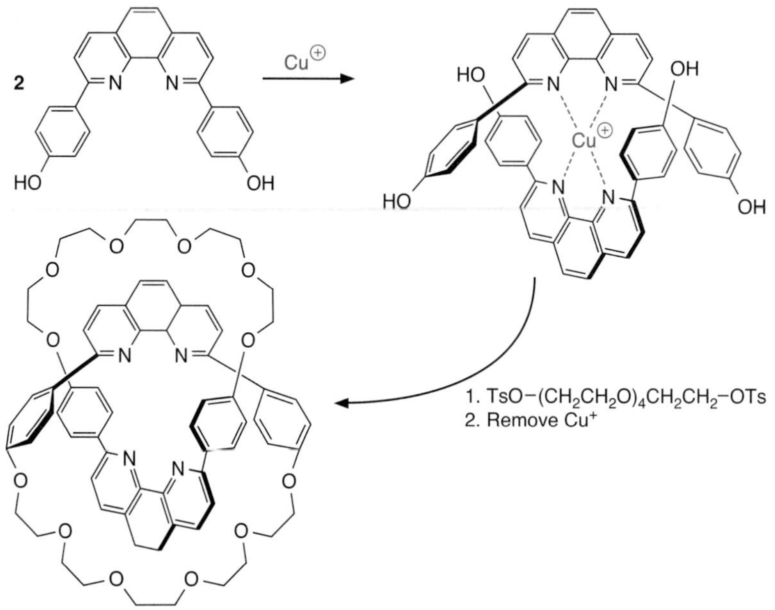

Figure 4.17
The Cu$^+$/phenanthroline strategy to prepare novel supramolecular architectures, as exemplified by the synthesis of a simple catenane. Sauvage, J.-P. "Transition Metal-Containing Rotaxanes and Catenanes in Motion: Toward Molecular Machines and Motors." *Acc. Chem. Res.*, **31**, 611–619 (1998).

a common herbicide). In an important intellectual leap, Stoddart reasoned that the same forces that led to such binding could be used to preorganize a system to favor catenane formation. Indeed, when the cyclophane crown, dication, and the dibromide shown in Figure 4.16 are mixed, the catenane is formed with a spectacular 70% yield! The likely mechanism is as follows. After the first alkylation by the dication, a bipyridinium is formed which binds in the cavity of the cyclophane crown in a threading process. Now, when the ring closure occurs, a catenane naturally forms. The key advance was that instead of just "hoping" that threading would occur, the forces of molecular recognition were used to strongly encourage it. This motif has been expanded all the way to the synthesis of "olympiadane" a structure with five catenated rings that is assembled from eight components in just two chemical steps! Also, similar designs can be used for the synthesis of rotaxanes. One of the most important aspects of this work is that structures like catenanes and rotaxanes represent physical entities that can be exploited to control movement of one object relative to another.

The second strategy for making supramolecular structures with novel topologies was developed by Sauvage, and it is shown in Figure 4.17. Quite simply, component pieces are first held together in a desired orientation using metal coordination. Then, when the system is assembled, the metal is removed, leaving a topologically interesting system. This templating strategy has considerable potential for the design of novel architectures, and systems based on Cu$^+$/phenanthroline have been especially successful. Highlights include the preparation of a trefoil knot, and other topologically unique systems that will be discussed in Section 6.6 ("Topological and Supramolecular Stereochemistry").

Nanotechnology

Molecular recognition and supramolecular chemistry are extremely important for building complex molecular entities for use in the new and emerging field called **nanotechnology**. Nanotechnology seeks to create devices, such as switches, gates, machines, etc., on the nm scale. The notions of self-assembly of chemical structures from the "bottom up" (construction by assembly of molecules) will ultimately meet and merge with the construction of devices from the "top down" (such as microlithography; see Section 17.6). In this quest, entities such as catenanes, rotaxanes, and other as yet undiscovered supramolecular motifs, along with self-assembly, will be key design principles (the following Connections highlight shows one of the most exploited self-assembly tools).

Connections

Biotin/Avidin: A Molecular Recognition/ Self-Assembly Tool from Nature

The insights from molecular recognition studies are now commonly being used to design and create molecule-based devices, and the discussion of supramolecular chemistry given above has shown some examples. When two molecules need to be non-covalently brought together in the creation of a device, one can exploit the natural interaction between biotin and avidin. Biotin is a cofactor that binds to the protein receptor avidin (or strep-avidin), and the binding constant has been measured to be near 10^{15} M^{-1}. This is one of the largest affinities associated with any molecular recognition event, and can be considered irreversible. Avidin is a tetrameric protein that binds four biotins independently. The fact that the protein binds four biotins and has such a large affinity for each, makes this system a powerful tool for assembly processes.

Strategies for coupling two different structures (A and B) together using biotin/avidin proceed as follows (see below). Both A and B are first covalently attached (**conjugated**) to biotin, a process known as **biotinylation**. Biotin-conjugated A is exposed to avidin, leading to association (we show two binding events below, but the binding is really statistical). Any open sites on avidin are now available for binding biotin-conjugated B. The "sandwich" thus created attaches one or more A to one or more B through the intermediacy of the avidin protein. Yet, no covalent bonds are required. Instead, the components are simply mixed in a specific order, and the final structure self-assembles. The biotin/avidin interaction is likely the most widely used self-assembly process exploited by chemists and biochemists, and we will see examples throughout this text.

Biotin

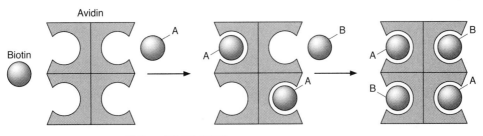

Biotin–avidin interaction

4.3.3 Container Compounds—Molecules within Molecules

One of the most fascinating new supramolecular structures has been the general motif of the **container compound**, a closed surface structure that completely encapsulates an inner void. Under appropriate conditions, this inner void can be filled by another molecule (nature abhors a vacuum), producing a novel supramolecular structure. The inner molecule is not held in the interior by any chemical bonding, as is an ion in a cryptand. And there is no linkage between the two components, as in a catenane. A molecule is simply trapped inside another molecule, with no viable escape route. These are intrinsically interesting types of structures, and it has even been proposed that the interior constitutes a "new state of matter", in that it is not obviously gas, liquid, or solid. In addition, we can imagine novel chemis-

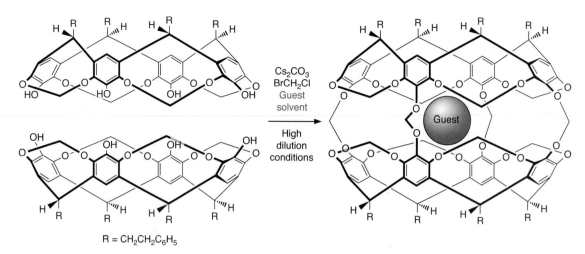

Figure 4.18
Synthesis of a carcerand. Cram, D. J. "Molecular Container Compounds." *Nature,* **356**, 29–36 (1992).

try occurring inside a container compound, and even the possibility of drug delivery systems, if controlled release could be developed.

We will consider two different types of container compounds. In the first, the container is formed fully by covalent bonds, and a molecule trapped in the interior can only get out if a bond breaks or if it can squeeze through a hole in the container. In the second form, the container itself is a supramolecule, formed by the spontaneous assembly of parts. Since this step can be reversible, so can encapsulation.

An important series of "covalent" container compounds has been developed by Cram and termed **carcerands**. Figure 4.18 shows the synthesis of the prototype system. It begins with a calixarene that has been modified to include bridges that lock the molecule into the cone form while leaving four phenols for further elaboration. When two such units are coupled through a one-carbon bridge, it is generally observed that a small molecule, typically a solvent such as DMF or DMSO, is trapped within, making a **carceplex**. Yields of the container complex are in the 50–60% range, quite impressive for a process in which eight new bonds are formed. Interestingly, if the reaction is attempted in a solvent that is too large to be contained in the cavity, such as $(CH_2)_5NCHO$, the assembly completely fails. This indicates that the entrapped guest is actually templating the synthesis of the carcerand, much like the template effect discussed earlier in the context of crown ethers.

Once inside, most molecules never get out. However, the entrapped molecules "communicate" with the outside world, as evidenced by the fact that carceplexes that differ only by their internal contents separate easily on thin layer chromatography. Many variants of the basic system have been prepared, including structures in which one of the links formed in the final synthetic step is missing (see the Connections highlight on the next page) or in which the link is longer. In these systems, termed **hemicarcerands** and **hemicarceplexes**, molecules still get trapped inside, but conditions can be found in which the guests move in and out of the cavity, allowing very novel dynamics studies.

Another intriguing class of compounds with well-defined interior cavities are the **cryptophanes**. These are analogous to the carcerands except, instead of a calixarene as the basic building block, cryptophanes are derived from a structure called cyclotriveratrylene. The phenols provide a valuable "hook" for linking units together. A typical structure is shown in the margin. It contains a C_3 axis, but no other symmetry elements, and so the cryptophanes are chiral. These structures form well-defined complexes with the guest completely encapsulated. Tetramethylammonium ion is very tightly held in the "crypt", presumably because of constrictions that impede escape and attractive cation–π interactions.

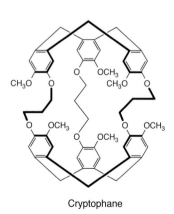

Cryptophane

Connections

Taming Cyclobutadiene—A Remarkable Use of Supramolecular Chemistry

The isolation and characterization of cyclobutadiene has been one of the holy grails of physical organic chemistry. It is the prototype antiaromatic compound, it has been the subject of innumerable experimental and theoretical studies, and it is discussed in several locations in this text. Cyclobutadiene did eventually succumb to experimental characterization, primarily involving spectroscopic studies in cryogenic matrices at very low temperatures that we will discuss later in this book. However, using supramolecular chemistry, cyclobutadiene can be characterized by NMR and IR spectroscopies *at room temperature!*

The key sequence is shown below. The cyclobutadiene precursor α-pyrone was inserted into a hemicarcerand. In a hemicarcerand, guests can shuttle in and out at elevated temperatures, and then be trapped in when the temperature is lowered. Photolysis leads to several reactions, but under appropriate conditions, CO_2 is expelled from the hemicarcerand, and cyclobutadiene is formed. The incarcerated cyclobutadiene shows a singlet in its NMR spectrum at δ 2.27 (a confirmation of its antiaromaticity). Thus, cyclobutadiene, a molecule that has always been viewed as extremely reactive, is indefinitely stable at room temperature—as long as it is encapsulated in a supramolecular container.

Cram, D. J., Tanner, M. E., and Thomas, R. "The Taming of Cyclobutadiene." *Angew. Chem. Int. Ed. Eng.*, **30**, 1024 (1991).

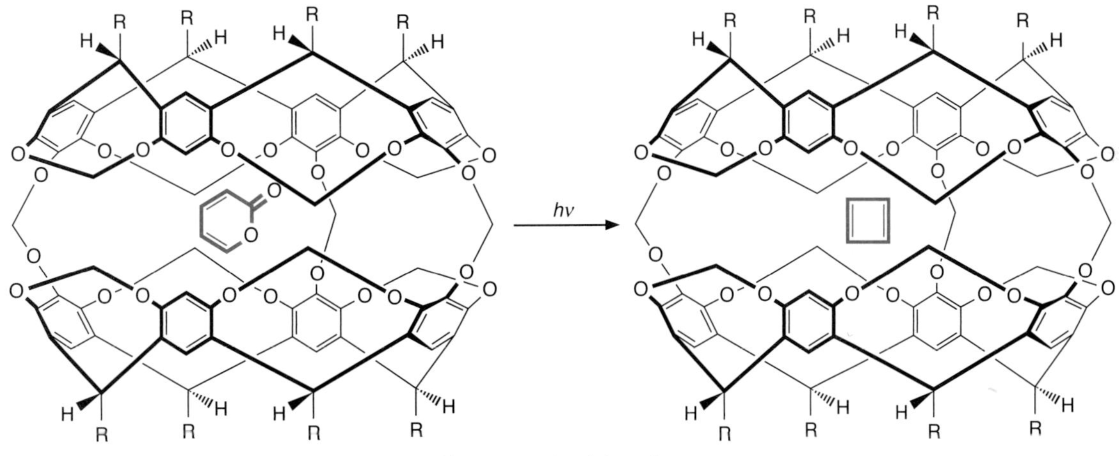

Encarcerated cyclobutadiene

An intriguing addition to the molecular container field has been the development of self-assembling containers. An early example is shown in Figure 4.19. The building block diphenylglycoluril is readily available. Rebek recognized that when two such units are linked to an aromatic ring, a system that is concave and self-complementary forms. That is, the molecule has both hydrogen bond donors and acceptors on it, and if two of these come together in the desired perpendicular arrangement, an effective dimerization is possible. Indeed, when the bisglycoluril is dissolved in chloroform, it dimerizes efficiently. In dimerizing, it forms an interior cavity that is large enough to encapsulate small molecules such as methane. This system resembles a tennis ball, which is similarly made by the union of two identical pieces.

When the basic hydrogen bonding unit is greatly expanded (Figure 4.19 **C**), dimerization still occurs, but a much larger cavity is formed. Now structures such as benzene or tetramethyladamantane can fit into the cavity. In fact, prototype chemical reactions such as a Diels–Alder reaction can be induced to take place in the larger cavity. This illustrates a potential benefit of such systems. Since formation of the container compound is a reversible, supramolecular process, one could in principle design systems that would function as unique catalysts that can exhibit true turnover. This is not very practical for the covalent containers, because guests are pretty much trapped in the interiors. We anticipate many other exciting new container molecules in the near future.

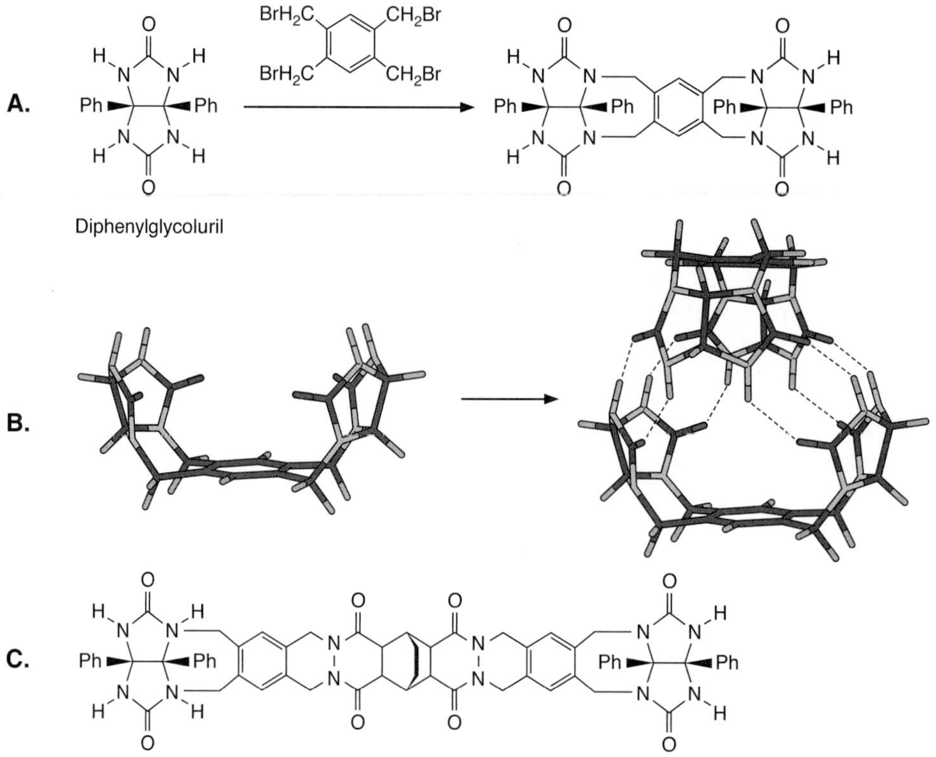

Figure 4.19
The "tennis ball" strategy. **A.** Synthesis of the monomer from the diphenyl glycoluril building block.
B. Dimerization of the monomer through eight hydrogen bonds produces a cavity-containing
supramolecule. Note that the phenyl rings are omitted for clarity. **C.** A larger building block that
produces larger cavities. Wyler, R., DeMendoza, J., and Rebek, J. "A Synthetic Cavity Assembles
Through Self-Complementary Hydrogen-Bonds." *Angew. Chem. Int. Ed. Eng.*, **32**, 1699–1701 (1993).

Summary and Outlook

This chapter has covered the thermodynamics of binding, analytical methods to determine
binding constants, several classes of receptors used in molecular recognition studies, and
supramolecular chemistry. Along with weak forces, the topics of this chapter could easily be
expanded into a complete textbook, and an entire one- or two-semester course. Therefore,
we have really only briefly introduced each of these topics. However, our discussion has
been deep enough that we can take the concepts and apply them to many of the topics of the
second part of this textbook. Examples of topics in the second part that use concepts from
this chapter include measuring thermodynamic parameters, solvent effects on reaction kinetics, and the use of weak forces to control reaction stereochemistry.

Yet, before analyzing kinetics, mechanisms, and catalysis, there are two remaining topics that must be covered. The first is acid–base chemistry. Since it is predominately a subject
that deals with thermodynamics, it belongs in the first part of this book. In fact, in this chapter we already alluded to the fact that a base can be considered a host for a proton. Hence,
topics in acid–base chemistry naturally evolve from the kind of understanding of the thermodynamics of complexation processes described in the last two chapters, and the mathematical development given in this chapter can be used to describe acid–base chemistry. The
second topic that must be covered is stereochemistry, the final chapter of Part I of this book.

Exercises

1. Discuss the kinds of forces that could be involved in the binding of paraquat to the polyether cyclophane of Stoddart shown in Figure 4.16.

2. Thermodynamic measurements establish that the binding of the artificial polyamide heterocycles to DNA (Figure 4.9) shows a large *favorable* entropy in aqueous media. Explain this, and be as specific as possible.

3. The following drawing shows the association constants of three different complexes in chloroform, all three of which have three hydrogen bonds. What is the difference between the three complexes that makes the binding constants so different? Explain in detail.

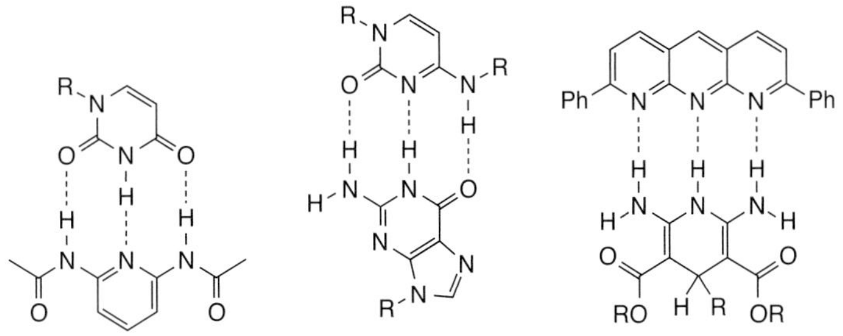

K_a near 10^2 M^{-1} $\qquad$ K_a near 10^4 M^{-1} $\qquad$ K_a above 10^5 M^{-1}

4. The cyclophane of Figure 4.10 **D** has been shown to be a general receptor for a wide range of organic cationic guests through cation–π interactions in water. Generally, guests of the sort RNMe$_3^+$ are preferred over RNH$_3^+$, with the latter generally not showing strong binding. Suggest a reason for this.

5. Derive the equation for ΔG° (NH–π) given in the Connections highlight entitled "A Thermodynamic Cycle to Determine the Strength of a Polar–π Interaction".

6. Show all the mathematics that leads to the double reciprocal plot used in the Benesi–Hildebrand method. How much does the host need to be kept in excess over the guest so we can safely assume that [H] = [H]$_o$?

7. In the Benesi–Hildebrand method, one component needs to be kept in large excess over another. In our Going Deeper highlight, we kept the host concentration much larger than the guest. Under such circumstances, one often finds that the guest is saturated upon the very first addition of host. Hence, no change in [H•G] occurs during the experiment. What is the simplest solution to this problem that you would implement in your second attempt to measure the K_a for this system?

8. Derive Eq. 4.19 starting from Eqs. 4.15, 4.18, and 3.22. When would you expect a significant heat capacity difference between the reactant and products, such that this analysis is important?

9. The following molecular tweezers bind 2,4,5,7-tetranitrofluorenone (TENF). The association constants for **A**, **B**, and **C** are 3.4×10^3, 7.0×10^2, and 1.7×10^2 M^{-1}, respectively. Explain this trend.

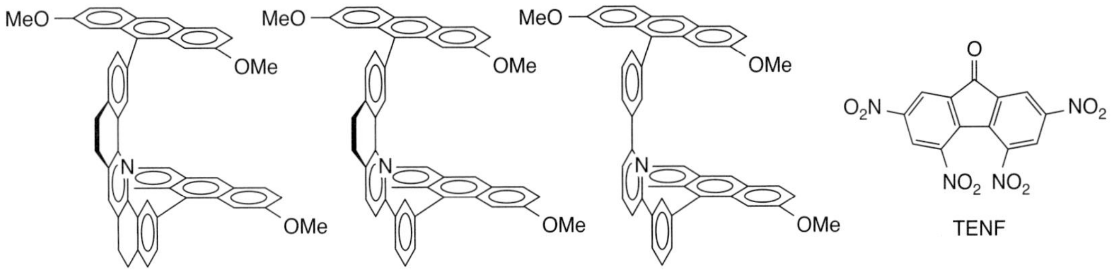

A. $\qquad$ **B.** $\qquad$ **C.**

10. The azacrown ether discussed in Section 4.2.1 was said to have reasonable selectivity for binding ammonium over K$^+$. Recently, however, the following structure, developed by Chin and Kim, was found to have approximately 400:1 selectivity for ammonium relative to K$^+$. Propose a binding geometry for ammonium, and explain why the selectivity

is so good. Also, explain the role of the three ethyl groups.

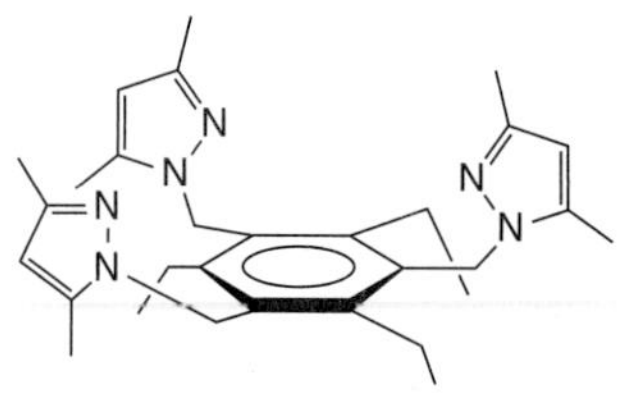

11. Explain why it is best to determine an equilibrium constant for any binding scenario when one of the component's concentrations is near K_d for the complex.

12. In Section 4.1.1 (on changing the standard state) we showed how an exergonic reaction could be viewed as an endergonic reaction simply by changing the standard state. Examine this discussion again, and decide how you could define the standard state for a binding reaction such that the GFE of that reaction is zero (that is, neither exergonic nor endergonic).

13. The following series of molecular recognition reactions were studied and the association constants given were determined. Based upon these values, and the definitions of cooperativity given in this chapter, does the binding of *cis*-1,3-cyclohexanediol in the last equation show positive cooperativity? By how many kcal/mol is the binding cooperative in a positive or negative sense?

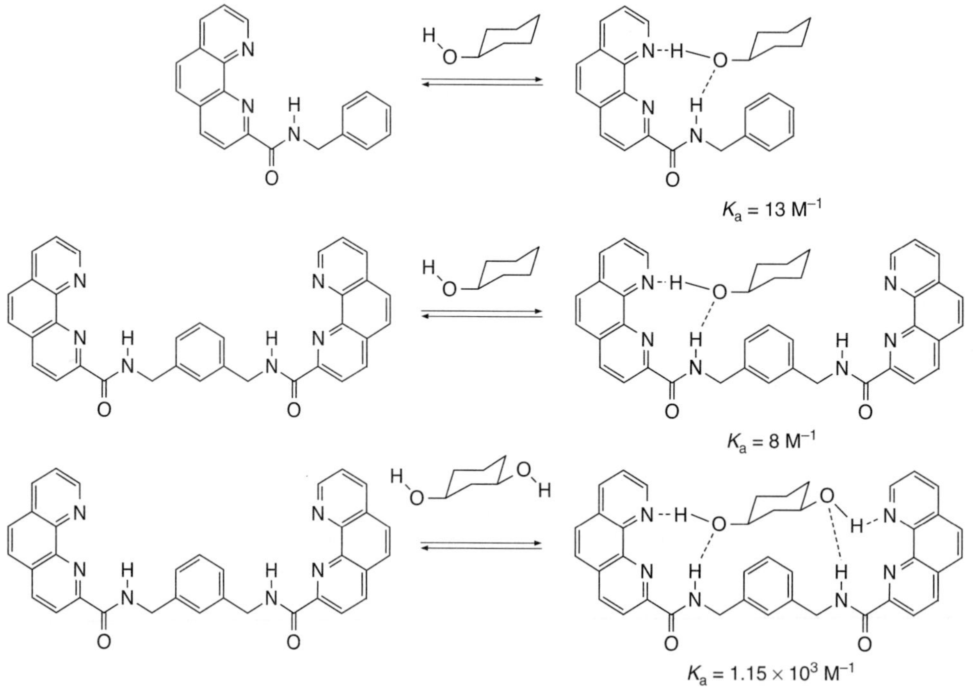

$K_a = 13 \, M^{-1}$

$K_a = 8 \, M^{-1}$

$K_a = 1.15 \times 10^3 \, M^{-1}$

14. Why is it a good general rule to perform a binding titration experiment so that one achieves at least 70% saturation of the component that is being kept relatively constant?

15. One of the simplest ways to make rotaxanes has been recently introduced by Vögtle. It involves the alkylation of a phenoxide or alkoxide anion with an alkyl halide. By examining the barbiturate receptor of Figure 4.8 **A**, write down a reaction sequence that could create a rotaxane.

16. In the following polymerization reaction, explain why one might expect a faster polymerization rate by lowering the temperature.

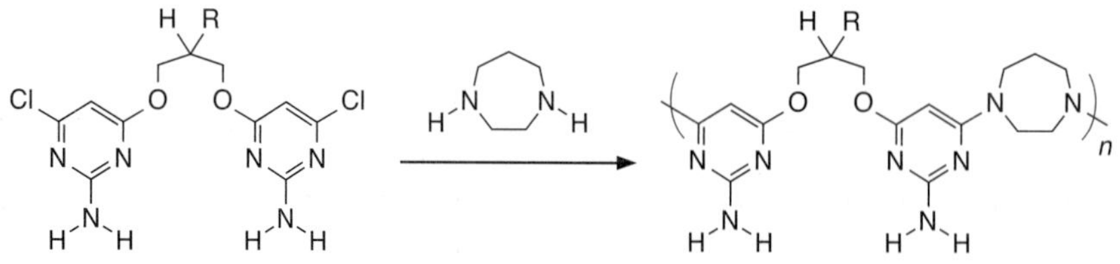

17. Explain the trend in binding constants (in CH_2Cl_2) for the complexation of phenol derivatives by the synthetic receptors shown.

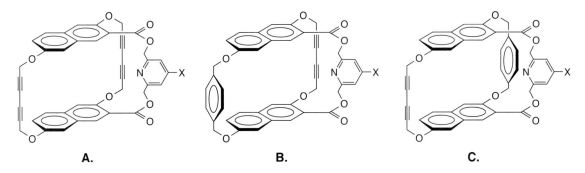

X	Y	K_a (M^{-1})
H	4–NO$_2$	2340
H	4–CN	815
H	3–NO$_2$	115
Cl	4–NO$_2$	700
NMe$_2$	4–NO$_2$	16000

18. The following cyclophanes bind *p*-nitrophenol in the manner shown in Exercise 17. The binding constants for **A**, **B**, and **C** are 1.6×10^4, 3.5×10^4, and 1.0×10^5 M^{-1}, respectively. Give a possible explanation for this trend.

A.   **B.**   **C.**

19. If two receptors bind the same guest, which one has the stronger affinity, the one with a K_d of 10 μM or the one with a K_d of 100 μM?

20. Boronic acids form reversible complexes with vicinal diols in aqueous media as shown below. This strategy has been incorporated into receptors for saccharides. It typically takes a few minutes for these equilibria to be achieved in solution. Given this simple observation, do you expect the study of these binding events to be on the slow or fast exchange time scale on a 300 MHz ^{1}H NMR spectrometer? Do you expect to see different resonances for the boronic acid and the boronate ester, or an average of the signals? Explain your answer.

21. Derive Eqs. 4.24 and 4.25.

22. Why is it that with isothermal calorimetry we can determine two thermodynamic parameters (K_a and $\Delta H°$), while titrations involving other methods give only one thermodynamic parameter (K_a)?

23. A useful graph of a binding isotherm plots the mole fraction of H•G with respect to H on the *y* axis and log[G]$_o$ on the *x* axis with [H] held constant and is called a semilogarithmic plot. Make such a plot where [H]$_o$ = K_d using log [G]$_o$/[H]$_o$ on the *x* axis. Span a concentration range for [G]$_o$ from 100 times lower than [H]$_o$ to 100 times greater than [H]$_o$. At what [G]$_o$ is the mole fraction of H•G = 0.5, and what are the mole fractions of H•G for each unit change in log[G]$_o$? What does this plot tell you about the extent of binding as a function of each decade change in guest concentration? (*Hint:* You will need to solve for [G] in Eq. 4.25, and then use this information in Eq. 4.24 to calculate the mole fraction of H•G.)

24. An interesting binding phenomenon is sometimes found between nitro groups and halogens on aryl rings, especially iodo. Further, amine groups also have an affinity for halogens on aryl rings, as well as iodos on fluorinated alkanes. Speculate as to the origins of these kinds of molecular recognition events.

25. In the text, we analyzed a scenario with a binding constant of 10 M^{-1} and initial concentration of H and G at 10 mM. Verify that at equilibrium we have $[\text{H}] = [\text{G}] = 9.2$ mM and $[\text{H} \bullet \text{G}] = 0.84$ mM.

Further Reading

Thermodynamics of Protein Ligand Interactions

Stites, W. E. "Protein–Protein Interactions: Interface Structure, Binding Thermodynamics, and Mutational Analysis." *Chem. Rev.,* **97**, 1233–1250 (1997).

Gill, S. J. "Thermodynamics of Ligand Binding to Proteins." *Pure Appl. Chem.,* **61**, 1009–1020 (1989).

Freire, E. "Statistical Thermodynamic Linkage Between Conformational and Binding Equilibria." *Adv. Protein Chem.,* **51**, 255–279 (1998).

Cooperativity in Binding

Jencks, W. P. "On the Attribution and Additivity of Binding Energies." *Proc. Natl. Acad. Sci. U.S.A.,* **78**, 4046 (1981).

Williams, D., and Westwell, M. S. "Aspects of Weak Interactions." *Chem. Soc. Rev.,* **27**, 57–63 (1998).

Isothermal Calorimetry

Freire, E., Mayorga, O. L., and Straume, M. "Isothermal Titration Calorimetry." *Anal. Chem.,* **62**, 950A–959A (1990).

Wadsoe, I. "Trends in Isothermal Microcalorimetry." *Chem. Soc. Rev.,* **26**, 78–86 (1997).

Complementarity and Preorganization

Cram, D. J. "The Design of Molecular Hosts, Guests, and Their Complexes." *Science,* **240**, 760–767 (1988).

Crown Ethers

Bradshaw, J. S., and Izatt, R. M. "Crown Ethers: The Search for Selective Ion Ligating Agents." *Acc. Chem. Res.,* **30**, 338–345 (1997).

Gokel, G. W. "Lariat Ethers: From Simple Sidearms to Supramolecular Systems." *Chem. Soc. Rev.,* **21**, 39–47 (1992).

An, H., Bradshaw, J. S., and Izatt, R. M. "Macropolycyclic Polyethers (Cages) and Related Compounds." *Chem. Rev.,* **92**, 543–572 (1992).

Tweezers and Clefts

Zimmerman, S. C., Zeng, Z., Wu, W., and Reichert, D. E. "Synthesis and Structure of Molecular Tweezers Containing Active Site Functionality." *J. Am. Chem. Soc.,* **113**, 183–196 (1991).

Molecular Recognition with a Large Hydrogen-Bonding Component

Schneider, H.-J. "Linear Free Energy Relationships and Pairwise Interactions in Supramolecular Chemistry." *Chem. Soc. Rev.*, **23**, 227–234 (1994).

Kato, Y., Conn, M. M., and Rebek, J., Jr. "Hydrogen Bonding in Water Using Synthetic Receptors." *Proc. Natl. Acad. Sci. U.S.A.*, **92**, 1208–1212 (1995).

Perrault, D. M., Chen, X., and Anslyn, E. V. "The Advantages of Using Rigid Polyaza-Clefts for Hydrogen-Bonding Molecular Recognition." *Tetrahedron*, **51**, 353–362 (1995).

DNA Binding with Cooperative Hydrogen Bonds

Distefano, M. D., and Dervan, P. B. "Energetics of Cooperative Binding of Oligonucleotides with Discrete Dimerization Domains to DNA by Triple Helix Formation." *Proc. Natl. Acad. Sci. U.S.A.*, **90**, 1179–1183 (1993).

Cyclodextrins

Breslow, R., and Dong, S. D. "Biomimetic Reactions Catalyzed by Cyclodextrins and Their Derivatives." *Chem. Rev.*, **98**, 1997–2011 (1998).

Takahashi, K. "Organic Reactions Mediated by Cyclodextrins." *Chem. Rev.*, **98**, 2013–2033 (1998).

Saenger, W., Jacob, J., Gessler, K., Steiner, T., Hoffmann, D., Sanbe, H., Koizumi, K., Smith, S. M., and Takaha, T. "Structures of the Common Cyclodextrins and Their Larger Analogs—Beyond the Doughnut." *Chem. Rev.*, **98**, 1787–1802 (1998).

Cyclophanes

Diederich, F. (1991). *Cyclophanes*. Monographs in Supramolecular Chemistry, J. F. Stoddart (ed.), The Royal Society of Chemistry, Cambridge, UK.

Sutherland, I. O. "Cyclophanes as Synthetic Receptors." *Pure Appl. Chem.*, **62**, 499–504 (1990).

Molecular Recognition Emphasizing π Systems

Meyer, E. A., Castellano, R. K., and Diederich, F. "Interactions with Aromatic Rings in Chemical and Biological Recognition." *Angew. Chem. Int. Ed. Eng.*, **42**, 1210–1250 (2003).

Supramolecular Chemistry

Lehn, J. M. "Perspectives in Supramolecular Chemistry: From Molecular Recognition Towards Self-Organization." *Pure Appl. Chem.*, **66**, 1961–1966 (1994).

Zeng, F., and Zimmerman, S. C. "Dendrimers in Supramolecular Chemistry: From Molecular Recognition to Self-Assembly." *Chem. Rev.*, **97**, 1681–1721 (1997).

Van Veggel, F. C. J. M., Verboom, W., and Reinhoudt, D. N. "Metallomacrocycles: Supramolecular Chemistry With Hard and Soft Metal Cations in Action." *Chem. Rev.*, **94**, 279–299 (1994).

Lawrence, D. S., Jiang, T., and Levett, M. "Self-Assembling Supramolecular Complexes." *Chem. Rev.*, **95**, 2229–2260 (1995).

Catenanes and Rotaxanes

Vogtle, F., Safarowsky, O., Heim, C., Affeld, A., Braun, O., and Mohry, A. "Catenanes, Rotaxanes and Pretzelanes—Template Synthesis and Chirality." *Pure Appl. Chem.*, **71**, 247–251 (1999).

Stoddart, F. "Making Molecules to Order." *Chem. Br.*, **27**, 714–718 (1991).

Balzani, V., Gomez-Lopez, M., and Stoddart, J. F. "Molecular Machines." *Acc. Chem. Res.*, **31**, 405–414 (1998).

CHAPTER 5

Acid–Base Chemistry

Intent and Purpose

Depending upon your definition of an acid and base, chemists can characterize almost all chemical reactions as some form of an acid–base reaction. Because proton transfers are ubiquitous in chemistry and biochemistry, we start this chapter focusing upon this single kind of acid–base reaction. However, a later portion of this chapter concentrates on filled and empty orbitals as our base and acid equivalents (Lewis definitions), thereby broadening our acid–base reactions considerably. Hence, the reason for discussing acid–base chemistry thoroughly, and even devoting an entire chapter to this one reaction type in Part I of this book, is that it plays a central role in all chemical disciplines.

One of the most important insights we can gain about acid–base chemistry is the ability to predict what is the stronger acid or base when confronted with a comparison. Here, this will be a completely thermodynamic analysis, and we leave it until Chapter 9 to discuss the kinetics of proton transfers. In order to be able to make a sound prediction, we will cover correlations between gas phase and solution acidities. Then, numerous factors that control acidity will be covered—namely, solvation, resonance, electronegativity, inductive effects, etc. These are all topics that we have covered in Chapters 1–4, and hence this chapter serves as a nice recap.

Related to predicting the stronger acid is being able to decide the protonation state of an acid or base at a particular pH. This is of paramount importance for understanding many types of reactions, as well as catalysis and enzymology. Although all introductory chemistry courses cover the mathematics of quantifying the extent of protonation of an acid at a particular pH, surprisingly, most students cannot make a qualitative estimate of the protonation state of a particular acid at a specific pH. We will show you how to do so here. Our analysis may seem a bit elementary at times, but acid–base chemistry is a topic that many students have difficulty understanding. Therefore, much of the beginning of this chapter will be a review.

In organic chemistry, the acids and bases are widely varied in structure, and are used under a variety of experimental conditions, including different solvents and temperatures. Hence, although we are used to considering acid–base chemistry with water as a reference, we should also become proficient at predicting acid–base chemistry under unusual circumstances. Therefore, discussions of nonaqueous solvents are included herein, as well.

5.1 Brønsted Acid–Base Chemistry

In Chapters 2, 3, and 4 we covered thermodynamics and equilibria in general. Based on the discussion of molecular recognition in Chapter 4, we can now think of a base as a host for a proton. Recall, too, that for reactions at equilibrium, the equilibrium ratio of concentrations can be expressed by equations such as Eq. 5.1.

$$K_{eq} = \frac{[P]}{[R]}$$

(Eq. 5.1)

The equilibrium constant (K_{eq}) gives quick insight into whether the reactant or product is more stable, and the extent to which the reactant or product is preferred.

Probably the most widely studied reactions, for many thousands of equilibrium constants have been measured, are Brønsted acid–base reactions (also called Brønsted–Lowry). A **Brønsted acid** is defined as a proton donor, and a **Brønsted base** is a proton acceptor (a more narrow definition was given by Arrehenius, where an acid is a proton donor and a base is a hydroxide donor). More precisely, a Brønsted acid is a hydron donor. A **hydron** is a general term for H^+, which does not imply an isotope; it includes proton, deuteron ($^2H^+$), and triton ($^3H^+$). Proton and hydron should not be confused with **protium**, which is the specific isotope for a hydrogen atom ^1H. Since most chemists realize these distinctions, but do not use these terms in everyday discussions, we will consistently just use the term proton to imply all the isotopes involved in natural abundance. Thus, an acid and a base under the Brønsted definition are substances that react as HA and B, respectively, in Eq. 5.2. We have arbitrarily chosen the acid and base to be neutral. However, each can have any charge: neutral, positive, or negative. Yet, independent of their initial charges, after proton transfer, the acid gains one unit of negative charge and the base gains one unit of positive charge.

$$\text{HA} + \text{B} \rightleftharpoons \text{A}^{\ominus} + \text{BH}^{\oplus}$$

$$\begin{array}{cccc} \text{Acid} & \text{Base} & \text{Conjugate base} & \text{Conjugate acid} \end{array}$$

(Eq. 5.2)

In the reverse reaction, the A and B entities change roles (Eq. 5.2). Now A^- is the base and BH^+ is the acid. To keep track of the side of the reaction we are referring to, we designate the substances in the reverse reaction as the **conjugate acid** and the **conjugate base**. The side of the equilibrium that is preferred depends upon the relative acid–base strength of all species. Hence, we need to be proficient at predicting the relative acid–base strengths of common chemicals to rapidly predict the outcome of such equilibria.

Inherently, the acid must have a labile bond to a proton, and the base must have two electrons that can accept the proton. In principle, any compound with an H can be an acid. Typically, the base has a lone pair of electrons, but sometimes even bonds (π or σ) can accept the proton. Therefore, in principle, any compound can be a base.

Since in principle any compound is a base, compounds we normally view as acids can also act as bases. For example, sulfuric acid can protonate acetic acid; now acetic acid is a base (Eq. 5.3). Further, compounds we normally consider as bases can be acids. Diphenylacetate can be deprotonated by *n*-butyllithium; now diphenylacetate is an acid (Eq. 5.4). In both of these examples, we needed an acid or base that is stronger than the acid or base we were considering in order to invert the expected reactivity. This further emphasizes that relative acid–base strength is a key factor we need to become proficient at predicting.

(Eq. 5.3)

(Eq. 5.4)

The solvent is not considered in Eq. 5.2, and indeed acid–base reactions can be performed in many different solvents or in the gas phase. When in solution, all four species in Eq. 5.2 will be solvated, and solvation is one of the major factors controlling which side of the equilibrium is preferred. We do not show H^+ as a free entity, because it does not exist to any appreciable extent when in a solution. It is always attached to an A, B, or the solvent (S).

Many solvents are capable of picking up a proton. For example, putting an acid into pyridine creates pyridinium, or an acid in an alcohol ROH creates ROH_2^+. The protonated form of the solvent is referred to as the **lyonium ion** of that solvent. The most well known example of this is hydronium ion (H_3O^+), the protonated form of water. The conjugate base of any solvent is called the **lyate ion**. For water this is hydroxide, and for DMSO this is $CH_3SOCH_2^-$. The lyonium and lyate ions are also solvated. In fact, hydronium ion is thought to exist in water with several waters strongly associated, and hence a formula such as $H^+(H_2O)_n$ is most appropriate. However, we will always simply write H_3O^+. Overall, the ability of a solvent to accept a proton is influenced both by its own electronic ability to accept the proton, but also by how well it solvates the lyonium ion.

5.2 Aqueous Solutions

If an acid is added to water, Eq. 5.5 describes the reaction, because the base in solution is water. Further, if a base is added to water, Eq. 5.6 describes the reaction, because now water is the acid. These acid–base reactions are critical to life itself, since nature's solvent is water. Having a good understanding of the thermodynamics of these reactions is not only important for understanding organic reactions in water, but is of the upper most importance in understanding biochemical reactions, almost all of which have acid–base dependencies. The factors that control the thermodynamics of acid–base reactions are the strengths of the acids or bases and the pH of the solution, so these measurements of acidity need to be examined in detail.

$$HA + H_2O \rightleftharpoons A^{\ominus} + H_3O^{\oplus} \tag{Eq. 5.5}$$

$$B + H_2O \rightleftharpoons BH^{\oplus} + HO^{\ominus} \tag{Eq. 5.6}$$

5.2.1   pK_a

The thermodynamics of Eq. 5.5 can be expressed in the typical equilibrium expression, Eq. 5.7, with the key quantity being K_a. Water is missing in the expression for K_a because it is the solvent and is in large excess (essentially 55.5 M in dilute acid solutions). Its concentration does not change significantly during the reaction, and therefore it is incorporated into K_a. It is important to remember this, though, when using K_a values to determine equilibrium concentrations of species. As with any K_{eq}, large values indicate that the products are more stable, and values less than 1 indicate that the reactants are more stable. Here, we are discussing **thermodynamic acidities**, where the relative free energies ($\Delta G_{rxn}°$) of the reactants and products determine the K_a values. This is in contrast to **kinetic acidities**, where the relative rates of deprotonation of various acids are used to determine the acid strengths. We will examine rates of proton transfer reactions in Chapter 9.

$$K_{eq} = \frac{[H_3O^+][A^-]}{[HA][H_2O]} \quad \text{or}$$

$$K_{eq} = \frac{[H_3O^+][A^-]}{[HA][55.5]} \quad \text{or}$$

$$[55.5]K_{eq} = \frac{[H_3O^+][A^-]}{[HA]}$$

$$K_a = \frac{[H_3O^+][A^-]}{[HA]}$$

(Eq. 5.7)

Since the K_a values reflect the relative stabilities of the species on the different sides of Eq. 5.5, we can use the K_a values to draw conclusions about the acid strength of HA relative to the strength of H_3O^+. For example, we can first conclude that for K_a values larger than 1, HA is a stronger acid than H_3O^+. This is because HA gives up its proton more efficiently than H_3O^+ does. Second, we can conclude that A^- is a weaker base than H_2O, because a K_a greater than 1 means that the proton prefers to be on water rather than on A^-. These two correlated conclusions about acid and base strengths on the different sides of the equilibrium are always coupled. The exact opposite conclusions are reached for K_a values less than 1; HA is a weaker acid than H_3O^+ and A^- is a stronger base than H_2O.

As we will see, K_a values range from very large (10^{12}) to extremely small (10^{-50}). Hence, it is not particularly convenient to write these numbers with a ridiculously large number of zeros before or after the decimal point. Hence, an acidity scale called **pK_a** is defined, which makes for a convenient way to tabulate the K_a values. The pK_a is given by Eq. 5.8. The negative sign in this formula often leads to confusion. The ramification of this sign is that strong acids have small or negative pK_as, whereas the weaker the acid the larger its pK_a. However, because this is a $\log_{10}$ scale, it makes for a particularly convenient way to compare acidities. All the tables of acid strength in solution given herein will include pK_a values.

$$pK_a = -\log_{10}(K_a) \qquad \text{(Eq. 5.8)}$$

Just as we expressed Eq. 5.5 in an equilibrium expression (Eq. 5.7), we can express Eq. 5.6 similarly (Eq. 5.9). The K_b value gives insight into whether the base is a stronger base than hydroxide ion. K_b values greater than 1 tell us that the added base is indeed stronger than hydroxide ion, while values less than 1 mean that the base is weaker.

$$K_b = \frac{[HO^-][BH^+]}{[B]} \qquad \text{(Eq. 5.9)}$$

The K_a of an acid and the K_b of its conjugate base are related to one another via Eq. 5.10, where K_w is the heterolytic dissociation constant of water ($10^{-14}\,M^2 = K_w = [H_3O^+][OH^-]$). Because of this relationship, the pK_a values for acids are also indicative of the base strengths of their conjugate bases. Hence, we do not tabulate either K_b or pK_b values, but instead only pK_as. As an example, an acid whose pK_a is 1 unit larger than another acid means that its conjugate base is 10 times stronger than the other acid's conjugate base. Hence, larger pK_a values directly correlate with larger basicities of the conjugate bases. This is a convenient concept to remember.

$$K_w = K_a K_b \qquad \text{(Eq. 5.10)}$$

Since the pK_a scale tells us the relative strengths of acids and bases, we can use it to predict the outcome of mixing two or more acids and bases, such as which side of Eq. 5.2 will be preferred. When the pK_a of HA is larger than that of HB^+, this means that BH^+ is the stronger acid. Hence, HB^+ is more reactive and the equilibrium shifts away from the more reactive acid, thereby preferring the reactants. Conversely, when the pK_a of HA is smaller than that of HB^+, now HA is the stronger acid, and the reaction prefers products. The guiding principle is that *the reaction will always prefer the side of the equilibrium that possesses the weaker acid*. This trend is also readily explained by considering base strength. Eq. 5.2 represents a competition between B and A^- for the proton. Whatever base is strongest will win the competition for the proton, thereby pulling the equilibrium towards its protonated form.

We do not want to rely on our own memory of pK_as, nor do we want to always go find pK_a values in books to make our predictions about reactions such as Eqs. 5.5 and 5.6. Instead, our chemical intuition should be good enough such that the structures of HA and BH^+, or the structures of A and B^-, will lead us to our predictions. Developing such a predictive ability is a large fraction of what this chapter addresses. However, we should also be able to predict, without resorting to calculations, what the protonation state of an acid or base will be when dissolved in water. This is determined by examining the pH and the pK_a.

5.2.2   pH

The positions of acid–base equilibria in water are not only controlled by the relative acid–base strengths, but also by the pH. The pH of a solution is given by Eq. 5.11. It tells us the concentration of H_3O^+ in solution. Since it is a $\log_{10}$ scale, a single unit pH difference reflects a 10-fold difference in H_3O^+ concentration. Further, once again, the negative sign can be confusing. More acidic solutions have a lower pH, whereas basic solutions have a high pH.

$$pH = -\log_{10}[H_3O^+] \qquad \text{(Eq. 5.11)}$$

Unlike the intrinsic strength of an acid (pK_a), which we can only manipulate by changing the acid, the pH of a solution is a quantity that we can experimentally adjust and manipulate. Changing the pH actually changes the extent of protonation of the acids and bases in the solution (the **protonation state**). To see this, it is convenient to take the $\log_{10}$ of both sides of Eq. 5.7 to acheive Eq. 5.12, known as the **Henderson–Hasselbalch equation**.

$$pH = pK_a + \log_{10}\left(\frac{[A^-]}{[HA]}\right) \qquad \text{(Eq. 5.12)}$$

If we take an HA with a pK_a of 4.0 off the shelf, make a dilute solution in water (see our discussion of "activity" below to understand why the solution must be dilute), and adjust the pH of the solution to 3, Eq. 5.12 tells us that the $[A^-]/[HA]$ ratio is $1/10$. This means there is 10 times more HA in solution than A^-. As another example, if the pH is adjusted to 7, the $[A^-]/[HA]$ ratio is $1000/1$. Now A^- has a 1000-fold preference over HA. Lastly, if the pH equals the pK_a, then $[A^-] = [HA]$.

In these examples we control the extent of protonation of A^- by physically adjusting the pH. The important point is that the pH is adjusted relative to the pK_a of the conjugate acid of the base A^- to change the protonation state. The pH therefore tells us the **proton donor ability** of the solution toward species A^-. In other words, it tells us the power of the solution to donate a proton to a particular base. In the next section we introduce a similar concept, acidity functions, which give the proton donor ability of extremely concentrated acid solutions.

Using the insight that the pH gives the proton donor ability of the solution, and the pK_a gives the proton donor ability of an acid, we are ready to make predictions as to the protonation states of acids based solely on their pK_as and the solution's pH. The following three rules apply:

1. *When the pH is the same as the acid's pK_a, the solution has the same ability to protonate the conjugate base A^- as the acid HA has the ability to protonate the solvent, and hence the acid exists as a 1:1 ratio of its HA and A^- forms.*

2. *When the pH is above the pK_a, the solution does not have enough donor ability to protonate the conjugate base, and therefore the acid exists mostly in its conjugate base form A^-.*

3. *When the pH is below the pK_a, the donor ability of the solution is strong enough to protonate A^-, and mostly HA exists.*

Therefore, all that is needed to predict protonation state is knowledge of the acid's pK_a and the pH of the solution. This is an extremely important insight to remember when looking at any acid–base chemistry that occurs in solution. These predictions are also easy to remember if we recall the general shape of a pH titration curve for a weak acid. Figure 5.1 shows experimental points for the titration of an acid, and a theoretical fit for that acid using a pK_a of 8.35. The plot was generated by using Eq. 5.12 (for a more realistic titration curve, see the Connections highlight on page 265). The units on the x axis are relatively arbitrary, since they represent the concentration of added base, and their values depend upon the concentration of the acid added at the start. However, the pH scale (y axis) is not arbitrary. The inflection point in the curve along the y axis is the point at which the measured pH is the pK_a of the acid. At concentrations of added base below the inflection point (low pH relative to the

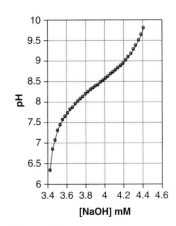

Figure 5.1
The pH titration curve of 3-nitrophenol with a pH range near the pK_a.

pK_a), the acid is mostly protonated, but at higher concentrations of added base the acid converts predominately to its conjugate base form (high pH relative to the pK_a), just as was discussed above.

The change in protonation state of an acid that occurs around its pK_a is exploited in the use of **pH indicators**, molecules that change color as a function of their protonation state. One drop of a pH indicator to a solution is often enough to visually signal a change in the pH of that solution during a reaction or titration. In this case the indicator is either protonated or deprotonated during the pH change, and the color changes. Such indicators are used in a wide variety of applications, and the next Connections highlight shows how they can be used in sensing applications.

This discussion of controlling the pH relative to a pK_a may sound familiar. In the last chapter, we examined the saturation of a host with a guest. If the host concentration is near K_d of the H•G complex and the guest is in excess, we say that the host is being saturated with guest. This is identical to the discussion of acids and bases given here. Since the pH controls the amount of hydronium ion, if it is lower than the pK_a of the acid, then the conjugate base is "saturated" with hydronium ion, meaning that the acid form dominates the equilibrium. This is true as long as the concentration of the conjugate base in solution is at or higher than the K_a for the acid.

Connections

Using a pH Indicator to Sense Species Other Than the Hydronium Ion

One very active subfield of molecular recognition is the creation of sensors using synthetic receptors. In Chapter 3 we discussed the use of cyclodextrins as entities for the binding of hydrophobic organic molecules. As we explore with multiple examples in this chapter, this being only the first, pK_a values are quite sensitive to solvation and charges near the acidic group. The attachment of the pH-sensitive indicator alizarin yellow to cyclodextrin leads to a sensor for various hydrophobic groups, because the microenvironment near the phenol changes upon binding various compounds to the cyclodextrin. In other words, the pK_a of the phenol is different for the free host and a host–guest complex. This leads to a differing extent of protonation of the indicator for the host and host–guest complex, and in turn a different color. In particular, the binding of 1-adamantanol to the cyclodextrin host shown to the right leads to a yellow-to-red transition, allowing one to quantify 1-adamantanol if desired. Hence, the

attachment of pH indicators to synthetic receptors allows one to exploit the response of the indicators to hydronium ion, while coupling that response to the addition of other analytes.

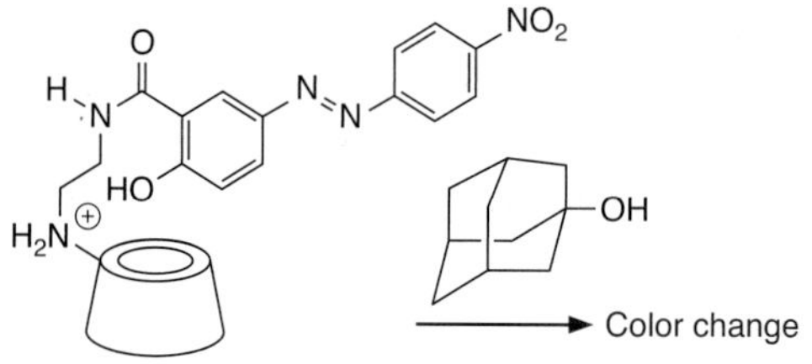

Phenol pK_a shift leads to color change

Aoyagi, T., Nakamura, A., Ikeda, H., Ikeda, T., Mihara, H., and Ueno, A. "Alizarin Yellow-Modified β-Cyclodextrin as a Guest-Responsive Absorption Change Sensor." *Anal. Chem.*, **69**, 659–663 (1997). Wiskur, S. L., Ait-Haddou, H., Lavigne, J. J., and Anslyn, E. V. "Teaching Old Indicators New Tricks." *Acc. Chem. Res.*, **34**, 963 (2001).

5.2.3 The Leveling Effect

In the discussion of K_a values given above, it was noted that conclusions can be drawn about the relative acid strength of acid HA and the acid H_3O^+ by examining the K_a (or the pK_a) of HA. For those acids where the K_a was greater than 1, the conclusion was that HA is a stronger acid than H_3O^+. We can also compare the relative pK_as. The pK_a of H_3O^+ is –1.74. Any acid whose pK_a is less than –1.74 must therefore be a stronger acid, and the equilibrium in Eq. 5.5 favors products. This logic leads us to the conclusion that H_3O^+ must be the strongest acid that can exist to any appreciable extent in dilute aqueous solutions (we will see below what happens in concentrated solutions of acids). The reasoning is as follows. All acids with pK_a values smaller that –1.74 create H_3O^+ because the equilibrium of Eq. 5.5 is shifted toward products. For example, HCl is a stronger acid than H_3O^+ because it has a pK_a of around −6. However, after adding HCl to water, it is essentially fully dissociated, and the real acid in such a solution is simply H_3O^+.

A similar situation holds for base strength. For a dilute solution, no base stronger than HO^- can exist to any appreciable extent in water. Adding a base whose conjugate acid has a pK_a above 15.7 will cause the products in Eq. 5.6 to dominate. For example, if one makes up a solution of the base $Na^+NH_2^-$ (the pK_a of NH_3 is 35) in water, the NH_2^- will completely deprotonate water, and therefore the real base in such a solution is simply HO^-. These phenomena have been discussed in terms of water, but are general to any solvent, leading to what is known as the **leveling effect**.

The leveling effect states that:

- An acid stronger than the conjugate acid of the solvent cannot exist in any appreciable concentration in that solvent, and

- A base stronger than the conjugate base of the solvent cannot exist in any appreciable concentration in that solvent.

The leveling effect creates a limitation on the strengths of acids and bases that can be determined in particular solvents:

- The pK_as of acids stronger than the conjugate acid of the solvent cannot be measured in that solvent, and

- The pK_as of acids whose conjugate bases are stronger than the conjugate base of the solvent cannot be measured in that solvent.

Given the leveling effect, the range of pK_a values that can be determined in water roughly spans −1.74 to 15.7. Moreover, in practice, those acids whose pK_as approach −1.74 and 15.7 become very difficult to measure. Throughout this chapter, however, pK_a values outside of the range of −1.74 and 15.7 are given. There is no single solvent that all these pK_as are determined in, but relative acidities of the acids can be determined in various solvents and scaled to a consistent set of values for a single solvent. Water is taken as the standard solvent for setting up the acidity scales for all compounds. Therefore, with the exception of our discussion of nonaqueous systems below, we can assume that the pK_as we give are relative to an aqueous solution.

Connections

Realistic Titrations in Water

The titration curve given in Figure 5.1 may not look familiar to you. The shape you may recall from an introductory chemistry or quantitative analysis course is given to the right. There are three parts that are near level. The one in the center is indicative of titrating the acid added to solution, such as the 3-nitrophenol of Figure 5.1. The other two level regions are indicative of the leveling effect. When one plots a large range of pH values, the graph is near level at low and high pH. The addition of a base to the low pH solutions leads to titration of hydronium ion, the strongest possible acid in water. The addition of a base to solutions at high pH leads to titration of water to create hydroxide ion, the strongest possible base in the medium. Hence, on the extremes of any titration curve the leveling effect takes over.

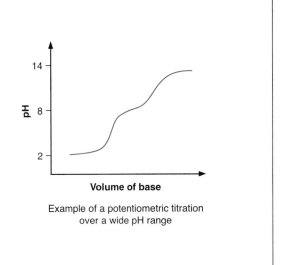

Example of a potentiometric titration over a wide pH range

5.2.4   Activity vs. Concentration

The discussion above focused upon the use of concentrations in the equilibrium expressions. Experimentally, the typical manner in which these concentrations are determined is related to measuring the amount of acid or base added to the solution by weight or by use of a syringe. The concentration of H_3O^+ is most often measured using a pH meter. The concentrations of HA and/or B⁻ are assumed to be directly related to the amount of each entity added to the solution. For dilute solutions, this is acceptable. However, for increasingly concentrated solutions these methods have increasing error, leading to what is known as **nonideal behavior**. In dilute solutions the acid, base, hydronium ion, and/or hydroxide ion all are well separated, and association of ion pairs is minimal. However, as solutions become concentrated, aggregation of these entities can occur, and ion pairing becomes significant. The concentrations of individual solutes surrounded only by solvent begins to drop because the solutes become involved in solute–solute interactions (recall our discussion of ion pairing in Chapter 3). Therefore, the concentration of free entities that you might expect from your laboratory measurements may not actually be what exists in the solution. For reactions involving ions, such as acid–base reactions, water is quite good at mediating charges, and hence these aggregation effects are not as severe as with other solvents. Yet, these effects need to be considered when working in concentrated solutions in water.

Hence, strictly speaking, the equilibrium expressions in Eqs. 5.7 and 5.9 should be expressed in terms of the activities (a) of the entities, not their concentrations as given. Likewise, the definition of pH in Eq. 5.11 should really include the activity of H_3O^+, not its concentration, because activity is what the pH meter really measures.

Recall that activity was used in our development of the thermodynamic driving force for a reaction in Section 3.1.5. It is the number of free solute particles in a solution that affects the entropy of that solution, and hence the activity has to be used in thermodynamic relationships.

The activity is related to the concentration (moles/liter) of the compound (X) by Eq. 5.13, where γ is the **activity coefficient**. Gamma is always less than one, and is an empirically determined factor that mediates the concentration of the compound to reflect the amount of compound free in the solution. The activity coefficient of ions can be estimated from **Debye–Hückel theory**. Consult any quantitative analysis textbook to get a thorough discussion of activities and activity coefficients. It is particularly important to use activities in place of concentrations for concentrated solutions in water, and for nonaqueous solvents, which are the next two topics.

$$a_x = \gamma_x \frac{[X]}{[X]_0}$$

(Eq. 5.13)

5.2.5   Acidity Functions: Acidity Scales for Highly Concentrated Acidic Solutions

For many organic reactions catalyzed by acids, such as alcohol eliminations, alkene hydrations, acetal hydrolysis, and ester hydrolysis, one needs to add high concentrations of very strong acids to water. Mineral acids such as sulfuric acid, hydrochloric acid, and phosphoric acid are typically used, often in concentrations ranging from 1 M to nearly neat solutions. These high concentrations of strong acids are required in order to create significant amounts of the protonated forms of the organic substances undergoing reaction. The nature of the solvent changes significantly in these strongly acidic solutions, such that we are no longer dealing with normal aqueous solutions. The mineral acids can be non-dissociated, aggregated with themselves, and aggregated with hydronium ion. Activities are definitely required in the equilibrium expressions.

The notion of pH given in Eq. 5.11 is not a sufficient measurement of the acidity of such solutions, even when activities are used instead of concentrations. Remember the leveling effect, which limits the acid strength one can achieve in a dilute solution. Therefore, now, new acidity scales are needed, giving a measurement of the effective ability of the concentrated solution to donate protons to an organic compound, just as the pH is the ability of a dilute acid solution to donate protons. These scales are known as **acidity functions**. Several acidity functions have been established to measure the proton donating abilities of highly concentrated acid solutions, all of which are determined by examining the extent of protonation of a series of weaker and weaker bases in more and more acidic solutions. Here, only one acidity function is analyzed.

The most commonly used acidity function is based upon the extent of protonation of a series of anilines with electron withdrawing groups (EWG) attached (Eq. 5.14). It is called the **Hammett acidity function**. Other acidity functions have been developed based upon nitroanilines, indoles, and amides by several individuals, such as Arnett, Cox, Katritzky, Yates, and Stevens. Since these functions are not as common as the Hammett function, they are not discussed here, but other physical organic textbooks referenced at the end of this chapter have good descriptions.

With the Hammett function, as more and more electron withdrawing groups are attached to an aniline, the basicity of the NH_2 group continues to drop. Anilines were chosen because UV/vis spectroscopy allows one to conveniently measure the extent of protonation in the different solutions of acids.

(Eq. 5.14)

The first aniline (B_1) that is examined has a basicity such that its extent of protonation can be measured in a normal dilute aqueous acidic solution. Eq. 5.15 gives the equilibrium expression for its conjugate acid, where $K_{a\text{-}B_1H^+}$ and the activity coefficients γ_{B_1} and $\gamma_{B_1H^+}$ can be determined under these normal conditions. $K_{a\text{-}B_1H^+}$ is a constant reflecting the acidity of the conjugate acid of B_1, and does not change in different concentrations of acids. However, the activity coefficients will change.

$$K_{a\text{-}B_1H^+} = \frac{a_{H_3O^+}[B_1]\gamma_{B_1}}{[B_1H^+]\gamma_{B_1H^+}}$$

(Eq. 5.15)

The second aniline (B_2) is too weak of a base to be protonated under normal dilute acid solutions in water. In a several percent solution of the strong acid (HB_1^+), however, it can be protonated. Importantly, B_1 and B_2 were chosen such that significant concentrations of both the protonated and deprotonated forms of both anilines will be present in this more acidic solution. Dividing Eq. 5.15 by an identical expression for B_2 leads to Eq. 5.16. The activity of H_3O^+ drops out because the measurements of B_1 and B_2 are made in the same acidic solution.

$$\frac{K_{a-B_1H^+}}{K_{a-B_2H^+}} = \frac{[B_1][B_2H^+]\gamma_{B_1}\gamma_{B_2H^+}}{[B_1H^+][B_2]\gamma_{B_1H^+}\gamma_{B_2}}$$

(Eq. 5.16)

Since B_1 and B_2 were chosen so that all the relevant concentrations can be determined by UV/vis spectroscopy, the unknowns in Eq. 5.16 are $K_{a\text{-}B_2}$ and the activity coefficients. A reasonable assumption is that the activity coefficients for the B_2 entities are the same as for the B_1 entities in the new solution, because aggregation and non-ideal behavior are related to molecular structure, and both B_1 and B_2 are anilines. Given this assumption, Eq. 5.16 reduces to Eq. 5.17. Now $K_{a\text{-}B_2H^+}$ can be determined. The next step is to choose an even weaker aniline base, B_3, which requires a higher percent acid for appreciable protonation, but for which the protonation state of B_3 and B_2 can both be measured in the same acidic solution. This allows calculation of $K_{a\text{-}B_3H^+}$, and so on.

$$\frac{K_{a-B_1H^+}}{K_{a-B_2H^+}} = \frac{[B_1][B_2H^+]}{[B_1H^+][B_2]} \qquad \text{(Eq. 5.17)}$$

After determining the K_a values for the series of conjugate acids of aniline bases, we are ready to establish the acidity function. Rearrangement of Eq. 5.15 gives 5.18, where h_o is defined as $a_{H_3O^+}\gamma_B / \gamma_{BH^+}$. Taking the $\log_{10}$ of both sides gives Eq. 5.19, similar to the Hendersen–Hasselbalch equation (Eq. 5.12), where the acidity function is defined as $H_o = -\log_{10}h_o$.

$$K_a\left(\frac{[BH^+]}{[B]}\right) = a_{H_3O^+}\left(\frac{\gamma_B}{\gamma_{BH^+}}\right) = h_0 \qquad \text{(Eq. 5.18)}$$

$$H_0 = pK_a + \log_{10}\left(\frac{[B]}{[BH^+]}\right) \qquad \text{(Eq. 5.19)}$$

Note that as the concentration of the strong acid approaches zero, the solutions become more like normal water, and the activity coefficients for B and BH$^+$ will approach 1. When the acid is sufficiently dilute, H_o will become pH. Hence, H_o can be viewed as a pH surrogate for very very acidic solutions, reflecting "pH" values as low as –10. Remember that for a dilute solution of acid, we could never achieve a pH of –10 in water (due to the leveling effect). For an example of the use of acidity scales in an industrially important reaction, see the next Connections highlight.

Since H_o is our pH surrogate, we need to know how to measure H_o for any acid solution that we might prepare in order to perform a reaction. pH is simply measured using a pH meter. However, when checking the ability of a concentrated acid solution to donate a proton (H_o), one can make up a solution of the acid in water, find the proper aniline base that is a mixture of protonated and deprotonated forms in that solution, and put those concentrations (determined via UV/vis spectroscopy) along with the pK_a into Eq. 5.19. Fortunately, for most acids, scales have already been developed, and hence we can just refer to graphs such as Figure 5.2. This graph shows H_o for several different mole fractions of strong acids in water. The acid that makes solutions with the largest proton donating ability is HClO$_4$. HCl and H$_2$SO$_4$ are similar, and HF becomes a very strong acid when neat. The organic acids HCO$_2$H and CF$_3$CO$_2$H both achieve H_o values near –2 when neat, making for very acidic solutions.

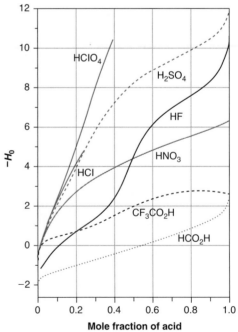

Figure 5.2
H_o values for mixtures of several different acids in water as a function of their mole fractions. The data come from Cox, R. A., and Yates, K. "Acidity Functions: An Update." *Can. J. Chem.*, **61**, 2225 (1983).

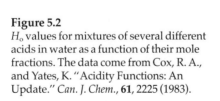

Connections

An Extremely Acidic Medium is Formed During Photo-Initiated Cationic Polymerization in Photolithography

In Chapter 17 we will describe the photolithography process, in which 0.25 μm or smaller features in silicon chips can be created. The process involves a photo-initiated cationic polymerization, and/or cross-linking reaction. In some cases, photo-initiated deprotection of phenol groups is used to change the polarity of certain regions of the chip. In all of these applications, the transformations are initiated by very strong, photo-generated acids.

As we explore later in this chapter, one way to produce an exceptionally strong protic acid is to have a non-aqueous environment and a very non-coordinating conjugate base, such as BF_4^-, PF_6^-, and AsF_6^-. In fact, one reason that computer chips are manufactured in clean rooms is to remove all sources of bases that quench the acid-catalyzed processes. Water is one such base, but so are the various amine bases that we have in our breath and fingerprints.

The photo-initiated production of catalytic amounts of HBF_4, HPF_6, or $HAsF_6$ most commonly derives from irradiation of diaryliodonium salts, where the counter ion MtX_n^- is any one of the special non-coordinating anions. Irradiation produces mixtures of cation radicals, cations, and radicals of the aryl groups, which are extremely reactive species. These compounds oxidize, abstract hydrogen atoms, or abstract hydrides from most monomers or the polymer backbone (RH). The resulting cationic monomers or polymers lose a proton to form unsaturated positions, yielding the $HMtX_n$ acid that initiates polymerization, cross-linking, or deprotection.

$$Ar-\overset{\oplus}{I}-Ar \ MtX_n^{\ominus} \xrightarrow{h\nu} Ar-\overset{\cdot\oplus}{I} \ MtX_n^{\ominus} + Ar\cdot \xrightarrow{RH} HMtX_n$$
$$\text{or } ArI + Ar^{\oplus} \ MtX_n^{\ominus}$$

Studies on the experimental conditions created by the photo-produced acidic media reveal H_o values from –15 all the way to –30. Thus, the modern day process of computer chip manufacturing relies heavily on the production of extremely strong acids, which generate acidities that can only be measured using acidity function scales.

Crivello, J. V. "The Discovery and Development of Onium Salt Cationic Photoinitiators." *J. Poly. Sci. Part A, Poly. Chem.*, **37**, 4241–4254 (1999).

The use of acidity functions to determine the proton donating ability of a solvent is useful for the analysis of acid-catalyzed reactions. Many reactions, however, are base-catalyzed, and strongly basic solutions can be created by adding strong bases to water. Therefore, acidity function methods have also been developed for basic solutions, establishing scales designated for the ability of the solution to remove a proton from a reactant. One such scale, H_-, is given by Eq. 5.20, and correlates mixtures of DMSO and water with added tetramethylammonium hydroxide. Again the A^- and AH are aniline derivatives that can be analyzed in the UV/vis spectra, but now the neutral forms are acting as acids. As shown in Figure 5.3, the solutions are made more basic by the addition of increasing percentages of DMSO. In almost

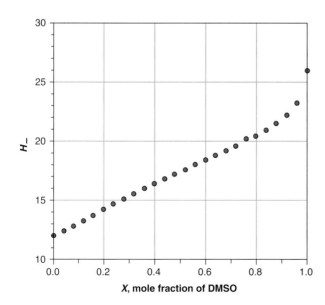

Figure 5.3
A plot of H_- values for solutions of 0.011 M $(CH_3)_4NOH$ in water/DMSO solutions as a function of the mole fraction of DMSO. The data come from Cox, R. A., and Stewart, R. "The Ionization of Feeble Organic Acids in DMSO-Water Mixtures. Acidity Constants Derived by Extrapolation to the Aqueous State." *J. Am. Chem. Soc.*, **98**, 488 (1976).

pure DMSO, the ability of the solution to remove a proton increases dramatically. Under these conditions the hydroxide ion is not solvated by DMSO (see the discussion in Chapter 3 of DMSO solvating anions) and the $(CH_3)_4N^+$ counter ion does not coordinate. Therefore the base is becoming essentially naked. Once again the acidity function is a pH surrogate, and solutions approaching a basicity value of 26 can be achieved.

$$H_- = pK_w + \log_{10}[OH^-] - \log_{10}(a_{H_2O}) - \log_{10}\left(\frac{\gamma_{A^-}}{a_{HA}\,\gamma_{OH^-}}\right) = pK_a + \log_{10}\left(\frac{[A^-]}{[HA]}\right)$$

(Eq. 5.20)

5.2.6   Super Acids

The extreme of strongly acidic solutions are those formed from what are called **super acids**. Our Connections highlight on the acids used in photolithography (see above) introduced these acids $(HMtX_n)$. Some of the strongest proton donating solutions known are those formed from BF_3, PF_5, AsF_5, and especially SbF_5 in liquid HF, often diluted with FSO_3H and SO_2ClF. This creates acids of the extremely non-nucleophilic and non-coordinating anions BF_4^-, PF_6^-, AsF_6^-, and SbF_6^-. Such solutions can protonate bases as weak as benzene to create persistent carbenium ions. This means that benzene is actually a stronger base than these anions! Even more impressive, they can be used to protonate alkanes, as the following Connections highlight shows.

Connections

Super Acids Used to Activate Hydrocarbons

If acidity functions of –20 to –30 are accessible with super acids, can one protonate alkanes? The answer is "yes", and in fact, the 1994 Nobel Prize in Chemistry presented to George Olah was partially in acknowledgement of this discovery. The realization of σ bond basicity led to the ability to activate hydrocarbons at low temperature toward various reactions.

One recent example of this is the carbonylation of methylcyclopentane in HF–SbF$_5$ solutions. An array of fascinating reactions occurs upon protonation of an alkane, and they are illustrated in the following scheme. The relative basicity of σ bonds in alkanes is 3° C–H > C–C > 2° C–H >> 1° C–H. Hence, the protonation of methylcyclo-

pentane leads to the two cations shown. Protonation of a C–H bond leads to loss of H$_2$ and formation of a carbenium ion (which undergoes rearrangements not shown), whereas protonation of a C–C bond leads to a linear carbenium ion after a rearrangement (along with other isomers not shown). With added CO, trapping of the carbenium ions occurs to make persistent acylium ions, which undergo a reaction with ethanol upon workup to make ethyl esters. This is just one example of an amazing reaction—the activation of a hydrocarbon with an acid.

Sommer, J., Hashoumy, M., Culmann, J.-C., and Bukala, J. "Activation and Carbonylation of Methylcyclopentane in HF–SbF$_5$." *New J. Chem.*, **21**, 939–944 (1997).

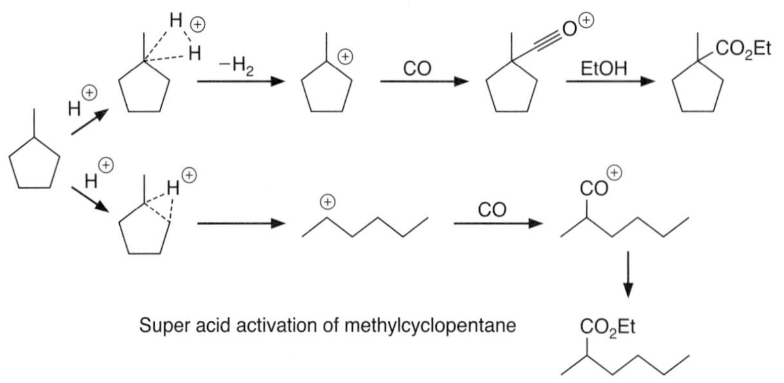

Super acid activation of methylcyclopentane

The previous sections have analyzed the scales used to tabulate the thermodynamics of acid–base reactions, the leveling effect, and how one determines the acidity of dilute and highly concentrated solutions. Much of what was discussed is applicable to any solvent, but water is the solvent most often referred to when one is discussing acid–base chemistry. Yet, chemists do not perform all of their acid–base reactions in water, and so an understanding of acid strengths in nonaqueous systems is very important also. Further, the interiors of enzymes and proteins are quite organic in nature, and therefore correlations between organic solvents and the microenvironments created by nature for catalyzing reactions are often made. Hence, we also want to explore acid–base chemistry in nonaqueous media.

5.3   Nonaqueous Systems

We noted earlier that all standard pK_a values are relative to those that can be determined in water. Most of the time pK_a values change significantly when measured in different solvents, although there are examples where acidities are nearly identical in different solvents. For example, picric acid (2,4,6-trinitrophenol) is just as strong an acid in DMSO as it is in water (see discussion below). Moreover, the relative ordering of acid strength can change from one solvent to another. For instance, in water HCN is a stronger acid than malononitrile $[CH_2(CN)_2]$, but in DMSO malononitrile is the stronger acid. Thus, solvation has a large influence in altering the intrinsic ability of a compound to act as a proton donor.

There are two general changes that occur for pK_a values in organic solvents relative to water. First, the pK_a values are almost always larger in the organic solvent than in water. Second, pK_a differences are accentuated in the organic solvent, meaning that they tend to spread out over a larger pK_a range. The reason pK_a values are larger is that organic solvents are not as effective at supporting the charges that develop upon creating the lyonium ion. Table 5.1 shows a few pK_a values in different organic solvents. Note that the lower the polarity of the solvent, the higher the pK_as of the various acids. The most common solvent, other than water, that pK_as have been measured in is DMSO, and Bordwell has developed an extensive listing.

Table 5.1
pK_as of Various Acids in Differing Solvents

Acid	Solvent				
	H_2O	CH_3OH	DMSO	DMF	CH_3CN
CH_3CO_2H	4.76	9.5	12.6	13.5	
p-$NO_2C_6H_4OH$	7.15	11.4	11.0	12.6	21
$PhNH_3^+$	4.6		3.2	4.2	

There are some severe limitiations to measuring pK_a values in organic solvents. Often salts are strongly ion paired so that the acidity of a cationic acid really reflects that of its salt. Similarly, for neutral acids, the resulting lyonium ion has to be ion paired with the conjugate base of the acid. These factors strongly affect the acid strength because the strength of an ion pair will influence the measured acidity. For example, when using an anionic base to deprotonate an acid, Eq. 5.21 describes the reaction. This kind of ion pairing often causes an acid–base titration to appear non-ideal. Figure 5.4 shows titrations of the sodium 15-crown-5 salts of 1,3-cyclohexanedionate and cyanonitromethide in acetonitrile, which clearly do not conform to the shape of the Henderson–Hasselbalch equation (Eq. 5.12 and Figure 5.1).

$$A_1H + A_2^{\ominus} M^{\oplus} \rightleftharpoons A_1^{\ominus} M^{\oplus} + A_2H \qquad \text{(Eq. 5.21)}$$

The second trend, that of accentuating the differences between acids, again derives from the fact that the organic solvents are not very effective at stabilizing the charges created

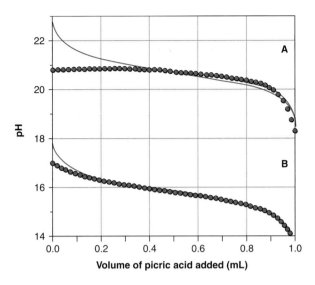

Figure 5.4
Titration of sodium 15-crown-5 salts of 1,3-cyclohexanedionate (**A**) and cyanonitromethide (**B**) with picric acid in acetonitrile. The line drawn through the data is a Henderson–Hasselbalch fit. The data come from Kelly-Rowley, A. M., Lynch, V. M., and Anslyn, E. V. "Molecular Recognition of Enolates of Active Methylene Compounds in Acetonitrile. The Interplay between Complementarity and Basicity, and the Use of Hydrogen Bonding to Lower Guest pK_as." *J. Am. Chem. Soc.*, **117**, 3438 (1995).

when the acid donates its proton to the solvent. Due to the relative lack of ion pairing, many pK_as of organic acids have been determined in DMSO, and hence this solvent makes for a nice comparison between acid strengths in an organic solvent. Examine the pK_as of the acids given in Table 5.2. The differences are generally greater in DMSO than in water. For example, the difference between the acidities of phenol and methanol is 5.5 pK_a units in water, but it is 11 in DMSO. Since the organic solvent cannot as effectively stabilize the anionic conjugate base of the acid, the intrinsic electronic factors that make one acid more acidic than another are accentuated. In essence, the trends in acidity in organic solvents and the differences between one acid and another become more similar to what is observed in the gas phase (see below). From another vantage point, the strong solvating nature of water very effectively stabilizes the anionic conjugate bases, thereby making the intrinsic stability of these bases less important, and the corresponding acidity differences between acids becomes smaller. When the anionic charge of the conjugate base is highly delocalized, solvation effects would be expected to be less important, and indeed with picric acid and malononitrile, the pK_as are essentially the same in water and DMSO.

Table 5.2
pK_a Values for Various Acids in Water and DMSO*

Acid	pK_a (water)	pK_a (DMSO)
HBr	−9	0.9
HCl	−8	1.8
HF	3.2	15
Picric acid	0.4	0.0
Acetic acid	4.75	12.3
Phenol	10.0	18.0
Methanol	15.5	29.0
Water	15.57	32
HCN	9.1	12.9
$CH_2(CN)_2$	11.0	11.0

*Bordwell, F. G. "Equilibrium Acidities in Dimethyl Sulfoxide Solution." *Acc. Chem. Res.*, **21**, 456 (1988).

5.3.1 pK_a Shifts at Enzyme Active Sites

Our analysis of nonaqueous pK_as makes it clear that the environment of an acid strongly influences its strength. As stated above, organic solvents generally lower acidity, whereas polar solvents increase acidity. Similarly, electrostatic interactions influence acidity. As we will explore later, the second pK_a of a dicarboxylic acid is higher than the first pK_a due to the formation of a dianion with the associated electrostatic repulsion. Hence, we must examine all the aspects of the microenvironment around an acid to fully understand or predict its strength.

The active sites of enzymes are not identical to water, and in fact are often considered quite organic in nature. However, they also can possess charges. Therefore, acids at enzyme active sites often have pK_as that are quite different than the normal pK_as for those acids in water. This is because the active site of an enzyme presents a significantly different solvation environment than just water. Table 5.3 lists a few enzymes and the pK_as of the side chains in solution and within the enzyme. For example, if a neutral acid, such as a carboxylic acid, is placed in proximity to a positive charge, the acid will become more acidic because the resulting anionic conjugate base is stabilized due to electrostatic attraction. Conversely, when a positive acid, such as ammonium, is placed near a negative charge, it will become less acidic. Now the electrostatic attraction is lost when the acid donates its proton and becomes neutral. These effects can be quite large, and changes in acidity strength of four to five orders of magnitude are common.

Table 5.3
pK_a Values of Side Chain Residues in Three Enzymes
Relative to the Solution pK_a Values*

Enzyme	Residue	pK_a in solution	pK_a in enzyme
Lysozyme	Glu-35	4.2	6.5
Acetoacetate decarboxylase	Lys-43	10.0	5.9
Papain	His-150	6.0	3.4

*Fersht, A. (1999). *Structure and Mechanism in Protein Science*, W. H. Freeman and Company, New York. Urry, D. W., Gowda, D. C., Peng, S. Q., Parker, T. M., and Harris, R. D. "Design at Nanometric Dimensions to Enhance Hydrophobicity-Induced pK_a Shifts." *J. Am. Chem. Soc.*, **114**, 8716 (1992).

5.3.2 Solution Phase vs. Gas Phase

Before launching into an analysis of pK_as of representative acids and a discussion of those factors that influence acidities, we examine **gas phase acidities**. In the gas phase, the intrinsic proton donor abilities between various acids can be determined without influence from solvents. Hence, an examination of gas phase acidities makes a nice introduction to determining relative solution acidities, where we can take the lessons from the gas phase and perturb them with knowledge of how the solvent influences acidities. For an example of how gas phase acidities have been used to interpret pK_a changes in solution, see the Connections highlight given at the end of this discussion.

How can we measure the acidity of an HA bond in the gas phase (it would require the heterolysis of the bond to create naked H^+ and A^-)? Such a reaction is quite unreasonable in the gas phase. For example, the heterolytic bond dissociation energy for methane is 417 kcal/mol (see Table 2.9), but the homolytic bond dissociation energy is only 105 kcal/mol (see Table 2.2). Hence, thermolysis would simply lead to homolysis. Moreover, most alkyl anions are not stable in the gas phase, possessing a negative or low positive ionization potential, and therefore spontaneously lead to a radical and an electron. Hence, in these cases, one must consider three different reactions. Adding together the bond dissociation energy of the A–H bond, the ionization potential of the hydrogen atom, and the electron affinity of the A radical (Eqs. 5.22–5.24) gives the acidity reaction (Eq. 5.25).

$$\text{A–H} \longrightarrow \text{A}\bullet + \text{H}\bullet \qquad \Delta H^\circ = \text{BDE(A–H)} \qquad \text{(Eq. 5.22)}$$

$$\text{H}\bullet \longrightarrow e^\ominus + \text{H}^\oplus \quad \Delta H^\circ = \text{IP(H)} \qquad \text{(Eq. 5.23)}$$

$$\text{A}\bullet + e^\ominus \longrightarrow \text{A}^\ominus \quad \Delta H^\circ = \text{EA(A)} \qquad \text{(Eq. 5.24)}$$

$$\text{A–H} \longrightarrow \text{A}^\ominus + \text{H}^\oplus \quad \Delta H^\circ(\text{HA}) = \text{BDE(A–H)} + \text{IP(H)} + \text{EA(A)} \qquad \text{(Eq. 5.25)}$$

We must measure the ΔH° of these three reactions to derive the acidity of HA in the gas phase. The acidity is reported as a ΔH° value, not a pK_a, because it does not reflect the donation of the proton to any solvent. Instead, the value is a heterolytic bond dissociation energy. In those cases where the base (anionic or neutral) is stable in the gas phase, one can determine the proton affinity of this conjugate base of the acid, which would be the negative of the heterolytic bond dissociation energy (Eq. 5.26). The most pertinent number for any reaction is ΔG° not ΔH°, which would require knowledge of ΔS°. ΔS° can often be estimated, but for most gas phase acidities the entropies of the reactions are all similar, such that relative ΔG° is barely different than relative ΔH°. Therefore, Table 5.4 lists gas phase acidities solely as ΔH° values.

$$\text{A}^\ominus + \text{H}^\oplus \longrightarrow \text{HA} \quad \text{PA(A}^\ominus) = -\Delta H^\circ (\text{HA}) \qquad \text{(Eq. 5.26)}$$

There are many points of interest in Table 5.4. First is the proton affinity of water in the gas phase, which is –167 kcal/mol. Interestingly, coordination of more waters continues to release energy. A second water releases 32 kcal/mol, and up to eight waters in the gas phase continues to release energy. This is part of the evidence for our earlier statement that H_3O^+ in water is better described as $H^+(H_2O)_n$. The increasing release of heat upon the addition of several water molecules to H_3O^+ has been interpreted to mean that the proton is equally bound to a few waters, rather than being most strongly associated with a single molecule of water.

One quite interesting trend is that the interchange of atoms with differing electronegativities does not necessarily yield the expected results. For example, examine the acidities of the various monohalogenated acetic acids. The most acidic is iodoacetic acid, and the least acidic is fluoroacetic acid. Similarly, of the substituted methanes, the most acidic is iodomethane and the least acidic is fluoromethane. This can be understood on the grounds that iodine is more polarizable than fluorine, being better able to accept and spread out the increase in electron density than the smaller harder fluorine in the absence of any solvent to mediate the charge. In fact, although surprising at first glance, chloroform is more acidic than fluoroform in water by a factor of 10^7 for the same reason. Although we will discuss electronegativity below as a major factor influencing acidity, we cannot always rely on simple electronegativity arguments to predict acidities.

Another interesting and important trend is the relative acidities of methanol, ethanol, isopropanol, and t-butyl alcohol. Acidity increases as the size of the alkyl group increases. Once again polarization is found to be important. The larger the alkyl group, the better it can accept the increase in electron density upon heterolysis of the O–H bond. We normally think of alkyl groups as electron donating, but whether they are electron donating or electron accepting depends upon the context. Interestingly, the exact opposite trend in acidity is found in solution, where methanol is the strongest acid. We examine the reasoning for this inversion in the order of acidities in the next section.

Finally, examine the trend for the decreasing acidities of ammonium species with increasing alkyl substitution, as revealed by studies from Brauman and Blair. Ammonium is more acidic than methyl ammonium, which in turn is more acidic than dimethylammonium, and so on. The charge is being stablized by dispersion with more alkyl groups.

Table 5.4
Selected Values of Gas Phase Acidities ($\Delta H°$) and Proton Affinities (PA) of Organic Compounds, in kcal/mol*

Compounds	$\Delta H°$	PA of conjugate base
Carbon acids		
Methane	416.6	
Ethane	420.1	
Ethylene	407.5	
Propene	390.8	−390.7
Benzene	400.7	
Fluoromethane	409	
Chloromethane	396	
Bromomethane	392	
Iodomethane	386	
Acetonitrile	373.5	
Acetone	370.0	−369.1
Toluene	377.0	−380.8
Nitromethane	357.6	−356.4
Methane–H⁺		−130.2
Ethane–H⁺		−142.7
Ethylene–H⁺		−162.6
Oxygen acids		
Diethyl ether–H⁺		−200.2
Tetrahydrofuran–H⁺		−198.8
Acetone–H⁺		−197.2
Propyl acetate–H⁺		−202.0
t-Butyl alcohol–H⁺		−195.0
Methanol–H⁺		−182.5
Water–H⁺		−167.3
Water	390.8	−390.7
Methanol	381.7	−380.6
Ethanol	378.6	
Isopropyl alcohol	376.7	
t-Butyl alcohol	375.9	
Trifluoroethanol	361.0	
Phenol	349.8	
Formic acid	345.3	
Acetic acid	341.5	
Benzoic acid	340.1	
Fluoroacetic acid	338.6	
Chloroacetic acid	336.0	
Bromoacetic acid	335.2	
Iodoacetic acid	334.7	
Trifluoroacetic acid	324.4	
Nitrogen acids		
Ammonia		−403.6
Ammonium		−204.0
Methylammonium		−214.1
Dimethylammonium		−220.6
Trimethylammonium		−224.3
Pyridinium		−220.8
Anilinium		−211.5

*Note how closely the two values agree for those cases where both values have been determined experimentally. See references at the back of the chapter for more extensive lists.

Connections

The Intrinsic Acidity Increase of a Carbon Acid by Coordination of BF₃

As discussed above, measuring acidities in the gas phase allows a determination of intrinsic acidity without solvation effects. In many synthetic reactions, Lewis acids are used to increase the enolization of an aldehyde or ketone as well as to enhance electrophilicity of the carbonyl carbon. Although the acidity of the α carbon to a carbonyl will undoubtedly increase with carbonyl oxygen coordination to an electrophile, quantitative information has been lacking.

Recently, in the gas phase, the increase in the acidity of acetaldehyde by the addition of BF₃ was determined. This was done using mass spectrometry. The acidity of acetaldehyde was found to be 365.8 kcal/mol, while the complex with BF₃ was found to be 316 kcal/mol, an astounding drop of 50 kcal/mol (equivalent to 36 pK_a units in solution). This makes the Lewis acid–base complex approximately as acidic as HI (314 kcal/mol) in the gas phase.

pK_a shift induced by BF₃

The effect of Lewis acid coordination on the acidity of acetaldehyde should be dramatically reduced in solution, because the solvation energy of the free enolate will be larger than the solvation of the Lewis acid–base complex. Using computational methods, estimates were made for this attenuation. It was calculated that the pK_a of the complex in water should be near –7, and it is known that the pK_a of acetaldehyde alone in water is 17. Hence, a 24-unit pK_a shift is predicted. This makes the Lewis acid–base complex as acidic as HBr in water—quite astounding!

Ren, J., Cramer, C. J., and Squires, R. R. "Superacidity and Superelectrophilicity of BF₃—Carbonyl Complexes." *J. Am. Chem. Soc.,* **121**, 2633–2634 (1999).

5.4 Predicting Acid Strength in Solution

As a practicing organic chemist performing reactions every day in the laboratory, it is not often that one needs to perform acid–base calculations, measure a pK_a in an organic solvent, or determine exact H_o values. Instead, we are most commonly confronted with comparisons between acids or bases, forcing us to make judgements about what base to use for a base initiated reaction, or possibly what acid to use to remove a protecting group. Further, we commonly need to choose experimental conditions to generate a reactive carbanion intermediate, such as an enolate or alkyl anion. In all these analyses, it is particularly useful to have a small handful of pK_a values of common structures set to memory, and when necessary, to have the ability to predict relative acid or base strengths. These are the items that we want to discuss in the next several sections.

5.4.1 Methods Used to Measure Weak Acid Strength

Before examining methods to predict acid strength, it is useful to have an understanding of how acid strengths are determined experimentally. For acids whose pK_as lie between the boundaries of the leveling effect of water, simple pH titrations in water suffice. For very strong acids, acidity function methods are used. For bases whose conjugate acid pK_as are around 14 to around 24, H_- acidity functions can be used. However, many organic structures are much weaker acids than water, and therefore their pK_as are far too high to be determined by any method that has water present. The pK_as of these very weak organic acids can often be determined in organic solvents, such as DMSO, giving a nonaqueous pK_a. Yet, as we have mentioned, all standard pK_as are referenced to water, and are referred to as "aqueous" pK_as. How are the "aqueous" pK_as of extremely weak acids determined?

Many methods have been employed to determine the pK_as of very weak organic acids. In fact, pK_as of alkanes near 45 and 50 have even been determined. The techniques are all built upon incremental comparisons. That is, the pK_a of a new compound under analysis is always determined using a compound whose pK_a is known. The structure whose pK_a is known was at some point analyzed relative to another compound whose pK_a was initially known. These comparisons are repeatedly done until finally a comparison can be made to a structure whose pK_a was determined in water.

In this approach, compounds that are sufficiently basic that they can be used to depro-tonate weak acids just outside of the water limit, but whose conjugate acids are just acidic enough to be measurable in water, are the key to linking all organic acids to a water pK_a scale. Two of the most important compounds used in this way are the acids cyclopentadiene and 9-phenylfluorene, whose pK_as are 16.0 and 18.5, respectively. A carbon acidity scale based upon 9-phenylfluorene was developed by Cram, and named the MSAD scale (named after McEwen, Streitwieser, Applequist, and Dessy). This scale is the one commonly used to de-fine pK_a values of near 18 to near 40. Values above 40 can be determined using an electro-chemical method developed by Breslow.

Using either 9-phenylfluorene or cyclopentadiene, the Cs^+ salt is created and allowed to equilibrate with other hydrocarbon acids (RH) in the novel solvent cyclohexylamine (Eq. 5.27). The extent of deprotonation of the hydrocarbon acid by the salt at equilibrium gives the pK_a of RH. The conjugate base of RH can then be used to establish an equilibrium with another less acidic RH', giving the pK_a of RH', and so on. Since the pK_a of cyclohexylamine is 41.6, pK_as of organic acids up to about 39 can be determined. The problem with such an analysis is similar to that encountered in nonaqueous titrations—namely, aggregation ef-fects. In cyclohexylamine, salts are strongly aggregated, and therefore the pK_as determined are really **ion pair acidities**, as noted by Streitwieser.

(Eq. 5.27)

There are cases when the acid–base equilibrium required to get a true thermodynamic difference between the compounds being compared is difficult to establish. The acid–base equilibrium is not established rapidly, and instead one determines a **kinetic acidity**. Kinetic acidities are relative rates for proton transfer, which can in some cases be related to thermo-dynamic acidities.

5.4.2 Two Guiding Principles for Predicting Relative Acidities

The most important goal in this section is to give the student a sufficient background to be able to make rational predictions as to the relative acidities of a series of HAs. All "aque-ous" pK_a values are based upon the acid HA reacting in the manner shown as Eq. 5.5. To pre-dict the equilibrium constant for this reaction, we call on principles introduced in Chapters 1–4. Recall that an equilibrium constant reflects the relative stability of the reactants and products, including factors of bond strength, any strains, and entropy factors—but also the entropy of mixing. With regards to predicting the relative strengths of several acids, entropy factors make minor differences. This is because all the acids act as given in Eq. 5.5, and hence the relative translational and rotational degrees of freedom between reactants and products are similar for all acids being compared. Therefore, we need to concentrate on enthalpy fac-tors to predict the relative acidities of acids. The enthalpy that we need to be concerned with is that of the different HAs and A^-s being compared.

In Eq. 5.5, all acids are acting as proton donors to water, and the conjugate acid for all acids being compared is H_3O^+. Therefore, in predicting the relative acidities we do not need to concern ourselves with the stability of H_2O or H_3O^+ because it is the same for all the acids.

Hence, we have two choices in predicting relative acidities; we can focus on the relative stabilities of all the HA structures being compared, or we can focus on the relative stabilities of all the A^-s created by the various HAs. Since chemists are well versed in those factors that stabilize charges (induction, resonance, etc.), it is easiest for us to rationalize acidity differ-ences by comparing charged species. Hence, two guiding principles emerge:

• When the acids being compared are neutral (HA) and create negative conjugate bases (A^-), it is most convenient to predict the relative acidities by examining the relative stabilities of the anionic conjugate bases. The acid with the most stable conjugate base A^- will be the strongest acid.

• When the acids being compared are cationic (HA$^+$) and create neutral conjugate bases (A), it is often most convenient to predict relative acidities by examining the relative stabilities of the acids themselves. The acid HA$^+$ that is most stable will be the weakest acid.

Our ultimate goal is to have a series of factors that we can use to understand and ultimately predict relative acidities. Several chemical phenomena are invoked when discussing acidities. They include electronegativity effects, inductive effects, electrostatic effects, resonance, aromaticity/antiaromaticity, solvation, hybridization effects, polarizability, bond strengths, and steric effects. Hence, almost all common factors that chemists use to rationalize experimental observations are used to explain acid–base chemistry. Thus, acid–base chemistry is an excellent topic that ties together many of the phenomena discussed to this point in the text. Below, we examine each of these effects in the context of rationalizing pK_a trends within sets of specific compounds. Often, pK_a ranges are given, but also many specific pK_as are listed. We take each factor individually, but often more than one will need to be considered to fully comprehend the acidity trends. Moreover, we combine our discussion of carbon acids along with other kinds of acids, because the exact same phenomena are involved. We simply place the carbon acid pK_as in a separate table for convenience of reference. Lastly, for purposes of discussion, a few pK_as are repeated in Tables 5.5, 5.6, and 5.7, when they are good examples of differing factors that influence acidities.

5.4.3   Electronegativity and Induction

The first factor we discuss that influences acidities is the electronegativity of the atom to which the acidic proton is attached. Table 5.5A shows the pK_a ranges of several acids of first row compounds. The acidity decreases in the following order: HF > ROH > R$_2$NH > R$_3$CH. The trend follows the stability of the conjugate base, since F is more stable when possessing a negative charge than is O, which is more stable than N possessing a negative charge, and finally C is least stable with a negative charge. This electronegativity effect is dramatic. Since pK_a is a logarithmic scale, we find that HF is approximately a trillion times more acidic than an alcohol, and an amazing factor of 10^{20} separates the acid strengths of alcohols and amines. There is a smaller separation between amines and alkanes—only on the order of a million to a billion.

Electronegativity effects are not just associated with elements directly bonded to the acid hydrogen. In fact, differences in the electronegativity of atoms attached to a molecule can affect acidity of a hydrogen from quite a far distance. As we noted in Chapter 1, when electronegativity effects arise from electron withdrawal by a remote group via sigma bonds, it is referred to as an **inductive effect**. We find such an effect on the various monohalogen substituted carboxylic acids (Table 5.5B). Fluoroacetic acid (pK_a 2.59) is more acidic than chloroacetic acid (pK_a 2.87), which is more acidic than bromoacetic acid (pK_a 2.90), and so forth. The more halogens, the more dramatic their effect, with trifluoroacetic acid (pK_a 0.52) being essentially just as acidic as picric acid (pK_a 0.4). Although the effects are small, even para substitution of a halogen on benzoic acid increases the acidity (*p*-fluorobenzoic acid has a pK_a of 4.14 while that of benzoic acid is 4.21; see Table 5.5C).

5.4.4   Resonance

In keeping with the concept that the stability of the anionic charge on the conjugate base of the acid dictates the acidity of the acid, resonance stabilization of the anion has a powerful influence on acidity. This is most dramatically seen when the acidity of methanol (pK_a 16.0) is compared to phenol (pK_a 10.02) and acetic acid (pK_a 4.76). These are all acids where the conjugate base possesses negative charge on an oxygen atom, yet the increasing ability to have resonance stabilization increases the acidity approximately one million-fold each time in going from an alcohol to a phenol to a carboxylic acid. However, the increased acidity of

Table 5.5
pK_a Values (or Ranges) of Many Different
Kinds of Acids (R = Alkyl or H)*

Compound	pK_a	Compound	pK_a
A. First row compounds and mineral acids		*D. Phenol derivatives*	
H*F*	3.18	Ph*OH*	10.02
RO*H*	16–18	*p*-ClC$_6$H$_4$O*H*	9.38
ArO*H*	–1 to 11.0	*p*-NO$_2$C$_6$H$_4$O*H*	7.15
R$_2$N*H*	38–42	2,4-Dinitrophenol	4.0
ArN*H*$_2$	18–28	Picric acid	0.4
R$_3$C*H* (simple R)	45–50	*E. Thiols and related compounds*	
H*Cl*	–6.1	CH$_3$S*H*	10.33
H*Br*	–8	PhS*H*	6.52
H*I*	–9	H$_2$S	7.0
HNO$_3$	–1.44	H$_2$Se	4.0
HNO$_2$	3.29	H$_2$Te	3.0
H$_2$SO$_4$	–9 (pK_{a1}); 1.92 (pK_{a2})	*F. Hybridization effects on RCO$_2$H*	
H$_2$SO$_3$	1.82	CH$_3$CH$_2$CH$_2$CO$_2$*H*	4.82
HClO$_4$	–10	*cis*-CH$_3$CH=CHCO$_2$*H*	4.42
H$_3$PO$_4$	2.12 (pK_{a1}); 7.21 (pK_{a2}); 12.67 (pK_{a3})	CH$_3$C≡CCO$_2$*H*	2.59
B. Acetic acid derivatives		*G. Hybridization effects on alcohols*	
CH$_3$CO$_2$*H*	4.76	CH$_3$CH$_2$CH$_2$O*H*	16.1
ClCH$_2$CO$_2$*H*	2.87	CH$_2$=CHCH$_2$O*H*	15.5
Cl$_2$CHCO$_2$*H*	1.30	HC≡CCH$_2$O*H*	13.6
Cl$_3$CCO$_2$*H*	0.64	*H. Solvation effects on alcohols*	
FCH$_2$CO$_2$*H*	2.59	CH$_3$O*H*	15.5
F$_2$CH$_2$CO$_2$*H*	1.34	CH$_3$CH$_2$O*H*	15.9
F$_3$CCO$_2$*H*	0.52	(CH$_3$)$_2$CHO*H*	17.1
BrCH$_2$CO$_2$*H*	2.90	(CH$_3$)$_3$CO*H*	19.2
ICH$_2$CO$_2$*H*	3.18	*I. Solvation effects on carboxylic acids*	
NO$_2$CH$_2$CO$_2$*H*	1.48	HCO$_2$*H*	3.75
NCCH$_2$CO$_2$*H*	2.46	CH$_3$CO$_2$*H*	4.76
PhCH$_2$CO$_2$*H*	4.31	CH$_3$CH$_2$CO$_2$*H*	4.88
H$_3$N$^+$CH$_2$CO$_2$*H*	2.31	(CH$_3$)$_2$CHCO$_2$*H*	4.85
HO$_2$CCO$_2$*H*	1.27	CH$_3$CH$_2$CH$_2$CO$_2$*H*	4.82
$^-$O$_2$CCO$_2$*H*	4.27	(CH$_3$)$_3$CCO$_2$*H*	5.03
HO$_2$CCH$_2$CO$_2$*H*	2.83		
$^-$O$_2$CCH$_2$CO$_2$*H*	5.69		
C. Benzoic acid derivatives			
PhCO$_2$*H*	4.21		
p-FC$_6$H$_4$CO$_2$*H*	4.14		
p-ClC$_6$H$_4$CO$_2$*H*	3.99		
p-BrC$_6$H$_4$CO$_2$*H*	4.00		
p-IC$_6$H$_4$CO$_2$*H*	4.00		
o-CH$_3$C$_6$H$_4$CO$_2$*H*	3.91		
m-CH$_3$C$_6$H$_4$CO$_2$*H*	4.27		
p-CH$_3$C$_6$H$_4$CO$_2$*H*	4.34		
o-CNC$_6$H$_4$CO$_2$*H*	3.14		
m-CNC$_6$H$_4$CO$_2$*H*	3.60		
p-CNC$_6$H$_4$CO$_2$*H*	3.55		
o-NO$_2$C$_6$H$_4$CO$_2$*H*	2.21		
m-NO$_2$C$_6$H$_4$CO$_2$*H*	3.49		
p-NO$_2$C$_6$H$_4$CO$_2$*H*	3.44		

*See references at the back of the chapter for more extensive lists.

Table 5.6
pK_as of Carbon Acids*

Compound	pK_a	Compound	pK$_a$
A. Carbonyl derivatives		*F. Phenyl stabilization*	
CH_3COCH_3	20.0	$PhCH_3$	41.2
CH_3COCH_2Cl	16.0	Ph_2CH_2	33.0
$CH_3COCHCl_2$	14.9	Ph_3CH	31.5
$CH_3COCH_2COCH_3$	8.84	*G. Ylides and electrostatic effects*	
CH_3CONH_2	25		
2-Acetylcyclopentanone	8	$Ph_3P^+–CH_3$	22.4
1-Ethylcarboxy-2-		$(CH_3)_3P^+CH_2Ph$	17.4
oxocyclopentane	10.5	$(CH_3)_3N^+CH_2Ph$	31.9
$CF_3COCH_2COCF_3$	5.35	$(CH_3)_3As^+CH_2Ph$	22.3
$CH_3CH_2O_2CCH_2CO_2CH_2CH_3$	13.3	$(CH_3)_2S^+–CH_2COPh$	8.3
$NCCH_2CO_2CH_2CH_3$	9	Thiamin	17.6
$CH_3CO_2^-$	24	*H. Hybridization effects*	
B. Nitro derivatives		Ethane	50
CH_3NO_2	10.2	Ethylene	44
$NO_2CH_2NO_2$	3.63	Benzene	37
$CH(NO_2)_3$	0.14	Acetylene	24
C. Cyano derivatives		Phenylacetylene	19.9
CH_3CN	25.0	*I. Aromaticity effects*	
$CH_2(CN)_2$	11.2	Cyclopentadiene	16.0
$CH(CN)_3$	5.13	Cycloheptatriene	38.8
HCN	9.21	1,2,3-Triphenylcyclopropene	50
D. Sulfones and sulfoxides		*J. Alkyl groups*	
CH_3SOCH_3	28.5	Cyclohexane	45
$EtSO_2CH(CH_3)SO_2Et$	14.6	$(CH_3)_3CH$	71
E. Cyclopentadienes		$CH_2=CHCH_3$	43
Cyclopentadiene	16.0		
1-Cyanocyclopentadiene	9.78		
2,5-Dicyanocyclopentadiene	2.52		
Indene	20.2		
Fluorene	22.7		
9-Phenylfluorene	18.5		

*See references at the back of the chapter for more extensive lists.

carboxylic acids has recently been noted to arise from an inductive effect also. The carbonyl group is a strongly electron withdrawing group, which stabilizes the negative charge on a carboxylate via induction (a similar effect rises for enolates). This may be a more important factor for the acidity of a carboxylic acid than resonance. It is often difficult to separate inductive effects from resonance effects.

Resonance structures

Table 5.7
pK_a Values of Cationic Heteroatom Acids*

Compound	pK$_{a1}$	pK$_{a2}$	pK$_{a3}$
A. Various effects on N acids (resonance, hybridization, sterics, and induction)			
Guanidinium	13.5		
1,8-Bis(dimethylamino)napthalene–H^+	12.37		
CH$_3$NH$_3^+$	10.6		
Imidazole–H^+	7.1		
Purine–H^+	2.39		
Pyridine–H^+	5.23		
Aniline–H^+	4.87		
Piperidine–H^+	11.1		
Pyrimidine–H^+	0.65		
Adenine	4.17	9.75	
Cytidine	4.08	12.24	
Guanine	3.3	9.2	12.3
NH$_4^+$	9.24		
CH$_3$NH$_3^+$	10.6		
(CH$_3$)$_2$NH$_2^+$	10.8		
(CH$_3$)$_3$NH$^+$	9.80		
RC≡NH$^+$	–12		
B. Protonated cationic oxygen			
RCO(H^+)X (X = H, R, OH, OR)	–2 to –8		
RCO(H^+)NR$_2$	0 to –4		
PhOH_2^+	–7		
THF–H^+	–2.08		
H$_3$O$^+$	–1.74		
DMSO–H^+	–1.5		
NH$_2$CO(H^+)NH$_2$	0.10		
PhO(H^+)CH$_3$	–6.54		
CH$_3$CO(H^+)CH$_3$	–7.5		
FCH$_2$CO(H^+)CH$_3$	–10.8		
F$_3$CCO(H^+)CH$_3$	–14.9		
RNO$_2$H$^+$	–12		

*See references at the back of the chapter for more extensive lists.

Another series of structures in which resonance plays a major role is the set of substituted phenols. The pK_a of p-nitrophenol is 7.15, compared to the 10.02 for phenol, whereas 2,4-dinitrophenol has a pK$_a$ of 4.0, and finally picric acid has a pK$_a$ of 0.4 (Table 5.5D). Certainly a large fraction of the effect of the nitro group arises from induction due to the electronegative character of the group, but the ability to delocalize the negative charge from the phenoxide anion to the nitro group also contributes to the stability. This is seen by comparing the pK$_a$ of p- and o-nitrophenol (7.15 and 7.22, respectively) to m-nitrophenol (8.36) to phenol (10.02). The nitro group in the conjugate base of m-nitrophenol cannot stabilize the negative charge via resonance, only via induction, which leads to over a two-unit pK$_a$ lowering relative to phenol. However, over another full unit of pK$_a$ lowering results from resonance.

If a benzene ring can stabilize a negative charge on oxygen via resonance, it should also stabilize a negative charge on N or C. In fact, the pK$_a$s of various anilines are between 18 and 28, whereas typical alkyl amines are in the range of 38 to 42 (Table 5.5A).

Resonance structures

Carbon acid acidity is dramatically influenced by resonance stabilization from an electron withdrawing group. Just the presence of one carbonyl directly adjacent to a C–H bond drops the pK_a to approximately 20—a full 10^{25}- to 10^{30}-fold change in acidity from an alkane! Whereas in an alkane the negative charge must reside on a carbon, with a ketone, the negative charge can largely reside on an oxygen. Two flanking ketones, such as in 2,4-hexanedione, produce a pK_a of 8.84, two flanking nitriles produce a pK_a of 11.2 (as in malononitrile), two sulfones produce a pK_a of 14.6, and the two nitro groups (as in dinitromethane) produce a pK_a of 3.63 (Tables 5.6A, B, C, and D). Hence, some of these carbon acids actually act as acids in water. When a carbon is flanked by two strongly electron withdrawing groups such as carbonyl, nitriles, and nitros, these compounds are known as **active methylene compounds**, reflecting their high acidity and use in various synthetic schemes requiring enolates (see Chapters 10 and 11).

Resonance structures for enolates

Another case where resonance influences carbon acidities is the comparison of toluene to diphenylmethane and lastly triphenylmethane (pK_as = 41.2, 33.0, and 31.5, respectively). The large shift between toluene and diphenylmethane is due to additional resonance stabilization of the conjugate base. However, the third additional phenyl ring has little effect. Several factors are involved to account for this small change. One is that the phenyl rings cannot all be planar with the anionic carbon, which is required for full resonance stabilization (examine the trityl radical discussed in Chapter 2), and instead a propeller twist develops in the anion. This is an example of a **steric inhibition of resonance**. A second factor is called a **resonance saturation effect**. Once the charge on the conjugate base is stabilized via resonance, the additional resonance is not as effective at stabilization.

Competitive resonance effects can be found. In general, C–H bonds alpha to ketones are more acidic than when alpha to ester carbonyls, which are more acidic relative to amide carbonyls, all three of which are more acidic than C–H bonds near carboxylates. The resonance stabilization gained by ionization of the C–H bond is increasingly lower in this series because the O, N, or O^- heteroatom on the ester, amide, or carboxylate, respectively, is increasingly involved in resonance with the carbonyl in the acid prior to ionization of the C–H bond. In such cases, we must consider the role of resonance in stabilizing the HA compound as well as A^-.

Resonance structures for enolates

Similar resonance effects influence the acidities of ammonium and amine acids. The most interesting examples are relevant to biological chemistry, and a discussion of these is left to the next section.

When contrasting the acidities of protonated amines, it is often convenient to analyze the stabilities of the acids themselves rather than the stabilities of the conjugate bases (see solvation effects below). However, the nitrogen lone pairs created upon ionization of an N–H bond are often involved in resonance interactions, and hence examination of the conjugate base is instructive. For example, the pK_a of anilinium is 4.87 while that of a simple primary ammonium such as methylammonium is 10.66. This is a dramatic difference. To explain this, it is important to note that the positive charge on the anilinium cannot be resonance delocalized. Yet, upon ionization of an N–H$^+$ bond in these structures, aniline or methylamine is obtained, and the lone pair on the nitrogen of aniline can delocalize into the phenyl ring whereas that on methylamine cannot. The delocalization stabilizes the conjugate base aniline relative to methylamine making anilinium more acidic.

5.4.5   Bond Strengths

The acidities of halogen acids follow the trend: HI > HBr > HCl > HF (Table 5.5A). A similar trend is found for group VIA acids: H_2Te > H_2Se > H_2S > H_2O (Table 5.5F). The major factor that contributes to this trend is bond strengths. The bond strength decreases in the following order: H–F > H–Cl > H–Br > H–I and H–OH > H–SH > H–SeH > H–TeH. Although these are trends in homolytic bond strengths, not heterolytic bond strengths, the trends certainly contribute to the increasing acidity with lower bond strength. However, when other factors are at play, homolytic bond strengths do not dominate trends in acidities. For example, when one moves away from a homologous series of structures all in the same row of the periodic table, bond strength trends do not correlate with acidities. R_2N–H bond strengths (R = alkyl) are weaker than both R_3C–H and RO–H bonds (see Table 2.2), yet N–H bonds are more acidic than C–H bonds.

If we remove an electron from a hydrocarbon we weaken the bonds. Therefore, radical cations of weak acids can become very acidic. Radical cations (R–H$^{\bullet+}$) can lose a proton by heterolytic cleavage to give a neutral radical (R$^\bullet$ + H$^+$), or undergo homolytic cleavage to give a carbocation by loss of a hydrogen atom (R$^+$ + H$^\bullet$). In polar solvents the loss of a proton is favored due to the favorable solvation energy. For example, in DMSO the pK_a of the methyl group in toluene is 43, but the pK_a of the radical cation of toluene is –20. This makes an amazing difference of 63 in pK_a values.

5.4.6   Electrostatic Effects

Neighboring charges greatly influence the ability to create further charges in close proximity. Hence, electrostatic effects can have quite large influences on acidities. Most chemists are familiar with the differences in the pK_{a1} and pK_{a2} of dicarboxylic acids (Table 5.5B). The closer the anionic charge of the first carboxylate to the second carboxylic acid, the higher the second pK_a relative to the first. Conversely, neighboring cationic charges increase the acidities of carboxylic acids. The pK_a of $^+H_3NCH_2CO_2H$ is 2.31, compared to a pK_a of 4.82 for $CH_3CH_2CO_2H$. Similarly, the partial charges in dipoles and quadrupoles will affect the strength of neighboring acidic sites.

Electrostatic effects have been capitalized upon in the creation of nucleophiles in common synthetic transformations. For example, alkyl anions can be readily created in direct proximity to positive phosphorus, nitrogen, and sulfur centers, creating what are known as **ylides**. The pK_a of $Ph_3P^+CH_3$ is 22.4, deprotonation of which gives the common phosphorus ylide exploited in the Wittig reaction. Table 5.6G shows a handful of other examples, where pK_as as low as 8 can be obtained when in conjunction with other effects.

Phosphorus ylide

5.4.7   Hybridization

A fascinating effect taught in beginning organic chemistry courses is the influence of the hybridization state of carbon on C–H acidities. Recall that alkynes are more acidic than

$$R-C\equiv C-H \quad > \quad \underset{R}{\overset{R}{>}}=\underset{H}{\overset{R}{<}} \quad > \quad \underset{R}{\overset{R}{>}}-H$$

Decreasing acidity with decreasing
s character in carbon hybridization

alkenes, which in turn are more acidic than alkanes (Table 5.6H). The principle influence is the hybridization of the carbon. An *sp* hybridized carbon bonds to the hydrogen atom with the most amount of *s* character. Since *s* orbitals feel the positive charge of the nucleus more effectively than *p* orbitals (*p* orbitals have a node at the nucleus), higher *s* character can better stabilize the resulting negative charge upon ionization of the C–H bond. Thus, the electronegativity of various hybridized forms of carbon follows the following trend: $sp > sp^2 > sp^3$.

This electronegativity difference due to the hybridization states of carbon should manifest itself in other acidity trends. Indeed, it does. Examine the trends in acidities of alcohols and carboxylic acids given in Tables 5.5F and G, respectively. Proximity of the acidic group to an alkynyl group results in the strongest acid, whereas a simple alkyl group gives the weakest acid.

Hybridization effects are evident in the acidities of protonated amines also. The pK_a of a protonated nitrile is around –12 (an *sp* nitrogen), while that of pyridinium is 5.23 (*sp²* nitrogen), and alkyl ammoniums are around 10 to 11 (*sp³* nitrogen).

$$R-C\equiv \overset{\oplus}{N}-H \quad > \quad \underset{R}{\overset{R}{>}}=\overset{\oplus}{N}\overset{R}{\underset{H}{<}} \quad > \quad \underset{R}{\overset{R}{>}}-\overset{\oplus}{N}-H$$

Decreasing acidity with decreasing
s character in nitrogen hybridization

5.4.8 Aromaticity

In Chapter 2 we showed that aromaticity strongly stabilizes organic structures. Accordingly, large effects are prevalent in acid–base chemistry. Tables 5.6E and I highlight some examples. The most well known is the acidity of cyclopentadiene (pK_a 16.0), which is similar to that of water. The resulting anion is aromatic. However, when the resulting anion is antiaromatic the compounds are dramatically less acidic, as expected; the cyclopropene pK_a is 61 and the pK_a of cycloheptatriene is 38.8.

Aromatic   Antiaromatic

Differences in aromaticity of
cyclic carbanions

5.4.9 Solvation

When examining the acidities of different functional groups, large differences derive from the interactions discussed above—namely, induction, resonance, polarizability, electrostatics, aromaticity, and hybridization. Solvation effects are often invoked to explain the small differences in acidities that exist within similar functional groups. For example, consider alcohols (ROH). The larger the R group the lower the acidity of the alcohol (Table 5.5H). A factor of approximately 3 pK_a units separates the stronger acid methanol from the weaker acid *t*-butanol. Recall that this is the opposite of the gas phase acidities, where the larger polarizability of the *t*-butyl group leads to the greater acidity of *t*-butanol. Solvation effects have completely reversed the intrinsic acidity order of the compounds. The most common explanation is that the bulkier the alkyl group, the more difficult it is for the solvent molecules to make close contact with and thereby stabilize the O⁻ in the alkoxide.

$$CH_3-OH \quad > \quad \diagdown-OH \quad > \quad \diagup-OH \quad > \quad \diagup-OH$$

Order of decreasing acidity of alcohols in solution

A similar solvation effect is found for carboxylic acids (Table 5.5I). Formic acid is the most acidic (pK_a 3.75) and pivalic acid is the least acidic (pK_a 5.03). In both the alcohol and carboxylic acid series, the smaller anion can be better solvated, leading to a larger acidity of the corresponding conjugate acid.

Order of decreasing acidity in solution

The notion that the smaller anion is better solvated suggests an enthalpy effect. However, the origin of the effect is entropy. The larger R group in acetic acid (R = CH_3) relative to formic acid (R = H) leads us to predict that formic acid would be the stronger acid because the resulting anion is better solvated. However, the enthalpies ($\Delta H_{rxn}°$) of formic acid and acetic acid donating a proton to water are –0.01 and –0.02 kcal/mol, respectively—hardly different from each other and in the opposite direction from that predicted. However, the $\Delta S°$ values for formic acid and acetic acid are –17.1 and –21.9 eu, respectively, making formic acid the stronger acid. The larger the R group, the more unfavorable the entropy of reaction. The reason for this has its roots in steric hindrance to solvation. In order for the water to form strong interactions and solvate the conjugate base of the acid, water must become more ordered when the R group is larger.

An interesting trend exists for the acidity of the various ammonium ions (Table 5.7A):

$$NH_4^+ > (CH_3)_3NH^+ > CH_3NH_3^+ > (CH_3)_2NH_2^+$$

It is most convenient to analyze the stabilities of these acids themselves to predict acidity instead of analyzing the conjugate bases, because the acids possess the charge. Further, to understand this trend, we must look at two conflicting effects—polarizability and solvation, as has been discussed by Arnett. As presented above in the section on gas phase acidities, the intrinsic acidities of the various ammonium ions follows the following trend:

$$NH_4^+ > CH_3NH_3^+ > (CH_3)_2NH_2^+ > (CH_3)_3NH^+$$

because more alkyl groups can spread out the positive charge better (see Table 5.4). However, solvation effects favor $(CH_3)_3NH^+$ as the most acidic, because its large size makes it the least well solvated. The solvation effect on the acidities leads to the following trend:

$$(CH_3)_3NH^+ > (CH_3)_2NH_2^+ > CH_3NH_3^+ > NH_4^+$$

Apparently, a combination of these two effects leads to the observed trend given at the beginning of this paragraph.

5.4.10 Cationic Organic Structures

The protonation of neutral oxygen-containing compounds creates exceptionally strong acids. The protonation of the oxygens of the carbonyl groups of aldehydes, ketones, acids, esters, and amides results in acids with pK_as ranging from –15 to around 0, depending upon the stability of the cationic charge created upon protonation. Protonation of a nitro group benefits from no inductive, resonance, or other effects, and in fact creates a positive charge on a very electronegative group, and hence the pK_a of a protonated alkyl nitro is around –12. Similarly, protonation of a nitrile leads to acids with pK_a values around –12.

Very strong acids

5.5 Acids and Bases of Biological Interest

Since proton transfers are fundamental chemical reactions required in numerous transformations, biology has its own set of acids and bases. For example, two of the classes of amino acid side chains have Brønsted definitions: basic and acidic. Both glutamic acid and aspartic acid have simple carboxylates as side chains. At pHs near neutral, these side chains are in their conjugate base carboxylate forms, and hence the aspartic and glutamic acid side chains are actually most often found to act as bases in enzymatic reactions.

Glutamic acid   Aspartic acid

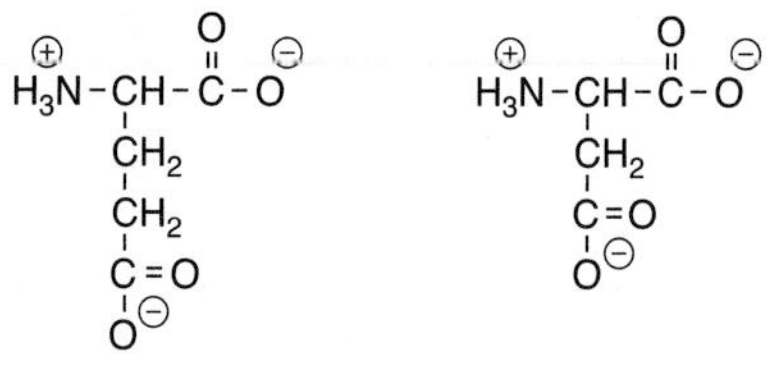

The basic amino acids are histidine, lysine, and arginine, possessing side chains incorporating an imidazole, a primary amine, and a guanidine, respectively. At pHs near neutral the primary amine and the guanidine are protonated, because the pK_as of the conjugate acids are significantly higher than 7. The high pK_a of guanidinium arises from the delocalization of the positive charge onto all three nitrogens of the functional group, greatly stabilizing the structure due to resonance. Due to their protonation state, the side chains of lysine and arginine are most commonly found to act as acids in enzymatic reactions. In contrast, because the pK_a of imidazolium is near 7, imidazole exists in both its basic and acidic forms at neutral pH. Hence, the side chain of histidine is found to be involved in both acid and base reactions in enzymes.

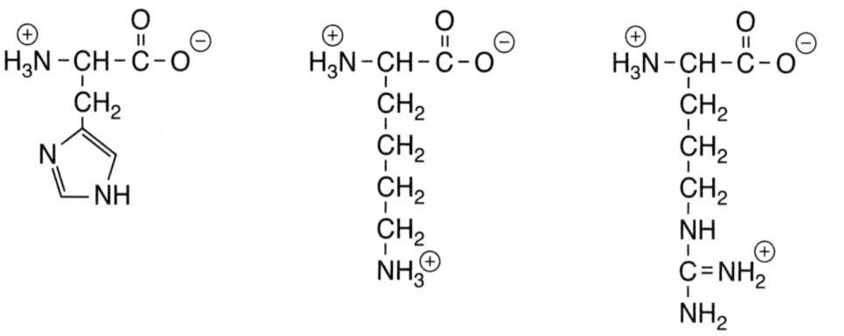

Histidine   Lysine   Arginine

One particularly interesting biological acid is the thiazolium group, part of the cofactor thiamine pyrophosphate. Thiazolium is a carbon acid with a pK_a near 21. This pK_a is low enough that enzymes can deprotonate thiazolium, creating a *carbon nucleophile* that commonly is found to transfer an acetyl group. The reason for the relatively high acidity is an electrostatic effect similar to that discussed for ylides. Due to the positive quaternary nitrogen, the neighboring hydrogen is acidic.

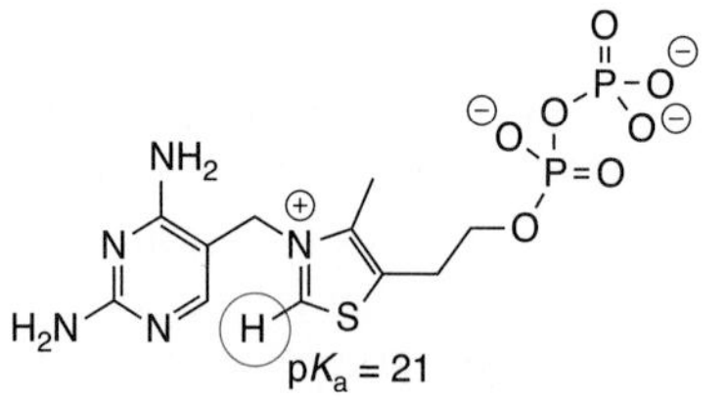

Thiamine pyrophosphate

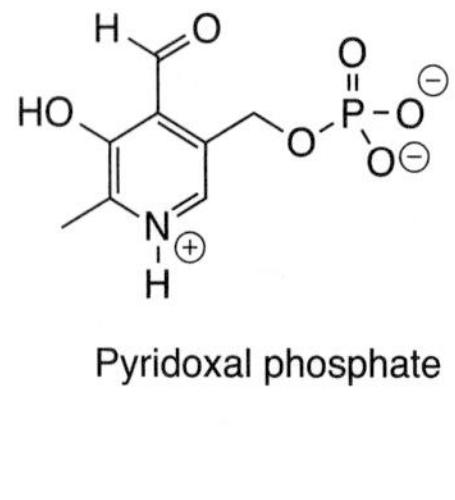

Pyridoxal phosphate

Another interesting cofactor whose activity is related to its pK_a is pyridoxal phosphate. Whereas pyridinium has a pK_a of 5.23, alkyl pyridiniums have pK_as approaching or exceeding 6. Hence, this cofactor is predominately protonated at physiological pH. In a protonated state the nitrogen of the pyridinium ring makes a good electron sink. In fact, this is exactly the manner in which the cofactor is most commonly thought to act, simply as a good two electron accepting entity.

The acid–base chemistry of amino acids and various cofactors are certainly not the only biological structures whose activity depends upon the protonation state. Even DNA responds to pH effects due to acidic and basic sites within its structure. At physiological pH the four heterocyclic bases of nucleotides are neutral, but as the pH is raised, thymine and uridine will undergo deprotonation near pH 10. As the pH is lowered, cytidine will become protonated around pH 6. This protonation has been used to control DNA triple helix formation, as shown in the following Connections highlight.

Connections

Direct Observation of Cytosine Protonation During Triple Helix Formation

DNA forms double helical structures using the **Watson–Crick hydrogen bonding** motif (Appendix 4). However, DNA can also form triple helices when a third strand forms what are termed **Hoogsteen hydrogen bonds**. The Hoogsteen hydrogen bonding motif between cytidine and adenine (shown below) requires that the cytidine be protonated, and in fact as the pH is lowered below 7 to between 5 and 6, triple helical DNA formation is enhanced.

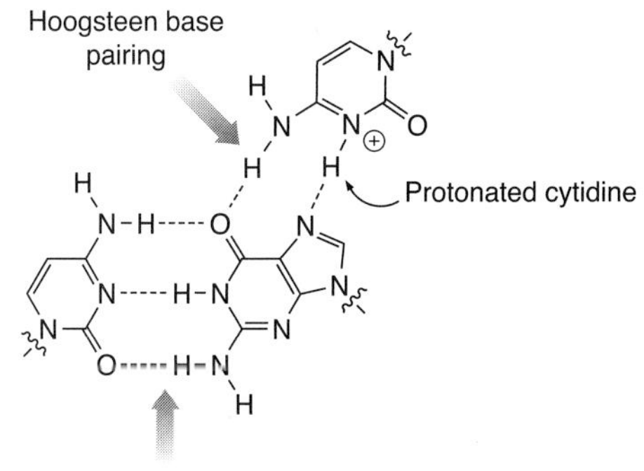

In one recent study the ^{15}N NMR signals of the three cytidines shown in bold in the following DNA strand were followed as a function of pH. The spectra clearly showed the presence of double stranded DNA at higher pHs and the corresponding triple helix at lower pH. By varying the pH and following the interconversion between double and triple stranded forms, apparent pK_a values for the various cytidines could be determined as 5.5, 6.0, and 6.7 for C_{15}, C_{20}, and C_{18}, respectively. This is a nice demonstration of a biological conformational change as a function of pH.

Leitner, D., Schröder, W., and Weisz, K. "Direct Monitoring of Cytosine Protonation in an Intramolecular DNA Triple Helix." *J. Am. Chem. Soc.*, **120**, 7123–7124 (1998).

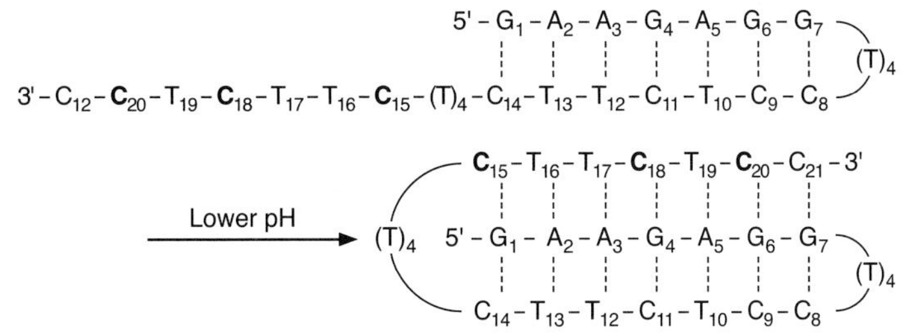

Hoogsteen base pairing

Protonated cytidine

Watson–Crick base pairing

$$5' - G_1 - A_2 - A_3 - G_4 - A_5 - G_6 - G_7 \smallfrown (T)_4$$
$$3' - C_{12} - \mathbf{C_{20}} - T_{19} - \mathbf{C_{18}} - T_{17} - T_{16} - \mathbf{C_{15}} - (T)_4 - C_{14} - T_{13} - T_{12} - C_{11} - T_{10} - C_9 - C_8$$

Lower pH →

$$\mathbf{C_{15}} - T_{16} - T_{17} - \mathbf{C_{18}} - T_{19} - \mathbf{C_{20}} - C_{21} - 3'$$
$$(T)_4 \quad 5' - G_1 - A_2 - A_3 - G_4 - A_5 - G_6 - G_7 \smallfrown (T)_4$$
$$C_{14} - T_{13} - T_{12} - C_{11} - T_{10} - C_9 - C_8$$

Double to triple helix formation

A recent debate in the enzymology community has centered around how carbon acids whose pK_as are far above neutral pH can be deprotonated by enzymes at physiological pH. For example, the enzyme mandelate racemase deprotonates mandelate (see in the margin), even though the α hydrogen of mandelic acid has a pK_a of 22. The pK_a of mandelate would be significantly higher. As the base, the enzyme uses an amine from lysine, which has a conjugate acid pK_a of 10. Therefore, the success of this deprotonation indicates that the enzyme has lowered the pK_a of mandelate by significantly more than 12 pK_a units—quite a feat!

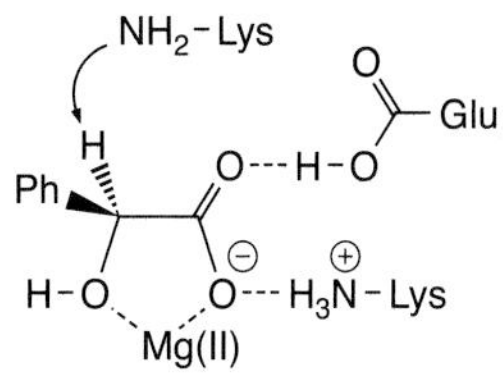

Mandelate racemase active site

As discussed in this chapter, there are two strategies an enzyme could employ to increase the acidity at a particular site. One is to raise the pK_a of the conjugate acid of the base by changes in the environment around this base, making it more basic. The second approach would be to present a microenvironment around the anionic conjugate base of the mandelate that is so stabilizing that it induces the large increase in acidity. It would appear that this latter strategy is required in mandelate racemase, and the stabilization forces have been the focus of the debate. Hydrogen bonds, short–strong (low barrier) hydrogen bonds (see Section 3.2.3), ion pairing effects, and metal coordination have been discussed as the possible stabilizing forces. The following Connections highlight shows that the proximity of metals can significantly increase the acidity of neighboring groups.

Connections

A Shift of the Acidity of an N–H Bond in Water Due to the Proximity of an Ammonium or Metal Cation

As a means of understanding enzymatic reactions, model studies have been undertaken to measure the extent to which certain microenvironments can influence acidities. In some cases, dramatic effects have been found. A case in point is the shift in acidity of N_1 of uracil when proximate to cationic centers. The pK_a of the N_1–H site in uracil alone is 9.9. However, when present in the following two complexes, the pK_a drops to 7.1 and below 5, as shown. Thus, electrostatic effects in water can be quite dramatic, lowering a pK_a in the case of a proximal zinc by at least five units.

Kimura, E., Kitamura, H., Koike, T., and Shiro, M. "Facile and Selective Electrostate Stabilization of Uracil N(1)-Anion by a Proximate Protonated Amine: A Chemical Implication for Why Uracil N(1) Is Chosen for Glycosylation Site." *J. Am. Chem. Soc.*, **119**, 10909–10919 (1997).

pK_a values change as a function of local charges

Nature has found many clever ways to alter the pK_a of a specific side chain by creating precise microenvironments. However, we should remember that the most common alteration of pK_as seen in proteins results from a relatively less specific type of interaction. As we noted in earlier chapters, the cores of most proteins are relatively hydrophobic. Such environments are much different than bulk water, and they generically favor neutral species over charged species. As such, if an ionizable side chain finds itself buried in the core of a protein (a less common, but certainly not impossible situation), it will more likely exist in the neutral form than if it were on the surface of the protein, exposed to water.

5.6   Lewis Acids/Bases and Electrophiles/Nucleophiles

While the Brønsted acid/base terms specifically refer to proton donors and acceptors, respectively, the Lewis approach (named after G. N. Lewis, who introduced the idea in 1923) greatly broadens the definitions of what is an acid and what is a base. Recall that a **Lewis acid** is an electron pair acceptor and a **Lewis base** is an electron pair donor. All common organic reactions that do not involve radicals or concerted pericyclic processes can in some manner be discussed as Lewis acid–base reactions. Similarly, all these reactions can be considered to be occurring between electrophiles and nucleophiles. Recall that an **electrophile** is any species seeking electrons and a **nucleophile** is any species seeking a nucleus (or positive charge) toward which it can donate its electrons. In this context, a Lewis base is synonymous with a nucleophile, and a Lewis acid is synonymous with an electrophile; it just depends upon your favorite choice of definitions. However, the terms are commonly used in different contexts. The Lewis acid/base terms are most often associated with thermodynamic discussions, while the terms nucleophile and electrophile are used when discussing reactivity and kinetics. Since we will explore the concepts of Lewis acids/bases and electrophiles/nucleophiles extensively in future chapters, our goal here is to draw analogies with Brønsted acid–base chemistry and to define hard and soft interactions.

Just as with the pK_a and pK_b scales for Brønsted acids and bases, other scales have been developed to relate the strength of one Lewis acid (electrophile) to another, or one Lewis base (nucleophile) to another. For example, scales based upon the relative reactivities of various nucleophiles and electrophiles in common organic reactions, such as S_N2 transformations, have been defined. We will cover these scales in Chapter 8 after we examine linear free energy relationships. Further, after examining the scales, we will discuss what factors make for a good nucleophile and electrophile.

For now, just remember that the more electron rich and the more accessible those electrons are (higher energy and sterically uncrowded), the better the nucleophile. In addition, increased polarizability implies increased nucleophilicity. Conversely, the more electron poor and the lower the energy of the empty orbitals that electrons can be donated toward, the better the electrophile.

Connections

The Notion of Superelectrophiles Produced by Super Acids

You've heard of supermodels, super-sized fries, and the Super Bowl; why not superelectrophiles, too? Superelectrophilic intermediates are typically generated when a cationic species is protonated or coordinated to a Lewis acid, creating a dication. The dications often undergo reactions with what we would consider to be among the weakest of nucleophiles, such as benzene. For example, benzaldehyde reacts with two equivalents of benzene in solutions of triflic acid (CF_3SO_3H, $H_o = -11$ to -14) to give triphenylmethane.

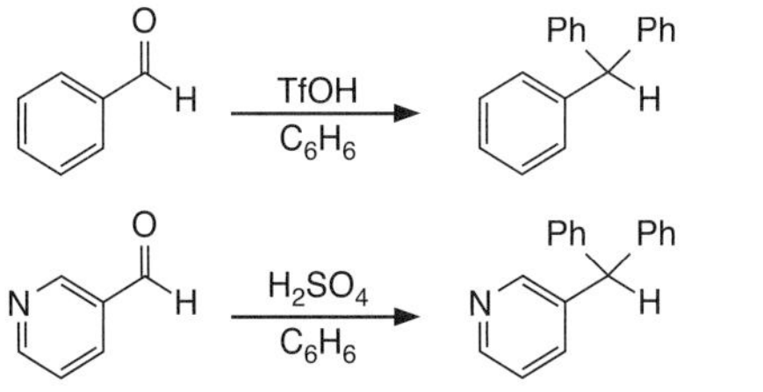

Recently, it has been shown that 3-pyridinecarboxaldehyde undergoes the same reaction in weaker acid solutions of sulfuric acid with H_o values around -9. NMR analysis definitively established that the dication shown below is generated under the reaction conditions, and the two positive charges enhance the electrophilic aromatic substitution with benzene.

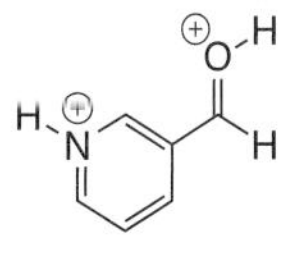

A superelectrophile

Klumpp, D. A., and Lau, S. "3-Pyridinecarboxaldehyde: A Model System for Superelectrophilic Activation and the Observation of a Diprotonated Electrophile." *J. Org. Chem.*, **64**, 7309–7311 (1999).

Reactions of superelectrophiles

5.6.1 The Concept of Hard and Soft Acids and Bases, General Lessons for Lewis Acid–Base Interactions, and Relative Nucleophilicity and Electrophilicity

In the brief guidelines given above for what makes a good nucleophile and electrophile, we touched on the energy and accessibility of the electrophilic and nucleophilic orbitals. This brings us to another related concept, that of "hard" and "soft" acids and bases. In this definition, the acids and bases are best viewed as being of the Lewis type. Here we examine the "hardness" and "softness" of the acid and base to predict reactivity. In this analysis, the character of a nucleophile or electrophile is most often correlated with the polarizability of the species; **hard reactants** are non-polarizable, whereas **soft reactants** are polarizable. The

Table 5.8
Hard, Moderate, and Soft Lewis Bases (Nucleophiles)
and Lewis Acids (Electrophiles)

Lewis bases	Lewis acids
Hard	
H_2O, OH^-, F^-, RCO_2^-, Cl^-, ROH RO^-, R_2O, NH_3, RNH_2, N_2H_4	H^+, Li^+, Na^+, K^+, Al^{3+}, Mg^{2+}, Ca^{2+}, BF_3, $B(OR)_3$, $Al(CH_3)_3$, $AlCl_3$, RCO^+
Moderate	
$PhNH_2$, C_5H_5N, N_3^-, Br^-	Fe^{2+}, Co^{2+}, Cu^{2+}, Zn^{2+}, Pb^{2+}, Sn^{2+}, $B(CH_3)_3$, R_3C^+, $C_6H_5^+$
Soft	
R_2S, RSH, RS^-, I^-, SCN^-, R_3P $(RO)_3P$, CN^-, C_2H_4, C_6H_6, R^-	Cu^+, Ag^+, Hg^+, Pd^{2+}, I_2, Br_2, carbenes, radicals

concepts of hard and soft, however, can also be related to orbital energies. The lower the energy of the orbitals containing the nucleophilic electrons, and the higher the energy of the empty orbital that can accept the electrons, the "harder" the respective base and acid. Lewis acids and bases with low energy empty and high energy filled orbitals are considered "soft". Table 5.8 lists examples of several hard and soft Lewis acids and bases. These definitions are quite qualitative, but the correlations found for reactivity can be quite insightful.

In general, interactions (defined as simple coordinations between the acid and base, or as reactions such as substitutions) are most facile between hard acids and hard bases and between soft acids and soft bases (a trend called the **principle of hard and soft acids and bases**, HSAB). In other words, interactions between acids and bases of the same type are more facile than between a hard acid and a soft base, and vice versa. For example, an interaction between lithium and hydroxide is predicted to be more favorable than between lithium and iodide, while iodide will interact well with Cu^{2+} (see Table 5.8). A well known example of this is the strong interaction between mercury metal and sulfur (leading to the old fashioned name mercaptan for a thiol).

Why is such a trend observed? Actually, the reason that hard acids and bases prefer to interact with each other is different than the reason that soft acids and bases prefer to interact. To see this, let's examine some mathematics that is meant to model the interaction between Lewis acids and bases in an early stage of their interaction. The analysis derives from perturbational molecular orbital theory (PMOT), which was briefly introduced in Chapter 1, and is explored in more depth in Chapter 14. In essence, three forces are considered to mediate the energy of interaction (E_i) between the acid and base as they approach each other in space (Eq. 5.28). One is the electrostatic repulsion between the electron clouds of the two entities, referred to as E_{core}, a positive destabilizing term. The second and third factors are both attractive and stabilizing. An electrostatic attraction between an acid and base occurs due to opposite charges on the acid and base; this is called E_{ES}. Lastly, a term called $E_{overlap}$, which is related to the net overlap of the nucleophilic and electrophilic orbitals, is found to lower the energy of the system as the nucleophilic electrons delocalize into the empty electrophilic orbital.

$$E_i = E_{core} + E_{ES} + E_{overlap} \qquad \text{(Eq. 5.28)}$$

For a series of similar acids and bases with similar atoms and structures, the E_{core} values are essentially the same because the repulsion due to their electrons clouds will be close to identical. Hence, the difference between the energy of interaction for one Lewis acid and base with another Lewis acid and base of similar structures is controlled by the E_{ES} and

$E_{overlap}$ terms. The E_{ES} and $E_{overlap}$ terms for each individual acid and base interaction are given by Eqs. 5.29 and 5.30, respectively. As expected, the electrostatic term looks like Coulomb's law, where r is the distance between the charges on the Lewis acid and base (q_a and q_b, respectively), and ε is the dielectric constant. In Eq. 5.30 c_a and c_b stand for the coefficients of the atomic orbitals, in the molecular orbitals, on atoms a and b that are interacting for the acid and base, respectively, while β is a measure of their interaction. E_a and E_b are the energies of the electrophilic and nucleophilic orbitals, respectively. The $E_{overlap}$ term tells us that the interaction is directly proportional to the size of the electrophilic and nucleophilic orbitals (related to the c's) and the extent to which they can interact (β), but it is inversely proportional to the initial energy separation between them. This means that orbitals of roughly the same energy experience a strong, stablizing interaction while orbitals of disparate energies do not interact well.

$$E_{ES} = \frac{q_b q_a}{\varepsilon \varepsilon_0 r} \qquad \text{(Eq. 5.29)}$$

$$E_{overlap} = \frac{2(c_b c_a \beta)^2}{E_b - E_a} \qquad \text{(Eq. 5.30)}$$

Eq. 5.30 is a general relationship for the interactions of electrophiles and nucleophiles, and is not restricted to definitions and discussions of hard and soft acids and bases. It tells us that the relative nucleophilicity of several Lewis bases will depend upon which electrophile is used, because the c's and β values will change for each different electrophile. Similarly, the relative electrophilicities of several Lewis acids will depend upon what nucleophile is used. We will see exactly such results when we explore quantitative scales for various nucleophiles and electrophiles, where the scales are highly dependent upon the particular reaction that is chosen to analyze relative reactivities (see Chapter 8). Eq. 5.30 nicely explains the reactivity trends for soft acids and bases. It predicts that the $E_{overlap}$ will be best for Lewis acids and bases that have electrophilic and nuclephilic orbitals of roughly the same energy, which is the cases for the soft acids and bases of Table 5.8.

The preferential interaction between hard acids and hard bases is actually mostly controlled by the electrostatic term. This term is large when the charges are highly localized on the acids and bases. Such is the case for the hard species (examine Table 5.8 for yourself to see this). This is also true for neutral examples, which have charged regions due to dipoles or quadrupoles. Hence, we see a fundamental difference between hard–hard vs. soft–soft interactions, in that the former are controlled primarily by electrostatic attractions (Eq. 5.29), while the latter depend on orbital mixing interactions (Eq. 5.30). These are descriptive terms, and there will always be intermediate cases. Still, the hard/soft classification has proven to be quite successful in predicting or rationalizing reactivity.

Several individuals have worked to quantify these trends, making scales of Lewis acidity and basicity. Gutmann has created a series of donor numbers (DN) and acceptor numbers (AN) for various solvents, while Drago and Wayland have assigned parameters E and C, which measure electrostatic interactions and covalent bonding potential, respectively. Lastly, Pearson treats each Lewis acid and base with two parameters, relating what is called the strength of the acid/base and the softness/hardness of the acid/base. In fact, the HSAB trend discussed above is primarily a concept developed by Pearson.

In summary, the concepts of electrophilies and nucleophiles are very similar to those of Lewis acids and bases. A more thorough discussion of what makes good electrophiles and nucleophiles is left to Chapter 8. Until then, it is instructive to simply realize that trends of preferential reactivity fall into classes defined as hard and soft species, where nucleophiles and electrophiles within these individual classes prefer to react. The reactivity of the soft species is primarily due to better overlap of the orbitals, while for the hard species the electrostatic attraction dominates.

Summary and Outlook

One goal of this chapter was to develop tools for comparing the thermodynamics of one reaction to another, where acid–base chemistry was our setting. In that regard, this chapter allowed us to analyze again many of the chemical effects that chemists routinely call upon to explain trends—namely, induction, resonance, aromaticity, solvation, hybridization, polarizability, and electrostatics. Although none of these effects were new, this chapter showed for the first time how their interplay affects the thermodynamics of reactions. Really, the only common effect that was not focused upon in this chapter is that of sterics, because a proton is so small that steric effects are minimal.

A second goal of the chapter was to delineate many of the experimental techniques used to analyze the thermodynamics of the acid–base reactions, and to explore some of the scales used to determine the strengths of various acidic and basic solutions. As we move beyond acid–base chemistry into future chapters, we will return to similar methods of analysis for other reactions, and various other scales that relate the strengths of nucleophiles and electrophiles to perform chemical reactions other than Brønsted acid–base reactions. Hence, this chapter included some review, but also launches us into the second section of the book, which deals with the kinetics and mechanisms of organic reactions other than acid–base chemistry. There is only one last topic that we must cover prior to analyzing organic mechanisms, and that is stereochemistry. Hence, hold on for just one more chapter to set the stage for our discussions of organic, bioorganic, and organometallic reactivity.

Exercises

1. Why is the pK_a of hydronium -1.74, not 0?

2. The pK_a of methane is near 45, acetonitrile's pK_a is 25.0, whereas that of malononitrile is 11.2, and the pK_a of tricyanomethane is 5.1. Why does each additional cyano group have less of an effect at lowering the pK_a?

3. For an acid HA with a pK_a of 9.2, what is the ratio of the conjugate base to the acid at pH 7.2 for the following two conditions: total concentrations of HA being 0.0042 M and 0.00042 M?

4. 1,3-Cyclopentanedione is more acidic than 1,3-cyclohexanedione. What is the reason for this?

5. Explain the statement and give the reason why it is true that "an acid–base equilibrium will always prefer the side with the weaker acid".

6. Explain the reasoning behind the statement that "pH is best thought of as the proton donating ability of a solution".

7. If the pH of an aqueous solution is adjusted by a researcher to be 7.1, and p-nitrophenol, triethylamine, and acetic acid are all present in that solution, what are the dominant protonation states of these species?

8. Rationalize the better interaction between Hg^{2+} and S^{2-} than between Hg^{2+} and O^{2-} using a molecular orbital mixing diagram.

9. In the following pairs of compounds, predict which is the stronger acid and explain why.

E.

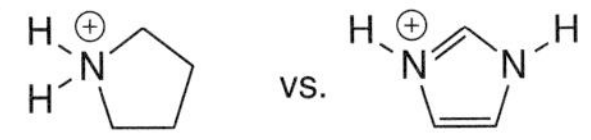

vs.

10. Give an example of how each of the following effects may influence the acidity of various alcohols: hybridization, electro-statics, induction, resonance, and solvation.

11. The pK_as of the conjugate acids of 2-aminopyridine and 4-aminopyridine are as shown. Which nitrogen is the more basic in these structures and why? Further, why are the acidities of the conjugate acids of these structures different by about two and a half orders of magnitude? Why are these structures so much more basic than pyridine?

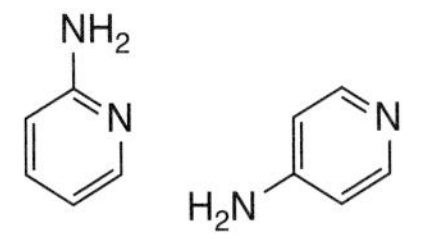

6.71   9.11

12. The pK_a of 2-aminopyridine is approximately 24, and the ratio of tautomers is approximately as shown below. Calculate the pK_a of compound B.

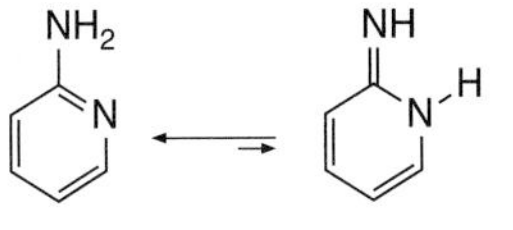

A (92%)   B (8%)

13. The following table shows the pK_as of a series of phenyl–tetramethyl–guanidinium structures in water and acetonitrile. Can you find a correlation between the pK_as in the different solvents? Rationalize why you might expect a correlation.

X	Acetonitrile	H_2O
H	20.6	12.2
NO_2	17.8	9.8
CN	18.4	10.9
CF_3	19.0	11.4
Cl	19.7	11.7
Br	19.6	11.6
CH_3	20.9	12.4
OCH_3	21.0	12.6

14. The pK_as of alkanols are typically in the range of 16–17. However, the pK_as of vicinal diols are commonly 14–16, and those for geminal diols are in the range of 13–14. Explain this trend.

15. The three pK_as for citric acid are 3.1, 4.7, and 5.4. Explain why the first pK_a is lower than for normal carboxylic acids, and why the third pK_a is higher than normal.

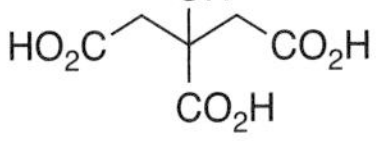

16. The three pK_as for phosphoric acid (H_3PO_4) are 2.1, 7.2, and 12.0. Hence, each ionization is separated by about 5 pK_a units. One typical explanation for the acidity of phosphoric acid is the resonance that is achieved in the anions. However, recent calculations find that the P=O in phosphoric acid is better thought of as P^+-O^-. Given this, can you think of another expla-nation for the high acidity of phosphoric acid? Further, why is there such a large difference between each pK_a value?

17. We have explained in this chapter that one needs acidity functions to measure pK_as in water that are below the pK_a of hydronium. Using acidity functions, the pK_as of F_3CSO_3H, HBr, HCl, CH_3SO_3H were determined to be −14, −9, −8, and −0.6, respectively. However, in DMSO, when acidity functions are not used, the pK_as of these acids are 0.3, 0.9, 1.8, and 1.6, respectively. Explain why there is very little difference between these acids in DMSO.

18. The preferred conformation of carboxylic acids is called s-*trans*, where the hydrogen on the OH is trans to the R group, as shown for acetic acid below. *Ab initio* computational modeling finds a 5.9 kcal/mol preference in the gas phase for acetic acid. However, further computational work modeling aqueous solvation found only a 1.7 kcal/mol preference. First, state

which lone pair in a carboxylate anion is the more basic, the syn or anti lone pair. Second, explain why there is a difference in the equilibria between the gas phase and water, and why the preference is smaller in water.

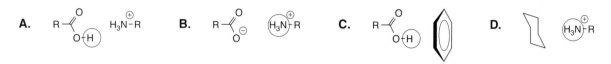

19. Predict whether you would expect the pK_a values for the circled acids to shift up or down from their normal aqueous solution values when in proximity to the species shown. Explain your answers.

A.   **B.**   **C.**   **D.**

20. Using Figure 5.2, predict the percent protonation of the following aniline in 0.5 mole fraction H_2SO_4. The pK_a of the conjugate acid of this aniline is –6.2.

21. Explain the following trend in acidities.

22. The traditional reasoning for why phenol is more acidic than cyclohexanol resides in a resonance analysis. What is another reason that phenol is more acidic?

Further Reading

Brønsted Acid–Base Chemistry

Muller, P. "Glossary of Terms Used in Physical Organic Chemistry." *Pure Appl. Chem.*, **66**, 1077 (1994).
Bell, R. P. (1973). *The Proton in Chemistry*, 2d ed., Cornell University Press, Ithaca, NY.
Albert, A., and Serjeant, E. P. (1984). *The Determination of Ionization Constants: A Laboratory Manual*, 3d ed., Chapman and Hall, London, p. 203.
Bates, R. G. (1964). *Determination of pH: Theory and Practice*, John Wiley & Sons, Inc., New York.
Stewart, R. (1985). *The Proton: Applications to Organic Chemistry*, Academic Press, New York.

Techniques to Study Acid–Base Reactions in the Gas Phase

Van Doren, J. M., Barlow, S. E., DePuy, C. H., and BierBaum, V. M. "The Tandem Flowing Afterglow–Sift–Drift." *Int. J. Mass Spectron, Ion Proc.*, **81**, 85 (1987).
Pellerite, M. J., and Brauman, J. I. in *Comprehensive Carbanion Chemistry, Part A: Structure and Reactivity*, E. Buncel and T. Durst (eds.), Elsevier Scientific Publishing Company, Amsterdam, 1980, pp. 55 *ff*.
Aue, D., and Bowers, M. T. in *Gas Phase Ion Chemistry*, M. T. Bowers (ed.), Academic Press, New York, 1979, Vol. 2, pp. 1–51.

Bartness, J. E., and McIver, R. T., Jr. in *Gas Phase Ion Chemistry*, M. J. Bowers (ed.), Academic Press, New York, 1979, Vol. 2, pp. 87–121.

Gas Phase Acidities

Siggel, M. R. F., and Thomas, T. D. "The Anomalous Gas-Phase Acidity of Formic Acid. Importance of Initial State Polarization." *J. Am. Chem. Soc.*, **114**, 5795 (1992).

Rodriquez, C. F., Sirois, S., and Hopkinson, A. C. "Effect of Multiple Halide Substituents on the Acidity of Methanes and Methyl Radicals. Electron Affinities of Chloro and Fluoromethyl Radicals." *J. Org. Chem.*, **57**, 4869 (1992).

Solvation Effects on Acid–Base Chemistry

Tunon, I., Silla, E., and Pascual-Ahuir, J.-L. "Theoretical Study of the Inversion of the Alcohol Acidity Scale in Aqueous Solution. Toward an Interpretation of the Acid–Base Behavior of Organic Compounds in Solution." *J. Am. Chem. Soc.*, **115**, 2226 (1993).

Brauman, J. I., and Blair, L. K. "Alkyl Substitute Effects on Gas-Phase Acidities. The Influence of Hybridization." *J. Am. Chem. Soc.*, **93**, 4315 (1971).

Brinck, T., Murray, J. S., and Politzer, P. "Relationships Between the Aqueous Acidities of Some Carbon, Oxygen, and Nitrogen Acids and the Calculated Surface Local Ionization Energies of Their Conjugate Bases." *J. Org, Chem.*, **56**, 5012 (1991).

Acid Dissociation Constants

Serjeant, E. P., and Dempsey, B. (1979). *Ionization Constants of Organic Acids in Aqueous Solution*, Pergamon Press, Oxford, England.

Kortum, G., Vogel, W., and Andrussow, K. (1961). *Dissociation Constants of Organic Acids in Aqueous Solution*, Butterworths, London.

Perrin, D. D., Dempsey, B., and Serjeant, E. P. (1981). *pKa Prediction for Organic Acids and Bases*, Chapman and Hall, London.

Barlin, G. B., and Perrin, D. D. "Prediction of the Strengths of Organic Acids." *Quart. Rev. Chem. Soc.*, **20**, 75 (1966).

Albert, A., and Serjeant, E. P. (1962). *Ionization Constants of Acids and Bases, A Laboratory Manual*, Methuen and Co., Ltd., London.

The Leveling Effect

Hammett, L. P. (1970). *Physical Organic Chemistry: Reaction Rates, Equilibria and Mechanisms*, 2d ed., McGraw–Hill Book Company, New York, pp. 272–273.

Hammett Acidity Functions

Cox, R. A., and Yates, K. "Excess Acidities. A Generalized Method for the Determination of Basicities in Aqueous Mixtures." *J. Am. Chem. Soc.*, **100**, 3861 (1978).

Cox, R. A., and Yates, K. "The Excess Acidity of Aqueous HCl and HBr Media. An Improved Method For the Calculation of x-Function and H_o Scales." *Can. J. Chem.*, **59**, 2116 (1980).

Stewart, R. (1985). *The Proton: Applications To Organic Chemistry*, Academic Press, New York.

H_ Scale

Paul, M. A., and Long, F. A. "H_o and Related Indicator Acidity Functions." *Chem. Rev.*, **57**, 1 (1957).

Long, F. A., and Paul, M. A. "Application of the H_o Acidity Function to Kinetic and Mechanisms of Acid Catalysis." *Chem. Rev.*, **57**, 935 (1957).

Hammett, L. P. (1970). *Physical Organic Chemistry: Reaction Rates, Equilibria and Mechanisms*, 2d ed., McGraw–Hill Book Company, New York, p. 228.

Activity Coefficient

Albert, A., and Serjeant, E. P. (1984). *The Determination of Ionization Constants: A Laboratory Manual*, 3d ed., Chapman and Hall, London, p. 4.

Super Acids

Olah, G. A., Prakash, G. K. S., and Sommer, J. (1985). *Super Acids*, John Wiley & Sons, Inc., New York.

DMSO vs. Water and pK_a Value

Bordwell, F. G. "Equilibrium Acidities in Dimethyl Sulfoxide Solution." *Acc. Chem. Res.*, **21**, 456 (1988).

Taft, R. W., and Bordwell, F. G. "Structural and Solvent Effects Evaluated From Acidities Measured in Dimethyl Sulfoxide and in the Gas Phase." *Acc. Chem. Res.*, **21**, 463 (1988).

Ion Pair Acidity in Nonaqueous Solvent

Kaufman, M. J., Gronert, S., and Streitwieser, A., Jr. "Carbon Acidity. 73. Conductimetric Study of Lithium and Cesium Salts of Hydrocarbon Acids. A Scale of Free Ion Acidities in Tetrahydrofuran. Revision of the Ion Pair Scales." *J. Am. Chem. Soc.,* **110**, 2829 (1988).

Kaufman, M. J., and Streitwieser, A., Jr. "Carbon Acidity. 72. Ion Pair Acidities of Phenyl Alkyl Ketones. Aggregation Effects in Ion Pair Acidities." *J. Am. Chem. Soc.,* **109**, 6092 (1987).

Gronert, S., and Streitwieser, A., Jr. "Carbon Acidity. 74. The Effects of Hetero-Substituted Pendant Groups on Carbanion Reactivity. Solvent-Separated-Conduct Ion Pair Equilibria and Relative pKLi/THF's for 9-Substitued Fluorenyllithiums in Tetrahydrofuran. The Importance of Internal Chelation." *J. Am. Chem. Soc.,* **110**, 2836 (1988).

Lewis Acid–Base Chemistry

Jensen, W. B. (1980). *The Lewis Acid–Base Concepts: An Overview,* Wiley–Interscience, New York.

CHAPTER 6

Stereochemistry

Intent and Purpose

Stereochemistry is the study of the static and dynamic aspects of the three-dimensional shapes of molecules. It has long provided a foundation for understanding structure and reactivity. At the same time, stereochemistry constitutes an intrinsically interesting research field in its own right. Many chemists find this area of study fascinating due simply to the aesthetic beauty associated with chemical structures, and the intriguing ability to combine the fields of geometry, topology, and chemistry in the study of three-dimensional shapes. In addition, there are extremely important practical ramifications of stereochemistry. Nature is inherently chiral because the building blocks of life (α-amino acids, nucleotides, and sugars) are chiral and appear in nature in enantiomerically pure forms. Hence, any substances created by humankind to interact with or modify nature are interacting with a chiral environment. This is an important issue for bioorganic chemists, and a practical issue for pharmaceutical chemists. The Food and Drug Administration (FDA) now requires that drugs be produced in enantiomerically pure forms, or that rigorous tests be performed to ensure that both enantiomers are safe.

In addition, stereochemistry is highly relevant to unnatural systems. As we will describe herein, the properties of synthetic polymers are extremely dependent upon the stereochemistry of the repeating units. Finally, the study of stereochemistry can be used to probe reaction mechanisms, and we will explore the stereochemical outcome of reactions throughout the chapters in parts II and III of this text. Hence, understanding stereochemistry is necessary for most fields of chemistry, making this chapter one of paramount importance.

All introductory organic chemistry courses teach the fundamentals of stereoisomerism, and we will only briefly review that information here. We also take a slightly more modern viewpoint, emphasizing newer terminology and concepts. The goal is for the student to gain a fundamental understanding of the basic principles of stereochemistry and the associated terminology, and then to present some of the modern problems and research topics in this area.

6.1 Stereogenicity and Stereoisomerism

Stereochemistry is a field that has often been especially challenging for students. No doubt one reason for this is the difficulty of visualizing three-dimensional objects, given two-dimensional representations on paper. Physical models and 3-D computer models can be of great help here, and the student is encouraged to use them as much as possible when working through this chapter. However, only simple wedges and dashes are given in most of our drawings. It is these kinds of simple representations that one must master, because attractive, computer generated pictures are not routinely available at the work bench. The most common convention is the familiar "wedge-and-dash" notation. Note that there is some variability in the symbolism used in the literature. Commonly, a dashed wedge that gets larger as it emanates from the point of attachment is used for a receding group. However, considering the art of perspective drawing, it makes no sense that the wedge gets bigger as

297

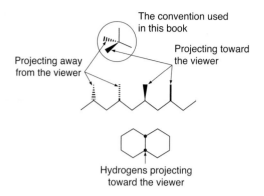

The convention used
in this book

Projecting toward
the viewer

Projecting away
from the viewer

Hydrogens projecting
toward the viewer

it moves further away. Yet, this is the most common convention used, and it is the convention we adopt in this book. Many workers have turned to a simple dashed line instead (see above), or a dash that does get smaller. Similarly, both a bold wedge and a bold line are used to represent forward-projecting substituents. Another common convention is the bold "dot" on a carbon at a ring junction, representing a hydrogen that projects toward the viewer.

The challenge of seeing, thinking, and drawing in three dimensions is not the only cause for confusion in the study of stereochemistry. Another major cause is the terminology used. Hence, we start this chapter off with a review of basic terminology, the problems associated with this terminology, and then an extension into more modern terminology.

6.1.1   Basic Concepts and Terminology

There was considerable ambiguity and imprecision in the terminology of stereochemistry as it developed during the 20th century. In recent years, stereochemical terminology has clarified. We present here a discussion of the basics, not focused solely on carbon. However, in Section 6.2.4 we will examine carbon specifically. While most of this should be review, perhaps the perspective and some of the terminology will be new.

Let's start by delineating the difference between a stereoisomer and other kinds of isomers. Recall that **stereoisomers** are molecules that have the same connectivity but differ in the arrangement of atoms in space, such as *cis*- and *trans*-2-butene. Even gauche and anti butane are therefore stereoisomers. This is in contrast to **constitutional isomers**, which are molecules with the same molecular formula but different connectivity between the atoms, such as 1-bromo- and 2-bromobutane. The **constitution** of a molecule is defined by the number and types of atoms and their connectivity, including bond multiplicity. These definitions are straightforward and clear (as long as we can agree on the definition of connectivity—see the Going Deeper highlight on page 300).

An historical distinction, but one that is not entirely clear cut, is that between **configurational isomers** and **conformational isomers**. Conformational isomers are interconvertible by rotations about single bonds, and the **conformation** of a molecule concerns features related to rotations about single bonds (see Chapter 2). There is some fuzziness to this distinction, attendant with the definition of a "single" bond. Is the C–N bond of an amide a single bond, even though resonance arguments imply a significant amount of double bond character and the rotation barrier is fairly large? Also, some olefinic "double" bonds can have quite low rotation barriers if the appropriate mix of substituents if present. Because of these examples, as well as other issues concerning stereochemistry, we simply have to live with a certain amount of terminological ambiguity. A related term is **atropisomers**, which are stereoisomers that can be interconverted by rotation about single bonds but for which the barrier to rotation is large enough that the stereoisomers can be separated and do not interconvert readily at room temperature (examples are given in Section 6.5).

The term configurational isomer is a historic one that has no real value in modern stereochemistry. It is generally used to encompass enantiomers and diastereomers as isomers (see definitions for these below), but stereochemical isomers is a better term. The term **con-**

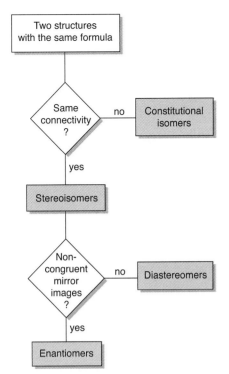

Figure 6.1
Simple flowchart for classifying various kinds of isomers.

figuration is still useful. Mislow defines configuration as "the relative position or order of the arrangement of atoms in space which characterizes a particular stereoisomer". A related term is **absolute configuration**, which relates the configuration of a structure to an agreed upon stereochemical standard. For example, later in this chapter we discuss the D and L nomenclature system, where the arrangement of atoms in space is related to that of (+)-glyceraldehyde. If the arrangement of atoms in space in a molecule can be related to (+)-glyceraldehyde, or some other standard, we state that we know that molecule's absolute configuration.

When two stereoisomers are nonsuperposable mirror images of each other, they are known as **enantiomers** (see the schematic examples in the margin). To achieve the mirror image of a molecule, simply imagine a sheet of glass placed alongside the molecule of interest, then pass each atom through the glass such that each atom ends up the same distance from the sheet of glass as in the original structure. Stereoisomers that are not enantiomers are known as **diastereomers**. Figure 6.1 shows a simple flow chart for classifying isomers.

Any object that is nonsuperposable (noncongruent) with its mirror image is **chiral**. If an object is not chiral—that is, if its mirror image is congruent with the original—it is **achiral**.

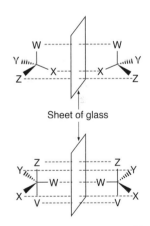

Enantiomers: non-superimposable
mirror images

Classic Terminology

There are a series of terms used in the context of stereochemistry that are ingrained in the literature, and several you are likely familiar with from beginning organic chemistry. We define many of these terms here, and examine how they can be misleading. After a look at this classic terminology, more modern and concise terms are given.

Confusion with respect to terminology arises with terms such as "optically active" and "chiral center", which often mislead as much as they inform. **Optically active** refers to the ability of a collection of molecules to rotate plane polarized light (a phenomenon that we explore in detail in Section 6.1.3). In order for a sample to be optically active, it must have an excess of one enantiomer. Now comes the confusion. Optically active was generally used as a synonym for chiral in the earlier literature, and unfortunately this usage continues at times even today. We discourage this use. The problem is that there are many examples of chemical

Connections

Stereoisomerism and Connectivity

A crucial concept in the definition of stereoisomers given above is "connectivity". In methane or 2,3-dichlorobutane, there is no doubt as to the connectivity of the system. However, there is an innate arbitrariness to the term, and this can lead to some ambiguity about stereoisomerism. For example, do hydrogen bonds count in our list of connectivity? No, but consider the implications of this. If hydrogen bonds "don't count", then how do we think about isomerism in double-helical DNA? Do we just ignore the interaction of the two strands? As a simpler example, in a solution of a racemic carboxylic acid, does dimerization create true diastereomers?

Do diastereomers exist in a solution of enantiomeric carboxylic acids?

Further, what about metal coordination? We are comfortable with a clear connectivity pattern in inorganic complexes such as iron pentacarbonyl or a porphyrin complex. But what about Mg^{2+} ions complexing a carbonyl? When is a bond too weak to be considered relevant for stereoisomerism?

Stereoisomers?

Finally, there has been a modern emphasis on "topological isomerism", structures with loops or interlocking rings in which large parts of the molecule are not connected to each other in any conventional way. This can produce novel stereochemical situations, as we will see in Section 6.6.

In the end, there is no universally agreed upon convention for connectivity as it relates to stereoisomerism. Usually, the connectivity of a system is clear. When there is the potential for ambiguity, though, a clear statement of the ground rules should be made.

---

samples that contain chiral molecules, but the samples themselves are not optically active. A **racemic** mixture, a 50:50 mixture of enantiomers, is not optically active, but every molecule in the sample is chiral. It is important to distinguish between a sample that is optically inactive because it contains a racemic mixture and a sample that is optically inactive because it contains achiral molecules, and the earlier terminology made this difficult.

Also, it is easy to imagine molecules, even when enantiomerically pure, that would not rotate plane polarized light to any *measurable* extent. The extent of rotation of plane polarized light depends upon differences in the refractive indices with respect to right and left circularly polarized light as it passes through the sample. Enantiomers that do not have dramatically different refractive indices would not result in measurable rotations. Examples would be a carbon with four different *n*-alkyl chains attached, with chain lengths of maybe 10, 11, 12, and 13 carbons; or one with four C_{10} chains, but terminating in $-CH_3$, $-CH_2D$, $-CHD_2$, and $-CD_3$. In each case the molecule is chiral, but any rotation of plane polarized light would be immeasurably small. Operationally, they are optically inactive. Finally, even an enantiomerically pure sample of a chiral molecule will show zero rotation at certain wavelengths of light, as we move from (+) rotation to (−) rotation in the optical rotatory dispersion (ORD) curve (see Section 6.1.3). "Optically active" is an ambiguous description.

More confusion arises with terms that are meant to focus on the chirality at a particular point in a molecule. The prototype is the **chiral center** or **chiral carbon**, which is defined as an atom or specifically carbon, respectively, that has four different ligands attached. Here, the term "ligand" refers to any group attached to the carbon, such as H, R, Ar, OH, etc. The particular case of a carbon with four different ligands has also been termed an **asymmetric carbon**. One problem with such terms, as we will show below, is that "asymmetric carbons" and "chiral centers/carbons" exist in molecules that are neither asymmetric nor chiral. In addition, many molecules can exist in enantiomeric forms without having a "chiral center". Classic examples include dimethylallene and the twisted biphenyl shown in the margin—we'll see more below. Given all this, although the terms may already be part of your vocabulary, we discourage their use.

Chiral molecules without a "chiral center"

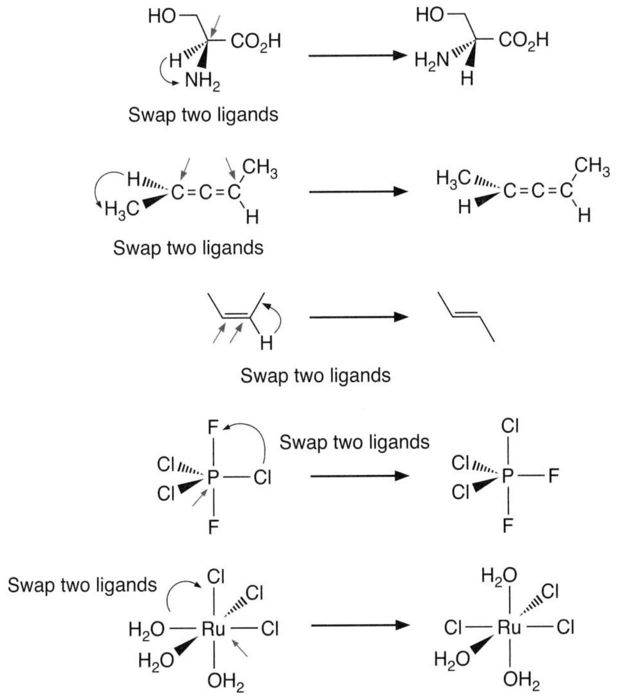

Figure 6.2
Molecules with stereogenic centers. The stereogenic centers are marked with colored arrows, and a curved black arrow is used to show how ligand interchange at a stereogenic center produces a new stereoisomer.

More Modern Terminology

Much of the confusion that can be generated with the terms given above was eliminated with the introduction of the **stereogenic center** (or, equivalently, **stereocenter**) as an organizing principle in stereochemistry. An atom, or a grouping of atoms, is considered to be a stereogenic center if the interchange of two ligands attached to it can produce a new stereoisomer. Not all interchanges have to give a new stereoisomer, but if one does, then the center is stereogenic. The center therefore "generates" stereochemistry. A **non-stereogenic center** is one in which exchange of any pair of ligands does not produce a stereoisomer. The term "stereogenic center" is, in a sense, broader than the term "chiral center". It implies nothing about the molecule being chiral, only that stereoisomerism is possible. The structures in Figure 6.2 show several stereogenic centers. Note that in more complex geometries, such as pentacoordinate or hexacoordinate atoms, we do not need all the ligands to be inequivalent in order to have a stereogenic center. Given these new terms, we strongly encourage students to abandon the term "chiral center" and to reserve "optically active" as a description of an experimental measurement.

A related and more encompassing concept is that of a **stereogenic unit**. A stereogenic unit is an atom or grouping of atoms such that interchange of a pair of ligands attached to an atom of the grouping produces a new stereoisomer. For example, the $C=C$ group of *trans*-2-butene is a stereogenic unit because swapping a CH_3/H pair at one carbon produces *cis*-2-butene. A tetrahedral atom is a stereogenic unit, where swapping the positions of any two of four different ligands gives a stereoisomer (see below).

In the examples of chiral molecules without "chiral centers" noted above, the $C=C=C$ unit of the allene and the biphenyl itself are stereogenic units. Many workers have adopted terms such as **planar chirality** and **axial chirality** to describe systems such as chiral biphenyl and allene based structures, respectively. The justification for these terms is that such molecules do not have stereogenic centers, but rather stereogenic units. Admittedly, terms that address chirality without stereogenic centers could be useful. However, since a molecule that is truly planar (i.e., has a plane of symmetry) must be achiral, planar chirality is an odd use of the word "planar". Developing precise, unambiguous definitions of these terms is a challenge that, in our view, has not yet been met. Currently, the best term is "stereogenic unit", where the biphenyl or allene groups have the ability to create chirality, just as a tetrahedral atom has the ability to generate chirality.

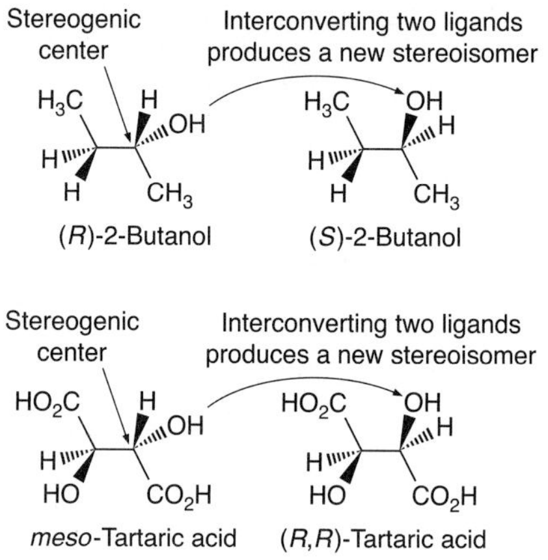

Figure 6.3
Illustration of the concept of the stereogenic center in the context of carbon. Whether in a chiral molecule like 2-butanol or an achiral molecule like *meso*-tartaric acid, interconversion of two ligands at a stereocenter produces a new stereoisomer.

To illustrate the value of the newer terminology, let's review two prototypes of organic stereochemistry. First, consider a molecule that has a carbon with four different ligands, a carbon we will describe as CWXYZ. A specific example is 2-butanol (Figure 6.3). If we interchange any two ligands at carbon 2, we obtain a stereoisomer—the enantiomer—of the original structure. Thus, C2 of 2-butanol is a stereogenic center. The analysis can get more complicated in systems with more than one CWXYZ center. Let's consider such a case.

Figure 6.3 also shows tartaric acid. Beginning with the structure labeled "meso", if we interchange two ligands at either C2 or C3, we obtain a new structure, such as (R,R)-tartaric acid. (If you do not recall the R and S notation, look ahead to Section 6.1.2.) This structure has the same connectivity as *meso*-tartaric acid, but the two are not congruent (verify for yourself), and so the new structure is a stereoisomer of the original. However, (R,R)- and *meso*-tartaric acid are *not* mirror images, so they are not enantiomers. They are diastereomers.

Note that the meso form of tartaric acid is achiral; verify for yourself that it is congruent with its mirror image. However, C2 and C3 of *meso*-tartaric acid are stereogenic centers; that is, swapping any two ligands at either center produces a new stereoisomer. This is one value of the stereogenic center concept. As we noted above, in earlier literature a CWXYZ center such as C2 or C3 was called a chiral center, but it seems odd to say we have *two* chiral centers in an achiral molecule! A CWXYZ center does not guarantee a chiral molecule. However, *a CWXYZ group is always a stereogenic center.*

Tartaric acid has two stereogenic centers and exists as three possible stereoisomers. This is an exception to the norm. Typically, a molecule with n stereogenic, tetracoordinate carbons will have 2^n stereoisomers–2^{n-1} diastereomers that each exist as a pair of enantiomers. For example, a structure with two stereogenic centers will exist as RR, SS, RS, and SR forms. In tartaric acid the RS and SR forms are identical—they are both the meso form—because C2 and C3 have the same ligands.

The 2^n rule quickly creates complexity in molecules with multiple stereogenic centers. In complex natural products that are often targets of total synthesis efforts, it is conventional to note the number of possible stereoisomers (for example, 10 stereogenic centers implies 1024 stereoisomers), with only one combination defining the proper target (see the Following Connections highlight). Polymers, both natural and synthetic, can produce extraordinary stereochemical diversity when each monomer carries a stereogenic center. We'll return to this issue below.

When many stereogenic centers are present in a molecule, it becomes difficult to refer to all the possible stereoisomers. It is often useful to consider only two different isomers, called epimers. **Epimers** are diastereomers that differ in configuration at only one of the several stereogenic centers. Imagine taking any one of the many stereogenic centers in everninomicin (shown in the next Connections highlight) and changing the stereochemistry at only that one stereogenic center. This creates an epimer of the original structure. Another example is the difference between the α- and β-anomers of glucose, which are epimeric forms of the sugar (look ahead to Figure 6.18 for definitions of α- and β-anomers).

Connections

Total Synthesis of an Antibiotic with a Staggering Number of Stereocenters

Synthetic chemists are continually in search of new methods to control the stereochemical outcome of synthetic transformations. Although the exact methods used are best described in textbooks with a focus upon asymmetric synthesis, it is worth mentioning here how sophisticated the field is becoming. By analyzing how the topicity relationships within reactants will influence enantiomeric and disasteromeric selectivities, a multitude of reactions with good stereochemical control have been developed. One particular example that highlights just how far advanced

these techniques have become is the total synthesis of everninomicin 13,384–1. This compound contains 13 rings and 35 stereocenters (3.4×10^{10} possible stereoisomers). Although many of the stereocenters were derived from the "chiral pool" (see Section 6.8.3), several stereocenters associated with the ring connections and ring-fusions were set with reactions that proceed with varying degrees of stereoselectivity and specificity.

Nicolaou, K. C., Mitchell, H. J., Suzuki, H., Rodriguez, R. M., Baudoin, O., and Fylaktakidou, C. "Total Synthesis or Everninomicin 13,384–1—Part 1: Synthesis of A1B(A)C Fragment." *Angew. Chem. Int. Ed. Eng.*, **38**, 3334–3339 (1999), and subsequent communications.

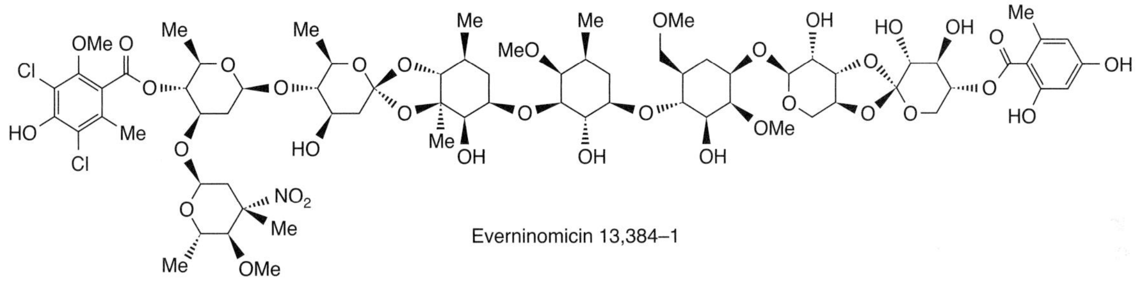

Everninomicin 13,384–1

6.1.2 Stereochemical Descriptors

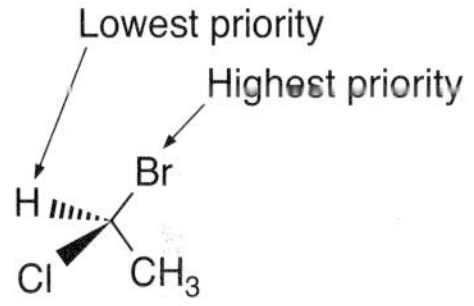

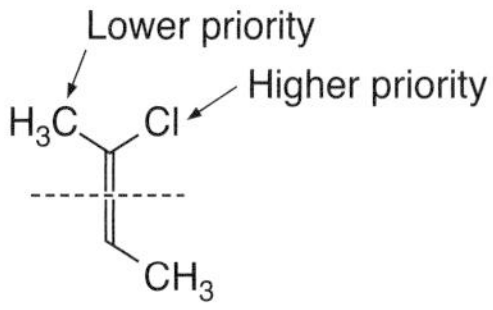

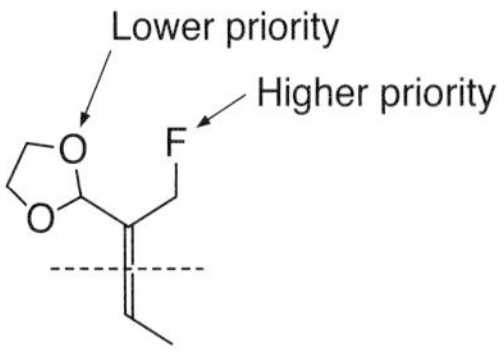

All introductory organic chemistry texts provide a detailed presentation of the various rules for assigning descriptors to stereocenters. Here we provide a brief review of the terminology to remind the student of the basics.

Many of the descriptors for stereogenic units begin with assigning priorities to the attached ligands. Higher atomic number gets higher priority. If two atoms under comparison are isotopes, the one with higher mass is assigned the higher priority. Ties are settled by moving out from the stereocenter until a distinction is made. In other words, when two attached atoms are the same, one examines the next atoms in the group, only looking for a winner by examining individual atomic numbers (do not add atomic numbers of several atoms).

Multiple bonds are treated as multiple ligands; that is, C=O is treated as a C that is singly bonded to two oxygens with one oxygen bound to a C. For example, the priorities shown below for the substituted alkene are obtained, giving an *E*-stereochemistry.

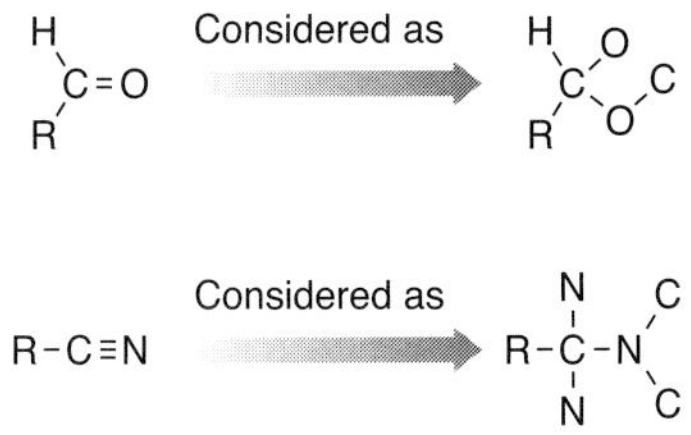

Considered as

Considered as

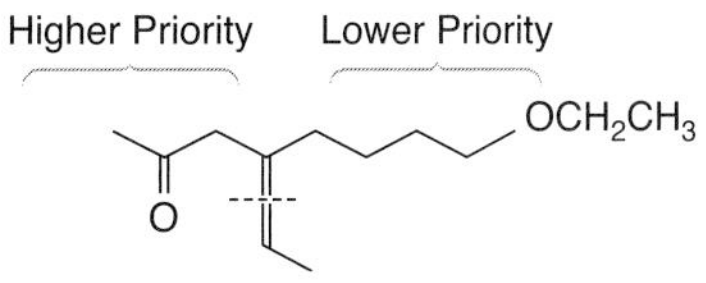

Higher Priority   Lower Priority

An *E*-alkene

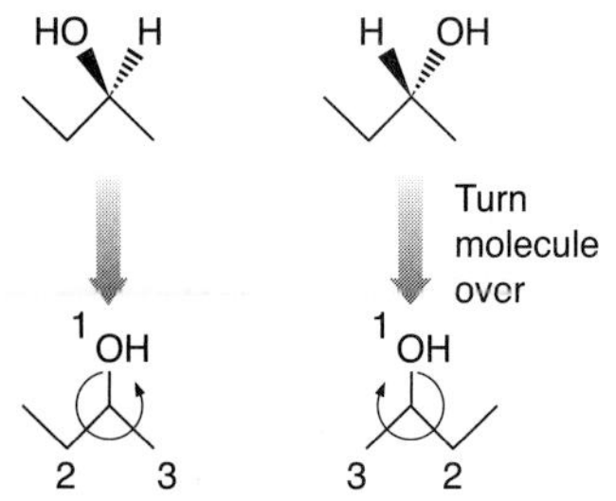

R,S System

For tetracoordinate carbon and related structures we use the Cahn–Ingold–Prelog system. The highest priority group is given number 1, whereas the lowest priority group is given number 4. Sight down the bond from the stereocenter to the ligand of lowest priority behind. If moving from the highest (#1), to the second (#2), to the third (#3) priority ligand involves a clockwise direction, the center is termed *R*. A counterclockwise direction implies *S*.

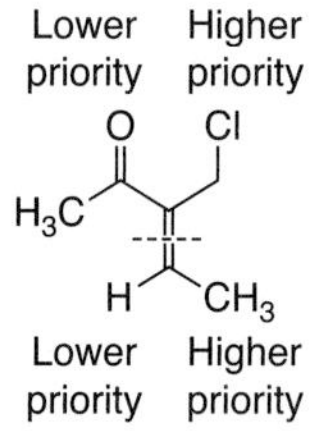

E,Z System

For olefins and related structures we use the same priority rules, but we divide the double bond in half and compare the two sides. For each carbon of an olefin, assign one ligand high priority and one low priority according to the rules above. If the two high priority ligands lie on the same side of the double bond, the system is *Z* (zusammen); if they are on opposite sides, the system is *E* (entgegen). If an H atom is on each carbon of the double bond, however, we can also use the traditional "cis" and "trans" descriptors.

D and L

The descriptors D and L represent an older system for distinguishing enantiomers, relating the sense of chirality of any molecule to that of D- and L-glyceraldehyde. D- and L-glyceraldehyde are shown below in **Fischer projection** form. In a Fischer projection, the horizontal lines represent bonds coming out of the plane of the paper, while the vertical lines represent bonds projecting behind the plane of the paper. You may want to review an introductory text if you are unfamiliar with Fischer projections. The isomer of glyceraldehyde that rotates plane polarized light to the right (*d*) was labelled D, while the isomer that rotates plane polarized light to the left (*l*) was labelled L.

To name more complex carbohydrates or amino acids, one draws a similar Fischer projection where the CH_2OH or R is on the bottom and the carbonyl group (aldehyde, ketone, or carboxylic acid) is on the top. The D descriptor is used when the OH or NH_2 on the penultimate (second from the bottom) carbon points to the right, as in D-glyceraldehyde, and L is used when the OH or NH_2 points to the left. See the following examples.

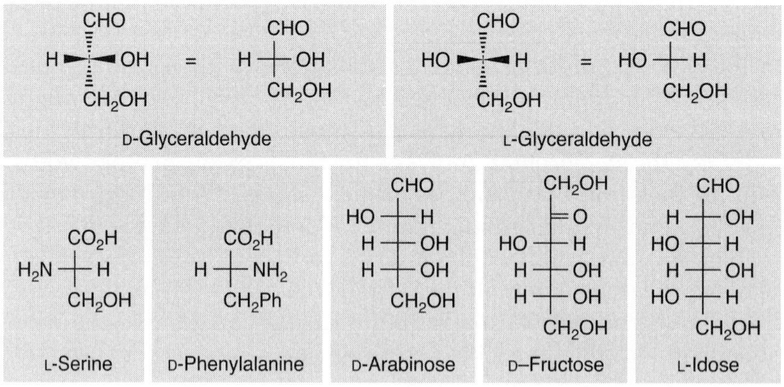

The D and L nomenclature system is fundamentally different than the *R/S* or *E/Z* systems. The D and L descriptors derive from only one stereogenic center in the molecule and are used to name the entire molecule. The name of the sugar defines the stereochemistry of all the other stereogenic centers. Each sugar has a different arrangement of the stereogenic centers along the carbon backbone. In contrast, normally a separate *R/S* or *E/Z* descriptor is used to name each individual stereogenic unit in a molecule. The D/L nomenclature is a carry over from very early carbohydrate chemistry. The terms are now reserved primarily for sugars and amino acids. Thus, it is commonly stated that all natural amino acids are L, while natural sugars are D.

Erythro and Threo

Another set of terms that derive from the stereochemistry of saccharides are erythro and threo. The sugars shown below are D-erythrose and D-threose, which are the basis of a nomenclature system for compounds with two stereogenic centers. If the two stereogenic centers have two groups in common, we can assign the terms erythro and threo. To determine the use of the erythro and threo descriptors, draw the compound in a Fischer projection with the distinguishing groups on the top and bottom. If the groups that are the same are both on the right or left side, the compound is called **erythro**; if they are on opposite sides, the compound is called **threo**. See the examples given below. Note that these structures have enantiomers, and hence require R and S descriptors to distinguish the specific enantiomer. The erythro/threo system distinguishes diastereomers.

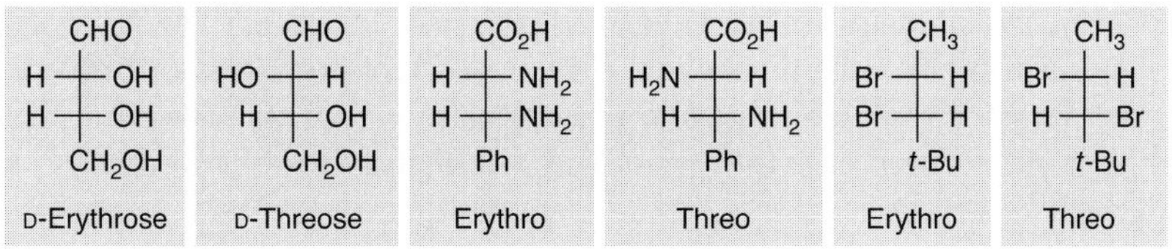

Helical Descriptors—M and P

Many chiral molecules lack a conventional center that can be described by the R/S or E/Z nomenclature system. Typically these molecules can be viewed as helical, and may have propeller, or screw-shaped structures. To assign a descriptor to the sense of twist of such structures, we sight down an axis that can be associated with the helix, and consider separately the "near" and "far" substituents, with the near groups taking priority. We then determine the highest priority near group and the highest priority far group. Sighting down the axis, if moving from the near group of highest priority to the corresponding far group requires a clockwise rotation, the helix is a right-handed helix and is described as P (or plus). A counterclockwise rotation implies a left-handed helix and is designated as M (or minus). As in all issues related to helicity, it does not matter what direction we sight down the axis, because we will arrive at the same descriptor. Three examples of molecules with M/P descriptors are shown below.

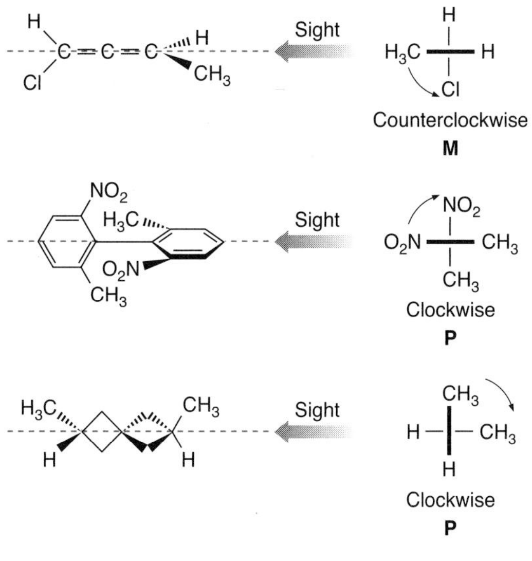

As another example, consider triphenylborane (Eq. 6.1, where a, b, and c are just labels of hydrogens so that you can keep track of the rotations shown). Triphenylborane cannot be fully planar because of steric crowding, and so it adopts a conformation with all three rings twisted in the same direction, making a right- or left-handed propeller. The M or P descriptors are most easily assigned by making an analogy to a common screw or bolt. Common screws or bolts are right-handed ("reverse thread" screws and bolts are left-handed). If the sense of twist is the same as a screw or bolt, it is assigned the P descriptor (check the P and M descriptors for yourself in Eq. 6.1).

$$(\text{Eq. } 6.1)$$

Rotation about the C–B bonds of triphenylborane is relatively facile, and the motions of the rings are correlated in the sense shown (Eq. 6.1). In Eq. 6.1 the arrows denote the direction of bond rotation, not the helical direction. Two rings rotate through a perpendicular conformation while one moves in the opposite way. This "two-ring flip" reverses helicity and, in a substituted case (now a, b, and c in Eq. 6.1 are substituents), creates a new diastereomer.

Ent and Epi

Because of the stereochemical complexity of many natural products, short and simple descriptors have come into common use to relate various stereochemical relationships. For example, the enantiomer of a structure with many stereogenic centers has the prefix **ent-**. Ent-everninomicin is a trivial name that can be given to the enantiomer of everninomicin. Similarly, due to the stereochemical complexity of many natural products, the prefix **epi-** has become a convenient way to name structures where only one stereogenic center has undergone a change in configuration. For example, any epimer of everninomicin can be called epi-everninomicin. Usually, a number precedes "epi-" to distinguish which center has changed configuration.

Using Descriptors to Compare Structures

Compounds that have the same sense of chirality at their individual stereogenic centers are called **homochiral**. Homochiral molecules are not identical—they just have the same sense of chirality, much like all people's right hands are distinct but of the same chirality. As a chemical example, the amino acids L-alanine and L-leucine are homochiral. Those molecules with a differing sense of chirality at their stereogenic centers are called **heterochiral**. The same sense of chirality can often, but not always, be analyzed by examining whether the different kinds of stereochemical descriptors at the stereogenic centers are the same. For example, (R)-2-butanol and (R)-2-aminobutane are homochiral. Further, all the naturally occurring amino acids are L, so they are all homochiral (see the next Connections highlight).

Homochiral has been used by some as a synonym for "enantiomerically pure". This is another usage of a term that should be discouraged, as homochiral already had a clear and useful definition, and using the same term to signify two completely different concepts can only lead to confusion. A better term for designating an enantiomerically pure sample is simply **enantiopure**.

6.1.3   Distinguishing Enantiomers

Enantiomers are distinguishable if and only if they are placed in a chiral environment, and all methods to separate or characterize enantiomers are based on this principle. Suppose, for example, that we have a collection of right- and left-handed gloves, and we want to retrieve only the right-handed ones. Using a simple hook to reach into the pile cannot succeed because a hook is achiral—it cannot distinguish handedness. A chiral object, however, like a right hand, can distinguish between the gloves just by trying them on.

Connections

The Descriptors for the Amino Acids Can Lead to Confusion

As just noted, all amino acids have the same sense of chirality in that they are all L in the D/L terminology system. Yet, in the more modern Cahn–Ingold–Prelog system, they do not all have the same designators. All have the S stereochemisty, except cysteine, which has the same sense of chirality but is R because the sulfur makes the sidechain have a higher priority than the carbonyl carbon. In addition, the amino acids threonine and isoleucine have two stereocenters and can exist as diastereomers. In the natural amino acids, the sidechain is R for threonine and S for isoleucine. The diastereomers obtained by reversing the stereocenter at the sidechain only are termed allo-threonine and allo-isoleucine.

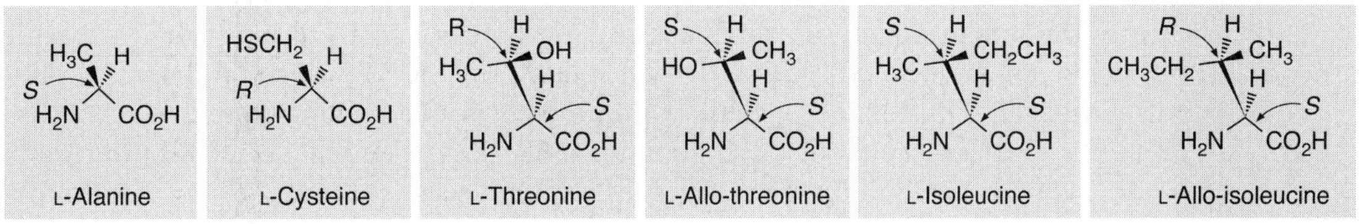

L-Alanine L-Cysteine L-Threonine L-Allo-threonine L-Isoleucine L-Allo-isoleucine

Figure 6.4 shows some chemical examples of this. If a racemic mixture of 2-aminobutane is allowed to react with an enantiomerically pure sample of mandelic acid, the two amides that are produced are diastereomers. The two diastereomers can be separated by any conventional method (such as crystallization or chromatography), and subsequent hydrolysis of a pure diastereomer gives enantiomerically pure 2-aminobutane.

The interaction that creates diastereomers out of enantiomers need not be covalent. Weaker, non-covalent complexes are often discriminating enough to allow separation of enantiomers. The most classical way to separate enantiomeric amines is to form salts with a

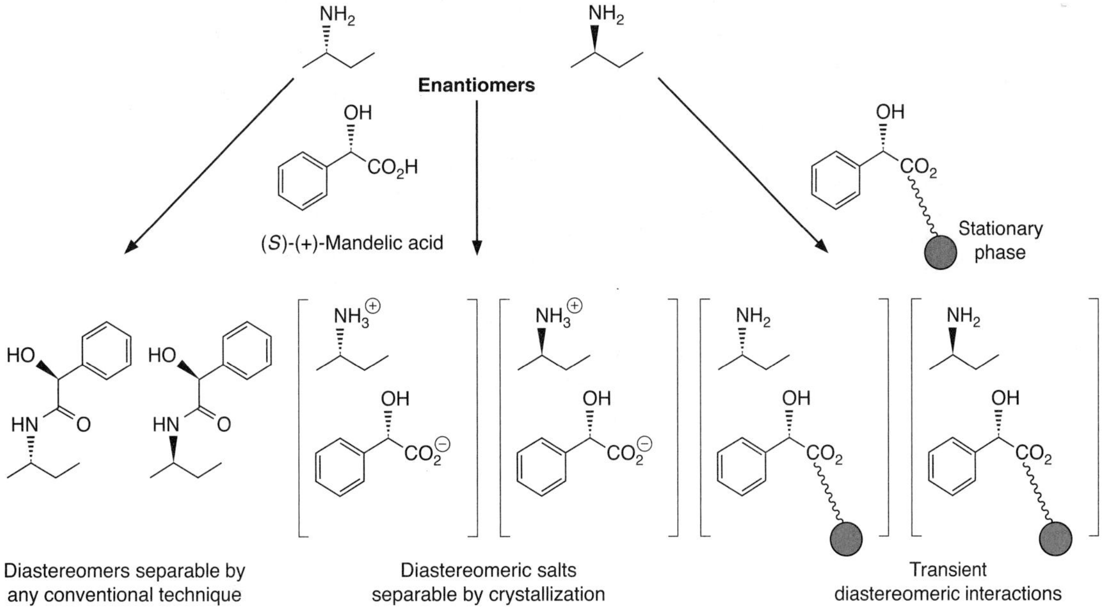

Diastereomers separable by any conventional technique

Diastereomeric salts separable by crystallization

Transient diastereomeric interactions

Figure 6.4
Strategies for separating enantiomers, using 2-aminobutane as an example. Left: Forming diastereomeric derivatives—in this case, amides of mandelic acid. Center: Forming diastereomeric salts that can be separated by crystallization. Right: Chiral chromatography, making use of transient, diastereomeric interactions between the enantiomers of 2-aminobutane and the chiral stationary phase.

chiral acid and use crystallization to separate the diastereomeric salts. There are many variations on this theme, and this traditional approach is still very commonly used, especially for large scale, industrial applications.

For the smaller scales associated with the research laboratory, chiral chromatography is increasingly becoming the method of choice for analyzing and separating mixtures of enantiomers. We show in Figure 6.4 a hypothetical system in which the mandelic acid we have used in the previous examples is attached to a stationary phase. Now, transient, diastereomeric interactions between the 2-aminobutane and the stationary phase lead to different retention times and thus to separation of the enantiomers. Both gas chromatography and liquid chromatography are commonly used to separate enantiomers.

With a tool to discriminate enantiomers in hand, we can determine the **enantiomeric excess (ee)** of a sample. This commonly used metric is defined as $X_a - X_b$, where X_a and X_b represent the mole fraction of enantiomers a and b, respectively. Usually ee is expressed as a percentage, which is $100\%(X_a - X_b)$. Analogous terms such as **diastereomeric excess (de)** are also used. The traditional tools for evaluating ee are the chiroptical methods discussed below. However, methods such as high field NMR spectroscopy with chiral shift reagents (see the Going Deeper highlight below), NMR spectroscopy of derivatives that are diastereomeric, and chromatography (HPLC and GC) with chiral stationary phases, are becoming ever more powerful and popular.

Going Deeper

Chiral Shift Reagents

A convenient technique to measure the ratio of enantiomers in a solution is to differentiate them in the NMR spectrum using what is known as a **chiral shift reagent**. These reagents are typically paramagnetic, enantiomerically pure metal compounds that associate with the enantiomers to form complexes. The complexes formed between the chiral shift reagent and the enantiomers are diastereomeric, and thus can be resolved in NMR spectroscopy. The paramagnetic nature of the reagents induces large chemical shifts, further assisting with the resolution of the spectral peaks associated with the diastereomeric complexes.

For example, the enantiomeric forms of 2-deuterio-2-phenylethanol can be readily distinguished in the NMR using a complex known as Eu(dcm). Coordination of the alcohol to the Eu center leads to diastereomers. The ^{1}H NMR spectrum shown to the side of the H on the stereogenic center of 2-deuterio-2-phenylethanol indicates that the two enantiomers (in a 50:50 ratio) are easily distinguished.

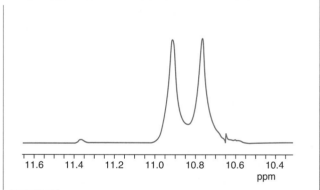

McCreary, M. D., Lewis, D. W., Wernick, D. L., and Whitesides, G. M. "Determination of Enantiomeric Purity Using Chiral Lanthanide Shift Reagents." *J. Am. Chem. Soc.*, **96**, 1038 (1974). Buchwald, S. L., Anslyn, E. V., and Grubbs, R. H. "Reaction of Dicyclopentadienylmethylenetitanium with Organic Halides: Evidence for a Radical Mechanism." *J. Am. Chem. Soc.*, **107**, 1766 (1985).

Eu(dcm)   Eu$_3$ +   50/50

Optical Activity and Chirality

Historically, the most common technique used to detect chirality and to distinguish enantiomers has been to determine whether a sample rotates plane polarized light. **Optical activity** and other **chiroptical properties** that can be measured using ORD and CD (see below) have long been essential for characterizing enantiomers. Their importance has lessened somewhat with the development of powerful NMR methods and chiral chromatographic methods, but their historical importance justifies a brief discussion of the methodology.

All introductory organic chemistry textbooks cover the notion of optical activity—the ability of a sample to rotate a plane of polarized light. We check to see if the plane in which the polarized light is oscillating has changed by some angle relative to the original plane of oscillation on passing through the sample. A solution consisting of a mixture of enantiomers at a ratio other than 50:50 can rotate plane polarized light to either the right (clockwise) or the left (counterclockwise). A rotation to the right is designated (+); a rotation to the left is designated (–). Earlier nomenclature used **dextrorotatory** (designated as *d*) or **levorotatory** (designated as *l*) instead of (+) or (–), respectively. Typically, light of one particular wavelength, the Na "D-line" emission, is used in such studies. However, we can in principle use any wavelength, and a plot of optical rotation vs. wavelength is called an **optical rotatory dispersion** (ORD) curve. Note that as we scan over a range of wavelengths, any sample will have some wavelength regions with + rotation and others with – rotation. Since the rotation must pass through zero rotation as it changes from + to –, any chiral sample will be optically inactive at some wavelengths. If one of those unique wavelengths happens to be at (or near) the Na D line, we could be seriously misled by simple optical activity measurements. Furthermore, at the Na D line, rotation is often small for conventional organic molecules. In addition, we previously discussed instances in which a chiral sample might be expected to fail to rotate plane polarized light. Thus, optical activity establishes that a sample is chiral, but a lack of optical activity does not prove a lack of chirality.

Why is Plane Polarized Light Rotated by a Chiral Medium?

We have said that we need a chiral environment to distinguish enantiomers, and so it may seem odd that **plane polarized light** can do so. To understand this, we must recall that electromagnetic radiation consists of electric and magnetic fields that oscillate at right angles to each other and to the direction of propagation (see Figure 6.5 **A**). In normal light (such as that coming from a light bulb or the sun), the electric fields are oscillating at all possible angles when viewing the radiation propagating toward you (Figure 6.5 **B**). Plane polarized light has all the electric fields oscillating in the same plane (Figure 6.5 **B** and **C**), and can be viewed as the single oscillation shown in Figure 6.5 **A**. The representation in Figure 6.5 **A**

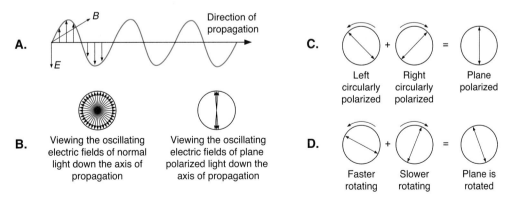

Figure 6.5
The phenomenon of optical activity. **A.** Oscillating electric and magnetic fields. **B.** The difference between normal (non-polarized) light and plane polarized light, viewing the oscillating electric fields down the axis of propagation. **C.** Plane polarized light is a combination of right and left circularly polarized light. **D.** If the differential index of refraction causes one form to "rotate" faster than the other, the effect is to rotate the plane of polarization.

does not look chiral, yet plane polarized light can be used to distinguish enantiomers. To reconcile this, we must appreciate that plane polarized light can be considered to be created by two circularly polarized beams of light, one rotating clockwise and one counterclockwise. **Circular polarization** means that the plane of the oscillating electric field does not remain steady, but instead twists to the right or the left, referred to as right or left **circularly polarized light**. In other words, the linear vector that traces out the plane polarized wave is formed from two circularly polarized waves, one rotating clockwise and one rotating counterclockwise (Figure 6.5 **D**). Taken separately, these circularly polarized beams are rotating in a helical fashion, and hence are chiral. The right and left polarized beams of light are therefore enantiomers of each other. So, indeed, we again find that it takes chiral entities to distinguish between chiral chemical structures.

As the plane polarized light passes through a chiral sample, several different kinds of interactions between the light and the material are possible. One is actual absorption of the light, which we explore below when circular dichroism is discussed. However, another is simple refraction. The indices of refraction of the chiral material for the right and left polarized light are expected to be different, which means that the speed of light through the medium is different for the two polarizations, a phenomena called **circular birefringence**. Therefore, one of the light components will lag behind the other. "Lagging behind" means a slower rate of propagation due to a different refractive index for that form of light (Figure 6.5 **D**). The result is that right- and left-handed twists no longer have the same phase matching to cancel along the original plane, but instead they cancel along a slightly different plane, rotated away from the original plane.

Circular Dichroism

In the discussion above, plane polarized light was described as a combination of right and left circularly polarized light. Just as a chiral medium must refract left and right circularly polarized light differently, chiral molecules must have different absorptions of the left and right circularly polarized light. **Circular dichroism** (CD) spectroscopy measures this differential absorption. This technique involves the same absorption phenomenon that occurs in UV/vis spectroscopy, which is discussed in Chapter 16.

One collects a CD spectrum by measuring the difference in absorption of right and left circularly polarized light as a function of the wavelength of the light. At certain wavelengths of circularly polarized light, the right-handed form is absorbed more (defined as a positive value) than the left-handed form, and vice versa at other wavelengths. There are specific rules related to **exciton coupling** (coupling of electronic states between two or more chromophores) that dictate which form of light is absorbed the most at various wavelengths. This is beyond the scope of this chapter, but extensive discussions of this phenomenon are available in the more specialized texts cited at the end of this chapter.

Because of the predictability of CD spectra, in earlier times, CD was frequently used as a means of establishing the absolute configuration of chiral molecules, and extensive correlations of CD spectra with molecular structure were developed based upon empirical rules. The shapes of the curves, called either plain curves or curves possessing positive and/or negative **Cotton effects**, can be correlated with structure. In more recent times, x-ray crystallography has become the most common way to establish absolute configuration (see below). One area in which CD has remained quite a powerful and commonly used tool is in studies of protein secondary structure. We will discuss this application of CD later in this chapter.

X-Ray Crystallography

If we have a crystal of an enantiomerically pure compound, and we determine its crystal structure, you might think that we would then know its absolute configuration. Actually, this is typically not the case. Nothing in the data collection or analysis of x-ray crystallography is inherently chiral, and so we cannot tell which enantiomer we are imaging in a typical crystallography study. There are two ways around this. One is an advanced crystallographic technique called **anomalous dispersion**. Anomalous dispersion occurs when the x-ray wavelength is very close to the absorption edge of one of the atoms in the structure. This leads to an unusual scattering interaction that contains the necessary phase information to allow enantiomer discrimination. Originally a somewhat exotic technique, the method has become more common as more diverse and brighter x-ray sources have become available.

The alternative approach to determine absolute configuration by x-ray crystallography is to functionalize the molecule of interest with a chiral reagent of known absolute configuration. Returning to the example of Figure 6.4, if we determine the crystal structure of one of the separated amide diastereomers, crystallography will unambiguously establish the *relative* configurations of the original molecule and the appended carboxylic acid. Since we independently know the absolute configuration of the the (S)-(+)-mandelic acid that we used, we know the absolute configuration of the 2-aminobutane.

6.2 Symmetry and Stereochemistry

Stereochemistry and symmetry are intimately connected, and in developing some more advanced aspects of modern stereochemistry, it is convenient to be able to invoke certain symmetry operations. A proper understanding of symmetry can greatly clarify a number of concepts in stereochemistry that can sometimes seem confusing. One operation that we have already used extensively is that of reflection through a mirror plane, and simple guidelines using imaginary sheets of glass were given. We will not need to develop the entire concept of point group symmetries in this textbook. For those who are familiar with point groups and irreducible representations, we will occasionally mention them where appropriate, but they are not required. However, for those students not well versed in symmetry operations, we now give a very short summary of some of the basics.

6.2.1 Basic Symmetry Operations

A **symmetry operation** is a transformation of a system that leaves an object in an indistinguishable position. For molecular systems, we need be concerned with only two types of symmetry operations: **proper rotations** (C_n) and **improper rotations** (S_n). A C_n is a rotation around an axis by $(360/n)°$ that has the net effect of leaving the position of the object unchanged. Thus, a C_2 is a 180° rotation, a C_3 a 120° rotation, and so on. These are termed "proper" rotations, because it is actually physically possible to rotate an object by 180° or 120°. Some examples are shown below, with the atoms labeled only to highlight the operation.

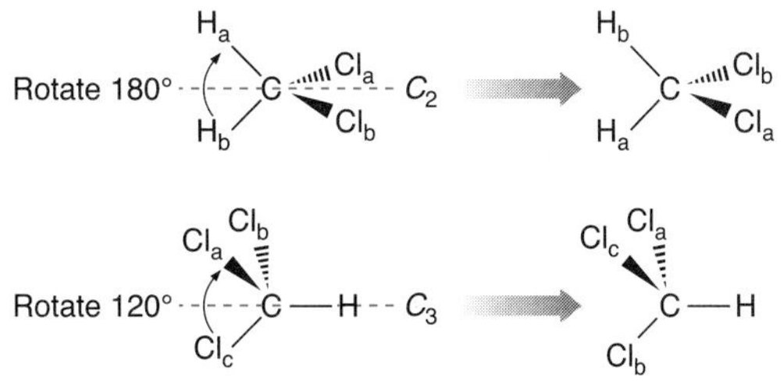

In contrast, improper rotations are not physically possible. An S_n involves a rotation of $(360/n)°$, combined with a reflection across a mirror plane that is perpendicular to the rotation axis (see examples on the next page). Note that S_1 is equivalent to just a mirror reflection (denoted with a σ), while S_2 is equivalent to a center of inversion (denoted with an i). The C_1 operation also exists. It leaves an object completely unmoved and is also termed the **identity operation**, sometimes symbolized as E. An internal σ plane that includes a C_2 axis is designated a σ_v, while a σ plane perpendicular to a C_2 axis is designated σ_h.

6.2.2 Chirality and Symmetry

Now we can further refine the connection between symmetry and chirality. Quite simply, for a rigid molecule (or object of any sort), *a necessary and sufficient criterion for chirality is*

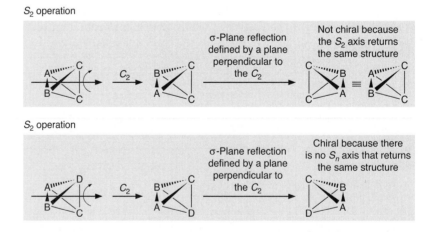

an absence of S_n axes; the existence of any S_n axis renders an object achiral. For example, consider the two structures shown below. The first object has an S_2 axis and is not chiral, while the second object does not have an S_2 axis, let alone any S_n axis, and so the structure is chiral.

In addition, when a chiral molecule is subjected to any improper rotation, it is converted into its enantiomer. Since the simplest improper axis to use is an S_1, the σ plane (see many of our examples above), most chemists first look for an internal mirror plane in a molecule to decide if it is chiral or not. If the molecule possesses an internal mirror plane in any readily accessible conformation, then the molecule is achiral. For those familiar with point groups, it is a simple matter to show that all chiral molecules fall into one of five point groups: C_n, D_n, T, O, or I. All other point groups contain an S_n axis.

Chiral molecules need not be asymmetric. **Asymmetric** is defined as the complete absence of symmetry. However, many chiral molecules have one or more *proper* rotation axes —just no improper axes are present. These compounds can be referred to as **dissymmetric**, essentially a synonym for chiral. Thus, while all asymmetric (point group C_1) molecules are chiral, not all chiral molecules are asymmetric. Importantly, high symmetry chiral molecules play a special role in many processes, especially in efforts to influence the stereochemistry of synthetic reactions (see the following Connections highlight).

Connections

C_2 Ligands in Asymmetric Synthesis

The use of C_2 symmetric ligands in catalytic asymmetric induction is a common design motif. Below are shown a series of chiral Lewis acid catalysts that have been used for Diels–Alder reactions. In every case a C_2 axis exists in the structures. Also, in every case the metal is non-stereogenic. Most catalytic processes involve weak interactions between substrate and catalyst, and this often leads to a situation in which several different binding interactions between substrate and catalyst are possible. Each different binding interaction might produce different stereoselectivity, making it difficult to achieve high enantio-

meric excess. Since the metal is non-stereogenic in a C_2 symmetric complex, coordination of the Diels–Alder reactants to either face of the metal produces identical complexes. We ask that you show this in an Exercise at the end of the chapter. The environment around the metal is still chiral, however, and so asymmetric induction is possible. This same motif will be seen in a Going Deeper highlight on polymerization reactions given in Section 6.7.

Evans, D. A., Miller, S. J., Lectka, T., and von Matt, P. "Chiral Bis(oxazoline)copper(II) Complexes as Lewis Acid Catalysts for the Enantioselective Diels–Alder Reaction." *J. Am. Chem. Soc.*, **121**, 7559–7573 (1999).

C_2 symmetric catalysts

6.2.3 Symmetry Arguments

We argued above that any rigid molecule lacking an S_n axis is chiral. We don't need to know anything else about the molecule to reach this conclusion with confidence. This is an example of a **symmetry argument**—a statement from first principles that depends only on the symmetry, not on the precise nature, of the system under consideration.

Two important features of symmetry arguments must always be remembered. First, the most compelling symmetry arguments are based on an *absence* of symmetry. If we can be sure that a certain kind of symmetry is lacking, then firm conclusions can be reached. Stated differently, two objects (molecules or parts of molecules in our context) are equivalent if and only if they are interconvertible by a symmetry operation of the system. On the other hand, if two objects are not interconvertible by a symmetry operation, they are expected to be different, and they are different in essentially all ways. We cannot rule out the possibility of accidental equivalence. However, we expect that, in most instances, if the precision of our measurement is high, objects that are not symmetry equivalent will be measurably different. We will generally use a phrase such as "are expected to be different" to acknowledge the possibility that in some systems the differences between two symmetry inequivalent objects may be too small to be detected at the present level of precision.

For example, consider the C1–C2 vs. the C2–C3 bonds of *n*-butane. We can be certain that there can never be a symmetry operation of butane that will interconvert these two bonds. As such, they are different, and they are different in all ways. They will have different bond lengths, different IR stretching frequencies, and different reactivities.

The absence of symmetry can be unambiguous—we know for sure that the two C–C bonds discussed above cannot be interconverted by symmetry. On the other hand, we must be careful about using a symmetry argument to declare two objects to be equivalent, because that can be a cyclic argument. For example, consider a CH_2 group in cyclobutane. It is tempt-

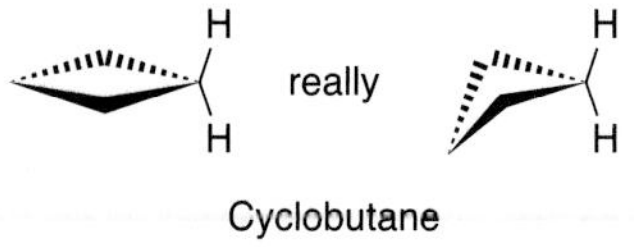

Cyclobutane

ing to conclude that the two hydrogens are equivalent. If we draw the molecule as square and planar, there are symmetry operations that interconvert them (a C_2 axis and a σ plane). We had to assume a structure for the system, and we chose a high symmetry structure. However, there is no law that molecules will adopt the highest possible symmetry, and in the particular case of cyclobutane, the molecule indeed adopts a lower symmetry form, as we saw in Section 2.3.2. Cyclobutane is nonplanar, and the hydrogens of a given CH_2 are inequivalent (the time scale is of importance in this argument, as we discuss later in the chapter). Thus, in the absence of independent information about the symmetry of a system, it is risky to simply look at a structure and say two parts are equivalent.

On the other hand, if we have independent evidence that a molecule has certain symmetry elements—for example, from an x-ray structure—then we can use those symmetry elements to make statements about equivalence. Restating, two objects are equivalent if and only if they are interconverted by a symmetry operation of the system, and if they are not interconverted by a symmetry operation of the system, they are expected to be different.

Another important aspect of symmetry arguments is that they tell us *nothing about magnitudes*. We can conclude that two angles are expected to be different, but they may differ by 10° or by 0.0000000001°. Symmetry arguments are oblivious to such distinctions. Objects are either different or not; that is all we can conclude.

6.2.4   Focusing on Carbon

While most chemists are justifiably enamored of symmetry, in a sense it is the absence of symmetry that makes things happen. Let's illustrate this by considering the desymmetrization of methane. The carbon in methane is not a stereogenic center—that is, interchanging the positions of two hydrogens does not produce a new stereoisomer in this high symmetry structure. We often say that a carbon atom with four covalent ligands has "tetrahedral" symmetry. What does that mean? It means that in CH_4 the four hydrogens lie at the vertices of a regular tetrahedron, with the C at the center (Figure 6.6). Every H–C–H angle is arc cos($-\frac{1}{3}$) ~ 109.47°, and every bond length is the same. These two descriptors (one length, one angle) are enough to fully describe such a system, and the same geometry holds for most CX_4 systems.

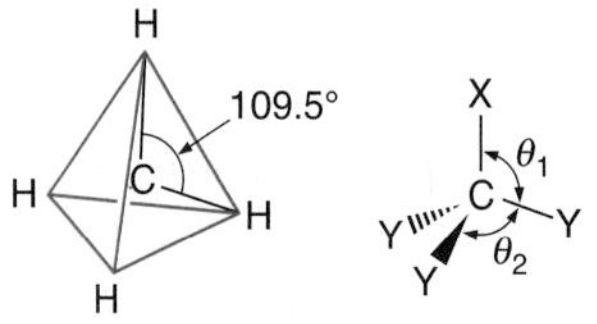

Figure 6.6
Left: The "tetrahedral" carbon atom.
Right: Differing angles in a CXY_3 molecule.

Things get more interesting when all four ligands are different. As first appreciated by Pierre Curie, it is the *lack of symmetry* that gives rise to observable phenomena. For example, in CXY_3, a desymmetrized CX_4, there are now two different valence angles (X–C–Y and Y–C–Y) (Figure 6.6) and two bond lengths, so there was an increase in the number of observables on lowering the symmetry. Desymmetrization to produce a CXY_3 structure also leads to a new molecular property that is not possible for CX_4—a dipole moment (Chapter 2). With further desymmetrization to CX_2Y_2, three angles are now possible, and so on. These systems no longer correspond to a perfect, regular tetrahedron, but we still tend to refer to them as "tetrahedral". They just happen to be irregular tetrahedrons.

Full desymmetrization to produce CWXYZ gives four different bond lengths and six different angles. As already discussed, this complete desymmetrization also leads to chirality. We noted in Chapters 1 and 2 that most organic molecules do not have perfect tetrahedral angles, and that all C–C bonds lengths are not the same. In that context, we focused on the quantitative deviations from the standard norms, and how specific bonding theories could rationalize them. Here, we are arriving at similar conclusions, but from a different perspective. Our argument that a CXY_3 molecule has two different angles can be made with confidence and without any knowledge of what X and Y are, as long as they are different. It is a symmetry argument, and so it is incontrovertible, but qualitative in nature.

6.3   Topicity Relationships

Thus far we have focused on terminology appropriate for describing the stereochemical relationships between molecules. As we will see, it is also convenient to describe relationships between regions of molecules such as two different methyl groups or two faces of a π system. In such cases we are considering the **topicity** of the system. The topicity nomenclature is derived from the same roots as topography and topology, relating to the spatial position of an object.

6.3.1   Homotopic, Enantiotopic, and Diastereotopic

If two objects cannot be interconverted by a symmetry operation, they are expected to be different. This reasoning applies not only to entire molecules, but also to differing regions within a molecule. When the groups can be interconverted by a symmetry operation, they are chemically identical. Yet, depending upon the symmetry operation, they can act differently. The terms we introduce here have the suffix *-topic*, which is from the Greek for "place". When identical groups or atoms are in inequivalent environments, they are termed **heterotopic**. They can be either constitutionally heterotopic or stereoheterotopic. **Constitutionally heterotopic** means that the connectivity of the groups or atoms is different in the molecule. **Stereoheterotopic** means the groups or atoms have different stereochemical relationships in the molecule under analysis.

Consider the CH_2 group of 2-butanol. There are no symmetry operations in 2-butanol, and as such the two hydrogens of the CH_2 cannot be interconverted by a symmetry operation. Therefore, these two hydrogens are expected to be different from one another in all meaningful ways, such as NMR shift, acidity, C–H bond length, bond dissociation energy, reactivity, etc. They have the same connectivity, but there is no symmetry operation that interconverts them in any conformation. They are stereoheterotopic, and defined specifically as **diastereotopic**.

Now consider the CH_2 group of propane. There is, or more properly can be, a C_2 operation that interconverts the two hydrogens, and so they are considered to be equivalent. The modern terminology is **homotopic**, and is defined as interconvertable by a C_n axis of the molecule. These hydrogens are equivalent in all ways.

We have one more case to consider, exemplified by the CH_2 group in ethyl chloride. There is a symmetry element that interconverts the two hydrogens—a mirror plane. Here is where the distinction between proper and improper symmetry elements becomes important. These hydrogens are equivalent because they are interconverted by a symmetry element. However, just as with two enantiomers, such an equivalence based upon a mirror plane will be destroyed by any chiral influence. As such, these hydrogens are termed **enantiotopic**—that is, interconverted by an S_n axis of the molecule ($n = 1$ in this case). Enantiotopic groups, when exposed to a chiral influence, become distinguishable, as if they were diastereotopic. The example of the use of a chiral shift reagent given on page 308 illustrates this point.

Homotopic groups remain equivalent even in the presence of a chiral influence. Since chiral molecules need not be asymmetric (they can have C_n axes), groups can be homotopic even though they are part of a chiral molecule. Consider the chiral acetal shown in the margin. The methyl groups are homotopic because they are interconvertable by a C_2 operation. A chiral influence cannot distinguish these methyl groups.

Another common situation where topicity issues become important is at trigonal centers, such as carbonyls and alkenes. As some examples, let's focus on carbonyl groups. The two *faces* of the carbonyl are homotopic in a ketone substituted by the same groups [R(C=O)R], such as acetone, because the molecule contains a C_2 axis (see below). The faces are enantiotopic in an unsymmetrically substituted ketone, such as 2-butanone, because they are interconverted by a σ plane. The faces are diastereotopic in a structure such as either enantiomer of 3-chloro-2-butanone, because there are no symmetry elements that interconvert the faces.

Different in all ways

H_a H_b

HO   H

Diastereotopic hydrogens

Equivalent in all ways

H_a H_b

Homotopic hydrogens

Equivalent unless within
a chiral environment

H_a H_b

Cl

Enantiotopic hydrogens

Chiral molecule with
homotopic methyl groups

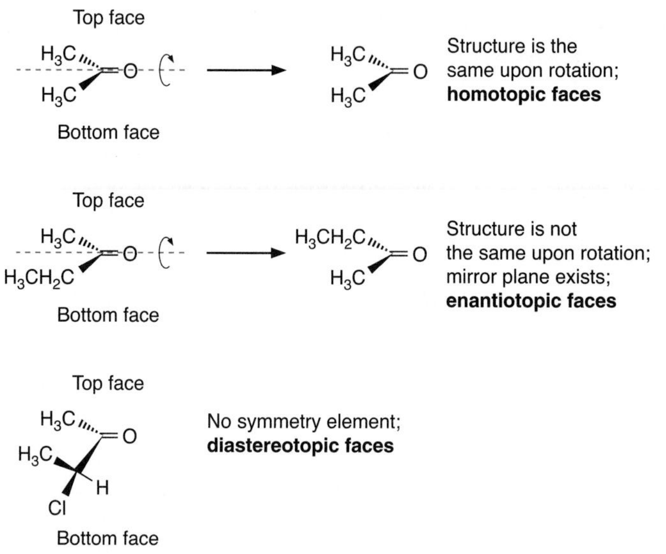

Top face

H₃C⁗∙∙∙=O

Bottom face

Structure is the same upon rotation; **homotopic faces**

Top face

H₃C⁗∙∙∙=O / H₃CH₂C

Bottom face

Structure is not the same upon rotation; mirror plane exists; **enantiotopic faces**

Top face

H₃C⁗∙∙∙=O / H₃C / Cl / H

Bottom face

No symmetry element; **diastereotopic faces**

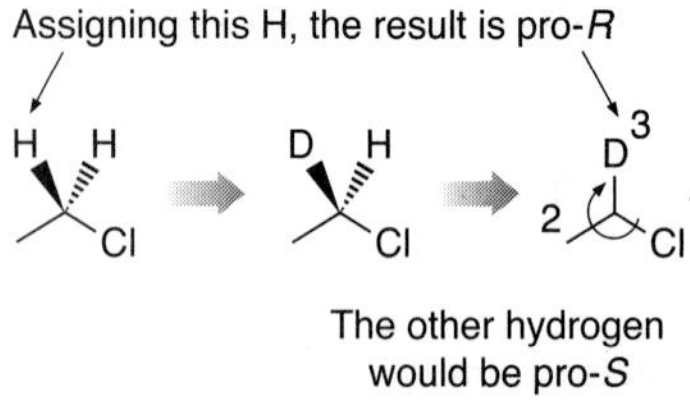

6.3.2 Topicity Descriptors—Pro-*R*/ Pro-*S* and *Re/Si*

Assigning this H, the result is pro-*R*

The other hydrogen would be pro-*S*

Just as it was convenient to have descriptors to distinguish enantiomeric molecules, it is also useful to be able to identify enantiotopic hydrogens. To do so, we use something similar to the *R* / *S* notation. For a CH₂ group, first take the hydrogen that is being assigned a descriptor and mentally promote it to a deuterium. Now assign priorities in the normal way. If the result is that the newly formed stereogenic center is *R*, the hydrogen that we mentally replaced by deuterium is denoted **pro-*R***, and if the new stereocenter is *S*, the hydrogen is denoted **pro-*S***. An example using chloroethane is given in the margin. The same nomenclature convention can be used with diastereotopic hydrogens.

The "pro" terminology is meant to imply that the center would become stereogenic (and hence worthy of an *R* / *S* descriptor) if the substitution were made. For this reason, the carbon containing the enantiotopic hydrogens is also referred to as a **prochiral** center. While some find this term useful, it can lead to confusion, and as such, describing the situation in terms of enantiotopic groups is preferable. It should be apparent that the enantiotopic groups need not be hydrogens. For example, two methyl groups or two chlorines can be enantiotopic. The pro-*R* / *S* distinction would be made by converting the methyl to be named to a CD₃ group, and the Cl to be named to a higher isotope (see below).

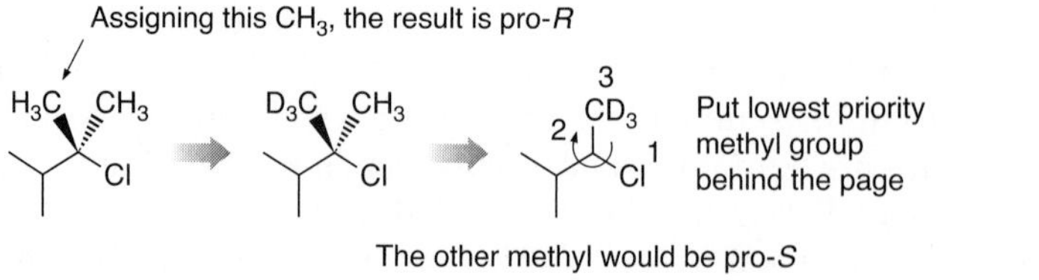

Assigning this CH₃, the result is pro-*R*

The other methyl would be pro-*S*

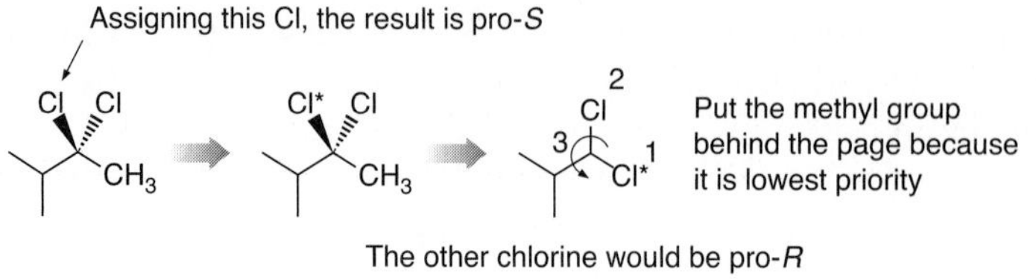

Put lowest priority methyl group behind the page

Assigning this Cl, the result is pro-*S*

The other chlorine would be pro-*R*

Put the methyl group behind the page because it is lowest priority

When assigning a descriptor to the enantiotopic faces of a trigonal structure, start by simply placing the molecule in the plane of the paper. Next assign priorities to the groups using the same methods for R/S and E/Z. If the result is a clockwise rotation, the face we are looking at is referred to as **Re**; if it is a counterclockwise rotation, the face is **Si**. An example using 2-butanone is given in the margin. Once again, it is common to refer to the carbon of the carbonyl as prochiral, because attachment of a different fourth ligand will create a stereogenic center and possibly a chiral molecule.

The *Re* face   The *Si* face

6.3.3 Chirotopicity

The terms enantiotopic and diastereotopic describe the relationship between a pair of atoms or groups in a molecule. Sometimes it is also useful to describe the local environment of a single atom, group, or location in a molecule (even if it does not coincide with an atomic center) as chiral or not. A **chirotopic** atom or point in a molecule is one that resides in a chiral environment, whereas an **achirotopic** atom or point does not. All atoms and all points associated with a chiral molecule are chirotopic. In achiral molecules, achirotopic points are those that remain unchanged (are invariant) upon execution of an S_n that is a symmetry operation of the molecule. For most situations, this means that the point either lies on a mirror plane or is coincident with the center of inversion of the molecule. Importantly, there will generally be chirotopic points even in achiral molecules.

These terms can be clarified by looking at some specific examples. In the following rotamers of *meso*-1,2-dichloro-1,2-dibromoethane, the only achirotopic site in rotamer A is the point of inversion in the middle of the structure. Every atom is in a locally chiral environment, and so is chirotopic. For rotamer B, all points in the mirror plane (a plane perpendicular to the page of the paper) are achirotopic. All other points in these conformers are chirotopic, existing at sites of no symmetry. In other words, all other points in these conformers feel a chiral environment, even though the molecule is achiral.

As another example, consider once again the chiral acetal shown in the margin. The C atom indicated resides on a C_2 axis but not on any type of S_n axis, and so it is chirotopic. Note, however, that the C is non-stereogenic. Hence, non-stereogenic atoms can reside in chiral environments. Refer back to the first Connections highlight in Section 6.2.2. In this highlight all the metals are chirotopic but nonstereogenic. The term "chirotopic" focuses us on the points in a molecule that are under a chiral influence, which is the most important factor for using stereochemical principles to understand spectroscopy and reactivity.

A.

Achirotopic point

B.

Achirotopic plane

Carbon is chirotopic

6.4 Reaction Stereochemistry: Stereoselectivity and Stereospecificity

Topicity relationships and symmetry arguments provide a powerful approach to anticipating reactivity patterns. Whether by habit, intuition, or full realization, it is the topicity relationships discussed above that synthetic chemists use to develop chemical transformations that yield asymmetric induction.

6.4.1 Simple Guidelines for Reaction Stereochemistry

Consider the three ketones in Figure 6.7 and the topicities of their carbonyl faces. In acetone, the two faces of the carbonyl are homotopic—interconverted by a C_2 rotation. In 2-butanone, the faces are enantiotopic (prochiral)—interconverted only by a mirror plane. In (R)-3-chloro-2-butanone, the two faces are diastereotopic. This molecule is asymmetric, and so there can be no symmetry operation that interconverts the two faces of the carbonyl. A consequence of this lack of symmetry in (R)-3-chloro-2-butanone is that the carbonyl group is expected to be nonplanar—that is, O, C2, C1, and C3 *will not* all lie in a plane. The point is that because the two faces of the carbonyl are inequivalent, the carbonyl cannot be planar. This is a symmetry argument of the sort mentioned previously, and as with all symmetry arguments, we cannot predict how large the deviation from planarity must be, only that it is expected to be there. As such, if we obtain a crystal structure of (R)-3-chloro-2-butanone, we should not be surprised to find a nonplanar carbonyl.

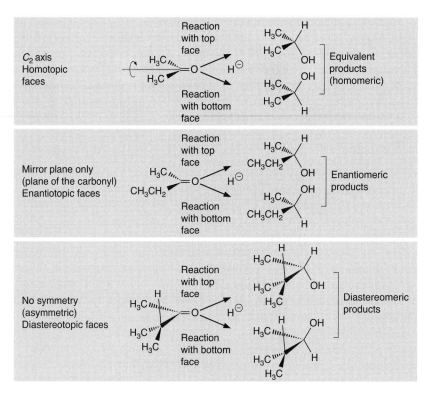

Figure 6.7
Stereochemical consequences of reacting three different types of
carbonyls with a hydride reducing agent.

Let's consider the reactivity of the three carbonyls shown in Figure 6.7. For acetone, reaction with an achiral reagent such as LiAlH₄ produces the same product regardless of which carbonyl face reacts. This will always be the case for homotopic faces. For 2-butanone, reaction with LiAlH₄ at enantiotopic faces gives enantiomeric products, (R)- and (S)-2-butanol. For (R)-3-chloro-2-butanone, the two carbonyl faces are different. They will give different products from the reaction with LiAlH₄—namely, (R,S)- and (R,R)-2-chloro-3-butanol, which are diastereomers.

As we can anticipate the stereochemical relationships among the products, we can also evaluate the symmetry properties of the transition states of the hydride addition reactions. For acetone, there is only one possible transition state and only one product. For 2-butanone, the transition states derived from "top" and "bottom" attack are enantiomeric. As such they will have equal energies, and so $\Delta G^{\ddagger}$ will be the same for the formation of the two enantiomeric products. As a result, a racemic mixture must form. Finally, in the reduction of (R)-3-chloro-2-butanone, the two transition states are diastereomeric, and so they are expected to have different energies (diastereomers differ in all ways). Since the starting point for the two reactions is the same, $\Delta G^{\ddagger}$ is expected to be different for the two, and therefore the rates for formation of the two diastereomeric products cannot be the same. Since the rates of formation of the two products are not the same, we can state with certainty that the reduction of (R)-3-chloro-2-butanone *is expected to not produce a 50:50 mixture of the two products* in the initial reaction. This can be anticipated from first principles. When we start from a single reactant and produce two diastereomeric products, we do not expect to get exactly a 50:50 mixture of products. However, as is always true of a symmetry argument, we cannot anticipate how large the deviation from 50:50 will be—it may be 50.1:49.9 or 90:10. We can only say that it is not 50:50.

Let's examine what happens if we use a single enantiomer of a chiral hydride reducing agent. Acetone still gives only one product—isopropanol. However, we would now expect the two enantiotopic faces of 2-butanone to be distinguished. The transition states corresponding to attack from opposite faces of the carbonyl are now diastereomeric, and something other than a 50:50 mixture of the enantiomeric products (a non-racemic sample) is expected to result from such a reaction. Achieving **asymmetric induction** is therefore anticipated by simple symmetry arguments. The only issue is whether the magnitude of the effect is small or large. To visualize how a chiral environment can distinguish enantiotopic groups, see the Connections highlight below that describes enzyme catalysis and molecular imprints.

Lastly, in the reduction of (R)-3-chloro-2-butanone, the different faces of the ketone were already diastereotopic due to the presence of the stereogenic center. Hence, even an achiral reducing agent such as LiAlH$_4$ will give something other than a 50:50 ratio of R- and S-centers at the newly formed alcohol. Interestingly, switching from LiAlH$_4$ to a chiral hydride agent has no impact (from a symmetry standpoint) on the reduction of (R)-3-chloro-2-butanone; we still expect something other than a 50:50 mixture of two diastereomers.

In summary:

1. *Homotopic groups cannot be differentiated by chiral reagents.*

2. *Enantiotopic groups can be differentiated by chiral reagents.*

3. *Diastereotopic groups are differentiated by achiral and chiral reagents.*

6.4.2 Stereospecific and Stereoselective Reactions

The terms stereospecific and stereoselective describe the stereochemical outcomes of the sort we have been discussing. Even these terms, though, are sometimes used in confusing ways. Figure 6.8 illustrates the definitions of these terms as originally presented. In a **stereospecific** reaction, one stereoisomer of the reactant gives one stereoisomer of the product, while a different stereoisomer of the reactant gives a different stereoisomer of product. Hence, to determine whether a reaction is stereospecific, one has to examine the product ratio from the different stereoisomers of the reactant. An example would be the epoxidation of 2-butene by mCPBA. The trans olefin gives the trans epoxide and the cis olefin gives the cis epoxide (Figure 6.8 **A**). S$_N$2 reactions are also stereospecific, in that inversion of the stereo-

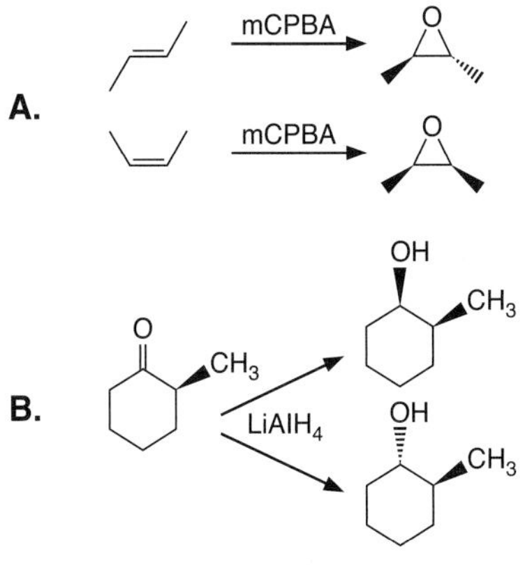

Figure 6.8
A. An example of a stereospecific reaction (mCPBA is *meta-*chloroperbenzoic acid). **B.** An example of a stereoselective reaction. If the enantiomer were analyzed, the reaction would also be stereospecific.

Connections

Enzymatic Reactions, Molecular Imprints, and Enantiotopic Discrimination

The general concept that enantiotopic groups can be distinguished chemically by a chiral environment is of paramount importance to enzymatic catalysis. Since enzymes are constructed from chiral entities—α-amino acids—they are themselves chiral. Enzymes are well known for their stereoselectivity. The fact that enzymatic reactions are diastereoselective or enantioselective is not surprising; this is expected to happen when the reagent (the enzyme) is chiral and enantiomerically pure. The remarkable feature of enzymatic reactions is the high *degree* of stereoselectivity they generally display.

Enzymes possess binding sites that are complementary to their substrates using the same principles of complementarity and preorganization introduced for

synthetic receptors in Chapter 4. As a simplification of the notion of complementarity, we can consider an enzyme binding site as an imprint of the substrate, similar to the imprint of an object in wet sand. The analogy leads to a very simple visual image of how an enzyme can distinguish enantiotopic groups. Consider the picture of the molecular model of ethyl chloride sitting in wet sand shown below with one enantiotopic hydrogen of the CH_2 group embedded in the sand (**A**). After removing the plastic model, an impression is left in the sand (**B**). We cannot pick up and place the ethyl chloride back into the impression in any way besides the original placement (**A**). Hence, this impression in the sand leads to only one of the two enantiotopic hydrogens buried in the sand, thus clearly differentiating among these two hydrogens.

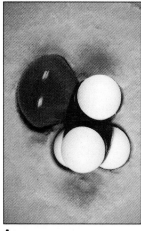

A.

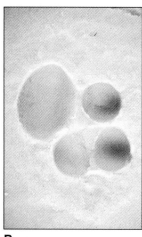

B.

C.

chemistry on stereogenic centers is consistently observed, so that enantiomers of reactants must give different enantiomers of the products. For a few other examples, see Table 6.1 **A**. A reaction need not be perfectly stereospecific. If an 80:20 mixture of stereoisomers is produced, we could call the reaction 80% stereospecific.

Whether a reaction is or is not stereospecific has significant mechanistic implications, and we will look at stereochemical analyses of this sort in future chapters. In essence, when a reaction is stereospecific, a common intermediate *cannot* be involved in the mechanisms of reaction of the two stereoisomeric reactants.

A **stereoselective** reaction is one in which a single reactant can give two or more stereoisomeric products, and one or more of these products is preferred over the others—even if the preference is very small. Now we only need to examine one stereoisomer of the reactant to make this determination for a reaction. In fact, the reactant may not even exist as stereoisomers, yet the reaction can be stereoselective. See the example in Table 6.1 **B**.

A reaction is also stereoselective when two stereoisomers of the starting material give the same ratio of stereoisomeric products, as long as the ratio is not 50:50. This just means the reaction is not stereospecific. For example, this may occur if the mechanisms of reaction for

Table 6.1
Stereospecific Reactions (A), a Stereoselective Reaction (B), and Stereoselective but Not Stereospecific Reactions (C)

the two stereoisomeric reactants proceed through a common intermediate, and that intermediate gives two stereoisomeric products with one in excess. However, there are also reactions where the different stereoisomeric reactants give the same ratio of stereoisomeric products, even when a common intermediate is not formed (Table 6.1 **C**). All stereospecific reactions are stereoselective, but the converse is not true.

Another example of a stereoselective reaction is the previously discussed reduction of (*R*)-3-chloro-2-butanone (see Figure 6.7). In this case the two products are diastereomers, and the reaction is referred to as **diastereoselective**. This reaction is also stereospecific, in that (*S*)-3-chloro-2-butanone will give a different ratio of products with the same reducing agent. If the two products are enantiomers [as in the reduction of 2-butanone (Figure 6.7)], the reaction is **enantioselective** if one enantiomer is formed preferentially.

Unfortunately, an alternative usage of these terms exists. Often in the organic synthesis literature, stereospecific is taken to mean 100% stereoselective. This is a necessarily vague distinction, because it depends on the tools used to measure the product ratios. A reaction that appears "stereospecific" by a relatively crude measure such as optical activity, may turn into a "stereoselective" reaction when chiral HPLC reveals a 99:1 product ratio. Also, the mechanistic implications of stereospecificity are lost in this alternative usage. However, it seems likely that both usages will exist side-by-side for some time, and the student needs to be aware of the distinction.

Terminology aside, the reaction of a chemical sample composed of only achiral molecules (such as 2-butanone) cannot give rise to products with any chiral bias (i.e., any enantiomeric excess) without the intervention of an external chiral influence. This observation has significant implications for discussions of such topics as the origin of chirality in natural systems (see Section 6.8.3).

A term similar to stereoselective is regioselective. "Regio" in this context is defined as a site in a molecule where a reaction can occur, and the difference in the reactivity of various sites is called **regiochemistry**. When more than one site reacts, a **regioselective reaction** is one where an excess of one of the possible products results. A common example is the Markovnikov addition of HCl to a double bond (see Chapter 10), where the chloride preferentially adds to the more substituted carbon (Eq. 6.2). Hence, this is a regioselective reaction. Here, the two carbons of the alkene are considered to be the two "regions" or sites in the molecule that can react. Once again, there are varying degrees of regioselectivity, ranging from 100% (completely selective) to 0% (completely unselective).

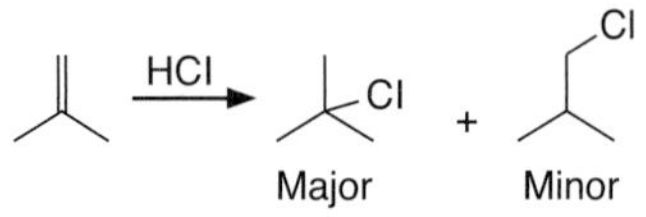

(Eq. 6.2)

6.5   Symmetry and Time Scale

2-Butanol is asymmetric, so the two hydrogens of the CH_2 group are diastereotopic. Shouldn't the three hydrogens of the CH_3 group at C1 (or C4) be diastereotopic also? It depends. In particular, it depends on the time scale of our observation of the molecule.

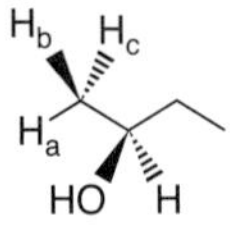

When considering the symmetry of any system, we must *always* include a time scale. In Section 6.2.2 when we gave a symmetry argument for predicting chirality, we explicitly limited ourselves to rigid molecules. Symmetry arguments and stereochemistry are much simpler if we treat all molecules as rigid, geometric objects. However, real molecules are in motion, and if the motion is fast compared to the time scale of observation, we have to include the motion in our analysis of symmetry. If we are considering 2-butanol at room temperature, the rotation of the CH_3 groups will be fast under most time scales of observation. Since that rotation interconverts the three hydrogens, they become equivalent; they are not diastereotopic under these conditions. However, there is no rotation that ever interconverts the hydrogens on the methylene group, and therefore the methylene hydrogens are *always* diastereotopic, regardless of the time scale.

If we lower the temperature or greatly increase our speed of observation, rotation will appear to be slow, and the hydrogens of the CH_3 groups will be different. In practice this is difficult. However, computational methods typically produce static structures. Look carefully at the output of a computed structure of even a simple asymmetric molecule using molecular mechanics or quantum mechanics. In the particular case of 2-butanol, there are three different C–H bond lengths calculated for both of the methyl groups.

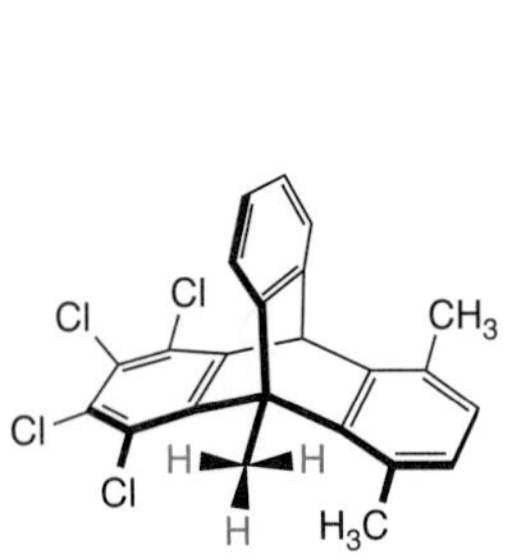

Alternatively, in very crowded systems we can slow methyl rotation enough to see individual hydrogens of a CH_3. The structure shown in the margin, a triptycene derivative of the kind we have seen before (Section 2.5.3, Figure 2.22), gives three unique NMR signals for the colored hydrogens at $-90\,°C$. Nevertheless, under most experimental circumstances it is safe to treat the three hydrogens of a methyl group as equivalent.

The H's of the methyl group are actually inequivalent

Slowed rotation in a methyl group

Symmetry and time scale are always tightly coupled. For example, we discussed in Chapter 2 that cyclobutane is not planar, but rather adopts a lower symmetry, puckered geometry. The methylene hydrogens are diastereotopic in this geometry. However, the interconversion of the puckered forms is rapid on most time scales, and so for most analyses of cyclobutane, all hydrogens appear equivalent. In fact, a planar representation (the time average of two interconverting puckered forms) is acceptable for many analyses. The molecule is only planar when the fleeting transition state between the puckered forms is achieved, but on most time scales it behaves *as if* it were planar. The same analysis can be made for the CH$_2$ groups in cyclohexane. However, the time scale must be considerably more leisurely for the averaging of the axial and equatorial hydrogens of cyclohexane to occur, because of the much higher barrier (and therefore slower rate) for ring inversion in cyclohexane compared to other cyclic hydrocarbons.

Typically, if a flexible molecule can achieve a reasonable conformation that contains a symmetry element, the molecule will behave as if it has that symmetry element. The classic example is an amine with three different substituents. The pyramidal form is chiral, but the two enantiomers interconvert rapidly by pyramidal inversion (Eq. 6.3). That rapid inversion leads to an effectively achiral system is appreciated when we consider that the transition state for inversion is a planar, achiral structure.

Planar, achiral transition state

$$R_3\text{''''}\ddot{N} \rightleftharpoons R_3\text{''''}N \qquad \text{(Eq. 6.3)}$$

Time scale is important for *all* stereochemical concepts. Even our most cherished stereochemical concept, the stereogenic tetracoordinate carbon, is undone if we are at high enough temperatures and long enough time scales that inversion of the center is possible through bond cleavage reactions.

There are many chiral molecules for which enantiomeric forms can be interconverted by a rotation about a single bond. The enantiomeric conformations of gauche butane provide an example, where rapid rotation interconverts the two under most conditions. If the rotation that interconverts a pair of such enantiomers is slow at ambient temperature, however, the two enantiomers can be separated and used. Recall from our first introduction of isomer terminology (Section 6.1) that stereoisomers that can be interconverted by rotation about single bonds, and for which the barrier to rotation about the bond is so large that the stereoisomers do not interconvert readily at room temperature and can be separated, are called **atropisomers**. One example is the binaphthol derivative shown in the margin. It is a more sterically crowded derivative of the biphenyl compound discussed previously as an example of a chiral molecule with no "chiral center". A second example is *trans*-cyclooctene, where the hydrocarbon chain must loop over either face of the double bond (Eq. 6.4). This creates a chiral structure, and the enantiomers interconvert by moving the loop to the other side of the double bond.

Binaphthol

$$\text{(Eq. 6.4)}$$

Facile rotation does not guarantee interconversion of conformational isomers. One of the most fascinating dynamic stereochemistry systems is exemplified by the triarylborane shown in Eq. 6.1. Correlated rotation of the rings, the "two-ring flip", is facile at room temperature. There are three different two-ring flips possible, depending on which ring does the "non-flip". All two-ring flips are fast, but in a highly substituted system, not all possible conformations can interconvert. As long as only two-ring flips can occur, we have two sets of rapidly interconverting isomers, but no way to go from one set to the other. This has been termed **residual stereoisomerism**. We have two separate stereoisomers, each of which is a collection of rapidly interconverting isomers. Clearly, stereoisomerism and time scale are intimately coupled in such systems.

6.6   Topological and Supramolecular Stereochemistry

One of the more interesting aspects of modern stereochemistry is the preparation and characterization of molecules with novel topological features. As we indicated in Chapter 4, supramolecular chemistry has produced a number of structures with novel topologies such as catenanes and rotaxanes. "Simple" molecules (i.e., not supramolecules) can also have novel topological features such as knots or Möbius strips. Here we will introduce some current topics in this fascinating area, emphasizing the aspects that relate to stereoisomerism. But first, we must agree upon a definition of "topology".

The mathematical definition of **topology**, and the one that is best suited to stereochemistry, concerns studies of the features of geometrical objects that derive solely from their connectivity patterns. Metric issues—that is, those associated with numerical values (such as bond lengths and bond angles)—are unimportant in topology. The easiest way to see this is to consider two-dimensional topology as the study of geometric figures that have been drawn on a rubber sheet. You can stretch and bend and flex the sheet as much as you like without changing the topology of a figure on the sheet (Figure 6.9 **A**). Thus, a circle, a triangle, and a square are topologically equivalent because we can deform one to the other. Topologically, all three are just a closed loop. In three dimensions the same concept applies, with the additional requirements that you cannot break a line or allow any lines to cross, and you cannot destroy a vertex. In a dictionary, one will often see a second definition of topology that does include metric issues, so it is a synonym for topography. In topography (i.e., map making), it matters how high the mountain is, but in the mathematical definition of topology we will use here, it does not (in fact, the mountain can be "stretched flat").

With the very few special exceptions discussed below, all stereoisomers are, perhaps surprisingly, topologically equivalent. If you are allowed to stretch and bend bonds at will, it is a simple matter (Figure 6.9 **B**) to interconvert the enantiomers of 2-butanol without crossing any bonds (simple mathematically, but not chemically!). Similar distortions are possible with almost any molecule, allowing stereoisomers to interconvert. This is consistent with our definition of stereoisomers as molecules with the same connectivities (topologies) but different arrangements of atoms in space. Since topology concerns only issues that derive from the connectivity of the system, structures with the same connectivity have the same topology. There are stereoisomers that have different topologies, however, and that is the topic of this section.

We should first make explicit the natural connection between chemistry and mathematics that allows us to discuss topology. Topology deals with graphs—objects that consist of edges and vertices (points where two or more edges meet). In considering chemical topology, we are considering a **chemical graph**, in which the edges are bonds and the vertices are atoms. The ambiguity concerning connectivity still applies (see the Going Deeper highlight in Section 6.1.1), but once we agree on a definition we can consider topological issues.

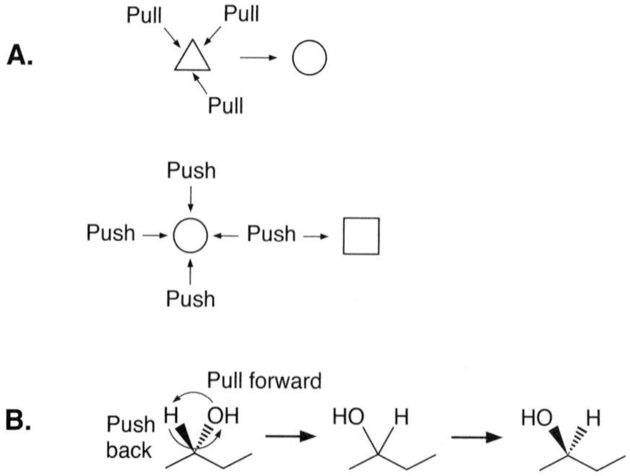

Figure 6.9
The interconversion of topologically equivalent structures.
A. Topologically, the triangle, circle, and square are all just closed loops (as long as we do not consider the "corners" of the triangle and square to be vertices).
B. Interconversion of the enantiomers of 2-butanol can be accomplished by flexing and bending without crossing any bonds, and so the two enantiomers are topologically equivalent.

6.6.1 Loops and Knots

What are the simplest systems that can produce topological stereoisomers? All we need is a cyclic structure. Figure 6.10 **A** shows a circle and a classic trefoil knot. Both structures are simply a single, closed loop (which is the definition of a knot, with a circle being the simplest knot or the "unknot"). It is not possible to interconvert the two structures without crossing edges—they are topologically different. Molecular realizations of the circle and the trefoil knot would be examples of **topological stereoisomers**. Since they are not non-congruent mirror images, it is sensible to call them **topological diastereomers**. To create a chemical version of this situation, a structure as simple as $(CH_2)_n$ could serve the purpose. Interestingly, knots are actually relatively common in biochemistry, as the next Connections highlight describes.

A.

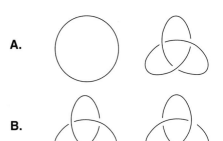

Figure 6.10
A. Topological stereoisomers—a circle and a trefoil knot.
B. Enantiomorphous trefoil knots.

B.

How are these stereoisomers different from conventional diastereomers? The circle and the knot can be infinitely deformed—bent, twisted, stretched, and compressed—but they will never be interconverted (as long as we don't cross any bonds). Conventional isomers can be interconverted by deformation, as in the case of 2-butanol in Figure 6.9. Conventional stereoisomerism depends on the precise location of the atoms in space, leading to the terms **geometric** or **Euclidian isomerism**. With topological stereoisomers, we can move the atoms all around, and retain our isomerism.

Going Deeper

Biological Knots—DNA and Proteins

All we need to make a knot is a cyclic structure. If the ring is large enough to allow the necessary twisting, a knotted structure could form. While it may seem fanciful to consider such structures, and we might expect their preparation to depend on exotic synthetic methods, knotted structures turn out to be common in nature. Circular, double-stranded DNA molecules have been known for some time, with very large "ring sizes" (thousands of nucleotides). Indeed, these large cycles do form knots, which are in fact fairly common structures that can be directly observed by electron microscopy. Catenated circular DNAs have also been observed.

What about proteins? Typically, naturally occurring proteins are not closed circles as in cyclic DNA; the C and N termini are not connected. However, cycles are intro-duced when crosslinks occur between separate regions of the backbone, most typically via disulfide bonds. Rare examples of unique topologies in such systems are known. However, it was recently realized that when the analysis includes cofactors and prosthetic groups such as seen in quinoproteins or iron–sulfur cluster proteins, interesting topologies including knots and catenanes are in fact more common than previously realized. As always, in considering stereochemical phenomena, our definition of connectivity is crucial. Earlier studies had counted only the amino acids as contributing to the connectivity of the system. When cofactors are included, more complex connectivities result.

Liang, C., and Mislow, K. "Knots in Proteins." *J. Am. Chem. Soc.*, **116**, 11189 (1994).

6.6.2   Topological Chirality

If we can have topological diastereomers, can we have topological enantiomers—that is, is there **topological chirality**? There is, and the trefoil knot is a simple example. Figure 6.10 **B** shows two trefoil knots, and these two knots are enantiomorphs. The term enantiomers is reserved for molecules; **enantiomorphs** applies to geometrical objects. How do we know, however, that we could not just deform one structure into the other by stretching and pulling? If we could, the two forms would be topologically equivalent and thus not enantiomers, and the trefoil knot would be topologically achiral. Perhaps surprisingly, there is no general way to prove a knot is chiral. One can prove it is achiral by just finding one way to draw the knot (called a presentation) that is itself achiral. However, if you fail to find an achiral presentation, that doesn't prove the knot is chiral; maybe you just weren't able to find the achiral presentation. In the case of the trefoil knot, however, the structure is indeed chiral.

6.6.3   Nonplanar Graphs

We mention briefly here another topological issue that has fascinated chemists. For the overwhelming majority of organic molecules, we can draw a two-dimensional representation with no bonds crossing each other. This is called a **planar graph**. If you cannot represent the connectivity of a system without some crossing lines, you have a **nonplanar graph**. It may seem surprising, but most molecules have planar graphs. Figure 6.11 **A** shows some examples that illustrate that this is so. Remember, we are doing topology, so we can stretch and bend bonds at will.

Graph theory is a mature branch of mathematics, and graph theorists have established that all nonplanar graphs will conform to one of two prototypes, called K_5 and $K_{3,3}$ in graph theory terminology (Figure 6.11 **B**). K_5 is simply five vertices, maximally connected. Every vertex is connected to every other. $K_{3,3}$ contains two sets of three vertices, with every vertex of one set connected to every vertex of the other set. The fact that $K_{3,3}$ is nonplanar is proof of

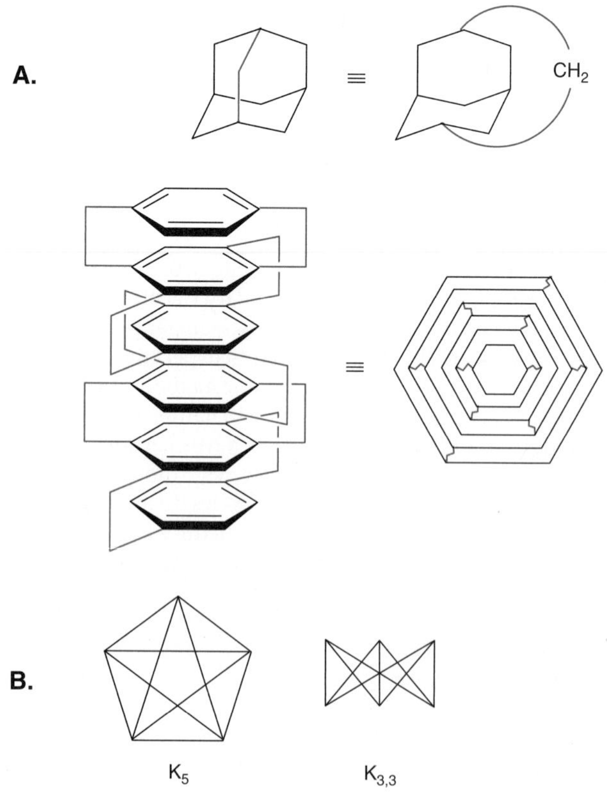

Figure 6.11
A. Examples of how most chemical structures can be represented as planar graphs.
B. K_5 and $K_{3,3}$ nonplanar graphs.

the architectural conundrum, "three houses, three utilities". It is impossible to have three houses, each connected to three utilities (such as water, electric, and phone) without at least one instance of "lines" crossing. We will see molecular versions of these nonplanar graphs below.

6.6.4   Achievements in Topological and Supramolecular Stereochemistry

Recent efforts have produced chemical structures that successfully realize many interesting and novel topologies. A landmark was certainly the synthesis of a trefoil knot using Sauvage's Cu^+/phenanthroline templating strategy described in Section 4.3.2. This nonplanar, topologically chiral structure is a benchmark for the field. Other more complicated knots have also been prepared by this strategy. Vögtle and co-workers have described an "all organic" approach to amide-containing trefoil knots, and have been able to separate the two enantiomeric knots using chiral chromatography.

Another seminal advance in the field was the synthesis and characterization of a "Möbius strip" molecule (Figure 6.12). A **Möbius strip** can be thought of as a closed ribbon with a twist, and it has long fascinated mathematicians and the general public. Although the concept behind the Möbius strategy for preparing novel topologies was enunciated in the late 1950s, it was not chemically realized until the 1980s. A clever strategy based on tetrahydroxymethylethylene (THYME) ethers was developed by Walba. Ring closure could proceed with or without a twist, and when the reaction is performed, the two are formed in roughly equal amounts. An important design feature was that the "rungs" of the ladder system were olefins, which could be selectively cleaved by ozonolysis. Cleavage of the untwisted product produced two small rings, but cleavage of the Möbius product gives a single, larger macrocycle, thereby differentiating the two topological stereoisomers.

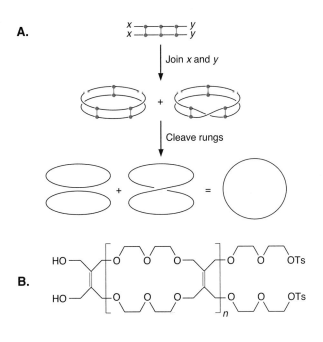

Figure 6.12
A. The synthetic strategy for the preparation of a molecular Möbius strip, and the results of rung cleavage. **B.** A THYME polyether that can ring close to make a Möbius strip.

Even without the twist, the three-rung Möbius ladder compound is a molecular realization of an interesting topology. It is a simple example of a nonplanar graph with the $K_{3,3}$ topology. Another example of a recently prepared molecule with a $K_{3,3}$ topology is given in Figure 6.13 **A**. A structure with the K_5 nonplanar graph has also been prepared, and it is shown in Figure 6.13 **B**.

As suggested in our discussion of supramolecular chemistry in Section 4.3, the facile preparation of complex catenanes and rotaxanes using the various preorganization strategies has led to the consideration of a number of novel stereochemical situations. Topolog-

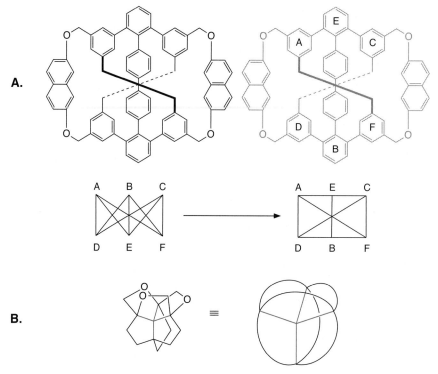

Figure 6.13
Examples of structures with nonplanar graphs. **A.** A $K_{3,3}$ molecule. To see this as a $K_{3,3}$, begin with the schematic graph as presented in Figure 6.11 **B**, and move the vertices B and E. This is topology, so that is legal because all the connectivities stay the same. The structure on the right, then, is labeled in the same way. See also the three-rung ladder molecule of Figure 6.12 **A** for another example of a $K_{3,3}$ molecule. **B.** A K_5 molecule, and a schematic showing the sense that it has the K_5 connectivity.

ical stereoisomers have become commonplace. In addition, other types of isomerism that really do not fit any pre-existing categories are perhaps best regarded as **supramolecular stereoisomerism**.

For example, rotaxanes and catenanes can often exist in different forms that are stereoisomers, but with some unique properties. Figure 6.14 shows several examples. The rotaxane of Figure 6.14 **A** has been studied using electrochemistry, which drives the macrocycle from one "station" to the other. However, without oxidation or reduction of the paraquat, we expect an equilibrium between two forms that are differentiated solely by the position of the macrocycle along the rotaxane axle. Likewise, a catenane with two different building blocks in one of the rings will exist in two different forms (Figure 6.14 **B**). A similar form of supramolecular stereoisomerism arises in the "container compounds" discussed in Section 4.3.3. As shown in the schematic of Figure 6.14 **C**, when the container has two distinguishable "poles", an unsymmetrical guest can lie in isomeric positions. Such isomerism has been observed for both covalent and non-covalent container compounds.

For each case in Figure 6.14, we have stereoisomers—structures with the same connectivities but differing arrangements of the atoms in space. They are not enantiomers, so they must be diastereomers. The novelty lies in the fact that these stereoisomers interconvert by a *translation or reorientation* of one component relative to the other. In some ways these structures resemble conformers or atropisomers, which involve stereoisomers that interconvert by rotation about a bond. For the supramolecular stereoisomers, however, interconversion involves rotation or translation of an entire molecular unit, rather than rotation around a bond. Note that for none of the situations of Figure 6.14 do we have topological stereoisomers. In each case we can interconvert stereoisomers without breaking and reforming bonds.

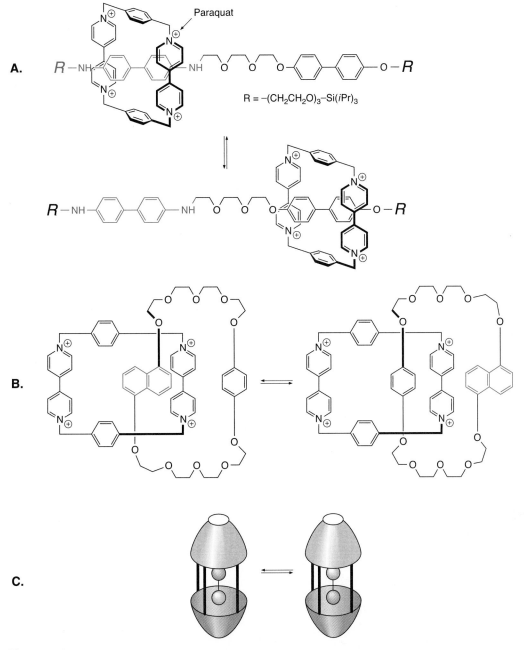

Figure 6.14
Supramolecular isomerism in rotaxanes, catenanes, and "container" compounds. **A.** Moving along a rotaxane axle can lead to isomerism if there are two different "docking stations". **B.** Similarly, catenanes can exist in isomeric forms if there is structural diversity in one of the rings. **C.** A conceptualization of isomerism in a container compound.

More complex catenanes can produce topological stereoisomers. Consider a [3]cate-nane with two types of rings, symbolized in Figure 6.15 **A**. Having the unique ring in the outer position vs. the inner position defines two stereochemical possibilities. These structures are now topological diastereomers. They cannot be interconverted without breaking bonds. A large number of stereoisomers becomes possible with [*n*]catenanes as *n* gets larger and each ring is different.

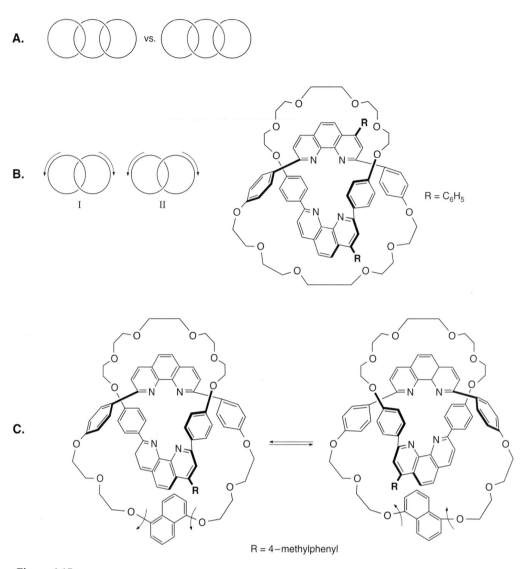

Figure 6.15
Topological isomerism in simple catenanes. **A.** A kind of "positional" isomerism that can occur in a [3]catenane with two different kinds of ring. **B.** A more subtle isomerism that involves "oriented" rings, and a chemical example. **C.** The "toplogical rubber glove", a pair of enantiomers that can interconvert readily without ever going through an achiral conformation.

A more subtle case of topological isomerism arises in a [2]catenane in which the two rings are not simple, symmetrical circles, but rather have a sense of direction (Figure 6.15 **B**). Now, topological enantiomers (I vs. II) are possible. This may be easier to see with a real chemical example (Figure 6.15 **B**). Again, the Sauvage Cu$^+$/phenanthroline templating strategy was used to assemble two directional rings, producing a topologically chiral [2]catenane. You should convince yourself that the catenane shown can exist as a pair of enantiomers, and that no amount of spinning the rings can interconvert them.

If one ring has a sense of direction, but the other does not, an even more subtle phenomenon occurs. Figure 6.15 **C** shows such a case. The molecule is chiral. The two enantiomers, however, can interconvert readily by simply rotating the 1,5-dioxynaphthyl ring and translating the other macrocycle. Sauvage and Mislow realized, however, that at no point during this process does an achiral conformation appear. In fact, it is impossible to create an achiral representation of this structure. The molecule has been referred to as a "topological rubber glove", referring to the fact that a rubber glove can be converted from right-handed to left-handed by pulling it inside out, but at no point in the process does an achiral form appear.

6.7   Stereochemical Issues in Polymer Chemistry

Many unnatural polymers of considerable commercial importance have one stereocenter per monomer, such as in polypropylene and polystyrene (Figure 6.16). Unlike the "polymerization" involved in forming a protein or nucleic acid (see the next section), these unnatural systems typically start with a simple, achiral monomer (propene or styrene), and the polymerization generates the stereogenic centers. Control over the sense of chirality for each polymerization step is often absent. As a result, considerable stereochemical complexity can be expected for synthetic polymers. For example, molecular weight 100,000 polypropylene has approximately 2400 monomers, and so 2400 stereogenic centers (look at the next Going Deeper highlight for an interesting ramification of this). There are thus 2^{2400} or approximately 10^{720} stereoisomers! The R,S system is not very useful here. Hence, polymer stereochemistry is denoted by a different criterion called tacticity.

Tacticity describes only local, relative configurations of stereocenters. The terms are best defined pictorially, as in Figure 6.16. Thus, **isotactic** polypropylene has the same configuration at all stereocenters. Recall the two faces of propylene are enantiotopic, and the isotactic polymer forms when all new bonds are formed on the same face of the olefin. If, instead, there is an alternation of reactive faces, the polymer stereocenters alternate, and a **syndiotactic** polymer is produced. Finally, a random mixture of stereocenters produces **atactic** polymer.

Control of polymer stereochemistry is a major research area in academic and industrial laboratories. This is because polymers with different stereochemistries often have very different properties. For example, atactic polypropylene is a gummy, sticky paste sometimes used as a binder, while isotactic polypropylene is a rugged plastic used for bottle caps. Recent advances (see the Going Deeper highlight on the next page and Chapter 13) have greatly improved the ability to control polymer stereochemistry, leading to commercial production of new families of polymers with unprecedented properties.

Another stereochemical issue is helicity, as some simple polymers can adopt a helical shape. We defer discussion of this to Section 6.8.2, in which we discuss helicity in general.

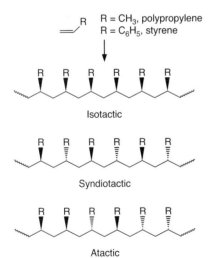

Figure 6.16
Different forms of polypropylene and polystyrene.

<hr/>

Going Deeper

Polypropylene Structure and the Mass of the Universe

Just for fun, calculate the mass of a sample of molecular weight 100,000 polypropylene that has just one molecule of each of the 10^{720} possible stereoisomers. In doing so, you will exceed the entire mass of the universe by a large margin. In fact, even though millions of tons of polypropylene are made every year, every possible stereoisomer of a polypropylene sample of molecular weight 100,000 has never been made and never will be!

Going Deeper

Controlling Polymer Tacticity—The Metallocenes

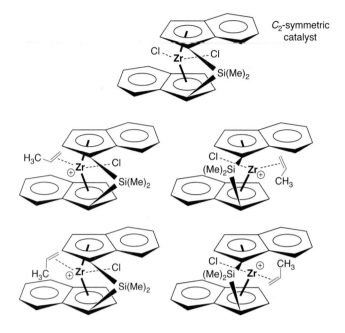

The C_2-symmetric Zr metallocene catalyst (top) and a highly schematic view of propylene complexing to it. The middle two structures use the same face of the propylene, and lead to the same tacticity because of the C_2 symmetry of the catalyst. The bottom two structures use the opposite face of the olefin. The adverse steric interaction of the CH_3 with the aromatic ring disfavors these structures.

One of the most exciting recent advances in organic and organometallic chemistry has been the development of new catalysts that produce polypropylene with high stereochemical purity. Both isotactic and syndiotactic polypropylene are now made commercially with a new class of metallocene catalysts, prototypes of which are shown below. The mechanism of the polymerization reaction is discussed in Chapter 17. Here we will focus on the stereochemistry, because symmetry principles of the sort we discussed above were crucial in the design of this chemistry.

A key step in metal-induced olefin polymerization has the olefin π face complexing to the metal center. The two faces of the propylene double bond are enantiotopic. Isotactic polypropylene forms when only one face of the propylene monomer consistently reacts to make polymer. Thus, a chiral catalyst is needed to distinguish enantiotopic faces of an olefin. But, how do we ensure that only one face reacts? It is a complicated problem, because when an olefin like propylene complexes to a metal center in a typical chiral environment, not only will both faces complex to some extent, but many orientations are possible for each complex. This leads to many different reaction rates, and a mixture of stereochemistries. A key to the solution, then, was to develop a catalyst that is chiral but not asymmetric. In particular, the C_2-symmetric metallocene shown below was prepared. The metal is chirotopic but non-

stereogenic. Hence, the chlorines are homotopic and either can be replaced with propylene, giving identical structures. By making one side of the coordination site much more bulky than the other, the propylene will complex to the metal (the first step in the reaction) with the methyl group away from the crowded side. There are two different ways to do this, but they are symmetry equivalent, and both involve the same face of the propylene. If the catalyst is enantiomerically pure, stereochemical control becomes possible.

The production of pure, syndiotactic polypropylene was even more challenging, but again symmetry notions played a key role. Syndiotactic polypropylene requires an alternation of stereochemistry at the catalyst center. Formally, a syndiotactic polymer is like a meso compound, and so a chiral catalyst is not required. To achieve the desired stereochemistry, a catalyst with a mirror plane of symmetry (C_s) was developed (see next page). The idea was that the growing polymer would move back and forth between mirror-image (enantiotopic) sites of the catalyst (caused by steric influences of the growing chain), and this alternating behavior would lead to an alternation in the stereochemistry of monomer incorporation. This was a bold suggestion, but this strategy has been successfully implemented into commercially viable processes.

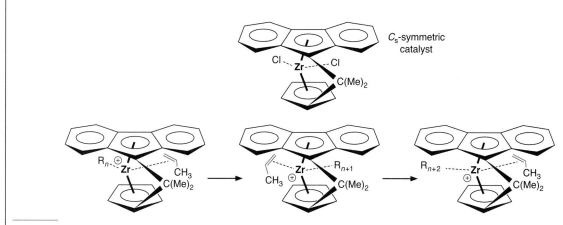

Coates, G. W., "Precise Control of Polyolefin Stereochemistry Using Single-Site Metal Catalysts." *Chem. Rev.,* **100**, 1223–1252 (2000); Resconi, L., Cavallo, L., Fait, A., and Piemontesi, F. "Selectivity in Propene Polymerization with Metallocene Catalysts." *Chem. Rev.,* **100**, 1253–1345 (2000).

6.8 Stereochemical Issues in Chemical Biology

Molecular shape is a crucial concept in chemical biology. The "lock-and-key" metaphor of enzyme–substrate or antigen–antibody interactions is useful for understanding biological phenomena, and it depends crucially on molecular shape. Despite the marvelous diversity and apparent complexity of biomolecules, at a fundamental level, biopolymers are built up from really fairly simple monomers and connecting units. The structural complexity arises from an accumulation of a large number of individually straightforward interactions. As such, only a few basic stereochemical notions are necessary for dealing with biopolymers. Since many of the complex chemical structures that make up life (proteins, nucleic acids, and polysaccharides) are biopolymers, our current understanding of small molecule stereochemistry and polymer topology allows us to explore the stereochemistry of these biological structures.

6.8.1 The Linkages of Proteins, Nucleic Acids, and Polysaccharides

As stated previously, polymer stereochemistry depends critically upon the structures of the monomers and how they are assembled. No new stereocenters are produced when amino acids are combined to make proteins, or nucleotides are combined to make nucleic acids. This is because the linkages created in forming the polymers are not stereogenic. The same is not true for polysaccharides, where the newly formed anomeric center is stereogenic. We will consider these three types of biopolymers separately.

Proteins

Proteins are polymers built from a concatenation of α-amino acid monomers. There are twenty common amino acids, and all but one (glycine) are chiral. Thus, a protein—a poly(α-amino acid)—could have a huge number of stereoisomers. This is no way to build a living organism. As such, living systems contain only one enantiomer of each amino acid. Polymerization then produces only one stereoisomer, an isotactic polymer (Figure 6.17 **A**). The polymerization itself—the peptide bond formation—does not create a new stereogenic center. As a result, unlike polypropylene, the polymerization of amino acids does not require any special stereochemical control of the bond forming reaction.

The newly formed peptide bond is not a stereogenic unit, so amino acid polymerization is in some ways different than propylene polymerization. However, as we noted earlier in

A.

B.

s-trans   *s*-cis

Figure 6.17
Basic stereochemical issues in protein structures. **A.** The conventional representation of a protein chain, and an alternative representation that emphasizes the isotactic nature of the polymer. **B.** *S*-cis and *s*-trans geometries in a conventional peptide bond and in a peptide bond involving proline.

Chapter 1, the peptide bond does have significant conformational preferences. The group is planar, and in secondary amides of the sort found in most peptide bonds, there is a significant preference for what is termed the *s*-trans or the *Z* stereochemistry (Figure 6.17 **B**). This preference is typically on the order of 4 kcal/mol, and it has a profound effect on the potential shapes that proteins can adopt. The difference in this system from the polypropylene system is that the barrier separating the two forms of the peptide bond (~19 kcal/mol) is such that they equilibrate readily at conventional temperatures. Thus, exerting stereochemical control over the formation of the peptide bond would be futile, because the system would quickly adjust to the thermodynamic equilibrium. Still, this highlights the inherent ambiguity of many stereochemical concepts. If the rotation barrier in amides was 29 kcal/mol (or we lived at –78 °C!), the peptide bond would be a stereogenic center, and tacticity would be a key issue in protein chemistry. The conformational preference of the peptide bond results from several factors, including adverse steric interactions in the *s*-cis and a favorable alignment of bond dipoles in the *s*-trans form (Chapter 1). An exception arises when proline contributes the N to an amide bond (Figure 6.17 **B**). Now the N has two alkyl substituents, and the cis–trans energy difference is much smaller. As such, proteins often adopt unique conformations in the vicinity of a proline.

Nucleic Acids

The only stereogenic centers of DNA and RNA are found at the sugar carbons, and because the ribose or deoxyribose are enantiomerically pure, natural nucleic acids are isotactic. The P of the phosphodiester backbone of a nucleic acid is not a stereogenic center, but the two O$^-$ groups of a connecting phosphate are diastereotopic. The phosphorus is thus prochiral. This has led to the use of labeled phosphates in mechanistic studies, as described with one example in a Connections highlight on the next page.

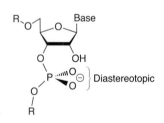

Polysaccharides

In contrast to proteins and nucleic acids, the linkages formed between saccharide monomers are made at stereogenic centers, and so stereochemical control of the polymerization step is critical. The crucial carbon, the **anomeric center**, is highlighted in Figure 6.18, which defines the nomenclature convention for this stereogenic center. This stereochemical distinc-

Going Deeper

CD Used to Distinguish α-Helices from β-Sheets

The two most prominent secondary structural features of protein chemistry are the α-helix and the β-sheet (the basic structures are described in Appendix 4). As mentioned earlier, all helices have an inherent chirality. In contrast, sheets are in a sense flat, and therefore, they are not *inherently* chiral even though the peptide building blocks themselves are chiral. In addition to the α-helix and the β-sheet, peptides and proteins can lack any defined shape, called a random coil. Once again, no inherent chirality would be associated with this structure, although the building blocks are chiral. This suggests that spectroscopic methods that probe chirality could be used to probe protein secondary structure. Circular dichroism is by far the one most commonly employed.

The most useful region of the spectrum is from 190–240 nm. Absorbances in this region are dominated by the amide backbone rather than the sidechains, making them more sensitive to secondary structure. In a CD spectrum, two negative peaks of similar magnitude at 222 and 208 nm are indicative of an α-helix. A β-sheet is revealed by a negative band at 216 nm and a positive one of similar magnitude near 195 nm. Lastly, a strong negative band near 200 nm and often a positive one at 218 nm is indicative of a lack of well-defined structure (the random coil). These are empirical observations that have been confirmed in many systems. The figure shows prototype spectra of each structural type in black, and the experimental CD spectrum of myoglobin in color. Fitting the experimental spectrum as a linear combination of the three prototype curves leads to an estimate of 80% α-helix, with the rest mostly random coil. This is in good agreement with the value of 77% α-helix derived from the x-ray structure of myoglobin.

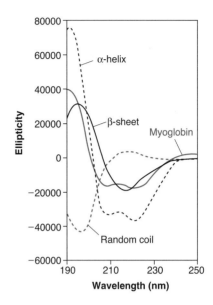

tion is crucial. For example, when glucose is polymerized exclusively with α-1,4-glycoside bonds, a helical structure called amylose (a starch) is obtained. Conversely, all β-1,4 linkages leads to a "rigid-rod" linear structure called cellulose. As with the stereoisomers of polypropylene, these two stereoisomeric polymers have distinctly different properties. Starch is formed by animals and is primarily used for energy storage, while cellulose is a structural material found in plants. Thus, the enzymes that make glycosidic bonds are well developed to control the stereochemistry of the coupling.

Connections

Creating Chiral Phosphates for Use as Mechanistic Probes

When one O^- in a phosphodiester of DNA or RNA is replaced by, for example, a specific isotope or by S^-, two stereoisomers are possible. This allows one to follow the stereochemistry of the reactions that take place at the phosphorus center, potentially revealing the mechanisms of these reactions. For example, RNase A (an enzyme) catalyzes ring opening of the specific diastereomer of the cyclic phosphodiester shown to the right, giving only a single product in methanol. This corresponds to what is known as an **in-line attack**, because the leaving group is in line with the nucleophilic attack (similar to an S_N2 reaction). We will examine the use of stereochemical analyses to probe mechanisms many times in the context of organic reactions in part II of this book.

Stereogenic phosphodiester group

Usher, D. A., Erenrich, E. S., and Eckstein, F. "Geometry of the First Step of Reaction of Ribonuclease A." *Proc. Natl. Acad. Sci. USA*, **69**, 116 (1972).

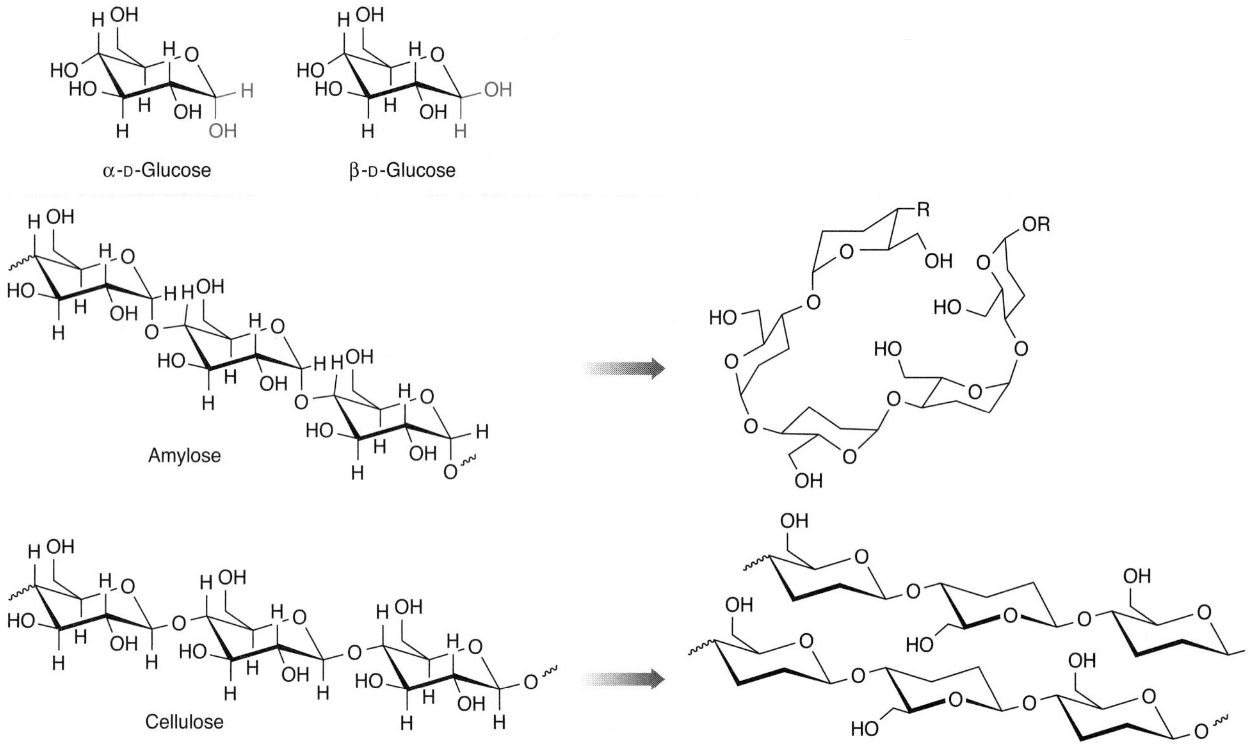

Figure 6.18
α- vs. β-D-Glucose with the key anomeric carbon highlighted, along with the structures of amylose and cellulose. For amylose and cellulose, the structures shown to the right of the arrows represent their structures in solution (with many hydroxyls eliminated for clarity).

6.8.2 Helicity

While helicity can be associated with many kinds of molecules, it is most frequently associated with polymers (especially biopolymers). Here we briefly cover the helix as a general stereochemical element. All helices are chiral, as evidenced by the fact that we refer to helices as right- or left-handed. Typically, with molecular helices the right- and left-handed forms are topologically equivalent—that is, we can interconvert the two without breaking or crossing bonds. A helix is a stereogenic unit, but it is not the interchange of ligands that interconverts opposite helices, but rather just the unwinding and rewinding of the helix.

In structural biology helices are associated with both DNA and proteins. Some polysaccharides adopt helical structures (see amylose in Figure 6.18), but this is not common. The double helix of DNA is right-handed. There is also a left-handed helical form of DNA termed Z-DNA. It is *not* the enantiomer of the much more common right-handed DNA. To make the enantiomer we would have to invert all the stereocenters of the deoxyribose sugars, which does not happen in nature. Z-DNA is a diastereomeric conformer, and it is favored by certain sequences and salt conformations, although its relevance to biology is debated. Thus, while in simple, prototype helices the right- and left-handed forms are enantiomers, in a system with enantiomerically pure, homochiral building blocks, reversing the sense of helicity produces a diastereomer.

In proteins, the most common structural motif is the α-helix discussed in Chapter 3 and depicted in Appendix 4. Again, because the building blocks (amino acids) are chiral and enantiomerically pure, right- and left-handed α-helices are diastereomers. In nature only the right-handed form is seen. A second, much less common helix, termed 3_{10} is also right-handed, and is just a conformer of the α-helix with different hydrogen bonding arrangements.

Synthetic Helical Polymers

Synthetic polymers that are isotactic are similar to biological building blocks in that all the stereocenters are homochiral. As such, it should not be surprising to learn that helical structures can show up in synthetic polymers, but usually not with the well-defined structural integrity of DNA or protein α-helices. In nucleic acids and proteins, there are strong stereochemical biases built into the monomers, and these lead to strong preferences for one helical form over the other. In synthetic polymers, such strong biases are often absent. However, in certain cases substantial helical biases can be seen in synthetic polymers (see the next Connections highlight for an example).

A truly remarkable example of a helical synthetic polymer is the series of polyisocyanates studied by Green and co-workers and summarized in Figure 6.19. The polyisocyanate backbone contains contiguous amide groupings reminiscent of a peptide or a nylon derivative [nylon-6 is $-C(O)(CH_2)_5NH-$; polyisocyanates have been termed nylon-1; see Chapter 13 for further discussion of nylons]. The structure shown describes the basic layout of the backbone, but steric clashing between the carbonyl oxygen and the R group precludes a planar geometry. A trade-off between conjugation and sterics produces a helical structure, but in a simple polyisocyanate we expect no particular bias for the right- or left-handed helix, as the two are enantiomers.

One way to produce a helical bias is to convert the enantiomeric helices into diastereomers by incorporating stereogenic centers into the sidechains (R), much as with natural biopolymers. This strategy works spectacularly well with polyisocyanates. As shown in Figure 6.19, making the sidechain stereogenic simply by virtue of isotopic substitution leads to a huge helical bias. That this is so is seen by the tremendous increase in optical activity and the reversal in sign on polymerizing the monomer. Both the magnitude and the change in sign establish that the inherent optical activity of the monomer is not responsible for the optical activity of the polymer. With a helical backbone, now the chromophoric amide units contribute to the optical rotation. Full CD studies support this analysis.

What is the cause of this effect? It has been estimated that the bias for one helical handedness over the other induced by the isotopic substitution is on the order of 1 cal/mol per subunit—a miniscule amount. Thus, we are seeing an extreme example of **cooperativity**. Once a tiny bias is established, it propagates down the chain, each successive monomer being more

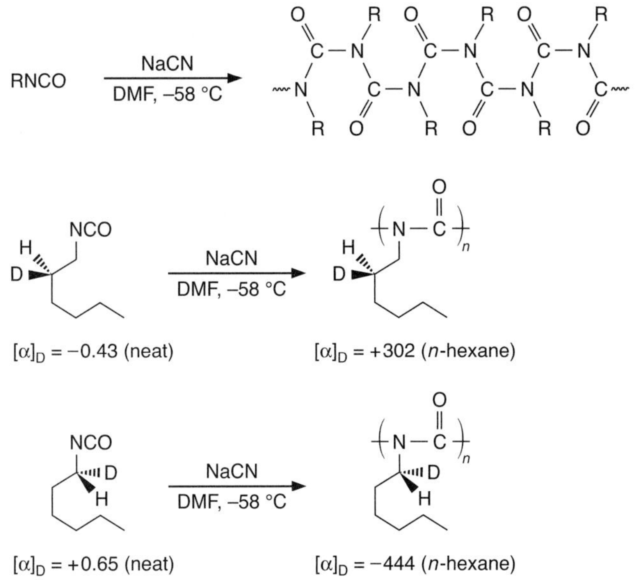

Figure 6.19
Examples of helicity in simple, non-natural polymers. Note that the optical rotation values given are on a per monomer basis, so the large increase in absolute value on polymerization is meaningful.

inclined to adopt the currently accepted chirality. It is truly amazing, though, that such a trivial inherent bias can ultimately lead to such an obvious effect. The detailed analysis of this sort of cooperativity involves some fairly complex math and physics, so we direct the interested student to the references at the end of the chapter.

The amplification of chirality inherent in the polyisocyanates described is an example of a phenomenon wherein a small initial chirality leads to a bias resulting in high enantiomeric excesses. This phenomenon has been termed the **sergeants and soldiers principle**, implying that the initial chiral influence is the "sergeant" that aligns all the "soldiers". This is a phenomenon that has been observed not only in polymer chemistry, but also with self-assembled supramolecular complexes driven by π interactions and hydrogen-bonded systems.

The optical rotations given in Figure 6.19 are extraordinarily large. The reason is not that these helical structures are somehow "more chiral" than typical molecules. Rather, the large rotations are due to the fact that with the polyisocyanates we are probing an **intrinsically chiral chromophore**. The feature of the molecule that is interacting most strongly with the light, the amide group, is itself distorted into a chiral shape. A more typical situation is a **chirally perturbed, intrinsically achiral chromophore**, such as a carbonyl group (intrinsically achiral, as in acetone) with a nearby stereogenic carbon. In such cases, much smaller rotations and differential absorptions are typically seen.

Connections

A Molecular Helix Created from Highly Twisted Building Blocks

The creation of helices using synthetic structures has attracted considerable attention due to the common helical motif in peptides and nucleic acids. Achieving a synthetic polymer with a complete right- or left-handed twist is difficult. One approach to helical molecules has been to make compounds known as **helicenes**, highly conjugated aromatic structures that naturally possess a twist due to the physical overlap of benzene rings. Convince yourself that if the [6]helicene shown were planar, unacceptable steric clashes would occur. The shapes of these structures are akin to that which one would get if one segment of a spring were cut off. Many helicenes have been made, including the [6]helicene shown and higher homologues. Not surprisingly, these structures show high optical rotations, because they are very much intrinsically chiral chromophores.

More recently, a polymer based on the helicene motif has been prepared. The key step in the synthesis of a helical polymer based upon a helicene is the condensation of a chiral [6]helicene that has salicylaldehyde functionality at each end with 1,2-phenylenediamine in the presence of a Ni salt. This gives the chemical structure shown to the right (bonds enormously stretched for clarity of presentation). The ORD spectra of structures of this kind display extraordinarily large rotations, and the circular dichroism

spectra reveal comparably large differential extinction coefficients for right- and left-handed circular polarized light, confirming the helical nature of the polymers.

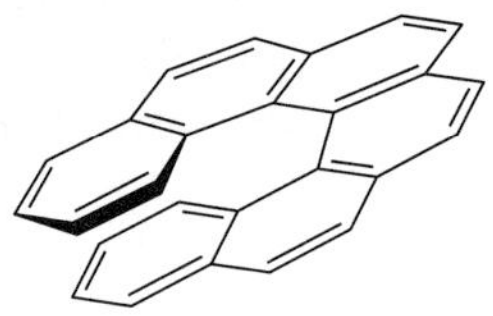

[6]Helicene

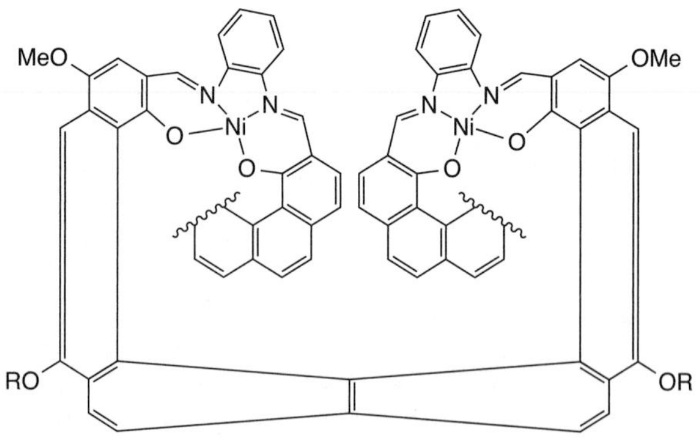

Dai, Y., Katz, T. J., and Nichols, D. A. "Synthesis of a Helical Conjugated Ladder Polymer." *Angew. Chem. Int. Ed. Eng.*, **35**, 2109 (1996).

6.8.3   The Origin of Chirality in Nature

The molecules of life are for the most part chiral, and in living systems they are almost always enantiomerically pure. In addition, groups of biomolecules are generally homochiral —all amino acids have the same sense of chirality and all sugars have the same sense of chirality. As already discussed, the chirality of the amino acids leads to chiral enzymes, which in turn produce chiral natural products. All the chiral compounds found in nature that are readily accessible to synthetic chemists for the construction of more complex molecules are referred to as the **chiral pool**.

What is the origin of the chirality of the molecules of life, and the reason for the homochirality? We cannot distinguish enantiomers unless we have a chiral environment. Further, in a reaction that forms a stereocenter, we cannot create an excess of one enantiomer over another without some chirality to start with. In the laboratory today, all enantiomeric excesses that we exploit ultimately derive from natural materials. Whether it is the interaction with an enantiomerically pure amino acid from a natural source, or an individual manually separating enantiomorphous crystals (first achieved by Pasteur), the source of enantiomeric excess in the modern chemistry laboratory is always a living system. But how was this achieved in the absence of life? This is a fascinating, complex, and controversial topic that we can touch on only briefly here. This question is often phrased as the quest for the origin of chirality in nature, but more correctly it is the origin of enantiomeric excess and homochirality we seek.

Models for the origin of life generally begin with simple chemical systems that, in time, evolve to more complex, self-organizing, and self-replicating systems. It is easy to imagine prebiotic conditions in which simple condensation reactions produce amino acids or molecules that closely resemble them, and indeed experiments intended to model conditions on the primitive earth verify such a possibility. However, it is difficult to imagine such conditions producing anything other than a racemic mixture.

Essentially, there are two limiting models for the emergence of enantiomeric excess in biological systems. They differ by whether enantiomeric excess arose naturally out of the evolutionary process or whether an abiotic, external influence created a (presumably slight) initial enantiomeric excess that was then amplified by evolutionary pressure (maybe a type of sergeant–soldier effect). The first scheme is a kind of selection model. The building blocks (let's consider only amino acids here) are initially racemic. However, there is considerable advantage for an early self-replicating chemical system to use only one enantiomer. For example, consider a simple polymer of a single amino acid. If both enantiomers are used, the likely result is an atactic polymer, which may well have variable and ill-defined properties. However, if only a single enantiomer is used, only the isotactic polymer results. This kind of specificity could be self-reinforcing, such that eventually, only the single amino acid is used. The homochirality of nature could result because addition of a second amino acid to the mix might be less disruptive if the new one has the same handedness as the original. The details of how all this could happen are unknown, but the basic concept seems plausible. Certainly, the remarkable cooperativity seen in polyisocyanates provides an interesting precedent.

While we begin with racemic materials, there will never be *exactly* identical numbers of right- and left-handed molecules in a sample of significant size. This is a simple statistical argument. For example, earlier we considered the reduction of 2-butanone with lithium aluminum hydride under strictly achiral conditions (Figure 6.7), and stated that we expect a racemic mixture without a *significant* enantiomeric excess. However, if we start with 10^{23} molecules of ketone, the probability that we will produce *exactly* 0.5×10^{23} molecules of (R)- and 0.5×10^{23} molecules of (S)-alcohol is essentially nil. There will always be statistical fluctuations. For example, for a relatively small sample of 10^7 molecules there is an even chance that one will obtain a $\geq 0.021\%$ excess of one enantiomer over the other (we cannot anticipate which enantiomer will dominate in any given reaction). Perhaps such a small excess from a prebiotic reaction, or a significantly larger excess from a statistical fluke, got amplified through selective pressure, and ultimately led to the chirality of the natural world.

The alternative type of model emphasizes the possible role of an inherently chiral bias of external origin. One possibility for this bias is the inherent asymmetry of our universe reflected in the charge–parity (CP) violation of the weak nuclear force. In particular, β decay of ^{60}Co nuclei produces polarized electrons with a slight excess of the left- over the right-handed form. From this point, several mechanisms that translate the chirality of the emission to a molecular enantiomeric excess can be envisioned. Unfortunately, all attempts to measure such enantiomeric enrichment in the laboratory have produced at best extremely small enrichments that have proven difficult to reproduce. An alternative proposal for an external chiral influence is an enantioselective photochemical process involving circularly polarized light, which is well established in the laboratory to give significant enantiomeric excesses. At present, however, no clear mechanism for creating circularly polarized light with an excess of one handedness in the prebiotic world has been convincingly demonstrated, although models have been proposed. Only further experimentation in the lab, or perhaps examination of the chirality of extraterrestrial life forms, will resolve this issue.

6.9   Stereochemical Terminology

Stereochemistry has engendered a sometimes confusing terminology, with several terms that are frequently misused. Here we provide definitions of the most common terms. This collection is based in large measure on a much more extensive listing in the following book: Eliel, E. L., Wilen, S. H., and Mander, L. N. (1994). *Stereochemistry of Organic Compounds*, John Wiley & Sons, New York.

Absolute configuration.   A designation of the position or order of arrangement of the ligands of a stereogenic unit in reference to an agreed upon stereochemical standard.

Achiral.   Not chiral. A necessary and sufficient criterion for achirality in a rigid molecule is the presence of any improper symmetry element (S_n, including σ and i).

Achirotopic.   The opposite of chirotopic. See "chirotopic" below.

Anomers.   Diastereomers of glycosides or related cyclic forms of sugars that are specifically epimers at the anomeric carbon (C_1 of an aldose, or C_2, C_3, etc., of a ketose).

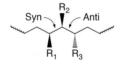

Anti.   Modern usage is to describe relative configuration of two stereogenic centers along a chain. The chain is drawn in zigzag form, and if two substituents are on opposite sides of the plane of the paper, they are designated anti. See also "syn", "antiperiplanar", and "anticlinal".

Anticlinal.   A term describing a conformation about a single bond. In A–B–C–D, A and D are anticlinal if the torsion angle between them is between 90 and 150 or –90 and –150. See Figure 2.7.

Antiperiplanar.   A term describing a conformation about a single bond. In A–B–C–D, A and D are antiperiplanar if the torsion angle between them is between +150° to –150°. See Figure 2.7.

Apical, axial, basal, and equatorial.   Terms associated with the bonds and positions of ligands in trigonal bipyramidal structures.

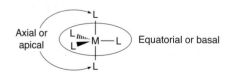

Asymmetric.   Lacking *all* symmetry elements (point group C_1). All asymmetric molecules are chiral.

Asymmetric carbon atom. Traditional term used to describe a carbon with four different ligands attached. Not recommended in modern usage.

Atactic.   A term describing the relative configuration along a polymer backbone. In an atactic polymer, the stereochemistry is random—no particular pattern or bias is seen.

Atropisomers.   Stereoisomers (can be either enantiomers or diastereomers) that can be interconverted by rotation about single bonds and for which the barrier to rotation is large enough that the stereoisomers can be separated and do not interconvert readily at room temperature.

Chiral.   Existing in two forms that are related as non-congruent mirror images. A necessary and sufficient criterion for chirality in a rigid molecule is the absence of any improper symmetry elements (S_n, including σ and i).

Chiral center.   Older term for a tetracoordinate carbon or similar atom with four different substituents. More modern, and preferable, terminology is "stereogenic center" (or "stereocenter").

Chirotopic.   The term used to denote that an atom, point, group, face, or line resides in a chiral environment.

Cis.   Describing the stereochemical relationship between two ligands that are on the same side of a double bond or a ring system. For alkenes only, Z is preferred.

Configuration.   The relative position or order of the arrangement of atoms in space that characterizes a particular stereoisomer.

Conformers or conformational isomers.   Stereoisomers that are interconverted by rapid rotation about a single bond.

Constitutionally heterotopic.   The same groups or atoms with different connectivities.

D and L.   An older system for identifying enantiomers, relating all stereocenters to the sense of chirality of D- or L-glyceraldehyde. See discussion in the text. Generally not used anymore, except for biological structures such as amino acids and sugars.

Diastereomers.   Stereoisomers that are not enantiomers.

Diastereomeric excess (de).   In a reaction that produces two diastereomeric products in amounts A and B, de = $100\%(|A - B|)/(A + B)$.

Diastereotopic.   The relationship between two regions of a molecule that have the same connectivity but are not related by any kind of symmetry operation.

Dissymmetric.   Lacking improper symmetry operations. A synonym for "chiral", but not the same as "asymmetric".

Eclipsed.   A term describing a conformation about a single bond. In A–B–C–D, A and D are eclipsed if the torsion angle between them is approximately 0°.

Enantiomers.   Molecules that are related as non-congruent mirror images.

Enantiomeric excess (ee). In a reaction that produces two enantiomeric products in amounts A and A', ee = 100%(|A − A'|)/(A + A').

Enantiotopic. The relationship between two regions of a molecule that are related *only* by an improper symmetry operation, typically a mirror plane.

Endo. In a bicyclic system, a substituent that is on a bridge is endo if it points toward the larger of the two remaining bridges. See also "exo".

Epimerization. The interconversion of epimers.

Epimers. Diastereomers that have the opposite configuration at only one of two or more stereogenic centers.

Erythro and threo. Descriptors used to distinguish between diastereomers of an acyclic structure having two stereogenic centers. When placed in a Fischer projection using the convention proper for carbohydrates, erythro has the higher priority groups on the same side of the Fischer projection, and threo has them on opposite sides.

Exo. In a bicyclic system, a substituent that is on a bridge is exo if it points toward the smaller of the two remaining bridges. See also "endo".

E, Z. Stereodescriptors for alkenes (see discussion in the text).

Gauche. A term describing a conformation about a single bond. In A–B–C–D, A and D are gauche if the torsion angle between them is approximately 60° (or −60°). See section 2.3.1.

Geminal. Attached to the same atoms. The two chlorines of 1,1-dichloro-2,2-difluoroethane are geminal. See also "vicinal".

Helicity. The sense of chirality of a helical or screw shaped entity; right (P) or left (M).

Heterochiral. Having an opposite sense of chirality. For example, D-alanine and L-leucine are heterochiral. See also "homochiral".

Heterotopic. The same groups or atoms in inequivalent constitutional or stereochemical environments.

Homochiral. Having the same sense of chirality. For example, the 20 natural amino acids are homochiral—they have the same arrangement of amino, carboxylate, and side-chain groups. Has also been used as a synonym for "enantiomerically pure", but this is not recommended, because homochiral already was a well-defined term before this alternative usage became fashionable.

Homotopic. The relationship between two regions of a molecule that are related by a *proper* symmetry operation.

Isotactic. A term describing the relative configuration along a polymer backbone. In an isotactic polymer, all stereogenic centers of the polymer backbone have the same sense of chirality.

Meso. A term describing an achiral member of a collection of diastereomers that also includes at least one chiral member.

Optically active. Rotating plane polarized light. Formerly used as a synonym for "chiral", but this is not recommended.

Prochiral.   A group is prochiral if it contains enantiotopic or diastereotopic ligands or faces, such that replacement of one ligand or addition to one face produces a stereocenter. See Section 6.3.2.

R, S. The designations for absolute stereochemistry (see earlier discussion in the text).

Racemic mixture or **racemate.**   Comprised of a 50:50 mixture of enantiomers.

Relative configuration.   This refers to the configuration of any stereogenic center with respect to another stereogenic center. If one center in a molecule is known as R, then other centers can be compared to it using the descriptors R* or S*, indicating the same or opposite stereochemistry, respectively.

Resolution.   The separation of a racemic mixture into its individual component enantiomers.

Scalemic.   A synonym for "non-racemic" or "enantiomerically enriched". It has not found general acceptance, but is used occasionally.

S-cis and s-trans.   Descriptors for the conformation about a single bond, such as the C2–C3 bond in 1,3–butadiene, or the C–N bond of an amide. If the substituents are synperiplanar, they are termed s-cis ("s" for "single"); if they are antiperiplanar, they are termed s-trans.

Stereocenter.   See "stereogenic center".

Stereogenic center.   An atom at which interchange of any two ligands produces a new stereoisomer. A synonym for "stereocenter".

Stereogenic unit.   An atom or grouping of atoms at which interchange of any two ligands produces a new stereoisomer.

Stereoisomers.   Molecules that have the same connectivity, but a different arrangement of atoms in space.

Stereoselective.   A term describing the stereochemical consequences of certain types of reactions. A stereoselective reaction is one for which reactant A can give two or more stereoisomeric products, B and B', and one or more product is preferred. There can be degrees of stereoselectivity. All stereospecific reactions are stereoselective, but the converse is not true.

Stereospecific.   A term describing the stereochemical consequences of certain types of reactions. A stereospecific reaction is one for which reactant A gives product B, and stereoisomeric reactant A' gives stereoisomeric product B'. There can be degrees of stereospecificity. Stereospecific does *not* mean 100% stereoselective.

Syn.   Modern usage is to describe the relative configuration of two stereogenic centers along a chain. The chain is drawn in zigzag form, and if two substituents are on the same side of the plane of the paper, they are syn. See also "anti", "synperiplanar", and "synclinal".

Synclinal.   A term describing a conformation about a single bond. In A–B–C–D, A and D are synclinal if the torsion angle between them is between 30° and 90° (or –30° and –90°). See Figure 2.7.

Syndiotactic.   A term describing the relative configuration along a polymer backbone. In a syndiotactic polymer, the relative configurations of backbone stereogenic centers alternate along the chain.

Synperiplanar.   A term describing a conformation about a single bond. In A–B–C–D, A and D are synperiplanar if the torsion angle between them is between +30° and –30°. See Figure 2.7.

Tacticity.   A generic term describing the stereochemistry along a polymer backbone. See "atactic", "isotactic", and "syndiotactic".

Trans.   A term describing the stereochemical relationship between two ligands that are on opposite sides of a double bond or a ring system. For alkenes only, *E* is preferred.

Vicinal.   Attached to adjacent atoms. In 1,1-dichloro-2,2-difluoroethane, the relationship of either chlorine to either fluorine is vicinal. See also "geminal".

Summary and Outlook

The excitement that chemists feel for the area of stereochemistry has hopefully rubbed off during your reading of this chapter. From simple enantiomers and diastereomers, to rotaxanes, catenanes, and knots, stereochemistry continues to challenge organic chemists to create molecules of increasing complexity, which inevitably leads to molecules with intriguing properties and simple aesthetic beauty.

Furthermore, stereochemical concepts shed important light on the study of reaction mechanisms. It is this topic that we still need to develop further. In our analyses of reaction mechanisms we will rely heavily upon the concepts and terminology introduced in this chapter. Further, in textbooks and journal articles related to chemical synthesis, the control of stereochemistry during chemical transformations is a topic of paramount importance. Now that we have a firm background on the fundamentals of stereochemistry, it is time to launch into the practical applications.

Exercises

1. We have stated that the stereogenic center in L-cysteine is *R*, while all other L-amino acids are *S*. Show this.

2. State whether the following sugars are L or D.

3. Label the following alkenes as either *Z* or *E*.

4. We have stated that the preferred conformation of a peptide bond is Z, also known as *s*-trans (referring to a trans arrangement of the single bond between C–N). Show that Z is the appropriate descriptor.

5. Show that propylene and styrene are prochiral, and label the faces of propylene as *Re* or *Si*.

6. How many diastereomers are there for the following compound? Draw them all with chair cyclohexane representations. Also, draw them flat in the page as shown below, except with solid dots on the bridgehead hydrogens to represent the cases where the hydrogens project up.

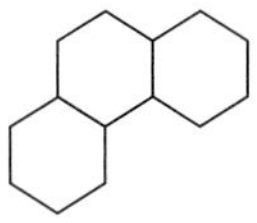

7. Draw enantiomers of the following compounds.

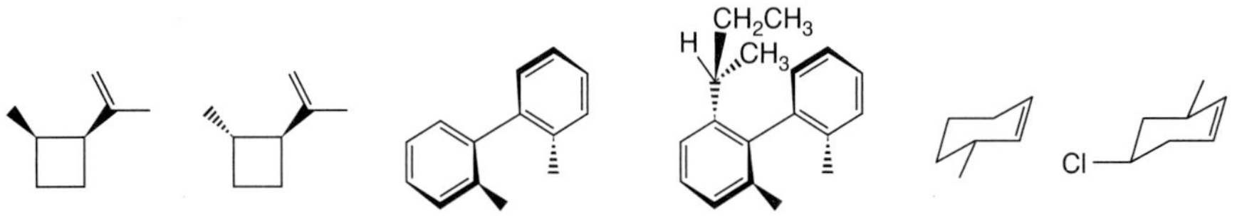

8. Identify the stereogenic centers or units in the following compounds.

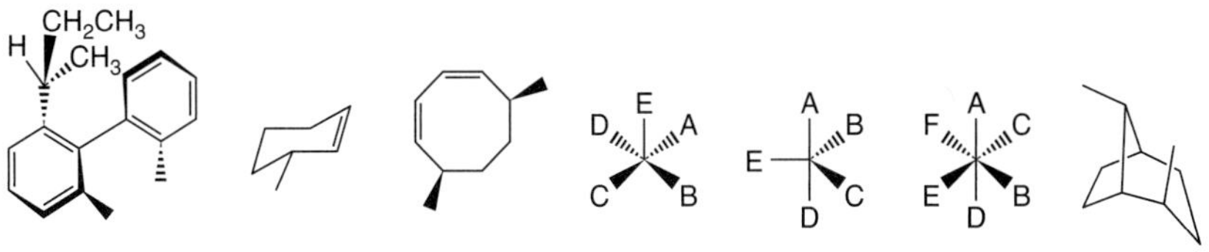

9. For each structure shown, label the pair of methyls as homotopic, enantiotopic, diastereotopic, or constitutionally heterotopic.

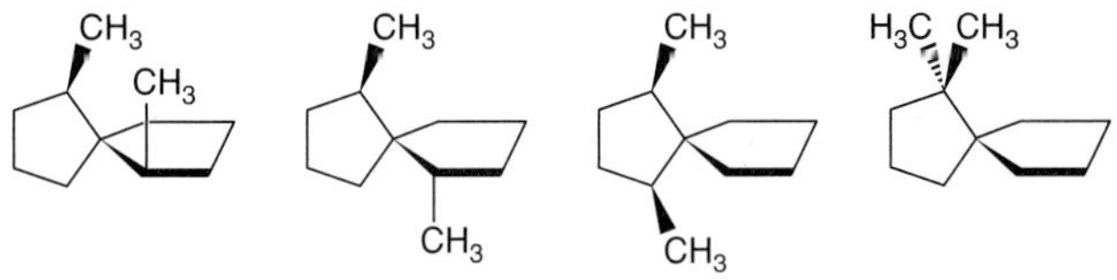

10. Is the structure shown chiral? Is it asymmetric?

11. Find the achirotopic points in the following compounds. If there are no achirotopic points, state this. If all points are achirotopic, state this also.

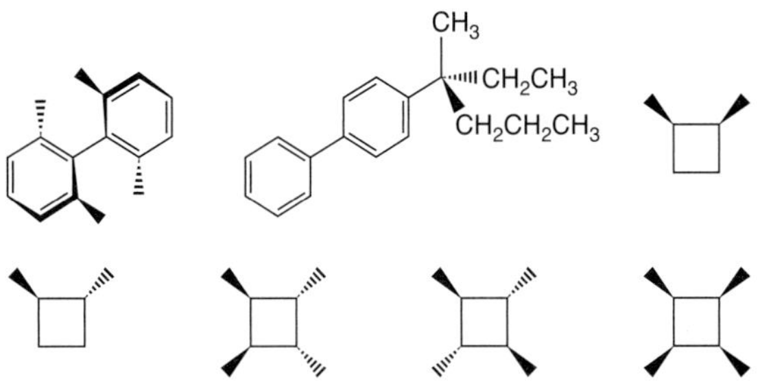

12. Label any C_n or S_n axes (including mirror planes) in the molecules in Exercise 11.

13. Draw a molecule that contains a C_3 axis and a single mirror plane.

14. Solutions of the molecule shown are optically active. However, upon reaction with itself, all optical activity vanishes. Explain this phenomenon. In addition, generalize the result. That is, describe the stereochemical features necessary for such a situation to occur.

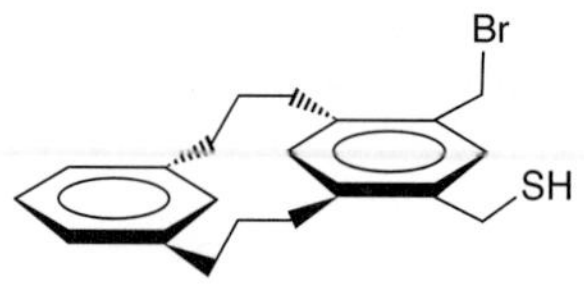

15. Draw a diastereomer of the following molecule that is not an epimer.

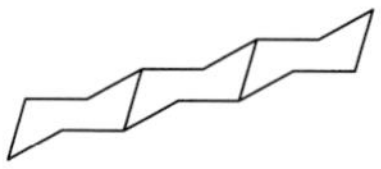

16. Find the prochiral hydrogens in the following molecules, and circle any pro-S hydrogens. If there are no prochiral hydrogens, state this.

17. Predict whether the product ratio of the following reactions will be 50:50 or a number other than 50:50.

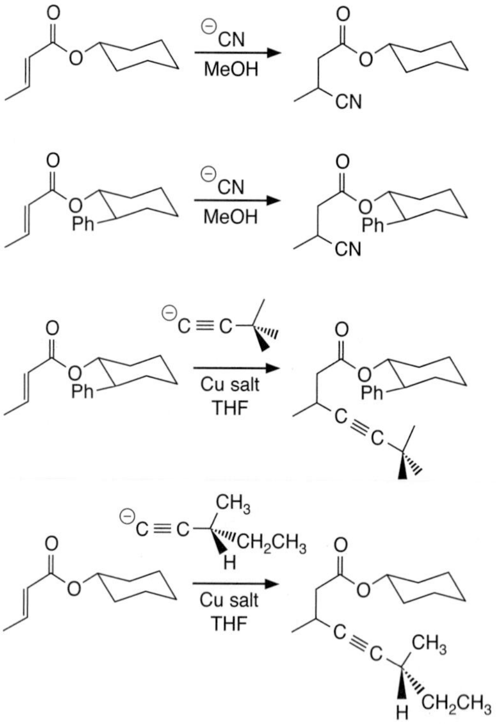

18. The following polymerization catalyst produces blocks of isotactic polypropylene with alternating stereochemistry for each block. Explain how this happens.

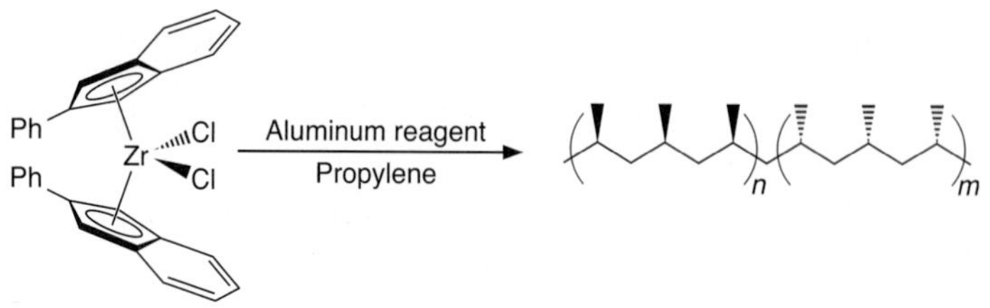

19. Show that the hydrogens of the CH$_2$ groups of the following molecules are never equivalent in any conformation.

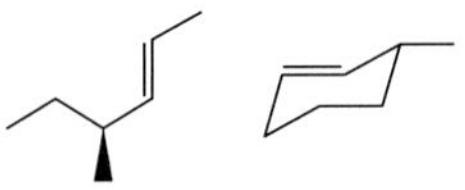

20. For each molecule shown, determine whether the two faces of the olefin or carbonyl are homotopic, enantiotopic, or diastereotopic. For ethyl phenyl ketone, designate the *Re* and *Si* faces.

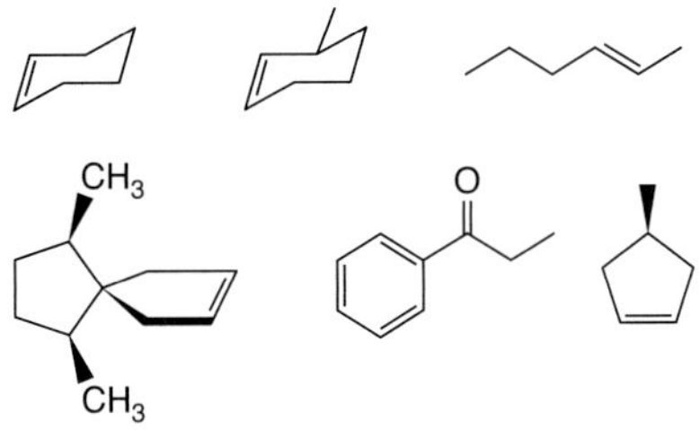

21. Show that the hydrogens of the CH$_3$ group of the following molecules are not equivalent in the conformation shown, but average due to bond rotation.

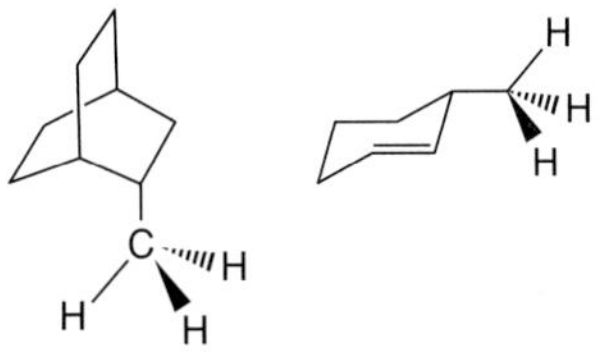

22. Define the following reactions as stereoselective and/or stereospecific, and if so, determine the percent stereoselectivity and/or stereospecificity. The products in **A**, **D**, **E**, and **F** are as shown. The product ratios in **B** and **C** are hypothetical for purposes of this question.

A.

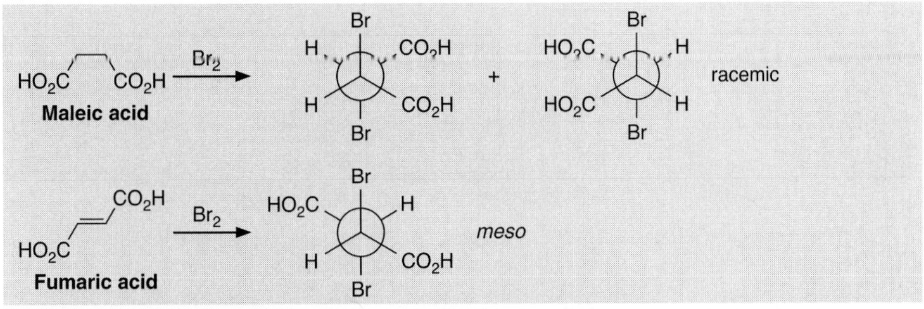

B.

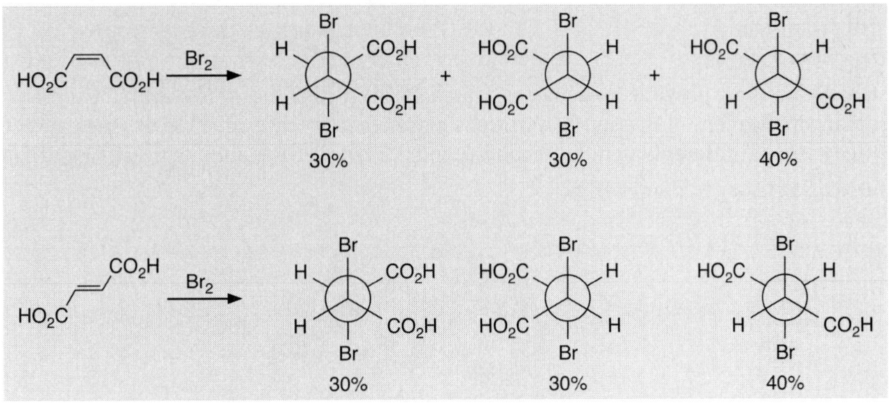

C.

HO₂C—CH=CH—CO₂H $\xrightarrow{Br_2}$

Br / H / CO₂H / H / CO₂H / Br (30%)   +   Br / HO₂C / H / HO₂C / H / Br (30%)   +   Br / HO₂C / H / H / CO₂H / Br (40%)

HO₂C—CH=CH—CO₂H $\xrightarrow{Br_2}$

Br / H / CO₂H / H / CO₂H / Br (40%)   +   Br / HO₂C / H / HO₂C / H / Br (40%)   +   Br / HO₂C / H / H / CO₂H / Br (20%)

D.

Ph, H — N(CH₃)₂⁺—O⁻ $\xrightarrow{\text{Heat}}$ Ph—C(CH₃)=CH₂

Ph, H — N(CH₃)₂⁺—O⁻ $\xrightarrow{\text{Heat}}$ Ph—C(CH₃)=CH₂

E.

N(CH₃)₂⁺—O⁻ $\xrightarrow{\text{Heat}}$ (71%) + (29%)

N(CH₃)₂⁺—O⁻ $\xrightarrow{\text{Heat}}$ (71%) + (29%)

F.

cyclohexene $\xrightarrow[n\text{-BuLi}]{\text{PhOCH}_2\text{Cl}}$ PhO— (26%) + —OPh (14%)

23. Draw any molecule that contains an enantiotopic pair of hydrogens that are not attached to the same atom.

24. We showed that rapid rotation about the C1–C2 bond of 2-butanol makes the three hydrogens at C1 symmetry equivalent. Why is it that rapid rotation about the C2–C3 bond (or any other bond) does not make the two hydrogens at C3 equivalent?

25. How many stereoisomers are possible for a linear [3]catenane? Which of these are chiral (presume that the individual rings have a mirror plane in the plane of the ring)? Consider separately three cases: a. all three rings are equivalent and directional, b. all three rings are different but not directional, and c. all three rings are inequivalent and directional.

26. Convince yourself that C₆₀ has a planar graph.

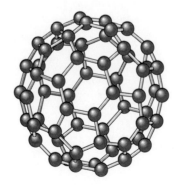

27. The THYME polyether of Figure 6.12 could also close with two twists. If it does, what would be the product of ozonolysis?

28. In the section on "Helical Descriptors" (part of Section 6.1.2), we showed an allene and two related structures and gave M/P assignments. Show that the same assignments are obtained if you sight down the opposite end of the axis shown.

29. Recall the [5]catenene olympiadane of Chapter 4. How many stereoisomers would be possible if each ring of the system were different, while maintaining the Olympic ring motif? Assume that all the rings are non-directional.

30. Ferrocene has two limiting conformations, an eclipsed form and a staggered form. Each has an S_n axis. What is n for each?

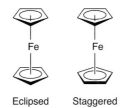

Eclipsed   Staggered

31. We discussed the "toplogical rubber glove", a system in which two enantiomers can interconvert without ever going through an achiral form. A related phenomenon was observed much earlier with the biphenyl derivative shown, first prepared by Mislow. The nitro groups are large enough that the biphenyls cannot rotate past one another on any meaningful time scale. Convince yourself that a. this molecule is chiral, b. the enantiomers can readily interconvert by rotations about single bonds, and c. at no time during the enantiomerization is a structure that is achiral involved.

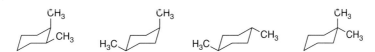

32. For each structure shown, determine whether the two methyl groups are homotopic, enantiotopic, diastereotopic, or constitutionally heterotopic, both on a time scale where ring inversion is slow and on a time scale where ring inversion is fast.

33. We saw in a Going Deeper highlight in Section 2.5.3 that hexaisopropylbenzene adopts a geared conformation. Consider a structure in which two adjacent isopropyl groups are replaced by 1-bromoethyl groups (that is, one CH_3 of an isopropyl is replaced by Br in two adjacent groups). Maintaining the rigorously geared structure, sketch all possible stereoisomers for this compound, and describe them as chiral or not and establish pair-wise relationships as enantiomeric or diastereomeric. Consider especially the consequences of reversing the direction around the ring of the geared array.

34. Convince yourself that the metals in the complexes shown in the Connections highlight entitled "C_2 Ligands in Asymmetric Synthesis" are indeed chirotopic but non-stereogenic. Also show that the coordination to either face of the metal in these complexes produces identical structures.

35. For the mathematically inclined, calculate the probability of obtaining an *exactly* 50:50 ratio of enantiomers from the LAH reduction of 2-butanone when the amount of starting material is a. 10 molecules, b. 10^3 molecules, and c. 10^{21} molecules.

36. In Section 6.8.1, a [6]helicene is shown in a Connections highlight. Assign an M or P descriptor to this helicene. Furthermore, what is the appropriate M or P descriptor for the binaphthol compound show in the margin of Section 6.5?

37. Draw the stereoisomers of tris(*o*-tolyl)borane. What bond rotations are required to interconvert diastereomers, and which are required to inconvert enantiomers?

38. The reaction of phenylacetylene with Br_2 only gives (Z)-1,2-dibromo-1-phenylethene, and therefore the reaction is 100% stereoselective. Is the reaction also stereospecific? Explain your answer.

39. A famous topological construct is the **Borromean rings**, shown below. At first they appear to be just three interlocking rings, but look more closely. No two rings are interlocked. If we break any one ring, the entire construct falls apart. These rings hold together only if all three are intact. The symbolic significance of such a structure has been appreciated for centuries in many diverse cultures. Chemically, the challenge is clear. We cannot build up the Borromean rings by first linking a pair of rings and then adding another, because there are no pairwise linkages. Alternative strategies are required, and several have been suggested. For the synthetically intrepid, design a synthesis of the Borromean rings using the general metal templating strategies that Sauvage applied to the creation of catenanes. Focus on strategic and topological issues rather than detailed chemical issues. Very recently, a molecular realization of the Borromean rings has been brilliantly synthesized by Stoddart and coworkers. See Chickak, K. S., Cantrill, S., Pease, A. R., Sheng-Hsien, C., Cave, G. W. C., Atwood, J. L., and Stoddart, J. F. "Molecular Borromean Ring." *Science,* **304,** 1308 (2004).

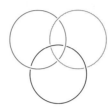

Further Reading

Classic Review Articles and Textbooks on Stereochemistry

Mislow, K. (1966). *Introduction to Stereochemistry,* W. A. Benjamin, Inc., New York.

Eliel, E. L., Wilen, S. H., and Mander, L. N. (1994). *Stereochemistry of Organic Compounds,* Wiley, New York. An extensive compilation of all topics related to organic stereochemistry. Also includes a comprehensive glossary of stereochemical terminology.

Mislow, K. "Molecular Chirality." *Top. Stereochem.,* **22,** 1 (1999).

Juaristi, E. (1991). *Introduction to Stereochemistry and Conformational Analysis,* Wiley-Interscience, New York.

Klyne, W., and Buckingham, J. (1978). *Atlas of Stereochemistry,* 2d ed., Oxford University Press, New York.

Three-Dimensional Drawing of Chemical Structures

Hoffmann, R., and Laszlo, P. "Representation in Chemistry." *Angew. Chem. Int. Ed. Eng.,* **30,** 1 (1991).

Chiral Molecules with High Symmetry

Farina, M., and Morandi, C. "High Symmetry Chiral Molecules." *Tetrahedron,* **30,** 1819 (1974).

Stereogenic and Chirotopic

Mislow, K., and Siegel, J. "Stereoisomerism and Local Chirality." *J. Am. Chem. Soc.,* **106,** 3319 (1984).

Symmetry and Point Groups

Cotton, F. A. (1971). *Chemical Applications of Group Theory,* 2nd ed., Wiley-Interscience, New York.

Heilbronner, E., and Dunitz, J. D. (1993). *Reflections on Symmetry,* Verlag Helvetica Chimica Acta, Basel.

Stereochemical Nomenclature and Terminology

http://www.chem.qmw.ac.uk/iupac/stereo/

Cahn, R. S., Ingold, C. K., and Prelog, V. "Specification of Molecular Chirality." *Angew. Chem. Int. Ed. Eng.,* **5,** 385 (1966).

Rigaudy, J., and Klesney, S. (1979). *Nomenclature of Organic Chemistry,* Pergamon Press, Oxford, England.

Hirschmann, H., and Hanson, K. R. "On Factoring Chirality and Stereoisomerism." *Top. Stereochem.,* **14,** 183 (1983).

Mislow, K., and Raban, M. "Stereoisomeric Relations of Groups in Molecules." *Top. Stereochem.,* **1,** 1 (1967).

Nicolaou, K. C., Boddy, C. N. C., and Siegel, J. S. "Does CIP Nomenclature Adequately Handle Molecules with Multiple Stereoelements? A Case Study of Vancomycin and Cognates." *Angew. Chem. Int. Ed. Eng.*, **40**, 701 (2001).

Prochiral Nomenclature

Hanson, K. R. "Applications of the Sequence Rule. I. Naming the Paired Ligands g,g at the Tetrahedral Atom Xggij. II. Naming the Two Faces of a Trigonal Atom Yghi." *J. Am. Chem. Soc.*, **88**, 2731 (1996).

Stereoselective and Stereospecific Reactions

Zimmerman, H. E., Singer, L., and Thyagarajan, B. S. "Overlap Control of Carbanionoid Reactions. I. Stereoselectivity in Alkaline Epoxidation." *J. Am. Chem. Soc.*, **81**, 108 (1959).
Adams, D. L. "Toward the Consistent Use of Regiochemical and Stereochemical Terms in Introductory Organic Chemistry." *J. Chem. Educ.*, **69**, 451 (1992).

Optical Activity and Chiroptical Methods

Hill, R. R., and Whatley, B. G. "Rotation of Plane-Polarized Light. A Simple Model." *J. Chem. Educ.*, **57**, 306 (1980).
Brewster, J. H. "Helix Models of Optical Activity." *Top. Stereochem.*, **2**, 1 (1967).
Snatzke, E., ed. (1967). *Optical Rotary Dispersion and Circular Dichroism in Organic Chemistry*, Heyden and Son, London.
Crabbe, P. "Optical Rotary Dispersion and Optical Circular Dichroism in Organic Chemistry." *Top. Stereochem.*, **1**, 93 (1967).

Atropisomers

Oki, M. (1993). *The Chemistry of Rotational Isomers*, Springer-Verlag, Berlin.

Molecular Propellers and Residual Stereoisomerism

Mislow, K. M. "Stereochemical Consequences of Correlated Rotation in Molecular Propellers." *Acc. Chem. Res.*, **9**, 26 (1976).

Polymer Stereochemistry

Goodman, M. "Concepts of Polymer Stereochemistry." *Top. Stereochem.*, **2**, 73 (1967).

Helical Isocyanates

Green, M. M., Park, J.-W., Sato, T., Teramoto, A., Lifson, S., Selinger, R. L. B., and Selinger, J. V. "The Macromolecular Route to Chiral Amplification." *Angew. Chem. Int. Ed. Eng.*, **38**, 3138 (1999).

Topological Issues

Schill, G. in *Catenanes, Rotaxanes, and Knots*, J. Boeckmann (ed.), Academic Press, New York, 1971.
Sauvage, J. P. "Interlacing Molecular Threads on Transition Metals: Catenands, Catenates, and Knots." *Acc. Chem. Res.*, **23**, 319 (1990).
Liang, C., and Mislow, K. "Topological Features of Protein Structures: Knots and Links." *J. Am. Chem. Soc.*, **117**, 4201 (1995).
Walba, D. M. "Topological Stereochemistry." *Tetrahedron*, **41**, 3161 (1985).
Merrifield, R. E., and Simmons, H. E. (1989). *Topological Methods in Chemistry*, Wiley, New York.
Chambron, J.-C., Sauvage, J.-P., and Mislow, K. "A Chemically Achiral Molecule with No Rigidly Achiral Presentations." *J. Am. Chem. Soc.*, **119**, 9558 (1997). The "topological rubber glove".

Life's Handedness

Sevice, R. F. "Does Life's Handedness Come from Within?" *Science*, **286**, 1282–1283 (1999).
Bonner, W. A. "Origins of Chiral Homogeneity in Nature." *Top. Stereochem.*, **18**, 1 (1998).

REACTIVITY, KINETICS, AND MECHANISMS

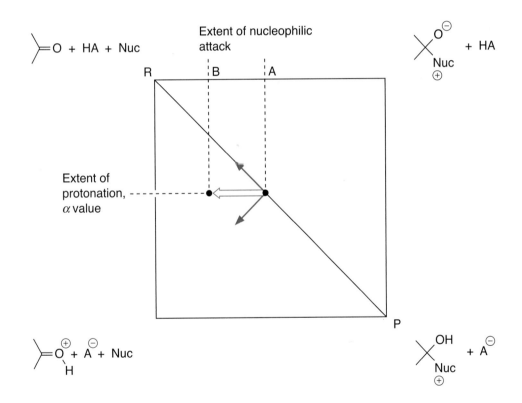

Energy Surfaces and Kinetic Analyses

Intent and Purpose

Chapter 7 is the first of two chapters that focus on teaching the "tools of the trade" of traditional physical organic chemistry as used in deciphering reaction mechanisms. We will consider several of the methods used to characterize reaction coordinates and reveal the nature of intermediates and activated complexes. The student should leave these next two chapters with the ability to anticipate experiments that we will describe in Chapters 9–11, and to use the "tools" to his or her own advantage when solving problems encountered in everyday chemical research. Along with the methods discussed here and in Chapter 8, we note the ever increasing contributions of electronic structure theory (Chapter 14) in deciphering mechanisms, and we leave spectroscopy to a dedicated textbook.

It is important to realize throughout all the discussions in these two chapters that a mechanism can never really be proven. Mechanisms become "well accepted" or "established". On the other hand, a well-designed experiment can definitively rule out one or more of several possible mechanisms. Many have argued that good experiments should be designed to disprove or falsify a model, and mechanistic chemistry is an area that is especially suited to this view. Since we can never prove a mechanism, we should always be open to the possibility of a different mechanism, and treat the well accepted mechanisms simply as good models and not necessarily reality.

In this chapter we explore the pathways that connect reactants and products. The main focus will be on kinetics and dynamics—that is, understanding energy surfaces, delving into molecular motions and collisions, measuring activation parameters, measuring rate constants, determining the sequence of chemical steps in a mechanism, and defining the rate-limiting (rate-determining) steps. We want to leave the student with a good sense of the complexity of a chemical reaction, and the manner in which we simplify our view of chemical reactivity. Several examples from organometallic chemistry, bioorganic chemistry, and enzymology are used to highlight the utility of the techniques in different fields. However, in many of the examples of this chapter and the next, we focus on S_N2 and S_N1 chemistry. Our assumption is that these reactions are among those that students are the most familiar with coming out of introductory organic chemistry, and therefore we can discuss them freely. If you are not "up-to-speed" with these reactions, you may want to review them in an introductory textbook before looking over this chapter, or skim the section in Chapter 11 where they are covered in detail.

After getting a good grasp of kinetics, we will return to energy surfaces and give guidelines on how to draw and think about these surfaces in both two and three dimensions. After finishing this chapter, the student should be able to design a kinetic experiment that leads to conclusions about the molecularity of a reaction, that gives insight into the enthalpy and entropy changes upon achieving the rate-determining transition state, and that indicates the sequence of steps in a mechanism.

Deciphering a reaction mechanism is the most enabling knowledge that a chemist has to control the outcome of a reaction. In industry this is of paramount importance. It allows the bench chemist to make rational predictions as to how to change experimental parameters such as the solvent, the reactant structure, and the temperature, in order to maximize the yield. This is particularly true in process chemistry within pharmaceutical firms, where optimization of yields is a primary goal. In organometallic chemistry, the knowledge of a mechanism enables subtle catalyst manipulations that can lead to different regio- and stereochemical outcomes. In enzymology, knowledge of the mechanism can give insights into rational drug design and inhibitor design. The bottom line is that knowledge of the reaction mechanism gives the quickest insight into how to manipulate matter at the atomic/molecular level, without making haphazard changes that rely on serendipity to give the desired outcome. Due to the central role that kinetics play in discerning mechanisms, kinetics is one of the most important disciplines within all of chemistry.

7.1   Energy Surfaces and Related Concepts

When we study the **kinetics** of a reaction, we are making experimental measurments on a reaction mixture, determining how quickly product forms as a function of concentrations, temperature, and other variables. Our goal is to relate these experimental observations to molecular scale concepts, such as molecular motions, molecular collisions, and molecular vibrations, as well as energy concepts such as free energy, enthalpy, and entropy. **Reaction dynamics**, on the other hand, is the molecular scale analysis of reaction rates. For example, in Section 3.3.4 we discussed the molecular dynamics method of computationally simulating a reaction. When we are talking about individual (or small groups of) molecules traversing a well-defined surface (see below), we will tend to call such analyses "dynamics studies". Discussion of macroscopic measurements of real reacting systems will be termed "kinetics".

An analysis of kinetics can arguably be considered the most informative study we can perform to delineate a reaction mechanism. However, the data obtained cannot give a complete picture of a mechanism, because the data do not give us information about which bonds are broken or formed. The greatest value of a kinetic analysis is that it often provides a framework for designing experiments to test a proposed mechanism.

The study of kinetics is concerned with the details of how one molecule is transformed into another and the time scale for this transformation. This is in stark contrast to thermodynamics. In our analysis of thermodynamics (Chapters 2–5), we were solely concerned with the initial and final states of a system; for chemical reactions, this means the reactant and product (often an intermediate), respectively. The mechanism involved in the transformation is not considered in thermodynamics, and therefore, time is not a factor. Yet, the two disciplines, kinetics and thermodynamics, are highly interrelated. In Section 7.1.3, for example, the most widely accepted theory for understanding rate constants (transition state theory) is based upon a thermodynamic analysis. Moreover, at equilibrium, the rate of the overall forward transformation equals the rate of the overall reverse transformation.

Before starting an in-depth discussion of how to perform a kinetics experiment and how to interpret the data, we must first have a good understanding of chemical reactions, as well as the energetic relationships among all the different chemical species involved in a mechanism. To do this we will turn to transition state theory (TST). TST is an outgrowth of a postulate first put forth by Arrhenius, that most reactions involve an energy barrier that needs to be surmounted during the transformation. The notion of an energy barrier implies a surface that the molecules are traveling upon. This surface gives the energy of the molecules as a function of their structure. We have already examined energy surfaces in Chapter 2 when we looked at conformational analysis. Those surfaces related conformation to energy. In this chapter we significantly expand upon energy surfaces, and now use them to explain chemical reactions, not just conformational changes. Having a good grasp of how to think about an energy surface is a prerequisite for understanding TST, and thereby kinetics, rates, and rate constants. Therefore, we start our analysis of kinetics by delving deeper into the notion of energy surfaces.

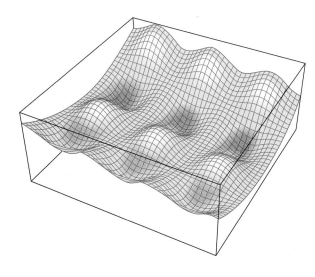

Figure 7.1
Energy surface showing multiple valleys, passes for interconverting between valleys, and saddle points.

7.1.1 Energy Surfaces

To start our analysis of energy surfaces, let's imagine hiking a mountain range that has multiple hills and valleys along with mountain passes that lead from one valley to another. This hypothetical mountain range is devoid of trees and boulders and is instead perfectly smooth. An example of what you should be imagining is given in Figure 7.1. While hiking from one valley to another, you are traversing a trail that takes you to and from each valley, with mountains rising up on your sides. At the peak of a mountain pass, the view of the mountain range beneath your feet resembles the center of a horseback riding saddle. Ahead and behind you the trail goes downhill, but to your left and right sides the mountains rise up, and hence this point along the trail is referred to as a **saddle point**.

In chemistry, such a three-dimensional topological map is called an **energy surface**. In hiking, the vertical axis is altitude, but in kinetics it is energy. In our analogy between hiking and kinetics, the valleys represent molecules with relatively lower energies, so this is where we expect to find reactants, products, and intermediates. You, as the hiker exploring the mountain range, represent a molecule traveling over the energy surface. The very tops of the mountains represent extremely high energy chemical structures that are seldom if ever achieved, and the tops of the mountain passes (saddle points) represent the highest energy points that must be traversed for the transformation of one molecule into another. Importantly, for a molecule to get from one valley to another valley, an elevated pass must be traversed. If you want to expend the least amount of energy, the trail you would choose to hike from one valley to the next would be the one with the lowest mountain passes. Although it is common for chemists to speak of molecules traversing the energy surface, and we will do so herein, what really traverses the surface is a hypothetical point that summarizes the geometry of the molecule.

Furthermore, the surface is not just three-dimensional. There is a single energy axis, but the structural coordinates are in $3N - 6$ space, where N is the number of degrees of freedom of the molecule. This leads to a **hypersurface** that is very difficult to imagine, let alone draw. For purposes of discussion, a one- or two-geometrical-coordinate version for the structural axes is the one we will use.

Chemically speaking, as a reaction occurs, the internal energy of the molecules rises to a peak at a saddle point and then drops off as the molecules fall into a neighboring valley. The molecular structure associated with this peak on the energy surface is called an **activated complex**. The physical point on the curve that represents the structure of the activated complex is called the **transition state**. The terms "activated complex" and "transition state" are often used interchangeably, although strictly speaking, they are different. The height of the mountain pass is called the **barrier to the reaction**, or the **activation barrier** for the reaction. The amount of energy required to attain the transition state is the **activation energy**.

What do we mean by the statement "the internal energies of the molecules rise to a peak at the saddle point and then drop off", and what kind of energy are we considering? In such discussions we are always speaking in terms of interchanging kinetic and potential energy of some sort, and internal energy can be both potential and kinetic. Both kinetic and potential energy (PE) can take many forms. Recall our analogy to a spring in Chapter 2. PE is associated with the tension in a compressed spring, or the gravitational pull on a brick that is being held off the ground. When considering energy surfaces, we think about the Gibbs free energy of the molecules rising and falling during the reaction. Often, however, we simply plot enthalpy, one of the components that goes into making up the Gibbs free energy. This is because we have good methods for determining enthalpy, and it is related to bond strength (see Chapters 2–5 and below). Therefore, the z axis in Figure 7.1 is best considered as potential energy in the form of the Gibbs free energy of the entire solution, consisting of reactants, intermediates, products, and solvent. Frequently, though, the z axis is considered to be the enthalpy of the reactants, products, transition states, and intermediates.

Before considering how a molecule can increase its internal energy, we should have an idea of how chemical reactions occur. Most chemical reactions are the result of molecular collisions, either between two molecules undergoing the reaction or between solvent molecules and reactants. The kinetic energy from the collision is taken up by the colliding molecules into increased molecular strains (distortions). As the molecules climb the activation barrier, their translational kinetic energies decrease; that is, they slow down due to the collision as would billiard balls at the point of impact. At the same time their potential energies increase in the form of weaker bonds due to structural strains or distortions, or losses in degrees of freedom (see below). The increase in potential energy can be associated with excitation of vibrational modes such as torsions, bends, and stretches.

Recall from the discussion in Section 3.1.5 that the standard change in Gibbs free energy between structures can be considered analogous to a driving force for their interconversion. Hence, you will often hear chemists speak of the **driving force for a reaction**, which is the degree of exergonicity of that reaction. Spontaneous changes to a solution will occur in order to lower the Gibbs free energy of the entire solution. In this chapter we consider the pathways that lead to a lowering of the Gibbs free energy—the mountain passes shown in Figure 7.1. Barriers to the lowering of the Gibbs free energy are present in almost all chemical reactions, and the ability to traverse a barrier is related to the kinetic energy of the molecules, and therefore the temperature of the solution. In essence, the molecules move around on the energy surface, driven to lower the Gibbs free energy of the total solution, but thermal motion is what allows them to traverse the barriers.

In Chapters 2–4 we mostly examined enthalpy as the energy associated with changes in conformations and changes in bond strengths that derive from collisions or the redistribution of energy between different vibrational states. However, changes in entropy also affect Gibbs free energy. For example, an S_N2 reaction not only has weaker bonds at the transition state, but the entropy becomes less favorable, too, because two molecules are combining. Conversely, in a fragmentation reaction the entropy increases and is thus favorable. In the case of a fragmentation, the increased entropy makes the reaction more favorable than would be predicted based upon bond strength changes alone. These changes in degrees of freedom can often dominate the Gibbs free energy along the reaction coordinate. **Degrees of freedom** is an all encompassing term that refers to the different forms of motions that molecules can possess (see Chapter 2). There are translational motions ("flying around" in the reaction vessel), rotational motions (tumbling motions—not bond rotations), and internal motions (bond stretches, torsions, bond angle distortions, etc.). The more degrees of freedom a molecule possesses, the more favorable its entropy. We will show later in this chapter that entropy is designated on enthalpy energy surfaces as the width of the wells (valleys) and saddle points.

Since the energy taken up by the molecules to the point of the transition state is potential energy, the energy released back to the solution as the molecules proceed to product or reactant is kinetic energy. In other words, molecular velocities increase. This notion is quite similar to the collision of any two particles, such as rubber balls. Prior to the collision they are moving with their respective velocities. At the point of collision the balls slow, distort, and then speed away from each other along different trajectories. In chemistry, if the energy of the collision is high enough and the distortion is severe enough to achieve the activated complex, the collision can lead to a chemical reaction. Otherwise the molecules just glance off each other without reacting.

Let's return to the three-dimensional image of an energy surface. In our analysis the transition states were likened to mountain passes. In a real mountain pass, the trail at the top can be narrow and hard to squeeze through, or the trail can be wide open and easy to traverse. Therefore, the **entry** and **exit channels** associated with different transition states can have differing shapes. Further, the pass can be relatively low and easy to climb, or it can be quite high and require substantial energy to climb. All these parameters are complex functions of the structures of the molecules undergoing the reactions, which vibrational modes of the molecules are changing along the reaction coordinate, and the amount of change in the organization or disorganization of the molecular entities as they traverse the pass (each of these items will be discussed in this and the next chapter). However, there is another dimension to these energy surfaces not shown at all in Figure 7.1 that has no analogy to mountain ranges. This has to do with the vibrational states of the molecules. Each valley and pass has rungs associated with it, which means that there are many layers of vibrational states associated with the surface shown, each layer having a certain fraction of the molecules residing in it (recall the discussion of quantized energy states for vibrations in Chapter 2). Therefore, real energy surfaces are not nearly as simple as that shown in Figure 7.1.

For now, it is sufficient to note that many trails or trajectories are possible for molecules to take when traversing a particular pass. Depending upon the initial velocities or momenta of the molecules, whether they are in their ground state or an excited state, which of several conformations are present, and the individual vibrational states of the molecules, each molecule will take a slightly different path to product. Many of these paths will not be the absolutely lowest energy path. This means that there is no single chemical path that every reactant takes to get to the product; instead, there is an ensemble of paths (quickly glance at Figure 7.6 on page 369 for a graphical representation of this notion).

The idea that there is no single path implies that the rate one observes for a chemical reaction is a weighted average of all the rates for all the different possible pathways. This picture of kinetics represents a very difficult scenario from a theoretical standpoint because there are so many variables. Instead, we concentrate on the minimum energy pathway from the reactant to the transition state, or a single pathway that represents the weighted average of all the pathways. In this manner we are able to simplify our picture of kinetics to a manageable level. This leads to reaction coordinate diagrams. These diagrams, therefore, represent a composite picture of all the pathways that the molecules take during the reaction mechanism. Interestingly, techniques to follow the kinetics of single molecules have been developed recently (see the next Going Deeper highlight), allowing one to consider reaction coordinates of single molecules.

7.1.2 Reaction Coordinate Diagrams

The minimum energy pathway, or the pathway we depict as the weighted average of all the pathways, is called the **reaction coordinate**. Plotting this pathway in two dimensions, where one variable is the energy, gives us what we call a **reaction coordinate diagram**. To draw a reaction coordinate diagram, we focus on a cross-section of Figure 7.1 that shows a single geometrical coordinate as the x axis and energy as the y axis. Such two-dimensional plots give a curve that represents the lowest energy pathway, and they are extremely useful for qualitative discussions of reaction mechanisms (Figure 7.2). The **transition state** is the highest point on the lowest energy path interconverting reactant and product.

Reaction coordinate diagrams show clear distinctions between transition states, stable structures, and transient intermediates. A relatively stable structure is given by a low energy depression (well) in the curve, reactive intermediates are in high energy shallow wells, and transition states are peaks. Any chemical structures that last longer than the time for a

Going Deeper

Single-Molecule Kinetics

In recent years it is becoming more and more common to observe single molecules, and in some instances we can see one molecule undergo a chemical transformation. A variety of techniques allow this, many involving sophisticated laser techniques. What do we expect from single-molecule kinetics? When we determine the experimental rate constant for a reaction under conventional conditions, we obtain a single value for the rate constant, k. However, we just noted that this is actually the average of an ensemble of values, with each individual reaction in the flask occurring at its own individual rate. If we watched reactions occur just one molecule at a time, we might expect to measure many different rate constants, and indeed this is the case.

As an example, we consider the oldest and most powerful method for single-molecule kinetic analysis, the **patch clamp** analysis of ion-channel proteins. Ion channels regulate the flow of simple ions such as Na^+, K^+, and Ca^{2+} across cell membranes. Movement of ions is equivalent to electrical current, and we have very sensitive methods for detecting electrical currents. If we take a small glass pipette and insert it into a membrane just right, we can electrically insulate the patch of membrane inside the pipette from the rest of the world. If done properly, a single ion channel will be in the patch, and we can watch

it open and close in real time. Typically the channel is closed, and no ions (current) flow. However, in response to a stimulus the channel can "gate", opening to allow ionic flow. The left figure below shows a typical measurement for such an experiment. We see only two states: closed (no current) and open (a set current). The kinetic parameter is the open time—how long the channel is open in any given opening event. It is a lifetime, and so is the reciprocal of a rate constant (see below for definition of "lifetime"). As can be seen by visual inspection, there is a considerable variation of lifetimes. However, the current value is always the same; the channel is either open or it is closed. To report a single lifetime value, we prepare the histogram shown, allowing a definition of a mean lifetime (in this case 2 ms).

Biologists have had the ability to do such single-molecule kinetics for over 20 years, and the 1991 Nobel Prize for Medicine or Physiology was presented to Neher and Sakmann for the development of the patch clamp. The statistical analysis methods for single-molecule systems were mostly worked out in this context, and now these methods are being applied in many new ways to analyze single-molecule systems.

Sakmann, B., and Neher, E. (eds.). *Single Channel Recording*, Plenum, New York, 1983, p. 503.

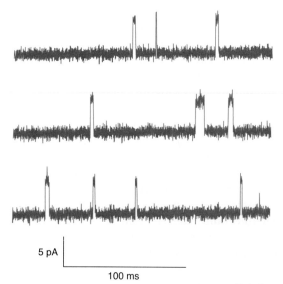

5 pA

100 ms

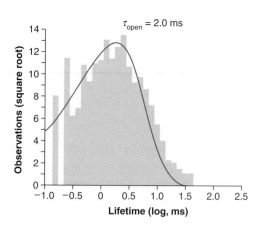

τ_{open} = 2.0 ms

Data for a patch clamp experiment

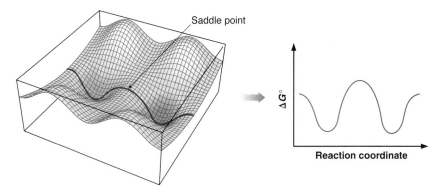

Saddle point

$\Delta G°$

Reaction coordinate

Figure 7.2
An energy surface with a reaction coordinate shown, and its two-dimensional projection. The saddle point is shown on the three-dimensional surface.

typical bond vibration (10^{-13} to 10^{-14} s) can be considered an intermediate. The larger the barriers leading to and from each well, the longer the lifetime of the structure represented by that depression in the curve.

When a reaction involves more than one elementary chemical step, one or more intermediates are formed. This means that there is more than one energy barrier that must be traversed during the reaction, and that there is more than one transition state (T.S.). The step of the reaction whose rate determines the observed rate of product formation is called the **rate-determining step (rds)**. It is commonly the chemical step that involves the highest energy transition state. When there is only one step to a reaction, it is obviously rate determining (Figure 7.3 **A**). However, in multistep reactions the observed reaction rate is related to the overall barrier height between the reactant and the highest energy T.S. It is important to note that the highest energy T.S. is not necessarily associated with the microscopic step that has the highest barrier. For example, in Figure 7.3 **B**, **C**, and **D**, the rate-determining step is indeed the step with the highest barrier. In part **E** the barriers from R to I and from I to P are of similar height, while in part **F** the barrier from R to I is the largest. In cases **E** and **F** the second step (I to P) is still the rate-determining step. Sometimes, the term **rate-limiting step (rls)** is used in such cases. To avoid confusion, we feel rds and rls are best considered as synonyms, along with the term **rate-controlling step**. We use rds and rls interchangeably in this book.

More information can be gleaned from Figure 7.3. In parts **D** and **E** there is an equilibrium formed between the reactant (R) and the intermediate (I). This conclusion is based upon the fact that the energy of the intermediate is equal or similar to that of the reactant, and the barrier for conversion of the intermediate back to reactant is much lower than that for formation of the product. However, in part **F**, the intermediate is of high energy, and the reverse and forward barriers for the intermediate are similar in energy. A fully established equilibrium with such an intermediate is not likely. This is a case where the steady-state approximation will be used to derive kinetic expressions (see Section 7.5.1).

Another aspect of reaction coordinate diagrams that is often overlooked is that the curves shown in Figure 7.3 do not necessarily represent hills and valleys on an energy surface that are in a straight pathway. The curves instead represent a pathway on an energy surface where often the molecules need to "change direction" to proceed from valley to valley to valley (see Figure 7.1 to envision this). The two-dimensional diagrams shown in Figure 7.3 show a flattened representation. The molecules typically can only make turns on an energy surface from valleys and not from saddle points. Once in a valley, the molecules have a lifetime and are free to climb all available activation barriers, meaning that different prod-

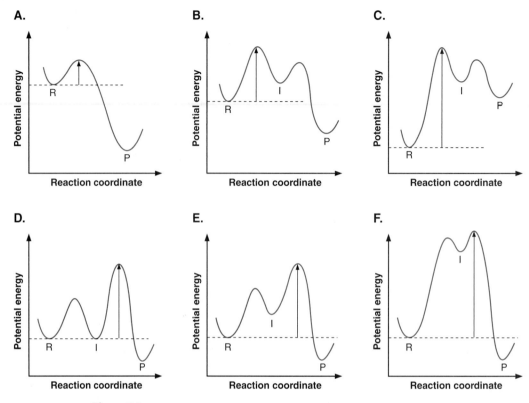

Figure 7.3
Several reaction coordinate diagrams used to define "rate-determining step" and "rate-limiting step".

ucts or intermediates can be formed by heading in different directions on the energy surface. From a saddle point, however, only moving forward or backwards leads to valleys, and hence turns on the surface from a transition state lead to very high energy structures. Furthermore, the trajectory of the movements of atoms within the activated complex is along a direction that is leading to product and cannot be suddenly changed at the point of a transition state to a different direction. Hence, when viewing a reaction coordinate diagram, remember that the paths from valley to valley may actually be different directions on the energy surface.

The manner by which a chemist first sketches a reaction coordinate diagram comes primarily from intuition, knowledge of the free energy or enthalpy of the reaction, and whether or not there are intermediates. Some guidelines assist the process (such as the Hammond postulate, discussed later in this chapter). One guideline used for drawing diagrams for complex reactions is called the **principle of least motion**. The favored reactions proceeding from reactant to intermediate, or from intermediate to another intermediate or product, are those that have the least change in nuclear position or electronic configuration. In other words, although many chemical reactions involve dramatic changes in positions of nuclei, these changes commonly occur by a series of simple reactions (most of these simple reactions are covered in Chapters 10 and 11).

Rigorous methods for creating reaction coordinate diagrams also exist. High-level computational methods such as we present in Chapter 14 can be used. Further, as described later in this chapter, the method developed by Marcus predicts both the position along the x axis and the energy associated with the transition state for simple one-step reactions.

7.1.3 What is the Nature of the Activated Complex/Transition State?

The activated complex is a molecular entity that has a lifetime no longer than a vibration, meaning that there is a particular movement of atoms that tilts the activated complex in the direction of product or reactant. Like all structures along the reaction coordinate, except for the end points, the activated complex is a species that exhibits some structural characteristics of both the reactant and the product. What makes the transition state unique is that it represents a point along the reaction coordinate that reflects the most strained or unstable structure involved in the reaction.

Recall that in Chapter 2 we discussed the molecular mechanics method and how it provides analytical expressions for the variation of energy of a system (∂E) as a function of a geometry change (∂x). Those derivatives can provide another way to think about energy surfaces, reaction coordinate diagrams, and transition states. A **stationary point** on the surface is one that has no forces on it. The force corresponds to the first derivative, $\partial E / \partial x$, and so first derivatives with respect to all dimensions are zero at a stationary point. Recalling basic analytical geometry, a point on a surface with a zero first derivative is either a minimum or a maximum, based on whether the second derivative ($\partial^2 E / \partial x^2$) is positive or negative, respectively. This is matrix algebra, and so the terms **eigenvectors** (forces are vector quantities) for the first derivative and **eigenvalues** (force constants are just numbers) for the second derivative are used. For example, consider our definition of a transition state. It is a minimum in all dimensions except along the reaction coordinate, where it is a maximum (consider the picture of a mountain range and saddle point given in Figure 7.2 to see this). Thus, the mathematical definition of a transition state is a stationary point (all zero first derivatives) with one, and only one, negative eigenvalue. A true minimum (a stable structure) is a stationary state with all positive eigenvalues. These definitions are very helpful in efforts to characterize transition state structures using computational methods, whether they involve molecular mechanics (Chapter 2) or quantum mechanics (Chapter 14).

We have spent considerable time analyzing how reactions occur, the energy surfaces that the molecules may traverse, and the kinds of energies involved in chemical transformations. In Chapters 2–4 we discussed methods for placing numerical values on the thermodynamics of reactions. However, we have not yet discussed any numerical values for barriers. In order to give numerical values to the relative energies of the peaks and valleys on a reaction coordinate diagram, chemists typically turn to two related methods: transition state theory (TST) and the Arrhenius rate law. Rate constants are used in rate expressions with both these theories, and hence we need to briefly review rates, rate constants, and rate expressions.

7.1.4 Rates and Rate Constants

Given our examination of energy surfaces, the rate of a reaction should depend upon the barrier height that needs to be surmounted and the temperature. Mathematically, this dependence is represented by a proportionality constant between concentration of reactants ([R]) and the reaction rate known as the **rate constant** (see Eq. 7.1 for an example). The rate constant is represented by the letter k, sometimes with a subscript as in k_n, where n tells the order of the reaction or which step the rate constant refers to in a multistep reaction. We will use the latter numbering system in this book. The reaction rate should also depend upon the amount of reactants present (i.e. their concentration). If the concentration of a reactant is zero, then no reaction can occur and the rate is zero. Conversely, a large concentration of reactant should lead to a large rate.

$$\text{rate} = k[\text{R}] \tag{Eq. 7.1}$$

The dependence of rate upon concentration means that reaction rates are time dependent, because the reactant's concentration changes as a function of time. The rate decreases as the concentration of reactant(s) decreases. Commonly, the rate of a reaction is at its highest at inception (those reactions without an induction period) and slows down to zero at the end of reaction or when the system has reached equilibrium. Plotting the concentration of product as a function of time gives plots similar to those shown in Figure 7.4.

Since the rate is the change in concentration per unit time, the tangent to the line at each time point is the rate of the reaction at that time point. Tangents are mathematically defined

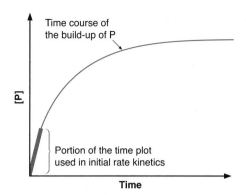

Figure 7.4
The build-up of product as a function of time
for a typical kinetic analysis.

by derivatives. Hence, we write the rate as a derivative reflecting the change in reactant (R) or product (P) concentration as a function of time (Eq. 7.2). The units are concentration per time, similar to the units involved in the rate of driving an automobile (distance per time = miles per hour).

$$\text{rate} = -\frac{d[R]}{dt} = \frac{d[P]}{dt} \tag{Eq. 7.2}$$

7.1.5 Reaction Order and Rate Laws

Each reactant may or may not actually influence the rate of the reaction. Further, the extent to which the concentration of each reactant influences the rate may be different. To explain these statements, let's picture a hypothetical reaction of species A, B, and C to give P (Eq. 7.3). If the reaction rate has a linear dependence on a reactant's concentration, then we define the reaction as **first order** with respect to this reactant. For example, if the concentration of A is doubled and the rate doubles, then the reaction is first order with respect to A. If the reaction rate quadruples when the concentration of B is doubled, the reaction is **second order** with respect to B. Sometimes changing the concentration of a reactant (let's say C in this case) does not affect the rate at all, in which case the reaction is **zero order** with respect to that reactant.

$$A + B + C \longrightarrow P \tag{Eq. 7.3}$$

Combining the logic that the rate of a reaction should depend upon both rate constants and concentrations leads to **differential rate equations**, also known as **rate laws**. A generalized rate law is given in Eq. 7.4 for the reaction of three molecules A, B, and C to give P. The concentration of each reactant has an exponent that is the **order** of the reaction for that reactant (0 for zero order, 1 for first order, 2 for second order, etc.). Integral exponents are very common, but fractional exponential dependence can be seen in reactions with complex mechanisms. Summing the reaction order for each species gives what is known as the overall order of the reaction. For example, if our hypothetical reaction is first order in A, second order in B, and zero order in C, we say the overall reaction is third order. In this particular case, the rate equation of Eq. 7.4 would reduce to Eq. 7.5. Integration of these equations lead to what are called the **integrated rate equations**. We derive a few integrated rate equations in Section 7.4.2 and give several in Table 7.2.

$$\frac{d[P]}{dt} = k[A]^a[B]^b[C]^c \tag{Eq. 7.4}$$

$$\frac{d[P]}{dt} = k[A]^1[B]^2[C]^0 = k[A][B]^2 \tag{Eq. 7.5}$$

The rate laws for many reactions can involve the sum of terms similar to Eq. 7.4. This means that there are two or more routes (mechanisms) for transformation of the reactant(s) into the product(s). Recall from your introductory organic chemistry class that an ester can undergo acid- or base-catalyzed hydrolysis reactions; each has a distinctly different mechanism. Therefore, the rate of hydrolysis of an ester will involve a term that has acid dependence and a term that has base dependence (Eq. 7.6).

$$\frac{d[P]}{dt} = k_a[\text{ester}][H_3O^+] + k_b[\text{ester}][OH^-]$$

(Eq. 7.6)

The **molecularity** of a chemical reaction is the number of molecules involved in the transition state of the reaction. The term can only be applied to single-step reactions, also known as **elementary** reactions. If only a single molecule is involved in the transition state, the reaction is **unimolecular**. For example, a thermal rearrangement such as a Cope rearrangement is typically unimolecular. If two molecules are involved the reaction is **bimolecular**, with the S_N2 reaction being the prototype. **Termolecular** processes involve three molecules and are rare, but not unprecedented (we show a few in Chapter 10).

It is important to keep the distinction between kinetic order and molecularity clear. Kinetic order is determined by experimental measurements of the rates of reaction. We can know the order of a reaction, but know nothing about its mechanism. Also, the kinetic order is meaningful for both elementary reactions and those reactions involving more than one step, known as **complex** reactions. Molecularity applies only to elementary reactions and is basically a statement about the mechanism of the reaction; we are saying we know something about the nature of the transition state. In complex reactions, each step will have its own molecularity. Only for elementary, single-step processes, do we expect the kinetic order and the molecularity to track with one another. We will delineate the difference between order and molecularity again in Section 7.4 when we examine how kinetic experiments are performed. Now that we have a grasp of rate constants, we can start to put energy values on the barrier heights of energy surfaces. The most common way to do this is transition state theory.

7.2 Transition State Theory (TST) and Related Topics

In the derivation of TST, one approach assumes that the reactants and activated complex are in pre-equilibrium. Statistical mechanics then is used to calculate the concentration of the activated complex. This concentration is used, along with the rate at which the activated complex proceeds to the product, to give a rate constant for the reaction. The statistical mechanics analysis can be given in terms of the common thermodynamic parameters of enthalpy, entropy, and Gibbs free energy. When associated with chemical kinetics and rates, these parameters are called the **activation parameters** ($\Delta G^{\ddagger}$, $\Delta H^{\ddagger}$, and $\Delta S^{\ddagger}$). The ability to measure activation parameters gives us information about the manner in which the transformation to the activated complex occurs—the kinds of structural changes that are occurring, the entropy changes, and the changes in solvation. The activation parameters are also the numerical values that give the relative energies of the reactants and activated complex. However, before learning how to experimentally measure the activation parameters and how to interpret them, we should examine some of the mathematics behind TST.

7.2.1 The Mathematics of Transition State Theory

Our analysis is not going to use statistical mechanics. Instead, we focus immediately upon the use of the common thermodynamic parameters (ΔG, ΔH, and ΔS). We start our analysis by considering a bimolecular reaction of A and B giving C in a single step (Eq. 7.7), where the activated complex is represented by $AB^{\ddagger}$ (a unimolecular reaction is also perfectly amenable to this analysis). The rate of a bimolecular reaction can be expressed as the rate of change of the concentration of the product, $d[C]/dt$, as given by Eq. 7.8.

In TST analysis, the rate by which C would be produced is directly proportional to the concentration of the activated complex $AB^{\ddagger}$ and the intrinsic rate constant $k^{\ddagger}$ for its decomposition to C (Eq. 7.9).

$$A + B \longrightarrow C \qquad \text{(Eq. 7.7)}$$

$$\frac{d[C]}{dt} = k[A][B] \qquad \text{(Eq. 7.8)}$$

$$\frac{d[C]}{dt} = k^{\ddagger}[AB^{\ddagger}] \qquad \text{(Eq. 7.9)}$$

Given the thermodynamic relationships introduced in Section 2.1, we can write an equilibrium expression for the formation of $AB^{\ddagger}$, where $K^{\ddagger}$ represents the equilibrium constant (Eq. 7.10). This is the primary postulate of TST—that the reactants are in equilibrium with the activated complex.

$$[AB^{\ddagger}] = K^{\ddagger}[A][B] \qquad \text{(Eq. 7.10)}$$

Combining Eqs. 7.8, 7.9 and 7.10 gives Eq. 7.11.

$$k = k^{\ddagger}K^{\ddagger} \qquad \text{(Eq. 7.11)}$$

The rate constant, $k^{\ddagger}$, for the activated complex converting to products is related to a vibration in the activated complex that makes it more resemble the product, thus tipping it along the reaction coordinate toward the product. Therefore, passage of the activated complex over the transition state can be identified by a vibrational mode (this is why we stated in Section 7.1.1 that an activated complex has a lifetime no longer than that of a vibration). We define the frequency of vibration to be v. However, not every oscillation associated with v will convert the activated complex to product. This is because other atoms in the molecule may not be properly arranged for a transition to product, or because the rotational state of the molecule interferes with the transition to product. TST takes these factors into account by using a factor κ, called the **transmission coefficient**. We assume that the rate of passage of the activated complex over the transition state to product is proportional to the vibrational frequency of relevance (v) and the transmission coefficient (κ), which in most cases is near unity (Eq. 7.12).

$$k^{\ddagger} = \kappa v \qquad \text{(Eq. 7.12)}$$

Now we need to solve for $K^{\ddagger}$. This is the point at which TST turns into a statistical mechanics analysis. We invoke statistical mechanics because a transition state does not have a Boltzmann distribution of states (see the next section), because its lifetime is so fleeting. Using statistical mechanics, it is found that $K^{\ddagger}$ is proportional to a new equilibrium constant ($K^{\ddagger\prime}$) that can be viewed in the same manner as the simple equilibrium constants given in Chapter 2. Thus, this $K^{\ddagger\prime}$ is equal to $\exp(-\Delta G^{\ddagger}/RT)$. The exact expression found for $K^{\ddagger}$ is Eq. 7.13, where k_B, h, v, and T are the Boltzmann constant, Planck's constant, vibrational frequency, and absolute temperature, respectively (see Appendix 1 for values).

$$K^{\ddagger} = \left(\frac{k_B T}{hv}\right) K^{\ddagger\prime} \qquad \text{(Eq. 7.13)}$$

Substituting Eqs. 7.12 and 7.13 into Eq. 7.11 gives the **Eyring equation**, named after Henry Eyring (Eq. 7.14). The equation consists of the term $\kappa k_B T/h$, which is near $10^{12}\,s^{-1}$ and is similar in magnitude to the frequency of many bond vibrations, as well as the $K^{\ddagger\prime}$ term.

$$k = \kappa \left(\frac{k_B T}{h}\right) K^{\ddagger\prime} \qquad \text{(Eq. 7.14)}$$

Finally, although Eq. 7.14 is the form put forth by Eyring in his TST analysis, it is most useful to chemists when transformed in terms of $\Delta G^{\ddagger}$ (Eq. 7.15 shows several ways to do this), and further into $\Delta H^{\ddagger}$ and $\Delta S^{\ddagger}$ using the equality $\Delta G^{\ddagger} = \Delta H^{\ddagger} - T\Delta S^{\ddagger}$ (Eq. 7.16).

$$k = \kappa \left(\frac{k_B T}{h} \right) e^{(-\Delta G^{\ddagger}/RT)}$$

$$k = 2.083 \times 10^{10}\, T e^{(-\Delta G^{\ddagger}/RT)} \qquad \text{(Eq. 7.15)}$$

$$\Delta G^{\ddagger} = 4.576\, T\, [10.319 + \log (T/k)] \text{ kcal/mol}$$

$$\text{(assuming } \kappa = 1)$$

$$k = \kappa \left(\frac{k_B T}{h} \right) e^{[(-\Delta H^{\ddagger}/RT) + (\Delta S^{\ddagger}/R)]}$$

$$= \kappa \left(\frac{k_B T}{h} \right) e^{(\Delta S^{\ddagger}/R)}\, e^{(-\Delta H^{\ddagger}/RT)} \qquad \text{(Eq. 7.16)}$$

Eqs. 7.15 and 7.16 are the ones used to experimentally measure the activation parameters $\Delta G^{\ddagger}$, $\Delta H^{\ddagger}$, and $\Delta S^{\ddagger}$, and therefore put values on the barriers of energy surfaces, as discussed later in this chapter.

7.2.2 Relationship to the Arrhenius Rate Law

There is another analysis often used to experimentally determine energies for the reaction barrier. This is the **Arrhenius rate law** (Eq. 7.17). This law was derived empirically by Arrhenius long before the development of TST. Arrhenius observed that the rates of reactions increased exponentially as the absolute temperature increased, producing the simple relationship of Eq. 7.17. There are two parameters associated with this law, the **pre-exponential factor (A)** and the **activation energy (E_a)**. In this method the barrier to the reaction is associated with the activation energy.

$$k = A e^{(-E_a/RT)} \qquad \text{(Eq. 7.17)}$$

When comparing the Arrhenius rate law to the form of the Eyring equation given in Eq. 7.16, it is tempting to let the pre-exponential factor A equal $\kappa(k_B T/h)\exp(\Delta S^{\ddagger}/R)$ and to assume that E_a and $\Delta H^{\ddagger}$ are the same. However, this is incorrect. The reason for this is the differing origins of these two treatments. The Arrhenius equation arises from empirical observations of the **macroscopic rate constants** for a particular conversion, such as A going to B by various paths. It is ignorant of any mechanistic considerations, such as whether one or more reactive intermediates are involved in the overall conversion of A to B. In contrast, the Eyring equation analyzes a **microscopic rate constant** for a single-step conversion of a reactant to a product. In a multistep process involving reactive intermediates, there is an Eyring equation and thus a $\Delta G^{\ddagger}$ for each and every step. In contrast, E_a describes the overall transformation. The two equations are actually addressing fundamentally different phenomena.

Still, connections can be made. For example, for a single-step unimolecular or bimolecular reaction, Eq. 7.18 holds. In addition, the pre-exponential factor A is in fact associated with entropy, and with single-step reactions the relationship is as in Eq. 7.19. As such, comparing one A value to another can provide insight into the relative activation entropies involved in different chemical reactions. A result worth remembering from Eq. 7.19 is that a simple reaction with $\Delta S^{\ddagger} = 0$ will have $\log A = 13.23$ at 25 °C. Larger values of $\log A$ imply favorable $\Delta S^{\ddagger}$, while smaller values of $\log A$ imply unfavorable (negative) values of $\Delta S^{\ddagger}$. Again, these relationships between the Eyring and Arrhenius equations are only viable if we are considering a single-step process. Given the fact that chemists have been taught to think in terms of ΔG, ΔH, and ΔS values, we will most often use the Eyring equation and the corresponding activation parameters.

$$E_a = \Delta H^{\ddagger} + RT \qquad \text{(Eq. 7.18)}$$

$$\Delta S^{\ddagger} = 4.576(\log A - 10.753 - \log T) \qquad \text{(Eq. 7.19)}$$
$$= 4.576(\log A - 13.23) \quad \text{at } 25\,°C$$

7.2.3 Boltzmann Distributions and Temperature Dependence

Both the Arrhenius rate law and the Eyring equation tell us that rate constants are temperature dependent. However, the potential energy surface is generally treated as being temperature independent. The barrier heights and the heat of reaction are determined solely by the structures of the molecules undergoing reaction. Sometimes, a heat capacity difference between individual species on the surface can lead to a temperature dependence of the surface (see Chapter 3 for a discussion of heat capacity). However, this is rare, and we will not consider this possibility further here.

Reaction kinetics depend upon molecular motions and collisions. As we raise the temperature, molecules will have more kinetic energy, and therefore more frequent collisions of the kind that will allow them to traverse the activation barriers. As a result, the rate at which the molecules can traverse the energy barriers increases. Stated in a slightly different manner, higher temperature gives molecules a higher average kinetic energy such that a larger fraction of them can traverse the barrier. This is shown in Figure 7.5 by plotting the distribution of molecular kinetic energies at two different temperatures. If the activation energy for a reaction is as noted in this figure, the higher temperature reaction proceeds faster due to the larger area under the curve past the activation energy.

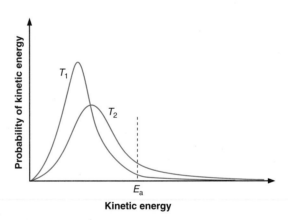

Figure 7.5
Diagram showing the Boltzmann distribution of molecules as a function of kinetic energy for two different temperatures. T_1 is the lower temperature and T_2 is a higher temperature.

Boltzmann derived a mathematical expression that relates the population of molecules at a particular energy to the temperature. In this expression is it assumed that thermal energy is distributed throughout the various energy levels of a molecule. This includes vibrational, rotational, electronic, and translational energies. As expressed in Eq. 7.20, the **Boltzmann distribution** gives the ratio of the populations of molecules, N_i/N_j, as a function of their energy difference, $E_i - E_j$, where R is the gas constant and T is the absolute temperature. Thus, this equation is convenient for comparing discrete energy levels, such as ground and excited rotational, vibrational, or electronic states.

$$\frac{N_i}{N_j} = e^{[-(E_i - E_j)/RT]} \qquad \text{(Eq. 7.20)}$$

7.2.4 Revisiting "What is the Nature of the Activated Complex?" and Why Does TST Work?

When chemists consider an equilibrium between two molecules, these molecules reflect a Boltzmann distribution of energies. The equation $K_{eq} = \exp(-\Delta G^\circ / RT)$ derives from analyzing systems that have Boltzmann distributions of energies. Therefore, our notion of an equilibrium between a transition state and a reactant implies that the transition state itself is populated by a significant number of molecules that are in equilibrium with each other and that are long lived enough for full exchange of energy among the various rotations and vibrations available to them. This, however, is *untrue*. The transition state does not have a sufficient lifetime for redistribution of energy among rotations and vibrations. Hence, we do not take the idea of an equilibrium with a transition state literally. Recall that in the derivation of the Eyring equation we needed to rely on statistical mechanics to make a substitution for the equilibrium constant for achieving the transition state. In that substitution, a distribution of states for the transition state is indeed assumed, but it is not a Boltzmann distribution.

In Figure 7.6, we pictorially represent an energy surface showing multiple trajectories that the molecules may take when passing over the saddle near the exact point we refer to as the transition state. In this analysis we are ignoring the rungs associated with various vibrational states for simplicity of presentation, even though these multiple trajectories may in part include different energy levels of various vibrational modes. Each trajectory represents a slightly different path a molecule may take depending upon its starting energy, conformations, and atomic motions. For example, not all S_N2 reactions proceed by a perfect 180° alignment of the nucleophile with the departing leaving group. Each different angle of attack would have a different energy transition state. The curved line that connects all the highest energy points along each trajectory is called the **col** in the surface. The col separates the energy surface into regions that represent molecules that are on the upward track (entry channel) toward the transition state or on a downward track (exit channel) to products. Each transit point on the col leading over the saddle gets less probable as its distance from the lowest energy transition state increases. Hence, there is a distribution of ways to pass over the col, not just one way.

The insert in Figure 7.6 shows a distribution of states for the activated complexes as a function of their distance from the lowest energy transition state along the reaction coordinate. This is important because now the transition state can be considered to represent molecules with a distribution of various energies. Thus, we can rationalize the idea of an equilibrium of states for the activated complex, allowing for a thermodynamic analysis that connects the energy of the reactant with the products, just as is done in transition state theory.

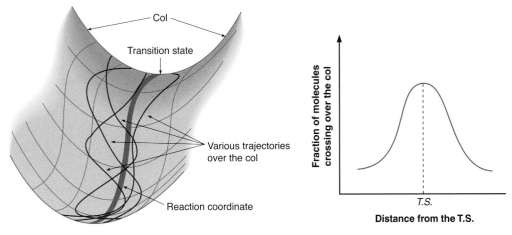

Figure 7.6
Diagram showing multiple hypothetical trajectories that take the molecules over the region of the energy surface near the transition state. The insert shows the distribution of molecules passing over the col as a function of distance from the transition state.

7.2.5 Experimental Determinations of Activation Parameters and Arrhenius Parameters

In the analysis of transition state theory and the Boltzmann distributions, we emphasized that the rate constant for a reaction is temperature dependent. If it is assumed that $\Delta H^{\ddagger}$ and $\Delta S^{\ddagger}$ do not vary with temperature, we can use the Eyring equation to solve for these parameters. If we take the natural log of both sides of Eq. 7.16, we obtain Eq. 7.21. Rearranging Eq. 7.21 gives Eq. 7.22. This now conforms to the equation of a straight line. Plotting $\ln(kh/\kappa k_B T)$ versus $1/T$ gives a line whose slope is $-\Delta H^{\ddagger}/R$ and intercept is $\Delta S^{\ddagger}/R$. Such a plot is called an **Eyring plot**. A similar rearrangement of the Arrhenius equation (Eq. 7.17) gives Eq. 7.23, which can be fit to a line (an example is given in the Connections highlight below). Lastly, a plot of $\ln k$ vs. pressure (P) can measure what is called the **volume of activation**, $\Delta V^{\ddagger} = -RT[\partial \ln(k)/\partial P]$, at constant temperature. This is the change in molar volume occupied by the sum of the reactants versus the transition state.

$$\ln(k) = \ln\left(\frac{\kappa k_B T}{h}\right) - \frac{\Delta H^{\ddagger}}{RT} + \frac{\Delta S^{\ddagger}}{R} \qquad \text{(Eq. 7.21)}$$

$$\ln\left(\frac{kh}{\kappa k_B T}\right) = -\left(\frac{\Delta H^{\ddagger}}{R}\right)\left(\frac{1}{T}\right) + \frac{\Delta S^{\ddagger}}{R} \qquad \text{(Eq. 7.22)}$$

$$\ln(k) = -\left(\frac{E_a}{R}\right)\left(\frac{1}{T}\right) + \ln(A) \qquad \text{(Eq. 7.23)}$$

Connections

Using the Arrhenius Equation to Determine Differences in Activation Parameters for Two Competing Pathways

Let's say we would like to know the difference in activation energies between two different pathways that branch from a common intermediate. We could measure the rate constants for formation of the two different products, and determine the ratio of these two rate constants as a function of temperature. Dividing the Arrhenius equation that describes the formation of one of the two products by the Arrhenius equation for the other product would lead to a plot of $\ln(k/k')$ versus $1/T$.

This is the approach that was taken for the analysis of the products derived from the photolysis of compound *i*. The photolysis leads to the reactive carbene intermediate *ii*, which in turn gives two products, *iii* and *iv*. Plotting $\ln(k_a/k_r)$ as a function of temperature (see below) gave a straight line revealing an activation energy difference of 1.6 kcal/mol and a ratio of A^a/A^r of $10^{-4.3}$.

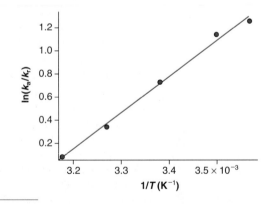

Reaction sequence for the Arrhenius equation example

Nigam, M., Platz, M. S., Showalter, B. M., Toscano, J. P., Johnson, R., Abbot, S. C., and Kirchoff, M. M. "Generation and Study of Benzylchlorocarbene from a Phenanthrene Precursor." *J. Am. Chem. Soc.*, **120**, 8055–8059 (1998).

Depending upon the form of the gas constant R used, the $\Delta H^{\ddagger}$ term can be obtained in kcal/mol or kJ/mol. The $\Delta S^{\ddagger}$ term is in units for entropy, J/(K)(mol) or cal/(K)(mol), the latter being known as entropy units (eu; see Chapter 2). Entropy values need to be multiplied by the temperature to be converted to energy values. Hence, the entropy contribution to the energy of activation is temperature dependent. At lower temperatures the energy associated with changes in entropy affects reaction rates less than at higher temperatures. This is logical given the general expression, $\Delta G^{\ddagger} = \Delta H^{\ddagger} - T\Delta S^{\ddagger}$. $\Delta G^{\ddagger}$ is therefore temperature dependent, and so often it will be given with a subscript denoting the temperature.

Due to the fact that the influence of entropy will diminish as one lowers the temperature, a reaction can switch from being entropy controlled to enthalpy controlled as the temperature is lowered. **Enthalpy control** occurs when $\Delta H^{\ddagger}$ is the major contributor to $\Delta G^{\ddagger}$, and **entropy control** occurs when $T\Delta S^{\ddagger}$ is the major contributor to $\Delta G^{\ddagger}$.

Just as in our treatment of the van't Hoff plot in Chapter 4, curvature in an Eyring plot can indicate that $\Delta H^{\ddagger}$ is temperature dependent, so heat capacity effects are important. Again, this is not common, but it can be most significant if the rates of a given reaction are measured over a large temperature range. On the other hand, a clear break in an Eyring plot most commonly indicates a change in mechanism or rate-determining step (see the Connections highlight below for an example). If a conversion has two possible paths, one with a favorable $\Delta H^{\ddagger}$ but an unfavorable $\Delta S^{\ddagger}$, while the other has a favorable $\Delta S^{\ddagger}$ but an unfavorable $\Delta H^{\ddagger}$, it is easy to imagine a situation where the mechanism, and hence the slope of the Eyring plot, would change with temperature (you are asked to examine this scenario in Exercise 9 at the end of the chapter).

Connections

Curvature in an Eyring Plot is Used as Evidence for an Enzyme Conformational Change in the Catalysis of the Cleavage of the Co–C Bond of Vitamin B$_{12}$

Vitamin B$_{12}$ (5'-deoxyadenosylcobalamin, AdoCbl) possesses a bond between a Co and a carbon, making it one of the few organometallic reagents used by nature. The homolysis of the Co–C bond leads to radical intermediates, which are postulated to undergo skeletal rearrangements catalyzed by various enzymes requiring B$_{12}$ as a cofactor.

The manner in which the Co–C bond is cleaved in the enzyme ribonucleotide triphosphate reductase (RTPR) was studied in part using an Eyring analysis. At approximately 30 °C a break was found in the Eyring plot (see to the right). This break was interpreted as indicating a conformational change to an inactive form of the enzyme below 30 °C, thereby requiring isomerization to the active form for catalysis. The required isomerization imposes the thermodynamics of the isomerization on the observed rate constant. This creates a mechanism change, where the new mechanism possesses a conformational switch in the catalyst prior to the catalytic steps.

Interestingly, the Eyring analysis revealed that the entropy of activation of the enzyme catalyzed pathway is nearly identical to the entropy of activation for an uncatalyzed cleavage of the Co–C bond. The catalysis, therefore, completely arises from a change in the enthalpy of activation. A full 13 kcal/mol drop in the enthalpy of activation

for the enzyme catalyzed pathway was found, a truly impressive decrease in this activation parameter, leading to a very large rate enhancement of 1.6×10^9.

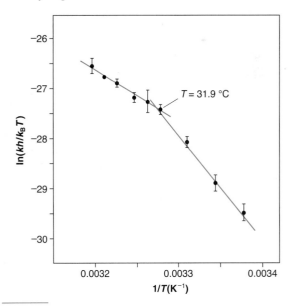

Brown, K. L., and Li, J. "Activation Parameters for the Carbon–Cobalt Bond Homolysis of Coenzyme B$_{12}$ Induced by the B$_{12}$-Dependent Ribonucleotide Reductase from *Lactobacillus leichmannii.*" *J. Am. Chem. Soc.,* **120,** 9466–9474 (1998).

7.2.6 Examples of Activation Parameters and Their Interpretations

The magnitudes of $\Delta H^\ddagger$ and $\Delta S^\ddagger$ give us information about how the enthalpy and entropy of the transition state differ from those of the reactants. In the activated complex some bonds will be partially broken, meaning that the bond strengths of the activated complex are lower than in the starting materials. Thus, achieving the transition state is almost always endothermic. However, $\Delta H^\ddagger$ can also be influenced by factors other than just bond strengths, such as solvation effects. A solvent that can interact favorably with the transition state will lower $\Delta H^\ddagger$ even though it may not significantly influence the extent of the bond breaking or bond forming. On rare occasions $\Delta H^\ddagger$ is near zero or negative, such as with the combination of radicals.

The $\Delta S^\ddagger$ term is more complex. It is a measure of the degree of order produced or lost when comparing the transition state to the reactant. If translational, vibrational, or rotational degrees of freedom are lost or gained in going to the transition state, the $\Delta S^\ddagger$ term can become negative or positive, respectively. Solvation can also affect the $\Delta S^\ddagger$ value. If a significant amount of ordering is required in the solvent to make favorable enthalpy interactions with the transition state, the $\Delta S^\ddagger$ value can become more negative than you might expect. A further complication arises from the fact that the Eyring equation commonly used (as given herein) is really only applicable to elementary reactions. When it is applied to complex reactions, as is commonly done, the resulting $\Delta S^\ddagger$ term depends upon the implied standard state, and the interpretation of absolute values of $\Delta S^\ddagger$ is incorrect. Comparison of $\Delta S^\ddagger$ for reactions of the same molecularity is fine, while comparison for reactions of different molecularity should be done with caution.

Table 7.1 gives several reactions and their corresponding $\Delta S^\ddagger$ values, along with the rate-determining step in each mechanism. Negative values of $\Delta S^\ddagger$ imply decreasing translational and rotational degrees of freedom, sometimes indicating the combination of two molecules, whereas a positive $\Delta S^\ddagger$ implies an increase in the translational and rotational degrees of freedom, often indicating the creation of two molecules. Note that entries 1, 2, 4, and 6 in Table 7.1 all have large, negative entropies of activation, suggesting steps involving molecules combining. In contrast, entries 3, 5, and 7 are all positive, suggesting activated complexes with a significant amount of bond cleavage to form two molecules.

It is worth considering the potential influence of these terms on reaction rates. The forms of both the Eyring and the Arrhenius equations are similar to our discussion in Chapter 2 of equilibrium constants. The rate constants depend in an exponential way on an energy ($\Delta G^\ddagger$ or E_a), just as an equilibrium constant K_{eq} varies with reaction free energy ($\Delta G°$). In fact, this parallel is the reason the empirically observed E_a was viewed as an activation energy. Also in parallel with our discussion of equilibria, every change of 1.36 kcal/mol in $\Delta G^\ddagger$ or E_a is worth a factor of ten in a rate constant at 298 K. Similarly, a change of 1.36 kcal/mol in $\Delta H^\ddagger$ or $T\Delta S^\ddagger$ is worth a factor of 10. In the first reaction of Table 7.1, the –26 eu for $\Delta S^\ddagger$ corresponds to 26 cal/mol•K × 298 K = 7,750 cal/mol in $\Delta G^\ddagger$, implying a decrease in the rate constant by a factor of 500,000, relative to a reaction with a $\Delta S^\ddagger = 0$.

7.2.7 Is TST Completely Correct? The Dynamic Behavior of Organic Reactive Intermediates

Research over the last several years has revealed that TST may not always be adequate to describe the behavior of reactive intermediates generated in the course of traditional organic chemistry experiments. We stated in our discussion of TST that the equations used ultimately derive from statistical mechanics (statistical kinetic models). This approach requires that any excess internal energy that a molecule may have after it has traversed the transition state and arrived at the position of a reactive intermediate be redistributed within the molecule and the solvent at a rate faster than chemical events such as bond cleavage or formation. In other words, the molecules must "rattle" around in the high energy valley, achieving a Boltzmann distribution before proceeding on to the next reaction. In this sce-

Table 7.1
Examples of Common Reactions
and Representative $\Delta S^{\ddagger}$ Values

Entry	Reaction	ΔS of activation	Rate-determining/ rate-limiting step
1	Acid-catalyzed hydrolysis of ethyl acetate in water	−26 eu	
2	Acid-catalyzed ring opening of ethylene oxide in water	−6 eu	
3	Acid-catalyzed hydrolysis of α-methylglucopyranoside in water	+4.5 eu	
4	Displacement of iodide from methyliodide by pyridine	−31 eu	
5	Hydrolysis of *t*-butyl chloride in water	+10 eu	
6	A conjugate addition reaction	−17 eu	
7	Peroxide homolysis	+11 eu	

nario, if there is more than one possible reaction for the intermediate, that with the lowest barrier would dominate. Stated another way, the intermediate has a lifetime long enough to change its trajectory on the energy surface from the trajectory that landed it at the intermediate. However, a high energy intermediate that has very low barriers to subsequent reactions may undergo those reactions at a rate that is comparable to or faster than the redistribution of internal energy. In other words, if the energy involved in the vibrational motions that lead to the reaction is not redistributed within an intermediate, the molecular motions involved in the reaction can continue to direct the molecule along a particular pathway in ways that are inconsistent with TST. The momentum of the atoms carries them in a particular direction during the reaction, and Newton's laws of motion end up playing a role in dictating the products. Such dynamic effects in chemical reactions are a topic of current research interest, as discussed in the next Going Deeper highlight.

Going Deeper

Where TST May be Insufficient

Let's examine one reaction where it has been postulated that the reaction of the intermediate occurs faster than the redistribution of internal energy. Azo compound *i* undergoes reaction to give two bicyclopentanes, *ii* and *iii*, in a ratio that favors the exo isomer *iii* by about 3 to 1 in the gas phase. One mechanism for explaining the dominance of *iii* would be the formation of biradical *iv*, which reacts in a concerted fashion with inversion of configuration at the carbon from which N_2 is departing. In addition, there would be a competing reaction that forms the biradical *v*, which can give either product in a 1 to 1 ratio. However, molecular dynamics calculations (an advanced version of the type of calculations discussed in Section 3.3.4) give a different view.

Detailed theoretical analysis suggests that cleavage of the second C–N bond always precedes formation of the C–C bond, meaning that an intermediate such as *v* is always present. However, some fraction of the population of *v* closes to *iii* before it has time to randomize its internal energy. The preference for *iii* over *ii* comes from the trajectory of the atoms involved in the expulsion of N_2. The momentum of the CH_2 group as the hydrocarbon recoils from the expelled N_2 is in a direction that directs it down past the plane formed by a planar symmetric diradical *v*, leading preferentially to *iii*. Stated in more sophisticated language, the entrance channel to *iii* has better dynamic matching to the exit channel from *iv*. In contrast, if *v* had been formed and had sufficient lifetime to allow all internal energy to randomize, both exit channels to *ii* and *iii* would have been traversed equally.

Reaction sequence for dynamic control

Studies such as that briefly described here reveal that the simple kinetic and structural models that organic chemists have traditionally relied upon may not always be adequate. Typically, when we see a mixture of products we propose two different mechanistic pathways. In the present case, however, we have a single path but with molecular dynamics influencing the choice of exit channels. How many other reaction mechanisms may have been similarly incorrectly delineated due to our reliance on TST?

Carpenter, B. "Dynamic Behavior of Organic Reactive Intermediates." *Angew. Chem. Int. Ed. Eng.*, **37**, 3341 (1999). Reyes, M. B., and Carpenter, B. K. "Mechanism of Thermal Deazetization of 2,3-Diazabicyclo[2.2.1]-hept-2-ene and Its Reaction Dynamics in Supercritical Fluids." *J. Am. Chem. Soc.*, **122**, 10163 (2000).

7.3 Postulates and Principles Related to Kinetic Analysis

Based upon many years of empirical evidence, chemists have developed a series of guiding principles that give us insight into chemical reactivity. Among other things, these principles allow us to quickly decide the shape of a reaction coordinate diagram, predict a product ratio, and determine how the stability of a molecule would affect its reactivity. Here we introduce these time honored tools of the physical organic chemist.

7.3.1 The Hammond Postulate

The Hammond postulate is likely the most widely used principle for estimating the structures of activated complexes. In fact, it is so useful that all introductory organic chemistry textbooks cover it, and many chemists use it intuitively. The value of the Hammond postulate stems from the fact that transition states are transient in nature and generally cannot

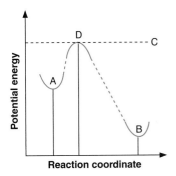

Figure 7.7
Diagram showing how the placement of the transition state is
determined using the Hammond postulate. Draw the placement of A
and B with knowledge of the heat or Gibbs free energy of the reaction.
Place the energy of the transition state (C) so that it is above the energy of
A and B, and then align the transition state along the reaction coordinate
to be closer in energy to structure A (intersection D), then fill in the
curve.

be directly characterized by experimental means. Therefore, any guiding principle that can
provide some insight as to the chemical structure of a transition state is quite useful. According to Hammond, "If two states, as for example, a transition state and an unstable intermediate, occur consecutively during a reaction process and have nearly the same energy
content, their interconversion will involve only a small reorganization of the molecular
structures".

In essence, the postulate tells us that *the activated complex most resembles the adjacent reactant, intermediate, or product that it is closest in energy to*, as long as the energy difference between the transition state and the adjacent structure is not too large. Hammond originally
used statements such as "in highly exothermic steps it will be expected that the transition
states will resemble reactants closely and in endothermic steps the products will provide
the best models for the transition states". This postulate allows us to accurately predict the
shape of a reaction coordinate diagram, and it gives insight into the structures of activated
complexes. The postulate ties the structure of the activated complex to the structures of reactants, intermediates, and products.

To put this postulate to use, let's examine an exothermic reaction (Figure 7.7). We place
the peak that represents the transition state higher on the diagram than the reactant and
product. But where on the x axis do we show the transition state? The Hammond postulate
tells us that because the transition state is closer in *energy* to the reactant than the product, it
is also closer in *structure* to the reactant than the product. Remember, the x axis in these diagrams is related to the progressive change in molecular geometry as we transform reactants to products. The fundamental assumption of Hammond is that molecules do not undergo rapid, discontinuous structure changes along a reaction coordinate. Rather, structural
changes are generally smooth and continuous along the reaction pathway. Therefore, we
draw the peak corresponding to the transition state closer to the reactant valley than to the
product valley. Connecting the valleys and peaks with curves completes the diagram. For
endothermic reactions, on the other hand, the transition state would be placed closer to the
product valley.

Now let's consider various energy surfaces. In Figure 7.8 the rear left corner of each energy surface is the reactant and the front right corner is the product. As the reaction varies
between exothermic and endothermic the shape of the surface varies as is shown going from
A to **B** to **C** in this figure. Plotting the cross-section of the surface that is the reaction coordinate on two-dimensional surfaces gives the reaction coordinate diagrams shown. We draw
the transition state shifting toward the product as the reaction moves from exothermic to
endothermic.

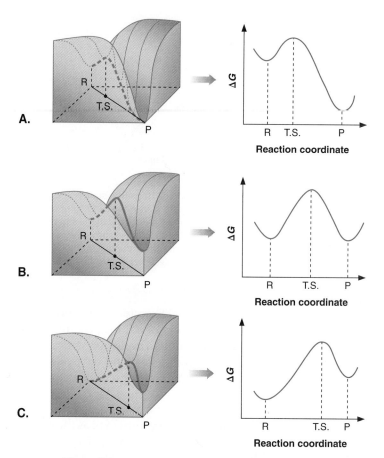

Figure 7.8
Various energy surfaces and their two-dimensional
projections that obey the Hammond postulate.

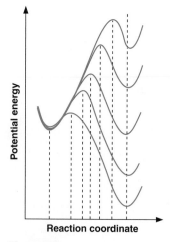

Figure 7.9
Several reaction coordinate
diagrams that obey Hammond's
postulate and where the
activation energy has a direct
correlation with the energy
change of the reaction.

The movement of the transition state is denoted in Figure 7.8 as a dot on a plane that is a projection of the energy surface. The diagonal across this plane is the actual reaction coordinate axis (x axis) that we plot in the reaction coordinate diagrams on the right of the figure. Note that the dot moves across the diagonal toward the product corner as the reaction becomes more endothermic. In general, the transition state moves along this diagonal toward the corner that is higher in energy. Such projections of the energy surface are going to become quite useful when we consider increasingly complex reactions (see Section 7.8).

Based partly upon the Hammond postulate, chemists typically write a continuum of reaction coordinate diagrams as shown in Figure 7.9 for similar reactions. The shape of each of these curves indicates a smooth shifting of the transition state structure from resembling the product to resembling the reactant as the reaction becomes increasingly exothermic. This predicts that a thermoneutral reaction has a transition state that is close to a one-to-one mixture of the structure of the reactants and products.

The Hammond postulate does not predict the height of the barrier compared to the reactant and product, only its position along the reaction coordinate. For example, we are not forced to draw the continuum as shown in Figure 7.9 in order to obey the Hammond postulate. In fact, Figure 7.10 shows a different overlay of reaction coordinate diagrams, each of which also conforms to the Hammond postulate, and you may occasionally encounter a system that exhibits this kind of behavior. However, most reactions that are of similar type but vary in their thermodynamics will have reaction coordinates that resemble those drawn in Figure 7.9.

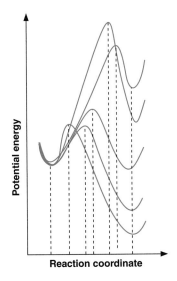

Figure 7.10
Several reaction coordinate diagrams that obey Hammond's postulate, but for which the activation energies do not necessarily correlate with the reaction free energies.

Connections

The Transition States for S_N1 Reactions

One of the most well known examples of the application of the Hammond postulate is the comparison of the structures of the various carbocations in an S_N1 reaction. This use is covered in all introductory organic chemistry textbooks. The relative stabilities of carbocations decrease in the following order: $3° > 2° > 1° > CH_3$. The Hammond postulate leads us to draw the reaction coordinate diagrams for the various heterolysis reactions as shown to the right with the transition state shifting toward reactant with increasing carbocation stability. Coupled with this shift in the position of the transition state is a lowering in energy of the transition state. Since the mechanism of the heterolysis of the bond between a carbon and a leaving group is similar, regardless of how substituted the carbon is, the transition state becomes more stable as the reaction becomes less endothermic, because it has character of the more stable carbocation. Don't forget, however, that even though the transition state to the most stable carbocation may have the character of the more stable carbocation, it

actually resembles the reactant more than any of the other transition states.

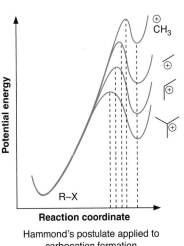

Hammond's postulate applied to carbocation formation

7.3.2 The Reactivity vs. Selectivity Principle

It is often stated, "The more reactive a compound is, the less selective it will be". In order to understand such statements, we must first determine what is meant by "more (or less) reactive" and by "selective". Both terms imply a comparison. Concerning selectivity, the comparison is the yield of two or more products as a function of the reactivity of two or more reactants. A reaction is **selective** if one product is formed in a higher yield than another.

More reactive molecules can be viewed either as being higher in energy or having more exothermic reactions. We can anticipate that the more reactive species will produce a transition state that more resembles the reactant. Hence, the transition state is not very sensitive to the structure of other components involved in the reaction, and it is affected little by the structure of the product. If the reaction is not sensitive to the structure of the product, it cannot select between different products and hence is not selective. Although this principle is quite logical and useful, there are *many* exceptions, and it should be applied with caution. The following Connections highlight describes a classic example of the reactivity vs. selectivity principle.

Connections

Comparing Reactivity to Selectivity in Free Radical Halogenation

A classic example of how reactivity is related to selectivity is concerned with the free radical halogenation of alkanes by Cl_2 and Br_2. In this free radical chain reaction, the step that sets the position of the halogen in the alkane is a hydrogen atom abstraction step. The carbon based radical created in this first propagation step then abstracts a halogen atom from the Cl_2 or Br_2, giving the alkyl halide (see below). In free radical halogenation by either Cl_2 or Br_2, tertiary alkyl halides are created preferentially to secondary, which in turn are formed preferentially to primary alkyl halides. This reflects the fact that the order of radical stability decreases from tertiary to secondary to primary. Yet, the extent of the selectivity for tertiary over secondary over primary is quite different for chlorination and bromination.

The relative rates for abstraction of a tertiary, secondary, and primary hydrogen by chlorine radical are 6:4:1. In contrast, the relative rates for abstraction by bromine radical are 19000:200:1; hence, the bromination is much more selective. This is explained by the fact that the chlorine radical is much more reactive than the bromine radical (recall the H–Cl vs. H–Br bond strengths in Chapter 2). In this case the hydrogen atom abstraction by the chlorine radical is an exothermic reaction and quite indiscriminate about which hydrogen is abstracted. By the Hammond postulate, there is very little radical character on carbon developing in the transition state during the hydrogen abstraction (called an **early transition state**). On the other hand, hydrogen atom abstraction by the bromine radical is an endothermic reaction and quite selective to which hydrogen is abstracted. Now, there is a lot of radical character on carbon in the transition state (a **late transition state**), allowing differences in carbon radical stability to control the reaction.

X = Cl or Br

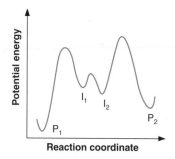

Hydrogen and halogen radical abstractions

7.3.3 The Curtin–Hammett Principle

The Curtin–Hammett principle is concerned with product ratios when there are two or more competing pathways commencing from interconverting isomers, conformers, or intermediates. In its most common implementation, the principle simply states that the ratio of products is determined by the relative heights of the highest energy barriers leading to the different products, and is not significantly influenced by the relative energies of any isomers, conformers, or intermediates formed prior to the highest energy transition states. This seems straightforward, but it is surprising how often chemists will fail to appreciate its implications.

We can understand the Curtin–Hammett postulate by examining the reaction coordinate diagram given in Figure 7.11. We have a reaction in which we generate an equilibrating pair of intermediates, I_1 and I_2. They can equilibrate readily because the barrier between them is much smaller than either exit barrier to P_1 or P_2. The pathway to P_2 derives from a relatively stable structure I_2, whereas the pathway to P_1 derives from a less stable structure I_1, but has a lower energy transition state. Given this scenario, the majority of the product arises from the less stable structure I_1. It is the barrier heights, not the relative stabilities of the intermediates, that determines the product ratio. The Curtin–Hammett principle is most commonly considered in synthetic organic chemistry, and the following Connections highlight gives just one example.

Our analysis above is a descriptive one. However, we can solve for the product ratio by using rate laws. The product ratio is given by Eq. 7.24 (you are asked to derive this expres-

Figure 7.11
Diagram demonstrating the Curtin–Hammett principle, where the product ratio is determined by the relative heights of the barriers and not the relative stabilities of the intermediates.

Connections

Using the Curtin–Hammett Principle to Predict the Stereochemistry of an Addition Reaction

There have been several empirical models for predicting the diastereoselectivity of nucleophilic addition to carbonyl groups adjacent to stereogenic centers (see Chapter 10). These models rely upon placing one group eclipsing the carbonyl and predicting that the favored approach of the nucleophile to the carbonyl is from the side with the smaller of the two remaining groups (see to the right). The addition of hydride and Grignard reagents to carbonyl groups have activation energies between 8 and 15 kcal/mol, whereas the rotation barriers along single bonds between sp^3 and sp^2 carbons are *much* lower. Hence, the Curtin–Hammett principle tells us that the diastereomeric product ratio should depend strongly upon the free energy difference between the diastereomeric transition states for nucleophilic attack, and not upon the relative energies of the different rotamers. To predict the relative energies of the diastereomeric transition states we examine steric interactions between the incoming nucleophile and substituents on the stereogenic carbon, and eclipsing interactions between the carbonyl group and the stereocenter at the transition state.

For example, of the three different trajectories for reaction shown at right, we need not at all consider which rotamer is more stable. Instead, we see that the first trajectory is the lowest in energy and would give the most product. This is because the nucleophile approaches the carbonyl from the side with the smallest group, and the

eclipsing between the carbonyl oxygen and the medium-sized group is more favorable than in any of the other possibilities. See Chapter 10 for more analyses of how diastereomers can be predicted by the analysis of conformations in the activated complex.

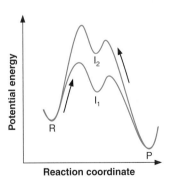

Various attack trajectories on carbonyls

Karabatsos, G. J. "Asymmetric Induction. A Model for Additions to Carbonyls Directly Bonded to Asymmetric Carbons." *J. Am. Chem. Soc.*, **89**, 1367 (1967).

sion in Exercise 1). The equation contains both the equilibrium constant (K_{eq}) between the interconverting intermediates and the relative rate constants (k_1, k_2) for formation of the products from the respective intermediates. The Curtin–Hammett principle is applicable when the barrier interconverting the different intermediates is much lower than the barriers to form products. The Curtin–Hammett principle is *not* applicable if this is not the case.

$$\frac{[P_1]}{[P_2]} = K_{eq}\left(\frac{k_1}{k_2}\right) \qquad \text{(Eq. 7.24)}$$

7.3.4 Microscopic Reversibility

This principle is concerned with an analysis of the individual pathways that chemical reactions take when the reactant is transformed to the product, and when the product is transformed back to the reactant. It states that the pathway for conversion of the product back to the reactant is the exact microscopic reverse of the forward pathway. The same intermediates and transition states are achieved in either direction. An example of how useful this notion becomes is given in the next Connections highlight.

Let's consider the implications of microscopic reversibility. Suppose we have two structures, R and P, and we propose that R is converted to P via intermediate I_1, and that R can also be converted to P via intermediate I_2. Since they are different, I_1 and I_2 must have different energies. Consider the reaction coordinate diagram shown in Figure 7.12, which describes this situation. Both R to P and P to R conversions will occur, in a ratio dictated by the rel-

Figure 7.12
Reaction coordinate diagrams that show two paths interconverting reactant to product.

Connections

Applying the Principle of Microscopic Reversibility to Phosphate Ester Chemistry

The hydrolysis of cyclic phosphotriesters involves an addition–elimination mechanism with the formation of trigonal bipyramidal intermediates (see below). These intermediates can undergo pseudorotation processes that exchange the axial and equatorial groups. To explain the ratio of products formed from exocyclic cleavage (cleavage of the ligand that is not part of the ring) and endocyclic cleavage (cleavage of a ligand within the ring) one must invoke leaving group departure commencing from an axial position solely (see reasoning given in the reference). If leaving group departure is from the axial posi-

tion, the principle of microscopic reversibility tells us that nucleophilic attack (the opposite of leaving group departure) must occur on the phosphotriester with the incoming nucleophile approaching along an axial position. Indeed, relating these two reactions in this manner, along with knowledge about which groups prefer axial over equatorial positions, is sufficient to explain the products obtained from hydrolysis of such phosphoesters.

Kluger, R., Covitz, F., Dennis, E., Williams, D., and Westheimer, F. H. "pH-Product and pH-Rate Profile for the Hydrolysis of Methyl Ethylenephosphate. Rate limiting Pseudorotation." *J. Am. Chem. Soc.*, **91**, 6066–6072 (1969).

Hydrolysis of methyl ethylene phosphate

ative barrier heights. The important point in the principle of microscopic reversibility is that along the path containing I_1, the same transition states will be formed in the forward and reverse directions. The same statement can be made for the path containing I_2. The trajectories of the atoms will be moving in opposite directions at each transition state in the forward and reverse reactions, but the chemical structures of the activated complexes formed at the transition states are the same. Further, if the lowest energy path to product uses I_1, the lowest energy path back to reactants must also pass through I_1.

7.3.5   Kinetic vs. Thermodynamic Control

The terms kinetic control and thermodynamic control are concerned with the manner in which the ratios of products of a reaction are determined. When a reaction is under **kinetic control**, the ratio of two or more products is determined by the relative energies of the transition states leading to these products. The relative stabilities of the products do not matter. Under **thermodynamic control**, the ratio of the products is determined solely by the relative energies of the products. In this case the energies of the pathways leading to the products do not matter. In our discussion of the Curtin–Hammett principle, we only analyzed the energies of the transition states and not the energies of the products, and thus this principle applies to kinetic control. Thermodynamic control, which ultimately produces the equilibrium (thermodynamic) mixture of products, can only be achieved when it is possible for the products to interconvert (equilibrate) under the reaction conditions. The simplest way for the

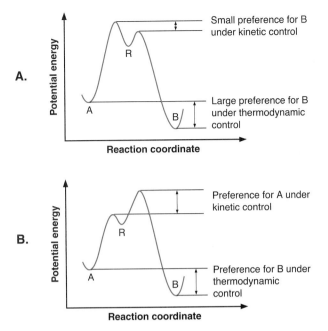

Figure 7.13
Diagrams used to explain kinetic control versus thermodynamic control.
A. The kinetic product and thermodynamic product are both B.
B. The kinetic product is A and the thermodynamic product is B.

products to interconvert is just by reversal of the reaction that formed them. Thus, thermodynamic control is often associated with readily reversible reactions. However, we can also get a thermodynamic product ratio if the products equilibrate after their formation through some other chemical pathway.

Often, the lower energy product is formed via the lower energy transition state. In this case the thermodynamic product is the same as the kinetic product. The only question is the exact amounts of the two products. Let's consider the exothermic reactions available to structure R shown in Figure 7.13 **A**. When working at a temperature where the reaction is irreversible, the product ratio would be determined by the relative activation energies of the two possible pathways. In this case there would be a small preference for product B. However, if the reaction is performed at a temperature where the reactions are reversible, the relative energies of the pathways no longer matter because both transition states can be traversed readily. Under reversible reaction conditions, the thermodynamic equilibrium between A and B will be established, and so it is the relative stability of the products that matters, producing in this case a large preference for product B. This is actually the most common scenario; the kinetic and thermodynamic products are the same.

The more interesting scenario occurs when the thermodynamic product is different from the kinetic product. Consider Figure 7.13 **B**. At low temperatures product A is favored because the barriers for formation of the products dictate the product ratio. But at higher temperatures product B is favored because all barriers on the energy surface are surmountable, allowing all the molecules to find the most stable point on the surface. This is a case where the heights of the barriers do not follow the continuum shown in Figure 7.9, but instead are as in Figure 7.10. It will always be true that whether we are seeing so-called thermodynamic or kinetic products will be determined by whether the products are or are not able to equilibrate under the reaction conditions. If so, we get the thermodynamic ratio; if not, we see the kinetic ratio. Running reactions at lower temperatures will often disfavor equilibration and thereby lead to the kinetic product ratio. A common example of exploiting this difference in product ratios in a synthetically important reaction is given in the following Connections highlight.

Connections

Kinetic vs. Thermodynamic Enolates

One of the most common uses of kinetic vs. thermodynamic control is the ability to manipulate the regiochemistry of the alkylation of enolates. As a simple example, consider forming the two possible enolates from 2-methylcyclohexanone (see to the right). Deprotonation is easiest at the α-carbon lacking the methyl group because those hydrogens are less sterically hindered. Yet, deprotonation of the α-carbon bearing the methyl group gives the more stable enolate. This enolate is more stable because sp^2 hydridized carbons are stabilized by higher alkyl group substitution levels (Chapter 2). We can take advantage of these differences to direct the regiochemistry of alkylation. Deprotonation of the unsymmetrical ketone and alkylation at low temperatures gives the product from the

kinetic enolate. However, if we expose the enolate to higher temperatures and thus equilibrating conditions, the product derived from the thermodynamic enolate is favored.

Two possible enolates

7.4 Kinetic Experiments

We now turn our attention to the goals of studying kinetics, how kinetic experiments are performed, and how they are analyzed. The study of the kinetics of a reaction is often one of the first experiments that chemists consider when setting out to determine the mechanism of a reaction. However, it is usually just one of the experiments performed, and the information obtained is combined with data collected from many other observations. It is important to note that kinetics does not give a complete picture of a mechanism. The information can only be used to support or refute a particular mechanism, and cannot be used to prove a mechanism.

7.4.1 How Kinetic Experiments are Performed

In Sections 7.1.4 and 7.1.5 we introduced rates, rate laws, rate constants, kinetic order, and molecularity. You may want to quickly review these sections before proceeding if you are uncomfortable with these terms and concepts. Our discussion here assumes you have a good understanding of these concepts.

The goal of kinetics is to establish a quantitative relationship between the concentrations of reactants and / or products, and the rate of the reaction. As we will see, this usually entails a separate kinetic study at several different concentrations of a reactant, with the goal of discovering the kinetic order of the reaction with respect to that reactant. A thorough analysis gives the kinetic order for each reactant, which in turn gives the rate law for the reaction. Once the rate law is in hand, one of the several analyses given below is used to determine a rate constant.

One obvious piece of information that a kinetic experiment yields is the actual rate of the reaction. Is the reaction relatively fast (over in microseconds) or relatively slow (takes days to weeks to go to completion)? The rate is expressed as the time dependence of the appearance of product or disappearance of reactant and is a positive number (Eq. 7.2). Determining the concentration of the reactant and / or product as a function of time is the essence of every kinetic experiment.

What does the order of the reaction tell us about the mechanism? How do we determine the rate law of a reaction, and thereby the order of the reaction with respect to each reactant? In addition, how do we determine rate constants? These are the questions that are of paramount importance in our analysis. In order to answer these questions we must first realize that the kinetic order of a complex reaction is highly dependent upon the sequential position of the rate-determining or rate-limiting step in the mechanism. Only reaction steps that occur before and at the rate-determining step can be detected during a routine kinetic study; steps after the rate-determining step cannot be detected. This means that the kinetic order of the different reactants only tells us about their involvement in the mechanism prior to or during the rate-determining step.

Given the above caveat about the position of the rate-determining step, we can examine the relevance of kinetic order. For example, a first order reaction means that only one species reacts during the rate-determining step. Intramolecular rearrangements or fragmentations are common examples (Eqs. 7.25 and 7.26). A second order reaction can involve the reaction of two identical species during the rate-determining step, in which case the reaction is also second order with respect to this species (Eq. 7.27). However, a second order reaction can also be represented by the reaction of two different species, where the reaction is first order in each of these species (Eq. 7.28). Either the two species can react together in the rate-determining step, or one of the two can react in a step prior to the rate-determining step, which requires the second species. However, we cannot tell in what order they react. These are called **intermolecular reactions**, because the reaction occurs between two different molecules. The following Going Deeper highlight gives an example relating intramolecular and intermolecular reactions. Finally, a zero order dependence typically means that the involvement of that reactant occurs past the rate-determining step (such as a zero order dependence of B in Eq. 7.29). Therefore, the order of the reaction and the order with respect to each reactant tells us about whether a given species reacts prior to or at the rate-determining step, and can give insight into which reactants react in a step after the rate-determining step.

$$A \longrightarrow P \qquad \text{(Eq. 7.25)}$$

$$A \longrightarrow P_1 + P_2 \qquad \text{(Eq. 7.26)}$$

$$A + A \longrightarrow P \qquad \text{(Eq. 7.27)}$$

$$A + B \longrightarrow P \qquad \text{(Eq. 7.28)}$$

$$A \longrightarrow I + B \longrightarrow P \qquad \text{(Eq. 7.29)}$$

To deduce the rate law for a reaction, we vary the concentrations of each individual reactant and measure the rate. From the experimental observations we discover if the reaction is zero, first, second, etc., or a fractional order in each species. This is often done by plotting the rate as a function of starting concentration of each species (Figure 7.14) while holding every other reactant concentration constant (also solvent, temperature, etc.). A slope of zero indicates zero order dependence, a slope of one indicates first order behavior, and a curved plot that fits an exponent of two indicates second order kinetics in that reactant.

Note that we are suggesting you plot several rates versus concentration, while other reactant concentrations are kept constant. However, reaction rates depend upon the concentration of the species involved and they change with time. Hence, in this experiment, we need to measure the rates at equivalent points. One must measure the various rates starting at points in time where the concentrations of those species whose concentrations are being kept constant are relatively invariant. Initial-rate kinetics, which is described later in this chapter, is particularly useful in this regard. In contrast, if we algebraically incorporate concentration into the rate constant, we can plot rate constants as a function of concentration to determine kinetic orders (wait until the section on pseudo-first order kinetics to understand this fully).

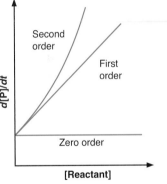

Figure 7.14
The rate of a reaction as a function of reactant concentration for second, first, and zero order kinetic dependence upon the reactant. The y-axis is also commonly a k_{obs} value generated by pseudo-first order conditions or an initial rate analysis (see text for discussion).

Going Deeper

Molecularity vs. Mechanism. Cyclization Reactions and Effective Molarity

A useful illustration of the distinctions between mechanism, molecularity, and order arises in the analysis of intramolecular versions of typically intermolecular reactions. Consider a classic S_N2 reaction of an amine and an alkyl iodide. The reaction is second order (first order in both amine and alkyl iodide) and bimolecular (two molecules involved in the transition state; that's what the "2" in "S_N2" stands for). The mechanism involves the backside attack of the nucleophilic amine on the C, displacing the iodide in a single step. Now consider a long chain molecule i that terminates in an amine on one end and an alkyl iodide on the other. Now two types of S_N2 reactions are possible. If two different molecules react, we still have a second order, bimolecular, **intermolecular** reaction. The product would ultimately be a polymer, ii, and we will investigate this type of system further in Chapter 13. Alternatively, an **intramolecular** reaction could occur, in which the amine reacts with the iodide on the same molecule producing a cyclic product, iii. This is still called an S_N2 reaction, even though it will be first order and unimolecular.

The differing kinetic orders of these two reaction pathways provide a simple means to select one product over another. The polymerization reaction depends on the square of the concentration of i, while the cyclization is first order in [i]. This states mathematically what we know intuitively. High concentrations favor the polymerization while low concentrations favor cyclization. When the cyclic product is desired, we simply run the reaction at high enough dilution that the polymerization reaction becomes implausibly slow. What is that concentration? Let's set up the rate equations for the two reactions. Cyclization will be favored when reaction I is slower than reaction II—that is, when the ratio of rates is less than 1. The ratio of rate constants has the units of molarity, and it is a characteristic of the particular system known as the **effective molarity (EM)** (see Chapter 9 for another analysis). If [i] is significantly less than the effective molarity of the reaction system, then the major product will be the cyclized molecule.

Reaction I: $d[i]/dt = k_{inter}[i]^2$; Reaction II: $d[i]/dt = k_{intra}[i]$

Ratio I/II $= (k_{inter}[i]^2)/(k_{intra}[i]) = (k_{inter}/k_{intra})[i] = [i]/EM$

Polymerization versus cyclization varies as a function of concentration

7.4.2 Kinetic Analyses for Simple Mechanisms

For elementary reactions the kinetics are relatively simple, and there are straightforward mathematical expressions that allow us to solve for rate constants. These simple mechanisms are those we analyze first. They involve first and second order kinetics, along with variations including pseudo-first order and equilibrium kinetics. We also look at a method to measure rate constants known as initial-rate kinetics. We analyze complex reactions only under the simplifying assumption of the steady state approximation (Section 7.5.1), and show how kinetic orders can change with concentration. More advanced methods for analyzing complex reactions are left to texts that specialize in kinetics.

The ways that kinetic data are used to assist us in deciphering a mechanism are quite varied depending upon the case. First, we can examine the kinetic data and decide in what manner the reactants combine, and then write a mechanism that is consistent with the data. Second, we can already have a potential mechanism or mechanisms in mind, which will predict particular kinetic observations. We test our hypotheses by performing the experiments designed to display the predicted observations. Both approaches require that we have the ability to derive the rate laws that predict the experimental observations. Thus, we now give a discussion of generic mechanisms and their particular rate laws. In Chapters 9, 10, 11, and 12, rate laws will be examined to support particular mechanisms.

Inherent in kinetic experiments is the determination of one or more rate constants. Many of these examples we now give have more than one step, and we number the rate constants reflecting the sequence of the steps and not the kinetic order.

First Order Kinetics

The rate law for a simple first order transformation (Eq. 7.25) can be written as in Eq. 7.30. Rearranging this gives Eq. 7.31 which, upon integration, gives the integrated rate law 7.32 where $[A]_o$ is the starting concentration of A. Another common way of writing this is Eq. 7.33, which shows an exponential dependence of the concentration of A upon the rate constant.

$$\frac{d[P]}{dt} = -\frac{d[A]}{dt} = k[A] \qquad \text{(Eq. 7.30)}$$

$$\frac{d[A]}{[A]} = -kdt \qquad \text{(Eq. 7.31)}$$

$$\ln[A] = \ln[A]_o - kt \qquad \text{(Eq. 7.32)}$$

$$[A] = [A]_o\, e^{-kt} \qquad \text{(Eq. 7.33)}$$

Therefore, if one uses a spectroscopic or chromatographic method to monitor the concentration of A at various time points t, a plot of $\ln[A]$ vs. time will give a straight line with a slope that is the negative of the rate constant. If the data can be modeled with this equation, this suggests that the reaction is first order (see the next Connections highlight for an example).

The **half-life** $(t_{1/2})$ of a reaction is the time required for 50% of the starting material to be consumed. A rule of thumb is to follow the reaction to five or more half-lives to obtain an accurate first order rate constant. For a first order reaction the half-life is $t_{1/2} = \ln(2)/k = 0.693/k$. A related term is the **lifetime** of a species, defined as $1/k$, where k is the rate constant for the first order disappearance of the species. For a first order reaction, the time required for reaction does not depend upon how much reactant one starts with. For higher order reactions the half-life and lifetime do depend upon the concentrations of the reactants.

We can also work with Eq. 7.30, using $d[P]/dt$. However, we need to have [P] in terms of [A]. Since the total concentration of A and P must always equal the initial amount of A ($[A]_o = [A] + [P]$, assuming no side reactions), we derive Eq. 7.34. Now a plot of $\ln\{[A]_o/([A]_o - [P])\}$ versus t gives the rate constant as the slope.

$$\ln\left(\frac{[A]_o}{[A]_o - [P]}\right) = kt \qquad \text{(Eq. 7.34)}$$

Connections

First Order Kinetics: Delineating Between a Unimolecular and a Bimolecular Reaction of Cyclopentyne and Dienes

Due to the extremely strained nature of cyclopentyne, its characterization has been challenging, and data even supporting its existence are slim. One proposed way to generate cyclopentyne is through the elimination of LiBr from compound *i*. In fact, if *i* is allowed to react with *ii*, both compounds *iii* and *iv* are formed, suggesting cycloaddition reactions. But, is the precursor to *iii* and *iv* truly cyclopentyne as shown in path A below? One can envision a mechanism leading to *iii* and *iv* without the intermediacy of cyclopentyne, as depicted in path B. This mechanism is bimolecular, involving a collision between *i* and *ii* in a rate-limiting step.

To distinguish between these two possibilities, the time course of the disappearance of *i* (at mM concentra-

tion) was followed as a function of several concentrations of *ii* varying from 1 M to near 4 M. A first order analysis was used with ln[*i*] as the *y* axis, as indicated by Eq. 7.32, and straight lines were obtained. The rate constants increased only by a factor of 50% for a four fold change in concentration of *ii*, indicating that *ii* is not involved in the rate-limiting step (the slight increase in rate with increasing *ii* was attributed to a solvent effect). Thus, the plots support a rate-determining unimolecular decomposition of *i* to give cyclopentyne, which subsequently reacts with *ii* to yield *iii* and *iv*. This is a good example of the use of Eq. 7.32, and the logic that goes along with using a kinetic analysis to support or refute particular mechanistic alternatives.

Gilbert, J. C., McKinley, E. G., and Hou, D.-R. "The Nature of Cyclopentyne from Different Precursors." *Tetrahedron*, **53**, 9891–9902 (1997).

Creation and reaction of cyclopentyne

Second Order Kinetics

The differential rate law describing the reaction in Eq. 7.28 is given in Eq. 7.35. Rearranging and integrating this equation leads to Eq. 7.36. The mathematical manipulations along with integrating the rate expression are left to the Exercises at the end of the chapter, as is solving the kinetics for the type of reaction depicted in Eq. 7.27. Using Eq. 7.36 and plotting $[1/([B]_o - [A]_o)] \ln([A]_o[B]/[B]_o[A])$ versus t gives the rate constant from the slope. This requires a spectroscopic or chromatographic method for monitoring both [A] and [B], unless we express these concentrations in terms of each other. If the data can be modeled with this equation it suggests that the reaction is second order. The following Connections highlight shows an example of how the observation of second order kinetics was used to support one mechanism versus another.

$$\frac{d[P]}{dt} = k[A][B] \qquad \text{(Eq. 7.35)}$$

$$\left(\frac{1}{[B]_o - [A]_o}\right) \ln\left(\frac{[A]_o[B]}{[B]_o[A]}\right) = kt \qquad \text{(Eq. 7.36)}$$

Connections

The Observation of Second Order Kinetics to Support a Multistep Displacement Mechanism for a Vitamin Analog

Thiamine is a vitamin used during the construction of β-keto acids. The kinetics of displacement of a leaving group (LG) from thiamine analog *i* was used to provide insight into the analogous enzyme-catalyzed reaction (the enzyme is known as thiaminase I). The reaction of sulfite ion with thiamine analog *i* to give *ii* appears at first glance to be a simple S_N2 displacement, for which we would predict first order behavior for sulfite. However, investigation of the kinetics showed the reaction to be second order in sulfite. This observation of second order kinetics

prompted further experimentation. The data supported the first sulfite adding adjacent to the imminium, expulsion of the leaving group, addition of a second sulfite, and then elimination of the first sulfite. Here, the sulfite was not used in a simple elementary reaction with second order behavior, but instead used twice in a multistep mechanism. This study is an excellent example of how a kinetic analysis may be used to demonstrate that the most obvious route is not necessarily the route used in nature.

Zoltewicz, J. A., Uray, G., and Kauffman, G. M. "Intermediate in Nucleophilic Substitution of a Thiamin Analogue. Change from First-Order to Second-Order Kinetics in Sulfite Ion." *J. Am. Chem. Soc.*, **102**, 3653 (1980).

A reaction sequence showing second order kinetics in sulfite

Pseudo-First Order Kinetics

While Eq. 7.36 does provide a way to analyze second order kinetics, it is not always feasible to simultaneously monitor the concentrations of both reactants. Very often, though, experimental conditions can be developed in which the relatively complex second order kinetics can be reduced to first order kinetic behavior. Let's assume a scenario as in Eq. 7.28 and employ a large excess (usually 10 equivalents or more) of B. Now the concentration of B will change very little during the course of the reaction, and [B] can be approximated by $[B]_o$. Because $[B]_o$ is a constant, it can be incorporated into the rate constant to give a new rate constant and a rate law that appears first order (Eq. 7.37). This process and the associated experimental conditions lead to what are called **pseudo-first order** kinetics (where $k' = k_{obs} = k[B]_o$), because a reaction that is in fact second order is being analyzed with first order reasoning. The application of pseudo-first order kinetics is very common. This is because the integrated rate law (either Eq. 7.32 or 7.33) is much simpler than that given in Eq. 7.36, and it is usually quite easy to execute the required experimental conditions.

$$\frac{d[P]}{dt} = k[A][B]_o = k'[A] \qquad \text{(Eq. 7.37)}$$

Using the integrated rate law given in Eq. 7.32 gives k', a number that depends upon the starting concentration of B. This new rate constant is often referred to as k_{obs}. Dividing k' by $[B]_o$ gives k. An even better method of determining k is to plot k_{obs} versus $[B]_o$ for several concentrations of B. The slope of the plot is k.

Another case in which pseudo-first order kinetics is often observed is a catalytic reaction. If one species is a catalyst, it is not consumed, so its concentration does not change (A +

B gives P + B). If we let B be the catalyst, then [B] always equals $[B]_o$, even if B is not in excess. Applying the same concepts expressed with regard to Eq. 7.37 allows us to solve for k_{obs} and therefore k. We will encounter examples of this kind of kinetic behavior when we examine general acid–base catalysis in Chapter 9.

Pseudo-first order kinetics is actually the most common form of kinetics used when determining the kinetic order of a reactant. For example, a linear dependence of k_{obs} upon $[B]_o$ indicates first order kinetics in B. A curved plot that conforms to a quadratic relationship would indicate second order kinetics in B. Hence, plotting k_{obs} vs. $[B]_o$ when $[B]_o >> [A]_o$ is an alternative method to plotting rates as a function of concentration (see the discussion of saturation kinetics later in the chapter) as a means of determining kinetic order in the reactants. The next Connections highlight demonstrates the use of pseudo-first order kinetics.

Connections

Pseudo-First Order Kinetics: Revisiting the Cyclopentyne Example

With regard to the reaction of cyclopentyne and dienes, we showed in a Connections highlight on page 386 how a first order kinetic analysis of the time course data gave straight lines revealing rate constants for the reaction that had little dependence upon the concentration of diene. This supported Path A involving cyclopentyne. What would be expected for this reaction if Path B were operating?

Since large excesses of *ii* were used relative to *i*, a pseudo-first order kinetic treatment was appropriate.

If path B were operative, the k_{obs} values would include $[ii]_o$, and we would expect a four-fold increase in the concentration of *ii* to directly give a four-fold increase in the k_{obs} value. Hence, the experiment given to distinguish Path A from Path B was actually designed to distinguish first order kinetics from pseudo-first order kinetics, thereby supporting either a unimolecular or bimolecular mechanism, respectively. It is important to stress that pseudo-first order kinetics, despite the "first order" in the name, will usually represent bimolecular or even more complex reaction mechanisms.

Equilibrium Kinetics

Although in principle all reactions are reversible, with the forward and reverse reactions proceeding through the same transition state (microscopic reversibility), often the forward reaction is sufficiently exothermic that the rate of the reverse reaction is so slow as to be negligible. We then consider the forward reaction to be irreversible. For many reactions, though, the forward and reverse reactions are comparable in rate, in which case an equilibrium between reactant(s) and product(s) is established. At equilibrium the concentrations of the reactant(s) and product(s) are not changing, but the system is dynamic. In fact, at equilibrium, reactions are going on all the time. It is just that at equilibrium the rate of the forward reaction is equal to the rate of the reverse reaction, and so there is no net change in concentrations of any species (Eq. 7.38, where k_f and k_r are the rate constants for the forward and reverse reactions, respectively). However, when we initiate such a reaction with only reactants present, a certain amount of time is required before equilibrium is established, so we can calculate a rate for the approach to equilibrium. This is the scenario that we now want to examine.

$$\frac{d[\text{reactant(s)}]}{dt} = \frac{d[\text{product(s)}]}{dt} = k_f[\text{reactant(s)}] = k_r[\text{product(s)}] \quad \text{(Eq. 7.38)}$$

Consider A in equilibrium with B (Eq. 7.39), where we allow the reaction to start with some A and B present ($[A]_o$ and $[B]_o$). We define a variable x that represents the extent of reaction. The variable x is simply the amount of A that has been depleted at a certain time t, which in this case would also be equal to the amount of B that has increased. This leads to Eq. 7.40. Rearranging and integrating leads to Eq. 7.41.

$$A \rightleftharpoons B \qquad\qquad \text{(Eq. 7.39)}$$

$$\frac{d[x]}{dt} = k_f([A]_o - [x]) - k_r([B]_o + [x]) \qquad\qquad \text{(Eq. 7.40)}$$

$$\ln\left(\frac{k_f[A]_o - k_r[B]_o}{k_f[A]_o - k_r[B]_o - (k_f + k_r)[x]}\right) = (k_f + k_r)t \qquad\qquad \text{(Eq. 7.41)}$$

This is quite a complex integrated rate equation. However, if we study the kinetics of the reaction at points in time near the establishment of equilibrium, we make the assumption that the forward and reverse rates are becoming equal (as when equilibrium is really established). At equilibrium we define $[x]$ as $[x]_e$, where the extent of reaction is as far as it is going to go, which leads to $k_f([A]_o - [x]_e) = k_r([B]_o + [x]_e)$. Solving this equality for $k_f[A]_o - k_r[B]_o$, and substituting the result into Eq. 7.41, leads to Eq. 7.42. This tells us that as one approaches equilibrium, the rate appears first order with an effective rate constant that is the sum of the forward and reverse rate constants. This is an approximation because we defined $[x]$ as $[x]_e$ to obtain this answer, but it is a very common way to analyze equilibrium kinetics. Chemists qualitatively estimate that the rate to equilibrium is the sum of the rates of the forward and reverse reactions.

$$\ln\left(\frac{[x]_e}{[x]_e - [x]}\right) = (k_f + k_r)t \qquad\qquad \text{(Eq. 7.42)}$$

Initial-Rate Kinetics

We commonly encounter reactions that are slow enough that it is difficult to follow them to several half-lives in order to obtain a reliable rate constant. Further, many reactions start to have significant competing pathways as the reaction proceeds, causing deviations from the ideal behaviors discussed above. In these cases we often turn to initial-rate kinetics. In this procedure we only follow the reaction to 5% or 10% completion, thereby avoiding complications that may arise later in the reaction and/or allowing us to solve for rate constants in a reasonable time period. This approach is inherently less accurate than a full monitoring of a reaction over several half-lives, but often it is the best we can do.

We start with Eq. 7.30 and note that very early on in the reaction the change in concentration of A is small. As a result we can approximate [A] with $[A]_o$. This means that $d[P]/dt$ is essentially a constant equal to $k[A]_o$. Plotting [P] versus t over the first few percent of the reaction gives a line whose slope is $k[A]_o$ (Figure 7.4). Dividing the slope by $[A]_o$ gives k. This is a simple way to measure first order rate constants as long as you can accurately measure the low concentrations of product produced early in a reaction, or accurately measure the small decrease in reactant concentration early in a reaction. It is also a good method for generating plots such as Figure 7.14, which give kinetic orders, by measuring initial rates for different $[A]_o$.

Tabulating a Series of Common Kinetic Scenarios

Now that we have looked at some of the most common scenarios, it is useful to tabulate these along with more complex ones. This should serve as a simple reference table for you to apply when implementing kinetic experiments. Table 7.2 shows several reaction stoichiometries along with the rate laws and the integrated rate laws. Almost all of these scenarios are amenable to reduction to simpler forms when a large excess of one reagent is used or an initial-rate kinetic treatment is applied.

Table 7.2
Reactions, Rate Laws, and Integrated Rate Laws

Reaction	Rate law	Integrated rate law
$A \rightarrow P$	$d[P]/dt = k[A]$	$\ln([A]/[A]_o) = -kt$
$2A \rightarrow P$	$d[P]/dt = k[A]^2$	$1/[A] = kt + 1/[A]_o$
$3A \rightarrow P$	$d[P]/dt = k[A]^3$	$0.5(1/[A]^2 - 1/[A]_o^2) = kt$
$A + B \rightarrow P$	$d[P]/dt = k[A][B]$	$[1/([B]_o - [A]_o)] \ln([A]_o[B]/[B]_o[A]) = kt$
$A + B + C \rightarrow P$	$d[P]/dt = k[A][B][C]$	$\{1/[([A]_o-[B]_o)([B]_o-[C]_o)([C]_o-[A]_o)]\ln\{([A]/[A]_o)^{([B]_o-[C]_o)}([B]/[B]_o)^{([C]_o-[A]_o)}([C]/[C]_o)^{([A]_o-[B]_o)}\} = kt$
$2A + B \rightarrow P$	$d[P]/dt = k[A]^2[B]$	$\{2/(2[B]_o - [A]_o)\}\{(1/[A]) - (1/[A]_o)\} + \{2/(2[B]_o - [A]_o)^2\} \ln([B]_o[A]/[A]_o[B]) = kt$

7.5   Complex Reactions—Deciphering Mechanisms

7.5.1   Steady State Kinetics

Thus far we have considered reactions with relatively simple mechanisms that can be assigned well-defined rate expressions. However, most organic, bioorganic, and organometallic mechanisms have more than one step, and frequently involve reactive intermediates. Such mechanisms can produce complex rate laws that are difficult to analyze. Fortunately, we can take advantage of the reactive nature of the intermediate(s) to simplify our analysis. In the course of a reaction with a transient intermediate, we do not expect large concentrations of that intermediate to ever accumulate in the reaction medium. There should always be only a very small concentration of the intermediate, and to a good approximation, *the concentration of the reactive intermediate is constant during the reaction*. This is the basis of the **steady state approximation (SSA)**, and we will develop its uses here.

It might seem odd that the concentration of an intermediate can be treated as effectively constant during a reaction if it is absent at the beginning of the reaction, is there in the middle, and is absent again at the end. What is important for the SSA to be valid is that the absolute changes in concentration of the intermediate be small with respect to changes in the concentrations of the reactants and products. It is therefore easy to realize that if the concentration of the intermediate is always very small, the absolute changes in its concentration must also be very small.

The SSA is best explained by working through an example. The SSA can be applied to reactions involving one or more intermediates, but our first example involves only a single intermediate. Consider the first order formation of an intermediate that is followed by a second order reaction for the formation of the product (Eq. 7.43). Eq. 7.44 gives the rate of the reaction, which depends on the concentration of I. Intermediate I is created from A and depleted by reversion to A and by reaction with B, so its concentration change as a function of time is given by Eq. 7.45. The SSA allows us to set Eq. 7.45 equal to zero; that is, the concentration of I does not change with time. Upon rearranging Eq. 7.45 to give Eq. 7.46 and substituting the resulting expression for [I] into Eq. 7.44, we arrive at Eq. 7.47. Importantly, Eq. 7.47 predicts first order behavior in the concentration of A, but a more complicated dependence on the concentration of B.

$$A \underset{k_{-1}}{\overset{k_1}{\rightleftharpoons}} I \xrightarrow{k_2B} P \qquad \text{(Eq. 7.43)}$$

$$\frac{d[P]}{dt} = k_2[I][B] \qquad \text{(Eq. 7.44)}$$

$$\frac{d[I]}{dt} = k_1[A] - k_{-1}[I] - k_2[I][B] = 0 \qquad \text{(Eq. 7.45)}$$

$$[I] = \left(\frac{k_1[A]}{k_{-1} + k_2[B]} \right) \qquad \text{(Eq. 7.46)}$$

$$\frac{d[P]}{dt} = \frac{k_1 k_2 [A][B]}{k_{-1} + k_2[B]} \qquad \text{(Eq. 7.47)}$$

The application of the SSA is extremely common, but before going further we should explicitly consider the implications of the approximation. To be valid, the intermediate(s) must be very reactive and therefore present at very small concentrations. To satisfy this, the intermediate must exist in a shallow potential energy well with small barriers to formation of product and/or reversion to reactant relative to the barriers for formation of I from R or P (see Figure 7.3 **F**). It is because of these small barriers that the concentration of the intermediate never builds up, but remains at a small level—essentially constant. This is mathematically expressed by setting the rate of change of the intermediate's concentration equal to zero ($d[I]/dt = 0$; see Eq. 7.45). The notion that the intermediate can rapidly revert to reactant implies that a pre-equilibrium is part of the mechanism. However, there is nothing in the SSA that requires the equilibrium concentration of the intermediate to be fully established, and in fact the equilibrium usually is not established due to the intermediate's transient nature. Finally, it is generally true that the product will be much more stable than the intermediate(s), making the last step in the sequence substantially exothermic. In practice, we usually apply the SSA to reactions where the last step is irreversible. This is not required in the SSA analysis, but by not establishing an equilibrium between the reactant and product, we do not have to use any of the complexities associated with the analysis of equilibrium kinetics (see above). If the intermediate approaches the energy of the reactant and/or product, the SSA is invalid because the concentration of the intermediate will change with time, and the approximation is certainly invalid when the energy of the intermediate drops below either the reactant or product.

Many organic, organometallic, and bioorganic reactions involve intermediates that are indeed reactive and transient. Reactive intermediates such as carbocations, radicals, carbanions, and carbenes are common to organic and bioorganic transformations, whereas coordinatively unsaturated transition metals, and low and high oxidation state metals are common to organometallic reactions (see Chapter 12).

Let's now consider a scenario where the same number of reactants are used as in the above example, but a different sequence is involved. Here the intermediate is formed in a second order reaction, and the intermediate converts to product in a first order reaction (Eq. 7.48). Eq. 7.49 expresses the rate of the reaction, and Eq. 7.50 expresses the SSA. Solving Eq. 7.50 for [I] leads to Eq. 7.51, which upon substitution into Eq. 7.49 gives Eq. 7.52. Eq. 7.52 has several rate constants incorporated into a product and quotient, which taken together is a constant that we call k_{obs}. This mechanistic scenario predicts that the reaction is first order in A and B, distinctly different than that presented in the last mechanistic scenario. This comparison reveals the power of a kinetic analysis when deciphering complex reaction mechanisms, because we are able to predict the order of the reaction with respect to different reactants for different possible mechanisms. However, this analysis also shows that we could not distinguish the mechanism of Eq. 7.48 from a simple elementary second order reaction of A and B, because both rate laws have a single rate constant, k or k_{obs}. We cannot decipher whether a rate constant represents a single elementary step or a combination of several rate constants for individual elementary steps.

$$A + B \underset{k_{-1}}{\overset{k_1}{\rightleftarrows}} I \overset{k_2}{\longrightarrow} P \qquad \text{(Eq. 7.48)}$$

$$\frac{d[P]}{dt} = k_2[I] \qquad \text{(Eq. 7.49)}$$

$$\frac{d[I]}{dt} = k_1[A][B] - k_{-1}[I] - k_2[I] = 0 \qquad \text{(Eq. 7.50)}$$

$$[I] = \frac{k_1[A][B]}{k_{-1} + k_2} \qquad \text{(Eq. 7.51)}$$

$$\frac{d[P]}{dt} = \frac{k_1 k_2[A][B]}{k_{-1} + k_2} = k_{obs}[A][B], \qquad \text{(Eq. 7.52)}$$

$$\text{where } k_{obs} = \frac{k_1 k_2}{k_{-1} + k_2}$$

Finally, let's increase the complexity just one step further. Consider a mechanism in which a reactive intermediate is formed in a first order reaction along with a stable product (P_1), followed by a second order reaction with a second reactant that converts the intermediate to product P_2 (Eq. 7.53). We go through the same mathematical process of analyzing the rate (Eq. 7.54), applying the SSA (Eq. 7.55), and performing algebraic manipulation (Eq. 7.56), to arrive at the result (Eq. 7.57). The prediction is that the reaction is first order in A, less than first order in B, and is retarded by P_1 (since its concentration is only in the denominator). Such predictions are important in guiding the experiments used to test if a reaction fits this mechanistic scenario.

$$A \underset{k_{-1}}{\overset{k_1}{\rightleftarrows}} P_1 + I$$

$$I + B \xrightarrow{k_2} P_2 \qquad \text{(Eq. 7.53)}$$

$$\frac{d[P_2]}{dt} = k_2[I][B] \qquad \text{(Eq. 7.54)}$$

$$\frac{d[I]}{dt} = k_1[A] - k_{-1}[I][P_1] - k_2[I][B] = 0 \qquad \text{(Eq. 7.55)}$$

$$[I] = \frac{k_1[A]}{k_{-1}[P_1] + k_2[B]} \qquad \text{(Eq. 7.56)}$$

$$\frac{d[P_2]}{dt} = \frac{k_1 k_2[A][B]}{k_{-1}[P_1] + k_2[B]} \qquad \text{(Eq. 7.57)}$$

There are three items that you should note about the form that the rate laws take when the SSA analysis is used. The first one leads to a useful "rule of thumb" for writing out the resulting rate law without doing all the mathematical manipulations given above. Note that the numerator of the final rate law is a *product* of all the forward rate constants (those consecutive steps that take the reactant to the product), and the concentrations of all the reactants required to give the product. In the three examples given above, k_1 and k_2 represent the two steps leading to product, and both A and B were needed to form the product. Therefore, the numerator of each rate law had these terms (Eqs. 7.47, 7.52, and 7.57). In contrast, the denominator is a *sum* of terms, each of which reflects a different route by which the intermediate can react. There is a term reflecting reversion to starting material, and a term that is representative of conversion to product. For example, consider the denominator of the rate law for the last mechanistic possibility given (Eq. 7.57). The intermediate can react with P_1 to give the reactant A or it can react with B to give P_2 with the rate constants k_{-1} or k_2, respectively. Finally, the concentration of the intermediate is never involved in either the numerator or the denominator. Therefore, the "rule of thumb" is the following: Write out all the rate constants and concentrations for the forward steps in the numerator as a product, and write out all the rate constants and concentrations for the ways that the intermediate can branch as a sum in the denominator. This generalization cannot be used when there is more than one intermediate that the SSA is applied to.

The second item that we note about the form of the rate expressions is that the terms in the denominator either involve rate constants alone or products of rate constants and concentrations. The terms that represent first order chemical reactions of the intermediate are represented by rate constants alone, whereas the terms that represent second order chemical reactions of the intermediate involve the product of rate constants and concentrations. When the SSA is applied to more than one intermediate, these terms become more complex. The magnitudes of the terms that are only rate constants are not amenable to manipulation by the chemist performing the experiment, except by changing the temperature or solvent properties. However, the magnitude of the terms that involve both rate constants and concentrations are easily adjusted by a chemist by simply increasing or decreasing initial concentrations. This is because these terms are representative of second order reactions, whose rates and half-lives are concentration dependent. The ability to change the magnitude of the terms in the denominator of such rate laws is critical to using kinetic experiments to decipher mechanisms.

The terms in the denominator are not necessarily comparable in magnitude. Very often one term is negligible compared to another and can be dropped, resulting in a considerably simplified rate law. In fact, when the terms in the denominator involve species whose concentrations can be controlled by manipulating their initial concentrations, we can use this ability to simplify the rate law when testing for a particular mechanism (see below). The simplification of rate laws often leads to reduction in the kinetic order for a reactant, where zero order becomes the result for that reactant. Before analyzing this, we show in the next Going Deeper highlight how zero order kinetics can arise for a reactant regardless of the concentration of that reactant. Then, in the following Connections highlight, we give an example from organometallic chemistry to show the power of using the SSA.

Going Deeper

Zero Order Kinetics

Often a reactant will not appear in the final rate law for a reaction, making the reaction formally zero order in that reactant. No reaction can be zero order in all species and therefore be zero order overall. This would mean that the reaction has no concentration dependence upon any of the reactants, which is impossible. But a reaction can be zero order in a single component, and the way this can occur involves a reaction where the species displaying zero order kinetics reacts after the rate-determining step.

Let's consider the scenario given below as an example. In this reaction, two irreversible steps and one reactive intermediate (I) are involved. If the first step is irreversible, it has to be rate-limiting. Application of the SSA follows, and thus the reaction is zero order with respect to B. This is a clear example of how a step beyond the rate-determining step does not influence the kinetics.

A case of zero order kinetics has been reported for the polymerization of norbornene by titanacyclobutanes. Plots of the concentration of norbornene (the alkene shown over the arrow of the second step) during polymerization, for three different initial concentrations of catalyst and norbornene, as a function of time, are shown below. The straight line plots establish that the rate of consumption (polymerization) of the norbornene is constant, indicating that the concentration of norbornene is not affecting the rate. Remember that if the species affects the rate, then an exponential concentration versus time plot should be found (as in Figure 7.4). The observation of zero order behavior in norbornene was interpreted as supporting a rate-determining step involving repetitive ring openings of the catalyst, with each ring opening followed by a rapid reaction with norbornene. Thus, the polymerization is first order in catalyst, but because the reaction with norbornene occurs after each rate-determining ring opening, one sees no kinetic dependence upon norbornene.

$$A \xrightarrow{k_1} I \xrightarrow{k_2 B} P$$

$$\frac{d[P]}{dt} = k_2[I][B]$$

$$\frac{d[I]}{dt} = k_1[A] - k_2[I][B] = 0$$

$$k_1[A] = k_2[I][B]$$

$$\frac{d[P]}{dt} = k_1[A]$$

Gilliom, L. R., and Grubbs, R. H. "Titanacyclobutanes Derived from Strained Cyclic Olefins: The Living Polymerization of Norborene." *J. Am. Chem. Soc.*, **108**, 733 (1986).

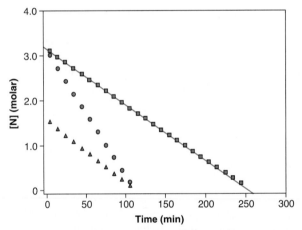

Ring-opening metathesis polymerization

Kinetic data for the polymerization of norbornene at 73 °C with the first metallacyclobutane shown above.
■, [catalyst] = 0.035 M and [norborene] = 3.2 M.
●, [catalyst] = 0.07 M and [norborene] = 3.2 M.
▲, [catalyst] = 0.035 M and [norborene] = 1.6 M.

Connections

An Organometallic Example of Using the SSA to Delineate Mechanisms

Complex i has been shown to react with allyl chloride (All–Cl) to give ii. The original investigators proposed two possible scenarios for this transformation. The first involved the sulfur acting as a nucleophile to displace the chloride followed by exchange of pyridine and chloride anion—an associative mechanism. The second involved loss of pyridine (pyr) in a first step, with subsequent reaction with allyl choride—a dissociative mechanism.

Associative and dissociative mechanisms
have different kinetic outcomes

These mechanisms should show distinctly different kinetics. Using the SSA on both pathways gives Eq. A for the associative mechanism and Eq. B for the dissociative mechanism. Eq. A predicts first order behavior in allyl chloride under all conditions with zero order behavior for pyridine. Interestingly, because chloride is used in both the forward and reverse steps branching from the inter-

mediate iii, it cancels in the numerator and denominator. In contrast, Eq. B shows that the kinetics would lose dependence upon allyl chloride at high concentration of allyl chloride (see the discussion of saturation kinetics given below) and the reaction would be slowed with addition of pyridine.

$$\text{rate} = k_1k_2[i][\text{All–Cl}][\text{Cl}^-]/(k_{-1}[\text{Cl}^-] + k_2[\text{Cl}^-])$$
$$= k_{\text{obs}}[i][\text{All–Cl}]$$

(Eq. A)

$$\text{rate} = k_3k_4[i][\text{All–Cl}]/(k_{-3}[\text{pyr}] + k_4[\text{All–Cl}]) = k_{\text{obs}}[i]$$

(Eq. B)

The original investigators, however, used a different kinetic "trick" to distinguish between the two mechanisms. Taking the reciprocal of k_{obs} from Eq. B gives Eq. C, which predicts a linear plot for $1/k_{\text{obs}}$ versus $([\text{pyr}]_o/[\text{All–Cl}]_o)$, which was found to be consistent with the experimental findings, thus supporting the dissociative mechanism over the associative one.

$$1/k_{\text{obs}} = 1/k_3 + (k_{-3}/k_3k_4)([\text{pyr}]_o/[\text{All–Cl}]_o)$$

(Eq. C)

Sweeney, Z. K., Polse, J. L., Andersen, R. A., and Bergman, R. G. "Cycloaddition and Nucleophilic Substitution Reactions of the Monomeric Titanocene Sulfide Complex (η^5-C$_5$Me$_5$)$_2$(C$_5$H$_5$N)Ti=S." *J. Am. Chem. Soc.*, **120**, 7825–7834 (1988).

7.5.2 Using the SSA to Predict Changes in Kinetic Order

Returning to the mechanism given in Eq. 7.53 and the corresponding rate law, Eq. 7.57, we now analyze the two terms in the denominator. The discussion given here is focused upon this single example, but it is meant to show how chemists in general intentionally look for changes in kinetic order to support or refute a mechanism by physically changing concentrations of reactants.

Let's examine the result obtained when $k_{-1} \gg k_2$, and the concentrations of P_1 and B are comparable and in excess at the beginning of the study. Since P_1 is a product of the first reaction, its concentration necessarily increases during the reaction. However, we are free to add this product in advance of performing the reaction as a means to keep its concentration nearly constant. With excess P_1 and B present from the start, their concentrations do not change appreciably during the reaction (hence $[P_1] = [P_1]_o$, $[B] = [B]_o$, and $k_{-1}[P_1] \gg k_2[B]$). Under these conditions the term $k_2[B]$ in Eq. 7.57 can be neglected, and the rate law reduces to Eq. 7.58. Now the reaction is found to be second order overall with respect to the reactants; that is, first order in both A and B. The rate of the reaction would be inversely proportional to the concentration of $[P_1]$ that was added at the beginning.

$$\frac{d[P_2]}{dt} = \frac{k_1k_2[A][B]}{k_{-1}[P_1]}$$

(Eq. 7.58)

This same result can also be achieved regardless of the relative sizes of k_{-1} and k_2 as long as we can add enough excess of P_1 so that $k_{-1}[P_1] \gg k_2[B]$. In this case we have manipulated the rates of the individual steps by controlling the concentrations of the reactants. This is only possible with second order or higher reactions where at least one of the terms in the denominator is concentration dependent.

In contrast, when the $k_2[B]$ term is much greater than the $k_{-1}[P_1]$ term, we can neglect the $k_{-1}[P_1]$ term. Neglecting $k_{-1}[P_1]$ in Eq. 7.57 cancels the $k_2[B]$ term in the numerator and denominator, giving Eq. 7.59. This scenario occurs when $k_2 \gg k_{-1}$ and the concentrations of P_1 and B are comparable, or when B is added in such a large excess that its associated term overwhelms the P_1 term. Under any of these conditions the reaction is found to be first order in A and zero order in B, and the concentration of P_1 does not affect the rate.

$$\frac{d[P_2]}{dt} = k_1[A] \qquad \text{(Eq. 7.59)}$$

The reduction of Eq. 7.57 to Eqs. 7.58 or 7.59, depending upon the experimental conditions, gives us the ability to test if the kinetics of the reaction under study responds in the manner predicted by these mathematical analyses. If the kinetics gives the behavior discussed, the data tell us that a mechanism involving A in the first step to give P_1 and a reactive intermediate that then reacts with B is plausible. This shows how one can experimentally control the form of the rate law and thereby test mechanistic postulates, such as the sequence of steps and the existence of an intermediate.

7.5.3 Saturation Kinetics

In the above hypothetical scenarios we showed that a mechanism that appears less than first order in B (Eq. 7.57) can be made to be either first order in B (Eq. 7.58) or zero order in B (Eq. 7.59) by manipulating the concentrations of P_1 or B, respectively. Therefore, depending upon the concentrations of B and P_1, as well as the relative sizes of k_{-1} and k_2, the reaction will vary between first and zero order in B. This leads to a graph like the one shown in Figure 7.15 for different concentrations of $[B]_o$. The pseudo-first order rate constant k_{obs} increases with increasing $[B]_o$, but levels off when $[B]_o$ is in a very large excess. At the lower concentrations of B the graph is indicative of first order behavior in B, while the plateau shows zero order dependence upon B. The rate has been saturated at a maximum level equivalent to $k_1[A]$. Such kinetic behavior is called **saturation kinetics**. An example using very common chemistry is given in the next Connections highlight.

Most mechanisms that have rate laws with concentrations of a reactant in both the numerator and the denominator will show saturation kinetics. It is always indicative of a pre-equilibrium in the mechanism, where the step involving this reactant (B in this case) is after the equilibrium. For example, the mechanism given in Eq. 7.43 will also show saturation kinetics in [B] (convince yourself by examining Eq. 7.47), whereas the mechanism given in Eq. 7.48 will not (convince yourself by examining Eq. 7.52).

Figure 7.15
The variation in k_{obs} as a function of the starting concentration of B for a reaction mechanism that shows saturation in B.

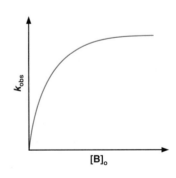

Saturation Kinetics That We Take for Granted—S$_N$1 Reactions

In introductory organic chemistry we often teach that the substitution of a tertiary alkyl halide with a nucleophile in ionizing solvents proceeds via a first order process (an S$_N$1 pathway). The reaction rate has no dependence upon the concentration of nucleophile. Does this really make sense? If we add no nucleophile there would be no product formation, so how can the rate really have no dependence upon nucleophile? The answer lies in the fact that under most experimental conditions for S$_N$1 reactions, the kinetics of the reaction are already in the saturation region.

Consider the kinetic expression for the reaction below. This substitution reaction has to be an S$_N$1 process because the alkyl halide is tertiary (remember that S$_N$2 reactions are prohibitively slow on tertiary centers). The kinetic expression indicates a kinetic order in cyanide. However, under most reaction conditions we would choose to perform this reaction (excess cyanide and an ionizing solvent), the kinetic expression would reduce to that shown. This is due to the carbocation formation being the highest barrier on the energy surface, k_2 being much larger that k_{-1}, and [CN$^-$] being greater than [Br$^-$]. We are already in the saturation regime.

$$\frac{d[P]}{dt} = \frac{k_1 k_2[R][NC^-]}{k_{-1}[Br^-] + k_2[NC^-]} = k_1[R] \text{ if } k_{-1}[Br] << k_2[NC^-]$$

An S$_N$1 mechanism and the associated rate law

7.5.4 Prior Rapid Equilibria

There is one last mechanistic scenario that we want to cover. It relates to a fully established equilibrium with an intermediate. Remember that by definition the equilibrium constant for a reaction is the quotient of the forward and reverse rate constants. Therefore, the k_1/k_{-1} term in Eq. 7.58 can be replaced by K_{eq}, which represents the equilibrium constant for the formation of the intermediate. This leads to Eq. 7.60 for such a mechanism (see the Connections highlight on the next page for an example).

$$\frac{d[P_2]}{dt} = \frac{K_{eq}k_2[A][B]}{[P_1]} \qquad \text{(Eq. 7.60)}$$

In summary, we have demonstrated how the kinetic behavior of a system can be manipulated by changing the concentrations of reactants. This leads us to conclusions about the sequence of the steps in the mechanism. Therefore, kinetic analyses are extremely informative for supporting or refuting mechanisms.

7.6 Methods for Following Kinetics

All rate laws express the concentration of either the reactant or the product as a function of time. In every case, a plot of one of these concentrations as a function of time needs to be generated. The data are plotted in the manner dictated by the integrated rate law so as to solve for the rate constant for the reaction. In all cases, some experimental technique must be used to monitor concentration as a function of time. Because time is a factor, the technique must be amenable to the time scale of the reaction and the ability of the chemist to introduce and mix the sample to initiate the reaction. Experimental techniques have been developed for the analysis of relatively slow reactions as well as exceedingly fast reactions.

Connections

Prior Equilibrium in an S$_N$1 Reaction

The reaction of *t*-butyl alcohol with HI in water proceeds via protonation of the alcohol, water departure in a rate-determining step, and trapping of the carbocation by iodide (see below). The kinetic expression for this reaction is shown below, where we have treated the protonation of the alcohol as a fully established equilibrium prior to the heterolysis reaction. We have not treated the protonated

alcohol as an unstable intermediate using the SSA because its energy is still significantly below that of the carbocation. Hence, during the time required for the formation of P, the equilibrium would be fully established. The kinetic expression simplifies when $k_2 > k_{-1}$ or [I$^-$] is high enough, and predicts that the rate depends only upon the pH and the concentration of alcohol.

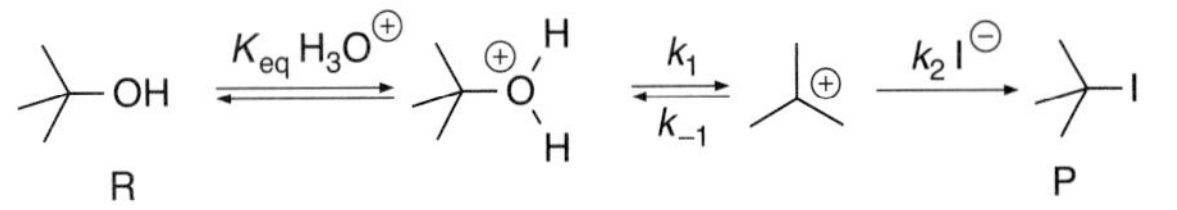

$$\frac{d[P]}{dt} = \frac{K_{eq}k_1k_2[R][H_3O^+][I^-]}{k_{-1}[H_2O] + k_2[I^-]} = K_{eq}k_1[R][H_3O^+] \quad \text{if } k_{-1}[H_2O] \ll k_2[I^-]$$

An S$_N$1 mechanism and the associated rate law

7.6.1 Reactions with Half-Lives Greater than a Few Seconds

When a reaction has a half-life greater than several seconds to minutes, we can usually mix the sample and record enough concentration data before the reaction is over such that a reliable time plot can be generated. Therefore, special techniques for generating the reactants and/or mixing the reactants are not required (see the next section). With these slower reactions, the kinetic analyses can be performed with chromatographic and spectroscopic methods. Most common is some form of spectroscopy. Spectroscopic techniques allow for continuous monitoring of changes in concentration. If the reactant and product have differing absorption or emission spectra, this is particularly convenient because all modern versions of these spectrometers are automated. Often we can directly use the absorbance or emission values in place of the concentrations when following first order kinetics, because concentrations are generally directly proportional to absorbance or emission via Beer's law (see Chapter 16). Nuclear magnetic resonance is also commonly used to follow the progress of a reaction.

Numerous other methods are also applicable. Chromatographic analysis with high performance liquid chromatography (HPLC) or gas chromatography (GC) is quite common. These methods are usually accompanied by an analysis that measures the response of the detector to the concentration of the species being analyzed. The peaks recorded in the chromatograms at each point in time are integrated in reference to an added standard, and then translated into concentrations using the analysis of the detector response. If the reaction generates a species that changes the pH, continuous pH measurements or acid–base titrations can be used to monitor the reaction. As a last example, if the reaction involves the interchange of optically active species, then polarimetry is a very convenient method to use. Essentially any method that can be related to concentration and followed as a function of time can be used to determine the rate of a chemical reaction.

7.6.2 Fast Kinetics Techniques

If the half-life for a reaction is just a few seconds or less, the reaction is typically completed within the time required to introduce and mix the reactants within the reaction vessel. The lifetime (τ) of a transient intermediate is the inverse sum of all the rate constants for its disappearance [$\tau = 1/(\Sigma k_i)$]. With very short lifetimes, conventional methods for following the time course of the reaction cannot be used. This is especially true for **diffusion limited reactions**, where every collision results in a reaction (k_{obs} is near 10^{10} M^{-1} s^{-1}; look back at the discussion of diffusion in Chapter 3). Here, we need especially fast mixing and kinetic methods. Increasing the viscosity of the media can slow the reaction and assist the analysis.

Further, if we want to measure the rate constants for reactions of high energy reactive intermediates, we cannot follow these reactions without some method of generating the intermediates at a rate faster than their subsequent reactions. In these cases we need to turn to what are known as **fast kinetic techniques**. The first one we discuss deals with a method for following a reaction that occurs faster than the time it takes to mix the reagents in a conventional manner, while the later two techniques are methods for the rapid generation of reactive intermediates.

Flow Techniques

When the rate of a reaction is faster than the time it takes us to introduce and mix reagents, we must turn to a method that achieves rapid mixing and has the ability to monitor the extent of reaction at various times. The most common methods involve flow tubes (Figure 7.16). The reactants are mixed at the point where the two tubes intersect, and then the extent of reaction is followed by an analysis of the mixture at various points along the observation tube. The flow rate of the tube is known, so that the time from the point of mixing is known at each analysis point along the tube. In the **stopped-flow** method, the flow is suddenly stopped at various times, and the analysis is performed at the same point in the observation tube at each different time. Several commercial versions of this apparatus are available for stopped-flow analyses, and it is a particularly common method for the analysis of enzyme-catalyzed reactions. With this technique reactions with half-lives as short as milliseconds can be measured.

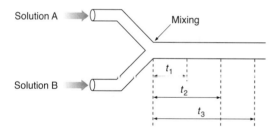

Figure 7.16
Diagram showing the general scheme for a flow analysis for measuring the kinetics of fast reactions.

Flash Photolysis

Flash photolysis is one of several methods used to create a non-equilibrium mixture of reactants within a very short time scale. It is most commonly used to generate a highly reactive intermediate that can be monitored on very short time scales. It can be used to prepare relatively high concentrations of electronically excited molecules and a variety of reactive intermediates. The most typical protocol is the **pump–probe** approach. An intense flash of light, the pump, is used to initiate a photochemical reaction that produces the reactive intermediate of interest. The duration of the light pulse must be shorter than the lifetime of the intermediate being probed. The intensity must be sufficient to generate enough of the reactive intermediate to be observable using some spectroscopic technique. Most typically the observation technique is optical spectroscopy (absorbance or fluorescence) using a second light, either as a continuous source or as a **probe pulse**.

By varying the time delay between the pump and probe pulses, information about the time it takes to form the intermediate can be gained. The actual lifetime of the intermediate can then be obtained by continuously monitoring the reactive intermediate over its short lifetime. This leads to decay traces such as that shown in Figure 7.17 **A**. These traces can be fit to the standard integrated rate laws to extract the appropriate rate constants. Some decay traces have more than one component. Figure 7.17 **B**, for example, shows a decay trace that indicates both a short and a long component. Hence, fast kinetic techniques can often be used to analyze multiple reactions of reactive intermediates. It may seem that the need to ini-

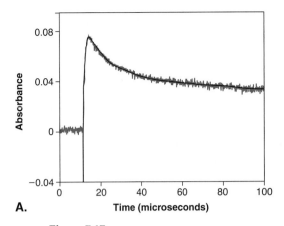
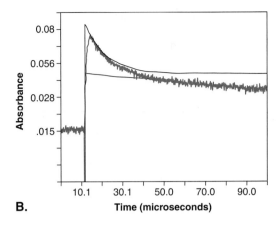

Figure 7.17
Fast kinetic traces. **A.** A single kinetic trace fits the data well, shown as a line through the data.
B. Two kinetic terms, a fast and slow component, are required to fit the data; both are shown as lines.

tiate the process with a photochemical reaction would severely limit the usefulness of flash photolysis techniques. In reality, though, clever chemistries have been developed that have allowed the photochemical generation of representative examples of all the major types of reactive intermediates. Some of our most detailed mechanistic insights have come from such studies.

A continuing theme of modern physical organic chemistry has been the quest for ever shorter pulses of light, allowing transients with ever decreasing lifetimes to be probed. Pioneering studies using flash lamps achieved millisecond and microsecond time resolution. The advent of the laser allowed shorter pulses that still delivered enough photons to generate sufficient quantities of intermediate for observation. Beginning with nanosecond (ns) pulses and moving on to picosecond (ps) pulses, the time scale of physical organic chemistry has been growing ever shorter. Now, as the Going Deeper highlight explains, even femtosecond chemistry is becoming a reality.

Going Deeper

Femtochemistry: Direct Characterization of Transition States, Part I

Throughout the 20th century it was a cornerstone of mechanistic analysis that we could never "see" a transition state. By definition, a transition state is not a stable point on a potential energy surface. Its lifetime is comparable to or less than that of a vibration, so how could we ever hope to see a transition state? Well, it turns out that vibrational times are on the order of ps (10^{-12} s); convince yourself of this by converting a typical IR stretch (3000–1000 cm^{-1}) to a time. What if we could do transient spectroscopy on a time scale that is faster than ps, namely in the femtosecond (10^{-15} s) time domain? What would we see?

Starting in the late 1980s, fs lasers became available, and it became possible to watch a reaction evolve on this time scale. What emerges is very much like a real-time movie of a chemical reaction, in which the transition state is just one "frame" of the film. Using pump–probe techniques with delays on the order of fs, we can watch as the starting material evolves into the transition state, and then follow the reaction further on to products. Early examples were for simple, prototype reactions such as the gas phase dissociation of NaI. However, as the methodology developed, more complex organic reactions have been studied, allowing direct insights into the transition states of the reactions. Perhaps with direct observation of a reaction transition state, we can claim to actually "prove" the mechanism of a specific reaction.

The 1999 Nobel Prize in Chemistry was awarded to Zewail for pioneering the field of femtochemistry. In its citation for the award the Nobel committee noted that in our quest for ever faster probes, "we have reached the end of the road: no chemical reactions take place faster than this."

Zewail, A. H. "Laser Femtochemistry." *Science*, **242**, 1641–1653 (1988); Pedersen, S., Herek, J. L., and Zewail, A. H. "The Validity of the 'Diradical' Hypothesis: Direct Femtosecond Studies of the Transition-State Structures." *Science*, **266**, 1359–1364 (1994).

Going Deeper

"Seeing" Transition States, Part II: The Role of Computation

In cases where femtochemistry is not readily applicable, we remain in the classical situation in which it is experimentally impossible to characterize a transition state. In this light, the development of reliable quantum mechanical methods for evaluating structures and energies of molecules, the so-called *ab initio* methods, has provided a real boon to mechanistic chemistry. These methods are discussed in detail in Chapter 14. Here we simply note that it is increasingly common to perform very high-level calculations on reactions of interest and to obtain valuable information. The calculations are certainly not infallible, and some caution is always in order when interpreting the results, but more and more, theory is providing important

mechanistic insights. The beauty of computational chemistry is that it is just as easy to determine the energy and geometry of a transition state as a ground state. We simply optimize to a structure with one and only one negative second derivative (eigenvalue—recall Section 7.1.3). We can then be sure we have a true transition state, and looking at the first derivative (eigenvector) can provide insights into the molecular motions along the reaction trajectory. Early computational models could not be trusted to perform such demanding tasks, but in the 21st century, the computational methods are becoming increasingly powerful. In the foreseeable future, we may see computation as the primary tool for evaluating the transition states of simple prototype reactions. More complex systems are likely to require experimental analysis for some time.

Pulse Radiolysis

One method of making and studying the reactions of radicals is pulse radiolysis. In this technique, a picosecond pulse of electrons is impinged into a solution, reducing various components therein. This leads to radical anions (Eq. 7.61) that can be monitored as a function of time. Decay traces similar to that shown in Figure 7.17 are generated, leading to rate constants for the reactions of these unstable species. It is another method for the rapid generation of highly reactive intermediates, and an example is given in the Connections highlight on the next page.

$$A \xrightarrow{\;e^{\ominus}\;} A^{\ominus}{}^{\cdot}$$

(Eq. 7.61)

7.6.3 Relaxation Methods

Whenever a chemical equilibrium is subjected to a perturbation, most commonly a change in temperature, pressure, pH, or other concentrations, the system will start to relax back to a new equilibrium state. The kinetics of this relaxation can be followed. Methods for quickly inducing a perturbation followed by monitoring the relaxation are referred to as **jump techniques**. Changes in temperature, pH, and pressure can often be done fast enough that reactions with half-lives in the microsecond range can be followed. For example, the equilibrium positions of Brønsted acid–base reactions are controlled by the pH, and therefore pH jump experiments are particularly useful with these reactions.

When the temperature of a system at equilibrium is changed by ΔT, the equilibrium concentrations will change, because $\ln(K_{eq}) = -\Delta G / RT$. The extent of change of the equilibrium constant as a function of the change in T is given by Eq. 7.62. The change in temperature can be induced quickly by a pulse of microwave radiation, by electric discharge in a conducting medium, or by the use of an iodine laser that emits at 1315 nm and excites O–H bond vibration overtones leading to heating.

$$\frac{\partial \ln K_{eq}}{\partial T} = -\frac{-\Delta H_{rxn}{}^{\circ}}{RT^2}$$

(Eq. 7.62)

Connections

The Use of Pulse Radiolysis to Measure the pK$_a$s of Protonated Ketyl Anions

The pK$_a$s of protonated ketyl anions can be determined using pulse radiolysis. For example, the addition of electrons into a solution of acetophenone causes the creation of the ketyl radical anion from the carbonyl. This radical anion can pick up a proton from solution, creating a ketyl radical that is in equilibrium with the ketyl anion. Monitoring the changes in absorbance spectra between the ketyl and ketyl anion (UV/vis spectra shown to the right) as a function of pH allows one to calculate the pK$_a$ of the ketyl (insert in the figure to the right). In this particular case, the pK$_a$ was found to be 10.5, significantly more acidic than a typical alcohol (pK$_a$ = 16–18). See Exercise 25 at the end of this chapter to consider the interpretation of the data shown in the graph.

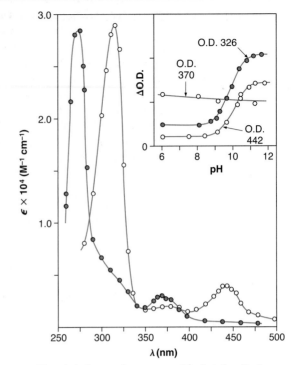

Acidity of ketyl radical to give ketyl anion

Closed circles are the spectra of the ketyl radical, while open circles are the spectra of the ketyl anion.

Hayon, E., Ibata, T., Lichtin, N. N., and Simic, M. "Electron and Hydrogen Atom Attachment to Aromatic Carbonyl Compounds in Aqueous Solution. Absorption Spectra and Dissociation Constants of Ketyl Radicals." *J. Phys. Chem.*, **76**, 2072 (1972).

7.6.4   Summary of Kinetic Analyses

In the preceding discussions we presented several mechanistic scenarios with the appropriate rate laws. Examples of data collected and the manner in which the data were used to support a particular mechanism were then discussed. Although kinetics is a useful mechanistic tool, several other pieces of experimental data typically go into analyzing a mechanism. For example, the kinetic data really only tell us what molecules are involved in a mechanism prior to and/or during a rate-determining step. As with some of the examples given above, the order of steps and the manner in which reactants combine could be determined. Importantly, kinetic results can be used to exclude mechanisms that are found experimentally to not conform to the predicted rate law. However, the kinetic results do not tell us the nature of any intermediates, nor do they indicate which bonds have been broken or formed during the reaction. To get this kind of information, we need to turn to other tools of physical organic chemistry, such as isotope effects and linear free energy relationships, as well as a myriad of other common techniques (see Chapter 8). However, to analyze in detail some of these other techniques, we need an increased understanding of energy surfaces and reaction coordinate diagrams. Therefore, at this stage we want to delve even deeper into rate constants and energy surfaces.

7.7 Calculating Rate Constants

Chemists have always had an intuitive sense that highly exothermic reactions should proceed more readily than only slightly exothermic reactions. It just seems to make sense that if it is energetically way downhill to products, it should be easier to get there. However, we have emphasized several times in this text that it is important to keep thermodynamic and kinetic distinctions clear. Detailed theoretical analysis of many reactions has revealed that the simple intuitive view is risky. In fact, in one of the most stunning predictions in the history of chemistry, Marcus predicted that in some cases, increasing the exothermicity of a process should *slow the reaction!* Decades of investigation by experimentalists finally verified this now famous prediction of Marcus theory, leading to Rudy Marcus receiving the 1992 Nobel Prize in Chemistry. Here we introduce the conceptual basis of Marcus theory and describe one simple example of its application. In practice, though, Marcus theory has had its largest impact in reactions involving electron transfer processes, as briefly described below.

7.7.1 Marcus Theory

Remarkably, one of the most profound theories of chemistry can be appreciated by simply considering certain geometrical features associated with parabolas. Why parabolas? Recall that in Section 2.1.4 we described the Morse potentials for bond stretching, and we noted that near the equilibrium bond position, these look very much like parabolas. In the molecular mechanics method these Morse potentials are often modeled using the equation for the potential energy involved in stretching and compressing a spring, which is a parabolic function. Furthermore, in Chapter 2 we examined the potential energy diagrams for rotamers. The depressions in these diagrams have a resemblance to parabolas, too. Taking this to its extreme, we consider that the depressions in energy surfaces such as Figure 7.1 can be approximated in two dimensions by parabolas. We consider a chemical reaction to be a transit from one surface minimum to another, or in two dimensions from one parabola to another. We show this in Figure 7.18. Note that the intersection of the two parabolas, then, is a good approximation to the transition state for the reaction.

The reaction we choose for demonstration purposes is the displacement of bromide from methyl bromide by hydroxide to form methanol—a classic S_N2 reaction. When using parabolic functions to mimic the Morse potentials for the C–Br or C–O bond stretches in methyl bromide or methanol, we would draw the respective potential energy diagrams as shown in Figure 7.19 **A**. Now let's consider that a chemical reaction depends upon the overlap of vibrational modes, and that these modes become coupled and interconvert from one to another during a reaction. Using Marcus theory, the shapes and intersection of the parabolas are determined by the free energy change and the intrinsic barrier ($\Delta G_{int}^{\ddagger}$).

The **intrinsic barrier** is an average of the barriers for the **self-exchange reactions**. For example, the activation free energy for hydroxide displacing hydroxide from methanol is 41.8 kcal/mol, and the activation free energy for bromide displacing bromide from methylbromide is 23.7 kcal/mol. These reactions are defined as the self-exchange reactions, and they give the $\Delta G_{int}^{\ddagger}$ to be 32.8 kcal/mol [(41.8 + 23.7)/2 = 32.8—note that these values are for reactions in the gas phase].

We start our parabola analysis at this point by drawing two parabolas that intersect at a barrier of 32.8 kcal/mol (Figure 7.19 **B**). Now the experimentally measured free energy change is brought in to play (in this case, $\Delta G° = -23.4$ kcal/mol). The parabola on the right is lowered by this free energy change, and the new point of intersection of the parabolas is defined as the transition state (Figure 7.19 **C**). The intersection point is calculated using the Marcus equation (Eq. 7.63), which was derived from the geometry of intersection of two parabolas. In the case of hydroxide displacing bromide from methyl bromide, an activation free energy ($\Delta G^{\ddagger}$) of 22.1 kcal/mol is found, extremely close to the experimentally determined value of 22.7 kcal/mol. Substituting the calculated free energy change into the Eyring equation (Eq. 7.15) allows us to calculate a rate constant for this reaction at different temperatures without a kinetic experiment.

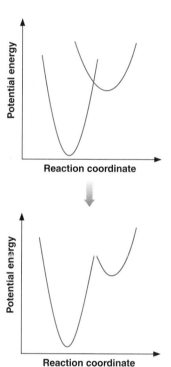

Figure 7.18
The intersection of two parabolas representing Morse potentials defines the shape of the reaction coordinate.

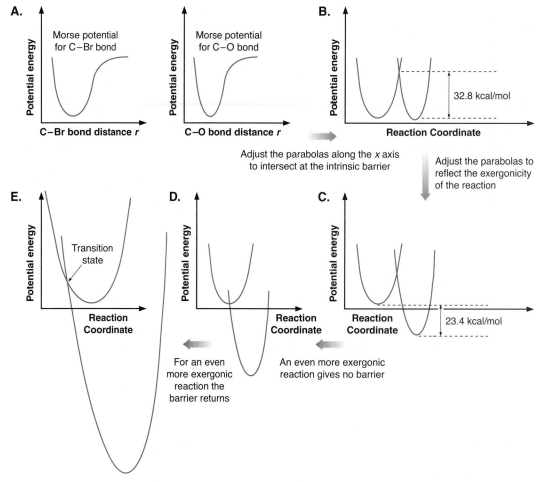

Figure 7.19
Diagram of the procedure used in Marcus theory to calculate rate constants (A–C is for the S_N2 displacement of bromide from methyl bromide by hydroxide; D and E are hypothetical curves that show how an inverted region would be predicted). **A.** Just the Morse potentials themselves for the two bonds, one in the starting material (C–Br) and one in the reactant (C–O). **B.** The overlay of the two Morse potentials, now modeled as parabolas, so that they intersect at the intrinsic barrier. **C.** This curve shows how the real barrier is found by lowering the product to reflect the free energy of the reaction. **D.** A general case where the exergonicity of the reaction leads to no barrier for the reaction. **E.** Further increasing the exergonicity of the reaction again introduces a barrier.

$$\Delta G^{\ddagger} = \Delta G_{int}^{\ddagger} + \frac{\Delta G^{\circ}}{2} + \frac{(\Delta G^{\circ})^2}{16\Delta G_{int}^{\ddagger}} \qquad (\text{Eq. } 7.63)$$

Besides solving for the intersection point of the two parabolas along the energy axis (the y axis) of the reaction coordinate diagrams, we can also find the intersection along the structure axis (the x axis). This allows us to follow the position of the transition state as a function of the free energy change for the reaction. The intersection point is given by Eq. 7.64. Making the reaction increasingly exergonic (a larger negative ΔG°) moves the position of the transition state toward the reactants, while making the reaction increasingly endergonic (a larger positive ΔG°) moves the transition state toward the products. These are exactly the same conclusions given by the Hammond postulate.

$$x^{\ddagger} = \frac{1}{2} + \frac{\Delta G^{\circ}}{8\Delta G_{int}^{\ddagger}} \qquad (\text{Eq. } 7.64)$$

Thus far, everything seems sensible. We make the reaction more exergonic and it goes faster. However, consider the consequences of making the reaction still more exergonic. At some point, we would have no barrier at all (Figure 7.19 **D**), but then going more exergonic would reintroduce a barrier! From here on, greater exergonicity implies a *slower* reaction (Figure 7.19 **E**). The novel finding is called the **Marcus inverted region**, the regime in which increasingly negative free energy slows the reaction. This is the signature finding of Marcus theory, and it launched a wide range of experimental efforts to find the inverted region and verify this prediction. As discussed in the Connections highlight below, it was in the study of electron-transfer reactions that convincing evidence for an inverted region, and thus confirmation of Marcus theory, appeared.

More advanced theoretical treatments have refined the version of Marcus theory presented here. Most important has been the inclusion of a tunneling term. It should be clear from Figure 7.19 that as the barrier height gets smaller, the barrier width also decreases. As discussed in the next chapter, tunneling probabilities increase dramatically as the barrier width decreases, and so a tunneling correction is generally included in a full treatment of such systems. The basic conclusions, including the existence of an inverted region, remain the same.

There are not many systems for which the self-exchange reactions are easily defined, or for which the thermodynamics of the self-exchange reactions have been well studied. Therefore, for more complex reactions of bioorganic, synthetic organic, and organometallic importance, Marcus theory is not directly applicable. However, one point of the analysis that we want to stress is the resulting view of an energy surface. To this point in the text we have shown that the depth of each valley represents the stability of the respective structure, and the height of the activation barrier represents the lowest energy pathway for interconverting the individual structures. Now we are adding the idea that the shapes of the valley are related to the bond vibrations of the molecules represented by the valley. The shape of the valley (flat, shallow, steep, or deep) is a composite of the potential energy curves for each degree of freedom the molecule possesses.

7.7.2 Marcus Theory Applied to Electron Transfer

Without a doubt, the discipline where Marcus theory has had the most impact is in the study of electron transfer, both in chemistry and biology. Electron transfers are ubiquitous in chemistry, being involved in electrochemistry, redox reactions, many enzymatic reactions, and photosynthesis. Furthermore, many classic organic reactions have now been shown to have an electron-transfer component (see SET reactions in Chapter 11 and exciplexes in Chapter 16).

Electron transfers occur from electron donors (D) to electron acceptors (A). These transfers occur much faster (10^{-16} s) than nuclear vibrations (10^{-13} s). Therefore, the nuclei do not change position during the time of the electron transfer. During the transfer, the electron does not change energy. In other words, the energy of the donor and acceptor orbitals must be the same prior to transfer. We have noted in Chapters 2 and 3 that reactants and the solvent are dynamic in structure. The energy levels of the donor and acceptor orbitals in the reactant and product are in continual flux due to internal nuclear movements and the solvent motions. For transfer, the donor and acceptor molecules must simultaneously achieve particular geometries and solvation arrangements that give matched energy levels between the donor and acceptor orbitals. After electron transfer, the nuclei of the donor and acceptor molecules relax to their optimum positions. The energy required to change the solvation sphere and internal structures bringing the donor and acceptor orbitals to the same energy is called the **reorganization energy**. This energy is what creates a barrier to the electron transfer.

In the calculation of rate constants for electron transfers, the Marcus equation takes a slightly different form than given in Eq. 7.63. As shown in Eq. 7.65, there is still a quadratic dependence upon the free energy change for the reaction, but now there is a new term, λ, which is a value that reflects the required reorganization energy. This term takes into account the rearrangement of the system of reactants and solvents discussed above that are neces-

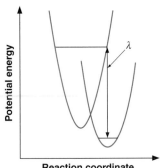

Reaction coordinate

Reorganization energy λ is the difference in energy between donor and acceptor orbitals that must be achieved by molecular distortions and/or solvent reorganization so that the orbitals become equal in energy

sary for the electron transfer to commence. The reorganization energy is defined as the energy that needs to be put into distorting either the reactant–solvent or product–solvent ensemble, or into both, to make the energy of the donor and acceptor orbitals the same. Hence, λ has two components (Eq. 7.66). The first is the internal reorganizational energy λ_i that measures the energy difference due to changes in bond lengths, angles, etc., which occur upon electron transfer. The second term, λ_o, measures the energy involved in reorganization of the solvent shell required for electron transfer. The quadratic dependence of $\Delta G^\ddagger$ on ΔG° still leads to the prediction of the Marcus inverted region. It is in the area of electron transfer that the predicted inverted region has been definitively discovered. One of the pioneering examples of this is given in the next Connections highlight.

$$\Delta G^\ddagger = \frac{\lambda}{4}\left[1 + \frac{(\Delta G^\circ)^2}{\lambda}\right]$$
(Eq. 7.65)

$$\lambda = \lambda_i + \lambda_o$$
(Eq. 7.66)

Connections

Discovery of the Marcus Inverted Region

One of the earliest confirmations that a reaction barrier can increase as the reaction becomes more exergonic came from a study of electron transfer. Shown to the right is a plot of rate constants for electron transfer from the biphenyl radical anion to a variety of acceptors (each electron acceptor is shown across the plot near their respective rate constants). In this study, the electron transfer occurs across a rigid steriod scaffold. In all cases the donor and acceptor are held at the same distance, but the reduction potential of the acceptor decreases across the x axis, making the reaction increasingly exergonic from left-to-right in the graph. The biphenyl radical anion was generated by pulse radiolysis (see discussion earlier in this chapter), and the rate of electron transfer to the acceptor was measured using fast kinetics techniques. As can be seen, the rate constant increases, and then actually decreases, even though the reaction is becoming increasingly exergonic.

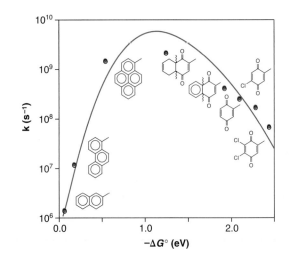

Miller, J. R., Calcaterra, L. T., and Closs, G. L. "Intramolecular Long-Distance Electron Transfer in Radical Anions. The Effects of Free Energy and Solvent on the Reaction Rates." *J. Am. Chem. Soc.*, **106**, 3047 (1984). Closs, G. L., and Miller, J. R. "Intramolecular Long-Distance Electron Transfer in Organic Molecules." *Science*, **240**, 440 (1988).

Electron-transfer reaction used to uncover the inverted region

7.8 Considering Multiple Reaction Coordinates

For predicting and discussing chemical reactivity, energy surfaces as in Figure 7.1 are hopelessly complex. Therefore, as previously discussed, we resort to drawing cross-sections of the surface that are two-dimensional, and we refer to these cross-sections as reaction coordinate diagrams (RCDs). One axis represents energy, and the other is a geometrical parameter that defines the progress of the reaction, called the reaction coordinate. When sketching an RCD without prior knowledge of the energy surface, which is by the far the most common situation, we draw upon the Hammond postulate to predict the position of the transition state along the reaction coordinate axis.

In more complicated mechanistic situations, however, such an approach is often unsatisfactory. Frequently, there are two feasible paths by which a reaction might occur. We could just plot two separate RCDs and see which has the higher barrier. However, this will often miss important features that arise because of the *interaction* of the two separate reaction paths. It is then necessary to consider three-dimensional RCDs, containing two reaction coordinates plus the energy. Again, of paramount importance is an analysis of where the transition state lies on the energy surface and its corresponding projection along the reaction coordinate. As we will see, such an analysis can reveal structural changes in the transition state that cannot be appreciated from simple two-dimensional RCDs and the Hammond postulate.

7.8.1 Variation in Transition State Structures Across a Series of Related Reactions—An Example Using Substitution Reactions

We stated above that insight into chemical reactions can be gained by considering multiple reaction paths and their potential interactions. This is easiest to see using an example. Consider the energy surfaces and reaction coordinates for an exergonic nucleophilic substitution reaction for a series of substituted alkyl groups (Figure 7.20). In this figure one axis represents the breaking of the bond between the carbon atom and the leaving group (x axis), and a second axis represents the formation of the bond between the carbon atom and the nucleophile (y axis). The third axis corresponds to free energy (z axis). A reaction path that involves first movement along the x axis and then movement along the y axis would describe an S_N1 reaction. Because the S_N2 reaction involves both of these bonding changes simultaneously, its reaction coordinate is diagonal to these two axes. To simplify our example, we assume there is no change in $\Delta G°$ as a function of the alkyl groups. To consider how alkyl substitution affects the reaction surface, we consider only variation in the heights of the transition states for the S_N2 reaction, and the relative stabilities of the carbocation intermediates in the S_N1 reaction. Recall that a second order nucleophilic substitution becomes more difficult as the steric bulk around the bond between the carbon atom and the leaving group increases. On the other hand, we lower the energy of the carbocation involved in the S_N1 reaction as we progress from methyl to primary to secondary to tertiary (see Chapter 11 for further discussion). Making these two incremental changes leads to the energy surfaces shown in Figures 7.20 **A**, **B**, **C**, and **D**.

How do the changes in these two parameters affect the possible reaction coordinates and the structures of the transition states? Diagonally across the diagram from the rear left point to the front right point is the idealized reaction coordinate for the S_N2 pathway. The reaction coordinate for the S_N1 pathway starts from the rear left point and proceeds to the front left point and then over to the right front point. In Figure 7.20 **A** the lowest energy pathway is the S_N2 reaction, because the methyl carbocation is so high in energy. Examining the energy surface for the departure of a leaving group (LG) from ethyl leads to a similar conclusion (Figure 7.20 **B**). The energy of ethyl cation, and the transition state leading to it, are still higher than the transition state for the S_N2 displacement. Importantly, however, the lowering of the energy of the structure at the front left corner has affected the overall shape of the potential energy surface. In particular, the S_N2 transition state has been "pulled" toward the front left corner. This means that the S_N2 activated complex has developed more carbocation character with ethyl–LG relative to methyl–LG. Now consider Figure 7.20 **C**. The energy of

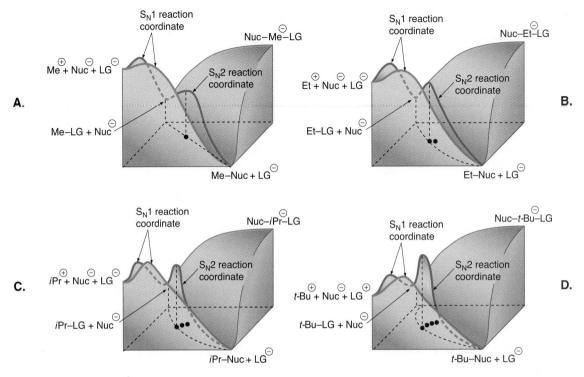

Figure 7.20
Energy surfaces for S_N1 and S_N2 reaction coordinates as a function of different R groups. X-axis (in and out of paper) represents R-LG bond breaking, y-axis (horizontal) represents R-Nuc bond formation, and z-axis (vertical) represents free energy. The contours on the energy surface are simplified from reality for sake of clarity. The reaction coordinates for both the S_N1 and S_N2 pathways are really in troughs that lead from one energy well to the next.

the S_N2 transition state has gone up again, and the energy of the carbocation has decreased again. At this point the transition state for the S_N2 reaction is comparable in energy to the transition state leading to the secondary carbocation. We predict two competing mechanisms for substitution; either the S_N2 or S_N1 pathway is possible. However, the S_N2 pathway has developed considerable carbocation character at its transition state. Lastly, the tertiary carbocation and the transition state leading to it are lower in energy than the transition state for the S_N2 reaction. The mechanism has completely changed to S_N1, involving a fully formed carbocation intermediate. At the same time, the transition state for the S_N2 pathway has considerable carbocation character.

The prediction of this analysis is that as the alkyl group changes from methyl to primary to secondary to tertiary, the activated complex for the S_N2 pathway develops increasing carbocation character. This is because perpendicular to the reaction coordinate for the S_N2 reaction, the S_N1 pathway is becoming progressively stabilized. This drop in energy warps the entire potential energy surface, and the S_N2 transition state feels this. This analysis is consistent with our chemical intuition, and we will see in Chapter 11 that it is also consistent with experiment.

The three-dimensional reaction coordinate diagram also provides insight into another important concept in reaction mechanisms, that of an **enforced mechanism**. The mechanism that a reaction uses is often dictated by the stability of possible intermediates. The interplay between S_N2 and S_N1 reactions depends upon the stability of the carbocation involved in the S_N1 reaction. When the stability of the carbocation and the transition state leading to it are lower in energy than the transition state for the S_N2 pathway, the S_N1 pathway dominates, as shown in the three-dimensional energy surface. In contrast, envision a case where the carbocation is so unstable that it has no lifetime; it reacts faster than the bond vibration that leads to leaving group departure. If so, the nucleophile must add while the leaving group is departing. In this case, the mechanism is forced to be S_N2. Enforced mechanisms are involved in additions to carbonyls, additions to alkenes, and eliminations to form alkenes. Some of these will be examined in Chapters 10 and 11.

In our comparison of hypothetical S_N2 and S_N1 reactions, the structure of the activated complex for the S_N2 reaction changed as we varied the alkyl group in a manner not immediately apparent by examining a two-dimensional RCD. In fact, the structures of activated complexes for most chemical reactions vary in a manner not immediately apparent by looking at just the structures of the reactant and product. To get a full picture of how the structure of the activated complex changes as we change the structures of the reactants and products, we need to examine the energies of any possible competing pathways, and how our changes in reactant and product structure affect these competing pathways. We do this by examining the projections of the three-dimensional RCDs onto a two-dimensional surface. Now, however, the two axes are not one reaction coordinate plus an energy, but rather are two competing reaction coordinates. This may sound complicated, but we can succeed by applying two simple rules described in the following section.

7.8.2 More O'Ferrall–Jencks Plots

In Figure 7.21 we show projections of the surfaces in Figure 7.20 of the sort we just described; both axes represent an individual reaction coordinate. In other words, these are what one would see if the surfaces of Figure 7.20 were viewed directly from the top with no topographical features included. The diagonal lines represent the S_N2 reaction paths, and the positions of the transition states are marked with a heavy dot. These two-dimensional projections are known as **More O'Ferrall–Jencks plots** (also known as More O'Ferrall plots, or More O'Ferrall–Albery–Jencks plots). As substitution on the alkyl group increases, the S_N2 reaction paths develop more curvature, and the transition state moves toward the carbocation quadrant of the plot. Each solid line (linear and curved) in these diagrams would be

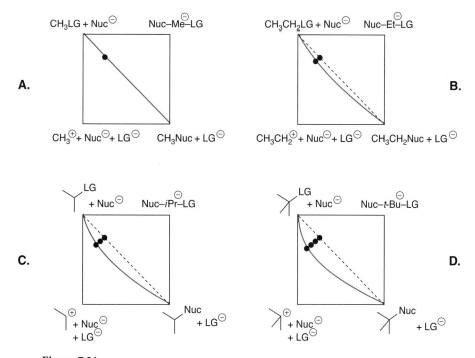

Figure 7.21
Projections of the energy surfaces of Figure 7.20, showing how the transition state for the S_N2 reaction shifts toward carbocation character with larger R groups.

the reaction coordinate axis (*x* axis) of a conventional two-dimensional RCD that has energy as the *y* axis. Conventional RCDs can not reveal the curvature representing the increase of carbocation character at the transition state. Examining the placement of the transition state in these projections along each axis, we find that the carbon–leaving group and the carbon–nucleophile bond distances both increase as the alkyl group becomes more sterically crowded.

If we draw a perpendicular line from the diagonal to each new transition state, all such lines will intersect the diagonal at the same place. This is because in our hypothetical series of substitution reactions we did not allow $\Delta G°$ to change as a function of the alkyl group. In our discussion of the Hammond postulate and of Marcus theory, we showed that the transition state moves along the reaction coordinate as a function of $\Delta G°$. Therefore, the transition state moves along the diagonal as a function of the free energy of the reaction in a More O'Ferrall–Jencks plot, and moves perpendicular to the diagonal as a function of the energy of competing pathways.

The movements of the transition state along the diagonal and perpendicular to the diagonal can be predicted by two rules. Here, the diagonal represents a reaction coordinate where we have not yet considered perturbation due to changes in structure that influence the competing pathways and $\Delta G°$. The rules are as follows:

1. *Along the diagonal the transition state shifts toward the corner that is raised in energy and away from the corner that is lowered in energy (referred to as a* **Hammond effect**).

2. *Perpendicular to the diagonal the transition state shifts toward the corner that is lowered and away from the corner that is raised (referred to as an* **anti-Hammond effect**).

Quite often we find that a structural change in the reactant affects more than just one corner of a More O'Ferrall–Jencks plot. Often an entire side of a plot is considered to be raised or lowered in energy. When this occurs the vector sum of the two changes in transition state structure derived from our two rules gives the real change of the position of the transition state on the diagram. For example, consider the two scenarios shown in Figure 7.22. With Figure 7.22 **A** the reactant and the upper right corner are raised in energy. The result is that the position of the transition state along the *y* axis (vertical) does not change appreciably, but instead changes only along the *x* axis (horizontal). Similarly, Figure 7.22 **B** shows a scenario where the reactant is made more stable, as is the structure represented by the lower left corner. Now the position of the transition state along the *x* axis does not change appreciably, but instead changes along the *y* axis. This combination of two changes in transition state structure is quite common, and we will return to it in Chapter 8 when we discuss linear free energy relationships, in Chapter 9 when we discuss acid–base catalysis, and in Chapter 10 when we discuss elimination reactions. In the next Connections highlight, an example of using these plots is given in the context of catalysis.

Raise corner   Raise corner

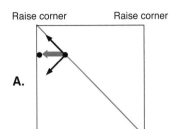

A.

Lower corner

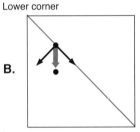

B.

Lower corner

Figure 7.22
How movement along a More O'Ferrall–Jencks plot can be the composite of two movements, one perpendicular and one parallel to the reaction coordinate. The arrows show the directions predicted by the discussion in the text, while the colored arrow shows the composite response.

Connections

Using a More O'Ferrall–Jencks Plot in Catalysis

The catalytic cleavage of phosphodiesters is an area of intense investigation due to the potential for the site-directed cleavage of these esters in DNA and RNA. Dinuclear metal complexes are currently among the best catalysts for phosphodiester hydrolysis. One motif that has been found to be successful is the bridging of two metals with hydroxide ligands, such as in *i*. This bis(cobalt) complex has *X*-phenylmethylphosphate in its coordination sphere, where *X* can be different substituents. A

proposed mechanism for cleavage of this phosphoester involves intramolecular nucleophilic attack of a bridging oxide on the phosphorus center with departure of a phenoxide leaving group. The corresponding uncatalyzed reaction involves nucleophilic attack of free hydroxide in solution on the phosphorus center with phenoxide departure.

It was found by experimental methods that the uncatalyzed reaction had about 37% leaving group departure. This means that the bond between P and the leaving group is 37% broken at the transition state. Further, there was a significant amount of bond formation between the nucleophile and phosphorus at the transition state. However, the catalyzed reaction was found to have about 77% leaving group departure at the transition state. The percentage of leaving group departure in the transition states was measured using a linear free energy relationship called β_{LG}, which is covered in Chapter 8.

It seems odd that the catalyzed pathway involving the better nucleophile would have a greater extent of leaving group departure in the transition state. One might anticipate that the better nucleophile would be able to attack the phosphorus with less requirement for simultaneous leaving group departure.

To understand why the activated complex structure has more bond cleavage to the leaving group in the catalyzed pathway than in the uncatalzyed pathway, a More O'Ferrall–Jencks analysis was employed (note that the diagonal is from bottom left to top right, in contrast to that previously shown in Figures 7.21 and 7.22). A dot was placed on the diagram as shown to designate the uncatalyzed reaction. This diagram was then applied to the catalyzed reaction, with the y axis representing nucleophilic attack of the oxide on the phosphorus and the x axis representing leaving group departure. Nucleophilic attack results in formation of two four-membered rings, thereby introducing significant strain into the structures. The structures of the compounds designated by the two top corners thus become higher in energy. Using the two rules for predicting movement of the transition state (previously discussed in this section), it was concluded that along the reaction coordinate the transition state moves toward the upper right, while perpendicular to the reaction coordinate the transition state moves toward the lower right. The vector sum of these two effects keeps the

extent of nucleophilic attack similar in the catalyzed and uncatalyzed reactions, but increases the extent of leaving group departure at the transition state for the catalyzed reaction. Thus, the use of the More O'Ferrall–Jencks analysis shows how leaving group departure can become greater even though a better nucleophile is involved.

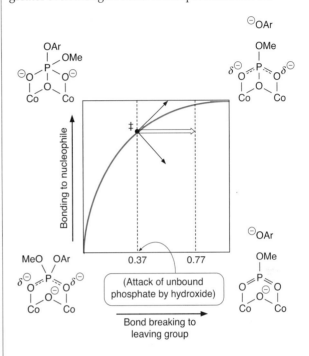

Williams, N. H., Cheung, W., and Chin, J. "Reactivity of Phosphate Diesters Doubly Coordinated to a Dinuclear Cobalt(III) Complex: Dependence of the Reactivity on the Basicity of the Leaving Group." *J. Am. Chem. Soc.*, **120**, 8079–8087 (1998).

Catalyzed reaction sequence

Uncatalyzed reaction sequence

In summary, we find that the structures of activated complexes are actually influenced by factors other than just the structures of the reactants and products. We need to keep this in mind when designing experiments to probe mechanistic postulates, especially in cases where we are considering the potential roles of two competing mechanisms. We may make a structural change in our reactant in order to test a theory, and thereby affect the transition state structure in a manner that we did not predict, leading to incorrect conclusions. In the extreme, we may even change the mechanism, such as with our example of comparing S_N2 and S_N1 reactions as a function of changing the alkyl group structure. We will give you examples to practice your predictive powers using More O'Ferrall–Jencks plots in the Exercises at the end of this and other chapters.

7.8.3   Changes in Vibrational State Along the Reaction Coordinate— Relating the Third Coordinate to Entropy

Up to this point our analysis of energy surfaces has concentrated on the relative energies of the reactant, transition state, intermediates, and product. Further, we analyzed the position of the transition state along the reaction coordinate relative to the reactant and product and relative to the structures of intermediates in competing pathways. We have not examined in detail the shapes of the potential energy wells that represent the reactant, intermediate, and product. However, an important factor in understanding kinetics, activation parameters, and isotope effects (see Chapter 8) is the manner in which the shapes of the potential energy wells change as the reaction proceeds.

In our analysis of Marcus theory, the potential energy wells that represent the structures of reactants and products on a reaction coordinate diagram were modeled using parabolas. An analogy between parabolas and the shapes of various potential energy plots for bond vibrations was drawn. The actual shape of each potential energy valley can be best considered as a composite of all the different vibrational modes of the molecule represented by the valley. The idea of a composite of vibrational states is hard to visualize; it implies a multidimensional surface where each potential energy well has a dimension representative of each degree of freedom for the molecule.

When using an energy surface to describe a reaction, however, the surface can be considerably simplified. Most vibrational modes for a molecule do not change significantly during a reaction. Only the modes associated with the stretches, bends, wags, etc., of the bonds that are actually changing have a significant influence on the reaction. Therefore, we concentrate on the vibrational modes that are undergoing changes along the reaction coordinate when analyzing the different shapes of the potential energy wells on the energy surface.

To understand how changes in the shapes of the valleys affect the kinetics of a reaction, let's return to the analogy we drew at the very beginning of the chapter between energy surfaces and mountain ranges. Imagine that you are guiding a herd of cattle from one valley to another over a mountain pass. A series of possible passes is shown in Figure 7.23. In Figure 7.23 **A** the mountain pass has a shape similar to the valley in which the cattle are starting out, and hence the cattle do not need to be more or less corralled and organized to climb the mountain. Equating "corralling cattle" with diminishing degrees of freedom, chemically speaking, the entropy of the reactants in Figure 7.23 **A** is similar to the entropy of the activated complex, and $\Delta S^\ddagger$ is near zero. In Figure 7.23 **B**, the shape of the mountain pass becomes wider perpendicular to the trail as the cattle climb the mountain. Thus, the cattle can become more spread out as they ascend the mountain range. In this analogy, the entropy of the transition state is increasing as the molecules traverse the barrier, and $\Delta S^\ddagger$ is positive. Finally, in Figure 7.23 **C**, the mountain pass is quite narrow at the top and therefore the cattle must become lined up more in a single-file fashion to climb the mountain. Analogously, the molecules are becoming more organized, and the entropy is decreasing as the transition state is achieved, so $\Delta S^\ddagger$ is negative.

Hence, the shape of the mountain pass is representative of the entropy changes along the reaction coordinate. The entropy changes are in turn related to changes in vibrational states (bond stretches, bends, wags, etc.) of the molecular structures. If vibrations (degrees of

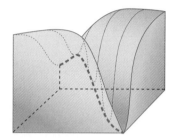

A.

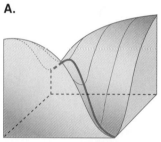

B.

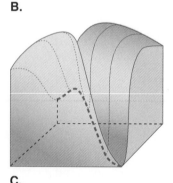

C.

Figure 7.23
Various energy surfaces showing different shapes of the valleys at the transition state relative to the reactant.
A. Relative shape does not change between reactant and transition state, so the entropy of activation is near zero.
B. The valley opens up at the transition state, so the entropy of activation is positive.
C. The valley constricts at the transition state, so the entropy of activation is negative.

freedom) are becoming more restricted in the activated complex relative to the reactant, $\Delta S^{\ddagger}$ will be negative and the energy surface will be similar to that of Figure 7.23 **C**. If the degrees of freedom of the molecules are increasing or becoming looser, then $\Delta S^{\ddagger}$ will be positive and the energy surface will resemble that of Figure 7.23 **B**.

Pushing our cattle analogy a bit further (perhaps a bit too far), imagine a case in which a narrow pass is significantly lower than an alternative broad pass. Under most circumstances, our leisurely ranchers would take their cattle over the lower pass, as this expends less energy. Imagine, however, a situation in which the herd is suddenly very excited, and it is imperative to get all the herd over the mountains as fast as possible—presumably some rustlers are approaching. Now, it may be worth it to spend the extra energy to go through the higher pass, because the greater width will allow more cattle to get through at any one time. This "excitation" of the herd is the analogue of temperature, and as it is elevated the broader pass becomes more attractive. In much the same way, at elevated temperatures $\Delta S^{\ddagger}$ plays an increasingly important role ($\Delta G^{\ddagger} = \Delta H^{\ddagger} - T\Delta S^{\ddagger}$), and so transition states in broader wells are more accessible at elevated temperatures.

Summary and Outlook

This chapter has concentrated on energy surfaces and kinetic analyses. The energy surfaces were presented as a means to understand the kinetics of reactions and how transition state structures change as a function of reactant, product, intermediates, and competing pathways. Further, now that the shapes of the valleys and passes in an energy surface have been tied to vibrational states of the molecules, we are ready to examine isotope effects, an experimental tool that has its origin in these changes in vibrational states. Moreover, now that our chapters on both thermodynamics and kinetics are behind us, we can describe experimental methods that rely on both thermodynamic and kinetic measurements—namely, linear free energy relationships. Hence, Chapter 8 continues to build up our repertoire of methods for deciphering reaction mechanisms.

Exercises

1. Using rate equations, derive Eq. 7.24 for the Curtin–Hammett principle. Discuss under what experimental conditions the ratio of products would be dominated by the activation barriers to the two reactions, and under what experimental conditions the ratio of products would be dominated by the stabilities of the respective products.

2. Can you envision a modification of the curve shown in Figure 7.11 that would cause the relative energies of the intermediates to dictate the product ratio, and not the relative sizes of the barriers for formation of the products? Redraw Figure 7.11 in a fashion that would indicate such a case, and discuss whether the reaction would have to be under kinetic or thermodynamic control for the relative energies of the intermediates to dictate the product ratio.

3. Show that $t_{1/2}$ for a first order reaction equals $0.693/k$.

4. Derive the integrated rate expressions for the reactions given in Eqs. 7.27 and 7.28.

5. The decomposition of nitrogen dioxide is a second order reaction with rate constants as follows: $522\,M^{-1}\,s^{-1}$ at 592 K, $755\,M^{-1}\,s^{-1}$ at 603 K, $1700\,M^{-1}\,s^{-1}$ at 627 K, and $4020\,M^{-1}\,s^{-1}$ at 652 K. Calculate E_a and $\log A$. What do these values tell you about the reaction?

6. Reduce Eq. 7.41 to Eq. 7.42 using the assumption that we are following the reaction close in time to the formation of equilibrium.

7. What is the form of Eq. 7.41 if we start the reaction initially with no B and the k_r rate constant is much smaller than the k_f rate constant? What form does the rate expression best resemble?

8. Let's examine E2 eliminations with More O'Ferrall–Jencks plots. We also examine similar scenarios in Chapter 10.

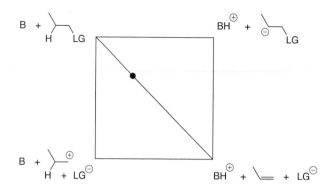

a. How would the position of the transition state change as the reactant changes from *n*-propyl-LG to isopropyl-LG?
b. How would the position of the transition state change as the base switches from ethoxide to diisopropylamide?
c. How would the position of the transition state change as the leaving group changes from chloride to iodide?

9. Imagine a hypothetical reaction with two possible mechanistic paths, one with a less positive $\Delta H^\ddagger$ and a negative $\Delta S^\ddagger$, and the other with a positive $\Delta S^\ddagger$ but a larger positive $\Delta H^\ddagger$. How could this produce a "kink" in an Eyring plot? Sketch what the Eyring plot would look like for such a case.

10. In a typical pump–probe flash photolysis experiment, the time delay between the pump and probe pulses is set by some type of electronic trigger device. However, in femtochemistry such a strategy is not possible—femtosecond electronic triggers don't exist. Instead, the delay is set by splitting a single laser pulse into two beams, and then using mirrors to have the two beams travel different distances before arriving at the reaction region. The one arriving first is the pump pulse, the one arriving second is the probe pulse. Consider an experimental set-up in which the pump pulse travels exactly 1 meter from the laser source to the reaction vessel. How long would the path for the probe pulse have to be in order to have a 10 fs delay between pump and probe?

11. The decomposition of dinitrogen pentoxide shows the following decrease in concentration as a function of time. What are the reaction order and the rate constant?

Time (s)	$[N_2O_5]$ (M)
0	0.0165
600	0.0124
1200	0.0093
1800	0.0071
2400	0.0053
3000	0.0039
3600	0.0029

12. What are the rate law and integrated rate law for the ring-opening metathesis polymerization of norbornene given in the Going Deeper highlight on page 393?

13. Consider the following data:

Recall the mechanism of aromatic sulfonation from your introductory organic chemistry course, or look at Chapter 10. Now draw two reaction coordinate diagrams on the same plot that show the relative energies of organic intermediates and products. Place the reactant at the center and the products at the left and right. Explain why the ratio is different at different temperatures, why one transition state is more stable than another, and why one product is more stable than another.

14. Recall from introductory organic chemistry that R groups are ortho–para directing in electrophilic aromatic substitution. Why, then, is the product of the reaction of excess EtBr and $AlCl_3$ with benzene the 1,3,5-triethylbenzene derivative after extended reaction times?

15. The reaction of the epoxide shown below with *cis*-stilbene gives the product shown. There are two reasonable mechanisms. Each involves different intermediates that have substituted benzyl cations and alkoxide anions. Write these two mechanisms, derive rate laws for the mechanisms, and describe how they can be distinguished using kinetics. What exact experiments would you perform?

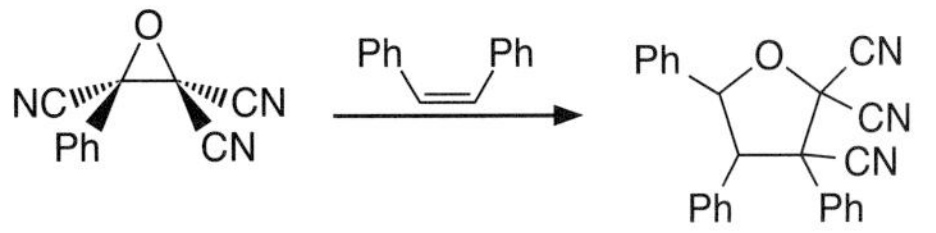

16. The reaction of benzoyl chloride with ethanol and pyridine to form ethyl benzoate and pyridinium chloride can proceed either by nucleophilic attack of the ethanol or pyridine in the first step. With nucleophilic attack by pyridine, an acyl–pyridinium species forms, so this kind of reaction amounts to nucleophilic catalysis by this added base. If ethanol attacks first, the pyridine only acts to scavenge the HCl formed in the reaction. Write these two possible mechanisms, and derive rate laws that distinguish these possibilities.

17. In the following reaction, it was found by isotope scrambling experiments that the acid and the water add to the alkene in a single step. Hence, the More O'Ferrall–Jencks plot shows a diagonal that represents this single-step process. The other possible competing mechanisms involve intermediates placed on the other corners of this diagram. Explain what happens to the extent of protonation in the transition state of the single-step process if the strength of the acid is increased. What happens to the extent of nucleophilic attack at the transition state as the acid strength is increased? Does your result seem reasonable? Explain.

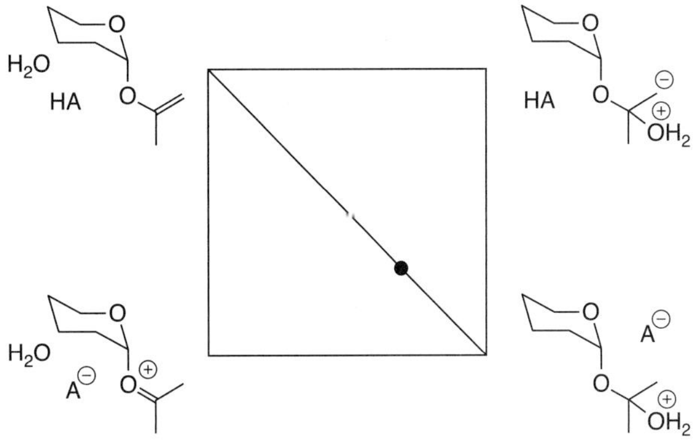

18. The following reaction is first order in the metallocyclobutane. The reaction is first order in diphenylacetylene at low concentrations, but becomes zero order in diphenylacetylene when 20 or more equivalents are used. When isobutylene is added the rate slows down. Derive a rate law that conforms to these data, and write a mechanism that is consistent with this information.

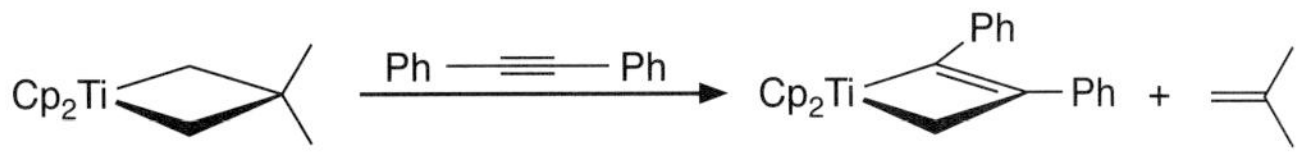

19. A very rare but interesting experimental observation is to see no temperature dependence on the ratio of two products. One's initial intuition might be that the entropies of the two reactions are the same. However, is this quick analysis correct?
   a. Determine what coincidence is required in enthalpies and entropies of activation for the two reactions. (*Hint:* Assume kinetic control and think about the ratios of rate constants in terms of the Eyring equation.)
   b. What is another possible explanation for why there may be no temperature dependence on a ratio of products? (*Hint:* Look back at Section 7.2.7.)

20. Why is it particularly easy to apply Marcus theory to electron transfers in redox reactions such as that shown below?

$$Fe^{2+} + Ce^{4+} \longrightarrow Fe^{3+} + Ce^{3+}$$

21. The effect of solvation is not specifically included in the Marcus equation (Eq. 7.63). However, one expects a solvent effect on the reaction of hydroxide with methyl bromide (the example used in this chapter when Marcus theory was discussed). In what manner does solvation come into this equation, such that it works in a variety of solvents?

22. In the electrocyclic ring closure of butadiene shown below, what torsional mode significantly affects the entropy of activation? Draw the energy surface for this torsion in both the reactant and the product. Does the change impart positive or negative entropy of activation to the reaction?

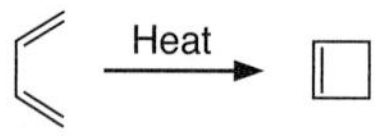

23. The bromination of styrene in carbon tetrachloride was followed by UV spectrophotometry. The disappearance of styrene was exponential when a 10-fold excess of bromine was used. When a 15-fold excess was used, the styrene loss was still exponential, but the observed rate was 2.25 times higher. What conclusions can you make about the mechanism?

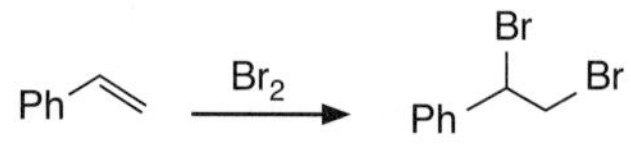

24. The thermal [2+2] cycloreversion reaction of bicyclo[1.1.1]pentane has been proposed to proceed by successive bond cleavage steps, while the analogous transformation in bicyclo[1.1.1]pentanone has been proposed to involve a concerted mechanism. How do the activation parameters support or contradict this assessment? Draw potential energy diagrams for the two processes as proposed. What are the half-lives of the two reactants at 120 °C?

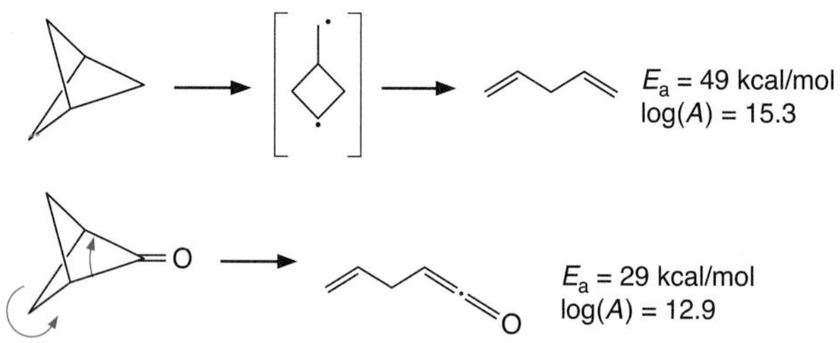

E_a = 49 kcal/mol
log(A) = 15.3

E_a = 29 kcal/mol
log(A) = 12.9

25. The Connections highlight on pulse radiolysis on page 402 used acetophenone as an example. Explain exactly how you anticipate this experiment was performed (you may need to consider issues described in Chapter 5). Why is the pK_a of the resulting alcohol so much lower than a normal alcohol?

26. Using Eq. 7.64, show how Marcus theory nicely reproduces the predictions of the Hammond postulate by considering values of $\Delta G°$ that represent exergonic and endergonic reactions. What kinds of reactions would you expect would not conform to the Hammond postulate?

Further Reading

Transition State Theory (TST)

Boudart, M. (1968). *Kinetics of Chemical Processes*, Prentice-Hall, Englewood Cliffs, NJ, pp. 35–46.

Amdur, I., and Hammes, G. G. (1966). *Chemical Kinetics, Principles and Selected Topics*, McGraw-Hill, New York, pp. 43–58.

Moore, J. W., and Pearson, R. G. (1981). *Kinetics and Mechanism*, John Wiley & Sons, New York, pp. 159–169.

Kreevoy, M. M., and Truhlar, D. G. in *Investigation of Rates and Mechanisms of Reaction*, C. F. Bernasconi (ed.), *Techniques of Chemistry*, 4th ed., Wiley-Interscience, New York, 1986, Vol. VI, Part 1.

Bamford, C. H., and Tipper, C. F. H. (eds.). *Comprehensive Chemical Kinetics*, Elsevier, Amsterdam, 1969–1980; 22 volumes.

William, I. H. "Interplay of Theory and Experiment in the Determination of Transition-State Structure." *Chem. Soc. Rev.*, **22**, 277 (1993).

Bunker, D. L. "Simple Kinetic Models From Arrhenius to the Computer." *Acc. Chem. Res.*, **7**, 195 (1974).

Fong, F. K. "A Successor to Transition-State Theory." *Acc. Chem. Res.*, **9**, 433 (1976).

Laidler, K. J., and King, M. C. "The Development of Transition-State Theory." *J. Phys. Chem.*, **87**, 2657 (1983).

Truhlar, D. G., Hase, W. L., and Hynes, J. T. "Current Status of Transition-State Theory." *J. Phys. Chem.*, **87**, 2664 (1983).

Albery, W. "Transition-State Theory Revisited." *J. Adv. Phys. Org. Chem.*, **28**, 139 (1993).

Saddle Point

Leffler, J. E., and Grunwald, E. (1963). *Rates and Equilibria of Organic Reactions*, John Wiley & Sons, New York, p. 65.

Weakness of TST

Bell, R. P. "Recent Advances in the Study of Kinetic Hydrogen Isotope Effects." *Chem. Soc. Rev.*, **3**, 513 (1974)

Miller, W. H. "Importance of Nonseparability in Quantum Mechanical Transition-State Theory." *Acc. Chem. Res.*, **9**, 306 (1976).

Carpenter, B. K. "Trajectories Through an Intermediate at a Fourfold Branch Point. Implications of the Stereochemistry of Biradical Reactions." *J. Am. Chem. Soc.*, **107**, 5730 (1985).

Gibbs Free Energy

Caldin, E. F. (1961). *An Introduction of Chemical Thermodynamics*, Oxford University Press, Oxford.

Guggenheim, E. A. (1967). *Thermodynamics*, North Holland, Amsterdam.

Mahan, B. H. (1963). *Elementary Chemical Thermodynamics*, Benjamin, New York.

Prigogene, I., and Defuy, R. (1954). *Chemical Thermodynamics*, Longmans, London.

Arrhenius Theory

Hulett, J. R. "Deviations From the Arrhenius Equation." *Quart. Rev. Chem. Soc.*, **18**, 227 (1964).

Gowenlock, B. G. "Arrhenius Factors (Frequency Factors) in Unimolecular Reactions." *Quart. Rev. Chem. Soc.*, **14**, 133 (1960).

Menzinger, M., and Wolfgang, R. "The Meaning and Use of the Arrhenius Activation Energy." *Angew. Chem. Int. Ed. Eng.*, **8**, 438 (1969).

Boltzmann Distribution

Wiberg, K. B. (1964). *Physical Organic Chemistry*, John Wiley & Sons, New York, p. 211.

Moore, W. J. (1972). *Physical Chemistry*, 4th ed., Prentice-Hall, Englewood Cliffs, NJ, p. 180.

Heat Capacity

Singh, S., and Robertson, R. "The Hydrolysis of Substituted Cyclopropyl Bromides in Water. IV. The Effect of Vinyl and Methyl Substitution on $\Delta C_p^{\circ\ddagger}$." *Can. J. Chem.*, **55**, 2582 (1977).

Robertson, R. E. "The Interpretation of $\Delta C_p^{\circ\ddagger}$ for S_N Displacement Reactions in Water." *Tetrahedron Let.*, **17**, 1489 (1979).

Enthalpy, Entropy, and Temperature

Benson, S. W., Cruickshank, F. R., Golden, D. M., Hugen, G. R., O'Neal, H. E., Rodgers, A. S., Shaw, R., and Walsh, R. "Additivity Rules for the Estimation of Thermodynamical Properties." *Chem. Rev.,* **69**, 279 (1969).

Moore, J. W., and Pearson, R. G. (1981). *Kinetics and Mechanism*, John Wiley & Sons, New York, pp. 159–169.

Solution Effects

March, J. (1977). *Advanced Organic Chemistry*, McGraw-Hill, New York.

Amis, E. S. (1966). *Solvent Effects on Reaction Rates and Mechanisms*, Academic Press, New York.

Amis, E. S., and Hinton, J. F. (1973). *Solvent Effects on Chemical Phenomena*, Academic Press, New York, Vol. 1.

Hammond Postulate

Hammond, G. S. "A Correlation of Reaction Rates." *J. Am. Chem. Soc.,* **77**, 334 (1955).

Leffler, J. E. "Parameters for the Description of Transition States." *Science*, **117**, 340 (1953).

Jencks, W. P. "A Primer for the Bema Hapothle. An Empirical Approach to the Characterization of Changing Transition-State Structures." *Chem. Rev.,* **85**, 511 (1985).

Le Nobel, W. J., Miller, A. R., and Hamann, S. D. "A Simple, Empirical Function Describing the Reaction Profile, and Some Applications." *J. Org. Chem.,* **42**, 338 (1977).

Miller, A. R. "A Theoretical Relation for the Position of the Energy Barrier Between Initial and Final States of Chemical Reactions." *J. Am. Chem. Soc.,* **100**, 1984 (1978).

Curtin–Hammett Principle

Shorter, J. "Hammett Memorial Lectures." *Prog. Phys. Org. Chem.,* **17**, 1 (1990).

Eliel, E. L. (1962). *Stereochemistry of Carbon Compounds*, McGraw-Hill, New York.

Seeman, J. I. "Effect of Conformational Change on Reactivity in Organic Chemistry. Evaluations, Applications, and Extensions of Curtin–Hammett / Winstein–Holness Kinetics." *Chem. Rev.,* **83**, 83 (1983).

Dauben, W. G., and Pitzer, K. S. in *Steric Effects in Organic Chemistry*, M. S. Newman (ed.), John Wiley & Sons, New York, 1956.

Principle of Microscopic Reversibility

Cope, A. C., Haven, A. C., Jr., Ramp, F. L., and Trumbull, E. R. "Cyclic Polyolefins XXIII. Valence Tautomerism of 1,3,5-Cyclooctatriene and Bicyclo[4,2,0]octa-2,4-diene." *J. Am. Chem. Soc.,* **74**, 4867 (1952).

Reactivity and Selectivity

Leffler, J. E., and Grunwald, E. (1963). *Rates and Equilibria of Organic Reactions*, John Wiley & Sons, New York.

Hine, J. "The Principle of Least Nuclear Motion." *Adv. Phys. Org. Chem.,* **15**, 1 (1977).

Kinetic vs. Thermodynamic Enolate

House, H. O. (1972). *Modern Synthetic Reactions*, 2d ed., W. A. Benjamin, Menlo Park, CA.

Caine, D. in *Carbon–Carbon Bond Formation*, R. L. Augusine (ed.), Marcel Dekker, New York, 1979.

Reaction Kinetics

Connors, K. A. (1990). *Chemical Kinetics*, VCH Publishers, Inc., New York, pp. 3–4.

Noyes, R. M. in *Investigation of Rates and Mechanisms of Reactions*, E. S. Lewis (ed.), Wiley-Interscience, New York, 1974, 3d ed., Part I, pp. 489–538.

Laidler, K. J. (1965). *Chemical Kinetics*, 2d ed., McGraw-Hill, New York, p. 4.

Moore, J. W., and Pearson, R. G. (1981). *Kinetics and Mechanism*, John Wiley & Sons, New York, pp. 159–169.

Kreevoy, M. M., and Truhlar, D. G. in *Investigation of Rates and Mechanisms of Reactions*, C. F. Bernasconi (ed.), *Techniques of Chemistry*, 4th ed., Wiley-Interscience, New York, 1986, Vol. VI, Part 1.

Capellos, C., and Bielski, B. H. J. (1972). *Kinetic Systems: Mathematical Description of Chemical Kinetics in Solution*, Wiley-Interscience, New York.

Steady State Approximation

Laidler, K. J. "Lessons From the History of Chemistry." *Acc. Chem. Res.,* **28**, 187 (1995).

Bunnett, J. F. in *Investigation of Rates and Mechanisms of Reactions*, 4th ed., C. F. Burnasconi (ed.), John Wiley & Sons, New York, 1986, Vol. VI, Part I, p. 251.

Flash Photolysis

Simon, J. D., and Peters, K. S. "Picosecond Studies of Organic Photoreactions." *Acc. Chem. Res.*, **17**, 277 (1984).

Scaiano, J. C. "Laser Flash Photolysis Studies of Reactions of Some 1,4-Biradicals." *Acc. Chem. Res.*, **15**, 252 (1982).

Hilinski, E. F., and Rentzepis, P. M. "Chemical Applications of Picosecond Spectroscopy." *Acc. Chem. Res.*, **16**, 224 (1983).

Michl, J., and Bonzcic-Koustecky, V. (1990). *Electronic Aspects of Organic Photochemistry,* Wiley-Interscience, New York, p. 74.

Pulse Radiolysis

Adams, G. E. "Pulse Radiolysis Studies on Reactive Intermediates in Organic Chemical Process." *Ann. Rep. Prog. Chem.*, **65B**, 223 (1968).

Marcus Theory

Marcus, R. A. "Chemical and Electrochemical Electron-Transfer Theory." *Ann. Rev. Phys. Chem.*, **15**, 155 (1964).

Albery, W. J. "The Application of the Marcus Relation to Reactions in Solution." *Ann. Rev. Phys. Chem.*, **31**, 227 (1980).

Albery, J., and Kreevoy, M. "Methyl Transfer Reactions." *Adv. Phys. Org. Chem.*, **16**, 87 (1978).

Marcus, R. A. "Electron Transfer Reactions in Chemistry: Theory and Experiment (Nobel Lecture)." *Angew. Chem. Int. Ed. Eng.*, **32**, 1111 (1993).

More O'Ferrall–Jencks Plots

Thornton, E. R. "A Simple Theory of Predicting the Effects of Substituent Changes on Transition-State Geometry." *J. Am. Chem. Soc.*, **89**, 2915 (1967).

More O'Ferrall, R. A. "Relationships Between E_2 and E1cB Mechanisms of β-Elimination." *J. Chem. Soc. B*, 274 (1970).

Jencks, W. P. "General Acid–Base Catalysis of Complex Reaction in Water." *Chem. Rev.*, **72**, 705 (1972).

More O'Ferrall, R. A. in *The Chemistry of the Carbon–Halogen Bond*, S. Patai (ed), John Wiley & Sons, New York, 1973, Vol. 2.

More O'Ferrall–Jencks Analysis of S_N2 and S_N1 Mechanisms

Harris, J. M., Shafer, S. G., Moffatt, J. E., and Becker, A. R. "Prediction of S_N2 Transition State Variation by the Use of More O'Ferrall Plots." *J. Am. Chem. Soc.*, **101**, 3295 (1979).

Potential Energy Hypersurface

Mezey, P. G. (1987). *Potential Energy Hypersurfaces*, Elsevier, Amsterdam.

CHAPTER 8

Experiments Related to
Thermodynamics and Kinetics

Intent and Purpose

The intent of this chapter is to teach several of the experimental tools routinely applied by physical organic chemists in the study of reaction mechanisms. The majority of these methods build upon an analysis of kinetics and thermodynamics, providing a natural progression from the previous chapter. Both the theory behind the methods and their application are covered, and many methods are illustrated by example.

The first topic is isotope effects. We consider the origin of isotope effects, what information they provide, and how they are analyzed. This includes solvent isotope effects, which are very relevant to the analysis of enzymatic reactions. After isotope effects, we delve into linear free energy relationships (LFERs), and show how structural changes can be used in a systematic way to gain insight into the nature of reactive intermediates. Electronic substituent effects (Hammett plots) are discussed in detail, but we also examine how structural changes in the solvent, nucleophile, and nucleofuge (a synonym for leaving group) can be used to probe a reaction mechanism. It is particularly instructive within the context of linear free energy relationships to examine trends in reactivity. Hence, in this chapter we also discuss the general effects of electron withdrawing and donating groups, changes in the ionizing power of the solvent, steric effects, and changes in nucleophilicity and leaving group ability (nucleofugality). The notions are not specific to individual reaction types, so they are discussed here instead of being examined only in the context of specific mechanisms in the next chapters. Finally, a variety of experiments that do not fit under any single heading are examined, such as cross-over experiments, scrambling experiments, clocks, etc. These experiments can be extremely important. We delineate the general approach used in each experimental type, but teach the methods mostly by example.

In all the sections of this chapter, examples are drawn from the literature. In doing so, we show that the tools of physical organic chemistry are used in most chemical subdisciplines. Our examples come from enzymology, bioorganic chemistry, organometallic chemistry, and traditional small molecule organic chemistry. Our intention is for the student to learn how general the experiments discussed herein are, so that he or she will immediately incorporate them into his or her repertoire for studying chemical reactivity.

8.1 Isotope Effects

In our analysis of kinetics in the last chapter, we emphasized that the information gained is limited, and that most studies of mechanisms involve other techniques. One piece of information that cannot be gained from a kinetic study is what bonds have been broken, formed, or rehybridized during the rate-determining step. Isotope effects can provide just this kind of information. Substituting one isotope for another at or near an atom at which bonds are breaking or rehybridizing typically leads to a change in the rate of the reaction. When the bonds being broken or formed involve those to hydrogen, the effect of replacing H with D often is relatively large and can be measured routinely. Isotope effects with other atoms have also been studied, but the effects are typically small and sometimes difficult to quantify. Therefore, our discussion of isotope effects will focus on hydrogen, although we give a brief introduction to heavy atom isotope effects later.

8.1.1   The Experiment

An isotope effect is measured to determine if the bond at which the isotopic substitution is being made changes in some manner during the rate-determining step. We express an isotope effect as a ratio of rate constants, where the numerator is the rate constant for the reaction with the natural abundance isotope, and the denominator is the rate constant for the reaction with the altered isotope. For example, when measuring isotope effects for reactions involving a substitution of hydrogen with deuterium, the isotope effect would be expressed as k_H/k_D. Measuring an isotope effect, therefore, typically requires us to run two kinetic analyses or to design a clever competition experiment.

For a hydrogen isotope effect, typically a first order (or pseudo-first order) rate constant for the reaction of interest is determined with the bond that is being analyzed (X–H) having a natural abundance distribution of isotopes. This is because the natural abundance of deuterium and tritium (0.015% and 1×10^{-4}%, respectively) is so low that their contribution to the rate is negligible. Next, the rate constant for the same reaction is determined with a version of the molecule in which synthesis (or solvent exchange; see Section 8.1.6) has introduced nearly 100% deuterium in place of the hydrogen.

The magnitude of the isotope effect (the variation of k_H/k_D from unity) gives us information about the reaction mechanism. If k_H/k_D is 1, one conclusion would be that the bond where the substitution occurred is not changing during the rate-determining step. However, it may just be that the isotope effect is too small to be measured accurately. If the ratio of k_H/k_D is different from one, more solid conclusions can be drawn. When k_H/k_D is greater than one, we call the isotope effect **normal**. When k_H/k_D is less than one, we call the isotope effect **inverse**. When the isotope effect can be attributed to a bond breaking event at the X–H/X–D bond, it is referred to as a **primary isotope effect**. When the effect is attributed to a rehybridization or arises from isotopic substitution remote from the bonds undergoing reaction, it is referred to as a **secondary isotope effect**. The situation we have discussed thus far, in which the isotope substitution changes the rate of the reaction, is called a **kinetic isotope effect** (KIE). When the interchange of isotopes alters the position of an equilibrium, we call it an **equilibrium isotope effect**.

8.1.2   The Origin of Primary Kinetic Isotope Effects

The origin of all isotope effects is the difference in frequencies of various vibrational modes of a molecule that arise when one isotope is substituted with another. Let's first analyze a bond that is breaking during the rate-determining step of a reaction (a primary kinetic isotope effect). To a good approximation, the potential energy of the system does not change with substitution of one isotope for another. In other words, the relative energies of the minima and maxima on the energy surface do not change with isotopic substitution. However, we noted in Chapter 7 that the *shapes* of the potential wells on an energy surface are composites of the various vibrational states of the molecule. When considering a reaction, it is sufficient to consider the shapes of these wells as being dominated by the vibrational modes that are undergoing the most change during the reaction.

Recall from our discussion of IR spectroscopy (Section 2.1.4) that vibrational states are quantized, and that each potential energy well has several rungs that represent different energies for that vibrational mode (Figure 2.2). The formula for the quantized energies (e_n) of the vibrational modes is given in Eq. 8.1, where v is the frequency of the vibrational mode being considered. These energies are measured from the lowest point in the potential energy well. At ambient temperature, the vibrational modes for bond stretches are dominated by $n = 0$, with $e_0 = \frac{1}{2}hv$. This energy is referred to as the **zero-point energy** (ZPE).

$$e_n = \left(n + \frac{1}{2}\right)hv \quad n = 0,1,2, \cdots \qquad \text{(Eq. 8.1)}$$

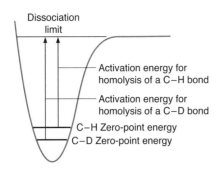

Figure 8.1
A Morse potential for a C–H bond showing that the activation energy
for homolysis of a C–D bond is larger than for a C–H bond.

For a bond breaking event, the stretching vibration of that bond is defined as the reaction coordinate. As was discussed in Chapter 2, the frequency of a stretching vibration is modeled by the classic equation for the stretching of a spring with a mass attached at both ends (Eq. 8.2). In Eq. 8.2 v is expressed in s^{-1}, whereas in IR spectroscopy the frequency is expressed in cm^{-1}, and is called $\bar{v}$ (where $\bar{v} = v/c$ and c, the speed of light, is given in cm/s).

The frequency is directly proportional to the square root of the force constant for the bond, and inversely proportional to the square root of the reduced mass. The reduced mass for a bond between a heavy atom such as C, N, or O with a light atom such as H is significantly affected when the H atom is changed to D. The stretching frequency of a bond with deuterium is lower due to the heavier mass, and hence the zero-point energy for the bond is lower also. We denote this on a potential energy well by showing a second rung that is lower in energy for the deuterium-containing bond (Figure 8.1).

$$v = \frac{1}{2\pi} \sqrt{\frac{k}{m_r}} \quad \text{where } m_r = \frac{m_1 m_2}{m_1 + m_2} \qquad \text{(Eq. 8.2)}$$

In a homolysis reaction that forms two radicals, the stretching vibration in the reactant is converted to a translational degree of freedom. Comparing a C–H and C–D bond (as done in Figure 8.1) shows a larger activation energy for the C–D bond. In this case, the full difference in ZPE establishes the magnitude of the isotope effect, and k_H/k_D is greater than 1. There are no vibrations for the bond that is breaking when the reaction is done, and therefore the force constant associated with this bond has gone to zero, after full bond cleavage. By assuming that the bond of interest is 100% broken at the transition state (typically not the case), we can calculate the maximum possible isotope effect using Eqs. 8.1 and 8.2 (left as an Exercise at the end of this chapter). The expectation is that an isotope effect measured at 298 K for a homolysis reaction involving a C–H bond, whose IR stretch appears at 3000 cm^{-1}, should be approximately $k_H/k_D = 6.5$. This is a relatively large primary kinetic isotope effect, because we assumed that the bond was completely broken.

Most isotope effects are attenuated from this value because reactions typically do not involve bonds that are completely broken in the transition state. An example of a reaction with a relatively large isotope effect is the hydroxylation reaction given in the Connections highlight on page 425. To understand any kinetic phenomenon, one always compares reactant with transition state. For isotope effects we compare the ZPEs of the various vibrations of the reactant and the activated complex. Usually the bond is only partially broken at the transition state, or another bond is starting to form at the transition state. Both of these will attenuate the isotope effect from that of total homolysis. To visualize this attenuation, we need to examine reaction coordinate diagrams and the associated vibrational modes.

A.

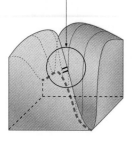

A potential energy well
perpendicular to the reaction
coordinate with the associated
C–H and C–D vibrational states

B.

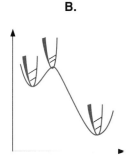

C.

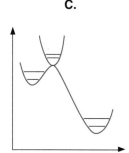

Figure 8.2
Various images of three- and
two-dimensional reaction
coordinate diagrams.
A. Showing the C–H and C–D
ZPE levels within the troughs
on the energy surface. **B.** Picture
of the troughs and rungs at the
reactant, T. S., and product,
perpendicular to the reaction
coordinate. **C.** Drawing the
troughs and rungs in the same
plane as the reaction coordinate.

Reaction Coordinate Diagrams and Isotope Effects

The energy surfaces discussed in Chapter 7 depicted energy wells from which troughs are traversed to achieve the saddle points. Each point along a reaction coordinate is associated with a potential energy well that is perpendicular to the reaction coordinate (Figure 8.2 **A**). Each well has vibrational states associated with it. To simplify these pictures for rapid drawing, we draw the reaction coordinate in the usual two-dimensional sense, and place a well perpendicular to the page at each point we want to discuss on the reaction coordinate (Figure 8.2 **B**). We then draw in the associated rungs representing the vibrational states. We now draw these same wells as if they were in the same plane as the reaction coordinate (Figure 8.2 **C**), fully realizing that they are really perpendicular.

Drawings such as Figure 8.2 **C** are the starting point for understanding the magnitudes of conventional isotope effects. In this figure we are achieving a transition state where there is still some bonding between the H or D and a heavy atom, such as carbon. Activated complexes have vibrational modes with ZPEs, just like any other molecule. For every vibrational mode there is a ZPE difference for C–H vs. C–D. However, the only vibrational modes we need consider are those that are undergoing a *change* along the reaction coordinate (see below). These will give a *differential* ZPE difference between C–H and C–D at the transition state vs. the reactant, and can thereby produce a kinetic isotope effect.

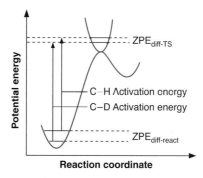

Figure 8.3
Diagram showing that a residual ZPE difference between C–H and
C–D bonds at the transition state diminishes the isotope effect.

With primary kinetic isotope effects (Figure 8.3) the ZPE difference in the activated complex ($ZPE_{diff-TS}$) is smaller than in the reactants ($ZPE_{diff-react}$). The activation free energy for the deuterium case is still larger than the hydrogen case, but the isotope effect is reduced from that associated with a full homolysis reaction by the amount of ZPE difference that is residual in the transition state. In other words, the difference in free energy of activation for the C–H and C–D reactions ($\Delta\Delta G_{CH/CD}^{o\ddagger}$) is the difference between the ZPE differences in the reactant and the activated complex (Eq. 8.3).

$$\Delta\Delta G_{CH/CD}^{o\ddagger} = ZPE_{diff-react} - ZPE_{diff-TS} \qquad \text{(Eq. 8.3)}$$

In summary, then, the magnitude of a primary kinetic isotope effect depends upon differences in the ZPEs in the reactant and activated complex for all the vibrational modes of the reactant and activated complex. The changes in vibrations for the bonds undergoing transformations during the reaction dominate the isotope effect. We need to identify these vibrational modes and understand how their force constants change during the reaction. We focus on the force constants because the frequencies of the vibrational modes are related to the force constants via equations such as Eq. 8.2. As is discussed below, the proper vibrational modes to consider can be a function of the angle of collision between the reactants. Further, the extent of change in the force constants is a function of the extent of bond breaking and forming in the transition state.

Connections

The Use of Primary Kinetic Isotope Effects to Probe the Mechanism of Aliphatic Hydroxylation by Iron(III) Porphyrins

One reaction that is uncommon in organic chemistry but is common in biological systems in the hydroxylation of alkanes to form alcohols. Cytochrome P-450, a heme containing enzyme, catalyzes this reaction. Large primary isotope effects are found, as well as an absence of carbocation-like skeletal rearrangements and loss of stereochemistry. These observations led researchers to conclude that a **radical cage mechanism** is operative. A radical cage mechanism is one in which radicals are created and react together before diffusing apart.

Several models of this heme enzyme are based upon substituted porphyrins. For example, compound *i* (oxidized Fe in porphyrin ligand shown) catalyzes the oxidation of norbornane to norboran-2-ol with iodosobenzene

as oxidant. With the deuterionorbornane shown (*ii*), a mixture of endo- and exo-norboran-2-ols is produced, and a primary kinetic isotope effect of 5 is measured. This is similar to the isotope effect found for the enzyme-catalyzed reaction, suggesting that *i* is a good model for the natural reaction. The large isotope effect supports a mechanism involving hydrogen atom abstraction in the rate-determining step followed by a rapid transfer of the hydroxyl from the iron to the carbon-based radical (hence the radical cage description). Furthermore, note that no carbon skeletal rearrangements were found, as would be expected for a norbornyl cation (see Chapter 11). This supports a mechanism that does not involve carbocations.

Traylor, T. G., Hill, K. W., Fann, W.-P., Tsuchiya, S., and Dunlap, B. E. "Aliphatic Hydroxylation Catalyzed by Iron(III) Porphyrins." *J. Am. Chem. Soc.*, **114**, 1308–1312 (1992).

Ar = 2,6-dichlorophenyl

Primary Kinetic Isotope Effects for Linear Transition States as a Function of Exothermicity and Endothermicity

To analyze which vibrational modes are relevant to an isotope effect, we must have a reference point to compare to the transition state, and we must identify the reaction coordinate. The Hammond postulate tells us that the structure of the activated complex most resembles the structure of the molecule to which it is closest in energy. This means that we can estimate the vibrational modes in the activated complex by making comparisons to the vibrational modes for the reactants and products (remembering that in this context the "product" might actually be a reactive intermediate). The force constants for the transition state will be most similar to the force constants in the structure that the transition state is closest to in energy. We must also identify the vibrational mode that is the reaction coordinate, because this vibration will not contribute to the isotope effect at the transition state. This is because the

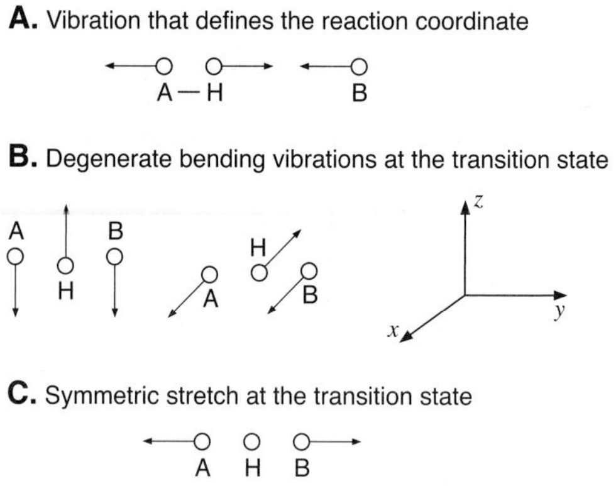

A. Vibration that defines the reaction coordinate

Figure 8.4
Vibrational modes relevant to kinetic primary isotope effects. **A.** Vibration that defines the reaction coordinate. **B.** In-plane and out-of-plane bends at the transition state. **C.** Symmetric stretch that forms at the transition state.

B. Degenerate bending vibrations at the transition state

C. Symmetric stretch at the transition state

reaction coordinate is not a vibration present in the transition state, but instead defines the reaction.

The normal modes we must pay the most attention to are those that are changing in the reactant and developing in the activated complex. Consider a deprotonation reaction of an acid (A–H) by a base (B⁻) that occurs with a linear transition state (Eq. 8.4). The vibration that is the reaction coordinate for this transformation is depicted in Figure 8.4 **A**, leading to A–H bond cleavage and B–H bond formation. This vibration does not contribute to the isotope effect, but all other vibrations that are changing do. As discussed above, the A–H stretching mode in the reactant is changing because this bond is breaking, and a simple analysis would lead to the conclusion that an isotope effect around 6.5 would result. However, because there is still some bonding in the transition state, there are other vibrational modes in the transition state that diminish this effect. One such mode is a set of degenerate bends that occur in perpendicular planes (Figure 8.4 **B**). However, a bending mode also exists in the reactant, and bending modes have significantly lower force constants than stretches. Therefore, differences in bends between reactants and transition states typically do not contribute greatly to primary isotope effects. In contrast, a symmetric stretching vibration develops at the transition state that has no analog in the reactant (Figure 8.4 **C**). This new vibration has a large force constant because it is a stretch, and hence it will have a large effect on the magnitude of the isotope effect. This is the mode that we want to examine in detail as a function of the free energy of the cleavage reaction.

$$\text{A–H} \;+\; \text{B}^{\ominus} \;\longrightarrow\; \left[\overset{\delta\ominus}{\text{A}}\text{----H----}\overset{\delta\ominus}{\text{B}} \right]^{\text{T.S.}} \;\longrightarrow\; \text{A}^{\ominus} \;+\; \text{H–B} \qquad\qquad \text{(Eq. 8.4)}$$

If the deprotonation is very exothermic, the transition state resembles the reactants; that is, very little bond breaking has occurred at the acid and very little bond formation has occurred at the base (Figure 8.5 **A**). Hence, the bond holding the hydrogen to the complex still greatly resembles the bond in the reactant. The newly forming symmetric stretch involves almost as much movement of the hydrogen as does the bond stretch in the reactant. The magnitude of the isotope effect is therefore predicted to be small, because the *difference* in ZPE H/D differences for the reactant and transition state is small.

When the reaction is very endothermic, the bond between the proton and the base has been almost completely formed at the transition state (Figure 8.5 **B**). Hence, the symmetric stretch at the transition state greatly resembles the normal bond stretch in the product. When examining a reaction where the force constant of the B–H bond is similar to that of the A–H bond, the force constant for the symmetric stretch at the transition state would therefore be similar to that for the A–H bond. Again, we find little difference in ZPE differences between the reactant and the activated complex, leading to a small isotope effect.

The situation is much more interesting when the reaction is thermoneutral or close to it (Figure 8.5 **C**). Now the bond that holds the hydrogen in the activated complex has the hy-

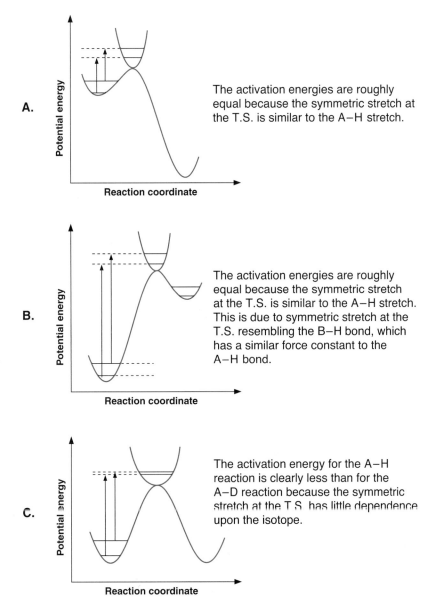

Figure 8.5
Diagrams showing how the ZPE difference between C–H and C–D bonds between the reactant and transition state changes as a function of the free energy of the reaction. **A.** Exothermic reaction, **B.** Endothermic reaction, and **C.** Thermoneutral reaction. The symmetric stretch referred to in the figure is that of Figure 8.4 **C.**

drogen almost equally shared between A and B, and the symmetric stretch (Figure 8.4 **C**) is centered around the hydrogen with little or no movement of the hydrogen. Therefore, the frequency of this stretch does not depend strongly upon the isotope. Now the ZPE difference at the transition state is very small. The resulting isotope effect is therefore predicted to be large, possibly approaching 6.5. Note that this analysis tells us that the magnitude of the isotope effect is a rough measure of the position of the transition state along the reaction coordinate; that is, large isotope effects are found for reactions with nearly symmetrical transition states, and smaller effects are observed as the transition state moves toward reactant or product. An example of this in a proton transfer reaction is described in the following Connections highlight.

Connections

An Example of Changes in the Isotope Effect with Varying Reaction Free Energies

In our discussion of primary kinetic isotope effects, we used a hypothetical acid–base reaction to explain why the magnitude of the isotope effect would be a maximum when the reaction is thermoneutral. Indeed, this kind of effect has often been observed in acid–base chemistry. As one example, the deprotonation of nitroethane ($CH_3CH_2NO_2$ or $CH_3CD_2NO_2$) by amines has a maximum isotope effect when the pK_a of the conjugate acid of the amine base matches the pK_a of nitroethane. In other words, when the acid strength on both sides of the reaction is the same, the isotope effect is maximized. The graph to the right shows various amines, and when the acidity of the ammonium and the nitroethane are close, the largest KIE is found.

Dixon, J. E., and Bruice, T. C. "Dependence of the Primary Isotope Effect (k^H/k^D) on Base Strength for the Primary Amine Catalyzed Ionization of Nitroethane." *J. Am. Chem. Soc.*, **92**, 905 (1970).

Nitroethane deprotonation

Isotope Effects for Linear vs. Non-Linear Transition States

Our analysis of the relative magnitudes of isotope effects has focused upon changes in ZPEs for various vibrations that are transforming during a reaction. When there are large changes in vibrational modes that are affected by mass differences, large isotope effects are found. One needs vibrational modes with relatively large force constants in the reactant in order for significant differences to arise between reactants and transition states.

When the hydrogen/deuterium transfer proceeds through an activated complex with bent bonds, bending modes become more significant (Eq. 8.5). The force constants for scissoring and bending motions (see Section 2.1.4) are significantly lower than for stretching motions, and therefore the changes in ZPEs between reactant and transition state are not as large as for reactions that involve changes in stretches. This leads to lower magnitudes for primary kinetic isotope effects. Moreover, the hydrogen in the symmetric stretch (the bent analog to Figure 8.4 **C**) still has a significant movement. Therefore, this vibration still has a mass dependence, and will diminish the magnitude of the isotope effect relative to a linear transition state. Hence, hydrogen transfers that proceeds through non-linear transition states have lower KIEs.

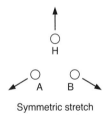

Symmetric stretch

$$A^H + B^{\ominus} \longrightarrow \left[\begin{smallmatrix} H \\ A \diagup \diagdown B \\ {\delta\ominus} \quad {\delta\ominus} \end{smallmatrix} \right]^{T.S.} \longrightarrow A^{\ominus} + \overset{H}{\diagdown}B \qquad \text{(Eq. 8.5)}$$

8.1.3 The Origin of Secondary Kinetic Isotope Effects

Secondary kinetic isotope effects arise from isotopic substitution at a bond that is not being broken, and typically involve a change in bond hybridization or involvement of the bond in hyperconjugation. They are defined as α or β **secondary isotope effects**. The terms refer to whether the isotope is on a position α or β to the bond that is changing. An α effect occurs when the atom undergoing reaction has the associated isotope, whereas a β effect occurs when the neighboring atom has the isotope. An example of how to use a secondary isotope effect in deciphering a mechanism is given in the Connections highlight on page 431.

Hybridization Changes

As with any kinetic isotope effect, a difference in ZPE differences between the reactant and the transition state is necessary for the isotope effect to be manifest. To understand a secondary effect, we need to consider all the changes in vibrational modes that occur when an atom (or atoms) associated with a bond undergoes rehybridization. The vibrational modes that have the largest force constants and those that undergo the largest changes will have the greatest influence on the isotope effect.

When a C–H bond involving an sp^3 hybridized carbon is changing to a bond involving an sp^2 hybridized carbon, there are only a limited number of vibrational modes that are undergoing large changes. These modes include stretches, as well as in-plane and out-of-plane bending motions. Similar vibrational modes change when an sp^2 hybrid changes to sp. Let's consider the stretches first. In Chapter 2 (Table 2.2), we showed that C–H bond strengths decrease in the order $sp > sp^2 > sp^3$. Similarly, the force constants for the stretching vibrations follow this trend. The trend is also reflected in the IR spectra, where stretching frequencies have the same order. Therefore, there is a change in force constant for stretches of a bond undergoing rehybridization, and we would predict an associated isotope effect. Yet, the change in force constant is not nearly as large as when the bond is breaking, as in a primary kinetic isotope effect. In fact, the change in force constant due to rehybridization is not large enough to create significant isotope effects. Hence, we must examine other vibrations to understand the origin of a secondary isotope effect.

Figure 8.6 shows the in-plane and out-of-plane bending motions for sp^3 and sp^2 hybridized carbons, along with the associated IR frequencies. The in-plane bend has essentially the same frequency in the sp^3 and sp^2 hybridized carbons, indicating there is little difference in force constants for these motions. The in-plane and out-of-plane bends for an sp^3 hybridized carbon are degenerate. However, the in-plane bend is a much stiffer motion for the sp^2 hybridized carbon than is the out-of-plane bend. This is because there is little steric hindrance for the out-of-plane bend of an sp^2 hybridized carbon. This large difference in force constant for the out-of-plane bend of an sp^3 hybrid versus an sp^2 hybrid means that there will be a significant difference in ZPE differences between C–H and C–D bonds in reactions that involve rehybridization between sp^3 and sp^2. Therefore, it is this bending mode that leads to a measurable secondary isotope effect. We can calculate the isotope effect expected from this frequency difference (left as an Exercise at the end of the chapter), and find a theoretical maximum value of 1.4. Typical secondary effects of around 1.1 to 1.2 are found, because the full difference between an sp^3 and sp^2 carbon is not felt at the transition state. Similarly, a large difference in the frequency of the in-plane bend exists between sp^2 and sp hybridized carbons, leading to secondary isotope effects. Note that these effects, even at their largest, are *much* smaller than typical primary KIEs, presenting a more significant challenge to the experimentalist. A Going Deeper highlight on page 432 describes an ingenious method for measuring very small isotope effects.

We have just explained that a secondary kinetic isotope effect arises from differences in bending vibrations. One draws this difference using the typical reaction coordinate diagrams. In Figure 8.7, we plot the potential energy wells for the vibrational states undergoing change, but now we are plotting bending motions. Since the transition state is developing sp^2 character at the carbon where the isotopic substitution has been made, the force constant is weaker at the transition state than for the reactant (Figure 8.7 **A**). We find that the reaction is slower when the reaction has a deuterium on the carbon undergoing rehybridization. This is a normal secondary kinetic isotopic effect.

Consider now a reaction that is the opposite from above—that is, one that involves rehybridization from sp^2 to sp^3 (Figure 8.7 **B**). Now the force constant for the bending motion is getting larger at the transition state because the vibration is becoming stiffer. In this scenario the ZPE difference is larger at the transition state than at the reactant, which means that the reaction actually proceeds faster with deuterium than with hydrogen. This is an inverse kinetic isotope effect. Isotope effect values of around 0.8 to 0.9 are common in these cases.

A secondary kinetic isotope effect can also arise from the involvement of a C–H(D) bond in hyperconjugation in a rate-determining step. For example, in an S_N1 reaction a carbocat-

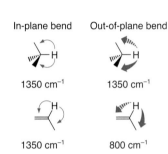

In-plane bend   Out-of-plane bend

1350 cm^{-1} 1350 cm^{-1}

1350 cm^{-1} 800 cm^{-1}

Figure 8.6
In-plane and out-of-plane bending vibrations for C–H bonds on sp^3 and sp^2 hybridized carbons.

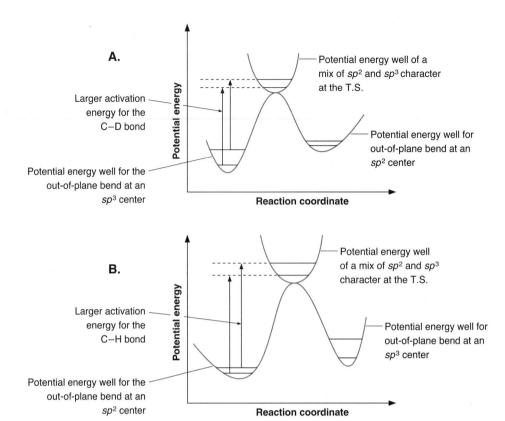

Figure 8.7
Diagrams showing the origin of normal (**A**) and inverse (**B**) secondary kinetic isotope effects. In **A** the C–H activation energy is smaller, but in **B** the C–D bond activation energy is smaller.

ion is created in the slow step, and C–H(D) bonds β to the cationic center stabilize the cation by hyperconjugation (see Section 1.4.1). This weakens the C–H(D) bonds, which will lead to a normal secondary kinetic isotope effect.

Steric Isotope Effects

There is an alternative mechanism by which we can see isotope effects in cases that do not involve breaking or making C–H bonds. This is a steric effect, in which the differing effective steric sizes of H vs. D come into play. In a C–H vs. a C–D bond, not only is the ZPE less for D, but we should also anticipate that the vibrational amplitude should be less for D than for H. This in turn should make D appear to be smaller than H in some contexts.

A classic demonstration of this effect is in the racemization of the chiral biphenyl compound characterized by Mislow and shown in Eq. 8.6. Rotation about the central bond racemizes the material and forces a severe steric clash between the two methyl groups in the transition state. Indeed, it is found that the deuterio compound racemizes faster than the protio, consistent with the notion that D is effectively smaller than H. The effect is certainly not a large one, as only a 15% difference is seen in a system with a really severe steric interaction and with multiple H/D substitutions. Nevertheless, steric isotope effects can be comparable to other secondary isotope effects, and so should be considered when evaluating experimental data.

(Eq. 8.6)

$$\frac{k\,(X=D)}{k\,(X=H)} = 1.15$$

Connections

The Use of an Inverse Isotope Effect to Delineate an Enzyme Mechanism

D-Amino acid oxidase catalyzes the oxidation of amino acids to imino acids via the transfer of a hydride to the coenzyme flavin adenine dinucleotide (FAD). The mechanism first involves deprotonation of the amino acid to create a carbanion (see below). This carbanion can then undergo either a nucleophilic addition to the flavin of FAD (Path A) or an electron transfer to the flavin, creating radicals that combine to give the same product as the nucleophilic addition (Path B). Expulsion of the flavin as a leaving group, concomitant with some proton transfers, gives the oxidized imino acid product.

To distinguish whether the reaction involves nucleophilic attack or radical intermediates, the conjugate base of nitroethane was used as a substrate for the enzyme (see to the right). In this case the enzyme does not need to deprotonate the substrate because it is a carbanion to begin with. This carbanion is sp^2 hydridized at the anionic carbon, and nucleophilic attack on the flavin would transform this carbon to sp^3 (Path A). Since the hybridization changes from sp^2 to sp^3, an inverse isotope effect is predicted. However, the alternative wherein a radical is formed from the carbanion keeps the hybridization state of the α-carbon roughly the same (Path B), giving little if any isotope effect. Following the kinetics of the reaction of nitroethane anion (α-H or D) with FAD and D-amino acid oxidase revealed a secondary inverse kinetic isotope effect of 0.84. This is relatively large for a secondary isotope effect, and so the data were interpreted to support the nucleophilic mechanism (Path A).

Probing radical vs. nucleophilic mechanisms

Kurtz, K. A., and Fitzpatrick, P. F. "pH and Secondary Kinetic Isotope Effects on the Reaction of D-Amino Acid Oxidase with Nitroalkane Anions: Evidence for Direct Attack on the Flavin by Carbanions." *J. Am. Chem. Soc.*, **119**, 1155 (1997).

Going Deeper

An Ingenious Method for Measuring Very Small Isotope Effects

Heavy atom and secondary hydrogen kinetic isotope effects are often quite small, so they can be difficult to measure due to the error values often associated with rate constants. However, as any reaction occurs, the reactants are incrementally enriched in the slower reacting components. Thus, for reactants with the natural abundance of heavy isotopes, near the end of the reaction the proportion of heavy isotopes in the reactants has increased relative to the proportion present at the beginning of the reaction. The isotope content of the recovered starting material relative to the original starting material (R/R_o) is related to the extent of reaction (F) and the kinetic isotope effect (KIE) via the following equation:

$$R/R_o = (1-F)^{(1/\text{KIE})-1}$$

As the reaction approaches completion (F approaches 1), the ratio of R/R_o becomes very sensitive to the value of the KIE.

Using this method, the kinetic isotope effects of ^{2}H and ^{13}C at each atom in isoprene for the Diels–Alder reaction with maleic anhydride were determined. NMR spectroscopy was used to measure the ^{2}H and ^{13}C content at each atom in the initial starting material, and in recovered starting material after the reaction had proceeded to 98.9% completion. The ratio of isotopic abundance at each carbon and hydrogen was used in the above equation to give the isotope effect at each center. The ^{13}C isotope effects for each carbon are shown below. Using this very clever method, one can now routinely measure very small isotope effects, as long as the reaction is sufficiently scalable to allow for recovery of enough starting material for NMR analysis.

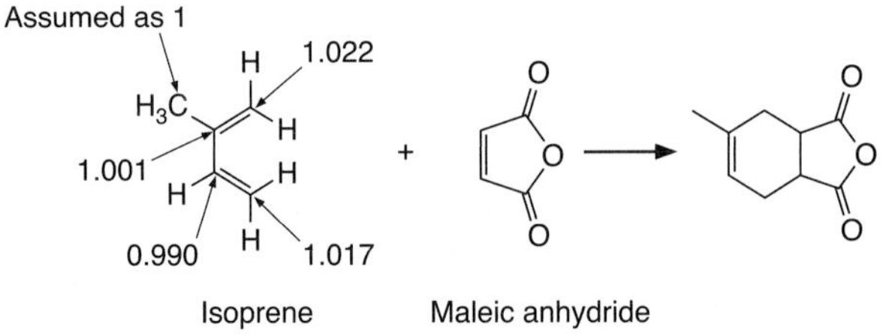

Isoprene   Maleic anhydride

Singleton, D. A., and Thomas, A. A. "High-Precision Simultaneous Determination of Multiple Small Kinetic Isotope Effects at Natural Abundance." *J. Am. Chem. Soc.*, **117**, 9357–9358 (1995).

8.1.4 Equilibrium Isotope Effects

Our analysis of kinetic isotope effects compared the difference in ZPE differences between C–H and C–D bonds in the reactants and transition states. There are also many reactions that establish equilibrium between two structures where rehybridization occurs or bond strengths change. For example, Figure 8.8 shows two cases, one where the ZPE difference is larger in the reactant and one where the ZPE difference is larger in the product. In the first case the equilibrium lies further to the right for the structure with hydrogen, and in the second case the equilibrium lies further to the right for the deuterated structure. Shifts in equilibria upon isotopic substitution are called **thermodynamic** or **equilibrium isotope effects**. We have been intentionally vague about which vibrational modes lead to these isotope effects, because any large change in force constant for any mode can yield such effects.

Note that in both examples of Figure 8.8 the deuterium prefers the bond with the larger force constant. This is often generalized into the statement "deuterium prefers the stronger bond". However, this implies bond strength, or bond dissociation energy (BDE), and it is not always true that the deuterium will concentrate in the bond with the larger BDE. Rather, deuterium will concentrate at the center with the larger force constants overall. Usually, but not always, this will be the center with the stronger bond to H/D.

Isotopic Perturbation of Equilibrium—Applications to Carbocations

The study of equilibrium isotope effects has found widespread application in analyzing near degenerate equilibria, and in this context is referred to as the **isotopic perturbation of equilibrium**. This technique, as developed by Saunders, has proven to be a powerful tool for studying carbocations. It provides a very clever way to distinguish a rapid equilibrium from a single, symmetrical structure. The method is best described by considering some specific examples, shown in Figure 8.9.

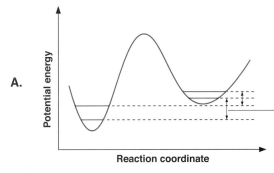

A.

Larger energy difference for
deuterated case; equilibrium lies
further toward products for the
hydrogen-substituted compound

B.

Larger energy difference for
hydrogenated case; equilibrium lies
further toward products for the
deuterium-substituted compound

Figure 8.8
Diagrams showing equilibrium isotope
effects. **A.** Increasingly favoring product
with H, and **B.** Increasingly favoring
product with D.

A.

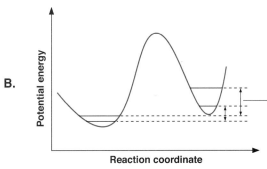
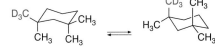

B.

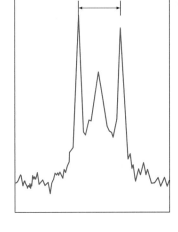

Figure 8.9
Isotopic perturbation of equilibrium. **A.** The 1,1,3,3-tetramethylcyclo-
hexane system. See text for discussion. **B.** ^{13}C NMR spectra used to study
equilibria given in part **A. C.** An isotopic perturbation of equilibrium
experiment for a prototype carbenium ion. **D.** Isotopic perturbation of
resonance for a substituted allyl cation. The equilibrium arrows are
hypothetical for sake of discussion in the text. It is really resonance,
not an equilibrium.

C.

D.

First, consider the chair–flip reaction of 1,1,3,3-tetramethylcyclohexane (Figure 8.9 **A**). For the purpose of discussion we have labeled the four methyl groups, which give a single line in the ^{13}C NMR at room temperature due to a rapid chair–flip. At $-110\,°C$, the ring flip is slow, and we see two methyl signals (axial and equatorial) separated by 9.03 ppm, a value we shall call Δ, with the axial methyls being upfield from the equatorials.

Now consider the situation in which one of the methyl groups is replaced by CD_3, let's say position 4 for discussion. The equilibrium is no longer degenerate; either the structure with the CD_3 axial or the one with it equatorial is preferred. The ^{13}C spectrum *at room temperature* allows us to choose. As shown in Figure 8.9 **B**, we see two sharp lines in the methyl region separated by 0.184 ppm, a value we shall call δ, with a broader line in between them. To assign these peaks, we consider first that C_4, the deuterated carbon, is not seen, because of extensive C–D coupling and diminished nuclear Overhauser effects (NOEs). C_3 is expected to be broadened due to longer range C–D coupling, allowing us to assign the broad, central peak to C_3. Of the sharper peaks, one is due to C_1 and one is due to C_2. Experience tells us that C_3 should be shifted *upfield* by the nearby deuteriums. Since C_3 would otherwise show the same chemical shift as C_1, this allows us to assign C_1 as the downfield member of the two sharp peaks. Thus, C_1 spends more time in the equatorial (recall axial methyls are upfield) position than C_2, establishing that the preferred conformer has the CD_3 axial, as shown in Figure 8.9 **A**. One expects CD_3 to preferentially occupy an axial position relative to CH_3, because as established previously in Eq. 8.6, the CD_3 group is smaller.

Eq. 8.7 allows us to relate the chemical shift separations we have discussed to the equilibrium constant for the process, K. For this case, inserting $\Delta = 9.03$ ppm and $\delta = 0.184$ ppm gives $K = 1.042 \pm 0.001$. This corresponds to an energy difference of 0.024 kcal/mol! The isotopic perturbation of equilibrium method has allowed us to measure a very small deviation of an equilibrium constant from a value of 1.

$$K = \frac{\Delta + \delta}{\Delta - \delta}$$
(Eq. 8.7)

The reason the isotopic perturbation method is so powerful for carbocations is that Δ is usually *very* large in such systems. Since the quantity we measure, δ, is dependent on Δ and K, we should see large values of δ if we are perturbing an equilibrium, no matter how small the perturbation. For example, Figure 8.9 **C** shows the 1,2-dimethylcyclopentyl cation set up for an isotopic perturbation of equilibrium experiment. This is a conventional carbenium ion that experiences the equilibration shown through a hydride shift. Now Δ between C_1 and C_2 is estimated to be 261 ppm. Even a K value only slightly different from 1.0 should produce a large δ. In this case, $\delta = 82$ ppm, and $K = 1.9$. This shows that the carbenium ion prefers to be adjacent to a CH_3 rather than a CD_3. In Section 8.1.3 we noted that hyperconjugation is preferred to a C–H bond relative to a C–D bond, and this experiment is supportive evidence.

Now consider the effects of isotopic perturbation on a static, symmetrical system. We show an isotopic perturbation of resonance in Figure 8.9 **D**. We know that an allyl cation such as that shown is a single symmetrical structure; it is *not* a rapidly equilibrating pair of structures. For the sake of argument, though, let's test what we would expect if we indeed had a rapidly equilibrating pair of structures as shown with equilibrium arrows in Figure 8.9 **D**. Again, we would expect Δ to be very large, and so δ should also be quite significant, regardless of how small the deviation of K is from 1.0. In the system shown, however, δ is only 0.33 ppm. This is clearly just a direct isotope effect on a chemical shift. There has been no isotopic perturbation of equilibrium, because there is no equilibrium; only a single, static structure is involved.

We can see, then, that the isotopic perturbation of equilibrium method offers a way to distinguish a situation in which we have a rapidly equilibrating pair of cations from one in which we have a single static structure. As discussed in Chapters 1, 2, 11, and 14, this is a recurring issue in carbocation chemistry, and so the method has proven to be very useful.

8.1.5 Tunneling

Sometimes isotope effects of 50 or larger are observed for comparisons of H vs. D. Such dramatic isotope effects are most often attributed to quantum mechanical tunneling. Originally viewed as an exotic consequence of the nuances of quantum mechanics, detailed study has shown that tunneling, while certainly not common, is frequently involved in organic reactions under conventional conditions. At the end of this section, a Connections highlight gives an example with a common synthetic transformation. Tunneling is very common under cryogenic conditions. Since tunneling will be relevant to topics discussed in subsequent chapters, we make a slight digression here to lay out the basic concepts involved. This will allow us to explain why very large isotope effects are a hallmark of tunneling.

Tunneling is a quantum mechanical phenomenon involving penetration of the wavefunction for the molecule *through* the barrier for the reaction rather than over it. Figure 8.10 illustrates the basic idea. A full description of the physics of tunneling is inappropriate here, but we can gain some valuable insights from analyzing the results of such treatments. One of the simplest analyses of tunneling is due to Bell, and it leads to a simple modification of the Arrhenius equation by a tunneling correction factor, Q, as in Eq. 8.8. Here A is the normal Arrhenius parameter, E is the height of the barrier, m is the mass of the tunneling particle, and $2a$ is the width of the barrier. Through the β term, we can see that the tunneling correction factor is very sensitive to the mass of the tunneling particle. The influence of β is greatest when it is in the exponential, and so tunneling is much more likely for a light particle than a heavy particle. In a proton or hydrogen atom transfer or abstraction, it is reasonable to consider the hydrogen as the tunneling particle, and so we have a very light tunneling particle with a mass of 1. This is why tunneling is usually associated with proton or hydrogen transfers or electron transfers, but there are important exceptions. On going to deuterium, we double the mass, and the exponential dependence on m can produce very large isotope effects. Hence, when tunneling is involved, our arguments concerning zero point energies and vibrational levels are less relevant. Instead, we focus on a generic mass effect, with an increase in mass greatly slowing a tunneling reaction.

$$k = QAe^{-E/RT}$$
$$\text{where, } Q = \frac{e^{\alpha}}{\beta - \alpha}\left(\beta e^{-\alpha} - \alpha e^{-\beta}\right)$$
$$\alpha = E/RT$$
$$\beta = 2a\pi^2(2mE)^{1/2}/h$$

(Eq. 8.8)

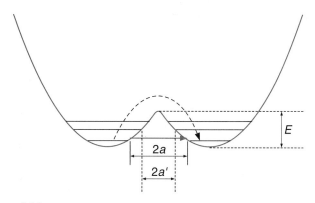

Figure 8.10
Diagram illustrating the key terms in a tunneling reaction. The gray, dashed arrow is the conventional, activated process, while the colored arrow is the tunneling path. Also shown are the barrier width, $2a$, and the width $2a'$, which would apply to tunneling out of the first excited vibrational state.

Eq. 8.8 reveals another interesting feature of tunneling. As with m, tunneling will be most effective when the barrier width, $2a$, is small (that is, the barrier is narrow). However, while the β term varies as the square root of m, it depends linearly on the barrier width, $2a$. As such, it is generally true that tunneling is much more sensitive to changes in barrier width than to mass. In fact, we will see that in certain cases particles much heavier than H/D can tunnel, as long as they are tunneling through a narrow barrier.

How do we think about the barrier width? Remember, the x axis in Figure 8.10 is a geometrical coordinate. A narrow barrier implies that nuclei must move only short distances to tunnel from one well to another. So, if we have a reaction in which only small nuclear movements can get us from one minimum to another, then tunneling might be important.

The tunneling correction term Q is *much* less sensitive to temperature than the Arrhenius component of Eq. 8.8. Thus, while conventional reactions slow down dramatically as we lower the temperature, tunneling reaction rates hardly change at all. As a result, tunneling is often much more important at very low temperatures, where even the smallest barrier could not be overcome, but it can be tunneled through.

Where does the generally small temperature dependence of tunneling come from? A particle tunneling through a barrier is a quantum mechanical effect, resulting from the fact that the true wavefunction of the reactant is not confined to the energy well of Figure 8.10, but can leak out into the next minimum. In this view, hydrogen tunnels well because its light mass gives it a very large de Broglie wavelength. Such a process should not be temperature dependent. However, consider the consequences of adding thermal energy to Figure 8.10. This could lead to a slight population of the second vibrational level ($n = 1$; remember that the lowest level is $n = 0$, the zero point). In a conventional, activated process, this excitation would have only a small influence on the rate, and that increase will be comparable to the Boltzmann factor disfavoring population of $n = 1$. In other words, this is not an issue for conventional reactions. However, for a tunneling reaction, the difference between $n = 1$ and $n = 0$ could be immense. This is because the *barrier width* is significantly diminished on going to the higher vibrational state; note how much smaller $2a'$ is than $2a$ in Figure 8.10. The shrinking of the barrier width can dramatically increase the tunneling rate, such that even a very small population of an excited vibrational state would dominate the tunneling process. It is in this way that tunneling reactions are sometimes found to have shallow, but non-zero temperature dependencies.

There are certain criteria used to determine if a large H/D isotope reflects a reaction that involves tunneling. First, the difference in the activation energies for the hydrogen- and deuterium-containing compound must be greater than the difference in their ZPEs. Second, tunneling reactions still proceed at low temperatures, as discussed above. This odd temperature dependence is supported by a deviation from linearity in an Eyring plot, where the slope becomes less negative at lower temperatures. Third, anomalously large differences in

---

Connections

An Example of Tunneling in a Common Synthetic Organic Reaction

The elimination of a selenoxide is a common, mild method for the formation of a carbon–carbon double bond, occuring via a syn-elimination pathway as shown to the right. The activation energy difference between the H- and D-substituted compounds was found to be 2.52 kcal/mol, giving an isotope effect of 74. Other experimental parameters were also found to support tunneling. For example, the A_H/A_D is 0.092, considerably different from 1. Furthermore, a barrier width for the reaction was calculated to be 0.82 Å, definitely less than the length of a common C–H bond (1.1 Å).

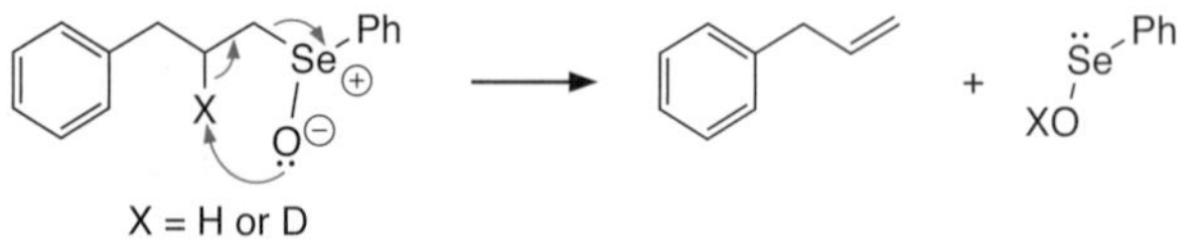

X = H or D

Selenoxide elimination

Kwart, L. D., Horgan, A. G., and Kwart, H. "Structure of the Reaction Barrier in the Selenoxide-Mediated Formation of Olefins." *J. Am. Chem. Soc.*, **103**, 1232–1234 (1981).

the Arrhenius pre-exponential factors are found, meaning A_D/A_H does not equal 1 as predicted by collisional theory. Remember, in an experimental determination, the product of Q and A (Eq. 8.8) will be the multiplier of the exponential. Since Q varies considerably for D vs. H, the empirically determined A factor will also vary considerably. Similarly, a large negative entropy of activation is indicative of barrier tunneling, recalling the connection between $\Delta S^\ddagger$ and the pre-exponential term. A large negative entropy of activation is interpreted to mean that a precise arrangement of the H must occur to permit tunneling through the barrier. Finally, the geometries of the reactants and products are usually very similar, meaning that the heavy atoms do not significantly change position during the reaction; instead, only the hydrogen moves.

8.1.6 Solvent Isotope Effects

The introduction of an isotope into a specific position (or positions) in a molecule in order to measure an isotope effect can sometimes be a synthetically challenging task. However, when the position to be deuterated is readily exchangeable with the solvent, stirring the reactant in a deuterated protic solvent will deuterate the reactant. For example, alcohol OHs, amine NHs, amide NHs, thiol SHs, etc., will all quite rapidly scramble the hydrogen atoms for deuterium atoms when placed in D_2O or deuterated alcohols such as CH_3OD. Reactions such as these are also excellent candidates for observing **solvent isotope effects**, which are changes in rates or equilibria that are seen in comparing normal vs. deuterated solvents.

Often the solvent is actually involved in the mechanism of the reaction, in which case isotope effects clearly can arise. Furthermore, one sometimes finds a solvent isotope effect when solvation of a transition state is different with the labeled and unlabeled solvent. The isotope effects can be primary, secondary, normal, or inverse depending upon the role that the solvent plays. As expected, direct bond cleavage of a hydrogen or deuterium atom in the solvent or of an exchangeable position in the rate-determining step will lead to a primary kinetic isotope effect. A protonation in an equilibrium prior to a rate-determining step can give either a normal or inverse effect depending upon how the equilibrium is shifted due to the isotopic substitution.

Fractionation Factors

As mentioned above, reactions involving a scrambling of the protons and deuterons between the reactant and solvent are good candidates for significant KIEs. This scrambling does not necessarily lead to a statistical distribution of the exchangeable entities; instead, equilibrium isotope effects are common. In these kinds of reactions, it is generally true that deuterium will prefer the site with the larger force constants. Measurable equilibrium isotope effects most often occur when the two basic centers exchanging the proton (or deuteron) are of different chemical types, such as oxygen or nitrogen. We measure what is known as a **fractionation factor** to determine whether the exchangeable site prefers hydrogen or deuterium relative to the preference of the solvent to have hydrogen or deuterium. The fractionation factors, therefore, tell us about the sum of all the different force constants in the exchangeable site [X–H(D)] relative to the sum in the solvent [S–H(D)]. At the end of the discussion of fractionation factors, an example of their use to characterize the strengths of hydrogen bonds is given in a Connections highlight.

Consider the equilibration of a solute molecule having an exchangeable X–H bond in the exchangeable solvent S–H (Eq. 8.9). To measure the fractionation factor, denoted as ϕ, we perform a competition experiment. We dissolve X–H in a mixture of S–H and S–D and measure the equilibrium constant for scrambling (Eq. 8.10).

$$X\text{–}H + S\text{–}D \; \rightleftharpoons \; X\text{–}D + S\text{–}H \qquad \text{(Eq. 8.9)}$$

$$K_{eq} = \phi = \frac{[S\text{–}H][X\text{–}D]}{[S\text{–}D][X\text{–}H]} = \frac{\dfrac{[X\text{–}D]}{[X\text{–}H]}}{\dfrac{[S\text{–}D]}{[S\text{–}H]}} \qquad \text{(Eq. 8.10)}$$

Table 8.1
Fractionation Factors*

Bond type	ϕ
RO–H(D)	1.0
HO⁻	0.5
RO–H₂(D₂)⁺	0.69
R₃C–H(D)	0.69
R₂N–H(D)	0.92
R₃N–H(D)⁺	0.97
RS–H(D)	0.42

*Gold, V. "Proteolytic Processes in H₂O–D₂O Mixtures." *Adv. Phys. Org. Chem.*, **7**, 259 (1969).

Hydrogen bonding to hydroxide

A fractionation factor greater than 1 means that the deuterium prefers site X over site S. Conversely, values less than 1 mean the deuterium prefers the solvent. Table 8.1 lists a few fractionation factors where the solvent is water. Alcohols statistically distribute deuterium with water ($\phi = 1.0$) because the vibrational modes in alcohols and water are all nearly identical. Carbon, nitrogen, and sulfur show a preference for hydrogen, meaning that the overall force constants are larger in water. Hydronium and hydroxide both show a preference for hydrogen. The bond in hydronium is predicted to be weaker than in water due to the positive charge, and since deuterium prefers sites with stronger bonds it makes good sense that the fractionation factor is less than 1.

Yet, it may seem odd that deuterium also prefers water over hydroxide. To understand this, remember that the deuterium concentrates where the overall force constants are larger, including the force constants for all intramolecular and intermolecular vibrations. Hence, we must examine hydroxide and its interactions with water. Approximately three hydrogen bonds are formed to the oxygen of hydroxide in water. These hydrogen bonds from water to hydroxide lower the force constants for the O–H bonds within the three respective waters. Apparently, this results in an overall preference for hydrogen in hydroxide. This means that the force constants for water hydrogen bonded to just water are larger overall than those for the hydroxide hydrogen bonded to water. In fact, if we assume a fractionation factor of 1.2 for the O–H bond in hydroxide and three fractionation factors of 0.7 for each hydrogen bond from water, the product of these four ($1.2 \times 0.7 \times 0.7 \times 0.7$) leads to the observed fractionation factor. Hence, we always use the product of fractionation factors for all the individual exchangeable sites to get the overall fractionation factor.

We can use the ratio of fractionation factor products to predict relative equilibrium constants for reactions in H₂O or D₂O. Consider the dissociation of an acid in H₂O or D₂O. Eq. 8.11 tells us that the ratio of equilibrium constants for a reaction in pure D₂O (K_D) to that in pure H₂O (K_H) is the ratio of the mathematical products of fractionation factors for the products (ϕ_i^P) and reactants (ϕ_j^r). Letters i and j are running tags of each exchangeable site in the product and reactant, respectively. In essence, the ratio between the equilibrium constants in these two solvents reflects the relative ability of the products and reactants to accept deuterium over hydrogen. We ask you to derive Eq. 8.11 in Exercise 4 at the end of the chapter.

$$\frac{K_D}{K_H} = \frac{\Pi_i^P \phi_i^P}{\Pi_j^r \phi_j^r} \qquad \text{(Eq. 8.11)}$$

Extending the concept to kinetics requires us to remember the tenets of transition state theory (TST). With TST we assume that there is an equilibrium between the transition state and the reactants. Therefore, we write Eq. 8.12 analogous to Eq. 8.11, where k_D is for the reaction with deuterium and k_H is for the reaction with hydrogen. As a result, we obtain the inverse of the kinetic isotope effect, because KIE $= k_H/k_D$.

$$\frac{k_D}{k_H} = \frac{\Pi_i^{\ddagger} \phi_i^{\ddagger}}{\Pi_i^r \phi_i^r} \qquad \text{(Eq. 8.12)}$$

Proton Inventories

Competition experiments between S–H and S–D for scrambling of an exchangeable site on a reactant X–H can be used to determine how many protons are being transferred or associated with rehybridation during the rate-determining step of a reaction. We demonstrate this below by examining an enzymatic reaction in a Connections highlight. To understand this experiment, we must consider measuring the rate constant for a reaction involving a change in the X–H(D) bond, where the molecules in solution are actually a mixture of X–H and X–D. The rate constant would be a weighted mean of the rate constants for the X–H and X–D molecules (Eq. 8.13). Letter n in Eq. 8.13 is the mole fraction of D at the exchangeable site, established both by the mole fraction of D in the solvent and the equilibrium fractionation factor for placing D at this site. We define a kinetic fractionation factor for the reaction as

---

Connections

Using Fractionation Factors to Characterize Very Strong Hydrogen Bonds

Recall from Chapter 3 the debate in enzymology about the relevance of a particular type of hydrogen bond known as a "low-barrier hydrogen bond" (LBHB) or "short–strong hydrogen bond". These hydrogen bonds can possess bond strengths on the order of 40 kcal/mol in the gas phase. In these bonds the barrier to hydrogen exchange between the donor (Dn) and acceptor atom (:Ac) is close to or lower than the ZPE of the bonds to the H. The force constant for the stretch involving H is smaller for the LBHB compared to a normal hydrogen bond, because the potential well that describes that bond vibration is wider with a LBHB.

One way that is used to characterize these hydrogen bonds is to measure a fractionation factor. Given the above analysis, a low isotope fractionation factor is predicted, because the force constant is smaller for the LBHB (look back at the potential wells for LBHBs in Chapter 3). For example, the observation that the fractionation factors for various substituted phthalate monoanions were between

0.5 to 0.6 was used to support the existence of LBHBs in these structures.

Strong hydrogen bond system

This example highlights the fact that we want to consider the force constants for vibrations to predict the site that the deuterium will prefer. LBHBs are very strong hydrogen bonds, but the force constant for the vibration of the hydrogen between the heavy atoms is low. Hence, fractionation factors are less than 1. The deuterium actually prefers the solvent.

Shan, S.-O., Loh, S., and Herschlag, D. "The Energetics of Hydrogen Bonds in Model Systems: Implications for Enzymatic Catalysis." *Science*, **222**, 97–101 (1996).

$\phi = k_D / k_H$. Here again the ratio is the inverse of the definition of the isotope effect. This leads to Eq. 8.14 and subsequently to Eq. 8.15 upon rearrangement. Eq. 8.15 makes good sense; when no D_2O is added, $n = 0$ and $k_n = k_H$. Conversely, when the reaction is run in pure D_2O, $n = 1$ and $k_n = k_D$.

$$k_n = k_H(1 - n) + k_D(n) \qquad \text{(Eq. 8.13)}$$

$$k_n = k_H(1 - n) + \phi k_H(n) \qquad \text{(Eq. 8.14)}$$

$$k_n = k_H(1 - n + n\phi) \qquad \text{(Eq. 8.15)}$$

Using analogous logic, we rewrite Eqs. 8.11 and 8.12, giving Eqs. 8.16 and 8.17, respectively.

$$\frac{K_n}{K_H} = \frac{\Pi_i^P(1 - n + n\phi_i^P)}{\Pi_j^r(1 - n + n\phi_j^r)} \qquad \text{(Eq. 8.16)}$$

$$\frac{k_n}{k_H} = \frac{\Pi_i^\ddagger(1 - n + n\phi_i^\ddagger)}{\Pi_i^r(1 - n + n\phi_i^r)} \qquad \text{(Eq. 8.17)}$$

Eq. 8.17 tells us that ratio of rate constants in mixtures of protio and deuterio solvents relative to a pure protio solvent depends upon each exchangeable site that has a fractionation factor different from unity between the transition state and the reactant. In practice, only those exchangeable sites that are undergoing either a bond cleavage or a rehybridization will have fractionation factors that differ between reactant and transition state. Hence, they are the only sites that typically need to be considered as contributing to Eq. 8.17. We call these protons ones that are "moving" or are "in flight" during the reaction. Very often it is assumed that the fractionation factors for the exchangeable sites on the reactant are close to unity because the respective bond strengths relative to the bond strengths in the solvent are similar, an assumption that is good for simple O–H and N–H bonds (see Table 8.1). This reasoning allows us to neglect the denominator of Eq. 8.17, thus leading to Eq. 8.18, where each term represents one of the protons that is moving in the transition state.

$$\frac{k_n}{k_H} = (1 - n + n\phi_1)(1 - n + n\phi_2)(1 - n + n\phi_3) \ldots \qquad \text{(Eq. 8.18)}$$

Each ϕ is the fractionation factor for a given proton that is moving—that is, the inverse of the isotope effect for each proton that is moving. If only one proton moves, a plot of n versus k_n/k_H is linear; if two protons move, the plot is quadratic in n; and if three protons move, the plot is cubic in n, etc. Hence, the term **proton inventory** is applied to such an experiment and plot. The shape of the curve allows us to inventory the number of protons that are moving, and fitting the curve can extract the individual fractionation factors.

Many reaction mechanisms involve the movement of one or more protons, so this technique is applicable to a vast number of reactions, particularly so in enzymology, where catalysis often involves the coupling of several proton transfers in a single step (see Chapter 9). Therefore, the proton inventory is an extremely powerful physical organic chemistry technique that has found use in an untraditional physical organic field.

Connections

The Use of a Proton Inventory to Explore the Mechanism of Ribonuclease Catalysis

The enzyme ribonuclease A cleaves single-stranded RNA and opens up 2′,3′-cyclic phosphodiesters. The classic mechanism for the ring opening involves an imidazole acting as a base to deprotonate water during nucleophilic attack, and an imidazolium protonating the leaving group (see below). If such a mechanism is operative, two protons are moving in the rate-determining step, and hence a two proton inventory should be found. Indeed, when plotting the catalytic rate constant ($k_{E,n}$) for this ring opening in various mixtures of D_2O and H_2O relative to the rate constant in pure H_2O ($k_{E,O}$), a curved plot was found, as shown to the right. The plot was fit using two ϕ values, both of 0.58. The inverse of the ϕ values can be considered as the isotope effect for each proton that is moving, giving in this case two isotope effects of 1.75. The results clearly support two simultaneous proton transfers in the rate-determining step.

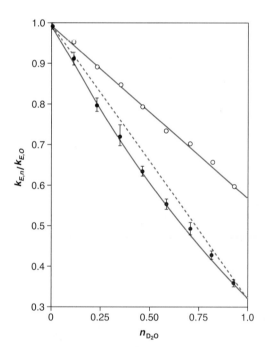

Closed circles represent the data used to fit a proton inventory of two protons moving in the rate-determining step. The open circles are the square root of the values for the closed circles. The dashed line simply shows a linear connection between the first and last points, thereby accentuating the curvature in the plot of the closed circles.

RNAse mechanism

Matta, M. S., and Vo, D. T. "Proton Inventory of the Second Step of Ribonuclease Catalysis." *J. Am. Chem. Soc.*, **108**, 5316 (1986).

8.1.7 Heavy Atom Isotope Effects

Our discussion of isotope effects has focused upon hydrogen versus deuterium. Isotope effects for other elements are possible, however, and experiments involving isotopes of C, O, N, and Cl are somewhat common. The problem becomes one of the sensitivity of the technique used to measure the isotope effects, because they are very small (see the earlier Going Deeper highlight entitled "An Ingenious Method for Measuring Very Small Isotope Effects", page 432). Recall that the stretching frequency of the bond undergoing cleavage in an experiment involving a primary kinetic isotope effect is given by Eq. 8.2, where the reduced mass of the bond is in the denominator. The difference in frequency of bonds with different isotopes arises from differences in the reduced masses. When comparing hydrogen to deuterium, there is a 100% increase in mass, which has a reasonable effect on the size of the two reduced masses. However, when comparing ^{13}C to ^{12}C, there is only an 8% change in mass, which has little effect on the reduced masses, and therefore very small primary isotope effects are found. Furthermore, secondary isotope effects are almost completely negligible.

Consider two examples—a decarboxylation and an S_N2 reaction. The primary kinetic isotope effects (PKIE, k_{12}/k_{14} for ^{14}C relative to the natural abundance C, which is mostly ^{12}C) at the methylene and carboxyl carbons in the decarboxylation of malonic acid are 1.076 and 1.065, respectively (Eq. 8.19). The values are very close, indicating that a bond to each of these carbons breaks in the rate-determining step. The ^{37}Cl isotope effect (k_{35}/k_{37}) in the displacement of chloride from benzyl chloride by cyanide is 1.0057, indicating that the bond to chlorine breaks in the rate-determining step (Eq. 8.20).

(Eq. 8.19)

(Eq. 8.20)

8.1.8 Summary

The study of isotope effects is extremely valuable in deciphering mechanisms. Such studies are necessary whenever we suspect that a bond to a hydrogen is moving or rehybridizing during a rate-determining step. A kinetic isotope effect greater than 1 is good evidence of a bond cleavage or rehybridization. A kinetic isotope effect less than 1 is good evidence of a rehybridization to a bond that has one or more stiffer vibrational modes.

8.2 Substituent Effects

So far, we have considered two types of experiments for studying reaction mechanisms. First was an analysis of kinetics (Chapter 7), which can give the order of reactants involved in a mechanism prior to or during the rate-determining step. Second, we now have a method to determine if a bond to hydrogen has been broken or rehybridized. This latter experiment gives us some limited structural information about the activated complex.

More in-depth analysis of the structure of the transition state is obtained from studies of substituent effects. A **substituent** is any group on a molecule, such as a methyl, nitro, hydroxy, etc. A **substituent effect** is the manner in which the reactivity of the molecule changes when substituents are changed. Note the conceptual difference here from the two methods presented thus far. In order to investigate the mechanism of a reaction using substituent effects, we are going to study a *different* reactant, one in which a substituent has been added or changed. It should be appreciated from the outset that this is a fundamentally different approach (we don't consider the H/D substitution associated with isotope effect studies to be a substituent effect), and one that is considerably less direct than kinetic or isotope effect studies. As such, studies of substituent effects are susceptible to misinterpretation and abuse; caution is in order. Nevertheless, studies of substituent effects represent a key pillar of physical organic chemistry. When they are carefully applied, substituent effects are used to determine how the free energies of reaction and activation vary as a function of chemical structure. In either a kinetic or thermodynamic analysis, we change the structure of the reactant or solvent and see how this changes the function (reactivity) of the molecule; hence the terms **structure–function relationship** or **structure–reactivity relationship**. In the context of pharmaceutical studies, where activity, bioavailability, and other medicinally related data are collected as a function of the chemical structure of the drug, the substituent effect studies are referred to as **structure–activity relationships (SARs)**. Most important for physical organic chemistry, the nature of the structure–reactivity relationship is often informative about the mechanism of the reaction.

The logic of conventional structure–function relationships should be familiar. Experiments can be designed to test for changes in charges along a reaction coordinate by interchanging functional groups, such as switching an electron donating group to an electron withdrawing group. If a positive charge is being created in a rate–determining step, then adjacent electron donating groups should stabilize the transition state and the reaction should speed up. Conversely, adjacent electron withdrawing groups should destabilize the transition state and therefore retard the reaction. Similar effects can be observed for equilibria.

As we will see below, there are linear relationships (called **linear free energy relationships** or **LFERs**) between the activation free energy or reaction free energy change induced by a substituent and a parameter that describes the electron donating or electron withdrawing characteristics of the substituent.

Changing substituents in the reactants or solvent should also influence the steric congestion, solvation, leaving group ability, nucleophilicity, acidity or basicity, and a variety of other chemical attributes. Just as with studying the change in charge along a reaction coordinate, we can design experiments where substituents are manipulated in order to determine the influence of these other chemical effects on the mechanism. We will show that there are LFERs for activation free energies and reaction free energies with parameters that describe all these chemical effects. In each analysis, the goal of changing substituents is to determine how that change affects the activation free energy (and thus the structure of the activated complex), or affects the equilibrium for the reaction.

As mentioned above, there are LFERs that correlate free energies with parameters that describe the character of substituent effects. It has been experimentally determined that in most cases, individual substituents influence reactions in a consistent manner. For example, electron withdrawing groups are consistently electron withdrawing, no matter what reaction they are involved in. What varies from one reaction to another is the magnitude of this influence. Sometimes a reaction is very sensitive to changes in substituents, and sometimes there is little or no effect.

When performing structure–function experiments we must remember that any structural change results in a perturbation of the energy surface. You can't do a structure–function study without changing structures, and hence the energy surface. Such perturbations for S_N2 and S_N1 reactions were analyzed in Section 7.8.1. In that example, the change of hydrogen substituents to alkyl substituents shifted the position of the S_N2 transition state on a three-dimensional reaction coordinate diagram. There was a point where additional alkyl substituents retarded the S_N2 route enough, and sped up the S_N1 route to such an extent that the mechanism changed. The interchange of substituents is always susceptible to such changes to the energy surface, and we must always be alert to a possible mechanism change.

In order to design successful experiments by varying substituents, we should start with a good understanding of how different groups influence structure and bonding and of the various mechanisms by which substituents can influence the rate of a reaction.

8.2.1 The Origin of Substituent Effects

In Chapter 1 two different views of bonding were described: valence bond theory (VBT) and molecular orbital theory (MOT). In MO theory there are orbitals that are spread out over all atoms in a molecule. When adding or changing substituents, new molecular orbitals are created that involve the atomic orbitals on the substituents. Although this theory can be used to analyze substituent effects successfully, the concept of delocalized molecular orbitals presents difficulties in visualizing localized changes in a molecule brought about by a substituent change. Thus, substituent effects are almost always discussed in terms of our hybrid VBT/MOT approach, wherein localized bonding effects and isolated bonds between atoms are considered.

Field Effects

When one substituent is changed to another, bond dipoles can change in magnitude and direction, or formal charges can change. Localized charges resulting from bond dipoles and formal charges present electric fields that can have an effect on a remote site in a molecule. A **field effect** is one that originates from such a through-space interaction. For example, a quaternary ammonium with a full positive charge produces an electric field that can influence distant atoms in the same or neighboring molecules. The quaternary ammonium group is also a strongly electron withdrawing group, and hence field effects are often very hard to separate from inductive effects (see below).

Usually, however, the field effect is associated with a bond dipole that is aligned so as to stabilize or destabilize a charge forming or diminishing during crossing of the transition state. Except in cases where full charges are present, field effects are typically small and rapidly diminish with distance, and when inductive or resonance effects are also present, the latter dominate.

Field effects

Dipole field effects

Inductive Effects

An **inductive effect** (introduced in Chapter 1) results from the ability of an atom or group of atoms to withdraw or donate electrons through σ bonds. Strongly electronegative atoms or groups are best at drawing electrons to themselves (see Section 1.1.8). Conversely, a group can donate electrons via the σ bond framework. The further away the group from the site of reaction, the lower its ability to affect the reaction via induction. For example, chloroacetic acid is substantially more acidic than acetic acid. However, the increase in acidity induced by the electronegative chlorine diminishes the further away from the carboxyl it is positioned (see other examples in Chapter 5, and recall from Chapter 5 that entropy effects are likely important in this example, too). An interesting example of an inductive effect is given in the Connections highlight below, where an important biomaterial is examined.

Increasing acidity due to an inductive effect

Convenient scales have been developed to indicate the relative inductive effects of substituents (see Tables 1.1 and 1.2 for atom and group electronegativities). In the present chapter, however, we will define a new scale called sigma (σ), to relate the different electron donating or electron withdrawing abilities of atoms and groups.

Connections

A Substituent Effect Study to Decipher the Reason for the High Stability of Collagen

Collagen is the most abundant protein in animals. It forms connective tissue of high tensile strength and thermal stability, such as cartilage. Collagen consists of three protein strands wrapped together to form a tight triple helix (see below). The sequences of the individual strands have glycine as every third amino acid with intervening prolines and hydroxyprolines. For some time it was thought that the hydroxyprolines led to the high structural stability by being involved in a network of hydrogen bonds between strands. However, to test whether the inductive nature of the hydroxyl substituent influenced the stability, this substituent was switched to fluorine, a group that is poor at making hydrogen bonds (see Chapter 3) but is still electronegative.

It was found that the circular dichroism spectra (see Chapter 6) are identical whether the strands possess fluoroproline or hydroxyproline. In fact, it was found that the fluoroproline-containing strands created even more stable collagen. The conclusion was that the trans conformation of the hydroxyprolyl peptide bond (see the Connections highlight in Section 1.1.10 for definitions of cis / trans peptide bonds), required in collagen formation, is preferred with increasing electron withdrawing groups, thereby stabilizing the overall peptide structure. This simple structure–function study greatly increased our knowledge of a very important biomaterial.

Fluoroproline   Hydroxyproline

Holmgren, S. K., Taylor, K. M., Bretscher, L. E., and Raines, R. T. "Code for Collagen's Stability Deciphered." *Nature*, **392**, 666 (1998).

Hydroxyproline

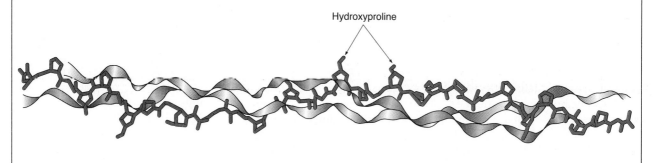

A common resonance effect

Resonance Effects

A **resonance effect** reflects the ability of an atom or group of atoms to withdraw or donate electrons through π bonds. This is also sometimes referred to as a **mesomeric effect** in older literature. This effect is solely a valence bond concept that describes the electronic structure of a molecule when more than one canonical structure of a molecule can be drawn (see Chapter 1 again). Resonance effects can be important for understanding charge distributions in molecules that result from substituent changes quite remote from the site of reaction. A typical example is shown in the margin, in which methoxy induces increased negative charge on a carbonyl group remote on a benzene ring. Scales have been developed that relate the ability of an atom or group to participate in resonance (see below).

Polarizability Effects

Another change that results when substituents are interchanged is the polarizability of the molecule. The polarizability of a molecule is defined as the extent to which the electron cloud of the structure can undergo distortion, and the polarizability of molecules and groups was extensively discussed in Section 1.1.12. A hard compound, in which the electron cloud is tightly held, is not very polarizable (see the discussion of hard and soft acids and bases in Section 5.6.1). Conversely, a soft compound is one in which the electron cloud is more diffuse, and therefore polarizable. Differences in polarizability influence nucleophile and nucleofuge properties, and relative solvation. For example, a sulfide is more polarizable than an alkoxide, making the sulfide a better nucleophile and a better leaving group, because in both leaving group departure and nucleophilic attack the electron clouds are distorting.

Steric Effects

Steric effects can also have a dramatic influence on the rate of a reaction, as well as conformations (see Chapter 2). Large atoms or groups influence the manner in which molecules collide, often deflecting the reactants away from the angle or depth of collision necessary for the reaction to occur. For example, the S_N2 reaction becomes slower due to steric effects as the carbon with the leaving group is more highly substituted with alkyl groups (see Chapter 11). The nucleophile cannot penetrate to the carbon with the leaving group when larger groups are attached. Again, scales have been developed to measure steric effects. You have already been introduced to one such scale; the A values given in Section 2.3.2 indicated the bias for equatorial versus axial substitution on a cyclohexane ring. A new scale, called the Taft parameters, is given below.

Solvation Effects

Finally, experiments involving changes in substituents are not just confined to the reactants. Changing groups in the solvent or completely switching solvents can also be considered to be a substituent effect. The five parameters—field, inductive, resonance, polarizability, and steric—all still play roles. However, now field effects can become quite important because the solvent dipoles can become directly involved in the reaction. The effects that changing the solvent have on the activation or reaction free energies are called **solvent effects**, to specifically distinguish them from substituent effects.

8.3 Hammett Plots—The Most Common LFER. A General Method for Examining Changes in Charges During a Reaction

Substituent effects come in the five kinds mentioned previously: field, resonance, inductive, polarizability, and steric. The first four can all be considered as electronic effects, whereas steric effects largely depend upon the size of the substituent. Hence, there is the often quoted notion that "trends in chemistry can be explained by either electronic or steric effects". However, even steric effects are electronic in origin. They are repulsions brought about by atoms approaching within their respective van der Waals contact distances, where the electron clouds of the groups involved repel each other. Most chemists, however, separate the concepts of sterics and electronics.

8.3.1 Sigma (σ)

To picture how reaction mechanisms vary as a function of the electronic changes induced by substituents, chemists use **Hammett plots**. Hammett defined a scale that measured the ability of substituents to influence the acidity of benzoic acid (Eq. 8.21). The substituents are placed meta or para to the carboxylic acid to eliminate any possible steric effects associated with an ortho substituent, and therefore only field, polarizability, inductive, and resonance effects should be operative.

$$\text{(Eq. 8.21)}$$

Eq. 8.22 was used to define a **substituent parameter** σ_X for each substituent X. Hydrogen is the reference substituent. Thus, all acidity equilibrium constants for the substituted benzoic acids are compared to the equilibrium constant for benzoic acid itself ($\sigma_H = 0$ by defi-

Table 8.2
σ Values for Several Commonly Encountered Substituents[*‡]

Substituent	σ_{meta}	σ_{para}	σ^+	σ^-
–NH$_2$	–0.09	–0.66	–1.3	
–OH	0.13	–0.38	–0.92	
–OCH$_3$	0.10	–0.27	–0.78	
–C(CH$_3$)$_3$	–0.09	–0.15	–0.26	
–CH$_3$	–0.06	–0.14	–0.31	
–Si(CH$_3$)$_3$	–0.04	–0.17		
–NHC(O)CH$_3$	0.14	0.0	–0.6	0.47
–Ph	0.05	0.05	–0.18	0.08
–I	0.35	0.18	0.13	
–Br	0.37	0.26	0.15	
–Cl	0.37	0.24	0.11	
–F	0.34	0.15	–0.07	
–C(O)CH$_3$	0.36	0.47		0.82
–OC(O)CH$_3$	0.39	0.31	0.18	
–C(O)OH	0.35	0.44		0.73
–CF$_3$	0.46	0.53		0.74
–CN	0.62	0.70		0.99
–NO$_2$	0.71	0.81		1.23
–N(CH$_3$)$_3{}^+$	0.99	0.96		

[*]Ritchie, C. D., and Sager, W. F. "An Examination of Structure–Reactivity Relationships."
Prog. Phys. Org. Chem., **2**, 323 (1964).
[‡]σ^+ and σ^- are for para substitution.

nition). Table 8.2 gives a number of σ values (σ^+ and σ^- values will be discussed later in this chapter). More extensive compilations are given in references at the end of the chapter. A different set of σ values is necessary for each different position on the benzoic acid, because the ability of a substituent to influence the acidity of benzoic acid depends upon its position relative to the carboxyl group. When σ is negative, the substituted benzoic acid is less acidic than benzoic acid itself, and when σ is positive, the substituted benzoic acid is more acidic. Note that electron donating groups have negative σ values and electron withdrawing groups have positive σ values. This trend is exactly as predicted, because electron withdrawing groups should stabilize the negative charge of the carboxylate and electron donating groups should destabilize this charge.

$$\log\left(\frac{K_X}{K_H}\right) = \sigma_X \qquad \text{(Eq. 8.22)}$$

Pulling electrons via induction

One interesting feature about the ionization of benzoic acids becomes apparent upon studying Table 8.2. The σ_{para} values generally reflect a larger influence of the substituent at this position than do the σ_{meta} values (the absolute value of $\sigma_{para} > \sigma_{meta}$), even though the meta position is closer to the ionizing group than is the para position. This difference in part reflects the ability of the para position to influence charge at the starred carbon (see margin) via resonance, an influence that is not possible for the meta position. This difference is clearly evident with the hydroxy and methoxy groups. In the meta position these groups are found to be electron withdrawing toward the starred carbon, an inductive effect. In the para position, these groups are electron donating, a resonance effect. Note that while this is a reso-

Donation of electrons via resonance

nance effect, it is *not* resonance with the carboxylate anion. The negative charge of benzoate anion is not in conjugation with the aromatic ring, and so cannot be stabilized by resonance.

8.3.2   Rho (ρ)

Now that a scale for substituent effects has been established, we can determine if other reactions respond to substituents the way benzoic acid does. The goal is to use benzoic acid ionization as a reference reaction that creates a negative charge and compare other reactions to it as a means to see if they also create a negative charge, or conversely, a positive charge. Furthermore, we want to determine if different reactions are more or less sensitive to the substituents than are the acidities of benzoic acid derivatives. To do this, we use the Hammett relationships given in Eqs. 8.23 and 8.24 for thermodynamic and kinetic analyses, respectively. To determine ρ, plot $\log(K_X/K_H)$ or $\log(k_X/k_H)$ versus σ_X for the new reaction under study. Rho (ρ) is simply the slope of this plot.

$$\log\left(\frac{K_X}{K_H}\right) = \rho\sigma_X \qquad \text{(Eq. 8.23)}$$

$$\log\left(\frac{k_X}{k_H}\right) = \rho\sigma_X \qquad \text{(Eq. 8.24)}$$

Rho describes the sensitivity of the new reaction to substituent effects relative to the influence of the substituent on the ionization of benzoic acid. It is called the **reaction constant** or **sensitivity constant** for each new reaction under study. The following values of ρ lead to the associated conclusions:

a. When $\rho > 1$, the reaction under study is more sensitive to substituents than benzoic acid, and negative charge is building during the reaction.

b. When $0 < \rho < 1$, the reaction is less sensitive to substituents than benzoic acid, but negative charge is still building.

c. When ρ is equal to or close to 0, the reaction shows no substituent effects.
    This can mean no change in charge occurs in the equilibrium or rate-determining step.

d. When ρ is negative, the reaction is creating positive charge.

Figure 8.11, for example, shows a plot of the log of ionization constants of substituted phenylacetic acids in water, and the ionization of substituted benzoic acids in ethanol. The ρ for phenylacetic acid derivatives is 0.56, while that for benzoic acid derivatives in ethanol is

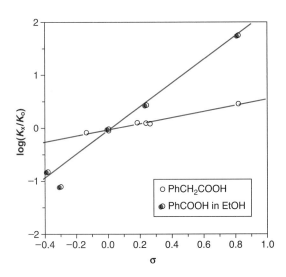

Figure 8.11
Hammett plots. For phenylacetic acid ionization constants and for benzoic acid in ethanol. Data to generate these plots were taken from Bright, W. L., and Briscoe, H. T. "The Acidity of Organic Acids in Methyl and Ethyl Alcohols." *J. Phys. Chem.*, **37**, 787 (1933), and Dippy, J. F., and Williams, F. R. "Chemical Composition and Dissociation Constants of Mono-Carboxylic Acids. Part I. Some Substituted Phenylacetic Acids." *J. Chem. Soc.*, 161 (1934).

2.25. These indicate much lower and higher sensitivities of the acidities to the substituent effects, respectively. The smaller ρ for phenylacetic acid derives from the more remote position of the substituents to the carboxyl group relative to that in benzoic acid. The larger ρ value for the ionization of benzoic acid in ethanol reflects reduced stabilization of the negative charge in the carboxylate product in ethanol relative to that in water (remember $\rho = 1$ in H_2O). Hence, the substituents become more important in stabilizing or destabilizing the negative charge in the product. Interestingly, the lines are linear even though the ρ values are derived from σ-values of benzoic acid ionization in water. The linear behavior is indicative of the general mathematics of LFERs, which we discuss in Section 8.6.1.

Very often the magnitude of ρ is used as a guide to the amount of charge that has developed in a transition state or in the product. Such an interpretation must be made with caution, because ρ really only relates the sensitivity of ionization to the substituents. In the examples just discussed, the amount of charge on the products is the same in all three reactions (phenylacetic acid, and benzoic acid in ethanol or water), but the ρ values are significantly different.

8.3.3   The Power of Hammett Plots for Deciphering Mechanisms

More interesting than the ionization of carboxylic acids is the application of Hammett plots to reactions that have no resemblance to an acid–base reaction. In fact, as has been alluded to, the σ values can be used to analyze the kinetics of reactions, even though they are based upon the analysis of the thermodynamics of an acid dissociation reaction. When a linear Hammett correlation is found for the kinetics of a reaction, the ρ value gives us information about the change in charge during the rate-determining step. We use the four scenarios labeled a–d in the preceding section to draw conclusions about the new reaction under study. Let's consider a few organic chemistry examples in detail.

The rate constants for the base-induced hydrolysis of methyl benzoate in water produce a ρ value of 2.23. This value tells us that the rate-determining step for this hydrolysis creates negative charge. Mechanisms involving nucleophilic displacement on the methyl group or addition to the carbonyl (Figures 8.12 **A** and **B**) are both consistent with this ρ value, because a full negative charge is created on the product. Here, the ρ value supports either mechanism, and other techniques are needed to distinguish the two possibilities.

The ρ value for the reaction of diphenylmethyl chloride with ethanol is –5.09. A large and negative ρ value indicates significant positive charge building in the rate-determining step. Both an S_N1 and S_N2 pathway create positive charge on the organic structure containing the substituents (Figures 8.12 **C** and **D**). It is the magnitude of the ρ value in this case that is informative. Such a large negative ρ is consistent with an S_N1-like mechanism, wherein a carbocation is forming. It is inconsistent with an S_N2 mechanism involving little or no change in charge on the carbon undergoing substitution.

You might be wondering what are considered large, medium, and small ρ values. As with many parameters in chemistry, there is no certain cut-off, but instead we consider a ρ value large or small relative to other ρ values for similar reactions. Thus, the conclusions drawn above are partially based upon knowledge of ρ values for other reactions.

As a final example, consider the ring opening of ethylene oxide by substituted phenoxides (Eq. 8.25). The ρ value for this reaction is –0.95. This value indicates the formation of positive charge during the rate-determining step on the organic structure containing the substituents. At first this interpretation might seem confusing, because no positive charges are involved in this reaction at all. In this case the diminution of negative charge is tantamount to the creation of positive charge. This value is perfectly logical, because the anionic oxygen of the nucleophile is adding to the electrophilic carbon of the epoxide. We learn from this example that we must be aware that decreases in negative charge will give negative ρ values, and correspondingly decreases in positive charge will give positive ρ values.

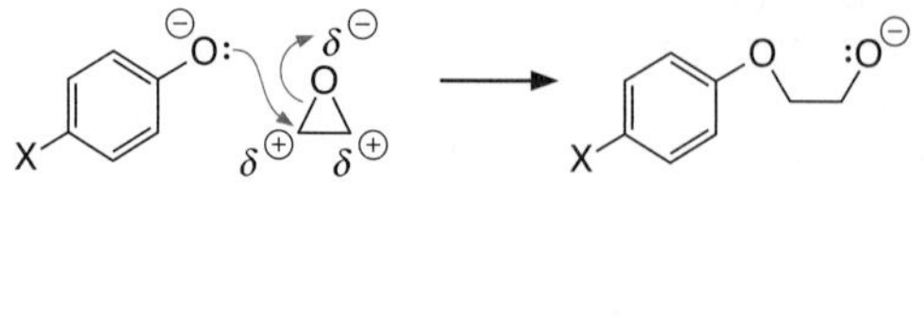

(Eq. 8.25)

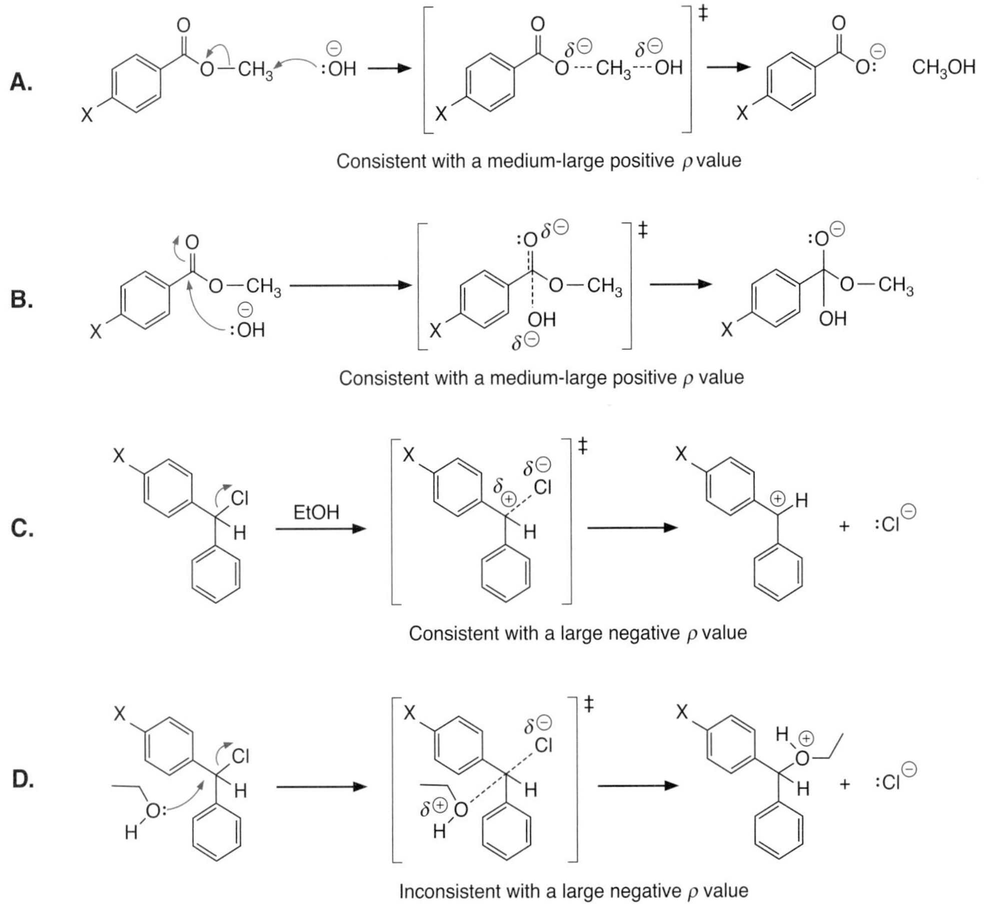

A. Consistent with a medium-large positive ρ value

B. Consistent with a medium-large positive ρ value

C. Consistent with a large negative ρ value

D. Inconsistent with a large negative ρ value

Figure 8.12
Examples of reactions and expected ρ values. A ρ value of 2.23 for the reactions shown in **A** and **B** does not distinguish between these two mechanistic alternatives. A ρ value of –5.09 for the first step of the substitution reaction given in **C** and **D** supports alternative **C**.

These examples highlight the power of LFERs, particularly Hammett plots. They indicate the change in charge when comparing the reactant to the transition state or product. This physical organic technique has often been applied (or in some cases "rediscovered") in fields well outside of traditional organic chemistry. The following Connections highlight presents an example from bioorganic chemistry, but numerous others can be found in the literature.

8.3.4 Deviations from Linearity

Recall that ρ, the parameter that characterizes a particular reaction in a Hammett analysis, is the slope obtained from the plot of k_X / k_H vs. σ. The plot must give a straight line to get a ρ value. What does it mean if a Hammett plot is not linear?

If a reaction creates a consistent amount of charge in its transition state regardless of the substituents attached, then we should find a linear Hammett plot. Therefore, if we define the reaction coordinate as the extent of creation of charge on the transition state, then the transition state must not move as substituents are interchanged. When the transition state does change significantly with regard to the extent of charge development, however, the Hammett plot will not be linear. For example, if more and more negative charge resides on the activated complex as the substituent becomes increasingly electron withdrawing, the Hammett plot will not be linear. This case is often associated with a gradual curve in the Hammett

Connections

Using a Hammett Plot to Explore the Behavior of a Catalytic Antibody

There are two competing mechanisms for the hydrolysis of aryl carbamate esters. One involves prior deprotonation followed by expulsion of the phenoxide leaving group to give an isocyanate that is trapped by water (Path A shown below). The second involves nucleophilic attack at the carbonyl followed by leaving group departure (Path B). With an H on the N, Path A is favored. When the carbamate N is missing a hydrogen, Path B dominates. The ρ values for Paths A and B are near 3 and 1, respectively (X = substituents).

Recently, a new way to develop catalysts has emerged known as the study of **catalytic antibodies**. Herein, transition state analogs for various reactions are created, and antibodies are isolated that bind these analogs. As we will see in Chapter 9, the binding of the transition state of a reaction is the major factor by which catalysis is achieved with enzymes. Thus, the antibodies created to bind a transition state analog are often found to catalyze the reaction involving the transition state the analog was patterned after.

To catalyze the hydrolysis of aryl carbamate esters such as i via Path B, transition state analog ii was synthesized (see above right). The tetrahedral phosphonate anion mimics the anionic tetrahedral intermediate formed along Path B. Carbamates i have a hydrogen on the N, and therefore are expected to hydrolyze via Path A in solution, and indeed a ρ value of 2.628 was found. Yet, a Hammett plot for the hydrolysis of carbamates catalyzed by antibodies that bind ii gave a ρ value of 0.526 (see plot at right). This value supports Path B, indicating that the use of transition state analog ii redirected the antibody-catalyzed mechanism to the less favored pathway. This study shows that antibody catalysis can be used to effect disfavored transformations, and highlights how ρ values can be used to distinguish between mechanisms.

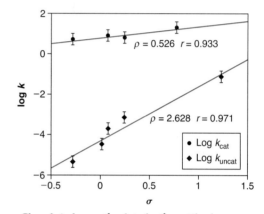

Transition state analog

Closed circles are the data for the antibody-catalyzed pathway. The diamonds are the data for the reaction in solution.

Wentworth, P., Jr., Datta, A., Smith, S., Marshall, A., Partridge, L. J., and Blackburn, G. M. "Antibody Catalysis of $B_{Ac}2$ Aryl Carbamate Ester Hydrolysis: A Highly Disfavored Chemical Process." *J. Am. Chem. Soc.,* **119**, 2315–2316 (1997).

Competing hydrolysis mechanisms

plot, indicating that the position of the transition state is incrementally changing as the substituents are interchanged.

An abrupt change in the slope of the plot with two intersecting lines can indicate a sudden change in the rate-determining step as a function of substituents. Such a break can also often indicate a change in mechanism for the reaction. After examining the σ^+ scale below, an example of an abrupt change in the slope of a Hammett plot is given in a Connections highlight. A change in the rate-determining step or mechanism is always of concern when examining LFERs, because changing substituents can change the relative energies of intermediates on the energy surface.

8.3.5 Separating Resonance from Induction

As noted above, we cannot draw resonance structures that delocalize the negative charge of benzoate onto the benzene ring via the π electron system. Therefore, the Hammett σ_{meta} and σ_{para} values do not include effects for direct resonance stabilization of the carboxylate negative charge. Yet, many reactions of interest create negative or positive charges that can be stabilized by delocalization via resonance with the substituent. For these reactions, we find that Hammett plots using σ values have considerable scatter. Therefore, two new substituent effect scales were produced, one for groups that stabilize negative charges via resonance (σ^-), and one for groups that stabilized positive charges via resonance (σ^+). The σ^- scale is based upon the ionization of para-substituted phenols (Eq. 8.26), for which groups like nitro can stabilize the negative charge via resonance (see margin). The σ^+ scale is based upon the heterolysis (S_N1) reaction of para-substituted phenyldimethyl chloromethanes (Eq. 8.27), for which groups like amino can stabilize the positive charge via resonance (see margin). Several σ^+ and σ^- values are given in Table 8.2. Note that the σ^+ values are defined so that negative ρ values correspond to the creation of positive charge, just as with normal Hammett plots. The electron withdrawing groups have positive σ^+ values, and the electron donating groups have negative σ^+ values, just as with σ values. One example of the use of the σ^+ scale is shown in the next Connections highlight.

p-Nitrophenolate

p-Aminobenzylic cation

(Eq. 8.26)

(Eq. 8.27)

When examining a brand new reaction, we often do not know which set of σ values is the most appropriate to use. However, the experimental measurements we make (rates or equilibria with varying substituents) are the same no matter what σ values are applied. Therefore, one develops plots based upon the various σ values and determines which has the least amount of scatter, thereby indicating if the stabilization of the charges involved is dominated by inductive or resonance effects.

A method for separating inductive effects from resonance effects was proposed by Yukawa and Tsuno (Eq. 8.28). Here the "r" value expresses the influence of resonance on the new reaction relative to the influence with benzoic acid. Since r is multiplied by the difference between the resonance and inductive substituent constants (σ^+ and σ, respectively), it describes the sensitivity of the reaction to resonance. When r is equal to zero, there is no difference in resonance effects for the new reaction compared to benzoic acid, but when $r > 0$,

Connections

An Example of a Change in Mechanism in a Solvolysis Reaction Studied Using σ^+

The substitution of various leaving groups from 1-phenyl-ethyl by azide or solvent (a 50:50 mixture of trifluoroethanol and water) was studied as a function of σ^+, and the results obtained are shown at right.

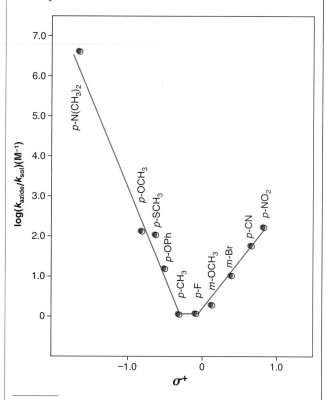

This plot is a slight variation from that described for a basic σ^+ plot, in that the reaction with solvent is used as the reference reaction rather than a comparison to one single substituent.

A very distinct break occurs. Those substituents with near zero or negative σ^+ values show rapidly increasing azide substitution relative to solvent. Those substituents with σ^+ values near zero produce nearly equal mixtures of products from azide and solvent, and finally, as σ^+ becomes more positive, the azide substitution again dominates but with a lower absolute value of the slope than with the negative σ^+ value substituents.

The data are consistent with a change in the mechanism of the substitution by azide from S_N1 to S_N2, which occurs as the stability of the carbocation is decreased. Those phenylethyl derivatives with negative σ^+ values (electron donating groups) undergo the substitution via an S_N1 mechanism, and those substituents with positive σ^+ values proceed via an S_N2 pathway. Furthermore, the data support a very clear reactivity–selectivity relationship. As the carbocation becomes more stable (the left side of the graph), it becomes more selective for the better nucleophile (azide).

Richard, J. P., and Jencks, W. P. "A Simple Relationship between Carbocation Lifetime and Reactivity–Selectivity Relationships for the Solvolysis of Ring–Substituted 1-Phenylethyl Derivatives." *J. Am. Chem. Soc.*, **104**, 4689–4691 (1982).

the new reaction is more sensitive than benzoic acid to resonance effects, and when $r < 0$, the new reaction is less sensitive.

$$\log\left(\frac{k_X}{k_H}\right) = \rho[\sigma + r(\sigma^+ - \sigma)] \qquad \text{(Eq. 8.28)}$$

The concept of separating the influence of resonance and induction on a reaction is quite appealing, because it tells us how the charge on the transition state is distributed. In fact, breaking down a substituent effect into all the different individual effects (resonance, induction, field, polarizability, steric) can be quite informative. For example, Swain and Lupton separated electronic LFERs into two substituent parameters: F (field or inductive) and R (resonance), with sensitivity factors f and r (Eq. 8.29), respectively. The F and R parameters can be found in references given at the end of the chapter. In this analysis, one can let the $fF + rR$ term equal the traditional $\rho\sigma$ values, thereby deciphering what fraction of the ρ value derives from resonance or induction. The following Connections highlight shows the use of this approach in an organometallic example.

$$\log\left(\frac{k_X}{k_H}\right) = fF + rR, \text{ where } fF + rR = \rho\sigma \qquad \text{(Eq. 8.29)}$$

Connections

A Swain–Lupton Correlation for Tungsten-Bipyridine-Catalyzed Allylic Alkylation

One very useful reaction in organic synthesis is the allylic alkylation of allyl carbonates catalyzed by various metal complexes (see Chapter 12 for mechanistic details). In the case of the alkylation of carbonates such as *i* and *ii*, catalyzed by the tungsten complex shown, the ratio of the two products was found to depend upon the para-substituent on the phenyl ring of the reactants. To understand the influence of the substitutent, a Hammett plot was derived based upon the Swain–Lupton parameters *F* and *R*. In this case, a plot of the ratio of *iii* to *iv* for the various substituents (with the reference substituent being hydrogen) versus *F* and *R* was made. The overall slope of the line (ρ) was –2.5 with the *R* constant being dominant (giving 80% of the slope). These data lead to the conclusion that there is a significant amount of positive charge on the benzylic carbon that is stabilized by resonance during the formation of the transition state that sets the regiochemistry.

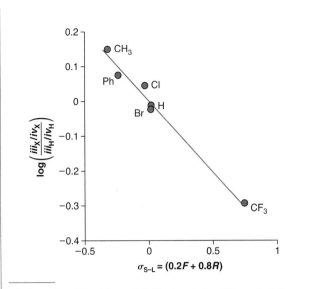

Lehman, J., and Lloyd-Jones, G. C. "Regiocontrol and Stereoselectivity in Tungsten-Bipyridine Catalyzed Allylic Alkylation." *Tetrahedron*, 8863–8874 (1995).

Allylic alkylation

Taft and Topsom broke the substituent constants into four parameters (Eq. 8.30), reflecting contributions from field (*F*), induction (*x*), polarizability (*a*), and *R* (resonance), along with appropriate sensitivity factors (these substituent constants are also given in references at the end of the chapter).

$$\log\left(\frac{k_X}{k_H}\right) = \rho_F \sigma_F + \rho_x \sigma_x + \rho_a \sigma_a + \rho_R \sigma_R \qquad \text{(Eq. 8.30)}$$

Although these multi-parameter approaches are intriguing, the overwhelming majority of electronic LFERs used in modern physical organic chemistry rely simply on σ, σ^+, and σ^- values, and therefore we do not explore the multi-parameter approaches to separating inductive and resonance in any more depth. However, the notion of separating reactivity into the various components remains appealing, and examples involving steric effects and solvent effects are given below.

8.4   Other Linear Free Energy Relationships

Changes in sterics, solvation, nucleophilicity, etc., can all be viewed as substituent effects. These changes nicely fit the concept of a structure–reactivity relationship, where we are probing the effect of changing steric size, solvating ability, nucleophilicity, etc., upon the function or reactivity of the groups undergoing change. Therefore, linear free energy relationships for these kinds of substituent effects have also been developed, and several are presented below.

8.4.1   Steric and Polar Effects—Taft Parameters

The σ values presented to this point include field, inductive, and resonance contributions for the substituents; they do not give the influence of sterics. Taft, however, developed a scale for LFERs that reflects the steric influence of substituents on various reactions. A parameter that reflects the polar nature (i.e., field and induction) of the substituent (σ^*) and a steric parameter (E_s) are both defined. The defining reaction is the acid-catalyzed hydrolysis of RCO_2Me, and the reference R group is methyl (Eq. 8.31). After examining this LFER, we give an example of its use in a Connections highlight concerned with structural inorganic chemistry.

$$\text{(Eq. 8.31)}$$

To measure the polar and steric substituent constants for the R groups in Eq. 8.31, the hydrolysis was performed in both acid and base, with the assumption that polar effects would only influence the base-catalyzed hydrolysis. This assumption was made due to the fact that the basic pathway takes a neutral reactant to a negatively charged intermediate in the rate-determining step (Eq. 8.32), whereas the acid-catalyzed pathway takes a positive reactant to a positive intermediate in the rate-determining step (Eq. 8.33). A further assumption was that the steric effects influence the acid and base pathways equally, because the intermediates differ simply by two small protons. Hence, the steric substituent constant E_s was determined solely from the acid-catalyzed pathway via Eq. 8.34. The σ^* values are determined from Eq. 8.35, where the subscripts A and B refer to the acid and base pathways, and the reference reaction is the hydrolysis of $CH_3CO_2CH_3$. The factor of 2.48 is introduced just to make the magnitude of these new σ^* values similar to the Hammett σ values. The general Taft expression combines the steric and polar substituent scales into one equation (Eq. 8.36), where ρ^* and δ are the sensitivity factors for a new reaction under study to polar and steric effects, respectively. Table 8.3 gives several E_s and σ^* values.

$$\text{(Eq. 8.32)}$$

$$\text{(Eq. 8.33)}$$

$$\log\left(\frac{k_s}{k_{CH_3}}\right) = E_s \qquad \text{(Eq. 8.34)}$$

$$\sigma^* = \frac{1}{2.48}\left[\log\left(\frac{k}{k_o}\right)_B - \log\left(\frac{k}{k_o}\right)_A\right] \qquad \text{(Eq. 8.35)}$$

Table 8.3
Selected Taft Parameters[‡]

R group	E_s	σ^*
–H	1.24	0.49
–Et	–0.07	–0.10
–iPr	–0.47	–0.19
–t-Bu	–1.54	–0.30
–CH₂Ph	–0.38	0.22
–Ph	–2.55	0.60

[‡]Taft, R. W. in *Steric Effects in Organic Chemistry,* M. S. Newman (ed.), John Wiley & Sons, New York, 1956.

$$\log\left(\frac{k_s}{k_{CH_3}}\right) = \rho^*\sigma^* + \delta E_s \qquad \text{(Eq. 8.36)}$$

These parameters show that the rate of the hydrolysis reaction is faster with hydrogen than methyl, and slows as the R group incrementally increases in size to *t*-butyl (that is, E_s becomes more negative). Phenyl, however, comes out surprisingly large in this analysis ($E_s = -2.55$). In Chapter 2 we noted that efforts to characterize the steric size of a group will be strongly context dependent, and this is an example. In an alternative measure of the steric size of substituents—the cyclohexane *A* value (Table 2.14)—phenyl is much smaller than *t*-butyl.

Connections

Using Taft Parameters to Understand the Structures of Cobaloximes; Vitamin B₁₂ Mimics

Cobaloximes (see right) are a class of compounds that have been used extensively to mimic the organometallic cofactor vitamin B₁₂. Crystal structures of many cobaloximes with different R groups have been obtained showing various Co–R bond distances, and various displacements of the Co atom from the plane formed by the four nitrogen ligands. Correlations of the Co–R bond distance and the displacement of the Co atom were created using a multiparameter approach. The Co–R bond distances gave a linear correlation to σ values describing the electron donating and withdrawing ability of the R group, with varying influences from resonance and induction. However, the displacement of the Co from the plane formed by the four nitrogens correlated best with the Taft E_s parameters, indicating that the steric bulk of the R group primarily determined the positioning of the Co atom within the ligand.

This study is an interesting application of LFERs,

because correlations with crystal structure information and not thermodynamic or kinetic data were made. Hence, the power of LFERs far exceeds just the analysis of reaction mechanisms, but can also be used to provide insight into molecular structure.

A cobaloxime

Randaccio, L., Geremia, S., Zangrando, E., and Ebert, C. "Quantitative Rationalization of Solution and Solid State Properties in Cobaloximes, RCo(DH)₂L, as a Function of the Electronic and Steric Properties of R." *Inorg. Chem.,* **33**, 4641–4650 (1994).

8.4.2 Solvent Effects—Grunwald–Winstein Plots

The Hammett LFER gives us a powerful tool for analyzing the changes in charge during a reaction. However, if we vary the structure of the solvent instead of varying the structure of the reactant, and follow the kinetics or thermodynamics of a reaction, we can also gain insight into changes in charge. We can determine whether charges form along the reaction coordinate, and how the solvent influences the formation of these charges. Such an LFER was developed by Grunwald and Winstein.

Many S_N1 reactions involve the formation of positive and negative species in the rate-determining step. Greater solvent polarity accelerates the rate of such reactions by stabilizing the transition states, which are highly charged. You might predict that the rates of these reactions would nicely correlate with the dipole moment of the solvent molecules (μ), the dielectric constant of the solvent (ε), or any one of the other solvent scales [such as Z, π^*, or $E_T(30)$; see Table 3.1]. However, none of these scales correlates well with chemical reactions. This lack of correlation reflects the fact that solvents do not just stabilize or destabilize point charges in a remote or "innocent" manner. Instead, the solvent often directly assists the reaction. With regard to an ionization such as the S_N1 reaction, an electrophilic solvent can directly coordinate with the leaving group as it departs. Furthermore, a nucleophilic solvent can directly coordinate in a loose manner with the carbocation as it forms, a process called **nucleophilic assistance**. None of these factors is directly taken into account with the scales of solvent polarity, dielectric constant, or polarizability. Thus, Grunwald and Winstein defined a new scale that was directly based on an ionization reaction.

The reference reaction chosen was the S_N1 reaction of t-BuCl (Eq. 8.37). This reaction was chosen because it was assumed that very little if any S_N2 pathway occurred. The reference solvent was 80% ethanol/20% water, both of which could act as a nucleophile to add to the incipient t-butyl carbocation. Hence, the LFER given in Eq. 8.38 was defined. Here, $k_{t\text{-BuCl, sol}}$ is the rate of the reaction in the solvent being analyzed, and $k_{t\text{-BuCl, 80\% EtOH}}$ is the rate of the reaction in 80% ethanol. Thus, Y reflects the ability of the new solvent to influence the rate of the reference reaction compared to 80:20 ethanol:water. Table 8.4 lists a variety of Y values.

$$\text{(structure)} \quad (\text{Eq. 8.37})$$

$$\log\left(\frac{k_{t\text{-BuCl, sol}}}{k_{t\text{-BuCl, 80\% EtOH}}}\right) = Y \qquad (\text{Eq. 8.38})$$

Several interesting trends are revealed in Table 8.4. First, pure water is the most ionizing solvent of all, with the ability to enhance the rate of the S_N1 reaction of t-BuCl by a factor of $10^{3.49}$ relative to 80% ethanol. Formic acid is a good ionizing solvent, too—better than ethanol and acetic acid. Acetic acid, ethanol, and 90% acetone, on the other hand, slow the reaction compared to 80% ethanol. There are two contributions to these trends. The first is that polar solvents better solvate charges, and hence speed the ionization. The second is that lower polarity solvents solvate neutral organic species, but a significant amount of this solvation is lost upon formation of a cation and halide anion, hence retarding ionization.

Our goal is not to study just the S_N1 reaction of t-BuCl, but instead to apply the LFER to many different reactions. Hence, an LFER with a sensitivity factor m for any new reaction under study is used (Eq. 8.39). Here, k_{new} is the rate constant for the new reaction in various solvents. The reference solvent is still 80% ethanol. Although the majority of the studies using this equation have been S_N1 and S_N2 reactions (see examples in Chapter 11), it is in principle applicable to any reaction.

$$\log\left(\frac{k_{new, sol}}{k_{new, 80\% EtOH}}\right) = mY \qquad (\text{Eq. 8.39})$$

Since the Y parameter was based upon a reaction that has little nucleophilic assistance, those reactions that have m values near 1 reflect nearly full ionization in the rate-determining step. For those reactions that have an m value less than 1, the reaction is not as sensitive to the ionizing ability of the solvent as is t-BuCl. This means less charge has been created in the transition state, which is most often accomplished by some degree of nucleophilic assistance (Chapter 11 discusses the shades of grey between pure S_N1 and pure S_N2 reaction mechanisms). Hence, a reaction with some S_N2 character will have a reduced extent of charge development in the transition state and therefore an m value less than unity. More-

Table 8.4
Values of Solvent Ionizing Ability as Defined
by LFERs Eqs. 8.38, 8.39 and 8.41*

Solvent	Y	Y_{OTs}	N_{OTs}
CF_3CO_2H	1.84	4.57	–5.56
H_2O	3.49	4.0	–0.41
CF_3CH_2OH	1.04	1.80	–3.0
HCO_2H	2.05	3.04	–2.35
CH_3CO_2H	–1.68	–0.61	–2.35
EtOH	–2.03	–1.75	0.00

Mixed solvent systems (all expressed as percent co-solvent in water)			
98% EtOH	–1.68		0.00
80% EtOH	0.00	0.00	0.00
50% EtOH	1.66	1.29	–0.09
80% HCO_2H	2.32		
50% HCO_2H	2.64		
50% CH_3CO_2H	1.94		
25% CH_3CO_2H	2.84		
90% Acetone	–1.86		
50% Acetone	1.40		
20% Acetone	2.91		

*Fainberg, A. H., and Winstein, S. "Correlation of Solvolysis Rates. III. *t*-Butyl Chloride in a Wide Range of Solvent Mixtures." *J. Am. Chem. Soc.*, **78**, 2770 (1956). Schadt, F. L., Bentley, T. W., and Schleyer, P. v. R. "The S_N2–S_N1 Spectrum. 2. Quantitative Treatments of Nucleophilic Solvent Assistance. A Scale of Solvent Nucleophilicities." *J. Am. Chem. Soc.*, **98**, 7667 (1976).

over, the *m* values of two different reactions in solvents of the same Y value but different nucleophilicities should be the same if ionization is rate-determining, but different if nucleophilic attack or assistance is involved.

8.4.3 Schleyer Adaptation

As stated above, many solvents can directly participate in an ionization reaction by coordination to the leaving group or carbocation. Schleyer noted that the ionization and rearrangement of norbornane derivatives did not fit the Grunwald–Winstein Y scale very well, and that nucleophilic assistance was likely occurring (see margin). In fact, extensive studies of solvolysis reactions began to indicate that nucleophilic assistance was even occurring to some extent in the Grunwald–Winstein reference reaction of *t*-BuCl. Therefore, Schleyer defined a scale (Eq. 8.40) based upon the ionization of 1-adamantyl chloride, a reactant for which no nucleophilic assistance can occur because the nucleophile cannot penetrate the adamantyl group to reach the backside of the C–Cl bond. Here, the Y_{Cl} parameter gives a measure of the solvent ionizing power on 1-adamantylchloride with no nucleophilic assistance, relative to 80% ethanol. Similarly, Y_{OTs} values have been measured for 1-adamantyl tosylate. These Y scales reflect only ionization.

Norbornane derivatives   1-Adamantyl system

$$\log\left(\frac{k_{1\text{-AdaCl, sol}}}{k_{1\text{-AdaCl, 80\% EtOH}}}\right) = Y_{Cl} \qquad \text{(Eq. 8.40)}$$

Although the 1-adamantyl system seems ideal, the ionization of 2-adamantyl systems has become the most widely accepted system for separating the effects of ionizing power and nucleophilic assistance. Again, the reference solvent is 80% ethanol. It may not be imme-

2-Adamantyl tosylate

diately apparent, but backside attack is also impossible in the 2-adamantyl system due to the axial hydrogens shown in color. We know that backside attack is not possible because the Y_{OTs} values based upon 2-adamantyl tosylate correlate very well with the Y_{OTs} values based upon 1-adamantyl tosylate. Furthermore, in the solvents 98:2 ethanol/water and acetic acid, the rates of ionization of 2-adamantyl tosylate are the same even though the ethanol/water mixture is more nucleophilic. In addition, the 2-adamantyl tosylate system gives a very large secondary isotope effect (1.23), indicating a large extent of rehybridization to sp^2 in the transition state. The reason 2-adamantyl is preferred over 1-adamantyl is primarily due to the fact that 2-adamantyl derivatives are more synthetically accessible than 1-adamantyl derivatives. Table 8.4 lists some Y_{OTs} values based upon 2-adamantyl tosylate.

Once again, the goal is to study the charge build-up on reactions other than the reference reaction. We compare the rate constants for the new reaction under study in a variety of solvents ($k_{new, sol}$) to the parameters developed from an analysis of the adamantyl systems. Most reactants, however, are not as restricted as the 1- or 2-adamantyl systems to only an ionization. Therefore, a substituent constant, called N_{OTs}, is also included. N_{OTs} gives the ability of the solvent to participate in nucleophilic assistance. N_{OTs} is determined from a reaction that is solely nucleophilic in nature, the S_N2 displacement of tosylate from methyltosylate (Eq. 8.41). In Eq. 8.41 the $0.3Y$ corrects the sensitivity of the rate of reaction of methyltosylate to the solvent ionizing power (Y in this case comes from the Grunwald–Winstein equation). Eq. 8.42 thus reflects both a factor for pure ionization (mY_{Cl}, where Y_{Cl} comes from the ionization of 1- or 2-adamantyl chloride) and a factor for pure nucleophilic assistance (lN_{OTs}, where N_{OTs} comes from S_N2 displacement on MeOTs). Table 8.4 lists some N_{OTs} values. We will look at various l and m sensitivity constants in Chapters 10 and 11. For now, the Connections highlight below describes one use of the Schleyer approach to solvent ionizing scales.

$$N_{OTs} = \log\left(\frac{k_{sol}}{k_{80\% \; EtOH}}\right) - 0.3Y \tag{Eq. 8.41}$$

$$\log\left(\frac{k_{new, sol}}{k_{new, 80\% \; EtOH}}\right) = lN_{OTs} + mY_{Cl} \tag{Eq. 8.42}$$

We have separated the ability of the solvent to assist in carbocation formation by electrostatics and nucleophilic assistance, but we have not considered the ability of the solvent to stabilize the leaving group. In fact, it has been found that the Y scales do not correlate well with reactions when the leaving group is different from that upon which the scale is based. Solvents that can directly assist the leaving group as it departs can do so to varying extents depending upon the structure of the leaving group. For example, electrophilic solvents can stabilize a small chloride anion better than a large tosylate anion. As a result, many scales based upon different leaving groups have been developed (such as Y_{Br}, Y_I, and Y_{OTs}).

In summary, we have defined another tool for studying the charges that change during a mechanism. Several LFERs were discussed that give one the ability to determine both the extent of ionization at the transition state of a specific reaction and how much nucleophilic assistance has occurred, all just by changing the structure of the solvent.

8.4.4 Nucleophilicity and Nucleofugality

In several of the reactions that you learned in introductory organic chemistry, and many of those that are discussed in Chapters 10 and 11, nucleophilic attack on an electron deficient center such as a carbocation or a polarized σ or π bond occurs. The ability of nucleophiles to participate in these reactions depends upon their molecular structure. A structural dependence suggests that there should be scales of relative nucleophilicity that depend upon the sterics, polarizability, and inductive/resonance properties of the nucleophile. Hence, changing the nucleophile becomes one of the tools that we have to study reactions. As with solvent and electronic substituent effects, LFERs provide insight into how the change in nucleophile structure affects the reaction. Similarly, the ability of a leaving group to depart

Connections

The Use of the Schleyer Method to Determine the Extent of Nucleophilic Assistance in the Solvolysis of Arylvinyl Tosylates

The solvolysis of vinyl tosylate i proceeds via an sp hybridized carbocation. The Grunwald–Winstein m value for this reaction was found to be small (0.62) for such an ionization reaction, indicating much less stabilization from the solvent than occurs in the ionization of t-butylchloride. This small m value could be due to increased nucleophilic assistance, or stabilization of the carbocation by some other factor, making the role of the solvent less important.

Solvolysis of a vinyl tosylate

Typically, $C(sp^2)$–X bonds as in i are very sluggish in solvolysis reactions compared to analogous $C(sp^3)$–X bonds. This sluggish behavior can be ascribed to less effective backside solvation of the developing carbocation with the sp^2 system, and the higher energy of an sp cation compared with an analogous trisubstituted "normal" sp^2 carbenium ion. Yet, with i, the steric congestion presented by the methyl groups on the phenyl ring results in a conformation where the vinyl group is initially perpendicular to

the benzene, while the vinyl cation becomes increasingly planar, allowing for resonance stabilization from the benzene. This stabilization by the phenyl ring could account for the low m values, because less stabilization by the solvent is required.

Stabilization of a vinyl cation via conjugation with phenyl

To discover if any nucleophilic assistance occurs with i, the Schleyer method was used. Instead of plotting an entire series of solvents with different Y and N values, followed by the application of Eq. 8.42, the investigators simply chose two solvent systems with the same Y values, but different nucleophilicities. Ethanol/water (98:2) and acetic acid both have Y values of –1.68, but N_{OTs} values of 0.0 and –2.35, respectively (see Table 8.4). Following the kinetics of solvolysis of i revealed a difference in rate constants in the two solvents of only 25%. Given the large difference in N values, the small difference in rate constants indicates that nucleophilic assistance does *not* occur in the solvolysis of i. Therefore, the stabilization imparted by the benzene ring is the reason for the low m values.

Yates, K., and Périé, J.-J. "Solvolysis of Arylvinyl Bromides and Tosylates." *J. Org. Chem.*, **39**, 1902 (1974).

---

from a structure should depend upon the electronics of the leaving group. In other words, the nucleofugalities of different groups should correlate with parameters that describe the properties of the leaving groups. Therefore, changing leaving groups is another tool for studying reactions.

Once again, as with all LFERs, the goal is to change the structure of the reactants, in this case the nucleophile and nucleofuge, and have parameters for relative nucleophilicity and nucleofugality that can be applied to many different reactions. Before describing the scales that have been developed for analyzing the ability of different atoms or molecules to act as nucleophiles or leaving groups, it is instructive to discuss what makes a good nucleophile or leaving group. The attributes of reactivity discussed below can be applied to all reactions involving nucleophiles, and hence we discuss them here as part of our tools for studying mechanisms, rather than discussing them in the context of specific reactions in future chapters.

Basicity/Acidity

First of all, there should be a relationship between nucleophilicity and basicity, and a correlation between nucleofugality and the acidity of the conjugate acid of the leaving group. Both basicity and nucleophilicity involve the donation of electrons to an electrophile. Likewise, a leaving group and an acid both accept electrons as they depart or lose a proton, respectively. Since nucleophiles and nucleofuges behave oppositely with regard to electron donation and acceptance, good nucleophiles are often poor leaving groups and vice versa. However, there are many exceptions, which we will discuss below.

Nucleophiles that have the same nucleophilic atom are better when the nucleophilic atom is negative compared to neutral. This charge dependence is the same for the relative basicity of the nucleophile. The opposite is true for leaving groups; that is, neutral is better than negative when the same atom is being compared.

When proceeding from left-to-right across a row in the periodic table, the atoms further to the left are more nucleophilic ($H_3C^- > H_2N^- > HO^- > F^-$). This trend correlates again with the basicity of the group, and the opposite trend is found for leaving group ability.

The more stable the negative charge of the nucleophilic atom, the less nucleophilic it is. For example, due to differing resonance stabilization, an alkoxide is more nucleophilic than a phenoxide, which in turn is more nucleophilic than a carboxylate. This trend reflects the relative basicities, as well. Below we will describe LFERs for nucleophilicity and nucleofugality that correlate the pK_as of the conjugate acids of the nucleophiles and nucleofuges to sensitivity parameters (β_{nuc} and β_{LG}).

Solvation

Another factor that has a dramatic influence on nucleophilicity is solvation. Solvation forces are stronger for small nucleophilic ions than for larger ones, an effect that is especially true in polar protic solvents where hydrogen bond donation from the solvent to the nucleophile is stabilizing. Because the solvent must be stripped from the nucleophile during the attack on the electrophile, the well solvated smaller ions are poor nucleophiles. Therefore, the more "naked" (less solvated) the nucleophile, the better it becomes. Hence, nucleophiles are consistently more reactive in aprotic solvents of low polarity. However, very often the nucleophile will not be soluble in these solvents because many nucleophiles are negatively charged (they are part of salts). Hence, solvents such as DMF and DMSO are commonly used. These are aprotic but polar enough to solubilize the nucleophile, and high reactivity is found.

Similar to the solvation effect, counterion effects can influence nucleophilicity. A tight ion pair between a negative nucleophile and a positive counterion will decrease the nucleophilicity (unless the counterion is involved in the reaction; see examples in Chapters 10 and 11). Conversely, if one can effectively solvate and perhaps even fully encapsulate the counterion, so that no coordination to the nucleophile is possible, the nucleophilicity will increase (see Section 4.2.1 on crown ethers).

Another interesting effect on nucleophilicity that arises in part from a solvation effect is called the α-effect. The **α-effect** arises when two nucleophilic atoms are adjacent to one another. This dramatically increases the nucleophilicity. For example, although hydroxide is about 16,000 times more basic than hydroperoxide (HOO⁻), hydroperoxide is approximately 200 times more nucleophilic. Similarly, the amine groups in hydroxylamines ($HONR_2$) or hydrazines (NR_2NR_2) are more nucleophilc than simple amines. There are two explanations for this effect. In one explanation the existence of lone pair electrons directly adjacent to the nucleophilic site is believed to destabilize the ground state, making the nucleophile more reactive. However, whereas the α-effect trend is seen in solution with hydroxide and hydroperoxide, hydroxide is the better nucleophile in the gas phase. Apparently, the inductive withdrawal of the negative charge by the adjacent electronegative oxygen in hydroperoxide makes this nucleophile less solvated, and hence more reactive in solution.

Polarizability, Basicity, and Solvation Interplay

A comparison of nucleophiles in which the nucleophilic atoms bear the same charge, and are in the same column of the periodic table, indicates that the nucleophilicity increases as one proceeds down the column. For example, the order of nucleophilicities for the halide anions is $I^- > Br^- > Cl^- > F^-$ in polar protic solvents. As discussed above, in these solvents, the anions are well solvated, and fluoride makes stronger bonds with the solvent because the anionic charge is more localized. This trend is opposite of that expected if there was a strict correlation between basicity and nucleophilicity (the order of basicity is $F^- > Cl^- > Br^- > I^-$). Similarly, thiolate anions are consistently better nucleophiles than analogous alkoxides, even though the alkoxides are significantly more basic. The increased nucleophilicity upon going down a column in the periodic table is due, in part, to a polarizability effect. The atom with the higher polarizability (see Chapter 1) forms the incipient covalent bond to the electrophile more easily because its electron cloud more readily accommodates changes in shape.

When polar aprotic solvents are used, anionic nucleophiles are not well solvated. In these cases, the nucleophilicity is more often found to track basicity. For example, in DMSO it can be found that $F^- > Cl^- > Br^- > I^-$ is the order of nucleophilicity for various reactions.

Shape

As a last example of a factor that influences nucleophilicity, we examine molecular shape. A nucleophile is required to penetrate the solvent shell around an electrophile, and often to hone in on an electrophilic center that may be sterically congested. In this regard, large bulky nucleophiles are not very effective, but small, "bullet" (linear) shaped nucleophiles can be quite effective. For example, the nucleophiles cyanide (NC^-) and azide (N_3^-) are particularly good. They are much more nucleophilic than would be predicted based upon their basicity.

In summary, shape, charge, basicity, polarizability, and solvation are all factors that determine the nucleophilicity of compounds. Understanding these factors allows us to use the differences to our advantage when probing mechanisms.

The LFERs that have been developed for structure–function relationships for nucleophiles do not universally correlate reactions of different types and in different solvents. As one changes from one electrophile to another, changes solvent, or changes counterions, the nucleophilicity parameters based upon a different type of reference reaction do not apply as uniformly as, for example, the Hammett parameters apply to reactions that have no resemblance to the ionization of benzoic acid. However, once armed with these caveats, there are nucleophilicity parameters that have found general use, and they are good tools for you to consider when studying any reaction that involves a nucleophilic attack prior to or during the rate-determining step.

8.4.5 Swain–Scott Parameters—Nucleophilicity Parameters

To create an LFER for nucleophilicity, Swain and Scott defined the reference reaction to be the S_N2 nucleophilic displacement of bromide from methyl bromide by the reference nucleophile water. A more widely accepted scale based upon methyl iodide is now used (Eq. 8.43). Eq. 8.44 shows the LFER, with n_X as the parameter for nucleophile X, and s is the sensitivity parameter for any new reaction under study. Additionally, k_{new, Nuc_X} is the rate constant for the new reaction under study with each nucleophile X, while k_{new, H_2O} is the rate constant for the same reaction with water as the nucleophile. Table 8.5 gives a series of n_X values for various nucleophiles (based upon CH_3I), and the pK_as of the conjugate acids. Note, as discussed above, that the pK_as do not necessarily correlate with the nucleophilicities, and that the effects of resonance on nucleophilicity are well accounted for with the n values (compare the n values for acetate, phenoxide, and methoxide).

$$\log\left(\frac{k_{CH_3-I, Nuc_X}}{k_{CH_3-I, H_2O}}\right) = n_X \qquad \text{(Eq. 8.43)}$$

$$\log\left(\frac{k_{new, Nuc_X}}{k_{new, H_2O}}\right) = s n_X \qquad \text{(Eq. 8.44)}$$

Recall that the substituent parameters in LFERs define a logarithmic scale. Hence, in general, methoxide is a better nucleophile than methanol by a factor of $10^{6.29}$, and iodide is a better nucleophile than fluoride by a factor of $10^{4.72}$. We will look at a few s values in Chapter 11 when substitution reactions are considered. Here, we simply give one example in the context of an acetal substitution reaction (see the Connections highlight on the following pages).

Table 8.5
Swain–Scott Parameters (n_X Values)
for the S_N2 Reaction of CH_3I*

Nucleophile	n_X	pK$_a$ of conjugate acid
F$^-$	2.7	3.45
Cl$^-$	4.37	−5.7
Br$^-$	5.79	−7.77
I$^-$	7.42	−10.7
N$_3^-$	5.78	4.74
NC$^-$	6.70	9.3
CH$_3$OH	near 0.0	−1.7
H$_2$O	0.0	−1.7
CH$_3$CO$_2^-$	4.3	4.75
PhO$^-$	5.75	9.89
CH$_3$O$^-$	6.29	15.7
Pyridine	5.23	5.23
Aniline	5.70	4.58
(Et)$_3$N	6.66	10.7
PhSH	5.70	—
PhS$^-$	9.92	6.52

*Pearson, R. G., Sobel, H., and Songstad, J. "Nucleophilic Reactivity Constants Toward Methyl Iodide and *trans*-Dichlorodi(pyridine)platinum(II)." *J. Am. Chem. Soc.*, **90**, 319 (1968).

Connections

The Use of Swain–Scott Parameters to Determine the Mechanism of Some Acetal Substitution Reactions

Compounds *i* and *ii* (PNP = *p*-nitrophenyl) undergo reactions with neutral and anionic nucleophiles giving substitution products. The reactions are second order with a first-order dependence on nucleophile, consistent with an S_N2 mechanism. However, reasonable S_N1 mechanisms can be written, given that the leaving groups are very good.

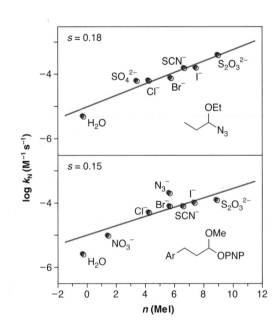

To determine the extent to which a nucleophile is involved in the transition state for S_N2 reactions of *i* and *ii*, Swain–Scott plots were generated (see right). The slopes of the plots (*s* values) for *i* and *ii* are only 0.15 and 0.18, respectively, meaning that the rate of the reaction does not depend very much upon the reactivity of the nucleophile.

The clear first-order kinetic dependence upon the nucleophile means that it has to be involved in the rate-determining step, as in an S_N2 mechanism. Yet, the very low s values indicate an early transition state with little bonding to the nucleophile at the transition state. The transition state has almost full loss of the leaving group with little nucleophilic attack (see right), analogous to an S_N1 reaction. This interpretation shows that the s value can provide insight into the extent of nucleophilic attack at a transition state relative to the extent of attack during the reference reaction.

Amyes, T. L., and Jencks, W. P. "Concerted Bimolecular Substitution Reactions of Acetal Derivatives of Propionaldehyde and Benzaldehyde." *J. Am. Chem. Soc.*, **111**, 7900 (1989).

8.4.6 Edwards and Ritchie Correlations

Although there have been many attempts to provide a multi-parameter approach for correlating relative nucleophilicity, only one is examined here. We examine only this one because it gives insight into how oxidation potential, a molecular property not yet discussed, can influence nucleophilicity. Edwards and Ritchie considered that nucleophilicity arises from two parameters that reflect the ability of the nucleophile to donate electrons: basicity and oxidation potential. To the extent that a nucleophilic attack on an electrophile resembles the loss of electrons from the nucleophile to the electrophile, then the oxidation potential of the nucleophile should correlate with its reactivity. Thus, two substituent parameters were defined—E_n for oxidation potential and H for basicity, with corresponding sensitivity factors α and β (Eq. 8.45). Because Eq. 8.45 incorporates oxidation potential and basicity, it is referred to as the **oxibase scale**. The parameters E_n and H are based upon the pK_a of the conjugate acid (Eq. 8.46) and the oxidation potential of the nucleophile, as given in Eq. 8.47 (where E_o is the standard oxidation potential). A reference reaction and nucleophile are still needed, and S_N2 attack on CH_3I with water were used. This method has been found to nicely correlate a variety of different S_N2 reactions.

$$\log\left(\frac{k_{Nuc_X}}{k_{H_2O}}\right) = \alpha E_n + \beta H \qquad \text{(Eq. 8.45)}$$

$$H = pK_a + 1.74 \qquad \text{(Eq. 8.46)}$$

$$E_n = E_o + 2.60 \qquad \text{(Eq. 8.47)}$$

8.5 Acid–Base Related Effects—Brønsted Relationships

A final linear free energy relationship that we cover is one that is still extensively used in modern research—almost as much as Hammett plots. It is called a Brønsted plot, and it relates acidity or basicity to other kinds of reactions. The reason that this LFER is still actively used is in part because this relationship is essential to an understanding of acid–base catalysis in enzymology. However, the same relationship can be used to study nucleophilicity and leaving group ability. We describe all three uses below, but we leave a more detailed analysis of the use of this relationship in studies of acid–base catalysis to Chapter 9.

8.5.1 β_{Nuc}

Some of the earliest correlations between nucleophilicity and reactivity focused solely on the pK_a of the conjugate acid of the nucleophile. If one keeps the electronic and steric properties of a series of nucleophiles similar, and uses the same solvent for the analyses, reactivity does nicely correlate with the basicity of the nucleophile (K_b; see Chapter 5). An LFER as defined in Eq. 8.48 can be used to correlate the data. This structure–function relationship is called a **Brønsted relationship**.

$$\log(k) = \beta_{Nuc} \log(K_b) + \log(C)$$
$$\text{or } \log(k) = \beta_{Nuc} pK_a + \log(C')$$

(Eq. 8.48)

In this equation k is the rate constant for the reaction under study. In place of a substituent parameter, we use $\log(K_b)$. Therefore, the reference reaction is the deprotonation of water by the nucleophile (Eq. 8.49). Log(C) is the y intercept of the plot, and has no physical meaning. Finally, β_{Nuc} is the sensitivity factor. Plotting the pK_a of the conjugate acid of the nucleophile ($\log K_b = pK_a + \log K_w$) versus $\log(k)$ should give a straight line when the dominant factor in determining how the structure of the nucleophile affects the reaction is basicity. When $\beta_{Nuc} = 1$, the full effect of the change in basicity is reflected in the nucleophilicity, meaning that a nucleophile that is 100 times more basic than another will speed up a reaction by a factor of 100. When $\beta_{Nuc} < 1$, then the sensitivity of the nucleophilic attack on the basicity of the nucleophile is diminished.

$$\text{Nuc}^{\ominus} + H_2O \rightleftharpoons \text{NucH} + \text{OH}^{\ominus}$$

(Eq. 8.49)

8.5.2 β_{LG}

Just as nucleophilicity can be found to correlate directly with basicity, the ability of a leaving group to depart is often found to correlate directly with the acidity of its conjugate acid. The same kind of equation used to correlate nucleophilicity with pK_a is used to correlate nucleofugality with pK_a (Eq. 8.50). In this case a β_{LG} is measured, which gives the sensitivity of the reaction to the acidity of the conjugate acid of the leaving group. Since the leaving group ability should increase with lower pK_as of the conjugate acid, Eq. 8.50 tells us that the β_{LG} values will be negative. It is very common to measure both β_{LG} and β_{Nuc} values when implementing a mechanistic analysis, since their comparison gives good insight into the relative amounts of nucleophilic attack and leaving group departure at the rate-determining step. An example is given in the next Connections highlight in the context of an important biochemical reaction, the hydrolysis of ATP.

$$\log(k) = \beta_{LG}(pK_a) + \log(C)$$

(Eq. 8.50)

Connections

ATP Hydrolysis—How β_{LG} and β_{Nuc} Values Have Given Insight into Transition State Structures

The phosphoryl transfer reaction from ATP is one of the most common biological reactions, being used to store energy, generate concentration gradients, and to transduce signals. The extent of bond breaking and forming that occurs at the transition state for phosphoryl transfer to alcohols has been studied to give insight into transition state structures during enzyme-catalyzed processes.

Shown below is the Brønsted plot for the reaction of ATP^{4-}, and the complex between Mg^{2+} and ATP^{4-}, as a function of the pK_a of primary alcohol nucleophiles (closed and open circles, respectively). The values of β_{Nuc} are 0.07 and 0.06, respectively, reflecting that the nucleophilicity of the alcohols has little effect on the rate of the reaction. This indicates very little bond formation to the nucleophiles in the transition state.

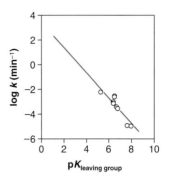

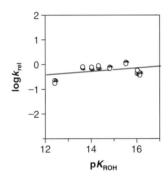

Shown above is a Brønsted plot for the hydrolysis reaction (phosphoryl transfer to water) of various phosphoanhydrides as a function of leaving group pK_a. The slope of the line (β_{LG}) is near –1.1. This value is large, indicating that there is a significant amount of bond cleavage to the leaving group at the transition state. In fact, because the absolute value is greater than 1, this leaving group departure is actually more sensitive to the structure and nature of the anion than is the reference deprotonation reaction of water. The small β_{Nuc} and large β_{LG} were interpreted to support a dissociative mechanism for the hydrolysis of ATP, with a transition state involving considerable metaphosphate-like character (see below).

Herschlag, D., and Admiraal, S. J. "Mapping the Transition State for ATP Hydrolysis: Implications for Enzymatic Catalysis." *Curr. Biol.*, **2**, 729–739 (1995).

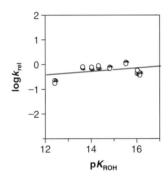

Going Deeper

How Can Some Groups be Both Good Nucleophiles and Good Leaving Groups?

As noted earlier, nucleophilicity in polar protic solvents increases as one goes down a column in the periodic table. Hence, the order of nucleophilicity for halide anions is $I^- > Br^- > Cl^- > F^-$. Iodide is the better nucleophile because it is more polarizable and is not as well solvated. We also stated that the better leaving groups are those that have the more acidic conjugate acids. Therefore, the order of leaving group ability for the halogens is $I^- > Br^- > Cl^- > F^-$, also. To many chemists, it seems counterintuitive that a group can be both a good leaving group and a good nucleophile.

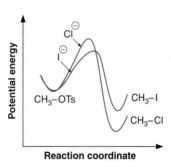

To understand this dichotomy we must understand that by "good" we mean faster—the groups are kinetically "good". It does not mean that the bond being formed or broken is particularly strong or weak. To see this, let's examine the hypothetical reaction coordinate diagrams for the S_N2 substitution reactions of methyl tosylate with iodide and chloride shown to the side. A C–Cl bond is stronger than a C–I bond, so methyl chloride is placed lower in energy on the diagram. We find from experiment that the barrier to the reaction is lower with iodide than with chloride even though the product is more stable with chloride. Iodide is therefore a better nucleophile.

Now consider the reverse reaction—that of tosylate displacing chloride or iodide. Due to microscopic reversibility, the reverse reactions have to proceed through the same transition states. Therefore, the barrier to the reverse reaction is also lower for iodide, and hence, iodide is the better leaving group. Once you tie the concept of microscopic reversibility to both nucleophilicity and leaving group ability, it is clear that in going down a column in the periodic table, the better nucleophiles are also the better leaving groups. (Before just accepting the explanation given above, you should ask yourself why hydroxide is a good nucleophile and a poor leaving group. Alternatively, try working Exercise 9 at the end of the chapter.)

8.5.3   Acid–Base Catalysis

Recall that Eqs. 8.48 and 8.50 are called Brønsted linear free energy relationships. If an acid or base is involved in the rate-determining step of a reaction, the rate of that reaction should depend upon the strength of the acid or base. Hence, a Brønsted correlation is often found. Eqs. 8.51 and 8.52 relate the rate constants for an acid- or base-catalyzed reaction, respectively, to the pK_a of the acid or conjugate acid of the base. The sensitivity of an acid-catalyzed reaction to the strength of the acid is α, whereas the sensitivity of a base-catalyzed reaction to the strength of the base is β. The α and β reaction constants indicate the extent of proton transfer in the transition state. In Chapter 9 we explore the use of these two equations in much more detail, and we apply them in Chapters 10 and 11.

$$\log(k) = -\alpha pK_a + \log(C) \qquad \text{(Eq. 8.51)}$$

$$\log(k) = \beta pK_a + \log(C) \qquad \text{(Eq. 8.52)}$$

8.6   Why do Linear Free Energy Relationships Work?

All the LFERs given above included a reaction constant that gave the sensitivity of the reaction to the change in substituent, and each substituent behaved the same in each reaction. It seems reasonable that an electron donating or electron withdrawing group should be electron donating or electron withdrawing regardless of the reaction. Similarly, a good leaving group is always a good leaving group, regardless of the reaction. Yet, why is the *relative* influence of the groups the same in each reaction, and why is their degree of influence the same? For example, why shouldn't nitro be more electron withdrawing than a cyano in one reaction, and vice versa in a different reaction? Further, why shouldn't one reaction be very sensitive to the addition of electron withdrawing groups and not electron donating groups, while another reaction is very sensitive to electron donating groups but not electron withdrawing groups? In general, we do not find this behavior. An LFER, such as a linear Hammett plot, Swain–Scott plot, or Brønsted plot, means that the relative influence of the substituents is always the same. To understand this behavior, we must first analyze some general mathematics that is the basis of all LFERs. After deriving the general form of all LFERs, we can return to this issue.

8.6.1 General Mathematics of LFERs

We first need to understand why the effect of changing substituents on one reactant will be similar to the effect on another reactant, even when the sensitivity of two different reactions to a change in substituents is quite different. Let's start by analyzing the effect one would expect on the free energy of a reaction due to a change in a substituent. The same analysis can be applied to the activation energy of a reaction. To determine how substituents affect equilibria, we need two reference points. First, we need a reference substituent, and second, we need a reference reaction. For example, in a Hammett analysis, the substituent that all other substituents are compared to is hydrogen, and the reaction that all reactions are compared to is the dissociation of benzoic acid. Here, we keep the analysis general.

The method used to determine the influence of new substituents compared to the reference substituent starts with Eq. 8.53, which gives the relationship of an equilibrium constant to the reaction free energy. Eq. 8.54 gives the difference in reaction free energies for the same reaction with two different substituents where ΔG_o is for the reference substituent and ΔG_X is for the new substituent X. In the following analysis, for simplicity of notation, we have dropped the "nought" superscript on $\Delta G°$ that indicates standard conditions.

$$\Delta G_o = -2.303RT \log_{10}(K_o) \tag{Eq. 8.53}$$

$$\Delta G_o - \Delta G_X = 2.303RT \log_{10}\left(\frac{K_X}{K_o}\right) \tag{Eq. 8.54}$$

Eq. 8.54 tells us that the difference in free energies of two reactions with different substituents is directly proportional to the log of the ratio of equilibrium constants (K_X/K_o). This quantity is the **substituent constant** for each X substituent (C_X in Eq. 8.55), and the reaction we use to measure each C_X is the **reference reaction**. The substituent constant tells us if substituent X shifts the equilibrium toward reactant or product compared to the reference substituent, and by how much. If C_X is negative, then substituent X shifts the equilibrium toward reactant, and if it is positive, substituent X shifts the equilibrium toward product. Substituting Eq. 8.55 into Eq. 8.54 yields a relationship between the substituent constant and the difference in reaction free energies between the reaction with the new substituent and the reaction with the reference substituent (Eq. 8.56).

$$C_X = \log_{10}\left(\frac{K_X}{K_o}\right) \tag{Eq. 8.55}$$

$$\Delta G_o - \Delta G_X = 2.303RT(C_X) \tag{Eq. 8.56}$$

The next step is to see if the various substituents influence a different reaction in the same manner as they influence the reference reaction. Another equation identical to Eq. 8.56 is written for the new reaction under study, but we designate it with primes on the variables (Eq. 8.57). As discussed earlier, this reaction can be totally unrelated to the reference reaction. Because we want to know the relative effect of the substituents on the two reactions, we take a ratio of Eqs. 8.56 and 8.57, giving Eq. 8.58. This equation states that the ratio of the free energy differences for the two reactions under comparison is the ratio of the log of the equilibrium constants of the new reaction to the log of the equilibrium constants for the reference reaction (C_X' / C_X). In our analysis, this ratio is defined as the **reaction constant** (Q in this generalized example). The ratio Q directly tells us if the reaction under study is more or less sensitive to a change in substituents than the reference reaction. If Q is greater than one, the substituents affect the thermodynamics of the new reaction to a greater extent than the substituents affect the reference reaction. Conversely, if Q is less than one, the reaction is less sensitive to changes in substituents than the reference reaction. Hence, we have also called Q the **sensitivity factor**. Rearranging Eq. 8.58 to 8.59 and simplifying to Eq. 8.60 gives further insight into LFERs. We now see that the extent to which the free energy of the new reaction is changed, via a change in substituent, is proportional to the extent to which the reference reaction was changed by the same substitution, but enhanced or diminished by Q.

$$\Delta G_o' - \Delta G_X' = 2.303RT(C_X') \tag{Eq. 8.57}$$

$$\frac{\Delta G_o' - \Delta G_X'}{\Delta G_o - \Delta G_X} = \frac{C_X'}{C_X} = Q \tag{Eq. 8.58}$$

$$(\Delta G_o' - \Delta G_X') = Q(\Delta G_o - \Delta G_X) \tag{Eq. 8.59}$$

$$\Delta\Delta G' = Q\Delta\Delta G$$
$$\text{where } (\Delta G_o' - \Delta G_X') = \Delta\Delta G' \text{ and } (\Delta G_o - \Delta G_X) = \Delta\Delta G \tag{Eq. 8.60}$$

The LFERs given in Eqs. 8.57 and 8.59 are not those routinely applied to experiments, such as a Hammett plot. Instead, a form that relates the sensitivity factor Q and the substituent constant C_X to a number that can be experimentally determined is used. To derive this form of the LFER, we perform some algebraic manipulations (left as an Exercise at the end of the chapter) to derive Eq. 8.61. This final form of an LFER is the one used in most such analyses (see Sections 8.3–8.5). An identical equation can be written for rate constants (Eq. 8.62).

$$\log_{10}\left(\frac{K_X'}{K_o'}\right) = QC_X \tag{Eq. 8.61}$$

$$\log_{10}\left(\frac{k_X'}{k_o'}\right) = QC_X \tag{Eq. 8.62}$$

8.6.2 Conditions to Create an LFER

In principle, the reaction constant Q does not have to be a constant for all substituents, because LFERs do not derive from any thermodynamic law. However, chemists find empirically that it most often is. When Q is constant, it means that two different reactions respond to all substituents in the same manner, but just to differing extents. To understand how this can be possible, we must substitute $\Delta G = \Delta H - T\Delta S$ into Eq. 8.60, giving Eq. 8.63.

$$(\Delta\Delta H' - T\Delta\Delta S') = Q(\Delta\Delta H - T\Delta\Delta S) \tag{Eq. 8.63}$$

Eq. 8.63 has two independent variables, $\Delta\Delta H'$ and $\Delta\Delta S'$ ($\Delta\Delta H$ and $\Delta\Delta S$ are already determined) and therefore can only hold when one of three conditions are met.

- First, the $\Delta\Delta H$s are coincidentally the same for both the new reaction under study and the reference reaction, and the $\Delta\Delta S$s are linearly proportional for the two reactions being compared.

- Second, the $\Delta\Delta S$s are coincidentally the same for both reactions and the $\Delta\Delta H$s are linearly proportional.

• Or third, the $\Delta\Delta H$s and $\Delta\Delta S$s are linearly related to each other for both the reference reaction and the new reaction. That is, for each reaction, as ΔH goes up ΔS increases also. In other words, they scale proportionately. In this scenario, the enthalpy and entropy compensate for each other, because in the Gibbs free energy equation ($\Delta G = \Delta H - T\Delta S$) the enthalpy and entropy terms have opposite algebraic signs. This is actually the most common situation, and is referred to as the **enthalpy–entropy compensation effect** (see Chapter 4).

Before exploring why enthalpy–entropy compensation effects exist, it is instructive to delineate the temperature dependence of LFERs. Because temperature dependence has been mostly explored with regard to kinetics, we present this discussion below in terms of activation parameters, but it also applies to reaction free energies.

8.6.3 The Isokinetic or Isoequilibrium Temperature

If we assume a linear relationship between $\Delta H^{\ddagger}$ and $\Delta S^{\ddagger}$ for the same reaction but with different substituents we can obtain Eq. 8.64. For most reactions, the $\Delta H^{\ddagger}$ and $\Delta S^{\ddagger}$ values are unknown. However, for those cases where these parameters have been measured and an LFER was found to hold, Eq. 8.64 does indeed correlate the $\Delta H^{\ddagger}$ and $\Delta S^{\ddagger}$ parameters. The slope β is a proportionality constant between $\Delta H^{\ddagger}$ and $\Delta S^{\ddagger}$, telling us how sensitive the compensation effect is between $\Delta H^{\ddagger}$ and $\Delta S^{\ddagger}$. Eq. 8.64 tells us that as the enthalpy of activation becomes larger and less favorable, the entropy of activation becomes larger, too, but is more favorable, leading to a compensating effect.

$$\Delta H^{\ddagger} = \beta\Delta S^{\ddagger} + \Delta H_o^{\ddagger} \qquad \text{(Eq. 8.64)}$$

Using the Gibbs free energy equation ($\Delta G^{\ddagger} = \Delta H^{\ddagger} - T\Delta S^{\ddagger}$) in combination with Eq. 8.64 leads to Eq. 8.65. Now the proportionality constant β is seen to have units of temperature, and is referred to as the **isokinetic** or **isoequilibrium temperature** for kinetic or thermodynamic LFERs, respectively.

$$\Delta G^{\ddagger} = \Delta H_o^{\ddagger} - (T - \beta)\Delta S^{\ddagger} \qquad \text{(Eq. 8.65)}$$

Eq. 8.65 states that the Gibbs free energies of activation for the same reaction with different substituents can be considered to be made up of two terms. One is constant, regardless of the substituent ($\Delta H_o^{\ddagger}$), and one is an entropy term that is different for each substituent. The extent that the entropy influences the reaction depends upon the difference between the experimental temperature T and the isotemperature β. A corollary of Eq. 8.65 is that LFERs should be interpreted with care when measured at temperatures near the isokinetic or isoequilibrium temperature. In other words, for the same reaction with varying substituents or in different solvents, if the enthalpy is linearly related to the entropy via a relationship such as Eq. 8.64, there will be a temperature where changes in substituents or solvent will not affect the rate or equilibrium at all. Therefore, when one measures any reaction constant and that value is near zero, it could simply be that the temperature of analysis is coincidentally near or at the isokinetic or isoequilibrium temperature. Hence, small reaction constants (such as ρ, s, m, β_{LG}, etc.) should be interpreted with care when drawing conclusions about reaction mechanisms. However, having stated this caveat related to LFERs, it is true that for most common organic reactions the isokinetic or isoequilibrium temperature is outside of experimentally accessible conditions.

8.6.4 Why does Enthalpy–Entropy Compensation Occur?

We have presented the fact that most reactions for which an LFER can correlate rate constants or equilibrium constants have linear relationships between ΔH and ΔS as described by Eq. 8.64. Because increases in $\Delta H^{\ddagger}$ impede a reaction while increases in $\Delta S^{\ddagger}$ assist a reaction, the two affects are compensating. This compensation is a general phenomenon, one that is often neglected when chemists attempt to improve the rate or yield of a reaction by changing substituents or reaction conditions. The most common rationalizations of enthalpy–entropy compensation invoke steric or solvation effects, and we discuss both briefly here.

Steric Effects

Consider what happens when you increase the steric hindrance to attack of a nucleophile on a carbonyl by increasing the size of the R group (Eq. 8.66). A larger R group would lead to a more crowded, less solvated transition state, thereby raising the $\Delta H^{\ddagger}$. However, the nucleophile is further away from the carbonyl carbon at the transition state, and therefore it and the solvent would be less tightly held at the transition state than the reaction with the smaller R group. These structural issues manifest themselves as a lower loss in entropy and therefore a more favorable entropy of activation.

$$\text{(Eq. 8.66)}$$

Solvation

Let's examine an S_N2 reaction between an amine and an alkyl halide to illustrate enthalpy–entropy compensation effects due to differential solvation (Eq. 8.67). The transition state is polar relative to the reactants, so the enthalpy of activation should be lower in increasingly polar solvents. In addition, the developing charges in the transition state will cause the solvent to be more constrained than when solvating the neutral amine or alkyl halide, and this effect will be greater with the more polar solvent. Hence, a higher polarity solvent leads to a lower $\Delta H^{\ddagger}$, but also makes $\Delta S^{\ddagger}$ more unfavorable; as a result, these are compensating effects.

$$\text{(Eq. 8.67)}$$

These two examples are indicative of a general phenomenon. When enthalpy becomes more favorable (stronger bonding) the system typically becomes more organized (meaning less entropy). Conversely, when the enthalpy change is slight there is often more disorganization at the transition state. We encountered this same phenomenon when we discussed binding in Chapter 4. Stronger intramolecular forces lead to stronger binding, but also more restricted movement of the compounds involved in the binding event, therefore compensating for each other with respect to the Gibbs free energy.

8.7   Summary of Linear Free Energy Relationships

A large number of LFERs were discussed above; many more can be found in the literature or other physical organic chemistry texts. This is especially true for Hammett-related sigma constants. For example, here is an abbreviated list of some you may encounter in the literature: σ^0, σ^n, and σ_I for the ionization of carboxylic acids other than benzoic acid; σ_I^q for inductive effects; σ^l and σ' for the effects of acidities in bicyclic compounds; and σ_R, σ_R°, $\sigma_{R(BA)}$, σ_R^+, and σ_R^- for various resonance effects. The number of these scales and their uses quickly become confusing.

Our intent here was to focus only on those that are still commonly used, and on the conceptual foundation of the methodology. The student should now be prepared to understand any new LFERs encountered in the literature. At this point, though, it may be helpful to tabulate the LFERs we have discussed in one place, so that their uses can be quickly deciphered

Table 8.6
A Summary of the Most Common LFERs Used*

LFER	Substituent constant	Reference reaction	Used to study	Reaction constant and its meaning relative to the reference reaction
Hammett	σ	Ionization of benzoic acid	Inductive effects	$\rho > 1$, more sensitive $0 < \rho < 1$, less sensitive $\rho = 0$, not sensitive $\rho < 0$, positive charge created
Hammett	σ^-	Ionization of phenol	Resonance in addition to induction	Same as above
Hammett	σ^+	Ionization of phenyldimethyl chloromethane	Resonance in addition to induction	Same as above
Taft	E_s	Hydrolysis of methyl esters	Steric size	$\delta > 1$, more sensitive $\delta < 1$, less sensitive
Grunwald–Winstein	Y	Ionization of t-BuCl in 80:20 EtOH/H_2O	Ionizing power of solvent	$m > 1$, more sensitive $m < 1$, less sensitive
Swain–Scott	n	S_N2 reaction of methyl iodide in water	Nucleophilicity	$s > 1$, more sensitive $s < 1$, less sensitive
Brønsted	pK_a	Acidity in water	Nucleophilicity	$\beta_{Nuc} > 1$, more sensitive $\beta_{Nuc} < 1$, less sensitive
Brønsted	pK_a	Acidity in water	Leaving group departure	$|\beta_{LG}| > 1$, more sensitive $|\beta_{LG}| < 1$, less sensitive
Brønsted	pK_a	Acidity in water	Acid catalysis	$\alpha > 1$, more sensitive $\alpha < 1$, less sensitive
Brønsted	pK_a	Acidity in water	Basic catalysis	$\beta > 1$, more sensitive $\beta < 1$, less sensitive

*Not all of the LFERs discussed in the text are included in Table 8.6. Only those most likely to be encountered in modern research are included.

when one is designing experiments. Table 8.6 gives the names of the major LFERs, the substituent constant, the reference reaction, the kinds of effects they are used to explore, and the meaning of the reaction constant.

8.8 Miscellaneous Experiments for Studying Mechanisms

A large number of experiments have been presented in Chapters 7 and 8 for the study of reaction mechanisms—namely, kinetic analyses, isotope effects, and structure–function analyses. These are standard experimental tools associated with physical organic chemistry. However, the experimentalist can often devise studies that are completely different from any of those presented thus far, depending upon his or her imagination and chemical ingenuity. In this section we summarize some of the more common approaches. We first analyze the most obvious experiments to perform, such as product and intermediate identification, but also more subtle experiments such as cross-over and isotope scrambling. Unlike our analyses of the isotope effect and substituent effects, the following techniques are given only brief general discussions, followed by one or more examples in Connections highlights. For most of the techniques mentioned, we give more examples in Chapters 9–12.

8.8.1   Product Identification

The experiment that should be performed first in the analysis of any mechanism is the determination of the products of the reaction. This experiment may seem obvious, but it is worth stating because failure to do so can cause problems (see the Connections highlight below). In particular, as much as possible, it is important to identify *all* the products of a reaction. Doing so is equivalent to ensuring that the **mass balance** of the reaction is complete; that is, everything that goes in as a reactant comes out as a product. Sometimes, identification of a minor product will provide a valuable clue as to the kind(s) of mechanistic pathways that are going on in the reaction. Alternatively, it is dangerous to make sweeping mechanistic conclusions based on a product that is present in only very low yield if we don't know the other products of the reaction.

Connections

An Example of an Unexpected Product

In the reaction of hydroxide with *p*-nitrobenzyl chloride, you might expect *p*-nitrobenzyl alcohol to be the major product via a simple S_N2 reaction. However, *p*-nitrobenzyl chloride is acidic enough to be deprotonated by hydroxide, leading to a carbon nucleophile that reacts with a second equivalent of *p*-nitrobenzyl chloride, subsequently

giving the stilbene via elimination. Thus, without identification of the product one might have interpreted the kinetic data incorrectly.

Tewfik, R., Fouad, F. M., and Farrell, P. G. "Elimination Reactions. I. Reaction of Bis(4-nitrophenyl)methyl Chloride with Sodium Hydroxide in Aqueous Dioxane." *J. Chem. Soc., Perkin Trans.*, **2**, 31 (1974).

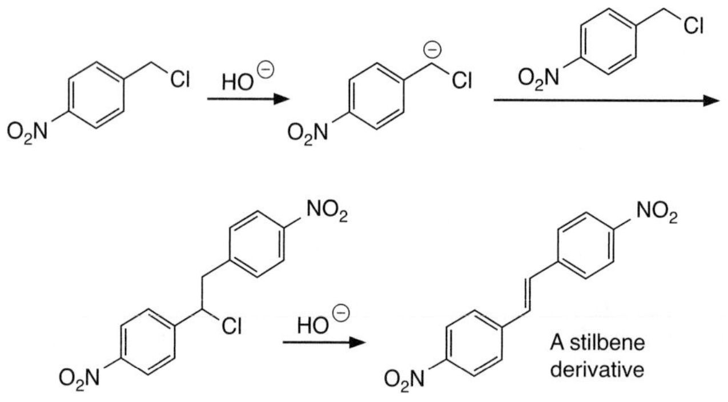

8.8.2 Changing the Reactant Structure to Divert or Trap a Proposed Intermediate

The nature of an intermediate can sometimes be deciphered by synthesizing a new reactant that is similar to the actual reactant under study, but for which the intermediate being proposed can react in a new and predictable manner. There are no standard ways to approach this kind of experiment, and you must design your own experiments on a case-by-case basis. We draw upon an enzymatic reaction as an example of the application of this technique.

Connections

Designing a Method to Divert the Intermediate

The enzyme mandelate racemase interconverts the enantiomers of mandelate ion (2-hydroxyphenylacetic acid). A carbanion alpha to the carboxylate was proposed as the intermediate formed. To test for the presence of this intermediate, an analog of mandelate ion, *i*, was synthesized and subjected to racemization by the enzyme. The product isolated was *ii*, which can result from formation of a carbanion and a subsequent 1,6-elimination of bromide followed by tautomerization. The result supported the intermediacy of carbanion *iii* in the mandelate ion pathway.

Lin, D. T., Powers, V. M., Reynolds, L. J., Whitman, C. P., Kozarich, J. W., and Kenyon, G. L. "Evidence for the Generation of .Alpha.-carboxy-.alpha.-hydroxy-*p*-xylylene from *p*-(Bromomethyl)mandelate by Mandelate Racemase." *J. Am. Chem. Soc.*, **110**, 323 (1988).

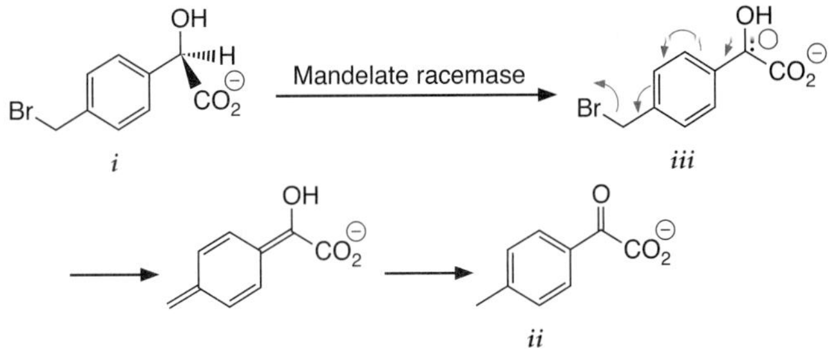

8.8.3   Trapping and Competition Experiments

A common method for intermediate identification is trapping of the intermediate with an added reagent. Several radical traps exist (see Section 8.8.8), and many good nucleophiles make viable traps for transient electrophiles such as carbocations. You should use your own chemical insight to devise traps for intermediates such as carbanions, carbenes, etc. Reactive intermediates are short lived, though, so the trap must be very reactive to compete with the standard reaction path of the reactive intermediate. Also, because the trapping reaction will typically be bimolecular, high concentrations of trap will often be required. Alternatively, the trap could be covalently tethered to the reactant, facilitating capture of the reactive intermediate.

A variant of a trapping experiment is a **competition experiment**. In our analysis of kinetic experiments in Chapter 7, we said that steps beyond the rate-determining step do not affect the kinetics, and thus information about them cannot be obtained. This lack of a kinetic dependence often leaves a large portion of the mechanism invisible to kinetic analysis. One way to probe chemical steps past the rate-limiting step is to use competition experiments. A competition experiment involves the addition of two or more reagents that compete for one or more intermediates. It is a variant on the trapping experiment, where now more than one trap is used (Eq. 8.68). The ratio of products derived from the different traps tells the ratio of rate constants for the reaction of the traps with the intermediate. From this ratio, some insight into the nature of the intermediate can be gained. The experiment is viable only when the trapping reaction is under kinetic control.

$$\text{Reactant} \underset{}{\overset{\text{Slow}}{\rightleftharpoons}} \text{Intermediate} \overset{\text{Trap 1}}{\underset{\text{Trap 2}}{<}} \begin{array}{l} \text{Product 1} \\ \text{Product 2} \end{array} \qquad \text{(Eq. 8.68)}$$

Connections

Trapping a Phosphorane Legitimizes Its Existence

Pentacoordinate species (phosphoranes) are proposed intermediates in the hydrolysis of RNA and DNA. Before such species were well accepted, chemists examined the chemistry of phosphoesters such as *i* as model systems. Compound *i* can cyclize to give phosphorane *ii*, although *ii* was never seen at room temperature. However, upon adding acetyl chloride to a solution of *i*, both *iii* and *iv* are isolated. Trapping experiments such as this one give good evidence that *ii* is present in a solution of *i*, showing that phosphoranes are legitimate species and possible intermediates in chemical reactions.

Sarma, R., Ramirez, F., McKeever, B., Nowakowski, M., and Marecek, J. F. "Crystal and Molecular Structure of Phosphate Esters. 9. Crystal and Molecular Structure of *o*-Hydroxyphenyl-*o*-phenylene Phosphate, (*o*-HOC$_6$H$_4$)(C$_6$H$_4$)PO$_4$. Equilibrium Between Pentavalent and Tetravalent Phosphorus in Solutions." *J. Am. Chem. Soc.*, **100**, 5391 (1978).

Phosphorane trapping

8.8.4 Checking for a Common Intermediate

Often similar reactions proceed via the same intermediate. For example, the S_N1 solvolysis of *t*-butyl bromide and *t*-butyl iodide in water would both be presumed to proceed via the *t*-butyl cation. We could easily verify this conclusion by performing a competition experiment where the addition of two nucleophilic traps would give two products resulting from the same intermediate in the same ratio. Any deviation in products and ratios would indicate different intermediates. This is exactly the same concept that was used in our example of competition experiments, except now we compare the product ratio from two different reactants.

Connections

Checking for a Common Intermediate in Rhodium-Catalyzed Allylic Alkylations

A very common structure in organometallic transformations is the π-allyl complex (see Chapter 12). Such a structure is a resonance hybrid of two forms with σ and π bond character. These structures are formed from allylcarbonates and can undergo attack by various carbon-based nucleophiles to give extended allylic systems.

π-Allyl complex   Individually known as σ + π complexes

A rhodium-based system that gives excellent control of stereochemistry has recently been reported, and it was shown by testing for a common intermediate that the putative allyl–Rh species is in fact unsymmetrical. The experiment consisted of subjecting the unsymmetrical secondary allylic carbonates *i* and *ii* to reaction with catalytic Rh(PPh₃)₃Cl. Conventional wisdom predicted the two would form the same π-allyl complex *iii*. If so, compound *iii* should give upon reaction with the nucleophile the two products shown in the same ratio regardless of whether *i* or *ii* is the starting material. Instead, the reactions of *i* and *ii* retained regiochemistry, with the nucleophile in both cases attached to the carbon that possessed the leaving group in the starting material. Hence, a common intermediate such as *iii* is *not* formed. The authors interpreted this result to support distinct σ + π intermediates instead of symmetric π-allyl complexes.

Evans, P. A., and Nelson, J. D. "Conservation of Absolute Configuration in the Acyclic Rhodium-Catalyzed Allylic Alkylation Reaction: Evidence for an Enyl (σ + π) Organorhodium Intermediate." *J. Am. Chem. Soc.*, **120**, 5581 (1998).

i: R₁ = Me, R₂ = *i*Pr   97   3
ii: R₁ = *i*Pr, R₂ = Me   3   97

Symmetrical π-allyl intermediate

8.8.5   Cross-Over Experiments

A cross-over experiment is used to determine if a reactant breaks apart to form intermediates that are released to solution before they recombine to give product. To accomplish this kind of experiment, we use two similar reactants, one that is labeled in some manner to distinguish it from the other. The label can be a different substituent or an isotope; the more subtle the variation the better. Consider a simple reaction in which A–B reacts to give C–D. We want to know whether the A and B fragments, which give rise to C and D, respectively, are ever free in the flask. The reactant (A–B) and its labeled variant (A*–B* in Eq. 8.69) are mixed together, and we analyze the products. No scrambling of the labeled and unlabeled portions of the reactant indicates no free intermediates to cross over between the two reactions (Eq. 8.69). When intermediates are released to solution, scrambling can occur (Eq. 8.70). However, sometimes the observation of no scrambling can still be consistent with fragmentation, where recombination of the intermediates occurs faster than diffusion of the intermediates into bulk solution. As always, a negative result should be interpreted with caution. In favorable cases, however, a cross-over experiment gives us information about whether a reaction occurs by an intramolecular or intermolecular process (look at Section 12.2.3 for examples).

$$\text{A–B} + \text{A*–B*} \longrightarrow \text{C–D} + \text{C*–D*} \qquad \text{(Eq. 8.69)}$$

$$\text{A–B} + \text{A*–B*} \longrightarrow \text{C–D} + \text{C*–D*} + \text{C*–D} + \text{C–D*} \qquad \text{(Eq. 8.70)}$$

8.8.6   Stereochemical Analysis

Very often the existence of an intermediate and/or the nature of a reaction can be deciphered from an analysis of the stereochemical outcome of an experiment. This type of experiment is covered in all introductory organic chemistry textbooks when the S_N2 and S_N1 reactions are first introduced. The observation of complete inversion of configuration in the product derived from an enantiomerically pure reactant is indicative of no intermediate and a concerted displacement of the leaving group by the nucleophile (Eq. 8.71, S_N2). On the other hand, complete or partial racemization is indicative of a planar intermediate (Eq. 8.72, S_N1). This simple but elegant probe into the existence and nature of an intermediate can be extended to much more complicated reactions, and stereochemical analysis is one of the most powerful mechanistic probes. We give a biochemical example in the Connections highlight, and we will see many examples of stereochemical analyses in subsequent chapters.

$$\text{(Eq. 8.71)}$$

$$\text{(Eq. 8.72)}$$

Connections

Pyranoside Hydrolysis by Lysozyme

The hydrolysis of β-pyranosides by lysozyme occurs strictly with retention of configuration at the anomeric center, giving a β-hemiacetal product. To explain the retention, the following mechanism is postulated. Protonation of the **aglycon group** (the leaving group attached to the anomeric center) by an enzyme carboxylic acid occurs during leaving group departure. Either backside nucleophilic attack or ion pairing with an enzyme carboxylate protects one face of the anomeric carbon from reaction with water. Thus, replacement of the departing saccharide with water leads to addition from the same face as the leaving group, giving a **double inversion** and hence overall retention. This study is a classic case of examining the stereochemical outcome of an enzyme-catalyzed reaction, providing support for a particular mechanistic scenario.

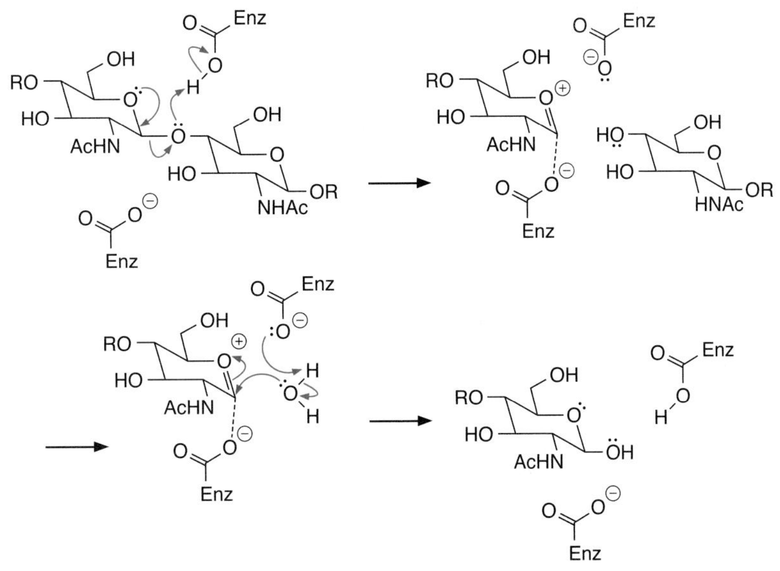

8.8.7   Isotope Scrambling

A related tool that can give us insight into the involvement of symmetrical intermediates is isotope scrambling. In this experiment one isotopically labels one portion of the reactant and checks to see where that label resides in the product. The position of the label in the product can tell us about the nature of the intermediate and/or whether dissociation has occurred to give a symmetrical intermediate. The classic check for a tetrahedral intermediate in an acyl transfer experiment is a good example (see Section 10.17.2). For another example, consider the thermal and photochemical Claisen rearrangements shown in Eqs. 8.73 and 8.74. Labeling the terminal position of the allyl group of the reactant and analyzing the products gives insight into the nature of the intermediates. The label cleanly migrates to only one place in the thermal rearrangement, suggesting a concerted mechanism with no free symmetric allyl fragment (see Section 15.5.4). However, the photochemical rearrangement gives a 50:50 scrambling of the label on both ends of the allyl fragment, suggesting the intermediacy of a symmetric allyl fragment somewhere during the mechanism, where both ends can react equally.

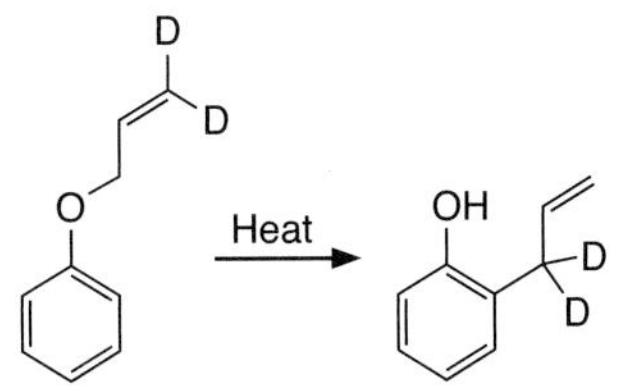

(Eq. 8.73)

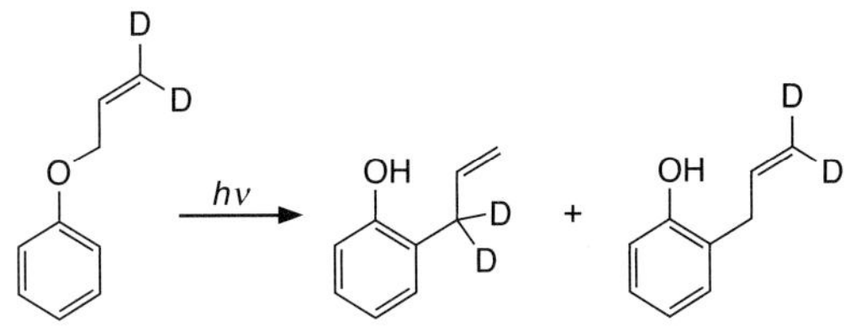

(Eq. 8.74)

Connections

Using Isotopic Scrambling to Distinguish Exocyclic vs. Endocyclic Cleavage Pathways for a Pyranoside

The hydrolysis of pyranosides is postulated to largely proceed via **exocyclic cleavage** to give a cyclic oxocarbenium ion (Path A). Yet, it has long been recognized that an alternative pathway is also possible—namely, **endocyclic cleavage** to give an acyclic oxocarbenium ion (Path B). In fact, molecular dynamics calculations have suggested that the endocyclic pathway may be operative in enzyme-catalyzed processes. In order to delineate the extent to which this alternative pathway occurs in solution, an isotopic scrambling experiment was performed.

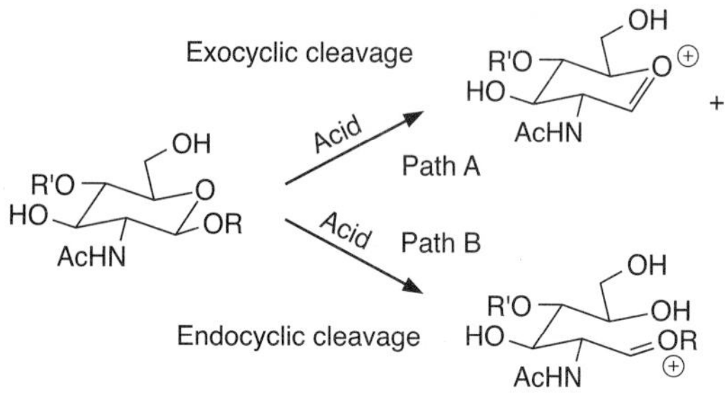

Compound *i* creates intermediates that either lack an internal mirror plane (*ii*) or possess an internal mirror plane (*iii*, where the symmetry is only broken due to deuterium incorporation) upon exocyclic or endocyclic cleavage, respectively. Analysis of whether the deuterium scrambles in the products leads to conclusions about the extent to which exocyclic or endocyclic cleavage occurs (see below). A lack of deuterium scrambling in the products (*iv* and *v*) indicates exocyclic cleavage, and the observation of deuterium scrambling in the products indicates endocyclic cleavage (*iv, v, vi,* and *vii*). Using this method, it was discovered that approximately 15% of the reaction proceeds via the endocyclic pathway, with the rest occurring by the exocyclic pathway.

Liras, J. L., Lynch, V. M., and Anslyn, E. V. "The Ratio between Endocyclic and Exocyclic Cleavage of Pyranoside Acetals is Dependent upon the Anomer, the Temperature, the Agylcon Group, and the Solvent." *J. Am. Chem. Soc.*, **119**, 8191–8200 (1997).

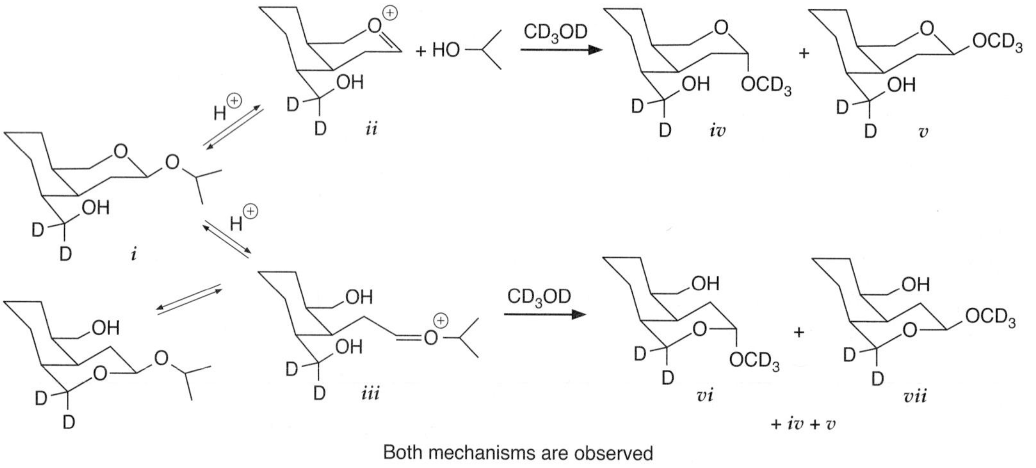

Both mechanisms are observed

8.8.8   Techniques to Study Radicals: Clocks and Traps

Radical clocks are one experimental technique that has received considerable use in the analysis of radical reactions. Most radical clocks involve an intramolecular free radical rearrangement that proceeds with a well-defined rate constant. The prototype is the rearrangement of 5-hexenyl radical to cyclopentylmethyl radical, which occurs with a unimolecular rate constant of $1.0 \times 10^5 \, \text{s}^{-1}$ at 25 °C (Eq. 8.75). The clock strategy is to embed a 5-hexenyl unit into the reactive system of interest. If a radical forms, and if its lifetime is comparable to or greater than 10^{-5} s, cyclopentylmethyl-derived products should form.

(Eq. 8.75)

Table 8.7
Various Radical Clocks and Their Rate Constants for Rearrangements*

Clock	Rate constant for rearrangement (s^{-1}), 25 °C	Clock	Rate constant for rearrangement (s^{-1}), 25 °C
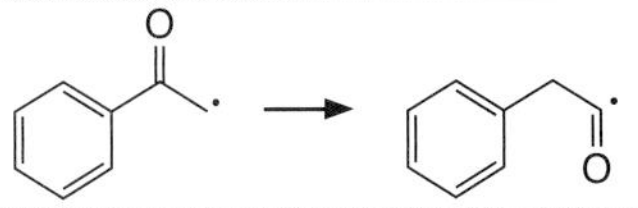	10	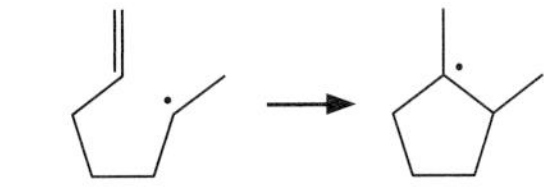	1.3×10^5
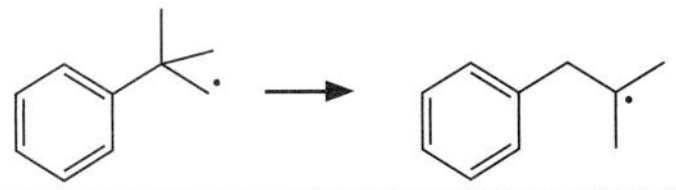	59	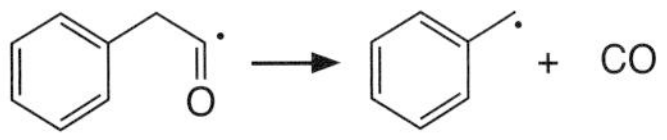	5.2×10^7
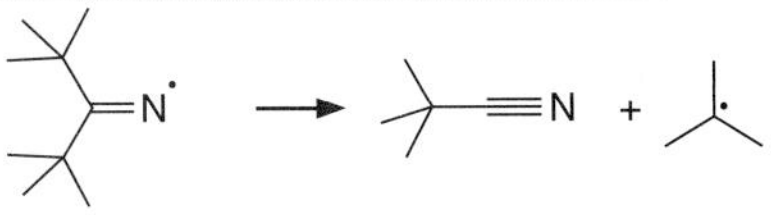	71		1.3×10^8
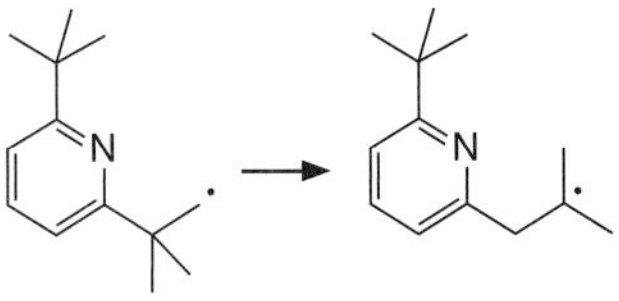	7.8×10^2	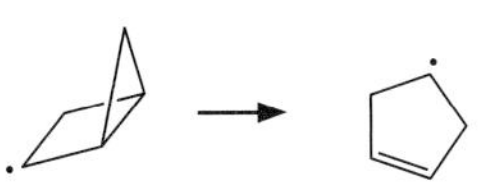	2×10^9
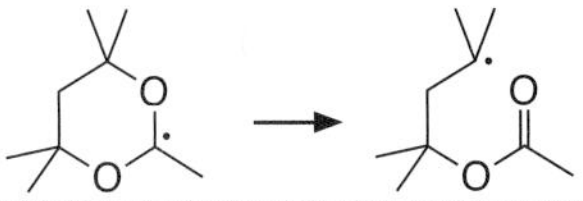	1.3×10^3		$(5-8) \times 10^{10}$
	9.8×10^3	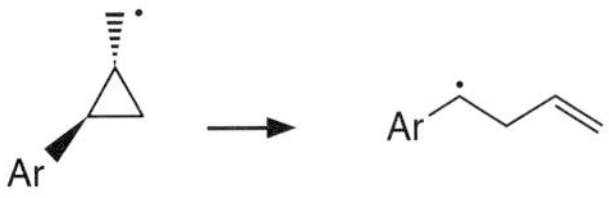	$(1-4) \times 10^{11}$
	3.3×10^4		

*Griller, D., and Ingold, K. U. "Free Radical Clocks." *Acc. Chem. Res.*, **13**, 317 (1980). Newkomb, M., and Toy, P. H. "Hypersensitive Radical Probes and the Mechanisms of Cytochrome P450-Catalyzed Hydroxylation Reactions." *Acc. Chem. Res.*, **33**, 449 (2000).

   In order to study the lifetimes of various radicals in new reactions, one requires several radical clocks with varying lifetimes. Incorporation of these clocks into the molecules under study is used both to show that radical intermediates do or do not exist, and if they do, their lifetimes relative to the clock. Several free radical clocks with their rate constants for re-arrangement are shown in Table 8.7. Such a collection has been termed an **horlogerie**, after a French term for a small shop that sells clocks. Seven orders of magnitude can be spanned by choosing the correct clocks.
   Another tool for studying radicals is the use of a **spin trap**, in an experiment called **spin trapping**. The addition of a free radical to a nitroso or nitrone group (the spin trap; see Eqs. 8.76 and 8.77) creates a **spin adduct**. The spin adduct is another radical, but it is typically a long lived radical that can often be studied using EPR spectroscopy. The EPR spectrum can be informative about the structure of the radical that added.

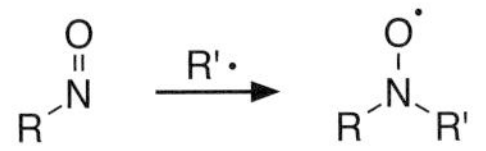

(Eq. 8.76)

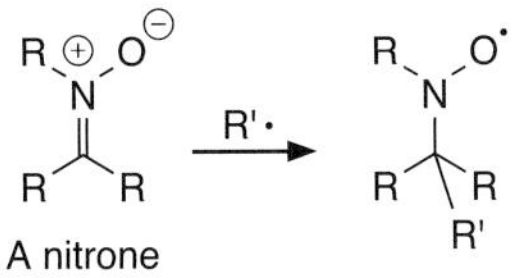

A nitroso

(Eq. 8.77)

A nitrone

Connections

Determination of 1,4-Biradical Lifetimes Using a Radical Clock

The photolysis of ketones leads to reactions that we will call Norrish type II in Chapter 16. The reaction involves carbonyl excitation followed by hydrogen atom abstraction from a γ-carbon. Imbedded carbon monoxide, in the form of an ester, placed between the ketone and the γ-carbon is known to accelerate the decay of the resulting 1,4-biradical by allowing a fragmentation pathway. To ascertain the lifetime of the radical intermediates formed from the photolysis of the following α-keto ester, the incorporation of a radical clock was performed. Upon photoly-

sis, three products are found, one that results from no opening of the radical clock, and two that result from the opening. Hence, the ring opening of the clock competes with the fragmentation. As seen in Table 8.7, the radical clocks of this type open with rate constants near 10^8. In this particular case, the rate constant is $9.4 \times 10^7 \, s^{-1}$. Hence, the lifetime for fragmentation of the kinds of biradical created upon photolysis of α-ketoesters is in the range of 1 to 4 ns.

Hu, S., and Neckers, D. C. "Lifetimes of the 1,4-Biradical Derived from Alkyl Phenylglyoxylate Triplets: An Estimation Using the Cyclopropylmethyl Radical Clock." *J. Org. Chem.*, **62**, 755–757 (1997).

Using a radical clock to probe the mechanism

8.8.9   Direct Isolation and Characterization of an Intermediate

If an intermediate is stable enough to be isolated, then common spectroscopic and other characterization techniques can be used to identify it. However, we must demonstrate that under the established reaction conditions the isolated intermediate proceeds to the same products as the reactants to be confident that it is an intermediate and not just a by-product. Conversely, if one suspects that a certain compound is an intermediate in a reaction, the suspected intermediate can be synthesized by an independent route and subjected to the experimental conditions of the reaction under study. If the correct product(s) is(are) produced, this gives good but not conclusive evidence (it may be coincidental that they give the same product) that the structure in question is an intermediate. If the proposed intermediate gives a different product, on the other hand, it is good evidence that it is not an intermediate in the reaction under study.

8.8.10   Transient Spectroscopy

For short lived reactive intermediates, simple isolation and characterization is not an option, so we must resort to other techniques. As we noted in Chapter 7, fast spectroscopy methods have allowed chemists to obtain real-time characterization of many types of reac-

Connections

The Identification of Intermediates from a Catalytic Cycle Needs to be Interpreted with Care

The mechanism of hydrogenation of alkenes using Wilkinson's catalyst (*i*) is shown below. The two other species outside the box have been either detected in, or isolated from, solutions undergoing catalysis (S = solvent). However, the actual catalytic cycle is given within the box. The species outside the box were found to be too sluggish to react as competent intermediates in the rapidly proceeding catalytic cycle. Hence, they are simply by-products of some of the catalytically active species. Their accumulation actually slows the rate of the catalytic reaction.

This example shows that the identification of a detectable species in a catalytic system can lead to misinterpretations of the catalytic cycle. Being able to detect an intermediate often means that it is stable, and therefore may not be active, especially in a catalytic system. We must always show that the species are kinetically competent to participate in the cycle.

Halpern, J. "Mechanism and Stereoselectivity of Asymmetric Hydrogenation." *Science,* **217**, 401 (1982).

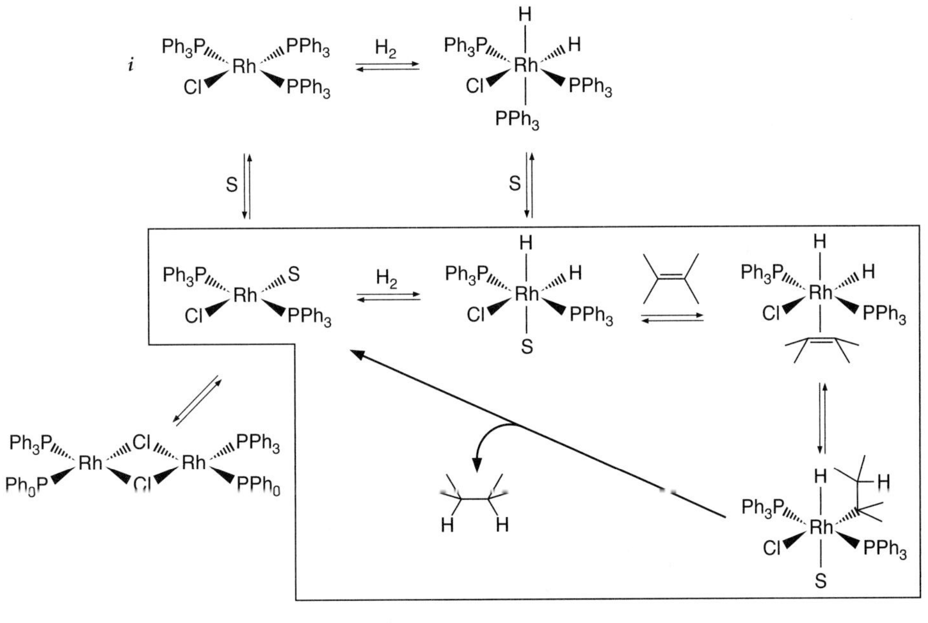

Hydrogenation mechanism

tive intermediates. Just as with isolating intermediates, we must demonstrate that the intermediate observed by the fast spectroscopy is indeed an intermediate involved in the mechanism under study. Furthermore, we should demonstrate that the spectroscopy is of the intermediate we are proposing and not of some other unknown species. These two proofs are often challenging, but careful experimental design can provide convincing support.

8.8.11 Stable Media

Another strategy for characterizing reactive intermediates that are short lived is to generate them in an environment in which there are no available reaction paths. If there is nothing the reactive intermediate can do, it will persist. There are two basic strategies. The first, typified by the **stable ion media** used so successfully to study carbocations, is to simply remove all possible species that could react with the reactive intermediate. For carbocations, this means removing all nucleophiles. This approach is only viable if the reactive intermediate cannot react with itself, either bimolecularly (dimerization) or unimolecularly (rearrangement). For example, because free radicals can usually dimerize readily, there is no such thing as "stable radical media". However, radicals can be observed in a frozen matrix (see below). Still, when a stable medium can be found for a reactive intermediate, the full array of spectroscopic tools becomes available, including NMR, IR, UV/vis, etc.

For extremely reactive structures for which no stable fluid medium can be envisioned, the technique of **matrix isolation** is useful. This approach combines two features to make the reactive intermediate observable. First, the reactive intermediate is generated in a solid (frozen) solvent. 2-Methyltetrahydrofuran forms a particularly useful glass at low temperature. The matrix suppresses all bimolecular reactions. Often this is enough to stabilize the intermediate, if the reactive intermediate has no unimolecular decomposition pathways. For example, methyl radical is incredibly reactive. But, if it is generated locked in a rigid glass, there is really nothing it can do. It can't rearrange, and it isn't photoreactive. Indeed, samples of methyl radical in glass are indefinitely stable at room temperature. This example highlights the distinction between stability and persistence given in Section 2.2.1.

More typically, highly reactive intermediates also have unimolecular decomposition pathways that need to be suppressed. For this purpose, matrix isolation experiments are typically carried out at extremely low temperatures. For example, temperatures as low as 4 K are not uncommon. Note that at these low temperatures, inert gasses such as Ar are rigid solids, and they make perfect inert matrices. You should convince yourself that at 4 K a barrier as small as 1 kcal/mol is completely insurmountable. The matrix isolation methodology is technically demanding, and it is best suited to IR, UV/vis, and EPR characterization.

Summary and Outlook

Kinetics, isotope effects, linear free energy relationships, and the varied experiments discussed in the last section of this chapter make up the vast majority of the methods chemists use to decipher reaction mechanisms. The level at which we discussed these techniques was aimed at setting the stage for a more detailed analysis of many of the common organic and organometallic reaction mechanisms. By having the chapters on experimental tools behind us, we are ready to examine how these tools are used to delineate the details of mechanisms. However, many organic reactions, as well as most biochemical reactions, are catalyzed. We still need to understand catalysis before examining common organic mechanisms. Therefore, we now examine general methods of catalysis (Chapter 9). After this, we look at the mechanisms of substitutions, eliminations, and additions (Chapters 10 and 11). Looking forward even further to Chapters 12 and 13, you will see the application of mechanistic tools in organometallic and polymerization reactions, respectively. Therefore, you may consider Chapters 7 and 8 the "bread and butter" of classical physical organic chemistry, and the subsequent chapters as the applications to classical and modern topics.

Exercises

1. One equation often used to calculate kinetic isotope effects is given below. Derive this equation.

$$k_H/k_D = \exp[hc(\bar{\nu}_H - \bar{\nu}_D)/2k_B T]$$

2. For a C–H bond with a stretching frequency of 3000 cm^{-1}, use the equation in Exercise 1 to calculate the expected isotope effect at 298 K for a full bond homolysis.

3. Given the frequencies of the out-of-plane bending modes in Figure 8.6, calculate the maximum secondary isotope effect that could arise during a reaction with a C–H(D) bond that rehybridizes from sp^3 to sp^2 at 298 K. (*Hint*: Although Eq. 8.2 was introduced in relation to stretching vibrations, it may also be used for bending vibrations.)

4. From our discussion of fractionation factors, derive Eq. 8.11. Here, assume two reactants and two products, each with only one exchangeable hydrogen.

5. Derive the Hammett equation in terms of rate constants instead of equilibrium constants. Use the discussion given in Section 8.6.1 as your guideline.

6. Steric effects are commonly dominated by differences in entropy. Taft parameters are dominated by entropy until the steric effect becomes very large, and then enthalpy effects start to dominate the steric effects.

   a. Using an entropy–enthalpy compensation analysis, explain why a linear free energy relationship for steric effects works at all.

   b. Explain how entropy–enthalpy compensation is involved in the Grunwald–Winstein LFER.

7. Consider a unimolecular reaction with an Arrhenius A value of 10^{13} s^{-1} and E_a of 0.5 kcal/mol. What is the half-life of this species at 4 K? At 8 K? Why is there such a large rate change in response to raising the temperature only 4 K?

8. A major research area is the consideration of the kinds of organic chemistry that might go on in interstellar space. It is possible that organic molecules formed in space could have been carried to earth on comets or meteorites, seeding prebiotic organic chemistry. Also, the organics in space can be a signature of star formation or collapse. In interstellar organic chemistry, tunnelling is often given much more consideration than in conventional, terrestrial chemistry. Why should this be so?

9. In the Going Deeper highlight on page 466 that discusses why a good leaving group can also be a good nucleophile, we tie the answer to the principle of microscopic reversibility. At first glance, reading this highlight may lead one to believe that all good nucleophiles are good leaving groups. This is not true, however, because hydroxide is a better nucleophile than water, but water is the better leaving group. Use reaction coordinate diagrams to explain when a good nucleophile is a good leaving group and when a good nucleophile is a poor leaving group.

10. Explain why the relief of ring strain or steric strain in an S_N1 reaction can often lead to small m values (reaction constants in Grunwald–Winstein plots).

11. In the nucleophilic addition to substituted alkenes, the ρ values are larger when the substituent is aryl than when the substituent is arylsulfonyl. This is also true if there is a second aryl group (denoted PhX). Why is this so?

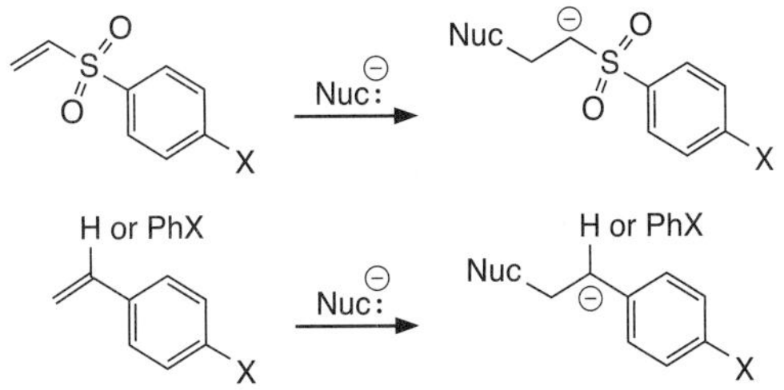

12. The acid-catalyzed hydrolysis of substituted ethylbenzoates has a ρ value of 0.14, whereas the base-catalyzed hydrolysis of the same series of compounds shows a ρ value of 2.19. Why is there such a difference?

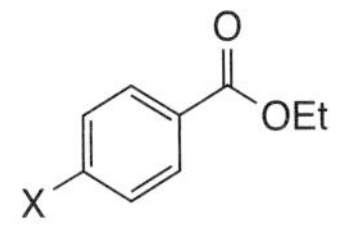

13. It is often stated that deuterium prefers the strongest bond. This implies that one should use bond strengths (meaning bond dissociation energies) to predict where deuterium will concentrate during a reaction that is in equilibrium and where the deuterium changes positions between two bonds. Explain what is not quite correct about this statement, and give an explanation of a better way to predict which bond deuterium would prefer.

14. Explain how one would experimentally perform a Hammett plot study for the following reaction. Do you expect a positive or negative ρ value?

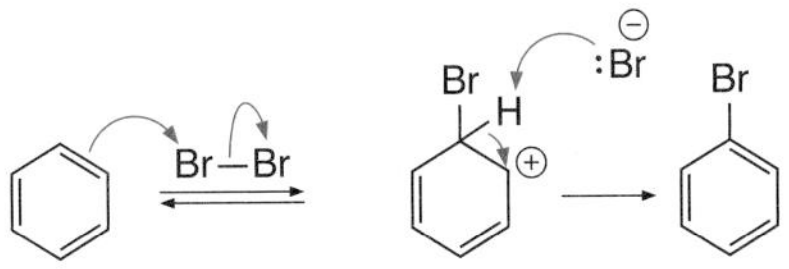

15. How can one distinguish experimentally if the ortho ester shown below hydrolyzes in acidic water by an S_N1 or S_N2 mechanism? There are actually two possible S_N2 mechanisms, but recall that an S_N2 at a tertiary carbon is uncommon, and therefore this is not one of the two choices that we are trying to distinguish.

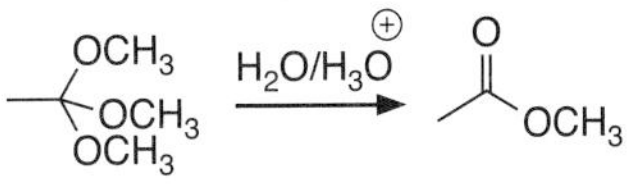

16. a. Given that a primary carbocation is less stable than a tertiary carbocation, would you expect that a reaction that creates a primary cation would be more or less responsive to solvent polarity than a reaction that creates a tertiary cation?
   b. Now give some reasons why the m value for ethyltosylate is less than the standard (t-butyl chloride) on which this free energy relationship is based. Does this experimental result agree with your answer to part a of this question?

17. In the following equilibrium the hydrogens (or deuteria) can be found in either terminal or bridging positions. It is harder to break the bond(s) to the H or D when they are bridging between the metals, and due to this, the reaction is slightly exothermic in the direction shown (from left-to-right) because the hydrogens are bound more strongly.

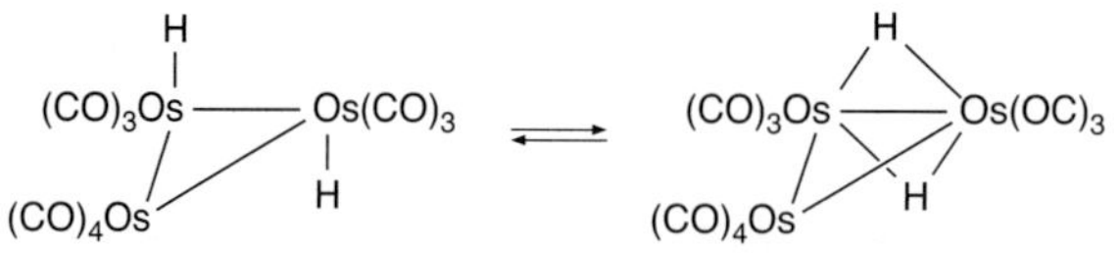

A. Terminal hydrogens   **B.** Bridging hydrogens

   a. Is the equilibrium constant for this reaction larger or smaller when one has D instead of H? Consider the equilibrium expression as $K_{eq} = [B]/[A]$.
   b. Draw a reaction coordinate diagram showing the relative shapes of the potential wells for the Os–H(D) bonds in the reactants and products, and the proper zero point energy levels within these wells. The Os–H bonds do not break and form in sequential steps. Use this diagram to explain your answer to part a.

18. In the following reactions substituent X was varied between electron donating and electron withdrawing. If you use the usual σ values, do you expect a positive or negative ρ value for each equilibrium? Should each reaction be more sensitive or less sensitive to the X substituent than benzoic acid? Explain your reasoning for each answer.

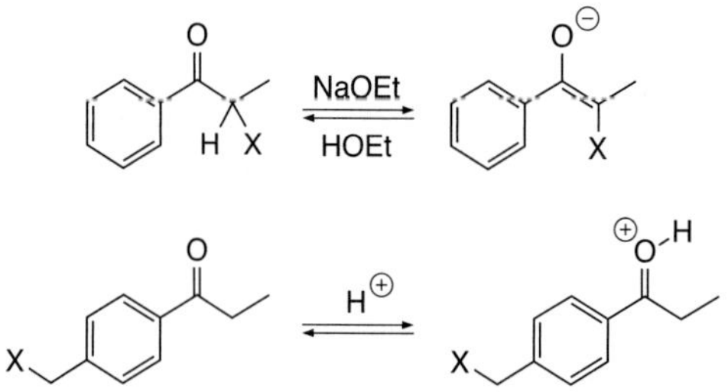

19. State whether the following reactions will show a normal or inverse, primary or secondary, kinetic isotope effect. Explain your reasoning.

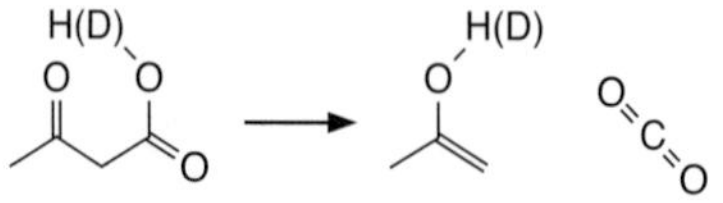

20. The following isotope effects are found for the ozonolysis of various deuterium-substituted propenes. What do these isotope effects tell you about the mechanism?

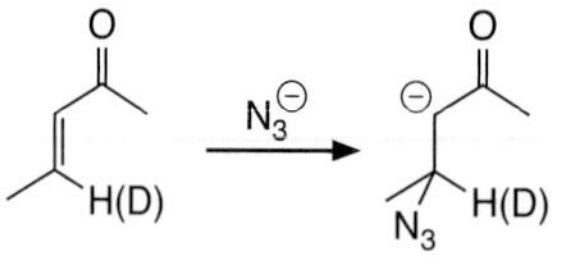

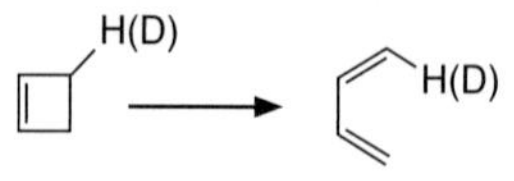

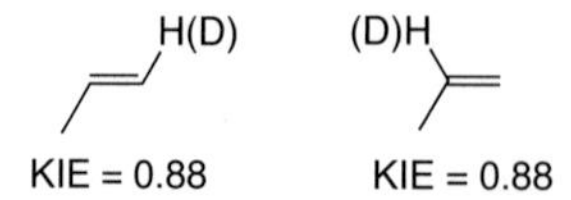

KIE = 0.88   KIE = 0.88

21. Two possible mechanisms for the Wittig rearrangement of benzyl ethers are shown below. Path 1 involves concerted intramolecular migration of CR₃, whereas Path 2 involves a heterolysis. In either mechanism, the deprotonation step is rate-determining. How would you apply the following experiments to distinguish between the two mechanisms: cross-over, trapping, and stereochemistry?

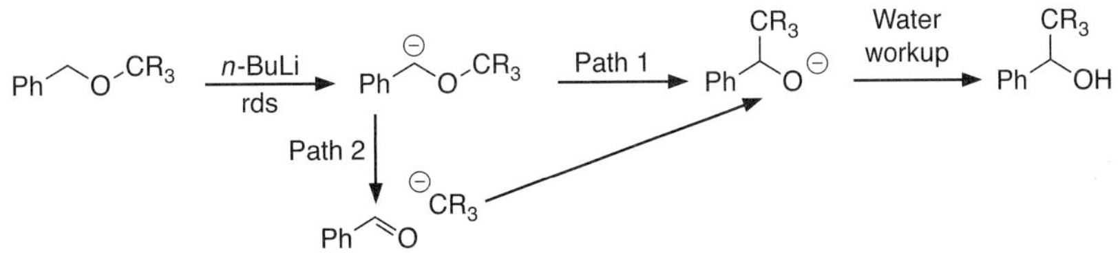

22. The treatment of chlorobenzene with potassium amide in liquid ammonia results in the formation of aniline. Propose four experiments discussed in this chapter, or others you can devise, that can be used to distinguish between the two mechanisms given below.

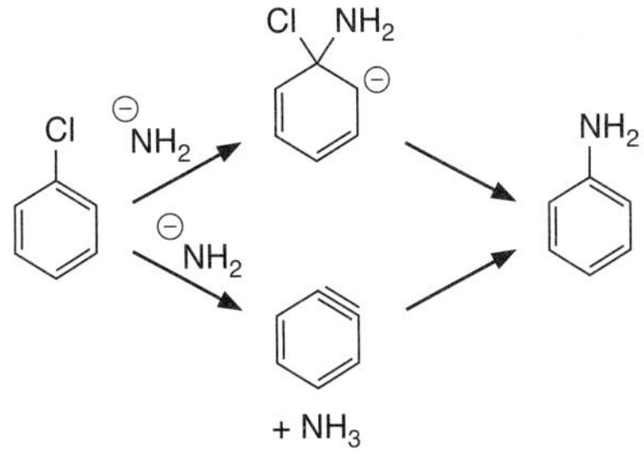

23. The hydrolysis of the following anhydrides gave the activation parameters listed. What is the isokinetic temperature for this reaction?

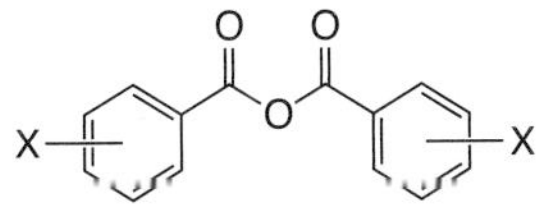

Substituent	$\Delta H^{\ddagger}$ (kcal/mol)	$\Delta S^{\ddagger}$ (eu)
m-Methyl	17.8	−31.4
p-Methoxy	20.1	−27.8
m-Nitro	11.6	−38.8
p-Nitro	10.7	−40.5

24. The acid-catalyzed Beckmann rearrangement of oximes to amides (see Chapter 11) has two possible rate-determining steps, the first one in each path shown below. Explain how you would use a Hammett plot analysis to distinguish these possibilities. It may be helpful to know that the R group trans to the departing water is the only one that migrates.

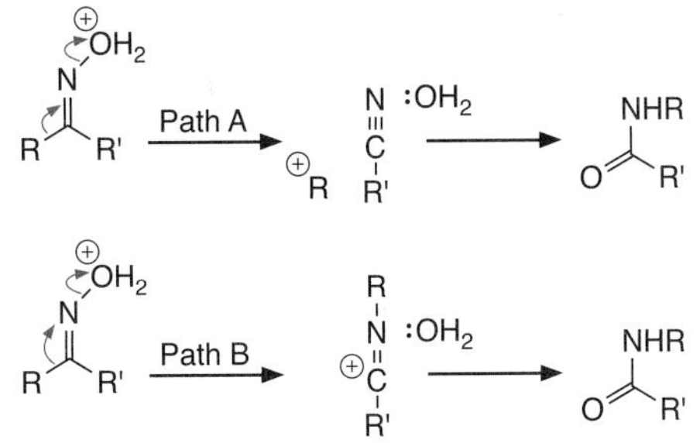

25. The relative rates of the S_N1 reactions of the following two alkyl p-nitrobenzoates (PNB) are as shown over the arrows below. The reason for the dramatic difference in the rate of departure of the nitrobenzoate leaving groups is the stabilization from the alkene to form a non-classical carbocation (recall non-classical carbocations from Chapters 1 and 2).

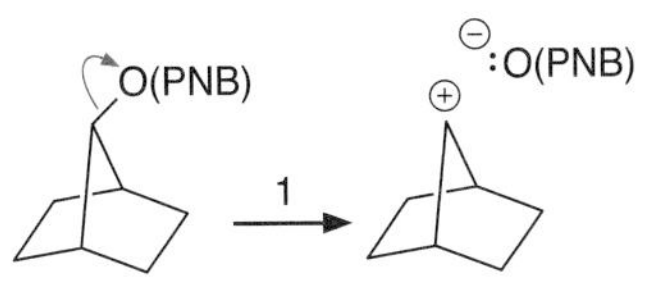

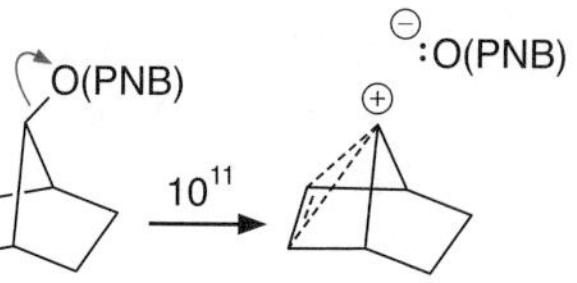

a. When *p*-anisyl is placed in the reactant as shown below, the 10^{11} rate enhancement by the double bond is reduced to only a factor of 3. Explain why introduction of this substituent changes the relative reactivity from the example given above.

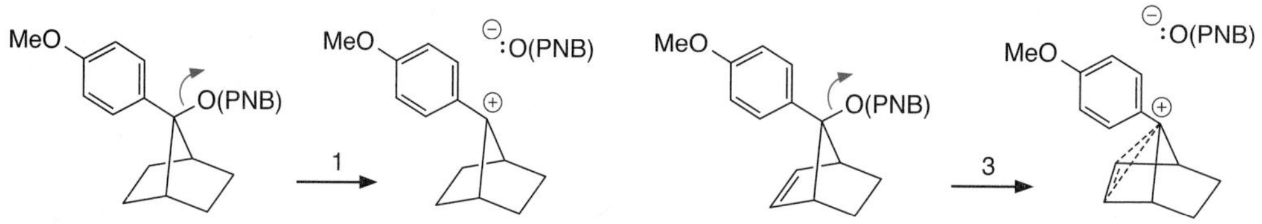

b. The ρ values for the S_N1 reaction of the following compounds were measured using σ^+ values. Interpret these values in light of the information given above. Do these ρ values support or contradict the conclusions based upon the difference in relative rates of the two comparisons given above?

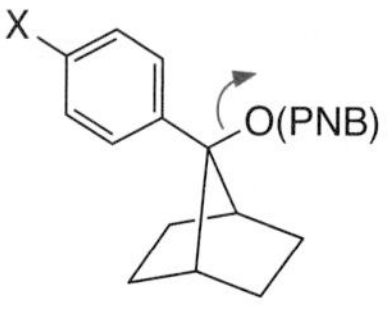

$\rho = -5.17$

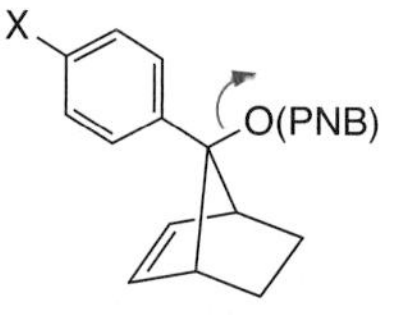

$\rho = -2.30$

c. When H and CF_3 are the X groups shown above, the relative rates of the S_N1 reaction of the norbornene structure compared to the norbornane analog are 41 and 35,000, respectively. Why are these relative rates higher than when X = OMe?

d. The data given in this problem have led to the **rule of increasing electron demand** for the extent of neighboring group participation. We will examine neighboring group participation in Chapter 11. Consider the alkene in the above reactants as a "neighboring group" that participates in the stabilization of the carbocation. Use all the data presented above to make a guess as to what the rule of increasing electron demand must state.

26. Draw two curves associated with proton inventory studies where two protons are "in flight" in the rate-determining step. In the first case each proton has an associated isotope effect of 2, while in the second case one isotope effect is 1.5 and the other is 2.5. Using a proton inventory analysis, can you differentiate the two possibilities?

27. The following reaction is an example of the Baeyer–Villager oxidation (see Chapter 11; *m*CPBA = *m*-chloroperbenzoic acid). A heavy atom isotope effect for the carbon labeled was found to be $k_{12}/k_{14} = 1.048$. What does this tell you about the mechanism (do not write the entire mechanism)?

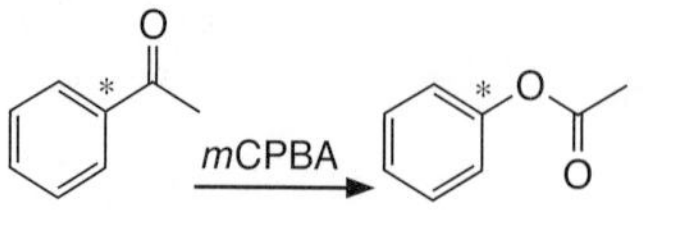

28. In general, how would you use radical clocks to measure the rate constant for the following radical rearrangement? Give one specific example.

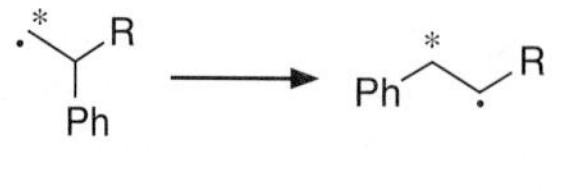

29. In an S_N2 reaction it is commonly observed that β_{Nuc} values are larger for reactions with poorer leaving groups. Similarly, β_{LG} values are larger for reactions involving weaker nucleophiles. Such interrelationships are called **Bema Hapothle** effects (an acronym of chemists' names). Explain the interdependence between β_{Nuc} and β_{LG} for S_N2 reactions.

Further Reading

Isotope Effects

Melander, L., and Saunders, W. A., Jr. (1980). *Reaction Rates of Isotopic Molecules*, John Wiley & Sons, New York.

Westheimer, F. H. "The Magnitude of the Primary Kinetic Isotope Effect for Compounds of Hydrogen and Deuterium." *Chem. Rev.,* **61**, 265 (1961).

Bigeleisen, J., and Wolfsberg, M. "Theoretical and Experimental Aspects of Isotope Effects in Chemical Kinetics." *Adv. Chem. Phys.,* **1**, 15 (1958).

Collins, C. J., and Bowman, N. S. (eds.), *Isotope Effects in Chemical Reactions*, ACS Monograph 167, Van Nostrand Reinhold, New York, 1970.

Shiner, V. J., Jr. in *Isotope Effects in Chemical Reactions*, C. J. Collins and N. S. Bowman (eds.), Van Nostrand Reinhold, New York, 1970.

Kresge, A. J. "Correlation of Kinetic Isotope Effects with Free Energies of Reaction." *J. Am. Chem. Soc.,* **102**, 7797 (1980).

Wiberg, K. B. "The Deuterium Isotope Effect." *Chem. Rev.,* **55**, 713 (1955).

Melander, L. (1960). *Isotope Effects on Reaction Rates*, Ronald Press, New York.

Linear vs. Non-Linear Transition State

More O'Ferrall, R. A. "Model Calculations of Hydrogen Isotope Effects for Non-linear Transition States." *J. Chem. Soc. B*, 785 (1970).

Kwart, H., Wilk, K. A., and Chatellier, D. J. "Verification and Characterization of the E_2C Mechanism. The Weak Base Catalyzed Elimination Reaction of β-Phenylethyl." *Org. Chem.,* **48**, 756 (1983).

Vitale, A. A., and San Filippo, J., Jr. "Four-Center Cyclic Transition States and Their Associated Deuterium Kinetic Isotope Effects: Hydrogenolysis of *n*-Octyllithium." *J. Am. Chem. Soc.,* **104**, 7341 (1982).

Anhede, B., and Bergman, N.-A. "Transition-State Structure and the Temperature Dependence of the Kinetic Isotope Effect." *J. Am. Chem. Soc.,* **106**, 7634 (1984).

Primary Isotope Effects

Bell, R. P., and Cox, B. G. "Primary Hydrogen Isotope Effects on the Rate of Ionization of Nitroethane in Mixtures of Water and Dimethyl Sulphoxide." *J. Chem. Soc. B*, 783 (1971).

More O'Ferrall, R. A. in *Proton Transfer Reactions*, E. Caldin and V. Gold (eds.), Chapman and Hall, London, 1975, p. 201.

Zero-Point Energy

Huskey, W. P. "Contributions of Internal Rotation to β-Deuterium Isotope Effects." *J. Phys. Chem.,* **96**, 1263 (1992).

Secondary Isotope Effects

Halevi, E. A. "Secondary Isotope Effects." *Prog. Phys. Org. Chem.,* **1**, 109 (1963).

Sunko, D. E., Hirsl-Starcevic, S., Pollack, S. K., and Hehre, W. J. "Hyperconjugation and Homohyperconjugation in the 1-Adamantyl Cation. Qualitative Models for *v*-Deuterium Isotope Effects." *J. Am. Chem. Soc.,* **101**, 6163 (1979).

Sunko, D. E., and Hehre, W. J. "Secondary Deuterium Isotope Effects on Reactions Proceeding Through Carbocations." *Prog. Phys. Org. Chem.,* **14**, 205 (1983).

Saunders, W. H., Jr. in *Investigation of Rates and Mechanisms of Reactions*, 4th ed., C. F. Bernasconi (ed.), John Wiley & Sons, New York, 1986, Vol. VI, Part I, p. 565.

Buncel, E., and Lee, C. C. (eds.), *Isotopes in Organic Chemistry*, Elsevier, Amsterdam, 1987, Vol. 7.

Tunneling

Lewis, E. S. in *Proton Transfer Reactions*, E. Caldin and V. Gold (eds.), Chapman and Hall, London, 1975, p. 317.

Saunders, W. H., Jr. "Contribution of Tunneling to Secondary Isotope Effects in Proton-Transfer Reactions." *J. Am. Chem. Soc.*, **106**, 2223 (1984).

Bell, R. P. (1980). *The Tunnel Effect in Chemistry*, Chapman and Hall, London.

Caldin, E. F. "Tunneling in Proton-Transfer Reactions in Solution." *Chem. Rev.*, **69**, 135 (1969).

Bell, R. P. (1960). *The Proton in Chemistry*, Methuen, London.

Solvent Isotope Effects

Schowen, R. L. "Mechanistic Deductions From Solvent Isotope Effects." *Prog. Phys. Org. Chem.*, **9**, 275 (1972).

Albery, W. J. in *Proton-Transfer Reactions*, E. Galdin and V. Gold (eds.), Chapman Hall, London, 1975.

Parker, A. J. "Rates of Bimolecular Substitution Reactions in Protic and Dipolar Solvents." *Adv. Phys. Org. Chem.*, **5**, 173 (1967).

Parker, A. J. "Protic–Dipolar Aprotic Solvent Effects on Rate of Bimolecular Reactions." *Chem. Rev.*, **69**, 1 (1969).

Waddington, T. C. (1969). *Non-Aqueous Solvents*, Thomas Nelson, London.

Kosower, E. M. (1968). *An Introduction to Physical Organic Chemistry*, John Wiley & Sons, New York, p. 259.

Mechanistic Analyses in General

Williams, A. "The Diagnosis of Concerted Organic Mechanisms." *Chem. Soc. Rev.*, 93 (1994).

Jencks, W. P. "A Primer for the Bema Hapothle. An Empirical Approach to the Characterization of Changing Transition-State Structures." *Chem. Rev.*, **85**, 511 (1985).

Linear Free Energy Relationships

Williams, A. (2003). *Free Energy Relationships in Organic and Bio-Organic Chemistry*, Royal Soc. of Chem., Cambridge, England.

Hammett Plots

Fuchs, R., and Lewis, E. S. in *Investigation of Rates and Mechanisms of Reactions*, 3rd ed., E. W. Lewis (ed.), Wiley-Interscience, New York, 1974, pp. 777–824.

Johnson, C. D. (1973). *The Hammett Equation*, Cambridge University Press, Cambridge, England.

Jaffé, H. H. "A Reexamination of the Hammett Equation." *Chem. Rev.*, **53**, 191 (1953).

McDaniel, D. H., and Brown, H. C. "An Extended Table of Hammett Substituent Constants Based on the Ionization of Substituted Benzoic Acid." *J. Org. Chem.*, **23**, 420 (1958).

Solvent Ionizing Plots

Bentley, T. W., and Llewellyn, G. "XY Scales of Solvent Ionizing Power." *Prog. Phys. Org. Chem.*, **17**, 121 (1990).

Brønsted Plots

Bender, M. L. "Mechanisms of Catalysis of Nucleophilic Reactions of Carboxylic Acid Derivatives." *Chem. Rev.*, **60**, 53 (1960).

Lewis, E. W. "Rate-Equilibrium LFER Characterization of Transition States. The Interpretation of α." *J. Phys. Org. Chem.*, **3**, 1 (1990).

Enthalpy–Entropy Compensation

Liu, L., Guo, Q.-X. "Isokinetic Relationship, Isoequilibrium Relationship, and Enthalpy–Entropy Compensation." *Chem. Rev.*, **101**, 673–695 (2001).

CHAPTER 9

Catalysis

Intent and Purpose

In this chapter on reactivity we cover catalysis, and how it is achieved. The study and development of catalytic reactions is currently of great interest in organic, organometallic, and bioorganic chemistry. Reactions as varied as the epoxidation of olefins, alkane C–H activation, DNA hydrolysis, cyclooctatetraene polymerization, and many more are currently being studied in an attempt to develop catalytic processes. In all cases, the principles of catalysis are essentially the same—to lower the energy barrier for the rate-determining step(s), thereby speeding up the reaction. There are two fundamental ways to do this. The first is to stabilize the transition state via binding interactions. The catalyzed reaction follows essentially the same mechanism as the uncatalyzed reaction, and occurs on a very similar energy surface but with lower barriers. This is the approach that many, but not all, enzymes utilize. The second approach to catalysis is to completely change the mechanism of the reaction. The catalyzed and uncatalyzed reactions occur on energy surfaces that are dramatically different. This is the approach to catalysis that is exploited with most, but not all, organometallic complexes. In this chapter we explore the multiple concepts that one employs to bind and stabilize transition states, along with some examples of change of mechanism. However, in Chapter 12, when catalysis using organometallic species is examined, the focus will be on changing the mechanism.

The chapter commences with a general overview of catalysis in the context of reaction coordinate diagrams and a simple thermodynamic cycle. Next, the most common factors invoked to explain transition state binding are explored: differential solvation, proximity, nucleophilic and electrophilic activation, and strain. We also look at covalent catalysis, which fundamentally involves a mechanism change.

After presenting general catalytic strategies we focus upon acid–base catalysis. The Brønsted linear free energy relationship briefly introduced in Chapter 8 is now a significant focus. Acid–base catalysis is exceedingly important in organic reaction mechanisms, and hence we provide a detailed treatment of this one form of catalysis. Since we have not yet discussed organic mechanisms, we develop acid–base catalysis in the context of carbonyl hydration, a mechanism that you should already be familiar with from introductory organic chemistry. Presenting the forms of acid–base catalysis in this chapter allows us to return to them repeatedly in Chapters 10 and 11.

Finally, we end the chapter with a discussion of nature's catalysts: enzymes. In fact, we allude to enzymes throughout the chapter. The general manner in which enzymes catalyze reactions is still a matter of debate, and so we present several theories. Our examination of enzymes is in preparation for a few specific enzymatic examples given in Chapters 10 and 11 as highlights for organic reaction mechanisms. Enzymes also provide an excellent setting in which to discuss Michaelis–Menton kinetics, the most common kinetic scenario used for catalysis. We also return to our analysis of the power of changing the thermodynamic reference state to examine reactivity, and show the manner in which an enzyme becomes "perfect".

9.1   General Principles of Catalysis

A **catalyst** is a compound that takes part in a reaction, resulting in an increased rate for that reaction, but is not consumed during the overall reaction (Eq. 9.1, where Cat = catalyst). Because a catalyst is not consumed, it can be added at sub-stoichiometric amounts. The catalyst then gives **turnover**, defined as the ability to act upon more than one reactant. The **turnover number** is the average number of reactants that a catalyst acts upon before the catalyst loses its activity. Because the catalyst affects the rate of the reaction, it is involved in the rate law for the reaction, but because it is regenerated, its concentration never changes.

The thermodynamics of a catalyzed reaction are unaffected by the catalyst, and hence catalysis falls solely within the realm of kinetics. A **catalytic reaction** is a reaction that is catalyzed. Some reactions are **promoted** by an additive. This definition is used when the additive speeds up the reaction but is converted in the reaction to another species. All reactions are in theory amenable to catalysis, meaning that there is some species, that when present, will speed up that reaction. However, not all reactions are easily catalyzed, because it is not always obvious what is the best strategy to accelerate the reaction or how to construct a catalyst that would yield a rate acceleration.

$$A \ + \ Cat \ \xrightarrow{\ k\ } \ B \ + \ Cat \qquad\qquad (\text{Eq. 9.1})$$

Why do we want to catalyze a reaction? Usually the goal is to make the rate fast enough for the reaction to be performed in a timely and practical manner. For example, C–H bonds in the presence of oxygen alone take an impossibly long time to be converted to alcohols, so a catalyst is needed if we are to efficiently oxidize unactivated C–H bonds.

Catalysts are classified as heterogeneous or homogeneous. A **heterogeneous catalyst** is one that does not dissolve in the solution, and hence the catalysis takes places in a phase separate from the solution (typically the surface of the catalyst). An example is the reduction of olefins with H_2 catalyzed by Pd/C, which we discussed in Chapter 2 as a method to measure heats of hydrogenation. A **homogeneous catalyst** is one that dissolves in the solution, and hence normal spectroscopic and chromatographic techniques can be employed to explore the mechanism and identify intermediates. Acids and bases typically fall into this category, as do most enzymes and most organometallic species.

Importantly, all catalysts operate by the same general principle—that is, the activation energy of the rate-determining step(s) must be lowered in order for a rate enhancement to occur. This is an obvious statement based upon our analysis in Chapter 7 of how rates are connected to energy surfaces. This leads to the typical potential energy diagram given in Figure 9.1 **B** to explain catalysis, where the transition state is placed lower in energy on the diagram relative to that shown in Figure 9.1 **A**.

Figure 9.1
Inherent in catalysis is the idea that the activation energies for any catalyzed reaction must be lower than the activation energies for the uncatalyzed reaction. **A.** The uncatalyzed path, **B.** A common way to view the catalyzed path, and **C.** A more realistic view of how catalysis can be achieved without a complete mechanism change. Binding of the reactant is required first. **D.** A totally new mechanism with a completely new reaction coordinate can also give catalysis.

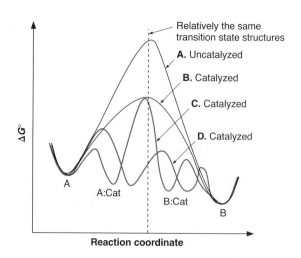

9.1.1   Binding the Transition State *Better* than the Ground State

Before examining general reaction coordinate diagrams for catalytic reactions, we need to define a few terms. First, a **substrate** is any starting material (reactant) used in a catalytic reaction. Second, the activated complex is simply referred to as the **transition state**, even though the strict differences given in Chapter 7 do exist. Since the transition state can be viewed as a high energy, strained form of the substrate, we often refer to the substrate as the **ground state**. We measure the success or efficiency of a catalytic reaction by determining the **rate enhancement**, which is the ratio of the rate constant for the catalyzed reaction to that for the uncatalyzed reaction.

Lastly, **binding** refers to any physical interaction that the catalyst has with the substrate and transition state, including a simple change in solvation, a coordination with a metal, a loose association with an acid or base, or any molecular recognition interaction. Binding in this context also includes proximity, where two functional groups may not have any physical attraction, but the simple fact that they are held close in space can lead to the enhancement of a chemical reaction. Let's look at reaction coordinate diagrams to examine how binding affects the potential energy surface.

The comparison of Figures 9.1 **A** and **B** gives the classic representation of catalysis found in introductory chemistry textbooks, and even some more advanced texts. The entire catalytic path is lower in energy than the uncatalyzed route. The substrate starts the reaction at the same energy level, with or without the catalyst, but the catalyzed route becomes increasingly lower and lower in energy relative to the uncatalyzed path until at the transition state the largest energy difference between the two paths is achieved. Is this a realistic view of how an energy surface can be changed by the addition of a catalyst?

To answer this question, let's consider what the catalyst is physically doing during the reaction given in Figure 9.1 **B**. First, the catalyst does not interact with the substrate, because the energy of the substrate is the same with or without the catalyst. However, the physical interaction between the substrate and catalyst (binding) becomes greater and greater as the substrate gradually changes structure until the transition state is achieved. Then, the binding of the catalyst to the product decreases until the product is fully formed, at which point there is no interaction between the catalyst and product. *Such a scenario is physically impossible.* It would mean that the catalyst has its highest affinity for the transition state of the reaction but has no affinity for the substrate or product. Moreover, because the catalyst binds the transition state best, the strongest binding would occur when the catalyst encounters the transition state in solution, which is impossible because of the fleeting nature of a transition state.

Since the activation energy for a catalyzed reaction must be lower than the uncatalyzed reaction, some interaction (binding) between the catalyst and the transition state is required. However, because it is always true that the structure of the transition state is a mixture of the substrate and product structures, it makes good sense that if the catalyst binds the transition state it should also bind the substrate and product. This leads to the reaction coordinate diagram shown in Figure 9.1 **C**. Here an equilibrium is established between the substrate and catalyst, followed by an activation barrier to the catalyzed reaction, and then an equilibrium dissociation of the product from the catalyst (Eq. 9.2). One can rightly view this as a mechanism change, because the energy surface has now changed and possesses two additional wells. However, this is not a full change in mechanism to achieve catalysis. The focus is the transition state, which has approximately the same structure in the catalyzed and uncatalyzed pathways (close along the dotted line in Figures 9.1 **A** and **C**). A full change in mechanism is shown in Figure 9.1 **D**. Now completely new transition states and intermediates are involved.

$$A \ + \ Cat \ \underset{}{\overset{K_1}{\rightleftharpoons}} \ A{:}Cat \ \underset{}{\overset{K_2}{\rightleftharpoons}} \ B{:}Cat \ \rightleftharpoons B \ + \ Cat \qquad (Eq. 9.2)$$

Let's examine the consequences of various energy levels for the transition states and valleys in diagrams such as Figure 9.1 **C**. $\Delta G°$ diagrams are used in this discussion. It is important to remember in this analysis that $\Delta G°$ has both enthalpy and entropy contributions.

Hence, the term "binding" is used to mean any kind of favorable enthalpic or entropic interaction between the catalyst and the substrate, transition state, or product.

Binding of the catalyst to the substrate and product, as well as the transition state, leads to several possible reaction coordinate diagrams, a few of which are given in Figure 9.2. Let's examine the four possibilities to explore the requirements for catalysis.

In Figure 9.2 **A**, the substrate binds to the catalyst, but the energy of the transition state has not been changed at all. Now the catalyst–substrate complex has fallen into an energetic hole, and the rate-determining barrier is actually larger than the uncatalyzed reaction. The reaction is slowed by the addition of the "catalyst". The term catalyst in this context is incorrect.

Now consider Figure 9.2 **B**. The substrate binds to the catalyst, but the complex is higher in energy than the substrate alone. This can be viewed as an interaction of the substrate with a very poor receptor ($K_a < 1$). Since the transition state is not lowered by the binding to the catalyst there is *still* no catalysis. Therefore, if strain were introduced in the substrate, that strain is not alleviated upon achieving the transition state.

In contrast, consider Figure 9.2 **C**. In this case, the interaction between the substrate and the catalyst stabilizes the substrate. However, because the transition state is stabilized to the same extent via its interaction with the catalyst, the rate of the reaction is identical with or without the catalyst.

In Figure 9.2 **D**, the catalyst binds to the substrate, and the complex is lower in energy than free substrate. In this scenario, however, the binding between the catalyst and transition state is even stronger. Now the activation energy is lowered relative to the uncatalyzed path, and the reaction rate increases. We show this schematically in Figure 9.3 with a binding pocket where more contacts with the catalyst are formed with the transition state than with the substrate or product.

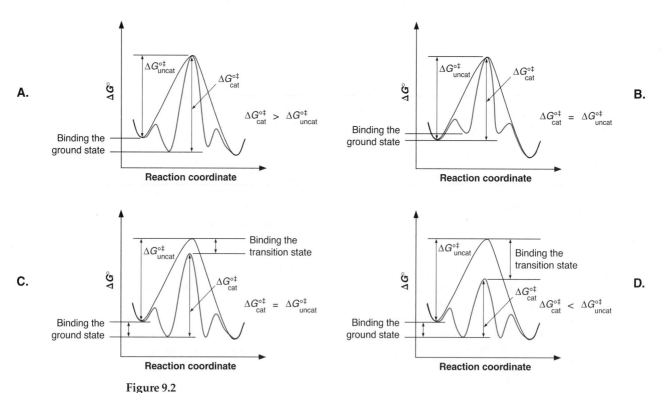

Figure 9.2
Reaction coordinate diagrams for various types of catalysts interacting with ground states and transition states. **A.** The substrate is bound by the catalyst, but there is no stabilization of the transition state and no catalysis occurs. The rate is actually slower than the background rate. **B.** The substrate is bound weakly, but there is no stabilization of the transition state. The rate is the same as the background. **C.** The substrate and transition state are bound to the same extent, and the uncatalyzed rate is the same as the catalyzed rate. **D.** The transition state is bound better than the substrate, and catalysis occurs.

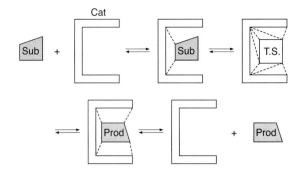

Figure 9.3
In many catalytic schemes (such as enzymatic, synthetic, or bioorganic), the substrate first binds to the catalyst, followed by interactions with the catalyst that stabilize the transition state, and then the catalyst releases the product. We show the catalyst here as having a pocket, but a surface, or even a single bond (such as to a proton), can act in a similar manner. Sub = substrate, Cat = catalyst, T.S. = transition state, and Prod = product.

In summary, *to achieve catalysis, the catalyst must stabilize the transition state more than it stabilizes the ground state*. That is, the transition state must be bound better than the ground state. This is one viewpoint that unifies a very large fraction of catalytic reactions.

What does it mean to bind a high energy structure such as a transition state (T.S.)? This is a bit of an esoteric notion, because transition states are highly strained and fleeting. No catalyst has ever stabilized a T.S. so well that the T.S. actually developed a long lifetime. Instead, the idea is that the catalyst is more complementary to the structure of the T.S. than the ground state (refer to Section 4.2.1 for a definition and examples of complementarity). The lowering of the T.S. energy arises from interactions between the catalyst and the substrate that release energy, but these interactions become stronger as the T.S. is achieved, releasing even more energy. The release of free energy due to binding interactions between the transition state and the catalyst, in part, counteracts the large uptake of free energy that is required to achieve the strained molecular structure at the transition state.

9.1.2 A Thermodynamic Cycle Analysis

The idea that catalysis arises from increased binding of the transition state relative to the ground state can be given a thermodynamic foundation by looking at a cycle. Recall that Gibbs free energy is a state function, and therefore is independent of the path that interconverts compounds. Figure 9.4 shows two pathways for interconversion of a substrate to the transition state. We consider the binding equilibrium between substrate and catalyst [$K_{a(Sub)}$], a hypothetical binding equilibrium between the transition state and the catalyst [$K_{a(Sub^\ddagger)}$], and then the two equilibria between the substrate and the transition state when bound to the catalyst ($K_1^\ddagger$) or free in solution ($K_2^\ddagger$). The notion of an equilibrium constant between a reactant and a transition state comes directly from transition state theory (recall Section 7.2). Thus, in the current analysis we can view the $K_1^\ddagger$ and $K_2^\ddagger$ values as being directly proportional to the catalyzed and uncatalyzed rate constants, respectively. Therefore, $K_1^\ddagger / K_2^\ddagger = k_{cat} / k_{uncat}$ (the rate enhancement). This cycle is similar to the one developed in the context of statistical perturbation theory in Section 3.3.5.

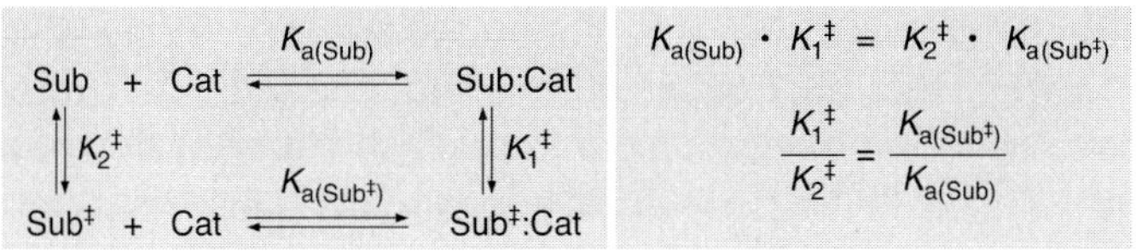

$$
\begin{array}{ccc}
\text{Sub} + \text{Cat} & \xrightleftharpoons{K_{a(Sub)}} & \text{Sub:Cat} \\
\Big\updownarrow K_2^\ddagger & & \Big\updownarrow K_1^\ddagger \\
\text{Sub}^\ddagger + \text{Cat} & \xrightleftharpoons{K_{a(Sub^\ddagger)}} & \text{Sub}^\ddagger\text{:Cat}
\end{array}
\qquad
\begin{array}{c}
K_{a(Sub)} \cdot K_1^\ddagger = K_2^\ddagger \cdot K_{a(Sub^\ddagger)} \\[1em]
\dfrac{K_1^\ddagger}{K_2^\ddagger} = \dfrac{K_{a(Sub^\ddagger)}}{K_{a(Sub)}}
\end{array}
$$

Figure 9.4
A thermodynamic cycle showing how catalysis can be directly related to the differential binding of the substrate and the transition state. Sub = substrate and Cat = catalyst.

Given these definitions and background, we can focus on why this thermodynamic cycle is informative. The goal is to solve for the relative rate constants for the catalyzed and uncatalyzed reaction, which is done using the mathematics shown to the right in Figure 9.4.

We want to know $K_1^{\ddagger}/K_2^{\ddagger}$, and it is equal to $K_{a(Sub\ddagger)}/K_{a(Sub)}$—the binding constant for the transition state relative to the substrate. Hence, we find again that one achieves catalysis by having a larger binding constant for the transition state than for the ground state. The larger this difference in binding, the greater the rate enhancement. This is a general strategy for achieving catalysis, but as the following Going Deeper highlight shows, it is the required rates for metabolism that dictate the extent of rate enhancement seen in biology.

Going Deeper

The Application of Figure 9.4 to Enzymes

The thermodynamic cycle given in Figure 9.4 has been criticized as being overinterpreted, especially when applied to enzymes. In an interesting study, Wolfenden found that k_{cat}/k_{uncat} ratios for enzymes were determined primarily by the k_{uncat} values, because the k_{cat} values were in a narrow range. He found that $\log(k_{cat})$ values did not give a good linear correlation with $\log[K_{a(Sub\ddagger)}]$, but instead they correlated well with $-\log[K_{a(Sub)}]$. In other words, the rate enhancement for many common enzymes is controlled by how slow the uncatalyzed (or background) reaction is. All the rate constants for the catalyzed reactions are similar.

The reason for this is that enzymes have evolved to execute their reactions within a specific time period appropriate for the metabolism of life. The reactions of life have to occur at a rate commensurate with metabolism and other life processes, and thus the rate constants for the enzymatic reactions fall within a fairly narrow range. For most biochemical processes, reactions that occur on the millisecond to microsecond time scale are fast enough.

Radzicka, A., and Wolfenden, R. "A Proficient Enzyme." *Science*, **265**, 90–93 (1995).

9.1.3   A Spatial Temporal Approach

Some researchers are critical of the analysis given above. As noted above, the notion of binding a transition state is a bit esoteric. Instead, another explanation for catalysis is given, called the **spatial temporal** postulate. This approach to catalysis is based on the fact that many intramolecular reactions are often much faster than corresponding intermolecular reactions (we will examine several in Section 9.2.2). The postulate states that *the rate of reaction between functionalities A and B is proportional to the time that A and B reside within a critical distance*. In this postulate, time and distance are the critical factors. This is a very intuitive notion, and therefore is very helpful for understanding catalysis. First, compounds A and B have to be at the proper distance to react, which is consistent with the idea from Chapter 7 that vibrational excitation and coupling between reactants leads to chemical transformations. Second, the reaction will occur when the vibration that defines the reaction coordinate is excited to a high enough level in the complex between A and B to surmount the barrier. The longer that A and B spend together in the correct geometry for reaction, the greater that probability.

The concept of having a spatial effect in catalysis can be related to the notion of differential binding of a transition state relative to a ground state. This is because distance is inherent in our notion of binding interactions. When bound to the catalyst, the distance between A and B is closer than when they are free in solution, and inherently the transition state brings A and B close because bonds are beginning to form. The temporal portion of this alternative view of catalysis can also be related to binding interactions. The rate of dissociation (off rate) of the catalyst–substrate complex determines, in part, the probability for catalysis. Therefore, the two views of catalysis are really very similar, but are distinctly different in their intuitive insights and terminology.

In summary, binding is the key element in the most widely accepted theory of how catalysis is achieved (without completely changing the mechanism). Greater binding of the transition state relative to the ground state is all that needs to be invoked to give a rate enhancement. Hence, to better understand specific molecular interactions that lead to catalysis, we need to understand all the different forms of chemical interactions that we are calling "binding".

9.2   Forms of Catalysis

Several specific chemical approaches are used to achieve increased transition state binding. The organic approaches include, but are not limited to, changing solvation, proximity, nucleophilic activation, electrophilic activation, introducing strain, acid–base chemistry, covalent catalysis, and supramolecular chemistry. Additionally, in some of these approaches there is a change in mechanism to achieve catalysis. All these methods are discussed below.

9.2.1   "Binding" is Akin to Solvation

We begin our analysis of specific molecular interactions that lead to catalysis by making an analogy to differential solvation. Recall from Chapter 4 that molecular recognition is often driven by differential solvation effects (Eq. 4.34). Solvation is inherently an intermolecular interaction, as is the binding between a receptor and a substrate. When a receptor and a substrate come together, they replace solvation interactions with their own intermolecular interactions and hence "solvate" one another.

As stated above, when considering catalysis, we must examine differential binding between the transition state and substrate. In essence, we require better molecular recognition of the transition state relative to the substrate. This is akin to saying we need better solvation of the transition state relative to the substrate. Imagine a catalyst where the dipoles, hydrogen bonds, and coordination interactions to metals, within its binding pocket (such as Eq. 9.3) are arranged to best solvate the structure of the transition state. This is a scenario that would lead to good catalysis.

Lastly, let's look ahead to our discussion of substitution reactions in Chapter 11 (see Section 11.5.6). There we will examine reaction coordinate diagrams for the use of different solvents in S_N2 and S_N1 reactions. Recall from introductory organic chemistry that the use of a more polar solvent speeds up an S_N1 reaction. During the heterolysis of an S_N1 reaction, the activated complex develops charge, and therefore a more polar solvent solvates the activated complex better; this leads to a faster rate. The extent to which the solvent affects the rates of these reactions can be dramatic. To see this, examine the order of magnitude differences predicted for rates given by the range of Grunwald–Winstein values listed in Table 8.4. Better solvation of the activated complex relative to the reactant is a very powerful way to speed reactions. It is perfectly akin to the discussion given here where better *binding* of the structure of a transition state within a catalyst's binding site leads to catalysis.

9.2.2   Proximity as a Binding Phenomenon

One form of "binding" that can occur between two reactants is simple proximity, a major tenet of the spatial temporal vision of catalysis. When a second order reaction occurs in solution, two reactants must collide in the rate-determining step. This causes a loss in translational degrees of freedom in the reactants, thereby increasing the Gibbs free energy at the transition state due to the increased order of the system (look back at Section 2.1.2 to review this idea). The translational and rotational entropies of a freely moving molecule in solution are both around 30 entropy units (eu).

If the reactants were bound together with a catalyst in a step prior to the rate-determining step, much of the energy cost in reducing the entropy of the reactants would be paid in the binding of the reactants, and not paid during the rate-determining step. Bringing together the reactants from dilute solution overcomes part of the entropy cost required in the rate-determining step. The transition state can be considered to be "bound better" than the ground state, even if no differential enthalpy interactions are present between the catalyst, the substrate, and the transition state. We would expect to see this reflected in the activation entropies of the catalyzed and uncatalyzed reactions.

To get an idea of how much the activation energy can be lowered by reducing the entropy of activation by tying two groups together, Benkovic and Bruice looked at a series of 40 reactions where the intramolecular version could be directly compared to intermolecular versions. On average, the $T\Delta S^{\ddagger}$ differed between the reactions by 4.6 kcal/mol at 25 °C. This correlates to a roughly 2×10^3 rate enhancement for an intermolecular reaction to an intramolecular reaction. Any rate enhancement beyond this value should be due to other forms of entropy or enthalpy factors.

Many examples of proximity effects are known. In general, whenever an intramolecular acid or base is invoked in acid–base catalysis, proximity effects can be a factor. Further, when any catalyst holds a substrate near a catalytic group at its active site, or holds two separate substrates next to each other, proximity effects can be relevant. Proximity effects are definitely prevalent in organometallic catalysis, as we will see in Chapter 12. Hence, proximity effects are key to many forms of catalysis.

Several attempts to measure the maximal rate enhancements that can be achieved through proximity effects alone have been performed. A typical example compares the intermolecular aminolysis of phenyl acetate by trimethylamine (Eq. 9.3) to the intramolecular cyclization of phenyl 4-(N,N-dimethylamino)butanoate (Eq. 9.4). The intermolecular reaction has a rate constant $k_2 = 1.3 \times 10^{-4}$ M^{-1} s^{-1}, while the intramolecular reaction has $k_1 = 0.17$ s^{-1}. This suggests a rate enhancement of 1200 for the intramolecular reaction. Even in pure trimethylamine, the pseudo-first order rate constant for the intermolecular reaction would be less than the intramolecular rate constant.

$$\text{(Eq. 9.3)}$$

$$\text{(Eq. 9.4)}$$

There is a slight problem with the rate comparison just made. First, it is difficult to make completely fair comparisons, since inherently the structures of the compounds undergoing the intra- and intermolecular reactions are different, suggesting possible intrinsic reactivity differences. More importantly, the rate constants for the two reactions have different units, typically M^{-1} s^{-1} and s^{-1} for the inter- and intramolecular reactions, respectively. What then is the meaning of the ratio in the previous paragraph?

When taking a ratio of the rate constants for intramolecular vs. intermolecular reactions, one is left with units of molarity (M). Hence, chemists have defined a term called the **effective molarity** (E.M.) or **intramolecularity**, which is the ratio of the first order to second order rate constants for the analogous reactions (Eq. 9.5). We examined some of these notions in a Going Deeper highlight in Section 7.4.1 that you may want to review. The E.M. tells us the concentration of one of the reactants that would need to be achieved to make the intermolecular reaction have a pseudo-first order rate constant identical to the first order rate constant for the intramolecular reaction. This is usually a concentration that is impossible to achieve (as with the example in Eqs. 9.3 and 9.4). In essence, the E.M. tells us the **effective concentration** of one of the components in the intramolecular reaction. Figure 9.5 shows a handful of relative cyclization rates, and Table 9.1 lists several interesting E.M.s.

$$\text{E.M.} = \frac{k_1}{k_2} \qquad \text{(Eq. 9.5)}$$

An interesting feature of Figure 9.5 is the energy difference between successive entries. The rate difference between the first and second reactions implies a $\Delta\Delta G^{\ddagger}$ of 2.8 kcal/mol. The difference between the second and third reactions is also 2.8 kcal/mol. We find that removing a freely rotating bond gives a consistent enhancement. These bonds are not freely rotating in the transition state for the cyclization, so that removing (or freezing) these bonds in the reactant creates a geometry that more resembles the transition state of the reaction. Also, note the further rate increase that occurs due to the addition of methyl groups (Table 9.1 **A**). This causes a **steric compression** that pushes the electrophile and nucleophile together. Since the nucleophile and electrophile come together to make a new bond in the

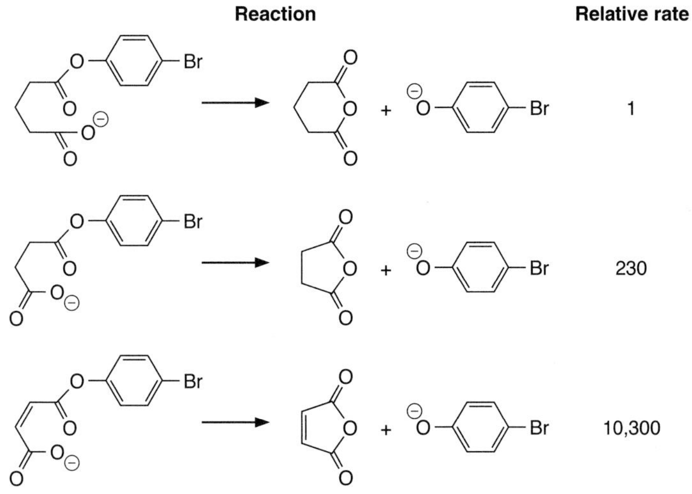

Reaction	Relative rate
	1
	230
	10,300

Figure 9.5
Relative rates for three similar intramolecular cyclization reactions.
Note how the freezing of rotamer distributions leads to increased rates.

Table 9.1
Effective Molarities (M) for a Variety of Cyclization Reactions*

A. Anhydride formation

3×10^9 6×10^{10} 4×10^{12} 5×10^{12}

B. Lactone formation

4×10^4 7×10^6 2.3×10^7 3.7×10^{11} 2×10^{13}

C. Ether formation

6.6×10^3 2.3×10^4 2.0×10^7 5.7×10^9

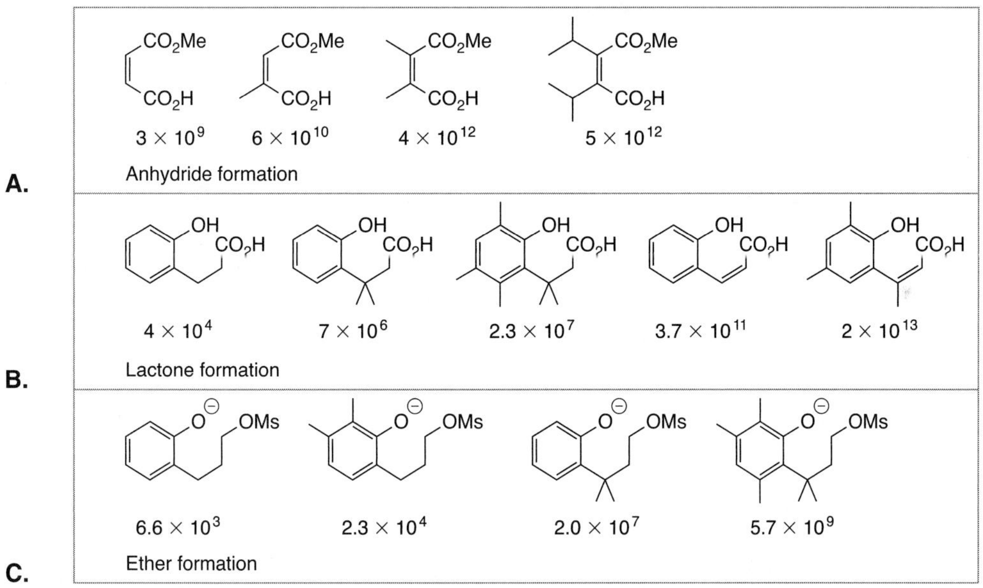

*Kirby, A. J. "Effective Molarities for Intramolecular Reactions." *Adv. Phys. Org. Chem.*, **17**, 183 (1980).

transition state, the increased compression in the reactant pushes its structure toward that of the transition state. Another common way to sterically compress two groups together and preorganize two reactants in proximity is the use of the gem–dimethyl effect (the **Thorpe–Ingold effect**; c.f. Exercise 44 in Chapter 2). Similar trends for losses of bond rotations and steric compression can be seen in the other examples listed in Table 9.1.

As already suggested, the first effect that goes into an E.M. is the freezing of translational, rotational, and vibrational degrees of freedom. The second is a steric compression between two reactants that is alleviated upon achieving the transition state. A general rule of thumb is that strain-free intramolecular reactions can achieve rate enhancements near 10^8 due to proximity alone. This number is based upon the average entropies of translation and rotation, and agree with the values from the Benkovic/Bruice study mentioned previously.

If steric strain is introduced in the ground state and alleviated in the transition state, values up to 10^{16} in rate enhancements can be obtained. In some cases, high proximity can even lead to the isolation of high energy intermediates, and the Connections highlight below shows an example from acyl transfer reactions (details of acyl transfers are given in Chapter 10).

A number of concepts have been developed to explain the large effective molarities given in Figure 9.5 and Table 9.1. We have already mentioned the expected enhancement in reactivity due to the prepayment of the translational and rotational entropy costs. One other manner in which reactivity can be enhanced is called **orbital steering**, an effect where a dependence of rate constants upon torsion and bond angles is postulated. Here, a very accurate alignment of orbitals is postulated to greatly facilitate reactions. A related notion is called **stereopopulation control**, where the freezing of one conformer into a productive geometry increases reactivity. A more recent concept related to having the appropriate geometry for attack is referred to as near attack conformations, as described in the Going Deeper highlight on the next page.

Many of the large effective molarity values given in Table 9.1 result from the release of strain when achieving the transition state. In these cases there exist steric, torsional, and/or bond angle strains that are relieved along the reaction coordinate. One can view the chemical synthesis route that produced the reactant as activating the reactant by the introduction of strain. In other words, the reactivities of the compounds that possess very high E.M.s ($> 10^8$) manifest themselves via a **transition state effect**. Relative to the control reaction, the intramolecular chelation results in stability imparted to the transition state and not the ground state (or, equivalently, instability in the ground state that is not present or lower in the transition state). Although certain bonds are breaking and forming in the transition states for the cyclizations given in Figure 9.5 and Table 9.1, the other bonds in the compounds are becoming stronger due to a relief of strain throughout the structure. Hence, there are a few different vantage points from which to understand effective molarities.

Connections

High Proximity Leads to the Isolation of a Tetrahedral Intermediate

No catalyst has ever stabilized a transition state so much that it became isolable. However, strong binding of a high energy *intermediate* can sometimes lead to enough stabilization that the intermediate can be isolated. Recall from introductory organic chemistry that the hydrolysis of esters and amides proceeds via a tetrahedral intermediate (see the equation below). Such intermediates are not isolated, but are inferred from kinetic studies and isotope scrambling experiments (see Section 10.17). However, as described here, a tetrahedral intermediate has been isolated and even characterized by crystallography.

Amide hydrolysis   Tetrahedral intermediate

Compound *i* rapidly converts in acid to compound *ii*, which is indefinitely stable. This is a tetrahedral intermediate analogous to those found in amide hydrolysis. The effective molarity of the amine in *i* is estimated to be 10^{11} to 10^{12} M. The effective molarity is so high that the amine leaving group is essentially permanently attached to the

carboxylate carbon, even when the amine of *ii* is protonated. However, at pH values lower than 4 or 5, *i* and *ii* are in a rapid equilibrium. This is an outstanding example of

Stable tetrahedral intermediate

the use of chemical synthesis to create a system where a normally very unstable and strained system becomes locked into a kinetically stable system (persistent).

You may be wondering, "Why doesn't a hydroxide act as a leaving group from *ii*?" Note that structure *iii* is a "twisted amide". It does not have the planar structure that imparts the stabilization normally associated with amides. Hence, compound *iii* is very reactive, and in water it immediately converts to *ii*.

Kirby, A. J., Komarov, I. V., and Feeder, N. "Spontaneous, Millisecond Formation of a Twisted Amide from the Amino Acid, and the Crystal Structure of a Tetrahedral Intermediate." *J. Am. Chem. Soc.*, **120**, 7101 (1998).

Going Deeper

The Notion of "Near Attack Conformations"

Another approach to understanding catalysis has been put forth recently by Bruice. This approach ties together orbital steering, stereopopulation control, spatial temporal effects, and transition state effects. The focus is upon **near attack conformations** (NACs). NACs are defined as conformations that achieve the required arrangement of juxtaposed reactants to achieve the transition state. For example, in the addition of a nucleophile to a carbonyl carbon (see Section 10.8.5), there is an optimum trajectory for attack called the Bürgi–Dunitz angle. A cone represents a set of attack trajectories that are roughly equally favorable, while attack angles outside this cone are less favorable. Furthermore, when the nucleophile is within the van der Waals radius (i.e., within about 3 Å) of the electrophile, the reaction is facilitated. Those intramolecular reactions where the reactant is largely in an NAC will be the most efficient. Importantly, an NAC is a ground state notion, not an interaction that is achieved in the transition state.

Nuc: ⊖   Cone for NAC

The researchers found that the rate constants for several intramolecular ester formations with a wide variety of E.M. values were directly correlated to the mole fraction of the reactants present as NACs. The mole fractions were calculated using molecular dynamics simulations. When the ground state resides naturally in an NAC, then an E.M. of around 10^8 was achieved (the same value that we introduced earlier as the upper limit to proximity effects). To achieve an NAC, the reactant must be placed in a conformation that also has a higher enthalpy, because the reactants are within van der Waals distances. Therefore, the rate enhancement obtained by NAC formation is postulated to be also enthalpy derived, not solely entropy derived.

In the NAC theory, the rate constants for catalyzed reactions depend upon the following: 1. the mole fraction of the reactant–catalyst complex that is in an NAC, 2. the difference in solvation between the reactant–complex in an NAC and the solvation of the NAC without catalyst, and 3. any electrostatic binding forces that can stabilize the transition state. Therefore, the theory considers both a ground state effect and increased binding of the transition state to fully cover the methods of catalysis.

Bruice, T. C., and Lightstone, F. C. "Ground State and Transition State Contributions to the Rates of Intramolecular and Enzymatic Reactions." *Acc. Chem. Res.*, **32**, 127–136 (1999).

How does all this relate to catalysis? The synthesis that went into the creation of the compounds that show high E.M.s is analogous to "binding" two reactants to a catalyst. In the intramolecular reactions we have looked at, covalent bonds provide the binding. In the synthetic formation of these bonds we have paid the entropy price for bringing A and B together, and furthermore, we have paid the enthalpy price associated with any steric strain of compressing A and B together.

9.2.3 Electrophilic Catalysis

Binding in catalytic reactions most often occurs in the form of some association of the reactant to an electrophile or nucleophile. This "association" or "binding" could include the formation of a covalent or non-covalent bond between the electrophile and the reactant. There are many examples of electrophilic catalysis, including simple electrostatics, hydrogen bonding, acid catalysis, and electrophilic metal coordination. Let's briefly look at a few examples.

Electrostatic Interactions

Proximal charges can create electric fields that catalyze reactions. These charges can be associated with full formal charges or dipoles/quadrupoles or hydrogen bond donating or accepting groups. Such electrostatic interactions are much more important in organic solvents with lower dielectric constants than in water (see Section 4.2.2). This means that the electrostatic stabilization of a charge that is developing along a reaction coordinate can be more effective when the reaction occurs in an organic solvent. It is postulated that enzymes (see Section 9.4) take advantage of this by creating catalytic pockets that are water-free after the reactant binds. Since an enzyme is essentially an organic molecule, its binding site can be very organic in character, thereby accentuating electrostatic effects.

As an example, dipoles and hydrogen bonds are often oriented in a manner that is roughly parallel, as a means to best solvate a developing charge. Consider the orientation of multiple hydrogen bond donors toward a carbonyl oxygen (Eq. 9.6). This creates a geometry that is routinely referred to as an **oxyanion hole**, because it will stabilize an oxyanion that would form upon nucleophilic attack at the carbonyl carbon. The positive ends of the hydrogen bond dipoles all converge, which is electrostatically disfavored in the catalyst, but is favorable for creation of the negatively charged oxygen. Some researchers have referred to the catalyst as being **electrostatically strained**. The nucleophilic attack is accelerated because stronger hydrogen bonds are formed to the negative charge that resides on the tetrahedral intermediate, and thus the transition state leading to it. A nice example of this kind of catalysis with a synthetic system is shown in the Connections highlight on the next page.

$$(Eq. 9.6)$$

Metal Ion Catalysis

If formal charges and dipoles can stabilize charges on transition states, charges on metals can act similarly. Metal coordination can polarize bonds, thereby enhancing their inherent reactivity. Metals are common parts of enzymes, where coordination of a substrate to a metal leads to activation toward nucleophilic attack. As an example of metal ion catalysis, the hydrolysis of glycine ethyl ester by hydroxide is increased 2×10^6-fold by coordination to (ethylenediamine)$_2$Co^{3+} (Eq. 9.7). In the next chapter, we will also see that metals are essential parts of many common organic reagents. For example, lithium aluminum hydride requires the lithium to activate a carbonyl group toward nucleophilic attack by hydride (see Section 10.8.4).

$$(Eq. 9.7)$$

Another aspect of metal catalysis is the ability to create high concentrations of hydroxide at neutral pH. Water is a common ligand to a metal, and the binding between the water and metal typically withdraws electrons from the water, making it more acidic. For example, in the complex shown in Eq. 9.8, the pK_a of the metal-bound water is 7.2, over 10^8 more acidic than water itself. Deprotonation gives a metal-bound hydroxide that can act as a base or a nucleophile. An example of using the electrophilic nature of metals coupled with the presence of a hydroxide ligand is given in the Connections highlight on page 502.

$$(Eq. 9.8)$$

Connections

Toward an Artificial Acetylcholinesterase

The use of hydrogen bonding to promote reaction rates is among the most well studied methods of catalysis. Several models have been created, taking nature as the inspiration. Fully synthetic systems, designed to catalyze a reaction in a manner analogous to an enzyme, are called **artificial enzymes**. Reviews of this field are given at the end of the chapter, and here we highlight one particular example.

Acetylcholine is a neurotransmitter released at a synapse as a means for one neuron to communicate with a neighboring neuron. The enzyme acetylcholinesterase rapidly hydrolyzes the ester to produce choline, terminating the signal.

To demonstrate that a similar reaction can be promoted by a synthetic receptor, the compound shown below was used to catalyze the methanolysis of *p*-nitrophenylcholinecarbonate (PNPCC) in 1% CH$_3$OH

in chloroform. The catalyst has a calix[6]arene subunit that binds to the quaternary ammonium end of the substrate, primarily via the cation–π effect (see Chapters 3 and 4). The binding constant for PNPCC to the catalyst is $6.0 \times 10^3 \, M^{-1}$.

Appended to the calixarene core is a bicyclic guanidinium group, which has the ability to hydrogen bond to and electrostatically stabilize the tetrahedral intermediate developed during the methanolysis of the carbonate functional group. This is shown in a schematic form below. This electrostatic/hydrogen bonding stabilization of the transition state leads to a 150-fold rate enhancement for the reaction relative to methanolysis in the absence of this artificial enzyme.

Cuevas, F., Stefano, S. D., Magrans, J. O., Prados, P., Mandolini, L., and de Mendoza, J. "Toward an Artificial Acetylcholinesterase." *Chem. Eur. J.*, **6**, 3228–3234 (2000).

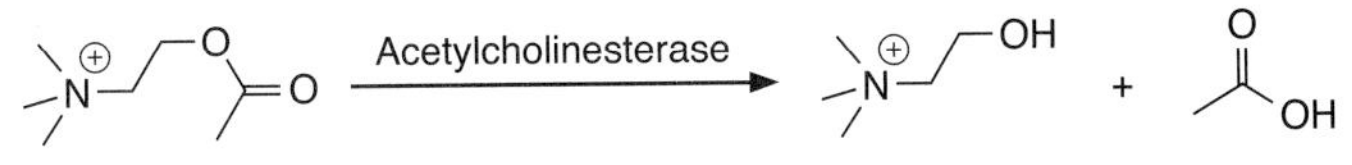

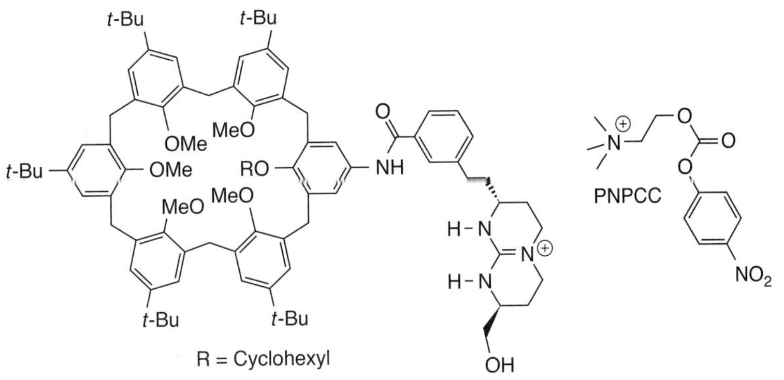

Calixarene catalyst

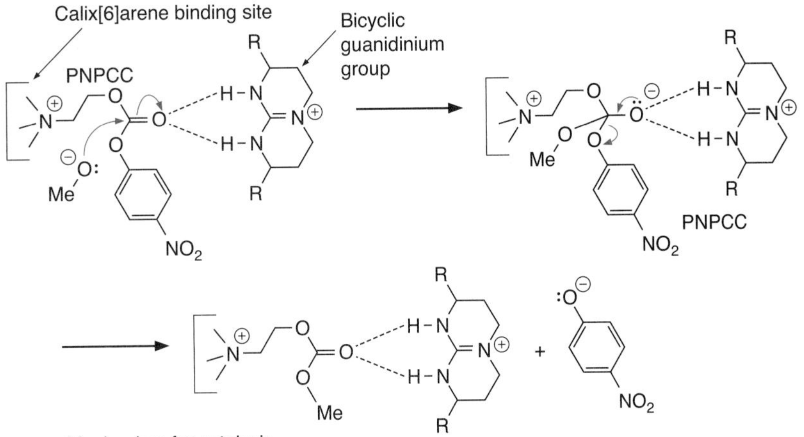

Mechanism for catalysis

Connections

Metal and Hydrogen Bonding Promoted Hydrolysis of 2′,3′-cAMP

Electrophilic activation can occur by both metal coordination and hydrogen bonding, and it is common for enzymes to combine these effects. Less common are synthetic systems that combine these, but one particularly simple example is the use of the diaminophenanthroline–Cu(II) complex shown below. This compound accelerates the hydrolysis of 2′,3′-cAMP (cAMP = cyclic adenosine monophosphate) approximately 20,000-fold compared to a Cu–phenanthroline complex with the amino groups replaced by methyls.

The amino groups are proposed to facilitate the reaction in two ways. First, they lead to a further lowering of the pK_a of the Cu-bound water due to hydrogen bonding. The bound hydroxide then acts as a nucleophile to attack the phosphodiester, leading to a Cu(II)-coordinated trigo-

nal bipyramidal phosphorane. The second effect of the amines is similar in origin, in that they should also lower the pK_a of the hydroxyl on the phosphorane intermediate. Deprotonation of this hydroxyl would further facilitate leaving group departure. Hence, in this system we find enhancement of the nucleophilic attack through metal-bound hydroxide, electrophilic activation of the phosphodiester substrate by coordination to the metal, and enhancement of the leaving group departure via facilitated deprotonation of the phosphorane intermediate. This very simple and elegant structure is one of the best artificial enzymes yet produced.

Wall, M., Linkletter, B., Williams, D., Lebuis, A-M., Hynes, R. C., and Chin, J. "Rapid Hydrolysis of 2′,3′-cAMP with a Cu(II) Complex: Effect of Intramolecular Hydrogen Bonding on the Basicity and Reactivity of a Metal-Bound Hydroxide." *J. Am. Chem. Soc.*, **121**, 4710–4711 (1999).

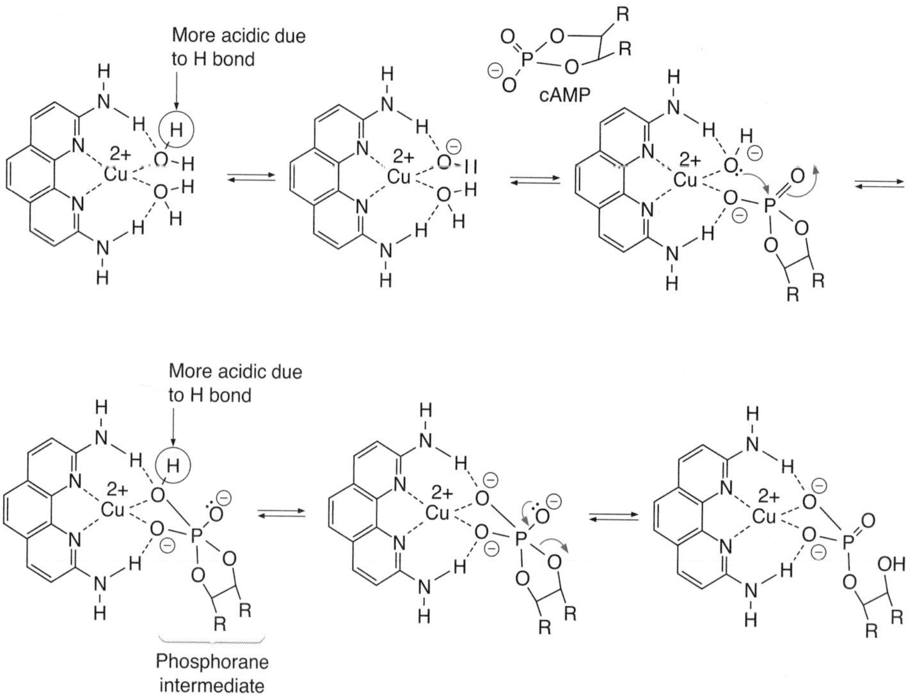

9.2.4 Acid–Base Catalysis

We will present an extensive discussion of acid–base catalysis in Section 9.3. It is mentioned here just for completeness, since it should always be listed as one of the primary interactions that can speed reaction rates. As we will see, the forms of acid–base catalysis can actually be quite complex.

9.2.5 Nucleophilic Catalysis

Nucleophilic catalysis arises when a nucleophile binds to a reactant and enhances its rate of reaction. Catalysis via nucleophilic activation is less common than electrostatic catalysis. Examples include reactions with bases (including hydroxide) and coordination of nucleophiles to electrophilic centers to improve their activity.

The most well known example is the addition of a tertiary amine or similar structure to an acid halide or anhydride (see Section 10.17.4). Here, the amine first acts as a nucleophile and adds to the carboxylic acid derivative (Eq. 9.9). The addition occurs faster than with water or an alcohol because an amine is more nucleophilic. However, the structure created is an even better electrophile and more reactive than the starting acid halide or anhydride, because a positive charge is proximal to the carbonyl carbon. Hence, the amine gives a faster rate of addition, and then creates a more reactive intermediate for further addition. The strategy is especially common in peptide synthesis, for which a variety of highly specialized nucleophilic catalysts has been developed. More recently, nucleophilic catalysis is being extended to much more complex reactions, and the next Connections highlight gives an example.

$$\text{(Eq. 9.9)}$$

Connections

Nucleophilic Catalysis of Electrophilic Reactions

Many organic reactions are catalyzed or promoted by strong electrophiles such as $AlCl_3$, $TiCl_4$, and $SiCl_4$. The electrophile often coordinates to the heteroatom of a polarized bond, activating that group toward nucleophilic attack, similar to the enhancements we showed in Section 9.2.3. Gutmann showed that some electrophiles become *more* electrophilic when Lewis bases (nucleophiles) coordinate to them. Coordination of a Lewis base to a Lewis acid sometimes induces ligand loss from the Lewis acid, leaving behind a positive charge that makes the Lewis acid an even more powerful electrophile, as shown in the reaction given below where the electrophilicity of $SiCl_4$ is enhanced.

Denmark has taken advantage of this nucleophilic enhancement of certain electrophiles to create enantioselective reactions. In one example, the coupling of allyl-

tributyltin with aldehydes using $SiCl_4$ was promoted by catalytic amounts of chiral phosphoramidates of varying structure. Coordination of the chiral phosphoramidate to $SiCl_4$ promotes chloride loss in a reversible reaction, giving a chiral Lewis acid. This Lewis acid activates aldehydes toward nucleophilic attack from tin reagents. The coupling of aldehydes and allyltributyltin catalyzed by the chiral activated Lewis acid gives enantiomeric excesses ranging from 50 to 95% depending upon the aldehyde R group (see next page). This is an excellent example of a relatively unexplored area of catalysis—nucleophilic enhancement of electrophilic reactions.

Gutmann, V. (1998). *The Donor–Acceptor Approach to Molecular Interactions*, Plenum Press, New York. Denmark, S. E., and Wynn, T. "Lewis Base Activation of Lewis Acids: Catalytic Enantioselective Allylation and Propargylation of Aldehydes." *J. Am. Chem. Soc.*, **123**, 6199–6200 (2001).

Lewis base   Lewis acid

Enhanced electrophilicity
of the Lewis acid

+ $SiCl_4$ ⇌

Chiral activated Lewis acid

9.2.6 Covalent Catalysis

Covalent catalysis is a general term applied when a catalyst forms full covalent bonds with the substrate, not just intermolecular non-covalent interactions. The nucleophilic catalysis by an amine given in Eq. 9.9 is a specific example where a nucleophile acts as a covalent catalyst.

As another example, consider the decarboxylation of acetoacetate. This compound will decarboxylate under acidic conditions with heating. However, the addition of aniline catalyzes the reaction, so it occurs at less acidic pH and ambient temperature (Eq. 9.10). Formation of the imine between aniline and the β-keto acid leads to a species that is protonated and can act as a good electron sink during the decarboxylation. Hydrolysis of the enamine product gives the ketone and regenerates the catalyst, thus leading to turnover.

(Eq. 9.10)

The examples of covalent and nucleophilic catalysis given above can be thought of as changes in the mechanism. However, this is a subtle difference from just transition state stabilization. Completely new chemical structures are created in the reactions of Eqs. 9.9 and 9.10 relative to the uncatalyzed pathways, although the general mechanistic considerations and arrow-pushing schemes are very similar in the catalyzed and uncatalyzed reactions. This is in contrast to the organometallic reactions we will discuss in Chapter 12, where the uncatalyzed reaction either does not occur at all, or the arrow-pushing and mechanistic schemes are completely and totally different for the catalyzed and uncatalyzed pathways.

Many nucleophilic catalytic schemes involve imine formation, as did the example above. Even though the amine adds as a nucleophile, it is the functional group properties of the imine that imparts the catalysis, and therefore the scenarios are considered covalent catalysis and not nucleophilic catalysis. The involvement of an imine is common in biology. Imines are used in decarboxylase and aldolase enzymes, and any enzymes using the cofactor pyridoxal phosphate. The following Connections highlight gives another example, but now from a relatively new field called organocatalysis.

Connections

Organocatalysis

In organic synthesis we usually think of catalysts as being based on transition metals, main group elements, and other elements besides C, H, N, and O. Recently, however, there have been significant advances in **organocatalysis**, the development of all organic systems that have desirable catalytic behaviors. Part of the motivation for such efforts is the air and water instability often associated with metal-based systems, the environmental benefit of avoiding toxic metals, and the ready availability of a large number of enantiomerically pure organic substances for developing chiral catalysts.

One especially successful example of organocatalysis is the chiral amine catalyst shown developed by MacMillan and co-workers. The key feature of such systems is that forming the iminium ion creates a much lower-lying LUMO relative to that found in the starting enone. This makes the structure much more susceptible to nucleo-

philic attack, ensuring that all reaction occurs via the iminium ion. The well-defined geometry of the system produces very high stereoselectivities. Turnover is achieved by hydrolysis of the product iminium ion, and often the water produced in the first step is sufficient for this purpose. While the turnover numbers generally seen with such organocatalysts are typically nowhere near those seen with organometallic catalysts, the benefits of inexpensive catalysts and the ability to run the reaction in the open air with wet solvents often more than compensates. Using this approach, a variety of reactions have succumbed to organocatalysis, including Diels–Alder reactions, Friedel–Crafts reactions, direct alkylations of heterocycles such as furan and indole, and a variety of Michael additions.

Austin, J. F., and MacMillan, D. W. C. "Enantioselective Organocatalytic Indole Alkylations. Design of a New and Highly Effective Chiral Amine for Iminium Catalysis." *J. Am. Chem. Soc.*, **124**, 1172 (2002).

Organocatalysis

9.2.7 Strain and Distortion

There is another vantage point from which one can consider catalysis. Our focus has been upon increased binding of the transition state relative to the ground state. But let's examine this notion a little deeper. When a substrate binds to a catalyst that is more complementary in structure or electronic characteristics to the transition state, the substrate may distort in order to optimize binding interactions. That is, because the catalyst is designed to optimally bind the transition state, it necessarily is not optimal for the most stable structure of the ground state. If the substrate distorts in order to compensate for this, we can speak of such distortion as a strain on the substrate. The strain can be slight or severe. Since the strain pushes the structure of the substrate toward a form closer to the transition state, we state that the substrate has been **activated**. The strain raises the energy of the substrate, and this can be thought of as an alternative way to diminish $\Delta G^{\ddagger}$. Therefore, the notions of increased binding of a transition state and activation of a substrate go "hand-in-hand".

Changes in the electronic structure of a substrate can also be considered an activation. For example, let's re-examine the oxyanion hole example given in Eq. 9.6. We first examined this example from the vantage point of increased transition state binding. As the nucleophile adds to the carbonyl carbon, the hydrogen bonds between the carbonyl oxygen and the amide hydrogen bond donors become stronger, thereby accelerating the nucleophilic attack.

However, it must also be true that binding of the carbonyl oxygen to the partially positive hydrogen bond donors increases the polarization in the carbonyl. This increased polarization is not present in the lowest energy structure of the substrate, and hence is an activation of the substrate. It is analogous to a structural strain, but instead is an electronic effect. This activation would also be introduced when coordinating a carbonyl oxygen to an electrophilic metal. The increased polarization makes the carbonyl carbon more electrophilic, thereby activating it toward nucleophilic attack. Hence, substrate activation and transition state stabilization derive from the same chemical phenomenon. The paradigmatic example of substrate strain comes from one particular enzymatic reaction, as detailed in the next Connections highlight.

Two important points are worth re-emphasizing here. In order for a strain placed in a substrate to facilitate a chemical reaction, that strain must be along the reaction coordinate. The strain must push the reactant toward the transition state on the energy surface, either structurally or electronically. Also, the strain must be partly or fully relieved upon achieving the transition state. A strain put into a reactant that remains in the transition state will not have any effect on the rate of the reaction (recall the discussion of Figure 9.2 **C**, where no catalysis is obtained).

Connections

Lysozyme

The most well known catalyst that introduces strain into its substrate is the enzyme lysozyme. This enzyme catalyzes the hydrolysis of a β-anomer linkage between saccharides. The enzyme strains the substrate by distorting a pyran ring from a chair to a half-chair. The most widely cited reason for this distortion is a stereoelectronic effect. Recall from Chapter 2 that the anomeric effect occurs for an α-anomer but not a β-anomer. Simply stated, the lone pair electrons on the endocyclic oxygen cannot be antiperiplanar with the leaving group in the β-anomer, but are instead gauche, and therefore they cannot assist the leaving group departure (also see our discussion of the stereoelectronics of additions and eliminations given in Sections 10.12.4 and 10.13.8).

When the chair is distorted into a conformation resembling a half-chair, the lone pairs on the endocyclic oxygen can become synperiplanar with the leaving group and thus facilitate leaving group departure. Lysozyme promotes this distortion toward a half-chair via a steric interaction between the CH$_2$OH group of the sugar undergoing hydrolysis and an enzyme side chain residue. This structural distortion facilitates the reaction by straining the substrate into a conformation that assists the formation of a lower energy transition state than would otherwise be obtained.

Strynadka, N. C. J., and James, M. N. G. "Lysozyme Revisited: Crystallographic Evidence for Distortion of an N-Acetylmuramic Acid Residue Bound in Site D." *J. Mol. Biol.*, **220**, 401–424 (1991). Kuroko, R., Weaver, L. H., and Matthews, B. W. "A Covalent Enzyme–Substrate Intermediate with Saccharide Distortion in a Mutant T4 Lysozyme." *Science*, **262**, 2030–2033 (1993).

Not periplanar with the leaving group

Steric clash with enzyme

Now one lone pair is synperiplanar with the leaving group

9.2.8   Phase Transfer Catalysis

Nucleophiles, bases, oxidizing/reducing agents, and other anionic species are particularly reactive when dissolved in aprotic low polarity solvents. Extraction of anions from aqueous media into these solvents can lead to large rate enhancements. **Phase transfer catalysis** involves enhancing the rates of reactions of ionic species such as nucleophiles and bases with organic molecules by the addition of a **phase transfer agent**. The experiments involve the use of a two-layer system: aqueous and organic. The organic reactant is placed in the nonpolar solvent, while the ionic species is dissolved in an aqueous phase, and stirring is used to enhance contact between the layers. The phase transfer agent is commonly a cationic surfactant or crown ether/cryptand (see Chapter 4). When using a surfactant, such as a tetralkylammonium or tetraalkylphosphonium, the cationic group ion pairs with the reactive anion and facilitates its transfer into the organic layer, where it is highly reactive. The use of crown ethers or cryptands leads to complexation of a Li^+, Na^+, or K^+ counterion of the anion, again facilitating its transfer into the organic layer. The phase transfer agents move between the two layers, extracting anions repeatedly into the organic phase.

9.3   Brønsted Acid–Base Catalysis

The most common kind of catalysis in organic chemistry is Brønsted acid–base catalysis, and therefore we are going to examine this kind of catalysis in considerable detail. In Brønsted acid–base catalysis a proton or hydroxide becomes involved in the reaction mechanism, lowers the energy of the transition state(s), accelerates the reaction, and is regenerated at the end of the reaction. Although strictly speaking a catalyst is regenerated at the end of any reaction, many reactions are said to be acid- or base-catalyzed even though they are really just acid- or base-promoted, meaning that the acid or base remains as part of the product.

To begin our understanding of this form of catalysis, we must first explore the different kinds of acid–base chemistry that can occur. As you will see, even this simple form of catalysis is really quite complex. Two classes have been identified, termed specific and general catalysis. We start with specific catalysis.

9.3.1   Specific Catalysis

The **specific acid** is defined as the protonated form of the solvent in which the reaction is being performed. For example, in water the specific acid is hydronium. In acetonitrile, the specific acid is CH_3CNH^+, and in DMSO the specific acid is $CH_3SO(H^+)CH_3$. The **specific base** is defined as the conjugate base of the solvent. As examples, in water, acetonitrile, and DMSO, the specific bases would be hydroxide, $^-CH_2CN$, and $CH_3SOCH_2^-$, respectively. These definitions lead to strict definitions for specific catalysis. **Specific-acid catalysis** refers to a process in which the reaction rate depends upon the specific acid, *not* upon other acids in the solution. **Specific-base catalysis** refers to a process in which the reaction rate depends upon the specific base, *not* upon other bases in the solution. To understand the kinds of reaction mechanisms that would depend only upon the specific acid or base, we need to examine some possible mechanisms and the associated kinetic analyses.

The Mathematics of Specific Catalysis

Let's start with an obvious example of specific-acid catalysis. We use water for demonstration purposes, but the kinetic development would be the same for other solvents. Eq. 9.11 shows a schematic mechanism in which hydronium protonates a reactant or substrate (R) prior to a reaction that is rate-determining. The slow step could be first order as shown in Eq. 9.11, or it commonly involves addition of a nucleophile that would then become part of the kinetic expression. Water is a common nucleophile. We keep the scenario simple here, because having additional reactants involved in the rate-determining step does not affect the conclusion we are leading to regarding the kinetic behavior found for the acid catalyst. After the reaction has proceeded, the proton is lost back to the solution, giving the product (P).

We start the kinetic analysis with Eq. 9.12, and substitute for RH⁺ using Eq. 9.13. Eq. 9.14 gives the kinetic expression for the mechanism of Eq. 9.11, assuming that the equilibrium between R and RH⁺ is completely established. The kinetic expression contains $[H_3O^+]$, as the definition of specific-acid catalysis implies. Hence, the reaction rate depends upon the pH. The expression also contains the acid dissociation constant (K_{aRH^+}) of RH⁺, which is an important factor that we will return to below. Note that k, $[H_3O^+]$, and K_{aRH^+} are constants during the reaction. Hence, we create a new rate constant, k_{obs}, showing that the reaction appears first order (Eq. 9.15, where $k_{obs} = k[H_3O^+]/K_{aRH^+}$).

$$R + H_3O^{\oplus} \rightleftharpoons RH^{\oplus} \xrightarrow{\text{Slow}} PH^{\oplus} \xrightarrow{\text{Fast}} P + H_3O^{\oplus} \qquad \text{(Eq. 9.11)}$$

$$\frac{d[P]}{dt} = k[RH^+] \qquad \text{(Eq. 9.12)}$$

$$[RH^+] = \frac{[H_3O^+][R]}{K_{aRH^+}} \qquad \text{(Eq. 9.13)}$$

$$\frac{d[P]}{dt} = \frac{k[R][H_3O^+]}{K_{aRH^+}} \qquad \text{(Eq. 9.14)}$$

$$\frac{d[P]}{dt} = k_{obs}[R] \qquad \text{(Eq. 9.15)}$$

Let's now examine the same reaction but under conditions for which it is not quite so obvious whether the reaction is catalyzed by the specific acid. If we add an acid such as acetic acid to water, small amounts of hydronium ion are produced, but the acid in highest concentration is acetic acid. Eq. 9.16 shows a mechanism in which the added acid protonates the reactant in an equilibrium prior to the rate-determining step. We designate the acid as HA with the implication that it could be acetic acid performing the protonation. If it were H_3O^+ performing the protonation, we would simply have the same scenario as presented in Eqs. 9.14 and 9.15.

$$R + HA \xrightleftharpoons{K_{eq}} RH^{\oplus} + A^{\ominus} \xrightarrow{\text{Slow}} PH^{\oplus} + A^{\ominus} \xrightarrow{\text{Fast}} P + HA$$
$$\text{(Eq. 9.16)}$$

Eq. 9.17 is the starting point for solving the kinetic analysis. The $[RH^+]$ can be derived from the expression for K_{eq} (Eq. 9.18) as shown in Eq. 9.19. However, Eq. 9.19 can be simplified by recognizing that $[HA]/[A^-]$ is equal to $[H_3O^+]/K_{aHA}$, leading to Eq. 9.20. Finally, K_{eq}/K_{aHA} is $1/K_{aRH^+}$ via Eq. 9.18. Hence, we end up with Eq. 9.21. This is exactly the same as Eq. 9.14, which we found for the mechanism shown in Eq. 9.11!

$$\frac{d[P]}{dt} = k[RH^+] \qquad \text{(Eq. 9.17)}$$

$$K_{eq} = \frac{[RH^+][A^-]}{[R][HA]} = \frac{K_{aHA}}{K_{aRH^+}} \qquad \text{(Eq. 9.18)}$$

$$[RH^+] = \frac{K_{eq}[HA][R]}{[A^-]} \qquad \text{(Eq. 9.19)}$$

$$\frac{d[P]}{dt} = \frac{kK_{eq}[R][H_3O^+]}{K_{aHA}} \qquad \text{(Eq. 9.20)}$$

$$\frac{d[P]}{dt} = \frac{k[R][H_3O^+]}{K_{aRH^+}} \qquad \text{(Eq. 9.21)}$$

This derivation teaches an important lesson. If the acid catalyst is involved in an equilibrium prior to the rate-determining step, and it is not involved in the rate-determining step, then the kinetics of the reaction will depend solely upon the concentration of the specific acid. This is true even if an added acid (such as acetic acid) is involved in protonating the reactant. The reason for this is that when a prior equilibrium is established, the concentration of RH$^+$ determines the rate of the reaction (Eqs. 9.12 and 9.17). The concentration of RH$^+$ depends solely upon the pH and the pK_a of RH$^+$, and does not depend upon the concentration of the acid HA that was added to solution.

A similar kinetic expression can be derived for the use of a catalytic base, B. When B is involved in an equilibrium with a reactant (RH) prior to a rate-determining step, Eqs. 9.22, 9.23, and 9.24 describe the situation (see Exercise 2). Now K_a and [H$_3$O$^+$] trade places in the numerator and denominator relative to the acid-catalyzed scenario (Eq. 9.21). The same expression will be derived if either the specific base or an added base is used in the equilibrium. Once again, because [R$^-$] controls the rate of the reaction, it is the pH of the solution and the pK_a of RH that are important, not the amount of B present in the solution. Finally, recognizing that $kK_{aRH}/[H_3O^+]$ is a constant during the reaction gives a kinetic expression that is first order in [RH] only (Eq. 9.25, where $k_{obs} = kK_{aRH}/[H_3O^+]$).

$$RH + B \rightleftharpoons R^{\ominus} + BH^{\oplus} \xrightarrow{\text{Slow}} P^{\ominus} + BH^{\oplus} \xrightarrow{\text{Fast}} PH + B \qquad \text{(Eq. 9.22)}$$

$$\frac{d[PH]}{dt} = k[R^-] \qquad \text{(Eq. 9.23)}$$

$$\frac{d[PH]}{dt} = \frac{kK_{aRH}[RH]}{[H_3O^+]} \qquad \text{(Eq. 9.24)}$$

$$\frac{d[PH]}{dt} = k_{obs}[RH] \qquad \text{(Eq. 9.25)}$$

The analysis given above did not show any particular reaction, because we wanted to develop the mathematics in a general fashion. To get a better feeling for specific catalysis, we show possible mechanisms for the hydration of a carbonyl using specific-acid and specific-base catalysis in Figure 9.6. Note in each case the equilibrium involving the acid or base is prior to the rate-determining step. It is important to note that we have designated the second step in Figure 9.6 **B** as rate-determining solely for discussion purposes, so that the mechanism corresponds to the definition of specific-base catalysis. In reality, nucleophilic attack by hydroxide would be rate-determining, because equilibria involving solely proton transfers are often established at diffusion controlled rates (see further discussion of this figure on page 512).

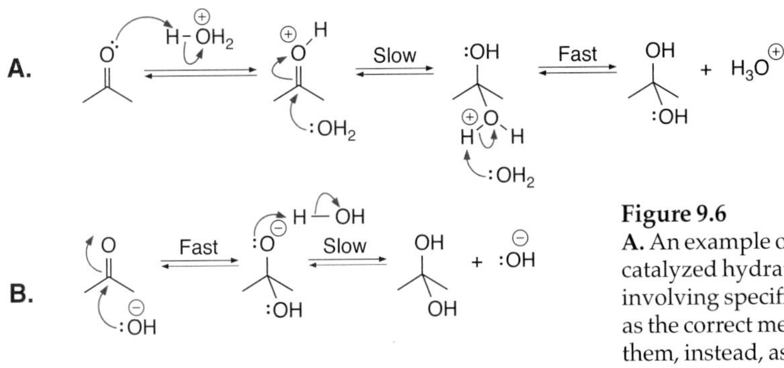

Figure 9.6
A. An example of a possible mechanism involving specific-acid-catalyzed hydration of acetone. **B.** An example of a possible mechanism involving specific-base-catalyzed hydration. Do not take these scenarios as the correct mechanisms for these addition reactions, but consider them, instead, as simply possibilities highlighting our discussion. Certainly, the order of relative rates in part **B** would be reversed.

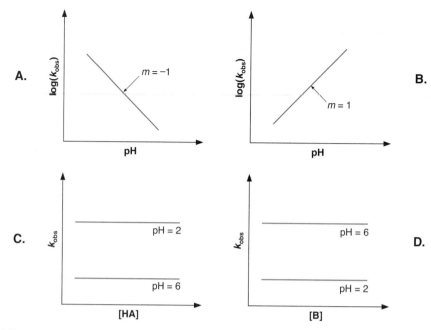

Figure 9.7
The distinctive kinetic plots for specific-acid and specific-base catalysis. **A.** The pH dependence of $\log(k_{obs})$ for a specific-acid-catalyzed reaction. **B.** The pH dependence of $\log(k_{obs})$ for a specific-base-catalyzed reaction. **C.** The dependence of k_{obs} for a specific-acid-catalyzed reaction on the concentration of an added acid HA at constant pH. **D.** The dependence of k_{obs} for a specific-base-catalyzed reaction on the concentration of an added base B at constant pH. The pH values 2 and 6 are just chosen as examples and are not indicative of any particular scenario.

Kinetic Plots

The hallmark of specific-acid or specific-base catalysis is that the rate depends upon the pH and not upon the concentration of various acids or bases. This always means that an equilibrium involving the acid or base occurs prior to the rate-determining step, and the acid or base is not involved in the rate-determining step itself. Experimentally, such reactions produce very distinctive kinetic plots. A plot of pH vs. the $\log(k_{obs})$ values produces a straight line whose slope is –1 for acid catalysis or 1 for base catalysis (Figures 9.7 **A** and **B**). The slopes of ±1 result because each one-unit change in pH changes the concentration of $[H_3O^+]$ or $[HO^-]$ by a factor of 10, and $[H_3O^+]$ or $[HO^-]$ has been incorporated into k_{obs}. Furthermore, if we keep the pH constant, and change the concentration of an added acid HA or base B, there is no change in k_{obs} (Figures 9.7 **C** and **D**). Recall that one can change the concentration of an acid HA or base B and keep the pH constant by keeping the ratios of $[HA]/[A^-]$ or $[B]/[BH^+]$ constant, respectively (see the Henderson–Hasselbach equation—Eq. 5.12).

9.3.2   General Catalysis

Now let's examine a scenario where the proton transfer *is* involved in the rate-determining step, not in a prior equilibrium. This leads to very different experimental observations, and a phenomenon called **general catalysis**. When an acid is involved in the rate-determining step, we have **general-acid catalysis**, and when a base is involved in the rate-determining step, we have **general-base catalysis**. The terms "general" and "specific" often get confused. Simply try to remember the following concepts. The term "general" refers to the fact that any acid or base we add to the solution will affect the rate of the reaction, and hence the catalysis is quite general. The term "specific" refers to the fact that just one acid or base, that from the solvent, affects the rate. The catalysis is therefore very specific.

The Mathematics of General Catalysis

Eq. 9.26 shows a scenario in which the first step involves an acid that facilitates a reaction of water with the substrate. This is the rate-determining step. An intermediate is formed that results from the addition of water to the reactant $[HR(H_2O)^+]$; the structural details are unimportant here. The intermediate then loses a proton in a second fast step to regenerate the acid HA and give the product. General catalysis does *not* require that water be one of the reactants. This just simplifies our current analysis because we assume that its concentration is constant and is incorporated into the rate constant k. The kinetic expression for this reaction is given in Eq. 9.27, which reflects only the first step of Eq. 9.26 because it is rate-determining. Alternatively, one can substitute the relationship $[HA] = [H_3O^+][A^-]/K_a$ to achieve Eq. 9.28. Since the acid or base is always regenerated after the reaction, its concentration never changes over the course of the reaction. Hence, the reaction is pseudo-first order, as in Eq. 9.29, where $k_{obs} = k[HA]$ or $k[H_3O^+][A^-]/K_a$.

$$\text{HA} + \text{R} + \text{H}_2\text{O} \xrightarrow{\text{Slow}} \text{HR(H}_2\text{O)}^{\oplus} + \text{A}^{\ominus} \xrightarrow{\text{Fast}} \text{P} + \text{HA} \qquad \text{(Eq. 9.26)}$$

$$\frac{d[\text{P}]}{dt} = k[\text{HA}][\text{R}] \qquad \text{(Eq. 9.27)}$$

$$\frac{d[\text{P}]}{dt} = \frac{k[\text{H}_3\text{O}^+][\text{A}^-][\text{R}]}{K_a} \qquad \text{(Eq. 9.28)}$$

$$\frac{d[\text{P}]}{dt} = k_{obs}[\text{R}] \qquad \text{(Eq. 9.29)}$$

The concentration of either HA or A^- is now in the rate expression, in contrast to that found for specific catalysis. This means that the concentration of the acid (or its conjugate base) that results from addition of an acid to the reaction vessel *does* affect the rate of the reaction. We should expect this because the acid is actually in the rate-determining step. Any acid in the solution will act as a catalyst.

When a base is involved in the rate-determining step to make an intermediate that then regenerates the base and gives the product, the scenario shown in Eq. 9.30 and the kinetic expression Eq. 9.31 are used. Once again, water could be involved in the rate-determining step, and this would not change the kinetic analysis. Here we show the base deprotonating the reactant or substrate, but it can also be used to deprotonate water while the water adds to the reactant. The concentration of base can be related to the concentration of hydroxide via $[B] = [OH^-][HB^+]/K_b$. One cannot eliminate both the concentration of the base and its conjugate acid from the rate expression. The reaction rate therefore depends upon the amount of base added to the solution, and any base in the solution will be active. Since the base is regenerated after the reaction, pseudo-first order kinetics holds, and Eq. 9.33 can be written, where $k_{obs} = k[B]$ or $k[HO^-][HB^+]/K_b$.

$$\text{R} + \text{B} \xrightarrow{\text{Slow}} \text{R}^{\ominus} + \text{HB}^{\oplus} \xrightarrow{\text{Fast}} \text{P} + \text{B} \qquad \text{(Eq. 9.30)}$$

$$\frac{d[\text{P}]}{dt} = k[\text{B}][\text{R}] \qquad \text{(Eq. 9.31)}$$

$$\frac{d[\text{P}]}{dt} = \frac{k[\text{HO}^-][\text{HB}^+][\text{R}]}{K_b} \qquad \text{(Eq. 9.32)}$$

$$\frac{d[\text{P}]}{dt} = k_{obs}[\text{R}] \qquad \text{(Eq. 9.33)}$$

Once again, as a means to better picture the kinds of mechanisms being discussed, we present the hydration of acetone as an example. Figure 9.8 shows possible general-acid- and general-base-catalyzed mechanisms. Note in each example the acid or base is involved in the slow step, which is the nucleophilic addition of water.

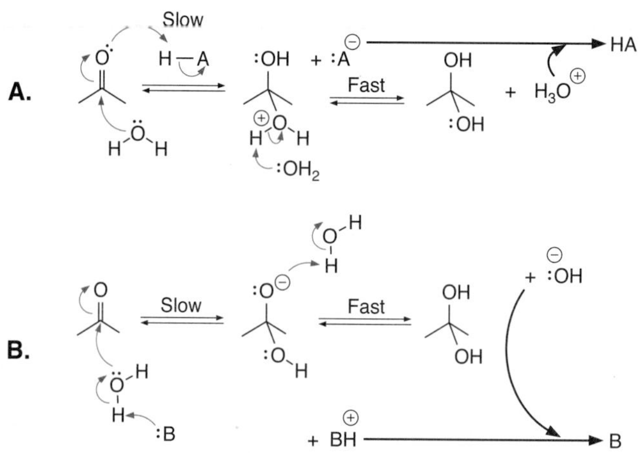

Figure 9.8
A. An example of a possible mechanism involving the general-acid-catalyzed hydration of acetone. **B.** An example of a possible mechanism involving a general-base-catalyzed hydration. Do not take these scenarios as the correct mechanisms for these addition reactions, but consider them, instead, as simply possibilities highlighting our discussion.

Before proceeding, it is important to make a note about the use of the term "specific". As stated, it is used to designate the protonated or deprotonated form of the solvent (hydronium or hydroxide for water, respectively), but it is also used to designate a mechanism involving an acid or base in an equilibrium prior to a rate-determining step. However, sometimes hydronium or hydroxide can be involved in the rate-determining step of a mechanism. When this occurs the specific acid and base are participating in general-catalysis. We noted above that any acid or base present in solution can function in general-catalysis, and so it should not be surprising that the specific acid and base can act in general-acid or base catalysis. To see this, let's reanalyze Figure 9.6 **B**. We discussed above that the mechanism is written to conform to the definition of specific-catalysis, but in reality the first step would be rate-determining. Therefore, in practice, hydroxide is acting as a general-base catalyst in this mechanism. If we alternatively write the first step of Figure 9.6 **B** as the slow step, the mechanism becomes the analog of Figure 9.8 **B**, except with hydroxide undergoing nucleophilic addition without assistance from an added base.

Kinetic Plots

Since the rate for general-acid or general-base catalysis always depends upon the concentration of acid or base added to the solution, and is not solely dictated by the pH, the experimental observations are quite different from specific-acid–specific-base catalysis. Figures 9.9 **A** and **B** show the typical plots observed for general catalysis as a function of added acid or base. Here we are plotting k_{obs} as the y axis, using Eqs. 9.29 and 9.33 to describe the reactions. Since the concentration of the acid or base has been incorporated into k_{obs}, this value is linearly related to the acid or base concentrations. The pH dependence, on the other hand, is more difficult to understand.

Figures 9.9 **C** and **D** show the pH dependence of $\log(k_{obs})$ for general-acid- and general-base-catalyzed reactions, respectively. Why is there curvature in these plots? Let's examine

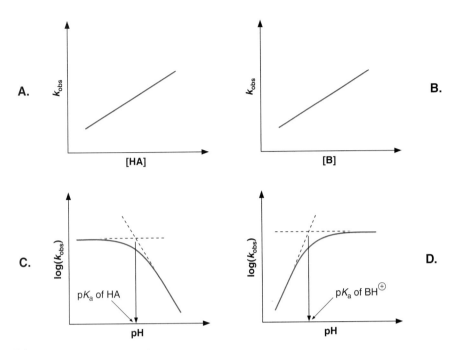

Figure 9.9
The distinctive kinetic plots for general-acid and general-base catalysis. **A.** The dependence of k_{obs} for a general-acid-catalyzed reaction on the concentration of an added acid HA at constant pH. **B.** The dependence of k_{obs} for a general-base-catalyzed reaction on the concentration of an added base B at constant pH. The pH values 4 and 7 are chosen simply for example purposes. **C.** The pH dependence of $\log(k_{obs})$ for a general-acid-catalyzed reaction. **D.** The pH dependence of $\log(k_{obs})$ for a general-base-catalyzed reaction.

general-acid catalysis first in order to answer this question. We have to go back to the Henderson–Hasselbach equation (Eq. 5.12). That equation tells us that the ratio $[A^-]/[HA]$ changes as a function of pH, and that when the pH is near the pK_a of the acid, this ratio is changing rapidly (review Sections 5.2.1, 5.2.2, and 5.2.3, if necessary). When the pH is one to two units below the pK_a of the acid, the acid is primarily protonated and further lowering of the pH does not create significantly more HA. Since $\log(k_{obs})$ for general-acid catalysis has [HA] incorporated into k_{obs}, Figure 9.9 **C** has a plateau for pH values below the pK_a. At pH values near the pK_a, HA is partially converted to A^-, a species that is not involved in the reaction. As the pH is increased the solution is being depleted in HA and therefore $\log(k_{obs})$ decreases. Finally, at pH values above the pK_a, the plot becomes linear with a slope of -1, because for each pH unit increase there is an order of magnitude decrease in the amount of HA relative to A^-. We can understand this statement by examining Eq. 9.28, which shows that the rate depends upon $[A^-]$. At pH values above the pK_a of the acid, this concentration is not changing significantly, whereas $[H_3O^+]$ is dropping by one order of magnitude for each pH unit increase.

The pH dependence of $\log(k_{obs})$ for a general-base-catalyzed reaction is the mirror image of the general-acid-catalyzed dependence. Now at pH values above the pK_a of the conjugate acid of the general base, the plot shows a plateau. At these pH values the catalyst is essentially completely converted to the base required in the reaction, and raising the pH does not give significantly more base. At pH values near the pK_a, there is a mixture of the base catalyst and the conjugate acid of the base that is not involved in the reaction. At pH values lower than the pK_a of the conjugate acid, the base is converted to the conjugate acid, and the pH dependence of $\log(k_{obs})$ is linear with a positive slope of 1. The positive slope can be understood by examining Eq. 9.32, which shows a dependence upon $[HO^-]$, which increases linearly while $[HB^+]$ is remaining relatively constant.

Comparing the kinetic plots for specific- vs. general-acid-catalyzed reactions, it is a simple matter to determine which mechanism is active (compare Figures 9.7 and 9.9). When examining the pH dependence, we find a constant linear dependence for the specific catalysis, whereas general catalysis changes slope near the pK_a of the acid or the conjugate acid of the base. The kinetic dependence upon acid or base is also very different. There is zero-order dependence for specific catalysis, and a first-order dependence for general catalysis.

Going Deeper

A Model for General-Acid–General-Base Catalysis

General catalysis can be conveniently thought of as a logical extension of the effects that stabilize high energy intermediates. Consider a molecule X–Y that can undergo a heterolysis reaction (reaction A). When the reaction occurs in the absence of solvent, full point charges are created. With the addition of water, these charges become dispersed as the anion and cation are solvated, where solvation of the anion is analogous to general-acid catalysis (reaction B). Addition of more water further disperses

the charges, and a reaction occurs where water acts as a general-base catalyst (reaction C). If the waters in reactions B and C are replaced by functional groups, most of which are better acids and bases than water, we arrive at the scenario given in reaction D. This reaction significantly stabilizes the enthalpy of heterolysis, but there are too many simultaneous collisions, such that the entropy cost for this reaction is too severe. However, we discuss at the end of this chapter that an enzyme places all the catalytic groups (acids and bases) in the right place when the substrate is bound, allowing such a scenario to occur.

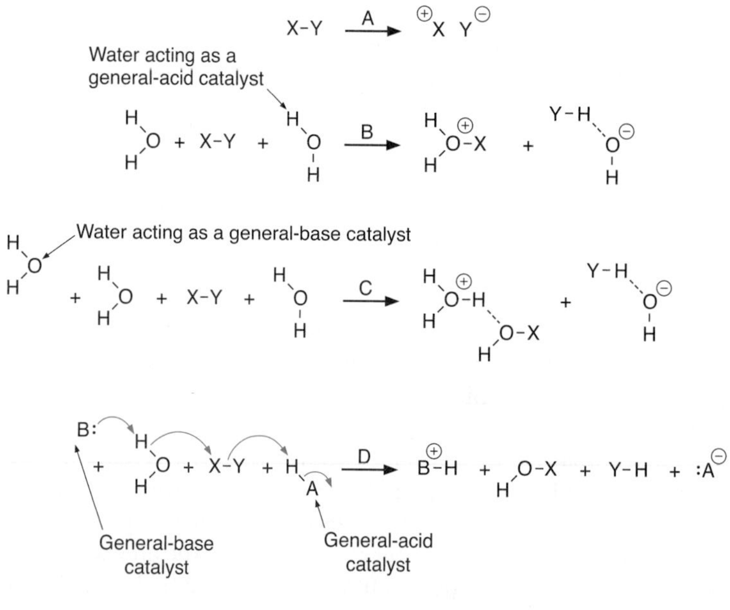

9.3.3 A Kinetic Equivalency

There is an interesting equivalency that arises from the analysis given above. To reiterate, the reason why the general-acid ([HA]) or general-base ([B]) catalyzed reactions had a negative or positive slope in Figure 9.9, above and below the pK_as, respectively, can be understood from an analysis of Eqs. 9.28 and 9.32, respectively. The [HA] is controlled by the ratio $[H_3O^+][A^-]/K_a$, and [B] is controlled by the ratio $[HO^-][HB^+]/K_b$. Therefore, one really doesn't know whether a reaction involves HA or a combination of H_3O^+ and A^-, because Eqs. 9.27 and 9.28 are equivalent ways to express the kinetic equation for general-acid catalysis. Similarly, for a base-catalyzed reaction, Eqs. 9.31 and 9.32 are equivalent. Therefore, a reaction that displays all the hallmarks of general-base catalysis could involve either B or

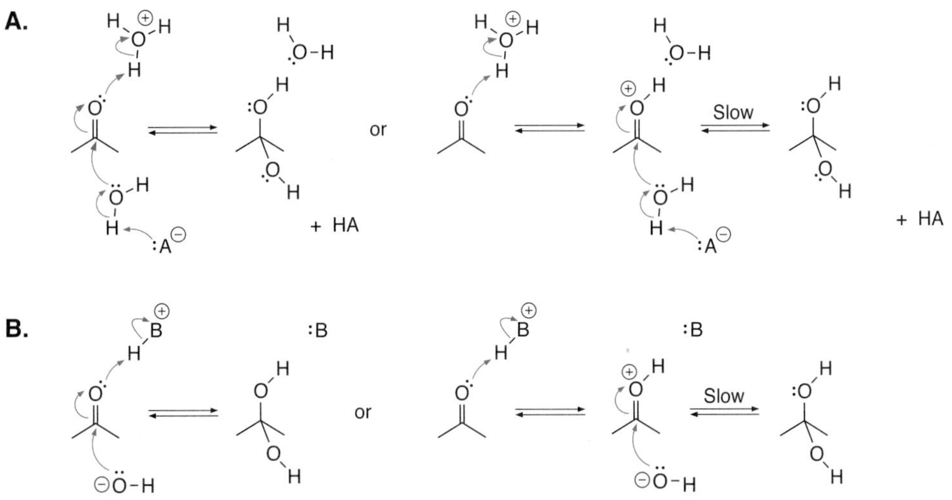

A.

B.

Figure 9.10
A. Examples of possible mechanisms involving the specific-acid–general-base-catalyzed hydration of acetone (type-n mechanisms). These are kinetically equivalent to the mechanism given in Figure 9.8 **A** (type-e mechanism). **B.** Examples of possible mechanisms involving the specific-base–general-acid-catalyzed hydration of acetone. These are kinetically equivalent to the mechanism given in Figure 9.8 **B**. Do not take these scenarios as the correct mechanisms for these hydration reactions, but consider them, instead, as simply possibilities highlighting our discussion. Equilibria prior to slow steps are assumed to be fast.

a combination of HO⁻ and HB⁺. The two possible mechanisms cannot be distinguished by a kinetic analysis. Examples of mechanisms that cannot be distinguished by pH and concentration experiments are shown in Figure 9.10. With acid catalysis, the standard general catalyzed path has been termed the **type-e** mechanism, while the kinetically equivalent combination of specific and general catalysis is called the **type-n** mechanism. In summary:

- A general-acid-catalyzed reaction is kinetically equivalent to a reaction that uses a specific acid and a general base in steps prior to or at the rate-determining step.

- A general-base-catalyzed reaction is kinetically equivalent to a reaction that uses a specific base and a general acid in steps prior to or at the rate-determining step.

9.3.4 Concerted or Sequential General-Acid–General-Base Catalysis

Sometimes both a general-acid and a general-base catalyst are required for a reaction. This is often the case with enzymes. In the following reaction scenario (Eq. 9.34) an acid HA and a base B are involved in the mechanism to make an intermediate that then forms the product and regenerates the two catalysts. Now both [HA] and [B] are present in the kinetic expression (Eq. 9.35). The k_{obs} rate constant would equal k[HA][B]. This means that the rate dependence upon pH is a combination of that observed for general-acid and general-base catalysis. Imagine combining the plots shown in Figures 9.9 **C** and **D**; a bell-shaped plot would result. The largest k_{obs} is found at a pH where the product of the concentrations of HA and B is at a maximum (Figure 9.11).

$$R + HA + B \xrightarrow{\text{Slow}} I \xrightarrow{\text{Fast}} P + HA + B \qquad (Eq.\ 9.34)$$

$$\frac{d[P]}{dt} = k[R][HA][B] \qquad (Eq.\ 9.35)$$

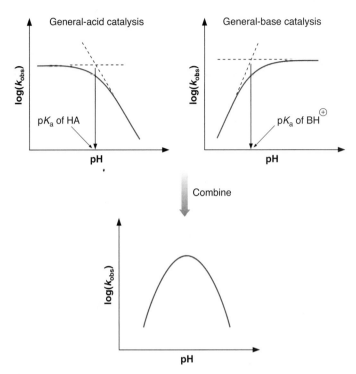

Figure 9.11
A bell-shaped pH vs. rate constant profile is found when general-acid-
and general-base-catalyzed processes are both involved in a mechanism
prior to or at the rate-determining step.

9.3.5 The Brønsted Catalysis Law and Its Ramifications

Since general-acid or general-base catalysis arises when the acid or base reacts in the rate-determining step, the rates of these reactions should depend upon the intrinsic reactivity of the acid and base being used as the catalyst. In other words, the k_{obs} values used in Eqs. 9.29 and 9.33 not only depend upon the concentrations of the catalysts, but also upon their structure. For example, it makes good sense that a general-acid-catalyzed reaction should have a larger k_{obs} when a stronger acid is used, because stronger acids would more easily donate their protons in the rate-determining step. Similarly, stronger bases should give larger k_{obs} values in general-base-catalyzed reactions. This analysis of rate constants is generally true, but when dealing with rates we must also take into account pH effects (see Exercise 3 at the end of the chapter).

A Linear Free Energy Relationship

The reasoning above tells us that there should be a relationship between the structure of the general-acid or general-base catalyst and the rate constant for the reaction. This means that a structure–activity relationship should exist, and a linear free energy relationship (LFER) may be operating (see Sections 8.3–8.6 for analyses of many other LFERs). Indeed, an LFER is often found for general catalysis, and the LFER is given by Eq. 9.36 and Eq. 9.37, both referred to as the **Brønsted catalysis law**. We briefly introduced these LFERs in Section 8.5.3. Eq. 9.36 tells us that the $\log(k)$ for a general-acid-catalyzed reaction is linearly dependent upon the pK_a of the general acid with a proportionality constant α. The negative sign in Eq. 9.36 reflects the fact that as the pK_a of the acid gets larger, the rate becomes smaller because the acid is weaker. The value of α gives the sensitivity of the reaction to the strength of the acid (see the reasoning below). Similarly, Eq. 9.37 tells us that the $\log(k)$ for a general-base-catalyzed reaction is linearly dependent upon the pK_a of the conjugate acid of the general base with a sensitivity constant β. As the pK_a of the conjugate acid becomes larger, the base becomes stronger, and the rate constant increases. The C and C' values are simply intercepts of the plots, and have no physical significance. These equations relate pK_a to k, not to the k_{obs} that includes the concentration of the catalyst.

$$\log(k) = -\alpha\, pK_a + C$$
(general-acid catalysis)

(Eq. 9.36)

$$\log(k) = \beta\, pK_a + C'$$
(general-base catalysis)

(Eq. 9.37)

Using the terminology defined in Chapter 8 to discuss linear free energy relationships, the pK_a values are the substituent constants and the α and β values are the sensitivity or reaction constants. These parameters are analogous to the σ and ρ values, respectively, for Hammett plots. All substituent constants are defined by some reference reaction. For example, the σ values of Hammett plots are defined by the ionization of various benzoic acids relative to benzoic acid itself. In the Brønsted analysis, the substituent constants are defined by the ability of each acid to protonate water (Eq. 5.5).

Let's set the foundation for the two Brønsted LFERs by establishing their relationship to free energies. Only the acid relationship is analyzed here, but the base dependence has an identical foundation. The free energy associated with the acidity of any acid HA is set by the free-energy difference between the conjugate base plus hydronium ion and the protonated acid—that is, $\Delta G° = G_{A^-}° + G_{H_3O^+}° - G_{HA}°$. It is reasonable to assume that the structural and electronic factors that lead to free-energy differences among acids for donating a proton to water will lead to a proportional difference in the free energies of activation for proton transfer from each acid to any reactant other than water. If this assumption is valid, the activation energy for proton transfer from any acid catalyst n, $\Delta G_n^{\ddagger}$, will be proportional to $\Delta G_n°$. Thus, Eq. 9.38 may be used if two acids (1 and n) are compared. This is the general form for a linear free energy relationship as given in Section 8.5.

$$\Delta G_1^{\ddagger} - \Delta G_n^{\ddagger} = \alpha\,(\Delta G_1° - \Delta G_n°)$$

(Eq. 9.38)

Using the relationship, $-\Delta G° = 2.303 RT \log K$, and the Eyring equation, $-\Delta G^{\ddagger} = 2.303\,(RT \log k - RT \log k_B T/h)$, gives Eq. 9.39.

$$\log k_n - \log k_1 = \alpha(\log K_n - \log K_1)$$

(Eq. 9.39)

Because HA_1 is the reference acid to which all evaluated general-acid catalysts are compared, $\log k_1$ and $\log K_1$ act as constants, leading to the Brønsted law (Eq. 9.36).

The Meaning of α and β

The magnitude of α or β gives mechanistic insight into general-acid- and general-base-catalyzed reactions, respectively. In particular, these reaction-specific constants reflect the extent of proton transfer in the transition state of the rate-determining step. Let's examine why this is so by analyzing the α values for a hypothetical general-acid-catalyzed reaction. If $\alpha = 1$, then the full difference in pK_a values between the various acids used in the reaction is reflected in the rate. In other words, for a one-unit pK_a change in acid strength between two acids, there is a 10-fold difference in the rate constants for catalysis by these two acids. This can most logically occur when the acid has completely performed the proton transfer to the reactant at the transition state, because the pK_a values themselves are defined by a complete proton transfer by the acid to water. In the other extreme, if $\alpha = 0$, there is no sensitivity to the strength of the acid. In this case, the acid must not be donating its proton at all in the rate-determining step. Hence, α values between 0 and 1 reveal intermediate extents of proton transfer at the transition state, from no proton transfer to full proton transfer. Similarly, a β value of 1 means that the base has completely deprotonated the reactant at the transition state, and a β value of 0 means there is no deprotonation of the reactant in the rate-determining step, with intermediate values of β reflecting intermediate extents of deprotonation.

We can reach the same conclusion by analyzing Eq. 9.38. Here, α tells how much of the thermodynamic difference between the strengths of the acids is reflected in the difference in the stabilities of the transition states derived from the two acids. The difference in the transition state structures must be the extent to which the proton has been transferred, and hence the Brønsted coefficient may be viewed as the extent of proton transfer in the activated complex.

Negative Brønsted coefficients should not be possible, because this would imply that increasing acid strength results in a decrease in catalytic activity, which violates the idea of acid catalysis as a proton transfer phenomenon. Likewise, coefficients greater than unity are rare, because this necessitates that $\log k$ increases faster than $\log K_a$ within a reaction series. When such behavior is seen, it simply implies that the reaction under study is actually more sensitive to the proton donor ability of the acid than is the protonation of water. The Going Deeper highlight on the next page gives a few examples of anomalous Brönsted values.

There is an informative extension of the Brønsted catalysis law that allows one to further analyze the structural differences between two transition states. Since the proton transfer at the transition state is intermediate between the amount of proton transfer in the reactants and products, a structural difference between two acids will lead to a difference in transition state free energies ($\Delta\Delta G^{\ddagger}$) as given in Eq. 9.40. Here, $\Delta\Delta G_b^{\circ}$ and $\Delta\Delta G_a^{\circ}$ are the differences in the standard free energy of the conjugate bases and acids, respectively, that result from structural changes within the acids. Eq. 9.40 is known as the **Leffler equation**. It tells us that the difference between the free energies of activation for two acid–base reactions is proportional to the difference in stability of the two conjugate bases minus the difference in stabilities of their acids, modulated by the sensitivity parameter α in the manner shown.

$$\Delta\Delta G^{\ddagger} = \alpha\,\Delta\Delta G_b^{\circ} - (1 - \alpha)\Delta\Delta G_a^{\circ} \qquad \text{(Eq. 9.40)}$$

$\alpha + \beta = 1$

When Brønsted relationships are applied to the forward and reverse steps of an equilibrium, a further restriction may be given to the coefficients. For a simple equilibrium, such as that given in Eq. 9.41, microscopic reversibility requires that the Brønsted relationship applies to both the forward (let's say acid-catalyzed) and reverse (therefore, base-catalyzed) steps. Rearranging Eqs. 9.36 and 9.37 to the form common to all linear free energy relationships and realizing that $K_b = 1/K_a$, gives Eqs. 9.42 and 9.43 (where the subscripted quantities refer to a reference acid A and it conjugate base B). Combining these equations results in Eq. 9.44. Since any equilibrium constant is the ratio of the forward and reverse rate constants, the only way for Eq. 9.44 to be satisfied is if $\alpha + \beta = 1$. Therefore, this equation allows one to measure an α value for the forward reaction and solve for the β-value of the reverse reaction.

$$A \underset{k_B\,\text{base}}{\overset{k_A\,\text{acid}}{\rightleftharpoons}} B \qquad \text{(Eq. 9.41)}$$

$$\log\left(\frac{k_A}{k_A^{\circ}}\right) = \alpha \log\left(\frac{K_A}{K_A^{\circ}}\right) \qquad \text{(Eq. 9.42)}$$

$$\log\left(\frac{k_B}{k_B^{\circ}}\right) = \beta \log\left[\frac{(1/K_A)}{(1/K_A^{\circ})}\right] \qquad \text{(Eq. 9.43)}$$

Going Deeper

Anomalous Brønsted Values

Deprotonation of acidic carbon atoms adjacent to groups capable of stabilizing the resulting carbanion through resonance stabilization often leads to odd Brønsted sensitivity parameters. In particular, when reactions of these so-called **pseudo-acids** are analyzed via a Brønsted plot, α values greater than one or less than zero can result. This is due to what is called an **imbalanced transition state**. In such a system, charge delocalization into the stabilizing group lags behind the deprotonation, causing the negative charge in the transition state to accumulate on the carbon rather than being delocalized into the π acceptor. This makes the rate constant for deprotonation more sensitive to structural changes than the thermodynamics of full deprotonation, because the developing anion is not "feeling" the stabilization that it achieves after full deprotonation. This can happen if the carbon acid is still primarily sp^3 hybridized at the transition state for deprotonation, and the rehybridization to sp^2 (allowing delocalization) primarily occurs after the transition state. This results in a profound disparity in the charge distribution of the transition state and the resulting resonance stabilized, delocalized carbanion.

A lot of negative charge build-up on the carbon in the transition state for deprotonation

Very little negative charge build-up on the oxygen in the transition state for deprotonation

Nitroalkane substrates, for example, are quite susceptible to this phenomenon, and odd α values for such substrates are termed **nitroalkane anomalies**. A study by Kresge on the deprotonation of simple nitroalkanes revealed a Brønsted α value of –0.5. The acidity of the nitroalkanes follows the order 2-nitropropane > nitroethane > nitromethane, while the rate constants for deprotonation follow the reverse order. The trend in the thermodynamic acidities is explained by hyperconjugative stabilization of the carbanion resonance structure with a C=N double bond. More substitution stabilizes double bonds in the anion, and hence makes the nitroalkane more acidic. However, the rate constants do not follow this trend because the lag in charge delocalization into the nitro group leads to little C=N character at the transition state. In fact, the inductive electron donating effect of the alkyl group(s) on the localized negative charge exacerbates the destabilization of the activated complex. These processes of delayed resonance stabilization have been described and quantified by Bernasconi, and called the **principle of non-perfect synchronization**.

R-group substitution destabilizes the creation of negative charge on the carbon in the transition state, yet stabilizes the final double bond

Kresge, A. J. "The Nitro Alkane Anomaly." *Can. J. Chem.,* **52**, 1897 (1974). Bernasconi, C. F. "The Principle of Non-Perfect Synchronization." *Adv. Phys. Org. Chem.,* **27**, 110–238 (1992).

$$\frac{\left(\dfrac{k_A}{k_B}\right)}{\left(\dfrac{k_A{}^\circ}{k_B{}^\circ}\right)} = \frac{(K_A)^{\alpha+\beta}}{(K_A{}^\circ)^{\alpha+\beta}}$$

(Eq. 9.44)

$$\text{i.e., } \alpha + \beta = 1$$

Deviations from Linearity

The rate of an acid-catalyzed reaction cannot continue to increase according to Eq. 9.36 indefinitely. As the acid increases in strength, the rate will increase only until the proton transfer process begins to approach the diffusion controlled limit. At that point, each acid catalyst encounter with reactant leads to a productive proton transfer, either coupled with another reaction or not. The proton transfer rate can no longer increase proportionally with the introduction of stronger acid catalysts, and α will therefore approach zero (Figure 9.12). Brønsted plots constructed using such acids will show curvature as the acid catalyst strength increases, then plateau at $\alpha = 0$. This reasoning implies that the Brønsted coefficient must actually change regularly between the values of zero and one, and cannot really provide a lin-

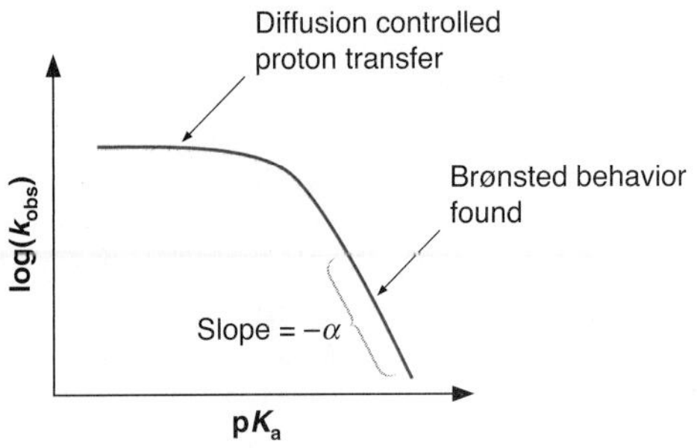

Figure 9.12
The rate of a proton transfer ultimately reaches a diffusion controlled plateau.

ear free energy relationship. Eqs. 9.36 and 9.37 are thus valid only over a limited range of catalyst pK_a values.

Diffusion controlled rates are expected for most proton transfers between certain acids and bases in water. As long as the pK_a of the conjugate acid of the proton acceptor B^- is greater than that of the donor HA by two or more units, the reaction is normally found to be diffusion controlled, and the rate becomes independent of the donor strength. When the pK_a of the conjugate acid of the acceptor drops below that of the donor, the forward reaction is endothermic, and hence the reverse reaction becomes diffusion controlled (proton transfer from BH^+ to A^-). We examine further the dynamics of proton transfers below.

9.3.6   Predicting General-Acid or General-Base Catalysis

The Libido Rule

When is general-acid or general-base catalysis feasible? As with any reaction, the driving force is the free-energy change, and for an acid–base reaction it is the free energy of the proton transfer reaction. The free energy of the transfer is determined by the relative pK_a values of the acid and the conjugate acid of the reactant/substrate that is accepting the proton. When the pK_a of the conjugate acid of the reactant/substrate is above that of the acid catalyst, the free energy for proton transfer will be negative and the reaction will spontaneously occur. In analyzing reaction rates we must consider activation energies, and this leads to what is known as the **libido rule**:

> *General-acid or general-base catalysis will occur when a thermodynamically unfavorable proton transfer in the ground state is converted to a thermodynamically favorable transfer in the transition state.*

Hence, if the substrate becomes more basic along the reaction coordinate of the rate-determining step and ultimately leads to a thermodynamically favorable proton transfer from a coordinated acid, general catalysis will occur. The proton transfer is thermodynamically viable only during the rate-determining step. This means that the pK_a of a general-acid catalyst must be between the pK_a values of a fully protonated reactant and the protonated product, with the assumption that the pK_a of the transition state and product are similar (Eq. 9.45). A similar statement can be made for a general-base-catalyzed reaction.

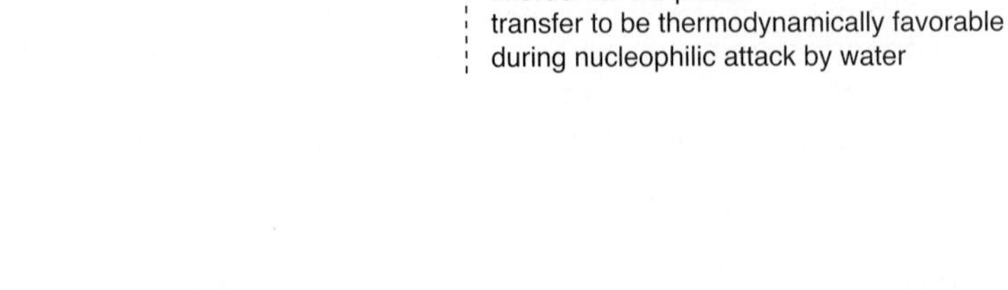

(Eq. 9.45)

Potential Energy Surfaces Dictate General or Specific Catalysis

Figure 9.13 depicts reaction coordinate diagrams for different mechanisms. Assume that each of these reactions involves a nucleophilic addition (e.g. to a carbonyl or acyl group) and a proton transfer. We will examine several such reactions in Chapter 10.

One of the important mechanistic considerations involved in addition and addition–elimination reactions of carbonyl compounds is the precise sequence of events. In particular, a major focus is on whether specific or general catalysis is involved in these reactions. In Chapter 10 we will consistently state whether the reactions are subject to general or specific catalysis. Let's examine the factors under which these various mechanisms operate. Figure 9.13 **A** shows a two-step process involving nucleophilic addition followed by protonation. The first step is rate-determining. The acid is not part of the kinetic equation, and therefore there is no acid catalysis of any kind, specific or general. This mechanism occurs for strong nucleophiles. As we will see in Chapter 10, the addition of cyanide to an aldehyde is one example.

When the nucleophile is not as good (Figure 9.13 **B**), k_{-1} is comparable to or larger than $k_2[HA]$, and the acid does enter the kinetic equation. We now have acid catalysis, and it is general-acid catalysis because protonation is involved in the rate-determining step. This particular scenario is called **enforced catalysis**, because the short lifetime of the intermediate requires that the acid trap it in order for the product to be formed. In other cases, the

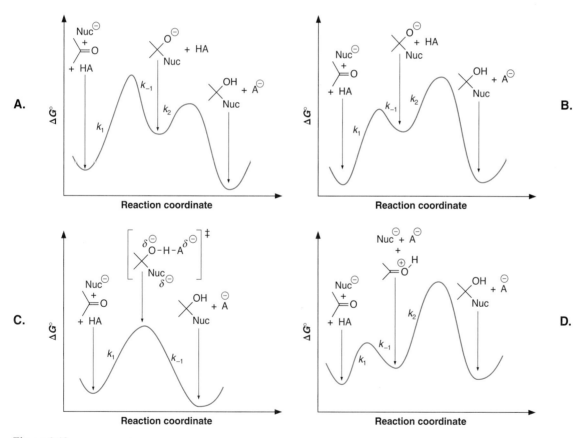

Figure 9.13
Reaction coordinate diagrams for various carbonyl addition possibilities. The steps defined as 1, –1, and 2 are designated by rate constants (k_n) for these barriers, and are referred to in the text. Remember that a larger barrier is associated with a smaller k. **A.** The rate-determining addition of a nucleophile followed by protonation. No acid catalysis is observed. **B.** When protonation is rate-limiting, we have enforced general-acid catalysis. **C.** Protonation simultaneous with nucleophilic attack also gives general-acid catalysis. **D.** Specific-acid catalysis results when nucleophilic attack cannot occur without full protonation of the reactant.

intermediate has no lifetime, and protonation is required simultaneously with nucleophilic attack, as in Figure 9.13 **C**. General-acid catalysis is again found. This is called **concerted proton transfer**. Finally, when the electrophile is so poor that it does not react without full prior protonation by the acid, we switch to specific-acid catalysis (Figure 9.13 **D**). In practice, all these alternatives can be found for addition to carbonyl compounds. The mechanism that dominates depends upon the reactivity of the nucleophile, the carbonyl compound, and the nature of the proton donor.

9.3.7 The Dynamics of Proton Transfers

Now that we have examined various forms of acid–base catalysis, and we have looked at how the thermodynamics of proton transfer are related to the kinetics via the Brønsted catalysis law, let's examine the mechanisms and rates of proton transfer in more detail. The transfer of a proton from an acid to a base is one of the simplest of all chemical reactions, and yet, even this reaction has been found to have several subtle mechanistic twists. The rate of the reaction generally depends upon the driving force (the thermodynamics) of the reaction, but there are cases where intrinsic barriers exist, making even very exothermic reactions slower than one might expect.

Proton transfers can be considered to occur in three stages (Eq 9.46). First, the acid and base diffuse together to form a hydrogen bond. Next, proton transfer occurs via a transition state with a bridging hydrogen between the acid and the base, forming a new complex with a hydrogen bond. Lastly, the new hydrogen bond complex must dissociate and diffuse apart.

$$
\begin{aligned}
&A\!-\!H \; + \; B^{\oplus} \longrightarrow A\!-\!H\cdots B^{\ominus} & \text{Diffusion together} \\
&A\!-\!H\cdots B^{\ominus} \longrightarrow [A^{\delta\ominus}\cdots H\cdots B^{\delta\ominus}]^{\ddagger} \rightarrow A^{\ominus}\cdots H\!-\!B & \text{Proton transfer} \\
&A^{\ominus}\cdots H\!-\!B \longrightarrow A^{\ominus} + \; H\!-\!B & \text{Diffusion apart}
\end{aligned}
\qquad \text{(Eq. 9.46)}
$$

The rates of diffusion were covered in Section 3.1.4. They are related to the radii of the diffusing species, in this case the acid and the base, and their diffusion coefficients. However, when charged entities are involved, an electrostatic term is incorporated that greatly increases the rate of diffusion together of oppositely charged molecules. As such, proton transfer rates between oppositely charged acids and bases are significantly greater than for neutrals.

In general, proton transfers to hydroxide from oxygen and nitrogen acids whose strengths are greater than water are diffusion controlled. Similarly, proton transfers from hydronium ion to oxygen and nitrogen bases whose base strengths are greater than water are diffusion controlled.

Marcus Analysis

The simple three-step process for understanding a proton transfer can be used to model these reactions using methods developed by Marcus (see Section 7.7.1). We focus solely upon achieving the transition state here, although the heat of reaction has also been modeled using this approach. The energy required for the reactants and products to diffuse together and apart, respectively, are referred to as the work functions w_r and w_p (Figure 9.14). Using this concept, the total free energy of activation is the sum of $\Delta G_{tr}^{\ddagger}$ (the barrier to the proton transfer step) and w_r. Using the Marcus method for calculating activation energies produces Eq. 9.47. As expected for a Marcus-type treatment, we find that the activation free energy is related to the free energy change that drives the reaction (ΔG_R°) and to a solvent reorganization term λ. Interestingly, ΔG_R° is not the standard free energy change of the reaction, but instead differs from this by the difference in the work required to bring together the reactants and to separate the products. The solvent reorganization term λ is related to the charges on the reactants and products. Plotting this equation as a function of ΔG_R° for a series of similar acid–base reactions results in a parabolic curve, meaning that the activation free energy does not linearly correlate with the reaction free energy change.

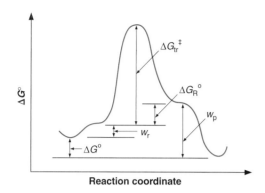

$\Delta G_{tr}{}^{\ddagger}$

$\Delta G_R{}^{\circ}$

w_p

w_r

ΔG°

ΔG°

Reaction coordinate

Figure 9.14
The energy surface for a proton transfer is treated as having three components. One work term is related to bringing together the acid and the base, forming a hydrogen bond complex. One term is the activation barrier, and the last work term is for the dissociation of the hydrogen bond complex with product.

$$\Delta G_{tr}{}^{\ddagger} = \left(1 + \frac{(\Delta G_R{}^{\circ})^2}{\lambda}\right)\left(\frac{\lambda}{4}\right) + w_r \qquad \text{(Eq. 9.47)}$$

A parabolic dependence of activation free energy on the reaction free energy is inconsistent with the Brønsted catalysis law, even when taking into account rates of proton transfer that are approaching diffusion control (Figure 9.12). We focused upon this LFER because experimentally one often finds a linear relationship between the rate constant and the strength of the acid or base for acid–base reactions. The inconsistency between the Brønsted laws (either Eq. 9.36 or 9.37) and the Marcus analysis (Eq. 9.47) can be resolved by realizing that the Brønsted plots for most acid–base reactions are experimentally evaluated over a fairly narrow pK$_a$ range, corresponding to a regime where the Marcus analysis gives only a slight curvature.

9.4 Enzymatic Catalysis

The masters of catalysis are enzymes. Enzymes are biomolecules typically based on proteins and often associated with small organic molecules or metal ions known as cofactors. In recent years it has become clear that RNA molecules can also catalyze important reactions, and such catalytic RNA molecules are referred to as **ribozymes**. Our focus here, however, will be on the more well known, protein-based enzymes, which mediate the overwhelming majority of biochemical transformations. These are nature's catalysts, and they can be incredibly efficient. As just one example, the hydrolysis of a phosphoester such as that used to link nucleotides together in DNA is estimated to have a half-life of hundreds of millions of years in water at neutral pH. Yet, the enzyme staphylococcal nuclease can catalyze this hydrolysis reaction with a half-life of a few minutes. Since this is a physical organic textbook, not a biochemistry textbook, we do not look at the structures of enzymes and how they are formed. Instead, we simply focus upon the mechanisms and kinetics of enzymatic catalysis.

While some details of how enzymes achieve such enormous rate enhancements are still being debated, and this debate is likely to continue for some time, one thing is for sure—enzymes are not *magical*. Their catalysis is achieved using the same principles of reactivity discussed in this chapter and elsewhere in this text. In a global sense, enzymatic catalysis arises from supramolecular and molecular recognition interactions. Here, again, binding of the transition state better than the ground state is the unifying principle. Hence, the following discussion nicely ties together Chapter 4 with concepts we have introduced over the last few chapters. We begin with the classic picture of enzyme kinetics. While some enzyme systems require more complex kinetic analyses, the vast majority of enzymes can be modeled with this scheme (for more complex scenarios, see the Further Reading at the end of this chapter).

9.4.1 Michaelis–Menten Kinetics

The simplest model for any catalytic reaction is the one most commonly used for enzymes—the **Michaelis–Menten** kinetic model. As previously shown in Figure 9.3, the substrate (S) and the catalyst (E, for enzyme) are in a reversible equilibrium with a non-covalent, enzyme–substrate complex, E:S. Once the complex is formed, a rate-determining step converts the substrate to the product, in the form of an analogous enzyme–product complex. Then, the product dissociates from the catalyst, setting up another equilibrium. The general picture of Figure 9.3 can be expressed equivalently as in Eq. 9.48, if chemical transformation of substrate to product is rate-determining (any steps past this step will not be evident in the kinetics). When association of the catalyst with the substrate or release of the product from the catalyst is rate-determining, the mathematical development we present here will be incorrect.

$$E + S \underset{k_{-1}}{\overset{k_1}{\rightleftharpoons}} E{:}S \xrightarrow{k_{cat}} E{:}P \rightleftharpoons E + P \qquad \text{(Eq. 9.48)}$$

Since the second step here is rate-determining, the rate of this reaction will be as given in Eq. 9.49. Using the steady state approximation for [E:S] gives Eq. 9.50, where we let [E] equal to $[E]_o - [E{:}S]$ and $[E]_o$ is the total concentration of enzyme. Solving for [E:S] and substituting the solution in Eq. 9.49 gives Eq. 9.51. Now, K_M is defined as $(k_{cat} + k_{-1})/k_1$ and is called the **Michaelis constant**, which leads to Eq. 9.52, which is called the **Michaelis–Menten equation**. This equation predicts a kinetic scenario that will show saturation behavior when $[S] \gg K_M$. Under this condition, the rate of the reaction is equal to $k_{cat}[E]_o$, which is called the **maximum velocity (V_{max})**. It is the fastest that the catalytic reaction can occur, because all the catalyst has been converted to the catalyst–substrate complex (E:S). The catalyst/enzyme is considered to be saturated with the substrate.

$$\frac{d[P]}{dt} = k_{cat}[E{:}S] \qquad \text{(Eq. 9.49)}$$

$$\frac{d[E{:}S]}{dt} = k_1([E]_o - [E{:}S])[S] - k_{-1}[E{:}S] - k_{cat}[E{:}S] = 0 \qquad \text{(Eq. 9.50)}$$

$$\frac{d[P]}{dt} = \frac{k_1 k_{cat}[E]_o[S]}{k_{-1} + k_{cat} + k_1[S]} \qquad \text{(Eq. 9.51)}$$

$$\frac{d[P]}{dt} = \frac{k_{cat}[E]_o[S]}{[S] + K_M}$$

$$K_M = \frac{k_{cat} + k_{-1}}{k_1} \qquad \text{(Eq. 9.52)}$$

There are several ways to measure k_{cat} and K_M, which we leave to a textbook devoted to biochemistry. Here, we want to understand the meaning of these constants, so that the student can easily read about enzyme kinetics, and also apply the notions to other forms of catalysis.

9.4.2   The Meaning of K_M, k_{cat}, and k_{cat}/K_M

The meaning of k_{cat} is the easiest to understand. This is the rate constant for the conversion of the substrate to the product within the active site of the catalyst, and is often called the **turnover number**. Note that it is a unimolecular rate constant, with units of s^{-1}. All the factors that we have examined in this chapter that can impart transition state stabilization will influence k_{cat}—namely, proximity, acid–base catalysis, electrostatic considerations, covalent catalysis, and the relief of strain. This rate constant is for the "chemical" step of catalysis, and it is thus the focus of efforts to interpret transition state binding relative to the substrate binding. Note that the scheme in Eq. 9.48 is for a simple, single-step conversion. As in other kinetic analyses, if multiple chemical steps are involved in converting the substrate to the product, the experimental kinetics can still be treated by a Michaelis–Menten scheme, but the derived k_{cat} will be a composite of all these steps, and is thus not amenable to easy mechanistic interpretation.

The meaning of the Michaelis constant (K_M) is more complex. There are two extremes. When $k_{-1} >> k_{cat}$, then $K_M = k_{-1}/k_1$, which is the dissociation constant (K_d) for the enzyme–substrate complex (see Section 4.1.1 for a discussion of dissociation constants). Many people therefore just consider the K_M value to be indicative of an *apparent* dissociation constant. Under these circumstances, K_M can provide insights into how good a receptor the catalyst is. A smaller K_M value means a better receptor. Note that in such scenarios, K_M reflects a physical process (binding) rather than a chemical transformation.

Under what circumstances would we expect K_M to be interpretable as K_d? Actually, this occurs when the catalyst is not particularly good. That is, if $k_{-1} >> k_{cat}$, the catalytic rate process is much slower than the off-rate associated with substrate binding. This is common for synthetic catalysts or "enzyme mimics" developed by chemists, but it is not true for many natural enzymes that have evolved to be very efficient (see Section 9.4.4).

The other extreme occurs when $k_{cat} >> k_{-1}$. Now K_M approaches k_{cat}/k_1, which is simply the ratio of the second to first rate constants in the sequence of steps in the reaction. Now K_M does not resemble the dissociation constant for the catalyst–substrate complex at all. The important point is that this is the situation for a very good catalyst.

The other very important value is the ratio of k_{cat} to K_M, called the **specificity constant**. When the substrate is in very low concentrations ($K_M >> [S]$), Eq. 9.52 reduces to Eq. 9.53. Here, $[E]_o$ has been set equal to $[E]$ because there is so little substrate that the total amount of enzyme is essentially the same as the amount of enzyme free in solution. The reaction appears to be second order, first order in both enzyme and substrate, where the apparent second order rate constant is k_{cat}/K_M.

$$\frac{d[P]}{dt} = \frac{k_{cat}[E][S]}{K_M}$$

(Eq. 9.53)

The usefulness of k_{cat}/K_M is in comparing competing substrates. One way to see this is to let K_M equal K_d, even though we know this is not strictly true. Since $K_d = 1/K_a$ (where K_a is the association constant for the catalyst and the substrate), the ratio k_{cat}/K_M is related to $k_{cat}K_a$. Thus, the specificity constant has a component related to how well the catalyst binds the substrate and how well the catalyst turns over the substrate to product. When comparing two substrates, we would like to know about both binding and catalysis. Increasing either should make a substrate more favorable for catalytic conversion by the enzyme. Often, however, it is difficult to measure k_{cat} and K_a separately. The specificity constant gives us one number that combines both.

9.4.3   Enzyme Active Sites

Enzymes are often very large molecules—much larger than the catalysts developed by chemists. However, the binding of the substrates and the catalytic transformation generally occur within a relatively small region of the enzyme called the **active site**. Often, active sites are associated with clefts or pockets of the enzyme. Let's look at just one enzyme to show how the principles of catalysis discussed in this chapter are put to use at an active site. We will look at several enzyme mechanisms in the next chapter, so this is just the first of many analyses.

The enzyme Ricin A is a potent cytotoxin isolated from seeds of the castor plant. It was used as a poison by the former Soviet Union's intelligence agency, the KGB, because it is essentially undetectable when administered. The enzyme attacks ribosomes, hydrolyzing the N-glycoside linkage of specific adenosine nucleotides in oligonucleotides. Figure 9.15 shows a model of the active site with a bound oligonucleotide based on crystal structures of the enzyme with substrate analogs. Multiple molecular recognition interactions hold the substrate in place for the stabilization of transition states and high energy intermediates.

A.

B.

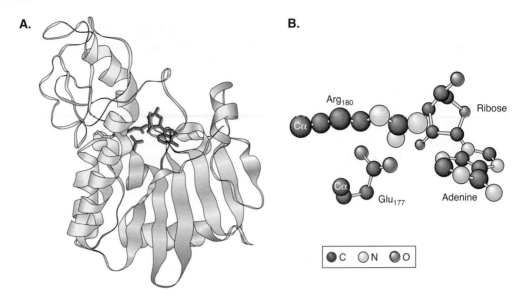

Figure 9.15
A. The enzyme Ricin A with a bound oligonucleotide (CGAGAG) modeled at the active site. **B.** Interactions involving Arg-180 and Glu-177 are noted. Monzingo, A. F., and Robertus, J. D. "X-ray Analysis of Substrate Analogs in the Ricin A-chain Active Site." *J. Mol. Biol.*, **227**, 1136–1145 (1992).

Figure 9.16 shows the proposed mechanism for the reaction catalyzed by Ricin A. The carboxylate from glutamate-177 ion pairs with the oxocarbenium ion created by the departure of the purine base, thereby stabilizing the transition state for the departure of this leaving group. Simultaneously, the leaving group is either protonated or ion paired with a neighboring guanidinium group from arginine-180. This further stabilizes the transition state for leaving group departure. A water that is held at the active site by hydrogen bonding to the same arginine and glutamate side chains then acts as a nucleophile in the second step. The nucleophilicity of the water is enhanced by general-base catalysis, as shown in the figure.

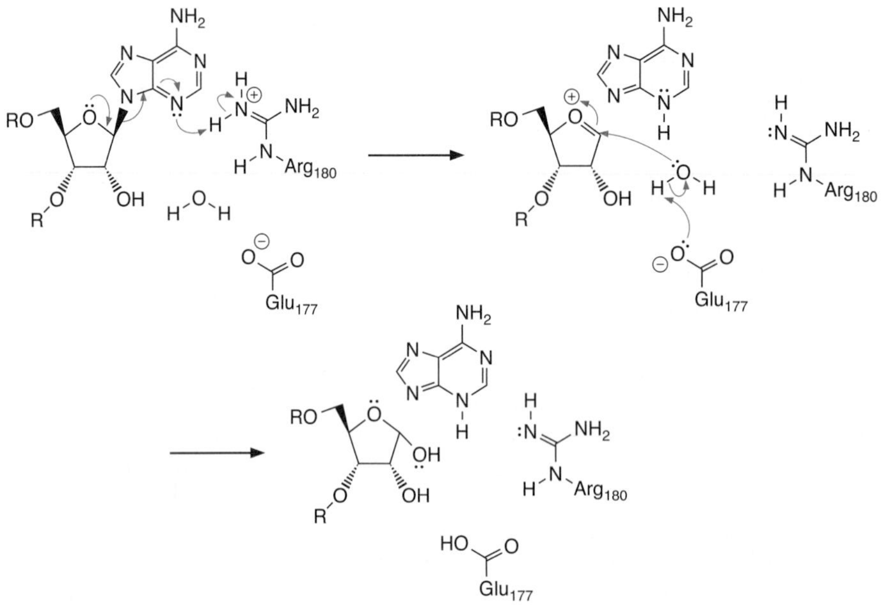

Figure 9.16
The proposed mechanism of depurination catalyzed by Ricin A. In the first step the leaving group departure is enhanced by ion pairing between the carboxylate anion of Glu_{177} and the oxocarbenium ion intermediate, as well as protonation of the leaving group. In the second step a water is delivered to the oxocarbenium ion using general-base catalysis.

The general base could be either the carboxylate, or the newly created guanidine base (we arbitrarily show the carboxylate). We show full deprotonation of the guanidinium by the adenine leaving group, although in reality only an ion pair may form. The pK_a of adenine is 9, while that of guanidinium is around 12, so unless these pK_a values are significantly perturbed by their microenvironment (a definite possibility), only an ion pair will form. The adenine base produced is protonated at N-3, but this proton would quickly tautomerize to the more stable protonation site of N-9. After loss of the adenine and the depurinated nucleotide from the enzyme active site, proton transfer and binding of another water molecule is all that is required to regenerate the catalyst for the next substrate molecule. This brief look at one enzyme should impress upon you the considerable sophistication of nature's catalysts.

9.4.4   [S] vs. K_M—Reaction Coordinate Diagrams

Let's tie together many of the thermodynamic concepts presented in this textbook to understand how an enzyme can improve. In other words, if there is evolutionary pressure on an organism to optimize a particular chemical pathway for survival, how could the enzyme that catalyzes that pathway become better at catalysis? The simple answer is to continue to increase the difference in binding between the transition state and the substrate. There is a limit to this, however, because the transition state does resemble the reactant. The enzyme must still bind the ground state in order to sequester its substrate from the cellular medium. Hence, the enzyme must effectively bind the substrate but still bind the transition state better. Recall from Section 4.1.1 that the extent to which a receptor binds a substrate depends upon both the K_d and the concentration of substrate. To maximize efficiency, an enzyme will adjust K_d in response to the natural concentration of the substrate. In particular, we expect K_d to be comparable to the substrate's physiological concentration, assuring a significant fraction will be bound. If K_d is much larger than the natural concentration of substrate, very little substrate will be bound at the active site. On the other hand, a very small K_d value implies binding "overkill"; there is no need to add more binding energy if the enzyme active site is already fully occupied, and very tight substrate binding makes it that much harder to bind the transition state even more tightly. How an enzyme can still efficiently bind the substrate but optimize preferential binding of the transition state is best understood by an analysis of reaction coordinate diagrams.

Examine Figure 9.2 once again. These are $\Delta G°$ diagrams, which means they reflect the relative energies of the various components in the reaction when present at one molar concentrations (see Section 3.1.5). For Figure 9.2 **B** the binding constants would be less than $1 M^{-1}$, making $K_d > 1 M$. The catalyst is therefore a *terrible* receptor, and would require a substrate near 1 M in concentration to efficiently sequester it (refer to Section 4.1.1 if you are not sure why this is). Many natural substrates are at μM or lower concentrations for common enzymatic reactions. Hence, the K_d or K_M values range from around 10^{-3} to 10^{-6} M for most enzymes, making enzymes good receptors for their substrates. With values such as these, all $\Delta G°$ reaction coordinate diagrams for enzymes should be drawn as in Figures 9.2 **A**, **C**, and **D**, indicating that the enzyme would be saturated if the substrate were 1 M. However, diagrams such as Figure 9.2 are difficult to use to see how an enzyme might evolve to optimize catalysis. Therefore, we define a new standard state for the analysis (ΔG^*), just as we showed can be done in Section 4.1.1. The new standard state is the natural abundance concentration of the substrate. For purposes of this analysis, let's say the concentration of a substrate for a particular enzymatic reaction is consistently maintained near 1 μM by a cell. Now examine Figure 9.17, where ΔG^* diagrams are shown.

Figure 9.17 **A** represents $[S] > K_d$ (when K_M and K_d are similar, we can use K_M here). This requires that the substrate drop in energy on binding to the enzyme, compared to its value in free solution, and so the substrate has to climb out of a well to reach the transition state. We can consider Figure 9.17 **A** as our starting point, and we want the enzyme to evolve into a more effective catalyst. The addition of another binding interaction that is equal in strength for both the substrate and the transition state (called a **uniform binding change**) does nothing to decrease the barrier associated with k_{cat}, and thus it does nothing to accelerate the reaction. An additional binding interaction that operates on the transition state but *not* on the

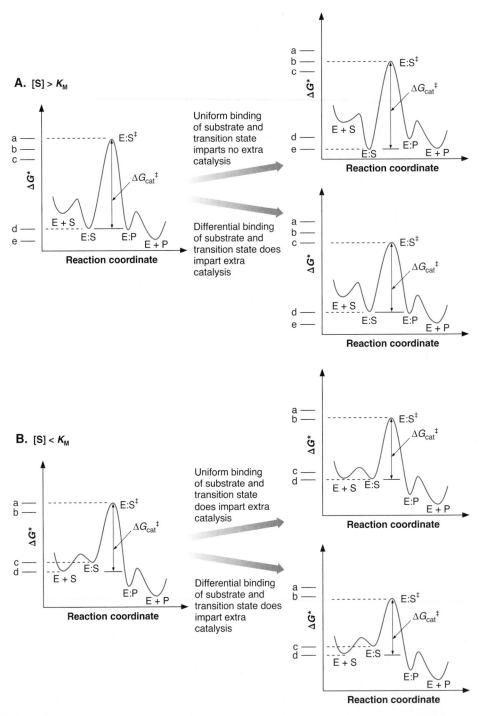

Figure 9.17

How an enzyme (or potentially any catalyst) can improve. **A.** When $[S] > K_d$, there is only one way to improve catalysis—focus more binding on the transition state relative to the ground state. Uniform binding drops the transition state energy level a to b, and the E:S complex the same amount, from d to e. Differential binding drops the transition state more, a to c, while the E:S complex either remains the same or drops less than the transition state. **B.** When $[S] < K_d$, there are two ways to improve catalysis—namely, focus binding upon both the ground state and transition state, or the transition state alone. Uniform binding of the transition state and the E:S complex (a to b and c to d, respectively) will give improved catalysis. Differential binding that drops the transition state from energy level a to b, while keeping the E:S complex at c also gives improved catalysis.

substrate (a **differential binding change**) necessarily improves catalysis. The differential binding need not be perfect; if there is some improved binding of substrate but a larger improvement in binding the transition state, the same effect will be seen. Hence, when $[S] > K_d$, an improvement in catalysis (i.e., an increase in k_{cat}/K_M) can only occur when additional binding interactions occur in the transition state.

Figure 9.17 **B** represents $[S] < K_d$. Again, a differential binding change favoring the transition state will improve catalysis. In addition, though, now a uniform binding change *will* improve catalysis. This occurs because, in this concentration regime, improving K_d will increase the concentration of the E:S complex significantly, thereby aiding catalysis. A series of uniform binding changes will improve catalysis until the binding of the substrate has become so good that K_d becomes lower than the natural $[S]$. Then, we return to the scenario depicted in Figure 9.17 **A**.

We find experimentally that many enzymes have K_M values nearly equal to or slightly greater than the natural concentration of their substrates. Often binding is just strong enough to have a high fraction of the E:S complex present at equilibrium.

The fastest that any catalyst can operate is at diffusion control. Bimolecular diffusion rates in water at conventional temperatures are in the range of 10^8 to $10^9\,M^{-1}\,s^{-1}$. We conclude that any enzyme that has a k_{cat}/K_M value in the range of 10^8 to $10^9\,M^{-1}\,s^{-1}$, and a $K_M > [S]$, cannot get any better. These are called **perfect enzymes**. Carbonic anhydrase and triosephosphate isomerase are two examples. For carbonic anhydrase, $k_{cat}/K_M = 8.3 \times 10^7$ and $K_M = 0.012\,M$, while for triosephosphate isomerase, $k_{cat}/K_M = 2.4 \times 10^8$ and $K_M = 2.5 \times 10^{-5}\,M$. Note that being a perfect enzyme does not imply the largest possible k_{cat}, because there is a significant difference in k_{cat} values for the two examples just given. The key to perfection is to have the enzyme properly tuned to the physiological conditions it is likely to experience.

9.4.5 Supramolecular Interactions

We have looked at methods for catalysis—namely, proximity, electrophilic catalysis, acid–base catalysis, and the interplay between strain and transition state stabilization. We have now also looked at the general features of enzyme catalysis. All of these are supramolecular interactions—beyond the molecule (see Chapter 4)—where interactions between the substrate and catalyst lead to the rate enhancement. If we think about tying all of the methods of catalysis together, we can get a good idea of how to catalyze a reaction. Not only can we design catalysts for ourselves, but we can make very good predictions as to the general features that an enzyme must possess. The Connections highlight on the next page discusses cyclodextrins, one of the most common building blocks chemists use to construct catalysts for various reactions.

Let's consider one particular reaction as a means of demonstrating the knowledge that you should have by this point. How would you catalyze an S_N2 reaction? The rate-determining step for this reaction has a transition state that is becoming planar at carbon, a nucleophile is adding to the carbon, a slightly positive charge may be developing on the carbon, and the leaving group is developing negative charge. Hence, a good catalyst would facilitate each of these features. In this hypothetical approach to the catalysis of this reaction, one might place point charges in a binding site that complement all the interactions occurring during the reaction, as shown below. We will discuss an enzymatic example for an S_N2 reaction in Chapter 11 that does indeed give all these anticipated interactions.

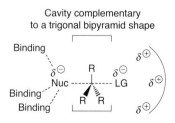

Cavity complementary
to a trigonal bipyramid shape

Connections

Artificial Enzymes: Cyclodextrins Lead the Way

Inspired by the remarkable efficiency of many enzymes, chemists have tried to prepare artificial systems that operate in form and function like enzymes do. Cyclodextrins are the most extensively used platforms for these efforts. The dominance of cyclodextrins stems from the pioneering observations of Breslow and Tabushi, who showed that simple organic compounds can display many of the hallmarks of enzymatic catalysis, such as binding, rate accelerations, and turnover. However, most artificial systems do not give the large rate enhancements that their natural counterparts impart. One example that does produce a large rate enhancement is based on a cyclodextrin dimer.

The dimer of β-cyclodextrin shown below (refer back to Section 4.2.4 to see the structure of β-cyclodextrin) presents a rigid binding cleft possessing a metal as an electrophilic catalyst. The spacing between the two cyclodextrins leads to binding of the substrate in a manner that places the ester functional group in direct proximity to a metal bound hydroxide, and coordinated to the copper metal as shown. Nucleophilic attack by the hydroxide on the ester carbonyl and leaving group departure completes the hydrolysis reaction. The rate acceleration of this hydrolysis reaction is approximately 10^4–10^5 over hydrolysis catalyzed by hydroxide itself. The *bis*-cyclodextrin catalyst shows enzyme-like behavior, in that saturation kinetics is observed that can be fit to a Michaelis–Menton model. A V_{max} value of 2.24×10^{-5} M s^{-1}, and a K_M value of 4.69×10^{-5} M were determined. This example is one of the very best artificial enzymes yet reported, and bodes well for continued efforts in this field.

Breslow, R., and Dong, S. D. "Biomimetic Reactions Catalyzed by Cyclodextrins and Their Derivatives." *Chem. Rev.*, **98**, 1997–2011 (1998). Zhang, B., and Breslow, R. "Ester Hydrolysis by a Catalytic Cyclodextrin Dimer Enzyme Mimic with a Metallobipyridyl Linking Group." *J. Am. Chem. Soc.*, **119**, 1676–1681 (1997).

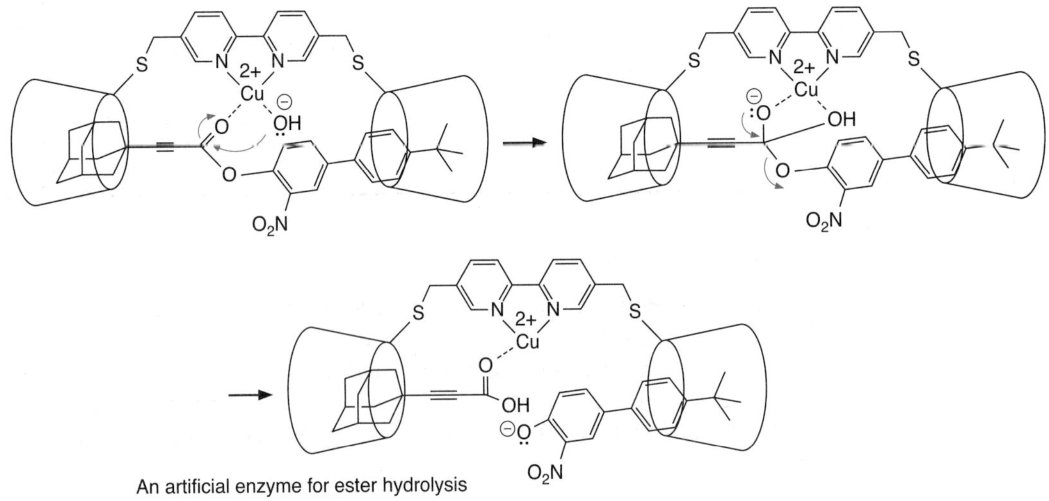

An artificial enzyme for ester hydrolysis

Summary and Outlook

The ability to catalyze a variety of chemical reactions is a frontier of physical organic chemistry. Whether the specific chemical approach involves heterogeneous or homogeneous organometallic species, bioorganic artificial enzymes, acid–base catalysts, or the use of enzymes, the principles involved are very similar. As stated in this chapter, one designs chemical structures that stabilize transition states more than they stabilize ground states, or create completely new mechanisms to transform the reactant into the product. Factors varying from proximity to supramolecular interactions are involved. Now that we have an understanding of the principles of catalysis, we can examine how catalysts are involved with specific reactions. This is the direction we turn to in the next chapter, where common organic reactions mechanisms are described. In this context, several highlights are given in the next chapter that cover either natural or artificial enzymes.

Exercises

1. The following reaction involves both a general-acid and a general-base catalyst in water. Yet, a bell-shaped $\log k_{obs}$ vs. pH profile is not found. Derive the rate expression and explain why the reaction does not lead to a bell-shaped pH dependence. How is this reaction fundamentally different from the reactions that would show a bell-shaped $\log k_{obs}$ vs. pH profile? What kind of catalysis would the kinetic expression you derived lead a researcher to propose for this reaction?

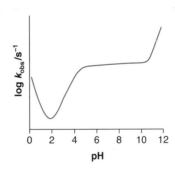

2. Show the mathematics that leads to Eq. 9.24.

3. The following hypothetical reaction is general-base-catalyzed. Let's examine the relative rates of this reaction for various possible bases and for various Brønsted β values, but all at the same pH of 7.0. For a β value of 0.2, calculate the relative rates for three different bases whose conjugate acid pK_a values are either 4, 7, or 10. For a β value of 0.5, calculate the relative rates for the same set of pK_a values. Finally, for a β value of 0.9, calculate the relative rates for the same pK_a values. For what β value and what pH do you find the largest relative rate? Is it always true that you want to use the strongest base possible to get the largest rate possible at any pH?

4. The following is a hypothetical example of a $\log(k_{obs})$ versus pH profile for the hydrolysis of an ester with an added acid such as acetic acid. Examine this curve and state which form of acid or base catalysis is occurring at the various pH regions. Write a rate law that would describe this curve.

5. Cast Eqs. 9.36 and 9.37 into the form normally written for LFERs, such as the Hammett equation.

6. Use the libido rule to predict which of the following acids could give specific catalysis or general catalysis in the following reaction (the pK_a of the conjugate acid of an imine is around 4 to 5): HCl, picric acid, acetic acid, ammonium, and phenol. It is important to remember that whether specific or general catalysis occurs is also related to the reactivity of the substrate, not just the pK_a of the acid.

7. Look at the following transformation, and state what features you would incorporate into a catalyst for this reaction.

8. Pyridoxal phosphate is used by enzymes to catalyze a reaction known as transamination. This cofactor and the reaction it catalyzes are shown below. Write a mechanism that shows how this cofactor can catalyze this reaction using hypothetical general-acid (BH^+) and general-base (B) catalysts present in an enzyme active site. (*Hint:* Imine functional groups are involved.)

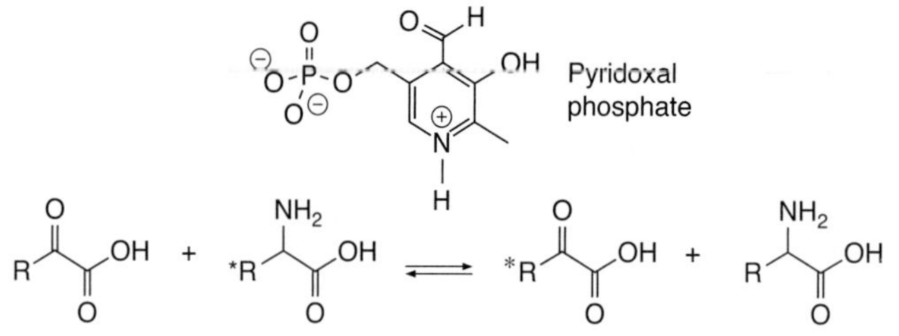

9. The compound *N*-hydroxybenzotriazole (HOBT) is often added to increase the rate of reaction of an ester with an amine to form an amide. What is the role of HOBT (see Chapter 10 for the answer)?

10. Exercises 10–13 are concerned with the enolization of acetone.

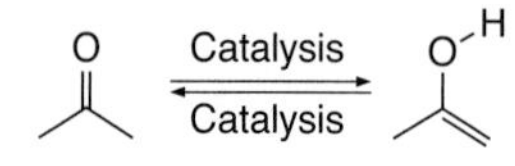

    a. Write a specific-acid-catalyzed mechanism for this reaction.
    b. Write a general-acid-catalyzed mechanism for this reaction.
    c. Write a specific-base-catalyzed mechanism for this reaction.
    d. Write a general-base-catalyzed mechanism for this reaction.

11. In a general-acid-catalyzed enolization of acetone, is the reverse mechanism also general-acid-catalyzed or some kinetic equivalent? Briefly explain.

12. Each of the mechanisms from Exercise 10 were supported by careful kinetic measurements performed for this enolization by Hegarty and Jencks in 1975. In other words, all these different forms of catalysis can operate simultaneously, the extent of each pathway being determined by the pH and the concentrations of the general acids and general bases. Correspondingly, the rate law looks like the following:

$$\text{Rate} = [\text{acetone}](k_{H^+}[H_3O^+] + k_{HA}[HA] + k_{OH^-}[OH^-] + k_B[B])$$

$$\text{term 1} \qquad \text{term 2} \qquad \text{term 3} \qquad \text{term 4}$$

Draw four separate pH versus $\log(k_{obs})$ kinetic plots, one for each term in this expression. To do this, incorporate the hydronium, hydroxide, general-acid (pK_a of 7.0), and general-base (conjugate acid pK_a of 7.0) concentration into the rate constants to give a k_{obs} for each of these four terms.

13. An additional kinetic term was found experimentally for the enolization of acetone. This fifth term involved both general acid and general base. Thus, it had the form, rate = $k_{AB}[HA][B][\text{acetone}]$. Furthermore, experiments revealed that for this kinetic term there were two isotope effects and the two different Brønsted relationships described below.
    i. When the reaction is run in D_2O, the k_{AB} is one-half the k_{AB} in H_2O. In other words, k_{H_2O}/k_{D_2O} is near 2.0 for the various acids. (*Hint:* In order to understand this isotope effect you need to remember that any acidic hydrogen on an acid in D_2O will become deuterated.)
    ii. Furthermore, if one deuterates the methyl protons of acetone, an isotope effect of $k_H/k_D = 5.8$ is found.
    iii. An α value of 0.2 and a β value of 0.88 were found for this third-order term.
        Write a mechanism for this enolization that is consistent with all these data (points i, ii, iii, and the rate law) for this fifth term. Explain each piece of data and how it is consistent with the mechanism you wrote.

14. The hydrolysis of isopropenyl glucopyranosides has recently been studied. One issue in question was the site of cleavage of these saccharides, whether it was the acetal linkage or the vinyl ether.

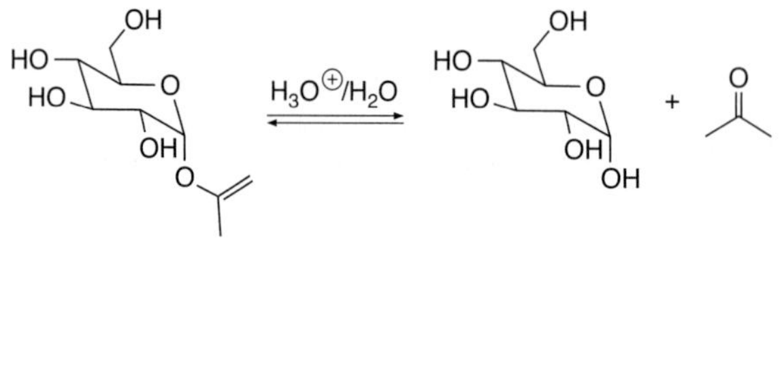

a. Using hydronium ion as the catalyst, write a mechanism for the hydrolytic cleavage of the acetal linkage (as shown). Show all arrow pushing.

b. Using hydronium ion as the catalyst, write a mechanism for the hydrolytic cleavage of the vinyl ether linkage (as shown). Show all arrow pushing.

15. The following two pieces of data were collected as a means to distinguish the site of cleavage of the glucopyranoside given in Exercise 14. The authors concluded that the site of cleavage was the vinyl ether linkage (part b of Exercise 14). Explain how the following data were used to support their choice of mechanism, and why the data do not support the alternative mechanism.

   i. The α-anomer gives exclusively α-hemiacetal products. The β-anomer gives exclusively β-hemiacetal products.

α-anomer β-anomer

   ii. When the reaction is run in ^{18}O water, the acetone product has heavy oxygen but none of the glucose products have heavy oxygen.

16. The following data were collected to elucidate the rate-determining step of the reaction given in Exercise 14.

   i. Electrophilic addition of $DOCD_3$ catalyzed with $D_2O^+CD_3$ resulted in the following product and no incorporation of deuterium was found in the starting material.

   ii. Similarly, when the hydrolysis was performed in D_2O, no deuterium was incorporated into the starting material, and only one atom of deuterium was found in the acetone produced.

   iii. An isotope effect of $k_H/k_D = 3.06$ was found when the reaction was catalyzed by H_3O^+ or D_3O^+.

   iv. An α value of 0.637 was found for this reaction.

   Do these four experiments support general-acid or specific-acid catalysis? Explain how each of the four pieces of data supports your conclusion as to the type of catalysis by which this reaction proceeds.

17. In Exercise 16, we found that the hydrolysis of the glucopyranoside had the same extent of proton transfer (63%) in the transition state regardless of the strength of the acid, because it followed the Brønsted relationship. Use the following More O'Ferrall–Jencks graph to show that the strength of the acid does not change the extent of proton transfer at the transition state. You may want to refer back to Exercise 17 in Chapter 7 for an earlier analysis of this same plot.

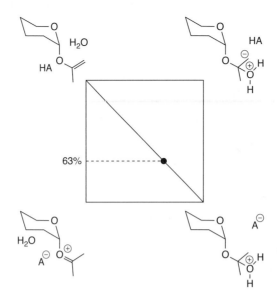

18. There are many ways in which catalysts can become reduced in activity, a general phenomenon known as **inhibition**. The inhibition of an enzyme is a common way to create a pharmaceutical that impedes a chemical reaction required in a bacterial or viral life cycle. In the following three examples, use the Michaelis–Menten kinetics model, and derive the kinetic expression for the rate of product formation when the inhibitor acts as shown [I = inhibitor, S = substrate, P = product, and E = enzyme (catalyst)]. Use your rate equations to explain how each form of inhibition causes the rate to move further from V_{max}.

a. In **competitive inhibition**, the inhibitor competes with the substrate for the binding site of the catalyst with binding constant K_I.

$$E + S \underset{k_{-1}}{\overset{k_1}{\rightleftharpoons}} ES \overset{k_{cat}}{\longrightarrow} P$$

$$\big\updownarrow K_I I$$

$$EI$$

b. In **non-competitive inhibition**, the inhibitor and the substrate both bind to the catalyst at the same time, but when the inhibitor is bound the catalyst is inactive. Assume that the binding constant K_I is the same for I binding to E and ES.

$$E + S \underset{k_{-1}}{\overset{k_1}{\rightleftharpoons}} ES \overset{k_{cat}}{\longrightarrow} P$$

$$\big\updownarrow K_I I \qquad \big\updownarrow K_I I$$

$$EI \underset{}{\overset{K_{eq} S}{\rightleftharpoons}} ESI$$

c. In **non-productive inhibition**, the substrate binds part of the time in an inactive mode (ES').

$$E + S \underset{k_{-1}}{\overset{k_1}{\rightleftharpoons}} ES \overset{k_{cat}}{\longrightarrow} P$$

$$\big\updownarrow K_{eq}$$

$$ES'$$

19. Explain the following trend in effective molarities (M) taken from Table 9.1.

2.3×10^7 3.7×10^{11} 2×10^{13}

Further Reading

Catalysis in General

Cornils, B., Herrmann, W. A., Schloegl, R., and Wong, C.-H. (eds.) (2000). *Catalysis from A to Z: A Concise Encyclopedia*, Wiley–VCH, Weinheim.

Ford, M. E. (eds.) (2000). *Catalysis of Organic Reactions*, Marcel Dekker, New York.

General and Specific Acid–Base Catalysis

Jencks, W. P. "Requirements for General Acid–Base Catalysis of Complex Reactions." *J. Am. Chem. Soc.*, **94**, 4731–4732 (1972).

Eigen, M. "Proton Transfer, Acid–Base Catalysis, and Enzymatic Hydrolysis." *Angew. Chem. Int. Ed. Eng.*, **3**, 1–19 (1964).

Jencks, W. P. "When is an Intermediate not an Intermediate? Enforced Mechanisms of General Acid–Base Catalyzed, Carbocation, Carbanion, and Ligand Exchange Reactions." *Acc. Chem. Res.*, **13**, 161–169 (1980).

Jencks, W. P. "General Acid–Base Catalysis of Complex Reactions in Water." *Chem. Rev.*, **72**, 705 (1972).

Jencks, W. P. "Enforced General Acid–Base Catalysis of Complex Reactions and Its Limitations." *Acc. Chem. Res.*, **9**, 425 (1976).

Theories of Enzyme Action

Bruice, T. C., and Benkovic, S. J. "Chemical Basis for Enzyme Catalysis." *Biochemistry,* **39**, 6267–6274 (2000).

Cleland, W. W. "What Limits the Rate of an Enzyme-Catalyzed Reaction?" *Acc. Chem. Res.*, **8**, 145–151 (1975).

Kraut, J. "How Do Enzymes Work?" *Science*, **242**, 533–540 (1988).

Neet, K. E. "Enzyme Catalytic Power Minireview Series." *J. Biol. Chem.*, 273, 25527–27038 (1998).

Gerlt, J. A. "Relationships Between Enzymatic Catalysis and Active Site Structure Revealed by Applications of Site-Directed Mutagenesis." *Chem. Rev.*, **87**, 1079 (1987).

Spatial Temporal Analysis

Menger, F. M. "On the Source of Intramolecular and Enzymatic Reactivity." *Acc. Chem. Res.*, **18**, 128–134 (1985).

Menger, F. M. "Enzyme Reactivity from an Organic Perspective." *Acc. Chem. Res.*, **26**, 206–212 (1993).

Notions of Intramolecularity and Orbital Steering

Bruice, T. C., and Benkovic, S. J. "The Compensation in $\Delta H^{\ddagger}$ and $\Delta S^{\ddagger}$ Accompanying the Conversion of Lower Order Nucleophilic Displacement Reactions to Higher Order Catalytic Processes. The Temperature Dependence of the Hydrazinolysis and Imidazole-Catalyzed Hydrolysis of Substituted Phenyl Acetates." *J. Am. Chem. Soc.*, **86**, 418–426 (1964).

Page, M. I., and Jencks, W. P. "Entropic Contributions to Rate Accelerations in Enzymic and Intramolecular Reactions and the Chelate Effect." *Proc. Natl. Acad. Sci.*, **68**, 1678–1683 (1971).

Bruice, T. C., Brown, A., and Harris, D. O. "On the Concept of Orbital Steering in Catalytic Reactions." *Proc. Natl. Acad. Sci.*, **68**, 658–661 (1971).

Mesecar, A. D., Stoddard, B. L., and Koshland, D. E., Jr. "Orbital Steering in the Catalytic Power of Enzymes: Small Structural Changes with Large Catalytic Consequences." *Science*, **277**, 202–206 (1997).

Port, G. N., and Richards, W. G. "Orbital Steering and the Catalytic Power of Enzymes." *Nature*, **231**, 312 (1971).

Dafforn, A., and Koshland, D. E., Jr. "Theoretical Aspects of Orbital Steering." *Proc. Natl. Acad. Sci.*, **68**, 2463 (1971).

Metal Ion Catalysis

Cacciapaglia, R., and Mandolini, L. "Catalysis by Metal Ions in Reactions of Crown Ether Substrates." *Chem. Soc. Rev.*, **22**, 221 (1993).

Straeter, N., Lipscomb, W. N., Klabunde, T., and Krebs, B. "Two-Metal Ion Catalysis in Enzymic Acyl- and Phosphoryl-Transfer Reactions." *Angew. Chem. Int. Ed. Eng.*, **35**, 2025 (1996).

Enzyme Mimics and Artificial Enzymes

Kirby, A. J. "Enzyme Mechanisms, Models, and Mimics." *Angew. Chem. Int. Ed. Eng.*, **35**, 707–724 (1996).

Murakami, Y., Kikuchi, J.-I., Hisaeda, Y., and Hayashida, O. "Artificial Enzymes." *Chem. Rev.*, **96**, 721–758 (1996).

Perfect Enzymes

Knowles, J. R., and Albery, W. "Efficiency and Evolution of Enzyme Catalysis." *Angew. Chem. Int. Ed. Eng.*, **16**, 285–293 (1977).

Chin, J. "Perfect Enzymes: Is the Equilibrium Constant Between the Enzyme's Bound Species Unity?" *J. Am. Chem. Soc.*, **105**, 6502–6503 (1983).

CHAPTER 10

Organic Reaction Mechanisms, Part 1: Reactions Involving Additions and/or Eliminations

Intent and Purpose

The preceding chapters have focused on structure and its relationship to energetics, as well as the tools of mechanistic physical organic chemistry. We have presented many examples where physical principles are used to explain trends in reactivity. However, it is not until this chapter and the next that we actually explore the steps involved in common organic reactions. The only major classes of organic reactions we do not cover in these two chapters are pericyclic reactions and photochemical reactions, which we leave until Chapters 15 and 16, respectively.

We start this chapter with a paradigm for predicting chemical reactivity that can guide you in writing reasonable mechanisms for almost all polar organic transformations. The paradigm can be understood from several vantage points: simple electrostatics, Lewis acid–base interactions, nucleophile–electrophile interactions, and orbital overlap. Teaching you how to use this paradigm to predict reactivity is an important goal of this and the next chapter, so we repeatedly show how various reactions conform to it. It is important to be able to predict reactivity—probably even more so than it is to know the "nit-picky" details of organic reaction mechanisms.

Another goal of these two chapters is to review the majority of the reaction mechanisms covered in introductory organic chemistry, and to significantly expand upon that knowledge base. The mechanisms are covered not only to reacquaint you with common organic transformations, but also to provide a setting for developing good electron pushing skills. It is often said, "Organic chemistry is the art of electron pushing." This chapter and the next should drive this point home firmly. However, it should also be appreciated that arrow pushing is a formalism akin to a bookkeeping method. It allows chemists to predict certain aspects of a reaction.

There is often more than one way to push electrons in writing an organic mechanism, and our paradigm for predicting reactivity can lead to many possible mechanisms. How do we know, therefore, if the predictions we make from our paradigm and electron-pushing rules are correct? At first we do *not* know if we are correct. To understand the mechanisms of organic transformations, it is not enough to just combine common steps and use correct electron-pushing techniques. We have to experimentally test whether the mechanism implied by the electron pushing is correct or not, and that is where we turn to the tools presented in Chapters 7 and 8. Deciphering the mechanisms of chemical reactions is a major focus of physical organic chemistry, and some of the key experiments that form the basis of our understanding of common organic reaction mechanisms are presented in this and the next chapter. It is important to appreciate that the successful application of our knowledge of a mechanism to control the reactivity of organic compounds in petrochemical, pharmaceutical, and other industrial applications represents a major triumph of physical organic chemistry.

Remember, a mechanism can never be completely proven. However, after having read Chapters 7 and 8, we are ready to present the experiments that are used to support a proposed mechanism. Furthermore, with the analysis of the structure and stability of reactive intermediates covered in Chapters 1 and 2, we are ready to explain their reactivity. Moreover, after having read Chapter 9, we have a foundation in catalysis. Therefore, Chapters 10 and 11 represent a culmination of much of what you have learned so far.

A general comment about reaction mechanisms is in order here. The early, heroic years of mechanistic organic chemistry sought to define the basic paradigms and assign reactions to mechanistic classes. Concepts like S_N2, tetrahedral intermediate, and chain mechanism were thoroughly developed. With further study, however, the lines separating mechanistic classes progressively blurred. This led to lengthy, often contentious, debates about the fine details of specific reaction mechanisms. With the benefit of hindsight, we can view this work in a different light. We will see that a modern perspective on mechanistic organic chemistry emphasizes a continuum of possibilities. There are limiting cases, reactions that perfectly follow a particular mechanistic paradigm. However, there are many reactions that fall in the middle, and there are endless variations of substrates and reaction conditions that push a reaction toward one mechanistic extreme or another (as implied by More O'Farrell–Jencks diagrams). It is important to learn the prototype, paradigmatic mechanisms, but keep in mind that many reactions fall in between the mechanistic extremes. Almost every variation imaginable can be realized with a cleverly designed system. This in no way invalidates our mechanistic paradigms. The mechanistic continuum is simply testimony to the incredible structural diversity of organic molecules and the facility with which molecules will always find the lowest energy path from starting material to product.

The intent of Chapters 10 and 11 is not to give all the details that some graduate-level physical organic textbooks present. We leave even more thorough analyses to dedicated textbooks referenced at the end of Chapter 11. Here, we give the information that we believe all organic chemists should know. If you finish these two chapters having absorbed all the information presented, you will be one of the more knowledgeable organic chemists in any lab. Therefore, in keeping with the spirit of this text, we present the fundamentals of each reaction type, and we emphasize modern research efforts when presenting examples.

10.1 Predicting Organic Reactivity

We have already looked at many reactions throughout this textbook, and it is now time to give a concrete definition. A **reaction** involves the redistribution of the connectivity of atoms within a chemical structure or between structures. One or more reactions constitutes the transformation of one molecule into another. The individual steps involved in a reaction are called the reaction mechanism when taken as a set. Understanding reaction mechanisms is a basic goal of the science of chemistry. In addition, there are also many practical consequences of having a good understanding of the mechanism of a reaction. For example, once a mechanism has been delineated, a chemist can choose the appropriate reaction conditions or modify the reactants so as to use the reaction to his or her advantage. Therefore, the understanding of a reaction mechanism is often the most important information a chemist can have to manipulate nature.

A **mechanism** describes, as a function of time, the chemical steps necessary for one molecule to be transformed into another. It gives the interrelationships among the molecules whose motions and collisions are necessary for the chemical transformation. Specifically, it provides the relative positions and energies of all nuclei and electrons in the reactants, intermediates, activated complexes, and products, as well as those of the solvent, at each stage of the transformation. Gould likened the mechanism to a motion picture of the chemical transformation. By analyzing each frame of the motion picture individually, we can analyze each step of the reaction in a sequential manner.

10.1.1   A Useful Paradigm for Polar Reactions

Since chemical reactions involve a change in the connectivity of atoms, and this connectivity is conveniently viewed as localized bonds consisting of two electrons, most chemical reactions can be described by considering the movement of two electrons. We will refer to these globally as **polar reactions**, emphasizing the importance of charges and/or partial charges in analyzing them. Some reactions involve the movement of one electron, and we will explore such radical reactions also. However, we first develop a paradigm for predicting two-electron movement.

Two electrons inherently have negative character, such as a π bond or a lone pair of electrons. Sometimes this negative charge is associated with a full negative charge on a molecule, as with hydroxide. It may be a partial negative charge on the negative end of a polar covalent bond, such as exists in a carbonyl. In order to have a positive region in a molecule, there must be either polar covalent bonds or a formal positive charge. Either way, the negative region of one molecule is inherently reactive toward a positive region of another molecule. This is based purely upon electrostatic considerations. Given this simple analysis, most polar organic reactions involve the combination of molecules with some regions of positive and negative charge, and these charges can be viewed as guiding the attraction between the molecules. In essence, the inherent polarity in bonds can be used as a guide to their reactivity.

We can use the above logic to state the following simple paradigm for making a first pass at predicting chemical reactivity:

Combine the positive region of one molecule with the negative region of the other molecule.

This sounds fairly obvious, but in fact many students do not appreciate that this is the first thing to look for in analyzing potential reaction pathways. We will refer to this as the electrostatic model. The electrostatic surface maps given in Appendix 2 are of great use in this regard, and we will refer to them in this and the next chapter. Real reactions, however, are not electro*static*. The redistribution of electrons is essential in all reactions. But often, the initial charge distributions in the reactants provide a good clue to the reactivity. Most of the polar reactions we will discuss in this and the next chapter follow this electrostatic paradigm. As we will discuss in detail below, this simple notion is typically couched in more sophisticated language, such as nucleophile–electrophile, or donor–acceptor interactions. In the end, though, a great deal of chemistry comes down to coulombic interactions—like charges repel, opposites attract. There are reactions that will turn out to be exceptions, but in most cases this simple predictive tool can guide us a long way in writing reasonable reactions that can be combined to create mechanisms for a large fraction of organic transformations.

The paradigm also guides our electron-pushing rules. Appendix 5 provides an overview of basic electron-pushing strategies. Unless you are quite comfortable with electron pushing, we strongly suggest that you read over Appendix 5 before studying the mechanisms we show in this and the next chapter.

There are at least three different definitions and vantage points that chemists use to expand upon the electrostatic paradigm. They are nucleophile–electrophile combinations, Lewis acid–base reactions, and donor–acceptor orbital interactions. We have used these terms repeatedly throughout this book because they are presented in introductory organic chemistry classes, but here we give them strict definitions. Each of these definitions is a subtle variation on the other, and often it is essentially a case of semantics to decide which best describes a particular reaction.

Nucleophiles and Electrophiles

Formally, a **nucleophile** is a compound that is seeking a nucleus, whereas an **electrophile** is a compound that is seeking electrons (Eq. 10.1). Since a nucleus is inherently positive and electrons are inherently negative, nucleophiles are considered to be seeking positive charge and electrophiles negative charge. Note that there is no mention of orbitals in this definition. Consistent with the paradigm presented above, we will generally find that nucleophiles have a negative charge or a substantial partial negative charge at their site of reactivity. Similarly, electrophiles generally have full or partial positive charges.

Nucleophile   Electrophile

(Eq. 10.1)

A **nucleofuge** is a nucleophile that is departing from a molecule instead of adding, and an **electrofuge** is an electrophile that is departing from a molecule instead of adding. For example, in the heterolysis of *t*-butylbromide to form *t*-butyl cation and bromide, the *t*-butyl cation is the electrofuge and the bromide is the nucleofuge (Eq. 10.2).

Electrofuge   Nucleofuge

(Eq. 10.2)

The terms nucleophile, electrophile, nucleofuge, and electrofuge are typically used in discussing reactivity and kinetics. Hence, when one refers to a good nucleophile or electrophile, this means that the nucleophile or electrophile reacts faster than some reference nucleophile or electrophile. The same can be said for a good electrofuge or nucleofuge.

Lewis Acids and Lewis Bases

The definitions of Lewis acids and Lewis bases also lead naturally to our paradigm (see Chapter 5 for another discussion of this topic). A **Lewis acid** is an electron pair acceptor and a **Lewis base** is an electron pair donor. An electron pair is any two paired electrons, such as an actual lone pair, but also σ or π bonds. The Lewis definitions imply an analysis of orbitals. Since lone pairs and bonds are partially negative due to the electron charges, so is the region of the Lewis base that is undergoing the reaction. Furthermore, because empty orbitals have some character near or on the nucleus, filling an empty orbital leads to donation of the electrons toward a positive center. Again, this leads to the conclusion that Lewis acid–base reactions are similar to electrophile–nucleophile combinations, which can be considered to be guided by electrostatic considerations.

A subtle distinction is that typically when we discuss strong versus weak Lewis acid-bases, we are making a thermodynamic distinction. This contrasts the inherently kinetic nature noted above when evaluating nucleophiles and the like.

Donor–Acceptor Orbital Interactions

Reactions can be understood by considering interactions between filled orbitals and empty orbitals. We emphasized in several places in this book that it is always favorable to mix a filled orbital with an empty orbital. The definitions of a Lewis acid and Lewis base imply an empty and a filled orbital, respectively, as do the definitions of an electrophile and a nucleophile. Most of the time this is obvious, such as in the reaction of water with a carbenium ion (Eq. 10.3). The water has the filled lone pair orbital (it is the nucleophile or Lewis base) that donates its electron density to the empty *p* orbital on the carbenium ion (the electrophile or Lewis acid).

(Eq. 10.3)

The emphasis on mixing filled with unfilled orbitals leads naturally to the notion of donors and acceptors. A **donor** is a structure with a filled orbital. An **acceptor** has an empty orbital. As discussed in Chapters 1 and 14, the closer in energy the donor orbital is to the acceptor orbital and the better the overlap between the donor and acceptor orbitals, the stronger the interaction (we gave an equation relating the two when we first introduced Lewis acids and Lewis bases in Section 5.6). As such, we should anticipate that reactions are most facile when the donor and acceptor orbitals that interact are close in energy. Thus, we should consider mixing the highest occupied molecular orbital (HOMO) of the donor/nucleophile/Lewis base and the lowest unoccupied molecular orbital (LUMO) of the acceptor/electrophile/Lewis acid. The donor–acceptor concept is a powerful one. It is a useful mnemonic with a firm foundation in bonding theory.

10.1.2   Predicting Radical Reactivity

The reactions of organic radicals do not easily fit into the paradigms discussed above. Radicals can have either electrophilic or nucleophilic character, or both. Furthermore, radical reactions often involve chain mechanisms that are of a different nature from what is seen in typical polar reactions. Radical chain reactions are extremely common in organic chemistry, but the reactions are often harder to control than polar reactions. Often a given radical will have several competing pathways available, and predicting which is preferred can be challenging. Nevertheless, we will give many examples of mechanisms involving radicals in this and the next chapter, and clear trends in the reactivity patterns will be evident. Do not discount the importance of radical reactions, which are commonly used in organic synthesis. They are also central to many industrial processes and are the dominant reaction types for the degradation of organic and biological chemicals due to exposure to O_2 and sunlight.

10.1.3   In Preparation for the Following Sections

In the following sections of this and the next chapter a large fraction of the common organic reaction mechanisms is covered. We start each section with specific examples of the reaction under consideration, along with the "standard" or "classic" electron-pushing procedure for the mechanism. The electron pushing refers to the reaction as written going from the left to the right, even though many of the mechanisms involve equilibria. This electron pushing should be a review of the mechanism as presented in introductory organic chemistry classes. Importantly, as you will see, what is traditionally presented in introductory organic classes is often not correct for all reactants, but instead represents a slice of a mechanistic continuum. Often the mechanism deviates from that given by the electron pushing due to variations in the structure of the reactants and the experimental conditions. We use Schemes to portray the standard electron pushing, and Figures to explore the more in-depth details of the mechanisms.

The sections in this chapter and the next are organized in terms of reaction classes: additions, eliminations, substitutions, and others. This organization stresses that almost all organic reactions involve various transformations of unsaturated systems (Chapter 10: C=C, C=O, Ar, etc.) or saturated systems (Chapter 11: C–O, C–N, C–X, etc.). For example, addition reactions to alkenes and carbonyls differ due to differing bond polarizations, but they are fundamentally similar because all π bonds undergo nucleophilic, electrophilic, and radical additions. The similarities are more unifying than the extent to which their differences separate them. Similarly, electrophilic and nucleophilic addition–elimination mechanisms for acyl derivatives are actually very similar to electrophilic and nucleophilic aromatic substitution mechanisms. This is because they are both transformations of a π system that involve additions and eliminations. Hence, it is logical to cover them together. Therefore, before starting the following sections, it is important to recognize the following unifying structure to organic mechanisms: *the key steps are typically addition to π bonds, eliminations to form π bonds, and substitution on σ bonds.*

Lastly, most mechanisms commence by the reaction of an electrophile or nucleophile on a reactant, or by a bond homolysis, and this in part forms a classification chemists give to each mechanistic possibility. For example, electrophilic aromatic substitution, or nucleophilic aliphatic substitution, refer to an electrophile adding to the aromatic structure, or a nucleophile adding to the aliphatic center, respectively. Besides such titles to describe reactions, chemists have developed a shorthand system to name reaction mechanisms based upon the following three aspects of the mechanism: 1. the kind of reaction that is occurring—addition (Ad), elimination (E), or substitution (S); 2. the character of the reactant that is interacting with the substrate—electrophilic (E), nucleophilic (N), radical (R or H for homolytic); and 3. the molecularity of the reaction—unimolecular (1), bimolecular (2), etc. S_N1 and S_N2 are classic examples of this nomenclature—that is, substitution, nucleophilic, unimolecular, or substitution, nucleophilic, bimolecular, respectively. A summary of this shorthand system is given in Appendix 6, and we use these kinds of acronyms where appropriate in various sections of Chapters 10 and 11.

Addition Reactions

An **addition reaction** involves the combination of two molecules to form a product containing atoms from both reactants. All addition reactions occur on unsaturated organic structures—namely, alkenes, alkynes, carbonyls, and arenes. Not surprisingly, three distinct mechanisms are observed—electrophilic, nucleophilic, and radical. In the electrophilic addition pathway, the unsaturated system acts as the nucleophile supplying electrons to the electrophilic reagent. The opposite is true in the nucleophilic pathway. In either of these two cases, simple electrostatic considerations assist in predicting what will occur.

At this point it may be useful to review our description of the bonding of simple carbonyls and olefins. In Chapter 1 we described the π and π^* molecular orbitals of both. Recall, especially, the polarizations in the orbitals of the carbonyl group, where the LUMO is polarized toward the carbon. Also, Appendix 3 has detailed drawings of orbitals associated with olefins and the carbonyls of many representative functional groups.

10.2   Hydration of Carbonyl Structures

We start our discussion by picking up where we left off in Chapter 9, with the topic of carbonyl hydrations. With all the background on acid–base catalysis and carbonyl hydration given in Chapter 9, we can now look at the experimental data relevant to Eq. 10.4. Since exactly the same pathways are followed in the formation of hemiacetals (Eq. 10.5), we briefly cover this reaction also. Figures 9.6 and 9.8 show multiple electron-pushing scenarios, so we do not need to open this discussion by showing electron-pushing possibilities. However, it should be noted that all the mechanisms given in Chapter 9 indicate that the carbonyl compound is electrophilic at the C of the C=O bond, just as the electrostatic potential surfaces (EPS) of Appendix 2 indicate.

Geminal diol

(Eq. 10.4)

Hemiacetal

(Eq. 10.5)

10.2.1 Acid–Base Catalysis

The acid-catalyzed hydration of most carbonyl structures involves alternative **C** of the reaction coordinates in Figure 9.13, where Nuc = H_2O or ROH. The mechanism we defined as a concerted proton transfer (Section 9.3.6) is operative, and it leads to a tetrahedral intermediate that rapidly loses a proton in a second step. This was the alternative shown in Figure 9.8 **A**, and it is shown again in Eq. 10.6. Furthermore, general-base-catalyzed hydration pathways are common, where Figure 9.8 **B** shows the mechanism. There are also specific-acid-catalyzed pathways that dominate at very low pH values. Hence, even this simplest of carbonyl reactions is subject to most of the forms of acid–base catalysis we discussed in Chapter 9.

(Eq. 10.6)

Brønsted α values for a few general-acid-catalyzed reactions are given in Table 10.1. Our discussion in Section 9.3.5 showed that the α value is interpreted as the extent of protonation of the reactant by the acid in the transition state. The fact that the Brønsted plots are linear for carbonyl hydrations means that the extent of protonation at the transition state remains the same regardless of the strength of the acid.

Table 10.1
A Few Examples of α and β Values for Hydration Reactions of Carbonyl Compounds*

Compound	α Value	β Value
1,3-Dichloro-2-propanone	0.27	0.50
Formaldehyde	0.24	0.40
3-Chlorobenzaldehyde	0.46	0.44

*McClelland, R. A., and Coe, M. "Structure–Reactivity Effects in the Hydration of Benzaldehydes." *J. Am. Chem. Soc.*, **105**, 2718 (1983).

A linear Brønsted plot, however, seems counter to the Hammond postulate. As the reaction becomes more exothermic with stronger acids, the transition state should shift toward the structure of the reactant. So, why is the extent of proton transfer in the transition state of a carbonyl hydration invariant? Actually, this question is relevant not just to hydration reactions, but also to all reactions where linear Brønsted plots are found.

We must consider More O'Ferrall–Jencks plots to find the answer. Figure 10.1 shows this plot for the hydration of a carbonyl (Nuc = H_2O), with the extent of protonation (Brønsted α value) and the extent of nucleophilic attack (β_{Nuc}) as the y and x axes, respectively. The diagonal line represents the concerted proton transfer mechanism with an α value similar to 3-chlorobenzaldehyde. A stronger acid raises the upper right and left corners of this plot, and therefore the resulting position of the transition state is along the composite arrow given (if you have forgotten how these plots are created and analyzed, review Section 7.8.2). Note that the extent of protonation remains the same. What changes is the extent of addition of the nucleophile. As the acid becomes stronger, there is less nucleophilic attack in the transition state. This makes sense, because with a stronger acid less nucleophilic attack is needed to induce proton transfer.

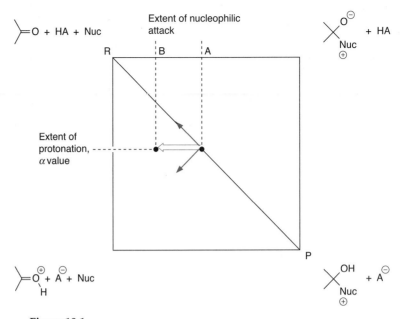

Figure 10.1
A More O'Ferrall–Jencks plot of hydration of a carbonyl compound.
As the acid becomes stronger, the transition state has less nucleophilic
attack (A to B), and the α value remains the same.

10.2.2   The Thermodynamics of the Formation of Geminal Diols and Hemiacetals

Table 10.2 **A** shows the association constants for the hydration of several carbonyl structures. For ketones and aryl aldehydes, the constants are less than unity, favoring the carbonyl. However, aliphatic aldehydes, carbonyl structures with electron withdrawing groups, and carbonyls in strained rings have equilibrium constants greater than unity. In

Table 10.2
A. Equilibrium Constants for the Hydration Reaction of Various Carbonyl Compounds B. Association Constants for Hemiacetal Formation in Methanol*

$$K_a = \frac{[\text{Geminal diol}]}{[\text{Carbonyl}]}$$

A. Examples highlighting steric and electronic differences

Reactant	K_a	Reactant	K_a
H₂C=O	2×10^3	(CH₃)(H)C=O	1.3
(CH₃)₂C=O	2×10^{-3}	ClCH₂(H)C=O	2.8×10^4
(ClCH₂)₂C=O	10	m-Cl-C₆H₄-CHO	2.2×10^{-2}

B. Examples highlighting angle strain

Reactant	K_a
Cyclobutanone	1
Cyclopentanone	$6.7 \cdot 10^{-2}$
Cyclohexanone	$4.5 \cdot 10^{-1}$
Cycloheptanone	$1.9 \cdot 10^{-2}$
Cyclooctanone	$3.7 \cdot 10^{-3}$

*Bell, R. P. "The Reversible Hydration of Carbonyl Compounds." *Adv. Phys. Org. Chem.*, **4**, 1 (1966). Wheeler, O. H. "Structure and Properties of Cyclic Compounds. IX. Hemiketal Formation of Cyclic Ketones." *J. Am. Chem. Soc.*, **79**, 4191 (1957).

general, aldehydes are more hydrated than ketones because steric congestion in the geminal diol is less. Furthermore, electron withdrawing groups destabilize the already electrophilic carbonyl, leading to greater hydration.

Table 10.2 **B** shows the association constants for various hemiacetals in methanol. Relative to large ring cyclic ketones, strained rings such as cyclobutanone prefer the sp^3 hybridization of a hemiacetal carbon over the sp^2 hybridization of a carbonyl carbon. This is because the smaller bond angle of an sp^3 center better matches the bond angles in small rings. Cyclohexanone also exists significantly as its hemiacetal, but the larger rings have increasingly lower extents of hemiacetal formation.

Connections

Cyclic Forms of Saccharides and Concerted Proton Transfers

Hemiacetals in five- and six-membered rings are considerably more stable than acyclic hemiacetals. They constitute the form in which many saccharides exist under biological conditions. Glucose exists primarily in its pyranose (six-membered ring) form, although furanose (five-membered ring) forms are present also. Both the pyranose and furanose forms have anomeric isomers, known as α and β. These are shown below for the pyranose form. The open chain form is present in neutral water at ambient temperature only to an extent of approximately 0.003% of the total equilibrium. In the context of saccharides, interconversion between the hemiacetals and the open chain forms occurs in a reaction called **mutarotation** (see below). This reac-

tion interconverts the α and β anomers, and it can be general-acid- and/or general-base-catalyzed. For glucose, Brønsted acid α and base β values of 0.27 and 0.36 are found, respectively.

The catalysis of mutarotation in organic solvents can be accomplished using simple compounds that are capable of tautomerization reactions. As shown below, simultaneous deprotonation and protonation of a hemiacetal with 2-pyridone leads to ring-opening. Bond rotation and closing of the ring interconverts the anomers. The 2-pyridone catalyzes this interconversion 7000 times faster than a mixture of phenol and pyridine.

Swain, C. G., and Brown, J. F. "Concerted Displacement Reactions. VII. The Mechanism of Acid–Base Catalysis in Non-Aqueous Solvents." *J. Am. Chem. Soc.*, **74**, 2534–2537 (1952).

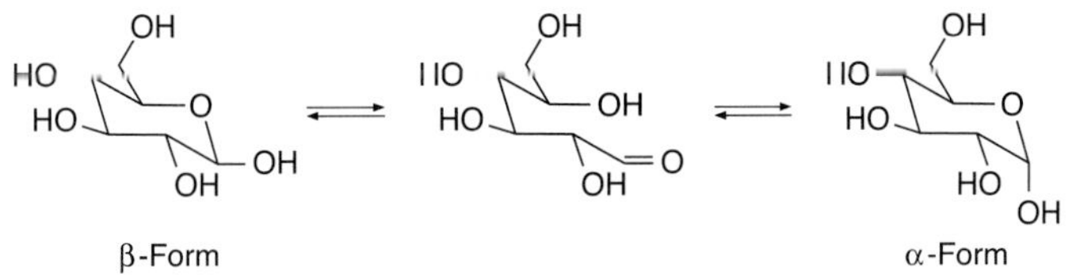

Mutarotation

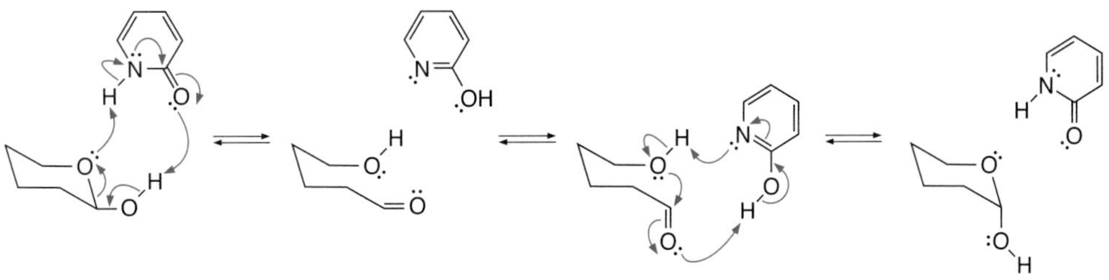

Catalysis of mutarotation

10.3 Electrophilic Addition of Water to Alkenes and Alkynes: Hydration

The mechanism of the hydration of alkenes is very similar to that of carbonyls. However, alkene protonation leads to a much more reactive species—a carbenium ion. As we will see in the following sections, many different electrophiles add to alkenes or alkynes, resulting in additions, all of which are called Ad$_E$2 reactions (addition, electrophilic, bimolecular). To start our analysis, let's briefly review the mechanism as given in introductory organic chemistry.

10.3.1   Electron Pushing

The π bond of an olefin can act as a nucleophile (see the red center in the EPS of ethylene and propene shown in Appendix 2) to remove a proton from hydronium ion (Scheme 10.1), and full proton transfer results in a carbenium ion. This highly electrophilic species reacts with water, and deprotonation gives the product, a simple alcohol. Note that every reaction in this mechanism involves the negative region of one molecule interacting with a positive region of the other molecule.

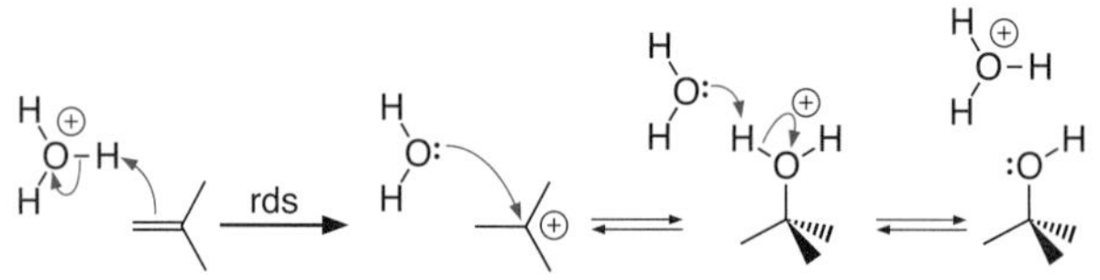

Scheme 10.1
The mechanism of the acid-catalyzed hydration of an alkene (rds = rate-determining step).

10.3.2   Acid-Catalyzed Aqueous Hydration

The electrophilic addition of water to an alkene shown in Scheme 10.1 occurs readily at ambient temperature only with extremely strong acids. Typically, acidic conditions that require the use of H_o scales (see Section 5.2.5) are necessary. Linear correlations between H_o and $\log(k_{obs})$ for these reactions are common. This means that the reaction is first-order in both acid (defined here as all the forms that contribute to the protonating power of the medium) and the alkene. Therefore, protonation is rate-determining. In support, primary kinetic isotope effects are common for the proton donating species, even up to a magnitude of 6. Moreover, isotope scrambling in the first step is not observed. For example, if β,β-dideuteriostyrene is treated with acid and water, the deuteriums are not scrambled with protons in the early stages of the reaction (scrambling at later times indicates a reaction of the alcohol product). The lack of scrambling means that the first step of the reaction is not reversible. Protonation as the rate-determining step makes sense in the hydration of an alkene, because the highest energy species shown in Scheme 10.1 is the carbenium ion, and its formation *should* therefore be rate-determining. This mechanism corresponds to general-acid catalysis; all forms of acid in the medium are reactive, and the protonation is rate-determining.

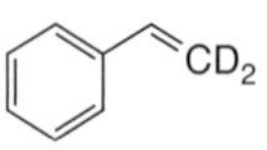

β,β-Dideuteriostyrene

10.3.3   Regiochemistry

When the double bond is substituted with alkyl groups or other electron donating groups, the effect of the substituents on the regiochemistry leads to the trend known as **Markovnikov addition**. Markovnikov addition predicts the dominant product to arise from nucleophilic addition to the more substituted carbon. Note that nucleophilic addition to the more substituted carbon forms the more sterically crowded and thus less stable product (Eq. 10.7). Therefore, Markovnikov addition predicts the kinetic product, not the thermodynamic product.

Major product
Markovnikov addition
Kinetic product

Minor product
Anti-Markovnikov addition
Thermodynamic product

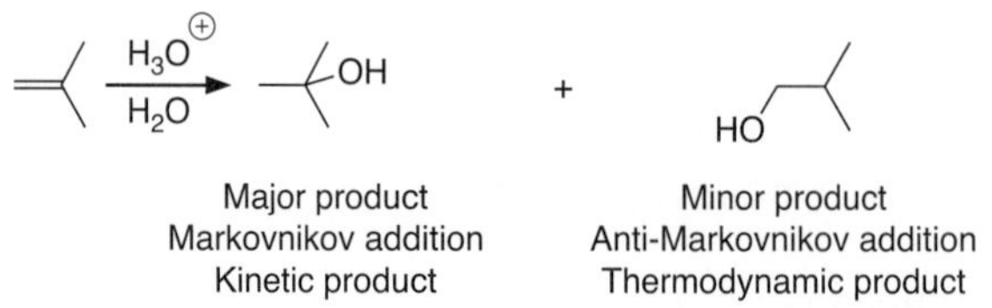

(Eq. 10.7)

Electron donating groups such as alkoxy and dialkylamino also stabilize carbenium ions, and so are expected to increase hydration rates. Furthermore, the protonation occurs on the carbon beta to these groups, giving Markovnikov addition. The protonation of ena-mines and vinyl ethers are the first steps in the hydrolysis of these species (Eqs. 10.8 and 10.9). In contrast, electron withdrawing groups such as cyano and chloride retard alkene hydration.

$$\text{(Eq. 10.8)}$$

$$\text{(Eq. 10.9)}$$

Figure 10.2 shows a Hammett plot for the hydration of 1,1-disubstituted alkenes. Here, the sum of the σ^+ substituents is used, and all the points, with very few exceptions, are near the line. This means that when the electron donating or withdrawing groups are on the same carbon their effect is additive. In contrast, when both carbons of the olefin carry the same substituent, little effect on the rate is observed relative to only a single substitution. For ex-ample, the rates of hydration of *cis*-2-butene and *trans*-2-butene are comparable to that of propene (see below right). Similarly, 2-methyl-2-butene hydrates four orders of magnitude faster than *cis*- or *trans*-2-butene, but no further increase is seen with 2,3-dimethyl-2-butene. These results support the involvement of a localized carbenium ion on the more substituted carbon, and not a carbonium ion possessing a hydrogen bridging both alkene carbons (Sec-tion 1.4.1).

Olefin hydration rates

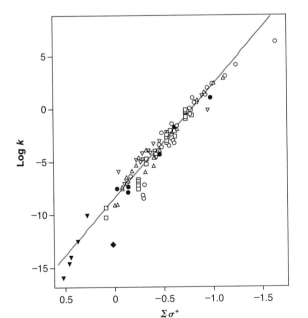

Figure 10.2
An LFER for the log of the rate constants for hydration of various alkenes as a function of the sum of the σ^+ values for the substituents on the double bond. The open and closed circles, boxes, and triangles represent different kinds of substituents including both EWGs and EDGs. Koshy, K. M., Roy, D., and Tidwell, R. R. "Substituent Effects of the Trifluoromethyl Group on Electrophilic Additions to Alkenes. Solvolysis of the Trifluoromethyl Groups. Protonation of Alkenes Less Basic than Ethylene, $\rho+$ Values of Deactivated Styrenes, and Reactivity–Selectivity Effects." *J. Am. Chem. Soc.*, **101**, 357 (1979).

10.3.4   Alkyne Hydration

Alkynes are similarly nucleophilic, as shown with the EPS for propyne in Appendix 2. Experimental details for alkynes are, therefore, very similar to those found for alkenes. The hydrations are general-acid-catalyzed, isotope scambling in the first step does not occur, and for substituted phenyl acetylenes, a large and negative ρ value (–3.83) was found. The data support rate-determining protonation of the alkyne.

With terminal alkynes, hydration gives the ketone instead of the aldehyde, consistent with Markovnikov addition (Eq. 10.10). When alkynes are treated with strong acids, vinyl cations are formed. Nucleophilic attack by water and deprotonation first give a vinyl alcohol (enol), which ultimately tautomerizes to a ketone. The tautomerization mechanism will be given in the next chapter.

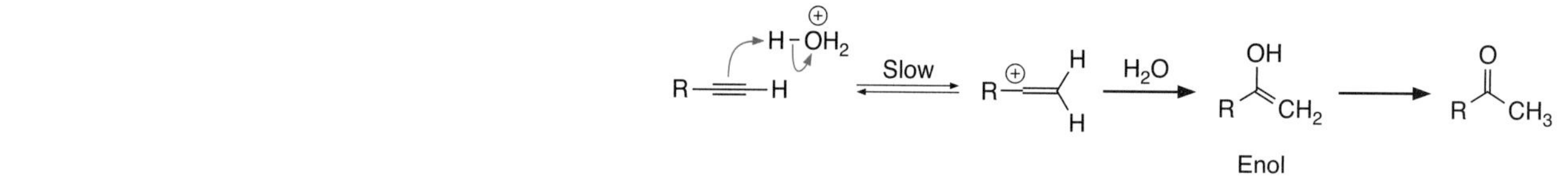

$$R\!-\!\!\equiv\!\!-\!H \xrightarrow{\text{Slow}} R\overset{\oplus}{-}\!\!=\!\!\overset{H}{\underset{H}{}} \xrightarrow{H_2O} \underset{\text{Enol}}{R\overset{OH}{=}CH_2} \longrightarrow R\overset{O}{\underset{CH_3}{}} \qquad \text{(Eq. 10.10)}$$

10.4 Electrophilic Addition of Hydrogen Halides to Alkenes and Alkynes

10.4.1 Electron Pushing

The standard mechanism written for electrophilic addition of HX to an alkene or alkyne is similar to acid-catalyzed hydration. As shown in Scheme 10.2, protonation is the first step, followed by nucleophilic attack of the halide on the resulting electrophilic carbenium ion. Just as with hydration, electrostatic effects guide our understanding of both steps of the reaction. As we'll show below, this standard electron pushing implies a mechanism that is much simpler than that often found experimentally.

Scheme 10.2
Electron pushing for the addition of HX species.

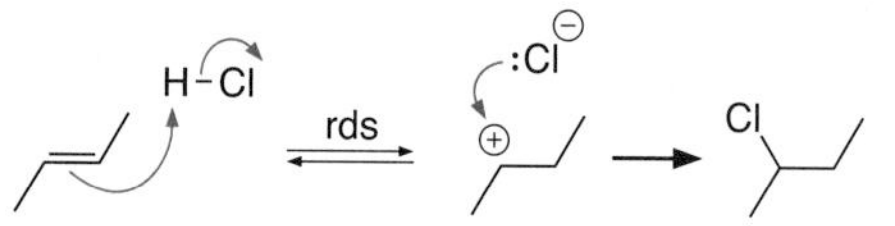

10.4.2 Experimental Observations Related to Regiochemistry and Stereochemistry

The rate-determining step for these reactions is protonation of the alkene to form the carbenium ion, so Markovnikov addition is observed. However, the timing of the addition of the halide can vary, depending upon the alkene and HX. Also, because the addition of HX is usually carried out in a solvent that is also nucleophilic (e.g., water, alcohol, or acetic acid), byproducts resulting from solvent addition are common. Very interesting kinetics and product distributions can be observed.

The addition of HCl to 3,3-dimethyl-1-butene in acetic acid gives the three products shown in Eq. 10.11 in the ratio indicated. Protonation is rate-determining as indicated by a kinetic analysis that is first order in both acid and alkene, and an isotope effect of 1.15 is found (comparing HCl in AcOH with DCl in AcOD). A carbenium ion rearrangement has occurred to give the second product. To verify that the rearrangement occurs during the reaction, the other two products were subjected to the reaction conditions, and they are stable.

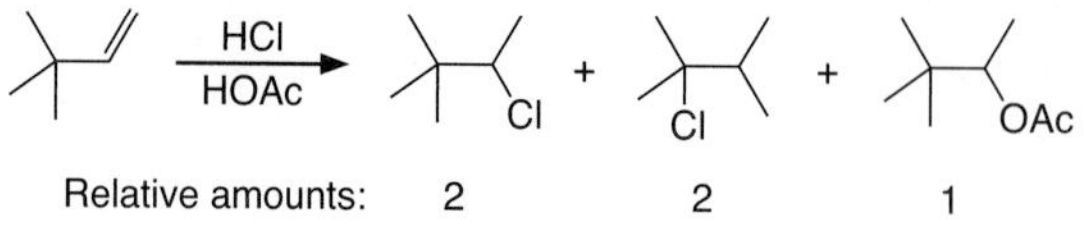

$$\xrightarrow[\text{HOAc}]{\text{HCl}} \quad + \quad + \quad \qquad \text{(Eq. 10.11)}$$

Relative amounts: 2   2   1

Interestingly, the product ratio depends upon neither the concentration of HCl nor the concentration of added chloride salts. If chloride and acetic acid (HOAc) compete for addition to a transient carbenium ion, one would expect increasing chloride concentrations to divert the intermediate to the formation of alkyl chloride products, but this is not seen. This

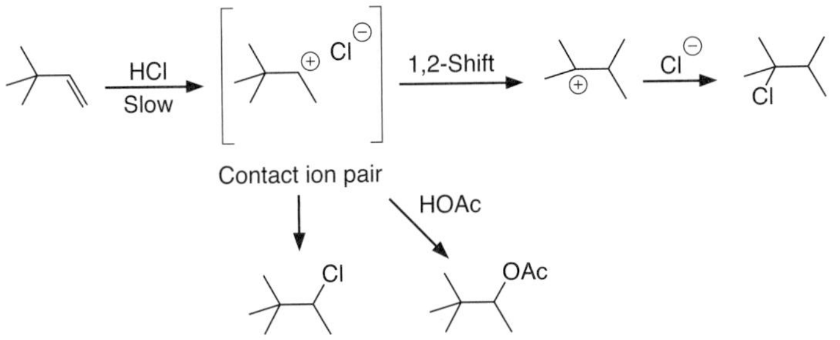

Figure 10.3
Formation of a contact ion pair that undergoes reaction with acetic acid and chloride, and that rearranges faster than chloride dissociation explains why the product ratio does not depend on chloride concentration.

means that the free chloride in solution does not influence the outcome of this reaction. How can this be explained?

The most logical explanation considers the microenvironment within which the reaction occurs, suggesting the formation of a contact ion pair between the carbenium ion and the chloride (Figure 10.3). A **contact ion pair** is a loosely held electrostatic complex between a cation and an anion. The two ions are considered to be trapped together for a very short time within a **solvent cage**. We will examine the nature of contact ion pairs extensively in our discussion of S_N1 reactions (see Section 11.5.3). In the reaction under consideration, rate-determining protonation leads to a reactive ion pair, not a free carbenium ion. The diffusion of the chloride into bulk solvent is slow relative to carbenium ion rearrangement, chloride addition, and acetic acid addition. In other words, the chloride concentration around the reactive species is never in equilibrium with the chloride concentration in solution, explaining why adding chloride does not affect the product ratios.

In contrast, the product ratio from the addition of HCl to cyclohexene in acetic acid (Eq. 10.12) *does* depend upon the concentration of added chloride salts. Addition of tetramethylammonium chloride increases the yield of chlorocyclohexane relative to cyclohexylacetate. Thus, very subtle structural changes can influence the mechanism of this reaction.

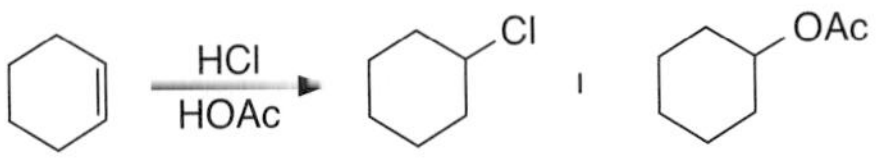

(Eq. 10.12)

A key observation in understanding the difference between these two systems is the stereochemistry of addition. As shown in Figure 10.4, five products are obtained from the reaction of 1,3,3-trideuteriocyclohexene with HCl in acetic acid. One product results from syn addition of HCl, while another from anti addition of HCl. A third product arises from anti addition of acetic acid, and the stereochemistry of addition of the fourth and fifth products cannot be ascertained. **Syn addition** is defined as the addition of both components (in this case H and Cl or H and OAc) to the same face of the alkene π bond. **Anti addition** is addition of the two components to opposite faces of the alkene.

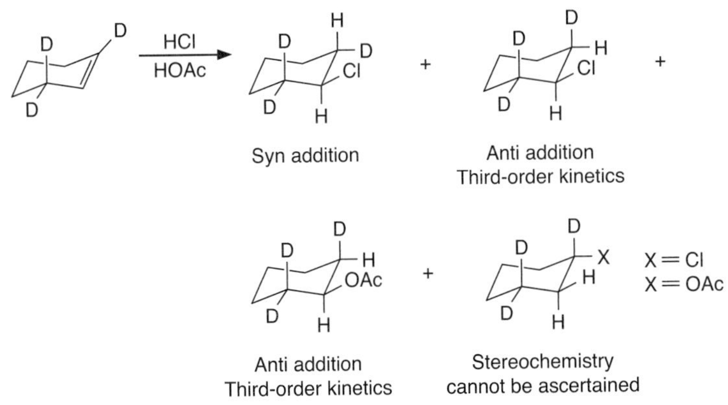

Figure 10.4
The stereochemistry of the products derived from the addition of HCl to 1,3,3-trideuteriocyclohexene in acetic acid.

The kinetics of the reaction reveals three competing pathways. The syn addition of HCl is found to be a second-order reaction, first order in alkene and HCl. Syn addition occurs via a contact ion pair, where the H⁺ and Cl⁻ necessarily add from the same face of the alkene. The kinetics of anti addition of chloride and acetic acid is *third* order—first order in alkene, first order in HCl, and first order in chloride or acetic acid, respectively. These additions are termolecular. This explains the dependence of the product distribution on chloride. Because the reaction is termolecular, electrophilic and nucleophilic attack on the alkene would be expected to occur on opposite faces of the alkene due to steric considerations, and anti addition products result. Termolecular collisions are rare, so that the actual reaction likely occurs by collision of the nucleophile with a weak complex between the alkene and the acid.

Now that we have examined two cases, cyclohexene and 3,3-dimethylpropene, it should be clear that the simple electron pushing of Scheme 10.2, although giving a general picture of the reaction, lacks insight into the subtleties of the mechanism.

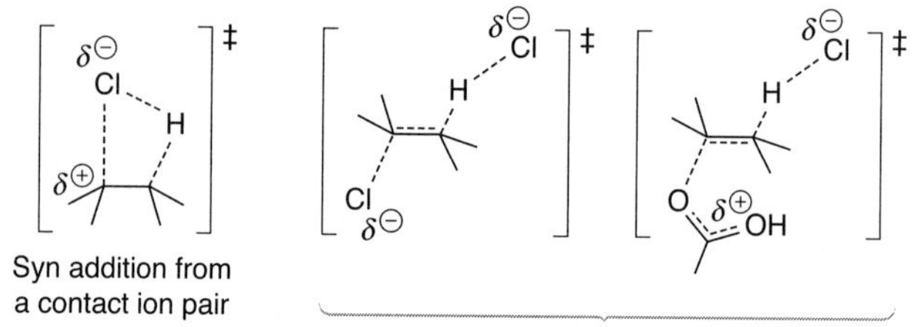

Syn addition from
a contact ion pair

Anti addition occurs from
a termolecular reaction

Connections

Squalene to Lanosterol

The electrophilic additions of water and hydrogen halides all involve nucleophilic attack on a carbenium ion intermediate. Many nucleophiles can add in a similar manner, and in fact, other alkenes can act as nucleophiles and add to carbenium ions, creating another carbenium ion. This is the basis of cationic polymerization (see Chapter 13). Nature exploits the addition of one alkene to another to create complex polycyclic natural products. As shown below, the conversion of squalene to lanosterol involves a cascade of alkene additions. The enzyme squalene oxi-

dase first epoxidizes the 2,3-double bond (step 1). Subsequently, there is an acid-catalyzed opening of the epoxide to initiate the cascade (step 2). After the generation of the final carbenium ion, a number of 1,2-shifts occur which are followed by a deprotonation to give lanosterol. Several more steps not shown ultimately give cholesterol. When squalene epoxide is exposed to acid under conventional reaction conditions, a complex mixture of products is obtained. The enzyme exhibits exquisite control over the carbocation rearrangements in order to obtain the desired product.

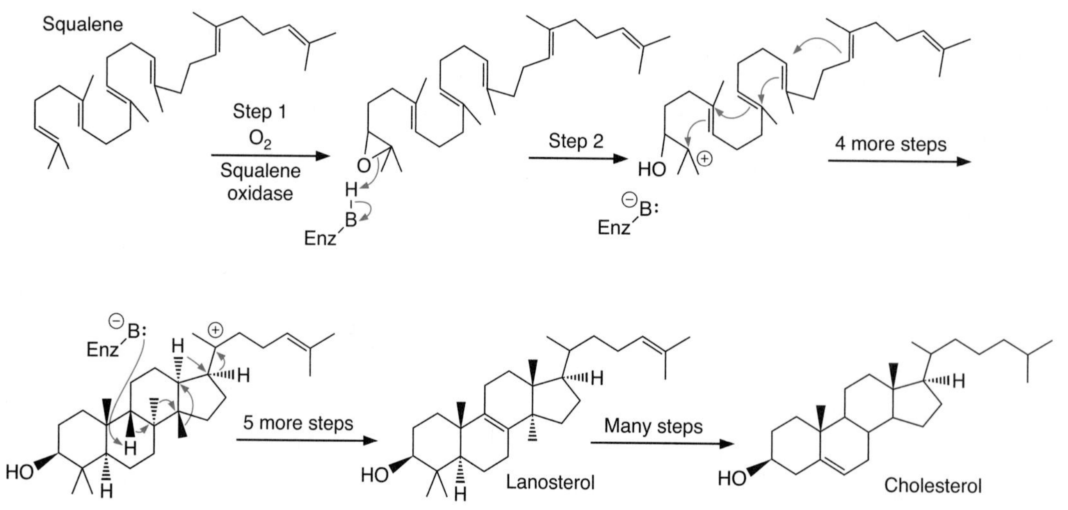

10.4.3   Addition to Alkynes

The addition of HX to alkynes follows a similar mechanism as that for addition to alkenes. The major difference is the fact that a vinyl carbenium ion is formed. The stability of these cations is lower than trigonal sp^2 carbenium ions, and thus the addition of HX is slower than with alkenes. The regiochemistry of addition is the same as with alkenes, in that the halogen attaches to the more substituted carbon. Since halogens can stabilize adjacent carbenium ions via resonance, the addition of a second equivalent of HX places the carbenium ion on the substituted carbon, and therefore geminal dihalides are the major products formed from the addition of two HX molecules (Eq. 10.13).

$$\text{(Eq. 10.13)}$$

Interestingly, because of the lower stability of vinyl cations relative to alkyl carbenium ions, these structures are sometimes not on the reaction's energy surface, and instead concerted reactions occur. For example, the electrophilic addition of HCl to 3-hexyne in acetic acid gives predominately the anti addition product, indicating that protonation occurs simultaneously with nucleophilic attack (Eq. 10.14).

$$\text{(Eq. 10.14)}$$

10.5   Electrophilic Addition of Halogens to Alkenes

10.5.1   Electron Pushing

The mechanism of the addition of halogens (X_2) and the standard electron pushing are again similar to the hydration of an alkene. However, in contrast to the addition of water or HX, which gives carbenium ions, the addition of halogen gives a bridging species, a **halonium ion**, also called a **σ complex**. This difference is due to the ability of a lone pair on a halogen atom adjacent to a carbenium ion to fill the empty carbon p orbital (see margin). Scheme 10.3 shows an example using Br_2. In this example, a relatively low energy empty σ^* orbital on Br_2 allows for the negative region of the alkene to add. The same mechanism is followed with alkynes, but commonly a second equivalent of X_2 adds, creating a tetrahalogenated product.

The reactions are exothermic for F_2, Cl_2, and Br_2. The reaction with F_2 is so exothermic that it is explosive, making it hard to study, but at –78 °C the reaction has been found to be manageable. The addition of Cl_2 to ethylene is exothermic by 44 kcal/mol, while that of Br_2 is exothermic by 29 kcal/mol. The reaction is near thermoneutral or endothermic for I_2, and therefore is readily reversible.

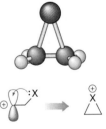

Halonium ion

Scheme 10.3
Electron pushing for the addition of X_2 species.

σ Complex

Vicinal dibromide

10.5.2 Stereochemistry

The classic analysis that implicates the formation of a cyclic halonium ion emphasizes the stereochemistry of the products. With unconjugated alkenes, anti addition is consistently seen. This requires backside attack of the nucleophile, indicating that the "frontside" has been blocked. For example, the addition of bromine to 3-*t*-butylcyclohexene gives trans-dibromo products (Eq. 10.15).

$$\text{(Eq. 10.15)}$$

Major   Minor

10.5.3 Other Evidence Supporting a σ Complex

Several other pieces of experimental data support a mechanism with a σ complex: kinetics, the isolation of certain bromonium ions, and the trapping of the bromonium ion. The kinetics is often first order in both bromine and alkene, with the rate increasing with more electron rich alkenes, and decreasing with electron poor alkenes. For example, vinyl bromide (an electron poor alkene) reacts approximately 10^4 times slower than ethylene. In contrast to the hydration reaction, and in further support of a bridging bromonium intermediate, electron donating groups on *both* carbons of the alkene lead to large rate enhancements, as shown in Table 10.3. If a non-bridging carbenium ion were formed, one would expect substituent effects similar to that seen for alkene hydration (see Section 10.3.3).

Table 10.3
Relative Rates for Addition of Bromine in Methanol with Added NaBr at 25 °C*

$=$	$\diagdown\!\!=$	$\diagdown\!\!=\diagdown$	$\diagdown\!\!=\!\!\diagup$	$\diagdown\!\!\!=\!\!\!\diagup$	$\diagdown\!\!\!=\!\!\!\diagup$	$\diagdown\!\!\!=\!\!\!\diagup$
1	61	1.7×10^3	2.6×10^3	5.4×10^3	1.3×10^5	1.8×10^6

*Dubois, J. E., and Mouvier, G. "Reactivité des Composes Ethyleniques Reaction de Bromation. XVI. Determination Quantitative des Influences Structures des Groups Alcoyles." *Bull. Chem. Soc. Fr.*, 1426 (1968).

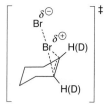

Bromination isotope effect

Isotope effects also support bromonium ion formation. An inverse isotope effect of 0.53 is obtained when comparing the bromination of cyclohexene to cyclohexene-d_{10} (all hydrogens replaced with deuterium, referred to as **perdeuterio**). The hydrogen/deuterium substitution that gives rise to the isotope effect is that on the alkene carbons of the cyclohexene. The inverse nature of the effect indicates rehybridization from sp^2 to sp^3 as expected, for conversion of an alkene to a bromonium ion, but the large value (normal inverse isotope effects are around 0.8) implies significant rehybridization of *both* alkene carbons in the rate-determining step (see margin).

Nucleophiles other than bromide, such as the solvent, can also add to the bromonium ion. The reaction of 1-hexene with bromine in methanol gives the product distribution shown in Eq. 10.16. There is more solvent attack at the secondary carbon than the primary carbon. Recall that secondary carbenium ions are more stable than primary carbenium ions, so a larger fraction of the positive charge in the bromonium ion is on the secondary carbon than the primary carbon.

$$\text{(Eq. 10.16)}$$

3   4.8   1.2

When the carbenium ion can be stabilized in a manner other than with a bridging bromine, the preference for anti addition diminishes. For example, *cis*- and *trans*-1-phenyl-propene give only 73% and 83% products from anti addition, respectively, while the remaining products are from syn addition (Eqs. 10.17 and 10.18). Here, there is an *equilibrium* between two forms of the carbenium ion: a cyclic bromonium ion and an open, resonance stabilized benzyl cation. Syn addition and anti addition products can arise from the open form due to bond rotation, but only anti addition products can arise from the cyclic form.

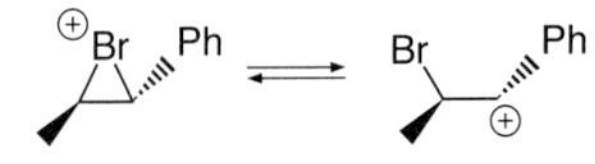

Bromonium–carbenium equilibrium

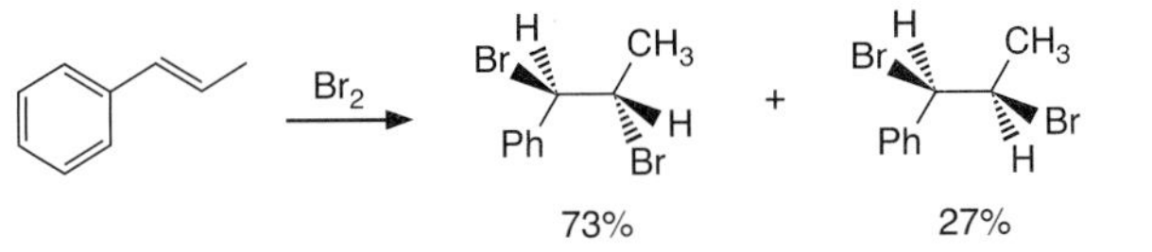

(Eq. 10.17)

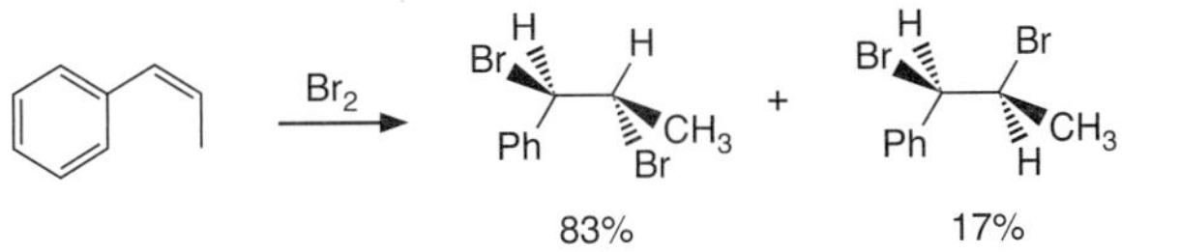

(Eq. 10.18)

Some bromonium ions have even been isolated and fully characterized. This is possible when steric hindrance impedes nucleophilic attack. For example, the structure shown in the margin has been characterized by x-ray crystallography.

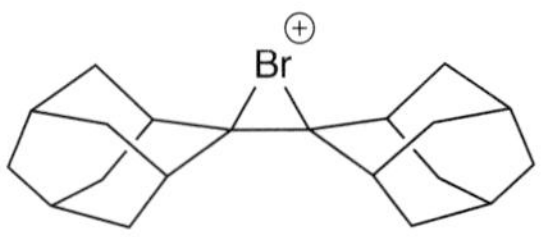

The σ complex is actually not the only intermediate formed in these reactions. The mixing of bromine with alkenes leads to a transient UV/vis absorption band, which indicates the formation of a charge–transfer complex (Eq. 10.19). In the case of the bromination of cyclohexene, the rate of decay of the absorption corresponds to the rate of formation of the bromine addition product. This indicates that the complex is formed rapidly prior to the rate-determining formation of the bromonium ion. Different alkenes have different stoichiometries for the charge–transfer complex, ranging from one Br_2 to as many as three. These higher order stoichiometries indicate that in many cases more than a single Br_2 is involved in creating the bromonium ion, as we discuss below.

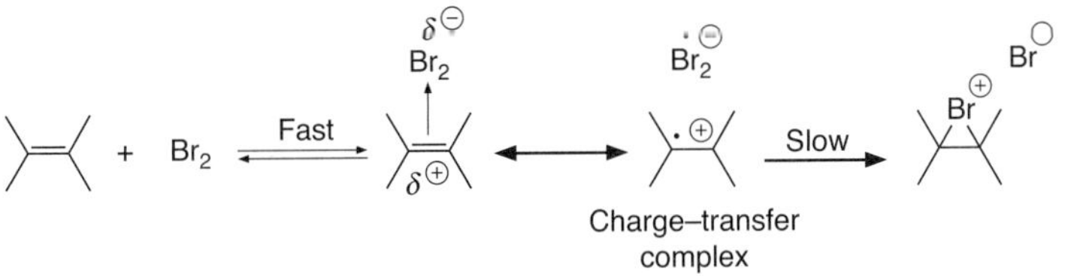

Stable bromonium ion

(Eq. 10.19)

10.5.4 Mechanistic Variants

For many experimental conditions, the mechanism in Scheme 10.3 is a simplification. This mechanism typically occurs only at low concentrations of bromine or in water and alcohol solvents. In solvents of lower polarity, even acetic acid, the reaction is second order in bromine. The second bromine assists the first step by polarizing the bromine that is adding to the alkene, creating Br_3^- instead of Br^- as a leaving group (Eq. 10.20). Similarly, iodinations of alkenes are commonly second order in I_2, and sometimes even third order. Kinetic expressions such as that given in Eq. 10.21 are often observed. The kinetic dependence on Br_3^- indicates a reaction between the alkene and Br_2, as well as Br^-, which is equivalent to Br_3^-.

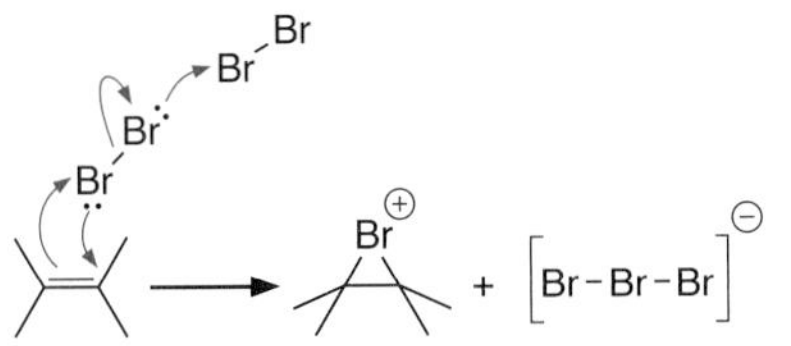

(Eq. 10.20)

$$\frac{d[P]}{dt} = (k[Br_2] + k'[Br_2]^2 + k''[Br_3^-])[alkene]$$

(Eq. 10.21)

Other mechanistic variants occur with the additions of F_2 and Cl_2. Although anti 1,2-addition products dominate the reactions with Cl_2 and Br_2, one also often finds allylic halogenation with Cl_2. Radical pathways occur in these reactions, and so the addition is not typically as clean as with Br_2. In contrast to both Cl_2 and Br_2, the addition of F_2 occurs via a syn addition mechanism. The prevailing data support the formation of a β-fluorocarbocation that rapidly combines with the fluoride counterion before dissociation of the ion pair (Eq. 10.22).

Ion pair

(Eq. 10.22)

10.5.5 Addition to Alkynes

Alkyne-derived bromonium ion

Vinyl cation

Alkyl-substituted alkynes show anti addition products with Br_2, making trans-dibromoalkenes, again supporting a bromonium ion intermediate. For example, 1-hexyne and 3-hexyne give only the trans-alkene products. However, due to the additional ring strain associated with the bromonium ion formed from an alkyne, and because a partial positive charge is placed on a carbon with a higher s-character hybridization than with alkenes, alkynes generally react 10^3 to 10^7 times slower than alkenes.

With substituents on the alkyne that can stabilize a non-bridging carbocation, mixed stereochemistry is found in the products. The reaction of 1-phenylpropyne with Br_2 in acetic acid gives the products shown in Eq. 10.23. The reaction obeys a kinetic expression such as Eq. 10.21 ([alkene] replaced by [alkyne]), and a ρ value of –5.17 for para substituents on the phenyl ring was found. The data support the intermediacy of a vinyl cation (see margin) that can combine with both bromide or solvent to form either E or Z products.

(Eq. 10.23)

Interestingly, the kinetic term in the reaction of 1-phenylpropyne with Br_2 that has a Br_3^- dependence is only observed when bromide anion is added. Furthermore, the reaction that corresponds to this term gives solely the anti addition product. This supports a concerted mechanism with nucleophilic attack by bromide simultaneous with electrophilic addition of bromine (Eq. 10.24).

(Eq. 10.24)

10.6 Hydroboration

In synthetic organic textbooks, many other electrophiles are covered as reagents for transforming alkenes into a variety of different functional groups. The acid-catalyzed addition of water or HX is seldom used in a synthetic sequence, because many functional groups undergo reactions in the presence of strong acids. Therefore, these reactions were covered primarily because they nicely highlight features of organic mechanisms and reactivity. The hydroboration of double bonds is a reaction that is both instructive for analyzing mechanisms and synthetically very relevant. This reaction leads to the addition of a hydride and a boron across a double bond, leaving the boron available for further derivitization, most often to an alcohol.

10.6.1 Electron Pushing

Borane (BH_3) and its organic derivatives (RBH_2 or R_2BH) will add to an alkene (Scheme 10.4). BH_3 dimerizes in the gas phase to make diborane. In solution reactions, however, this dimer is easily broken up (Eq. 10.25), especially if the reaction is occurring in a donor solvent such as THF, which produces a Lewis acid–base complex, $BH_3 \bullet THF$.

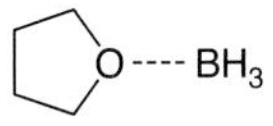

BH$_3$·THF complex

$$\text{H}_2\text{B}\text{-H-BH}_2\text{-H-BH}_2 \rightleftharpoons 2\,\text{H-BH}_2 \qquad \text{(Eq. 10.25)}$$

Scheme 10.4
Electron pushing for the addition of BH_3 to an alkene.

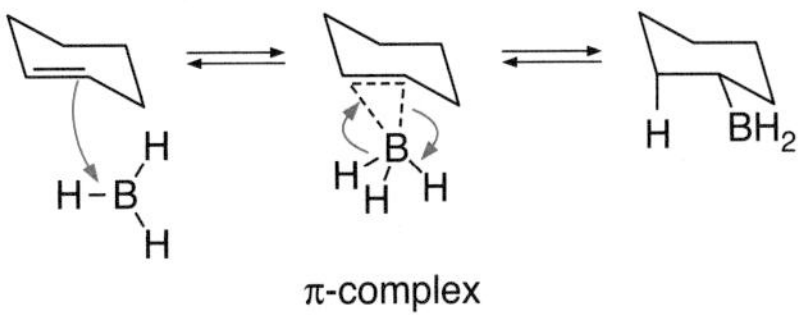

π-complex

The formation of a **π complex** between the olefin and BH_3 is the first step, which is viewed as a Lewis acid–base interaction. Concerted addition of the hydrogen and boron occurs in a second step, leading to syn addition. Commonly at this stage, addition of a peroxide such as H_2O_2 results in an alcohol at the position of the boron.

10.6.2 Experimental Observations

The addition of a borane to an alkene leads to an anti-Markovnikov product. The boron preferentially adds to the less hindered carbon, which occurs for both steric and electronic reasons. The adding hydrogen is the smallest group, so it makes sense that it would add to the more hindered position. Reinforcing this effect is the fact that the hydrogen is the nucleophilic portion of the B–H bond, so it adds to the carbon where the most positive charge resides within the initially formed π complex. This is also the more substituted carbon. One example is given in Eq. 10.26. Alkynes also undergo hydroboration, and again give syn addition and anti-Markovnikov regiochemistry.

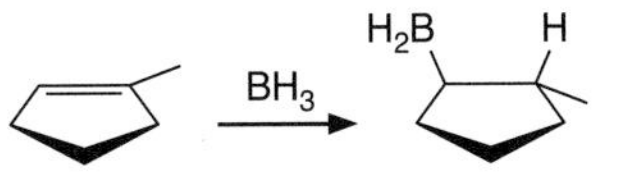

$$\qquad \text{(Eq. 10.26)}$$

Kinetic studies of hydroboration are difficult to perform because all three B–H bonds can add to an alkene. Despite this, studies have shown most alkenes react at very similar rates. When more than one alkene is present in a reactant, terminal alkenes are preferred, presumably due to steric considerations. In many cases the reaction is zero order in alkene. The lack of large substituent effects on the rate and the zero-order dependence on the alkene indicate that the dissociation of the dimer or the $BH_3 \bullet$ solvent complex to monomeric BH_3 is the rate-determining step. After the dissociation, the mechanism shown in Scheme 10.4 commences.

10.7 Epoxidation

Peroxides and peracids are electrophilic in a manner similar to halogens (X_2). This can be rationalized by appreciating that the highly electronegative oxygen typically has a partial negative charge when it is attached to any element except oxygen or fluorine. So, relative to a typical oxygen, the oxygens of a peroxide are positive, and thus electrophilic. However, direct nucleophilic attack on an O–O bond of H_2O_2 or ROOR seldom occurs, because hydroxide and alkoxides are poor leaving groups. A carboxylate or carboxylic acid is a much better leaving group, and so peracids (RCO_2OH) do react readily with nucleophiles, even nucleophiles as weak as alkenes. This leads to an epoxide product. Epoxidations are key reactions in organic synthesis, and asymmetric variants are especially important (see the Connections highlight on page 558).

10.7.1 Electron Pushing

The epoxidation of an alkene by a peracid is a single-step reaction. Several arrows are used to depict the reaction, as shown in Scheme 10.5. The alkene directly attacks the backside of the O–O bond in such a fashion that a carboxylic acid is the leaving group.

Scheme 10.5
Electron pushing for the single-step epoxidation reaction.

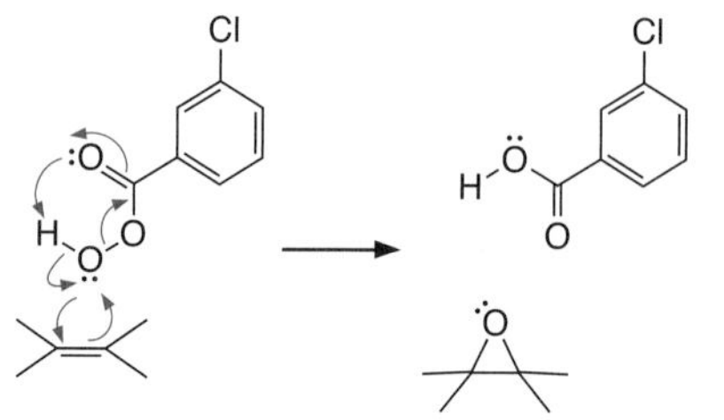

10.7.2 Experimental Observations

If there is only one step, the reaction has to be second order; first order in the peracid and first order in the alkene. The reaction rate has very little dependence upon the solvent, supporting a concerted mechanism with little charge developing at the transition state. The small charge development is also supported by the fact that the rates correlate with a Hammett parameter (σ^+), but the ρ value is only –1.1 for p-XArCH=CH$_2$. There are only small primary kinetic isotope effects. Values of k_H/k_D around 1.1 to 1.2 are found for the peracid [$RO_3H(D)$]. This means that the hydrogen atom transfer shown in the electron pushing of Scheme 10.5 has to be either minimal at the transition state or almost complete (see Section 8.1.2). Secondary deuterium isotope effects on the alkene carbons are inverse, as may be expected for an sp^2 to sp^3 transformation.

It is relatively straightforward to predict the stereochemistry and rates of these reactions. The more electron rich the double bond, the faster it will react with the peracid. Also, sterics are the primary factor directing the epoxidation stereochemistry. The least hindered face of a double bond is predominately epoxidized. The reaction is typically 100% stereospecific, with trans-olefins giving trans-epoxides and cis-olefins giving cis-epoxides.

When the electrophilic oxygen of the peracid is equally bonded to each alkene carbon at the transition state, the reaction follows what is known as the **butterfly mechanism**. Evidence suggests that this only occurs for symmetric alkenes like ethylene. When one carbon of the alkene is substituted, the bonding to the electrophilic oxygen is more advanced at the unsubstituted carbon, placing slight positive charge at the more substituted carbon. For example, the k_H/k_D values for styrene at the substituted and unsubstituted carbons are 0.99 and 0.82, respectively. This indicates that rehybridization at the transition state is minimal for the phenyl substituted carbon (see margin).

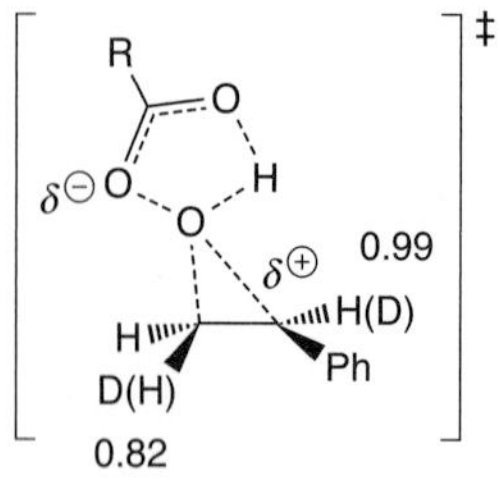

Unsymmetrical transition state

10.8   Nucleophilic Additions to Carbonyl Compounds

Much of the chemistry described above required electrophilic activation of the carbonyl or alkene in order for a nucleophile to add. However, many good nucleophiles will directly add to a carbonyl, and we commence this section by looking at the electron pushing for a few examples traditionally covered in introductory organic chemistry.

10.8.1 Electron Pushing for a Few Nucleophilic Additions

The classic and most straightforward example of a nucleophilic addition to a carbonyl is cyanohydrin formation (Scheme 10.6). Cyanide is a good nucleophile, and it can add directly to the carbonyl carbon, as predicted by the EPS of ketones given in Appendix 2. In a second step the resulting anion is protonated.

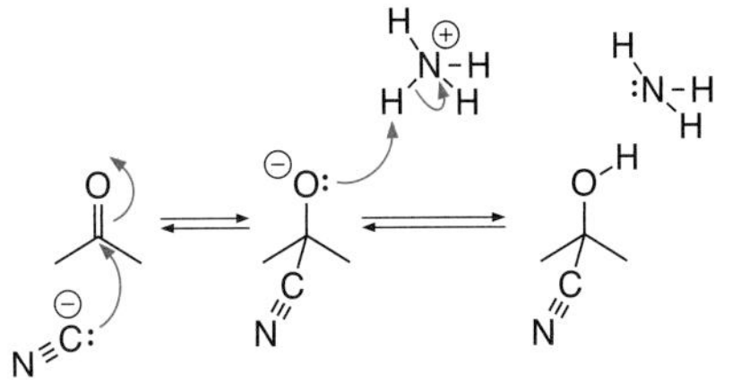

Scheme 10.6
Electron pushing for cyanohydrin formation.

The addition of a Grignard reagent to a ketone leads to attachment of the R group of the Grignard reagent (Scheme 10.7) to the carbonyl carbon. Although the exact structure of the reactive species in a Grignard addition is still debated and depends on the reaction conditions (see below), there is an R–Mg bond that is highly polarized, with considerable negative character on carbon, making it a very good nucleophile.

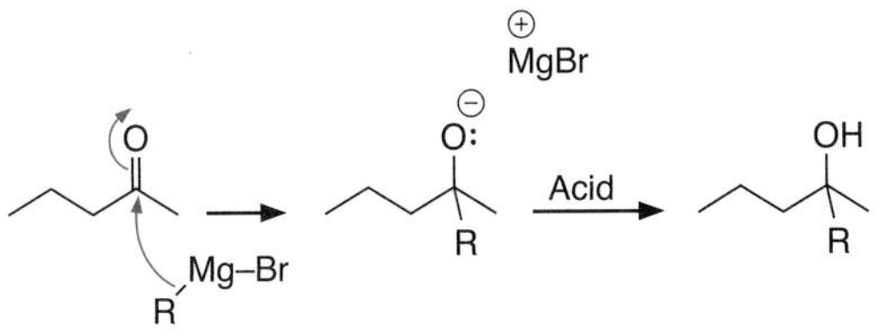

Scheme 10.7
Classic electron pushing for a Grignard addition to a ketone.

Another standard example is the nucleophilic attack by lithium aluminum hydride (LAH) on a ketone or aldehyde (Scheme 10.8). Each hydrogen in $LiAlH_4$ is partially negatively charged, and therefore the Al–H σ bonds are nucleophilic. After the nucleophilic attack by hydride, the resulting alkoxide anion coordinates with the Al species. These two steps are repeated three more times. Finally, dilute acid is added to supply a proton to the alkoxide anion (electron pushing not shown).

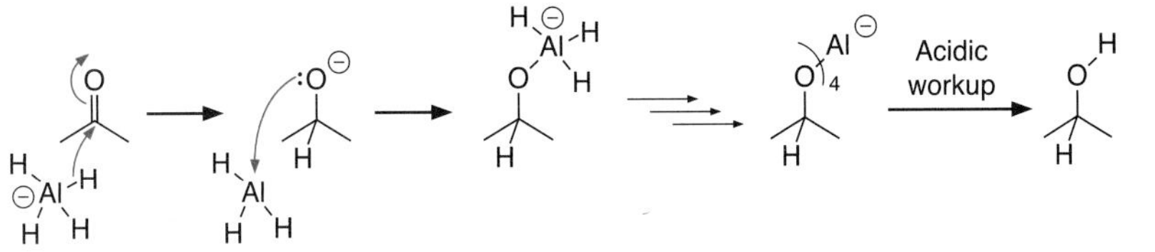

Scheme 10.8
Electron pushing for the LAH reduction of a ketone.

The overall effect is the addition of H_2 across the C–O π bond, and this is defined as reduction. Generally, oxidation and reduction are defined as the removal and addition of electrons to a compound, respectively. Here, **oxidation** and **reduction** are defined as elimination and addition of dihydrogen, respectively. Reduction using hydride reagents formally adds H_2 across the carbonyl, but it proceeds mechanistically as a nucleophilic addition. It is interesting to note that nature has an analog to LAH, called NADH. It also delivers a hydride to carbonyl groups (see the Connections highlight on page 566).

Connections

Mechanisms of Asymmetric Epoxidation Reactions

The epoxide functional group is often an intermediate in organic synthesis sequences because it can be readily opened by a variety of nucleophiles, leading to complex functionalization that starts with just an alkene. Asymmetric variants of epoxidation are important reactions, with the Sharpless and Jacobsen epoxidations being the most well known. Both these reactions involve metal catalysts.

The Sharpless epoxidation involves a dimeric titanium(IV) species with bridging tartrate ligands (**A**). The tartrates impart a chiral environment to the catalyst centers, leading to asymmetric induction. The dimeric species shown below, involving several alkoxide ligands, has been shown via an ^{17}O NMR study to be the active form of the catalyst. This structure undergoes rapid ligand exchange reactions, incorporating both an allylic alcohol and a peroxide onto a single metal center. Metallation of the peroxide (HOOR to L_nTiOOR) enhances the electrophilicity of the peroxide. The electron-pushing scheme shows how the OR group can become a good leaving group via transfer to the titanium. Small secondary inverse deuterium isotope effects (0.93 to 0.99) of roughly the same magnitude for both carbons of the double bond are found. This indicates nearly equal bond changes to the alkene carbons at the transition state, supporting the concerted epoxidation shown.

The Jacobsen epoxidation reaction uses an Mn–salen complex (**B**). The mechanism has been examined using many of the tools presented in Chapters 7 and 8. Two pathways have been proposed. Path A is a radical pathway and Path B involves a metallaoxetane. Path A is supported by several pieces of data. First, the reaction is enhanced by the addition of *N*-oxides, which supports the replacement of the chloride by these ligands and subsequent activation of the metal. A metallooxetane (Path B) would be extremely crowded at the metal center with a coordinated

N-oxide. Recent computational studies find the metallooxetane structure to be too high in energy to be a reasonable intermediate. In homogeneous solution linear Erying plots are found for styrene, indene, and cyclooctadiene. This supports Path A and argues against a mechanism with an equilibrium formation of a metallooxetane prior to rate-determining epoxide formation (as in Path B).

One might expect radical rearrangements in Path A, but they are not observed. In the epoxidation reaction (**C**), which incorporates the very rapid "phenylcyclopropyl clock" (see Section 8.8.8), no ring-opened products were observed, implying no free radical involvement. However, this observation falls into a class of experimental results known as **negative evidence**. Negative evidence is the failure to observe a possible result. It does not prove that an expected phenomenon is not occurring, just that it was not observed with the tools at hand. If one observed the expected rearrangement —positive evidence—it *would* support the existence of radicals. On the other hand, the lack of a rearrangement (or any observation in general) should not be used to negate a mechanism. In this case, the lack of a rearrangement can be explained if the radical closes to the epoxide with a rate constant faster than ring opening.

Finn, M. G., and Sharpless, K. B. "Mechanism of Asymmetric Epoxidation. 2. Catalyst Structures." *J. Am. Chem. Soc.*, **113**, 113–126 (1991). Finney, N. S., Pospisil, P. J., Chang, S., Palucki, M., Konsler, R. G., Hansen, K. B., and Jacobsen, E. N. "On the Viability of Oxametallacyclic Intermediates in the (Salen)Mn-Catalyzed Asymmetric Epoxidation." *Angew. Chem. Int. Ed. Eng.*, **36**, 1720 (1997). Linde, C., Arnold, M., Norrby, P.-O., and Akermark, B. "Is There a Radical Intermediate in the (Salen)Mn-Catalyzed Epoxidation of Alkenes?" *Angew. Chem. Int. Ed. Eng.*, **36**, 1723–1725 (1997). Linde, C., Akermark, B., Norrby, P.-O., and Svensson, M. "Timing Is Critical: Effect of Spin Changes on the Disastereoselectivity in Mn(salen)-Catalyzed Epoxidation." *J. Am. Chem. Soc.*, **121**, 5083 (1999). Cavallo, L., and Jacobsen, H. "Radical Intermediates in the Jacobsen–Katsuki Epoxidation." *Angew. Chem. Int. Ed. Eng.*, **39**, 589 (2000).

A.

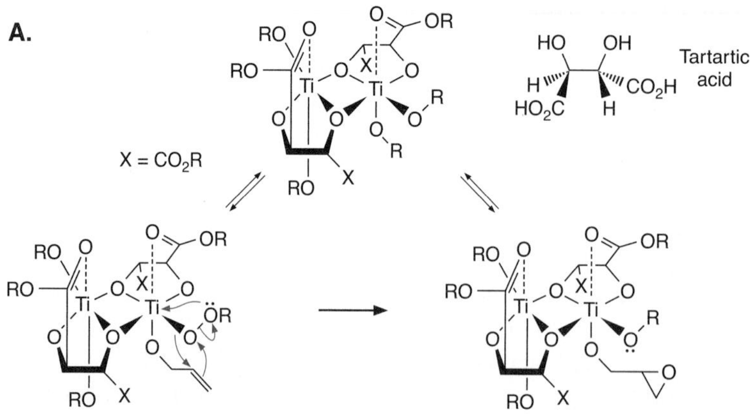

X = CO$_2$R

Tartaric acid

Sharpless epoxidation

B.

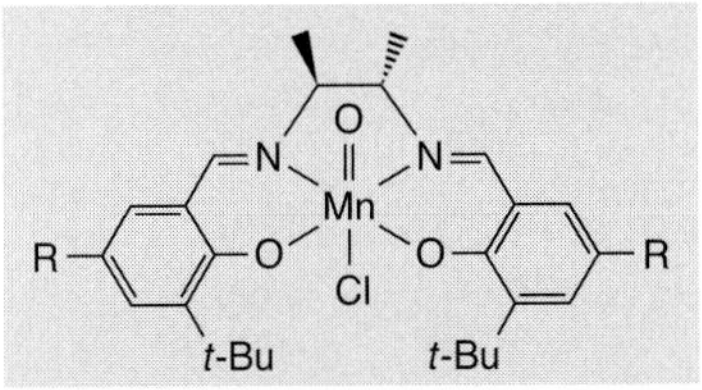

Mn-salen complex

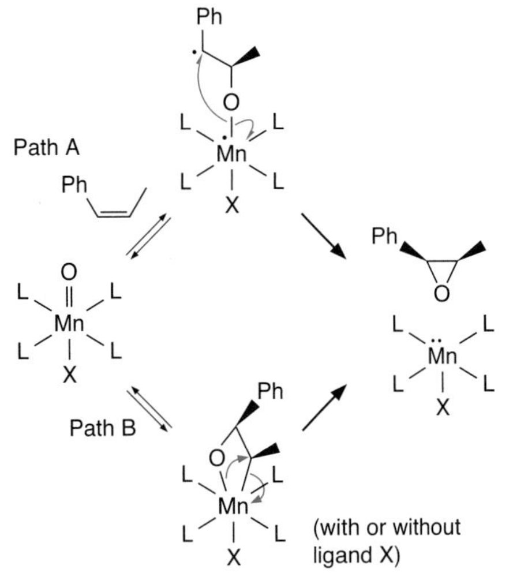

Jacobsen epoxidation

Metallaoxetane

(with or without ligand X)

Path A

Path B

C.

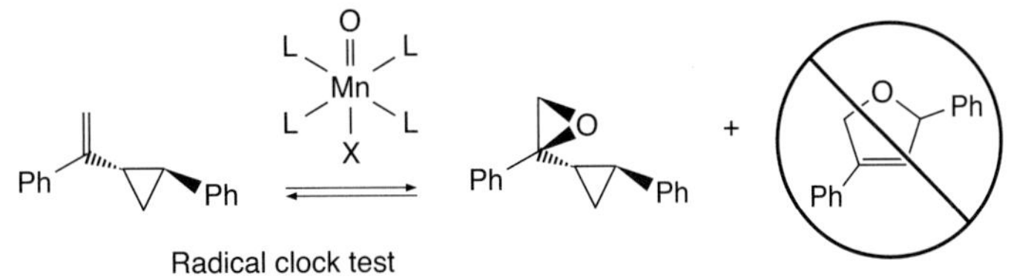

Radical clock test

10.8.2 Experimental Observations for Cyanohydrin Formation

Table 10.4 shows a few equilibrium constants for the addition of HCN to various carbonyl compounds. The trend is the same as for the extent of hydration (Table 10.2). One typically performs this reaction with a cyanide salt and an acid that is strong enough to protonate the tetrahedral intermediate, but not strong enough to protonate cyanide. In this manner, the lethal gas HCN is not produced.

**Table 10.4
Equilibrium Constants for Formation of Cyanohydrins***

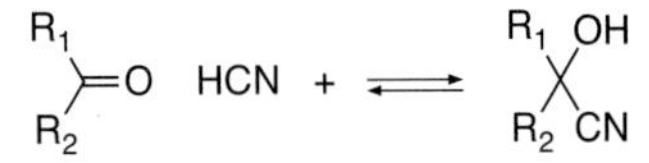

R_1	R_2	K
C_6H_5	H	2.2×10^5
C_6H_5	CH_3	0.77
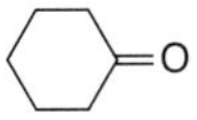 cyclohexanone		1.0×10^3

*Baker, J. W. "Substituent Effects on Hyperconjugation: Hyperconjugation Energy for Groups of the Type CH$_2$X." *Tetrahedron*, **5**, B5 (1959). Prelog, V., and Kobelt, M. "Über die Abhangigkeit der Dissoziationskonstanten der Ringhomologen Cyclanon-Cyanhydrine von der Ringgrösse." *Helv. Chim. Acta*, **32**, 1187 (1949).

The kinetics of HCN addition under these conditions is first order in cyanide anion and in the carbonyl-containing reactant, and the mechanism given in Eq. 10.27 is operating. The fact that cyanide is a very good nucleophile means that protonation is not required prior to or at the rate-determining step, and therefore no kinetic dependence on the acid is observed. Kinetic dependence upon the acid would only become evident with exceedingly low concentrations of acid. Hence, there is no specific-acid or general-acid catalysis.

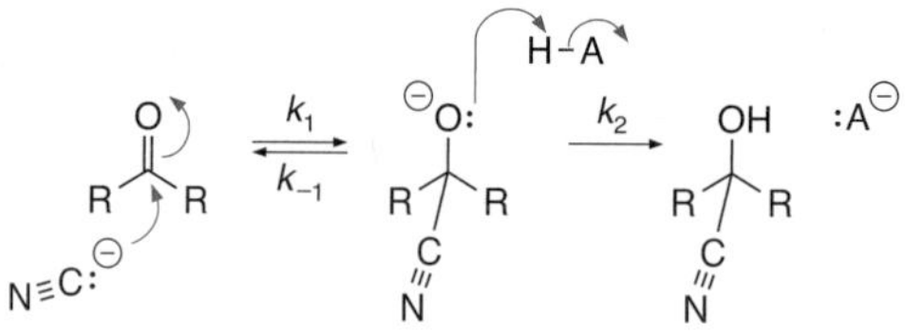

(Eq. 10.27)

10.8.3   Experimental Observations for Grignard Reactions

The addition of alkyl groups to carbonyl compounds is prototypically accomplished with the use of Grignard or organolithium reagents. The reactions are extremely fast, often over in milliseconds, even at –85 °C.

The mechanism of Grignard addition to carbonyls has been studied in detail, and Ashby is a leader in this regard. It is not always what you might expect. The electron pushing given in Scheme 10.7 indicates that the C–Mg bond acts as a two-electron nucleophile toward polarized bonds. This is acceptable electron pushing, in that it keeps track of the rearrangement of bonds, and in most cases, such as with the addition of CH_3MgBr to acetone, it is essentially correct. For some carbonyl compounds, however, Scheme 10.7 is a poor representation of the actual mechanism. Instead, an electron transfer mechanism operates (Figure 10.5 **A**). Here, the electron rich C–Mg bond acts to reduce the carbonyl group, giving a **ketyl anion**. Combination of the resulting radicals in a cage gives the product. Because a ketyl anion is an intermediate, carbonyl structures that lead to stabilized ketyl anions will favor this mechanism, such as conjugated enones, phenyl ketones, and phenyl aldehydes. As evidence for the mechanism, the radicals often escape from the cage, giving hydrogen abstraction products or radical coupling products (Figure 10.5 **B**). This is a case where detailed product analysis provides insight into the mechanism.

Other complications arise due to the fact that RMgX reagents are in equilibrium with R_2Mg and MgX_2 species, so that the exact entity that is adding to the carbonyl is sometimes ambiguous. Furthermore, kinetic studies indicate second-order dependence upon the Grignard or lithium reagent in certain cases, indicating that a mechanism involving a cyclic transition state as shown in the margin could be operative.

Proposed Grignard mechanism

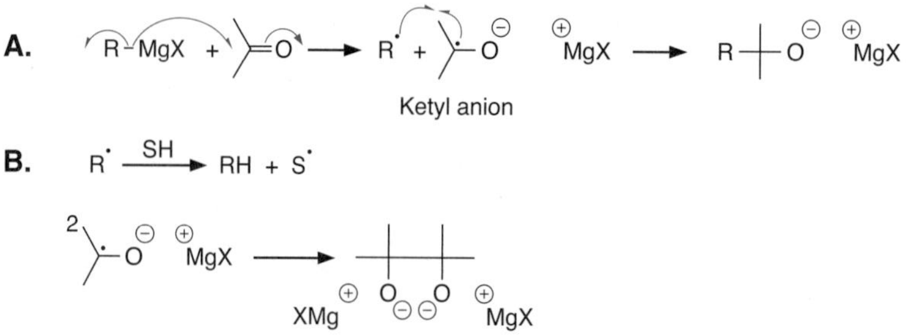

Figure 10.5
A. A common electron transfer mechanism for Grignard additions.
B. Side reactions that are indicative of this radical mechanism.

10.8.4 Experimental Observations in LAH Reductions

The mechanism given in Scheme 10.8 is a simplification for LAH reductions. In reality, complexities arise due to aggregation of the hydride reagents and ion pairing in ether solvents. Furthermore, participation by the counterion to the aluminum hydride is essential. For example, the rates of these reactions are different for LiAlH$_4$ and NaAlH$_4$. The $\Delta S^{\ddagger}$ values for the reactions indicate significantly more ordering in the transition state using LiAlH$_4$ than with NaAlH$_4$. Interestingly, if a Li-specific cryptand (see Section 4.2.1) is added to the reaction, no product is formed (Eq. 10.28). However, addition of a lithium salt will restore activity. These observations imply that the lithium cation is intimately involved in the reaction, a facet that is not displayed in the common arrow-pushing scenario given in Scheme 10.8. The lithium ion acts as a Lewis acid coordinating to the carbonyl oxygen prior to hydride addition. This polarizes the carbonyl, facilitating nucleophilic addition, and leads directly to an ion pair that stabilizes the resulting alkoxide anion (Eq. 10.29). This electrophilic activation of the carbonyl carbon by coordination to lithium is analogous to the action of a Brønsted acid in either specific-acid or general-acid catalysis.

$$\text{(Eq. 10.28)}$$

$$\text{(Eq. 10.29)}$$

10.8.5 Orbital Considerations

In Chapter 1 we emphasized that polar covalent bonds have an unequal distribution of electrons. This is reflected in unequal contributions of the atomic orbitals in the bonding and antibonding molecular orbitals, as predicted from perturbational molecular orbital theory. The bonding orbital has more character on the more electronegative atom, whereas the antibonding orbital has more character on the less electronegative atom. The acceptor orbital on the carbonyl of a ketone or aldehyde is the antibonding π^* orbital (Figure 1.17 and Appendix 3). The majority of its character is on the carbon, which is exactly where this molecule is electrophilic and slightly positive. Interaction of the nucleophile at the carbon rather than the oxygen leads to the greatest bond overlap in the combination of the nucleophile and electrophile.

The Bürgi–Dunitz Angle

What is the trajectory of nucleophilic attack on a carbonyl carbon? This trajectory will determine any steric effects that can influence stereochemistry or regiochemistry. The direction of attack is given by what is known as the **Bürgi–Dunitz angle**. In a creative use of experimental data, Dunitz and Bürgi studied the crystal structures of a large number of compounds that contained a carbonyl and a nucleophile. The geometric relationship between the two was not random, but instead showed a definite pattern, especially as the nucleophile got close to the carbonyl. By considering a large number of structures, the trajectory of attack was essentially mapped out by the crystallographic data.

The direction of attack was found not to be perpendicular to the plane formed by the carbonyl group, but rather to be approximately 105° (see margin). This approaches the angle for an sp^3 hybridized system. While the Bürgi–Dunitz angle was developed from a collection of experimental data, it has been completely confirmed by high level computational studies, which clearly show that nucleophiles approach carbonyls along an obtuse angle. As an ex-

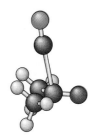

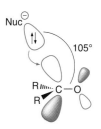

Bürgi–Dunitz angle

ample, a ball and stick model of the transition state for the addition of cyanide to acetone is shown in the margin on page 561, where the obtuse angle is clear.

The same angle of attack is seen for addition reactions to an alkene. The reason for the Bürgi–Dunitz angle is that the adding nucleophile needs to attack the empty π^* orbital of the C=O or C=C bond. Looking back at Sections 1.3.2 and 1.3.3, or in Appendix 3, we find that the character of this orbital on C points back and away from the bonding region.

Orbital Mixing

The stabilization obtained on mixing a filled and an empty orbital will be largest when the energy gap between the interacting orbitals is smallest (Rule 10 of Table 1.7; also see Chapter 14, Eq. 14.61). This predicts that the difference in reactivity between two compounds with the same reactant can be obtained by simply comparing the energies of the interacting orbitals. This approach can thus be used to predict relative reactivity. To do this, let's contrast nucleophilic addition to a C–C π bond vs. a C–O π bond. Figure 10.6 shows a QMOT model of nucleophilic addition.

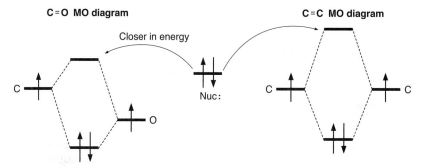

Figure 10.6
The molecular orbital diagrams for a C–C π bond and a C–O π bond on the same energy axis, along with the energy of a lone pair donor orbital. The donor orbital is closer in energy to the acceptor orbital of the C–O π bond.

Remember that the substitution of a heteroatom such as oxygen for carbon lowers the energy of *all* orbitals, both the π and π^*. In the case of the lone pair orbital on the nucleophile filling either the alkene or carbonyl π^* orbital, the interaction is best with the carbonyl, because the energies of the interacting orbitals are closer. The carbonyl is a better acceptor, because its empty orbital is lower in energy and closer to the energy of the donor orbital.

10.8.6   Conformational Effects in Additions to Carbonyl Compounds

The addition of nucleophiles to carbonyl compounds is often found to occur faster with six-membered ring cyclic ketones than with acyclic ketones or cyclopentanones. For example, the LAH reduction of cyclohexanone occurs nearly 300 times faster than for acyclic ketones. A conformational effect accounts for the increased reactivity (Figure 10.7). In cyclohexanone, the dihedral angle between the equatorial hydrogen and the carbonyl oxygen is only 4°. This near eclipsing interaction produces a conformational strain of around 4 kcal/mol that raises the ground state energy of cyclohexanones relative to acyclic systems. Upon nucleophilic attack, this near eclipsing interaction is relieved, but we introduce a 1,3-diaxial interaction with an oxygen anion. However, the diaxial interaction is estimated to be only 0.7 kcal/mol destabilizing (the *A* value for an OH group), and so the net effect is that a significant amount of strain has been released in this reaction. This makes the cyclohexanone reaction significantly faster.

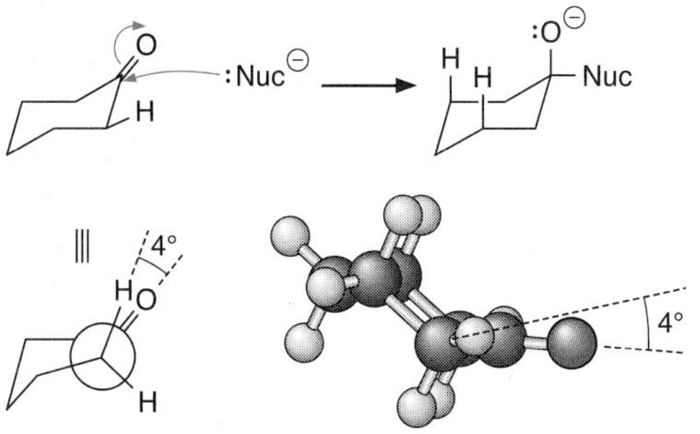

Figure 10.7
Torsional strain in cyclohexanone.

10.8.7 Stereochemistry of Nucleophilic Additions

One of the most important aspects of nucleophilic addition to carbonyl compounds is the stereochemistry of these additions. Controlling the stereochemistry can lead to diastereoselective syntheses. Although a complete analysis of this topic is more appropriate for a textbook on synthetic organic chemistry, a few examples are instructive here, because they nicely tie together several of our structure–reactivity principles.

Addition reactions of a nucleophile such as a hydride or Grignard reagent to a carbonyl adjacent to a stereocenter are stereoselective (see Section 6.9 for the definition). Many chemists have provided rationalizations for the observed products.

Three common models can be summarized as follows. One assigns the descriptors small (S), medium (M), and large (L) to the three different groups on the stereocenter adjacent to the carbonyl based upon the relative steric sizes of these entities. Recall that steric size is highly dependent upon the context, but we have given A values (Chapter 2) and Taft parameters (Chapter 8) as two examples, and usually a clear assignment can be made. The lowest energy transition state for nucleophilic addition has the lowest steric strain. Hence, we examine both the strain in the conformation around the carbonyl-to-stereocenter bond and the strain introduced during addition of the nucleophile.

In the **Cram model**, the only conformer considered is that which places the carbonyl oxygen gauche to the small group and medium group. Nucleophilic attack preferentially occurs on the diastereotopic face of the carbonyl that has the smaller group (Figure 10.8 **A**). An example is given in Eq. 10.30.

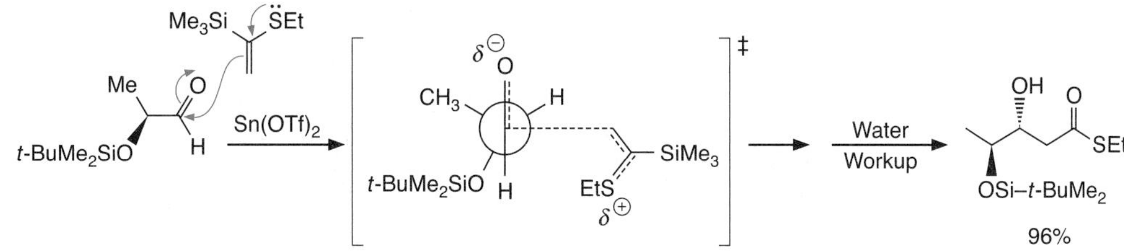

(Eq. 10.30)

In the **Karabatsos model**, the medium-sized group is placed eclipsed with the carbonyl oxygen. Nucleophilic attack is again predicted to preferentially occur from the diastereotopic face that has the smaller group (Figure 10.8 **B**). Note that the Cram and Karabatsos methods lead to the same predictions. However, the Karabatsos model has been found to more accurately predict product ratios.

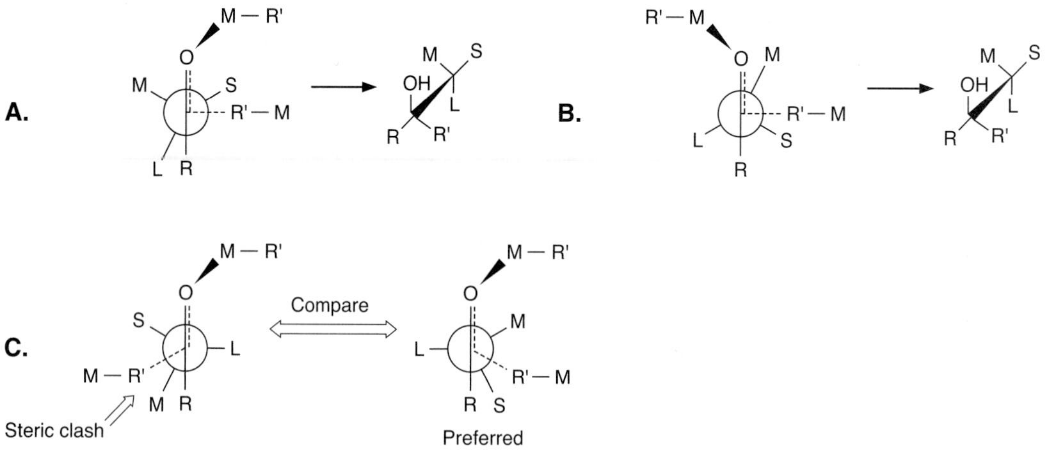

Figure 10.8
A. Cram's model, **B.** Karabatsos' model, and **C.** the Felkin–Ahn model for nucleophilic addition to a carbonyl.

The **Felkin–Ahn model** is the most sophisticated (Figure 10.8 **C**). Here the large group is placed approximately antiperiplanar to the approaching nucleophile. Of the two conformations with the L group antiperiplanar to the nucleophile, one has the S group near the carbonyl R group, whereas the other has the M group near the R group. Taking into account the Bürgi–Dunitz angle for attack, we can see that the former arrangement is preferred (Figure 10.8 **C**), again giving the same result as the other two models. In a variant of the model, if there is a low-lying σ^* orbital that can accept electrons (such as a C–X, C–N, or C–O bond) from the developing C–Nuc σ orbital, then that group is placed antiperiplanar to the approaching nucleophile. Then the same conclusions as to the preferred attack near an S or M group are drawn. Eq.10.31 shows an example of a reaction that obeys the Felkin–Ahn prediction. Here, the NBn$_2$ group (Bn=benzyl) is placed anti to the approach of the nucleophile, and the nucleophile is placed near the S group of the stereocenter (H in this example).

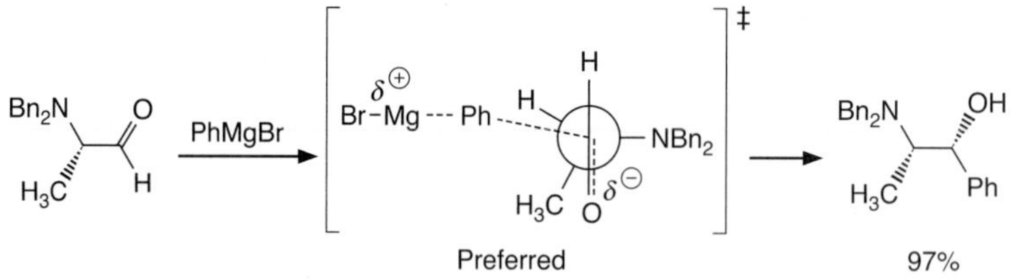

(Eq. 10.31)

Other stereochemical issues can arise with cyclic carbonyl structures. The analysis can be subtle, and significant variations are seen. Take 4-t-butylcyclohexanone, for example. Recall that the A value for t-butyl is 4.7–4.9 (Table 2.14), so 4-t-butylcyclohexanone is essentially 100% in the conformation shown in Eq. 10.32. Reduction using LiAlH$_4$ gives 90% of the trans isomer as shown, making the reaction stereoselective. Recalling the trajectory of approach implied by the Bürgi–Dunitz angle, we might expect that the axial hydrogens on the 3-positions would deflect the approach of any reducing agent or nucleophile. Therefore, the hydride nucleophile has added from the more sterically hindered face of the ketone. Indeed, when groups other than hydrogen are placed axial at the 3-position, the preferential face for attack of the reducing agent changes (the cis isomer dominates). In general, we also find that larger nucleophiles give preferential attack on the other diastereotopic face of the carbonyl; only small nucleophiles prefer the trajectory past the axial hydrogens.

(Eq. 10.32)

Developing gauche strain

To explain the experimental observations, recall that all strains introduced in a transition state need to be considered. With small groups such as hydrogens on the axial positions, little steric strain is introduced during formation of the trans isomer compared to when larger groups are present. However, there are other strains introduced during formation of the cis isomer in all cases. These are the gauche torsional strains between the approaching reducing agent and the axial hydrogens on the 2-positions. There are also developing interactions between the axial hydrogens on the 3-positions with the newly forming alkoxide anion. Thus, there is a delicate balance between steric effects, depending upon the trajectory of approach.

Developing 1,3-diaxial strain

Orbital reasoning has also been invoked to explain the stereochemistry of such addition reactions. The general term given to one particular orbital explanation is the **Cieplak effect**. The Cieplak effect predicts that nucleophilic addition to carbonyl groups or alkenes will occur syn to electron withdrawing groups and/or anti to electron donating groups. In the case of reduction of 4-t-butylcyclohexanone, the axial C–H's labeled (a) are the electron donating groups, and attack is preferentially anti to these bonds (Eq. 10.32).

But why is there a trend related to electron donating and withdrawing groups? This is best explained by first looking at a few more carbonyl addition reactions. Eq. 10.33 shows a schematic of an addition reaction to a substituted adamantanone. Nucleophilic attack occurs preferentially syn to the electron withdrawing fluorine. This is rationalized by invoking a donor–acceptor interaction of the sort discussed in Section 2.4.2. The newly forming C–Nuc bond in the transition state is creating a σ^* orbital that is very low in energy, because the bond is not yet fully formed and full orbital mixing has not occurred. It will be favorable to mix this low-lying empty orbital with a good donor, and recall that an anti relationship between the donor and the acceptor is preferred. In the adamantanone system, C–C bonds will be the donor (see center structure of Eq. 10.33). The electron withdrawing fluorine substituent makes the C–C bonds near it poorer donors by lowering their energies. So, the nucleophile adds anti to the C–C bonds that are further away from the fluorine, and hence syn to the fluorine. The general trend is shown in the margin. When X = an electron withdrawing group (EWG) or an electron donating group (EDG), the indicated stereochemistries of attack are observed.

Preferred if X = EWG

Preferred if X = EDG

(Eq. 10.33)

Let's now return to the attack on 4-t-butylcyclohexanone. The C–H bonds labeled (a) in Eq. 10.34 can similarly donate toward the developing σ^* orbital as the hydride nucleophile adds anti to them. The alternative approach can be stabilized by C–C bonds aligned with the developing σ^* orbital, but the alignment is not as favorable.

C–H bond aligned to donate
to the developing C–H σ^* orbital

(Eq. 10.34)

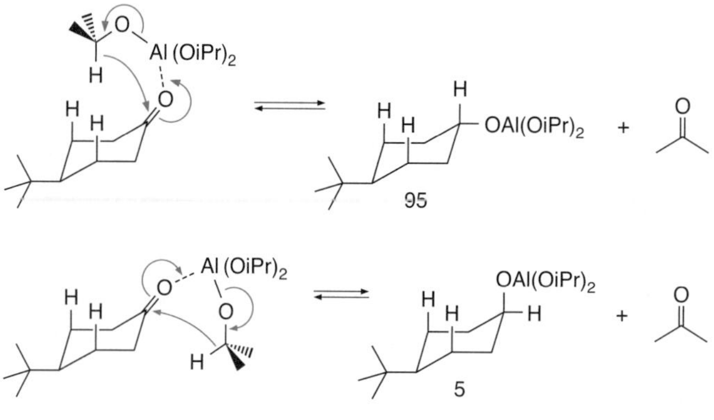

Figure 10.9
Reversible reductions lead to similar product ratios as irreversible
reactions on locked cyclohexanone rings.

The 9:1 ratio obtained in the reduction of 4-*t*-butylcyclohexanone with LAH (Eq. 10.32) is a kinetic result, because the hydride addition is irreversible. Yet, when the same reaction is run under reversible conditions, similar results are obtained. Figure 10.9 shows what is known as the **Meerwein–Pondorrf–Verley reduction** (the reverse reaction is called the **Oppennauer oxidation**), where the aluminum reagent is used in a large excess. Since this reduction is an equilibrium, only the relative energies of the two products that arise from hydride addition should control the product ratio. The cyclohexane *A* value of an OH predicts a 72:28 ratio (Table 2.14), but the *A* value needs to be bigger for an Al(O–iPr)$_2$ group, giving a product ratio of 95:5.

Connections

Nature's Hydride Reducing Agent

Lithium aluminum hydride reacts violently with water, so it cannot be used effectively in aqueous media. Hence, what does nature use as a reducing agent? The hydride reducing agent comes in the form of a cofactor called nicotinamide adenine dinucleotide (NADH, or NADPH for a phosphorylated version). The essential part of NADH is shown to the right, along with a hypothetical reduction reaction. A nitrogen lone pair can be viewed as the electron source. Electron pushing gives a picture of how a hydride can be transferred to a carbonyl carbon coupled with general-acid-catalyzed protonation of the carbonyl oxygen by the enzyme (Enz).

Nature has a variety of cofactors that it uses as reagents. **Cofactors** are organic compounds (or metal ions) that are bound within the active sites of enzymes and become reagents for various transformations that the 20 natural amino acids cannot catalyze on their own. It is fascinating to compare the reagents used in an organic synthesis laboratory to nature's "reagents". A nice compilation of such comparisons is given in the following reference.

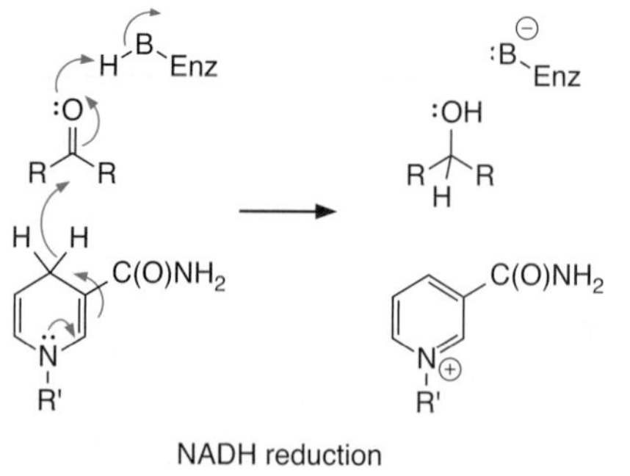

NADH reduction

Dugas, H. (1996). *Bioorganic Chemistry*, 3rd ed., Springer, New York, pp. 26–33.

10.9 Nucleophilic Additions to Olefins

Nucleophilic addition to an alkene is generally much less favorable than addition to a carbonyl, and the orbital analysis of Section 10.8.5 provides a nice rationalization. However, when strongly electron withdrawing groups are placed on an alkene, nucleophilic addition can occur, because the alkene LUMO is at lower energy. For example, addition of nucleophiles to perfluoroethylene and 2,2-dichloro-1,1-difluoroethylene does occur. Furthermore, as we will see in Chapter 12, when an alkene is coordinated to a transition metal, nucleophilic attack can be promoted. Even phenyl groups, on rare occasions, can act as electron withdrawing groups, allowing additions. However, by far the most common form of nucleophilic addition to an alkene occurs when a group capable of electron withdrawal via resonance (such as carbonyl, cyano, or imine) is placed on the alkene, termed Ad_N2 (addition, nucleophilic, bimolecular). This reaction is commonly referred to as a **1,4-addition** or **conjugate addition**. When the nucleophile is a carbanion, most commonly an enolate, the reaction is referred to as a **Michael addition**.

10.9.1 Electron Pushing

One specific example is given in Scheme 10.9, along with the classic electron pushing associated with this reaction. We draw arrows for the nucleophilic addition that place the negative charge on the enolate oxygen as the electron sink. In this example, protonation by the solvent gives the product.

Scheme 10.9
Electron pushing for a conjugate addition example.

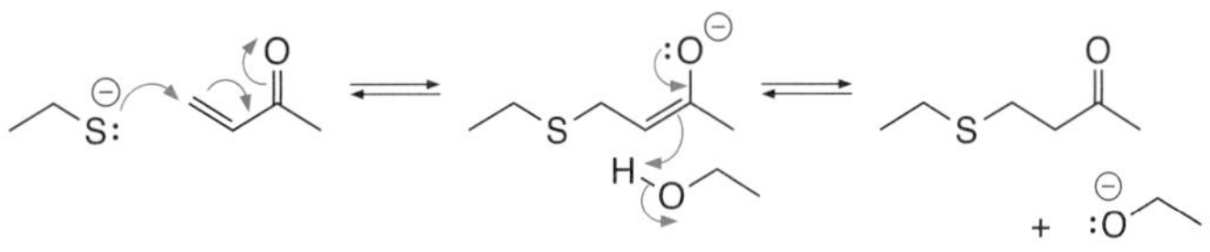

10.9.2 Experimental Observations

These reactions most commonly have nucleophilic attack as the rate-determining step. Supporting this bimolecular reaction as being rate-determining is a negative entropy of activation.

Recalling the discussion of the dynamics of proton transfer in Section 9.3.7, protonation and deprotonation of heteroatoms is faster than with carbon. Therefore, the mechanism given in Scheme 10.9 is, as with many classic electron-pushing schemes, a simplification. Protonation actually occurs first on the oxygen to make the enol form (Eq. 10.35), but the higher thermodynamic stability of the keto form relative to the enol form ultimately leads to the carbonyl product (see Section 11.1).

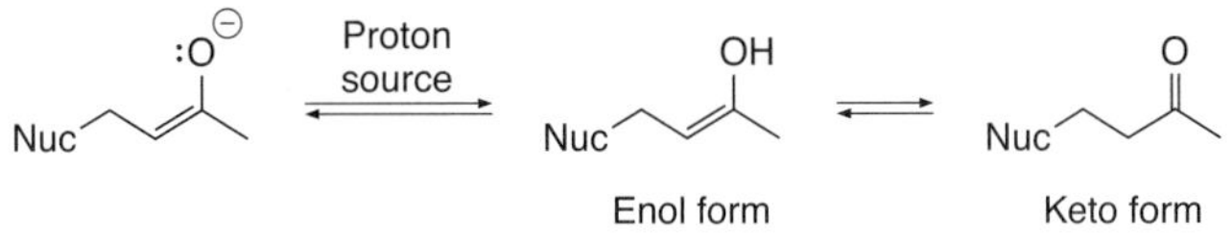

(Eq. 10.35)

10.9.3 Regiochemistry of Addition

We are discussing 1,4-additions, but a nucleophile can also add directly to the carbonyl in what is termed a **1,2-addition**, as discussed in Sections 10.2 and 10.8. The preferred regiochemistry of addition is in part controlled by the relative electrophilicity of the carbonyl carbon and the β-carbon in each specific case. Steric interactions can also influence the regiochemistry. Large groups on the carbonyl can direct the nucleophile away from attack at the carbonyl carbon.

However, the character of the nucleophile has been found to be the dominant factor in determining the regiochemistry of addition. Hard nucleophiles such as lithium reagents and Grignard reagents undergo preferential 1,2-addition, while softer nucleophiles such as amines, enolates, and thiolates undergo preferential 1,4-addition (the definition of hard and soft is covered in Section 5.6.1). The harder nucleophiles are directed to the carbonyl carbon because their counter ions coordinate to the carbonyl oxygen as part of the mechanism of attack, and these nucleophiles are closely ion-paired, thereby delivering the nucleophile to the carbon. The softer nucleophiles tend to add to the site where the largest orbital character in the LUMO exists, which is the β-carbon (see the LUMO of acrolein, Appendix 3).

10.9.4   Baldwin's Rules

The Bürgi–Dunitz angle of attack (Section 10.8.5) on an olefin or carbonyl leads to interesting trends in *intramolecular* additions that occur on sp^2 centers. The trends are relevant to additions to carbonyls, but also additions to alkenes and alkynes, so we discuss them all together here. Furthermore, the trends apply to nucleophilic, radical, or cationic additions. As a group, the trends are summarized as **Baldwin's rules**. These rules allow chemists to predict the ease of *ring closure* reactions. Three factors are considered: 1. ring size, 2. hybridization of the carbon undergoing attack, and 3. whether the bond undergoing attack will be endocyclic or exocyclic to the forming ring in the product.

Although reactivity depends upon the exact chemical reaction under study, there are clear trends related to ring size. The ease of intramolecular formation of a particular ring size generally follows the trend, 5 > 6 > 3 > 7 > 4 > 8–10. This holds for intramolecular nucleophilic, as well as radical and cationic ring closures.

As the ring size varies, however, the regiochemistry of attack will vary. To see this, let's first give some definitions. When ring closure occurs to place the cation, radical, or anion of the *product* exocyclic to the ring, this is called an **exo closure** (Eq. 10.36). When closure occurs placing the cation, radical, or anion within the ring, this is called **endo closure** (Eq. 10.37). Now we define the hybridization of the atom undergoing attack by a nucleophile, radical, or electrophile as follows: sp = dig, sp^2 = trig, sp^3 = tet. The terms are combined to describe a certain reaction. For example, Eq. 10.38 shows a reaction we would call a 5-exo-dig cyclization (5 denotes the size of the ring). If we used these terms to describe an intramolecular S_N2 reaction, the reaction would be considered an exo-tet cyclization, because the attack must be on an sp^3 center and the leaving group departs from the ring.

Exo closure   (Eq. 10.36)

Endo closure   (Eq. 10.37)

(Eq. 10.38)

Given these definitions, Table 10.5 shows the trends. We refer to certain closures as favored or unfavored, and you should be aware that the rules represent trends, not laws of nature. Notice it makes very little difference how six- or seven-membered rings are formed. The anion, radical, or cation can be exo or endo to the ring regardless of the hybridization of the reactant. Also, note that all exo-tet cyclizations are favored, meaning that any S_N2 reaction to form a ring is possible. However, it does matter how three-, four-, and five-membered rings are formed in other reactions. This is where the Bürgi–Dunitz angle of attack becomes important. Whereas the reaction in Eq. 10.38 is a favored reaction (5-exo-dig), the corresponding 5-endo-trig reaction given in Eq. 10.39 is not favored. Here, the nucleophile cannot achieve the Bürgi–Dunitz angle for the cyclization, and hence the reaction does not occur. For an analogous dig system, the reaction does proceed (Eq. 10.40).

Table 10.5
Ring-Closure Tendencies (Baldwin's Rules)*

Ring size	Exo			Endo	
	dig	trig	tet	dig	trig
3	unfav	fav	fav	fav	unfav
4	unfav	fav	fav	fav	unfav
5	fav	fav	fav	fav	unfav
6	fav	fav	fav	fav	fav
7	fav	fav	fav	fav	fav

*fav = favorable; unfav = unfavorable. Baldwin, J. E. "Rules for Ring Closure." *J. Chem. Soc. Chem. Comm.,* 734 (1976).

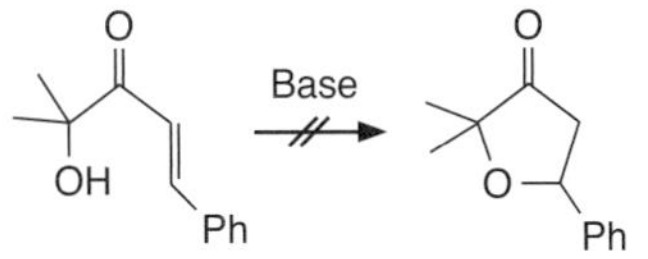

(Eq. 10.39)

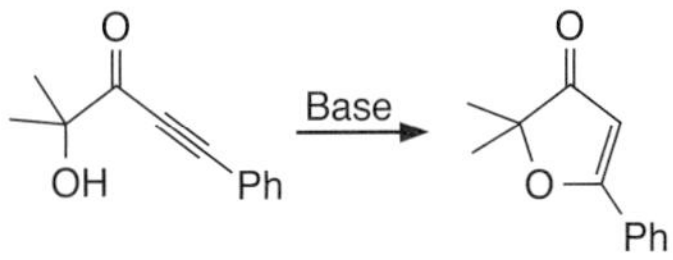

(Eq. 10.40)

10.10 Radical Additions to Unsaturated Systems

In the addition of electrophiles and nucleophiles to unsaturated systems (C=C and C=O), much of our focus has been upon the stability of the cationic or anionic intermediates. In the addition of radicals, the focus is again primarily upon the intermediate, but now there are no charges involved. Polarity effects will therefore influence the reactions to a much smaller extent, while steric factors will be more important. However, because increasing substitution with alkyl groups on the carbon bearing the radical center increases the stability of a radical via hyperconjugation, radicals most often add to double bonds at the less substituted center. If the adding radical is a halogen, this places the halogen at the less substituted center—**anti-Markovnikov addition**. Let's review the electron pushing for this reaction, called Ad_H2 (addition, homolytic, bimolecular).

10.10.1 Electron Pushing for Radical Additions

Radical additions, just like radical substitutions (see Section 11.7), most commonly occur via **chain reactions** (Scheme 10.10). The addition of HBr starts with the initiation of a rad-

Scheme 10.10
A radical chain mechanism for the addition of HBr to alkenes. I = initiator.

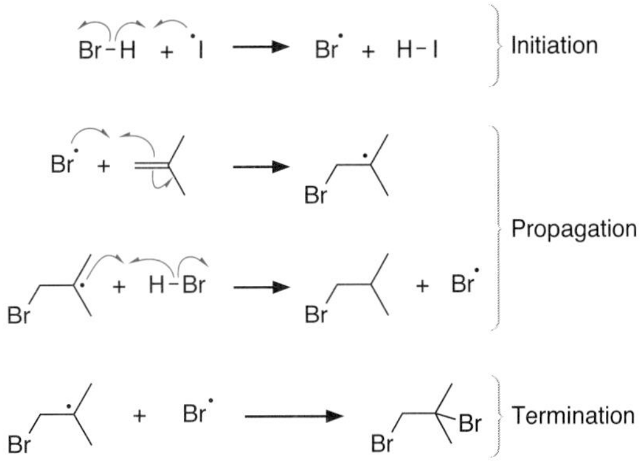

ical chain, sometimes created by an impurity in the solvent, or any of the numerous initiators discussed below. Abstraction of a hydrogen atom from HBr creates bromine radical, which adds to the alkene at the less substituted carbon, because this creates the more stable carbon-centered radical. Abstraction of a hydrogen atom from HBr by the carbon radical creates another bromine radical. The alternating creation of bromine and carbon radicals propagates the chain.

In a chain mechanism, three types of reactions occur: **initiation, propagation**, and **termination**. Initiation always creates a radical that starts the chain by creating a propagating species. Propagation involves an intermediate with an unpaired electron that undergoes a reaction to create a different species with an unpaired electron. The radical reactant in one step of the chain is necessarily a radical product in another step of the chain. Termination occurs whenever two radicals react with each other to produce closed-shell species. This removes some of the propagating species from the system, and hence terminates a chain. While initiation and termination are parts of a chain process, the propagation steps are what account for the conversion of reactants to products.

10.10.2   Radical Initiators

A variety of compounds have been created by chemists to initiate radical reactions. These **initiators** can be used not only in addition reactions, but also radical substitutions and polymerizations. The common design principle is the incorporation of a weak bond that undergoes homolysis thermally or with light. Organic peroxides and azo compounds are common examples that can decompose by both means (Eqs. 10.41–10.43). The decomposition of benzoyl peroxide (a common light-activated acne medicine; Eq. 10.42) leads to phenyl radical and carbon dioxide.

The $\Delta H^{\ddagger}$ for the homolysis of peroxides is near the bond dissociation energy (30 to 40 kcal/mol), and $\Delta S^{\ddagger}$ is positive for both peroxides and azo compounds. For example, the entropies of activation for homolysis of di-t-butylperoxide (Eq. 10.41) and AIBN (2,2′-azodiisobutyronitrile; Eq. 10.43) are 13.8 and 12.2 eu, respectively. These values are consistent with the increase in disorder at the transition state due to the breaking of the bonds.

$$\text{(Eq. 10.41)}$$

$$\text{(Eq. 10.42)}$$

$$\text{(Eq. 10.43)}$$

AIBN

Cage effects are common in initiation reactions. The radicals initially formed after photolysis or thermolysis are briefly held together in a solvent cage before they diffuse away from one another into free solution. Because recombination reactions of radicals are rapid, these reactions can compete with diffusion into solution. The terms **geminate recombination** and **internal return** (the reverse of Eq. 10.41 within a solvent cage) are applied to this phenomenon. The more viscous the solvent, the greater the internal return, and the lower the efficiency of initiation reactions, because a larger number of the first formed radicals simply recombine. This even occurs for azo compounds, where R–N=N–R initiators create R–R via dimerization of the first-formed radicals within a solvent cage.

Radicals can also be generated via the fragmentation of **radical anions**. Single electron reduction of various alkyl or aryl halides will lead to bond cleavage, as shown in Eqs. 10.44 and 10.45. The reductions commonly occur from dissolving metals such as Na and K, sodium naphthalenide ($Na^{+} Ar^{-}$), sodium benzophenone ketyl, or from pulse radiolysis (e^{-}).

(Eq. 10.44)

(Eq. 10.45)

In the addition of HBr, the **peroxide effect** is the historical initiation procedure. In the 1930s, the addition of HBr to alkenes was found to be both Markovnikov and anti-Markovnikov by various laboratories, and some laboratories found both results under seemingly identical conditions. Kharasch and Mayo found that in the presence of peroxides and often air, the anti-Markovnikov addition prevailed, while "clean" conditions gave Markovnikov addition. It is now clear that the anti-Markovnikov results arose from the presence of peroxides, often in trace amounts in ether solvents, which initiated radical chain reactions.

10.10.3 Chain Transfer vs. Polymerization

Chain transfer is the transfer of a radical from one propagating species to another, as in Eq. 10.46. Chain transfer leads to the kinds of products we have been considering. Alternatively, two species can combine as in Eq. 10.47, which is a different propagation process that ultimately leads to polymers. Depending upon the efficiency of the chain transfer in the radical process, we can get small molecule products or polymers. The mechanisms of the polymerizations are discussed in Chapter 13. Here we focus upon the factors that influence the extent of polymerization or chain transfer. If the concentrations of the alkene and H–X are similar, then we will get polymer when $k_p \gg k_{ct}$, and mostly small molecules when $k_{ct} \gg k_p$. When the two rate constants are similar, the products consist of a broad distribution of oligomers, sometimes called **telomers**.

(Eq. 10.46)

(Eq. 10.47)

In considering the addition of HX to olefins, HBr is more likely to create small molecules, while HCl gives significant amounts of oligomer. Let's see why by comparing the rates of Eqs. 10.46 and 10.47. The activation energy for the polymerization step (Eq. 10.47) is essentially independent of whether X is Cl or Br, and is typically between 6 and 10 kcal/mol. However, abstraction of a hydrogen atom (Eq. 10.46) from HCl is endothermic by 5 kcal/mol, while the same abstraction from HBr is exothermic by around 10 kcal/mol. The activation energies track these thermodynamic differences, and therefore chain transfer is much more likely with HBr, while HCl gives telomers. The radical addition of HI does not occur at all, because the addition of iodine radical to an alkene is too endothermic.

10.10.4 Termination

Termination can proceed via the combination of any two radical species to make a σ bond. A particular termination reaction is **disproportionation**, defined as a reaction between two identical species that leads to two different products. One possibility relevant to HBr addition is shown in Eq. 10.48. Here, hydrogen abstraction results in termination reactions due to the removal of two radical propagating species. In general, disproportionations are favored over radical dimerizations as the temperature is raised, even though dimerizations are more exothermic. The reason is that the entropy of activation for disproportionation is generally less negative.

$$\text{(Eq. 10.48)}$$

10.10.5   Regiochemistry of Radical Additions

As discussed, electrophilic addition of HBr occurs with Markovnikov regiochemistry, but radical addition of HBr occurs with *anti*-Markovnikov regiochemistry. The addition of the electrophilic bromine radical occurs at the less substituted carbon, resulting in the more stable carbon radical. This is analogous to the addition of an acid to an alkene, where the electrophilic proton attaches to the less substituted carbon to give the more stable carbenium ion. Since the definition of Markovnikov or *anti*-Markovnikov follows the placement of the halogen, the two additions have different regiochemistries "in name", although their reactivities are fundamentally similar.

Note that the addition of a radical to an alkene is often nearly thermoneutral or even strongly exothermic. Therefore, the transition state is early, and has less radical character than the extent of carbenium ion character in the transition state for the endothermic addition of a proton to an alkene. As such, electronic factors are less important in radical addition and steric factors can contribute to regioselectivity.

Substituents attached to an alkene direct radical additions in two ways, called **α effects** and **β effects** (Eq. 10.49). An α effect occurs when the radical adds to the carbon with the substituent, while the effect of a substituent on the other alkene carbon produces a β effect.

$$\text{(Eq. 10.49)}$$

Addition to a terminal alkene shows only a β effect. Both electron withdrawing and electron donating groups increase the rate of radical addition when in this position, because the orbital resulting after radical addition is lower in energy due to delocalization of the radical by either inductive, hyperconjugative, or resonance effects. The sensitivity to the Taft steric parameter is relatively small ($\delta = 0.28$), meaning that sterics play very little role in the β-position.

Electron withdrawing groups in the α-position speed radical addition, but the effect is less dramatic than the β effect. However, as might be expected, steric α effects are larger than β effects. The Taft sensitivity parameter δ is 1.4 for the α-position. In total, this means that when both double bond carbons have electron withdrawing or electron donating groups, it is sterics that dominate the position of attack.

10.11   Carbene Additions and Insertions

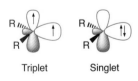

Triplet   Singlet

The last addition reactions we cover involve carbenes. In Chapter 1 we discussed the geometry of methylene (H$_2$C). Recall that methylene has σ(out) and *p* group orbitals that can each be singly occupied to create a triplet state, or the two nonbonding electrons can occupy the σ(out) orbital giving a singlet state (see margin). These two possibilities exist for all carbenes (R$_2$C: species), not just methylene. The products that result from carbene additions depend strongly upon the spin state of the carbene, as described below. Carbenes can undergo addition reactions with both alkenes and alkynes, and even with arenes in some instances. Another reaction of carbenes is **insertion** into a single bond, usually a C–H. Although not for-

Going Deeper

The Captodative Effect

In the discussion above about radical additions to alkenes, we noted that both electron donating and electron withdrawing groups increase the rate of addition. This is because the resulting radical is more stable. Interestingly, radical stability dramatically increases when both an electron donating group (EDG) *and* an electron withdrawing group (EWG) are placed on the radical center, an effect referred to as the **captodative effect**. For example, dicyano(dimethylamino)methyl radical and 1-ethyl-4-carbomethoxypyridinyl radical are remarkably persistent without any large steric hindrance to reactions.

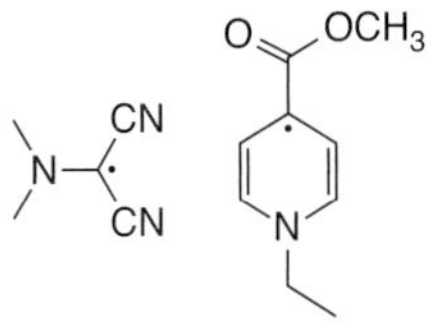

Captodative radicals

Molecular orbital calculations emphasize the importance of polar character in the structures of radicals that have both EWGs and EDGs. The stabilities of these structures can be readily understood using resonance theory, as shown to the right for an amino–cyano-substituted radical.

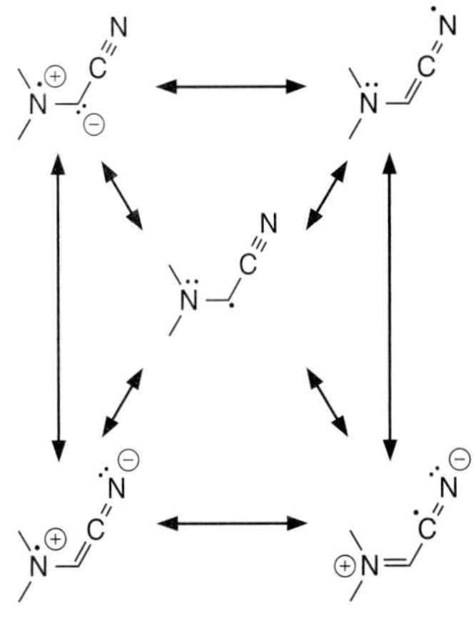

Resonance in a captodative radical

The consequence of the captodative effect on radical additions to alkenes is a strong directing β effect for addition of the radical to the center that gives the especially stabilized radical. The captodative effect has been used in a variety of synthetic schemes, many of which are given in the reference below.

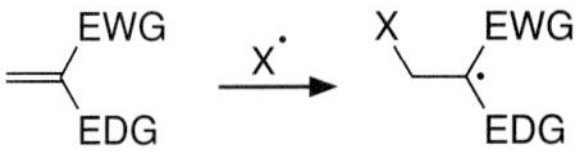

β-Effect

Viehe, H. G., Janousek, Z., and Merenyi, R. "The Captodative Effect." *Acc. Chem. Res.*, **18**, 148 (1985).

mally an addition reaction, we will consider it briefly here. Triplet carbene insertion is very similar to homolytic substitution at a saturated center, a reaction we will consider in detail in Chapter 11.

Methylene ($:CH_2$) is the prototypical carbene, and it has been the object of extensive experimental and theoretical scrutiny for decades (see Section 14.5.6). The interplay between theory and experiment over the years has been substantial, and at time, contentious. For the most part, high-level *ab initio* electronic structure theory has been quite successful in treating methylene. In fact, with regard to both the preferred geometry and the energy gap between the singlet and triplet states, theory has played a key role in rectifying erroneous interpretations of experimental data. Now, theory and experiment are in accord with regard to all essential properties of this crucial structure. Furthermore, the ability to predict spin states of carbenes using high-level theory is now well accepted, and it can be used in advance of studying reactivity to predict the various reactions we now describe for singlet and triplet carbenes.

10.11.1 Electron Pushing for Carbene Reactions

Scheme 10.11 shows the proper electron pushing for four different carbene reactions. The reactions of the singlet carbenes (**A** and **C**) are treated as concerted, whereas the reactions of triplet carbenes (**B** and **D**) are stepwise.

Scheme 10.11
Electron pushing for four different carbene reactions. **A.** Singlet carbene addition to alkenes, **B.** Triplet carbene addition to alkenes, and **C.** Singlet carbene insertion into a C–H bond. **D.** Triplet carbene insertions go via radical abstraction followed by recombination.

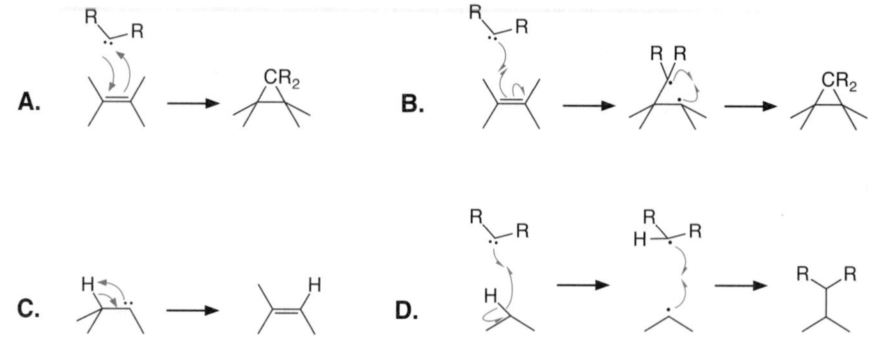

10.11.2 Carbene Generation

A variety of methods have been devised for the generation of carbenes. One common method is the thermal decomposition of diazoalkanes (Eq. 10.50). When R = H, the reactant is called diazomethane, and special glassware can be purchased for its formation (diazomethane explodes on contact with ground glass joints). A general protocol for the synthesis of a diazoalkane involves the base induced elimination of N-nitrosoureas (Eq. 10.51). Similarly, the base induced elimination of tosylhydrazones gives carbenes via a "diazo-like" intermediate (Eq. 10.52). A similar protocol to the use of diazo compounds for generating carbenes is the decomposition of diazirines (Eq. 10.53). Lastly, base induced α-eliminations from haloform reagents will generate dihalo carbenes (Eq. 10.54). The generation of carbenes from diazo and diazirine compounds can also be accomplished photochemically. Due to the extremely high reactivity of carbenes, photochemical generation using laser flash photolysis has been useful in the investigation of carbene reactivity and structure.

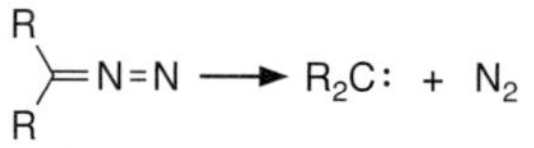

$$(\text{Eq. } 10.50)$$

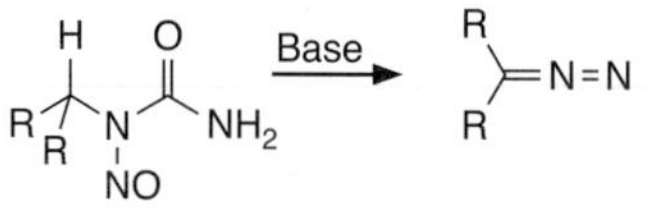

$$(\text{Eq. } 10.51)$$

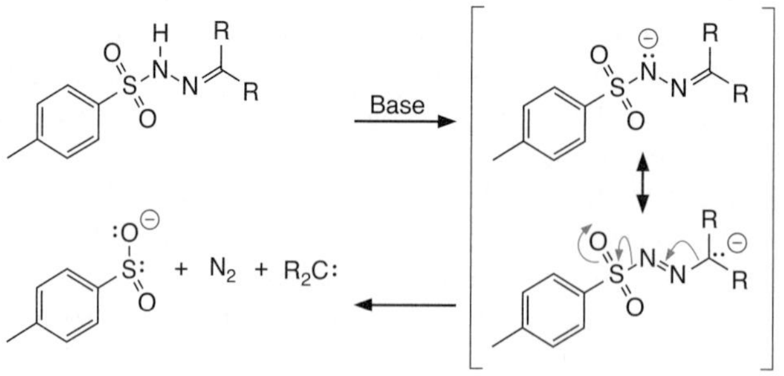

$$(\text{Eq. } 10.52)$$

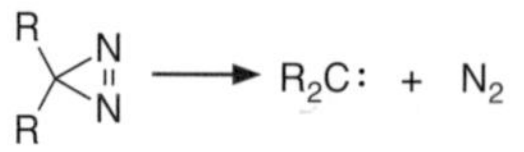

$$(\text{Eq. } 10.53)$$

$$(\text{Eq. } 10.54)$$

All the thermal protocols discussed above initially form singlet carbenes, as do the photolysis of diazo and diazirine compounds. For cases where the triplet is the ground state, relaxation to the triplet state can occur in competition with additions and insertions. In addition, it is possible to use a photochemical sensitizer (see Chapter 16) to produce a triplet state directly. Here, the **sensitizer** is a photo-excited triplet state of an additive that transfers its energy to the carbene precursor, creating the triplet carbene directly.

The production of carbenes from haloforms (Eq. 10.54) is an interesting reaction. The reaction sequence displays second-order kinetics, first order in both base and haloform. This supports a mechanism involving an equilibrium deprotonation prior to rate-determining α-halogen departure. The loss of an equivalent of HCl from $HCCl_3$ constitutes an elimination reaction, and is specifically called a **1,1-elimination** or an **α-elimination**.

Phase transfer catalysts (see Chapter 9 for a discussion) have been used to facilitate the reaction depicted in Eq. 10.54. The halomethyl anion is transported as an ion pair with a tetraalkylammonion counter cation into an organic phase, where reprotonation occurs at a lower rate. The α-elimination to form the carbene occurs in the organic medium where the alkene or alkyne targeted for addition resides.

The **Simmons–Smith reagent** (ICH_2ZnI) also acts as a carbene source. The reaction between CH_2I_2 and Zn does not generate a full-fledged free carbene, but instead a carbenoid (Eq. 10.55). A **carbenoid** is a carbene that is stabilized by complexation to a metal. Even the carbenes created by the reaction between strong bases and haloforms sometimes react as carbenoids, where the carbene is complexed to the counter cation of the strong base. We will examine more carbenoid species in Chapter 12, when alkylidenes and Fischer carbenes are discussed.

$$CH_2I_2 \xrightarrow{\text{Zn/Cu}} ICH_2ZnI \qquad \text{(Eq. 10.55)}$$

10.11.3   Experimental Observations for Carbene Reactions

As shown in Scheme 10.11, both singlet and triplet carbenes add to alkenes. The electronic structure of a singlet carbene (see the drawing in the margin on page 572) imparts reactivity that is analogous to both carbenium ions and carbanions. We will examine the orbital considerations for the singlet carbene addition to alkenes in Chapter 15, because this reaction is formally a pericyclic reaction. In contrast, triplet carbenes act as biradicals. This reactivity difference is most obvious with addition to alkenes, where singlet carbenes give 100% stereospecific reactions, and triplet carbenes give mixtures. With singlet carbenes, cis-alkenes give cis-cyclopropanes, and trans-alkenes give trans-cyclopropanes.

The reactions of triplet carbenes result from a stepwise addition involving the biradical intermediate shown in Scheme 10.11 **B**, which can undergo C–C bond rotation. Normally, radical combination is a very fast reaction, and closure to a cyclopropane is further facilitated because it is intramolecular. The reason that the rate of closure is slower than normal in this case is that an electron spin flip is required prior to closure (Eq. 10.56). The addition of a triplet carbene to an alkene first gives a triplet biradical that cannot cyclize. Instead, a spin flip must first occur, followed by radical combination to complete the ring closure. The stereochemistry of the alkene is typically not completely lost in the product, which indicates that the spin inversion and bond rotation rates must be comparable.

$$\text{(Eq. 10.56)}$$

Heteroatom stabilization

The reactivity of most singlet carbenes with alkenes is dominated by the electrophilic character of the carbene (the empty p orbital). Thus, the more electron rich the alkene, the faster the carbene addition. Increasing alkyl group substitution on alkenes increases the rate of addition. This trend parallels the reactivity for the addition of other electrophiles with alkenes, such as acids, X_2, and borane. Dialkylcarbenes are less selective than dihalocarbenes, whereas carbenes with neighboring O or N atoms are resonance stabilized (see margin) and are highly selective. This trend tracks the reactivity–selectivity principle (see Chapter 7), where the more stable carbenes are the more selective.

In contrast to the electrophilic nature of most carbenes, highly resonance-stabilized carbenes can be nucleophilic, as suggested by the right-hand resonance structure in the margin. Dimethoxycarbene is a good example of a nucleophilic carbene. Intermediate cases such as methoxychlorocarbene can be **ambiphilic**, showing increased reactivity to both electron rich and electron deficient olefins.

Carbenoid species, such as the Simmons–Smith reagent (Eq. 10.55), undergo facile additions to alkenes to create cyclopropanes. The reactions are stereospecific, making carbenoids synthetically useful versions of carbenes. Carbenoid compounds do not normally perform insertion reactions.

Carbenes are highly reactive species, and if an olefin or other addition partner is not available, carbenes will indiscriminately insert into C–H bonds. Singlet carbenes insert into C–H bonds in a concerted reaction that always proceeds with retention of configuration at the C–H bond undergoing insertion. Triplets undergo insertions via a stepwise process involving radical intermediates. Reactions of this type are covered in more detail in Chapter 11, where we discuss substitutions at saturated centers.

When additions and insertions are not possible, singlet carbenes react as either carbanions or carbocations. In protic solvents, for example, singlet carbenes give ethers. Singlet diphenylcarbene first abstracts a proton from the alcohol to give a carbenium ion, and then a nucleophile adds (Eq. 10.57). The alternative mechanism, in which the C–O bond is formed first, is seen for other carbenes. Either way, the overall transformation results in the insertion of the carbene into the O–H bond of the alcohol.

$$\text{(Eq. 10.57)}$$

Hindered carbenes

Very sterically hindered carbenes, such as di-*t*-butyl and diadamantyl carbene (see margin) can persist long enough to be observed spectroscopically at 40 K. Both are ground state triplets, but their reactions are indicative of both singlet and triplet states.

Eliminations

Elimination reactions are the opposite of addition reactions. In an elimination a single molecule splits into two or more different molecules. How the molecule splits apart can be predicted based upon the polarizations of the bonds, as we will see below. The prototype pathways for this reaction are denoted E1 and E2, standing for elimination unimolecular and elimination bimolecular, respectively. E1 and E2 refer to an overall reaction mechanism. As we will describe below, there are other mechanisms for eliminations, referred to as variants of E1cB (elimination, unimolecular, conjugate base). Other terms for elimination reactions are also common, including "1,2-elimination", "1,4-elimination", and "β-elimination". These latter terms refer to the relative position of the leaving group to the source of negative charge.

10.12   Eliminations to Form Carbonyls or "Carbonyl-Like" Intermediates

The addition reactions discussed in Section 10.2 create stable tetrahedral structures such as geminal diols and hemiacetals that are in equilibrium with their respective starting carbonyl compound. Hemiacetals can undergo further reactions in acid to create acetals. The first elimination mechanism we examine involves the conversion of acetals to carbonyl compounds.

10.12.1   Electron Pushing

The classic electron pushing for specific-acid-catalyzed hydrolysis of an acetal is given in Scheme 10.12. Many steps are involved. Protonation of an acetal oxygen creates a good leaving group (step 1), whose departure is assisted by the lone pairs on the other acetal oxygen (step 2). Leaving group departure creates a "carbonyl-like" intermediate that is susceptible to nucleophilic attack by water (step 3). Shuffling of the protons from the added water to the ether oxygen creates another good leaving group (steps 4 and 5). Leaving group departure (step 6), again assisted by an adjacent oxygen lone pair, creates a structure that simply needs to lose a proton to create the final ketone product (step 7).

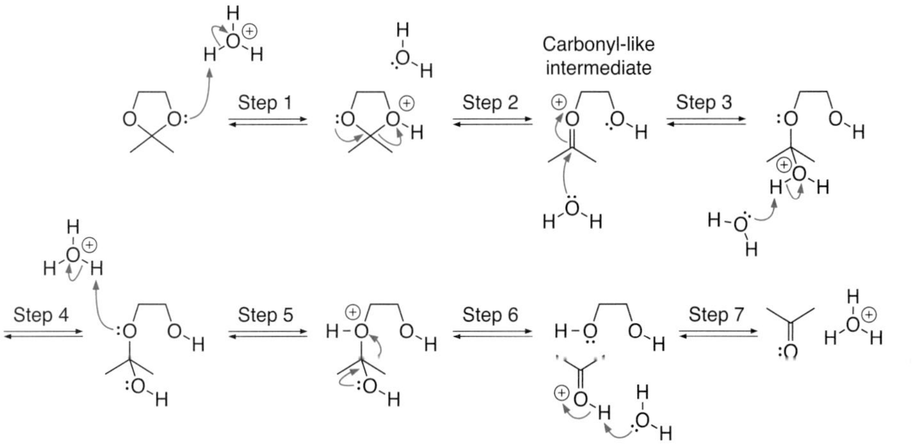

Scheme 10.12
Electron pushing for the conversion of an acetal to a carbonyl.

10.12.2   Stereochemical and Isotope Labeling Evidence

Although Scheme 10.12 shows one particular possibility, the initial cleavage of a C–O bond in an acetal can be envisioned to proceed via two possible paths, as shown in Eqs. 10.58 and 10.59. Either an oxocarbenium ion or a simple carbenium ion is created. Two kinds of experiments resolve which path is followed. When the R group is a stereocenter, retention is consistently found, even when R can make a reasonably stable carbenium ion. Furthermore, isotopic labeling of the acetal with ^{18}O shows loss of the label to solution. Both experiments support the pathway given in Eq. 10.58, where an oxocarbenium ion is the intermediate. This intermediate is "carbonyl-like", in that it can be viewed as a structure with an R group attached to a carbonyl oxygen. This greatly enhances the electrophilicity of the carbonyl carbon, similar to the way protonation of a carbonyl oxygen enhances the electrophilicity.

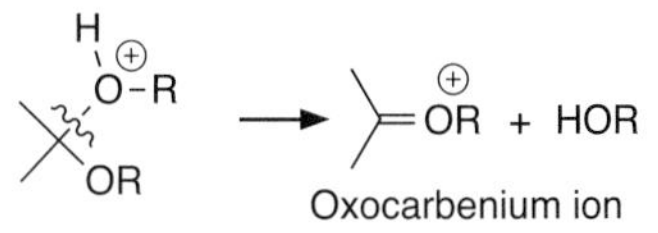

(Eq. 10.58)

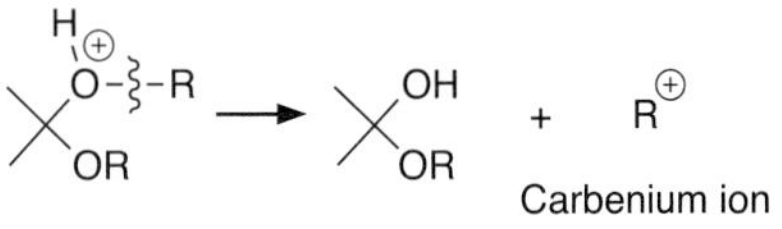

(Eq. 10.59)

10.12.3   Catalysis of the Hydrolysis of Acetals

Just as with the formation of geminal diols, there are several possibilities for the sequence of steps in the acid-catalyzed hydrolysis of acetals to hemiacetals and ultimately to carbonyl structures. Once again, the focus is upon specific vs. general catalysis. Figure 10.10 shows three possibilities. Mechanism **A** has an equilibrium involving acid prior to a rate-determining dissociative step, and therefore should show all the hallmarks of specific-acid catalysis and an S_N1 reaction. Mechanism **B** also has an equilibrium involving acid prior to the rate-determining step, but the nucleophilic attack is now an S_N2 reaction. The last mechanism (**C**) involves simultaneous protonation and expulsion of the leaving group, and therefore would give experimental results consistent with general-acid catalysis. As is often the case in organic chemistry, the mechanism that dominates depends upon the reaction conditions and the specific structures of the reactants.

For most acetals with poor OR leaving groups, specific catalysis is experimentally observed. The leaving groups are sufficiently poor that complete transfer of the proton to the leaving group is required prior to any further reaction. This indicates that the mechanisms of Figures 10.10 **A** and **B** could be operative. To distinguish between these two mechanisms, many of the methods covered in Chapter 8 have been employed. Hammett plots using meta-substituted benzaldehyde diethyl acetals show a ρ value of -3.3, supporting positive charge moving closer to the substituents in the transition state of the rate-determining step (rds in Eq. 10.60). Moreover, our chemical intuition should lead us to predict mechanism **A** over **B**. S_N2 reactions are retarded at centers with significant substitution, whereas this favors S_N1

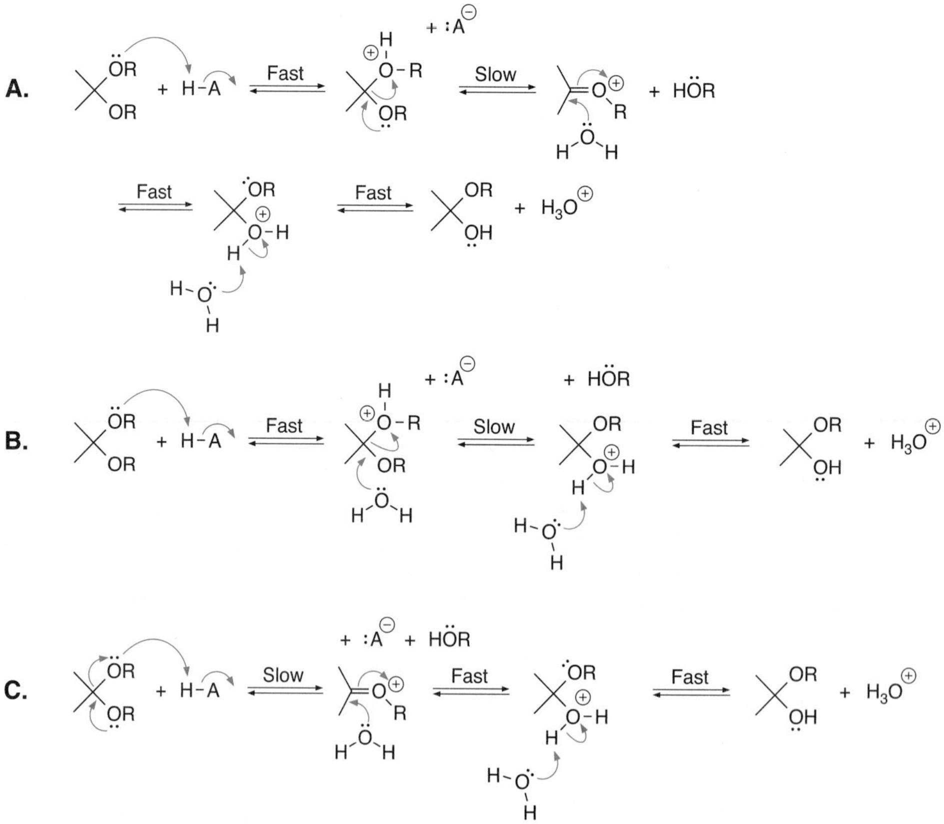

Figure 10.10
A. Specific-acid-catalyzed pathway for acetal hydrolysis that occurs with poor leaving groups. **B.** A pathway seldom if ever seen. **C.** General-acid-catalyzed pathway that occurs with good leaving groups.

mechanisms, especially when stabilization of the carbenium ion is possible via an adjacent heteroatom (see Section 11.5.11).

$$\rho = -3.3 \qquad \text{(Eq. 10.60)}$$

With good OR leaving groups, such as phenols, general-acid catalysis is often seen, supporting the mechanism shown in Figure 10.10 **C**. This leads us to conclude that there is a continuum of mechanisms between Figures 10.10 **A** and **C**. With the better leaving groups, only partial proton transfer is required to induce cleavage of the bond. As one example, 2-(*p*-nitrophenoxy)tetrahydropyran undergoes general-acid-catalyzed hydrolysis with a Brønsted α value of 0.5 (see margin).

$\alpha = 0.5$

General-acid catalysis

10.12.4   Stereoelectronic Effects

Interesting stereoelectronic effects have been observed in the reactivity of acetals and hemiacetals. Recall the discussion in Section 2.4.2 of the anomeric effect, in which the placement of electron lone pairs antiperiplanar to polarized bonds is found to be stabilizing. Similarly, lone pairs can enhance departure of a leaving group if they are arranged antiperiplanar to it. For example, Eq. 10.61 shows a strong preference for a reaction that benefits from two lone pairs antiperiplanar to the leaving group, leading to cleavage of the endocyclic C–N bond. The other products would derive from a leaving group departure that involves only one lone pair that is antiperiplanar to the departing group. Similar effects have been seen in models of enzymes (see the following Connections highlight).

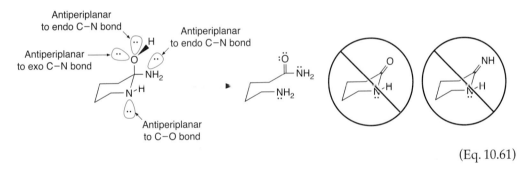

(Eq. 10.61)

Connections

Stereoelectronics in an Acyl Transfer Model

The notion of antiperiplanar lone pairs enhancing leaving group departure is a powerful one. We have seen it here, and we will return to it when eliminations to form alkenes are discussed. One particularly elegant example of this was shown by Groves with a mimic of Zn(II) peptidase enzymes. The appended amide of the compound shown to the right is cleaved one million times faster than background hydrolysis of the amide. The system combines two important factors for acyl transfer, facilitating both nucleophilic attack and leaving group departure. The first factor is the orientation of the hydroxide nucleophile to the amide carbonyl. The nucleophile is aimed at the π^* orbital, approximately as required by the Bürgi–Dunitz angle. The second facet of the design is the stereoelec-

tronic enhancement of the leaving group departure. Lone pair electrons on both oxygens of the tetrahedral intermediate can facilitate the departure.

Peptidase mimic

Groves, J. T., and Baron, L. A. "Models of Zinc-Containing Proteases. Catalysis of Cobalt(III)-Mediated Amide Hydrolysis of Pendant Carboxylate." *J. Am. Chem. Soc.*, **111**, 5442–5448 (1989).

10.12.5   CrO₃ Oxidation—The Jones Reagent

Oxidation is another form of elimination reaction. For example, the oxidation of a secondary alcohol to give a ketone formally involves the elimination of a molecule of dihydrogen, although the actual creation of H_2 is rare. Such a reaction is typically accomplished with some form of a metal-based oxidizing reagent, although some of the most useful oxidizing agents in organic synthesis are solely organic-based, such as DMSO (see the Swern oxidation in the Connections highlight below).

Electron Pushing

The sequence given in Scheme 10.13 shows the oxidation of an alcohol by CrO_3 in aqueous acid (**Jones reagent**). First, CrO_3 abstracts a proton from the acid (step 1). The alcohol oxygen then acts as a nucleophile and adds to the partially positive chromium (step 2). This reaction is followed by a proton abstraction by the solvent from the alcohol to give a chromate ester (step 3). These steps are analogous to the acid-catalyzed addition of an alcohol to a carbonyl. With CrO_3 oxidation, the O–Cr σ bond is drawn as a source of electrons in the last step for the reduction of the chromium metal (step 4).

Scheme 10.13
Electron pushing for the CrO_3 oxidation of a secondary alcohol.

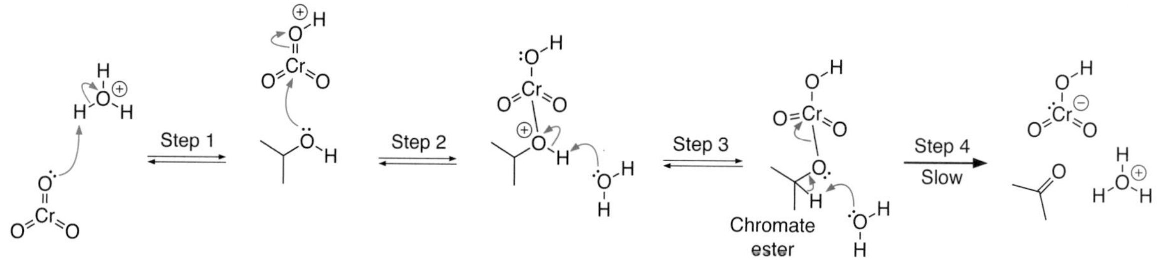

Connections

The Swern Oxidation

One of the most used alcohol oxidations in organic synthesis is the **Swern oxidation**. A large number of variants exist for this reaction, but a common one involves DMSO, oxalyl chloride, and a base (pyridine, dimethylaminopyridine, and triethylamine are common). The currently accepted mechanism is shown below along with electron pushing for some steps. The first part of the mechanism involves activation of DMSO by reaction with oxalyl chloride. This is followed by nucleophilic attack of the alcohol on this activated species, creating an alkoxysulfonium intermediate.

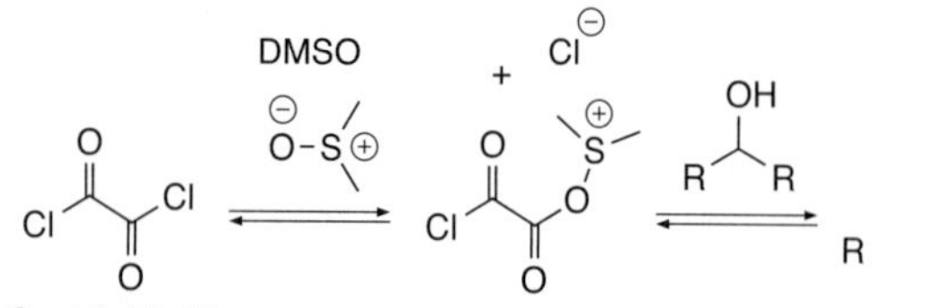

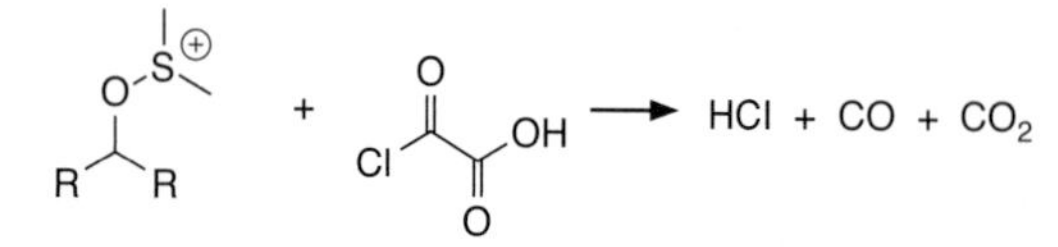

The alkoxysulfonium intermediate is next deprotonated to make an ylide. The ylide then undergoes an intramolecular proton abstraction, leading to bond cleavage and the creation of a carbonyl and dimethylsulfide. This mechanism was established by deuterium labeling experiments, where a deuterium on the α-carbon of the alcohol is incorporated into the dimethylsulfide product.

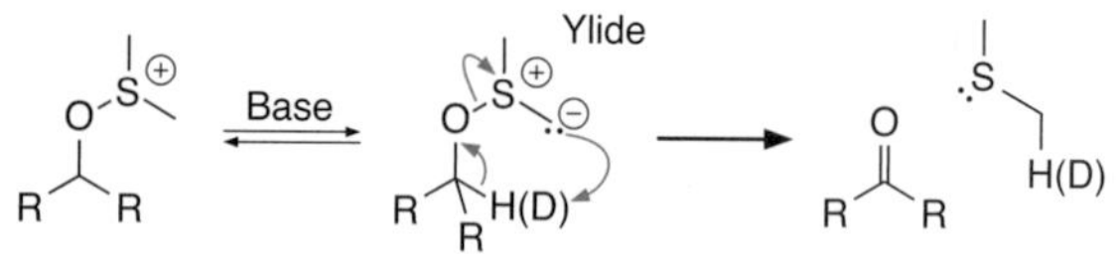

Torssell, K. "Mechanisms of Dimethylsulfoxide Oxidations." *Tet. Lett.*, 4445 (1966).

A Few Experimental Observations

It has been observed experimentally that the steps leading to the chromate ester are all fast. The rate-determining step is the final decomposition of the chromate ester to give the carbonyl compound. Factors that introduce strain into the chromate ester lead to faster rates of oxidation. This reaction can be considered as an E2 elimination (discussed in Section 10.13) in which an alcohol is first activated by forming a chromate ester, and the CrO_2OH^- is a good leaving group.

10.13   Elimination Reactions for Aliphatic Systems— Formation of Alkenes

Eliminations to create alkenes can occur from neutral structures, or they can involve carbenium ions, carbanions, or radical intermediates. We look at all these possibilities below within the context of aliphatic systems. Although eliminations can occur on alkenes to make alkynes and on arenes (see the benzyne reaction, Section 10.20), we concentrate our discussion on aliphatics, because this is the most common type of system to undergo elimination. Before looking at the details, let's review the E2 and E1 mechanisms and electron pushing, and define the terms "1,2-elimination", "1,4-elimination", and "β-elimination".

10.13.1   Electron Pushing and Definitions

The **E2** reaction (elimination, bimolecular) involves a base and an alkyl structure with a good nucleofuge, such as halide or tosylate. As shown in Scheme 10.14 for a simple alkyl bromide, the reaction involves three simultaneous bonding changes. A base removes a proton from the carbon adjacent to the C–Br bond; a new C–C π bond is formed; and the C–Br σ bond is cleaved. No intermediates are formed, and the reaction is completely concerted.

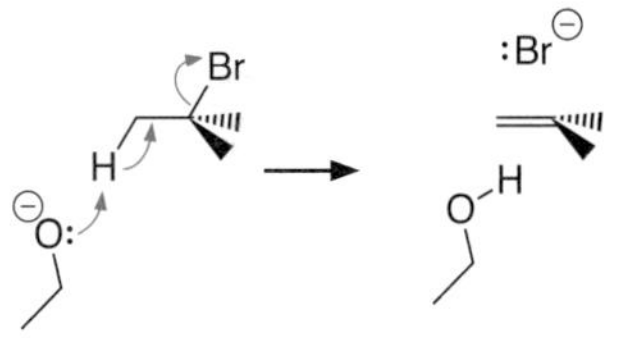

In the first step of an **E1** reaction (elimination, unimolecular; see Scheme 10.15) there is a heterolysis of a C–LG bond to form a carbenium ion. Carbenium ions are prone to rearrangement, and so rearranged products are always possible in an E1 reaction. The second step involves deprotonation of an adjacent hydrogen by any base, shown here as water. The initial heterolytic cleavage is as in an S_N1 reaction, but now an elimination occurs due to the deprotonation step after heterolysis.

Scheme 10.14
An example of an E2 mechanism. All bond breaking and formation occur in one single step.

Scheme 10.15
An example of an E1 mechanism. Two steps lead to the elimination.

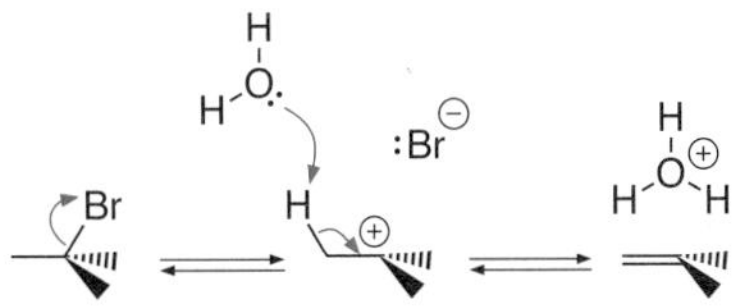

Both of the reactions shown in Schemes 10.14 and 10.15 are referred to as **β-eliminations** or **1,2-eliminations**. These generic terms apply to eliminations where the hydrogen is removed from the carbon beta to the carbon possessing the leaving group (the carbon possessing the leaving group is called the α-carbon).

A reaction known as **1,4-elimination** involves the expulsion of a leaving group on the 4-position relative to the hydrogen being removed (see Eq. 10.62). This reaction is an analog of the 1,2-elimination, except now the leaving group and the hydrogen have an intervening π bond. Continued insertion of π bonds can lead to 1,6-eliminations, 1,8-eliminations, etc.

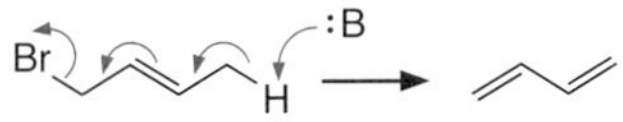

(Eq. 10.62)

The elimination reactions shown above occur under either neutral or basic conditions. 1,2- and 1,4-eliminations can also be acid-catalyzed. Consider the electron pushing for the elimination of water from a β-hydroxycarbonyl produced during an aldol addition (Scheme

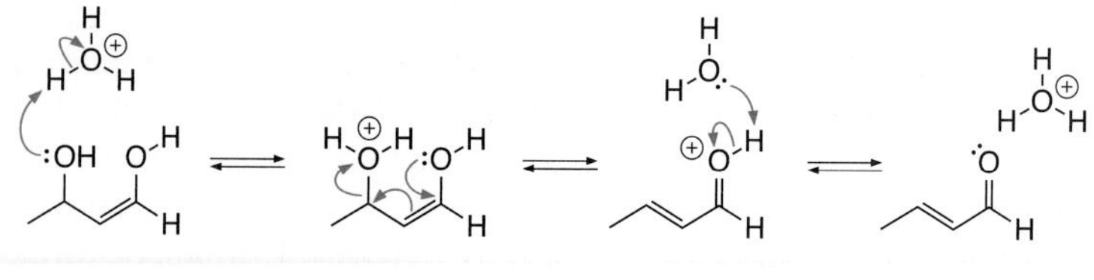

Scheme 10.16
An example of electron pushing for an acid-catalyzed 1,4-elimination.

10.16). Acid-catalyzed tautomerization of an aldehyde (see Chapter 11) gives an enol as the starting point. Protonation of the alcohol attached to the β-carbon makes it a good leaving group. In the second step, a lone pair on the enol OH is used as an electron source to expel the water. Deprotonation of the carbonyl yields the final product in a last step.

There are also a series of more complex eliminations that you should be aware of (Eqs. 10.63–10.67), although we are not going to look at these in any detail. One is the elimination of 1,2-dihaloalkanes and 1,4-dihaloalkanes (the **Grob fragmentation**) using Zn to create alkenes or dienes (Eqs. 10.63 and 10.64, respectively). The first step in both reactions involves the oxidative addition of Zn to a C–X bond, a reaction we will cover in detail in Chapter 12. Other eliminations involve γ-amino alkyl halides, which can spontaneously undergo elimination (Eq. 10.65), and the base-induced eliminations of both β-hydroxyketones (Eq. 10.66, the reverse aldol reaction) and δ-ketoketones (Eq. 10.67, the reverse Michael addition).

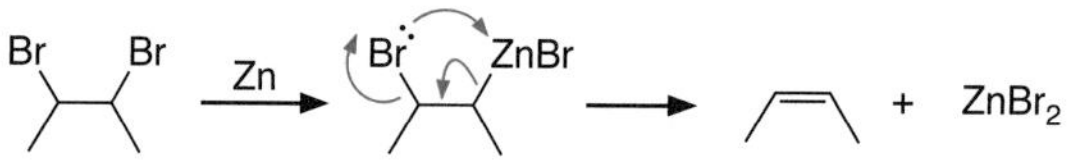

(Eq. 10.63)

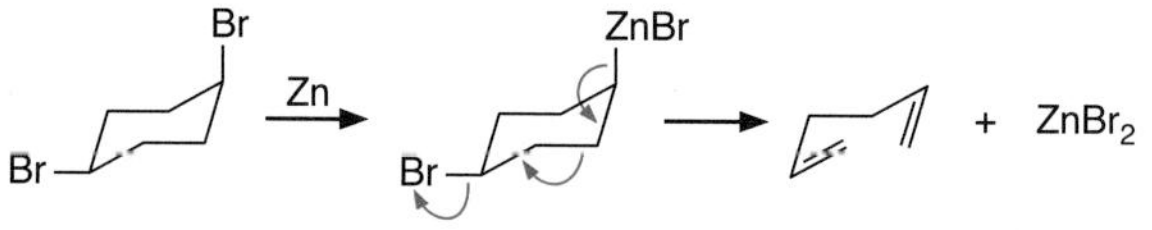

(Eq. 10.64)

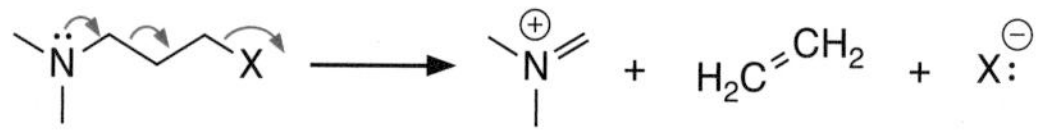

(Eq. 10.65)

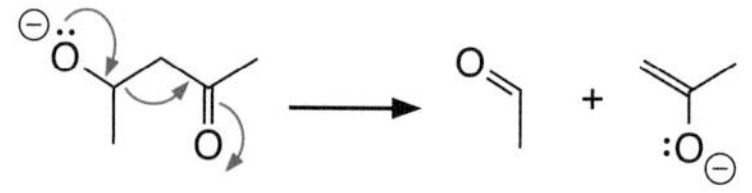

(Eq. 10.66)

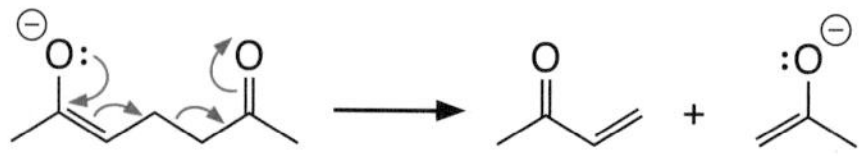

(Eq. 10.67)

10.13.2 Some Experimental Observations for E2 and E1 Reactions

Much of the work that delineated the details of E1 and E2 reactions came from the laboratories of Hughes and Ingold in the 1940s and 1950s. E2 reactions show second-order kinetics (Eq. 10.68 **A**), first order in base and first order in organic reactant. These reactions show large primary deuterium kinetic isotope effects at the site undergoing deprotonation, indicating significant deprotonation during the rate-determining step. In contrast, E1 reactions have no kinetic dependence upon the base, and are solely first order in the organic reactant (Eq. 10.68 **B**). Deuterium isotope effects at the site of deprotonation are minimal, as deprotonation is past the rate-determining step. E1 reactions do not need an added base, because the solvent or the departed leaving group can act as the base.

$$\text{A: } \frac{d[\text{P}]}{dt} = k[\text{reactant}][\text{base}]$$

$$\text{B: } \frac{d[\text{P}]}{dt} = k[\text{reactant}]$$

(Eq. 10.68)

10.13.3 Contrasting Elimination and Substitution

Almost all bases are also nucleophiles, and hence we expect competition between eliminations and substitutions. In both S_N2 and E2 reactions, the nucleophile or base reacts in a single rate-determining step with the reactant. In both S_N1 and E1 reactions, the nucleophile or base reacts in a step after the rate-determining heterolysis. Because the experimental observations for substitution and elimination reactions are so similar, we leave the discussion of kinetics to our discussion of substitutions in the next chapter. There are, however, some points that we should make about the factors that influence the extent of S_N2 versus E2 and S_N1 versus E1 reactions (Eq. 10.69).

(Eq. 10.69)

An E2 reaction requires the addition of a base, and can be performed in solvents of low or high ionizing power. Since most strong bases are also good nucleophiles, S_N2 reactions are important competitors. However, elimination will dominate if the carbon with the leaving group is not susceptible to nucleophilic attack, such as a tertiary R group. Examples of these trends are given in Table 10.6.

E1 reactions involve carbenium ion intermediates, and therefore are facilitated by all the factors that stabilize carbenium ions. These are the same factors that facilitate S_N1 reactions. Strongly ionizing solvents and substitution of electron donating groups on the carbon undergoing heterolysis are necessary.

In highly ionizing solvents and with R groups that readily form carbenium ions, the ratio of substitution to elimination products is typically independent of the leaving group. This evidence supports the notion that the substitution and elimination products are formed by branching from a common intermediate, and therefore the two reactions share a common rate-determining step. In contrast, in solvents of lower ionizing power, the ratio of substitution to elimination products does depend upon the leaving group, an indication that the two

Table 10.6
Percent Elimination Found for Various Alkyl Bromides in Two Different Solvents and with the Addition of the Very Weak Base Chloride or Strong Base Ethoxide*

Reactant	Solvent	Base	Elimination[‡]
⋎—Br	Ethanol	NaOEt	100%
⋎—Br	Acetone	NBu$_4$Cl	96%
⋋—Br	Ethanol	NaOEt	75%
⋋—Br	Acetone	NBu$_4$Cl	0%
⁄—Br	Ethanol	NaOEt	9%
⁄—Br	Acetone	NBu$_4$Cl	0%

*Biale, G., Cook, D. P., Lloyd, D. J., Parker, A. J., Stevens, I. D., Takahashi, J., and Winstein, S. "The E2C Mechanism in Elimination Reactions. II. Substituent Effects on Rates of Elimination from Acyclic Systems." *J. Am. Chem. Soc.*, **93**, 4735 (1971).
[‡]Any remaining product is due to substitution.

reactions do not share a common intermediate. Instead, contact ion pairs are formed, and either the leaving group or the solvent can remove the proton to give elimination products. The formation of a contact ion pair also has an influence on the stereochemistry of the elimination (see below). A much more thorough discussion of the factors influencing the balance between elimination and substitution reactions will be given in Chapter 11, after we have discussed substitution reactions.

10.13.4   Another Possibility—E1cB

When an electron withdrawing group is alpha to the hydrogen involved in an elimination, this hydrogen is relatively acidic compared to a hydrogen on a standard alkyl group involved in E2 and E1 mechanisms. Eliminations from these kinds of reactants occur with poor leaving groups and weaker bases than used with a E2 pathway. This means that a standard E2 mechanism is not occurring. Such reactants also do not support E1 mechanisms because poor leaving groups are viable in the reaction, and ionizing solvents are not required. Therefore, another mechanism is involved. Scheme 10.17 shows an example of deprotonation in the first step to give an enolate as an intermediate, and in a second step the leaving group departs. This elimination is the microscopic reverse of a conjugate addition. Any elimination that first forms the conjugate base of the reactant is referred to as **E1cB** (elimination, unimolecular, conjugate base).

Scheme 10.17
Electron pushing for a specific example of an E1cB mechanism.

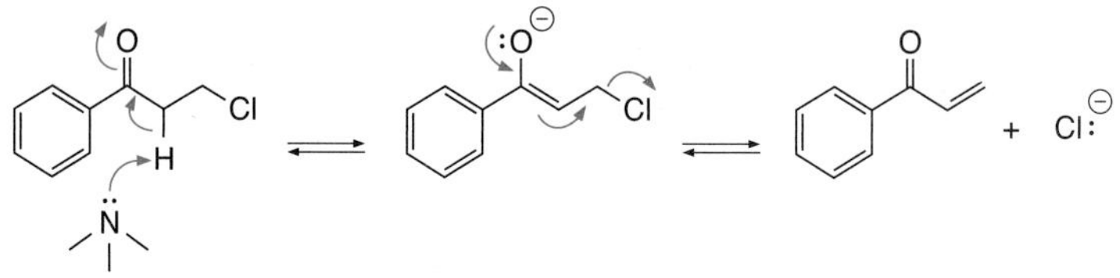

10.13.5   Kinetics and Experimental Observations for E1cB

Although at first glance the E1cB mechanism seems simple, the kinetics of the reaction can be complex. Depending upon the relative rate constants for the individual steps, the kinetics of the reaction can take three different forms. Let's see this by examining the rate expressions for the schematic mechanism given in Eq. 10.70. We treat the carbanion as a transient intermediate and use the steady state approximation (see Section 7.5). This gives Eq. 10.71.

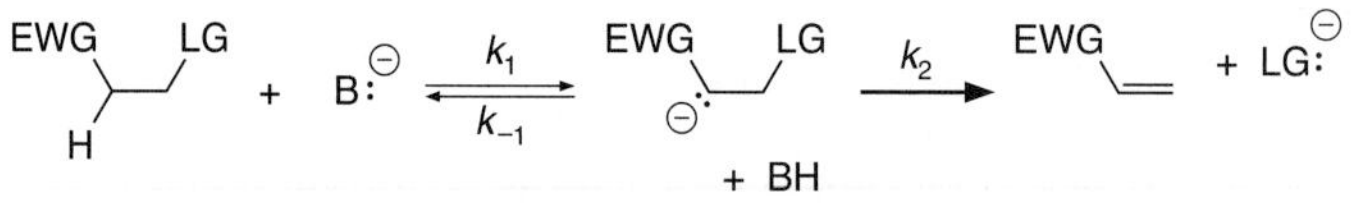

$$\frac{d[P]}{dt} = \frac{k_1 k_2 [\text{R–LG}][\text{B}^-]}{k_{-1}[\text{BH}] + k_2} \qquad \text{(Eq. 10.71)}$$

When the deprotonation occurs in a reversible step, leaving group departure is rate-limiting, $k_{-1} \gg k_2$, and Eq. 10.71 reduces to Eq. 10.72. Recall from Chapter 5 that the ratio of a base to its conjugate acid (here $[\text{B}^-]/[\text{HB}]$) sets the pH of a buffer, and is equal to $K_a/[\text{H}^+]$, where K_a is the acid dissociation constant of the conjugate acid of the base. We find that the rate of the elimination does not depend upon the concentration of the base, if the pH is kept constant. When the rate of a reaction that involves a base added to the solution depends only upon the pH, it is specific-base-catalyzed (see Section 9.3.2). Hence, many standard E1cB reactions are specific-base-catalyzed eliminations.

$$\frac{d[P]}{dt} = \frac{k_1 k_2}{k_{-1}} [\text{R–LG}] \left(\frac{[\text{B}^-]}{[\text{BH}]}\right) = \frac{K_a k_1 k_2}{k_{-1}[\text{H}^+]} [\text{R–LG}] \qquad \text{(Eq. 10.72)}$$

When the deprotonation occurs in a reversible step prior to leaving group departure, isotope scrambling from the solvent to the site of deprotonation will occur. With a carbonyl containing reactant, the α-hydrogens will become deuterated at a rate faster than elimination if the reaction is performed in a deuterated protic solvent. 4-Methoxy-2-butanone is an example of a reactant that shows all the elimination attributes discussed here (Eq. 10.73). Note, as we mentioned above, these eliminations can occur with poor leaving groups—in this case methoxide. Such reactions are called **E1cB$_R$**, where the R indicates a reversible first step.

$$\text{(Eq. 10.73)}$$

If in Eq. 10.70 $k_2 \gg k_{-1}$, Eq. 10.71 reduces to Eq. 10.74. Now the reaction is first order in both reactant and base under all experimental conditions, and the kinetics is identical to E2 (Eq. 10.68 A). However, because k_2 is not in the expression, changing the leaving group does not have as large an effect on the rate of the reaction as on the rate of an E2 reaction.

$$\frac{d[P]}{dt} = k_1 [R\text{–}LG][B^-] \qquad \text{(Eq. 10.74)}$$

As stated, the prediction is that one can distinguish this mechanism from that of a simple E2 reaction by changing the leaving group. 4-Benzoyl-2-butanone is a reactant that follows the experimental predictions discussed here, where substitutions on the benzoyl group do not affect the rate (Eq. 10.75). These reactions are called **E1cB$_{irr}$**, where the irr indicates an irreversible first step.

$$\text{(Eq. 10.75)}$$

One assumption in the derivation of Eq. 10.71 is that the deprotonation of the reactant forms a carbanion that can react with the BH that is present in solution to regenerate starting material. This is why [BH] is in the denominator. However, if the newly protonated base and the carbanion can react as a contact ion pair (common when the base is neutral to start and becomes positive) or as a hydrogen bonded complex (common when the base is anionic and becomes neutral), a different scenario can occur (see Eq. 10.76 for a general example). The subsequent reactions proceed within the solvent cage that surrounds the complex, and the reaction of the carbanion is not influenced by the bulk concentration of reactants. In this scenario the intermediate complex can be treated using the steady state approximation, and simple second-order kinetics overall is found, first order in reactant and first order in base (just like Eq. 10.74). Furthermore, no scrambling of hydrogens from the reactant with the solvent occurs. The reaction shows all the hallmarks of an E2 reaction except for one point. In many cases the deprotonation is an equilibrium prior to the rate-determining loss of the leaving group ($k_2 \ll k_{-1}$ in Eq. 10.76). We find *equilibrium* isotope effects on the first step, and these are normally small compared to a standard kinetic isotope effect that is found for a classic E2 elimination. Elimination of *cis*-1,2-dibromoethylene to give bromoacetylene shows all the experimental attributes discussed here (Eq. 10.77). Such reactions are called **E1cB$_{ip}$**, where ip stands for ion pair. In fact, because an sp^2 carbon is more acidic than an sp^3 carbon, many alkenes eliminate to give alkynes via an E1cB$_{ip}$ mechanism.

$$\text{(Eq. 10.76)}$$

$$\text{Br}\diagdown\diagup\text{Br} \ \underset{}{\overset{Et_3N}{\rightleftharpoons}} \ \left[\text{Br}\diagdown\diagup\overset{\ominus}{\text{Br}} \ \ H\overset{\oplus}{-}NEt_3 \right] \longrightarrow \ \equiv\!\!-Br \qquad\qquad \text{(Eq. 10.77)}$$

Ion pair

10.13.6   Contrasting E2, E1, and E1cB

There are several attributes of a reactant that one can look for when predicting whether elimination will proceed via E2, E1, or E1cB. First, most E1cB reactions need a strongly electron withdrawing group such as a cyano, carbonyl, or nitro adjacent to the proton being removed. However, as seen with the example involving 1,2-dibromoethylene (Eq. 10.77), an E1cB mechanism involving ion pairs does not necessarily meet this requirement. It is also usually quite clear when E2 and E1 mechanisms are operative because the experimental conditions used are very different. First, an E2 elimination is facilitated by a strong base, because the hydrogen undergoing deprotonation is not acidic. If no base is added to solution, E1 reactions are really the only possibility. Second, we need a good leaving group for E1 reactions, more so than with E2 reactions. Lastly, E1 eliminations require highly ionizing solvents, whereas E2 eliminations do not. Even though this seems clear at first glance, there is, as we have emphasized from the beginning of this chapter, a continuum of mechanisms between the extremes of E2, E1, and E1cB. The extent of deprotonation of the reactant at the transition state of the rate-determining step is the distinguishing feature of these reactions—complete in E1cB, partial in E2, and none in E1.

Since there is a continuum of mechanisms among the three prototypes we have discussed, either the deprotonation or the leaving group departure can be more advanced than the other at the transition state. In fact, the extents of deprotonation and leaving group departure in a simple E2 reaction depend upon each other. There are several experiments in the literature that show this dependence. For example, the deuterium isotope effect for the elimination reaction in Eq. 10.78 is 3.0 for $LG = N(CH_3)_3^+$ and 7.1 when $LG = Br$. The size of the isotope effect is indicative of a particular amount of deprotonation, which changed as the leaving group was changed. Similarly, Hammett ρ values that measure deprotonation also change as a function of the leaving group. For example, with $LG = I$, Br, Cl, and F for the reaction in Eq. 10.79, we find $\rho = 2.07$, 2.14, 2.61, and 3.12, respectively.

$$\text{(Eq. 10.78)}$$

$$\text{(Eq. 10.79)}$$

To understand the interdependence of the extent of leaving group departure and deprotonation, let's examine what makes a reaction such as E2 concerted. Jencks has described concerted reactions such as E2 as "forced" to be concerted. This term is used to describe a situation when no barrier separates a possible intermediate from the product. If there is no barrier to reaction of an intermediate, it has no lifetime, and does not represent a stable structure on the potential energy surface. For example, if a carbenium ion is too unstable to exist, an E1 reaction will become an E2 reaction. Furthermore, if a carbanion is too unstable to exist, an E1cB reaction will convert to E2. Stated another way, once we create a reactant that can achieve a carbanion or carbenium ion structure that is stable enough to exist, an E2 mechanism will convert to E1cB or E1, respectively.

As with all reactions that have possible competing pathways, we can examine the interdependence of the various mechanisms with More O'Ferrall–Jencks plots. In Figure 10.11, the diagonal shows the E2 reaction, which is a composite of the E1 and E1cB mechanisms shown proceeding at the bottom-left and top-right corners, respectively. We start our analy-

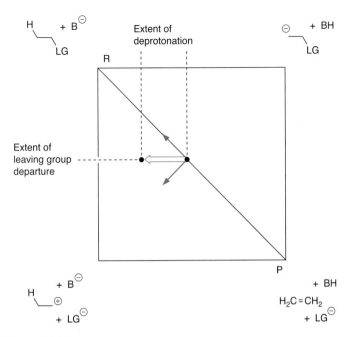

Figure 10.11
A More O'Ferrall–Jencks plot allows one to predict how deprotonation and leaving group departure in an E2 reaction change as a function of each other. We show what happens if the leaving group is made better.

sis with a dot halfway along the E2 path, corresponding to a reaction that is **synchronous**, both deprotonation and leaving group departure having proceeded to the same extent at the transition state. We can now consider how changes in leaving group structure, R group substitution, and base strength will affect the position of the E2 transition state. We only examine leaving group structure in detail here, while in Table 10.7 we summarize all the effects on transition state structure that this More O'Ferrall–Jencks plot predicts.

If we increase the leaving group ability in an E2 reaction, is there more leaving group departure in the transition state? Actually, there is not. Making the leaving group better lowers

Table 10.7
Effect on the E2 Transition State of Various Structural Changes

Change	Effect
Greater R group substitution on the carbon with the leaving group	a. Less deprotonation b. More leaving group departure c. More carbenium ion character
Better leaving group	a. Less deprotonation b. Same extent of leaving group departure
Placement of an electron withdrawing group on the carbon with the hydrogen	a. More deprotonation b. Less leaving group departure c. More carbanion character
Use of a stronger base	a. Same extent of deprotonation b. Less leaving group departure

the energy of both the bottom-left and bottom-right corners of the More O'Ferrall–Jencks plot. Lowering the bottom-right corner moves the transition state toward reactants along the reaction coordinate. Recall that this is a Hammond effect. Because the reaction is now more exothermic, the transition state more resembles the reactant. Lowering the bottom-left corner moves the transition state toward that corner, and this is perpendicular to the reaction coordinate (an anti-Hammond effect). The net result is the large arrow shown in Figure 10.11. The extent of leaving group departure at the transition state has not changed, but instead, the extent of deprotonation has decreased. The More O'Ferrall–Jencks plot leads to the conclusion that less deprotonation is necessary to expel a better leaving group. This makes sense, because expulsion of a better leaving group needs less neighboring negative charge to assist its expulsion. Such an effect has been observed by an examination of Hammett plots (Eq. 10.79). The ρ value was larger for fluoride, the worst leaving group. When a better leaving group is used, Hammett ρ values decrease, indicating less charge at the transition state. Conversely, more deprotonation would be necessary to expel a worse leaving group, but still the extent of leaving group departure would remain the same at the transition state. When the leaving group becomes so good that it departs prior to deprotonation, we have an E1 mechanism, and no deprotonation.

Going Deeper

Gas Phase Eliminations

A stark difference exists in the potential surfaces for solution phase and gas phase eliminations, giving very different kinetic observations. The addition of hydroxide to alkyl halides in the gas phase leads to spontaneous formation of an **ion–molecule complex**. The hydroxide associates with the alkyl halide with no activation barrier, because it is the only way the hydroxide can achieve any solvation at all, albeit only by ion–dipole and ion–induced-dipole interactions (see Section 3.2.2). Once the complex has been formed, either substitution reactions or elimination reactions can occur, both of which have significant barriers (see the Going Deeper highlight of Section 11.5.4). Elimination commonly dominates over

substitution, due to a lower entropy of activation. A substitution reaction requires backside alignment with the leaving group, whereas the geometries for elimination are less constrained. After the elimination, a complex between water, leaving group, and alkene is formed, which must dissociate to give product. When there is more than one possible regiochemistry for elimination, high selectivity is often seen because the ion–molecule complex is sufficiently stable and has a long enough lifetime to choose among the various barriers to reaction.

Pellerite, M. J., and Brauman, J. I. "Intrinsic Barriers in Nucleophilic Displacements." *J. Am. Chem. Soc.,* **102**, 5993 (1980). Wladkowski, B. D., and Brauman, J. I. "Substitution versus Elimination in Gas-Phase Ionic Reactions." *J. Am. Chem Soc.,* **114**, 10643–10644 (1992).

Ion–molecule complexes

10.13.7   Regiochemistry of Eliminations

When there are hydrogens on both carbons adjacent to the carbon with the leaving group, two possible double bonds can be formed in elimination reactions. **Saytzeff's rule** states that the more substituted double bond will dominate, a common observation for both E2 and E1 reactions. Here, the double bond is substituted by electron donating groups such as alkyls. Electron withdrawing groups can reverse the rule. The product with the more substituted alkene is referred to as arising from **Saytzeff elimination**, whereas the product with the less substituted double bond arises from what is called **Hofmann elimination**.

With E1 reactions, Saytzeff elimination dominates because the transition state for proton removal from the carbenium ion has double bond character. Since substitution by electron donating groups is known to stabilize double bonds (see Section 2.4.1), that stabilization is felt in the transition state, making the barrier to the more substituted double bond lower in energy. One example is given in Eq. 10.80. However, the formation of contact ion pairs can influence the regiochemistry.

$$\text{(Eq. 10.80)}$$

The rationalization for Saytzeff elimination in E2 reactions is similar to the reasoning for E1 reactions. As the base removes the proton and the leaving group is departing, the extent of double bond character developed at the transition state is large enough that the more substituted double bond will have a lower barrier to its formation. One example is given in Eq. 10.81. As one increases base strength, the base becomes more reactive and less selective for which hydrogen it removes.

$$\text{(Eq. 10.81)}$$

As the mechanism becomes more E1cB-like, the regiochemistry can start to change. With an E2 reaction that has a lot of carbanion character, the deprotonation of the more acidic hydrogen will dictate the elimination regiochemistry. In E1cB reactions, the regiochemistry is completely dictated by the relative acidity of the protons that can be removed. The double bond will form oriented to the carbon with the most acidic proton.

If there are severe steric factors that make the hydrogen on the more substituted carbon inaccessible, Hofmann elimination will dominate the product mixture (Eq. 10.82). The larger the base used in the elimination, the greater the extent of Hofmann elimination.

$$\text{(Eq. 10.82)}$$

It is also generally observed that elimination reactions with quaternary ammonium and sulfonium leaving groups give preferential Hofmann elimination. See Eqs. 10.83 and 10.84 for two examples. There is a steric bias toward deprotonation of the less hindered proton because the leaving groups are large, and there is also a statistical effect in that there are more hydrogens for deprotonation on the less hindered carbon. However, there is also an electronic effect operating in these eliminations. It has been theorized that a strongly electron withdrawing cationic leaving group (recall the group electronegativity of a quaternary ammonium ion, Table 1.2) creates a significant amount of positive charge on the neighboring hydrogens. However, electron donating alkyl groups diminish this charge on the neighboring hydrogens, and hence the most positive hydrogens are those on the less substituted carbon. This leads to preferential deprotonation of the less substituted carbon and thus formation of the less substituted double bond.

More positive charge here — $NR_3^{\oplus}$

Less positive charge here

$$\text{(Eq. 10.83)}$$

$$\text{(Eq. 10.84)}$$

Table 10.8 brings together some of the trends discussed above. By examination of this table, we can see leaving group effects, steric effects, and effects of the base strength. The stronger bases increase Hofmann elimination (see the first three entries), and the very large leaving groups give more Hofmann elimination.

Table 10.8
Relative Percentages of Hofmann and Saytzeff Elimination
Using Various Experimental Conditions*

Reactant	Base/solvent	1-Butene (Hofmann)	2-Butene (Saytzeff)
2-Iodobutane	Benzoate in DMSO	7%	93%
2-Iodobutane	Phenoxide in DMSO	17%	83%
2-Iodobutane	*t*-butoxide in DMSO	21%	79%
2-Bromobutane	*t*-butoxide in DMSO	33%	67%
2-Chlorobutane	*t*-butoxide in DMSO	43%	57%
2-Dimethylsulfoniumbutane	Ethoxide in ethanol	74%	26%
2-Trimethylammoniumbutane	Hydroxide in water	95%	5%

*Bartsch, R. A., Pruss, B. A., Bushaw, B. A., and Wiegers, K. E. "Effects of Base Strength and Size in Base-Promoted Elimination Reactions." *J. Am. Chem. Soc.*, **95**, 3405 (1973). Griffith, D. L., Meges, D. L., and Brown, H. C. "Reaction of 2-Butyl Halides with Potassium *t*-Butoxide, *t*-Heptoxide. Evidence for a Steric Effect of the Attacking Base in Influencing the Direction of Elimination in an E2 Reaction." *J. Chem. Soc., Chem. Comm.*, 90 (1968). Hughes, E. D., Ingold, C. K., Maw, G. A., and Woolf, L. I. "Mechanism of Elimination Reactions. Part XIII. Kinetics of Olefin Elimination from *iso*-Propyl, *sec*-Butyl, *m*- and 1-Phenyl-ethyldimethylsulfonium Salts in Alkaline Alcholic Media." *J. Chem. Soc.*, 2077 (1948). Cope, A. C., LeBel, N. A., Lee, H. H., and Moore, W. R. "Amine Oxides. III. Selective Formation of Olefins from Unsymmetrical Amine Oxides and Quaternary Ammonium Hydroxides." *J. Am. Chem. Soc.*, **79**, 4720 (1957).

10.13.8   Stereochemistry of Eliminations—Orbital Considerations

Let's start our examination of stereochemistry by looking at the requirements for orbital alignment in E1 reactions. If a π bond is to be created in a single step by removal of a proton adjacent to a carbenium ion, the C–H bond to this proton must be aligned with the empty *p* orbital to start. In fact, the proton removed by elimination is involved in hyperconjugation [see the π(CH$_3$) orbital interaction in the margin], making the hydrogen aligned with the empty *p* orbital more acidic.

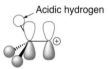

Acidic hydrogen

In a fully solvent-equilibrated carbenium ion, the stereoelectronic requirement just discussed would have no influence on the stereochemistry of the elimination, because the proton to be eliminated can be aligned with either the top or the bottom of the empty π orbital. However, a stereochemical consideration does arise when there is preferential removal of the proton from one of these two possibilities. This arises when contact ion pairs are formed in an E1 reaction, and the leaving group acts as the base to remove the proton. For example, the elimination of erythro-3-D-2-butyl tosylate gives only elimination products via a syn pathway in the low-ionizing solvent nitromethane (Figure 10.12). A contact ion pair is formed and the tosylate is the base that removes the proton. However, in more ionizing solvents, such as aqueous ethanol, all four products shown in Figure 10.12 are observed.

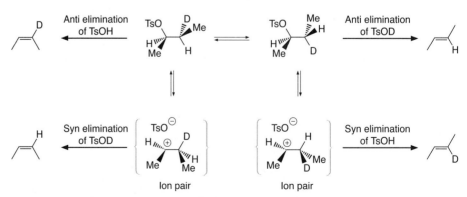

Figure 10.12
A contact ion pair formed in the E1 reaction of *erythro*-3-D-2-butyl tosylate results in products only from syn elimination.

A.

B.

Figure 10.13
Stereoelectronically-controlled E2 pathways.
A. Antiperiplanar, and **B.** Synperiplanar.

In contrast to E1 reactions, E2 reactions are always very strongly influenced by stereo-electronic factors. Since an E2 reaction has double bond character in the transition state, and the p orbitals that make the double bond must be coplanar in the product, the proton undergoing abstraction and the leaving group that is departing must be coplanar also. There are two conformations in which this is true: anti- and synperiplanar (see Figure 2.7). The orbitals are aligned as shown in Figure 10.13. In the antiperiplanar arrangement all the groups are staggered, and this is the favored arrangement. Elimination via the syn pathway involves a conformation that is actually a transition state along the potential energy curve for bond rotation.

Besides the conformational preferences, there are orbital reasons that cause anti elimination to dominate. If one aligns the σ bonding orbital of the C–H bond that is being deprotonated with the σ^* orbital of the C–X bond that is breaking, there is better overlap with the anti conformation. We discussed this point extensively in Chapter 2 (see Figure 2.19).

Let's look at a few examples of anti- and synperiplanar eliminations. As you might anticipate, anti elimination occurs for the vast majority of systems, even when the less stable isomer is produced, whereas syn elimination requires some special circumstances. When stereochemistry is possible in the product, the two paths typically give opposite results. For example, when the very weak base tetrabutylammonium chloride reacts with the brosylate compound (Bs = $SO_2C_6H_4Br$) shown in Eq. 10.85, only the less stable product is produced, that from anti elimination (see the Connections highlight on page 593 for another example).

$$\text{(Eq. 10.85)}$$

Less stable

Syn elimination can occur when one or more of the following circumstances occurs: 1. a synperiplanar arrangement can be achieved but an antiperiplanar one cannot; 2. the counterion of the base is ion paired with the base and the leaving group; and 3. strong steric factors favor the syn pathway. As an example of the first case, consider Eq. 10.86. Here, the deuterium is synperiplanar with the leaving group (see margin), but the hydrogen has a dihedral angle of 120°, not 180°, with the leaving group. We find only the syn elimination product.

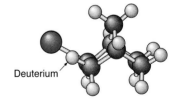

Deuterium

$$\text{(Eq. 10.86)}$$

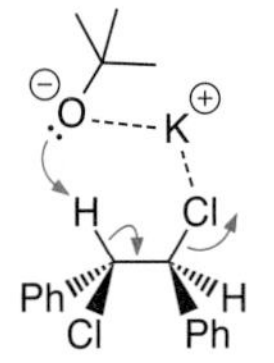

Ion pairing

Anti elimination

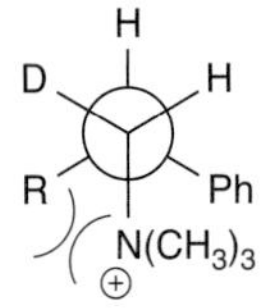

Disfavored by larger R group

Syn elimination

As an example of ion pairing, consider the elimination of meso-1,2-dichloro-1,2-diphenylethane in Eq. 10.87. Here, syn elimination occurs 13% of the time. Addition of 18-crown-6 completely wipes out any syn elimination. Apparently the small percent of syn elimination arises from an ion pair where the potassium cation bridges the base and the leaving group (see margin). Addition of the crown ether negates the ability of the K^+ to act in this manner.

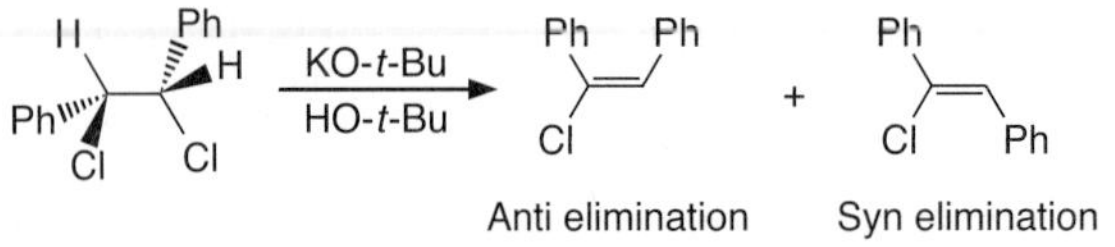

$$\text{(Eq. 10.87)}$$

Anti elimination   Syn elimination

As stated, one last way syn elimination can occur is via steric constraints. Consider the reaction shown in Eq. 10.88. When R = CH_3, the amount of syn elimination is 37%, and this increases to 69% when R is isopropyl. The large isopropyl group disfavors the conformation, and the resulting transition state, with the large leaving group gauche to this R group (as shown in the margin). Therefore, elimination occurs more often from the syn pathway.

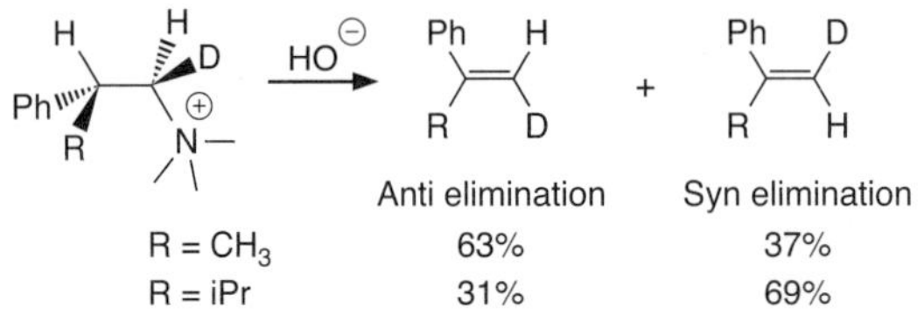

$$\text{(Eq. 10.88)}$$

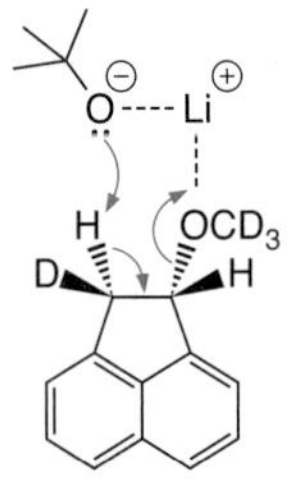

	Anti elimination	Syn elimination
R = CH_3	63%	37%
R = iPr	31%	69%

Eliminations involving E1cB mechanisms can occur by both syn and anti pathways. For example, the structure shown in Eq. 10.89 was found to undergo deuterium exchange with solvent in competition with elimination, thereby indicating an E1cB mechanism. When various salts of the base t-butoxide (M^+–$^-$O-t-Bu) are used, the syn/anti elimination ratio decreases in the order M^+ = Li^+ > Na^+ > K^+ > $(CH_3)_4N^+$. Syn elimination is favored with the Li^+ counterion because this cation is strongly ion paired with both the butoxide base and the departing methoxide leaving group. Hence, the base removes a proton from the same face of the newly forming double bond as the leaving group departs (see margin). As the counterion becomes less coordinating, anti elimination dominates because the reaction more resembles E2.

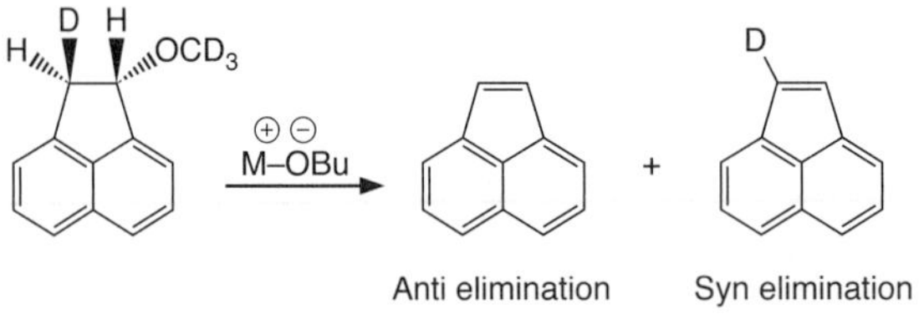

$$\text{(Eq. 10.89)}$$

Anti elimination   Syn elimination

10.13.9 Dehydration

One of the most common types of elimination is the dehydration of an alcohol. Formally, these reactions are the microscopic reverse of the hydration of an alkene, and therefore we have already covered these reactions (see Section 10.2). However, several points should be stressed here, because dehydrations are commonly used in synthesis. Furthermore, dehydrations and hydrations are important biosynthetic reactions, as many natural products possess alkenes and alcohols. The Connections highlight at the end of this section discusses how enzymes catalyze these reactions.

Electron Pushing

Scheme 10.18 shows the standard electron pushing for an acid-catalyzed dehydration reaction. Base-catalyzed reactions typically occur only when conjugated dienes are formed, and we do not cover these reactions here. Alcohols show specific-acid-catalyzed dehydrations, and thus the mechanism given starts with a reversible protonation of the alcohol.

Connections

Using the Curtin–Hammett Principle

For practice, let's predict the products and the relative rates of elimination of the following two alkyl chlorides.

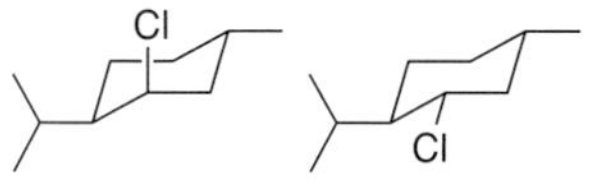

Isomeric substrates

The prediction of products is straightforward. We need to examine the two ring conformations of the reactants with a focus upon any antiperiplanar arrangements between the leaving group and a hydrogen. Note that an antiperiplanar arrangement requires the Cl to be in an axial position, so we only need to consider ring conformations that fulfill this requirement. The possible reactions are shown below with the hydrogens noted and the relative energies of the two ring conformations. This is the kind of situation where a molecular mechanics calculation would be quite useful for predicting the relative energies of the various ring conformations (not of the transition states!). In part **A** there are two possible products, and indeed both are observed. In part **B** there is only one possible product, and indeed only one is observed. This is

excellent experimental evidence in support of the stereo-electronic arguments presented above.

Predicting which reaction is faster is more challenging. It is actually the one shown in part **B**. In part **A** the conformation that is highest in concentration is the one that can eliminate, yet that is the slowest reaction. To understand this we need to recall the Curtin–Hammett principle (see Section 7.3.3). It is the relative barriers to the reactions that control the rates, not the relative populations of conformations. In part **B**, very little of the productive conformation exists in solution, but it is a highly reactive conformation. The compound is strained in a manner that facilitates the reaction. There is a considerable 1,3-diaxial interaction between the chlorine and the methyl group, and this strain is relieved upon elimination. No such strain exists in the productive conformation in part **A**. A strain that is relieved upon achieving the transition state facilitates the reaction. This is very similar to the concept we introduced in Chapter 9 related to catalysis. Binding of a substrate to a catalyst in a form that strains the substrate toward the geometry of the transition state will facilitate the reaction.

Boger, D. L. (1999). *Modern Organic Synthesis*, TSRI Press, La Jolla, CA, p. 28.

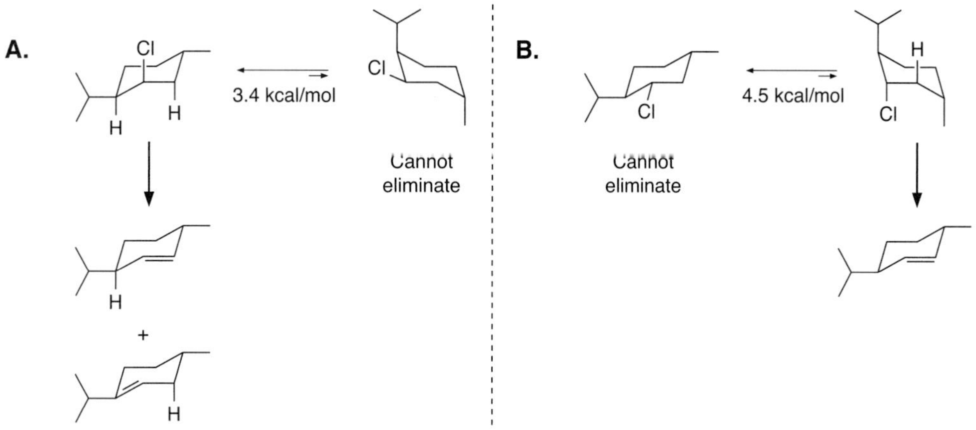

A.

3.4 kcal/mol

Cannot eliminate

B.

4.5 kcal/mol

Cannot eliminate

Acids such as HCl, H_2SO_4, and H_3PO_4 are effective in this regard. Departure of water as a leaving group creates a carbenium ion that undergoes deprotonation according to Saytzeff's rule. Since a carbenium ion is created, rearrangements are an important side reaction that must be anticipated.

Scheme 10.18
Standard electron pushing for a dehydration reaction, using cyclohexanol as an example.

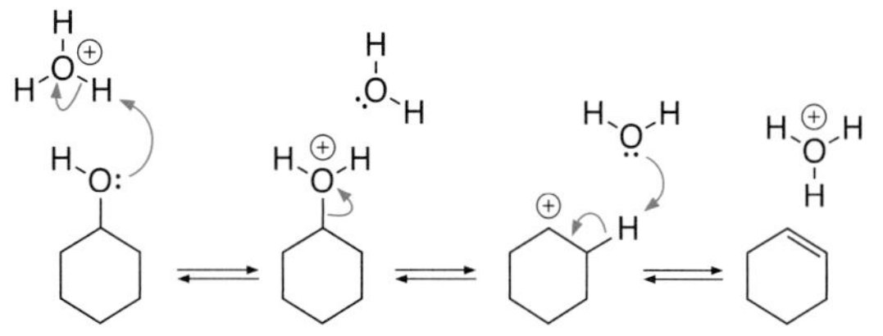

Other Mechanistic Possibilities

As with so many classic electron-pushing schemes, Scheme 10.18 is relevant to only certain reactants. The mechanism of the reaction depends upon the substitution pattern around the alcohol. For example, most alcohols actually exchange oxygen with the solvent faster than they eliminate. *t*-Butyl alcohol exchanges the natural abundance O with ^{18}O-labeled water about 30 times faster than it eliminates to make isobutylene in H_2SO_4. However, 1-butanol exchanges the O with ^{18}O-labeled water only about three times faster than elimination. With primary alcohols, water exchange occurs by an S_N2 process on the protonated alcohol. Even with some secondary alcohols, the water exchanges by an S_N2 process because inversion of stereochemistry is found. As an example, the rate of racemization of 2-butanol in sulfuric acid is twice the rate of oxygen exchange with the solvent (a classic test for S_N2 reactions; see Section 11.5.4). Since exchange with solvent is faster than deprotonation to create the alkene, the deprotonation must be the rate-determining step for these reactions.

The fact that primary carbenium ions are unstable suggests that the exchange with solvent is an S_N2 process with primary alcohols. If true, then are primary carbenium ions ever intermediates in dehydration reactions? Studies have shown that it depends upon the case. Neopentyl alcohol does form a primary carbenium ion, whereas 1-propanol does not. Acid-catalyzed elimination of 1-propanol to form propene occurs by a concerted E2 reaction (Eq. 10.90). Similarly, whether a secondary alcohol eliminates in acid via an E1 or E2 pathway depends on the case.

(Eq. 10.90)

10.13.10   Thermal Eliminations

In some cases, eliminations occur in non-ionizing solvents and without the addition of any base. In these cases the reactant itself has an internal base and a cyclic transition state leads to elimination. The symbolism for the reactions is Ei, standing for elimination, intramolecular. Only heat is required to induce the reaction, and hence these reactions are called **thermal eliminations** (the term **pyrolysis** is also sometimes used). Thioesters, xanthates, selenoxides, and *N*-oxides are common in these reactions. The **Cope elimination** involves the formation of an *N*-oxide and subsequent elimination via the pathway shown in Eq. 10.91, and the **Chugaev elimination** involves xanthate esters [ROC(S)SR]. The Chugaev elimination was shown to follow a syn elimination pathway based on the stereospecific nature of the reaction (Eqs. 10.92 and 10.93).

(Eq. 10.91)

(Eq. 10.92)

(Eq. 10.93)

Ester pyrolysis to form an alkene and a carboxylic acid is another common thermal elimination (Eq. 10.94). At temperatures between 400 and 450 °C, these reactions can be very efficient. The reactions are first order, and show only syn elimination products, indicating that at these high temperatures radicals are not involved. Deuterium isotope effects at the site undergoing deprotonation during the reaction are large, in the range of 2.0 to 2.5. This is actually about the maximum that an isotope effect can be at these high temperatures.

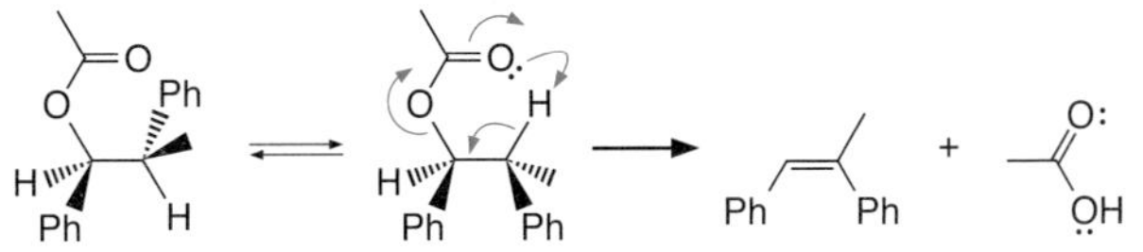

(Eq. 10.94)

Connections

Aconitase—An Enzyme that Catalyzes Dehydration and Rehydration

Given the acid-catalyzed mechanisms discussed above for dehydration and hydration, and our look at enzyme catalysis in Chapter 9, one might expect an enzyme to use a combination of general-acid and general-base catalysis for these reactions. Indeed, this is common, but nature can be surprising, too. The enzyme aconitase is the second enzyme in the citric acid cycle, and it catalyzes a dehydration–rehydration that interconverts citrate and isocitrate via the intermediate cis-aconitate. The enzyme uses an Fe–S cluster as a cofactor, a group normally associated with electron transfer chemistry. Therefore, deciphering the role of this cluster has been a challenge in enzymology.

The major tools used to probe the role of the Fe–S cluster were various forms of spectroscopy and x-ray crystallography. As shown below, the mechanism does involve

general-acid and general-base catalysis, but only for the removal and addition of the proton required in the reaction. The elimination of the hydroxide leaving group is facilitated by transfer to an Fe center (step 1). This transfer of electrons to the Fe–S cluster via coordination of a hydroxide leaving group is analogous to a reduction, a role this cluster often performs in redox chemistry. Apparently at this stage, the cis-aconitate flips in the enzyme active site, with the carboxylates changing positions (step 2). Rehydration then occurs by the reverse of the dehydration steps, producing the isomeric product (step 3). Therefore, the roles of the Fe–S cluster are to bind one carboxylate of the substrate, enhance hydroxide departure, and then redeliver hydroxide to the bound substrate.

Emptage, M. H. (1988). "Aconitase, Evolution of the Active-Site Picture" in *Metal Clusters in Proteins*, ACS Symposium Series #372, Chapter 17, pp. 343–371.

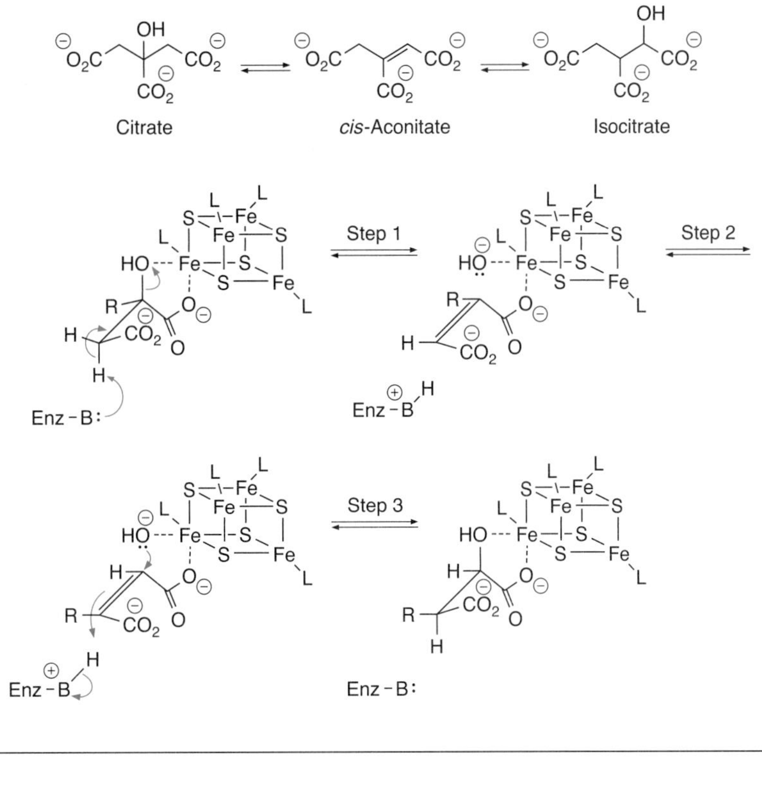

Citrate   cis-Aconitate   Isocitrate

10.14 Eliminations from Radical Intermediates

Radicals undergo various elimination reactions, sometimes as side reactions during the transformations we describe in this and the next chapter. These reactions are not part of standard radical mechanisms, so we simply group a few of them together here as examples of reactions you should know.

Just as deprotonation adjacent to a carbenium ion can form an olefin, similarly removal of H• adjacent to a free radical will form an olefin (Eq. 10.95). As we noted in Section 10.10.4, the process of Eq. 10.95 is referred to as radical disproportionation when the radicals are the same. Unimolecular elimination from a radical is the simple reverse of the addition of a radical to an alkene (Eq. 10.96). Since the addition is typically exothermic, it takes heat to reverse the addition. One example is the depolymerization of polystyrene, which will occur at temperatures of 300 °C (Eq. 10.97; see Chapter 13 for a discussion of the polymerization reactions). Strain in an adjacent ring will favor elimination, as shown in Eqs. 10.98 and 10.99. These two examples convert one radical to another, and such reactions will be discussed in more detail in Section 11.11.

$$\text{R•} + \quad \rightleftharpoons \quad \text{R–H} + \quad \tag{Eq. 10.95}$$

$$\rightleftharpoons \quad \text{R•} + \tag{Eq. 10.96}$$

$$\rightleftharpoons \quad + \quad m \tag{Eq. 10.97}$$

$$\longrightarrow \tag{Eq. 10.98}$$

$$\xrightarrow{35\ °C} \tag{Eq. 10.99}$$

Elimination from an alkoxy radical is a common occurrence at ambient temperature, making this reaction a competing process to other radical reactions (Eq. 10.100). This reaction forms a carbonyl-containing product, and is typically referred to as **β-scission**. It is more exothermic than elimination to form an alkene, because of the high bond strength of a carbonyl double bond, as well as the instability of an oxygen radical. As the departing radical becomes more stable, the rate of elimination increases. Similarly, when two or more R groups compete for elimination, the one that forms the most stable radical eliminates preferentially. Eliminations also readily occur from acyl radicals, leading to carbon monoxide and a carbon-based radical (Eq. 10.101).

$$\rightleftharpoons \quad \text{R•} + \tag{Eq. 10.100}$$

$$\longrightarrow \quad \text{R•} + \text{CO} \tag{Eq. 10.101}$$

Combining Addition and Elimination Reactions (Substitutions at sp^2 Centers)

Now that we have covered both additions and eliminations, we can combine them. Many functional group transformations involve such a combination. The result is a substitution reaction that occurs at an sp^2 hybridized carbon. They are best described as **addition–elimina-**

tion reactions or **elimination–addition reactions**, depending upon the sequence of the two steps. The more common addition–elimination sequences typically involve either addition to a polarized double bond with subsequent 1,2-elimination (Eq. 10.102) or conjugate addition followed by 1,4-elimination (Eq. 10.103). Some substitutions on aromatic rings and other reactants proceed via the alternative elimination–addition pathway (Eq. 10.104).

$$\text{(Eq. 10.102)}$$

$$\text{(Eq. 10.103)}$$

$$\text{(Eq. 10.104)}$$

For the next several sections we will focus upon addition–elimination reactions at carbonyl centers. All these reactions are easily understood using our paradigm of reactivity. In every case we will show how a nucleophile with a full or partial negative charge attacks the partially positive carbonyl carbon. It is the subtle details that make the examples interesting.

10.15 The Addition of Nitrogen Nucleophiles to Carbonyl Structures, Followed by Elimination

Many addition–elimination reactions at carbonyl centers involve a nucleophilic attack on the carbonyl carbon, followed by an elimination that restores the double bond. We first explore addition followed by 1,2-elimination, one of many types of reactions referred to as **condensations**. Strictly speaking, a condensation occurs when two large molecules combine to create a more complex molecule with the loss of a small molecule, such as water or an alcohol. Therefore, a condensation is a form of substitution. We will also examine condensation reactions that form polymers (see Chapter 13). One of the more complex condensations is the formation of an imine or enamine from a carbonyl and an amine. In both of these cases an oxygen is replaced by a nitrogen with loss of water (Eq 10.105 and 10.106).

$$\text{(Eq. 10.105)}$$

$$\text{(Eq. 10.106)}$$

Imines, in which the nitrogen R groups are alkyl and/or hydrogen, are commonly too unstable to be isolated from an aqueous medium, rapidly hydrolyzing back to the carbonyl structure. However, when aromatic groups are placed on either the C or N, the structures are more stable and can often be isolated. They are referred to as **Schiff bases** (Eq. 10.105; R or R′ = Ar). When an oxygen or nitrogen is attached to the imine N, the structures become quite stable. Such structures are often used for derivitization: oximes (R′ = OH), semicarbazones (R′ = NHCONH$_2$), and hydrazones (R′ = NHR).

10.15.1   Electron Pushing

Only the electron pushing for formation of an enamine is shown here (Scheme 10.19), because the formation of an imine follows the same pathway except for the last step. After combination of the electrophile and nucleophile (step 1), two proton transfers are required (steps 2 and 3). Such transfers do not typically occur as one intramolecular step, but instead involve two steps. The result is a tetrahedral intermediate, called a **carbinolamine**. The carbinolamines are sometimes stable enough to be isolated, thereby confirming their viability as intermediates in the formation of imines and enamines. A 1,2-elimination leads to hydroxide plus an iminium ion (step 4). To quench the positive charge on the nitrogen of the iminium, the hydroxide deprotonates an α-hydrogen (step 5). These reactions are often performed in benzene or a similar nonpolar solvent, so the proton transfers involve the added amine, as shown in the mechanism. The reaction can be driven to completion by the removal of water from the flask, often using a Dean–Stark apparatus.

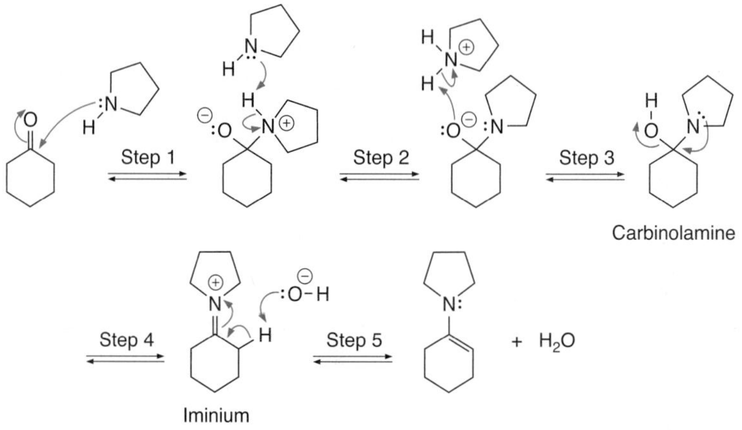

Scheme 10.19
Electron pushing for the formation of an enamine.

10.15.2   Acid–Base Catalysis

Commonly, bell-shaped pH versus rate profiles are found for imine and enamine formation. Figure 10.14 shows one example in the formation of an oxime using hydroxylamine. This amine is among the most nucleophilic of all amines. The bell-shaped profiles reflect acid catalysis superimposed upon the influence of pH on the protonation state of the amine nucleophile. Furthermore, the effects of nitrogen substituents on the reaction rate depend upon the pH. To understand the influence of substituents on the reaction, we group the amines into two classes. The first is strongly nucleophilic, with pK_a values of their conjugate acids between 6 and about 10 (hydroxylamines and alkylamines). The second is weakly nucleophilic amines, with pK_a values of the conjugate acids between 3 and 5 (aryl amines and semicarbazides).

With the strongly nucleophilic amines, spectral evidence is very revealing. At basic pH (beyond the pH maximum in curves such as Figure 10.14), the UV/vis absorbance band of the carbonyl rapidly disappears while the formation of the imine or enamine product takes much longer. This means the carbinolamine builds up in solution, and dehydration is the rate-determining step.

Figure 10.15 **A** shows the steps and relative rates involved in the addition with the strongly basic amines. The addition step has three possible pathways (step 1): direct addition, general-acid catalyzed addition, or specific-acid catalyzed addition. Because the amines are good nucleophiles, they add directly at all pHs, but below pHs around 4 this direct addition becomes rate-determining. This is because there is a low concentration of unprotonated amine present at low pHs. In some circumstances, enforced general-acid catalysis of the first step is found (see Section 9.3.6 for the definition of enforced catalysis). At the high pHs

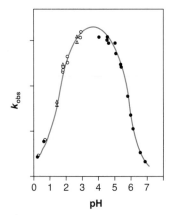

Figure 10.14
The k_{obs} values for the formation of the oxime of furfural as a function of pH. Jencks, W. P. "Studies on the Mechanism of Oxime and Semicarbazone Formation." *J. Am. Chem. Soc.,* **81**, 475–481 (1959).

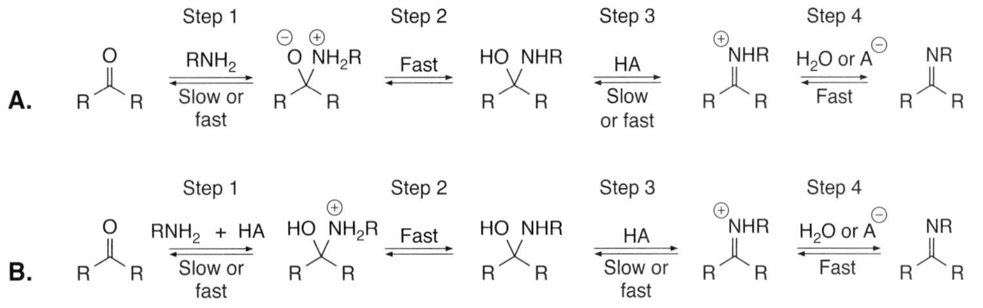

Figure 10.15
A. Imine formation with strongly nucleophilic amines. These amines add directly without acid catalysis, and dehydration can be rate-determining.
B. Imine formation with weakly nucleophilic amines. These amines require acid catalysis in both the nucleophilic addition and the dehydration.

where the carbinolamine breakdown is rate-determining (step 3), we find decreasing k_{obs} values as the pH is increased, indicating general-acid catalysis for the dehydration. The rate has a maximum where the amine is present in high enough concentrations as the free base form to react with a reasonable rate, but there is also enough acid present to catalyze the elimination of water from the carbinolamine, hence the bell-shaped pH–rate profile.

Different behavior is found for the weakly basic amines (Figure 10.15 **B**). The key difference is the addition step (step 1). Now the amines are not nucleophilic enough to directly add to the carbonyl, and general-acid catalysis is found for this step. Yet, superimposed upon the acid catalysis is the necessity that the amine be in its free base form, and therefore the rate still increases with increasing pH. At the higher pHs, the dehydration becomes rate-determining (step 3), and it involves general-acid catalysis, just as with the nucleophilic amines. Therefore, in this reaction both the addition and elimination steps are general-acid-catalyzed, but enough free base form of the amine still needs to be present to produce a reasonable rate.

10.16 The Addition of Carbon Nucleophiles, Followed by Elimination—The Wittig Reaction

One of the most useful C–C bond forming reactions is the Wittig reaction, which forms an olefin from a carbonyl compound and a phosphorus ylide (Eq. 10.107). Unlike most C–C bond forming reactions that make single bonds, this reaction creates a double bond in a single transformation. Many variants have been created, and they are covered in organic synthesis textbooks.

(Eq. 10.107)

The ylide is drawn as a **zwitterion** (a compound with a positive and a negative charge) in Eq. 10.107, as is triphenylphosphine oxide. It is common to see a P=C or P=O for these structures, instead. This is perfectly acceptable, although this requires the use of a d orbital on P to accommodate five bonds. Various forms of electronic structure theory calculations such as those discussed in Chapter 14 indicate that the d orbitals on P are too high in energy to participate in a significant manner in the bonding to phosphorus. Thus, the zwitterion forms are more representative of the true chemical structure.

10.16.1   Electron Pushing

The mechanism commences with a nucleophilic attack of the carbon from the ylide on the carbonyl carbon (Scheme 10.20). An intermediate is formed that exists in an equilibrium between a **betaine** (open form) and an **oxaphosphetane** (closed form). The alkene product is formed by direct elimination from the oxaphosphetane.

Scheme 10.20
Electron pushing for
the Wittig reaction.

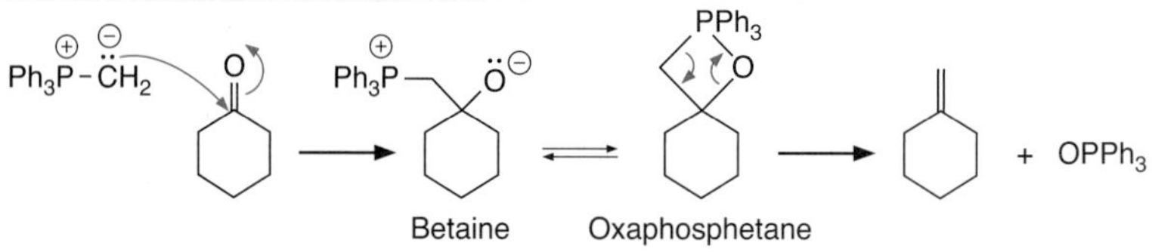

Betaine   Oxaphosphetane

10.17   Acyl Transfers

One of the most common substitutions at carbonyl centers is known as an **acyl transfer**. An **acyl group** is any RCO group (R = alkyl or aryl), where CO is a carbonyl. Acyl transfers form the basis of many biosynthetic and conventional organic synthesis procedures (several Connections highlights starting on page 604 discuss enzymatic examples, enzyme mimics, and efficient synthetic methods). In this reaction an acyl group attached to a leaving group is transferred to a nucleophile (Eq. 10.102). The leaving group has therefore been replaced with the nucleophile, commonly via an addition–elimination mechanism. When the addition–elimination occurs under basic conditions it is termed $B_{Ac}2$ (basic, acyl transfer, bimolecular), while under acidic conditions it is termed $A_{Ac}2$ (acidic, acyl transfer, bimolecular).

10.17.1   General Electron-Pushing Schemes

A simple example of an acyl transfer is the reaction of an acid chloride with an alcohol (Scheme 10.21). No catalysis is necessary due to the high reactivity of an acid chloride. The alcohol performs a direct nucleophilic attack on the electron deficient carbonyl carbon. The reaction is therefore very sensitive to the nucleophile. For example, the Swain–Scott s value for acyl transfer from benzoyl chloride is 1.43, indicating that the reaction is more sensitive to the nucleophile than is the S_N2 reaction on methyl iodide. Elimination of the chloride follows a typical 1,2-elimination pathway. The proton is removed from the intermediate either prior to or subsequent to chloride departure.

Scheme 10.21
Electron pushing for
formation of an ester
from an acid chloride.

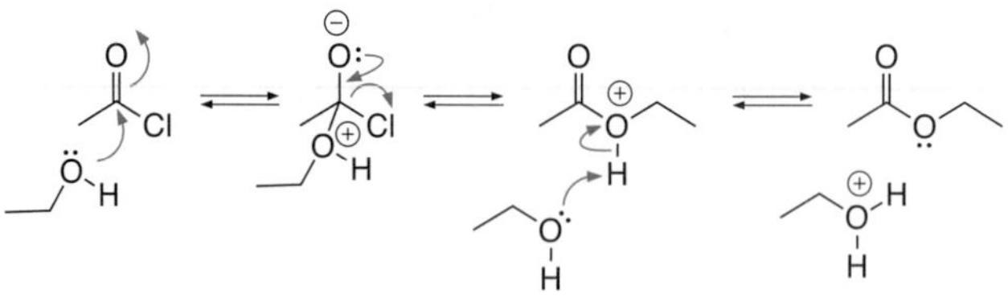

Acyl transfers are often performed under acidic conditions. Shown in Scheme 10.22 is the hydrolysis of an ester using specific-acid catalysis, which is normally taught in introductory organic chemistry. Here, the acyl group is transferred from an alcohol to water. The advantage to using acid catalysis is that the esters are activated toward nucleophilic attack. Protonation of the ester carbonyl by hydronium leads to a very electrophilic carbonyl (step 1) and the addition of water leads to cleavage of the carbonyl π bond (step 2). Subsequently a proton is removed (step 3). As we will see when we examine the experimental observations for these reactions, they proceed through what is called a **tetrahedral intermediate** (T.I.). To form the T.I. an acid-catalyzed addition reaction has occurred. However, given that a leaving group is attached to the central carbon, a 1,2-elimination reaction is possible. Protonation of the ethoxy group converts it into a good leaving group (step 4). Elimination is assisted by a lone pair of electrons on the OH to create a C–O π bond (step 5). Finally, deprotonation by water leads to the carboxylic acid product (step 6).

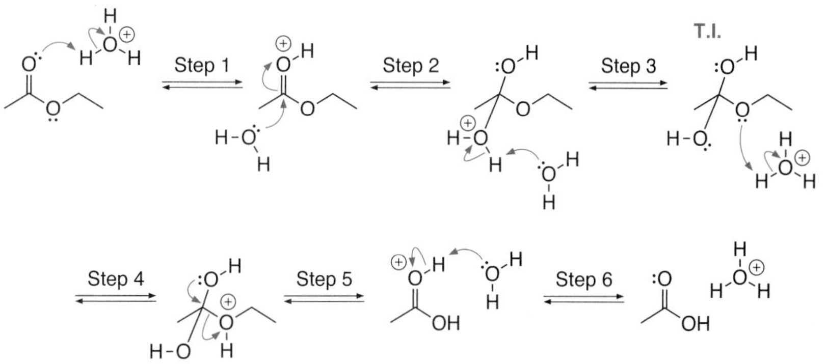

Scheme 10.22
Electron pushing for the acid
hydrolysis of an ester.

There are other possibilities besides a T.I., such as acylium ion formation, ketene formation, and direct displacement analogous to an S_N2 reaction (Figure 10.16). Such mechanisms are viable under certain conditions. Acylium ions are formed with acid halides, esters, and some amides under highly acidic conditions, and acid halides can form ketenes with base catalysis. However, by far the most common pathway is addition–elimination via a tetrahedral intermediate. Let's examine the evidence for this intermediate.

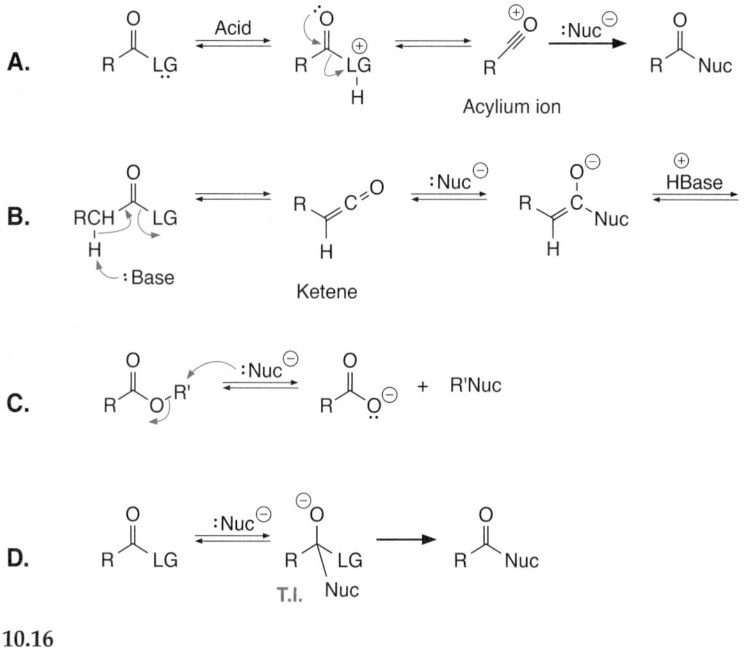

Figure 10.16
Acyl transfer mechanisms. **A.** Protonation of the leaving group rather than the carbonyl oxygen leads to an acylium ion intermediate. This is an S_N1-type substitution. **B.** Elimination of the leaving group using a base leads to a ketene intermediate. **C.** S_N2-type displacement. Strictly speaking, possibility **C** is not an acyl transfer because the acyl group remains attached to the original ester oxygen. **D.** T.I. formation (the most common mechanism).

10.17.2 Isotope Scrambling

Probably the most widely cited evidence for the existence of a tetrahedral intermediate comes from isotopic exchange reactions. For example, in the reaction of a carboxylic acid derivative, such as an ester, the two OH groups in the tetrahedral intermediate are equivalent (examine the T.I. in Scheme 10.22). If the reaction were performed in ^{18}O-labeled water, one of these OH groups would be isotopically labeled. Reversal of the nucleophilic addition step would exchange the ^{18}O into the carbonyl oxygen 50% of the time. As the reaction proceeds the reactant would build up ^{18}O substitution, which can be confirmed by isolation of the starting material after the reaction has been allowed to proceed to varying extents of completion. Such exchange has been observed in the hydrolysis of acid halides, anhydrides, esters, and amides. Apparently all these carboxylic acid derivatives can proceed through tetrahedral intermediates during acyl transfers.

A cautionary note is necessary at this stage. While the observation of isotope exchange is good evidence for a tetrahedral intermediate, the lack of isotope exchange is not necessarily evidence against a tetrahedral intermediate. If nucleophilic attack is rate-determining, so that $k_2 >> k_{-1}$ in Eq. 10.108, little exchange into the starting material will be seen. The tetrahedral intermediate predominately proceeds on to product instead of reverting to starting material. As we will discuss below, amides display such behavior under acidic conditions.

$$\text{(Eq. 10.108)}$$

10.17.3   Predicting the Site of Cleavage for Acyl Transfers from Esters

Ester hydrolysis is a paradigmatic acyl transfer. It is actually quite complicated. The reaction is susceptible to many forms of catalysis, and there are two possible mechanisms for cleavage: addition–elimination and S_N2 on the ester group (shown in Figures 10.16 **C** and **D**, respectively). Here, we discuss the manner in which the mechanism of cleavage is determined, while the forms of catalysis are examined below.

The most obvious way to distinguish the site of cleavage is with isotope labeling. If the ether oxygen of the ester group is ^{18}O-labeled, then the label is lost in the carboxylic acid product in an addition–elimination mechanism, but it is retained in the S_N2 mechanism. One finds that the addition–elimination mechanism is by far the most common, and the S_N2 mechanism is only viable for small ester groups that readily undergo nucleophilic attack, such as methyl and benzyl. Usually a very strong nucleophile such as RS^- or RSe^- is required also. The S_N2 path can also become favored if the acyl R group is sterically bulky.

Another method for analysis is to examine stereogenic centers at the ester group [RCO(OR')]. Addition–elimination leads to retention of the stereochemistry of the ester R' group, while S_N2 attack inverts the stereochemistry of the ester R' group.

10.17.4   Catalysis

Acyl transfer reactions are susceptible to many different forms of catalysis, including acid, base, and nucleophilic. We will look at each type here.

One of the most common methods to catalyze acyl transfers is the addition of a nucleophile, particularly when using acid halide reactants. For example, consider the reaction of an alcohol or water with an acid halide, but with the addition of triethyl amine. We add the triethylamine to neutralize the HX produced in this reaction, but it also significantly enhances the rate of the reaction (Eq. 10.109; also see Section 9.2.5 for a discussion of nucleophilic catalysis). Initial attack of the amine is faster than attack by the less nucleophilic alcohol or water, creating an intermediate with a positive charge. This is a highly reactive carbonyl that now adds nucleophiles faster than the starting acid halide.

$$\text{(Eq. 10.109)}$$

Several pieces of data support the role of certain amines as nucleophiles and others as general-base catalysts in this reaction. First, non-nucleophilic amines follow Brønsted relationships, but those amines that are small enough to be nucleophiles are markedly more active. Furthermore, solvent isotope effects are significantly different for the nucleophilic and general-base pathways. The small nucleophilic amines do not show solvent isotope effects, whereas the amines that act as general-base catalysts do show solvent isotope effects.

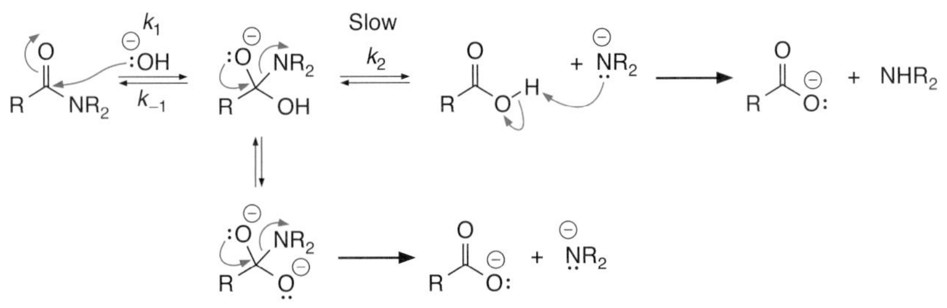

Figure 10.17
Hydrolysis of an amide under basic conditions.

Although we show an acid halide in our prototypical example of nucleophilic catalysis, other species such as anhydrides and esters are also susceptible to this form of catalysis. As the carboxylic acid derivative becomes less electrophilic, the catalyst needs to be more nucleophilic.

Let's examine a few cases. First, we look at amide hydrolysis under basic conditions (Figure 10.17). Nucleophilic attack by the hydroxide leads to a tetrahedral intermediate that can expel the amide anion leaving group, which subsequently removes a proton from the resulting carboxylic acid rapidly. Note that hydroxide is not regenerated in this reaction, so this is a base-initiated reaction, not a base-catalyzed reaction. Amide anions are very poor leaving groups, and hence k_2 represents the rate-determining step. Very rapid ^{18}O scrambling from water into the starting material relative to product formation is observed, supporting k_2 being rate-determining. Furthermore, there is some evidence that a dianionic form of the tetrahedral intermediate is required to produce an intermediate reactive enough to expel the leaving amide anion. Hence, base catalysis is not very effective for amide hydrolysis.

Base catalysis, however, is very effective for ester hydrolysis, and a mechanism similar to that for amides is operative (Figure 10.18). Now leaving group departure and reversion to starting materials have approximately the same rate constants ($k_2 \sim k_{-1}$). The balance between which rate constant is larger depends upon the ester R group. Hydroxide and alkoxide depart at similar rates, while a phenoxide departs much faster (recall what makes a good leaving group, discussed in Section 8.4.4).

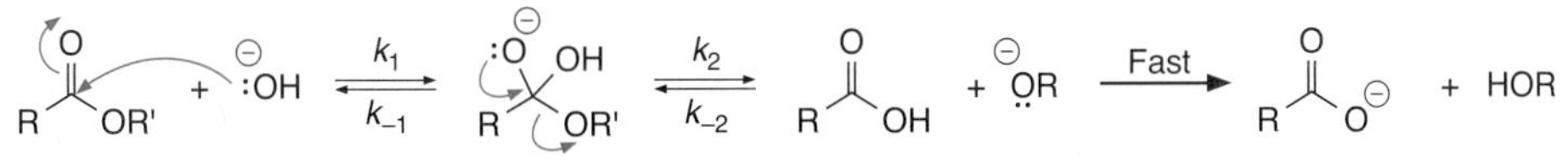

Figure 10.18
Hydrolysis of an ester under basic conditions.

The most unreactive carboxylic acid derivatives, esters and amides, usually require some form of catalysis to activate the carbonyl to nucleophilic attack. Acid catalysis for amide hydrolysis is quite effective. Since amides are so unreactive toward nucleophilic attack, specific-acid catalysis is most commonly observed. Here, full protonation of the amide carbonyl is necessary to activate the species enough that nucleophilic attack is possible (Figure 10.19 **A**). With acidic conditions up to about 80% acid (requiring H_o values; see Section 5.2.5), the carbonyl oxygen is the site of protonation, supporting an addition–elimination mechanism and the existence of a tetrahedral intermediate. Although there is no ^{18}O scrambling into the reactant, this does not establish the absence of such an intermediate. Leaving group departure from the tetrahedral intermediate is much faster than return to starting materials in acid ($k_2 \gg k_{-1}$ in Eq. 10.108), and hence little to no scrambling is observed. Above around 80% acid, the mechanism shifts to an S_N1-like process with an acylium ion intermediate (Figure 10.16 **A**). Here, protonation of the amine dominates the possible pathways, followed by ionization (Figure 10.19 **B**). Thus, in the acid-catalyzed hydrolysis of an amide, there is an additional issue of whether protonation occurs on the carbonyl oxygen or on the nitrogen.

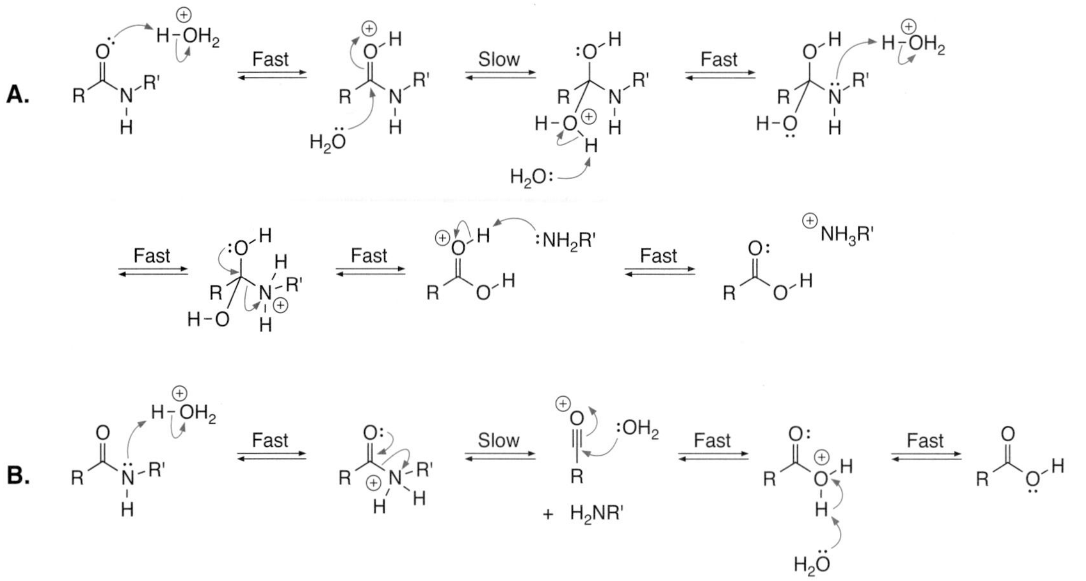

Figure 10.19
A. Amide hydrolysis under acidic conditions. **B.** Amide hydrolysis
under extremely acidic conditions.

Acid catalysis of ester hydrolysis is also very effective. Oxygen exchange from water is observed under most cases, supporting addition–elimination. Specific-acid catalysis is the most common mode of hydrolysis, although general-acid catalysis is observed with more electrophilic esters.

Connections

Enzymatic Acyl Transfers I: The Catalytic Triad

Now that we have an understanding of the organic reaction mechanisms for acyl transfers, and insight into acid and base catalysis, we can consider how an enzyme might approach catalyzing these reactions. One important biological acyl transfer is the hydrolysis of the amide linkage in a protein. There are three factors that are important in catalyzing this hydrolysis: 1. enhance the nucleophilic attack, 2. stabilize the tetrahedral intermediate, thereby stabilizing the transition state to it, and 3. enhance the leaving group departure. Based upon the mechanisms that have been presented here and the principles discussed in Chapter 9, the general aspects of the mechanism we present should not be too surprising, but we can see just how sophisticated nature is when it comes to catalysis.

Chymotrypsin is one of the most well studied **peptidases** (enzymes that cleave peptide bonds). Instead of using water as the nucleophile, the alcohol side chain of serine-195 is used. As shown on the next page, the nucleophilic attack by the serine is general-base-catalyzed by the side chain of histidine-57, an imidazole (step 1). Although general-base catalysis might be expected, it is even further enhanced by ion-pairing of the resulting imidazolium with the carboxylate from aspartate-102 at the active site. Hence, the nucleophilic attack is enhanced by the binding energy that develops with the formation of an ion pair.

The combination of the serine, histidine, and aspartate is a set of amino acids called the **catalytic triad**.

The tetrahedral intermediate is stabilized by hydrogen bonding in a pocket that is called the oxy–anion hole. Amide NHs from the peptide backbone contribute the hydrogen bonding donors.

Leaving group departure is general-acid-catalyzed by the imidazolium created in the first step (step 2). This forms an **acyl–enzyme intermediate**. Hence, this represents a form of covalent catalysis, where the high effective molarity of the serine nucleophile further enhances the rate. Water then comes into the active site, and then all the steps just discussed are simply repeated with water as the nucleophile.

Along with illustrating a strategy for peptide bond (amide) hydrolysis, the chymotrypsin structure established a number of other features of enzymatic catalysis. Although the protein might be large, the essential chemistry occurs in a relatively small region of the protein termed the active site (see Section 9.4.3). Also, the various residues that form the active site are physically close in space, but they are not close in peptide sequence (chymotrypsin, for example, uses His-57, Asp-102, and Ser-195.

Sigler, P. B., Blow, D. M., Matthews, B. W., and Henderson, R. "Structure of Crystalline α-Chymotrypsin II. A Preliminary Report Including a Hypothesis for the Activation Mechanism." *J. Mol. Biol.*, **35**, 243–264 (1968).

Connections

Enzymatic Acyl Transfers II: Zn(II) Catalysis

Another strategy for biological amide hydrolysis is exemplified by carboxypeptidase A. The electrophilic stabilization of the T.I. is now imparted by a Zn(II) ion and a guanidinium side chain from arginine-127. A zinc-bound water acts as the nucleophile, which is enhanced by general-base catalysis from the glutamate-270 side chain

(step 1). The newly generated carboxylic acid from the glutamate side chain is subsequently involved in general-acid-catalyzed leaving group departure (step 2). Thus, with this enzyme, we see proximity effects, electrophilic catalysis, and general-acid/general-base catalysis.

Christianson, D. W., and Lipscomb, W. N. "Carboxypeptidase A." *Acc. Chem. Res.*, **22**, 62–69 (1989).

Carboxypeptidase A

Connections

Enzyme Mimics for Acyl Transfers

There have been numerous synthetic supramolecular structures created to mimic the various aspects of the natural enzymes that catalyze acyl transfers. Here, we only show two with their respective binding geometries and electron pushing for the nucleophilic attack. The first shown below was developed by Lehn, and uses the well precedented binding between crown ethers and ammonium ions to form a complex between the catalyst and the substrate. The second example was developed by Breslow, where cyclodextrin (the toroid; see Chapter 4) is used to drive hydrophobic binding of the substrate to the catalyst. These two examples do indeed catalyze their respective acyl transfers, but with orders of magnitude lower

activity than the natural enzymes. Note that the majority of the catalysis in these systems arises from proximity, which results from binding the substrate near the nucleophile. It is still an important challenge in the field of physical organic chemistry to create an artificial system that catalyzes nucleophilic attack, stabilizes the tetrahedral intermediate, and catalyzes leaving group departure, thereby mimicking all aspects of the natural enzymes.

Lehn, J.-M., and Sirlin, C. "Catalyse Supramoléculaire: Conpure des Esters Activés d'Aminoacides Liés à un Récepteur Macrocyclique Portant des Résidus Cysteinyles." *Nouveau J. Chimie*, **11**, 683–702 (1987). Breslow, R., and Overman, L. E. "An Artificial Enyzme Combining a Metal Catalytic Group and a Hydrophobic Binding Cavity." *J. Am. Chem. Soc.*, **92**, 1075–1077 (1970).

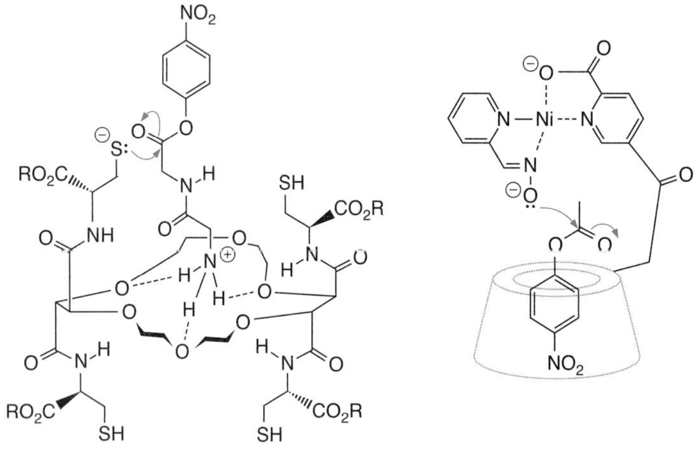

Two enzyme mimics

Connections

Peptide Synthesis—Optimizing Acyl Transfer

Peptides and proteins are constructed of amide bonds linking together a string of α-amino acids. There is a large industry centered around the synthesis of peptides (a **peptide** is a relatively short string of amino acids, roughly ≤50; longer stretches are called proteins) for use in research and as pharmaceuticals. For such efforts, it is important to maximize the coupling yields for the formation of each new peptide bond. For example, if we have a 95% coupling yield, but we have to perform the reaction 40 times to make a peptide with 41 amino acids, our overall yield will be $(0.95)^{40}$, which is less than 13%. Also, we will have a terrible mixture of products from coupling failures at each individual step of the sequence. We need yields that are 99% or better, and simple reactions like reacting an amine with an acid chloride will not suffice.

The key advance in this effort was Merrifield's development of solid phase protocols for peptide synthesis on resin beads, which earned him the 1984 Nobel Prize in Chemistry. One large advantage of solid phase synthesis is the ability to purify the growing peptide from reactants by simply washing the resin. This allows a large excess of the soluble reagent to be used, improving coupling yields. Now it is common for organic chemists to run all sorts of reactions, not just peptide couplings, on solid, polymeric beads (the solid circle in the drawing below). While this helped, there was still a need to optimize the acyl transfer chemistry. Acid chlorides react too indiscriminately. One reagent that has seen much use is dicyclohexylcarbodiimide (DCC). This reacts with the carboxylic acid, forming a highly reactive species. Nucleophilic addition of the amine from the bead produces an amide and dicyclohexylurea as the byproduct. The direct coupling of amines and carboxylic acids with DCC is a generally useful reaction.

In some cases in peptide synthesis, even DCC does not produce an active enough acyl transfer species, and special reagents are employed. A typical example is 1-H-hydroxybenzotriazole (HOBt). Usually, DCC is

involved in making the HOBt adduct, and the latter is highly efficient in peptide coupling reactions.

Hurby, V. J., and Meyer, J.-P. "Chemical Synthesis of Peptides" in *Bioorganic Chemistry: Peptides and Proteins*, S. M. Hecht (ed.), Oxford University Press, New York, 1988, pp. 27–64.

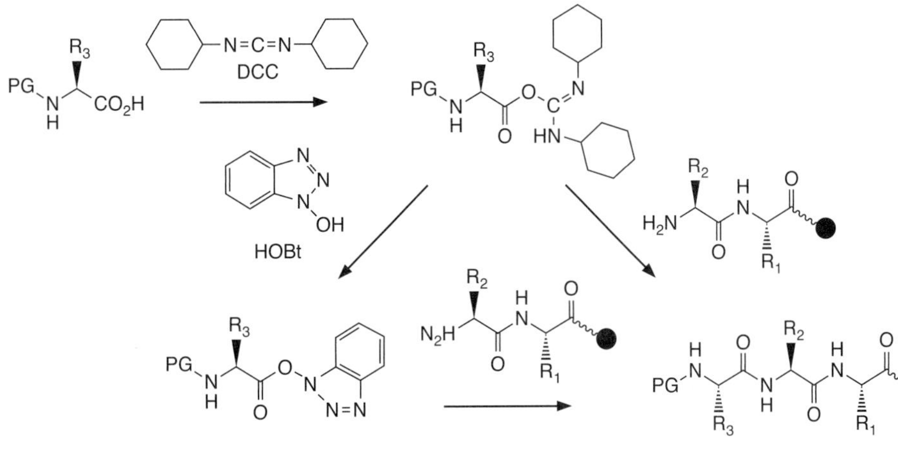

Peptide coupling chemistry

10.18   Electrophilic Aromatic Substitution

By far, aromatic substitution most commonly occurs via an electrophilic route. It conforms to our paradigm of reactivity, where the electron rich aromatic π system interacts with a partial or full positive charge on an electrophile (see the EPS in Appendix 2). Electrophilic aromatic substitution involves the substitution of a hydrogen on an aromatic ring with an electrophile (E) while giving off a proton (Eq. 10.110). All introductory organic chemistry textbooks cover this reaction, and the reaction mechanism commonly follows an addition–elimination sequence. A rare analog to this reaction is called **ipso substitution**, where a group other than hydrogen, most often iodine, undergoes the substitution by the electrophile.

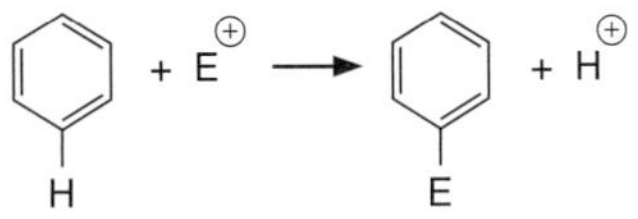

(Eq. 10.110)

10.18.1   Electron Pushing for Electrophilic Aromatic Substitutions

You may recall nitration, sulfonation, bromination, and Friedel–Crafts alkylation and acylation of aromatic rings from introductory organic chemistry classes. Our goal here is not to review each of these reactions, but to review the general S_EAr mechanism (substitution, electrophilic, aromatic), regiochemistry, and reaction rates, drawing on the principles of chemical reactivity discussed in this text to explain the experimental observations. Since the goal here is only to re-familiarize you with simple mechanisms and to teach good electron-pushing skills, one example of electron pushing will suffice—namely, Friedel–Crafts alkylation.

Friedel–Crafts alkylation is initiated by a complex formed by a Lewis acid–base interaction of a lone pair of electrons on the halogen of an alkyl halide with a strong electrophile such as $AlCl_3$ (Scheme 10.23). The complexation turns the halogen into a better leaving group, hence inducing a heterolysis of the C–Cl bond (step 2). The carbenium ion formed is a potent electrophile, and it consequently reacts with the π system of the aromatic ring (step 3). Subsequent deprotonation leads to the product (step 4).

Scheme 10.23
A Friedel–Crafts alkylation—
an example of the mechanism
of electrophilic aromatic
substitution.

10.18.2   Kinetics and Isotope Effects

In general, the mechanism involves two steps (Eq. 10.111), the first being addition of the electrophile to the aromatic ring to create a highly delocalized carbenium ion. The second step is loss of a proton to a base, most often the solvent.

$$(\text{Eq. } 10.111)$$

σ Complex

The kinetic expression for Eq. 10.111 is given in Eq. 10.112, where the σ complex has been treated using the steady state approximation. Commonly, the first step is rate-determining. This means that $k_2[B] \gg k_{-1}$, leading to the reaction being second order ($d[P]/dt = k_1[Ar][E^+]$). However, this is not always correct, because small isotope effects on the deprotonation step can be observed. A small isotope effect requires some k_2 influence on the reaction, which means for these cases $k_2[B]$ is of the same order of magnitude as k_{-1}. When the second step of a reaction is kinetically significant, although the first step is rate-determining, we say there is a **partitioning effect**. Here, the observation of a small isotope effect is due to the fact that the intermediate can partition back to starting material, making the second step kinetically observable. This simply means that the heights of the barriers on the energy surface for the k_{-1} and k_2 steps are comparable.

$$\frac{d[P]}{dt} = \frac{k_1 k_2 [Ar][E^+][B]}{k_{-1} + k_2 [B]} \qquad (\text{Eq. } 10.112)$$

10.18.3   Intermediate Complexes

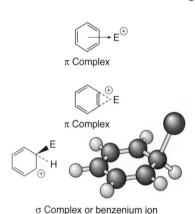

π Complex

π Complex

σ Complex or benzenium ion

The steps given in Eq. 10.111 ignore the generation of the electrophile, which is commonly a multistep process (see Scheme 10.23). Nucleophilic attack by benzene on the electrophile creates what is called the **σ complex**, also referred to as the **Wheland intermediate** or **benzenium ion** (see a structure with bromine added to a benzene ring in the margin). This intermediate is simply a carbenium ion structure in which the charge is delocalized around the ring. Simple resonance arguments predict that the positive charge will be on alternate atoms around the ring. The charge pattern can also be understood by recognizing that this is simply a pentadienyl cation. Figure 14.14 shows the LUMO of pentadienyl; it has nonzero coefficients only at carbons 1, 3, and 5.

There is NMR spectroscopic evidence that the electrophile does not always add to the aromatic nucleophile directly, as shown in Scheme 10.23 and implied by Eq. 10.111, but rather a **π complex** (also called an **encounter complex**) is formed prior to formation of the σ complex. We examined the cation–π effect in Chapters 3 and 4, and therefore it should not be surprising that an electrophile will make a weak complex with an aromatic ring prior to a reaction. Depending upon the experimental conditions and the aromatic compound used, either the formation of the electrophile, the encounter complex with the aromatic ring, or the σ complex will be rate-determining.

Sigma complexes can be observed spectroscopically. The reaction of HF and BF_3 with an aromatic ring leads to an observable σ complex at –140 °C, in which the electrophile is H^+. The stabilities of these complexes correlate with the rates of electrophilic substitutions, supporting the existence of these structures as intermediates in the mechanism. Table 10.9 shows several examples of how the two numbers are in agreement. The relative stability numbers derive from measurements of equilibrium constants for the protonation of the aromatic ring.

Table 10.9
Relative Stabilities of σ Complexes Formed with Various Aromatics and HF–BF₃, and Relative Rates of Bromination in 85% Acetic Acid*

Substituent	Relative σ complex stability	Relative rate of bromination
H	1	1
Methyl	790	605
1,2-Dimethyl	7,900	5,300
1,3-Dimethyl	1,000,000	514,000
1,4-Dimethyl	3,200	2,500
1,2,3-Trimethyl	2,000,000	1,670,000
1,3,5-Trimethyl	630,000,000	189,000,000

Olah, G. A. "Mechanism of Electrophilic Aromatic Substitutions." *Acc. Chem. Res.,* **4,** 240 (1971).

10.18.4 Regiochemistry and Relative Rates of Aromatic Substitution

Substitution on benzene raises no regiochemical issues because every hydrogen is equivalent, but as soon as there is one substituent on the ring, isomeric products can result—namely, ortho, meta, and para. The reactivities of the different sites on substituted aromatic rings are quantified by what are known as **partial rate factors** (f_n^R, where n = o, m, or p for ortho, meta, or para, respectively, and R = substituent). These numbers reflect the rate constants (k_n') for reaction of the individual ortho, meta, or para sites with an electrophile compared to the rate constant (k) for addition to benzene itself (Eqs. 10.113 **A**, **B**, and **C**). The rate constants for ortho and meta are divided by two because there are two ortho and meta hydrogens, and the rate constant for benzene is divided by six due to the six hydrogens.

$$\textbf{A.} \quad f_o^R = [(k_o'/2)/(k/6)]$$
$$\textbf{B.} \quad f_m^R = [(k_m'/2)/(k/6)] \qquad \text{(Eq. 10.113)}$$
$$\textbf{C.} \quad f_p^R = [(k_p')/(k/6)]$$

Table 10.10 shows some partial rate factors for various reactants. Notice that the para position in toluene undergoes nitration about 46 times faster than benzene itself, while the ortho position reacts about 39 times faster. Actually, all the sites on toluene, including meta, react faster than benzene. The methyl group on toluene is therefore described as **activating**. All sites on chlorobenzene react slower than benzene itself, and hence the chloro group is **deactivating**.

Aromatic substituents are broken into two categories regarding electrophilic aromatic substitution reactions: activating or deactivating, and ortho/para directing or meta directing. When $f_n^R > 1$, we say that substituent R is activating, and when $f_n^R < 1$, we classify that

Table 10.10
The Partial Rate Factors for Different Reactions*

Reactant	f_o^R	f_m^R	f_p^R	Reaction
Toluene (R = CH$_3$)	38.9	1.3	45.8	Nitration
t-Butylbenzene (R = t-Bu)	5.5	3.7	71.6	Nitration
Chlorobenzene (R = Cl)	0.028	0.00084	0.13	Nitration
Bromobenzene (R = Br)	0.030	0.00098	0.103	Nitration
Toluene (R = CH$_3$)	617	5	829	Halogenation (Cl$_2$)
Toluene (R = CH$_3$)	600	5.5	2420	Halogenation (Br$_2$)
Toluene (R = CH$_3$)	32.6	5.0	831	Acylation (PhCOCl)
Toluene (R = CH$_3$)	4.5	4.8	749	Acylation (MeCOCl)
Toluene (R = CH$_3$)	4.2	0.4	10.0	Alkylation (BnCl, AlCl$_3$)

*Stock, L. M. "A Classic Mechanism for Aromatic Nitration." *Prog. Phys. Org. Chem.*, **12**, 21 (1976). Stock, L. M., and Brown, H. C. "A Quantitative Treatment of Directive Effects in Aromatic Substitution." *Adv. Phys. Org. Chem.*, **1**, 35 (1963).

substituent as deactivating. All activators are ortho/para directing, while almost all deactivators are meta directing. The halogens are exceptions to this correlation, in that they are deactivating but ortho/para directing. Let's recall the reasoning given for these patterns of reactivity. Here we use resonance and the concept of localized bonds between atoms, although the concept of fully delocalized molecular orbitals over the entire benzene ring very nicely explains the trends also.

To rationalize substitution patterns, one examines the relative stabilities of the various regioisomeric σ complexes. For example, Figure 10.20 shows the resonance structures of the carbenium ion generated after addition of an arbitrary electrophile (E$^+$) to a benzene substituted in the ortho, meta, and para positions. If we envision X = CH$_3$, ortho and para substitution leads to carbenium ions that have some tertiary carbenium ion character, whereas meta substitution does not. Because 3° carbenium ions are more stable than 2° carbenium ions, they are formed faster in this reaction, and hence, ortho and para substitution is preferred. Since alkyl groups are stabilizing to carbenium ions via hyperconjugation, they not only direct the regiochemistry as described, but also enhance the rate of the reaction relative to benzene because the formation of the carbenium ion is rate-determining. Thus, alkyl groups are activating and ortho/para directing.

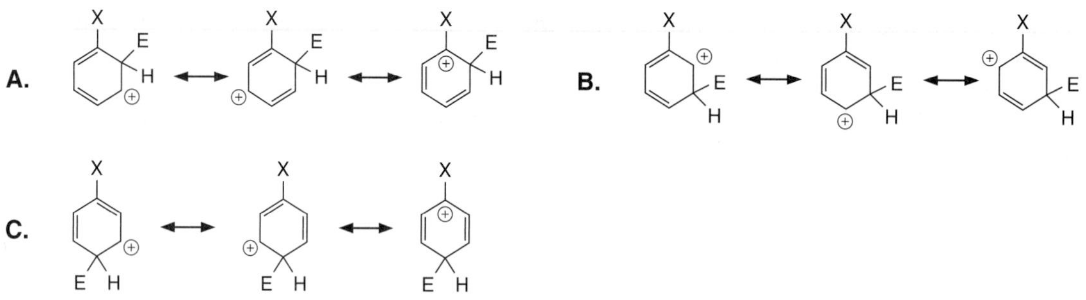

Figure 10.20
Resonance structures resulting from electrophilic addition to a substituted benzene ring. **A.** Ortho, **B.** Meta, and **C.** Para.

Other groups that can donate electrons are also activating. Heteroatoms with lone pairs that donate electrons to the carbenium ion via resonance, such as OR, NR$_2$, and SR, are examples (Figure 10.21).

Now imagine X = CN, an electron withdrawing group. Once again the σ complex derived from ortho and para addition of the electrophile has a resonance structure in which the

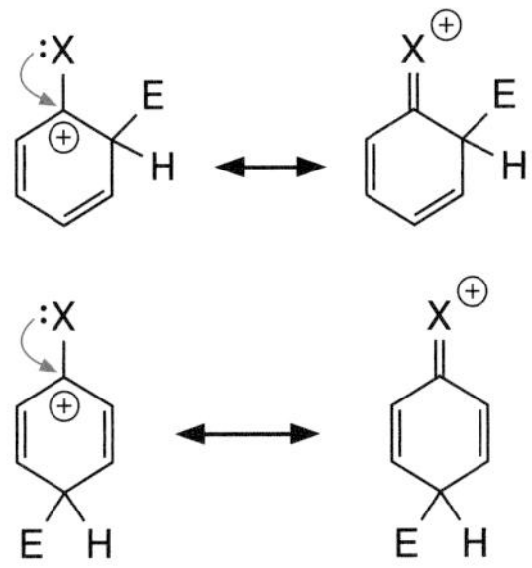

Figure 10.21
Groups with heteroatoms lead to activation via resonance when ortho or para to the site of addition.

Table 10.11
Relative Rates for Nitration of a Few Benzene Derivatives*

Substituent	Relative rate
OH	1000
CH_3	25
H	1
Cl	0.033
NO_2	6×10^{-8}
$N(CH_3)_3^+$	1×10^{-8}

*Ingold, C. K. (1969). *Structure and Mechanism in Organic Chemistry*, 2nd ed., Cornell University Press, Ithaca, NY.

positive charge is directly adjacent to the group, but now the positive charge would be destabilized. In fact, the creation of a positive charge, regardless of the regiochemistry of addition, is retarded, and hence the group is deactivating. However, attack at the meta position gives the least deactivation, because there are no resonance structures where the positive charge is directly adjacent to the EWG. Hence, such a group is meta directing.

Table 10.11 shows the relative rates of nitration of a few benzene derivatives, and these demonstrate the electron donating (activating) and withdrawing (deactivating) effect of several substituents. In fact, most chemist's intuition as to what groups are electron donating and withdrawing is derived from rates of electrophilic aromatic substitution, as well as the σ constants associated with Hammett plots.

10.19 Nucleophilic Aromatic Substitution

Nucleophilic aromatic substitution (called S_N2Ar) is less common than electrophilic aromatic substitution, and it usually occurs on an aromatic ring where a leaving group is ortho or para to one or more strongly electron withdrawing groups (commonly nitro or cyano).

10.19.1 Electron Pushing for Nucleophilic Aromatic Substitution

Consider the substitution of chloride by hydroxide in *p*-chloronitrobenzene (Scheme 10.24). Nucleophilic attack by the hydroxide places increased negative character on the nitro group, which is evident by the electron pushing. If you count the number of atoms from the site of nucleophilic attack to where the negative charge is deposited, you can see why this is called a 1,6-addition. Reversing these arrows leads to expulsion of the leaving group (a 1,6-elimination).

Scheme 10.24
Electron pushing for nucleophilic aromatic substitution.

Meisenheimer complex

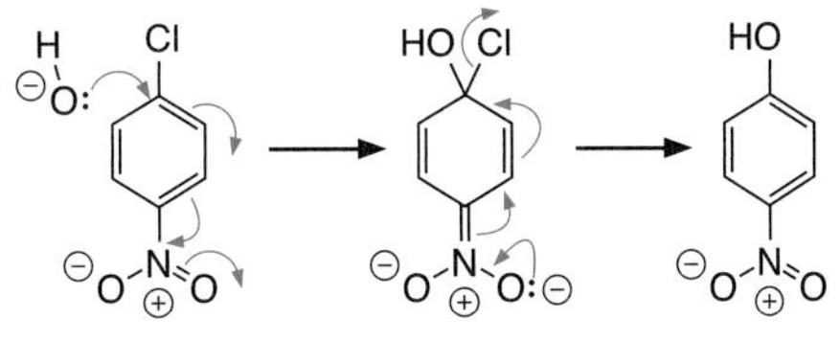

10.19.2 Experimental Observations

The kinetics of nucleophilic aromatic substitution is almost always second order—first order in nucleophile and first-order in the aromatic electrophile. The intermediate structure is called the **Meisenheimer complex** (or Jackson–Meisenheimer complex). The Meisenheimer complex can sometimes be directly observed at low temperatures. In aprotic solvents the loss of the leaving group is often rate-determining, because nucleophiles are very active in such solvents (see Section 8.4.4). In protic solvents either the nucleophilic attack or leaving group departure can determine the rate. Since the leaving group is on the carbon undergoing nucleophilic attack, it naturally influences the rate of the reaction regardless of whether the leaving group departure has any effect on the rate. Hence, we must be careful in interpreting leaving group effects.

The site of attack of the nucleophile is not influenced by the location of the other groups on the aromatic ring. This site of attack *has* to be the carbon with the leaving group for the attack to lead to a product. With electrophilic aromatic substitution, the groups on the ring influence the regiochemistry of the reaction, being either ortho/para or meta directing. This is an important distinction between nucleophilic and electrophilic aromatic substitution mechanisms.

As an important caution at this stage, we note that some aromatic substitutions involve single electron transfer (SET) pathways. Because the aromatic ring is necessarily electron deficient, and the nucleophile is electron rich, a single electron transfer from the nucleophile to the ring is increasingly probable as the aromatic ring becomes increasingly electron poor. For example, in the reaction of 1-chloro-2,4,6-trinitrobenzene with hydroxide (Eq. 10.114), the expected radical anion intermediate was directly observed using fast spectroscopy.

(Eq. 10.114)

On rare occasions, the reaction is first order in the aromatic structure and zero order in the nucleophile. This is reminiscent of an S_N1 reaction. This mechanism occurs with diazonium salts, where the leaving group is so good (N_2) that it can depart without assistance, leaving behind an aryl cation that is trapped by a nucleophile (Eq. 10.115). The nucleophile can be water to make a phenol, or CuX salts that place the X group on the ring (the **Sandmeyer reaction**). The Sandmeyer reaction actually involves electron transfer, as we describe in a Connections highlight in Section 12.2.3.

(Eq. 10.115)

10.20   Reactions Involving Benzyne

Alkyne   Cumulene   Biradical

Another substitution reaction that occurs on aromatic rings in the presence of nucleophiles involves the intermediacy of **benzyne**. Benzyne is benzene minus two adjacent hydrogens, producing a formal triple bond (C_6H_4). The structure of benzyne has been examined both experimentally and theoretically, and the alkyne representation is most widely accepted, although the cummulene and biradical structures are significant resonance contributors.

10.20.1   Electron Pushing for Benzyne Reactions

Instead of performing an addition reaction, the nucleophile acts as a base to deprotonate a position adjacent to a leaving group, resulting in an elimination reaction that generates benzyne (Scheme 10.25). When the leaving group is good, elimination is concerted as shown. Benzyne is highly strained and rapidly reacts as an electrophile by adding a nucleophile. This involves the formation of an sp^2 carbanion that very rapidly deprotonates the solvent. Hence, the sequence of reactions used in this mechanism is elimination–addition (termed E–Ad). Although amide anion is shown as the base in Scheme 10.25, oxygen bases such as hydroxide and alkoxides are also effective.

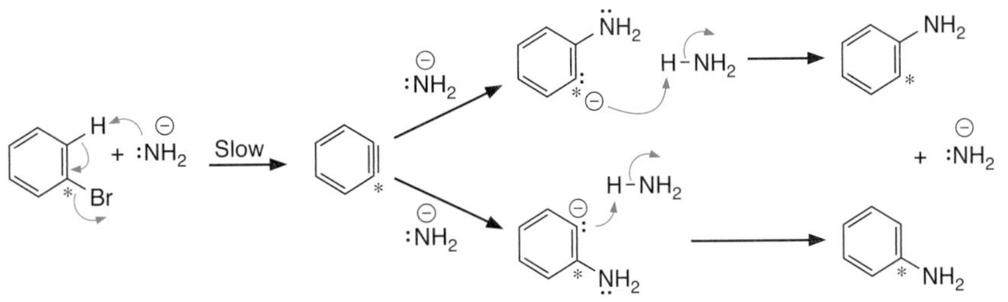

Scheme 10.25
Electron pushing for reactions involving benzyne.

10.20.2   Experimental Observations

The reaction is first order in the aromatic ring and the base. The formation of aniline from 2,6-dideuteriobromobenzene shows an isotope effect of 5.5, supporting rate-determining elimination via deprotonation. In a pioneering experiment, Roberts showed that when the carbon possessing the leaving group on the aromatic ring is ^{14}C-labeled (shown as an asterisk in Scheme 10.25), the label in the product is distributed between the carbons attached to and adjacent to the nucleophile, supporting a symmetric intermediate such as benzyne. The distribution of this label is independent of which leaving group is on the ring, supporting a common intermediacy of benzyne from different reactants. One finds the standard order of leaving group activity, where the reaction decreases in rate within the following series: I > Br > Cl > F. In fact, when the leaving group is electron withdrawing but very poor at departure, expulsion of the leaving group can become rate-determining. This means that a carbanion is created prior to leaving group departure. For example, fluorobenzene mixed with amide in ammonia scrambles a deuterium adjacent to fluorine but no elimination occurs (Eq. 10.116). Chlorobenzene also shows a small percentage of scrambling, meaning that the elimination to form benzyne is stepwise also. In this case the elimination component of the elimination–addition sequence has the character of an E1cB reaction. In contrast, with Br and I no scrambling is observed, and the elimination is a concerted E2 as written in Scheme 10.25.

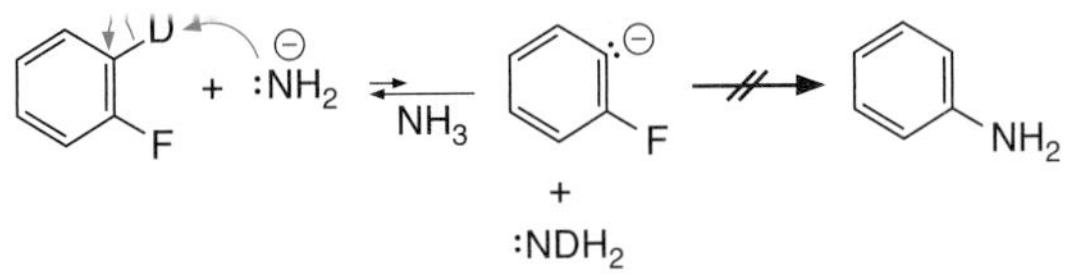

$$\text{(Eq. 10.116)}$$

10.20.3   Substituent Effects

When one examines the mechanism given in Scheme 10.25, two ways in which substituents could effect the regiochemistry of the products become apparent. The first is the regiochemistry of formation of the triple bond relative to the substituent (Y) when the leaving group (X) is meta to the substituent (Eq. 10.117). This concern is only with a meta arrangement in the reactant, since ortho and para arrangements of the substituent and leaving group can only generate one regiochemical arrangement in the benzyne intermediate (see Figure 10.22).

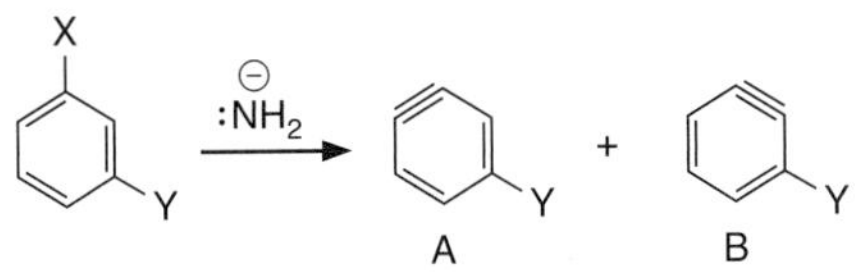

$$\text{(Eq. 10.117)}$$

Since the rate-determining step that sets the regiochemistry of the triple bond involves deprotonation, we expect the acidity of the hydrogen to influence the regiochemistry. When the substituent Y is electron withdrawing, regioisomer B (Eq. 10.117) is preferred, because the hydrogen ortho to the substituent is most acidic. If the substituent is electron donating, regioisomer A is preferred, because now the more remote hydrogen is most acidic.

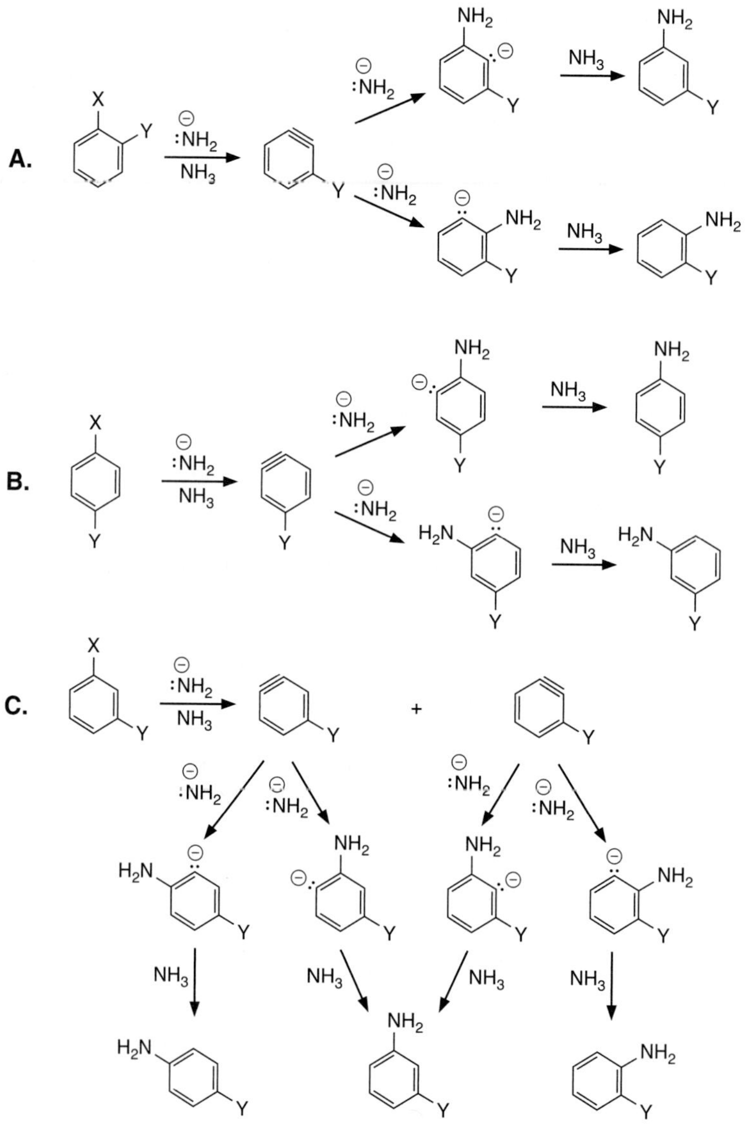

Figure 10.22
Regiochemical possibilities for reactions involving benzyne. The
regiochemical preferences can be predicted by an analysis of whether
Y is an EWG or EDG. **A.** Ortho reactants, **B.** Para reactants, and **C.** Meta
reactants.

The second expected substituent effect is the regiochemistry of addition of a nucleophile
to the triple bond. With ortho substitution in the reactant, the triple bond can undergo nucleophilic addition to give an ortho or meta product (Figure 10.22 **A**). Electron withdrawing
groups (EWGs) direct addition of the nucleophile to place the carbanion as close as possible
to the EWG. When the substituent is CF_3 and the reaction involves amide anion in ammonia,
only the meta product is found. With an electron donating group such as Y = CH_3, there is almost an equal mix of ortho and meta products.

When the leaving group is para to the substituent (Figure 10.22 **B**), the effect of the substituent on nucleophilic addition to the triple bond is lower, and many substituents give a
nearly equal mixture of para and meta products. The exact same reasoning predicts the regiochemistry of nucleophilic addition to the two different triple bond isomers derived from a
meta substituted reactant (Figure 10.22 **C**). In this case, if Y = CF_3, only the meta product is
found due to regioselectivity in both steps, but with Y = CH_3, all three possible products are
found, with meta dominating.

10.21 The $S_{RN}1$ Reaction on Aromatic Rings

Besides nucleophilic and electrophilic pathways for aromatic substitutions, there are also radical pathways. With aromatic rings that are easily reduced, this is a common mechanism, because the benzene ring can delocalize the radical anion. The radical chain mechanism is referred to as $S_{RN}1$, (substitution, radical–nucleophilic, unimolecular). An example is shown in Eq. 10.118.

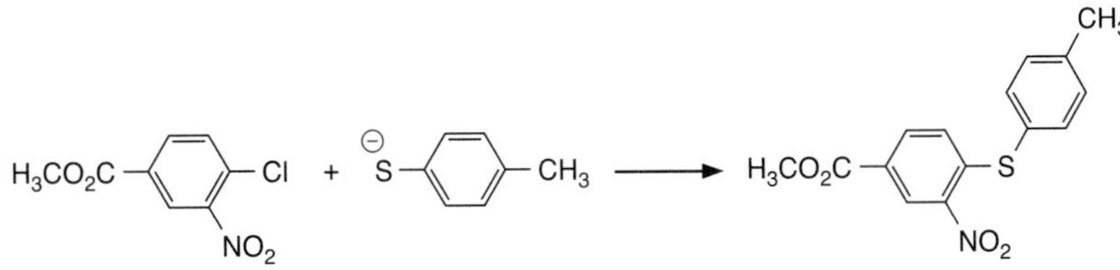

(Eq. 10.118)

10.21.1 Electron Pushing

An electron source donates an electron to the aromatic ring, producing the radical anion of the reactant. This leads to heterolytic departure of the leaving group (Scheme 10.26). Reaction of the resulting aryl radical with the nucleophile gives another radical anion, which transfers an electron to another reactant to propagate the chain. These reactions are surprisingly common, and often explain unusual products in aromatic substitution reactions.

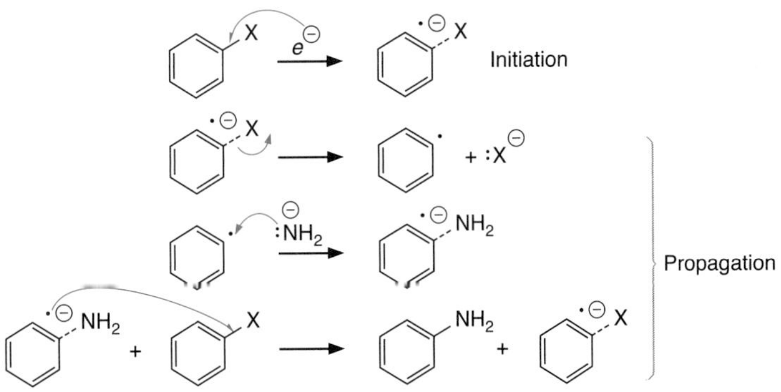

Scheme 10.26
Electron pushing for the $S_{RN}1$ mechanism for aromatic substitution.

10.21.2 A Few Experimental Observations

Polynuclear aromatic rings are more susceptible to this reaction than simple benzene, because the first formed radical anion of the reactant is more stable due to increased delocalization. Many leaving groups participate in the $S_{RN}1$ reaction of aromatic rings, including F, Cl, Br, I, SPh, and NR_3^+. One also requires a nucleophile with a low enough oxidation potential to give a favorable electron transfer to the aromatic ring. This mechanism can also be initiated with solvated electrons in ammonia, electrochemically, or with photoinduced electron transfer.

10.22 Radical Aromatic Substitutions

There are also aromatic substitution reactions involving radicals that do not proceed via chain reactions. Instead, the radical adds followed by radical elimination.

10.22.1 Electron Pushing

Radical aromatic substitution is stepwise, involving a radical intermediate that has several resonance structures. Abstraction of the hydrogen on the carbon where the radical added requires either another radical or an oxidizing agent. The reaction and proper electron pushing are shown in Scheme 10.27. When the hydrogen abstraction is by another radical it terminates a potential chain, and indeed, radical aromatic substitutions are typically *not* chain reactions.

The mechanism shown is really quite a simplification of what normally occurs in these reactions. There are many competing pathways. Dimerization of the intermediate radical is common, leading to regioisomeric products. Disproportionation and coupling with another radical are also possibilities.

Scheme 10.27
A radical pathway for aromatic substitution.

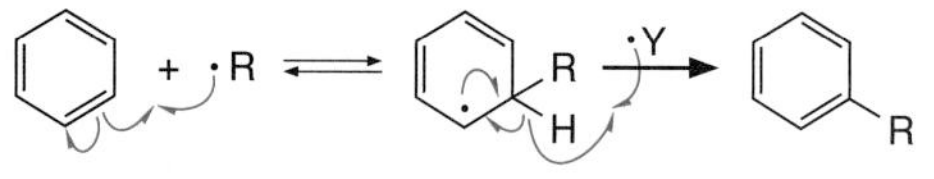

10.22.2   Isotope Effects

Either of the two steps shown in Scheme 10.27 can be rate-determining, depending upon the nature of the adding radical and the substituents on the benzene ring. The addition of phenyl radical to perdeuterio benzene shows no measurable isotope effect (Eq. 10.119), whereas the addition of benzoyl radical to perdeuterio benzene does have an isotope effect (Eq. 10.120). This means that the radical addition is rate-determining and irreversible for phenyl, whereas for benzoyl radical the first step is reversible, and the second step enters the rate expression.

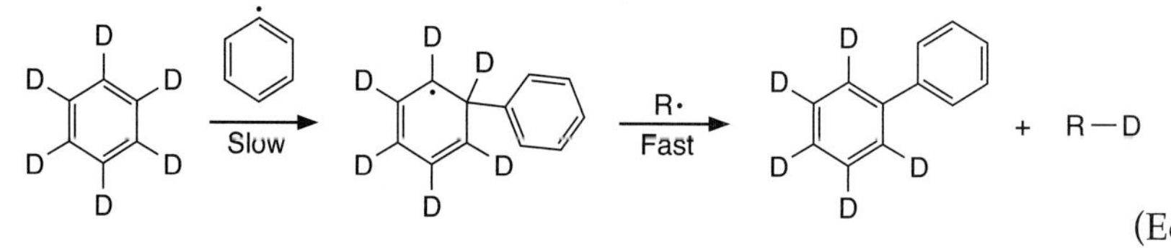

(Eq. 10.119)

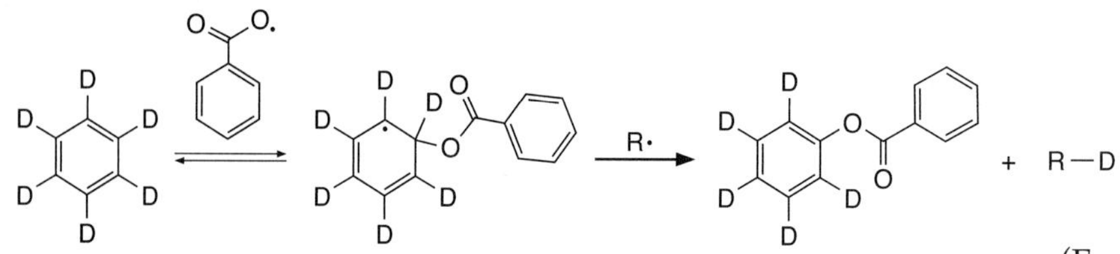

(Eq. 10.120)

10.22.3   Regiochemistry

Just as with nucleophilic and electrophilic aromatic substitutions, when substituents are present on the starting benzene regioisomeric products are possible. Almost any substituent on the benzene ring enhances the rate of radical addition, reflecting the fact that both electron donating and withdrawing groups stabilize radical intermediates. However, with radicals, the preferences for ortho, meta, or para substitution are significantly lower than with electrophilic substitution. For example, the addition of phenyl radical to toluene gives partial rate factors for ortho, meta, and para of 3.3, 1.1, and 1.3, respectively. For phenyl radical addition to nitrobenzene the numbers are 9.4, 1.2, and 9.1, respectively. As a last example, the addition of benzoyl radical to methoxybenzene gives only slightly larger numbers, having partial rate factors of 20.7, 0.3, and 20.4, respectively. Looking back at Table 10.10 to see the partial rate factors for electrophilic substitution, it is clear that the factors there are much larger.

We can see that the polar effects found for electrophilic aromatic additions are much less pronounced with radical additions. This is in part due to the fact that the influence on radical stability of electron donating or electron withdrawing groups is lower than the effect on carbenium ions. However, the lower sensitivity is also due to an early transition state for radical addition because the addition is exothermic, and a late transition state for electrophilic addition because this addition is endothermic. By the Hammond postulate, the early transition state for radical addition would have little radical character on the aromatic ring, and therefore substituents on the ring have a small effect. In contrast, electrophilic addition has a late transition state, and carbenium ion character has developed to a large extent in the transition state, and so substituents have a larger effect.

Summary and Outlook

This chapter focused upon reactions at unsaturated centers: alkenes, alkynes, arenes, and carbonyls. A large variety of additions and eliminations was examined. The complexity in the examples showed how diverse the possibilities are for additions and eliminations to such centers. One can have electrophilic, nucleophilic, and radical pathways for almost all reaction classes, and the dominant product depends upon the structures of the reactants and the experimental conditions. You should now be able to write the electron pushing for the standard reaction mechanisms (E1, E2, S_EAr, etc.), as well as be able to predict and examine experimentally the variations that are possible. Given this background on unsaturated centers, we turn our attention to reactions that occur at saturated centers, which is the first major topic of the next chapter.

Exercises

1. Identify any atoms, bonds, or lone pairs in the following molecules that are nucleophilic or electrophilic. In each case define the site on the molecule that can be considered as Lewis acidic or Lewis basic. Finally, draw a picture of what would be the donor (HOMO) and acceptor (LUMO) orbitals. Comment on how all these approaches to predicting reactivity relate.

A. **B.** **C.** **D.** **E.** **F.** **G.**

2. The equilibrium constants for the addition of thiols to carbonyl compounds are significantly larger than those given in Table 10.2 for the addition of water and alcohols. For aldehydes the equilibrium constants are approximately 10^3 to 10^4 larger, and for ketones they are 10 to 10^3 larger. Give a qualitative explanation for this. For more quantitive reasoning, consider that the average BDEs for an S–H and S–C bond are 90 and 60 kcal/mol, respectively, and explain the phenomenon again with a focus upon which bond formation drives the reaction to prefer thiol addition.

3. The addition of thiols to ketones shows very interesting kinetics. One finds specific-base catalysis of the addition step, whereas there is also general-acid catalysis for the reaction. Given this, propose a mechanism for the addition of ethane thiol to acetone in aqueous media. Explain why thiols may be expected to show specific-base catalysis in their addition, but that the nature of a thiol also leads to general-acid catalysis of the reaction.

4. Write mechanisms with full electron pushing that explain the following two reactions. Here we show that *d*- or *l*-erythro-3-bromo-2-butanol forms meso-2,3-dibromobutane upon treatment with HBr. However, *d*- or *l*-threo-3-bromo-2-butanol forms the corresponding chiral 2,3-dibromobutanes. Are these reactions stereospecific?

5. Bromination of 4-*t*-butylcyclohexene gives two products (Eq. 10.15). Why is the diaxial-dibromide the major product?

6. Bromination of *cis*- and *trans*-2-butene gives different products. Which gives meso product and which gives a mixture of *d* and *l* products?

7. Recall from the Connections highlight on page 550 that the cationic cyclization of squalene is a key step in steroid biosynthesis. It has been found that the active site of enzymes that catalyze this type of transformation is unusually rich in aromatic amino acids (Phe, Tyr, and Trp). Why might this be so?

8. The following is called the **Cannizaro reaction**. This is a base-catalyzed disproportionation of aldehydes that have no α-hydrogens (e.g., R = phenyl). Write a mechanism for this reaction showing all electron pushing. Derive a kinetic expression for your mechanism and predict the kinetic order in aldehyde and hydroxide. What experiments would you perform to test the validity of your mechanism? Can you write an alternative mechanism that would give different experimental results?

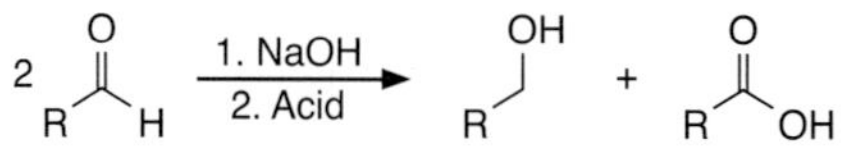

9. Write the steps involved in the acid-catalyzed hydrolysis of an oxime. Predict the relative rates of the primary steps in this mechanism.

10. We noted in this chapter that nucleophilic addition to alkenes occurs when groups capable of electron withdrawal by both induction and resonance are attached to the alkene. The addition of alkoxide anions to 2,2-dichloro-1,1-difluoroethylene occurs on the carbon with the fluorines. Why does the addition occur to place the negative charge of the carbanion intermediate on the carbon with the chlorines, rather than on the carbon with the more electronegative fluorines?

11. Write a detailed mechanism for the addition of $CXCl_3$ to cyclooctene. Why do the product distributions depend upon X?

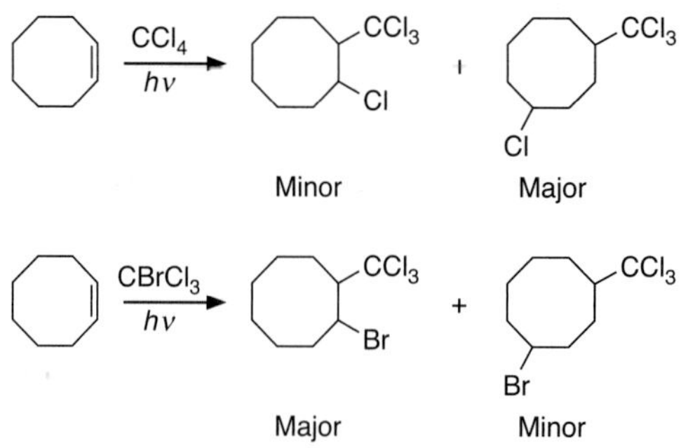

12. The addition of HCl to 1-phenylpropyne gives predominately the syn product with the regiochemistry shown below. Give an explanation of what the regiochemistry and syn addition indicate about the mechanism of this reaction.

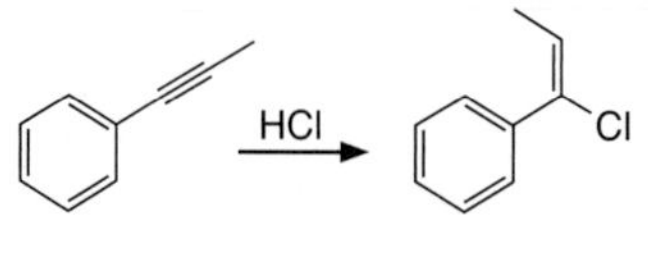

13. The **Bamford–Stevens reaction** involves the addition of tosylhydrazine to an aliphatic ketone, followed by treatment with a strong base and heating. The product is an alkene. Write a mechanism for this reaction, and describe experiments you would perform to confirm the presence of any reactive intermediate you postulate.

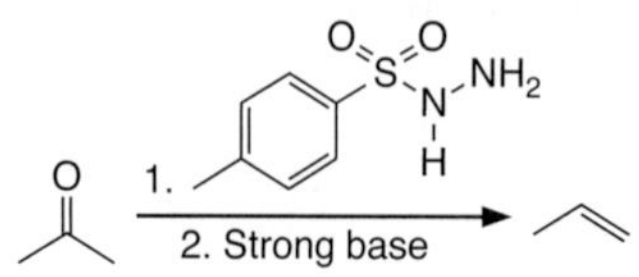

14. Explain how the following products arise.

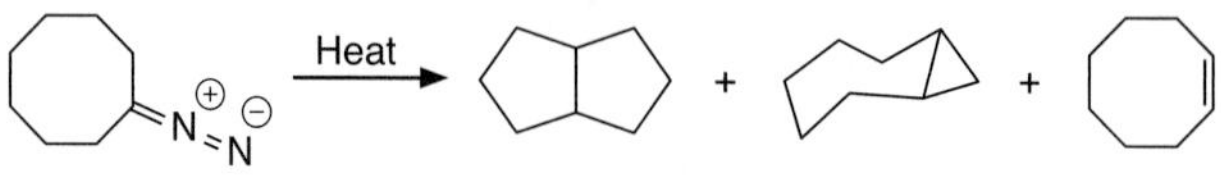

15. Derive the kinetic expression for the following nucleophilic aromatic substitution. Note that the elimination of the leaving group is rate-determining, and that this step is general-acid-catalyzed. Show why there is always a kinetic dependence upon nucleophile in an addition–elimination reaction even when leaving group departure is rate-determining.

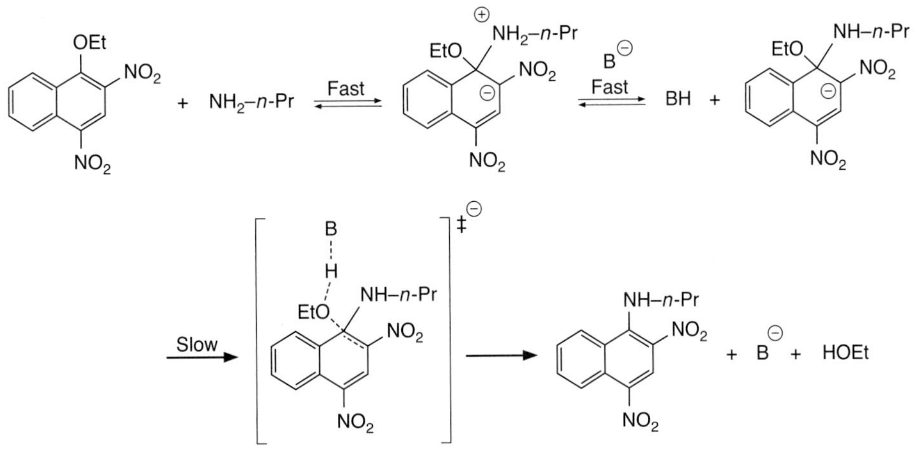

16. Write a mechanism for the substitution reaction shown below, given the following information. The nucleophile is found only on the carbon that possessed the leaving group. However, with the addition of a radical scavenger, the amine in the product is found either on the carbon with the leaving group or on the adjacent carbon.

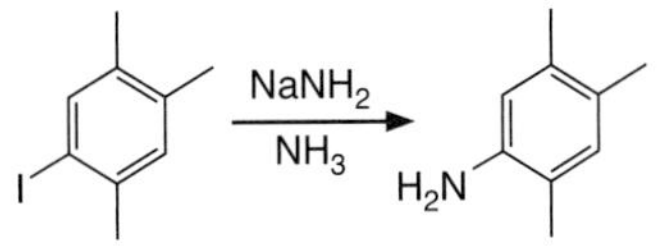

17. Write a mechanism for the following substitution reaction that is initiated by a single-electron reduction of the aromatic ring.

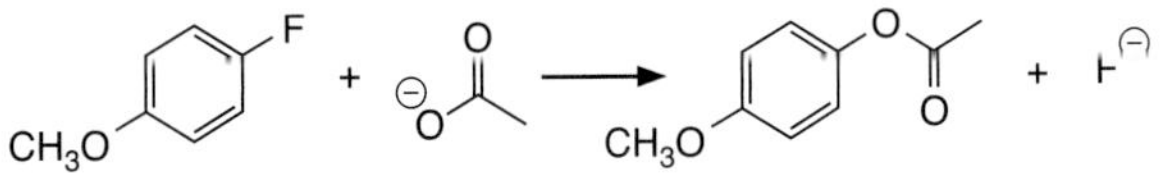

18. Sketch the reaction coordinate diagram for the electrophilic substitution reaction of benzene by molecular bromine. Be sure your barrier heights are appropriate to indicate the correct rate-determining step.

19. Explain how the products from the following reaction arise. Then explain the differences in product ratios as a function of temperature.

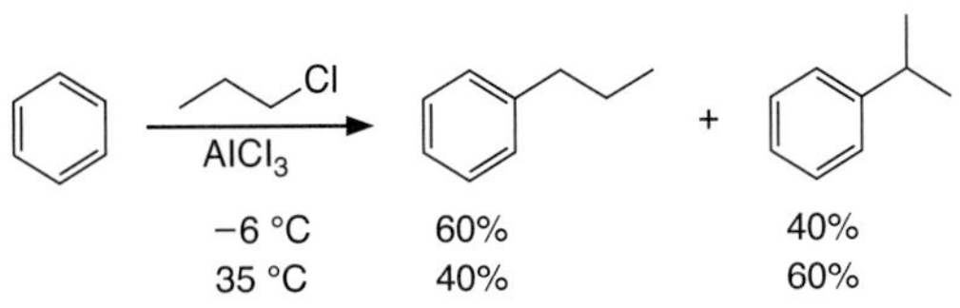

−6 °C	60%	40%
35 °C	40%	60%

20. What alkyl bromide would give the following as the predominant product upon treatment with KO-*t*-Bu / *t*-BuOH? Be sure to show stereochemistry.

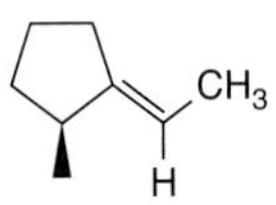

21. An interesting nucleophilic aromatic substitution occurs with cyanide on nitrobenzene, resulting in benzoic acid if the reaction is performed in water. This is called the **von Richter reaction**. Experiments involving isotope substitution confirm that the carboxylic acid functional group in the product is *ortho* to the position of the nitro group in the reactant. This has been termed **cine substitution** (Greek for "to move"). **Tele substitution** means that the nucleophile is placed *para* to the leaving group. Write a mechanism showing how nitrobenzene can be converted via cine substitution to benzoic acid with NaCN in water. (*Hint:* A traditional hydrolysis of a cyano group is *not* involved.)

22. The reaction of *m*-bromoanisole with amide anion in liquid ammonia gives solely *m*-methoxyaniline. We normally think of methoxy as an electron donating group, so this may seem confusing. Rationalize this result with respect to the two possible benzyne intermediates and the relative preferences for nucleophilic addition to benzyne.

23. One of the following acetals undergoes hydrolysis 10 trillion times faster than the other. Which is the faster and why?

24. The hydrolysis of acetals normally involves a specific-acid catalysis mechanism. However, the hydrolysis of tropone diethylketal is found to use a general-acid catalysis mechanism. Why is there this difference?

25. The following compound displays no acid or base catalysis for hydrolysis between pH values of 1.5 and near 13. Why?

26. In the addition of HBr to alkenes, it is often found that alkenes that undergo anti addition follow the rate law, rate = k[alkene][HBr]^2, while alkenes that undergo syn addition follow the rate law, rate = k[alkene][HBr]. Explain why this is so.

27. Table 10.7 lists changes to the transition state structure for E2 reactions that occur when structural changes are made to the reactants and solvent. Confirm each of these by looking at More O'Ferrall–Jencks plots.

28. We have discussed in this chapter that it is sometimes difficult to distinguish E2 and E1cB mechanisms by examining kinetics, isotope effects, and leaving group effects. An extremely ingenious method to distinguish these possibilities takes into account the relative rates of elimination in H_2O and D_2O as a function of the concentration of the base in a buffer. Examine the plot given for the elimination reaction shown. As one increases the base concentration (AcNHO$^-$ is acetoxyhydroxamate) in a buffer, keeping the pH constant, the k_{obs} levels off faster in H_2O than in D_2O, yet the k_{obs} in D_2O is also slowly leveling off. First, explain why this supports an E1cB mechanism and not an E2 mechanism. Second, explain why the k_{obs} values for an E1cB reaction would level off faster in H_2O than in D_2O. (*Hint:* You will have to examine the relative sizes of the k_{-1} and k_2 terms in the denominator of the rate expression in the two different solvents.)

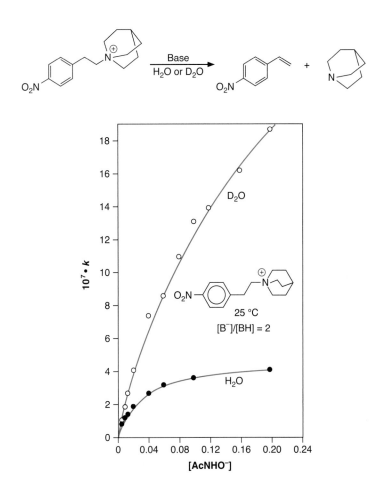

Keefe, J. R., and Jencks, W. P. "Large Inverse Solvent Isotope Effects: A Simple Test for the E1cB Mechanism." *J. Am. Chem. Soc.*, **103**, 2457 (1981).

29. In the elimination of 2-halohexanes by methoxide in methanol, the ratio of the 2-hexene / 1-hexene products is 0.43, 2.0, 2.6, and 4.2 for F, Cl, Br, and I, respectively. Explain the trend.

30. In the following eliminations, the percent Hofmann elimination increases as the leaving group becomes worse. Explain why this occurs using the Hammond postulate and the relative acidities of the hydrogen adjacent to the carbocation center.

LG = Cl	23%	77%
LG = OAc	45%	55%
LG = NHNH₂	60%	40%

31. Explain why there is such a high percentage of product from a syn elimination pathway in the following reaction.

32. The acid-catalyzed dehydration of 2-phenylcyclohexanol gives very different product ratios depending upon whether the reactant is cis or trans. Write a mechanism for these dehydration reactions that explains the product ratios.

33. There are two kinetically indistinguishable mechanisms for the general-base-catalyzed hydrolysis of esters, both of which would show acyl carbon to oxygen bond cleavage. What are these?

34. We noted in this chapter that specific-base catalysis is not very effective for the hydrolysis of amides. However, in solutions of water, ether, and large excesses of potassium *tert*-butoxide, catalysis by the basic medium can be quite effective. Propose a mechanism for this and a possible intermediate that would explain why these conditions are useful.

35. In the hydrolysis of amides, we stated that at very low pHs, especially when H_o scales are necessary, the pathway converts to protonation of the amide nitrogen followed by rate-determining acyl carbon to nitrogen bond heterolysis. Why does this mechanism become favored at very high acid concentrations?

36. The gas phase protonation energies of amides do not track well with the protonation energies of amines. This indicates that the amides are protonated on the carbonyl oxygen and not the nitrogen. Why are the carbonyl oxygens of esters and amides the more basic sites, and not the leaving group oxygen and nitrogen, respectively?

37. Rationalize why esters are most effectively formed with acid catalysis from carboxylic acids and alcohols, but more effectively hydrolyzed under basic conditions.

38. The initial steps of acid-catalyzed hydrolysis of the following orthoester can give two different dioxocarbenium ion intermediates. Show what these cations are and predict which will dominate.

39. The acid-catalyzed hydrolysis of *tert*-butyl esters proceeds via a less common mechanism that differs from any given in Figure 10.16. Predict what this mechanism is, rationalize why this new mechanism is viable, and propose one or two experiments to demonstrate that this new mechanism is operative.

40. The acid-catalyzed hydrolysis of methyl 2,4,6-trimethylbenzoate proceeds via a less common mechanism, neither the addition–elimination nor the S_N2 pathway. Predict what this mechanism is, rationalize why this new mechanism is viable, and propose one or two experiments to demonstrate that this new mechanism is operative.

41. The following mechanism shows a series of steps for the formation of a hemiacetal catalyzed by an added base. What is the kinetic expression for this mechanism if the first step, the second step, or the third step of the reaction is rate-determining? With the second and third possibilities, assume the first step achieves equilibrium. What experiments would you perform to figure out which step is rate-determining?

42. Draw a More O'Ferrall–Jencks plot for general-acid and specific-acid catalysis of the first step in acetal hydrolysis (shown below). For the general-acid-catalyzed pathway, does the extent of protonation in the transition state increase, decrease, or stay the same when the added acid is stronger, and what happens to the extent of leaving group departure? For the general-acid-catalyzed pathway, does the extent of leaving group departure increase, decrease, or stay the same as the leaving group becomes better, and what happens to the extent of protonation?

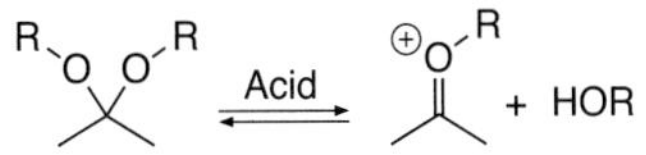

43. The nitration of very electron rich aromatic rings is diffusion controlled. Although the reactivity of NO_2^+ is extremely high, these reactions sometimes show good selectivity in their regiochemistry of addition. Propose relative rates for the various steps in a plausible mechanism, and predict which steps are in equilibrium in order to account for the seeming lack of correlation with the reactivity–selectivity principle.

44. In the following vinyl substitution reaction, retention of configuration at the double bond is observed. What must this mean about the mechanism?

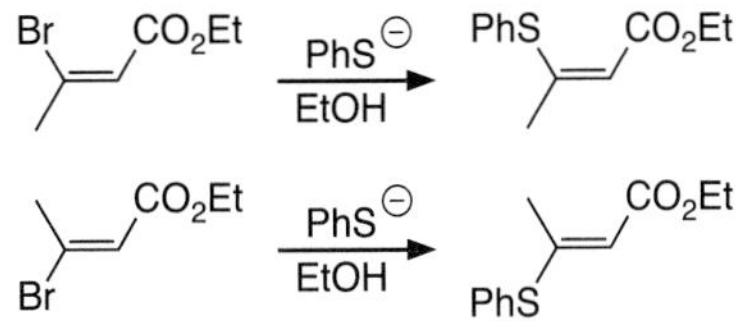

45. The following are name reactions or sequences that are often covered in introductory organic chemistry courses. Give the mechanisms of these transformations, showing all intermediates and all electron flow. (For more challenging electron-pushing exercises, see Appendix 5.)

A. Reformatsky

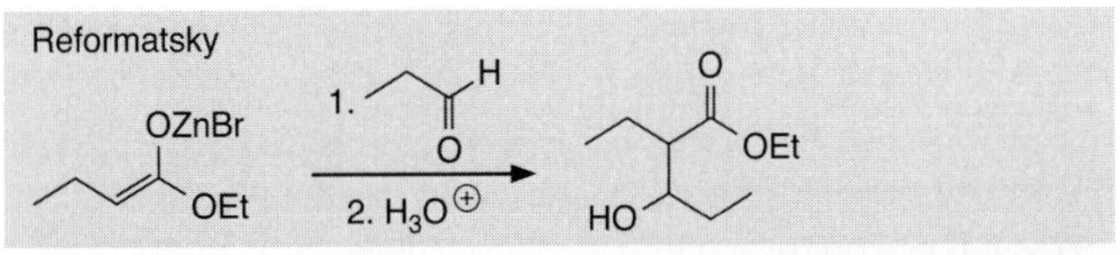

B. Dieckman condensation

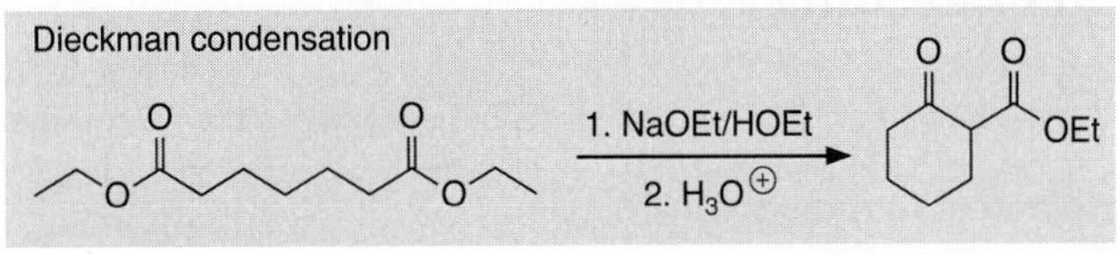

C. Mannich reaction

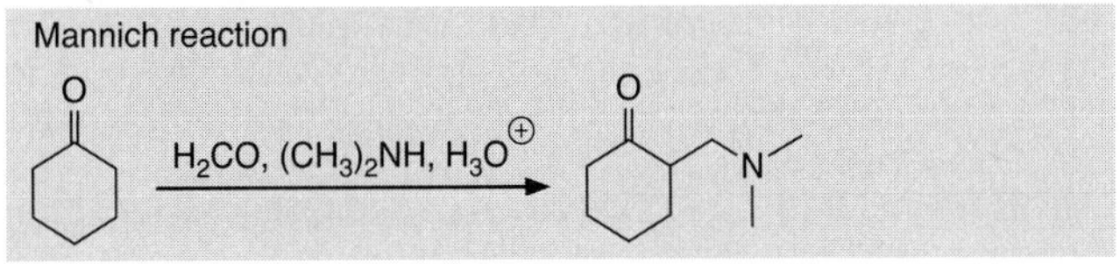

46. Show that the E2 reaction of 2-bromobutane is stereoselective, but not stereospecific.

47. The following alkyl bromide gives both *cis-* and *trans-*2-butene upon reaction with sodium ethoxide. Only one of these alkenes retains the deuterium label. Draw both products and explain why only one is deuterated.

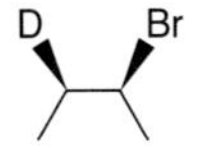

48. Write a complete mechanism with all electron pushing for the following reaction.

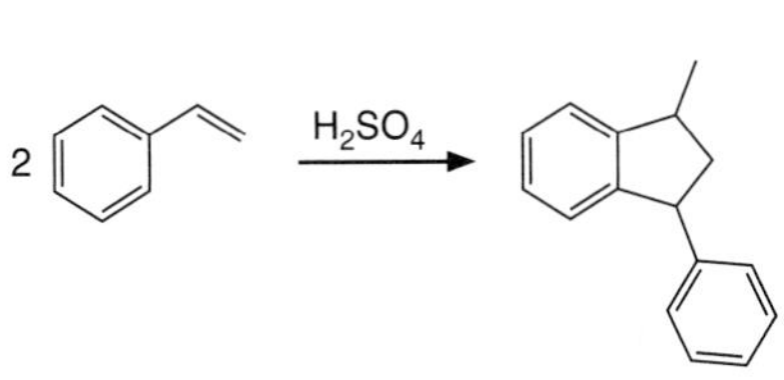

49. Provide a mechanism for the following reaction. Explain why this reaction occurs readily, given that 1-chloronaphthalene is unreactive under the same conditions.

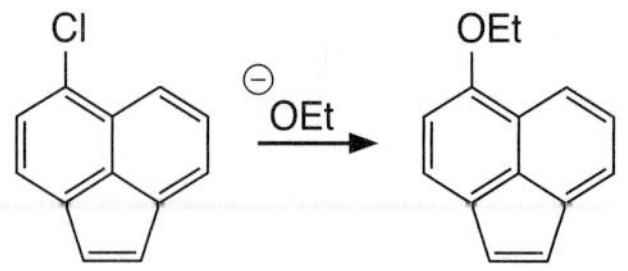

50. Predict the product of the following reaction.

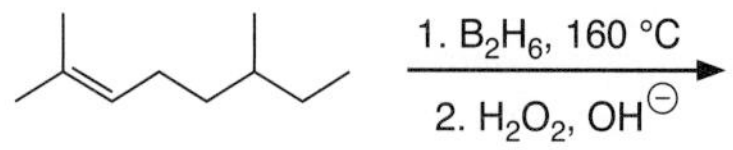

51. Give a mechanism showing all of the electron flow for the following reaction. Predict which product dominates.

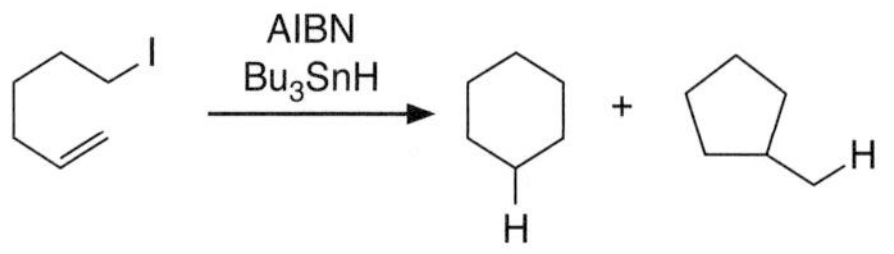

The following reaction also occurs, with the product distribution shown. Is this product distribution the same that you predicted above? Regardless of whether it is or not, give an explanation.

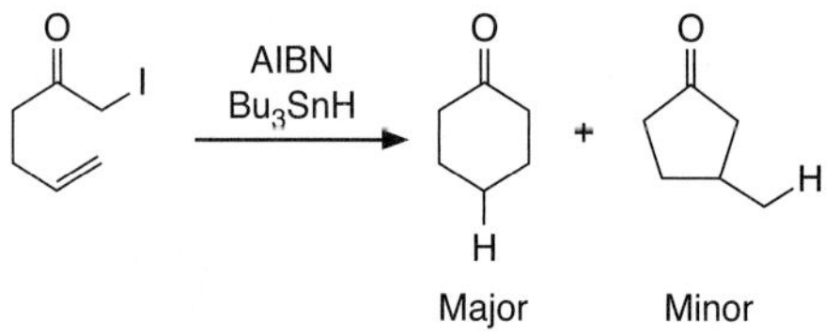

Further Reading

For textbooks on organic mechanisms in general, see the references in Chapter 11.

Hydration of Carbonyl Structures

Patai, S. (ed.) (1966). *The Chemistry of the Carbonyl Group*, Vol. 1, John Wiley & Sons, London.
Zabicky, J. (ed.) (1970). *The Chemistry of the Carbonyl Group*, Vol. 2, John Wiley & Sons, London.
Fife, T. H. "Physical Organic Model Systems and the Problem of Enzymatic Catalysis." *Adv. Phys. Org. Chem.*, **11**, 1 (1975).
Cordes, E. H., and Bull, H. G. "Mechanisms and Catalysis for Hydrolysis of Acetals, Ketals, and Ortho Esters." *Chem. Rev.*, **74**, 581 (1974).
Toullec, J. "Enolisation of Simple Carbonyl Compounds and Related Reactions." *Adv. Phys. Org. Chem.*, **18**, 1 (1982).
Bell, R. P. "The Reversible Hydration of Carbonyl Compounds." *Adv. Phys. Org. Chem.*, **4**, 1–29 (1966).

Electrophilic Addition of Water to Alkenes and Alkynes

de la Mare, P. B. D., and Bolton, R. (1982). *Electrophilic Additions to Unsaturated Systems*, 2nd ed., Elsevier, New York.
Schmid, G. H. in *The Chemistry of the Carbon–Carbon Triple Bond*, S. Patai (ed.), John Wiley & Sons, New York, 1978, p. 275.

Halonium Ion or Sigma Complex

Olah, G. A. "Carbocations and Electrophilic Reactions." *Angew. Chem. Int. Ed. Eng.*, **12**, 173 (1973).
Farcasiu, D. "Protonation of Simple Aromatics in Superacids. A Reexamination." *Acc. Chem. Res.*, **15**, 46 (1982).

π Complex

Banthorpe, D. V. "π Complexes as Reaction Intermediates." *Chem. Rev.*, **70**, 295 (1970).

Nucleophilic Additions to Olefins

Patai, S., and Rappoport, Z. in *The Chemistry of Alkenes*, S. Patai (ed.), John Wiley & Sons, New York, 1964.
Winterfeldt, E. "Additions to the Activated C–C Triple Bond." *Angew. Chem. Int. Ed. Eng.*, **6**, 423 (1967).
Rappoport, Z. "Nucleophilic Vinylic Substitution." *Adv. Phys. Org. Chem.*, **7**, 1 (1969).
Modena, G. "Reactions of Nucleophiles with Ethylenic Substrates." *Acc. Chem. Res.*, **4**, 73 (1971).
Dickstein, J. I., and Millers, S. I. in *The Chemistry of the Carbon–Carbon Triple Bond*, Part 2, S. Patai (ed.), John Wiley & Sons, New York, 1978, p. 813.

Baldwin's Rules

Johnson, C. D. "Stereoelectronic Effects in the Formation of 5- and 6-Membered Rings: The Role of Baldwin's Rules." *Acc. Chem. Res.*, **26(9)**, 476–482 (1993).
Juaristi, E., and Cuevas, G. "A 'Neumonic' for Baldwin's Rules for Ring Closure." *Rev. Soc. Quim. Mex.*, **36(1)**, 48 (1992).

Radical Additions to Unsaturated Systems

Cadogan, J. I. G., and Hey, D. H. "Free-Radical Addition Reactions of Olefinic Systems." *Quart. Rev. (London)*, **8**, 308–329 (1954).

Elimination to Form Carbonyls

Ogata, Y., and Kawasaki, A. in *The Chemistry of the Carbonyl Group*, Vol. 2, J. Aabicky (ed.), John Wiley & Sons, London, 1970, p. 1.

E1cB Mechanism

Banthorpe, D. V. (1963). *Elimination Reactions*, Elsevier, Amsterdam.
Saunders, W. H., Jr., and Cockerill, A. (1973). *Mechanisms of Elimination Reactions*, John Wiley & Sons, New York.

Tetrahedral Intermediate

Perrin, C. L. "Is There Stereoelectronic Control in Formation and Cleavage of Tetrahedral Intermediates?" *Acc. Chem. Res.*, **35(1)**, 28–34 (2002).

Electrophilic Aromatic Substitution

Stock, L. (1968). *Aromatic Substitution Reactions*, Prentice–Hall, Englewood Cliffs, NJ.
Norman, R. O. C., and Taylor, R. (1965). *Electrophilic Substitution in Benzenoid Compounds*, Elsevier, Amsterdam.
Stock, L. M., and Brown, H. C. "A Quantitative Treatment of Directive Effects in Aromatic Substitution." *Adv. Phys. Org. Chem.*, **1**, 35 (1963).
Taylor, R. "Nucleophilic Substitution." *Aromatic and Heteroaromatic Chem.*, **2**, 271 (1974).
Taylor, R. "Electrophilic Substitution on Carbon." *Aromatic and Heteroaromatic Chem.*, **3**, 220 (1975).

Friedel–Crafts Alkylation

Olah, G. A. (ed.) (1963–1965). *Friedel–Crafts and Related Reactions*, Vols. 1–4, John Wiley & Sons, New York.

Nucleophilic Aromatic Substitution

Miller, J. (1968). *Aromatic Nucleophilic Substitution*, Elsevier, Amsterdam.
Bernasconi, C. F. "Mechanisms and Reactivity in Aromatic Nucleophilic Substitution Reactions." *MTP Int. Rev. Sci. Org. Chem., Ser. One.*, **3**, 33 (1973).
Peitra, F. "Mechanisms for Nucleophilic and Photonucleophilic Aromatic Substitution Reactions." *Quart. Rev.*, **23**, 504 (1969).

Radical Aromatic Substitution

Perkins, M. J. in *Free Radicals*, Vol. II, J. K. Kochi (ed.), John Wiley & Sons, New York, 1973, p. 231.
Hey, D. H. "Arylation of Aromatic Compounds." *Adv. Free Radical Chem.*, **2**, 47 (1967).
Tiecco, M., and Testaferri, L. in *Reactive Intermediates*, Vol. 3, R. A. Abramovich (ed.), Plenum Press, New York, 1983.

Wittig Reaction

Zimmerman, H. E. in *Molecular Rearrangements*, Part 1, P. de Mayo (ed.), John Wiley & Sons, New York, 1963.

CHAPTER 11

Organic Reaction Mechanisms, Part 2: Substitutions at Aliphatic Centers and Thermal Isomerizations/Rearrangements

Intent and Purpose

In this second of two chapters on organic reaction mechanisms we focus on substitutions at aliphatic (saturated) centers. We start this chapter with an examination of substitution α to carbonyl groups, which is a logical continuation of the mechanisms that centered on these types of unsaturated structures covered in Chapter 10. Next, substitutions on aliphatic centers possessing leaving groups are covered. The classic mechanisms of substitution, S_N2 and S_N1, have been referred to frequently throughout this textbook. Now it is time to look at them in detail and show that, although they look relatively simple at first glance, there are complexities that reveal a continuum of mechanisms between these two extremes. In fact, there is even radical character to some of these mechanisms. Lastly, we wrap up our analysis of organic reaction mechanisms by examining isomerizations and rearrangements. These are reactions that either involve intramolecular migration of groups to electrophilic centers, or involve biradicals leading to stereochemical isomerizations or skeletal rearrangements. Other thermal rearrangements that can be classified as pericyclic are covered in Chapter 15.

As with Chapter 10, we introduce each section with the standard electron pushing for the mechanism under analysis. In many cases we show that this electron pushing only represents one particular variant of the mechanism, and that all reactants do not conform to the standard. Yet, the electron pushing shown will once again serve to accentuate the paradigm of reactivity given in the beginning of Chapter 10, where electrophile/nucleophile, Lewis acid/base, and donor/acceptor orbitals can be used to predict reactivity and guide one in writing proper electron-pushing schemes. We advise that you keep your focus on these schemes as the "big picture" for each mechanism, and then examine differences and subtleties that arise as the structures of the reactants vary.

Substitution α to a Carbonyl Center: Enol and Enolate Chemistry

Carbonyl-containing compounds undergo a variety of reactions apart from additions and addition–eliminations. The hydrogens on carbons adjacent to carbonyls, the **α-hydrogens**, are relatively acidic, with pK_a values in the range of 18–22. This acidity enables a broad class of substitution reactions, and such reactions are the topic of this section. The deprotonation of the α-hydrogens gives carbanions called **enolates**. Many reactions that are traditionally covered in introductory organic chemistry courses involve enolates, such as the aldol, Claisen, Michael addition, acetoacetic ester synthesis, malonic ester synthesis, and Robinson annulation. These reactions can also occur via enols, and hence we cover both enols and enolates together. We do not look at all these reactions here, but instead we focus upon the mechanistic aspects that are common to the reactions of enolates.

11.1 Tautomerization

Tautomerizations involve the shift of a hydrogen atom across a π system. The most typical tautomerization is a 1,3-shift, and the focus of this section is the interconversion of a ketone (or aldehyde) and an enol, often termed **keto–enol tautomerization**. The reaction can be catalyzed by acid or base, and it is technically an isomerization, a class of reactions we will cover later in this chapter. However, knowledge of the mechanism of keto–enol tautomerizations is crucial to understanding enol and enolate chemistry, and therefore we cover it here.

11.1.1 Electron Pushing for Keto–Enol Tautomerizations

Let's examine the electron pushing for an acid-catalyzed tautomerization (Scheme 11.1). Protonation of the carbonyl oxygen of a ketone or aldehyde renders the α-hydrogens even more acidic than normal. This leads to facile deprotonation and the formation of an enol.

Scheme 11.1
Acid-catalyzed formation of an enol from acetone.

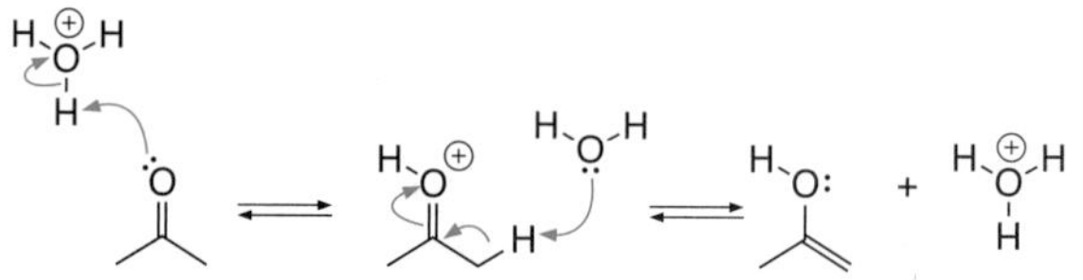

11.1.2 The Thermodynamics of Enol Formation

Based on bond dissociation energies (BDEs, Table 2.2), the keto form will dominate keto–enol equilibria for common carbonyl-containing structures. However, in many cases, enols can be spectroscopically observed or isolated, and in some cases they even dominate the equilibrium.

Table 11.1 shows a handful of keto–enol equilibrium constants. These values are not completely consistent, because varying experimental conditions were used, but they illus-

Table 11.1
Equilibrium Constants for Keto–Enol Tautomerization Reactions (K = [enol]/[keto])*

Compound	Equilibrium constant	Compound	Equilibrium constant
$CH_3-CO-CH_3$	6.3×10^{-8}	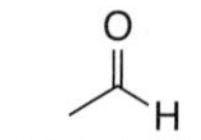	5.0×10^{-6}
$CH_3CH_2COCH_3$	5.0×10^{-9}	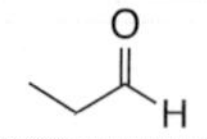	1.2×10^{-4}
$CH_3CH_2COCH_2CH_3$	1.6×10^{-8}	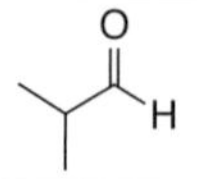	1.6×10^{-3}
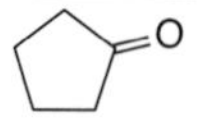	6.3×10^{-8}	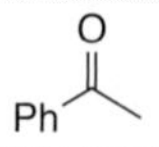	2.0×10^{-7}
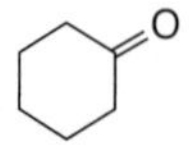	2.0×10^{-6}	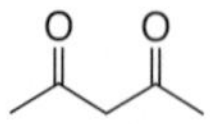	3.2
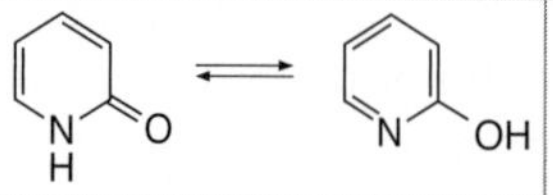	1.1×10^{-3}	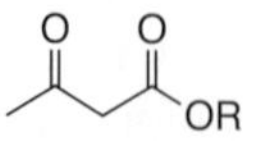	0.09
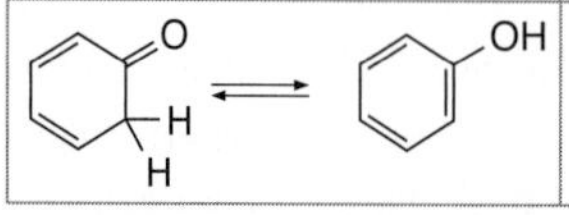	4×10^{13}		

*Guthrie, J. P., and Cullimore, P. A. "The Enol Content of Simply Carbonyl Compounds, a Thermochemical Approach." *Can. J. Chem.*, **57**, 240 (1979).

trate the key trends. For normal alkyl ketones and aldehydes, there is a miniscule fraction of the enol present at equilibrium. However, in β-diketones the enol form becomes a significant fraction of the equilibrium. This is because the creation of a conjugated π system and the formation of an intramolecular hydrogen bond stabilizes the enol form. Phenol and other enols within aromatic rings exist nearly exclusively in their enol form.

11.1.3 Catalysis of Enolizations

Enolization can be either acid- or base-catalyzed, and general catalysis mechanisms have been observed. With general-base catalysis, deprotonation of the α-carbon in the first step is rate-determining. Protonation of the enolate oxygen gives the enol (Eq. 11.1). Both reactions in the second equilibrium of Eq. 11.1 are faster than the initial deprotonation of the α-carbon of the keto form.

$$\text{(Eq. 11.1)}$$

In contrast, in the acid-catalyzed pathway the second step is rate-determining (Eq. 11.2). General-acid catalysis is found, but is commonly the combination of specific-acid and general-base catalysis (recall the kinetic equivalency given in Section 9.3.3). This combination of specific-acid and general-base catalysis is seen by examining the path given in Eq. 11.2. Protonation of the carbonyl oxygen occurs in a rapid equilibrium prior to rate-determining deprotonation. Since the protonation is not in the rate-determining step, that step is subject to specific-catalysis, while the rate-determining deprotonation is performed by the conjugate base of the added acid, thereby being general-base-catalyzed. Chemists find that the protonation or deprotonation at oxygen is faster than protonation or deprotonation at carbon.

$$\text{(Eq. 11.2)}$$

Evidence that deprotonation is rate-determining in the general-acid-catalyzed enolization of a carbonyl comes from studying the reverse reaction. The rate of conversion of the enol of cyclohexanone to cyclohexanone has been measured (Eq. 11.3). It occurs at the same rate as the hydrolysis of 1-methoxycyclohexene (Eq. 11.4), where protonation of the double bond of the enol ether is known to be rate-determining. The coincidence is interpreted to mean that protonation is rate-determining in both reactions of Eqs. 11.3 and 11.4. Hence, by microscopic reversibility, if protonation of the double bond of the enol is rate-determining in the formation of the ketone, it follows that deprotonation of the α-carbon is rate-determining in the formation of the enol (Eq. 11.2).

$$\text{(Eq. 11.3)}$$

$$\text{(Eq. 11.4)}$$

11.1.4 Kinetic vs. Thermodynamic Control in Enolate and Enol Formation

When a carbonyl has two different R groups with α-hydrogens, there are two possible enols and enolates. Furthermore, deprotonation of an acyclic structure with α-hydrogens can lead to enol and enolate isomers (E and Z). By judicious choices of reaction conditions (primarily temperature control) and base, it is usually possible to control which enolates are formed. Because enolates are used more often than enols synthetically, our focus is now on enolate formation. However, the kinetic and thermodynamic effects we discuss are also relevant to enols.

By analogy to olefin chemistry, we expect that the more stable enolate, the **thermodynamic enolate**, will be the one with the more heavily substituted C=C double bond. Deprotonation under equilibrating conditions will produce the thermodynamic enolate. The same trend is seen for enols; that is, the more substituted double bond gives the more stable enol.

If we use a bulky base at low temperatures, however, it will abstract the least sterically hindered α-hydrogen, which leads to the less substituted double bond in the enolate (the **kinetic enolate**). Examples of thermodynamic and kinetic enolate formation are given in Eqs. 11.5 and 11.6, respectively. Here, the use of the strong base LDA (lithium diisopropylamide) at low temperature gives 84% of the less stable enolate (kinetic control), while the use of triethylamine at ambient temperature gives 87% of the more stable enolate (thermodynamic control). Trimethylsilyl (TMS) chloride was used to trap the enolates, leading to silyl enol ether functional groups (defined as O-silylated enolates). The ratio of the two silyl enol ethers is indicative of the enolate ratio in solution, presuming that the trapping rates for the two enolates are the same.

A word of caution is in order when discussing issues such as the most stable enolate. Under many conditions of the sort commonly used in organic synthesis, enolates are not simple structures. Extensive aggregation is common, producing dimers, tetramers, and even octamers. Often the reactive species in an enolate reaction is one of these aggregated species, rather than a simple structure (see the next Going Deeper highlight). Thus, it may well be that some of the rationalizations we present here for enolate chemistry are more like mnemonics that get the answer right rather than detailed mechanistic insights.

$$\text{(Eq. 11.5)}$$

13%   87% (E and Z)

$$\text{(Eq. 11.6)}$$

84%   16% (E and Z)

Another issue is the ratio of enolate E and Z isomers that can form in appropriate systems. In general, Z-enolates are more stable than E-enolates due to lower steric interactions with the R group on the carbonyl carbon. For example, in the deprotonation of 3-pentanone with lithium tetramethylpiperidide (LTMP) at low temperature, the E-enolate is formed preferentially (Eq. 11.7). In contrast, the use of LTMP with added HMPA (hexamethylphosphoramide, which binds cations) at ambient temperature preferentially gives the thermodynamic enolate (Eq. 11.8). The HMPA assists in breaking up the aggregates mentioned above, and in the exchange of enolate deprotonation sites.

$$\text{(Eq. 11.7)}$$

R = ethyl   14%   86%
          Z-Enolate   E-Enolate

$$\text{(Eq. 11.8)}$$

R = ethyl   92%   8%
          Z-Enolate   E-Enolate

Going Deeper

Enolate Aggregation

We noted above the tendency for enolates to aggregate. The aggregates are organized in large part by the counterion of the base, Li$^+$ in the case of LDA or LTMP. To the side, prototypical dimers and tetramers are shown. In the absence of ligands such as HMPA, diamines, and THF, even larger aggregates exist.

Williard, P. G., and Carpenter, G. B. "X-Ray Crystal Structure of an Unsolvated Lithium Enolate Anion." *J. Am. Chem. Soc.*, **107**, 3345 (1985). Amstutz, R., Schweizer, W. B., Seebach, D., and Dunitz, J. D. "Tetrameric Cubic Structures of Two Solvated Lithium Enolates." *Helv. Chim. Acta*, **64**, 2617 (1981). Williard, P. G., and Hintze, M. J. "The First Structural Characterization of a Lithium Ketone Enolate-LDA Complex." *J. Am. Chem. Soc.*, **109**, 5539 (1987).

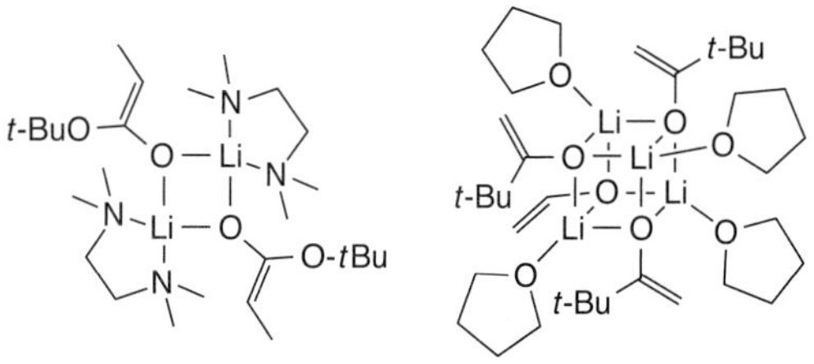

Lithium enolate aggregates

11.2 α-Halogenation

After tautomerization, an enol will act as a nucleophile in a variety of reactions. One is halogenation. Under acidic conditions, the acid catalyzes formation of the enol, which then undergoes a reaction with molecular halogen (X_2). Under basic conditions in protic solvents, the base catalyzes formation of the enol also, but now via the enolate. Either the enol or enolate can react with molecular halogen.

11.2.1 Electron Pushing

Scheme 11.2 shows the electron pushing schemes for α-bromination under both acidic and basic conditions. In base deprotonation creates an enolate. The enolate can rapidly attack electrophilic bromine. A similar electron-pushing scheme commences from the enol in acid.

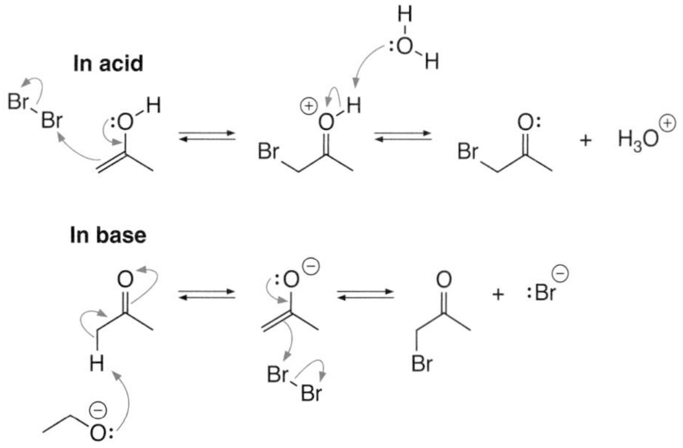

Scheme 11.2
α-Halogenation electron pushing via an enol or via an enolate.

11.2.2 A Few Experimental Observations

Under acidic or basic conditions, α-halogenation gives similar experimental observations. The rate is typically first order in the carbonyl structure and zero order in molecular halogen (Cl_2, Br_2, or I_2) . Racemization of a stereogenic center undergoing halogenation occurs competitively with the halogenation. Furthermore, exchange with a deuterium-labeled protic solvent proceeds at rates similar to halogenation. In addition, large primary kinetic isotope effects are found. For example, a value of 6.1 is obtained for the bromination of methyl cyclohexyl ketone in NaOMe/HOMe. All these experimental observations support a rate-determining formation of the enolate or enol, followed by rapid trapping of the intermediate by the molecular halogen.

Under acidic conditions, α-halogenation can be monitored and stopped after monohalogenation. However, because one equivalent of HX is a product of this reaction, the solution becomes more acidic as the reaction proceeds. Since acid catalyzes the reaction, it becomes faster as it proceeds, a phenomenon called **autocatalysis**.

Under basic conditions, multiple halogenations occur readily. As halogens are added to the α-carbon, the α-hydrogens become more acidic. Because deprotonation is rate-determining, subsequent halogenations become faster. This leads to either mixtures or replacement of all the α-hydrogens with X.

When different groups with α-hydrogens are attached to a carbonyl, halogenation can occur at either group. The regiochemistry of halogenation is different under acidic and basic conditions. In acid the halogenation occurs at the more substituted group, because the reaction proceeds via the most stable enol (Eq. 11.9). With base, one most commonly finds that the least substituted R group is preferentially halogenated, due to kinetic control.

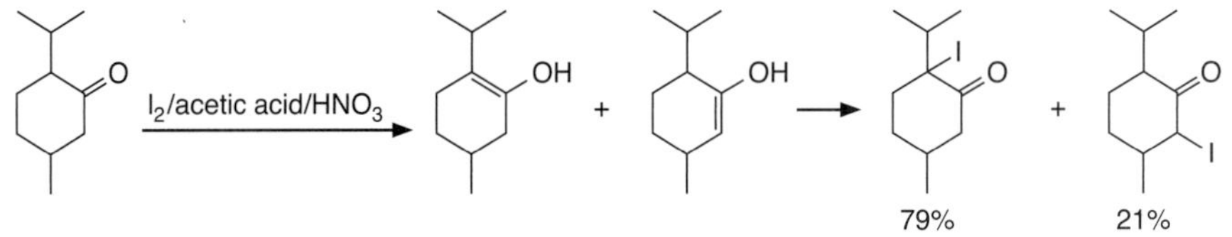

$$\text{(Eq. 11.9)}$$

If a methyl ketone is allowed to undergo multiple halogenations under basic conditions, a carboxylic acid will ultimately be produced via the **haloform reaction** (Eq. 11.10). This reaction involves three sequential halogenations, followed by an addition-elimination. It only works with a methyl ketone, because three halogens are required to create a good enough nucleofuge to depart ($^-CX_3$).

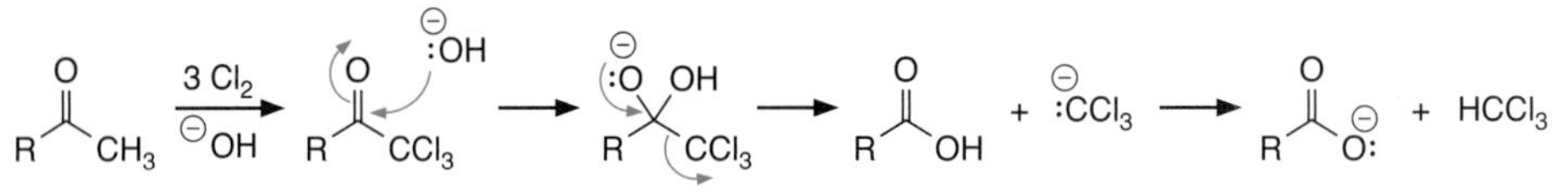

$$\text{(Eq. 11.10)}$$

11.3   α-Alkylations

The halogenation of the α-carbon of carbonyl structures is just one example of a vast field of substitution reactions that commence from enols and enolates. This field encompasses a large portion of organic synthetic procedures, and it is best left to a text devoted to that discipline. However, a few points about enolate alkylations are worth mentioning here, because the common rationalizations for many experimental observations are strongly derived from physical organic chemistry principles.

11.3.1   Electron Pushing

The alkylation of an enolate follows well-established procedures (Scheme 11.3). Irreversible deprotonation of an α-hydrogen by a strong base creates the enolate. The enolate is nucleophilic at both the C and the O. With strongly basic enolates and standard alkyl halides, alkylation at the carbon dominates.

Scheme 11.3
Electron pushing for
enolate alkylations.

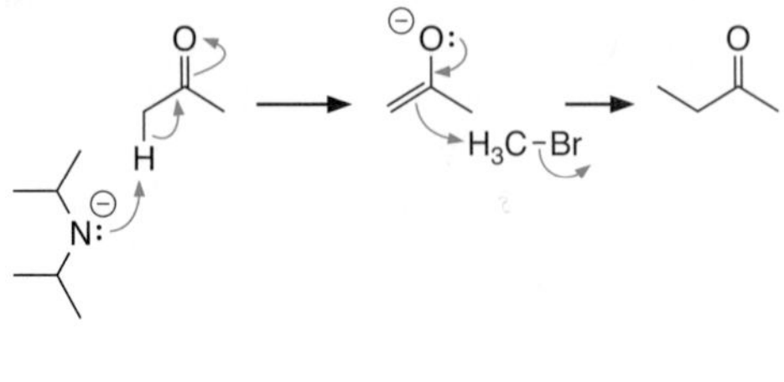

11.3.2 Stereochemistry: Conformational Effects

As with our discussion of the reduction of carbonyl compounds in Chapter 10, *t*-butyl-cyclohexanone makes an interesting structure to probe stereochemical preferences in carbonyl chemistry. Let's look at two specific cases of enolate alkylation: the alkylation of 4-*t*-butylcyclohexanone, and the alkylation of 4-*t*-butylcyclohexyl methyl ketone.

The major product obtained upon alkylation of the enolate formed by the reaction of lithium diisopropyl amide with 4-*t*-butylcyclohexanone is the trans isomer (Figure 11.1 **A**). The cis product would be the thermodynamically more stable structure, because the R group would be equatorial instead of axial. Hence, we must have kinetic control operating for this alkylation. The product stereochemistry can be explained by examining the structure of the enolate, as is done in Figure 11.1 **B**. From a "chair-like" conformation of the enolate, attack on the "top face" of the enolate leads to a transition state with the ring still in a chair conformation, while attack from the "bottom face" leads the ring into a boat-like conformation. Hence, the trans product dominates due to the lower energy "chair-like" transition state.

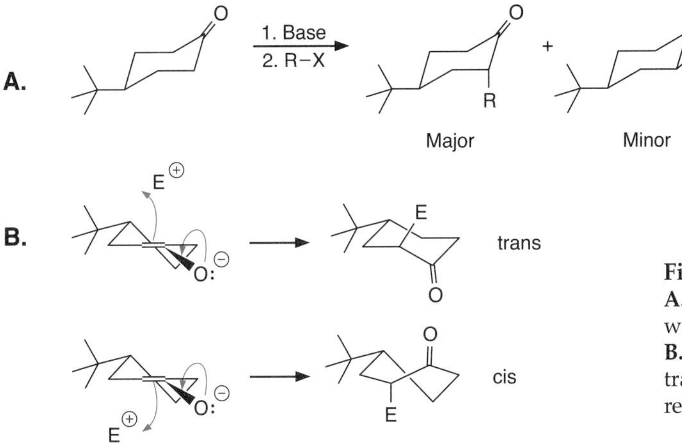

Figure 11.1
A. The reaction of 4-*t*-butylcyclohexanone enolate with alkyl halides gives the trans product predominately.
B. Two different trajectories lead to chair- and boat-like transition states giving trans and cis products, respectively.

In the alkylation of the exocyclic enolates of cyclohexane rings, the reaction occurs predominately to place the electrophile in the equatorial position (Eq. 11.11, path b). The electrophile avoids the hydrogens placed 1,3 to the nucleophilic carbon. However, because these alkylations are exothermic, the transition states are early, and the stereochemistry is not very sensitive to the alkylating agent.

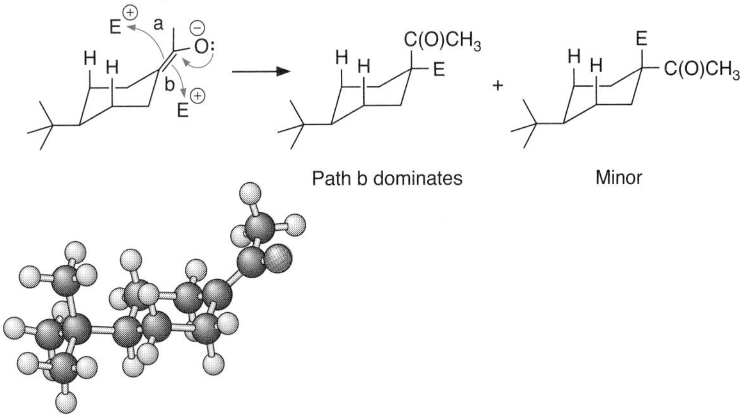

(Eq. 11.11)

Even in those cases where both chair forms of the cyclohexane ring are present, interesting stereochemical results for alkylation can be observed. For example, Eq. 11.12 shows another system where conformational analysis provides the explanation for the stereochemical outcome. In the alkylation of a 2-substituted exocyclic enolate, the major product places the 2-substituent and the electrophile trans, giving a stereoselective reaction. The enolate has two conformations (see margin), one of which suffers a 1,3-diaxial strain (typically around 2 kcal/mol) while the other suffers an allylic strain (around 3 to 4 kcal/mol). The π–facial selectivity of the alkylation is as in Eq. 11.11, where alkylation of the lower energy enolate gives the major product. This is a case where the lower energy conformation leads to the major product, but as the Curtin–Hammett principle states, this is not always the case.

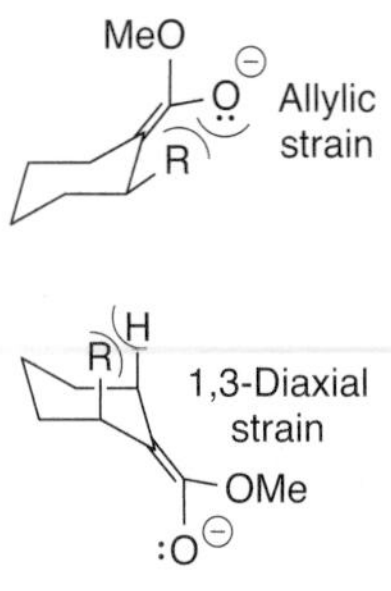

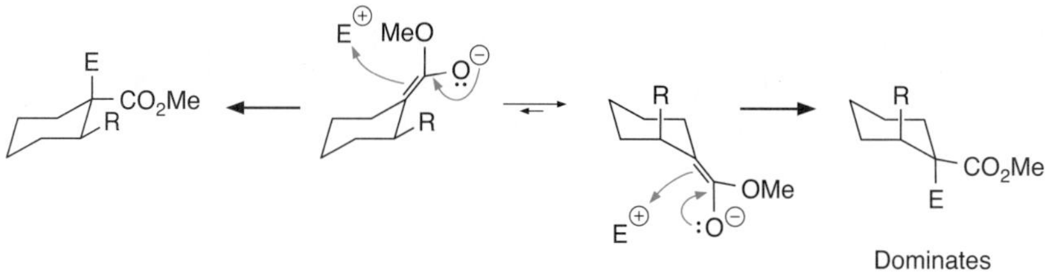

$$\text{(Eq. 11.12)}$$

11.4 The Aldol Reaction

The aldol reaction involves the substitution of an α-hydrogen by the carbonyl carbon of another carbonyl compound, thereby creating a β-hydroxycarbonyl product. Eq. 11.13 shows the coupling of two aldehydes, while Eq. 11.14 shows the coupling of two ketones. The reaction is similar to the alkylation and halogenation of an enolate, except that now the electrophile is another carbonyl compound, and so we have nucleophilic addition to a carbonyl as well as substitution on an α-carbon. As with other reactions we have seen in this section, the aldol reaction can proceed via an enol or an enolate. However, the most common pathway, and the one we will emphasize here, makes use of an enolate and so is base-catalyzed. With prolonged treatment in acid or base, the β-hydroxycarbonyl products will dehydrate to form α,β-unsaturated carbonyl structures.

The aldol reaction is reversible; cleavage of a β-hydroxycarbonyl is called the **reverse aldol** or **retro-aldol**. Equilibria between the β-hydroxycarbonyl products and the carbonyl reactants lie toward the products with alkyl aldehydes and toward the reactants with alkyl and aryl ketones.

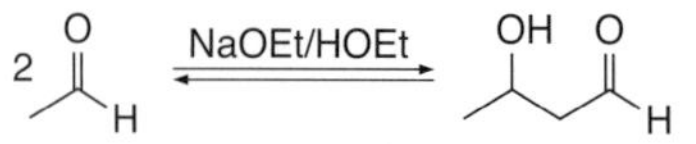

$$\text{(Eq. 11.13)}$$

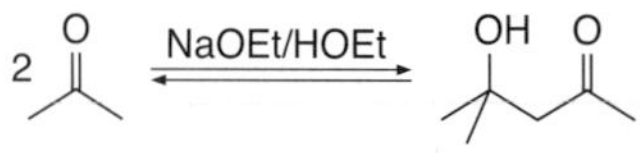

$$\text{(Eq. 11.14)}$$

11.4.1 Electron Pushing

As shown in Scheme 11.4, reversible deprotonation of the α-carbon of a carbonyl creates an enolate that can undergo nucleophilic addition to another carbonyl. Protonation of the resulting adduct leads to the product. The electron pushing is similar to other carbonyl additions performed under basic conditions.

Scheme 11.4
The electron-pushing scheme for the aldol reaction.

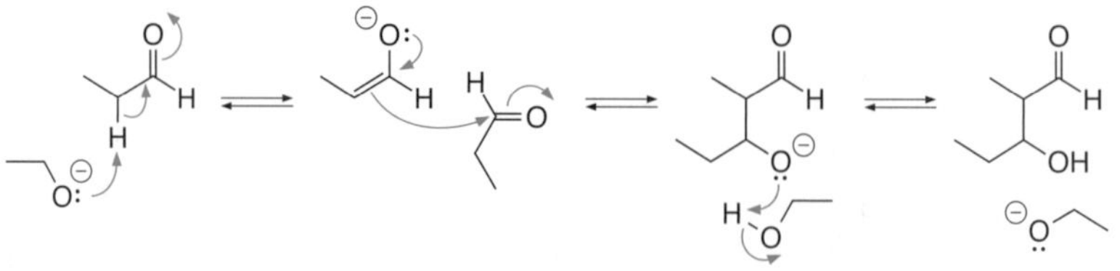

11.4.2 Conformational Effects on the Aldol Reaction

The importance of the aldol reaction in synthetic organic chemistry cannot be overstated. In this regard, controlling the stereochemistry of the reaction has received extensive investigation. Entire chapters have been written about just this one facet of synthetic organic chemistry, and we cannot hope to do the topic justice here. However, a few interesting features of the methods for controlling aldol stereochemistry should be mentioned here.

When the two carbonyl-containing species of the aldol reaction are the same, then a simple reflux in basic ethanol will lead to the reaction (Eq. 11.13). However, when two different carbonyls are involved, special precautions must be taken to avoid an intractable mixture of products. Enolate formation is achieved first, and the reaction conditions should promote quantitative and irreversible enolate formation. Typically, an aldehyde is required as the electrophilic carbonyl component, so that the addition reaction is fast. In this way, complications involving proton transfer from the α-carbon of the carbonyl to the enolate, making a new enolate, are avoided.

For example, consider the reaction of the enolate of cyclohexanone with 1-butanal (Eq. 11.15). Here, the nucleophile is first created by treatment with a strong base, followed by addition to the electrophile. Under proper conditions, no enolate from 1-butanal is formed, and a single product is isolated (ignoring stereoisomers). This is referred to as a **mixed aldol** reaction.

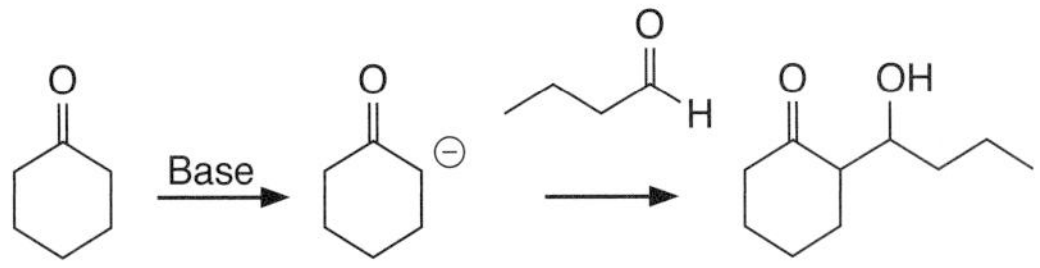

(Eq. 11.15)

A reaction such as that in Eq. 11.15 forms two new stereocenters, and so diastereomeric products are possible. A general trend for the diastereoselectivity is found in mixed aldol reactions. Z-Enolates give predominately syn (or threo) aldol products, while E-enolates give predominately anti (or erythro) products (Figure 11.2 shows the definitions we are discussing). These trends are accentuated when R_1 and R_3 (the substituent on the electrophilic aldehyde) are sterically demanding, while the trends are counteracted when R_2 is sterically demanding. Figure 11.2 gives a rationale for these observations based upon a conformational analysis that is called the **Zimmerman–Traxler model**. Transition states with "chair-like" conformations are proposed, where chelation of the counter cation to the enolate with the oxygens of the nucleophile and electrophile creates six-membered rings.

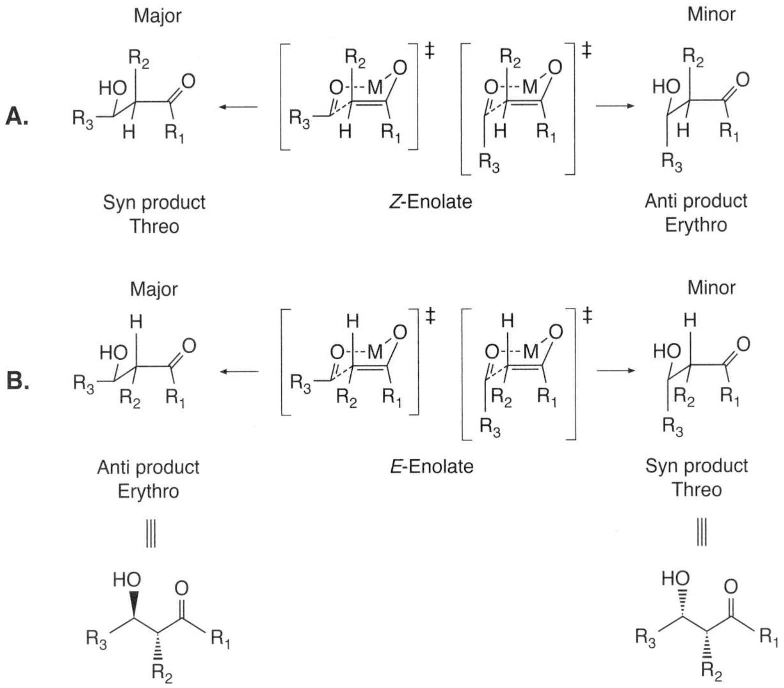

Figure 11.2
Analysis of conformational effects on aldol reactions. **A.** The equilibria and products from Z-enolates. **B.** The equilibria and products from E-enolates.

Let's first examine the Z-enolates in Figure 11.2 **A**. Placement of the R_3 group of the electrophile in an equatorial position is preferred, and the syn product therefore dominates. With E-enolates, the R_3 group again prefers the equatorial position, but now this leads to anti products. The preference for R_3 being equatorial is increased as the size of R_1 or R_3 increases due to larger 1,3-diaxial strain.

While we have just described the simplest of aldol reactions, the analysis is quite representative. Chelation control and application of the rules of conformational analysis to reaction transition states are common aspects of the analysis of more advanced aldol systems.

Going Deeper

Control of Stereochemistry in Enolate Reactions

Typically, the products of an aldol or other reactions of enolates are chiral molecules, and often the starting materials are not. Thus, another stereochemical issue arises—namely, the preferential formation of one enantiomeric product over the other. This again is a large research area, with many clever and effective solutions. One especially useful approach has been developed by Evans.

The Evans approach shown below makes use of enantiomerically pure oxazolidinones known as **chiral auxiliaries**. A chiral auxiliary is a fragment that is appended to a reactive species in order to influence the stereochemistry of subsequent reactions. After the desired effect is achieved, the auxiliary is removed and (hopefully) recycled for further use.

As shown in the scheme below, starting from the amino acid valine, reduction of the carboxylic acid with borane and reaction with diethyl carbonate produces the oxazolidinone with a bulky isopropyl group strategically positioned. The nitrogen is acylated, and then enolate formation with LDA produces a well-defined structure thanks to chelation of the lithium ion by the enolate oxygen and the carbonyl of the oxazolidinone. When an electrophile (E) approaches, the isopropyl group blocks the bottom face, and so attack occurs preferentially from the top face, producing excellent control of stereochemistry. Removal of the auxiliary produces an enantiomerically enriched product that formally arises from the reaction of an electrophile with the enolate of a propionic acid derivative. Many variations are possible, and a wide range of electrophiles is compatible with this useful reaction.

Evans, D. A., Ennis, M. D., and Mathre, D. J. "Asymmetric Alkylation Reactions of Chiral Imide Enolates. A Practical Approach to the Enantioselective Synthesis of α-Substituted Carboxylic Acid Derivatives." *J. Am. Chem. Soc.*, **104**, 1737–1739 (1982).

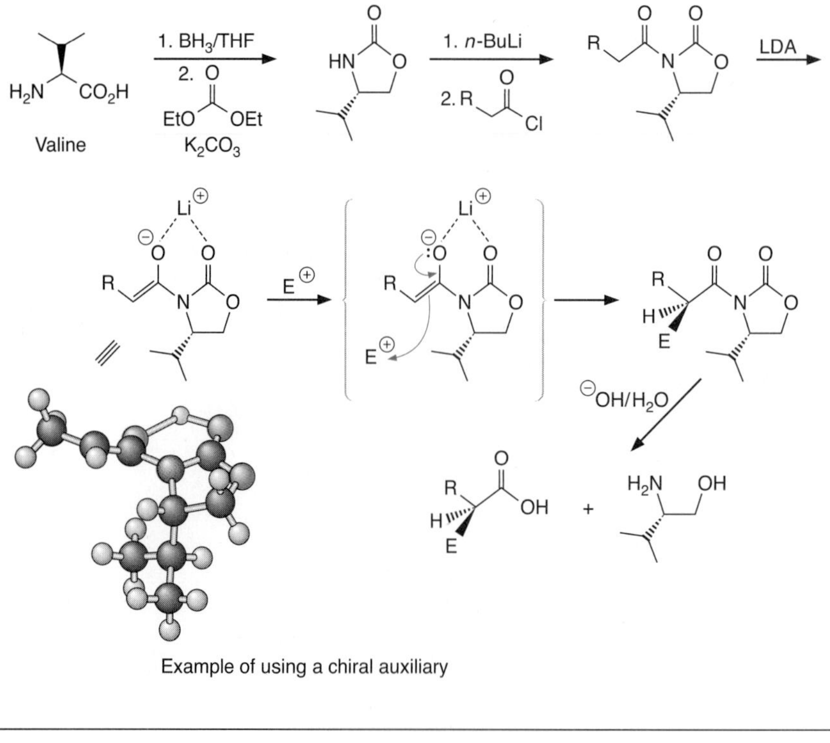

Example of using a chiral auxiliary

Substitutions on Aliphatic Centers

11.5 Nucleophilic Aliphatic Substitution Reactions

Interconversion of groups at aliphatic centers can occur via nucleophilic, electrophilic, and radical pathways. Nucleophilic mechanisms are the most common, and are given the classic S_N2 and S_N1 descriptors first suggested by Hughes and Ingold. Many of the pioneers of physical organic chemistry honed their skills on these reaction mechanisms. We will mention many below, but we note from the beginning the overarching contributions of Winstein to this area of organic chemistry.

11.5.1 S_N2 and S_N1 Electron-Pushing Examples

Three examples of S_N2 (substitution, nucleophilic, bimolecular) reactions are shown in Scheme 11.5. These are simple reactions from a mechanistic standpoint. They are concerted, and there is very little that can go wrong when considering the proper electron-pushing notation. These reactions fit our paradigm for predicting reactivity, because they are combinations of nucleophiles and electrophiles, whose reactivity can be predicted solely based upon electrostatic considerations. The specific reaction shown in Scheme 11.5 **B** is an example of the **Menschutkin reaction**, defined as the reaction between an amine nucleophile and an alkyl halide. Scheme 11.5 **C** shows the second step of the enolate alkylation reaction we described in Section 11.3.

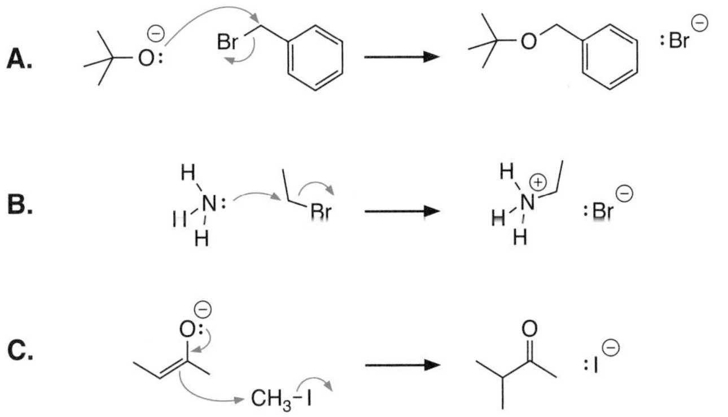

Scheme 11.5
S_N2 examples. **A.** Williamson ether synthesis, **B.** Alkylation of an amine, and **C.** Alkylation of an enolate.

A **concerted reaction** is one in which the bond breaking and bond forming steps all occur at the same time in one single elementary step. There is only a single transition state in a concerted reaction. A **synchronous reaction** is one in which the breaking and forming of bonds has occurred to the same extent at the transition state, whereas an **asynchronous reaction** has one of these two parts of the reaction lagging behind the other at the transition state. Most S_N2 reactions that we will consider in this chapter are asynchronous, meaning that the extents of nucleophilic attack and leaving group departure at the transition state are unequal.

A more complex substitution pathway is known as S_N1 (substitution, nucleophilic, unimolecular). Two steps are involved. The first is the formation of a carbocation, which is then followed by nucleophilic attack (Scheme 11.6). The overall conversion is not concerted, in that at least two steps are involved. Each individual step, however, is concerted; in fact, all elementary steps in any reaction mechanism are concerted. In the examples of Scheme 11.6, cleavage of the R–LG bond leads to the formation of a carbenium ion that is trapped by solvent. This kind of reaction is referred to as a **solvolysis**. A solvolysis is any substitution reaction where the nucleophile is the solvent. When the solvent is water, acetic acid, formic acid, or methanol, the reactions are known as **hydrolysis, acetolysis, formolysis**, or **methanolysis**, respectively.

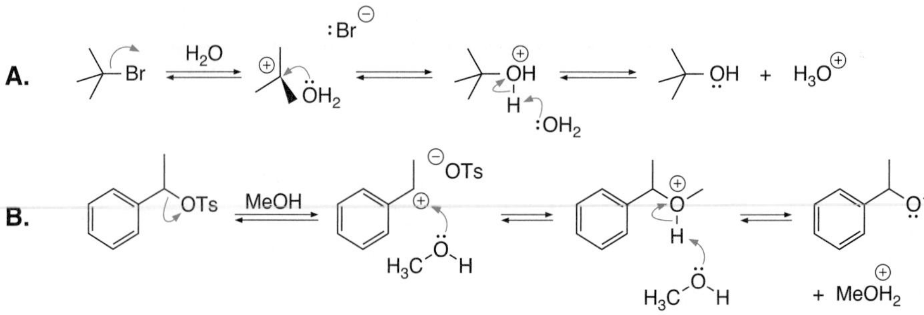

Scheme 11.6
S_N1 examples. **A.** Heterolysis of a tertiary carbon–bromine bond. **B.** Heterolysis of a benzyl–tosylate bond.

Far less common aliphatic substitution pathways include S_N2' and S_N1'. Here, nucleophilic attack occurs on the alkene of a vinylic group, leading to migration of the double bond (Scheme 11.7). We do not discuss these reactions any further here.

Let's examine the subtle details of these simple substitution reactions. As with most chemical reactions, they appear deceptively simple at first glance, but upon digging deeper we find quite a bit of complexity. Complexity arises when we examine the effects of various factors upon nucleophilic substitution—namely, the R group, the leaving group (X), the nucleophile, and the solvent.

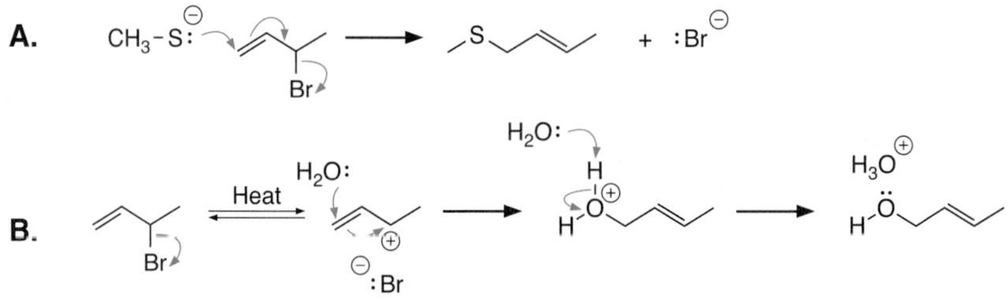

Scheme 11.7
A. An S_N2' reaction.
B. An S_N1' reaction.

11.5.2   Kinetics

Some of the earliest observations of substitution reactions were performed with hydroxide as the nucleophile and various quaternary ammonium salts of the form $RNMe_3^+$ (see two examples in Eqs. 11.16 and 11.17). The kinetics of these reactions were always found to be first order in electrophile, but first order in nucleophile for certain R groups, while zero order in nucleophile for other R groups.

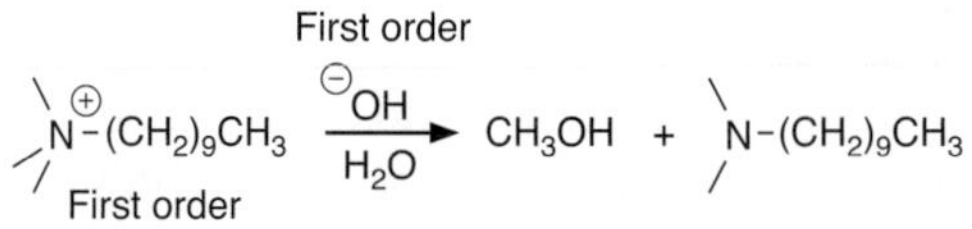

(Eq. 11.16)

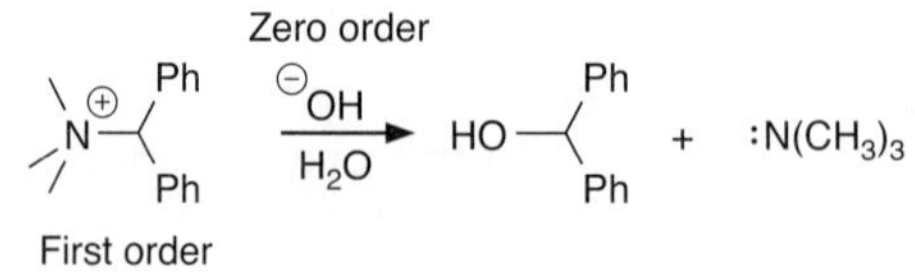

(Eq. 11.17)

The kinetic observations led to predictions about the mechanisms. The single-step reaction in Eq. 11.18 would display second-order kinetics, whereas the sequence of reactions in Eq. 11.19 could display first-order kinetics, being zero order in nucleophile if the first step were rate-determining. The major difference between the two mechanisms is the timing of the leaving group departure.

$$Nuc^{\ominus} + RX \longrightarrow Nuc{-}R + X^{\ominus}$$

(Eq. 11.18)

$$R{-}X \underset{k_{-1}}{\overset{k_1}{\rightleftharpoons}} R^{\oplus} + X^{\ominus} \overset{k_2\ Nuc^{\ominus}}{\longrightarrow} R{-}Nuc$$

(Eq. 11.19)

How do we test these hypotheses? The kinetic expression (Eq. 11.20) of the S_N2 reaction in Eq. 11.18 predicts that the nucleophile always affects the rate, even when it is in large excess. When it is in large excess, however, its concentration would hardly change during the reaction, and pseudo-first-order kinetics in electrophile alone should be found (Eq. 11.21). A linear dependence of k_{obs} on [Nuc⁻] with a non-zero slope is predicted, and this is indeed found for such reactions. Furthermore, the S_N2 reaction shows behavior predicted by simple Marcus theory (see Section 7.7.1). The more exothermic reactions typically have larger rate constants, and the transition states appear more like reactants.

$$\frac{d[P]}{dt} = k[RX][Nuc^-] \qquad \text{(Eq. 11.20)}$$

$$\frac{d[P]}{dt} = k_{obs}[RX], \text{ where } k_{obs} = k[Nuc^-] \qquad \text{(Eq. 11.21)}$$

Now let's delve into the kinetics of the S_N1 reaction. Does it really make sense that the reaction is zero order in nucleophile? After all, if you do not add the nucleophile, the reaction will not occur and the rate will be zero. There *has* to be some dependence upon nucleophile. If we write the full kinetic expression for the mechanism as given in Eq. 11.19, using the steady state approximation for the carbocation intermediate, we obtain Eq. 11.22. The concentration of the nucleophile is indeed in the kinetic expression. However, using the type of reasoning we presented in Chapter 7 for analyzing kinetic expressions, we can predict the reaction will show zero-order kinetics in nucleophile when either the nucleophile is in large excess or when $k_{-1} \ll k_2$. In a solvolysis reaction, the nucleophile is indeed in very large excess because it is the solvent.

$$\frac{d[P]}{dt} = \frac{k_1 k_2 [RX][Nuc^-]}{k_{-1}[X^-] + k_2[Nuc^-]} \qquad \text{(Eq. 11.22)}$$

The heights of the barriers that correspond to k_{-1} and k_2 on the energy surface for a substitution reaction will influence the kinetic dependence of an S_N1 reaction on the nucleophile. Since the reactions are normally observed to be zero order in nucleophile, it must be that $k_{-1}[X^-] \ll k_2[Nuc^-]$. This means that the energy barrier to recombination of the leaving group and the carbocation is higher than the barrier for reaction of the carbocation with the added nucleophile. This is indeed the common situation, because chemists normally add a reagent with a higher nucleophilicity than the leaving group.

One simple test of this mechanism is to add an alkali metal salt corresponding to the departed leaving group at the inception of the reaction. For example, if we are studying the solvolysis of an alkyl bromide, we can add sodium bromide to the reaction flask. One repeatedly analyzes the kinetics with increasing leaving group concentration until the $k_{-1}[X^-]$ term becomes comparable in magnitude to the $k_2[Nuc^-]$ term in Eq. 11.22. Under these circumstances the reaction will show a kinetic dependence on nucleophile, and the reaction will slow down. This is called the **common ion effect**, and it is the first and simplest of several salt effects we will be discussing. It is a classic, but not foolproof, test of an S_N1 mechanism.

11.5.3 Competition Experiments and Product Analyses

In Section 8.8 we examined several tools for deciphering reaction mechanisms, one of which was competition experiments, and another was examining the products. In a reaction that produces two or more products, a single reactive intermediate may be involved, with the two products arising from different reaction paths emanating from that intermediate. Chemists check for this possibility by generating the same intermediate from two different reactants. If we see the same product ratio from the two different reactants, a common intermediate is likely involved.

Because eliminations are in competition with substitutions (see Sections 10.13.3 and 11.5.16), these two paths can be used to check for a common intermediate in the two different reactions. S_N1 reactions and E1 reactions are often proposed to branch from a common carbenium ion intermediate, as shown in Eq. 11.23. Given this, let's examine some data. Eq. 11.24 shows the product distribution derived from three sources of *t*-butyl cation, all in the same solvent. The percent of alkene product changes as a function of leaving group. This means that the intermediates created by each *t*-butyl–X must be different. How can this be possible given our image of the S_N1 mechanism? The mechanism as portrayed in Scheme 11.6 and Eq. 11.19 is clearly inadequate.

$$\text{(Eq. 11.23)}$$

$$\text{(Eq. 11.24)}$$

X = Cl	56%	44%
X = I	68%	32%
X = S(Me)$_2^{\oplus}$	78%	18%

Since product branching from the intermediate changes as a function of the leaving group, the intermediate must not be the same for each leaving group. The only difference between the reactions is the leaving group, and so there has to be a memory of the leaving group. The most logical way for this to occur is that the leaving group is still in proximity to the carbenium ion when it reacts with the nucleophile or base in the substitution or elimination reactions, respectively. The carbenium ion and leaving group form what is called a **contact ion pair** or an **intimate ion pair**. This occurs for the reaction of Eq. 11.24 when X = Cl and I, whereas for X = S(Me)$_2$ the leaving group is neutral and diffuses away more rapidly. In a contact ion pair there is no solvent separating the carbenium ion and the leaving group. The ion pair is trapped for a finite time period within a **solvent cage**, in which the carbenium ion and leaving group undoubtedly jostle around and reorient. To explain the data, the substitution and/or the elimination reactions must occur at a rate faster than, or comparable to, the rate for the dissociation of the ion pair into solvent separated ions, and that rate depends upon the leaving group.

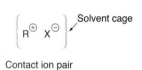

Contact ion pair

11.5.4 Stereochemistry

Different stereochemical outcomes are predicted for S_N2 and S_N1 mechanisms. This gives chemists another mechanistic test. In a famous experiment, Ingold considered the reaction in which one enantiomer of 2-iodooctane is allowed to react with radioactive iodide (I*−) in acetone solvent (Eq. 11.25). The substitution of I by I* in an S_N2 fashion could hypothetically occur from the frontside or the backside of the alkyl halide, or by an S_N1 reaction. In the Ingold experiment, racemization, as measured by the rate of loss of optical activity of the starting material, is twice as fast as radioactive iodide incorporation. This means that for each radioactive iodide that displaces a normal iodide, the enantiomer of the starting material must have formed. For example, if the starting material is the (+)-enantiomer, displacement produces the (−)-enantiomer. Not only is the (+) rotation of the starting material lost, an equal but oppositely signed rotation is associated with the new product. Another way to consider the observation is that if 50% of the reactants have become products with opposite chirality, the solution would be completely optically inactive.

$$\text{(Eq. 11.25)}$$

What mechanism does Ingold's observation support? First, it shows that a simple S_N1 mechanism is not operating. The creation of a carbenium ion that is not in a solvent cage with the leaving group would lead to an equal amount of both enantiomers of the product for each incorporation of radioactive iodide, and hence racemization at exactly the same rate as radioactive iodide incorporation. The experimental observation also rules out frontside attack by the nucleophile, because the stereochemistry would be retained when the radioactive iodide is incorporated. Finally, the result does support a mechanism with backside attack, because each radioactive iodide incorporation inverts the stereochemistry of the reactant. This was a landmark experiment that helped to solidify the notion of backside attack in S_N2 mechanisms. It has been validated repeatedly in a wide range of systems. S_N2 reactions always proceed with inversion of configuration. S_N2 reactions are 100% stereospecific. Interestingly, the inversion of stereochemistry in aliphatic substitution reactions was first reported in 1893 by Walden, and in honor of this discovery, the stereochemical outcome of S_N2 reactions is termed the **Walden inversion**.

S_N2 transition states

Going Deeper

Gas Phase S_N2 Reactions—A Stark Difference in Mechanism from Solution

Whereas rate constants for S_N2 reactions in solution commonly correlate with the exothermicity and endothermicity of the reaction, the rate constants for these reactions in the gas phase show little such correlation. Furthermore, the rate constants can be unexpectedly small. The mechanism for the substitution has been found to be very different than in solution. First, a complex between an anionic nucleophile and a neutral electrophile is formed due to strong ion–dipole interactions. The attraction drops the complex into a potential energy minimum for a loose **ion–molecule complex**. For this complex to react, proper orientation of the nucleophile and electrophile must be achieved, and a bond to the nucleophile must form while the bond to the leaving group breaks. Thus, both entropy and enthalpy terms contribute to creating a barrier to the reaction. Traversing the barrier leads to another loose ion–molecule complex, now between the product and the leaving group. Substantial energy must now be applied to break up the complex between the product and the leaving group.

As shown in the drawing, the barrier to the S_N2 reaction in the gas phase can be comparable to, or even below, the energy of the separated nucleophile and electrophile. For example, the transition state enthalpy for reaction of chloride with methylbromide in the gas phase is actually 2.5 kcal/mol *lower* than the separated reactants ($\Delta E = 2.5$). The reason that the rate constants can be very small, even though the enthalpy barrier is low relative to the reactants, is **entropy control**. The change in entropy going from the loose complex to the transition state is very large and negative.

The differences in the mechanisms in the gas and solution phases necessarily arise from differences in solvation. In the gas phase the nucleophile and electrophile solvate one another prior to reaction, and achieving the correct orientation between the nucleophile and electrophile for backside attack leads to a substantial entropy barrier. The rate constants therefore have little correlation with the enthalpy of the reaction. In solution, however, the electrophile and nucleophile are solvated prior to the reaction, the transition state is solvated during the reaction, and the product is solvated after the reaction. This yields rate constants that correlate with the heat of the reaction.

Olmstead, W. N., and Brauman, J. I. "Gas-Phase Nucleophilic Displacement Reactions." *J. Am. Chem. Soc.*, **99**, 4219 (1977).

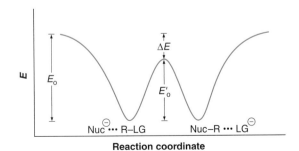

Reaction coordinate

Further support for backside attack in S_N2 mechanisms comes from compounds for which backside attack is blocked. The following structures show no reaction when treated with nucleophiles and experimental conditions that favor S_N2 chemistry (see the discussion of solvent effects below for an explanation of these conditions). The ring systems completely block access of a nucleophile.

Reactants with impossible S_N2
approach trajectories

If S_N2 reactions give stereochemical inversion, what do we expect for S_N1 reactions? A pure S_N1 mechanism with a simple carbenium ion that has a lifetime long enough to be fully dissociated from the leaving group will show complete racemization. Recall from Chapter 1 that simple carbenium ions are trigonal planar structures, and hence the plane forms a symmetry element within the structure. Ideally, addition of a nucleophile from either face of the carbenium ion will occur at the same rate, as shown in the margin.

Equal nucleophilic attack
from either face of the carbocation

More nucleophilic attack
from one face of the carbocation
in an ion pair

In reality, whereas full inversion of stereochemistry is always seen in S_N2 reactions, full racemization of stereochemistry in S_N1 reactions is rare. Instead, partial inversion of stereochemistry is commonly found. In the solvolysis of 1-phenylneopentyl tosylate in acetic acid (Eq. 11.26), the reaction goes with 10% excess inversion. This observation is consistent with the participation of a contact ion pair. We already concluded that such a structure had to be an intermediate due to the results of competition kinetics.

$$\text{Ph}-\text{OTs} \xrightarrow{\text{HOAc}} \text{Ph}-\text{OAc} + \text{Ph}-\text{OAc}$$
$$\qquad\qquad\qquad\qquad 45\% \qquad\quad 55\%$$

(Eq. 11.26)

It is also commonly observed that if we interrupt an S_N1 reaction and analyze the "unreacted" starting material, there will be some racemization. The most logical explanation is that the carbenium ion is formed, but it is in a contact ion pair with the leaving group. Rotation of the carbenium ion within the solvent cage leads to racemization after recombination with the leaving group. The following Going Deeper highlight discusses the consequences of contact ion pair formation on the kinetics of the reaction.

Going Deeper

A Potential Kinetic Quandary

Remember that first-order kinetics in an S_N1 reaction require $k_{-1}[X^-] \ll k_2[Nuc^-]$ (Eq. 11.22). Yet, racemization of the reactant sometimes occurs under conditions where the reaction is still first order. Do these results contradict each other? They don't, because racemization of the starting material can occur in two ways. The first involves a fully free carbenium ion that reacts with free X^- in solution. In this case, $k_{-1}[X^-]$ would have to be comparable to

$k_2[Nuc^-]$ for racemization to occur, and the reaction would *not* be first order. However, if the carbenium ion rotates as part of a contact ion pair within a solvent cage, after which the leaving group re-adds to the carbenium ion, this step is not part of the $k_{-1}[X^-]$ term. Re-addition of the leaving group within the solvent cage is referred to as **internal return**. The rate of internal return does not depend upon the free X^- concentration. One can assign a different rate constant (k_{-1}^*) for internal return of the leaving group within the ion pair, which now represents a first-order reaction, as shown to the right. This reaction can be very fast, much faster than the reaction with a nucleophile in solution (k_2), so it would be kinetically invisible. The only way we know that it occurs is racemization of a stereogenic center, or re-addition of the leaving group in a manner different than how it departed.

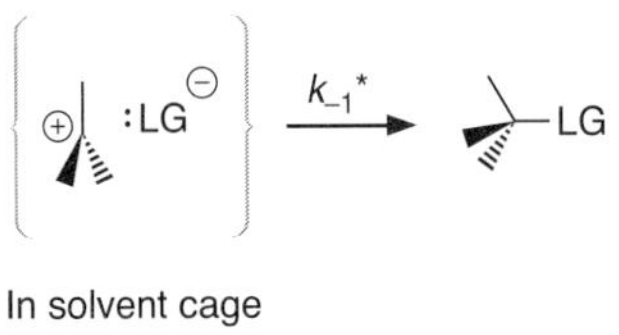

In solvent cage

11.5.5 Orbital Considerations

Whereas it is clear that a free carbenium ion should experience nucleophilic attack equally well from either face because it has a plane of symmetry, it may not be clear why an S_N2 reaction *has* to proceed via backside attack only. Leaving groups generally have electronegative atoms. This produces a polar C–LG bond, with a $\delta+$ on carbon. Simple electrostatic considerations thus predict a nucleophile should attack carbon, but they really cannot rationalize why the attack must be on the backside. Considering potential donor–acceptor interactions, however, nicely rationalizes the result.

The filled orbital on the nucleophile is usually obvious; it is commonly a lone pair (Eq. 11.27). The lowest-lying empty orbital on the electrophile is the antibonding σ^* C–X bond, which is polarized toward the carbon (look back at Section 1.3.5). Remember that electronegative elements lower the energies of orbitals, and so this empty orbital is lower in energy than, for example, a C–C σ^* molecular orbital. The largest character of this orbital lies on the backside of the carbon (see the LUMO of CH_3Cl in Appendix 3). Backside attack by the nucleophile is preferred because it maximizes the interaction of the LUMO on the electrophile with the HOMO on the nucleophile. Hence, using an orbital analysis nicely explains why nucleophilic attack occurs from the backside of the carbon undergoing displacement in an S_N2 reaction.

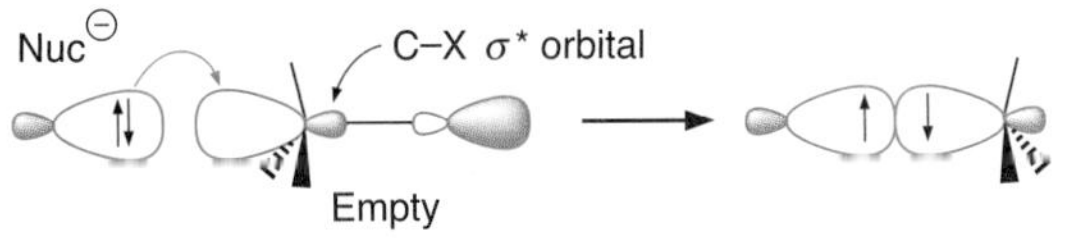

(Eq. 11.27)

11.5.6 Solvent Effects

Since charged species are often created or destroyed in nucleophilic substitution reactions, we can anticipate that solvent effects might be large. When examining the effect that the solvent can have on the rate of any reaction, it is important to compare the *relative* solvations of the reactants and the transition state. *Differences* in the solvation of the two affect the rate. In the case of nucleophilic aliphatic substitution, we need to compare the solvation of the alkyl–LG species and the separate nucleophile relative to the transition state. Often, the solvent itself is the nucleophile, and in these cases we are not as concerned with how it solvates itself.

To understand how the solvent affects nucleophilic aliphatic substitution reactions, we must first examine the possibilities for the creation or destruction of charge in S_N2 and S_N1 reactions. Table 11.2 shows the possibilities, considering charges on reactants and transition states for S_N2 and S_N1 reactions.

The guiding principle to understand how increasing the polarity of the solvent affects the rates of S_N2 and S_N1 reactions is to compare the charges on the reactants and the transition state. If the charges on the transition state are becoming larger or more localized, then increasing solvent polarity increases the rate of that reaction. This is because the polar solvent is better at solvating the transition state relative to the reactants. If charge is diminishing or dispersing when the reactants achieve the transition state, however, then increasing the solvent polarity decreases the rate. Now the solvent is better at stabilizing the reactants. This guiding principle can be seen in all the entries of Table 11.2. There are exceptions, and different reaction types show different magnitudes of solvent effects.

Table 11.2
Comparison of the Charge Distribution on Various Reactants and Transition States for a Series of A. S_N2 Reactions and B. S_N1 Reactions (LG = Leaving Group)

Reactants	Transition states	Effect of increasing the polarity of the solvent
A. S_N2 reactions		
$Nuc^- + R\text{–}LG$	$Nuc^{\delta-}\text{----}R\text{----}LG^{\delta-}$	Retards the reaction
$Nuc + R\text{–}LG$	$Nuc^{\delta+}\text{----}R\text{----}LG^{\delta-}$	Speeds the reaction
$Nuc^- + R\text{–}LG^+$	$Nuc^{\delta-}\text{----}R\text{----}LG^{\delta+}$	Retards the reaction
$Nuc + R\text{–}LG^+$	$Nuc^{\delta+}\text{----}R\text{----}LG^{\delta+}$	Retards the reaction
B. S_N1 reactions		
$R\text{–}LG$	$R^{\delta+}\text{----}LG^{\delta-}$	Speeds the reaction
$R\text{–}LG^+$	$R^{\delta+}\text{----}LG^{\delta+}$	Retards the reaction

Let's examine a few cases of Table 11.2 to understand the guiding principle. When a negatively charged nucleophile reacts with a neutral R–LG in an S_N2 process, the transition state disperses the negative charge between the nucleophile and the LG. Hence, the reaction slows down when the polarity of the solvent increases. This is because the anionic nucleophile with its full charge confined to a relatively small area is better solvated by a polar solvent than the transition state with dispersed charge. When a neutral R–LG undergoes a heterolysis in an S_N1 mechanism, charges are being created, and hence the reaction rate increases with a more polar solvent. However, when a positively charged reactant (R–LG$^+$) undergoes a heterolysis, the overall charge on the system remains the same, and in fact, is getting more dispersed in the transition state. This reaction typically shows a small solvent effect and may even slow down as the solvent becomes more polar.

We can examine the differential solvation of the reactant and the transition state by looking at reaction coordinate diagrams. Figure 11.3 shows two possibilities, corresponding to the first two entries of Table 11.2. To predict rate differences, we focus our attention on the relative sizes of the activation free energy barriers, not on the energy of any particular surface. Note that polar species such as nucleophiles and electrophiles are commonly better solvated in higher polarity solvents, and therefore the whole energy surface is lowered in energy.

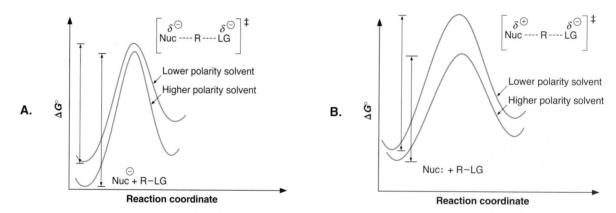

Figure 11.3
A. A reaction coordinate diagram where the higher polarity solvent leads to better solvation of all species, but preferentially solvates the reactant. Entry 1 in Table 11.2 would fit this scenario. **B.** A reaction coordinate diagram where the higher polarity solvent leads to better solvation of all species, but preferentially solvates the activated complex. Entry 2 in Table 11.2 would fit this scenario.

In Chapter 8 a handful of linear free energy relationships (LFERs) were discussed for studying solvent effects on reactions. In fact, much of what we know about solvation effects on reactions comes from studies of nucleophilic aliphatic substitution. The Grunwald–Winstein and Schleyer scales have been extensively used to study aliphatic substitutions that are solvolysis reactions. Table 8.4 shows several solvent ionizing constants using these scales. The values give quantitative comparisons for different solvents and mixtures of solvents for supporting the formation of charges. The ionizing power of the solvent influences the balance between S_N2 and S_N1 reactions, because there are differences in the extent to which charges change in these two reaction mechanisms. Furthermore, the sensitivity to a change in solvent as a function of the R group can be followed by the Grunwald–Winstein and Schleyer reaction constants.

When using the Grunwald–Winstein or Schleyer scales to compare solvent ionizing power, we must remember that the reference reactions are heterolysis reactions. The polarity and the protic/aprotic nature of the solvent are mixed in these scales as appropriate for the respective reference reactions. However, some substitution reactions may be influenced differently than the reference reactions by the protic/aprotic nature of the solvent, or its polarity. For example, S_N2 reactions are very sensitive to the protic/aprotic nature of the solvent, due to differential solvation of the nucleophiles. If the solvent is protic, it can hydrogen bond to the nucleophile and thereby substantially diminish its reactivity. The reference reactions for the Grunwald–Winstein or Schleyer scales reflect no effects on the nucleophiles because they are S_N1 reactions.

Many highly polar solvents, such as DMF and DMSO, are aprotic, and as discussed in Section 8.4.4, negative nucleophiles are very reactive in these solvents. Therefore, when changing from a polar protic solvent to a polar aprotic solvent, dramatic rate increases are observed for S_N2 reactions that involve negatively charged nucleophiles, such as ^-OH, $^-N_3$, ^-CN, etc. An example of this dramatic effect is shown in Table 11.3. The relative rates for the reaction of chloride anion with methyliodide span six orders of magnitude, with the aprotic, less polar solvents imparting the fastest rates.

Table 11.3
Relative Rate Constants (k_{rel}) for the Reaction of Chloride with Methyl Iodide*

Solvent	k_{rel}
Methanol	0.9
Water	1.0
Formamide	14.1
Nitromethane	1.41×10^4
Acetonitrile	3.58×10^4
DMF	7.08×10^5
Acetone	1.41×10^6

*Parker, A. "Protic–Dipolar, Aprotic Solvent Effects on Rates of Bimolecular Reactions." *Chem. Rev.*, **69**, 1 (1969).

However, one potential problem that can arise when running a substitution reaction in a polar aprotic solvent is that the nucleophile is not soluble. This becomes particularly problematic when the less polar of the aprotic solvents are used. The most common trick used to solve this problem is the addition of a crown ether (see Chapter 4), which solvates the counter cation well enough to induce the solubility of the corresponding anionic nucleophile. In these cases, the nucleophile is essentially "naked", and extremely large rates for S_N2 reactions are observed.

Our focus on solvent effects has been upon the polarity and the protic/aprotic nature of the solvent. However, the solvent can also be involved in **nucleophilic assistance** of leav-

Doering–Zeiss intermediate

ing group departure in an S_N1 reaction. Nucleophilic assistance involves solvation of the intermediate carbenium ion by lone pair electrons from the solvent. A loose association between the carbenium ion and solvent electron pairs, called a **Doering–Zeiss intermediate**, stabilizes the carbenium ion and its formation. Most polar solvents possess heteroatoms with lone pairs, and therefore their activity in assisting S_N1 reactions in part derives from some level of nucleophilic assistance.

The extent of solvation of the leaving group in both S_N2 and S_N1 reactions affects the rate, too. In both reactions, the charge on the leaving group is changing during the rate-determining step. Polar protic solvents enhance the departure of leaving groups that are developing negative charge. The addition of electrophilic reagents can enhance the leaving group departure even further by prior coordination to the leaving group. The most common example of this is the addition of a silver salt to an alkyl halide (R–X), where the metal enhances the departure of the halide and results in a precipitate of AgX.

An interesting way to increase the polarity of a solvent system is to add salt. Organic chemists add salt to water to increase the efficiency of extraction procedures by increasing the partition of organic structures from a water layer to an organic layer, an effect known as **salting out**. A related behavior is found for S_N1 reactions. Addition of a salt can lead to larger rate constants by increasing the polarity of the solvent. A relationship such as Eq. 11.28 is often found, where k_1 is the rate constant in the absence of an added salt, and b is a proportionality constant. However, some complications to this analysis may be expected, because salts can get involved in the formation of contact ion pairs (see the next Going Deeper highlight).

$$k_{obs} = k_1(1 + b[salt])$$

(Eq. 11.28)

11.5.7   Isotope Effect Data

During an S_N2 reaction there are hybridization changes at the reacting carbon. Simplistically, we view the carbon as going from sp^3 to sp^2 at the transition state. This would predict a normal secondary isotope effect. In practice, however, the isotope effects are typically minimal or zero. To understand this, we must recall the origin of a secondary isotope effect. The hybridization change from sp^3 to sp^2 normally leads to a looser out-of-plane bend motion for the hydrogens (see Section 8.1.3). At the transition state for an S_N2 reaction, however, the out-of-plane bend remains relatively stiff due to the presence of the incoming nucleophile and the outgoing leaving group. These groups block the out-of-plane bend. Hence, this vibration in the reactant and transition state has undergone little change, and therefore the isotope effect is small.

In an idealized S_N1 reaction, the transition state leading to the bond heterolysis step still has some bonding to the leaving group, but no bond with the nucleophile has been formed. Hence, the out-of-plane bend is a significantly weaker motion. When only one H is replaced by a D, isotope effects between 1.2 and 1.25 are common for sulfonate leaving groups, whereas chloride and bromide leaving groups show isotope effects of around 1.15 and 1.13, respectively.

Out-of-plane bend is stiff

S_N2

Bending is getting looser at the transition state

S_N1

11.5.8   An Overall Picture of S_N2 and S_N1 Reactions

Before exploring in depth the influence of the nature of the R group, the nucleophile, and the leaving group on the interplay between S_N2 and S_N1 reactions, it is useful to summarize the various aspects of these reactions we have discussed so far. Figure 11.4 summarizes the major pathways for aliphatic nucleophilic substitution, indicating the true complexity of these reactions. The S_N2 reaction is relatively straightforward, with full participation of the nucleophile in the rate-determining step. The S_N1 reaction is significantly more complex. Only a small fraction of S_N1 reactions proceed with a fully symmetric carbenium ion intermediate. In an S_N1 reaction, exchange of the leaving group and nucleophile can occur at the separated ion, solvent-separated ion, or contact ion pair stage. Taking all this complexity

Going Deeper

Contact Ion Pairs vs. Solvent-Separated Ion Pairs

When the salt added to increase the rate of an S_N1 reaction is not nucleophilic at all, unusual kinetic behavior can sometimes be observed. For example, the graph below shows the effect of added $LiClO_4$ on the acetolysis of the structure shown. As predicted by the discussion above, the rate constant for loss of stereochemistry at the leaving group carbon (top graph in the figure), and the rate constant for formation of diastereomeric products (bottom graph in the figure) both increase with increasing salt concentration. However, note the non-linear portion of the graph that shows product formation at low salt concentration. A rapid increase in k is seen at low salt concentrations, ultimately giving a linear and increasing plot at higher concentrations. Such behavior has been termed a **special-salt effect**.

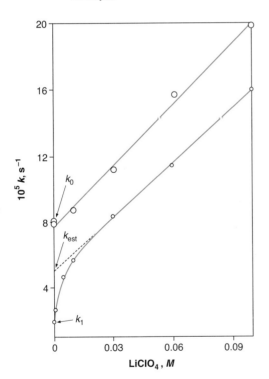

Acetolysis

How can such an unusual effect be explained? The currently accepted answer is that there are two kinds of ion pairs: **contact ion pairs** and **solvent-separated ion pairs**. A solvent-separated ion pair includes one (or perhaps several) solvent molecule between the cation and anion. It is distinct from a completely dissociated ion pair

in that the cation and anion still have a measurable electrostatic attraction, even though there is intervening solvent. The addition of any salt will increase the rate constants for the initial formation of the contact ion pair, the separation of that ion pair into the solvent-separated ions, and ultimately into completely dissociated ions. Return of the leaving group from the solvent-separated ion pair to form the contact pair and ultimately the covalent bond with the carbenium ion, will slow the overall substitution reaction. Differing terms have been given to the various possibilities: **ion pair return**, **external ion pair return**, and **external ion return**. The first two terms are different forms of internal return, referring to return of the leaving group from the contact ion pair and the solvent-separated ion pair, respectively. External ion return involves free ions in bulk solvent.

$$R^{\oplus} \ X^{\ominus}$$

Contact ion pair

$$R^{\oplus} \parallel X^{\ominus}$$

Symbolism for a solvent-separated ion pair

A very non-nucleophilic anion, such as perchlorate, can replace the leaving group at the solvent-separated ion pair stage. Upon reversion to a contact ion pair but with perchlorate, the perchlorate cannot form a stable compound with the carbenium ion. Hence, this salt becomes involved in the mechanism at a stage that blocks internal return, converting the carbenium ion to a more reactive form. This increases the rate of product formation more than an increase in the dielectric constant of the medium due to the polarity of the salt.

This postulate can be tested. If the rate enhancement from perchlorate at low concentrations is due to replacement of the leaving group at the stage of a solvent-separated ion pair, the extent of this replacement should depend upon the amount of leaving group present in the solution. Under conditions where the common ion effect is negligible, it has been observed that the effect of perchlorate can be counteracted by the addition of the leaving group ion. This is called an **induced common ion effect**. The added leaving group competes with perchlorate for replacement of the original leaving group in the solvent-separated ion pair. It should be increasingly clear that S_N1 mechanisms are not nearly as simple as portrayed in Scheme 11.6.

Winstein, S., and Robinson, G. C. "Salt Effects and Ion Pairs in Solvolysis and Related Reactions. IX. The *threo*-3-*p*-anisyl-2-butyl system." *J. Am. Chem. Soc.*, **80**, 169 (1958). Winstein, S., Klinedinst, P. R., Jr., and Robinson, G. C. "Salt Effects and Ion Pairs in Solvolysis and Related Reactions. XVII. Induced Common Ion Rate Depression and the Mechanism of the Special Salt Effect." *J. Am. Chem. Soc.*, **83**, 885 (1961).

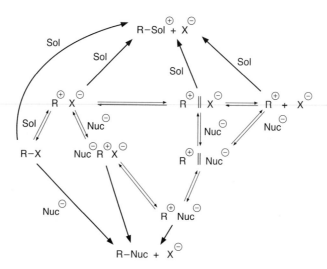

Figure 11.4
A variety of pathways for S_N2 and S_N1 reactions.

into account suggests that there is a continuum of reactivity between what we think of as "pure" S_N1 and "pure" S_N2, depending upon the structures of the reactants and solvent.

11.5.9   Structure–Function Correlations with the Nucleophile

The rates of S_N2 reactions correlate well with relative nucleophilicities, because the nucleophile is intimately involved in the rate-determining step. In S_N1 reactions, the nucleophile is involved after the rate-determining step. Furthermore, the product ratios observed for S_N1 reactions, when more than one nucleophile is present, do not correlate very well with relative nucleophilicities. This is because reactions with very unstable intermediates, such as carbenium ions, are not selective (see the reactivity–selectivity principle in Chapter 7). They occur at close to diffusion controlled rates. Therefore, in the following discussion, little needs to be said about the influence of the nucleophile on S_N1 reactions. Instead, our focus is on S_N2 reactions.

All else being equal, the better the nucleophile, the faster the S_N2 reaction. In Section 8.4.5 we examined a linear free energy relationship for nucleophilicity, called the Swain–Scott equation. The reference reaction for the Swain–Scott measure of nucleophilicities is an S_N2 reaction, that of methyl iodide and methanol. Therefore, the chemical intuition that most chemists rely upon for predicting relative nucleophilicity is actually based upon how the structure of the nucleophile influences S_N2 reactions (Table 8.5).

Recall also from Section 8.4.4 that several parameters influence nucleophilicities, including shape, donor atom, solvation, and the pK_a of the conjugate acid of the nucleophile. When the shape and donor atom of all the nucleophiles being compared are similar, and the same solvent is used, we find good correlations between the pK_a of the conjugate acid of the nucleophile and the rate of the reaction. This correlation is called a Brønsted linear free energy relationship, and the sensitivity parameter for an S_N2 reaction to the pK_a of the conjugate acid of the nucleophile is called β_{Nuc}. Figure 11.5 shows the Brønsted plots for the reaction of benzyl chloride with various nucleophiles of similar structure. Here, all the β_{Nuc} values are similar between the classes of nucleophiles, although they do not all reside on the same line. The carbon nucleophiles have the highest nucleophilicities, with oxygen second, and nitrogen last.

The similarity of the various β_{Nuc} values in Figure 11.5 is interesting. If β_{Nuc} represents the extent of nucleophilic attack at the transition state, this tells us that all the various nucleophiles have a similar extent of bonding to the benzyl carbon at their transition states. This is

Connections

An Enzymatic S_N2 Reaction: Haloalkane Dehydrogenase

Substitution reactions on aliphatic compounds are some of the most prevalent reactions known. Nature needs to perform these reactions, too, but at a rate appropriate for metabolism. Hence, enzymes are involved for many such reactions. As one example, the enzyme haloalkane dehydrogenase from the bacteria *Xanthobacter autotrophicus* catalyzes the detoxification of 1,2-dichloroethane. Many of the methods of catalysis noted in Chapter 9 are operative with this enzyme, such as proximity of the nucleophile and stabilization of the transition state.

Below we show a schematic drawing of the catalysis by this enzyme. Covalent catalysis is used, in that the nucleophile for the S_N2 displacement is an aspartate side chain. The nucleophile is in proximity to the polarized C–Cl bond. Leaving group departure is stabilized by hydrogen bonding of the developing charge on chloride by two NH hydrogen bond donors from the side chains of tryptophan residues (reminiscent of an oxy–anion hole; see Section 9.2.3). After leaving group departure, water held at the active site is used to hydrolyze the acyl–enzyme intermediate via general-base catalysis from a histidine side chain.

Verschueren, K. H. G., Franken, S. M., Rozeboom, H. J., Kalk, K. H., and Dijkstra, B. W. "Refined X-ray Structures of Haloalkane Dehalogenase at pH 6.2 and pH 8.2 and Implications for the Reaction Mechanism." *J. Mol. Biol.*, **232**, 856–872 (1993).

Dehalogenase mechanism

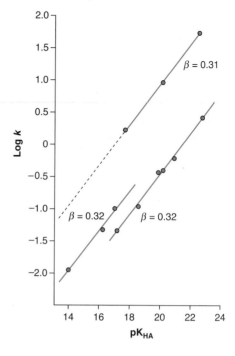

Figure 11.5
Brønsted plots for the reaction of benzyl chloride with three classes of nucleophiles in DMSO at 25 °C. Top: Carbanions, Middle: Oxy anions, and Bottom: Nitrogen anions. Bordwell, F. G., and Hughes, D. L. "S_N2 Reactions of Nitranion with Benzyl Chlorides." *J. Am. Chem. Soc.*, **106**, 3234 (1984).

not what one would predict from the Hammond postulate, which tells us that the transition state structure should change as a function of the exothermicity or endothermicity of the reaction, which depends upon the structure of the nucleophile. To understand how to properly correlate changes in nucleophile structure with transition state structure in S_N2 reactions, we turn to a More O'Ferrall–Jencks analysis (see Section 7.8.2, where we used the interplay between S_N2 and S_N1 chemistry to present this concept). Recall that a More O'Ferrall–Jencks plot is used to analyze any energy surface when two bonds are changing simultaneously; here they are the formation of a bond to the nucleophile and the cleavage of a bond to the leaving group. A change in the nucleophile also changes the energies of structures along competing pathways where these bond changes occur sequentially. This, in turn, affects the S_N2 pathway. Figure 11.6 shows a More O'Ferrall–Jencks plot that explains how the position of the transition state changes when the nucleophile becomes better. With a better nucleophile, the bottom left and right corners are lower in energy. Hence, the extent of nucleophilic attack does not change, but the extent of leaving group departure decreases (see reasoning in the figure caption). This makes sense because a better nucleophile needs less leaving group departure to attack the carbon. Stated another way, the better nucleophile makes the S_N2 reaction have less S_N1 character.

Figure 11.6
A More O'Ferrall–Jencks plot showing how the transition state for an S_N2 reaction changes when the nucleophile becomes better. The diagonal is the projection of the S_N2 reaction coordinate, while along the top and right side is the projection of the S_N1 path. Down the left and across the bottom represents a hypothetical mechanism involving full addition of the nucleophile with the formation of a negative pentacoordinate carbon. Along the reaction coordinate the transition state shifts toward reactants because the reaction is more exothermic with a better nucleophile. Perpendicular to the reaction coordinate, the transition state shifts to the lower left corner. The result is the same extent of nucleophilic attack, but less leaving group departure.

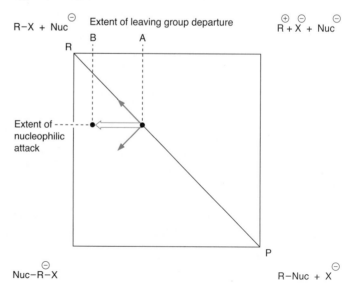

A variety of Swain–Scott sensitivity *s* values for different reactions are given in Table 11.4. These are based upon CH_3Br as the standard reactant. We find that the S_N2 reaction of ethyl tosylate is less sensitive to the power of the nucleophile than methyl bromide, as is benzyl chloride.

11.5.10 Structure–Function Correlations with the Leaving Group

In both S_N2 and S_N1 reactions the leaving group develops negative charge relative to its charge in the reactant during the rate-determining step. Therefore, the reactions proceed faster with leaving groups that are better at accepting this negative charge. Another reaction that correlates well with a group's ability to accept negative charge is the acidity of acids (see Chapter 5). Therefore, it may be expected that a good correlation would be found between leaving group ability and the acidity of the conjugate acid of that leaving group. However, just as with nucleophilicities, good correlation is only found within a class of leaving groups. In other words, the correlation between the affinity for a proton and the affinity for a carbon does not hold for all different classes of leaving groups. For example, sulfonates, such as tosylate, triflate, and mesylate, are much better leaving groups compared to halides than would be expected based upon the acidities of the conjugate acids of these leaving groups.

Table 11.4
Examples of S_N2 *s* Values*

Compound	*s*
Methyl bromide	1.00
Methyl iodide	1.15
Benzyl chloride	0.87
Ethyl tosylate	0.66

*Wells, P. R. "Linear Free Energy Relationships." *Chem. Rev.*, **63**, 171 (1963).

Going Deeper

The Meaning of β_{LG} Values

In Section 8.5.2 we examined a Brønsted LFER that yields a sensitivity parameter β_{LG}. This parameter measures the sensitivity of any reaction to leaving group departure by correlating rates of the reactions with the pK_a values of the conjugate acids of the leaving groups. The reference reaction for this LFER is the complete donation of the proton of the conjugate acid of the leaving group to water (H–LG + H_2O → LG^- + H_3O^+). Stated another way, the reference reaction represents a complete leaving group departure of LG^- from the proton, giving that proton to water. Therefore, chemists often view a reaction that has $\beta_{LG} = -1$ as indicating complete leaving group departure at the transition state of the new reaction under study. Any value between 0 and –1 indicates less than complete leaving group departure, and the value of β_{LG} can be correlated to a fraction of leaving group departure.

However, we must be wary. This analysis assumes that the new reaction under study behaves similarly to the reference reaction. Often β_{LG} values more negative than –1.0 are measured, which certainly cannot mean greater than 100% leaving group departure at the transition state. It simply means that the new reaction under study is actually more sensitive to charge build up on the leaving group than is the reaction where the leaving group departs from a proton.

11.5.11 Structure–Function Correlations with the R Group

The R group of R–LG has a very large effect on the rates of and the interplay between S_N2 and S_N1 reactions. One needs to focus upon several issues related to the structure of the R group to understand substitution reactions—namely, sterics, electronics, and adjacent groups. For simplicity, we split this discussion into a focus upon S_N2 reactions and then S_N1 reactions. However, keep in mind that there is a continuum of reactivity with varying decreases of nucleophilic attack and leaving group departure.

Effect of the R Group Structure on S_N2 Reactions

Since the nucleophile directly adds to the carbon with the leaving group in the rate-determining step of an S_N2 reaction, the more sterically encumbered the carbon under attack, the slower the reaction. This is a classic and reliable predictor of reactivity. Table 11.5 shows the average relative rate of displacement of leaving groups by nucleophiles for a series of R groups under experimental conditions that impede S_N1 pathways. As groups are attached to the carbon undergoing nucleophilic attack, the rate dramatically decreases. However, placing groups adjacent to the carbon under attack also slows the substitution, and to an even greater extent. In fact, the slowest substitution reported here is for R = neopentyl.

Table 11.5
Average Relative Rates
of S_N2 Substitution Reactions
on Various R–X Species*

R group	Relative rate
Methyl	1
Ethyl	3.3×10^{-1}
Propyl	1.3×10^{-2}
Isopropyl	8.3×10^{-4}
t-Butyl[a]	5.5×10^{-5}
Neopentyl	3.3×10^{-7}
Allyl	1.3
Benzyl	4.0

*Streitwieser, A., Jr. (1962). *Solvolytic Displacement Reactions*, McGraw–Hill, New York.
[a]Estimated from Cook, D., and Parker, A. J. "Halide Exchange at a Saturated Carbon Atom in Dimethylformamide Solvent. Comparison of Experimental Rates and Arrhenius Parameters with Values Calculated by Ingold." *J. Chem. Soc. B*, 142 (1968).

S_N2 transition state

Let's examine the reason behind these trends. Increasing substitution on the carbon undergoing nucleophilic attack results in larger steric repulsions at the transition state. At the transition state, the steric congestion between the R groups attached to the carbon undergoing nucleophilic attack is somewhat relieved, because the R groups change from being separated by 109° angles to nearly 120° angles. However, these groups are now approximately 90° from the nucleophile and the leaving group. The more groups placed 90° from the nucleophile and leaving group, the more strained the transition state. In addition, the degrees of freedom of the activated complex are more restricted with groups attached, because the nucleophile and leaving group bending vibrations are stiffer due to the steric influence of the attached groups. Hence, overall, the transition state is destabilized with increasing substitution at the carbon undergoing attack.

A related interpretation of the steric effects on S_N2 reactions focuses upon entropy. In general, there are larger negative entropies of activation for reactions where many heavy nuclei must move simultaneously on going from the reactant to the transition state. This occurs in an S_N2 reaction with larger and more R groups attached to the carbon undergoing nucleophilic attack. This interpretation of the R group structure–reactivity relationship is focused upon the mass of the groups attached to the carbon undergoing nucleophilic attack rather than on their steric size, and is called the **ponderal effect**.

In the case of adjacent groups, these effects are even more severe. For example, in the extreme, a neopentyl halide achieves a transition state where the nucleophile and leaving group are proximal to a *t*-butyl group. This leads to a substantial bending of the trajectory of the nucleophile away from the optimal linear approach, and therefore creates a very strained transition state.

Electronic effects do not play nearly as large a role in S_N2 reactions as do steric effects. However, electronic effects can be important. For example, Table 11.5 shows that a rate acceleration occurs due to a neighboring alkenyl or phenyl group. Benzyl, allyl, and methyl groups all have similar reactivities. This can be viewed as a resonance effect or an orbital mixing effect. On the next page we show the developing σ bond to the incoming nucleophile as essentially a *p* orbital on the allylic carbon. This orbital results from addition of the lone pair on the nucleophile to the antibonding R–X bond. It can mix with π* orbitals of the alkenyl or phenyl group. This developing σ bond must be aligned with the π bond to experience this orbital mixing/resonance stabilization.

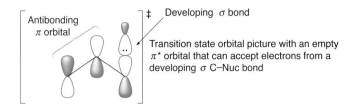

Alternatively, the rate acceleration found for allyl and benzyl groups in S_N2 reactions can be viewed as an inductive effect. The electron withdrawing nature of the sp^2 hybridized carbons of a vinyl or phenyl group makes the carbon more electrophilic, and therefore more reactive toward nucleophilic attack. Both effects are likely involved.

Other groups adjacent to the leaving group can also be expected to stabilize an S_N2 transition state by resonance, including cyano and carbonyl. Indeed, this is a common observation. However, the effect actually depends upon the nucleophile. For example, the reaction of α-bromoacetophenone is faster than that of benzyl bromide when the nucleophile is iodide, but is slower when the nucleophile is an amine. Furthermore, an additional effect arises to enhance substitution on a carbon with carbonyl or cyano substituents. Pearson has suggested that the increased reactivity of α-halo carbonyl structures with negatively charged nucleophiles results from a dual electrostatic attraction of the nucleophile to the positive end of both the carbonyl and the C–X bond, as shown in the margin.

To probe the effect of changing the R group on the interplay between S_N2 and S_N1 reactions, we can vary the solvent. The competition between S_N2 and S_N1 reactions can be deciphered by looking at the Grunwald–Winstein and Schleyer linear free energy relationships. Table 11.6 lists some l and m values (look back at Section 8.4.2 for definitions if necessary). For example, the primary systems ethyl and benzyl have large l values, indicating that they react with a significant fraction of S_N2 character. The secondary systems have large m values, indicating more carbenium ion and S_N1 character.

Nucleophile attracted to two sites

Table 11.6
l and m Values for the Substitution Reactions of Alkyl Tosylates*

Compound	l	m
Ethyl–OTs	0.89	0.40
Benzyl–OTs	0.75	0.64
Isopropyl–OTs	0.49	0.62
Cyclohexyl–OTs	0.32	0.78

*Schadt, F. L., Bentley, T. W., and Schleyer, P. v. R. "The S_N2–S_N1 Spectrum. 2. Quantitative Treatments of Nucleophilic Solvent Assistance. A Scale of Solvent Nucleophilicities." *J. Am. Chem. Soc.*, **98**, 7667 (1976).

Effect of the R Group Structure on S_N1 Reactions

The effect of R group structure in S_N1 reactions is of paramount importance. Since the rate-determining step in S_N1 reactions is ionization to give the carbenium ion, the rate directly depends upon R group structure. To make predictions about reactivity, we need to recall all the factors discussed in Chapters 1 and 2 that are found to stabilize carbenium ions. Even though we do not normally observe the nucleophilic addition step kinetically, the R group structure also affects this step. The R group influences the lifetime of the carbenium ion and whether it reacts as part of a contact ion pair, a solvent-separated ion pair, or a completely free carbenium ion.

We will not dwell again upon the factors that stabilize carbenium ions. Suffice it to say that increasing the number of alkyl and electron donating groups stabilizes these structures,

Two cartoon orbital interactions showing hyperconjugation

$k_{CH_3}/k_H = 5.5 \times 10^4$
$k_{Ph}/k_{CH_3} = 4.6 \times 10^3$

Relative rates of solvolysis

and hence speeds up S_N1 reactions. The increase in rate that results from alkyl group substitution is due to hyperconjugation (also called **σ conjugation**), which stabilizes the cation by electron donation to the empty p orbital [shown schematically in the margin using a $\pi(CH_3)$ group orbital on a methyl, or alternatively with an sp^3 hybrid].

Shown in the margin are some dramatic rate effects that were measured under conditions that assist S_N1 pathways, with only a small fraction of an S_N2 pathway. Substitution of a hydrogen by a methyl on isopropyl chloride gave a 5.5×10^4 increase in rate. Substitution of a methyl by a phenyl gives a 4.6×10^3 enhancement. As another series of examples, the relative rates of solvolysis of $PhCH_2Cl$, Ph_2CHCl, and Ph_3CCl in mixtures of diethylether/ethanol are 1, 1.75×10^3, and 2.5×10^7, respectively. Thus, the addition of phenyl rings dramatically stabilizes conjugated carbocations.

Table 11.7 shows a series of Hammett sensitivity parameters for various S_N1 reactions using σ^+ substituent constants. The ρ value is indicative of the amount of positive charge present in the transition state. The larger the magnitude of ρ, the more positive the charge. Some trends are noteworthy. The smallest negative number (entry 1) is for triphenylmethyl cation, the most stable carbenium ion. This is as expected, because the reaction would be the least endothermic, and by the Hammond postulate the transition state would be most reactant-like, thereby having the least cationic character.

Table 11.7
Hammett ρ Values for S_N1 Solvolysis Reactions of Various R–X Structures*

Entry	Reactant[‡]	ρ Value	Experimental conditions
1		−2.68	40% Ethanol/60% diethylether, 0 °C
2		−3.82	80% Aqueous acetone, 25 °C
3		−4.06	2-Propanol, 25 °C
4		−4.54	90% Aqueous acetone, 25 °C
5		−4.60	80% Aqueous acetone, 25 °C
6		−5.15	80% Aqueous acetone, 25 °C

*Brown, H. C., Ravindranathan, M., Peters, E. N., Rao, C. G., and Rho, M. M. "Structural Effects in Solvolytic Reactions. 22. Effect of Ring Size on the Stabilization of Developing Carbocations as Revealed by the Tool of Increasing Electron Demand." *J. Am. Chem. Soc.*, **99**, 5373 (1977).
[‡]OpNB = *p*-nitrobenzoate.

Table 11.8
Relative Rate Constants for the Solvolysis Reactions of Various Bridgehead Reactants*

Reactant	Approximate relative rate
LG	1
LG	10^{-3}
LG	10^{-7}
LG	10^{-15}
LG	10^{-19}

*Bingham, R. C., and Schleyer, P. R. "Calculation of Bridgehead Reactivities." *J. Am. Chem. Soc.*, **93**, 3189 (1971).

Strain in the formation of the carbenium ion will impede an S_N1 reaction (see entry 6 in Table 11.7). The strain can either be in the form of a carbon skeleton that does not allow the carbenium ion center to achieve a planar sp^2 hybridization, or a carbon skeleton where hyperconjugation by neighboring groups cannot occur due to a lack of orbital alignment. Bridgehead effects are the most common examples of these types of strain. Table 11.8 shows the relative solvolysis rates of a series of structures with bridgehead leaving groups. As might be expected, the rate of solvolysis decreases as the ring sizes in the bicyclic systems decrease, because the carbenium ion increasingly becomes more strongly pyramidal.

One must analyze all the characteristics of a substituent to predict its behavior. For example, a cyano or carbonyl group is electron withdrawing via resonance when stabilizing a negative charge. Furthermore, both are electron withdrawing via induction. However, they can also donate electrons via resonance, as shown in the margin. Although these resonance structures are not particularly good, they do stabilize the cation somewhat. Therefore, the electron withdrawing cyano and carbonyl groups actually impede S_N1 reactions more when they are one carbon away from the carbenium ion.

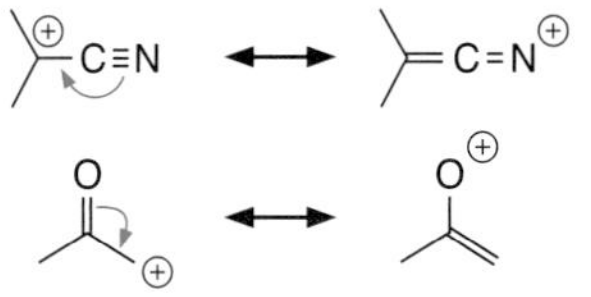

As an example, consider the relative rates of solvolysis shown below. The all-alkyl system is indeed the fastest, yet when the electron withdrawing cyano is placed directly on the site of carbenium ion formation, the rate is faster than for an adjacent cyano.

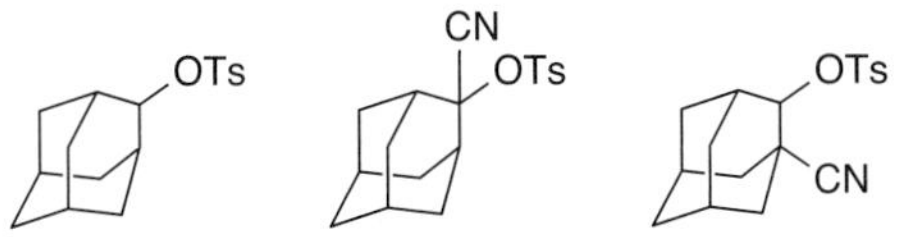

Relative rates of solvolysis

Another case in which the different characteristics of a group need to be considered is the effect of heteroatoms. When they are not directly attached to the carbon undergoing ionization, an electronegative atom such as O, N, or S destabilizes carbenium ions and slows S_N1 reactions due to inductive electron withdrawal. However, when they are directly attached, they accelerate S_N1 reactions due to resonance stabilization. Nitrogen is the best at stabilizing a carbenium ion and thereby accelerating S_N1 processes. Sulfur seems to have a variable effect in this resonance stabilization; its orbitals are larger than those of C and hence the sizes are mismatched, but S is very polarizable.

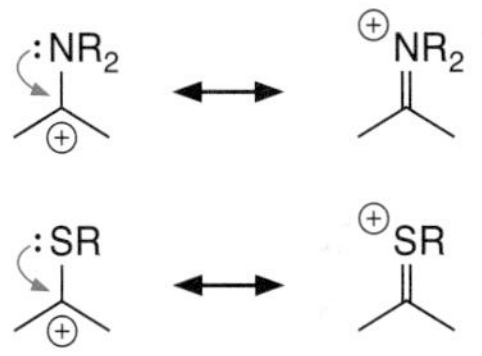

Heteroatoms are not the only groups that will facilitate an S_N1 reaction when in proximity to the cationic center. Resonance effects in allyl or benzyl carbenium ions stabilize the cationic center and hence facilitate the substitution reaction. Yet, π systems further away from the cationic center can also get involved if their geometry is such that they are oriented toward the carbenium ion's empty p orbital. Consider the solvolysis of cholesterol tosylate in Figure 11.7. Two products are formed, and the solvolysis rate is approximately 100 times

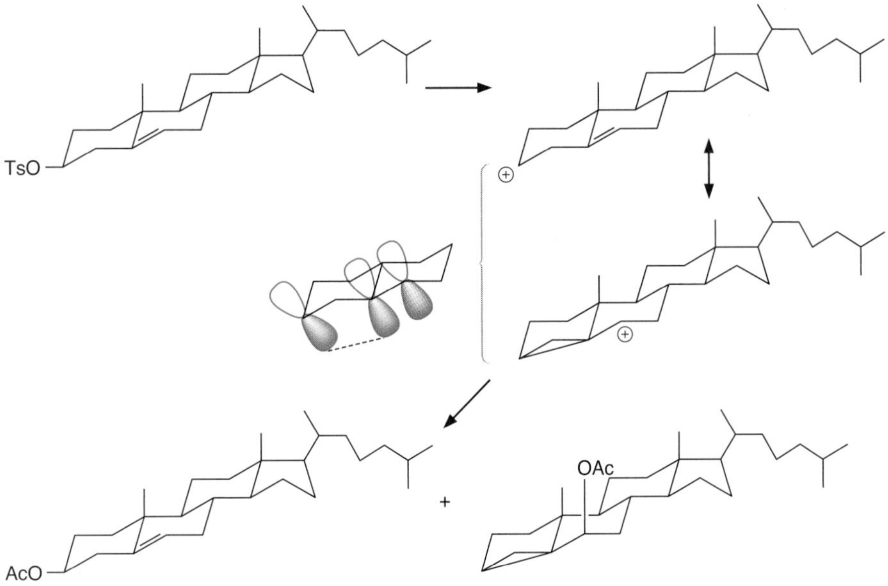

Figure 11.7
Solvolysis of cholesteryl tosylate enjoys an extra stabilization due to the geometry of the adjacent double bond.

faster than without the double bond. The starting material is referred to as **homoallylic**. This term is used when one carbon is between the leaving group and the double bond versus a normal allylic system. Correspondingly, the resulting carbenium ion is considered to be **homoconjugated** (see Section 2.4.1).

Hybridization affects the rates of S_N2 and S_N1 reactions, too. Until this point, our discussions of S_N2 and S_N1 reactions have focused upon alkyl R groups. This is because S_N2 reactions can *only* occur when the leaving group is attached to an sp^3 carbon; the rate is zero for an sp^2 hybridized carbon. However, S_N1 reactions, on *rare* occasions, will occur with alkenyl–LG compounds. Such reactions require a *very* good leaving group, such as triflate, but they can occur (Eq. 11.29).

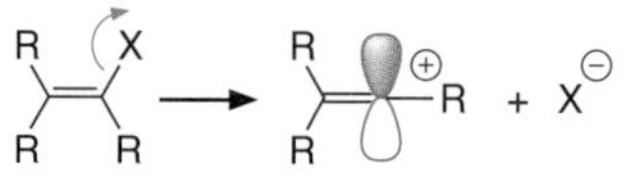

$$\text{(Eq. 11.29)}$$

11.5.12 Carbocation Rearrangements

Whenever a carbocation is an intermediate in a mechanism, rearrangements are possible. Besides the study of carbocation structures and reactivity in stable ion media, the majority of the information chemists have on carbocation rearrangements comes from S_N1 solvolysis reactions. A hydrogen, alkyl, or aryl group on a carbon adjacent (β) to the cationic carbon can shift to form a different carbocation. This is called a **Wagner–Meerwein shift**. Eq. 11.30 shows a thermoneutral example.

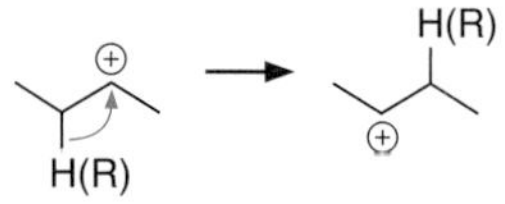

$$\text{(Eq. 11.30)}$$

Eq. 11.31 shows one example where the only product found in the S_N1 reaction derives from a carbocation rearrangement. In this example, a primary carbenium ion is formed first, but a methyl migration creates the more stable tertiary carbenium ion. However, the precise timing of such migrations is debated, as we will see below. Since S_N2 reactions are concerted, rearrangements are impossible (Eq. 11.32).

Interestingly, carbocation rearrangements occur even when the shift does not create a more stable cation. Eq. 11.33 gives a case where two products are obtained, both of which result from secondary carbocations that are expected to be comparably stable. Lastly, we note that rearrangement can be driven by a relief of ring strain. Eq. 11.34 shows an example in which the Wagner–Meerwein shift expands a cyclobutane to a cyclopentane ring, even though the shift converts a tertiary carbenium ion to a secondary carbenium ion.

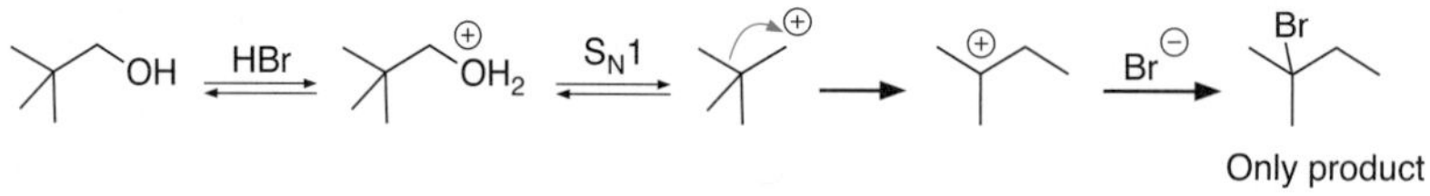

$$\text{(Eq. 11.31)}$$

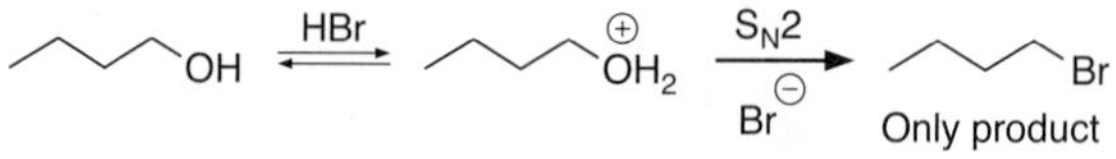

Only product

$$\text{(Eq. 11.32)}$$

Only product

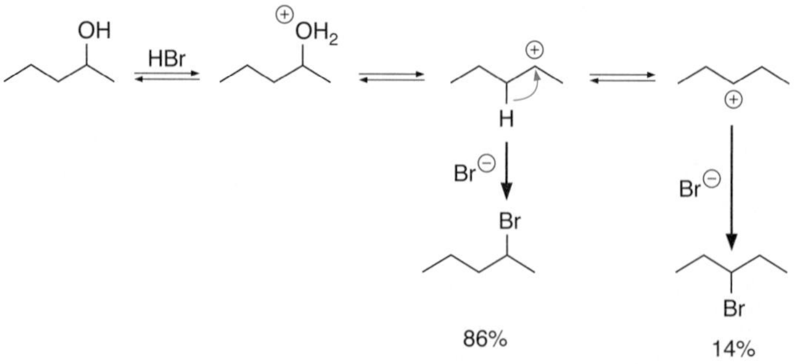

$$\text{(Eq. 11.33)}$$

86%   14%

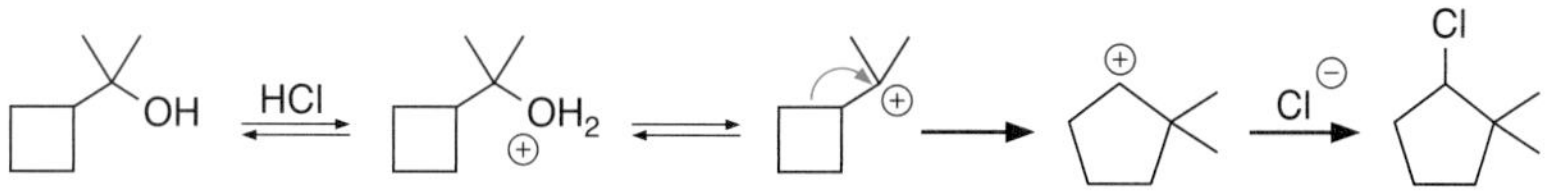

(Eq. 11.34)

There are stereochemical requirements for carbocation rearrangements. The group that migrates must be involved in hyperconjugation prior to shifting. Hyperconjugation requires that the bond to the migrating group be aligned with the empty p orbital on the carbocation center. A shift where the potential migrating group is orthogonal to the empty p orbital will not occur. For example, 2-adamantyl cation will not undergo an intramolecular migration of the adjacent hydrogen even though it would create a tertiary carbocation (see below). In the drawing on the left, the hydrogen is in the plane of the paper, as drawn, while the p orbital on the cationic carbon points in and out of the plane of the paper; the C–H bond and the p orbital are not aligned. The ball and stick figure shows a different orientation.

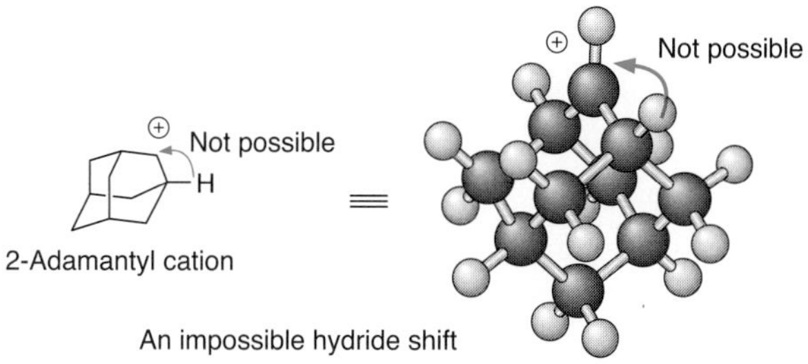

Not possible

Not possible

2-Adamantyl cation

An impossible hydride shift

It should be appreciated that carbocations are in general fluxional molecules. As discussed in Chapters 1 and 14, the barriers to hydride shifts are very low, and in many cases bridged structures are the lowest in energy. This means that carbenium ions formed from systems that at first glance do not appear to have any driving force for rearrangement often do, as in Eq. 11.33. As another example, cyclopentyl carbocation undergoes 1,2-hydride shifts, even though more stable structures are not created (Eq. 11.35). The rate of these degenerate shifts at $-139\,°C$ is $3.1 \times 10^7\ s^{-1}$, such that all five carbons become equivalent in the ^{13}C NMR spectrum.

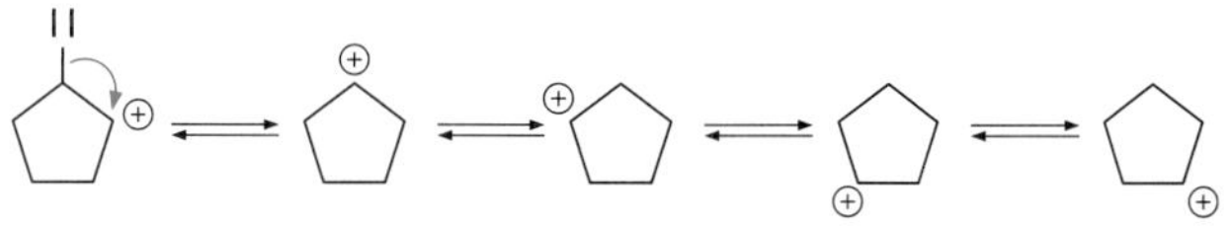

(Eq. 11.35)

Even rearrangements that create a less stable carbenium ion can occur, if the energy difference between the two cations is not too large. Isopropyl cation ($CH_3CH^+CH_3$) scrambles its hydrogens via formation of n-propyl cation ($CH_3CH_2CH_2^+$), although the thermodynamically more stable cation is clearly secondary. Interestingly, even the *carbons* in isopropyl carbocation scramble their positions in stable ion media. The proposed mechanism involves rearrangement to protonated cyclopropane (Eq. 11.36). This scrambling of the carbons is slower than the hydride shifts at $-78\,°C$, taking an hour for a radiolabel at the secondary carbon in isopropyl cation to be completely scrambled.

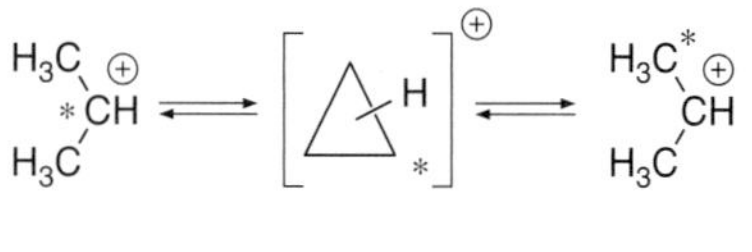

(Eq. 11.36)

Because carbocations are so fluxional, the timing of the migration during a substitution reaction has been questioned. Returning to a prototype reaction, such as Eq. 11.31, the question is whether the proposed, first-formed primary cation actually has a finite lifetime, or whether the methyl migration shown as a discrete step actually occurs simultaneously with leaving group departure (see Section 11.5.13). Note that simultaneous migration is well-established for phenyl groups and intramolecular nucleophiles. The question is whether hydrogens and alkyl groups migrate simultaneously with leaving group departure, and the debate has been contentious.

The timing of migration is a subtle issue. Extensive studies have shown that there is an entire spectrum of reactivity. In some systems migration of the electrons from the neighboring group is simultaneous with leaving group departure; at the other extreme lies the phenomenon of hyperconjugation. The two can be viewed as just different degrees of the same delocalization process. These subtle differences are part of the complexity in chemical reactivity that naturally arises when the potential surface for a structure is relatively flat, with small energy differences between different structures and the transition states that interconvert them. Such is almost always the case with a carbocation. When there is a direct interaction through space between the electrons in the σ bond that migrates and a developing cationic center, the term **σ participation** is used.

Going Deeper

Carbocation Rearrangements in Rings

We have presented carbocation rearrangements in the context of substitution reactions. Yet, rearrangements and issues that are relevant to carbocation structure are applicable to any reaction involving carbocations—namely, additions, eliminations, rearrangements, etc. One example is the acid-catalyzed ring opening of epoxides.

Unusual rearrangements occur in the acid-catalyzed ring opening of large-ring epoxides, and with other reactions that create carbocations within large rings. The acid-catalyzed ring opening of epoxycyclooctene in 90% formic acid / 10% water gives the products shown below, in addition to the expected *trans*-1,2-cyclooctanediol.

A hydrogen from across the ring has migrated to yield the products. This kind of a migration is referred to as a **transannular shift**. In Chapter 2 transannular strain (Figure 2.13) was discussed. It arises from steric repulsions between hydrogens across from each other in medium-sized rings. The hydrogens in position 5 from the cationic site shown below are in close proximity to the cationic carbon and they can shift across the ring.

The resulting 1,4-diol is exclusively cis, although the carbenium ion should lead to both cis and trans products. Therefore, the mechanism involves some extent of migration of the hydrogen during the ring opening, creating a bridged carbonium ion similar to those discussed with regards to non-classical carbocations (see Section 11.5.14). The hydrogen will migrate to the backside of the epoxide as it opens, and attack by water occurs from the top side, resulting in cis stereochemistry.

Roberts, A. A., and Anderson, C. R. "Ion Pairs and Hydride Participation in the Acetolysis of Cyclooctyl Tosylate." *Tetrahedron Lett.*, **44**, 3885 (1969), and references therein.

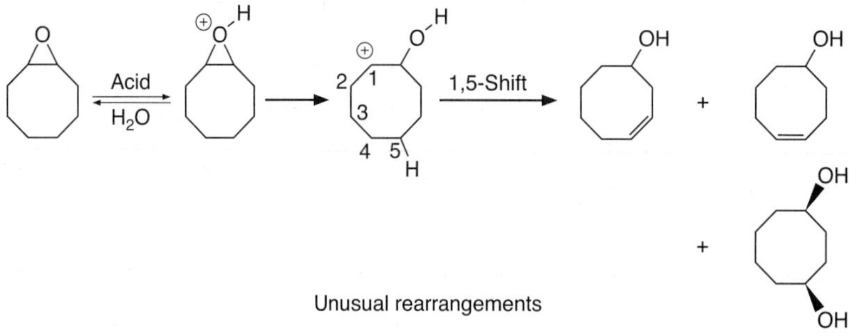

Unusual rearrangements

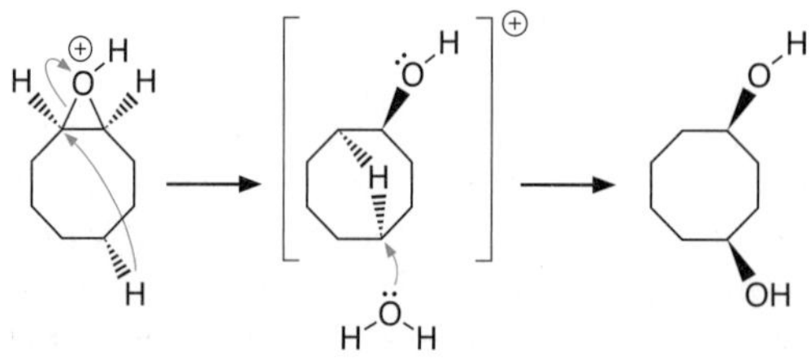

Stereochemistry of addition

11.5.13   Anchimeric Assistance in S$_N$1 Reactions

We noted previously that a N, O, or S on the same carbon as a leaving group can facilitate an S$_N$1 substitution reaction because of donation of the heteroatom's lone pair toward the carbenium ion center. However, an electron donating group does not have to be bonded to the same carbon as the leaving group to participate in this manner. When the donating group is not on the same carbon, yet facilitates formation of the carbenium ion, this phenomenon is known as **anchimeric assistance** or **neighboring group participation**. Even homoallylic interactions (see Figure 11.7) can be described by these terms. Eq. 11.37 shows a schematic example, indicating that this interaction is really just a form of intramolecular nucleophilic attack.

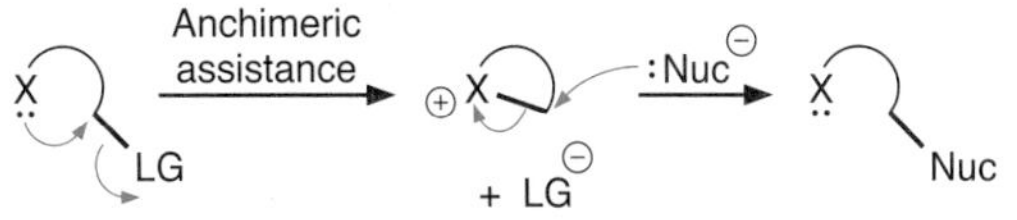

(Eq. 11.37)

One specific example involving anchimeric assistance is given in Figure 11.8, in which an enantiomerically pure reactant with two stereogenic centers undergoes solvolysis to give a racemic mixture of products. This can be explained by the intermediacy of the cation shown in part **B** of the figure. Here, the appended acetate "leaned" over to fill the developing empty *p* orbital of the cation. This creates a fully symmetric intermediate, and hence racemic products. Anchimeric assistance speeds the reaction, in that the reaction depicted in Figure 11.8 occurs approximately 10^3 times faster than the reaction of the analogous cis reactant. Therefore, the anchimeric assistance occurs in the rate-determining step involving leaving group departure, not after formation of the carbenium ion. In addition, the entropies of activation of the trans structures are typically more negative than the cis analogs, indicating a more ordered transition state.

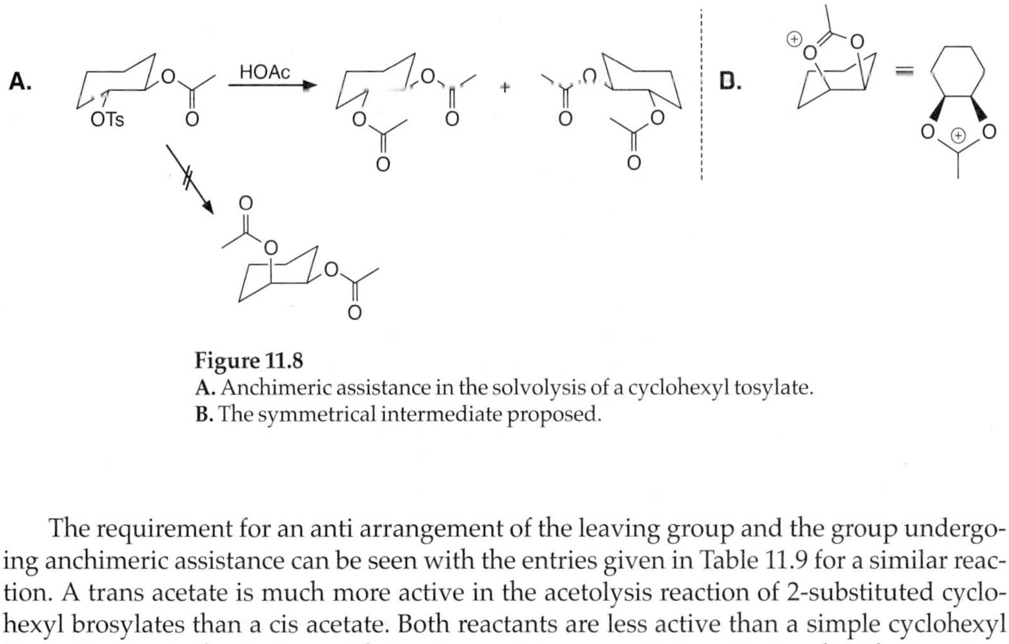

Figure 11.8
A. Anchimeric assistance in the solvolysis of a cyclohexyl tosylate.
B. The symmetrical intermediate proposed.

The requirement for an anti arrangement of the leaving group and the group undergoing anchimeric assistance can be seen with the entries given in Table 11.9 for a similar reaction. A trans acetate is much more active in the acetolysis reaction of 2-substituted cyclohexyl brosylates than a cis acetate. Both reactants are less active than a simple cyclohexyl derivative due to the electron withdrawing acetate groups. Interestingly, while the trans isomer is more active, its entropy of activation is more negative, supporting the notion of a more organized transition state due to the anchimeric assistance. A trans bromine or methoxy group adjacent to the leaving group are also enhancing relative to the cis acetate, both groups also being capable of anchimeric assistance.

Table 11.9
Relative Rate Constants and Activation Parameters for the Acetolysis of 2-Substituted Cyclohexyl Brosylates*

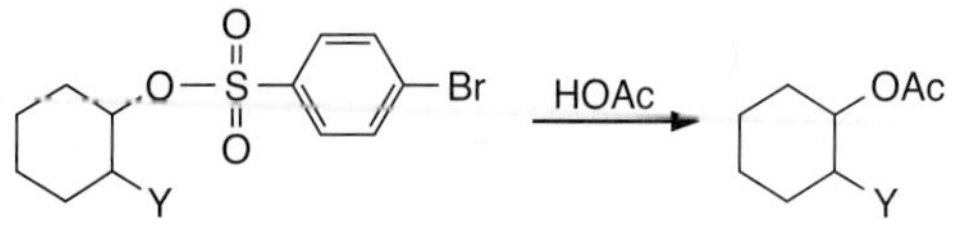

Y	k_{rel}	$\Delta H^{\ddagger}$ (kcal/mol)	$\Delta S^{\ddagger}$ (eu)
H	1	27	1.5
trans–OAc	0.24	26	−4.2
cis–OAc	3.8×10^{-4}	31	−3.5
trans–Br	0.1	28	0.8
trans–OMe	0.06	27	−3.4

*Winstein, S., Grunwald, E., and Ingraham, L. L. "The Role of Neighboring Groups in Replacement Reactions. XII. Rates of Acetolysis of 2-Substituted Cyclohexyl Benzene Sulfonates." *J. Am. Chem. Soc.*, **70**, 821 (1948).

Connections

Anchimeric Assistance in War

Although chemistry has played a large role in improving human life, it can also be put to negative uses. In World War I, chemical weapons were commonly used, and mustard gas was the most prevalent. The solvolysis of mustard gas occurs with no kinetic dependence upon the nucleophile. The reason for the S_N1 behavior is anchimeric assistance by the neighboring sulfur, creating a cyclic sulfonium ion. This cyclic structure rapidly reacts with a variety of nucleophiles, including water, but also proteins and other biomolecules. These reactions create HCl, which severely burns and blisters the skin.

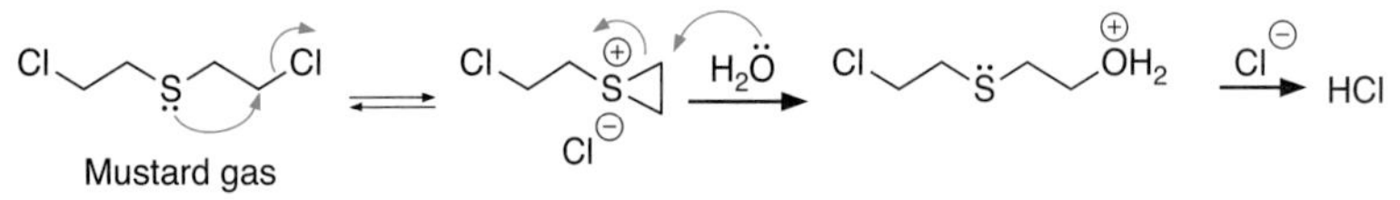

Mustard gas

Although anchimeric assistance primarily occurs from groups possessing lone pair electrons, other forms of electron rich groups can also get involved, including π bonds. Probably the most common is a phenyl ring. Cram, for example, studied the solvolysis of *erythro-* and *threo*-3-phenyl-2-butyltosylate in acetic acid and found products indicating retention and scrambling, respectively. The mechanisms given in Figure 11.9 are consistent with these observations, giving the proper retention and scrambling. Departure of the tosylate is facilitated by the neighboring phenyl ring to yield these results. An intermediate called a **phenonium ion** is formed, where the phenyl bridges between two carbons. The phenonium ion derived from the erythro or threo reactants contains a C_2 axis or a mirror plane of symmetry (C_s), respectively (see Figure 11.9 **C**). Equal attack at the two symmetry-related carbons leads to the observed stereochemical outcomes.

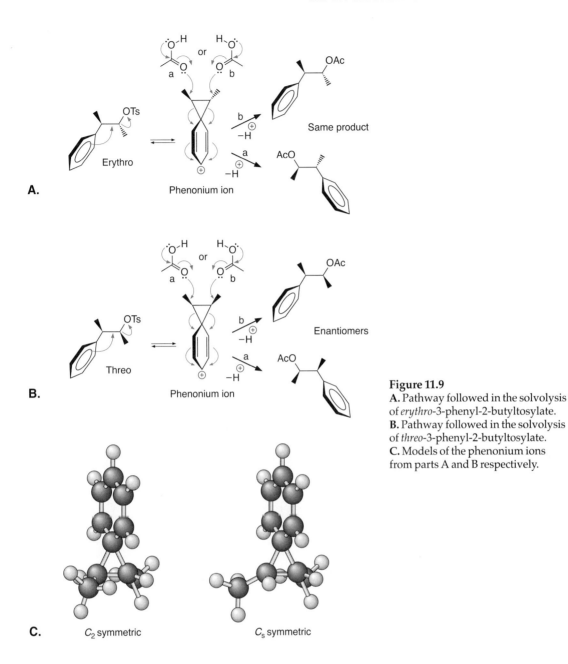

Figure 11.9
A. Pathway followed in the solvolysis of *erythro*-3-phenyl-2-butyltosylate.
B. Pathway followed in the solvolysis of *threo*-3-phenyl-2-butyltosylate.
C. Models of the phenonium ions from parts A and B respectively.

11.5.14 S$_N$1 Reactions Involving Non-Classical Carbocations

When carbons are attached to more than four ligands, they are said to be **hypervalent**. Because carbocations are so electron deficient, they often become hypervalent as a means of gaining more electrons, and the associated cation is called a **carbonium ion**. While this might seem like an unusual situation, a number of structures with hypervalent carbons are known, including some very stable molecules (see the Connections highlight given on page 666). We noted in Chapter 1 that such structures employ 3-center 2-electron bonds, a common motif in electron deficient molecules. When the hypervalent interaction can be viewed as the donation of a C–C σ bond to a cationic center, we have what are called **non-classical carbocations**. Initially, solvolysis reactions that produced rearranged products were the primary testing ground for non-classical carbocations. In time, however, essentially every major tool of physical organic chemistry was brought to bear on this challenging, and sometimes frustrating, problem.

Endo product   Exo product

Norbornyl Cation

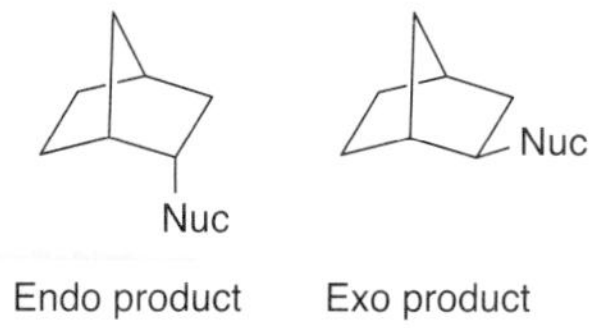

Figure 11.10 shows the products from the solvolysis of *exo*-2-chloronorbornane (one carbon is labeled with an asterisk in order to follow the carbon skeleton rearrangement). Only exo products are obtained, which means that the endo face of the bicyclic system is protected by the bonding geometry of the intermediate carbocation (see the definition of "endo" and "exo" in the margin and in Chapter 6). In addition, the product is racemic, suggesting an intermediate with a plane of symmetry.

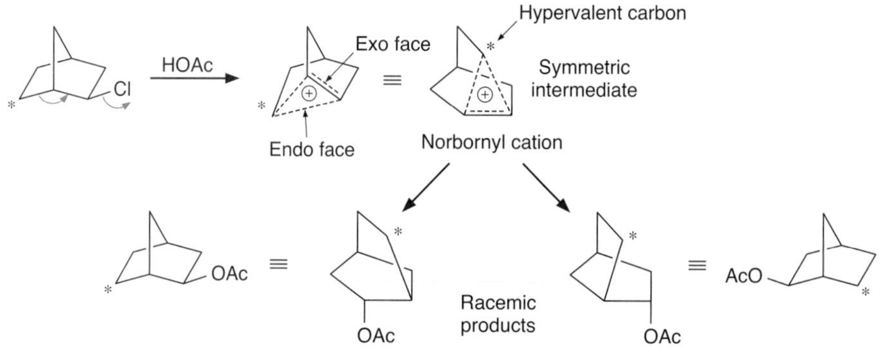

Figure 11.10
The accepted solvolysis pathway for 2-norbornyl systems.

The experimental observations are consistent with the intermediacy of a non-classical carbocation called the **norbornyl cation** (see Figure 11.10), a structure primarily attributed to Winstein. The non-classical structure possesses an internal mirror plane of symmetry, and the endo face of the compound is protected by the bridging interaction. This, at the time, was a highly novel proposal, and it was not universally accepted. The experimental observations could alternatively be explained by invoking two classical carbocations rapidly equilibrating via carbon shifts (Eq. 11.38). The biggest proponent of this explanation was Brown. The distinction between these two possibilities is an important issue to discuss.

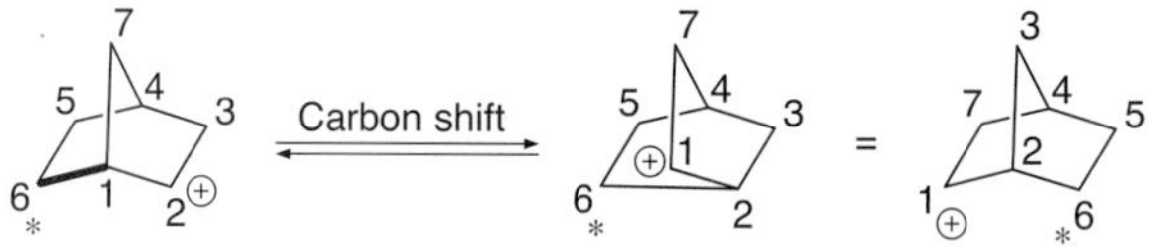

$$\text{(Eq. 11.38)}$$

Let's first examine some of the hydride shifts of the norbornyl cation as a means of exploring the once contentious issues associated with this structure. The most conventional reaction of the norbornyl cation is the simple hydride shift seen in many cations, and called here a 3,2-shift to designate the originating and terminating carbons (Eq. 11.39). Less conventional is the 6,2-hydride shift (Eq. 11.40). This is a facile process because the endo C6–H bond aligns well with the empty p orbital at C2, greatly facilitating migration (see in the margin). These hydride shifts alone cannot explain the products found from the solvolysis of 2-norbornyl systems. For example, consider Eq. 11.41. The predicted carbon label positions are not in the same positions as experimentally observed (Figure 11.10).

$$\text{(Eq. 11.39)}$$

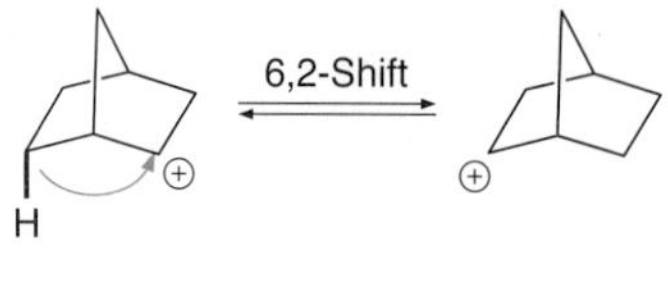

$$\text{(Eq. 11.40)}$$

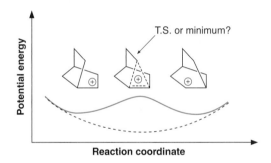

(Eq. 11.41)

Let's return to the conventional Wagner–Meerwein carbon shift as proposed by Brown, in which C6 migrates to the cationic center, making C1 the new cationic center (Eq. 11.38). The unique symmetry of the norbornyl system is such that this novel rearrangement is degenerate, creating a new 2-norbornyl cation. Trapping of the two cations given in Eq. 11.38 with solvent would give racemic products with the label in the correct positions. In fact, if the hydride shifts of Eqs. 11.39 and 11.40, as well as this carbon shift of Eq. 11.38 are all facile, all the carbons and all the hydrogens of 2-norbornyl cation will become equivalent. Indeed, both the ^{1}H and ^{13}C NMR spectra show only one line at room temperature in stable ion media.

The hydride shifts of norbornyl cation are well accepted; it is the carbon shift that has generated so much controversy. The issue is whether Eq. 11.38 is very facile, or whether we have a single symmetrical structure (see the drawing below). A large fraction of physical or-

ganic chemistry studies in the 1950s and 1960s focused on the norbornyl carbocation. It is fair to state that the evidence strongly supports a non-classical cation, and it is now the most well accepted interpretation. Importantly, the studies on the structure of norbornyl cation caused chemists to re-think many of their paradigms of reactivity and structure.

As just stated, a good deal of evidence indicates that the bridged structure is the most stable, both in the gas phase and in stable ion media. The highest level electronic structure theory calculations find the bridged form more stable by 2–4 kcal/mol. Furthermore, recall from Chapter 2 that the gas phase hydride ion affinity (HIA) data also support the non-classical view. As shown in Table 2.8, the 2-methyl-2-norbornyl cation has an HIA of 225 kcal/mol, a typical value for an eight-carbon 3° cation, and it is generally accepted that this is a conventional carbenium ion. However, the HIA of 2-norbornyl is 231 kcal/mol, only 6 kcal/mol higher than the analogous 3° system. We discussed in Chapter 2 that a typical HIA 2°/3° energy difference is 17 kcal/mol. The substantially smaller value in this case indicates a special stabilization of this 2° ion, consistent with the notion that there is something unique about its structure.

Even though the 2-methyl-2-norbornyl cation is viewed as a carbenium ion, there is still a strong interaction between the C1–C6 bond and the cationic center formally at C2. This is illustrated by an x-ray structure of the 1,2,4,7-tetramethyl-2-norbornyl cation, key features of which are shown in the margin (some methyl groups have been deleted for clarity). The substantial lengthening of the C1–C6 bond (see Eq. 11.38 for the numbering scheme) and the shortening of C1–C2 show the strong hyperconjugative interaction in this "classical" system. As we have seen many times before, there are rarely clear-cut dividing lines in physical

HIA = 225   HIA = 231

Hydride affinities

1.71 Å   1.41 Å

2.11 Å

Geometry of methylnorbornyl cation

organic chemistry. This 3° carbocation lies at neither extreme of the classical–non-classical ion continuum; it is an intermediate structure.

Returning to the parent 2-norbornyl ion, several studies in stable ion media support the bridged non-classical structure. At low temperatures, the 3,2- and 6,2-hydride shifts are slow on the NMR timescale. However, at no temperature have the C1 and C2 carbons become inequivalent. This could either mean that, in fact, the stable structure is the symmetrical bridged ion, or that the barrier to the reaction given in Eq. 11.38 is so small that the reaction is fast even at the lowest temperatures that are feasible for solution NMR.

In order to go to even lower temperatures to try to freeze out the putative Wagner–Meerwein shift of Eq. 11.38, several studies in *frozen* stable ion media have been performed. The most impressive is the solid state NMR spectrum of 2-norbornyl cation taken at 5 K! Under these conditions, the system still appears to be a single, symmetrical ion as shown in Figure 11.10. If there is a rapid equilibration of two structures over a finite barrier, that barrier must be ≤ 0.2 kcal/mol. It is generally considered that, if anything, the solid state should artificially increase barriers due to steric hindrance to atom movement, so if the structure is not symmetrical, the barrier is *very* low.

In summary, the prevailing wisdom now finds that the 2-norbornyl cation is non-classical, and that this should not be all that surprising. Donation of filled orbitals to empty orbitals is always stabilizing, and when the filled orbitals are aligned with the empty p orbital of a carbocation, this donation can and will occur.

Cyclopropylcarbinyl Carbocation

Another one of the more extensively investigated carbocation systems derives from S_N1 reactions on homoallyl, cyclobutyl, or cyclopropylcarbinyl derivatives. Solvolyses of all three of these systems give very interesting product mixtures. Let's examine just a few examples.

The solvolyses of cyclopropylcarbinyl and cyclobutyl derivatives often give exactly the same products, in close to the same ratios, as observed by Roberts in 1951. The reaction of certain homoallyl derivatives will also give these products, but homoallyl structures are also very susceptible to S_N2 reactions and will therefore sometimes deviate in the product ratios. A comparison where all three derivatives give the same products in similar ratios is shown in Eq. 11.42. The data indicate that there is likely a common intermediate in all three of these reactions.

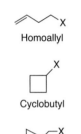

Homoallyl

Cyclobutyl

Cyclopropylcarbinyl

$$\text{(Eq. 11.42)}$$

48%   48%   4%

Furthermore, radiolabeling studies have revealed that the solvolysis of cyclopropylcarbinyl diazonium (the first reactant given in Eq. 11.42) results in partial scrambling of the carbon attached to the diazonium group into all of the carbons of all three products. Hence, there must be an intermediate where several of the carbons become equivalent. Another facet to this puzzle is that cyclopropylcarbinyl, cyclobutyl, and homoallyl derivatives all undergo solvolysis faster than analogous structures. For example, cyclopropylcarbinyl tosylate undergoes solvolysis approximately 10^6 times faster than isobutyl tosylate.

Below we lay out the key carbocationic intermediates that could be involved in these solvolysis reactions. Cyclobutyl, cyclopropylcarbinyl, and 3-butenyl structures are conventional carbenium ions. Less conventional are the bicyclobutonium and tricyclobutonium ions. Although part of the earlier analyses of the solvolysis reactions, the tricyclobutonium ion is no longer considered a likely contributor to most reactions, and will not be discussed further here.

| Cyclobutyl cation | Cyclopropylcarbinyl cation | 3-Butenyl cation | Bicyclobutonium cation | Tricyclobutonium cation |

Roberts originally proposed a series of rapidly equilibrating bicyclobutonium ions to explain these results. We can view bicyclobutonium ion as a cyclopropylcarbinyl system in which the CH_2^+ group is "leaning over" to interact with one edge of the cyclopropyl ring (see below). In Chapter 14 we will analyze a model of bonding for cyclopropane (Walsh orbitals) that emphasizes substantial p character in the bonding orbitals of the ring. The C–C bonding MOs do not lie along the line connecting carbons of the ring, but instead bulge out and away from the ring. The electrons in these cyclopropane C–C bonds are higher in energy than in standard alkanes (the ring is strained). The use of these orbitals in interactions with an

Ways to draw the delocalization
in cyclopropylcarbinyl cation

empty p orbital seems quite sensible. The CH_2^+ can "lean" to the left or the right, so two different bicyclobutoniums can form (Eq. 11.43). The interconversion of the two bicyclobutonium ions places partial positive charge on each carbon of the cyclopropane ring.

(Eq. 11.43)

Another way to envision bicyclobutonium is to start from the homoallyl cation and write an interaction between the cationic carbon and both carbons of the double bond. We already analyzed the stabilization of cations by a homoallyl interaction in reference to Figure 11.7, and so the bicyclobutonium ion should not be too surprising.

Ways to draw the delocalization
in homoallyl cation

Despite having made these analogies to other systems, the exact nature of the $C_4H_7^+$ system is less clear than that of norbornyl, and it is still under investigation. Considerable evidence exists in support of *both* the bicyclobutonium and cyclopropylcarbinyl carbenium ions as important contributors to $C_4H_7^+$. Under stable ion conditions these two appear to be in equilibrium. As with so many carbocations, a very flat potential energy surface is implied, with structures of similar energy and low barriers to interconversion.

Connections

Further Examples of Hypervalent Carbon

In Chapter 1 we explicitly noted the isoelectronic relationship between BH_3 and CH_3^+, and we used the three-center–two-electron bonding in boranes as a springboard to rationalize bonding features of carbocations. Another hallmark of borane chemistry is the existence of highly-bridged, polyhedral structures such as the dodecaborane dianion, $B_{12}H_{12}^{2-}$, which has a perfect icosahedral structure (see below). Since CH^+ is isoelectronic with BH, disubstitution into dodecaborane produces the neutral **carborane** structures, $C_2B_{10}H_{12}$. As shown below, three isomeric carboranes are possible, and they are called ortho, meta, and para by analogy to aromatic chemistry. The conclusion is unavoidable that the carbon atoms of the carboranes are *hexacoordinate* (remember, • = CH). While this suggests an exotic bonding scheme, the molecules themselves are quite robust. At very high temperatures ($> 500\,°C$), the ortho/meta/para isomers interconvert, but there is no decomposition. It is generally accepted that carboranes can be viewed as three-dimensionally aromatic molecules. These are certainly electron deficient molecules (convince yourself that there are not enough valence electrons to describe all the bonds drawn in the carborane structures as conventional two-electron bonds). Again, we see that under electron deficient conditions carbon, like boron, can form structures that are stable but hypervalent.

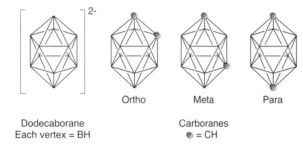

Dodecaborane
Each vertex = BH

Ortho   Meta   Para

Carboranes
● = CH

The analogy to polyhedral boranes led to the consideration of other exotic carbocation structures. For example, the B_5H_9 structure is common. Recognizing that C_2H_2 is isoelectronic with B_2H_4 produces the stable $B_3C_2H_7$. A second similar replacement, plus replacing the remaining B by C^+ gives the square pyramidal structure for $(CH)_5^+$, first proposed by Williams in 1971. It soon received theoretical support from Hoffmann, and a dimethyl deriva-

tive was observed experimentally by Masamune. The molecule $(CH)_5^+$ can be thought of as a complex between cyclobutadiene and CH^+.

B_5H_9

$B_3C_2H_7$

$C_5H_5^{\oplus}$

Hypervalent structures

Another structure of this type is $(CCH_3)_6^{2+}$. Both ^{13}C and ^1H NMR spectroscopy support a pentagonal pyramidal structure for this ion. For this and other proposed polyhedral cationic structures, recent methods for accurate calculation of NMR chemical shifts have proven to be especially valuable. A number of complex systems have recently succumbed to detailed analysis using a combined theoretical/experimental approach (see further discussion in Section 14.5.5).

$[C(CH_3)]_6$ dication

Grimes, R. N. (1970). *Carboranes*, Academic Press, New York.

11.5.15 Summary of Carbocation Stabilization in Various Reactions

In summary, neighboring groups that can stabilize a developing carbenium ion will assist an S_N1 mechanism and other mechanisms involving the intermediacy of carbocations. The stabilization can come from many sources (Figure 11.11), including hyperconjugation from alkyl groups, resonance with electron donating groups possessing lone pairs (N, O, and S), resonance with π systems (allyl, benzyl, cyano, and carbonyl), homoconjugation, hypervalency via donation of aligned σ bonds (non-classical carbocations), bridging in halonium ions, and anchimeric assistance. There are subtle differences in all the effects mentioned here, yet they all have a common origin—filled orbitals aligned with the p orbital on the highly electrophilic cationic carbon will donate electrons toward this center, either after or concurrent with carbocation formation.

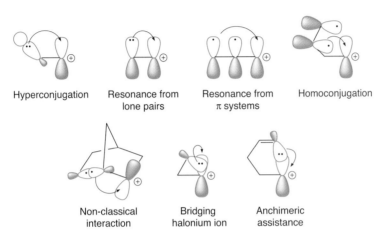

Hyperconjugation   Resonance from lone pairs   Resonance from π systems   Homoconjugation

Non-classical interaction   Bridging halonium ion   Anchimeric assistance

Figure 11.11
Various ways to stabilize carbenium ions. All are similar in that filled orbitals aligned with the empty p orbital on the cation center will donate electrons toward that center.

11.5.16 The Interplay Between Substitution and Elimination

Many of the reagents commonly used as nucleophiles in substitution reactions are also bases, and they can thus promote elimination reactions. Since both elimination and substitution reactions commence from alkyl–LG species, it is not surprising that substitutions and eliminations are often competitive, and we gave some examples in Section 10.13.3. Now that we have considered both substitutions and eliminations, let's analyze the competition.

Figure 11.12 summarizes the interplay between nucleophilic aliphatic substitution and elimination reactions. Beginning with R–LG, where LG is a good leaving group, and then adding B:, a structure that is a base and/or nucleophile, many options arise. The figure shows conclusions for 1°, 2°, and 3° systems separately.

There are some special systems that do not fit neatly into the chart in Figure 11.12, but for the most part they follow expectations based on previous discussions. For example, 1° systems that are highly hindered exhibit quite slow S_N2 reactions. If possible, as in a *sec*-butyl system, elimination may become much more competitive. In neopentyl systems, which cannot undergo elimination, rearrangement reactions can be seen under S_N1 conditions. Also, systems that can form resonance stabilized cations, such as benzyl or allyl, generally undergo rapid substitution reactions under either S_N1 or S_N2 conditions.

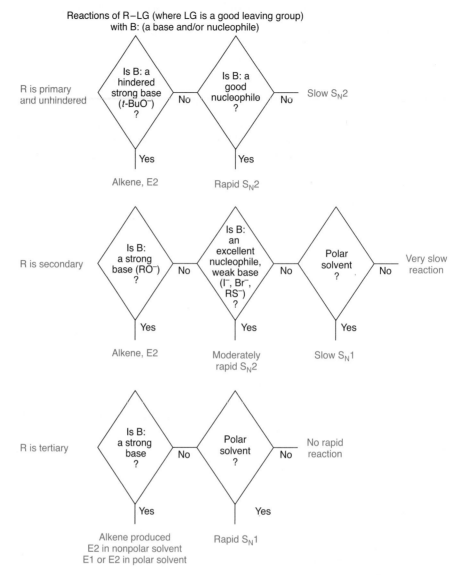

**Reactions of R–LG (where LG is a good leaving group)
with B: (a base and/or nucleophile)**

Figure 11.12
The interplay between nucleophilic substitution and elimination in 1°,
2°, and 3° systems. See the text for a discussion of certain exceptions and
special cases. Adapted from a scheme presented in Kemp, D. S., and
Vellaccio, F. (1980). *Organic Chemistry*, Worth Publishers, New York,
pp. 227–228.

11.6   Substitution, Radical, Nucleophilic

Some S_N2 reactions have electron transfer characteristics. Such **single-electron transfer
(SET)** reactions start with an electron transfer from the nucleophile to the electrophile.

11.6.1   The SET Reaction—Electron Pushing

Scheme 11.8 shows a series of steps involved in an SET mechanism. The SET from the
nucleophile leads to reduction of the R–X bond, producing a radical anion, R–X$^{\bullet-}$, and a radical form of the nucleophile. The radical anion may be a true intermediate with a finite lifetime, or it may undergo immediate cleavage. Either way, we produce the carbon radical with
departure of the leaving group. Combination of the R and Nuc radicals gives the product.

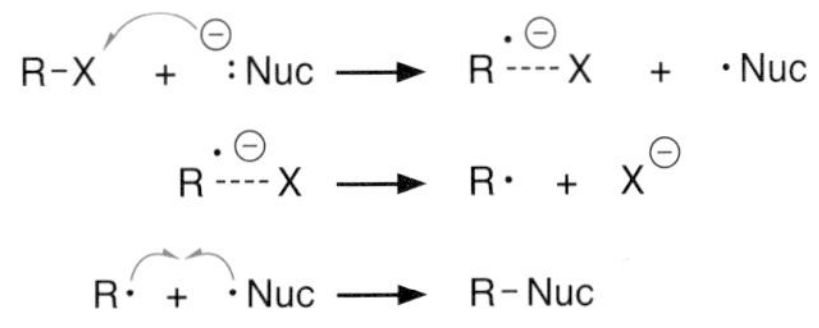

Scheme 11.8
An SET mechanism for substitution reactions.

11.6.2 The Nature of the Intermediate in an SET Mechanism

Pross and Shaik have examined this reaction in detail, proposing a **single-electron shift** model. In this model, the nucleophile transfers an electron to the electrophile as they approach each other. The reaction coordinate involves several valence bond structures such as [Nuc$^-$---R$^\bullet$---X$^\bullet$] and [Nuc$^\bullet$---R$^\bullet$---X$^-$] and some zwitterionic forms. The point is that a free radical may never be formed, but instead a complex between the electrophile and nucleophile is formed wherein an electron shifts between the two reactants. In essence, the mechanism of Scheme 11.8 is reduced to a single step that occurs smoothly along one reaction coordinate with no intermediates (Eq. 11.44).

$$ R{-}X \xrightarrow{\quad \overset{\ominus}{Nuc} \quad} \left[\overset{\delta \cdot \ominus}{Nuc} {-} R {-} \overset{\delta \cdot \ominus}{X} \right]^{\ddagger} \longrightarrow R{-}Nuc \qquad \text{(Eq. 11.44)} $$

You may be wondering how the reaction of Eq. 11.44 differs in practice from an S_N2 reaction, and indeed Pross and Shaik propose that many "normal" S_N2 reactions are instead single-step SET reactions. Therefore, Scheme 11.8 represents a limiting case for a pure SET reaction, while Eq. 11.44 shows an alternative view of a classic S_N2 reaction. These are subtle distinctions, as we are considering exactly how an electron moves from one center to another. This may be a risky proposition for an inherently quantum mechanical object like an electron, but it does explain the results found for some reactions. We can imagine a continuum of possible intermediate mechanisms depending upon the structure of the electrophile, nucleophile, and solvent.

11.6.3 Radical Rearrangements as Evidence

The most convincing evidence that a radical is involved in some substitution reactions is the observation of rearrangements (recall our discussion of free radical clocks in Section 8.8.8). One example is given in Figure 11.13, and several others are known. Lithium aluminum hydride can be used to reduce a C–X bond, which is formally a substitution. The kinetics and stereochemistry of these reduction reactions indicate that they commonly proceed via S_N2 mechanisms, where AlH_4^- is the nucleophile that donates a hydride. Yet, look at the products formed from the reduction of 6-iodo-5,5-dimethyl-1-hexene by $LiAlD_4$ (Figure 11.13). Radicals have been formed, as evidenced by the classic 5-hexenyl radical ring closure rearrangement. The formation of cyclic products is indicative of radical formation, and the position where the deuterium should be has some protium, indicating hydrogen abstraction from the solvent.

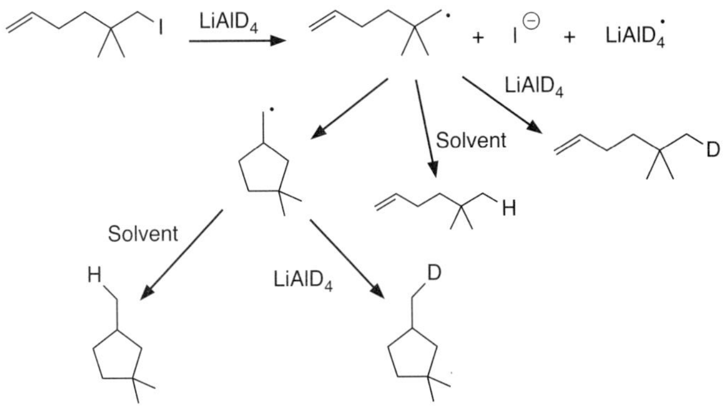

Figure 11.13
Pathway undertaken in the reduction of 6-iodo-5,5-dimethyl-1-hexene with $LiAlD_4$.

11.6.4   Structure–Function Correlations with the Leaving Group

Leaving group effects follow a slightly different order with SET reactions than with S_N2 reactions. For S_N2 reactions of alkyl halides, I^- is similar to ^-OTs in leaving group ability, followed by Br^- and Cl^-. In an SET mechanism, ^-OTs is the worst leaving group, while the order of halogen leaving group ability is the same. For example, $LiAlH_4$ reductions of C–X bonds are thought to proceed primarily via SET mechanisms for X = I, but S_N2 for the other leaving groups. This distinction is evident from the observation of racemization of stereogenic alkyl iodide centers, but inversion with other alkyl halides. This trend is consistent with the general observation that radical anion formation is easier as we move down a column of the periodic table, presumably in part because of the much greater polarizability of the heavier elements. With more electron rich nucleophiles, such as $^-Sn(CH_3)_3$, even alkyl chlorides proceed via the SET mechanism.

11.6.5   The $S_{RN}1$ Reaction—Electron Pushing

There is still another substitution mechanism to consider, although it is much less common. It is called $S_{RN}1$ (substitution, radical, nucleophilic, unimolecular) and involves a radical chain mechanism, unlike the SET mechanism just described. We have seen a radical chain substitution mechanism in Chapter 10 when we considered radical aromatic substitution (Section 10.22).

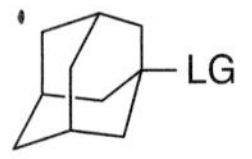

Eq. 11.45 shows an example of an $S_{RN}1$ reaction. It does not seem to fit the criteria of either S_N2 or S_N1. Substitution takes place on a tertiary center that has a very electron withdrawing group on it. The tertiary center impedes an S_N2 reaction, while the electron withdrawing group impedes an S_N1 reaction. In addition, the nucleophile seems too bulky for a standard S_N2 reaction. The $S_{RN}1$ mechanism allows otherwise unfavorable substitutions to occur. Even 1-adamantyl systems can undergo facile substitution, and the leaving groups can be NO_2, N_3, and sometimes even phenyl!

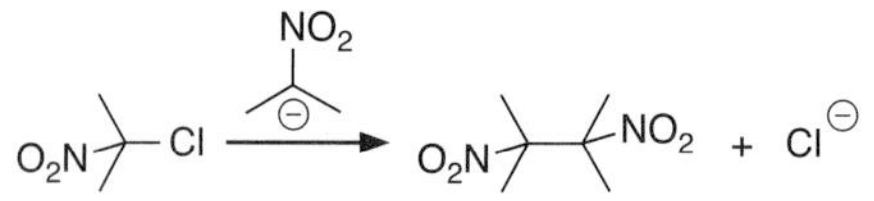

1-Adamantyl–leaving group

$$O_2N \!-\!\!\!\!\!\!\triangleleft\!\!-Cl \;\xrightarrow{\overset{NO_2}{\underset{\ominus}{\triangle}}}\; O_2N\!-\!\!\!\!\!\!\triangleleft\!\!-NO_2 \;+\; Cl^{\ominus} \qquad\qquad \text{(Eq. 11.45)}$$

Scheme 11.9 shows a general example with all the proper one-electron pushing for an $S_{RN}1$ reaction, clearly showing the chain process. In an $S_{RN}1$ mechanism, an initiator (I) first transfers an electron into the carbon–leaving-group bond. The electron goes into the antibonding C–LG σ^* orbital. In many cases the reduction is simultaneous with leaving group departure and the radical anion is never really present. This is called **dissociative electron transfer**. The reaction is most facile when there are one or more electron withdrawing groups on the same carbon as the leaving group, or on a phenyl ring attached to the carbon with the leaving group. These electron withdrawing groups can accept the electron first, prior to transfer into the C–LG σ bond. After bond cleavage, the resulting carbon-based radical combines with the nucleophile to form another radical anion, which subsequently transfers its electron to the starting material. This forms the product and propagates the chain by transferring an electron to the starting material.

Scheme 11.9
An electron-pushing scheme for the $S_{RN}1$ reaction. In$^\bullet$ = initiator.

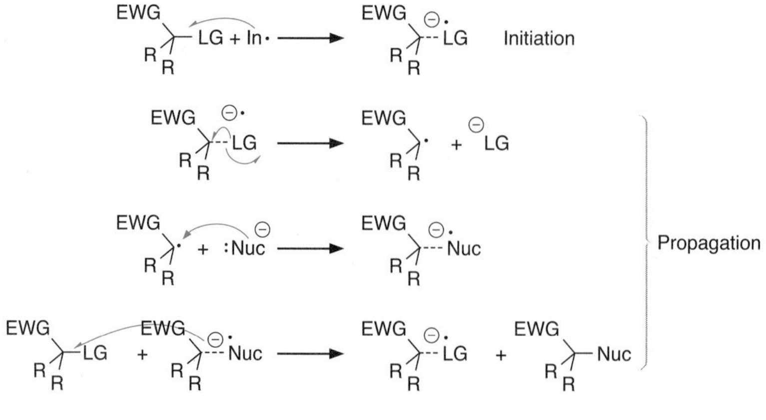

11.7 Radical Aliphatic Substitutions

In the preceding section we saw that nucleophilic substitutions on R–X reactants can, under some circumstances, occur via a radical chain mechanism. A much more common substitution pathway involves substitution of a hydrogen on an R–H reactant by a halogen. This involves conventional free radicals and a chain mechanism. The term given to these kinds of substitutions is S_H2 (substitution, homolytic, bimolecular). The most synthetically useful free radical substitution involves halogenation, and *N*-bromosuccinimide (NBS) is a common reagent (covered in the Connections highlight at the end of this section).

11.7.1 Electron Pushing

We start our analysis with a review of the chain mechanism for free radical halogenation, which is typically given in introductory organic chemistry texts. Scheme 11.10 shows the chlorination of methane as an instructive example. The process is initiated by homolysis of a bond, typically that of the molecular halogen (X_2). Light is a common source of energy for initiation. The halogen radical abstracts a hydrogen atom from the alkane creating a carbon-centered radical, which in turn abstracts a halogen atom from molecular halogen. These two steps turn over many times to propagate the reaction. Both atom abstraction steps occur on a σ bond; if an alkene were present in the reactant, addition to the double bond would often be more favorable (see Section 10.10).

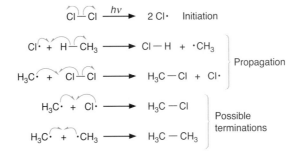

Scheme 11.10
Mechanism and electron pushing for the free radical halogenation of methane.

11.7.2 Heats of Reaction

If one looks at the average bond dissociation energies for X_2, C–X, H–X, and C–H bonds (Table 2.2), an average heat of reaction for the halogenation of alkanes can be calculated. The results in kcal/mol are as follows: F = –101, Cl = –22, Br = –4, and I = 16. The variation in these numbers comes from a continual decrease in H–X and C–X bond strengths in the series F, Cl, Br, and I. These heats of reaction reflect a dramatic change in reactivity. Free radical fluorination is so exothermic that it occurs spontaneously and very explosively. Chlorination and bromination can be controlled and are useful reactions. Free radical iodination rarely occurs.

11.7.3 Regiochemistry of Free Radical Halogenation

With alkanes that have more than one kind of hydrogen, free radical halogenation will usually lead to a mixture of constitutional isomers. Which isomer dominates will depend upon the relative reactivity and number of different hydrogens that can be abstracted. While the structure of the substrate determines which sites are the more reactive, the quantitative selectivity of a reaction will depend upon the reactivity of the halogen radical atom.

The relative reactivities of various radicals for abstracting hydrogens attached to 1°, 2°, and 3° carbons are given in Table 11.10. These numbers have been corrected for statistical factors, meaning that to use these numbers to predict product ratios, you need to multiply each number by the actual number of 1°, 2°, and 3° hydrogens present in your reactant.

As seen in Table 11.10, abstraction of 3° hydrogens is always more favorable than 2° hydrogens, which is more favorable than 1° hydrogens. Recall that the order of stability of radicals is 3° > 2° > 1°. Correspondingly, the order of C–H bond strengths for various substituted

Table 11.10
Relative Rates of Hydrogen Abstraction by Various Radicals*

Radical	Temperature (°C)	1° C–H	2° C–H	3° C–H
F•	25	1	1.2	1.4
Cl•	25	1	4	6
Br•	40	1	200	19000
H•	35	1	5	40
CH$_3$•	110	1	4	46
Ph•	60	1	9	47

*Poutsma, M. L. in *Free Radicals*, J. K. Kochi (ed.), John Wiley & Sons, New York, 1973, Vol. II. Russell, G. A., and de Boer, C. "Substitutions at Saturated Carbon–Hydrogen Bonds Utilizing Molecular Bromine or Bromotrichloromethane." *J. Am. Chem. Soc.*, **85**, 3136 (1963).

carbons is 1° > 2° > 3° (Table 2.2). Therefore, the regiochemistry of free radical substitutions is apparently controlled by thermodynamics and is always the same—tertiary substitution dominates over secondary which dominates over primary.

Quantitatively, though, the relative reactivities of the C–H bonds is very different depending upon the nature of the radical that is doing the abstraction. The more reactive radicals are less selective. "More reactive" in this case refers to a more exothermic hydrogen atom abstraction step. The more exothermic this step, the more the transition state resembles the reactants, with a correspondingly lower character of the carbon-centered radical product. This makes the reaction less sensitive to the stability of the carbon-centered radical. This is a classic example of the reactivity–selectivity principle, and was examined in a Connections highlight in Section 7.3.2.

Table 11.11 provides further insight into the origin of the selectivity in free radical halogenations. The E_a is essentially zero for all hydrogen atom abstractions by fluorine atom, and the free energy barrier arises solely from the log A term of the Arrhenius equation, which is near 13 for all the halogenations given in Table 11.11. The activation energies for abstraction by chlorine atoms are also exceedingly small but in the direction of the trends discussed. Lastly, the activation energies for abstraction by bromine atoms are substantial, and they clearly produce the differential reactivities of 3°, 2°, and 1° C–H bonds. Because of these differences the relative selectivities of Table 11.10 are temperature dependent.

Table 11.11
Approximate Activation Energies for Free Radical Halogenations (kcal/mol)*

	1° C–H	2° C–H	3° C–H
Radical	E_a	E_a	E_a
F•	0	0	0
Cl•	1	0.5	0
Br•	13	10	7

*Trotman-Dickenson, A. F. "The Abstraction of Hydrogen Atoms by Free Radicals." *Adv. Free Radical Chem.*, **1**, 1 (1965).

The relative reactivities of hydrogens on 3°, 2°, and 1° carbons are a reflection of radical stability, and so it is not surprising that the addition of groups that can stabilize a radical via resonance, such as phenyl, favors halogenation at that site. Selectivities can change, in that the relative rates for Br• abstracting hydrogen from C–H bonds in PhCH$_3$, Ph$_2$CH$_2$, and Ph$_3$CH are 1:10:17. Compare this to the trend given for bromine in Table 11.10. The lower selectivity in the benzylic series arises because the carbon-centered radical has a source of stability other than just substitution level—resonance. Stated another way, upon addition of phenyl rings, the C–H bonds become more reactive than simple alkane C–H bonds, and so their reactions are less selective.

Connections

Brominations Using *N*-Bromosuccinimide

Although free radical halogenation as described above is used industrially in certain contexts, it is not a method commonly used in synthesis laboratories. One method that is routinely used involves *N*-bromosuccimide (NBS). NBS will cleanly brominate allylic and benzylic positions. The mechanism, with electron pushing, is shown to the right. Studies have shown that the halogenating agent is actually molecular bromine, which is maintained at a low concentration throughout the reaction. After initiation, a bromine or succinimidyl radical is reactive enough to abstract a hydrogen atom from the weaker benzylic and allylic C–H bonds in the reactants.

Hydrogen abstraction by bromine radical gives HBr, which in turn reacts with NBS to give molecular bromine. Br_2 will react with the carbon-centered radical to give the product and produce bromine radical to propagate the chain.

Incremona, J. H., and Martin, J. C. "*N*-Bromosuccinimide. Mechanisms of Allylic Bromination and Related Reactions." *J. Am. Chem. Soc.*, **92**, 627–634 (1970).

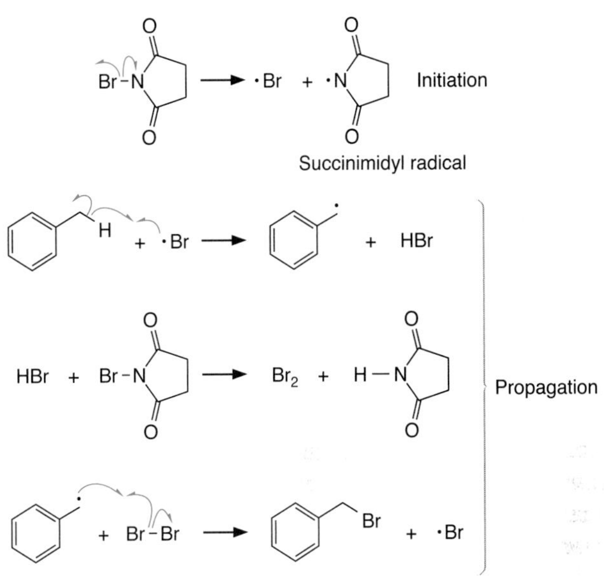

11.7.4 Autoxidation: Addition of O_2 into C–H Bonds

The addition of molecular oxygen across a C–H bond (also called an insertion) to create a peroxide (ROOH) is called **autoxidation**. It is called autoxidation because it forms products that can initiate the same reaction, thus accelerating the reaction as it goes. The reaction is generally slow for standard C–H bonds, but it does represent an important reaction in the degradation of organic molecules and materials that are exposed to the atmosphere for extended periods. You have likely been taught to never pick up an old bottle of diethyl ether that has crystals in it. The crystals are a peroxide of the ether formed from autoxidation, and they are highly explosive. As a result, many organic substances are sold with small percentages of inhibitors of autoxidation. Sterically-encumbered phenols can undergo hydrogen atom abstraction, leading to a relatively unreactive radical. Such compounds make good inhibitors. An example is butylated hydroxytoluene (BHT), a common food preservative.

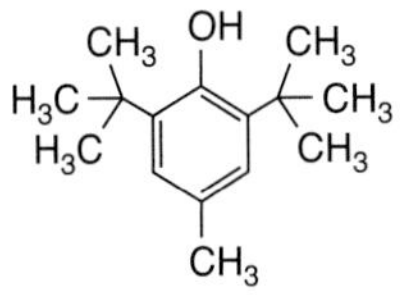

BHT

Electron Pushing for Autoxidation

Autoxidation is a consequence of the triplet nature of molecular oxygen (O_2), which by virtue of its radical character will react with other radicals to form a σ bond (Scheme 11.11). This leads to a radical on the terminal oxygen of a peroxide, called a **peroxyl radical**. As with so many radical reactions, autoxidation involves a chain sequence incorporating initiation, propagation, and termination steps. Here, some species (often an impurity) initiates the process by formation of a radical, which subsequently reacts with molecular oxygen to create a peroxyl radical. The creation of the peroxyl radical leads to a propagation step wherein a hydrogen atom is abstracted to form a hydroperoxide and a new carbon radical. Termination usually involves the formation of tetroxides. These species decompose in one of two ways, depending upon whether the R group on the tetroxide is 1°, 2°, or 3°. Tertiary R groups

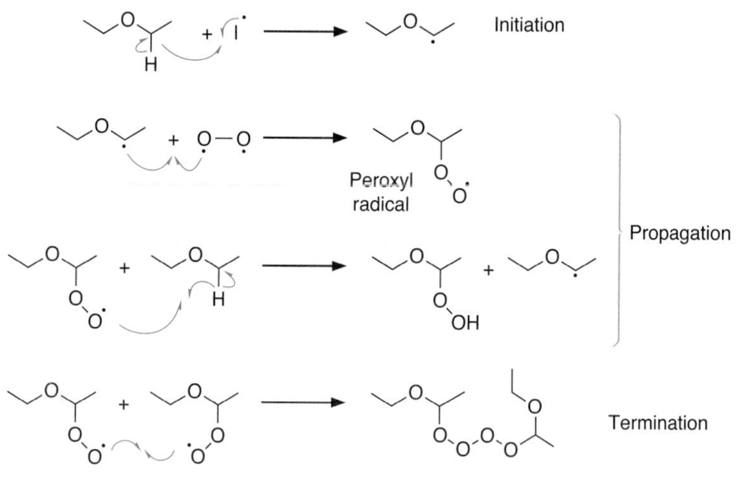

Scheme 11.11
The radical chain mechanism for autoxidation, using diethyl ether as the example.

give elimination of O_2 and $2RO^\bullet$. If any $RO^\bullet$ formed from one of these termination steps escapes before dimerization, it can initiate more autoxidation and other new products. Primary and secondary R groups eliminate O_2 to give an alcohol and a carbonyl.

Isomerizations and Rearrangements

An **isomerization** is a reaction that interconverts compounds with the same molecular formula but with different structures. Just as there are two fundamental types of isomers—stereoisomers and constitutional isomers—so we have **stereoisomerizations** and **constitutional isomerizations**. In earlier literature, these are sometimes described as **geometrical isomerizations** and **positional isomerizations**, respectively, but the alternatives are clearer in their meaning and more consistent with modern usage. In a **rearrangement**, the connectivity of the heavy atoms (C, N, O, S, etc.) making up the molecular skeleton changes. A rearrangement generally involves the conversion of one functional group to another.

The steps in a rearrangement are often intramolecular combinations of nucleophiles and electrophiles, and hence the reactions can still be understood by the guiding paradigm of reactivity given in Section 10.1. A large number of rearrangements involve the movement of an electron rich center to an electron poor center, as we will see in the next six sections. Recall that we previously discussed one rearrangement in Section 11.1.

11.8   Migrations to Electrophilic Carbons

As described in Section 11.5.12 with reference to S_N1 reactions, a hydride or alkyl shift in a carbenium ion is a common rearrangement. It entails the movement of a hydrogen, alkyl, or vinyl/aryl group from a carbon adjacent to a carbenium ion to the electrophilic center in order to create a more stable carbenium ion. Many rearrangements have a similar migration as a key step, with the additional feature of a heteroatom on the β-carbon that stabilizes the newly formed electron deficient center (Eq. 11.46). Two prototype examples are the **pinacol rearrangement** and the **benzilic acid rearrangement**.

(Eq. 11.46)

11.8.1 Electron Pushing for the Pinacol Rearrangement

In the pinacol rearrangement, acid is added to a solution of a vicinal diol. Scheme 11.12 shows the steps using the compound whose common name is pinacol. Protonation of a hydroxyl group allows for a facile heterolysis because water is a good leaving group. After leaving group departure, the resulting carbenium ion undergoes a 1,2-alkyl shift to produce a more stable oxycarbenium ion. Deprotonation yields the carbonyl product.

Scheme 11.12
The pinacol rearrangement of the compound pinacol.

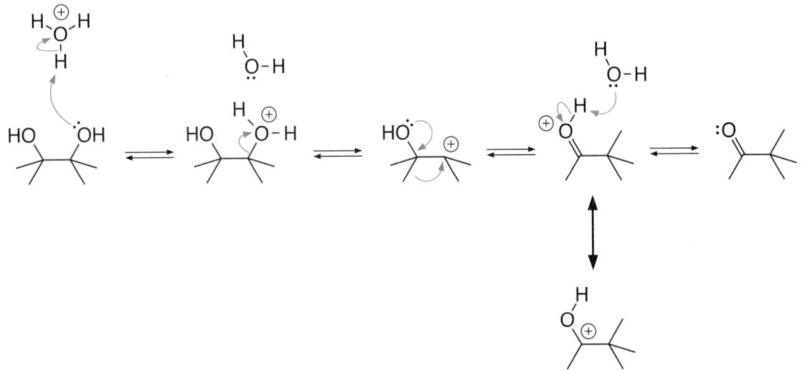

11.8.2 Electron Pushing in the Benzilic Acid Rearrangement

Another well known rearrangement involving migration to an electrophilic carbon is the benzilic acid rearrangement. As shown in Scheme 11.13, three steps are involved. The first step is nucleophilic attack by hydroxide on a carbonyl carbon. The resulting anionic oxygen assists phenyl migration to the adjacent electron poor carbonyl carbon. Rapid proton transfer gives the product.

Scheme 11.13
Electron pushing for the benzilic acid rearrangement.

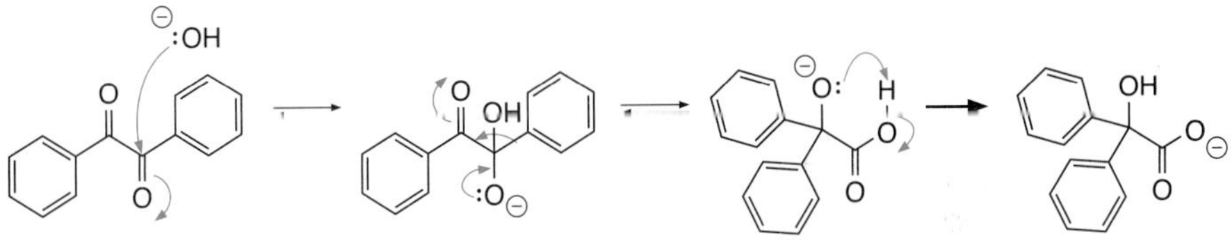

11.8.3 Migratory Aptitudes in the Pinacol Rearrangement

The pinacol rearrangement is a very good reaction for studying **migratory aptitudes**—that is, the relative tendency for a group to undergo a 1,2-shift to an electron deficient center. In essence, we use an experiment described in Chapter 8 as a competition, but the competition is now intramolecular. Symmetrical pinacol structures such as those shown in Eq. 11.47 are used. Regardless of which alcohol group departs, the same carbenium ion is formed. Then, R_1 and R_2 compete for migration to the cationic center as shown.

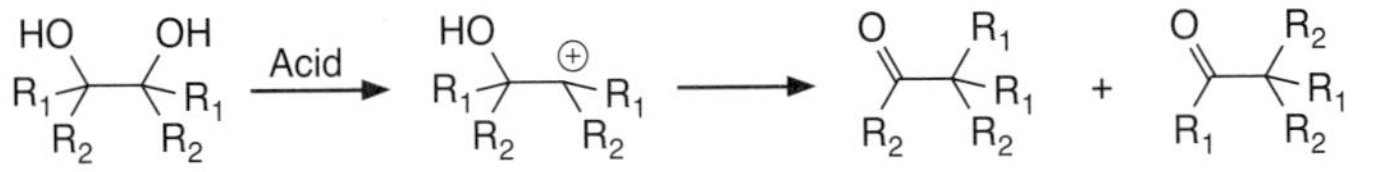

(Eq. 11.47)

Some caution is necessary in interpreting the results of Eq. 11.47. The relative migratory aptitude of R_1 and R_2 does not solely depend upon which group intrinsically migrates the best. The relative ability of R_1 and R_2 to migrate also depends upon the group that does not migrate, because the group left behind stabilizes the carbocation created at the center from which the migrating group departs. Therefore, the group that migrates may not do so because it has a better inherent ability to migrate, but because the group left behind is better at stabilizing carbocations. Furthermore, if leaving group departure is simultaneous with migration, the migration will depend upon the group's ability to give anchimeric assistance to the departing nucleofuge. Nevertheless, some general trends emerge from these kinds of studies. Aryl groups usually migrate better than alkyl groups. Hydrogens give mixed results relative to aryl and alkyl groups depending upon both the structure of the reactant and the experimental conditions. Furthermore, due to the similarity between methyl and ethyl, neither is found to consistently migrate better.

Comparisons among aryl groups give expected trends. Thus, electron donating groups ortho and para speed migration (see Table 11.12). Electron withdrawing groups retard migration in all positions. The data shown in Table 11.12 correlate well with electrophilic aromatic substitution and can form the basis of a Hammett plot (try Exercise 37).

Table 11.12
Relative Migratory Aptitudes in the Pinacol Rearrangement*

R group	Migratory aptitude
p-Methoxyphenyl	500
p-Tolyl	15.7
p-Biphenyl	11.5
m-Tolyl	1.95
m-Methoxyphenyl	1.6
Phenyl	1.0
p-Chlorophenyl	0.66

*Bachmann, W. E., and Ferguson, J. W. "The Pinacol–Pinacolone Rearrangement. VI. The Rearrangement of Symmetrical Aromatic Pinacols." *J. Am. Chem. Soc.*, **56**, 2081 (1934).

11.8.4 Stereoelectronic and Stereochemical Considerations in the Pinacol Rearrangement

The pinacol rearrangement can be initiated from either alcohol, which can complicate mechanistic analysis. Therefore, many studies have investigated a related reaction called the **semi-pinacol rearrangement**. This reaction is initiated by the deamination of a β-amino alcohol. The amine is first converted to a diazonium group, and N_2 is the leaving group. The reaction results again in a carbonyl-containing product. This reaction has been studied extensively to evaluate the stereochemistry of the migrating group. As shown in Eq. 11.48, retention of configuration of the migrating group is found.

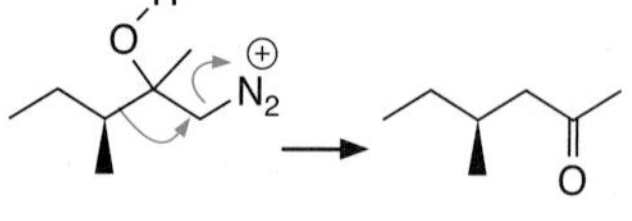

88% retained

(Eq. 11.48)

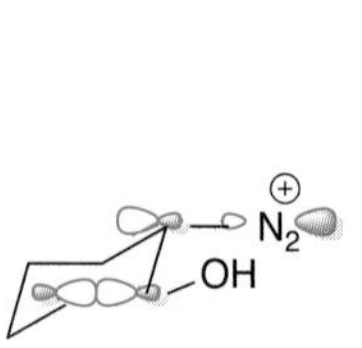

In favorable cases, migration of the R group in the pinacol rearrangement can occur simultaneously with leaving group departure. Such a process would simply be an intramolecular backside displacement of the leaving group by the migrating group. Hence, only groups that are antiperiplanar to the leaving group can migrate in such a scenario. Eq. 11.49 shows an example. Only ring contraction is observed, because the antiperiplanar C–C σ bond is properly aligned to migrate, as shown in the margin. Other groups, such as axial hydrogens, are not antiperiplanar and cannot migrate (Eq. 11.50).

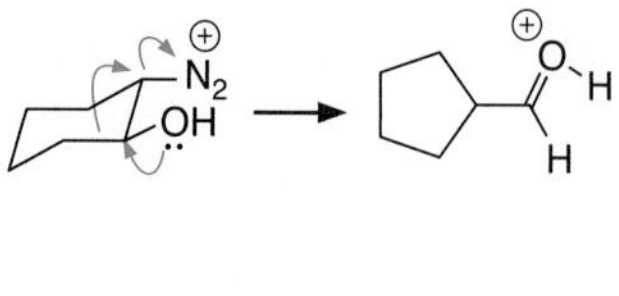

Orbital alignment

(Eq. 11.49)

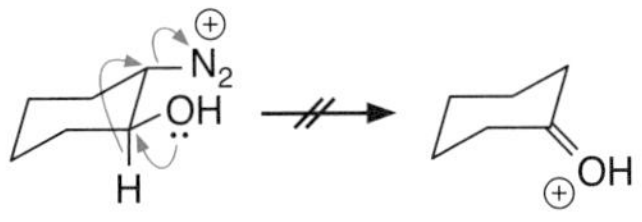

(Eq. 11.50)

Connections

An Enzymatic Analog to the Benzilic Acid Rearrangement: Acetohydroxy-Acid Isomeroreductase

The pathway involved in the biosynthesis of branched-chain hydrophobic amino acids such as valine, isoleucine, and leucine involves several unusual reactions and fascinating enzymes. One in particular, acetohydroxy-acid isomeroreductase, catalyzes the sequence shown below (the R group varies according to the amino acid). The first step involves a rearrangement that is analogous to the second step of the benzilic acid rearrangement, which is followed by a reduction using NADPH (see the Connections highlight in Section 10.8.7).

We can anticipate many features of the enzyme-catalyzed rearrangement mechanism by looking at Scheme 11.13. Deprotonation of the hydroxyl group alpha to a carbonyl needs to be facilitated by a basic resi-

due. In addition, the alkoxide anion resulting from R group migration should be stabilized by an electrophile, and the resulting alkoxide needs to be protonated. In the enzyme, two magnesium ions and their ligands are involved in all these anticipated steps. The deprotonation of the substrate occurs via a Mg(II)-bound hydroxide in the first step. Migration of the R group to the carbonyl is enhanced by ion pair formation between the resulting alkoxide anion and a Mg(II). Protonation then occurs from a Mg(II)-bound water. This completes the rearrangement portion of the mechanism. In the reduction step, the newly formed carbonyl is highly polarized because it is flanked by two Mg(II) ions, which facilitates hydride attack from NADPH. This is a fascinating mechanism for the catalysis of a classic organic rearrangement reaction.

Dumas, R., Biou, V., Halgand, F., Douce, R., and Duggleby, R. G. "Enzymology, Structure, and Dynamics of Acetohydroxy Acid Isomeroreductase." *Acc. Chem. Res.*, **34**, 399–408 (2001).

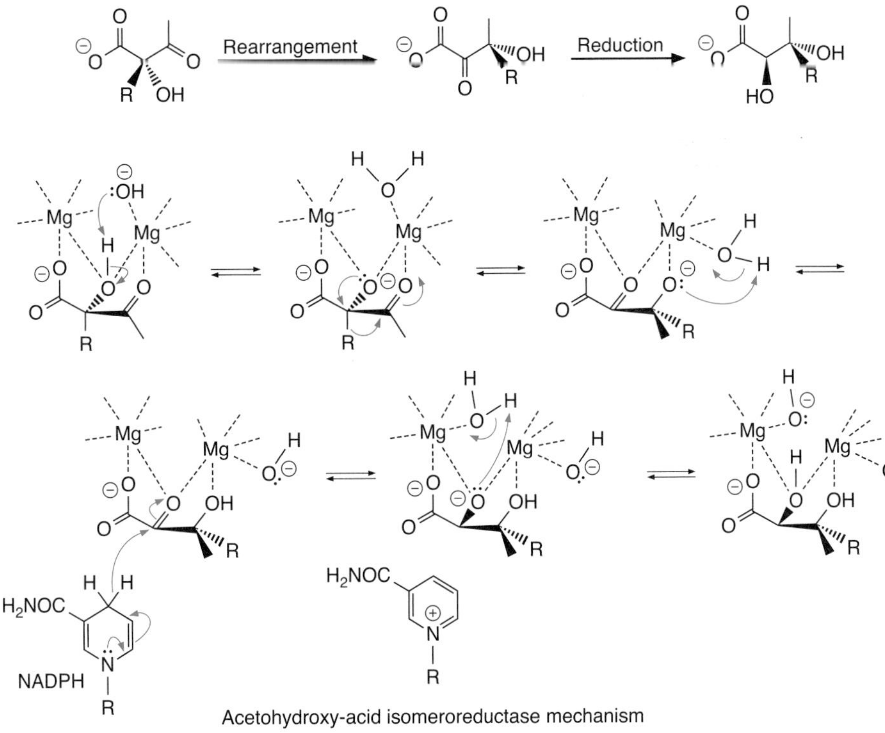

Acetohydroxy-acid isomeroreductase mechanism

11.8.5   A Few Experimental Observations for the Benzilic Acid Rearrangement

The common name of the diketone shown in Eq. 11.51 is benzil. As noted above, re-arrangement of this α-diketo structure to give a carboxylate anion occurs upon addition of hydroxide, and acid workup gives the carboxylic acid product—benzilic acid. The reaction occurs through an anionic intermediate, but involves migration to an electron deficient car-bon just as with the pinacol rearrangement. There is an enzymatic analog to this reaction, which is described in the Connections highlight on the previous page.

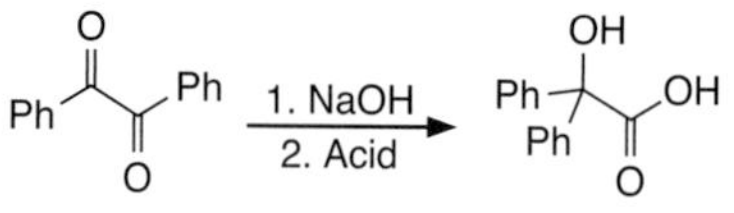

Benzil

(Eq. 11.51)

The reaction is second order, first order in both hydroxide and benzil. However, this does not delineate which step of Scheme 11.13 is rate-determining, because all steps prior to the rate-determining step would be evident in the kinetics. Therefore, isotope scram-bling and a kinetic isotope effect experiment were used to discover the rate-determining step. If one adds ^{18}O-labeled hydroxide, the label scrambles into the carbonyl oxygens of benzil faster than product formation. Therefore, the first step is reversible and is not rate-determining. If one uses OD^- instead of OH^-, no isotope effect is found, so the third step must be past the rate-determining step. Hence, the rearrangement step must be rate-determining.

11.9   Migrations to Electrophilic Heteroatoms

Many rearrangements involve migrations to electrophilic heteroatoms, such as N and O. We examine four here, and several are also presented in the Exercises at the end of the chapter. Electron pushing is presented for each first, followed by some experimental observations.

11.9.1   Electron Pushing in the Beckmann Rearrangement

The Beckmann rearrangement involves the transformation of an oxime to an amide (Eq. 11.52). In this case, migration is to an electron poor nitrogen, rather than an electron poor car-bon, as in the last two rearrangement reactions.

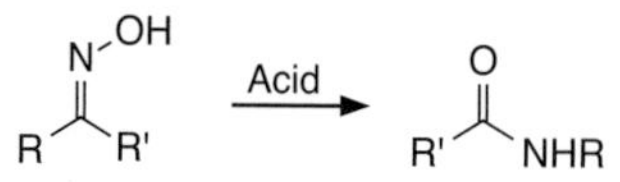

(Eq. 11.52)

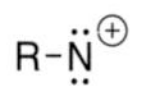

$R–\ddot{N}^{\oplus}$

A nitrenium ion

In the first step (Scheme 11.14), acid protonates the oxime OH group, making it a good leaving group. Direct heterolysis of the N–O bond, however, is not chemically reasonable, because it would place a positive charge on a nitrogen lacking an octet of electrons (a **ni-trenium ion**). Instead, the evidence supports a concerted migration of an alkyl group to the electron deficient nitrogen, assisted by the nitrogen lone pair, and loss of water. In so doing, a relatively stable **nitrilium ion** is formed. An addition reaction involving nucleophilic at-tack of water on the nitrilium carbon leads to a tautomeric form of an amide after deprotona-tion, called an **imidate**. Tautomerization gives the product.

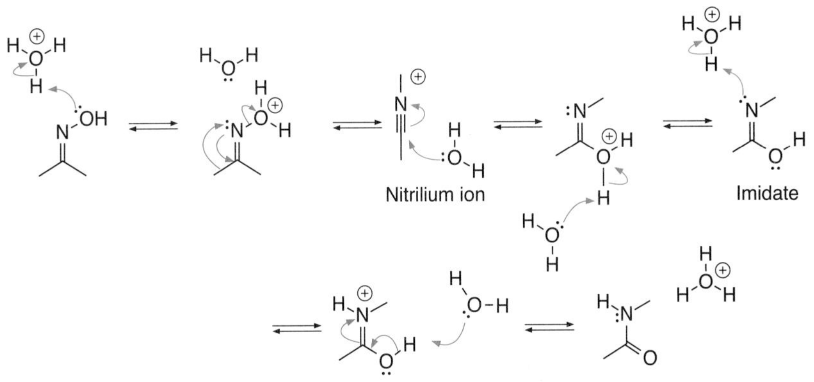

Scheme 11.14
An example of the Beckmann rearrangement.

11.9.2 Electron Pushing for the Hofmann Rearrangement

The Hofmann rearrangement involves the treatment of primary amides (those with an NH_2 group) with bromine in basic media. Amides are converted to amines and CO_2 is eliminated (Eq. 11.53). The mechanism involves an N-bromo amide and an isocyanate, both of which have been isolated as intermediates.

$$\text{(Eq. 11.53)}$$

Just as the treatment of a ketone containing an α C–H with a strong base will lead to an enolate, the reaction of an amide N–H with a strong base will create an anion of the amide. Here, the anionic charge is now delocalized over the amide oxygen and nitrogen (see Scheme 11.15). Continuing with the analogy to carbonyl chemistry, treatment of an enolate with bromine gives α-bromo carbonyls (see Section 11.2), and treatment of the amide anion with bromine yields an N-bromo amide. Furthermore, just as in α-halogenations under basic conditions, a second deprotonation is faster. With an N-bromo amide, this creates an anionic amide with an attached bromine leaving group. Migration of the R group to the amide nitrogen expels a bromide, creating an isocyanate (RNCO). Nucleophilic attack by hydroxide on the isocyanate ultimately leads to a decarboxylation.

Just as with the Beckmann rearrangement, the migration step in the Hofmann rearrangement is to an electron poor nitrogen. Although the nitrogen formally has a partial negative charge, it is attached to an electron withdrawing carbonyl and a bromine atom. An N–Br bond is also relatively weak, and hence susceptible to nucleophilic attack, similar to a Br–Br bond. Thus, this rearrangement is another example of migration to an electrophilic center.

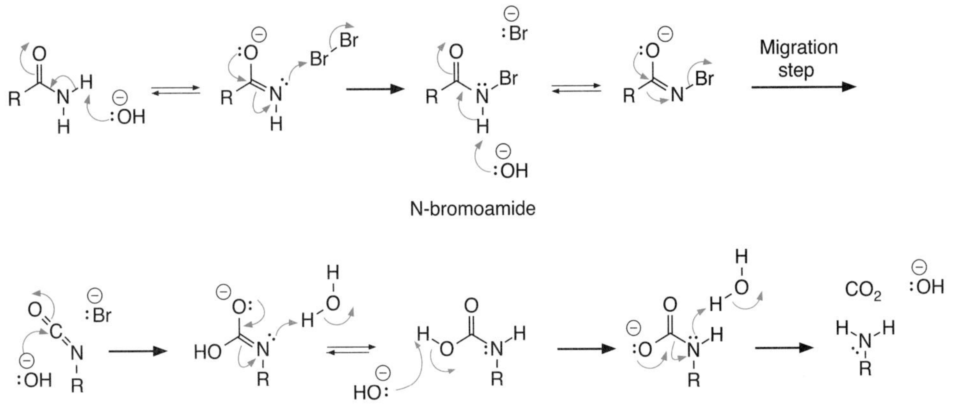

Scheme 11.15
Proper electron pushing for the Hofmann rearrangement.

11.9.3 Electron Pushing for the Schmidt Rearrangement

When an aldehyde or ketone is treated with hydrogen azide, amides are formed. This is our third example of a rearrangement to an electron deficient nitrogen. However, now the leaving group is N_2, rather than water or bromide.

Hydrogen azide (or another acid) first protonates the carbonyl oxygen, followed by nucleophilic addition of azide (Scheme 11.16). After protonation and loss of water, rearrangement via migration of an R group to the electron poor nitrogen, concomitant with loss of N_2, gives a nitrilium ion. This rearrangement is plausible because a lone pair on nitrogen can be used as an electron source to stabilize the resulting cation. After the rearrangement, nucleophilic attack by water followed by tautomerization gives the amide.

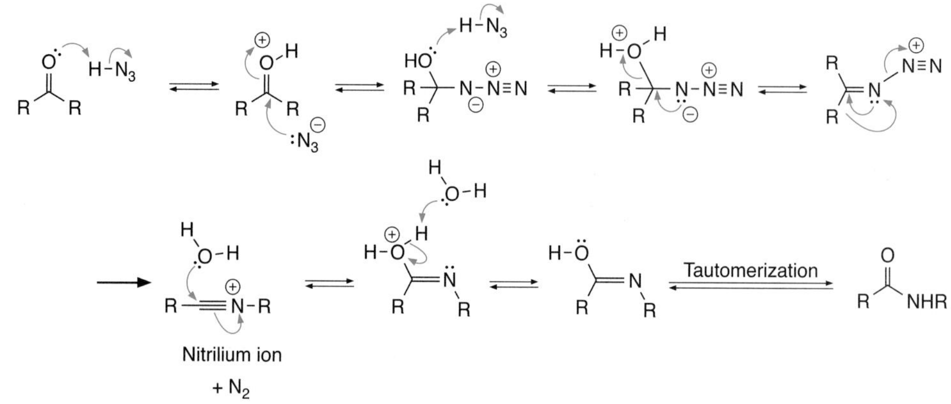

Scheme 11.16
An example of a Schmidt rearrangement.

Nitrilium ion
+ N_2

11.9.4 Electron Pushing for the Baeyer–Villiger Oxidation

When aldehydes and ketones are treated with peracids, esters are formed (Eq. 11.54). Once again, the mechanism involves migration to an electron deficient center, but now it is to an oxygen, not a nitrogen or carbon.

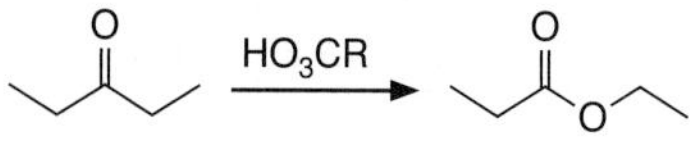

(Eq. 11.54)

Addition of a peroxyacid to the carbonyl forms a tetrahedral intermediate that undergoes rearrangement as shown in Scheme 11.17. An R group migrates to an electron poor oxygen in a peroxy linkage. Addition of acid can speed up the reaction by protonating the leaving group prior to departure.

Scheme 11.17
An example of a Baeyer–Villiger oxidation, in this case making a lactone.

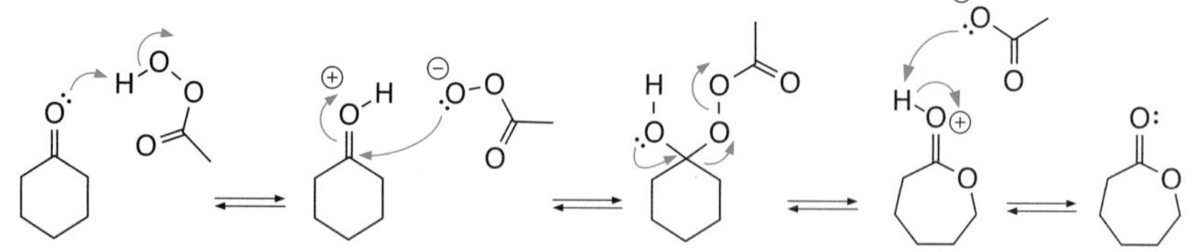

11.9.5 A Few Experimental Observations for the Beckmann Rearrangement

As shown in Scheme 11.14, the role of the acid is to protonate the OH group, thus allowing loss of water. In a subsequent step, water adds back. Therefore, oxygen transfer in this mechanism is an intermolecular process. In support, addition of varying percentages of ^{18}O-labeled water gives exactly the same percentage of ^{18}O incorporation in the product.

Sometimes an acid can act to esterify the leaving group, hence activating it. For example, sulfuric acid has been suggested to create sulfonate esters of the oxime oxygen. When the leaving group is very good to start, acid is not needed at all. For example, picric acid oximes react with no acid activation (Eqs. 11.55 and 11.56).

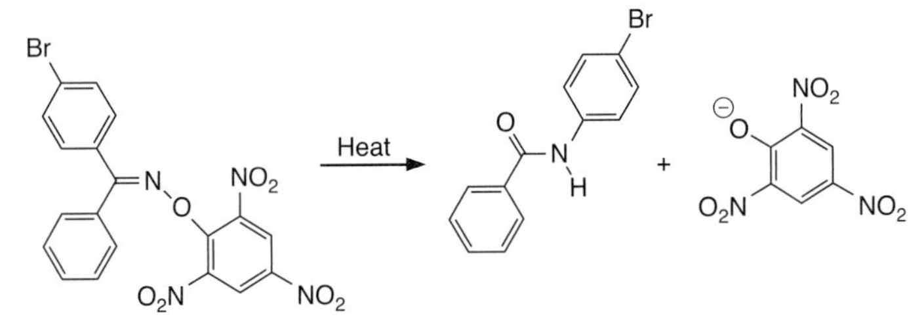

(Eq. 11.55)

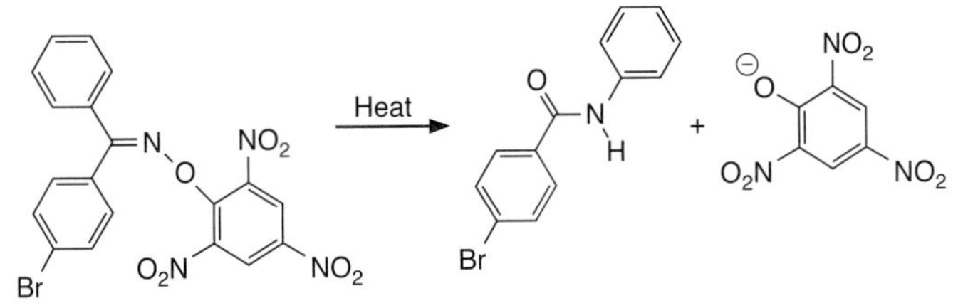

(Eq. 11.56)

Because the heterolytic cleavage of the N–O bond in the activated oxime occurs simultaneously with R group migration, there are stereochemical requirements. Oximes exist as E and Z isomers that are configurationally stable at ambient temperatures. Under conditions where these forms cannot interconvert (acid can catalyze the isomerization), the group anti to the leaving group is the one that migrates (Eqs. 11.55 and 11.56). This supports a mechanism involving simultaneous migration, because stereoelectronic considerations dictate that the anti C–C bond is aligned to fill the antibonding nitrogen–leaving group orbital.

Studies of the rearrangement of anti acetophenone oximes (Eq. 11.57) indicate that the rearrangement is rate-determining. The reaction correlates well with Hammett σ^+ values, supporting a phenonium-like transition state.

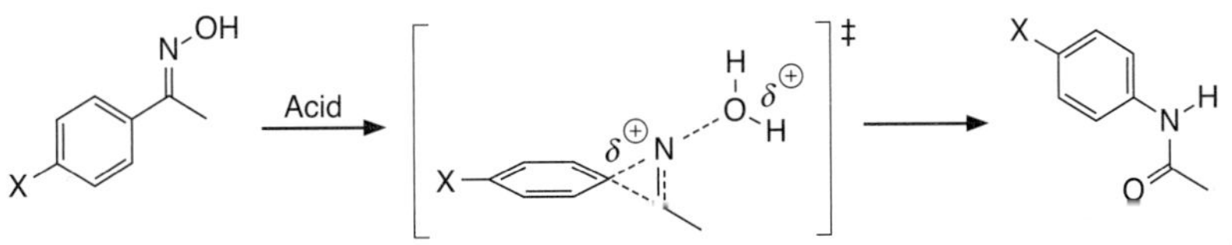

(Eq. 11.57)

11.9.6 A Few Experimental Observations for the Schmidt Rearrangement

Several experiments support the general picture given in Scheme 11.16. First, the same intermediates can be generated from vinyl azides by protonation (Eq. 11.58), and from there the same products as from a Schmidt rearrangement arise. Furthermore, in support of the intramolecular nature of the rearrangement, stereogenic R groups migrate with retention of configuration. Moreover, just as with the Beckmann rearrangement, stereoelectronic considerations require that the R group anti to the leaving group migrates. Typically, it is found that the largest R group is the one that migrates. This suggests that the migration occurs from a structure where the nitrogen leaving group is anti to the largest R group.

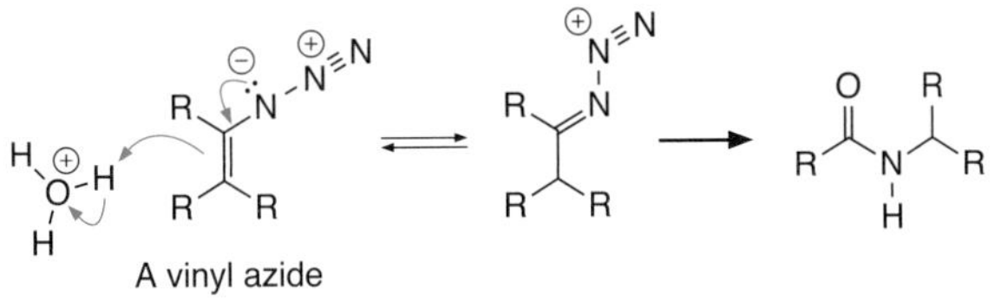

A vinyl azide

(Eq. 11.58)

11.9.7 A Few Experimental Observations for the Baeyer–Villiger Oxidation

Although a tetrahedral intermediate such as that shown in Scheme 11.17 has not been observed or isolated, migration from such an intermediate is supported by the fact that ^{18}O-labeling of the ketone or aldehyde remains in the carbonyl oxygen of the ester product.

Furthermore, stereogenic R groups migrate with retention of configuration.

For most ketones and aldehydes, the rate-determining step is the rearrangement, because significant substituent effects for the migrating group are found. Hammett plots for the migration of substituted phenyl groups from acetophenones (Eq. 11.59) are linear with negative slopes, as would be expected for a group migrating to an electron deficient center. If formation of the tetrahedral intermediate were rate-determining, there should be no substituent effect on the rate, because the migration would be after the rate-determining step. However, when the migrating group is extremely active, the formation of the tetrahedral intermediate can be rate-determining.

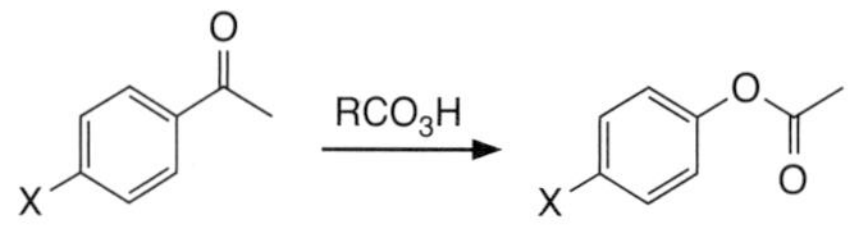

(Eq. 11.59)

11.10   The Favorskii Rearrangement and Other Carbanion Rearrangements

Another common and useful rearrangement is the **Favorskii rearrangement**, which involves a carbanion intermediate. An example is shown in Eq. 11.60. It involves the base-induced conversion of an α-halo carbonyl to a carboxylate. If the starting material is a cyclic ketone, a ring contraction results, and this is one of the most useful applications of the Favorskii rearrangement.

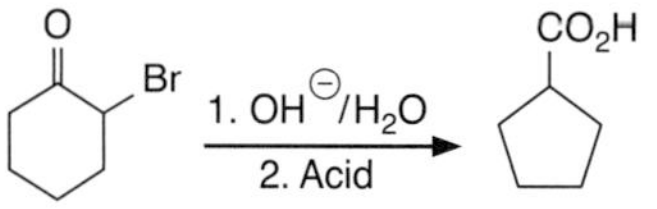

(Eq. 11.60)

11.10.1   Electron Pushing

The standard mechanism for the Favorskii rearrangement is given in Scheme 11.18. Creation of an enolate with base is followed by loss of bromide to give the novel **oxallyl** species. This is a neutral species that can be thought of as a zwitterion comprised of an alkoxide and an allyl cation. The oxallyl species is in equilibrium with a cyclopropanone, and the carbonyl of the cyclopropanone can undergo nucleophilic attack by hydroxide. A standard addition–elimination sequence then produces the product. Release of strain in the cyclopropane provides the driving force for the C–C bond cleavage. The leaving group is drawn as a carbanion, but in protic solvent this likely is not on the energy surface and is protonated during explusion.

Scheme 11.18
Standard electron pushing for the Favorskii rearrangement.

Oxallyl species

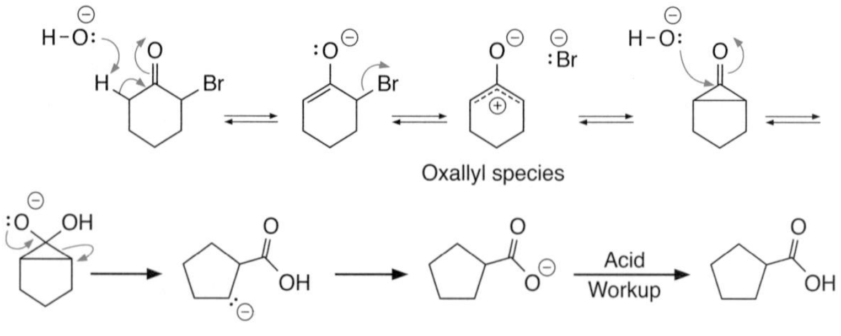

In some cases an alternative mechanism is involved in what has been called a **pseudo-Favorskii rearrangement** (Eq. 11.61). The reaction involves nucleophilic addition to the carbonyl, followed by breakdown of the tetrahedral intermediate with concomitant migration of the C–C bond to give the product. No oxyallyl intermediate is involved. In this variant, the R group migrates to an electron poor center, assisted by the negative alkoxide. This is similar to the rearrangement step in the Hofmann rearrangement (see Section 11.9.2).

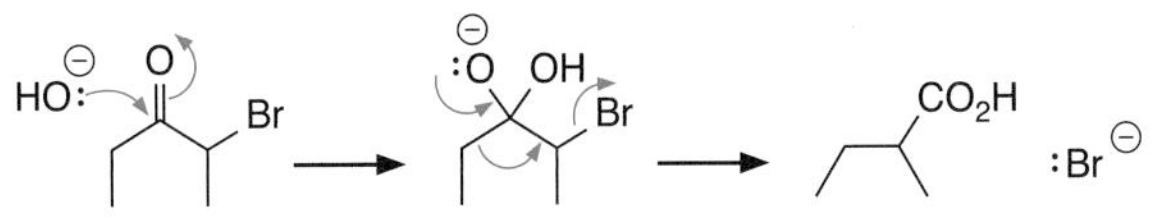

(Eq. 11.61)

11.10.2 Other Carbanion Rearrangements

Carbanions in general are prone to rearrange if a more stable structure can be obtained, but carbanion rearrangements are not as common as carbocation rearrangements. Eqs. 11.62 and 11.63 give two examples. The rearrangement generally does not occur unless the group moving is unsaturated. In each case the rearrangement yields a more delocalized, and hence, more stable carbanion. As shown in Eq. 11.64, phenyl group migration involves the intermediacy of a bridging phenyl ring, the carbanion analog of the phenonium ion. The bridging intermediate can be trapped with CO_2, giving a carboxylic acid product.

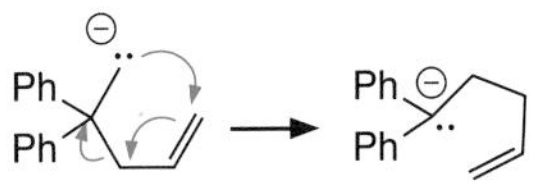

(Eq. 11.62)

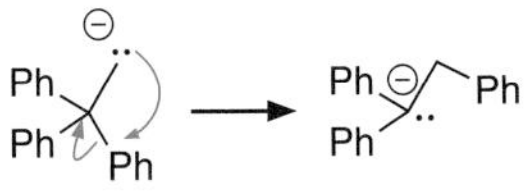

(Eq. 11.63)

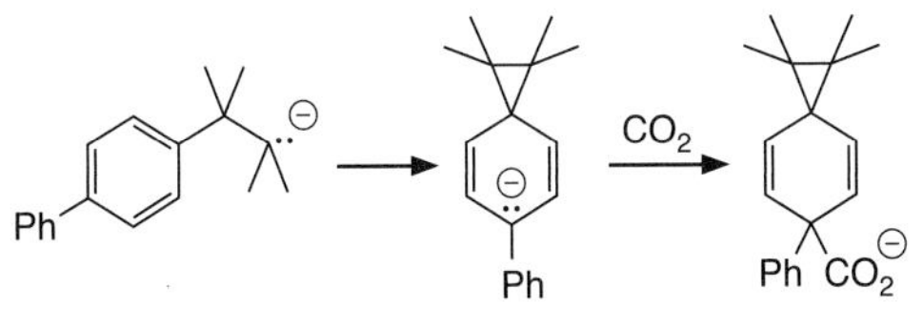

(Eq. 11.64)

11.11 Rearrangements Involving Radicals

Intramolecular rearrangements of free radicals are not nearly so common as those of carbocations. In fact, the most important rearrangements of free radicals are those associated with free radical "clocks", as discussed in Section 8.8.8 and listed in Table 8.7. Here we describe a few other rearrangements of radical systems.

11.11.1 Hydrogen Shifts

Hydrogen shifts in radicals are simply intramolecular variants of hydrogen atom abstraction reactions. The arrow pushing is straightforward, as shown in Eq. 11.65. The defining feature is the distance between the newly formed and the original radical centers, and we classify radical hydrogen shifts accordingly.

(Eq. 11.65)

The 1,2-hydrogen shift that is the hallmark of carbocation chemistry is completely absent in free radical chemistry. There are no documented cases at temperatures below 600 °C. The 1,3-hydrogen shift is rare, but it will occur if the reaction is substantially exothermic. One of the few unequivocal cases is shown in Eq. 11.66.

(Eq. 11.66)

Similarly, a 1,4-hydrogen shift is uncommon in solution and must be exothermic. An interesting case is shown in Eq. 11.67, involving one of the persistent radicals of Table 2.6. For the 1,4-shift shown, the kinetic isotope effect, k_H/k_D, is 80 at −30 °C and 13,000 at −105 °C. These huge values indicate that, at low temperatures, quantum mechanical tunneling is involved.

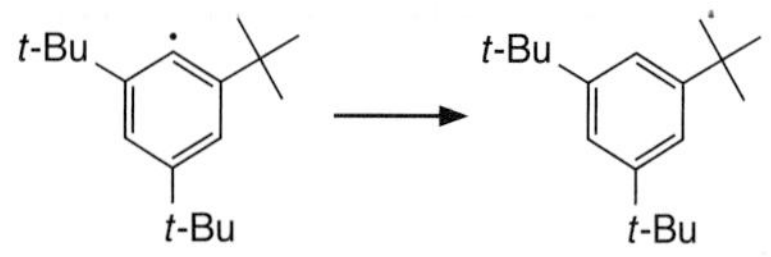

(Eq. 11.67)

Both 1,5- and 1,6-hydrogen shifts are common if the reaction is exothermic, especially the 1,5-shift. Eqs. 11.68 and 11.69 show two examples. A very important version of such a shift is the **Barton reaction**, in which an oxygen-based radical is generated by photolysis of a nitrite (Eq. 11.70). We have discussed before how oxygen radicals are especially reactive, so a 1,5-hydrogen abstraction will occur if it is available. Given the high reactivity of oxygen radicals, even abstraction geometries that are not ideal are viable.

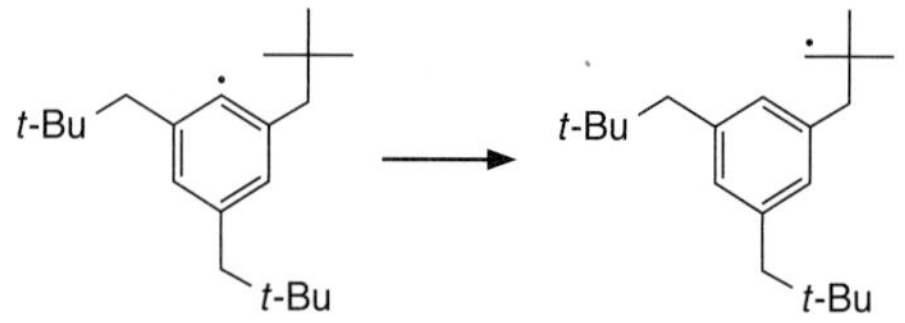

(Eq. 11.68)

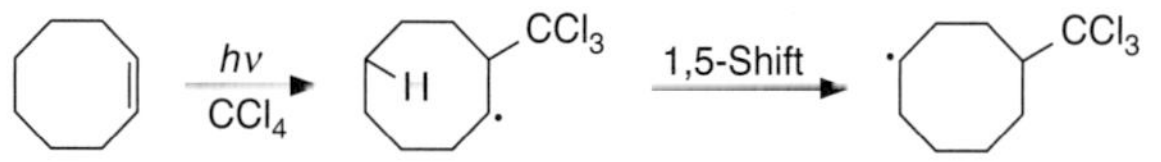

(Eq. 11.69)

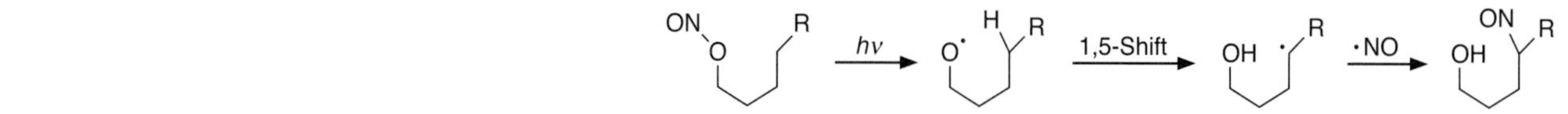

(Eq. 11.70)

Breslow exploited the Barton reaction to selectively functionalize a steroid. Eq. 11.71 shows the target steroid, and the goal was to functionalize just the C18 methyl (shown in color). The trick was to use the nearby OH to deliver a radical properly positioned to react with C18. After the carbon radical is formed, it reacts with the NO that was generated to ultimately produce an oxime, a useful functional group.

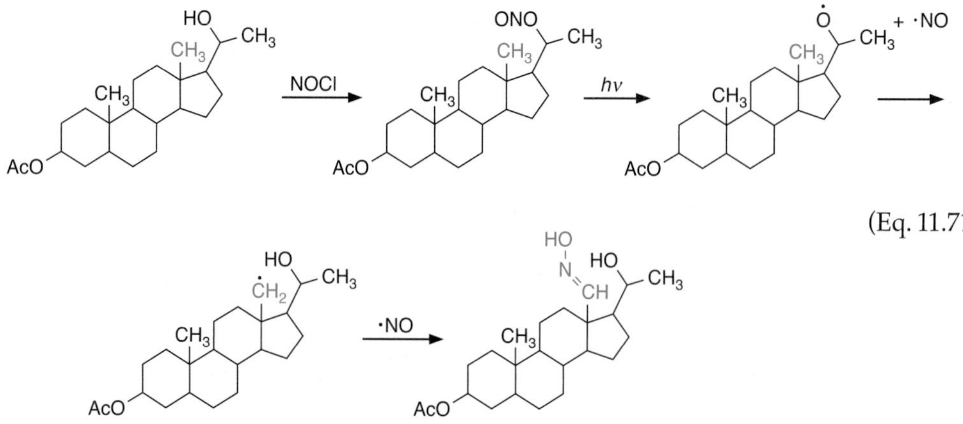

(Eq. 11.71)

11.11.2   Aryl and Vinyl Shifts

In contrast to hydrogen shifts, 1,2-shifts of aryl and vinyl groups are quite facile. Eq. 11.72 shows the arrow pushing for a 1,2-vinyl shift. The reaction is fundamentally radical addition to an olefin, and then the reverse reaction.

(Eq. 11.72)

Cyclohexadienyl radical
intermediate

A similar sequence is involved in the 1,2-aryl shift, and Eq. 11.73 shows a prototypical example. While we expect the reaction to be reversible, the case shown is driven to the right to form the more stable 3° radical. Such shifts presumably involve a cyclohexadienyl radical of the sort shown in the margin, but such a species is not detected and cannot be trapped.

(Eq. 11.73)

The 1,2-vinyl shift shown in Eq. 11.72 proceeds via a familiar structure, the cyclopropylcarbinyl radical we introduced in the context of free radical clocks (Table 8.7). In this case, the two species involved, the allylcarbinyl and cyclopropylcarbinyl radicals, are both discrete chemical entities that have been thoroughly characterized by EPR spectroscopy (Eq. 11.74). The equilibrium very strongly favors the ring-opened form, making the clock reaction, the opening of cyclopropylcarbinyl, essentially irreversible.

	log A	E_a (kcal/mol)	k (25 °C, s^{-1})
	12.5	5.9	1.3×10^8
	9.1	10.4	4.9×10^3

$$K (25\ °C) = \frac{1.3 \times 10^8}{4.9 \times 10^3} = 2.4 \times 10^4$$

$$\Delta G_{298K} = -RT \ln K = -6.03\ \text{kcal/mol}$$

(Eq. 11.74)

11.11.3 Ring-Opening Reactions

We've noted often that radical chemistry is strongly influenced by thermodynamics, and all the rearrangements we have shown follow that pattern. However, just because a thermodynamically favorable rearrangement is available to a system, that does not mean it will occur. As examples, consider the hypothetical ring-opening reactions of cyclopropyl and cyclobutyl radicals (Eqs. 11.75 and 11.76). Both reactions are exothermic, by about 23 kcal/mol for the cyclopropyl and roughly 5 kcal/mol for the cyclobutyl. However, these are not facile rearrangements, such that when cyclopropyl or cyclobutyl radicals are generated in a reaction sequence, the products generally still contain the small ring.

(Eq. 11.75)

(Eq. 11.76)

11.12   Rearrangements and Isomerizations Involving Biradicals

We've seen a great number of reactions in this chapter and Chapter 10 in which two species combine, resulting in an addition or substitution, or in which a "reagent" such as acid or base initiates a process. There are some reactions, however, that are inherently unimolecular, not requiring any kind of reagent. Chapter 16 describes unimolecular reactions initiated by the absorption of a photon. Chapter 15 considers unimolecular thermal reactions and rearrangements that involve a cyclic array of orbitals, called pericyclic reactions.

For the vast majority of organic molecules, however, heating does not lead to an elegant, concerted process involving a cyclic array of orbitals. By far the most common consequence of heating a molecule is simple bond cleavage. For polar molecules in polar solvents, heterolytic cleavage is possible. This is how S_N1 reactions begin. Now we consider a different scenario. For hydrocarbons and other nonpolar molecules, bond homolysis is much more likely than bond heterolysis. In the gas phase, homolysis will always be preferred over heterolysis, because a polar solvent is required to stabilize the ionic structures that result from heterolysis. For simple molecules like ethane or peroxides, **thermolysis**—thermally induced bond cleavage—simply leads to a pair of radicals that then go on to do conventional radical chemistry. However, for cyclic molecules, homolytic cleavage of a C–C bond produces a new species with two radical centers, a **biradical** (the term **diradical** is also used). This is a new class of reactive intermediates with a range of intrinsic properties, much like with carbocations or carbanions. In appropriate cases, the biradical can go on to do unique chemistry, with the net outcome being the production of a new product that is an isomer of the starting material. Such thermal rearrangements are the focus of this section. Keep in mind that some issues in biradical chemistry are the subject of ongoing debate. These are fleeting species that are difficult to characterize, and often very small energy differences are important. Our goal here is not to resolve these issues, but rather to familiarize you with the major systems and the types of questions involved.

Before discussing the chemistry of biradicals, we must consider spin state. Just as we saw with carbenes, the two weakly interacting electrons of a biradical can exist as either a low spin, singlet state or a high spin, triplet state. The two spin states will display noticeably different chemistries. Thermolysis generally produces singlet biradicals initially, because the starting material is a closed shell cyclic molecule that is necessarily a singlet. For simple biradicals, the singlet and triplet states are often close in energy and can interconvert fairly rapidly. At this point, we are not ready to discuss the subtle electronic features that favor the singlet or the triplet as the ground state. We need some more advanced electronic theory concepts, and these are developed in Section 14.5.6. At this point, we simply need to keep in mind that both the singlet and the triplet are viable for simple biradicals.

11.12.1   Electron Pushing Involving Biradicals

In Scheme 11.19 we show the general case for thermolysis of a cycloalkane to produce a biradical. For large rings this process is difficult to distinguish from homolysis in an acyclic structure. We produce two essentially independent radicals, which then go on to do conventional radical chemistry. What processes are available to a pair of radicals? One possibility is the recombination of the radicals, which in this case corresponds to a ring closure, reforming the original ring system. We will see this is an important process in biradical chemistry, making the original thermolysis readily reversible. Bond rotations in the biradical could lead to scrambling of any stereochemistry associated with the cleaved bond, and so the ring closure could lead to a stereoisomer of the initial reactant. Disproportionation is another common radical process, and with a cyclic biradical the analogous process involves intramolecular hydrogen abstraction to produce a terminal olefin.

Scheme 11.19
The possible pathways
available to a biradical.

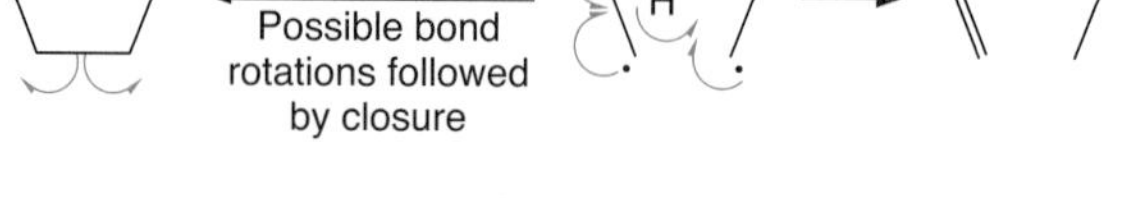

In smaller ring systems, other interesting processes become feasible. Eqs. 11.77 and 11.78 show the biradicals formed from C–C bond homolysis in cyclopropane and cyclobutane, respectively. Cyclopropane homolysis produces a 1,3-biradical, and the parent system is called **trimethylene**. Cyclobutane homolysis produces a 1,4-biradical, and the parent system is called **tetramethylene**. In both of these biradicals, the two radical centers are close enough to experience some interaction. As such, it is best to think of these systems not as two radicals, but as a single, unique reactive intermediate, a biradical, just as we consider a carbene to be a single reactive intermediate rather than two separate radicals.

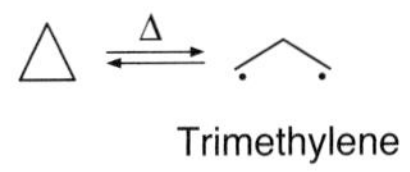

$$\triangle \;\rightleftharpoons\; \curlywedge \qquad\qquad (Eq.\ 11.77)$$

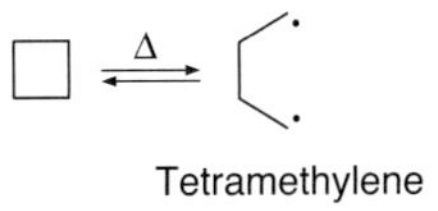

Trimethylene

$$\square \;\rightleftharpoons\; \text{biradical} \qquad\qquad (Eq.\ 11.78)$$

Tetramethylene

We are now ready to consider some specific examples of biradical chemistry. Rather than organize the following sections according to a specific reaction, as we have done so far in this chapter, we will use the biradicals themselves as the organizing theme. The reason for this is that often the same biradical can be generated from several different precursors, and so it is the unifying structure. We begin with tetramethylene, a prototypical biradical, and then consider the intriguing case of trimethylene.

11.12.2 Tetramethylene

Thermolysis of cyclobutanes leads to cis–trans isomerization of substituents, and the tetramethylene biradical is a sensible intermediate. As shown in Eq. 11.79, there are other options available for this structure. The arrow pushing is shown for reactions moving from left to right. Cleavage to two ethylenes is a quite facile process, and in simple systems it is faster than ring closure (E_a for the conversion of cyclobutane to ethylene is ~62 kcal/mol). The reverse process is also known; that is, under conditions of high temperature and pressure, olefins can dimerize to make a cyclobutane. Thus, a 1,4-biradical has three routes available: bond rotation, which scrambles stereochemistry; ring closure to make a cyclobutane; and cleavage to make two olefins. Much study has gone into determining the relative rate constants of these processes.

A powerful tool for the study of biradicals has been the use of cyclic 1,2-diazenes (azo compounds) as precursors (Eq. 11.80). Thermolysis or photolysis of a diazene generally leads to the extrusion of N_2 and the production of the biradical. Photolysis with a sensitizer (see Section 16.2.3 for a discussion of sensitized photolysis) allows a direct route to the triplet biradical. This provides the most convenient way to probe the reactivity differences of singlet and triplet biradicals.

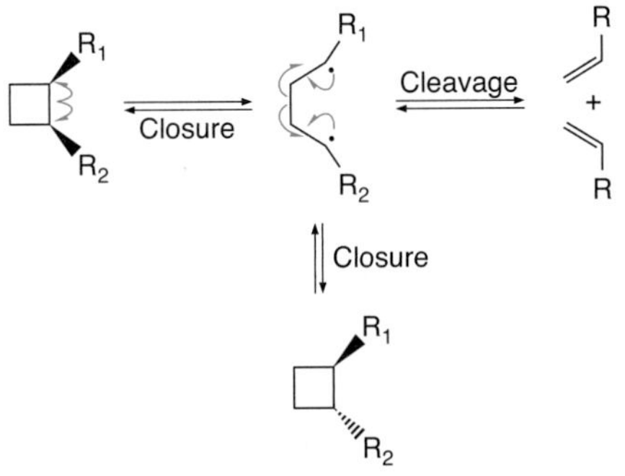

(Eq. 11.79)

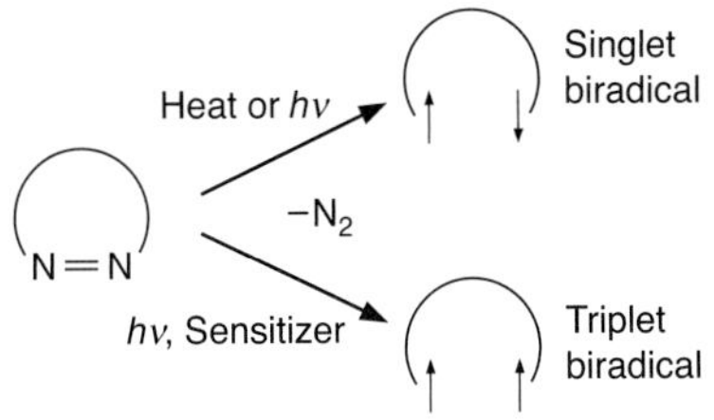

(Eq. 11.80)

As shown in Figure 11.14, a diazene with appropriate stereochemical labeling can provide information on the relative values of C–C bond rotation (k_{rot}) and cleavage (k_{cleav}) by examining appropriate product ratios. Such studies have been conducted by Dervan for differ-

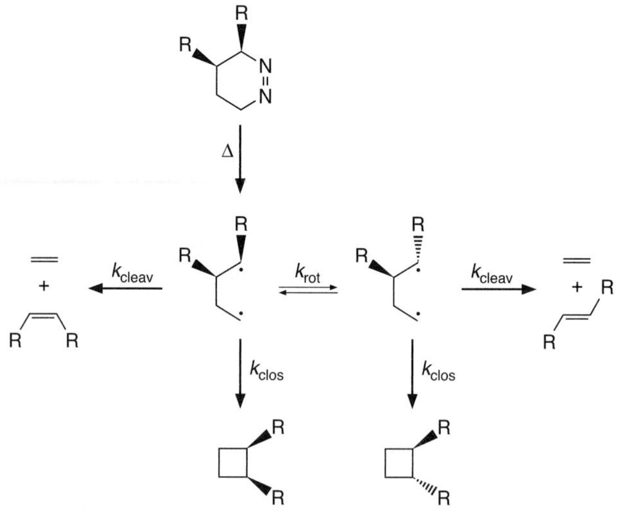

Figure 11.14
Pathways for the thermolysis of diazene precursors to tetramethylene
derivatives.

Table 11.13
Rotation vs. Cleavage
in Tetramethylenes
(Gas Phase, 425–440 °C)***

Tetramethylene	k_{rot}/k_{cleav}
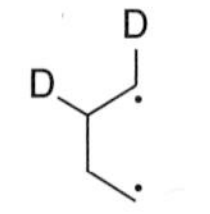	5.5
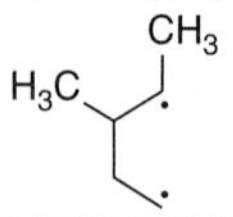	cis: 0.8 trans: 0.3
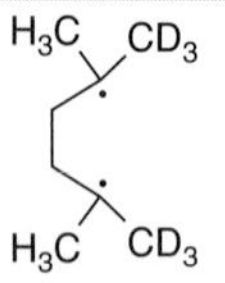	0.02

*Dervan, P. B., and Dougherty,
D. A. "Nonconjugated Diradicals
as Reactive Intermediates" in *Diradicals*, W. T. Borden (ed.) Wiley–
Interscience, New York, 1982.

ent substituents, and the results are summarized in Table 11.13. Interestingly, the value of k_{rot}/k_{cleav} varies as the substituent is changed. The trend makes sense, though. As the substitution at the radical center increases, k_{rot}/k_{cleav} decreases. This is almost certainly a consequence of k_{rot} slowing down, and it teaches us that bonds with more substituents rotate more slowly, an example of the ponderal effect.

The thermolysis studies of Figure 11.14 involve singlet biradicals. We noted above that we would expect different behaviors for singlet vs. triplet biradicals, and many types of experiments bear this out. A classic experiment is summarized in Eq. 11.81. The dimethyl-diethyldiazene was decomposed by three different methods: heat, direct photolysis, and sensitized photolysis. Thermolysis or direct photolysis produces a considerable preponderance of the cyclobutane with retention of stereochemistry. However, the triplet biradical produced by sensitized photolysis leads to an almost complete scrambling of the stereochemistry. The triplet biradical cannot directly close to cyclobutane, because that reaction is spin forbidden. This increases the lifetime of the biradical and allows much more bond rotation prior to closure. Cleavage products (olefins) are also observed in these reactions, and it is generally observed that more olefin product relative to cyclobutane is observed in the triplet manifold.

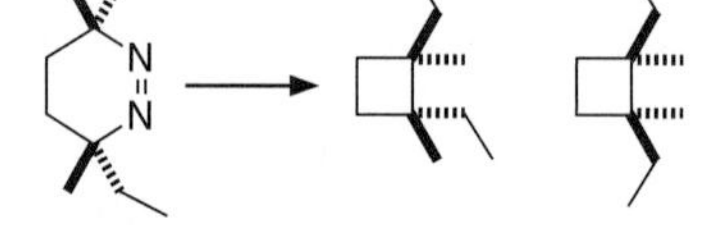

Method	A/B
Heat	49
hv	32
hv, Sensitized	2

(Eq. 11.81)

An important aspect of tetramethylene chemistry is that similar results are seen whether the biradical is prepared by cyclobutane thermolysis, by ethylene dimerization, or by diazene photolysis. Seeing the same product ratios from different modes of preparation is one of the most stringent tests for the existence of a common reactive intermediate.

11.12.3 Trimethylene

Superficially, the thermolysis of cyclopropanes resembles that of cyclobutanes. Several processes are observed, and a biradical is a reasonable intermediate. However, we will see some subtleties that will require considering other options.

The essential reactions that occur upon heating cyclopropane in the gas phase are shown in Eqs. 11.82 and 11.83, along with conventional electron pushing. The stereochemical isomerization and 1,2-shifts have been observed with a range of R groups.

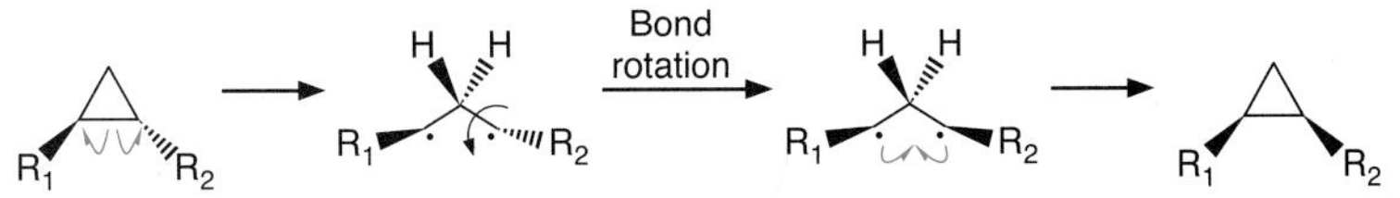

(Eq. 11.82)

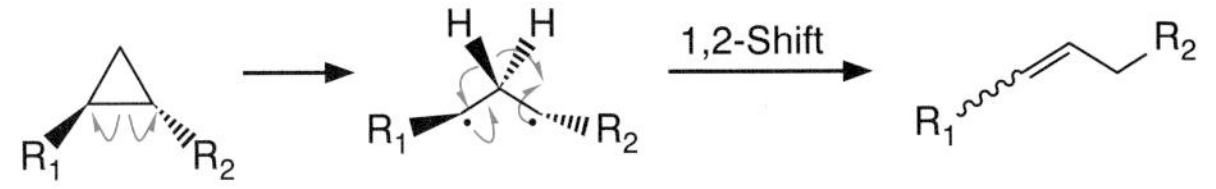

(Eq. 11.83)

The stereochemical isomerization implies a competition between ring closure and bond rotation in the biradical reactive intermediate. As with tetramethylene, a competition is supported by studies of appropriately substituted diazenes which, on thermolysis, lose N_2 and presumably produce the biradical (Eq. 11.84). However, in an early indication that things are not as simple as with tetramethylene, the diazene experiments show a "cross-over" effect. The cis diazene preferably produces trans cyclopropane, and vice versa.

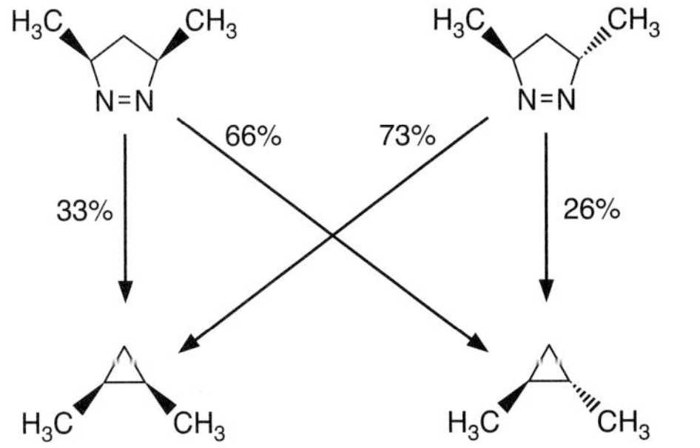

(Eq. 11.84)

While the biradical process is appealing, proving that this is indeed the mechanism is not so straightforward. Is a singlet, 1,3-biradical a viable reactive intermediate with a lifetime? That is, can you really have two carbon radicals that are very close to each other in space, with a roughly 60 kcal/mol exothermic process available that requires very little atomic movement? If the biradical does exist, is its lifetime long enough for bond rotation to occur, leading to a scrambling of the stereochemistry? Is the 1,2-hydrogen shift proposed in alkene formation a viable reaction, given that 1,2-hydrogen shifts do not occur in monoradicals? Does a common intermediate lie on the paths for stereoisomerization and alkene formation?

Thermochemistry is a useful tool in considering thermal rearrangements. Recall the group increments methodology introduced in Chapter 2. The heat of formation of cyclopropane is +12.7 kcal/mol. Using the data of Table 2.7, we can estimate that the heat of formation of trimethylene is $2(35.82) + (-4.95) = 66.7$ kcal/mol. The other useful type of information is the set of activation energies for rearrangement. These are 64.2 kcal/mol for cis–trans isomerization and 65.6 kcal/mol for propene formation. These lead, by simple addition, to heats of formation of the transition states for isomerization and propene formation of 76.9 kcal/mol and 78.3 kcal/mol, respectively. These data can be used to assemble a potential energy surface for cyclopropane thermolysis, as shown in Figure 11.15.

The key feature of this surface is the clear prediction that the heats of formation of the two transition states lie above the heat of formation of the trimethylene biradical. This

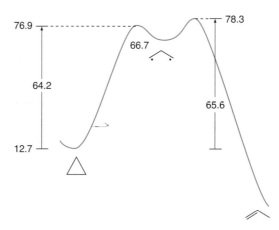

Figure 11.15
Hypothetical potential energy surface for cyclopropane isomerization based on the group increments method.

means that the biradical is a viable structure along the reaction pathway. In fact, the trimethylene biradical is predicted to lie in a significant potential well, with barriers to escape on the order of 10 kcal/mol (referred to as a **Benson barrier** since it derives from group increments). Since we expect rotation barriers around the C–C bonds of trimethylene to be much less than 10 kcal/mol, the model predicts loss of stereochemistry if trimethylene is formed.

The energy surface in Figure 11.15 comes with two large caveats, one methodological and one conceptual. First, the group increments of Table 2.7 were developed in the 1970s based on the best available data. We present these Benson increments because they are the most comprehensive set of self-consistent data. However, later experimental determinations of radical heats of formation suggest that some revision is necessary. These revisions, although not universally accepted, lead to a reduction in the Benson barrier.

The conceptual problem stems from a fundamental feature of the group increments method. By using the value from Table 2.7 for a $CH_2^\bullet$ group, we are assuming that we can use the increments associated with simple monoradicals to study biradicals. Implicitly, we are assuming there is no interaction between the radical centers in trimethylene. While this is no doubt an excellent assumption for large-ring biradicals, the whole point of trimethylene is that the radical centers are close and that they certainly interact. This must affect the heat of formation of the molecule, and so the group increments method may not be applicable.

We could now reproduce a lengthy and ongoing literature debate on the best way to refine the potential energy surface of Figure 11.15 and the reality of the Benson barrier, but that would miss the point. It will *always* be true that thermochemical approaches can provide guidelines only; there will always be uncertainties, especially for reactive intermediates. In the present case, we can say that the group increments method does not rule out trimethylene as a viable reactive intermediate on the path of cyclopropane isomerization. It certainly does not prove its involvement. In other cases, group increments can provide a fairly strong case that a proposed reactive intermediate is simply too high in energy to be seriously considered; we will see an example of this in Chapter 15. Thus, thermochemistry can provide fairly strong evidence ruling out a particular reactive intermediate, but only permissive evidence for the involvement of a reactive intermediate.

Given its small size, trimethylene was one of the first reactive intermediates to be explored using modern quantum mechanical computational techniques (Chapter 14). The calculations established that there was indeed a significant interaction between the two radical centers, giving good reason to be suspicious of the thermochemical data. As shown in Figure 11.16 **A**, in appropriate geometries the $\pi(CH_2)$ orbital of the central methylene can be aligned

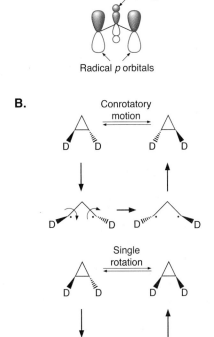

Figure 11.16
A. Mixing of the $\pi(CH_2)$ orbital of C2 of trimethylene with the radical centers.
B. Stereochemical test of coupled, conrotatory motions vs. single-rotation mechanisms in cyclopropane.

with the p orbitals on the radical centers. This results in an orbital mixing phenomenon termed **through-bond coupling**, which we will discuss in detail in Chapter 14. For the present purposes, we simply note that this mixing of the $\pi(CH_2)$ orbital and the radical p orbitals led Hoffmann to an interesting theoretical prediction. Most chemists would expect the cis–trans isomerization to arise from simple, independent rotations around the C1–C2 and C2–C3 bonds of trimethylene, leading to scrambling of stereochemistry. Theory suggested, however, that because of the through-bond coupling, such rotations in trimethylene should be coupled, with a bias toward both bond rotations occurring in the same direction, called a **conrotatory** motion. This is illustrated in Figure 11.16 **B**, using deuterium substitution as a stereochemical label. If the bond rotations that lead to the product from trimethylene are conrotatory, by microscropic reversibility, the bond rotations that create trimethylene from cyclopropane must also be conrotatory. This is not at all the behavior we saw with tetramethylene. If the Hoffmann prediction were true, trimethylene would have unique properties, associated specifically with the fact that it is a 1,3-biradical.

An elegant test of the Hoffmann prediction was executed by Berson and co-workers. They prepared the *trans*-1,2-dideuterocyclopropane of Figure 11.16 **B** *enantiomerically enriched*—this simple molecule is chiral by virtue of isotopic substitution. If, after ring opening at the C1–C2 bond, there is a coupled, conrotatory rotation about the C–C bonds, we will directly produce the enantiomer of the original trans starting material. Alternatively, if the C–C bonds of the biradical rotate independently, we expect formation of the *cis-*

dideuterocyclopropane (an achiral molecule) to compete with the racemization of the trans form of the reactant. The experiment is a challenging one, but the result was clear: racemization of the trans reactant was measurably faster than cis–trans isomerization. This supports coupled conrotatory rotations in the stereoisomerization of cyclopropane, with no trimethylene intermediate.

The coupled rotation seen in the Berson experiment is fascinating. However, it is pretty much exclusive to the parent system. When radical-stabilizing substituents such as phenyl or cyano are placed on the ring, simple, independent bond rotations are observed, much as in tetramethylene.

For the parent cyclopropane, the path to propene must be different than the concerted double rotation. For substituted cyclopropanes, however, it seems reasonable to propose that propene products arise from a 1,2-hydrogen shift of the biradical. We have noted earlier that 1,2-shifts are essentially never seen in simple radicals. Thus, the substantial driving force of forming a new C–C π bond makes this usually unfavorable process feasible.

Several simple 1,3-biradicals have been directly observed spectroscopically. Photolysis of 2,3-diazabicyclo[2.2.1]heptene in a frozen solvent at temperatures below 10 K produced a simple biradical, 1,3-cyclopentanediyl, the triplet state of which was directly observed by Closs using EPR spectroscopy (Eq. 11.85). Similarly, a series of substituted 1,3-cyclobutanediyls was produced (Eq. 11.86), showing that substituent effects on simple, triplet biradicals mirror those of conventional free radicals.

$$\text{(Eq. 11.85)}$$

1,3-Cyclopentanediyl

$$\text{(Eq. 11.86)}$$

An intriguing feature of the reactions in Eqs. 11.85 and 11.86 is that, even at cryogenic temperatures, ring closure to the bicycloalkane could be observed. The reaction is spin forbidden, converting a triplet biradical to a singlet bicyclic product. The spin forbiddenness shows up in the Arrhenius pre-exponential terms, with log A values in the range of 10^8 s^{-1}. This is very low for a unimolecular reaction. The thermal barrier to ring closure is on the order of 1–2 kcal/mol for both systems. At 5 K, a 1 kcal/mol barrier is essentially insurmountable. This leads to the conclusion that quantum mechanical tunneling is involved in these ring-closure reactions. This is a novel result, because tunneling is normally associated with the movement of light particles like hydrogens or electrons; **heavy-atom tunneling** of the sort that must be involved in these ring-closure reactions is much less common.

In recent years, several other reactions have been observed to exhibit heavy-atom tunneling. An intriguing example is the interconversion of the equivalent rectangular forms of singlet cyclobutadiene (Eq. 11.87—see Chapter 14 for a detailed discussion of cyclobutadiene). In 2003, several groups described the rearrangement of the fluorocarbene of Eq. 11.88 at 8 K. Through a combination of experiment and theory, the workers concluded that the reaction occurred faster by a factor of 10^{152} over what would be expected from a conventional, thermally activated path. Thus, tunneling can strongly influence the rates of reactions at cryogenic temperatures!

$$\text{(Eq. 11.87)}$$

$$\text{(Eq. 11.88)}$$

Going Deeper

Femtochemistry and Singlet Biradicals

We have described several examples of the direct observation of triplet biradicals. However, the fleeting nature of singlet biradicals has made it very difficult to obtain direct information on their structure and reactivity. With the advent of femtosecond lasers, however, this has become possible. For example, Zewail and co-workers directly characterized simple tetramethylene and trimethylene biradicals. The approach involved a molecular beam of either cyclopentanone or cyclobutanone with crossed laser beams. Photolysis led to extrusion of CO and formation of the biradical. For tetramethylene, a clear biradical structure is seen, with a lifetime of 700 fs. However, for trimethylene, no such "long-lived" structure was observed. Instead, a transient with a lifetime of only 120 fs was seen. This is so short that it is really not meaningful to distinguish this transient from a transition state. These time-resolved results are considered to be consistent with the more conventional thermolysis studies. Tetramethylene is a well behaved biradical, a true reactive intermediate. However, trimethylene is better thought of as a transition state.

Pederson, S., Herek, J. L., and Zewail, A. H. "The Validity of the 'Diradical' Hypothesis: Direct Femtosecond Studies of the Transition-State Structures." *Science*, **266**, 1359–1364 (1994).

11.12.4 Trimethylenemethane

Methylenecyclopropane derivatives are prone to thermal rearrangement, as shown in Eq. 11.89. Along with cis–trans isomerization (as in cyclopropane), a structural rearrangement is observed. The rational mechanistic proposal is that the C2–C3 bond cleaves to produce the biradical trimethylenemethane (TMM; Eq. 11.90). This biradical is fundamentally different from the biradicals we have seen so far. First, TMM is substantially resonance stabilized, with three equivalent resonance forms, such that the structure has true, three-fold symmetry. Ring closure can occur in any of three equivalent ways, leading to the structural isomerization in Eq. 11.89. Second, the two unpaired electrons are part of the same π system, and so they interact with each other strongly. In this case that interaction leads to a *triplet* ground state, and the preference is quite strong (~14 kcal/mol; see Section 14.5.6). As a result, TMM shows two well-defined reactive modes, one for the singlet and one for the triplet.

(Eq. 11.89)

Trimethylenemethane (TMM)

(Eq. 11.90)

TMM was the first biradical to be directly characterized by low temperature, matrix isolation EPR in pioneering work by Dowd (Eq. 11.91). The spectroscopy confirmed the three-fold symmetry and the triplet ground state that theory had predicted, and allowed kinetic studies under matrix isolation conditions.

(Eq. 11.91)

When the diazene in Eq. 11.91 is employed to generate TMM in solution, irreversible ring closure to methylenecyclopropane is too rapid to allow characterization of the biradical. In order to increase the lifetime of TMM and make it more amenable to direct characterization, Berson introduced the modification shown in Eq. 11.92. By incorporating the TMM into a five-membered ring, the two possible ring-closure paths produce *very* highly strained molecules.

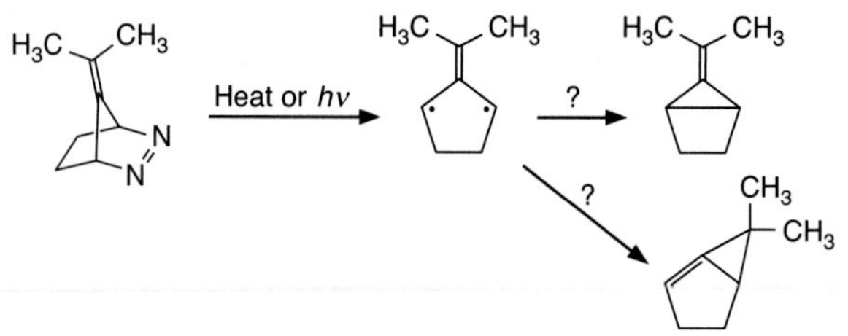

(Eq. 11.92)

As shown in Figure 11.17, the ring constraints were successful in suppressing the ring-closure reaction, such that bimolecular chemistry is possible. A **cascade mechanism** is evident, which means structures of progressively lower energy are produced. The singlet diazene produces the singlet TMM, which produces the triplet TMM, that goes on to dimers. In the presence of appropriate trapping agents, the cascade can be diverted. An interesting feature is the differing trapping reactivities of the singlet and triplet biradicals, the former producing bicyclo[3.3.0] products, and the latter producing bicyclo[2.2.1] products. This is one of the clearest demonstrations of the different natures of singlet and triplet biradicals.

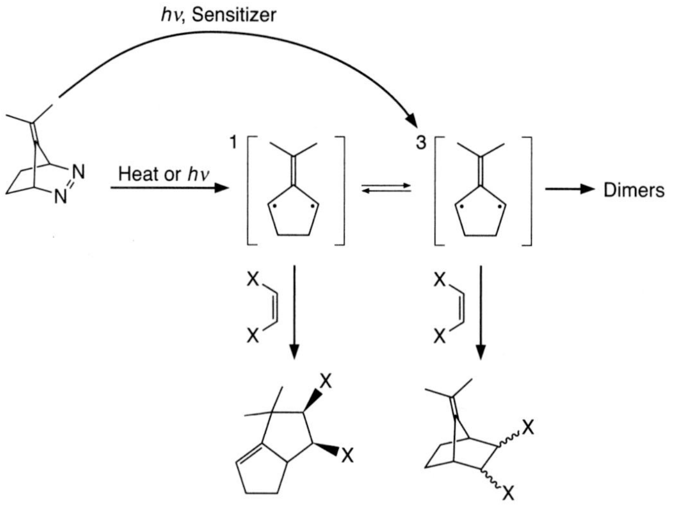

Figure 11.17
The cascade mechanism associated with the chemistry of the "Berson TMM".

What about the ring-closed forms of Eq. 11.92? Are they actually formed? Further study has characterized derivatives of each. From a combination of mechanistic and computational studies, it has been concluded that the ring openings of both of the ring-closed products shown in Eq. 11.92 to triplet TMM are *exothermic*, but especially so for the bicyclo[3.1.0]-hexene derivative. Thus, we have a system that is more stable when it is missing a C–C bond than when the bond is present (Eq. 11.93). Such structures have a *negative bond dissociation energy*! This observation is a consequence of the special stabilization of the TMM biradical, and the high strain of the ring-closed systems.

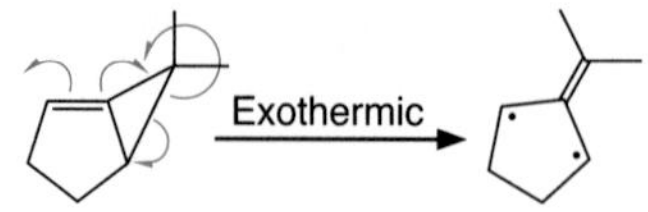

(Eq. 11.93)

Summary and Outlook

The 20th century was a triumphant one for mechanistic organic chemistry. The notions of a reaction mechanism and of a reactive intermediate were conceived, and their importance to chemistry unambiguously established. The fundamental mechanistic paradigms—E2, E1, S_N2, S_N1, addition–elimination, etc.—were established. In the second half of the century, all the key reactive intermediates that had been proposed to explain various kinetic and stereochemical experiments were directly characterized through stable media and fast kinetic techniques. Even some reaction transition states were directly observed! In addition, mechanistic enzymology provided many examples of nature's use of the mechanistic paradigms developed on small organic molecules. Thus, the development of mechanistic organic chemistry stands as one of the great scientific achievements of the 20th century.

So where are we now? While novel structures and new reactions continue to be developed, it seems unlikely that fundamentally new mechanistic pathways for organic molecules will emerge. New organic reactions will likely follow some combination of the various fundamental mechanisms discussed herein.

A recurrent theme of this chapter and Chapter 10 has been that the traditional characterizations of a reaction as either E2 or E1, or as S_N2 or S_N1, are fuzzy. We see a continuum of mechanisms, where many (if not most) systems are best thought of as intermediate cases that lie more toward one end of the mechanistic spectrum than the other, or even lie directly in the middle. The limitless diversity of organic structures and reactions ensures that almost all mechanistic nuances are feasible. Thus, each new case you confront in your research must be considered individually. Simply looking at a reaction and declaring it to be S_N2, or any other traditional mechanism, seems risky. Such declarations, instead, should be your starting points for analysis, analogous to how the standard electron pushing was the starting point for each mechanism given herein. Because the tools for studying mechanisms have been covered, and the nuances explained, you should now know what to look for in a possible mechanism and how to test that prediction. Nothing is more enabling if a chemist wants to control or modulate a chemical reaction than knowledge of the mechanism. As such, it seems certain that the tools and concepts developed here and in the last several chapters will forever be a key component of organic chemical research.

We now have the essentials of mechanistic organic chemistry in hand. We are ready to move on to some more advanced topics, beginning with organometallic chemistry, followed by the chemistry and materials properties of polymers.

Exercises

1. The rate of nucleophilic attack in an S_N2 reaction often depends upon the counterion of the nucleophile. For example, when the counterion is Li^+, the order of reactivity for the reaction of halogen nucleophiles with methyl brosylate is $I^- > Br^- > Cl^-$. The order of reactivity is reversed when the counterion is tetrabutylammonium. Why?

2. The secondary isotope effect for the S_N2 reaction of bromide anion with methyl iodide in the gas phase is 0.76. Why is there any isotope effect at all, and why is it inverse?

3. The S_N1 solvolysis of one of the following alkyl tosylates is 10^5 times faster than the other. Which is faster and why?

4. During the S_N1 solvolysis of allyl chloride, some scrambling of the position of the chloride often occurs. The extent of scrambling has no dependence upon the concentration of added chloride salts. Why is this?

5. Using a More O'Ferrall–Jencks plot, analyze how the position of the S_N2 transition state changes for the reaction of *n*-butylchloride with azide under the following conditions.
a. The nucleophile azide is changed to phenoxide.
b. The leaving group chloride is changed to iodide.

6. The secondary isotope effects for the solvolyses of the following compounds are listed. Explain why there is such a difference between the two examples.

$$\frac{k_H}{k_D} = 1.44 \qquad \frac{k_H}{k_D} = 1.10$$

$$\left(OBs = O\text{-}SO_2\text{---}\text{---}\!\!Br \right)$$

7. The S_N1 solvolysis of one of the following alkenyl bromides is approximately 10^5 times faster than the other. Which is faster and why?

8. What is the mechanism of the following substitution reaction? Show all electron pushing.

9. Why is there a difference in the isotope effects in the solvolysis reactions of the following alkenyl tosylates? Explain why the second example has the larger isotope effect.

$$\frac{k_H}{k_D} = 1.20$$

$$\frac{k_H}{k_D} = 1.31$$

10. In the solvolysis of stereogenic structures, stereochemical scrambling is often faster than product formation. To test whether such scrambling occurs within a contact ion pair or via reaction of the carbenium ion with free leaving group, radiolabeled leaving groups can be added to the reaction. What does it mean when the reactant undergoes stereochemical scrambling faster than incorporation of a radiolabeled leaving group?

11. Goering studied the solvolysis of structures similar to the following. He followed both the racemization of the stereogenic center in the reactant, and the scrambling of the isotopically-labeled oxygen in the reactant between the ether and carbonyl positions. The scrambling of the isotopic label was faster than the scrambling of the stereochemistry during solvolysis reactions. What step in the solvolysis mechanism does measuring the rate for isotopic scrambling give a rate constant for, and mathematically how is the rate of this scrambling related to the rate constant for this particular step?

12. We stated in this chapter that the observation of a rearrangement reaction in what appears to be an S_N2 reaction is evidence for an SET mechanism. But is the opposite true? For example, the reaction of 6-bromo-1-hexene with $NaSN(CH_3)_2$ sometimes gives a cyclized product. What is this product? How do you explain the fact that this reaction gives no cyclization in THF, but in 1:1 THF/pentane the cyclized product does form?

13. One of the following structures undergoes acetolysis 10^{11} times faster than the other. Which one is faster and why? What stereochemistry in the products do you expect for both these reactants?

14. The following graph shows a correlation between the logarithm of the rate constant for the solvolysis of the various reactants in 80% aqueous acetone with added azide (y axis) to the logarithm of the relative rates of substitution by azide and water (x axis). Does this graph follow the reactivity–selectivity principle discussed in Chapter 7? Also, comment upon the factors that stabilize an intervening carbocation in each reactant shown in this graph (Ad = adamantyl).

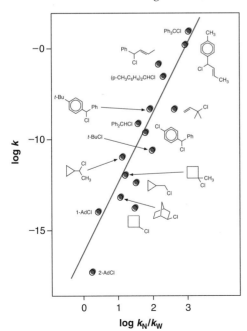

Raber, D. J., Harris, J. M., and Schleyer, P. v. R. "The Use of Added Sodium Azide as a Mechanistic Probe for Solvolysis Reactions." *J. Am. Chem. Soc.*, **93**, 4821 (1971).

15. The scrambling of hydrogens in *sec*-butyl carbocation ($CH_3CH^+CH_2CH_3$) between the two secondary carbons occurs at –110 °C. At –40 °C or slightly lower temperatures, a radio-label (^{14}C) originally at the cationic carbon becomes statistically distributed over all four carbons. Furthermore, at –40 °C all nine hydrogens in the structure become NMR equivalent. At temperatures above –40 °C, the cation rearranges to *t*-butyl carbocation. Give mechanisms for all the hydrogen and carbon shifts described here.

16. Show which carbons of the bicyclobutonium ion (Section 11.5.14) need to be attacked by a nucleophile to give homoallyl, cyclobutyl, and cyclopropyl carbinyl derivatives as the final products.

17. Calculate the exothermicity or endothermicity of the following hydrogen abstraction reactions involved in free radical halogenation of alkanes (the BDEs of HF, HCl, HBr, and HI are 136, 103, 87, and 71 kcal/mol, respectively). Which reaction would you predict is the least selective?

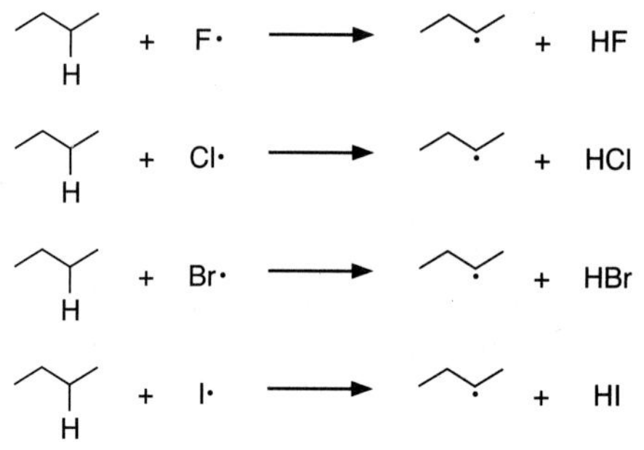

18. Explain why placement of an oxygen in the chlorocyclooctane ring slows down the solvolysis reaction in one case and speeds it up in the other.

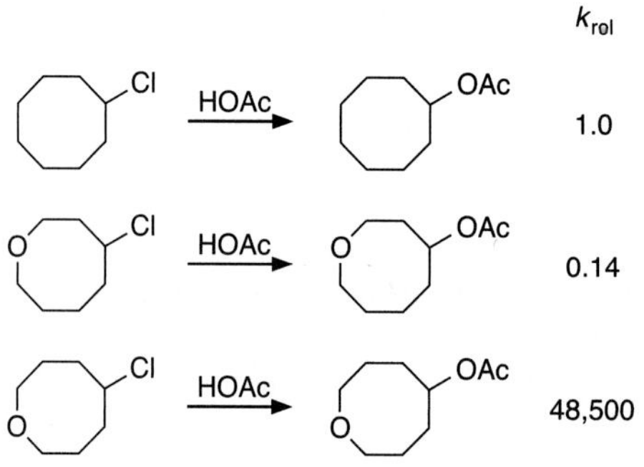

19. Below are shown relative rates for the solvolysis of haloalcohols and relative rates for solvolysis of haloamines, both in water. Explain the trends seen.

2-Chloroethanol	2000
3-Chloro-1-propanol	1
4-Chloro-1-butanol	5700
5-Chloro-1-hexanol	20
1-Amino-2-bromoethane	72
1-Amino-3-bromopropane	1
1-Amino-4-bromobutane	60000
1-Amino-5-bromopentane	1000
1-Amino-6-bromohexane	2

20. What is the flaw in the experimental design in using a pinacol structure such as that shown below for determining the relative migratory aptitudes of R_1 and R_2?

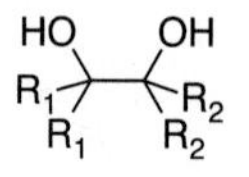

21. The thermal Criegee rearrangement is shown below. Write a mechanism for this reaction with all electron pushing. Explain the fact that when the carbonyl oxygen of the perester is labeled with ^{18}O, it is almost entirely retained in the product carbonyl oxygen position. Propose another experiment that might help support the mechanism you wrote.

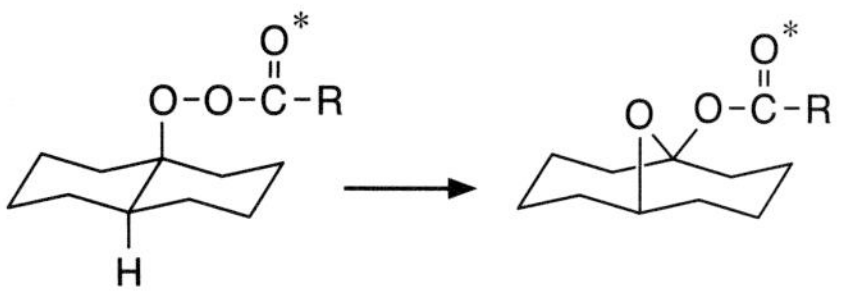

22. Explain the products from the following reactions. Why is only one product found in each case?

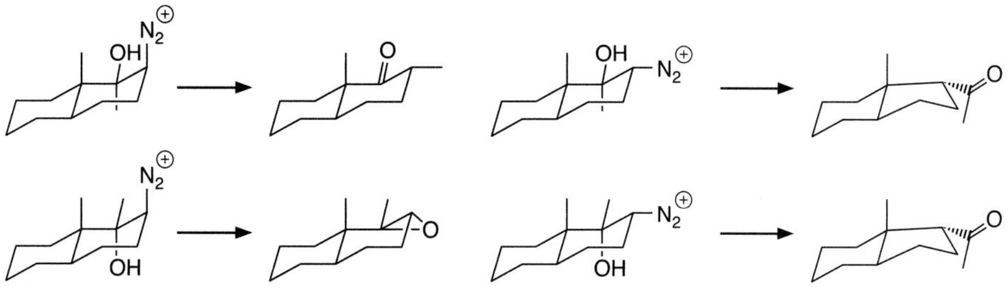

23. The benzilic acid rearrangement can be catalyzed with cyanide, and the conditions do not require the addition of a strong base. Write a mechanism for this catalyzed reaction with the appropriate electron pushing.

24. The following are examples of several rearrangements. Rearrangements are excellent reactions for applying multiple experimental tools given in this book to explore the mechanisms. Propose mechanisms for these reactions along with electron pushing. Describe isotope scrambling experiments to test where the oxygens or nitrogens in the products arise from in each case. In addition, where appropriate, propose Hammett plots to test for intermediates and changes in charge that occur in the rate-determining steps. Finally, for appropriate cases, propose cross-over experiments to determine if the reactions are intramolecular or intermolecular.

Neber rearrangement

A.

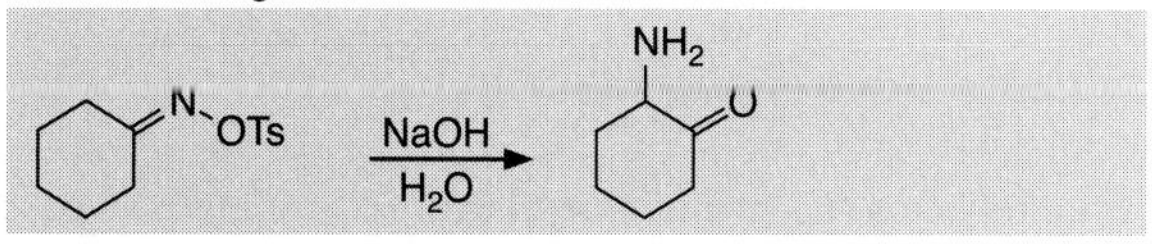

Curtius rearrangement

B.

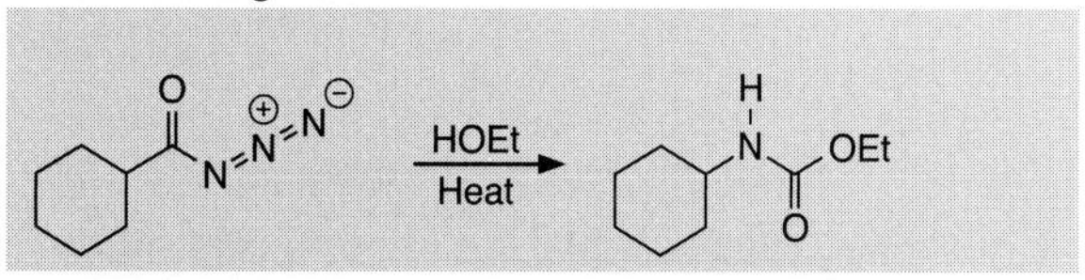

Lossen rearrangement

C.

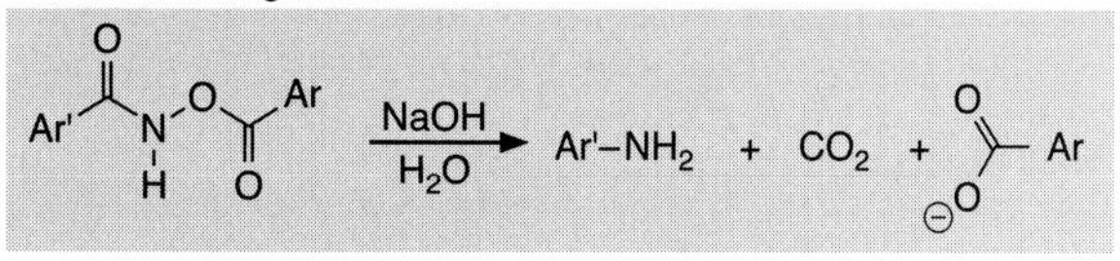

Sommelet–Hauser rearrangement

D.

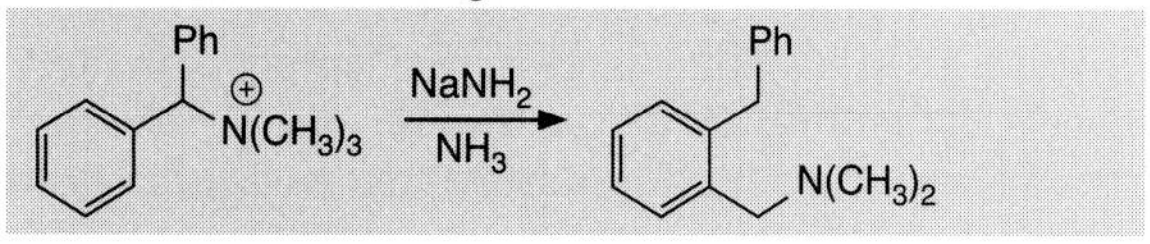

Smiles rearrangement

E.

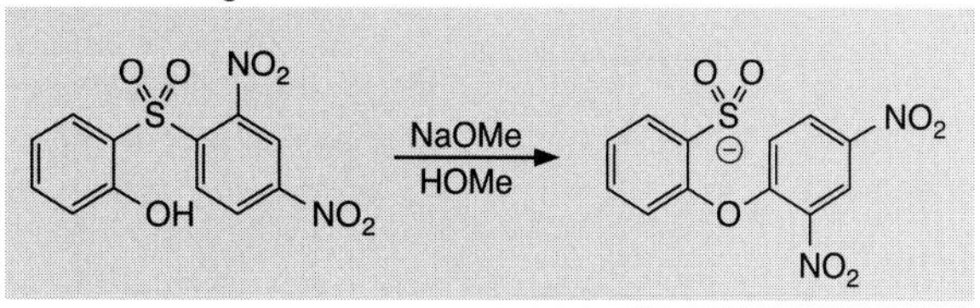

Wallach rearrangement

F.

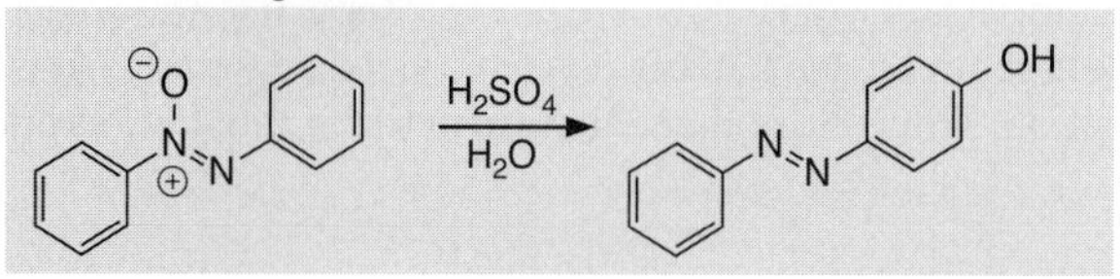

25. Explain the products and stereochemistry of the following two reactions.

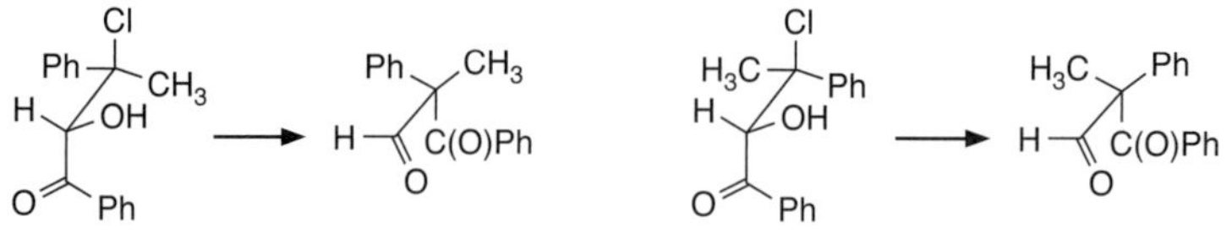

26. How well do group increments reproduce the thermodynamics for the cyclopropylcarbinyl–allylcarbinyl system of Eq. 11.74?

27. Convince yourself that at 5 K a 1 kcal/mol barrier is essentially insurmountable.

28. Perform an analysis of the thermochemistry for cyclobutane cleavage to two ethylenes. Does it predict that the tetramethylene biradical (Section 11.12.2) is a viable reactive intermediate?

29. Although the slopes of the lines in Figure 11.5 are all the same, the vertical placements of the lines are not what one would predict. For a general series of C, N, and O nucleophiles, what order of reactivity would you predict? What factors could change this order of reactivity?

30. Consider the dimethyl diazene results of Eq. 11.84. Are these results consistent with the proposal of a preferential conrotatory motion of the trimethylene biradical? Can you propose another explanation?

31. Briefly answer the following questions about aldol reactions.
   a. Why are equilibrium constants for the formation of β-hydroxyl carbonyl compounds generally higher for aldehydes than ketones?
   b. Given that the equilibrium constant for the formation of the β-hydroxyl compound from acetone is only 0.02, why does it sometimes complicate reactions that are performed in acetone?
   c. Why is the problem from part b exacerbated at elevated temperatures?

32. Isotope effects for β-hydrogens in S_N1 reactions are commonly in the range of 1.1 per deuterium substitution. How does this isotope effect arise?

33. Give a mechanism for the conversion shown. Predict an activation energy for the process.

34. Why is the autoxidation of diethyl ether so facile? Why is it even more facile for diisopropyl ether?

35. A common reaction that has shown some synthetic utility is the vinylcyclopropane to cyclopentene rearrangement. E_a for the parent system is 49.6 kcal/mol. Write a mechanism for this reaction, propose an intermediate, and include appropriate electron pushing. How might you test your mechanistic postulate?

36. Spiropentanes undergo a thermal rearrangement to methylenecyclobutanes; E_a for the parent system is 54.5 kcal/mol. In appropriately substituted systems cis–trans isomerization of substituents on the spiropentane is seen. Furthermore, heating a substituted methylenecyclobutane can lead to scrambling of any substituents on the ring. Write a mechanism for this reaction, propose an intermediate, and include appropriate electron pushing. How might you test your mechanistic postulate?

37. Make a Hammett plot out of Table 11.12. What is the best substituent constant to use, and how sensitive is the pinacol rearrangement to the substituents?

38. The isotope effect on the hydrogen atom abstraction from *t*-butylthiol by various radicals has been studied. Using various radicals to do the abstraction, the enthalpies of abstraction varied between –24 and 18 kcal/mol. The largest isotope effect was found when the enthalpy of abstraction was as close to zero as possible, in this case 5 kcal/mol. Why is this true, and what does this tell you about the reaction?

39. We have already analyzed the following reaction, where a small percentage of stereoselectivity is found. If the enantiomer of the reactant gives the same products, with the ratio 55 to 45, is the reaction stereospecific? Explain.

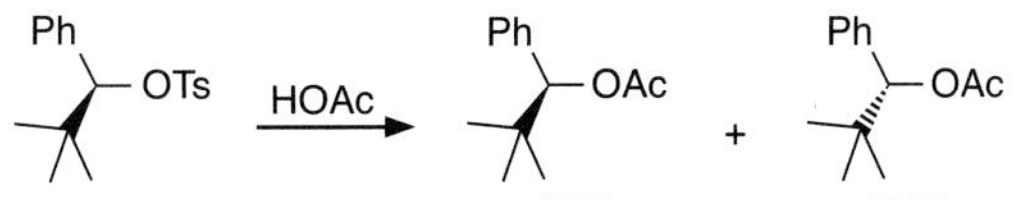

40. When *trans*-2-deuterio-4-*t*-butylcyclohexanone is treated with a strong base, only the deuterium on C2 can be deprotonated by the base. Explain why.

41. Provide a mechanism for the following reaction, showing the electron pushing. Rationalize the observed activation parameters: $\Delta H^{\ddagger} = 23.8$ kcal/mol and $\Delta S^{\ddagger} = -6.0$ eu.

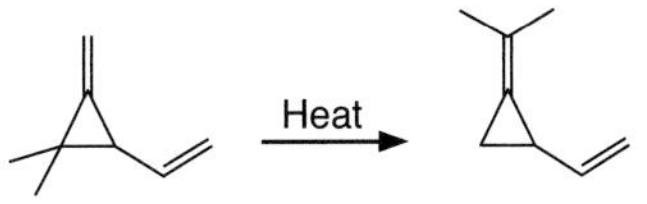

42. Draw two different alkyl bromides that might efficiently produce the following carbonium ion.

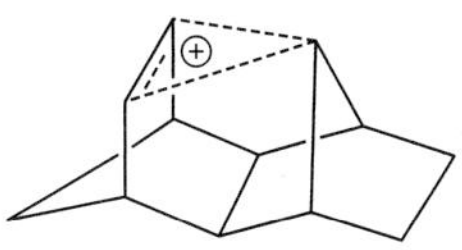

43. Consider the alkylation of acetone shown using methoxide or diisopropylamide as the base. The pK_a of acetone is 19, while the pK_a values of methanol and diisopropylamine are 16 and 36, respectively. Discuss how the selection of the base affects the strategy behind the reaction. It is possible that either base will work?

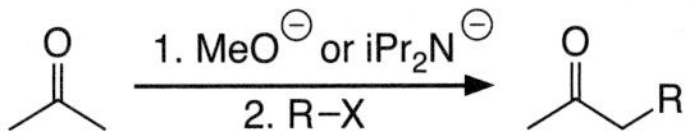

44. Write a mechanism for the following reaction, showing all intermediates and electron-flow arrows.

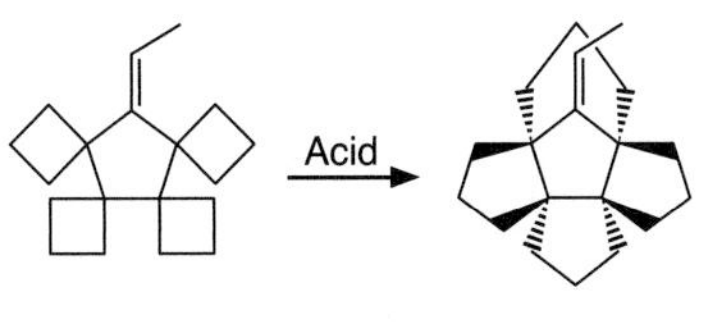

45. Experiments designed to study the effects of substituents, solvents, nucleophiles/bases, and leaving groups on the balance between S_N2 and S_N1 reactions most often involve secondary substrates. Explain why this is so.

46. Solvolysis of the reactants shown in aqueous solvent give essentially identical ratios of the same two alcohols (7:3). One explanation is that two carbocations are formed that rapidly interconvert and have essentially the same energy. Draw these two carbocations and comment on whether this explanation is reasonable. An alternative explanation is that the same non-classical carbocation is formed from both reactants. Draw this carbonium ion.

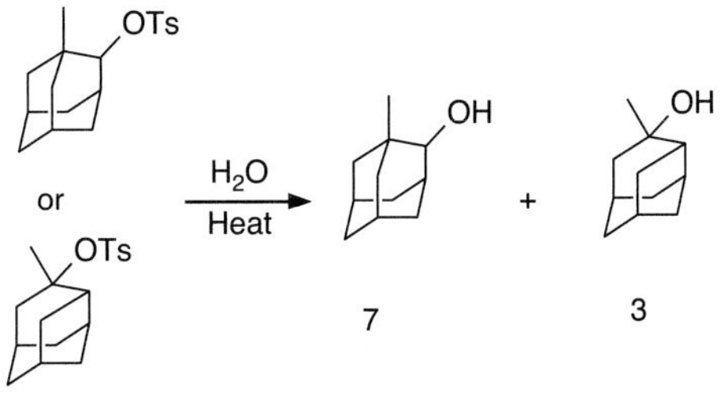

47. Which of the following carbanion rearrangements might reasonably be expected to occur? Give a mechanism for each and explain why they would or would not be expected to occur.

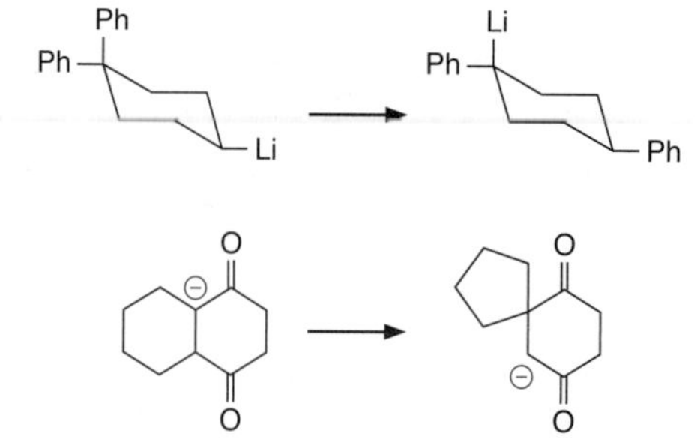

48. The relative rates for the reaction of three benzyl bromides with potassium thiocyanate are given below. Give a mechanistic interpretation of the observation that both the methoxy- and nitro-substituted substrates undergo faster reactions than the parent compound.

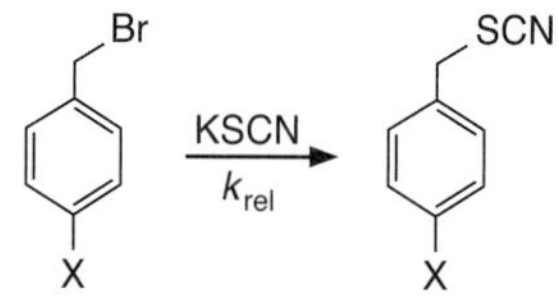

X	k_{rel}
OMe	11.2
H	1.0
NO$_2$	2.94

49. Write a mechanism with all intermediates and electron-flow arrows for the following reaction.

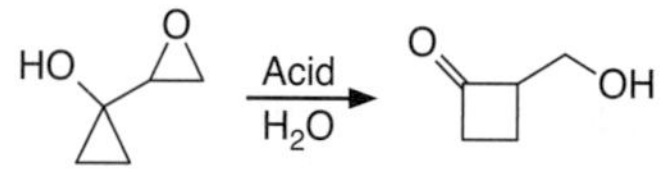

50. a. Give an example of a good nucleophile that is also a weak base and a poor nucleophile that is also a strong base. In each case, point out the properties of the compound that lead to this disparity.
b. Is the identity of the leaving group typically more important in nucleophilic aliphatic substitution or nucleophilic aromatic substitution? Explain.

51. Choose between the following options to maximize the probability of an S$_N$2 mechanism and to minimize the probability of an S$_N$1 mechanism. Explain each of your choices.

A. 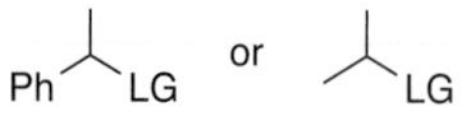 Ph—LG   or   —LG
B. Leaving group: Br$^{\ominus}$ or F$_3$CSO$_3^{\ominus}$
C. Nucleophile: Me$_3$N or N$_3^{\ominus}$
D. Solvent: CF$_3$CH$_2$OH or CH$_3$OH
E. Concentration of nucleophile: High or low

52. Draw mechanisms for each of the following reactions involving rearrangements. In the last example predict the stereochemistry.

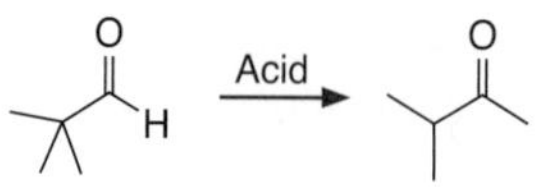

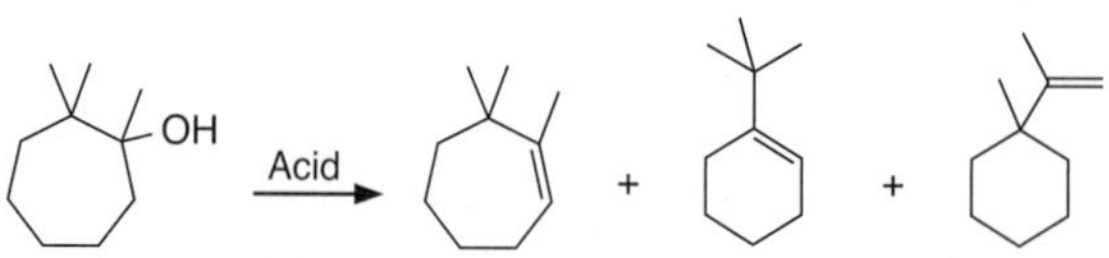

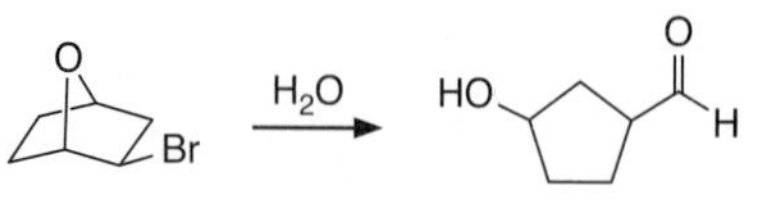

D.

53. The reaction of the following compound with KO-*t*-Bu in warm *t*-BuOH provides a product of formula $C_{10}H_{16}O$ in 90% yield. A different $C_{10}H_{16}O$ product is obtained in 80% yield when the same compound is treated with LDA in Et_2O at −60 °C. Draw structures of the two products and explain the selectivity.

54. Predict the product of the following aldol reaction at −78 °C. Explain your choice.

Further Reading

Classic Physical Organic Textbooks (for more examples and analysis of the material presented in Chapters 10 and 11)

Carey, F. A., and Sundberg, R. J. (2000). *Advanced Organic Chemistry; Part A*, 4th ed., Kluwer Academic, New York.

Smith, M. B., and March, J. (2001). *Advanced Organic Chemistry*, Wiley–Interscience, New York.

Lowry, T. H., and Richardson, K. S. (1987). *Mechanism and Theory in Organic Chemistry*, 3rd ed., Harper Collins, New York.

Carroll, F. A. (1998). *Perspectives on Structure and Mechanism in Organic Chemistry*, Brooks Cole, Pacific Grove, CA.

Isaacs, N. S. (1995). *Physical Organic Chemistry*, 2nd ed., Longman Scientific and Technical, Essex.

Textbooks Solely Focused on Organic Reaction Mechanisms

Miller, B. (1998). *Advanced Organic Chemistry*, Prentice–Hall, London.

Edenborough, M. (1999). *Organic Reaction Mechanisms*, 2nd ed., Taylor and Francis, London.

Bruckner, R. (1996). *Reaktionsmechanismen*, Spektrum, Berlin.

Bioorganic Examples of Organic Reaction Mechanisms with a Focus on Physical Organic Principles

Page, M., and Williams, A. (1997). *Organic and Bio-Organic Mechanisms*, Longman, Essex.

S_N2 Reactions

Shaik, S. S. "The Collage of S_N2 Reactivity Patterns: A State Correlation Diagram Model." *Prog. Phys. Org. Chem.*, **15**, 197–337 (1985).

Alpha-Halogenations

Freeman, F. "Possible Criteria for Distinguishing between Cyclic and Acyclic Activated Complexes and Among Cyclic Activated Complexes in Addition Reactions." *Chem. Rev.*, **75**, 439 (1975).

Enolates (Kinetic and Thermodynamic)

Bell, R. P. (1973). *The Proton in Chemistry*, 2nd ed., Cornell University Press, Ithaca, NY, pp. 141 and 171.

Anchimeric Assistance or Neighboring Group Participation

Capon, B., and McManus, S. P. (1976). *Neighboring Group Participation*, Plenum, New York.

Phenonium Ion

Lancelot, C. J., Cram, D. J., and Schleyer, P. v. R. in *Carbonium Ions*, Vol. III, G. A. Olah, and P. v. R. Schleyer (eds.), John Wiley & Sons, New York, 1972.

Single-Electron Transfer

Eberson, L. "Electron-Transfer Reactions in Organic Chemistry." *Adv. Phys. Org. Chem.*, **18**, 79 (1982).
Cannon, R. D. (1980). *Electron Transfer Reactions*, Butterworth, London.
Masuhara, H., and Mataga, N. "Inorganic Photodissociation of Electron Donor–Aceptor Systems in Solution." *Acc. Chem. Res.*, **14**, 312 (1981).

Benzilic Acid Rearrangement

Selman, S., and Eastham, J. F. "Benzilic Acid and Related Rearrangements." *Quart. Rev. (London)*, **14**, 221 (1960).

Beckmann Rearrangement

Donaruma, L. G., and Heldt, W. Z. "The Beckmann Reaarrangement." *Org. Reactions*, **11**, 1 (1960).

Nitrenium Ion

Gassman, P. G. "Nitrenium Ions." *Acc. Chem. Res.*, **3**, 26 (1970).
Abramovitch, R. A. in *Organic Reactive Intermediates*, S. McManus (ed.), Academic Press, New York, 1973, Chapter 3, p. 181.
Lwowski, W. in *Reactive Intermediates* Vol. 2, M. Jones, Jr., and R. A. Moss (eds.), Wiley–Interscience, New York, 1981, Chapter 8, p. 327.

Contact Ion Pair or Intimate Ion Pair

Franks, F. (ed.) (1975). *Water, A Comprehensive Treatise*, Plenum Press, New York, p. 1.
Szwarc, M. (ed.) *Ions and Ion Pairs in Organic Reactions*, John Wiley & Sons, New York, Vol. 1, 1972; Vol. 2, 1974.

Radical Rearrangements

Beckwith, L. J., and Ingold, K. U. (1980). "Free Radical Rearrangements" in *Rearrangements in Ground and Excited States*, Vol. 1, P. de Mayo (ed.), Academic Press, New York.

Thermal Rearrangements and Biradicals

Gajewski, J. J. (1981). *Hydrocarbon Thermal Isomerizations*, Academic Press, New York.
Berson, J. A. (1980). "Hypothetical Biradical Pathways in Thermal Unimolecular Rearrangements" in *Rearrangements in Ground and Excited States*, Vol. 1, P. de Mayo (ed.), Academic Press, New York.
Borden, W. T. (ed.) (1982). *Diradicals*, John Wiley & Sons, New York.

Heavy Atom Tunneling

Buchwalter, S. L., and Closs, G. L. "Electron Spin Resonance and CIDNP Studies on 1,3-Cyclo-pentadiyls. A Localized 1,3 Carbon Biradical System with a Triplet Ground State. Tunneling in Carbon–Carbon Bond Formation." *J. Am. Chem. Soc.*, **101**, 4688–4694 (1979).
Whitman, D. W., and Carpenter, B. K. "Limits on the Activation Parameters for Automerization of Cyclobutadiene-1,2-d2." *J. Am. Chem. Soc.*, **104**, 6473–6474 (1982).
Sponsler, M. B., Jain, R., Coms, F. D., and Dougherty, D. A. "Matrix-Isolation Decay Kinetics of Triplet Cyclobutanediyls. Observation of Both Arrhenius Behavior and Heavy-Atom Tunneling in Carbon–Carbon Bond-Forming Reactions." *J. Am. Chem. Soc.*, **111**, 2240–2252 (1989).
Zuev, P. S., Sheridan, R. S., Albu, T. V., Truhlar, D. G., Hrovat, D. A., and Borden, W. T. "Carbon Tunneling from a Single Quantum State." *Science*, **299**, 867–870 (2003).

CHAPTER 12

Organotransition Metal Reaction Mechanisms and Catalysis

Intent and Purpose

This chapter is intended to extend our concepts of catalysis and our knowledge of organic mechanisms to organometallic chemistry. **Organometallic chemistry** is the chemistry of compounds containing metals with bonds to carbon. It is specifically referred to as **organotransition metal chemistry** when transition metals are involved. This field of chemistry has developed rapidly in the last 50 years, with a strong emphasis on synthetic transformations and industrially useful catalytic cycles. Our emphasis here is on the steps and mechanisms involved in fundamental organometallic transformations. We will then focus on how these basic reactions are combined to create catalytic cycles. The fundamental steps involve ligand exchanges, insertions, oxidative additions, reductive eliminations, and reactions of π systems. Before examining these reaction types, however, we must first learn a little about structural organometallic chemistry and electron counting, and this is how the chapter commences. Our goal is to give an overview of the field and impart familiarity with the terms and topics. Advanced treatments are referenced at the end of the chapter.

Why put a chapter on organometallic chemistry in a book on physical organic chemistry? Partly because we would like to further emphasize how the concepts of physical organic chemistry are used in many other fields of chemistry. The techniques discussed in Chapters 7 and 8 are routinely used to decipher organometallic reaction mechanisms, not just organic mechanisms. Indeed, we look at substituent effects, isotope effects, and kinetics quite extensively in this chapter. However, most important is the fact that organometallic chemistry is now so common and pervasive that all organic chemists should be versed in the terminology and reaction types, and be able to propose a mechanism for an organometallic reaction as readily as for an organic reaction. Hence, our primary goal is to teach you how to predict and appreciate organometallic mechanisms and catalytic cycles.

12.1 The Basics of Organometallic Complexes

The large variety of structural motifs, ligands, and metals can make organometallic chemistry seem like a more daunting topic to learn than organic chemistry, where carbon is the focus. In order to organize organometallic chemistry into a comprehensible discipline, several methods of analyzing structure and reactivity have been developed, along with nomenclature rules. We must have at least a rudimentary understanding of these methods and the terminology used in order to comprehend basic reactivity. The brief introduction to the topics given here is sufficient to make organometallic chemistry accessible. In this section of the chapter, we analyze electron counting, common ligand types, common structural motifs, and a little nomenclature.

12.1.1   Electron Counting and Oxidation State

One of the first things one does when looking at an organometallic reaction is to count the number of electrons the metal has in the reactant and product. This is called the **electron count** for the metal. Next, the **oxidation state** of the metal is determined. Lastly, it is often useful to determine how many electrons are in the *d* orbitals of the metal, a number referred to as the *d* **electron count**. The result of these determinations gives insights into both the reactivity of the metal complexes and what has occurred during the reaction.

Electron Counting

Although there is some difference of opinion as to the "correct" way to count electrons, here we avoid that debate and present only one of the possible methods. This is just a formalism; electrons don't "belong" to any one atom. In the method we will use, ligands to the metal that make covalent bonds (σ and / or π) are considered to donate one electron for each bond. In essence, these bonds are considered to arise from a radical ligand making a bond with one electron from the metal. Alkyl groups, hydrogen, halogens, and oxygen or nitrogen (singly-bonded) are common examples of the kinds of ligands considered to donate one electron to the metal during σ bond formation.

$$\text{M}-\text{H} \quad \text{M}-\text{R} \quad \text{M}-\text{X} \quad \text{M}-\text{OR} \quad \text{M}-\text{NR}_2$$

Covalent (one-electron donating) bonds to metals

Dative bonds are formed by donation of a pair of electrons from the ligand to the metal. Ligands such as carbon monoxide (CO), phosphines, and amines all have lone pairs prior to donation, and they therefore contribute two electrons. These ligands can be considered as Lewis bases that are making a complex with a Lewis acidic metal. Upon coordination, the formal charge of a phosphine or amine would be positive. Similarly, the metal would formally have a negative charge, but neither of these formal charges are normally drawn. The dative bonding electrons are still formally assigned to P or N, just as they are in a nonbonded lone pair. Dative bonds are often drawn as arrows, but they are also often drawn the same way as covalent bonds, requiring chemists to recognize them as dative from a knowledge of ligand types.

$$\text{M}-\text{PR}_3 \quad \text{M}-\text{NH}_3 \quad \text{M}-\text{CO} \quad \text{or} \quad \text{M}\leftarrow\text{PR}_3 \quad \text{M}\leftarrow\text{NH}_3 \quad \text{M}\leftarrow\text{CO}$$

Dative (two-electron donating) bonds to metals

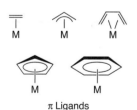

π Ligands

Alkenes also contribute two electrons to a metal, by coordination of the π bond. To show this a line is drawn between the metal and the center of the double bond. As the π system becomes larger, more electrons can be donated to the metal. Allyl can donate three electrons, butadiene can donate four, cyclopentadienyl (Cp) can donate five, and benzene can donate six electrons.

Organometallic chemistry is full of complex ligands and bonding arrangements. This is part of the utility of organometallic chemistry. One often has the ability to complex and stabilize what would otherwise be a quite reactive organic species. In the examples of ligands given above, the allyl group, normally a very reactive structure as either a cation, radical, or anion, complexes to the metal and makes a stable structure. A neutral three-carbon allyl ligand is a radical, and so it could be considered to make one covalent σ bond with one electron on the metal as well as to donate a π bond to the metal, contributing a total of three electrons. Since allyl is symmetric, either terminal carbon can be considered to make the σ bond, and therefore two resonance structures are possible.

π Allyl

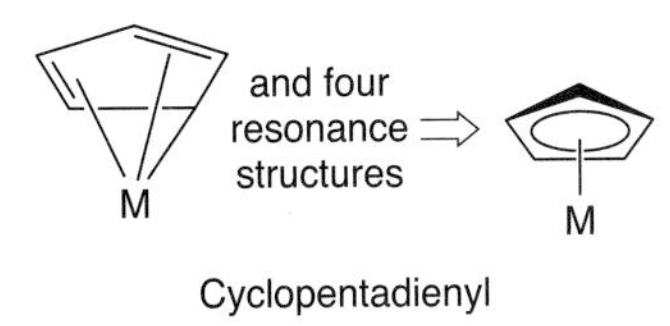

A similar situation arises in the cyclopentadienyl ligand. A neutral cyclopentadienyl ligand contributes five electrons. This is because a neutral ligand would be a radical that can make one σ bond, and donate two π bonds. The resonance structures lead us to represent the ligand as a pentagon with a circle in the middle.

Another common set of ligands are **alkylidenes** and **alkylidynes**. These are defined as carbon ligands that make double and triple bonds to metals, respectively. They can be viewed as involving the bonding of carbenes ($:CR_2$) and methines ($:\dot{C}R$), respectively, to metals. An alkylidene contributes two electrons to the metal count, one each from the σ and π bonds. An alkylidyne contributes three electrons to the metal count.

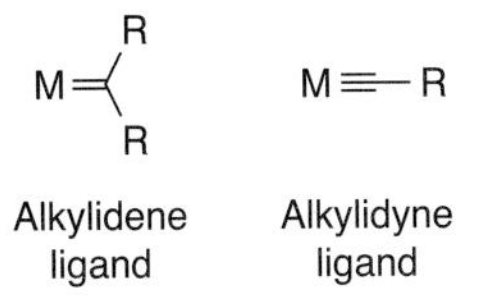

The metal contributes to the overall electron count according to its position in the periodic table. Figure 12.1 shows a periodic table for just the rows that contain the transition metals. One simply counts from the left-hand side of each row to the metal to find the number of electrons that the metal is considered to bring to a complex.

Cyclopentadienyl

Alkylidene   Alkylidyne
ligand   ligand

Figure 12.1
A section of the periodic table showing the transition metals and the number of electrons they contribute to a complex.

Lastly, any positive charge on the metal complex is considered to remove electrons from the metal, and a negative charge on the metal complex adds electrons to the metal. Using these rules, the electron counts for a series of organometallic complexes are shown in Figure 12.2. We suggest that you look over each to confirm for yourself how the electron count is determined, and then try Exercise 1 at the end of the chapter.

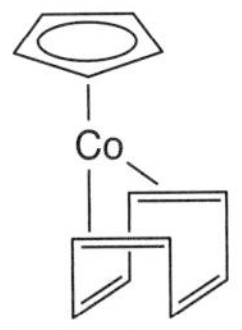

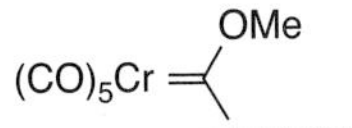

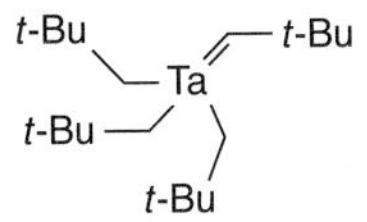

Figure 12.2
A number of organometallic complexes, along with an analysis of the electron counting and the oxidation state. Note that the NO ligand is a three-electron donor (think about its Lewis dot structure).

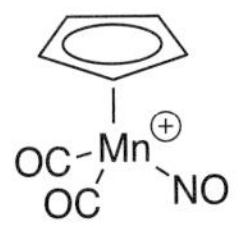

Oxidation State

The oxidation state of the metal is now easy to determine. Ligands making σ or π covalent bonds (not dative) are considered to completely take an electron from the metal when making that bond; both electrons of the bond are considered to be associated with the ligand (which is almost always more electronegative than the metal). This imparts a positive charge to the metal for each of these bonds. On the other hand, dative bonds from amines and phosphines, and π bonds from alkenes and the like do not alter the oxidation state of the metal. The number of covalent σ bonds or π bonds to the metal, along with a consideration of the overall charge on the complex, leads to the oxidation state of the metal. Note that for "radical" π ligands, such as allyl and pentadienyl, the model that involves one σ bond and the rest of the system as a π complex is used. Thus, allyl and pentadienyl oxidize the metal by one electron, but a butadiene ligand does not oxidize the metal.

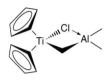

Tebbe's reagent

Let's do one example in detail. The complex shown in the margin is called Tebbe's reagent. Each Cp ligand interacts with the metal by donating one σ bond and two π bonds. The single σ bond inherent in the Cp ligand is considered to take one electron from the Ti. The chloride and the bridging CH_2 group also take one electron each from the Ti. This is a total of four electrons that are defined as being removed from the Ti, and therefore the oxidation state of Ti is plus four, denoted as Ti(IV). Look at Figure 12.2 to see other examples.

The oxidation state is best considered as a bookkeeping notation. A metal with an oxidation state of two does not have a full plus two charge. In reality, the charge is significantly lower than the oxidation state would imply. This is because the method views the ligands as fully withdrawing an electron, even if the ligand is not very electronegative. The ligands *do not* fully remove an electron from the metal. That would create a completely ionic bond. Most bonds between transition metals and common ligands are highly covalent; even bonds to electronegative ligands such as halides and alkoxides.

d Electron Count

The last determination to make is the *d* electron count. The number of electrons in the metal's *d* orbitals is the number that the metal starts with, based on its position in the periodic table, minus the number of electrons that are considered to be removed by oxidation. The count is designated as d^n, where *n* is the number of electrons in the *d* orbitals. This number is not used as often as the full metal electron count and the oxidation state when predicting reactivity, but it is quite useful when coupled with crystal field theory to predict the spectroscopy and spin state of the organometallic complex. Once again, Figure 12.2 shows a series of examples.

Ambiguities

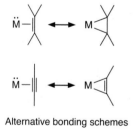

Alternative bonding schemes

There are certain ligands that present different possible electron counting and oxidation state options. Only a few are considered here. For example, an alkene and an alkyne can be drawn as bonding to a metal in two different ways (see margin). These are just resonance structures, but they impact the electron counting and oxidation states differently. In the first resonance structure of each example in the margin the ligands act as Lewis bases and donate two electrons to the metal with no oxidation state change. In the second structures, the picture is of a metallacyclopropane or metallacyclopropene. Now each σ bond is formed from a single electron from the ligands, so the ligands are still overall two-electron donors. However, the metal oxidation state has increased by two in these forms. Hence, we find the same electron count but differing views of the oxidation state. In general, **early transition metals** (to the left in the periodic table) are considered to have more metallacyclopropane or propene character if they have *d* electrons. Complexes of the **late transition metals** (toward the right in the periodic table) are better described by the simple donation picture. This guideline has been established by examining crystal structures of organometallic compounds with alkenes and alkynes as the ligand. More pyramidalization of the alkene carbons occurs with the early metals, while the alkenes remain closer to planar when coordinated to late metals.

Another example involves carbon monoxide as a ligand. As shown in the margin, two different resonance structures can be drawn with this ligand also. In the first, the ligand acts as a two-electron donor via donation of a lone pair from carbon to the metal. In the second structure, electrons from the metal donate to the ligand, creating a double bond to carbon. Electron donation from any metal to a ligand is referred to as **back-bonding**. Just as with ethylene, CO donates the same number of electrons in each case, but the resonance structure involving back-bonding increases the formal oxidation state of the metal. However, the standard procedure for determining oxidation states always views CO in the manner shown in the first resonance structure.

$$\ddot{M}-C≡O: \longleftrightarrow M=C=\ddot{O}:$$

Back-bonding

We have noted several times in this book that resonance structures are inherently a valence bond theory (VBT) concept. Molecular orbital theory (MOT) does not require such structures. Hence, there are MOT bonding concepts that describe the bonding pictures given above for alkenes, alkynes, and CO. A simple MOT picture is given in the following Going Deeper highlight.

Going Deeper

Bonding Models

The resonance structures shown above for the coordination of alkenes and CO to metals either involve no back-bonding or complete back-bonding from the metal. These two extremes are rarely found. Instead, the bonds formed between the metals and the ligands are created by various extents of mixing between the metal and ligand orbitals, giving differing contributions of resonance structures to the real electronic structure. The metal and ligand orbitals that give rise to the bonding with alkenes and CO are easy to understand based upon the symmetries of the metal and ligand orbitals.

The bonding picture for the alkene is called the **Dewar–Chatt–Duncanson model**. Here the HOMO of the π bond donates to a d_{z^2} orbital (look back at Figure 1.26 to see the d orbitals), while the LUMO interacts with a d_{yz} orbital. If there are electrons in the d_{yz} orbital, the complex can have back-bonding. The extent of back-bonding is determined by the relative energies of the alkene LUMO and the metal d_{yz} orbital. If they are close in energy, there is a lot of back-bonding, but if they have a large energy difference, little back-bonding occurs.

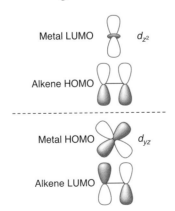

The bonding picture for CO is similar. A d_{z^2} orbital mixes with the carbon lone pair orbital, while a LUMO of C–O π mixes with a d_{yz} orbital. The number of electrons on the metal and the energy difference between the metal and ligand orbitals dictates the extent of back-bonding.

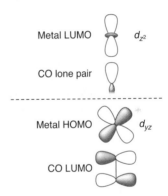

The bonding picture presented here placed the ligand along the z axis, but this was completely arbitrary. It is important to note that a "d_{z^2}-like" orbital can be formed along any axis, and then the proper "π symmetry" d orbital can be identified to create a molecular orbital that yields a back-bonding interaction. It is also important to note that when many σ bonds must be formed to a metal, such as six of them in $Cr(CO)_6$, $d_{x^2-y^2}$ orbitals and p orbitals can be used along with the d_{z^2} orbital. A useful model is to consider d^2sp^3 hybrid orbitals (see Chapter 1). However, it is still the "π symmetry" orbitals, d_{xy}, d_{yz}, and d_{xz}, that are used in back-bonding.

12.1.2 The 18-Electron Rule

A unifying principle in organic chemistry is the octet rule. There is a similar rule for organometallic species, called the **18-electron rule**. As can be concluded by counting the number of electrons in row 4 of Figure 12.1, krypton has 18 electrons in its valence shell. To achieve a noble-gas configuration, all the transition metals need 18 valence electrons, analogous to elements in rows 2 and 3 requiring 8 valence electrons. Therefore, in general, organometallic complexes wherein the metal has 18 electrons are relatively stable compared to species with lower electron counts. In the electron counting method discussed above, the goal is often to find out if the metal has 18 electrons, or the extent to which it deviates from an 18 electron count. This gives immediate insight into whether a complex can react with ligands that are added to solution, or if dissociation of a ligand from the metal is required before other reactions can commence.

The 18-electron rule is not nearly as "hard and fast" as the octet rule. Many organometallic species do not have 18 electrons on the metal, and yet they are perfectly stable and / or persistent. Metals that are early in the *d* block elements commonly have 16 electrons, as do complexes from group VIII metals, where square planar geometries dominate. However, as the number of electrons drops significantly below 16, transition metal complexes become increasingly electrophilic. Many are unstable in air and react violently with oxygen and water. This is due to the large exothermicity of reactions that give the metals more electrons.

When the metals have 18 electrons, they are referred to as **coordinatively saturated**. They cannot easily take on another ligand. Metals with lower electron counts are **coordinatively unsaturated**. If their geometries allow, these metals can react with additional ligands to bring the electron count up to 18. These metals are considered to have **open coordination sites**, defined as positions on the metal where another ligand can add. Attachment of another ligand increases the **coordination number** (the number of ligands attached) of the metal by one.

You might be wondering why chemists consider oxidation states and *d* electron counts. The goal is to predict reactivity. All metals have characteristic oxidation states that have distinct spin states, and deviations from these oxidation states tell us that the complex will be reactive in a manner that brings it back to one of the common oxidation states. We do not discuss in length the origin of these preferred oxidation states, but you may remember from an inorganic chemistry course that the origin derives in part from crystal field theory and depends upon geometry (tetrahedral, trigonal bipyramidal, square planar, or octahedral). Similarly, chemists count *d* electrons to understand reactivity and spin state. The *d* electron count allows us to predict if back-bonding can occur (see above), and the number of electrons populating the various molecular orbitals that are mostly metal-*d*-orbital in character (see Section 12.1.6 on crystal field theory). Furthermore, the spin state of an organometallic complex depends upon the number of electrons in these *d* orbitals, and the strength of the crystal field.

12.1.3 Standard Geometries

Organometallic complexes normally adopt one of five different structures: octahedral, tetrahedral, square planar, trigonal bipyramidal, or square pyramidal, which can be predicted based upon the simple valence-shell electron-pair repulsion rule (see Chapter 1). As with organic compounds, when the groups attached to the central atom are not the same, geometries with considerable deviations from the standard bond angles can arise. The reason is the shapes of the ligands. For example, although the Cp ligand makes five contacts to the

| Octahedral | Tetrahedral | Square planar | Trigonal bipyramidal | Square pyramidal |

metal, it can be considered as one group in a single position of any one of these geometries. Hence, the complex Cp_2TiCl_2 is predicted to be tetrahedral due to four groups attached to Ti. However, due to sterics, the complex has a Cl–Ti–Cl bond angle of 94.5°, and a Cp–Ti–Cp angle of 131.0°. It therefore has a distorted tetrahedral geometry. Similar deviations from the ideal structures occur for most common organometallic complexes. These subtle deviations from standard geometries are not dominant factors influencing reactivity, and therefore we do not delve into them here.

12.1.4 Terminology

The nomenclature for organometallic complexes can become very complicated. This might be expected based upon the number of metals there are, the number and variety of ligands, the different ways that ligands can be attached to a metal, and multiple oxidation states. Therefore, we are not going to discuss nomenclature in enough depth to learn all the rules for naming organometallic species. Instead, only the terminology applied to various ligands is explored here. This is because the manner in which the ligands are attached to the metal(s) influences the reactivity of the ligands, and this is the issue chemists are most focused upon. Figure 12.3 shows a variety of common metal–ligand motifs and the terminology applied. As you can see, the term "eta^n" (η^n) is given for ligands that make multiple contacts to a metal, where n = the number of contacts to the metal(s). The term "mu$_m$" (μ_m) is given to indicate that the ligand bridges two or more metals, where m indicates the number of metals that the ligand contacts (when $m = 2$ it is not written). Lastly, any complex with two Cp ligands is called a **metallocene**.

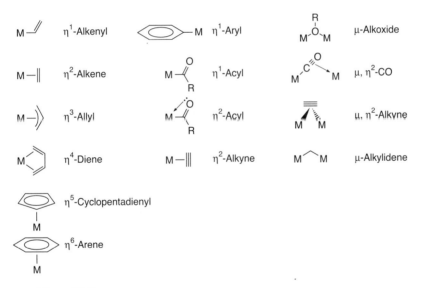

Figure 12.3
Terminology given to various coordination modes of different ligands.

12.1.5 Electron Pushing with Organometallic Structures

The electron-pushing conventions used in organometallic chemistry are identical to those used in organic chemistry (see Appendix 5). The difficulty arises when trying to portray reactions with ligands that are simply shown as coordinating. For example, consider a nucleophilic attack on an alkene ligand that is coordinated to a metal (Eq. 12.1). Since the ligand is shown with a line (a bond) between the metal and the center of the alkene, where do we place the electron arrows? Several conventions are acceptable. First, in your mind, keep the two electrons of the alkene π bond associated with the alkene and ignore the line drawn to the metal. The electron pushing of Eq. 12.2 ensues, which creates a σ bond to the metal. In the other convention, recall that we can write a resonance structure of the dative alkene interaction as a metallacyclopropane, which involves considerable back-bonding from the metal. As shown in Eq. 12.3, this picture of the bonding allows one to easily write arrow pushing for the nucleophilic attack and the formation of a lone pair on the metal.

$$M-\| \xrightarrow{\overset{\ominus}{:}Nuc} \overset{\ominus}{M}\diagup^{Nuc}$$

(Eq. 12.1)

$$M\overset{\frown}{\triangleleft}\| \xrightarrow{\overset{\ominus}{:}Nuc} \overset{\ominus}{M}\diagup^{Nuc}$$

(Eq. 12.2)

$$M\overset{\frown}{\triangleleft} \overset{\frown}{}\overset{\ominus}{:}Nuc \longrightarrow \overset{\ominus}{M}\diagup^{Nuc}$$

(Eq. 12.3)

Let's also examine arrow pushing for the nucleophilic attack on a carbon monoxide ligand. Once again, the electron pushing can be considered to start with either of the two resonance structures drawn for coordinated CO (Eqs. 12.4 and 12.5). In Eq. 12.4, we give the formal charges on the metal and CO that we alluded to in our earlier discussion, because it makes the electron pushing clear. Considerations similar to those given here, along with common organic conventions, should be sufficient to write the electron pushing for any organometallic reactions and mechanisms.

$$\overset{\overset{\ominus}{:}Nuc}{\overset{\curvearrowleft}{\underset{M-C\equiv O}{\ominus}}} \longrightarrow \overset{\ominus}{M}\diagdown\diagup^{Nuc}_{\underset{:\!\underset{O}{}}{}}$$

(Eq. 12.4)

$$\overset{\overset{\ominus}{:}Nuc}{\overset{\curvearrowleft}{\underset{M=C=O}{}}} \longrightarrow \overset{\ominus}{M}\diagdown\diagup^{Nuc}_{\underset{O}{}}$$

(Eq. 12.5)

12.1.6   *d* Orbital Splitting Patterns

One reason that chemists evaluate the *d* orbital count is to determine if the metal complex is diamagnetic or paramagnetic. A **diamagnetic** complex has all electrons paired. NMR spectra with chemical shifts similar to organic compounds can be obtained for these structures. A **paramagnetic** complex has one or more unpaired electrons. Very often, standard NMR spectra for these structures cannot be obtained. Either the resonances are so broad that they cannot be detected, or the chemical shifts are spread over an unusually wide ppm range.

Due to the unpaired electrons, paramagnetic compounds often react as radicals. In the section on oxidative addition below, we'll see that radical mechanisms can be involved. In fact, many organometallic mechanisms involve radicals, and having a knowledge of which orbitals these radicals reside in is informative. The identity of the singly occupied orbitals is predicted by an examination of the *d* orbital splitting.

The *d* **orbital splitting** is defined as the pattern of energies of the *d* orbitals, and it results from the geometry of the complex. You might recall from a class in inorganic chemistry that **crystal field theory** predicts this splitting. The reasons behind the splitting and the magnitude of the separation between the *d* orbitals are best discussed in a textbook focused on inorganic or organometallic chemistry. Suffice it to say, a general rule for predicting the splitting is that *d* orbitals that lie along the bonds to the ligands are raised in energy. The *d* orbitals that are partially aligned along the bond axes are not raised in energy as much. The *d* orbitals that do not lie along the bond axes are relatively unperturbed in energy. Given these simple guidelines, Figure 12.4 shows the various *d* orbital splitting patterns for most common inorganic and organometallic geometries. These are for idealized structures where all the ligands are of the same type. Different ligands split the orbital energies to different

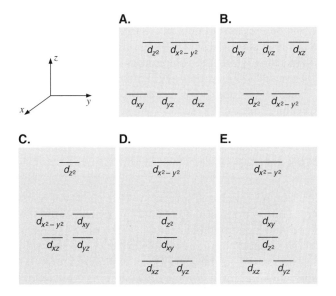

Figure 12.4
d Orbital splitting patterns. **A.** Octahedral complexes, **B.** Tetrahedral complexes, **C.** Trigonal bipyramidal complexes, **D.** Square pyramidal complexes, and **E.** Square planar complexes. Refer to the pictures in Section 12.1.3 to see the various geometries.

extents. Strongly donating ligands (such as phosphines) give larger energy differences, whereas weakly donating ligands (such as an alkene) give smaller differences. When different ligands are on the same metal, the orbitals that are degenerate in Figure 12.4 can split.

A few uses of the *d* orbital splitting patterns are noteworthy. One is the population of these orbitals with electrons. If the ligands are strongly donating, the electrons are placed in the *d* orbitals according to the Aufbau principle with paired spins if necessary. This occurs because strongly donating ligands create large energy splittings between the *d* orbitals, and the energy difference between the orbitals is larger than the energy required to pair the electrons. This creates **low spin complexes**. Alternatively, when the ligands are not strongly donating, the electrons are placed in the *d* orbitals with aligned spins (for d^2 up to d^5), now giving **high spin complexes**. Due to the smaller energy differences between the various *d* orbitals, it is more energetically costly to pair them than to put them in higher energy orbitals.

Most organometallic complexes that we examine in this chapter are either 16- or 18-electron species. Furthermore, low spin complexes dominate because carbon ligands are generally strongly donating. For example, $Cr(CO)_6$ is an octahedral complex with a d^6 metal, and hence the three lower energy orbitals in Figure 12.4 **A** contain all six of the *d* electrons. With $NiCl_2(PMe_3)_2$, the metal is d^8, and the complex is square planar. Consequently, only the lowest four *d* orbitals in Figure 12.4 **E** are populated. As a last example, $Pd(PPh_3)_4$ has a d^{10} metal, and therefore all the *d* orbitals must be fully occupied.

Another use of the *d* electron count is the determination of whether a metal has a nucleophilic orbital. Square planar complexes with *d* orbital counts above four have electrons in a d_{z^2} orbital. This orbital is completely accessible because there are no ligands along the *z* axis. The orbital can act as a nucleophilic lone pair, much like a lone pair on a nitrogen atom. For example, in $Rh(Cl)(CO)(PPh_3)_2$, the metal is d^8 and the complex is square planar. Hence, the structure is nucleophilic at Rh. However, the complex only has 16 electrons, and thus it is also electrophilic. This is not uncommon; some inorganic and organometallic complexes can often accept and donate electrons, and the reactivity patterns reflect this.

12.1.7 Stabilizing Reactive Ligands

In the various analyses given above, we considered reactive carbon species such as carbenes and allyl radicals coordinated to transition metals. This is a hallmark of organotransition metal chemistry: the stabilization of reactive species by coordination to a metal. Sometimes these ligands retain reactivity patterns analogous to their standard organic reactivity, but often completely new reaction pathways are induced due to the metal coordination.

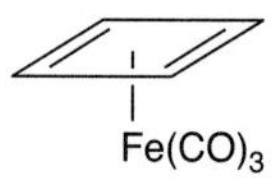

Fe(CO)₃

A cyclobutadiene complex

One of the most striking examples of the stabilization of a highly reactive organic compound involves organometallic complexes of cyclobutadiene. In Chapter 2 we addressed the concept of antiaromaticity, and noted that cyclobutadiene is highly unstable due to this phenomenon. In Chapter 14 we will look in more detail at the electronic structure of this unusual molecule. However, several complexes of cyclobutadiene with metals have been isolated, and as a ligand cyclobutadiene is completely stable. We show one example in the margin involving iron, making an 18-electron complex.

12.2   Common Organometallic Reactions

In the preceding chapters on organic reaction mechanisms, several specific classes of reactions were discussed, including additions, eliminations, substitutions, and rearrangements. The same classes of reactions are involved in organometallic chemistry, but now, the reactions occur at one or more metal atoms. The nomenclature used for the reaction classes is slightly different, as it reflects what has happened to the metal(s) as well as the organic reactant. In this section we describe several of the most commonly encountered reaction classes in organometallic chemistry, including ligand exchange, oxidative additions, reductive eliminations, α- and β-eliminations, insertions, and electrophilic/nucleophilic attack on ligands.

12.2.1   Ligand Exchange Reactions

Organic reactions mediated by organometallic complexes often begin with coordination to a metal or metals, where subsequent reactions ensue. This binding leads to activation, reminiscent of the activation steps we examined when catalysis was discussed in Chapter 10. Therefore our first discussion is the manner in which ligands coordinate to a metal and exchange between metals.

Reaction Types

When an organometallic species has 18 electrons, the complex is coordinatively saturated and typically unreactive. To initiate a reaction from such structures, a ligand needs to first dissociate, as shown for $Cr(CO)_6$ in Eq. 12.6. Most such reactions involve loss of ligands that are two-electron donors, and hence are most commonly followed by reactions that replace these two electrons.

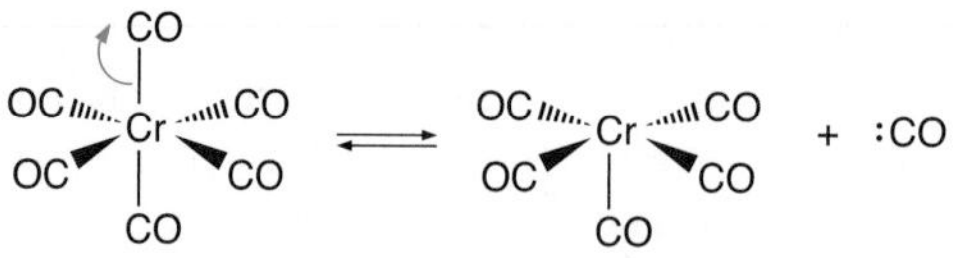

$$\text{(Eq. 12.6)}$$

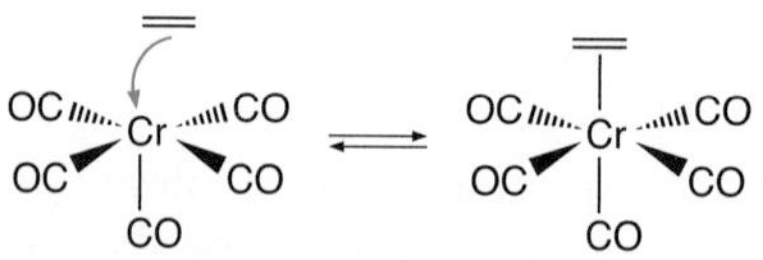

$$\text{(Eq. 12.7)}$$

Once there is an open coordination site, new ligands can attach to the metal. Eq. 12.7 shows an example where an olefin coordinates. These **association** reactions are also referred to as **ligation** reactions. Most involve coordination of a two-electron donor to an open site on the metal. These reactions bring organic reagents onto the metal center, where they can undergo subsequent reactions. Coupling a dissociative step with a subsequent association leads to a ligand exchange, referred to as a **dissociative mechanism** (Eq. 12.8 gives another example).

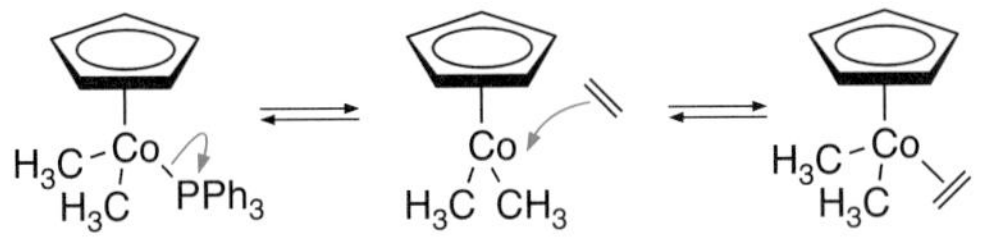

(Eq. 12.8)

The dissociative path is the most common mechanism for ligand exchange, especially when the metal has 18 electrons before and after ligand exchange. However, some organometallic species undergo ligand exchange via an **associative mechanism**. Eq. 12.9 shows an example. With this mechanism, the metal typically has 16 electrons to start, and is commonly square planar.

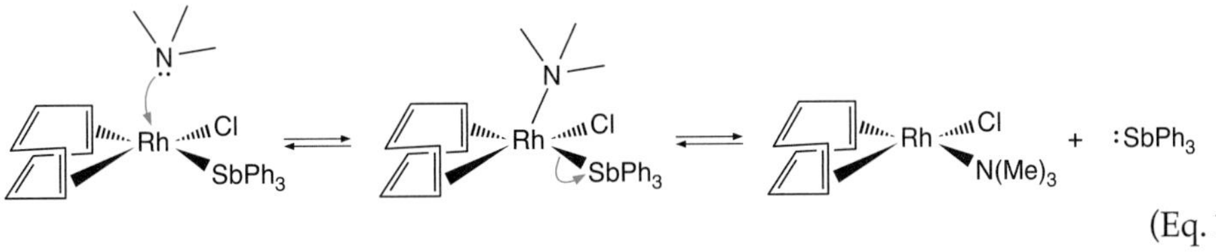

(Eq. 12.9)

Another ligand exchange mechanism is called **transmetallation** (Eq. 12.10). Here, one metal–R species directly exchanges a ligand with an organometallic species. These reactions commonly require open coordination sites on the two metals involved, and can occur via a mechanism known as electrophilic aliphatic substitution (S_E2; see the next Going Deeper highlight). The hallmark of a transmetallation is the metal–R reagent required—namely, an R group attached to Mg, Zn, Zr, Sn, B, Al, or Li. A halogen or other electronegative ligand is simultaneously exchanged with the R-group. A specific example is given in Eq. 12.11. This is called a **σ bond metathesis** reaction. The term **metathesis** is a common one in organometallic chemistry, and means the pair-wise interchange of two ends of two bonds. Exchange of an alkyl group for a halide or other one-electron donating group via a reaction that is first order in each metal species is an example. It is postulated to occur via a four-centered transition state that involves simultaneous transfer of the groups undergoing exchange between the two metals (see margin).

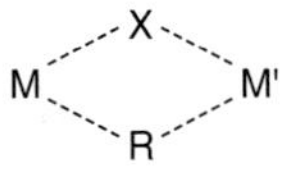

Four-centered transition state for σ bond metathesis

$$M-R + M'-X \rightleftharpoons M-X + M'-R$$ (Eq. 12.10)

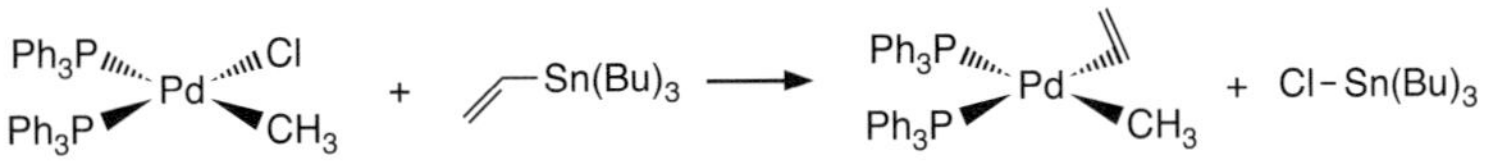

(Eq. 12.11)

Going Deeper

Electrophilic Aliphatic Substitutions (S_E2 and S_E1)

As we saw in Chapter 11, the vast majority of aliphatic substitutions proceed by nucleophilic S_N2, S_N1, SET, or $S_{RN}1$ mechanisms. However, there are less common reactions that proceed via exchange of electrophiles. Shown below are mechanisms called S_E1 and S_E2. These mechanisms almost always occur with organometallic reagents where metals are the electrophiles (E), such as with alkyl magnesium and mercury compounds.

One of the most interesting aspects of these reactions is stereochemistry. Varying results are obtained depending upon the metal, the solvent, and the R group. Transfer between alkyl mercury compounds most often occurs with retention. This is opposite of S_N2, and implies that the exchange of attached electrophiles occurs by frontside attack on the R group. Other metals, such as Co and Sn, give mixtures of retention or inversion, depending upon the case.

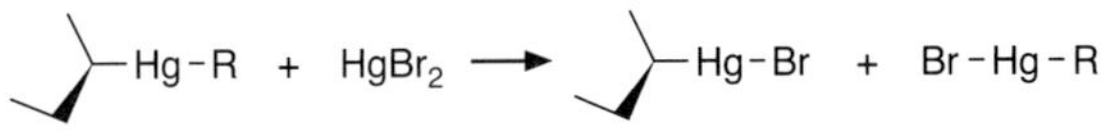

$$R-E_1 \rightleftharpoons E_1^{\oplus} + R^{\ominus} \xrightarrow{E_2^{\oplus}} R-E_2 \quad S_E1$$

$$E_2^{\oplus} + R-E_1 \rightleftharpoons R-E_2 + E_1^{\oplus} \quad S_E2$$

Electrophilic substitution

Kinetics

Although many ligands undergo exchange reactions, we primarily look at the kinetics of CO here, which is indicative of the reactivity patterns found for many other ligands. The kinetics of exchanging a CO ligand for another ligand, such as a phosphine or amine, is normally zero order in the adding ligand. Specific examples where this is found are the substitution of CO ligands in $Ni(CO)_4$ and $Mn(CO)_5Br$. This supports a mechanism where dissociation of CO is rate-determining, followed by a fast addition of the new ligand (Eq. 12.12), thereby conforming to the dissociative mechanism. This scenario is analogous to the S_N1 mechanism for substitutions at alkyl centers. In fact, the same kinetic equations describe the two mechanisms (compare Eqs. 12.13 and 11.22). Generally, $k_2[L] \gg k_{-1}[CO]$, and Eq. 12.13 reduces to $d[P]/dt = k_1[M(CO)_n]$. In further support of ligand dissociation being rate-determining, the entropy of activation for these reactions is large and positive. For example, the exchange of ^{18}O-labeled CO or PPh_3 with CO ligands on $Ni(CO)_4$ has a $\Delta S^{\ddagger}$ of $+13$ eu.

$$M(CO)_n \underset{k_{-1}}{\overset{\overset{k_1}{-CO}}{\rightleftharpoons}} M(CO)_{n-1} \xrightarrow{k_2[L]} M(CO)_{n-1}L \qquad \text{(Eq. 12.12)}$$

$$\frac{d[P]}{dt} = \frac{k_1 k_2 [M(CO)_n][L]}{k_{-1}[CO] + k_2[L]} \qquad \text{(Eq. 12.13)}$$

To measure the relative rates of various ligands to replace CO, chemists must turn to competition experiments (see Section 8.8.3). Competition experiments using the coordinately unsaturated 16-electron species $Ni(CO)_3$, $Fe(CO)_4$, and $Mo(CO)_5$, and alkenes, phosphines, and amines as ligands show only small differences in product ratios (between 1 and 10). Because these three 16-electron coordinately unsaturated species are not very selective in their reactions, they must all be very reactive. They react with almost any nucleophile with a very low energy of activation. In support of this view, the rate constants for the addition of the new ligands to such unsaturated systems have been estimated to be near 10^6 M^{-1} s^{-1}; that is, they are approaching diffusion control.

Structure–Function Relationships with the Metal

Within the same column of the period table, dissociative substitution reactions are generally fastest for metals in the second row of the d block. For example, the rate of CO substitution by other ligands with $M(CO)_6$ complexes is 14 times faster for Mo compared to Cr, and 500 times faster compared to W. The effect is even more dramatic with Ni, Pd, and Pt complexes. The rate of substitution of the phosphite ligands in $M[P(OEt)_3]_4$ complexes is 3×10^8 faster for Pd relative to Ni, and 4×10^3 faster for Pd relative to Pt. Trends going across a row in the periodic table are not as ordered, but the fastest CO substitution reactions occur with $Co(CO)_4$ and $Mn(CO)_5$ (both 17-electron complexes), and the slowest are for $Cr(CO)_6$ and $Fe(CO)_5$ (both 18-electron complexes). The reasons for these variations are not totally understood, but are related to the total d electron count and what is called the **crystal field activation energy**. This predicts activation energies by considering the change in geometry and how the crystal field thereby changes.

Structure–Function Relationships with the Ligand

Ligands other than the departing carbon monoxide will influence the rates of substitution reactions. Good electron donating ligands such as amines and phosphines speed dissociative substitution reactions, because they compensate for the loss of electrons on the metal that occurs in the rate-determining step. This is an important factor to consider when one wants to increase the rate of ligand loss in a catalytic cycle.

The CO ligands that most rapidly exchange are those that are cis to the donating group (Eq. 12.14). This is because the donating ligands increase the extent of back-bonding mostly to the CO ligands trans to them, thereby increasing the bond strengths of the trans ligands to the metals. This is opposite to the classic trans effect in inorganic chemistry, where strongly donating ligands accelerate the dissociation of ligands trans to them by weakening those bonds. The reason for the difference is the back-bonding capability of CO, which is not possible with most ligands.

$$PPh_3-\overset{CO}{\underset{OC}{\overset{CO}{\underset{CO}{Cr}}}}-CO \longrightarrow PPh_3-\overset{CO}{\underset{OC}{\overset{CO}{\underset{CO}{Cr}}}}-CO + :CO \qquad \text{(Eq. 12.14)}$$

Substitutions of Other Ligands

Besides the dissociation of CO, the most common other ligand dissociation is that of a phosphine (PR_3) or phosphite [$P(OR)_3$]. There are two factors that influence the rate of dissociation of a phosphine—its donor ability and its cone angle. Both of these factors affect the M–P bond strength. Phosphines are better donors than phosphites because the P atom is more electron rich. For the same reason, alkyl phosphines are better donors than aryl phosphines (i.e., $PMe_3 > PPh_3$). Better donors create stronger Lewis acid–Lewis base interactions with transition metals, and therefore dissociate slower.

The cone angle is also very important, and sometimes even dominates the reactivity. The **cone angle** is the angle at the bonded metal to the phosphorus substituents (as shown in the margin). It is specifically defined from a point that is 2.28 Å from the phosphorus atom and forms lines that just touch the van der Waals radii of the outermost atoms of the R groups, such that it is defined as a property of the phosphine ligand itself. The larger the cone angle, the faster the dissociation due to steric strain. Table 12.1 gives the cone angles for several phosphorus ligands (L), along with the rate constants for the dissociation of that ligand from cis-$Mo(CO)_4L_2$. The rates reflect a balance between the electronic and steric effects of the ligands.

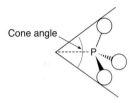

Cone angle

Table 12.1
Cone Angles and the Rate Constants for Dissociation of Ligand (L) from cis-$Mo(CO)_4L_2$ at 70 °C in Carbon Tetrachloride

Ligand	Cone angle	Rate constant (s^{-1})
PMe_2Ph	122°	$<1.0 \times 10^{-6}$
$PMePh_2$	136°	1.3×10^{-5}
PPh_3	145°	3.2×10^{-3}
$P(OPh)_3$	128°	$<1.0 \times 10^{-5}$

Now that we have examined some of the aspects of ligand dissociation—the first step involved in many organometallic reactions—we can shift to those reactions that actually manipulate bonds within organic structures.

12.2.2 Oxidative Addition

Oxidative addition is an addition reaction that occurs on a metal and raises its oxidation state (Eq. 12.15). The reaction is similar to that of an insertion (discussed in Section 10.11), and it expands the coordination sphere of the metal. A number of mechanisms for oxidative addition exist, and which one occurs is a function of the metal and the adding group. Although experimental studies of these reactions have been performed on many different kinds of complexes, we only examine a few prototypical examples here.

$$M + X-Y \longrightarrow X-M-Y \qquad \text{(Eq. 12.15)}$$

Stereochemistry of the Metal Complex

Several experimental observations, many from the research groups of Collman, Halpern, Kubota, and Osborn, have been combined to present a picture for the mechanisms of oxidative addition to square planar 16-electron species. One is the stereochemistry of the resulting metal complex. For example, Eq. 12.16 shows the addition of H_2 to *trans*-$Ir(CO)(PPh_2Me)_2Cl$, which gives only a cis arrangement of the added ligands. Addition of C_3F_7I also gives only cis addition of the organic fragment (Eq. 12.17).

$$\text{(Eq. 12.16)}$$

$$\text{(Eq. 12.17)}$$

Other reactants, however, give trans products. The oxidative addition of methyl iodide or acetylbromide to *trans*-$Ir(CO)(PPh_2Me)_2Cl$ results in a trans arrangement of the groups that added to the metal (Eqs. 12.18 and 12.19). Therefore, there must be different mechanisms for the addition of different organic structures.

$$\text{(Eq. 12.18)}$$

$$\text{(Eq. 12.19)}$$

The most obvious difference between those organic groups that give cis and trans addition is their polarity. Highly polarized bonds, or those that are known to be susceptible to nucleophilic attack, result in trans stereochemistry at the metal. Bonds possessing no polarity, or those that cannot undergo nucleophilic attack, give cis stereochemistry. This difference in reactivity correlates to other experimental observations, such as the kinetics and the stereochemistry of the R group.

Kinetics

The kinetics as a function of ligands, R group, X group, and solvent can be very informative as to the mechanism. For example, the reaction given in Eq. 12.18 has been explored in detail. The reaction is second order, first order in both the metal complex and CH_3X for all concentrations of these reactants. Hence, both reactants are involved at or prior to the rate-determining step. The $\Delta H^{\ddagger}$ and $\Delta S^{\ddagger}$ values are 5–9 kcal/mol (depending upon X) and –50 eu, respectively. The very negative entropy of activation supports a highly ordered transition state with a loss of translational freedom, and is in the same range as that for S_N2 reactions. The reaction is found to be faster in increasingly polar solvents, implying an increase in charge in the rate-determining step. Furthermore, the reaction shows the common trend in leaving group ability: $CH_3I > CH_3Br > CH_3Cl$. Lastly, the more electron donating the phosphine ligand, the faster the rate, suggesting that the metal is acting as a nucleophile. All of this evidence supports the mechanism given in Eq. 12.20.

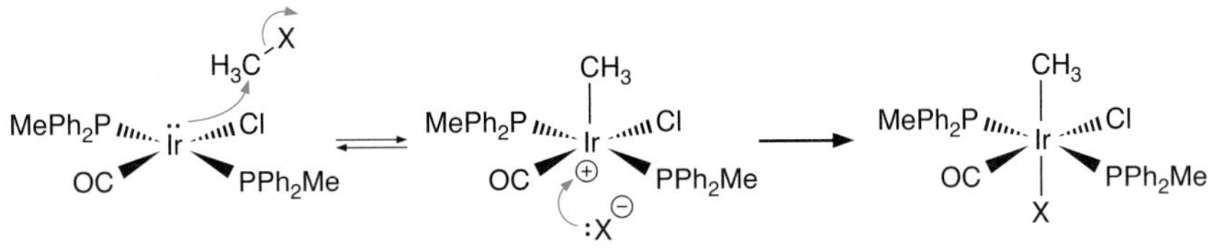

(Eq. 12.20)

The kinetics of oxidative addition of H_2 as a function of various structural parameters also sheds light on this mechanism. The reaction is first order in both metal complex and H_2. The $\Delta H^\ddagger$ and $\Delta S^\ddagger$ values for addition to Ir complexes such as those discussed here are 11–12 kcal/mol and around –23 eu. Solvent effects are small. There is very little H–H bond breaking at the transition state because isotope effects are small (k_H/k_D is typically around 1.2). For square planar 16-electron complexes, the addition of H_2 is rapid and reversible. Furthermore, d^8 complexes generally work well but d^{10} complexes do not. Therefore, one needs an empty d orbital for this reaction to proceed. The reason for this is shown in Figure 12.5, where we present a drawing of how the HOMOs and LUMOs of the metal and H_2 can mix to form two M–H bonds. In one mixing interaction H_2 acts as a nucleophile to fill the required empty orbital on the metal. The evidence suggests a concerted reaction, as shown in the margin.

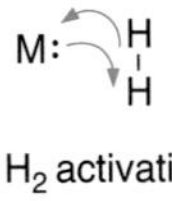

$$M: \begin{matrix} H \\ | \\ H \end{matrix}$$

H_2 activation

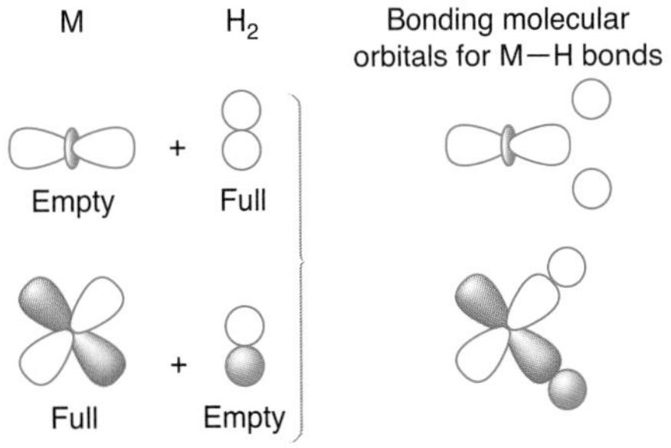

Figure 12.5
Orbital interactions that describe the oxidative addition of H_2.

Stereochemistry of the R Group

The stereochemistry of the R group during oxidative addition is another informative piece of experimental data for deciphering the mechanism. With R groups that can readily undergo nucleophilic attack, inversion of the stereochemistry is seen. Eq. 12.21 shows one example. This supports an S_N2-like mechanism, where the metal is the nucleophile. With electron rich late metal complexes, it is generally found that S_N2 mechanisms dominate oxidative addition for R groups that are very susceptible to S_N2 attack. When the R group is less susceptible to nucleophilic attack, mixed stereochemical results are commonly found, as shown in Eq. 12.22.

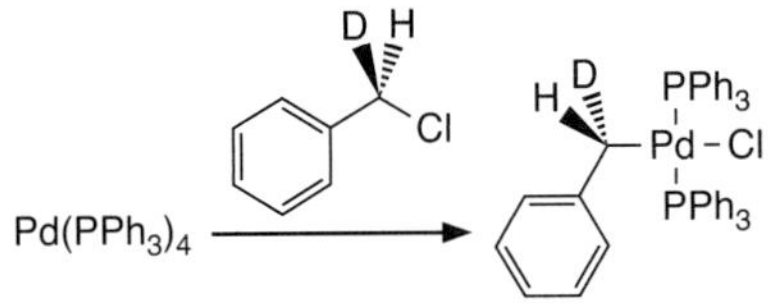

(Eq. 12.21)

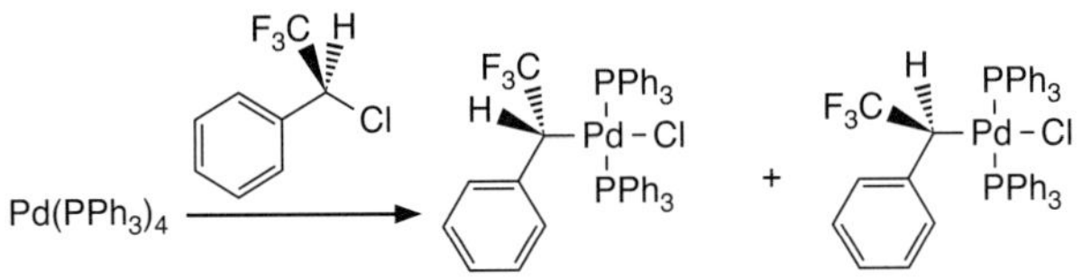

(Eq. 12.22)

Structure–Function Relationship for the R Group

As stated, S_N2 reactions are common for electron rich metals and R groups susceptible to nucleophilic attack. The general order of reactivity for R groups is similar to that found for organic mechanisms: $CH_3 > 1° > 2° >> 3°$. Benzyl and allyl are particularly good.

However, what kind of mechanism can explain mixed stereochemistry such as that given in Eq. 12.22? Two interesting examples that shed light on the mechanism are given in Eqs. 12.23 and 12.24. The reaction displayed in Eq. 12.23 yields products from both inversion and retention. Importantly, the reaction is very slow unless a radical initiator, such as benzoyl peroxide, is added, and the rate slows if a radical trap is added. Further support for radicals in some oxidative addition mechanisms can be seen in Eq. 12.24. Here, the common radical clock, cyclopropylcarbinyl (see Section 8.8.8), was used to probe the possible involvement of radical intermediates. Indeed, ring opening occurs, supporting the intermediacy of radicals.

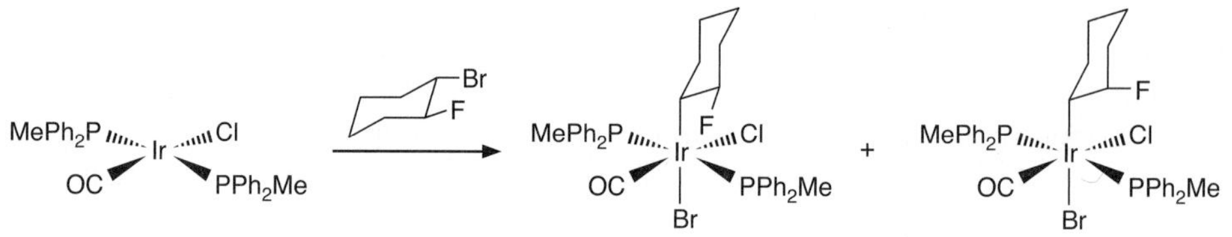

(Eq. 12.23)

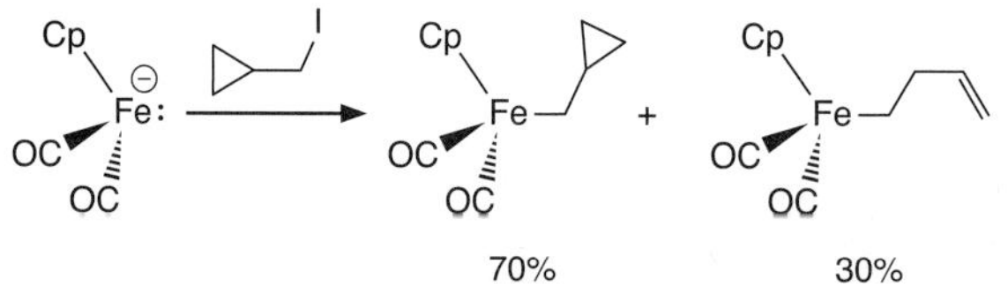

(Eq. 12.24)

The radical-based mechanism for oxidative addition, as delineated by Hill and Puddephatt, is shown in Figure 12.6. It is a typical radical chain process. A radical species, commonly from an initiator, adds to the metal. The resulting metal-based radical abstracts a halogen from the organic reactant, thereby generating a carbon radical. These two steps represent the initiation. Once the carbon radical is formed, it can add to another metal, creating a new metal-based radical. Abstraction of a halogen gives the oxidative addition product and another carbon radical that propagates the chain. Termination involves the combination of any two radicals.

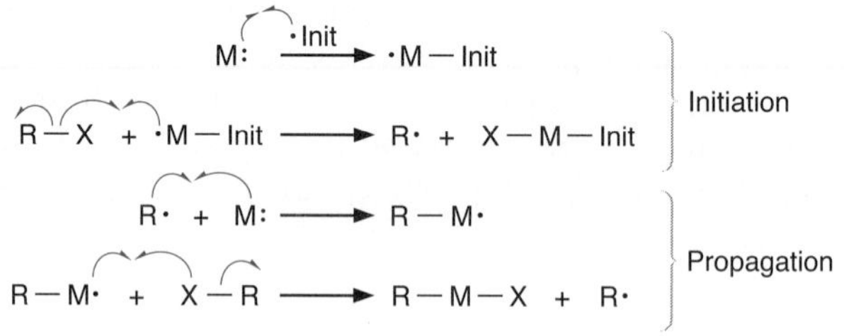

Figure 12.6
Arrow pushing for the radical mechanism involved in oxidative addition.

Structure–Function Relationships for the Ligands

The rates for oxidative addition to $Ir(CO)(PPh_2Me)_2X$ species display only a small sensitivity to X. However, the larger the donor capability of the phosphine, the faster the rate. Negative Hammett ρ values are found for phenyl-substituted phosphines in the oxidative addition of both methyl iodide and benzyl chloride (Figure 12.7), as might be expected because positive charge is building on the metal center. These data further support the Ir complex acting as a nucleophile in the oxidative addition of CH_3I. In the oxidative addition of H_2 to Ir complexes, the reactivity increases in the order $Cl < Br < I$ as ligands, and there is very little dependence upon the donor ability of the phosphine.

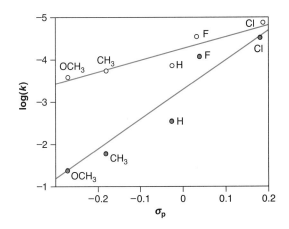

Figure 12.7
A Hammett plot for the oxidative addition of CH_3I (closed circles) and benzyl chloride (open circles) to Vaska's complex [*trans*-IrCl(CO)(p-X–PPh$_3$)$_2$]. Ugo, R., Pasini, A., Fusi, A., and Cenini, S. "A Kinetic Investigation of Some Electronic and Steric Factors in Oxidative Addition Reactions to Vaska's Complex." *J. Am. Chem. Soc.*, **94**, 7374 (1972).

Oxidative Addition at sp^2 Centers

Aliphatic R groups are not the only ones that undergo oxidative addition. Vinyl and phenyl groups work also; Eqs. 12.25 and 12.26 give examples. S_N2 mechanisms cannot be operative, because nucleophilic substitution does not occur on sp^2 hybridized carbons. With alkenes, the stereochemistry of the double bond is almost always retained. This rules out radical intermediates because a vinyl radical would undergo rapid inversion, thereby scrambling the stereochemistry of the alkene in the product.

$$\text{Pd(PPh}_3)_4 \longrightarrow \quad \text{(Eq. 12.25)}$$

$$\text{Pd(PPh}_3)_4 \longrightarrow \quad \text{(Eq. 12.26)}$$

Therefore, the mechanism must be concerted. However, in contrast to the concerted oxidative addition of H_2, the addition of vinyl and phenyl C–X bonds occurs more rapidly with d^{10} complexes than with d^8 complexes. The reactions are still second order, and the normal order for leaving group reactivity is found.

The best way to view the mechanism of these reactions is to postulate prior formation of a metal–olefin π complex, and to view the olefin complex with the metal as a metallacyclopropane. From this vantage point, it is a simple matter of electron pushing to expel the leaving group and retain the stereochemistry (see the examples in the margin).

Oxidative addition at an sp^2 center

Summary of the Mechanisms for Oxidative Addition

The stereochemistry of both the metal center and the R group, the kinetics, and the ligand dependence, can be combined to support three mechanisms for oxidative addition. For apolar bonds, such as H_2, and for vinyl and aryl C–X bonds, there is a concerted single-step reaction. The H_2 or π system coordinates to the metal center, and a single transition state exists involving both bond making to the metal and bond breaking within the H_2 or organic structure.

For polarized bonds where the R group is susceptible to nucleophilic attack, oxidative addition commonly proceeds via an S_N2 mechanism. The metal is nucleophilic by virtue of having a lone pair of electrons in a d orbital. For polarized bonds where the R group is not susceptible to nucleophilic attack, radical mechanisms dominate. Propagation occurs via metal radicals that extract leaving groups from the organic reactant resulting in carbon based radicals that subsequently add to another metal.

Before analyzing the next organometallic reaction type, it should be noted that oxidative addition is the most common manner that σ bonds are cleaved in organic compounds, leading subsequently to other transformations. In this regard, it is still a very actively studied reaction, particularly so with bonds that are not very susceptible to this reaction. The following two Connections highlights discuss mechanisms for oxidative addition of C–H bonds, one of the most inert bonds of this reaction type.

Connections

C–H Activation, Part I

Currently, one of the most sought after procedures in organometallic chemistry is a homogeneous catalytic cycle that involves the activation of a C–H bond in an alkane. We have just examined the oxidative addition of C–X bonds, and in Section 12.3 we will show how this activation of the C–X bond leads to further chemistry. Regardless of how useful this may be, one still needs a C–X bond to start. It would be very useful to directly activate a C–H bond in an alkane, without needing to first create a C–X bond. However, the oxidative addition of a standard C–H bond is a rare reaction. A handful of systems have been developed, and we discuss two. No homogeneous catalytic cycles have yet to be developed, so the examples are simply stoichiometric reactions.

A classic example derives from the Bergman laboratories. Irradiation of $[C_5(CH_3)_5]Ir[P(CH_3)_3]H_2$ in various solvents leads to C–H oxidative addition. Even neopentane and cyclohexane work well.

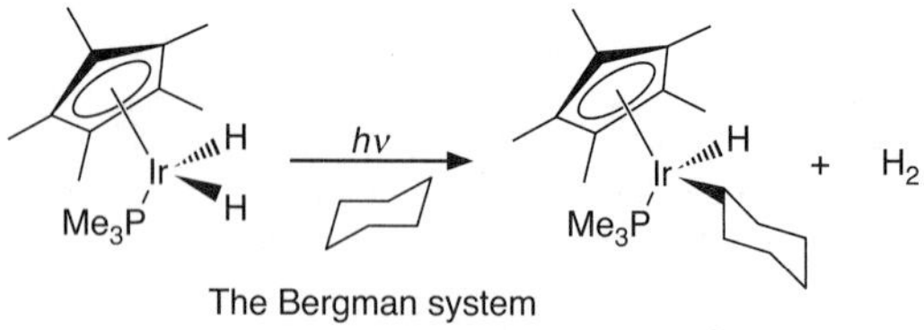

The Bergman system

Three possible mechanisms (**A**, **B**, and **C**) can be envisioned, as shown below [$Cp^* = C_5(CH_3)_5$]. The first is homolysis of an Ir–H bond to create hydrogen and Ir radicals, abstraction of a hydrogen radical from the alkane, and then radical combination (path **A**). A similar radical pathway starts with reductive elimination of H_2 to form $[C_5(CH_3)_5]Ir[P(CH_3)_3]$, followed by hydrogen atom abstraction by Ir, and then radical combination (path **B**). A last mechanism involves a concerted C–H insertion by

Ir after H_2 elimination (path **C**). These last two pathways convert an 18-electron compound to an electron deficient 16-electron structure in the first step.

Path **A** was ruled out using an isotope labeling experiment. This mechanism would retain one of the hydrogens from the reactant in the product. However, when completely deuterated cyclohexane is used as solvent, the product has deuterium and not hydrogen. The plausibility of path **B** was explored using a competition experiment. When p-xylene is the solvent, the Ir intermediate can react with either the benzylic C–H bonds in the methyl groups of xylene, or the C–H bonds on the aromatic ring. Hydrogen abstraction from C–H bonds is well known to preferentially occur with the weaker C–H bonds (see Section 11.7.3). Therefore, this mechanism would predict abstraction from the benzylic C–H bond (see Table 2.2). However, the aromatic C–H bonds are preferentially activated instead, and therefore mechanism **B** is incorrect. Therefore, mechanism **C** seemed most likely.

To test this mechanism, a cross over experiment was performed. If oxidative addition occurs in a concerted mechanism, there should be no cross-over of hydrogens and deuteriums in the product when a mixture of cyclohexanes (C_6H_{12} and C_6D_{12}) is the solvent. Indeed, only a very low level of cross-over (around 7%) was found. This little bit of cross-over was postulated to be due to a hydrido exchange reaction that also occurs during photolysis. These experiments—isotope labeling, bond dissociation energy analysis, and cross-over—highlight how physical organic chemistry can be applied to a very important organometallic reaction.

Janowicz, A. H., and Bergman, R. G. "Activation of C–H Bonds in Saturated Hydrocarbons on Photolysis of (μ^5-C$_5$Me$_5$)(PMe$_3$)IrH$_2$. Relative Rates of Reaction of the Intermediate with Different Types of C–H Bonds and Functionalization of the Metal-Bound Alkyl Groups." *J. Am. Chem. Soc.*, **105**, 3929 (1983).

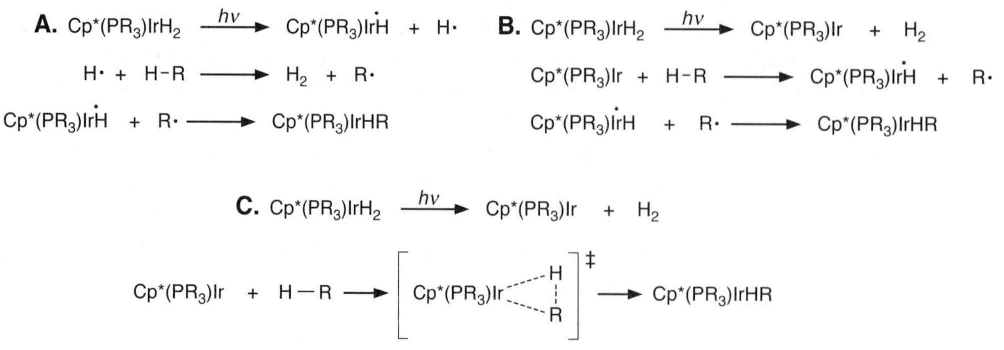

Connections

C–H Activation, Part II

One of the most recent examples of C–H activation comes from the Hartwig laboratories. The complex [C₅(CH₃)₅]W(CO)₃[B(1,2-O₂C₆H₂-3,5-dimethyl)] undergoes a photochemically induced reaction with alkane solvents to form alkylboronate esters, thus activating the C–H and functionalizing the alkyl groups. Various metal-carbonyl–boryl complexes undergo this reaction.

Once again, a variety of possible mechanisms can be envisioned. As mentioned in the previous Connections highlight, any mechanism involving radical abstraction of a hydrogen from an alkane would give preferential abstraction of the weakest C–H bond: 3° > 2° > 1°. However, with all the metal complexes studied, Hartwig found that the terminal carbon of the alkane was the one functionalized. This argues against a radical mechanism, and supports a direct concerted insertion into the least sterically hindered C–H bond.

Since the starting boryl complexes are 18-electron species, one would predict that CO dissociation is required to start the reaction (step 1). However, the presence of excess CO does not slow the C–H activation. Yet,

the addition of a phosphine does stop the C–H activation, and an organometallic product where one CO is replaced with the phosphine is found. This supports a mechanism where the C–H activation of the alkane occurs faster than CO re-association, but slower than association of a phosphine.

Once a CO has been lost, there are two possible mechanisms for the alkyl group transfer to boron (see below, where Bcat′ = the boryl ligand). One is a direct insertion of the metal into R–H (step 2), and the other is a σ bond metathesis (step 3). The first possibility would require a reductive elimination (see Section 12.2.3) to form the alkylboronate ester (step 4). Both pathways would re-associate a CO to give the final organometallic product (step 5). The authors prefer the direct C–H activation, but this has yet to be proven. In summary, C–H activation and functionalization is an elusive process, one that hopefully will have further breakthroughs in the near future.

Waltz, K. M., and Hartwig, J. F. "Functionalization of Alkanes by Isolated Transition Metal Boryl Complexes." *J. Am. Chem. Soc.*, **122**, 11358 (2000).

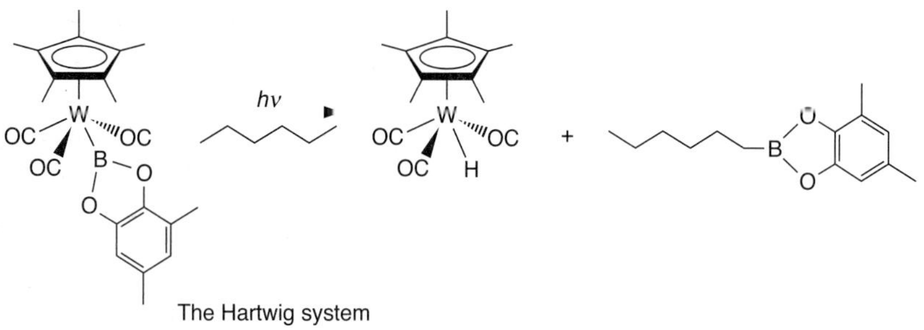

The Hartwig system

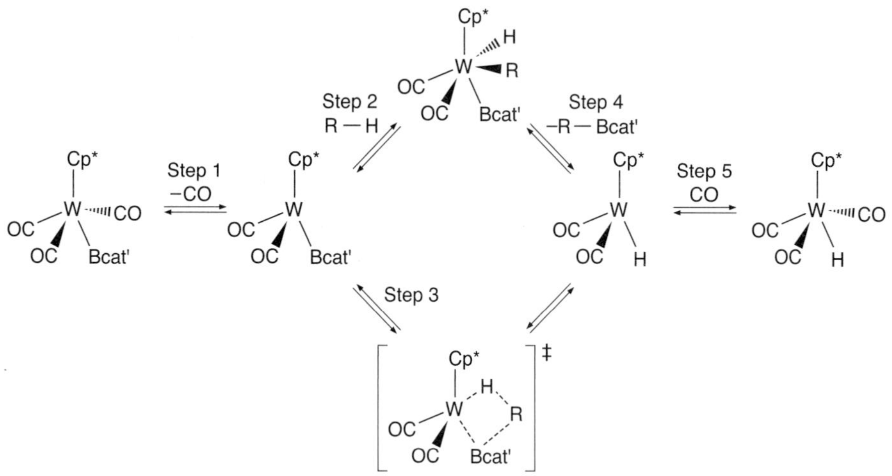

Proposed mechanism

12.2.3   Reductive Elimination

A **reductive elimination** is the reverse of an oxidative addition. The coordination sphere of the metal is diminished, an organic molecule or other structure is eliminated, and the metal is reduced (Eq. 12.27). The study of these reactions is typically done independently of oxidative addition, and therefore these reactions have their own structure–function relationships.

$$\begin{array}{c} A \\ | \\ M-B \end{array} \longrightarrow M \ + \ A-B \qquad\qquad\text{(Eq. 12.27)}$$

Reductive eliminations do not always lead to stable metal products, because the organo-metallic complex is losing electrons and therefore is typically dropping below 18 electrons. These reactions are normally very fast in catalytic cycles, and therefore difficult to observe. Hence, the study of reductive eliminations has not been as extensive as that of oxidative additions. However, this reaction is certainly just as important as oxidative addition in catalysis, because it represents the manner in which organic products are often released from the metal center.

Structure–Function Relationship for the R Group and the Ligands

The kinetics of reductive elimination is first order in the metal complex. One reaction that has been studied extensively is shown in Eq. 12.28. This reductive elimination of methane has to be studied at low temperature because it is too fast at ambient temperature. The following reaction rate order is found: phenyl > ethyl > methyl > allyl. The trend represents a balance between the steric bulk of the R group and the strength of the C–H bond formed. In general, the larger R groups undergo reductive elimination faster, and the stronger the bond that is being formed, the faster the reaction.

$$\underset{Ph_3P}{\overset{Ph_3P\prime\prime\prime\prime}{\diagdown}}Pt\overset{\prime\prime\prime\prime H}{\underset{CH_3}{\diagdown}} \longrightarrow \underset{Ph_3P}{\diagup}Pt\overset{\prime\prime\prime\prime PPh_3}{} \ + \ CH_4 \qquad\text{(Eq. 12.28)}$$

Reductive eliminations are exothermic because the bond strengths are greater in the products. An average bond dissociation energy (BDE) for an M–R bond is 30 kcal/mol, while that of an M–H bond is 60 kcal/mol. Recall from Chapter 2 that the BDE for a C–H bond is around 100 kcal/mol, making that one single bond stronger than the combination of two bonds to the metal. This is also true for elimination of two R groups, where each M–R BDE is around 30 kcal/mol, while a C–C BDE is 80–90 kcal/mol.

The deuterium isotope effect for the reaction shown in Eq. 12.28 is 3.3 (comparing a hydride ligand to a deuteride). This is a fairly large isotope effect, supporting cleavage of the M–H(D) in the rate-determining step. Cross-over studies have shown that there is little to no scrambling of isotopes between metal centers (Eq. 12.29), indicating that the elimination occurs in a monomeric fashion, and does not involve bridging metallic species. These two pieces of data support a concerted mechanism. In support of this, the stereochemistry is retained if the R groups are stereogenic centers.

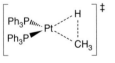

Reductive elimination

$$\underset{Ph_3P}{\overset{Ph_3P\prime\prime\prime\prime}{\diagdown}}Pt\overset{\prime\prime\prime\prime H}{\underset{CH_3}{\diagdown}} \ + \ \underset{Ph_3P}{\overset{Ph_3P\prime\prime\prime\prime}{\diagdown}}Pt\overset{\prime\prime\prime\prime D}{\underset{CD_3}{\diagdown}} \longrightarrow \underset{40}{CH_4} \ + \ \underset{1}{CH_3D} \ + \ \underset{1}{CD_3H} \ + \ \underset{40}{CD_4}$$

$$\text{(Eq. 12.29)}$$

The steric bulk of the ligands also influences the rates of the reactions. The order of reactivity found for reductive elimination of ethane from Pd dimethyl complexes is shown below. The larger triphenylphosphine ligand gives the faster rate. Larger ligands often enhance the rate of elimination due to more steric strain in the reactants. However, for this particular reaction, an additional interesting effect is observed. The addition of free ligand to solution impedes the reaction. Apparently, ligand dissociation occurs prior to reductive elimination. With a *bis*-phosphine ligand the chelate effect greatly impedes full dissociation of a phosphine, and the reaction is much slower. The **chelate effect** is a term used to describe the lower dissociation rate or greater binding constant of a ligand that forms two or more co-ordinations to a metal. Therefore, although an open coordination site is not required for a re-ductive elimination, the electron donating nature of a phosphine impedes the reduction. The Pd complexes must first lose a ligand, lowering the electron count, before reduction of the metal can occur.

Trend in reductive elimination

Since the metal is formally reduced in these reactions, making the ligands more electron withdrawing or better π acceptors will commonly speed reductive eliminations. The elimi-nation of methane shown in Eq. 12.28 has such a dependence, although it is small. Here, para substitution on the triphenyl phosphine ligand with the following groups gives the relative rate constants listed: Cl (9.2) > H (4.5) > CH_3 (1.4) > OCH_3 (0.47).

The reductive elimination of ethane from analogous Pt complexes does not occur. Even heating the complex shown in Eq. 12.30 to 180 °C does not produce ethane, but instead makes methane. One of the hydrogens in the methane comes from the solvent. The reason for this is discussed below when osmium complexes are examined.

$$\text{(Eq. 12.30)}$$

Stereochemistry at the Metal Center

When the groups that eliminate are held trans on the metal center, often no reduc-tive elimination is found. For example, *trans*-$(CH_3)_2Pd(II)(PPh_3)_2$ will not eliminate even at 100 °C. This means that the two groups undergoing elimination have to be in proximity, which further supports a concerted mechanism for this particular reaction.

Other Mechanisms

Not all reductive eliminations are concerted. Many involve radicals. This makes good sense, because M–C bonds are weak. This means that warming above 150 °C will start to give significant amounts of M–C bond homolysis. Let's examine an interesting series of ex-amples that shows a mixture of reductive elimination mechanisms.

The reductive eliminations and the organic products observed for three similar cases are given in Eqs. 12.31–12.33. Dramatically different temperatures are required to make the reac-tions occur at similar rates. For the reaction depicted in Eq. 12.31, one of the hydrogens in the methyl group derives from the solvent. The high temperature required and the abstraction of a hydrogen from the solvent indicates a radical mechanism. The rate-determining step is simply bond homolysis, where the resulting methyl radical abstracts a hydrogen from the solvent and the remaining Os-centered radical goes on to form Os clusters with other Os rad-icals (an analogous mechanism is postulated for Eq. 12.30).

$$\text{(Eq. 12.31)}$$

$$\text{(Eq. 12.32)}$$

$$\text{(Eq. 12.33)}$$

In contrast, the experimental observations for the reactions given in Eqs. 12.32 and 12.33 do not support radical mechanisms. With respect to Eq. 12.32, an M–H bond strength averages about 60 kcal/mol, and therefore there is no homolysis at 125 °C. Interestingly, Norton found that when a mixture of $(CO)_4OsH_2$ and $(CO)_4OsD_2$ is heated together, a statistical mixture of H_2, HD, and D_2 is produced. Yet, addition of D_2 gas to $(CO)_4OsH_2$ gives no HD product. This means that the reductive elimination is not concerted from each individual $(CO)_4OsH_2$ and $(CO)_4OsD_2$ complex, but instead must involve a bridging structure. This is further supported by the formation of $H_2Os_2(CO)_8$ as the single organometallic product. However, the kinetic order in $(CO)_4OsH_2$ is only first order. Therefore, the rate-determining step must not involve the formation of a bridging structure. The only plausible reaction which occurs at above 100 °C and is first order in an Os complex is the loss of a CO ligand. A mechanism that accomodates all the observations is given in Figure 12.8. Ligand loss opens a coordination site (step 1) that can result in a bridging structure from which reductive elimination occurs (step 3). An Os dimer is formed that captures another CO ligand (step 4). Because methyl groups do not easily bridge, this mechanism is not available to the $(CO)_4Os(CH_3)_2$ analog.

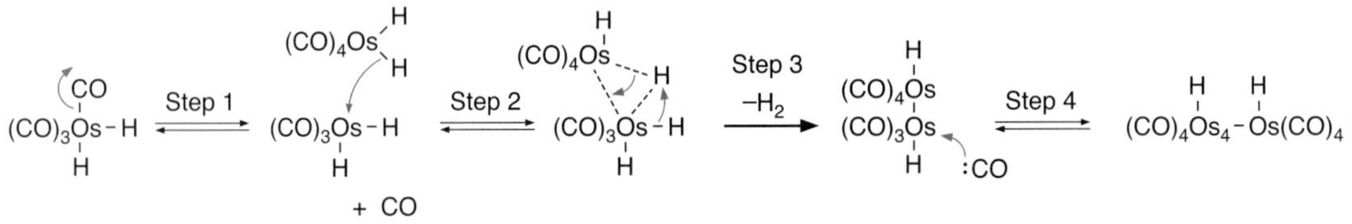

Figure 12.8
Mechanism for the elimination of H_2 from $(CO)_4OsH_2$.

A different mechanism must be operative for the reaction given in Eq. 12.33. This should be immediately obvious because M–H or M–C bond homolysis cannot occur at only 40 °C, and neither does CO dissociation. However, there is a similarity to the mechanism given in Figure 12.8, in that bridging structures are involved. This was demonstrated once again with a cross-over study. When $(CO)_4OsH(CD_3)$ and $(CO)_4OsD(CH_3)$ are mixed together, a statistical mixture of CD_3H, CD_4, CH_3D, and CH_4 is formed. Because the creation of a bridging structure requires an open coordination site on one of the two bridging Os complexes, the first step must now be a migratory insertion (see Section 12.2.5), a reaction known to proceed at lower temperatures than ligand loss or bond homolysis.

Summary of the Mechanisms for Reductive Elimination

As with oxidative addition, we find several mechanisms for reductive elimination. The most common, and the one we will invoke for all the catalytic cycles discussed at the end of this chapter, is a concerted elimination. However, radical mechanisms and those involving bridging structures are possible. In all cases, the loss of a ligand or the use of electron withdrawing ligands facilitates the reactions due to the formal reduction at the metal center.

Connections

The Sandmeyer Reaction

The Sandmeyer reaction is usually covered in introductory organic chemistry in a chapter on electrophilic aromatic substitution, along with diazonium ion chemistry and the formation of diazo dyes. The Sandmeyer reaction involves the substitution on an aryl diazonium compound with Cu(I) salts, as shown to the right. The mechanism of these transformations is sketchy, but now that we know

some common organometallic and organic reactions, we can look over the current thoughts about this reaction.

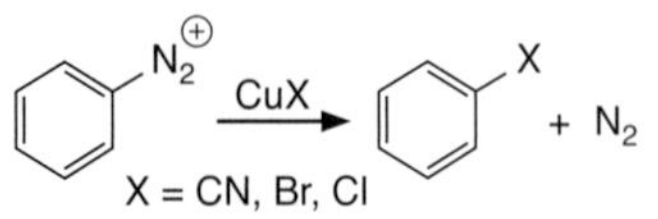

Sandmeyer reaction

Let's examine the possibilities with CuCl. In HCl (typically present under the reaction conditions), cuprous chloride is in equilibrium with dichlorocuprate ion ($CuCl_2^-$). This electron rich anion can transfer an electron to the diazonium ion (step 1), which then eliminates N_2, generating phenyl radical (step 2). At this point, the mechanism is not clear, but the phenyl radical can either abstract a chlorine atom from the $CuCl_2$ radical, or it can couple with this radical (this is the mechanism we show below; step 3). After coupling, reductive elimination gives phenyl chloride and CuCl (step 4).

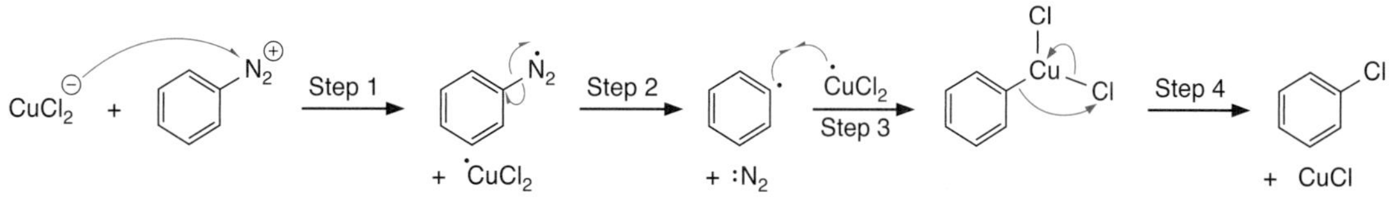

Proposed mechanism

12.2.4   α- and β-Eliminations

There is an entire class of eliminations that occur at metal centers where a group or atom from the ligand migrates to the metal while at the same time an unsaturated system is produced (Eq. 12.34). These reactions are reminiscent of elimination reactions in organic chemistry that were covered in Chapter 11.

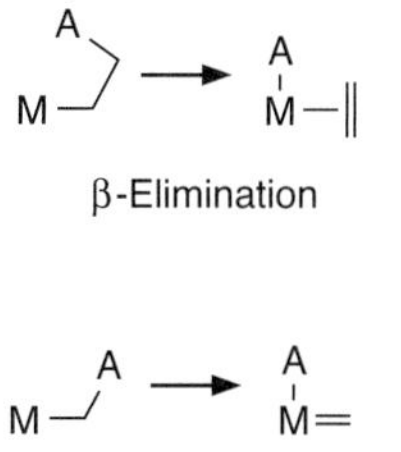

β-Elimination

(Eq. 12.34)

α-Elimination

General Trends for α- and β-Eliminations

By far the most common elimination is **β-hydride elimination**. One example is given in Eq. 12.35. These reactions are very exothermic because a weak M–C bond and a C–H bond are converted to two stronger bonds overall: an M–H bond and a C–C π bond, along with coordination of the π bond to the metal. This reaction can occur whenever a hydrogen is β to the metal and the metal has an open coordination site.

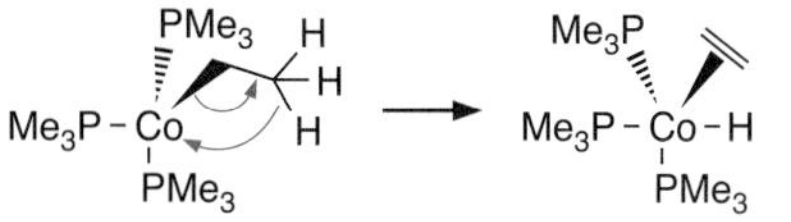

(Eq. 12.35)

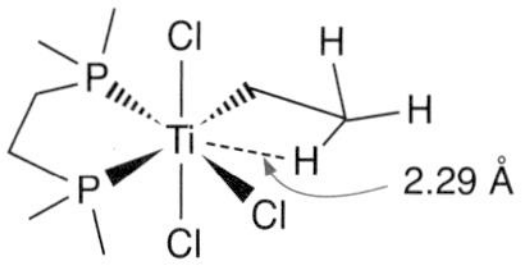

β-Hydride elimination is preceded by a bridging complex where the hydrogen to be eliminated is weakly coordinated to the metal. This coordination is referred to as an **agostic interaction**. Sometimes these bridging species are actually isolated, and one example is shown in the margin. In this complex the Ti---H distance is only 2.29 Å, and the Ti–C–C bond angle in the ethyl group is only 86°.

β-Alkyl elimination is much less common than hydride elimination. This is because it is difficult to have an alkyl group form an intramolecular interaction with the metal, as is required prior to the elimination, and the reaction is now sometimes endothermic because an M–C and a C–H bond are traded for another M–C bond and a C–C π bond. However, β-alkyl elimination can occur. For example, Eq. 12.36 shows an example involving Lu, where

2.29 Å

An agostic interaction

methyl and hydride elimination are competitive. In addition, Eq. 12.37 shows a case where the elimination is only that of an alkyl group, because the hydrogens on the β-position cannot achieve the proper orientation to form an agostic interaction.

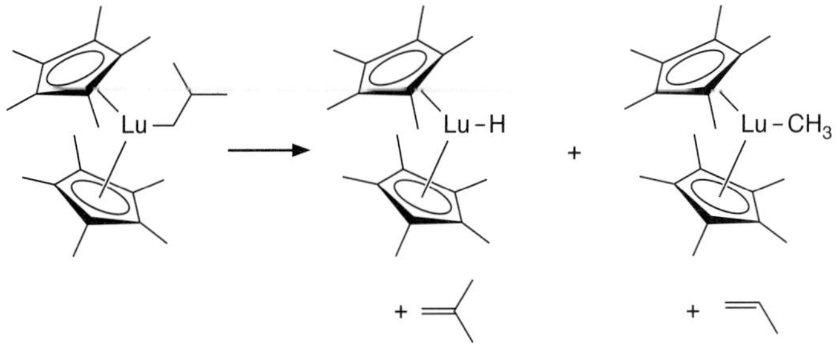

(Eq. 12.36)

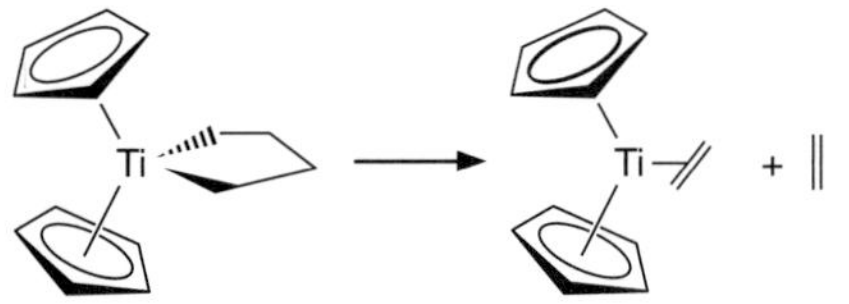

(Eq. 12.37)

With highly electrophilic early transition metal complexes, **α-hydride elimination** is a common reaction. If other alkyl groups are present on the metal center besides the group undergoing elimination, the facile nature of reductive elimination will lead to formation of an alkane. An example of this is given in Eq. 12.38. Whether α- or β-hydride elimination occurs, neopentane is the product that arises from reductive elimination of a neopentyl ligand with a hydride ligand.

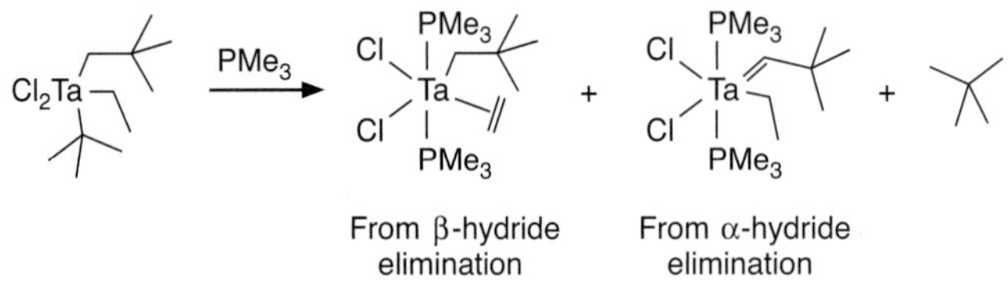

From β-hydride
elimination

From α-hydride
elimination

(Eq. 12.38)

Kinetics

Because β-hydride elimination leads to the coordination of an alkene, the metal undergoes a coordination sphere expansion. This means that 18-electron complexes must first lose a ligand before β-hydride elimination can occur. As a result, ligand loss is often rate-determining for β-hydride elimination. The kinetics of the reaction shown in Eq. 12.39 are first order in the Pt complex. It is inhibited by the addition of free phosphine, thereby indicating the formation of an intermediate resulting from phosphine loss prior to the hydride shift and reductive elimination. When hydrogens and deuteriums can be competitively abstracted, isotope effects ranging from 2 to 4 are seen.

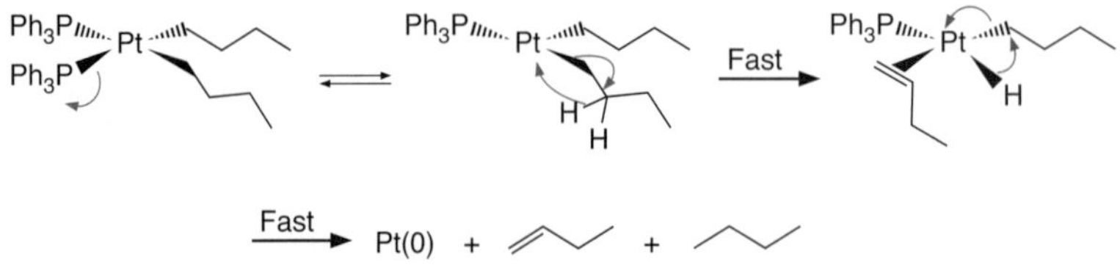

(Eq. 12.39)

One often wants to avoid a β-hydride elimination during catalytic cycles, because it can give unwanted side products. The existence of a β-hydrogen on an alkyl group is a requirement for β-hydride elimination. In fact, it is generally observed that alkyl complexes that lack hydrogens on the β-carbon are thermally more stable than those with β-hydrogens. For example, $W(CH_3)_6$ is much more stable than $W(CH_2CH_3)_6$. In addition, due to the requirement of an open coordination site for β-hydride elimination, coordinatively-saturated complexes or those with chelated ligands are more thermally stable. Any electronic or steric effects that hinder an increase in coordination number will impede β-hydride elimination.

Stereochemistry of β-Hydride Elimination

There are stereoelectronic requirements for β-hydride elimination, just as there are for E2 reactions in organic chemistry. In Section 10.13.8, we showed that E2 reactions require either synperiplanar or antiperiplanar arrangements of the hydrogen and leaving group that are being eliminated. Similar requirements are necessary in β-hydride elimination, but because the elimination occurs intramolecularly, only the synperiplanar arrangement facilitates the reaction. As shown in the Newman projection in the margin, the hydrogen must be aligned with the metal to directly migrate to it, and the other groups on the alkane must be aligned in order to form the π bond. This creates a four-centered transition state with a completely concerted mechanism.

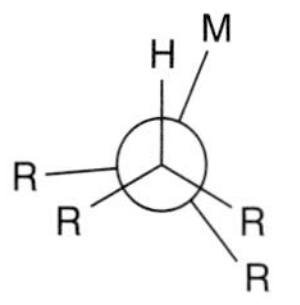

As support for this stereoelectronic requirement in the elimination, the relative rates for the β-hydride elimination shown below have been found. The β-hydride elimination from the metallacycloheptane occurs at approximately the same rate as the dibutyl complex, but the metallacyclopentane eliminates 10^4 times slower. The β-hydrogens cannot achieve a synperiplanar arrangement with the metal, and the elimination is greatly retarded.

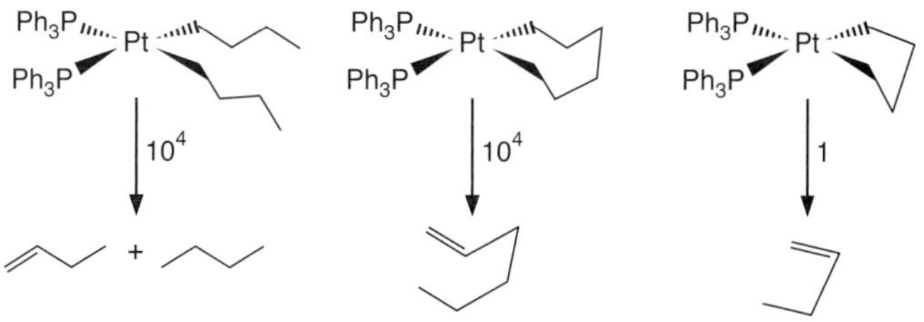

Syn elimination

β-Hydride eliminations

Further support for the synperiplanar arrangement in β hydride eliminations comes from the stereochemistry of the alkene created. Eq. 12.40 shows a case where the less stable product is found, due to the required synperiplanar arrangement of the metal and the hydrogen.

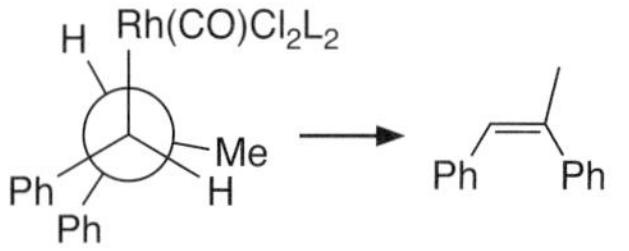

(Eq. 12.40)

12.2.5 Migratory Insertions

Another reaction class in organometallic chemistry involves insertions. These are the reverse of eliminations. The term "insertion" is not one that was used extensively in Chapters 10 and 11, although it does have analogs in organic chemistry (recall carbene reactions and additions). In organometallic chemistry, however, insertions play an important role. These reactions involve insertion of a ligand into another metal–ligand bond (Eq. 12.41). Here, one group (A) migrates to the other (B–C), and hence this is known as a **migratory insertion**. The oxidation state of the metal has not changed, but a coordination site to the metal has opened up.

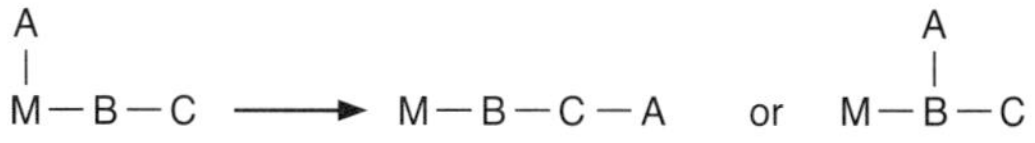

(Eq. 12.41)

A very large variety of insertion reactions are known. While the A group of Eq. 12.41 is typically a hydride, alkyl, or aryl, the B–C ligand takes many forms. Essentially any unsaturated compound that can act as a ligand to a metal can undergo an insertion. Just a few examples include CO, NO, alkenes, alkynes, SO_2, CO_2, carbonyls, imines, cyanides, and isocyanides. Most mechanistic studies have focused on migration to CO and alkenes, and these are the only examples we look at below. However, in Section 13.2.7, we will return to migratory insertions when we examine the industrially important process of alkene polymerization.

Kinetics

Migration of CO to alkyl, aryl, or hydride ligands is generally induced by the addition of a ligand, such as a phosphine or additional CO. This is because the insertion opens up a coordination site on the metal, which needs to be filled by another ligand, otherwise the insertion is reversible. Hence, the mechanism can be schematically given as in Eq. 12.42, and the rate expression using the steady state approximation for the intermediate is given in Eq. 12.43. This equation tells us that there can be a kinetic dependence upon added ligand. However, when $CH_3Mn(CO)_5$ is studied, the rate of insertion of CO into the methyl–Mn bond is independent of the concentration of the ligand that is adding in the second step: L = CO, PR_3, or NH_2R. This means that migratory insertion is rate-determining, and $k_{-1} \ll k_2[L]$.

$$\text{(Eq. 12.42)}$$

$$\frac{d[P]}{dt} = \frac{k_1 k_2 [MR(CO)][L]}{k_{-1} + k_2[L]} \qquad \text{(Eq. 12.43)}$$

We have noted in our discussion of previous reaction classes that relative rates can be correlated with bond dissociation energies. The same is true of migratory insertions. As shown in the reactions grouped under Eq. 12.44, migratory insertions of CO or alkenes into M–alkyl bonds are typically much faster than insertions into M–hydride bonds. This is due to the much stronger M–H bonds relative to M–R bonds.

$$\text{(Eq. 12.44)}$$

The rate of migration of alkyl groups to CO ligands can be dramatically increased by the addition of a Lewis acid. This can be viewed as electrophilic activation of the CO group toward nucleophilic attack by a metal–R bond. One example is given in Eq. 12.45.

$$\text{(Eq. 12.45)}$$

Studies to Decipher the Mechanism of Migratory Insertion Involving CO

There are three possible mechanisms for migratory insertion involving a CO ligand and an M–alkyl bond. These possibilities are shown in Figure 12.9 for $(CO)_5MnCH_3$. One mechanism involves direct insertion of free CO into R–M (possibility A, Figure 12.9). The second involves the movement of a CO ligand to a cis R group (possibility B), while the third involves movement of the R group to a cis CO ligand (possibility C). The first possibility is easy

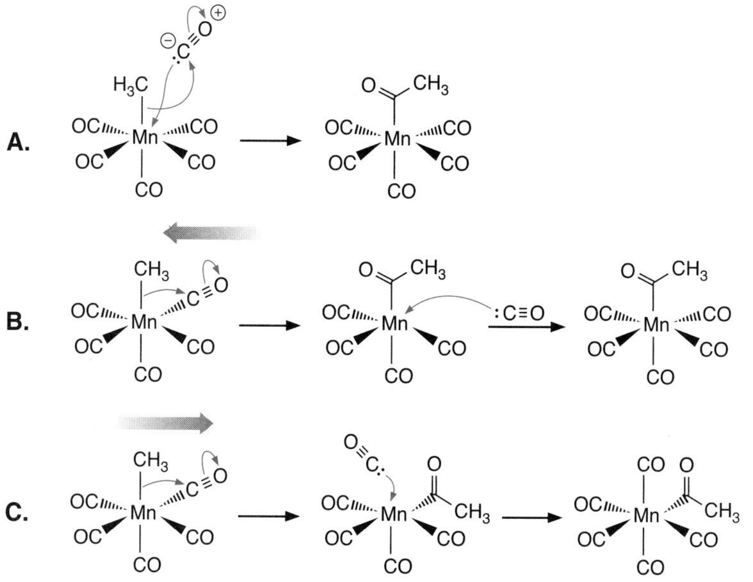

Figure 12.9
Three possible mechanisms for CO insertion reactions. **A.** Direct insertion into free CO, **B.** Migration of the CO to the R group, and **C.** Migration of the R group to a CO ligand. The colored arrows denote the direction for flow of two electrons, while the bold arrow shows the direction of movement of groups.

to test for. As shown in Eq. 12.46, when isotopically-labeled CO gas is added to the reaction vessel, only the product where the label is in a cis CO ligand is isolated. Hence, the CO gas did not directly insert into the R group, but instead coordinated to the metal, so mechanism **A** of Figure 12.9 must be incorrect.

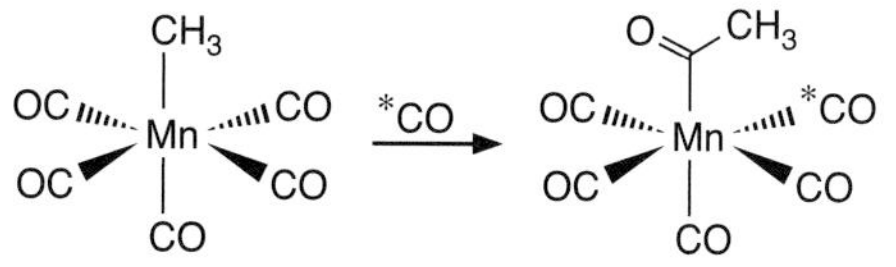

(Eq. 12.46)

To differentiate between mechanisms **B** and **C** in Figure 12.9, one needs a stereochemical analysis. Such an experiment was performed on the reverse of the migratory insertion. By examining the de-insertion reaction, the product ratio given in Eq. 12.47 was found for the products incorporating the label. This product ratio supports movement of the R group to the CO (that is, mechanism **C**). Let's see the reasoning by examining Figure 12.10.

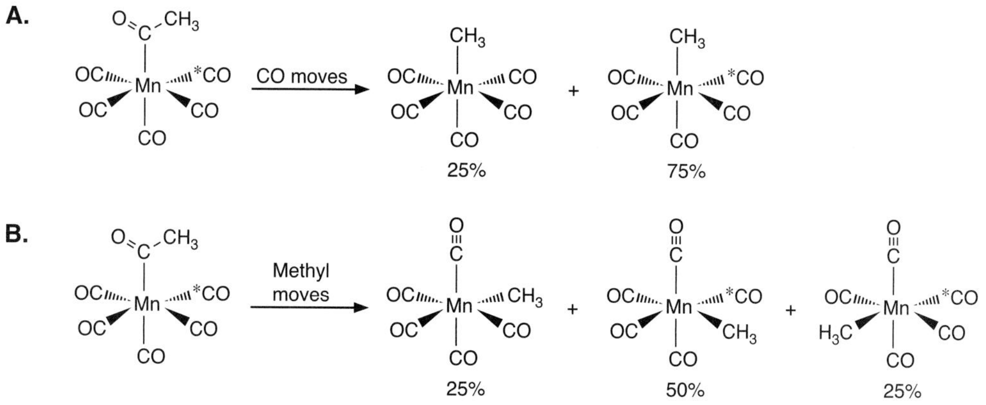

Figure 12.10
Reverse of the mechanisms given in Figure 12.9 leads to different stereochemistry in the products. **A.** Movement of the CO group in the de-insertion, and **B.** Movement of the methyl group in the de-insertion.

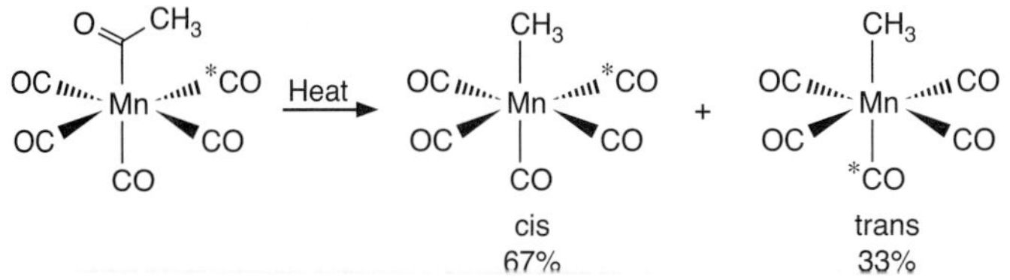

(Eq. 12.47)

If the CO portion of the acyl group moves during the de-insertion (Figure 12.10 **A**), it will replace the cis-labeled CO 25% of the time, and never give any trans product. In contrast, if the methyl group moves during the de-insertion, it will replace the labeled CO 25% of the time, but there is a 2-to-1 chance of moving to a position that is cis or trans to the labeled CO ligand (Figure 12.10 **B**). Hence, the 2-to-1 ratio given in Eq. 12.47 supports migration of the R group.

Another set of studies similarly supports R group migration in Pd complexes. As shown in Figure 12.11, the products are different for the reaction of *cis*- and *trans*-Pd(PPh$_3$)$_2$(Et)$_2$ with CO gas. The trans reactant gives 3-pentanone as the product, whereas the cis reactant gives ethylene and propanal. If the R group moves, then as shown in Figure 12.11 **A** for the trans reactant, there is never an open coordination site cis to an ethyl group (after trapping with CO), and hence no β-hydride elimination can occur. With the cis reactant, however, β-hydride elimination after CO coordination can occur if the R group moves. Importantly, studies on complexes other than the Mn and Pd compounds discussed here have shown different results. Therefore, although the R group is commonly the species found to be migrating, this is not always the case.

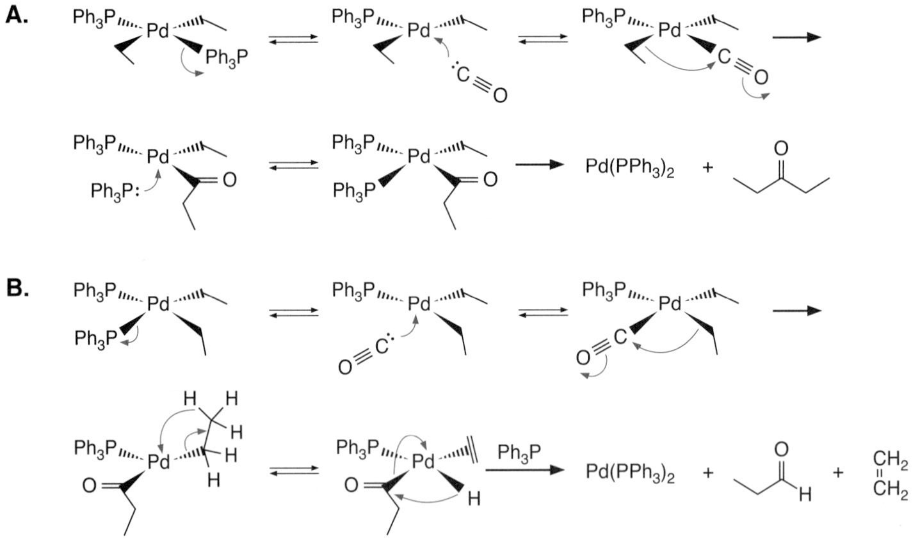

Figure 12.11
Different mechanistic pathways followed for the reaction of Pd(PPh$_3$)$_2$(Et)$_2$ with CO gas, depending upon the stereochemistry of the reactant. The products support movement of the ethyl group in the migratory insertion step for both reactants.

Other Stereochemical Considerations

The group that migrates routinely does so with retention of stereochemistry. Two examples are given in Eqs. 12.48 and 12.49. These two examples show stereogenic centers directly attached to the metal, which do not change their sense of chirality upon migration. Therefore, when the CO ligand inserts into the M–R bond, the R group adds via frontside attack, not backside attack as in an S$_N$2 reaction.

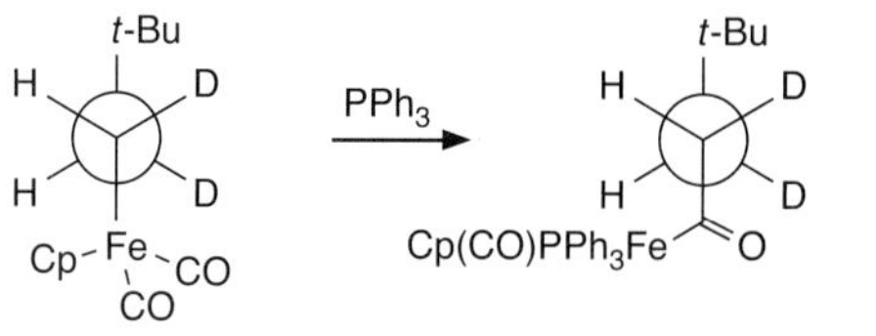

(Eq. 12.48)

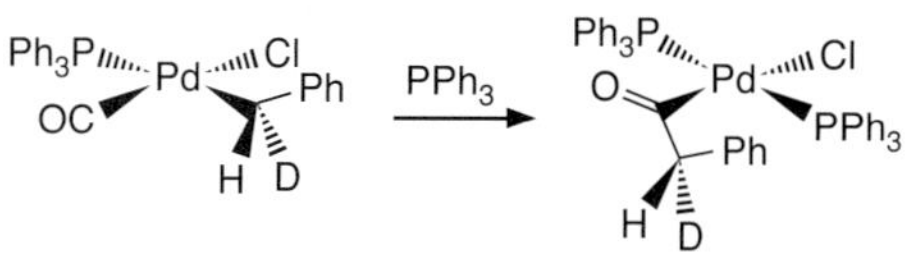

(Eq. 12.49)

Migration to stereogenic centers in alkenes also routinely occurs with retention of stereochemistry. Eq. 12.50 shows the insertion of an isotopically-labeled E-alkene into a M–D bond from Cp_2ZrDCl. The product retains the stereochemistry of the alkene, thereby supporting a syn addition of the Zr–D bond. The syn addition leads to a picture of the insertion involving attack of the Zr and D on the same face of the alkene—that which is coordinated to the metal.

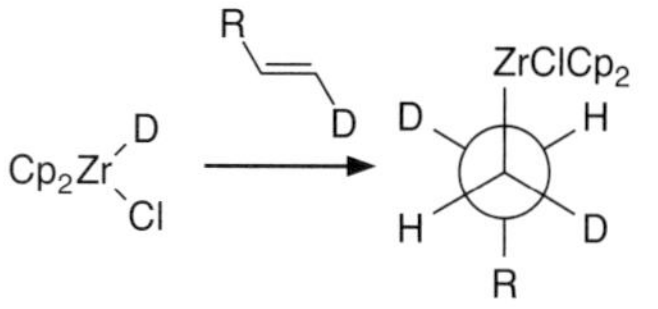

(Eq. 12.50)

The stereochemistry for migratory insertion of alkynes has led to effective methods for creating each of the different stereoisomers of deuterated terminal alkenes. Eq. 12.51 shows a sequence of reactions involving t-butylacetylene. Due to the 100% stereoselective syn addition of the Zr–D bond, only one product is obtained.

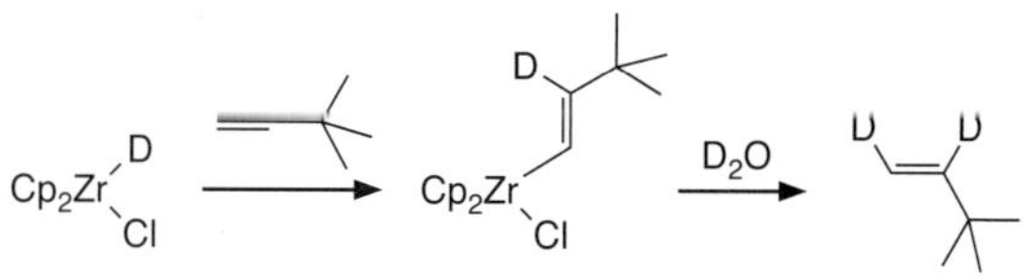

(Eq. 12.51)

12.2.6 Electrophilic Addition to Ligands

Electrophilic addition involves the addition of an electrophile to a metal-bound ligand. These additions can release the newly formed structure from the metal, or they can simply alter the ligand. They fit the general reaction type given in Eq. 12.52.

$$M-A \xrightarrow{E^{\oplus}} M-\overset{\oplus}{A}-E \quad \text{or} \quad \overset{\oplus}{M} + A-E$$

(Eq. 12.52)

Reaction Types

The most common electrophilic reagents for attack on ligands are protic acids, alkylating and acylating reagents, halogens (X_2), and Lewis acids (often B, Al, or Hg reagents). Below, a few examples are shown with an emphasis on variety. Eq. 12.53 gives an example using a protic acid that releases an organic structure. Here the M–R bond was protonated to release the R group. Eq. 12.54 shows an example using Hg, which constitutes a transmetallation (see Section 12.2.1). Eq. 12.55 shows an alkylation of a ligand that is nucleophilic by virtue of possessing free lone pair electrons.

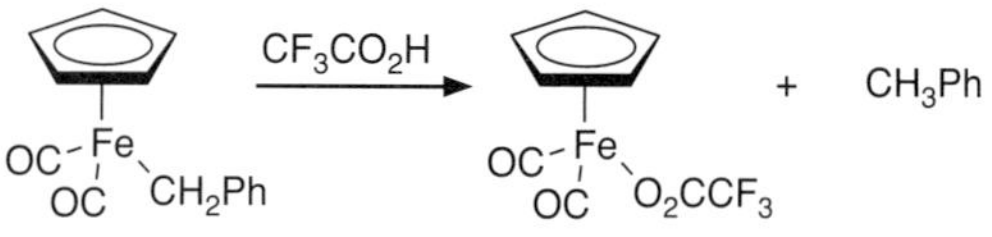

$+ \quad CH_3Ph$

(Eq. 12.53)

$$\text{(Eq. 12.54)}$$

$$\text{(Eq. 12.55)}$$

There are several reactions that have more direct analogs to organic reactions. Aromatic rings coordinated to metals can undergo Friedel–Crafts alkylations and acylations (see Eq. 12.56 for an example). Alkylation of enolate analogs is also a common reaction (Eq. 12.57). In this case, the alkylation leads to a complex with a heteroatom directly attached to an alkylidene carbon, creating what is known as a **Fischer carbene**.

$$\text{(Eq. 12.56)}$$

$$\text{(Eq. 12.57)}$$

Common Mechanisms Deduced from Stereochemical Analyses

There are two mechanisms found for most electrophilic reactions that remove alkyl groups from metal centers. The first is the S_E2 reaction, which we examined in the Going Deeper highlight on page 715. These reactions can involve either frontside or backside attack on the carbon, giving retention or inversion of stereochemistry, respectively. Protonation is commonly frontside with early transition metals, but reactions with Hg, Al, and B reagents give differing results based upon the organometallic species and the electrophile. The second common mechanism involves radical intermediates, and it is most often involved in reactions with halogens and late transition metal elements. These reactions occur via diverse pathways, giving retention, inversion, or racemization depending upon the case.

One example of a stereochemical analysis that gives clear insight into the mechanism of electrophilic attack involves the reaction shown in Eq. 12.58. With early transition metals, the addition of X_2 commonly gives retention of configuration. The X atom is delivered to the ligand from the frontside in a concerted four-centered transition state.

$$\text{(Eq. 12.58)}$$

12.2.7 Nucleophilic Addition to Ligands

Nucleophilic attack on coordinated ligands is also a common reaction in organometallic chemistry (Eq. 12.59). With nucleophilic attack, the metal often acts as an electron sink. Hence, cationic complexes undergo nucleophilic attack more readily than do neutral or anionic complexes. In addition, many organic structures are activated to nucleophilic attack by coordination to transition metals. The metals withdraw electrons from the organic molecule, and thereby act as electrophilic catalysts. Common nucleophiles include Grignard reagents, organolithiums, alkoxides, amines, phosphines, and hydride.

$$\overset{\oplus}{M}-A \xrightarrow{\;:\overset{\ominus}{N}\;} M-A-N \quad \text{or} \quad M \;+\; A-N \qquad \text{(Eq. 12.59)}$$

The site of nucleophilic attack is commonly one of the atoms that donates directly to the metal, because it is the most electrophilic. However, the LUMO of the overall complex will dictate the site of attack, where the atom that contributes most to the LUMO will be the preferred site of attack.

Reaction Types

There are a very large number of ligand types that are susceptible to nucleophilic attack when coordinated to transition metals, including CO, alkenes, alkynes, arenes, alkylidenes, and alkylidynes. Here we show several examples to impart an appreciation for the tremendous variety. Eqs. 12.60 and 12.61 involve attack on carbon monoxide, which is likely the most common nucleophilic attack. Eq. 12.62 shows nucleophilic addition to an alkylidene, which can lead to leaving group departure and substitution on the alkylidene carbon. This is analogous to the transformation of an ester to a ketone, where the metal acts in place of oxygen as the electron sink for addition of the nucleophile. Eqs. 12.63, 12.64, and 12.65 show addition to a coordinated alkene, alkyne, and allyl group, respectively. Such additions to unsaturated systems are common reactions in synthetic sequences. Finally, Eq. 12.66 shows nucleophilic aromatic substitution on a coordinated arene. The metal acts as an electron withdrawing group that activates the arene ring toward the nucleophilic attack. In all of these nucleophilic additions, the metal is an electron sink that accepts, via either resonance or inductive effects, the additional negative charge.

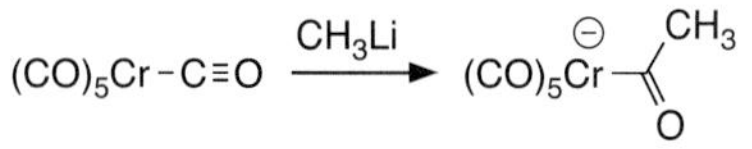

(Eq. 12.60)

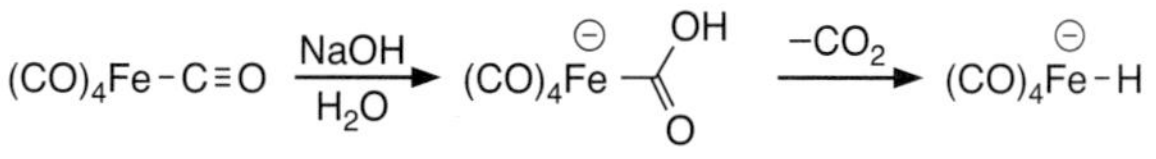

(Eq. 12.61)

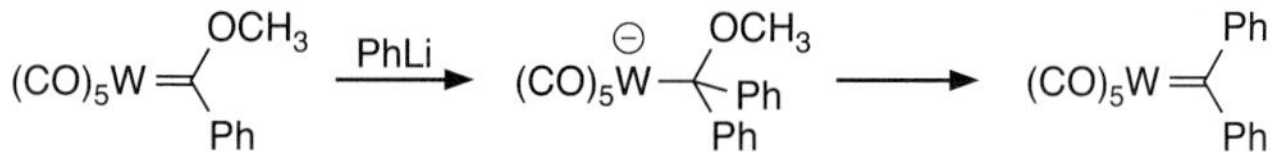

(Eq. 12.62)

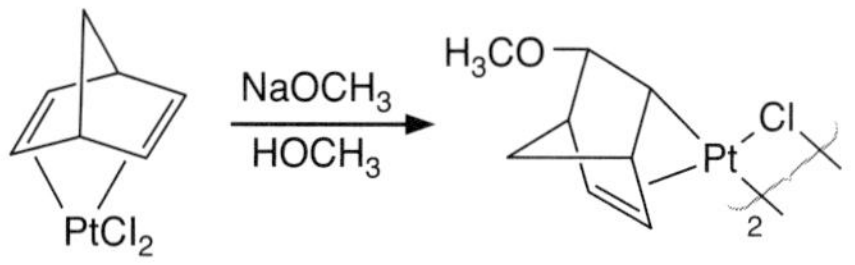

(Eq. 12.63)

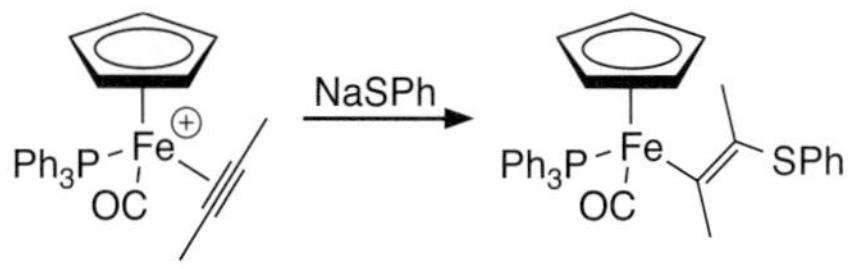

(Eq. 12.64)

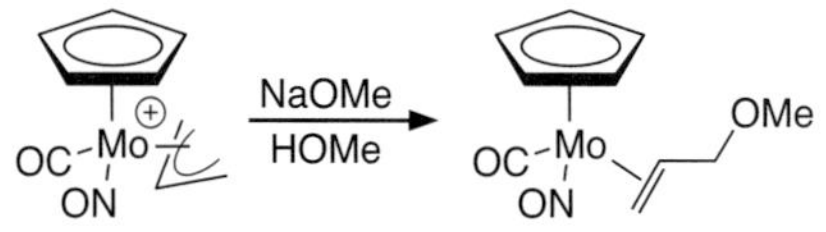

(Eq. 12.65)

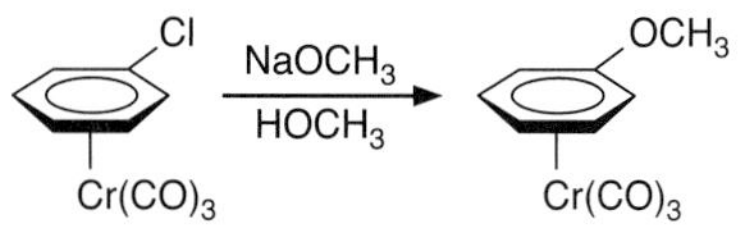

(Eq. 12.66)

Stereochemical and Regiochemical Analyses

There is not much to say about the mechanism of nucleophilic addition to ligands. The nucleophile typically adds in the rate-determining step, and in some cases a leaving group departs in a subsequent step. The most interesting aspects of the reactions are the stereo-chemistry and regiochemistry.

The reaction shown in Eq. 12.67 gives complete inversion at the stereogenic center attached to Fe. This requires a backside attack, and the most logical way for this to occur is to start with an oxidative addition of the Br_2, and then nucleophilic attack by bromide with Fe as the leaving group. The sequence of electrophilic addition of X_2 to the metal followed by nucleophilic attack on the ligand is common for middle-to-late transition metals. Interest-ingly, when phenyl is in the β-position, the reaction proceeds with retention. Retention is best explained by a double inversion, and the phenonium ion has been substantiated as the intermediate formed (Eq. 12.68).

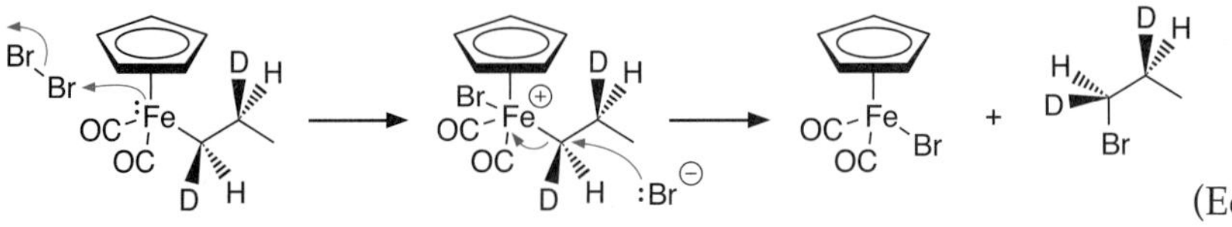

(Eq. 12.67)

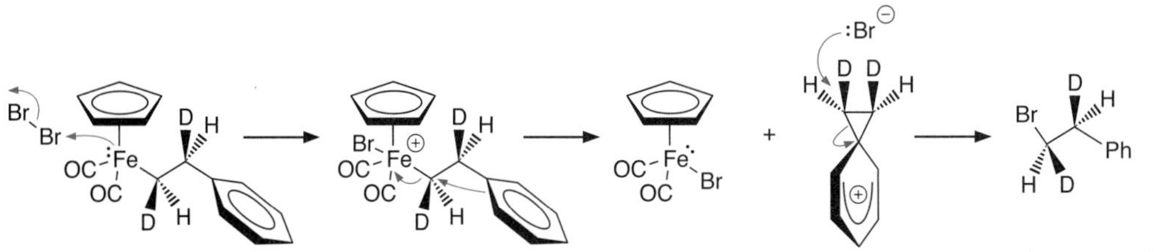

(Eq. 12.68)

Nucleophilic attack on alkenes proceeds almost always by an anti addition pathway. Eq. 12.69 shows an example. The nucleophile has added to the face of the π system opposite to the metal. This makes good sense, in that sterics would preclude attack on the alkene from the face where the metal is coordinated. However, there are some examples of the syn addi-tion of a nucleophile, and these reactions presumably occur via coordination of the nucleo-phile first to the metal, followed by addition to the alkene. Coordination followed by addi-tion formally constitutes a migratory insertion, and such reactions were covered in Section 12.2.5.

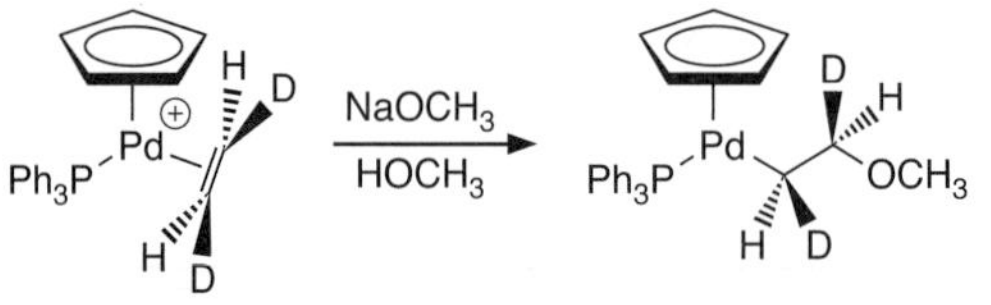

(Eq. 12.69)

The regiochemistry of addition to conjugated dienes normally occurs at the ends of the extended π system. One of the most well studied examples involves the η^5-dienyl cationic iron complexes shown in Eqs. 12.70 and 12.71. Addition to either end of the extended π sys-tem gives regioisomeric products, which are difficult compounds to obtain using other syn-thetic routes.

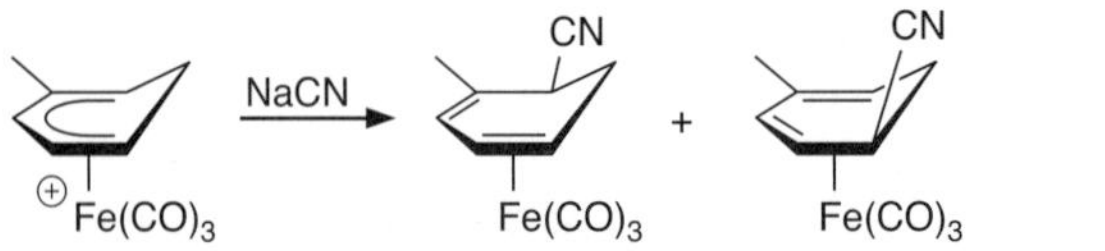

(Eq. 12.70)

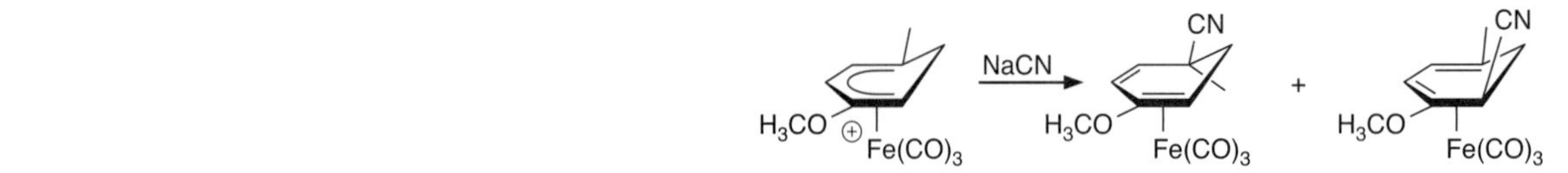

(Eq. 12.71)

Going Deeper

Olefin Slippage During Nucleophilic Addition to Alkenes

In a Going Deeper highlight on page 709, we discussed the Dewar–Chatt–Duncanson model of metal–olefin bonding. This bonding model involves both donation of electrons from the alkene to the metal and back-bonding from the metal to the alkene. Normally, the donation is greater than the back-bonding, and therefore the alkene is somewhat electrophilically activated toward nucleophilic attack. However, activation is actually best when the olefin is slightly displaced from a perfectly symmetrical η^2-alkene.

The displacement places more positive charge on the carbon that is further from the metal. In other words, the displacement shifts the structure of the alkene complex toward the structure of an alkyl complex with a cat-

ionic β-carbon. Such slippage of the alkene structure naturally occurs along the reaction coordinate of the nucleophilic addition, and it is proposed to occur early during the addition leading to a more facile reaction.

Some experimental results support this notion. The crystal structure of the following complex reveals the Pd–C bond lengths shown. The bond lengths to carbons C and D are nearly identical, so this alkene has a symmetrical η^2-structure. However, the bond lengths to carbons A and B are significantly different. Nucleophilic attack by alkoxide anions, enolates, and amines occurs at carbon A only, supporting the notion that slipping the alkene enhances its susceptibility to nucleophilic attack.

Eisenstein, O., and Hoffmann, R. "Activation of a Coordinate Olefin Toward Nucleophilic Attack." *J. Am. Chem. Soc.,* **102**, 6148 (1980).

12.3 Combining the Individual Reactions into Overall Transformations and Cycles

Now that most of the common steps involved in organometallic transformations have been covered, we can consider combining them. It is the combination of several individual organometallic reactions that makes new organic compounds, and hence gives organometallic chemistry an important role in organic synthesis and catalysis.

One of the main goals of this chapter is to provide you with enough information to predict a mechanism for an organometallic transformation, and then, based upon the information presented in previous chapters, propose experiments to test the prediction. Here, we go through the logic for proposing a complete mechanism. This is done in the context of several catalytic cycles. These are cycles that we feel all organic chemists should be exposed to at some time in their education, and even better, these mechanisms should become just as familiar as the standard S_N2, S_N1, acyl transfer, and other common organic mechanisms that were covered in Chapters 10 and 11.

Nearly all organometallic sequences used in synthesis or catalytic cycles require coordinating the organic compound with the metal. There are two ways that we have covered for this to occur. The first is coordination to an empty coordination site on the metal, exploiting a simple Lewis acid–base interaction. The second reaction is an oxidative addition, using any one of the number of mechanisms presented in Section 12.2. As may be expected, most catalytic transformations end with reactions that release the products. Ligand loss, β-hydride elimination, attack on ligands, and/or reductive elimination can expel the organic product from the metal. Since ligand exchange reactions and oxidative addition/reductive elimination are necessary in almost all organometallic reactions, we will encounter these over and over again in subsequent discussions. The various reactions in between ligand addition and loss are what create the synthetically and industrially useful compounds. Insertions, hydride eliminations, nucleophilic additions, and other reactions are combined to transform the organic reactants into products.

12.3.1   The Nature of Organometallic Catalysis—Change in Mechanism

Chapter 9 examined methods of catalysis. We have seen that acid–base catalysis and enzymatic catalysis can all be considered to occur due to an increased binding of the transition state of the reaction relative to binding of the substrate of that reaction. We explained catalysis by examining the potential surface of the uncatalyzed reaction, and then contrasted it to the surface of the catalyzed reaction (see Figures 9.1 and 9.2). In these comparisons, very similar transition states are involved. With organometallic catalysis, there are completely new transition states, completely new intermediates—essentially a completely new mechanism.

In the next several sections, we examine a series of catalytic cycles. When examining these cycles, determine for yourself whether there is an organic chemistry mechanism that will lead to the products without the metal complex being present. Try to combine steps from Chapters 10 and 11. For example, the hydroformylation reaction is covered. Here, a terminal alkene, carbon monoxide, and hydrogen gas are transformed into an aldehyde. There is no reasonable organic mechanism available for this transformation in the absence of the catalyst. Instead, catalysis derives from the fact that combining organometallic reactions does give a reasonable path to the product.

12.3.2   The Monsanto Acetic Acid Synthesis

The Monsanto acetic acid synthesis is a classic example of an industrially useful transformation based upon organometallic reactions. In this synthetic procedure, carbon monoxide and methanol are coupled to form acetic acid in water. Two catalysts are used. The first is HI, playing the role of a strong acid with a nucleophilic counterion, and the second is the 16-electron anionic species $Rh(CO)_2I_2^-$ (Eq. 12.72).

$$CH_3OH + CO \xrightarrow[HI]{Rh(CO)_2I_2^-} CH_3CO_2H \qquad \text{(Eq. 12.72)}$$

Let's begin our analysis of this cycle by examining what is required in the reaction. Logically, the C=O in acetic acid comes from carbon monoxide, while the methyl group comes from the methanol. Therefore, CO has inserted into the C–O bond in methanol. This, in turn, requires that Rh has inserted into the methanol C–O bond, or an equivalent bond, at some point in the mechanism. With these ideas in mind, we can start to analyze a logical pathway for this cycle (Figure 12.12).

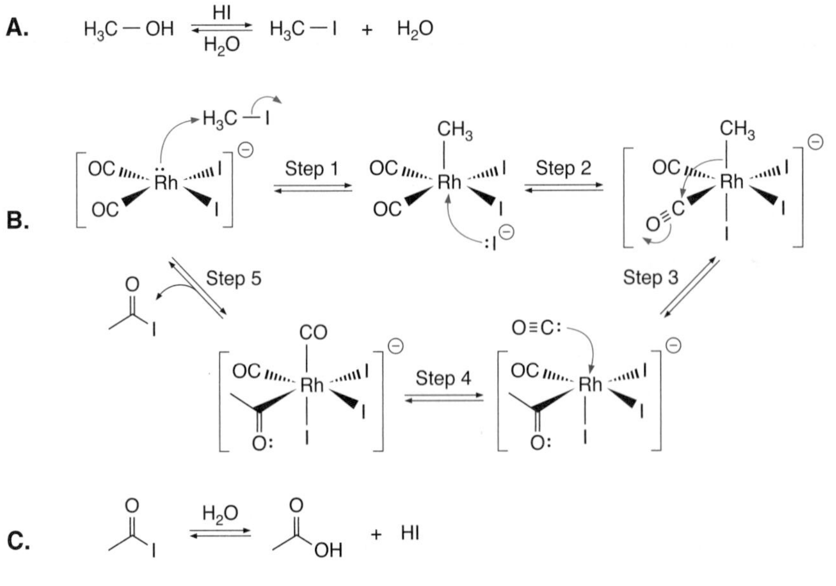

Figure 12.12
The catalytic cycle involved in the Monsanto acetic acid synthesis. **A.** An equilibrium established in solution that supplies CH_3I.
B. The organometallic portion of the reaction.
C. Conversion of the acetyl iodide to acetic acid.

Since CO has inserted into the methanol C–O bond or some equivalent bond, this bond needs to oxidatively add to the Rh. The Rh has 16 electrons in the catalyst, and it is an example of a complex that is nucleophilic due to a lone pair and a negative charge. Therefore, of the mechanisms used for oxidative addition (Section 12.2.2), the nucleophilic route seems most logical. However, nucleophilic attack on methanol is illogical because hydroxide is a very poor leaving group. Therefore, methanol must be transformed to a compound that can undergo the oxidative addition. This is the role of HI. As shown in Figure 12.12 **A**, the HI converts methanol to methyl iodide, which is readily amenable to oxidative addition. The oxidative addition places the methyl group cis to a CO ligand (step 1, Figure 12.12 **B**), which is appropriate for the necessary migratory insertion.

Figure 12.12 **B** shows the organometallic steps, where an insertion drops the electron count back to 16 (step 3). Coordination of another CO ligand from solution brings it back up to 18 (step 4). This coordination of a sixth ligand places the acetyl and iodide groups closer together on the metal, assisting their coupling in a reductive elimination step (step 5). The reductive elimination has to occur from the 18-electron complex, because this brings the metal back to 16 electrons. Alternatively, elimination from a 16-electron complex would lead to a 14-electron complex, which is too unstable. The reductive elimination regenerates the catalyst, but it does not give acetic acid. Instead, acetyl iodide is produced. In acidic water, the acyl iodide is rapidly transformed to acetic acid, thereby regenerating the HI catalyst.

12.3.3 Hydroformylation

Hydroformylation involves the addition of both H_2 and CO to a terminal alkene to create an aldehyde. It can be catalyzed by many transition metal complexes, but a classic example is the use of $HCo(CO)_4$ (Eq. 12.73).

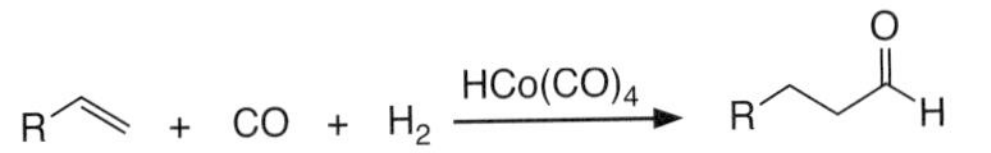

$$R \diagup\!\!\!\diagdown + \; CO \; + \; H_2 \xrightarrow{\;HCo(CO)_4\;} R \diagdown\!\!\diagup\!\!\diagdown\!\!\underset{H}{\overset{O}{\diagup}} \qquad\qquad \text{(Eq. 12.73)}$$

In hydroformylation the alkene has added both CO and H_2. This means that the alkene must at some point in the cycle become an alkyl group in order to insert into CO, which requires that the insertion of the alkene into a hydride occurs first. At the end of the cycle, the most straightforward manner to form an aldehyde is reductive elimination of an acyl group and a hydride. Using this logic, we can write the mechanism.

$HCo(CO)_4$ is an 18-electron complex, so ligand loss is required in the first step (Figure 12.13). The resulting open coordination site can facilitate either oxidative addition of H_2, or coordination of the alkene. You can write a mechanism starting with either, but oxidative addition of H_2 would require a second loss of CO to again open a coordination site for the coordination of the alkene (Eq. 12.74). Since CO loss is more likely with a Co(I) oxidation state than Co(III) because the metal is more electron rich, coordination of the alkene occurs first, as shown in Figure 12.13 (step 2). Hydride insertion to give the most sterically assessible linear alkyl group (step 3), followed by addition of another CO ligand generates $RCH_2CH_2Co(CO)_4$ (step 4), an 18-electron species. The intermediate $RCH_2CH_2Co(CO)_4$ is

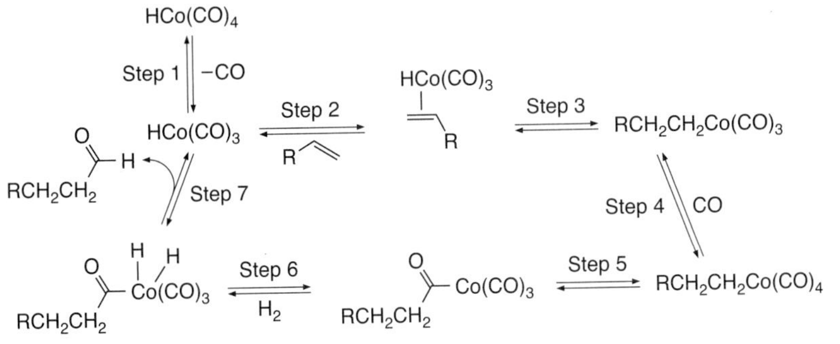

Figure 12.13
The catalytic cycle for hydroformylation using $HCo(CO)_4$.

isoelectric with the catalyst $HCo(CO)_4$. Insertion of this alkyl group into a CO ligand creates the acyl group necessary in the product (step 5), and opens up a coordination site. Oxidative addition of H_2 at this stage (step 6) gives a Co(III) species with 18 electrons. Reductive elimination of the organic product finalizes the cycle (step 7).

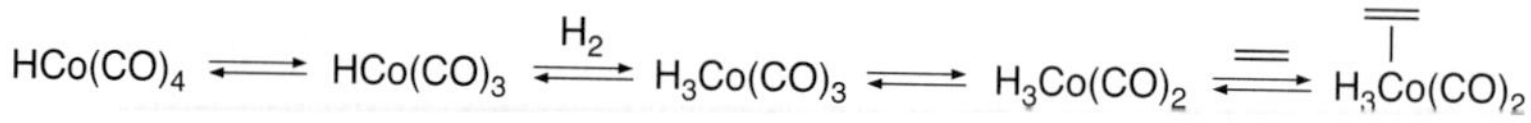

$$HCo(CO)_4 \rightleftharpoons HCo(CO)_3 \xrightarrow{H_2} H_3Co(CO)_3 \rightleftharpoons H_3Co(CO)_2 \rightleftharpoons H_3Co(CO)_2$$

$$\text{(Eq. 12.74)}$$

A few aspects of this cycle are worth mentioning at this stage. The first is that the reductive elimination of the aldehyde product likely occurs via the formation of a π complex with the metal, followed by dissociation. Also, there is debate as to whether the $RCH_2CH_2Co(CO)_3$ complex can actually fully oxidize to Co(III) upon interaction with H_2. It may be that the H_2 merely forms a complex with the Co, simply lengthening the H–H bond but not fully breaking it.

12.3.4 The Water-Gas Shift Reaction

In the previous catalytic cycles, and many others, CO and H_2 are used as reactants. In the petrochemical industry, an equimolar mixture of CO and H_2 gas can be generated by the high temperature reaction of graphite ("coke") and steam (Eq. 12.75). Similarly, the high temperature reaction of methane with water gives CO and H_2 (Eq. 12.76). A mixture of CO and H_2 is referred to as **water gas** or **synthesis gas**. In basic water, the CO can be converted to CO_2 and more H_2 gas, using the **water-gas shift** reaction (Eq. 12.77). After conversion of the "coke" or methane to a mixture of CO, CO_2, and H_2, the gases are "scrubbed" by passing through a solution of KOH, thereby removing the CO_2, and then over a nickel catalyst that uses some of the H_2 to reduce the CO, leaving 99% or greater pure H_2 gas. It is the water-gas shift reaction that is of interest here, because it can be catalyzed using hydroxide and standard organometallic reactions that involve $Fe(CO)_5$ (Figure 12.14).

$$C \text{ (graphite)} + H_2O \xrightarrow[\text{Pressure}]{\text{Heat}} CO + H_2 \qquad \text{(Eq. 12.75)}$$

$$CH_4 + H_2O \xrightarrow[\text{1000 °C, 10 atm}]{\text{NiO}} CO + 3\,H_2 \qquad \text{(Eq. 12.76)}$$

$$CO + H_2O \xrightarrow[\text{Metal oxide}]{\text{375 °C}} CO_2 + H_2 \qquad \text{(Eq. 12.77)}$$

Analysis of the reactants and products, and considering standard organometallic reactions, can lead to the proper mechanism. The water-gas shift reaction transforms the hydrogens in water to H_2 gas, and places the water oxygen into CO, making CO_2. Therefore, the mechanism must involve the breaking of the H–O bonds in water, and attack of a water oxygen on the C of a CO. Since the reaction is run in basic media, hydroxide acts as a nucleophile and adds to a coordinated CO ligand, thus attaching an oxygen to the carbon of the CO ligand (Figure 12.14, step 1). We show the electron pushing for this reaction with an M=C=O

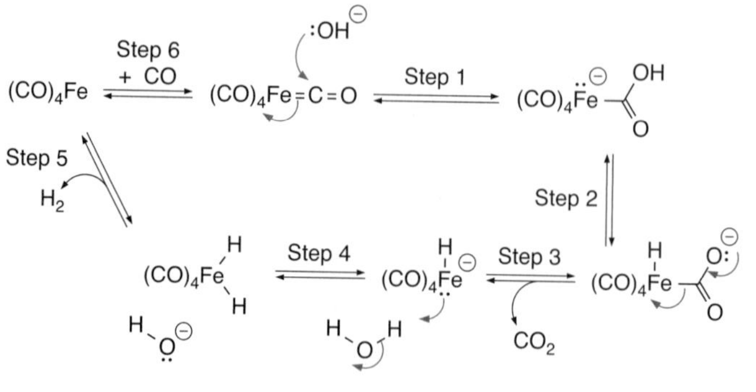

Figure 12.14
Catalytic cycle for the water-gas shift reaction.

resonance structure, and the metal as the electron sink. This is because the resulting Fe anion then picks up the proton from the coordinated carboxylic acid (step 2), because the metal is more basic than a carboxylate. All these reactions keep the Fe at 18 electrons. At this stage it is logical to lose CO_2 with $(CO)_4FeH^-$ as a leaving group (step 3). The Fe can readily accept an anionic charge because this creates another 18-electron species. Now all that is left is the formation of H_2, which requires another hydrogen to add to the Fe center. The Fe is basic, and it can remove a proton from water, thus regenerating the hydroxide catalyst (step 4). The resulting Fe–dihydride complex undergoes reductive elimination to furnish H_2 gas (step 5). The reductive elimination gives the only complex that is 16 electrons in the entire cycle $[Fe(CO)_4]$, while coordination of a CO ligand to this species regenerates the 18-electron $Fe(CO)_5$ and starts the cycle again (step 6).

12.3.5 Olefin Oxidation—The Wacker Process

The oxidation of an olefin to a carbonyl can be performed using the organometallic compound $PdCl_4^{2-}$ with water as the oxygen source (Eq. 12.78). In this reaction, HCl is also produced. Since the olefin is oxidized, some species has to be reduced. In this case it is the metal, which goes from Pd(II) to Pd(0). Regeneration of the $PdCl_4^{2-}$ is done with a catalytic amount of $CuCl_2$, which is reduced to CuCl (Eq. 12.79). In turn, the CuCl is oxidized back to $CuCl_2$ with O_2 and the HCl produced during the Pd cycle (Eq. 12.80). This process is used commercially to produce acetaldehyde from ethylene, and acetone from propene. It was discovered in the 1950s by a research group at Wacker Chemie, and bears the name of this company.

$$PdCl_4^{2-} + C_2H_4 + H_2O \longrightarrow CH_3CHO + Pd^0 + 2\,HCl + 2\,Cl^- \qquad \text{(Eq. 12.78)}$$

$$Pd^0 + 2\,CuCl_2 + 2\,Cl^- \longrightarrow PdCl_4^{2-} + 2\,CuCl \qquad \text{(Eq. 12.79)}$$

$$2\,CuCl + 0.5\,O_2 + 2\,HCl \longrightarrow 2\,CuCl_2 + H_2O \qquad \text{(Eq. 12.80)}$$

We once again use logic and our knowledge of organometallic reactions to predict a plausible mechanism (Figure 12.15). $PdCl_4^{2-}$ is a 16-electron species, and it can therefore associate another ligand, either water or the alkene. Since the oxygen from water needs to add to the alkene, it is logical to consider the alkene coordinating to the metal (step 1), which activates it toward nucleophilic attack by water. The attack will occur on the more substituted carbon because this makes the least sterically crowded attachment to the Pd (step 5). It also places the oxygen on the internal carbon in a terminal alkene, which is what is observed in the product. Our logic can bring us to this point, but it is difficult to conclude exactly what ligands are on the metal during the nucleophilic attack.

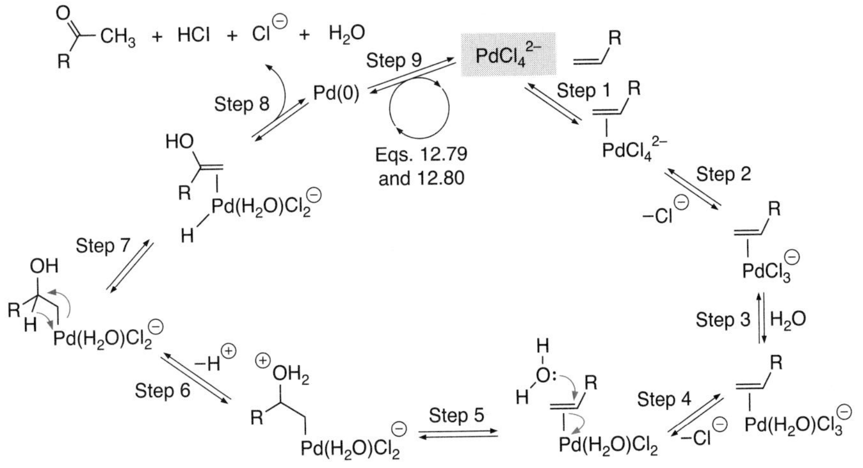

Figure 12.15
The catalytic cycle for the Wacker process.

The ligands on the Pd during the nucleophilic attack have been investigated, and it is generally agreed that the complex that undergoes nucleophilic addition is neutral with both a single water and a single alkene coordinated to the Pd. Presumably a neutral complex renders the alkene most susceptible to nucleophilic attack. The nucleophilic attack occurs from solution, and not via migration of the Pd-coordinated water to the alkene. That cannot be predicted *a priori*, but analysis of the stereochemistry of attack on *cis*-CHD=CHD shows anti addition, supporting nucleophilic addition from solution.

At this stage logic can again guide us to the end of the mechanism. We need to remove the hydrogen attached to the same carbon as the added oxygen, which is readily accomplished by a β-hydride elimination to give a coordinated enol (step 7). Dissociation of the enol and tautomerization to the carbonyl gives the organic product (step 8). The metal is now Pd(0), and can be written with or without ligands. Regeneration of $PdCl_4^{2-}$ occurs using the redox reactions given in Eqs. 12.79 and 12.80 (step 9).

12.3.6 Palladium Coupling Reactions

There are a wide variety of coupling reactions catalyzed by Pd that follow the general scheme given in Figure 12.16. Instead of giving a specific example, Figure 12.16 shows a generalized mechanism. The Pd(0) always has ligands, commonly phosphines, that keep it soluble. A series of reactions fit this mechanism, and they are very synthetically useful (see the Connections highlight below). In these reactions, oxidative addition of an R–X bond (where R is alkyl or aryl) first occurs to a zero-oxidation-state metal. The halogen ligand then undergoes a transmetallation with an R'M' (where R' is also alkyl or aryl), replacing the halogen with the R' group. Reductive elimination gives R–R', and regenerates the zero-oxidation-state metal. The cycles oscillate between Pd(0) and Pd(II), although in some cases with very electron donating ligands, cycles involving Pd(II) and Pd(IV) species occur.

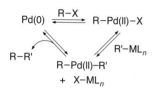

Figure 12.16
A common catalytic cycle for coupling reactions involving transmetallations.

Connections

Pd(0) Coupling Reactions in Organic Synthesis

The great utility of coupling reactions in organic synthesis has lead to a large number of variations, and several "name reactions". Here, we simply want to accentuate the diversity, and show the kinds of structures that can be formed based upon some variant of the mechanism given in Figure 12.16. Note that in all these reactions, R and R' are ultimately placed upon the same metal, leading to their coupling via a reductive elimination.

Stille coupling The hallmark of a Stille coupling is the use of a tin reagent for transmetallation. After oxidative addition of an R–X (R is alkyl, aryl, vinyl, or alkynyl, and X is halogen or triflate) to Pd(0), the Sn transmetallates its R group to Pd. Alkyl groups will transmetallate slowly, but vinyl, aryl, and alkynyl transmetallate quickly, while benzyl and allyl are intermediate. Reductive elimination couples the two groups as shown in Figure 12.16. In principle, any group that will oxidatively add to the Pd(0) can be coupled to any organotin group that will transfer.

$$R-X \ + \ R_3SnR' \ \xrightarrow{Pd(0)} \ R-R' \ + \ R_3SnX$$

Stille coupling

Suzuki coupling The hallmark of a Suzuki coupling is the use of an alkyl boronate with a base in the mechanism of Figure 12.16. Since boron is electrophilic, the attached alkyl groups are not nucleophilic enough to transmetallate to Pd. However, the addition of a base enhances the nucleophilicity, thus allowing the transmetallation. The boron can carry an alkyl, alkenyl, or aryl group. In fact, the Suzuki coupling is the method of choice to form biaryls.

$$R-X \ + \ R'B(OH)_2 \ \xrightarrow{Pd(0)} \ R-R' \ + \ XB(OH)_2$$

Suzuki coupling

Sonagashira coupling This coupling involves the use of terminal alkynes, Cu(I), and a base. Here, the base assists metallation of the Cu(I) with the terminal alkyne. The copper complex then transmetallates the alkyne to Pd as shown in Figure 12.16. Reductive elimination gives an alkyne.

$$RC{\equiv}CH \ \xrightarrow[\text{Base}]{Cu(I)} \ RC{\equiv}CCu(I) \ \xrightarrow{Pd(0)/R'-X} \ RC{\equiv}CR'$$

Sonagashira coupling

Negishi coupling Another named coupling that follows the mechanism of Figure 12.16 involves the use of vinyl zirconium and aluminum species with vinyl halides. This efficiently gives conjugated alkenes, and has been extensively used in the creation of oligomeric alkenes.

Negishi coupling

Kumata coupling The last variant we give here that follows the mechanism in Figure 12.16 involves a Grignard reagent and an aryl halide. Many variants that use Ni(II) in this reaction are also effective.

$$R-MgX \; + \; Ar-X \xrightarrow{\;Pd(0)\;} Ar-R$$

Kumata coupling

Heck reaction The Heck reaction is a variant on the mechanism given in Figure 12.16. There is no transmetallation step. Instead, an R–X reagent and an alkene are coupled via Pd(0) catalysts. All R–X reactants susceptible to oxidative addition can participate.

Heck reaction

The Heck reaction is carried out by heating the R–X, the alkene, catalytic amounts of Pd(II)acetate, and an excess of a tertiary amine. The mechanism given below is well accepted. Oxidative addition of R–X to Pd(0) occurs (step 1). The alkene coordinates (step 2), and alkyl group migration subsequently ensues (step 3). A β-hydride elimination furnishes the coupled product (step 4), and reductive elimination regenerates the Pd(0) species (step 6). In this reaction, the tertiary amine plays two roles. It reduces the Pd(II) acetate to Pd(0), and it neutralizes the HX produced in the last step.

Proposed mechanism of the Heck reaction

In summary, the discovery of the catalytic cycle shown in Figure 12.16 has had one of the largest impacts on organic synthesis of any organometallic breakthrough. Hopefully, given our discussion here, you will not view these reactions as "black boxes", but instead be familiar with the steps, and even create your own variants.

12.3.7 Allylic Alkylation

Another synthetically useful reaction is catalytic allylic alkylation (Eq. 12.81). With this reaction, the X group can be a halogen, as well as other commonly available substrates such as acetate or carbamate. The synthetic utility derives mostly from stereochemical control, which is briefly introduced in the Connections highlight below. This reaction is widespread in organometallic chemistry and has been found to be catalyzed by a variety of metals, including nickel, palladium, platinum, rhodium, iron, ruthenium, molybdenum, and tungsten.

(Eq. 12.81)

We show the mechanism using a Pd(0)L$_4$ complex, where L is commonly triphenylphosphine (Figure 12.17). A Pd(0)L$_4$ complex is an 18-electron species, so the first step will be ligand dissociation (step 1). The olefin can then coordinate (step 2). An oxidative addition of the allylic R–X group to the metal center is then required. Another phosphine ligand must first dissociate, however, in order to create a 16-electron species (step 3). Oxidative addition will occur via the nucleophilic mechanism (step 4, Section 12.2.2). In this case, the oxidative

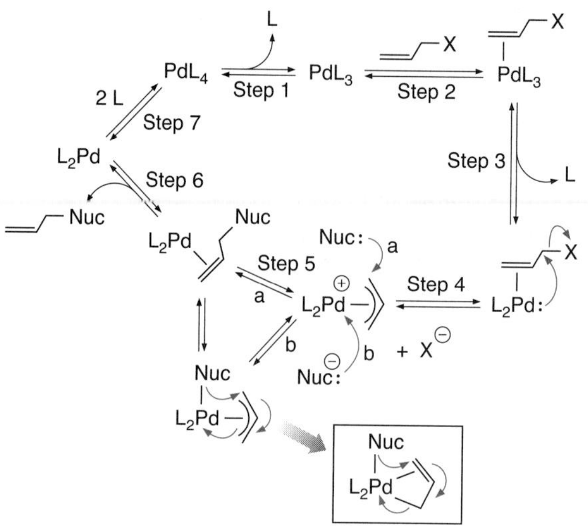

Figure 12.17
Catalytic cycle for allylic alkylations. In the box is shown the arrow pushing for a single resonance structure of the π-allyl complex.

addition creates another 16-electron species because coordination of the counterion does not occur. Instead, there is nucleophilic attack on the coordinated ligand (step 5), followed by dissociation of the product from the metal (step 6). There are two possible ways that the nucleophilic attack can occur. One is via direct attack from solution (arrow a and equilibrium path a in Figure 12.17), and the second is via coordination of the ligand to the metal first, followed by insertion of the nucleophile into the bound allyl group (arrow b and equilibrium path b in Figure 12.17). In general, nucleophiles whose pK_a values of their conjugate acids are above 25 (hard nucleophiles) will first coordinate to the metal, while softer nucleophiles add directly from solution. The reaction has been heavily used to form C–C bonds by using enolates as the nucleophiles. This is one of the simplest of organometallic reactions, yet as is shown in the following Connections highlight, the opportunities for stereochemical control are immense.

12.3.8   Olefin Metathesis

The last catalytic scheme we examine is olefin metathesis. This reaction has been widely used to create small and large rings, as well as to open up rings to make polymers. Olefin metathesis involves the pairwise exchange of the alkene carbons in two olefins (Eq. 12.82). These reactions were reported as early as the 1950s, but it was not until the early 1970s that the currently accepted mechanism was proposed by Chauvin.

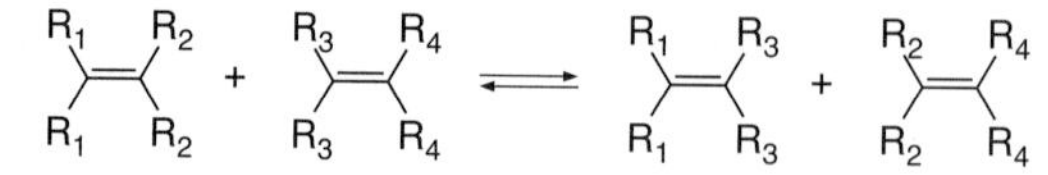

$$(Eq.\ 12.82)$$

The mechanism of olefin metathesis does not involve the classic reactions we have covered—namely, oxidative addition, reductive elimination, β-hydride elimination, etc. Instead, it simply involves a [2+2] cycloaddition and a [2+2] retrocycloaddition. The [2+2] terminology derives from pericyclic reaction theory, and we will analyze this theory and the orbitals involved in this reaction in Chapter 15. In an organometallic [2+2] cycloaddition, a metal alkylidene (M=CR$_2$) and an olefin react to create a metallacyclobutane. The metallacyclobutane then splits apart in a reverse of the first step, but in a manner that places the alkylidene carbon into the newly formed olefin (Eq. 12.83). Depending upon the organometallic system used, either the alkylidene or the metallacycle can be the resting state of the

Connections

Stereocontrol at Every Step in Asymmetric Allylic Alkylations

The catalytic asymmetric alkylation of allylic groups is an excellent reaction to highlight various approaches to enantiomeric control. Catalytic asymmetric induction is a major focus of current organometallic studies oriented toward synthetic applications. The vast majority of these reactions involve addition to double bonds. In contrast, the allylic alkylation reaction represents one of the few examples where achiral reactants are efficiently transformed to chiral products with high enantiomeric excesses, and where the chirality in the product results from reactions at sp^3 centers in the reactants. The enantiomeric control results from the use of chiral ligands on the metal center, making the metal a chirotopic center. Depending upon which ligand is used, the metal can be either stereogenic or non-stereogenic (see Chapter 6 to review this terminology).

The schematic structures shown to the right summarize the sources of enantio-discrimination during the reaction. Each picture shows enantiomeric reactions, which can be under enantiocontrol. As shown in **A**, there is enantiotopic face discrimination possible during the olefin coordination. There is also discrimination possible in the nucleophilic displacement of enantiotopic leaving groups (**B**). Both external and internal nucleophilic attack on the allyl ligand can be controlled by chirality on the metal center (**C** and **D**). Lastly, enantiotopic faces of the nucleophile, as with an enolate, can be discriminated by a chiral metal complex (**E**). Hence, each step of the mechanism (Figure 12.17) that involves a chemical reaction on the allyl group can have enantiocontrol. This high level of control has made the catalytic asymmetric allylic alkylation one of the most useful organometallic reactions yet produced.

Allylic alkylation

Trost, B. M., and Van Vranken, D. L. "Asymmetric Transition Metal-Catalyzed Allylic Alkylations." *Chem. Rev.*, **96**, 395 (1996).

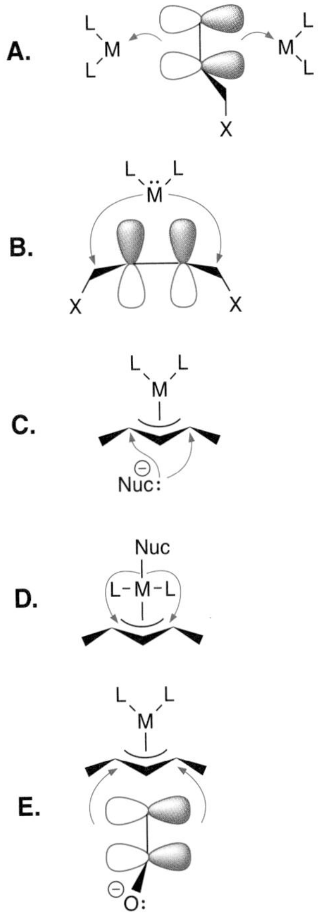

catalyst, with the corresponding metallacycle or alkylidene being a high energy intermediate, respectively. Let's see how the reaction of Eq. 12.82 can be used to make rings and create polymers.

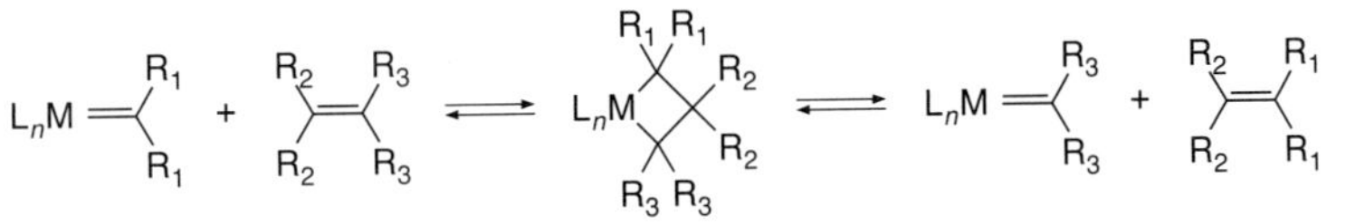

(Eq. 12.83)

Ring-closing metathesis (RCM) creates rings by the sequence of reactions given in Eq. 12.84. A diene reacts with the metal alkylidene via the cycloaddition–retrocycloaddition sequence, placing the alkylidene group into a released olefin, and creating an alkylidene with a pendant olefin. Next, the intramolecularly attached olefin undergoes the cycloaddition–retrocycloaddition sequence, giving back the starting alkylidene catalyst and the carbocyclic ring.

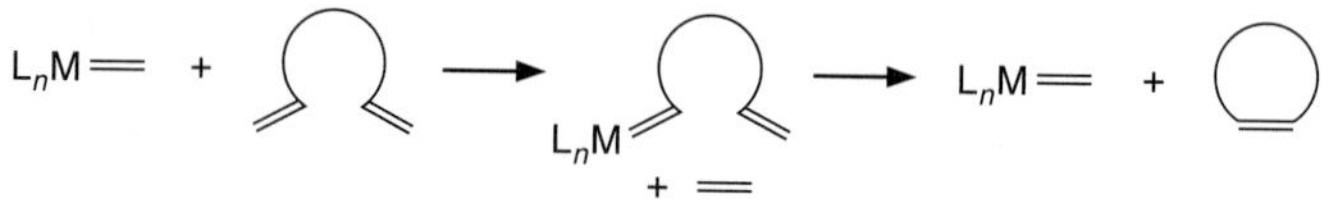

$$\text{(Eq. 12.84)}$$

In **ring-opening metathesis polymerization** (ROMP), an olefin that is part of a strained ring undergoes the cycloaddition with the metal alkylidene. The strain in the ring is relieved by opening the newly created metallacyclobutane (see the example in a Connections high-light in Section 7.5.1). The strain drives the ring opening, where each subsequent ring open-ing adds a monomer to a growing polymer (Eq. 12.85).

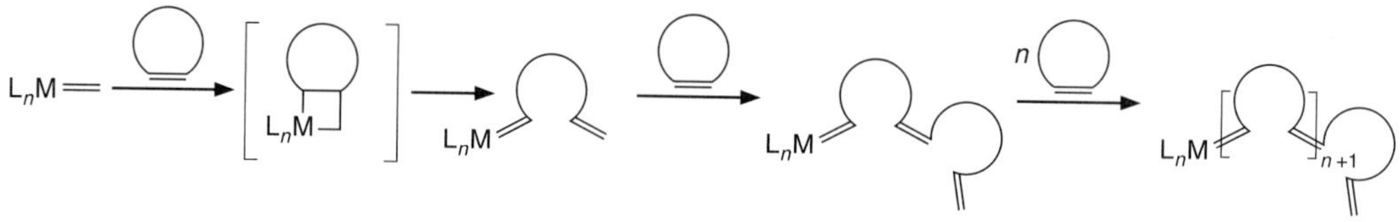

$$\text{(Eq. 12.85)}$$

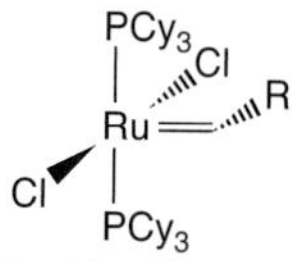

Grubbs, first-generation catalyst

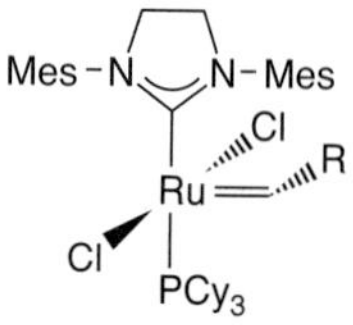

Grubbs, second-generation catalyst

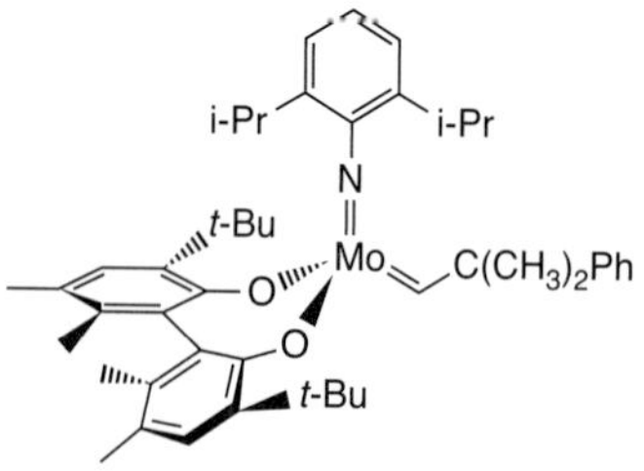

Schrock–Hoveyda catalyst

We have not yet discussed the actual organometallic catalysts that perform olefin me-tathesis. This is because the catalysts used are varied. In industry, multi-component ho-mogeneous and heterogeneous catalysts have been used, such as WCl_6/Bu_4Sn, $WOCl_3/EtAlCl_2$, and Re_2O_7/Al_2O_3. However, there is no doubt that the reason olefin metathesis is now widely used is the creation of the air-stable Grubbs catalysts shown in the margin (where PCy_3 = tricyclohexylphosphine, and Mes = 2,4,6-trimethylphenyl). The Grubbs second-generation catalyst replaces one of the phosphines with a carbene ligand, so that the Ru–C bond shown in actuality involves only a trivalent carbon. In addition, Schrock and Hoveyda have worked extensively with chiral metathesis catalysts. In 2005, Grubbs, Schrock, and Chauvin shared the Nobel Prize in chemistry for their work on olefin meta-thesis.

Kinetic studies on the Grubbs catalysts show that the mechanism involves prior coor-dination of the alkene to the Ru center, creating an 18-electron species, followed by loss of phosphine (step 1, Figure 12.18). Metallacyclobutane formation commences (step 2), fol-lowed by retrocycloaddition (step 3), to generate the new olefin coordinated to the Ru center. The rest of the steps in the cycle are the reverse of the steps leading to this point. The Grubbs first- and second-generation catalysts have been widely adopted by the chemical commu-nity, and have found uses in natural product synthesis, supramolecular chemistry, and ma-terials chemistry. The following Connections highlight shows a spectacular case of ROMP that leads to the largest of all carbon macrocycles yet created.

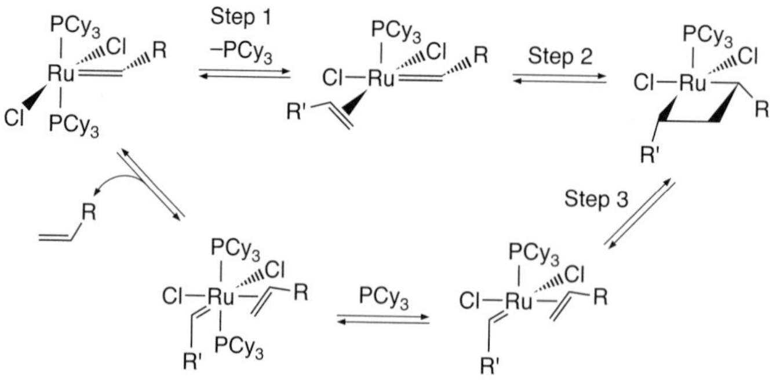

Figure 12.18
Catalytic cycle for the Grubbs catalyst in olefin metathesis.

Connections

Cyclic Rings Possessing Over 100,000 Carbons!

The Grubbs second-generation catalyst will efficiently ring open cyclooctene to make a linear polymer as shown below. The resulting polyoctenamer can be hydrogenated to make polyethylene.

In a very ingenious modification of this procedure, ROMP and RCM were combined into a single procedure to make cyclic polyethylene. Catalyst *i*, shown below, represents a version of Grubbs' second-generation catalyst where the alkylidene is conjugated to the carbene ligand. Ring opening of cyclooctene gives a polymer attached to the carbene ligand. After consumption of the cyclooctene, RCM occurs to liberate the cyclic polymer. Hydrogena-

tion gives cyclic polyethylene. Using end-group analysis, differential scanning calorimetry, molecular weight determinations, and the observation of phase separation with linear polyethylene (see such techniques in Chapter 13), it was established that the polymers are cyclic rather than linear. The molecular weights created by this method exceeded 10^3 kD (kiloDaltons), which is greater than 100,000 carbons per ring on average. This Connection highlight shows the power of the olefin metathesis reaction for both ring opening and ring closing, and makes a nice tie to the next chapter on organic materials chemistry.

Bielawski, C. W., Benitez, D., and Grubbs, R. H. "An 'Endless' Route to Cyclic Polymers." *Science,* **297**, 2041 (2002).

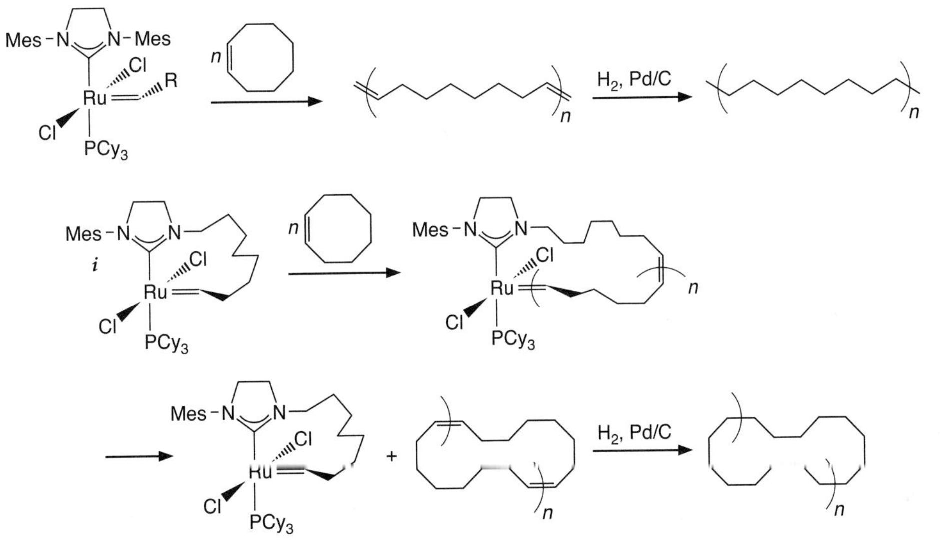

Summary and Outlook

The main goals of this chapter were to familiarize you with the reaction types found in organometallic chemistry and to provide enough information and examples so you can predict a plausible mechanism for new reactions. Since organic chemists now routinely exploit organometallic reactions in the synthesis of natural products, materials, polymers, and bioorganic structures, the mechanisms and reactions given herein should naturally be part of your knowledge base. At this point, we turn to another topic that is becoming increasingly important in all facets of organic chemistry—namely, polymers and other organic materials.

Exercises

1. Assign the number of electrons in the valence shell of the metal, the oxidation state, and the d electron count to the five complexes shown below.

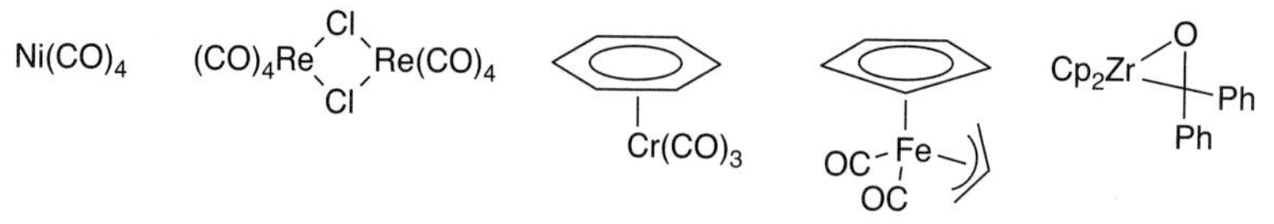

2. Discuss the reason for the trends in IR C–O stretching frequencies for the following *trans*-$Cr(CO)_4L_2$ complexes.

L	ν (cm^{-1})
P(OMe)$_3$	1916
PPh$_3$	1884
PMe$_3$	1873

3. The general order of bond strength for ligands bound to metal carbonyl compounds is M–PMe$_3$ > M–P(OMe)$_3$ > M–PPh$_3$ > M–NMe$_3$. Discuss how this trend is a balance between donating ability and sterics.

4. The insertion of CO into the methyl–Mn bond in CH$_3$Mn(CO)$_5$ is driven by the addition of gasous CO, and reverses under a vacuum. Explain this. Also, draw a reaction coordinate diagram for this reaction.

5. Discuss how the replacement of a CO by a phosphine ligand in CH$_3$Mn(CO)$_5$ will affect the rate of CO insertion into the methyl–Mn bond. How would the addition of a Lewis acid to CH$_3$Mn(CO)$_5$ affect the rate of CO insertion?

6. In the Heck reaction, we stated that almost all R–X groups that can undergo an oxidative addition to Pd(0) will undergo coupling. The Heck reaction works well for R = methyl, aryl, and vinyl. However, the reaction fails for R = ethyl, propyl, and butyl. Why is this so, and what would be expected to happen with these reactants?

7. The Stille, Sonagashira, and Heck reactions can often be catalyzed by the addition of Pd(II) salts. However, we noted that Pd(0) is required to start these cycles. Triethylamine is often added to reduce the Pd(II). How does this reduction occur, and what is the mechanism?

8. On average, how exothermic is a β-hydride elimination? In addition to bond strengths given in this chapter and Chapter 2, you will need to know that the BDEs for coordinated alkenes and metals are typically about 20 kcal/mol.

9. Write a full mechanism with all electron pushing for the following transformation.

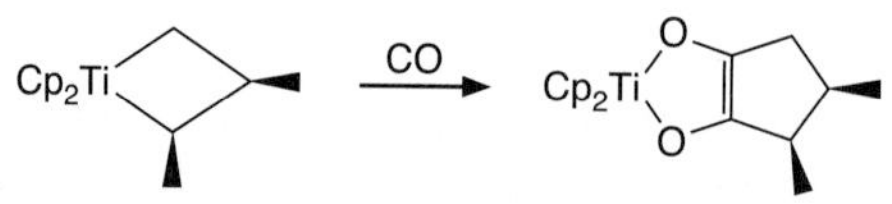

10. The following reaction occurs with the addition of Br$_2$.
   a. What experiment would you perform to determine the stereochemistry of this reaction at the carbon attached to the metal?
   b. When R = *t*-Bu, the stereochemistry is inverted, while when R = phenyl the stereochemistry is retained. How do you explain this, and what experiment would you perform to test your hypothesis?

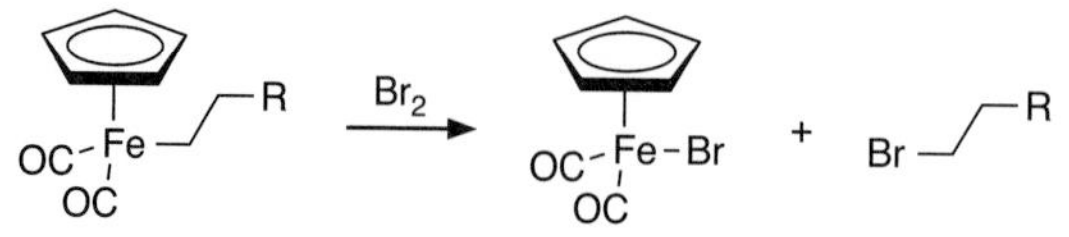

11. Propose mechanisms with electron pushing for the following reactions, the first of which is catalytic.

A.

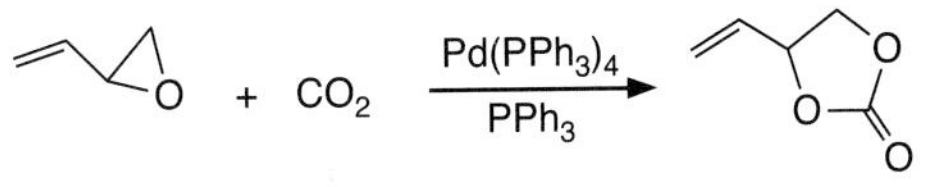

B.

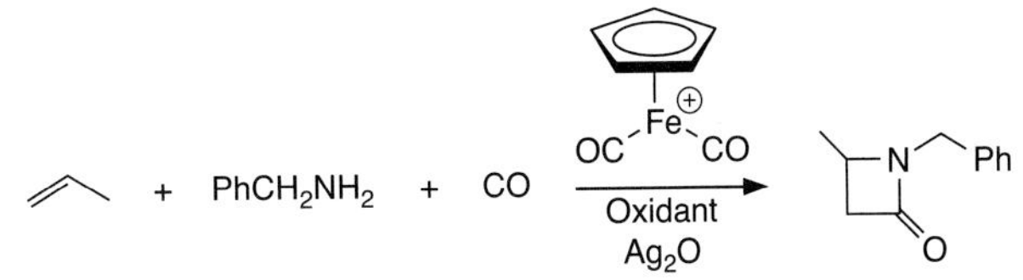

12. Decipher what the data supplied with each mechanism means, and write a mechanism consistent with these data.

A.

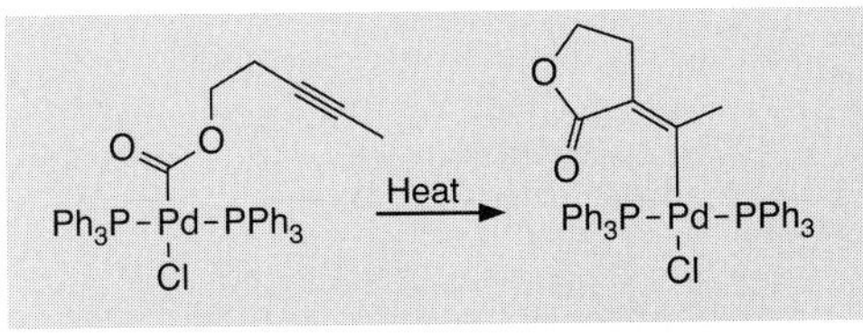

The rate of the reaction slows if PPh₃ is added to the solution

B.

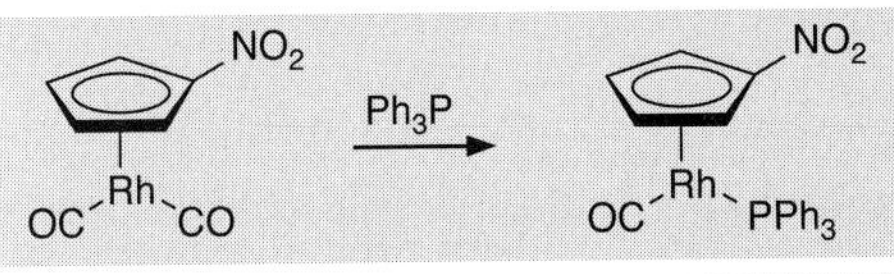

The reaction is first order in both the Rh complex and PPh₃, and $\Delta S^{\ddagger} = -17.9$ eu

C.

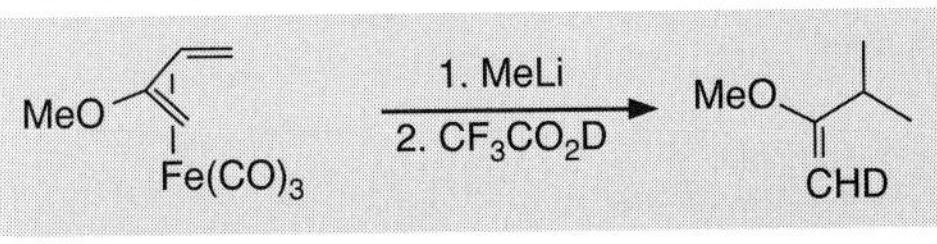

The deuterium is found both cis and trans to the isopropyl group in the product

D.

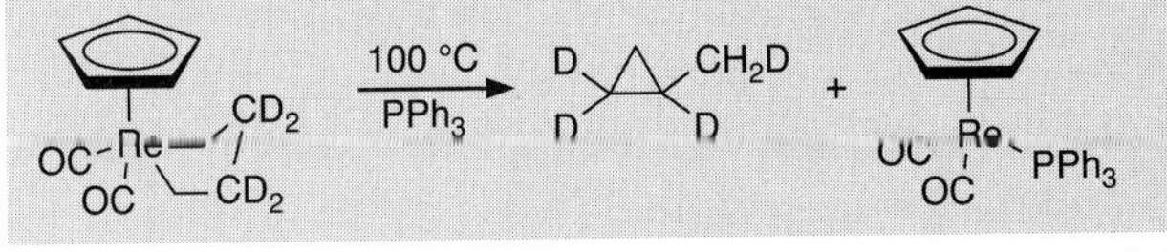

The reaction is zero order in phosphine and CO does *not* dissociate in the first step

E.

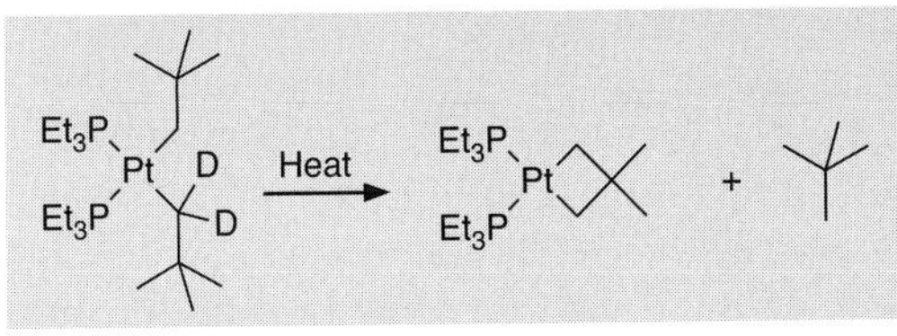

The reaction slows with added phosphine, and gives the Pt and neopentane products as mixtures of d_0, d_1, and d_2.

13. Why, among alkene complexes, do those of early transition metals consistently have a greater extent of metallacyclopropane vs. η²-alkene character in cases where the metal has *d* electrons before alkene coordination?

14. When a mixture of the following two rhodium complexes (which are in equilibrium) is added to *p*-methylbenzoyl chloride in *o*-xylene at 144 °C, conversion to *p*-chlorotoluene occurs. Write a catalytic cycle for this reaction.

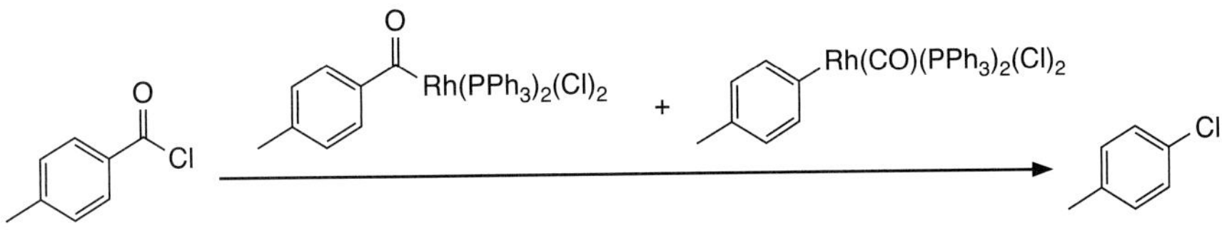

15. Write probable mechanisms for the following oxidative addition reactions. Evaluate the stereochemistry when appropriate.

A. $Ni(CO)_4 \xrightarrow{\quad \diagup\diagdown Cl\quad} (\eta^1\text{-}C_3H_5)Ni(CO)_3Cl$

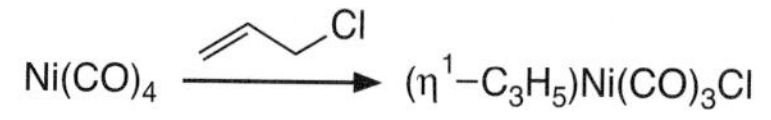

B. $Os(CO)_3(PPh_3)_2 \xrightarrow{\quad Br_2 \quad} Os(CO)_2(Br)_2(PPh_3)_2$
Phosphines in   Predict the
axial positions   stereochemistry

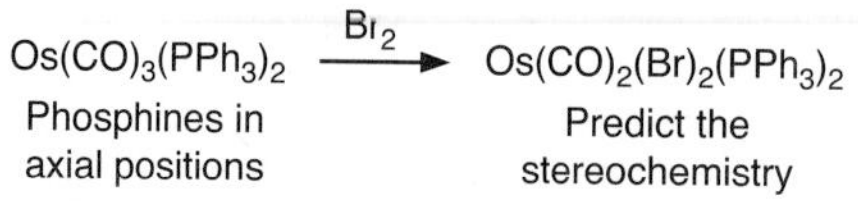

C. $Os(CO)_5 \xrightarrow{\quad H_2 \quad} Os(CO)_4H_2$
                                Predict the
                                stereochemistry

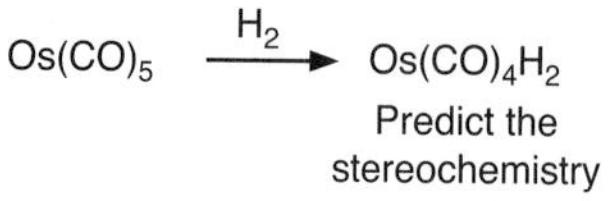

D.   *trans*-$IrCl(CO)(PMe_3)_2 \xrightarrow[\substack{\text{Trace } O_2 \\ \text{required}}]{\quad Br\diagup F \diagdown \quad}$ Product from oxidative
                                                                                addition of the C—Br bond

                                                                                Predict the stereochemistry at
                                                                                the metal and organic ligand

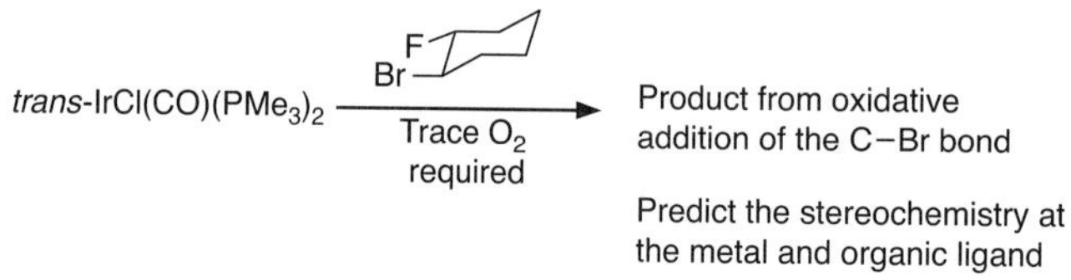

16. What is the product of the following reaction? Explain why the stereochemistry at the stereogenic center indicated is lost in the product, but is retained at the other stereogenic center.

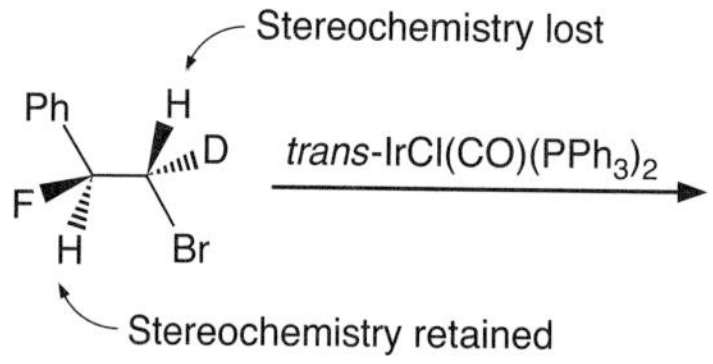

Stereochemistry lost

$\xrightarrow{\quad trans\text{-}IrCl(CO)(PPh_3)_2 \quad}$

Stereochemistry retained

Further Reading

General Organometallic Textbooks

Lukehart, C. M. (1985). *Fundamental Transition Metal Organometallic Chemistry*, Brooks Cole, Belmont, CA.

Jordan, R. B. (1998). *Reaction Mechanisms of Inorganic and Organometallic Systems*, 2d ed., Oxford University Press, New York.

Hegedus, L. (1999). *Transition Metals in the Synthesis of Complex Organic Molecules*, 2d ed., University Science Books, Mill Valley, CA.

Collman, J. P., Hegedus, L. S., Norton, J. R., and Finke, R. G. (1987). *Principles and Applications of Organotransition Metal Chemistry*, University Science Books, Mill Valley, CA.

Kochi, J. (1978). *Organometallic Mechanisms and Catalysis*, Academic Press, New York.

Heck, R. F. (1974). *Organotransition Metal Chemistry*, Academic Press, New York.

Crabtree, R. H. (2000). *The Organometallic Chemistry of the Transition Metals*, 3d ed., John Wiley & Sons, New York.

Yamamoto, A. (1986). *Organotransition Metal Chemistry*, Wiley–Interscience, New York.

Transmetallation

Davies, G., El-Sayed, M. A., and El-Toukhy, A. "Transmetallation and Its Applications." *Chem. Soc. Rev.*, **21**, 101–104 (1992).

General Mechanisms of Inorganic Reactions

Basolo, F., and Pearson, R. G. (1967). *Mechanisms of Inorganic Reactions*, 2nd ed., John Wiley & Sons, New York.

Pearson, R. G., and Ellgin, P. C. "Mechanisms of Inorganic Reactions in Solution" in *Physical Chemistry, An Advanced Treatise*, Vol. VII, H. Eyring (ed.), Academic Press, New York, 1975.

Oxidative Addition/Reductive Elimination

Vaska, L. "Reversible Activation of Covalent Molecules by Transition Metal Complexes." *Acc. Chem. Res.*, **1**, 335 (1968).

Halpern, J. "Oxidative-Addition Reactions of Transition Metal Complexes." *Acc. Chem. Res.*, **3**, 386 (1968).

Stille, J. K., and Lau, K. S. Y. "Mechanisms of Oxidative Addition of Organic Halides to Group 8 Transition-Metal Complexes." *Acc. Chem. Res.*, **10**, 434 (1977).

Schrock, R. R., and Parshall, G. W. "σ-Alkyl and Aryl Complexes of the Group 4–7 Transition Metals." *Chem. Rev.*, **76**, 243 (1976).

Norton, J. "Organometallic Elimination Mechanisms: Studies on Osmium Alkyls and Hydrides." *Acc. Chem. Res.*, **12**, 139 (1979).

Crabtree, R. "Iridium Complexes in Catalysis." *Acc. Chem. Res.*, **12**, 331 (1979).

Parshall, G. W. "Intramolecular Aromatic Substitution in Transition Metal Complexes." *Acc. Chem. Res.*, **3**, 139 (1970).

Insertion Reactions

Flood, R. C. "Stereochemistry of Reactions of Transition Metal–Carbon Sigma Bonds." *Top. Stereochem.*, **12**, 37 (1981).

Heck, R. F. "Addition Reactions of Transition Metal Compounds." *Acc. Chem. Res.*, **2**, 10 (1969).

Elimination Reactions

Braterman, P. S., and Cross, R. J. "Organo-Transition-Metal Complexes, Stability, Reactivity, and Orbital Considerations." *Chem. Soc. Rev.*, **2**, 271 (1973).

Davidson, P. J., Lappert, M. F., and Pearce, R. "Metal Sigma-Hydrocarbyls. MRn, Stoichiometry, Structures, Stabilities, and Thermal Decomposition Pathways." *Chem. Rev.*, **76**, 219 (1976).

Schrock, R. R. "Alkylidene Complexes of Niobium and Tantalum." *Acc. Chem. Res.*, **12**, 98 (1979).

Electrophilic and Nucleophilic Addition to Ligands

Candlin, J. P., Taylor, K. A., and Thompson, D. T. (1968). *Reactions of Transition Metal Complexes*, Elsevier, New York.

C–H Activation

Labinger, J. A., and Bercaw, J. E. "Understanding and Exploiting C–H Bond Activation." *Nature*, **417**, 507 (2002).

Olefin Metathesis

Trnka, T. M., and Grubbs, R. H. "The Development of $L_2X_2Ru=CHR$ Olefin Metathesis Catalysts: An Organometallic Success Story." *Acc. Chem. Res.*, **34**, 18 (2001).

Grubbs, R. H. (ed.). (2003). *Handbook of Metathesis*, Vols. 1–3, Wiley–VCH, Weinheim.

CHAPTER 13

Organic Polymer and Materials Chemistry

Intent and Purpose

Training in physical organic chemistry is an excellent preparation for a career in developing and characterizing new materials. Beginning with revolutions such as the preparation of nylon (introduced in the 1939 World's Fair), and seminal discoveries such as Ziegler–Natta polymerization of ethylene and propylene, the latter half of the 20th century was dominated by the materials revolution, which continues into the early 21st century. In this context, physical organic analyses have directly led to improvements in everyday lives.

Throughout this book we have highlighted several advances related to polymers and other types of materials that used physical organic concepts. Now it is time to pursue these topics in greater depth. This chapter is the last one of Part II of this book, where kinetics and mechanisms of reactions as well as the tools of physical organic chemistry are the emphasis. In this regard, much of the focus of this chapter is on various mechanisms of polymer synthesis, the kinetics of polymerization, and the methods used to characterize polymers. However, we also define some of the basic structural motifs common to polymers. We will also use the section on mechanisms of synthesis to introduce many of the more common polymers of modern materials science.

We also consider several other novel structures that are important in modern materials chemistry. These include dendrimers, liquid crystals, and fullerenes. Although polymers and these other structures are key components of organic materials chemistry, these topics alone do not encompass all of organic materials chemistry. The impact of organic chemistry on materials chemistry is far broader, including but not limited to the control of interactions between surfaces, the construction of nano-devices, and recently the creation of supramolecular materials that self-assemble. You should appreciate that we just cannot give a comprehensive treatment of all aspects of organic materials herein.

A quick glance at Table 13.1 should convince you that polymers represent a *big* business! All signs point to sustained growth in this industry, and therefore many of you will find employment performing research with materials and polymers after graduation. As an ex-

Table 13.1
1999 U.S. Output of Common Polymers*

Polymer	Output (millions of pounds)
Polyethylene	29,500
Polypropylene	15,000
Polystyrene	6,500
Polyvinyl chloride[1]	15,000
Synthetic fibers[2]	10,000

*Chem. Eng. News, June 26, **2000**, pp. 50–56.
1. Includes polyvinyl chloride copolymers.
2. Includes nylon, acrylic, olefin, and polyester fibers.

ample of the impact of polymers on the field of organic chemistry, almost 30 billion pounds of polyethylene are produced annually in the U.S. alone. Not surprisingly, then, a great deal of effort has gone into understanding and optimizing every aspect of polymer production, such that every section and subsection of this chapter merits a large textbook of its own. References to such works are listed at the end of the chapter. Our goal here is to introduce key concepts and illustrate some of the triumphs and remaining challenges of materials chemistry from a physical organic chemistry perspective. The topic also provides many opportunities to revisit key concepts introduced previously in the text. With this level of background, you should be able to appreciate advances described in the current literature, and you should be well prepared to delve into specific topics in greater detail.

Another facet of modern organic materials chemistry is the group of so-called "electronic materials". These include conducting polymers, organic magnetic materials, nonlinear optical materials, organic superconductors, and the like. These, too, are part of the materials revolution, and will have a strong impact on 21st century technology. However, to appreciate these structures, we will need some advanced concepts in molecular orbital theory and detailed insights into the interaction of light with organic molecules. As such, we defer discussion of these types of materials until Part III of this text.

13.1   Structural Issues in Materials Chemistry

One important issue that you should appreciate from the start is the molecular diversity associated with synthetic polymeric materials. When we say "cyclohexane" there is no ambiguity as to what we mean, because cyclohexane represents a single kind of molecule. But, when we say "polypropylene", things are not that simple. First, a sample of polypropylene is a mixture of similar molecules, all formed by polymerizing propylene to different chain lengths and extent of variation of chain length. Second, there are in fact a very large number of *materials* that are composed exclusively of the *polymer* polypropylene. They range in properties from rubbers to tough durable thermoplastics. These materials are processed differently, and hence have different physical properties. There is also a stereochemistry issue (Chapter 6): we can have isotactic, syndiotactic, and atactic polypropylene, and in fact varying degrees of tacticity (or **stereoregularity**). The material properties vary considerably depending on polymer stereochemistry. Also, all real materials will have **defects** such as impurities or deviations from perfect linearity. Sometimes these defects are desirable, if they modify the material properties in a favorable way. Thus, polypropylene can be many different things. Hence, one goal of this chapter is to give you a feeling for some of the key issues that define the properties of a particular material.

13.1.1   Molecular Weight Analysis of Polymers

We are now talking about very large molecules, structures with molecular weights from the thousands to the millions. You should not be surprised to hear that the materials properties of a polymer, features such as solubility, flexibility, melting temperature, and so on, depend strongly on the molecular weight of the polymer. However, molecular weight is a slightly more complicated concept for polymers than for small molecules.

Historically, artificial polymers such as polypropylene and nylon have been made in a relatively uncontrolled fashion. A solution of monomers is prepared, and then a reaction is initiated which links monomers together to produce an ever growing chain. In such systems, it will *always* be true that the final material will consist of a range of molecular weights. We need a way to describe this distribution of molecular weights.

Number Average and Weight Average Molecular Weights—M_n and M_w

Two common representations of the molecular weight of a polymer are the **number average molecular weight (M_n)** and the **weight average molecular weight (M_w)**, which are defined by Eqs. 13.1 and 13.2. In these equations, the counter i corresponds to the **degree of polymerization (DP)**, and it is simply the number of monomer units in a given chain. The other terms are N_i, the number of molecules with a DP of i; M_i, the molecular weight of a molecule with a DP of i; and w_i, the weight in the sample of all the molecules with a DP of i. Therefore, $w_i = N_i M_i$ and $\sum_i w_i = \sum_i N_i M_i$.

$$M_n = \frac{\sum_i N_i M_i}{\sum_i N_i}$$ (Eq. 13.1)

$$M_w = \frac{\sum_i N_i M_i^2}{\sum_i N_i M_i} = \frac{\sum_i w_i M_i}{\sum_i w_i}$$ (Eq. 13.2)

Why should we be concerned with two different definitions of molecular weight? It is easiest to see with an example. Consider a polymer sample that consists of 10 molecules (a small sample!) with molecular weights of 10,000, 11,000, 12,000, 13,000, 14,000, 15,000, 16,000, 17,000, 18,000, and 19,000. M_n is just the conventional average of the molecular weights, and so M_n for this sample is 14,500. According to Eq. 13.2, M_w is 15,070; the two values differ very little. Now consider a different polymer sample in which half the time the polymerization failed, producing a polymer of molecular weight only 10. We might have 10 molecules with molecular weights of 10, 10, 10, 10, 10, 15,000, 16,000, 17,000, 18,000, and 19,000. M_n is now 8,500, yet M_w is 17,100. The value of M_w has changed much less than that of M_n, and in fact M_w has increased while M_n has decreased. What is going on here? Often in polymer science the important consideration is not the average of all the molecules in the sample (M_n). Rather, we are concerned with the molecular weight of the largest fraction of the *weight* of a sample; M_w provides this measurement. In our second sample, the molecule with molecular weight 15,000 contributes *much* more to the final *weight* of the sample than one of the molecules with a molecular weight of 10. Indeed, the overwhelming majority of the weight of the sample is made up of the five high molecular weight polymer molecules, with the low molecular weight molecules contributing very little. The M_w value correctly reflects this, while M_n does not. M_n is an averaging method based on counting each molecule equally, while M_w is an averaging method based on counting molecules in a way that is biased by their weight. Accordingly, many material properties depend more strongly on M_w than M_n.

Essentially all polymer samples consist of molecules with a range of molecular weights, and there can be variation in this spread. Of the two samples discussed above, the first is more nearly homogeneous. All the individual molecular weights lie near a central value. The range of molecular weights in a polymer sample is called the **polydispersity**, and it is designated by a number called the **polydispersity index (PDI)** which is defined by Eq. 13.3 (an example is given in the Connections highlight on page 757). For the two samples discussed above the PDI values are 15,070/14,500 = 1.04 and 17,100/8,500 = 2.01. The PDI correctly reflects that the second sample has a broader range of molecular weights. One thinks of a **molecular weight distribution** based upon the PDI, with larger PDIs indicating a larger distribution of molecular weights. The PDI is a key factor in determining the materials properties of a sample, and efforts to control the polydispersity of synthesized polymers remain at the forefront of modern polymer science. The PDI of 1.04 in our first hypothetical sample is *extremely* low, and synthetic polymers usually do not have PDIs in this range. In fact, a PDI of 2.01 is also relatively low, and many commercial polymers have PDIs in the range of 5–10.

$$\text{PDI} = \frac{M_w}{M_n}$$ (Eq. 13.3)

Polymer samples in which every molecule has exactly the same molecular weight are called **monodisperse**. The macromolecules of biology are usually monodisperse. A protein is simply a polymer made from a concatenation of similar monomers, but the number (and sequence) of monomers in the chain is precisely defined. That is because in protein synthesis each monomer is added individually to the growing chain in a precisely controlled fashion by a remarkable molecular machine, the ribosome. Protein synthesis is really not a "polymerization" in the usual sense. Inspired by the monodispersity of biopolymers, chemists have long sought ways to make truly monodisperse artificial polymers. The key is to have precise control over the chemistry, so monomers are added one at a time, as in the solid phase synthesis of proteins and oligonucleotides. The problem is that this is incredibly tedious, and takes a huge number of chemical steps to make even a modestly-sized polymer (automated machines greatly assist the process). We will see below that the dendrimer strategy offers one way around this problem. The following Going Deeper highlight provides another.

Going Deeper

Monodisperse Materials Prepared Biosynthetically

Nature prepares truly monodisperse polymers using the biosynthetic machinery of the ribosome. In recent years, Tirrell and co-workers have shown that it is possible to hijack the synthetic machinery of bacteria to prepare "non-natural" amino acid-based polymers that are monodisperse and that have useful materials properties. Using standard molecular biology protocols, the exact amino acid sequence of a protein can be designed, allowing a rational modulation of properties. Perhaps most surprising is that with modern biotechnology techniques, large quantities of these artificial proteins can be made, allowing the development of useful new materials with novel properties. For example, the polymer shown in schematic form to the right was designed and produced in substantial quantities by *E. coli*. The helical domains are based on the **leucine zipper** helix motif, an α-helical structure found in many proteins that tends to self-associate at neutral pH but not at elevated pH. The middle segment (colored line) is unstructured but quite water soluble. At high pH the polymer is freely soluble in water. However, if the pH is dropped to 7, association of the helical regions cross-links the polymer (see below for definitions of these terms). In a typical polymer such cross-linking would cause precipitation. However, the water soluble segment of the polymer traps a great deal of water, discouraging precipitation. The net effect is the formation of a gel. Hence, the entire sample "solidifies" at low pH, but the process can be reversed simply by raising the pH.

Unnatural amino acids can also be incorporated into materials. This is done by taking advantage of mutant bacteria that cannot biosynthesize a particular amino acid (such bacteria are known as **auxotrophs**). If auxotrophs are fed a similar, unnatural amino acid, they will incorporate it biosynthetically in place of the missing natural amino acid. For example, a thiophene ring can be substi-

tuted for the benzene of all phenylalanine residues in the peptide. These thiophenes can then be coupled to make oligothiophenes that have novel and interesting electronic properties (see the discussion of polythiophenes in Chapter 17).

pH = 10
↓
pH = 7

Peptides aggregate to polymerize at lower pH

Petka, W. A., Harden, J. L., McGrath, K. P., Wirtz, D., and Tirrell, D. A. "Reversible Hydrogels from Self-Assembling Artifical Proteins." *Science*, **281**, 389 (1998).

13.1.2 Thermal Transitions—Thermoplastics and Elastomers

A key feature of any polymer is how it responds to heating. For example, a polymer that on heating becomes much more flexible, perhaps even fluid, can be shaped or molded into a particular form at high temperatures and then rigidified upon cooling so as to maintain the new shape. Processability issues such as these are crucial in polymer science. As with most topics in this field, we cannot hope to cover all aspects of polymer thermodynamics. Here we simply introduce a few key terms that are commonly used and provide valuable initial insights into a polymer's characteristics.

Generally, polymers are classified as either **amorphous** or **crystalline**. A solid amorphous material that has no significant order on the molecular scale is characterized as a **glass**. Other polymers are characterized as crystalline, but this does *not* mean that the entire sample is crystalline, as implied by a crystalline sample of a molecular solid. Rather, in polymer science "crystalline" implies that some small regions of the material, termed **domains**, are crystalline and have the ability to diffract x rays. Polymers are rarely fully crystalline

---

Connections

An Analysis of Dispersity and Molecular Weight

The manner in which molecular weights and polydispersities are calculated most often derives from a form of size exclusion chromatography known as **gel permeation chromatography (GPC)**. Here, the polymer is eluted through a gel that contains pores of a defined average size. Large polymers are not retained by the pores and so are more readily eluted, while the smaller polymers become temporarily entrained in the cavities of the gel and so are better retained. Typically, polystyrene standards of known molecular weight are used to form an elution volume versus molecular weight standardization curve for each particular chromatography column.

As an example of how molecular weight and polydispersity are apparent from such a method, we show to the right the overlay of four GPC traces from various samples of poly(methylmethacrylate). Here the *x* axis is the elution volume and the *y* axis is the response of a refractive index detector. The four samples were formed by continued irradiation of the polymer by γ rays. Trace 0 was prior to irradiation, and trace 3 was after the highest dose of irradiation.

Irradiation of the polymer leads to lower molecular weight species as the retention on the column increases. However, we also see that the width of the chromatography traces becomes increasingly narrow with continued irradiation. This indicates more uniform lengths among the polymer strands upon irradiation, meaning a lower polydispersity. How is this explained?

As shown below, irradiation leads to homolysis of the bond between the acyl and tertiary carbon along the polymer backbone. The tertiary radical eliminates to break the chain into smaller segments. This leads to lower molecular weight material. Larger chains are more likely to be cleaved than shorter chains, just because they have more cleavage linkages. This decreases the polydispersity of the starting polymer because the smaller molecular weight polymers formed have a more consistent size.

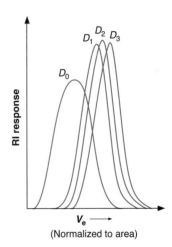

Thompson, L. F., Willson, C. G., and Bowden, M. J. "Introduction to Microlithography." ACS Symposium Series 219, ACS, Washington DC, 1983.

Photolysis leads to homolysis

+ CO, CO_2, $\cdot CH_3$, $CH_3O\cdot$

(proteins can be fully crystalline). The crystalline domains, instead, are intermixed with (or embedded in) glassy domains. Often, a **percent crystallinity** is assigned to polymers.

At low temperatures, both amorphous and crystalline polymers behave as glasses and are completely rigid. Many polymers, though, are **thermoplastic**, meaning they soften when they are heated. Upon cooling, **thermosets** are formed—that is, structures that conform to their container. As the temperature is raised, a critical temperature is reached called the **glass transition temperature (T_g)**. We transit from an essentially rigid glass to a more flexible, rubbery material (an example analysis is given in the next Connections highlight). The transition is fairly sharp, often occurring over a range of 2–5°. As the temperature is raised further above T_g, an amorphous polymer goes through a *gradual* series of changes from a solid to a rubber to a gum and finally to an actual liquid. These are fairly vague terms, and no further sharp transitions are seen.

A crystalline polymer above T_g is a flexible thermoplastic. On raising the temperature further, a second relatively sharp transition occurs at the **melting temperature (T_m)**. It is at this point that the crystalline domains melt, and above T_m the material is a liquid. By definition, an **elastomer** is a polymer in the temperature range between T_g and T_m.

The glass transition temperature is one of the most important properties of a polymer. It sets that temperature range over which the polymer has useful structural properties and also defines the conditions under which the material can be thermally processed. As illustrated in Table 13.2, there is a considerable variation in the transition temperatures for common polymers. Because of differing synthesis and processing strategies, as well as variations in stereoregularity and levels of defects, there are variations in T_g and T_m for any given polymer.

Table 13.2
T_g and T_m Values (°C) for Some Common Polymers*

Polymer	T_g	T_m
Polystyrene (isotactic)	100	240
Poly(*m*-methylstyrene) (isotactic)	70	215
Poly(methyl methacrylate) (isotactic)	48	160
Poly(methyl methacrylate) (syndiotactic)	126	200
Poly(*cis*-1,4-isoprene) (natural rubber)	−70	36
Poly(*trans*-1,4-isoprene) (gutta percha)	−68	74
Poly(ethylene oxide) (carbowax)	−67	66
Poly(dimethylsiloxane) (silicone rubber)	−123	−29
Polyacrylonitirle (Orlon, Creslan)	85	317
Nylon-66	45	267
Poly(ethylene terephthalate) (Mylar/Dacron)	69	270
Polyethylene	−20	141
Polypropylene (isotactic)	−35	130
Poly(vinyl chloride) (atactic, PVC)	80	285

*Allcock, H. R., and Lampe, F. W. (1990). *Contemporary Polymer Chemistry*, Prentice Hall, Englewood Cliffs, NJ., Lewis, O. G. (1968). *Physical Constants of Linear Homopolymers*, Springer–Verlag, Berlin.

What structural features influence T_g? Along with molecular weight, issues such as the precise connectivity pattern of the polymer, its topology, stereochemistry, and functional group composition are crucial. Stereoregular polymers tend to be able to pack better, leading to greater crystallinity and thus higher transition temperatures. Polar and polarizable groups favor intermolecular interactions and so also tend to raise transition temperatures.

Connections

A Melting Analysis

As an example of a melting curve analysis, we show to the right the calorimetry data for a sample of crystalline Nylon-6. The x axis is the temperature of the sample, and the y axis is the heat absorbed or released as a function of the change in temperature. The process involves a scan where the sample does not come to complete thermodynamic equilibrium for each change in temperature. The top curve shows a scan starting near 0 °C and ramping to 250 °C at a rate of 20.0 °C / min. Two peaks are evident. The first is a very small negative peak at around 45 °C. This is a glass transition corresponding to a very small portion of the polymer that was not crystalline, but was glassy, becoming crystalline. The onset of the second peak at 210 °C defines the melting temperature. The integrated area under this peak is the **heat of melting** for the polymer. The lower curve was generated after letting the sample rest at 250 °C for 0.2 min, and then cooling to near 0.0 °C at 20.0 °C / min. The crystallization temperature is the onset of the peak at about 170 °C, and the integrated area is the **heat of fusion** for the polymer sample.

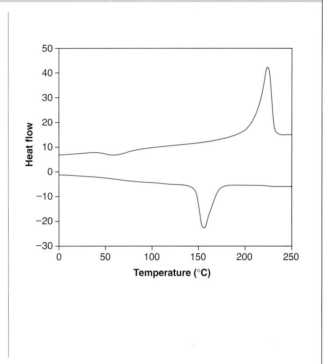

We thank Professors Don Paul and Kris Haggard of the University of Texas for supplying these data.

13.1.3 Basic Polymer Topologies

In your mind you might envision a polymer as an indefinitely long, linear chain. In reality, this is rarely the case. Whether by design or by accident, most polymers deviate from perfect linearity, and these deviations profoundly affect a polymer's properties. Here we review some of the basic issues in polymer topology (recall from Chapter 6 that for most structures the topology of a molecule is established solely by its connectivity).

Linear polymers are, not surprisingly, truly linear with respect to connectivity, consisting of long chains of the polymer backbone. Figure 13.1 provides an illustration. Note

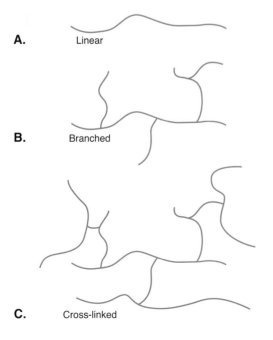

A. Linear

B. Branched

C. Cross-linked

Figure 13.1
Basic polymer topologies.
A. A linear polymer.
B. A branched polymer.
C. A cross-linked polymer.

that there may be side chains hanging off the main chain. A **branched** polymer is a linear polymer that also contains branches of the same basic structure emanating from the main chain (see Figure 13.1 **B**). A **cross-linked** (or network) polymer is one in which linkages occur between the main chains, as in Figure 13.1 **C**. One other issue is stereochemistry. As discussed in Chapter 6, polymer stereochemistry focuses on tacticity, with linear polymers that contain side chains being termed syndiotactic, isotactic, or atactic (see Section 6.7).

While it is always risky to make sweeping generalizations, we can make some general comments about the relationships between polymer structure and properties. We noted above that polymers can be characterized as crystalline or amorphous, with the former involving two well-defined phase transitions (T_g and T_m) and a polymer lying between these two being an elastomer. To form a crystalline domain, polymer chains must be able to pack together well. Fewer branches encourage better packing and thus more crystallinity. This in turn raises transition temperatures. Highly linear polymers are generally much more crystalline than highly branched polymers. A similar result holds with stereochemistry. A more regular pattern, whether syndiotactic or isotactic, promotes efficient packing and thus higher crystallinity. Stereoregular polymers are generally more crystalline than stereorandom polymers.

A special case arises for cross-linked polymers. Consider a polymer with an average DP of ~10,000, and with one cross-link for every 100 monomers (termed 1% cross-linking). In such a sample, every chain has many cross-links, and it is inevitable that a covalent path exists between every chain. That is, for all practical purposes, every chain is ultimately connected to every other chain—*the polymer is a single molecule!* Rubber is a highly cross-linked polymer of isoprene (Table 13.2), so a rubber band may very well be a single molecule.

Cross-linking profoundly affects the properties of a polymer. If you've ever tried to dissolve a rubber band in anything, you've noticed it is pretty much impossible (without decomposing the material). This is typical of highly cross-linked polymers. Instead of dissolving, exposure to "solvents" causes the cross-linked polymer to swell, and the lighter the amount of cross-linking, the more the swelling.

Another facet to polymer topology arises when you consider what happens if more than one monomer is used. A single monomer gives a **homopolymer** (. . . AAAA . . .), whereas a **random copolymer** results if two equally reactive monomers (A and B) are added together at the outset of the polymerization reaction (such as . . . ABAABBABBAABABA . . .). If the monomers are added sequentially, then a **diblock copolymer** can result (. . . AAAAA– BBBBB . . .). Continuing with this logic, a **triblock copolymer** can be created by adding the monomers one after another (. . . AAAA– . . . BBBB . . . – . . . AAAA . . .), leading finally to a **multiblock copolymer** [(A)$_n$–(B)$_m$)$_q$]. One of the most interesting issues that arises with block copolymers is their phase behavior, because different bulk morphologies can result. This is the next issue we turn to.

13.1.4   Polymer–Polymer Phase Behavior

The easiest to analyze, and the most well understood phase behavior, arises from a binary mixture of homopolymers, or diblock copolymers. These solutions consist of either one or two phases (neglecting crystallization) at equilibrium. The simplest phase behavior is a completely homogeneous solution where the polymers are completely miscible with one another, and only one phase is present. A **macrophase separation**, defined as two or more phases, arises when the two polymers are immiscible (Figure 13.2 **A**). The surface tension within each phase of homopolymer leads to a minimization of the surface area of contact between the phases. Alternatively, the use of a diblock copolymer, where the two monomers A and B are incompatible, commonly leads to striation, giving what is known as a **microphase segregation** (Figure 13.2 **B**). Lastly, mixing together homopolymers of A and B with A–B diblock copolymers can result in microdomains, and **intermediate-scale phase separation** (Figure 13.2 **C**). Given these possibilities, the most important factor determining phase behavior is the introduction of the covalent bond between blocks in copolymers, which prevents macroscopic phase separation.

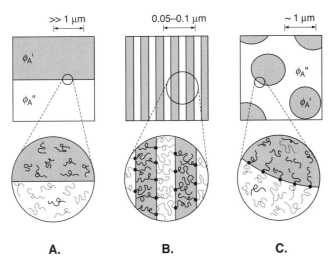

Figure 13.2
A. Macrophase separation into two layers, which commonly arises when two homopolymers are mixed. **B.** Microphase segregation can arise with diblock copolymers. Here, striated layers appear. **C.** Mixing homopolymers with the corresponding diblock copolymer gives intermediate-scale phase separation. A surfactant-like mixture occurs. Figure modified from Bates, F. S. "Polymer–Polymer Phase Behavior." *Science,* **251**, 898 (1991).

The phase behavior that results is controlled by the Gibbs free energy (GFE) of the entire solution. When mixing polymers a spontaneous phase change results, lowering the GFE. We can rely on many of the principles already explored in depth throughout this book to understand the interactions that contribute to the GFE. Four factors are involved: molecular architecture, choice of monomers, composition, and degree of polymerization. When mixing homopolymers or homopolymers with diblock copolymers, the **composition** is defined as the fraction of the volume of a component (ϕ_n, where n = A or B). Experimentally, ϕ is controlled by changing the percentage of the different homopolymers, and/or the stoichiometry of the A and B blocks in a copolymer. The molecular architecture and the choice of monomer establish what are called the **contact energies** between the different polymer segments. When mixing two homopolymers, or mixing two homopolymers with a diblock copolymer, three different contact energies are established: ε_{AA}, ε_{BB}, and ε_{AB}, representing interactions of mostly A and B with themselves, or A and B together, respectively. Favorable contact interactions are attractive, and therefore associated ε contact energies are negative values. The enthalpy of mixing is related to the Flory–Huggins segment–segment interaction parameter χ (Eq. 13.4; k = Boltzmann constant).

$$\chi = \frac{1}{kT}\left[\varepsilon_{AB} - 0.5(\varepsilon_{AA} + \varepsilon_{BB})\right] \qquad \text{(Eq. 13.4)}$$

A negative χ means there is a favorable enthalpy for mixing segments of A and B. The A–B interactions (hydrogen bonding, dispersive forces, etc.) produce a lower energy system than the A–A and B–B interactions. A positive value means that interaction between segments A and B is unfavorable. The weak forces discussed in Chapter 3 are the dominant factors leading to either positive or negative χ values.

However, enthalpy alone does not control the solution; the GFE is the controlling factor. Eq. 13.5 gives the **Flory–Huggins equation** for estimating the change in free energy (ΔG_m) due to mixing polymer chains of A and B. Here, N is the average number of units of A and B per polymer molecule.

$$\frac{\Delta G_m}{kT} = \frac{\phi_A \ln \phi_A}{N_A} + \frac{\phi_B \ln \phi_B}{N_B} + \phi_A (1 - \phi_A) \chi \qquad \text{(Eq. 13.5)}$$

There are three terms in the Flory–Huggins model. The first two account for the entropy of mixing. Mixing increases the randomness of the system, and naturally increases $\Delta S°$, and hence leads to a decrease in the GFE of mixing. On a per mass basis, long polymers have a fewer number of statistical ways to mix than smaller polymers, and therefore the entropy of mixing decreases with increasing values of N. In other words, phase separation increases in likelihood with longer polymers. The $\Delta H°$ of mixing is given by the third term, which involves the χ parameter discussed above. The Flory–Huggins approach is called a **mean-field theory**, meaning that it ignores fluctuations between phases and only looks at the mean of the interactions involved between the polymers. When N_A and N_B are equal to 1, the equation describes the GFE for mixing two liquids. An important insight given by Eqs. 13.4 and 13.5 is that the standard factors associated with the entropy and enthalpy of solutions control polymer phase behavior. Weak intermolecular forces determine the factor χ, while the entropy of mixing, which we found in Section 3.1.5 is so crucial to solubility and reactivity, in part controls phase separation. The lessons given by the Flory–Huggins model are far reaching, and have even been used to model protein dynamics, where folding is analogous to phase transitions (see the following Connections highlight).

Connections

Protein Folding Modeled by a Two-State Polymer Phase Transition

Proteins are heteropolymers, and therefore application of the simple analysis given above for mixing homopolymers with diblock copolymers would seem to be a stretch for proteins. Yet, one of the simplest models for proteins divides the twenty natural amino acids into two categories: hydrophobic (H) and ionic/polar (P). The attraction between H-type monomers models the tendency of hydrophobic polymers in water to collapse in order to minimize their exposed surface area, but also to minimize their interactions with P-type monomers. The minimiza-

tion of the H–P interactions derives from a positive χ value. The collapse of the H-type monomers leads to compact states for proteins with nonpolar cores and exposed P-type monomers. The collapse of the polymer is not entropically favored because the entropy of mixing is decreased, even though solvent release is entropically favored. Importantly, the more compact, collapsed structure is thought to be a prerequisite to and favors the development of complex secondary structure in the protein.

Dill, K. A., Bromberg, S., Yue, K., Fiebig, K. M., Yee, D. P., Thomas, P. D., and Chan, H. S. "Principles of Protein Folding —A Perspective from Simple Exact Models." *Prot. Sci.*, **4**, 561 (1995).

13.1.5 Polymer Processing

An incredibly important issue in polymer chemistry is **processing**, the manner in which the crude polymer product is transformed into the final material used in applications. This is a topic best reserved for a chemical engineering textbook, but it should not be dismissed here, either. As an example, poly(ethylene terephthalate), which is considered the workhorse of the polyester industry, is used as both a fiber and a plaster under the trade names of Dacron and Mylar, respectively (look ahead to Figure 13.15 to see the structures). The processing procedure leads to the different properties and the ultimate applications.

Similarly, virtually all polymers in commercial use contain **additives**. Their purpose is two-fold: to alter the polymer properties and to enhance ease of processing. Plasticizers are used to modify mechanical properties. Modifiers can be lubricants, cross-linking agents, emulsifiers, and thickening agents, just to name a few. Pigments and odorants are common additives used for aesthetic reasons.

13.1.6 Novel Topologies—Dendrimers and Hyperbranched Polymers

Dendrimers

We mentioned above the challenge of making truly monodisperse artificial polymers, and that the various synthesis methods outlined in Section 13.2 generally fail in this regard. However, in the early 1980s a strategy was developed for overcoming this limitation. The approach involves highly branched molecules, termed **arborols** in early work by Newkome or **dendrimers** in early work by Tomalia, the latter having become accepted in the literature (from the Greek "dendron" meaning tree). These systems are also sometimes referred to as **starburst** polymers. The structures are best described with reference to Figure 13.3. In Figure 13.3 we represent two functional groups that can couple together, one with a filled colored box and the other with an open or white box. We start with a "core", which in this example contains three colored boxes. The key molecule is referred to in Figure 13.3 as the "branch", which contains one white box and two colored boxes, the latter being protected (P) or otherwise unreactive. Reaction of excess branch with the core, followed by deprotection produces a new structure that now has six colored boxes on its periphery. This is an iterative process, and so this material is called the first generation dendrimer, symbolized as **G1**. Now we simply repeat this two-step addition–deprotection sequence and successively make **G2**, **G3**, and so on.

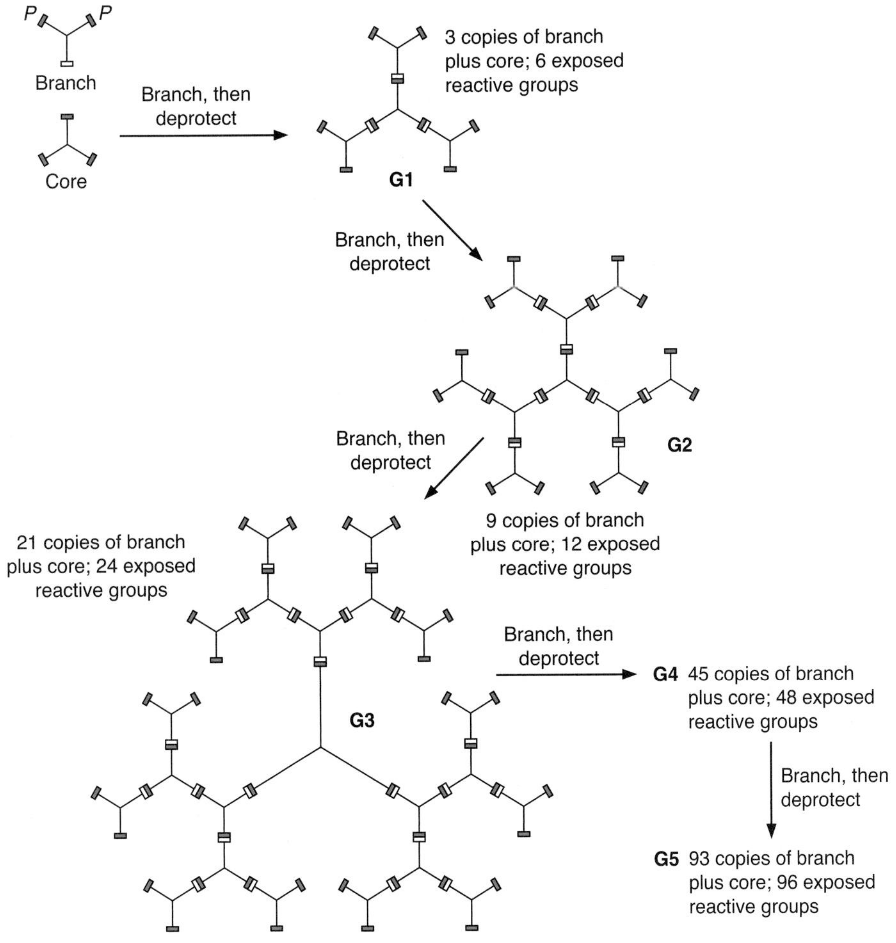

Figure 13.3
The dendrimer strategy.

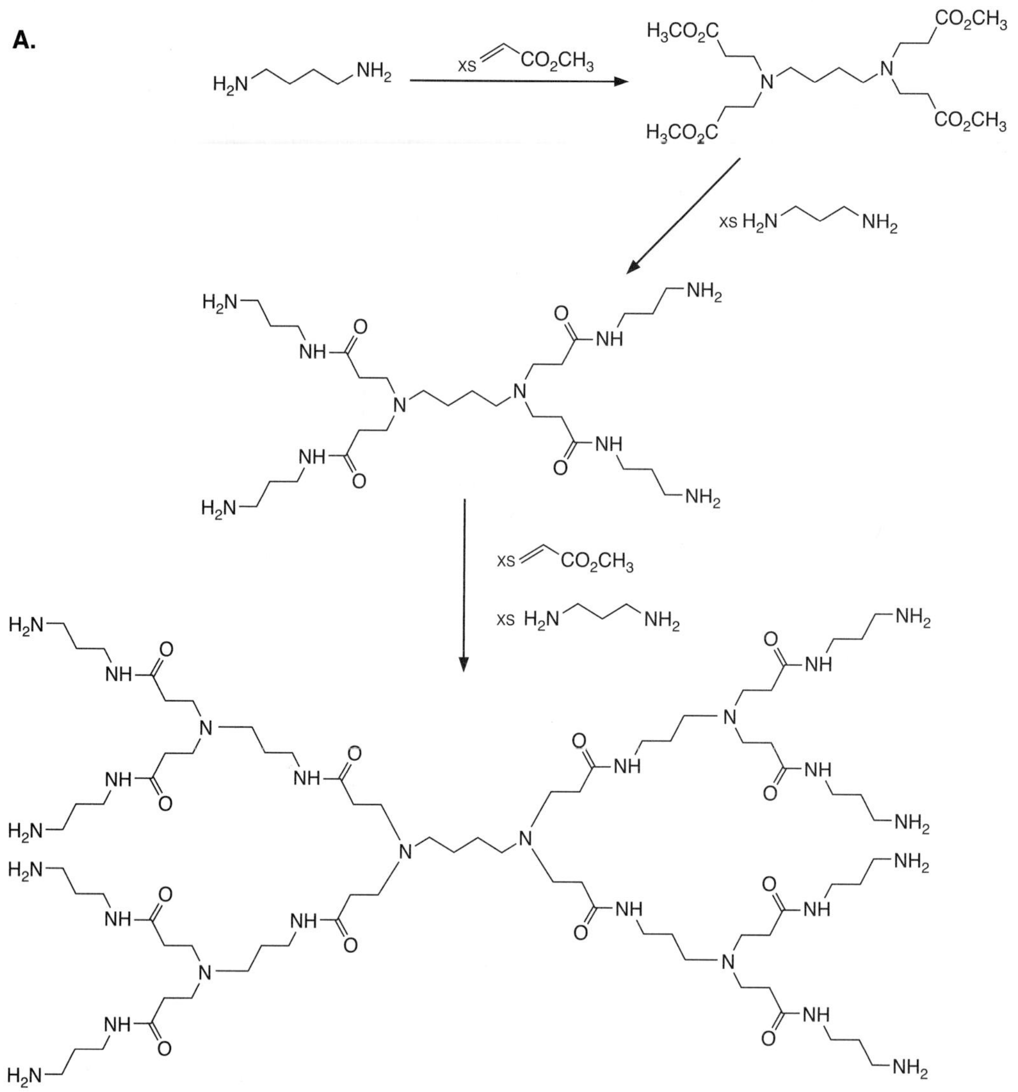

Figure 13.4
Two examples of specific dendrimer systems. **A.** The PAMAM system.

The increasingly branched nature of these systems as we go to higher generations is reminiscent of a tree, and hence the term dendrimer for structures with this novel type of topology. Note how quickly the system increases in size. By **G5** we have run only 10 simple reactions, but we have assembled 94 monomers into a precise structure. To build a monodisperse linear polymer of this size would take many more reactions. For example, if the monomer weighs 100 g/mol, we are already near molecular weight 10,000 g/mol by **G5**, and the next step will more than double the weight! In principle, each generation is a single pure molecule; we really should have a monodisperse polymer.

Figure 13.4 shows some specific dendrimer examples. Very early examples of this "cascade synthesis" approach were reported by Vogtle. Here we show two especially well characterized systems. The first has been termed PAMAM [poly(amido amine)]. The second structure, the poly(propylene imine), has been especially well optimized and characterized. Both systems are now available commercially and are produced on the tens of kilogram scale. Many variations of the structures in Figure 13.4 can be obtained by starting with a different core or choosing different reagents.

B. The poly(propylene imine) system. Both syntheses are of the divergent type.

In practice, absolute monodispersity is difficult to achieve. Detailed studies by Meijer on the poly(propylene imine) dendrimer show that the fifth generation has an experimental PDI (determined by mass spectrometry) of 1.002, with 23% of the material being the desired fully elaborated dendrimer. This is one of the best studied, most highly optimized dendrimer systems, and so this may be the best that is practically achievable.

The syntheses of Figure 13.4 are typical and have been termed **divergent**, in that one starts from the core and branches out further and further with each successive generation. In 1990, Frechet and co-workers introduced the **convergent** approach to dendrimer synthesis, in which conceptually one starts at the periphery and builds to the core. Again, this is best seen with an example, and Figure 13.5 shows an early example from the Frechet group. The key structure is the diphenol branch. Only the more readily deprotonated phenols react with a benzylic bromide, and two bromides can react with one equivalent of the branch, producing the necessary branching for a dendrimer. The alcohol is then converted to a benzylic bromide, and this new bromide is again allowed to react with the diphenol branch. This protocol can be repeated several times (shown only twice in Figure 13.5). Finally, the bromide is

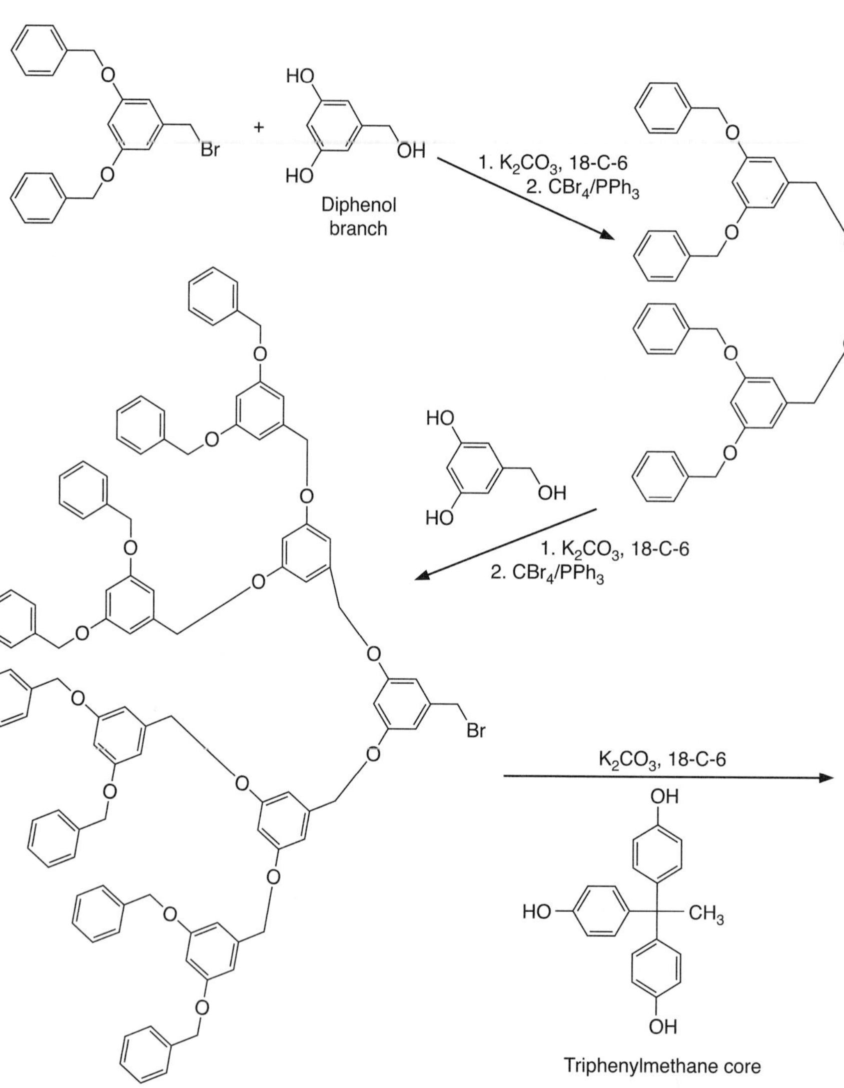

Figure 13.5
Convergent synthesis of dendrimers.

stirred with a "core" molecule, in this case a triphenylmethane core. The core really is the last part of the molecule added, and the first units are those on the outside of the final product. The convergent synthesis is the divergent synthesis run backwards. Again, many variations are possible at each step.

There are some practical advantages to the convergent approach. At each step, no matter how far along in the sequence, the reaction is simply two equivalents of a benzylic bromide reacting with a bisphenol. Contrast this with later stages of a divergent synthesis, in which, for example, 48 surface functional groups must be modified. Even with a very efficient reaction, there are sure to be failures, and these defects will propagate through the successive generations. Another potential advantage of the convergent approach involves purification of the final product. For example, in the final step of Figure 13.5, the major impurity might be expected to be the system in which only two equivalents of bromide reacted with the trisphenol. It should be possible to separate this byproduct from the desired product because of the substantial difference in molecular weights. In the divergent approach, we may be attempting a separation between two species that differ only by the difference of 47 or 48 copies of a relatively small molecule being added, a much more difficult task. At present, both divergent and convergent approaches are in use. Which to choose depends on the desired structure, the chemistry involved, and the availability of starting materials.

We noted above that both the PAMAM and poly(propylene imine) dendrimers are commercially available on a large scale, and no doubt others will follow. Why would anyone want a kilogram of a dendrimer? Monodispersity alone does not make dendrimers useful. Certainly, dendrimers have a certain aesthetic appeal, but again this is not enough to produce an article of commerce. However, the unique topologies of these systems influence their properties in many ways. First, for almost any molecular weight, dendrimers tend to be quite soluble. Apparently, the highly branched structure disfavors the kind of ordered packing arrangements that favor crystallinity. However, for applications commonly reserved for polymers, dendrimers have poor material properties, such as low viscosity.

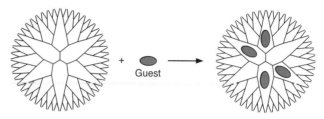

Figure 13.6
Encapsulation of a guest molecule in the interior of a dendrimer.

The most intriguing feature of dendrimers is only evident at very high generations. Figure 13.6 shows another idealized view of a dendrimer. First, it must be appreciated that dendrimers are certainly not flat, disk-shaped molecules. Steric interactions are significant for the higher generations, and to avoid end group clashes, the system adopts a more nearly spherical shape. The image of Figure 13.6 should be viewed as a two-dimensional representation of a three-dimensional object. The figure does nicely illustrate a key feature of dendrimers. While there is substantial crowding on the surface of the dendrimer, there are generally holes or voids toward the interior. It is these voids that have especially intrigued chemists. One can envision encapsulating other molecules in the voids for use in drug delivery or developing novel catalytic sites.

A number of workers have established that encapsulation can be quite efficient. Meijer has shown that if, after encapsulation, the surface groups of the dendrimer are functionalized with a bulky group, the steric interactions on the surface become so severe that the guest is permanently entrapped in the dendrimer. This is another example of a supramolecular container compound of the sort discussed in Chapter 4. Hence, dendrimers have been proposed as receptors for use in sensing and catalysis, due to their interior pockets and tailorability.

In addition, the huge number of functionalities on the surface of a dendrimer can be exploited. For example, attaching multiple copies of a substrate for an enzyme, or a recognition molecule for an antibody or cell adhesion protein can lead to novel effects. MRI image-enhancement molecules (complexes of Gd^{3+}) can decorate the surface of a dendrimer, producing an especially effective contrast agent. Studies thus far suggest that dendrimers are often quite nontoxic, and we can anticipate that more medical applications will appear.

Another novel feature of dendrimers is that, because of their rational, stepwise synthesis, the surface can be modified in very specific ways. For example, you could build up a very hydrophobic dendrimer, but then at the last step decorate the surface with a large number of hydrophilic species such as carboxylates or sulfonates. Such a system would then be water soluble, and would very much resemble a unimolecular micelle.

A great many very clever dendrimer systems have been designed and synthesized over the past twenty years. Initially, the emphasis was on developing appropriate synthetic approaches and characterization tools. That stage is nearly complete, and future work on dendrimers will emphasize the useful, functional properties of these novel materials. It will be very interesting to see what new systems are developed by physical and synthetic organic chemists.

Hyperbranched Polymers

Let's return to Figure 13.3. Consider what would happen if there were no protecting groups on the "branch" molecule. Now, under appropriate conditions, it could react with itself to make a polymer. It may be valuable to stop right now and write out for yourself what the product of such a self-condensation would be. As you continue adding monomers, you should see regions with definite dendritic character emerging. It is not a dendrimer, however, because there is no precise generational control of the developing polymer. The polymer structure is not well defined, but there are clearly similarities to a dendritic system. Such

Going Deeper

Dendrimers, Fractals, Neurons, and Trees

One of the fascinating aspects of dendrimers is that they can be analyzed as **fractals**, geometrical objects that exhibit "self-similarity". Fractals are built up from ever increasing numbers of similar pieces, and in many cases the size of the building blocks shrink as their number grows. As a result, whether we are looking at a fractal up close or from afar, we see the same basic pattern. Stated differently, in a fractal each building block resembles the object as a whole. The mathematics of fractals, pioneered by Mandelbrot, is a fascinating field of its own that has implications across broad fields of science. This mathematics applies directly to analyses of dendrimers with regard

to, for example, the number of end groups after a given number of generations. Fractal patterns are common in nature, where the self-similarity is often imperfect but is clearly evident. Examples include the branching patterns in trees (recall the origin of the word dendrimer), blood vessels, river basins, and the dendritic outgrowths of neurons. Fractals are a very popular topic on the world wide web; just type "fractals" into your favorite search engine and you'll find hundreds of sites. Some with high quality color pictures include the following: http://spanky.triumf.ca; http://www.techlar.com/fractals; and http://sprott.physics.wisc.edu/fractals.html.

structures are termed **hyperbranched polymers**, and they can be expected to arise from any AB_2 monomer if A and B can react with each other.

In fact, many systems can produce a hyperbranched topology. An AB_3 monomer will produce a hyperbranch, as will an A_2B_3 monomer. Yet, the A_2B_3 monomer also leads to a **network** (a highly cross-linked macromolecule). An alternative strategy is to react two different monomers, such as an A with a B_2 or B_3 monomer. Again, a hyperbranched system will result.

What features can we expect from a hyperbranched polymer? It will certainly not be monodisperse. This is a typical polymerization, and significant PDIs are expected. However, some of the desirable features of dendrimers are retained. For example, hyperbranched polymers tend to be quite soluble, an apparent consequence of their dendritic topology. This is much more evident for systems made from a single monomer. That is, the product from an AB_3 polymerization will tend to be more soluble than that which results from the reaction of an A with a B_3. In addition, for their molecular weight, the undesirable low viscosity associated with a dendrimer is retained, and hyperbranched polymers have poor mechanical properties.

Hyperbranched polymers have been used to make unimolecular micelles and novel liquid crystals (see below for a discussion of liquid crystals). A major advantage of hyperbranched systems is that their synthesis is a single step! It seems likely that for some applications, hyperbranched systems will perform well, but for others, the precision of a dendritic system will be required.

13.1.7 Liquid Crystals

We introduce here another fascinating aspect of organic materials chemistry—the liquid crystal phase. Along with highlighting their intrinsic importance, a discussion of liquid crystals allows us to introduce some important concepts such as cooperativity and domain formation. In addition, liquid crystals are the first example of systems in which *dimensionality* plays a key role. This is another large and active field of physical organic research, and certainly we cannot hope to present a comprehensive treatment here. As before, our goal is to introduce some of the basic terminology, concepts, and applications of liquid crystals. References to more detailed discussions of this topic are given at the end of the chapter.

Liquid crystals represent a state of matter that is intermediate between the crystalline and the liquid state. Because of this intermediate nature, the liquid crystal phase is sometimes referred to as a **mesophase**, and molecules that form liquid crystals are called **mesogens**. Mesogens can be small molecules or polymers. The crystalline state is associated with complete molecular order and extremely high barriers to reorienting any molecule in the crystal, as long as we are well below the melting temperature. The liquid state involves molecules that are randomly oriented and rapidly reorienting. In liquid crystals there is molecu-

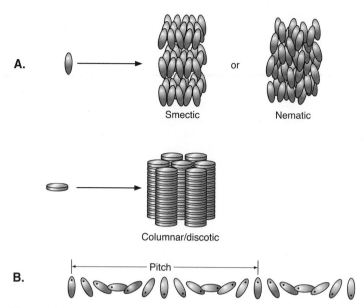

Figure 13.7
Liquid crystals. **A.** Generic description of select phases. A rod-shaped mesogen can form a smectic or a nematic phase. A disk-shaped mesogen can form a columnar/discotic phase. **B.** Illustration of how the director of a cholesteric phase rotates as one moves through the material, giving rise to a helical pitch. Note that the oval in this part of the figure is *not* a single mesogenic molecule, but rather represents the director of a thin section of the cholesteric phase.

lar order, but it is not nearly so precise as in a molecular crystal. Liquid crystalline phases not only possess orientational order, but the direction of orientation can be induced by the application of very modest external shear forces. This ability to reorient the liquid crystalline phase with an external force is a hallmark of liquid crystals and the foundation of much of the technology built around liquid crystals.

Temperature plays a critical role here. The mesophases occupy an intermediate temperature regime. At low temperatures we have crystals; at high temperatures we have liquids. In the intermediate temperature range, liquid crystalline behavior is possible. Tuning this intermediate temperature range, so that it is appropriately broad or narrow and does or does not encompass room temperature, is a key component of liquid crystal research.

In fact, a liquid crystalline phase may be one of dozens of phases with exotic names, like smectic C*. What differentiates these is the degree and nature of the molecular order in the phase. As we steadily raise the temperature, the same material may pass from the crystalline phase through several different liquid crystalline phases before becoming an isotropic liquid. An **isotropic liquid** is one where there is no structure along any direction, although there may be local order due to alignment of dipoles or other weak forces (see Chapter 3). A key feature of liquid crystals is that they are **anisotropic**, having a definite *directionality* over long distances (Figure 13.7 provides examples). The temperature at which an anisotropic liquid turns isotropic is called the **clearing temperature** (T_c).

The bulk anisotropy of liquid crystals depends upon the structures of the mesogens, the molecules that form liquid crystalline phases. Not all molecules can form liquid crystalline phases, and the ones that do tend to have certain shapes. More than in most other fields of chemistry, molecular shape rather than the precise nature of the functionality is a key determinant of liquid crystal behavior. Mesogens most often have a significant elongation in one direction (rod-shaped), leading to the common symbolism of an oval, as in Figure 13.7.

The most ordered type of liquid crystalline phase is the **smectic** phase, an example of which is shown in Figure 13.7 **A**. The ordering of the mesogenic molecules in this phase points the monomers in the same direction, along the long axis of the mesogens. The alignment is imperfect, but it is clearly there. The unique direction defined by this orientational preference is called the **director**. In Figure 13.7 **A** the director runs along the "north–south" direction. A second type of order exists in a smectic phase; the molecules tend to form layers or planes. As such, a smectic liquid crystal is not too different from a molecular crystal. The main difference is that there is only partial order in the orientation and position of the molecules within the layers, with no register between layers.

Nematic liquid crystalline phases can be thought of as smectic phases absent the layering effect (Figure 13.7 **A**). There is clearly still a preferred orientation, a director, but that is the only type of order. Another type of phase termed the **columnar** or **discotic** liquid crystalline phase can form when the mesogen is more disk-shaped rather than cigar-shaped (Figure 13.7 **A**).

An important new feature of liquid crystal order appears when the mesogen is a chiral molecule and is enantiomerically pure, or the liquid crystal is doped with a small amount of an enantiomerically pure compound. Chiral nematic liquid crystals are also known as **cholesteric** phases, recognizing that cholesterol derivatives provided early examples of chiral mesogens that form such a phase (recall the discussion in Chapter 2 of the distinctive shape of steroids such as cholesterol). The innate handedness of the chiral mesogen exerts an additional bias on the material in the form of a subtle twist of the director. That is, as we move through the material in a direction perpendicular to the director, the tilt of the director varys continuously, tracing out a helical shape (Figure 13.7 **B**). As with all helices there is a direction and a pitch associated with a cholesteric phase, defined as the distance it takes for the director to achieve one full turn. Light with a wavelength that matches the pitch is reflected by the liquid crystal, making the material colored if the pitch is of a distance comparable to the wavelength of visible light. The pitch of many cholesterics expands or contracts along the helix axis as the temperature is raised or lowered. As such, these materials change color with changing temperature, and this is the basis for liquid crystalline thermometers.

Given this brief introduction to liquid crystals, we can describe an important and familiar technological application of liquid crystals, the **liquid crystal display** or **LCD**. The interaction of light with liquid crystals is distinctive, a key issue being the polarization of light induced by interaction with a liquid crystal. In Chapter 6 we discussed plane polarized light. Plane polarized light can be produced by passing light through a polarizer. Often a material as simple as a piece of plastic that has been stretched in one direction can act as a polarizer; light that passes through the plastic emerges as polarized light. The reason the plastic can do this is that the stretching has rendered the material anisotropic. In particular, polymer molecules will tend to be aligned along the stretching direction.

Anistropy is also a property associated with liquid crystals, and indeed this property is crucial to the use of liquid crystals in LCDs and other devices. Because of their anisotropy, liquid crystals are **birefringent**. This means the material has two refractive indices, one for light whose electric field vector is parallel to the director of the liquid crystal and one for light whose vector is perpendicular to the director. Recall that the refractive index of a material is simply the speed of light in the material relative to vacuum. When placed between crossed polarizers, a birefringent material (in most orientations) can rotate the plane of polarization, allowing light to come through the second polarizer.

The most common type of LCD is the **twisted nematic** display (Figure 13.8). It consists of a nematic liquid crystal sandwiched between two plates of glass. The glass is treated such that it can orient the director of the liquid crystal in contact with it. This can be achieved by simply coating the glass with a thin film of polymer (often a polyimide) and then literally rubbing the glass in a particular direction to create anisotropy on the glass surface. This anisotropy is then transferred to the liquid crystal that contacts the glass, presumably through a kind of epitaxial growth process. If the "rubbing directions" of the two glass plates are rotated 90° relative to each other, the directors of the liquid crystals touching the top and bottom pieces of glass will also be rotated by 90°. This rotation occurs gradually throughout

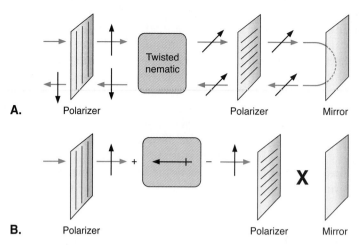

Figure 13.8
Schematic description of a simple LCD display. The colored arrows represent the direction of propagation of the light, while the black arrows represent the direction of polarization. **A.** With the twisted nematic between the two perpendicular polarizers, the polarization induced by the liquid crystal allows the light to pass. **B.** When an external electric field aligns the liquid crystal parallel to the first polarizer (indicated by the dipole arrow) the light cannot pass through the second polarizer and so there is no reflected light.

the liquid crystal layer, hence the term "twisted nematic". This is similar to a cholesteric phase (see Figure 13.7 **B**), and sometimes a small amount of a cholesteric mesogen is added to facilitate the twisting. The thickness of the liquid crystal layer is adjusted such that as light passes through a twisted nematic, it will emerge with its polarization rotated 90° relative to its value as it entered the liquid crystal.

We need to discuss only one more effect to understand an LCD. Consider the case where each mesogen molecule of our liquid crystal phase has a permanent dipole moment. If we apply an electric field across the liquid crystal, a force will be generated that will align the molecules appropriately with the field. This force exists regardless of the phase the molecules are in. In a solid, the crystal packing forces are much larger than this force, and no alignment can occur. In a liquid, thermal energy causes constant reorientation of the molecules and so disfavors bulk orientation. The solid is too rigid and the liquid too flexible, but the liquid crystal is just right. The director of the liquid crystal will be realigned by the external field so as to make a favorable interaction of the molecular dipoles with the field. When the applied field surpasses a certain critical value, the desire to align with the field overcomes and destroys the twisting induced by the glass plates, and the mesogens reorient.

So how does a simple LCD work? Figure 13.8 shows a highly schematic design. In Figure 13.8 **A** light passes through a polarizer then through a twisted nematic that causes a 90° rotation of the plane of polarization. As a result, the light can pass through the second polarizer, which is rotated 90° relative to the first. It then hits a mirror, bounces back through the back polarizer, through the liquid crystal and gets its plane of polarization rotated 90° back. The net rotation is 0° and so the light can emerge through the original polarizer. Light of this sort is the familiar grey of LCDs on digital watches. If, however, a voltage is applied across the liquid crystal (Figure 13.8 **B**), the twisting effect is lost. Then the light is unable to pass through the second polarizer, and so is not reflected back by the mirror, giving a black appearance. We now have our on and off states. LCDs involve patterned regions of twisted nematic liquid crystalline phases and appropriate circuitry to allow the controlled application of voltages across these regions.

More advanced LCDs take advantage of so-called **super-twisted nematic** displays, in which the director rotates 270° between the two polarizers. This leads to a crisper distinction between on and off states. Color LCDs simply include appropriate dyes to make red, green, and blue (RGB) pixels—simple in principle but a fairly complex technology.

An important aspect of LCDs and other liquid crystal-based devices is that while they can switch between two states, that switching depends on the bulk reorientation of molecules. As such, the switching is not very fast. This is not a problem for video displays, where rates under 100 Hz are typical.

A few more comments about reorientation of liquid crystal phases are in order. We noted that if the mesogen has a permanent dipole moment, an applied external electrical field can reorient the director of the liquid crystal. Actually, a permanent dipole moment is not essential. Absent a permanent moment, an external field will produce an induced dipole moment (recall the discussion of induced dipoles in Chapter 3). This is enough to cause orientation of the director, although the effect is typically smaller than for cases with a permanent dipole moment. Amazingly, even an external *magnetic* field can cause alignment of a liquid crystal. The forces exerted by a magnetic field are extremely small. For magnetic alignment, we are counting on the anisotropy of the diamagnetic moment of the mesogen (see Chapter 17 for a brief discussion of diamagnetism). Because the effect is so small, magnetic alignment is typically slow, sometimes requiring hours or even days to reach its maximal effect, although strong fields can give fast alignment.

Why can such small forces produce such noticeable effects? This is an excellent example of **cooperativity**. As soon as one molecule starts to align with an external field, the strong shape effects inherent in liquid crystals cause neighboring molecules to want to align. The effects feed off of each other, and in this way a force that is intrinsically quite weak can be amplified to ultimately produce a very big effect. For example, addition of a small percentage of a chiral dopant can induce a nematic phase to become cholesteric.

We can take this opportunity to introduce one more important concept for materials chemistry, and that is the notion of a **domain**, or **microdomain**. The kinds of alignment effects we are discussing are essentially enthalpy driven, relying on steric and electronic interactions. Alignment is entropically unfavorable. Entropy favors randomness, not alignment. Therefore, small regions of the material align, and these are termed domains. But, in the absence of fields or surface effects, the bulk alignment (the sum of the domain alignments) is zero. Hence, the domains are randomly aligned relative to each other. A bulk unaligned LC looks milky due to light scattering from the domain boundaries, while a bulk aligned LC is clear.

The birefringence inherent to liquid crystals has very interesting consequences for unaligned, polydomain samples placed between crossed polarizers. Differently oriented domains rotate the polarized light different amounts, such that some areas appear dark and others light. Furthermore, since the amount of rotation depends on the wavelength of light, fascinating, colorful, and beautiful patterns emerge, providing one of the most attractive aspects of working with liquid crystals. Remarkably, the patterns in such images are informative about the nature of the liquid crystalline phase. The pattern produced by a nematic phase is visibly different from a smectic phase when viewed through crossed polarizers, primarily due to differences in the structure of the domain boundaries. This carries through to a wide variety of phases, and though it requires some training and experience, visual observation is the most common tool for categorizing liquid crystalline phases.

Now that we have the basic concepts in place, lets look at some molecules. Figure 13.9 shows some typical mesogens. Not surprisingly, given the very small magnitude of the effects involved, design principles are derived from empirical observations rather than theoretical insights. Rigid rod molecules are common, and so aromatic rings are usually involved. However, true crystallinity must be avoided, so mesogens typically also have flexible portions that disfavor crystalline order. Polar groups such as cyano and nitro are common, especially in switching applications such as LCDs. Not surprisingly, discotic phases tend to arise from disk-shaped molecules. Thousands of mesogens are known, and the structures in Figure 13.9 should only be viewed as representative.

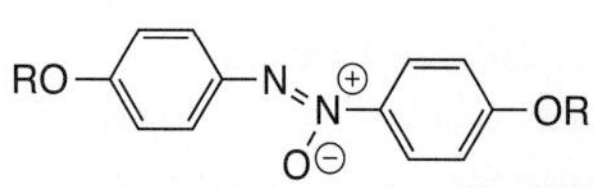

Series of PAA (*p*-azoxyanisole) derivatives

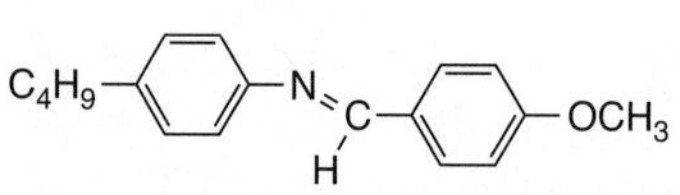

p-Methoxybenzylidene-*p*-*n*-butylaniline (MBBA),
the first stable room temperature liquid crystal

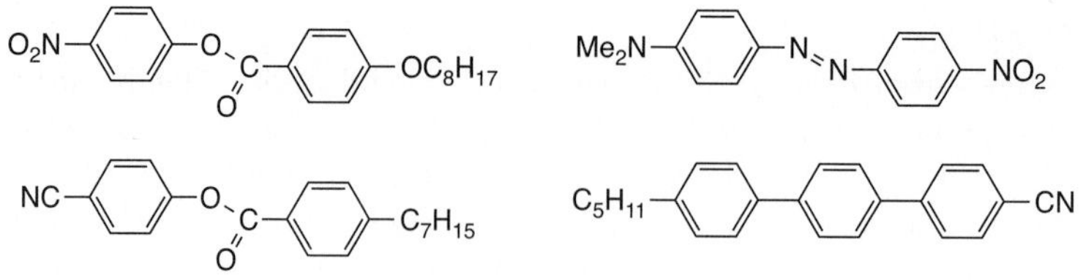

R = CH$_3$, C$_2$H$_5$, C$_3$H$_7$

Structures with polar end groups;
favorable for LCD applications

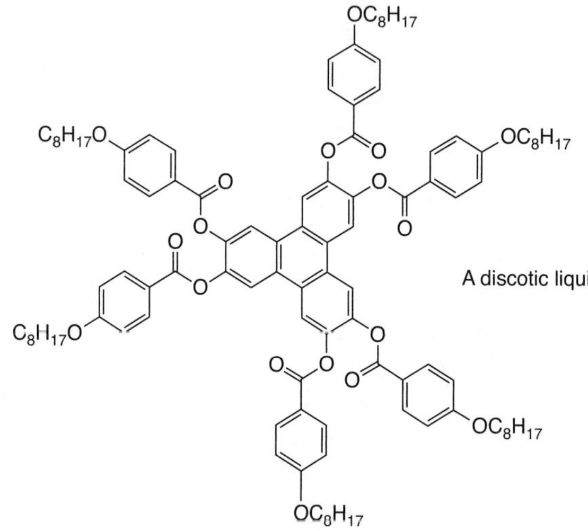

A discotic liquid crystal

Figure 13.9
Representative structures of mesogens.

Connections

Lyotropic Liquid Crystals: From Soap Scum to Biological Membranes

The liquid crystal systems we have discussed so far are termed **thermotropic**, in that temperature is the key factor in determining what phase we see. Another way to regulate phase behavior is by varying the relative concentrations in a two-component system. Such systems are termed **lyotropic**, the most common of which involves the hydrophobic effect. Under certain conditions, very high concentrations of amphiphilic molecules in water form structures that are much larger than micelles or vesicles. For a typical soap or a phospholipid, when the concentration approaches 50% the micelles or vesicles aggregate into higher order structures. Liquid crystalline phases will develop, the exact nature of which depends on the concentration and the structure of the amphiphile in water. Examples include the **hexagonal phase**, in which

long cylindrical rods of amphiphiles assemble into hexagonal close-packed domains; and the **lamellar phase**, which involves flat bilayers that are separated by thin layers of water. The "scum" that forms at the bottom of your soap dish is a lyotropic liquid crystal.

A perhaps more significant example of a lyotropic liquid crystal is the cell membrane. The phospholipid bilayers of all cells behave as liquid crystals, maintaining a definite director and displaying only limited fluidity. Interestingly, most membranes are not pure phospholipids, but contain additives whose function is presumably to tune the structural and dynamic properties of the membrane. In animals, the most common additive is cholesterol, while plants use a similar steroid. As such, cell membranes can truly be thought of as cholesteric liquid crystalline phases.

13.1.8 Fullerenes and Carbon Nanotubes

In 1985 Kroto, Smalley, and Curl reported that laser-induced vaporization of graphite under appropriate conditions produced in the gas phase an abundance of a species with mass 720. They correctly deduced that this C_{60} species had the soccer ball or "Buckminsterfullerene" geometry (Figure 13.10), work for which they earned the 1996 Nobel Prize in Chemistry. In 1990 Krätschmer and Huffman reported that this gas phase novelty could be produced in useful quantities by simply running large electrical currents through graphite rods and extracting C_{60} and related compounds from the soot that forms on the side of the flask. Fullerene chemistry was born.

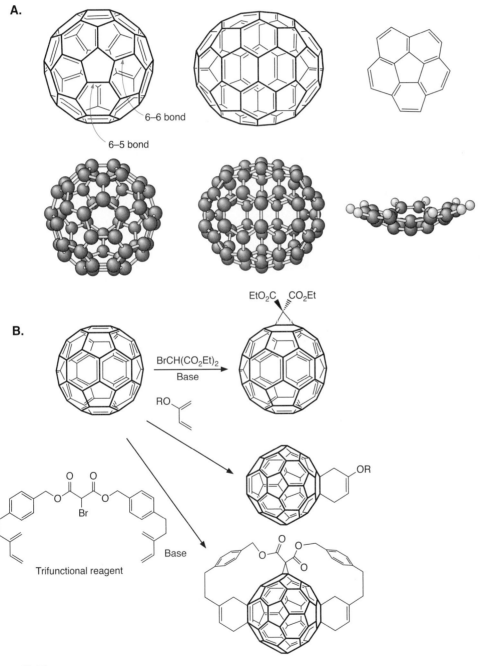

Figure 13.10
Fullerene chemistry. **A.** The structures of C_{60}, C_{70}, and corannulene are shown. **B.** Representative reactions of fullerenes. Shown are prototypical carbene and Diels–Alder reactions. Also shown is a trifunctional reagent used to regioselectively provide a multifunctional fullerene.

Fullerenes are discrete C_n molecules. The fullerenes constitute a new allotrope of elemental carbon, adding to the previously well known forms of diamond and graphite. It was really quite stunning to appreciate that after centuries of investigation a new form of carbon could be discovered. This class of molecules includes C_{60} and C_{70} (the second most abundant product in the soot), as well as their derivatives (Figure 13.10). These structures are based on the motif of hexagons and pentagons to close the sphere or ellipsoid, with the C_{60} structure completely analogous to that of a soccer ball (with apologies to our friends outside the United States who consider this object to be a football). The name is a tribute to Buckminster Fuller, the architect who developed geodesic domes based on the same geometric principles that allow C_{60} to form a closed space. As Fuller appreciated, to completely enclose space a fullerene must have exactly 12 pentagons. For a C_n fullerene, there will be $(n/2-10)$ hexagons, and so C_{60} has 20 hexagons. A structure with only hexagons is a flat sheet, as in graphite, or cylindrical, as in nanotubes (see below). It is the deviation from perfect hexagonal angles in the pentagons that forces the materials to buckle and deviate from planarity. The truncated icosahedron of C_{60} is the smallest system for which no two pentagons are adjacent; every five-membered ring of C_{60} is completely surrounded by six-membered rings. This bonding pattern contributes to the special stability of C_{60}. In C_{70}, the nearly spherical soccer ball of C_{60} is elongated to an ellipsoidal, "rugby ball" shape with still no pentagons sharing an edge.

The core structure of C_{60} is contained in corannulene (Figure 13.10), a single five-membered ring surrounded by hexagons which is also nonplanar. Like C_{60}, corannulene was initially available in only minute quantities, but clever syntheses developed by Scott and Siegel made corannulene and derivatives readily available. This allowed detailed investigations into their properties. For example, the bowl inversion of corannulene was determined to have a barrier of 10.2 kcal/mol.

There are two types of bonds in C_{60}, the 6–6 bonds, which lie at the junctions between two six-membered rings, and the 6–5 bonds, which lie at the junctions of six- and five-membered rings. Considerable alternation in bond lengths is seen, with the 6–6 bonds being $\sim$1.38 Å and the 6–5 bonds being $\sim$1.45 Å. Based on these observations, and the reactivity patterns discussed below, the 6–6 bonds are considered to have much more double bond character than the 6–5.

The predominant reaction mode of C_{60} is that of a strained, electron deficient alkene. Addition reactions typically occur in a 1,2 fashion across the 6–6 bonds, adding across the olefin and not opening up the basic C_{60} ring system. Representative of such reactions are the carbene addition and the Diels–Alder reaction shown in the first two examples of Figure 13.10 **B**. Radical additions also occur, but they are typically more difficult to control and generally lead to multiple adducts.

More challenging has been the regioselective and stereoselective synthesis of multifunctionalized fullerenes. Several strategies have emerged. One clever approach developed by Hirsch makes use of a *reversible* Diels–Alder addition of 9,10-dimethylanthracene across 6–6 bonds. When an additional, irreversible reaction (such as a carbene addition) is run in the presence of dimethylanthracene, the steric bulk of the anthracene adduct directs further additions to more remote locations. The reversibility of the reaction ensures that multiple functionalizations will occur. With this strategy, good yields of symmetrical, multifunctionalized fullerenes can be obtained. An alternative strategy for regiocontrol uses a tethering approach. For example, Diederich designed the multifunctional reagent shown in Figure 13.10 **B**, which first undergoes a carbene addition. Heating the adduct promotes a double Diels–Alder reaction, and the constraints provided by the tethers direct the reactions such that a 60% yield of the single adduct shown can be obtained. Many variants of this strategy have been developed.

A good deal of the excitement surrounding the discovery of the fullerenes was their perceived potential as novel materials for a number of applications. Early speculations suggested that the essentially spherical structure of C_{60} might be the basis of novel lubricants, providing "molecular ball bearings". However, the first hint of special properties came in 1991 when a group at Bell Labs reported superconductivity at 18 K in a sample of C_{60} that

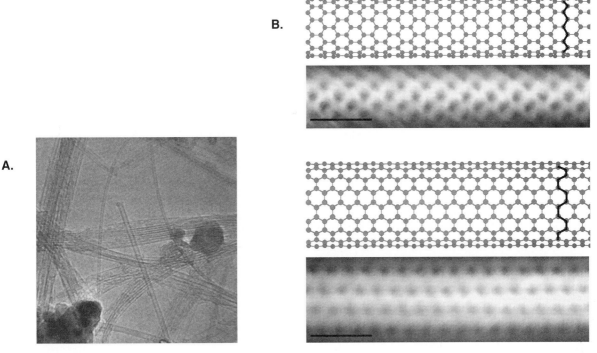

Figure 13.11

Carbon nanotubes. **A.** Transmission electron micrograph (TEM) of 1–2.5 nm diameter SWNT both isolated and in bundles. **B.** Molecular models and scanning tunneling microscopy images of SWNT. The scale bar in both images is 1 nm. The two forms [termed (12,0) (top) and (9,9) (bottom) in reference to crystallographic indices of graphite] differ in the orientation of the graphite sheet prior to "rolling up" the sheet into a nanotube. We thank Professor Charles Lieber, Harvard University, for these images.

had been "doped" with K. Doping is discussed extensively in Chapter 17. Here, we are simply referring to the reaction of K metal with C_{60}. Not surprisingly, C_{60} has a low-lying LUMO. It is in fact three-fold degenerate, a consequence of the very high symmetry of the molecule (point group I_h). This orbital is a good acceptor of electrons, and under appropriate conditions reaction with K metal produces C_{60}^{3-} with three K^+ counterions. Films of this material are superconducting. When first discovered, 18 K was the highest critical temperature yet seen for a molecular superconductor, and it launched a race to achieve still higher temperature superconductors based on fullerenes. However, very high temperature superconductivity has not yet been seen with fullerenes. While the theoretical importance of superconductivity in doped fullerenes cannot be denied, in the end these materials seem unlikely to ever be of technical or commercial importance as conductors, having been eclipsed by the inorganic, copper oxide, high temperature superconductors (see Chapter 17).

The efficient production of fullerenes through electrical treatment of graphite prompted further investigations of this novel procedure. Under certain conditions, it was noticed that long, needle-like structures formed. We noted above that a closed surface requires 12 pentagons, but there is no limit to the number of hexagons that can be accommodated. Imagine taking a graphite sheet, comprised entirely of hexagons, and rolling it up to make a cylinder (Figure 13.11). The ends of the cylinder could be capped off with appropriate "half-fullerenes", creating a long tube of carbon. Extensive investigations have shown that such **carbon nanotubes** do in fact form. While their diameter is typically on the order of 1–2 nm, their length can be on the order of microns, leading to a huge aspect ratio for these novel structures. We have described a **single-walled nanotube (SWNT)**, but there are also **multi-walled nanotubes (MWNT)**, which contain concentric cylinders of graphite. In recent years it has become apparent that new technological advances are more likely to be associated with SWNT materials than with molecular fullerenes.

Two features of SWNT have generated considerable excitement. The first is their remarkable strength. Carbon fibers with much less well-defined structures have been used for some time to make strong, lightweight materials. In a sense, SWNT have the potential to perform as super carbon fibers. As produced, SWNT typically self-assemble into rope-like bundles. These carbon fibers have superior mechanical properties, placing them among the strongest and stiffest materials known. In addition, because of the hollow nature of the cylinders, these very strong materials are very lightweight. If uniform SWNT can be prepared in large quantities, a new generation of lightweight, very strong materials should emerge.

SWNT also have novel electronic properties. Depending on their mode of preparation, these tubes can be semiconductors or electrical conductors (see Chapter 17 for more details on conductivity). These molecular-scale wires show unique properties, including true metal-like electrical conductivity. It seems likely that the "molecular electronics" of the future could involve heavy use of these molecular wires.

The combination of high strength and electrical conductivity can be combined to produce highly novel phenomena. For example, in 1999 Lieber and co-workers prepared "nanotube nanotweezers". Two nanotubes were attached to independent electrodes and positioned near each other. When a voltage bias was placed across the two, the two nanotubes reversibly bent toward each other, fully mimicking the motions of a tweezer. These electrically activated tweezers can grab and manipulate polystyrene microspheres or silicon carbide nanoclusters. This early work could presage the development of true nanomachines that could perform unprecedented operations on a near-atomic distance scale.

Connections

Organic Surfaces: Self-Assembling Monolayers and Langmuir–Blodgett Films

A common theme in modern chemistry is the study of so-called **low-dimensional materials**. All materials are three-dimensional, but in some systems a material can have quite large dimensions in one or two directions, but very small dimensions in the other two or one dimensions. In this sense, the SWNT just discussed are examples of one-dimensional materials, and some of their unique electronic properties arise from this feature.

Two-dimensional materials—surfaces or films—are also of considerable interest. Surface science is a huge field that is primarily concerned with metallic materials and related crystalline solids. However, some novel and interesting "organic" surfaces can be prepared with special techniques. These are intrinsically interesting and perhaps technologically useful structures that have attracted much research interest in the past two decades.

The simplest structure to understand is that formed when a gold surface is exposed to a simple, linear chain alkanethiol, $CH_3(CH_2)_nSH$. The attraction of thiols for gold has long been appreciated. When a thiol and metallic gold interact, a covalent, Au–S bond forms. As we add more thiol, more absorbs to the surface until it has been passivated—that is, all of the exposed surface area has reacted. Weakly attractive van der Waals interactions induce the alkyl chains to align into two-dimensional ordered structures, commonly known as **self-assembled monolayers (SAMs)**. By varying the thiol structure, the surface properties can be systematically controlled. For example, using a thiol such as $X(CH_2)_nSH$, where X is OH, produces a hydrophilic surface, while $X = CH_3$ produces a hydrophobic surface (see below).

Another class of monolayers are known as **Langmuir–Blodgett films**. Instead of gold, the initial surface is liquid water. Long chain molecules that terminate in a very polar or charged group tend to put the polar head group in the water, and align the hydrophobic tail away from the water. The spontaneous formation of a well-defined monolayer does not usually occur on this liquid surface. However, if a movable trough (see the drawing, below) is used to compress the monolayer, well-defined surfaces will form. If you dip a glass slide or other solid structure (called the substrate) into the solution and then pull the substrate out while applying pressure in the trough, the monolayer on the surface of the solution will transfer to the substrate, creating a film. As a bonus, by measuring the force required to compress the monolayer in the trough, remarkably detailed insights into the forces involved in forming the monolayer can be obtained.

Ulman, A. "Formation and Structure of Self-Assembled Monolayers." *Chem. Rev.*, **96**, 1533 (1996). Bryce, M. R., and Petty, M. C. "Electrically Conductive Langmuir–Blodgett Films of Charge-Transfer Materials." *Nature*, **374**, 771 (1995).

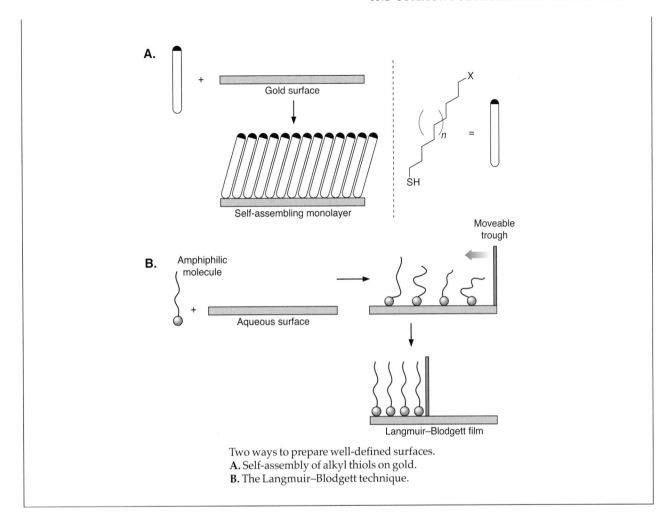

Two ways to prepare well-defined surfaces.
A. Self-assembly of alkyl thiols on gold.
B. The Langmuir–Blodgett technique.

In summary, much of the current research in organic materials is focused on the creation of macromolecules or interfaces with controlled properties. To control the bulk material properties, chemists must be able to control molecular weights and polydispersities, stereochemistry, sequence, and even macromolecular shape and topology. Many new kinds of materials have emerged in this effort, such as dendrimers, liquid crystals, and fullerenes. Achieving the kinds of control we are discussing often has a basis in the knowledge of the mechanisms of polymerizations, and that is the topic we now consider.

13.2 Common Polymerization Mechanisms

13.2.1 General Issues

In many ways the types of polymers available and the properties expected for them are defined by the synthesis strategies used to prepare them. Different synthesis strategies produce different trends with regard to issues such as molecular weight distributions and the role of defects. As with common functional group interconversions, classes of reactions emerge. For the most part, polymerization reactions are of three types: condensation reactions, addition reactions, and ring-opening reactions. Before discussing particular polymers, we briefly discuss the generic reaction types and some general differences among them. Schematics of the various polymerization types are shown in Figure 13.12.

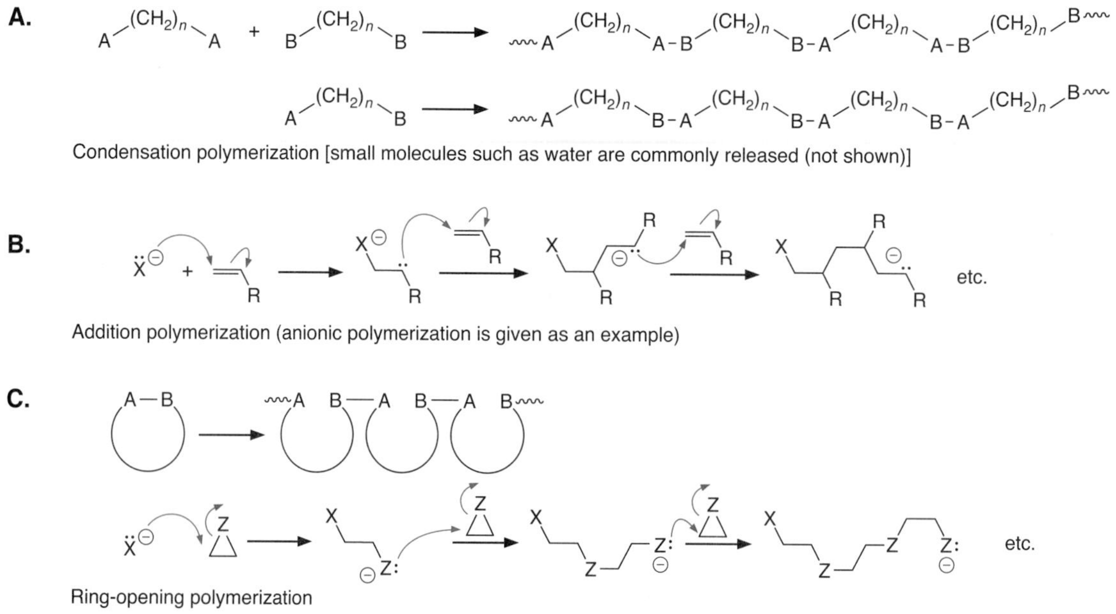

A.

Condensation polymerization [small molecules such as water are commonly released (not shown)]

B.

Addition polymerization (anionic polymerization is given as an example)

C.

Ring-opening polymerization

Figure 13.12
Schematics of the basic types of polymerization reactions. **A.** Condensation polymerizations, showing both A_2/B_2 and AB systems. The small molecule released (often water) is not shown. **B.** Addition polymerization. **C.** Ring-opening polymerizations, showing both a path that has some relationship to condensation polymers and one that is driven by ring strain and looks more like an addition polymerization.

A **condensation polymerization** is one in which two or more molecules combine with the loss of a small molecule, typically water or ammonia. We can anticipate that reactions such as alcohols combining with carboxylic acids to make esters with the loss of water and comparable amide-forming reactions will be important. The reactive groups, which we'll call A and B, can either be in differing, necessarily difunctional monomers, or they can be part of a single, difunctional AB molecule.

Addition polymerization involves the addition of a reactive functionality such as a radical or an ion to a double or triple bond such as an olefin or a carbonyl (Figure 13.12 **B**). Many strategies exist, mostly based on chemistries we have described earlier in Chapter 10. A number of well known polymers can be formed by addition reactions, including polyethylene, polypropylene, polystyrene, polyvinylchloride (PVC), polyacrylonitrile (Orlon), polytetrafluoroethylene (Teflon), and poly(methyl methacrylate) (Plexiglas, Lucite).

Ring-opening polymerizations start with a cyclic monomer that, in the presence of a catalyst, opens up and polymerizes (Figure 13.12 **C**). Often a ring-opening polymerization produces a polymer similar or identical to what would be prepared from a condensation polymerization of an AB monomer. The difference is that, instead of being part of the polymerization reaction, the condensation reaction occurred in the synthesis of the monomer, with the AB functionalities reacting to make a ring with extrusion of a small molecule. Release of ring strain provides a valuable thermodynamic driving force for many ring-opening polymerizations.

The three classifications just described are essentially based on monomer structure. Condensation polymerizations arise when the two monomers are stable but have functionalities that can react with each other. Addition polymerizations require unsaturation in the monomer that is vulnerable to attack by radicals or ions, and ring-opening polymerizations require cyclic monomers. An alternative and more modern classification emphasizes the differing *mechanisms* of polymerization, producing two classes: step-growth reactions and chain-growth reactions. Figure 13.13 contrasts the two mechanisms.

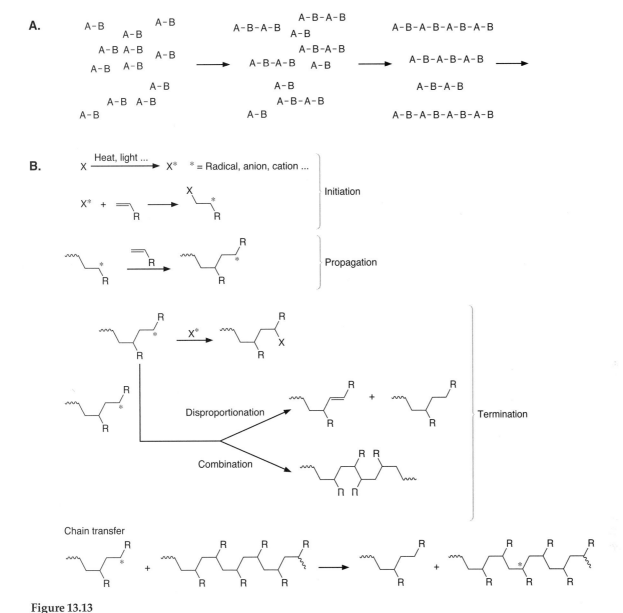

Figure 13.13
A. Step-growth polymerization, illustrated with a typical condensation polymerization (the small molecule that is eliminated is not shown). Note that essentially all monomer is gone after only one or two "reaction cycles" and that many reactive groups (A and B) are present in the flask at all times. **B.** Chain polymerization. The familiar steps of initiation, propagation, and termination are shown. Not all termination pathways are available for all mechanisms. Also shown is a chain transfer process, whereby the reactive functionality (*) is transferred from the end of the growing chain to any position, including internal positions, of another chain.

Step-growth reactions involve a polymer chain that grows in a stepwise fashion. Two monomers react to make a dimer that has the same functionality at its two ends as the original monomers. The dimer can react with another monomer to make a trimer or with a dimer to make a tetramer. Condensation reactions generally fall into this category.

Chain-growth reactions are familiar from the radical mechanisms discussed in Chapter 10. The standard steps of initiation, propagation, and termination are all involved. Prototypically, chain polymerizations involve the rapid addition of olefin molecules to a growing chain end. At the end of the polymer chain is a reactive functionality not part of the basic monomer structure, such as a radical or an ion. Although our previous exposure to chain reactions was restricted to radical chemistry, there are important ionic polymerizations that proceed by a chain mechanism. Ring-opening polymerizations also generally follow a chain mechanism.

13.2.2   Polymerization Kinetics

In efforts to control polymer synthesis and thus polymer composition, kinetics plays a key role. Building off our development of the essentials of kinetics in Chapter 7, here we develop the basics of polymerization kinetics. A key issue is the difference between step-growth/condensation polymerization and the chain mechanism. We will see that after a few basic assumptions, the kinetics of polymerization are not too different from those of conventional reactions.

Much of what we will describe in this section is based on the pioneering work of Flory, who earned the 1974 Nobel Prize in Chemistry for his work in polymer chemistry. Flory was hired by DuPont in 1934, where he was assigned to the labs of Carothers. The brilliant Carothers is certainly one of the fathers of polymer chemistry, and his DuPont labs first produced neoprene, polyesters, and nylon, among others. Carothers would have certainly earned a Nobel Prize himself, but he committed suicide in 1937 at the age of 41. Beginning in the environment that created modern polymer science and later in an academic career, Flory developed the foundations of the physical organic chemistry of polymerization.

A key simplifying assumption made by Flory was that the reactivity of a monomer with a growing chain was independent of chain length. This may seem counterintuitive at first, and indeed in the early stages of a polymerization when the chains are short, it is incorrect. However, as the reaction progresses and the chains get longer, "end-group" effects disappear and the assumption is valid, as we will show below. Now we can consider the differing kinetics of step-growth vs. chain polymerizations. Our treatment closely follows that of the excellent polymer text by Painter and Coleman, cited at the end of the chapter.

Step-Growth Kinetics

Consider the condensation of two monomers, A–A and B–B, such as a dicarboxylic acid and a diol to make a polyester. We assume that a trace amount of acid catalyst is present, and the concentration of the catalyst is a constant and so can be absorbed into the rate constant, giving the rate law of Eq. 13.6. This is just the familiar second-order rate law (see Eq. 7.32). Note that [A] and [B] are the concentrations of *functional groups*, not of monomers. We assume that the starting concentrations of each functional group are the same, and define c_0 as the initial concentration and c for the concentration at a particular time in the reaction. This makes it possible to simplify Eq. 13.6. We define the extent of reaction as p, and express c in terms of p as in Eq. 13.7. The treatment is then as in Eq. 13.8, giving the final rate law in Eq. 13.9.

$$\frac{-d[A]}{dt} = k[A][B] \quad \frac{-dc}{dt} = kc^2 \qquad \text{(Eq. 13.6)}$$

$$c = c_0\,(1-p)$$
$$p = \frac{\text{Number of A groups reacted}}{\text{Number of A groups initially present}} \qquad \text{(Eq. 13.7)}$$

$$\int_{c_0}^{c} \frac{dc}{c^2} = k \int_{t=0}^{t} dt$$
$$\frac{1}{c} - \frac{1}{c_0} = kt \qquad \text{(Eq. 13.8)}$$
$$\frac{1}{c_0\,(1-p)} - \frac{1}{c_0} = kt$$

$$\frac{1}{1-p} - 1 = c_0 kt \qquad\qquad\text{(Eq. 13.9)}$$

If our assumptions are correct, a plot of $1/(1-p)$ vs. time should be linear, and indeed this is empirically observed for step-growth polymerizations. To determine p you need a method that can follow the conversion of all functional groups in the monomers to all the associated functional groups in the polymer, such as carboxylic acids to esters in the formation of polyester.

The **number average degree of polymerization** (average number of monomers per chain), $\overline{DP}$, is M_n divided by the molecular weight of the monomers. It varies as a function of p and thus evolves during the course of the reaction. In fact, it can be shown that $\overline{DP}$ equals the leading term in Eq. 13.9, as restated in Eq. 13.10 (called the **Carothers equation**). This means that in a step-growth polymerization, the average molecular weight of the system gradually increases during the process. This tends to produce a relatively broad molecular weight distribution in the final product. In addition, monomer is generally depleted very early in the reaction, because a large number of reactive groups are always present. We can expect to see high molecular weight polymer only toward the end of the reaction, as p approaches its maximum value.

$$\overline{DP} = \frac{1}{(1-p)} \qquad\qquad\text{(Eq. 13.10)}$$

Now consider the situation at the end of the polymerization reaction. The value of p is hypothetically 1, but in a real system this is never so. At the end of the reaction, p reflects the chemical yield of the condensation reaction. High molecular weights are achieved only if p is very nearly 1. For example, inserting a p of 0.90 (corresponding to a 90% chemical yield) into Eq. 13.10 gives a $\overline{DP}$ of 10; that is, the typical polymer will have only 10 monomer units! Even with a p of 0.99 we get a $\overline{DP}$ of 100, not an especially large number for synthetic polymers. As such, in a step-growth polymerization, which includes all condensation polymerizations, only extraordinarily efficient processes can produce high molecular weight polymer. Massive industrial research efforts have focused on improving a reaction conversion from 99.9% to 99.99%. This is a sobering thought for chemists. Many reactions that we consider to be quite good, and for which we could imagine bifunctional substrates that would produce polymers, will never produce high molecular weight material.

The **weight average degree of polymerization**, $\overline{DP}_w$, is given by Eq. 13.11. Given this and Eq. 13.10, the theoretical PDI for a step-growth polymerization is given by Eq. 13.12. In the limit of $p = 1$, we see that the PDI must equal 2. Less efficient reactions will produce narrower molecular weight distributions, but shorter polymers.

$$\overline{DP}_w = \frac{(1+p)}{(1-p)} \qquad\qquad\text{(Eq. 13.11)}$$

$$\frac{\overline{DP}_w}{\overline{DP}} = \text{PDI} = (1+p) \qquad\qquad\text{(Eq. 13.12)}$$

Free-Radical Chain Polymerization

For a chain process we have initiation, propagation, and termination to consider. We will frame our discussion in terms of a radical polymerization, but the analysis applies to any chain reaction. We begin with initiation.

We will describe the initiator as X_2, recognizing that many initiators give rise to two radicals, each of which can launch a polymerization. We will symbolize the monomer as M. We consider the initiation process to be the combination of the two reactions in Eq. 13.13. The subscript 1 on the M indicates it is the first monomer of the chain. Typically the first step, decomposition of the initiator, is rate-determining. In addition, not every radical formed will successfully initiate a polymerization, so we introduce a factor f equal to the proportion of radicals generated that do so. Putting this all together then leads to the expression in Eq. 13.14 for r_i, the rate of initiation. The factor of two reflects the formation of two radicals from each initiator.

$$X_2 \xrightarrow{k_d} 2X\bullet$$

$$X\bullet \ + \ M \xrightarrow{k_i} M_1\bullet \tag{Eq. 13.13}$$

$$r_i = \frac{d[M_1\bullet]}{dt} = 2fk_d[X_2] \tag{Eq. 13.14}$$

Propagation involves the addition of monomer to the growing polymer chain and leads to the expression for r_p shown in Eq. 13.15, where $M\bullet$ represents a polymer chain that terminates in a radical. This analysis implicitly assumes that the propagation rate is independent of chain length, as discussed above.

Both of the major termination reactions, disproportionation and recombination, involve two radicals. The termination rate, r_t, can thus be written as in Eq. 13.16, where both the exponent and the factor of two reflect this feature of termination.

$$r_p = \frac{-d[M]}{dt} = k_p[M\bullet][M] \tag{Eq. 13.15}$$

$$r_t = \frac{-d[M\bullet]}{dt} = 2k_t[M\bullet]^2 \tag{Eq. 13.16}$$

In order to obtain the rate of polymerization, r_p, apply the steady state approximation to $M\bullet$. That is, assume the rate of formation of $M\bullet$ equals the rate of destruction, or $r_i = r_t$. This produces Eq. 13.17. Inserting this result back into the equation for r_p (Eq. 13.15) gives Eq. 13.18 for the rate of polymerization. This is an important result. It tells us that the rate of polymerization is proportional to the concentration of monomer, but it will also be proportional to the *square root* of the concentration of initiator. The initiator concentration is changing constantly during the polymerization, but it is a first-order decomposition. Thus, we can write the first-order rate behavior as Eq. 13.19 (recall Eq. 7.34) and substitute into Eq. 13.18 to get Eq 13.20.

$$2fk_d[X_2] = 2k_t[M\bullet]^2$$

$$[M\bullet] = \left(\frac{fk_d[X_2]}{k_t} \right)^{\frac{1}{2}} \tag{Eq. 13.17}$$

$$r_p = k_p \left(\frac{fk_d[X_2]}{k_t} \right)^{\frac{1}{2}} [M] \tag{Eq. 13.18}$$

$$[X_2] = [X_2]_o \, e^{-k_d t} \tag{Eq. 13.19}$$

$$r_p = \left\{ k_p \left(\frac{fk_d}{k_t} \right)^{\frac{1}{2}} \right\} \left\{ [X_2]_o^{\frac{1}{2}} [M] \right\} e^{-k_d t/2} \tag{Eq. 13.20}$$

Eq. 13.20 is broken into three parts. From the first term, the rate is proportional to k_p but inversely proportional to the square root of k_t. This can have some interesting consequences on, for example, the temperature dependence of polymerization. Raising the temperature will increase both these individual rate constants, but not with the same effect on r_p. Since the square root of the k_t term is in the denominator, r_p will increase more rapidly than you might expect. This can produce an intriguing, and potentially disastrous effect. Sometimes, as a polymerization progresses, the sample becomes progressively more viscous, as polymer forms. This increased viscosity has little impact on propagation, because the monomer is still mobile. However, termination requires that two macromolecules come together, and so k_t can be substantially lowered by viscosity. This leads to continued propagation, typically an exothermic reaction, but less termination, and can lead, literally, to an explosive situation termed **autoacceleration**.

The second term of Eq. 13.20 shows that the polymerization still depends on [M], but also on the square root of the *initial* concentration of the initiator. It is really a poor strategy to try to speed up the reaction by increasing the initiator concentration, because the square root term will attenuate the effect, and you will generally see a decrease in polymer chain length (see below).

The third term shows that the polymerization slows down in an exponential fashion as the initiator is consumed. To keep the polymerization going at a steady rate, more initiator may have to be added.

What chain lengths do we expect from a radical polymerization? A thorough analysis is complicated by the fact that the two possible termination reactions, disproportionation and recombination, affect the final molecular weight differently. Disproportionation does not affect molecular weight, but recombination doubles it. If we set aside this issue for the moment, we can define a **kinetic chain length**, v, as simply the ratio of the propagation and initiation rates. Substituting in appropriate values for r_p and r_i gives Eq. 13.21. Substituting our value for [M•] from the steady state approximation gives Eq. 13.22. The key result here, which holds up in a more detailed analysis that includes the different kinds of termination, is that the chain length varies as the *inverse* of the square root of the initiator concentration. We thus see in Eq. 13.22 the rationale for the statement above that adding initiator may speed up a reaction, but it will produce shorter chains.

$$v = \frac{r_p}{r_i} = \frac{k_p[M][M•]}{2fk_d[X_2]} \qquad \text{(Eq. 13.21)}$$

$$v = \left\{ \frac{k_p}{2(fk_dk_t)^{1/2}} \right\} \left(\frac{[M]}{[X_2]^{1/2}} \right) \qquad \text{(Eq. 13.22)}$$

In a chain reaction, only a relatively few growing chains are present at any one time, because the lifetime of the radicals, or whatever reactive species, is typically low relative to the time scale of polymerization. At any point in the reaction, there are only a few reactive centers with which the monomer can react. As such, the monomer concentration decreases steadily during the reaction, and at any time there will be almost exclusively polymer and monomer in the reaction vessel. As a consequence, polymer molecular weight is not steadily increasing during the course of a chain polymerization. The major factor that controls molecular weight is the competition between propagation and termination reactions.

Living Polymerizations

One other general aspect of polymer synthesis is a novel type of system referred to as a **living polymer**. A living polymer is one for which, after initiation, monomer (typically an olefin or diene) successively adds to the reactive terminus of the growing chain until all the monomer is exhausted (Figure 13.14). At this point the polymerization stops, but a reactive moiety (typically an anion or a metal complex) is still attached to the end of each chain. Termination has not occurred. The polymer is still "alive", in that if we now add more monomer, the polymerization starts up again.

Living polymerizations have several useful properties. For example, we could start the reaction with monomer A and run it to completion, and then we could add monomer B and append a totally new polymer to the original chains, giving a **diblock copolymer** (Figure 13.14), with each chain containing two distinct regions. This is in contrast to a ran-

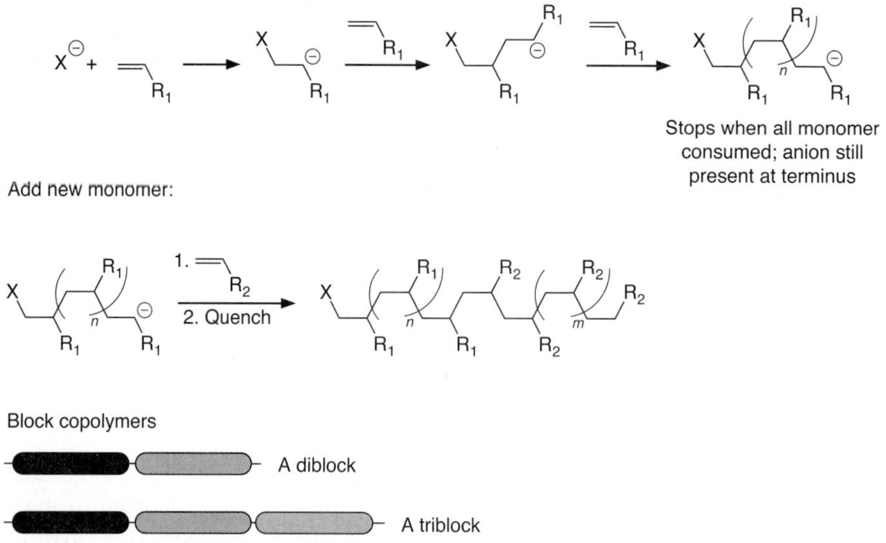

Stops when all monomer
consumed; anion still
present at terminus

Add new monomer:

Block copolymers

A diblock

A triblock

Figure 13.14
Living polymerization and block copolymers. The example is an anionic polymerization, but other types of
polymerization can also be living. Also shown are schematics of diblock and triblock polymers.

dom copolymer, which typically results from initiating a polymerization in the presence
of two different monomers (a **copolymerization**), or an **alternating copolymer**, a perfect
... ABABAB... system that can be prepared in special cases. Concerning block copolymers,
one can imagine that if the two blocks had very different properties, such as one being crystalline and the other very flexible, novel materials would result, and indeed many important
commercial polymers are block copolymers.

The equations for the kinetics of living polymerization reactions take different forms,
depending upon whether the reactive end group is formed during each propagation step directly, or if the end group is formed in a rate-determining step after each propagation step.
When the reactive end group is formed in each propagation step directly, then we have a
first-order dependence upon both the monomer and end group. For an example of a case
where the end group is formed in a step subsequent to each propagation step, see the Going
Deeper highlight in Section 7.5.1 which shows zero-order kinetics in monomer. Furthermore, the kinetic expression depends upon both the initiation and propagation rates, which
will take limiting forms depending upon which of these steps is slower. We leave the mathematics to a more specialized text.

An interesting feature of living polymerizations is that under optimal conditions they
can produce very narrow molecular weight distributions. If the initiation step is faster than
the propagation step, from the very beginning every catalyst site will have a polymer chain
growing off of it. Since the propagation rate is the same for all of these, they will all grow together until monomer runs out, and all should be approximately the same length. PDI values as low as 1.06 are often seen. The analogy has been made that such a situation is like the
start of a horse race, where all the horses start in the gate, then all the gates open simultaneously, and all the horses start together. Such as situation is much more likely to have all the
horses finish together, having run the same distance, than a situation where different horses
start at different times. The latter situation is akin to a polymerization in which propagation
steps are comparable to or faster than initiation. Also, in a living polymerization the molecular weight is relatively easy to control. In optimal systems the final molecular weight of the
product will be precisely that predicted by the initial ratio of monomer to catalyst/initiator.
That is, if we start with 10,000 monomer molecules for every single molecule of catalyst, then
$\overline{DP}$ will be 10,000, and the PDI will be very low. This can be extremely useful for designing
highly tailored materials. We will see several examples of living polymerizations below.

Going Deeper

Free-Radical Living Polymerizations

We will discuss later in this chapter that it is quite common to have anionic polymerizations and ring-opening metathesis polymerizations that are living. These polymerizations require that a reactive end-group persist on the polymer chains, living and ready to react when more monomer is added. It seems counterintuitive that radical polymerizations could be living, because radicals are just too reactive to persist. Recently, however, free-radical living polymerizations have become a reality. The key is having a free-radical trap (see Chapter 8), which can cap the polymer end but also allows regeneration of the radical. This technique is referred to as **atom transfer radical polymerization**.

One example is the polymerization of styrene using 1-chloro-1-phenylethane as an initiator in the presence of Cu(I)(bpy) (bpy = bipyridine). Chlorine atom abstraction by the Cu(I) complex gives a radical that initiates polymerization. The chlorine atom can be transferred back to the chain at any time in an equilibrium process. This becomes particularly important when monomer runs low because the propagating radical is "capped" in this form. Addition of more monomer leads to further polymerization due to reaction again with radical generated from chlorine atom abstraction.

A second example uses a more standard radical trap called TEMPO (2,2,6,6-tetramethylpiperidinyl-1-oxy). This nitroxide radical trap is not a reactive enough radical to initiate polymerization, but it does form adducts with the propagating radical end group, protecting the chain end from termination reactions. Homolysis of the bond regenerates reactive radical end groups, and polymerization commences again.

TEMPO

Capping the radical with TEMPO

Wang, J.-S., and Matyjaszewski, K. "Controlled 'Living' Radical Polymerization. Atom Transfer Radical Polymerization in the Presence of Transition-Metal Complex." *J. Am. Chem. Soc.*, **117**, 5614 (1995). Georges, M. K., Veregin, R. P. N., Kazmaier, P. K., and Hamer, S. K. "Narrow Molecular Weight Resins by a Free-Radical Polymerization Process." *Macromolecules*, **26**, 2987 (1993).

Atom transfer polymerization

Thermodynamics of Polymerizations

Now that we have looked at the kinetics of polymerizations, let's briefly examine the thermodynamics. Table 13.3 provides some quantitative data. Recall that the pathway (mechanism) does not affect the thermodynamics. Instead, the relative energies of the monomers and polymers set the thermodynamics. As we noted in discussing autoacceleration, the propagation step of a radical chain reaction is typically quite exothermic. This is also seen in the thermodynamics of many addition polymerizations (see Table 13.3), because in each step a π bond is converted to a σ bond (which is stronger). Note, however, that the entropies of polymerizations are quite negative. This is due to the fact that the free translation of individual monomers is lost for each additional propagation step.

An important issue for polymers that derives from their thermodynamics is the sealing temperature. Because all chemical reactions are in equilibrium, polymerization reactions

Table 13.3
Thermodynamic Parameters for Various Addition
Polymerizations to Form Amorphous Materials*

Monomer	$\Delta H°$ (kcal/mol)	$\Delta S°$ (eu)
Ethylene	−24.7	−37.0
Propylene	−22.5	−43.3
Styrene	−17.4	−24.8
Methyl methacrylate	−13.4	−27.8

*Allen, P. E. M., and Patrick, C. R. (1974). *Kinetics and Mechanisms of Polymerization Reactions*, John Wiley & Sons, New York.

can run in reverse—a process called **depolymerization**. As with all chemical reactions, the position of equilibrium for polymerization is controlled by the Gibbs free energy change ($\Delta G°$). The Gibbs–Helmholtz equation ($\Delta G° = \Delta H° - T\Delta S°$) tells us that there will be a particular temperature where the difference between the GFE of the monomers and polymers is zero ($\Delta G° = 0$). This is the **sealing temperature**. The fact that the entropies of polymerizations are routinely negative means that temperatures lower than the sealing temperature favor polymer, while higher temperatures favor monomer. In some special polymerizations, the enthalpies of the monomer and polymer are similar. An example is the ring opening polymerization of cyclohexene. In such cases the negative entropy of the reaction can actually lead to depolymerization to monomer, if the proper conditions of temperature, catalyst, and concentrations are achieved.

13.2.3   Condensation Polymerization

Condensation polymerizations produce some of the most versatile and structurally interesting of the synthetic polymers. Condensation polymerizations are commonly step-growth polymerizations. As illustrated in Figure 13.15, the chemistry is often fairly ordinary, involving reactions such as amide or ester formation. However, given the constraints of Eq. 13.10, we can imagine that a great deal of effort has gone into optimizing these simple reactions to produce high molecular weight polymer. We should not underestimate the complexities of synthesizing high molecular weight polymers from even a simple reaction, especially when one needs to make fibers or sheets or molded shapes. Each commercial polymer has a huge base of research behind it involving both synthesis and processing. Here we can only illustrate the kinds of materials that are made by condensation polymerizations. Several excellent polymer texts that address these additional, important features are cited at the end of the chapter.

Just as simple esters and amides can be made by condensation reactions, polyesters such as Dacron and Mylar are made by reacting diols with dicarboxylic acids (Figure 13.15 **A**). Similarly, polyamides such as nylon form from amines and carboxylic acids. Starting from anhydrides and amines, polyimides can be formed. These structures tend to be extremely stable, and so are valuable in high temperature applications.

A related family of materials is shown in Figure 13.15 **B**, along with a highly efficient synthetic route. Diisocyanates are common, especially attractive monomers. Reaction with diols produces polyurethanes [the –NH–C(O)–O– functional group is also known as a carbamate], while reaction with diamines produces polyureas. Technically, no small molecule such as water or ammonia is given off, but such reactions are generally viewed as condensation polymerizations. More accurately, they are step-growth polymerizations.

Polyurethanes find many uses, but the most common is as the soft foams found, for example, in car seats, cushions, and bedding. Along with the structural properties of the polymer, polyurethane foams take advantage of another important reaction of isocyanates.

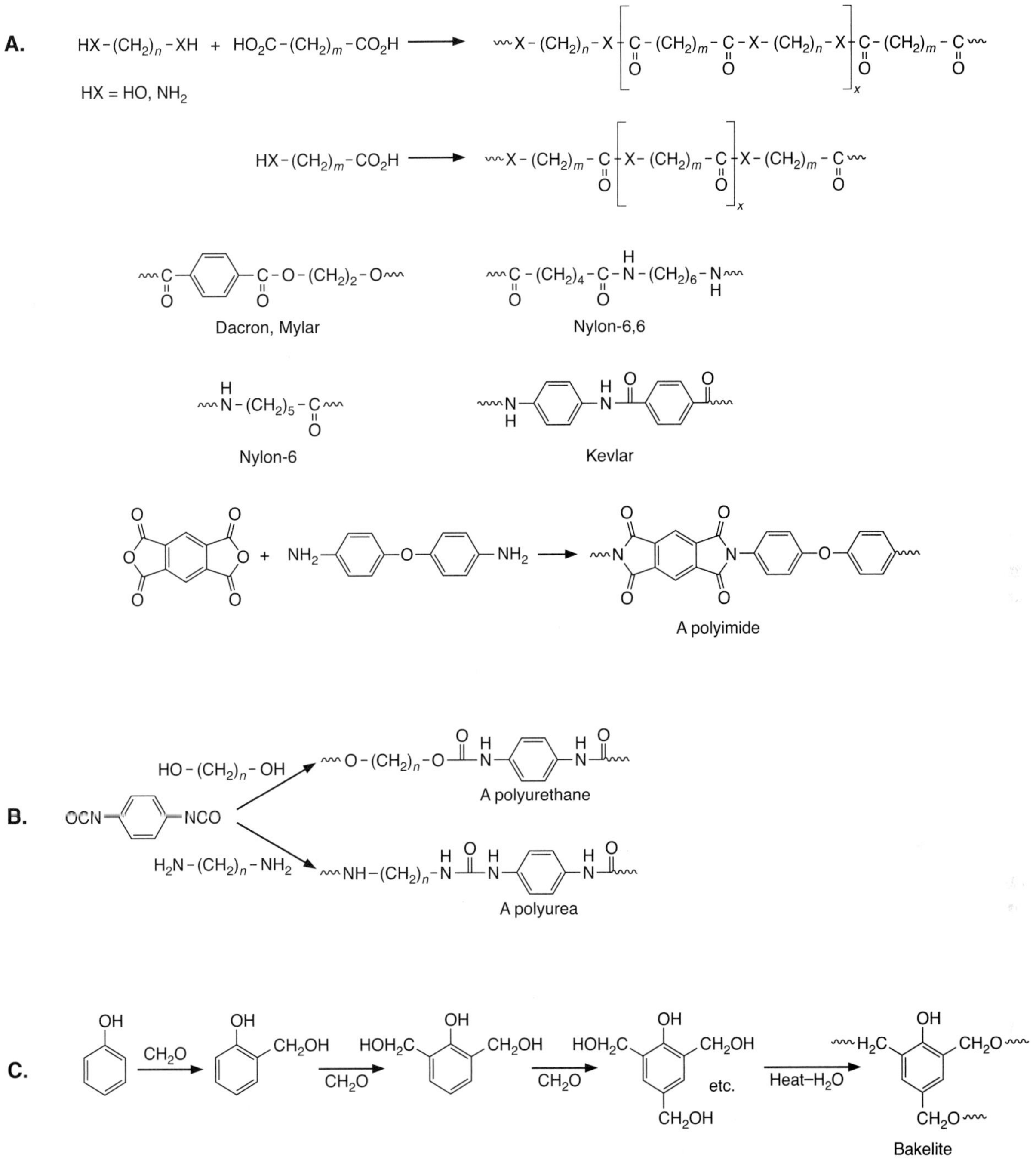

A.

$HX-(CH_2)_n-XH$ + $HO_2C-(CH_2)_m-CO_2H$ $\longrightarrow$

$HX = HO, NH_2$

$HX-(CH_2)_m-CO_2H$ $\longrightarrow$

Dacron, Mylar

Nylon-6,6

Nylon-6

Kevlar

$+$ NH_2—◯—O—◯—NH_2 $\longrightarrow$

A polyimide

B.

$HO-(CH_2)_n-OH$

OCN—◯—NCO

$H_2N-(CH_2)_n-NH_2$

A polyurethane

A polyurea

C.

OH

CH_2O

CH_2O

CH_2O

etc.

Heat–H_2O

Bakelite

Figure 13.15
Condensation polymers. **A.** Polyester/polyamide/polyimide formation
by combining A_2 and B_2 monomers or from a single AB monomer.
Several common polymers of this class are shown. **B.** The diisocyanate
route to polyurethanes and polyureas. **C.** Bakelite synthesis. Note that
the final product is heavily cross-linked because of the presence of bis
and tris adducts in the initial reaction with formaldehyde.

When an isocyanate reacts with water, CO_2 is given off. When just the right amount of water is added to the polymerization, the CO_2 gas that is generated *in situ* acts as a "blowing agent" and creates the cavities in the final material that make it a foam.

$$R-N=C=O + H_2O \longrightarrow R-NH_2 + CO_2\uparrow$$

Connections

Lycra/Spandex

The development of nylon, Dacron, and a host of other synthetic fibers revolutionized many industries, including clothing, carpeting, and the like. One of the most remarkable members of this family is Lycra, the trademark ascribed by DuPont to a family of fibers known generically as spandex. In the 1960s, Lycra was introduced as a great improvement over natural rubber in "ladies foundation garments".

The modern market for spandex includes "active wear" and a wide range of light-weight, highly flexible materials. DuPont's market alone is $1.5 billion per year. Spandex is always blended with other fibers such as cotton or wool. It is lighter in weight than threads made from rubber. An important advantage that you might not immediately consider is that, unlike rubber, spandex does not break down on exposure to body oils, perspiration, lotions, detergents, or ozone.

The structure of a typical spandex is shown. It consists of a quite rigid urea component, and a highly flexible segment. The rigid component makes spandex viable as a fiber in fabrics, but the flexible segment provides the elastic properties we most associate with such materials. Remarkably, spandex can expand as much as 600% on stretching and then return to its original shape.

"What's That Stuff: Spandex." *Chem. Eng. News*, February 15, **77**, 70 (1999).

Spandex

Other well known condensation polymers include phenol–formaldehyde resins, the prototype of which is Bakelite (Figure 13.15 **C**). Such structures were known as early as the 1870s, and in the early 20th century these tough, durable thermosets were among the first synthetic polymers of commerical importance. More modern versions of this type of polymer are known as Novolac. This chemistry is also that which makes calixarenes (Chapter 4), which are cyclic tetramers rather than linear polymers.

Proteins are condensation polymers, formed by coupling amines and carboxylic acids. In proteins, the monomers (amino acids) contain both the amine and the carboxylic acid, and so they are an example of the AB type monomer shown in Figure 13.15. In fact a protein, or more specifically polyglycine, could be described as Nylon-2. Recall that Nylon-1 is the polyisocyanate structure discussed in Section 6.8.1.

The persistence of the reactive functionalities of the monomers throughout the entire process at the ends of the growing polymer chains can lead to some novel features in a condensation polymerization. For example, consider a polyester made by reacting a dicarboxylic acid with a diol. After reaction is complete, if the final product is heated, the alcohol at the terminus of one chain could react with an ester in the middle of another chain, leading to a transesterification. This would not change the total number of polymer molecules in the sample, but the average molecular weight could change. Similarly, if we mix two samples of the same polymer that have very different molecular weight distributions, such cross-reactivity will scramble the system. One way to prevent such problems is to quench the product. For example, a polyester could be quenched by adding an excess of a *monofunctional* acid and/or alcohol.

Another potential complication of condensation polymerizations is ring formation. If an AB monomer is used, the growing chain will be A–[polymer]–B. This molecule can cyclize on itself, effectively terminating the polymerization process. Since polymerization is a bimolecular process and cyclization is unimolecular, polymerization is favored by high monomer concentrations. As such, condensation polymerizations typically involve very concentrated, perhaps even neat, monomer solutions. All the usual factors that favor or disfavor ring formation also apply here, such that once the polymer chain gets longer than 15–20 atoms, cyclization usually cannot compete with polymerization.

13.2.4 Radical Polymerization

Radical polymerization is also extremely common. Many of the most common polymers are polyolefins made by radical polymerization, with the monomer very often having the formula $CH_2=CHR$. Figure 13.16 gives representative examples. Typically, radical polymerizations give atactic polymers, and thus polymers of low crystallinity.

Mechanistically, radical polymerizations are addition polymerizations, and they follow the exact same radical chain process discussed in Chapters 10 and 11. The key steps of initiation, propagation, and termination are again involved, as recapitulated in Figure 13.13 **B**. Based upon studies of substituents on the carbon beta to the site of radical addition, it has been concluded that the adding radical has almost completely formed a bond at the transition state. Secondary isotope effects support this notion of a late transition state, with values between 1.05 to 1.17 for deuterium at the site beta to addition.

As noted above, the key factor influencing polymer length in these chain reactions is the competition between propagation and termination. One can expect that it will be crucial to exclude any impurities that can react with the growing radical chain and terminate the polymerization, if you want high MW polymer. On the other hand, if smaller molecular weights are desired, judicious addition of inhibitors and chain transfer agents could be useful.

A wide range of initiators are employed in radical polymerization chemistry. These include organic peroxides such as benzoyl peroxide, azo compounds such as azoisobutyronitrile [AIBN, $Me_2C(CN)N=NC(CN)Me_2$], redox agents such as persulfates or soluble metal ions, heat, radiation, and electrolysis.

Monomer	Polymer
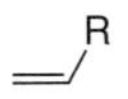 R / = Monosubstituted olefins	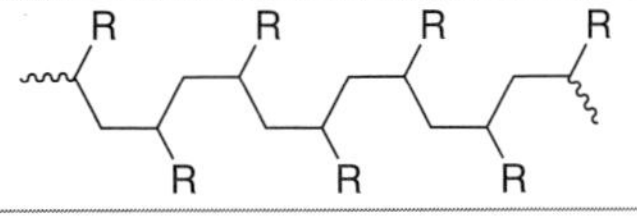
R = H	Polyethylene
CH$_3$	Polypropylene
Cl	Poly(vinyl chloride) (PVC)
Ph	Polystyrene
CN	Polyacrylonitrile (acrylic fiber)—Orlon and Creslan
OAc	Poly(vinyl acetate)
Others	
$CF_2=CF_2$	Poly(tetrafluoroethylene) (PTFE)—Teflon
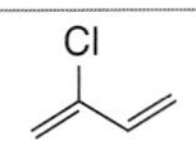 Cl Chloroprene	*trans*-1,4-Polychloroprene (neoprene rubber)
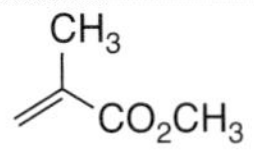 CH$_3$ CO$_2$CH$_3$ Methyl methacrylate	Poly(methyl methacrylate) (PMMA)—Lucite, Plexiglas, and Perspex

Figure 13.16
Examples of polymers made by radical polymerization.

Radical Copolymerization—Not as Random as You Might Think

While living polymerizations can be exploited to produce block copolymers, a copolymerization should give polymer chains that contain both monomers distributed throughout. You might expect that a radical chain polymerization would give a truly random copolymer. Radicals are quite reactive and not known for their selectivity. In fact, though, radical copolymerizations are not totally random, and some quite distinctive polymer compositions can be achieved. Consider a radical polymerization progressing in the presence of two different monomers. As shown in the drawing, there are now four possible propagation rate constants, where k_{12} is the rate constant for a chain terminating in M_1 reacting with monomer M_2, etc. A good estimate of the likely composition of a copolymer can be obtained by consideration of just these rate constants. If there is absolutely no selectivity in the radical reactions, all four rate constants are the same and a truly random copolymer results. Actually, this is rarely the case. Instead, there is some selectivity that can be defined by **reactivity ratios**: $r_1 = (k_{11}/k_{12})$ and $r_2 = (k_{22}/k_{21})$. The reactivity ratio represents the preference of a given radical to react with its own monomer rather than the other monomer in the mixture.

Four possible propagation reactions

Reactivity ratios have a large influence on copolymer compositions. Let's consider some limiting cases. The case of $r_1 = r_2 = 1$ is the one we just discussed above. There is absolutely no selectivity, and a truly random copolymer results. On the other hand, if $r_1 = r_2 = \infty$, each radical is perfectly selective. One chain should be nothing but M_1 and the other chain exclusively M_2. This is fairly rare. An interesting case is $r_1 = r_2 = 0$, in which each radical would rather add to the other monomer. A perfectly alternating copolymer should be produced.

Given the commercial importance of the polymer industry, you should not be surprised to learn that a number of reactivity ratios have been determined. Each pair of monomers has its own pair of r values, and the r value for a given monomer is not transferrable from one monomer pair to another. We list r values for several monomer pairings below. There is considerable variation, and some interesting pairings can be found. It is worth contemplating these pairings. Are they consistent with the reactivities of free radicals discussed in Chapter 10? You will be given an opportunity in Exercise 10 to decide.

M_1	M_2	r_1	r_2
Methyl methacrylate	Acrylonitrile	1.22	0.15
Styrene	Methyl methacrylate	0.52	0.46
Styrene	Vinyl acetate	55	0.01
Vinyl acetate	Methyl methacrylate	0.015	20

Often, free radical polymerizations involve relatively inexpensive monomers, and they are run neat, with the olefin serving as the solvent. However, if the reaction is very exothermic, heating can get out of hand and the reaction can become too violent. In such cases a solvent is added, but care must be taken as many organic solvents will react with radicals. Even if the reaction with solvent is not especially efficient, an occasional termination by abstraction of a hydrogen from solvent will be quite deleterious.

Given this situation, it makes good sense that water should be a very good solvent for radical polymerizations. Recall from Chapter 2 that the OH bond strength of water is very high (the bond dissociation energy is large), such that organic radicals are highly unlikely to abstract a hydrogen from water. This makes water an ideal solvent for radical polymerizations except for one small thing—most organics are insoluble in water! To get around this problem, a common technique is **emulsion polymerization**. The monomer is mixed with water, a soluble initiator such as a redox agent (for example, persulfate), and a detergent such as SDS. The detergent forms micelles, the monomer goes into the micelles, and polymerization occurs in the novel environment of the interior of a micelle. The micelles contain most of the monomers, and once the initiation takes place the monomers are in high concen-

---

Connections

PMMA—One Polymer with a Remarkable Range of Uses

Poly(methyl methacrylate), PMMA, is readily formed by a variety of polymerization procedures and it is one of the foundations of the modern polymer industry. First developed by Rohm and Haas as a clear plastic known as Plexiglas, it is a very tough, highly transparent material. You can easily see through sections of Plexiglas that are over a foot thick, making it ideal as a protective glass substitute (for example, protecting hockey fans from errant pucks). Another interesting use is in the largest single window in the world at California's Monterey Bay Aquarium. It is a single piece of PMMA, 54 feet long, 18 feet high,

and 13 inches thick. Imperial Chemical Industries markets PMMA as Lucite. In this form, it is used in hot tubs and single piece showers. In addition, PMMA can be made by emulsion polymerization to form a latex that is a staple of the water-based household paint industry. PMMA is an excellent example of how one polymer can in fact represent a wide range of materials.

Methyl methacrylate   PMMA

---

tration for propagation. This allows monomers that are typically rather sluggish in the propagation rates, such as butadiene, to polymerize efficiently. The sizes of the micelles are such that they contain, on average, only one growing chain, so the termination reactions are impeded. As monomer is consumed, more diffuses into the micelles from bulk solution. The resulting aqueous suspension of an organic polymer in a micelle is generically called a **latex**. Both natural polymers, such as natural rubber, and synthetic polymers can form latexes. Water-based paints are generally formed by this technology. There are many ways to control emulsion polymerization. Choice of detergent, concentrations, initiator, etc., should all influence the final nature of the product.

Along with termination, another important side reaction in radical polymerizations is **chain transfer** (Figure 13.13 **B**). This happens when the terminal radical of a growing chain reacts not with another monomer, but rather with another polymer chain by hydrogen abstraction. This terminates the original chain, but puts a reactive center on the new chain. The number of growing chains is not different, but chain transfer can affect the molecular weight distribution.

An important feature of chain transfer is that the hydrogen abstraction can occur anywhere along a polymer chain. As such, chain transfer generally leads to **branching**. So, contrary to initial expectation, radical polymerization of a simple monomer like styrene can lead to a branched polymer. Branching substantially affects the material properties of a polymer, discouraging crystallinity and thus lowering T_m.

13.2.5 Anionic Polymerization

Conceptually, anionic polymerizations are related to free radical polymerizations in that they are chain processes with a reactive end group on the growing chain, in this case an anion rather than a radical. However, the reaction conditions and ultimate polymer properties are much different for the two types of polymerization.

The basic scheme for anionic polymerization is a chain mechanism, such as that shown in Figure 13.13 **B**, with the reactive functionality being an anion of some sort. In fact, a variety of initiators have been employed. These include dissolved metals such as Na in liquid ammonia, alkyllithium and Grignard reagents, and organic radical anions such as sodium naphthalenide. The key step is simply to create the carbanion of the monomer, and then the propagation process occurs. In anionic (and cationic) polymerizations the initiator is also sometimes referred to as a catalyst, although it is never recovered unchanged from the reaction.

The prototype monomer is a monosubstituted olefin with a substituent that can stabilize a negative charge. Typical monomers include butadiene, styrene, acrylonitrile, and methyl methacrylate. Mixtures of styrene and butadiene are used in running shoes, and described in the next Connections highlight. Certain ring-opening polymerizations are also initiated by anions, and we will discuss those in Section 13.2.8.

The list of monomers compatible with anionic polymerization overlaps the radical list (Figure 13.16) considerably, but the unique features of anionic polymerization mean that the same polymer can be a different material. For example, one important feature of anionic polymerizations is that they show a tendency to produce stereoregular polymers, in contrast to radical polymerizations. As such, polystyrene produced by anionic polymerization is more crystalline than polystyrene produced by radical polymerization. This is just another example of how one polymer—polystyrene—can represent several different materials. Note also that because carbanions generally do not abstract protons from C–H bonds, chain transfer (and branching) is typically not a problem in anionic polymerizations.

A common material made by anionic polymerization is the random copolymer composed of a roughly 1:3 mixture of polystyrene and polybutadiene known as **SBR** (for styrene butadiene rubber). Both of these monomers are compatible with anionic polymerization, and indeed SBR is made this way as a popular substitute for natural rubber, often used in the automotive industry. Styrene and butadiene are also both compatible with radical polymerization, and SBR can be made under emulsion polymerization conditions using radical initiators.

A monomer such as butadiene opens up another stereochemical option—that is, whether the remaining double bond is cis or trans. Not surprisingly, the differing stereochemistries have different properties, with the trans tending to have higher transition temperatures. Conditions have been found that allow some, but not complete, control over the cis/trans content of the polymer.

Another feature of anionic polymerizations is that they are often living. Under appropriate conditions carbanions can be much more stable (actually, persistent is a better term) than radicals, and so it is not unreasonable that once all the monomer is exhausted, the polymer chains can retain a reactive, anionic group at their termini. As noted above, having a living polymerization opens up the possibility of preparing block copolymers.

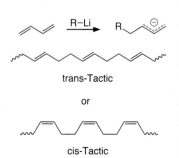

trans-Tactic

or

cis-Tactic

13.2.6   Cationic Polymerization

Cationic polymerization is the direct analogue of anionic polymerization but with a cation at the terminus of the growing polymer chain. Initiators are either strong acids such as sulfuric, perchloric, or hydrochloric acids, or Lewis acids such as BF_3 or $TiCl_4$. Usually the Lewis acids require water or methanol as a "cocatalyst", suggesting that the reactive initiator is actually a protic acid. The list of viable monomers is somewhat limited, including isobutene, methyl vinyl ether, and butadiene—all structures that can give stabilized cations. Cationic polymerizations are the least common of the types discussed here.

13.2.7   Ziegler–Natta and Related Polymerizations

A major advance in polymer science occurred in the 1950s when Ziegler and Natta both established that aluminum alkyls could polymerize ethylene under high pressure. They discovered that addition of transition metal compounds such as $TiCl_4$ or VCl_5 accelerated the reaction, so that ethylene could be polymerized at atmospheric pressure and room temperature (Eq. 13.23). Propylene could also be polymerized by these systems. While these very simple, and readily available, olefins can be polymerized under radical conditions with difficulty, such reactions were far from optimal. The discovery of Ziegler–Natta polymerization launched a huge effort in polyethylene and polypropylene science that continues to this day, with tens of millions of pounds of each being produced every year.

$$CH_2\!=\!CH_2 \xrightarrow[\text{catalyst}]{\text{Ziegler–Natta}} \text{polyethylene}_n \qquad \text{(Eq. 13.23)}$$

Connections

Living Polymers for Better Running Shoes

One of the most common implementations of living anionic polymerization is a triblock polymer of styrene–butadiene–styrene called **SBS rubber** (there is also SIS rubber from styrene–isoprene–styrene). The polystyrene has a T_g of 100 °C while the polybutadiene has a T_g of –63 °C. Under normal conditions, the polystyrene gives the material a rigid structure, resisting the rubbery properties of the polybutadiene. However, under high impact the polybutadiene can "give", preventing the material from shattering. The key is that because this is a block copolymer, the polystyrene sections can aggregate to form tough domains that act like pure polystyrene, but the rubbery polybutadiene is still around. There is a good chance that SBS rubber is a component of the soles of your running shoes.

The synthesis of SBS rubber is also interesting. First, a block of polystyrene (**A**) is created by initiating styrene polymerization with, for example, *n*-butyllithium. Next, butadiene is added and a polybutadiene block forms. You might think that the next step would be to add styrene to create the third block, but unfortunately, the polybutadiene terminus is not reactive enough to initiate styrene polymerization. A clever way around this is to "cap" all the chains by reacting with dichlorodimethylsilane. Next, we add another living chain of pure polystyrene (**A** again, although we could instead choose a different block polymer). The polystyrene terminus reacts with the chlorosilane, and we have our triblock.

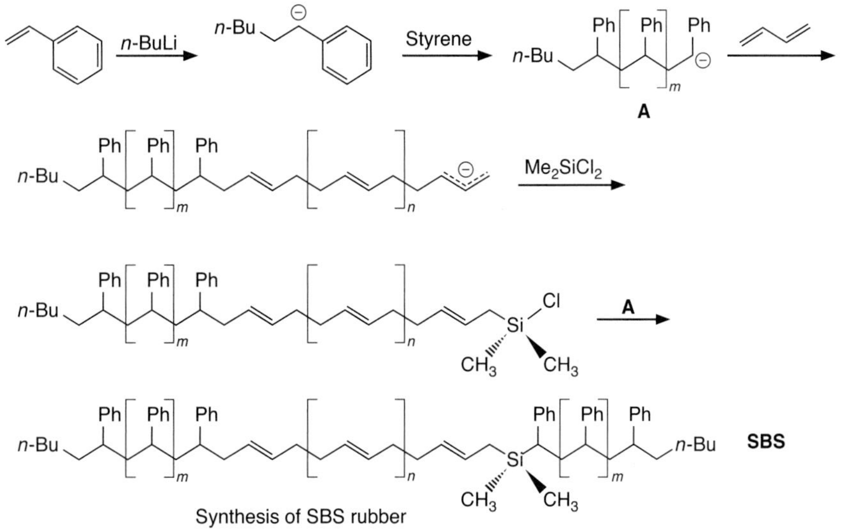

Synthesis of SBS rubber

A great range of Ziegler–Natta catalysts has been developed, and all consist of at least two components. Usually, a trialkylaluminum compound is involved, but alkyl aluminum halides, alkyl sodiums, and dialkyl zincs have also been used. The second component is typified by $TiCl_4$ or $TiCl_3$, although other transition metal compounds have been used. Other additives (a third component) have included NaF, amines, and HMPA. These catalysts are not especially well-defined or well characterized chemical entities. They are empirically obtained mixtures that produce desirable results. We return to this issue below.

Along with the mild reaction conditions, there are two other key features of Ziegler–Natta polymerization. First, chain transfer to the polymer, a common side reaction in radical polymerization (Figure 13.13), is essentially absent in such systems. As such, highly linear polymers can be made using Ziegler–Natta systems. The polyethylene made by a Ziegler–Natta reaction has high density and is more crystalline than polyethylene made by a radical polymerization.

Second, Ziegler–Natta systems can give highly stereoregular polymers. Depending on the monomer and the catalyst, stereoregularity from 20% to >99% can be achieved. The isotactic polypropylene made by Ziegler–Natta catalysis again has much different material properties than the atactic polypropylene prepared by radical polymerization, and it has become an important article of commerce.

Mechanistic studies of Ziegler–Natta polymerizations are challenging because of the ill-defined nature of the catalyst. Much debate has focused on the roles of the various metals and on the exact nature of the active species. In fact, it seems likely that more than one active catalyst is present in the reaction mixture, and that these different catalysts have differing reactivities and differing stereoselectivites. This limits the usefulness of the traditional, heterogeneous, Ziegler–Natta catalysts. Nevertheless, the spectacular success of Ziegler–Natta systems got organometallic chemists thinking about plausible intermediates along the polymerization pathway, especially those involving metal olefin complexes and insertion reactions involving metal alkyls (recall Section 12.2.5). This fundamental organometallic research led to the next big advance in polymerization chemistry.

Single-Site Catalysts

Beginning in the mid-1980s, another revolution in olefin polymerization occurred with the development of metallocene-based catalysts for Ziegler–Natta type polymerizations. The key catalysts typically involve a group IV metallocene (Ti or Zr) mixed with methylaluminoxane (MAO), which is prepared by partial hydrolysis of trimethylaluminum. A bit of the old Ziegler–Natta magic remains. The essential mechanism of olefin polymerization by metallocene catalysts is given in Figure 13.17. With metallocenes, we now have a pure, well-defined, single catalyst in the reaction mixture. This is in stark contrast to the ill-defined, complex mixture in more conventional Ziegler–Natta preparations. With every polymer chain growing off the same kind of catalyst site, even greater control is possible. The polyethylene and polypropylene so produced are now more linear and more stereoregular than ever before.

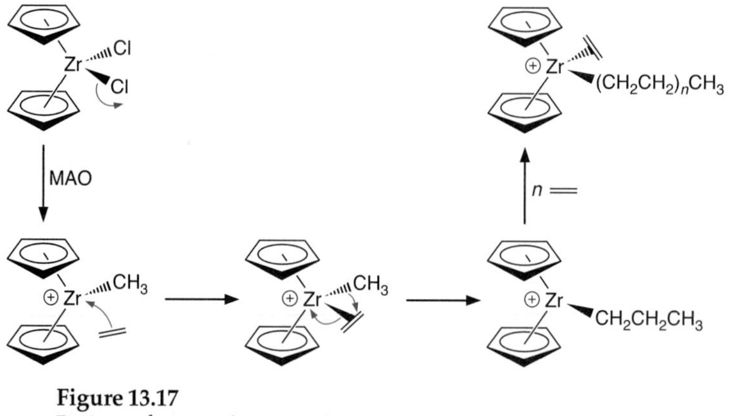

Figure 13.17
Basic mechanistic features of metallocene-based, Ziegler–Natta polymerization.

Recall from Chapter 6 that the other key feature of metallocene catalysts for Ziegler–Natta polymerization is enhanced control of stereochemistry (see the Going Deeper highlight in Section 6.7). Beautifully designed, chiral, C_2-symmetric metallocenes have been developed for the synthesis of isotactic polypropylene, while C_s-symmetric metallocenes lead to syndiotactic polypropylene. Throughout the world, billions of pounds of polyethylene and polypropylene are now made using these catalysts, the result of modern organometallic chemistry and mechanistic insight.

Going Deeper

Using ^{13}C NMR Spectroscopy to Evaluate Polymer Stereochemistry

Once it was appreciated that metallocenes could be potent catalysts for olefin polymerizations, a huge number of systems were designed and synthesized, with the goal of maximizing, among other things, the stereoregularity of the polymer. This required a simple, quantitative way to evaluate the tacticity of polymers, and modern ^{13}C NMR spectroscopy provided a perfect solution. For polypropylene we can describe any pairwise relationship between stereocenters as "meso", m (local mirror symmetry), or "racemic", r (local C_2 symmetry). ^{13}C NMR spectroscopy can easily distinguish an m diad from an r diad, where the

resonance of the central C of the five carbon unit (pentad) is indicative of the stereochemistry. In fact, modern high field machines can distinguish an mmmm pentad (five stereocenters, four pairwise relationships) from an rrrr pentad or any other pentad. The ^{13}C NMR spectrum of atactic polypropylene shows ten peaks in the methyl region, corresponding to the ten possible pentads. You are asked in Exercise 4 to list the ten. All the peaks have been assigned to particular pentads, and so now a simple ^{13}C NMR spectrum provides a thorough analysis of polypropylene tacticity. Comparable analyses are available for other polymers.

Polypropylene pentads

With the advent of metallocene catalysts, polyethylene and polypropylene production has reached new heights (Table 13.1). Differing synthesis conditions produce noticeably different materials, leading to new classifications. The high temperature, radical polymerization to give polyethylene produces a material with extensive branching due to chain transfer. This material cannot pack well as a solid, and so it is not crystalline. It is termed **low-density polyethylene (LDPE)**. Because of the low crystallinity, the material is highly flexible, and it is used in packaging film and in plastic grocery bags.

Since Ziegler–Natta polymerization does not suffer chain transfer, the polyethylene made by this route is highly linear (no branching). If desired, a small amount of controlled branching can be introduced by copolymerizing ethylene with 1-butene and perhaps small quantities of 1-hexene and/or 1-octene. Now the basic backbone is linear, but small alkyl side chains emanate from it. This material is also not highly crystalline, and it is termed **linear, low-density polyethylene (LLDPE)**. Its properties are similar to those of LDPE, but the convenience (low temperatures and pressures) and control of Ziegler–Natta polymerization have made LLDPE a strong competitor for LDPE.

When branching and side chains are avoided altogether by polymerizing ethylene alone with Ziegler–Natta or metallocene catalysts, a highly crystalline, higher density material emerges that is called **high-density polyethylene (HDPE)**. This is a much stronger, more rigid material that is used in plastic milk bottles, bottle caps, and other tough plastic applications. Roughly half the polyethylene made is HDPE, with the remainder being evenly split between LDPE and LLDPE.

13.2.8   Ring-Opening Polymerization

Classical ring-opening polymerizations represent a diverse group of reactions and polymer types. Figure 13.18 shows several prototypes. Often a ring-opening strategy simply provides an alternative approach to a polymer that can be made by another route. This is certainly the case for the caprolactam-based synthesis of Nylon-6. The formation of polyformaldehyde (also known as Delrin) by ring-opening polymerization of the cyclic formaldehyde trimer also falls into this category.

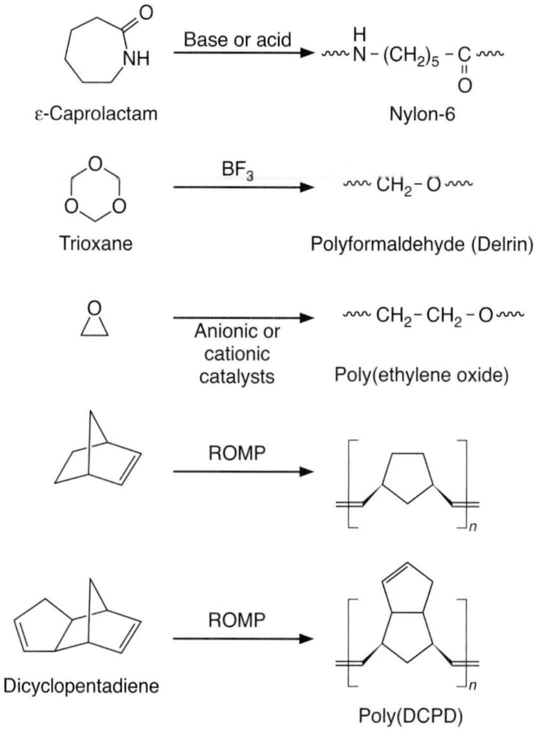

Figure 13.18
Examples of ring-opening polymerization.

Relief of ring strain is often a driving force for ring-opening polymerizations. The anionic polymerization of ethylene oxide [making poly(ethylene oxide)] is certainly an example of this. Because of the substantial relief of strain, a wide range of initiators exists, including hydroxide, alkoxides, organometallic compounds, Lewis acids, and protic acids. Propylene oxide can also be induced to polymerize, as can analogous thiiranes.

A novel bit of chemistry is involved in the formation of **epoxy resins** (Figure 13.19). It involves ring-opening and other reactions. It also is an example of cross-linking being designed into a material to increase toughness. Epoxy resins see commercial use in the surface coatings of electronic circuit boards. The common two-part epoxy adhesive that you can get at a hardware store is also a member of this family.

Most epoxy resins are made by combining two well known monomers, epichlorohydrin and bisphenol A (so named because it is formed by the reaction of two equivalents of phenol with acetone). The initial linear polymer (middle structure in Figure 13.19), arises from ring opening and then reclosure, to ultimately produce bisphenols linked by CH_2–$CH(OH)$–CH_2– groups. An excess of epichlorhydrin is used so that every chain terminates with an epoxide on both ends. Because of the reactive termini, cross-linking of the chains is possible. While a variety of cross-linkers (also known as **curing agents**) are used, amines such as diethylenetriamine are common. Nucleophilic ring opening provides extensive opportunities for cross-linking, ultimately producing a very tough material. The two components in an epoxy adhesive system are the bis-epoxide capped polymer in one tube and the polyamine curing agent in the other. When they are mixed, cross-linking occurs, and an adhesive is formed. For this application, n in the capped polymer is usually small, perhaps 1–5. In other applications, such as surface coatings, n is larger.

A new addition to the array of polymer synthesis methodologies is **ring-opening metathesis polymerization (ROMP)**. Although the basic reaction had been known for some time, it was not until the development of new transition metal catalysts based on work by Schrock and Grubbs (Chapter 12), that the methodology became practical (Figure 13.18). Typically the reaction requires strained olefins such as norbornene so that relief of strain

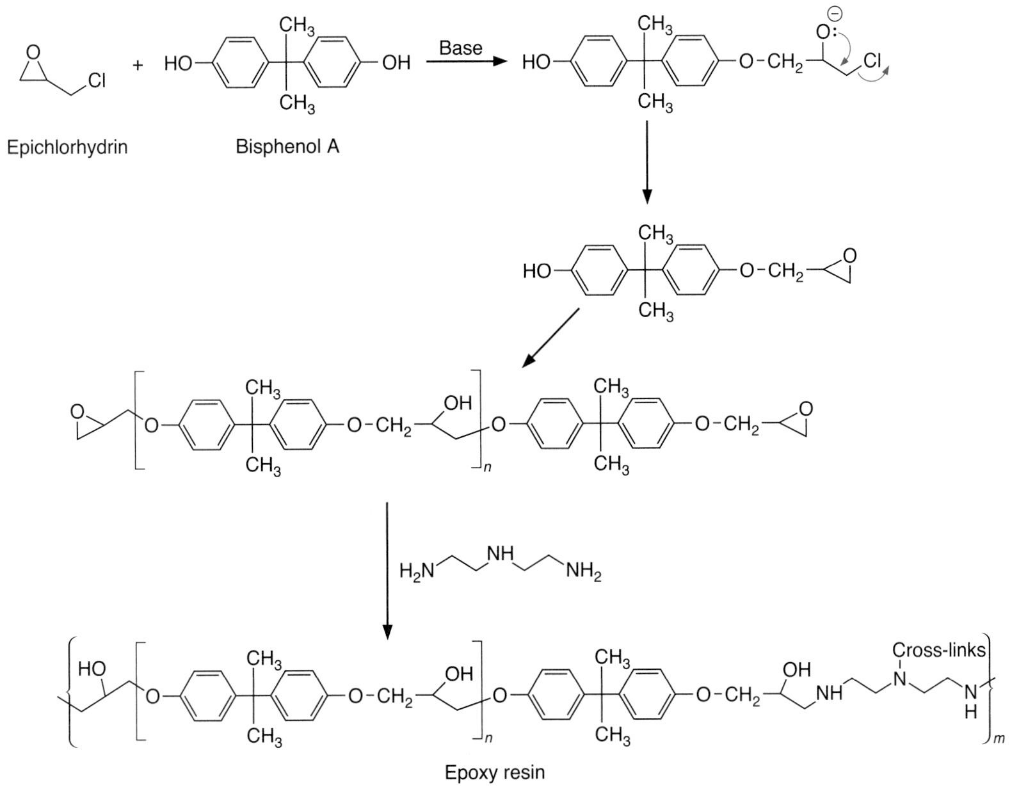

Figure 13.19
Formation of an epoxy resin. The "cross-links" shown in the last
structure indicate bonding arising from nucleophilic opening of
an epoxide on another bis-epoxide.

on ring opening can push the reversible metathesis reaction toward polymerization. The
readily available and inexpensive dimer of cyclopentadiene is an ideal substrate, such
that poly(dicyclopentadiene) (polyDCPD) is now a common material. Recently developed
much more reactive Ru-based catalysts can even polymerize essentially unstrained olefins
such as 1,5-cyclooctadiene, and this could expand the range of usefulness of ROMP. Remarkably, these transition metal-based catalysts are quite tolerant of heteroatom functionality, allowing polar and even proton-donating functionalities to be part of the monomers. In
addition, living polymers are often produced from ROMP, providing new strategies for
diblock and triblock polymer synthesis.

13.2.9 Group Transfer Polymerization (GTP)

Another relatively new addition to polymer synthesis methodology, **group transfer polymerization** (GTP), uses some fairly sophisticated chemistry and a novel polymerization
mechanism. The prototype reaction, as introduced by Webster and co-workers at DuPont in
1983, is shown in Figure 13.20. It involves a vinyl monomer such as methyl methacrylate, a
silyl ketene acetal as an initiator, and a nucleophilic catalyst such as bifluoride, cyanide, or
azide. It is proposed that a hypervalent silicon species is involved, and that the silyl group is
directly transferred from one oxygen to another, hence the name group transfer polymerization. The transfer results in the formation of a new silyl ketene acetal at the end of the growing polymer chain, allowing the process to continue. Consistent with this mechanism is the
observation that only monomers with a potentially coordinating side chain such as methacrylate or acrylonitrile are compatible with GTP.

GTP produces a living polymer, such that each chain is terminated by a silyl ketene acetal that will be stable once all monomer is exhausted. Thus, all the advantages of living poly-

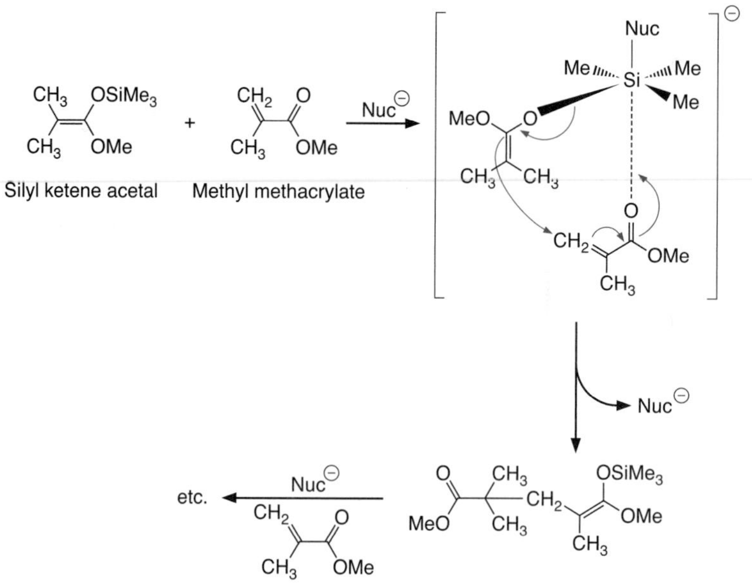

Figure 13.20
Group transfer polymerization.

merization discussed above, including simple control of molecular weight, narrow molecular weight distributions, and facile formation of block copolymers, apply to GTP. Because of its unique mechanism, GTP offers some new options for polymerization, involving side chains that may not be compatible with conventional radical polymerizations. This feature, and the fact that it produces living polymers, has generated considerable excitement about this relatively newer polymerization strategy, and commercial products based on GTP have already appeared.

Summary and Outlook

In this brief overview of a vast field, we have introduced many of the basic concepts of polymer science. Differing structural types and polymerization strategies have been described, and the importance of molecular weight analysis emphasized. In this regard, classic polymers such as polyethylene and polypropylene were examined. Furthermore, polymer phase behavior was found to rely primarily on the same principles given in Chapter 3 that dictate the thermodynamics of solutions. We have also introduced novel types of organic materials, such as dendrimers, fullerenes, and liquid crystals. You will no doubt be seeing many studies on these kinds of structures in the modern chemical literature. The polymer chemistry we have discussed builds upon many of the types of reaction mechanisms introduced earlier, and it is clearly an area where mechanistic insight can rapidly lead to practical ramifications and applications.

Some of the most exciting new types of organic polymers have applications in the electronics industry. In Chapter 17 we will return to polymers again. There, it will be the electronic structure of the polymers rather than the functionality and topology that will be key. However, in order to understand the electronic structure of these polymers, we need a more in-depth understanding of molecular orbital theory. Thus, we now turn to Part III of this book, where molecular orbital theory is the starting point from which several topics are launched, including pericyclic reactions, photochemistry, and electronic materials.

Exercises

1. Draw G1 and G2 poly(propylene imine) dendrimers starting from a tetraaminoadamantane core. How many exposed amines will be present at G5?

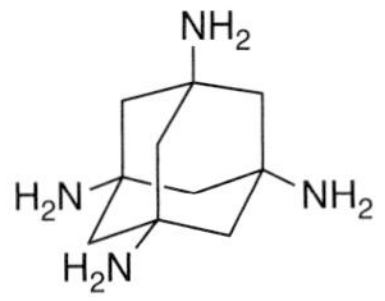

2. Calculate the yield for an intact G5 dendrimer of the PAMAM type if each individual linkage in the sequence occurs with a 99.5% yield.

3. Why does PVC have a higher T_m than polypropylene?

4. Show that there are 10 and only 10 stereochemical pentads for a polypropylene-type polymer and list them all.

5. Polymer theory tends to emphasize simple systems, such as polyethylene. Unfortunately, ethylene polymerization, even using advanced Ziegler–Natta methods, is not controlled enough to allow production of, for example, a series of polyethylenes with M_n of 10,000, 15,000, 20,000, etc. for comparison with theory. Suggest how olefin metathesis chemistry could provide a solution to this problem.

6. Suggest a synthesis of the Spandex molecule shown in Section 13.2.3. (*Hint:* MDI is a commonly available molecule.)

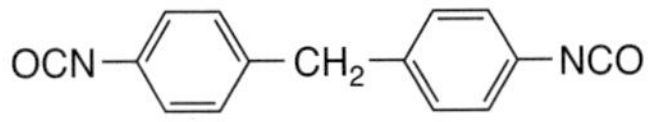

7. Figure 13.20 shows the proposed mechanism for group transfer polymerization, involving a hypervalent silicon intermediate. An alternative proposal invokes full displacement of the silyl group from the initiator, forming, for example, fluorotrimethylsilane if a fluoride-based catalyst is used. Suggest an experiment that would distinguish these two mechanisms.

8. Methyl methacrylate can be polymerized both by free radical addition and GTP methods. Compare and contrast the polymer structure and the ways in which one would manipulate the molecular weight distributions.

9. A recent addition to the condensation polymerization field has been acyclic diene metathesis polymerization (ADMET). When an acyclic diene is reacted with a modern metathesis catalyst, a condensation polymerization occurs with the loss of the small molecule ethylene. In ADMET the loss of ethylene pushes the reaction toward the polymer product. Making high molecular weight polymers through ADMET has only recently become feasible, and it remains to be seen how useful this approach will be. With ADMET and ring-opening metathesis polymerization (ROMP; see Figure 13.18) it is possible to make the "same" polymer by two different routes. For example, ROMP of 1,5-cyclooctadiene might be expected to give the same polymer as ADMET of 1,5-hexadiene. Discuss how the products of these two reactions might be expected to differ, both with regard to structure and molecular weight distribution.

10. Consider the four pairs of reactivity ratios given in the Going Deeper highlight on copolymerization on page 792. Does each number make sense given what we know of radical reactivity? Why are both numbers for vinyl acetate so low? Briefly discuss the implications of each pairing for the composition of the copolymer.

11. Propose an experimental approach to measure the inversion barrier in corannulene. (*Hint:* Consider the symmetry of a monosubstituted corannulene.)

12. At least two factors are considered to contribute to the inversion barrier in corannulene, one favoring and one disfavoring planarity. Discuss these and any other factors that might contribute to the 10.2 kcal/mol barrier.

13. How many 6–6 and how many 6–5 bonds are there in C_{60}? (*Hint:* If you have trouble seeing this, get your hands on a soccer ball.)

14. As suggested in the figure in the Connections highlight on page 778, alkanethiols self-assembled onto gold surfaces do not line up perfectly perpendicular to the surface, but rather tilt all in the same direction by about 30°. Suggest why this might be so.

15. Given the polymer molecular weight distributions A and B shown below, rank order the following four quantities: $M_n(A)$, $M_n(B)$, $M_w(A)$, and $M_w(B)$. Just to be clear, both distributions are symmetrical and peak at the same molecular weight.

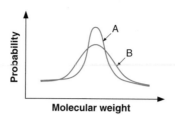

16. In one polymerization reaction it was found that termination processes were much slower for longer polymer chains. The molecular weight distribution was found to be as shown below.
a. Explain how the molecular weight distribution is consistent with the termination rate data noted.
b. Will the smaller peak have a larger effect on M_n or M_w? Explain.

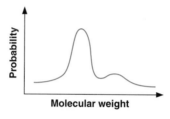

17. In one sense, cross-links in a polymer are a type of branching. However, addition of cross-links tends to increase the melting point of a polymer, while addition of branches tends to decrease the melting point. Explain.

18. Though its general shape is cigar-like, the following compound does not form any liquid crystalline phases. Propose both an explanation for this observation and structural modifications that might result in liquid crystallinity.

19. Liquid crystals have been used for many years as solvents in NMR studies. Though tumbling of solute molecules in isotropic solvents has the great advantage of simplifying NMR spectra, additional information about the structures of solute molecules can be gained from analysis of the more complicated spectra obtained for oriented solute molecules in liquid crystalline solvents.
a. Fortunately, liquid crystal domains with random orientations are generally not encountered in these experiments. Instead, oriented phases are obtained. Explain the reason for this fact.
b. Some molecules are better aligned than others in an oriented liquid crystalline phase. What characteristics of a molecule should lead to good alignment?

20. It is often found that for high molecular weight polymers, the melting temperature (T_m) and the degree of polymerization (DP) have the following relationship: $1/T_m - 1/T_{m(infinity)} = 2R/(\Delta H_{fus} \cdot DP)$. Here R is the gas constant and ΔH_{fus} is the energy of interaction between each individual monomer unit with a neighboring monomer unit. Explain why this occurs, and interpret the y intercept.

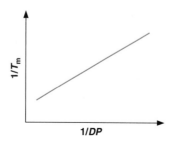

21. Interestingly, a variety of polymers become insoluble upon heating and are soluble again upon cooling. This is opposite to the behavior commonly found for small molecules. The following polymer is soluble in water below 30 °C, while above that temperature it precipitates. Explain the general phenomenon that leads to this behavior, with a focus upon entropy and enthalpy considerations of solubility, rather than on the specific structure of this polymer.

22. Often there is a slight deviation from Eq. 13.9 at the beginning of the polymerization. Why would there be a deviation from linearity?

23. Through the course of a free radical polymerization, the termination rate typically varies much more than the propagation rate. Give two reasons for this phenomenon.

Further Reading

Polymer Chemistry

Flory, P. J. (1953). *Principles of Polymer Chemistry*, Cornell University Press, Ithaca, NY.

Allen, P. E. M., and Patrick, C. R. (1974). *Kinetics and Mechanisms of Polymerization Reactions*, John Wiley & Sons, New York.

Bovey, F. A., and Winslow, F. H. (1979). *Macromolecules: An Introduction to Polymer Science*, Academic Press, Inc., New York.

Saunders, K. J. (1973). *Organic Polymer Chemistry*, Chapman and Hall, New York.

Stevens, M. P. (1999). *Polymer Chemistry, An Introduction*, Oxford University Press, Oxford.

Dendrimers

Newkome, G. R., Moorefield, C. N., and Vogtle, F. (2001). *Dendrimers and Dendrons*, 2d ed., John Wiley & Sons, New York.

Fischer, M., and Vögtle, F. "Dendrimers: From Design to Application—A Progress Report." *Angew. Chem. Int. Ed. Eng.*, **38**, 884–905 (1999).

Jansen, J. F. G. A., de Brabander-van der Berg, E. M. M., and Meijer, E. W. "Encapsulation of Guest Molecules into a Dendritic Box." *Science*, **266**, 1226–1229 (1994).

Tomalia, D. A., and Durst, H. D. in *Genealogically Directed Synthesis: Starburst/Cascade Dendrimers and Hyberbranched Structures*, E. Weber (ed.), Springer–Verlag, New York, 1993, pp. 193–313.

Bosman, A. W., Janssen, H. M., and Meijer, E. W. "About Dendrimers: Structure, Physical Properties, and Applications." *Chem. Rev.*, **99**, 1665–1688 (1999).

Hecht, S., and Frechet, J. M. J. "Dendritic Encapsulation of Function: Applying Nature's Site Isolation Principle from Biomimetics to Materials Science.". *Angew. Chem. Int. Ed. Eng.*, **40**, 74–91 (2001).

Zeng, F., and Zimmerman, S. C. "Dendrimers in Surpamolecular Chemistry: From Molecular Recognition to Self-Assembly." *Chem. Rev.*, **97**, 1681–1712 (1997).

Liquid Crystals

Collings, P. J. (2001). *Liquid Crystals: Nature's Delicate Phase of Matter*, 2d ed., Princeton University Press, Princeton, NJ.

Kirsch, P., and Bremer, M. "Nematic Liquid Crystals for Active Matrix Displays: Molecular Design and Synthesis." *Angew. Chem. Int. Ed. Eng.*, **39**, 4216–4235 (2000).

Fullerenes and Carbon Nanotubes

Diederich, F., and Thilgen, C. "Covalent Fullerene Chemistry." *Science,* **271,** 317–323 (1996).

Ajayan, P. M. "Nanotubes from Carbon." *Chem. Rev,* **99,** 1787–1799 (1999).

Hu, J., Odom, T. W., and Lieber, C. M. "Chemistry and Physics in One Dimension: Synthesis and Properties of Nanowires and Nanotubes." *Acc. Chem. Res.,* **32,** 435–445 (1999).

Kim, P., and Lieber, C. M. "Nanotube Nanotweezers." *Science,* **286,** 2148–2150 (1999).

Wanda, A. "Computational Approach to the Physical Chemistry of Fullerenes and Their Derivatives." *Ann. Rev. Phys. Chem.,* **49,** 405 (1998).

Ouyang, M., Huang, J.-L., and Lieber, C. M. "Scanning Tunneling Microscropy Studies of the One-Dimensional Electronic Properties of Single-Walled Carbon Nanotuges." *Ann. Rev. Phys. Chem.,* **53,** 201 (2002).

Self-Assembled Monolayers

Prime, K. L., and Whitesides, G. M. "Self-Assembled Organic Monolayers: Model Systems for Studying Adsorption of Proteins at Surfaces." *Science,* **252,** 1164–1167 (1991).

Ulman, A. "Formation and Structure of Self-Assembled Monolayers." *Chem. Rev.,* **96,** 1533–1554 (1996).

Polymer Phase Behavior

Bates, F. S. "Polymer–Polymer Phase Behavior." *Science,* **251,** 898 (1991).

Painter, P. C., and Colemen, M. M. (1997). *Fundamentals of Polymer Science,* Technomic Publishing Co., Inc., Lancaster, PA, Chapter 9.

Ring-Opening Metathesis Polymerization

Schlueter, A.-D. "Ring Opening Metathesis Polymerization—Recent Developments." *Angew. Chem.,* **100,** 1259 (1988).

Schrock, R. R. "Living Ring-Opening Metathesis Polymerization Catalyzed by Well-Characterized Transition-Metal Alkylidene Complexes." *Acc. Chem. Res.,* **23,** 158–165 (1990).

Group Transfer Polymerization

Senkler, C. A. "The Chemistry and Process of Group Transfer Polymerization." *Org. Coat.,* **8,** 1–10 (1986).

Webster, O. W. "Group-Transfer Polymerization." *Encycl. Polym. Sci. Eng.,* **7,** 580–588 (1987).

Bywater, S. "Group-Transfer Polymerization—A Critical Overview." *Makromol. Chem., Macromol. Symp.,* **67,** 339 (1993).

Step-Growth Polymerization

So, Y. H. "The Effect of Limited Monomer Solubility in Heterogeneous Step-Growth Polymerization." *Acc. Chem. Res.,* **34,** 753 (2001).

Condensation Polymerization

Flory, P. J. "Fundamental Principles of Condensation Polymerization." *Chem. Rev.,* **39,** 137 (1946).

Olefin Polymerization

Angermund, K., Fink, G., Jensen, V. R., and Kleinschmidt, R. "Toward Quantitative Prediction of Stereospecificity of Metallocene-Based Catalysts for α-Olefin Polymerization." *Chem. Rev.,* **100,** 1457 (2000).

Hlatky, G. G. "Heterogeneous Single-Site Catalysts for Olefin Polymerization." *Chem. Rev.,* **100,** 1347 (2000).

Rappé, A. K., Skiff, W. M., and Casewit C. J. "Modeling Metal-Catalyzed Olefin Polymerization." *Chem. Rev.,* **100,** 1435 (2000).

Emulsion Polymerization

Richards, D. H. "The Polymerization and Copolymerization of Butadiene" *Chem. Soc. Rev.,* **6,** 235 (1977).

ELECTRONIC STRUCTURE: THEORY AND APPLICATIONS

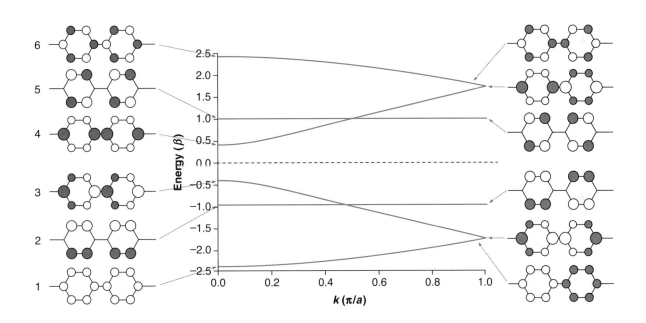

CHAPTER 14

Advanced Concepts in
Electronic Structure Theory

Intent and Purpose

This chapter presents a substantially more advanced analysis of electronic structure theory than does Chapter 1. We begin with a brief introduction to quantum mechanics. The goal is to make you comfortable with the notion of wavefunctions, the Schrödinger equation, and the form of the Hamiltonian for atoms and molecules. We also describe why a chemical bond forms. Most chemists think about bonding improperly, even though it is a fundamental concept in chemistry.

We then present **ab initio molecular orbital theory**. This is a well-defined approximation to the full quantum mechanical analysis of a molecular system, and also the basis of an array of powerful and popular computational approaches. Molecular orbital theory relies upon the linear combination of atomic orbitals, and we introduce the mathematics and results of such an approach. Then we discuss the implementation of *ab initio* molecular orbital theory in modern computational chemistry. We also describe a number of more approximate approaches, which derive from *ab initio* theory, but make numerous simplifications that allow larger systems to be addressed. Next, we provide an overview of the theory of organic π systems, primarily at the level of Hückel theory. Despite its dramatic approximations, Hückel theory provides many useful insights. It lies at the core of our intuition about the electronic structure of organic π systems, and it will be key to the analysis of pericyclic reactions given in Chapter 15.

We also present a brief introduction to density functional theory (DFT). This is a newer method that is steadily gaining popularity. We do not develop DFT as extensively as conventional *ab initio* theory, in part because the latter is the rigorous implementation of our qualitative notions of structure and bonding. So, while DFT may in time become the computational method of choice, *ab initio* molecular orbital theory is a pedagogically more useful introduction to advanced molecular quantum mechanics.

With a more developed view of MO theory than given in Chapter 1, we can consider several advanced topics for which MO theory is especially useful. These include aromaticity, the bonding in cyclopropane, planar methane, through-bond coupling, the novel bonding possibilities of carbocations and other electron deficient systems, and the spin preferences of organic biradicals and carbenes. We conclude the chapter by tying together the MO theory of organic molecules and the MO theory of organometallic complexes, an insight into bonding known as the isolobal analogy.

The concepts developed in this chapter set the stage for the study of several topics for which orbital concepts are crucial, including pericyclic reactions, photochemistry, and organic electronic materials. In addition, computational methods play an increasingly important role in all facets of modern chemistry. We firmly believe that knowledge of the basics of these approaches is essential for any practicing physical organic chemist, but also for all synthetic, organometallic, and bioorganic chemists. All desktop computers can now routinely run electronic structure theory calculations, and all practicing chemists can avail themselves of this powerful tool. Yet, the calculations should not be simply a "black box". Developing a deep understanding of these methodologies is beyond the scope of this text, but knowing the basic assumptions and methodologies can be very valuable. This familiarity should allow you to read and appreciate the modern computational literature, and to perform the calculations themselves with a degree of understanding such that you can interpret the results in an meaningful manner for your own research needs.

14.1   Introductory Quantum Mechanics

The goal of this first section of the chapter is to cover some of the basics of quantum mechanics that are relevant to calculational methods. This introduction is likely a review for most of you. However, if the only encounter you have had with quantum mechanics is in undergraduate physical chemistry classes, it is likely that we present the material in a different order. The intent is to present quantum mechanics in a manner that leads step-by-step to the modern view of orbitals and bonding. This presentation is rigorous, but is meant to be only sufficient for a good understanding of organic molecular structure and reactions, and it is in no way as thorough as you would find in an advanced text on quantum mechanics. In several places we just give the result, and leave it to the interested reader to delve into a more advanced text if they desire.

14.1.1   The Nature of Wavefunctions

One of the starting points for quantum mechanics was the observation that matter exhibits interference phenomena, as does light. De Broglie showed in the 1920s that electrons diffract just as does electromagnetic radiation. Hence, electrons exhibit the same "wave–particle duality" that is necessary to describe the properties of light. In classical mechanics, waves are mathematically described by an amplitude function, or **wavefunction** $[\psi(r,t)]$, which has a positional (r) and a time (t) dependence. Recall from trigonometry that the wave nature of the simple time independent function is $\psi(x) = \cos(x)$. To create a time dependent wavefunction, imagine any wave that moves along in space, meaning that the wave propagates at a particular rate (Figure 14.1). Waves in swimming pools are time dependent, moving across the surface of the pool and dying off to zero amplitude with time.

The wavefunction $\psi(r,t)$ describes the position of a wave in three-dimensional space at coordinates r at time t. For waves, this describes all their observable properties. Due to their wave-like nature, it was postulated in the 1920s that matter such as electrons and molecules could be described with wavefunctions also. These wavefunctions contain all the observable information about the system. Therefore, producing the wavefunctions for electrons, and for atoms and molecules as a whole, is of paramount importance for understanding the nature of chemical structure and reactivity.

Given the importance of wavefunctions, it is worthwhile to consider some of their properties. First of all, the probability (P) of the amplitude of a wave being a particular value at a certain coordinate r and time t is given by the square of the wavefunction evaluated at position r and time t $[P = \psi(r,t)^2]$. With respect to the wave-like nature of an electron, the exact same expression holds, giving the probability of finding an electron at a particular position at a particular time.

An important insight into wavefunctions is that they are additive. The superposition of two wavefunctions leads to the addition of the individual amplitudes at coordinates r and time t, resulting in a new wavefunction (Eq. 14.1). For example, when you jump into one end of a pool and your friend jumps in the other end, the waves collide in the center and add together. The wave that results from collision can be mathematically described by a wavefunction that is the sum of the wavefunctions for the two individual waves. The addition is either constructive or destructive, meaning that the amplitudes either increase or decrease, respectively, at positions r and times t (Figure 14.2). The exact same properties hold for electron wavefunctions. In fact, the ability to add wavefunctions is one of the prime tenets of molecular orbital theory, and we will use it extensively.

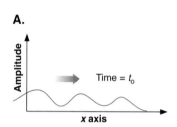

A.

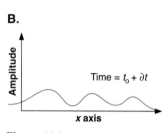

B.

Figure 14.1
A simple wave that is a function of x and t. **A.** The wave at time t_o, and **B.** The wave at time $t_o + \partial t$. A physical wave propagates as the result of an applied force.

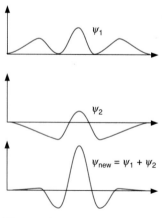

Figure 14.2
Two waves are additive, creating a new wave.

$$\psi_{new} = \psi_1 + \psi_2 \qquad\qquad \text{(Eq. 14.1)}$$

14.1.2 The Schrödinger Equation

Everything that we want to know about electrons can be extracted from their wavefunctions. Hence, we need to solve for these functions. This is done using the **Schrödinger equation** (Eq. 14.2). Here H has the dimensions of energy and is called the **Hamiltonian operator**, representing all the forces on the system. The forces are a function of the various positions of the electrons and nuclei. An **operator** indicates a mathematical operation (such as differentiation, multiplication, etc.) that acts on a function. Eq. 14.2 represents a particular situation in which an operator (H) acts on a function (ψ) and gives back the same function (ψ) multiplied by a constant (E). This is called an **eigenvalue** problem, with ψ being the **eigenfunction** and E the eigenvalue.

$$H\psi = E\psi \qquad \text{(Eq. 14.2)}$$

One of the most confusing aspects of quantum mechanics can be the notation. We will do our best here to keep it clear and simple. From this point forward in the chapter, wavefunctions (the eigenfunctions) Ψ, ψ, and ϕ will be defined as follows. The symbol Ψ will represent the *total* wavefunction for the system under consideration (usually a molecule); ψ will represent molecular orbitals; and ϕ will represent atomic orbitals. In addition, E will represent the total energy of a system while ε will be an individual orbital energy (both are eigenvalues); H is a Hamiltonian operator and h is Planck's constant. The choice of the Hamiltonian and how the wavefunctions are expressed is much of what the next few sections are about.

14.1.3 The Hamiltonian

The Schrödinger equation tells us that the forces operating on a wavefunction will generate a wave with energy E. To solve for the wavefunction and the energy we need to know the forces operating on the system. In classical mechanics the forces on a system create two kinds of energy—kinetic and potential. The Hamiltonian operator H is therefore broken into two operators (Eq. 14.3), one that represents kinetic energy (T) and one that represents potential energy (V).

$$H = T + V \qquad \text{(Eq. 14.3)}$$

The kinetic energy of any particle is generated by the potential field the particle inhabits. When the potential field has no dependence upon the velocity of the particle and is simply related to the coordinates of the particle, the kinetic energy of the particle is only related to its momentum (p) and its mass (m) (Eq. 14.4). The operator that generates the momentum of a three-dimensional wave is given by Eq. 14.5, where $\nabla = (\partial/\partial x)i + (\partial/\partial y)j + (\partial/\partial z)k$ (where i, j, and k are orthogonality operators; consult a vector calculus text if you are unfamiliar with these). Substituting Eq. 14.5 into Eq. 14.4 gives the kinetic energy operator (Eq. 14.6). The nature of the ∇^2 term (Eq. 14.7) is extremely important to understanding the energies of electrons and the nature of bonding, and we will return to it below. For each and every particle, whether an electron or nucleus, in an atom or molecule, we will need to write a term such as Eq. 14.6 representing its kinetic energy.

$$T = \left(\frac{1}{2m}\right)p^2 \qquad \text{(Eq. 14.4)}$$

$$p = \left(\frac{\hbar}{i}\right)\nabla \qquad \text{(Eq. 14.5)}$$

$$T = \left(\frac{-\hbar^2}{2m}\right)\nabla^2 \qquad \text{(Eq. 14.6)}$$

$$\nabla^2 = \frac{\partial^2}{\partial x^2} + \frac{\partial^2}{\partial y^2} + \frac{\partial^2}{\partial z^2} \qquad \text{(Eq. 14.7)}$$

$$H = -\frac{\hbar^2}{2} \sum_{A}^{N} M_A^{-1} \nabla_A^2 + \sum_{A<B} e^2 Z_A Z_B \, r_{AB}^{-1} - \frac{\hbar^2}{2m} \sum_{i}^{n} \nabla^2$$

❶   ❷   ❸

$$-\sum_{A}\sum_{i} e^2 Z_A \, r_{Ai}^{-1} + \sum_{i<j} e^2 \, r_{ij}^{-1}$$

❹   ❺

UPPER CASE = Nuclei, lower case = electrons; m or M = mass;

Z = Nuclear charge; e = electron charge;

r_{ij} = Distance between electrons i and j;

r_{Ai} = Distance between nucleus A and electron i.

∇^2 = Kinetic energy operator = $\dfrac{\partial^2}{\partial x^2} \cdots$

The terms are:
❶ = Kinetic energy of nuclei
❷ = Nuclear–nuclear repulsions
❸ = Kinetic energy of electrons
❹ = Nuclear–electron attraction
❺ = Electron–electron repulsion

Figure 14.3
The general form of the Hamiltonian for any molecule, along with the meaning of each term.

Now we need to consider the form of the potential energy—that is, the potential field that created the kinetic energy described above. For isolated atoms or molecules not subjected to any external influence, there is no time dependence to the potential field. Furthermore, when dealing with electrons and nuclei, the potential field experienced by the particles depends solely upon their charge and position, and hence Coulomb's law is all we need. The repulsive interaction between two electrons (1 and 2) would be given by Eq. 14.8 **A**; the attraction between an electron (1) and a nucleus (A) would be given by Eq. 14.8 **B**; and the repulsion between two nuclei (A and B) would be given by Eq. 14.8 **C**. Here e is the charge on an electron or proton, Z is the atomic number for the particular nucleus being considered, and r is the distance between the particles. Given these equations, we can now write the Hamiltonian for any atom or molecule (Figure 14.3).

A. $V = \dfrac{e^2}{r_{12}}$ **B.** $V = \dfrac{-Ze^2}{r_{1A}}$ **C.** $V = \dfrac{Z_A Z_B e^2}{r_{AB}}$ (Eq. 14.8)

We will describe in Section 14.2 several techniques for solving the Schrödinger equation for complex systems using Hamiltonians of the form given in Figure 14.3. For now, we simply want to introduce a simplifying notation. To solve the Schrödinger equation, we first rewrite Eq. 14.2 in the form shown in Eq. 14.9, multiplying both sides by ψ^* and integrating over all space. Because some wavefunctions contain the imaginary number i ($i = \sqrt{-1}$), in integrals such as these we must use the complex conjugate, ψ^*, where i is replaced by $-i$. The energy E is simply a number and can be written outside the integral. Since the integral sign is cumbersome to keep writing, a simpler notation is adopted known as **bracket notation** (also called bra-ket, Eq. 14.10). Here the "$< x|$" term is called the "bra" and the "$|x >$" term is called the "ket" (who ever said quantum chemists and physicists don't have a sense of humor?). Together they mean the integral over all space, and the complex conjugate of the first wavefunction in the bra is implicit in the notation.

$$\int \psi^* H \psi \, d\tau = E \int \psi^* \psi \, d\tau \qquad \text{(Eq. 14.9)}$$

$$<\psi | H | \psi> = E <\psi | \psi> \qquad \text{(Eq. 14.10)}$$

Once the Hamiltonian has been identified, substitution into Eq. 14.10, and some extensive calculus produce the desired wavefunctions, each with a distinct energy. The **expectation value** of the energy (E) can be obtained from Eq. 14.11. An example is given in the following Connections highlight, where we look at a very simple system, the hydrogen atom.

$$E = \frac{<\psi | H | \psi>}{<\psi | \psi>} \qquad \text{(Eq. 14.11)}$$

Connections

The Hydrogen Atom

In introductory chemistry courses, the solutions to the Schrödinger equation for the hydrogen atom are commonly presented. Let's review how these solutions arise. The Hamiltonian for a hydrogen atom is given below. Only two terms are necessary, one reflecting the kinetic energy of the electron and the other the attraction between electron 1 and the nucleus A. We do not need a kinetic energy term for the hydrogen atom nucleus, because we will let it be the point of reference.

$$H = (-h^2 / 2m)\nabla^2 + e^2 / r_{1A}$$

When this Hamiltonian is used, the solutions are the wavefunctions for the H atom (ϕ's). In this specific case, the Hamiltonian can be directly substituted into $H\phi = E\phi$, and the Cartesian coordinate system (x, y, and z) is converted to spherical coordinates (r, θ, and ϕ). The solutions to the resulting differential equations can be directly obtained and are found to have a radial function $R(r)$ that is separate from an angular portion [$Y(\theta,\phi)$]. The radial portions of several of these wavefunctions are given in the table shown to the right [the angular portion, $Y(\theta,\phi)$, is not shown].

There are an infinite number of solutions, each with a distinct energy (they are quantized). The solutions contain numbers that come in particular sets, which we defined as the principle, azimuthal, and magnetic quantum numbers in Chapter 1. Note that the solutions are nothing more than equations describing standing waves in three-dimensional space. The equations indicate an amplitude

of the wave at a particular coordinate in space relative to the nucleus. Some of these wavefunctions have different algebraic signs in different regions of three-dimensional space. The different signs represent different phases, shown by the shading or color difference drawn for the lobes of certain orbitals.

$1s$	$R_{10}(r) = (Z/a_o)^{3/2} \, 2 \exp(-\rho/2)$
$2s$	$R_{20}(r) = (Z/a_o)^{3/2} \, (\frac{1}{2}\sqrt{2}) \, (2-\rho) \exp(-\rho/2)$
$2p$	$R_{21}(r) = (Z/a_o)^{3/2} \, (\frac{1}{2}\sqrt{6}) \, \rho \exp(-\rho/2)$
$3s$	$R_{30}(r) = (Z/a_o)^{3/2} \, (\frac{1}{9}\sqrt{3}) \, (6 - 6\rho + \rho^2) \exp(-\rho/2)$
$3p$	$R_{31}(r) = (Z/a_o)^{3/2} \, (\frac{1}{9}\sqrt{6}) \, (4-\rho) \, \rho \exp(-\rho/2)$
$3d$	$R_{32}(r) = (Z/a_o)^{3/2} \, (\frac{1}{9}\sqrt{30}) \, \rho^2 \exp(-\rho/2)$

The radial dependence (R_{nl}) of hydrogen wavefunctions take the form $\phi_{AO} = R_n(r) Y(\theta,\phi)$, where $\rho = 2Zr/na_o$, $a_o = 52.92$ pm, Z is the atomic number, n is the principal quantum number, and l is the azimuthal quantum number.

The angular functions $Y(\theta,\phi)$ modulate amplitudes in various directions along the x, y, and z axes. Three-dimensional pictures of the solutions are the familiar shapes we call the $1s$, $2s$, $2p$, $3s$, $3p$, $3d$, etc., orbitals. Interestingly, solving $H\phi = E\phi$ gives the p_z orbital in the manner we normally present it, but what we normally draw as p_x and p_y orbitals are actually linear combinations of the $l = \pm 1$ solutions, meaning that the pictures we draw are just convenient representations of the orbitals.

14.1.4 The Nature of the ∇^2 Operator

We calculate the kinetic energy of a wave by using Eq. 14.6, which involves the ∇^2 operator. Because ∇^2 is a second derivative, this operator gives the rate of change of the gradient (first derivative) of the wavefunction at a particular point r. However, in our calculation of the energy E (Eq. 14.11), we take the integral over all space. Hence, the kinetic energy of the wave is the average kinetic energy over all space—that is, the average change of the gradient of the wavefunction.

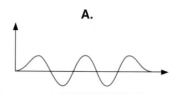

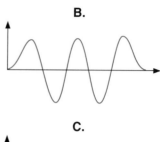

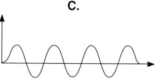

Figure 14.4
In the comparison of waves **A** and **B**, **B** has the higher
kinetic energy due to the larger amplitude. In the comparison
of waves **A** and **C**, **C** has the higher kinetic energy due to the
larger frequency. The fact that the change in the slope of the
waves is larger for **B** and **C** is the mathematical way of stating
that they have the higher kinetic energies.

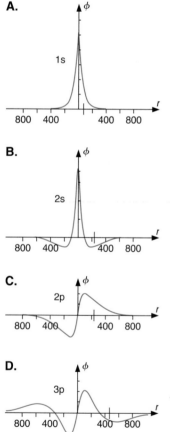

We want to apply the ∇^2 operator in a qualitative sense to get a feeling for the relative kinetic energies of different waves. In Figure 14.4, wave B has higher kinetic energy than wave A even though they have the same frequency. This can be appreciated by visual inspection, in that the change in slope of B has to be more rapid than for A because of its higher amplitude. Due to its higher frequency, wave C has higher kinetic energy than wave A. Integrated over the entire wave, the rate of change of the slope of B is larger than A, and likewise for C relative to A.

Now apply this reasoning to the wavefunctions for the hydrogen atom shown in the margin. These plots represent amplitude as a function of one dimension in space. We find that the kinetic energy is lowest for wave A (1s), because it has the least amount of change in slope relative to B (2s), C (2p), and D (3p). However, it is harder to determine in a qualitative fashion whether B or C has the lower kinetic energy, and a calculation is actually needed. Yet, it is a clear call for comparing wavefunctions C (2p) and D (3p); D has the higher kinetic energy due to the greater change in gradients over all space. This kinetic energy analysis also explains the notion that the more **nodes** an orbital has, the higher is its energy. Recall from Chapter 1 that nodes are points of zero electron density, where the wavefunction changes sign. Having more nodes in an orbital is analogous to a higher frequency wave in a swimming pool.

Extending these concepts, the kinetic energy of an orbital drops as it spreads out in space (Figure 14.5 **A**). In the extreme, a wave with a constant amplitude spread out over all space has a kinetic energy of zero. Conversely, if the electron collapsed into the nucleus, such as is being approximated in Figure 14.5 **C**, the kinetic energy would be infinite. Therefore, the kinetic energy of an electron becomes lower the more diffuse the wavefunction becomes. This is the basic reason we are taught that electrons preferentially delocalize, and that the more resonance structures a compound has, the lower its energy.

If the kinetic energy drops as the orbital spreads out in space, why don't all electrons spread out to infinity? The answer is that the electrostatic attraction between electrons and protons favors a collapse of electrons into the nucleus. There is a balance between the electron wanting to spread out in space to diminish its kinetic energy versus being attracted to the nucleus due to electrostatics (potential energy). A balance between kinetic and potential energies also occurs in bonding, as described below.

14.1.5 Why do Bonds Form?

It is now time to analyze our first chemical bond. Why does a bond form? A typical answer is that the energy decreases because the electrons primarily reside in between the

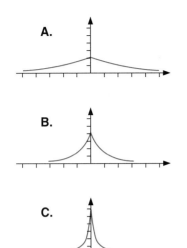

Figure 14.5
Kinetic energy is lowest for a diffuse wavefunction.
A. Lowest energy wavefunction, **B.** Intermediate
energy, and **C.** Highest energy wavefunction.

nuclei of the two atoms in the bond, shielding the internuclear repulsions and holding the
atoms together. Actually, such an explanation is simply untrue, because it is based solely
upon an electrostatic argument. In reality, the total electrostatic repulsions between nuclei
and electrons increase when a bond forms. Think of it this way. The internuclear repulsion
has to be higher for a bond relative to an infinite distance between the atoms, even if the electrons shield the nuclear charges. Also, if the electrons concentrate between the nuclei, the
repulsion between them would increase. Hence, there must be a kinetic energy reason that
bonds form.

To understand the changes in kinetic energy and potential energy that occur when a
bond forms, let's analyze the simplest of bonds, that in H_2^+. The Hamiltonian for H_2^+ is given
in Eq. 14.12, where A and B represent the individual nuclei. Here we leave out the internuclear repulsion term, drawing on a simplification known as the Born–Oppenheimer approximation (discussed more in Section 14.2.1).

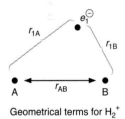

Geometrical terms for H_2^+

$$H = \left(\frac{-\hbar^2}{2m}\right)\nabla^2 + \frac{e^2}{r_{1A}} + \frac{e^2}{r_{1B}} \qquad \text{(Eq. 14.12)}$$

The Schrödinger equation for H_2^+ can be solved in an alternative coordinate system called
confocal elliptical coordinates. Just as with the hydrogen atom, certain wavefunctions result, each of which has a distinct energy.

In Figure 14.6 we show cross-sections of the two lowest energy solutions, along with a
cross-section through a single $1s$ orbital. One solution for H_2^+ has no node (g, referred to as
gerade, symmetric, bonding) and one solution has a node (u, **ungerade**, antisymmetric, antibonding). Now that we have the result of the ∇^2 operator upon a wavefunction, it is obvious
that the u wavefunction is higher in kinetic energy.

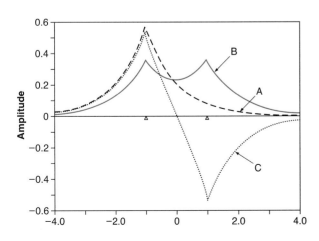

Figure 14.6
Cross-sections along the x axis showing amplitude
versus distance for **A.** $1s$ orbital, and for the two lowest
energy wavefunctions **B.** ψ_g and **C.** ψ_u for H_2^+ relative
to the center of the bond. The x axis is internuclear
distance (Bohr radii). Δ represents H positions.
Adapted from course notes by W. A. Goddard III,
"Nature of the Chemical Bond."

One important feature of these exact wavefunctions for a simple molecule is that they resemble $1s$ orbitals on the individual hydrogen atoms that were either added (the g state) or subtracted (the u state). This observation is one of the justifications for linearly combining atomic orbitals on atoms to create molecular orbitals.

We can use our knowledge of the ∇^2 operator to examine the kinetic energy of the g state relative to two $1s$ orbitals. The average change in the gradient of two isolated $1s$ wavefunctions is larger than that of the g state because in the region where the $1s$ AOs overlap, the bonding orbital is lower in average kinetic energy. The wavefunction is "flatter" in this area with the g state. This is the fundamental reason that a bond forms; the kinetic energy of the electrons in the bonding region is lower than the kinetic energy of the electrons in isolated atomic orbitals.

Recall that the potential energy increases due to increasing electrostatic repulsions during bonding. Hence, there must be a balance between the increased potential energy and the decreased kinetic energy, with a net lower energy overall winning out for certain internuclear distances. We can see this by examining Figure 14.7, where kinetic and potential energy terms are plotted relative to an energy of zero for an infinite separation distance between the atoms. For H_2^+, T_g (kinetic energy of the gerade state) has a minimum at a particular interatomic distance, while V_g (potential energy of the gerade state) is always repulsive and increases steeply at small distances. The greater the overlap between the orbitals at the point of the minimum in the T_g curve, the lower the kinetic energy. This is why bonds are stronger with larger orbital overlap. Figure 14.7 also shows the T_u and V_u (ungerade state) values as a function of internuclear distance. Note that T_u is always repulsive due to the node, but that V_u is actually attractive for certain internuclear distances. Also note that kinetic energy (T_g) is never actually negative, but is just negative relative to two separate s-orbitals.

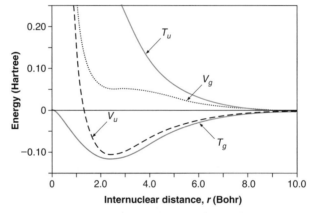

Figure 14.7
The kinetic (T) and potential (V) energies of the g and u states for H_2^+ as a function of internuclear distance. The addition of the T_g and V_g energies leads to the typical Morse potential curve, while the addition of the T_u and V_u energies leads to a function that is always repulsive. Adapted from course notes by W. A. Goddard III, "Nature of the Chemical Bond."

Since the real energy of the system is T plus V, we must add the T and V curves for both the g and u states. Upon addition for the g state, you find there are certain internuclear distances that the T term dominates, causing a bond to form. At shorter distances the V term dominates, giving the typical shape found for a Morse potential curve (see Sections 2.1.4 and 8.1.2). For all internuclear distances, the T term dominates in the u state, and this state is always repulsive. This is the origin of the repulsion between atoms that arises when populating an antibonding orbital with electrons.

In summary, bonds form because the kinetic energies of electrons are lower in bonding molecular orbitals than they are in isolated atomic orbitals. The potential energy of the system is always higher. Addition of the kinetic and potential energies results in an optimum bond distance and the common Morse potential diagrams describing bond lengths and strengths.

14.2 Calculational Methods—Solving the Schrödinger Equation for Complex Systems

Now that we have reviewed some fundamentals of quantum mechanics, and laid a foundation for why bonds form, we want to turn our attention to calculating the electronic structures of atoms and molecules that are more interesting than H and H_2^+. As always, we have to use the Schrödinger equation, but now the mathematics is much more complicated. Instead of describing all the math in detail, we touch on the fundamental math required, and we describe several of the modern techniques used in such an analysis.

Because the Schrödinger equation cannot be solved for any but the simplest systems, for years after its development the equation sat idle, with only heroic efforts on mechanical calculators producing any useful results. However, with the advent of high powered computers, it became feasible to calculate *approximate* solutions to the Schrödinger equation for molecular systems of moderate complexity. The approximations follow two strategies. In one, terms that are computationally intensive are replaced with adjustable parameters, often related to experimental quantities (empirical parameters); or, they are simply ignored. As we will see, these terms are one- and two-electron integrals of various kinds. Alternatively, one can reject all recourse to adjustable parameters, and simply make well-defined approximations and solve all necessary integrals, the so-called ***ab initio*** approach. Here we will present an overview of both strategies.

14.2.1 *Ab Initio* Molecular Orbital Theory

The term "*ab initio*" means "from first principles"; it does not mean "exact" or "true". In *ab initio* molecular orbital theory, we develop a series of well-defined approximations that allow an approximate solution to the Schrödinger equation. We calculate a total wavefunction and individual molecular orbitals and their respective energies, without any empirical parameters. Below, we outline the necessary approximations and some of the elements and principles of quantum mechanics that we must use in our calculations, and then provide a summary of the entire process. Along with defining an important computational protocol, this approach will allow us to develop certain concepts that will be useful in later chapters, such as spin and the Born–Oppenheimer approximation.

Born–Oppenheimer Approximation

The mass of the nucleus (M) is more than a thousand times the mass of an electron (m). Therefore, electrons move faster than nuclei and can rapidly adjust to changes in nuclear position. Thus, one can keep the nuclei fixed and simply consider electron motions, giving an electronic Hamiltonian as the following terms from Figure 14.3: $H_e = ③ + ④ + ⑤$. This is an approximation that is almost universally made. Breakdowns of the Born–Oppenheimer approximation occur only when nuclei are moving very rapidly, as can happen when two surfaces meet at a conical intersection (see Section 16.3.1).

If we calculate the wavefunction for a molecule using the Born–Oppenheimer approximation, we must choose the coordinates of each nucleus beforehand. However, using the Born–Oppenheimer approximation we can determine molecular geometries in the following way. First, evaluate H_e, then add on ② from Figure 14.3 (which is just Coulomb's law between nuclei) to give E_{tot}. Term ① from Figure 14.3 is typically not included because the nuclei are not allowed to move, and therefore have no kinetic energy. Next, we pick a new geometry (move the nuclei), repeat the calculation, and continue until geometrical changes no longer lower the energy.

The Orbital Approximation

When we are calculating the total wavefunction for an atom or molecule, the orbital approximation is used (Eq. 14.13). Ψ, the total wavefunction, is considered as a product of one electron wavefunctions known as **orbitals**—that is, $\psi_a(1)$, $\psi_b(2)$, $\psi_c(3)$, etc. We usually think of orbitals as holding two electrons, but the proper definition of an orbital is as a one-electron wavefunction. This is a subtle terminology issue that we will address fully below. The number in parentheses indicates a different electron, simply allowing us to keep track of them, and the letter designates a particular orbital. When we are evaluating a molecule, the ψ's are molecular orbitals. When we are evaluating an atom, we use atomic orbitals (ϕ's).

$$\Psi = \psi_a(1)\,\psi_b(2)\,\psi_c(3)\,\bullet\bullet\bullet\,\psi_n(m) \qquad \text{(Eq. 14.13)}$$

Since both Ψ and all the ψ's can be related to probabilities (refer back to Section 14.1.1), the orbital approximation defines a total probability as a product of individual probabilities. This is only true in probability theory if the individual events (ψ's) are independent, such as each flipping of a coin to get "heads" or "tails" is an independent event. With respect to electrons, this means that the probability that electron 1 will exist at a certain position in space is completely independent of the positions of electrons 2, 3, 4, etc. The electrons' movements are therefore not correlated, a severe approximation (see the discussion of electron correlation given later). This is therefore called an **independent electron theory**.

Since the probability of the electron being somewhere in the universe is 1, it must be true that the integral over all space of the square of a one-electron wavefunction is 1. Hence, one tenet of quantum mechanics is that all orbitals must be **normalized**, defined as satisfying Eq. 14.14.

$$<\psi_a|\psi_a> = 1 \qquad \text{(Eq. 14.14)}$$

Another important attribute is that the orbitals are **orthogonal**, meaning that any pair of orbitals have zero overlap. Orthogonality is expresssed by Eq. 14.15. This greatly simplifies the calculational process, but with no cost in accuracy, as we will see later in this section.

$$<\psi_a|\psi_b> = 0 \qquad \text{(Eq. 14.15)}$$

Spin

Our ψ and ϕ provide a complete description of the spatial distribution of a given electron. However, electrons also have an internal structure that we refer to as **spin**. Each electron can have an up or down spin (designated as α or β), the important point being that the two states are opposite. Mathematically speaking, the two states are orthogonal. The integral of $\alpha(1)\alpha(1)$ over all space is 1, but the integral of $\alpha(1)\beta(1)$ is zero.

The total wavefunction of a molecule needs to take spin into account. The way to do this is to simply split our one-electron wavefunctions (orbitals) into a product of two parts—a space part and a spin part: ψ (MO) or ϕ (AO) for the space part, and α or β for the spin. The actual orbital is the product of the two and is termed a **spin orbital**. For example, the function $\psi_a(1)\alpha(1)$ means that electron 1 has spatial distribution ψ_a and is spin α.

The Pauli Principle and Determinantal Wavefunctions

Electrons are **fermions**—that is, indistinguishable particles. Thus, no observable physical property of a system can change if we simply rename or renumber the electrons. The manner in which this statement manifests itself mathematically is that the total wavefunction of an atom or molecule (Ψ) must be antisymmetric with respect to an exchange of the coordinates of any two electrons. This is a rule called the **Pauli principle** (you must just accept this rule, unless you want to delve into a textbook on advanced quantum mechanics). Antisymmetric means that the exchange of two electrons creates the negative of the original function. Consider an operator $P_{1,2}$ that permutes the position of any two electrons, arbitrarily denoted as electron 1 and 2, in a multi-electron system. Ψ is an acceptable wavefunction if $P_{1,2}$ operating on Ψ produces $-\Psi$ (Eq. 14.16).

$$P_{1,2}\Psi = -\Psi \qquad \text{(Eq. 14.16)}$$

Let's start checking for adherence to the Pauli principle by considering the simplest wavefunction for a two-electron system (Eq. 14.17), where $\psi_a(1)$ is read as electron 1 in orbital ψ_a. Here, we use molecular orbitals (ψ's) to analyze molecules, but we could just as easily have used atomic orbitals (ϕ's) for atoms.

$$\Psi = \psi_a(1)\alpha(1)\,\psi_b(2)\beta(2) \qquad \text{(Eq. 14.17)}$$

The exchange of the two electrons creates a new system where $P_{1,2}\,\Psi = \psi_a(2)\alpha(2)\psi_b(1)\beta(1)$, which is not equal to $-\Psi$, and hence the Pauli principle is not satisfied.

Now consider an alternative wavefunction (Eq. 14.18). Ψ' is an acceptable wavefunction because exchanging the positions of electrons 1 and 2 does produce $-\Psi'$ (Eq. 14.19).

$$\Psi' = \psi_a(1)\,\alpha(1)\,\psi_b(2)\,\beta(2) - \psi_a(2)\,\alpha(2)\,\psi_b(1)\,\beta(1) \qquad \text{(Eq. 14.18)}$$

$$P_{1,2}\Psi' = \psi_a(2)\,\alpha(2)\,\psi_b(1)\,\beta(1) - \psi_a(1)\,\alpha(1)\,\psi_b(2)\,\beta(2) = -\Psi' \qquad \text{(Eq. 14.19)}$$

Why is it that the exchange of two electrons creating the negative of the wavefunction does not change any physical observables? Remember that any physical observable relates to Ψ^2, which is completely invariant to electron interchange if Ψ is antisymmetric, because $(-\Psi)^2 = (\Psi)^2$. Note that Ψ' is in the form of a determinant:

$$\Psi' = \begin{vmatrix} \psi_a(1)\,\alpha(1) & \psi_b(1)\,\beta(1) \\ \psi_a(2)\,\alpha(2) & \psi_b(2)\,\beta(2) \end{vmatrix}$$

A spin–orbit wavefunction of this form is called a **Slater determinant**, after the pioneering physicist who developed the concept. Any Ψ can be written as a Slater determinant using the general form shown, creating a total wavefunction that obeys the Pauli principle. Note that we have now collapsed the spatial and spin parts into a single symbolism, such that $\psi_n(m)$ contains both space and spin components.

$$\Psi = \begin{vmatrix} \psi_a(1) & \psi_b(1) & \psi_c(1) & \dots & \psi_n(1) \\ \psi_a(2) & \psi_b(2) & \psi_c(2) & \dots & \psi_n(2) \\ \psi_a(3) & \psi_b(3) & \psi_c(3) & \dots & \psi_n(3) \\ \vdots & & & & \\ \psi_a(n) & \psi_b(n) & \psi_c(n) & \dots & \psi_n(n) \end{vmatrix}$$

It is a mathematical property of determinants that exchange of any two rows produces the negative of the determinant. Thus, with a **determinantal wavefunction**, you can be sure that exchange of any two electrons will produce the negative of the original wavefunction, satisfying the Pauli principle. When a Slater determinant is formed from a product of individual wavefunctions, we say that the wavefunction has been **antisymmetrized**. An operator A is designed to refer to this process. Thus, when a total wavefunction is written, we normally write the A operator out front to simplify the notation (Eq. 14.20), and to signify that the function is really a Slater determinant.

$$A\Psi = A[\psi_a(1)\,\psi_b(2)\,\psi_c(3)\bullet\bullet\bullet\psi_m(n)] \qquad \text{(Eq. 14.20)}$$

Note also that if any two columns of a determinant are identical, the value of the determinant is zero. Thus, in order to have a non-zero wavefunction, no two electrons can occupy the same spin orbital, meaning that they cannot have the same spin when in the same spatial orbital—a statement of the Pauli principle in a more familiar form.

Actually there are two different ways to build our Slater determinant with spin orbitals. The most obvious way is to give each electron its own spatial function and a spin of α or β.

This is termed an **unrestricted** wavefunction, and it would have the form of Eq. 14.21. Here, the total number of orbitals—last of which is Ψ_n—equals the total number of electrons (n).

$$\Psi = A\{[\psi_a(1)\alpha(1)][\psi_b(2)\beta(2)][\psi_c(3)\alpha(3)][\psi_d(4)\beta(4)] \bullet\bullet\bullet [\psi_n(n)\beta(n)]\} \quad \text{(Eq. 14.21)}$$

More typically, however, a simplification is made. We can have two electrons share the same spatial wavefunction and achieve orthogonality through the spin part. We would have $\psi_a(1)\alpha(1)\psi_a(2)\beta(2)$ for a two-electron system. Writing out the entire wavefunction for an n electron system we obtain Eq. 14.22. This is termed a **restricted** total wavefunction—electrons are required to line up in pairs in spatial orbitals, with one α and the other β. Now the total number of orbitals is half the number of electrons, and the subscript on the last ψ reflects this. This is the orbital theory with which we are all familiar. Molecular and atomic orbitals contain two electrons, with the electrons having opposite spins. Often, this is viewed as the definition of the Pauli principle. Actually, the Pauli principle is expressed in Eq. 14.16. We satisfy it by using a determinantal wavefunction, and if we have a determinantal wavefunction, two electrons with the same spin cannot occupy the same orbital.

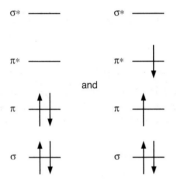

Two different electronic configurations of ethylene

$$\Psi = A\{[\psi_a(1)\alpha(1)][\psi_a(2)\beta(2)][\psi_b(3)\alpha(3)][\psi_b(4)\beta(4)] \dots [\psi_{n/2}(n-1)\alpha(n-1)][(\psi_{n/2}(n)\beta(n)]\}$$
$$\text{(Eq. 14.22)}$$

Since we designated the population of only particular molecular orbitals with electrons, the Slater determinant only describes a single electronic configuration. By **electronic configuration** we refer to the exact placement of the electrons into specific orbitals. For example, the molecular orbital diagrams shown to the side represent different electronic configurations for ethylene, using a wavefunction that considers only the σ and π bonds of the C=C bond. A different Slater determinant would be used for these two systems, because electrons are in different orbitals in each configuration.

The Hartree–Fock Equation and the Variational Theorem

The **variational theorem** states that the energy, E, of any approximate solution to the Schrödinger equation will always be higher than the true energy. Thus, we minimize the total energy of any wavefunction to get as close to the true energy as possible. The manner in which the minimization is performed will be touched upon when we discuss secular determinants below.

When we apply the variational theorem to a determinantal wave function, we obtain the Hartree–Fock equations (Eq. 14.23).

$$\left[H_{ii} + \sum_j \left(2J_{ij} - K_{ij} \right) \right] \psi_i(m) = \varepsilon_i \psi_i(m)$$
$$\text{(Eq. 14.23)}$$
$$\underbrace{\qquad\qquad}_{\text{Fock operator}}$$

There are n such equations, one for each electron in the atom or molecule, where the specific electron is designated by its number (1, 2, 3, . . .), here written as the letter m. The subscript i stands for each one-electron wavefunction from the Slater determinant, running from values of a, b, c, . . . , where again the number of letters equals the number of electrons (n) in the system under analysis. For notational convenience, we will not always explicitly write out the spin terms α and β, but they are always there. The subscript j stands for all the other one-electron wavefunctions. For example, to evaluate the energy for the orbital with i = b, there will be a single H_{bb} term, and n J and K terms ($J_{ba}, J_{bb}, J_{bc} \dots$). The reason for the factor of 2 in front of the J term will be explained when we examine the terms of this equation and the influence of spin.

As noted above there are two ways to build our Slater determinant, depending on whether we do not allow or do allow two electrons to share the same spatial orbital. In this context, these are termed **unrestricted Hartree–Fock (UHF)** and **restricted Hartree–Fock (RHF)** theories, respectively. For most systems RHF is adequate, such that the "R" is usually dropped. However, for radicals, polyradicals and other **open-shell** systems in which there is an imbalance in the number of α vs. β electrons, it is often better to use separate orbitals for separate spins. A UHF calculation is then more appropriate.

The H, J, and K terms are themselves operators that create specific kinds of integrals upon multiplying out Eq. 14.23. It is the sum of all these integrals that produces the energy (ε_i) for each molecular orbital ψ_i. In our notation for these integrals given below, we have dropped the physical constants h, e, and m normally associated with Hamiltonians for simplicity.

H_{ii} is referred to as the **core integral**, and is a **one-electron integral** (Eq. 14.24)—it depends only on the coordinates of one electron. Here we use a specific number to designate the electron, in this case "1", because we will only need to keep track of the specific electrons when we consider the J and K integrals. The origin of H_{ii} is terms ③ and ④ from Figure 14.3, and hence it reflects the kinetic energy of an electron in orbital ψ_i and its attraction to the various nuclei (A, B, C, ... with different charges Z). This term simply gives the energy that an electron in the molecular orbital ψ_i would have if there were no other electrons in the molecule.

$$H_{ii} = \int \psi_i(1)\left[-\frac{1}{2}\nabla_i^2 - \sum_A Z_A r_{Ai}^{-1}\right]\psi_i(1)\,d\tau \qquad \text{(Eq. 14.24)}$$

The J and K integrals derive from the electron–electron repulsion term ⑤ in Figure 14.3. J is the called the **Coulomb integral** and K is the **exchange integral**. It is because of the determinantal nature of the wave function that these two types of electron–electron interaction terms arise. In order to understand this, let's consider a molecule with only two electrons. We first introduce some new symbolism to simplify the notation (Eq. 14.25).

$$\Psi = \psi_i(1)\,\psi_j(2) - \psi_i(2)\,\psi_j(1) \equiv i(1)j(2) - j(1)i(2) \equiv ij - ji \qquad \text{(Eq. 14.25)}$$

Recall that we solve the Schrödinger equation using Eq. 14.11. Substituting Eq. 14.25 into 14.11 gives $E = <(ij - ji) \,|H|\, (ij - ji)>$ if the total wavefunction is normalized. After multiplying out the terms within the bracket, you get the **two-electron integrals** shown in the margin, all involving a $1/r$ term coming from ⑤ of Figure 14.3. ❶ and ❹ are **Coulomb integrals** (such as Eq. 14.26). These are double integrals, one for each electron of the pair over all space. These integrals correspond to the classical Coulomb repulsion between two electron clouds, one given by ψ_i and the other by ψ_j. This term is thus destabilizing, and enters with a positive sign. Terms ❷ and ❸ are the **exchange integrals** (Eq. 14.27). They are non-classical terms, meaning that non-quantum mechanical models will not produce exchange integrals. They clearly arise from the use of a determinantal wavefunction. That is, exchange integrals are a direct consequence of the Pauli principle.

❶ $\left\langle ij\,\middle|\frac{1}{r}\middle|\,ij \right\rangle$

❷ $-\left\langle ij\,\middle|\frac{1}{r}\middle|\,ji \right\rangle$

❸ $-\left\langle ji\,\middle|\frac{1}{r}\middle|\,ij \right\rangle$

❹ $\left\langle ji\,\middle|\frac{1}{r}\middle|\,ji \right\rangle$

$$J_{ij} = \iint \psi_i(1)\,\psi_j(2)\,r_{1,2}^{-1}\,\psi_i(1)\,\psi_j(2)\,d\tau\,d\tau \qquad \text{(Eq. 14.26)}$$

$$K_{ij} = \iint \psi_i(1)\,\psi_j(2)\,r_{1,2}^{-1}\,\psi_j(1)\,\psi_i(2)\,d\tau\,d\tau \qquad \text{(Eq. 14.27)}$$

It is now useful to consider the role of the spin functions in Hartree–Fock theory. The spin functions α and β have no effect on the one-electron core integrals (H_{ii}). They are factored outside of the spatial portion of the integral (Eq. 14.28) and their integral equals 1.

$$\int \alpha(1)\alpha(1)\,d\tau \int \psi_i(1)\left[-\frac{1}{2}\nabla_i^2 - \sum_A Z_A r_{Ai}^{-1}\right]\psi_i(1)\,d\tau \qquad \text{(Eq. 14.28)}$$

Unlike the core integrals, however, spin affects the two-electron integrals significantly, and differently for J and K. To understand this, we return to a two-electron system. In the case of the Coulomb integrals, all four possible spin combinations [$\alpha(1)\alpha(2)$, $\alpha(1)\beta(2)$, $\beta(1)\alpha(2)$, $\beta(1)\beta(2)$] for electrons 1 and 2 are allowed. Eq. 14.29 shows the kind of integral we are referring to with the $\alpha(1)\beta(2)$ state as the example. Such integrals are allowed because any given electron has the same spin on both sides of the $1/r$ term. This is a quadruple integral, one integral over each electron's spin and spatial functions.

$$J_{ij} = \iiint \psi_i(1)\alpha(1)\,\psi_j(2)\,\beta(2)\,r_{1,2}^{-1}\,\psi_i(1)\alpha(1)\,\psi_j(2)\beta(2)\,d\tau d\tau d\tau d\tau$$

$$= \int\alpha(1)\alpha(1)d\tau \int\beta(2)\,\beta(2)\,d\tau \iint\psi_i(1)\,\psi_j(2)\,r_{1,2}^{-1}\,\psi_i(1)\,\psi_j(2)\,d\tau d\tau \qquad \text{(Eq. 14.29)}$$

$$= \iint\psi_i(1)\,\psi_j(2)\,r_{1,2}^{-1}\,\psi_i(1)\,\psi_j(2)\,d\tau d\tau$$

The situation is more complicated and more interesting with the exchange integral. Both the spatial and spin functions for the electrons are exchanged. Consider again the $\alpha(1)\beta(2)$ spin combination, but now within an exchange integral as given in Eq. 14.30. Here, electron 1 is α on the left side of the integral, whereas it is β on right. Similarly, electron 2 is β on the left and α on the right. Because α and β spin functions are orthogonal (see above), when either the $\alpha(1)\beta(1)$ or the $\beta(2)\alpha(2)$ functions are integrated over all space they give zero, causing this entire integral to go to zero. Such a cancellation will not arise if the two electrons have the same spin.

$$K_{ij} = \iiint \psi_i(1)\alpha(1)\,\psi_j(2)\beta(2)\,r_{1,2}^{-1}\,\psi_j(1)\beta(1)\,\psi_i(2)\alpha(2)\,d\tau d\tau d\tau d\tau$$

$$= \int\alpha(1)\beta(1)d\tau \int\beta(2)\alpha(2)\,d\tau \iint\psi_i(1)\,\psi_j(2)\,r_{1,2}^{-1}\,\psi_j(1)\psi_i(2)\,d\tau d\tau$$

$$= 0$$

$$\text{(Eq. 14.30)}$$

Given this discussion, there are two important consequences of the introduction of spin into the wavefunction.

1. *All four spin combinations are acceptable for* J.
   However, only $\alpha(1)\alpha(2)$ and $\beta(1)\beta(2)$ are viable for K. This produces the factor of two with J seen in the Hartree–Fock equation (Eq. 14.23).

2. *K is non-zero only for $\alpha(1)\alpha(2)$ and $\beta(1)\beta(2)$—that is, only for parallel electron spins.*
   Now we can understand the origin of **Hund's rule** (see Chapter 1). A wavefunction with parallel spins will be more stable by a value related to the magnitude of their K integral. This will become important in our discussion of high-spin molecules (see below).

We can see now that exchange integrals are in a sense correction terms. They correct the wavefunction for the fact that electrons of the same spin do not move independently. The Pauli principle prevents two electrons of the same spin from occupying the same region of space.

Given all this, let's remember the goals of this analysis. One goal is to derive the molecular orbitals and their energies (ε_i's). To do this we turn to a method called SCF theory, discussed directly below. Once we have these orbitals and energies, we can use them to determine the total energy of the molecule. To calculate the total energy of the molecule, it might seem sensible to simply add together the energies of the orbitals that are populated with electrons to get a total energy, including a factor of two because each orbital has two electrons in a closed-shell molecule. However, when using the Hartree–Fock equation, the total energy is *not* just the sum of the electron orbital energies, per Eq. 14.31. Rather, the total energy is expressed by Eq. 14.32, because each electron–electron repulsion is counted twice if we simply sum the orbital energies (that is, ij and ji are both counted).

$$E_{\text{tot}} \neq 2\Sigma\,\varepsilon_i \qquad \text{(Eq. 14.31)}$$

$$E_{\text{tot}} = 2\Sigma\,\varepsilon_i - \Sigma\Sigma\,(2J_{ij} - K_{ij}) \qquad \text{(Eq. 14.32)}$$

SCF Theory

In order to find a molecule's MOs, ψ_i, we solve Eq. 14.23 *m* times. However, in order to evaluate the two electron integrals *J* and *K*, we must already know ψ for each and every filled orbital. This is because we must know the orbitals that electrons occupy in order to calculate the repulsion each electron feels from all the other electrons. Therefore, we need to perform an iterative calculation. We guess the ψ's, evaluate all *J* and *K* integrals, solve Eq. 14.23 and obtain the E_{tot} value (Eq. 14.32). We then use new ψ's that are generated by this procedure and repeat the entire process until the calculation converges to the lowest energy, meaning that a new cycle leaves the energy unchanged. The orbitals are then said to be self-consistent with the field generated by the electrons, hence the term **self-consistent field (SCF)** calculations. In an SCF wavefunction, each electron moves in a potential field that is generated by all the other electrons. This field does not respond to the movement of the individual electron. We still have an independent electron model.

Linear Combination of Atomic Orbitals—Molecular Orbitals (LCAO–MO)

We must represent the molecular orbitals in our calculations in a manner such that they can be incrementally changed in the SCF process. A common approach is to represent a molecular orbital as a linear combination of atomic orbitals (Eq. 14.33). Here, the subscript *i* is the same as in the HF equation, where *i* is a, b, c, etc. The subscript *k* stands for all different atomic orbitals that are included on the atoms in the molecule. Hence, every MO can potentially have a contribution from every AO in the molecule.

$$\psi_i = \sum_k c_{ik}\,\phi_k \qquad \text{(Eq. 14.33)}$$

This is called the linear combination of atomic orbitals method for creating molecular orbitals (the **LCAO–MO** method). As long as each AO is represented by a single mathematical function (corresponding to a minimal basis set; see below), the calculations produce the same number of ψ's as ϕ's included. This corresponds to the statement that one gets as many molecular orbitals for a molecule as there are atomic orbitals on the individual atoms. Note that LCAO–MO is just one of many possible ways to computationally develop MOs. It is computationally expedient, and it is consistent with our notion that molecules are built up from combinations of atomic orbitals, a conceptual advantage over other possible ways of building up MOs.

Every molecular orbital ψ_i starts by including every atomic orbital ϕ_k on every atom in the molecule. The SCF iterative process changes the coefficients (c_{ik}) for each ψ_i, creating new ψ_i's until a minimum energy for the molecule is achieved.

The atomic orbitals used in the LCAO–MO procedure are represented by what is known as the **basis set**. Typically, you must distinguish three types of basis sets:

1. *Valence: Only valence orbitals; for example, 2s and 2p on carbon; 1s on H.*

2. *Minimal: One function is used for each orbital; all orbitals up to and including the valence shell are included; that is, 1s, 2s, and 2p on C; 1s on H.*

3. *Extended: Additional functions beyond the minimal basis set are added; typically more than one mathematical function is used to represent each AO, with each individual function receiving its own coefficient in the wavefunction.*

We do not use hybrid orbitals in these calculations. Hence, as we discussed in Chapter 1, MOT does not directly produce the familiar *sp*³, *sp*², and *sp* hybrids. Furthermore, because each molecular orbital can potentially be made up from all the atomic orbitals on every atom in the molecule, "bonds" can be spread out among many different atoms in a molecule. The concept of individual localized bonds between adjacent atoms is gone, although mathematical transformations of the wavefunctions that produce localized bonding orbitals can be developed.

Common Basis Sets—Modeling Atomic Orbitals

Once we have decided to let the molecular orbitals be linear combinations of atomic orbitals, we need to decide the mathematical form we will use for the atomic orbitals. One choice would be to simply use the hydrogenic wavefunctions adapted for other atoms. Such a function is called a **Slater-type orbital** (**STO**; Figure 14.8 **A**). These wavefunctions have radial forms possessing terms such as $r^{n-1} e^{-\zeta r}$ ($\zeta = Z/n$). Although such functions are used in some kinds of semi-empirical calculations (see below), they generally are not used in *ab initio* quantum mechanics. The reason for this is that it is very difficult to evaluate the complex two-electron integrals (J and K) with STOs. To address this issue, other types of basis sets have been developed.

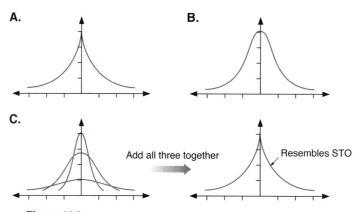

Figure 14.8
A. Shape of STOs. **B.** Shape of GTOs. **C.** Addition of three GTOs resembles an STO.

A popular alternative to STOs is a **Gaussian-type orbital** (**GTO**). These orbitals have a different spatial function, $X^l Y^m Z^n e^{-\zeta r^2}$. The GTOs are not a good match for atomic orbitals (Figure 14.8 **B**), but the integrals that must be evaluated are much easier to compute with GTOs. The computational advantage is so substantial that it is more efficient to represent a single AO as a combination of several GTOs rather than as a single STO.

If we add a few GTOs with differing shapes, the result is a function that resembles an STO (Figure 14.8 **C**). Now the basis set is not just the atomic orbitals, but is instead all the GTOs that are used to make up the atomic orbitals. Hence, the terms that have evolved in the literature to represent a basis set are acronyms for understanding how the STO mimics are created.

Generally, basis sets are developed by optimizing the appropriate coefficients to minimize the energy or to fit experimental data. The latter approach would constitute a violation of our definition of *ab initio* given at the beginning of Section 14.2.1. Nevertheless, there may be some empirical parameters hidden in the basis sets used to describe the atomic orbitals.

A common basis set used in early *ab initio* calculations is **STO-3G**, an STO mimic created by a linear combination of three GTOs (Eq. 14.34). Here the a's are coefficients that simply reflect the extent to which each G is added to create the atomic orbital ϕ. These a's are not changed during the SCF calculation, but rather they define the basis set. The a's are therefore completely different than the c_{ik}'s used in the LCAO–MO method (Eq. 14.33), which are optimized during the SCF method. The a's were optimized to fit experimental data by computational chemist Pople, who won the 1998 Nobel Prize in Chemistry for the development of many of the essential components of modern *ab initio* theory.

$$\phi_i = a_1 G_1 + a_2 G_2 + a_3 G_3 \tag{Eq. 14.34}$$

The STO-3G basis set is a minimal basis set, which lacks the flexibility required for high quality calculations. As computer power grows, STO-3G sees less and less use.

Another common basis set is **4–31G**, an extended basis set. For each core orbital ($1s$ on C; $1s$, $2s$, and $2p$ on Si; . . .), a linear combination of four Gaussians is used instead of the three used for STO-3G. More importantly, each *valence* AO is split into two functions ($\varnothing_i$ and $\varnothing_i'$). The functions $\varnothing_i$ and $\varnothing_i'$ are a linear combination of three Gaussians and a single diffuse Gaussian, respectively. This is thus referred to as a **split valence** basis set, denoted by the symbolism 31G.

$$\varnothing_i = a_1 G_1 + a_2 G_2 + a_3 G_3$$

$$\varnothing_i' = a_4 G_4 \text{—usually a more diffuse function}$$

The two basis functions, $\varnothing_i$ and $\varnothing_i'$ for each AO, are treated independently in the SCF procedure, and each has its own coefficient for each atomic orbital in each MO (Eq. 14.35). This gives greater flexibility to the wavefunction, and this is what makes this basis set "extended" beyond a minimal basis set like STO-3G.

$$\phi_i = c_{i1} \varnothing_i + c_{i2} \varnothing_i' \qquad \text{(Eq. 14.35)}$$

3–21G is a smaller basis set than 4–31G, but it is still split valence. Its performance rivals that of 4–31G, but at less computational cost, and so 3–21G has largely replaced 4–31G in most *ab initio* calculations.

A basis set that sees more extensive use than 3–21G in current *ab initio* calculations is **6–31G***. It is a split valence basis set that uses six Gaussians for the core orbitals and a 31G split valence shell. In addition, the * indicates that a full set of d orbitals is included on all "heavy" atoms (which in this context means all atoms other than H and He). Finally, **6–31G**** is similar to 6–31G*, but includes p orbitals on H as well as d orbitals on heavy atoms.

What is the purpose of the extra functions included in the split valence and further extended basis sets (* and **)? Their role is to increase the flexibility of the basis set and thereby produce more realistic wavefunctions. The split valence approach allows different carbon $2p$ orbitals to have different sizes. Consider, for example, the ethyl cation. The $2p_z$ orbital on the charged carbon should be contracted relative to the $2p_z$ orbital on the methyl carbon because of the larger effective nuclear charge of the former carbon. With a minimal basis set this cannot be done, because all C $2p$ orbitals are the same. With a split valence basis set, however, the SCF calculation can adjust the coefficients of the $\varnothing_i$ and $\varnothing_i'$ in the $2p_z$ valence orbitals differently for the two carbons, effectively giving them different sizes.

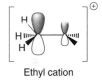

Ethyl cation

The d orbitals added to C and the p orbitals added to H in a basis set such as 6–31G** serve a different purpose. Just as the basis AOs need not be all the same size, they need not all be centered on a particular atom. Returning to the example of the ethyl cation, you can imagine that the $2p_z$ orbital on the methyl carbon might "lean" toward the cationic center, attempting to add some electron density to a very electron deficient site. As shown in the margin, an appropriate mixing of a d orbital with a p can produce a new orbital that still basically looks like a p but is "leaning", or alternatively, is centered off the nucleus. This is called polarization of the orbital, and the d orbitals that are mixed in to allow this are often called **polarization functions**.

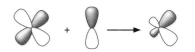

Polarization of orbitals

The value of adding polarization functions is just that; they allow basis AOs to polarize. They should not be thought of as literally C $3d$ orbitals. That is to say, if the quality of a calculation improves substantially by including polarization functions, that does *not* mean that the actual molecule recruits C d orbitals in its bonding. Such orbitals are typically much too high in energy to be meaningfully involved in bonding. The same is true for elements such as S and Si, and discussions that invoke d orbitals in the bonding of sulfoxides or sulfones, for example, are controversial at best, and just plain wrong in the opinion of many. The polarization functions are simply mathematical tools that allow you to give the basis set more flexibility, and thus produce a better calculation.

Extension Beyond HF—Correlation Energy

When we first introduced the "orbital approximation" and noted that, at least for electrons of opposite spin, it is an independent electron theory, we mentioned that the electrons are therefore not correlated. Hence, in the HF–SCF method each electron moves in a general field created by all the other electrons. In reality, however, the motions of electrons are pairwise correlated to keep the electrons apart; when one electron "moves left", a nearby electron will move out of its way, because like charges repel. This correlation of electron motions is a stabilizing effect, and thus E_{SCF} is always too high, even at the so-called **Hartree–Fock limit** (the result you would get with a basis set of infinite size). The error associated with the Hartree–Fock limit is called the correlation energy (Eq. 14.36).

$$E_{correlation} = E_{reality} - E_{SCF} \qquad \text{(Eq. 14.36)}$$

This equation ignores relativistic effects, which are very small for typical organic molecules, but can be significant for heavier elements.

An estimate of the correlation energy can be obtained by considering the total wave function, not as a single Slater determinant (configuration), but as a linear combination of Slater determinants (Eq. 14.37), each of which represents a different electronic configuration of the molecule. Typically, Ψ_1 is the lowest energy electronic configuration, often referred to as the reference configuration, while the other Ψ's represent excited configurations. In essence, small fractions of excited states are mixed into the ground state to create Ψ_{CI}, which approximates a correlated wavefunction. The c_i's are incrementally changed until Ψ_{CI} has the lowest energy. This is termed **configuration interaction** (CI). It is computationally quite expensive, but the results can be quite good.

$$\Psi_{CI} = c_1 \Psi_1 + c_2 \Psi_2 + c_3 \Psi_3 \dots \qquad \text{(Eq. 14.37)}$$

Let's consider why mixing an excited configuration into the ground state would approximate a correlated system. Recall the idea that a Slater determinant represents only one electronic configuration. With ethylene, the ground state has two electrons in the same π bonding orbital, but the excited state has the electrons in two different orbitals. To the extent that the π^* orbital has a different shape than the π orbital, this helps the two electrons avoid each other. Since correlation arises from the tendency of the electrons to avoid each other, the admixture of a small fraction of this excited state into the ground state can be stabilizing. Configuration interaction calculations offer a significant improvement over Hartree–Fock for determining total energies.

Except for small molecules and small basis sets, complete CI (allowing all possible excited configurations) is prohibitively expensive. Usually the size of a CI calculation is restricted by limiting the number of excited configurations that are involved. For example, **CISD** (configuration interaction, singles and doubles), a commonly used method, allows only configurations that can be derived from the reference configuration by single or double excitation to contribute to the final wavefunction.

Alternatively, there are perturbation methods to estimate $E_{correlation}$. Briefly, in these methods, you take the HF wavefunction and add a correction—a perturbation—that better mimics a multi-body problem. Møller–Plesset theory is a common perturbative approach. It is called **MP2** when perturbations up to second order are considered, **MP3** for third order, **MP4**, etc. MP2 calculations are commonly used. Like CISD, MP2 allows single and double excitations, but the effects of their inclusion are evaluated using second-order perturbation theory rather than variationally as in CISD. An even more accurate type of perturbation theory is called **coupled-cluster theory. CCSD** (coupled-cluster theory, singles and doubles) includes single and double excitations, but their effects are evaluated at a much higher level of perturbation theory than in an MP2 calculation.

Note that correlation effects present a more physically sensible way to think about the van der Waals or dispersion interactions that we discussed in Chapter 3. That is, the induced dipole–induced dipole interaction really implies that electrons on one molecule are correlating their motions with electrons on another molecule. Since these weak interactions are essentially a result of correlation effects, simple MO theory (HF theory) is unable to address them. Indeed, we find that conventional MO theory is unable to model van der Waals complexes and other weak non-covalent interactions that do not have a strong electrostatic component. However, van der Waals forces are included in calculations at the CISD, MP2, or CCSD levels of theory.

Solvation

Thus far, all the calculations we have been discussing are performed on isolated molecules—that is, molecules in the gas phase. While this is appropriate for evaluating fundamental molecular properties such as ionization potentials and orbital patterns, most reaction chemistry occurs in solution. In Chapter 3 we noted that computational approaches to solvation follow one of two paths: explicit or implicit solvation models. While this is technically true of quantum mechanical approaches to solvation, explicit solvation models are severely limited. It is simply not possible to evaluate millions of configurations for a solute plus 267 water molecules using *ab initio* methods! Occasionally, you might add a small number (such as one to ten) of solvent molecules (typically water) to the gas phase calculation to get some feeling for the impact of solvation, but this can only provide a hint of the true impact of solvent. As such, for the forseeable future, quantum mechanical solvation models will primarily involve implicit solvent. We also noted in Chapter 3 that the most promising of these is a hybrid method, in which empirically derived parameters related to the solvent accessible surface area (SASA) are combined with electrostatic terms derived from an accurate quantum mechanical wavefunction. We can anticipate continued development in this area, making the inclusion of solvation effects in *ab initio* calculations more and more common.

General Considerations

We note here a few general issues associated with *ab initio* MO theory. Typically the error in E_{tot} for various *ab initio* methods is $\ll 1\%$, a very impressive feat. However, E_{tot} is a very large number, and so chemically significant errors can remain.

Presently, several program packages are available for performing *ab initio* calculations. Examples include SPARTAN and the GAUSSIAN package developed by Pople and co-workers, the current version being denoted by a year, such as GAUSSIAN03. These user friendly systems include many basis sets, such as STO-3G, 3–21G, 6–31G**, etc.; full geometry optimization; and post-HF methods, such as CI, MP2, MP4, etc. Often geometry optimization is done at a lower level, then single point energies are produced at a higher level. This is symbolized as 6–31G**//6–31G*–MP2, meaning the geometry optimization was done using HF theory (the default) with the 6–31G** basis set, then the energy was evaluated by doing a single calculation at the optimized geometry using MP2 theory and the 6–31G* basis set.

With current high-level computer workstations, geometry optimization is feasible for molecules with 20–30 heavy atoms and an extended basis set. Similar constraints apply to very high-level single-point energies. With supercomputers, bigger systems can be considered, and computers keep getting faster. However, it must be appreciated that for an HF *ab initio* calculation, the computer time goes up as n^4 where n is the number of basis functions. So, to double the size of the molecule, you need a factor of 2^4 or 16 in computing power. Thus, it will be some time before we are doing *ab initio* geometry optimizations on proteins!

Summary

What takes place in an *ab initio* calculation is pretty complicated. However, because all the mathematics is done by a computer, *ab initio* calculations are not difficult to perform. So, let's briefly review what is done in *ab initio* methods. This approach is generally called the **Roothaan procedure**, or Roothaan's formulation of the Hartree–Fock method (minus step 7, below).

1. *Total wavefunctions for molecules are written as products of one-electron wavefunctions, termed orbitals, and the total wavefunction is antisymmetrized so that it obeys the Pauli principle.*

2. *Using an antisymmetrized wavefunction, the Hartree–Fock equations are solved self-consistently, giving an optimal set of MOs for the molecule at the chosen geometry.*

3. *When solving the Hartree–Fock equations, a series of integrals over all space must be evaluated.*
   The one-electron integrals give the kinetic energy of an electron and the nuclear attraction it feels in each MO. The Coulomb integrals give the classical electron–electron repulsion between an electron in a given one-electron orbital and an average field created by all the other electrons. The exchange integrals are non-classical terms that serve as corrections to the Coulomb integrals. They result from the antisymmetrization of the wavefunction, which embodies the Pauli principle. Antisymmetrization keeps electrons of the same spin from occupying the same region of space, and thus from occupying the same AO.

4. *Typically a molecular orbital is treated as a linear combination of all the atomic orbitals on every atom in the molecule (LCAO–MO).*

5. *To evaluate the integrals required in the Hartree–Fock equation you need to know the molecular orbitals.*
   Finding the MOs thus requires an iterative process known as an SCF calculation, where the coefficients in the LCAO–MO are optimized to lower the total energy of the molecule.

6. *To facilitate integral evaluation, the atomic orbitals themselves are usually expressed as a linear combination of Gaussian functions.*
   The number and type of Gaussians in the basis set play an important role in the accuracy of the calculation.

7. *For energies that are quantitatively reliable, electron correlation must be included, either variationally (for example, CISD) or via perturbation theory (for example, MP2 or CCSD).*

Remember, *ab initio* means "derived from first principles", *not* "without approximation". All the approximations noted above (Born–Oppenheimer, orbital, SCF, LCAO–MO, finite basis set, and GTOs) are in place. However, to implement the calculations, we do draw upon rigorously sound quantum mechanical principles (Hamiltonians, the Pauli principle, and spin). Coupling these principles with the approximations, especially with large basis sets and inclusion of correlation effects, can produce very impressive results.

We now turn our attention to a method of deriving molecular orbitals that does not require evaluation of two-electron integrals. This is the method used with the approximate techniques known as extended Hückel and Hückel theory. It is also used in perturbational molecular orbital theory, which we present later in this chapter. This method takes advantage of secular determinants, a way in which to represent the Schrödinger equation as a diagonalizable matrix.

Connections

Methane—Molecular Orbitals or Discrete Single Bonds with sp^3 Hybrids?

Chemists typically view methane as containing four equivalent C–H σ bonds, each resulting from the union of an sp^3 hybrid at carbon and a hydrogen $1s$ orbital. The molecular orbitals obtained using Hartree–Fock theory describe a different picture. Which view is "correct"? Both models are valid and each has strengths and weaknesses. However, only the MO model just described can explain the experimentally observed ionization potentials of methane measured by **photoelectron spectroscopy**. Photoelectron spectroscopy uses x-ray and γ-ray irradiation to eject electrons from a molecule, and measures the energies of the liberated electrons, producing measured ionization potentials. Methane shows two valence ionization potentials, at 14.2 and 22.9 eV, not one as the sp^3 model would predict. Using Koopman's theorem (see the very next Going Deeper highlight), we can assign these ionizations based on our knowledge of the energies of atomic orbitals. The $2s$ AO of C (IP = 21.4 eV) is much lower that the $2p$ AO (IP = 11.4 eV), and this energy gap carries over to the MOs built from these AOs ($2a_1$ and $1t_2$ in graph, respectively).

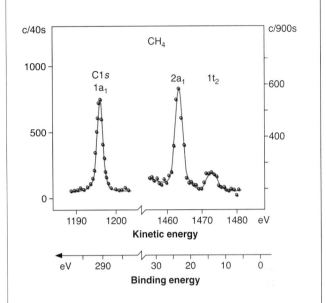

Hamrin, K., Johansson, G., Gelius, U., Fahlman, A., Nordling, C., and Siegbahn, K. "Ionization Energies in Methane and Ethane by Means of ESCA (Electron Spectroscopy for Chemical Analysis)." *Chem. Phys. Lett.*, **1**, 613 (1968).

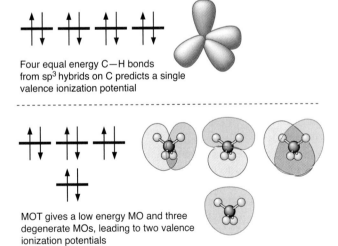

Four equal energy C—H bonds from sp³ hybrids on C predicts a single valence ionization potential

MOT gives a low energy MO and three degenerate MOs, leading to two valence ionization potentials

Going Deeper

Koopmans' Theorem—A Connection Between *Ab Initio* Calculations and Experiment

How can we judge whether the results of *ab initio* calculations are "correct"? The ultimate test of a theoretical model must always be a comparison with experiment. One direct link that can be made between the results of a calculation and an experimental measurement is a comparison of the calculated orbital energies with the measured ionization potentials (IP) of a molecule.

In our discussions of methane above and water below, we use patterns in IPs to support the results of an *ab initio* calculation. In fact, the connection can be made more quantitatively. **Koopmans' theorem** states that the valence state IPs of a molecule can be directly related to the MO energies obtained from a HF calculation on the system:

$$IP_i = -\varepsilon_i$$

Although this agrees with intuition, it actually works because of a fortuitous cancellation of errors. Recall that IP is the energy required to remove an electron from a molecule. In computational terms, it is the difference between the energy of the molecule, M, and the energy of the radical ion produced on ionization, $M^{+\bullet}$. In relating the IP to an orbital (one electron) energy, we are ignoring the fact that the electronic structure of $M^{+\bullet}$ rearranges after ion-

ization, while Koopmans' theorem uses the same orbitals for both states. Thus, the true energy of $M^{+\bullet}$ is lower than our model suggests, making $IP_i < \varepsilon_i$. The second error is the neglect of correlation energy. The energies of both M and $M^{+\bullet}$ will be overestimated in the HF calculation, but the error should be greater for M, because there is one more electron to correlate than there is for $M^{+\bullet}$. Thus, this error makes $IP_i > \varepsilon_i$. In practice, the two errors tend to cancel.

Koopmans' theorem is routinely invoked to relate IP measurements to computational results. The development of photoelectron spectroscopy as a tool to measure the first few IPs of molecules greatly increased the opportunity for such correlations. Typically, Koopmans' theorem will not get the IPs exactly correct, but a simple correlation between IPs and ε_i will be linear and quite frequently reliable.

Just as an aside, it is interesting to note that Koopmans won the Nobel Prize in 1975, not for chemistry, but for economics for a "theory of optimal allocation of resources". He used the same kind of procedures that minimize the solutions of multiple simultaneous functions in atomic and molecular electronic structure theory to analyze functions that lead to a minimization of costs in economics. You just never know where insights from chemical analyses will be applicable!

14.2.2 Secular Determinants—A Bridge Between *Ab Initio*, Semi-Empirical/Approximate, and Perturbational Molecular Orbital Theory Methods

The HF–SCF methods that have been discussed to this point are very intensive computationally, primarily because of the huge number of integrals that are required to treat electron–electron repulsion. However, there is an alternative approach to the problem of calculating molecular orbitals. We can set up the problem in terms of an undefined Hamiltonian, *H*. We can then solve for MOs and their energies in terms of *H*. While exact energy values are not obtained, trends and patterns are evident that provide very useful insights into bonding. In addition, we can substitute in empirical parameters for *H* that provide useful semiquantitative results.

The key concept here is the **secular determinant**. Secular determinants are central to Hückel theory. Hückel theory is the simplest electronic structure theory tool that organic chemists use, and we will use it later to explore many different facets of organic chemistry. Moreover, secular determinants are central to perturbational molecular orbital theory, which we also cover later in this chapter. In fact, what we are showing here can be described as first-order perturbational molecular orbital theory.

To see what a secular determinant is and how it arises, we present the mathematics of the simplest orbital mixing problem, the mixing of two orbitals of equal energy. We can visualize these two orbitals any way we like, because the forms of the resulting MOs are going to be very general. For example, they could be the two 1s orbitals for the formation of H_2, or two sp^3 hybrids for the formation of the central σ bond in ethane, or two *p* orbitals for the formation of the π bond in ethylene. We will see that the concepts introduced previously concerning the *g* and *u* states of H_2^+ in Section 14.1.2, as well as the general shapes of molecular orbital diagrams that organic chemists routinely draw, follow naturally from this analysis.

The "Two-Orbital Mixing Problem"

Let's mix two atomic orbitals on different atoms to form an MO, as in Eq. 14.38. This is the LCAO method of forming molecular orbitals. We let c_A and c_B represent coefficients that tell us the proper proportion of the starting orbitals (ϕ_A and ϕ_B) that should be present in the resulting MO. The ϕ's are known quantities (possibly from the solution of the hydrogen atom, or are basis functions). We do not know what the c's should be (they are the unknowns). We will solve for the c's that create the lowest energy wavefunctions, just as with HF–SCF methods.

$$\psi = c_A\phi_A + c_B\phi_B \qquad \text{(Eq. 14.38)}$$

Besides finding the c's that give the lowest energy wavefunctions, we also would like to know the energy of any new wavefunctions. In fact, solving for the energy is our starting point. We use Eq. 14.11, with Eq. 14.38 as the wavefunction, thus giving Eq. 14.39.

$$E = \frac{<c_A\phi_A + c_B\phi_B|H|c_A\phi_A + c_B\phi_B>}{<c_A\phi_A + c_B\phi_B|c_A\phi_A + c_B\phi_B>} \qquad \text{(Eq. 14.39)}$$

After multiplying out the numerator and denominator of Eq. 14.39, the following substitutions are made as a means of further simplifying our notation:

Let $<\phi_A|H|\phi_A> = H_{AA}$,

$<\phi_B|H|\phi_B> = H_{BB}$,

$<\phi_A|H|\phi_B> = <\phi_B|H|\phi_A> = H_{BA} = H_{AB}$,

$<\phi_A|\phi_A> = <\phi_B|\phi_B> = 1$ because the atomic orbitals are normalized, and

$<\phi_A|\phi_B> = <\phi_B|\phi_A> = S$

The terms represented by these integrals have physical meaning. The H_{AA} and H_{BB} terms are called the **core integrals**. They are simply the energy of an electron in the starting atomic orbitals, ϕ_A and ϕ_B, respectively, including both the kinetic and potential energy. They are not the same as the Hartree–Fock core integrals. H_{AA} equals H_{BB} when the atomic orbitals A and B (or basis sets) are the same, but we will not make this substitution until later. The H_{AB} and H_{BA} terms are called the **resonance integrals**. These integrals are stabilizing, and reflect the energy gained by the fact that the electrons can be in the regions where the orbitals overlap. Recall from Section 14.1.5 that this is particularly favorable, because in this region the electron has lower kinetic energy. Also, recall that ϕ_A and ϕ_B are not orthogonal because they are on different atoms, and therefore the S term is called the **overlap integral**. It measures the extent to which the ϕ's overlap in space. It is typically given subscripts also, S_{ij}, but we drop these here because only two orbitals are overlapping. Using all these substitutions gives Eq. 14.40, where we let the numerator be referred to as N and the denominator as D.

$$E = \frac{c_A{}^2H_{AA} + c_B{}^2H_{BB} + 2c_Ac_BH_{AB}}{c_A{}^2 + c_B{}^2 + 2c_Ac_BS} = \frac{N}{D} \qquad \text{(Eq. 14.40)}$$

We are now ready to solve for the c's that give the lowest energy ψ's. We differentiate Eq. 14.40 with respect to the c's, setting the result equal to zero, as a means of finding the minimum energy for the wavefunction. This is our best possible wavefunction according to the variational theorem. However, Eq. 14.40 has two unknown c's, and hence the "quotient rule" from calculus must be used (Eq. 14.41). Eq. 14.41 can only be satisfied if the partial derivatives are both equal to zero ($\partial E/\partial c_A$ and $\partial E/\partial c_B = 0$). These derivatives are of the quotient N/D from Eq. 14.41 and can be written as the set given in Eq. 14.42. These two equations can only be satisfied when the numerators are equal to zero. Therefore the set in Eq. 14.43 results. Lastly, rearranging Eqs. 14.43 **A** and 14.43 **B** results in the set of Eqs. 14.44, called the **secular equations**.

$$dE = \left(\frac{\partial E}{\partial c_A}\right) dc_A + \left(\frac{\partial E}{\partial c_B}\right) dc_B = 0 \qquad \text{(Eq. 14.41)}$$

A. $\quad \dfrac{\partial E}{\partial c_A} = \dfrac{\dfrac{\partial N}{\partial c_A} - E \dfrac{\partial D}{\partial c_A}}{D} = 0 \qquad$ **B.** $\quad \dfrac{\partial E}{\partial c_B} = \dfrac{\dfrac{\partial N}{\partial c_B} - E \dfrac{\partial D}{\partial c_B}}{D} = 0$

(Eq. 14.42)

A. $\quad \dfrac{\partial N}{\partial c_A} - E\left(\dfrac{\partial D}{\partial c_A}\right) = 2c_A H_{AA} + 2c_B H_{AB} - E(2c_A + 2c_B S) = 0$

(Eq. 14.43)

B. $\quad \dfrac{\partial N}{\partial c_B} - E\left(\dfrac{\partial D}{\partial c_B}\right) = 2c_B H_{BB} + 2c_A H_{AB} - E(2c_B + 2c_A S) = 0$

A. $\quad c_A(H_{AA} - E) + c_B(H_{AB} - SE) = 0 \qquad$ **B.** $\quad c_A(H_{AB} - SE) + c_B(H_{BB} - E) = 0$

(Eq. 14.44)

The secular equations play a pivotal role in this analysis. They allow us to solve for E, c_A, and c_B. We will learn how to write the secular equations simply by looking at the molecule that we want to analyze, and without going through all the steps that brought us to this point.

When mixing two orbitals, the secular equations are two equations in three unknowns: c_A, c_B and E. The resonance, core, and overlap integrals are all numbers that we could calculate by identifying the Hamiltonian and solving the respective integrals, as is done in HF theory. However, as mentioned above, for qualitative results, we can simply carry them through as parameters and solve for c_A, c_B, and E using linear algebra, in particular by calling on Kramer's rule. This rule states that the determinant of the coefficients of the unknowns (the c's) must be equal to zero. This is a bit confusing, because we call the c's the coefficients for the orbitals, but now "coefficient" refers to the terms in the parentheses multiplied by the c's. Kramer's rule leads to the determinant shown below, called the **secular determinant**, which gives the polynomial in E listed in Eq. 14.45. To solve for the roots, we use the substitution specific for degenerate identical orbitals that $H_{AA} = H_{BB}$. The roots of this polynomial (Eqs. 14.46 **A** and 14.46 **B**) are the energies of the ψ's resulting from our initial linear combination (Eq. 14.38). Hence, the roots are the energies of individual molecular orbitals, and as such, we change our notation (E to ε) to remain consistent with the rest of the chapter. Just as with the secular equations, one of our ultimate goals is to write the secular determinant for any system simply by inspection, rather than by deriving all the math we have given here.

$$\begin{vmatrix} H_{AA} - E & H_{AB} - SE \\ H_{AB} - SE & H_{BB} - E \end{vmatrix} = 0$$

The secular determinant
for two-orbital mixing

$$(H_{AA} - E)(H_{BB} - E) - (H_{AB} - SE)^2 = 0 \qquad \text{(Eq. 14.45)}$$

A. $\quad \varepsilon_1 = \dfrac{H_{AA} + H_{AB}}{1 + S} \qquad$ **B.** $\quad \varepsilon_2 = \dfrac{H_{AA} - H_{AB}}{1 - S} \qquad$ (Eq. 14.46)

Let's examine ε_1 and ε_2. First of all, because H_{AB} is a stabilizing term, it is a negative number. Second, since there is some overlap between ϕ_A and ϕ_B when the bond is formed, S is a positive number. Therefore ε_1 (bonding) is lower in energy than ε_2 (antibonding). Third, since H_{AA} is the energy of the starting orbitals, ε_1 is lower than H_{AA} by the magnitude of the resonance integral (H_{AB}) but attenuated by 1 plus the amount of overlap (S) in the denominator. Similarly, ε_2 is higher in energy than H_{AA} by an amount proportional to the exchange integral, but the energy increase is accentuated by $1 - S$ in the denominator. Therefore, the antibonding MO is destabilized more than the bonding MO is stabilized. In other words, *antibonding orbitals are more antibonding than bonding orbitals are bonding*. This is one of the most important lessons of the analysis. All this leads to the typical molecular orbital diagram, shown in Figure 14.9 and referred to earlier when we discussed perturbational molecular orbital theory in Chapter 1 (see Figure 1.11 also).

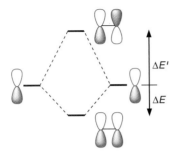

Figure 14.9
A typical MO diagram for mixing two degenerate orbitals, in this case p orbitals.

Lastly, the coefficients (c's) for the molecular orbitals need to be derived. To accomplish this task, ε_1 and ε_2 are separately substituted into the secular equations. For both ε_1 and ε_2, this results in two equations in two unknowns, and relative relationships between the c's can be found. For ε_1, we find that $c_A = c_B$, and for ε_2 we find that $c_A = -c_B$. Therefore, we find two molecular orbitals, ψ_1 and ψ_2, corresponding to energies ε_1 and ε_2, respectively. Note that the lower energy orbital is the positive mixture of the starting atomic orbitals (bonding), whereas the higher energy orbital is the negative mixture (antibonding). If the starting orbitals were $1s$ orbitals on H, then Eq. 14.47 would correspond to Figure 14.6 **B** for H_2^+, whereas Eq. 14.48 would correspond to Figure 14.6 **C** for H_2^+. The notion of adding and subtracting atomic orbitals to achieve the bonding and antibonding molecular orbitals is exactly the lesson taught in introductory organic chemistry, but now we see that it is a simple ramification of arbitrarily mixing two orbitals and solving the linear algebra associated with the Schrödinger equation.

$$\psi_1 = c_A \phi_A + c_A \phi_B \tag{Eq. 14.47}$$

$$\psi_2 = c_A \phi_A - c_A \phi_B \tag{Eq. 14.48}$$

We are still not quite done. The orbitals given in Eqs. 14.47 and 14.48 are not normalized, and we do not have an exact value for c_A. To solve for c_A we need to set the brackets $< \psi_1 | \psi_1 >$ and $< \psi_2 | \psi_2 >$ both to equal 1. We substitute Eqs. 14.47 and 14.48 into these brackets, multiply out the terms, and find that c_A equals the square root of $1/(2 + 2S)$ for ψ_1 and $1/(2 - 2S)$ for ψ_2. Hence, the final form of the molecular orbitals are as given in Eqs. 14.49 and 14.50.

$$\psi_1 = \sqrt{\frac{1}{2 + 2S}}\, \phi_A + \sqrt{\frac{1}{2 + 2S}}\, \phi_B \tag{Eq. 14.49}$$

$$\psi_2 = \sqrt{\frac{1}{2 - 2S}}\, \phi_A - \sqrt{\frac{1}{2 - 2S}}\, \phi_B \tag{Eq. 14.50}$$

Note that we have "solved" the Schrödinger equation and produced valuable insights into bonding without ever actually evaluating H_{AA}, H_{AB}, or S! To get quantitative data these terms could be evaluated, but useful qualitative insights come from just considering the form of the mathematics involved. The method given here for two-orbital mixing can be generalized to larger molecules, and the secular equations can be written by inspection of the molecule of interest. We do this just below and later in the context of Hückel theory. However, the next Going Deeper highlight shows that matrix algebra can be used to set up and solve the Schrödinger equation. This approach can also be generalized to larger molecules.

Going Deeper

A Matrix Approach to Setting Up the LCAO Method

There is another way to solve the two-orbital mixing problem, which does not involve all the calculus shown above. It is a general approach that results from the variational theorem, and can be applied to *ab initio*, semi-empirical, or Hückel theory.

Eqs. 14.44 **A** and 14.44 **B** can be rewritten as follows:

$$c_A H_{AA} + c_B H_{AB} = E(c_A + c_B S)$$

$$c_A H_{AB} + c_B H_{BB} = E(c_A S + c_B)$$

These equations can be expressed as the product of the matrices shown below. The matrix of H_{ii} and H_{ij} values is called the **Hamiltonian matrix**. Although the matrix elements H_{ii}, H_{jj}, H_{ij}, S_{ij}, etc., will be different depending on whether we do a Hückel, semi-empirical, or *ab initio* calculation, the Hamiltonian matrix and other matrices will have this generalizable form. This series of matrices is simply another way of writing $H\psi = E\psi$ when the ψ's are formed by the LCAO method.

$$\begin{vmatrix} H_{AA} & H_{AB} \\ H_{AB} & H_{BB} \end{vmatrix} \begin{vmatrix} c_A \\ c_B \end{vmatrix} = \begin{vmatrix} E & 0 \\ 0 & E \end{vmatrix} \begin{vmatrix} 1 & S \\ S & 1 \end{vmatrix} \begin{vmatrix} c_A \\ c_B \end{vmatrix}$$

This series of matrices tells us what happens when the Hamiltonian matrix is multiplied by a matrix of coefficients from the LCAO–MO wavefunction. The Hamiltonian matrix is transformed into a matrix whose diagonal elements are the orbital energies and whose off-diagonal elements are all zero, times a matrix that represents the overlap between the atomic orbitals—namely, S_{ij} values, where $S_{ii} = 1$ (here $S_{AB} = S_{BA}$ and is simply called S). The matrix of LCAO coefficients is thus said to **diagonalize** the Hamiltonian matrix. Moreover, the elements of the Hamiltonian matrix always have the same physical meaning. The diagonal terms are the energies of the electrons in the individual atomic orbitals, while the off-diagonal elements represent the energies due to overlap. When two atomic orbitals do not overlap because they are orthogonal, they are of different symmetries, or are on atoms far from each other, both the H_{ij} and S_{ij} terms can be ignored when qualitative results are desired.

Linear algebra on this series of matrices will lead to the following:

$$\begin{vmatrix} H_{AA} - E & H_{AB} - SE \\ H_{AB} - SE & H_{BB} - E \end{vmatrix} \begin{vmatrix} c_A \\ c_B \end{vmatrix} = 0$$

In order for this product of matrices not to have the trivial solution (that $c_A = c_B = 0$), the first matrix must equal zero. This directly gives the secular determinant discussed above, and the logic from there flows as we have already discussed.

Writing the Secular Equations and Determinant for Any Molecule

In the two-orbital mixing problem, we showed that when molecular orbitals are defined as linear combinations of atomic orbitals and are put into the Schrödinger equation, followed by differentiation to minimize E, a series of simultaneous equations in the c's and E results. When mixing two orbitals, only two energies result along with two molecular orbitals. It is not much of a stretch to realize that there will be as many molecular orbitals with distinct energies as the number of atomic orbitals (or basis functions) we use to create the molecular orbitals. Moreover, there will be the same number of secular equations as the number of starting atomic orbitals or basis functions (n), and hence the secular determinant will be n by n.

It is actually easier to write the secular determinant first, and then create the secular equations (although we saw in the "Two-Orbital Mixing Problem" that they are derived in the opposite order). All secular determinants have the following form:

$$\begin{vmatrix} H_{AA} - E & H_{AB} - S_{AB}E & H_{AC} - S_{AC}E & ------ & H_{An} - S_{An}E \\ H_{BA} - S_{BA}E & H_{BB} - E & H_{BC} - S_{BC}E & ------ & H_{Bn} - S_{Bn}E \\ H_{CA} - S_{CA}E & H_{CB} - S_{CB}E & H_{CC} - E & ------ & H_{Cn} - S_{Cn}E \\ ------ & ------ & ------ & ------ & ------ \\ H_{nA} - S_{nA}E & H_{nB} - S_{nB}E & H_{nC} - S_{nC}E & ------ & H_{nn} - E \end{vmatrix}$$

The diagonal elements represent the energies of the atomic orbitals in the molecule minus E. The off-diagonal elements consist of the resonance integrals between the various atomic orbitals minus their respective overlaps times E.

The secular equations can be written by inspection of the secular determinant. Each row in the determinant corresponds to a separate secular equation. To generate all the secular equations, we take the first term in each row and multiply it by c_1, the second term is multiplied by c_2, the third term by c_3, and so on. Equation 14.51 shows a generalized form of the equations that result from the first row of the secular determinant. The result is n equations in n c_{ik}'s and a single ε_i.

$$c_{i1}(H_{AA} - \varepsilon_i) + c_{i2}(H_{AB} - S_{AB}\varepsilon_i) + c_{i3}(H_{AC} - S_{AC}\varepsilon_i) + \dots c_{in}(H_{An} - S_{An}\varepsilon_i) = 0$$

(Eq. 14.51)

To calculate the wavefunctions using this method, we have to evaluate each H_{nn}, H_{nm}, and S_{nm} integral (or carry them through as parameters), substitute them into the secular determinant, and solve for the roots (ε_i's) of the resulting polynomial. Just as with the mixing of two orbitals, each root ε_i is substituted back into the secular equations, and the series of simultaneous equations in the c_{ik}'s are solved to find the relative relationships between the c_{ik}'s for each ε_i. Exact values for the c_{ik}'s are obtained after normalizing. This whole procedure is probably best understood by looking at some examples, and we will do so by setting up specific secular equations and determinants in our analysis of Hückel molecular orbital theory below.

Now that we have examined *ab initio* methods and secular determinants, let's turn our attention to semi-empirical and approximate methods. Here, many of the integrals that are evaluated in the *ab initio* approach are parameterized, or in some cases simply ignored.

14.2.3   Semi-Empirical and Approximate Methods

Neglect of Differential Overlap (NDO) Methods

The computationally expensive part of an SCF–HF calculation is the evaluation of the two-electron J and K integrals. Therefore, approximate methods tend to emphasize ways to shorten the time it takes to evaluate these integrals. To understand these simplifications, let's examine a J integral such as Eq. 14.26, but now include the fact that our MOs are modeled as linear combinations of atomic orbitals, giving integrals with the form of Eq. 14.52.

$$J_{ij} = \iint \sum_k c_{ik}\phi_k(1)\sum_k c_{jk}\phi_k(2)r_{1,2}^{-1}\sum_k c_{ik}\phi_k(1)\sum_k c_{jk}\phi_k(2)\,d\tau d\tau \qquad \text{(Eq. 14.52)}$$

If we write out the summations in this integral, and then multiply out the terms, we will generate a sum of many integrals, such as $<c_{ia}\phi_a(1)c_{ib}\phi_b(2)|1/r_{12}|c_{id}\phi_d(1)c_{ie}\phi_e(2)>$, where the subscripts a, b, d, and e stand for particular atomic orbitals on potentially different atoms in the molecule. There are as many of these subscripts as the number of atomic orbitals in the basis set, and so there are a great number of such integrals. The same situation holds for the K integrals. We abbreviate the notation for these J and K integrals as follows: $<\phi_a\phi_b|r_{1,2}^{-1}|\phi_d\phi_e>$.

$\phi_a\phi_b d\tau$ is called **differential overlap**. $<\phi_a\phi_b|r_{1,2}^{-1}|\phi_d\phi_e>$ is therefore referred to as a **differential overlap integral**. For neglect of differential overlap (NDO) methods, these integrals are assumed to be zero unless a = b and d = e. This *greatly* reduces the number of integrals. The remaining integrals are not directly evaluated, but instead are *parameterized*—that is, fit to experimental data.

Generally, such methods incorporate the following features from *ab initio* methods: LCAO–MO, SCF, single Slater determinant, STO, and a valence basis set. The reason we can use STOs in this method instead of GTOs, is that the difficult two-electron integrals are either neglected or parameterized. In addition, overlap is neglected when normalizing the orbitals. This latter point is a subtle issue to which we will return below.

i. CNDO, INDO, PNDO (C = Complete, I = Intermediate, P = Partial)

These methods are parameterized to reproduce *ab initio* results. As such, they are **approximate molecular orbital** theories—they approximate a well-defined theoretical model. They were very important in the early development of computational MO theory, but they are not extensively used now, except for certain specialized versions for particular applications, such as CNDO/S for spectroscopy.

ii. The Semi-Empirical Methods: MNDO, AM1, and PM3

An alternative strategy was to develop methods wherein the two-electron integrals are parameterized to reproduce *experimental* heats of formation. As such, these are **semi-empirical molecular orbital** methods—they make use of experimental data. Beginning first with modified INDO (MINDO/1, MINDO/2, and MINDO/3, early methods that are now little used), the methodological development moved on to modified neglect of diatomic differential overlap (MNDO). A second MNDO parameterization was created by Dewar and termed Austin method 1 (AM1), and finally, an "optimized" parametrization termed PM3 (for MNDO, parametric method 3) was formulated. These methods include very efficient and fairly accurate geometry optimization. The results they produce are in many respects comparable to low-level *ab initio* calculations (such as HF and STO-3G), but the calculations are *much* less expensive.

It is an area of ongoing debate as to the merits of MNDO/AM1/PM3 vs. *ab initio* methods. Proponents of semi-empirical methods claim that including experimental results in the parameterization increases accuracy, even to the point of including some correlation energy, even though a single Slater determinant calculation is made. Others dispute this view. As with all parameterized methods, the parameterizations are based on a set of standard structures, in a way similar to the manner in which a molecular mechanics parameterization is established. This means that some caution is required in applying the method to unusual structures not represented in the parameterization. Nevertheless, there will always be systems that are simply too large for *ab initio* theory. In such cases, only the semi-empirical methods will be computationally feasible. When applied to appropriate systems, and considered with a skeptical eye, these methods can be quite useful.

Extended Hückel Theory (EHT)

This method *completely ignores* electron–electron repulsion (term ⑤ in Figure 14.3). As such, there are no J and K integrals to evaluate, and so this is not an SCF method. The Schrödinger equation is just set up as a secular determinant as we have already discussed. The diagonal elements, H_{ii}, are set equal to the valence state ionization energies for the AOs. The off-diagonal element's H_{ij}'s are given by Eq. 14.53, in which the parameter k is a constant that is usually set equal to 1.75.

$$H_{ij} = k \frac{(H_{ii} + H_{jj})}{2} S_{ij} \qquad \text{(Eq. 14.53)}$$

This method is *very* fast, though ignoring electron–electron repulsion is a serious disadvantage. However, unlike NDO methods, *overlap is not neglected* in normalizing and evaluating the wavefunction. Furthermore, since the complex two electron integrals are not present, the method uses STOs as the basis set.

As the name implies, EHT is pretty much Hückel theory but with π and σ electrons both included, whereas Hückel theory only includes π electrons (see below). EHT is quite good for visualizing the shapes of orbitals, and for constructing orbital correlation diagrams and Walsh diagrams. However, the geometries and absolute energies obtained from EHT are unreliable. It may seem like an unacceptably crude method, but it was the primary method used by Hoffmann (Nobel Prize in Chemistry, 1981) in the earliest efforts to visualize molecular orbitals, leading to many of the basic ideas about orbital interactions that are now standard tools for chemists. In addition, the fact that it is not a neglect of overlap method has some advantages, as we will show below.

Hückel Molecular Orbital Theory (HMOT)

Hückel theory is the electronic structure method that most chemists have as the basis of their understanding of molecular orbitals. It is the simplest of all the methods, but can be amazingly insightful. In the present context, Hückel theory can be summarized as follows: only valence π electrons are considered; only nearest neighbor interactions are included; the orbital overlap S is set equal to zero; and electron–electron repulsion is neglected. Basically, HMOT analyzes connectivity (topology) of a π system in a planar molecule.

Despite its monumental approximations, HMOT produces qualitatively correct orbitals and orbital energy patterns. The proof that HMOT has been a very useful theory is the fact that it has inspired a great deal of beautiful experimental physical organic chemistry. As Pauling noted in the introduction to *The Nature of the Chemical Bond*, "The principal contribution of quantum mechanics to chemistry has been the suggestion of new ideas." Thus, even a very crude level of theory can be extremely useful when it is applied in an intelligent way and it is not pushed into areas that are beyond its capabilities. Likewise, very high-level theory can be misused and misinterpreted. Even with all the computer power in the world, you still have to think about what is being calculated!

We will examine the implementation of HMOT in more detail below, learning how the method is used, the kinds of results it gives, and some of the insights obtained, all of which will be of paramount importance in understanding the following three chapters.

14.2.4 Some General Comments on Computational Quantum Mechanics

We have introduced an array of computational methods that spans the range from highly sophisticated to quite simple, and we have spent considerable time analyzing each method. This is because these methods are now routinely available to organic chemists for everyday implementation in their research via "canned programs" such as GAUSSIAN and SPARTAN. Solving for the energies and orbitals of organic compounds allows us to predict reactivity. For example, we can predict reactivity by solving for the molecule's frontier orbitals—those orbitals that act electrophilic and nucleophilic. Solving for the orbitals and energies also allows us to anticipate relative reactivities.

Given the power of these methods, organic chemists should avail themselves of them for a more complete approach to their research interests, and we encourage you to use these whenever they seem appropriate. There are challenges. For example, how do we know what is the right level of the theory to use? One possible answer is to use the highest level that our computational resources will allow. However, even the highest levels are often insufficient and frequently they are unnecessary. How do we decide when and how to use theory?

Knowledge of the underlying theory for a given method—such as that provided here—is a good starting point. We wouldn't use HMO for a nonplanar system, and it's well established that EHT is poor for optimizing bond lengths. When considering a particular application, another important step is to consult the chemical literature to see if a similar system has been addressed, and what levels of theory performed well.

An important, but often neglected strategy, is to perform a "control experiment". For example, suppose we want to calculate the first ionization potential (IP) of a new structure with novel bonding properties. A useful control would be to find a known molecule that is as close as possible in structure to our new molecule and whose IP has been *experimentally* determined. We then determine which level of theory correctly reproduces the experimental IP. The more experimental data we have to test our calculations on the better. The key is to make sure that the experimental data we are reproducing are as close as possible to the quantity we want to calculate. If we can find a level of theory that successfully models a range of relevant experimental data, we can proceed with some confidence to our new system. Often it is true that extremely high levels of theory are unnecessary—a more modest model will suffice. This is especially true if we are looking for *trends* across a series of related molecules, rather than quantitative accuracy. For all levels of theory we are more likely to get trends right than we are to *a priori* predict some effect quantitatively, but often trends are what most interest the experimentalist.

The computational methods are extremely useful, but we are not often armed with a computer when making quick judgments about reactivity at the bench, or when discussing reactions in a group meeting. Under these circumstances, qualitative insights and acquired intuition concerning electronic structure are essential. We will build such intuition in this chapter by presenting the results of Hückel theory in more detail, and by developing perturbational molecular orbital theory as a simple way to analyze complex systems.

14.2.5   An Alternative: Density Functional Theory (DFT)

In recent years, an alternative approach to implementing the Schrödinger equation for quantitative electronic structure calculations has appeared. Instead of calculating wavefunctions of the sort we have described, these methods focus on the **electron density (ρ)** across the entire molecule. It has been shown that if ρ is known precisely, one can in principle determine the total energy (and all other properties of the system) precisely. In addition, ρ is certainly simpler than the complicated total wavefunction used in the orbital approximation (Eq. 14.13).

Unfortunately, we do not know the mathematical function that relates ρ to the energy. Therefore, the function must be guessed at rather than derived. In fact, because ρ is a function of the coordinates of the system, the function we need to relate ρ to energy is actually not a function, but rather a **functional**. A functional is a mathematical function that has a mathematical function (in this case ρ) as its argument. For example, if $g(x) = \rho$, then $f(g(x))$ is the functional that gives the energy. Thus, this theoretical approach is called **density functional theory (DFT)**.

The basics of DFT are embodied in Eq. 14.54. The total energy is partitioned into several terms. Each term is itself a functional of the electron density. E^T is the electron kinetic energy term (the Born–Oppenheimer approximation is in place, so nuclear kinetic energy is neglected). The E^V potential energy term includes both nuclear–electron attraction and nuclear–nuclear repulsion. The E^J term is sometimes called the Coulomb self-interaction term, and it evaluates electron–electron repulsions. It has the form of Coulomb's law. The sum of the first three terms ($E^T + E^V + E^J$) corresponds to the classical energy of the charge distribution.

$$E = E^T + E^V + E^J + E^{XC} \qquad \text{(Eq. 14.54)}$$

The key term in DFT is E^{XC}, the **exchange–correlation** term. As the name implies, this functional is intended to account for *both* the exchange interaction arising from the Pauli principle and electron–electron correlation. To the extent that this term is successful in accounting for correlation and exchange, DFT has an advantage over wavefunction-based methods in that the correlation energy is included from the start. Since E^{XC} is not significantly more difficult to calculate than the other terms of Eq. 14.54, the potential for a considerable savings in computation time exists.

It is not obvious what the optimal form for the E^{XC} functional is. Usually, it is broken into two parts, the exchange functional and the correlation functional (Eq. 14.55). A number of different forms have been suggested for each. A currently popular form for the exchange functional is one developed by Becke in 1988. While the exact form of this functional is not given here, it is important to realize that it contains a parameter (γ) that is chosen to fit experimental data related to atomic exchange energies. As such, DFT methods have a semiempirical flavor, in that there is a fit to experimental data (see Section 14.2.3). DFT as typically implemented is therefore *not* an *ab initio* method.

$$E^{XC} = E^X(\rho) + E^C(\rho) \qquad \text{(Eq. 14.55)}$$

Popular forms for the correlation functional have been developed by Perdew and Wang and by Lee, Yang, and Parr. In recent years a method known as Becke3LYP (or equivalently, B3LYP) is emerging as a favorite of DFT practitioners. This is actually a hybrid HF/DFT method, in which E^{XC} is composed of both HF and DFT exchange terms (recall HF does include exchange interactions) and DFT electron correlation functionals. These various terms have weighting coefficients, and Becke developed an optimal three parameter fit to a body of experimental data, hence the "3" in the acronym.

One other feature of DFT calculations worth mentioning concerns the manner in which the E^{XC} term is calculated. While all the other terms are evaluated in a way similar to HF theory, E^{XC} cannot be evaluated analytically. Instead, a numerical integration is performed to evaluate E^{XC}. Basically, a grid of points is placed around the molecule, and E^{XC} is evaluated at each point. A key issue is how fine this grid is. More points will produce better results, but at a cost of more computing time. A balance must be struck, therefore, and this is something the user needs to keep in mind.

An advantage of DFT is that electron correlation is intrinsically part of the basic method via the E^{XC} term, as we noted above. The interesting feature, though, is that the time it takes to execute a DFT calculation is not very different from that required for an HF calculation. This is especially true for larger molecules, because DFT computer time scales formally as the third power of the molecular size, vs. formal fourth power scaling for HF methods. In many cases, DFT gives results that are comparable to MP2, but in times that are closer to HF.

One disadvantage is that DFT is not variationally correct. One cannot assume that any energy obtained is always too high, and so it cannot be assumed that anything that lowers the total energy constitutes an improvement in the wavefunction. HF and CI methods are variationally correct, but MP2 and related methods are not. In addition, unlike the case with *ab initio* calculations, DFT cannot be systematically improved (for example, by expanding the basis set or including more configurations in a CI calculation). Using a different functional may give a different result, but there is no way to know if it is a better one.

As typically implemented, DFT calculations are similar to HF calculations. A set of MOs called the **Kohn–Sham orbitals** are iteratively improved until they converge on self-consistency. The number, shape, and symmetry properties of the Kohn–Sham orbitals are similar to the HF orbitals. However, the orbital energies are generally not in good agreement with those from either HF theory or experiment, and so there is no analogue to Koopmans' theorem in DFT.

DFT is still an evolving method, but it is seeing more and more use in the modern literature. At present a somewhat bewildering array of functionals is in use, with B3LYP emerging as the consensus method of choice, at least for organic molecules. DFT is definitely more applicable to larger systems than conventional *ab initio* methods, and it has seen many considerable successes. However, like most semi-empirical methods, DFT can, unexpectedly, fail to give good results in certain cases. Thus, some caution is still in order when applying even a well-developed DFT functional such as B3LYP to new kinds of systems. Many theoretical chemists are working to further improve DFT functionals, and it seems certain that their efforts will continue to make DFT methods an even more attractive alternative to wavefunction-based methods in the coming years.

14.3 A Brief Overview of the Implementation and Results of HMOT

We now turn our attention to *qualitative* applications of MO theory. Our goal is to develop general guidelines for predicting and understanding the electronic structures of organic molecules in ways that are most useful to experimentalists. To start this off, we analyze Hückel molecular orbital theory (HMOT), because it is very useful in giving quick insights into electronic structures and qualitative insights into pericyclic reactions (Chapter 15). In the 1950s and 1960s, prior to the advent of powerful computers, a large amount of work was done on Hückel theory. Many strategies were developed for simplifying the calculations (a useful exercise because they were done by hand). In addition, many extended analyses of the results were developed, evaluating spin densities, bond orders, and the like. With the ready access to powerful computers, however, it is now a simple matter to obtain the Hückel results for almost any system. As such, there really is no need to dwell on the tricks of the trade for doing HMO calculations by hand; one simply uses a desktop computer. In addition, the detailed analysis of Hückel wavefunctions is much less common, because much more accurate wavefunctions can now be obtained.

Why worry about HMOT at all then? There are at least two good reasons. A great deal of organic chemistry involves molecules with π systems, and essentially all of our intuition about the electronic structure theory of π systems is based on notions grounded in HMOT. While it may be less commonly used now, its historical reach is long. Second, the ease with which we can relate the results to simple notions of bonding engenders a level of insight and understanding that is sometimes difficult to obtain looking at the complex output of a higher level calculation. We will develop several rules or patterns of MO theory that lead to a good qualitative feel for how orbitals are formed. These rules hold strictly only at the level of HMOT, but the remarkable reality is, the *qualitative* conclusions we reach from HMOT are almost universally supported by more advanced levels of theory. The facts are, the most important thing about a π system is how atoms are connected to each other, and HMOT treats this issue well. From our discussion in Chapter 6, you should appreciate that fundamentally HMOT treats the *topology* of a molecule. For this reason, the results of HMOT can be understood using the mathematics of graph theory and other topology-related fields.

Although the results of HMOT are easy to obtain, and many excellent treatments of the quantitative methodology are available, computer programs should not be treated as "black boxes". Hence, it is useful to understand how HMOT is implemented. Here we will provide a brief overview of the manner in which Hückel theory is applied, the results obtained from HMOT, and some examples of how it provides useful and general qualitative insights into the electronic structures of organic molecules.

14.3.1   Implementing Hückel Theory

Recall the basic assumptions of HMOT (Section 14.2.3). The method is restricted to planar molecules and evaluates only the π electrons—the σ framework is ignored. This is not as drastic an assumption as it may seem, because σ systems and π systems are of opposite symmetry in a planar molecule, and so σ and π orbitals do not mix. The σ electrons can be viewed as simply providing part of the potential field experienced by the π electrons.

Our starting point for understanding how HMOT is practiced is the analysis of the generalized secular determinant given in Section 14.2.2. In HMOT, the energy of an electron in an isolated C $2p$ orbital is designated as α (a negative number). The energy of any carbon p orbital anywhere in the molecule is the same value, α. Hence, all H_{ii} values in the secular determinant of a hydrocarbon are now α. The energy of interaction between any two adjacent C $2p$ orbitals (that is, those on atoms linked by a σ bond) is called the resonance integral, β. This is a stabilizing interaction, and hence β is negative. All H_{ij} integrals are now β when atoms i and j are adjacent. Otherwise, H_{ij} is 0. Since overlap is totally ignored, all S_{ij} integrals are assigned a value of 0. For heteroatoms such as N and O, α is replaced by α + a constant, and β may or may not be changed.

Putting this all in the context of only two adjacent carbon p orbitals gives the Hückel secular determinant for ethylene shown below. Compare this to the secular determinant for the "Two-Orbital Mixing Problem" given in Section 14.2.2.

$$\begin{vmatrix} \alpha - E & \beta \\ \beta & \alpha - E \end{vmatrix} = 0$$

Since we have already examined the mathematics of mixing two orbitals, we can simply take the energies (Eqs. 14.46 **A** and **B**) and substitute in α for H_{AA} and β for H_{AB}, letting $S = 0$, thereby giving the Hückel energy diagram for ethylene shown in Figure 14.10. Since overlap is ignored in Hückel theory, the antibonding MO is *not* destabilized more than the bonding MO is stabilized. Because HMOT is so ingrained in many chemist's minds, the more realistic way to write such diagrams (Figure 14.9) is often forgotten. Always remember that when two filled orbitals interact, the result is actually net destabilization, because the out-of-phase combination is really raised more than the in-phase combination is lowered.

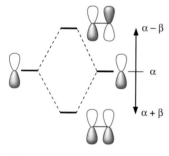

$\alpha - \beta$

α

$\alpha + \beta$

Figure 14.10
A molecular orbital diagram for ethylene at the Hückel level. An isolated p orbital has an energy of α, and the interaction energy for an adjacent carbon p orbital is β.

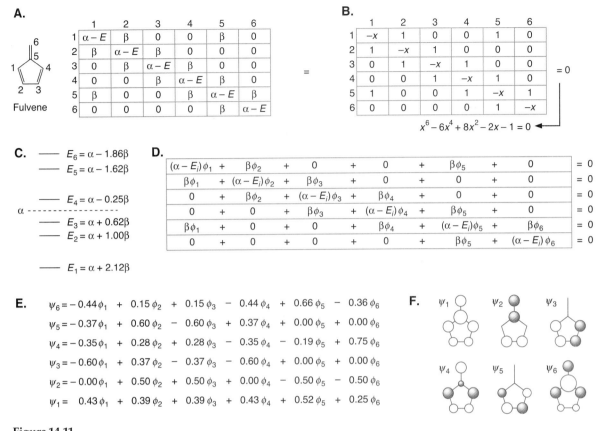

Figure 14.11
All the parts of the HMOT analysis of fulvene. **A.** The secular determinant with numbering as given in the picture of fulvene. **B.** The secular determinant after dividing by β and substituting $(\alpha - E)/\beta = -x$. **C.** The energy diagrams derived from solving the sixth-order polynomial from the secular determinant. **D.** The secular equations for fulvene. **E.** The wavefunctions for the molecular orbitals. **F.** Pictures of all the Hückel MOs.

Now let's consider a substantially more complicated system than ethylene, the molecule fulvene. The Hückel secular determinant for fulvene is given in Figure 14.11 **A**. Note that all the off-diagonal terms that do not correspond to adjacent atoms are zero. After dividing the entire determinant by β, we let $(\alpha - E)/\beta = -x$, giving the associated determinant (Figure 14.11 **B**). This determinant gives a sixth-order polynomial that has the following roots (x_i's): 2.115, 1.00, 0.618, –0.254, –1.618, and –1.861. Since $\varepsilon_i = \alpha + x_i\beta$, the energy diagram for the molecular orbitals shown in Figure 14.11 **C** arises. To generate the molecular orbitals, each energy ε_i is substituted into the secular equations (see Figure 14.11 **D**), and for each, the six simultaneous equations in the six unknowns (c_{ik}) are solved. The wavefunctions are given in Figure 14.11 **E**, and pictures of all the orbitals for the π system of fulvene are given in Figure 14.11 **F**. These pictures simply show the tops of the p orbitals on each carbon with the planar molecule placed in the page. They nicely indicate the nodal patterns.

Using Hückel theory it was relatively easy to arrive at the energies and the wavefunctions of the π MOs of fulvene. Even with all the assumptions of HMOT, the energy splitting pattern and the contributions from the p orbitals on each carbon in the molecular orbitals are similar to those obtained with higher level calculations. In the Exercises at the end of the chapter you will be asked to solve for the energies and molecular orbitals of allyl and butadiene (we give the answers in Figure 14.15) using the associated secular determinants and equations. However, the molecular orbitals for these simple structures can also be derived from some trends that Hückel theory produces. These trends are the next items to be considered.

14.3.2 HMOT of Cyclic π Systems

For simple, monocyclic hydrocarbon π systems, the results of HMOT can be readily predicted. The HMO energy of orbital i, ε_i, is given by Eq. 14.56, where N = number of vertices (atoms in the cyclic structure). The i is the orbital numbering and runs from $i = 0, \pm 1, \pm 2, \ldots,$ up to $\pm N/2$ for even N or $\pm(N-1)/2$ for odd N. This equation leads to the observation that the orbital splitting patterns for cyclic π systems can be predicted by the circle mnemonic shown in Figure 14.12. One inscribes the appropriate polygon into a circle of radius 2β, and the orbital energies are $\alpha + x\beta$, where x is the point of intersection along the vertical axis of the circle. This produces a number of degenerate orbitals for the cyclic hydrocarbons. The lowest MO is always $E = \alpha + 2\beta$ for these systems.

$$\varepsilon_i = \alpha + 2\beta \cos\left(\frac{2i\pi}{N}\right) \qquad \text{(Eq. 14.56)}$$

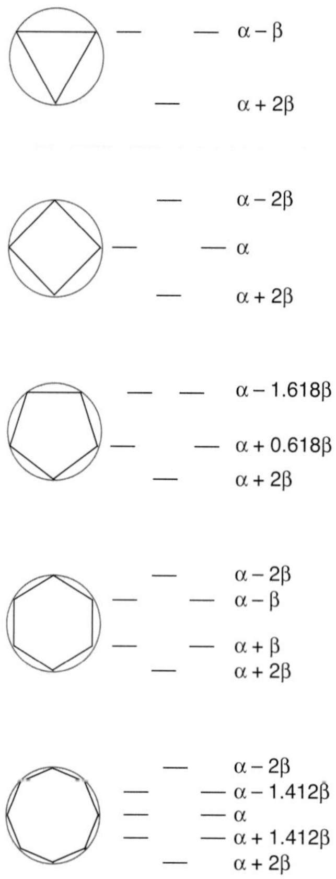

Figure 14.12
The circle mnemonic for determining the π system molecular orbital diagram for cyclic hydrocarbons at the Hückel level.

The nodal properties of the MOs follow the familiar patterns with the orbital of lowest kinetic energy at the bottom (no nodes) and highest kinetic energy at the top (greatest number of nodes).

The coefficients of the MOs can be easily obtained from four simple rules (these rules are for any molecule, although we are presenting them in the context of cyclic hydrocarbons). Our symbolism will continue to be

$$\text{MO} = \psi_i, \quad \text{AO} = \phi_k, \quad \text{and} \quad \psi_i = \sum_k c_{ik}\phi_{ik}$$

Rule 1. For each ψ_i:

$$\sum_k c_{ik}^2 = 1$$

This is the normalization criterion. The sum of the squares of the coefficients for any MO must be 1, when no overlap integral S is considered.

Rule 2. For each ϕ_k:

$$\sum_i c_{ik}^2 = 1$$

That is, the sum of the squares of the coefficients for any AO over all MOs must be 1. This means that the p orbital on each carbon must be used up completely over all the different molecular orbitals.

Rule 3. For a degenerate pair of molecular orbitals, each atom must contribute equally to the pair. Each atom's contribution is just the sum of the squares of its coefficients for the pair.

Rule 4. The coefficients of the AOs must yield MOs that are either symmetric or antisymmetric with respect to any symmetry operation of the molecule.

Consider, for example, the MOs of benzene (Figure 14.13). For MO number 1 (counting from lowest to highest energy), all coefficients are $1/\sqrt{6}$ (Rules 1 and 4). Similarly, for MO number 6, all coefficients are $\pm 1/\sqrt{6}$. For the first degenerate pair we can say that the c's must all be the same by symmetry, and that $c = \frac{1}{2}$ by Rule 1 because $4(\frac{1}{2})^2 = 1$. For coefficients a and b, we set up two equations.

First, $a^2 = b^2 + c^2 = b^2 + (\frac{1}{2})^2$ (Rule 3). The square of coefficient a is set equal to the sum of the squares of b and c because coefficient a only contributes to one of the orbitals in the degenerate set.

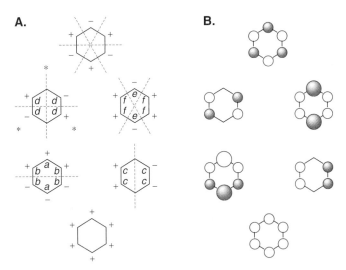

Figure 14.13
A. The nodal patterns for the π MOs of benzene. **B.** Pictures of the π MOs. Letters a, b, c, d, e, f, and the * refer to discussions given in Sections 14.3.2 and 14.3.4.

Second, $2a^2 + 4b^2 = 1$ (Rule 1).

Solving these gives $b = 1/\sqrt{12}$ and $a = 2/\sqrt{12}$. Note that the atom that is on the node in one member of the degenerate pair has a larger coefficient in the other orbital. This is a general result. The coefficients for the other degenerate pair can be obtained equivalently or by the pairing theorem (see below).

14.3.3 HMOT of Linear π Systems

For linear polyenes the HMO energy of orbital i, ε_i, can be obtained from Eq. 14.57, where N = the number of vertices (atoms in the chain) and i is the orbital numbering starting with 1 for the lowest energy orbital and proceeding up to N.

$$\varepsilon_i = \alpha + 2\beta \cos\left(\frac{i\pi}{N+1}\right) \qquad \text{(Eq. 14.57)}$$

The nodal pattern follows a familiar form (Figure 14.14 **A**). We start with no nodes, and then progressively add nodes as the energy increases. For chains with an odd number of atoms, there is an MO that resides at the energy level of a p orbital (α). Before noting some obvious patterns to these orbitals, this is a convenient place to review some definitions. The orbital at the α level, that just below the α level, and that just above the α level, are known as the **frontier orbitals** for the individual hydrocarbons. When adding electrons to these π systems to create neutral structures, the hydrocarbons with even numbers of electrons have only the orbitals below the α level populated. The highest-occupied molecular orbital for each structure is referred to as the **HOMO**. The lowest-unoccupied molecular orbital is the **LUMO**. For chains with an odd number of carbons, neutral structures have one electron in the MO at the α level. This orbital is therefore referred to as a **SOMO**, a singly-occupied molecular orbital. Furthermore, in these specific cases, because the SOMO is at the α level, it is not bonding or antibonding. Hence it is referred to as an **NBMO**, a nonbonding molecular orbital.

There are several patterns to the MOs of Figure 14.14 **A** that are important. First, the HOMO of even-electron systems is always a collection of what appears to be isolated π bonding pairs. Second, there is a pattern to the symmetry of the HOMOs and LUMOs. The HOMO of ethylene is symmetric (S) with respect to a C_2 axis in the molecule. The HOMO of butadiene is antisymmetric (A); the HOMO of hexatriene is S; the HOMO of octatetraene

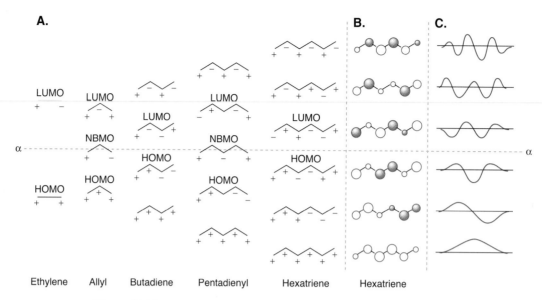

Figure 14.14
A. The nodal pattern for the linear polyenes. **B.** Relative contributions of the p orbitals to each MO of hexatriene. **C.** Analogous two-dimensional waves showing increasing kinetic energy.

would therefore be A; etc. The LUMOs follow a similar alternating pattern, starting from the A LUMO of ethylene. Furthermore, the NBMOs of linear hydrocarbons with odd numbers of carbons have coefficients on every other atom, consistently changing phase on every other atom.

We focus upon hexatriene to note two other trends. First, the nodal patterns of Figure 14.14 **A** do not give insight into the relative contribution of each p orbital to the MOs, and as seen in Figure 14.14 **B**, the relative contributions of the p orbitals are quite different for each MO. Second, you can take the nodal patterns along with the relative contributions of the p orbitals in each MO and sketch a picture of an analogous two-dimensional wave for each MO wavefunction (Figure 14.14 **C**). These pictures nicely highlight the increasing kinetic energies of the wavefunctions, from the bottom to the top of the figure.

Two important structures in HMOT are allyl and butadiene, because we will use these in Chapter 15. As such, Figure 14.15 shows pictures of the π MOs for these structures, their Hückel energies, and the coefficients of the atomic orbitals for each MO.

An analysis of the NBMO of allyl nicely ties together the valence bond concept of resonance (see Chapter 1) with the MO concepts presented in this chapter. The NBMO only has coefficients on the end carbons, meaning that allyl radical does not have radical character on the central carbon at this level of theory. (See, however, the discussion of negative spin density in Section 17.3.2.) This is exactly what the resonance structures for allyl radical tell us.

14.3.4   Alternant Hydrocarbons

For moderately complex polyenes, the nodal patterns of the higher molecular orbitals can be difficult to decipher. Fortunately, things are greatly simplified because many systems of interest are members of a class called **alternant hydrocarbons** (AH). An AH is a system for which all the atoms can be divided into two classes, termed * and non-* (starred and non-starred), such that no two members of the same class are adjacent. Further, AH are termed "even" or "odd" depending on whether the number of carbon atoms is even or odd, respectively. Figure 14.16 gives examples. Neutral odd AH are radicals. A necessary and sufficient criterion for a non-AH is the presence of an odd-membered ring.

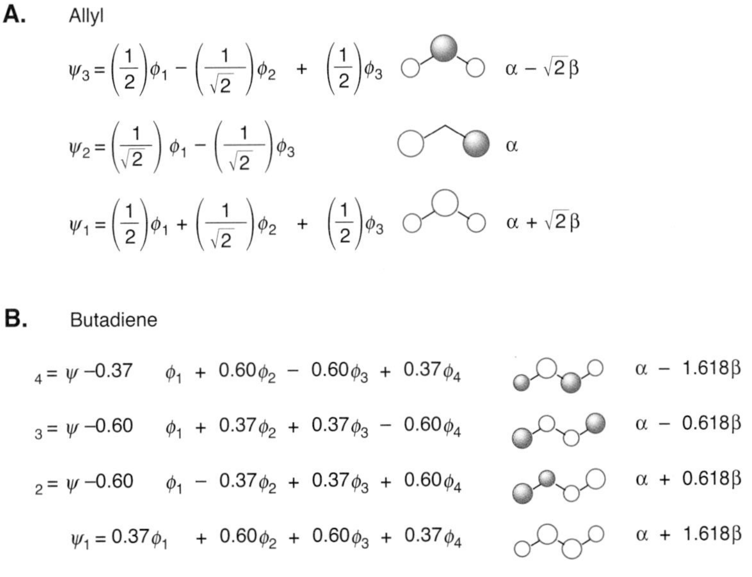

A. Allyl

$$\psi_3 = \left(\frac{1}{2}\right)\phi_1 - \left(\frac{1}{\sqrt{2}}\right)\phi_2 + \left(\frac{1}{2}\right)\phi_3 \qquad \alpha - \sqrt{2}\beta$$

$$\psi_2 = \left(\frac{1}{\sqrt{2}}\right)\phi_1 - \left(\frac{1}{\sqrt{2}}\right)\phi_3 \qquad \alpha$$

$$\psi_1 = \left(\frac{1}{2}\right)\phi_1 + \left(\frac{1}{\sqrt{2}}\right)\phi_2 + \left(\frac{1}{2}\right)\phi_3 \qquad \alpha + \sqrt{2}\beta$$

B. Butadiene

$$\psi_4 = -0.37\phi_1 + 0.60\phi_2 - 0.60\phi_3 + 0.37\phi_4 \qquad \alpha - 1.618\beta$$

$$\psi_3 = -0.60\phi_1 + 0.37\phi_2 + 0.37\phi_3 - 0.60\phi_4 \qquad \alpha - 0.618\beta$$

$$\psi_2 = -0.60\phi_1 - 0.37\phi_2 + 0.37\phi_3 + 0.60\phi_4 \qquad \alpha + 0.618\beta$$

$$\psi_1 = 0.37\phi_1 + 0.60\phi_2 + 0.60\phi_3 + 0.37\phi_4 \qquad \alpha + 1.618\beta$$

Figure 14.15
The Hückel molecular orbitals of **A.** Allyl, and **B.** Butadiene.

The usefulness of AH is in the following two **pairing theorems**, which apply only to AHs:

Rule 1. All MOs come in pairs, ψ_i^+ and ψ_i^-, with energies $\alpha + x_i\beta$ and $\alpha - x_i\beta$.

Rule 2. For all atoms, the coefficients in ψ_i^+ and ψ_i^- are the same in magnitude and differ solely by changing the signs of the starred atoms only.

With these two rules it is a simple matter to determine the nodal patterns of the higher MOs of the polyenes. Let's use these rules to work out the coefficients of the higher energy degenerate pair of orbitals for benzene (Figure 14.13). We start by placing a star on every other carbon in benzene; it does not matter where the first star is placed (refer to Figure 14.13). We showed above that $c = \frac{1}{2}$, $b = 1/\sqrt{12}$, and $a = 2/\sqrt{12}$. To create the partner to the MO with c coefficients, we simply change the sign at the starred carbons, giving d as $\frac{1}{2}$ also. For the partner to the MO with coefficients a and b, we again just switch the sign on the starred carbons. Hence, $f = 1/\sqrt{12}$ and $e = 1/\sqrt{12}$. Therefore, it is a simple procedure to derive the π MOs for benzene, but also for any other alternant polyene.

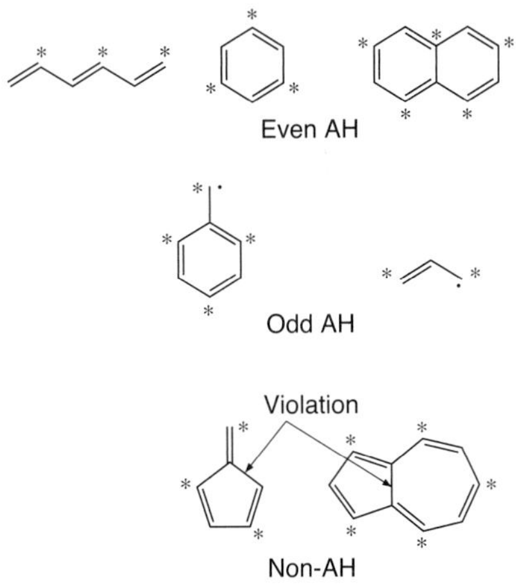

Even AH

Odd AH

Figure 14.16
Examples of alternant and non-alternant hydrocarbons.

Violation

Non-AH

The rules above apply strictly to even AH. For odd AH the two rules still apply, but there is one additional MO. It has $\varepsilon = \alpha$, and it has coefficients only on the starred atoms. Here, we write the stars to create the largest number of starred atoms possible, and this NBMO represents the radical orbital for a neutral molecule. To write the coefficients for this orbital, we simply find a set of coefficients for which the coefficients for atoms around each non-starred position sum to zero. See the examples in Figure 14.17. Note that this NBMO does, in a sense, still satisfy the pairing theorem, in that it is "self-paired". Since it only has coefficients on the * atoms, changing the signs of all the * atoms gives the same MO. Since its energy is α, it can be viewed as paired with itself energetically, with $E = \alpha \pm 0\beta$.

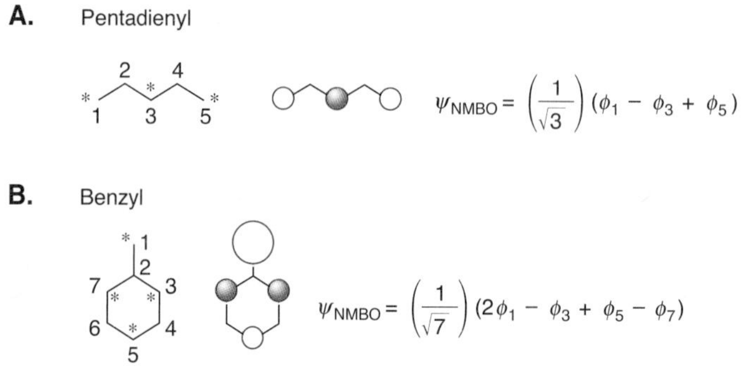

A.   Pentadienyl

$$\psi_{NMBO} = \left(\frac{1}{\sqrt{3}}\right)(\phi_1 - \phi_3 + \phi_5)$$

B.   Benzyl

$$\psi_{NMBO} = \left(\frac{1}{\sqrt{7}}\right)(2\phi_1 - \phi_3 + \phi_5 - \phi_7)$$

Figure 14.17
Examples of the NBMOs of two odd alternant hydrocarbons. The sum of the coefficients around each non-starred center must be zero.

An interesting example given in Figure 14.17 is that of benzyl. Once again, valence bond theory concepts agree nicely with MOT. Resonance structures for benzyl radical place the radical on the benzylic carbon and also on the ortho and para carbons. Note that the NBMO for benzyl covers only these carbons (the * atoms). Additionally, the resonance structure that does not disrupt the aromaticity of the ring is most stable, and hence contributes the most to the overall structure of the molecule. We find that the coefficient on the benzylic carbon is twice that of the coefficient on the ring carbons, supporting our VBT analysis.

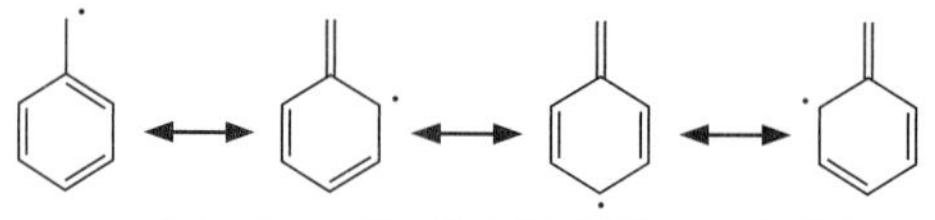

Benzyl resonance

14.4 Perturbation Theory—Orbital Mixing Rules

The basic concept of perturbation theory is to begin with a Hamiltonian for a system, and then consider any new interactions as a perturbation to that Hamiltonian. Precise solutions to problems of this type are available, but our goal is to summarize the salient results of these treatments. The important point is that you need not actually know the correct wavefunctions for the unperturbed Hamiltonian in order to appreciate the effects of a perturbation on the wavefunctions and their energies.

We will be concerned with combining the orbitals of two fragments or molecules. When mixing orbitals, three factors must be considered:

1. Symmetry: Only orbitals of like symmetry can interact.

2. Overlap: Orbitals must overlap in order to interact; the greater the overlap, the stronger the interaction. The symmetry restriction is a special case of the overlap rule.

3. Relative energy: The closer in energy the interacting fragment orbitals, the stronger the interaction.

These general ideas can be put on a quantitative basis by examining the results of first-order and second-order perturbations.

14.4.1 Mixing of Degenerate Orbitals—First-Order Perturbations

When a pair of degenerate orbitals (AOs or MOs of molecular fragments) interact, we obtain the mixing diagram in Figure 14.9, identical to the results from the "Two-Orbital Mixing Problem". The initial orbital energies are H_{AA} and the interaction matrix element is H_{AB}. The lower orbital is the bonding orbital and is comprised of the in-phase combination of the original orbitals, while the higher lying, antibonding orbital is the out-of-phase combination. The changes in energy from the H_{AA} level are given by Eqs. 14.58 and 14.59.

$$\Delta E = \frac{H_{AA}S + H_{AB}}{1 + S} \qquad \text{(Eq. 14.58)}$$

$$\Delta E' = \frac{H_{AA}S - H_{AB}}{1 - S} \qquad \text{(Eq. 14.59)}$$

This is termed a **first-order perturbation** because H_{AB} is only raised to the first power. A two-center, two-electron interaction is stabilizing, and the stabilization energy is $2\Delta E$ (the "2" is present because there are two electrons). *It is always favorable to mix a filled orbital with an empty orbital or to mix two singly occupied orbitals.* Importantly, because S is a positive number, it is always true that $|\Delta E'| > |\Delta E|$. That is, as already discussed, the antibonding orbital is destabilized more than the bonding orbital is stabilized. As such, the mixing of two filled orbitals, a two-center, four-electron interaction, is inherently destabilizing. This is called **closed-shell repulsion**, and it is an important result of orbital mixing.

Now, if we assume that $S = 0$, then $|\Delta E'| = |\Delta E|$ and closed-shell repulsion is undone. This may seem an unreasonable situation, but in fact many levels of theory make this approximation. As noted earlier in this chapter, most semi-empirical methods, such as AM1 and MNDO, neglect overlap. As such, there is no closed-shell repulsion at these levels of theory, a serious deficiency in some situations. In contrast, for all its limitations, extended Hückel theory (EHT) does not neglect overlap, and so closed-shell repulsion survives. As discussed, classical Hückel theory does neglect overlap. In fact, as previously noted, HMOT can be viewed as perturbation theory with $S = 0$, $H_{AA} = H_{BB} = \alpha$, and $H_{AB} = \beta$.

14.4.2 Mixing of Non-Degenerate Orbitals—Second-Order Perturbations

When rigorous perturbational methods are applied to the mixing of non-degenerate orbitals, the results shown in Figure 14.18 are obtained. Again, the lower orbital is the in-phase combination and the higher orbital the out-of-phase. A key observation is that the mixing coefficient, λ, is always less than 1. Thus, the lower orbital is primarily ϕ_A with some fraction of ϕ_B mixed in. Likewise, the higher orbital is primarily ϕ_B with some fraction of ϕ_A mixed in (recall that we already examined this briefly in Chapter 1). The greater the initial energy separation between ϕ_A and ϕ_B, the smaller λ is—that is, the more the lower combination orbital looks like the original lower orbital and the higher combination orbital looks like the original higher orbital.

As always, the two-center, two-electron interaction is stabilizing. Assuming the two electrons began in the lower energy starting orbital, the stabilization is as given in Eq. 14.60.

$$\Delta E_2 = \frac{2(H_{AB} - E_A S_{AB})^2}{E_A - E_B} \qquad \text{(Eq. 14.60)}$$

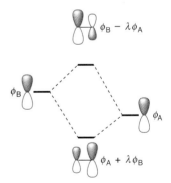

Figure 14.18
The non-degenerate, two-orbital mixing diagram. We show the fragments as p orbitals, but they could be any type of AO or MO.

This is a second-order perturbation because of the square in the numerator, and typically second-order orbital interactions are smaller than those between degenerate orbitals. In the situation where overlap is neglected, we have Equation 14.61.

$$\Delta E_2 = \frac{2H_{AB}^{\,2}}{E_A - E_B}$$

(Eq. 14.61)

In either case, the stabilization scales as the square of the interaction matrix element H_{AB}, which is roughly proportional to overlap. In addition, stabilization scales inversely with the initial energy separation of the interacting orbitals. The larger the initial energy gap, the weaker the interaction. This is why we state that mixing orbitals of equal energy gives the most stabilization.

As in the first-order perturbation, the two-center, four-electron interaction is destabilizing, because $|\Delta E'| > |\Delta E|$ (Figure 14.18). Eq. 14.62 gives the destabilization that results when Figure 14.18 is populated with four electrons,

$$\Delta E_4 = \frac{4(E_o S_{AB}^{\,2} - H_{AB} S_{AB})}{1 - S_{AB}^{\,2}}$$

(Eq. 14.62)

where E_o is the average energy of the original interacting orbitals. Again, if $S_{AB} = 0$, then $\Delta E_4 = 0$, and closed-shell repulsion is neglected.

14.5   Some Topics in Organic Chemistry for Which Molecular Orbital Theory Lends Important Insights

There are several phenomena that have been discussed throughout this book that can be better understood now that we have a stronger background in MOT. These include the unusual stability and instability imparted by aromaticity and antiaromaticity, respectively, and the unusual bonding and high donor ability of cyclopropane. We now reexamine these and other phenomena using molecular orbitals, and we introduce a new concept called through-bond coupling.

14.5.1   Arenes: Aromaticity and Antiaromaticity

Aromaticity is extensively discussed in every introductory organic text, and aspects of it were given in Chapter 2, such as the $4n+2$ rule. Key features that are diagnostic of an aromatic system include the following:

• Special stability,

• Equalization of bond lengths,

• Ring current effects in the NMR, and other special magnetic properties, and

• Reactivity that is different from that of olefins and polyenes, such as electrophilic substitution rather than addition.

For all of these, benzene is the prototype. We discussed the special thermodynamic stability of benzene in Chapter 2, and it is well accepted that all C–C bonds in benzene are equal in length, and that length is intermediate between a single and double bond (Table 1.1). A diamagnetic ring current associated with the π electrons deshields nuclei on the edge of the system, and shields nuclei over the π face. As such, the NMR chemical shifts of protons around the periphery of aromatic rings are downfield from analogous alkene protons, while those above the ring are shifted upfield. Antiaromaticity—the destabilization associated with $4n$ systems—was introduced much later than aromaticity and was considered controversial at first. The low thermodynamic stablity of cyclobutadiene discussed in Chapter 2, however, gives strong support to the notion.

It is commonplace to state that MO theory, including HMOT, "explains" aromaticity and antiaromaticity. Certainly, the results of Hückel theory, as summarized by the circle mnemonic of Figure 14.12, make it possible to distinguish between systems with $4n+2$ electrons and those with $4n$ electrons. The antiaromatic systems are predicted by this level of theory to be biradicals (put four electrons into the energy diagram for square cyclobutadiene in Figure 14.12). The aromatic systems are more stable relative to analogous linear conjugated systems because their β values are larger for the bonding orbitals. In other words, relative to an analogous acyclic π system, the **HMOT energy**—which is just the sum of the one electron energies—is always lower for an aromatic cycle, again suggesting stability for the aromatic system. However, we've discussed above how the energy of a system does not equal the sum of the one-electron energies, and this might be especially so for such a crude theoretical model as HMOT. As a result, some caution is in order in interpreting HMOT energies, but the trends are sensible. Apart from the stability of benzene and the instability of cyclobutadiene, notable successes of this approach include the stability of cyclopropenyl cation and cyclopentadienide, and the nonplanarity plus relatively large inversion barrier of cyclooctatetraene because it would be $4n$ when planar (Figure 14.19, **A**, **B**, and **C** respectively).

Further insights into the $4n+2$ rule can be gained from the 10-electron systems (Figure 14.19). For example, the planarity and relative stability of the dianion of cyclooctatetraene (Figure 14.19 **B**) is generally considered a major success of the $4n+2$ rule. However, it is not obvious that Hückel theory predicts this dianion to be especially stable. Referring to Figure 14.12, we see that in making this dianion we must put four electrons into two degenerate, *nonbonding* molecular orbitals. This hardly seems like a stable situation, but the dianion is aromatic because it fits the features listed above. Again, the qualitative arguments that arise from HMOT are sound, but caution is appropriate for quantitative analyses.

For neutral 10-electron systems, the story is more complex. Naphthalene is often viewed as a 10-electron system, but really it is better viewed as two six-electron cycles that share an edge. Note that six-electron cycles are especially stable even within the context of aromatic systems. For more complex ring systems, it is generally observed that the larger the number of benzene-like rings, the more stable the structure.

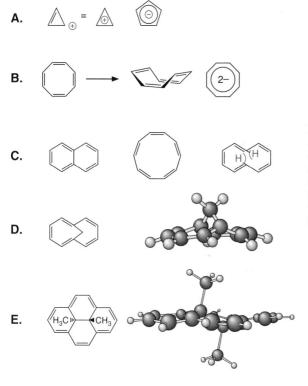

Figure 14.19
A. The aromatic ions cyclopropenyl cation and cyclopentadienyl anion. **B.** Cyclooctatetraene is nonplanar, but the dianion is planar and aromatic. **C.** 10 π electron systems: naphthalene, all-cis [10]annulene, and a possibly planar [10]annulene with severe transannular interactions. **D.** A bridged [10]annulene. **E.** Dimethyldihydrophenanthrene.

The monocyclic 10-π electron system, cyclodecapentaene, also known as [10]annulene, has long been an experimental challenge. The all-cis cyclodecapentaene cannot be planar; if it were, the C–C–C angles would be 144°. Introducing trans bonds would allow the system to be planar, but severe transannular repulsions ensue (see Section 2.2.2). This can be seen by imagining breaking the central bond of naphthalene to make [10]annulene—the hydrogens we would have to add will point right at each other. The system will have to become nonplanar. The issue of the preferred conformation of [10]annulene is still open. The molecule has been observed at low temperatures by Masamune, and its NMR suggested a highly nonplanar structure with primarily olefinic character. Actually, two separate conformers of [10]annulene were observed, and it is an issue of ongoing debate as to the nature of these conformations.

A recent computational study finds five low energy conformations for [10]annulene, with the most stable form being a twisted structure that has a C_2 axis. Interestingly, only a very high level of theory, involving extensive configuration interaction, produced this result. Both DFT and MP2 theories predict different, and presumably incorrect structures, emphasizing the challenging nature of this system.

A clever strategy of introducing a bridging saturated carbon relieves the transannular strain, and produces a more nearly planar and much more stable system. Vogel was the first to synthesize this structure, and despite the considerably reduced π overlap at the bridgehead carbons (Figure 14.19 **D**) this 10-π system is aromatic by the conventional criteria.

Higher annulenes have also been prepared, including [14]- and [18]annulene. One of the best characterized of this type of molecule is the [14]annulene derivative dimethyldihydropyrene (Figure 14.19 **E**). The outer periphery is essentially planar, and a high quality x-ray structure shows extensive bond equalization, with an average C–C bond length of 1.393 Å. This is clearly a bond length associated with an aromatic system. In a remarkable display of a ring current effect, the two methyl groups of dimethyldihydropyrene appear at δ −4.25 ppm in the ^{1}H NMR, placing them clearly in the shielding region of the ring current associated with the peripheral π system. The reactivity patterns of this system are also benzene-like. Therefore, the $4n+2$ rule holds up for larger n.

Finally, recent calculations have called into question one of the most sacred tenets of aromaticity theory. Hiberty and Shaik have come to the conclusion that the reason benzene has all equal bond lengths in not due to the π system. Rather, they propose that *it is the σ framework that enforces six equivalent bond lengths in benzene*! We will see in Chapter 17 that benzene is indeed an exception when it comes to extended π systems, in that bond alternation is generally the rule. By separating full *ab initio* wavefunctions for benzene into σ and π components and analyzing them separately, it appears that the benzene π system also prefers bond alternation. However, a cyclic array of six σ bonds (that is, the σ framework underlying the π system) prefers equal lengths, and it wins out in benzene. This is a controversial notion, and references for further reading are given at the end of this chapter. We mention this here simply to highlight the fact that even after more than a century of work, aromaticity remains a topic of discussion and research.

14.5.2   Cyclopropane and Cyclopropylcarbinyl—Walsh Orbitals

One of the most challenging molecules for organic electronic structure theory is cyclopropane. It is difficult to understand C–C–C angles of 60° with conventional bonding models. Also, in Section 11.5.14 we examined the unusual reactivity of the cyclopropylcarbinyl cation. Much of the rearrangement chemistry of this cation, as well as its special thermodynamic stability (Chapter 2), can be understood to arise from the uniquely strong donor character of the σ bonds in cyclopropane. Here we provide a more detailed description of the bonding in cyclopropane that nicely rationalizes these observations.

Cyclopropane is very well treated by a combination of group orbitals and the perturbation theory analysis described in this chapter and Chapter 1. First, however, we must consider a new aspect of orbital mixing, the general problem of combining three, equivalent orbitals in a cyclic array. We now have two choices on how to proceed in solving this problem.

We can perform all the analysis in a manner similar to how the two-orbital mixing problem was addressed (Section 14.2.2). Alternatively, we can use some of the logic we have developed to this point in the chapter, and some of the rules of HMOT, to derive the answer. We choose the latter method here.

The Cyclic Three-Orbital Mixing Problem

We begin with the simplest possible three-orbital mixing, that of three $1s$ atomic orbitals located at the vertices of an equilateral triangle. We will operate at the zero overlap level, and we will use the perturbation rules derived in previous sections. The basic approach is outlined in Figure 14.20. First, we mix two of the atomic orbitals. This is a degenerate, first-order mixing, and so the in-phase combination drops in energy by H_{AB}, and the out-of-phase rises by H_{AB}, just as with the two-orbital mixing described in Section 14.4.1 with $S = 0$.

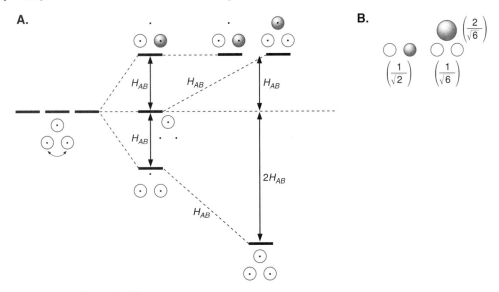

Figure 14.20
The mixing of three s orbital in a cyclic array. **A.** The mixing pattern and orbital energies. **B.** Coefficients for the degenerate pair of orbitals.

Next, we add the third of the original orbitals, allowing it to mix with the newly formed MOs from the two-orbital mixing. The third orbital cannot mix with the upper combination, because it lies on the node of that molecular orbital. So, the upper orbital of the original two-orbital mixing comes across unchanged in the final molecular system. The third AO and the lower MO from the two-orbital mixing can interact. This is now a non-degenerate, second-order mixing. For such a situation, we use Eq. 14.61. The factor of two given in Eq. 14.61, which simply accounted for the fact that there are two electrons, has been dropped, because here we wish to discuss only the drop of the orbital energy, not the total stabilization energy. The denominator of Eq. 14.61 is precisely H_{AB}. Concerning the numerator, the interaction of the new AO with the MO can be treated as a sum of AO–AO interactions (because the MO is an LCAO–MO). The interaction energy for each AO–AO interaction is H_{AB}, because these are still interactions between $1s$ AOs, except now the interaction energy has to be scaled by the coefficient $(1/\sqrt{2})$ of the AOs in the MO. That is, because the AO contributed by the MO is not a "full" AO, we have to scale the interaction energy. Thus, each AO–AO interaction is worth $[(1/\sqrt{2})H_{AB}]^2$. There are two such interactions, and so the total interaction is given in Eq. 14.63.

$$\Delta E = \frac{2[(1/\sqrt{2})\,H_{AB}]^2}{H_{AB}} = H_{AB} \qquad \text{(Eq. 14.63)}$$

That is, the fully in-phase MO is lowered by H_{AB} from the position of the original lower orbital, placing the completely in-phase orbital at an energy of $2H_{AB}$. The out-of-phase combination of the AO and the MO is raised from the initial AO position ($E = 0$) by H_{AB}, making it degenerate with the higher orbital from the original two-orbital mixing.

We can derive the orbital coefficients from the rules outlined in Section 14.3.2. In the lowest MO all coefficients are $1/\sqrt{3}$, by symmetry. For the degenerate pair, the MO with a node through one atom has coefficients of $\pm 1/\sqrt{2}$. For the other MO there will be two different coefficients. The two AOs corresponding to the AOs involved in the other member of the degenerate pair will have one coefficient, a, while the third AO will have a different coefficient, b. It is a simple matter to set up two equations using the rules of Section 14.4.2.

$$2a^2 + b^2 = 1 \tag{Rule 1}$$

$$a^2 + (1/\sqrt{2})^2 = 0^2 + b^2 \tag{Rule 3}$$

Solving gives $a = 1/\sqrt{6}$ and $b = 2/\sqrt{6}$. As before, the AO that lies on the node in one member of a degenerate pair has a larger coefficient in the other member.

The results of the three-orbital mixing problem can be summarized as follows:

- There is one, low-lying bonding MO and a degenerate pair of higher-lying, antibonding MOs;

- The bonding MO is stabilized twice as much as the antibonding pair is destabilized;

- The lowest MO is fully symmetrical; and

- In the degenerate pair, one MO has a node through one atom; in the other MO, this unique atom has a larger coefficient and is out-of-phase with the other two.

These rules are general for the mixing of three degenerate orbitals in a cyclic arrangement at the level of Hückel theory.

The MOs of Cyclopropane

We can now build the MOs of cyclopropane by combining three CH_2 groups positioned at the corners of an equilateral triangle, with the H–C–H plane perpendicular to the C–C–C plane. The patterns of one down and two up found above will be evident in all the orbital mixing given here. We use the CH_2 group orbitals developed in Section 1.2.4, and the basic scheme is shown in Figure 14.21. We emphasize only mixings between originally degenerate group orbitals. Beginning with the predominantly C–H bonding orbitals, both $\sigma(CH_2)$ and $\pi(CH_2)$ group orbitals combine in the usual three-orbital mixing way (one down, two up, as described above) to produce a total of six C–H bonding cyclopropane MOs (convenient, because there are six C–H bonds, Parts A, B, C, and E in Figures 14.21 and 14.22). The splittings are not large, because these group orbitals are more focused toward the hydrogens.

In contrast, the mixing of the three $\sigma(out)$ group orbitals is quite large. These group orbitals have a major component that points directly toward the center of the cyclopropane ring, and they overlap substantially (Figure 14.21 **D** and **G**). The in-phase combination produces a unique cyclopropane MO with a large density in the center of the ring (Figure 14.22 **D**). Since in this MO a great deal of electron density lies in a region that does not correspond to where one usually thinks there are C–H or C–C bonds, it would be quite difficult to explain this orbital using conventional σ bonding models using sp^3 hybridized carbons. However, this MO, which is obtained from any type of calculation, is easily rationalized by the group orbital concept.

Considering the remaining CH_2 group orbital, p in character, introduces a new concept. Note that this p orbital lies, not along any C–C bond, but rather is "tangent" to the ring. In the conventional three-orbital mixing problem, the lowest-lying MO is the completely in-phase combination of the basis orbitals. However, for a cyclic array of p orbitals of the type involved here, completely "in-phase" is difficult to define. There is no way to arrange three p orbitals in a circle so that all interactions are bonding (try it).

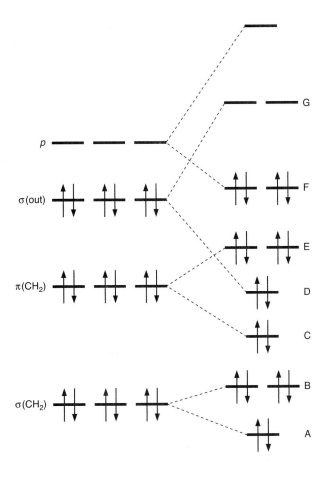

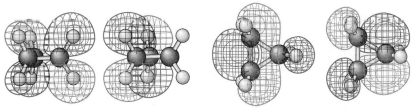

Figure 14.21
The orbital mixing diagram for cyclopropane formed from three CH₂ groups.

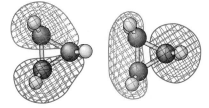

π (CH₃) degenerate pair (E)

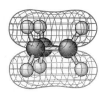

"The Walsh orbitals" formed from
C p orbitals; degenerate HOMO (F)

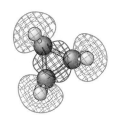

σ (CH₂) degenerate pair (B)

In-phase combination
of π (CH₂) (C)

C–C bonding from σ (out) (D)

Figure 14.22
The molecular orbitals of cyclopropane from an *ab initio* calculation. The letters correspond to the lettering from Figure 14.21.

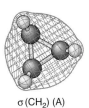

σ (CH₂) (A)

A.

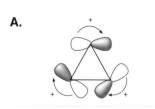

B.

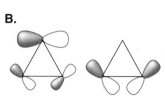

Figure 14.23
A. The Möbius interaction of *p* orbitals placed around a cyclopropane. **B.** Cartoons of Walsh orbitals of cyclopropane.

To understand the MOs derived from the CH_2 *p*-like orbital, we need to define some new conventions. We adopt a convention that a *p* orbital has a "positive" coefficient if, moving around the ring in a clockwise direction, we go from negative to positive in the *p* orbital. Now, we can examine the $+/+/+$ MO—the "in-phase" combination of *p* orbitals. It is completely antibonding (Figure 14.23). The MOs of the degenerate pair ($+$/node/$-$ and $+/2-/+$) are more bonding than the $+/+/+$ MO. *The energy ordering in the three-orbital mixing has been reversed;* it is now two down and one up. This is because of the unusual nodal properties of the basis orbitals. For simple *s* orbitals (as in the discussion above) or for *p* orbitals oriented perpendicular to the plane [as in the $\pi(CH_2)$ orbitals of cyclopropane], the usual mixing results. For *p* orbitals in the plane of the ring, however, a different pattern is seen. This unique arrangement is termed a **Möbius topology**. It has been studied extensively, and we will consider it more completely in Chapter 15, where we consider pericyclic reactions. In the present context, the important issue is that the mixing pattern is reversed for the *p* group orbitals.

Given this analysis, the HOMO of cyclopropane is the degenerate pair arising from the *p* orbitals of the CH_2 groups (Figures 14.21 **F** and 14.22 **F**). These F orbitals are called the **Walsh orbitals** of cyclopropane. Their electron density does not lie along the direct lines connecting carbon atoms. These C–C bonding orbitals can be considered to define **bent bonds**, or to have considerable π character. Either way, they are relatively high-lying in energy for C–C bonding orbitals in a saturated hydrocarbon.

The *p*-character of the F orbitals is responsible for the unique reactivity of cyclopropane. These Walsh orbitals are excellent "donors", interacting well with low-lying empty orbitals of other systems (see Section 11.5.14). The prototype example is the cyclopropylcarbinyl cation (Figure 14.24). The empty *p* orbital of the cationic center can interact very well with the Walsh orbitals of the cyclopropane. The mixing is shown in Figure 14.24 **A**. It splits the degeneracy of the Walsh orbitals, as only the orbital with significant density on the substituted carbon interacts. This mixing of a filled orbital with an empty orbital is clearly stabilizing, and this accounts for the unusual thermodynamic stability of the cyclopropylcarbinyl cat-

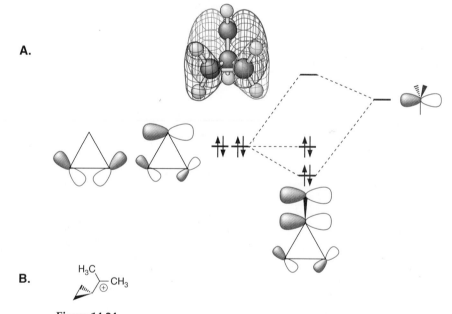

Figure 14.24
A. Orbital mixing diagram for cyclopropylcarbinyl. Also shown is a filled MO of cyclopropylcarbinyl cation, showing the mixing of a cyclopropyl Walsh orbital with the empty *p* orbital of the cationic center. **B.** The preferred geometry of a cyclopropylcarbinyl cation.

ion. It also explains a unique geometrical preference of the system. In order to mix with the Walsh orbital, the cationic center must adopt a specific geometry, illustrated for the dimethyl derivative in Figure 14.24 **B**. Using NMR spectroscopy under stable ion conditions, Olah was able to establish that there are indeed two inequivalent methyl groups in the cation, with one methyl over the cyclopropane ring and one pointing away. Rotation around the $CH–CMe_2$ bond makes the two methyls equivalent, allowing NMR to determine the activation energy for the process as 13.7 kcal/mol. This is a reasonable estimate of the stabilization in the cyclopropylcarbinyl cation, establishing it to be a very significant effect.

It is important to remember that the major electron density is on the edges of the cyclopropane rings, not in the middle of the ring. The cyclopropane MO with large density in the middle of the ring formed from the σ(out) group orbitals is too low-lying to be important in cyclopropane reactivity.

While the HOMO of cyclopropane is relatively high-lying and has "π" character, the LUMO (orbital G of Figure 14.21) is *not* especially low-lying and is not of π character. The degenerate LUMO arises from the σ(out) group orbitals. Thus, cyclopropane is a good π donor, but not a good π acceptor.

14.5.3   Planar Methane

A severe angle distortion that has long been a goal of physical organic chemistry has been planar, tetracoordinate carbon. Two strategies have been applied. The first is "brute force" distortion, using ring constraints to twist the normally tetrahedral carbon into a planar array. Molecules such as the **fenestranes** are examples of this approach. The [4.4.4.4]fenestrane shown has yet to be prepared.

The second strategy is more subtle, and it takes advantage of the unique electronic structure of a planar, tetracoordinate carbon. We begin by considering the MOs shown in Figure 14.25. This analysis is very similar to that of developing the group orbitals for a CH_3 group from Chapter 1. We can understand the lowest energy orbital as arising from the in-phase combination of the carbon 2s orbital and the hydrogens. Next, we have a degenerate pair of MOs arising from the interactions of the carbon p_x and p_y orbitals and the hydrogen atomic orbitals. What remains, then, is the carbon p_z orbital, and this orbital cannot interact with the hydrogens. We now add eight valence electrons, four from carbon and one from each hydrogen, and we see a very distinctive feature of planar methane. The p orbital is doubly occupied. In effect, *planar methane contains a lone pair of electrons occupying a carbon p orbital.* Thus, only six valence electrons are used to make the four C–H bonds, not unlike the situation with three center–two electron bonding. This nicely explains why tetrahedral methane is more

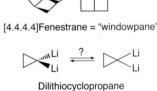

[4.4.4.4]Fenestrane = "windowpane"

Dilithiocyclopropane

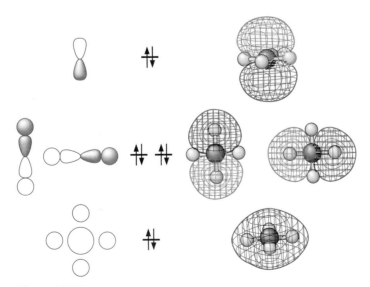

Figure 14.25
The MOs of planar methane. Shown are qualitative descriptions of the MOs along with the actual MOs from an *ab initio* calculation.

stable than planar methane. The tetrahedral structure involves all eight valence electrons in C–H bonding, while the planar form uses only six.

Along with providing an MO explanation of why tetracoordinate carbon is tetrahedral rather than planar, the orbitals of Figure 14.25 suggest a novel way to stabilize a planar carbon. We can anticipate that any factors that stabilize the lone pair orbital will stabilize planar carbon. We want substituents that can act as π acceptors, with low-lying, empty p/π orbitals. Another factor is the smaller C–C–C angles in the planar form vs. the tetrahedral (90° vs. 109.5°). Considering these factors led Schleyer and co-workers to propose candidate structures such as 1,1-dilithiocyclopropane. The small ring favors small angles and the empty orbitals on Li can stabilize the lone pair. Indeed, *ab initio* computational studies predict that 1,1-dilithiocyclopropane and related molecules should prefer a planar geometry. Poly-lithiated molecules have been prepared in the gas phase, but to date no structural information that could provide support for such structures has been forthcoming.

14.5.4   Through-Bond Coupling

We next consider an "effect" that when first observed seemed quite anomalous, but with modern perspectives on bonding is now seen to be quite reasonable. The problem can be set up (Figure 14.26) by considering 1,4-cyclohexadiene (1,4-dihydrobenzene). For the purposes of discussion we will consider the molecule to have a planar carbon framework.

The question is the nature of the π orbitals associated with the two double bonds. A molecular orbital analysis would begin with linear combinations of the π orbitals: an in-phase (symmetric, S) and an out-of-phase (antisymmetric, A). To the extent that there is any direct interaction between the double bonds, one would expect S to lie below A. Such a direct mixing is termed **through space**. The problem is that a number of observations lead to the conclusion that the actual orbital ordering in 1,4-cyclohexadiene is A < S. What has gone wrong? This is a planar molecule, so there is a π system that can be confidently treated as separate from the σ framework. Isn't it always better to mix orbitals in-phase?

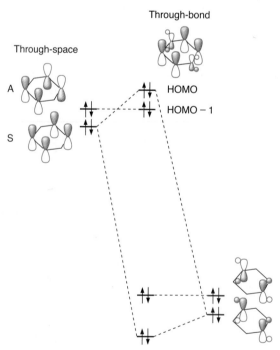

Figure 14.26
Through-bond coupling in planar cyclohexadiene.

The problem is that we have not considered all the π orbitals! A central theme of Chapter 1 was that CH_2 groups also have MOs of π symmetry, and cyclohexadiene is a perfect system for them to be involved. As shown in Figure 14.26, the $\pi(CH_2)$ orbitals lie on the nodal plane of the A combination of conventional π orbitals, and so they cannot mix. However, the $\pi(CH_2)$ orbitals can mix with the S combination. This mixing lowers the energy of the $\pi(CH_2)$ orbitals, but it raises the energy of the S combination of π orbitals. The interaction is strong enough that the final ordering is A < S. The actual molecular orbitals of cyclohexadiene confirm expectations (Figure 14.27).

As originally described, this mixing of conventional π molecular orbitals with $\pi(CH_2)$ orbitals was termed **through-bond coupling**, to distinguish it from the direct, through-space interaction. The π orbitals were described as mixing via the C–H bonding orbitals of the CH_2 groups. This is simply different terminology for what we now call a mixing of conventional π bonding orbitals with $\pi(CH_2)$ orbitals.

In analyzing cyclohexadiene, we emphasized the mixing of filled orbitals. We have noted frequently that the orbital that goes up in energy always goes up more than the other orbital drops in energy. Hence, if all the orbitals in Figure 14.26 are filled as shown, this mixing should be net destabilizing. Why, then, does it occur? A simple rationalization is to remember that the energy of a system is not influenced *only* by the orbitals we are examining. Overall stability or instability is related to all the interactions within the molecule. If filled orbitals are aligned, of the correct symmetry, and within the proper distance to overlap, they *will* mix. The only way to avoid the mixture is to introduce a distortion into the structure to remove the overlap. In this case the distortion must be more destabilizing than the closed-shell mixing. We confronted a similar analysis in Chapter 1 when examining why alkyl groups raise the energy of alkene bonding π orbitals (Section 1.3.5).

14.5.5 Unique Bonding Capabilities of Carbocations— Non-Classical Ions and Hypervalent Carbon

The structural diversity of carbocations is unparalleled in organic chemistry. Along with a large collection of conventional carbenium ions, a vast array of unusual and exotic bridged species, carbonium ions, have been observed or proposed. In Chapter 1 we discussed some aspects of the novel bonding possibilities for carbocations, emphasizing the role of three center–two electron bonding in producing bridged structures. We also discussed a simple prototype system, ethyl cation, which exists in two forms: a carbenium ion that is fairly conventional but experiences a strong hyperconjugative interaction, and a fully bridged carbonium ion. In Chapter 2, we noted how the bridging nature of some carbocations gives rise to extra stability, as evidenced by lower hydride affinity values. In Chapter 11 we examined substitution chemistry that proceeds via non-classical carbocations, and the interesting scrambling that occurs in the C–C skeletons of such structures. Here we expand the discussion of these cations, considering the electronic structures of a wide range of carbocations and some of the methodologies used to investigate them. As we will see, carbocations provide a number of excellent examples of the bonding principles and orbital mixing approaches we have developed in this chapter.

Inevitably, such discussions lead to the non-classical ion problem. Few subjects have generated as much experimental and computational effort—as well as debate, controversy, and animosity—as the non-classical ion problem. At one time it *appeared* that almost all of physical organic chemistry was focused on this single issue, and many have argued that this intensity, bordering on obsession, was ultimately detrimental to physical organic chemistry. With the benefit of hindsight, we can put this era into perspective and emphasize the positive outcomes of the diverse efforts to address the problem.

The non-classical ion problem focussed attention on a general issue in mechanistic/reactive intermediates chemistry: how do we distinguish an intermediate in a very shallow energy well from a transition state? In addition, the non-classical ion problem inspired the development of a number of new experimental and computational approaches that continue to serve physical organic chemistry well. We cannot hope to cover all the nuances and complexities of this issue in this text, especially when entire volumes have been written on

HOMO

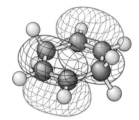

HOMO – 1

Figure 14.27
The HOMO and next lowest MO (HOMO – 1) of planar cyclohexadiene from an *ab initio* calculation. Note there is no contribution from $\pi(CH_2)$ orbitals in the HOMO – 1, but that they mix out-of-phase in the HOMO.

the subject. Here, we only consider structural issues—predictions about the structures of various ions based on theory and on experimental methods. A key to addressing this problem computationally was a method to address transition states. Hence, we must first take a sojourn into the methods used to calculate transition state structures.

Transition State Structure Calculations

It is convenient to phrase the key issues in terms of potential energy surfaces, as in Figure 14.28. We are dealing with a continuum of possibilities (a common theme in physical organic chemistry). We could have the "conventional" situation (Figure 14.28 **A**), in which there are two carbenium ions that interconvert via a single transition state that is a bridged structure (a carbonium ion). In prototype systems, this carbonium ion is a symmetrical structure relating two less symmetrical forms. Alternatively, the carbonium ion may itself be a stable structure also, and it could lie higher than (Figure 14.28 **B**) or lower than (Figure 14.28 **C**) the carbenium ion. Finally, we could have the situation in which the only stable structure on the surface is the symmetrical carbonium ion (Figure 14.28 **D**). The "non-classical ion problem" boils down to a question of whether the potential energy surfaces of Figures 14.28 **B**, 14.28 **C**, and, most especially, 14.28 **D** are viable.

Two major advances in theoretical organic chemistry have proven especially useful in addressing problems in carbocation chemistry. The first breakthrough occurred when analytical expressions for the first and second derivatives of the *ab initio* energy expressions with respect to normal coordinates of the molecule became available. As we noted in our discussion of geometry optimization using the molecular mechanics method (Section 2.6), having analytical first and second derivatives of the energy expression allows efficient, complete geometry optimization of molecules. While such expressions are straightforward to derive for molecular mechanics, it took some effort to develop them for *ab initio* wavefunctions. Prior to this advance, only approximate/partial geometry optimizations were possible. Given the subtlety of the non-classical ion issue, this was unsatisfactory. We discuss the application of *ab initio* geometry optimizations here in the context of carbocations, but they are applicable to all molecules.

An important aspect of *ab initio* geometry optimizations is the analysis of the derivatives of energy with respect to atom positions. It allows us to distinguish a true minimum from a transition state. As shown in Figure 14.28 **E**, when all the internal forces on a molecule are zero—that is, when all the first derivatives dE/dx (where x is a geometrical coordinate) are zero—we have a structure that is called a **stationary point**. By internal forces, we mean the driving forces to change bond length, bond angles, torsion angles, etc. Figure 14.28 **E** illustrates this for a single geometrical coordinate. In a calculation, the geometrical coordinates are the **normal coordinates** of the molecule (discussed in Chapters 2 and 7). The normal coordinates are all the vibrational modes available to the molecule. There are $3N - 6$ normal coordinates to most organic molecules, where N is the number of atoms. The first derivatives of energy with respect to each normal coordinate, called components of the **gradient**, are slopes, and so a stationary point corresponds to a point where are all the slopes are zero. The sign of the second derivative of the energy expression, d^2E/dx^2, tells us whether a stationary point is a minimum or a maximum, the former corresponding to a positive second derivative, the latter a negative.

A stable molecule is a structure with all first derivatives of the energy expression with respect to the normal coordinates equal to zero, and with all second derivatives positive. The molecule is at a minimum in energy with respect to all normal modes. Now, recall the definition of a transition state as the highest energy point on the lowest energy path between reactants and products (Section 7.1.2). A transition state is at a maximum along one and only one normal coordinate, the reaction coordinate. The second derivatives, also known as **eigenvalues**, thus provide an unambiguous way to distinguish a stable molecule from a transition state (these are not the same eigenvalues that represent energy discussed in Section 14.1.2). A stable molecule has all positive eigenvalues, whereas a transition state has one and only one negative eigenvalue. Theory is thus able to unambiguously determine which of the potential energy surfaces of Figure 14.28 fits a particular system. One seeks a single negative eigenvalue in an *ab initio* geometry optimization to find a transition state.

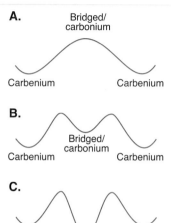

A.   Bridged/carbonium

Carbenium   Carbenium

B.

Bridged/carbonium

Carbenium   Carbenium

C.

Carbenium   Carbenium

Bridged/carbonium

D.

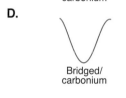

Bridged/carbonium

E. $dE/dx = 0$

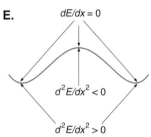

$d^2E/dx^2 < 0$

$d^2E/dx^2 > 0$

Figure 14.28
A–D. Potential energy surfaces for hypothetical carbocation rearrangements, highlighting the continuum of possibilities. **E.** Definitions of stationary points on a potential energy surface.

The second important theoretical advance for the study of carbocations was the development of reliable *ab initio* methods to calculate NMR shifts of organic molecules. The most popular of these is **IGLO** (individual gauge for localized orbitals), developed by Kutzelnigg and Schindler. The theory and mathematics behind IGLO are beyond the scope of this text, but lead references are given at the end of the chapter. Extensive studies have shown that with extended basis sets and geometries optimized at fairly high levels of theory (such as MP2), excellent agreement between calculated and observed NMR chemical shifts can be obtained.

Application of These Methods to Carbocations

The interaction between theory and experiment has been especially fruitful for carbocation chemistry for several reasons. First, a large quantity of gas phase thermodynamic data on carbocation stabilities is available (Table 2.8), allowing direct comparisons with computed values, which typically refer to the gas phase. Second, a large amount of structural data, mostly by NMR, is available under conditions of stable ion media. Importantly, as we noted in Chapter 2, compelling thermodynamic data establish that direct comparisons between gas phase and stable ion media results are possible, allowing theory to address stable ion media results. In addition, the carbocations of interest are fairly small molecules, and so very high levels of theory are applicable. Finally, compared to other reactive intermediates, carbocations benefit from their electron deficiency. That is, carbocations put one fewer electron into the given basis set. This leads to less of a contribution from correlation energy. This may seem like a small effect, but it does make cations much easier to treat than anions. The open-shell nature of radicals, carbenes, and biradicals adds additional problems, such that carbocations are the easiest of the reactive intermediates to treat theoretically. For all these reasons, in more recent years theory has played an increasingly important role in resolving questions concerning non-classical ions.

NMR Effects in Carbocations

The presence of a cationic center has a profound effect on chemical shifts. Shown in the margin are the ^{13}C NMR shifts for two typical ions. The cationic carbon is shifted *very* far downfield, while adjacent carbons are also shifted downfield, but not to such a large extent. This downfield shifting has been found to be fairly general, and it can be quantified according to Eq. 14.64, where δ is the ^{13}C chemical shift of a carbon. Thus, Δ_Σ is the sum of the downfield shifts of *all* carbons of the cation, compared to the analogous hydrocarbon (addition of hydride to the ion). Study of a number of typical carbenium ions shows that Δ_Σ is fairly constant, generally in the range of 350–390 ppm. In some instances even larger values have been observed.

$$\Delta_\Sigma = \sum \delta(R^+) - \sum \delta(RH) \qquad \text{(Eq. 14.64)}$$

We might anticipate, however, that the three center–two electron bonding of carbonium ions might lead to atypical values of Δ_Σ. In particular, we might expect the chemical shifts of hypervalent carbons to be unusual, and indeed they are shifted far upfield from what is seen for carbenium centers. This leads to greatly reduced values of Δ_Σ.

There are cations that all workers agree have unconventional structures. Figure 14.29 shows a simple example, the 7-norbornenyl cation. The value of Δ_Σ is clearly outside the standard range, indicating that the interaction between C7, which carries the formal cationic center, and the C2–C3 double bond is so strong that a new bonding situation arises. In effect, these three carbons are linked together by two "π" electrons, a typical three center–two electron bonding situation. We also show a calculated structure for this ion, along with the HOMO and the LUMO. Note how C7 "leans" very strongly toward the olefinic carbons. The HOMO and LUMO are exactly as we would expect from orbital mixing arguments. We have the in-phase and out-of-phase combinations of the olefin π bond and the empty *p* orbital at C7. As expected, the C2–C3 double bond is longer than a typical double bond, because π electron density has been removed from the C2–C3 bonding region and delivered to C7.

$\delta = 4.15$
$\delta = 335$
H_3C
CH_3
H_3C
$\delta = 47.5$

$\delta = 4.5$
$\delta = 321$
H_3C
$H \longleftarrow \delta = 13$
H_3C
$\delta = 51.5$

^1H and ^{13}C NMR of carbocations

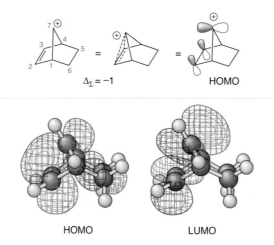

Figure 14.29
The norbornenyl cation. Shown in schematic form are the limiting structures. Also shown is a calculated geometry along with the HOMO and LUMO.

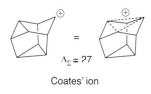

Coates' ion

If an appropriately positioned olefin can interact with a cationic center to make a carbonium ion, we might expect that the *edge* of a cyclopropane ring could do the same. An example of this is the ion derived from pentacyclo[4.3.0.0^{2,4}.0^{3,8}.0^{5,7}0]nonane, which is known as **Coates' ion** (shown in the margin). Even the most ardent opponents of the non-classical ion concept accept that this is a bridged, symmetrical carbonium ion. The value of Δ_Σ fully supports this interpretation.

The unusual chemical shifts associated with carbonium ions carry over to ^{1}H NMR as well. Eq. 14.65 shows the novel bridged structure obtained when cyclodecanol is exposed to the highly acidic environment associated with stable ion media. The remarkable upfield shift seen for one proton confirms its bridging nature. This system is a beautiful example of the types of transannular effects discussed in Section 2.3.2.

$$\text{(Eq. 14.65)}$$

The Norbornyl Cation

Both the conventional (carbenium ion) and non-classical (carbonium ion) representations of the 2-norbornyl cation are shown in Figure 14.30. If, for the moment, we assume a traditional structure for the ion, the system is very well set up for rearrangements during substitution reactions. Yet, the non-classical bridged structure also explains the myriad of products found in substitution reactions (see Section 11.5.14). The structures of Figure 14.30 (obtained from quantum mechanical calculations) illustrate how subtle this distinction is. It is just a matter of whether a CH_2 group leans a bit to the right or not. Such a subtle difference is what constitutes the core of the non-classical ion problem, especially with regard to norbornyl cation.

The NMR data support the bridged structure. Under stable ion conditions, the value of Δ_Σ (Eq. 14.64) is 175 ppm. This is far below the usual range seen for conventional carbenium ions, although the deviation is not as dramatic as that seen for the Coates' ion. The isotopic perturbation of equilibrium technique (Chapter 8) has also been applied. At very low temperatures, the C1 and C2 of norbornyl cation show a single, merged line in the ^{13}C NMR spectrum. When a 2-deuterio precursor is used in an effort to break this degeneracy, a single,

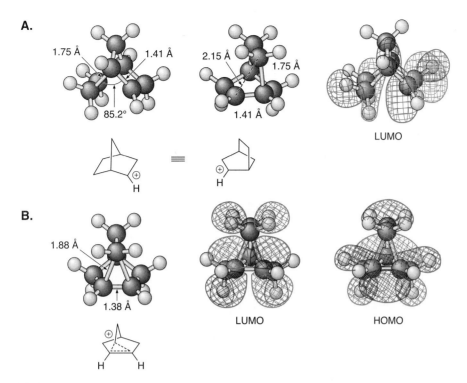

Figure 14.30
Structures for the norbornyl cation. **A.** The classical carbenium ion form. Two views of the computed structure are given along with the LUMO. Note the extensive contribution of the C1–C6 bond to the LUMO. **B.** The bridged non-classical structure. A structure is given, along with the LUMO and HOMO. Note the similarity of these MOs to those of the bridged ethyl cation (Figure 1.25).

broad line is still seen, indicating a value of $\delta \leq 2.3$ ppm. Again, this is much smaller than is seen for conventional cations, supporting the bridged structure. It seems undeniable that, under stable ion conditions, the 2-norbornyl cation has the symmetrical, bridged, carbonium ion structure. This proves that such a structure can be stable, and so it must be considered as a plausible reactive intermediate under conventional conditions. The remaining issue, then, is whether the symmetrical, bridged norbornyl cation is indeed involved in solvolysis reactions discussed in Chapter 11. While it is ultimately very difficult to *prove* that the symmetrical ion plays a key role in these reactions, all the data are consistent with its involvement.

14.5.6 Spin Preferences

In Chapter 1 we briefly mentioned the issue of differing spin states in carbenes. Chapters 10 and 15 also highlight differing reactivities of singlet vs. triplet states. Here we consider the electronic structure factors that control spin states in organic molecules. Not surprisingly, a thorough understanding of spin states requires consideration of electron–electron repulsion at a bare minimum, and often electron correlation effects are important. We are now armed with the necessary theoretical background to attack this intriguing problem. We also note that because electron spin is at the heart of magnetism, an understanding of molecular spin preferences is crucial in efforts to design organic magnetic materials, as discussed in Chapter 17.

Two Weakly Interacting Electrons: H_2 vs. Atomic C

In essence, we are talking about systems that have two weakly interacting electrons and asking what is the preferred spin state. Carbenes and nitrenes are examples of reactive intermediates that contain two weakly interacting electrons, as are biradicals. Figure 14.31 reviews the basics of spin states and shows two prototypes for the two-electron system: H_2 at

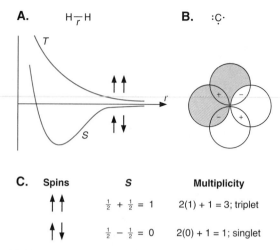

C.

Spins	S	Multiplicity
↑ ↑	$\frac{1}{2} + \frac{1}{2} = 1$	2(1) + 1 = 3; triplet
↑ ↓	$\frac{1}{2} - \frac{1}{2} = 0$	2(0) + 1 = 1; singlet

Figure 14.31
Prototypical spin preferences. **A.** At all separations, r, molecular hydrogen shows a singlet ground state. **B.** Atomic carbon shows a triplet ground state because of a cancellation of regions of positive and negative overlap of the $2p$ atomic orbitals. **C.** Possible spin states for the two-electron system.

long bond lengths and atomic carbon. Both have two weakly interacting electrons, but their spin preferences are quite different.

At bonding distances H_2 is, of course, a singlet, with the triplet as a very high-lying excited state. In order to make the bond in H_2, we must put two electrons into the bonding MO, and to do this and satisfy the Pauli principle, these electrons must be spin paired. In the triplet state, one electron is in the bonding MO and one is in the antibonding MO, and there is no H–H bonding. As we stretch the H–H bond, the energy gap between the bonding and antibonding MOs collapses, and the singlet and triplet states converge. However, high-level calculations consistently establish that at no point does the triplet drop below the singlet. Absent any special effects, two weakly interacting electrons will form a singlet, because a little bit of bonding is better than no bonding at all. This is an important and general result. It tells us that if we bring two radicals together whose orbitals can overlap, whether they are part of the same molecule (a biradical) or two monoradicals near each other in solution or in a crystal, the most common result is a singlet (low-spin) state.

Atomic carbon has four valence electrons, two paired in a carbon $2s$ orbital and two that occupy the degenerate trio of $2p$ orbitals. The two electrons in the p orbitals are high-spin paired, and atomic carbon has a triplet ground state. Why is it that these two weakly interacting electrons produce a triplet ground state, in contrast to the situation in stretched H_2? It turns out that exchange integrals play a key role here. The two p orbitals that contain the electrons on carbon are orthogonal due to their nodal properties. However, this does not mean that the electrons do not interact. Orthogonal means that the overlap *integral* between the orbitals is zero. Overlap favors bonding, and bonding requires a low-spin state, so one requirement for a high-spin state is zero or near zero overlap. As the H_2 result shows, even very weak overlap can produce a preference for a singlet state. The reason the AOs in H_2 have minimal overlap is that they are very far apart. However, as shown in Figure 14.31, the reason the AOs in atomic carbon have zero overlap is that regions of positive overlap are canceled by regions of negative overlap. The two orbitals interact substantially by partially occupying the same space, defined as **co-extensive**, but the *net* interaction is zero. This is an underappreciated feature of AOs on the same atomic center—they are very much co-extensive in space. The reason this is important is that the exchange integral, K, does not follow the same kinds of nodal properties as the overlap integral. To a good approximation, the values of exchange integrals track with the overlap of the *squares* of the orbitals. There are no sign differences once we square the orbitals, and so regions of positive and negative overlap do not cancel. As such, in atomic carbon the overlap integral is zero, but the exchange integral is not. Recall our discussion in Section 14.2.1 that the exchange integrals are stabilizing, but only for parallel spins. If there is no bonding opportunity, and thus no strong driving force for spin pairing, the high-spin state is preferentially stabilized. This is the origin of Hund's rule. The situation with carbon is a perfect recipe for a high-spin state, and atomic carbon has a triplet ground state. When we stretch H_2, we diminish both overlap and exchange integrals, and overlap wins out to produce a singlet at all distances.

The results of the discussion of H_2 and atomic carbon can be generalized. Typically, weakly interacting radicals will produce a weak preference for the singlet state. However, when the nodal properties of the system are such that the orbitals that contain the single electrons are orthogonal but co-extensive in space, significant exchange interactions result, and this can produce a triplet ground state.

With this analysis in hand, we can now look at two prototype biradical systems: cyclobutadiene (CBD) and trimethylenemethane (TMM). As shown in Figure 14.32, CBD and TMM have qualitatively equivalent MO diagrams. Both have a degenerate pair of nonbonding molecular orbitals (NBMO) occupied by two electrons. However, the spin preferences of the two systems are completely opposite: CBD is a singlet and TMM is a triplet. The magnitudes of these preferences are comparable: ~10 kcal/mol favoring the singlet in square CBD, and ~14 kcal/mol favoring the triplet in TMM.

Closer consideration of the NBMOs of these systems explains the differing spin preferences. For a degenerate pair, any linear combination of the MOs is acceptable. For CBD, we can make a linear combination of the NBMOs such that they have no atoms in common (Figure 14.32 **A**). One NBMO is confined to atoms 1 and 3, while the other is confined to atoms 2 and 4. Such orbitals are said to be **disjoint**; they occupy different sets of atoms. In contrast, it is not possible to find a linear combination of the NBMOs of TMM that are disjoint (Figure 14.32 **B**). No matter what you do, there will always be atoms in common, and the NBMOs are termed **non-disjoint**.

In light of our discussion of exchange and overlap, we can see that whether the NBMOs are disjoint or not should be important. In both CBD and TMM the NBMOs are orthogonal. However, for disjoint NBMOs such as seen in CBD, the exchange integral is also very nearly zero. The two nonbonding electrons are not co-extensive in space. It is as if one electron is on atoms 1 and 3 and the other is on atoms 2 and 4. Now, both to first order (one-electron theory) and to second order (including electron–electron repulsion), the singlet and triplet are degenerate. Technically this is all we can say at this stage (the Going Deeper highlight on the next page looks at a higher level of theory). However, experience has shown that when electron correlation effects are included in the analysis of a disjoint system, the singlet state is preferentially stabilized and becomes the ground state.

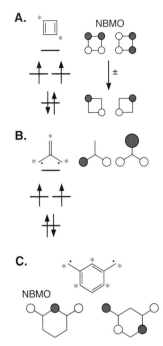

Figure 14.32
The similarities and differences between CBD and TMM.
A. CBD. Note that taking plus and minus combinations of the familiar NBMOs produces a pair of disjoint NBMOs. **B.** TMM. No mixture of the NBMOs can produce a disjoint pair.
C. Another prototype biradical, *m*-xylylene, which has an excess of * atoms and thus nondisjoint NBMO and a triplet ground state.

Going Deeper

Through-Bond Coupling and Spin Preferences

A subtle effect of through-bond coupling is an influence on the spin preferences of certain simple biradicals. Both 1,3-cyclobutanediyl and 1,3-cyclopentanediyl have triplet ground states. A simple analysis of such systems would model them as two *p* orbitals that are only weakly interacting, each containing one electron. The two radical centers are roughly 2.10 Å and 2.37 Å apart in 1,3-cyclobutanediyl and 1,3-cyclopentanediyl, respectively. This looks like just a "π" version of stretched H_2, and so we should expect a singlet ground state. Indeed, calculational studies of two singly occupied *p* orbitals at these distances do produce singlet ground states. However, we now should suspect that the π(CH₂) orbitals will be

involved, and they have a profound effect. Briefly, in both systems the direct, through-space interaction produces an S < A ordering (analogous to the S and A orbital symmetries of 1,4–hexadiene discussed in Section 14.5.4), and the gap is larger in the cyclobutanediyl because of the shorter distance. However, mixing in the π(CH₂) orbitals (the CH₂CH₂ bridge of the five-ring can be ignored) raises the S orbital, and in each case this produces a near degeneracy of the S and A NBMO. Thus, these simple localized biradicals have triplet ground states.

Cyclobutanediyl   Cyclopentanediyl

For the non-disjoint NBMOs of TMM the situation is quite different. Now the exchange interactions are significant, because the degenerate but orthogonal orbitals have character on the same atoms (co-extensive). A triplet ground state results. We can generalize these observations as follows. When the NBMOs of a biradical are disjoint, we expect the singlet and triplet to be close in energy, and often higher order effects will produce a singlet ground state. When the NBMOs are non-disjoint, exchange interactions are significant, and we expect a triplet ground state.

Earlier in this chapter we introduced the concept of alternate hydrocarbons (AH). TMM and CBD are both even AHs, but there is a significant difference. While CBD has two * and two non-* atoms, TMM has three * and one non-*. It is not a coincidence that these two systems with differing spin preferences also appear different in the */non-* analysis. The */non-* rule is basically a topology rule—it reflects the connectivity of the system. The topology of the system also influences the nature of the NBMOs. In fact, it can be shown that a system with an excess of * over non-* atoms will in general have non-disjoint NBMOs, while a system with equal numbers of * and non-* atoms will have disjoint NBMOs.

Figure 14.32 **C** shows the NBMOs of *m*-xylylene, another prototype high-spin system. This structure also has two more * than non-* atoms. As such, its NBMO are non-disjoint, and a triplet state for the biradical is preferred. In our discussion of organic magnetic materials in Section 17.3 we will see that this */non-* approach to predicting spin states can be extended to a remarkable degree, allowing the rational design of very high-spin organic molecules.

Going Deeper

Cyclobutadiene at the Two-Electron Level of Theory

As we just discussed, the HMO triplet state for cyclobutadiene is modified significantly once we include exchange interactions. Now cyclobutadiene has a singlet ground state. Here we consider one additional feature of the singlet state. It is not square (look back at Eq. 2.34). Rather, it is rectangular, with alternating long (1.54 Å) and short (1.37 Å) bonds. It really is more like a diene than a resonance hybrid of two equal forms. Why does this distortion occur? Consider the MOs of cyclobutadiene, and the consequences of a distortion from square to rectangular (now it is more convenient to use the linear combination of NBMOs shown). On going to the rectangular form, one MO will be significantly stabilized, and one will be destabilized, opening up a significant HOMO–LUMO gap. At higher levels of theory, this is definitely stabilizing. The square form of cyclobutadiene is actually

the transition state for the interconversion of equivalent rectangular forms, and it lies 5–10 kcal/mol above the ground state. Such a change from a high symmetry form to a lower symmetry form is called a **pseudo Jahn–Teller distortion**. It is an effect that is only seen at higher levels of theory, and it has a significant impact on the molecular and electronic structures of many systems.

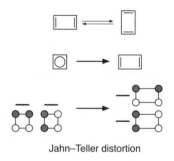

Jahn–Teller distortion

14.6   Organometallic Complexes

We now introduce simple concepts in organometallic and inorganic bonding. This continues our theme of treating organometallic complexes alongside organic complexes. Fortunately, once one is knowledgeable of organic structure and bonding, there is a very simple extension that can be applied to organometallic and inorganic structure and bonding called the isolobal analogy. This notion of bonding is essentially a continuation of our group orbital and molecular fragment concepts of Chapter 1, except in the setting of organometallic and inorganic complexes. Before looking at this, we examine a simple molecular orbital diagram for octahedral complexes, because this is a launching point for examining bonding in other structures.

14.6.1 Group Orbitals for Metals

Organic structures can be viewed as being constructed from groups of methyls, methylenes, carbonyls, alcohols, etc. The group orbital concepts from Chapter 1 form a good theoretical vantage point for this notion, because to achieve a qualitative feeling for the bonding in organic molecules, all we need to know is the bonding within the fragments.

Now let's consider organometallic structures. We once again find distinctive groups that are incorporated repeatedly in the structures—carbonyls, cyclopentadienyls, dienes, alkylidenes, etc., and metals. Just as with organic molecules, we can consider these groups to contribute a specific set of orbitals in the construction of their respective metal complexes. In other words, once again, for a simple qualitative understanding of the bonding in these complexes, all we need to know is the electronic structures of the fragments and the metal.

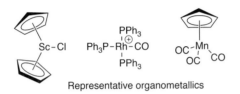

Representative organometallics

Let's start by examining the electronic structures of metals. We begin by recalling the d orbitals given in Figure 1.26. These atomic orbitals can be used to make bonds, house lone pair electrons, or remain empty, depending upon the particular organometallic complex. In the LCAO–MO method, the basis sets for metals contain functions that represent the shape and energies of the atomic d orbitals. Hence, *ab initio*, semi-empirical, or approximate methods can all be used to calculate the electronic structures of metal-containing complexes. The results of the calculations show specific patterns of orbitals, much as with organic structures. Many of these patterns can be understood by considering metals as adopting, at least initially, octahedral-like coordination geometries.

As discussed in Chapter 1 with regard to octahedral complexes, the s, all three p orbitals, and the d_{z^2} and $d_{x^2-y^2}$ orbitals can all mix to create six identical d^2sp^3 hybrid orbitals (Figure 14.33 **A**). On the left-hand side of this diagram are shown lobes representing these hybrids, along with what is known as the t_{2g} set. The t_{2g} set contains the d_{xy}, d_{xz}, and d_{yz} orbitals. They are lower in energy than the hybrid orbitals because the hybrids contain character from s and p orbitals of a principle quantum number one higher than the d orbitals. (Recall that the $3d$ orbitals are in the same row of the period table as the $4s$ and $4p$ orbitals.) On the right-hand side of Figure 14.33 **A** are six orbitals representing six identical ligands. Mixture of these ligand orbitals with the d^2sp^3 hybrids gives a very simple molecular orbital diagram with six degenerate bonding orbitals and six degenerate antibonding orbitals. One can envision each of the degenerate orbitals as a discrete and localized σ bond between the metal and each ligand; a valence bond model.

As with organic structures, molecular orbital theory calculations do not give hybrid orbitals and discrete localized bonds. Figure 14.33 **B** shows a molecular orbital treatment of the bonding in an octahedral complex. Instead of six equivalent hybrid orbitals plus the t_{2g} set, we now have symmetry adapted orbitals, according to octahedral symmetry. To create the mixing diagram, prior to mixing the ligand orbitals with the metal orbitals, the ligand orbitals are mixed with themselves. This is done with an eye towards creating mixtures of proper symmetry to mix with the d orbitals. The resulting molecular orbitals are now more complex than shown in part **A**, but they better mimic what is found in *ab initio* calculations. One of the essential features of inorganic and organometallic octahedral complexes is the same in the two diagrams—namely, the t_{2g} set is unperturbed. However, with the MO ap-

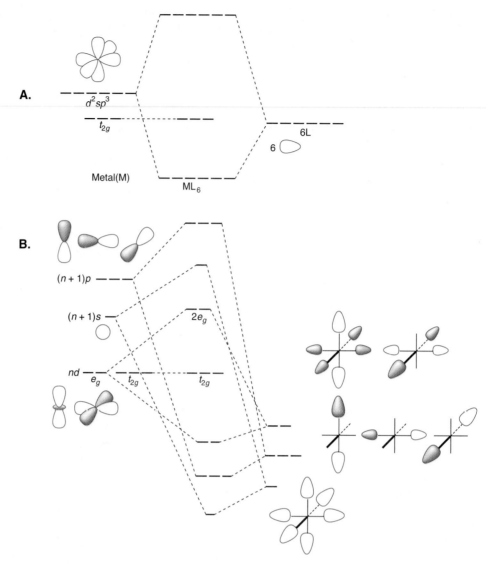

Figure 14.33
Molecular orbital diagrams for octahedral complexes. **A.** The mixing diagram for six equivalent ligands that bond to a transition metal possessing d^2sp^3 hybrid orbitals. **B.** The MO diagram for six equivalent ligands that bond to an unhybridized metal.

proach, another set of orbitals, called $2e_g$, is found near in energy to the t_{2g} set. These molecular orbitals have signficant d_{z^2} and $d_{x^2-y^2}$ character.

We can use either model given in Figure 14.33 to predict d orbital population. An octahedral molecule with six identical ligands is $Cr(CO)_6$, which is known for its stability. Each CO ligand contributes two electrons while the metal contributes six electrons (see the discussion of electron counting for metals given in Section 12.1.1). After placing 18 electrons into the diagram, a completely closed-shell arrangement is found in either model. The t_{2g} set contains six electrons. You can use the diagrams in Figure 14.33 for various electron counts to predict the orbital population for octahedral complexes.

The molecular orbital diagrams for square planar, tetrahedral, and trigonal bipyramidal complexes become quite complicated. However, our goal here is to develop a model that yields a qualitative "feeling" for the bonding. In this regard, it is actually the hybridization picture of bonding that serves as a useful starting point. To see this, we continue to analyze

A.   **B.**

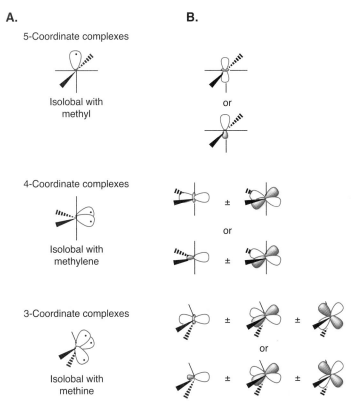

Figure 14.34
The group orbitals for ML_5, ML_4, and ML_3 complexes available for
making bonds to other fragments. **A.** Hybridization view of the group
orbitals. **B.** Some of the possibilities for combining atomic orbitals to
construct the group orbitals. Alternatively, these orbitals can also be
considered the group orbitals of the fragments.

octahedral geometries, with various ligands removed as a means of creating molecular fragments that can be used to make new complexes.

Let's consider a basic octahedral core for systems that contain only five, four, and three
ligands (Figure 14.34 **A**), with each ligand contributing two electrons. The bonding MOs between the metal and these ligands drop in energy and are completely filled with electrons.
We do not need to consider these. The interesting orbitals are those that are left over. Using
the d^2sp^3 hybrid orbitals, we have one, two, or three unused lobe orbitals in the five-, four-,
and three-coordinate systems, respectively. These left over orbitals, along with the t_{2g} set, are
a representation of the "group orbitals" that can be used to construct bonds to other fragments. The t_{2g} set is often not used in bonding, but it is available to donate or accept electrons
depending upon the electron count of the complex.

The orbitals shown in Figure 14.34 **A** are just lobes. However, the atomic orbitals that are
the origin of these lobe orbitals can be very useful in creating molecular orbital diagrams. As
shown in Figure 14.34 **B**, there are different possibilities for viewing the origin of the lobes in
part **A** of the figure. Only a few of the possibilities are shown. As an example, let's consider
the four-ligand case shown in the middle of the figure. The two lobes can be considered as
arising from the mixture of a d orbital with π symmetry with either a σ(out) hybrid or a d orbital possessing σ symmetry. The origin of the other lobe orbitals can likewise be viewed in
various ways (see the figure). These lobe orbitals constitute a set of "group orbitals" that can
make bonds to other molecular fragments. We consider the orbitals in both Figures 14.34 **A**
and 14.34 **B** as the frontier orbitals for the metal fragments.

14.6.2 The Isolobal Analogy

Now that we have a picture of the frontier orbitals for several metal fragments, we can consider combining these fragments with other fragments, and making correlations with organic structures. One way to do this is known as the **isolobal analogy**. It is a thought process whereby we make comparisons between the group orbitals of organic fragments and metal fragments, correlating their structures, the structures of the molecules created from them, and their reactivities. Two fragments are known as **isolobal** if the number, symmetry properties, approximate energies, and shapes of the frontier orbitals are similar, and the number of electrons in these orbitals is the same. Importantly, before deciding what groups are isolobal, we need to know the electron count on the organometallic group.

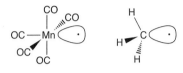

Consider the $Mn(CO)_5$ fragment and the methyl group. Methyl radical has a lone radical electron. Similarly, $Mn(CO)_5$ has a radical electron, because the metal has seven electrons from the metal, and the CO ligands contribute a total of 10 electrons, making a 17-electron complex. Both have a single electron in an orbital protruding away from the other atoms in the molecule. The orbital on methyl is the $\sigma(out)$ group orbital (see Chapter 1), while the radical orbital for $Mn(CO)_5$ can be viewed as shown in Figure 14.33 **A**. The $Mn(CO)_5$ and methyl fragments are considered isolobal, because their frontier orbitals are similar. In fact, they behave similarly. $Mn(CO)_5$ dimerizes to form an Mn–Mn bond, and similarly methyl dimerizes to make ethane. Furthermore, they can react with each other to make $CH_3Mn(CO)_5$.

Using the isolobal analogy, we can understand a number of different organometallic complexes. Any species with a single electron should be isolobal with methyl. This not only includes $d^7 ML_5$ species, but also $d^9 ML_4$ species (17-electron complexes). As an example of a $d^9 ML_4$ species, $Co(CO)_4$ is isolobal with methyl. It combines with alkyl groups and itself. Furthermore, $CpFe(CO)_2$ is isolobal (17-electron species). Indeed, $CpFe(CO)_2$ dimerizes and makes single bonds to alkyl fragments.

As another example, let's examine $Fe(CO)_4$, a $d^8 ML_4$ fragment. This is a 16-electron complex, and therefore has two radical electrons. The frontier orbitals are those shown in Figure 14.34 for the four ligand case. Both $Fe(CO)_4$ and CH_2 have two electrons in their frontier orbitals. Hence, $Fe(CO)_4$ is isolobal with CH_2. If we consider the middle of Figure 14.34 **B**, the group orbitals resemble the $\sigma(out)$ and p orbitals of CH_2 given in Chapter 1. Indeed, the following structures are known to form, although the last one has only been isolated in a frozen matrix.

$$CH_2=CH_2$$

Organometallic analogues of ethylene

In summary, the isolobal concept makes meaningful connections between inorganic fragments and the common organic groups. The lobe orbitals of Figure 14.34 give the fastest insight into which metal fragments are isolobal with methyl, methylene, and methine. Table 14.1 shows the isolobal relationships, where each ligand (L) contributes two electrons. This table makes a convenient reference point when considering the kinds of structures possible when combining organometallic fragments with organic fragments.

Table 14.1
Isolobal Relationships*

Organic groups	Metal fragments				
Methyl	$d^9 ML_4$	$d^7 ML_5$	$d^5 ML_6$	$d^3 ML_7$	$d^8 CpML_2$
Methylene	$d^8 ML_4$	$d^6 ML_5$	$d^4 ML_6$	$d^2 ML_7$	$d^9 CpML$
Methine	$d^9 ML_3$	$d^7 ML_4$	$d^5 ML_5$	$d^3 ML_6$	$d^1 ML_7$	$d^{10} CpM$

*Each neutral ligand (L) contributes two electrons. The d count refers to the metal atom without ligands.

14.6.3 Using the Group Orbitals to Construct Organometallic Complexes

As examined above, the isolobal analogy not only leads to ties between inorganic or organometallic complexes and organic structures, but it also gives a convenient manner to visualize the group orbitals for metal fragments. We can use these group orbitals to build up the bonding in complex molecules. Let's do this for one case. We examine how the group orbitals that derive from the isolobal analogy can be used to predict the bonding in a complicated organometallic complex that has no organic analogy.

Consider the MO diagram given in Figure 14.35. Here, $Fe(CO)_3$ is coordinated to trimethylenemethane (TMM), which at first glance seems like a difficult structure to analyze. Yet, using the group orbitals for an ML_3 complex from Figure 14.34 **B** immediately reveals how the MOs are constructed. Note how linear combination of the three $Fe(CO)_3$ lobes produces the familiar "one down, two up" pattern of three-orbital mixing. The phasing and symmetry of the two degenerate MOs of TMM match the phasing and shape of the two ML_3 π symmetry orbitals, while the low-lying TMM MO properly matches the ML_3 σ symmetry orbital.

Without great experience or a calculation, it would be difficult to predict bonding for $(TMM)Fe(CO)_3$. Yet, the combination of the metal group orbitals and the π orbitals of TMM naturally leads to the appropriate molecular orbitals. The important point is that the orbitals for the three-ligand case of Figure 14.34 match the symmetry of the TMM orbitals, allowing one to clearly predict that such a complex should be stable.

In summary, the isolobal analogy ties the group orbital concepts from organic chemistry to organometallic and inorganic chemistry. Fragments with similar frontier orbitals create similar structures. The method is particularly insightful for organic chemists, who are naturally adept at considering the kinds of structures that combinations of organic fragments can create. By making the isolobal connection to an inorganic fragment, we can have similar insights into the vast possibilities for structures created from metal-containing fragments.

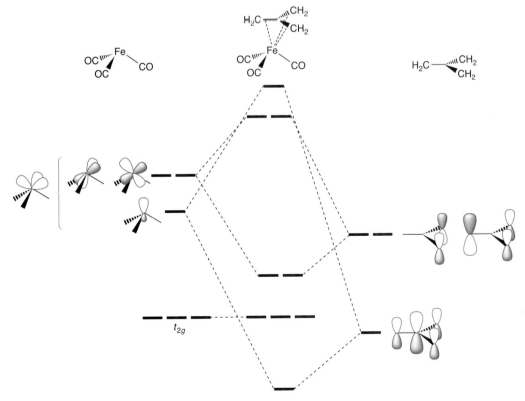

Figure 14.35
The MO diagram for the formation of $(TMM)Fe(CO)_3$.

Summary and Outlook

Many of the methods for electronic structure theory calculations were described in this chapter. We found that for small molecules, *ab initio* calculations can give great insight into bonding and common effects that we routinely use to rationalize chemical reactivity, such as hyperconjugation. We highly advise students of organic chemistry to learn how to use the common programs for *ab initio*, semi-empirical, and approximate methods and employ the results in your research. Hopefully, this chapter has given a foundation for these theories, such that the programs can be approached with less trepidation. However, these more sophisticated methods are not always necessary. Even methods as simple as Hückel theory can be extremely useful, explaining the bonding properties of π systems quite nicely.

Up to this point in the textbook, the valence bond concept of discrete and localized bonds between adjacent atoms and the use of group orbitals and QMOT were sufficient to understand reactivity and mechanism. Hence, we left the detailed analysis of full MOT until this late point in the book. The next few chapters, however, greatly benefit from an understanding of MOT. In Chapter 15—pericyclic reactions and theory—MOT is the launching point. The methods of predicting the regiochemistry and stereochemistry of pericyclic reactions begin with correlation diagrams and frontier orbitals—purely MOT concepts. Following this chapter is an analysis of photochemistry. Again, to understand electronic transitions in molecules, the molecular orbitals need to be understood. Finally, materials chemistry is strongly dependent upon the results of MOT. In the last chapter of this book, band structure, electrical conductivity, and magnetism, will all be explained using MOT. Hence, although our model of bonds between atoms using sp^3, sp^2, and sp hybridization states has taken us a long way, now that we have a foundation in molecular orbitals, we are ready to launch into new and exciting territory.

Exercises

1. Show all the math used to derive Eqs. 14.49 and 14.50, along with the energies of these wavefunctions, starting with Eq. 14.38.

2. a. Look at the following wavefunctions for molecular orbitals of methanol, all of which are filled. Considering the ∇^2 function, rank them from lowest energy to highest energy. What orbital(s) do you find as the HOMO? Does this lead you to any prediction as far as reactivity? Some orbitals will be very close in energy, and hence simply consider them as degenerate.

   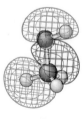 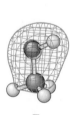 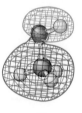 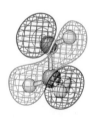

A.   B.   C.   D.   E.   F.   G.

b. Shown below is the LUMO of methanol. Does this suggest anything about the reactivity of methanol?

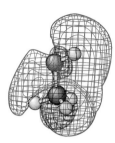

3. Using Figure 14.3 as a guide, write the full Hamiltonian for methane. Assume completely tetrahedral symmetry with all C–H bond lengths identical.

4. Using T_g and V_g plots (Figure 14.7), along with the resulting Morse potentials, show in a schematic fashion that bond strengths can be increased with lower electrostatic repulsions (between electrons or between nuclei), and with lower kinetic energies in the molecular orbitals relative to the atomic orbitals.

5. Explain in detail why a bond between two He atoms cannot form.

6. Read the following experimental descriptions and briefly discuss all the underlined terms.

Electronic structure calculations were carried out with the Gaussian 94 suite of programs at the levels of <u>second-order Møller–Plesset</u> (MP2) and <u>hybrid density functional theory</u> (B3LYP) with <u>basis sets</u> developed by Pople, McLean, and co-workers. Geometry optimizations with the <u>6–31G*</u> basis set were followed by single-point calculations with the <u>6–311G*</u> basis set.

Krogh-Jespersen, K., Yan, S., and Moss, R. A. "Ab Initio Electronic Structure Calculations on Chlorocarbene-Ethylene and Chlorocarbene-Benzene Complexes." *J. Am. Chem. Soc.*, **121**, 6269–6274 (1999).

All molecular orbital calculations were carried out using Spartan 4.1. Geometries and conformer studies were carried out with the <u>AM1</u> and <u>PM3</u> Hamiltonians using the default geometry optimizer and convergence criteria. Single-point <u>SCF</u> *ab initio* calculations were then done on those structures using one of several standard Gaussian basis sets.

Lipkowitz, K. B., Gao, D., and Katzenelson, O. "Computation of Physical Chirality. An Assessment of Orbital Desymmetrization Induced by Common Chiral Auxiliaries." *J. Am. Chem. Soc.*, **121**, 5559–5564 (1999).

7. Using Eq. 14.57, show that the energies in Figure 14.15 for allyl and butadiene are correct.

8. Using full Hückel theory with a secular determinant, derive the wavefunctions and molecular orbital energies for allyl and butadiene given in Figure 14.15. For allyl, you will derive a cubic equation whose roots x are $\sqrt{2}$, 0, and $-\sqrt{2}$. For butadiene, you will have to solve a fourth-order polynomial, the roots of which are $x = 0.618, -0.618, 1.618$, and -1.618.

9. Use the rules of Section 14.3.4 to draw the MOs of the π system for pentadienyl.

10. Provide coefficients for the NBMO of the odd alternate hydrocarbons given below.

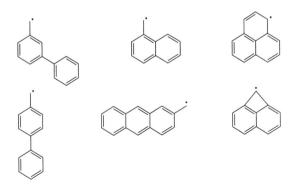

11. Derive the HMOs for the π system of cyclobutadiene.

12. Below is the photoelectron spectrum for ethane. Does this spectrum better agree with the MO diagram given in Figure 1.13, or does it best agree with six equivalent C–H bonds and a single C–C bond? Explain your answer.

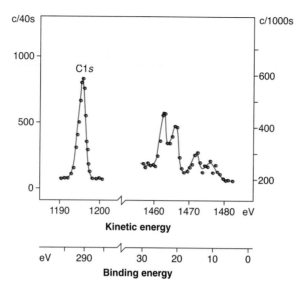

13. Show that $HFe(PR_3)_4$ and $CpFe(CO)_2$ are isolobal with methyl.

14. Use the methods of secular determinants to derive the MOs and energies for the cyclic three-orbital mixing problem. Let the three orbitals be simple s orbitals. As with Hückel theory, let the overlap integrals equal zero. The answer should be the same as given in Figure 14.20.

15. For dienes **1** and **4**, there is an issue as to the sequence of the two MOs associated with the olefinic π bonds. The "natural" order would have the in-phase, symmetric (S) combination lower in energy than the out-of-phase, antisymmetric (A) combination. Using the PES data below, build orbital mixing diagrams (**1** + **3** to make **2**; **4** + **6** to make **5**) to determine whether the orbital ordering is the "natural" one (S < A) or is reversed. For each system, draw an orbital mixing diagram and sketch all relevant orbitals.

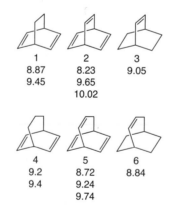

1	2	3
8.87	8.23	9.05
9.45	9.65	
	10.02	

4	5	6
9.2	8.72	8.84
9.4	9.24	
	9.74	

16. Why does it not matter whether 1,4-cyclohexadiene is planar for the through-bond analysis to hold?

17. Write the Hückel secular determinants for the following two polyenes.

18. Examine the mixing of the NBMOs of allyl and benzyl fragments to predict which of the following polyenes have the most stable π systems. Draw pictures of the bonding molecular orbitals that result for linear combination of the NBMOs. Rank the stability of the π systems based upon your analysis.

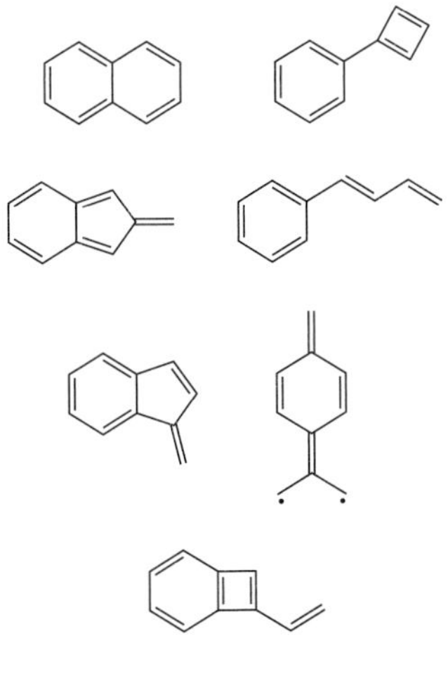

19. Using Eq. 14.56, confirm the π MO energies of cyclopentadienyl as given in Figure 14.12.

20. Use the cyclic HMOT rules to write the MOs for the high-energy degenerate pair of benzene. The answer should be the same as in Figure 14.13.

21. Treat tetramethyleneethane with qualitative first-order perturbational MOT using the π MOs of allyl to derive the entire π MO diagram of this molecule.

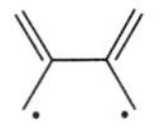

22. Using the energies for the orbitals of butadiene and allyl given in Figure 14.15, calculate in terms of α and β the activation energy for the bond rotation shown below. Assume that the transition state for the reaction is biradical-like in which the plane of the terminal CHD group and the allyl group are perpendicular.

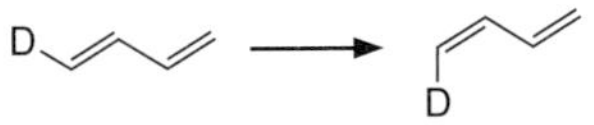

23. Explain the following statement: "The reason we can only put two electrons into an MO is that electrons are fermions."

24. Semi-empirical methods such as AM1 are much faster than *ab initio* methods for standard organic molecules involving C, H, N, and O. The difference in speed is even more pronounced when heavier elements such as Si and P are involved. Why?

25. Show that if k is set equal to 1 in Eq. 14.53, mixing two orbitals produces no bonding (refer to Eq. 14.46). Discuss how this result emphasizes the fact that bonding results from the lowering of energies in the overlap region relative to that of the isolated orbitals.

26. The C–C bond length in ethane is 1.531 Å, while that of the ethyl cation is 1.438 Å. Rationalize this difference, citing two reasons.

27. According to *ab initio* HF calculations (3–21G* basis set), the C–C bond of cyclopropane is 1.513 Å long, in good agreement with experiment. A comparable calculation on cyclopropylcarbinyl cation gives the bond lengths shown. Rationalize the differences.

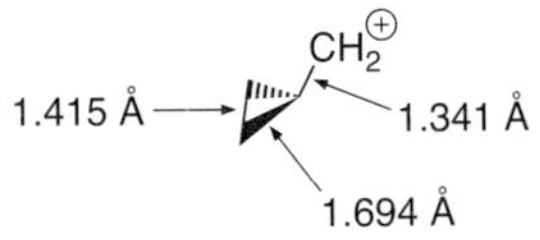

28. Shown below are the absolute values of the Hückel coefficients for selected atoms in the Hückel MOs for vinyl alcohol ($CH_2=CH-OH$). The MOs are in order of increasing energy, with oxygen as the atom on the right. Determine magnitudes of the remaining coefficients using the rules presented in Section 14.3.2. To determine the signs, build up the Hückel orbitals of vinyl alcohol by mixing ethylene and hydroxyl (OH). The end result should be a complete Hückel MO diagram for vinyl alcohol, with coefficients and signs clearly indicated (orbital energies not required).

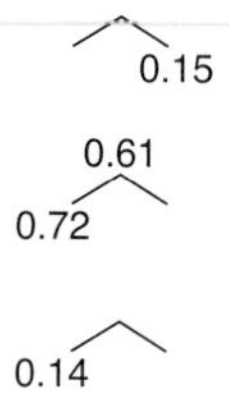

0.15

0.61
0.72

0.14

29. Shown below is the structure of norcaradiene. The C1–C6 bond is unusually long (cyclopropane itself has a C–C bond length of 1.51 Å). Use an orbital mixing diagram to rationalize this result. Sketch the orbital that is most responsible for the effect.

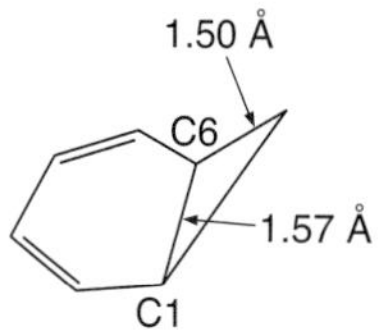

1.50 Å

C6

—1.57 Å

C1

30. While methylene (CH_2) is a ground state triplet (T), substituents can have a large effect on the spin preference. Aminomethylene $CHNH_2$, for example, prefers the singlet (S) ground state by 42 kcal/mol!
 a. Use an orbital mixing diagram to rationalize this result. (Note: $CHNH_2$ is a *planar* molecule.) Sketch the molecular orbitals that are most crucial for this argument.
 b. Based on the results for part a, explain the following trend in S–T gaps for CHX molecules:

X	$E_T - E_S$ (negative value implies T ground state)
H	−9.2
F	12.2
OH	27
NH_2	42

(A whole new mixing diagram is not necessary—just relate comments to part a.)

31. The cation CH_5^+ is a well known gas phase species. We will consider here two possible geometries for this interesting structure.
 a. Build the molecular orbitals for trigonal bipyramidal CH_5^+ (shown on the left) by combining the orbitals of *planar* CH_3^+ with those of two hydrogens that are far apart (the apical hydrogens).
 b. The alternative structure for CH_5^+ is less symmetrical (shown on the right), and is best thought of as a complex between *pyramidal* CH_3^+ with molecular H_2. Build the molecular orbitals of this species from those two fragments.
 c. Based on your two diagrams, rationalize the fact that the experimental structure corresponds to the structure on the right.

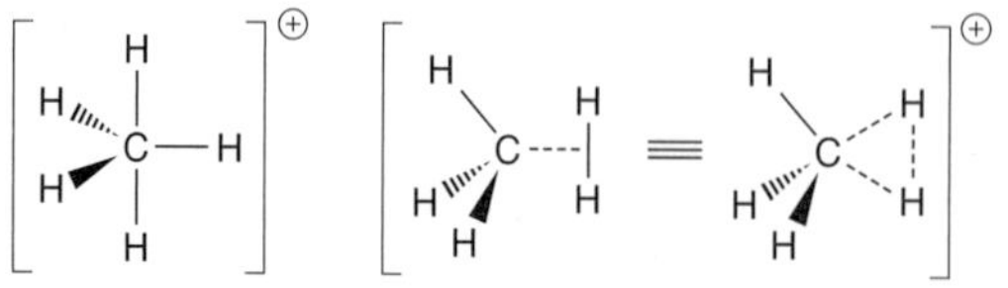

32. The photoelectron spectrum of tetraene **1** gives only the three π ionizations shown. Rationalize this result based on the data given. Draw an orbital mixing diagram and sketch all relevant orbitals.

PES data: 8.35   9.09
              9.67   11.55
           13.10

33. Derive the valence MOs for fully planar ethane (all eight atoms lie in one plane) by first constructing planar CH_3 in the normal geometry, distorting to a T-shaped structure, and then combining two such fragments to give planar ethane. Show an orbital mixing diagram and sketch all relevant orbitals. Discuss briefly the nature of the C–C bond and the overall C–H bonding situation in planar ethane.

34. The first two ionizations in bicyclo[2.1.0]pentane (**1**) are believed to arise from the Walsh-type orbitals of the three-membered ring, the degeneracy of which has been split by substitution. Using the PES data shown, assign the first two ionizations of **1**. Sketch the MOs responsible for the three ionizations of **2**.

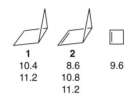

   1   **2**
  10.4   8.6   9.6
  11.2   10.8
         11.2

35. Ethylene (C_2H_4) is planar, but B_2H_4 is perpendicular. Using the orbitals of CH_2 and BH_2 groups, build up the molecular orbitals of both molecules, and rationalize the structural difference. Sketch all relevant orbitals.

36. Your goal in this problem is to rationalize the photoelectron spectrum of [1.1]paracyclophane (**1**). The first four ionizations are found at 7.03, 7.77, 8.63, and 9.82 eV, all of which can be traced to the HOMOs of benzene (IP = 8.9 eV*). Proceed as follows:

a. Start with *p*-xylene (*p*-dimethylbenzene) and rationalize its IPs (8.33 and 8.86 eV). Sketch the relevant orbitals.

b. Distort *p*-xylene to the boat-like structure contained in **1**. Rationalize the calculated IPs of this hypothetical structure (7.64 and 8.82 eV).

c. Combine two such boat structures to make **1** and explain the ionizations of **1**. Sketch the relevant orbitals.
   You need consider only contributions from the aromatic rings.

*For consistency, all IPs are the results of *ab initio* calculations, and so may differ slightly from known experimental data.

  1

37. Shown below are two π MOs of benzene. At HMO, $a = b = 1/\sqrt{6} = 0.41$. In an EHT calculation, $a = 0.33$ and $b = 0.54$. Explain the difference between the HMO and EHT results and between the values of a and b in EHT.

$$\psi_1 = \qquad \psi_2 = $$

38. The structure of spiro[2.4]heptadiene (**1**) has been accurately determined by microwave spectroscopy. Use an orbital mixing diagram to rationalize the structural differences between **1** and the model compounds cyclopropane and cyclopentadiene. Include sketches of all relevant orbitals. (Numbers shown are bond lengths in Å.)

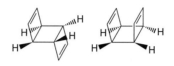

1

39. Rationalize the variation in the ΔIP values seen in the photoelectron spectra of the two isomeric structures shown below.

IPs (eV)	IPs (eV)
8.96	9.08
9.93	9.44
ΔIP = 0.97	**ΔIP = 0.36**

40. Considering the Going Deeper highlight on 1,3-cyclobutanediyl and 1,3-cyclopentanediyl (Section 14.5.6), draw an orbital mixing diagram for each showing how through-bond coupling can produce nearly degenerate NBMOs. Also show that the final NBMOs are non-disjoint.

41. Make an orbital interaction diagram between a cyclopropane and a methyl cation that mimics the bonding in "Coates' ion" (Section 14.5.5). Sketch all relevant orbitals.

42. Show an orbital mixing diagram that combines CH^+ with square cyclobutadiene to make $(CH)_5^+$. Use this to explain why CH^+ is stabilized by this interaction.

43. While CH^+ is stabilized by coordination to cyclobutadiene (see Exercise 42), the analogous structure in which a CH^+ is complexed to benzene, $(CH)_7^+$, is *not* stabilized. Explain why this would be so.

44. What do the $2e_g$ orbitals look like in Figure 14.33? Does population of these orbitals with electrons stabilize or destabilize octahedral complexes?

45. By inspection of the bonding and antibonding π orbitals of benzene and the s, p, and d orbitals of a metal, draw the general molecular orbital diagram for the sandwich complex $M(aryl)_2$, where M = Cr, Mo, and W, and aryl = benzene. (*Hint:* Mix the orbitals on the two eclipsed benzenes first to derive linear combinations of the benzene orbitals. Because the benzenes are so far from one another, the energies of the mixtures only change slightly from the energies of the starting benzene molecular orbitals. Then examine the symmetries and energies of these mixtures, and match them with the appropriate s, p, and d orbitals on the metals to write the entire orbital diagram.) Which d orbital(s) is (are) not used significantly in the bonding?

46. Using Table 14.1, design organometallic compounds that you suspect should make stable complexes with triple bonds to methine (CH).

47. Write a similar table to that of Table 14.1 where one ligand, X, is a covalent ligand (contributing only one electron to the M–X bond), while the other ligands, L, are neutral, dative ligands (contributing two electrons each).

48. Why can't a stationary point with two negative eigenvalues be a transition state?

49. When you have three, four, five, or n electrons in a molecule, the Slater determinant gives a total wavefunction that is very complicated. It consists of n terms, each of which is a product of n orbital wavefunctions. Yet, all the integrals we considered in the Hartree–Fock equation are only one- and two-electron integrals. What happened to the n electron integrals you should get if such a determinantal wavefunction is used in the Schrödinger equation? In addition, why do the one-electron integrals only involve the same orbital; in other words, why don't H_{ij} terms arise? Furthermore, why do the two-electron terms always involve the same two orbitals (i and j in our notation)—why not a $<ij|1/r|kl>$ type integral? Answer these three questions with a mathematical analysis.

Further Reading

Textbooks on Molecular Orbital Theory

Albright, T. A., Burdett, J. K., and Whangbo, M.-H. (1985). *Orbital Interactions in Chemistry,* John Wiley & Sons, New York.

Borden, W. T. (1975). *Modern Molecular Orbital Theory for Organic Chemists,* Prentice–Hall, Englewood Cliffs, NJ.

Gimarc, B. M. (1979). *Molecular Structure and Bonding: The Qualitative Molecular Orbital Approach,* Academic Press, New York.

Hehre, W. J., Radom, L., Schleyer, P. V., and Pople, J. A. (1986). *Ab Initio Molecular Orbital Theory,* John Wiley & Sons, New York.

Pople, J. A., and Beveridge, D. L. (1970). *Approximate Molecular Orbital Theory,* McGraw–Hill Series in Advanced Chemistry, McGraw–Hill, New York.

Salem, L. (1982). *Electrons in Chemical Reactions: First Principles,* John Wiley & Sons, New York.

Zimmerman, H. E. (1975). *Quantum Mechanics for Organic Chemists,* Academic Press, New York.

Roberts, J. D. (1961). *Notes on Molecular Orbital Calculations,* W. A. Benjamin, New York.

Molecular Orbital Pictures

Jorgensen, W. L., and Salem, L. (1973). *The Organic Chemist's Book of Orbitals,* Academic Press, New York.

Early Pictures of Bonding

Pauling, L. (1960). *The Nature of the Chemical Bond and the Structure of Molecules and Crystals: An Introduction to Modern Structural Chemistry,* 3d ed., George Fisher Baker Non-Resident Lectureship in Chemistry at Cornell University (v. 18), Cornell University Press, Ithaca, NY.

Annulenes

For a recent overview of the [10]annulene problem see: King, R. A., Crawford, D., Stanton, J. F., and Schaefer III, H. F. "Conformations of [10]Annulene: More Bad News for Density Functional Theory and Second-Order Perturbation Theory." *J. Am. Chem. Soc.,* **121,** 10788 (1999), and references therein.

Aromaticity and a Recent Re-Examination

Jug, K., Hiberty, P. C., and Shaik, S. "Sigma–Pi Energy Separation in Modern Electronic Theory for Ground States of Conjugated Systems." *Chem. Rev.,* **101,** 1477 (2001).

The Isolobal Analogy

Hoffmann, R. "Building Bridges Between Inorganic and Organic Chemistry." *Angew. Chem. Int. Ed. Eng.,* **21,** 711 (1982).

CHAPTER 15

Thermal Pericyclic Reactions

Intent and Purpose

In this chapter we discuss a special subset of organic reactions that have transition states that are characterized by a cyclic array of interacting orbitals, called pericyclic reactions. Throughout the first half of the 20th century, several interesting reactions were found to be quite reliable and synthetically important, but they involved no reactive intermediates, and were generally insensitive to solvent effects. These useful but mysterious transformations were sometimes described as "no-mechanism" reactions. As we will see, there are indeed no intermediates, but there is of course a mechanism. A unifying feature of these reactions is that an analysis of the electronic structures of starting materials and products is crucial to developing an understanding of the mechanisms.

The announcement of theoretical models emphasizing molecular orbitals to explain these reactions inspired experimentalists to test and expand upon the theorists' work. Some of the most elegant and insightful mechanistic studies ever executed were designed to probe pericyclic reactions. As such, this chapter will also give us the opportunity to apply many of the mechanistic tools we have learned in earlier chapters.

A number of distinguished workers developed differing analyses of these "no-mechanism" reactions. Seminal papers in 1965 by R. B. Woodward and R. Hoffmann described the "conservation of orbital symmetry" as the controlling feature. The orbital correlation diagrams associated with orbital symmetry arguments can be related to state correlation diagrams developed earlier by Longuet-Higgins and Abramson. K. Fukui established the importance of the HOMO and LUMO, the frontier orbitals, in controlling such reactions. Zimmerman developed a novel approach based on aromatic transition states, and similarly important contributions were made by Dewar, Salem, Oosterhoff, and van der Lugt. Hoffmann and Fukui were awarded the 1981 Nobel Prize in Chemistry for this work. Woodward would almost certainly have shared the 1981 prize (his second), but he died several years earlier.

One goal of this chapter is to familiarize you with the several models developed to treat pericyclic reactions. No one model is "right"; which you use depends upon the particular reaction under consideration and your own particular preferences. Sometimes orbital symmetry is most useful; sometimes frontier orbitals are easier to analyze. Another goal is to analyze these reactions at a level beyond that typically given in introductory organic chemistry texts. In particular, state correlation diagrams will be introduced, as these provide a compelling analysis of the reactions, and they also introduce several terms and concepts that will be useful in subsequent chapters. Actually, only the state correlation diagram model needs the rigor of the previous chapter, where the notion of "electronic states" as single determinant wavefunctions was introduced. It is the *elegant simplicity* of the Woodward–Hoffmann model, and the other models, that makes them so powerful. In fact, as you'll see, all the models can be quickly analyzed without computational approaches. Our approach will not be to present a separate section on each model. Rather, we will analyze in detail two prototype reactions, the [2+2] and [4+2] cycloadditions, and we will show how all the models treat these two. We then compare all the models together when examining other pericyclic reaction classes.

In addition, we present a brief survey of the types of reactions that adhere to the various models developed. In reality, the reactions considered here constitute a relatively small subset of organic chemistry, with the Diels–Alder reaction, the Claisen rearrangement, and the oxy-Cope rearrangement providing the majority of reactions that are used "day-to-day" in the chemistry lab. However, the importance of the development of the molecular orbital models described here is not so much that they rationalize a huge body of literature. Rather, the work described in this chapter profoundly influenced the interaction between molecular orbital theory and organic chemistry, forever changing the education and mind-set of organic chemists with regard to electronic structure theory.

15.1   Background

Apart from stereochemistry, perhaps no field has spawned more terminology than pericyclic reactions, some of which can be confusing. We will do our best to make sure you have a clear understanding of the various terms in this field. We begin with "pericyclic", the defining term for this chapter. A **pericyclic reaction** is one that involves a transition state with a cyclic array of atoms and an associated cyclic array of interacting orbitals. A reorganization of σ and π bonds occurs within this cyclic array. The meaning will become more clear as we present examples.

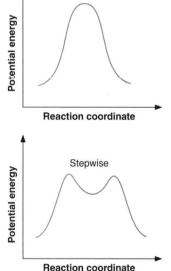

The reactions we are studying are also concerted. A **concerted** reaction is one that occurs in a single step without any intermediate, while a **stepwise** process has one or more intermediates (carbocations, radicals, carbenes, or carbanions), and are thus distinct from pericyclic reactions. Pericyclic and concerted, however, are not inextricably linked. Not all concerted reactions are pericyclic. For example, the S_N2 reaction is concerted, but it is not pericyclic. Pericyclic and concerted really pertain to two very different aspects of a reaction. Pericyclic refers to the geometry of the transition state; a cyclic array is required. Concerted refers to a particular kind of reaction coordinate diagram—one without intermediates. The reactions of this chapter are indeed both concerted and pericyclic, but keep clear in your mind the differing implications of these two terms.

A more recent addition to the lexicon of pericyclic chemistry is the term **synchronous**. In most pericyclic reactions, two (or more) bonds are being formed or broken. Even if the reaction is concerted, that does not necessarily mean that all bond making and bond breaking have occurred to the same extent at the transition state. For example, one bond might be nearly completely formed and one only barely formed at the transition state. Alternatively, the bonds may form and break in synchrony, producing a truly symmetrical transition state. Although this is a fairly subtle distinction, the debate over whether specific pericyclic reactions are synchronous or not has at times been contentious, and elegant experiments have been developed to evaluate synchronicity, some of which will be presented below.

15.2   A Detailed Analysis of Two Simple Cycloadditions

To develop the concepts related to understanding all pericyclic reactions, we will study two prototype reactions, shown in Eqs. 15.1 and 15.2. Both are **cycloadditions**, reactions in which two (or more) molecules combine to make a new ring system. We will develop the nomenclature more fully below, but for now it is convenient to refer to the dimerization of ethylene to give cyclobutane as a [2+2] cycloaddition, and the combination of butadiene and ethylene to give cyclohexene as a [4+2] cycloaddition. We assume a **pericyclic transition state**, and that the two partners approach each other in a symmetrical fashion, forming a symmetrical cyclic transition state, and then go on to product. "Symmetrical" means that the trajectory for approach of the reactants and the geometry of the transition state have a particular symmetry element, such as a σ plane, C_n axis, S_n axis, etc. We will refer to this symmetry element in many of the theoretical models used to analyze pericyclic reactions.

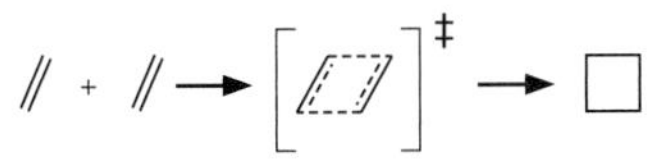

$$// \;+\; // \longrightarrow \left[\; \diagbox{} \; \right]^{\ddagger} \longrightarrow \square \qquad\qquad \text{(Eq. 15.1)}$$

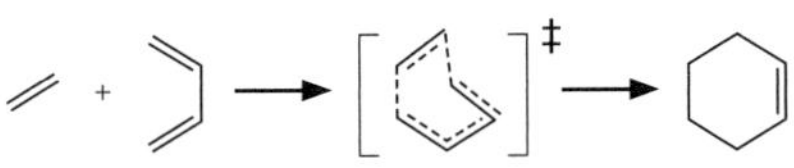

$$\diagup\!\diagup \;+\; \diagdown\!\!\diagup \longrightarrow \left[\; \right]^{\ddagger} \longrightarrow \bigcirc \qquad\qquad \text{(Eq. 15.2)}$$

The starting point for analyzing all pericyclic reactions is to predict a geometry for the reactants during the collision that leads to products. In other words, *we assume a geometry for the transition state and analyze its bonding properties.* In this chapter we are examining thermal pericyclic reactions, which means that the activation barrier is surmounted by energy that comes from collisions. The theoretical models we develop in this section will lead us to a conclusion as to whether the reaction under analysis can occur readily with the geometry predicted. All pericyclic reactions are allowed for a particular geometric arrangement of the reactants. It is a matter of deciding what is the correct geometry.

Just because we analyze a given reaction as involving a cyclic, symmetrical transition state does not mean that the reaction actually goes that way. Alternative paths are usually available, and frequently a stepwise reaction that does not involve a pericyclic transition state will be preferred. In this section especially, we are performing a theoretical analysis of hypothetical reactions. In the lab, neither of the reactions in Eqs. 15.1 and 15.2 are especially efficient. However, they are prototypes for reactions that are important. This is a recurrent theme of this chapter. We will usually do our theoretical analyses on simplified systems where the symmetry is obvious, and then assume that the conclusions apply to the more complicated reactions employed routinely in the laboratory.

15.2.1 Orbital Symmetry Diagrams

The bonding changes in Eqs. 15.1 and 15.2 are straightforward. In the [2+2] reaction, two π bonds are breaking and two σ bonds are forming. In the [4+2] reaction, three π bonds are breaking and a new π bond and two σ bonds are forming in the product. We want to analyze these bonding changes in some detail, and a key principle will be to consider orbitals and bonding changes in the context of the symmetry of the system. The prototype reactions given in Eqs. 15.1 and 15.2 can proceed along high symmetry paths, facilitating such analysis. The essence of an orbital symmetry analysis is to draw the MOs of the starting material and the product. One then determines whether the MOs are symmetric or antisymmetric with respect to the symmetry assumed for the reaction, and then one checks to see how the symmetries of the orbitals relate to each other in the reactants and products. This is done in Figure 15.1 for the [2+2] reaction of Eq. 15.1.

[2+2]

In the method given here, we are creating what are called **orbital symmetry diagrams**, also known as **orbital correlation diagrams**. The starting point for the [2+2] cycloaddition is a pair of ethylene molecules interacting only weakly. The symmetry leads to in-phase and out-of-phase combinations of the HOMOs and LUMOs. The two sets of orbitals in boxes on the bottom left of Figure 15.1 **B** are the combinations of ethylene HOMOs, while the top left are combinations of LUMOs. The weak interaction breaks the inherent degeneracy of the energy levels of the HOMOs and LUMOs that would exist for non-interacting ethylenes. Taking the combinations of the HOMOs and LUMOs also facilitates the analysis of the symmetries of these orbitals during their approach to each other. Although the interaction is small, the in-phase combination of π MOs is placed lower in energy than the out-of-phase. Similarly with the π^* orbitals, the in-phase is below the out-of-phase combination.

The product is cyclobutane. We only draw the MOs relevant to the bonds being formed. These bonds are drawn as C–C σ bond group orbitals of the sort we have seen before in Chapter 1. Each individual bond is created from combinations of hybrid orbitals on the carbons where bonding changes have occurred. The molecular orbitals are created by combining the σ bond group orbitals. In-phase combinations of the σ bonds are lower in energy than out-of-phase combinations.

A.

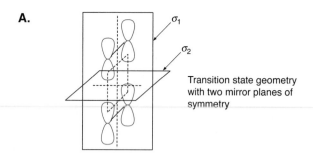

Transition state geometry
with two mirror planes of
symmetry

B.

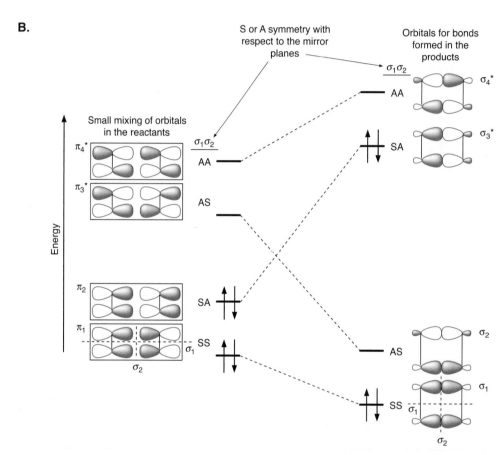

Figure 15.1
The orbital correlation diagram for the prototype [2+2] cycloaddition.

After reading Chapter 14, you might be wondering why we can represent the σ bonds in such a simple manner, because fully delocalized MOs without carbon hybridization are the results obtained from sophisticated electronic structure theory. Experience has shown that in pericyclic reactions there is no need to include "spectator bonds", bonds that remain intact throughout the reaction, in the analysis. Remember that orbitals and our notions of bonding are just models. Any model that explains experimental results and can predict them is useful, and the simplest is always the best. Here we show the simplest model that works. We could complicate our model by including the spectator bonds, but the results would be the same.

Crucial to the analysis is the two mirror planes of symmetry, σ_1 and σ_2, *which are maintained throughout the entire reaction* (Figure 15.1 **A**). Symmetry is crucial here, but only symmetry elements that are present in reactants, products, and transition state are useful. In addition, only symmetry elements that relate regions of the molecule that are undergoing bonding changes are relevant. We designate each molecular orbital as symmetric (S) or antisymmetric (A) with respect to each mirror plane. The designation SS means symmetric with respect to both planes, while SA means symmetric with respect to σ_1 and antisymmetric with respect to σ_2, and so on.

We are now ready for the central tenet, which is that during a pericyclic process there is a **conservation of orbital symmetry**, a notion introduced by Woodward and Hoffmann. That is, an orbital of a given symmetry (for example, SA) in the starting material is smoothly converted to an orbital of the *same* symmetry in the product. The two orbitals are said to **correlate** to one another. In the particular example of Figure 15.1, this means that the SS orbital (π_1) of the two ethylenes correlates (is transformed into) the SS orbital (σ_1) of cyclobutane; the SA to SA; the AS to AS; and the AA to AA. This transformation of one molecular orbital to another can actually be followed if the MOs are calculated for several geometries along the reaction coordinate using the techniques given in Chapter 14. The intermediate geometries will have molecular orbitals that are combinations of the reactant and product orbitals that correlate in symmetry in these diagrams. The orbital symmetry diagram gives a simple way to predict how the reactant orbitals mix and finally result in the product molecular orbitals without such a calculation.

Having drawn all our orbital correlations, we can now follow the electrons from starting material to product. We have a problem, though. *The ground state of two ethylenes correlates to a doubly excited state of cyclobutane*. That is, if orbital symmetry is conserved, and filled orbitals stay filled while empty orbitals stay empty, two ethylenes will combine to make cyclobutane with two electrons in a bonding, SS orbital (σ_1), and two electrons in an antibonding SA orbital (σ_3^*). This is an undesirable outcome in that it creates a very high energy excited state of cyclobutane. Such a reaction is termed **forbidden** in orbital symmetry parlance, and indeed the concerted pericyclic dimerization of ethylene to give cyclobutane is an inefficient reaction (we will have more to say about the meaning of "forbidden" below). More precisely, we should conclude that the trajectory of reaction where the two ethylenes approach each other face-to-face as shown in Figure 15.1 **A** is forbidden. Another approach may allow the reaction to occur, and later we will show this is true.

[4+2]

Now let's consider the [4+2] cycloaddition of Eq. 15.2. The relevant orbital correlation diagram is in Figure 15.2. We again start by predicting a trajectory for collision of butadiene and ethylene, and the face-to-face one shown is most logical. Because the orbitals of butadiene and ethylene are not degenerate, we do not need to start with a weak mixing as was done for the [2+2] reaction. Furthermore, with the [4+2] we have only one persistent symmetry element, a mirror plane, so all molecular orbitals are either just S or A. The molecular orbitals of the starting materials are straightforward: the π MOs of butadiene and ethylene (see Figure 14.14). Those of the product again involve C–C σ bond group orbitals, as well as a conventional π/π^* pair. The conservation of orbital symmetry is required as with the [2+2] reaction, but now the outcome is qualitatively different. All filled molecular orbitals of the reactants correlate with filled molecular orbitals of the products, while empty orbitals correlate with empty orbitals. We do not form an excited state of the product, but instead the ground state of the reactant correlates with the ground state of the product. The unfavorable nature of the [2+2] reaction is gone, and the [4+2] reaction is termed **allowed** in an orbital symmetry sense. It is "allowed" with the geometry shown in the figure. This is the prototype of the Diels–Alder reaction, which is a common and efficient reaction. As we noted above, the exact reaction studied here, ethylene plus butadiene, is an inefficient reaction in the lab. We use the simple prototype to perform the orbital analyses that are applicable to the real world examples.

The two examples just discussed illustrate the conservation of orbital symmetry approach. Orbitals of reactants correlate with product orbitals of like symmetry. Whether a reaction is allowed or forbidden in the geometry assumed is determined by whether the ground state of the starting material correlates to the ground state or an excited state of the products. The key to the method is to recognize symmetry elements that persist throughout the reaction and to correlate molecular orbitals of like symmetry. In subsequent sections we

A.

B.

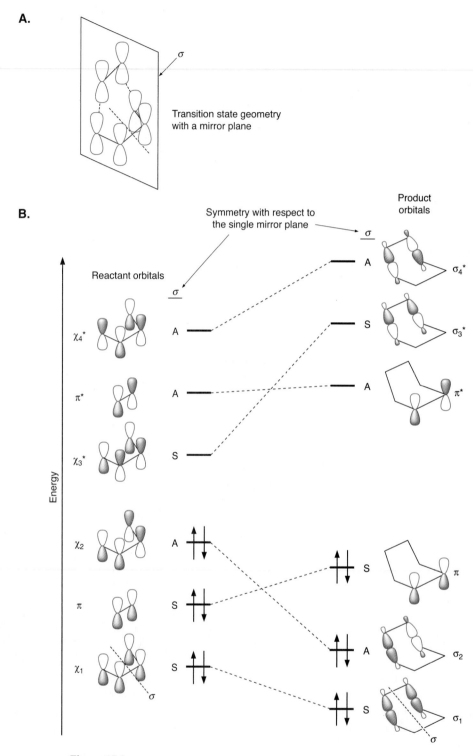

Figure 15.2
The orbital correlation diagram for the prototype [4+2] cycloaddition.
Energy levels for the various orbitals are only qualitatively meaningful.

will present other, more descriptive ways to analyze pericyclic reactions. While these other approaches are always reliable, in some rare instances their application may not be totally straightforward. When in doubt, construct an orbital correlation diagram; this will always give a clear result.

Why does symmetry matter? Why must there be a conservation of orbital symmetry? Recall from Chapter 14 that molecular orbitals are just convenient creations of an approximation to the Schrödinger equation. Yet, orbital wavefunctions do have inherent symmetries. Wavefunctions of different symmetry *cannot* mix, or stated more accurately, their mixture does not give any net interaction. As we will explain further below, it is fruitless to mix them. To see this better, we now turn to an analysis of the mixture of total wavefunctions, where symmetry again is of paramount importance. Here, the theoretical basis for why symmetry matters is more clear.

15.2.2 State Correlation Diagrams

The conservation of orbital symmetry, as developed by Woodward and Hoffmann, launched a flurry of experimental and theoretical organic chemistry. In its time, it was much more "theoretical" than was typical for organic chemistry, as most practicing organic chemists usually did not worry too much about the molecular orbitals of their starting materials and products. The approach has a rigor to it, and its predictions are powerful. However, as stated above, it is natural to wonder what the fundamental, theoretical justification of the method might be.

There is a more fundamental principle at work, underlying the conservation of orbital symmetry, and it is embodied in **state correlation diagrams**. Before looking at these diagrams, we must define a state. **Quantum mechanical states** are well-defined solutions to the Schrödinger equation. Unlike orbitals, which result from approximations we must make in order to calculate a wavefunction, states are true solutions to the Schrödinger equation, and they have a definite correspondence to reality. A state is not the same as an electron configuration, which we defined in Chapters 1 and 14 as corresponding to a particular placement of electrons into a collection of orbitals. A state can be made up of a linear combination of configurations, as in configuration interaction, defined in Section 14.2.1. Often, however, a single electron configuration dominates a particular state, and it is convenient to write a single configuration to describe a state. Importantly, only configurations of like symmetry can combine to make a given state. Thus, by considering only the dominant configuration of a state, we capture the essential symmetry properties of that state. Here we will be writing states as single configuration wavefunctions, but you should remember that the true wavefunction of the state is more complicated. Recall also from Chapter 14 that when the dominant electron configuration of a state has all the electrons in the lowest possible orbitals, we speak of the ground state, while excited states are formed by promotion of electrons to higher-lying orbitals.

A state correlation diagram is really no different from the reaction coordinate diagrams we have seen throughout this text. However, instead of just following the progressive conversion of reactant to product for one electronic state, the ground state, we also follow many excited states along with the ground state. The y axis is an energy scale as usual, and the x axis is the reaction coordinate.

State correlation diagrams provide an extension of orbital correlation and a more rigorous analysis for evaluating pericyclic reactions. After all, it makes sense that the state of the molecule, not just its individual molecular orbitals, should dictate its reactivity. In addition, state correlation diagrams introduce several useful concepts that will be of value for the study of photochemistry in Chapter 16. Conveniently, the starting point for a state correlation diagram is an orbital correlation diagram of the sort we constructed in Figures 15.1 and 15.2.

[2+2]

We begin our analysis of state symmetry by examining the [2+2] cycloaddition of two ethylenes. To start a state correlation diagram, we must know how to assign symmetries to

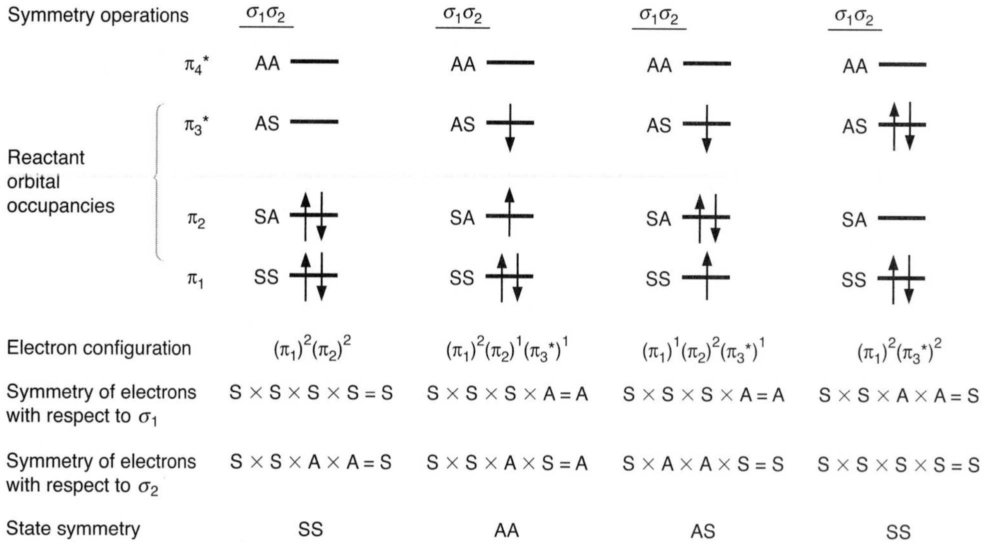

Symmetry operations		$\sigma_1\sigma_2$	$\sigma_1\sigma_2$	$\sigma_1\sigma_2$	$\sigma_1\sigma_2$
	π_4^*	AA ⎯	AA ⎯	AA ⎯	AA ⎯
Reactant orbital occupancies	π_3^*	AS ⎯	AS ↕	AS ↕	AS ↑↓
	π_2	SA ↑↓	SA ↑	SA ↑↓	SA ⎯
	π_1	SS ↑↓	SS ↑↓	SS ↑	SS ↑↓
Electron configuration		$(\pi_1)^2(\pi_2)^2$	$(\pi_1)^2(\pi_2)^1(\pi_3^*)^1$	$(\pi_1)^1(\pi_2)^2(\pi_3^*)^1$	$(\pi_1)^2(\pi_3^*)^2$
Symmetry of electrons with respect to σ_1		S × S × S × S = S	S × S × S × A = A	S × S × S × A = A	S × S × A × A = S
Symmetry of electrons with respect to σ_2		S × S × A × A = S	S × S × A × S = A	S × A × A × S = S	S × S × S × S = S
State symmetry		SS	AA	AS	SS

Figure 15.3
Examples of how to calculate state symmetries based on electron configurations.
The diagram uses the [2+2] cycloaddition as a representative example.

electronic states. We use the symmetries of the molecular orbitals when two ethylenes are allowed to slightly interact. These are shown on the reactant side of Figure 15.1. Recall that the symmetries of the reactant orbitals were assigned with respect to the two mirror planes inherent in the face-to-face approach of two ethylenes. Therefore, we are correlating states for this single approach trajectory, and no other. There are several ways to populate these four orbitals with electrons, and four different electronic states are shown in Figure 15.3 (more are certainly possible). To determine the symmetry of each of these states, one first writes the electron configuration for each state. For example, the ground state with all electrons paired and in the two lowest energy orbitals (first state in Figure 15.3) has the electron configuration $(\pi_1)^2(\pi_2)^2$.

The total symmetry of each state is determined by the symmetries of the populated molecular orbitals in that state. We need not concern ourselves with molecular orbitals that are not populated with electrons. In the analysis of the [2+2] cycloaddition, there were two mirror planes defined as σ_1 and σ_2. We need to define a symmetry for each state with respect to each of these mirror planes. To do this, one creates a multiplication product of the symmetries of the electrons with respect to each symmetry operation. Let's start with the ground state of the mixture of ethylene orbitals. With respect to σ_1, we have four electrons in orbitals that are S, so the state symmetry is S × S × S × S. With respect to σ_2, two electrons are in an orbital that is S, and two electrons are in an orbital that is A. Therefore the product is S × S × A × A.

Next we need to know how to determine the product of symmetries. The "multiplication" rules are given in Eq. 15.3.

$$S \times S = A \times A = S$$
$$S \times A = A \times S = A$$

(Eq. 15.3)

Hence, the result of S × S × S × S is S, while S × S × A × A = S also. So, the ground state symmetry is SS. It will always be true that any closed shell state will be completely symmetric. As we promote electrons to make excited states, new symmetries are possible, as illustrated further in Figure 15.3. Check for yourself that the symmetries for the states are as listed in this figure.

The symmetries of the states of the product also have to be assigned. Once again reference to Figure 15.1 starts the process. The symmetries of the molecular orbitals with respect

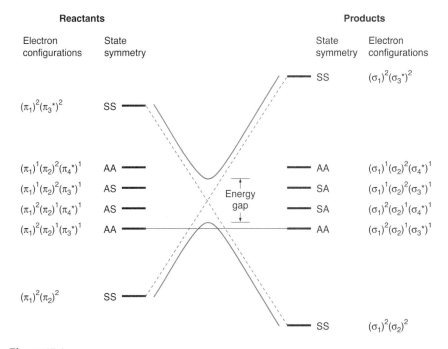

Figure 15.4
The state correlation diagram for the [2+2] cycloaddition of Eq. 15.1. The intended crossing of states with like symmetry is shown with dotted lines, while the avoided crossing is shown as a solid colored line.

to the two mirror planes are again used to define the symmetries of the electrons, and multiplication of the symmetries of the electrons in each state gives the symmetry of that state with respect to each symmetry element. The result is given on the right-hand side of Figure 15.4.

We can now construct the state correlation diagram for the [2+2] cycloaddition, as shown in Figure 15.4. On the left are six electronic states of the starting materials. We generated the symmetries of four of these in Figure 15.3. The ground state is lowest in energy, and then there is a significant energy gap to the first excited state, which we characterize as an excitation from the HOMO to the LUMO. Other states formed by excitation of a single electron from a filled to an empty molecular orbital, termed **single excitations**, will be nearby in energy. The precise energy ordering of these singly excited states is not crucial to the analysis. Then there will be another energy gap to reach **doubly excited states**, formed by promotion of two electrons from filled to empty orbitals. We only show one doubly excited state of the reactant, because it will be especially relevant to our analysis. For the products, the excitations are necessarily σ to σ^*, and therefore the energy gaps should be larger. However, such quantitative issues are not essential here. We need only get the basic, qualitative layout of the diagram correct. What is most important is to have the proper symmetries.

Now we are ready to correlate states, rather than orbitals. The state correlation is achieved by connecting states of the reactant and product that have the same orbital occupancies. Hence, we have to refer back to the orbital correlation diagrams. For example, the electron configuration of the reactants in Figure 15.1 is $(\pi_1)^2(\pi_2)^2$, and if we follow those four electrons to the orbitals in the reactants, we get the $(\sigma_1)^2(\sigma_3^*)^2$ configuration. Of necessity, the two correlated states will have the same symmetry with respect to each symmetry element. As shown in Figure 15.4, the $(\pi_1)^2(\pi_2)^2$ state of the reactant correlates to a doubly excited state of the products (the dashed line). Just as we saw in the orbital correlation diagram, the ground state of the reactants correlates to an excited state of the product. The reverse is also true. That is, the ground state of the product correlates to a doubly excited state of the reactants.

The only other state correlation of interest here is the first excited state of reactants to the first excited state of products, the importance of which we will return to later. However, let's look here at how the states are correlated. In this first excited state, the electron configuration of the reactants is $(\pi_1)^2(\pi_2)^1(\pi_3{}^*)^1$, as deduced from Figure 15.1. If we follow those electrons to the orbitals in the products, we get $(\sigma_1)^2(\sigma_3{}^*)^1(\sigma_2)^1$ as the orbital occupancy. The state symmetries are the same because the orbitals have the same symmetries. However, one always writes an electron configuration with the orbitals listed in order of increasing energy, so the electron configuration of the product is $(\sigma_1)^2(\sigma_2)^1(\sigma_3{}^*)^1$. We find in Figure 15.4 that the AA state of the reactant correlates to an AA state of the product without having a significant increase or decrease in energy.

Thus far, not much in the state correlation diagram is different from the orbital correlation diagram. However, now we introduce a very important concept, termed the **noncrossing rule**. Simply put, states of the same symmetry may not cross in a state correlation diagram. Instead, as they approach each other in energy, they mix and diverge; the lower energy state drops in energy and the higher state rises. This mixing of states is precisely analogous to the kinds of orbital mixings we have seen throughout this book. Orbitals of the same symmetry can mix, and the closer they are in energy the more they can mix (see perturbational molecular orbital theory in Section 14.4.2). The SS states in Figure 15.4 have the same symmetry, and when they are close in energy we expect linear combinations to develop.

Thus, when two states of the same symmetry approach each other in a state correlation diagram, they do not cross, but instead there is an intended but **avoided crossing**, shown by the solid colored lines in Figure 15.4. The size of the gap that forms in the region of the avoided crossing is determined by the extent of interaction between the states. Just as with orbital mixing, the degree to which the two states overlap will be an important factor in determining the size of the gap. The more the two states overlap, the larger the gap.

With Figure 15.4 in hand, let's consider the thermal cycloaddition reaction of two ethylenes to produce cyclobutane. When we take into account the avoided crossing, the ground state of reactants correlates via the solid colored line with the ground state of product. However, because the state correlation diagram represents a reaction coordinate diagram, we find a significant electronic barrier to the reaction. This is defined as a **symmetry imposed barrier** to the reaction. We can go from the reactants to the products, but there is a substantial barrier to overcome. This electronic barrier is in addition to any barrier that might arise from steric effects or from the introduction of strain in the products, as we would expect from a reaction that forms a small ring product such as cyclobutane.

[4+2]

We can now make a comparable state correlation for the [4+2] cycloaddition, and the results are shown in Figure 15.5. A series of states (described as single configurations) for the reactants and products are shown in Figure 15.5, placed at approximate energy levels. Our starting point is Figure 15.2, where the molecular orbitals of the reactants and products are given descriptors such as π, σ, and χ. We need to correlate states with the same orbital occupancies. Let's do both the ground state and the first excited state as examples. The ground state electron configuration, as deduced from Figure 15.2, is $(\chi_1)^2(\pi)^2(\chi_2)^2$. The electrons flow into orbitals σ^1, σ^2, and π in the products, giving the electron configuration $(\sigma_1)^2(\sigma_2)^2(\pi)^2$ and a direct correlation in Figure 15.5 for the ground S states. In contrast, the electron configuration of the first excited state of the reactants is $(\chi_1)^2(\pi)^2(\chi_2)^1(\chi_3)^1$. As seen in Figure 15.5, these electrons flow into the σ_1, π, σ_2, and σ_3 orbitals of the products. This results in a very highly excited state of the products: $(\sigma_1)^2(\sigma_2)^1(\pi)^2(\sigma{}^*_3)^1$.

As we would anticipate from the orbital correlation diagram, the ground states correlate. The first excited states do not, and we see several avoided crossings among the various excited states. Focusing on the ground state, however, we conclude that, unlike the [2+2] cycloaddition, the [4+2] cycloaddition does not experience a symmetry imposed barrier. All other things being equal, we expect the [4+2] cycloaddition to be more favorable than the [2+2] cycloaddition, consistent with observation.

It can be seen that state correlations provide, in a sense, a justification for the simpler orbital symmetry analysis. Clearly, the origin of the avoided crossing seen in the state correlation of the [2+2] cycloaddition can be traced back to the original orbital correlation diagram. It will generally be true that the predictions from an orbital correlation will carry over to the state correlation diagram. Thus, it is rare that practicing chemists construct a state correla-

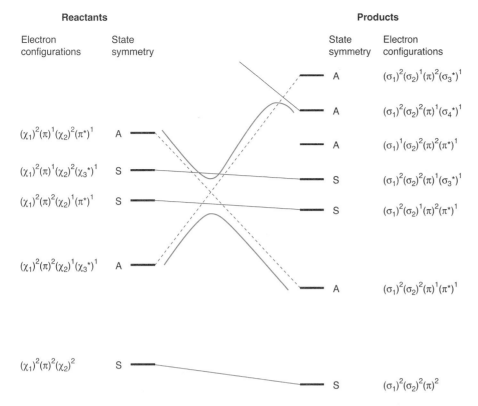

Figure 15.5
The state correlation diagram for the [4+2] cycloaddition of Eq. 15.2. The intended crossing of states with like symmetry is shown with dotted lines, while the avoided crossing is shown as a solid colored line.

tion diagram, except, as we will see in Chapter 16, when considering photochemistry. For thermal pericyclic reactions the main value of state correlation diagrams is that they provide a reassuring theoretical underpinning to the more descriptive methods.

Interestingly, both the orbital and state correlation diagrams use symmetry as their underpinning. Symmetry is perceived by chemists; the molecules do not "know" about symmetry. Yet, there is a mathematical reason that chemists can qualitatively use symmetry in their analysis. The following Going Deeper highlight discusses the fundamental reason behind this.

Going Deeper

Symmetry Does Matter

In our discussions of state and orbital correlations we have noted that: 1. orbitals with different symmetries do not mix, 2. only states with the same symmetry mix in configuration interaction, and 3. states of the same symmetry do not cross. Therefore, symmetry plays a major role. Symmetry matters because the Hamiltonian for a molecule is a totally **symmetric operator**. This means that when it operates on a wavefunction, the resulting wavefunction has the same symmetry. Hence, the result of the operation $|H|\Psi>$ has the same symmetry as Ψ. Further, the Hamiltonian does mix electronic states of like symmetry, while states of different symmetry lead to "no mixing". In other words, integrals of the sort $<\Psi_a|H|\Psi_b>$ and $<\Psi_a|\Psi_b>$ must equal zero when Ψ_a and Ψ_b have different symmetries. This implies that the energy of interaction is zero, and that

the total wavefunctions have no overlap. If their symmetries are the same, they have overlap and a net interaction energy does result. This is the basis of why we can use symmetry elements to examine interactions and correlations between quantum mechanical states.

In reactions of more realistic substrates, typically there will be no symmetry elements because substituents have been attached to the ethylene or butadiene structures. Now we cannot say that $<\Psi_a|H|\Psi_b>$ must be zero, because Ψ_a and Ψ_b cannot have different symmetries if they have no symmetry. However, substituents usually do not completely alter the wavefunction for a system. Thus, if $<\Psi_a|H|\Psi_b>$ is exactly zero in the high symmetry, prototype case, it will likely be small in the substituted case, and the same conclusions will arise.

15.2.3   Frontier Molecular Orbital (FMO) Theory

Throughout this book we have often referred to the importance of the HOMO and LUMO in understanding a molecule's properties. Fukui reasoned that these **frontier molecular orbitals** might play an especially important role in concerted, pericyclic reactions. We have also noted before that it is always favorable to mix filled molecular orbitals with empty molecular orbitals. Combining these ideas led to frontier molecular orbital (FMO) theory, an elegant analysis of the types of reactions of importance here. Quite simply, frontier molecular orbital theory states that if we can achieve a favorable mixing between the HOMO of one reactant and the LUMO of another reactant, a reaction is allowed. If we cannot, the reaction is forbidden.

Contrasting the [2+2] and [4+2]

FMO theory is quite easy to implement. For the [2+2] cycloaddition we simply examine the HOMO and LUMO of ethylene (Figure 15.6 **A**). We can see that the HOMO of one ethylene cannot mix with the LUMO of another during a face-to-face approach because the phasing does not match. Because they are of opposite symmetry, there can be no constructive interaction. On the other hand, for the [4+2] cycloaddition the HOMO of ethylene is of the same symmetry as the LUMO of butadiene and vice versa (Figure 15.6 **B**). In this case the phasing of each matches on the carbons that are beginning to bond together. The same conclusion will be drawn regardless of which reactant's HOMO is considered, as long as the other reactant's LUMO is paired with it.

In summary, all pericyclic reactions can be examined simply by writing the HOMO of one component, the LUMO of the other component (where "component" is defined as "separate interacting orbitals"; see below), and determining whether, at the geometry being assumed, the orbitals can produce a mixing that is in-phase. Often, as we'll explore below, the HOMO and LUMO to be analyzed are within the same molecule, and maybe even in conjugation.

FMO theory makes good sense. As two molecules approach each other, mixing between filled and empty molecular orbitals has to be stabilizing. We know that HOMO–LUMO mixings will be especially favorable because these orbitals will have the smallest energy gap. If such stabilizing interactions can occur as the reactants approach each other, they will cer-

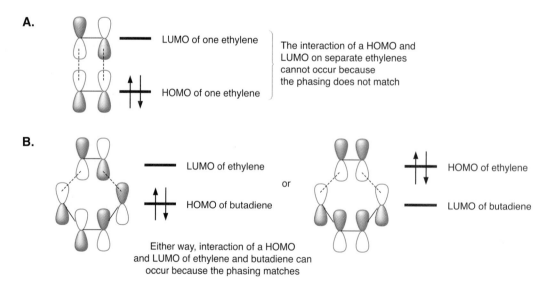

Figure 15.6
Frontier molecular orbital theory. **A.** Analysis of a [2+2] cycloaddition, showing unfavorable mixing between the HOMO and LUMO of the two ethylene fragments. **B.** Analysis of a [4+2] cycloaddition, showing the favorable HOMO–LUMO mixings.

tainly persist in the transition state, lowering its energy and favoring the reaction. Unlike an orbital or state correlation diagram, we don't need to know all the molecular orbitals of the reactants and products—only the HOMO and LUMO of the reactants are required. We have already discussed the properties of these orbitals in simple π systems (Chapters 1 and 14), and so the analysis is fairly straightforward.

15.2.4 Aromatic Transition State Theory/Topology

Another approach to analyzing concerted pericyclic reactions is based on the observation that the forbidden [2+2] cycloaddition involves a cyclic array of four electrons in the transition state, while the allowed [4+2] cycloaddition involves a cyclic array of six electrons. This is a familiar pattern that immediately calls to mind aromaticity, in which the ground states of molecules with four π electrons in a cycle are destabilized and termed antiaromatic, while molecules with six π electrons are stabilized and aromatic. Building off an earlier analysis by Evans, Zimmerman developed **aromatic transition state theory**. Simply put, reactions with a simple cyclic array of $4n + 2$ electrons (commonly six) in a pericyclic transition state will be stabilized by aromaticity, making the reactions favorable. Note that the relevant electrons need not be exclusively in π orbitals; a mixture of σ and π bonds in a cyclic array is acceptable.

In order to expand the range of reactions that can be analyzed using aromatic transition state theory, we need to expand our definition of aromaticity. Figure 15.7 shows how to do this. In a conventional cyclic array, we can always arrange for all the constituent orbitals to be in-phase, in which case all neighboring interactions are favorable. This is evident in the lowest-lying π molecular orbital of any planar, cyclic π system, and in the first orbital array of Figure 15.7 **A**.

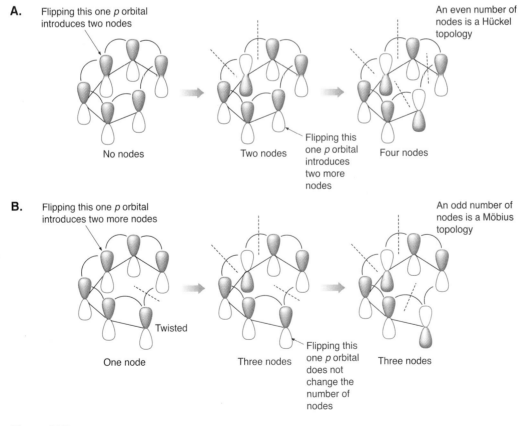

A.
Flipping this one *p* orbital introduces two nodes

An even number of nodes is a Hückel topology

No nodes   Two nodes

Flipping this one *p* orbital introduces two more nodes

Four nodes

B.
Flipping this one *p* orbital introduces two more nodes

An odd number of nodes is a Möbius topology

Twisted

One node   Three nodes

Flipping this one *p* orbital does not change the number of nodes

Three nodes

Figure 15.7
A. A Hückel topology with no nodes, or an even number of nodes. **B.** A Möbius topology. The twist in the system alters the topology, such that the "top" of one *p* orbital must interact with the "bottom" of the other. Such a system must have at least one node (dotted line), and will always have an odd number of nodes.

However, some arrays of orbitals cannot achieve such a perfect phase matching. These are necessarily nonplanar systems, such as the twisted array of Figure 15.7 **B**. Here we have moved one end of the array down, so that the "top" lobe of its p orbital overlaps with the "bottom" lobe of the orbital at the other end of the array. When this happens, a node has to be introduced into the system, as shown by the dotted line. There is *no way* to color the orbitals of the cyclic array in Figure 15.7 **B** without having at least one node. More generally, a system such as this must have an odd number of nodes (also referred to in the literature as **sign inversions**). We can see from the figure that reversing any p orbital introduces two new sign inversions, bringing the total to three (still an odd number). With this analysis, it is important to note that we do not count the nodes within the p orbitals themselves, only nodes between orbitals. It is also important to note that the alignment of p orbitals used does not have to correspond to any conventional molecular orbital. The signs of the p orbitals are chosen arbitrarily. The conventional system of Figure 15.7 **A** has zero nodes. More generally, a system like this will always have an even number of nodes, as seen when any p orbital is flipped.

Because of the twisted ribbon nature of the system in Figure 15.7 **B**, these systems are described as having a **Möbius topology**. Systems with an odd number of nodes will always have a Möbius topology. The conventional topology, which must have an even number of nodes, is described as having a **Hückel topology**. For Hückel systems the rule for aromaticity is familiar. Systems with $4n + 2$ electrons in a cyclic array are aromatic; systems with $4n$ electrons are antiaromatic. These rules are exactly reversed for a Möbius topology; that is, $4n$ is aromatic and $4n + 2$ is antiaromatic. Therefore, there are two pieces of information that must be obtained to use this method to analyze pericyclic reactions: the topology (Hückel or Möbius) and electron count ($4n + 2$ or $4n$).

The application of aromatic transition state theory proceeds as follows. We sketch the cyclic array of orbitals that is necessarily associated with the assumed transition state of any pericyclic reaction. Although not necessary, it is convenient to shade the orbitals to produce the minimum number of nodes. We then assign the system as having either Hückel or Möbius topology, depending on whether the number of nodes is even or odd, respectively. Once the topology is assigned, we follow the appropriate aromaticity rule to determine whether the reaction transition state is aromatic or antiaromatic, with the former corresponding to an allowed reaction, and the latter corresponding to a forbidden reaction. This method is easily applied to the [2+2] and [4+2] cycloadditions, as seen in Figure 15.8.

Once you get used to analyzing cyclic arrays of orbitals, aromatic transition state theory can be a simple and rapid way to analyze pericyclic reactions. As with the other approaches, it is well suited to some types of reactions, but less well suited to others. With practice, you will develop some instincts as to which model to apply to a given reaction.

4-electron
Hückel system:
forbidden

6-electron
Hückel system:
allowed

Figure 15.8
The application of aromatic transition state theory to cycloadditions.

15.2.5 The Generalized Orbital Symmetry Rule

The field of pericyclic chemistry has spawned a large amount of terminology. We've already introduced "pericyclic", "concerted", "stepwise", "allowed", and "forbidden". We need to introduce a few more terms here to further facilitate our analysis of pericyclic reactions, and then give a rule that can be used to analyze all pericyclic reactions.

An important pair of terms is **suprafacial** and **antarafacial**. These describe the topology of interaction of a given system in a pericyclic transition state, and they are best defined with reference to Figure 15.9. Here, lines which appear as "loops" or "arcs" show where interactions occur on the orbitals. These lines define a geometry for interaction of the orbitals shown with other orbitals. This should become clearer below with examples.

For π systems and lone pairs the distinction is simple: suprafacial interactions involve the same face of the system, while antarafacial interactions are on opposite faces. Although we have only examined π systems thus far, we will examine pericyclic reactions below that involve σ bonds. Therefore, we need a similar definition for these kinds of bonds. For σ bonds, the distinction is less obvious, but is consistent with the other systems. With suprafacial interactions, the two loops are drawn to either the inner lobes or outer lobes, while with an antarafacial interaction one loop is to an inner and one is to an outer lobe (see Figure 15.9 for this to make sense). Note that a suprafacial interaction at a σ bond involving two sp^3 hy-

A. π Systems

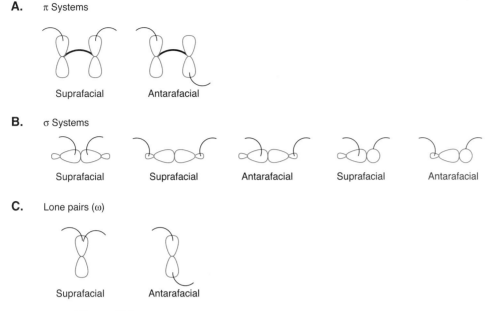

B. σ Systems

C. Lone pairs (ω)

Figure 15.9
Definitions of suprafacial and antarafacial for various types of orbitals.

bridized carbons can either lead to double inversion or double retention of stereochemistry. An antarafacial process will always lead to inversion at one center and retention at the other.

With these definitions, a symbolism to describe pericyclic processes can be developed. The **components** for a reaction are the parts of a molecule (separate bonds or conjugated bonds) that undergo a change during a pericyclic reaction. We then assign an electron count to the component, a suprafacial or antarafacial descriptor, and identify if the component involves π or σ bonds. For example, if a component of the transition state (or the entire transition state) contributes four electrons to the pericyclic array, the electrons are contained in a π system, and the interaction of the π system is suprafacial, the designation is $_\pi 4_s$. Alternatively, a two-electron, sigma system with an antarafacial interaction is designated $_\sigma 2_a$. Lone pairs are designated by the Greek letter ω. In this symbolism, the reactions of Eqs. 15.1 and 15.2 would be $_\pi 2_s + _\pi 2_s$ and $_\pi 4_s + _\pi 2_s$, respectively. With this symbolism, and in all analyses of pericyclic reactions, keep in mind that *the crucial, distinguishing feature is the number of electrons involved in the process, not the number of orbitals.* An allyl anion is a $_\pi 4$ system, while an allyl cation is a $_\pi 2$ system, even though both contain three orbitals (see margin).

With these descriptors in hand, we can look at the **generalized orbital symmetry rule**. There is a definite binary nature to the theory of pericyclic reactions. For cycloadditions, [2+2] is forbidden (all suprafacial), whereas [4+2] is allowed (all suprafacial). Continuing with the series, [6+2] is forbidden, and [8+2] is allowed. We will also encounter patterns in the other kinds of pericyclic reactions presented: electrocyclic reactions, sigmatropic shifts, etc. Based on patterns such as these, Woodward and Hoffmann proposed the following rule for all pericyclic reactions:

A pericyclic reaction is allowed if the number of 4q + 2 suprafacial plus 4r antarafacial components is odd.

Here *q* and *r* are integers. This means that any 2-, 6-, 10-, 14-electron, etc., suprafacial component is considered, and any 0-, 4-, 8-, 12-electron, etc., antarafacial component is considered when determining if there are an odd number of components for the reaction under consideration. If the number is even, the reaction is forbidden.

$\diagup\diagup \ominus \quad _\pi 4$

$\diagup\diagup \oplus \quad _\pi 2$

Electron counts for allyl groups

Let's once again look at the cycloaddition reactions to implement this method of analyzing pericyclic reactions. The $_\pi 2_s + {_\pi 2_s}$ components given in Eq. 15.1 are both suprafacial, and they both fit the formula $4q + 2$, and hence there are two components to this reaction that conform to the generalized orbital symmetry rule. This is an even number, and therefore the reaction is forbidden. However, with the $_\pi 4_s + {_\pi 2_s}$ reaction, only the $_\pi 2_s$ component fits the $4q + 2$ pattern. One is an odd number, and the reaction is allowed.

This is a very powerful rule, and it is especially useful when there are several components to a pericyclic reaction. With several components it is often difficult to identify the appropriate HOMOs and LUMOs for an FMO analysis, and difficult to quickly write an orbital or state correlation diagram. In such cases, aromatic transition state theory, or the generalized orbital symmetry rule, are the easiest approaches for analyzing the reaction. It is your decision as to which works best for you.

15.2.6　Some Comments on "Forbidden" and "Allowed" Reactions

When Woodward and Hoffmann developed the conservation of orbital symmetry, they introduced the terms "allowed" and "forbidden" to describe reactions such as the [4+2] and [2+2] cycloadditions, respectively. This terminology caught on, and has become fairly standard in the field. With the benefit of a historical perspective, though, we can now see that these terms are too definitive.

When we say a reaction is **forbidden**, what we really mean is that there is expected to be a barrier on the reaction path that results from the unfavorable orbital properties of the system, what we call an **electronic barrier**. We can arrive at this conclusion based on orbital correlations or state correlations or other models, but the basic concept is the same. The electronic structure of the system is not especially favorable for the reaction to proceed in the geometry considered. The reaction isn't really *forbidden*. It is just that if the reaction is going to occur, it has to overcome an electronic barrier in addition to any other "barriers" that are intrinsic to the system, such as steric effects. Alternatively, the reaction has to proceed along a different geometry (through a transition state with a significantly different structure) from the one we used in the orbital analysis, or the reaction occurs stepwise. In reality, all forbidden pericyclic reactions are allowed in some alternative geometry, though the allowed path might be unfavorable due to strain or poor orbital overlap.

Similarly, a reaction that is **allowed** is simply one that does not have such an electronic barrier. This does not automatically mean, however, that the reaction will be favorable. Steric interactions or other factors could make the reaction quite slow. All we know is that no additional barrier due to electronic factors contributes to the overall activation energy.

The two cycloaddition reactions we have discussed so far illustrate some of these points. The [2+2] cycloaddition is forbidden. However, olefins can dimerize to make cyclobutanes. It is just that the reaction is not concerted, but rather involves a biradical intermediate. The [4+2] cycloaddition is allowed, but in fact the concerted cycloaddition of ethylene and butadiene requires high temperature and pressure. It does occur by a concerted allowed path, but the activation barrier is high.

With this perspective the terms "forbidden" and "allowed" seem overly sweeping. Something like "disfavored" and "not disfavored" would be less dramatic (and less grammatical) but closer to the reality of the science. However, "forbidden" and "allowed" are firmly entrenched in the pericyclic literature, and we will use them here. We hope this discussion, however, will provide you with the proper perspective and will discourage any tendency to interpret the terminology too literally.

15.2.7　Photochemical Pericyclic Reactions

An oft-cited dichotomy is that if a reaction is thermally forbidden, it is photochemically allowed and vice versa. In fact, photochemical [2+2] cycloadditions are well known (see Chapter 16), and other examples of thermally forbidden processes that proceed photochemically can be found. A justification for this binary aspect of pericyclic reactions can be gleaned from the orbital or state correlation diagrams. Figure 15.4 shows a direct correlation between the first excited states of reactants, produced by photolysis, and products for the thermally forbidden [2+2] cycloaddition. In contrast, the first excited state of the [4+2] reaction correlates to an even higher excited state of the products (Figure 15.5). This seems consistent with the idea of the [2+2] reaction being photochemically "allowed" and the [4+2] being photochemically "forbidden". However, as discussed in detail in Chapter 16, photochemical processes are intrinsically more complicated than thermal reactions, and a large number of factors determine whether a given photochemical process is or is not favorable. Also, both singlet and triplet excited states must be considered in a discussion of photochemistry, and none of the tools developed in this chapter are appropriate for such distinctions. A detailed analysis of photochemical mechanisms indicates that in some instances the barrier on the ground state surface associated with a thermally forbidden process can facilitate a photochemical process by forming a funnel (see Section 16.3.1). However, this is not always the case, and even when it is, it is not clear that orbital symmetry issues are controlling the photochemical process.

As such, we will not consider photochemical processes in this chapter, deferring such topics to Chapter 16, which is devoted entirely to photochemistry. When we make tables to present rules for various types of reactions, describing them as allowed or forbidden, we will only be addressing thermal conversions. The photochemical part of such tables has always been redundant; you just reverse the thermal predictions. However, on a more basic level we feel that predictions about photochemical reactions based on the level of analysis presented in this chapter are risky and fail to take into account the many subtleties of photochemistry. If you want to consider a photochemical pericyclic reaction, it is best to consider it in the context of the entire field of photochemistry, rather than as the opposite of a thermal process.

15.2.8 Summary of the Various Methods

The discussion above has given five approaches to analyzing pericyclic reactions: orbital correlation diagrams, state correlation diagrams, frontier molecular orbital analysis, aromatic transition state theory, and the generalized orbital symmetry rule. Each approach gave the same answer, and this must always be true. When working out a problem in your research, if you find different answers from the different approaches, you have done something wrong. Keep in mind that all the approaches are only models that lead to predictions that explain and predict the experimental observations. One is not necessarily any better than the other.

With this introduction to the theories and terminology of pericyclic reactions, we are now ready to begin our survey of the various types of reactions involved and the manners in which the differing theories are used to analyze them. Traditionally, pericyclic reactions are categorized as cycloadditions, electrocyclic reactions, sigmatropic rearrangements, or cheletropic reactions. While each class is analyzed in slightly different ways, the fundamental issues are the same for all thermal pericyclic reactions. Transition states that experience stabilizing electronic interactions in a cyclic array of orbitals will be favored.

In anticipation of considering a variety of reactions, you may wish to re-examine Figure 14.14, which shows the nodal patterns of basic linear systems. These are the building blocks for much of what we will be studying in the following sections.

15.3 Cycloadditions

The discussion given above for the analysis of pericyclic reactions has set the stage for using these predictive methods for all pericyclic reactions. The following sections analyze each reaction type in detail, referring back to these methods for each case. We start with cycloadditions, where our analysis of the theory will not be as in-depth as with the other reaction types, since we have already discussed the theory above.

Two examples of efficient cycloadditions are given in Eqs. 15.4 and 15.5, along with the appropriate electron pushing. Both reactions are examples of dienes that are locked into an s-cis conformation (see below) with electron poor 2π components and electron rich dienes.

The 2π component is called the **dienophile**, because it is seeking the diene.

$$\text{(Eq. 15.4)}$$

$$\text{(Eq. 15.5)}$$

As with many of the examples given in Chapters 10 and 11, the electron pushing does not reflect how the reaction actually occurs. It is simply a bookkeeping method that allows chemists to keep track of bonds and lone pairs. The mechanisms are best described by the orbital and state correlation diagrams given above. The lack of insight from the electron pushing into the real mechanisms will hold for all the examples of the pericyclic reaction classes given throughout this chapter. Yet, the electron pushing shows how to distinguish between the bonding in the reactants and products, and is a common procedure for organic chemists, even with pericyclic reactions.

In cycloaddition reactions, two π systems come together to make a new ring. The pericyclic nature of the reaction is easy to see, as we clearly have a cyclic array of atoms (orbitals) in the transition state. These are among the most useful of pericyclic reactions, with Diels–Alder reactions such as those of Eqs. 15.4 and 15.5 being especially common. Cycloadditions are, as we noted above, denoted by an $[m + n]$ symbolism, where m and n are the *number of electrons* contributed by each reacting partner. In the overwhelming majority of cases, the cycloaddition transition states are as we depicted in Figures 15.1 and 15.2—that is, suprafacial on both partners. In this geometry, the $[_\pi 4_s + _\pi 2_s]$ cycloaddition is allowed, and the $[_\pi 2_s + _\pi 2_s]$ cycloaddition is forbidden. Yet, pericyclic reactions are always allowed with some geometry. Let's see what geometry for a [2+2] cycloaddition would be allowed by theoretical approaches.

15.3.1   An Allowed Geometry for [2+2] Cycloadditions

We have already mentioned the binary nature of the pericyclic reaction rules. If we change the mode of interaction of one of the reactants, we will reverse the allowed/forbidden nature of the reaction. For example, if we change the interaction mode of *one* of the reaction partners in the cycloaddition from suprafacial to antarafacial, now the [4+2] cycloaddition is forbidden, and the [2+2] cycloaddition is actually allowed. The [2+2] cycloaddition is now designated as $[_\pi 2_s + _\pi 2_a]$. In Figure 15.10 **A** we define the $[_\pi 2_s + _\pi 2_a]$ reaction, and in Figure 15.10 **B** we show a realistic geometry that could achieve the necessary orbital interactions. The two π systems approach in a perpendicular orientation, and the lines define the suprafacial and antarafacial interactions.

How do the various models rationalize this geometry of the [2+2] reaction as allowed? Orbital and state correlation diagrams are left as an Exercise at the end of the chapter. In day-

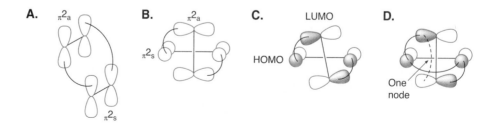

Figure 15.10
The $[_\pi 2_s + _\pi 2_a]$ cycloaddition reaction. **A.** A schematic of the reaction. **B.** A representation of a realistic transition state geometry for the reaction. **C.** Illustration of the favorable HOMO–LUMO mixing for the reaction. **D.** The Möbius topology of the transition state.

to-day practice, chemists rarely have to construct a correlation diagram to understand a reaction. The simpler rules typically provide a clear prediction.

The FMO analysis is as shown in Figure 15.10 **C**. The HOMO–LUMO interaction is now favorable and leads naturally to the formation of the two new bonds. Figure 15.10 **D** shows the aromatic transition state analysis. Using the looped lines, we have designated the full cyclic array of interactions. As shown, there is one node in the system, so this is a Möbius system. Since there are four electrons in the cyclic array, the reaction is allowed. By the generalized orbital symmetry rule, this approach trajectory ($[_\pi 2_s + _\pi 2_a]$) is thermally allowed [only the $_\pi 2_s$ component fits the $(4q + 2)$s and $(4r)$a formulas]. In summary, it is incorrect to say that a [2+2] cycloaddition is forbidden. It is a suprafacial/suprafacial approach that is forbidden. A suprafacial/antarafacial approach is allowed. Similarly, the forbidden nature of a [2+2] cannot be uniformly applied to systems where the symmetry of orbitals are different than with ethylene plus ethylene, as noted in the following Going Deeper highlight that examines an organometallic [2+2] cycloaddition.

Going Deeper

Allowed Organometallic [2+2] Cycloadditions

The [2+2] cycloaddition of alkenes with metal alkylidene species is very facile in many cases. With early transition metal complexes, evidence points to the reaction being pericyclic. The stereochemistry of the alkene is preserved, and the sterics of the organometallic species are usually such that only an aligned approach (not twisted as in Figure 15.10) seems viable. Hence, these systems would be classified as allowed $[_\pi 2_s + _\pi 2_s]$ reactions. Therefore, the generalized orbital symmetry rule given in this chapter must *only* relate to organic systems, but the other methods such as correlation diagrams, FMO analyses, and aromatic transition state analysis should be applicable to all molecular systems, including organometallic species.

$$M=CR_2 \quad + \quad R'_2C=CR'_2 \rightleftharpoons$$

[2+2] for metallacycle formation

How do we rationalize this allowed reaction? Both FMO and aromatic transition state theory are easy to apply. As shown below, the extra node in the d orbital used in the alkylidene π bond allows the HOMO of the M=C bond to interact with the LUMO of the C=C bond constructively. Similarly, the extra node in the d orbital makes the four-electron system Möbius (remember we do not count nodes in the atomic orbitals themselves), and therefore allowed.

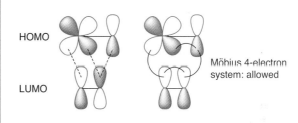

HOMO

LUMO

Möbius 4-electron system: allowed

Upton, T. H., and Rappé, A. K. "A Theoretical Basis for the Low Barriers in Transition-Metal Complex $2_\pi + 2_\pi$ Reactions: The Isomerization of $Cp_2TiC_3H_6$ to $Cp_2TiCH_2(CH_2CH_2)$." *J. Am. Chem. Soc.*, **107**, 1206 (1985).

15.3.2   Summarizing Cycloadditions

We are now ready to summarize the trends expected for cycloadditions. This is done in Table 15.1. The key is the sum of $m + n$, where once again we emphasize that m and n represent a number of electrons, not atoms. The rules are reminiscent of aromaticity, in that for the conventional s + s geometry, a total of $4q + 2$ electrons is allowed, while a total of $4q$ is forbidden. The less common s + a geometry is Möbius, so $4q$ is preferred in that case.

15.3.3   General Experimental Observations

We can also anticipate how cycloadditions will be characterized by the various "tools" of physical organic chemistry. The reactions are bimolecular, and one sees second-order kinetics. The transition state is highly organized and brings two molecules together. As such, highly negative entropies of activation are found. Since ionic species are not involved, there is not a very strong dependence of reaction rates on solvent polarity in organic solvents.

Table 15.1
Rules for Thermal $[m + n]$ Cycloaddition Reactions

$m + n$	Allowed	Forbidden
$4q$	s + a a + s	s + s a + a
$4q + 2$	s + s a + a	s + a a + s

However, water dramatically enhances Diels–Alder rates because the hydrophobic effect brings the two reactants together (for a further discussion see the Going Deeper highlight on page 923). Since four sp^2 centers are being converted to sp^3 centers, inverse kinetic isotope effects are the norm. Lastly, because two reactants combine to create a product that is smaller in volume than the separate reactants, high pressure will facilitate cycloadditions.

There are substantial substituent effects on cycloaddition reaction rates. We noted earlier that the prototype Diels–Alder reaction of Eq. 15.2, with butadiene as the diene and ethylene as the dienophile, does not actually proceed efficiently. In reality, some substituents are always necessary. The most common pattern is to have one or more electron withdrawing groups (EWG) on the dienophile, and electron donating groups (EDG) on the diene. Hence, the diene is made electron rich (i.e., add alkyl groups, amino groups, ethers), and the dienophile is made electron poor (i.e., add cyano, esters, nitro). We saw this with the examples given in Eqs. 15.4 and 15.5. While both HOMO–LUMO interactions between the diene and dienophile are favorable (Figure 15.6), one of these interactions can be accentuated by substitution. The EWG(s) on the dienophile lower its LUMO and the EDG(s) on the diene raise its HOMO. The reason for the effect on the dienophile LUMO is that EWGs have electronegative elements, and as we noted in Chapters 1 and 14, electronegative elements lower the energies of all orbitals in which they are involved. The reason for the effect on the diene HOMO was discussed in Chapter 1, where we showed that the mixing of a $\pi(CH_3)$ group orbital with the π bond of ethylene raises the HOMO. In a Diels–Alder reaction, therefore, the diene HOMO and dienophile LUMO are brought closer in energy, lowering the energy gap and thus making this interaction more favorable. In support of this analysis, the $\log(k)$ for Diels–Alder reactions correlate quite well with the inverse of the difference in energy between the ionization potential of the diene (related to the energy of the HOMO) and the electron affinity of the dienophile (related to the energy of the LUMO). The more electron rich the diene and the more electron poor the dienophile, the faster the cycloaddition. To further enhance the electrophilicity of the dienophile, it is common to add Lewis acids that can complex the electron withdrawing groups on the dienophile, further lowering the LUMO.

Another common strategy to facilitate the reaction is to incorporate the diene into a ring, thereby favoring the *s*-cis geometry required for reaction. The *s*-cis geometry is required in the cycloaddition for the dienophile to reach both ends of the diene (see margin).

15.3.4 Stereochemistry and Regiochemistry of the Diels–Alder Reaction

Synthetically, the Diels–Alder reaction is the most important cycloaddition and arguably the most important pericyclic reaction. Because of this, we will consider several additional features of this reaction here. It is a $[_\pi4_s + _\pi2_s]$ cycloaddition, and it best illustrates a key feature of pericyclic reactions that we have yet to touch on. Since, by definition, pericyclic reactions involve a well controlled array of atoms / orbitals in the transition state, well-defined stereochemistry is a hallmark of pericyclic reactions. In general, *a high degree of control of stereochemistry is associated with pericyclic reactions, and this is one of their most valuable features.* Eq. 15.6 illustrates this aspect of the Diels–Alder reaction. As many as four new stereocenters are created, and the control is often complete.

s-trans   *s*-cis

(Eq. 15.6)

An Orbital Approach to Predicting Regiochemistry

When multiple substituents are involved, a new issue arises, that of the regiochemistry of the Diels–Alder reaction. When both the diene and dienophile have a substituent, we can speak in terms of pseudo-ortho, meta, and para patterns for the product, as shown in Figure 15.11. The nomenclature is imperfect, as two different pseudo-meta forms are shown, but usually it is clear in context which isomers are being discussed. Houk and co-workers have

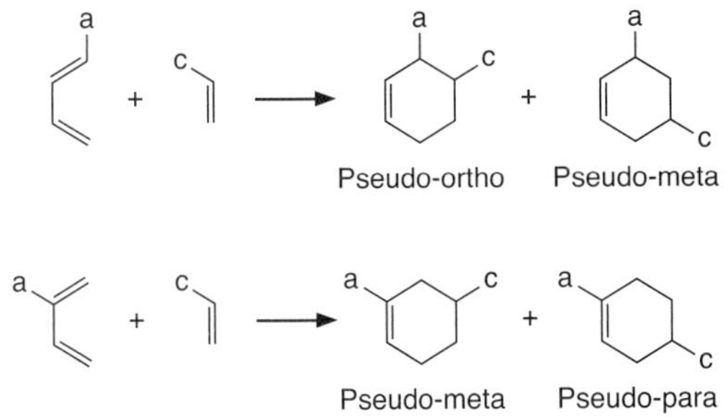

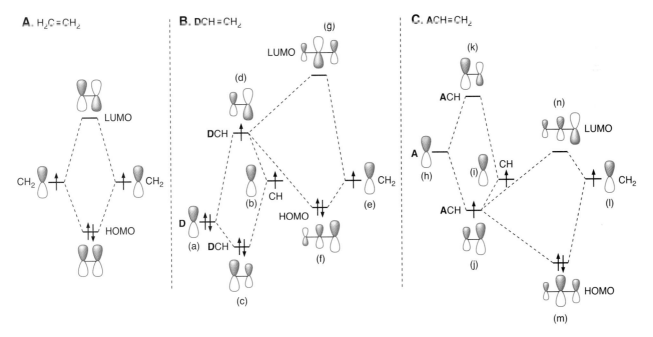

Figure 15.11
Regiochemical issues that arise in the Diels–Alder reaction when substituted dienes and dienophiles are involved.

developed a model for such systems that provides a nice example of the kinds of qualitative molecular orbital arguments we have developed throughout this text. Figure 15.12 describes the model.

Our starting point is the mixing of two p orbitals to make the π and π^* molecular orbitals of ethylene, originally given as Figure 1.14, and recapitulated here as Figure 15.12 **A**. Now we will consider the effects of substituents on this scheme. As before, we will consider two kinds of substituents, donor (**D**) and acceptor (**A**). In this scheme, a donor is characterized as having a high-lying, doubly-occupied orbital, whereas an acceptor has a low-lying empty orbital. Our goal is to construct orbitals for $CH_2=CHD$ and $CH_2=CHA$ that are analogous to the π and π^* molecular orbitals of $CH_2=CH_2$.

We consider the donor first, and begin at the left of Figure 15.12 **B**, where we mix the high-lying filled orbital of **D** (a) with a carbon p orbital (b). Note the polarizations in the resulting orbitals: the lower doubly-occupied orbital (c) is polarized toward **D**, while the higher singly-occupied orbital (d) is polarized toward carbon. This is exactly as we would expect from our earlier discussions of perturbation theory in Chapter 14. Now, the newly formed singly-occupied orbital (d) mixes with another singly-occupied p orbital (e) of the unsubstituted carbon to make the π orbitals (f) and (g). Note that the addition of the **D** atom makes both the HOMO and LUMO higher in energy than in ethylene.

Figure 15.12
Analysis of olefin substituent effects relevant to the Diels–Alder reaction. **A.** The familiar mixing diagram for the formation of the π and π^* orbitals of ethylene, which serves as a reference for the rest of the figure. **B.** Mixing diagram to develop the molecular orbitals of an olefin substituted with a donor substituent (**D**). **C.** Mixing diagram to develop the molecular orbitals of an olefin substituted with an acceptor substituent (**A**). See text for a discussion of parts **B** and **C**.

The new HOMO is different from the ethylene HOMO in two ways. First, as stated above, it is higher in energy. This is because the first-order (degenerate) mixing seen with ethylene is now a second-order mixing, resulting in a smaller energy lowering. Second, the HOMO is polarized toward the *unsubstituted* carbon (CH_2). This is because the carbon p orbital (e) starts lower in energy than the CH**D** orbital (d). The LUMO is raised in energy and polarized toward the substituted carbon (CH). However, because the coefficient on the substituted carbon was reduced by mixing with the **D** orbital, the polarization seen in the LUMO is less than in the HOMO. A secondary interaction, not shown in Figure 5.12 **B**, is also part of the analysis. The newly formed HOMO (f) can interact with the lower-lying CH**D** orbital (c). This will further raise the energy of the HOMO, accentuating the primary effect of the orbital mixing.

The polarizations shown for the HOMO and LUMO of a CH_2=CH**D** system can be anticipated using resonance theory. As shown in Eq. 15.7 with an enamine as the example, the unsubstituted carbon is nucleophilic. This leads one to predict that the HOMO will be polarized toward this carbon, and this is what we found with orbital (f) in Figure 15.12 **B**. Furthermore, it is generally true that the polarization in the HOMO is opposite to that in the LUMO, and therefore the LUMO will be polarized toward the substituted carbon [orbital (g) in Figure 5.12 **B**].

$$(Eq. 15.7)$$

A similar analysis of the CH_2=CH**A** system is given in Figure 15.12 **C**. The acceptor orbital (h) is higher in energy than the CH orbital (i). Mixing these orbitals and populating with only one electron (from the CH group) gives a low-lying **A**CH orbital (j) for mixing with the other carbon orbital (l). Now, the resulting HOMO (m) and LUMO (n) are both lower in energy than the ethylene reference HOMO and LUMO. We conclude that the important effect is the lowering of the energy of the LUMO, along with a polarization toward the unsubstituted carbon. Effects on the HOMO will be less. Again, a secondary interaction between the LUMO and the higher-lying CH**A** orbital (k) further lowers the LUMO energy.

Once again these perturbations of coefficients and energies of the ethylene orbitals can be anticipated using resonance. Eq. 15.8 shows the expectations for substitution of an electron withdrawing group on an ethylene. The β-carbon is electrophilic. Therefore, the LUMO should be polarized to the unsubstituted carbon as seen in orbital (n). The reverse polarization is predicted for the HOMO, and this is seen in orbital (m).

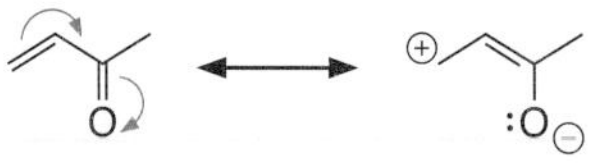

$$(Eq. 15.8)$$

The basic conclusions of this analysis are that a donor substituent, **D**, raises the energy of the HOMO and polarizes it toward the unsubstituted carbon. Furthermore, an acceptor substituent, **A**, lowers the energy of the LUMO and also polarizes it toward the unsubstituted carbon. A third kind of substituent is an unsaturated group (**U**), such as a phenyl or vinyl group. From our previous analysis of π systems, we know that extended conjugation will raise the HOMO and lower the LUMO (Section 14.3.3, Figure 14.14). We find polarization of the HOMO and LUMO toward the unsubstituted carbon, but the effect is much less dramatic than with a **D** or **A** substituent.

To predict regiochemistry, we emphasize two features of the Diels–Alder reaction. First, as noted above, the key interaction is usually the HOMO of the diene with the LUMO of the dienophile. Second, we line up the reacting partners so that the larger coefficient at a terminus of the diene reacts with the larger coefficient of the dienophile. The reason that we align the larger coefficients of the HOMO and LUMO is that this has been shown to give the largest overlap ($<\psi_{HOMO}|\psi_{LUMO}>$) in the transition state, leading to the most facile reaction pathway.

In Figure 15.13 **A** we show an example of such an analysis, with the reaction of isoprene and acrylonitrile. The coefficient values and orbital energies are obtained from a HF–SCF

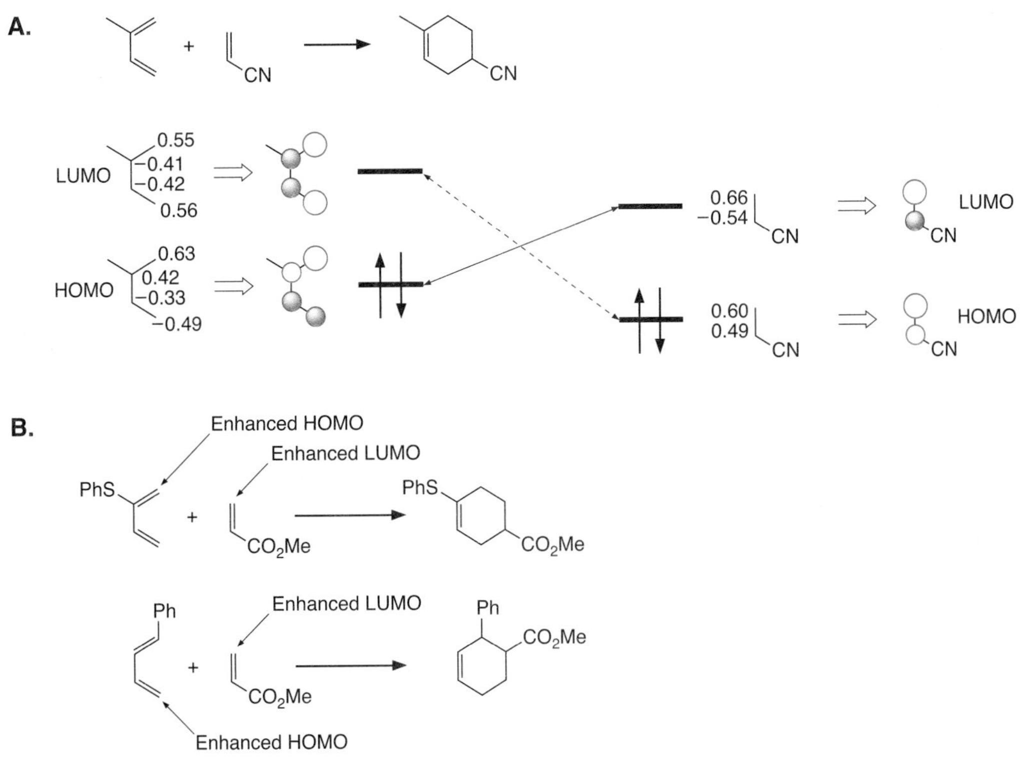

A.

B.

Figure 15.13
Substituent effects in the Diels–Alder reaction. **A.** Reaction of acrylonitrile with isoprene. Molecular orbital coefficients for the HOMO and LUMO of each reactant are shown, along with the major mixing (solid arrow) and the minor mixing (dashed arrow). **B.** Further examples of regioselectivity in the Diels–Alder reaction.

calculation of the sort described in Chapter 14. In this case, both the HOMO and the LUMO of acrylonitrile are polarized toward the unsubstituted carbon, but the effect is more pronounced in the LUMO, as predicted above. The methyl group of isoprene acts as a weak donor. As anticipated from Figure 15.12, the larger polarization in the diene is in the HOMO while the larger polarization of the dienophile is in the LUMO. To rationalize the regiochemistry, we line up the largest coefficient of the diene HOMO with the larger coefficient of the dienophile LUMO. Doing so produces the pseudo-para product, as is indeed observed. Other examples of regioselective Diels–Alder reactions are given in Figure 15.13, all of which are consistent with the analysis we have presented.

The Diels–Alder reactions we have considered thus far are typical and are described as having **normal electron demand**. Acceptor substituents on the dienophile accelerate the reaction, as do donor substituents on the diene. The primary interaction is diene HOMO and dienophile LUMO. However, in cases where there is a donor group on the dienophile and an acceptor on the diene, the result is an **inverse electron demand** Diels–Alder reaction. Now the key interaction is dienophile HOMO and diene LUMO. These reactions are less common, but are sometimes useful.

The Endo Effect

Another common feature of Diels–Alder stereochemistry is the so-called **endo effect**. Phenomenologically, this is a stereochemical effect whereby an acceptor substituent on the dienophile ends up in the endo position of the product, as shown in Eqs. 15.9 and 15.10. This is a useful and fairly general feature of the Diels–Alder reaction.

$$\text{(Eq. 15.9)}$$

Endo 74%   Exo 26%

$$\text{(Eq. 15.10)}$$

Endo 99%   Exo 1%

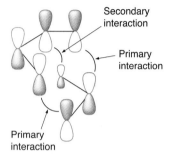

Secondary interaction

Primary interaction

Primary interaction

The longest standing model to rationalize this effect emphasizes so-called **secondary orbital interactions**. These are molecular orbital interactions other than those primarily used to define whether the reaction is allowed or forbidden, and that are frequently invoked to describe additional features of pericyclic reactions. Consider the LUMO of the $CH_2=CHA$ fragment in Figure 15.12 **C** (orbital n). The p orbital associated with **A** is in-phase with the adjacent carbon. Now consider the HOMO of a diene. C2 and C3 are similarly in-phase with C1 and C4, respectively. Thus, if we tuck the substituent **A** under the diene in the transition state, an additional interaction between the HOMO and LUMO can occur, and this must be stabilizing (see margin).

This model has provided a sensible rationalization of the endo-effect. However, recent studies suggest that additional factors such as steric and electrostatic effects also play an important role. This has led some to conclude that secondary orbital interactions do not play a determining role in the Diels–Alder reaction nor in other pericyclic reactions.

Going Deeper

Semi-Empirical vs. *Ab Initio* Treatments of Pericyclic Transition States

A recurring conflict during the development of computational quantum mechanics was the differing conclusions about pericyclic transition state geometries reached by semi-empirical vs. *ab initio* methods. Consistently, semi-empirical methods such as MINDO, MNDO, and AM1 favored highly unsymmetrical transition states, often involving biradical intermediates. In contrast, *ab initio* methods always favored more symmetrical transition states and concerted processes. It is now clear from very high-level *ab initio* studies and extensive experimental work that the earlier semi-empirical work was flawed. What was the cause of this error? An interesting analysis by Houk provides an explanation.

We noted in Chapter 14 that the semi-empirical methods neglect overlap in normalizing the wavefunction. One consequence of this is that closed-shell repulsion is absent in these methods (recall Section 14.4). It is not destabilizing to mix filled orbitals in a semi-empirical wavefunction. Consider a simple Diels–Alder reaction. In the symmetrical approach favored by *ab initio* methods (and by the molecules themselves!), the HOMO of the diene and the HOMO of the dienophile are necessarily of opposite symmetry. As such, they cannot mix, and there is no

closed-shell repulsion. However, in a highly unsymmetrical (i.e., low symmetry) approach, this restriction is relaxed, and a significant HOMO–HOMO closed shell repulsion develops. This should destabilize the unsymmetrical approach relative to the symmetrical approach. However, because the semi-empirical methods do not include closed-shell repulsion, they miss this effect. As such, the semi-empirical methods predict that the unsymmetrical transition states are more stable than they really are.

In support of this analysis, it was demonstrated that extended Hückel theory (EHT) correctly predicts a symmetrical transition state. Despite the substantial approximations of EHT, it *does* include overlap and thus closed-shell repulsion. In a clever "control experiment", a modified EHT code that does not include overlap produced an unsymmetrical transition state. It appears that the case of pericyclic transition states is one in which the approximations necessary to develop a rapid, semi-empirical computational model are too severe, and the semi-empirical methods are not applicable to such reactions.

Caramella, P., Houk, K. N., and Domelsmith, L. N. "On the Dichotomy between Cycloaddition Transition States Calculated by Semiempirical and Ab Initio Techniques." *J. Am. Chem. Soc.*, **99**, 4511–4514 (1977).

15.3.5 Experimental Observations for [2+2] Cycloadditions

According to the various theories of pericyclic reactions, thermal [2+2] cycloadditions must occur in an [s + a] fashion. When both partners are alkenes, this geometry is unfavorable, suffering adverse steric interactions. As such, all [2+2] cycloadditions of olefins to produce cyclobutanes occur by a stepwise, biradical pathway.

However, when one of the reacting partners is a ketene or an allene, the steric problems are less severe, and a concerted [2+2] cycloaddition does occur, with the ketene reaction being much more common (Eq. 15.11). Evidence for the concerted nature of the reaction includes retention of stereochemistry in the olefin component, large negative entropies of activation, and small solvent effects on the rate. Ketene cycloadditions are favored by electron donating groups on the alkene; vinyl ethers are especially favorable. The reaction works with stable ketenes such as diphenyl ketene and with transiently generated ketenes, dichloroketene being a common, easily produced substrate (Eq. 15.12).

$$\text{(Eq. 15.11)}$$

$$\text{(Eq. 15.12)}$$

A novel feature of these reactions is that they tend to give the sterically most crowded product (Eq. 15.13). This result is nicely rationalized by the required $[_\pi 2_s + {}_\pi 2_a]$ interaction geometry (Eq. 15.14). Bringing the reactants together so as to put the larger substituent of the ketene away from the olefin ultimately produces the more crowded product.

$$\text{(Eq. 15.13)}$$

$$\text{(Eq. 15.14)}$$

L = Large substituent
S = Small substituent

15.3.6 Experimental Observations for 1,3-Dipolar Cycloadditions

An important and interesting class of cycloadditions is the broad group of [4+2] cycloadditions in which the four-electron component is a three-atom system. The resulting product is thus a five-membered ring. Remember, though, these are six-electron systems, and so the favorable [s + s] pathway is allowed.

Pioneering work by Huisgen established that a huge range of three-atom, four-electron systems are viable in this reaction, with the unifying themes being the presence of heteroatoms and the existence of at least one dipolar resonance form. The latter gives rise to the **1,3-dipolar cycloaddition** terminology. Figure 15.14 lists some dipolar molecules, though many other variants exist. Retention of stereochemistry on the olefinic component and relative insensitivity to solvent polarity support the assignment of these as concerted, pericyclic reactions.

Regiochemistry is an issue in these reactions. With the advent of the orbital symmetry rules (Huisgen's definitive overview of the field was published two years before Woodward

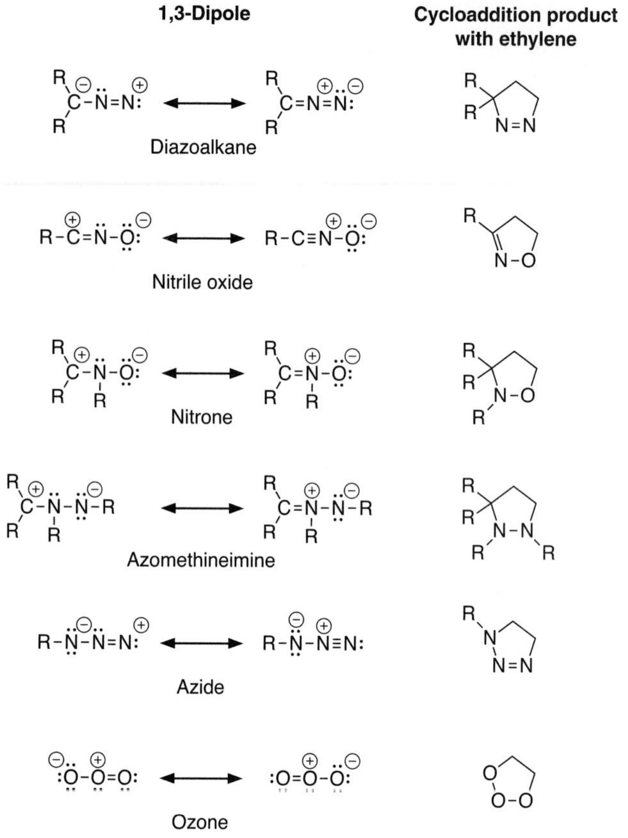

1,3-Dipole   **Cycloaddition product with ethylene**

Figure 15.14
Representative 1,3-dipoles and their cycloadducts with ethylene.

and Hoffmann's first papers), a rational analysis of this issue became feasible. Despite the great diversity of the reactions, analyses along the lines presented above for the Diels–Alder reaction are generally successful. That is, we examine the coefficients of the frontier orbitals and match the larger coefficient values. Nowadays, any desktop *ab initio* software package can provide this information with an advanced level of theory. As an example, Eq. 15.15 shows the addition of a nitrile oxide to a **dipolarophile** (an alkene seeking a dipolar molecule) with a substituent that can be viewed as either a donor or a unit of unsaturation (**D** or **U**). In this situation, the LUMO of the dipole is the controlling orbital, although this issue must be addressed on a case-by-case basis. The dipoles are isoelectronic with allyl anion, and so the phase properties of the π systems follow the familiar allyl pattern. Eq. 15.15 also shows the LUMO of the parent nitrile oxide, with the largest coefficient being on the carbon. We predict a large coefficient on the carbon by analysis of a resonance structure, which places positive charge on the carbon. As described in Figure 15.12, the HOMO of the dipolarophile with a **D** or **U** substituent will be polarized toward the unsubstituted carbon. Matching the largest coefficients produces the 5-substituted heterocycle. Indeed, the benzonitrile oxide with styrene gives 100% of the 5-substituted product.

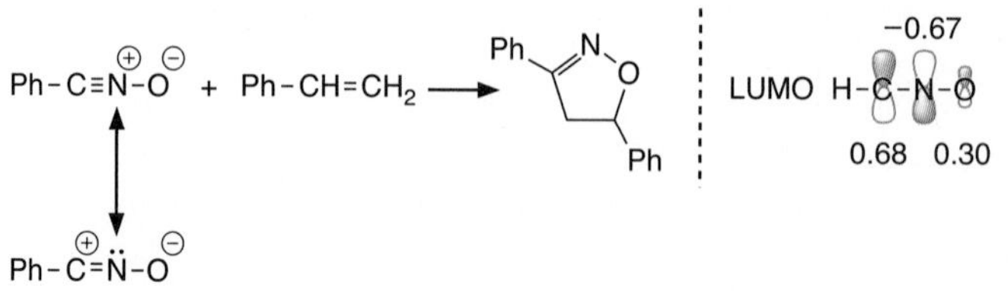

(Eq. 15.15)

15.3.7   Retrocycloadditions

If a reaction is deemed allowed, its reverse is also allowed. In principle, any allowed cycloaddition we have discussed can also be run in the reverse direction, producing two fragments. Since retrocycloadditions (also called **cycloreversions**) are entropically favored relative to cycloadditions, we can imagine that at high enough temperatures cycloreversion would be common. Indeed, **retro Diels–Alder** reactions are common, as are reversals of other reactions we have discussed. A common example is the cracking of dicyclopentadiene (Eq. 15.16). Note also that there is no need to do another orbital or state correlation diagram to analyze the retrocycloaddition. The principle of microscopic reversibility ensures that the analysis of the reverse direction uses the same diagram as the forward reaction. An important message is that when encountering a new reaction, sometimes it might be easier to analyze it in the reverse direction; the results are just as applicable.

$$\text{(Eq. 15.16)}$$

15.4 Electrocyclic Reactions

An **electrocyclic reaction** involves the conversion of a π system with n electrons to a cyclic system with $n-2$ π electrons and a σ bond, or the reverse. Eqs. 15.17 and 15.18 show two prototype reactions, the butadiene–cyclobutene interconversion and the hexatriene–cyclohexadiene interconversion. Once again, the arrow pushing does not reflect the mechanism of the reaction.

$$\text{(Eq. 15.17)}$$

$$\text{(Eq. 15.18)}$$

In each case the ring closure involves the rotation of the terminal carbons so that the p orbitals of the π system, which are necessarily parallel to each other in the polyene, point toward each other and make a new bond. It is the direction of rotation (clockwise or counterclockwise) that changes depending upon electron count and orbital interactions. This direction of rotation influences the stereochemical outcome of the reactions, and therefore our analysis really focuses on rationalizing and predicting stereochemistry. Before examining which rotations are allowed or forbidden, we must introduce some new terminology.

15.4.1 Terminology

The nomenclature we need to examine electrocyclic reactions (and cheletropic reactions, see below) is one that describes how p orbitals at the termini of a π system rotate. As can be seen in Figure 15.15, if the two p orbitals rotate in the same direction (both clockwise or both counterclockwise) the process is termed **conrotatory**. If they rotate in opposite directions (one clockwise and one counterclockwise), the process is termed **disrotatory**. The same terms are used to describe the direction of rotation of atoms involved in σ bonds. Figure 15.15 also shows their use in this context.

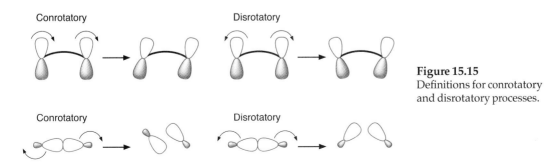

Conrotatory   Disrotatory

Conrotatory   Disrotatory

Figure 15.15
Definitions for conrotatory and disrotatory processes.

15.4.2   Theoretical Analyses

Let's start our theoretical analysis with an orbital correlation diagram for the four-electron case. These diagrams are very straightforward to apply to electrocyclic reactions. If the two ends of the π system rotate in the same direction—a conrotatory motion—a C_2 symmetry axis is maintained throughout the process (Figure 15.16 **A**). If the two termini rotate in opposite directions—that is, disrotatory—a mirror plane (σ) is maintained. Figure 15.16 **B** shows the orbital symmetry correlation diagrams for the butadiene–cyclobutene interconversion based on the conserved symmetry elements. The butadiene molecular orbitals are the familiar ones from Figure 14.14. The cyclobutene molecular orbitals are simple π/π* and σ/σ* pairs. As with the analyses given in Figures 15.1 and 15.2, we only need to consider molecular orbitals for the bonds undergoing changes. For the two different processes (conrotatory or disrotatory), the molecular orbitals are the same; it is just the symmetry designations that change. The black labeling shows the symmetry of the orbitals for the conrotatory process, while the colored labels give the orbital symmetries for the disrotatory process.

If we place four electrons in the lowest energy orbitals of the butadiene, it is clear from Figure 15.16 **B** that the conservation of orbital symmetry predicts that the conrotatory process (black lines) is preferred. The disrotatory (colored dashed lines) process leads from butadiene to an excited state of cyclobutene. The same conclusions are reached by considering the reaction in the reverse direction.

A.

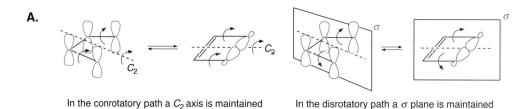

In the conrotatory path a C_2 axis is maintained   In the disrotatory path a σ plane is maintained

B.

Figure 15.16
A. Orbital symmetry analysis of the butadiene–cyclobutene interconversion. **B.** The lines connecting the reactant and product orbitals for the conrotatory process are given in black, while the correlation for the disrotatory process is given in dashed color.

By now you should anticipate that exactly the opposite conclusions are reached for the hexatriene–cyclohexadiene interconversion. Indeed, a full orbital symmetry analysis, given as an Exercise at the end of the chapter, leads to the conclusion that the disrotatory process is orbital symmetry allowed, whereas the conrotatory process is forbidden. Furthermore, a state correlation analysis constructed along the lines of Figures 15.4 and 15.5 supports the conclusions of the orbital symmetry analysis. Again, we leave this as an Exercise at the end of the chapter.

The FMO analysis of the electrocyclic reactions in Eqs. 15.17 and 15.18 is quite straight-forward also, and is given in Figure 15.17 **A**. One typically looks at the ring-opening not the closure, although there are FMO approaches to the closure (covered in the Exercises at the end of the chapter). With the opening, the HOMO of the σ bond is analyzed with regard to how it would correlate with the LUMO of the π system. The black loops given in Figure 15.17 **A** show that conrotatory rotation of the σ bond gives in-phase interactions for cyclobutene, while disrotatory rotation would give out-of-phase interactions. The exact opposite FMO conclusions are reached for 1,3-cyclohexadiene.

Figure 15.17 **B** shows the aromatic transition state analysis of these reactions. We draw a picture of an opening pathway with the minimum number of phase changes and examine the number of nodes. The four-electron butadiene–cyclobutene system should follow the Möbius/conrotatory path, and the six-electron hexatriene–cyclohexadiene system should follow the Hückel/disrotatory path. As such, aromatic transition state theory provides a simple analysis of electrocyclic reactions. The disrotatory motion is always of Hückel topology, and the conrotatory motion is always of Möbius topology.

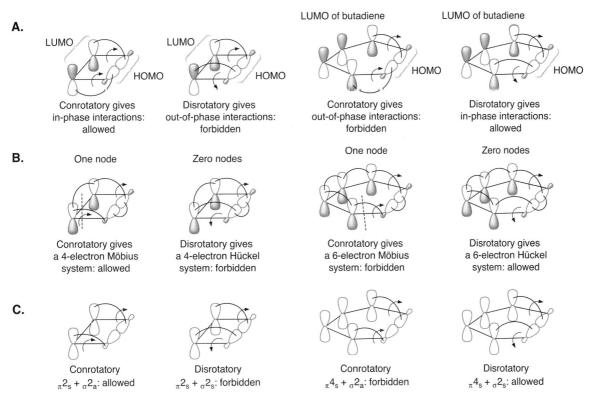

Figure 15.17
Analysis of electrocyclic reactions using a variety of methods and the various conclusions that are drawn. **A.** FMO theory for ring-opening. The LUMOs of the π systems are compared to the HOMO of the C–C σ bond in cyclobutene and 1,3-cyclohexadiene. **B.** The Hückel/Möbius approach. **C.** Using the generalized orbital symmetry rule. Note, as always, that all the methods predict the same outcome.

n	Allowed	Forbidden
$4q$	conrotatory	disrotatory
$4q + 2$	disrotatory	conrotatory

The analysis of the generalized orbital symmetry rule nicely follows from the above discussion (Figure 15.17 **C**). The conrotatory opening of cyclobutene is a $[_\pi 2_s + _\sigma 2_a]$ reaction and therefore allowed, while the disrotatory reaction is $[_\pi 2_s + _\sigma 2_s]$ and forbidden. The conrotatory opening of 1,3-cyclohexadiene is a $[_\pi 4_s + _\sigma 2_a]$ path that is forbidden, while the disrotatory $[_\pi 4_s + _\sigma 2_s]$ reaction is allowed. Summarizing all these analyses, we come up with the selection rules listed in Table 15.2.

15.4.3 Experimental Observations: Stereochemistry

Figure 15.18 shows several examples of electrocyclic processes. Since the reactions are always allowed in either a conrotatory or disrotatory manner, the key issue is the control of stereochemistry. Electrocyclic reactions provide a good example of the power of pericyclic reactions in this regard. In all cases, the reaction proceeds as predicted from the various theoretical approaches. The restrictions placed by the orbital analysis on the reaction pathway are nicely demonstrated by examples **D** and **E** in Figure 15.18; only the stereochemistry given is found. An instructive example of the fact that it is the number of electrons that controls the process, not the number of atoms or orbitals, is the conrotatory ring closure of the four-electron pentadienyl cation prepared by protonation of a divinyl ketone (example **G**).

Just how selective are pericyclic reactions? That is, how large is the preference for the allowed path over the forbidden path? Brauman and Archie found that the electrocyclic ring-opening of *cis*-3,4-dimethylcyclobutene was 99.995% stereospecific, corresponding to an energy difference of 11 kcal/mol between the activation energies of allowed and forbidden paths (Figure 15.18, example **A**). Another indication of the magnitude of the preference is given in the tetraphenyldimethylbutadiene reaction shown as example **H**. This system was kept at 124 °C for 51 days, during which time each molecule underwent almost 3 million ring-openings and closings. However, no products other than those shown were seen, indicating that in this system the preference for conrotatory motions is on the order of 15 kcal/mol.

An intriguing and famous application of the orbital symmetry rules concerns Dewar benzene (Eq. 15.19). This valence isomer of benzene is highly strained and *much* less stable than benzene. It would appear that a very simple process of just cleaving the central bond would relieve a great deal of strain and produce a highly stable, aromatic product. Nevertheless, Dewar benzene is a relatively persistent molecule.

Dewar benzene

(Eq. 15.19)

The standard analysis ascribes the surprising persistence of Dewar benzene to the fact that the ring-opening to give benzene is a forbidden process. Conversion of Dewar benzene to benzene must be disrotatory, but because the reaction is essentially a cyclobutene ring-opening, it prefers the conrotatory path. Conrotatory opening creates a trans double bond (Eq. 15.20), which is far too highly strained to occur. The Dewar benzene isomer is trapped in a kinetic prison whose origin is orbital symmetry. You might ask why this is a four-electron, and not a six-electron process. This illustrates a general feature of such analyses. If a bond or π system is really just a spectator (meaning it does not change position) in the reaction, then we do not consider it in the analysis. One of the π bonds of Dewar benzene really just goes along for the ride, and so we consider this a four-electron process. An orbital correlation diagram makes this clear, and an Exercise at the end of the chapter gives you the opportunity to show this.

(Eq. 15.20)

The selection rules for electrocyclic ring-openings can also explain rate differences in the solvolysis of cyclopropyl halides as a function of stereochemistry. For example, shown in

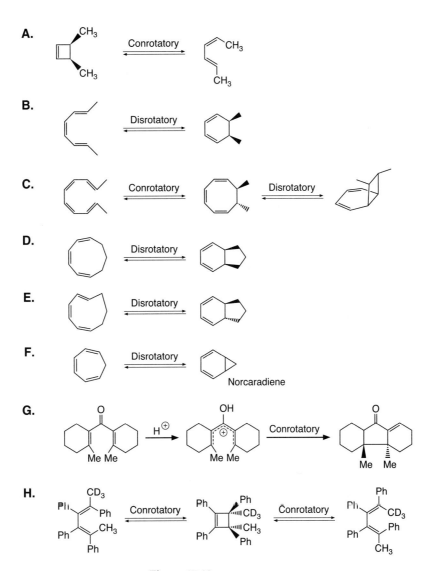

Figure 15.18
Examples of electrocyclic reactions.

the margin are the relative rates for solvolysis of the reactants given. Although the all-cis arrangement of groups is the least stable, the other isomer reacts over 1000 times faster at 150 °C in acetic acid. The reason is that the solvolysis involves ring-opening to form an allylic cation. The ring-opening occurs in a disrotatory fashion because of the FMO analysis shown in the margin. For this motion to occur in the all-cis isomer, the methyl groups necessarily will collide during the bond rotations, dramatically impeding the reaction.

15.4.4   Torquoselectivity

We have shown that we can easily choose whether the conrotatory or disrotatory path is preferred for a given system. However, this is not the whole story. Both paths can occur in two different "directions", and in many systems, the two give different products. This is illustrated in Figure 15.19. The conrotatory openings of cyclobutenes have been extensively investigated in this regard, leading to a novel theory that nicely illustrates some qualitative molecular orbital reasoning.

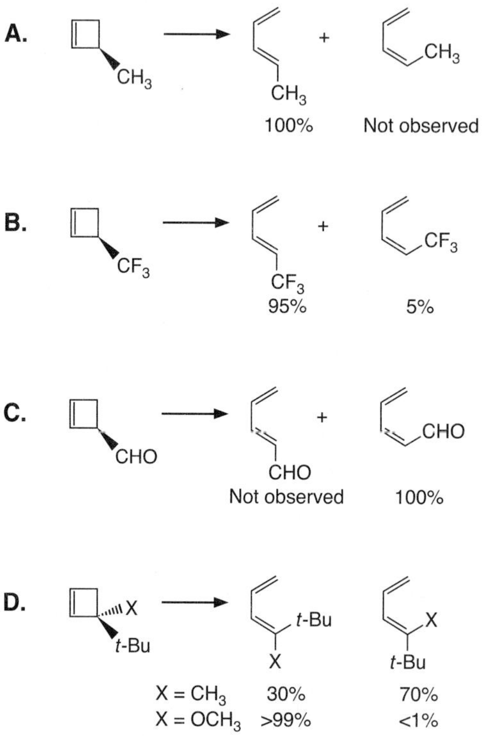

A.

100%   Not observed

B.

95%   5%

C.

Not observed   100%

D.

| X = CH$_3$ | 30% | 70% |
| X = OCH$_3$ | >99% | <1% |

Figure 15.19
Torquoselectivity in electrocyclic openings.

Early observations were interpreted in terms of steric interactions. For example, 3-methylcyclobutene opens exclusively to the less crowded trans-pentadiene (example **A**, Figure 15.19). However, further study revealed some intriguing non-steric effects. For example, the trifluoromethyl group, generally considered larger than a methyl (recall, for example, Table 2.14), gives measurable amounts of the cis product (example **B**). Remarkably, 3-formylcyclobutene gives 100% of the cis product (example **C**). In another stunning example, 3-methoxy-3-*tert*-butylcyclobutene rotates the bulky *tert*-butyl group inward (example **D**). All these observations suggested that electronic, rather than steric, effects were controlling these reactions.

At the same time that these stereochemical results were being studied, some interesting substituent effects on the reaction *rates* were noticed. For example, 3-formylcyclobutene has an activation energy that is over 5 kcal/mol smaller than the parent hydrocarbon. The effects are substantial, and they have been the topic of considerable investigation. Dolbier,

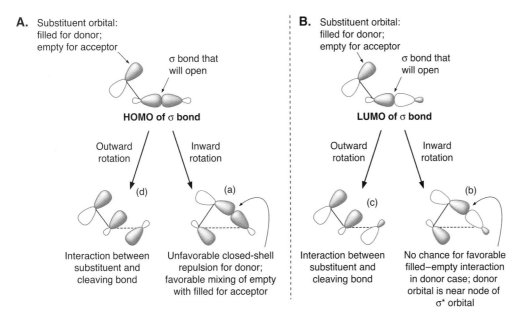

A. Substituent orbital:
filled for donor;
empty for acceptor

σ bond that
will open

HOMO of σ bond

Outward
rotation

Inward
rotation

(d)

(a)

Interaction between
substituent and
cleaving bond

Unfavorable closed-shell
repulsion for donor;
favorable mixing of empty
with filled for acceptor

B. Substituent orbital:
filled for donor;
empty for acceptor

σ bond that
will open

LUMO of σ bond

Outward
rotation

Inward
rotation

(c)

(b)

Interaction between
substituent and
cleaving bond

No chance for favorable
filled–empty interaction
in donor case; donor
orbital is near node of
σ* orbital

Figure 15.20
Orbital interactions explaining inward or outward rotation in electrocyclic ring-opening.

Houk, and co-workers have developed a model for these systems that explains the reactivity trends and also rationalizes the stereochemical issues discussed. The term **torquoselectivity** was coined to describe these rotational preferences.

The essence of the analysis is given in Figure 15.20. *Ab initio* calculations on the conrotatory ring-opening of cyclobutene revealed that in the *transition state* the HOMO is the cleaving σ bond, and the LUMO is the corresponding σ* orbital. We now consider the effects of a substituent on the cleaving bond. In Figure 15.20 we show the substituent as contributing a *p*-type orbital to the HOMO and the LUMO. For a donor substituent, this will be a filled orbital. On inward rotation, this filled orbital interacts strongly with the HOMO of the σ bond. This is a mixing of filled orbitals (a), and so it is destabilizing. Now let's examine the effect of inward rotation of a donor on the σ* LUMO. Because the substituent orbital is directed toward the middle of the cleaving bond, near the node of σ*, the mixing of the filled orbital with the transition state LUMO is minimal (b). As such, a donor substituent should cause outward rotation. In the outward rotation, the mixing of the filled donor orbital with the transition state LUMO (σ*) is favorable (c), and so a donor substituent should accelerate the outward rotation reaction relative to the parent. Donors such as methyl therefore rotate outward.

For an acceptor substituent, the substituent orbital is empty, and so now the mixing of this orbital with the transition state HOMO is quite favorable (a). We predict a substantial preference for inward rotation. The inward rotation reaction should be accelerated, consistent with the above observations with the formyl acceptor placed on cyclobutene (Figure 15.19, example **C**). For substituents that are somewhat in between in π donor and acceptor ability (CF_3, example **B**) or when two donors compete (methyl vs. *tert*-butyl, example **D**), mixed results are observed. However, the observation that donor substituents rotate outward and acceptor substituents rotate inward is quite general. In fact, an excellent linear correlation is seen between calculated differences in activation energies for inward vs. outward rotation and a Hammett-type parameter ($\sigma_R°$) that measures the π donating/accepting ability of a substituent.

Connections

Pericyclics in Cancer Therapeutics

One fascinating example of a pericyclic ring closure involves a 1,4-biradical intermediate. Bergman first studied this reaction in detail, and it is now generally termed the **Bergman rearrangement**. The enediyne shown undergoes a thermally induced, **cycloaromatization** reaction. The product is 1,4-dehydrobenzene, an isomer of benzyne. The deuterium labeling experiment shown, along with quantitative formation of benzene when hydrogen-atom donors are present, provided compelling evidence for the viability of such a 1,4-biradical as a reactive intermediate.

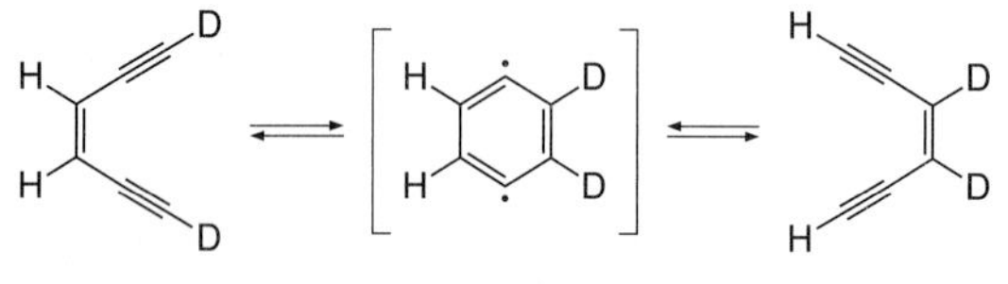

Enediyne   1,4-dehydrobenzene

Interest in this reaction exploded when, in 1987, several novel antitumor antibiotics containing enediynes or related structures were reported. Ultimately many such compounds were found, and representative structures

include calicheamicin, dynemycin, and neocarzinostatin ("enediyne" unit shown in color). Extensive work has shown that these antibiotics target DNA and undergo a Bergman rearrangement. When the rearrangement occurs, the biradical intermediate forms, and this reacts with the DNA leading to scission of the double helix and cell death. These enediyne-based antibiotics are very efficient cell killers, and they have been the targets of extensive synthetic and pharmaceutical research.

The parent Bergman rearrangement occurs around 200 °C, but nature has tuned the antibiotics such that they can cycloaromatize under physiological conditions. The enediyne group is strained in the antibiotics, facilitating electrocyclization. Also, in each case a "triggering" event is required. For example, conjugate addition of a thiolate to calicheamicin and neocarzinostatin (at the position marked with a *), or reduction followed by epoxide ring-opening for dynemicin induce the Bergman rearrangement. The trigger launches the biradical formation when the antibiotic is in the vicinity of DNA, enhancing the efficiency of scission.

Bergman, R. G. "Reactive 1,4-Dehydroaromatics", *Accts. Chem. Res.* **6**, 25–31 (1973).Smith, A. L., and Nicolaou, K. C. "The Enediyne Antibiotics", *J. Med. Chem.*, **39**, 2103–2117 (1996).

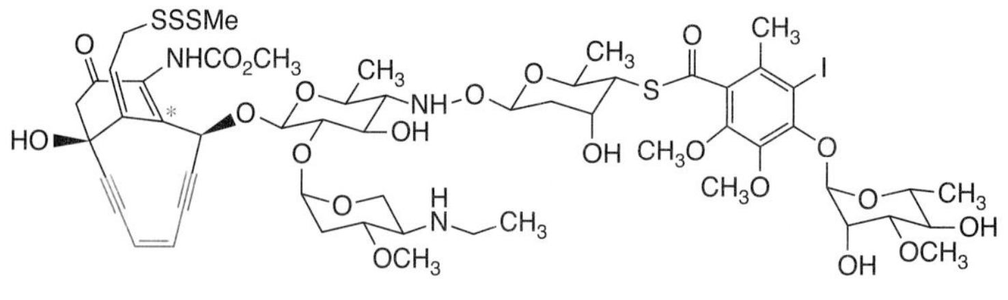

Calicheamicin

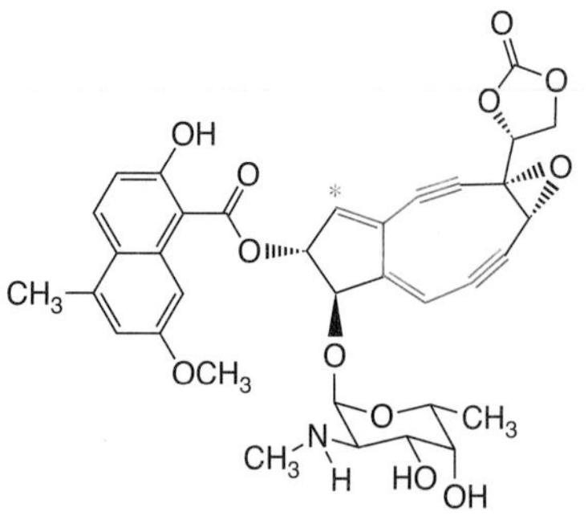

Neocarzinostatin

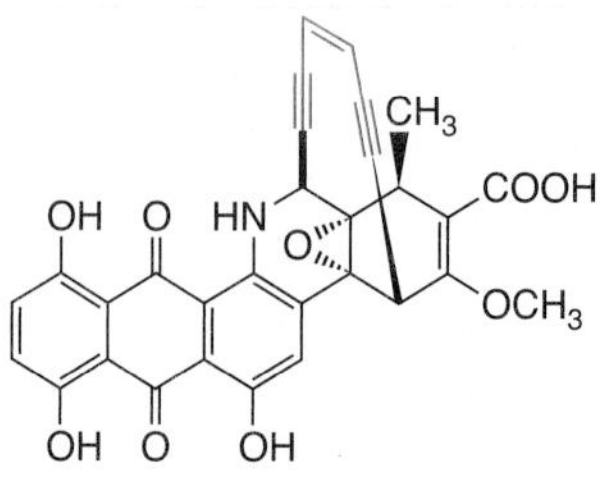

Dynemicin

15.5   Sigmatropic Rearrangements

A **sigmatropic shift** is defined as a reaction wherein a σ bond migrates over one or more π systems. Numbers in brackets are used to define the movement of the σ bond. An [*i,j*] sigmatropic rearrangement is the migration of a σ bond flanked by one or more π systems to a new location *i-1* and *j-1* atoms away. Like many concepts in this chapter, sigmatropic shifts are easier to explain graphically than with words, and Eqs. 15.21 and 15.22 do just that. We show a [3,3] sigmatropic shift and a [1,5] sigmatropic shift, the most important members of this reaction class. As with electrocyclic reactions, all sigmatropic shifts are allowed with some geometry. It is the stereochemistry of products that is dictated by the orbital analysis and that changes with the number of electrons.

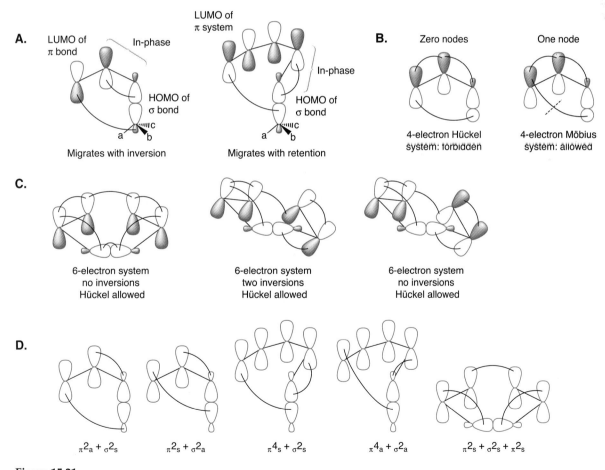

(Eq. 15.21)

(Eq. 15.22)

15.5.1 Theory

The FMO analysis gives quick insight into these reactions. We show the analysis in Figure 15.21 **A** for a [1,3] and [1,5] shift of a pyramidal group. Although the σ bond can be viewed as the HOMO or LUMO, it is common to draw it as the HOMO. The LUMO of the π

Figure 15.21
Theoretical predictions for sigmatropic rearrangements. **A.** FMO analyses of [1,3] and [1,5] carbon shifts. **B.** Aromatic transition state theory analysis of [1,3] hydrogen shifts. **C.** Aromatic transition state theory analysis of a [3,3] shift. This is a Cope Rearrangement—see Section 15.5.3, Figure 15.23. Note the left image corresponds to the boat transition state and is doubly disrotatory, while the right two images correspond to the chair transition state and are doubly conrotatory. **D.** Using the generalized orbital symmetry rule for various examples showing the allowed possibilities.

system is then drawn such that at the point at which the HOMO and LUMO interact, the orbitals are in-phase. In order for the group to perform a [1,3] shift, it does so by using the rear end of the σ bond in order to keep the interactions in-phase (Figure 15.21 **A**). This leads to inversion of stereochemistry for the migrating group (Eq. 15.23). An odd-looking transition state is obtained where the migrating group is planar and bonding to both ends of the π system (Eq. 15.23). In contrast, the [1,5] shift proceeds by directly moving the migrating group to the end of the π system (Figure 15.21 **A**). This leads to retention of stereochemistry of the group that is migrating, because at the transition state this group is still pyramidal (Eq. 15.24). It is important to note that the orbitals drawn at the transition states of Eqs. 15.23 and 15.24 are not the real MOs of these structures, but are simply constructs that show the orbital phasing leading to the observed stereochemistry. This analysis highlights one of the important issues with sigmatropic shifts—whether the migration occurs with retention or inversion of stereochemistry.

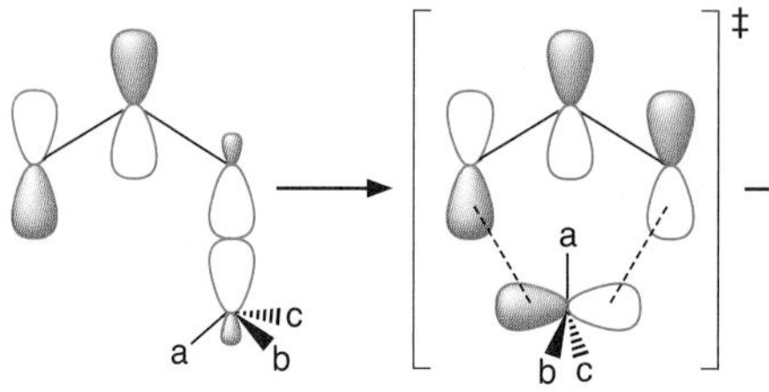

(Eq. 15.23)

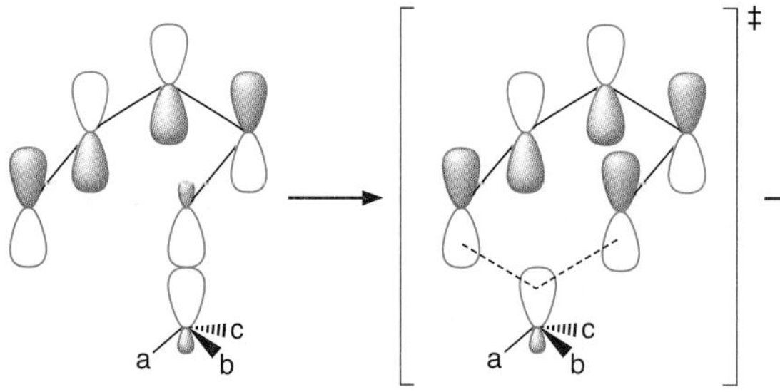

(Eq. 15.24)

Another important issue with sigmatropic shifts is whether the group that migrates does so to the same face or opposite face of the π system from where it started. FMO analysis can be used to predict this, but for demonstration purposes we look here at aromatic transition state theory. This theory provides a simple analysis of sigmatropic rearrangements. Consider a [1,3] sigmatropic shift of a hydrogen (Figure 15.21 **B**). Two pathways are feasible. In one, the hydrogen simply migrates across one face of the π system. In the alternative pathway, the hydrogen migrates to the opposite face of the π system from where it started. Migration across the same face gives a Hückel topology, and so is unfavorable for this four-electron process. In contrast, migration to the opposite face has a Möbius topology, making that migration allowed (although difficult because the hydrogen must bridge both faces of the π system in the transition state). Following now familiar reasoning, the [1,5] sigmatropic shift should be allowed via migration across the same face, and forbidden with migration to the other face.

We also show the aromatic transition state analysis of a [3,3] sigmatropic shift (Figure 15.21 **C**). There are no nodes. This is a six-electron, Hückel system, and so is allowed. A more realistic representation of the [3,3] sigmatropic shift is given in Eq 15.25, showing a chair-like transition state with all orbitals in-phase.

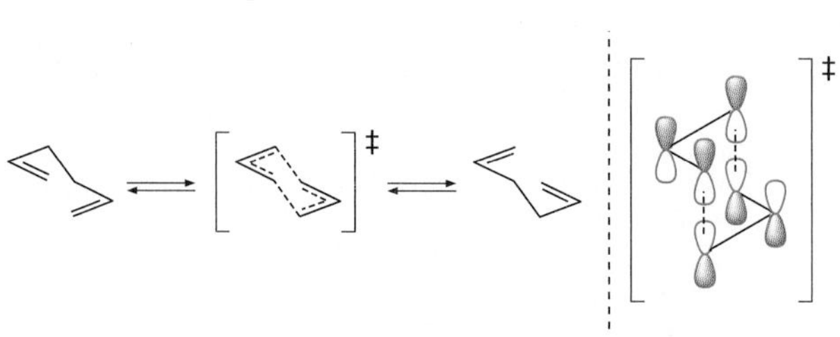

(Eq. 15.25)

Lastly, we can examine all these different shifts using the generalized orbital symmetry rule. Various examples of allowed combinations are shown in Figure 15.21 **D**. The [3,3] shift is best viewed as a three-component reaction, where all three act in a suprafacial manner in order to be allowed. Generalizing all these approaches for two-component reactions yields Table 15.3, which summarizes sigmatropic shifts.

Table 15.3
Rules for Thermal [*i,j*] Sigmatropic Rearrangements

i + j	Allowed	Forbidden
4*q*	s + a	s + s
	a + s	a + a
4*q* + 2	s + s	s + a
	a + a	a + s

As we have noted several times in this chapter, just because a pericyclic reaction is allowed does not necessarily mean that it is facile. However, one particularly facile reaction is a [3,3] shift involving a σ bond that migrates from and to a cyclopropane ring. A prototypical example is the rearrangement of homotropylidene (shown in the margin). Even more complex examples of this kind of are known, and the most famous involves bullvalene, which is described in the next Going Deeper highlight.

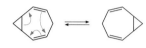

Homotropylidene

Going Deeper

Fluxional Molecules

A novel class of structures that attracted considerable attention in the years following the announcement of the orbital symmetry rules are the so-called **fluxional molecules**. These are structures that can undergo facile, unimolecular, pericyclic rearrangements that are degenerate, producing the same molecule, but with the atoms rearranged. The prototype is certainly bullvalene, the three-fold symmetric structure (point group, C_{3v}) shown to the right. This $(CH)_{10}$ structure undergoes a series of especially facile [3,3] sigmatropic shifts; $\Delta H^{\ddagger}$ for the process is ~11 kcal/mol. Remarkably, at room temperature both the ^1H and ^{13}C NMR spectra of bullvalene show a single line! All the carbons (and all the hydrogens) are made equivalent by the sigmatropic shifts. If every carbon were uniquely labeled, there would be 10!/3 = 1,209,600 different isomers, all of which are interconverting.

Degenerate interconversions can be considered a special case of another class of rearrangements termed **valence isomerizations** or **valence tautomerizations**.

These are interconversions of $(CH)_n$ molecules, in which no hydrogens move, only skeletal rearrangements occur. This may seem a fairly narrow class, but it is remarkably rich. For example, cyclooctatetraene is a $(CH)_8$ molecule. However, there are at least 21 such structures, cubane being another example. Typically, a large number of pericyclic reactions can be envisioned that would interconvert all the $(CH)_8$ valence isomers. For an example of such an analysis, see the article by L. R. Smith.

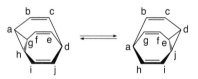

A fluxional molecule

Smith, L. R. "Schemes and Transformations in the $(CH)_8$ Series." *J. Chem. Ed.*, **55**, 569 (1978).

15.5.2   Experimental Observations: A Focus on Stereochemistry

As we have already mentioned, two of the important factors in sigmatropic shifts are the stereochemistry of the migrating group and whether migration occurs to the same or opposite faces of the π system. Here, we look at some examples that verify the predictions given above.

One of the most common shifts is the [1,5] hydrogen shift. The reaction typically shows a large kinetic isotope effect of ~5 at 200 °C for the migrating hydrogen, indicating considerable bond breaking in the transition state. That the reaction does indeed occur in a suprafacial manner across the π system was shown with elegant stereochemical studies by Roth. As shown in Eqs. 15.26 and 15.27, the *S/E* stereoisomer (*S* stereo descriptor for the tetrahedral carbon, *E* descriptor for the double bond) can produce two products via the allowed suprafacial migration of the hydrogen shown, giving *S/Z* and *R/E* stereochemistry. This predicted stereospecificity was seen, confirming the suprafacial nature of the rearrangement. You should convince yourself that the forbidden reaction (antarafacial on the π system) would produce the *S/E* and the *R/Z* products.

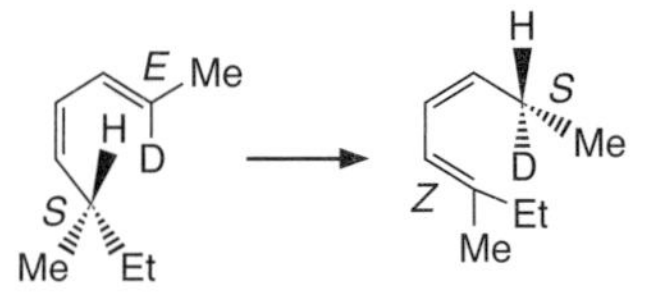

(Eq. 15.26)

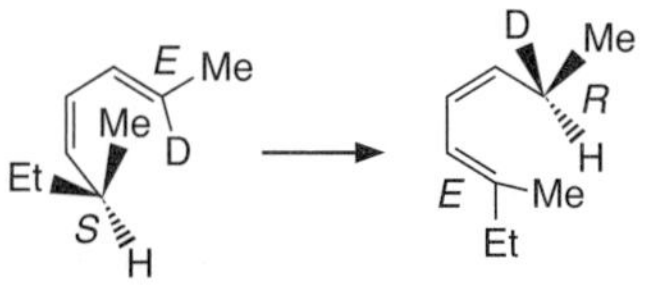

(Eq. 15.27)

The [1,5] hydrogen shift is extremely facile in 1,3-cyclopentadiene, where the ring locks the π system into the perfect geometry for the hydrogen migration (Eq. 15.28). The shift is so rapid that it is not possible to isolate single isomers of simple substituted cyclopentadienes, such as the methylcyclopentadiene shown, because the isomers equilibrate via the [1,5] hydrogen shift. The reaction looks like a [1,2] shift (as implied by the arrow pushing), but it is a six-electron process, and so is better described as a [1,5] shift. The preference for the allowed path is strong enough that the apparent [1,3] sigmatropic shift of indene has been established to be a pair of [1,5] shifts (Eq. 15.29). This occurs even though the first step involves the formation of the highly unstable isoindene, where the aromaticity of the benzene ring has been broken.

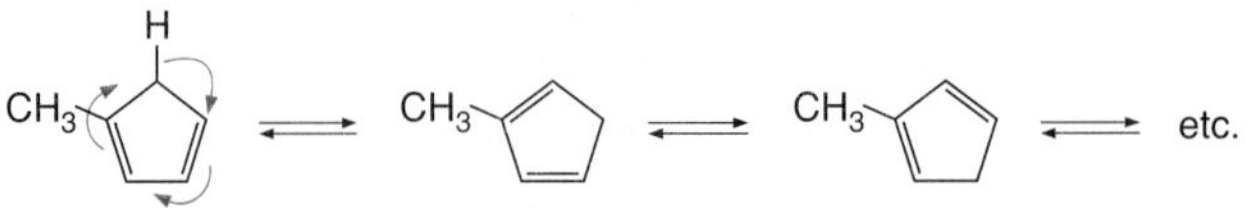

(Eq. 15.28)

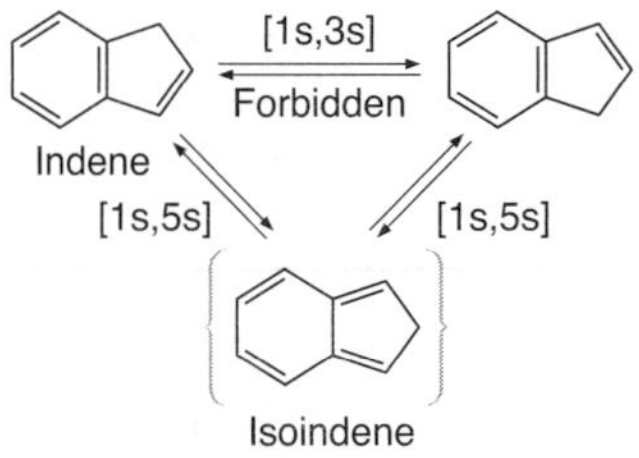

(Eq. 15.29)

We noted above that a [1,3] hydrogen shift must be antarafacial on the π system, and that this is a difficult geometry to achieve. However, when the migrating group is a carbon, it can act as the required antarafacial component. An elegant series of experiments by Berson set out to test this concept. As shown in Figure 15.22, a bicyclo[3.2.0]heptene system can thermally rearrange to a norbornene system. The clean inversion of stereochemistry at the migrating group suggests an antarafacial interaction, consistent with expectation. Interestingly, when methyl groups are placed in the reactant, only the reactant that can migrate where the methyl swings away from the carbon skeleton does so with clean inversion (Figure 15.22 **B**). With the opposite stereochemistry in the reactant, the methyl group must swing up toward the carbon skeleton during an antarafacial migration. This cannot occur due to steric reasons, and therefore the reaction is much slower, giving both inversion and retention of stereochemistry. Theoretical studies indicate that dynamic effects of the sort discussed in Section 7.2.7 likely have a strong influence on reaction stereochemistry in this system.

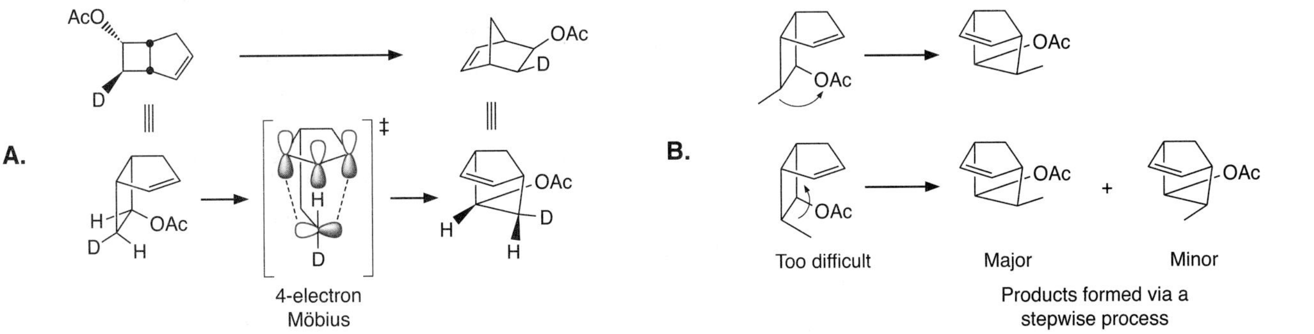

Figure 15.22
A. Demonstration of the $[_\sigma 2_a + _\pi 2_s]$ reaction. **B.** Differences when methyl groups are used.

Another very common sigmatropic shift is the [1,2] hydride shift associated with car-bocations (Eq. 15.30). Although not usually analyzed this way, these migrations involve a cyclic, two-electron system and are Hückel aromatic (see margin). Viewing these reactions this way nicely rationalizes why comparable [1,2] shifts are not seen in carbanions; that is, they would involve a four-electron Hückel antiaromatic transition state.

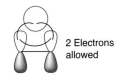

2 Electrons
allowed

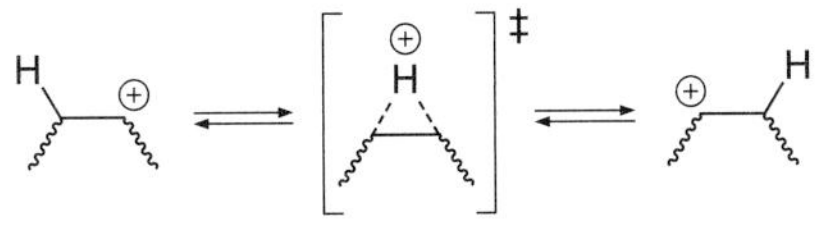

$$(Eq. 15.30)$$

Nevertheless, formal [1,2] shifts involving anions can be observed in systems involving heteroatoms, the prototype being the Wittig and Stevens rearrangements given in Eqs. 15.31 and 15.32, respectively. These appear to be pericyclic reactions, but are they? The Wittig and Stevens rearrangements are formally four-electron systems and so should proceed with inversion of configuration at the migrating carbon. However, using the deuterium-labeled ylide shown in Eq. 15.33, Baldwin showed that migration occurs with *retention* of configura-tion. When the reaction was run in an EPR spectrometer, evidence was obtained that radical pairs are involved in the reaction, at least to some extent. Although it is difficult to quantify the EPR effect, the observations suggest that the Stevens rearrangement, and by extension the Wittig, might not be concerted pericyclic processes.

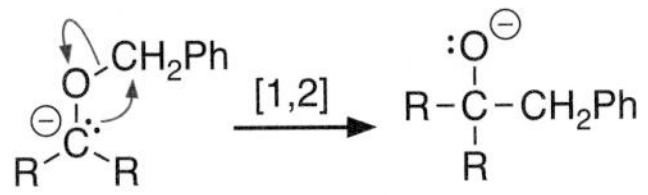

$$(Eq. 15.31)$$

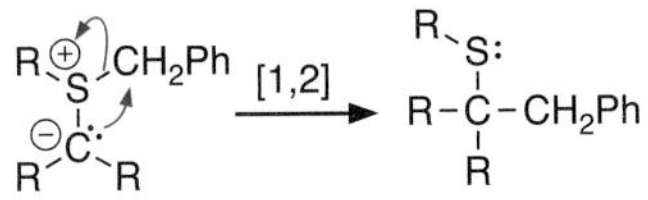

$$(Eq. 15.32)$$

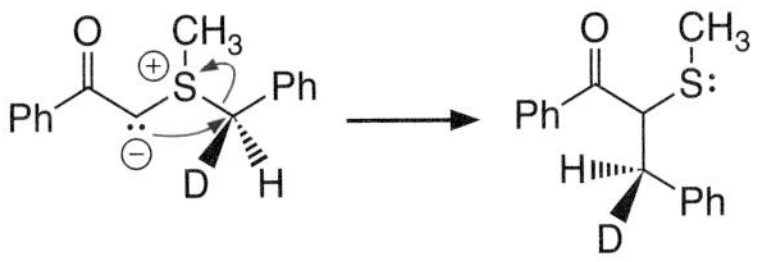

$$(Eq. 15.33)$$

Another interesting and useful reaction is the [2,3] sigmatropic shift (Eq. 15.34). It is a six-electron process, where two of the electrons are contributed by a lone pair, and thus only five atoms are involved. Sulfoxides and various types of allylic ylides are common substrates (Eq. 15.35). Note that the [2,3] shift of allyl sulfonium ylides is a pericyclic, concerted process, unlike the [1,2] Stevens shift of simple sulfonium ylides. The [2,3] sigmatropic shift has proven to be synthetically useful, as suggested in the examples given in Eqs. 15.36 and 15.37.

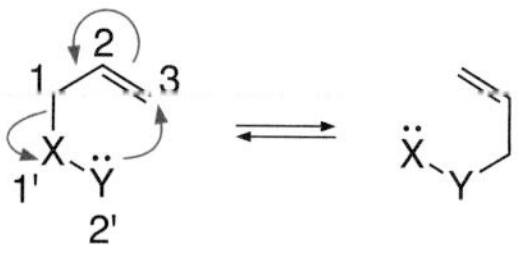

(Eq. 15.34)

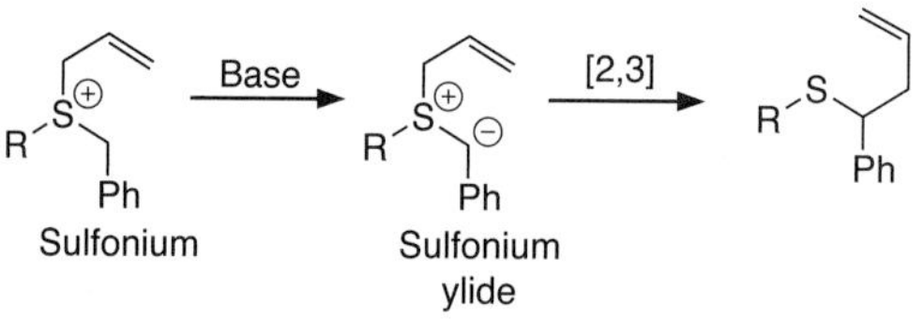

(Eq. 15.35)

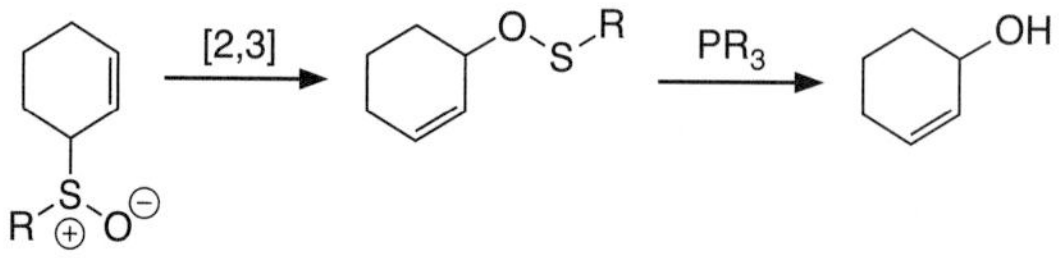

(Eq. 15.36)

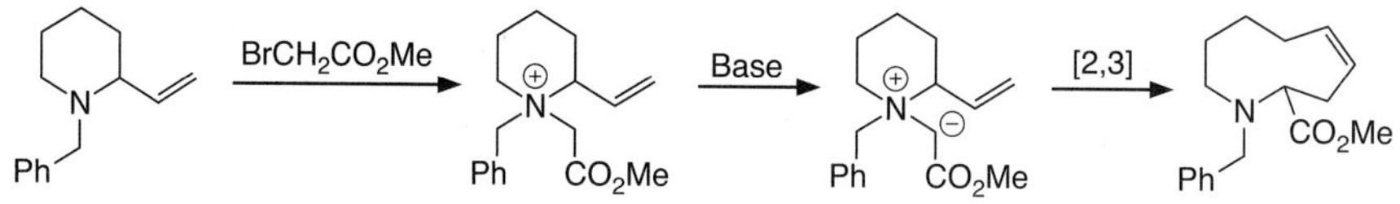

(Eq. 15.37)

15.5.3 The Mechanism of the Cope Rearrangement

To this point in the chapter, we have provided theoretical analyses, presented rules, and given examples of successful applications of the rules. You might have the impression, therefore, that the field of pericyclic reactions has been devoid of controversy, with everyone agreeing that all can be understood by just analyzing orbital symmetry. In fact, there have been times when the situation was very different. Many of the basic approaches and assumptions of this chapter have been called into question and subjected to substantial experimental and theoretical challenges. We illustrate this by considering the mechanistic investigations of the **Cope rearrangement** in greater detail. As you will see, many of the mechanistic tools and concepts developed throughout this book have been brought to bear on this important reaction.

In 1940 Arthur Cope discovered what is now considered the prototypical [3,3] sigmatropic rearrangement that bears his name (Eq. 15.38). The first papers on the conservation of orbital symmetry did not appear until 1965, and so, like the Diels–Alder reaction, the Cope rearrangement was known experimentally long before the theory of these reactions was developed.

(Eq. 15.38)

One of the most elegant and telling mechanistic investigations of the Cope rearrangement was reported in 1962 by Doering and Roth. The experiment involves the conversion of the two stereoisomers of 3,4-dimethyl-1,5-hexadiene to 2,6-octadiene (Figure 15.23). This experiment was designed to probe the geometry of the Cope transition state. Already in 1962 it was assumed that a cyclic array of six carbons was involved in the transition state. As such it seemed reasonable to analyze the reaction in terms of two limiting transition state structures, one resembling chair cyclohexane and one resembling boat cyclohexane. As summarized in Figure 15.23, both diastereomers of the reactants can react via either a chair-like transition state (one possibility for the meso, two for the *d,l*) or a boat-like transition state (two for

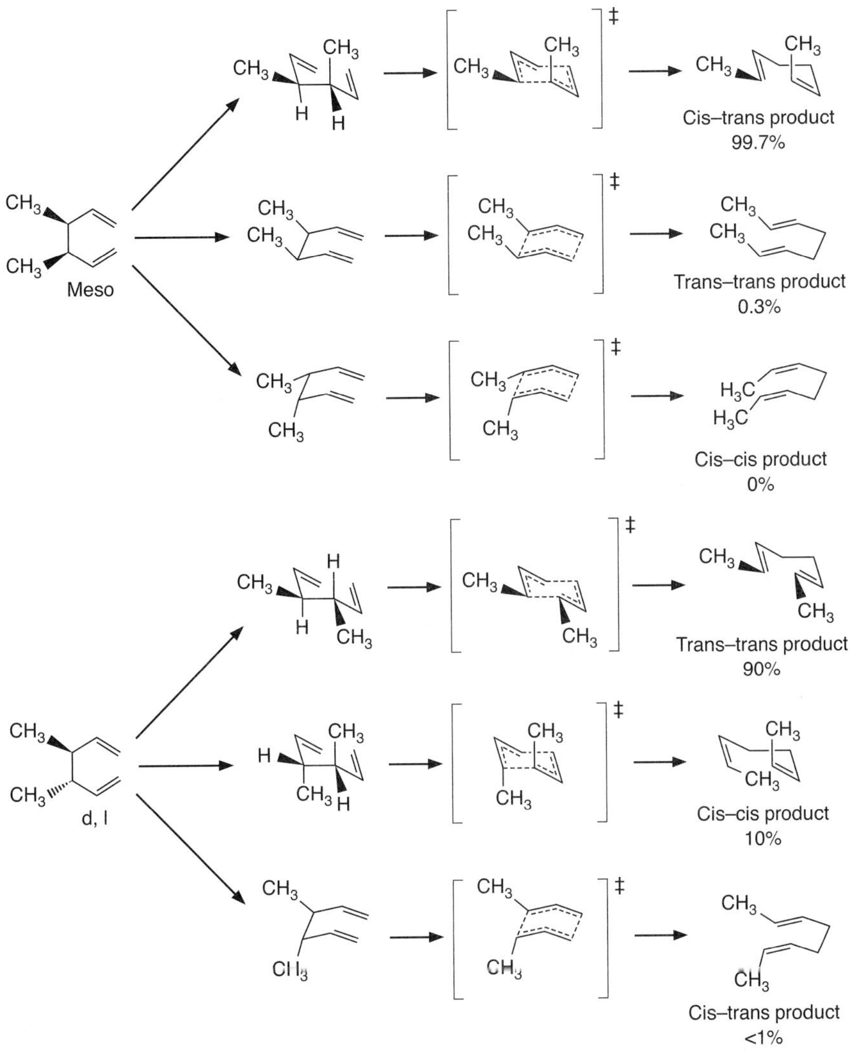

Figure 15.23
The Doering–Roth experiment, which established the chair-like nature
of the transition state for the reaction.

the meso, one for the *d,l*). Different products are seen from the chair and boat transition states. Nothing in the orbital symmetry analyses would predict that one is preferred over the other. However, based on the product ratios shown in Figure 15.23, the chair transition state is substantially preferred. Presumably this is due to the more stable chair conformation relative to a boat conformation. A later study that utilized deuterium instead of CH_3 as the stereochemical marker reached the same conclusion and estimated $\Delta\Delta G^{\ddagger}$ (boat–chair) as 5.8 kcal/mol.

A recurring controversy concerning the Cope rearrangement is whether the reaction really is a concerted, pericyclic process. Especially in heavily substituted systems, one can imagine stepwise processes involving biradical intermediates. In addition, whether the reaction is synchronous or not has been a topic of discussion. We can imagine three limiting reaction paths (Figure 15.24). First is the synchronous, concerted path that produces a symmetrical transition state (as we have assumed throughout our discussion). Second is a stepwise mechanism in which σ bond making precedes σ bond breaking, involving a 1,4-cyclohexanediyl biradical as a true intermediate. The third mechanism has complete σ bond breaking preceding σ bond making, yielding a pair of allyl radicals as intermediates in the

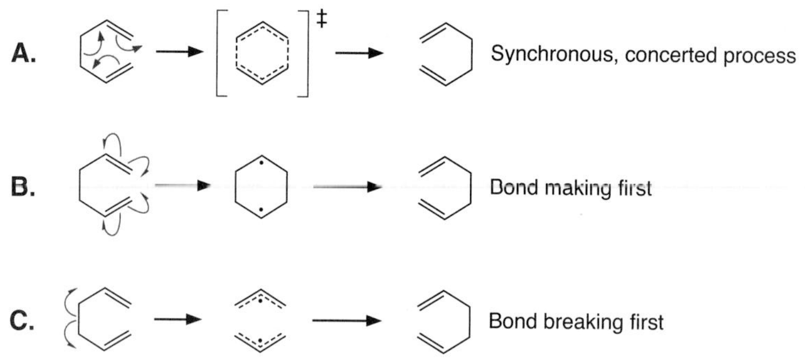

Figure 15.24
Limiting transition state models for the Cope rearrangement.

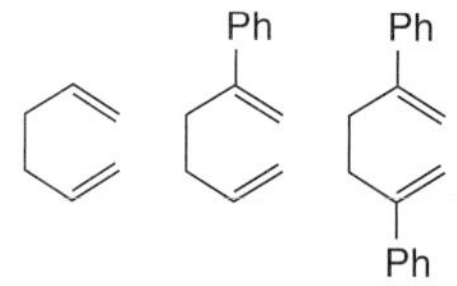

process. If biradical intermediates are involved, they must be short lived and must have properties consistent with the high stereospecificity of these reactions, but this is not inconsistent with singlet biradical chemistry.

An excellent starting point in the analysis of the three mechanisms in Figure 15.24 is an evaluation of the thermochemistry of the various paths. That is, are the proposed biradical intermediates viable species along the reaction path? This issue can be addressed by using the thermochemical arguments developed in Chapter 2. Using group increments, the heat of formation of 1,5-hexadiene is found to be 20.2 kcal/mol. Also, the activation energy of the parent Cope rearrangement is 34.3 kcal/mol. Thus, the highest energy point along the reaction pathway of the Cope rearrangement has a heat of formation of approximately 20.2 + 34.3 = 54.5 kcal/mol (approximate because we have not considered E_a vs. $\Delta H^{\ddagger}$ differences). If the heat of formation of the proposed biradical intermediates is substantially higher than that number, we can rule out the biradical pathway. Using the radical group increments of Chapter 2, we can estimate that the heat of formation of two allyl radicals is ~76 kcal/mol. As such, this biradical intermediate is not viable in the parent Cope rearrangement. Perhaps with highly radical stabilizing substituents at appropriate sites, such a structure could become important, but not in the parent system.

The situation is much different with the cyclohexanediyl. Group increments estimate ΔH_f° to be ~55 kcal/mol. Given the uncertainties in group increments for radicals, let alone their application to biradicals, we must conclude that this biradical is a viable possibility in the Cope rearrangement. Substituent effects on the rate of the Cope rearrangement also seem consistent with the biradical mechanism. As shown in the margin, phenyl groups at the 2 and 5 positions substantially accelerate the reaction in a way that is essentially additive ($4900 \approx 69^2$). These are fairly substantial effects, especially when we consider that the substituents are on carbons that are not undergoing bond making or bond breaking during the rearrangement. It is not obvious why the concerted pericyclic reaction would show such large effects (another large substituent effect is discussed in the Connections highlight on page 921). However, once again, the large substituent effects surely rule out the involvement of two allylic radicals, as the phenyls are located on what would be nodes in allyl radicals. Therefore, cyclohexanediyl is definitely a possible intermediate.

Gajewski and co-workers have conducted an elegant series of experiments designed to probe the nature of the transition state of the Cope rearrangement. The key tool has been secondary kinetic isotope effects. With appropriate deuterium substitution, Gajewski defined a bond breaking kinetic isotope effect (BBKIE, Eq. 15.39) and a bond making kinetic isotope effect (BMKIE, Eq. 15.40). The ratio of these two, **R** as defined in Eq. 15.41, was considered to be a good indicator of the relative degree to which bond making and bond breaking have occurred at the transition state. When **R** is measured for a series of hexadienes (see below), considerable variation is seen. For the parent system (with a methyl substituent), **R** is 1.8, suggesting a bit more bond making than bond breaking at the transition state. But this small

k (relative)   1   69   4900

value certainly does not support a cyclohexanediyl-like transition state. Phenyl substituents at carbons 2 and 5 have the expected effect of increasing **R**. They lead to substantially more bond making than bond breaking at the transition state. Thus, we are increasing cyclohexanediyl character at the transition state because phenyls at this position stabilize such an intermediate (option **B**, Figure 15.24). Interestingly, with two strongly radical stabilizing cyano groups on C3, we see **R** < 1. This indicates that now bond breaking has progressed further than bond making in the transition state, and the transition state has increased bis(allyl) biradical character (option **C**, Figure 15.24) relative to the parent system.

Gives bond breaking kinetic
isotope effect (BBKIE)

(Eq. 15.39)

Gives bond making kinetic isotope effect (BMKIE).
For this case only, defined as k_D/k_H.

(Eq. 15.40)

$$\mathbf{R} = \frac{BMKIE - 1.0}{BBKIE - 1.0}$$

(Eq. 15.41)

Clearly, the nature of the transition state is varying as we introduce substituents. This should remind you of our discussion of such effects in Chapters 7, 10, and 11, and in particular this would appear to be an excellent system for a More O'Ferrall–Jencks plot. Indeed, such a plot has been used to analyze this reaction, and one is given in Figure 15.25.

| **R** | 1.8 | 3.3 | 8.1 | 0.3 |

The two axes for the More O'Ferrall–Jencks plot were developed as follows. Along with the kinetic isotope effect, the equilibrium isotope effect was measured for each system. We wish to have one axis gauge the extent of bond making in the transition state, and the other to gauge the extent of bond breaking. We can choose the ratio of the appropriate kinetic isotope effect to the equilibrium isotope effect to serve this purpose. For example, the bond making axis represents the ratio of the BMKIE to the thermodynamic isotope effect for the whole reaction. This is taken to be a fair indicator of the extent of bond making in the transition state. Similar reasoning holds for the bond breaking axis.

The More O'Ferrall–Jencks plot illustrates the variation in transition state structure as a function of substituents. The lower left to upper right diagonal represents a perfectly synchronous reference case. The transition state would be the dot in the center where the percent bond making equals the percent bond breaking. Interestingly, the structures probed do not lie along this line. For all the structures analyzed, bond making and bond breaking have occurred to differing degrees in the transition state. However, the transition state structures all lie along the diagonal from the upper left to the lower right. The fact that they all lie along this line shows that the extents of bond making and bond breaking at the transition state are tightly coupled, such that % bond making + % bond breaking ≅ 100%. This tight coupling of

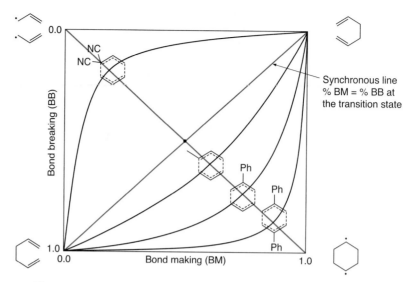

Figure 15.25
More O'Ferrall–Jencks plot for the Cope rearrangement. The diagonal from the lower left to the upper right is the synchronous pathway, with a transition state right in the center that is an equal mixture of bond making and bond breaking. The diagonal from the upper left to the lower right contains a series of possible transition states that vary from full bond breaking with no bond formation to full bond formation with no bond breaking. Note that both axes run from no bond present (0.0) to a full bond present (1.0).

bond making and bond breaking is consistent with a pericyclic concerted process that is varying its structure in response to substituents. It also seems consistent with the expectation that, for the parent thermoneutral reaction, we have a symmetrical transition state with extensive bond making and bond breaking. This is very much in line with the model assumed for a pericyclic concerted reaction presented numerous times in this chapter.

Ultimately, these experimental studies cannot completely rule out biradical intermediates in the Cope rearrangement, especially for the heavily substituted systems. In the end, only theory can make a definitive statement about the structure of a transition state. The interplay between theory and experiment in the Cope rearrangement and other pericyclic processes has at times been contentious. In addition, differing theoretical models have often made diametrically opposed predictions (see the Going Deeper highlight on page 900). This further fueled the debate over the true nature of pericyclic transition states.

In recent years, however, theory and experiment have arrived at the same conclusions. It is now possible to probe pericyclic reaction pathways with very high levels of *ab initio* quantum mechanics. These studies consistently and unambiguously favor a concerted mechanism, with a single aromatic transition state for the Cope rearrangement and other prototype pericyclic reactions such as the Diels–Alder. Most impressively, theory can now completely rationalize the extensive kinetic isotope effect data that have been collected on these systems. For example, the measured bond making and bond breaking kinetic isotope effects for the tetradeuterio systems of Eqs. 15.39 and 15.40 are 1.07 ± 0.025 and 0.89 ± 0.018, respectively. *Ab initio*, DFT-calculated values, based on a concerted, aromatic transition state, are 1.07 and 0.88, respectively. Similar results have been reported for Diels–Alder reactions. The stunning agreement between a subtle experimental measurement and the *a priori* calculation of precisely the same quantity provides compelling evidence that the reactions are concerted pericyclic processes.

Connections

A Remarkable Substituent Effect: The Oxy-Cope Rearrangement

A potentially quite useful variation of the Cope rearrangement would be to have an OH substituent on C3. The rearrangement product would be an enol that would quickly tautomerize to the δ,ε-unsaturated carbonyl. However, like most Cope rearrangements of simple systems, this reaction requires considerable heating, making it not very useful synthetically. For this reason, the discovery by Evans that simply deprotonating the alcohol allowed the reaction to proceed under much milder conditions generated considerable interest. Further study showed that, depending on the exact system, the rate enhancement associated with deprotonation ranged from a factor of 10^{10} to 10^{17}! This variation, now known as the **oxy-Cope rearrangement**, has provided a useful new way to make δ,ε-unsaturated carbonyls.

Oxy-Cope rearrangement

The origin of this rate enhancement has been investigated theoretically. In this case there is significant bond cleavage in the transition state, with option **C** of Figure 15.24 playing a dominant role. Given that there is significant bond cleavage in the transition state, the substituent effect on the radical that would be produced by complete cleavage was studied using *ab initio* theory, calculating appropriate bond dissociation energies (BDEs). The effect of an HO on an adjacent C–H BDE is not large. However, deprotonation resulted in a 17 kcal/mol drop in BDE of a neighboring C–H.

	BDE (calculated, kcal/mol)
$HO-CH_2-H \longrightarrow HO-CH_2\cdot + H\cdot$	91
$^{\ominus}O-CH_2-H \longrightarrow {}^{\ominus}O-CH_2\cdot + H\cdot$	74

Bond dissociation energies as a function of protonation state

If a significant fraction of this kind of stabilization of a radical is felt in the transition state of the oxy-Cope rearrangement, the rate acceleration is understandable. It is simply attributed to the fact that a radical is stabilized adjacent to an oxy-anion more than when adjacent to an OH. We can view this as a resonance effect.

Ketyl anion resonance structures

Steigerwald, M. L., Goddard III, W. A., and Evans, D. A. "Theoretical Studies of the Oxy Anionic Substituent Effect." *J. Am. Chem. Soc.*, **101**, 1994 (1979).

15.5.4 The Claisen Rearrangement

Uses in Synthesis

The most synthetically useful sigmatropic shift is the Claisen rearrangement, shown in Eq. 15.42. The reaction involves the conversion of an allyl vinyl ether to a γ,δ-unsaturated carbonyl, and it is a [3,3] sigmatropic shift closely related to the Cope rearrangement. Unlike the Cope, however, the Claisen is essentially irreversible. This is because of the substantially greater stability of a carbonyl double bond vs. an olefinic double bond (Table 2.2). An interesting variant is the conversion of an O-allylphenol to an ortho allylphenol by way of a carbonyl intermediate that quickly tautomerizes to restore aromaticity (Eq. 15.43). For a biological example of this reaction, see the Connections highlight below.

(Eq. 15.42)

(Eq. 15.43)

Connections

A Biological Claisen Rearrangement— The Chorismate Mutase Reaction

It appears that nature rarely uses concerted, pericyclic reactions in biosynthesis or metabolism. However, in at least one instance a Claisen rearrangement is key to a biosynthetic pathway. It is, in fact, a crucial pathway, the one that biosynthesizes the amino acids phenylalanine (Phe) and tyrosine (Tyr). The pathway only exists in plants— our bodies cannot synthesize Phe and Tyr, making Phe and Tyr so-called **essential amino acids**.

Beginning with erythrose, shikimic acid is prepared, and it is converted to chorismic acid (a.k.a., chorismate).

This then does a Claisen rearrangement to produce prephenate, which is taken on to both Phe and Tyr. The Claisen rearrangement is catalyzed by the enzyme chorismate mutase. The enzyme catalyzes the reaction by as much as a factor of 10^6. Extensive mechanistic studies have established that the enzyme-mediated reaction has all the hallmarks of a Claisen rearrangement, including appropriate isotope effects. A major role of the enzyme is to pre-organize chorismate into the proper conformation for rearrangement. However, the enzyme also uses substantial electrostatic interactions to stabilize partial charges in what is a fairly polar pericyclic transition state.

Shikimate   Chorismate   Prephenate

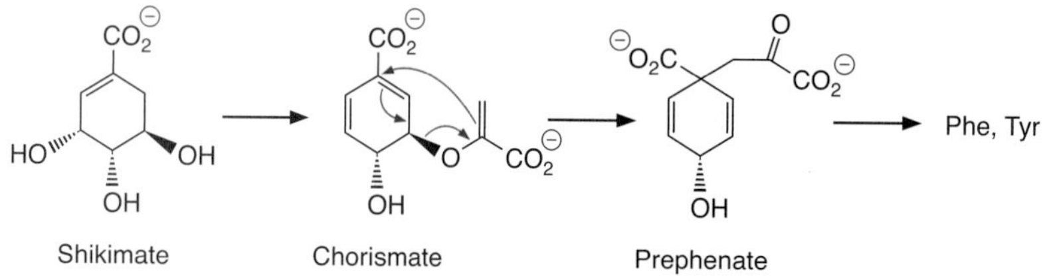

An important variant of the Claisen rearrangement is the ester enolate Claisen developed by Ireland (Eq. 15.44). Starting with a simple allyl ester, the ester enolate is prepared and trapped as, for example, a trimethylsilyl ether. Mild heating then produces the Claisen rearrangement. The beauty of the sequence is that allyl esters are easy to make, much more so than allyl vinyl ethers. The reaction produces a new carbon–carbon bond, and because of the highly structured nature of this (and other pericyclic) transition states, excellent control of stereochemistry is seen, as shown in Eq. 15.45. A surprise is that the reaction proceeds under very mild conditions. The trimethylsiloxy substituent at C2 that is naturally part of the ester enolate Claisen lowers the activation free energy by approximately 9 kcal / mol, making the reaction proceed at much lower temperatures than most Claisen reactions. The mild conditions and stereospecific formation of a new C–C bond have made the **Ireland–Claisen** rearrangement one of the most powerful of the pericyclic reactions. The reaction can also be preformed on the *in situ* generated enolate, without the silyl trapping step.

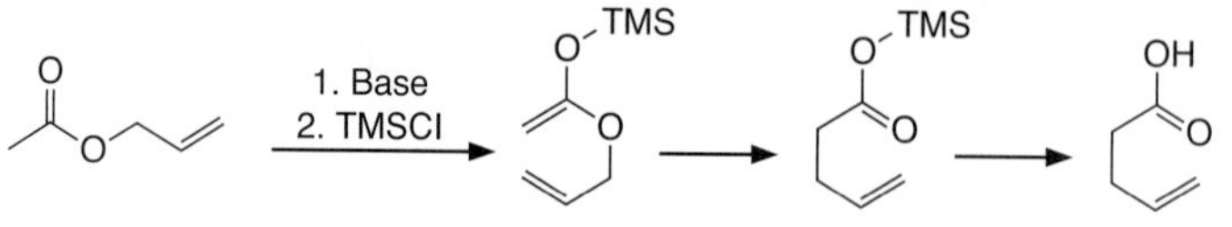

(Eq. 15.44)

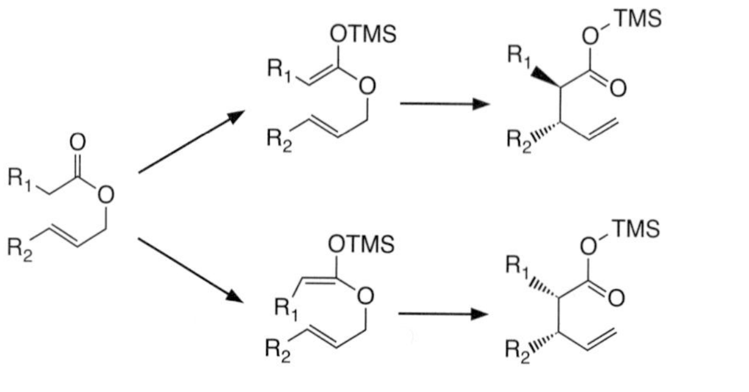

(Eq. 15.45)

Mechanistic Studies

Just as with the Cope rearrangement, mechanistic studies of the Claisen rearrangement have been extensive. Again, kinetic isotope effects play a central role, with typical k_H / k_D data shown in Figure 15.26 **A** (for H_2 to D_2 substitution). The sum of the evidence points to a concerted, pericyclic mechanism for the Claisen rearrangement, but not surprisingly the reaction is not perfectly synchronous. The kinetic isotope effects suggest there is significantly more bond breaking than bond making in the transition state $[1.09 > (1/0.976)]$. For the ester enolate Claisen, both isotope effects are more significant (larger normal and smaller inverse effects), indicating that while bond breaking still proceeds further than bond making, the transition state is much further along in the process with the siloxy substituent. As with the oxy-Cope, this substituent effect is rationalized in terms of the stabilization of the oxyallyl radical afforded by the siloxy substituent.

A.   1.09   0.976   1.48   0.917

B. k_{rel} at 100 °C   1   0.9   111   270   16   0.11

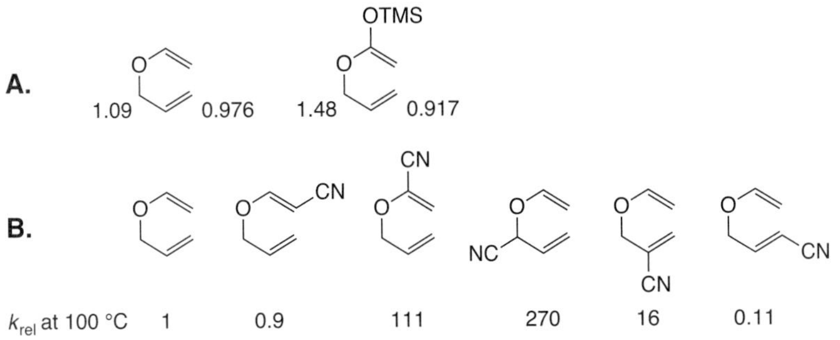

Figure 15.26
A. Kinetic isotope effects and **B.** Substituent effects in the Claisen rearrangement.

Other substituent effects can also be significant. For example, Carpenter has evaluated the effect of a cyano group at every position around the parent system (Figure 15.26 **B**). That the largest effect is seen at C4 (fourth entry in Figure 15.26 **B**) is consistent with the idea that there is more bond breaking than bond making in the transition state, as are the relatively smaller values for C1 and C6 (second and sixth entries), although the effects here are surprisingly small. The origin of the substantial substituent effects at C2 and C5 has been debated (third and fifth entries), with models emphasizing the electron withdrawing ability or the radical stabilizing ability of the substituent being proposed.

Going Deeper

Hydrophobic Effects in Pericyclic Reactions

A characteristic of pericyclic reactions that contributed to them being described as "no-mechanism" reactions is their relative insensitivity to solvent effects. These are typically not highly polar reactions, and so the usual solvent effects of the sort discussed in Chapter 3 are small. Given this, it was surprising when it was observed that both the Diels–Alder reaction and the Claisen rearrangement proceed significantly faster in aqueous media than in organic solvents. Efforts to describe this as a solvent polarity effect or a hydrogen bonding effect were unsuccessful. Ultimately, it was concluded that, while polarity and hydrogen bonding effects may contribute, a substantial and often dominant contributor to the rate accelerations was the hydrophobic effect. Recall from Section 3.2.6 that the primary feature of the hydrophobic effect is the minimization of solvent exposed surface area of the hydrocarbon portions of molecules. Now consider a pericyclic transition state. It will generally be true that a pericyclic transi-

tion state is more compact than the starting material(s). Certainly, in the Diels–Alder we can see that two molecules that were free in solution and fully exposed to solvent come together in a transition state that will have some buried hydrocarbon surface area. Similarly, the Claisen transition state will be "coiled" and more compact than the extended conformations available to the ground state. Thus, on going from starting material(s) to the transition state, there will be less solvent exposure. In water, then, this will selectively destabilize the ground state relative to the transition state, lowering the activation free energy. The hydrophobic rate acceleration seen in such systems can be quite substantial and is synthetically useful in many contexts.

Breslow, R. "Hydrophobic Effects on Simple Organic Reactions in Water." *Acc. Chem. Res.*, **24**, 159 (1991). Gajewski, J. J. "The Claisen Rearrangement. Response to Solvents and Substituents: The Case for Both Hydrophobic and Hydrogen Bond Acceleration in Water and for a Variable Transition State." *Acc. Chem. Res.*, **30**, 219 (1997).

15.5.5   The Ene Reaction

The **ene reaction** is an important pericyclic reaction that is in some ways difficult to classify (Eq. 15.46). The reaction involves the migration of an allylic C–H bond to one end of an olefin, while the other end of the olefin forms a C–C bond to the opposite end of the allylic system. There are several different ways to think about the ene reaction. It resembles a sigmatropic shift, where the hydrogen shifts through space to the **enophile** (the hydrogen acceptor). If we draw "virtual" bonds between the partners (shown as dotted lines in the two reactions of Eq. 15.47), then we can view the reaction as either a [1,5] or a [3,3] shift. Alternatively, the reaction can be viewed as a cycloaddition, in which the allylic C–H bond plays the role of the second double bond in the diene. Either way, it is a six-electron cyclic process that can be drawn to conform to any of the theories given in this chapter.

$$\text{(Eq. 15.46)}$$

$$\text{(Eq. 15.47)}$$

An especially important type of ene reaction uses singlet oxygen in place of the olefin partner (see Section 16.5 for a detailed discussion of singlet oxygen). The product is an allylic hydroperoxide (Eq. 15.48). We have looked at some of the details of the chain reactions that ensue from a hydroperoxide when autooxidation using triplet oxygen was discussed in Section 11.7.4. This method of oxygenating a hydrocarbon can be synthetically useful, and is also seen in a biological context, where the reactions of singlet oxygen play deleterious roles.

$$\text{(Eq. 15.48)}$$

15.6   Cheletropic Reactions

Although **cheletropic reactions** are viewed as a separate class of pericyclic reactions, they are simply cycloadditions in which one partner interacts through a single atom, making two new bonds to one center. Examples of the formation and breaking of bonds are shown in Eqs. 15.49–15.52. In the bonding breaking direction, in which a small molecule is eliminated, these reactions are often referred to as **extrusions**. Frequently the reaction involves the addition, or even more likely, the expulsion of a small stable fragment such as CO, N_2, or SO_2. Stereospecificity is seen, as with all pericyclic reactions (Eqs. 15.51 and 15.52).

$$\text{(Eq. 15.49)}$$

$$\text{(Eq. 15.50)}$$

$$\text{(Eq. 15.51)}$$

$$\text{(Eq. 15.52)}$$

Going Deeper

Pericyclic Reactions of Radical Cations

For some time it has been known that many pericyclic reactions can be greatly accelerated if they are run under single electron transfer (SET) conditions (also known as **electron transfer catalysis**, ETC). Examples include Diels–Alder reactions, electrocyclic openings of cyclobutenes, and retro [2+2] cycloadditions. From the beginning it has been debated whether these SET reactions really are concerted processes with aromatic transition states, or whether they are better thought of as stepwise processes involving radical cation intermediates.

As always, stereochemistry has proven to be a crucial indicator of mechanism. Many examples of highly stereospecific SET reactions have been found. An example is the Diels–Alder reaction of the 1,2-di(aryloxy)-ethylenes shown below. Mixing the dienophile with cyclopentadiene and the very convenient SET reagent tris(p-bromophenyl)aminium (**1**$^{+\bullet}$) gives the cycloaddition adducts with high stereospecificity (first two examples below). The observation of several cases like this led many to conclude that the SET reactions really were concerted, pericyclic processes. However, more recent work has found clear exceptions. The deuterated 4-methoxystyrene shown adds to cyclopentadiene under the same conditions with extensive loss of stereochemistry (third example). These systems are more complicated than conventional pericyclic reactions.

Once again, theory may provide the best way to understand these observations. As we have discussed elsewhere, computational studies of open shell systems such as the radical ions involved here are challenging, but in recent years the computational power and theoretical methods have developed to the point where very informative computational studies of SET reactions can be performed. Summarizing a number of such calculations, several trends emerge. First, the potential energy surfaces for these reactions are very flat; that is, small barriers separate a number of species with similar energies. This is reminiscent of what is seen with biradicals, excited states, and carbocations. As such, these reactions are also much more sensitive to solvent and substituent effects than conventional pericyclic reactions. It is therefore not surprising that one system might be stereospecific and the other not. Stepwise pathways are definitely viable for these systems, and they often resemble analogous biradical paths that are considered for the neutral systems. Finally, the unpaired electron can often lead to a disfavoring of symmetrical pathways through a Jahn–Teller type of distortion.

The overall conclusion is that, while these reactions superficially resemble the pericyclic processes we have been considering in this chapter, they are mechanistically quite different. Each system has to be considered individually as to whether stereospecific reactions can be expected or not.

Bauld, N. L., and Yang, J. "Stereospecificity and Mechanism in Cation Radical Diels–Alder and Cyclobutanation Reactions." *Org. Lett.*, **1**, 773–774 (1999). Gao, D., and Bauld, N. L. "Mechanistic Implications of the Stereochemistry of the Cation Radical Diels–Alder Cycloaddition of 4-(cis-2-Deuteriovinyl)anisole to 1,3-Cyclopentadiene." *J. Org. Chem.*, **65**, 6276–6277 (2000). Saettel, N. J., Oxsgaard, J., and Wiest, O. "Pericyclic Reactions of Radical Cations." *Eur. J. Chem.*, 1429–1439 (2001).

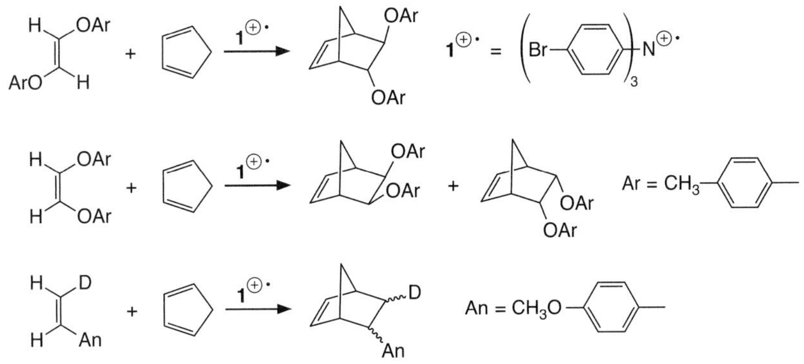

Radial cation Diels–Alder reactions

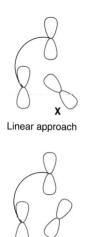

X

Linear approach

X

Non-linear approach

Figure 15.27
Definitions of linear and non-linear approaches in chelotropic reactions.

15.6.1 Theoretical Analyses

The analysis of cheletropic reactions is typically done in the addition direction. The small fragment, designated **X** in Figure 15.27, is considered to contribute two electrons to the pericyclic transition state. As always, the reaction is allowed with some specific geometry. The key issues with these reactions are whether the approach of the small fragment is **linear** or **non-linear**, and if the π system rotates conrotatory or disrotatory. Figure 15.27 defines the linear or non-linear approaches. In the linear approach, the occupied orbital of **X** points directly at the π system it is adding to. In a non-linear approach, this orbital approaches at a skew angle.

The π system must rotate to form the new bonds. The direction of rotation is different for different electron counts. Figure 15.28 shows the various theoretical approaches to understanding the linear or non-linear approaches. For approach of the two-electron single fragment to a four-electron π system, the rotation is disrotatory for linear, and conrotatory for non-linear. This can be seen using an FMO analysis (Figure 15.28 **A**). Furthermore, the linear approach makes a six-electron Hückel aromatic transition state with a disrotatory motion (Figure 15.28 **B**), whereas the non-linear path requires a conrotatory motion to create a six-electron array with no nodes. Lastly, the generalized orbital symmetry rule is in agreement (Figure 15.28 **C**). These directions of rotation are the pattern for $4n + 2$ systems: linear approach requires disrotatory motion of the π system, whereas non-linear approach requires

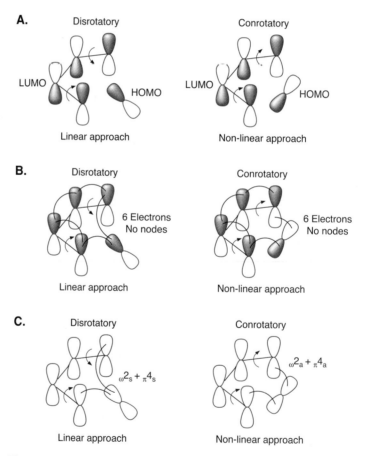

A.

Disrotatory — LUMO / HOMO — Linear approach

Conrotatory — LUMO / HOMO — Non-linear approach

B.

Disrotatory — 6 Electrons No nodes — Linear approach

Conrotatory — 6 Electrons No nodes — Non-linear approach

C.

Disrotatory — $_\omega 2_s + {}_\pi 4_s$ — Linear approach

Conrotatory — $_\omega 2_a + {}_\pi 4_a$ — Non-linear approach

Figure 15.28
Theoretical approaches to cheletropic reactions, showing only allowed reaction paths. **A.** An FMO analysis. **B.** Hückel vs. Möbius transition states. **C.** Use of the generalized orbital symmetry rule.

conrotatory. As always, the rules are reversed for $4n$ systems. The sulfur dioxide additions to the two hexadienes shown in Eqs. 15.51 and 15.52 illustrate how the selection rules control the stereochemistry, both being disrotatory, suggesting that the approach of SO_2 is linear.

15.6.2   Carbene Additions

The most important cheletropic reaction is the addition of singlet carbenes to olefins to make cyclopropanes. Only singlet carbenes will be considered here; the pericyclic selection rules cannot be applied to triplet states. The electronic structure of a singlet carbene involves an empty p orbital and a roughly sp^2 hybrid that has two electrons (see, for example, the two lone pair orbitals of the water molecule in Appendix 3). We know from Chapter 10 that singlet carbenes add stereospecifically to olefins, and that the olefin stereochemistry is retained in the cyclopropane product. As such, in the present context, the reaction would be described as suprafacial on the olefin.

The conventional analysis of carbene additions emphasizes the interaction of the filled carbene orbital with the olefin π system. This makes the system a four-electron system, and so a non-linear approach should be preferred. The drawings in Figure 15.29 show why this should be so. A linear approach creates a four-electron Hückel system that is antiaromatic. A non-linear approach creates a four-electron Möbius system, and so it is allowed.

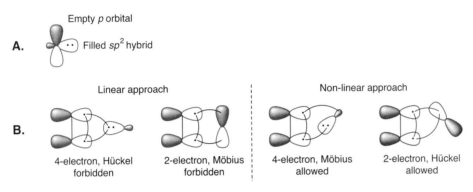

Figure 15.29
Using a Hückel/Möbius analysis to understand the addition of singlet carbenes to alkenes. **A.** The orbitals for singlet carbenes. **B.** Linear vs. non-linear approaches.

This analysis, like those for many other cheletropic processes, is complicated by the fact that there is a second low-lying orbital on the carbene center, the empty p orbital. It would be favorable to mix this empty orbital with the filled olefin π orbital. The linear approach aligns the empty p orbital in a manner that it is symmetry forbidden to mix with the olefin filled π orbital, and Figure 15.29 **B** shows that this is also forbidden because it is a two-electron Möbius system. Tilting the carbene on approach—that is, making a non-linear approach (Figure 29 **B**)—would allow a favorable mixing of filled and empty orbitals. We can alternatively view the right-hand drawings of each approach in Figure 15.29 **B** as FMO analyses where the olefin HOMO mixes with the carbene empty p-orbital LUMO (see Exercise 20 at the end of the chapter).

Regardless of our analysis, the carbene should approach the olefin in a non-linear manner. Frankly, the experimental significance of this is not completely clear. There are no obvious experimental implications for a linear vs. a non-linear approach. Theory clearly favors the non-linear path, but in this instance that prediction does not strongly influence experimental observations.

15.7 In Summary—Applying the Rules

We have introduced several different ways to analyze thermal pericyclic reactions. The highly formalized orbital symmetry analyses produce explicit rules, the conclusions of which are summarized in several tables. Other methods, such as FMO and aromatic transition state theories can be applied on a case by case basis, although they can be generalized in the same way. The realities are, in day-to-day chemistry, most pericyclic reactions will be familiar types or straightforward extensions of reactions we have covered here.

Occasionally, though, you will run across a more exotic pericyclic process, and will want to decide if it is allowed. In a complex case, a reaction that is not a simple electrocyclic ring-opening or cycloaddition, often the basic orbital symmetry rules or FMO analyses are not easily applied. In contrast, aromatic transition state theory and the generalized orbital symmetry rule are easy to apply to any reaction. With aromatic transition state theory, we simply draw the cyclic array of orbitals, establish whether we have a Möbius or Hückel topology, and then count electrons. Also, the generalized orbital symmetry rule is easy to apply. We simply break the reaction into two or more components and analyze the number of electrons and the ability of the components to react in a suprafacial or antarafacial manner.

The best way to develop a feel for which model is best suited to any given reaction is practice. The Exercises give you several opportunities to do so. Also, extensive books and reviews, cited at the end of the chapter, provide many examples of both simple and complex pericyclic reactions to analyze.

Summary and Outlook

In this chapter we have introduced the essential theories used to analyze thermal pericyclic reactions. We have also provided a number of examples. The conservation of orbital symmetry generated a great deal of experimental and theoretical work, with hundreds of often exotic tests and examples of the theory. We cite several excellent monographs below that provide many further examples of the concepts developed here.

Perhaps most importantly, though, the analysis of thermal pericyclic reactions introduced to organic chemists the power of orbital-based arguments in analyzing reaction pathways. Prior to this work, many experimentalists were justifiably skeptical that full molecular orbital theory could seriously impact their research. Those days are now long gone.

The remaining two chapters in this book build further on the knowledge of electronic structure that we have developed thus far. Chapter 16 discusses photochemistry. As we will see, photochemistry is complex, and requires detailed consideration of molecular orbitals and states. In several instances we will be able to make direct connections to reactions or concepts developed in this chapter. Chapter 17 extends our knowledge of molecular electronic structure to infinite systems, allowing us to consider conducting polymers and many other novel systems.

Exercises

1. Draw both an orbital correlation and a state correlation diagram for the $[_\pi 2_s + _\pi 2_a]$ reaction of Figure 15.10. (*Hint:* The only conserved symmetry element is a C_2 axis.)

2. Use FMO and aromatic transition state theories to show that the $[_\pi 4_s + _\pi 2_a]$ and $[_\pi 4_a + _\pi 2_s]$ cycloaddition reactions are forbidden.

3. When a Diels–Alder reaction such as that in Eq. 15.10 is run in water rather than the conventional organic solvents, the preference for the endo product *increases*. Explain this observation.

4. Draw an orbital correlation diagram for the suprafacial-suprafacial reaction of an allyl species with an olefin to give the corresponding cyclopentyl species, and establish whether the reaction is allowed or forbidden for the cation, anion, and radical.

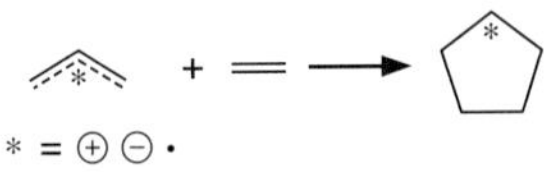

5. Draw the orbital correlation diagram for the addition of ozone to ethylene to give a five-membered ring ozonide. Is the reaction allowed or forbidden?

6. Predict the product of the following reaction. The reaction involves an intermediate (I) which sets the stereochemistry in the product. What is this intermediate, and what is the stereochemistry of the product? (*Hint:* The first reaction involves a pericyclic C–C bond cleavage.)

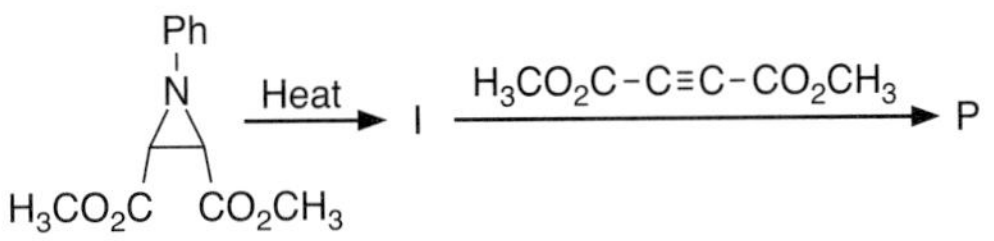

7. Develop an orbital symmetry correlation diagram for the hexatriene–cyclohexadiene interconversion of Eq. 15.18. Consider both conrotatory and disrotatory processes.

8. Develop state correlation diagrams for the conrotatory and disrotatory butadiene–cyclobutene interconversions and discuss their implications

9. Develop state correlation diagrams for the conrotatory and disrotatory hexatriene–cyclohexadiene interconversion and discuss their implications.

10. Rationalize the following transformation.

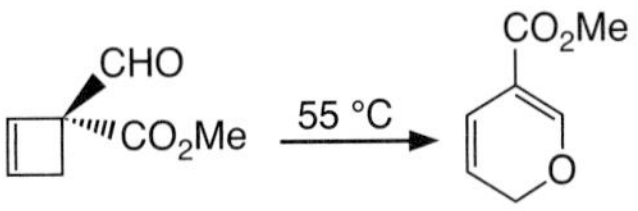

11. Predict the preferred product of the thermolysis shown.

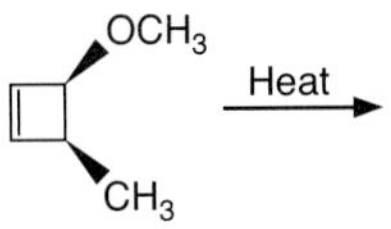

12. Explain why the solvolysis of the endo chloride to give the cyclohexenyl product is rapid, while the exo chloride reacts quite slowly. (*Hint:* This reaction involves carbocation intermediates.)

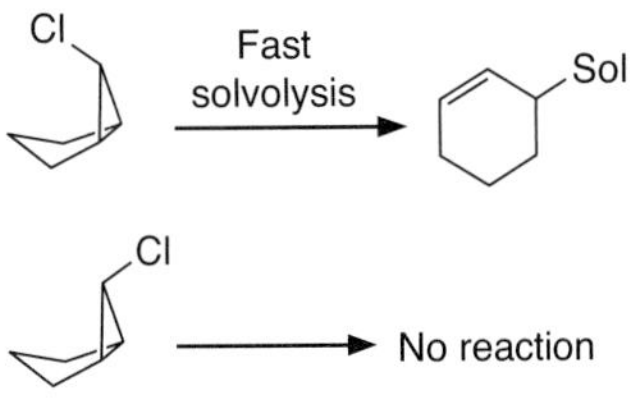

13. On treatment with strong base, compound **1** readily opens to an allylic anion, but compound **2** does not. Rationalize this difference.

14. Draw an orbital correlation diagram to show that the conversion of Dewar benzene to benzene is forbidden.

15. Thermolysis of the diene shown gives two products, with one substantially favored. Rationalize the stereoselectivity of this process.

16. Use the group increments of Chapter 2 to derive the heat of formation values for various structures considered in the discussion of the Cope rearrangement (Figure 15.24).

17. The oxy-Cope rearrangement is faster when the counterion is K^+ than when it is Na^+. Rationalize this effect.

18. Draw transition state structures that rationalize the stereochemistry of the two Ireland–Claisen rearrangements shown in Eq. 15.45.

19. Sulfoxides are pyramidal molecules, and typically they are stably chiral, with racemization barriers on the order of 40 kcal/mol. However, the sulfoxide shown has a racemization barrier of only 21 kcal/mol. Explain this observation.

20. Present an FMO analysis of the cheletropic carbene addition to ethylene using the LUMO of ethylene and the HOMO of the carbene. Does it lead to the same conclusions given in the chapter?

21. *cis*-Divinylcyclopropane rearranges to cycloheptatriene with an activation energy of roughly 20 kcal/mol. This is an example of what familiar reaction? How is it different from the prototype? The dimethyl compound has a substantially higher activation energy. Why?

22. Characterize the reaction shown as thermally allowed or forbidden and rationalize your conclusion.

23. Examine the following complex pericyclic reactions, and designate them using the electron count and suprafacial/antarafacial terminology. State if they are allowed or forbidden based upon the generalized orbital symmetry rule.

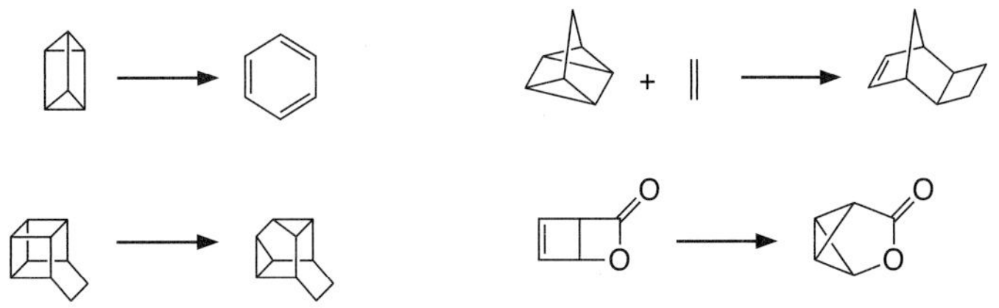

24. Use the aromatic transition state theory method to determine whether the reactions given in Exercise 23 are allowed or forbidden as written.

25. Give series of allowed pericyclic reactions that explain these overall transformations.

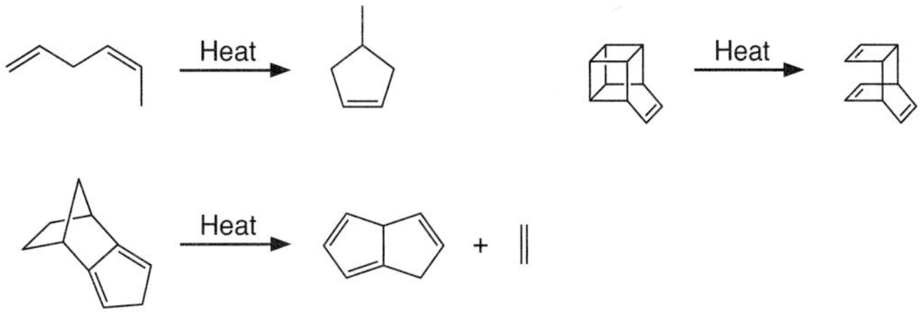

26. Examine the sigmatropic shifts of the following compounds, and state whether the a and b substituents exchange positions relative to the ring. The rearrangements indicated are called the **walk rearrangements**, because the group appears to be walking around the ring.

27. By examining FMO interactions, predict the regiochemistry of the following Diels–Alder reactions.

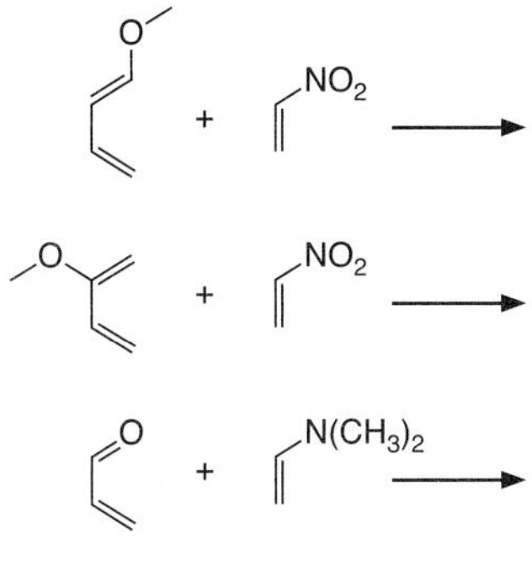

28. The addition of Lewis acids such as $AlCl_3$ tends to speed up reactions such as that shown below. Give an explanation.

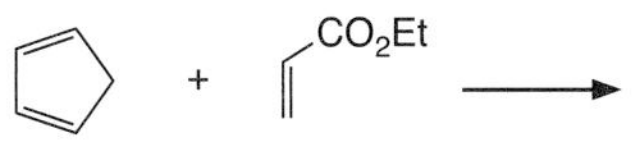

29. We noted in the chapter that if we can characterize a reaction as allowed in one direction, it is necessarily allowed in the reverse reaction. Therefore, we expect the same conclusion for the electrocyclic ring-closing reaction as with the opening. To apply FMO theory to an electrocyclic ring closure, we divide the π system into two parts, and write the HOMO for one part and the LUMO for the other part. This is done so that the HOMO and LUMO are in-phase where they interact with each other in the reactant, therefore leading to a constructive interaction between these carbons in the transition state. Using this method, show that for the closure of butadiene a conrotatory process is found when this system is analyzed as conjugated ethylenes, while for hexatriene a disrotatory process occurs when this system is considered to be a butadiene conjugated to an ethylene.

30. There is another approach to predicting the stereochemistry of electrocyclic ring closures. One simply looks at the ends of the HOMO to conclude the proper direction for rotation of the bond by creating in-phase interactions during closure. Show that this method also predicts conrotatory closure for butadiene and disrotatory closure for hexatriene.

31. Predict the thermally allowed products of the following two reactions.

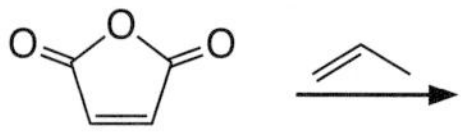

32. The following reaction is known as the **Carroll rearrangement**. Write a mechanism for this reaction. What pericyclic reaction is involved?

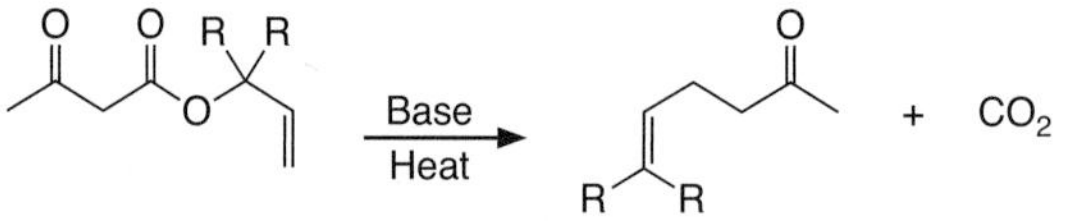

33. At first glance, the following reaction appears to be a [2+2] cycloaddition of a single double bond from cyclopentadiene with the C=C double bond of diphenylketene. However, it has been experimentally shown that the product actually results from two more conventional pericyclic reactions. Show by writing an alternative mechanism what these two reactions must be.

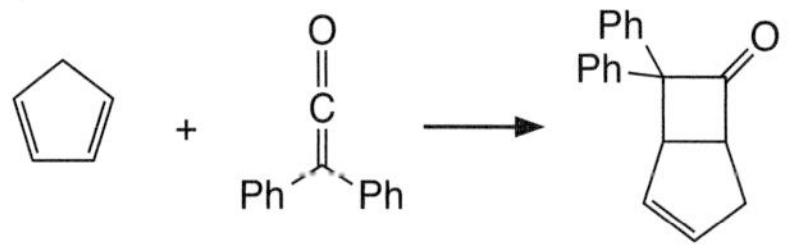

34. The **retroene** reaction is, as the name implies, the reverse of an ene reaction. Show how the following reaction conforms to this name. Explain the cis stereochemistry in the product.

35. A rare, but sometimes observed, class of pericyclic reactions is called **atom transfer** or **group transfer** reactions. A single example is given below, where two hydrogens are transferred to an alkene in a concerted single-step reaction. Using ethylene and ethane as models for the alkene and alkane portion of this reaction, draw an orbital correlation diagram for this reaction. Is it allowed or forbidden? Similarly, analyze the reaction using an FMO approach, and lastly examine aromatic transition state theory and the generalized pericyclic selection rule.

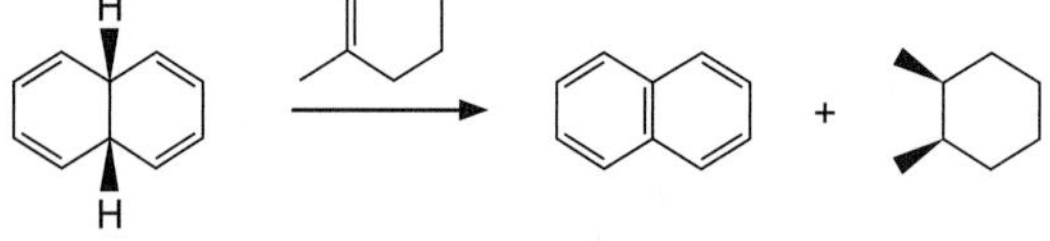

36. Figure 15.9 defined suprafacial and antarafacial interactions of a lone pair in a p orbital (called an ω component). Using these definitions, predict if the ring-opening of the cyclopropyl anion shown below will occur in a conrotatory or disrotatory fashion. What will be the stereochemistry of the product?

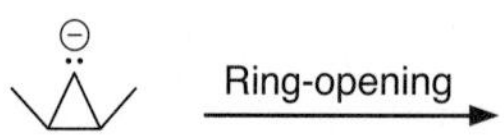

37. Pseudo-pericyclic reactions are reactions that appear at first glance to be pericyclic, in that arrow pushing shows a cyclic array of electrons at the transition state. Several of the other facets of pericyclic reactions are also commonly evident, such as low barriers if the geometries are favorable. However, these reactions are never forbidden based upon a symmetry argument. Examples of such reactions are given below. Write the proper arrow pushing for these reactions, and identify the one facet of these reactions that does not properly conform to standard pericyclic reactions, thereby making them "pseudo-pericyclic". (*Hint:* In your arrow pushing, it is important to identify the correct source of each arrow.)

Further Reading

General References on Pericyclic Reactions

Woodward, R. B., and Hoffmann, R. (1970). *The Conservation of Orbital Symmetry,* Verlag Chemie International, Deerfield Beach, FL.

Fukui, K. (1975). *Theory of Orientation and Stereoselection,* Spinger–Verlag, Berlin.

Zimmerman, H. E. "The Möbius–Hückel Concept in Organic Chemistry. Application to Organic Molecules and Reactions." *Acc. Chem. Res.,* **4**, 272–280 (1971).

Fleming, I. (1976). *Frontier Orbitals and Chemical Reactions,* John Wiley & Sons, London.

Fleming, I. (1999). *Pericyclic Reactions,* Oxford University Press, Oxford.

Lehr, R. E., and Marchand, A. P. *Orbital Symmetry; A Problem-Solving Approach,* Academic Press, New York.

Epiotis, N. D. "Theory of Pericyclic Reactions." *Angew. Chem. Int. Ed. Eng.,* **86**, 825 (1974).

Mango, F. D. "Transition Metal Catalysis of Pericyclic Reactions." *Coord. Chem. Rev,* **15(2–3)**, 109–205 (1975).

Bauld, N. L., Bellville, D. J., Harirchian, B., Lorenz, K. T., Pabon, R. A., Jr., Reynolds, D. W., Wirth, D. D., Chiou, H. S., and Marsh, B. K. "Cation Radical Pericyclic Reactions." *Acc. Chem. Res.,* **20**, 371 (1987).

Detailed Experimental and Computational Studies

Gajewski, J. J. "Energy Surfaces of Sigmatropic Shifts." *Acc. Chem. Res.,* **13**, 142–148 (1980).

Houk, K. N., Li, Y., and Evanseck, J. D. "Transition Structures of Hydrocarbon Pericyclic Reactions." *Angew. Chem. Int. Ed. Eng.,* **31**, 682 (1992).

Houk, K. N., Gonzalez, J., and Li, Y. "Pericyclic Reaction Transition States: Passions and Punctilios, 1935–1995." *Acc. Chem. Res.,* **28**, 81 (1995).

Wiest, O., and Houk, K. N. "Density Functional Theory Calculations of Pericylic Reaction Transition Structures." *Topics Curr. Chem.,* **183**, 1 (1996).

Wiest, O., Montiel, D. C., and Houk, K. N. "Quantum Mechanical Methods and the Interpretation and Prediction of Pericyclic Reaction Mechanisms." *J. Phys. Chem. A,* **101**, 8378 (1997).

Torquoselectivity

Dolbier, W. R., Jr., Koroniak, H., Houk, K. N., and Sheu, C. "Electronic Control of Stereoselectivities of Electrocyclic Reactions of Cyclobutenes: A Triumph of Theory in the Prediction of Organic Reactions." *Acc. Chem. Res.,* **29**, 471–477 (1996).

Dipolar Cycloadditions

Huisgen, R. "1,3-Dipolar Cycloadditions Past and Future." *Angew. Chem. Int. Ed. Eng.*, **2**, 565 (1963).

Houk, K. N., Sims, J., Duke, J., Strozier, R. W., and George, J. K. "Frontier Molecular Orbitals of 1,3 Dipoles and Dipolarophiles." *J. Am. Chem. Soc.*, **95**, 7287 (1973).

Houk, K. N., Sims, J., Watts, C. R., and Luskus, L. J. "The Origin of Reactivity, Regioselectivity, and Periselectivity in 1,3-Dipolar Cycloadditions." *J. Am. Chem. Soc.*, **95**, 7301 (1973).

Padwa, A. "Intramolecular 1,3-Dipolar Cycloaddition Reactions." *Angew. Chem. Int. Ed. Eng.*, **88**, 131 (1976).

Gothelf, K. V., and Jorgensen, K. A. "Asymmetric 1,3-Dipolar Cycloaddition Reactions." *Chem. Rev.*, **98**, 863 (1998).

Claisen Rearrangement

Tarbell, D. S. "The Claisen Rearrangement." *Chem. Rev.*, **27**, 495 (1940).

Jefferson, A., and Scheinmann, F. "Molecular Rearrangements Related to the Claisen Rearrangement." *Quart. Rev. Chem. Soc.*, **22**, 391 (1968).

Ziegler, F. E. "Stereo- and Regiochemistry of the Claisen Rearrangement: Applications to Natural Products Synthesis." *Acc. Chem. Res.*, **10**, 227 (1977).

Lutz, R. P. "Catalysis of the Cope and Claisen Rearrangements." *Chem. Rev.*, **84**, 205–247 (1984).

Ziegler, F. E. "The Thermal, Aliphatic Claisen Rearrangement." *Chem. Rev.*, **88**, 1423 (1988).

Frontier Molecular Orbital Theory

Dewar, M. J. S. "A Critique of Frontier Orbital Theory." *THEOCHEM*, **59**, 301 (1989).

Kato, S. "Perspective on 'A Molecular Orbital Theory of Reactivity in Aromatic Hydrocarbons'." *Theo. Chem. Acc.*, **103**, 219 (2000).

Dannenberg J. J. "Using Perturbation and Frontier Molecular Orbital Theory to Predict Diastereofacial Selectivity." *Chem. Rev.*, **99**, 1225 (1999).

Houk, K. N. "Frontier Molecular Orbital Theory of Cycloaddition Reactions." *Acc. Chem. Res.*, **8**, 361 (1975).

Cycloadditions

Dilling, W. L. "Photochemical Cycloaddition Reactions of Nonaromatic Conjugated Hydrocarbon Dienes and Polyenes." *Chem. Rev.*, **69**, 845 (1969).

Bartlett, P. D. "Mechanisms of Cycloaddition." *Quart. Rev., Chem. Soc.*, **24**, 473 (1970).

Herndon, W. C. "Theory of Cycloaddition Reactions." *Chem. Rev.*, **72**, 157 (1972).

Schore, N. E. "Transition Metal-Mediated Cycloaddition Reactions of Alkynes in Organic Synthesis." *Chem. Rev.*, **88**, 1081 (1988).

Katritzky, A. R., and Dennis, N. "Cycloaddition Reactions of Heteroaromatic Six-Membered Rings." *Chem. Rev.*, **84**, 827 (1989).

Secondary Orbital Interactions

Garcia, J. I., Mayoral, J. A., and Salvatella, L. "Do Secondary Orbital Interactions Really Exist?" *Acc. Chem. Res.*, **33**, 658 (2000).

Cope Rearrangement

Rhoads, S. J., and Raulins, N. R. "Claisen and Cope Rearrangements." *Org. React.*, **22**, 1 (1975).

Lutz, R. P. "Catalysis of the Cope and Claisen Rearrangements." *Chem. Rev.*, **84**, 205 (1984).

CHAPTER 16

Photochemistry

Intent and Purpose

The reactions of organic molecules initiated by the absorption of light—**photochemistry**—have always played an important role in physical organic chemistry. The mechanistic issues associated with photochemistry are intrinsically interesting and important. In addition, photochemistry often plays a crucial role in the characterization of reactive intermediates, as in studies involving laser flash photolysis or the generation of reactive intermediates under matrix isolation conditions.

Many modern aspects of organic chemistry also involve photochemistry. New synthetic methods based on photochemistry continue to be developed, although photochemistry is still a relatively minor component of synthetic organic chemistry. In contrast, photochemistry has blossomed into a crucial facet of many other active research areas. In bioorganic chemistry, photochemical methods are increasingly prominent. Examples we will consider below include the use of "caged" substrates to provide temporal and spatial control of biochemical processes; fluorescent dyes and sensors that play a central role in various imaging strategies; and fluorescence resonance energy transfer (FRET), a powerful approach to obtaining geometric information on complex systems. In materials chemistry, the photoresist strategy is a crucial tool of the modern electronics industry (as will be discussed in Chapter 17). Also, luminescent materials associated with LEDs and other types of displays are of central importance to modern media.

The present chapter thus has two major goals. First, we will introduce the conceptual and mechanistic foundations of photochemistry. We will see that before we can understand photochemistry, we must understand photophysics. Indeed, photochemistry is controlled by a competition among a variety of rate constants associated with fundamental physical processes, and our goal is to help you develop a sense of the relative values of these rate constants. Second, we will survey the basic photochemical processes of organic chemistry. In reality, a relatively small number of basic reaction types dominate organic photochemistry.

You should appreciate from the outset that a thorough mechanistic understanding of a photochemical process presents an intrinsically more severe challenge than an effort to understand a thermal process. We have seen throughout this book how difficult it can be to develop a complete understanding of the potential energy surface associated with a thermal reaction. In a photochemical process, we must consider two or more such surfaces, their intrinsic natures and the interactions between (among) them. This can indeed be challenging. Nevertheless, the essential issues associated with such an analysis are well within the grasp of a modern physical organic chemist, and this chapter should provide a basic understanding of most photochemical processes.

16.1   Photophysical Processes—The Jablonski Diagram

When light interacts with matter, any of a number of things can happen. Most are **photophysical processes**—that is, the chemical structure of the molecule is unchanged in the end. Others, however, are **photochemical processes**—that is, the interaction of light with matter leads to any of a large array of chemical structural changes. Before we can understand photochemical processes, we must understand photophysical events, and we begin this chapter with a brief discussion of light and energy.

16.1.1   Electromagnetic Radiation

Table 16.1 shows the familiar electromagnetic spectrum and provides useful reference points for future discussions. Recall the basics of Eqs. 16.1 and 16.2. The energy of **monochromatic light** (that of one wavelength) is proportional to its frequency (ν), and inversely proportional to the wavelength (λ). In other words, longer wavelength light has a lower energy. We show wavelength in the condoned unit of nanometer (10^{-9} m), but another common unit is the angstrom (Å, 10^{-10} m); for example, 200 nm equals 2000 Å. Another common measurement is the reciprocal of the wavelength, $\bar{\nu}$, typically measured in units of cm^{-1} or wavenumbers. Table 16.1 also shows the energy equivalent of one mole of photons, known as an **einstein**, which allows us to relate the energy associated with a given wavelength of light to the measure of energy that chemists use for reactions: kcal/mol. A useful relationship is E (kcal/mol) $= 2.86 \times 10^4 / \lambda$ (nm).

$$E = h\nu \quad \text{with } \nu \text{ in s}^{-1} \text{ and } h = 6.63 \times 10^{-27} \text{ erg} \bullet \text{s} \qquad \text{(Eq. 16.1)}$$

$$\nu = \frac{c}{\lambda} \qquad \text{(Eq. 16.2)}$$

Keep in mind the difference between the energy of the light used, its power, and the intensity of the light. The energy is given by Eq. 16.1, while the **power** depends both upon the energy per photon and the number of photons in a given time. The unit of radiant power is the watt (W), which is one Joule per second. The intensity of light is also called the **power density**, indicating that it is the power per unit area of light (flux). The intensity depends upon the light source. For example, a laser can deliver a high intensity of either low or high energy radiation.

Table 16.1
Energy and Time Scales

	λ (nm)	$\bar{\nu}$ (cm^{-1})	ν (s^{-1})	E (kcal/mol)[a]	
Ultraviolet	200	50,000	15×10^{15}	143.0	Electronic absorption
	300	33,333	1×10^{15}	95.3	
Visible	400	25,000	7.5×10^{14}	71.5	
	500	20,000	6×10^{14}	57.2	
	600	16,666	5×10^{14}	47.7	
	700	14,286	4.2×10^{14}	40.8	
Infrared	1,000	10,000	3×10^{14}	28.6	Nuclear vibrational motion
	5,000	2,000	6×10^{13}	5.8	
	10,000	1,000	3×10^{13}	2.9	
Microwave	10^7	10	3×10^{11}	3×10^{-2}	Electron spin precession (EPR)
	10^9	0.1	3×10^9	3×10^{-4}	
Radio wave	10^{11}	0.001	3×10^7	3×10^{-6}	Nuclear spin precession (NMR)

[a]Energy of "1 mole" of photons (1 einstein).

Photochemistry is associated with electronic excitation—that is, the promotion of an electron from one orbital to a higher–lying orbital, thereby creating an **excited electronic state** (see below). The energies required to do this are in the range of ultraviolet or visible light, and so this is the range we will consider in this chapter. Table 16.1 does not list x rays and γ rays, which are so high in energy that they lead to complete ejection of electrons from atoms and molecules, called **ionization**. As we move from UV/vis radiation to longer wavelengths, corresponding to lower energies, we progress to infrared radiation, which excites nuclear vibrations (hence, IR spectroscopy); microwaves, which excite electron precessions in a magnetic field (EPR), molecular tumbling, and some lower energy torsional motions; and radiowaves, which excite nuclear precessions in a magnetic field (NMR).

Consider the energy range spanned by UV/visible light, roughly 40–140 kcal/mol. Now recall Table 2.2, which lists bond dissociation energies (BDE) for typical bonds found in organic molecules. The range of energies associated with single bonds is nicely spanned by UV/visible light. It is no coincidence, then, that photochemistry is associated with UV/visible light. It is this type of light that has sufficient energy to break bonds and thus do chemistry.

Multiple Energy Surfaces Exist

Throughout this textbook we have discussed reactions and conformational analyses that occur on a single energy surface. We explored these surfaces in some depth in Chapter 2 for bond rotations and ring interconversions, while in Chapter 7 the surfaces were examined in the context of chemical reactions. In other chapters the surfaces were used to explain phenomena related to reactivity, product distributions, etc. Although we usually did not explicitly state it, all these surfaces represented the **ground electronic state** of the molecule(s). This means that the electrons are in the lowest energy molecular orbitals of the molecule(s) as dictated by the Aufbau principle. However, we have noted many times, especially in Chapters 1 and 14, that many other molecular orbitals in a molecule exist. Within a **closed shell molecule** (all electrons paired and two electrons per orbital), the unoccupied molecular orbitals are most commonly antibonding.

The absorption of a photon of UV/vis light leads to an **excited electronic state** of the molecule (more on this below when we discuss absorption). Chemists state that the molecule is in an excited state, and this commonly means that an electron is promoted from a bonding to an antibonding orbital. For many organic molecules, the orbital that is losing an electron is the HOMO or an orbital near it in energy, while the orbital receiving the electron is the LUMO. The next Going Deeper highlight describes simple ways to envision excited states. The excited electronic state has a completely different potential energy surface that dictates the conformations of the molecule and any possible chemical reactions. Since there are many bonding and antibonding orbitals in common organic molecules, there are in theory a very large number of possible excited states and corresponding energy surfaces for every molecule. However, only a handful (two or three) normally need to be considered for photochemical discussions, since with the energy of UV/vis light only certain electronic transitions can be accessed.

Going Deeper

Excited State Wavefunctions

The typical notation for a ground state and its first excited state is that shown in the orbital energy diagram below. The ground state has a closed-shell configuration, and the excited state is formed by exciting an electron from the HOMO to the LUMO. While this is often a good starting point, it is in reality a massive oversimplification. Recall from our discussions in Chapter 14 that such representations correspond to single determinant or single config-

uration descriptions of the total wavefunction. While often adequate for ground states, single configuration wavefunctions are *never* adequate for an electronic excited state. Even a moderately adequate wavefunction will have to go beyond Hartree–Fock theory, most typically by including a significant amount of configuration interaction (CI). This is because, for the excited state, there are other configurations than the one shown that are of the appropriate symmetry to mix and are not very much higher in energy. Thus, they will contribute significantly to the true wavefunction. For a closed-shell ground state configuration, the energy gap to the first alternative configuration worth considering is often large enough that CI can be safely ignored. Because of the complexities of CI and other approaches to correlated wavefunctions, computational photochemistry remains a very challenging research area. Although excited states are represented throughout this chapter by relatively simple diagrams and symbolisms of the kind shown here, the real situation is much more complex.

Another useful model for thinking about excited states can be applied to systems where electron donating and withdrawing groups are placed in conjugation via alkenes, alkynes, and/or aryl groups. The ground states of such systems are best represented by a standard resonance structure with no charges, while the excited state can be considered to have considerable character of the resonance structure possessing charges. See below for an example. This model is another vast simplification, but it can be useful when considering the polar character of excited states and designing non-linear optical materials (see Section 17.5).

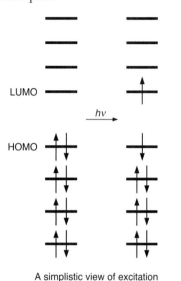

A simplistic view of excitation

A highly polar excited state

Figure 16.1 shows three energy surfaces and identifies the key photophysical processes we must consider when irradiating molecules and materials with UV/vis light. This figure summarizes a great deal of information. It is a slightly modified form of a diagram first proposed by Jablonski in 1935, and hence is called a **Jablonski diagram**. Each surface is represented by a Morse potential, just as we did for bonds in molecules discussed in Chapter 2. Remember, however, that the energy surface for a molecule is a composite of all its vibrational modes and possible interconversions between chemical structures. As such, what we show in Figure 16.1 is a simplification for each energy surface. In figures given later in this chapter, when photochemistry is discussed, the ground and excited state energy surfaces will appear more like those given in Chapter 7, representing reaction coordinates. Yet, Figure 16.1 is very useful for understanding the photophysical phenomena that are preludes to photochemistry, and in this section we will walk step-by-step through the various processes shown in Figure 16.1.

As stated above, each electronic excited state of the molecule has its own energy surface, placed at a higher energy than the ground state surface on the Jablonski diagram. It is very common that the chemical and structural properties of excited states are quite different than the ground state. The preferred geometries of atoms—tetrahedral, trigonal planar, etc.—can change. The preferred bond rotomers and ring conformations can be different on the different surfaces. When the lowest energy structures of the ground state and excited state are different, we place the Morse potentials on the Jablonski diagram at different positions along the x axis (only done for the triplet state in Figure 16.1). Hence, while the y axis represents the typical notion of potential energy, the x axis in this diagram represents a structural axis much like in reaction coordinate diagrams.

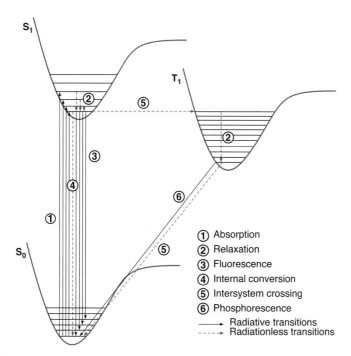

(1) Absorption
(2) Relaxation
(3) Fluorescence
(4) Internal conversion
(5) Intersystem crossing
(6) Phosphorescence
———▸ Radiative transitions
- - - -▸ Radiationless transitions

Figure 16.1
The Jablonski diagram. See text for a discussion of the various features.

16.1.2 Absorption

Photochemistry begins in the ground electronic state of the molecule of interest. In all but a few exceptional structures, the ground state of an organic molecule is a closed shell, singlet state, and so it is referred to as S_0. When a molecule is exposed to light of an appropriate energy, there is a chance that a single photon will be absorbed, promoting the molecule to an excited electronic state. According to the **Stark–Einstein law**, a molecule will only absorb light to bring about a single electronic transition, and the energy of the light must match the difference between the ground state and some excited state. Absorption of two photons or multiple photons to give a single electronic transition does not occur, except with special equipment that allows high laser intensity and flux within a small volume of space.

In quantum mechanical terms, the transition from one state to another is controlled by an operator (more on this below) and must obey certain symmetry restrictions. In particular, high probability transitions must be spin conservative. If we start out in a singlet state, we will end up in a singlet state. Thus, absorption from S_0 will be to an excited singlet state, the first of which is termed S_1 (Figure 16.1). Alternative spin states are possible for excited states, and the most common one is a triplet (designated T_n). Here, the spins of the electrons in the different orbitals are aligned.

What if the photon is energetic enough to excite the molecule to a higher-lying singlet state, such as S_2 or S_3? This can and does happen. However, experience has shown that, with very few exceptions, all photochemistry is initiated from the lowest excited states, S_1 and T_1. This is termed **Kasha's rule**. It is a consequence of the fact that relaxation for S_n or T_n states to S_1 and T_1, respectively (a form of internal conversion, see below), is always very fast. The terms **singlet manifold** and **triplet manifold** are used to refer to the different energy levels associated with singlet and triplet states, respectively.

We mentioned above that much of photochemistry is an exercise in relative rates, and so we need to consider the time scale of absorption. The transition from S_0 to S_1 is essentially instantaneous—it occurs over 10^{-16}–10^{-14} s. No chemical-scale process can compete with absorption. Absorption happens so rapidly because only electron movement is involved. Because of the substantially greater mass of nuclei vs. electrons, nuclear motions are considerably slower. They occur on a time scale of 10^{-13}–10^{-12} s. This realization leads to a key concept in photochemistry/photophysics, the **Franck–Condon principle**. The principle states that because electronic transitions occur faster than nuclear motion, absorption is vertical on an axis that represents nuclear motion (the x axis of a Jabonski diagram). This means that electronic transitions are most favorable when the geometries (nuclear positions) of the initial and final states are the same. Consider a situation in which the preferred geometry of the excited state is very different from that of the ground state. When electronic excitation occurs, the sluggish nuclei can't possibly keep up with the fleet electrons, and so initially a very unfavorable geometry of the excited state will be produced. This disfavors excitation.

Along with having similar geometries, the ground and excited states must also have their electronic structures matched in an appropriate way. In anticipation of later discussions, it is worthwhile to briefly describe the quantum mechanical treatment of absorption. When a molecule absorbs a photon of light, it undergoes a transition from one electronic state to another. Not surprisingly, this process is described by a quantum mechanical integral that connects the initial state, Ψ_i, to the final state, Ψ_f via an appropriate operator. The oscillating electric field of the light induces a **transition dipole**, D, in the molecule, and the transition probability is proportional to this dipole strength (Eq. 16.3). The integral, m, is called the **transition moment integral**, and the operator, M, is called the **dipole moment operator**. This operator has the form of a dipole moment, a charge (e) multiplied by a distance (r). The sum $j = 1$ to n is over the distance r_j that each electron j moves in the induced dipole associated with the transition. It reflects the extent to which charges move on going from the initial to the final state.

$$D = \left(\int \Psi_i M \Psi_f \, d\tau \right)^2 = m^2$$

$$M = \sum_{j=1}^{n} er_j$$

(Eq. 16.3)

Experimentally, the efficiency of absorption is expressed by the **molar extinction coefficient**, ε, in **Beer's law**, Eq. 16.4. The quantity $\log[I_0/I]$, or A, is termed the **optical density (OD)** or **absorbance** of the sample. Note that it is a logarithmic quantity. An OD of 1 implies that 90% of the incident light is absorbed (10% is transmitted), and an OD of 2 indicates that 99% of the light is absorbed (1% is transmitted). There is often a lack of sensitivity in experimentally measuring OD values above 2, because their differences are associated with less than 1% of transmitted light, which is difficult to measure accurately. Beer's law often breaks down as the OD values get much larger than 2, especially approaching 3 to 4.

The extinction coefficient is the most convenient way to express the efficiency of light absorption, with a larger extinction coefficient corresponding to more efficient absorption. For each wavelength of light, there is a different ε value (there are coincidental equalities). A plot of absorption as a function of wavelength is called the **absorbance spectrum**, and with light associated with electronic transitions, the plots are often referred to as **UV/vis spectra**. The spectra are generated with instruments referred to as UV/vis spectrophotometers.

$$\log[I_0/I] = A = \varepsilon bc$$

I_0 = Intensity of incident light

I = Intensity of transmitted light

b = Path length (cm)

c = Concentration (mol/L)

ε = Molar extinction coefficient (L/mol•cm)

(Eq. 16.4)

The most obvious factor influencing absorption efficiency, shown in Figure 16.1, is that there must be a precise matching between the energy of the incoming photon and the gap between the initial and final states. Figure 16.1 shows that there is more than one way to achieve this matching. We assume that all our molecules begin in the ground vibrational level of S_0. However, absorption can occur to any of a number of vibrational levels in S_1. In fa-

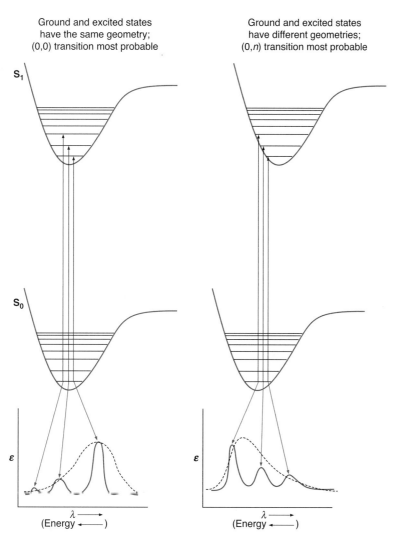

Ground and excited states have the same geometry; (0,0) transition most probable

Ground and excited states have different geometries; (0,n) transition most probable

Figure 16.2
Absorption of light causes conversion from S_0 to S_1. On the left the two states are of very similar geometries so the (0,0) transition is most likely. The case on the right involves a geometry change on excitation, causing a (0,n) transition to be more likely. Below each potential energy diagram is the corresponding absorption spectrum. The solid line is for the case where vibrational fine structure can be resolved, while the dotted line is for the more typical case where it cannot.

vorable cases, the individual transitions from S_0 to the various vibrational levels of S_1 can be observed, as shown in Figure 16.2. Such instances are especially informative.

Figure 16.2 shows vertical transitions for two different cases. If the geometries of S_0 and S_1 are nearly the same, the highest probability transition will be from the zero vibrational level of S_0 to the zero vibrational level of S_1, termed the (0,0) transition. The (0,0) transition is necessarily the longest wavelength (lowest energy) transition possible. On the other hand, if the geometry of S_1 is significantly different from that of S_0, designated in Figure 16.2 by a shift of the S_1 surface, then the transition with the smallest geometry change will be to a higher vibrational level of S_1.

Consider the various absorption transitions shown in Figures 16.1 and 16.2. The energy spacings separating the various absorptions correspond to a vibrational spacing in the *excited state*. Thus, in favorable cases, an absorption spectrum can provide geometrical information about the excited state of a molecule, revealing the nature of its vibrational states.

Typically, detailed vibrational spacings are not seen in an absorption spectrum. Instead, a single, broad transition is seen (called an **absorption band**), as shown by the dotted lines in

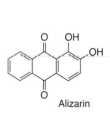

Alizarin

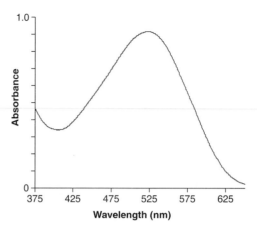

Figure 16.3
The absorbance spectrum of alizarin in 3:1 methanol/water at pH 7.4.

Figure 16.2. A specific example is given in Figure 16.3, where the UV/vis spectrum of alizarin is shown. This spectrum shows a single very broad absorption band. In typical molecules under conventional conditions, we must consider more than just a single, simple vibrational level implied by Figure 16.2. Most molecules will have many vibrational and rotational levels available, so that a continuum of absorption is seen, as in Figure 16.3.

Vibrational detail is typically seen only for very rigid molecules. For example, Figure 16.4 shows the absorption spectrum of anthracene, with the associated fine structure. Each peak is associated with a narrow absorption band. Also, very low temperatures and rigid media can favor the observation of vibrational fine structure.

We designate the various surfaces of Jablonski diagrams as S_n and T_n where n is the energy level. We have also stated that the formation of an S or T state, where n does not equal 0, involves excitation of one electron into an empty orbital. This leaves one electron behind in the starting molecular orbital. Chemists have accepted descriptors that define which orbital the electron departed from and where it landed. For example, when an electron is excited from a π molecular orbital to a π^* molecular orbital, there is a single electron in the original π orbital and now an electron in the π^* orbital. We call this a **π,π^* transition**, and the resulting state of the molecule is called a **π,π^* state**. Often an excitation derives from a nonbonding orbital (lone pair, designated n) to an antibonding π^* orbital, and such a transition is designated as an **n,π^* transition** which gives an **n,π^* state**. We tie these definitions back to S_n and T_n by using an $^m(x,y)$ notation, where m tells the spin state (1 for S and 3 for T), and x and y are the orbitals involved as described here.

Anthracene

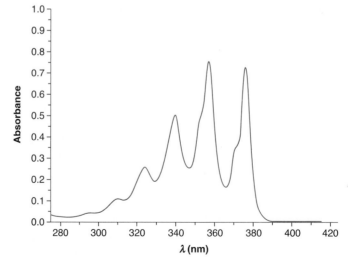

Figure 16.4
Absorption spectrum of anthracene.

Table 16.2
Typical Organic Chromophores

Chromophore	λ_{max} (mn)	ε_{max} (L/mol•cm)	Type
C−C and C−H	<180	1000	$\sigma \rightarrow \sigma^*$
C−O (alcohol)	~180	200	$n \rightarrow \sigma^*$
C−S (alkyl sulfide)	210	1200	$n \rightarrow \sigma^*$
C−Br (alkyl bromide)	208	300	$n \rightarrow \sigma^*$
C−I (alkyl iodide)	260	400	$n \rightarrow \sigma^*$
C=C	180	10,000	π,π^*
C=C−C=C	220	20,000	π,π^*
C=C−C=C−C=C	260	50,000	π,π^*
C_6H_6	260	200	π,π^*
Styrene	282	450	π,π^*
Acetophenone	278	1100	n,π^*
Naphthalene	310	200	π,π^*
Anthracene	380	10,000	π,π^*
Nitrobenzene	330	125	n,π^*
C=O	280	20	n,π^*
N=N	350	100	n,π^*
−NO$_2$	270	20	n,π^*
N=O	660	200	n,π^*
C=C−C=O	350	30	n,π^*
C=C−C=O	220	20,000	π,π^*

At what wavelengths do organic molecules absorb to promote electrons? Table 16.2 lists some typical organic **chromophores**. A chromophore is a functional group or combination of functional groups that absorbs UV/vis light. There are two general types of chromophores, π,π^* and n,π^*, and these differ in several characteristic ways, as summarized in Table 16.3. Further, the trends in λ_{max} (the wavelength of maximal absorbance) are predictable. The more extended a π system, the longer the wavelength of the π,π^* absorption. For a given chromophore such as a carbonyl, the n,π^* transition will be at a lower energy (longer wavelength) than the π,π^* transition, consistent with the notion that a nonbonding MO should be higher lying in energy than a bonding π MO.

Table 16.3
Characteristics of π,π^* and n,π^* Transitions

π,π^*	n,π^*
"Allowed" $\varepsilon > 10^3$	"Forbidden" $\varepsilon < 10^2$
Increased polarity of solvent → increased λ_{max} (red shift)	Increased polarity of solvent → decreased λ_{max} (blue shift)
ΔE (S$_1$ − T$_1$) > 20 kcal/mol	ΔE (S$_1$ − T$_1$) < 10 kcal/mol
Typical reactions: Pericyclic rearrangements Cycloadditions/cycloeliminations	**Typical reactions:** Atom abstraction Radical addition Electron transfer Cleavage reactions

In general, π,π^* transitions are more intense than n,π^* transitions, with ε_{max}, the extinction coefficient of the absorption at λ_{max}, typically being $> 10^3$ for the former and $< 10^2$ for the latter. We describe π,π^* transitions as "spatially allowed" and n,π^* transitions as "spatially forbidden". While the origin of these terms is in the strict quantum mechanics of absorption, in practice they are not hard and fast rules. Instead, "forbidden" n,π^* transitions are observed, just with lower probabilities. The basis for the distinction relates back to Eq. 16.3. In order for the transition moment integral to have a large value, Ψ_i and Ψ_f must be co-extensive (see definition in Section 14.5.6) in space. This is not a problem when Ψ_f is a π,π^* state, since all electrons that began in π orbitals stay in π orbitals. However, on excitation to produce an n,π^* state, an electron that began in a lone pair orbital, which is typically of σ symmetry, is promoted to a π-type orbital. This makes the ground state and the n,π^* state less co-extensive, and so the transition is less favorable. Two common molecules in Table 16.2, benzene and naphthalene, represent exceptions to this rule. Because of the very high symmetries of these systems, the ground state and the first π,π^* states are of the wrong symmetries to interact, and so the absorptions have low extinction coefficients.

Empirically, it has generally been observed that π,π^* and n,π^* transitions can be distinguished by how they respond to changes in solvent polarity. The trends shown in Table 16.3 are generally observed. "Explanations" for these trends vary, often invoking differential hydrogen bonding to ground vs. excited states. However, the same trends are seen with solvents that are polar but not hydrogen bonding. The best way to view these trends is as a series of experimental observations that generally hold true, but have complex origins that may differ from system to system.

The λ_{max} of absorption depends upon the microenvironment of the chromophore, as does the intensity of absorbance. For example, as just discussed, solvation can affect the energies of transitions. Binding of the chromophore to a receptor, natural or synthetic, can also often influence λ_{max}, as can a variety of other effects. When the shift in absorption is to shorter or longer wavelengths it is defined as a **hypsochromic** (blue shift) or **bathochromic shift** (red shift), respectively. When the intensity of absorption increases or decreases it is called a **hyperchromic** or **hypochromic effect**, respectively.

Connections

Physical Properties of Excited States

As we have discussed in this chapter, excited states represent chemical structures of molecules, just as do ground states. The physical properties of an electronic excited state of a molecule can be significantly different from those of the ground state. For example, the dipole moment of formaldehyde is 2.3 D for the ground state, but 1.5 D and 1.3 D in the S_1 and T_1 states, respectively, both of which are n,π^* states. An electron from an oxygen lone pair orbital has been excited to a π^* MO, which is primarily on carbon (Figures 1.16 and 1.17). This decreases the charge on oxygen and so diminishes the dipole moment. There are also geometry changes associated with excitation; both the S and T excited states are pyramidal at the carbonyl carbon, while the ground state is planar.

Carbonyl excited state

16.1.3 Radiationless Vibrational Relaxation

Step 2 in Figure 16.1 involves conversion of the vibrationally excited states of S_1 created by absorption to the ground vibrational state of S_1. No photons are emitted. The excess energy is given off as heat to the surrounding medium. Such a process is fast, but not instantaneous. Typical rates are 10^{11}–10^{12} s^{-1}. Generally, the relaxation occurs fully before any other process can compete. However, fast lasers can easily deliver a second photon long before relaxation occurs, allowing excitation to still higher energy levels. This is a common approach in chemical physics, but in preparative organic photochemistry, vibrational relaxation is generally complete.

16.1.4 Fluorescence

Conversion of S_1 back to S_0 with concomitant emission of a photon is termed **fluorescence** (Figure 16.1). Spectra that measure fluorescence are typically referred to as **fluorescence spectra** or **emission spectra**. Fluorescence can produce the ground or vibrationally excited states of S_0. In optimal cases a spacing can be seen in the emission spectrum that corresponds to vibrational structure in the *ground* state. For example, examine the emission spectrum of anthracene given in Figure 16.5, which shows several bands corresponding to different vibrational levels of the ground state. As shown with anthracene, in most cases fluorescence spectra are observed at longer wavelengths (lower energy) than the absorption spectrum (compare to Figure 16.4), and the absorption and emission spectra display a mirror image relationship. Figure 16.1 should make it clear why this is so. Molecules that fluoresce have a portion of their structure that is called a **fluorophore**, analogous to the term chromophore in molecules that absorb light. More commonly the entire molecule is referred to as the fluorophore.

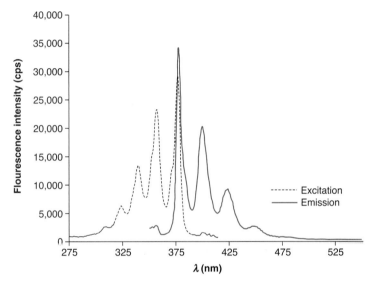

Figure 16.5
Overlay of the emission and excitation spectra of anthracene. Both are taken in toluene.

As stated above, fluorescence involves relaxation of S_1 to S_0 via emission of a photon. This is by far the most common form of fluorescence. However, some molecules emit light from higher singlet states, before relaxation to S_1 occurs. When this occurs we have what is called **anomolous fluorescence**. Azulene gives anomolous fluorescence.

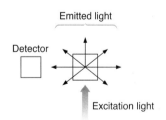

Azulene

A fluorescence spectrum is obtained by irradiating the sample at a wavelength where the molecule absorbs light, the **excitation wavelength** (commonly λ_{max} of the UV/vis spectrum). The spectrum gives the wavelength dependence of the intensity of the emitted light. Instruments that measure this dependence are called fluorimeters or spectrofluorimeters. The light emitted from the sample is commonly detected at 90° to the incident light, in order to avoid detecting residual incident light (see in the margin).

The intensity of the emitted light (I_e) is related to the path length of the cell (b), the concentration of the fluorophore (c), and the likelihood of fluorescence at each wavelength (Φ_f, called the quantum yield; see Section 16.1.8). However, because the fluorophore must first absorb a photon, there is a dependence of the emitted light upon the irradiation intensity (I_o) and the extinction coefficient for the fluorophore absorption (ε). Hence, an equation similar to Beer's law (Eq. 16.4) can be written for fluorescence (Eq. 16.5, valid only at low concentrations), but it contains more constants than the simple Beer's law.

$$I_e = 2.3\, I_o\, \varepsilon\, \Phi_f\, bc \qquad \text{(Eq. 16.5)}$$

Recall the fluorescence spectrum of anthracene that is shown in Figure 16.5 (solid line). A related measurement is an **excitation spectrum**. Here, one varies the excitation wavelength and measures the fluorescence intensity at a fixed wavelength. Because the fluorescence intensity is directly proportional to the extinction coefficient (ε) for absorption (Eq. 16.5), an excitation spectrum should faithfully reproduce the *absorption spectrum* of the material. As an example, look at Figure 16.5 (dashed line), which shows the excitation spectrum of anthracene, and compare it to Figure 16.4. Seeing such an excitation spectrum provides compelling evidence that the fluorescence being observed is indeed the result of excitation of the desired molecule rather than, for example, some impurity in the sample. As described in the next Going Deeper highlight, impurities can often hamper analysis of fluorescence.

The difference between the λ_{max} of the longest wavelength absorbance band and the shortest wavelength emission band is called the **Stokes shift**, which is often simply the difference between the absolute λ_{max} for each spectrum. With molecules that experience a significant geometry change on excitation, or a significant reorganization of solvent, the Stokes shift will be larger and the mirror image relationship will be less apparent. The analysis of Figure 16.5 shows no Stokes shift. Figure 16.6, however, shows an overlay of the absorption spectrum and the emission spectrum for fluorescein. The mirror image relationship is still apparent as in Figure 16.5, but the λ_{max} values are separated by approximately 45 nm (the Stokes shift). With fluorescein, the structure is very rigid, so the Stokes shift primarily arises due to different solvation of the excited state relative to the ground state.

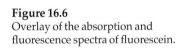

Fluorescein

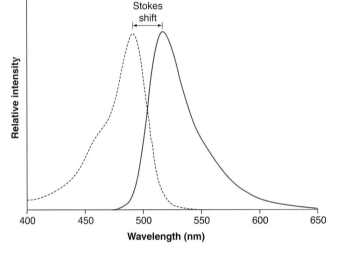

Figure 16.6
Overlay of the absorption and fluorescence spectra of fluorescein.

In rigid structures for which S_1 and S_0 have very similar geometries, and the solvation of the excited state is very similar to that of the ground state, the Stokes shift will be small. For example, look again at Figures 16.4 and 16.5, where the longest wavelength absorption band or excitation band for anthracene coincides with the shortest wavelength emission band. When this correspondence occurs, it is referred to as **resonance fluorescence**. Resonance fluorescence is actually not very common.

On very rare occasions, the initial excitation event can occur from excited vibrational states of S_0, leading to fluoresence from S_1 that occurs at a shorter wavelength than absorption. In such instances, the difference in the maxima on the absorption and emission curve is called an **anti-Stokes shift**.

ground technique. In an absorbance experiment, we irradiate the sample at a given wavelength and then observe the intensity of transmitted light at that same wavelength. We get a signal at the detector even if there is no chromophore that absorbs at that wavelength. This is because we are monitoring the decrease in a signal. In contrast, in a fluorescence experiment we irradiate at one wavelength and detect at another; if there is no fluorophore there is no signal. As long as we set up the optics of the experiment properly, there is no background irradiation reaching the detector. This allows us to use very sensitive detectors designed to amplify very weak signals. This is not possible with a subtractive measurement like an absorbance spectrum.

High sensitivity is a good thing. It is now possible, with fairly routine optics, to see fluorescence from *single molecules*, an exciting research area with considerable

potential. We will soon see an example of this. However, high sensitivity can also produce problems. Fluorescence measurements are especially susceptible to spurious signals from impurities. Even a very small amount of a highly fluorescent material can give an intense fluorescent signal that could be misinterpreted as due to the major species in the sample. Care must be taken to purify samples for fluorescence measurements as much as possible. In this regard, an excitation spectrum can be especially helpful. As noted above, if the shape of the excitation spectrum exactly follows the absorption spectrum of the species of interest, then we can be sure that the observed fluorescence is due to that species.

Connections

GFP, Part I: Nature's Fluorophore

The Pacific jellyfish *Aequorea victoria* has proven to be a real boon to chemical biologists. This beautiful creature lives in the deep dark waters of the Pacific Ocean, where it glows a brilliant green. The source of this color is a remarkable molecule known as **green fluorescent protein (GFP)**. This protein is 238 amino acids long and contains a chromophore that produces an intense green fluorescence. The chromophore is formed from consecutive amino acids Ser65–Tyr66–Gly67. Remarkably, the conversion of these amino acids to a fluorophore occurs spontaneously without any enzymatic assistance, as long as oxygen is available. This unique structure (*i*) absorbs at 397 nm and emits at 509 nm. The structure of the full protein is quite elegant. A robust barrel structure formed from β sheets completely encapsulates the chromophore, preventing quenching from outside influences and creating a favorable environment for fluorescence. We show an image of the structure with the outside "barrel" in color, and the chromophore as a white, space-filling model.

for GFP to the gene for the protein of interest. When the desired protein is expressed, it will be tagged with a bright green marker. In this way the protein can be followed. It can be determined if the protein is associated with the cell membrane, if it is ever transferred from one cell to another, what its fate is during embryonic development, and so on. The key is that the chromophore forms with no assistance, and so the fluorescence develops in almost any environment. In fact, green fluorescent animals have been prepared by transgenically incorporating GFP into the animal's genome. GFP-based assays for various cellular processes have been developed because the fluorescence is so easy to detect. Entire volumes have been written on the diverse ways this structure from a jellyfish has been put to good use. The 2008 Nobel Prize in Chemistry was awarded to Osamu Shimomura, Martin Chalfie, and Roger Tsien for their discovery and development of GFP.

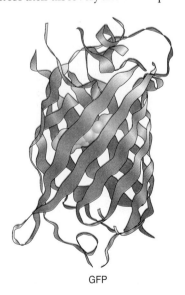

GFP

Biologists love this structure so much because it can be used to fluorescently tag a wide range of proteins expressed in living cells. One simply appends the gene

Cubitt, A. B., Heim, R., Adams, S. R., Boyd, A. E., Gross, L. A., and Tsien, R. Y. "Understanding, Improving and Using Green Fluorescent Proteins." *TIBS*, **20**, 448–455 (1995). Ormo, M., Cubitt, A. B., Kallio, K., Gross, L. A., Tsien, R. Y., and Remington, S. J. "Crystal Structure of the *Aequorea victoria* Green Fluorescent Protein." *Science*, **273**, 1392 (1996).

Table 16.4
Fluorescence-Related Properties of Selected Molecules*

Compound	k_f	k_{ST}	Nature of S_1
Naphthalene	2×10^6	5×10^6	$\pi,\pi*$
9,10-Diphenylanthracene	5×10^8	$<10^7$	$\pi,\pi*$
Stilbene	10^8	10^9	$\pi,\pi*$
1-Bromonaphthalene	10^6	10^9	$\pi,\pi*$
1-Iodonaphthalene	10^6	10^{10}	$\pi,\pi*$
Acetone	10^5	10^9	$n,\pi*$
Benzophenone	10^6	10^{11}	$n,\pi*$

*Adapted from an analogous table in Turro, N. J. (1978). *Modern Molecular Photochemistry*, The Benjamin/Cummings Publishing Company, Inc., Menlo Park, CA. k_f and k_{ST} are the rates in s^{-1} of fluorescence and singlet-to-triplet conversion, respectively.

Fluorescence is generally a fairly fast process. Table 16.4 shows properties of selected molecules, and we can see that fluorescence rate constants, k_f, are in the range of 10^8–10^5 s^{-1}, giving **fluorescence lifetimes** ($1/k_f = \tau_f$) in the range of 10^{-8}–10^{-5} s. These fluorescence lifetimes are the so-called **radiative lifetimes**, which are in the 100 μs to 10 ns range for most organic structures. **Inherent fluorescence lifetimes** are defined as the lifetime of the excited state if fluorescence were the only relaxation pathway. Just as the absorption process ($S_0 \rightarrow S_1$) is allowed for $\pi,\pi*$ but not for $n,\pi*$, the reverse process, fluorescence, should show a comparable rate dependence, and this is evident in Table 16.4. Emission from $\pi,\pi*$ states is more facile, producing k_f values typically in the 10^7–10^8 s^{-1} range, while emission from $n,\pi*$ states is typically slower, in the 10^5–10^6 range. Looking ahead, the longer lived an excited state is, the better the chance it can do productive photochemistry. For this reason, $n,\pi*$ states are often more photochemically reactive.

More so than absorption, fluorescence is sensitive to the chromophore's environment. Many fluorophores can show very substantial shifts in both λ_{max} of emission and fluorescence lifetime in response to changes in solvent or other environmental factors. When the change results from a variation in solvent, this is termed **solvatochromism** (the term is applicable to both absorption and emission). This can often be put to good use, in which λ_{max} of a fluorophore can be used to probe whether the local environment is, for example, relatively hydrophobic or hydrophilic. Look back at Table 3.1 to see several examples of solvent scales based upon solvatochromism. Examples of two typical compounds that show large solvatochromic effects are shown.

	Solvent	λ_{max} (emission, nm)
	Hexane	410
	Water	606
	Cyclohexane	455
	Water	549

Solvatochromic molecules

A clever and very powerful variant of this has been pioneered by Tsien and co-workers. A wide array of **fluorescence imaging** techniques are now available that allow one to directly monitor chemical processes in cells in real time. A good example of this is the collection of commercially available dyes exemplified by BAPTA and Fura-2. These are both derivatives of EGTA, a common chelating ligand. However, when BAPTA or Fura-2 bind Ca^{2+} they undergo a large increase in their fluorescence intensity. Since Ca^{2+} is a major signaling species in biology, these structures provide powerful tools for monitoring cellular activity.

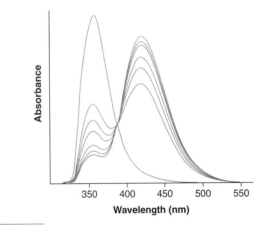

EGTA   BAPTA   Fura-2

Connections

Isosbestic Points—Hallmarks of One-to-One Stoichiometric Conversions

A change in absorption or emission spectra is often monitored as a function of time during a chemical reaction as a means of measuring kinetics. Furthermore, a change in absorption or emission is routinely monitored as a function of the concentration of a reactant to determine thermodynamic parameters, such as the equilibrium constant of a process. When one compound converts to another as a function of time, or converts to another as a function of concentration, the stoichiometry of the reaction is one to one, such as A goes to B. With such scenarios, it is often found that the probability of absorption or emission is coincidentally the same for the reactant and product at one or more wavelengths (with absorption this means the ε values are coincidentally the same). If the probability is the same at a particular wavelength, the absorption or emission at that wavelength will not change during the reaction. Such a point on the spectrum is called an **isosbestic point**. The observation of such a point is often used to support a one-to-one interconversion of reactant and product.

For example, shown to the right are the UV/vis spectra of 2-naphthol as a function of pH, where the spectrum with λ_{max} at 360 nm corresponds to the phenol and the spectrum with λ_{max} at 430 nm corresponds to the phenoxide anion. The isosbestic point at approximately 390 nm indicates that the phenol and phenoxide coincidentally have the same extinction coefficient at this wavelength, and that the deprotonation involves an interconversion between two separate species, the phenol and phenoxide.

Lawrence, M., Marzzacco, C. J., Morton, C., Schwab, C., and Halpern, A. M. "Excited-State Deprotonation of 2-Naphthol by Anions." *J. Phys. Chem.*, **95**, 10294 (1991).

16.1.5 Internal Conversion (IC)

The excited state S_1 can also return to S_0 without the emission of a photon. The excess energy is usually given off as heat to the medium. This energy wasting process is termed **internal conversion (IC)**. Conceptually, it is similar to a productive photochemical reaction, and so we delay further discussion of IC until we are ready to discuss reactions.

16.1.6   Intersystem Crossing (ISC)

A third way to exit S_1 is to form the triplet of the excited state, which is termed $\mathbf{T_1}$ (Figure 16.1). This process is **intersystem crossing (ISC)**, a general term for the interconversion of differing spin states. Before discussing ISC, a few comments about T_1 are in order. Just as with S_1, to the extent that T_1 is formed in a vibrationally excited state, radiationless vibrational relaxation occurs quickly. Note that T_1 is drawn as being at lower energy than S_1, and this is generally the case. The triplet state is lower in energy because of Hund's rule, reflecting a greater number of exchange interactions (see Section 14.5.6). The gap is typically larger for π,π^* states than for n,π^* states, meaning that the triplet is more stable relative to the singlet state for a π,π^* state. We can rationalize this as follows. In a π,π^* state, the two electrons are in the same π system, and so coulombic repulsions should be quite large in either the singlet or triplet state. However, the exchange interactions in the triplet state are quite large because the orbitals are very co-extensive. This leads to a large singlet–triplet gap. On the other hand, in an n,π^* state, the two electrons are in orbitals that are much less co-extensive in space, and so both the coulombic replusions and exchange interactions are smaller, making the singlet–triplet gap smaller. This is precisely analogous to our analysis of disjoint and non-disjoint biradical systems (Section 14.5.6).

Spin flipping in an organic molecule is typically not a rapid process. Unlike the $\pi,\pi^*/n,\pi^*$ distinction made above, ISC truly is quantum mechanically forbidden to first order because singlet and triplet wavefunctions are orthogonal. Given this, how do these transitions occur? One useful way to think about this is to remember that spin is associated with a spin angular momentum. Converting S_1 to T_1 involves a change in spin angular momentum, but nature desires a conservation of angular momentum in all processes. It is thus desirable to couple the spin flip with a compensating change in some other kind of angular momentum, and the most reasonable partner is orbital angular momentum. That is, ISC is favored by **spin–orbit coupling**, an interaction between orbital angular momentum and spin angular momentum. There are two ways to think about favoring a coupling of spin and orbital angular momenta in organic molecules.

The first gives rise to the **heavy atom effect** in organic photochemistry. As we move down the periodic table the mixing of spin and orbital quantum numbers for atoms is common. In fact, as we move down the periodic table, the distinction between spin and orbital quantum numbers becomes less and less meaningful, such that it is eventually better to just think in terms of a single spin–orbit quantum number. A practical consequence of this is that adding a heavy atom to an organic molecule can substantially enhance ISC rates. As shown in Table 16.4, the effects on the ISC rate, k_{ST}, can be significant, such that 1-bromo- and 1-iodonaphthalene undergo ISC much more rapidly than the parent naphthalene.

In molecules comprised only of C, H, N, and O, spin–orbit coupling is more difficult. There is, however, a strategy for increasing the efficiency of ISC, and it is embodied in **El-Sayed's rules**. These rules state that ISC is slow ("forbidden") when the two states being interconverted are both π,π^* or when both are n,π^*, since orbital angular momentum does not change. However, when we are converting a π,π^* state to an n,π^* state, or vice versa, ISC is more favorable ("allowed") because now the orbital angular momentum has changed.

Figure 16.7 **A** provides a rationalization of this effect. We start with a carbonyl chromophore and undergo an n,π^* excitation. We then convert S_1 to T_1 by flipping a spin and simultaneously moving the electron from a p_y orbital to a hybrid that has substantial p_x character. It is this change from p_y to p_x that constitutes the change in orbital angular momentum (recall the magnetic quantum numbers from Chapter 1: -1, 0, and 1). Thus, the change in spin angular momentum can at least to some extent be compensated by a change in orbital angular momentum, making the process more favorable. This is only possible, though, if our process is $^1(\pi,\pi^*) \rightarrow {}^3(n,\pi^*)$ or $^1(n,\pi^*) \rightarrow {}^3(\pi,\pi^*)$.

An important example of El-Sayed's rules in action concerns the ISC efficiencies of simple ketones. Figure 16.7 **B** summarizes the effect. For a typical ketone, S_1 and T_1 are n,π^*, while T_2 is π,π^*. Therefore, S_1 and T_1 interconversion is inefficient, but conversion to T_2 may be efficient. Let's examine some cases. For a dialkyl ketone, T_2 lies above S_1, and so ISC is inefficient (conversion to a higher energy state almost never occurs). However, aryl substitu-

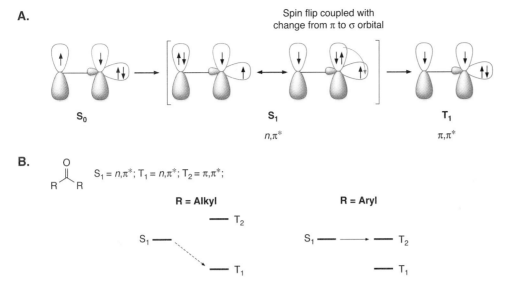

Figure 16.7
Intersystem crossing. **A.** An example of coupling a change in spin angular momentum with a change in orbital angular momentum on converting a $^1(n,\pi^*)$ state to a $^3(\pi,\pi)$ state. **B.** Dialkyl vs. diaryl ketones. ISC is more efficient for diaryl ketones because T_2, which is π,π^*, is comparable in energy to S_1, allowing ISC, consistent with El-Sayed's rules.

tion preferentially stabilizes the π,π^* T_2 state. As a result the S_1 to T_2 ISC is plausible, and indeed efficient, due to El-Sayed's rules. Typical ISC rates are 10^8 s^{-1} for dialkyl ketones and 10^{11} for diaryl ketones; two specific examples are given in Table 16.4. Once T_2 is formed, it quickly relaxes to T_1 (Kasha's rule). We will see below that the photochemistry of benzophenone and related structures is completely dominated by triplet states because of the efficiency of ISC. This feature has frequently been exploited in both chemical and biological contexts.

16.1.7 Phosphorescence

The conversion of an excited triplet state to the ground singlet state with the concomitant emission of a photon is termed **phosphorescence**. As with fluorescence, the phosphorescence spectrum of a molecule will generally be shifted to longer wavelengths relative to the absorbance spectrum. And, as Figure 16.1 suggests, the effect should be larger for phosphorescence than for fluorescence because T_1 is lower in energy than S_1. Also, vibrational fine structure can be seen in a phosphorescence spectrum that is informative of the ground state of the molecule (not explicitly shown in Figure 16.1). Note that if chemists wish to discuss emission in general terms, without distinguishing fluorescence vs. phosphorescence, they refer to **luminescence**.

Since phosphorescence is spin forbidden, just like ISC, it is much slower than fluorescence. What is remarkable is just how slow it can be. Reasoning similar to El-Sayed's rules applies, and so phosphorescence rates from n,π^* excited states are faster than from π,π^* states. Typical rates for the former are 10^2–10^3 s^{-1}; those for the latter are 10^{-1}–10 s^{-1}. That's right—phosphorescence lifetimes (inverse of the rate) can be on the order of seconds, even minutes! As a result, a phosphorescent sample can continue to glow even after the excitation source is turned off, something never seen with fluorescence. This phenomenon is sometimes seen in nature, and it is the basis for glow-in-the-dark Frisbees, toy dinosaurs, and the like. Note that "glow sticks" and other objects that can emit a bright light for quite a long time do not involve phosphorescence, but rather chemiluminescence, which we will discuss below.

In Figure 16.1, we also show radiationless conversion of T_1 to S_0, but this is not common. Similarly uncommon is direct absorption from S_0 to T_1. It is technically not impossible, just extremely inefficient, with ε values on the order of 10^{-2}–10^{-3}. In a few cases it has been useful for mechanistic studies, but it is generally not preparatively useful.

16.1.8   Quantum Yield

A key concept in photochemistry is that of the **quantum yield** for process n, Φ_n, defined as in Eq. 16.6. For example, the fluorescence quantum yield, Φ_f, is equal to the number of photons emitted divided by the number of photons absorbed. If all possible processes are considered, then the sum of all quantum yields should be 1.0 in typical systems. A formal exception to this rule is the case of a photochemically initiated polymerization, **photopolymerization**. In this case, if we define the quantum yield as the number of couplings between monomers divided by the number of photons absorbed, we expect to see $\Phi \gg 1$.

$$\Phi_n = \frac{\text{Number of molecules that undergo process } n}{\text{Total number of photons of light absorbed}} \qquad \text{(Eq. 16.6)}$$

Much of photochemistry is an exercise in relative rates, and that is certainly true when considering quantum yields. As we show in Eq. 16.7, Eq. 16.6 could be rewritten as a ratio of rate constants, where k_n is the rate constant for the process under consideration, and k_i are the rate constants for all possible processes from the excited state. In either case, the key is that we must consider all possible processes in the denominator in order to get the proper quantum yield for the process in the numerator. As the next Going Deeper highlight describes, internal conversion is a particularly important process competing with fluorescence.

$$\Phi_n = \frac{k_n}{\sum\limits_i k_i} \qquad \text{(Eq. 16.7)}$$

Pyrene 3-carboxaldehyde

The quantum yield is an indicator of how efficient a particular process is. However, some care must be taken in comparing quantum yields for different systems, because the quantum yield is always measured relative to other processes in the molecule. For example, Φ_f (f for fluorescence) is roughly 0.2 for both benzene and pyrene-3-carboxaldehyde. This might lead to the conclusion that fluorescence is equally efficient for the two compounds. However, the fluorescence *rates*, k_f, are 2×10^6 s^{-1} and 1×10^8 for benzene and pyrene-3-carboxaldehyde, respectively. Fluorescence is in a sense more efficient for the pyrene derivative. The reason this is not reflected in the quantum yields is that we must also consider competing processes. The ISC rate for pyrene-3-carboxaldehyde is also much faster than that for benzene due to the ability to access a $^3(n,\pi^*)$ state that is not available for benzene. This competing ISC process limits the amount of fluorescence, and by coincidence the two compounds end up with the same fluorescence quantum yield. Thus, while the quantum yield tells you about the efficiency of a process for a given molecule, it alone cannot tell you why the process is or is not efficient.

16.1.9   Summary of Photophysical Processes

We have discussed a number of factors that influence the transitions between various states in the Jablonski diagram, and it may be useful to summarize them here. We introduced the notion of "allowed" vs. "forbidden" transitions, but quickly added that these are not hard and fast rules, but rather guidelines for the relative efficiencies of various processes.

The factor that most strongly influences allowedness of a transition is spin. Transitions between states with different spins, such as singlet and triplet, are very inefficient. Direct S_0 to T_1 absorption is rarely important, and the reverse emissive process, phosphorescence, occurs quite slowly. The heavy atom effect can relax this selection rule through a spin–orbit effect. As a result, S–T interconversions are much more facile in molecules that contain atoms such as Br or I. The second major factor that influences the efficiency of transitions is the general spatial overlap of the wavefunctions for the two states. This term favors π,π^* transitions over n,π^* transitions, for reasons discussed above. Another important factor is a general **energy gap law**. For processes such as intersystem crossing, the smaller the gap between the

Going Deeper

The "Free Rotor" or "Loose Bolt" Effect on Quantum Yields

One interesting trend in fluorescence quantum yields nicely illustrates how all quantum yields reflect a competition among various rates. As shown to the right, the quantum yield for fluorescence is much higher for 9-methylanthracene than for 9-(t-butyl)anthracene. The electronic structures of these two molecules should be very similar for both the ground and excited states, so it is difficult to imagine a large difference in the fluorescence rates for the two. Instead, the big difference between the two is in the rate of internal conversion. Recall that internal conversion involves a radiationless transition from S_1 to S_0, and that the excitation energy must be "used up" in some manner. One way to do this is to convert the electronic excitation to vibrational/rotational excitation. We saw in Chapter 2 that bond rotations are quantized. In t-butyl-

anthracene there are many more low-lying vibrational/rotational modes to absorb the excess energy. As such, internal conversion is much more efficient than for the methylanthracene, and the fluorescence quantum yield is correspondingly lower. This is a fairly general phenomenon termed the **free rotor** or **loose bolt** effect. It is often seen that as substituents that have many degrees of vibrational/rotational freedom are added to a structure, the energy wasting internal conversion becomes more efficient at the expense of emissive or photochemical processes.

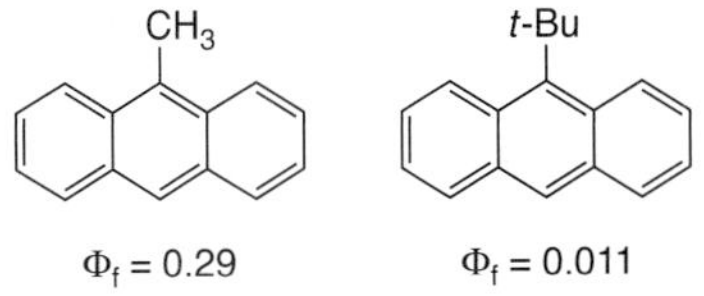

$\Phi_f = 0.29$ $\Phi_f = 0.011$

two states, the more rapid the process will be. We saw examples of this in the discussion of El-Sayed's rules, where factors that adjust the energy of the $^3(\pi,\pi^*)$ state of carbonyl compounds strongly influence intersystem crossing rates.

Finally, it is worth repeating that following absorption of a photon, many processes are possible. What we observe in the lab is a result of a competition of rates for these many processes. We can enhance the probability of observing a given process by either increasing its intrinsic rate constant, or by decreasing the rate constant of a competing process.

16.2 Bimolecular Photophysical Processes

16.2.1 General Considerations

We are now ready to consider more complex processes induced by the absorption of light. Before considering photochemical reactions, however, we will address the several things that can happen when a molecule in an excited electronic state encounters another molecule that is in its ground state, and related phenomena. These bimolecular photophysical processes are interesting and sometimes useful, and they provide further insights into the nature of the excited state.

First, though, let's consider the likelihood of a collision between an excited state molecule and another molecule. For such a collision to occur, the rate of collision must be competitive with the rate of decay of the excited state. The collision is a second-order process that depends upon the concentration of the molecule that the excited state is colliding with ([X]), while the decay is a first-order process. In a typical solvent at room temperature, the bimolecular diffusion rate constant, k_{diff}, is roughly 10^{10} M^{-1} s^{-1}. If we consider excited state lifetimes to be comparable to fluorescence lifetimes, we have unimolecular excited state decay rate constants, k_f, on the order of 10^8–10^5 s^{-1}. Thus, to compete with decay of the excited state (k_f similar to k_{diff}[X]), the colliding molecule must have a concentration in the range of 10^{-2}–10^{-5} M, depending on the precise excited state lifetime. More generally, the efficiency of such encounters will be concentration dependent.

16.2.2 Quenching, Excimers, and Exciplexes

When a molecule in an excited state collides with another molecule, many things can happen. For example, our excited state is constantly colliding with solvent molecules. These collisional processes facilitate the several radiationless relaxation processes of Figure 16.1.

Here, though, we are considering collisions with solute molecules that have chromophores of their own.

Quenching

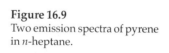

The most common outcome from a collision between a molecule in its excited state and another chromophoric or reactive molecule (apart from a photochemical reaction) is **quenching**, the collision-induced, radiationless relaxation of an excited state to the ground state. As noted above, the quenching process implies an interesting kinetic competition, the treatment of which is referred to as a **Stern–Volmer analysis** (Figure 16.8). Since the excited state lifetime can be determined both in the absence (τ_1) and presence (τ_2) of quencher (Q), the latter as a function of [Q], we can easily determine the quenching rate constant, k_q. We simply plot $1/\tau_2$ as a function of [Q], and the slope is k_q. The maximum possible value for k_q is the bimolecular diffusion rate constant, and any deviation from that value gives an indication of the efficiency of the quenching process. Often it is observed that k_q values for a given quencher with a given class of chromophores do not vary extensively. If so, more complex kinetic schemes can be analyzed because one potential unknown, k_q, is now "known". An example is given in the Exercises at the end of the chapter.

In general, quenching requires an intimate contact between the excited state molecule and the quencher. Extensive compilations of k_q values for various quenchers interacting with carbonyl (n,π^*) states have been produced. Many olefins and simple dienes are efficient quenchers, and an interesting relationship between the ionization potential (IP) of the quencher and k_q has been observed. Both electron rich (low IP) and electron poor (high IP) π systems are effective quenchers, while "normal" alkenes and dienes are less effective. As we will see below, the mechanism of quenching can be viewed as a photochemical reaction that, instead of producing product, returns to starting material.

Excimers and Exciplexes

Quenching is a radiationless process involving two molecules. There are also radiative processes involving two molecules. These are necessarily cooperative processes involving non-covalent molecular complexes. If two molecules act cooperatively to *absorb* a photon, an **absorption complex** is involved. If two molecules act together to *emit* a photon, an **exciplex** (electronically excited complex) is involved. The particular case of an exciplex in which the two molecules are the same is termed an **excimer** (electronically excited dimer).

With either an absorption complex or an excimer/exciplex the absorption/emission band associated with the complex will be at a different and *longer* wavelength than that associated with either individual molecule. The intensity of the new absorption/emission will necessarily depend on the concentrations of both species involved. For example, two emission spectra for pyrene are shown in Figure 16.9. In dilute solution the complex between

Figure 16.8
Stern–Volmer quenching kinetics.

$$A^* \xrightarrow{k_1} A$$
Lifetime of A^* without Q = $\tau_1 = 1/k_1$

$$Q + A^* \xrightarrow{k_q} A$$
Lifetime of A^* with Q = τ_2

$$1/\tau_2 = k_1 + k_q[Q] = 1/\tau_1 + k_q[Q]$$

Pyrene

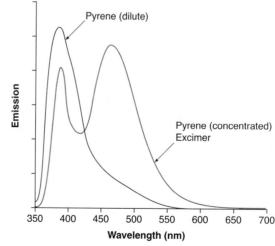

Figure 16.9
Two emission spectra of pyrene in *n*-heptane.

two pyrene molecules is not formed. At higher concentration, a dimer of pyrene molecules forms, and the UV/vis emission spectrum shows a new emission band associated with the excimer. Note that absorption complexes, when they absorb a photon, do not necessarily produce an exciplex, and exciplexes, when they emit, do not necessarily produce an absorption complex.

An excited state of a molecule (A*) can collide with another molecule (B), giving rise to quenching, or the formation of a stable complex that can emit a photon (exciplex or excimer). Excited states are expected to be highly polarizable species, because excitation, in a sense, breaks a bond (excitation from a bonding to a nonbonding orbital). As such, two electrons are not tightly held in bonds. This increased polarizability promotes the formation of a weak complex via dipole–induced-dipole interactions or London dispersion forces. Additionally, donor–acceptor interactions of the sort discussed in Section 3.2.4 are possible, and these should also stabilize the exciplex or excimer. Usually the excited state complex remains as a singlet before relaxation, but sometimes an excited state will undergo ISC faster than back electron transfer. **Back electron transfer** is a general term given to any system where relaxation occurs by electron transfer from the acceptor back to the original donor. After back electron transfer occurs in a triplet exciplex or excimer, the first excited state triplet of one of the components is created. Hence, the formation of a excimer or exciplex leads to preferential stabilization of the excited state, but when the exciplex or excimer emits, both fluorescence and phosphorescence are possible, and the emission will be at considerably longer wavelengths than for the isolated excited molecule.

The most common type of absorption complex is associated with **charge-transfer absorption**, also know as electron **donor–acceptor absorption**. The donor (D) is a molecule with a low ionization potential; the acceptor (A) has a high electron affinity. The two molecules can associate in the ground state, and this donor–acceptor complex gives a broad, intense absorption band shifted far from the individual absorbances (look back to Section 4.2 to review this phenomenon in the context of molecular recognition). In fact, the charge-transfer absorption is often in the visible range of the spectrum, making the charge-transfer complexes colored. One way to think of these systems is that the ground state is a weak, intermolecular complex, (D•A). Often such molecules are polar and/or quite polarizable, facilitating complex formation. Because one partner is intrinsically a good electron donor (low ionization potential) and the other is a good acceptor (high electron affinity), the excited state will have a large contribution from states such as (D⁺•A⁻) that involve electron transfer, a configuration not possible for either isolated molecule. This preferentially stabilizes the excited state, leading to a lower energy (longer wavelength) absorption. Because of this highly polar character, charge-transfer absorptions are extremely sensitive to solvent polarity, moving to longer wavelengths as the solvent polarity increases. Such complexes most commonly undergo relaxation by back electron transfer.

Photoinduced Electron Transfer

Photoinduced electron transfer (PET) is a common process that can lead to quenching of fluorescence. Because an excited state has an electron in an antibonding orbital (often the LUMO), this state has a significantly different oxidation potential than the ground state. It is easier to oxidize an excited state because the electron to be removed is in a higher energy orbital, and the excited state is considered to be a good donor. Furthermore, because there is a vacancy in the lower energy orbital from which the excited electron was removed, the excited state is easier to reduce and thus a better acceptor than the ground state. Therefore, the excited state has both a lower oxidation and a lower reduction potential. This means that excited states can undergo electron transfers, accepting electrons from ground states that have a higher HOMO and transferring electrons to ground states that have a lower LUMO. Both or either process would lead to quenching of the fluorescence. This is shown schematically in Figure 16.10. In part **A** the excited state transfers an electron to the ground state of another molecule, while in part **B** a ground state molecule transfers an electron to the excited state. Both scenarios are examples of photoinduced electron transfers.

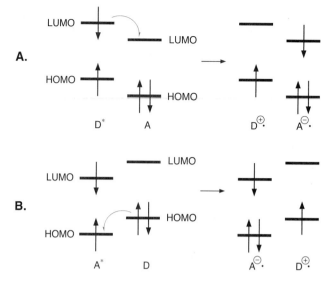

Figure 16.10
Photoinduced electron transfer scenarios. **A.** Example where the LUMO of the excited state is above the LUMO of the ground state of the acceptor. **B.** Example where the HOMO of the excited state is below the HOMO of the ground state of the donor.

16.2.3   Energy Transfer I. The Dexter Mechanism—Sensitization

We've seen that bimolecular processes involving excited states can take many forms. Collision can facilitate relaxation to the ground state (quenching) or formation of an excited state complex (exciplex or excimer). Alternatively, bimolecular association can occur prior to excitation, leading to an absorption complex. In this and the next two sections we consider a new outcome for the interactions of an excited state, D*, with another molecule, A (Eq. 16.8). Now the result is energy transfer from one molecule to another, producing electronically excited A (A*). Different mechanisms are possible, and these energy transfer processes are very important in photochemistry and other fields.

$$D^* + A \longrightarrow D + A^* \tag{Eq. 16.8}$$

The most trivial form of energy transfer is called **radiative transfer**, and occurs when one molecule emits a photon and another absorbs that photon. This can occur when there is a substantial overlap of the emission spectrum of D* and the absorption spectrum of A. This is not a particularly common event, given that the intensity of emission is low. Another form of energy transfer occurs in the solid state, called **exciton migration**. Here, the excited state energy migrates rapidly between nearest neighbors because of the close promixity of reacting D* and A partners in a packed organized medium.

The first mechanism we consider in detail is a relatively direct process involving collision between D* and A, and it is called **collisional energy transfer**. Such a collision produces a direct interaction of the wavefunctions of D* and A, and it leads to the notion of an **electron exchange** formalism (see Figure 16.11). Dexter developed a detailed model for the process, producing Eq. 16.9 to describe the electron exchange energy transfer rate, k_{ee}. Here, K is related to the magnitude of a specific orbital interaction that promotes the electron exchange. The quantity J is a spectral overlap integral, describing the extent to which the absorption spectrum of the donor (D) overlaps the absorption spectrum of the acceptor (A). A subtlety is that J is normalized for the extinction coefficient of A (ε_A). As such, k_{ee} does not depend on ε_A, a point we will return to below. The key feature of Eq. 16.9 is that k_{ee} depends exponentially on the distance between D and A, r_{DA}, with L being the sum of the van der Waals radii.

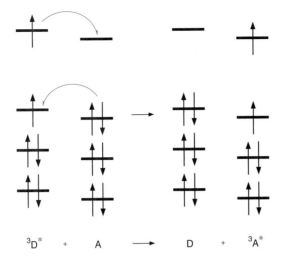

$$^3D^* \quad + \quad A \quad \longrightarrow \quad D \quad + \quad ^3A^*$$

Figure 16.11
The Dexter or electron exchange mechanism of energy transfer.
As discussed in the text, this is a highly schematic diagram.

As such, the efficiency of electron exchange energy transfer falls off steeply as the separation between D and A increases.

$$k_{ee} = KJe^{\frac{-2r_{DA}}{L}}$$

(Eq. 16.9)

Given the steep distance dependence, in fluid media this mechanism effectively requires collision between D* and A. However, energy transfer is also often employed in systems in which D* and A are held in a fixed position relative to each other. This can be achieved by embedding the two in a rigid environment such as a frozen solvent or a polymer matrix. Alternatively, the two chromophores could be attached to the same large molecule at a fixed distance. In either case, the efficiency of energy transfer will strongly depend on distance. For the electron exchange mechanism considered here, the exponential dependence of the efficiency on r ensures that energy transfer is only efficient over fairly short distances, on the order of $r = 5$–10 Å. If not a direct collision, we need the two structures to be very near each other.

The most useful form of electron exchange energy transfer involves the triplet state of D*. Since the energy transfer must conserve spin, the product is the triplet state of A*. Figure 16.11 shows a *highly* schematic way of thinking about such a process. It can be seen that the idea of electron exchange does allow a natural conversion of $^3D^*$ to $^3A^*$. However, these are quantum mechanical phenomena, and care should be taken not to interpret Figure 16.11 too literally; it is just a useful mnemonic.

When $^3D^*$ is used to produce $^3A^*$ through energy transfer, the process is known as **sensitization**. Compound D is a sensitizer used to produce the triplet state of A. This can be a very useful process. As discussed above, there is considerable variation in the efficiency of ISC for different molecules. For many compounds, excitation does not produce any significant quantity of T_1, because other processes such as fluorescence are too fast. But what if we want to investigate the photochemistry of the triplet state of such a system? As we will see later in this chapter, singlet and triplet excited states often lead to quite different photoproducts, and perhaps the products we want are those expected from triplet photochemistry. This is where sensitization comes in handy. We begin with a molecule with a very efficient ISC, such as benzophenone. We excite the benzophenone (the **sensitizer**), and allow it to collide with the molecule whose triplet state we wish to produce. Energy transfer occurs, and the desired triplet state is produced. This is a common and very useful technique in photochemistry.

Table 16.5
Triplet Energies of Common Sensitizers

Sensitizer	Triplet energy (kcal/mol)
Acetophenone	73.6
Benzophenone	68.5
Anthraquinone	62.2
Biacetyl	54.9

Table 16.5 shows some common sensitizers along with their **triplet energies** (the energy gap between T_1 and S_0). It must be true that the triplet energy of the sensitizer is greater than or equal to the triplet energy of the acceptor for the energy transfer to be efficient. Thus, we cannot use biacetyl to sensitize a molecule that has a triplet energy of 70 kcal/mol.

Any sensitization must obey what is called **Wigner's spin conservation rule**. This rule states that total spin must be conserved, which can be envisioned by fixing the spin quantum numbers of electrons as $^1/_2$ and $-^1/_2$ while they exchange positions. Singlet sensitization by energy transfer from another singlet is certainly acceptable, as is triplet sensitization from another triplet. However, triplet energy transfer to another triplet can produce two singlets (we will see this later with energy pooling). Try Exercise 34 at the end of the chapter to see how this occurs.

16.2.4 Energy Transfer II. The Förster Mechanism

We now consider an alternative and very important energy transfer mechanism. In this mechanism no collision is necessary between D* and A. Instead, a surprisingly long distance interaction develops, leading to a novel and very useful effect. The theoretical analysis of this mechanism was developed by Förster, and it is referred to as the **Förster energy exchange** or **coulombic energy exchange** mechanism.

We begin with Figure 16.12, another highly schematic diagram, but one that highlights the difference between the Dexter and Förster mechanisms. The Förster mechanism is more typically associated with singlet states, and so Figure 16.12 is shown with singlet energy transfer. Note that this is *not* radiative transfer. The Förster mechanism does not operate this way; no photon is emitted.

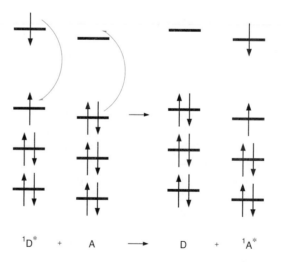

$$^1D^* \quad + \quad A \quad \longrightarrow \quad D \quad + \quad ^1A^*$$

Figure 16.12
A schematic of the Förster energy transfer mechanism. The colored arrows are meant to represent the coupled transitions that occur simultaneously.

Just as the absorption of a photon requires a quantum mechanical coupling of the initial and final states induced by light, so the Förster mechanism requires a coupling of two *transitions*, from D* to D and A to A*. Recall that a transition probability is proportional to a transition dipole, D (Eq. 16.3). The Förster mechanism is proportional to the interaction of two transition dipoles; it is therefore a **dipole–dipole coupling**. We have discussed dipole–dipole interactions and couplings before in the context of molecular dipoles and molecular recognition (see Section 3.2.2 and Eq. 3.25). As with any dipole–dipole coupling, the interaction energy, E, is proportional to the product of the dipoles divided by the distance cubed, r^3 (Eq. 16.10). We are also interested in the rate constant of the energy transfer, and that is proportional to E^2 (Eq. 16.11).

$$E \propto \frac{D_D D_A}{r_{DA}^{\,3}}$$
(Eq. 16.10)

$$k \propto E^2 \propto \frac{D_D^{\,2} D_A^{\,2}}{r_{DA}^{\,6}}$$
(Eq. 16.11)

The important difference between the Förster and the Dexter mechanisms is that the Förster mechanism does *not* require a direct overlap of the wavefunctions of D* and A. It is simply a coupling of transition dipoles. There is, though, a requirement for an energy matching between the two states. A will get excited by an energy equal to the energy lost on D* going to D. This energy matching is similar to that required between a photon to be absorbed and the gap between initial and final states, and so is referred to as a **resonance condition** (as in nuclear magnetic resonance, for example). This requires some overlap between the emission spectrum of D* and the absorption spectrum of A; the larger the overlap the more likely the energy transfer (see Figure 16.13). Because the Förster mechanism does not require physical overlap of the wavefunctions, energy transfer by this mechanism can occur over much larger distances than by the Dexter mechanism. Indeed, energy transfer by the Förster mechanism can occur over quite long distances. This is the foundation of FRET, the most important implementation of Förster energy transfer.

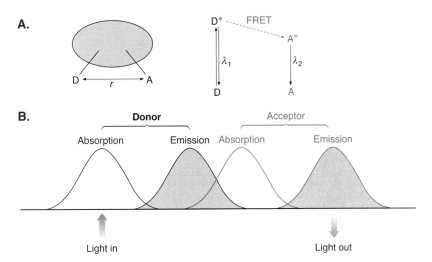

Figure 16.13
Schematic of the FRET process, an example of Förster energy transfer. **A.** In one application a large biomolecule is represented as a gray oval. Attached to it are the donor (D) and acceptor (A) chromophores. The goal is to measure the distance r. On the right the competing emission processes and the energy transfer are shown. As described in the text, the relative intensities of emission at λ_1 and λ_2 can lead to a determination of r. **B.** The necessary relationship between the absorption/emission characteristics of the donor and those of the acceptor. The emission spectrum of the donor must overlap the absorption spectrum of the acceptor.

16.2.5   FRET

A powerful application of Förster energy transfer is **fluorescence resonance energy transfer (FRET)**. This phenomenon is used extensively in probing biological structures, in studying interactions at interfaces, and in creating sensors. Although rarely described this way in the literature, FRET is simply an application of Förster energy transfer. The basic approach is summarized in Figure 16.13 **A**. A large biological structure is labeled with two fluorescent groups, which we will refer to as D and A. D is excited, and it can fluoresce at its characteristic wavelength, λ_1. However, if A is close enough to D, Förster energy transfer can occur, producing the excited state of A. Now A* can emit at its characteristic wavelength, λ_2. Recalling the distance dependence of Förster energy transfer, we can anticipate that the closer together D and A are, the more efficient FRET will be. Typically we consider the ratio of donor fluorescence in the presence or absence of acceptor, and 1 minus that ratio is the FRET efficiency, E_f. In fact, E_f will depend on a ratio of *rates*—the k_f value for D*, which is a constant for a given system, and the FRET rate constant. As noted above, the rate constant for FRET is proportional to $(1/r)^6$, and so E_f also depends on $(1/r)^6$. As such, the fluorescence intensity ratio for the two chromophores is a very sensitive measure of the distance between the two. For typical chromophores E_f can serve to distinguish distances in the 20–80 Å range, a very convenient range for biomolecules. Shorter separations lead to essentially complete FRET, while longer distances lead to no observable FRET. Note that, as shown in Figure 16.13 **B**, FRET can produce a very large gap between the wavelengths of absorption and emission.

There are, in fact, several complications associated with all FRET measurements. These are summarized by Eqs. 16.12 and 16.13, which express the distance dependence in terms of a parameter r_0. r_0 is the **Förster radius**, and it can be thought of as the distance at which 50% of the donor is deactivated by FRET. It is an invariant characteristic of a given pair of chromophores. For example, the fluorescein•••tetramethylrhodamine pair shown in the margin, which is commonly used in labeling biomolecules, has an r_0 value of 55 Å. The value of r_0 is given by Eq. 16.13. The various terms are as follows. The term Φ_D is the fluorescence quantum yield of D in the absence of A. It makes sense that the more efficient the fluorescence of D, the better it can act as a FRET donor. The $J(\lambda)$ is the spectral overlap of the emission of D* and the absorption of A, an important parameter we noted above in our discussion of Förster energy transfer. This J is different than that used in Dexter energy transfer (Eq. 16.9). The Dexter J is the overlap of two absorption spectra; the Förster J is the overlap of an emission spectrum with an absorption spectrum. The Φ_D and $J(\lambda)$ terms usually do not present severe problems in interpreting FRET results, especially when one is comparing a series of FRET efficiencies at varying distances for the same pair of fluorophores.

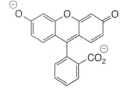

Fluorescein

Tetramethylrhodamine

$$E_f = \frac{1}{\left[1 + \left(\dfrac{r}{r_0}\right)^6\right]} \qquad \text{(Eq. 16.12)}$$

$$r_0^6 = 8.8 \times 10^{-28} \bullet \Phi_D \bullet \kappa^2 \bullet n^{-4} \bullet J(\lambda) \qquad \text{(Eq. 16.13)}$$

The other two terms in Eq. 16.13, κ and n, provide a more serious challenge in interpreting FRET data. The n is the refractive index of the medium. This is a simple matter when only bulk solvent intervenes between D and A, but consider the more relevant situation symbolized in Figure 16.13 **A**. The intervening medium can be a protein or nucleic acid, or a material such as a polymer or dendrimer. It is not at all obvious what to assign as the refractive index of the intervening medium in such a case. Again, for a series of measurements all involving the same biomolecule or the same material, this may not be a problem, but it is an important issue.

The final term, κ, is the orientation term. As discussed in Section 3.2.2, the interaction of two dipoles shows a $3\cos^2\theta - 1$ orientation dependence, and that is just what is contained in the κ term. Using simple math, one can establish that κ^2 ranges from 0 to 4, depending on the orientation of the two transition dipoles, and so this can have a significant effect on FRET efficiency. In cases where the fluorophores experience flexible reorientation during the fluorescence lifetime of D*, κ^2 averages to $\frac{2}{3}$, and absent any other information this is the value that is often assumed. In many applications of FRET we are interested only in relative values in closely related systems, and such simplifications are acceptable. However, in other cases in which quantitative interpretation of FRET efficiencies is desired, the uncertainties in n and especially in κ can cloud the analysis.

Connections

Single-Molecule FRET

Recent advances in optics and light sources have made it increasingly possible to observe fluorescence spectra on *single molecules*. In fact, for simply observing the presence of a fluorescent molecule, relatively simple optics can detect a single molecule. With sophisticated optics, real-time kinetics using FRET can be performed on single molecules. We highlight an example here.

Shown below is a schematic of an enzyme and its substrate. In this particular example, the enzyme is a ribozyme and the substrate is an oligonucleotide, but that is not essential here. After binding of the substrate, the enzyme undergoes a conformational change, converting from the so-called undocked to the docked state, and that is what is being probed by FRET. In this study, the enzyme was first immobilized by attaching it to a surface through a biotin–streptavidin complex. Such immobilization is very helpful in single-molecule studies, as it literally prevents the molecule from moving out of the field of visualization during the experiment. Second, a donor chromophore was attached to the region that must move in the docking, while an acceptor chromophore was placed near the site of immobilization. Structural studies indicated that the distance between the two chromophores changes

from ~70 Å to 10–20 Å on going from undocked to docked, the ideal distance range for FRET. Finally, the donor was excited, and the fraction of emission from the acceptor (indicating FRET was occurring) was monitored as a function of time. The plot of emission intensity from A (I_A) relative to intensity of both A and D versus time shows the *real-time* observations from a single molecule. Only two states are observed: the undocked state in which ~30% of the emission comes from the acceptor, and the docked state in which ~90% comes from the acceptor. In this trace, we are watching a single molecule as it undergoes a conformational change, interconverting two forms. We can directly measure the lifetimes of the two conformations. Averaging over many observations gives the average lifetime of each state, and the inverses of the lifetimes give the rate constants k_{dock} and k_{undock} given below. The ratio of these two rates (~7) is the equilibrium constant between the docked and undocked states. Thus, very valuable information can be obtained from such single-molecule studies.

Zhuang, X., Bartley, L. E., Babcock, H. P., Russel, R., Ha, T., Herschlag, D., and Chu, S. "A Single-Molecule Study of RNA Catalysis and Folding." *Science*, **288**, 2048 (2000).

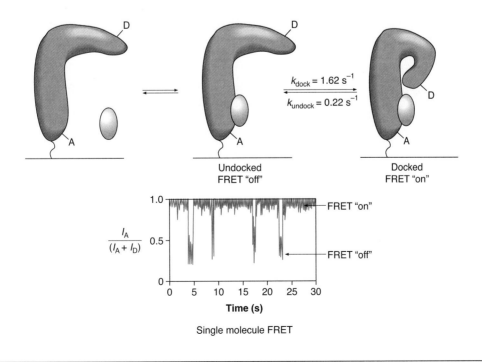

Single molecule FRET

16.2.6 Energy Pooling

Energy transfer can also occur when two excited states collide, and this is known as **energy pooling**. The term "pooling" is used because the excited state energy in one molecule is transferred to the other, pooling into only one molecule. However, the probability of a bimolecular reaction between two excited states is low, because these states are typically in very low concentrations. This makes the process essentially impossible for singlet states because their lifetimes are so short. However, for triplet states bimolecular processes can be important. When two triplets collide, one is the energy acceptor (A_t^*) and one the donor (D_t^*), and the collision can give an excited singlet state and a singlet ground state (A_s^* and D_s in Eq. 16.14), which obeys Wigner's spin conservation rule. Because all the energy is transferred to one of the reactants, it will be formed in a singlet excited state typically higher than S_1 ($S_{n>1}$ states are created). Kasha's rule predicts relaxation to S_1, and very often emission from there is observed. The resulting fluorescence occurs much later from the time of initial absorption than normal fluorescence due to the fact that it resulted from collision of two long lived triplet states, and hence it is called **delayed fluorescence**.

$$A_t^* + D_t^* \longrightarrow A_s^* + D_s \qquad \text{(Eq. 16.14)}$$

16.2.7 An Overview of Bimolecular Photophysical Processes

We have seen a number of processes in this section, and here we briefly summarize and contrast them. Collision between an excited state molecule and a ground state molecule frequently leads to quenching, a nonradiative relaxation to the ground state. Alternatively, collision between an excited state molecule and a ground state molecule can produce an exciplex or excimer, a non-covalent complex that emits light at a longer wavelength than is possible for either individual molecule. A less common scenario involves prior association of two molecules, with the resulting absorption complex absorbing light at longer wavelengths than either individual molecule.

Energy transfer is an important bimolecular photophysical process, and it can occur by one of two very different mechanisms. The Dexter or electron exchange mechanism is typically associated with a collision between the energy donor, D, and the acceptor, A. In its most useful form, the triplet state of D is used to produce the triplet state of A by energy transfer, a process known as sensitization. The Förster or coulombic energy exchange mechanism requires a coupling of transition dipoles and operates over longer distances. It is the basis of the powerful tool of FRET, which has been used to obtain geometric information about complex molecular systems.

The key differences between the Dexter and Förster mechanisms are as follows. The efficiency of the Dexter mechanism depends exponentially on the separation between D and A, and it is thus only valuable over relatively short distances (up to 10 Å). The efficiency of the Förster mechanism (FRET) varies as r^{-6}, and it is thus viable up to 80 Å or so, a very useful range for studies of macromolecular systems. Also, the efficiency of the Dexter mechanism does not depend on the efficiencies of the transitions involved. In contrast, the efficiency of the Förster mechanism depends on the efficiencies of both the $D^* \rightarrow D$ and the $A \rightarrow A^*$ transitions, with the latter generally being the more important term.

16.3 Photochemical Reactions

16.3.1 Theoretical Considerations—Funnels

A thorough theoretical analysis of photochemistry is a challenging task. Excited states are intrinsically complicated, and treatment of a photochemical reaction requires the consideration of the *interactions* of at least two electronic surfaces, the ground and excited states. It is unfortunately beyond the scope of this text to fully develop the present understanding of this area, which is considerable but still incomplete. Here, we highlight some of the key concepts and provide a sense of what the important issues are. For the student wishing to delve deeply into the present state of theoretical organic photochemistry, the text by Michl and Bonacic-Koutecky cited at the end of the chapter is an excellent but challenging starting point.

Diabatic Photoreactions

The most common scenario for a photochemical reaction is schematized in Figure 16.14. The reaction is termed **diabatic** because two potential energy surfaces are involved in the photoreaction (we will see an adiabatic, single surface reaction below). We are taking lessons from the Jablonski diagram, which simply shows geometrical differences between ground and excited states, and introducing reactions for the ground and excited states. We, therefore, show two reaction coordinates in this diagram, one for the ground state and one for the excited state. In Figure 16.14, excitation from a stable structure on the S_0 energy surface leads to an unstable structure on the S_1 energy surface. The first formed structure on the S_1 surface moves toward a minimum on this excited state surface. In favorable cases for photochemistry, the geometry that corresponds to a minimum on S_1 corresponds to a maximum on S_0, bringing the two surfaces very close together. This small energy gap between surfaces favors crossing from one surface to another, and so the molecule "hops" from S_1 to S_0. That is, the electronic state of the system changes (from S_1 to S_0), but the geometry stays the same. This is not too different from the other types of transitions between states that we have seen. In this case, the energy gap law comes into play, such that the smaller the energy gap between the two states, the more efficient the "hop".

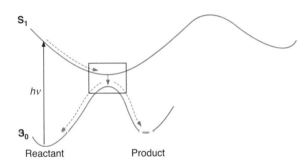

Figure 16.14
Schematic surface diagram for a diabatic photoreaction. Excitation is followed by a geometry change on S_1 toward the funnel region, designated by a box in the figure. At the minimum on S_1, relaxation to S_0 occurs. Depending on the precise nature of this jump from one surface to another, the photochemistry can be productive (producing product) or non-productive (reforming starting material).

What happens next depends on the precise geometry of the system. The imagery of Figure 16.14 is that if the system, once it returns to S_0, evolves to the "left", we return to the reactant structure; no photochemistry has occurred. This process is identical to internal conversion, number 4 in the Jablonski diagram of Figure 16.1. On the other hand, if the system evolves to the "right", a new product is formed; photochemistry has happened. Thus, the key to photochemistry is the precise conformation or geometry of the molecules in the region where S_0 and S_1 are close in energy. Because of its shape in the two-dimensional representation of Figure 16.14, this region has been termed a **funnel**. To see this, envision a three-dimensional plot of the potential energy surface, as described in Chapter 7, where one dimension is the reaction coordinate, the other dimension is a vibration, and the last dimension is energy. When the excited state surface approaches the ground state surface in energy, a funnel that empties on to the ground state surface is formed. The term **conical intersection** is also used. Note that photochemistry is one area where the Born-Oppenheimer approximation can break down. Nuclei can be moving rapidly as S_1 approaches the funnel, and dynamic effects of the sort discussed in Section 7.2.7 can be important.

Funnels or conical intersections play a crucial role in photochemistry. They provide an efficient exit point from the excited state to the ground state, and all photochemical reactions must end up back on the ground state surface. In addition, the precise geometry of the funnel determines whether the photochemistry is efficient—that is, whether it tends to produce product (exits to the "right"). What kind of geometries should be conducive to funnel formation? That is, what structures have a very small gap between the S_0 and S_1 surfaces? This is best addressed by considering a very simple, but prototypical photochemical reaction, the cis–trans isomerization of olefins.

Figure 16.15 presents a model for the photochemical cis–trans isomerization of an olefin. First consider thermal cis–trans isomerization. We start at the trans form, rotate 90° to a transition state structure with a great deal of biradical character, and then continue on to the cis form. Now consider the π,π^* state of a simple olefin. The π bond is now "broken", and there is no reason for it to stay planar. In fact, steric effects and electron repulsion effects favor the twisted form, and that is the minimum on S_1 and T_1. We have a situation in which a geometry that is a maximum on S_0 is a minimum in the excited state, and this is ideal for funnel formation. For direct irradiation of olefins, the S_1 and S_0 states are close enough that fairly efficient hopping from S_1 to S_0 can occur. If we add a sensitizer to the mix, T_1 of the olefin is formed, and this can actually be lower in energy than S_0 at the twisted geometry. Now a relatively stable, triplet biradical intermediate may lie on the cis–trans isomerization pathway.

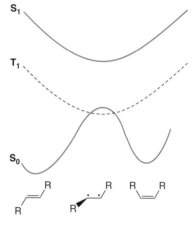

Figure 16.15
Schematic potential energy surface for olefin cis–trans isomerization. The biradical structure that is a maximum on S_0 is a minimum on S_1 and T_1.

We can generalize the scheme of Figure 16.15 to some extent. First, biradicals and biradical-like structures will often be quite unstable on S_0—after all, they represent bond-broken structures. However, excited states often have considerable biradical character. This is especially true of triplet states, but it is also true of some excited singlet states. We will see examples of both in the coming sections when we look at specific reactions. Thus, funnels will often be associated with structures that have considerable biradical character.

There are variations on the diabatic process. One important feature is that some reactions will have a small barrier on S_1 that separates the initial excited state geometry from the funnel geometry. This can adversely affect photochemical efficiency and produce temperature dependent quantum yields. Still, the basic idea of finding geometries in which the excited state and ground state are close in energy is central to photochemistry.

Other Mechanisms

While the diabatic mechanism we just discussed is typical of photochemical processes, there are some other less common yet interesting paths, summarized in Figure 16.16. In an **adiabatic** reaction (Figure 16.16 **A**), the conversion from reactant geometry to product geometry occurs on just one surface, the excited state surface. This is then followed by relaxation back down to the ground state. This relaxation could in principle be emissive, such that we would excite the reactant and see fluorescence from the product.

Another possible path is a so-called **hot ground state** reaction (Figure 16.16 **B**). Excitation is followed by internal conversion back to the ground state surface. However, before the energy deposited into vibrational excited states can be lost via collisions with the solvent, it is used to initiate a *thermal* reaction on the ground surface. The products and all other effects are just as in the thermal reaction. We have simply used light to deliver the heat. Again, this mechanism is uncommon, but it should be considered in mechanistic investigations.

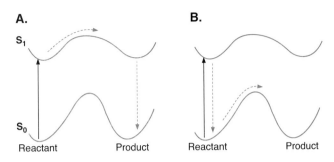

Figure 16.16
Examples of alternative photochemical mechanisms. **A.** An adiabatic photoreaction. **B.** A hot molecule reaction.

We now present an overview of the basic types of organic photochemical reactions. We begin with acid–base reactions, and then turn to reactions of hydrocarbon π systems, such as olefin isomerizations, cycloaddition reactions, and the di-π-methane rearrangement. We then study "heteroatom" photochemistry, the photoreactions of carbonyls and nitrogen-containing chromophores.

16.3.2 Acid–Base Chemistry

One of the simplest processes that can be induced photochemically is acid–base chemistry. The pK_a values of excited states often differ considerably from those of the ground state. For example, 2-naphthol is a much stronger acid in the excited singlet state than in the ground state (pK_a values are 3 and 9, respectively). In other words, upon excitation 2-naphthol will ionize to a greater extent in a pH range between 3 and 9, where normally this compound does not ionize significantly in this pH range (Eq. 16.15). In contrast, 2-naphthylamine is a much poorer base in the excited state than in the ground state (pK_a values of the conjugate acids are –2 and 4, respectively; Eq. 16.16). Upon excitation, protonated 2-napthylamine will lose its proton to a greater extent at pH values between –2 and 4 than will the ground state. These are dramatic effects that can substantially influence excited state reactivity patterns. In the Exercises, you are given the opportunity to rationalize these trends.

$$\left[\begin{array}{c} \text{OH} \end{array} \right]^* \xrightarrow{H_2O} \left[\begin{array}{c} O^{\ominus} \end{array} \right]^* + H_3O^{\oplus} \qquad \text{(Eq. 16.15)}$$

$$\left[\begin{array}{c} \overset{\oplus}{N}H_3 \end{array} \right]^* \xrightarrow{H_2O} \left[\begin{array}{c} NH_2 \end{array} \right]^* + H_3O^{\oplus} \qquad \text{(Eq. 16.16)}$$

16.3.3 Olefin Isomerization

A prototypical photochemical reaction that was alluded to above is cis–trans olefin isomerization. While in principle simple olefins like 2-butene are amenable to photochemical isomerization, inspection of Table 16.2 should convince you that this will be possible only with very short wavelengths and thus very high energy light. Most labs are not set up to handle irradiations using light of $\lambda < 200$ nm, a region of the electromagnetic spectrum called the far UV or vacuum UV, and so this is not a viable reaction. With simple alkenes, excitation below 200 nm produces a π,π^* state. However, there is also a weak absorption called a **Rydberg transition**, leading to what is known as a **Rydberg state**. Rydberg states have very weakly held electrons in an exceedingly high energy orbital. With alkenes, the Rydberg state has considerable $3s$ character on the carbons, denoted as $(\pi,3s)Ry$.

To make olefin isomerization a practical reaction that can be studied with conventional excitation sources, one must add additional groups in conjugation with the olefin to make a useful chromophore, as shown in Eq. 16.17. Addition of a phenyl ring is also a common strategy. Furthermore, oligo-alkenes absorb visible light and isomerize, which is the molecular basis of vision (see the Connections highlight on page 968).

$$\text{(Eq. 16.17)}$$

An important factor in alkene photoisomerization is the fact that both the starting alkene (for example, the cis olefin) and the product (the trans olefin) are likely to be photochemically reactive. We might start with pure cis, irradiate for while to generate trans, but then the trans will start absorbing light and revert to cis. In rare cases it may be possible to find a wavelength of light that is absorbed by only one isomer, allowing us to drive the reaction in one direction. More typically, though, the absorption spectra of the isomers overlap extensively, but they certainly are not identical. Extensive irradiation will produce the **photostationary state**, a particular proportion of isomers that does not vary upon further irradiation. It is similar to an equilibrium mixture, but the proportions at the photostationary state are *not* those expected at thermodynamic (thermal) equilibrium.

What sets the photostationary state? Let's consider the hypothetical interconversion of a cis compound, **C**, and a trans compound, **T**, using irradiation at a single wavelength with a laser. Two features will distinguish the isomers: their efficiency of absorption at the particular wavelength used, ε, and their quantum yield for conversion to the other isomer. The photostationary state will be as given in Eq. 16.18. In the more typical instance of irradiating with a broad range of wavelengths, we must consider the relative absorbance efficiencies of both isomers at all wavelengths. Often it is possible to find a wavelength range in which one isomer absorbs more strongly than the other, allowing us to push the photostationary state toward the other isomer. This is an important way in which the photostationary state differs from the equilibrium distribution.

$$\frac{[T]}{[C]} = \frac{\varepsilon_C \Phi_C}{\varepsilon_T \Phi_T}$$

$$\text{(Eq. 16.18)}$$

Both direct irradiation and sensitized photolysis can lead to isomerization, often with differing results. An extensively studied system is stilbene (1,2-diphenylethylene) and its substituted derivatives (Eq. 16.19). Studies involving dozens of different stilbene derivatives and dozens of different sensitizers have produced a detailed view of the excited state structures of these systems. Indeed, the strongest experimental support for a scheme such as shown in Figure 16.15 comes from studies of stilbenes.

$$\text{(Eq. 16.19)}$$

Hochstrasser used picosecond spectroscopy to analyze stilbene isomerization. Irradiation of the trans isomer with a pulse at 265 nm produces an intermediate with a λ_{max} of 584 nm and a lifetime of 68 ps. As described above, these isomerization reactions proceed via twisting of the double bond in the excited state, and in this case the intermediate with the 68-ps lifetime is the $^1(\pi,\pi^*)$ state. In contrast, the lifetime of the intermediate formed from *cis*-stilbene is less than 1 ps. Apparently there is a barrier to rotation on the excited state for the trans isomer, with a much lower barrier or no barrier for the cis isomer.

The isomerization of phenyl-*t*-butyl ethylene (Eq. 16.20) illustrates a very powerful feature of photochemistry, seen throughout the field, not just in cis–trans isomerization. The cis compound is much less stable than the trans due to steric repulsions between the phenyl and the *t*-butyl groups. Nevertheless, with photochemistry we can efficiently prepare the less stable isomer. The reason this is easily possible with photochemistry is made clear in Figure 16.14. The key issue in determining the product is the manner in which the system exits from S_1 (or T_1) onto S_0. The relative energy of the two minima on S_0 plays no direct role in determining the product ratio. If the arrangement of the excited and ground state surfaces is right, we can end up with a very high energy structure on S_0. If the reaction is done at low temperatures, *very* high energy species with very small thermal barriers to more stable products can be observed. This is the basis of most matrix isolation experiments. The *trans*-cycloheptenone of Eq. 16.21 is an example of this, as is *trans*-cyclohexene (see the Going Deeper highlight below).

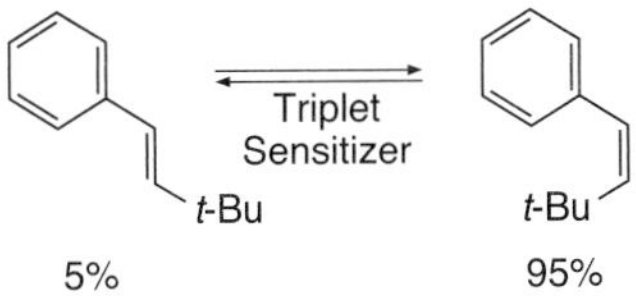

(Eq. 16.20)

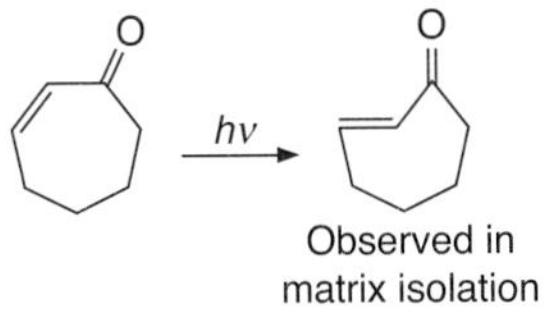

(Eq. 16.21)

Going Deeper

Trans-Cyclohexene?

Laser flash photolysis of phenylcyclohexene produces a new species that can be observed by fast absorption spectroscopy. The new species shows $\lambda_{max} = 380$ nm and has a lifetime of 9 μs. Even at −75 °C it quickly dimerizes. On the basis of these properties, the short-lived species was tentatively assigned the structure of *trans*-phenylcyclohexene, which should be very strained. The assignment was confirmed when the x-ray structure of the final dimer product was solved. The dimer arises from a [4+2] cycloaddition followed by a [1,3] hydrogen shift. The key evidence is the stereochemical relationship of the two groups shown in color. They are trans in the final product, establishing

that one partner in the cycloaddition was *trans*-phenylcyclohexene.

In later work, Peters combined this synthetic approach to *trans*-phenylcyclohexene with photoacoustic calorimetry to determine the strain energy of the transient species. It is 48.6 kcal/mol, so it is indeed a very highly strained olefin!

Dauben, W. G., van Riel, H. C. H. A., Hauw, C., Leroy, F., Joussot-Dubien, J., and Bonneau, R. "Photochemical Formation of *trans*-Phenylcyclohexene. Chemical Proof of Structure." *J. Am. Chem. Soc.*, **101**, 1901–1903 (1979); Goodman, J. L., Peters, K. S., Misawa, H., and Caldwell, R. A. "Use of Pulsed Time-Resolved Photoacoustic Calorimetry to Determine the Strain Energy of *trans*-1-Phenylcyclohexene and the Energy of the Relaxed 1-Phenylcyclohexene Triplet." *J. Am. Chem. Soc.*, **108**, 6803 (1986).

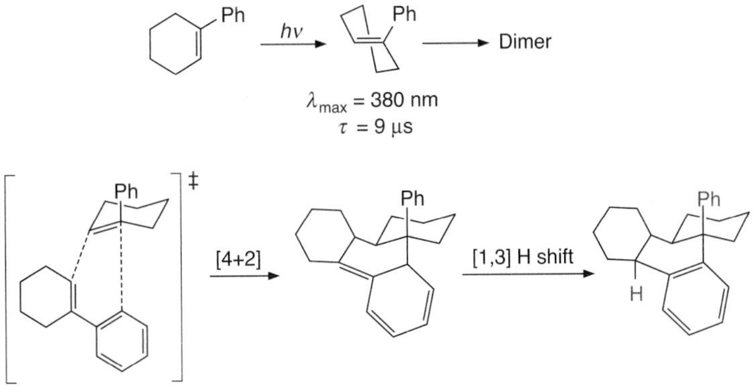

Trapping a trans cyclohexene

Connections

Retinal and Rhodopsin—
The Photochemistry of Vision

The photochemical cis–trans isomerization of an olefin in rhodopsin leads to the primary event in vision. The olefin comes from 11-*cis*-retinal. Rhodopsin is a membrane-bound protein that contains a cavity for the retinal and a lysine side chain that can react with the aldehyde. These two form an imine that is the fundamental chromophore of vision. This system is found in rod cells, primary photo-receptors that function in dim light and do not distinguish color—they are black and white receptors. As such, the newly formed chromophore shows a relatively broad absorption with a λ_{max} of 500 nm, a good match for the solar spectrum, as well as a very large extinction coefficient of 40,000 l/(mol•cm). Absorption of a photon causes an isomerization to the all-trans retinal. The isomerization is very rapid, occurring on the ps time scale, but the thermal back isomerization to the cis form is very slow. The structural change associated with photoisomerization is substantial. Within the confines of the binding cavity, this structural change is sensed by the rest of the rhodopsin

protein, launching a signaling cascade in the cell that ultimately reaches the visual cortex of the brain.

What about color vision? This occurs in the other photoreceptors, the cone cells. Just as with a computer monitor, cone cells are RGB: there are specific red, green, and blue receptors. Remarkably, *the chromophore is the same for all three colors!* It is the same 11-*cis*-retinal bound to a protein through an imine. The three color proteins and the "black-and-white" protein discussed above are all members of the same structural family and differ by less than 10% in their amino acid sequences. Color selectivity is achieved by precise positioning along the chromophore of specific amino acid side chains, especially those with strong local dipoles. These perturb the absorption characteristics of the chromophore, producing red (λ_{max} = 560 nm), green (λ_{max} = 530 nm), and blue (λ_{max} = 410 nm) photoreceptors. This is one of many examples of nature's conservative approach. Once you get a chromophore that works, stick with it. There's no need to invent three chromophores, just tune the one that works well.

Retinal isomerization

16.3.4 Reversal of Pericyclic Selection Rules

We noted in Chapter 15 that, for the most part, the orbital symmetry rules are not directly applicable to photochemistry. However, some photochemical reactions of simple π systems do give products that are consistent with expectations based on orbital symmetry, although this does not prove that these are concerted, pericyclic processes. The photochemical selection rules for pericyclic reactions are opposite of those for thermal pericyclic reactions. For example, there are many examples of [1,3] and [1,7] sigmatropic shifts that appear to go by the photochemically "allowed" suprafacial–suprafacial pathway; Eqs. 16.22 and 16.23 show two (recall that the thermal reactions would be suprafacial–antarafacial). These reactions occur upon direct irradation, while sensitized photolysis produces products more consistent with biradical-type reactions.

(Eq. 16.22)

Connections

Photochromism

The rhodopsin system just discussed is an example of **photochromism**, a light-induced, reversible change of color. Many simple organic molecules exist in two forms that can be interconverted photochemically and have different absorption spectra. These are interesting molecules that can form the basis for "switches". A photochromic system is found in eyeglasses that darken when exposed to sunlight, but lose their color in a non-photochemical reaction when exposed to less intense light. It has been proposed that optical computer memory systems could be based on photochromic organic molecules.

Several common photochromic organic molecules are shown to the right. The spiropyrans and spirooxazines are quite effective, and are used in photochromic eyeglasses. The second example is representative of a broad class of diarylethylenes that have been considered for possible electronic switching applications. With the spiropyrans, the photochemical process not only changes the optical properties of the system, but it also interconverts conjugated and unconjugated π systems. In the proper context, this could interconvert electrically conducting vs. insulating systems. Finally we show the well studied azobenzene system. The large geometrical change associated with cis–trans isomerization has made this the foundation for a large number of interesting systems whose properties can be photochemically switched.

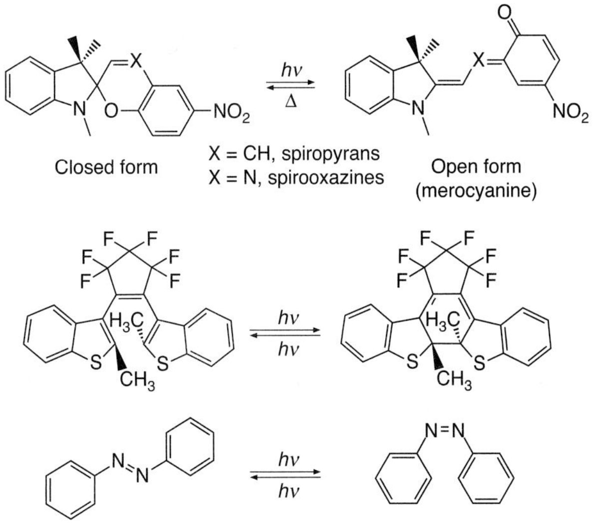

Closed form   X = CH, spiropyrans   X = N, spirooxazines   Open form (merocyanine)

Photochromic systems

For an excellent discussion of photochromism, see the IUPAC Technical Report on Organic Photochromism prepared by H. Bouas-Laurent and H. Dürr, available at http://www.iupac.org/publications/pac/2001/pdf/7304×0639.pdf.

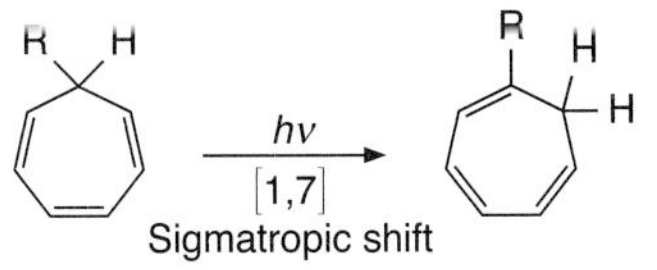

$$\text{(Eq. 16.23)}$$

Sigmatropic shift [1,7]

There are also examples of electrocyclic reactions that follow the stereochemical outcomes (conrotatory vs. disrotatory) expected for reactions under orbital symmetry control. For example, the photochemical ring opening of Eq. 16.24 should be a six-electron, conrotatory process, and indeed the product has the predicted trans double bond. An important biological example of such a process is the photochemical conversion of ergosterol to previtamin D (Eq. 16.25), a key event in the synthesis of vitamin D.

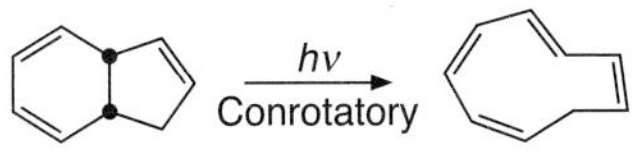

$$\text{(Eq. 16.24)}$$

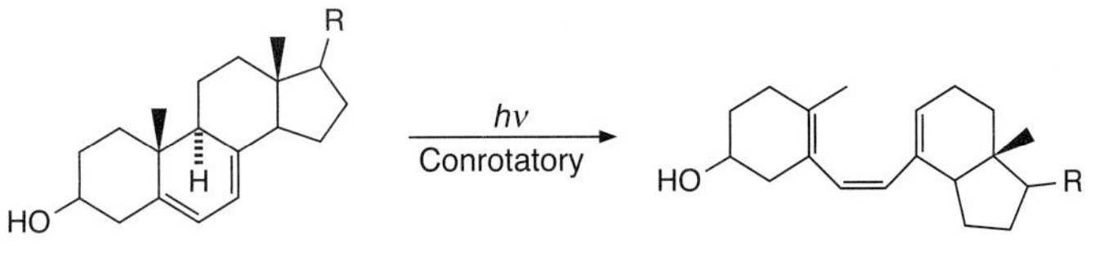

Conrotatory

$$\text{(Eq. 16.25)}$$

Ergosterol   Pre-Vitamin D

An example of introducing strain using photochemistry is the synthesis of **Dewar benzene**. Dewar benzene represents a classic strained ring system that many chemists have studied. One convenient synthesis is shown in Eq. 16.26, the key step being a photochemical disrotatory electrocyclic ring closure.

$$\text{(Eq. 16.26)}$$

16.3.5 Photocycloaddition Reactions

Cycloadditions represent an important class of photochemical reactions. We discussed thermal cycloadditions extensively in Chapter 15, with the prototype being the [4+2] cyclo-addition of the Diels–Alder reaction. Orbital symmetry reasoning would lead us to expect that photochemical cycloadditions should be typified by a [2+2] reaction. Indeed, formal [2+2] photocycloadditions are common. However, *most photochemical cycloadditions involve triplet states and biradical intermediates.* Concerted photochemical cycloadditions are rare. As such, orbital symmetry arguments are not directly relevant, and instead we must focus on potential biradical intermediates and possible funnels and other surface crossing points. Some photochemical cycloadditions do proceed via singlet states, and usually these involve the formation of exciplexes.

While photocycloadditions are typically not concerted, pericyclic processes, our analy-sis of the thermal [2+2] reaction from Chapter 15 is instructive. Recall that suprafacial–suprafacial [2+2] cycloaddition reactions are thermally forbidden. Such reactions typically lead to an avoided crossing in the state correlation diagram, and that presents a perfect situa-tion for funnel formation. This can be seen in Figure 16.17, where a portion of Figure 15.4 is reproduced using the symmetry and state definitions explained in detail in Section 15.2.2. The barrier to the thermal process is substantial, but the first excited state has a surface that comes close to the thermal barrier. At this point a funnel will form allowing the photochemi-cal process to proceed. It is for this reason that reactions that are thermally forbidden are often efficient photochemical processes. It is debatable, however, whether to consider the [2+2] photochemical reactions orbital symmetry "allowed". Rather, the thermal forbid-denness tends to produce energy surface features that are conducive to efficient photochem-ical processes. As we will see below, even systems that could react via a photochemically "al-lowed" concerted pathway, often choose a stepwise mechanism instead.

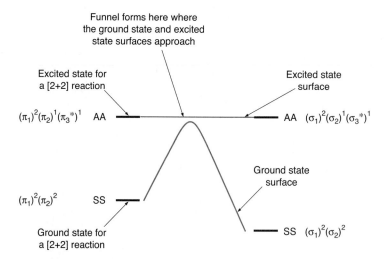

Figure 16.17
State correlation diagram for a [2+2] cycloaddition. There is a substantial barrier on the ground state energy surface, but the first excited state surface approaches the ground state surface, and a funnel forms that allows the excited state to exit to the ground state, facilitating the reaction.

Formal [2+2] cycloadditions can be conveniently divided into three classes: addition of two simple olefins to form a cyclobutane; addition of an olefin to a carbonyl to form an oxetane; and reaction of an α,β-unsaturated carbonyl with an olefin to form a cyclobutane. We will briefly discuss each reaction here.

Olefin dimerization (Eq. 16.27) necessarily involves π,π^* states. The reaction generally proceeds via triplet states, and since intersystem crossing is very slow for hydrocarbon π,π^* states, a sensitizer is generally required. Note that the triplet states of simple olefins are typically fairly high in energy, so only high energy sensitizers such as acetophenone are viable; even benzophenone is often not enough (recall Table 16.5). For simple olefins, these reactions are often not overly efficient, since cis–trans isomerization can interfere as an energy-wasting step. Given this, we should anticipate the possibility of scrambling of stereochemistry in these reactions. The biradical nature of the process is emphasized by the product distribution in the sensitized photolysis of butadiene shown in Eq. 16.27. The triplet state $[^3(\pi,\pi^*)]$ is represented as a biradical that can add to a ground state diene to produce a doubly allylic biradical. This intermediate can close to form a cyclobutane (both stereoisomers) or a cyclohexene.

(Eq. 16.27)

In contrast to acyclic olefins, sensitized photodimerization of cyclic olefins is often fairly efficient (Eq. 16.28). A major reason for this is that cis–trans isomerization is no longer an efficient side reaction. As shown with the example of cyclohexene, this is an excellent way to make cyclobutanes, but control of stereochemistry is problematical.

23%   27%   42%

(Eq. 16.28)

Connections

UV Damage of DNA—A [2+2] Photoreaction

A notorious photochemical olefin dimerization occurs on exposure of DNA to UV radiation. The primary event involves the dimerization of adjacent thymine (T) residues to produce the thymine dimer shown. Needless to say, this is not a favorable event for a living system, and extensive repair systems involving enzymes termed photolyases exist to excise the dimer and repair the DNA lesion. If not adequately repaired, however, such thymine dimers can lead to cell death (our skin peels when we get a sunburn) or can convert a healthy cell into a cancer cell. An inherited disorder involving defects in the repair system can lead to xeroderma pigmentosum, which involves hypersensitivity to UV irradiation and increased risk of skin cancer. An interesting consequence of this photochemistry is that species that live at very high altitudes,

where UV exposure is more intense, have evolved to have a lower proportion of T residues in their genome. This lowers the probability of adjacent T residues in their DNA, and thus minimizes the impact of this potentially lethal photoreaction.

Thymine dimer

A similar photochemical cycloaddition is the addition of a carbonyl across an olefin to produce an oxetane, a four-membered ring containing an oxygen (Eq. 16.29). This reaction is sometimes referred to as the **Paternò–Büchi reaction**. Reaction occurs from either the singlet or triplet n,π^* states of the carbonyl and involves a stepwise process with a biradical intermediate. The spin multiplicity of the initially formed biradical must be the same as that of the n,π^* state of the carbonyl. For aryl carbonyls the excited state is a triplet, while for alkyl carbonyls the excited state is a singlet. From here on in it is just standard biradical chemistry. The product is a 1,4-biradical of the sort we have seen in Section 11.12.2, and that we will see below in our discussion of Norrish II chemistry. Cleavage and closure reactions are possible, the former regenerating starting materials, the latter producing the oxetane product. Since a biradical intermediate is involved, we can expect some loss of stereochemistry, especially in the triplet manifold, while stereochemistry would be more retained in the singlet manifold. Eqs. 16.30, 16.31, and 16.32 show a non-stereospecific reaction and two reactions that demonstrate stereospecificity, respectively.

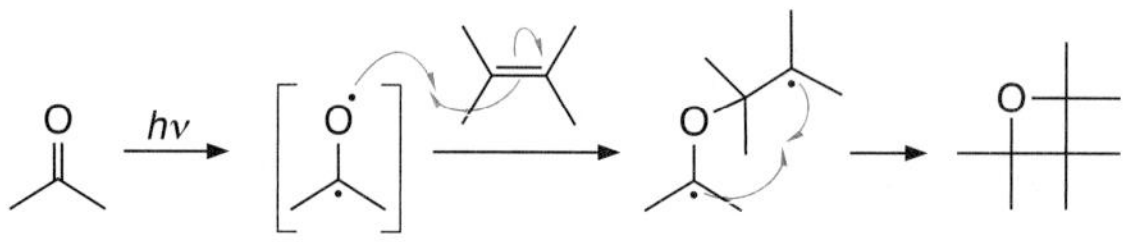

(Eq. 16.29)

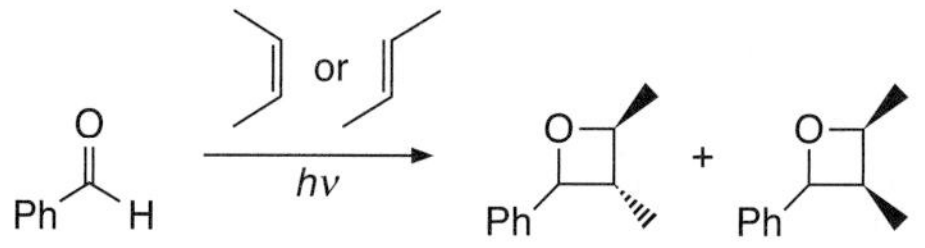

(Eq. 16.30)

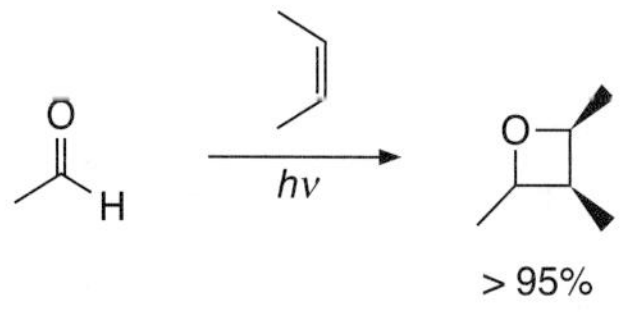

> 95%

(Eq. 16.31)

> 90%

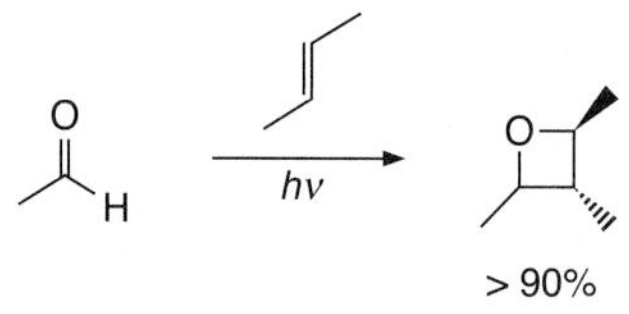

(Eq. 16.32)

If the biradical cleaves to produce starting material, we have a bimolecular process that returns an excited state to its ground state. This is an example of quenching, which we discussed in Section 16.2.2. Indeed, olefins and especially polyenes are efficient quenchers of carbonyl excited states. This illustrates the statement we made in Section 16.2.2 that quenching is often closely related to a photochemical process.

The photocycloaddition of α,β-unsaturated carbonyls (enones) with an olefin to form cyclobutanes is an especially efficient reaction (Eqs. 16.33 and 16.34). This reaction has proven to be of considerable synthetic utility. In principle these types of reactions can proceed from n,π^* or π,π^* states of the carbonyl, and for enones these two states can often be quite close in energy, making detailed analysis complicated.

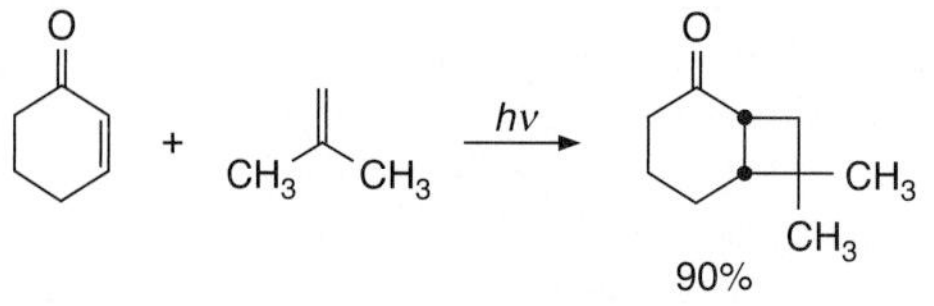

90%

(Eq. 16.33)

~50:50

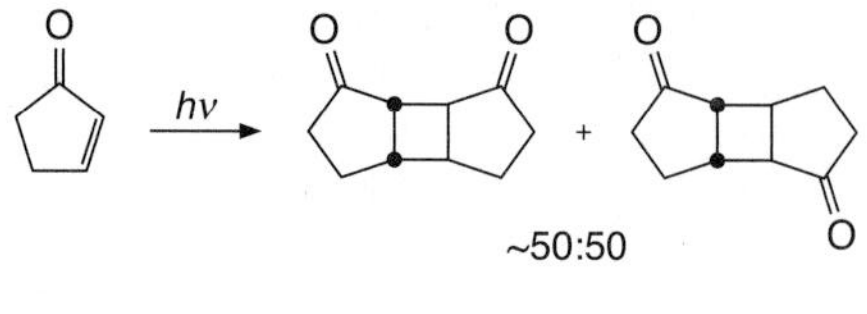

(Eq. 16.34)

The most studied case is that of a cyclic enone reacting with an olefin. In these systems several trends are observed. The reaction proceeds from T_1, which may be either n,π^* or π,π^*, and triplet biradicals are likely involved. Electron-rich olefins react more rapidly and add with predictable regiochemistry. Cis–trans stereochemistry in the olefin is lost in the cycloaddition, and trans fused rings can form.

The rationalization of the regiochemistry is given in Eq. 16.35. It is convenient to think of the T_1 state of the enone as being polarized in the manner shown. If we consider the excited state to have biradical character, a radical next to a carbonyl should be relatively electrophilic. When a polarized olefin approaches, the product arises by connecting the nucleophilic end of the olefin to the electrophilic carbon α to the carbonyl. Whether this is a mechanistic insight or a useful mnemonic, it does allow the prediction of cycloaddition regiochemistry in many, but not all, cases.

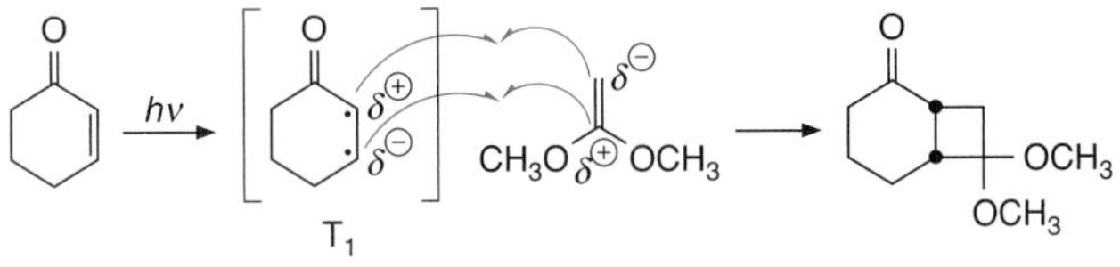

(Eq. 16.35)

Making Highly Strained Ring Systems

The [2+2] reaction of olefins is especially efficient in rigid polycyclic systems that hold the two π systems in close proximity. This has proven to be an excellent way to make exotic, highly strained systems. A classic example is Eaton's original synthesis of cubane (Eq. 16.36; recall our discussions of cubane in Section 2.5.2). The Diels–Alder reaction proceeds with endo selectivity (Section 15.3.4). This places the two olefins in fairly close proximity (making models of the product will help to make this more clear). As a result, the photocycloaddition is quite efficient, and it produces, in one step, a large portion of the polycyclic ring system. The other key step in the synthesis is a double Favorskii rearrangement (Section 11.10), which contracts two five-membered rings to make the cubane skeleton.

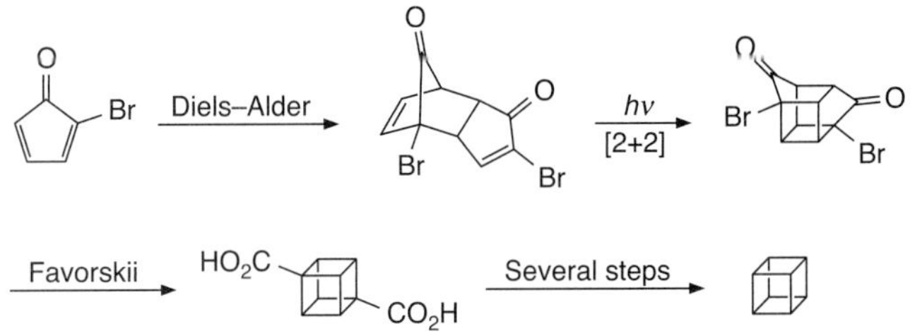

(Eq. 16.36)

It has been proposed that the large amount of strain energy built into a system upon intramolecular photocycloaddition could be put to good use. Consider the simple system of Eq. 16.37. The photochemical reaction converts norbornadiene to the highly strained quadricyclane structure. The reverse reaction would release a great deal of strain energy, which would be given off as heat. In this way, the energy of the absorbed photon has been stored (as molecular strain energy), and it can subsequently be released when so desired by exposure to a catalyst. This is one approach to using solar energy; light can be absorbed during the day, to be released as heat in the cooler evening. One needs to design efficient sensitizers with absorption spectra that match the solar spectrum and to identify appropriate catalysts for the reverse reaction. Much effort has been put into both steps, but more importantly, Eq. 16.37 illustrates a general strategy.

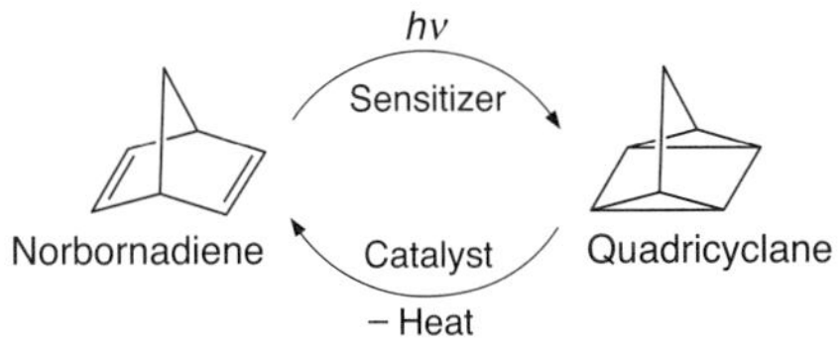

(Eq. 16.37)

Breaking Aromaticity

The high reactivity of excited states allows structures that are usually very unreactive, such as aromatic rings, to react. Photochemical cycloaddition to aromatics is a common reaction. For an olefin adding to a simple aromatic, the full range of [2+2], [3+2], and [4+2] cycloadditions has been seen (Eqs. 16.38–16.40). The [4+4] photodimerization of anthracene is one of the oldest known photochemical reactions. Concerning the [n+2] cycloadditions, strongly electron donating or withdrawing groups on the olefin favor the [2+2] path, while simpler alkyl olefins tend to give the [3+2] path. No single mechanistic scheme can rationalize the variations seen, but it is generally considered that the aromatic S_1 state is involved, not T_1.

$$\text{(Eq. 16.38)}$$

$$\text{(Eq. 16.39)}$$

$$\text{(Eq. 16.40)}$$

16.3.6   The Di-π-Methane Rearrangement

One of the most interesting and general photochemical reactions is the di-π-methane rearrangement. It has been extensively investigated, most notably by Zimmerman and coworkers, leading some to refer to it as the **Zimmerman rearrangement**.

The basic reaction is shown in Eq. 16.41. The starting material contains a pair of π systems separated by a single saturated carbon, making a di-π-methane. The product is a vinyl cyclopropane, and the numbering shows how the new bonds are formed. Also shown is a *schematic* mechanism, proposing a key intermediate that is a cyclopropyldicarbinyl biradical. This then fragments in the standard cyclopropylcarbinyl way to give a new biradical. Alternatively, the reaction can be viewed as concerted, with the pericyclic array of the transition state as shown in the margin. The easiest way to analyze this is as a Möbius system with six electrons, making the process thermally forbidden and thus photochemically allowed. As we will see below, both mechanistic schemes are useful in interpreting results. The "reality" of the mechanism is debated. Certainly, reactions involving $^3(\pi,\pi^*)$ states are not concerted. On the other hand, $^1(\pi,\pi^*)$ reactions can be, and singlet states can lead to high levels of stereospecificity (such as Eq. 16.42).

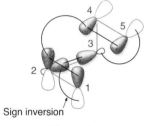

Sign inversion

Di-π-methane
Pericyclic analysis

$$\text{(Eq. 16.41)}$$

Cyclopropyldicarbinyl
biradical

$$\text{(Eq. 16.42)}$$

The di-π-methane rearrangement also shows interesting multiplicity effects, typified by the reactions in Eqs. 16.43–16.46. For acyclic π systems, direct irradiation, producing a $^1\pi,\pi^*$ state, gives the di-π-methane rearrangement. In contrast, sensitized photolysis, producing a $^3\pi,\pi^*$ state, gives no reaction (Eq. 14.44). For rigid systems such as the benzobarrelene shown in Eq. 16.45, the di-π-methane rearrangement is usually seen for the sensitized photolysis. Direct irradiation often produces alternative reaction paths, such as the one shown in Eq. 16.46.

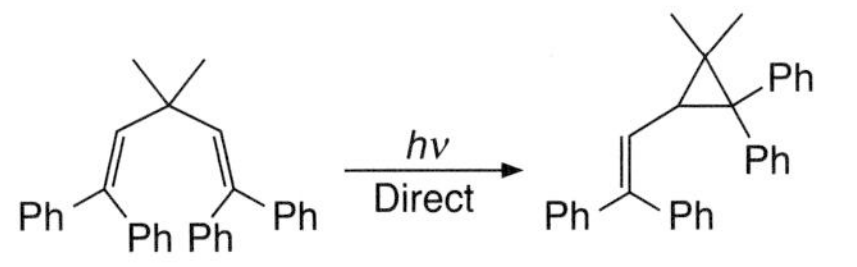

$$\text{(Eq. 16.43)}$$

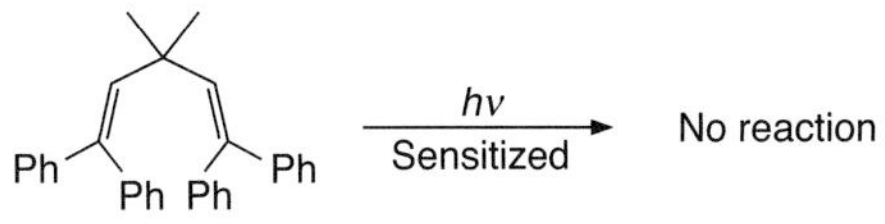

$$\text{(Eq. 16.44)}$$

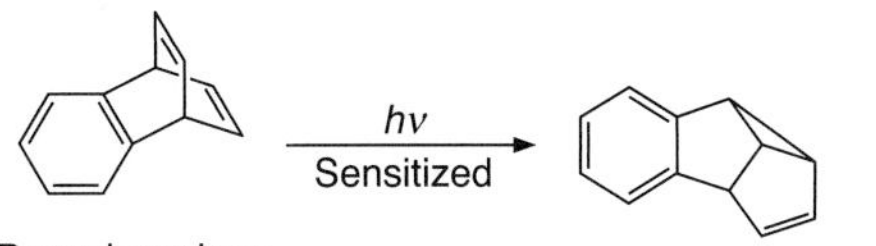

$$\text{(Eq. 16.45)}$$

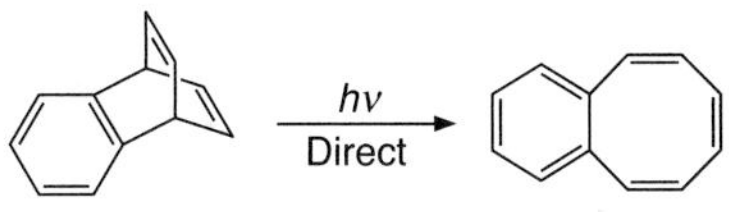

$$\text{(Eq. 16.46)}$$

The observations of Eqs. 16.42–16.46 are typical, and a consistent explanation for all possible systems is challenging. Nevertheless, the standard rationalization goes as follows. For acyclic systems, the $^3(\pi,\pi^*)$ state will typically lead to cis–trans olefin isomerization. For the tetraphenylpentadiene system shown this is an undetectable process (Eq. 16.44), but in other less heavily substituted structures it is often seen. Recall that cis–trans isomerization is favored by a biradical-like excited state (Figure 16.15), and the $^3(\pi,\pi^*)$ state has much more biradical character than the $^1(\pi,\pi^*)$. The $^1(\pi,\pi^*)$ is less susceptible to this effect, and the di-π-methane rearrangement proceeds, perhaps via a concerted process (Eq. 16.43).

In rigid bicyclic systems cis–trans isomerization is not a factor, and so the $^3(\pi,\pi^*)$ state displays efficient di-π-methane rearrangement photochemistry (Eq. 16.45). What about the $^1(\pi,\pi^*)$ state? The reason the di-π-methane rearrangement is often not seen is that other, more conventional reactions such as [2+2] photocycloadditions predominate. It is not that the di-π-methane rearrangement is not feasible, but rather that other reactions are more facile.

The regiochemistry of the di-π-methane rearrangement is best understood by referring to the biradical mechanism of Eq. 16.41. Consider an unsymmetrically substituted system of the sort shown in Eq. 16.47. The second step, the cleavage of the cyclopropylcarbinyl moiety, occurs so as to put the better radical stabilizing substituents on the radical center. Thus, these substituents become part of the cyclopropane ring in the product.

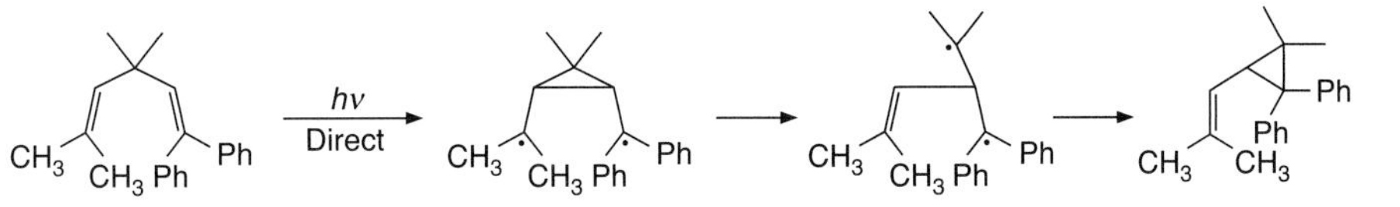

$$\text{(Eq. 16.47)}$$

In contrast, the stereochemistry of the di-π-methane rearrangement is best understood with reference to the pericyclic transition state shown above. As such, we see retention of stereochemistry at C1 (Eq. 16.48), inversion at C3 (Eq. 16.49), and retention at C5 (Eq. 16.50).

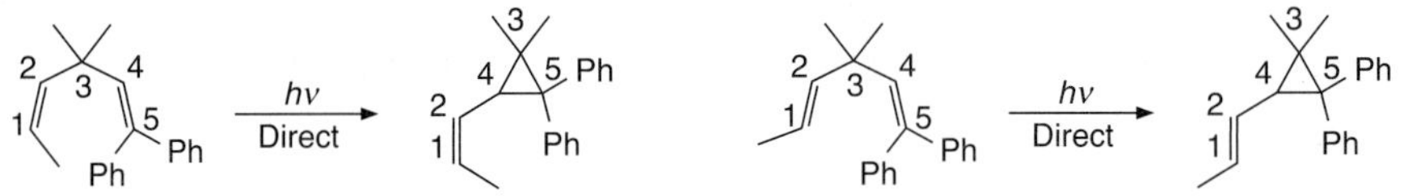

(Eq. 16.48)

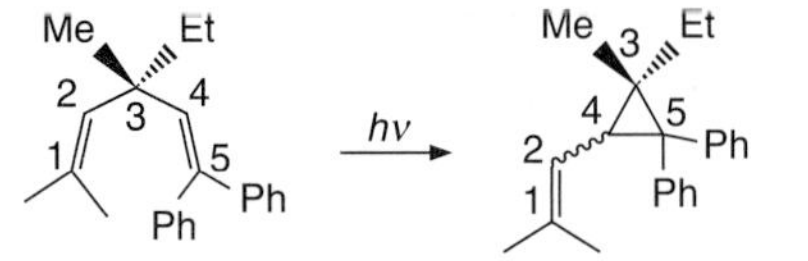

(Eq. 16.49)

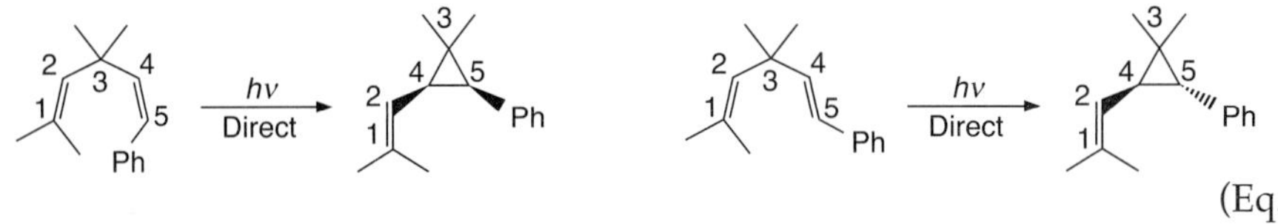

(Eq. 16.50)

A related reaction is seen with β,γ-unsaturated ketones, termed the **oxadi-π-methane rearrangement**. Eq. 16.51 shows the basic reaction and a specific example is given in Eq. 16.52. Since carbonyl double bonds are much stronger than olefinic double bonds, the product of an oxadi-π-methane rearrangement is always an acyl cyclopropane, not a vinyl oxirane. The specific example in Eq. 16.52 highlights a fairly general feature of the reaction. Triplet states favor the oxadi-π-methane rearrangement, while singlet states often do other types of photochemistry (Eq. 16.53). In the particular case shown (Eq. 16.53), the product of direct irradiation results from an initial α cleavage, as in the Norrish I reaction (see below). In this case, however, an allyl radical is formed, and ring closure at the other end of the allyl group gives the product. Given the considerable variation in intersystem crossing rates seen for carbonyls as a function of substituents, we can expect the relative proportions of oxadi-π-methane rearrangement vs. other reactions to vary considerably from system to system.

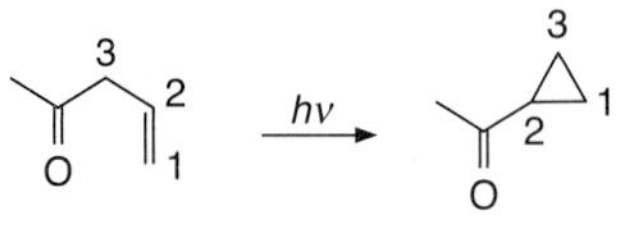

(Eq. 16.51)

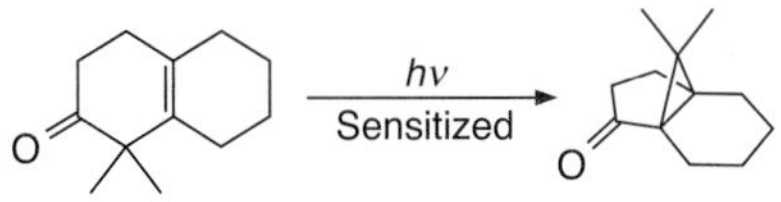

(Eq. 16.52)

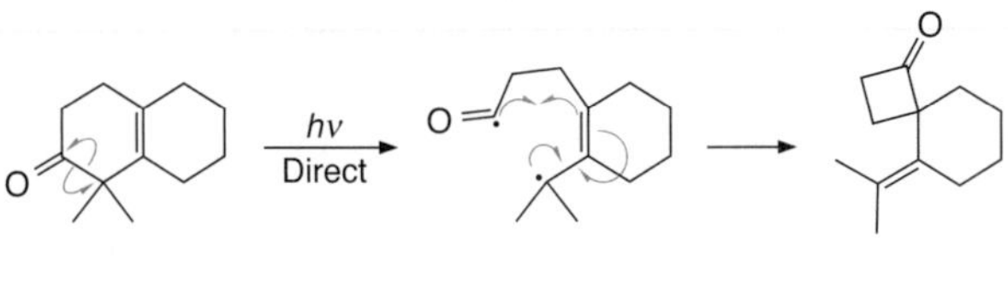

(Eq. 16.53)

16.3.7   Carbonyls Part I: The Norrish I Reaction

A dominant chromophore in organic photochemistry is the carbonyl group. Its n,π^* absorption is readily accessible and can lead to either singlet or triplet excited states. Also, fluorescence rates are relatively slow (Table 16.4), allowing plenty of time for reactions to occur. Most photoreactions of carbonyls follow one of two basic paths, the first of which we probe here.

Before a discussion of specific reactions, let's briefly consider the nature of the carbonyl excited state. *For n,π^* states*, a good model for the carbonyl excited state is the biradical-like structure shown in Eq. 16.54, especially for the triplet state. What should we expect from such a structure? There are two reactive centers, a carbon-based radical and an oxygen radical. To first order they are independent, one electron in the plane and the other in the π system (see Figure 16.7). The oxygen radical should be *extremely* reactive. Recall the high bond strength of the O–H bond discussed in Chapter 2, and how this makes hydroxyl radical a highly reactive species. Similar things might be expected here. The nature of the carbon radical will depend on the substituents R_1 and R_2. If they can delocalize and thereby stabilize the radical, it may be relatively unreactive for a radical. On the other hand, if these substituents are in no way stabilizing, we will have a fairly reactive center. One other issue to consider in anticipating carbonyl photochemistry is the strength of the carbonyl bond. Recall from Table 2.2 that a C=O double bond is quite strong, stronger than a C=C by a substantial margin. Thus, there should also be a considerable driving force to reform the C=O double bond, if possible.

$$\text{(Eq. 16.54)}$$

With that background, we are now ready to discuss the first class of carbonyl photochemical reactions. Eq. 16.55 shows the **α cleavage** or **Norrish type I** photoreaction. After excitation, the reaction involves reformation of the carbonyl double bond with concomitant cleavage of a C–C bond α to the carbonyl and expulsion of a carbon radical. Also formed is an acyl radical that, based on the bond dissociation energies of Table 2.2, is a fairly stable radical.

$$\text{(Eq. 16.55)}$$

Acyl radical

This is the dominant reaction pathway for acyclic, dialkyl ketones. It occurs entirely from the $^3(n,\pi^*)$ state. In contrast, bond cleavage in aryl alkyl ketones [ArC(O)R] is 10^2–10^4 times slower, because the triplet state of such molecules is π,π^* or has considerable π,π^* character. With either kind of ketone, cleavage occurs to give the more stable carbon radical. In fact, cleavage rates have been directly measured. For R_1 equal to methyl, 1° alkyl, or benzyl, the rates of bond cleavage of the $^3(n,\pi^*)$ states are 10^3, 10^7, and 10^{10} s^{-1}, respectively, reflecting the stabilities of these radicals. Ring strain can also promote the reaction, if it is relieved by the α cleavage.

What happens after the α cleavage depends on the particular system. For acyclic ketones, it is pretty much conventional free radical chemistry. All the reactions discussed in Chapters 10 and 11 are viable (atom abstraction, addition to olefins, etc.), and precisely what happens depends on the structure of the radicals generated and the presence of any additional species. Really the "photochemistry" is over once the α cleavage has occurred, and it is just regular free radical chemistry the rest of the way.

In cyclic ketones, there are some interesting possibilities, as summarized with the case of 2-methylcyclohexanone in Eq. 16.56. Now the immediate product of the α cleavage is a *biradical*, which opens up new reaction paths. Note that cleavage occurs preferentially to give the more stable radical. Initially we expect the biradical to be formed in a triplet state, but since the radical centers are not strongly interacting once cleavage has occurred, no strong spin preference is expected to be retained in the biradical. Like any pair of radicals, the biradical can undergo disproportionation reactions. If the acyl radical abstracts a hydrogen, the final product is a δ,ε-unsaturated aldehyde (Eq. 16.57). If the carbon radical abstracts a hydrogen, a ketene is formed, which is usually trapped by an alcohol or similar compound (Eq. 16.58). Typically, both of these formal [1,5] hydrogen shifts occur.

$$\text{(Eq. 16.56)}$$

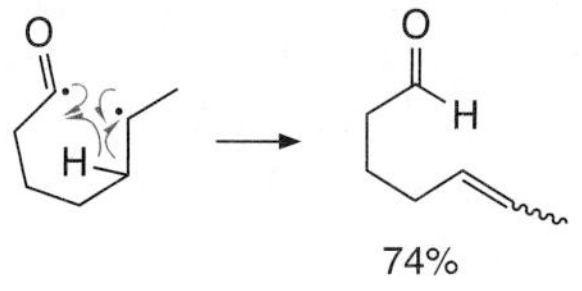

(Eq. 16.57)

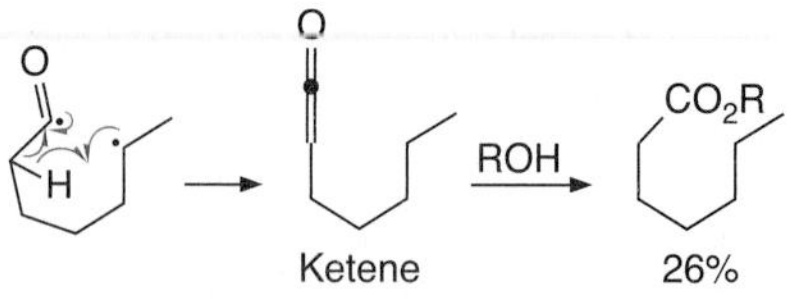

74%

(Eq. 16.58)

Ketene   26%

Eqs. 16.59–16.62 show some examples of variations on the theme of α cleavage. For β,γ-unsaturated ketones, the equivalent of a [1,3] sigmatropic shift can occur (Eq. 16.59), but this is an allylic rearrangement of a biradical, not a concerted reaction. Not surprisingly, if a free radical rearrangement is available, it will occur, as in the cyclopropylcarbinyl system shown in Eq. 16.60. If a stabilized biradical, such as trimethylenemethane (TMM), can form, loss of CO is a possibility (Eq. 16.61). Cyclobutanones can undergo a novel rearrangement to form an oxacarbene, which then goes on to do further chemistry (Eq. 16.62).

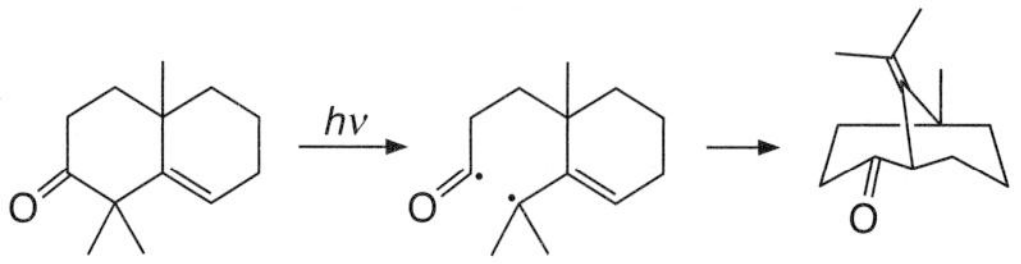

(Eq. 16.59)

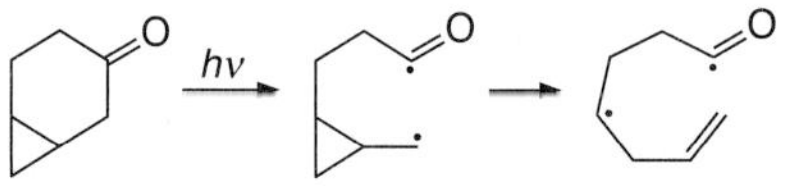

(Eq. 16.60)

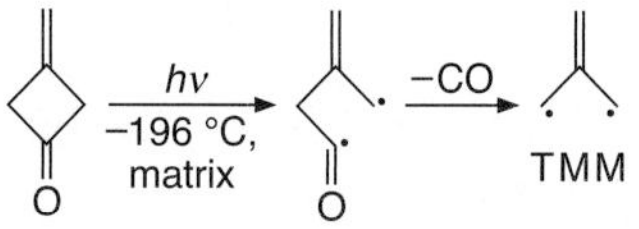

(Eq. 16.61)

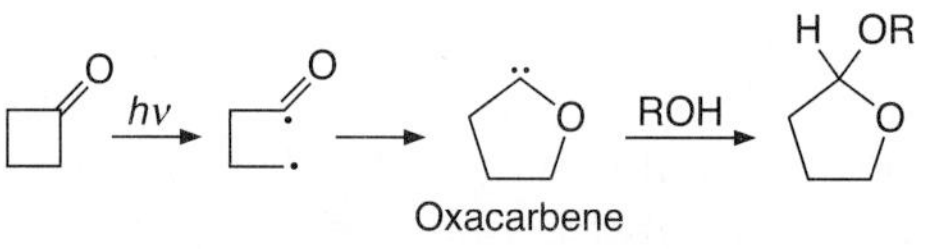

TMM

(Eq. 16.62)

Oxacarbene

Finally, a rearrangement of aryl esters and amides known as the **photo-Fries rearrangement** has occasionally found synthetic utility (Eq. 16.63). Although these reactions are superficially diverse, they all are initiated by α cleavage at the carbonyl, followed by conventional radical chemistry.

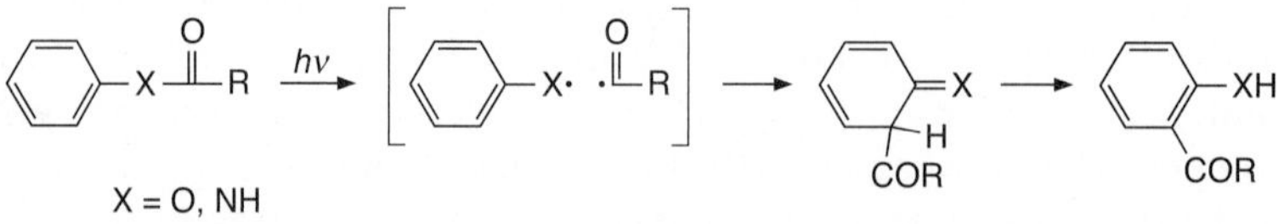

(Eq. 16.63)

X = O, NH

16.3.8   Carbonyls Part II: Photoreduction and the Norrish II Reaction

We discussed how the n,π^* states of carbonyls, especially the triplets, have considerable radical character. In addition, oxygen-centered radicals are very reactive because of the high strengths of O–H and O–C bonds. Hydrogen atom abstraction should be expected, and indeed it is a common reaction. A typical reaction involves a carbonyl n,π^* state undergoing a bimolecular hydrogen atom abstraction, and that is the basis for the common process called **photoreduction**.

The prototype is the reduction of benzophenone by isopropanol, as in Eq. 16.64. Irradiation of many ketones in hydroxylic solvents leads to reduction to the alcohol. Note that it is *not* the OH of the alcohol that is first abstracted—that is a very strong bond. Rather, the oxygen of the ketone n,π^* state abstracts H from the C–H bond adjacent to the alcohol OH. In the particular case shown, this forms a pair of **ketyl radicals**, one from benzophenone, and one from isopropanol. The reaction can then proceed along two paths. Simple reduction is possible via a second hydrogen atom abstraction, producing the alcohol (in this case benzhydrol). Alternatively, the ketyl radical from the original carbonyl can dimerize, giving a pinacol-type product. Reaction conditions can be adjusted to favor one path over the other. In particular, use of alkaline isopropanol favors reduction over pinacol formation. In alkaline isopropanol, the ketyl radical is mostly deprotonated and present as the ketyl anion, which does not readily dimerize due to electrostatic repulsions. Photoreduction is synthetically useful in simple cases, where isopropanol is used as solvent. In other cases, it is better thought of as a potential side reaction when carbonyl photochemistry is conducted in hydroxylic solvents.

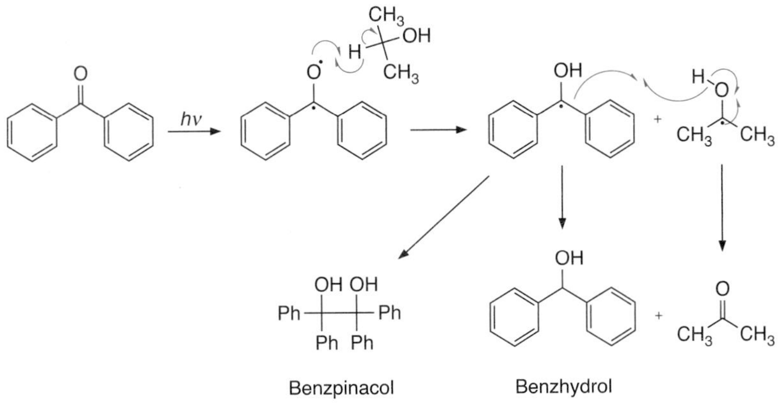

Benzpinacol   Benzhydrol

(Eq. 16.64)

The hydrogen atom abstraction just described can also occur in an intramolecular fashion, and this leads to the second major reaction path for carbonyls, the **β cleavage** or **Norrish type II** photoreaction. It is summarized in Eq. 16.65. This reaction begins with a γ hydrogen abstraction, to produce a 1,4-biradical in a cyclic process involving six atoms. This can also be viewed as a [1,5] hydrogen shift akin to the shift we saw after Norrish I cleavage of a cyclic ketone. We will return to this shift below, but first let's finish the Norrish II reaction. We have already discussed 1,4-biradicals in Section 11.12.2, and we know that there are two reaction paths available: cleavage to a pair of olefins or closure to a cyclobutane. The 1,4-biradical obtained from a Norrish II reaction can undergo both. In this case one of the cleavage products is an enol that rapidly isomerizes to a ketone; the closure product is a cyclobutanol. Thus, the possible products from a Norrish II process are a ketone plus an olefin and/or a cyclobutanol.

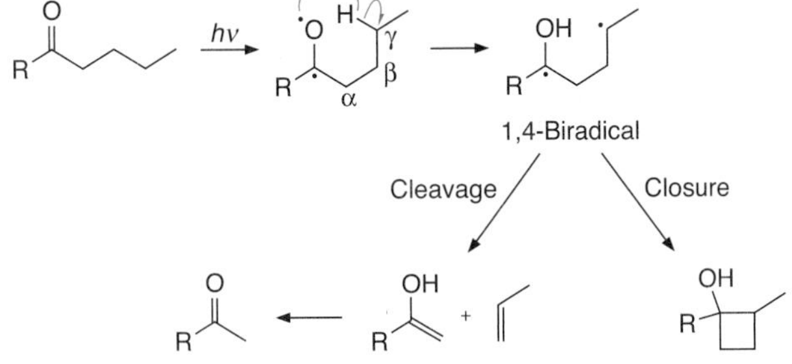

1,4-Biradical

Cleavage   Closure

(Eq. 16.65)

A number of factors influence the efficiency and mechanistic detail of the Norrish II reaction. Hydrogen abstraction by oxygen is faster if a more stable carbon radical is formed, typical rates being 10^7, 10^8, and $5 \times 10^8 \ s^{-1}$ for abstraction to produce primary, secondary, and tertiary radicals, respectively. Since aryl alkyl ketones undergo intersystem crossing very rapidly ($k_{ISC} \approx 10^{11} \ s^{-1}$), all reactions from such substrates arise from the triplet state. In fact, all reaction comes from the $^3(n,\pi^*)$ state, even if the $^3(\pi,\pi^*)$ state is lower in energy. This is consistent with our notion that it is n,π^* states that are more likely to be involved in hydrogen abstraction reactions.

For aliphatic ketones, with $k_{ISC} \approx 10^8$ s^{-1}, reaction from both $^1(n,\pi^*)$ and $^3(n,\pi^*)$ states is possible. Often it is possible to favor one over the other by using specific quenchers. For example, dienes and molecular oxygen preferentially quench triplet states by a spin-allowed conversion to a singlet state, and so their presence favors the $^1(n,\pi^*)$ pathways for the Norrish II reaction. There are significant differences between the singlet and triplet pathways. Typically, singlet states favor the cleavage pathway over the cyclization pathway, and give retention of stereochemistry in the cleavage products. This has led some to propose a concerted mechanism for the singlet cleavage per the structure shown in the margin, a view that has some experimental support but that has not been unambiguously established. In contrast, there is strong evidence that a true biradical is involved in the triplet reaction. In fact, the triplet 1,4-biradical has been observed by flash photolysis and has been trapped by conventional radical trapping reagents.

There is good evidence that the initial hydrogen abstraction is reversible. This is tantamount to a radiationless decay for the n,π^* state. The best evidence for this comes from studies of $^3(n,\pi^*)$ states with a stereogenic center at the γ carbon, as in Eq. 16.66. The long lived triplet biradical has time to lose its stereochemical integrity before returning to a now racemized starting material. In contrast, racemization is often not seen with $^1(n,\pi^*)$ states, consistent with the much shorter lifetimes expected for the singlet biradical.

No racemization

Racemization

(Eq. 16.66)

16.3.9 Nitrobenzyl Photochemistry: "Caged" Compounds

A photoreaction of increasing importance involves, in effect, the use of a nitrobenzyl unit as a photochemically removable protecting group. The basic scheme is shown in Eq. 16.67. A nitrobenzyl group is attached to a heteroatom, shown as an oxygen. The overall transformation is fairly simple. Photolysis cleaves the nitrobenzyl, to produce o-nitrosobenzaldehyde and the deprotected heteroatom. The mechanism shown is speculative in spots, but the so-called aci-nitro anion intermediate has been observed in flash photolysis studies, with an absorption maximum of 408 nm. The decay of the intermediate is pH dependent, with its lifetime varying from the ms range at pH 5–6 to the seconds range at pH > 8. We have chosen to write the n,π^* excited state of the nitrobenzyl moiety as having radical character. In this way we can see that the abstraction of the benzylic hydrogen that initiates the cleavage is quite similar to the initial step of the Norrish II process.

Acinitro intermediate

(Eq. 16.67)

The nitrophenyl chromophore absorbs in the 300–360 nm region. This makes the group useful in a biological context, because proteins and nucleic acids are transparent at the long wavelength end of this region. Also, several laser sources are potent in this region of the spectrum. Most uses of the nitrobenzyl group have been to address biological questions, making use of so-called **caged** groups. This refers to a biological species that is protected as a nitrobenzyl or similar group, making it impotent in the biological system under investigation. A flash of light then liberates the molecule from its "cage", allowing temporal and spatial control over a biological process. The terminology is in some ways unfortunate, because in physical organic chemistry a cage is more typically associated with solvent cage effects, especially in free radical chemistry. However, the terminology is now firmly entrenched in the chemical biology literature.

Below are shown examples of caged structures that have been developed. Caged ATP has been used to regulate muscle contraction. Calcium is a universal signaling ion, and so the caged calcium ion has seen extensive use. Many caged neurotransmitters, including the glutamic acid shown, have been used to probe synaptic signaling, while caged tyrosine has been incorporated as an unnatural amino acid into proteins. Common variants of the nitrobenzyl strategy are also shown. Adding a methyl to the benzylic carbon often increases photochemical efficiency, while a carboxylate favorably affects efficiency and solubility.

Caged ATP

Caged glutamic acid
Neurotransmitter

Caged Ca^{2+}
Photolysis releases Ca^{2+}

Caged tyrosine
Incorporated into protein

16.3.10 Elimination of N$_2$: Azo Compounds, Diazo Compounds, Diazirines, and Azides

We've seen loss of CO from ketones under certain circumstances (Eq. 16.61), and expulsion of a small stable molecule is a common photochemical reaction. Dinitrogen, N$_2$, is an extremely stable fragment, and photochemical expulsion of dinitrogen can occur from several different types of structures. Photochemical elimination of N$_2$ has been especially useful in a wide array of studies of reactive intermediates. Whether under conventional conditions or cryogenic, matrix isolation conditions, photolysis with loss of N$_2$ has been used to generate many types of reactive intermediates. We saw examples of this in Chapter 10 when radical additions were discussed and in Chapter 13 in the context of radical polymerizations.

Azoalkanes (1,2-Diazenes)

Eqs. 16.68–16.69 show prototype azoalkanes, also known as 1,2-diazenes, or just azo compounds. A primary use of cyclic diazenes has been to provide an alternative entry into biradical structures that have been postulated as intermediates in other thermal or photochemical reactions. For example, photolysis of 2,3-diazabicyclo[2.2.1]heptene produces 1,3-

cyclopentanediyl (Eq. 16.68 and Eq. 11.85), a postulated intermediate in the thermal isomerization of bicyclo[2.1.0]pentane. When the photolysis is carried out under cyrogenic, matrix-isolation conditions, the triplet biradical can be directly observed by EPR spectroscopy.

$$\text{(Eq. 16.68)}$$

Similarly, the polycyclic diazene shown in Eq. 16.69 provides an approach to the cyclopropyldicarbinyl biradical that has been proposed to be involved in the di-π-methane rearrangement (Eq. 16.41). Recall our discussion of benzobarrelene above (Eq. 16.45). The di-π-methane rearrangement was favored in the triplet state, not the singlet. Consistent with these observations, sensitized photolysis of the analogous diazene strongly favors the di-π-methane product, while direct photolysis also regenerates the benzobarrelene.

$$\text{(Eq. 16.69)}$$

	24%	73%
Direct		
Sensitized	0%	100%

An alternative type of diazene is typified by azobenzene (Eq. 16.70). Photolysis now does not lead to N_2 extrusion, because the phenyl radicals that would be formed are too unstable. Instead, a fairly efficient cis–trans isomerization occurs. This process can be repeated many times, and since the cis and trans diazenes usually have substantially different absorption spectra, wavelengths can be chosen that strongly favor the cis or the trans form in the photostationary state. As discussed in a Connections highlight on photochromism on page 969, these factors have made azobenzene and derivatives favorites for the development of systems that are photochemically switchable between two forms.

$$\text{(Eq. 16.70)}$$

trans-Azobenzene   *cis*-Azobenzene

Diazo Compounds and Diazirines

Diazo compounds have long been exploited as excellent sources of carbenes (see Section 10.11.2). A valuable feature of this route to carbenes is that the spin multiplicity of the carbene can be controlled. Experiments such as those shown in Eqs. 16.71 and 16.72 were among the first to provide definitive evidence of the differing reactivity patterns of singlet vs. triplet carbenes.

$$\text{(Eq. 16.71)}$$

$$\text{(Eq. 16.72)}$$

Photolysis of α-diazoketones produces acyl carbenes, which efficiently rearrange to ketenes that are subsequently trapped (Eq. 16.73). This photochemical version of the **Wolff re-**

arrangement has found synthetic utility, especially in the ring contraction version shown, where it has been used to produce highly strained rings.

(Eq. 16.73)

Wolff rearrangement

Diazirines are simply cyclic isomers of diazo compounds (Eq. 16.74). In fact, the two forms can sometimes be interconverted photochemically. The photochemistry of a diazirine mirrors that of the analogous diazo compound—that is, loss of N_2 to produce a carbene. In some circumstances the diazirine form is more desirable, especially in photoaffinity labeling applications.

(Eq. 16.74)

A diazirine

Azides

Photolysis of azides again leads to loss of N_2, but in this case what is left behind is a nitrene (Eq. 16.75). Photolysis of azides is the best route to nitrenes. Because of the high reactivity of nitrenes, aryl azides have been popular reagents for photoaffinity labeling (see the Connections highlight on the next page).

(Eq. 16.75)

Connections

Using Photochemistry to Generate Reactive Intermediates: Strategies Fast and Slow

In our discussions of reactive intermediates throughout this text, we have alluded to and relied upon information obtained from studies involving direct observations of reactive intermediates. Usually, a photochemical reaction has been used to generate the reactive intermediate. The unique features of photochemistry have been used in two different ways in studies of reactive intermediates. First, since most photochemical reactions do not require any thermal activation, photochemistry can be conducted at very low temperatures, making it ideal for the matrix isolation technique. With matrix isolation, a precursor to a reactive intermediate is frozen in a matrix and cooled to very low temperatures, often in the 2–4 K range. Photolysis then produces the desired reactive intermediate. Because the system is very cold, the newly formed structure cannot cross any substantive thermal barriers. Furthermore, the medium is rigid, and bimolecular chemistry is ruled

out. Under these conditions, even very highly reactive species become long lived. An environment is created where a normally transient species is persistent.

A prominent strategy in matrix isolation photochemistry has been the expulsion of N_2 from stable precursors. Both diazo/diazirine precursors and 1,2-diazene precursors have been useful. The former generate carbenes and related species, while the latter are useful precursors to biradicals.

The second photochemical strategy for the study of reactive intermediates is flash photolysis (see Section 7.6.2). A very brief pulse of light generates the reactive intermediate under more conventional conditions, typically fluid media at ambient temperatures. Then a fast spectroscopic technique is used to directly monitor the ensuing reactions of the photochemically generated reactive intermediate. As laser pulses have gotten progressively shorter in duration, ever faster processes have been probed.

Connections

Photoaffinity Labeling—A Powerful Tool for Chemical Biology

A general problem in chemical biology goes as follows. A new drug or peptide has been discovered to have a potent biological effect. However, the target for the active compound, perhaps a complex protein, is unknown. Often, photoaffinity labeling has come to the rescue in such instances, with diazirines, azides, and related structures playing a prominent role. The basic concept is outlined in the drawing. The ligand is modified by a photoreactive group (**P**) in a way that does not disrupt the binding of ligand to its receptor. Although a mixture of proteins is present, the ligand will associate with its receptor. The system is then photolyzed, generating a reactive species that reacts with the protein. The ligand and receptor are now covalently linked. From here, several strategies are possible. The ligand could also be radioactively tagged with ^{125}I. Then, chromatography with isolation of only the radioactive fraction would allow the target protein to be isolated. Alternatively, the ligand could, in addition to the photoreactive group, have a biotin attached. Recall our discussion of the very potent biotin•••streptavidin interaction in Chapter 4. In this scenario, running the mixture down a column with streptavidin attached would isolate the desired protein.

Several types of photoreactive groups have been successfully employed in such studies. Substituted phenyl rings are popular, since aromatic rings are common components of drugs, and the amino acid phenylalanine is easily modified for incorporation into peptides. The basic strategy is to generate a species that will undergo insertion into X–H bonds, forming a covalent link (C–H bonds are useful, but so are O–H, N–H, etc.). It is helpful if the reaction is relatively indiscriminate, so that labeling will not be sensitive to the particular protein or environment being probed.

Both azide and diazirine groups are popular. In the latter case, the trifluormethyl group has been found to be valuable because it stabilizes the diazirine. Also, a side reaction is formation of the analogous diazo compound, and the trifluoromethyl group minimizes interference from this route. Benzophenone derivatives are also popular. The carbonyl abstracts a hydrogen, and then radical recombination creates the cross-link.

Brunner, J. "New Photolabeling and Crosslinking Methods." *Ann. Rev. Biochem.*, **62**, 483 (1993). Kotzyba-Hibert, F., Kapfer, I., and Goeldner, M. "Recent Trends in Photaffinity Labeling." *Angew. Chem. Int. Ed. Eng.*, **34**, 1296 (1995).

Mixture of proteins   Ligand with photoreactive group   Covalent link of ligand and receptor

Precursor   Reactive species   Protein   Cross-linked product

16.4 Chemiluminescence

Chemiluminescent reactions are *thermal* reactions that produce a product in an electronic excited state. The product then emits a photon. These are fascinating and aesthetically pleasing processes that have found use in both mechanistic and commercial arenas. Here we lay out the basic mechanistic schemes for chemiluminescence, and we give several examples. As we will see, chemiluminescence is most frequently associated with oxidation chemistry, and often explicitly with molecular oxygen or other species with an O–O bond. It is thus conventional to discuss certain aspects of oxygen chemistry in connection with chemiluminescence, and we will segue to a discussion of the first electronic excited state of molecular oxygen—singlet oxygen.

16.4.1 Potential Energy Surface for a Chemiluminescent Reaction

Figure 16.18 shows the potential energy diagram for a typical chemiluminescent reaction. A thermal reaction begins in the usual way, progressing uphill energetically toward a transition state. However, near that transition state the ground and excited states become close in energy, and there is a finite probability that the system will "hop" onto the excited state surface. The reaction progresses on to a minimum on the excited state surface, which then relaxes back to the ground state with emission of a photon. The emission is essentially fluorescence from the product.

In some ways a chemiluminescent process is just a typical photochemical reaction run "in reverse". That is, if we take the reaction diagram of Figure 16.18 and instead start at the right and progress to the left, and change the emission of light to absorption, we have a typical photochemical reaction. The value of this analogy is that it tells us that the same kinds of features that favor the surface crossing ultimately required of all photochemical reactions will also favor chemiluminescent reactions. In chemiluminescence, we have to cross from one surface to another, and this will be more favorable if the surfaces are close in energy. As such, chemiluminescent reactions require a funnel-like structure on the potential energy surface just as much as photochemical reactions do. Again, biradical-like structures will be important in chemiluminescent reactions, just as they are important in photochemistry.

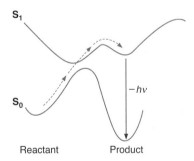

Figure 16.18
A generic potential energy surface for a chemiluminescent reaction. Transit from a ground state to an excited state must occur producing the product in the excited state leading to emission of a photon.

A related mechanism for chemiluminescence is called **chemically initiated electron exchange luminescence (CIEEL)**. Here, electron transfer from a donor to an acceptor initiates a thermal reaction that releases enough energy to place one of the reactants in an excited state. One case studied by Schuster is shown in Eq. 16.76, where electron transfer initiates CO_2 loss. This is followed by back electron transfer, giving the excited state of 9,10-diphenylanthracene, which then emits.

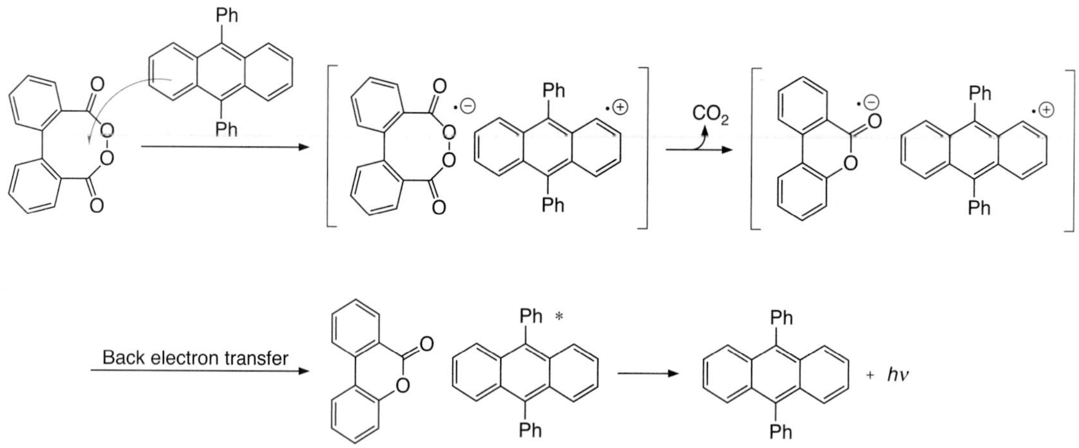

(Eq. 16.76)

16.4.2 Typical Chemiluminescent Reactions

Eq. 16.77 shows the reaction catalyzed by the enzyme luciferase that produces the characteristic light of fireflies, while Eq. 16.78, a chemical analogue, shows the exposure of the common substance luminol to oxygen under basic conditions.

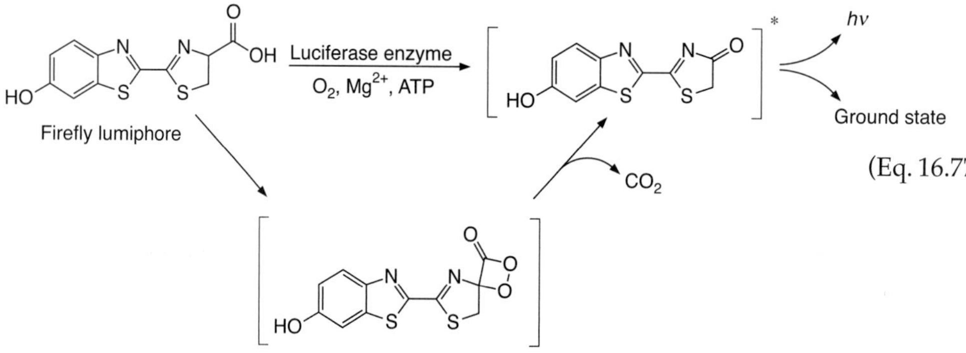

(Eq. 16.77)

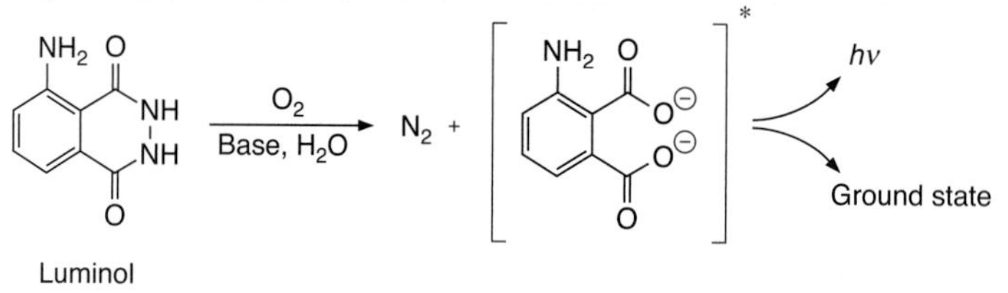

(Eq. 16.78)

The third example, Eq. 16.79, is a simple prototype, the thermolysis of tetramethyldioxetane to produce two equivalents of acetone with the emission of light. The high efficiency of this last reaction and the fact that it is induced simply by heating have made it a favorite of mechanistic investigations, as we will see below.

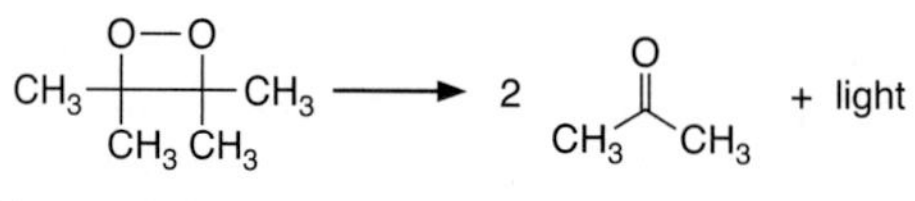

(Eq. 16.79)

Connections

Light Sticks

A fascinating and entertaining bit of organic chemistry gives rise to the glowing light sticks and necklaces that are popular at concerts and amusement parks. The prototype is based on the reaction of diphenyl oxalate with hydrogen peroxide to produce the highly strained dioxetane shown below. This decomposes with emission of a pho-

ton. Dyes are added to absorb the photon (radiative energy transfer) and then emit colored light. Standard dyes include 9,10-bis(phenylethynyl)anthracene for the classic green color and 9,10-diphenylanthracene for blue. These systems were developed by chemists at American Cyanamide and dubbed Cyalume.

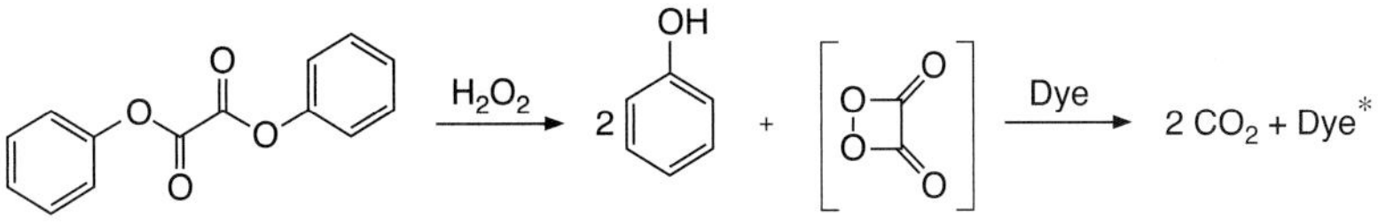

Diphenyl oxalate   A dioxetane derivative

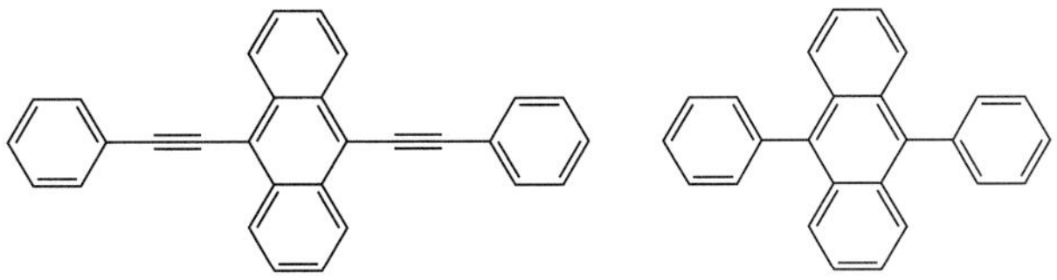

9,10-Bis(diphenylethynyl)anthracene   9,10-Diphenylanthracene

16.4.3 Dioxetane Thermolysis

Most chemiluminescent reactions involve cleavage of a strained O–O bond in the key step that generates an excited state. For this reason the thermolysis of tetramethyldioxetane has been extensively investigated. We will summarize the results of those studies here.

In looking at tetramethyldioxetane one should immediately focus on two key features. First, the molecule is a cyclobutane analogue, and so it is expected to be significantly strained (the strain energy of cyclobutane is 26.5 kcal/mol). Second, O–O bonds are very weak, with typical bond dissociation energies on the order of 34 kcal/mol. These two features, plus the fact that the product, acetone, is quite a stable molecule (C=O bonds are very strong) means that the thermolysis of tetramethyldioxetane to produce two equivalents of ground state acetone is very exothermic. Indeed, $\Delta H° \approx -63$ kcal/mol. In addition, the $\Delta H^{\ddagger}$ for the thermal reaction is ~ 27 kcal/mol.

With these numbers in hand, we can determine whether it is thermodynamically feasible to produce excited state acetone from thermolysis of tetramethyldioxetane. The S_1 state of acetone lies 85 kcal/mol above S_0, while the T_1 state is 78 kcal/mol above S_0. This is the amount of energy that needs to be produced in the thermal reaction of tetramethyldioxetane to create an excited state of the acetone. The energy released from the point of the transition state of the thermal reaction to the ground state of the products is the sum of $\Delta H_{rxn}°$ plus $\Delta H^{\ddagger}$, or $63 + 27 = 90$ kcal/mol. Hence, there is enough energy released from the point of the transition state on the thermal reaction, such that if the energy is placed into *one* of the two acetone product molecules, it can be in an excited state, and either S_1 or T_1 is possible.

The experimental result is somewhat surprising. The yields of the S_1 and T_1 states of acetone are 0.5% and 50%, respectively. The major pathway is the spin-forbidden production of the triplet state of acetone. This is nicely shown in an experiment of the following sort. When tetramethyldioxetane is thermolyzed in rigorously deoxygenated solutions, the emission is dominated by acetone *phosphorescence*, which occurs at 430 nm. However, if O_2 is added, the overall emission intensity drops by a factor of around 100, and we see only acetone *fluorescence*. This is because molecular oxygen is an efficient, selective quencher of triplet states. Remember, O_2 is a ground state triplet. Two triplet states can react to produce two singlets in a spin-allowed fashion, and that is why O_2 quenches triplets effectively.

Why is T_1 of acetone formed in preference to S_1? We can provide rationalizations based on several aspects of the process. First, the conversion of tetramethyldioxetane to two acetones is a retro-[2+2] cycloaddition. It is, therefore, a thermally-forbidden process with a substantial activation barrier. As discussed above, such reactions are very good for forming funnels that allow the efficient crossing from one electronic surface to another. Second, the reaction is initiated by cleavage of the very weak O–O bond, and at the transition state there should be considerable biradical character where singlet and triplet states are similar in energy. Third, since the biradical is oxygen-centered, spin–orbit coupling should be much stronger than for hydrocarbon biradicals. Recall our discussion of El-Sayed's rules, in which the presence of lone pairs at a radical center facilitates a coupling of a spin flip—that is, a change in spin angular momentum with a change in orbital angular momentum.

Combining these features suggests a potential energy surface like that of Figure 16.19. The S_0 and S_1 surfaces are arranged as in a typical forbidden reaction. There is an avoided crossing, and ultimately S_0 correlates to S_1. However, as we climb S_0, we first hit a crossing point with T_1. The transition to a triplet state occurs here, and we proceed on to T_1 of acetone.

Let's summarize the key features of the process. A forbidden thermal pericyclic reaction leads to biradical character in the transition state. Spin–orbit coupling facilities intersystem crossing. The thermodynamics are such that both T_1 and S_1 of the product lie below the transition state for the thermal process. Put this all together and we have a remarkably efficient chemiluminescent process.

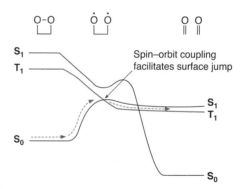

Figure 16.19
A surface diagram for the thermal decomposition of tetramethyldioxetane (methyl groups omitted for clarity).

Tetramethyldioxetane is the prototype for all chemiluminescent processes. It will generally be the case that a dioxetane or similar structure will be formed. Thermal decomposition of this high energy structure then produces an excited state product. Details vary, but many of the basics of Figure 16.19 will be involved. Thus, species containing a strained O–O bond play a special role in chemiluminescent mechanisms. For that reason, we now discuss some aspects of O_2 chemistry that are relevant to the formation of dioxetanes and related species.

Connections

GFP, Part II: Aequorin

Earlier we noted the remarkable utility of the simple protein GFP (green fluorescent protein) derived from a jellyfish. Because of this highly fluorescent protein, the jellyfish glows in the dark ocean waters. If it is dark, though, what is the excitation source for GFP? The answer is a chemiluminescent protein called **aequorin**. This remarkable protein contains a hydroperoxide that is bound to the enzyme. By a mechanism that is not well understood, external Ca^{2+} ions induce a structural change in the protein. This structural change launches the chemical cascade shown. As expected for a chemiluminscent process,

a dioxetane is involved. The penultimate product of the reaction sequence is the excited state of the anion shown in brackets. This structure emits blue light, producing the ground state. Because of this property, aequorin has found extensive use as a fluorescent sensor for biological Ca^{2+}.

In the jellyfish, aequorin is closely associated with GFP. In this environment, no photon is emitted but efficient FRET occurs, instead, exciting GFP and producing the green glow of the jellyfish. Along with providing two very useful biological tools, the *Aequorea* family of jellyfish has provided lovely examples of several of the concepts developed in this chapter.

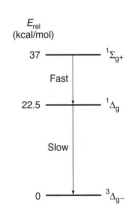

Aequorin chemistry

16.5 Singlet Oxygen

Throughout this book we have often mentioned how the remarkable properties of water play such an important role in establishing the chemistry and biology of our planet. Arguably, the other key player is molecular oxygen—O_2 or dioxygen. In its ground state, O_2 is a highly reactive, radical-like species. However, there is a very accessible excited state of dioxygen, termed singlet oxygen, that has entirely different properties. The chemistry of this excited state of dioxygen is the focus of this section.

The ground state of O_2 is a triplet, and a simple molecular orbital treatment of the bonding in dioxygen nicely rationalizes this result. The molecule contains two electrons in a degenerate pair of NBMOs, and the triplet ground state results because these NBMOs are orthogonal but co-extensive in space. Just as with the triplet n,π^* state of a carbonyl, the considerable radical character centered on oxygen atoms makes dioxygen a very reactive species. Recall that dioxygen in our atmosphere is mixed with dinitrogen (the atmosphere is 21% O_2 and 78% N_2)—an atmosphere of pure dioxygen would be quite reactive. The chemistry of ground state dioxygen is very radical-like, dominated by hydrogen abstractions and additions to double bonds.

Because of its biradical character, there are two low-lying singlet excited states for dioxygen, as shown in Figure 16.20. Relaxation from the higher singlet state to the lower singlet state is very fast, such that all reactivity involving a singlet dioxygen is from the $^1\Delta_g$ state. It lies only 22.5 kcal/mol above the ground triplet state, corresponding to light of 1270 nm in the near infrared.

E_{rel}
(kcal/mol)

37 —————— $^1\Sigma_{g^+}$

Fast

22.5 —————— $^1\Delta_g$

Slow

0 —————— $^3\Delta_{g^-}$

Figure 16.20
The triplet ground state and two low-lying excited singlet states for dioxygen.

Table 16.6
Singlet Dioxygen Lifetimes in Various Solvents*

Solvent	Lifetime (μs)	Solvent	Lifetime (μs)
H_2O	3.8	$(CH_3)_2C=O$	50
D_2O	62.0	$(CD_3)_2C=O$	723
CH_3OH	10.0	C_6H_6	30
$CHCl_3$	264.0	C_6D_6	630
$CDCl_3$	740.0	Freon-113	15,800

*Values are averages of several determinations, as compiled by Foote and Clennan: Foote, C. S., and Clennan, E. L. in *Active Oxygen in Chemistry*, SEARCH Series, Volume 2, C. S. Foote, J. S. Valentine, A. Greenberg, and J. F. Liebman (eds.), Blackie Academic, London, 1995. A very extensive compilation of singlet oxygen data is presented in Wilkinson, F., Helman, W. P., and Ross, A. B. *J. Phys. Chem. Ref. Data*, **24**, 663–1021 (1995); available on the web at http://www.rcdc.nd.edu/compilations/SingOx/SingOx.htm.

The relaxation from the excited singlet to the ground triplet state in dioxygen is quite slow. Table 16.6 shows some representative lifetimes. The lifetimes are relatively long, in the μs range, and are strongly solvent dependent. However, the most remarkable feature of Table 16.6 is the solvent isotope effect on singlet oxygen lifetimes. Deuteration of solvent always increases the lifetime, often by an order of magnitude or more. The origin of this effect can be understood as follows. The excitation energy associated with singlet oxygen can really only be given up as vibrational energy; at 1270 nm (~8000 cm⁻¹) there is no real chance for Dexter or Förster type energy transfer. Solvents with higher vibrational frequencies are more efficient quenchers. Thus, H_2O, with a vibrational frequency around 3600 cm⁻¹, is a very efficient quencher. Structures with C–H bonds (~3000 cm⁻¹) are next most efficient, while structures with no O–H or C–H bonds such as a freon are very poor quenchers. The solvent isotope effect can also be understood in this light. C–D bonds vibrate at lower frequencies than C–H bonds (see Chapter 8), and so perdeuterated solvents are much poorer quenchers.

As we have noted often, photochemistry is a game of relative rates. The longer the intrinsic lifetime of a species, the more likely it is to react. Thus, the efficiency of singlet oxygen chemistry can be profoundly influenced by just the judicious choice of solvent. Note that the relatively short lifetime of singlet oxygen in water means that under biological conditions this potentially reactive species is fairly short lived.

For the preparation of singlet oxygen, direct irradiation of the triplet ground state is not an option, because it is a spin forbidden process. However, singlet oxygen is readily prepared by one of two approaches. The first is sensitized irradiation, albeit in the reverse direction from what is usually seen in sensitization. We now have a triplet ground state to singlet excited state sensitization. Because of the low energy required to achieve the singlet state, any of the sensitizers of Table 16.5 can be used to prepare singlet oxygen. However, since all that energy is not required and could lead to unwanted side reactions, lower energy sensitizers are typically used. These are molecules that absorb in the visible region of the electromagnetic spectrum, and include classic dyes such as rose bengal and methylene blue. In addition, a variety of porphyrins and porphyrin derivatives can sensitize the formation of singlet oxygen. These are typically efficient processes, with quantum yields for singlet oxygen formation ranging from 0.5 to 0.9.

In a biological context, singlet oxygen production can have deleterious, even lethal, consequences. This process is generically referred to as a **photodynamic effect**. For example, hypericin, a pigment found in the weed St. John's wort, causes photodynamic damage to

Rose bengal

Methylene blue

Hypericin   α-Terthienyl

grazing livestock. Remarkably, plants such as marigolds produce α-terthienyl, a molecule that looks more suited to use as a conducting polymer (see Chapter 17), and it serves as a naturally produced photodynamic insecticide. Thus, if you are a gardener, it is wise to plant marigolds interspersed with your vegetables to cut down on bugs.

Connections

Photodynamic Therapy

The photodynamic effect we just discussed has been put to use in the clinic, where it is known as **photodynamic therapy (PDT)**. The basic idea is to give a patient a photosensitizer that, when exposed to light, produces singlet oxygen. This reactive oxygen species then kills cells in the immediate vicinity of its creation. Singlet oxygen has a short lifetime in water, so it cannot diffuse very far from where it is generated. Remarkably, many of these agents tend to accumulate in tumor cells, by mechanisms that are not yet certain, thus selectivity killing those cells.

Typical photodynamic agents are porphyrins and related compounds. The earliest successes came with Photofrin, a complex mixture of porphyrins that is presently approved for treatment of esophageal and lung cancer. Other examples include verteporfin (Visudyne) and SnET$_2$ (Puryltin), which are used to treat age-related macular degeneration. These compounds absorb in the 630–690 nm range, well out in the visible and a part of the spectrum that is not at all directly harmful to cells. Since light at these wavelengths only penetrates a few centimeters on direct exposure to skin, the light is usually delivered by small fiber optic devices, giving access to many but not all areas of the body where tumors could develop.

A recent addition to the PDT field has been the expanded porphyrin derivatives known as texaphyrins (that is, Texas-sized porphyrins). Sessler has developed a number of these molecules. Appropriate derivatives absorb as far out as 732 nm, and provide better penetration of the light to tissues. The lutetium derivative, termed motexafin lutetium (Lutrin), is under investigation for the treatment of breast cancer. In addition, some derivatives appear to accumulate in atherosclerotic plaques that can line blood vessels, making the notion of photoangioplasty a real possibility.

Mody, T. D., Fu, L., and Sessler, J. L. "Texaphyrins: Synthesis and Development of a Novel Class of Therapeutic Agents." *Prog. Inorg. Chem.*, **49**, 551 (2001).

Verteporfin (R$_1$, R$_2$ = H, Me)   SnET$_2$ Motexfin lutetium

The alternative approach to preparing singlet oxygen is to use a chemical process. The classic approach is the reaction of sodium hypochlorite with H_2O_2, producing singlet oxygen in nearly quantitative yield (Eq. 16.80). Reaction of ozone with phosphites produces phosphite ozonides, which can thermally produce singlet oxygen (Eq. 16.81). Another approach is a retrocycloaddition, as illustrated with the anthracene derivative shown in Eq. 16.82. While these approaches are generally not as convenient as photosensitization for preparative singlet oxygen chemistry, they have sometimes been convenient for mechanistic studies.

$$H_2O_2 + NaOCl \longrightarrow {}^1O_2 + H_2O + NaCl \qquad \text{(Eq. 16.80)}$$

$$(RO)_3PO_3 \longrightarrow {}^1O_2 + (RO)_3PO \qquad \text{(Eq. 16.81)}$$
Phosphite ozonide

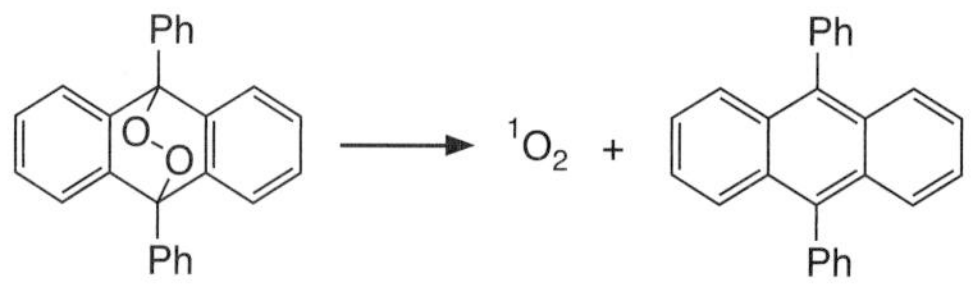

(Eq. 16.82)

The reactivity of singlet oxygen is much different from that of the ground state triplet. Instead of radical-like reactivity, we see something much more like conventional, closed-shell reactivity. In the most significant reaction, singlet oxygen behaves as a potent *dienophile* in [4+2] cycloadditions. A classic example is given in Eq. 16.83. The cyclic product from these reactions is an **endoperoxide**, and it can be easily reduced to a diol. Another synthetically useful protocol is based on the singlet oxygen cycloaddition to the so-called Danishefsky diene (Eq. 16.84). Another common and sometimes useful reaction of singlet oxygen is the ene reaction with a simple olefin (Eq. 16.85). Dioxygen inserts into an allylic C–H bond, with concomitant migration to the double bond. The high reactivity of singlet oxygen is shown by its addition to diphenylfuran, a diene species generally considered to be quite unreactive (Eq. 16.86). The diphenylanthracene adduct shown in Eq. 16.82 as a thermal precursor to singlet oxygen is in fact synthesized by cycloaddition of singlet oxygen to the anthracene in the first place.

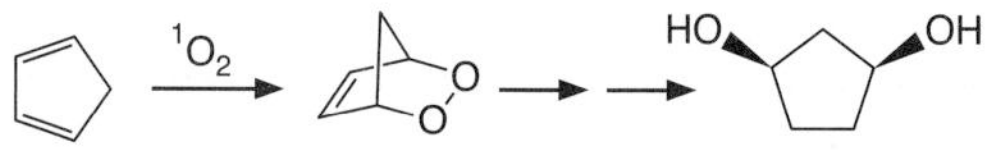

(Eq. 16.83)

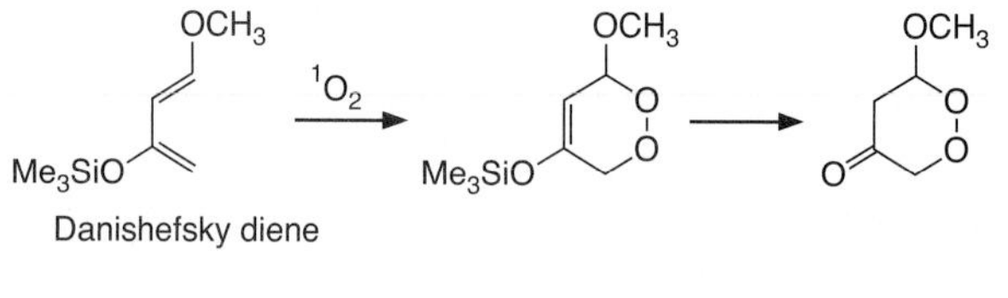

(Eq. 16.84)

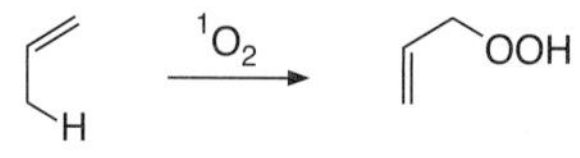

Danishefsky diene

(Eq. 16.85)

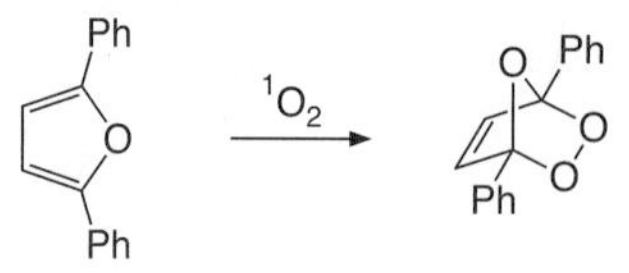

(Eq. 16.86)

Summary and Outlook

In this chapter we have shown how producing excited electronic states leads to interesting and important new reactivity patterns. We extended our picture of a single energy surface that dictates the kinetics and thermodynamics of chemical reactions to include the surfaces of electronic excited states. Multiple photophysical events can occur from these excited state surfaces—namely, internal conversion, fluorescence, intersystem crossing, and phosphorescence. When an excited state surface approaches the ground state surface in energy, it can lead to the production of a funnel, giving a path for the excited state to convert to the ground state. This will lead to chemistry in many circumstances. Due to the fact that the funnel can lead to the ground state surface at a point where the ground state structure is highly strained, unusual ring systems and twisted alkenes are common products of photochemical reactions. Furthermore, the spin state of the excited state will dictate the photochemical outcome, where a focus on the singlet/triplet character and the n or π nature gives trends as to the kinds of products expected.

In Chapter 17, we introduce "electronic organic materials". The goal there is to describe new structures with novel electronic properties. These structures can be put to use as conducting or magnetic materials, and even organic superconductors. We will return to the interaction of light with matter in Chapter 17 when we describe photoresists and non-linear optics.

Exercises

1. Consider a Stern–Volmer-type analysis of a system such as in Figure 16.8, but with one additional process, the conversion of A* to photochemical product B with rate constant k_{rxn}. Show that a plot of relative quantum yield for product formation vs. [Q] is linear and can give a value for the lifetime of A* if we assume a value for k_q. The definition of relative quantum yield is the quantum yield in the absence of quencher divided by the quantum yield in the presence of quencher.

$$A^* \xrightarrow{k_1} A$$
$$Q + A^* \xrightarrow{k_q} A$$
$$A^* \xrightarrow{k_{rxn}} B$$

2. Draw a simple MO energy diagram for a (D•••A) charge-transfer absorption complex that shows how the (D⁺•••A⁻) configuration should be a strong contributor to the excited state wavefunction.

3. Explain why charge-transfer absorptions are extremely sensitive to solvent polarity, moving to longer wavelengths as the solvent polarity increases.

4. Consider a 10^{-3} M solution of pyrene. If it is excited with a laser pulse, and then the emission is measured 1 ns after absorption, a well structured emission with a small Stokes shift is observed. However, if a delay of 100 ns is introduced between excitation and observation of emission, a broad featureless emission with a much larger Stoke's shift is observed. Explain these observations.

5. Chandross and Dempster studied the emission of linked naphthalenes. Compared to naphthalene, 1,3-bis(1-naphthyl)-propane fluoresces at a much longer wavelength. Why? Furthermore, attachment of the naphthalene groups with regiochemistry other than that shown leads to much less intense long wavelength emission. Speculate as to why.

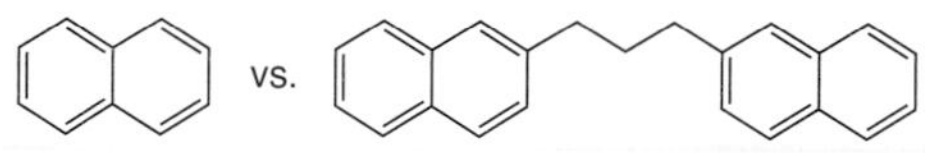

6. Why are the absorptions associated with charge-transfer complexes and the emissions associated with exciplexes always so broad and featureless?

7. In our discussion of dipole coupling in Section 3.2.2, in particular in the Going Deeper highlight on the "Magic Angle", we noted that the $3\cos^2\theta - 1$ dependence is relevant in NMR only for fixed samples, not for fluid samples in which the molecules are tumbling freely in solution. Most FRET studies of biomolecules are done in solution. Why, then, do we still need to consider the $3\cos^2\theta - 1$ term?

8. The photostationary state for stilbene photoisomerization on irradiation at 313 nm is 93% cis, 7% trans. Explain why cis is preferred.

9. The photoisomerizations of 2,4-hexadiene shown in Eq. 16.17 reflect a general trend. Singlet states often lead to isomerization of only one double bond, while triplets lead to both one- and two-bond rotations. A rationalization of this is that the singlet states have more zwitterionic character, while the triplets are biradical in nature. Show how this reasoning explains the observations.

10. Rationalize the differences in n,π^* λ_{max} values shown.

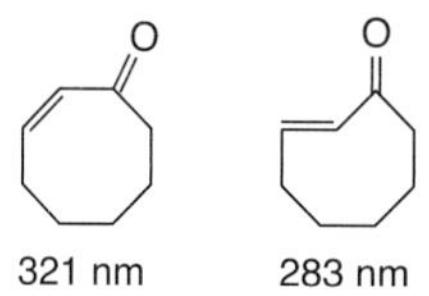

11. On photolysis, **A** undergoes primarily Norrish II chemistry, but **B** undergoes primarily Norrish I chemistry. Rationalize the difference and draw the major products expected from each reaction.

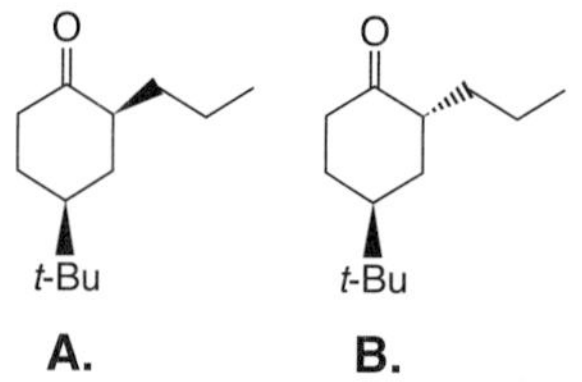

12. Discuss why the rate of decay of the acinitro species of Eq. 16.67 should be slower at higher pH values.

13. In the caged ATP structure of Section 16.3.9, a methyl group is added to the benzylic carbon of the nitrobenzyl group. Why should this increase photochemical efficiency?

14. The *p*-hydroxyphenacyl group has been proposed as a new "caging" group, shown here in color caging a simple phosphate. Photolytic release of the group is very efficient, and involves a complex rearrangment of the phenacyl group. If you want a substantial challenge, propose a mechanism for the reaction shown.

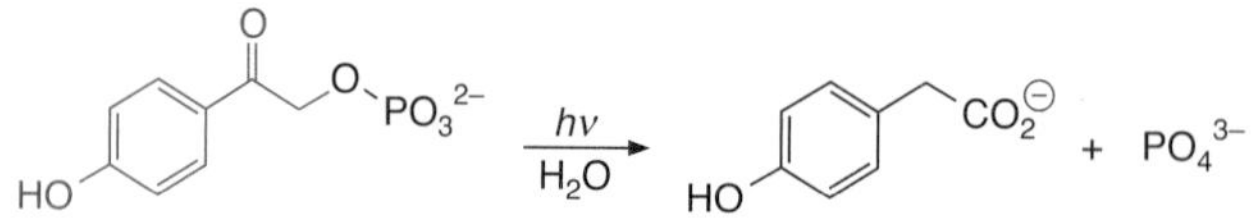

15. Draw a mechanism for the $^1\pi,\pi^*$ conversion of benzobarrelene to benzocyclooctatetraene shown in Eq. 16.46.

16. Write a mechanism for the following conversion.

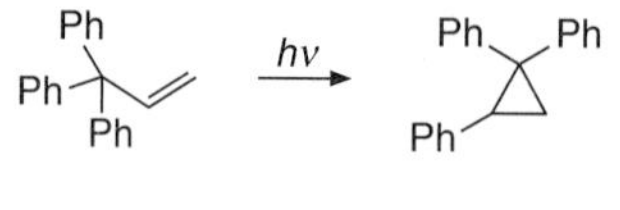

17. A student attempting the sensitized photodimerization of norbornene made the following observations. Rationalize the results.

18. Consider the three cyclohexene dimers shown in Eq. 16.28. Do the relative yields reflect the thermodynamic stabilities of the products?

19. A student wished to prepare of series of cyclobutanes that can be considered to be dimers of cycloalkenes. Sensitized photolysis of the cycloalkenes proceeded efficiently for cyclopropene, cyclobutene, cyclopentene, cyclohexene, and cycloheptene. However, cyclooctene produced little or no dimer. Rationalize this result.

20. If the photocycloaddition of acetone to *cis*-dimethoxyethylene is stopped before completion, the recovered olefin is a mixture of cis and trans isomers. Rationalize this result.

21. Rationalize the relative yields of the two isomeric products shown.

22. In the photochemical cycloaddition to form an oxetane, the biradical intermediate is considered to arise from formation of an O–C bond, as in Eq. 16.29. An alternative path would be to form the C–C bond first, followed by ring closure to the oxetane. Why is this path disfavored?

23. Provide a mechanism for the following conversion.

[plus other products]

24. Suggest a mechanism for the generation of chemiluminescence in the luminol reaction of Eq. 16.78.

25. Suggest an explanation for the relative lifetimes of singlet oxygen in benzene vs. acetone.

26. In the cellular medium, singlet oxygen has a very short lifetime. However, it is generally considered that singlet oxygen has a significantly longer lifetime in cell membranes than in the cytosol. Why would this be so?

27. Rationalize the excited pK_a values of compounds in Eqs. 16.15 and 16.16. (*Hint:* Singlet states sometimes have zwitterionic character.)

28. We noted in Section 16.3.3 that photochemical cis–trans isomerization is often complicated by the fact that both isomers absorb in similar wavelength regions, producing ultimately a photostationary state. However, in Section 16.3.8 we noted that for azo compounds, the cis and trans forms generally have substantially different absorption spectra, with the cis form always absorbing at longer wavelengths. Rationalize this result. (*Hint:* Think carefully about the structural differences between an olefin and a diazene. This is not a simple steric interaction, because both azobenzenes and simple dialkyl diazenes show the effect.)

29. a. Photolysis of a mixture of *m*-xylene and cyclopentene gives the two products shown. At least three possible mechanisms can be envisioned: 1. a concerted pathway that proceeds from an exciplex, 2. a stepwise pathway that proceeds from an exciplex to form a biradical, and 3. a stepwise pathway that involves a biradical formed from *m*-xylene, followed by reaction with cyclopentene. Sketch pathways 2 and 3.

b. In an effort to probe the mechanisms of the photoreaction, Sheridan and co-workers prepared the two azoalkanes shown. Discuss how these might be informative of the arene photoaddition reaction, including a clear explanation of what kind of results you might see and how they would or would not support mechanisms 1–3.

30. Provide a mechanism for the reaction shown.

31. In a general case, explain how a paramagnetic impurity can assist ISC.

32. Make a molecular orbital diagram for O_2, and rationalize why it is a ground state triplet.

33. Explain how a thermal barrier on S_1 prior to a funnel point to S_0 can lead to a temperature dependent quantum yield for a photochemical reaction.

34. Show that the collision between two triplet states can produce two singlet ground states, obeying Wigner's spin conservation rule. (*Hint:* Keep the spin quantum number of the electrons the same, but exchange them so that the products are singlets.)

35. Direct irradiation of cyclobutene with 185-nm light gives the following products. Rationalize the origin of each product with a mechanism that gives each. (*Hint:* Carbene intermediates are involved in some pathways, and the products can derive from common intermediates.)

36. Benzene undergoes some unusual photochemistry from its S_1 and T_1 states. Write intermediates that are consistent with the products found from these two states.

37. The photo-Claisen and photo-Fries reactions are shown below. Propose an experiment for each to test whether these reactions are concerted or stepwise, involving initial bond homolysis.

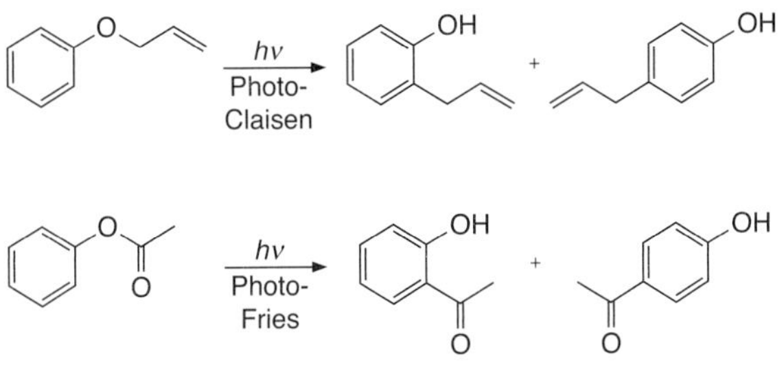

38. Give a mechanism by which the following compound undergoes deconjugation upon photolysis.

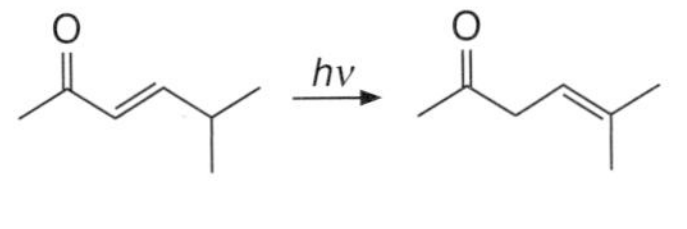

39. The irradiation of cyclic alkenes in alcohols forms ethers, or in water forms alcohols (called **photohydration**). For example, photolysis of 1-methylcyclohexene in methanol with *p*-xylene as a sensitizer gives the ether shown. Write a mechanism for this reaction. Does the addition to the double bond occur from an excited state or ground state structure?

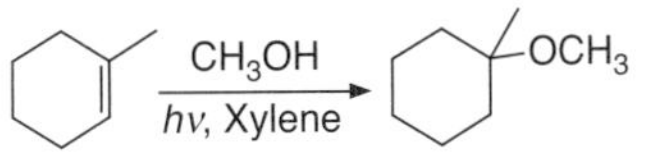

40. As mentioned in the chapter, excited states of molecules with electron donating and accepting substituents in conjugation often are well represented by resonance structures that show this character of the substituents. What is a good representation of the excited state of the following reactant? With this picture in mind, propose a mechanism for the reaction given. Propose an isotope labelling experiment to test your mechanism.

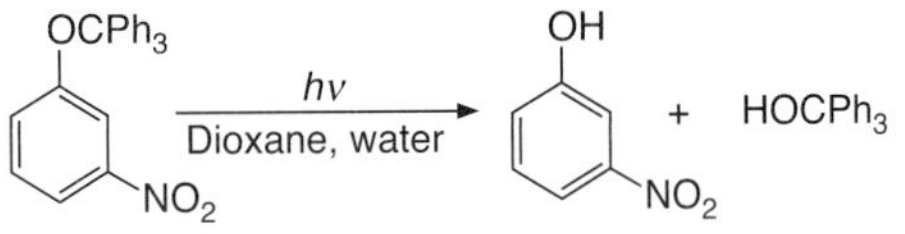

41. Draw a Jablonski diagram showing the appropriate absorption and emission energy gaps associated with an anti-Stokes shift. Similarly, draw a Jablonski diagram that shows the appropriate transitions associated with anomalous fluorescence.

42. In the discussion of solvatochromism, we noted a large dependence of the λ_{max} of emission of some fluorophores as a function of solvent (one example is given below). A larger Stokes shift in more polar solvents is often found for fluorophores where the excited state has a significant separation of charge. Give an explanation for the larger Stokes shift in the higher polarity solvents. (*Hint:* Consider the Frank–Condon factor, and then the solvation.)

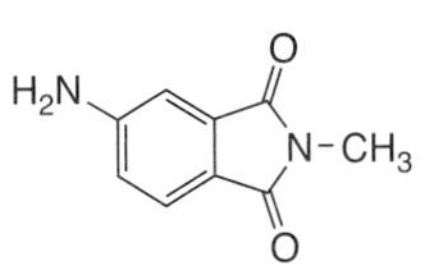

Solvent	λ_{max} (emission, nm)
Hexane	410
Water	606

43. Give short answers to the following questions.
   a. Why is the n,π^* absorption ε_{max} higher for benzophenone (115 L/mol • cm) than for acetone (15 L/mol • cm)?
   b. In an adiabatic photoreaction, the sum of three quantum yields, $\Phi_f + \Phi_{f'} + \Phi_{prod}$, was found to be equal to 1.2, where "f" represents reactant fluorescence, "f'" represents product fluorescence, and "prod" represents product formation. Why is this sum greater than 1.0?
   c. How would you classify the S_0 to S_1 absorption process in a 1,2-diazene (azo compound)? Is it spatially allowed or forbidden?
   d. Given that internal conversion is a very difficult process to study directly, how might you estimate Φ_{IC}?

44. Which of the following processes are spin-allowed? Spatially-allowed? For each, name the photophysical process involved.
   a. Benzene, S_0 to T_1, radiative.
   b. Formaldehyde, S_1 to S_0, nonradiative.
   c. Formaldehyde, T_1 to S_0, radiative.

45. The measured lifetime for the T_1 state of benzophenone is markedly lower in the presence of 1,3-pentadiene, but the T_1 state of 1,3-pentadiene cannot be detected by any of several techniques successfully used to detect the benzophenone triplet. What is happening?

46. Under appropriate conditions, the quantum yield for the disappearance of anthracene approaches 2 during the following reaction. Propose a mechanism that is consistent with this result.

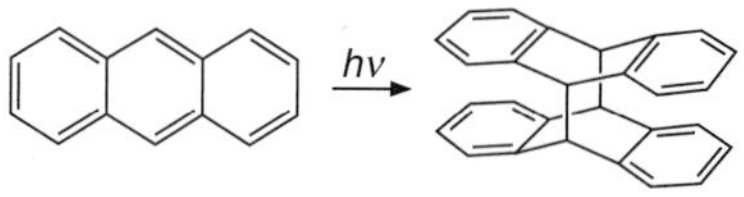

47. Most photophysical processes begin with absorption by the reactant S_0 state to give the reactant S_1 state. Another possible process could begin with absorption by the reactant S_0 state to directly produce the S_1 state of the product. Why is this generally less favorable?

48. The following quantum yield data have been obtained for the photochemical reactions of (S)-4-methyl-1-phenyl-1-hexanone.

Process	Quantum yield
Type II elimination	0.23
Cyclobutanol formation	0.03
Racemization	0.78

   a. Provide the structure of each product, along with a mechanism for its formation.
   b. Which, if any, of these products could be isolated in greater than 25% yield by this reaction?

49. Provide mechanisms for the formation of products in the following photoreactions. Why is no bicyclohexenyl product observed in the second case?

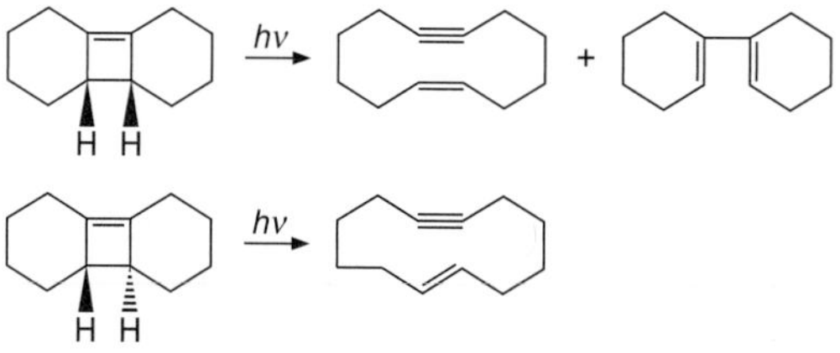

50. Fluorescence and phosphorescence spectra, while often displaying near mirror symmetry with the corresponding absorption spectra, often have significantly larger spacings between maxima (or shoulders). Why?

51. The following cyclobutanone is inert to both direct and sensitized irradiation, while cyclobutanone itself is quite photoreactive. Rationalize this difference.

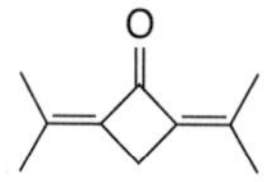

52. Provide a mechanism for the following photoreaction.

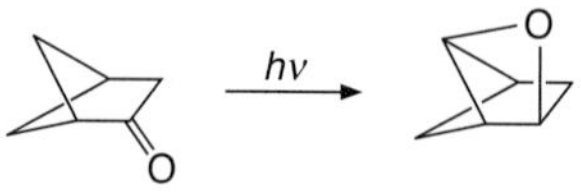

Further Reading

Textbooks on Photochemistry

Turro, N. J. (1978). *Modern Molecular Photochemistry*, The Benjamin/Cummings Publishing Company, Inc., Menlo Park, CA. The definitive and most accessible treatment of organic photochemistry.
Michl, J., and Bonacic-Koutecky, V. (1990). *Electronic Aspects of Organic Photochemistry*, John Wiley & Sons, Inc., New York. An advanced treatise on theoretical organic photochemistry.

FRET and Fluorescence Imaging

Tsien, R. Y. "Fluorescence Imaging Creates A Window on the Cell." *Chem. Eng. News*, 34–44 (1994).
Reichardt, C. "Solvatochromic Dyes as Solvent Polarity Indicators." *Chem. Rev.*, **94**, 2319–2358 (1994).
Lilley, D. M. J., and Wilson, T. J. "Fluorescence Resonance Energy Transfer as a Structural Tool for Nucleic Acids." *Curr. Opin. Chem. Biol.*, **4**, 507–517 (2000).

Funnels

Robb, M. A., Bernardi, F., and Olivucci, M. "Conical Intersections as a Mechanistic Feature of Organic Photochemistry." *Pure Appl. Chem.*, **67**, 783 (1995).

Solar Energy Storage

Schwendiman, D. P., and Kutal, C. "Transition Metal Photoassisted Valence Isomerization of Norbornadiene. An Attractive Energy-Storage Reaction." *Inorg. Chem.*, **16**, 719–721 (1977).
Franceschi, F., Guardigli, M., Solari, E., Floriani, C., Chiesi-Villa, A., and Rizzoli, C. *Inorg. Chem.*, **36**, 4099–4107 (1997).

Chemistry of Vision

Chabre, M., and Deterre, P. "Molecular Mechanism of Visual Transduction." *Eur. J. Biochem.*, **179**, 255 (1989).
Nakanishi, K., Chen, A. H., Derguini, F., Franklin, P., Hu, S., and Wang, J. "Rhodopsins Containing 6- to 9-Membered Rings. The Triggering Process of Visual Transduction." *Pure Appl. Chem.*, **66**, 981 (1994).

Singlet Oxygen

Foote, C. S., Valentine, J. S., Greenberg, A., and Liebman, J. F. in *Active Oxygen in Chemistry*, J. F. Liebman and A. Greenberg (eds.), Chapman & Hall, New York, 1995, Vol. 2, p. 333.

Excimers and Exciplexes—A Materials Vantage Point

Jenekhe, S. A., and Osaheni, J. A. "Excimers and Exciplexes of Conjugated Polymers." *Science*, **265**, 765 (1994).

Olefin Photoisomerization

Waldeck, D. H. "Photoisomerization Dynamics of Stilbenes." *Chem. Rev.*, **91**, 415 (1991).
Goerner, H., and Kuhn, H. J. "Cis–Trans Photoisomerization of Stilbenes and Stilbene-Like Molecules." *Adv. Photochem.*, **19**, 1 (1995).

Photocycloadditions

Schuster, D. I., Lem, G., and Kaprinidis, N. A. "New Insights Into an Old Mechanism: [2+2] Photocycloaddition of Enones to Alkenes." *Chem. Rev.*, **93**, 3 (1993).
Winkler, J. D., Bowen, C. M., and Liotta, F. "[2+2] Photocycloaddition/Fragmentation Strategies for the Synthesis of Natural and Unnatural Products." *Chem. Rev.*, **95**, 2003 (1995).
Griesbeck, A. G., Mauder, H., and Stadtmueller, S. "Intersystem Crossing in Triplet 1,4-Biradicals: Conformational Memory Effects on the Stereoselectivity of Photocycloaddition Reactions." *Acc. Chem. Res.*, **27**, 70 (1994).

Di-π-Methane Rearrangement

Zimmerman, H. E., and Armesto, D. "Synthetic Aspects of the Di-π-Methane Rearrangement." *Chem. Rev.*, **96**, 3065 (1996).
Hixson, S. S., Mariano, P. S., and Zimmerman, H. E. "Di-π-Methane and Oxa-Di-π-Methane Rearrangements." *Chem. Rev.*, **73**, 531 (1973).

Norrish Chemistry

Swenton, J. S. "Photochemistry of Organic Compounds. II. Carbonyl Compounds." *J. Chem. Educ.*, **46**, 217 (1969).

Davidson, R. S. "[Organic] Photochemistry." *Org. React. Mech.*, 467 (1972).

Nitrobenzyl Photochemistry

Bayley, H., Chang, C.-Y., Miller, W. T., Niblack, B., and Pan, P. "Caged Peptides and Proteins by Targeted Chemical Modification." *Methods Enzymol.*, **291**, 117 (1998).

Azo, Diazo, and Azirene Photochemistry

Padwa, A. "Azirine Photochemistry." *Acc. Chem. Res.*, **9**, 371 (1976).

Ye, T., and McKervey, M. A. "Organic Synthesis with .alpha.-Diazo Carbonyl Compounds." *Chem. Rev.*, **94**, 1091 (1994).

Chemiluminescence

Kricka, L. J. "Chemiluminescence and Bioluminescence." *Anal. Chem.*, **71**, 305R (1999).

CHAPTER 17

Electronic Organic Materials

Intent and Purpose

The materials chemistry we discussed in Chapter 13 was in some ways "classical", in that we were making polymers from monomers that have fairly conventional functional groups. Nylon has been around a long time, and while new advances such as group transfer polymerization, dendrimers, and metallocene catalysts continue to vitalize the field, the production of polymers is a relatively mature industry.

More recently, a new branch of organic materials chemistry has emerged which we will refer to as "electronic organic materials". These are polymers/materials for which the electronic structure of the material is a key component in developing desirable properties. Often these are properties such as conductivity or magnetism that are more typically associated with metallic/inorganic materials. The possibility of combining the desirable materials properties of organic polymers, such as solubility, rational synthesis and redesign, and ease of processing, with electronic properties normally associated with metals spawned the field of conducting polymers. Great advances have been made in this field, such that conducting organic materials are now important articles of commerce. To understand the conductivity in such materials, we must expand upon the concepts of electronic structure theory introduced in Chapters 1 and 14 to include the notion of band structure. We will show how the common concepts of molecular electronic structure theory have precise analogues in the terminology of condensed matter physics that is more commonly associated with conducting materials. A related, but less well-developed, area involves the quest for molecular magnetic materials. The differing strategies in pursuing a conducting vs. a magnetic material provide a nice recapitulation of some of the key concepts of electronic structure theory that we have discussed previously.

Other novel electronic behaviors that are goals of current organic materials include superconductivity and non-linear optical behaviors, and we will provide brief introductions to these. Finally, we will mention a key bit of chemistry that has been crucial to the electronics revolution—the chemistry of photoresists. Our goal is not to give an exhaustive treatment, but instead to give the student enough of a knowledge base to read modern physical organic literature on these topics.

17.1 Theory

Historically, **electronic materials** (structures with novel electrical, magnetic, or optical properties) have been the domain of condensed matter physicists. The materials involved are metals or ionic solids, structures with infinite, three-dimensional lattices and no discrete building blocks. These are quite different from the molecular-based systems that dominate organic chemistry. Organic materials are "molecular", whether they are truly molecular solids with well-defined molecules held together in a molecular lattice by relatively weak interactions, or polymers, where the molecular origins are clear, and weak intermolecular interactions are still crucial. Since the fundamental terminology and conceptual foundation of the electronic materials field was developed by scientists with a quite different perspective from that of an organic chemist, there is a language barrier that modern organic chemists

1001

Table 17.1
Correspondence Between Various
Molecular and Solid State Terms

Molecular	Solid State
LCAO–MO	Tight binding
Molecular orbital	Crystal orbital or band orbital
HOMO	Valence band; top = Fermi level (E_F)
LUMO	Conduction band
HOMO–LUMO gap	Band gap (E_G)
Jahn–Teller distortion	Peierls distortion
High spin	Magnetic
Low spin	Non-magnetic

must overcome. In reality, many concepts from condensed matter physics translate quite directly to well known chemical concepts. One goal of this chapter is to provide a translation table, designed to familiarize chemists with some of the important ideas related to electronic materials. Such translation is given in Table 17.1, and our goal now is to show how the molecular terms relate to the solid state terms.

17.1.1 Infinite π Systems—An Introduction to Band Structures

The novel electronic properties we will consider here are almost always associated with extended, conjugated π systems. Our earlier analysis of the electronic structure theory of π systems will certainly be relevant here, but we must adapt it to materials science. We begin by considering the infinite linear polyene at the Hückel molecular orbital (HMO) level. Keep in mind that throughout Section 17.1 we are discussing hypothetical polymers that are infinitely long, perfectly linear, and defect free. Such systems are the best for developing key theoretical insights. However, real materials never satisfy these criteria, and beginning in Section 17.2 we will look at real materials.

First, we will model our infinite π system as a polymer with butadiene as the "monomer". Actually, it is the π system of butadiene that is the monomer; we are not actually polymerizing butadiene as discussed in Chapter 13. Conceptual coupling of two monomers here implies the loss of H_2 so as to keep the π system intact. As shown later in this section, the polymer we are considering is called **polyacetylene**. However, viewing this polymer initially as being formed from butadiene fragments is instructive for our theoretical development. We are operating at the HMO level, and so are considering only π orbitals. In solid state work, this level of theory is often referred to as a **tight binding** model. Figure 17.1 summarizes the results.

For butadiene ($n = 1$), we have the familiar four π orbitals. For $n = 2$ (octatetraene) the eight MOs can be considered to be the in-phase and the out-of-phase combinations of the four butadiene MOs—each butadiene MO gives rise to two octatetraene MOs. There is a slight split in energy, with the in-phase lying below the out-of-phase. The 12 MOs for $n = 3$ can similarly be thought of as four sets of three, each set containing the three linear combinations of one of the four original butadiene orbitals. The overall spread in energy for each set is larger than for $n = 2$. The pattern continues, and as we add more and more diene monomers, the set of MOs related to an individual butadiene MO contains more and more orbitals and spans a wider range of energies.

At the infinite polyene, each of the four sets of MOs contains an infinite number of MOs, each of which is a particular linear combination of one of the MOs of the monomer. All MOs in a given set trace their origin to the same monomer MO. Such a collection of orbitals is called a **band**. An individual MO within a band is called a **crystal orbital** or a **band orbital**. Just like MOs, crystal orbitals can hold two electrons, the spins of which must be paired. The energies of the individual crystal orbitals within a band are not equally spaced, although the

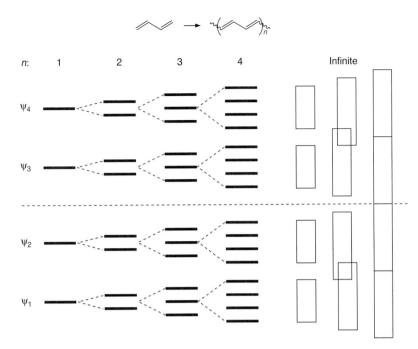

Figure 17.1
Band structures. Evolution from π MOs in finite polyenes to bands in an
infinite π system.

picture of a band might give this impression. Instead, the crystal orbitals may cluster around
certain energies within a band, such that there are different densities of states within a band.
The **density of states** refers to the number of orbitals around a given energy.

A priori, it is difficult to predict the range of energies for a given band, and Figure 17.1 illustrates several possibilities. The four bands might retain their discrete characters. Some
bands may spread so much that they overlap with their neighboring bands, as shown in the
second case in Figure 17.1. In some cases, bands might "touch", with the top of one band being degenerate with the bottom of another.

How can we develop a more quantitative model of these bands? One strategy is to
model the infinite linear polyene not as a linear combination of smaller polyenes, but rather
as an infinite cyclic π system, $-(CH)_n-$. If n really is infinity, there is no difference between a
line and a circle (from the viewpoint of electronic structure, that is, but not from the viewpoint of topology!). Recall Eq. 14.56, here restated as Eq. 17.1. This gives the HMO energies
for any cyclic polyene, where i is the orbital number and N is the number of vertices (atoms
or p orbitals in the cycle [infinity in this case]). In this equation i runs from $i = 0, \pm1, \pm2 \ldots$ up
to $\pm N/2$ for even N or $\pm(N-1)/2$ for odd N. As such, this index directly corresponds to the
number of nodal planes in the MO.

$$\varepsilon_i = \alpha + 2\beta\cos\left(\frac{2i\pi}{N}\right) \qquad \text{(Eq. 17.1)}$$

Figure 17.2 plots the results of Eq. 17.1 in a convenient format for $N = 5$ and $N = 15$. The
x axis represents the number of nodes in the MO; the plus and minus values are used to represent a degenerate pair. That is, for $N = 5$ (cyclopentadienyl) there is a single MO with no
node; two MOs with one node (denoted as ±1 in the figure); and two MOs with two nodes.
As noted above, the x axis has a total range of $\pm(N-1)/2$. The y axis is energy, and for a cyclic
polyene in HMO theory, its range is $\alpha \pm 2\beta$. With increasing N, the number of orbitals increases with an ever increasing number of nodes, but the basic pattern remains the same. As
N approaches infinity, the collection of isolated MO energies will become a continuous line
in the shape of a cosine curve.

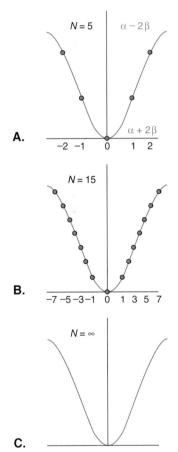

Figure 17.2
Schematic depicting HMO orbital energies for cyclic polyenes
with **A.** 5, **B.** 15, or **C.** an infinite
number of vertices. The x axis is
the number of nodes, and the y
axis is energy.

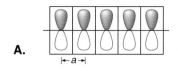

A.

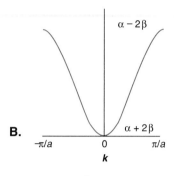

B.

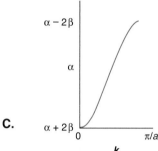

C.

Figure 17.3
A. Definition of a unit cell and the parameter a for polyacetylene with a one carbon unit cell.
B. Recasting the infinite cyclic polyene diagram of Figure 17.2 into a band structure format.
C. The conventional presentation of a band structure, showing only half of the Brillouin zone.

Let's recast this reasoning into solid state terms. Our reason for doing this is to learn the nomenclature normally used in solid state chemistry and physics, but also to make connections between molecular orbital schemes (how chemists think) and the model for electronic structure typically used for infinite solids. For an infinite system, it is useful to define a **unit cell**—the simplest pattern that repeats throughout the solid. For the "polybutadiene" discussed above, the unit cell would have four carbons. However, when modeling the same system as an infinite cyclic π system, the simplest unit cell is a single C atom (Figure 17.3 **A**). Recall that at the HMO level, all interactions between adjacent atoms are equivalent; there is no alternation between single and double bonds as implied by the polyene structure. We will return to this point later. The monomer thus has only one "MO" (it's actually a p orbital), and so we expect only one band for the polymer. We define the size of the unit cell as a (equivalent to the distance between adjacent atoms in this case), such that the position of unit cell p (R_p) is as given by Eq. 17.2. By position, we mean the distance of the unit cell p from the beginning of the polymer. Here, p is any number from one to infinity.

$$R_p = (p-1)a \qquad \text{(Eq. 17.2)}$$

Returning to Eq. 17.1, it is now convenient to define a parameter k as in Eq. 17.3. We will discuss the nature of k more fully below, but for now we can make a few comments. Returning to our cyclic polyene analysis, the lowest energy MO is the one with $i = 0$ and so it has a value of $k = 0$. The highest energy MO corresponds to $i = \pm(N/2)$, and plugging this into Eq. 17.3 shows that the maximum k value is $\pm\pi/a$. Inserting Eq. 17.3 into Eq. 17.1 then gives Eq. 17.4, the standard way to represent orbital energies in an infinite linear system.

$$k = \frac{2\pi i}{Na} \qquad \text{(Eq. 17.3)}$$

$$E_k = \alpha + 2\beta \cos(ka) \qquad \text{(Eq. 17.4)}$$
$$\text{where } k \text{ runs from 0 to } \pm\pi/a$$

In Figure 17.3 **B** we redraw Figure 17.2 **C**, but the x axis is now in terms of k, rather than the number of nodes. The total range is now $\pm\pi/a$ rather than $\pm(N-1)/2$. The curve is the same, we have just re-labeled the axes and cast things into more traditional solid state terms. The region of the plot of Figure 17.3 **B** from $k = 0$ to $k = \pm\pi/a$ is called the **first Brillouin zone**. The points $k = \pm\pi/a$ are called the **zone edges**, and the point $k = 0$ is called the **zone center**. Finally, it is convenient to consider only the right half of the plot in Figure 17.3 **B**; the left half is redundant. With these modifications, we can write the HMO band structure for the infinite polyene with a unit cell of a single atom as in Figure 17.3 **C**. As we noted above, the infinite linear polyene is a model for a real material—polyacetylene—and so Figure 17.3 **C** presents the HMO band structure for this known material.

A truism of solid state physics is that the electronic structure of a material should not change if we simply redefine the unit cell. To verify this, let's instead model polyacetylene with a two carbon unit cell, as in Figure 17.4 **A**. This creates a new parameter a', where $a' = 2a$. The monomer now has two MOs, and so we expect two bands, one from π and one from π^* of the ethylene monomer. The lower band should arise from π, the higher from π^*.

It is time to consider the meaning of k. It is an indication of the phase relationships of the band—in fact, the x axis of such figures is called a **phase space**. When $k = 0$, all the monomer MOs are in-phase, and when $k = \pi/a$, all the monomer MOs are out-of-phase. *However, the "phase" we are talking about is a solid state phase—an expression of how the unit cells are assembled, not the orbitals.* Two unit cells are "in-phase" if one can be obtained by a direct translation of the other. "Out-of-phase" implies a sign change accompanying the translation. Let's clear this up with an example.

Consider the band that arises from the π MO of ethylene (Figure 17.4 **B**). At $k = 0$, we put all the unit cells in-phase and obtain a crystal orbital that is completely in-phase in an electronic sense. Thus, all intercell overlaps (those between atoms in adjacent unit cells) are favorable and the energy is low. In fact, the energy of the band at this point is exactly $\alpha + 2\beta$, the

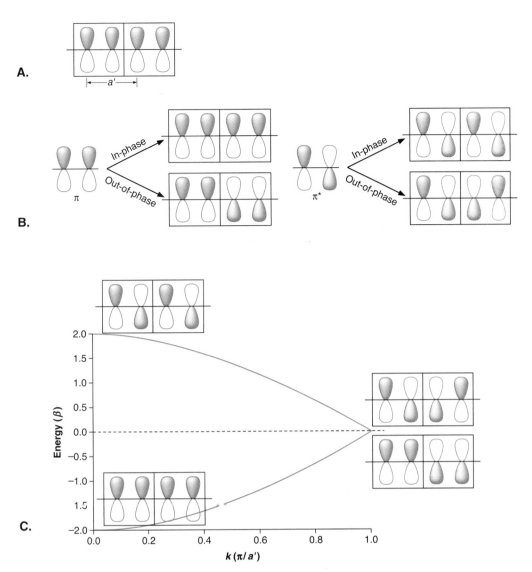

Figure 17.4
The band structure of polyacetylene using a two-carbon (ethylene) unit cell. **A.** A two-atom unit cell and the new unit cell parameter a'. **B.** The definition of "in-phase" and "out-of-phase" in infinite polymers. Note how for the band derived from the π MO, the intercell overlap is favorable for the in-phase combination and unfavorable for the out-of-phase combination, while the opposite is true for the band derived from the π^* MO. **C.** The band structure. The lower band arises from the π MO of ethylene and the higher band from the π^* MO. The x axis is now k in multiples of π/a', and so it ranges from 0 to 1. The y axis is β that would be used in the equation $E = \alpha + 2\beta$, so the 0.0 point is energy α.

energy of the lowest MO in every cyclic polyene. As we move along the x axis from left to right, marching through phase space, we are progressively adding more *intercell* nodes (the intracell interaction is always favorable in this band), and the energy is steadily rising.

Now, let's go to $k = \pi/a'$. This is the completely out-of-phase combination of ethylene π MOs, and it is drawn in Figure 17.4 **B**. For every intracell bonding interaction, there is a compensating intercell antibonding interaction. The crystal orbital is net nonbonding, and so the energy is exactly α.

Analysis of the the band arising from the ethylene π^* MO brings home the meaning of phase in such diagrams. At $k = 0$ we have the totally in-phase combination of ethylene π^* MOs, drawn in Figure 17.4 **B**. It can be seen that in-phase means *translationally* in-phase. That is, the MO in any unit cell can be obtained by just translating the MO in the adjacent cell with no adjustment. In contrast to what we saw with the lower lying band, at $k = 0$ all intercell interactions are antibonding. Since the intracell interactions are also antibonding, the crystal orbital is completely antibonding, and it has energy $\alpha - 2\beta$. Now as we march through phase space, making adjacent unit cells out-of-phase, we are introducing *bonding* intercell interactions, and the energy of the band drops. Jumping to $k = \pi/a'$, we now place each π^* ethylene orbital out-of-phase with its neighbor, maximizing the intercell bonding interaction. Inspection will reveal that at the zone edge, the crystal orbital derived from the ethylene π^* MO is completely nonbonding. It is equivalent to the crystal orbital derived from the out-of-phase interaction of the ethylene π MOs; the two are degenerate.

Putting this all together, we arrive at the band structure diagram of Figure 17.4 **C**. We have two bands. One starts at low energy and rises in energy as k increases; the other starts at high energy and falls in energy as k increases. The two meet at $k = \pi/a'$.

We said that we should get the same result regardless of how we define our unit cell. So, are the bands for polyacetylene shown in Figure 17.3 **C** and Figure 17.4 **C** the same? To address this, we need to consider where the electrons reside. Recall that a band is just a collection of an infinite number of crystal orbitals, each of which can hold two electrons. To fill the bands, we use the aufbau principle and consider how many electrons are in the MOs of the molecular building block. For Figure 17.3 **C**, with a one-atom building block, there is one electron in each p orbital before we begin mixing p orbitals to make bands. The unit cell MOs (actually AOs) are half full. The crystal orbitals formed by linear combinations of the p orbitals can each hold two electrons. As such, the single band formed for the polymer should be half full. There are enough crystal orbitals that if each p orbital started out with two electrons in it, we could accommodate them all. However, the p orbitals started out half full, so the band derived from them is half full. Where is the HOMO? Or, more correctly, where is the **Fermi level**, E_F, the energy of the highest occupied crystal orbital? It is halfway up the band, with an energy of precisely α.

Now let's put electrons into the band diagram of Figure 17.4 **C**. The building block MOs start out with two electrons in π and no electrons in π^*. Thus, we would expect the lower band, derived from π, to be completely filled, and the higher band, derived from π^*, to be completely empty, and that is the case. Where is the Fermi level? It is exactly at the top of the lower band, at energy α.

So, both models of polyacetylene lead to the same predictions as to electronic structure. We start at energy $\alpha + 2\beta$ and then progress through to an energy of precisely α. Note also that both models predict that the **band gap**, E_g, the energy gap between the highest occupied crystal orbital and the lowest unoccupied crystal orbital, should be precisely zero. This is an important finding that we will return to below.

Figure 17.5 illustrates an interesting way to show that the two analyses of polyacetylene band structure really are the same. On going from Figure 17.3 **C** to Figure 17.4 **C** we doubled the size of the unit cell. However, the x axis in a band diagram tracks k, which is proportional to $1/a$ (Eq. 17.3). Thus, phase space has been halved. We illustrate this by "folding" the Figure 17.3 **C** band structure around the midpoint of the x axis. This produces precisely the band structure for the larger unit cell. This is a general technique that will work for any system.

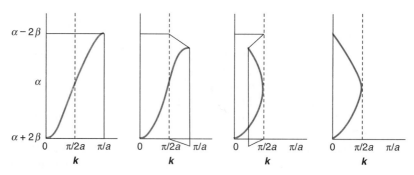

Figure 17.5
Illustration of how doubling a unit cell corresponds to folding a band diagram in half.

We have developed our model for the simplest possible system, one p orbital per unit cell, and then argued qualitatively about the system with two AOs per unit cell. In fact, there is a rigorous, analytical solution for the generalized problem of a unit cell with any number of MOs in an infinite polymer. This produces the **Bloch equations** that are the direct analogues of HMO theory as applied to infinite systems. To create computed band structures of the sort shown throughout this chapter, we simply solve the Bloch equations at several points throughout the Brillouin zone and then interpolate between these points to produce the smooth curves shown in band structures.

Throughout this chapter we will only consider one-dimensional band structures. Real materials, however, are three-dimensional. But, as we discussed in Chapter 13, in some materials there is one dimension that is much "stronger" than the others, and theoretical modeling can, to a good approximation, focus on that. The conducting polymers we will be considering here fall into that category. Three-dimensional band structures can be developed, and they are essential for many types of systems such as true metals (i.e., copper or silver). Conceptually, the process is identical, except now we have three k's to consider: k_x, k_y, and k_z. The only real problem is how to visualize a multidimensional band structure, and the agreed upon symbolism takes some getting used to. For those of you who are interested, the book by Hoffmann referenced at the end of the chapter has an especially lucid introduction to higher-dimensional band structures.

What is the value of these band structures? Just as an MO diagram is highly informative about the nature of a molecule, the band structure makes clear predictions about the nature of an electronic material. Figure 17.6 illustrates this by showing certain representative band diagrams. We return to the symbolism of Figure 17.1, showing a band as a block of crystal orbitals, with the understanding that this is equivalent to the representations of Figures 17.2 **C** and 17.4. Again, filled and empty bands are crucial. The highest-lying filled band is called the **valence band**, and the lowest-lying empty band is called the **conduction band**. Thus, the Fermi level is the top of the valence band, and the band gap is the energy separation between these two crucial bands.

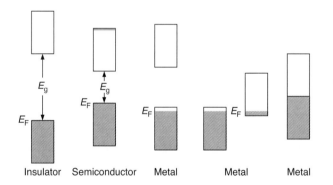

Figure 17.6
Relationships between various aspects of a band structure diagram and the expected electronic properties of a material. An insulator has a very large band gap, while a semiconductor has a small but finite band gap. A metal has a zero band gap, because of an incompletely filled band or because of the overlap of two bands.

As the name suggests, electrical conduction in a material is generally associated with population of the conduction band. If E_g is large, this is very unlikely, and we have an **insulator**. On the other hand, if E_g is finite but relatively small, we could have a **semiconductor**. An **intrinsic semiconductor** requires a finite thermal population of the conduction band, and so E_g cannot be much larger that kT. This is rare, and most semiconductors require doping to show any significant conductance (see Section 17.1.3).

Our polyacetylene band structure has an E_g of zero. Figure 17.6 shows different ways of achieving this. We can have a partially filled band, or we can have two bands that overlap or touch (as in Figure 17.4 **C**) at the Fermi level. Either way we have zero band gap, and such a system is a **metal** and is expected to be a **conductor**. Based on this, we conclude that *at the level of HMO theory, polyacetylene is an organic metal!* We hasten to add that the real situation, beyond HMO, is more complicated, such that real polyacetylene is a semiconductor. We will return to this below. Still, given the success of HMO theory in making interesting predictions about molecular properties, not always quantitatively correct but usually qualitatively cor-

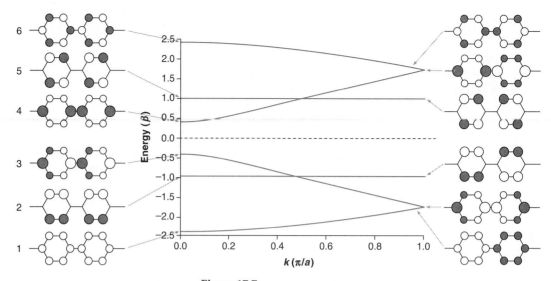

Figure 17.7
The band structure analysis for PPP.

rect, we can understand the excitement about highly conjugated polymers generated by the analysis of polyacetylene.

Knowing the band structure provides some interesting predictions about the potential properties of a conjugated organic polymer. As such, it is valuable to be able to predict or rationalize band structures. As always, computer programs are available to calculate band structures at the HMO level and beyond. There is considerable value in being able to predict prototype band structures, however, as this allows you to understand and interpret the results of sophisticated calculations. To develop this skill, we describe here the band structure of an infinite polymer of para-linked benzenes, poly(p-phenylene), PPP, another prototype and an important material in the conducting polymers field.

Figure 17.7 shows the calculated band structure for PPP at the HMO level, and here we will simply walk through the figure and discuss the origin of each band. We assume a coplanar structure, although real PPP is not totally coplanar. The MOs of the monomer are familiar (see Figure 14.13); note that because of the para substitution pattern in the polymer, we must represent the NBMOs in the way shown, which is consistent with the symmetry. We number the six bands **1–6**, starting with the lowest energy band. Band **1** arises from the fully symmetric benzene MO and rises in energy as we proceed from left to right. The starting energy of the band is now below $\alpha + 2\beta$, because the fully in-phase combination should be more stable than the starting MO, which itself began at $\alpha + 2\beta$. Bands **2** and **5** are unusual. There are no intercell interactions (remember we are at the level of HMO, so only interactions between connected atoms count). As such, these bands are completely flat. Band **3** is qualitatively similar to the higher band from Figure 17.4. Because the coefficients of the MO at the two connection points of the polymer are oppositely signed, the band starts at high energy and falls as we march though phase space. This system also provides an example of two bands crossing, a common feature in more complex systems. The higher-lying bands, **4–6** mirror the lower-lying bands, as reasoning based on the pairing theorems would predict.

Where are the electrons? We have six electrons per monomer to distribute, and these are used to fill bands **1–3**. Bands **4–6** are empty. Band **3**, therefore, is the valence band because it contains the highest occupied crystal orbital, and band **4** is the conduction band. The Fermi level E_F lies at the zone center in this system, while it is at the zone edge for polyacetylene. There is no real-world significance to this distinction. One very meaningful distinction between polyacetylene and PPP, however, is evident at HMO level. By inspection you can see that the highest occupied and lowest unoccupied crystals orbitals should not be degenerate in the case of PPP. PPP has a finite band gap in HMO theory, while polyacetylene does not.

The band structure of Figure 17.7 illustrates one more feature we need to discuss. Note that the range of energies spanned by a band on going from $k = 0$ to $k = \pi/a$ is smaller for bands **1** and **6** than it is for bands **3** and **4**. And, the spread for bands **2** and **5** is zero. The range of energies spanned by a band in the first Brillouin zone is called the **dispersion** of the band. What causes the dispersion? The difference in energy between $k = 0$ and $k = \pi/a$ is related to how strong the intercell interaction is, and that should be related to the size of the coefficients on the interacting atoms. For bands **1** and **6**, the coefficients of the monomers at the point of connection are all $1/\sqrt{6}$ (0.408); for bands **3** and **4** the value is $2/\sqrt{12}$ (0.577). The larger coefficient for bands **3** and **4** leads to larger intercell interactions and thus a greater dispersion in the band. The MOs that produce bands **2** and **5** have coefficients of zero at the key atoms, and so the dispersion is zero.

We are now ready to lay out a prescription for developing a qualitative band structure for any conjugated polymer. Remember, we are operating at the level of HMO theory. The simple steps are:

1. *Choose a unit cell.*

2. *Create one band for each MO of the unit cell molecule.*

3. *Whether a band rises or falls in energy on going from* k $= 0$ *to* k $= \pi/a$ *is determined by the relative signs of the coefficients at the connection points between unit cells.*
   If they have the same sign, the band rises; if they have the opposite sign, the band falls.

4. *The starting and ending energies of a band can be estimated from the starting energy of the MO and how strong the intercell interactions are.*

5. *The dispersion of a band is determined by the magnitude of the coefficients of the atoms that form intercell connections in the polymer.*

With this simple formula, quite useful band structures can be prepared for a wide range of conjugated materials. You will be given the opportunity for practice in the Exercises at the end of the chapter.

17.1.2 The Peierls Distortion

It's time for a reality check. We know well that HMO theory has some severe limitations, so we should be skeptical about the remarkable predictions it makes about polyacetylene and other systems. Qualitative trends should be correct, but is polyacetylene really a metal? We need to go to the next level of theory, and just as with molecular systems, the next step is to include the effects of electron–electron repulsion that are so ardently avoided in HMO theory.

At the level of HMO theory, polyacetylene is a metal. When we use a two-carbon unit cell, which we will do from now on, the band structure consists of a filled band and an empty band that are degenerate at the zone edge, producing a zero band gap. We have seen a similar situation before in the molecular world with the MO diagram of cyclobutadiene (Section 14.5.1). The HOMO and the LUMO are degenerate and contain a total of two electrons. We also saw (the Going Deeper highlight in Section 14.5.6) that when we include electron–electron repulsions, a Jahn–Teller distortion converts cyclobutadiene to a rectangular form, and this geometrical distortion opens up a HOMO–LUMO gap in the molecule.

A similar effect occurs in polyacetylene (Figure 17.8). Just as with cyclobutadiene, polyacetylene undergoes a geometrical distortion from a structure with all equal bond lengths (implicit in the HMO model), to a structure with alternating long and short bonds. In condensed matter physics this is called a **Peierls distortion**, but it is directly analogous to the Jahn–Teller distortion seen in cyclobutadiene. The actual bond lengths in polyacetylene are approximately 1.35 Å and 1.45 Å. And, just as a HOMO–LUMO gap opens up in cyclobutadiene, the geometrical distortion opens up a band gap in polyacetylene.

Figure 17.8 shows why this is so. By shortening the C–C distance within the unit cell (illustrated by drawing the intracell C–C bond as a double bond), the intracell interaction becomes stronger than the intercell interaction. At the zone edge, therefore, the band made

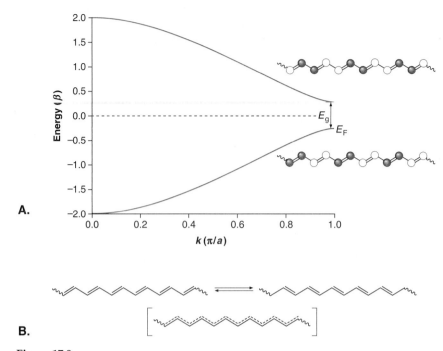

Figure 17.8
The Peierls distortion in polyacetylene. **A.** The band structure after the Peierls distortion. Note that now a finite band gap exists, and E_F lies below α. **B.** The two degenerate forms of polyacetylene along with the hypothetical, bond-equal form (in brackets).

from the ethylene π MO will be more stable than the band from π^* (recall the drawings in Figure 17.4 **C**). In the lower band at the zone edge, bonding interactions occur across the shorter bonds (double bonds) and antibonding interactions across the longer bonds; the opposite is true of the higher band. The resulting E_g in polyacetylene is 1.5 eV, a significant value.

Because of the Peierls distortion, polyacetylene shows alternating short and long bond lengths and has a significant band gap. As such, *pure polyacetylene is a semiconductor, not an organic metal*. The alternating bond length pattern in polyacetylene is worth contemplating. Organic students generally come away from introductory courses viewing benzene as the prototype conjugated π system, with all C–C lengths equal. Actually, benzene is more the exception than the rule; most neutral, closed-shell π systems show alternating bond lengths. Ions and radicals (such as allyl), though, do tend to equalize bond lengths. We noted in Chapter 14 that even for benzene, some have argued that the π system prefers alternating bond lengths as it does in polyacetylene, and it is the σ framework that produces equal bond lengths.

Physicists describe polyacetylene as having a **degenerate ground state**. This is a somewhat unfortunate terminology. In this context "degenerate" means there are two equivalent forms, as shown in Figure 17.8 **B**. Conceptually, there is a higher-lying, more symmetrical, form that is intermediate between the two, but this form may not exist in the real world. The two degenerate structures are *not resonance structures*. Remember, all resonance structures for a molecule have the same geometry—that is, nuclei do not move in resonance. The degenerate forms of polyacetylene have different geometries, in that nuclei would have to move in order to interconvert the pair. That said, there is no guarantee that the two forms can or will interconvert. There could be a very large barrier separating the two. As we will see below, the important issue is that the two forms exist.

Peierls distortions occur in other systems, and in fact they are fairly common in solid state physics. A common situation in which a Peierls distortion might be expected is in a crystal of relatively flat molecules that stack on each other along one crystal axis. Rather than spacing evenly along the axis, the molecules will often "dimerize", forming a closer contact with one neighbor and opening up a larger gap to the other neighbor in the stack. This, too, is a Peierls distortion, and it is ultimately driven by the same kinds of arguments as presented here.

17.1.3 Doping

We now know that polyacetylene is a semiconductor, with a degenerate ground state and a significant band gap caused by a Peierls distortion. The intriguing band structure of Figure 17.4 turns out to be an artifact of the low level of theory used to produce it. Given this, you might wonder why there is so much excitement about conducting polymers. One other obvious target, polyparaphenylene (PPP), has a significant band gap also, even at the level of HMO theory. In fact, this is the case for all other systems except some very exotic, strictly theoretical structures (see the Exercises at the end of the chapter). The facts are, linear conjugated polymers built from molecular π systems such as ethylene (to give polyacetylene) or benzene (to give PPP) or other conjugated molecules are at best semiconductors, and often are insulators.

The rescue from this situation requires a chemical treatment of the polymers known as **doping**. Doping is another term from condensed matter physics originating in studies of more conventional semiconductors such as silicon or gallium arsenide. To a chemist, doping is nothing more than the oxidation or reduction of a polymer system. Exposure of a material such as polyacetylene to oxidizing agents such as I_2 or AsF_5 will remove electrons from the polymer. This is termed **p-type doping**, because positive charges are introduced. Alternatively, reducing agents such as sodium metal or sodium naphthalenide will introduce negative charges, leading to **n-type doping**. Both p-type and n-type doping can also be achieved electrochemically. All semiconductors need to be doped in order to be useful, and conducting polymers are no exception. The focus of our discussion will be on p-type doping, but the terminology and reasoning are identical for n-type doping.

Let's consider the consequences of the p-type doping of polyacetylene (Figure 17.9). We remove one electron to make a radical cation. This radical cation will be delocalized, but to what extent will it be delocalized? At this point, the intuition that most organic chemists have developed fails. As noted above, we are explicitly and implicitly taught that delocalization is "intrinsically good", and the more delocalization, the better. Based on that view, removal of one electron from a chain of polyacetylene should produce an infinitely delocalized radical cation. However, this view is wrong. Theory and experiment both establish that *the radical ion obtained by doping an infinite conjugated polymer is only partially delocalized*. Figure 17.9 illustrates such structures. The partially delocalized radical cation (anion) obtained by oxidative (reductive) doping of a conjugated polymer is called a **polaron** (a different entity called a **soliton** also arises; see the Going Deeper highlight on page 1015). These local alterations in structure are often referred to generically as **defects** in the solid state literature.

Why doesn't a polaron fully delocalize? We can rationalize this by thinking a little deeper about the stabilization afforded by delocalization. For example, we know that allyl radical is much more stable than ethyl radical (as evidenced from bond dissociation energies), and we attribute that stabilization to delocalization. It is also true that pentadienyl is more stable than allyl, but the difference is much smaller than the difference between allyl

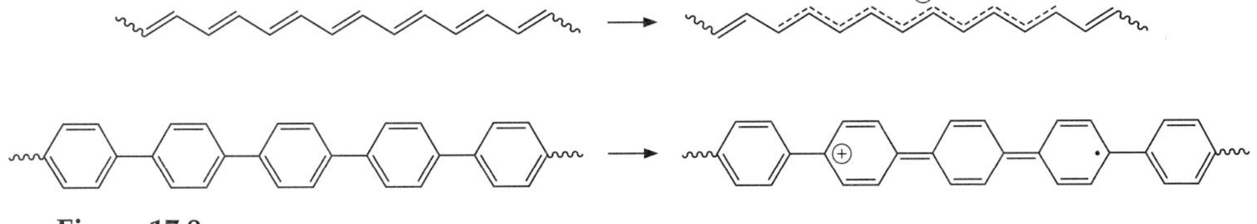

Figure 17.9
P-type doping of polyacetylene and PPP. Note that in the actual material the polarons are more extensively delocalized than shown.

and ethyl. As we continue to delocalize, producing heptatrienyl, nonatetraenyl, etc., stabilization energies grow, but by ever diminishing amounts. Eventually, very little is gained by further delocalization. On the other hand, the Peierls distortion tells us that there is energetic advantage to having alternating short and long bonds, and delocalizing the polaron disrupts this. For other polymers, such as PPP, the disruption caused by delocalization is even greater. In PPP, each time we delocalize the polaron into another ring, we are disrupting an aromatic system. Eventually, in any system, the small benefit of further delocalization is outweighed by the cost of disrupting the structure, and the polaron "stops delocalizing". Another factor disfavoring full delocalization is the counterion. We are introducing charges into the polymer, and there must be a counterion. The counterions are not mobile in the solid, and it is energetically beneficial for the charge on the polymer to stay in the general vicinity of its counterion.

Thus, the doping of polyacetylene or PPP produces partially delocalized radical ions called polarons. These polarons can migrate up and down the polymer chain, and this is what produces a conducting polymer. The migration takes the polarons from the vicinity of one counterion to that of another counterion. *All conducting organic polymers are in fact doped forms of otherwise semiconducting or insulating conjugated polymers.* As we continue to expose the polymer to oxidant, more polarons are introduced along the chain, and the conductivity goes up. This is the light doping phase of the process of doping, and because the doping can be stopped at any time, it is simple to prepare lightly doped samples. If we continue the doping reaction, though, producing heavily doped samples, a few more interesting things happen. To understand these, however, we need to look a little more deeply at the electronic consequences of doping.

A perhaps counterintuitive result of doping is the development of so-called **mid-gap states**. One way to think of the band gap is as an energy region where no orbitals are found. Doping, however, introduces orbitals into this region (Figure 17.10). The terminology is a bit sloppy here; "mid-gap states" is the conventional term, but we are really talking about mid-gap orbitals, and we will use that terminology here. When we remove an electron from the pristine polymer, we take it from the top of the valence band. If nothing else happens, we simply have a **hole** in the top of the band. "Hole" is a conventional term in this field, meaning a positive charge associated with the loss of an electron. To solid state physicists, there is no difference between a hole and an electron (except the sign of the charge), and systems are often referred to as hole conductors or electron conductors, depending on the nature of the mobile species. P-type doping introduces a hole at the top of the valence band, but doping leads to a geometrical change in the system. The radical ion delocalizes to make the polaron, and this relaxation of the structure stabilizes the system. What may not be immediately obvious is that this relaxation leads to the singly occupied crystal orbital lying in the band gap, isolated from the rest of the valence band.

One way to understand why this is so is to make a crude analogy to simple radical chemistry. At the level of HMO, what is the difference between methyl radical and allyl? To make the comparison balanced, we should consider methyl + ethylene vs. allyl, the former representing the polaron before it delocalizes. Allyl is more stable, but actually the singly occupied MO is at the same energy in both systems, an energy of precisely α (look back at Figure 14.15). Allyl is more stable because the *filled* orbital has dropped in energy (from $\alpha + \beta$ in ethylene to $\alpha + \sqrt{2}\beta$ in allyl), not the singly occupied orbital. In a sense this is what happened in the band structure. Several orbitals in the valence band have dropped in energy, leaving a singly occupied crystal orbital alone in the gap. A similar thing happens in the conduction band, producing an empty orbital (a hole) in the gap. Another explanation for the appearance of orbitals in the band gap is that the Peierls distortion is removed in the vicinity of the polaron. We saw above that the Peierls effect is the reason for the band gap, and without it the orbitals fill the gap and the bands meet at the Fermi level. More optical transitions are now possible involving these mid-gap orbitals, and some transitions will be of longer wavelength than was possible for the undoped polymer. As a result, doping generally introduces color into the material, and a heavily doped material is typically black.

Now we can consider the progression from light to heavy doping. As we just noted, at first doping just produces more polarons throughout the material. However, on continued

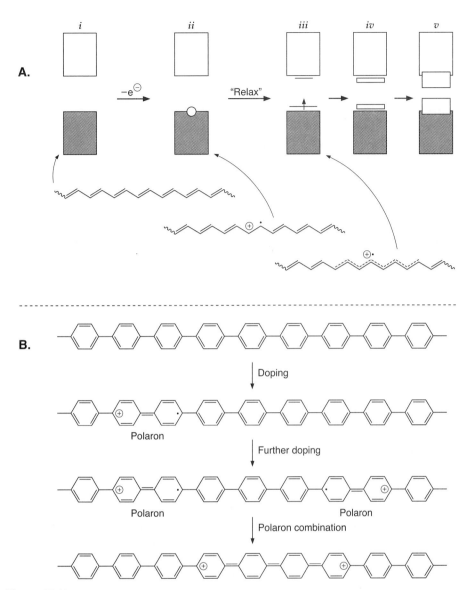

Figure 17.10
The consequences of doping on the band structure of a conjugated system. **A.** The evolution of the band structure. *i.* Pristine polymer. *ii.* Hypothetical structure after removal of one electron. *iii.* Structure in *ii* after relaxation to form the polaron. *iv.* Formation of polaron bands. *v.* Illustration of how in some cases the polaron/bipolaron bands can merge with the valence and conduction bands to produce a true metal. **B.** Formation of bipolarons on extended doping. In the actual materials the polarons are more extensively delocalized than shown.

doping an odd thing happens. EPR reveals that the number of spins starts *decreasing*. We are doping more, the conductivity is increasing, but we appear to be losing polarons. Indeed, at heavy doping we start to form a new species termed a **bipolaron**. A bipolaron in this context is a spin-free, doubly-charged, partially-localized structure, and Figure 17.10 **B** illustrates bipolaron formation. One way to think about bipolaron formation is as the direct combination of two polarons on the same chain. Polarons can migrate along the chain, and two could run into each other. The more polarons we make, the more likely they are to run into each other. Why would two polarons combine to make a bipolaron? Polarons are radicals, and two radicals should naturally combine, making a new bond in the process. Polarons are also ions, so shouldn't the coulombic repulsion associated with bringing two like charges together prevent bipolaron formation? This is certainly true in the gas phase, where radical cations never combine to make dications. We are in the solid state, though, not the gas phase. The solid surroundings provide a significantly higher dielectric constant than the gas phase (anything is better than the gas phase), and this diminishes the coulombic repulsion. Also, doping always introduces counterions into the polymer, and often these are large diffuse ions such as I_3^- or I_5^- from I_2 or $As_2F_{11}^-$ from AsF_5. These substantially shield the charges associated with polarons and facilitate bipolaron formation. With enough shielding, the energetic advantage of forming a new bond between two radicals outweighs the cost of bringing together like charges, and bipolarons form.

There is another way to think about bipolaron formation. Consider the band structure of Figure 17.10 right after removal of one electron and relaxation of the structure. If you are an oxidant, what is the next electron you would remove? The mid-gap electron is the highest-lying, so we might expect it to be the most easily removed. Doing so would diminish the number of spins, and therefore oxidation is just another way to think about bipolaron formation.

At light doping, therefore, we have polarons, and as the doping gets heavier we have bipolarons. Either of these can migrate along the chain, making the material conductive. At high doping levels we start to produce polaron/bipolaron bands, and these get broader (their dispersion increases) as the doping progresses (Figure 17.10 **A**). In some instances, at very heavy doping, the bipolaron band can merge into the valence band. At this point, we have a true metal, and indeed the properties of some heavily doped conjugated polymers do resemble those of a true, conductive metal. We'll discuss these issues further in Section 17.2.

UV/vis spectroscopy is a convenient tool to analyze aspects of the band structure of a material. In an undoped material, for example, a transition from the valence band to the conduction band is often seen, and the energy of this transition is one measure of E_g. Such spectroscopy is also very informative about the doping process. Even light doping leads to a substantial enrichment of the UV/vis spectrum of the polymer, as several new, longer wavelength bands are seen.

Figure 17.11 shows a series of absorption spectra as a function of doping for a conducting polymer, and these provide considerable experimental support for the theoretical model of Figure 17.10. The spectra are for a film of polypyrrole, but they are typical of the results seen for many conducting polymers. The spectra are interpreted as follows. The bottom spectrum is that of "pristine" polypyrrole, and the large absorption at 3.2 eV corresponds to the transition from the top of the valence band to the bottom of the conduction band—the "HOMO–LUMO" transition that defines the optical band gap. However, defects are present even in this sample, and the weak bands at 0.7, 1.4, and 2.1 eV involve polarons and bipolarons. As doping progresses (moving up Figure 17.11), the band at 1.4 eV first increases in intensity and then disappears. Monitoring of the number of spins in the sample by EPR spectroscopy shows a corresponding rise and then a fall in signal intensity. As such, the absorption at 1.4 eV involves polarons. On the other hand, the bands at 0.7 and 2.1 eV show a continual increase on doping, and so they are associated with bipolaron states.

There is an important consequence of the polaron model of conductivity in conducting polymers. In a metal such as copper or silver, conductivity increases as the temperature is lowered. This is because cooling reduces vibrations in the crystal lattice. These vibrations disrupt the free movement of electrons in conduction bands, and so damping them by lowering the temperature improves conductivity. In contrast, conductivity typically decreases as we lower the temperature in a doped, conjugated polymer. This is because there is a finite thermal activation energy for conduction in such a system. Remember, the geometry (C–C bond lengths and angles) of a polaron is different than that of the undoped polymer. Thus, as a polaron migrates along the chain, nuclei must move to interconvert between the doped and undoped geometries. This process requires energy, and so conductivity is better at higher temperatures. The measurement of conductivity as a function of temperature is straightforward, and it provides a key guide as to whether one is dealing with a true metallic system.

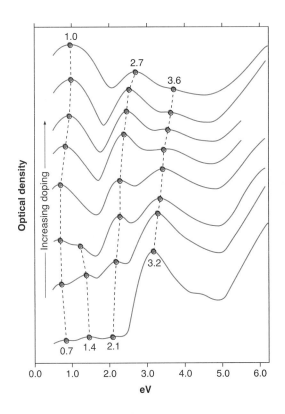

Figure 17.11
Optical absorption spectra of a thin film of polypyrrole as a function of degree of doping. See text for a description. Patil, A. O., Heeger, A. J., and Wudl, F. "Optical Properties of Conducting Polymers." *Chem. Rev.*, **88**, 183–200 (1988).

Going Deeper

Solitons in Polyacetylene

Our treatment of the doping of polyacetylene has in fact been a bit misleading. While the progression we have described, involving polarons and bipolarons, is appropriate for the overwhelming majority of conducting polymers, including PPP, the situation is subtly different in polyacetylene. Even in an undoped sample of polyacetylene, neutral defects analogous to a polaron can form. A partially delocalized radical defect called a **soliton** will exist in any real sample of polyacetylene.

A soliton

The physics of such a system are quite interesting, and the properties of a soliton resemble those of a solitary wave (hence the name). Such a neutral defect can only exist in systems with a degenerate ground state (two identical geometrical forms), and so solitons only form in polyacetylene and a few other, relatively exotic materials. A soliton forms a phase boundary between the two degenerate forms of polyacetylene. Note how in the structure shown, the double bonds go "southwest to northeast" to the left of the soliton and then go "southeast to northwest" to the right of the soliton. Any sample of polyacetylene will contain a significant number of spins detectable by EPR even before doping. Initial oxidative doping will lead to a *decrease* in the number of spins, as neutral solitons are converted to charged, spinless solitons. Extensive doping starts to introduce spins after all the neutral solitons are destroyed. The distinction between a soliton and a polaron is of little consequence for chemists interested in designing many kinds of better materials (such as a better battery). However, anyone who delves into the conducting polymer literature will certainly run across the term, and so we have provided a brief introduction here.

To summarize, all conjugated organic polymers are semiconductors or insulators in their pristine state. Doping produces polarons and then bipolarons, where polarons are spin-containing defects and bipolarons are spinless defects. These defects can migrate along the polymer chain, allowing the system to be conductive. This is the theory. The reality is much more complicated. In Section 17.2 we will discuss some real conducting polymers. This theoretical foundation, coupled with some practical, chemical aspects of the systems, will produce a good working model for how to understand conducting polymers.

17.2   Conducting Polymers

In this section we introduce the polymers that are actually seeing commercial use or are approaching that point. We will also discuss some of the practical aspects that substantially influence the extent to which a material can be conductive. First, though, we introduce the basic terminology of the field, and refresh your memory with some of the basics of electrical conductivity.

17.2.1   Conductivity

The essential equation for consideration of conductivity is **Ohm's law**, Eq. 17.5, where I is **current** in amps, R is **resistance** in ohms, and V is the **voltage** drop (in volts) across the material. The equation shows that for a given V, higher current requires lower resistance. The resistance of a wire (Eq. 17.6) depends on the length, l, the cross-sectional area, A, and the intrinsic ability of the material to conduct, which is defined in terms of the **resistivity**, ρ (with units of ohm•cm). The inverse of resistance is **conductance** (in units of Siemens, S, equivalent to ohm^{-1}), and the inverse of resistivity is the **conductivity**, σ [in units of S•cm^{-1} or (ohm•cm)$^{-1}$] (Eq. 17.7).

$$V = IR \qquad \text{(Eq. 17.5)}$$

$$R = \frac{\rho l}{A} \qquad \text{(Eq. 17.6)}$$

$$\sigma = \frac{1}{\rho} \qquad \text{(Eq. 17.7)}$$

Typically, we grade a material based on its conductivity, and a huge range is seen in this important parameter. Conventional metallic conductors such as copper, iron, or silver have conductivities on the order of 10^6 S•cm^{-1}, while intrinsic semiconductors have much lower values, such as 10^{-2} S•cm^{-1} for Ge and 10^{-5} S•cm^{-1} for Si. Insulators, naturally, have even smaller conductivities, with values such as 10^{-10} S•cm^{-1} for glass, 10^{-14} S•cm^{-1} for diamond, and 10^{-18} S•cm^{-1} for quartz or Teflon. Thus, conductivity spans a range of 24 orders of magnitude in common materials.

On the molecular level, conductivity can be related to the number of charge carriers (electrons and/or holes, n) and their intrinsic **mobility**, μ, as in Eq. 17.8 (where e is the charge on an electron). Thus, to increase conductivity, we want to increase the number of carriers (n), which implies increasing the doping level. Also, we want to maximize the mobility of the carriers (μ). Although it is not always simple to relate mobility to a molecular feature, it is reasonable to assume that increased delocalization and the absence of defects would both favor increased mobility.

$$\sigma = n\mu e \qquad \text{(Eq. 17.8)}$$

Before discussing real materials we must address one other issue. As noted above, our band structure analyses are one-dimensional, and everything we have discussed so far has emphasized conduction along the conjugated backbone of the polymer. In real materials, however, that is not enough. Real samples have to conduct in all directions, and because it is highly unlikely that every (or even one) polymer molecule will extend fully from electrode to electrode, conduction must occur *between* chains as well as along chains. Often this interchain conductivity is the limiting factor in the usefulness of a material. We will not go into the details here. We simply mention that interchain conductivity implies a "hopping" of polarons from one chain to another, in what is fundamentally just an electron transfer reaction. Factors such as packing density and the alignment of chains play important roles, so keep these in mind as different polymers are discussed. For example, adding long alkyl chains may increase the solubility of the polymer, but may also discourage the close packing of chains that would encourage hopping. Such trade-offs are common in efforts to design optimal materials.

17.2.2 Polyacetylene

Both theoretically and experimentally, the launching point for conducting polymers is polyacetylene. We've discussed the theory, and now we consider the actual material. Early efforts by Natta to polymerize acetylene produced a black, air-sensitive, insoluble powder. A key breakthrough was Shirakawa's discovery in the early 1970s, which showed that when gaseous acetylene was passed over a film of a classical Ziegler–Natta-type catalyst (prepared by evaporating a solution of the catalyst onto the walls of a flask), a shiny, copper-colored film of polyacetylene formed that could be easily peeled off the flask (Figure 17.12 **A**). Analysis revealed that this material had almost exclusively the cis configuration about the double bonds. Heating the material to 150 °C (or running the polymerization at higher temperatures) produced the thermodynamically more stable all-trans form, which is a silvery substance with a mirror-like surface. These appearances are remarkable for a hydrocarbon, but neither material was conductive. *cis*-Polyacetylene shows a conductivity of 10^{-10}–10^{-9} S•cm^{-1}, while *trans*-polyacetylene shows a conductivity of 10^{-5}–10^{-4} S•cm^{-1}, making the latter a reasonable semiconductor.

It remained for a chemist and a physicist, MacDiarmid and Heeger, respectively, to appreciate the potential of doping for improving the conductivity of pristine polyacetylene. In 1977, Shirakawa, MacDiarmid, and Heeger reported that exposure of polyacetylene to Cl_2, Br_2, or I_2 produced a material with conductivities as high as 10^3 S•cm^{-1}, and the field of conducting polymers was born. Suddenly, materials that always had been viewed as insulators had the potential to transform the electronics industry. For this work, Shirakawa, MacDiarmid, and Heeger shared the 2000 Nobel Prize in Chemistry.

While the intellectual importance of polyacetylene is undeniable, the technological importance is less so. The material is insoluble, brittle, and thus difficult to process. Recall that a key promise of electronic organic materials is to combine the facile synthesis and processability features of organic polymers with novel electronic properties. Polyacetylene does not quite meet the goal.

Many efforts to improve the solubility and processability of polyacetylene have been reported. A recurring theme in this field is the interplay between crystallinity and solubility. Planar conjugated polymers tend to be highly crystalline and thus notoriously insoluble, making them difficult to handle. The simplest way to increase the solubility is to break up

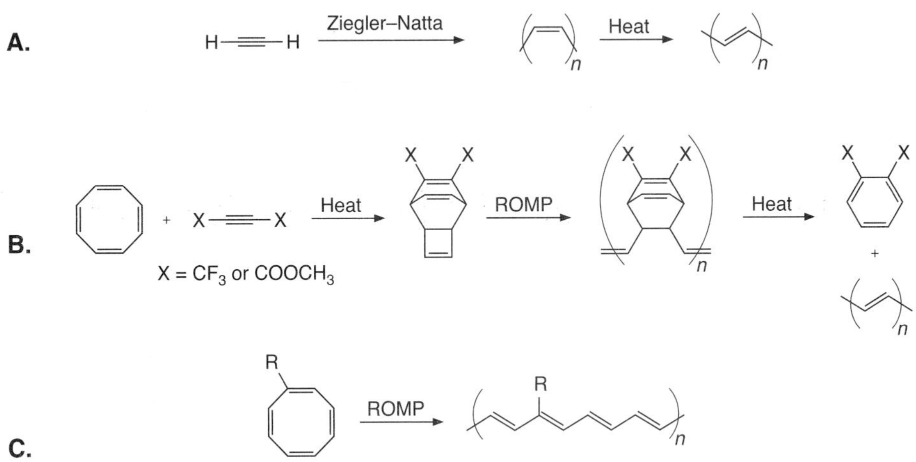

Figure 17.12
Various routes to polyacetylene and derivatives. **A.** Direct Ziegler–Natta polymerization of acetylene as developed by Shirakawa. **B.** Feast's retro Diels–Alder, precursor polymer route. **C.** Ring-opening metathesis polymerization (ROMP) of a substituted cyclooctatetraene as developed by Grubbs.

the crystallinity by adding flexible side chains, not unlike the strategy for preparing low-density polyethylene (Chapter 13). One could imagine that polymerizing a terminal acetylene (RC≡CH) would add an alkyl chain to every other carbon, and this would increase the solubility. Indeed, when R is alkyl or aryl the solubility of the derived polyacetylene is increased. However, the conductivity of the doped material disappears. The R group introduces significant twisting in the backbone, disrupting the conjugation. The trade-off between increasing solubility and disrupting planarity is a universal theme in the conducting polymers field (see the Connections highlight of Section 2.4.1).

Several clever strategies around this conundrum have been reported. Feast developed a retro-Diels–Alder approach to polyacetylene, summarized in Figure 17.12 **B**. Ring-opening metathesis polymerization (ROMP) of the Diels–Alder adduct of an acetylene and cyclooctatetraene gives a **precursor polymer** that is completely soluble and can be easily processed into films. Simple heating launches the retro-Diels–Alder reaction, producing true polyacetylene in film form. Although the retro-Diels–Alder directly produces a cis double bond, the temperature for the reaction is sufficient such that cis–trans isomerization occurs at the same time. This precursor polymer approach has been used in many systems to fight the solubility battle.

Another strategy involves the ring-opening metathesis polymerization (ROMP) of cyclooctatetraene. While ROMP of the strained cyclobutene in the Feast system could be achieved with traditional metathesis catalysts, only the more advanced catalysts developed by Schrock and Grubbs enabled the direct ROMP of cyclooctatetraene. Grubbs recognized this as a alternative route to polyacetylene, and the resulting material has a very high conductivity, perhaps because of fewer defects. This material is still insoluble and intractable. Solubility can be dramatically improved, however, by adding a *single* alkyl group to the COT monomer. Now, instead of an alkyl group on every other carbon, as is obtained from polymerizing an alkyl acetylene, there is one every eight carbons. The resulting polymer is soluble and processable, but it can still be doped to acceptable levels of conductivity. A good trade-off between solubility and twisting has been reached.

Another important advance in polyacetylene chemistry is the polymerization protocol developed by Naarman and co-workers. Using a specially prepared catalyst system and clever *in situ* processing conditions, acetylene could be polymerized to produce a near defect-free, highly crystalline polyacetylene. On doping, conductivities as high as $10^5\,\text{S} \bullet \text{cm}^{-1}$ were seen. An organic polymer was showing conductivites that rival those of copper metal!

Despite these more advanced efforts, polyacetylene has yet to become an article of commerce. Its importance lies in launching the conducting polymers revolution, and in establishing that very high conductivities can be seen with doped organic polymers. Its downfall is its insolubility. Other systems have been developed that have more desirable combinations of conductivity and processability, and we will review these below. Note that for many applications, conductivity values that rival those of copper are unnecessary. Often values of 10^2–$10^3\,\text{S} \bullet \text{cm}^{-1}$ are quite acceptable.

17.2.3   Polyarenes and Polyarenevinylenes

The most common variation on polyacetylene in efforts to produce more tractable conducting polymers is to replace the simple double bonds of polyacetylene with aromatic rings. This generally increases the stability of the system and offers several strategies for introducing solubilizing groups. Unfortunately, such systems intrinsically start with larger band gaps, as we showed previously in our discussion of PPP. However, the trade-off between solubility and conductivity is favorable for many of these systems.

Figure 17.13 shows prototype structures. While none of these is likely to replace copper wiring any time soon, the novel combinations of properties associated with these types of materials has led to an array of commercial applications that continues to grow in importance. These are typically applications in which very high conductivity is not essential, and features such as light weight, flexibility, and processability are more important. Examples include the use in fabrics for antistatic properties, radar blocking, electromagnetic interfer-

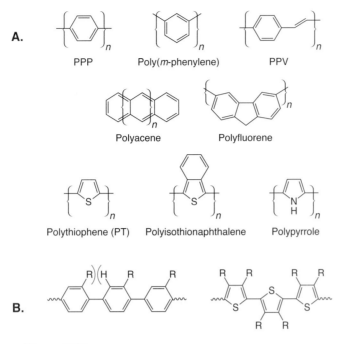

Figure 17.13
A. Representative conducting polymers. **B.** Steric interactions.

ence (EMI) shielding, and in certain types of batteries. Another advantage of organics is the ability to use substituents to tune the band gap (see below). This, in turn, influences the optical properties of the material, much like changing the HOMO–LUMO gap of a molecule affects its absorption and emission properties (recall the discussion in Section 16.1.2).

One of the most promising applications of conducting polymers is as thin film materials for use in light emitting diodes (LEDs) and flat panel displays. Organic light emitting devices (OLEDs) convert electrical energy into light, a phenomenon known as **electroluminescence**. When an electric current passes through some conducting polymers, light is emitted. In its simplest form, an OLED consists of a layer of the luminescent material placed between two electrodes. When an electric current is applied, light is emitted with a color that depends on the particular material used. At the anode, holes in the material are created, while at the cathode, electrons are added to the material. The conducting properties of the polymer allow the holes and electrons to migrate together, and when they meet a localized excited state of the polymer is created, analogous to the state created upon absorption of a photon. The excited state energy is released in the form of fluorescence, thereby producing light.

By varying substituents, the optical properties of PPV and PT derivatives can be tuned to a considerable degree, allowing a wide range of colors to be produced (see Figure 17.13 for structures). Displays based on organic polymer films are already in use (for example, the display for car stereo systems). Prototypes of displays that are highly flexible, able to be bent and twisted to a considerable degree, have also been prepared. This seems likely to be a high growth area for conducting polymers in coming years. Another promising application of conducting polymers in the near future is as organic transistors and field effect transistors.

Table 17.2 shows band gap values for representative systems; structures can be found in Figure 17.13. The E_g values given are computed using the valence effective Hamiltonian (VEH) approach, an *ab initio* protocol similar to HF theory. We show computed values so that consistent comparisons can be made. Experimental values are very close to those shown, but there is variation, depending on the method used to measure E_g and the condition of the par-

Table 17.2
**Computed Band Gaps for Representative
Conjugated Polymers**

Structure	E_g (eV)
Polyacetylene	1.5
Poly(p-phenylene) (PPP)	3.2
Poly(m-phenylene)	4.5
Poly(p-phenylenevinylene) (PPV)	2.5
Polyacene	0.0
Polythiophene (PT)	1.7
Polyisothionaphthalene	0.5
Polypyrrole	3.0

Source: Brédas, J. L., Chance, R. R., Baughman, R. H.,
and Silbey, R. "Ab initio Effective Hamiltonian Study
of the Electronic Properties of Conjugated Polymers."
J. Chem. Phys., **76**(7), 3673–3678 (1982); Brédas, J. L., Silbey,
R., Boudreaux, D. S., and Chance, R. R. "Chain-Length
Dependence of Electronic and Electrochemical Properties
of Conjugated Systems: Polyacetylene, Polyphenylene,
Polythiophene, and Polypyrrole." *J. Am. Chem. Soc.*,
105, 6555–6559 (1983).

ticular polymer (number of defects, cis / trans ratio, etc.). Several features of Table 17.2 bear comment. For example, PPP is predicted to have a significantly larger gap than polyacetylene, even at the HMO level, and indeed this is borne out. PPV has a significantly lower band gap than PPP, and this trend holds up for many other systems. Polyacene has been a topic of theoretical study for some time, but it has yet to be synthesized. Even at high levels of theory a negligible band gap is predicted, and some analyses even suggest the polymer could be a superconductor. You are asked in the Exercises at the end of the chapter to develop the band structure of polyacene.

There is a general sense that smaller band gaps are better for producing good conductors, although it is not immediately clear why this should be so for doped systems. With a small band gap, though, the merging of the polaron bands with the conduction and valence bands can happen more readily, and certainly true metallic conduction is more strongly associated with polythiophene and other small band gap systems.

An advantage of PPP, PT, and similar structures is that it is possible to introduce solubilizing groups without completely destroying conjugation and thus conductivity. Figure 17.13 **B** shows representative examples. In PPP, substitution does cause considerable twisting, but the effect is much less severe in PT. This makes sense, because the smaller angles associated with the five-membered ring pull the substituents back away from the conducting backbone.

Several basic synthetic strategies are applicable to these types of systems. Thiophene and related structures can be directly polymerized electrochemically. Oxidation gives a relatively stable radical cation which then reacts with neutral thiophene to launch the polymerization. A thin film of conducting, doped PT can be induced to form on the oxidizing electrode. Although this approach is less controlled than some methods, it is very convenient. For PPP and derivatives, early approaches involved a simple reaction of dihaloaromatics and various metal-mediated coupling reactions. However, these approaches often gave mixtures of regioisomers (meta vs. para coupling). A substantial advance was the development of the much better controlled Suzuki- and Stille-type couplings (Figure 17.14 **A**), for which there is no ambiguity as to the newly formed connection points. These allowed the design of more sophisticated structures, including rational design of polyarene dendrimers.

Several routes to arenevinylenes can be envisioned. We show in Figure 17.14 **B** a route to PPV derivatives that involves sulfonium chemistry. The final elimination can be thermally induced directly in a film of the polymer. As such, this route involves a soluble, precursor

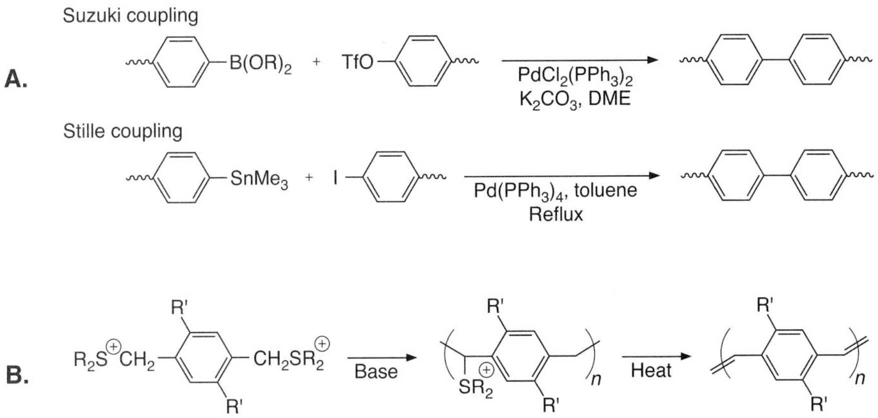

A. Suzuki coupling

Stille coupling

B.

Figure 17.14
Synthetic approaches to conducting polymers. **A.** The Suzuki and Stille couplings, which greatly improved the ability to control regiochemistry in the formation of polyarenes and related structures. **B.** A soluble precursor route to PPV compounds.

polymer, like we previously saw with polyacetylene. This has advantages in processing the material. You are given a chance in the Exercises at the end of the chapter to discuss the mechanism of the polymerization step.

A holy grail in the field has been the development of very low band gap organic polymers. Such systems could be intrinsic conductors, negating the need for doping. This would be an advantage because doping substantially degrades the materials properties of conjugated polymers. Once doping has occurred, a brittle, insoluble material is typically formed that cannot be processed any further. Thus, an organic polymer that conducts without doping would be very attractive. There is still a long way to go, but some progress has been made.

One example of a very low band gap material is polyisothianaphthalene (Figure 17.13). This material has been synthesized. Because of the low band gap, the material is blue in the undoped state. Doping, as usual, moves the absorption to longer wavelengths, but in this case that means out of the visible range. Doped polyisothianaphthalene is a transparent material, and there are many uses for thin film, transparent, conducting materials. You are asked in the Exercises to rationalize the low E_g of polyisothianaphthalene.

Recall the single-walled carbon nanotubes (SWNT) discussed in Section 13.1.8. Because of their highly extended π systems, they are conductive materials. In fact, depending on the precise structure, SWNT behave as semiconductors or true metallic conductors *even without doping*.

17.2.4 Polyaniline

Electrochemical polymerization of aniline produces an intriguing material known as polyaniline (Figure 17.15). Although drawn as all para, there are defects involving ortho and, to a lesser extent, meta coupling. Yet, recent coupling chemistry developed by Buchwald and Hartwig can yield pure para material. Polyaniline and several variants are produced commercially in significant quantity. Laboratory samples can achieve conductivities as high as 10^3 S/cm, while the commercial material more typically shows conductivities on the order of 5 S/cm. The material is soluble enough that transparent films can be cast from organic solvents or from water. Numerous applications are presently known or envisioned, including thin film coatings for corrosion protection and for printed circuit boards, transparent antistatic coatings, and "intelligent windows" that darken on exposure to sunlight but become clear at night.

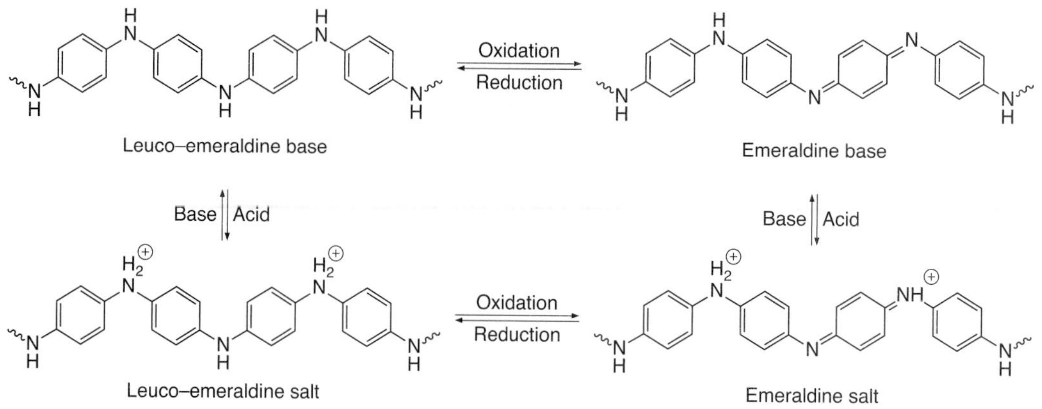

Figure 17.15
Polyaniline. Shown are the various forms, which differ by protonation state and oxidation state. These structures are representative only; degrees of oxidation and precise protonation states will vary. The conducting form is the emeraldine salt.

The most interesting feature of polyaniline is that it is, in a sense, an intrinsically conductive polymer. That is, the harsh chemical or electrochemical doping processes previously discussed are unnecessary to produce a conductive material. Instead, the state of the material can be modulated by pH. Figure 17.15 shows the various forms of polyaniline that are generally considered. These structures are representative only; the actual materials will be diverse, more random mixtures of the various components shown. The most intriguing state is the green, emeraldine salt form, which is the conductive form. In fact, by several (but not all) criteria, polyaniline in this form is a true organic metal. The emeraldine base is not conductive. Thus, the analogue to doping with AsF_5 or I_2 discussed previously for polyacetylene and the like is simple exposure to protonic acids—a clean, reversible process. In addition, the nonconducting forms are yellow or blue (depending on protonation state), and so simple acid–base treatments can modulate both conductivity and optical properties. These features, combined with the processability of polyaniline directly from aqueous solutions, make it one of the most promising of the conducting polymers.

With this general introduction to conducting polymers, you should be able to read the modern literature and understand the terminology and issues being presented. This is a rapidly moving field, and we can expect many more advances and applications in coming years.

17.3   Organic Magnetic Materials

The successes of conducting polymers are considerable, and more are certain to come. After decades of intense research, both theoretical and experimental, organic-based materials that are electrically conductive, a property usually associated with metals, have been prepared and are now articles of commerce. Can magnetism, another property not normally associated with organics, be coaxed onto an organic scaffold?

Along with being intrinsically interesting, a discussion of organic magnetic materials introduces several concepts that are of importance throughout materials science. These include notions such as cooperativity and the critical temperature. Superconductivity shares many similar features, and we will discuss it briefly below. It turns out, though, that we can build on a number of important ideas introduced earlier to develop a fairly rational description of magnetism. This is much more challenging with superconductivity, making magnetism a better platform to develop such phenomena.

As we will see, two fundamentally different approaches to preparing organic magnetic materials have emerged. Before we describe them, though, we present a very brief overview of some basic principles of magnetism.

17.3.1 Magnetism

Magnetism arises from unpaired electron spins. Any material with potentially interesting magnetic properties must have a large number of centers that individually contain unpaired electrons. The magnetic properties of the material are then determined by the interactions among these spins. This is best seen in Figure 17.16. A **ferromagnet** is a material for which all interactions are high-spin. Actually, in a bulk sample of a ferromagnet, it will not typically be true that every spin in the sample will be aligned with every other spin. The material will contain **domains**, microscopic regions throughout which all spins are aligned. The interactions of domains then determine the bulk properties of the material. The situation is similar to that discussed for liquid crystals in Chapter 13. For the most part we will not consider domains further, and will just focus on local interactions. The net alignment of spins in a ferromagnet produces a **magnetic moment**, which, just like a dipole moment in a molecule, implies a directionality.

In an **antiferromagnet** all neighboring spin–spin interactions are low-spin. There is a complete cancellation between spin up and spin down centers, and so there is no net moment on the sample. The distinction is subtle between an antiferromagnet and a **diamagnet**, a material that has no unpaired spins at all (i.e., most materials).

An interesting case is the **ferrimagnet**. A ferrimagnet contains two different types of spin centers that have different local magnetic moments (for definitions of spin states, see Chapter 14). For example, one could be spin $\frac{1}{2}$ (a radical) and one could be spin 1 (a triplet biradical). Locally, all pairwise spin–spin interactions in a ferrimagnet are low-spin. However, because the local moments are of different magnitude, there is an incomplete cancellation of spin. As such, there is still a net magnetic moment in a macroscopic sample of a ferrimagnet. In practice, it is difficult to distinguish a ferrimagnet from a ferromagnet. In fact, many common magnetic materials are ferrimagnets, including **magnetite**, the original "lode stone" or leading stone known to the ancients and the foundation of the compass, the first technological application of magnetism. Magnetite is a mixture of Fe^{2+} and Fe^{3+} in an ferrimagnetic lattice. Finally, Figure 17.16 shows a system in which there is no significant interaction at all among the magnetic centers; instead, all spins are randomly oriented and rapidly re-orienting. Such a structure is a **paramagnet**, and it possesses no net moment.

It should be appreciated from the start that *magnetism is an intrinsically solid state phenomenon—there is no such thing as a "magnetic molecule"*. Magnetism by definition involves the interaction of spin centers throughout large regions of a solid sample. A single molecule cannot do this. It has become common to refer to pairwise spin interactions, either within a molecule or between two molecules, as **ferromagnetic coupling** or **antiferromagnetic coupling**, which correspond to high-spin or low-spin states. However, keep in mind that ferromagnetic coupling is a long way from ferromagnetism. Also, the types of magnetism depicted in Figure 17.16 are just the tip of the iceberg. Magnetism is a very complicated phenomenon, with over 14 different kinds of magnetism known with exotic names like micto-magnetism and spin glass. In fact, despite the ancient history of magnetism and its pervasive importance in the real world, a comprehensive, predictive theory of magnetism remains a challenge for modern condensed matter physics. We mention this complexity because the unambiguous characterization of a magnetic material as a ferromagnet, for example, can be experimentally challenging, and claims that are not extensively documented should be viewed with caution.

Magnetism is a **critical phenomenon**. Like superconductivity (see below), all magnetic behaviors are associated with a **critical temperature**, T_c, below which the magnetic behavior exists, but above which it does not. In a ferromagnet, the critical temperature is called the **Curie temperature**. We can understand qualitatively why this is so. Consider the difference between a ferromagnet and a paramagnet. In the ferromagnet, there is a local interaction that makes it energetically favorable to align spins, and this alignment propagates throughout the material. No such interaction exists in the paramagnet. The energy of this coupling interaction is an enthalpy. *Entropically*, however, the paramagnet is much preferred over the ferromagnet. The paramagnet is completely random, existing in many different, ever changing, combinations of spin states. The ferromagnet is perfectly ordered, the lowest entropy state

Ferromagnet

Antiferromagnet

Ferrimagnet

Paramagnet

Figure 17.16
Schematic showing several of the basic forms of magnetism. The arrow represents an individual magnetic moment on an atom or a molecule, and the arrays are meant to symbolize three-dimensional packing of the individual moments.

imaginable. Since raising temperature always emphasizes entropy ($\Delta G = \Delta H - T\Delta S$), entropy wins out at high enough temperature and magnetic ordering is destroyed. *Above* T_c, *any ferromagnet, antiferromagnet, or ferrimagnet is converted to a paramagnet.*

With this minimal introduction to magnetism, we can now consider approaches to organic magnetic materials. These can be divided into two different basic strategies. One involves high-spin molecules that interact favorably in a molecular crystal lattice, which we will refer to as the molecular approach. The other involves polymers and related structures, such that differing spin centers are covalently linked, and we will refer to this as the polymer approach. These two strategies have also been referred to as the through-space and through-bond approaches, respectively. In each case, the crucial issue is to maximize the magnitude of the local spin–spin interaction that favors ferromagnetic coupling over antiferromagnetic coupling. The stronger this interaction, the higher we can expect T_c to be.

17.3.2 The Molecular Approach to Organic Magnetic Materials

Conceptually, this approach is straightforward. We make a stable, high-spin molecule, crystallize it into an ordered, solid state form, and, if the intermolecular interactions enforce net high-spin, a magnet can result. The challenge for this approach was laid out in Section 14.5.6. Barring special circumstances, whenever we allow two spins to experience a weak interaction, we expect a low-spin ground state. None of the tricks for obtaining high spins, such as the */non-* rule, are viable because the spins are not covalently linked.

There is one novel approach that can provide some guidance for the design of magnetic solids. The strategy, first laid out by McConnell, takes advantage of **negative spin densities** that are seen in most delocalized radicals. Figure 17.17 **A** illustrates the nature of negative spin densities, using allyl radical as an example. The effect is a consequence of electron correlation, and so is evident only in high-level wavefunctions. To first order, we expect all the spin density in allyl radical to be on carbons 1 and 3, because there is a node at C2 in the NBMO (see Figure 17.17 **A**). However, EPR studies clearly show a finite coupling between the electron spin and the hydrogen at C2; its value is smaller than analogous coupling at C1/3, but the coupling *is there*. More sophisticated measurements reveal that the sign (positive or negative) of the coupling constant at C2 is opposite to that of the main couplings at C1 and C3. The relative sign of the coupling constant tells us the relative quantum number ($\frac{1}{2}$ or $-\frac{1}{2}$) of the electron spin. For example, if this sign is positive at C1/3 and negative at C2, this means that the spins of the radical electrons on these centers are opposed.

To rationalize these observations, consider the consequences of electron correlation. We consider the unpaired electron to have spin α, and it resides exclusively on carbons 1 and 3. The lowest π MO has electrons of spin α and β. Correlation effects for these two electrons will be different. The α electron of the lowest MO will be naturally correlated with the unpaired electron because they have like spin; that is, both will never simultaneously occupy any atomic orbital due to the exclusion principle. As a consequence, the radical electron and the α electron of the lowest energy orbital cannot simultaneously occupy the p orbitals on C1 or C3. However, the β electron is not in the same situation, and there is nothing in the single determinant wavefunction that would prevent the β electron of the bonding MO and the α electron of the NBMO from simultaneously occupying the same p orbital. This will produce adverse electron–electron repulsions. However, this limitation of a two-electron wavefunction can be ameliorated in a correlated wavefunction. In particular, correlation will allow the β electron to avoid C1/3 and concentrate at C2. To compensate, the α electron of the lowest MO will concentrate at C1/3. The effects of this are two-fold. There will be an excess of β spin density at C2. This is manifest as the finite EPR coupling constant at C2 discussed above. Since C2 has β spin density, the sign of the coupling is negative. The second consequence is that the amount of α spin density at C1/3 is higher than what might be expected based on simple wavefunctions such as HMO. Indeed, efforts to correlate spin densities and EPR coupling constants show that this is so. Negative spin densities are actually common, and Figure 17.17 **A** shows the pattern for benzyl radical, too. Here an alternation of large positive and small negative spin densities is found, as is typical in radical systems.

A.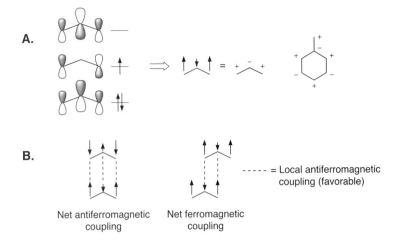

B.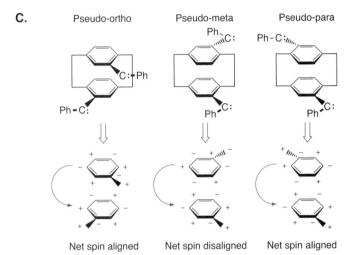

- - - - - = Local antiferromagnetic
coupling (favorable)

Net antiferromagnetic coupling   Net ferromagnetic coupling

C.   Pseudo-ortho   Pseudo-meta   Pseudo-para

Net spin aligned   Net spin disaligned   Net spin aligned

Figure 17.17
Negative spin densities in delocalized radicals.
A. The MOs of allyl, and the spin density pattern that arises from electron correlation effects. Also shown is the spin density pattern of the benzyl radical. **B.** Stacking two allyl radicals. Perfect stacking produces a net antiferromagnetic coupling, but a slipped stack can lead to ferromagnetic coupling. **C.** Iwamura's cyclophane biscarbenes designed to test the McConnell stacking model. Spin pairing of the spin densities on the carbons directly stacked on each other leads to the spin densities on the benzyl carbons aligned in the pseudo-ortho and pseudo-para examples, but not the pseudo-meta case. The curved arrows in the lower diagram of part **C** point out just one example of the six spin pairings of directly aligned and stacked benzene carbons.

How can negative spin densities be used to create a magnet? Consider stacking two allyl radicals on top of one another. Recall from Section 14.5.6 that bringing together two simple radicals, such as methyl, with only a through-space interaction will favor antiferromagnetic coupling. When we stack two allyls, the spins of the p orbitals that align will experience a local antiferromagnetic interaction. If the allyls are perfectly stacked, as shown in Figure 17.17 **B**, the local antiferromagnetic interactions produce an overall antiferromagnetic coupling of the two radicals.

Consider, however, a slipped arrangement, in which C1 of one allyl aligns with C2 of another. The antiferromagnetic coupling of the positive spin density at C1 with the negative spin density at C2 necessarily produces a ferromagnetic coupling of the α spins of the two systems. As such, the overall coupling of the two radicals is ferromagnetic. This is the McConnell recipe for high-spin coupling. That is, align regions of positive spin density with regions of negative spin density.

A brilliant test of the McConnell model was developed by Iwamura (Figure 17.17 **C**). Diphenylcarbene derivatives are well-established to have a strong preference for a triplet ground state (see the high-spin polycarbenes discussed below), and they can be readily generated by photolysis of appropriate diphenyldiazomethanes (Chapter 16). The spin polarization of the π systems of these diphenylcarbene derivatives follows that of the benzyl radical. Using a [2.2]paracyclophane scaffold to control geometries, pairs of triplet carbenes were placed in pseudo-ortho, pseudo-meta, and pseudo-para relationships. The spin densities pair for the carbons that are stacked upon each other (pairing denoted as + and – on top

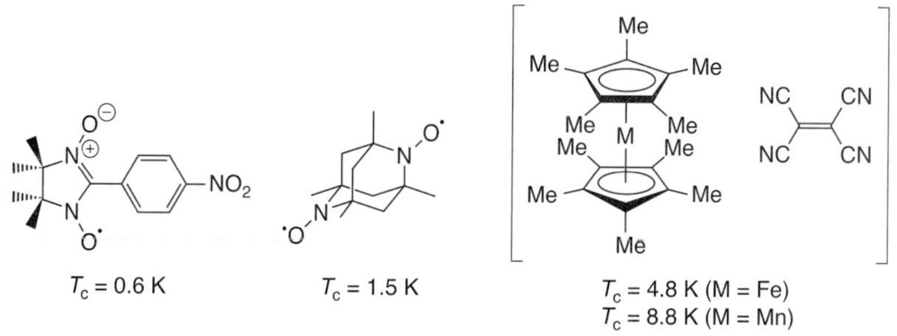

$T_c = 0.6$ K $T_c = 1.5$ K $T_c = 4.8$ K (M = Fe)
$T_c = 8.8$ K (M = Mn)

Figure 17.18
Molecular ferromagnets. Shown are several structures which, when crystallized, form solids that are ferromagnets. Curie temperatures for each system are also given.

of each other in Figure 17.17 **C**). This pairing directs the alignment of the spins on the benzyl carbene centers. As anticipated by the McConnell model, the pseudo-ortho and pseudo-para isomers showed high-spin coupling of the carbene centers to produce a quintet ground state, while the pseudo-meta compound was low-spin.

With this background we can now discuss molecular magnets. A key structure is the *p*-nitrophenyl nitronylnitroxide of Figure 17.18. The molecule crystallizes in at least four different phases. Kinoshita and co-workers have conducted an extensive interrogation of the so-called β phase using an array of condensed matter physics tools and have unambiguously established it to be a true organic ferromagnet. This is an important result. Early in the 20th century Heisenberg categorically stated that a ferromagnet that did not contain a transition metal or a rare earth element could never be made. Thus, the clear creation of a true ferromagnet comprised solely of C, H, N, and O represents an intellectually important landmark. Now the bad news. The Curie temperature for this solid is 0.6K! Magnets based on this material are not likely to be tacking notes to your refrigerator any time soon.

With such a weak magnetic coupling interaction (as evidenced by the low Curie temperature), it is risky to develop a detailed analysis of the coupling mechanism. Nevertheless, analysis of the crystal packing does support the notion that regions of negative spin density are interacting with regions of positive spin density. Thus, the McConnell model may be important in this system.

A large number of nitronyl nitroxides and other amine oxides have been found to display interesting magnetic behaviors in the solid state, including ferromagnetism at very low temperatures. Figure 17.18 shows a bis(nitroxyl) that appears to be a true ferromagnet. The 1.5 K critical temperature of this compound is the highest yet reported for an all-organic structure.

Another important system is the group of ferrocene–TCNE charge–transfer complexes developed by Miller and Epstein. The system forms stacks of alternating decamethylferrocene and TCNE with significant charge transfer between the two. The Mn system developed by Hoffman does have an elevated T_c, much higher than the all-organic systems. The exact mechanism of ferromagentic coupling is debated, and again, the critical temperatures of all these systems are very low, making convincing analyses of the coupling mechanisms quite challenging.

It has proven difficult to prepare a molecular magnet with a high critical temperature. The interactions *between* molecules in organic solids are typically small and always unpredictable. As shown in Figure 17.17, just a shift of a few tenths of an angstrom in the packing arrangement could convert a ferromagnetic stacking interaction to an antiferromagnetic one. At present, we do not have the ability to rationally design organic solids to meet precise packing specifications. The quest for organic magnets is one of several areas that is driving efforts to gain a better understanding of crystal packing forces and arrangements, so that the rational design of organic solids becomes possible.

17.3.3   The Polymer Approach to Organic Magnetic Materials— Very High-Spin Organic Molecules

The weakness of the molecular approach to organic magnetic materials is that the coupling among spin centers relies on relatively weak, intermolecular interactions. Commercial magnets are ionic solids in which all centers are linked through a network of strong, directly bonding interactions. This leads naturally to the question of whether an organic system with such stronger connections could be created. We know from biradical studies that very strong magnetic interactions can develop in organic systems. The singlet–triplet gap in trimethylenemethane is 14 kcal/mol—more than enough energy to make a room temperature magnet. Thus, organic chemists have sought to expand upon the lessons from studies of biradicals to design high-spin materials The initial steps in this effort focused on strategies to concatenate more and more spins that are all high-spin coupled. The goal was to prepare very high-spin organic molecules. Some quite spectacular successes have been achieved in this area, as summarized in Figure 17.19.

A landmark structure is the m-phenylene biscarbene (Figure 17.19 **A**), which was independently prepared by Itoh and Wasserman and reported as the first very high-spin hydrocarbon in 1967. Photolysis of the bis(diazo) precursor gives a system with four unpaired electrons, all of which are high-spin coupled. As such, the spin, S, is $4 \times \frac{1}{2} = 2$; the multiplicity, m_s, is $(2S + 1) = 5$, making the system a quintet state. Diphenyl carbene is a triplet ground state, with one electron in the delocalized π system and one in an orthogonal, in-plane, sp^2 orbital. We showed in Section 14.5.6 that two π electrons linked meta through a benzene are high-spin coupled. A combination of these two makes the overall system high-spin. Remarkably, this bis(carbene) provided the first experimental evidence that the m-xylylene motif was high-spin. It would not be until 1983 that Platz would report the direct observation of the m-xylylene biradical as a triplet ground state.

The polycarbene motif has been expanded by the groups of Itoh and Iwamura. Highlights are shown in Figure 17.19 **B**. Linear extension of the strategy has produced the tetracarbene shown with a nonet ground state. A branching strategy reminiscent of the dendrimer approach to materials yields the nonacarbene system shown. In this case, photolysis of the nona(diazo) precursor was apparently not 100% efficient, so that rather than the maximal value of $S = 9$, a slightly reduced value of $S = 7$ is seen for a bulk sample after photolysis. This suggests a mixture of partially photolyzed molecules.

A useful notion in this field has been the **ferromagnetic coupling unit (FC)**. This is a general structural motif that will take *any* pair of spin-containing moieties and cause them to experience a high-spin coupling. Figure 17.19 **C** illustrates this concept. By far the most common ferromagnetic coupling unit is m-phenylene. Coupling any two spins meta through a benzene produces a high-spin system. Along with the polycarbenes previously discussed, other examples are shown. 1,1-Ethylene is also expected to be an effective FC, because linking two simple radicals through it produces TMM, the prototypical high-spin biradical. Cyclobutanediyl appears to be a general FC, as the biradical is a triplet (Section 14.7.5), and like m-phenylene, it is able to couple two triplet TMMs to make an overall quintet.

An important variant of the polycarbene work is the development of analogous systems based on trityl radical rather than diphenylcarbene, as developed by Rajca (Figure 17.19 **D**). Topologically, the systems are identical, the difference being the trityl compounds use $S = \frac{1}{2}$ spin-containing units, while the polycarbenes use $S = 1$ units. An advantage of the trityl systems is that the trityl radical is much more stable than a diarylcarbene. Also, the synthetic routes to trityls are somewhat more flexible, allowing more complex structures to be developed. In particular, dendrimer type synthetic strategies have been developed that allow the preparation of large systems with very novel high-spin topologies.

Remarkably, the */non-* approach described in Section 14.5.6 for predicting singlet vs. triplet ground states of biradicals adapts very well to higher spin systems. If there are two excess *, a triplet is expected; four excess * implies a quintet, and so on. This was first confirmed by Berson, as illustrated in Figure 17.20 **A**. The starting point was m-quinomethane, a structure that is topologically equivalent to m-xylylene but more synthetically accessible. It

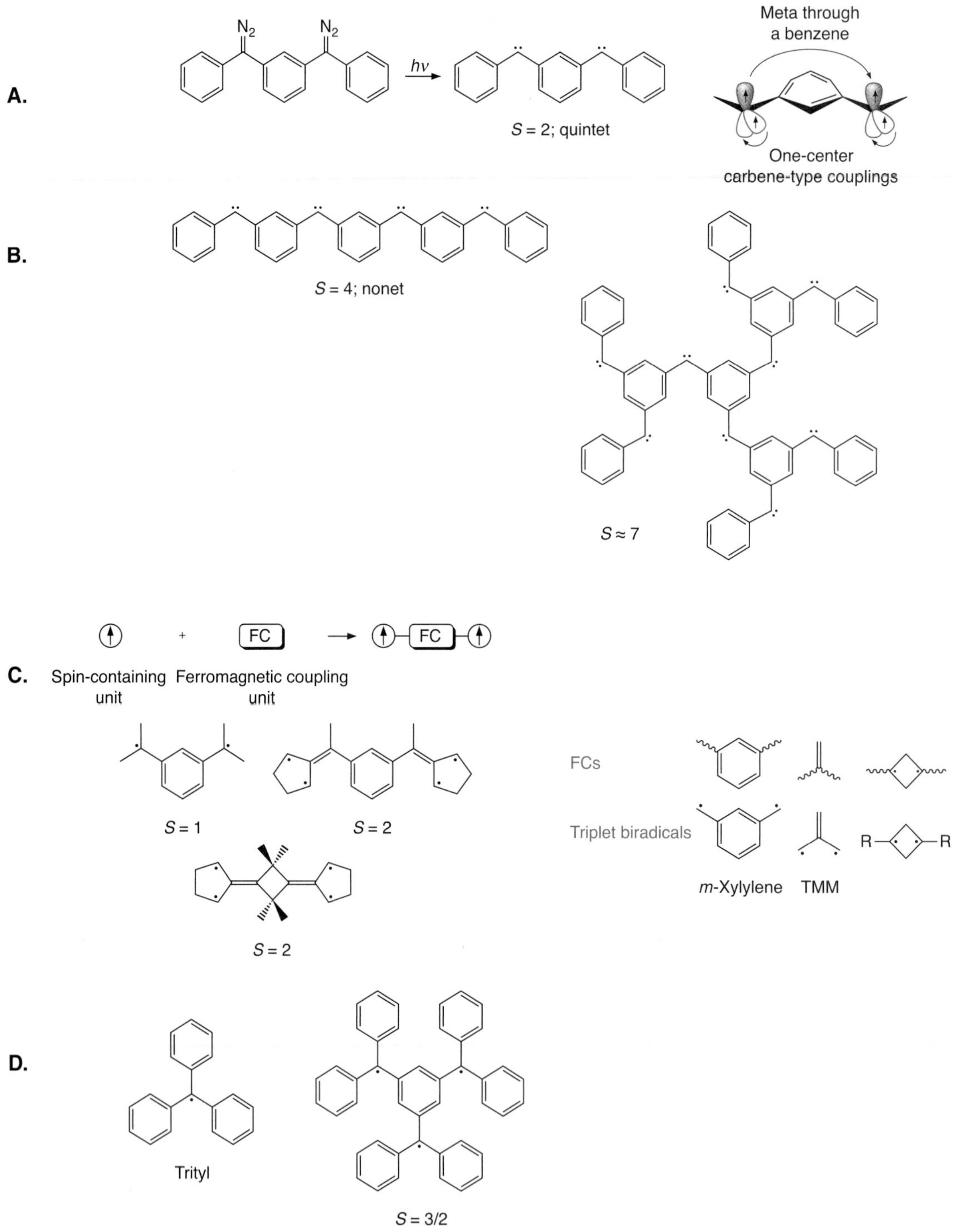

Figure 17.19
Selected very high-spin organic molecules.
A. The biscarbene that was prepared as the first very high-spin molecule.
B. Examples of polycarbenes that display very high S values.
C. The concept of the ferromagnetic coupling unit and several examples.
D. Trityl radical and a high-spin molecule based on it.

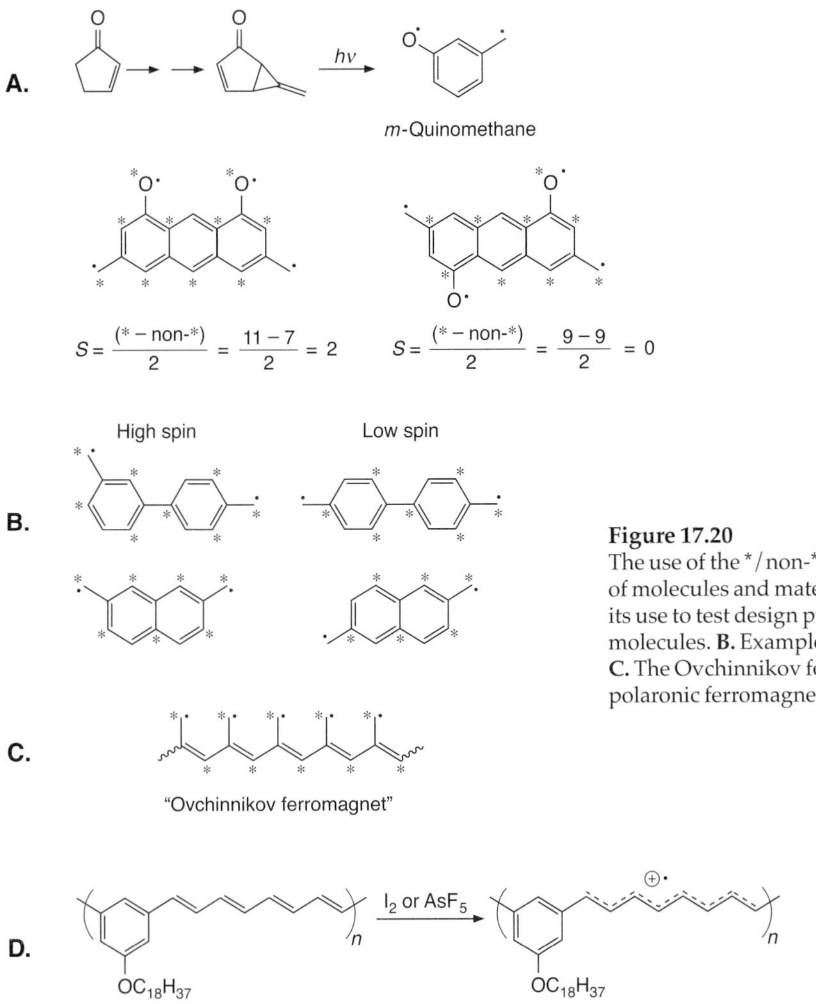

Figure 17.20
The use of the */non-* concept to predict spin states of molecules and materials. **A.** *m*-Quinomethane and its use to test design principles for very high-spin molecules. **B.** Examples of using the */non-* rule. **C.** The Ovchinnikov ferromagnet. **D.** A test of the polaronic ferromagnet concept.

has a definite triplet ground state. Two *m*-quinomethanes can be linked into an anthracene-like framework in different ways that have very different topologies, according to the */non-* rule. Indeed, the two isomers shown in Figure 17.20 **A** have very different spin preferences. The structure predicted to have $S = 2$ does indeed have a quintet ground state, while the other isomer very definitely has a lower spin (most likely triplet) ground state. Recall that the */non-* rule is very good at predicting high-spin states when there are excess *, but it produces ambiguous predictions in cases with equal * and non-*. Figure 17.20 **B** shows other examples of using the */non-* rules to predict the spin preferences of complex polyradicals.

Another key concept in this field is the so-called **Ovchinnikov ferromagnet**. This is a theoretical construct, shown in Figure 17.20 **C**, and is not really a ferromagnet. Rather, it is a polymer that shows an increase in spin multiplicity as each monomer is added. It is an innately one-dimensional system, although extensions to two and even three dimensions can be envisioned. The */non-* rule nicely rationalizes this system.

A novel strategy for magnetic materials is the so-called **polaronic ferromagnet**. The idea is to introduce spins by doping a conjugated polymer, the radical ions so produced being analogous to polarons. Figure 17.20 **D** shows a test system. Indeed, doping introduces stable spins, and unlike in a conducting topology, high-spin coupling of the polaron spins is seen. This is a one-dimensional system, so bulk ferromagnetism cannot be expected, but these early results are encouraging. A nice feature of the connectivity through the meta position of the benzene is that it creates a place to attach a solubilizing alkyl chain without causing any twisting of the polymer backbone. The polymer of Figure 17.20 **D** shows an example of this.

The polaronic strategy highlights the fundamental difference between magnetism and conductivity. In a conducting system, we want strong conjugation and maximal delocalization, so polarons can migrate along the chain. Thus, a para topology connecting monomers is preferred. In contrast, magnetism implies *localized* spins. If spins can delocalize, they will tend to spin pair. And so, the meta topology or the cross-conjugated topology of the 1,1-ethylene motif of TMM is preferred. In fact, magnetism doesn't require *any* delocalization, as evidenced by the fact that 1,3-cyclobutanediyl (Figure 17.19) is high-spin.

While the polymer strategy has been successful in producing some remarkable, very high-spin molecules and materials with interesting magnetic properties, no organic ferromagnet has been produced by this strategy. There are several reasons for this. First, the basic notion of using a ferromagnetic coupling unit to link spin systems is innately one-dimensional. Getting the second dimension is straightforward; a 1,3,5-substitution pattern on a benzene is completely high-spin. However, getting strong ferromagnetic coupling in three dimensions, a necessity for bulk magnetism, is challenging. Cross-linking polymers produces infinite three-dimensional arrays, and it could be used to make magnetic coupling pathways. However, as we noted in Chapter 13, cross-linked materials tend to be insoluble and intractable. At that point, we have lost the advantage of organic magnetic materials (we already have plenty of insoluble intractable magnets). In addition, all the strategies discussed here are vulnerable to defects. Any center that is designed to be, but is not spin-containing, breaks the chain of connectivity and so disrupts the entire system. Finally, there is the issue of the stability and the efficiency of preparation of organic spin centers. Inorganic chemists have a tremendous advantage here, because stable, high-spin metals are common. This is less true in the organic world.

In summary, some very interesting, very high-spin structures have been prepared. It has been demonstrated, moreover, that all-organic ferromagnets can exist, and many of the basic design principles have been enunciated and tested. What remains is to find the right balance of stability, synthetic accessibility, and desirable materials properties. As chemists become more and more adept at manipulating the solid state, we can anticipate that further advances in the field of organic magnetic materials will emerge.

17.4   Superconductivity

Superconductivity is a fascinating phenomenon that is the focus of much research worldwide. Superconductivity is, literally, conduction of electrons with zero resistance. If current is injected into a circle of superconducting wire, it will flow around the wire indefinitely. Electrical current propagating in a circle generates a magnetic field perpendicular to the plane of the ring, and this is the basis for superconducting magnets used in NMR spectrometers and other devices.

From a theoretical point of view, superconductivity has some parallels to magnetism. Both are critical phenomena, involving a critical temperature T_c. Both involve cooperative interactions among electrons. Nevertheless, there are also significant differences. Spin coupling, the fundamental local interaction that gives rise to magnetism, is directly connected to basic concepts in electronic structure theory we have discussed throughout this book (especially in Chapter 14). However, the basic interaction that is central to theories of superconductivity, formation of **Cooper pairs** through electron–phonon coupling, is not easily related to any topic we have discussed so far. Another difference is that there are "degrees" of magnetism. There are many kinds of magnetism, some "more magnetic" than the others. Also, we know we want high spin with magnetism, and as we progress to higher and higher spin states, we know we are making progress toward the ultimate goal. In contrast, superconductivity appears to be an all or nothing phenomenon. There is no intermediate level. Something either is or is not a superconductor. This makes rational design and systematic study of superconducting materials more challenging. It takes a good deal of condensed matter physics to get a solid understanding of superconductivity, and such detail is left to the references at the end of the chapter. Here, we only give a descriptive analysis of Cooper pairs and superconductivity.

Recall that conductivity arises from electrons that are not tightly associated with an atom or a bond, but rather move within a partially filled band. One-dimensional conductivity, as predicted by simple theory for polyacetylene, is disrupted by structural distortions such as the Pierels effect. Superconductivity also requires partially filled bands, but the Cooper pair carriers are influenced differently by distortions of the solid state structure. To see this, we first give a picture of Cooper pairs.

Cooper pairs are loosely bound electron pairs, which move at the same speed but in opposite directions. An analogy can be made to two balls on the ends of a spring, where the balls move together and apart at the same speed but always in opposite directions. Superconductivity arises from highly coordinated motion of all Cooper pairs within a solid, where these electron pairs are all propagating in the same direction—that of the current flow.

How does this "spring-like" motion arise? During any electrical conduction, the passage of electrons through the material will slightly deflect any positive entities (cations or nuclei) toward the moving electron. This causes a ripple in the lattice as the electron translates. Just behind the translating electron (in its wake) the positive ions are pulled together (see Figure 17.21 **A**). The positive ions then relax back to their equilibrium positions, driven

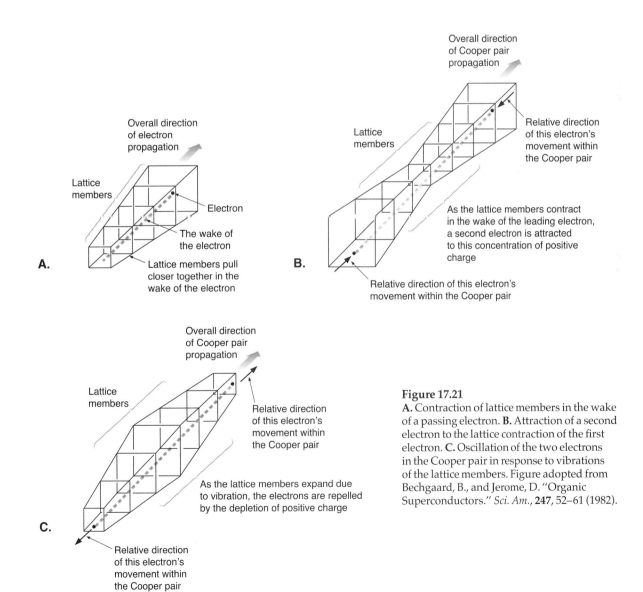

Figure 17.21
A. Contraction of lattice members in the wake of a passing electron. **B.** Attraction of a second electron to the lattice contraction of the first electron. **C.** Oscillation of the two electrons in the Cooper pair in response to vibrations of the lattice members. Figure adopted from Bechgaard, B., and Jerome, D. "Organic Superconductors." *Sci. Am.*, **247**, 52–61 (1982).

by what is called a **restoring force** on the lattice. The restoring force on the positively charged lattice members causes them to overshoot their equilibrium positions, and the lattice members begin to vibrate. Thus, the positive charges also vibrate in a manner analogous to being attached to the ends of springs. Since the mass of an electron is significantly lower than the mass of a nucleus, the vibration persists long after the electron is gone. Vibrations within a lattice are called **phonons**.

In superconducting materials, the increased concentration of positive charge created in the wake of the electron attracts a second electron, and these two electrons make up the Cooper pair. The second electron is being pulled toward the first electron as they both propagate through the lattice. However, the trailing electron moves faster than the leading electron, so the two move as balls on the end of a spring during compression (Figure 17.21 **B**). This is what occurs when the lattice members are closer together than their equilibrium positions. Keep in mind, however, that the lattice members also vibrate as described in the previous paragraph. When the positive charges obtain distances larger than their equilibrium positions, the electrons in the Cooper pair are now pulled apart rather than attracted together (Figure 17.21 **C**). Now the leading electron moves faster than the trailing electron, and the electrons resemble balls on the end of a spring under extension. In essence, the electrons' motions are correlated in response to vibrations in the lattice induced by their very own movement. This perfect coupling of electron motion with lattice motion, termed **electron–phonon coupling**, creates a situation with zero resistance—a true superconductor.

This correlation exists below the critical temperature, but above that temperature other thermal motions of the lattice break the correlations. Conduction in a superconductor results from movement of all the Cooper pairs throughout the material. Based on quantum mechanics, the energy of the system is minimized if all Cooper pairs have the same momentum. With no current the Cooper pairs vibrate in place but do not translate. With current they all move in concert. Once set in motion, their net momentum cannot decay, but scattering off of imperfections in the lattice can decrease conduction just as in normal conductors. In superconducting materials, scattering requires more energy than that which holds the pair together. Furthermore, changing the momentum of one pair would require that all pairs change their momentum, requiring much more energy than is in available in the lattice at low temperature. However, as stated previously, higher energy is available at temperatures above T_c, which therefore leads to breaking up of the pairs.

17.4.1   Organic Metals/Synthetic Metals

While both classical superconductors and the high temperature copper oxide superconductors that burst onto the world stage in the late 1980s are inorganic materials, there is a long-standing interest in preparing organic superconductors. Indeed, superconductivity has been seen in organic solids, although the critical temperatures do not at present approach those of the copper oxides. Progress toward raising the critical temperature with organic compounds has been difficult. As such, we only briefly explore these structures here.

A class of organic molecules known as "charge–transfer" or "donor–acceptor" complexes can display electrical conduction and superconductivity. These are molecular structures made from discrete single molecules, not polymers, and therefore are analogous to the molecular magnetic materials previously discussed. These materials have been referred to as **organic metals** or **synthetic metals**. More recently, these terms are also being applied to conducting polymers.

In order to have conduction in a molecular solid, the building blocks must fit together in such a manner that electrons can move from molecule to molecule. Furthermore, the energy cost to partially filling or opening a valance band must be low. This means that the ability to remove or add an electron must be low in energy, with little structure change to the packing. One common way to achieve this with organic molecules is to pair planar π electron rich molecules, most often incorporating N, S, Se, or Te, with some form of an acceptor. The planar molecules form one-dimensional stacks in the solid state, with orbital overlap allowing electron movement between monomers.

Probably the most well studied example is a co-crystal of tetrathiafulvalene (TTF) with 7,7,8,8-tetracyano-*p*-quinodimethane (TCNQ). Here, TTF is an electron rich donor, and TCNQ is an electron poor acceptor. The crystal has segregated homo-stacks of TCNQ and TTF. Approximately 0.59 electrons per molecule of TTF are transferred to TCNQ, thereby opening a valence band in TTF and partially filling a conduction band in TCNQ. The material is 500 times more conductive along the direction of the stacks than perpendicular to the stacks. However, as with all one-dimensional materials, a Peierls distortion occurs below some temperature. As with polyacetylene, the Peierls distortion involves a form of dimerization, with alternating short and long spacings between molecules along the stacks. With TTF–TCNQ, this occurs at 53 K, and the material is a semiconductor below this temperature.

Interactions among molecules beyond the first dimension (that is, perpenicular to the stacks) are generally weaker, and much as with molecular magnetic materials, it is the relative weakness of the interstack interaction that has made it difficult to achieve high critical temperatures. To increase intrastack and interstack interactions, larger heteroatoms are often used, as these can lead to stronger interactions in the other dimensions. Se and Te in the donor have become standards in this regard. Numerous donor–acceptor complexes have been studied and indeed many conducting materials have been created. However, the driving force in this endeavor is the creation of superconductors.

Some of the first donor–acceptor superconductors were tetramethyl tetraselenafulvalene hexafluorophosphate ($TMTSF_2 \cdot PF_6$) and perchlorate ($TMTSF_2 \cdot ClO_4$). A large number of these types of salts show superconductivity with different transition temperatures, and are commonly know as **Bechgaard salts**. It is thought that these salts are more two-dimensional than TTF–TCNQ, and not only suppress the Pierels distortion, but also other superconducting/semiconductor transitions. Nevertheless, these compounds still have transition temperatures in the single digits Kelvin.

An important step in increasing the transition temperature in superconducting charge–transfer salts was the development of the ET salts. The important component in ET salts is bis(ethylenedithia)tetrathiafulvalene. The ET salts can stack in dimer pairs, which are perpendicular to one another, thereby giving two-dimensional sheets. The increased dimensionality leads to transition temperatures in the low teens Kelvin.

Moving beyond two-dimensional materials is of paramount importance for the creation of a practical superconductor. In this regard C_{60} has attracted considerable attention. Since it is spherical, the band structure is identical in all three dimensions. Doping with K metal creates a molecular solid of formula K_3C_{60}, resulting in a superconducting material with a T_c of 18 to 19 K. In 1995 it was reported that Cs_3C_{60} was superconducting at temperatures as high as 40 K when under high pressures, although it is not at all superconducting at ambient pressure. The quest for higher superconducting temperatures appeared to stall here, but a report in late 2000 showed that films of C_{60} that were p-doped using a field effect transistor arrangement were superconducting at 52 K. Perhaps we have not yet seen the limits of C_{60} superconductivity.

17.5 Non-Linear Optics (NLO)

We discussed the basics of the interaction of light with matter in Chapter 16. A fundamental relation was Beer's law (Eq. 16.4), which shows a direct, linear relationship between the absorption of light and the extinction coefficient, ε, an intrinsic property of the absorbing material. Under conditions of low to moderate light intensity, Beer's law is quite adequate. Similarly, other common interactions of light with matter, such as refraction, reflection, and diffraction, increase in magnitude linearly with light intensity. However, the advent of lasers allowed studies of the interaction of matter with very intense light sources. Under these circumstances, an expanded description of the interaction of light with matter is required. Now, along with the first-order linear term, higher order terms that depend on the square, cube, or higher powers of the light intensity need to be considered. These non-linear terms, if significant, can produce very novel optical phenomena. Here we provide a brief overview

Tetrathiafulvalene
TTF

Tetracyanoquinodimethane
TCNQ

Tetramethyltetraselenafulvalene
TMTSF

Bis(ethylenedithio)tetrathiafulvalene
BEDT-TTF

of the concepts involved in non-linear optics. In addition, we discuss some of the design principles for developing organic materials with significant non-linear properties. As with conducting, magnetic, and superconducting materials, inorganic materials have a leg up on organic materials in the non-linear optics field, but the organics are closing fast.

First some terminology. Recall from Chapter 16 that we can view light as an oscillating electric field. As light passes through a material, that oscillating field induces a polarization in the medium. This effect is described by Eq. 17.9, where P is the polarization of the medium and E is the electric field component of the light. The P_o term is the intrinsic polarization of the medium, which is zero for an isotropic medium such as a fluid solution. The coefficients χ^n are termed the nth-order susceptibilities (first, second, third, etc.). These bulk properties also have direct *molecular* analogues, as given in Eq. 17.10 for describing molecular polarization, μ. In the first term, μ_o is simply the intrinsic dipole moment of the molecule. The coefficient α determines the response of the molecule to an external electric field, and so it is exactly the same molecular polarizability parameter that we discussed extensively in earlier chapters (see Sections 1.1.12 and 3.2.5). The coefficients β and γ are, respectively, the **hyperpolarizability** and the **second hyperpolarizability** of the molecule.

$$P = P_o + \chi^{(1)}E + \chi^{(2)}E^2 + \chi^{(3)}E^3 + \cdots \qquad \text{(Eq. 17.9)}$$

$$\mu = \mu_o + \alpha E + \beta E^2 + \gamma E^3 + \cdots \qquad \text{(Eq. 17.10)}$$

What is the significance of these non-linear terms? This is best seen with an example. When intense light is passed through a crystal with a large value of $\chi^{(2)}$, very interesting phenomena can occur. The frequency of the emergent light can be doubled due to scattering of the incident light at $\frac{1}{2}$ its wavelength. This **second harmonic generation (SHG)** is extremely valuable and commonly used. It is as if two photons interact with a molecule at the same time, and their energies add together to make a photon of double the energy. The chances of two photons impinging on the same molecule at the same time are slim with normal light, but become practical with lasers. An important example involves the output of a Nd–YAG laser, which is 1.06 μm (not a very useful wavelength for photochemistry). However, SHG cuts the wavelength in half to 532 nm, a much more convenient region of the spectrum. Another remarkable property of non-linear materials is **frequency mixing**. That is, if two intense beams of light of differing wavelength arrive at the same spot, the frequency of the resulting beam can be the sum *or the difference* of the two! This is an interesting way to make very long wavelength, monochromatic light. Such non-linear effects should be very valuable in optical switching and optical computing applications. Note that SHG and similar effects are never 100% efficient, because not all the light that goes into the crystal will be frequency doubled. The efficiency will depend on the magnitude of $\chi^{(2)}$. For chemists, it is the molecular term β that needs to be maximized. Before discussing how to do this, however, we must make one other important point.

The E in Eqs. 17.9 and 17.10 is an electric field *vector*. As such, the symmetry properties of the material will be important. In particular, it can be shown that the odd-order terms of these equations are independent of any symmetry considerations, but the even-order terms vanish in a centrosymmetric environment. That is, $\chi^{(2)}$ is zero for any sample that has a center of inversion. This is always the case for bulk liquids, and it is true for most solids. Thus, symmetry and orientation effects should be crucial for discussions of SHG and related phenomena. We have noted earlier that crystal engineering has significant implications for many types of applications, and NLO is definitely one of them. The ability to rationally design or predict non-centrosymmetric crystals would be very valuable. Similarly, at the molecular level, β will be zero for molecules with a center of symmetry. No such restrictions apply to the third-order terms ($\chi^{(3)}$ and γ).

There is really nothing unique or special about Eqs. 17.9 and 17.10. In most cases, when a simple linear equation models a physical phenomenon, it is really just an approximation to a more complete, power series of the sort described here. Non-linear behavior is simply the

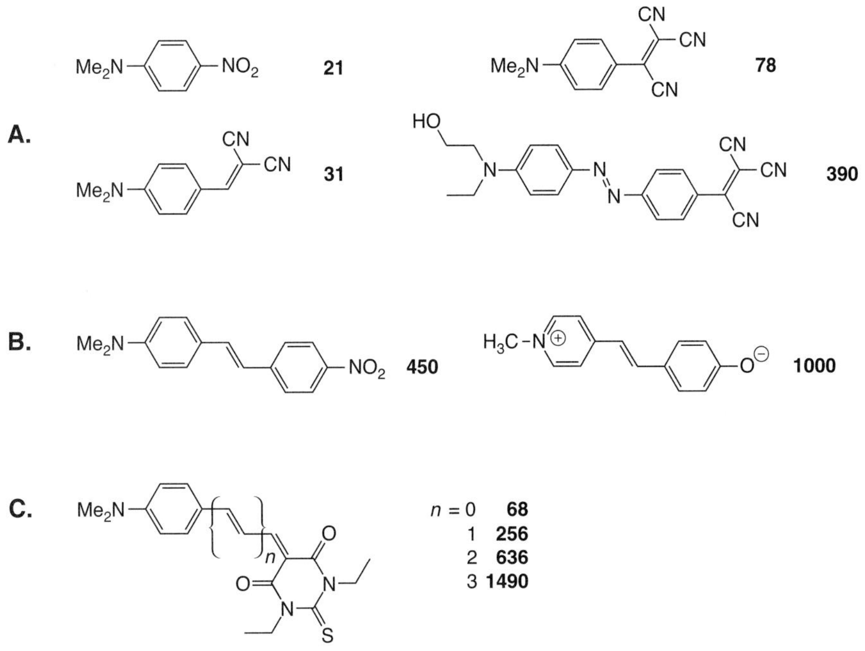

Figure 17.22
NLO β values for selected molecules. Units are 10^{-30} esu. Sources: **A.** Katz, H. E., et al. "Greatly Enhanced Second-Order Nonliner Optical Susceptibilities in Donor–Acceptor Organic Molecules." *J. Am. Chem. Soc.*, **109**, 6561–6563 (1987); **B.** Twieg, R. J., and Jain, K. "Organic Materials for Optical 2nd Harmonic Generation." *ACS Symposium Series*, **233**, 57–80 (1983); **C.** Marder, S. R., et al. "Large First Hyperpolarizabilities in Push–Pull Polyenes by Tuning of the Bond Length Alternation and Aromaticity." *Science*, **263**, 511–514 (1994).

consideration of the higher order terms, and it becomes important under any type of intense or extreme conditions.

SHG was first discovered in single-crystal quartz in 1961. Other inorganic crystals such as lithium niobate ($LiNbO_3$) were soon recognized as especially effective for SHG. While inorganic crystals remain important in the field, the promise of greater flexibility and diversity presented by organics has prompted intense development of non-linear organic materials. We discuss some of those structures here.

The most prominent motif for designing organic molecules with large β values has been to place π donor and acceptor substituents on opposite ends of conjugated π systems. There is a general trend that the stronger the donor–acceptor interaction, the larger the β. The series of compounds in Figure 17.22 **A** illustrates this point. As the acceptor is progressively improved from nitro to dicyanovinyl to tricyanovinyl, β steadily increases. The fourth member of the series couples the best acceptor with an extended donor, producing a further increase in β. The pair of compounds in Figure 17.22 **B** further illustrates the effect of improving the donor–acceptor characteristics.

A more systematic evaluation of such effects was carried out by Marder and co-workers. The series in Figure 17.22 **C** is representative of several studied. Theoretical work indicated that an additional key parameter to achieve large β values was the extent to which there was bond length alternation along the extended conjugated backbone. The relationship was not simple. It gave small β values when there was no bond length alternation, larger values at intermediate degrees of bond length alternation, and smaller β values again as the bond length alternation became maximal.

Designing and preparing a molecule with a large β value is only half the battle, and often it is the easy half. To actually be used for SHG or other effects, a macroscopic sample with a finite $\chi^{(2)}$ value must be prepared. All of the molecules of Figure 17.23 have large, permanent dipole moments. Molecules of this sort tend to crystallize into centrosymmetric space groups. This allows molecular dipoles to align in opposite directions, thereby canceling each other and producing a crystal with a zero value of $\chi^{(2)}$. For simple molecules there is no general solution to this often vexing problem. Of course, an enantiomerically pure sample of a chiral molecule cannot crystallize into a centrosymmetric space group, and this approach is being pursued.

An alternative approach is to prepare so-called **poled polymer** samples. An NLO compound is dispersed into a polymer at a temperature above T_g for the polymer. A large voltage (for example, 10^6 V/cm) is placed across the material. Since the NLO molecules have large dipole moments, they will tend to align with the field. After adequate alignment is achieved, the temperature is dropped below T_g, locking in the now non-centrosymmetric arrangement of NLO chromophores. This fairly simple procedure can be used to prepare samples with quite acceptable $\chi^{(2)}$ values.

The centrosymmetry problems are not an issue for $\chi^{(3)}$ materials. This makes design somewhat easier, but third-order NLO effects (such as frequency tripling and various combinations of frequency mixing) require quite substantial laser powers to be evident. It is generally assumed that molecules that have a large β will also have a large γ. This can be true, but there are subtleties. In particular, it has been proposed that variations of β and γ with the strength of the electric field induced by the incoming radiation have very different forms. At some field strengths both β and γ show maximal values, while at others one is at a maximum and the other at a minimum. Thus, some caution is in order in making predictions about $\chi^{(3)}$ materials based on analogous $\chi^{(2)}$ studies.

17.6   Photoresists

We conclude this chapter with a discussion of the photoresist, one other type of organic material that is crucial to the electronics revolution. Photoresist technology represents a beautiful amalgam of the polymer chemistry discussed in Chapter 13 and the photochemistry discussed in Chapter 16, all in service of the electronics industry alluded to throughout this chapter. We begin with a basic overview of the process, and then discuss separately the kinds of chemistries involved in negative and positive photoresists.

17.6.1   Photolithography

Photolithography involves an optical process that imprints a pattern onto a substrate. Lithography literally means "stone writing" and implies carving features manually. In photolithography light does the "carving". For our purposes, the key component is the **photoresist**. This is generally an organic polymer that can be coated onto a relevant surface. The polymer is photoreactive, such that exposure to light changes the properties of the material, typically making it more soluble or less soluble in an appropriate solvent. As we will see, the distinction between increasing or decreasing solubility on exposure to light is crucial and, in effect, defines the lithographic process. Photolithography is the basic process involved in patterning the integrated circuits that define modern computer processors and memory chips. As the demand for ever smaller feature sizes continues, the chemistry of the photoresist process is continually evolving. New chemistries and new photochemical processes are continually being developed. Here we give examples of classical photoresist technology that illustrate the principles and should allow you to anticipate future directions. Not all applications of photoresist technology are focused on creating the smallest possible "lines". The rapidly growing field of microdevices and micromachines also uses photolithographic techniques to create devices that are very small, although large on the scale of integrated circuits. At the end of this section we discuss in a Going Deeper highlight a new and promising technique that is an alternative to conventional photolithography, called soft lithography.

Figure 17.23 shows in schematic form some of the basics of photolithography. The desired pattern is created by irradiating through a **mask**, thus exposing only selected regions of the photoresist material. The key distinction is that between negative and positive photoresists. In a **negative photoresist**, exposure to light renders the photoresist insoluble. Treatment with solvent then removes the unexposed regions of the photoresist, revealing the substrate below. Since the substrate is actually the key electronic material here (typically silicon or gold, etc.), the process is considered "negative" because exposed regions of the substrate are those that were *not* exposed to light. In contrast, exposure of a **positive photoresist** to light increases its solubility, such that treatment with solvent reveals those regions of the substrate that *were* exposed to light. In early years of the semiconductor industry, the dominant approach involved negative photoresists. More recently, however, positive photoresists have gained in popularity because in some instances they can produce finer features. We will discuss the chemistry of each type below.

17.6.2 Negative Photoresists

A negative photoresist is typically a polymer that becomes less soluble upon exposure to light. Based on our discussions in Chapter 13, the most sensible way to do this is through cross-linking, and indeed that is the standard strategy. The prototype negative photoresist is based on synthetic rubbers such as polyisoprene (Figure 17.24 **A**). This is a soluble material that is easily cast into thin films. Dissolved in the film is a photoactivatable cross-linking agent such as the bis(azide) shown. Photolysis produces a bis(nitrene) that inserts into C–H bonds and cross-links the material. The surface is then treated with a "stripping agent" such as methyl ethyl ketone to dissolve away the polymer that has not been cross-linked. A similar strategy uses light to produce free radicals, which then initiate polymerization and cross-linking of a suitable acrylate monomer (Figure 17.24 **B**).

An alternative strategy makes use of polymers that can be induced to cross-link without the addition of an external reagent, a so-called single-component negative photoresist. The epoxide-containing polymer of Figure 17.24 **C** is a typical example. Cross-linking is initiated by "irradiation" with an electron beam through the mask, which generates anions that launch the cross-linking process.

Negative photoresists are still sometimes used in computer chip and microdevice manufacturing, but for many of the most demanding applications positive photoresists have become more popular. One reason for this is that the stripping step meant to remove the unexposed region will still swell the cross-linked, exposed region (Chapter 13). This swelling degrades the sharpness of the features that can be created by such technology, and was a major reason for the development of positive photoresists.

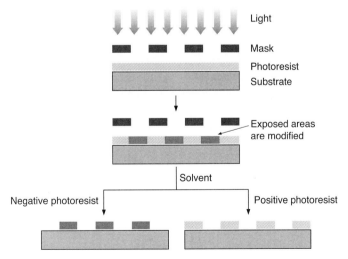

Figure 17.23
Photolithography. Passing light through a mask alters the properties of the polymeric photoresist, such that its solubility decreases (negative) or increases (positive). Treatment with solvent then reveals the pattern in the mask. In actual practice the situation may be more complex. For example, if the substrate is Si, there would likely be a layer of SiO_2 (an insulator) between the substrate and the photoresist. Treatment with solvent then reveals the SiO_2, which is then etched away in a subsequent step to reveal the Si surface.

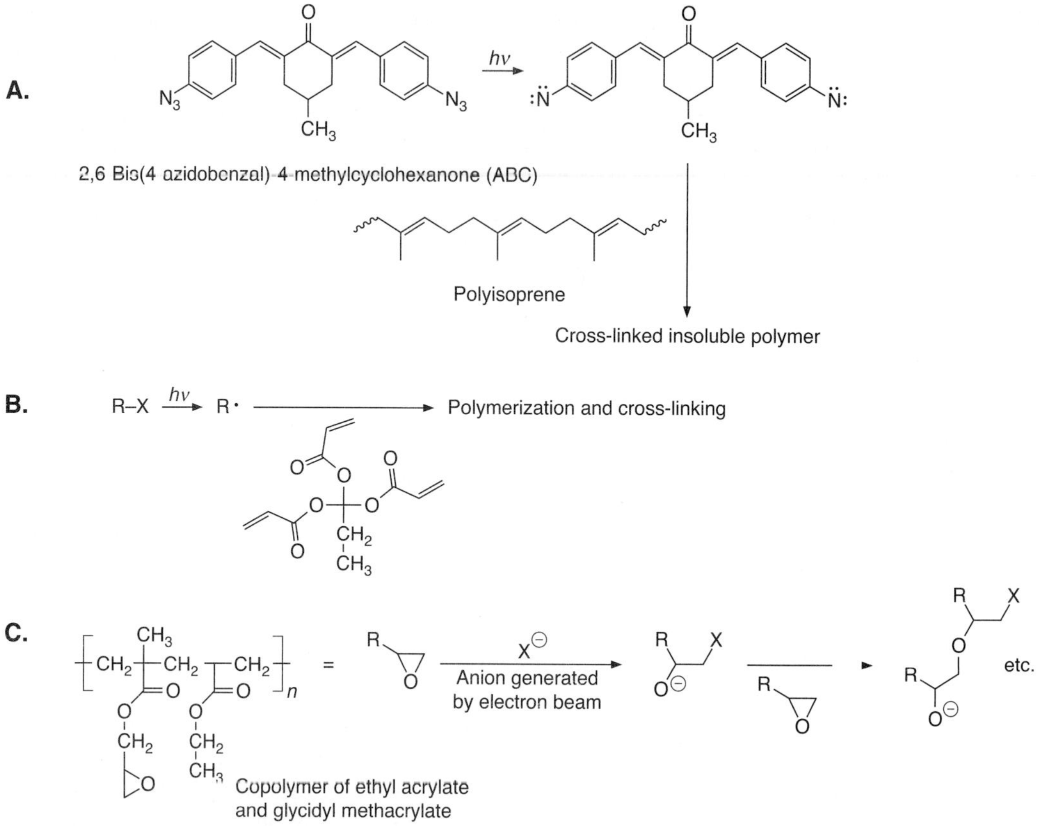

A.

2,6-Bis(4-azidobenzal)-4-methylcyclohexanone (ABC)

Polyisoprene

Cross-linked insoluble polymer

B.   R–X $\xrightarrow{h\nu}$ R· ⟶ Polymerization and cross-linking

C.   Copolymer of ethyl acrylate and glycidyl methacrylate

Anion generated by electron beam

etc.

Figure 17.24
Negative photoresist chemistry. **A.** Cross-linking of a polyisoprene rubber by a photoreactive bisazide. **B.** A radical induced polymerization and cross-linking of an acrylate monomer. **C.** A single component resist. Electron beam irradiation produces anions that launch a cascade of epoxide ring-opening reactions to cross-link the polymer.

17.6.3   Positive Photoresists

In a positive photoresist, exposure to light must make a material *more* soluble so that exposed regions can be directly washed away. This is a more challenging task than simply cross-linking to create insolubility, but clever approaches have been developed, as shown in Figure 17.25.

The classical strategy is surprisingly simple and begins with a Bakelite-type phenol–formaldehyde resin (Section 13.2.3; Figure 13.15; Figure 17.25 **A**). Such polymers are generally soluble in highly basic, aqueous solutions that deprotonate the phenol. This process is slowed in a film of the resin, because the polymer is somewhat hydrophobic and that slows the penetration of polar reactants. This effect is amplified when the polymer is impregnated with the hydrophobic diazonaphthoquinone shown (sometimes referred to as a **dissolution inhibitor**). However, photolysis of the diazoquinone, followed by Wolff rearrangement (see Section 16.3.10) and quenching with water, produces the indene carboxylic acid. This product is much more polar than the starting material, especially after ionization, and it facilitates the penetration of aqueous base into the polymer film, allowing the ultimate dissolution of the exposed region. The diazo compound absorbs in the near UV ($\sim$310 nm), and with optimization this approach allowed the development of circuits with 0.5 μm resolution. However, to achieve even smaller features, it became apparent that even shorter wave-

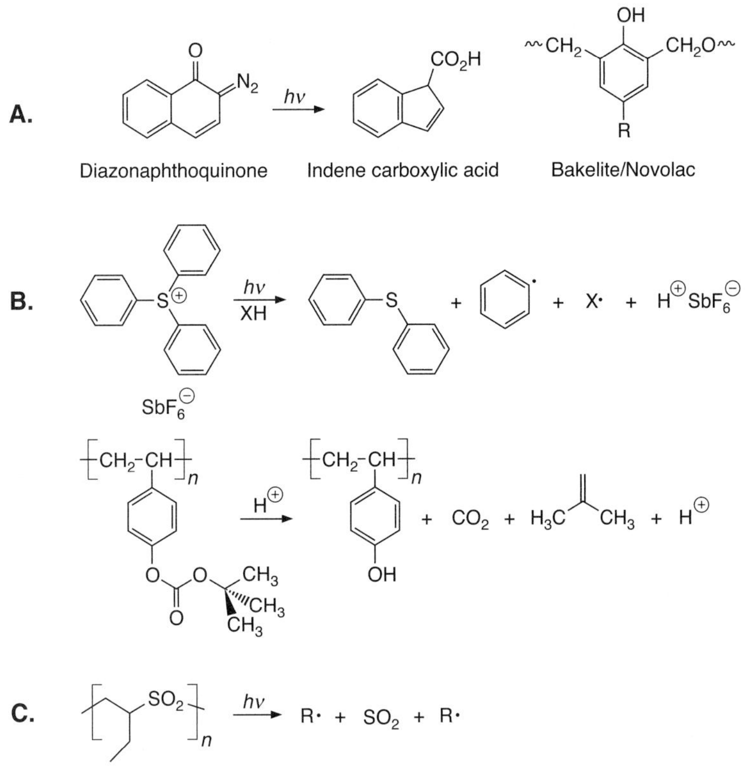

Figure 17.25
Positive photoresist chemistry. **A.** Photolysis of a diazoquinone creates a much more polar environment allowing aqueous base to dissolve a Bakelite-type polymer. **B.** Photogenerated acid and an example of a polymer whose solubility would change dramatically after acid is generated locally. **C.** A single component positive photoresist.

lengths of light, down into the deep UV (the 200-nm range) would be required. New types of photochemistry would have to be developed.

A very clever approach to deep UV photoresists is outlined in Figure 17.25 **B**. The key advance was the discovery of a family of "onium" salts that, on exposure to deep UV irradiation, give rise to protonic acids (see the Connections highlight in Section 5.2.5). Compared to conventional free radical initiators, these onium compounds are more thermally stable and less sensitive to oxygen. A typical strategy for using such compounds for photoresist technology begins with a *t*-butyloxycarbonyl-protected polystyrene derivative that is quite insoluble (*t*-butyloxycarbonyl, or *t*-BOC, is a common protecting group). Photogeneration of acid via the onium compound deprotects the phenol, making the polymer soluble in aqueous base. An interesting feature of this approach is that the deprotection of the *t*-BOC group is catalytic in acid. Thus, a single photon absorbed by the onium compound will generate a single equivalent of acid, but the acid can deprotect many equivalents of *t*-BOC. This is convenient from a photoefficiency point of view, but it opens up the possibility of acid diffusing out of the irradiated zone and thereby degrading the resolution.

Single-component positive photoresists have also been developed, and an example is shown in Figure 17.25 **C**. The polysulfone is sensitive to electron beam or deep UV irradiation, and it produces direct backbone scission of the polymer in a photoreaction similar to the Norrish type I process (see Section 16.3.7). This renders the exposed region soluble.

Going Deeper

Scanning Probe Microscopy

The advent of methods to create nanoscale structures such as SAMs, supramolecular assemblies, lithographic patterns, and the like, gave birth to a field called nanotechnology (see Section 4.3.2 for a brief introduction). Tools are required to characterize these structures, and the most widely used methods rely on various scanning probe microscopy techniques. All of these techniques rely on the use of a specialized "tip" that is brought into proximity to the surface to be visualized. As schematically shown below, the tip moves around the surface and maps out macroscopic features or electronic properties of the surface using a variety of techniques. In the scheme shown, the tip encounters a raised rectangle and moves over it. Rastoring the tip back and forth over the surface maps out the full length of the rectangle. Several techniques of this type have been developed, but we only describe four here.

The simplest example is **scanning tunneling microscopy** (STM). The tip is brought to the surface and an electric field is applied between the tip and the surface. Electrons will tunnel between the tip and the surface in either direction depending on the direction of the applied voltage. The energies of the tunneling electrons are recorded, giving information about the electronic spectra of the entities on the surface. Both filled and empty electronic states on the surface can be probed.

With **scanning electron microscopy** (SEM) the tip has an opening at the bottom, from which a flow of electrons exits. The electron beam is focused on the surface. The electrons hit the surface, and knock secondary electrons off the surface. The secondary electrons are collected by the recorder, giving an image of the surface.

Another technique is called **atomic force microscopy** (AFM). Here the tip approaches the surface until it is repelled. The force applied between the tip and the surface is recorded. As the probe encounters a "bump", it must move up to circumvent the bump and then move down as the bump diminishes in height. One of the most intriguing capabilities that this form of microscopy gives is the measurement of forces involved in molecular recognition events. If the tip is chemically derivatized with molecules that bind to structures on the surface, the force required to remove the tip from the surface is greater than without these binding forces. Examples have included complementary oligonucleotides on the tip and surface, as well as antigen–antibody interactions.

The last technique, called **near-field scanning optical microscopy** (NSOM), is gaining increased attention. This technique combines topographical information with information on electronic structure. Here the probe has a tiny opening at the bottom, through which light passes. The light is used to record electronic spectra of the entities on the surface, and the force applied to the probe maps out topology.

These techniques are incredibly powerful tools for probing nanostructures. The images shown in the next Going Deeper highlight would not have been possible without the advent of such techniques. These tools are increasingly becoming common in materials research and physical organic chemistry in general.

Weiss, P. S., and McCarty, G. S. "Scanning Probe Studies of Single Nanostructures." *Chem. Rev*, **99**, 1983–1990 (1999).

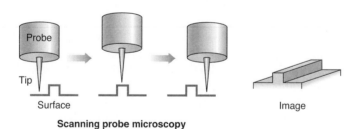

Scanning probe microscopy

Going Deeper

Soft Lithography

A rather new approach to creating two-dimensional patterns on the nanoscale has recently arisen, primarily through the efforts of the Whitesides group at Harvard University. This patterning technique is called soft lithography, and it relies on the use of various elastomeric stamps rather than rigid photomasks. The most common forms of soft lithography are called **microcontact printing** and **replica molding**. In each of these techniques a stamp or mold is created out of poly(dimethylsiloxane) that has relief structures on its surface. In replica molding, the stamp is a two-dimensional surface possessing the nanoscale relief structures, and it is used to make multiple copies of the pattern by pressing it repeatedly into moldable polymeric surfaces. With microcontacting printing, chemicals are applied to the raised structures on the surface of the stamp, and they are transferred to the surface being printed upon, much like an address stamp transfers ink to an envelope. The stamp can be a two-dimensional surface with relief structures, or it can be a roller with relief structures, similar to a decorative paint roller used to paint a repetitive flower or ivy trim on a house. Rolling this structure over a surface transfers the chemicals from the stamp to the surface.

One application of microcontact printing is to create patterns for etching as with standard lithography. The following structures were microcontact printed on a layer of silver with planar stamps that delivered hexadecane thiol self-assembled monolayers (SAMs; see the Connections highlight of Section 13.1.8). After deposition of the SAM, the silver was etched with ferricyanide. The images were created using scanning electron microscopy (see the previous Going Deeper highlight). Although soft lithography techniques are not used as widely as conventional lithography, it is clear that the techniques are growing rapidly in their capabilities and applications.

Xia, Y., and Whitesides, G. M. "Soft Lithography." *Angew. Chem. Int. Ed. Eng.*, **37**, 550–575 (1998).

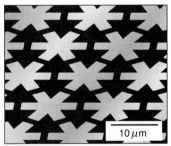

Summary and Outlook

We've covered a lot of territory in this chapter, and necessarily some topics have been more extensively covered than others. Our major goal has been to introduce you to some potentially unfamiliar research areas and terminology, and to illustrate basic concepts and strategies. These topics are at the forefront of modern physical organic chemistry, and it is certain that many of the areas described here will experience major advances in the coming years. It is hoped that with the background provided here, you will be able to read and understand current publications and, with a little more reading, be able to develop your own contributions to these exciting areas.

Exercises

1. Sketch the band structure (HMO) of polyacetylene using butadiene as the unit cell. Also, sketch the consequences of a Peierls distortion.

2. Rationalize why bands **1** and **3** of Figure 17.7 are degenerate at the zone edge. Be as quantitative as possible, remembering we are operating at the level of HMO theory.

3. Sketch the HMO-level band structure of poly(1,3-cyclobutadiene), being as quantitative as possible, and describe the origin of each band. Treat cyclobutadiene as a perfect square. Sketch the crystal orbital at the zone center and the zone edge for each band. What might you expect the properties of such a material to be?

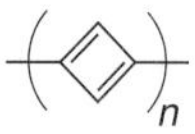

4. Consider the consequences of a Jahn–Teller distortion in each cyclobutadiene unit in poly(1,3-cyclobutadiene) that introduces long and short bonds into the ring (giving a Peierls distortion to the polymer). Sketch how this would change the band structure developed in Exercise 3.

5. Shown below are two different drawings of PPP. They have the same energy. Why, then, isn't PPP a degenerate system that can support solitons?

6. It has been argued that carrier mobility in a conducting polymer (μ) should be better the greater the dispersion of the valence (conduction) band for p-type (n-type) doping. Rationalize this proposal.

7. Draw a band structure for poly(*m*-phenylene) and discuss why E_g is so much larger for this polymer than for PPP. Describe the origin of each band, and sketch the crystal orbital at the zone center and the zone edge for each band.

8. Draw a band structure for polyacene, using butadiene as the unit cell. Describe the origin of each band, and sketch the crystal orbital at the zone center and the zone edge for each band. Rationalize in as quantitative a way as possible the zero band gap for the structure. Also, show that this system can support solitons.

9. Draw a band structure for polythiophene. Describe the origin of each band, and sketch the crystal orbital at the zone center and the zone edge for each band. HMO orbitals and coefficients for thiophene are given below.

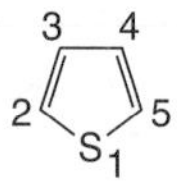

$$\varepsilon_5 = \alpha - 1.62\beta \quad \psi_5 = 0.000\phi_1 + 0.372\phi_2 - 0.602\phi_3 + 0.602\phi_4 - 0.372\phi_5$$
$$\varepsilon_4 = \alpha - 0.97\beta \quad \psi_4 = 0.386\phi_1 - 0.581\phi_2 + 0.296\phi_3 + 0.296\phi_4 - 0.581\phi_5$$
$$\varepsilon_3 = \alpha + 0.62\beta \quad \psi_3 = 0.000\phi_1 + 0.602\phi_2 + 0.372\phi_3 - 0.372\phi_4 - 0.602\phi_5$$
$$\varepsilon_2 = \alpha + 1.05\beta \quad \psi_2 = 0.694\phi_1 - 0.028\phi_2 - 0.508\phi_3 - 0.508\phi_4 - 0.028\phi_5$$
$$\varepsilon_1 = \alpha + 2.02\beta \quad \psi_1 = 0.607\phi_1 + 0.402\phi_2 + 0.393\phi_3 + 0.393\phi_4 + 0.402\phi_5$$

10. Rationalize the low band gap seen in polyisothianaphthalene. Begin with the band structure of polythiophene (PT) developed in Exercise 9. Next, create an orbital mixing diagram in which the appropriate MOs of butadiene are combined with the highest occupied and lowest unoccupied crystal orbitals of PT.

11. Sketch the HMO structure of polyfulvene (see Figure 14.11 for the MOs of the monomer). What band gap does your diagram predict? The computed value of E_g (analogous to those of Table 17.2) for the real material is 0.9 eV. Rationalize this observation.

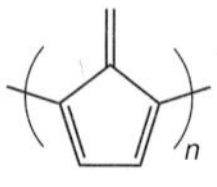

12. Crystallinity in conducting polymers is often a problem, leading to very insoluble materials. One strategy for circumventing this is to introduce variability or diversity in the polymer, thereby discouraging the ordered packing arrangements

associated with crystallinity. An example of this approach is the direct polymerization of racemic 3-(2-ethylhexyl)thiophene shown below. Give two ways in which polymerization of this monomer should introduce irregularity into the structure.

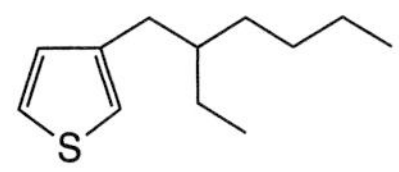

3-(2-Ethylhexyl)thiophene

13. When the sulfonium-based polymerization reaction of Figure 17.14 **B** is run at low temperatures, a short-lived intermediate with an absorption at 317 nm is observed. Propose a structure for this intermediate. From this intermediate, the mechanism could proceed along one of two different paths. One involves radical and biradical-type structures, and the other is a fundamentally anionic mechanism. Describe both and suggest a way to experimentally distinguish the two.

14. Suggest an experimental way to distinguish an antiferromagnet from a diamagnet.

15. Theory clearly establishes that it is not possible to have a one-dimensional ferromagnet (a linear chain with an infinite number of spins, all high-spin aligned) at any temperature above 0 K. Discuss why this might be so. (*Hint:* Consider entropy.)

16. Suggest ways to minimize the effects of defects in the high-spin material designs of Figure 17.19.

17. A common way to kinetically stabilize organic radicals is to add steric crowding. Perchlorotrityl is much more stable than trityl. Trityl units have been used as components of very high-spin molecules. Suggest why perchlorination may not be a good approach to making better organic magnets based on a trityl system.

18. Quartz is a very simple material, just a network of silicon and oxygen. Why, then, is SHG possible in single-crystal quartz?

19. Discuss what properties of polymers lead to good candidates for creating a poled material for NLO applications.

20. Referring to the description of bipolaron formation in Figure 17.10 **B**, show that a single bipolaron should have more bonding interactions than two isolated polarons.

21. The polytrityl strategy of Figure 17.19 **D** can be conveniently elaborated into a strategy for preparing dendrimeric polyradicals. For the present purposes, consider the standard synthetic approach to trityl systems as reaction of a diarylketone with an aryl Grignard reagent to make a triarylmethanol that can subsequently be elaborated to a trityl radical. Show an appropriate approach to a dendrimeric system based on this chemistry, and draw the second generation product in its alcohol form. Consider the core of the dendrimer to be a 1,3,5-substituted benzene.

22. In preparing very high-spin polyradicals, defects—that is, sites that are designed to be spin-containing but are not—are always a problem. Discuss the ways in which the impact of a single defect will be different in two different polycarbene systems, each containing 21 carbene centers. One is a completely linear system analogous to the nonet of Figure 17.19 **B**, and the second is the dendrimeric system obtained by elaborating the $S \approx 7$ system of Figure 17.19 **B** one "generation" further.

23. One of the most surprising confirmations of the importance of the primacy of topology in establishing spin preferences concerns the related biradicals *m*-xylylene and *m*-quinomethane. The surprise is that the magnitude of the singlet–triplet gap (favoring triplet) is very similar for the two. Simple reasoning had suggested that *m*-quinomethane should have a much smaller gap. Reproduce that reasoning here. (*Hint:* Think resonance.)

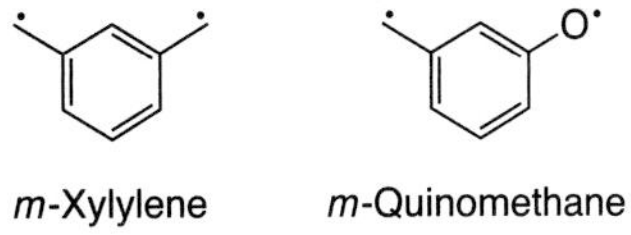

m-Xylylene   *m*-Quinomethane

24. Predict whether the following systems are high-spin or low-spin.

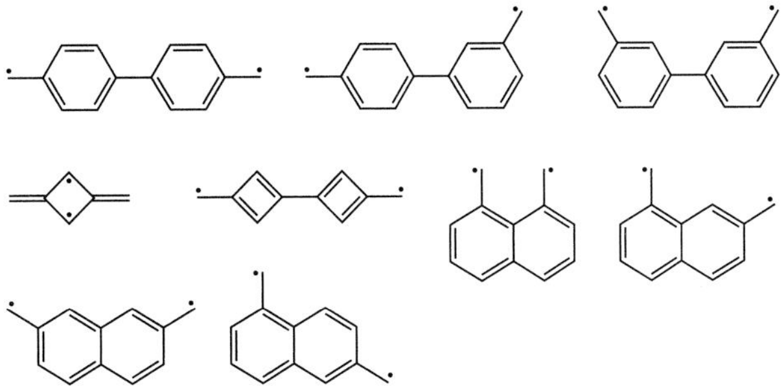

25. Suggest two-dimensional analogues of the Ovchinnikov ferromagnet that are expected to experience exclusively ferromagnetic coupling.

26. Convince yourself that the triradical of Figure 17.19 **D** is high-spin.

27. Since the formation of Cooper pairs is related to the deformation in the lattice, one might expect the "stiffness" of the lattice to affect T_c. Do you expect higher or lower stiffness to raise the T_c? Explain.

28. Show two monomers that could be combined to make a dendrimeric polyphenylene via a Suzuki coupling.

Further Reading

Band Structure Theory

Albright, T. A., Burdett, J. K., and Whangbo, M.-H. (1985). *Orbital Interactions in Chemistry*, John Wiley & Sons, New York.

Hoffmann, R. (1988). *Solids and Surfaces: A Chemist's View of Bonding in Extended Structures*, VCH Publishers, Inc., New York.

Hoffmann, R. "How Chemistry and Physics Meet in the Solid State." *Angew. Chem. Int. Ed. Eng.*, **26**, 846 (1987).

Burdett, J. K. "From Bonds to Bands and Molecules to Solids." *Prog. Solid St. Chem.*, **15**, 173 (1984).

Conducting Polymers

Skotheim, T. A. (1986). *Handbook of Conducting Polymers, Vol. 1*, Marcel Dekker, Inc., New York, p. 727.

Skotheim, T. A. (1986). *Handbook of Conducting Polymers, Vol. 2*, Marcel Dekker, Inc., New York, p. 676.

Patil, A. O., Heeger, A. J., and Wudl, F. "Optical Properties of Conducting Polymers." *Chem. Rev.*, **88**, 183–200 (1988).

Brédas, J. L., and Street, G. B. "Polarons, Bipolarons, and Solitons in Conducting Polymers." *Acc. Chem. Res.*, **18**, 309–315 (1985).

Adeloju, S. B., and Wallace, G. G. "Conducting Polymers and the Bioanalytical Sciences: New Tools for Biomolecular Communication. A Review." *ANALYST*, **121**, 699 (1996).

Mort, J. "Polymers as Electronic Materials." *Adv. Phys.*, **29**, 367 (1980).

Heeger, A. J., Kivelson, S., Schrieffer, J. R., and Su, W. P. "Solitons in Conducting Polymers." *Rev. Mod. Phys.*, **60**, 781 (1988).

Organic Magnetic Materials

Itoh, K., and Kinoshita, M. (2000). *Molecular Magnetism: New Magnetic Materials*, Gordon & Breach, Langhorne, PA.

Epstein, A. J., and Miller, J. S. "Molecule- and Polymer-Based Magnets, A New Frontier." *Synthetic Metals*, **80**, 231–237 (1996).

Rajca, A. "Organic Diradicals and Polyradicals: From Spin Coupling to Magnetism?" *Chem. Rev.*, **94**, 871–893 (1994).

Non-Linear Optical Materials

Eaton, D. F. "Nonlinear Optical Materials." *Science*, **253**, 281–287 (1991).

Marder, S. R., Cheng, L.-T., Tiemann, B. G., Friedli, A. C., Blanchard-Desce, M., Perry, J. W., and Skindhoej, J. "Large First Hyperpolarizabilities in Push–Pull Polyenes by Tuning of the Bond Length Alternation and Aromaticity." *Science*, **263**, 511–514 (1994).

Brédas, J. L., Chance, R. R., Silbey, R., Nicolas, G., and Durand, P. A. "Nonempirical Effective Hamiltonian Technique for Polymers: Application to Polyacetylene and Polydiacetylene." *J. Chem. Phys.*, **75**, 255–267 (1981).

Zyss, J., and Ledoux, I. "Nonlinear Optics in Multipolar Media: Theory and Experiments." *Chem. Rev.*, **94**, 77 (1994).

Charge–Transfer/Donor–Acceptor Complexes

Bechgaard, K., Jerome, D. "Organic Superconductors." *Sci. Am.*, **247**, 52–61 (1982).

Williams, J. M., Schultz, A. J., Geiser, U., Carlson, D., Kini, A. M., Wang, H., Kwok, W.-K., Whangbo, M.-H., and Schirber, J. E. "Organic Superconductors—New Benchmarks." *Science*, **252**, 1829–1830 (1991).

Wudl, F. "From Organic Metals to Superconductors: Managing Conduction Electrons in Organic Solids." *Acc. Chem. Res.*, **17**, 227–232 (1984).

Miller, L. L., and Mann, K. R. "π-Dimers and π-Stacks in Solution and in Conducting Polymers." *Acc. Chem. Res.*, **29**, 417–423 (1996).

Blythe, A. R. "Electrically Conducting Organic Materials." *Advan. Sci.*, **27**, 321 (1971).

Shirota, Y. "Syntheses, Properties and Applications of Photo- and Electro-Active Organic Materials." *New Funct. Mater.*, C 377 (1993).

Seo, D.-K., and Hoffmann, R. "Direct and Indirect Band Gap Types in One-Dimensional Conjugated or Stacked Organic Materials." *Theor. Chem. Acc.*, **102**, 23 (1999).

Torrance, J. B. "The Difference Between Metallic and Insulating Salts of Tetracyanoquinodimethone (TCNQ): How to Design an Organic Metal." *Acc. Chem. Res.*, **12**, 79 (1979).

Superconductivity

Khan, H. R., and Raub, C. J. "Ferromagnetism and Superconductivity." *Annu. Rev. Mater. Sci.*, **15**, 211 (1985).

Haddon, R. C. "Electronic Structure, Conductivity and Superconductivity of Alkali Metal Doped (C60)." *Acc. Chem. Res.*, **25**, 127 (1992).

Conversion Factors and Other Useful Data

Energy Conversion Factors (for example, 1 kcal/mol = 4.184 kJ/mol while 1 kJ/mol = 0.239006 kcal/mol)

	cm^{-1}	K	eV	au	kcal/mol	kJ/mol
cm^1	1	1.438769	1.240×10^{-4}	4.556×10^{-6}	2.859×10^{-3}	11.963×10^{-3}
K	0.69504	1	8.617×10^{-5}	3.167×10^{-6}	1.9872×10^{-3}	8.314×10^{-3}
eV	8065.54	1.160×10^{4}	1	3.675×10^{-2}	23.06035	96.4853
au	219474.7	3.1577×10^{5}	27.2116	1	627.5095	2625.500
kcal/mol	349.755	503.217	4.336×10^{-2}	1.594×10^{-3}	1	4.184
kJ/mol	83.5935	120.272	1.036×10^{-2}	3.809×10^{-4}	0.239006	1

au = atomic unit of energy (also known as the Hartree), eV = electron volts, K = thermal energy at 1 kelvin.

Useful Guidelines:

NMR and EPR

1 Tesla = 10,000 G; corresponds to ca. 0.93 cm^{-1} or 2.67 cal/mol or 1.34 K

1 cm^{-1} = 10,700 G

1 MHz = 9.53708×10^{-8} kcal/mol = 0.3573 G

100 MHz ^{1}H NMR implies H_o = 23,500 G = 2.35 T; 400 MHz implies 9.4 T; 900 MHz implies 21.2 T

Temperature

Thermal energy at 1K, (kT/hc) = 0.695 cm^{-1}, which corresponds to 1.987 cal/mol (that is, the value of the gas constant)

Thermal energy at 298 K = ca. 600 cal/mol

Eyring Equation (k = rate constant)

$k = 2.083 \times 10^{10}\, T \exp(-\Delta G^{\ddagger}/RT)\ \text{s}^{-1}$

$\Delta G^{\ddagger} = 4.576\, T\,[10.319 + \log(T/k)]\ \text{cal/mol}$

$E_a = \Delta H^{\ddagger} + RT$

$\Delta S^{\ddagger} = 4.576(\log A - 10.753 - \log T)$

$\quad = 4.576(\log A - 13.23)$ at 25 °C

1047

Miscellaneous

1 atm = 101,323 Pa = 760 torr

Boltzmann's constant (k): 1.3807×10^{-23} J/K = 0.69504 cm^{-1}/K

Planck's constant (h): 6.62697×10^{-34} J•s

Gas constant (R): 8.31451 J/K•mol = 1.987 cal/K•mol

Elementary charge (e): 1.602177×10^{-19} C

Avogadro's number (N_A): 6.02214×10^{23}/mol

Vacuum permittivity (ε_o): 8.85419×10^{-12} C^2/J•m

1 joule (J) = 10^7 erg

1 erg/molecule = 1.44×10^{13} kcal/mol

1 D = 3.33564×10^{-30} C•m

Electrostatic Potential Surfaces for Representative Organic Molecules

NOTE: *Color images of the following items are shown on the front and rear inside covers of this book.*

The first group consists of relatively polar molecules, and has an electrostatic potential range of ±50 kcal/mol. The second group contains less polar molecules, and is plotted with a ±25 kcal/mol range. The third group shows substituent effects on the electrostatic potential surfaces of simple aromatics. Surfaces are calculated at the HF level with a 3–21G* basis set. Some structures are given more than once, allowing for comparisons between groups.

Group Orbitals of Common Functional Groups: Representative Examples Using Simple Molecules

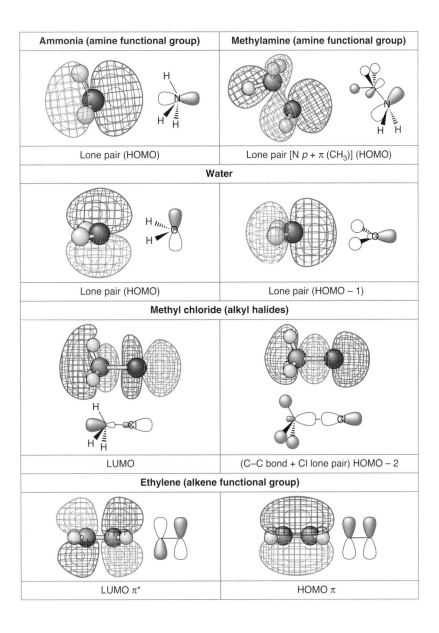

Ammonia (amine functional group)	Methylamine (amine functional group)
Lone pair (HOMO)	Lone pair [N p + π (CH$_3$)] (HOMO)

Water

Lone pair (HOMO)	Lone pair (HOMO – 1)

Methyl chloride (alkyl halides)

LUMO	(C–C bond + Cl lone pair) HOMO – 2

Ethylene (alkene functional group)

LUMO π^*	HOMO π

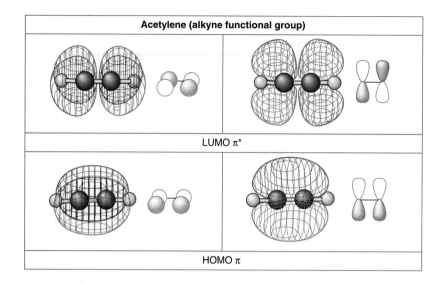

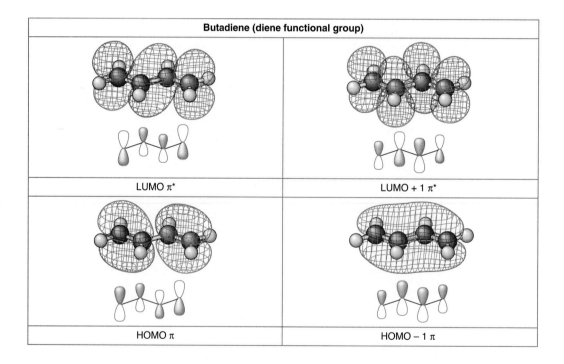

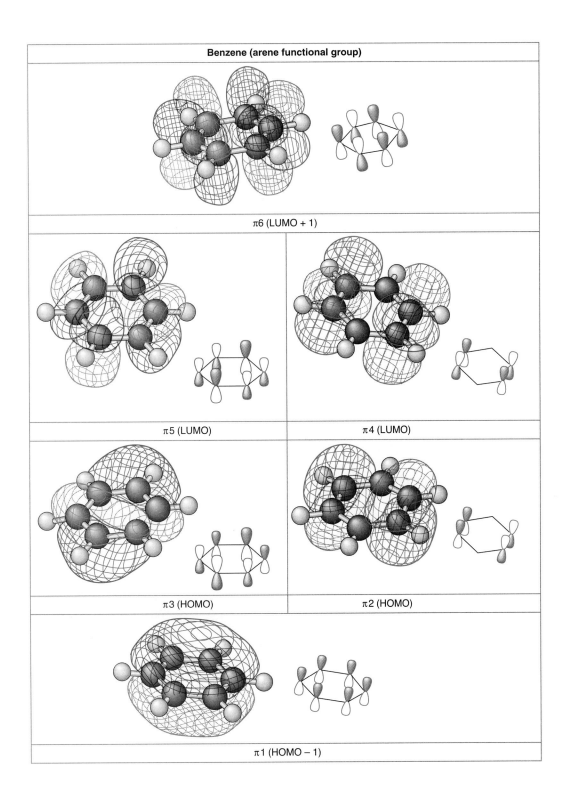

Benzene (arene functional group)

π6 (LUMO + 1)

π5 (LUMO)

π4 (LUMO)

π3 (HOMO)

π2 (HOMO)

π1 (HOMO − 1)

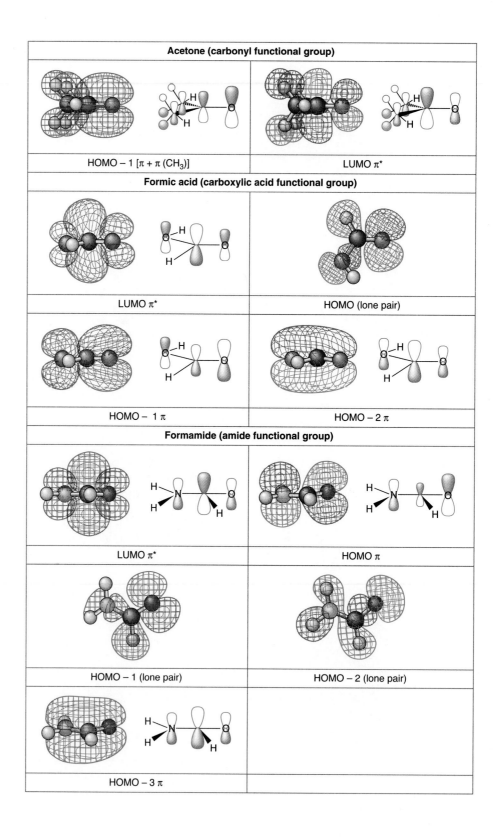

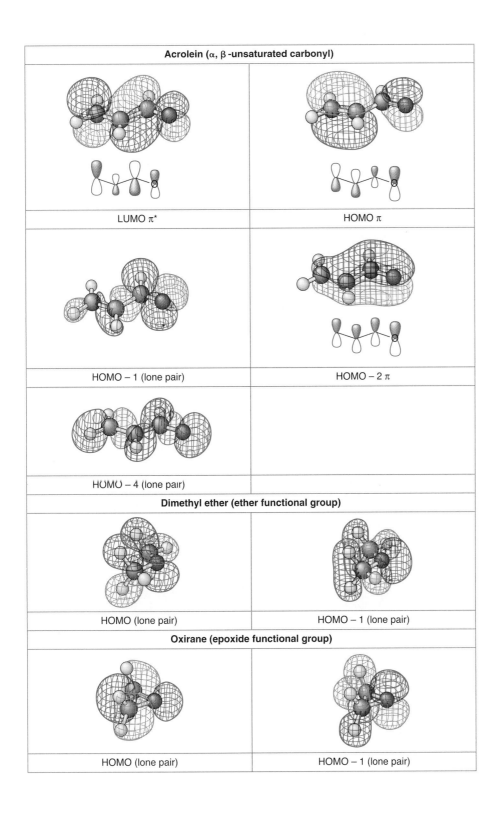

Acrolein (α, β -unsaturated carbonyl)

LUMO π*

HOMO π

HOMO − 1 (lone pair)

HOMO − 2 π

HOMO − 4 (lone pair)

Dimethyl ether (ether functional group)

HOMO (lone pair)

HOMO − 1 (lone pair)

Oxirane (epoxide functional group)

HOMO (lone pair)

HOMO − 1 (lone pair)

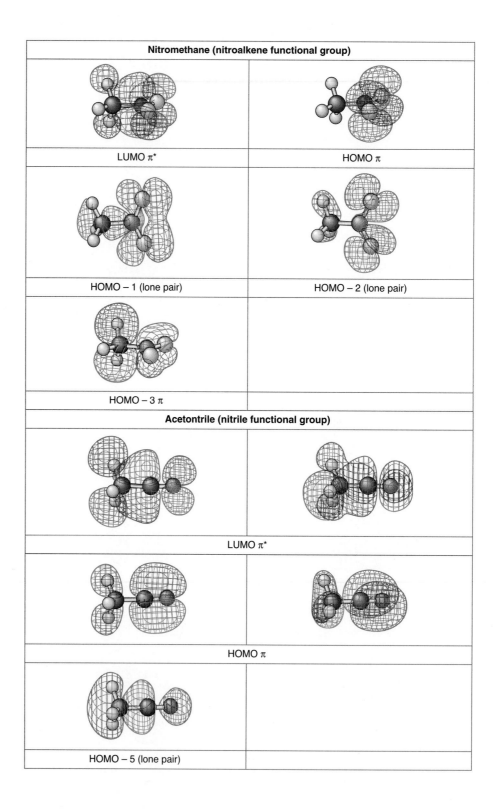

The Organic Structures of Biology

This is not a book about biology, biochemistry, bioorganic chemistry, or chemical biology—it is a book covering physical organic chemistry. However, since its inception, a major goal of physical organic chemistry has been to develop a chemical-scale understanding of the organic chemistry of life. Early efforts to understand ester/amide hydrolysis, substitutions at phosphorus centers, and the hydrolysis of acetals derived a strong motivation from the fact that the chemistry of peptides, nucleotides, and saccharides, respectively, involved analogous reactions. This is increasingly true, as studies of molecular recognition and supramolecular chemistry derive their inspiration from biochemistry, and our understanding of structural biology continues to grow.

Because of this, the structures of proteins, nucleic acids, and so on, are often part of the backdrop of discussions of modern physical organic chemistry. Many physical organic chemists (including the authors!) often forget the details of these structures, and so we compile them here. If the discussion in the text mentions tryptophan, and you should happen to forget its structure, there is a convenient place to look it up.

This appendix includes:

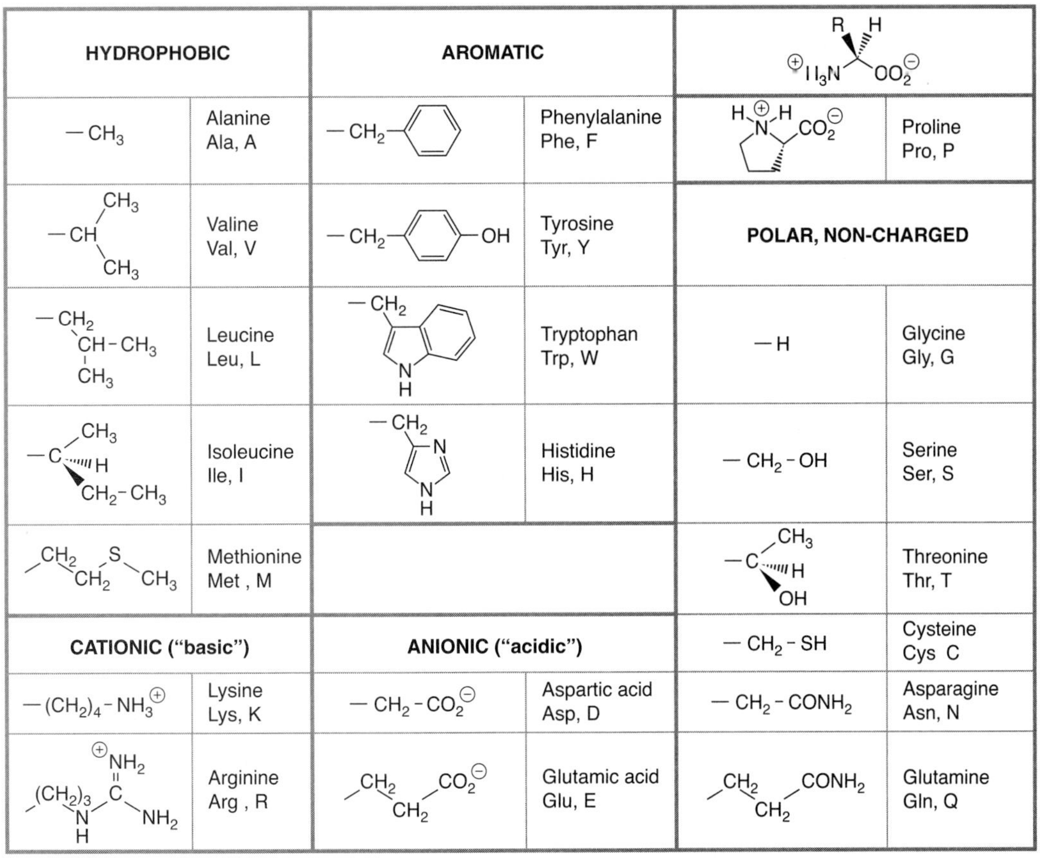

Figure A4.1
The 20 common amino acids.

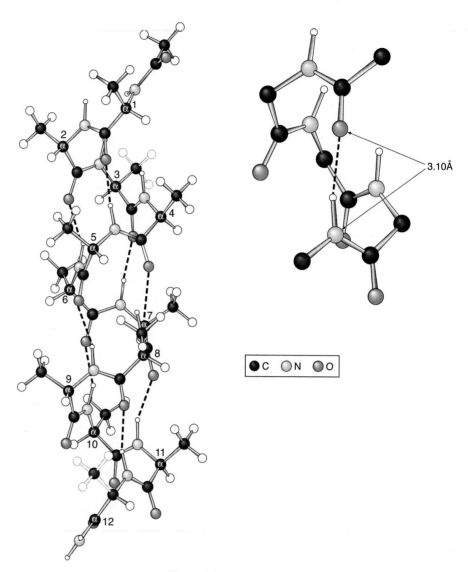

Figure A4.2
The protein α-helix.

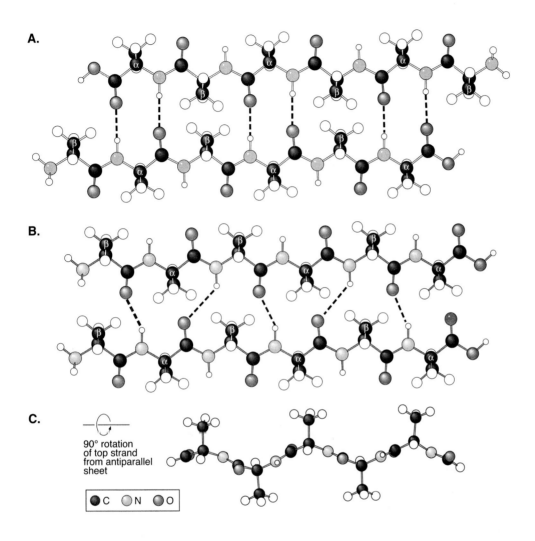

A.

B.

C.

90° rotation
of top strand
from antiparallel
sheet

● C ○ N ○ O

Figure A4.3
The protein β-sheet.

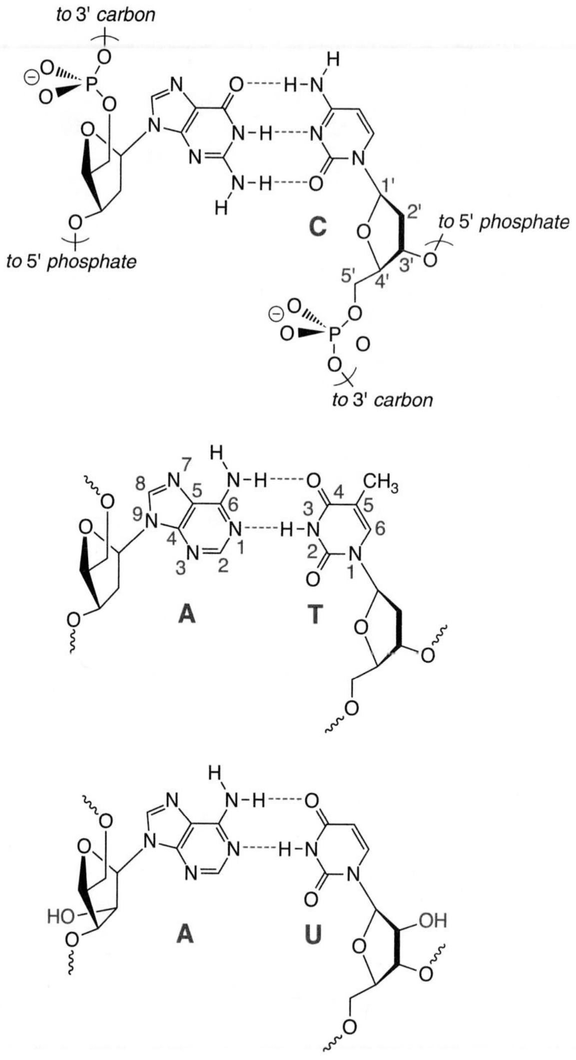

Figure A4.4
The fundamentals of nucleic acid structure.

Pushing Electrons

The manner in which organic chemists write mechanisms is known as "pushing electrons", "arrow pushing", or "electron flow". Becoming proficient at pushing electrons is extremely important, because it allows us to pictorially communicate mechanisms, can keep us from writing unreasonable mechanisms, and can guide us in the choice of experiments. Teaching the fundamentals of this notation is the goal of this appendix. Herein, the concepts of electron sources and electron sinks will be covered, along with resonance, and many examples of correct and incorrect arrow pushing.

A5.1   The Rudiments of Pushing Electrons

When drawing reaction mechanisms, chemists pictorially represent the movement of bonds, electrons, or electron pairs using an electron-pushing notation. Understanding the basic principles behind pushing electrons allows us to analyze the stability of compounds via resonance (see Chapter 1), and to communicate conceptually how reactions are occurring. The arrows used to denote how electrons are moving in chemical reactions are drawn within a chemical structure or between chemical structures. An arrow with a full head on one end denotes the flow of two electrons. An arrow with only one slash on one end is used to denote the flow of only one electron (see margin). Both of these arrows are often curved or S-shaped, and therefore are totally different than a reaction arrow that is straight. A reaction arrow connects a reactant to a product, and denotes a chemical step (or several chemical steps), and does not imply a mechanism. A fourth kind of arrow that chemists use is one with full heads on both sides. This is a resonance arrow, denoting two different ways in which to draw the bonding in a chemical structure. One other arrow notation involves two arrows pointing in opposite directions. This represents an equilibrium between the reactant and product.

The practice of electron pushing is solely a bookkeeping method—a notation. It does not represent the real movement of electrons. This means that the use of a double headed arrow to show the flow of two electrons does not literally mean that electrons are actually moving around, within and between molecules in the matter drawn. Nevertheless, the notation is useful within a valence bond theory (VBT) context, because it indicates how discrete bonds and lone pairs have been rearranged when comparing the reactant to the product.

As stated above, the full headed arrow represents the flow of two electrons. In the drawing of this notation the tail of the arrow is placed near the electron source and the head of the arrow is pointed to the electron sink (several examples of sources and sinks are given below). An **electron source** is always some form of two electrons, such as a lone pair of electrons or a σ/π bond. An **electron sink** is always an atom or bond within a molecule that can accept a pair of electrons in the form of a lone pair or an additional bond.

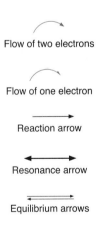

Flow of two electrons

Flow of one electron

Reaction arrow

Resonance arrow

Equilibrium arrows

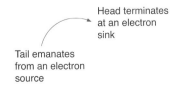

Head terminates
at an electron
sink

Tail emanates
from an electron
source

A5.2   Electron Sources and Sinks for Two-Electron Flow

Each and every arrow must start at an electron source and end at an electron sink. Therefore, before analyzing electron-flow procedures, a list of common electron sources and sinks is useful. Table A5.1 shows several electron sources and sinks. The sources are listed first. They often consist of lone pairs of electrons on heteroatoms, and the atoms can be either negative or neutral. The neutrals include, but are not limited to, alcohols, water, amines, and thiols. The anionic examples include, but are not limited to, alkoxides, amide anions, hydroxide, and thiolates. In all of these examples the lone pairs of electrons are the actual electron source, not the entire chemical structure itself.

Table A5.1
Common Electron Sources and Sinks

I. Sources (R = H, alkyl, aryl)
Nonbonding Electrons
A. The lone pairs on the heteroatoms on the following structures: X^-, RO^-, R_2O, NR_3, NR_2^-, RCO_2^-, RCO_2R, $R(CO)R$, $R(CO)NR_2$, R_2S, RS^-, CN^-, N_3^-, R_3P, R_2Se, and RSe^-
Electron Rich or Strained σ Bonds
A. Organometallics: $RMgX$, RLi, R_2CuLi, and R_2Zn B. Hydride reagents: $LiAlH_4$, $NaBH_4$, NaH, and BH_3 C. Cyclopropropyl or cyclobutyl
π Bonds
Alkenes, dienes, alkynes, allenes, and arenes
Electron Rich π Bonds
Enols, enamines, enolates, anilines, and phenols
II. Sinks (R = H, alkyl, aryl)
Species with Empty Orbitals
Carbocations, aluminum and boron containing Lewis acids, and transition metal (mercury, cadmium, and zinc) reagents
Acidic Hydrogens
Mineral acids, carboxylic acids, water, alcohols, amines, and terminal acetylenes
Weak Single Bonds
Peroxides (ROOR), molecular halogens (X_2), bleach (HOX), and disulfides (RSSR)
The Carbons in Polarized σ Bonds
Alkyl halides, alkyl tosylates, protonated alcohols (ROH_2^+), and protonated amines (RNH_3^+)
The Carbons in Polarized Multiple Bonds
$R_2C=O$, nitriles, α,β-unsaturated carbonyl compounds, acyl halides, anhydrides, esters, and amides

Sigma bonds between carbons are not normally electron sources, because the energy of the electrons in these orbitals is too low for them to be involved in common chemical reactions. Examples of σ bonds that can be electron sources are highly strained bonds such as in cyclopropane, and σ bonds that are highly polarized and have a partial or full negative charge on carbon. These include lithium and Grignard reagents. However, if the sink is electrophilic enough, an unstrained σ bond can be a source (see the carbenium ion arrangement below).

In contrast to σ bonds, simple π bonds from alkenes, alkynes, conjugated alkenes, allenes, and aromatic rings are often electron sources. Pi bonds are significantly better sources when they are electron rich due to an attached electron donating group. These donating groups are heteroatoms with lone pairs of electrons for which resonance structures can be written showing negative character on a carbon of the π bond (see the enamine below and see the next section on the details of denoting resonance structures).

Note that all the electron sources in Table A5.1 are also nucleophiles. As discussed in Chapter 10, nucleophiles can donate electrons to a positive or partially positive atom. All nucleophiles are electron sources.

However, not all electron sources are considered to be nucleophiles. For example, consider the deprotonation of hydronium by ammonia. One of the arrows used to depict this reaction involves the ammonia lone pair as an electron source and a hydrogen of hydronium as a sink. The other arrow denotes that the H–O σ bond is an electron source to create a lone pair on the positive oxygen, which is the sink for this arrow. We would not consider the H–O σ bond to be a nucleophile, although it is a source in this example.

Another example is the deprotonation of a carbon α to a carbonyl by ethoxide. In this reaction the ethoxide is an electron source, but so is the C–H σ bond and the C–O π bond. In this example the C–H bond would not be considered to be nucleophilic.

Chemists do not always draw all the lone pairs on heteroatoms, and it is important for the reader to realize that the lone pairs are there. When the lone pair is used as a source we always draw it, and when a lone pair is created we will draw it. This is the convention used in this textbook. Very often, however, chemists stop drawing all lone pairs, even when they are sources, and you should recognize what is meant by the arrows.

Electron sinks are more varied in nature than electron sources (several are shown in Table A5.1). Anything with a positive charge and/or an empty orbital is an electron sink. Carbocations, Lewis acids, and metal cations are good electron sinks. Atoms involved in σ bonds can also be electron sinks. The proton in a Brønsted acid or weak single bonds between pairs of heteroatoms often act as sinks. Atoms on the partial positive end of polarized bonds are common electron sinks—namely, the carbons in carbonyls, nitriles, conjugate acceptors, and alkyl halides. All of these sinks are also electrophiles.

However, sinks are not necessarily the same as electrophiles. The sink is always an atom that can accept a negative charge and still be relatively stable. To explain what is meant by this, consider the nucleophilic attack of cyanide on the carbonyl of acetone. The source for the first arrow is the lone pair of electrons on the carbon of cyanide, and the sink is the partially positive carbon of the carbonyl.

However, we cannot stop there. Since the drawing of the arrow from the cyanide source to the carbonyl sink implies that a bond is being formed between the source atom and the sink atom, in this example five bonds to carbon would be formed. Hence, another arrow is needed to show that some bond to this carbon must also be breaking. In this case the bond that is breaking is a π bond (one of the traditional sources), and the sink is the oxygen of the carbonyl, where a lone pair is being created. Chemists would not consider the oxygen as an electrophile, although it is an electron sink in this case. This dichotomy in the terminology is the reason that the terms "source" and "sink" are associated with electron pushing, whereas the terms "nucleophile" and "electrophile" are used to describe reactivity patterns.

The sink is always the place where the head of the arrow terminates. Confusion can arise as to the exact place that the arrow should terminate. When the sink is a heteroatom that is accepting a lone pair, place the head of the arrow near that heteroatom. Similarly, when the sink is an electrophilic atom, just place the arrow head near this atom. However, when the sink leads to the formation of a new bond, there are two acceptable conventions on where to place the end of the arrow. You can either place the arrow head between the two atoms where the new bond will be drawn (**A**, below), or place the arrow head pointing towards the atom where the bond is being formed (**B**, below). Both methods are acceptable.

In all mechanisms that we write, the total charge from one step to the next never changes. This is called the **conservation-of-charge** rule. If an anionic nucleophile adds to a neutral compound, the product must be negative. If an anionic intermediate expels a leaving group and becomes neutral, the leaving group must be departing with a negative charge. If a neutral compound fragments, it must do so to create either two neutral species or a negative and a positive one. This rule of maintaining the net charge from one set of structures to the next is very useful for checking if the steps being proposed in a mechanism are plausible or not. If the charges are not maintained, then something is wrong.

A5.3   How to Denote Resonance

Resonance was defined in Chapter 1. We commonly accompany a collection of resonance structures with electron flow arrows to describe how two electrons were moved around, thereby creating the new bonding arrangement. As an example of using arrows to depict resonance, consider acetate. An arrow starts at a lone pair on the single-bond oxygen and terminates near the carbonyl carbon. This arrow states that a second bond will be formed to the carbonyl carbon. A second arrow denotes that a bond in the C=O becomes a lone pair on oxygen.

We noted in Chapter 1 that drawing resonance structures often gives insight into the reactivity of a molecule. For example, consider the resonance structure of the cation shown below. In the first depiction there is a positive charge on the carbon and this carbon has an empty p orbital. Therefore, we predict nucleophilic attack at the carbon is possible.

However, directly adjacent to the empty orbital is an oxygen atom with lone pairs of electrons. Since the carbon is lacking an octet of electrons, a neighboring electron pair can donate into the adjacent orbital to give a π bond, thereby stabilizing this structure. We denote this donation using a double headed arrow as shown. The arrow starts at the lone pair and ends in between the carbon and oxygen, showing that a π bond is formed. Thus, the electron source was the lone pair and the sink was the empty carbon p orbital. Since no atomic movement has occurred and only electrons have been moved within the molecule, we have created a resonance structure.

Resonance structures most often involve p orbitals that are in conjugation. For example, the molecule acrolein has its π bonds in conjugation, and the resonance structure shown indicates that there is some π bond character between the two central atoms.

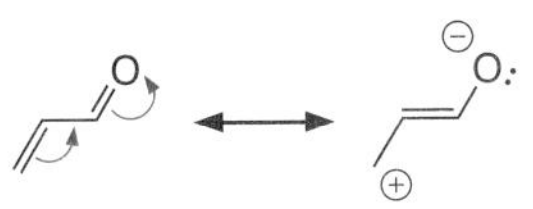

A5.4 Common Electron-Pushing Errors

There are several errors that are common for students when they are first learning to use electron-pushing notation. It is instructive to cover many of the common mistakes so as to spotlight them.

Backwards Arrow Pushing

Likely the most common mistake is pushing the arrow backwards. In other words, the arrow is started at a sink and ended at a source. Three examples are given below. The easiest way to avoid this mistake is to remember that the arrow must start from an electron rich region of a molecule. Most important, the arrow always starts with two electrons—namely, lone pairs, σ bonds, or π bonds. Do not use the positive regions of a molecule to start an arrow. The vast majority of the time the arrow will terminate at a center with some positive charge or a center that can accept a lone pair.

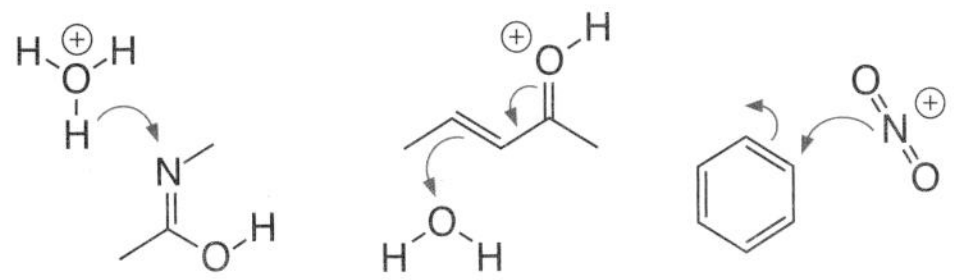

Common backwards electron flow (incorrect)

Not Enough Arrows

The second most common error is to not show enough arrows. As an example, consider the electron pushing shown below that is meant to indicate an E2 reaction. The base is abstracting the proton while the leaving group is departing, but there is no arrow to denote the formation of the double bond. The easiest way to avoid problems like this is to remember that each arrow leads to the formation of a bond or a lone pair of electrons. Hence, keeping track of where the arrow starts and terminates defines either a bond between the respective atoms or a lone pair localized on a specific atom. Using this analysis on the example given shows that the base is forming a bond to hydrogen, but since no bond to hydrogen is shown as breaking, the result is two bonds to hydrogen. Furthermore, since the bond to the leaving group is breaking to form a lone pair on bromide, and there is no arrow shown to fill the void left behind, a carbocation must be forming.

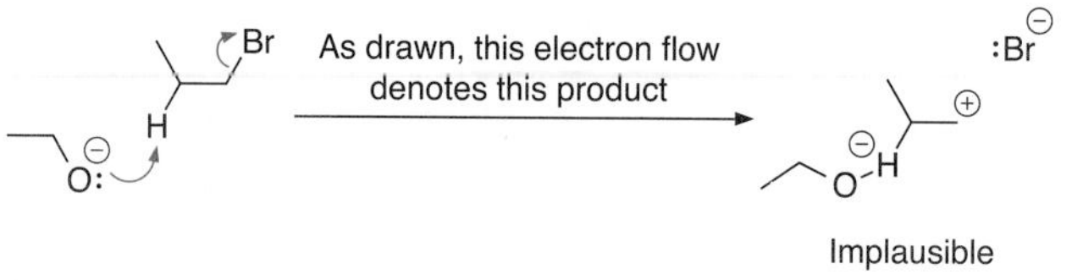

Implausible

Losing Track of the Octet Rule

Another common mistake is showing arrows that create atoms with electron counts above and beyond an octet. As examples of this, consider a resonance structure of nitrate and the attack of cyanide on the oxygen of an oxycarbonium ion. Both examples indicate that an electron source is quenching the positive charge of an electron sink. But, these positively charged atoms are not electron sinks capable of accepting two electrons in the form of an additional bond. In each case the number of electrons on the respective sink atoms increases to ten, two beyond an octet.

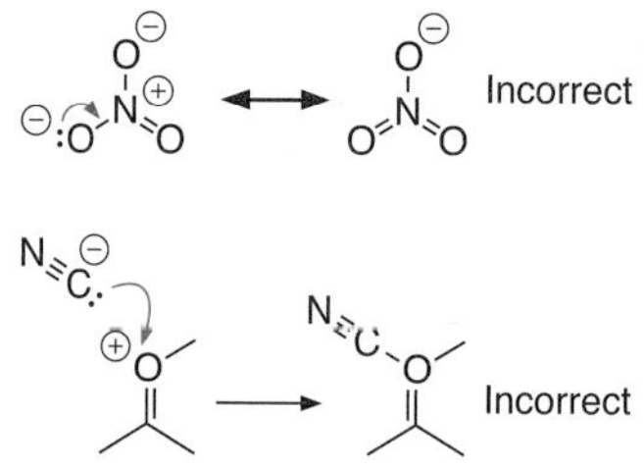

This problem is quickly remedied when we remember how many bonds each atom can form and that we do not routinely draw the lone pairs of electrons. In both examples given above, the number of bonds increases beyond what is allowed (that is, five to nitrogen and four to an oxygen possessing an undrawn lone pair). Therefore, although nitrogen and oxygen are positively charged in these examples, they are not electron sinks for formation of additional bonds.

Losing Track of Hydrogens and Lone Pairs

One of the easiest mistakes to make is to forget that hydrogens on carbons are not drawn in stick structures. This can often lead to the movement of an arrow to a carbon that already has four bonds, although at first glance the carbon may seem to be a reasonable electron sink. In the example shown below, the first resonance structure given is simply incorrect because a double bond is drawn to a carbon that already has four bonds. A second problem also exists. The oxygen is left without an octet, and should therefore have a plus-two charge; no lone pair was added to give the oxygen eight electrons. The real result of the electron pushing is also shown. It is an implausible structure. The only way to avoid these mistakes is to mentally take note of the hydrogens that are not drawn. Moreover, if you draw all the lone pairs when embarking on an electron-pushing exercise, the latter of these two problems will be much easier to spot and avoid.

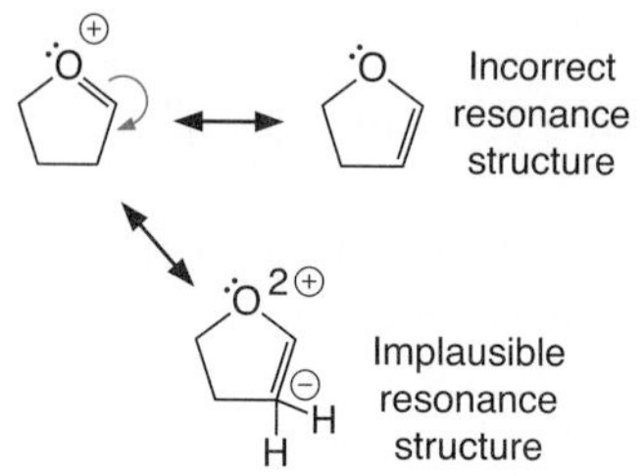

Not Using the Proper Source

A harder error to spot is the use of the wrong electrons as the source. As an example, in Chapter 9 general-base catalysis is covered. In some cases the base deprotonates water while the water simultaneously adds to a carbonyl carbon. The best source of electrons for showing the addition to the carbonyl sink is a lone pair of electrons on the oxygen of water. However, the electron flow could also be depicted as shown in Mechanism A5.1. As shown in this electron-pushing scheme, the H–O σ bond is the electron source for forming a bond to the carbonyl carbon sink. This is not so serious because electron pushing is just a bookkeeping notation, and the number of bonds and lone pairs is correctly portrayed in this scheme. However, it is not the best reflection of how one thinks about this nucleophilic attack.

In Mechanism A5.2, the lone pair on oxygen is correctly shown as the source for the carbonyl sink, but the deprotonation of the water is incomplete. Instead, two bonds are being formed to hydrogen because there is no arrow showing the breakage of the O–H σ bond. The best method is shown in Mechanism A5.3.

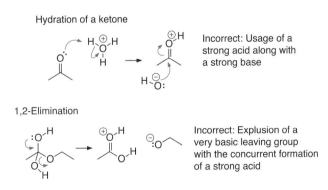

(Mechanism A5.1)

Incorrect source for the arrow
depicting the nucleophilic attack

Not enough arrows; this
indicates the formation
of two bonds to hydrogen

(Mechanism A5.2)

(Mechanism A5.3)

Best electron-pushing notation

Mixed Media Mistakes

An important rule to remember with arrow pushing, much like on the bench top, is to not mix strong acids and bases. If the reaction is performed in acidic media, it does not make sense to show the creation of a strong base in your mechanism. Similarly, if a reaction is run with added base, it is unreasonable to create a strong acid in the mechanism. Significant concentrations of strong acids and bases cannot co-exist in the same medium. For example, any medium that is acidic enough to protonate a carbonyl reactant would not have any appreciable hydroxide. Two examples of this mistake are given below.

Hydration of a ketone

Incorrect: Usage of a
strong acid along with
a strong base

1,2-Elimination

Incorrect: Explusion of a
very basic leaving group
with the concurrent formation
of a strong acid

Too Many Arrows—Short Cuts

The last error that we cover is the use of too many arrows in too complex of a scenario. It is sometimes tempting to combine several steps together into one step as a means of taking a short cut to the product. For example, consider the pinacol rearrangement, shown below.

The electron-pushing notation does indeed lead to the product, but this proposed pathway is chemically unreasonable. It involves the breaking and forming of several bonds simultaneously. The entropy disadvantage of such a reaction would be quite large. Such a combination of steps is not what chemists observe when studying these reactions (for the correct scheme see Section 11.8).

The second example involves an acyl transfer from chloride to water. The addition and elimination reactions are combined into one step. Although the arrows do keep track of the electrons involved in the reaction, such steps are known not to occur simultaneously, and thus the electron-pushing notation does not reflect what is known about the mechanism.

A5.5   Complex Reactions—Drawing a Chemically Reasonable Mechanism

There are many factors that go into writing a reasonable mechanism when a chemist is first confronted with the product of a reaction. All the chemical intuition that the chemist has, built upon past experience, is used to create the mechanism. Any available experimental data, any knowledge as to the feasibility of intermediates, and just a "gut" intuition are often the starting points. In actuality, for most chemists, the "gut" intuitive feeling is based on either a conscious or unconscious recognition of a logical electron-pushing pathway. In other words, when first considering a reaction, most chemists apply the rules of electron-pushing notation to visualize a logical sequence of chemical reactions that can lead to the observed product. These steps are then examined in light of the experimental data. If there are no experimental data, the electron-pushing analysis is used to create hypotheses that can be tested experimentally. Hence, being able to apply the electron-flow rules to completely new scenarios is one mark of a sophisticated organic chemist.

It is difficult to translate into words the mental process that a chemist goes through when coming up with a mechanism for a new reaction, especially since undoubtedly every chemist does it differently. However, we attempt to do just that here. A few simple rules will assist.

1. *Find a 1:1 correspondence between all atoms in the reactants and the products.*
    This may lead you to ask a question such as "Where does this oxygen come from?"

2. *Keep your mind on where you are going.*
    In other words, look for a path that will lead to the product. To do this, note which groups have added to or left from the reactant, and make sure that such steps are included in the mechanism.

3. *Measure your progress at intermediate stages based upon how many bonds still need to be formed or broken.*

4. *Note any rearrangement of atoms within the chemical structure and make sure appropriate steps are included.*

5. *Do not push too many arrows as a way to create a short cut to the product.*
   To do this, always stick to common reaction steps such as those presented in Chapters 10 and 11.

6. *Avoid the common electron-pushing mistakes.*

7. *Finally, do not form any intermediates of unreasonably high energy.*
   Similarly, do not form high energy intermediates when other intermediates of lower energy are possible. This is a more difficult analysis to make. However, the information and examples given throughout this text should provide excellent guidance.

If you can remember all of these items, you should be able to write out mechanisms that are both chemically sound and often prove to be correct after an experimental analysis. Two examples of complex chemical reactions, along with a discussion of how to proceed in writing a mechanism, are discussed here, and several practice problems are given at the end of this appendix.

A5.6 Two Case Studies of Predicting Reaction Mechanisms

Our first example is the acid-catalyzed hydrolysis of an enamine to give a ketone and a protonated amine.

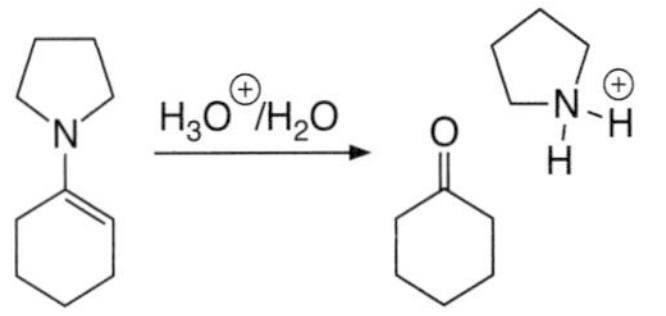

The second suggestion given in Section A5.5 is to note where you are going. In this case, the amine has to leave the molecule and an oxygen has to add. With regard to the departure of the amine, amide anions (R_2N^-) or neutral amines (NR_3) can be considered as possible leaving groups. However, since an amide anion is highly basic, it is a very poor leaving group. Moreover, we want to avoid the creation of a strong base in the presence of acid. Both of these points make it clear that the amine must depart as a neutral species. The nucleophile must be water and not hydroxide since the reaction is run in acidic conditions. The next point to note is that the double bond in the enamine is within the ring and between carbons, but the double bond in the product is exo to the ring and is to oxygen. Hence, the double bond must change position at some point in the mechanism.

Once it is clear what bonds must change positions and what groups must add and leave the reactant, the next step is to consider which reaction steps can be used to accomplish these tasks. In this case, because the reactions are being performed under acidic conditions, an acid can be used to move the position of the double bond in a manner similar to an acid-catalyzed tautomerization reaction. The acid can also be used to protonate the leaving group so that the amine can depart as a neutral species. Furthermore, the acid could be used to activate a polarized π bond toward nucleophilic attack, if necessary.

Once the likely reactions have been identified, it is a matter of putting them in the correct order. The reactant currently does not have a polarized π bond for nucleophilic attack by water, and neither is the leaving group ready for departure as a neutral amine. Therefore, we should start by either creating a polarized π bond, or making the leaving group ready for departure. Let's consider the latter first. Protonation of the amine can lead to its departure as shown below. However, this creates a vinyl cation, an unreasonably high energy intermediate. Hence, this is not plausible.

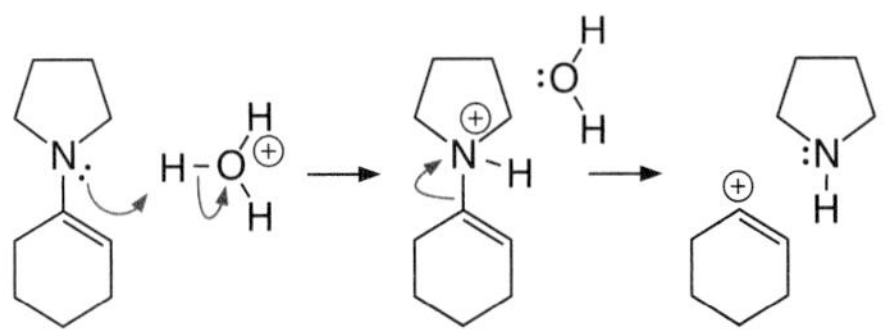

If protonation occurs on the β carbon, however, the cation formed is an imminium, which is much more stable than the vinyl cation shown above. In addition, this protonation leads to the migration of the double bond, as is required in the mechanism. Moreover, there is now a highly polarized π bond within the molecule. Due to the fact that this reaction contains several of the aspects of the mechanism that were identified as necessary, it is a good place to start.

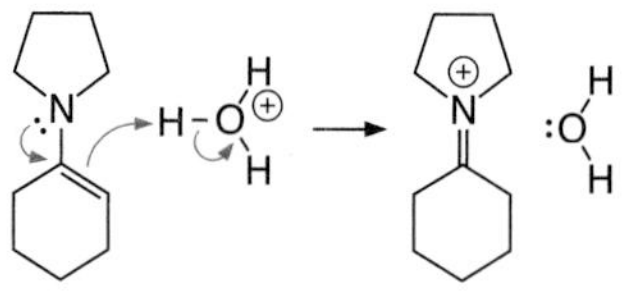

Keeping in mind where we are going, leaving group departure and nucleophilic attack are still required. The amine is not ready to depart since it is held to the structure via a double bond. Therefore, addition of water to the polarized π bond, followed by loss of a proton, is the next logical sequence of events.

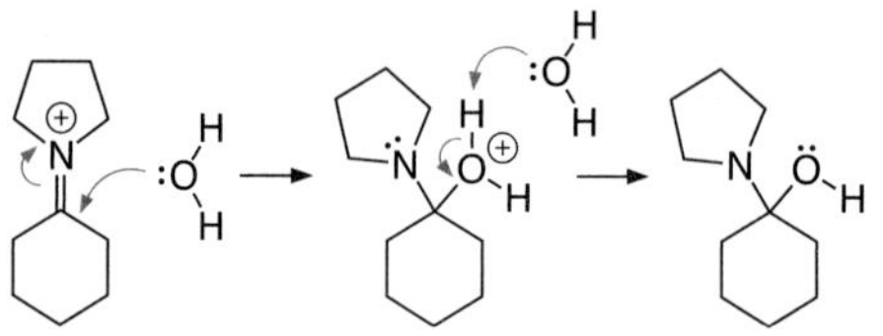

All that remains now is leaving group departure. Protonation of the amine followed by a 1,2-elimination assisted by the neighboring lone pairs on oxygen gives an intermediate that is simply a proton transfer away from the correct product.

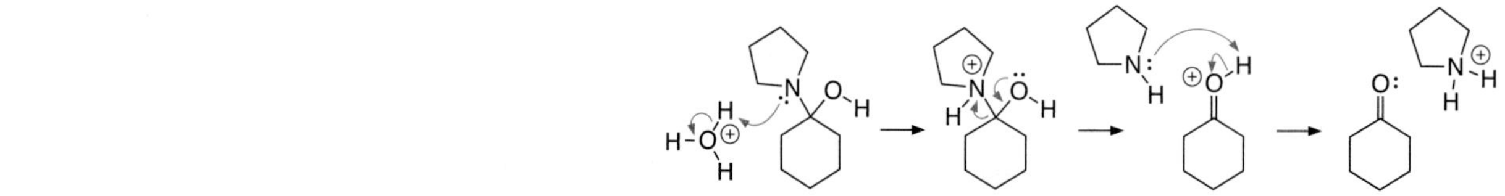

The next example involves base catalysis, and is known as the Robinson annulation. Once again, the first item to note is where we are going.

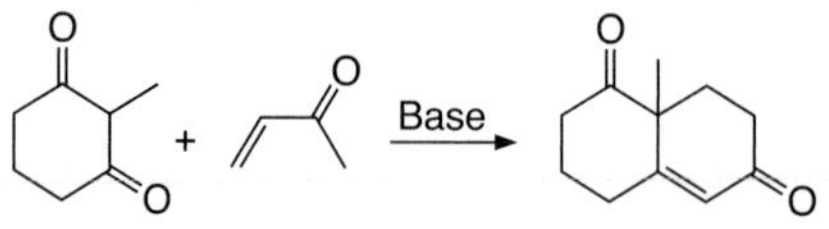

By counting carbons it is clear that no additional carbon atoms are required. To see how the two reactants are put together, it is useful to letter or number some or all of the carbons of the reactants and place those markers near the same carbons in the product. Note that in the starting material carbons a and c are identical due to symmetry in the molecule.

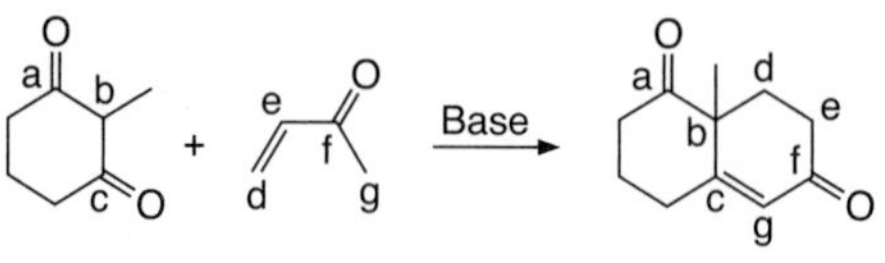

Once it is clear how the two pieces have to go together, we can make decisions as to the appropriate reactions to use to fuse the two reactants. First, note that carbon b has undergone a conjugate addition forming a bond to carbon d. Under basic conditions a conjugate addition would start via formation of an enolate nucleophile. Second, the bonds formed between carbons c and g require the loss of a molecule of water. The loss of water most likely arises from the elimination of water from an alcohol, since no other scheme seems reasonable, given the reactants and the experimental conditions. Elimination of water from an alcohol under basic conditions would require a 1,4-elimination reaction on a β-hydroxy carbonyl. Finally, β-hydroxy carbonyl structures are formed via aldol reactions.

Once the reactions required are known, it is a matter of stringing them together in the proper sequence. Since this reaction is performed under basic conditions, it is logical to have the base start by abstracting the most acidic proton. This would be the hydrogen that is alpha to two carbonyls. Furthermore, since the resulting enolate carbon was found by the previous analysis to be attached to carbon d, it makes sense to draw the conjugate addition as the first carbon–carbon bond forming sequence in the mechanism.

At this point the rest of the mechanism must consist of the aldol reaction followed by the loss of water. The electron pushing indicates the formation of an enolate followed by an aldol reaction. The loss of hydroxide is simply a 1,4-elimination.

In summary, the power of electron pushing is the ability to write chemically reasonable mechanisms by combining sources and sinks using reactions that are well precedented in organic chemistry. The electron pushing allows chemists to communicate their thoughts as to steps involving nucleophile and electrophile combinations. In all the discussion to this point, we have focused upon two-electron arrow pushing. None of these reactions involves radical intermediates. However, radicals are common intermediates in organic transformations, and therefore we also need an understanding of how to perform electron pushing for radical reactions.

A5.7   Pushing Electrons for Radical Reactions

To denote the movement of single electrons we use arrows with a single slash head. The arrow still starts at an electron source, but the source can now be any bond, any single electron, or any lone pair. Unlike two-electron arrow pushing, we do not consider the source for one-electron arrows to have any analogy to nucleophiles. The arrow still ends at an electron sink, but now the sink is defined as any site that can accept a single electron. These are not necessarily traditional electrophilic sites.

Two arrows starting at a bond and spreading apart are used to denote the homolytic cleavage of a bond. Two arrows starting on separate radicals and coming together represent the formation of a bond.

Other common reactions are hydrogen abstraction and radical additions to double bonds. These reactions are denoted with single-headed arrows as combinations of the two kinds of electron pushing given just above. Combining the kinds of steps shown here, with different radical reactants, will allow you to write the electron pushing for most radical reactions.

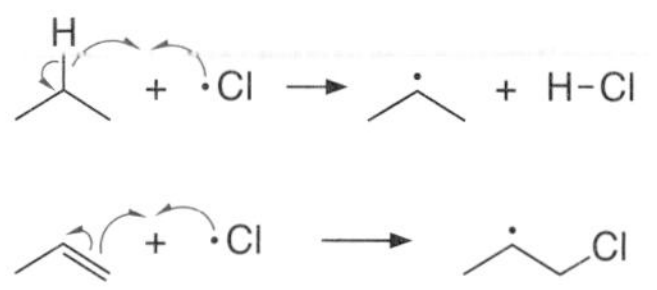

One last reaction type that is commonly encountered is electron transfer to a π or σ bond. This bond must be capable of receiving a negative charge, and it is common for these bonds to have electronegative atoms. Since the electron transfer is to an intact bond, the newly added electron goes into an antibonding orbital for that bond. In the case of an alkene or carbonyl, this can be denoted by the creation of a anion radical with the two species drawn on separate atoms. For a σ bond, however, it is difficult to designate where the electron went, and we simply draw a radical anion next to a dashed bond.

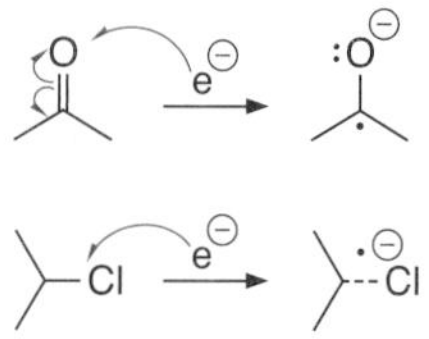

Given the above examples, we can write the mechanisms of common radical reactions that involve multiple steps. As our only example of single-electron arrow pushing, let's consider the mechanism of HBr addition to alkenes under radical conditions (see Section 10.10). Under radical conditions, there is an initiation of the reaction to create bromine radicals, often by a peroxide. The peroxide first homolyzes, and the resulting radical abstracts a hydrogen atom from HBr. These are the initiation steps. The electron pushing is a combination of homolysis and hydrogen abstraction steps. Propagation is a combination of radical addition to an alkene followed by hydrogen abstraction. Termination is the combination of any two radicals to create a σ bond. Note that the electron pushing is the same for each step as in the simple examples given above.

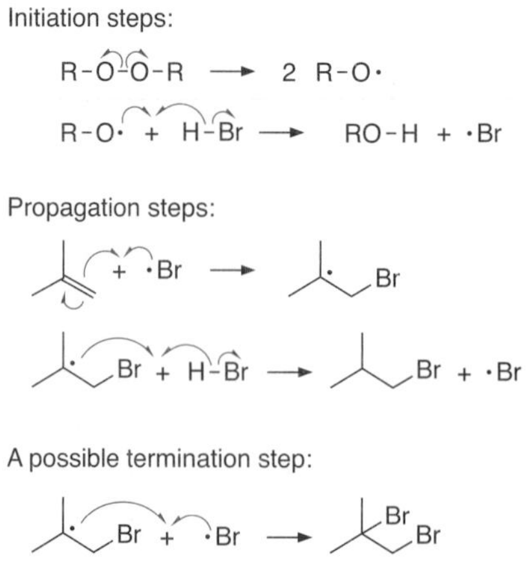

Initiation steps:

Propagation steps:

A possible termination step:

Practice Problems for Pushing Electrons

1. Identify any atoms, bonds, or lone pairs in the following molecules that could be considered as electron sources or sinks for two-electron arrows.

A.   B.   C.   D.   E.   F.   G.

2. Show mechanisms for the following transformations, along with all electron pushing. These reactions involve more than one step.

3. The following are more complicated reactions than those presented to this point. However, they all involve two-electron steps that are similar to those presented in this book, especially in Chapters 10 and 11. Use your best chemical intuition to write reasonable mechanisms for these transformations. Draw all intermediates and show all electron pushing. The mechanisms that you write may not actually be the ones that have been supported by experiments. You cannot be expected to know this. However, your mechanism should be reasonable, and the electron pushing should be correct.

I.

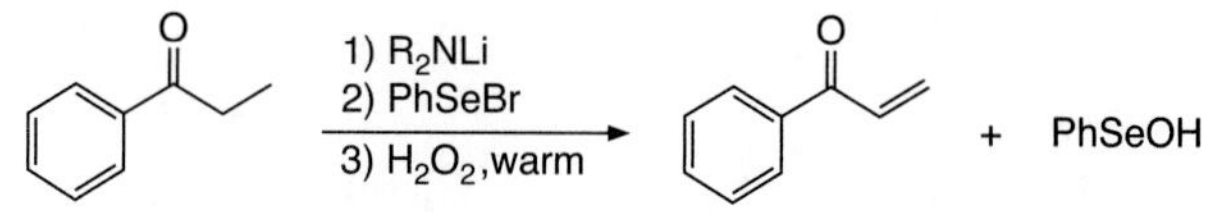
O
1) R₂NLi
2) PhSeBr
3) H₂O₂,warm
O
+ PhSeOH

J.

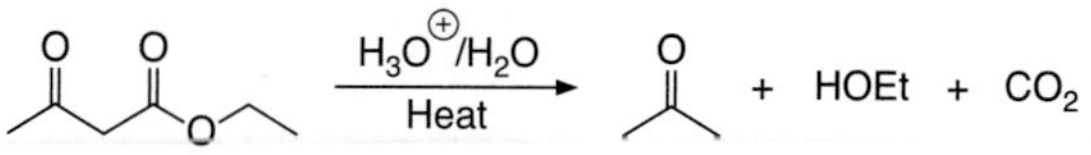
O
O
O
H₃O⁺/H₂O
Heat
O
+ HOEt + CO₂

K.

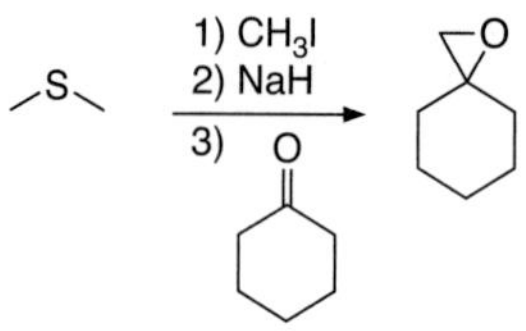
S
1) CH₃I
2) NaH
3)
O
O

L.

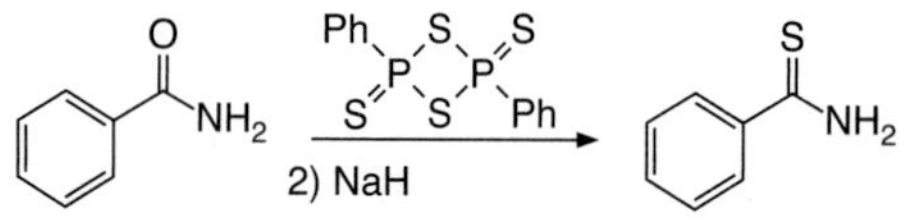
O
NH₂
Ph
S
S
P
P
S
S
Ph
2) NaH
S
NH₂

M.

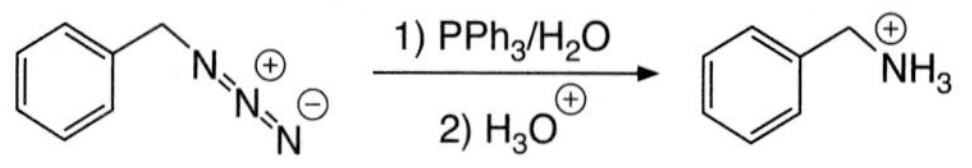
N
N
N
1) PPh₃/H₂O
2) H₃O⁺
NH₃

N.

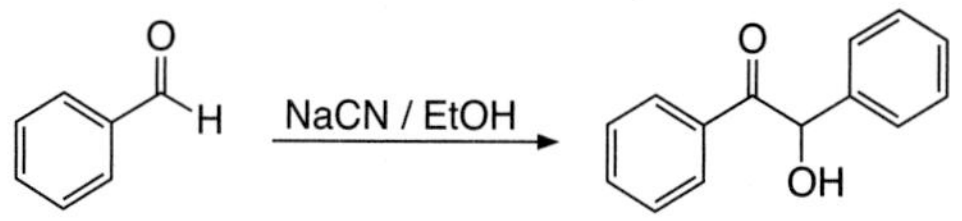
O
H
NaCN / EtOH
O
OH

O.

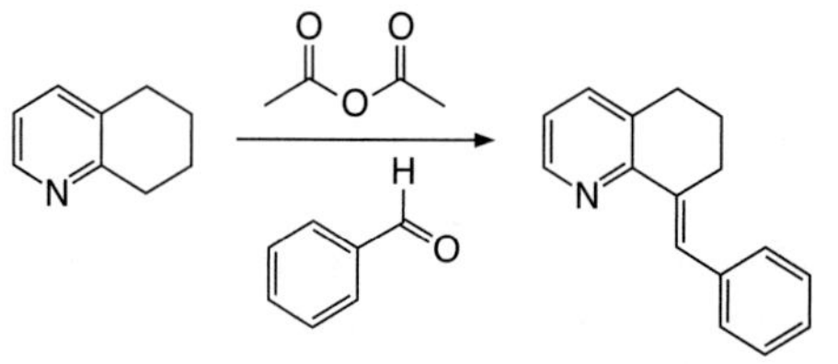
N
O
O
O
H
O
N

P.

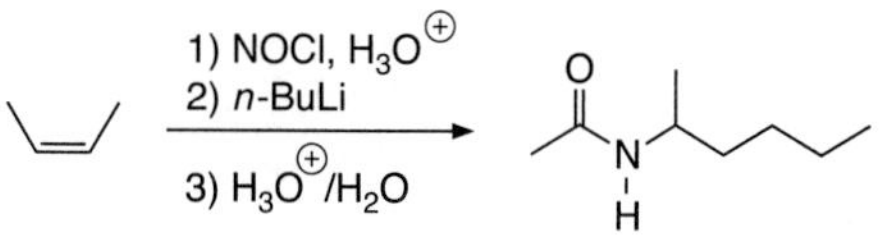
1) NOCl, H₃O⁺
2) n-BuLi
3) H₃O⁺/H₂O
O
N
H

4. Write reasonable steps for the following radical reactions, showing all the proper electron pushing.

A.

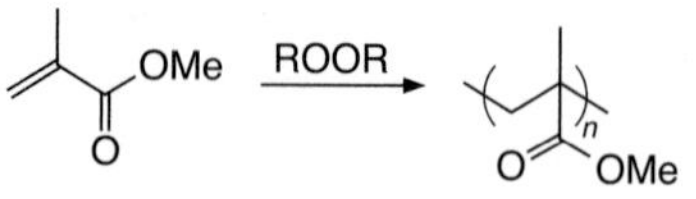
OMe
O
ROOR
n
O
OMe

B.

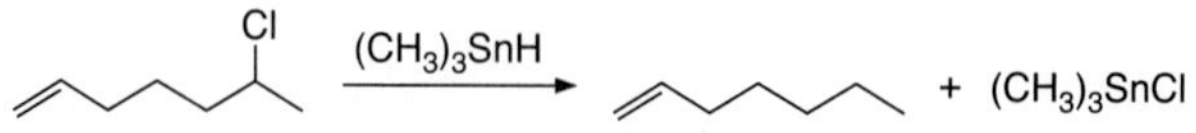
Cl
(CH₃)₃SnH
+ (CH₃)₃SnCl

C.

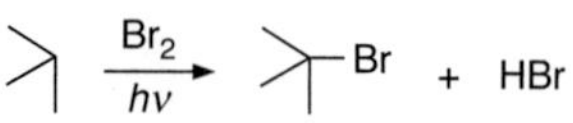
Br₂
hv
Br
+ HBr

Reaction Mechanism Nomenclature

Chemists have devised a nomenclature system to name reaction mechanisms. The nomenclature is given by an acronym that describes some of the essential features of the mechanism. This procedure has not been thoroughly adopted by organic chemists, although it is certainly used by many researchers. In this book, we have used only some of the more common examples—those which are used in everyday discourse between chemists: S_N1, S_N2, E2, E1, E1cB, and $S_{RN}1$. Although the method is not entirely systematic, it uses a symbol to represent the reaction as a substitution (S), addition (Ad), or elimination (E). This is followed by whether the reaction commences by a nucleophilic ($_N$), electrophilic ($_E$), or homolytic ($_H$) event, and finally the molecularity of the rate-determining step (1, 2, etc.). Other symbols are incorporated to communicate specific facets of the mechanism, such as A (acid-catalyzed), B (base-catalyzed), Ar (aromatic), R (reduction or radical), O (oxidation), i (intramolecular), EWG (electron withdrawing group), and cB (conjugate base). Lastly, Ac (acyl transfer) is used in substitutions on carboxylic acid derivatives when the acyl–leaving group bond is cleaved.

The following list of reactions shows the most common uses of the various acronyms. Many more exist, but we only show the ones that you are likely to encounter while reading the literature. N:⁻ and E⁺ represent generic nucleophiles and electrophiles, respectively, while N and E are nucleofuges and electrofuges, respectively. X is typically a halogen, while Y is a group that can support a radical. We also list the section where discussions of these mechanisms can be found.

S_N1 (substitution, nucleophilic, unimolecular), Section 11.5

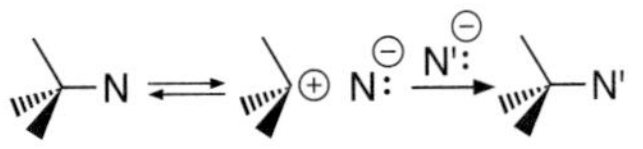

(Eq. A6.1)

S_N2 (substitution, nucleophilic, bimolecular), Section 11.5

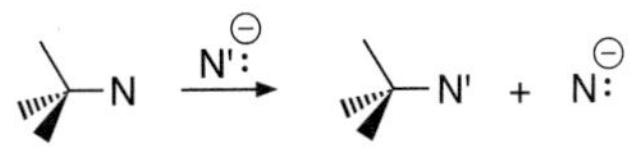

(Eq. A6.2)

S_N1' (substitution, nucleophilic, unimolecular, allylic rearrangement), Section 11.5

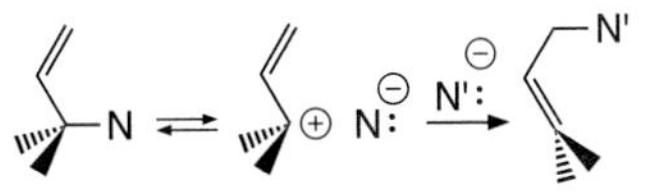

(Eq. A6.3)

S_N2' (substitution, nucleophilic, bimolecular, allylic rearrangement), Section 11.5

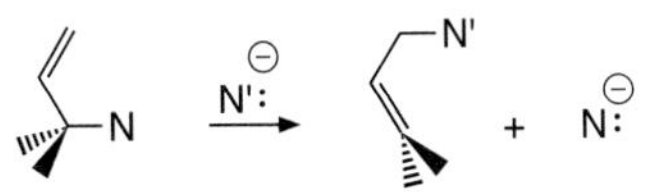

(Eq. A6.4)

S_E1 (substitution, electrophilic, unimolecular), Section 12.2

(Eq. A6.5)

S_E2 (substitution, electrophilic, bimolecular), Section 12.2

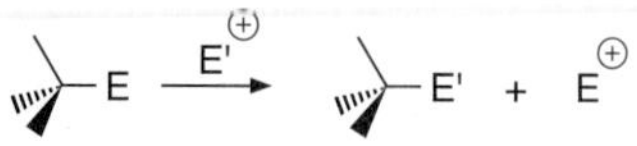

(Eq. A6.6)

$S_{RN}1$ (substitution, radical, nucleophilic, unimolecular), Section 11.6

(Eq. A6.7)

S_H1 (substitution, homolytic, unimolecular), Section 11.7

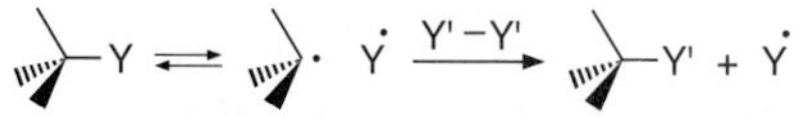

(Eq. A6.8)

S_H2 (substitution, homolytic, bimolecular), Section 11.7

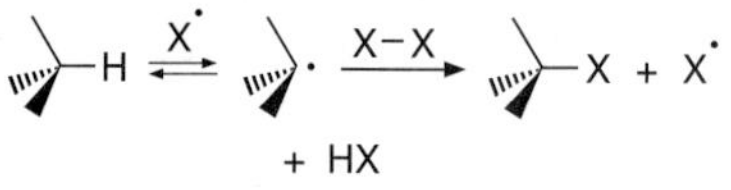

(Eq. A6.9)

S_N2Ar (substitution, nucleophilic, bimolecular, aromatic), Section 10.19

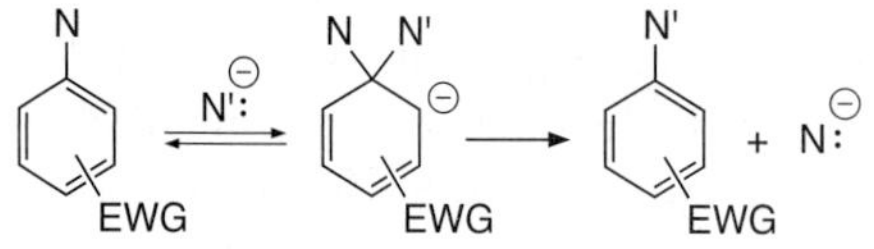

(Eq. A6.10)

S_E2Ar (substitution, electrophilic, bimolecular, aromatic), Section 10.18

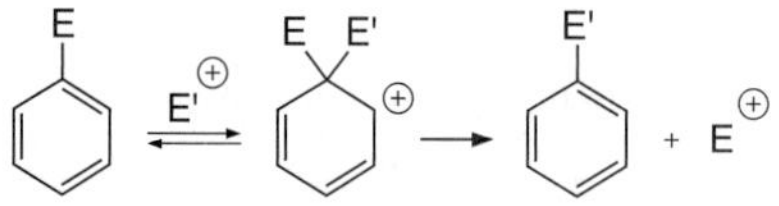

(Eq. A6.11)

E1 (elimination, unimolecular), Section 10.13

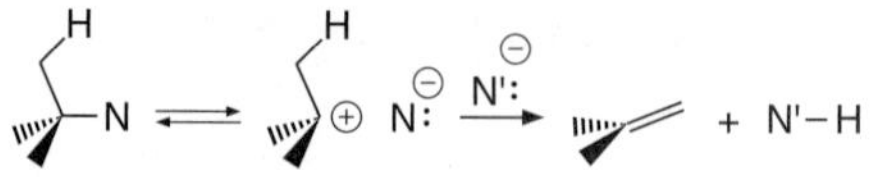

(Eq. A6.12)

E2 (elimination, bimolecular), Section 10.13

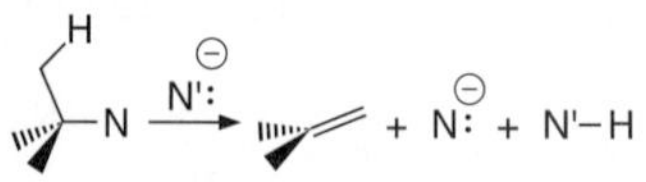

(Eq. A6.13)

E1cB (elimination, unimolecular, conjugate base), Section 10.13.4

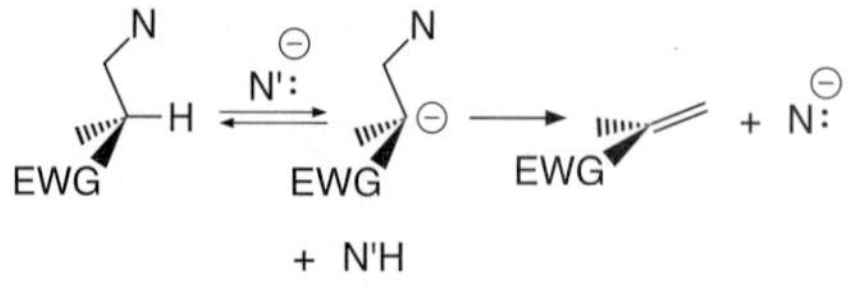

(Eq. A6.14)

Ei (elimination, intramolecular), Section 10.13.10

$$\text{(Eq. A6.15)}$$

Ad$_E$2 (addition, electrophilic, bimolecular), Section 10.3

$$\text{(Eq. A6.16)}$$

Ad$_N$2 (addition, nucleophilic, bimolecular), Sections 10.8 and 10.9

$$\text{(Eq. A6.17)}$$

Ad$_H$2 (addition, homolytic, bimolecular), Section 10.10

$$\text{(Eq. A6.18)}$$

E–Ad (elimination, addition), Section 10.20

$$\text{(Eq. A6.19)}$$

D$_{Ac}$2 (basic conditions, acyl transfer, bimolecular), Section 10.17

$$\text{(Eq. A6.20)}$$

A$_{Ac}$2 (acidic conditions, acyl transfer, bimolecular), Section 10.17

$$\text{(Eq. A6.21)}$$

Index